CHILTON'S TRACTOR REPAIR MANUAL

8 hp through 30 PTO hp – 1960 and later models

Editorial Director	Alan F. Turner
Managing Editor	Kerry A. Freeman, S.A.E.
Senior Editor	Richard J. Rivele, S.A.E.
Production Manager	Warren Owens
Assistant Production Manager	Timothy Frelick
Production Assistant	Nancy A. Hassler
Mechanical Pasteup	Dru Brown
	Donna P. Fisher
	Robin Miller
	Margaret A. Stoner

OFFICERS

President	William A. Barbour
Executive Vice President	James Miades
Vice President & General Manager	John P. Kushnerick

CHILTON BOOK COMPANY
Chilton Way, Radnor, Pa. 19089

Manufactured in USA
© 1981 by Chilton Book Company
ISBN 0-8019-7057-1
Library of Congress Catalog Card No. 80-70265

234567890 098765432

Notices

SAFETY NOTICE

Proper service and repair procedures are vital to the safe, reliable operation of all motor vehicles, as well as the personal safety of those performing repairs. This manual outlines procedures for servicing and repairing vehicles using safe effective methods. The procedures contain many NOTES, CAUTIONS and WARNINGS which should be followed along with standard safety procedures to eliminate the possibility of personal injury or improper service which could damage the vehicle or compromise its safety.

It is important to note that repair procedures and techniques, tools and parts for servicing motor vehicles, as well as the skill and experience of the individual performing the work vary widely. It is not possible to anticipate all of the conceivable ways or conditions under which vehicles may be serviced, or to provide cautions as to all of the possible hazards that may result. Standard and accepted safety precautions and equipment should be used when handling toxic or flammable fluids, and safety goggles or other protection should be used during cutting, grinding, chiseling, prying, or any other process that can cause material removal or projectiles.

Some procedures require the use of tools specially designed for a specific purpose. Before substituting another tool or procedure, you must be completely satisfied that neither your personal safety, nor the performance of the vehicle will be endangered.

PART NUMBERS

Part numbers listed in this book are not recommendations by Chilton for any product by brand name. They are references that can be used with interchange manuals and aftermarket supplier catalogs to locate each brand supplier's discrete part number.

NOTE: Although the information in this guide is based on industry sources and is as complete as possible at the time of publication, the possibility exists that the manufacturer made later changes which could not be included here. While striving for total accuracy, Chilton Book Company cannot assume responsibility for any errors, changes, or omissions that may occur in the compilation of this data.

Allis—Chalmers
INDEX

300 & 400 SERIES
- Front Axle 2
- Steering 3
- Clutch 4
- Brakes 4
- Manual Transmission 4
- Differential and Rear Axle 10
- Bevel Gear Box 10
- Hydrostatic Drive 11
- Variable Speed Pulley 12
- 400 Series Shuttle Drive 14
- Engine 14

MODELS 608 & 610
- Wheels 15
- Clutch and Brake 15
  - Drive Belt 15
  - Idler Pulley 15
  - Engine Pulleys 16
  - Clutch, Parking Brake, and Brake Rod 16
  - Brake Linings 17
- Power Take-Off 17
- Steering and Front Axle 18
- Transaxle 19
- Engine 23

MODELS 608LT & 611LT
- Wheels 23
- Sheet Metal 23
- Steering 23
- Clutch, Brake and Drive Pulley .27
- Transaxle 29
- Engine 31

MODELS 616 & 620
- Wheels, Bearings and Hubs 32
- Brakes 32
- Clutch and Driveshaft 33
- Front Axle and Linkage 34
- Steering 35
- Transaxle 36
  - Differential 36
  - Shift Assemblies, Bevel Gear, Bevel Pinion and Sliding Gears ..38
- Hydrostatic Drive 40
- Power Take-Off 44
- Engine 49

MODELS 710, 712, 716
- Wheels and Tires 49
- Clutch and Brake 51
- Front Axle and Steering 51
- Differential and Hub 55
  - Hubs 55
  - Differential 56
  - Axle 57
- Bevel Gear Drive 58
- Manual Transmission 60
- Shuttle Drive 67
- Hydrostatic Transmission 68
- Dual Range 6-Speed Pulley 70
- Engine 71

MODEL 720
- Wheel Bearings and Hubs 71
- Front Axle 72
- Steering 72
- Brakes 74
- Clutch and Driveshaft 76
- Transmission and Power Take-Off 77
  - Front PTO Electric Clutch ... 77
  - PTO Shaft 77
  - PTO Top Transmission Shaft .. 79
  - PTO Center Idler Gear 80
  - PTO Lower Shaft 80
- Hydrostatic Drive 82
  - Neutral Safety Switch 84
- Drop Housing 84
- Bevel Gear, Input Shaft and Gears 85
- Differential 87
- Engine 88

MODELS 808GT, 810GT
- Wheels 89
- Sheet Metal 89
- Steering and Front Axle 91
- Transmission and Differential . 92
- Engine 95

MODELS 910 6-SPEED, 912 HYDRO, 914 SHUTTLE, 916 HYDRO, 917 HYDRO
- Front Wheels and Axle 96
- Steering 99

- Clutch and Brake 101
- Electric Lift 103
- Power Take-Off 104
- Rear Wheels, Differential and Axle 106
  - Differential 108
  - Axle 109
- Bevel Gear Drive 110
- Transmission 112
- Shuttle Clutch 119
- Dual Range 6-Speed Unit 120
- Hydrostatic Transmission 121
  - Overhaul 121
  - Hydraulic Motor 124
- Engine 127

B SERIES
- Front Axle and Wheels 128
- Steering 128
- Brakes and Clutch 129
- Transaxle, except B-207, 208 .. 130
- Hydrostatic Transmission 135
- Transaxle, B-207, 208 139
- Two-Speed Pulley 142
- Engine, Driveshaft and Coupling 144

MODELS 5020 & 5030
- Front Axle 145
- Steering 145
- Clutch 145
- Torque Housing 146
- Transmission, 2-Wheel Drive ... 146
- Rear Axle, 2-Wheel Drive 150
- Transmission, 4-Wheel Drive ... 152
- Front Drive Axle 154
  - Axle 154
  - Hub and Gear Case Cover 155
  - Gear Case and Axle 155
  - Differential Carrier 156
  - Drive Unit 158
  - PTO Extension 159
- Hydraulic Pump 159
- Engine 160
  - Cooling System 160
  - Overhaul 160
  - Crankcase 169
  - R&R 170

ALLIS-CHALMERS
300 and 400 Series

MODELS: 310, 310D, 312, 312D, 312H, 314, 314D, 314H, 410 3-SPEED, 410S, 414S, 416S, 416H

FRONT AXLE ASSEMBLY

REMOVAL AND INSTALLATION

300 Series

1. Axle main member is mounted in the main frame and pivots on the large bolt and sleeve bearing.
2. To remove axle assembly, support front of the tractor frame and disconnect the tie rods from the steering plate.
3. Remove the pivot bolt and roll the axle assembly away from the tractor.
4. Check the sleeve and sleeve bearings for wear or damage and replace when needed.
5. Assemble unit in the reverse order.

400 Series

1. Axle main member is mounted in the main frame and pivots on a pivot pin.
2. To remove axle assembly, support front of the tractor frame and disconnect the tie rod ball joint ends from the steering knuckle arms.
3. Remove the bolt and drive out the pivot pin.
4. Raise the front of the tractor and roll the axle assembly from the tractor.
5. Check the pivot pin and main frame for wear and replace where needed.
6. Assemble unit in the reverse order.

Front Wheel Bearings

ALL MODELS

Every 100 hours of operation or once a year the front wheel bearings should be removed and repacked with grease. Proceed as follows:

1. Block or jack up front of tractor so the wheels are off the ground.
2. Remove cap A by prying it off with screwdriver.
3. Use an Allen wrench to loosen setscrew in collar B.
4. Remove the set collar B, washer E, outer bearing C, wheel D.

If seal G remains in hub instead of staying on the bearing shaft remove it and remove inner bearing. Wash the bearing shaft, bearings, wheel housing and seal with a suitable solvent and wipe dry.

NOTE: It is extremely important that bearings, all other parts, and the grease to be packed with them be kept clean.

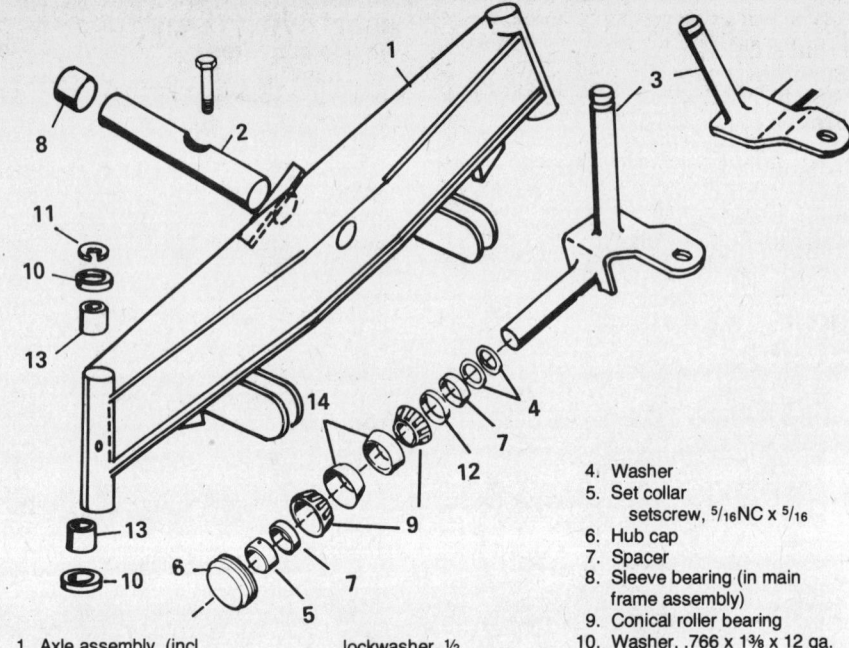

1. Axle assembly, (incl. bushings)
2. Axle pivot shaft capscrew, ½NC x 2¼, gr. 8
   lockwasher, ½
   nut, ½NC
3. Spindle assembly, lefthand
   Spindle assembly, righthand
4. Washer
5. Set collar
   setscrew, 5/16NC x 5/16
6. Hub cap
7. Spacer
8. Sleeve bearing (in main frame assembly)
9. Conical roller bearing
10. Washer, .766 x 1⅜ x 12 ga.
11. Retaining ring
12. Seal
13. Axle bushings
14. Cup bearing

Front axle assembly, 400 series

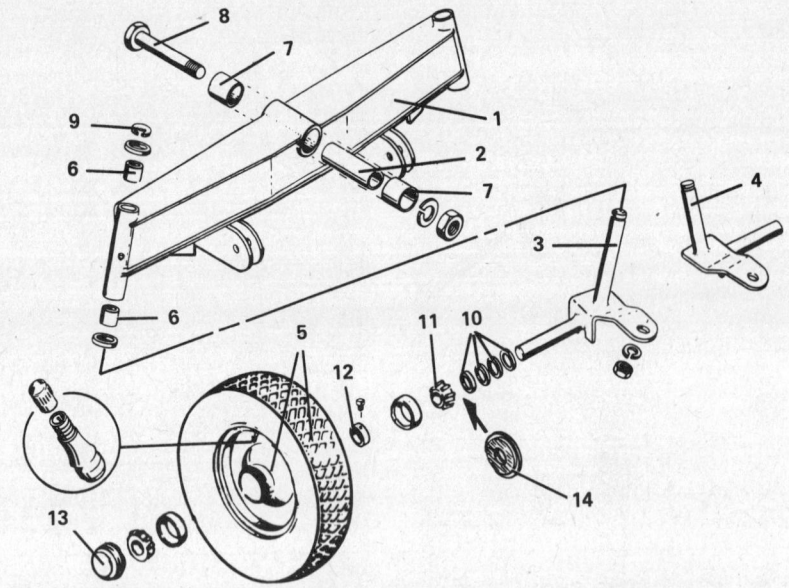

1. Axle assembly, front (incl. item 7)
   lube fitting (pkg. of 10)
2. Sleeve, axle
   Lockwasher, ¾
   Nut, ¾NC
3. Spindle assembly, lefthand
   washer, plain, 49/64 x 1⅜ x 23 ga.
4. Spindle assembly, righthand
   washer, plain, 49/64 x 1⅜ x 12 ga.
5. Wheel assembly
   Wheel assembly
   Bearing cup
   Valve assembly
   Tube
   Tube
6. Bushing
7. Sleeve bearing
8. Capscrew, ¾NC x 4½ (special thd. length)
9. Retaining ring
10. Washer
11. Bearing assembly, cone roller
12. Collar assembly, set (incl. setscrew)
    setscrew, 5/16NC x 5/16 sckt. hd.
13. Hub cap
14. Seal

Front Axle assembly, 300 series

Allis-Chalmers

Bearings should also be replaced in same position from which removed.

5. Using the palm of your hand force a good quality wheel bearing grease into the bearings. Place a coating of grease on seal where it turns in hub.

6. Replace the inner bearing and seal in hub. Make sure the five washers that were back of seal G are placed on the bearing shaft and then slide the wheel on the shaft.

7. Replace the outer bearing, washer and set collar. Spin the wheel slowly and press in on set collar to seat bearings. Be sure the seal on the inside of wheel is properly seated. Hold in on set collar and tighten Allen screw securely.

8. Replace cap A.

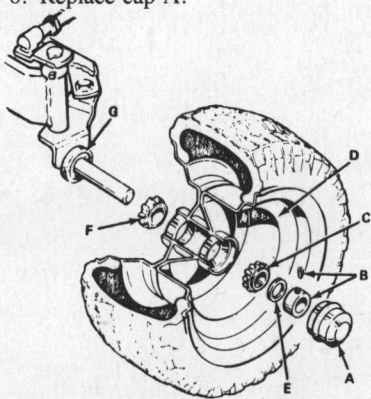

Front axle bearing assembly

STEERING

Steering Knuckles

REMOVAL AND INSTALLATION

300 Series

1. To remove the steering knuckles, support front of the tractor and remove the front wheels.

2. Disconnect the tie rod ends from the steering knuckles.

3. Remove the E-ring from the top of the steering knuckle and lower steering knuckles out of the axle main member.

4. Inspect the spindles and bushings for wear or damage and replace where needed.

5. Assemble unit in the reverse order.

400 Series

1. To remove the steering knuckles, support front of the tractor and remove the front wheels.

2. Disconnect the tie rods from the steering knuckles.

3. Remove the E-ring from the top of the steering knuckle and lower steering knuckle out of the axle main member.

4. Inspect the spindles and bushings for wear or damage and replace where needed.

5. Assemble unit in the reverse order.

Steering Gear

REMOVAL AND INSTALLATION

1. Remove steering wheel and loosen clamp bolts at instrument panel.

2. Disconnect drag link from steering gear arm and remove bolts securing steering gear housing to frame.

3. Raise tractor up and lower steering gear and shaft from underside of tractor.

DISASSEMBLY AND ASSEMBLY

1. Disassemble steering gear.

2. Check all parts for wear or damage and replace where needed.

3. To assemble, reverse the order in which it was disassembled.

4. Tighten plug (2) to 10-14 ft.lb. Make sure steering shaft turns freely after plug is installed.

5. Install lever and bolts assembly (1) with seal (12) and retainer (11).

6. Loosen locknut (10) and back off follower stud (9) two turns.

7. Adjust nuts (7) so that seal (12) is in full contact with housing (16), but do not compress seal.

8. Locate the steering lever in mid-position (half way between full left and full right turn).

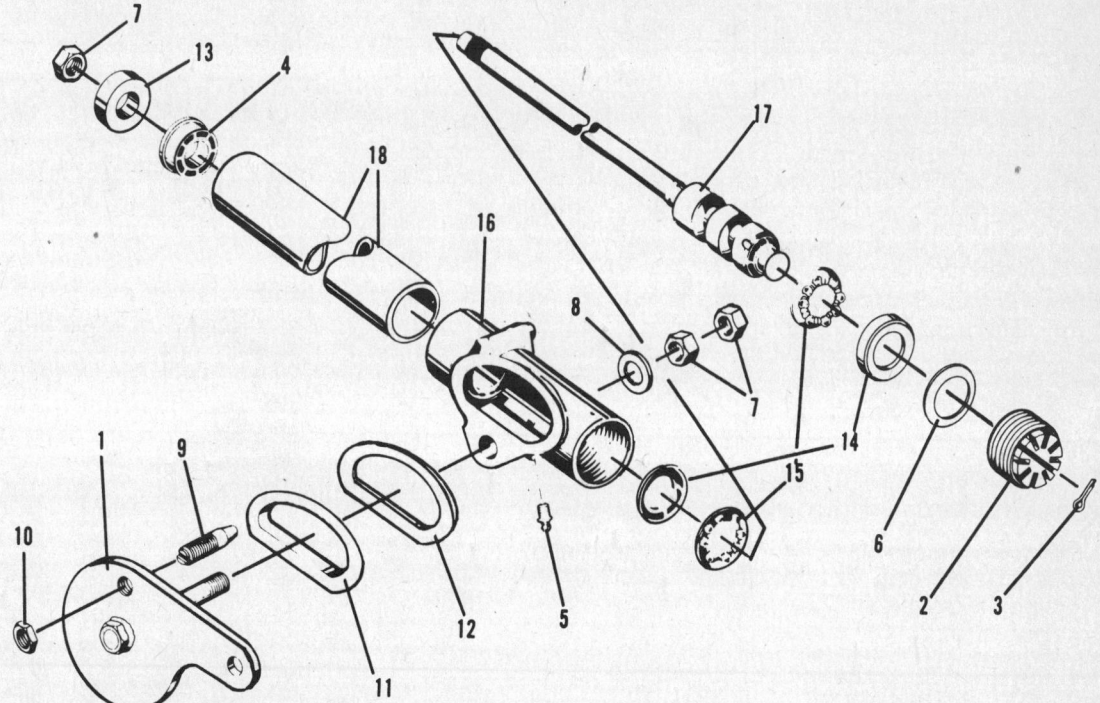

— Steering gear assembly (incl. item 1-18)
1. Lever and bolt assembly
2. Adjustment plug
3. Cotter pin, 1/8 x 1/2, pltd.
4. Bearing
5. Grease fitting
6. Belleville washer
7. Jam nut, 5/8-18, hex.
8. Washer
9. Stud
10. Adjusting nut
11. Seal retainer
12. Seal
13. Dust seal
14. Ball cup (incl. item 15)
15. Ball and retainer
16. Housing (not serviced)
17. Cam (not serviced)
18. Jacket tube (not serviced)

Steering gear components

Allis-Chalmers

9. Turn follower stud (9) in to obtain zero backlash and tighten locknut (10).

10. Steering shaft should turn from full right to full left turn position without binding.

11. Lubricate unit with approximately ¼ pound of No. 1 lithium base grease.

Steering Linkage

1. Inspect all ball joints for wear or damage and replace where needed.

2. Inspect sleeve bearing and replace if worn.

3. To adjust toe-in, disconnect tie rod ball joints from steering knuckles.

4. Loosen locknuts and turn tie rod ball joints in or out as required.

5. Tie rods should be adjusted so that there is approximately ⅛ in. toe-in.

CLUTCH

For clutch removal and installation procedures, see the applicable transmission section.

Adjustments

MANUAL TRANSMISSION MODELS

1. Place the variable speed control lever in the slow position.

2. Adjust the two locknuts on the clutch rod to give a spacing of ⅞ in. between the inside nut and the idler pulley pivot arm rod guide.

NOTE: The nut must not touch the rod guide when the control lever is in the FAST position.

HYDROSTATIC DRIVE MODELS

1. Place the control lever in NEUTRAL.

2. Adjust the two locknuts on the clutch rod to give a spacing of ⅝ in. between the inside nut and the idler pulley pivot arm rod guide.

BRAKES

For brake assembly removal and installation, see the applicable transmission section.

Adjustments

MANUAL TRANSMISSION MODELS

See Steps 4 and 6 in the Variable Speed Control Section.

HYDROSTATIC DRIVE MODELS

1. Place the control lever in NEUTRAL.

2. Depress the clutch-brake pedal completely and turn the brake band adjusting nuts until the brake band is tight against the brake drum.

TRANSMISSION

Manual Transmission

REMOVAL

1. Drain the oil from the transmission case.

2. Remove the ground cable from the battery.

3. Raise the tractor to remove the weight from the rear tires.

4. Place suitable blocks under the tractor frame to safely support the tractor.

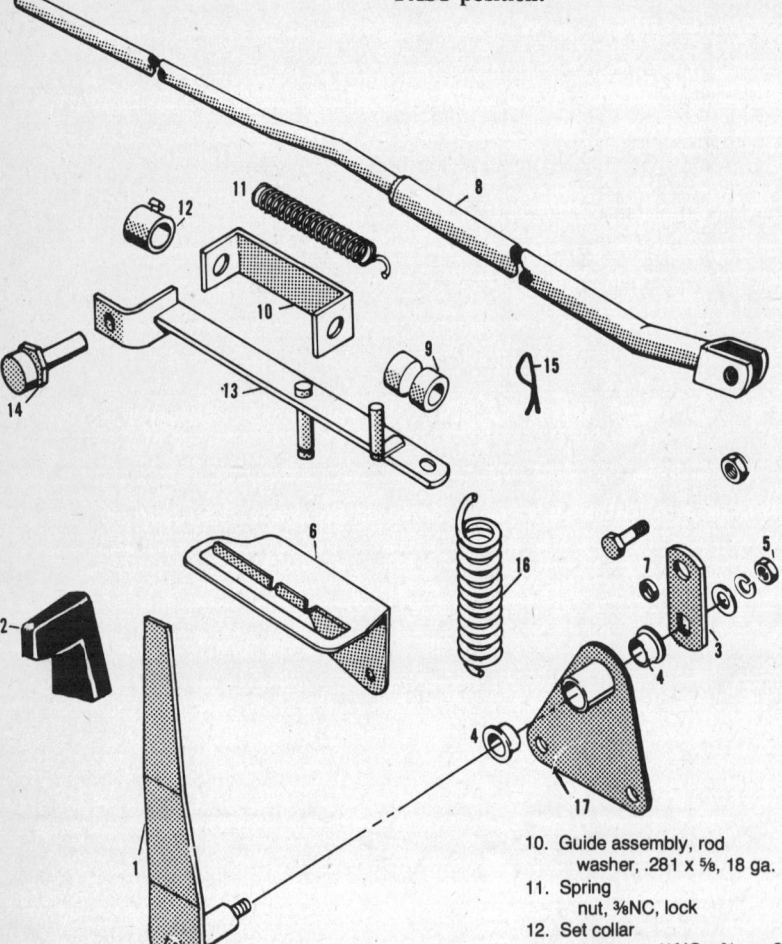

1. Control lever assembly screw, ¼NC x ¾ thd. forming
2. Knob
3. Link
4. Flanged bushing washer, ⅜ lockwasher
5. Nut, ⅜NC
6. Quadrant capscrew, ¼NC x ⅝ lockwasher, ¼ cage nut, ¼NC
7. Spacer capscrew, ⅜NC x 1 nut, ⅜NC, lock
8. Rod assembly
9. Brake, detent
10. Guide assembly, rod washer, .281 x ⅝, 18 ga.
11. Spring nut, ⅜NC, lock
12. Set collar setscrew, ¼NC x ⅜, gr. 8
13. Brake lever assembly
14. Brake pad assembly nut, 5/16NC
15. Pin, 3/32 x ¾, cotter
16. Spring washer, 5/16
17. Support bracket assembly setscrew, 5/16NC x ½ capscrew, 5/16NC x ¾ lockwasher, 5/16 washer, 5/16 nut, 5/16

Shuttle clutch system

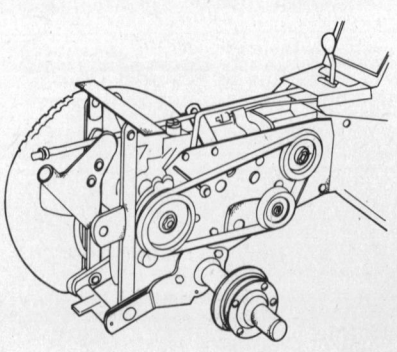

Transmission

Allis-Chalmers

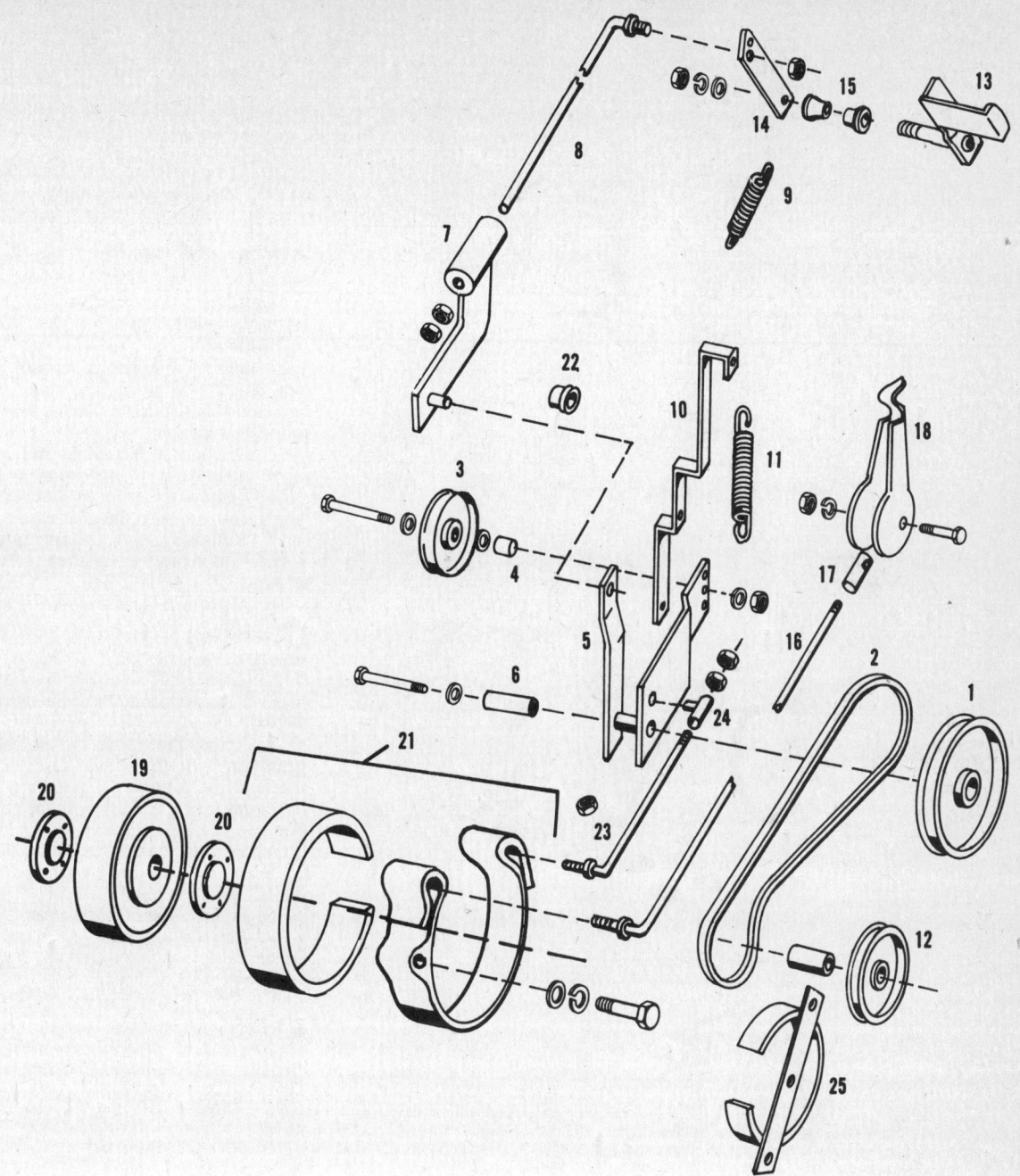

1. Pulley
2. Hydrostatic drive belt
3. Idler pulley
   washer, ⅜
4. Spacer
   capscrew, ⅜NC x 2¼
   lockwasher, ⅜
   nut, ⅜NC
   nut, ⅜NC-lock
5. Lever assembly
   washer, .781 x 1.25 x 7 ga.
   nut, ¾NF-jam lock
6. Spacer
   washer, ⁷/₁₆
   capscrew, ⁷/₁₆NC x 3¾,
7. Rod guide assembly
   nut, ⁷/₁₆NC
   nut, ⁵/₁₆NC-lock
8. Clutch and brake rod
   nut, ⁵/₁₆NC
9. Spring
10. Spring bracket
11. Extension spring
12. Driven pulley
13. Foot pedal assembly
14. Foot pedal arm
    washer, ⅜
    lockwasher, ⅜
    nut, ⅜NC
15. Flanged bushing
16. Parking brake rod
17. Rod end
18. Parking brake lever
    nut, ¼NC-lock
19. Drum, brake
20. Special washer
    washer, ⁷/₁₆
    nut, ⁷/₁₆NC-lock
21. Brake band assembly (incl. lining)
22. Bearing, flanged
23. Brake rod assembly
    nut, ⅜NC
    nut, ⅜NC-lock
24. Guide assembly
    nut, ⁵/₁₆NC-lock
25. Retainer assembly

Clutch and brake system with hydrostatic transmission

5

Allis-Chalmers

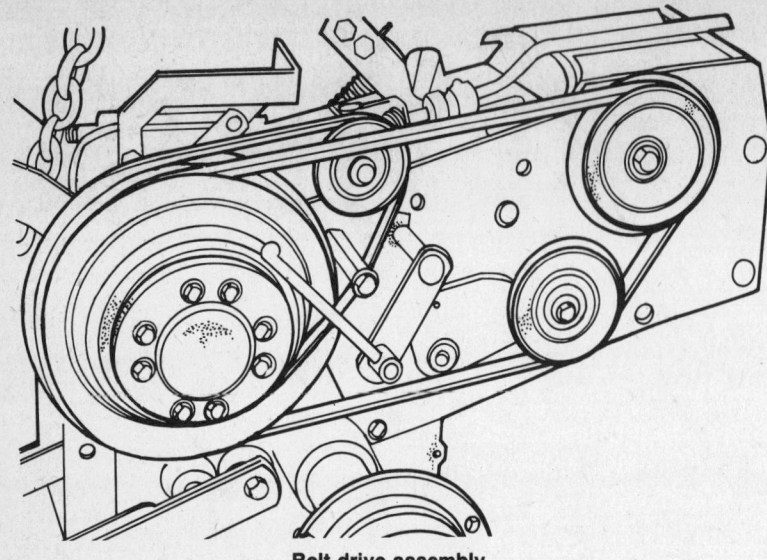
Belt drive assembly

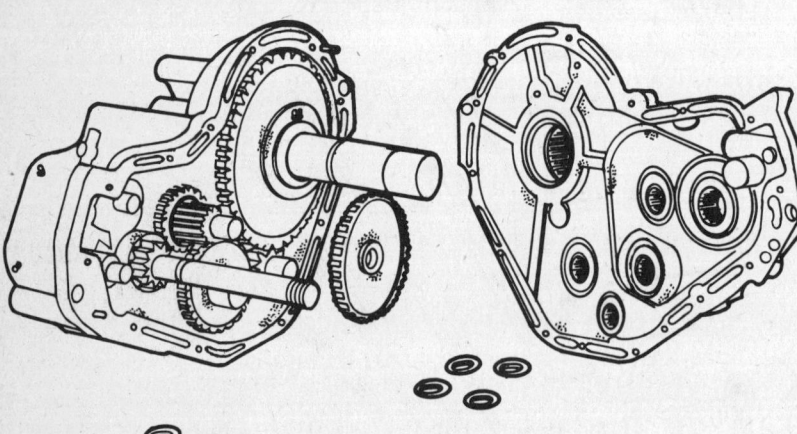
Transmission gear carrier

5. Place wedges between the front axle and tractor frame at the righthand and lefthand side for stability.

6. The seat and rear fenders may be removed for convenience.

7. Remove the rear wheels.

8. Remove the two nuts and spacer from the lift rod.

9. Remove 3 capscrews from the right and 3 capscrews from left side of the transmission to remove the drawbar and lift bracket assembly.

10. Refer to Differential and Axle Assembly Removal Section and remove.

11. Remove the drive belt, driven pulley or shuttle clutch assembly from the righthand side.

12. Support the gear case with a floor jack or suitable blocks and remove three capscrews that attach righthand rear frame to the gear case.

13. Disconnect the brake rod and remove the brake band and drum.

14. Remove 3 capscrews that attach left rear tractor frame to the gear case.

15. Remove transmission interlock switch. Disconnect the gear shift lever extension.

16. Move the gear case to a suitable work area.

DISASSEMBLY

1. Remove paint, rust and burrs from the axle tube.

2. Remove the capscrews attaching the gear case cover to the gear case.

3. Drive the two roll pins through the cover until they are flush with the gear case flange.

4. Remove the cover.

5. Remove the thrust washers from the shafts.

6. Remove the 50 tooth gear.

7. Remove axle tube, gear and thrust washer.

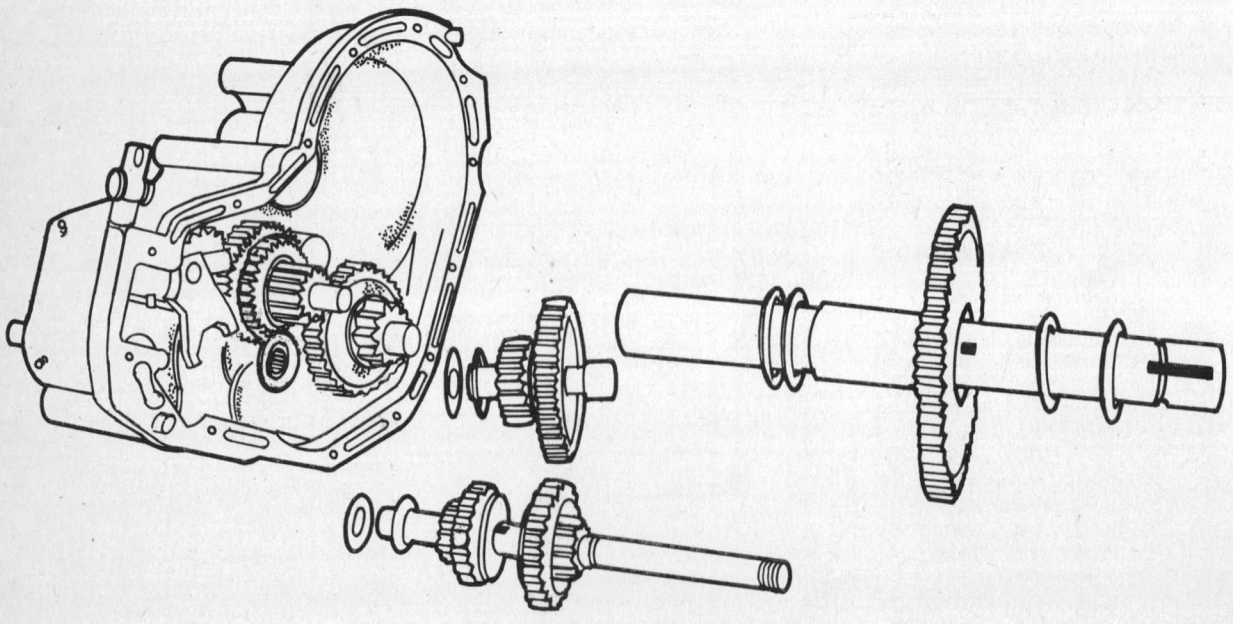
Transmission gear shafts

8. Remove the 2nd intermediate shaft, gears and spacer.

9. Remove the 3rd intermediate shaft gears and thrust washers.

10. Position the I-III shifter fork in neutral. The lower edge of the shifter stem slot will be even with the end of the roll pin.

11. Move the II-IV or II-Reverse fork out past the detent until the edge of the shifter is even with the short shifter shaft.

12. Remove the input shaft, retaining rings, gears and thrust washers.

13. It will be necessary to raise the brake drum shaft and gears slightly, to remove the input shaft.

14. The input shaft cam with gears can be removed with the shifter forks if the Allen screws at the top of the gear case are removed to allow shifter shaft removal.

15. Remove the brake drum shaft and gears.

16. Clean the case and cover with a suitable solvent.

17. If the needle bearings in the case are not to be replaced do not wash dirt and chips into the bearings.

18. The bearings should not be removed unless replacement is necessary.

19. Remove the 3/8 capscrew and nut from the reverse bracket assembly and remove the bracket and reverse gear.

ASSEMBLY AND INSTALLATION

1. To install needle bearings always press against the bearing end that has the bearing number and manufacturers name. Otherwise the bearing may be damaged. Use a tool that will contact the outside area of the bearing end to prevent distortion. Use a press or puller; do not drive bearing in place.

2. The needle bearings are pressed in the case so that the inside edge of the bearing is 1/8 in. below the machined surface of the case or cover.

3. The brake drum shaft bearing is open at both ends and an oil seal is to be installed after the case is assembled.

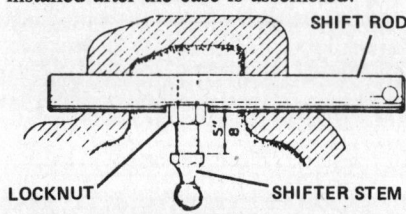

Locknut positioning

4. The input shaft bearing, second and third intermediate shaft bearings have one end closed and are installed with the closed end toward the outside.

5. The axle tube is pressed in the case until the end of the bearing is flush with the small diameter bore at the outside. This will provide space for the oil seal which will be installed after the case is assembled.

6. Check the shift stem guide (roll pin).

7. The pin should be pressed in the case until the end of the roll pin is $1^{39}/_{64}$ in. below the machined surface of the case.

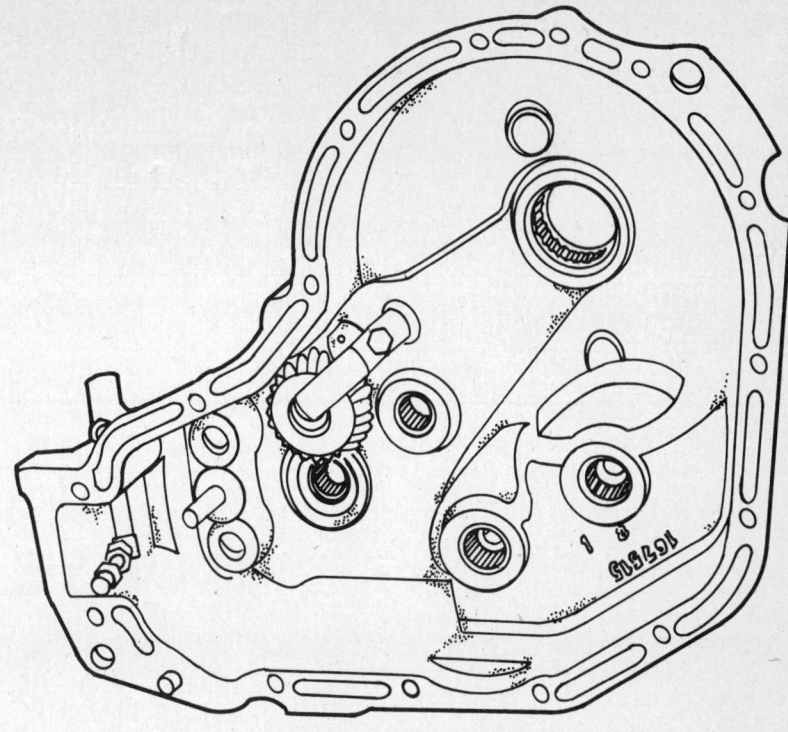

Transmission front case half

8. A steel ball is installed in the roll pin.

9. Lubricate the bearings with Lubriplate® before installing the shafts.

10. Press the input shaft ball bearing in the case and install retaining ring to hold bearing.

11. Press the brake drum shaft bearing, 2nd intermediate and 3rd intermediate shaft bearings in the case. The inner end of the bearing case should be 1/8 in. below the machined surface of the case.

12. Always press against the end of the bearing that has the bearing number and manufacturers identification. Use a suitable fixture so that the bearing case is not distorted during installation. Press the bearings in place.

13. Press the axle tube bearing (long) in the case. The outer end of this bearing should be flush with the small bore in the protrusion of the cover.

14. The cover is machined at the outside end of the protrusion for an oil seal which will be installed later.

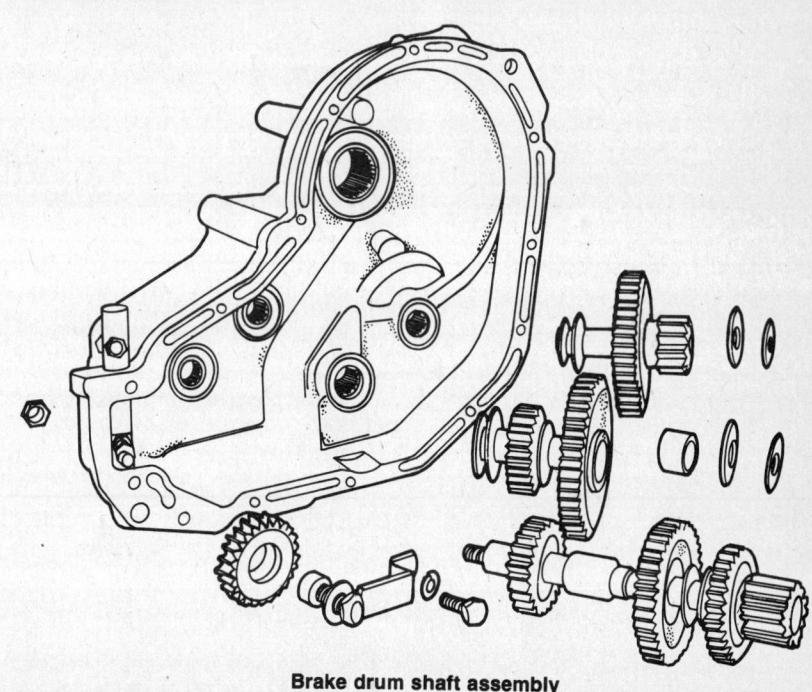

Brake drum shaft assembly

Allis-Chalmers

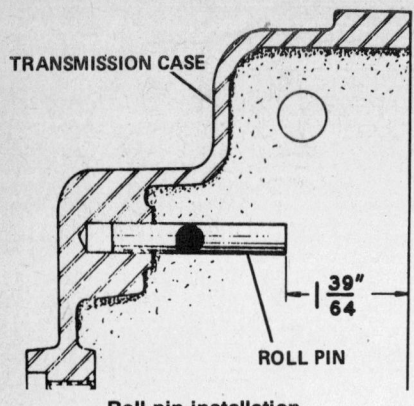

Roll pin installation

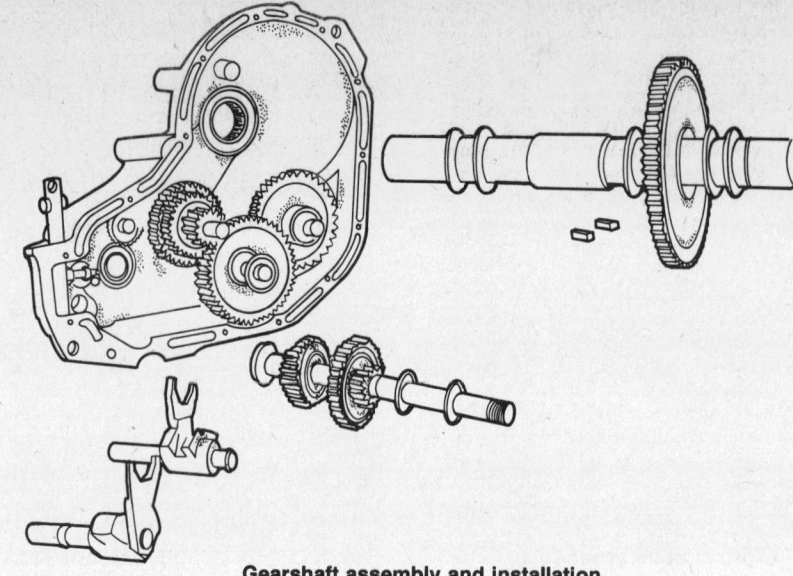

Gearshaft assembly and installation

15. If a new shift stem guide (roll pin) is installed in the cover press the roll pin in until the distance from the machined surface of the cover to the end of the roll pin is 1³/₁₆ in.

16. A steel ball is pressed inside the roll pin to lock it in place.

17. Lubricate the bearings before installing the shafts and bearings.

18. Install the shifter rod in the gear case and adjust shifter stem.

19. The distance from the round section of the shift rod to the shoulder of the shifter stem is ⅝ in. Use Loctite® on threads. The oil seals on the shift rod can be installed later.

20. The locknut has been eliminated on high serial number tractors. This shift stem is torqued to the shaft without a locknut.

21. Install on the reverse bracket one thrust washer, spacer and reverse pinion.

22. Install the pinion with the tooth chamfer and recess toward the bracket as shown.

23. Install the assembled pinion and bracket in the gear case. The bracket is secured with a capscrew and the locknut is torqued on the ½ in. diameter bracket capscrew. Torque to 55 ft.lb.

24. This reverse pinion is not used in shuttle clutch tractors with 4-speed transmissions. Plugs are used to close the capscrew holes when the reverse pinion is not used.

25. Press the long key in the key way.

26. Install the square retaining ring in the groove.

27. On the brake drum end of the shaft install the spacer and 25 tooth reverse or IV speed pinion.

28. Install the pinion with the chamfered end of the teeth toward the retaining ring.

29. Install the II driven gear, 38 tooth, with the chamfered end of the teeth toward the retaining ring.

30. Install two spacer washers on the shaft.

31. Install the III driven gear, 30 tooth, on the shaft with the chamfered end of the teeth away from the spacer washers.

32. Install one spacer washer.

33. Install the 15 tooth pinion on the shaft.

34. Install the assembled shaft and gears in the case with the threaded section of the shaft through bearing as shown.

NOTE: There are no thrust washers used between the gear case and gear or the gear and cover.

3rd Intermediate Shaft Assembly

35. Press two Woodruff keys in the 10 tooth pinion hub and install the 47 tooth gear on the hub. Install the tru-arc retaining ring in the groove on the shaft.

36. Install the gear and pinion on the shaft with the 49 tooth gear toward the retaining ring.

37. Install one steel thrust washer on the shaft next to the pinion and one organic thrust washer with the oil grooves toward case.

38. Install one steel washer next to the retaining ring and one organic thrust washer with oil grooves toward case as shown.

39. Coat the thrust washers with a thick chassis lube to retain them in location for installation.

40. Install the assembled shaft and gears in the case.

2nd Intermediate Shaft Assembly

41. Press the key in the key way in the shaft.

42. Install the retaining ring (tru-arc) in the groove.

43. Install the 25 tooth gear, one thrust washer, 50 tooth gear, spacer, one thrust washer and organic washer, with oil grooves toward case.

44. Install one thrust washer against the retaining ring followed by one organic washer with the oil grooves toward bearing.

45. Coat the washers with thick chassis lube to retain them in proper location during installation.

46. Install the assembled shaft and gears in the case.

47. Install the long key in the key way in the input shaft.

48. Install one retaining ring (wire type) in the groove near the center of the shaft (pulley end).

49. Install the I-III sliding gear, 15-35 tooth, on the shaft with the 15 tooth section toward the pulley end. The shifter collar will be near the center of the shaft.

Four Speed Shuttle Clutch Transmission

50. Install II-IV 27 and 40 tooth gear, with the 27 tooth section and shifter collar toward the I-III gear.

Three Speed Transmissions

51. Install the II-Reverse sliding gear, 15-27 tooth, with the 27 tooth and shifter collar toward the I-III gear. Install the retaining ring (wire type) in the groove. Install one thin recessed washer on the shaft so that the recess will encase the retaining ring.

52. Install one organic washer with oil grooves toward the gear case.

53. On the pulley end of the shaft install the thick recessed washer, with the recess toward the retaining ring. Install one thrust washer as shown.

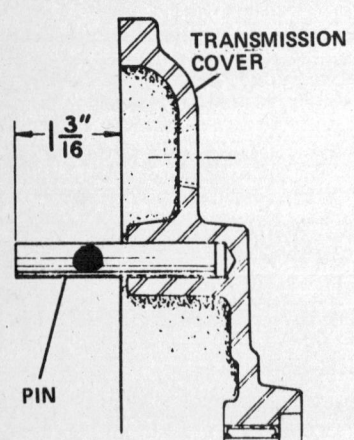

Shift stem guide installation

Allis-Chalmers

Shifter Forks and Shaft

54. Install the retaining ring in the groove at the end of the long shaft as shown.

55. Install the detent spring and ball in the I-III shifter fork.

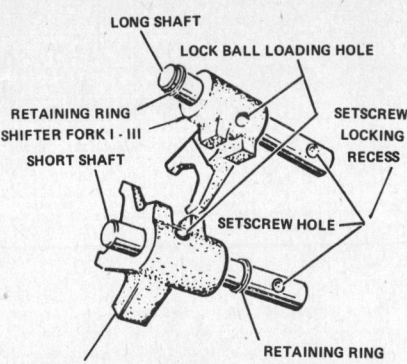

Shifter fork assembly

56. Install the shifter fork on the long shifter shaft with the long hub section toward the retaining ring. Use a small punch or bolt to force the ball and spring down in the loading hole so that the long shaft can be moved through the fork and trap detent ball.

57. Install the retaining ring on the short shaft as shown.

58. Insert the detent spring and ball in the loading hole in the II-Reverse shifter fork.

59. Install the shifter fork on the shaft with the long hub section toward the retaining ring. Use a small punch or long 5/16 bolt to force the ball and spring down so that the shaft can be pushed through the shifter fork.

60. The shifter shafts can be installed in the case as shown. Align the recesses in the end of the shaft with Allen setscrew holes in the case so that when the setscrews are torqued the end of the setscrew will lock the shifter shafts in the case.

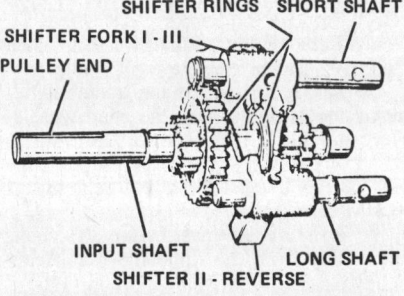

Shifter fork and input shaft installation

61. Install the input shaft in the case as shown.

62. It is more convenient to remove the organic washer and recessed washer from the shaft and position them on the machined area of the case at the input shaft bearing bore. Coat the washers with thick chassis lube to hold them in place. Position the recessed washer so that the recess will encase the retaining ring that is on the end of the input shaft.

63. It will be necessary to raise the brake drum shaft and gears slightly to allow the installation of the shaft and sliding gears and install the shifter forks in the collars of the gears.

64. Be sure the I-III shifter fork is in its neutral position. Move the I-Reverse fork out until the end of the fork is flush with the shaft, and shifter shaft stem is in the I-III shifter fork slot.

65. The shifter forks and input shaft assembly can be installed at the same time if they are assembled as shown.

66. Be sure and install the thrust washers as outlined previously and rotate the shifter shafts so that the Allen setscrews will be in the recesses in the shafts to lock the shafts in the case.

Axle Tube and Gears

67. If the axle tube bushings require replacement press the bushings in the tube until the end of the bushing is flush with the end of the tube.

68. Install the wire type retaining ring in the groove nearest the center of the tube.

69. Install the spacer with the recessed end toward the retaining ring.

70. Press two keys in the key ways near the center of the tube. Install one thrust washer with key slots over the keys.

71. Install the large driven gear on the shaft and over the keys. Install one thrust washer with key slots and one organic washer with oil grooves toward cover.

72. On the other end of the axle tube install one thrust washer and one organic washer with oil grooves toward the case.

73. Coat the thrust washers with chassis lube to hold washers on the shaft during installation.

74. Lubricate the bearing with Lubriplate® and install axle tube and gears as shown.

75. Install on the brake drum shaft the 50 tooth driven gear with the chamfered end of the teeth toward the 10 tooth pinion as shown.

76. Check the thrust washer installation. The oil grooves in the organic washers are toward the cover. The brake drum shaft does not have thrust washers.

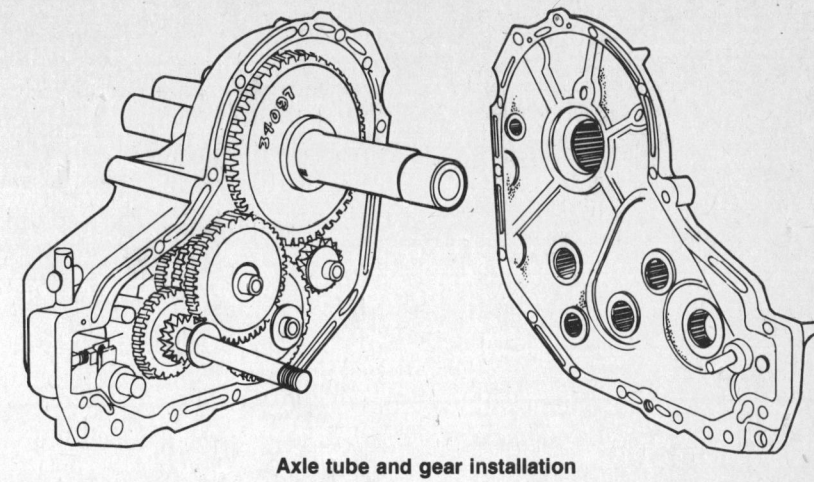

Axle tube and gear installation

77. Lubricate the bearing with Lubriplate®. Coat the gasket with a suitable sealer and install gasket and cover.

78. Install the capscrews finger tight. Drive the roll pins through the cover and gear case flange. Torque the capscrews. Install the oil seals.

79. Cover the key ways, retaining ring and threaded area with Scotch tape to prevent oil seal damage during installation. Lubricate the oil seal and shaft with Lubriplate® to aid installation. Use a suitable drive to install the seals to prevent distortion and seal damage. Install the seals with the lips or seal springs toward the oil sump. Install the assembled transmission in the tractor and fill with SAE 90 gear lube to the proper level.

80. Refer to the differential installation instructions and install axle assembly and differential.

81. Install lube fitting in axle tube and fill tube with chassis lube.

82. Replace driven pulley, clutch belt drawbar, lift assembly, wheels and all other items that were removed. Adjust brake and clutch.

ADJUSTMENTS

1. Adjust jam nuts on rear end of brake rod so that when foot clutch brake pedal is pushed firmly forward clutch arm will stop with its forward edge 5/8 in. to rear of rear corner of bevel gear box.

2. Loosen jam nut on parking brake rod and turn lever and rod end on parking brake rod so that brake is tight when parking brake lever is pulled up vertically as shown. Tighten jam nut against rod end.

3. With clutch pedal up in engaged position adjust jam nuts on clutch rod 1/2 in. away from clutch rod guide.

4. Rotate front belt guard around front guard bolt until guard has clearance of 3/32 to 1/8 in. to outside diameter of front pulley at the closest point. Be sure guard stays in proper location as front guard bolt is tightened.

5. Rotate the idler pulley belt retainer around pulley until rear edge of retainer is 7 in. from front side of axle shaft as shown. Securely tighten bolt through pulley hub.

Allis-Chalmers

6. Adjust rear pulley belt retainer to have ¼ in. clearance to top of belt.

7. The transmission safety interlock switch may be adjusted up or down by repositioning the large flat nuts that hold it in its mounting bracket. It should be just low enough that when switch button is resting centered on carriage bolt head as shown the switch is closed permitting the starter to operate. The switch must not be closed when the gear shift lever is engaged in any of the four gear positions.

DIFFERENTIAL AND REAR AXLE ASSEMBLY

REMOVAL AND INSTALLATION

1. Raise the tractor to remove the weight from the rear wheels and use jack stands to support the tractor safely.
2. Remove the rear wheels.

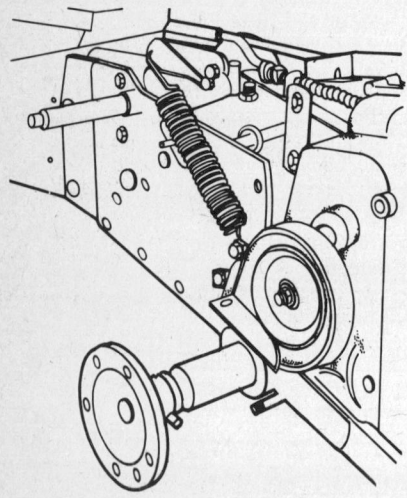

Rear axle shaft

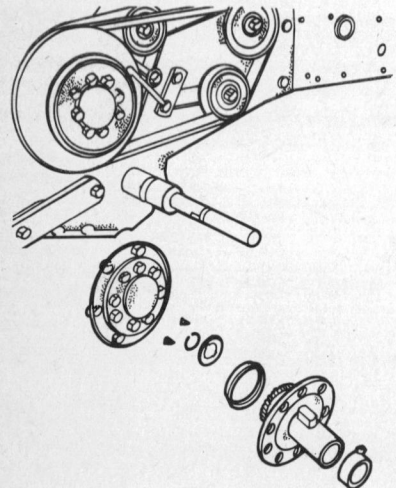

Axle shaft exploded view

Differential components

3. Remove the paint from the lefthand end of the axle shaft assembly. Loosen the lefthand retaining collar setscrew and move the axle toward the right as far as possible.
4. Remove the righthand retaining collar, righthand hub, seal, recessed washer, axle retaining ring and two keys from the axle assembly. The differential assembly can be removed from the axle tube.
5. Remove the burrs and smooth the axle assembly before removing the axle from the lefthand side.
6. If the tractors have the lefthand hub fastened to the axle with two setscrews remove hub, keys, set collar and thrust washer.
7. Install the differential retaining ring in the axle tube and two keys in the key ways.
8. Install one retaining collar and thrust washer on the axle shaft assembly lefthand end.
9. Install the axle assembly through the axle tube. Move the axle assembly to the right as far as possible.
10. Install the differential assembly on the axle tube. The differential hub will encase the two axle tube keys with the recess in the differential hub over the retaining ring. Push the differential gear to the left so that the thrust washers contact the axle tube. The two key ways will be exposed sufficient to permit installation of the two keys in the axle assembly.
11. Install the axle retaining ring.
12. Install the recessed washer on the axle with the recess toward the retaining ring.
13. Install the oil seal on the differential cover flange.
14. Install the righthand hub and retaining collar.
15. Move the hub on the axle shaft left as far as possible to position the recess in the washer over the retaining ring to remove all end play. Torque the collar setscrews to hold the hub in place.
16. Align the two keys in the axle with the key ways in the differential gear and move the axle to the left, until the recessed washer contacts the differential gear hub.
17. Remove all end play in the assembly, move the lefthand retaining collar and thrust washer against the axle tube and torque the setscrews.
18. Install the rear wheels.

BEVEL GEAR BOX

REMOVAL

1. Before removing the bevel gear box remove the battery ground cable.
2. Remove the rear fenders and seat.
3. Disconnect the drive shaft by removing the capscrews from the drive coupling. Remove the coupling and yoke.
4. Remove four capscrews that fasten the bevel gear box to the front frame.
5. Raise the rear of the tractor to remove the weight from the rear wheels. Place suitable blocks under the front frame ahead of the bevel gear box to support the rear of the tractor. Remove the lefthand rear wheel.
6. Remove the transmission drive belt, and pulley from the righthand end of the pulley shaft.
7. Remove the implement drive pulley or clutch from the lefthand end of the shaft if one has been installed for rear or center mounted implements.
8. Disconnect the brake spring and other items that are attached to the lefthand rear frame.
9. Remove the capscrews that attach the lefthand rear tractor frame to the transmission case, bevel gear box and front tractor frame.
10. Remove the lefthand frame.
11. Remove the capscrews that fasten

the bevel gear box to the righthand tractor frame and remove the gear box. Retain the bearing retaining shims that are installed between the righthand tractor frame and oil seal.

DISASSEMBLY

1. Remove all paint and rust from the pulley shaft and input shaft.
2. Drain the oil from the case.
3. Remove the capscrews and case cover.
4. All 300 series tractors and 400 series tractors prior to serial #25841001, 26041001, and 26141001 have a cover with street-L and pipe plug for oil level and filling. Tractors above serial #25841001 and 26141001 have the fill plug located on top of the gear case. The plug has a gauge attached that indicates fill level. The plug is vented and in the left rear portion of the case.
5. Move the pulley shaft left and remove the bevel pinion key and retaining ring as the shaft is removed from the left.
6. Remove driven bevel pinion.
7. Remove the input shaft bearing retaining capscrew and clamping plate.
8. Press the ball bearing and shaft from the gear case.
9. Remove the oil seals.
10. Use a suitable solvent to remove the oil and foreign material from the case.
11. Check the ball and needle bearings and replace if worn or damaged. Remove the needle bearings with a suitable drive if replacement is necessary. When replacing the bearings always use a suitable fixture and press against the end with the bearing manufacturer's name and number. Press or pull these bearings in place. The input shaft bearing is pressed into the case $1/16$ past the machined recess for the oil seal.
12. Lubricate the bearings with Lubriplate®.
13. If the input shaft ball bearing or bevel pinion require replacement remove the capscrew and locking washers from the end of the shaft to remove pinion or bearing.

ASSEMBLY

1. Install on the input shaft retaining ring, recessed washer with recessioner ring, bearing, key, bevel pinion, locking washers and capscrew. The locking washer has a flat section to fit the flat area of the shaft. Torque the capscrew and bend a section of the locking washer against one flat section of the capscrew head to lock the capscrew in place.
2. Install the assembled input shaft and pinion in the gear case. Install the locking plate and capscrew to retain the shaft in place.
3. Press the lefthand pulley shaft bearing in the case $1/16$ in. past the machined area for the oil seal.
4. Press the pulley shaft ball bearing in gear case as far as possible.
5. Start the pulley shaft through the lefthand needle bearing.

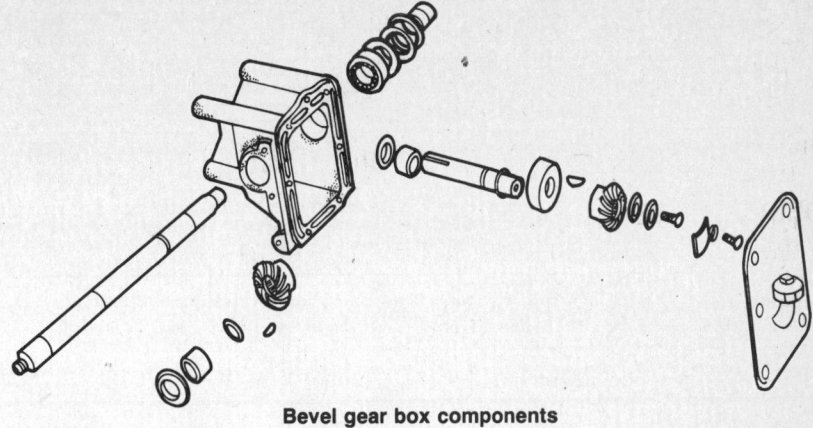
Bevel gear box components

6. Install the retaining ring on the shaft, key in the shaft and bevel gear on the shaft.
7. Press the shaft through until the retaining ring is in its groove with the gear against the ball bearing and retaining ring.
8. Cover the keyways and sharp areas of the shafts with cellophane tape. Lubricate the shaft and cellophane with Lubriplate® to prevent seal damage while installing seals.
9. Install the seals with the lip or seal spring toward the oil supply. Use a seal driver that will not bend the metal seal case. Press the righthand pulley shaft seal firmly against the ball bearing.

INSTALLATION

1. Before installing the assembled case in the tractor select the correct thickness of shims to hold the righthand pulley shaft bearing in place. The shims are installed when the bevel gear case is installed in the tractor frame. Install sufficient shims so that when they are against the oil seal a shim will extend .010 to .015 in. past the outside machined surface of the bevel gear case.
2. Apply a coat of gasket sealant to the cover edge of the gear case and install new gasket. Install the cover and torque the capscrews to hold the cover in place.
3. Use chassis lube or a heavy grease to keep the predetermined thickness shims while installing the case in the tractor.
4. Install the bevel gear case in the righthand tractor frame. Use the three righthand frame capscrews to hold the gear case in and shims in place. Do not torque these capscrews at this time.
5. Install the lefthand rear tractor frame and install the three bevel gear case capscrews but do not torque.
6. Install the four capscrews that fasten the bevel gear box to the front frame and torque.
7. Torque all the capscrews in the righthand and lefthand frames that were removed or loosened.
8. Fill the bevel gear box with SAE 90 oil to the proper level.
9. Replace the driveshaft coupling, key and yoke.
10. Install the pulleys, clutches, belts and other items that were removed to remove the rear left frame.
11. Adjust the clutch, brake etc.
12. Refer to the adjustment section for correct adjustment.
13. Replace the lefthand rear wheel and remove the blocks used to support the tractor.
14. Replace the seat, fenders and other items that were removed during disassembly.
15. Replace the battery ground cable.

HYDROSTATIC DRIVE

Both the 300 and 400 series use the Vickers T66 and TA6 Series 10 units. For Removal Overhaul and Installation see the B Series Tractor Section of this manual under HB-112 and 212 tractors. A gear reduction unit is used on all models.

Gear Reduction Unit

REMOVAL AND INSTALLATION

1. Raise and support the tractor with jack stands under the frame just ahead of the bevel gear housing.
2. Remove the rear seat deck and fender assembly.
3. Drain the reduction unit housing.
4. Remove the rear wheels, hubs and differential assembly and axle shaft.
5. Remove the bevel gear PTO belt pulley and disconnect the tension spring. Support the gear reduction housing and remove the left side plate.
6. Drain the transmission reservoir, disconnect the oil lines and control rod.
7. Unbolt and remove the reservoir, oil cooler, shroud and cooler fan.
8. Unbolt and remove the hydrostatic transmission and brake band.
9. Remove the capscrews securing the

Allis-Chalmers

gear reduction unit to the right side plate and lift the unit from the tractor.

10. Installation is the reverse of removal. Fill the reduction unit with SAE 90EP gear oil. Fill the transmission with Dexron® automatic transmission fluid.

ADJUSTMENTS

1. Loosen nut from rod end at front of parking brake rod. Turn parking brake handle and rod end until parking brake is fully tight when parking brake handle is pulled up against fender as shown.
2. With parking brake tight adjust jam nuts on end of foot brake rod to provide ½ in. clearance to rod guide.

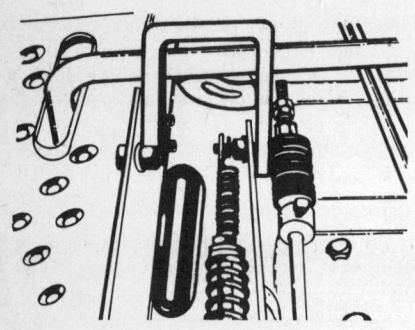
Clutch rod adjustment

3. Adjust jam nuts on clutch rod to have ⅝ in. clearance from clutch rod guide as seen.
4. Put the Hydrostatic Control Lever in the Neutral position. If the tractor "creeps" forward or backward, the turnbuckle on the control linkage should be adjusted. Loosen the locknuts at each end of the turnbuckle. Rotate the turnbuckle clockwise or counterclockwise, whichever is necessary to stop the creeping and retighten the locknuts using two wrenches so that the turnbuckle will not move. If the tractor still creeps, readjust by the same procedure.
5. Check that transmission interlock safety switch is properly located, that the switch button is depressed far enough to make contact only when the hydrostatic control is in neutral. Switch can be moved in its mounting bracket to align tip of control arm and switch button.

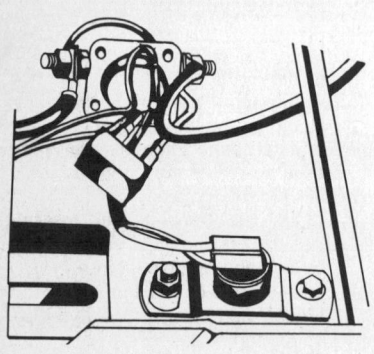
Safety interlock switch

VARIABLE SPEED PULLEY

OVERHAUL

1. Remove the drive belt.
2. Remove the locknut on the pulley shaft.
3. On the rear pulley it will be necessary to remove transmission fork assembly.
4. Remove pulley halves and bearing assembly.
5. Inspect and replace Teflon® bushing and bearing assembly if necessary.

NOTE: The bearing and bearing retainer are serviced assembled with Loctite®.

6. Clean all parts, lightly oil the Teflon® bushing and reassemble pulleys.
7. Do not exceed 50 lb. torque on the locknut when reassembling. Over-torquing will cause the pulley halves to bind.

Turnbuckle

REMOVAL AND INSTALLATION

1. Remove the ⅜ in. nut and bolt from the arm and rocker assembly, reference letter A, in the accompanying figure.
2. Remove the locknuts, reference letter B, and lift out bolt and rocker assembly.

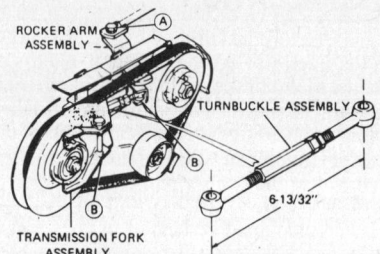
Turnbuckle adjustment

3. Remove the transmission fork assembly and turnbuckle as follows:
Adjust the turnbuckle so that the holes of the eyebolts are parallel. The distance between the holes is 6¹³⁄₃₂ in. Note that one end of the turnbuckle body is grooved to identify the righthand thread end. The grooved end should be placed forward when assembling in the tractor. Assemble both eyebolts with the same length of threads engaged in the turnbuckle.
4. Install the turnbuckle assembly.

NOTE: After tightening the locknuts, reference letter B, insure that there is free movement of the fork assemblies.

ADJUSTMENTS

The following adjustments are made with the variable speed control lever in the "Fast" position.
1. Check the locknut on the ball joint holding the rod assembly to the rocker arm.
2. Check for the proper clearance of ³⁄₁₆ in. between the belt and belt guard.

NOTE: Adjustments 3 through 6 are made with the variable speed control lever in the "Slow" position.

3. Check the proper clearance of 9¼ in. from the inside rear axle tube to the upper lip of the idler pulley belt stop. Also, check the ¹⁄₁₆-⅛ in. clearance between the belt and the belt stop.
4. Check for the proper clearance of ⅞ in. between the nuts on the clutch rod and the set collar. Check in the "Fast" position to see that the nuts do not touch the rod guide.
5. The wide belt in the large rear pulley should be approximately ⅛ in. below the top of the pulley.
6. Check the brake adjustment.

Belt Slippage Adjustments

If the belt is slipping check the following points:
1. See that the parking brake is fully disengaged.

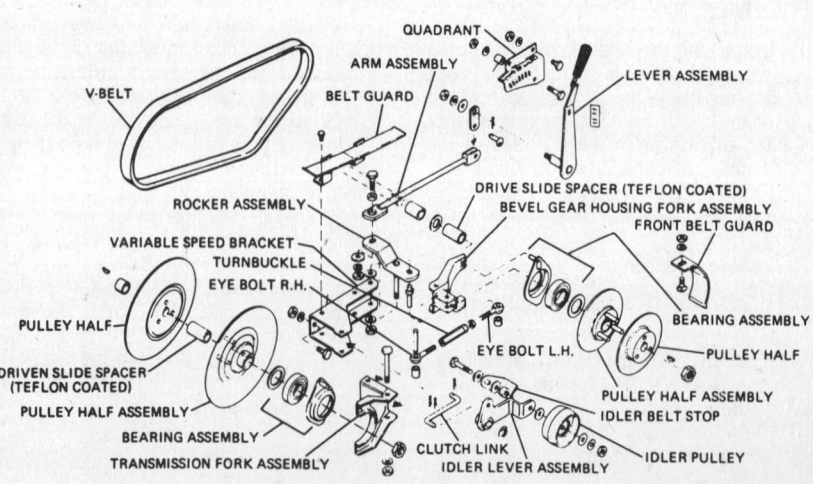
Variable speed pulley components

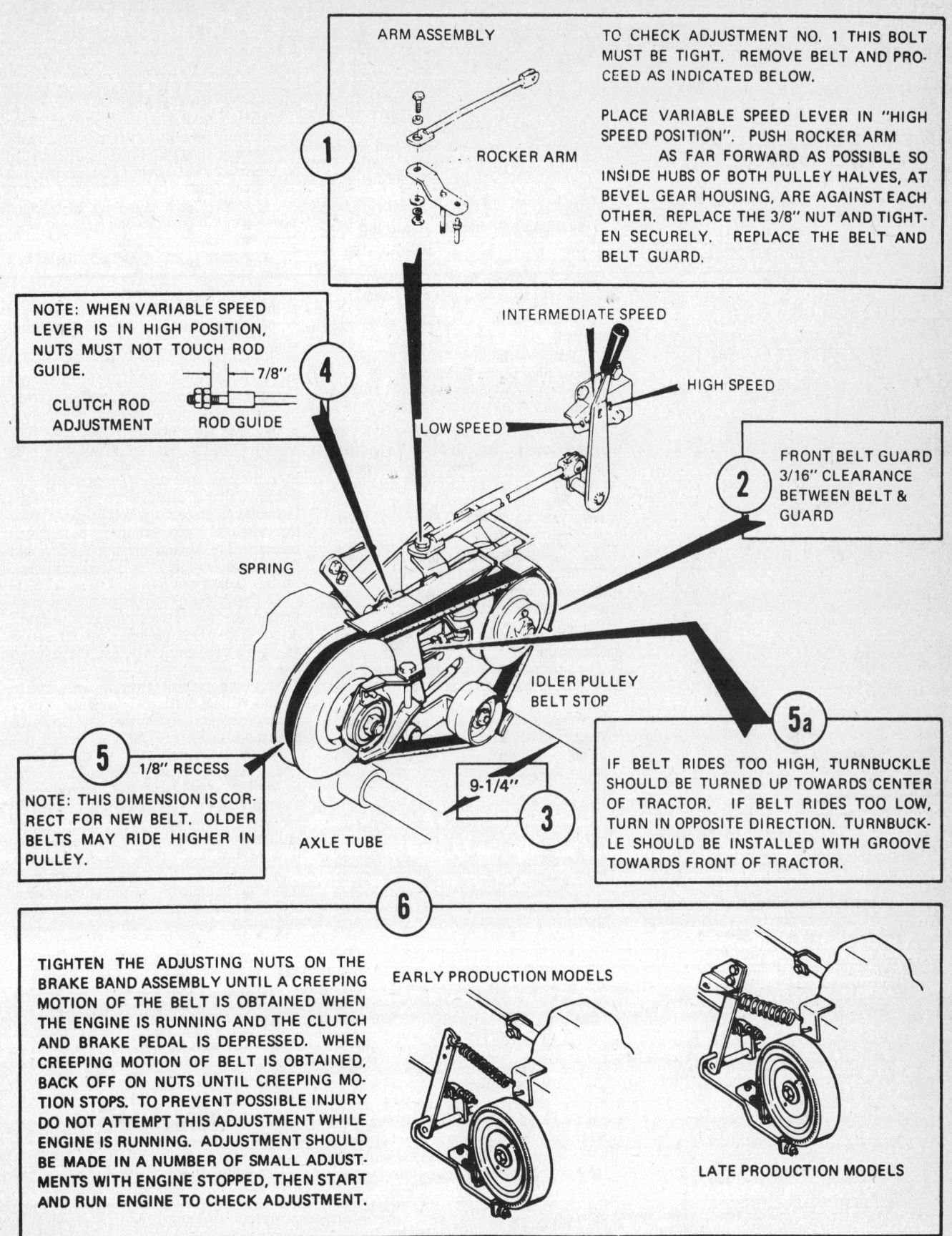

Variable speed pulley adjustments

Allis-Chalmers

2. Proper clearance of locknuts on clutch rod and rod guide.

3. Proper 1/8 in. clearance of the belt below the lip of the large pulley.

400 SERIES SHUTTLE DRIVE

Shuttle Clutch

REMOVAL AND INSTALLATION

1. Jack up the rear of the tractor so that the right rear wheel may be removed. Remove the wheel.

2. Raise the seat deck to allow ample work room.

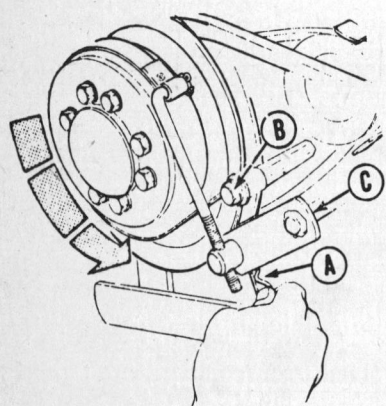

Brake band removal

3. Remove the spring clip A from the brake band assembly. Loosen the capscrew B two turns. The brake band may now be removed from the bracket C and rotated downwards.

4. Using a large screwdriver, pry the dust cap from the planetary.

5. Place a large screwdriver through the slots A in the two pulley sheaves to

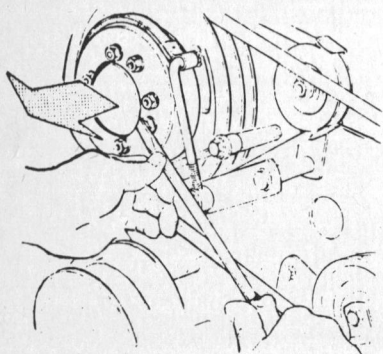

Prying dustcap from planetary

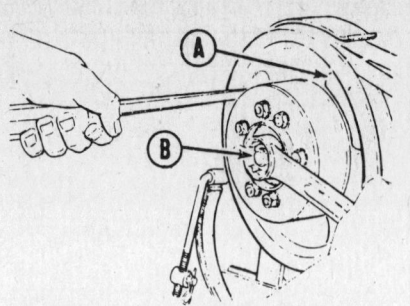

Preventing planetary from turning

prevent them from turning. Remove the locknut B securing the planetary assembly.

6. Remove the planetary assembly from shaft.

7. Remove the key A from the transmission pulley shaft. Be sure the key used is of the regular Woodruff design—**NOT** a high profile design.

8. Remove the thrust washer from the transmission pulley shaft. The pulley sheave may now be removed. This is accomplished by depressing the foot clutch and sliding the pulley sheave from the transmission input shaft. The sheave can then be removed from the belt.

9. Place the forward clutch belt down, away from the split sheave. Remove the split sheave which is keyed to the transmission shaft.

10. Disassembly and assembly of planetary unit.
   a. To disassemble the unit, remove the nuts from bolts (14) and the planetary shafts.
   b. Remove cover (6), seal ring (9), and bearing (10). The needle bearings (1), planetary pinions (2), sleeves (4), and planetary carrier (11) can now be removed.
   c. Clean all parts and replace any parts that show excessive wear or other damage.
   d. Use new seal rings, and reassemble by reversing the disassembly procedure.
   e. Tighten the nuts on capscrews (14) and planetary shafts (3) to a torque of 12-15 ft.lb.

11. The assembly process is the reverse of the disassembly.
   a. There are two sides to the planetary, one with eight hex nuts showing, and the other with four hex nuts and four capscrews showing. The side with the eight hex nuts faces outward from the tractor.
   b. The locknut securing the planetary is torqued to 75 ft.lb.
   c. The capscrew securing the brake band assembly is tightened to a torque of 30 ft.lb. When tightening the capscrew, use a wrench on the inside capscrew to prevent the assembly from turning.
   d. Pack the planetary unit with No. 2 lithium grease. Do not get any grease on the planetary brake band.
   e. Replace cover (5) and adjust shuttle drive linkage as outlined in "Adjustments" section.

ADJUSTMENTS

1. Adjust jam nuts on rear end of brake rod so that when foot clutch brake pedal is pushed firmly forward clutch arm will stop with its forward edge 5/8 in. to rear of rear corner of bevel gear box.

2. Loosen jam nut on parking brake rod and turn lever and rod end so that brake is tight when parking brake lever is pulled up vertically as shown. Tighten jam nut against rod end.

3. With clutch pedal up in engaged position adjust jam nuts on clutch rod 3/4 in. away from clutch rod guide.

4. Rotate front belt guard around belt guard bolt until guard has clearance of 3/32 to 1/8 in. to outside diameter of front pulley at the closest point. Be sure guard stays in proper location as belt guard bolt is tightened.

5. Rotate the idler pulley belt retainer around pulley until rear edge of retainer is 7 in. from front side of axle shaft as shown. Securely tighten bolts through pulley hub.

6. The transmission safety interlock switch may be adjusted up or down by repositioning the large flat nuts that hold it in its mounting bracket. It should be just low enough that when switch button is resting centered on carriage bolt head as shown the switch is closed permitting the starter to operate. The switch must not be closed when the gear shift lever is engaged in any of the 4 gear positions.

7. With control lever in neutral position, make sure center of notch in brake detent is centered on brake pin. If necessary loosen setscrew in detent, move it and retighten setscrew.

8. With control lever in full reverse position adjust brake pad assembly to have 1/8 in. clearance between pad and surface of pulley.

9. With control lever in neutral position loosen setscrew in set collar behind forward drive belt spring. Move rod guide assembly and spring forward until slack is taken out of forward drive belt. Tighten setscrew. Move control lever to the full forward position. There should now be 13/64 in. clearance between rear surface of set collar and rear leg of rod guide assembly. If necessary reset set collar to obtain the 13/64 in. dimension when control lever is fully forward.

10. With control lever in neutral position rotate pivot pin on threads on brake rod so that when pin is reinstalled in hole in brake lever all of the slack will be taken out of brake band. Fasten pivot pin in place in brake lever with lock pin.

ENGINE

The 300/400 series use Kohler K241, 301, 321, 341 engines. For removal and installation of the engine assembly, see the 710, 712, 716 tractor section. For overhaul of the engine, see the Kohler engine part of the Engine Unit Repair section of this book.

Allis-Chalmers

Models 608 and 610

WHEELS

Rear Wheel and Tire

REMOVAL AND INSTALLATION

1. Lock brake and support axle of tire being removed.
2. Pry off hub cap and button plug.
3. Remove cotter pin and pin and slide wheel and tire assembly from rear axle.
4. Clean all dirt, grease, oil, or other foreign material from tire and associated parts.
5. If tires are badly worn or have an excessive amount of breaks in them, they should be replaced. Tires which are worn unevenly or have sections of tread torn out will cause excessive vibration. Separate tire from tube if replacement required.
6. If tire is being replaced, also check condition of tube for leaks by holding under water and observing for bubbles. While under water note if valve leaks. Valve may be loose or valve seat worn. Tighten or replace as required.
7. Install tube in tire and tire assembly in wheel being careful not to bend valve stem in wheel.
8. Slide wheel and tire assembly on rear axle and secure with pin and cotter pin.
9. Lower tractor and evenly tap on button plug and hub cap. Remove blocks.

Front Wheel and Tire

REMOVAL AND INSTALLATION

1. Raise and block up front axle.
2. Pry off hub cap.
3. Loosen setscrew and slide set collar from front axle spindle.
4. Remove wheel assembly and two washers from spindle.

NOTE: Rear tires are 18 x 9.50, inflation pressure 6 to 8 psi. Front tires are 5.30 x 6.00, inflation pressure 12 to 15 psi. The front wheels of the tractor have sintered iron bearings lubricated by grease fittings.

5. See procedure for rear wheel.
6. If wheel bearing is nicked or damaged in any way, press it evenly from wheel. Using a rubber mallet or hammer and block of wood, press new bearing in place.
7. Install tube in tire and tire assembly in wheel, being careful not to bend valve stem in wheel.
8. Slide wheel and tire assembly and two washers on spindle.
9. Secure wheel and tire assembly on spindle with set collar and setscrew.
10. Lower tractor and tap on hub cap.

CLUTCH AND BRAKE

Description

The clutch and brake are controlled by one pedal located on the right front of the frame. The clutch is a soft action, touch-o-matic V-belt type. The brake is an external band type. A parking brake is standard equipment.

Drive Belt

REMOVAL AND INSTALLATION

1. Park tractor on level ground.
2. Put tractor in neutral and set the parking brake.
3. Remove mower if attached.
4. Depress the foot pedal to remove tension from the drive belt.
5. Loosen drive belt stops, and move them away from the engine pulley.
6. Remove drive belt.
7. Clean belt of all oil and grease and inspect for excessive wear or deterioration. Replace as required. An old belt will be inflexible or hard and should also be replaced.
8. Excessive belt wear may be caused by foreign matter in the pulley grooves. Clean the grooves thoroughly with a cloth moistened in alcohol.
9. Install drive belt insuring it is twisted in the same manner as removed. The part of the belt leaving the right side of the engine drive pulley goes to the bottom side of the transmission pulley.

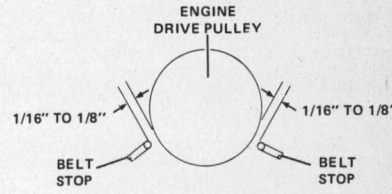

Drive belt stop adjustment

10. Adjust drive belt stop as shown.
11. Check belt for proper adjustment.

BELT ADJUSTMENTS

1. Belt stops are in proper adjustment when there is a clearance of 1/16 in. to 1/8 in. between belt and stop when the belt is engaged (foot pedal released). Loosen attaching bolts and readjust as required.
2. Adjust clutch rod as specified below.

Idler Pulley

REMOVAL AND INSTALLATION

1. Remove drive belt.
2. Remove capscrew, flatwasher, lockwasher, and hex nut securing idler pulley to bracket assembly. Remove idler pulley.
3. Clean pulley with cloth moistened with alcohol.
4. Inspect pulley groove for nicks, scratches, or breaks in the metal surface that may tear belt cords. Replace pulley if so damaged.
5. Secure idler pulley to bracket assembly with capscrew, flatwasher, lockwasher and hex nut.

Wheels and tires

ALL MODELS (FRONT WHEEL & TIRES)
NOTE: SEE PARTS BOOK FOR REPLACEMENT TIRES, TUBE, ETC.

Allis-Chalmers

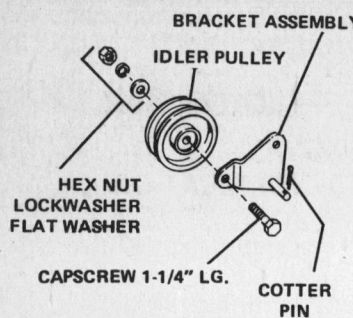

Idler pulley removal

Engine Pulleys

REMOVAL AND INSTALLATION

1. Remove drive belt.
2. Loosen setscrews and remove pulleys and keys from engine shaft. Separate key from engine shaft. Separate key from pulley key way.
3. Clean pulleys with cloth moistened with alcohol.
4. Inspect keys, setscrews, and thrust washers for damage. Replace as required.

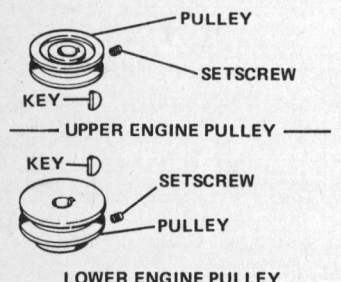

Engine pulley components

5. Inspect pulley groove for nicks, scratches, or breaks in the metal surface that may tear belt cords. Replace pulley if so damaged.
6. Be sure key ways are clean and edges not rounded. File edges as required.
7. Insert keys in pulleys completely and install pulleys on engine shaft. Tighten in position with setscrews.
8. Install drive belt.

Clutch, Parking Brake, and Brake Rod

REMOVAL AND INSTALLATION

NOTE: On some units the clutch and brake rod ends have locknuts instead of set collars.

1. Remove the clutch shaft assembly as follows:
   a. Remove two capscrews, lockwashers, hex nuts, and two E-rings securing the left foot rest to the tractor frame and clutch shaft assembly. Remove the foot rest.
   b. Remove the two locknuts holding two rod guide assemblies to the clutch shaft assembly. Remove the two rod guide assemblies with the attached clutch rod and brake rod from the clutch shaft assembly.
   c. Remove two setscrews, lockwashers and hex nut securing the clutch shaft assembly to the right side of the tractor frame. Remove clutch shaft assembly.
2. Remove clutch rod as follows:
   a. Remove two set collars, spring, and rod guide assembly.
   b. Remove cotter pin holding the clutch rod to the bracket assembly. Remove the clutch rod.
3. Remove the return spring from the idler pulley bracket assembly and lefthand transmission side plate.
4. Remove cotter pin holding the idler pulley bracket assembly to the righthand transmission side plate and remove the bracket assembly.
5. Remove brake rod as follows:
   a. Remove set collar, flat washer, spring, and rod guide assembly.
   b. Remove rod end, flat washers, spring, and hex nut.
   c. Remove capscrew, lockwasher, and hex nut securing brake rod to brake band assembly. These attaching parts hold both parking brake and brake rods. Remove brake rod.
6. Remove parking brake rod as follows:
   a. Remove capscrew, lockwasher, and hex nut securing parking brake lever to parking brake rod.
   b. Remaining end of parking brake rod already removed at brake rod removal. Remove parking brake rod.
7. Clean rods and attaching parts with an approved cleaning solvent.
8. Replace rods that are distorted and/or broken.
9. Install the rod or rods on the tractor in reverse order of disassembly.
   a. With the foot pedal in the released (up) position, secure the brake rod set collar against the tapered coil spring. Be sure the spring is not compressed. Depress the foot pedal all the way down. It should stop 2½ in. from the front edge of the foot rest. Adjust the set collar if necessary to obtain the 2½ in. distance.
   b. With foot pedal released (up) the clutch idler pulley is pressed firmly against the belt. Under this condition the distance between the clutch rod guide assembly should be ⅝ in.
   c. With the clutch rod set collar adjusted correctly on the clutch rod, depress the foot pedal all the way and

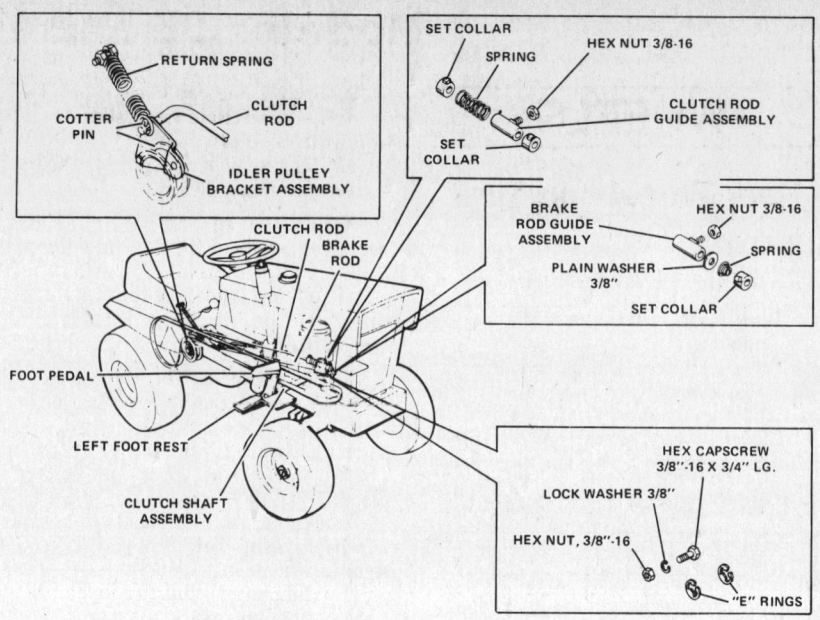

Clutch and brake rod removal

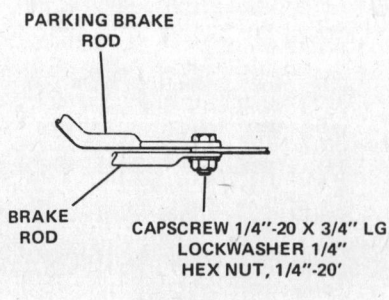

Brake rod

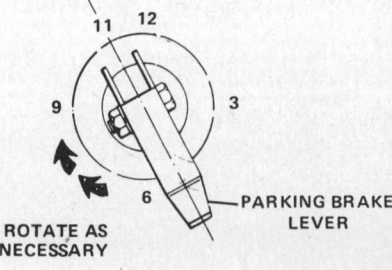

Parking brake lever rotation

Allis-Chalmers

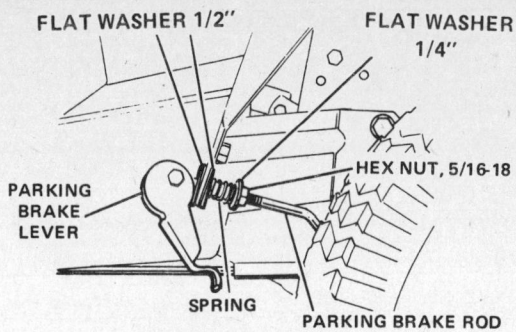
Brake rod installation

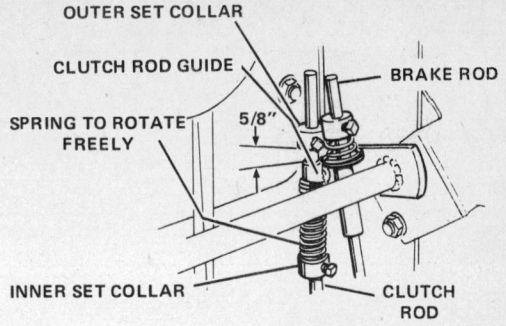
Clutch rod adjustment

adjust the inner set collar at the rear of the spring so that the clutch rod spring is just free to rotate on the rod.

10. To adjust the parking brake, loosen the nut located behind the seat deck on the parking brake rod. Rotate the parking brake lever clockwise to tighten and counter-clockwise to loosen the parking brake. The brake should be tight enough to hold in reverse, yet be easily placed in the "locked" position. Before securing the locking nut, be sure the parking brake lever is in the "11 o'clock" position when brake is engaged.

Brake Lining

REMOVAL AND INSTALLATION

1. Put tractor in gear and block the wheels, so that the tractor will not roll.
2. Remove capscrew and lockwasher holding brake band assembly to transmission cover.
3. Remove capscrew, lockwasher, hex nut securing brake rod to brake band assembly. This hardware holds both parking brake and brake rods to the brake band assembly.
4. Drill out three rivets from brake band and separate brake lining from brake band.
5. Clean brake band with an approved cleaning solvent.
6. Replace old brake lining with new.
7. Using new brake lining and rivets of the proper size, rivet lining to band.
8. Secure brake band and lining to brake rod or rods with capscrew, lockwasher, and hex nut.

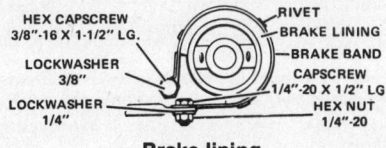
Brake lining

Brake drum-to-band alignment

9. Secure brake band assembly to transmission cover with capscrew and lockwasher.
10. Depress foot pedal and check contact of the brake band with the brake drum. The band must conform fully to the curvature of the drum. If it does not, loosen the capscrew as shown and align the band with the drum. Tighten capscrew.

POWER TAKE-OFF UNIT

Description

The power take-off is a direct belt drive from the engine controlled from the driver's position through an idler system at the front of the tractor. This system provides power to front and mid-mounted attachments.

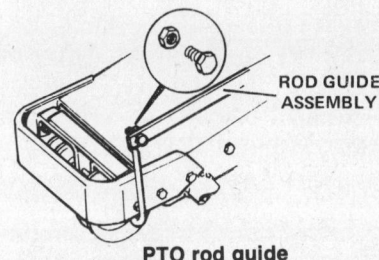
PTO rod guide

REMOVAL AND INSTALLATION

1. Remove PTO belt, if installed.
2. Remove the capscrew and locknut holding the rod guide assembly to the pulley clevis assembly.
3. Remove the two capscrews, spacers, flat washers, lockwashers and hex nuts attaching clevis to front bumper. Remove the pulley clevis assembly from the tractor frame.
4. Remove the 4¼ in. capscrew, flat washers, spacers, stop, pulleys, lockwashers, and hex nut from the pulley clevis assembly.
5. Loosen the setscrew holding the set collar and remove the PTO clutch rod and spring from the rod guide assembly.
6. a. Remove cotter pin holding clutch rod to lever. Remove clutch rod.

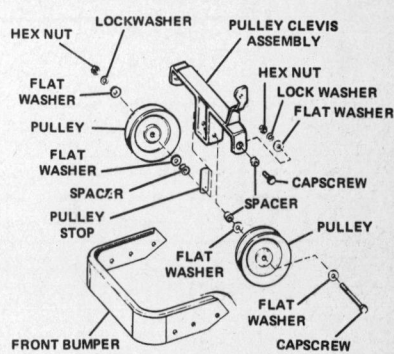
PTO pulley removal

b. Remove locknut, two flat washers, two nylon washers, stud, lockwasher, and hex nut attaching lever to tractor frame. Remove lever and ball knob.
7. On all models inspect all parts for damage and wear. Replace worn or damaged parts.
8. Clean all parts thoroughly and repaint as needed. Clean pulleys with cloth moistened with alcohol.
9. Install front PTO in reverse order of removal, being careful to position items as removed.
10. Apply a few drops of general lubricating oil at all pivot points.

ADJUSTMENTS

The PTO is normally in proper adjustment when there is ¼ in. clearance between the set collar and the end of the rod guide assembly with the PTO engaged. If more tension is needed disengage the PTO clutch lever. Loosen the collar setscrew and increase slightly the distance between the end of the rod guide assembly and the set collar. Engage the PTO and check tension. Only enough tension to drive the implement being used is necessary. Excessive tension may cause premature failure of belts and pulley bearings.

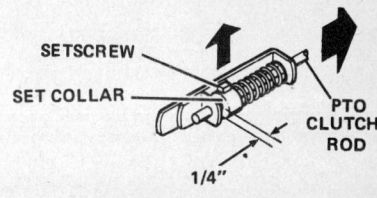
PTO clutch rod removal

17

Allis-Chalmers

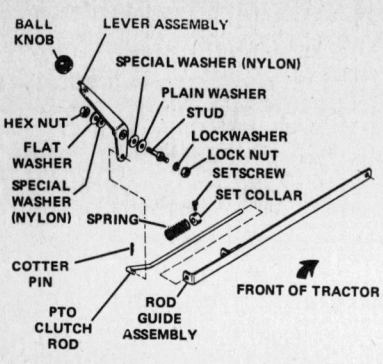

PTO clutch lever components

STEERING SYSTEM AND FRONT AXLE

Description

The steering system for both models consists of a gear and sector with a 3.3 to 1 ratio controlled by a steering wheel.

Steering Wheel, Shaft and Gear

REMOVAL AND DISASSEMBLY

1. Remove the capscrew and hex nut holding the steering wheel to the steering shaft assembly.
2. Remove hood (two thumbscrews).
3. Remove battery from dash assembly.
4. Drain and remove fuel tank from tractor.
5. Disconnect and mark for reassembly all electrical wiring under the dash assembly. Disconnect the choke and throttle control cables at the engine.
6. Remove the capscrews holding the dash assembly to the frame of the tractor. Remove the retaining ring from the bushing on the steering shaft assembly. Remove the dash assembly. Remove bushing and lower retaining ring from steering shaft.
7. Remove locknut (7), flat washer (8), and spacer (9) securing the fork and pinion assembly (10) to the frame. Remove shaft with fork and pinion from tractor.
8. Remove capscrew (2) and hex nut (13) securing universal joint pin (12) to steering shaft (1). Remove steering shaft.

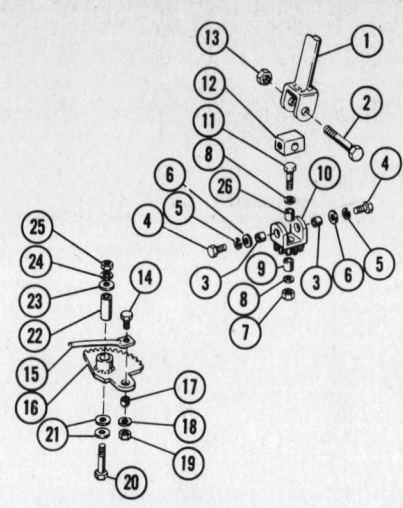

1. Steering shaft assembly
2. Hex capscrew, 5/16"-18 x 1½" lg.
3. Spacer
4. Hex capscrew, 3/8"-16 x ¾" lg.
5. Lock washer 3/8"
6. Flat washer 3/8"
7. Hex hut, jam, lock 3/8"-16
8. Flat washer
9. Spacer
10. Fork and pinion assembly
11. Hex capscrew, 3/8"-16 x 1½" lg.
12. Universal joint pin
13. Hex nut full lock, 5/16"-18
14. Hex capscrew, 3/8"-16 x 1" lg.
15. Drag link
16. Steering gear assembly
17. Spacer
18. Flat washer, 3/8"
19. Hex nut full lock, 3/8"-16
20. Hex capscrew, 3/8"-16 x 2½" lg.
21. Special washers
22. Spacer
23. Flat washer 3/8"
24. Lockwasher 3/8"
25. Hex nut, 3/8"-16
26. Bushing

Steering gear components

9. Remove two capscrews (4), lockwashers (5), flat washers (6), and spacers (3) attaching universal joint pin (12) to fork and pinion (10). Remove universal joint pin, capscrew (11), flat washer (8), and bushing (26) from fork and pinion assembly.
10. Remove locknut (19), flat washer (18), spacer (17) and capscrew (14), holding drag link (15) to steering gear (16).
11. Remove capscrew (20), two special washers (21), spacer (22), flat washer (23), lockwasher (24), and hex nut (25) attaching steering gear (16) to frame. Remove steering gear.

CLEANING, INSPECTION AND REPAIR

1. Clean all parts thoroughly using a suitable cleaning solvent to remove all grease and foreign matter.
2. Inspect teeth of steering gear and fork and pinion assembly. Small nicks and burrs

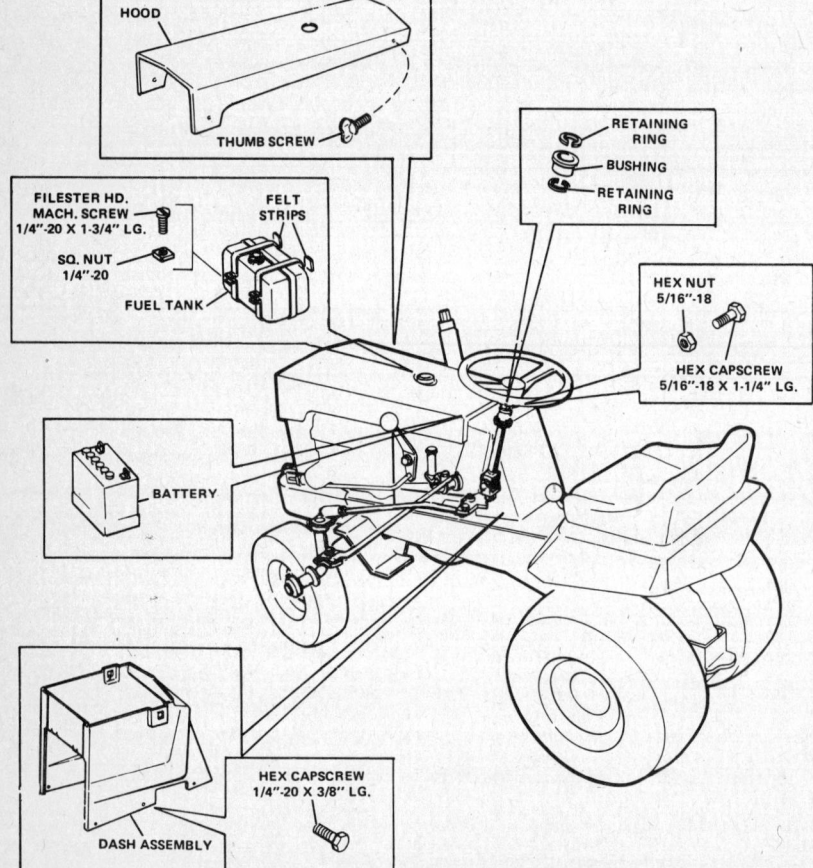

Steering wheel and shaft components

Allis-Chalmers

on gear teeth may be removed by using a slip stone or hone. If damage is severe, replace part.

3. Check bushing and spacers for damage and replace as required. Lubricate bushing with a light oil before assembly.

ASSEMBLY AND INSTALLATION

1. Be sure front wheels of tractor are straight ahead.

2. Secure steering gear (16) to frame with capscrew (20), two special washers (21), spacer (22), flat washer (23), lockwasher (24), and hex nut (25).

3. Attach drag link (15) to steering gear with locknut (19), flat washer (18), spacer (17), and capscrew (14).

4. Install capscrew (11), flat washer (8), and bushing (26) in fork and pinion assembly (10).

5. Install bushing (26), flat washer (8), capscrew (11) and universal joint pin (12) in fork and pinion assembly (10). Secure universal joint pin to fork and pinion with two capscrews (4), lockwashers (5), flat washers (6), and spacers (3).

6. Secure steering shaft (1) to universal joint pin (12) with capscrew (2) and hex nut (13).

7. Align fork and pinion assembly on tractor frame so that the holes securing the universal joint pin are positioned straight ahead. Secure fork and pinion to tractor frame with locknut (7), flat washer (8), and spacer (9) and previously installed bushing (26), flat washer (8), and capscrew (11).

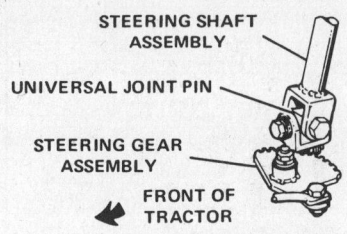

Steering gear alignment

8. Install bushing and lower retaining ring on steering shaft.

9. Apply a few drops of general lubricating oil on all pivot points of the steering system.

10. Install dash assembly over steering shaft and secure it with attaching capscrews. Install top retaining ring on steering shaft.

11. Connect choke and throttle cables and all electrical wiring under dash, noting wire locations as previously marked.

12. Fill the fuel tank.

13. Install battery, hood, and steering wheel.

Drag Link and Steering Lever Assembly

INSPECTION

Check drag link, ball joint, and steering lever for damage. If repair necessary, remove as follows:

REMOVAL AND INSTALLATION

1. Position front wheels straight forward.

2. Remove locknut, flatwasher, spacer and capscrew attaching drag link to steering gear.

3. Remove hex nut and lockwasher attaching ball joint to steering lever.

4. Remove drag link and ball joint from tractor.

5. Note location of jam nut on drag link and separate ball joint from drag link.

6. Loosen two setscrews on steering lever and pull lever straight up from spindle, being careful not to damage both keys in spindle key ways.

7. Remove two keys.

8. Clean all parts with a suitable cleaning solvent.

9. Replace drag link with a new one if it is broken or bent beyond normal straightening.

10. Inspect key ways in spindle and keys for damage. Replace bent or broken keys and attempt to file key ways straight if their corners are rounded. If key ways damaged severely, replace spindle per the procedure in this section.

11. Check setscrews for stripped threads or unusual wear. Replace as required.

12. Be sure grease fitting in axle is clean.

13. Install the steering lever and drag link in reverse order of removal noting the following:

   a. Keys must be inserted completely into key ways and setscrews must be secured tightly against them.

   b. Ball joint must be screwed on drag link and locked in position by the jam nut the same distance on drag link as when it was removed.

Tie Rod and Spindle

REMOVAL AND INSTALLATION

1. Raise tractor front wheels off ground and block up axle.

2. Remove hex nut and lockwasher attaching ball joint to steering lever. Remove ball joint from steering lever and move drag link away from tractor.

3. Loosen setscrews on steering arm and remove steering lever and keys from spindle.

4. Pry off hub caps from both front wheels.

5. Loosen set collar setscrews and slide set collar, washer, and wheel and tire assembly from both spindles. Remove remaining washer.

6. Remove both capscrews, flat washers, spacers, lockwashers, and hex nuts attaching tie rod to spindles. Remove tie rod from tractor.

7. Tap lefthand spindle assembly from front axle. Remove one washer and, using a screwdriver and hammer, tap two bushings from axle. Remaining washer should be removed from spindle.

8. Remove righthand spindle retaining ring and tap righthand spindle from front axle. Remove one washer and, using a screwdriver and hammer, tap two bushings from axle. Remaining washer should be removed from spindle.

9. Thoroughly clean all parts.

10. Inspect tie rod for unusual bending or damage. Replace as required.

11. Examine each of four bushings for nicks, scratches, or unusual damage. Replace as required.

12. Inspect spindles for wear, damage, or weak or broken weldments. Check key way in spindle to see if its sides are square with its bottom and not rounded. If possible, file key way square. Replace spindle damaged beyond immediate repair or if weldments are weak or broken.

13. Replace broken or bent keys.

14. Using a rubber mallet, tap four bushings in position in front axle.

15. Slide one washer on each key way end of spindle.

16. Insert righthand spindle (key way end) into its respective front axle and secure it in place with a washer and retaining ring.

17. Insert lefthand spindle (key way end) into front axle and slide washer over key way end of spindle.

18. Position keys in key ways. Install steering lever on lefthand spindle and secure it by tightening two setscrews on keys.

19. Secure drag link ball joint on steering lever with lockwasher and hex nut.

20. Install both wheel and tire assemblies on spindles, placing one washer on each side of wheel and securing the assembly with a set collar and setscrew. Be sure setscrew is tight and wheel turns freely.

21. Lower tractor and tap on hub caps.

22. Using a lithium base automotive type grease, lubricate both front wheels and spindles.

23. A few drops of engine oil should be applied to points indicated. Do not allow oil to get on belts and pulleys.

TRANSAXLE

Transmission Removal

1. Raise the rear of the tractor to lift the rear wheels from the ground and support just ahead of the frame swivel point. Remove the pins and spring clip holding wheels and tires to axles. Remove wheels. Remove drain plug and drain the transmission. Remove seat deck.

2. Unhook the idler pulley spring from the inside of the lefthand transmission side plate and allow the idler pulley to swing free of the drive belt. Remove drive belt from transmission pulley.

3. Remove transmission pulley by loosening setscrew on the hub of the transmission pulley and pull the pulley from the transmission shaft.

Allis-Chalmers

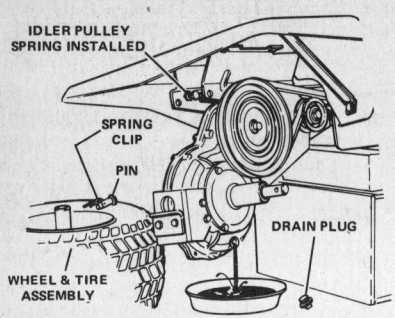

Draining transaxle

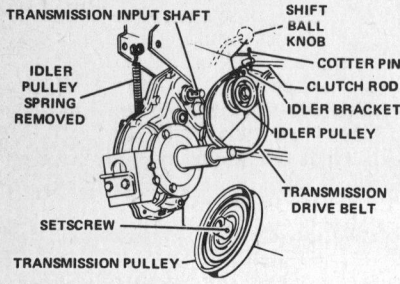

Idler spring removal

4. Remove the cotter pin from the clutch rod at the idler pulley bracket assembly. Remove the clutch rod from idler pulley bracket. Unscrew the shift knob from the shift lever.

5. Remove the electrical connection at the transmission safety interlock. Remove the support bracket and pivot support assembly of the safety interlock.

6. Remove the capscrew holding the brake band to the transmission case. On Model 608 and 610 the parking brake is removed by unscrewing the rod end from the parking brake rod. Remove the brake assembly from the transmission.

7. Loosen the two setscrews and remove the set collar holding the pivot rod. Slide the transmission to the rear until it is free of the tractor frame. Be sure not to lose the washer located between the frame and transmission.

8. Clean the outside of the case with a suitable solvent.

Transmission Disassembly

1. For ease of disassembly secure the transmission in a bench vise as illustrated. Grip the righthand side plate in the vise jaw. Remove the rear hitch assembly from

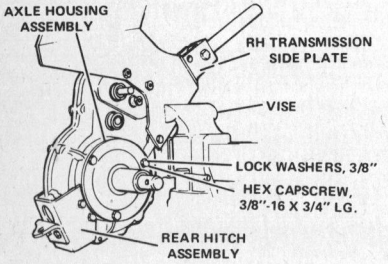

Transmission mounted in a bench vise

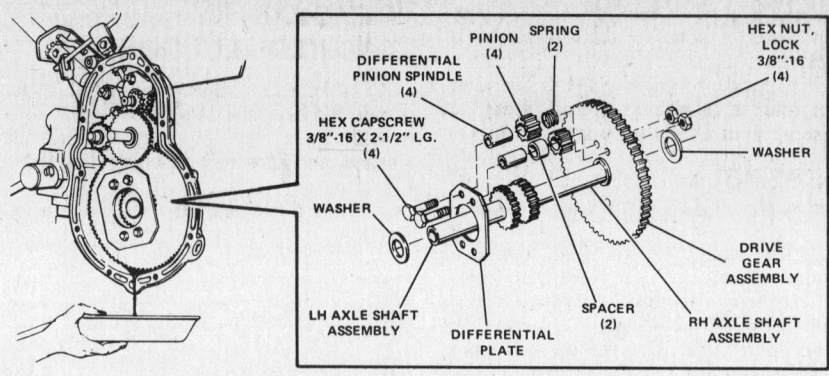

Right half axle shaft and driven gear removal

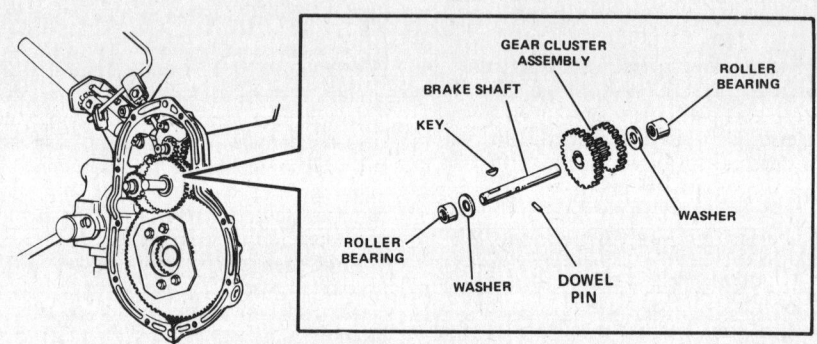

2nd/3rd gear assembly removal

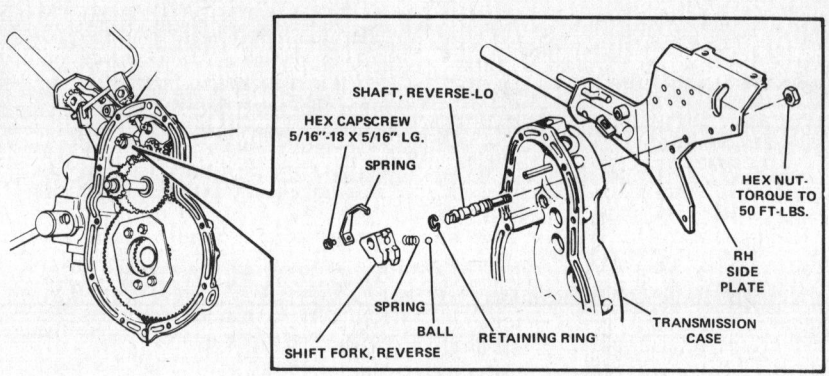

Reverse/low shaft fork removal

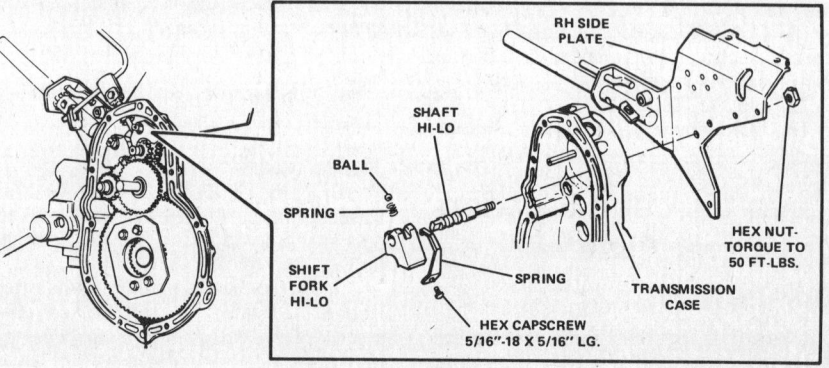

High/low shaft and fork removal

the transmission case by taking out the four capscrews and lockwashers.

2. Remove the brake drum by loosening the two setscrews and sliding the brake drum off the shaft and away from the transmission. Loosen and remove the four capscrews, lockwashers, and one hex nut holding the lefthand side plate in place and remove it from the transmission.

3. Using fine emery cloth, remove any burrs and paint from the shafts and axles, so that the bearings will not be damaged.

4. Drive out the two roll pins located at the top and bottom of the transmission. Be sure the pins are driven completely out of the transmission cover. Remove the eight capscrews and lockwashers securing the transmission case and separate cover from case. Place a pan beneath case to catch residue oil.

Differential Removal

1. Remove the right axle shaft and the driven gear assembly from the gear case.

2. DO NOT take the differential apart unless it is damaged.

2ND-3RD GEAR ASSEMBLY

1. Remove the 2nd-3rd gear assembly from the case, and the two washers. Tilt the gear assembly to clear the high-low pinion assembly.

2. Inspect the shaft and gear for wear. Replace any damaged parts.

LOW GEAR

1. Remove the low gear assembly from the case, and the two recessed washers.

2. Inspect the shaft and gear for wear. Replace any damaged parts.

REVERSE PINION

1. Remove the hex nut securing the reverse pinion shaft from the outside of the transmission case. Grasp the reverse pinion shaft and pull the reverse pinion shaft out of the transmission case.

2. Inspect the shaft and gear for wear. Replace any damaged parts.

REVERSE-LOW SHAFT AND FORK

1. Remove the hex nut securing the reverse-low shaft from outside the right side plate and transmission case. Remove the shaft with reverse shift fork from transmission case.

2. Inspect the shaft and fork assembly for wear. Replace any damaged parts.

HIGH-LOW SHAFT AND FORK

Remove the hex nut securing the high-low shaft from outside the right side plate and transmission case. This is done so that the high-low pinion, can be removed with the shaft and fork assembly.

HIGH-LOW PINION AND PTO SHAFT

Remove the shaft, fork, and PTO shaft from the case together.

AXLE HOUSING ASSEMBLY

1. Remove three capscrews and lockwashers securing the axle housing to the transmission case. Remove the axle housing assembly from transmission.

2. Remove the two oil seals and O-ring and, using a bearing puller, remove both roller bearings from the axle housing. Replace old oil seals with new. Refer to "Bearings and Oil Seals."

BEARINGS AND OIL SEALS

1. Remove and replace with new oil seals.

NOTE: Seals are to be replaced after shafts and axles are installed.

2. Remove all roller bearings from transmission cover, case, and, if not already removed, from axle housing.

3. Thoroughly clean by soaking in an approved solvent all bearings. Bearing must soak long enough to loosen all grease and dirt. After all visible dirt is removed, rinse bearing in a clean container of clean solvent, and immediately dip in oil or light grease.

4. Inspect bearings for broken or cracked races, dented seals or shields, cracked or broken separators, balls or rollers. If bearing has been overheated, it will be brownish blue or blue-black color. If bearing is found to have any of the above conditions, replace it with a new one.

5. Bearings are to be recessed 0.010 in. from the machined surface inside of the case and cover.

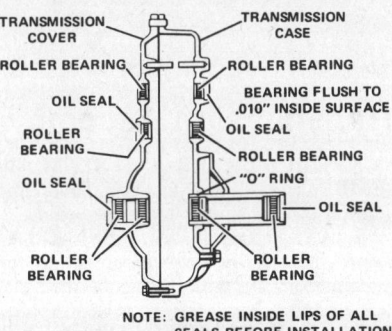

Bearing and oil seal positioning

SHIFT LEVER AND STEM ASSEMBLY

1. Remove capscrew and locknut securing the shift lever to the shift stem.

2. Shift knob already removed.

3. Remove the shift stem and the two half-ball bushings.

4. Inspect and replace worn or broken parts.

5. Install the two half-ball bushings and the shift stem in the transmission case.

6. Secure the shift lever to the shift stem with a capscrew and new locknut.

ROLL PINS

1. Check two roll pins for damage. One is located in the transmission case and the other in the transmission cover.

2. If replacement is necessary the proper installation is with the groove up and a distance of 0.808 in. (+0.000-0.010) from the face of the transmission cover to the end of the roll pin subassembly. The roll pin subassembly and the transmission case is properly installed with the groove up and a distance of 1.180 in. (+0.010-0.000) between the end of the roll pin subassembly and the face of the transmission case.

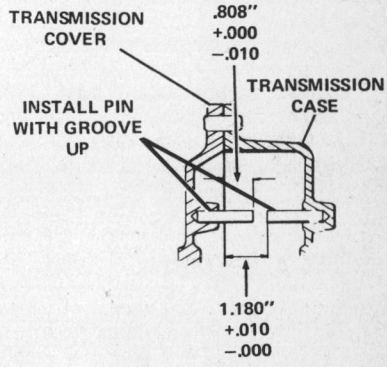

Roll pin positioning

CLEANING

Clean all transmission assemblies, cover, and case with an approved solvent. Dry thoroughly. Do not use high pressure air on bearings.

PIPE PLUGS

Check, clean, and replace with new, if necessary, the three pipe plugs in the transmission case.

Differential Assembly

1. Remove the four capscrews and locknuts securing the differential spindles, four differential pinions, two differential spacers, and two springs to the drive gear assembly.

2. Replace worn or broken parts.

3. Assemble the righthand axle assembly through the drive gear assembly and butt the gear of the lefthand axle assembly against the gear of the righthand axle assembly before mounting the differential pinions, spacers, springs and capscrews.

4. Assemble differential assembly with the righthand top and lefthand bottom capscrews holding two differential spindles, two differential pinions and two springs with the heads of the capscrews on the outside of the differential plate and the springs against the drive gear assembly. Assemble the lefthand top and righthand bottom capscrews through the differential plate, two spindles, two spacers against the

Allis-Chalmers

differential plate and two pinions. Secure all four capscrews with four new locknuts torqued to 20 ft.lb.

NOTE: Under torque will allow slippage and over torque will cause hard steering due to lack of differential action.

2ND—3RD GEAR ASSEMBLY

1. Remove the two washers.
2. The brake shaft and gear cluster assembly is best replaced as a complete assembly.
3. Replace broken or worn parts.
4. Slide gear cluster assembly over the roll pin and on the brake shaft.
5. Replace the two washers, one on each side of the gear cluster assembly.

LOW GEAR ASSEMBLY

1. Remove the two washers.
2. Remove the two retaining rings.
3. Replace broken or worn parts.
4. Secure the two retaining rings and washers to the gear and shaft assembly with the washers to the outside of the retaining rings.

HIGH-LOW PINION ASSEMBLY

1. Remove the two washers.
2. Remove the two retaining rings.
3. Slide the high-low pinion assembly over the key and off the shaft.
4. Remove the key.
5. Replace worn or broken parts. Check the key way carefully for burrs. Remove any burrs with a fine grit emery cloth. Replace broken or bent key.
6. Assemble the key and high-low pinion assembly on the pulley PTO shaft.

NOTE: The longer end of the shaft is facing outward (transmission cover side).

7. Secure the two retaining rings to the shaft and mount the two washers outboard of the retaining rings.

REVERSE PINION ASSEMBLY

1. Remove O-rings. Replace with new.
2. Remove roller bearing.
3. Remove reverse pinion.
4. Replace worn or broken parts.
5. Secure the roller bearing and reverse pinion to the reverse pinion shaft.

REVERSE-LOW SHAFT AND FORK ASSEMBLY

1. Remove the capscrew holding the spring to the shift fork. Remove the spring.
2. Slide reverse-low shifter fork off the reverse-low shift shaft.
3. Remove the ball and spring from the reverse-low shifter fork.
4. Clean and inspect all parts with special attention to the springs. Replace all worn or broken parts.

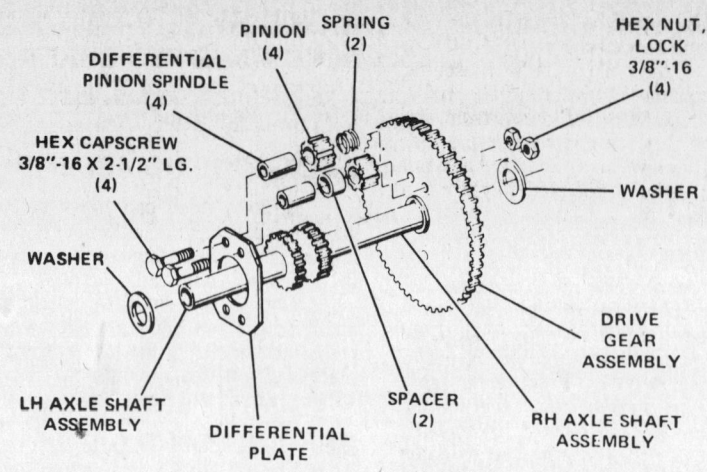

Differential components

5. Install the spring and the ball into the reverse-low shift fork. Slide the shift fork onto the reverse-low shifter shaft.
6. Install the spring and secure with capscrew.
7. Install the retaining ring on the reverse-low shifter shaft.

HIGH-LOW SHAFT AND FORK ASSEMBLY

1. Remove the capscrew holding the spring to the high-low shifter fork. Remove spring.
2. Slide the high-low shifter fork off the high-low shift shaft.
3. Clean and inspect all parts with special attention to the springs. Replace all worn or broken parts.
4. Install the spring, then install the ball in the high-low shift shaft.
5. Install spring and secure with capscrew.

Transmission Assembly

1. **SHIFT LEVER AND STEM ASSEMBLY:** Assemble the shift lever and stem on the transmission as instructed.
2. **AXLE HOUSING ASSEMBLY:** Secure the axle housing assembly to the transmission case with three capscrews and lockwashers.
3. **GEAR CLUSTER ASSEMBLY:** Move differential assembly towards you slightly and install gear cluster assembly, tilting it to clean other installed assemblies.
4. **HIGH-LOW SHAFT AND FORK ASSEMBLY:** Before installing the high-low shift fork be sure the flat spot on the shift shaft is turned so the spring rests in it. When securing the shaft to the transmission case it is very important that this shaft not be allowed to turn. Use a screwdriver on the end of the shaft to hold it from turning. Use a new hex nut torqued to 50 ft.lb. to secure the high-low shaft and fork assembly to the transmission case and righthand side plate.

5. **REVERSE-LOW SHAFT AND FORK ASSEMBLY:** Before installing the reverse-low shift fork be sure the flat spot on the shift shaft is turned so the spring rests in it. When securing the shaft to the transmission case it is very important that this shaft is not allowed to turn. Use a screwdriver on the end of the shaft to hold it from turning. Use a new hex nut torqued to 50 ft.lb. to secure the reverse-low shaft and fork assembly to the transmission case.

NOTE: After each fork is installed, check to see that there is clearance between it and the transmission case to allow easy movement in all positions. It may be necessary to grind away part of the case in some instances.

6. **REVERSE PINION ASSEMBLY:** Install the reverse pinion assembly in the transmission case and secure it with a hex nut, torqued to 50 ft.lb., on outside of case.
7. **HIGH-LOW PINION ASSEMBLY:** Mount the PTO pulley shaft through the transmission case carefully to prevent damage to bearing and oil seal. Tilt the high-low shifter fork out of the way for ease of removal.
8. **GEAR AND SHAFT ASSEMBLY:** Carefully insert gear and shaft assembly into transmission case bearing meshing its gears with the already installed gear assemblies. Be sure smaller gear faces out or away from case.
9. **DIFFERENTIAL ASSEMBLY:** Mount the differential assembly with the righthand axle assembly through the transmission case and axle housing assembly.
10. Carefully slide lefthand axle into transmission cover and install washer on axle.
11. Install a new gasket on transmission case. Drive two roll pins in transmission cover. Refer to accompanying illustration for roll pin location and positioning.
12. Secure the transmission cover on the case with eight capscrews and lockwashers.
13. Install the brake drum with key in place and secure with two setscrews.

14. Secure the lefthand side plate with four capscrews, lockwashers and one hex nut.
15. Install the rear hitch assembly on the transmission with four capscrews and lockwashers.

Transmission Installation

1. Coat the pivot rod with a good grade of automotive grease. Slide the washer on the pivot rod and install the transmission on the tractor frame. Secure the pivot rod to the frame with two setscrews and a set collar.
2. Remove transmission filler and level plug at rear of transmission. Fill transmission with one quart of SAE 90 transmission oil. Oil should be level with bottom of filler and level hole. Reinstall filler and level plug securely.
3. Secure the brake band assembly to the transmission cover with a capscrew and lockwasher.
4. Secure the parking brake to the seat deck. The ¼ in. flat washer is assembled against the hex nut on the parking brake rod. One large plain washer is placed on each side of the seat deck and the rod end is passed through both. The spring is placed over the rod end and the rod end is threaded onto the parking brake rod. (See Clutch and Brake Adjustments.)
5. Install the support bracket and pivot support assembly of the safety interlock system between the transmission side plates. Connect the electrical lead to the interlock switch.
6. Secure the idler pulley bracket assembly to the right side plate with a cotter pin.
7. Install shift knob.
8. Mount transmission pulley to the pulley shaft of the transmission with a key and setscrew. Transmission hub must be towards the transmission.
9. Secure the return spring with the capscrew and hex nuts to the lefthand side plate and the idler bracket assembly.
10. Install the drive belt. Push down on the idler pulley and slip the belt around the transmission pulley. Release the pressure on the idler pulley. (See Clutch and Brake Adjustments.)
11. Install the seat deck on the transmission side plates.
12. Before mounting wheel and tire assembly coat the axles with a good grade of automotive grease. Be sure transmission drain plugs are secure. Mount wheel and tire assembly and secure with pins and cotter pins.

Allis-Chalmers

ENGINE

For detailed engine overhaul, see the Briggs & Stratton section in the Engine Unit Repair Section of the book.

REMOVAL AND INSTALLATION

1. Remove the hood.
2. Disconnect the battery.
3. Disconnect the throttle and choke cables at the carburetor.
4. Tag and disconnect all wiring at the engine.
5. Two capscrews hold the grille to the frame. Remove them and lift off the grille. Some models have a screw holding the grille support to the engine.
6. Remove the regulator, pulley guard, generator belt, regulator bracket and generator.
7. Remove all belts and pulleys from the underside of the engine.
8. Remove all bolts and nuts securing the engine to the frame.
9. Lift the engine from the frame.
10. Installation is the reverse of removal. Clean and inspect all parts before installation.

Models 608LT and 611LT

WHEELS

Front Wheels

REMOVAL AND INSTALLATION

1. Raise and support the front end.
2. Remove the retaining ring and washer and remove the wheel.
3. Inspect the wheel bushings for damage or excessive wear. Replace if necessary.
4. Thoroughly grease the wheel inner hub surface using multipurpose chassis lube.
5. Slide the wheel onto the axle shaft and secure it with the flat washer and retaining ring.

Rear Wheels

REMOVAL AND INSTALLATION

1. Raise and support the rear of the tractor.
2. For right wheel removal, remove the retaining ring, wheel spacer, and washer. For the left side remove the retaining ring, washer, wheel and washer.
3. Remove the key from the axle and check it for damage or excessive wear. Replace it if necessary.
4. On the right side, install the washer, spacer, key, wheel and retaining ring.
5. On the left side install the washer, wheel, washer and retaining ring.

SHEET METAL

Hood

The hood on both the 608 and 611 is secured to the bumper by four carriage bolts. The hood may be replaced or aligned by removing or loosening these bolts.

Grille

608LT

The grille may be removed by removing the four screws from the hood. The top two screws are used for securing the lights, if so equipped. When tightening, torque these screws to 30 in.lb.

611LT

This grille is secured by six capscrews. For installation, torque these screws to 30 in.lb. for the bottom four, and 8 ft.lb. for the top two.

Dash Panel

Dash panel R&R is found under Fuel Tank in the Fuel System section.

STEERING

Upper Steering Assembly

REMOVAL AND INSTALLATION

1. Remove the upper dash, fuel tank and lower dash.
2. Remove the steering support and steering plate from the frame. NOTE: On 608LT serial #169026300001 and up and 611LT serial #169034500001 and up, there is no steering support.

Allis-Chalmers

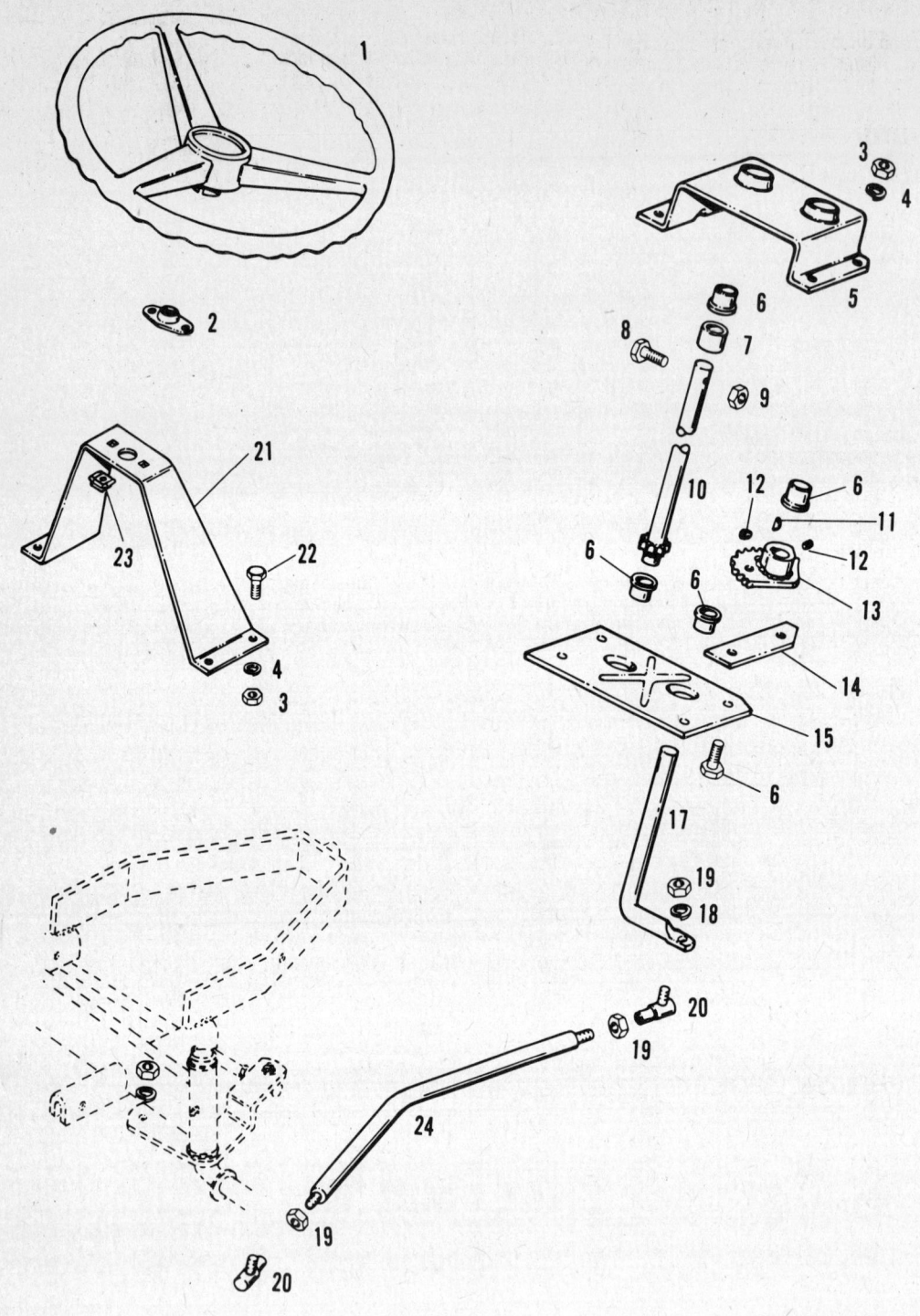

1. Steering wheel
2. Bearing
3. Nut, 5/16NC
4. Lockwasher, 5/16
5. Plate
6. Bushing
7. Spacer
8. Capscrew, 5/16NC x 1¼
9. Locknut, 5/16
10. Shaft assembly
11. Key
12. Setscrew, 5/16NC x ⅝
13. Steering gear
14. Stop spacer
15. Plate
16. Capscrew, hex, 5/16NC x 1
17. Steering rod
18. Lockwasher, ⅜
19. Nut, ⅜NC
20. Ball joint
21. Steering support
22. Capscrew, 5/16NC x ¾
23. Speed nut
24. Drag link

Upper steering components on tractors after serial # 169019000001

24

Allis-Chalmers

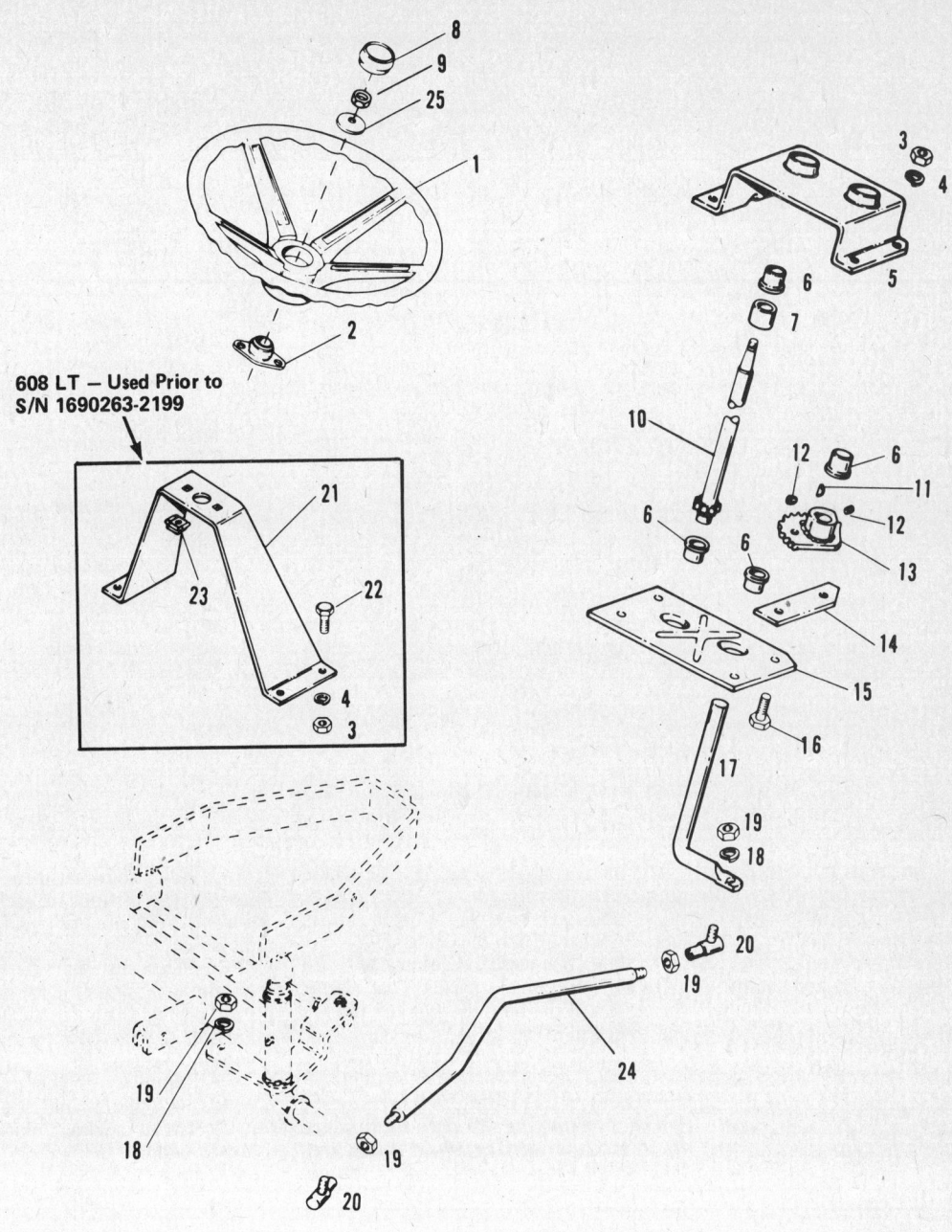

1. Steering wheel
2. Bearing
3. Nut, 5/16NC
4. Lockwasher, 5/16
5. Plate
6. Bushing
7. Spacer
8. Cap
9. Locknut, 5/16NC
10. Shaft assembly
11. Key
12. Setscrew, 5/16NC x 5/8
13. Steering gear
14. Stop spacer
15. Plate
16. Capscrew, 5/16NC x 1
17. Steering rod
18. Lockwasher, 3/8
19. Nut, 3/8NC
20. Ball joint
21. Steering support
22. Capscrew, 5/16NC x 3/4
23. Speed nut
24. Drag link
25. Compression washer

Upper steering assembly on tractors after serial #160345000001

Allis-Chalmers

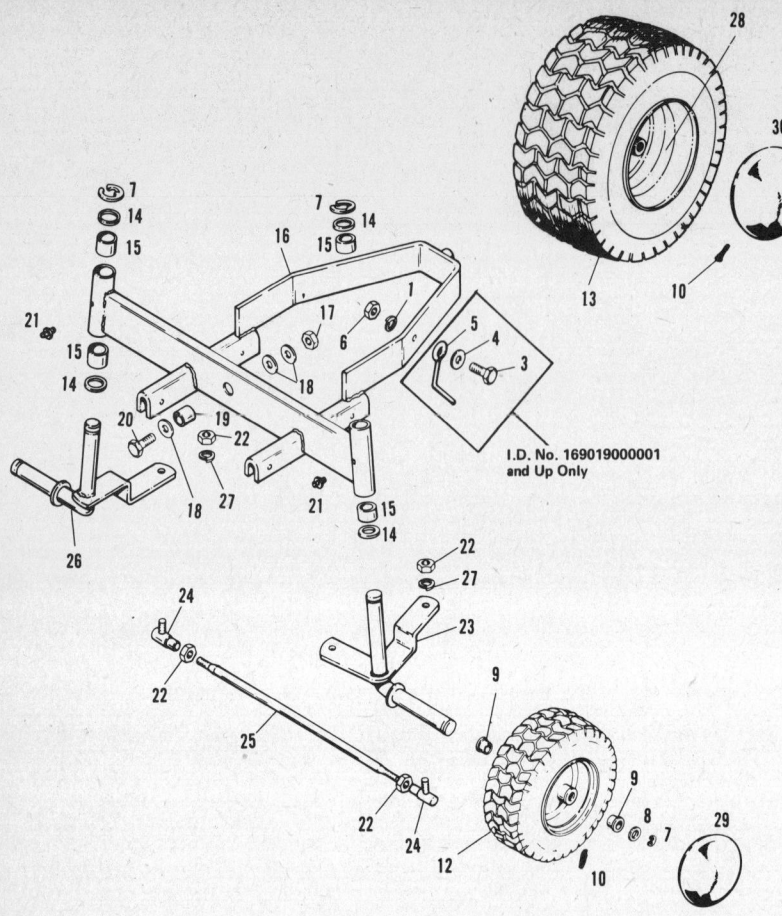

1. Lockwasher, 3/8
2. Nut, 3/8NC
3. Capscrew, 3/8NC x 1
4. Washer, 5/16
5. Belt stop
6. Nut, 3/8NC
7. Retaining ring
8. Washer
9. Bushing (Service Only)
10. Valve stem and cap
11. Wheel
12. Tire
13. Tire
    Tire
14. Washer
15. Bushing
16. Axle assembly
17. Locknut, 1/2NC
18. Washer
19. Spacer
20. Capscrew, 1/2NC x 1 1/2
21. Fitting
22. Nut, 3/8NF
23. Spindle, lefthand
24. Ball joint
25. Tie rod
26. Spindle, righthand
27. Lockwasher, 3/8
28. Wheel
29. Front hub cap
30. Rear hub cap

Lower steering components and front axle

3. Check the two bushings in the steering plate for damage or excessive wear.
4. Inspect the two bushings in the base plate for damage or excessive wear.
5. When reassembling the steering gear to the steering rod, press the gear on until it is flush with the key in the rod. The rod should extend approximately 1/4 in. above the gear. Torque the two setscrews to 15 ft.lb.

NOTE: The bevel on the spacer faces the gear. The hole in the steering shaft faces front and back. Center the gear.

Lower Steering Assembly

FRONT AXLE

Removal and Installation

1. Disconnect the battery ground.
2. Raise and support the front end.
3. Remove the front wheels.
4. Disconnect the drag link at the steering rod or left spindle.
5. Remove the bolt that secures the axle to the frame. There is a replaceable spacer inside the axle. If it is damaged or worn, replace it.
6. Pull the axle assembly out of the center pivot and remove it from the frame.

7. Inspect all parts and replace where necessary.
8. Coat the axle pivot stud with chassis lube and place it into the center pivot.
9. Place one washer on each side of the axle, and one on the back of the tractor frame. Install the bolt, lockwasher and nut and torque to 90 ft.lb.
10. Install the drag link and tie rod.

— CAUTION —
When reassembling the drag link, adjust it, so that the steering wheel is centered when the front wheels point straight ahead.

11. When assembling the tie rod adjust the toe-in to 1/8-1/4 in. measuring at the center of the tread at the front and rear of the tire.

Front Spindle

REMOVAL AND INSTALLATION

1. Disconnect the battery ground.
2. Raise and support the tractor.
3. Remove the wheel.
4. Disconnect the tie rod from the spindle being worked on and disconnect the drag link from the left spindle.

5. Take note that there is a washer at the top and bottom of each spindle. Remove the retaining ring and slide the spindle down.
6. When installing the bushings in the spindle tube, press the top bushing until it is firm and flush with the top of the tube. Press the bottom bushing in until it is flush with the bottom.
7. Install the spindle and washers and secure with the retaining ring.
8. Install the drag link and tie rod. Adjust as outlined below.
9. Install the wheel as outlined above.
10. Connect the battery ground.

Drag Link Adjustment

1. Assemble the ball joints loosely on the drag link by engaging about a half of the threads. Leave the jam nuts loose.
2. With the wheels pointed straight ahead, add the sector gear at the mid-point of its travel, measure between the centers of the steering arm holes where the ball joints attach. This measurement should be the same as the measurement between the centers of the ball joint studs.
3. Install the drag links assembly on the tractor using the lockwashers and hex nuts.

Allis-Chalmers

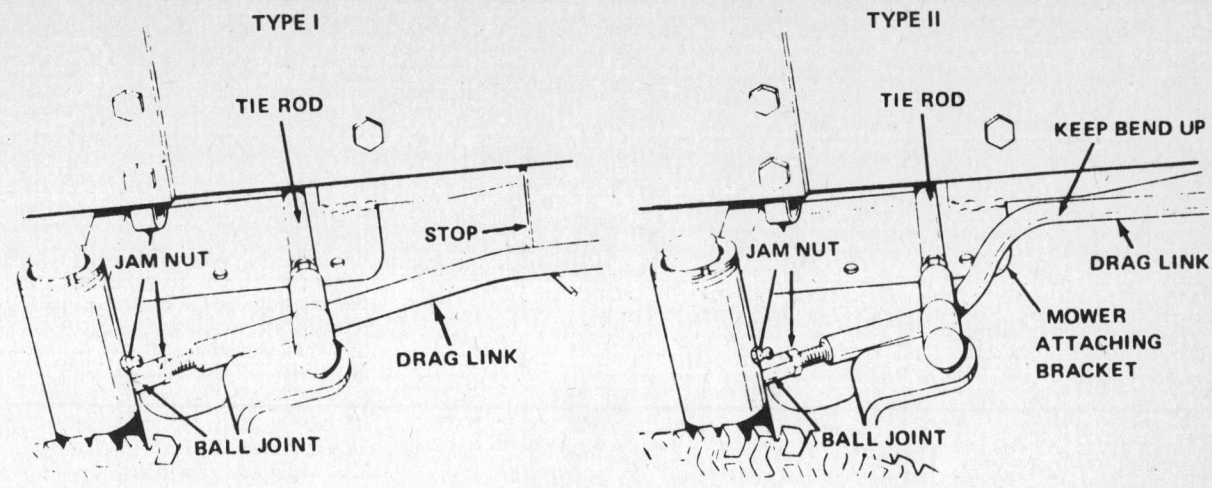

Drag link adjustments

CAUTION
To avoid damaging the ball joint, use a wrench on the flat surface of the ball to hold it when tightening the jam nut.

4. On 608LT models, the drag link should rest lightly on the stop on the left frame. On 611LT models it should be positioned to just clear the mower attaching bracket. Hold the flat surfaces on the ball joints parallel with the ground and, with the drag link in position, tighten the jam nuts.

5. Check to be sure that the stops on the spindle arms and the internal steering stops on the sector gear are reached at the same moment and both left and right full lock. No binding or bending in the linkage should occur at full lock.

6. When installing the tie rod, adjust the toe-in to 1/8 in. measured at the center tread at the front and rear of the tires.

CLUTCH, BRAKE AND DRIVE PULLEY

The clutch and brake are controlled by a single pulley on the right side. The clutch is engaged by tensioning the drive belt between the engine drive pulley and the transmission pulley by means of an idler pulley. The brake is engaged by fully depressing the clutch pedal or engaging the parking brake lever.

Adjustments

CLUTCH/BRAKE PEDAL ADJUSTMENTS

1. With the foot brake/clutch pedal in full up position and the brake lever fully forward, make sure that the jam nuts on the forward end of the foot brake rod and the

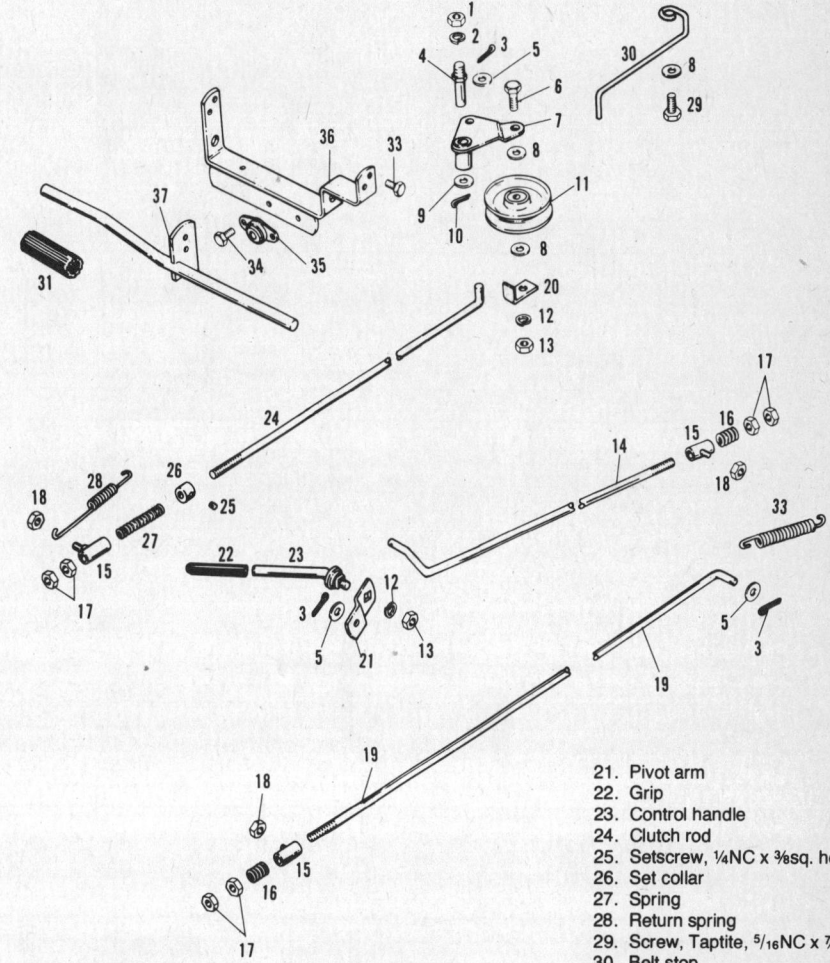

1. Hex nut, 7/16NC
2. Lockwasher, 7/16
3. Cotter pin, 3/32 x 3/4
4. Stud
5. Washer, 1/4
6. Capscrew, 3/8NC x 1 1/2
7. Idler assembly
8. Washer, 5/16
9. Washer, 3/8
10. Cotter pin, 1/8 x 3/4
11. Idler pulley
12. Lockwasher, 3/8
13. Hex nut, 3/8NC
14. Parking brake rod
15. Guide assembly
16. Spring
17. Hex nut, 5/16NC
18. Locknut, 5/16NC
19. Brake rod
20. Belt guide
21. Pivot arm
22. Grip
23. Control handle
24. Clutch rod
25. Setscrew, 1/4NC x 3/8 sq. hd.
26. Set collar
27. Spring
28. Return spring
29. Screw, Taptite, 5/16NC x 7/8
30. Belt stop
31. Brake pedal
32. Tension spring
33. Screw, Taptite, 5/16NC x 5/8
34. Screw, Taptite, 3/8NC x 1/2 washer hd.
35. Side flange bearing assembly
36. Brake pedal support assembly
37. Brake lever assembly

Typical clutch and brake assembly

27

Allis-Chalmers

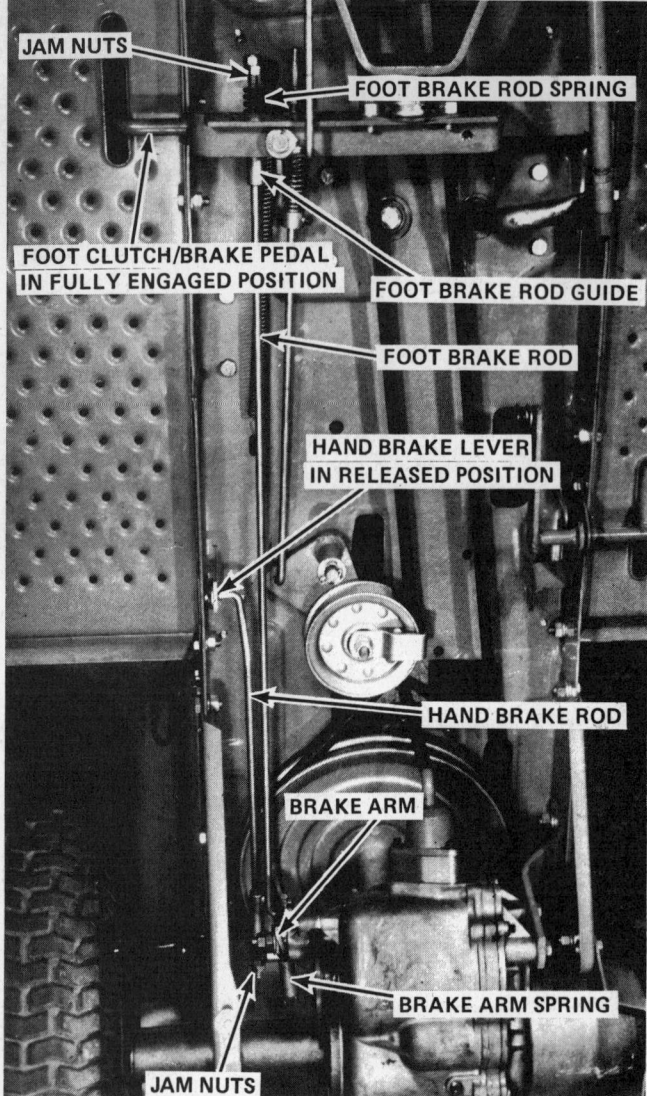

Foot and handbrake adjustments

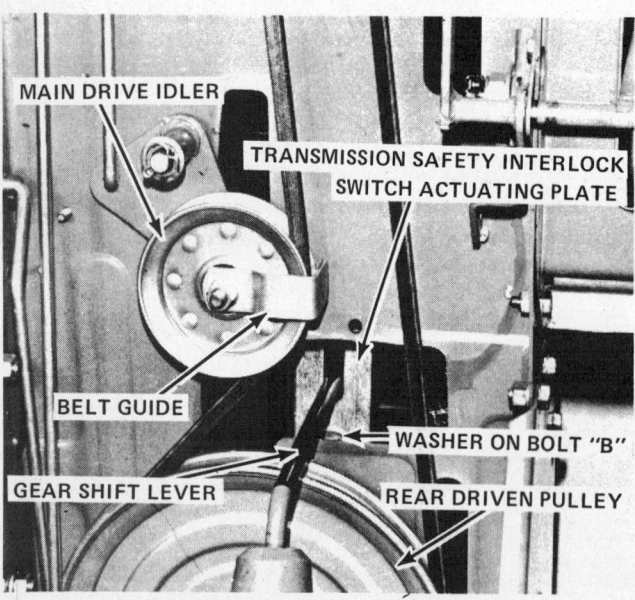

Main drive adjustments

jam nuts on the rear end of the hand brake rod are far enough away from the rod guides to allow the brake arm spring to rotate the brake arm until the rear edge of the brake arm is against the rear spacer.

2. Hold the upper end of the brake arm where the hand brake rod guide bolts to it and rotate the arm forward as tightly as possible. Hold it there until the clearance between the rear edge of the brake arm and the rear spacer is set at 1/8 in. The feeler gauge should be held against and parallel to the rear edge of the brake arm and should be a snug fit over the rear spacer while the arm is being rotated. To adjust the clearance, turn the brake arm adjusting nuts tighter to reduce the clearance and looser to increase it. Lock the nuts after the adjustment is made.

3. Adjust the jam nuts on the front of the foot brake rod to give a clearance of 5/8 in. to 3/4 in. between the rear nut and the foot brake rod spring when the foot brake rod is held tightly against the rod guide with the thumb and fingers of one hand. Lock the jam nuts together tightly after adjustment is made. For the final adjustment, move the pedal forward against its stop, and hold it there while measuring the length of the foot brake rod spring. The spring length must not be less than 7/8 in. and not more than 15/16 in. Adjust the jam nuts on the foot brake if necessary.

4. With the pedal in the full up position, and the hand brake in the full forward position, and the brake arm resting on the rear spacer, adjust the jam nuts on the rear of the hand brake rod to hold the hand brake rod spring against the hand brake rod guide without moving the brake arm away from contact with the rear spacer. Lock the jam nuts tightly after adjustment is made. For final adjustment, engage the hand brake fully. The length of the hand brake rod should be 7/8 - 15/16 in. If not, adjust the jam nuts on the end of the hand brake rod.

NOTE: As the brake rods wear, follow the above steps in order given to maintain braking action. NEVER attempt to adjust the brakes by adjusting the brake rod springs. ALWAYS adjust the brake arm clearance, then check the springs.

MAIN DRIVE CLUTCH ADJUSTMENT

1. With the pedal in the full up position, hold the main drive idler assembly and pull it firmly to the left into the belt with about 25 lbs. of force two or three times to seat the belt.

2. Position the set collar on the clutch rod to give a 2 1/4 in. spring length from the set collar to the rear side of the clutch rod guide pin against which the front end of the clutch rod spring rests as shown when the pedal is released.

3. With the clutch rod spring set at the 2 1/4 in. length, adjust the jam nuts on the end of the clutch rod to give a 3/8 in. clearance on the front end of the clutch rod guide. Lock the jam nuts.

4. Position the idler belt guide approximately at the forward strand of the belt. As

Allis-Chalmers

Clutch adjustments

the belt wears, check the clutch rod spring length frequently.

MAIN DRIVE BELT STOP ADJUSTMENTS

1. Loosen capscrews B in the accompanying illustration.
2. Hold the stops on place to give a clearance of 1/16-1/8 in. at point A.
3. Torque the capscrews to 25 ft.lb.

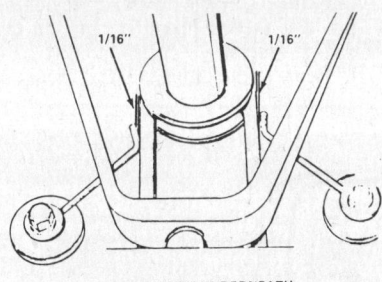

Drive belt stop adjustments

CLUTCH ROD ADJUSTMENT

1. Push the clutch idler pull against the belt firmly by hand.
2. The gap between the adjusting nut on the end of the clutch rod and the rod guide should be ½ in. If not, tighten the adjusting rod nuts to decrease the gap and loosen them to increase the gap. Lock the nuts after adjustment.
3. After the clutch rod length is properly adjusted, depress the pedal fully. Loosen the collar behind the spring on the clutch rod. Move the collar toward the spring so that it is just touching the spring. Push the collar ½ in. further toward the spring to place a slight tension on the spring and tighten the setscrew on the collar.

CAUTION
Never perform this adjustment with the engine running.

Drive Belt Replacement

1. Remove the drive belt stop from the idler pulley.
2. Remove the shift lever from the transaxle.
3. Press the clutch/brake pedal all the way down to release the tension. Remove the belt from the engine pulley and idler pulley.
4. Turn the belt sideways to remove it from the transaxle pulley. It may be necessary to loosen the transaxle pulley and press it down onto the shaft in order to get clearance between the frame and pulley.
5. Installation is the reverse of removal. Align the transaxle pulley and tighten the setscrew.

POWER TRAIN

Transaxle Removal and Installation

1. Remove the drain plug and drain all the oil from the case.
2. Thoroughly clean all the dirt from the case.
3. Place the shift lever in neutral and remove the three capscrews from the shift lever assembly. Remove the shift lever from the case.
4. Raise the rear of the tractor and support it on jack stands.
5. Remove the rear wheels by removing the retaining rings. The left axle has two washers outside the rim and one inside. The right axle has one washer outside the rim and a spacer inside. Both axles have keys.
6. Remove the drivebelt and brake linkage from the transaxle.
7. Remove the support strap from the side of the transaxle and the U-bolt from the right axle housing.

NOTE: On units with ID# 169019000001 and up, there are two brake rods hooked to the brake linkage. On units with ID# 169026300001 and up

Allis-Chalmers

and 169034500001 and up, there is one brake rod hooked to the linkage. See Brake Adjustments in the Brake Section.

8. Remove the left support plate from the tractor frame and remove the transaxle on a rolling jack.

9. Remove the drive pulley and key from the input shaft.

Shift Lever Overhaul

NOTE: Special tools are required for this job!

1. Place the shift lever assembly in a vise so that the lever housing is at least one inch above the jaws of the vise.
2. Using a snapring pliers, remove the snapring from the housing.
3. Loosen the vise and disassemble the parts.
4. Remove the lever from the housing. Check the roll pin in the ball of the shift lever, and if bent or worn, replace it. If the roll pin is replaced, position it so that equal lengths protrude from each side.

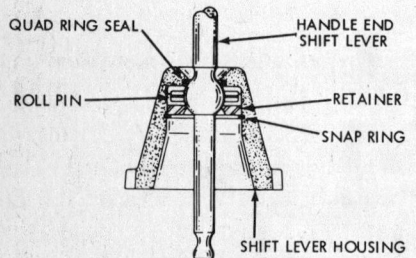

Shift lever assembly

Shift lever installation

5. Check the point at which the lever enters the housing. Oil leakage past this point requires replacement of the seal in the shift lever housing.
6. Before assembling, make sure that the bends in the lever correspond to the mounting on the tractor.
7. Assembly is the reverse of disassembly. Before installing the lever and housing in the transaxle, check that the forks are in the neutral position. Align the index marks on the shift lever assembly and case. Always use new gaskets between the lever housing and the case. Torque the capscrews to 85-108 in.lb.

Transaxle Overhaul

NOTE: Special tools are required for this job!

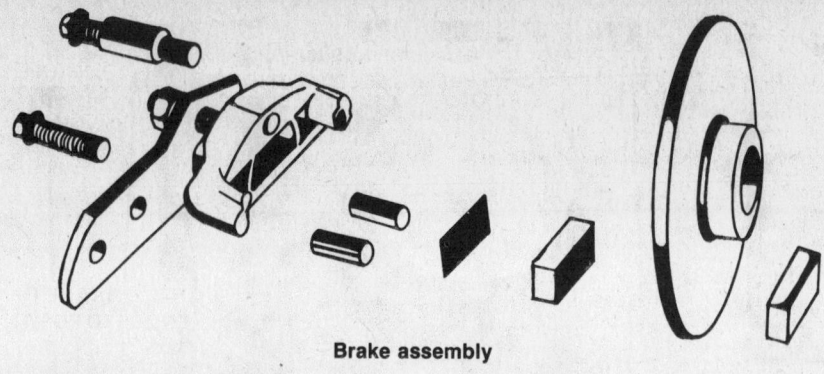

Brake assembly

1. Remove the two capscrews that hold the brake assembly to the case and remove all of the parts.
2. Remove all keys from key ways, and clean and deburr all shafts.
3. Remove and discard all seals and gaskets.
4. Remove both axle housings.
5. Place the case in a soft-jawed vise so that the socket-head capscrews are facing up.
6. Remove the capscrews and lift the cover off the case.
7. To remove the differential assembly, it may be necessary to replace two or three capscrews to hold the center plate assembly down. Pull the assembly straight up. If it is tight, tap on the lower axle with a wood or plastic mallet.
8. Drive the two dowel pins from the center section and case. Remove the temporary holding screws from the center section and lift off the center plate assembly.
9. If necessary, remove the bushing in the center section using bushing tool #670205.
10. Remove the complete shifter assembly by grasping the shifter gears, shaft and both shifter rods as a unit.
11. Remove the reverse idler shaft and spacer, cluster gear assembly and thrust washer.
12. Lift the idler gear assembly out of the case.
13. If the input shaft is not damaged, don't remove it.
14. Remove the input shaft oil seal to allow access to the snapring. Remove the snapring and the input shaft will slide out.
15. Remove and install the bushings with bushing installer #670207A. Install

Differential gear

the bushing by pressing it in from the top of the case until the tool is seated in the lip of the case. The bushing should be flush to 0.005 in. inside.

16. Replace all worn or damaged parts and install the input shaft assembly as shown in the accompanying illustrations.
17. Install a new seal using installer #670209. Use a seal protector #670102 to protect the seal while going over the key way in the input shaft.

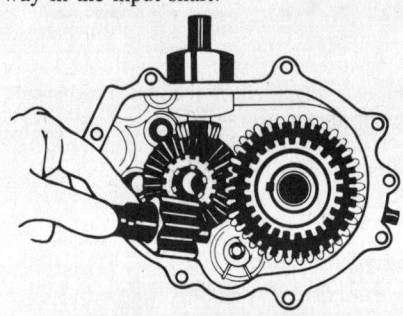

Reverse idler assembly

18. Using a suitable driver, drive out the roll pin that secures the drive pin, in the differential assembly.
19. Remove the thrust washers from the differential assembly.

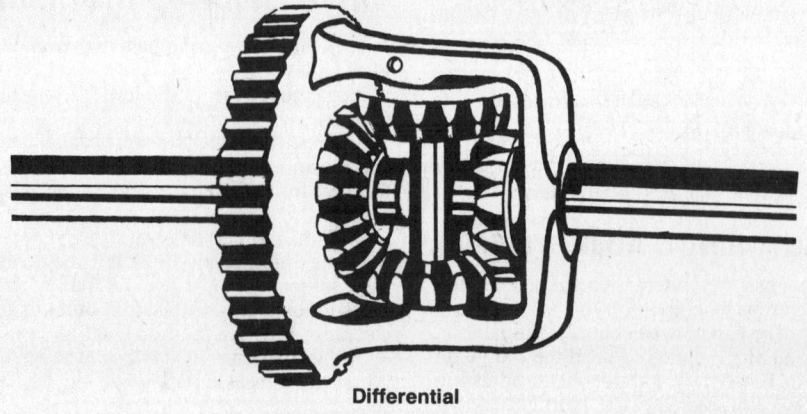

Differential

20. Remove the bevel pinions simultaneously by rotating the gears in opposite directions. This will move the gears out of position.
21. On double D type drives, remove the snapring, bevel gear and thrust washer. Slide the axle out.
22. Inspect the bushings and gears for wear and replace where necessary. Use bushing tool part #670204 to remove and install the bushing. On installation, bushings should be tightly recessed from the inside surface of the gear and brace.

Cluster Gear Subassembly—Type I

a. Remove the cluster gear from the case as a unit.
b. The cluster gear can be disassembled by pressing the key and gears from the shaft.
c. If the bushings in the shaft need to be replaced use tool #670204.
d. Install the gear with 34 teeth on the key with the bevel facing the short end of the key and press it on the shaft.
e. Press the gear with 25 teeth on the shaft where the long end of the key is with the bevel facing the 34 toothed gear.

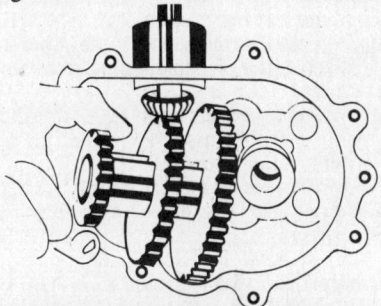

Cluster gear assembly

f. Press the gear with 39 teeth on the shaft where the short end of the key is. There is no bevel on this gear. The key edge ends must align with the shaft ends at the top and bottom.

Cluster Gear Subassembly—Type II

a. Remove the cluster gear as a complete unit from the transmission case.
b. Remove the three gears and two spacers from the splined shaft.
c. If the bushings in the shaft need to be replaced use tool #670204.
d. Install the gear with 39 teeth on the end of the shaft and then the small spacer. Install the gear with 34 teeth on the shaft with the bevel facing the large gear. Install the large spacer and the gear with 25 teeth on the shaft with the bevel facing the other two gears.
e. The two end gears should be flush with the end of the splined shaft.

NOTE: Types I and II cluster assemblies are interchangeable as a complete unit.

23. Place the left and right axles into the differential gear assembly. Install the thin thrust washers.

NOTE: The axles differ in length so make sure you don't mix them up.

24. On double D type drives, place the bevel gears on the shaft and install the snapring in the groove on the shaft.
25. IMPORTANT: Install the bevel pinions simultaneously by rotating the pinions in opposite directions while sliding them into position on the gear assembly. Check the alignment by inserting a finger into the drive pin holes. Take extra care in removing and replacing bevel pinions as just one tooth out of position will affect alignment.
26. After alignment is assured, insert the thick thrust washers behind each pinion. Insert the drive pin and secure with the roll pin.
27. To remove or install the shift assembly, squeeze the top end of the shifter rods. This will cause a binding that holds all parts in place during removal.
28. Disassemble the shift shafts according to the accompanying illustrations.
29. Assemble the shift shaft parts according to the accompanying illustrations. Be careful to follow the annular grooves in the shifter rods and snaprings. The shifter forks are interchangeable during installation. Slide the shifter fork onto the shifter rod until it comes in contact with the hole in the indexing ball and spring. With a standard screwdriver press the indexing ball into the hole and move the shifting fork completely onto the rod.
30. Referring to the accompanying illustration, move the shifter fork to the neutral position. The neutral groove is the center groove. The groove can be seen through a hole in the shifter fork.
31. When the shifter forks are properly installed, the ends of the notches in the shifter forks are in alignment.
32. Assemble the two flanged gears on the shifter shaft. The large gear goes on first with the flange side toward the bottom of the shaft. Slide the smaller gear on with the flange side toward the larger gear.
33. When assembling the shifter fork and rod to the flanged gears on the shifter shaft, the shifter fork which is on the rod always engages the flange in the larger gear. Always check the illustrations to make sure that the parts are assembled correctly.
34. Once the shifter fork and rod assemblies have been engaged with the flanged gears, allow the shifter rods to lay open in your hand and position the shifter stop. The notch in the shifter stop is the guide for correct positioning. This notch should be aligned with the corresponding notches in the shifter forks and insert the stop. Squeeze the ends of the shifter rods and insert the assembly into the transaxle.
35. The shifter assembly is correctly installed in the case if the notches in the shifter forks are in the center of the case opening.
36. Set the case assembly open side up. Insert the idler gear shaft assembly, thrust washers and bearing. Note the sequence of the washers and bearings in the accompanying illustration.

NOTE: Place the reverse idler shaft into the bearing to aid in holding the washers, thrust bearing idler shaft and gear assembly, before installing the shifter assembly.

NOTE: One thrust washer is thicker than the other.

37. Insert the thrust washer and then the three-gear cluster assembly.
38. Install the reverse idler gear making sure that the bevel is up with the spacer on top of the gear.
39. Place a new gasket on the case and install the center plate.
40. Install the gear case dowel pins, leaving the dowel pins slightly exposed on top to help locate the cover assembly.
41. Place a new gasket on the center plate and install the differential assembly with the longer axle in the down position. Be sure that the gear on the shifter shaft is on the shaft. Use seal tool #670143 to protect the seal in the case before installing the differential. Make sure the thrust washer behind the cluster gear does not slide out of place.
42. Install the transaxle cover assembly and secure with eight capscrews. Torque to 84-108 in.lb. Use seal tool #670143 to protect the seal in the cover.
43. Install the left housing support plate with the long flat edge down. Secure with four axle housing bolts and torque to 13-15 ft.lb.
44. Install the long axle housing with the tractor mounting groove up and the spring anchor bracket in the back with the large hole to the top. Install the seals flush with the end of the axle housing. Use tool #670143 to start the seals over the key way and tool #670204 to install them. Torque the housings to 13-15 ft.lb.
45. Install a new brake shaft seal flush with the case with tool #670209.
46. Install the complete brake assembly and torque the two cap-screws to 84-108 in.lb. See brake adjustments.
47. Fill the transaxle case with 1½ pints of SAE EP90 oil.
48. Install the transaxle in reverse order of removal.
49. The shift lever is installed after the transaxle is installed. Use new gaskets between the riser block and align the index marks on the shift lever and the case. Torque the three capscrews to 84-108 in.lb.

ENGINE

These tractors use the Briggs and Stratton models 191707 and 252707 respectively. For detailed repair and overhaul, see the Briggs and Stratton Engine part of the Engine Unit Repair section.

REMOVAL AND INSTALLATION

For removal and installation procedures, see the Engine section under Models 608 and 610 immediately preceding this section.

Allis-Chalmers

Models 616 and 620

WHEELS, BEARINGS AND HUBS

REMOVAL AND INSTALLATION

1. Raise the front of the tractor to remove the load from the front wheels. Place stands or blocks under the axle to safely support the weight.
2. Remove the hub cap. The wheel may be removed from the hub for convenience.
3. Remove the cotter pin, nut, washer and outside bearing cone. The hub can be removed with the inner bearing cone and oil seal.
4. Remove the oil seal and bearing cone from the hub. If the bearings are to be reused always install the same cup with the original cone.
5. Remove all the lubricant from the hub and bearings. Clean the hub and bearings in a suitable solvent.
6. Check the condition of the hub and bearings. Replace the hub if cracked or damaged. Replace chipped or worn cones and cups. If either cup or cone are to be replaced always replace both. The cups will have to be pulled or driven from the hub.
7. Press the bearing cups in the hub.
8. Use a bearing packer or work a good quality of wheel bearing lubricant in the inner bearing. Install the cone in the hub and press the oil seal in the hub.
9. Fill the hub with sufficient wheel bearing lubricant to provide additional lubrication for the bearings.
10. Install the hub on the axle.
11. Fill the outer bearing cone with wheel bearing lubricant, install the cone on the spindle, thrust washer and slotted nut.
12. Torque the nut until a slight preload is detectable as the wheel is rotated. Replace the cotter pin and hub cap. Replace the wheel if it was removed from the hub.

BRAKES

Brake Linkage

REMOVAL AND INSTALLATION

1. Raise the hood and remove the fuel tank.
2. Remove the six washer headed capscrews that attach the top cover to the tractor frame and remove the cover.
3. Unhook the brake return springs (4-5) and cotter pins to disconnect the front brake rods.
4. Remove setscrew (8) and move brake rod anchor assembly (9) left to permit removal of Woodruff key (10).
5. Remove Woodruff key (10) and remove righthand brake lever assembly from the tractor.
6. Remove lefthand brake lever assembly (15) and anchor assembly (9).
7. The front and rear rod assemblies can be removed by removing the pins (20) at the brake housing.
8. Inspect the bearings (12) in the lefthand brake lever assembly. If replacement is necessary remove the bearings with a suitable press or driver and install new bearings.
9. Inspect the righthand pivot bearings and replace if worn.

NOTE: These bearings are in the tractor frame but not in the illustration.

10. Check the brake rods and levers at the pivot areas and replace if worn.

The link plate (29) and rear brake levers can be inspected after the drop housing and axle extensions are removed and the brake assembly is exposed.

11. Replace the brake levers, rods and springs in the reverse order of removal.

Brake Assembly Removal and Installation

1. Position the tractor safely on suitable blocks or stands and remove the tires and wheels.

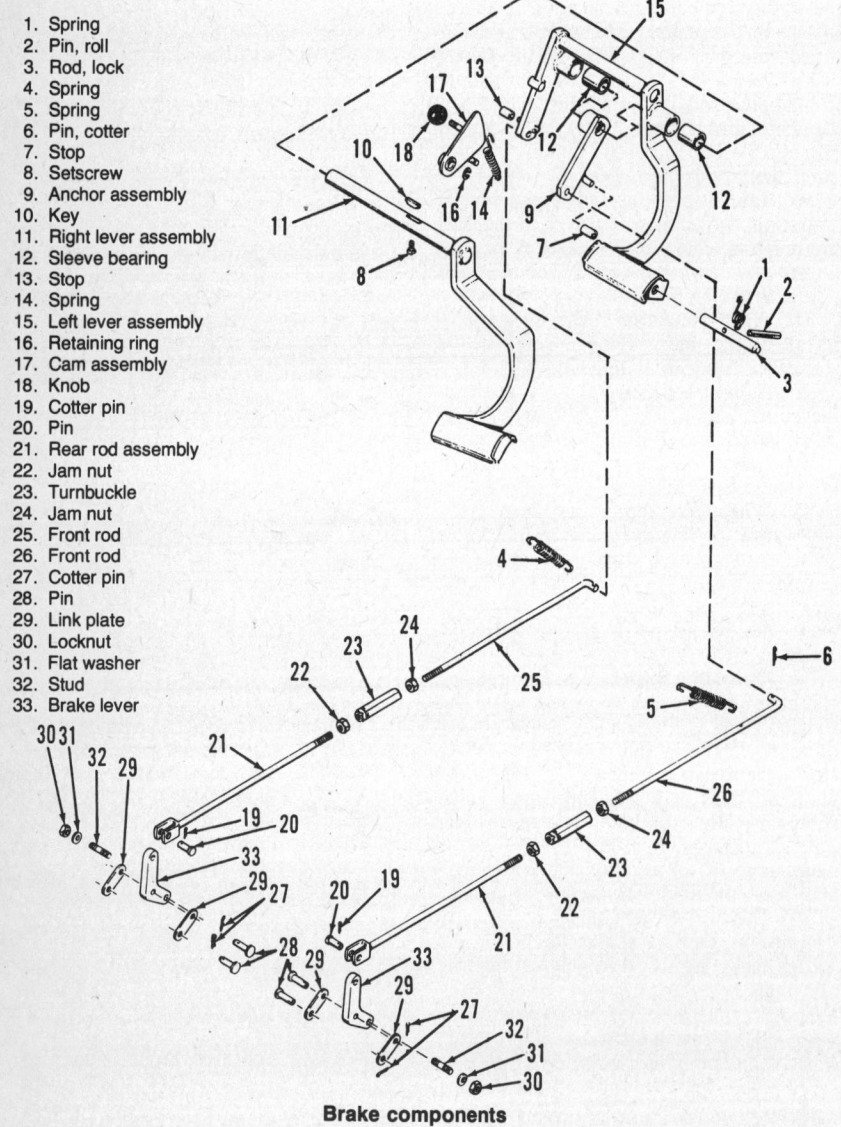

1. Spring
2. Pin, roll
3. Rod, lock
4. Spring
5. Spring
6. Pin, cotter
7. Stop
8. Setscrew
9. Anchor assembly
10. Key
11. Right lever assembly
12. Sleeve bearing
13. Stop
14. Spring
15. Left lever assembly
16. Retaining ring
17. Cam assembly
18. Knob
19. Cotter pin
20. Pin
21. Rear rod assembly
22. Jam nut
23. Turnbuckle
24. Jam nut
25. Front rod
26. Front rod
27. Cotter pin
28. Pin
29. Link plate
30. Locknut
31. Flat washer
32. Stud
33. Brake lever

Brake components

Allis-Chalmers

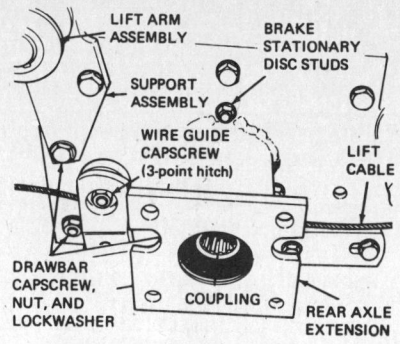

Brake assembly attaching points

2. Remove the four capscrews attaching the wheel guard to the drop housing.

3. Remove the four capscrews attaching the drop housing to the axle extension.

4. Remove the drop housing.

5. Disconnect the hydraulic lift cylinder from the lefthand axle extension and lift shaft and move out of the way.

6. Remove the capscrews and pins and remove the lift shaft assembly.

7. Remove capscrews and nuts attaching the drawbar to the axle extensions and remove the drawbar assembly.

8. Remove the transmission oil filter.

9. Remove the four capscrews attaching each axle extension to the transmission case.

10. Remove the hex nuts and lockwashers from the stationary brake disc studs.

11. Remove the rear axle extension and coupling.

12. Remove the outer stationary disc.

13. Remove differential output shaft with outer rotating disc assembly and retaining ring.

14. Remove the two flat edge spacers and one round edge spacer from the stationary studs.

15. Remove pin attaching link plate to yoke link assembly and remove disc actuating assembly.

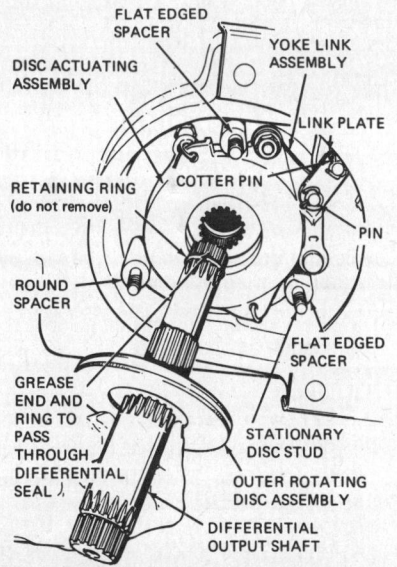

Differential output shaft removal

16. Clean all the brake parts and housing with suitable solvent to remove all grease and dirt from brake parts.

17. Inspect all linkage and pins for wear at the pivot areas and replace items that are worn excessively.

18. Check rotating disc assemblies for wear. Replace if necessary.

19. Check the outer stationary discs, actuating disc assembly and differential supports for scoring and wear. If new rotating discs are installed replace the other brake parts that are worn or scored.

20. Remove the oil seals from the differential supports.

21. Remove the springs from the actuating disc assemblies and replace if stretched or worn in the hook area.

NOTE: Check the three ¾ in. steel balls and ramps for wear or roughness. Replace worn or rough parts.

22. Install new oil seals in the differential supports and install the supports in the transmission case. Torque the capscrews.

23. Install the inner rotating disc assembly.

24. Install the two flat edge spacers on the stationary disc studs with the flat edge toward the yoke link assembly. Install the round spacer on the stationary disc stud.

25. Install the actuating disc assembly in the brake housing. Attach the yoke link assembly and actuating link to the link plate with a pin and cotter pin.

26. Install the retaining ring and outer rotating disc assembly on the differential output shaft. Lubricate the inner splines and retaining ring with Lubriplate® to prevent oil seal damage.

27. Install the outer stationary disc.

28. Install the axle extension and fasten with lockwashers, through hex nuts and four capscrews.

NOTE: Some 616 tractors may not have a drain hole drilled in the brake area of the axle extensions. If necessary drill a ⅜ hole in the bottom area to allow water and oil to drain from the brake housing (axle extension).

29. Install the coupling on the differential shaft, spring on the pinion shaft and install the drop housing.

30. Install the lift shaft assembly hydraulic ram, drawbar, wheel guard and accessories that were removed.

CLUTCH AND DRIVESHAFT

REMOVAL AND INSTALLATION

1. To remove the clutch and driveshaft remove the oil cooler, shrouds and fuel tank from the tractor.

2. Remove the bottom cover from the tractor frame.

3. Hold the clutch pedal forward and remove the clutch belts from the engine crankshaft pulley.

4. Remove two capscrews holding support (45) and pivot shaft (18) from the tractor frame.

5. Remove the capscrews (616) or

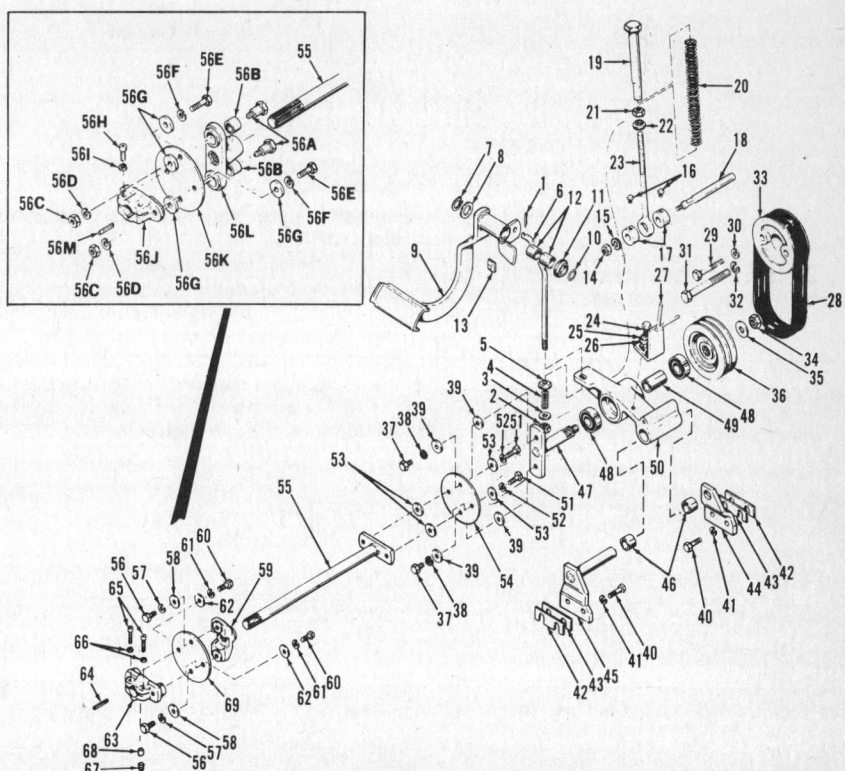

Clutch and driveshaft assembly

Allis-Chalmers

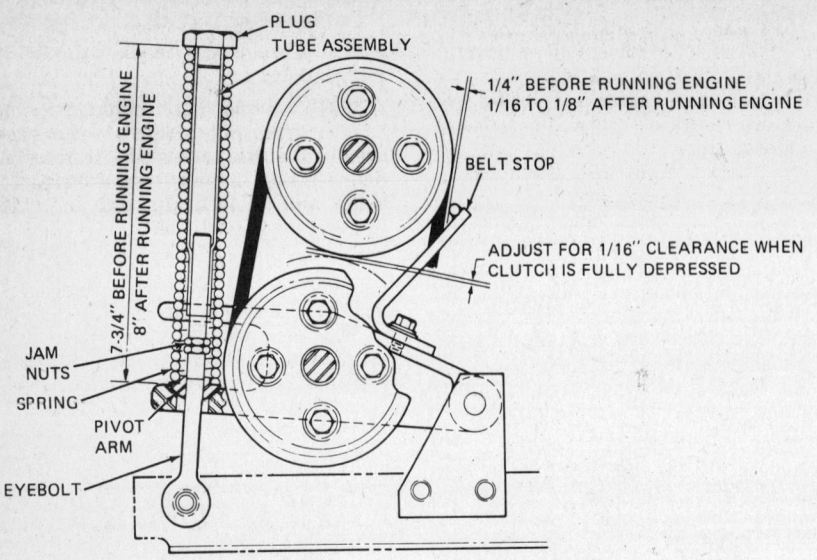

Spring tension and belt stop adjustment

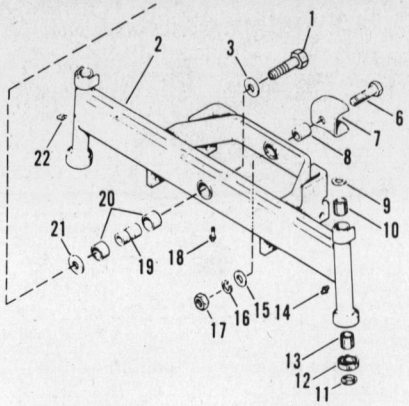

1. Capscrew
2. Axle assembly
3. Thrust washer
4. Lockwasher
5. Nut
6. Bolt
7. Clevis
8. Reach bushing
9. Thrust washer
10. Spindle bushing
11. Bearing spacer
12. Thrust bearing
13. Spindle bushing
14. Grease fitting
15. Flat washer
16. Lockwasher
17. Nut
18. Grease fitting
19. Axle spacer
20. Axle bushing
21. Thrust washer
22. Grease fitting

Front axle components

shoulder bolts (620) attaching rear driveshaft yoke to coupling disc.

6. Remove the drive shaft assembly from the tractor frame.

7. Inspect the shaft assembly bearings and the pivot bearings. Replace if necessary.

8. Check the splines on shaft assembly (47) and clutch pulley (36) and replace if they have excessive clearance.

9. Check the clutch belts and replace if cracked.

10. Check the rear drive shaft (55) and yoke splines. The 620 tractor will have a neoprene and felt washer on the driveshaft to form a grease seal at the coupling yoke. Clean all parts in non-flammable solvent.

11. Check clutch spring (20), Free height 9.75 in., compressed height 6.5 in., 62 to 68 pounds at compressed height. Clutch and spring (4), Free length 1.375 in., compressed under 57 pounds pressure.

INSTALLATION

1. Lubricate and install pivot bearings (46) in the pivot arm.

2. Install driveshaft bearings (48) and spacer (49) in the pivot arm.

3. Install shaft assembly (47) in the pivot assembly and pulley (36) on the shaft.

Use flange nut part #2028644 to secure pulley to the shaft and torque to 75 ft.lb.

4. Install the pivot assembly, pivot shaft (18), collars (17), eyebolt (23), pivot supports (44-45), shims (42-43) and clutch belts. Install the shims (42-43) in their original locations.

5. Install front driveshaft coupler (54) to the driveshafts (47-55) with hardware shown.

6. Install rod spacer (5) on the eyebolt.

7. Install the jam nuts on the eyebolt and adjust to maintain 1/16 in. clearance between the drive and driven sheaves when the pivot arm is raised as far as possible.

8. Install the spring, tube assembly, clutch rod, and belts on engine pulley.

9. Adjust spring tension and belt stop as shown.

10. Assemble the rear driveshaft coupling with hardware and parts as shown for the 620 and 616. Items 56 to 60 are 616 coupling items.

11. Replace bottom cover, engine shrouds, oil cooler and fuel tank.

FRONT AXLE AND LINKAGE ASSEMBLY

REMOVAL AND INSTALLATION

1. Raise the front of the tractor to remove the weight from the front wheels and support safely with suitable stands or blocks.

2. To remove the spindles remove the wheels from the hubs.

3. Remove the righthand and lefthand steering arms and Woodruff keys. Remove the spindles. The spindle bushings can be removed with a suitable puller or driver.

4. To remove the front axle from the frame remove the two capscrews (1-6).

5. Check the axle pivot bushings (20), reach pivot bushing (8) and spindle bushings. Replace bushings that have excess clearance to cause steering problems or inconvenience. Check the thrust bearings and replace if they do not rotate smoothly.

6. Clean all parts in suitable solvent. Press new bushings in where required.

7. The threaded block should be welded to the tractor frame and replace nut and lockwasher (16-17).

The early production 616 tractors used a nut and lockwasher to fasten the reach capscrew. This block part #2029577 should be ordered from your local Allis-Chalmers dealer and welded to the tractor frame as shown if the 616 tractor uses the nut and lockwasher and does not have the block welded in place.

8. In some cases when a 616 tractor was equipped with high flotation tires the tires would strike the tractor frame. This condition can be corrected by adding a thin weld to the stop as shown to limit the rotation of the spindle.

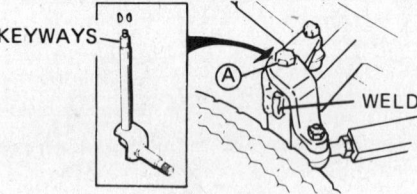

Correcting tire interference caused by floatation tires on 616 models

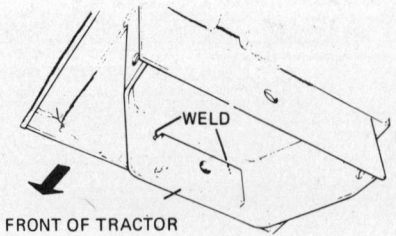

Part #2029577 welded in place

9. Press the spindle bushing in place.

10. Install the axle pivot bushings and spacer.

11. Install the axle in the tractor frame and fasten with the front and rear pivot capscrews (1-6).

NOTE: Tie rod clevis is fastened in place by capscrew (6). Torque the capscrews.

12. Install the spindles, linkage and items removed during disassembly.

Allis-Chalmers

13. The drag link, tie rod, and ball joints should be replaced if excess clearance is apparent.
14. Adjust the tie rod to have approximately 1/16 in. toe-in.

STEERING

Steering Gear

REMOVAL

1. Disconnect the battery.
2. Remove oil cooler, left engine shroud and fuel tank from the tractor.
3. Remove cotter pin and machine washer attaching the clutch rod to the lever assembly.
4. Remove tube assembly, spring, jam nuts and clutch spacer from the eyebolt.
5. Remove the clutch belts from the engine pulley.
6. Remove ten washer headed capscrews attaching the bottom cover to tractor frame and remove the cover.
7. Remove the capscrew securing the steering wheel to the shaft. Remove steering wheel and rubber washer from the steering shaft.
8. Remove the choke cable from the carburetor and the clips that hold the choke cable and throttle cable in place.
9. Remove the six capscrews attaching the instrument panel to the support. Raise the panel over the steering shaft without damaging the electrical wires.
10. Remove the roll pin from the U-joint and steering shaft and move the steering shaft up in the support tube.

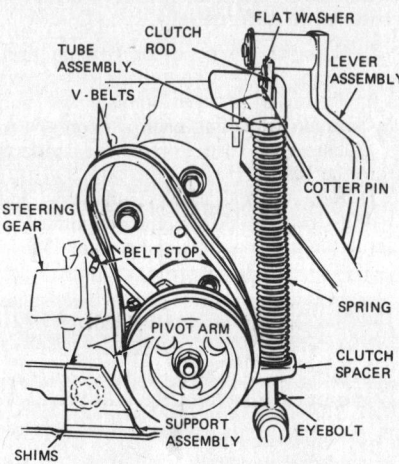

Clutch belt stop removal

11. Loosen the capscrew that holds the clutch belt stop in place.
12. Remove the four capscrews and lockwashers that attach the clutch pivot arm supports to the frame. Tilt the pivot arm and support assemblies away from the frame and remove support assembly shims. When the support is installed during assembly the shims must be placed in their original location.
13. Remove the nut and lockwasher from the Pitman shaft and remove the Pitman arm (20).
14. Remove the three capscrews and lockwashers attaching the steering gear to the frame to remove the steering gear assembly from the tractor.
15. Remove the U-joint from the steering gear.
16. Steering shaft assembly removal: remove the steering shaft from the support tube.

Remove the bearings from the panel support tube with a suitable bearing puller if bearing replacement is necessary. Clean the bearings and parts in a suitable solvent and check all parts. Replace worn or damaged parts.

DISASSEMBLY

1. Clamp the steering gear assembly in a vise with the wormshaft in a horizontal position. Clamp to one of the mounting tabs.
2. Rotate the worm from one stop to the other and count the revolutions of the wormshaft. Rotate the wormshaft away from a stop ½ the revolutions required to rotate from stop to stop.
3. Prepare to catch the lubricant from the gear case in a pan when the cover is removed.
4. Remove the three self locking capscrews that attach the side cover to the gear case. Tap on the end of the Pitman shaft with a plastic hammer to loosen the cover. Remove the Pitman shaft with the cover from the case.
5. Remove the locknut and adjuster plug with the lower bearing and race from the gear case.
6. Remove the worm nut with the ball nut assembly from the gear case. Do not permit the ball nut assembly to wedge at either end of the wormshaft. The assembly may be damaged if this occurs.
7. Remove the upper bearing from the wormshaft.
8. Inspect the bearing areas and worm shaft for chipped or worn areas and replace if necessary.
9. Remove the ball guide clamps and ball guides. The ball bearings can be re-

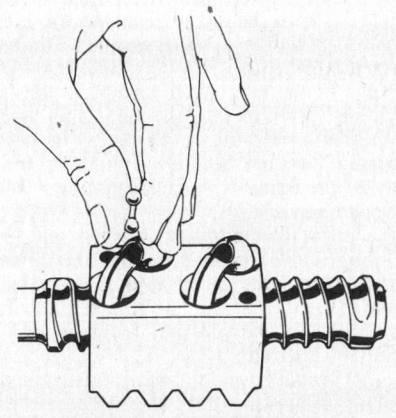

Removing ball bearings

moved from the ports. There should be a total of 48 ball bearings in the ball nut assembly. The nut can now be removed from the wormshaft.
10. Thoroughly clean all parts in a suitable solvent. Inspect the ball bearings for roughness or flat areas and replace damaged bearings. Check the ends of the ball guides for bent ends which could interfere with the ball bearing travel. This could cause steering problems.
11. Remove the wormshaft seal and upper wormshaft bearing race.

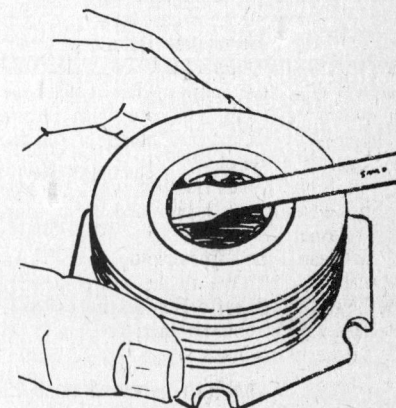

Prying out bearing retainer

ASSEMBLY AND INSTALLATION

1. Place the ball nut in a clean area and install the wormshaft in the ball nut. Block the wormshaft up to center it in the ball nut. Rotate so the grooves in the worm shaft align with the four holes in the ball nut.
2. Replace the ball guides in the ball nut.
3. Install 24 ball bearings in each circuit of the ball nut assembly. Install the ball guide clamps and secure with machine screws and washers.
4. Rotate the ball nut assembly on the wormshaft and lubricate with Lubriplate®.
5. Remove and inspect the lower bearing by using a suitable bar to pry out the bearing retainer. The bearing cup and plug are serviced as an assembly. If the bearing is worn or damaged replace the complete assembly.

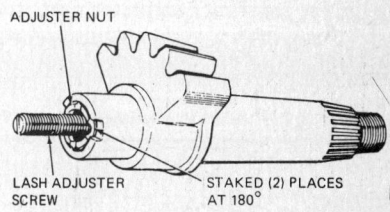

Staking lash adjuster nut

6. Pitman shaft and side cover. Remove the lash adjusting screw locknut and turn the lash adjusting screw clockwise to remove the cover. Clean the parts in a solvent.
7. Check the needle bearing in the gear case and if replacement is necessary remove and replace with a suitable puller.

35

Allis-Chalmers

When pressing a needle bearing in place always press against the end with the manufacturers name or identification. Do not press the bearing into the offset in the bore. To do so may distort the bearing. Replace the oil seal in the gear case.

8. Remove the wormshaft seal from the gear case. If the upper wormshaft bearing is to be replaced remove the bearing race and install a new one. Install a new wormshaft oil seal.

9. The lash adjusting screw and Pitman shaft are serviced as an assembly. If it is worn or damaged replace the assembly.

10. Clamp the gear case in a vise with the wormshaft horizontal. Install the upper bearing on the wormshaft. To protect the oil seal cover the splines with Scotch tape or cellophane and lubricate with Lubriplate®. Lubricate the bearing and install the wormshaft in the case.

11. If the lower bearing was removed from the adjuster plug replace the bearing and retainer.

12. Install the adjuster plug assembly in the case and remove all the end play from the bearings or shaft. The bearings should preload until it requires three to six in.lb. to rotate the wormshaft.

NOTE: To check the preload use a 12 point $11/16$ in. socket on the splined end of the wormshaft with an in.lb. torque wrench to rotate the shaft. Turn the adjuster plug in while rotating the wormshaft with the in.lb. torque wrench until a reading of between three and six pound torque is observed on the torque wrench. Tighten the adjuster plug locknut without changing the preload setting.

13. Pack nine oz. of automotive chassis type lubricant in the case. Rotate the worm shaft from one stop to the other to lubricate the bearings and ball nut assembly.

14. Rotate the wormshaft until the ball nut is at mid-point between the two stops. Lubricate Pitman shaft bearing.

15. Cover the splined end of the Pitman shaft with Scotch tape and lubricate with Lubriplate® to protect the oil seal. Install the Pitman shaft in the case so that the center tooth of the sector gear enters the center groove of the ball nut.

16. Coat the side cover gasket with a thin coat of Permatex® gasket sealer, lubricate the needle bearing in the side cover and install the side cover on the case. Rotate the lash adjuster screw counterclockwise until the Pitman shaft contacts the cover assembly. Turn the adjusting screw clockwise ½ turn for starting lash adjustment. Install the locknut and torque while holding adjusting screw in predetermined position. Install the self locking cover capscrews and torque.

17. Rotate the wormshaft to position the ball nut in the center of the sector. To locate the center rotate the wormshaft clockwise until it stops then count the shaft revolutions as you rotate the wormshaft counterclockwise to the opposite stop. Turn the wormshaft one-half this number of revolutions clockwise. This will center the ball nut assembly on the wormshaft.

18. Loosen the Pitman shaft lash adjuster locknut and turn the Pitman lash adjusting screw clockwise to remove all the ball nut to sector lash without binding. Torque the locknut.

NOTE: The ball nut to worm lash should be "0" when the ball nut is in the center of the wormshaft and will require about ten in.lb. torque to rotate the wormshaft through the center Pitman shaft arc area.

19. Check the torque required to rotate the wormshaft with an $11/16$ in. socket and in.lb. torque wrench. The torque should not exceed eleven in.lb. to rotate the wormshaft when the ball nut is in the center area but should not be less than three in.lb. Recheck wormshaft bearing adjustment or Pitman shaft lash if necessary.

20. Install the bearings in the support tube. Use a suitable puller and press the end of the needle bearings that have the manufacturers name or identification. Lubricate the bearings with wheel bearing lubricant. Install the roller bearing in the bottom section of the support tube.

21. Install the retaining ring and thrust washer on the steering shaft.

22. Place the shaft in the support tube bearings.

23. Install the universal joint, align the holes in the U-joint with the hole in the steering shaft and drive in the roll pin.

24. Install the gear case with the splined end of the wormshaft in the universal joint and secure to the tractor frame with three lockwashers and capscrews. Torque capscrews to 70-90 ft.lb.

25. Secure the steering arm to the Pitman shaft with lockwasher and nut. Torque nut 162-198 ft.lb.

26. Move the clutch assembly to its original position. Install the shims that were removed during disassembly in their original location. Secure the clutch assembly to the frame with four capscrews and lockwashers.

27. Install the thrust washer on the U-joint, move the shaft through the thrust washer and in the universal joint. Align the hole in the U-joint with the hole in the shaft and drive in the roll pin, retaining ring, thrust washer etc.

28. Install the following: instrument panel, choke and throttle cables, rubber washer and steering wheel.

29. Install the clutch V-belts. Install the clutch eyebolt spacer on the pivot arm.

30. Install the jam nuts on the clutch eyebolt and adjust so that $1/16$ in. clearance is maintained between the drive and driven sheaves when the pivot arm is moved up as far as possible to release the clutch.

31. Install the spring and tube assembly.

32. Attach clutch pedal lever to the clutch and install washer and cotter pin.

33. Adjust belt stop.

34. Replace oil cooler and shrouds. Replace fuel tank and other items that were removed.

TRANSAXLE

Differential

REMOVAL

1. Raise and support the tractor on jack stands.

2. To drain transmission lubricant remove the ten washer head capscrews attaching the bottom cover to the frame, place a pan under the sending unit, remove the electrical wire from the transmission oil temperature sending unit, and remove the sending unit.

3. Remove the wheels, drop housings, axle extensions, brake assemblies. Seat with seat support and transmission cover.

4. If the tractor is equipped with a rear PTO remove the top power take off shaft. Refer to PTO section for shaft removal.

NOTE: If the transmission case is to be removed from the tractor frame it is not necessary to remove the front PTO shaft. It is not necessary to remove transmission case from the tractor to remove the differential.

5. Remove the upper brake stationary disc studs.

6. Remove the two capscrews and lockwashers attaching each support (5) to the transmission case. Remove the supports from the case.

7. Remove and discard oil seals and O-rings from the differential supports. Examine the bearing cups and replace if they

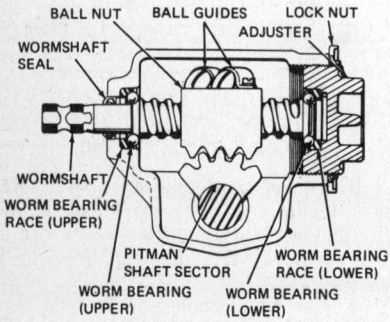

Pitman shaft installation

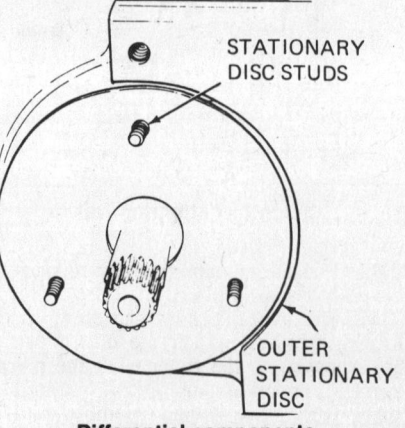

Differential components

are rough or pitted or if new bearing cones are to be used on the differential assembly.

8. Remove the differential assembly from the case.

DISASSEMBLY

1. Inspect bearing cones (9) on the differential gear assembly and gear carrier. If the cones require replacement remove with a suitable bearing puller.
2. Remove the pinion shaft retaining rings (12) and lock plates (13).
3. Remove six capscrews (10) and lockwashers to separate the gear carrier (24) from the gear assembly (15).
4. Dowel pins (14) are pressed into the gear assembly. Unless replacement is necessary do not remove.
5. Remove the pinions, spacers and shafts from the gear carrier.
6. Remove two thrust washers, gears, differential clutches and Belleville washers from the gear carrier.
7. Remove all the oil from the transmission case. Clean the transmission case and all parts in a suitable solvent. Inspect all parts and replace any that are worn excessively or damaged.

ASSEMBLY AND INSTALLATION

1. If bearing cones were removed from gear assembly hub (15) and gear carrier hub (24) press new bearing cones on the hubs.
2. Install pinion shafts (16) in the gear carrier and secure with lock plates (13) and retaining rings (12).
3. Install in the gear carrier in the following order, one thrust washer (19), one gear (20), one clutch plate (21) with the flat side toward carrier, one Belleville washer (22) with the concave side toward the clutch plate, another Belleville washer with the convex side toward the first Belleville washer, install the third Belleville washer

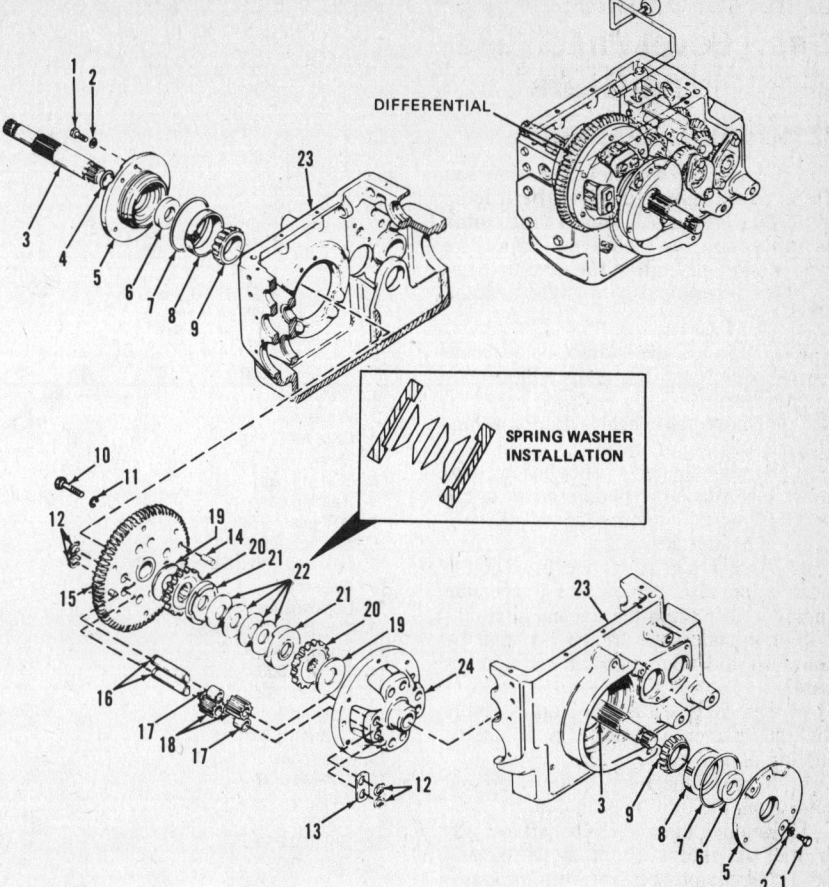

1. Capscrew
2. Lockwasher
3. Input shaft
4. Retaining ring
5. Support
6. Seal
7. O-ring
8. Bearing cup
9. Bearing cone
10. Capscrew
11. Lockwasher
12. Retaining ring
13. Lock plate
14. Pin
15. Gear assembly
16. Shaft, pinion
17. Spacer
18. Pinion gear
19. Thrust washer
20. Differential gear
21. Plate clutch
22. Spring washer
23. Differential case
24. Gear carrier

Upper brake stationary disc studs

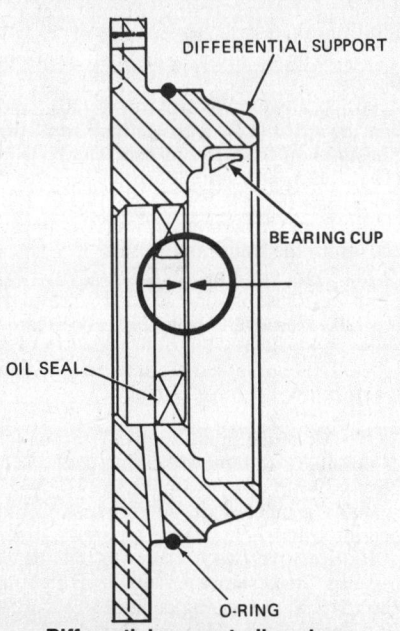

Differential support oil seals

with the concave side toward the second Belleville washer, install the fourth Belleville washer with the convex side toward the third Belleville washer, install the clutch plate with the flat side away from the Belleville washers, install the differential gear and thrust washer as shown.

4. Install one pinion gear (18) on four alternate shafts, then place one spacer on each shaft.
Install one pinion gear on each of the four remaining shafts that are without pinions.
5. Align the gear assembly (15) on the carrier assembly so that the dowel pins in the gear align with the holes in the carrier.
6. Secure the gear assembly to the carrier with the six capscrews.
7. Install the retaining rings (12) on the gear assembly ends of the pinion shafts.
8. Install the differential assembly in the transmission case.
9. Press bearing cup in the differential supports.
10. Install new oil seals in the differential supports as shown. Lubricate the O-rings and oil seal with Lubriplate®. Install differential supports in case.

11. Replace the brake stationary disc studs and secure each carrier with two capscrews and lockwashers.
12. Replace the top PTO shaft with new oil seal. Install the electric clutch hub. Use Loctite® on the hub attaching capscrew. Install the front PTO shaft.
13. Install the transmission assembly in the tractor frame if it was removed.
14. Install the brake assemblies, axle extensions, drop housings, drawbar, lift shaft, ram, oil filter, wheel shields, wheels, and temperature sending unit.
15. Replace the transmission cover.
16. Fill transmission to correct level with Allis-Chalmers Power Fluid 401 or its equivalent. Approximately 6 quarts (616) and 5 quarts (620) of 401 Power Fluid will be required to fill the transmission.
17. Adjust the hydrostatic transmission control lever so that the tractor does not move when the lever is in neutral.
18. Adjust the brakes for proper pedal free travel.
19. Replace the top cover, bottom cover, seat with support, connect clutch, electrical connections, light wire connectors, and any items removed.

Allis-Chalmers

Shift Assemblies, Bevel Gear, Bevel Pinion and Sliding Gears

REMOVAL

NOTE: If the complete transmission case is to be removed from the tractor frame the hydrostatic pump and motor assembly can be removed from the transmission case after the complete assembly is removed from the tractor frame.

1. Disconnect the hydrostatic oil lines, control linkage and the temperature sending unit wire.
2. Remove the wheels, drop housing, axle extensions and brakes.
3. Remove the ten washer headed capscrews that attach the bottom cover to the tractor frame and remove cover. Disconnect clutch shaft at U-joint.
4. Place a pan under the front of the transmission case, remove the temperatue sending unit to drain the transmission oil.
5. Disconnect the front PTO shaft by removing the four hub attaching capscrews.
6. Remove the transmission assembly from the tractor and place in a suitable working area.
7. Remove the hydrostatic pump and motor from the transmission case.
8. Remove the top PTO shaft and differential assembly. (Refer to Differential and Transmission section) from the transmission case.
9. Remove capscrew (21) and lockwasher attaching shift stem (2) to shift shaft assembly and remove shift stem.
10. Remove shift lever assembly (22) and seal (19). Discard oil seal.
11. Remove two capscrews (13) and washer seals (14) attaching the shift guide (3) to the differential case.
12. Remove shift guide and two spacers (18).
13. Remove two capscrews (11) and lockwashers attaching shift shaft assembly to transmission case.
14. Pull the shift shaft assembly (10) from the case and the two shifter forks (4-7). Remove the shaft from the forks carefully so that the detent springs (5) and balls (6) encased in the forks are not lost.
15. Remove the O-rings and gasket from the shift shaft assembly and discard.

ASSEMBLY

1. Install new O-rings (8-9) on shift shaft assembly (10) and lubricate. Apply a light coat of Permatex® to gasket (15) and install on shifter shaft plate. Start the shifter shaft assembly through the bore in left side of the transmission case.
2. Install the detent spring (5) and ball (6) in the third speed shifter fork (7) and place the groove in the shifter fork over the 20T gear. Hold the detent ball and spring far enough in the drilling in the shifter fork and allow the shifter shaft to pass through the bore in the fork (7). Move the shaft through the fork a sufficient distance to retain the detent ball and spring.
3. Install the detent spring (5) and ball (6) in the drilling in first and second shifter fork (4) and install the shifter fork in the groove around the first and second speed gear.
4. Hold the detent ball and spring in the shifter fork so that the shifter shaft can move through the fork until the plate and gasket contact the transmission case. Coat the capscrew threads with a sealer, install lockwashers and secure. Secure shifter shaft assembly to the transmission case with the two capscrews.
5. Install seal washers on two capscrews (13) and start them through the differential case. Install one spacer (18) on each capscrew. Install the shift guide in the case and secure with the two capscrews (13).
6. Install the differential assembly in the tractor.
7. Install top PTO shaft in transmission case. Install new PTO shaft oil seal. Use Loctite® on clutch hub capscrew.
8. Install hydrostatic pump and motor assembly on the transmission case.

Bevel Gear, Input Shaft and Gears

DISASSEMBLY

1. Remove two capscrews (1) lockwashers, and cap (3) from the transmission case. Remove O-ring and discard.
2. Remove bearing cone (6) from the sliding gear shaft (8) and store with bearing cap (3). If bearings are to be reused always reinstall with the original cup and in same location.
3. Remove 20T gear (7) from the shaft.
4. Move the shaft to the left and remove the bearing cone (17) and gear (9) from the shaft and case.
5. Remove sliding gear shaft (8) from the case.
6. Remove two capscrews (10), and bearing cap (12) with shims from the transmission case. Store cap, shims, cup (16) with right bearing cone (17) removed from shaft (8).
7. Remove two capscrews (18) and cap (20) from the transmission case.
8. Remove shims (22-25) and cup (26) from the transmission case and store with cap (20). Remove O-ring (21) and discard.

NOTE: Some early production 616 tractors do not have O-ring.

9. Move the input shaft (33) left as far as possible and remove bearing cone (42), gears (30-32) and two spacers (29-31).
Remove input shaft, bevel gear (28) and bearing cone (27) from the top of the case.
10. Remove two capscrews, (34) bearing cap (36), shims (37-40) and bearing cup (41). Store bearing cone (42) with cap (36) bearing cone (42) and shims.

Bevel gear components

1. Capscrew
2. Washer
3. Cap, retaining
4. O-ring
5. Bearing cup
6. Bearing cone
7. 20T gear
8. Sliding gear shaft
9. Low and 2nd gear assembly
10. Capscrew
11. Lockwasher
12. Cap
13. Shim, 0.005
14. Shim, 0.003
15. Shim, 0.002
16. Bearing cup
17. Bearing cone
18. Capscrew
19. Lockwasher
20. Cap
21. O-ring
22. Shim, 0.010
23. Shim, 0.005
24. Shim, 0.003
25. Shim, 0.002
26. Bearing cup
27. Bearing cone
28. Bevel gear assembly
29. Spacer
30. Gear, 23T
31. Spacer
32. Gear, 18T
33. Input shaft
34. Capscrew
35. Lockwasher
36. Cap
37. Shim, 0.010
38. Shim, 0.005
39. Shim, 0.003
40. Shim, 0.002
41. Bearing cup
42. Bearing cone
43. Differential case

11. Clean all parts and case with suitable cleaning solvent. Check all bearing cones, cups and other parts for wear or damage. If a bearing cone is replaced the original cup must be replaced also.

12. The lefthand cone is a press fit to the input shaft and should not be removed unless the bearing cone, bevel gear, or cup is to be replaced or the shim thickness (22-25) is to be changed. Use a suitable bearing puller to remove the cone from the shaft.

NOTE: DO NOT press cone (27) on the input shaft at this time.

13. Check the internal splines and gear teeth for wear. Check the shafts for wear or chipped splines.

ASSEMBLY

1. If the lefthand input shaft bearing cone (27), cup (26), cap (20), or transmission case is to be replaced the shim stack height will need to be corrected. The bevel gear to pinion lash is determined by the location of the bevel gear in the transmission case with these shims.

2. To determine the height of the shim stack refer to the chart.

3. Place the lefthand bearing cap with the hub up. Place the bearing cup on the cap hub and bearing cone in the cup. Rotate the cone to be sure cone is seated in the cup.

4. Measure the distance from the top surface of the cap to the top surface of the bearing cone inner race and record. This dimension will be 1.25+ in.

5. A case dimension is stamped on the transmission case to establish the bevel gear location. This dimension will be a four digit number 3.11+.

6. Refer to the chart to select correct

DIAL INDICATOR READ OUT	DIMENSION "A"										
		3.118	3.119	3.120	3.121	3.122	3.123	3.124	3.125	3.126	3.127	3.128
DIMENSION "B"	1.255	.012	.011	.010	.009	.008	.007	.006	.005	.004	.003	.002
	1.257	.014	.013	.012	.011	.010	.009	.008	.007	.006	.005	.004
	1.259	.016	.015	.014	.013	.012	.011	.010	.009	.008	.007	.006
	1.261	.018	.017	.016	.015	.014	.013	.012	.011	.010	.009	.008
	1.263	.020	.019	.018	.017	.016	.015	.014	.013	.012	.011	.010
	1.265	.022	.021	.020	.019	.018	.017	.016	.015	.014	.013	.012
	1.267	.024	.023	.022	.021	.020	.019	.018	.017	.016	.015	.014
	1.269	.026	.025	.024	.023	.022	.021	.020	.019	.018	.017	.016
	1.271	.028	.027	.026	.025	.024	.023	.022	.021	.020	.019	.018
	1.272	.030	.029	.028	.027	.026	.025	.024	.023	.022	.021	.020

Bevel gear pinion shim chart

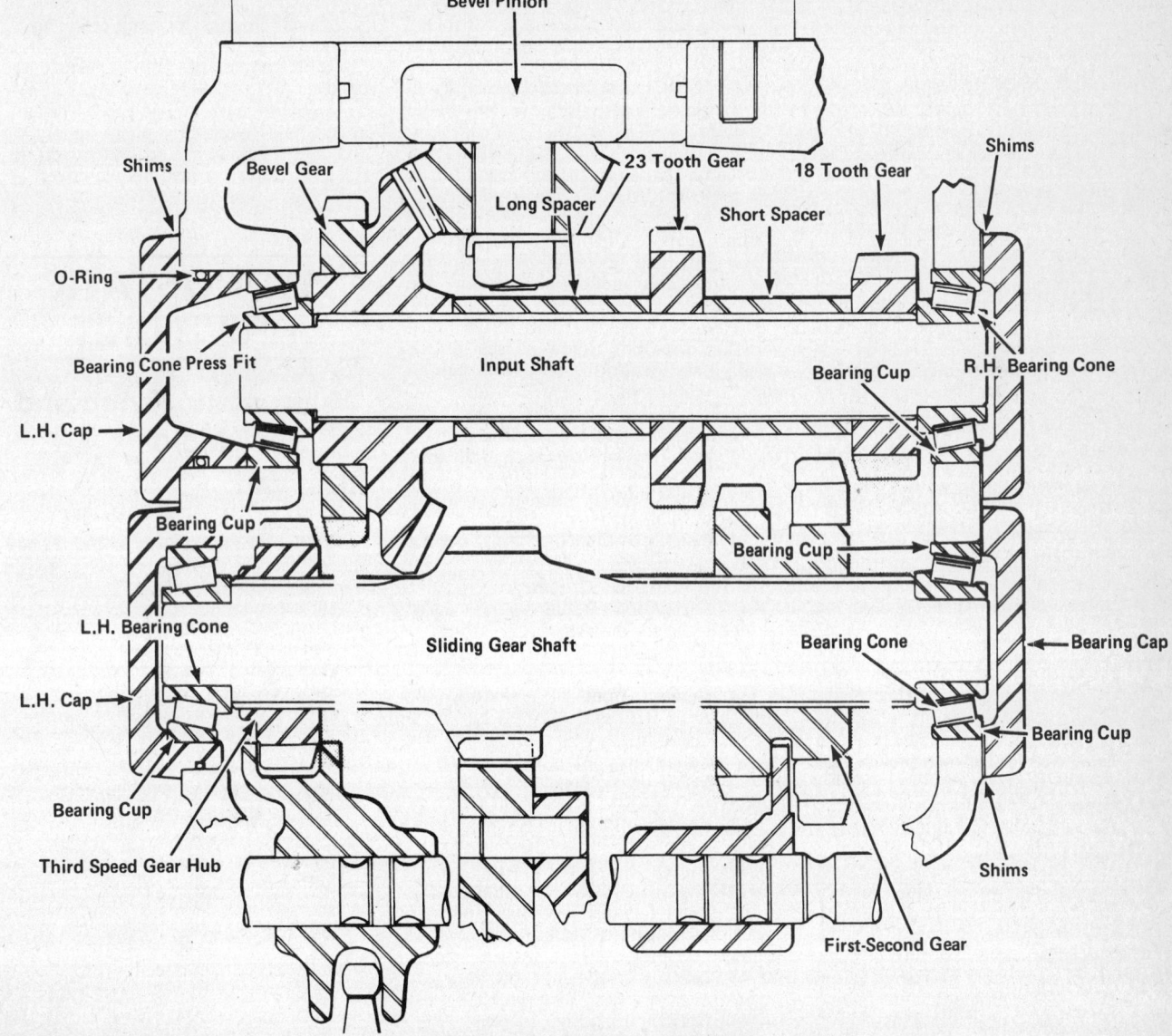

Bevel gear installation on input shaft

Allis-Chalmers

shim stack height. The top line numerals are stamped on the transmission case. The lefthand column of numerals is bearing and cap dimensions determined in steps 3 and 4.

7. **Example:** Case dimension is 3.119 Bearing and cap dimension is 1.267

Locate the numeral 1.267 in the lefthand column and move toward the right on the same line to the column directly under the case dimension 3.119. The numeral 0.023 is in this column and is the correct shim height required. Shims are supplied in .002, .003, .005 and .010 thickness.

Use one .003 and two .010 thickness shims for .023.

8. Install the bevel gear on the input shaft as shown. Press the lefthand bearing cone on the input shaft.

NOTE: This is the bearing cone used in steps 3 and 4 to determine shim stock thickness. This bearing will be on lefthand side of the transmission.

9. Install the assembled gear, bearing and input shaft in the case and hold to the left as far as possible.
10. Install the long spacer, second speed gear (23T) with the tooth chamfer away from the bevel gear, short spacer, first gear (18T) with the tooth chamfer toward the bevel gear.
11. Lubricate the bearing cup (used in paragraph 3-4) and install cup on cone. Install O-ring on bearing cap and lubricate. Apply a light coat of Permatex® to the shims determined in steps 3 and 4 and install bearing cap with shims in place on transmission case.
12. Apply a suitable sealer to the capscrew threads and torque the capscrews to 20 ft.lb.
13. Lubricate the righthand bearing cone and install on the input shaft. Install the bearing cup in the transmission case. Install the bearing cap with the shims that were removed during disassembly. If the shims were destroyed or lost, install approximately .040 shims of various thickness to start. Shims are available in .002, .003, .005 and .010 thickness.
14. Tap the caps to seat the bearing cones and cups.
15. Check the input shaft for end play and bearing adjustment.
16. The shaft should rotate freely without end play. If a vertical or lateral movement of the shaft or gears is apparent remove shims from between the righthand bearing cap and transmission case to obtain correct adjustment. Do not preload the bearings.
17. When the adjustment is correct the spacers can be rotated without the gears and shaft rotating.
18. After the correct bearing adjustment is made remove the cap and shims. Apply a light coat of Permatex® to the shims and a suitable sealer to the capscrew threads. Torque the capscrews to 20 ft.lb.
19. Install the sliding gear shaft in the transmission case with the first and second speed gear installed on the shaft as shown.
20. Install the third speed gear on the sliding gear shaft with the hub toward lefthand as shown. Lubricate and install the lefthand bearing cone on the shaft. Press the lefthand bearing cup in the cap. Lubricate a new O-ring and install on the cap. Install the cap on the transmission case.
21. Apply a sealant to the capscrew threads and torque to 20 ft.lb. Lubricate the righthand bearing cone on the sliding gear shaft. Install the bearing cup in the transmission case. Install the shims that were removed during disassembly on the cap and install the cap on the case. If the shims are destroyed and require new ones a shim stack of .040 of shims of various thickness could be used to start. Shims are available in .002, .003, .005 and .010 thickness.
22. Torque capscrews to 20 ft.lb. Tap the caps to seat the bearings.
23. The shaft should rotate freely without end play.
24. Check the shaft for vertical or lateral movement and remove shims to remove end play. If the bearings are preloaded add shims to adjust bearing to a free rolling fit.
25. After the bearings are adjusted correct remove cap and apply a light coat of Permatex® to the shims and sealer to the capscrew threads. Torque capscrews to 20 ft.lb.
26. Install the differential assembly in the transmission case. Refer to Differential Assembly.
27. Install new shifter shaft level oil seal (19) in transmission case and install shifter shafter lever (22) in the transmission case.
28. Move the shift forks (4-7) so that sliding gears are in the neutral position. Place shift stem (2) in the recesses of the shift forks and secure with capscrew and lockwasher.
29. Install top PTO shaft in the transmission case. Use a new oil seal on the shaft. Use Loctite® on the clutch hub attaching capscrews.
30. Install the hydrostatic pump and motor assembly on the transmission.
31. Install transmission case in the tractor.
32. Install brakes (refer to Brake Installation section).
33. Install the following:
- Axle extensions.
- Lift shaft, and lift ram.
- Transmission oil filter.
- The tractor drawbar.
- Drop housing.
- Wheels.
34. Connect front PTO shaft to top transmission PTO shaft.
35. Install hydrostatic transmission if it was not done in step 31.
36. Install oil temperature sending unit and wire.
37. Connect oil filter to charge pump oil line.
38. Install the transmission safety switch and hydrostatic control linkage.
39. Refer to the illustrations and connect clutch shaft to pump input shaft as shown.
40. Install top hydraulic oil lines and items removed during disassembly.
41. Install transmission cover.
42. Fill transmission with type A automatic transmission fluid. Adjust the brakes.
43. Start the tractor and check hydraulic system and hydrostatic pump and motor and adjust. Adjust neutral start switch. Install seat, top cover and bottom cover.

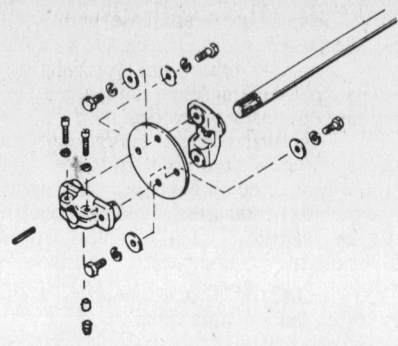

616 clutch shaft-to-pump coupling

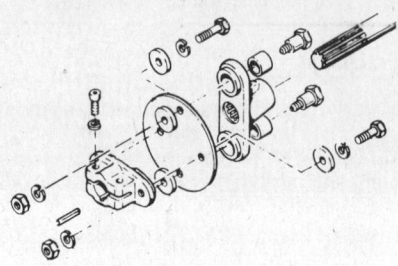

620 clutch shaft-to-pump coupling

HYDROSTATIC DRIVE UNIT

Hydrostatic Pump and Motor

REMOVAL

1. Remove the seat.
2. Remove the six washer headed capscrews that fasten the top cover to the frame and remove cover.

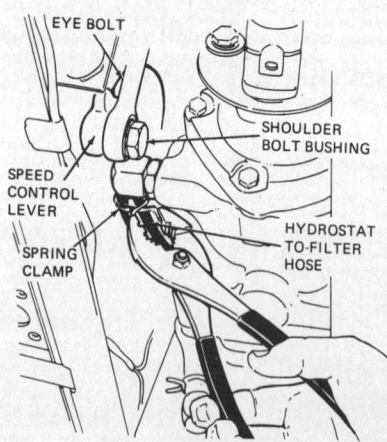

Removing filter hose

3. Disconnect the pump to spool valve oil tube (steel).

4. Remove hose clamp to disconnect the cooler oil return line.

5. Remove the hose clamp and disconnect oil filter hose at the charge pump elbow.

6. Disconnect the speed control lever by removing the shoulder bolt and bushing from the eyebolt.

7. Remove the capscrew, remove the speed control lever and locating pin from the hydrostatic pump. It will be necessary to remove the switch and bracket from pump case on 620.

8. Remove the ten washer headed capscrews that attach the bottom cover to the tractor frame and remove the cover.

9. Remove the wire from the oil temperature sending unit.

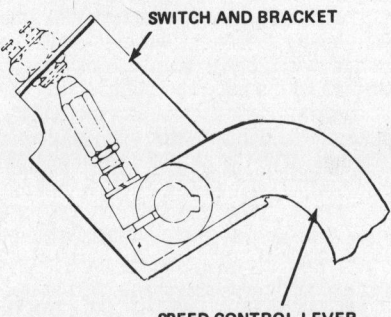

Disconnecting speed control lever

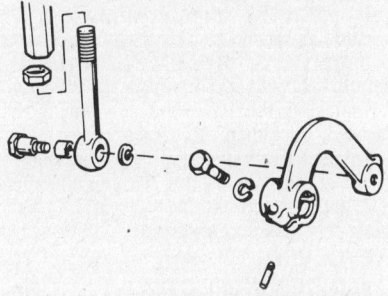

Removing switch and bracket from pump case

10. Place a drain pan under the transmission and remove the sending unit to drain the transmission oil.

11. Remove the capscrews, nuts, washers, and spacers that secure the driveshaft yokes to the coupling disc.

12. Loosen the two socket head capscrews in the hydrostatic pump input shaft yoke.

13. Slide the coupling from between the yokes.

14. Remove the yoke from the driveshaft.

NOTE: The 616 yoke is held in place with a setscrew. Loosen the setscrew before attempting to remove the yoke.

15. Remove the yoke and key from the hydrostatic pump input shaft.

16. Remove access cover in lefthand tractor frame.

17. Remove the two washer nuts attaching the hydrostatic pump and motor to the transmission case.

18. Move the hydrostatic pump and motor forward to remove it through the opening in the bottom of the tractor frame.

OVERHAUL
Charge Pump Relief Valve

The charge pump relief valve is located in righthand front section of the pump housing. This valve is adjusted to hold 70 to 150 psi in the hydrostatic unit. This oil will lubricate and cool the internal parts of the unit, and the excess oil from the charge pump will flow through the open center spool valve and cooler; and then return to sump.

REMOVAL AND INSPECTION

1. Remove the ⅝ in. recessed plug located on the righthand side of the pump housing. The plug may have shims in its recess.
2. Remove the spring and ball.
3. Inspect the ball for flat spots and erosion, and replace the ball if it is damaged.
4. Check the condition of the valve seat.

INSTALLATION

1. Lubricate parts during installation. Place the ball and spring in the housing.
2. Install a new O-ring on the plug.
3. Replace the shims that were in the plug when the plug was removed, and place the plug in the housing.
4. Start the engine and run long enough to fill hydrostatic unit, and oil lines with oil.
5. Check the charge pump pressure.
6. Add shims if the pressure is below 70 psi, or remove shims if the pressure exceeds 150 psi.

NOTE: Do not mix charge pump valve parts with implement lift valve parts.

Implement Lift Relief Valve

The implement lift relief valve is located in the lefthand front section of the pump housing. This valve is set to bypass the oil when the pressure exceeds 700 psi when the spool valve is moved to activate the hydraulic system.

REMOVAL

1. Remove the ⅝ in. recessed plug located on the lefthand side of the pump housing. This plug may have shims in its recess.
2. Remove the spring and cone shaped valve.
3. Inspect the valve for flat spots and erosion and replace the valve if it is scored or damaged.
4. Inspect the valve seat.

INSTALLATION

Lubricate parts during installation.

1. Place the cone shaped valve and the spring in the housing so the spring contacts the end of the valve.
2. Install a new O-ring on the plug.

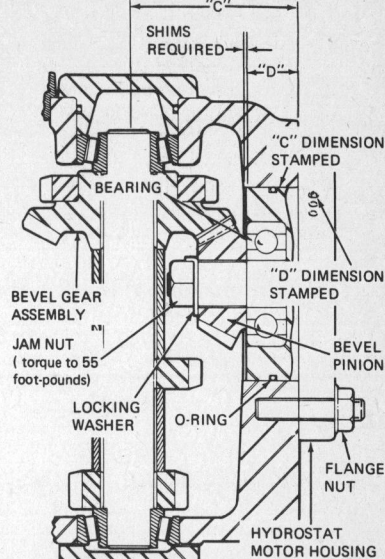

Shim positioning to regulate pump pressure

Allis-Chalmers

DIMENSION "D"
HYDROSTATIC PUMP MOTOR (On Motor Housing)

DIMENSION "C" TRANSMISSION NUMBER (On Case)	900	901	902	903	904	905	906	907	908	909	910	911	912	913	914	915
3000	.038	.037	.036	.035	.034	.033	.032	.031	.030	.029	.028	.027	.026	.025	.024	.023
3001	.039	.038	.037	.036	.035	.034	.033	.032	.031	.030	.029	.028	.027	.026	.025	.024
3002	.040	.039	.038	.037	.036	.035	.034	.033	.032	.031	.030	.029	.028	.027	.026	.025
3003	.041	.040	.039	.038	.037	.036	.035	.034	.033	.032	.031	.030	.029	.028	.027	.026
3004	.042	.041	.040	.039	.038	.037	.036	.035	.034	.033	.032	.031	.030	.029	.028	.027
3005	.043	.042	.041	.040	.039	.038	.037	.036	.035	.034	.033	.032	.031	.030	.029	.028
3006	.044	.043	.042	.041	.040	.039	.038	.037	.036	.035	.034	.033	.032	.031	.030	.029
3007	.045	.044	.043	.042	.041	.040	.039	.038	.037	.036	.035	.034	.033	.032	.031	.030
3008	.046	.045	.044	.043	.042	.041	.040	.039	.038	.037	.036	.035	.034	.033	.032	.031
3009	.047	.046	.045	.044	.043	.042	.041	.040	.039	.038	.037	.036	.035	.034	.033	.032
3010	.048	.047	.046	.045	.044	.043	.042	.041	.040	.039	.038	.037	.036	.035	.034	.033
3011	.049	.048	.047	.046	.045	.044	.043	.042	.041	.040	.039	.038	.037	.036	.035	.034
3012	.050	.049	.048	.047	.046	.045	.044	.043	.042	.041	.040	.039	.038	.037	.036	.035
3013	.051	.050	.049	.048	.047	.046	.045	.044	.043	.042	.041	.040	.039	.038	.037	.036
3014	.052	.051	.050	.049	.048	.047	.046	.045	.044	.043	.042	.041	.040	.039	.038	.037
3015	.053	.052	.051	.050	.049	.048	.047	.046	.045	.044	.043	.042	.041	.040	.039	.038
3016	.054	.053	.052	.051	.050	.049	.048	.047	.046	.045	.044	.043	.042	.041	.040	.039
3017	.055	.054	.053	.052	.051	.050	.049	.048	.047	.046	.045	.044	.043	.042	.041	.040
3018	.056	.055	.054	.053	.052	.051	.050	.049	.048	.047	.046	.045	.044	.043	.042	.041
3019	.057	.056	.055	.054	.053	.052	.051	.050	.049	.048	.047	.046	.045	.044	.043	.042
3020	.058	.057	.056	.055	.054	.053	.052	.051	.050	.049	.048	.047	.046	.045	.044	.043
3021	.059	.058	.057	.056	.055	.054	.053	.052	.051	.050	.049	.048	.047	.046	.045	.044
3022	.060	.059	.058	.057	.056	.055	.054	.053	.052	.051	.050	.049	.048	.047	.046	.045
3023	.061	.060	.059	.058	.057	.056	.055	.054	.053	.052	.051	.050	.049	.048	.047	.046
3024	.062	.061	.060	.059	.058	.057	.056	.055	.054	.053	.052	.051	.050	.049	.048	.047

Shim selection chart

Allis-Chalmers

3. Place the shims in the recess in the plug, and place the plug in the housing. (Start with four shims.)

4. Start the engine and run long enough to fill hydrostatic unit and oil lines with oil.

5. Move the spool valve to extend the ram and check the left system pressure. It should be between 550 and 700 psi. If the pressure is low, add shims to bring the pressure within this range. If the pressure is high, remove shims to bring the pressure within this range.

NOTE: Do not confuse this valve with the charge pump relief valve.

Inspection of Charge Pump

The charge pump is located over the exposed end of the pump input driveshaft.

REMOVAL—REMOVE DRIVE SHAFT AND UNIVERSAL JOINTS

1. Remove external couplings and keys. Inspect the driveshaft for sharp edges or raised portions that could damage the oil seal. Smooth the shaft if necessary. If damage is excessive, driveshaft replacement is required. Inspect for misalignment of the driveshaft.

2. Remove the four (4) capscrews which hold the charge pump housing to unit. Carefully pull the housing from the driveshaft. It is recommended that the splines or key ways be wrapped before the seal is pulled over surfaces. Inspect the seal for wear or damage. Replace if necessary.

3. Pull charge pump assembly, (gerotor type consisting of rotor and roller) from the drive pin and driveshaft. Inspect the pin for cracks or bends.

4. Inspect charge pump for scoring, pitting, pickup, or excessive wear. Replace as an assembly if damaged. A damaged pump can damage the charge pump housing.

5. Inspect the charge pump housing cavity which contains the charge pump for scoring, pitting, pickup, or excessive wear. A damaged housing can damage the charge pump and/or cause low oil flow and/or pressure. A new seal is to be used if a new housing is installed.

6. Seal may be removed by inserting tool in I.D. and roll-out. Do not distort housing in process. Discard seal.

7. Discard O-ring.

INSTALLATION

1. Lubricate all rotating parts, oil seals, etc., with a light coat of pressure gun grease.

2. Install a new seal on the charge pump housing until it seats by pressing squarely on outer edge. A light coating of lubrication aids installation. Check for garter spring on I.D.

3. Install the drive pin into the drive shaft and locate the charge pump over the pin on the drive shaft.

4. Install a new O-ring in the charge pump housing over charge pump. Torque screws to 19-21 ft.lb.

NOTE: Orientation of the charge pump housing to charge pump and unit housing must be only one way. Install the charge pump housing for counterclockwise or lefthand rotation.

The cavity in the charge pump housing which accepts the charge pump is offset from the driveshaft centerline. This offset creates the pumping action. The large offset side has ½ in. dimple cast $1/10$ in. deep on its outer wall. This dimple is to go to the top righthand corner of the pump case. (Looking at the input shaft.) Refer to the Figure for correct installation per input rotation and offset or cast circle.

Inspection of Check Valves

The check valves (2) are in the communicating oil passages between the charge pump and the pump to motor oil passages. They are ball and spring type valves without any psi specifications. The valves allow the passage of oil from the charge pump to the pump, motor and oil passages in the unit but prevent the return of oil from pump, or motor, to the charge pump. Each valve is located in an oil passage in the center section.

REMOVAL AND INSPECTION

1. Remove the four capscrews that hold the center section between the pump and motor housings.

2. Separate the center section from the pump and motor housings.

3. Remove the plugs, O-rings, springs, and balls.

4. Inspect the springs and balls. Replace if broken or damaged.

INSTALLATION

1. Place the balls and springs in the housing.

2. Put new O-rings on the plugs.

3. Put the plugs in the center section.

Inspection of Accelerating Valve Assemblies

There are two acceleration valves located in the center section. The reverse acceleration valve is located in this lefthand side behind a ⅞ in. plug. The reverse acceleration valve has a small orifice drilled in the first groove. The forward acceleration valve is in the righthand side of the center section. The valve is held in the body with a ⅞ in. plug. The accelerating valve spring is in the oil passage between the two valves and when it is installed properly it will be in the recess of each valve face.

REMOVAL AND INSPECTION

1. Remove the two ⅞ in. plugs in the lower center section.

2. Inspect the external area of each valve.

3. Each valve must move to and away from its seat without any bind. A quick check would be to install the spring in the center section and hold one valve flush with the housing then force the opposite valve against the spring until the valve contacts its seat. The spring should move the valve back from its seat until the spring returns to its free length. Check the other valve in the same manner. If the valves or bore are rough it may be possible to polish them with a fine polishing cloth. Check the bore in the housing and polish if necessary.

4. The valves each have a ball type relief valve that is held on its seat with a spring. This valve allows the oil that is trapped between the cap and plug to escape readily when the spring moves the valve against the plug. The cap can be removed to inspect the spring and ball. Replace the O-ring if the cap is removed.

REPLACEMENT

1. Install new O-rings on the plugs. Install the forward accelerating valve assembly in the righthand port and torque plug.

2. Install the accelerating valve spring in the oil passage. Be sure the spring enters the recess in the forward accelerating valve.

3. Install the reverse accelerating valve. Be sure spring enters the recess in the reverse accelerating valve. Install the plug and torque.

Hydrostatic Motor Inspection

The pump can be separated into three sections by removing four capscrews. For easier assembly, the sections can be marked with a chisel on pump.

The pump valve plate and motor valve plate are held in position with the center bearings. The valve plate retaining pins prevent the valve plates from rotating.

The motor valve plate has four hydrostatic oil relief areas. These are necessary because the revolving cylinder turns clockwise or counterclockwise.

The pump valve plate has only two hydrostatic oil relief areas. The revolving cylinder always rotates in the same direction. The revolving cylinder is held against the valve plate by an internal spring. The spring can be checked by applying pressure against the revolving cylinder face. It should be possible to push the revolving cylinder in the case and the spring will move it out when the pressure is released.

Inspect the valve plate for scratches or wear. It may be possible to use a face plate and lapping compound to put the valve plate in a usable condition.

REVOLVING CYLINDER INSPECTION

1. Remove the revolving cylinder from the output shaft.

2. The pistons can be removed from the cylinder.

3. Check the pistons for wear and score marks.

4. Check the piston shoes for wear.

5. The center area of the shoes is recessed for lubrication and oil pressure. The oil pressure in this area minimizes the pressure of the outer ring that contacts the swash plate. There is an orifice drilled through the shoe plate and bottom of the piston to allow oil to be pressurized in the recess in the piston shoe.

6. Check the shoe plate. A bent shoe plate will allow piston chatter which erodes the piston shoes and swash plate. Place shoe plate on a face plate or flat surface to check for distortion.

7. Check the bearing area of the revolving cylinder and shoe plate. The shoe plate must pivot freely on the revolving cylinder contact area.

8. Check the internal splines for damage. To remove the internal spring it will

Allis-Chalmers

be necessary to obtain a pilot with the O.D. small enough to go through the snapring. The spring will have to be compressed with a press or bolt for snapring removal.

Bearing and Shaft
REMOVAL
1. To prevent damage to the motor case during bearing removal, the bearing case should be heated to approximately 200 degrees to 250 degrees. This can be done by placing the bearing end of the case in oil that has been heated to approximately 250 degrees or careful use of a butane torch.
2. A soft hammer can be used to drive the shaft and bearing from the housing while the housing is hot.
3. If the bearing or shaft needs replacing, remove the bearing with a suitable press.

INSTALLATION
1. Press the bearing on the shaft.
2. Heat the housing as outlined in bearing and shaft removal and push bearing in the case.
3. Install the swash plate.
4. Install the revolving cylinder assembly.

NOTE: Lubricate all parts during assembly.

5. Place the motor valve plate on the center section bearing with the retaining pin in the groove.

NOTE: There are four hydrostatic oil relief areas on the motor valve plate.

6. Install new gaskets and O-rings in their proper location on the center section.
7. When installing the motor housing, note that there is a gusset cast in the housing that forms two depressions that must be on top when the assembly is bolted together. The bottom area of the motor housing is smooth. This positions the motor swash plate for proper pinion rotation.
8. Bolt the motor to the center section and pump case with the four capscrews.

NOTE: The internal pump and motor springs will have to be compressed while the capscrews are torqued.

9. Tighten the capscrews finger-tight and check the spring pressure. If the assembly is correct, both cases can be forced against the center section by hand.
10. Torque the capscrews evenly as the items are drawn together to prevent binding and damage.

Pump Inspection
The pump revolving cylinder is the same as the motor revolving cylinder. For detailed inspection information refer to the Motor Inspection Section.

The valve plate is different; it has only two hydrostatic oil relief areas.

REMOVAL
1. Remove the swash plate from the rotating trunnion.
2. To remove the rotating trunnion drive the two roll pins from the trunnion and control shaft.

NOTE: There are two short roll pins to allow ample space for removal.

3. Drive the single roll pin from the trunnion and stub shaft.
4. Remove the control shaft and stud shaft from the trunnion and pump case.
5. Remove the rotating trunnion.
6. Remove the input shaft.
7. Remove the stub shaft and the control shaft oil seals.
8. Inspect the control shaft and stub shaft bearings and replace them if necessary. If the bearings are to be replaced use a suitable fixture for replacing them to prevent damage or distortion to the bearing case.
9. Replace the oil seals. Apply a sealant to the outside surface of the seal to eliminate oil seepage. Use a suitable driver to prevent distortion of the seals.
10. Lubricate the shafts, bearings, and oil seals.
11. Check the input shaft for rough bearing surface, worn splines, etc. Check the condition of the bearing and replace if not suitable. Press the bearing on the shaft and install the bearing and shaft in the housing.
12. Place the trunnion in the pump housing and install the stub shaft and control shaft.
13. Align the hole in the trunnion with the hole in the control shaft so that the roll pins can be driven through the control shaft. Drive one roll pin down about ½ in. below the trunnion surface. Then drive the second roll pin in. Continue to drive the roll pins in until the second roll pin is approximately ¼ in. below the surface of the trunnion.
14. Align the hole in the trunnion with the hole in the end of the stub shaft. Drive one roll pin in until it is approximately ¼ in. below the surface of the trunnion.
15. Replace the washers and snaprings on the stub shaft and control shaft. Replace the swash plate.
16. Place the revolving cylinder on the driveshaft. Push back on the face of the revolving cylinder to check the internal spring. The revolving cylinder should move on the driveshaft.
17. Replace the center section gasket and charge pump oil passage O-rings.
18. Lubricate all parts during installation.
19. Install the pump valve plate on the center section bearing and retaining pin.

NOTE: There are two hydrostatic oil relief areas on the pump valve plate.

20. Install the pump on the center section and motor housing with the four capscrews. Turn the capscrews in finger-tight and check the assembly. The pump housing can be forced toward the center section with hand pressure and will be moved away with the internal spring force.
21. Torque the four capscrews evenly to prevent damage to the pump or motor.
22. Follow the pre-installation and start-up instructions before installing the unit in a tractor.
23. Install a new O-ring on the pinion.

INSTALLATION
1. If a new bearing is installed in the hydrostatic motor housing or a new bevel pinion is installed on the pump shaft always install the correct number of shims between the pinion hub and ball bearing. Torque the nut to 55 ft.lb.
2. If a complete pump and motor assembly is installed the bevel pinion location must be correct.
3. A four digit pinion locating number starting with 30-- is stamped on top of the tractor transmission.
4. A three digit number starting with 900 is stamped on the hydrostatic motor mounting flange. The numerals in steps 3 and 4 are used to determine shim thickness for pinion location.
5. One of the three digit numerals that will be stamped on the hydrostatic pump flange will be in the row of numerals across the top of the chart. The four digit numeral on top of the transmission case will be one of the numerals in the left column of the above chart.
6. **Example:** If the four digit dimension "C" stamped on the transmission case is 3011 and the three digit dimension stamped on the pump flange is 904 locate the numeral 3011 in the lefthand column of the chart and follow this row of figures to the right of the column directly under the pump dimension 904.

C Transmission Dimension   3011
D Motor Mounting Flange
    Dimension   904
Shim Thickness is   .045

Shims are available in .002, .003 and .005 thickness.

7. Install .045 thickness of shims between pinion hub and ball bearing. Install pinion and torque nut to 55 ft.lb.
8. Position the hydrostatic pump and motor on a bench with the charge pump inlet port upward. Rotate the pump shaft counterclockwise while pouring type A transmission lubricant in the charge pump inlet port. This should fill the charge pump and oil cavities in the hydrostatic pump and motor to provide lubrication for the unit's initial start.
9. Install a new O-ring on the pump case and install the assembly in the tractor frame. Torque the two flange nuts that fasten the hydrostatic unit to the transmission case.
10. Install the driveshaft coupling assembly, pump shaft key, and yoke. To use proper capscrews etc. refer to 616 diagram. Refer to 620 diagram, for correct material to install coupling.
11. Move the pump shaft yoke to obtain ⅛ to ¼ in. clearance between the oil seal and yoke. Torque the two Allen head capscrews.
12. Install the transmission oil temperature sending unit.
13. Install the alignment pin in the pump trunnion shaft with the head of the pin toward the front of the tractor. On 620 install the switch bracket. Install the speed control lever on the shaft and torque capscrew to hold the lever in place. The lever should be located so that a clearance of ⅛ in. is between the lever and oil seal.
14. The 620 safety switch actuating detent is attached to the speed control capscrew and will need to be adjusted to

Allis-Chalmers

actuate the safety switch for starting. Adjust detent so switch closes the circuit when the hydrostatic lever is in neutral. Adjust 616 switch to close the circuit when the hydrostatic lever is in neutral. Switch Kit 2087558 is available to improve 616 safety switch performance.

15. Install access hole cover in the tractor frame.
16. Connect the speed control lever to the eyebolt with the shoulder bolt and bushing.
17. Connect the transmission to oil filter hose.
18. Connect oil cooler to hydrostatic pump oil return line. Connect charge pump to spool valve oil line (Steel).
19. Remove the transmission filler plug and fill the transmission to the proper level with Power Fluid 401, or its equivalent.
20. Remove the spark plugs from the engine.
21. Remove one pipe Allen type pipe plug from the front section of the hydrostatic pump and pour some type A transmission oil in this port.
22. Rotate the charge pump with the engine starter until oil flows from the open port.
23. Install a pressure gauge in the open port so that charge pump pressure and implement relief valve pressure can be checked.
24. Move the transmission gear selector lever to have the transmission in neutral.
25. Replace the spark plugs in the engine and operate the engine at idle speed to fill the hydraulic system with oil.
26. Move the spool valve lever to extend and retract the ram to eliminate air pockets.
27. Check the charge pump pressure. Correct charge pump pressure is 70-150 psi.
28. Move the spool valve to extend the ram and activate implement lift relief valve. The pressure should be 550-700 psi.
29. Remove the pressure gauge and install the pipe plug. Use a sealer on the pipe plug threads.
30. Replace the top cover, bottom cover, seat etc.

Drop Housing
DISASSEMBLY

1. Position the tractor safely on suitable stands or blocks.
2. Remove the tire and wheel.
3. If the drop housing requires disassembly drain the lubricant from the housing before removing it from the tractor.
4. Remove the capscrews, nuts and lockwashers that attach the wheel guard to the drop housing.
5. Remove the lockwashers and capscrews attaching the drop housing to the axle extension.
6. Slide the drop housing away from the axle extension. When the pinion shaft clears the coupling the assembly can be moved to a suitable work area.
7. Remove the drop housing plate by removing the ⅜ x 1¼ in. capscrews and lockwashers. Tap plate to remove from housing.

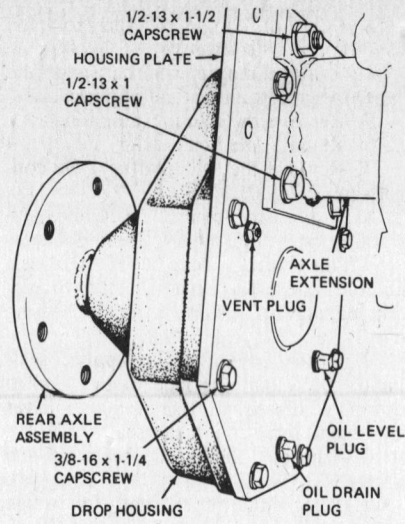
Drop housing

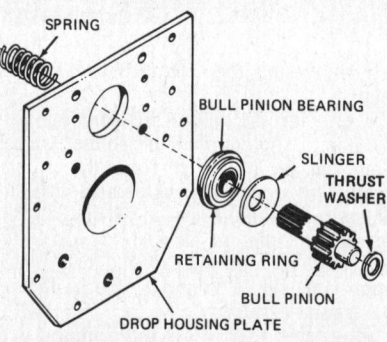
Bull pinion shaft and bearing removal

8. Remove the spring from the bull pinion shaft and press the shaft and bearing from the plate.
9. Remove the bull pinion thrust washer from the drop housing.
10. Check the vent and remove if damaged.
11. Press the pinion shaft from the bearing and remove the slinger.
12. To remove bull gear, bearings or axle shaft from the housing.
13. Remove locknut and flat washer from the rear axle assembly and remove bull gear.
14. Remove the rear axle assembly from the gear case.
15. Remove the inner and outer bearing cones and oil seal.
16. Clean the gear case and all parts in a suitable solvent and inspect the gear and bearings. Replace the gear if worn or chipped and bearings if they are rough and worn. Check bull pinion roller bearing and remove if replacement is necessary. Remove the two bearing cups if the axle bearing cones are to be replaced.
17. Use a brush to remove dirt or oil from the bull pinion shaft sealed bearings. Do not allow the sealed bearings to remain in the solvent. If EP90 oil enters the brake housing from the drop housing replace the sealed bull pinion shaft bearings.
18. Rotate the sealed bearings to check for damage or wear.

ASSEMBLY

1. Press the bull pinion roller bearing, inner and outer bearing cones in the drop housing. Replace the roll pins if they have been removed.
2. Lubricate the outer bearing cone with SAE 90EP and place in the housing.
3. Press a new oil seal in the housing. Lubricate the shaft contact area of the seal with Lubriplate®.
4. Install the rear axle assembly through the oil seal and outer bearing cone. Install the inner bearing cone.
5. Install the bull gear on the axle shaft.
6. Install the flat washer and locknut.
7. Torque the locknut to slightly preload the bearings. Tap each end of the axle to seat the cones and cups. Place a socket wrench on the nut and use an in.lb. torque wrench to check the effort required to rotate the axle assembly. Carefully continue to torque the nut until it requires twelve in.lb. to rotate the axle in the bearings.
8. Install the slinger on the bull pinion shaft and press the sealed bearing on the pinion shaft.

NOTE: The bearing must be installed on the shaft so that the narrowest area between the retaining ring and bearing edge is toward the pinion.

9. Press the sealed bearing in the drop housing plate.
10. Lubricate the bull pinion roller bearing and install the thrust washer on the pinion shaft.
11. Coat the gasket area of the plate and housing with a sealer and install a new gasket. Install new vent plug if necessary.
12. Install the assembled bull pinion and plate on the drop housing. Tighten the eight ⅜ x 1¼ in. capscrews.
13. Place the coupling on the differential output shaft, spring on the bull pinion shaft, and install the drop housing assembly on the axle extension. Torque the nuts and capscrews that attach the drop housing to the axle extension and to the wheel guards. Install the wheel and tire.
14. Fill the drop housing with SAE 90EP oil to the correct level.

POWER TAKE-OFF SYSTEM

PTO Shaft
REMOVAL

1. Raise the tractor hood and disconnect the negative battery terminal. Fasten battery cable to prevent it from swinging back and touching the negative terminal.
2. Remove the four capscrews that hold the seat support to the top of the transmission housing, and remove the seat and support.

**NOTE: Be sure that electric wires to the lights in the righthand and lefthand fenders are disconnected from the terminal if the wires are strung over the seat support structure.

3. Remove the four ¼ in. capscrews that fasten the frame cover to the frame and remove the cover.

4. Remove the two capscrews that hold the lefthand end of the fuel tank straps to the frame. Disconnect the fuel lines and remove the fuel tank.

5. Remove the upper rear oil cooler shroud.

6. Remove the righthand and lefthand oil cooler capscrews and retaining washers. Swing the oil cooler out of the way to righthand side and remove the righthand and lefthand oil cooler shrouds.

7. Remove the capscrews that fasten the righthand and lefthand side of the console to the tractor frame to permit raising the rear side of console enough to provide clearance to remove rear PTO shaft assembly.

8. Remove the four capscrews that attach the electric clutch to the hub (37). Disconnect electric connector. The 616 tractor has a center PTO shaft bearing (11). Remove the two capscrews that attach the center bearing support (15) to the tractor frame. Remove the capscrew attaching clutch bracket (11) to the tractor frame.

9. Remove the two capscrews, spacers and washers that attach the coupling disc (6) to the flywheel. The 616 yoke has an Allen setscrew that secures the yoke (8) to the shaft (9). Loosen Allen screw.

10. Allow the front section of the shaft with yoke to drop down and remove the electric clutch and PTO shaft from the rear.

616 DRIVE SHAFT DISASSEMBLY

1. Remove the two capscrews, washers, and lockwashers to remove the coupling disc (6) from yoke (8).

2. Loosen setscrew to remove yoke (8) from shaft (9).

3. Loosen bearing set collar and remove bearing (11) and support (15) from the shaft.

4. Remove capscrew (33) lockwasher, and collar (31).

5. Move the shaft forward to expose the retaining ring (20) on the bearing end of the clutch shaft. Remove retaining ring (20) and bearing retaining collar (21) from the bearing retainer. Remove bearing (23) from the shaft.

6. Remove clutch assembly from the shaft. Separate the clutch assembly and remove the metal washer (26) from between the field and armature. Remove key (19).

7. Inspect bearings, shafts and other components for wear or damage and replace if necessary.

ASSEMBLY AND INSTALLATION 616 DRIVE SHAFT

1. Install key (19) in clutch (39).

2. Assemble bearing (23) in retainer (22) with washer (26).

3. Align key (19) with key slot in shaft (9) and install clutch on shaft.

4. Move clutch on the shaft sufficient to expose ring groove and install a new retaining ring (20) in the groove.

5. Move clutch back on retaining ring (20) and secure clutch to the shaft (9) with collar (31), lockwasher, and ⅜-16 x 1¼ capscrew (33). Apply Loctite® to capscrew and torque to 45 ft.lb.

6. Place center bearing (11) and bearing support (15) on the shaft.

7. Fasten disc coupling (6) to yoke (8) with two capscrews, lockwashers, and flat washers.

8. Install yoke on the shaft.

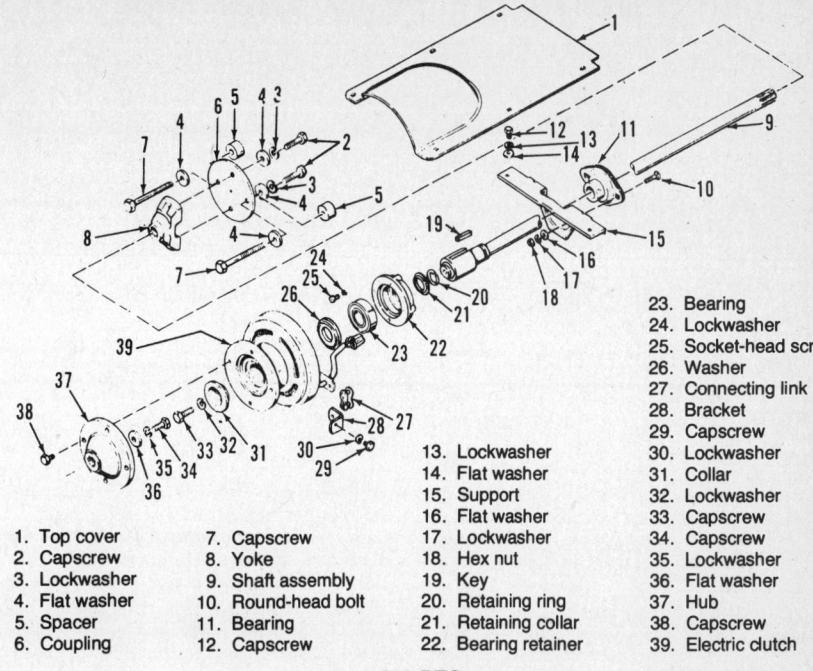

23. Bearing
24. Lockwasher
25. Socket-head screw
26. Washer
27. Connecting link
28. Bracket
29. Capscrew
30. Lockwasher
31. Collar
32. Lockwasher
33. Capscrew
34. Capscrew
35. Lockwasher
36. Flat washer
37. Hub
38. Capscrew
39. Electric clutch

1. Top cover
2. Capscrew
3. Lockwasher
4. Flat washer
5. Spacer
6. Coupling
7. Capscrew
8. Yoke
9. Shaft assembly
10. Round-head bolt
11. Bearing
12. Capscrew
13. Lockwasher
14. Flat washer
15. Support
16. Flat washer
17. Lockwasher
18. Hex nut
19. Key
20. Retaining ring
21. Retaining collar
22. Bearing retainer

616 PTO

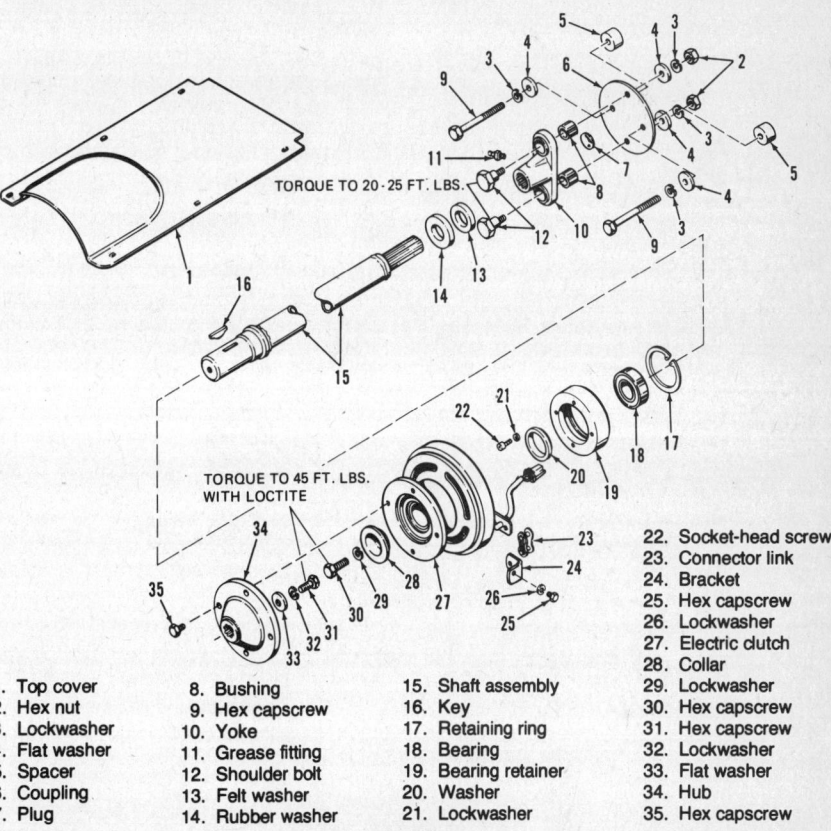

1. Top cover
2. Hex nut
3. Lockwasher
4. Flat washer
5. Spacer
6. Coupling
7. Plug
8. Bushing
9. Hex capscrew
10. Yoke
11. Grease fitting
12. Shoulder bolt
13. Felt washer
14. Rubber washer
15. Shaft assembly
16. Key
17. Retaining ring
18. Bearing
19. Bearing retainer
20. Washer
21. Lockwasher
22. Socket-head screw
23. Connector link
24. Bracket
25. Hex capscrew
26. Lockwasher
27. Electric clutch
28. Collar
29. Lockwasher
30. Hex capscrew
31. Hex capscrew
32. Lockwasher
33. Flat washer
34. Hub
35. Hex capscrew

620 PTO

Allis-Chalmers

NOTE: The complete assembly may be easier to install if the shaft assembly is installed under the console and with the yoke and disc coupling removed. Install the yoke and coupling on the driveshaft before attaching clutch to rear hub (34).

9. Place the driveshaft assembly in the tractor. Install the front yoke with coupler attached. Align the clutch bracket (28) with hole in the frame to install capscrew (29).

10. Apply Loctite® to the four capscrews (38) that secure hub to the clutch. Torque capscrews to 30 ft.lb.

11. Torque clutch bracket capscrew (29) to secure bracket to frame and connect electrical fitting.

12. Attach bearing support (15) and bearing to the frame with capscrews, lockwashers and flat washers.

13. Loosen the two carriage bolts (10) that attach bearing flange to the bearing support. Bearing support collar should be loose so that the bearing will be positioned correct when the shaft assembly installation is completed.

14. Secure the front yoke to the engine flywheel with two capscrews, spacers, flat washers and lockwashers. Apply Loctite® to the capscrew threads and torque the capscrews to keep yoke from moving freely.

15. Use a dial indicator and check the shaft for runout. Remove the spark plugs from the engine and rotate the engine by hand to check the runout. Tap the yoke to align the shaft and reduce the runout to less than .0025 in.

16. Torque the capscrews to 30 ft.lb.

17. Torque the capscrews that fasten the bearing support (15) to the frame and torque the two carriage bolts (10) to secure the bearing flange to the bearing support (15).

Rotate the bearing locking collar to secure bearing to the shaft and lock with setscrew.

620 PTO SHAFT DISASSEMBLY

1. Remove yoke (10) and coupling (6) from the shaft (15).
2. Remove the felt (13) and rubber washer (14) from the shaft.
3. Remove two shoulder bolts (12) flat washers, lockwashers attaching coupling (6) to yoke (10).
4. Check the yoke bushings (8) and replace if worn or loose in yoke.
5. Remove capscrew (30), lockwasher and clutch collar (28).
6. Remove the socket head screw (22), lockwasher (21) from bearing retainer (19).
7. Remove the field and clutch parts from the shaft.
8. Remove Woodruff key (16), washer (20).
9. Remove the bearing (18) and bearing retainer (19) from the shaft.
10. To remove bearing (18) from retainer (19) remove retaining ring (17).
11. Clean parts in suitable solvent. Do not wash lubricant from the sealed bearings. Check all parts and replace worn, damaged or parts that will not function properly.

620 PTO SHAFT ASSEMBLY

1. Install the bearing in the bearing retainer and secure with retaining ring.
2. Install the bearing and washer on the shaft.
3. Install Woodruff key in shaft.
4. Attach armature to the bearing retainer with four socket head capscrews.
5. Install the remaining clutch parts on the shaft.
6. Use Loctite® on the capscrew threads and torque to 45 ft.lb. to hold the collar and clutch assembly in place.
7. Install the rubber washer (14) and felt washer (13) on the shaft as shown.

Check yoke (10) and install plug (7) if plug was removed.

8. Assemble the coupling disc (6) and yoke (10) with the shoulder bolts, washers, lockwashers and nuts.
9. Place the shaft and clutch under the console between the tractor frame. Install the assembled front yoke on the shaft.
10. Align bracket (24) with the capscrew attaching hole in the frame.
11. Apply Loctite® to the four capscrews and attach the clutch to the hub (34). Torque capscrews to 30 ft.lb.
12. Secure the clutch bracket (24) to the frame with capscrew. Connect electrical connection.
13. Apply Loctite® to the two capscrews (9) and attach front disc coupling (6) to the flywheel. Torque capscrews to prevent the yoke from moving freely on the flywheel.
14. Use a dial indicator to check the PTO shaft runout. Remove the spark plugs from the engine and rotate the engine by

Note:

616 POWER TAKE-OFF SHAFT

There have been some failures of the front PTO yoke. The problem can be corrected by installing the following parts:

Part No.	Quantity	Description
923938	2	Capscrew
2029382	1	Coupling
917421	2	Nut ⅜ NC Jam Lock
2028584	2	Shoulder Bolt
2029380	1	Yoke
2029593	1	Plug
9002040	1	¼-28 Taper Thread Grease Fitting
2029387	1	Shaft Assembly
2029388	1	Key
2029389	1	Washer
2029390	1	Bearing
2029391	1	Bearing Retainer
2029392	1	Retaining Ring
2028628	1	Rubber Washer
2029386	1	Felt Washer
2028589	2	Bushing

These items replace the following:

⅜ NC x 3¾	2	Capscrew
2028769	1	Coupling
920415	2	Capscrews ⅜x16x1¼
2028770	1	Yoke
2028771	1	Support
2028772	1	Bearing
2028773	1	Shaft
2028571	1	Key
1124170	1	Bearing
2028775	1	Retainer Bearing
2028778	1	Ring
2029779	1	Collar
2028818	1	Washer

Refer to the 620 PTO shaft assembly when the above list of parts is used.

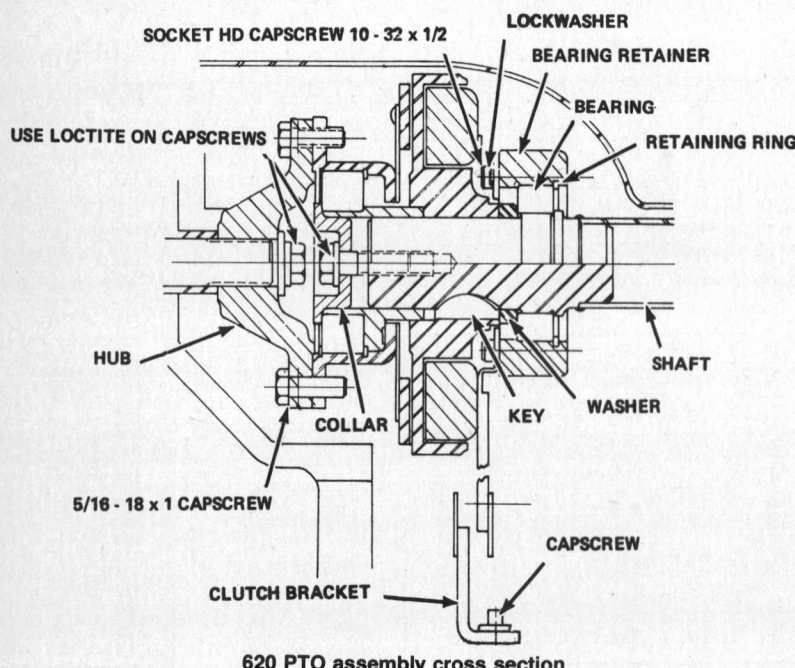

620 PTO assembly cross section

Allis-Chalmers

hand to check PTO shaft runout. Tap the shaft to center the shaft with the engine flywheel to reduce the runout to less than .020.

15. Torque the capscrews to 30 ft.lb. Lubricate the yoke splines thru lube fitting.

16. Move the felt washer (13) and rubber washer (14) against the yoke (10). Lubricate yoke splines.

17. Replace console capscrews, top cover, bottom cover, oil cooler and other items that were removed for shaft removal.

PTO Top Transmission Shaft

DISASSEMBLY

1. Remove the electric clutch and shaft assembly as outlined.

2. Remove the upper link supports and upper link from rear of transmission.

3. Remove the capscrews that fasten the transmission cover to the top of transmission.

NOTE: To remove the two rear capscrews first remove the two top bolts holding the 3 point hitch lift tube supports to the tractor frame and loosen the two lower bolts as shown. Rotate the lift tube out of way of the rear capscrews.

4. Remove the transmission cover.

5. If the gears, bearings or other items are worn and damaged drain the oil from the transmission case.

6. To drain the transmission case remove the ten washer head capscrews that attach the bottom cover to the tractor frame and remove the cover.

7. Place a pan under the front of the transmission case, remove the electrical wire and temperature sending unit to drain oil from case. Clean all abrasive and foreign material from the transmission case with a suitable solvent.

8. Remove oil seal (22) retaining ring (19) from the front bearing housing.

9. Move the shaft forward as shown and remove the snapring (14) and gear (15), then remove the shaft from the case.

10. If the top shaft is not equipped with a removable gear it will be necessary to remove the bearing, spacer and O-rings, then remove the shaft from the rear.

11. Check roller bearing (13) and remove if not satisfactory.

ASSEMBLY

1. Replace roller bearing (13) if it was removed. Press against the end of the bearing with the manufacturers identification. Use a suitable fixture so that the bearing case is not damaged.

2. Press bearing (18) on the shaft. Install two O-rings (20) in spacer (21). Install the spacer on the shaft with the O-rings toward the bearing (18).

3. Install the shaft through the front bearing housing and install tru-arc retaining ring (16) on the shaft. Move the snapring past the front ring groove.

4. Install gear (15) on the shaft with the recessed end toward the rear. Move the gear forward to expose rear snapring groove and install the round (wire type) snapring (14).

5. Move gear back so that snapring (14) is in the recess in the gear and move the front snapring (16) in the groove in the shaft to hold the gear in place.

6. Align the shaft so it will enter the roller bearing (13) and press shaft (17) with bearing (18) in front bearing housing.

7. Install the retaining ring (19) in the groove in bearing housing to hold the bearing and shaft in place.

8. Lubricate spacer (21) and oil seal (22) with Lubriplate® and press oil seal in the housing.

9. If top shaft (17) has gear (15) attached to the shaft permanently install as follows:

10. Install the shaft through the front bearing housing from the rear. Install the bearing (18) on the shaft (17). Install the two O-rings (20) in the recess in the spacer and install spacer on the shaft with the O-rings toward the bearing (18). Press the bearing in the housing and install the retaining ring (19) in the groove to hold the bearing in place.

11. Lubricate the oil seal and spacer with Lubriplate® and press the oil seal in the housing.

12. Install the front hub. Wedge a clean rag between the gears to prevent the shaft from turning while torquing hub retaining capscrew.

13. Apply Loctite® to capscrew threads and torque to 45 ft.lb.

Center Idler Gear

DISASSEMBLY

NOTE: The cap assembly (29) and thrust washers (24-26) can be removed without removing the top transmission PTO shaft (17), however, if the gear or bearing need to be removed from the case the top shaft will have to be removed.

1. Remove the two capscrews (30) and cap assembly (29). The steel washer, and fibre thrust washer will drop to the bottom of the case.

2. If the roller bearing is worn or damaged press from the gear.

3. Clean all parts in a suitable solvent.

4. Check the two thrust washer retaining roll pins in the cap assembly and replace if distorted.

5. Replace the thrust washers whenever the center idler is removed. Check the spring and replace if broken or worn.

ASSEMBLY

1. The accompanying figure shows an enlarged view of the parts that make up the center idler gear. To assemble: Coat one side of washer (23) with grease and stick washer against the front support surface, with the flat side of hole toward the left.

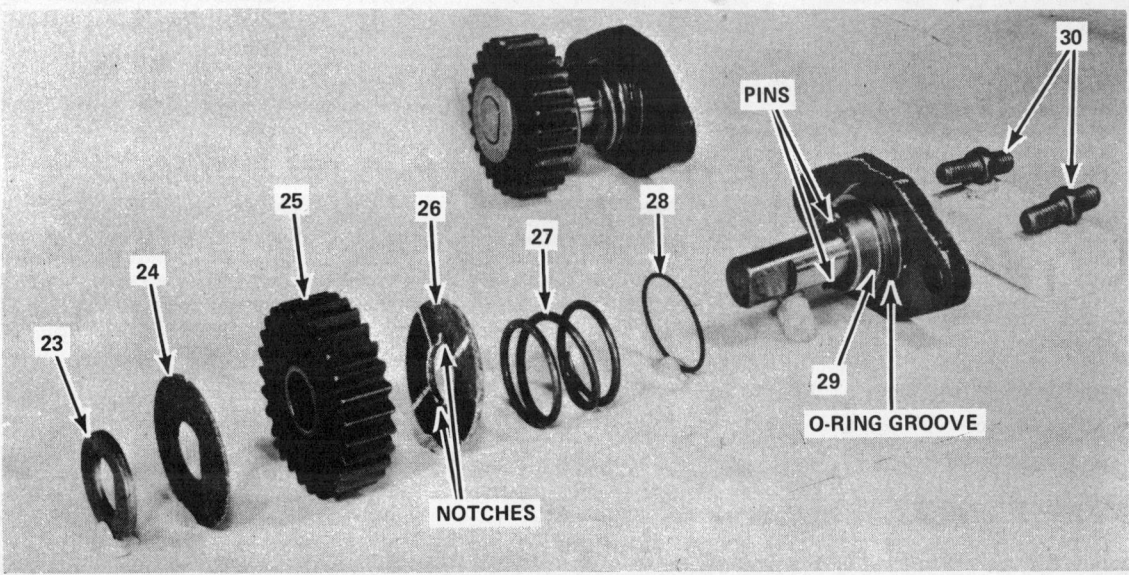

Central idler gear components

47

Allis-Chalmers

Using grease, stick the grooved side of thrust washer (24) against the forward face of gear (25) and stick the grooved side of thrust washer (26) to the rear face of gear with the notches in washer in vertical position.

2. Place O-ring (28) in O-ring groove in cap assembly (29). Place spring (27) over shaft of cap assembly. While holding the gear and thrust washers down inside the transmission, insert the cap assembly through the rear of transmission with the flat side of shaft to left. Pick up the thrust washers and gear with the shaft. Make certain that thrust washer (24) and washer (23) align with the flat side of shaft as shown. When washers are in place on shaft, install long end of capscrews (30) through cap assembly and into tapped holes in transmission. As cap is tightened down make certain that notches in thrust washer, (26), align with and slip over the 2 pins in cap assembly. Tighten capscrews securely.

3. Install shield assembly with decal (50) on the exposed threaded end of the cap assembly capscrews.

Power Take-Off (Lower) Hollow Shaft with Internal Splines Output Sleeve

DISASSEMBLY

NOTE: The output shaft (38) and rear bearing can be removed by removing the locking collar and items 44 to 49 and cap (42) without removing the center idler gear (25) and top PTO transmission shaft. To remove the gear (34) front bearing cone (33) and race it will be necessary to remove the PTO shaft and center idler gear and shaft.

1. Remove the top shaft and center idler gear and shaft.
2. 620 tractor S/N 1658 and up—Remove retaining ring (49), collar (48), spring (46), washer (45), retaining ring (44), and (3 locking balls).
3. 620 tractors prior to S/N 1658 do not have retaining ring (49) but use collar assembly part #2029885. It will be necessary to use a pair of thin tru-arc snapring pliers to spread the internal square snapring. A slot is provided in the collar to permit spreading the retaining ring to remove collar from output sleeve.
4. 616 tractors have three locking pins in place of the three balls otherwise the lock removal is similar.
5. Remove the bearing cap (42) and shims (51).
6. Pull the output shaft (38) with bearing cone and cup (40) out.
7. Remove the gear (34) and bearing cone (33) from the case.
8. Clean all the parts in suitable solvent and inspect for wear or damage. Check the bearing cups and cones for wear or damage. Replace all parts that are needed.
9. If bearing cone (33) requires replacement pull bearing cup (32) from the transmission case. Remove bearing cone from the output shaft with a suitable puller to prevent damage to the oil seal area of the output shaft.
10. The plug and O-ring are held in the hollow shaft with retaining ring (35). If oil leaks through the center of the shaft replace O-ring.

ASSEMBLY

1. Place bearing race (32) in front bore of lower hole in transmission with large end of cup facing the rear. Press race firmly in place against shoulder in casting with suitable adapter tool.

2. Place O-ring (37) in groove in front end of PTO output sleeve and carefully work plug (36) into place under O-ring. Install retaining ring (35) on outside of plug. Press rear bearing cone on output sleeve.

3. Coat bearing cone (33) with grease and place in bearing race (32). Lower gear (34) from top of transmission housing and while holding it at the rear of bearing cone (33), place front end of PTO output sleeve (38) through the center of gear and into bearing cone (33). Align splines in gear and sleeve and carefully press sleeve fully into bearing race (33).

4. Install bearing race (40) over rear of PTO output sleeve and press it into rear of housing. Place shims under bearing cap (42) and bolt cap in place tightly to finally seat the bearing races. Tap end of shaft with plastic hammer to seat bearing cones and cups.

5. Set up a dial indicator against the end of the PTO output sleeve and measure the end play of sleeve. Correct end play is .002 in. to .005 in. The dark red gaskets are .010 in. thick, the green ones .003 in. and the light red gaskets are .002 in. thick. Remove gaskets as required to reduce end play to the .002 in. to .005 in. specifications, or add to increase bearing clearance.

6. To determine an approximate shim thickness hold the cap to the bearing cup with a slight torque on the two capscrews, then measure the clearance between the cap and the machined surface of the case with feeler gauge then add approximately .005 to the distance determined with the feeler gauge. Then select the quantity of shims to equal this figure.

7. When correct end play is obtained, remove bearing cap and install O-ring (41) in groove in cap. Press the oil seal (43) into the bearing cap with the lip of seal facing inward and outside of seal flush with rear face of bearing cap. Lubricate O-ring and seal lip freely. Cover snapring grooves in PTO output sleeve with cellophane tape to protect the oil seal. Place bearing cap over sleeve and work it into place taking care to protect O-ring and lip of seal. Tighten cap securely.

8. Remove cellophane tape from snapring groove in PTO output sleeve and install snapring (44). Slide cupped washer (45) onto sleeve and against the snapring with cup lip to rear. Place spring (46) on sleeve against the cupped washer. Place spring retainer (47) in groove in collar (48). Place the 3 steel balls (39) in the three holes in PTO output sleeve and hold in place with grease, push the collar (48) with spring retainer (47) in it over the 3 balls until they snap in place. Install snapring (49) in groove in sleeve to rear of collar.

9. 620 tractors prior to S/N 1658 may not have retaining ring (49). These tractors use only two locking balls to retain the driveshaft in place. The collar assembly 2029885 has a square type internal snapring to lock the collar on the sleeve.

10. 616 tractors have a hex shaped collar that requires rotation to release the implement PTO shaft.

11. To replace, install retaining ring

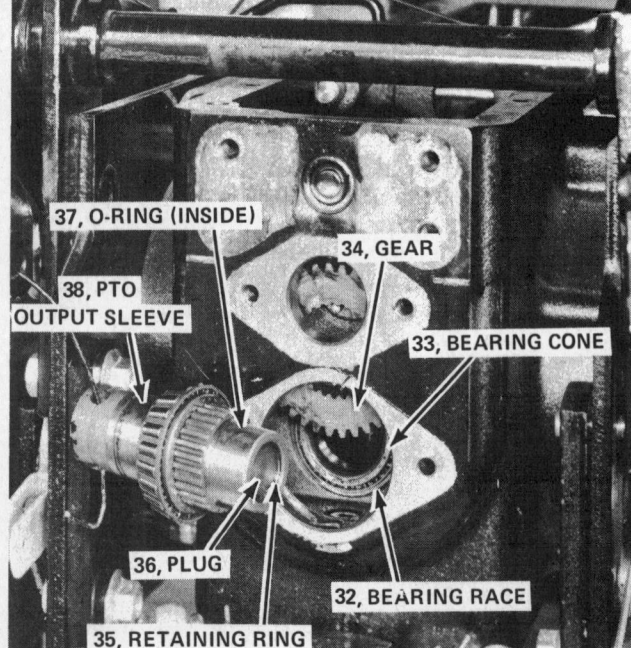

Pulling transmission bearing cone

Allis-Chalmers

(44), retainer, spring, three lock pins, cam, outer cover and retaining ring.

12. Replace center idler gear, top transmission PTO shaft, rear PTO shaft, oil sending unit and wire bottom cover, transmission cover and gasket, seat and accessories that were removed. Fill the transmission with Type A transmission fluid.

ENGINE

REMOVAL AND INSTALLATION

1. Remove the hood by removing the two shoulder capscrews from the hinge assembly. Remove the hood.

2. Remove the battery cover, disconnect the battery ground cable and the positive cable. Remove the battery clamp to remove the battery.

3. Disconnect the throttle and choke cable from the engine. Disconnect the electrical connections at the front of the engine. Disconnect the fuel lines at the fuel tank. The clamps should be moved off of the end of the hose to prevent overstretching the clamps.

4. Remove the capscrews and washers that fasten the oil cooler to the shrouds and move the oil cooler away from the shrouds. Remove the six capscrews that attach the righthand, lefthand and center shrouds to the engine and remove the shrouds.

5. Remove the two capscrews, washers, lockwashers and spacers that fasten the rear PTO yoke to the engine flywheel if the tractor is equipped with a PTO.

6. Hold the clutch pedal forward and remove the two clutch belts from the engine pulley.

7. Remove the cotter pin from the clutch rod and allow the clutch assembly to drop to the frame.
Remove the bottom cover from the tractor frame.

8. Remove the engine mounting capscrews from the tractor frame. Three capscrews are used on the 616. Four capscrews are used on the 620.

9. Attach a chain or web to the engine as shown and raise the engine with a hoist to move to a suitable work area.

NOTE: For engine service refer to the Onan Engine section of the Engine Unit Repair part of this book.

10. Check the condition of the clutch belts and sheaves and replace if necessary before installing the engine.

11. For 616 tractors equipped with a rear PTO:
If the rear PTO front shaft uses a center bearing remove the top cover from the tractor frame and loosen the nuts on the carriage bolts that fasten the bearing flanges to the bearing support.

12. Loosen the setscrew in the front PTO yoke and move the yoke rearward.

13. Position the electrical wires, throttle cable and choke cable so they will not be damaged during engine installation. Install the engine in the tractor frame and replace the engine mounting capscrews, lockwashers and nuts. Move the engine as far forward as possible and torque the mounting capscrews to 30 ft.lb.

14. Attach the front PTO yoke to the flywheel with the spacers, washers, lockwashers and capscrews. Apply Loctite® to the capscrews. Torque the capscrews sufficient to prevent the yoke from moving freely on the flywheel.

15. Rotate the engine and tap the coupling to align the shaft with the engine flywheel. A dial indicator should be used to align the shaft with the engine flywheel. Torque the capscrews to 25-30 ft.lb. Do not over torque.

16. If the yoke has a setscrew, torque the setscrew. Torque the center bearing carriage bolts after the shaft is aligned and replace the top cover. If the PTO yoke has a lube fitting lubricate with automotive type lube.

17. Attach the clutch rod to the clutch pedal arm. Hold the clutch pedal forward and install the two V-belts. Refer to the pperator manual for clutch spring adjustment and pedal free travel. Replace the bottom cover.

18. Install the righthand, lefthand and upper engine shrouds.

19. Install the oil cooler. A capscrew and special washer is used to fasten the righthand end of the oil cooler. For the lefthand end of the cooler use 2 neoprene washers and one special steel washer. Place one neoprene washer between the cooler tube and the engine shroud. Place a steel washer on the capscrew and one neoprene washer between the steel washer and the cooler tube. Torque the two capscrews.

20. Attach the fuel lines to the fuel tank and secure with clamps.

21. Connect the electrical wires at the right front of the engine. Install the battery and fasten with clamp at the front of the tractor frame. The long battery cable is attached to the starter solenoid and must be attached to the positive (+) battery terminal. The short cable is attached to the negative (−) battery terminal and tractor frames. Install the battery cover and secure with two springs.

22. Install the choke and throttle cables. Refer to the Engine service section in the Engine Unit Repair part of this book for correct cable installation and adjustments.

Models 710, 712, 716

WHEELS AND TIRES

Front Wheel

REMOVAL AND INSTALLATION

1. The dust caps are retained by friction and must be pryed out. Start separation by carefully driving a large, blunt chisel into the groove (A) between the end of the hub (B) and the flange of the dust cap (C). Hold the chisel against the wheel rim so the angle of entry will be as shallow as possible. Avoid cocking the cap. Turn the wheel between successive blows to distribute the pressure uniformly around the groove.

2. When the groove has been opened enough to admit the nib of a wedge bar (D), substitute the latter for the chisel. Rest the heel of the wedge bar on a block of wood (E) about 1¼ in. high, laid against the wheel web. Rotate the wheel between successive applications of the wedge bar so the cap will not become cocked.

3. Set dust cap (A) aside with open end up. With an Allen wrench, loosen the setscrew in the set collar (B). Slip off the collar and washer or washers (C). Pull the wheel out a half inch or so, and then push it back to initial position, leaving the outside roller bearing cone (D) accessible. Remove the cone, and place the collar, washers, and cone in the up-ended dust cap.

4. Pull off the wheel. Lay it aside with the dust cap and contents. Identify these parts as belonging to the right or left side of the tractor for reference at assembly.

5. With a rag, wipe the grease off of the outer end of the spindle (H). Wrap a clean cloth around the inner bearing cone (E) and seal (F). With a fine file, remove any burrs raised on the spindle around the score mark left by the collar setscrew. Wipe the filings off of the spindle and remove the cloth from the bearing and seal.

6. Using finger pressure only, try to remove both inner bearing cones (E) and seals (F). If they respond, lay them aside with the other parts for their respective

Allis-Chalmers

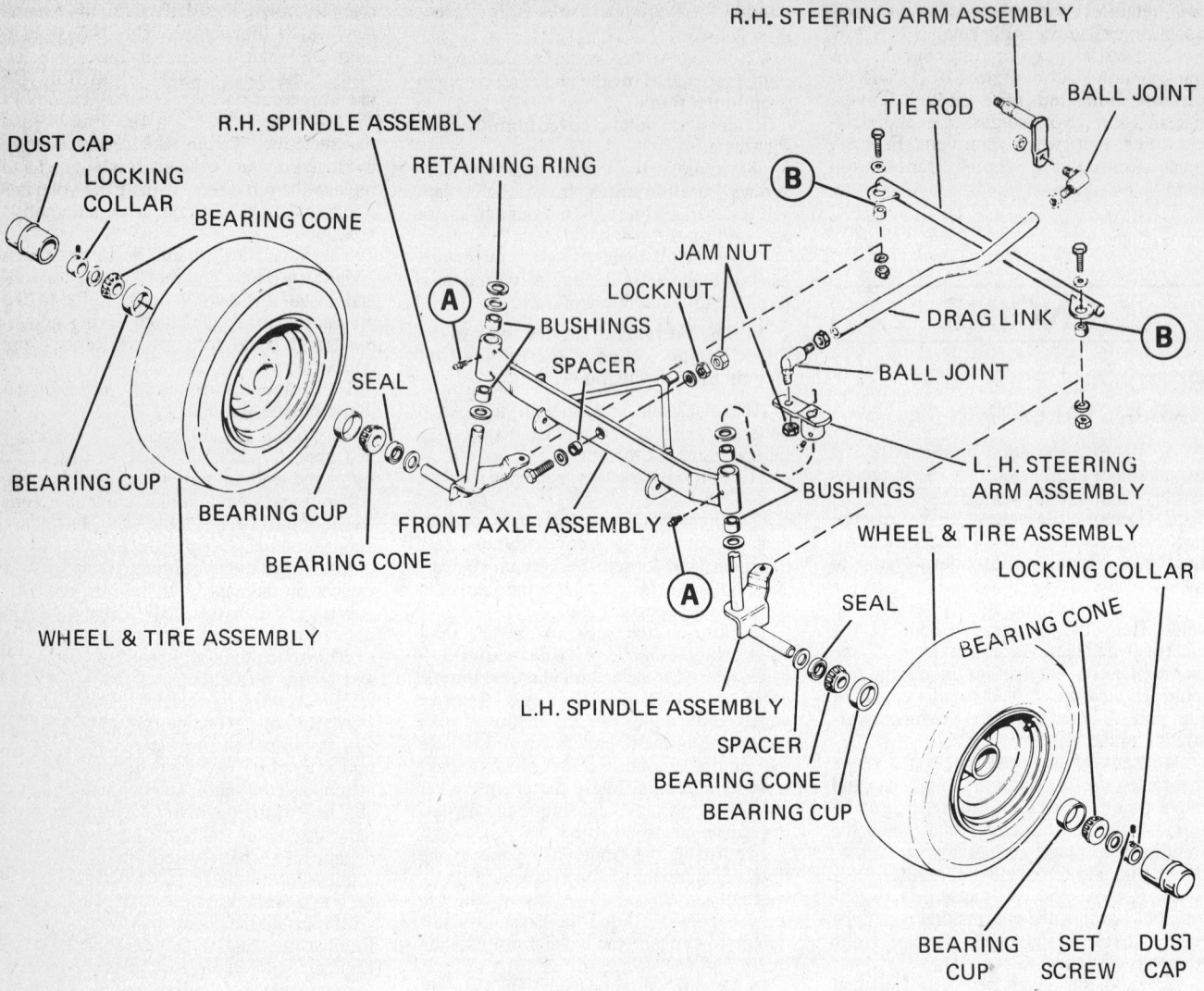

Front axle assembly

spindles. If not, remove them as described in steps 7 through 9.

7. Insert a screwdriver blade between the spacer collar (G) and the adjacent structure of the spindle assembly. Pry the collar outward so it forces the seal (F) and bearing cone (E) to move with it. Do not pry against the seal or bearing cone.

----- **CAUTION** -----
Do not touch either the seal or the cone with either the wooden block or the hammer.

8. When the spacer (G) has been well started by prying, drive it off, using a wooden block and hammer. Distribute the blows as uniformly as possible around the spacer collar, so it will not tend to cock and pinch.

9. If the movement becomes sufficiently free to respond to finger pressure, remove them manually. Lay the bearing cone, seal, and spacer aside with the other parts of the same wheel and tire assembly.

10. If required, repeat steps 1 through 9 for the other wheel.

11. Using a grease solvent, clean all of the parts individually. Remove any caked deposits of grease and dirt. Flush all of the grease out from between the rollers in the bearing cones. Flush all of the grease out of the interior of the wheel hubs, particularly the bearing cups.

12. When the parts are clean and dry, examine them carefully for wear and damage.

----- **CAUTION** -----
Do not spin bearings with compressed air.

13. Examine the rollers of the roller bearings and the mating surfaces of the cones and cups for discoloration due to overheating, scoring, spalling, cracking, or chipping. If any defects are present on one part of a bearing, discard the entire cone and cup assembly and replace them with a new set.

14. If a bearing cup (in a wheel hub) is to be removed, use a bearing puller. Do not try to remove the cups with a rod and hammer.

15. Because of the importance of the seals in retaining grease and excluding dirt, and the difficulty of detecting minor wear or damage, they should be replaced with new seals.

16. Inspect tire for wear and tire valve for leakage. Replace as required. Inflate tire to specified pressure.

17. If a wheel bearing cup is to be replaced, press or drive the replacements into the wheel hub using a rod slightly smaller than the seat bore. Do not continue forcing a cup if it becomes cocked. Remove it, and start over. The smaller diameter of the cup must be placed inward. Seat the cup solidly against the inner shoulder of the bore.

18. If not already installed, assemble the seals (F) and inner bearing cones (E) on both spindles. Place flanged side of seal next to the spacer collar (G) and the largest end of the bearing cone next to the seals.

NOTE: If force is required, use a hammer and a brass or aluminum tube against the outer face of the cone. The tube should have an inner diameter slightly larger than the spindle and a wall thickness no greater than 1/8 in. Drive the cone in until the seal is firmly confined between the spacer collar and the cone.

19. Lubricate the two inner bearing cones with prime quality wheel bearing grease. Work the grease thoroughly into the spaces between the rollers. Extend the grease a distance out on the spindles and spread a layer of grease over the portion of the seal that the wheel hub runs on.

20. Replace the wheels. The tire air valves should be on the inside nearest the tractor frame.

21. After placing each of the outer bearing cones partly on its spindle, lubricate it with grease as explained in step 19.

22. Assemble the spacer washers (C) and the set collars (B). Spin each wheel while urging the set collar inward, and then, with an Allen wrench, tighten the setscrew in each collar. Spin the wheels again, and then test the fit by urging the tire tread back and forth sideways in a wobbling motion. If side play is more than barely perceptible, reposition the set collar.

23. Replace the dust caps (A), driving them home with a rubber mallet applied near rim of the caps while spinning the wheels. If a cap becomes cocked, do not force it; remove it and start over.

24. Lower the front of the tractor.

Rear Wheel

REMOVAL AND INSTALLATION

1. Raise rear wheels and block up tractor as described in preparation procedure.
2. Remove five hub bolts. Then remove wheel and tire assembly from hub.
3. Repeat step 2 on other side of tractor.
4. Check tire for excessive wear on tread, weak sidewalls, or cracked beads.
5. Check tire valve for leakage.
6. Replace or repair defective tire or valve and inflate tire to specified pressure.
7. Place wheel and tire assembly against hub. Align holes in wheel and hub and loosely install the five hub bolts.
8. Tighten hub bolts evenly and snugly to secure wheel and tire assembly to hub.

CLUTCH AND BRAKE

Description

The brake and clutch group includes a foot pedal, an idler pulley, a brake drum and band, and a series of rods, links, and pivots that connect these items together. The connecting items change frequently. Separate illustrations show the clutch-brake group for standard 3 and 4-speed drives, the hydrostatic drive and the dual range 6-speed drive.

The basic operation of the clutch and brake group is the same for all models. Pressing the foot pedal pulls the clutch rod forward to turn a pivot. As the pivot turns, it moves a linkage to pull the idler pulley away from the drive belt between the bevel gear box and the transmission. Thus, the belt tension is reduced to disengage the transmission from the bevel gear box and engine.

The parking brake is a lever connected by a separate rod to the brake band. This lever can be operated independently of the foot pedal to lock the brake.

Pivot motion also moves a rod that passes through the brake band. If the pivot moves enough, a nut on this rod contacts the brake band and tightens it around the brake drum. The friction that results slows or locks a transmission gear shaft to brake the tractor.

Inspection

Before replacing items of the brake and clutch group, check the overall operation of both the clutch and brake. Be sure that the problem is not one that can be corrected simply by performing one of the adjustments described in the Adjustment Section.

If belt slippage is a problem, check both the belt and the pulleys. Be sure that both are clean of oil and grease. Use a clean rag soaked in solvent to remove any grease or oil.

If belt breakage or excessive wear is a problem, check the pulleys. The pulleys should be free of burrs and sharp edges that can cause belt breakage.

REMOVAL AND INSTALLATION

For clutch and brake removal and installation procedures, see Transmission Removal and Installation for the type of unit in your tractor.

FRONT AXLE AND STEERING

Right Front Spindle Assembly

REMOVAL AND INSTALLATION

1. Remove right front wheel and tire assembly.
2. Disconnect the tie rod (B) from the

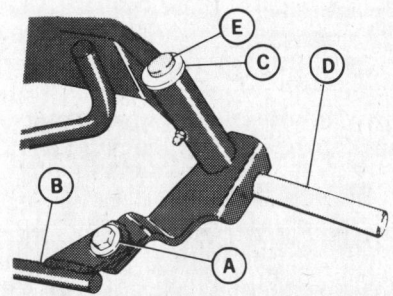

Front spindle assembly

right spindle assembly by removing the hex head capscrew (A) with its bushing, two washers, and nut.

3. Remove the retaining ring (C) and the plain washer immediately under it. Pull the pintle (D) of the spindle assembly out of its bushings in the axle assembly. Remove the lower plain washer from the pintle.

4. Place all of the parts removed from the right end of the axle assembly together and identify them for future reference.

5. Insert bushing into tie rod end lug and shake from side to side. If excessive play exists, discard bushing and tie rod.

6. Push the pintle of each spindle assembly temporarily back into its bushings in the end of the axle assembly. Test for wear by shaking the pintle sideways, noting the extent of the side play. The running fit allowance in these bushings is normally rather generous. The amount of shake, even with no wear, will be quite perceptible. If there is appreciable rattle, however, the bronze bushings should be replaced. If they are to be removed, press or drive them out using a rod slightly smaller than the seat bore.

7. Examine the pintle areas that run in the wheel bearings. If there is appreciable wear, or scoring indicative of seizing or galling, discard the spindle assembly.

8. Put a plain washer on the pintle (D) of the righthand spindle assembly, and insert the pintle upward into the bushings of the axle assembly. Put another plain washer on the pintle and then install the retaining ring (C).

9. Slip a plain washer and a spacer bushing on hex head capscrew (A). Insert capscrew with bushing and washer downward through the hole in the tie rod lug.

10. Lay the end lug of the tie rod (B) on top of the bent lever arm and push the capscrew assembly through its hole.

11. Place a lockwasher the capscrew and secure assembly with nut.

12. Install right front wheel and tire assembly.

13. Wipe grease fitting (A) clean. Using multi-purpose lithium base gun grease, lubricate the fitting until the grease is forced from the bushing.

14. Lubricate pivot points (B) with a film of grease.

Left Front Spindle Assembly

REMOVAL AND INSTALLATION

1. Remove left front wheel and tire assembly.
2. Disconnect tie rod end lug from spindle assembly by removing hex head capscrew (A) with its bushing, two washers, and nut.
3. Remove jam nut attaching ball joint (E) to lefthand steering arm assembly (C) and move drag link (F) away from steering arm.
4. Loosen the two square head set-

Allis-Chalmers

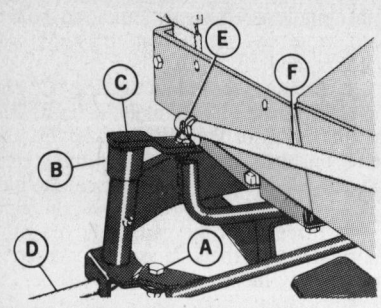

Left front spindle

screws (B) two turns each. Using a hammer and a piece of brass rod applied against the lower end of the steering arm hub (C), drive the steering arm up and off the pintle (D). Do not try to drive the pintle down, out of the steering arm. Lay the steering arm aside.

5. Remove the two keys (A) and washer (B) from pintle (C).

6. With a cloth, wipe all of the dirt and grease off of the exposed end of the pintle. With a fine file, remove any burrs that may exist around the two key ways. Wipe off all of the filings.

7. Pull the pintle out of its bushing in the axle assembly and remove the bottom plain washer (D).

8. Place all of the parts removed together and identify them for future reference.

9. See right front spindle assembly.

10. Examine keys for swelling caused by pressure of the setscrews when previously installed. Use a fine file to remove swollen material, but do not reduce size of keys. Replace keys that are worn, bent, or broken. Replace the bronze bushings if needed in the same manner as for the right front spindle.

11. Retrieve the spindle assembly having the longer pintle (C) with key ways.

12. Place a plain washer (D) on the pintle (C). Then insert the pintle upward through the bronze bushings in the left end of the axle. Place another plain washer (B) on the end of the pintle.

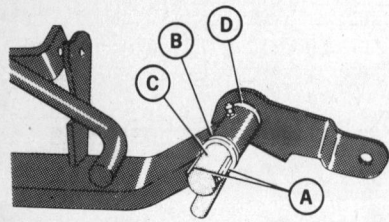

Removing the pintle

13. Assemble the keys, if necessary, driving them home in their key ways. Use a brass hammer, and back up the pintle with a lead mallet.

14. Place the steering arm (C) over the pintle with its hub toward, and its arm opposite, the spindle (D).

15. Engage the keys in the key ways, and drive the arm onto the pintle with a plastic mallet while backing up the lower end of the pintle with a lead mallet. Drive the arm down on the pintle until there is barely perceptible end play. Then permanently tighten the two square head setscrews (B).

16. Secure the drag link ball joint (E) to the steering arm (C) with a jam nut and torque nut to 40-60 ft.lb.

17. Retrieve the tie rod and identify the left end lug (marked or tagged during disassembly). Slip a plain washer, and then a steel spacer bushing, on the hex head capscrew (A). Insert the capscrew with bushing and washer downward through the hole in the tie rod lug. Place the lug on top of the bent lever arm of the spindle assembly, and push the capscrew through its hole. Place a lockwasher on the capscrew, and then install the nut, tightening it permanently.

18. Install front wheel and tire assemblies.

Tie Rod

DESCRIPTION

The tie rod functions to transmit the same degree of turn to each wheel assembly by connecting or tying the left and right spindle assemblies. It differs from the drag link, in that the drag link transmits turn commands from the steering gear to the left spindle assembly.

REPAIR

The tie rod can be removed and installed from the right and left spindle arms by removing the attaching nut, flat and lockwashers, capscrews and bushings. Bushings should be cleaned, inspected, and replaced if worn. The tie rod should be replaced if distorted beyond repair or if bushings do not fit securely in rod end lugs.

Drag Link

DESCRIPTION

The drag link connects and transmits turn commands from the steering gear to the left spindle assembly. The tie rod connects both left and right spindle assemblies, thus providing uniform turning of the wheel assemblies.

REMOVAL AND INSTALLATION

1. Disconnect the front ball joint (A) of the tie rod from the steering arm (B) by removing the nut (C). Prevent rotation of the stem by applying an end wrench to the hex flange of the stem located just above the arm.

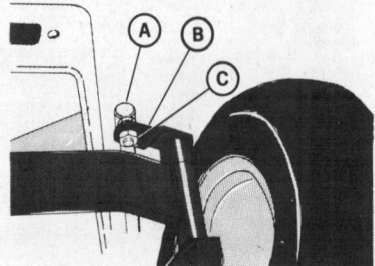

Front view of steering arm

2. Access for a subsequent disassembly operation requires that the left foot rest be partially separated from the frame. Remove the three carriage bolts (A) together with their nuts and lockwashers.

In addition to the three carriage bolts just removed, the foot rest has a fourth fastening, a hex head bolt located near the top of and behind the bent-up rear portion of the foot rest. Do not remove this fourth bolt. Using it as a hinge, gently swing the foot rest down until it comes to rest on the power take-off idler pulley.

3. The two hex capscrews (B) and (C) fasten the steering bearing casting to the tractor side frame. Remove the capscrew (B) and lay it and its two washers aside. Loosen the front capscrew (C), backing it out until only two or three threads remain engaged in the casting.

4. Repositioning the ball joint to the front end of the crescent slot (F) will provide access to the hex flange. While keeping the free end of the tie rod disengaged from the foot rest, turn the steering wheel shaft counterclockwise (left turn) as far as it will go.

5. With the ball joint (A) at the front end of the crescent slot (B), gently pull the tie rod outward, away from the tractor side frame, as far as it will go, and lay it on the front end (C) of the foot rest.

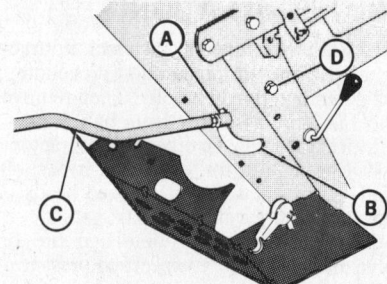

Left side foot rest

6. Shift the bearing casting (A) inward, away from the tractor side frame (B) as far as possible. Tilt it slightly, if necessary, so an end wrench (C) can be fitted onto the ball joint hex flange between the steering arm (D) and the side frame.

7. Holding the wrench (C) stationary to prevent rotation of the stem, apply a socket wrench to the nut (E) and remove the latter. Withdraw the stem and lay the drag link aside.

8. Examine drag link for unusual bends or distortion and ball joint threads for damage or wear. Replace parts as required.

9. If necessary, install ball joints at both ends of the tie rod. Select the ball joint nearest the bend in the rod and insert its stem through the crescent slot (B) into the hole in the steering arm.

10. Pull the tie rod out gently as far as it will go, and lay it on the front end of the foot rest.

11. Tilt the casting slightly, if necessary, and insert an end wrench (C) between the arm (D) and the tractor side frame (B) onto the hex flange of the ball joint stem.

12. Install and tighten the nut (E) onto the stem, preventing roatation of the stem

Allis-Chalmers

with the end wrench. Then remove end wrench.

13. Turn the front wheels and/or the steering shaft as necessary and insert the stem of the front tie rod ball joint through the steering arm. Install and torque nut (C) to 40-60 ft.lb., preventing rotation with an end wrench on hex flange of the ball joint stem.

Front Axle Assembly

REMOVAL AND INSTALLATION

1. Remove both wheel and tire assemblies.
2. Remove both spindle assemblies, tie rod and axle assembly end of drag link.
3. Remove the nut from the back end of the hex head bolt (C). Later models have a locknut, which has to be removed first.
4. While supporting the axle assembly manually, withdraw the bolt.
5. Still supporting the axle assembly, pull the latter toward the front of the tractor until the rear trunnion (D) is free of its bearing in the frame.
6. Remove and lay aside the bushing from the center hole in the axle assembly that receives the bolt removed in step 4.
7. Take the now separated axle assembly to the bench and grasp the main bar in a vise.
8. Using a grease solvent, clean entire axle assembly. Flush all of the grease out from between the pintle bushings in the axle pivot tubes.
9. Push the pintle of each spindle assembly temporarily back into the axle pivot tube bushings and shake the assembly noting the extent of side play. If there is appreciable rattle, the bushings must be replaced. If bushings are to be replaced, use a rod slightly smaller than the seat bore.
10. Make sure grease fittings are clean, with unobstructed hole and are secure in axle pivot tube.
11. Check the front axle and hitch bar for damage, distortion, or broken or weak weldments. Replace entire assembly if damaged excessively.
12. Check axle bushing for wear and replace if excessively worn.
13. Retrieve one of the two plain washers (E). Spread a layer of thick grease on its flat side. Using the grease as a temporary adhesive to hold the washer in place, center the washer over the bolt hole in the front face of the frame crossmember (F).

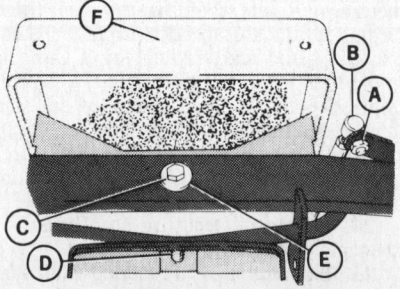

Lower front side of front axle

14. Slip the second plain washer (E) on the bolt (C), rounded face next to the bolt head. Lay the bolt, with washer assembled, at the front of the tractor where it can be easily reached.
15. Bring the axle assembly to the front of the tractor. Spread some thick grease on the steel bushing for bolt (C), and slip the bushing into its hole at the center of the axle main member.
16. Supporting the axle assembly, insert the trunnion (D) into its bushing in the frame, simultaneously aligning the holes for bolt (C). Insert bolt (C) with washer (E) through the axle main member, the second washer (E) pasted to the frame, and then through the frame crossmember (F). Finally, install and torque nut to 80-100 ft.lb. A jam nut is also required on later models.
17. Install both spindle assemblies, tie rod, and axle assembly end of drag link.
18. Install both wheel and tire assemblies.
19. Lubricate grease fittings (A) with multi-purpose lithium base grease. Wipe fittings clean and lubricate each until the grease is forced from the bushing.
20. Lubricate pivot points (B) with a film of grease.

Steering Shaft and Gear

REMOVAL AND INSTALLATION

1. Park tractor on a hard, level surface. Place park brake in "off" position and all other controls in "off" or "neutral" positions. Remove ignition switch key.
2. Remove steering wheel and battery.

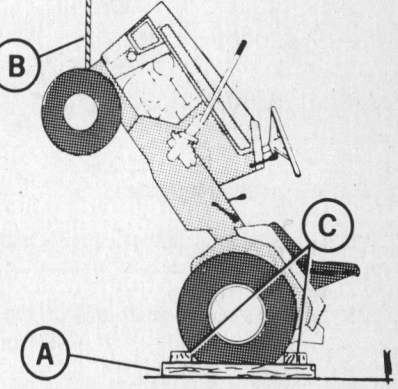

Blocking arrangement

3. Block the rear wheels at front and back. Use a hoist and raise tractor so that the steering gear assembly can be removed.
4. Remove drag link.
5. Turn steering arm to allow access to capscrew. Remove capscrew and lockwasher.

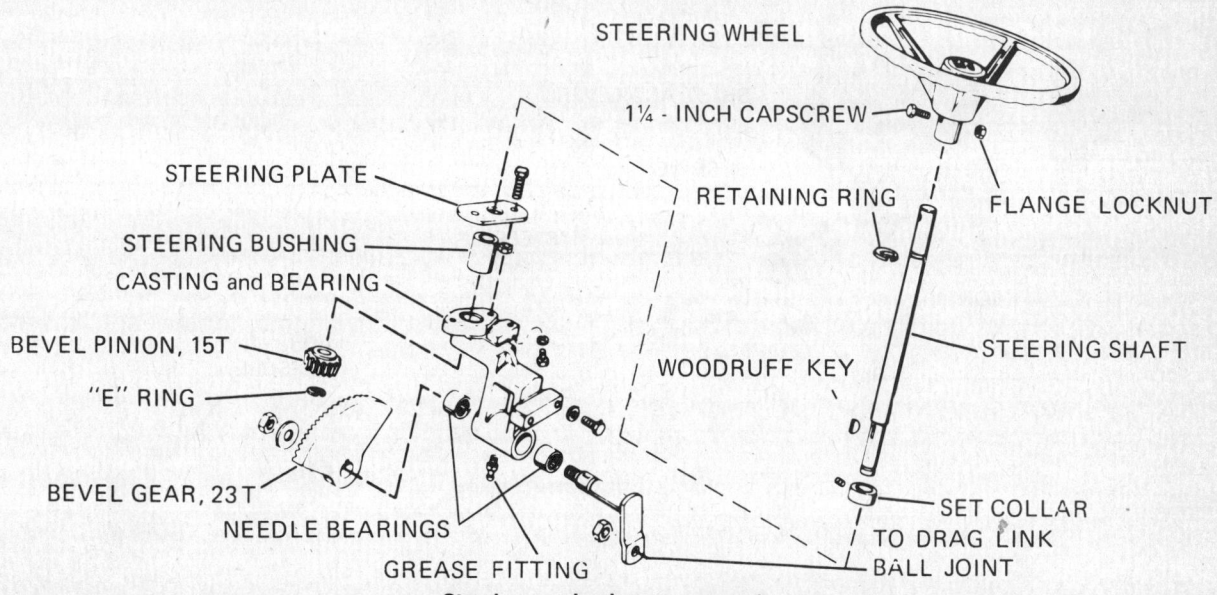

Steering mechanism components

53

Allis-Chalmers

6. Remove the two capscrews with lockwashers that attach steering gear casting to side of frame.
7. Move entire steering gear and shaft assembly forward until casting lug clears edge of frame opening. Lower assembly until shaft has passed completely out of opening.
8. Clean entire assembly with steam or with a grease solvent and stiff brush.
9. Secure casting in vise and remove locknut (B) and flat washer.
10. Using a plastic mallet, drive bevel gear (C) from steering arm shaft and casting. Remove gear by sliding it sideways and out of engagement with pinion gear (D).

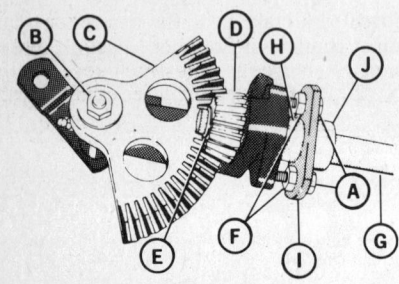

Steering gear

11. Using bearing puller, remove both needle bearings from casting.
12. Remove retaining ring (E).
13. Secure steering shaft (G) in vise. Using a hammer and a piece of hardwood, drive pinion gear (D) off shaft.
14. Remove Woodruff key.
15. Pull out shaft (G) from bearing bushing (H).
16. Loosen locknuts (F) and back out capscrews (A) to separate the plate (I) from the casting.

17. Remove bushing (H) from casting.
18. Remove collar (J) with setscrew from shaft (G).
19. Flush out casting bores and remove all grease from bearings.
20. Clean all teeth on both gears using a wire brush and solvent.
21. Examine steering arm shaft in areas that serve as journals for needle bearings. If there is excessive wear, or scoremarks showing evidence of seizing or galling, replace both the arm assembly and the needle bearings.
22. If no such defects are apparent, insert steering arm shaft into both needle bearings. Rotate shaft, at the same time urging it from side to side. If rotation is not perfectly free and smooth, or if there is more than a barely perceptible side shake, replace both the arm assembly and the needle bearings.
23. If needle bearings are to be replaced, press them out of the casting bore with a rod slightly smaller than the bore.
24. Inspect all teeth on both gears for damage. If damaged, replace gears.
25. Examine area of steering shaft that runs in casting bushing. If there is excessive wear or score marks showing evidence of seizing or galling, replace both shaft and bushing. Check shaft for bends or distortion.
26. Slip bushing on steering shaft to its normal location. Rotate shaft, at the same time urging it from side to side. If rotation is not smooth, or if there is a side shake, replace both shaft and bushing.
27. If old shaft is to be reassembled, file away any burrs around retaining ring groove and key way. Clean the shaft and remove the filings.
28. Press needle bearings, one from each end into the bore. Be sure the end of the bearing marked "Torrington" is facing outward—installation pressure must be applied on the marked end.

The bearing at the gear end of the bore must be seated 1/8 in. below the face of the casting on which the gear runs. Press the bearing in using a rod slightly smaller than the bore.

The bearing at the arm end of the bore must be seated flush with the face of the casting against which the arm runs. Press this bearing in, using a rod at least 1/4 in. larger than the bore.

29. Slip the collar (J) on the shaft, leaving the setscrew untightened. Spread a film of oil over the shaft from the key way back a distance of a couple of inches. Be sure that there is a retaining ring installed in the groove near the steering wheel end.
30. Slip the bushing (H) into the casting.
31. Assemble the plate (I) by running the two capscrews (A) into the casting until the end of the bushing opposite the plate protrudes about 1/64 in. from the adjacent face of the casting. Be sure that the plate is not cocked from one screw being run in farther than the other. Tightly run the locknuts back against the plate finger.
32. Insert the end of the shaft into the bushing until the shoulder emerges.
33. Assemble the Woodruff key by driving it with a brass hammer while backing up the shaft with a lead hammer.
34. Drive the shaft into the pinion gear (D) up to the shoulder using a plastic or wooden mallet against its free end while supporting the pinion on a wooden block.
35. Install the retaining ring (E).
36. With the casting gripped between lead jaws in a vise, use a plastic mallet to drive the end of the shaft protruding from the pinion backward, so the retaining ring bears snugly against the face of the pinion.
37. Check that the plate is perpendicular

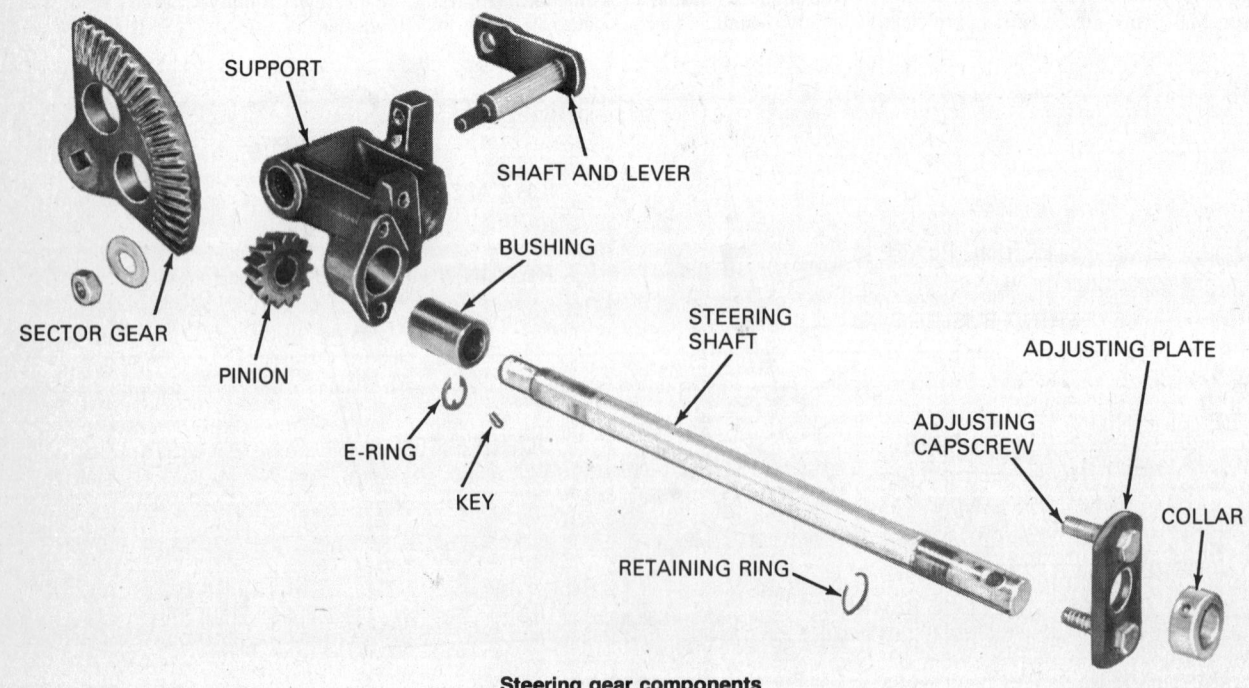

Steering gear components

to (square with) the shaft. Then move the collar (J) up against the plate and tentatively tighten its setscrew.

38. Test the fit by rotating the shaft. Any binding is probably due to the collar rubbing on the plate. Check that the locknuts are against the plate and then tap the free end of the shaft with a plastic mallet, shifting the shaft slightly through the collar.

39. Repeat the above procedure, resetting the collar if necessary, until the shaft rotates freely with only a barely perceptible end play.

40. Examine the face of the pinion at the small end. Note the rectangular depression embossed in the end of one tooth. This is the witness mark (A). Examine the top face of the sector gear (C). Note the circular dimple-like depression embossed at the center of the tooth arc, between the two large holes. This is the witness mark (B). In assembly, the two witness marks must coincide for proper timing.

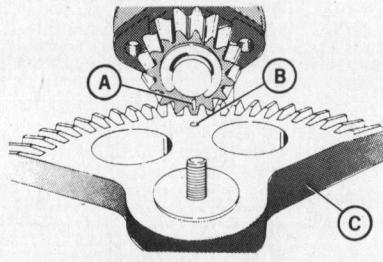

Witness marks

41. Turn the steering shaft so the witness mark (A) is down; that is, nearest the space to be occupied by the sector gear. Then, aligning the witness mark (B) with (A), slide the sector gear into engagement with the pinion, centering its double-flat hole over the arm shaft bore.

42. Oil the arm shaft, and insert it through its needle bearings toward the sector gear. Rotate the arm shaft until its flats are aligned with those in the gear. Then, urge the shaft into the gear, tapping the shaft, if necessary, with a plastic mallet.

43. Run on the nut (B) and tighten it permanently.

ADJUSTMENT

The steering gear mechanism must be adjusted for minimum play in the steering before the assembly is installed in the tractor. Minimum play is obtained when the gears are adjusted for the least amount of backlash that can be obtained without creating binding of the gears. This adjustment is accomplished by loosening or tightening the capscrews (A) as follows:

NOTE: To make an adjustment, first loosen the locknuts (F). Then turn the capscrews small, equal amounts in the same direction to keep the plate (I) perpendicular to the steering shaft. After completing the above, retighten the locknuts and rotate the gears to check for binding or excessive backlash.

1. Tighten capscrews (A) until gears begin to bind. Due to size variations, the initial binding may occur at one specific position of the gears. If so, concentrate all adjustments on this specific position.

2. When binding occurs, loosen the capscrews slightly and tap the end of the steering shaft in the same direction with a plastic mallet. Retest and repeat this step until the binding is just eliminated.

3. Check the backlash of the gears by urging the sector gear back and forth while holding the steering shaft stationary. Note the amount of sector gear movement that is possible without moving the pinion gear. This play or backlash should be no more than a few thousandths of an inch or the thickness of a sheet of paper as seen at the tip of the sector gear. If the backlash is excessive, repeat steps 1 and 2 above.

4. When the backlash is properly adjusted, tighten both locknuts (F) and the setscrew for the steering shaft set collar (J). Then repeat step 3 to determine if tightening of the locknuts and set collar affected the adjustment. If it did, loosen these items and repeat the adjustment steps.

5. Urge the steering shaft strongly toward the sector gear and retest the fit throughout the entire range of the gears. If binding occurs, repeat entire adjustment.

LUBRICATION

Using a grease gun, fill steering casting bore through grease fitting until grease seeps out at both arm and gear ends of the bore. Oil the gears and the shaft, the latter at three points: between the gear and the casting, between the bushing and the plate, and between the collar and the plate.

INSTALLATION

Prior to installation in the tractor, turn the steering shaft clockwise, as viewed from the steering wheel end, as far as it will go. Then install the steering gear mechanism into the tractor in the reverse order of removal. Replace the steering wheel and the battery.

DIFFERENTIAL AND HUB ASSEMBLY

NOTE: One differential gear is part of the right rear hub.

Right Hub Removal and Installation

1. Raise and support the rear of the tractor on jack stands.
2. Loosen the two setscrews and remove the rear axle collar from the end of the axle.
3. Use a fine file to remove any burrs left by the setscrews.
4. Slide the hub off the end of the axle.
5. Wipe the exposed end of the axle to remove any dirt and grease.
6. Using a thin grease solvent, clean entire right hub assembly. Be sure to flush all grease from differential gear and bearings within center bore of hub.
7. Check differential gear on hub for signs of excessive wear and broken or chipped teeth.
8. Check bearings within bore of hub for discoloration and other signs of overheating. Insert a finger into the bore and check bearings for grooving or cracks. If any bearing defects are noted, also check portion of axle that runs in the hub. Axle defects, other than minor surface discoloration or shallow scratches and pitting that are easily removed by polishing, are cause for axle replacement.
9. The right wheel hub assembly includes a pair of bearings and a grease fitting. Disassembly consists only of removing these items from the combination hub and differential gear.
10. Remove grease fitting from hub.
11. Remove the bushing type bearings from the bore of the hub. The removal is best done with an internal bearing puller. If a puller is not available, the bearings can also be driven out of the hub. Begin by inserting a long rod about half the size of the axle into the bore of the hub. Tilt this rod slightly so that its end bears against the inner face of the bearing at the far end of the bore. Then drive the far bearing out, being careful not to scratch or mar the bore. Drive the second bearing out in the normal manner, using a rod just slightly smaller than the bore of the hub.
12. Be sure that hub and bearings are clean.
13. Wipe a light coat of oil over the outer diameter of the bearings. Then press or drive the bearings into the hub until the outer faces are flush with the ends of the hub bore. When driving the bearings into place, use a driving rod slightly smaller in diameter than the hub bore. Never hammer directly on the bearings.
14. Install the grease fitting in the proper hole of the hub.
15. Check rubber seal in outer rim of differential assembly. Replace seal if it is worn or cracked.
16. Check axle collar. One face must be smooth and free of paint. The mating surface on the wheel hub must be in the same condition.
17. Spread a thin coat of grease over the exposed portion of the axle. Apply grease liberally to the differential gear on the wheel hub.
18. Slide wheel hub over end of axle with differential gear facing differential assembly. Rotate wheel hub as necessary to align teeth of its differential gear with teeth of pinion gears within differential assembly.
19. Grease paint-free surface of axle collar. Slide axle collar onto axle with greased surface nearest hub. If axle has not been replaced, axle collar setscrews should be aligned with old scoremarks in axle.
20. Press axle collar inward to firmly seat hub in differential assembly while tightening axle collar setscrews to secure hub and collar in place.

Allis-Chalmers

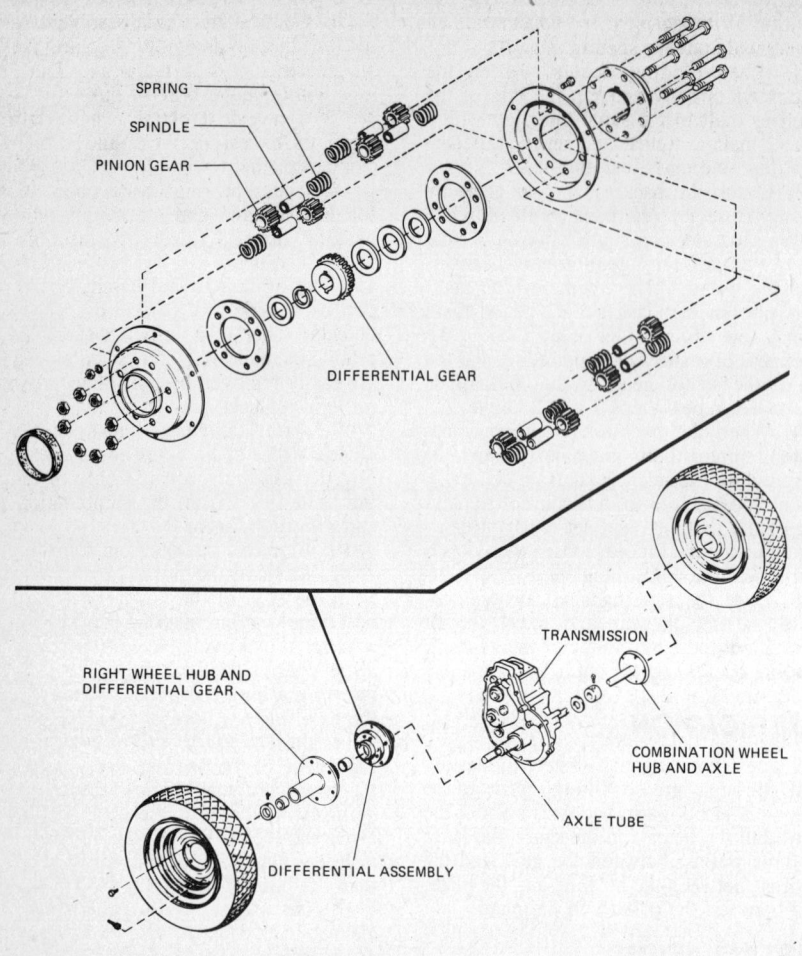

Differential and hub

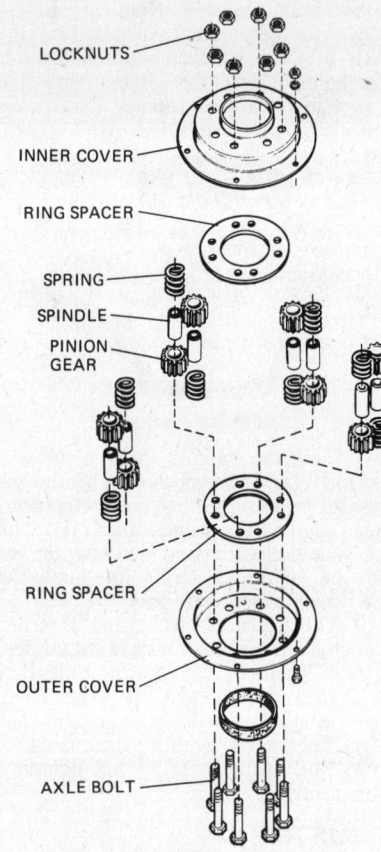

Later style differential components

and remove the axle shaft. Remove all burrs and sharp edges from the axle shaft, before removing.

8. Remove the axle tube keys and retaining ring if replacement is needed or if the gear case is to be disassembled.
9. Remove the six screws with lockwashers and nut around the outer rim of the differential covers.
10. Remove the eight locknuts from the pinion gear axle bolts. The locknuts should be loosened slowly in succession until the pressure of the internal springs is released.
11. Remove the inner differential cover and all internal parts.
12. Clean all parts using a thin grease solvent.

21. Check for end play of hub. If necessary, loosen axle collar setscrews and repeat step 6 until end play is eliminated.
22. Adjust and lubricate wheel hub.
23. Install right wheel on hub.

Differential

REMOVAL AND INSTALLATION

1. Raise and support the rear of the tractor on jack stands.
2. Remove both rear wheel and tire assemblies from wheel hubs.
3. Remove right wheel hub.
4. Remove the paint and burrs from the lefthand axle shaft.
5. Loosen the lefthand set collar and move the axle all the way to the right, so that the set collar is against the lefthand hub.
6. Remove the recessed washer, retaining ring and the two keys from the right end of the axle shaft.
7. Pull the differential assembly off

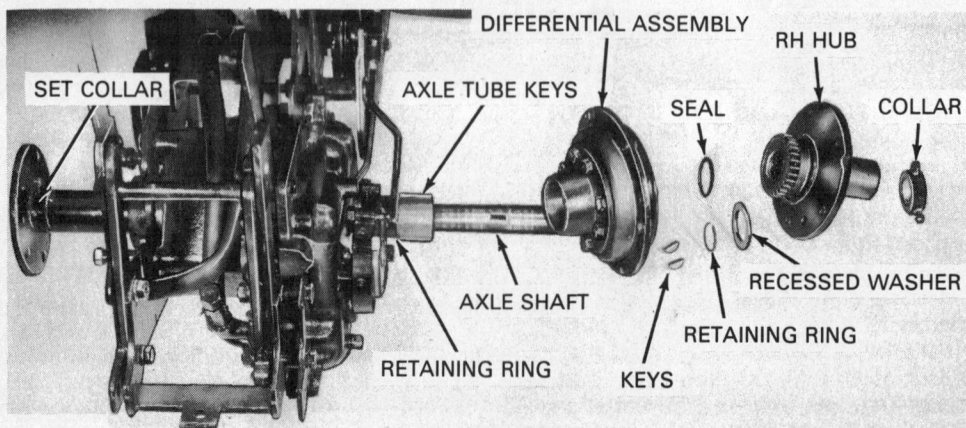

Differential components

13. Check pinion gear teeth for excessive wear, breaks or chips. Check gear faces for scoring. Discard or replace any damaged gears.

14. Check the pinion gear spindles for scoring and signs of overheating. Discard and replace any bad spindles.

15. Springs should be checked for broken or brittle ends as well as general strength. Discard and replace any springs with even questionable defects.

16. Replace the rubber seal in the lip of the cover and replace any broken, stripped or scored pinion gear axle bolts.

17. Scrape clean mating surfaces of the two covers. Replace both covers if either is bent or cracked.

Assembly

18. Insert eight bolts that serve as pinion gear axles through holes from outside to inside of outer differential cover. The outer differential cover has the lip for the rubber seal.

19. Set outer differential cover with lip and bolt heads down on a flat surface. Coat portion of bolts that will be within covers with grease. (Use a good grade of lithium base automotive chassis lubricant.)

20. Inspect one ring spacer and grease face that shows the least wear. Install this ring spacer over the eight pinion gear axle bolts with the greased side facing up.

21. Coat all eight spindles with grease and install one on each pinion gear axle bolt.

22. Coat one spring and one pinion gear with grease. Install these items, spring first, on a spindle. Repeat this process three additional times, installing a spring and pinion gear on every other spindle.

23. Place a differential gear between the four pinion gears that are already installed and rotate the latter until all four mesh with the differential gear at the same time. Then remove the differential gear. From this point on, exercise care to prevent rotation of pinion gears once they are installed.

24. Pinion gears and springs must still be installed on the four remaining spindles. Coat a pinion gear and spring with grease. Place pinion gear over an empty spindle, rotating it until its teeth are centered in gaps between teeth on adjacent previously installed pinion gear. Then allow the pinion gear to drop straight down on the spindle and place spring over spindle on top of pinion gear. Repeat this process three more times until a pinion gear and spring are installed on all eight spindles.

25. Inspect remaining ring spacer and grease face with least wear. Then install ring spacer over eight pinion gear bolts with greased side facing spindles.

26. There are six larger bolt holes and one smaller alignment hole around the outer rim of the inner cover. Matching this alignment hole to a similar hole in the outer cover, install inner cover over the eight pinion gear axle bolts.

27. Obtain six ¼-20 by 2 in. long capscrews and mating nuts. Install these items in the six holes around the rims of the cover halves. Do not attempt to draw the cover halves together with these capscrews; the capscrews are being used only to keep the covers aligned during the next step.

28. Install locknuts on the eight pinion gear axle bolts.

29. Tighten locknuts successively in small increments to draw covers together. To avoid cover damage, stop tightening of locknuts if resistance suddenly increases as pinion gears are drawn together. In such cases, it will be necessary to return to conditions of step 7 and realign the pinion gears.

29. When covers are drawn together, remove capscrews and nuts installed in step 10 and replace them, one at a time, with the normal ⅝ in. long capscrews, lockwashers and nuts.

30. Pack extra grease around all pinion gears within differential assembly.

NOTE: The installation procedure given in the steps below may be used whenever the axle is already installed in the tractor. If the axle, right hub and differential are to be installed as a unit, the differential carrier should be removed from the differential and installed on the transmission axle tube as detailed in the first four steps below. After cleaning and oiling the axle, the other parts can be installed by simply inserting and slipping the axle through the axle tube. To complete the installation, rotate unit until pinion gear axle bolts pass through differential carrier, install and tighten locknuts to secure carrier to differential, and then install the left wheel hub and the two wheel and tire assemblies.

31. Insure the retaining ring is installed on transmission axle tube.

32. Obtain keys (A) for differential carrier. File small chamfer along side and top edges at one end of keys. Then place keys in key ways with plain end against retaining ring (G). Tap keys home in key way with a brass hammer, being sure that keys remain butted against retaining ring.

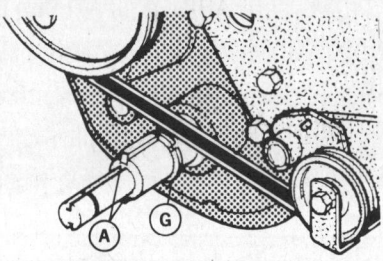

Installing gear

33. If axle is installed, pull it from left side of transmission to right.

34. Coat keys (A) and axle tube outboard of retaining ring (G) with a film of oil. Place differential over keys and tap it with a brass hammer until it sets against the retaining ring.

35. Install the two Woodruff keys into the axle shaft key way and install the retaining ring onto the axle shaft.

36. Align the two Woodruff keys in the axle shaft with the key ways in the differential gear and push the axle shaft toward the left so that the gear encases the keys and the retaining ring contacts the gear.

Move the lefthand set collar against the gear case and remove all the end play and tighten the setscrews.

37. Install the recessed thrust washer on the shaft with the recess toward the retaining ring, and install the hub seal.

38. Install the righthand hub and set collar, and remove all the end play from the axle shaft and tighten the setscrews.

39. Replace the drawbar and lift assembly.

40. Refer to the hydraulic section for the following instructions:
   a. Installation of the hydrostatic drive unit.
   b. Adjustment of the hydrostatic drive unit, brakes, clutch and other linkage.
   c. Filling of the gear case with oil. (Three quarts of Dexron® Hydraulic Fluid.)

41. Install the wheels, seat, springs and other items that were removed and adjust as outlined in the operators manual.

Axle

REMOVAL AND INSTALLATION

1. Raise and support the tractor on jack stands.

2. Remove right wheel hub.

3. Remove differential assembly including differential gear and its keys from axle.

4. Remove any rough edges and burrs from edges of differential gear key ways in axle using a file. Use polishing cloth to smooth out any scratches in the axle. Then use a clean cloth to remove grease, dirt, filings and abrasive particles from exposed righthand portion of axle. Polishing and cleaning axle in this manner is necessary to prevent damage to bearings within axle tube of transmission.

5. Grasp left wheel hub and pull axle out of axle tube from left side of tractor. Care must be taken to maintain axle alignment, particularly after axle slips through first bearing in transmission, to avoid binding within the bearings.

6. Check axle for signs of overheating and excessive wear in the areas that run in the bearings of the axle tube and right wheel hub.

7. Check key ways and retaining ring slot to be sure that edges are sharp and able to perform retaining functions.

8. If damage is evident in checks above, axle must be replaced. In such cases, inspect and replace as necessary parts that mate with damaged area on axle.

9. Use a clean rag to give the axle a final cleaning. Then spread a light film of oil over the axle.

10. Carefully insert right end of axle into left end of axle tube. The axle is a running fit in the axle tube bearings. Any binding as the axle is installed will be due to axle misalignment. With this in mind, carefully continue insertion of axle through axle tube. Be especially careful when axle reaches near end of bearing at right side of axle tube. Once axle is protruding from

Allis-Chalmers

right side, simply push axle through the rest of the way.

11. Install differential assembly.
12. Install right wheel hub.
13. Install wheel and tire assemblies.

LUBRICATION

Two grease fittings must be lubricated after performing repairs on the axle, differential or right wheel hub.

1. Wipe fitting on right end of axle tube with a clean rag. Use a grease gun to pump a good grade lithium based automotive chassis grease through the fitting until grease escapes at either end of axle tube.
2. Wipe fitting on right wheel hub with a clean rag. Apply four shots of a good grade lithium based automotive chassis grease through the fitting.
3. Wipe fittings, ends of axle tube and right end of axle to remove excess grease.

BEVEL GEAR DRIVE

Description

The bevel gear drive houses a driveshaft, driven shaft and their respective bevel gear, bearings and seals. This assembly transfers the engine drive to the transmission through a drive pulley and V-belt arrangement.

NOTE: If the bevel gear drive requires servicing, the complete drive unit must be removed from the tractor in order to gain access to the bevel drive gear.

OVERHAUL

1. Remove drive unit, drive belt, and pulleys:
   a. If tractor is equipped with 3-speed or shuttle transmission see that section.
   b. If tractor is equipped with a hydrostatic transmission, see that section.
2. Refer to PTO section for removal of power take-off, which has to be removed before the side plates can be removed from the bevel gear housing. Remove drive pulley and key.
3. Remove the three capscrews from each side plate.
4. Pull the side plates apart, separating them from the bevel gear housing, but leaving the linkage parts still assembled to each.
5. Remove the large, thin, spacer washers from the short end of the driven shaft.

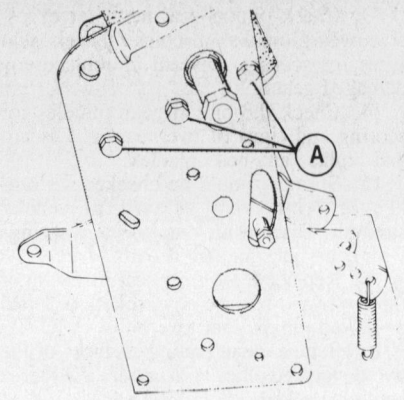
Removing capscrews

6. Examine the housing in the areas of the cover gasket and oil seals for signs of oil seepage. If so, note the location for future attention.
7. Remove all dirt and grease from the exterior of the housing and shafts. Steam cleaning is preferable, but a water soluble grease solvent may be used. Brush solvent on thickly, let the solvent act for a time, and then hose it off with hot water.
8. Spin the driver shaft several revolutions in both directions while applying light braking pressure to the driven shaft. Repeat this procedure, spinning the driven shaft.

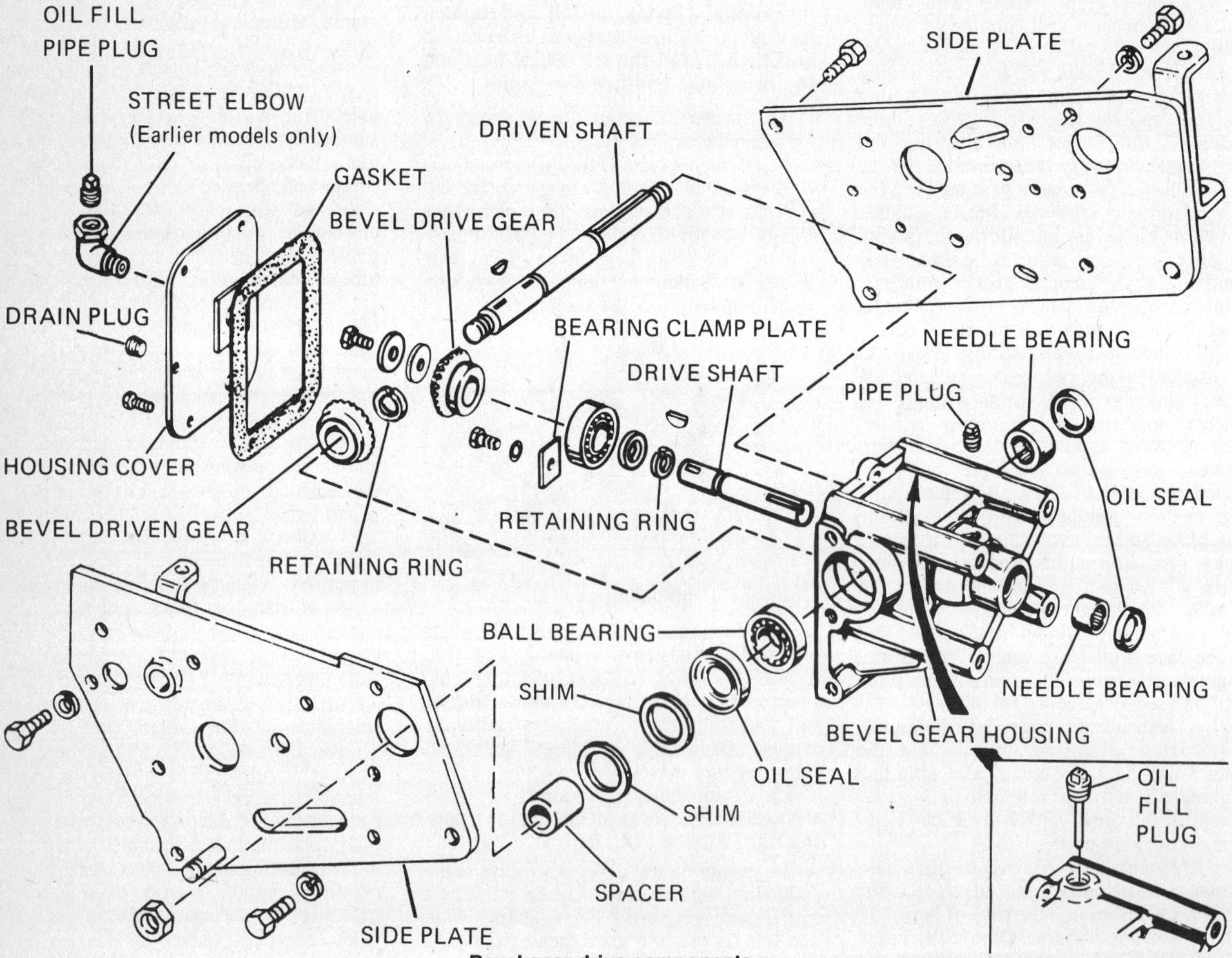

Bevel gear drive components

Allis-Chalmers

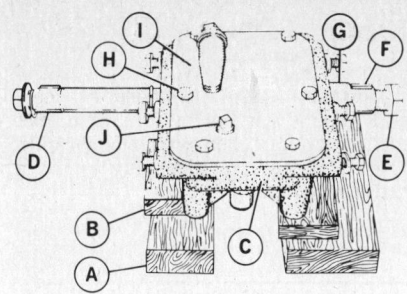

Blocking of the bevel gear drive assembly

Note for future attention any noticeable binding, roughness, unevenness in the motion, or excessive backlash (lost motion) between the gears.

9. Remove drain plug (J) and drain fluid from housing. Then replace and tighten drain plug (J). If housing has filler plug with dipstick on top of housing, remove filler plug during disassembly and assembly procedures.
10. Block the housing up on short pieces of wood on a bench.
11. Remove the nut (E) from the short end of the driven shaft.
12. Remove the Woodruff key (F).
13. Using a fine file, reduce any burrs around the key way to shaft surface level. Do not remove any surface material.
14. Remove the spacer sleeve (G).
15. With a strip of fine abrasive cloth, polish the exposed short end of the shaft to remove any file marks.
16. Using a clean rag, remove all abrasive dust from the shaft, the seal, and the key way. Spread a film of light oil over the entire short end of the shaft.
17. Remove the six capscrews (H).
18. Using a screwdriver, carefully pry off the cover (I).
19. Applying a lead hammer as a buck to back up the gear (A), drive the shaft (B) through the gear to expose the key (C).
20. Remove the key (C).
21. If it is present, disengage the retaining ring (D) from its groove and slide it over near the gear (A).

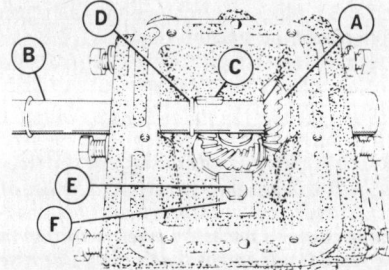

Bevel gear drive housing with cover removed

22. Stuff a clean rag into the housing under the driven shaft. Then use a fine file to remove any burrs from around the key way and ring groove. Reduce all protrusions to shaft surface level, but do not remove any surface material. Do not use any abrasives.
23. Remove the rag, being careful not to spill any filings into the gears or bearings.
24. Pull the shaft (B) out of the housing with one hand, while holding the gear (A) with the other hand.
25. Remove the capscrew (E) and lift out the clamp plate (F).
26. Turn the housing over as shown.
27. Place a piece of brass rod on the overhanging end of the key (B) and drive the key out of its key way.

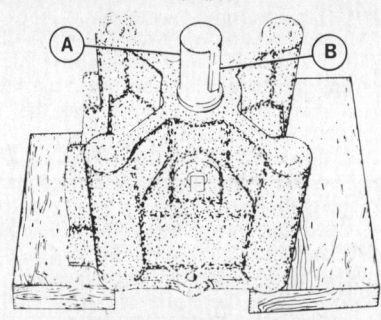

Removing driver shaft

28. Using a fine file, remove any burrs from around the key way. Using a strip of abrasive cloth, polish the shaft to smooth the file marks.
29. With a clean cloth, clean grit from the shaft, the key way, and around the seal. Spread a film of light oil over the entire exposed area of the shaft.
30. With a plastic mallet, drive the shaft downward until it begins to move freely. Be careful not to hit the seal with the mallet.
31. Turn the housing over and lift the shaft, gear, and ball bearing assembly out of the housing.
32. Hold the driver shaft with one hand and spin the outer race of the ball bearing with the other hand. Any binding or unevenness in the motion or perceptible side play of the outer race indicates the need for bearing replacement.
33. Examine the area of the driver shaft that serves as a journal of the needle bearing. Any scoring, galling, excessive wear, or discoloration due to overheating indicates the need for shaft and needle bearing replacement.
34. Examine the area of the driver shaft that turns in the oil seal. Any scratches or other surface defects indicate the need for shaft and oil seal replacement.
35. To disassemble the shaft from the gear and ball bearing for replacement of any of these items, proceed as follows:
    a. Lay the shaft on a wooden block. Drive the key at the end opposite the gear back into the key way. An end of the key should hang over the end of the shaft to permit easy removal.
    b. With gear facing upward, grasp portion of driver shaft below bearing between thick lead jaws of a vise. The shaft key should be against a jaw to prevent shaft rotation in the next step.
    c. With a socket wrench, back out capscrew (A) about ¼ in.
    d. Remove shaft from vise. Grasp the gear only with a leather gloved hand and strike head of capscrew (A) with a plastic mallet until capscrew bottoms out on gear.

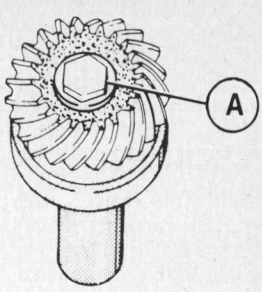

Driver shaft

    e. Remove capscrew (A) and washer(s). Then repeat step d using a piece of brass or plastic rod to drive shaft out of gear.
    f. Remove both keys from the driver shaft. Holding the shaft with the bearing above to keep filings out of bearing, remove any burrs from key way with a fine file.
    g. Use a clean rag to remove the filings and then apply a light coat of oil over the short end of the shaft.
    h. Reinstall the capscrew (A). Hold the bearing with a gloved hand and gently tap the capscrew with a plastic mallet to drive the shaft out of the bearing.
36. If initial inspection indicated fluid leaks around shafts, replace related seals. Pry seals out carefully; do not scratch or score the bore that the seal fits.
37. Examine the needle bearings. To remove a damaged needle bearing, it is first necessary to remove the adjacent seal. Press the bearing inward, using a smooth faced plug with a diameter slightly smaller than the bore.
38. Spin the inner race of the ball bearing in the housing. Any binding, unevenness in motion, or sideplay indicate the need for replacement of the ball bearing and adjacent seal. To remove the bearing, push it outward with a smooth faced plug slightly smaller than the bore.
39. Examine the driven shaft. Any evidence of scoring, galling, seizing, excessive wear, or discoloration due to overheating indicates the need for shaft, seal, and needle bearing replacement. If the shaft is all right except for scratches in the area that runs in a seal, replace the shaft and the seal.
40. Examine both gears. If gears are worn or damaged or if initial spin tests indicate problems and bearing problems are not evident, replace both gears.
41. Remove all foreign material from housing gasket flange and the mating surfaces of the cover. Scrape surfaces until clean, but do not remove any surface material.
42. Thoroughly clean inside of housing with a grease solvent.
43. Install the ball bearing for the driven shaft into the housing. Use a press and press the bearing inward until it seats against the shoulder of the bore. Do not press against the inner race.
44. Install the related ball bearing onto the driver shaft. Use a press and press the bearing until it seats against the shoulder of the shaft. The pressing plug should be hollow with a rim that bears only against the inner race.

Allis-Chalmers

45. Install the needle bearings into the housing with the end marked "Torrington" facing outward. Press the bearing inward until it is flush with the bottom of the counterbore. Use a smooth faced plug about ⅛ larger in diameter than the bearing.

46. Install the seals into the housing with the spring element facing inward. Use a smooth faced plug with an outer diameter slightly smaller than the counterbore for the seal. When properly seated, the seals other than the one for the driven shaft ball bearing will be about flush with the housing surface; the seal for the ball bearing will be slightly below the flush point.

47. Before installing the shafts, grease the sealing lip of each seal.

48. Install the Woodruff key (A) in the gear end of the driver shaft (B) and, resting the gear (C), face down, tap the shaft downward with a plastic mallet until its hub is within a quarter of an inch of the ball bearing.

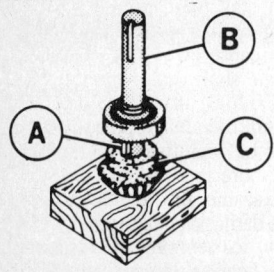

Assembling driver shaft

49. With the key in the outboard end of the shaft temporarily installed to prevent rotation, grasp the driver shaft between thick lead jaws in a vise. Replace the washer and screw, tighten screw securely, drawing the gear down against the inner race of the ball bearing.

50. Remove the outboard key, clean the shaft, and cover shaft with a light film of oil.

51. Block the housing up.

52. Install the driver shaft assembly (B) in the housing, pushing it gently downward from the inside until the ball bearing enters its seat. Do not strike the gear. Use a plastic rod held at an angle against the outer race of the bearing.

53. Install the clamp plate (C) with the arch upward. Then install and tighten the capscrew (D) until the clamp plate bends downward and is tightly clamped to the housing wall.

54. Clean the driven shaft (E) and cover it with a light coat of oil. Then insert the large threaded end of the driven shaft into the housing through the bore at the left side when the housing is viewed as shown.

55. Clean the driven shaft (E) and cover it with a light coat of oil. Then insert the large threaded end of the shaft into the housing through the bore at your left when viewing the housing as shown.

56. Slip retaining ring (F) onto the shaft. Then hold the gear (G) inside the housing while turning the shaft into the gear. When the slot appears, snap the retaining ring (F) into place.

57. Place Woodruff key (H) in its key way on the shaft. Double electrical tape around both the shaft and the key to protect them and squeeze the key home with a pair of Channel-Lock® pliers. Do not use a hammer to drive the key into place.

58. Remove the electrical tape from the key and turn the shaft to align the key to the key way slot in the gear (G).

59. Install the spacer (I) with the painted end outward on the short end of the driven shaft. Turn the spacer gently through the oil seal until it seats against the ball bearing.

60. Bore a ¹⁵/₁₆ in. hole through a block of soft lumber, stand the housing with driven shaft (E) vertical and the outer end of sleeve (I) centered over the bored hole. Drive the shaft downward with a plastic mallet until the retaining ring contacts the face of the gear.

61. Test operation of the assembly by spinning first the driver shaft (B) and then the driven shaft (E). If the motion is not completely free, isolate the cause of the difficulty and correct it before proceeding further.

62. Lay a new gasket on the flange of the housing. Place the cover over the gasket and align the six screw holes in the housing, gasket and cover. Then install and tighten the six capscrews (H).

63. Use a lead hammer to back up the driver shaft (A) and a brass hammer to drive the key (B) into its key way. For easy future removal, the key should overhang the end of the shaft by about ⅛ in.

INSTALLATION

Install bevel gear box in reverse order of removal. (Refer to removal procedures at beginning of this section.)

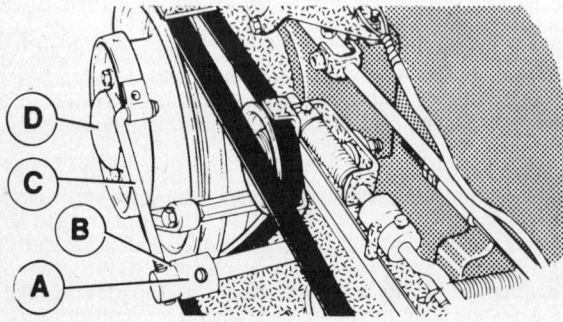

Shuttle gears and sheaves

LUBRICATION

Fill bevel gear box with SAE 90 transmission fluid. Pour fluid slowly and check level often. Gear box is full when level reaches top of filler pipe elbow (older models) or tip of filler plug—dipstick (newer models).

TRANSMISSION

Three Speed Manual and Shuttle Transmissions

REMOVAL

1. Select a good site for transmission removal. The site should have a level and hard floor with a vertical brick or block wall that can be used to prevent the tractor from rolling forward.

2. Park tractor with front wheels against the vertical wall to prevent forward motion of tractor. Shift transmission to first or second gear and set the parking brake.

3. Place wooden blocks or bricks behind front wheels to keep tractor from rolling backwards.

4. Remove seat deck. Seat deck is secured with nuts, lockwashers, and bolts.

5. Drain fluid from transmission. Place large pan under transmission and remove drain plug. Then remove vent plug to speed draining.

6. While fluid is draining from transmission, remove capscrew with its lockwasher and hex nut and disconnect the shifter rod (C) from the transmission. To prevent loss, reinstall capscrew with its lockwasher and hex nut on the shifter rod.

7. To remove the transmission safety switch (B), disconnect electrical plug from this switch at this time.

8. When fluid has finished draining from transmission, reinstall both the vent and drain plugs and remove the drain pan from under the tractor.

CAUTION

Use the jack in the following step only for initial raising of the tractor. Use the pair of jack (safety) stands to support the tractor before beginning any work. The use of jack stands is a good practice. This practice becomes a necessity when working on the raised rear end of the tractor. Because of the pivot between the front wheels and frame of the tractor, support must be provided at two additional points to assure side-to-side stability of the raised tractor.

9. Place a portable or floor jack under the rear hitch and raise rear of tractor off floor. Adjust jack (safety) stands to keep rear wheels off floor and place one stand under each end of seat deck holddown bar. Then release and lower jack to lower tractor weight onto jack stands.

Allis-Chalmers

10. Remove rear wheels, differential, and rear axle. See Differential and Axle.

3 Speed Tractors

11. Remove nut from pulley shaft.
12. Depress clutch-brake while pulling main drive belt off drive pulley.
13. Remove pulley from pulley shaft. If necessary, use a bearing puller or tap pulley lightly with a brass hammer.
14. To prevent its loss, wrap key to pulley shaft with one or two turns of electrical tape.

Shuttle Tractors

15. Remove cotter pin from pivot at lower end of shuttle brake rod. After disconnecting pivot from lever, reinstall cotter pin in pivot.

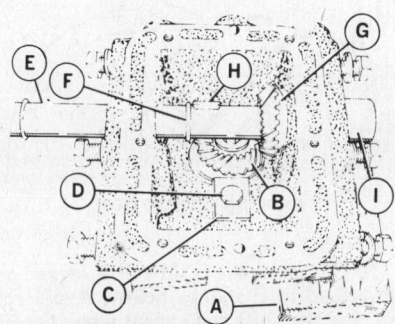

Installing shafts in housing

16. Using a screwdriver, pry grease cover off planetary gear assembly.
17. Insert a long screwdriver between the webs of the pulleys. Use this screwdriver to hold pulleys while using a wrench to remove nut from end of pulley shaft. Then reinstall grease cover.
18. Slip planetary gear assembly off end of pulley shaft.
19. Remove key from pulley shaft.
20. Use a fine file and polishing cloth to remove any burrs from edges of key way in pulley shaft. Then use a clean cloth to remove any grit or filings.
21. Remove brass washer from pulley shaft.
22. Depress clutch-brake pedal while pulling main drive belt from sheave and gear assembly.
23. Remove sheave and gear assembly from end of pulley shaft.
24. Remove spacer, second brass washer, and half-sheave from pulley shaft.
25. Prepare tractor for transmission removal.
26. Remove main drive belt and drive pulley or planetary gear assembly from drive shaft on transmission.
27. Remove three capscrews that secure hitch assembly to right side of transmission.
28. If tractor is equipped with a rear lift kit, remove cotter pin and disconnect rod (B) from lift kit. To prevent loss, reinstall cotter pin in rod.
29. Remove two capscrews and stud to separate hitch assembly from transmission. Then reinstall all hitch assembly fasteners in transmission holes.

30. With parking brake engaged to help prevent turning of brake drum, loosen and remove nut. Then disengage parking brake and slide brake drum off its shaft. To prevent loss, reinstall nut and any spacers on the brake drum shaft.

CAUTION

The transmission, which weighs approximately 140 pounds, will be separated from the tractor in the next step. Some form of support is required to prevent dropping of the transmission. Such support can be provided in many ways including:
a. Support the transmission from underneath using a floor jack.
b. Support the transmission from above using a sling attached to a chain hoist or to a portable crane.

31. Remove three capscrews on each side to separate the transmission from the side-plates. Use a crane, hoist or jack to hold the transmission up and to pull it clear from the tractor.
32. To prevent loss, temporarily reinstall the six capscrews in the transmission. The longer capscrew goes into the lower hole on the left side.

TRANSMISSION EXTERNAL CLEANING

To prevent the gears from being contaminated with dirt, it is necessary to clean the exterior of the transmission. While the need for cleanliness is at its greatest during assembly, the cleaning is best done prior to disassembly while the transmission is still on the floor.

Before beginning the cleaning, examine the case for signs of oil leakage. Note areas of suspected leaks so the related seal or gasket can be checked after disassembly.

Steam cleaning of the transmission case is preferred. If facilities for steam cleaning are not available, cap both ends of the axle tube with plastic bags held in place by rubber bands. Then brush a strong, water soluble grease solvent thickly over the entire case. Allow the solvent to stand for a time to act. Then hose the casing off with hot water and dry it using compressed air or towels.

DISASSEMBLY

Disassembly of the transmission is best done on a work bench. The bench used should be small enough to permit access to the transmission from at least three sides.

Before hoisting the transmission to the bench, check the brake shaft and the portion of the axle tube on the same side of the transmission as the brake shaft. Burrs or paint on either must be removed before removing the gears and shafts to prevent damage to the bearings. To clean the shafts, stuff a clean rag into the open end of the axle tube. Then smooth and clean the surfaces using abrasive cloth and, if necessary, a file. To complete this process, remove the rag from the axle tube. Whether or not you polished the shafts, wipe both

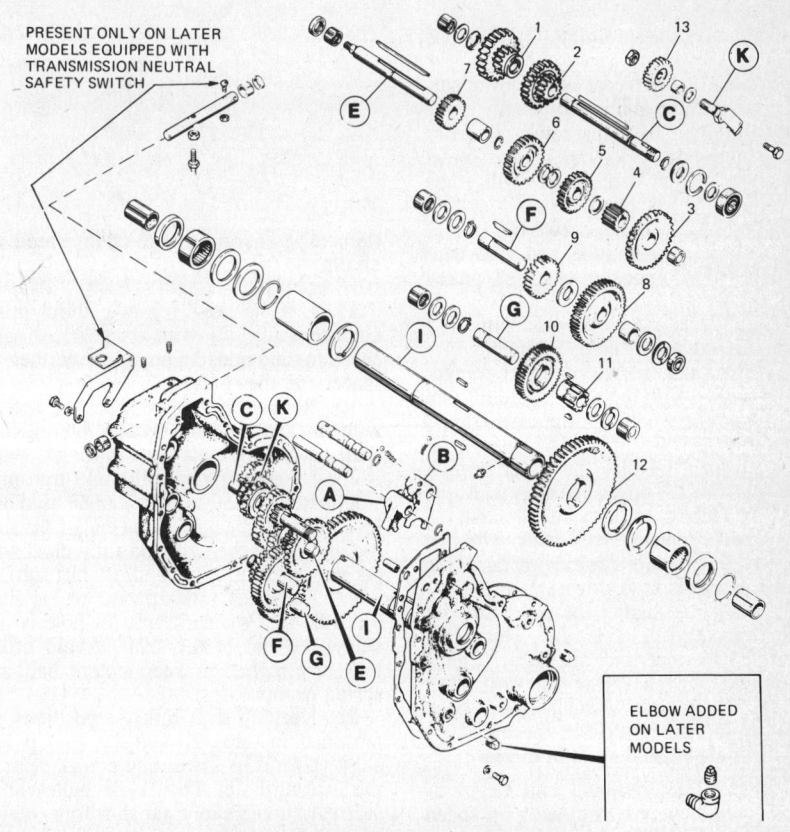

3-speed transmission components

Allis-Chalmers

clean with a rag before applying a light coat of oil to each.

Hoist the transmission to the bench top using a chain hoist or a portable crane. Then lay the transmission on wooden blocking with cover facing upward.

1. Remove twin keys from axle tube.
2. Remove retaining ring (B) from axle tube.

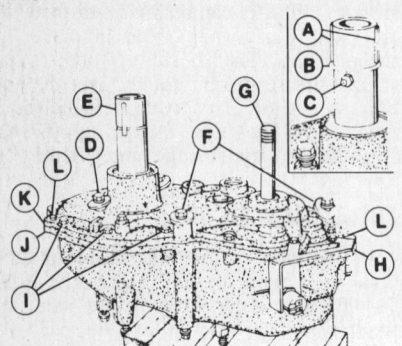

Removing transmission cover

3. Check axle tube (E) and pulley shaft (G) to be sure that both are free of paint and burrs. If so, omit steps 4 and 5.
4. Stuff a clean rag into the exposed open end of the axle tube (E).
5. Using abrasive cloth and, if necessary, a file, smooth and clean surfaces of axle tube (E) and pulley shaft (G) until all traces of paint and burrs that would otherwise damage bearings during cover removal are eliminated.
6. Use a clean cloth to remove grease, dirt, grit, and filings from axle tube and pulley shaft.
7. Remove grease fitting (C) from axle tube.
8. Wipe a light coat of oil over surfaces of axle tube and pulley shaft.
9. Remove 14 capscrews (I) from around outer edge of cover. If one is present, safety switch bracket (H) will come free and can be removed.
10. Remove two capscrews (D) on opposite sides of axle tube and two capscrews (F) on opposite sides of pulley shaft.
11. Drive the two dowel pins (L) downward out of the holes in the cover.
12. Use a screwdriver or other wedge-shaped item as a prying tool. Insert this tool at various points between the case (J) and its cover (K) and pry upward to separate the cover. Be sure to raise the cover straight up. Otherwise the cover bearings will bind on the axle tube or pulley shaft. If either the axle tube or pulley shaft rises with the cover, use a small rawhide hammer to pound the shafts back down.
13. Continue raising the cover straight up until it clears the axle tube. Then set it aside with the inner side facing up. Check the inside of the cover for any shaft washers that may have stuck to and been removed with the cover. Slip these washers back over the related shaft in the case.

NOTE: Gear removal can begin as soon as the cover is removed. To speed the process, gears are removed from the transmission still installed on the gear shafts for later disassembly. The shafts must be removed in a specific order as follows:

14. Remove washers above top gear on the pulley, second intermediate and third intermediate shafts (A, C and D).
15. Remove top gear (G) from first intermediate shaft (B).

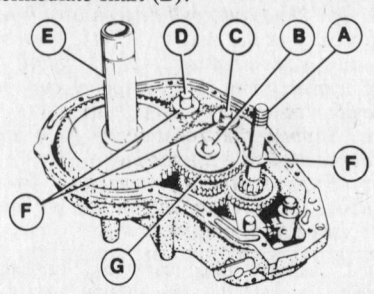

Components with cover removed

16. Lift axle tube (E) and its gear straight up and out of the transmission. Check bottom of case around axle tube hole for washers or spacers. If any are found, reinstall them on the axle tube before setting the latter aside.
17. Pull second intermediate shaft with its gears upward out of the transmission. Repeat check at bottom of case and reinstall any loose washers or spacers on shaft.

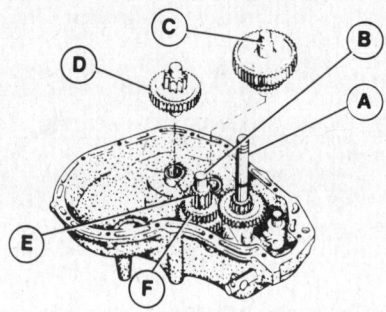

Removing second and third intermediate shafts

18. Lift up and remove third intermediate shaft (D) with its gears. Again, check for and reinstall any loose washers or spacers on the shaft.
19. Remove two upper gears (E and F) with any spacers from second intermediate shaft (B).
20. Push shift rod fully into transmission. Insert something that can be used as a lever through the hole in the end of the shift rod and rotate the shift rod fully clockwise (as viewed from lever end of shift rod).
21. Pull shift fork upward on its shaft until the fork's detent ball fill hole is just below the end of the shaft. Avoid lifting fork any higher to keep detent ball and spring in fork.
22. Keeping shift forks in positions established in steps 20 and 21, pull pulley shaft up until its lower end comes clear of the bearing hole. Then slide pulley shaft sideways until gears clear shift forks before completing removal of shaft from transmission. Be sure that any washers or spacers left in bottom of transmission are reinstalled on the shaft.
23. Remove first intermediate shaft (B) by pulling it straight up and out of the transmission.
24. On three-speed transmission only, remove locknut on opposite side of case from reverse gear. Then remove capscrew (A) and pull reverse gear (B) with its bracket out of the transmission.

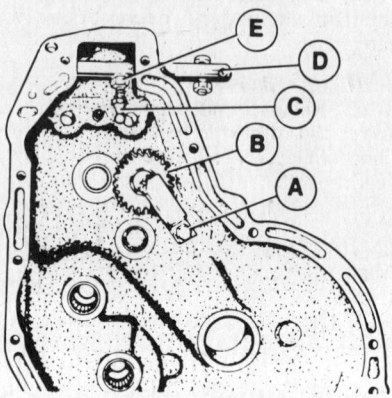

Removing reverse gear

NOTE: Remove a shift fork or rod as described in the steps below. Avoid removing the shift rod except when necessary to replace the rod, its bearings, or the bearing seals.

25. Pull shift fork and its shaft out after loosening related setscrew.
26. Once the shift fork and shaft are removed, separate them by sliding the work along the shaft away from the retaining ring. Cup a hand over the fork to prevent the loss of the detent ball or spring.
27. Before removing shift rod, loosen locknut and remove shift rod stem.
28. Remove shift rod by pulling it forward, out of the transmission.
29. Clean the inside of both the case and its cover. Use a thin grease solvent. Be sure to also remove any metal chips or filings.
30. Use a putty knife to scrape any remaining gasket material off flanges of case and cover.
31. Clean all internal gears, shafts, and spacers. Use a thin grease solvent, drying all parts thoroughly before use.
32. Inspect all gears. Replace any that have broken or chipped teeth or any that show signs of excessive wear.
33. Check all spacers. Replace any that show galling or deep scratches on the faces.
34. Check shafts for signs of excessive wear, over-heating, galling, scratching in the areas that run within bearings. Replace the shaft and the related bearing when such defects are found.
35. Check and replace any damaged keys on the gear shafts.

NOTE: There are a total of 12 bearings and bushings in the transmission case and cover and two more bushings in the axle tube. Remove a bearing only when it is necessary to replace that bearing. Note that there are oil seals on the

Allis-Chalmers

outer side of some bearings (C, G and H). These oil seals must be removed before removing the related bearing or when the initial exterior inspection indicated a possible leak from the seal. To remove an oil seal, pry it out using an awl or similar tool. Take care not to damage or scratch the seal seat in the case or cover. Once the seal is removed, discard and replace it. Never attempt to reuse an oil seal.

36. A ball bearing (A) is used in the pulley shaft hole of the cover. On later models, a retaining ring against the inside surface of the bearing must be removed first. Use special retaining ring pliers. Then drive or press the bearing inward from the outside of the cover using a rod slightly smaller in diameter than the bearing.

37. To remove a needle bearing (B through F) or the bearing (G) which can be either a bronze or needle type depending upon the tractor model, drive or press the bearing inward from the outside of the case or cover. Use a driving rod slightly smaller than the outer diameter of the bearing. Be careful to avoid damage to the bearing housing.

38. Shift rod bushings (H) can also be removed by driving or pressing them inward. Remember to remove the oil seals first.

39. Should it be necessary to remove a bushing inside the end of the axle tube, the bushing at the opposite end should also be removed. Use a bearing puller to remove these bushings.

ASSEMBLY

Gear Replacement

Once the gear shaft assemblies are removed from the transmission during disassembly, it is easy to replace individual gears. Simply disassemble one gear shaft assembly at a time, replacing parts as necessary. Use the exploded views to identify and properly locate parts on the shafts.

Bearing Installation

Bearing installation is done as the first step of transmission assembly. The location of all bearings except the two within the ends of the axle tube are shown. The table lists the descriptions of each of the bearings. It also lists the step below that should be used to install the particular bearing.

Normally, it is best to press the bearings into place. However, the bearings can be driven into the housing when a press is not available. When driving a bearing into a housing, do not hammer directly on the bearing. Use a smooth, flat faced driving rod slightly smaller than the diameter of the bearing housing. For the ball bearing, the driving rod should be hollow or have its end machined away so that the rod bears only against the outer race of the bearing.

1. The ball bearing is installed from the inside of the cover. Drive or press the bearing into the housing until it bottoms out on the shoulder within the housing. Then install the retaining ring in the groove within the housing to hold the bearing in place.

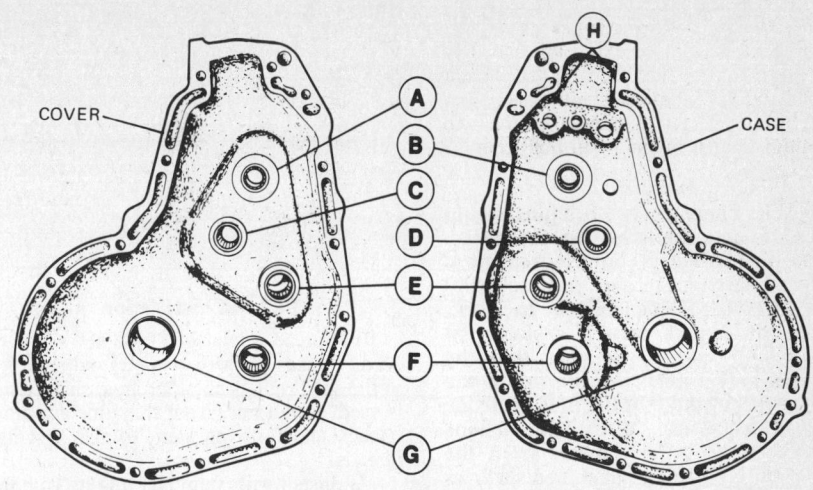

Transmission bearing locations

2. The needle bearing for the pulley shaft and the four needle bearings (E and F) for the second and third intermediate gear shafts are all sealed on one end and are installed the same way. Start these bearings, unsealed end first, into the bearing housing from the outside of the case or cover. Drive the bearings inward until the unsealed end is flush with the machined surface of the housing on the inside of the case or cover.

3. The needle bearing in the cover for the first intermediate gear shaft is also sealed on one end. Start this bearing, unsealed end first, into the bearing housing from the outside of the cover. Drive the bearing inward until its unsealed end is positioned as shown.

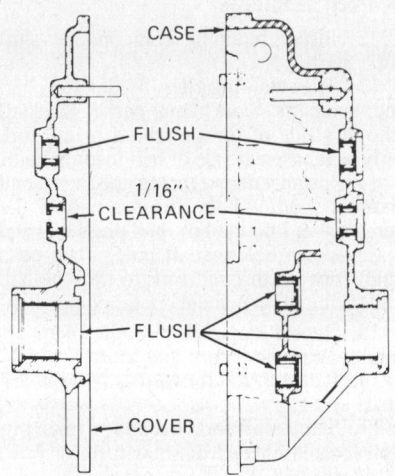

Bearing locations in housing

4. The needle bearing in the case for the first intermediate gear shaft is not sealed on either end. Start this bearing with the end marked "Torrington" facing outward into the bearing housing from the outside of the case. Drive the bearing inward until its unmarked end is positioned as shown.

5. Bearings for the axle tube may be bronze bushings or needle bearings depending upon the tractor model. These bearings are always replaced in pairs. Drive the bearings into the housings from the outside to the inside of the case or cover. If needle bearings are used, the end marked "Torrington" must face outward. The inner end of the bearing in the case must be flush with the machined surface on the bearing housing inside the case. The outer end of the bearing in the cover must be flush with the second (smaller) counterbore visible within the bearing housing from the outside of the cover. If bronze bushings are used, install the one in the case first. Then place the cover on the case using the dowel pins in the flanges for alignment and insert the axle tube through the case and cover. Now slide the other bushing over the axle tube and press or drive it into place in the cover. Using the axle tube as an alignment mandrel in this manner prevents misalignment and binding of the bearings.

TRANSMISSION BEARING DATA

Bearing Location	Type	Installation Step
A	Ball bearing	1
B	Needle bearing	2
C	Needle bearing	3
D	Needle bearing	4
E	Needle bearing	2
F	Needle bearing	2
G	Bronze bushing or needle bearing	5
H	Bronze bushing	6
Inside axle tube	Bronze bushing	7

6. The bronze bushings for the shift rod are driven into the case from the outside until the inner end is flush with the end of the bearing housing on the inside of the case.

7. The bronze sleeve bushings in the axle tube are to be driven in flush with the ends of the axle tube. Use a rod at least a quarter of an inch larger than the bore of the tube. Drive in the first bushing, and then insert the axle shaft through it and out

Allis-Chalmers

of the other end of the axle tube. Place the second bearing on the axle shaft and then center the bearing with the bore and drive it while guided by the axle shaft. This procedure insures exact alignment of the two bushings so that binding will not occur.

Roll Pin Installation

NOTE: There are two roll pins, one in the case and the other in the cover. These pins (A and B) limit movement of the shift mechanism and must be properly installed if the shift is to work. Check these pins even if replacement was not necessary. Be sure that the grooves in the pins face the top of the case or cover and that the pins protrude from the casting to the dimensions shown. Make measurements from the top of the pins to the machined surfaces on the flanges of the case or cover.

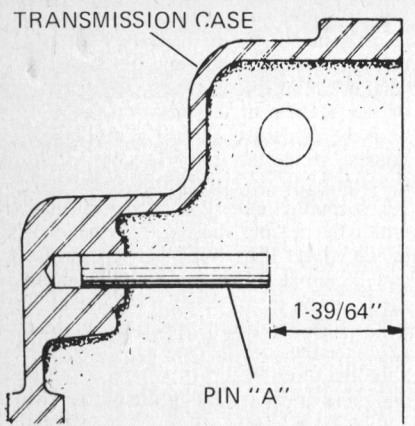

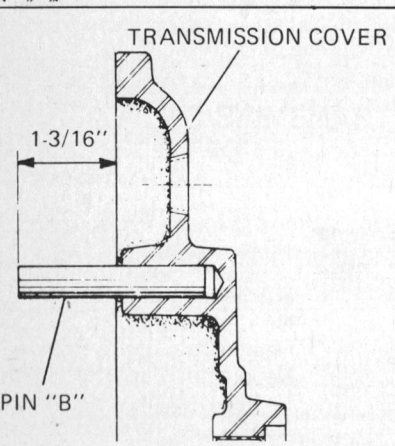

Roll pin locations

Shift Rod Installation

8. If seals (C and E) are installed, run a greased finger gently around the seal lips.

9. Coat rod with a film of light machine oil. Then insert end opposite clevis pin hole through front seal (C) and bearing and push the rod gently rearward through the rear bearing and seal (E). If extra resistance is felt as rod enters rear seal, do not force rod through. Instead, pull rod forward again and pry out and discard rear seal. Take care to avoid damage to seal seat and bearing. Then push rod rearward again until threaded hole is centered in housing. Seals, if removed, will be installed later.

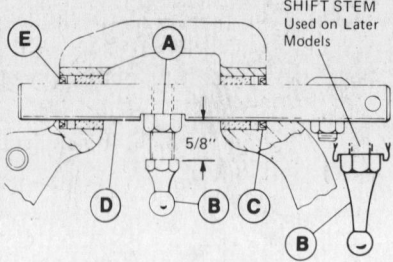

Shift rod installation

10. Spread a coat of Lok-Tite®, Grade A, on threads of shift stem (B) and its hole in the shift rod. If an older style shift stem is used, install nut (A) at this time and run it back and forth on stem to also wet its threads.

11. Install shift stem (B) into its hole in the shift rod. Tighten new style stem against flat of shift rod. Install earlier style stem so the shoulder is ⅝ in. below rounded surface of shift rod. Then hold shift stem with one wrench while using another wrench to tighten nut (A) against flat on shift rod. The ⅝ in. gap is important to proper shifting.

Shift Fork Installation

NOTE: The two shift forks are identical. But, the shafts for the forks are different. One shaft is slightly longer than the other and has its retaining ring groove near one end. The other shaft has its retaining ring groove near the setscrew hole. Note that both forks are installed on the shafts with the long portion of the hub facing the retaining ring. To assemble and install the shift forks, proceed as follows:

12. Install retaining ring on one shift fork shaft.

13. The retaining ring divides the shaft into two parts. Start longer part of the shaft into hub side of fork. Push shaft into fork only as far as near side of ball loading hole.

14. Install a detent spring and then a ball into ball loading hole. Insert a narrow flat screwdriver into the hole and press ball and spring down below shaft hole. Then push shaft forward through fork to trap the ball and spring in permanent operating position.

15. Repeat steps 12 through 14 to assemble other shift fork and shaft.

16. Pull shift rod forward as far as it will go.

17. Select shift fork that has its retaining ring located on end of shaft furthest from setscrew hole.

18. Install selected shift fork in case with the setscrew hole down and facing the setscrew.

19. Tighten setscrew while feeling for setscrew hole. When the two are in apparent alignment, loosen setscrew one turn and rotate shift fork shaft slightly. Then retighten setscrew and note whether or not setscrew forces shaft to return to prior position. If it does, setscrew is properly seated and can be tightened. If not, the entire step should be repeated.

20. Push shift rod as far rearward as it will go.

21. Install other shift fork assembly (A) into case as described in steps 18 and 19.

Gear Installation

NOTE: Once the shift rod and forks are installed in the case, gear installation can begin. To speed the process, all gears should be assembled on their shafts before installation.

22. On three-speed transmissions only, begin gear installation by installing reverse gear. Assemble gear with its spacer and washer on the bracket. Then install this assembly in case as shown. Torque capscrew (C) to 20 ft.lb. Complete assembly by installing locknut on pin at backside of case and torque it to 55 ft.lb.

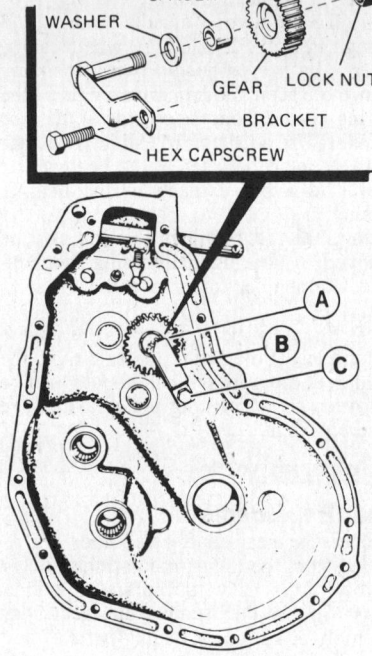

Installing reverse gear

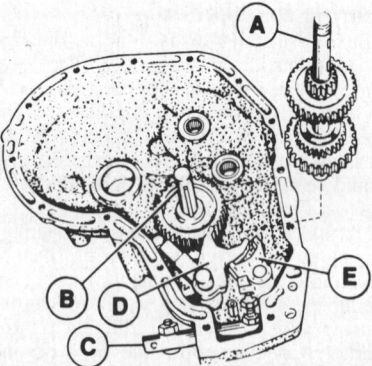

Removing first, intermediate and pulley shafts

23. Push shift rod fully into transmission. Insert something that can be used as a lever through hole in the end of the shift rod and rotate the shift rod fully clockwise (as viewed from lever end of shift rod).

24. Pull shift fork (D) upward on its shaft until the fork's detent ball fill hole is just below the end of the shaft. Avoid lift-

ing fork any higher to keep detent ball and spring in fork.

25. Remove the two washers from lower end of pulley shaft (A). Coat these washers with grease and place them over pulley shaft hole in bottom of case. The composition washer must be on the bottom with its grooved side down.

26. Install first intermediate gear shaft assembly (B) into its needle bearing at bottom gear case. Then remove the three top gears from this shaft and set them aside for later installation.

27. Lower pulley shaft assembly (A) with long end up into bottom of gear case away from shift forks (D and E). While holding large gear on first intermediate gear shaft up for clearance and tilting pulley shaft (A) as necessary, slide pulley shaft sideways toward forks. Engage rear fork (E) in groove above lower gear on pulley shaft and front fork (D) in groove below upper gear on pulley shaft.

28. With gears properly engaged in forks, insert lower end of pulley shaft through washers installed in step 25 and into bearing at bottom of case.

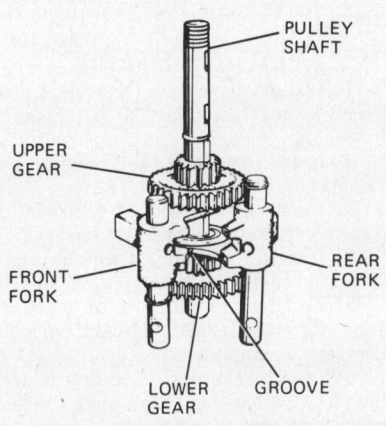

Engaging shift forks in the shift grooves

29. Carefully tap front shift fork (D) back down to its neutral position, where detent ball will snap into groove in shaft.

30. Turn shift rod (C) back to neutral position to lower rear shift fork (E).

31. Pull shift rod (C) back and forth several times and be sure that shift stem alternately engages both forks.

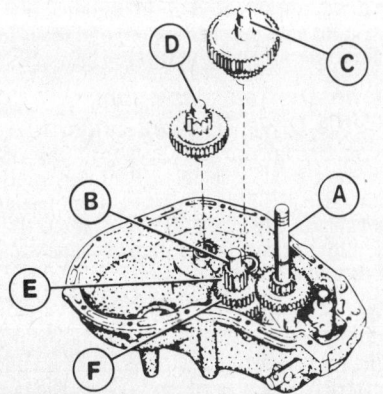

Installing second and third intermediate shafts

32. Be sure that two spacers are placed above the upper gear on first intermediate gear shaft. Then install in order the larger gear (F) with its beveled side up, another spacer, and the smaller gear (E).

33. Remove two washers from lower ends of both the second and third intermediate gear shaft assemblies (C and D). Coat these washers with grease and center them over respective bearings in bottom of case. The composition washers must be on the bottom with the groove sides down.

34. Install third intermediate gear shaft assembly (D) into its bearing in bottom of case. The larger gear on this shaft should be on the bottom.

35. Install second intermediate gear shaft assembly (C) into its bearing in bottom of case. The larger gear on this shaft should be at the top and a spacer should be above the gear.

36. Remove two washers from lower end of axle tube assembly. After coating these washers with grease, center them over the axle tube bearing in bottom of case. The composition washer must be on the bottom with the grooved side down.

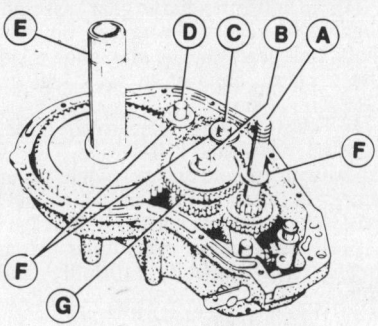

Final gear installation

37. Install axle tube assembly (E) into its bearing in bottom of case. Be sure that the two spacers above the gear on this shaft are installed.

38. Install top gear (G) with its beveled side down on first intermediate gear shaft (B).

39. Grease the two pairs of washers for the second and third intermediate gear shafts. Then install one pair of washers on each shaft (C and D). The composition washers must be on the top with the grooved side up.

40. Grease the two washers for the upper end of pulley shaft (A). Then install them on the shaft. The thicker washer must be on the bottom with its large flat face up.

Cover Installation

NOTE: When gear installation is completed, the cover can be installed on the transmission case. It is then advisable to check the gears for proper shifting and lack of binding before completing assembly of the transmission.

41. Drive the two dowel pins in flange of case upward. These pins must protrude above flange a distance equal to the thickness of the cover flange.

42. Place cover gasket on flange of case. Be sure that gasket is flat and properly centered over dowel pins and holes in flange.

43. Coat the upper ends of the gear shafts in the case with a light film of oil.

44. Install the cover. Begin by centering the proper bearings in the cover over the axle tube and pulley shaft. Then hold the cover level while lowering into place. Do not force the cover. If binding occurs, raise the cover slightly to relevel it before lowering it again. Continue until the dowel pins in the case flange enter the holes in the cover.

45. When the dowel pins begin entering the holes in the cover, it will be necessary to begin driving the cover into place. Be sure that the pins are centered in the cover holes. Then use a wooden mallet and strike the cover alternately over the two dowel pin holes.

46. When the cover is seated on the gasket, place lockwashers over the 14 capscrews and loosely install the capscrews through the cover and into the case. On later models, two of these capscrews also pass through the safety switch bracket which should be installed at this time.

47. Select four capscrews evenly spaced around and near the top, bottom and two sides of the cover. Alternately tighten these four capscrews snugly, but not completely, to draw the cover evenly down to the case. Then proceed around the cver to snug up the other ten capscrews. Finish by tightening all 14 capscrews a little at a time until the cover is firmly secured.

48. Check the gears for binding. Temporarily install a lathe dog or pulley on the pulley shaft. Insert a rod through the shift rod clevis and shift the gears into high. Then turn the pulley shaft while observing the axle tube. The tube should turn freely without binding in all gears. If binding occurs, try striking ends of pulley and first intermediate shafts moderately hard blows with a wooden hammer. It this treatment does not relieve binding immediately, it will be necessary to disassemble the transmission to locate and correct the trouble.

Oil Seal Installation

There are five oil seals over the bearings in the transmission case and cover. (See Table below.) If any of these seals were removed, new seals should be installed only after the transmission is reassembled and checked. Installation should be done with a pressing tool and thimble. If a thimble is not available, heavy paper lubricated with grease or oil can be used as a substitute. Wrap this paper in a spiral fashion over the shaft beginning near the seal seat. Be sure that the paper or thimble covers any key ways in the shaft. Specific steps for seal installation are as follows:

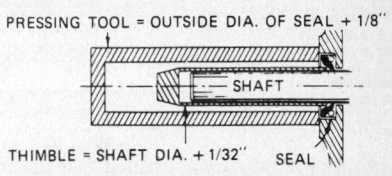

Seal installation tool

Allis-Chalmers

OIL SEAL DATA

Shaft	Qty.	Case	Cover
Shaft Rod	2	2	
1st Intermediate Shaft	1		
Axle Tube	2	1	1

49. Check shaft in area of seal seat. If shaft shows nicks, scratches, or burrs in this area it must be replaced. Attempts to file or remove flaws will only cause flat spots and oil leakage.

50. Check the seal seat or bore. Be sure to remove any nicks, scratches, burrs and foreign material.

51. Slide thimble over shaft and coat it lightly with clean grease.

52. Remove seal from package and gently lubricate it with clean grease. Do not run finger roughly around seal as it is easily damaged.

53. Place seal onto thimble with spring element facing case or cover.

54. Drive seal into seat in cover or case using the pressing tool. Avoid tilting the seal as it starts into seat and do not hammer directly on seal. When properly seated, outer surface of seal should be flush with shoulder of bore or seat in cover or case.

NOTE: All that remains to complete assembly of the transmission is to install parts on the axle tube. Do this as follows:

55. Install grease fitting in its hole in end of axle tube.

56. Before installing retaining ring (B) check it for damage. Discard it if it has bent out of round or if the ends are more than 1/8 in. apart.

57. Use retaining ring pliers to expand ring (B) and slide it over axle tube into its groove.

58. Install twin keys (A). Flat end of keys must butt up against retaining ring. Use a brass hammer to tap keys down into key ways. Be sure keys are fully bottomed along entire length.

INSTALLATION

1. Hoist transmission into place between side plates at rear of tractor.

2. Align holes and install capscrews. Installation is easiest if upper capscrew on each side is installed first. Note that one of the six capscrews is longer and must be installed into lower hole on left side.

3. Install brake drum on shaft.

4. Engage parking brake to help prevent turning of brake drum. Then install and tighten nut (G) to secure brake drum in place.

5. Lift rear hitch (E) into place on transmission. Secure hitch by installing two capscrews (A) and stud (F) on left side and three capscrews on right side.

6. If tractor is equipped with a rear lift kit connect rod (B) to lift kit and install cotter pin (C).

PLANETARY GEAR AND SHEAVE INSTALLATION, SHUTTLE TRACTORS

1. Use a clean rag to wipe all grease and oil off the sheave and half-sheave.

2. Install half-sheave on pulley shaft with belt groove facing outward. If necessary, use a plastic mallet to fully seat half-sheave.

3. Stretch shorter V-belt over pulley shaft and into groove of half-sheave. Belt must be seated properly for good drive.

4. Install brass washer and spacer on pulley shaft.

5. Install sheave and gear assembly on pulley shaft.

6. Depress and hold clutch pedal of tractor while stretching main drive V-belt over pulley shaft and into groove of sheave.

7. Install composition washer on pulley shaft.

8. Install key using a plastic hammer to seat it in key way.

9. Install planetary gear assembly aligning its key way with key on pulley shaft. Bull gear must be on outboard side. Use a plastic hammer to fully seat gear assembly on key.

10. Insert a long screwdriver through webs of sheave and half-sheave. Use this screwdriver to hold sheaves while using a wrench to install and tighten nut on end of pulley shaft. Torque nut to 50-55 ft.lb.

11. Install grease cap. Wipe off any grease from planetary assembly brake drum.

12. Install brake band. Hold hex shaft while tightening capscrew.

13. Connect pivot at lower end of brake rod to lever. Then install cotter pin through hole in pivot to secure pivot to lever.

DRIVE BELT AND PULLEY INSTALLATION, THREE-SPEED TRACTORS

1. Install pulley aligning its key way with key on pulley shaft. Tap pulley lightly with a brass hammer to seat it on key.

2. Depress clutch-brake pedal while stretching main drive belt over pulley shaft into groove of pulley.

3. Install nut on pulley shaft.

3-Speed Transmission Adjustments

1. Adjust jam nuts on rear end of brake rod so that when foot clutch brake pedal is pushed firmly forward clutch arm will stop with its forward edge 5/8 in. to rear of the rear lift cable guide near the bevel gear box when the brake is locked tight.

2. Loosen jam nut on parking brake rod and turn lever and rod end on parking brake rod so that brake is tight when parking brake lever is pulled fully upward against the fender opposite position. Tighten jam nut against rod end.

3. With clutch pedal up in engaged position adjust jam nuts on clutch rod 3/8 in. away from clutch rod guide.

4. Slide the front belt guard forward or back in slot at bolt C until guard has 1/8 in. clearance to the outside of the front pulley at the front side. Make sure that the belt guard does not touch the pulley or the belt at any point while clutch is fully engaged.

5. Check that the upper 90° formed leg of the idler pulley belt retainer is resting on and is parallel to the top edge of the idler arm as shown to properly locate the belt retainer.

6. Adjust rear pulley belt retainer to have 1/4 in. clearance to top of belt.

7. The transmission safety interlock switch is actuated by a special flat top carriage bolt screwed into the gear shift rod and held by a jam nut. The carriage bolt should be adjusted in or out of the rod to the correct height to actuate the switch when shift rod is in neutral but not high enough to drag tightly against the switch. The bolt must not touch the switch when shift is in any of the 4 gears.

FINAL INSTALLATION STEPS

1. Install drive belt and pulley or planetary gear assembly on transmission driveshaft.

2. Install rear axle, differential, and wheels.

3. Connect shift rod to transmission using a capscrew, lockwasher, and nut.

4. If tractor is equipped with transmission safety switch, connect plug to switch.

5. Complete lubrication of transmission, axle, and on shuttle tractors only, the planetary assembly.

6. Install seat deck. Secure it with nuts, lockwashers, and capscrews. On later models, be sure to install clip.

7. It is recommended that the tractor be operated at this time to check the repair as well as to identify any pre-existing conditions that might warrant adjustments. If adjustments of the clutch, brake or electrical parts are required, refer to the adjustments following this procedure.

8. The transmission fill plug is visible from right rear side of tractor. Remove plug. Be sure drain plug is tight. Then add SAE 90 transmission fluid through plug hole until level reaches bottom of threads in hole. Reinstall plug and wipe up any spills.

Planetary Gear Assembly, Shuttle Drive Tractors Only

9. Use a screwdriver to pry cap off planetary gear assembly. Pack assembly with clean #2 lithium grease. Then install the cap and wipe up any excess grease. Do not allow grease to get onto planetary assembly brake band or drum.

Rear Wheel and Axle

10. There is one grease fitting on the right rear wheel and another on the right rear axle. Wipe these fittings clean and use a grease gun to lubricate the axle through the fittings.

Allis-Chalmers

SHUTTLE DRIVE

The shuttle drive is a hand controlled differential type unit with a split sheave type V-belt forward clutch and band type reverse clutch. Driven by the shuttle shaft unit, the gear transmission has four forward and four reverse ranges.

REMOVING SHUTTLE CLUTCH UNIT FROM TRACTOR

1. Jack up the rear of the tractor so that the right rear wheel may be removed. Remove the wheel.
2. Raise the seat deck to allow ample work room.
3. Remove the spring clip from the brake band assembly. Loosen the capscrew two turns. The brake band may now be removed from the bracket and rotated downwards.
4. Using a large screwdriver, pry the dust cap from the planetary.

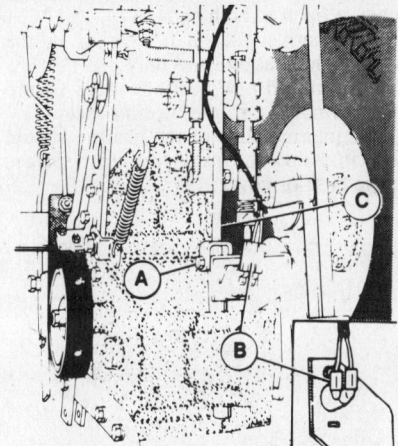

Connecting shift rod and safety switch

5. Place a large screwdriver through the slots in the two pulley sheaves to prevent them from turning. Remove the locknut securing the planetary assembly.
6. Remove the planetary assembly from shaft.
7. Remove the key from the transmission pulley shaft. Be sure the key used is of the regular Woodruff design—**NOT** a high profile design.
8. Remove the thrust washer from the transmission pulley shaft. The pulley sheave may now be removed. This is accomplished by depressing the foot clutch and sliding the pulley sheave from the transmission input shaft. The sheave can then be removed from the belt.
9. Place the forward clutch belt down, away from the split sheave. Remove the split sheave which is keyed to the transmission shaft.
10. Disassembly and assembly of planetary unit.
  a. To disassemble the unit, remove the nuts from bolts (14) and the planetary shafts.
  b. Remove cover (6), seal ring (9), and bearing (10). The needle bearings (1), planetary pinions (2), sleeves (4), and planetary carrier (11) can now be removed.
  c. Clean all parts and replace any parts that show excessive wear or other damage.
  d. Use new seal rings, and reassemble by reversing the disassembly procedure.
  e. Tighten the nuts on capscrew (14) and planetary shafts (3) to a torque of 12-15 ft.lb.
11. The assembly process is the reverse of the disassembly.
  a. There are two sides to the planetary, one with eight hex nuts showing, and the other with four hex nuts and four capscrews showing. The side with the eight hex nuts faces outward from the tractor.
  b. The locknut securing the planetary is torqued to 75 ft.lb.
  c. The capscrew securing the brake band assembly is tightened to a torque of 30 ft.lb. When tightening the capscrew, use a wrench on the inside capscrew to prevent the assembly from turning.
  d. Pack the planetary unit with No. 2 lithium grease. Do not get any grease on the planetary brake band.
  e. Replace cover (5) and adjust shuttle drive linkage as described in Adjustments, below.

Shuttle Drive Adjustments

1. Adjust jam nuts on rear end of brake rod so that when foot clutch brake pedal is pushed firmly forward clutch arm will stop with its forward edge ⅝ in. to rear of the rear lift cable guide near the bevel gear box when the brake is locked tight.

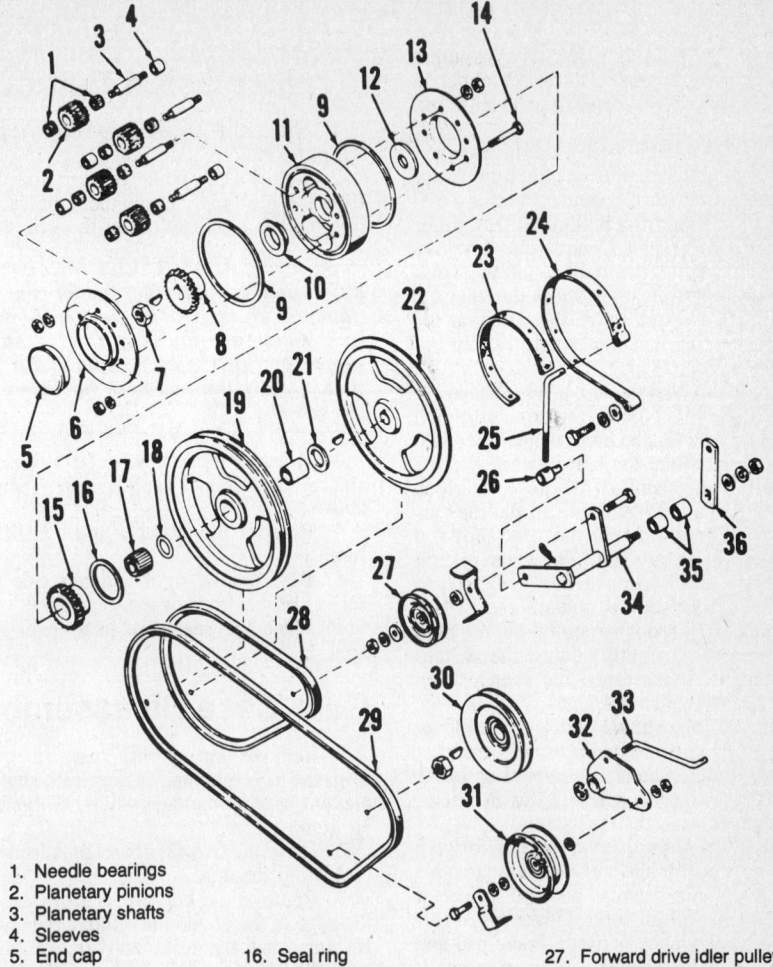

1. Needle bearings
2. Planetary pinions
3. Planetary shafts
4. Sleeves
5. End cap
6. Cover (outer)
7. Nut
8. Gear
9. Seal ring
10. Flange bearing
11. Planetary carrier
12. Thrust washer
13. Cover (inner)
14. Thru bolts (4 used)
15. Gear
16. Seal ring
17. Roller bearing
18. Seal ring
19. Pulley
20. Inner race
21. Thrust washer
22. Forward drive pulley half
23. Brake lining
24. Reverse drive brake band
25. Rod
26. Swivel
27. Forward drive idler pulley
28. Forward drive belt
29. Main drive belt
30. Pulley (bevel gear driven)
31. Clutch idler pulley
32. Idler arm
33. Clutch link
34. Shuttle drive lever and shaft assembly
35. Bushings
36. Lever

Forward/reverse shuttle drive unit components

Allis-Chalmers

2. Loosen jam nut on parking brake rod and turn lever and rod end so that brake is tight when parking brake lever is pulled fully upward against the fender opposite the position. Tighten jam nut against rod end.

3. With clutch pedal up in engaged position adjust jam nuts on clutch rod ¼ in. away from clutch rod guide.

4. Slide the front belt guard forward or back in slot at bolt C until guard has ⅛ in. clearance to the outside of the front pulley at the front side. Make sure that the belt guard does not touch the pulley or the belt in any point while clutch is fully engaged.

5. Check that the upper 90° formed leg of the idler pulley belt retainer is resting on and is parallel to the edge of the idler arm to properly locate the belt retainer.

6. The transmission safety interlock switch is actuated by a special flat top carriage bolt screwed into the gear shift rod and held by a jam nut. The carriage bolt should be adjusted in or out of the rod to the correct height to actuate the switch when the shift rod is in neutral but not high enough to drag tightly against the switch. The bolt must not touch the switch when shift is in any of the 4 gears.

7. With the shuttle clutch control lever in neutral position make sure center of notch in brake detent is centered on brake pin. If necessary loosen setscrew in detent, move it, and retighten setscrew.

8. With control lever in full forward position adjust brake pad assembly, to have ⅛ in. clearance between pad and surface of pulley. Check with control lever in full rear position, clearance between brake pad and pulley must be at least 1/16 in. or more. If less, readjust to a full 1/16 in. clearance. With the control lever in neutral there must be at least 1/16 in. clearance between brake pin and notch of brake detent.

9. With control lever in neutral position loosen setscrew in set collar behind forward drive belt spring. Move rod guide assembly and spring forward until slack is taken out of forward drive belt. Tighten setscrew. Move control lever to the full forward position. There should now be 3/16 in. clearance between rear surface of set collar and rear leg of rod guide assembly. If necessary reset set collar to obtain the 3/16 in. dimension when control lever is fully forward.

10. With control lever in neutral position rotate pivot pin on threads on brake rod so that when pin is reinstalled in hole in brake lever all of the slack will be taken out of brake band. Fasten pivot pin in place in brake lever with lock pin.

HYDROSTATIC TRANSMISSION

This is an automatic transmission which transmits power through the brake shaft and intermediate shaft to the axle tube. No shifting of gears is necessary.

NOTE: Removal of the transmission is necessary for most repairs.

Transmission Removal

1. Drain the transmission and replace the plug. Remove the seat deck.
2. Remove the brake linkage and drum from the gear case. Disconnect the oil suction line from the bottom of the gear case.

— CAUTION —
Raise and support the tractor on jack stands. With the type of work to be done, jack stands are the only safe means of supporting the tractor. Place the jack stands under the seat deck holddown bar at each end.

3. Remove the rear wheels, differential and axle. See the appropriate sections above.
4. Remove the drawbar and rear lift assembly.
5. Remove the right and left side capscrews at the front of the gear case.
6. Place the gear case in a clean work area.

Gear Case Disassembly

1. Remove any paint, rust, or burrs from the axle tube and brake drum shaft to prevent bearing damage when components are removed.
2. Drive the two (2) dowel pins from the cover into the case.
3. Remove the capscrews from around the edge of the cover and remove the cover. Be sure that the keys and grease fitting have been removed from the axle tube.
4. Remove the intermediate gear and shaft and two (2) thrust washers.
5. Remove the drive gear, axle tube and thrust washers.
6. Remove the brake shaft seal from the cover and axle tube seals from the case and cover.
7. Clean the case, cover and components that were removed with a suitable solvent and compressed air. Do not wash dirt or any foreign substance into the bearings.

BEARING INSTALLATION

1. Check the condition of the roller bearings. Do not remove bearing unless replacement is necessary. If the bearings are removed use a suitable fixture so that the bore in the gear case or cover is not scuffed.
2. When new bearings are installed always press against the end with the manufacturers name or number.
3. The bearings for the intermediate shaft and brake drum shaft in the gear case are open at one end only.
The brake drum shaft bearing in the cover is open at both ends.
The intermediate shaft bearing in the cover is open at one end only.
4. To install new bearings use a fixture that will not distort the bearing case.
The internal edge of the bearing should be ⅛ in. below the inside machined surface of the case or cover.

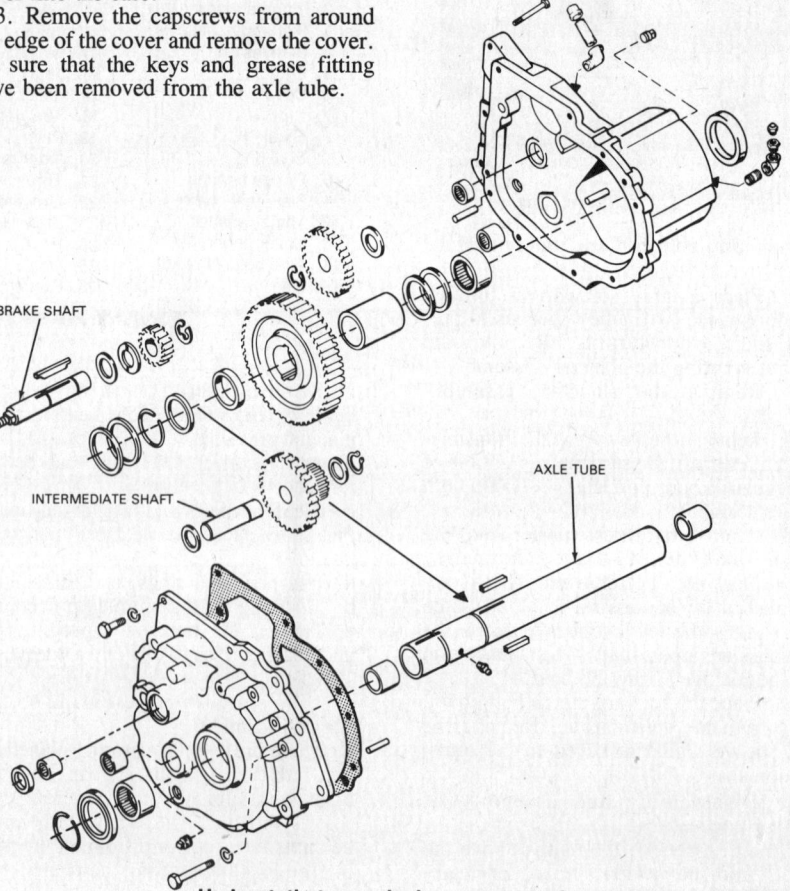

Hydrostatic transmission components

Allis-Chalmers

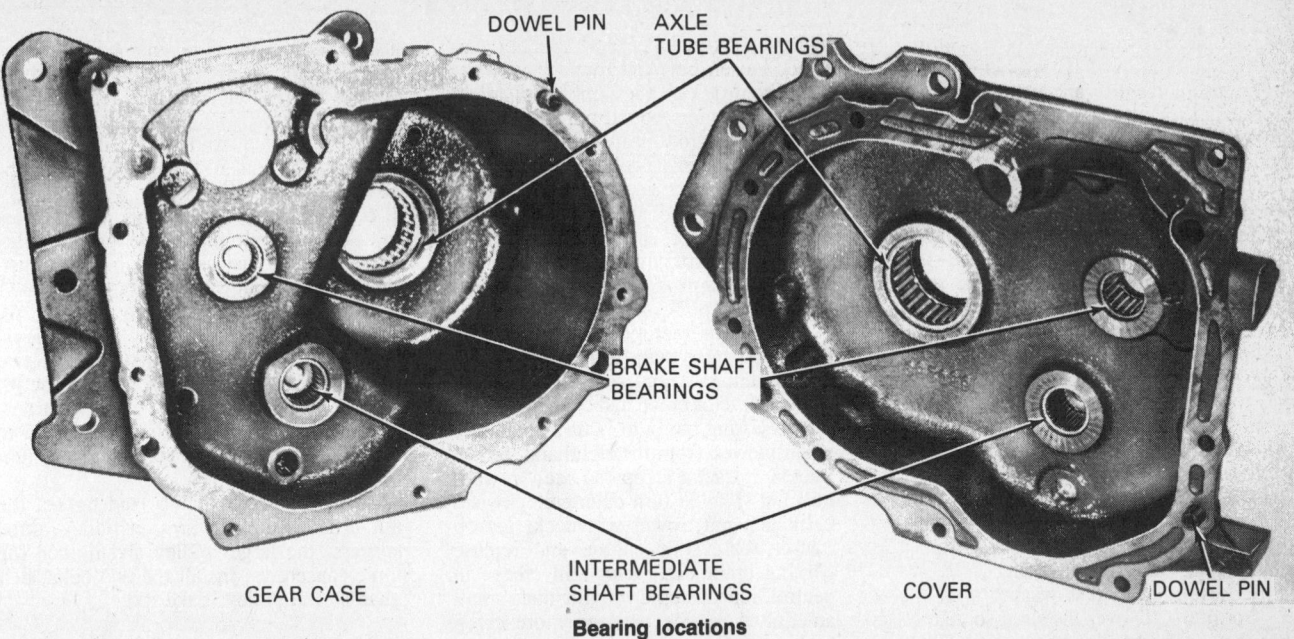

Bearing locations

Gear Case Reassembly

5. Press the axle tube bearing in the cover and gear case so that the outside end of the bearing is flush with the outside end of the small bore in the case or cover.
6. Lubricate the roller bearings in the cover and case with Lubriplate®.
7. Inspect the bushings in the axle tube and replace if worn or damaged.

1. Install two keys in the axle tube key way.
2. Install the short spacer and wire type retaining ring on the axle tube as shown.
3. Install the axle tube into the drive gear, with the two keys engaging the key ways in the drive gear.
4. Install the long spacer on the long end of the axle tube next to the gear.
5. Install two thrust washers next to the long spacer, and push the axle tube assembly through the large bearing in the case.

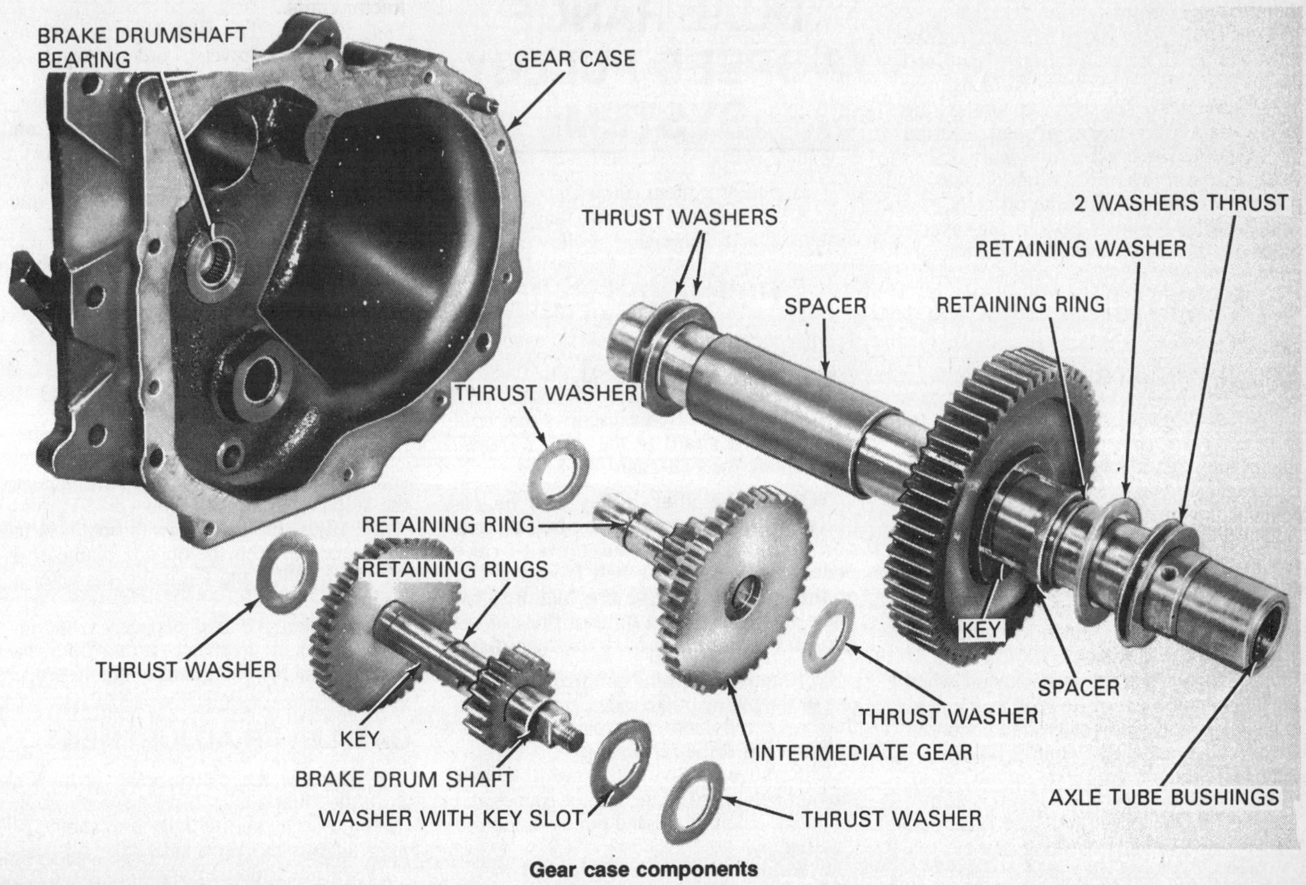

Gear case components

69

Allis-Chalmers

6. Install one large ID washer that will encase the retaining ring on the axle tube end and install two thrust washers on the shaft against the large ID washer.

7. Install the tru-arc snapring on the intermediate shaft, and install the pinion and gear assembly on the shaft with the small pinion toward the retaining ring.

8. Install one thrust washer on the shaft next to the retaining ring and one thrust washer against the large intermediate gear.

9. Install the assembled gear and shaft in the gear case bearing so that the small pinion will engage the large drive gear.

10. Install two wire type retaining rings on the brake drum shaft, and one long key in the shaft key way, between the retaining rings.

11. Install the large gear on the shaft end without the treads and a thrust washer as shown.

12. Install the small pinion on the shaft end with the treads and align the shaft key with the key way in the pinion and push pinion against the retaining ring.

13. Install the thrust washer with the key slot in the ID over the shaft so that the shaft key is in the slot and the washer is against the pinion, and install the second thrust washer.

14. Install the assembled brake drum shaft in the gear case, with the small pinion engaging the intermediate gear.

15. Drive the two roll pins into the case, leaving enough stick up to engage the cover. Install a new gasket on the case and install the cover.

16. Install the capscrews and torque them to 12-15 ft.lbs.

17. Install new axle tube and brake drum shaft oil seals in the gear case and cover. Protect the oil seals from damage by covering the key way, retaining ring grooves and other uneven or sharp areas on the axle tube or brake drum shaft. Use a suitable fixture when installing oil seals to prevent any distortion of the oil seals.

18. Install the gear case in the reverse order that it was removed from the tractor.

HYDROSTATIC TRANSMISSION ADJUSTMENTS

1. Loosen nut from rod end at front of parking brake rod. Turn parking brake handle and rod end until parking brake is fully tight when parking brake handle is pulled up against fender as shown.

2. With parking brake tight, adjust jam nuts on end of foot brake rod to provide ¾ in. clearance to rod guide.

3. Adjust jam nuts on clutch rod to have ¼ in. clearance from clutch rod guide.

4. *Hydrostatic Neutral Adjustment*. If tractor moves forward or backward when the hydrostatic control lever is in the neutral notch of the quadrant as shown proceed as follows to adjust the control linkage.

a. Park tractor on level ground, make sure hydrostatic control lever is firmly seated in the neutral notch of quadrant, stop engine and set parking brake.

b. Raise the seat deck and check if the pump control arm roller is exactly centered with the centering mark on the hydrostatic control cam. If it is not centered, loosen bolt and move the hydrostatic control cam assembly forward or backward in the slotted hole in the hydrostatic control strap until centering mark is centered on roller. Tighten bolt. Lower set deck, get in the operator's seat, start engine and release parking brake. If tractor still creeps with hydrostatic control lever in neutral, note which direction it creeps and proceed with next step.

c. Again stop tractor engine, set parking brake and raise seat deck. Loosen jam nut on lefthand end of the cam pivot shaft. If tractor creep had been in reverse turn adjusting nut ⅛ to ¼ turn clockwise when viewed from the righthand side of tractor. If tractor creep had been forward turn nut ⅛ to ¼ turn counterclockwise. Lock jam nut, lower seat deck, get on tractor seat, start engine and release parking brake. If tractor still creeps in neutral, repeat step C moving nut a small amount at a time until no more creep occurs.

CAUTION
ALWAYS stop tractor engine before leaving tractor seat. NEVER attempt to make adjustments while engine is running.

DUAL RANGE 6-SPEED PULLEY SYSTEM

This pulley system allows six forward speeds by means of a dual range pulley system. One reverse gear is allowed.

Pulley and Belt

REMOVAL AND INSTALLATION

1. Remove the two capscrews that hold the front belt guard to the tractor frame, and remove the belt guard.

NOTE: The idler pulley can be replaced without removing the belt guards and belts. Depress the clutch-brake pedal and remove the belt from the top of the pulley. Remove the locknut, replace the pulley, and tighten the nut securely. Place the belt back on the pulley.

2. Remove the cotter pin from the front end of the pulley brake rod. Do not remove the brake rod from the rear belt guard, unless it has to be replaced.

3. Remove the two capscrews that hold the rear belt guard to the tractor frame, and remove the belt guard and pulley brake rod together.

4. Replace the needed belt, and install the belt guards and pulley brake rod in the reverse order that they were removed. Adjust as seen in the Adjustment Section below. If either the drive pulley or driven pulleys have to be replaced, remove both belts.

5. To replace the drive pulleys, remove the nut in the center of the pulleys, from the bevel gear box cross shaft.

6. Replace drive pulleys and tighten nut. Replace belt guards, belts, and pulley brake rod. Adjust as noted in the adjustment section below.

7. To replace the driven pulleys, remove the three capscrew and pulleys. There are three spacers between the two pulleys. The driven pulley hub can be removed, by removing the nut on the transmission input shaft. This hub is keyed to the input shaft. When installing a new hub, place the key.

8. Install the key, hub, and tighten the nut. Then install the smaller pulley, three spacers, the larger pulley and tighten the three capscrews. Install the two belts, belt guards, and pulley brake rod.

Dual Range Controls

1. To replace the idler pivot assembly, depress the clutch-brake pedal and lift the belt off of the idler pulley.

2. Disconnect the control rod from the cam lever, and remove the setscrew from the cam lever.

3. Slide the pivot assembly to the right, remove the key and remove it from the tractor frame.

4. Replace the needed parts, and assemble in the reverse order. Adjust as noted in the adjustment section below.

5. If the bushings in the dual range lever need replacing, remove nut, washers, and control link. Slide the control lever to the right and replace one bushing from the outside and one from the inside of the tractor frame.

6. Install the control link, washers, and secure with nut.

6-Speed Adjustments

REAR BELT GUARD ADJUSTMENT

1. Loosen, but do remove the two capscrews which secure the rear belt guard to the side of the transmission.

2. Adjust the belt guard to obtain ⅛ in. clearance between the outside diameter of the large pulley and the inside diameter of the retainer.

3. Tighten the two capscrews and recheck for ⅛ in. clearance. It may be necessary to bend the belt guard to obtain proper clearance.

CAM LEVER ADJUSTMENT

1. Depress the clutch-brake pedal and shift the dual range lever to high range (forward). The sliding idler arm should be tight against the right side plate bushing. This will slide the shifting cam lever into

Allis-Chalmers

contact with the lefthand side of the gate finger.

2. Check to see if binding occurs between the cam lever and gate finger. To eliminate binding, loosen the setscrew in the hub on the shift cam lever and reposition the cam lever on the idler arm shaft, so that it is against the gate finger by sliding the cam lever right or left on the idler arm shaft. Tighten the setscrew securely.

IDLER HEIGHT ADJUSTMENT

1. Depress the clutch-brake pedal and move the dual range lever until the cam lever is under the gate finger. Release the pedal and let the spring tension hold the cam lever in position.

2. With the outside belt resting on both the front and rear belt guards, loosen the idler adjusting bolt and position the idler pulley so there is a minimum of 1/8 in. clearance between the belt and outside diameter of the idler pulley. Tighten the adjustment bolt securely.

CLUTCH ROD ADJUSTMENT

1. Depress the clutch-brake pedal and move the dual range lever in high-range.

2. With the pedal up in engaged position, there should be a clearance of 3/8 in. between the front jam nut and the rod guide. Pull back on the pedal to make sure it is completely in engaged position.

3. To make the proper adjustment, loosen the two jam nuts and turn the front jam nut in or out to obtain the 3/8 in. measurement. Hold the front jam nut in position and tighten the rear jam nut securely.

TRANSMISSION BRAKE AND PARKING BRAKE

1. Depress the clutch-brake pedal and move the dual range lever unit the cam lever is under the gate finger.

2. Place a .030 in. feeler gauge between the cam lever and the end of the gate finger. Release the pedal and let the spring tension hold the feeler gauge in place.

3. Loosen the jam nut behind the parking brake lever and turn the lever clockwise to increase tension or counterclockwise to decrease tension, until tension is felt when putting the lever into engaged position. Tighten locknut when properly adjusted. Clockwise and counterclockwise direction given is standing in front of the tractor looking at the parking brake lever.

4. With the parking brake engaged, loosen the brake rod jam nuts, and turn the front jam nut in or out, until it just comes in contact with the brake band plus 1/3 turn. Hold the front jam nut in position and tighten the rear adjusting nut securely.

PULLEY BRAKE

1. With the clutch-brake pedal engaged and the dual range lever in High or Low range, set the coil length of the return spring (the rear spring) to 1 3/8 in. Loosen the jam nuts and turn the front jam nut in or out to obtain this measurement.

2. Depress the clutch-brake pedal and move the dual range lever until the cam lever is under the gate finger.

3. Loosen the jam nuts behind the front spring and turn the front jam nut in or out until the spring is compressed to 1 3/4 in. between the rod guide and the front nut. Tighten the jam nuts securely.

NOTE: The idler spring should be in the bottom hole of the idler arm. If belt slippage occurs with a heavy drawbar load, move the spring into the next hole. When the spring is moved into a higher hole, there will be an increase in tension on the clutch-brake pedal, due to the increased tension on the idler arm. This will require more pressure to depress the clutch-brake pedal.

ENGINE

The 700 series is equipped with Kohler engines as follows: 710-Kohler K241S; 712-Kohler K301S; 716-Kohler K341S. For engine repairs, see the Kohler Engine section in the Unit Repair section of this book.

Engine Removal

1. Disconnect the fuel line from the fuel pump and plug the hose or drain the fuel tank of all gasoline.

CAUTION
DO NOT drain fuel from the tank when the engine is hot, while smoking, near an open flame, near anything that is hot, or in a closed building.

2. Disconnect the battery ground cable, all electrical wires, choke cable, and throttle cable from the engine.

3. Remove the two capscrews with the thick washers and spacers that fasten the drive shaft to the engine flywheel.

4. Remove the four capscrews that hold the engine to the tractor frame, and remove the engine.

Engine Installation

1. Place the engine on the tractor frame and align the mounting holes in the engine with the slotted holes in the tractor frame.

2. Install the capscrews into the holes, with the flat washers, lockwashers and nuts on the bottom side of the frame. Use a flat washer on each slotted hole or the engine will not hold properly. Do not tighten the capscrews at this time.

3. Fasten the drive coupling to the engine flywheel in the following way. On each capscrew place one lockwasher and one thick washer and place the capscrew through the coupling disc. Then install one thick washer next to the coupling disc with a spacer between the thick washer and the engine flywheel. Install the second capscrew in the same order and torque both capscrews.

4. Move the engine in the slotted holes so that the engine flywheel and the drive shaft are properly lined up. The drive coupling should be kept flat and not bowed or bent. The drive shaft has to be at right angles to the flywheel. Torque the four engine mounting capscrews.

Model 720

WHEEL BEARINGS AND HUBS

REMOVAL AND INSTALLATION

1. Raise the front of the tractor to remove the load from the front wheels. Place stands or blocks under the axle to safety support the weight.

2. Remove the hub cap. The wheel may be removed from the hub for convenience.

3. Remove the cotter pin, nut, washer and outside bearing cone. The hub can be removed with the inner bearing cone and oil seal.

4. Remove the oil seal and bearing cone from the hub. If the bearings are to be reused always install the same cup with the original cone.

5. Remove all the lubricant from the hub and bearings. Clean the hub and bearings in a suitable solvent.

6. Check the condition of the hub and bearings. Replace the hub if cracked or damaged. Replace chipped or worn cones and cups. If either cup or cone are to be replaced always replace both. The cups will have to be pulled or driven from the hub.

7. Press the bearing cups in the hub.

8. Use a bearing packer or work a good quality of wheel bearing lubricant in the

Allis-Chalmers

inner bearing. Install the cone in the hub and press the oil seal in the hub.

9. Fill the hub with sufficient wheel bearing lubricant to provide additional lubrication for the bearings.

10. Install the hub of the axle.

11. Fill the outer bearing cone with wheel bearing lubricant, install the cone on the spindle, thrust washer and slotted nut.

12. Torque the nut until a slight preload is detectable as the wheel is rotated. Replace the cotter pin and hub cap. Replace the wheel if it was removed from the hub.

Front Axle

DISASSEMBLY

1. Raise the front of the tractor to remove the weight from the front wheels and support safely with suitable stands or blocks.

2. To remove the spindles remove the wheels from the hubs.

3. Remove the righthand and lefthand steering arms and Woodruff keys. Remove the spindles. The spindle bushings can be removed with a suitable puller or driver.

4. To remove the front axle from the frame remove the two capscrews.

5. Check the axle pivot bushings, reach pivot bushing and spindle bushings. Replace bushings that have excess clearance to cause steering problems or inconvenience. Check the thrust bearings and replace if they do not rotate smoothly.

6. Clean all parts in suitable solvent. Press new bushings in where required.

7. The threaded block should be welded to the tractor frame and replace nut and lockwasher.

ASSEMBLY

1. Press the spindle bushings in place.

2. Install the axle pivot bushings and spacer.

3. Install the axle in the tractor frame and fasten with the front and rear pivot capscrews.

NOTE: Tie rod clevis is fastened in place by capscrew. Torque the capscrews.

4. Install the spindles, linkage and items removed during disassembly.

5. The drag link, tie rod and ball joints should be replaced if excess clearance is apparent.

6. Adjust the tie rod to have approximately $3/16$ in. toe-in.

STEERING

Steering Gear

REMOVAL

1. Disconnect the battery.

2. Remove oil cooler, lefthand engine shroud and fuel tank from the tractor.

3. Remove cotter pin and machine washer attaching the clutch rod to the lever assembly.

4. Remove tube assembly, spring, jam nuts and clutch spacer from the eyebolt.

5. Remove the clutch belts from the engine pulley.

6. Remove ten washer headed capscrews attaching the bottom cover to tractor frame and remove the cover.

7. Remove the capscrew securing the steering wheel to the shaft. Remove steering wheel and rubber washer from the steering shaft.

8. Remove the choke cable from the carburetor and the clips that hold the choke cable and throttle cable in place.

9. Remove the six capscrews attaching the instrument panel to the support. Raise the panel over the steering shaft without damaging the electrical wires.

10. Remove the roll pin from the U-joint and steering shaft and move the steering shaft up in the support tube.

11. Loosen the capscrew that holds the clutch belt stop in place.

12. Remove the four capscrews and lockwashers that attach the clutch pivot arm supports to the frame. Tilt the pivot arm and support assemblies away from the frame and and remove support assembly

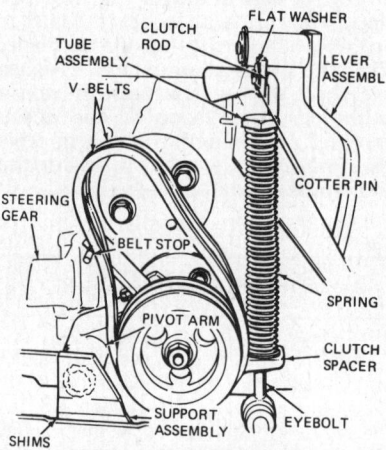
Removing clutch parts

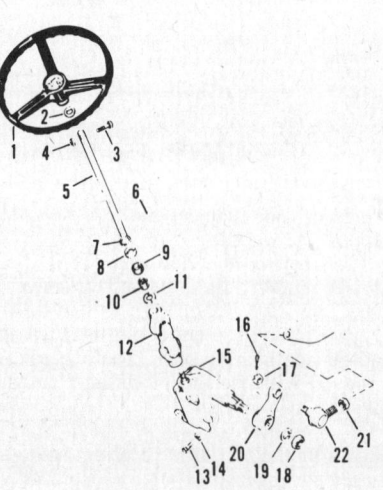
Steering components

shims. When the support is installed during assembly the shims must be placed in their original location.

13. Remove the nut and lockwasher from the Pitman shaft and remove the Pitman arm (20).

14. Remove the three capscrews and lockwashers attaching the steering gear to the frame to remove the steering gear assembly from the tractor.

15. Remove the U-joint from the steering gear.

16. Remove the steering shaft from the support tube.

17. Remove the bearings from the panel support tube with a suitable bearing puller if bearing replacement is necessary. Clean the bearings and parts in a suitable solvent and check all parts. Replace worn or damaged parts.

Steering Gear

DISASSEMBLY

1. Clamp the steering gear assembly in a vise with the wormshaft in a horizontal position. Clamp to one of the mounting tabs.

2. Rotate the worm from one stop to the other and count the revolutions of the wormshaft. Rotate the wormshaft away from a stop $1/2$ the revolutions required to rotate from stop to stop.

3. Prepare to catch the lubricant from the gear case in a pan when the cover is removed.

4. Remove the three self locking capscrews that attach the side cover to the gear case. Tap on the end of the Pitman shaft with a plastic hammer to loosen the cover. Remove the Pitman shaft with the cover from the case.

5. Remove the locknut and adjuster plug with the lower bearing and race from the gear case.

6. Remove the worm nut with the ball nut assembly from the gear case. Do not permit the ball nut assembly to wedge at either end of the wormshaft. The assembly may be damaged if this occurs.

7. Remove the upper bearing from the wormshaft.

8. Inspect the bearing areas and wormshaft for chipped or worn areas and replace if necessary.

9. Remove the ball guide clamps and ball guides. The ball bearings can be removed from the ports. There should be a total of forty-eight ball bearings in the ball

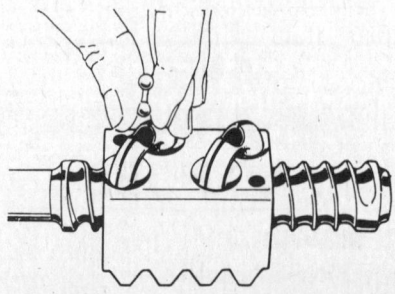
Removing the 48 ball bearings from the ball guide

nut assembly. The nut can now be removed from the wormshaft.

10. Thoroughly clean all parts in a suitable solvent. Inspect the ball bearings for roughness or flat areas and replace damaged bearings. Check the ends of the ball guides for bent ends which could interfere with the ball bearing travel. This could cause steering problems.

ASSEMBLY OF BALL NUT

1. Place the ball nut in clean area and install the wormshaft in the ball nut. Block the wormshaft up to center it in the ball nut. Rotate so the grooves in the wormshaft align with the four holes in the ball nut.

2. Replace the ball guides in the ball nut.

3. Install 24 ball bearings in each circuit of the ball nut assembly. Install the ball guide clamps and secure with machine screws and washers.

4. Rotate the ball nut assembly on the wormshaft and lubricate with Lubriplate®.

5. Remove and inspect the lower bearing by using a suitable bar to pry out the bearing retainer. The bearing cup and plug are serviced as an assembly. If the bearing is worn or damaged replace the complete assembly.

6. Pitman shaft and side cover. Remove the lash adjusting screw locknut and turn the lash adjusting screw clockwise to remove the cover. Clean the parts in a solvent.

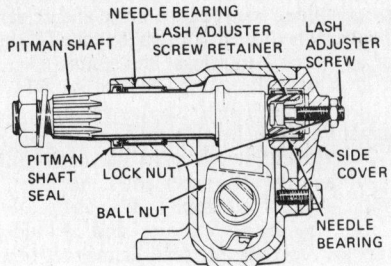

Pitman shaft installation

7. Check the needle bearing in the gear case and if replacement is necessary remove and replace with a suitable puller. When pressing a needle bearing in place always press against the end with the manufacturers name or identification.

Do not press the bearing into the offset in the bore. To do so may distort the bearing.

Replace the oil seal in the gear case.

8. Remove the wormshaft seal from the gear case. If the upper wormshaft bearing is to be replaced remove the bearing race and install a new one. Install a new wormshaft oil seal.

9. The lash adjusting screw and Pitman shaft are serviced as an assembly. If it is worn or damaged replace the assembly.

ASSEMBLY OF STEERING GEAR

1. Clamp the gear case in a vise with the wormshaft horizontal. Install the upper bearing on the wormshaft. To protect the oil seal cover the splines with Scotch tape or cellophane and lubricate with Lubriplate®. Lubricate the bearing and install the wormshaft in the case.

2. If the lower bearing was removed from the adjuster plug replace the bearing and retainer.

3. Install the adjuster plug assembly in the case and remove all the end play from the bearings and shaft. The bearings should preload until it requires three to six in.lb. to rotate the wormshaft. To check the preload use a 12 point $^{11}/_{16}$ in. socket on the splined end of the wormshaft with an in.lb. torque wrench to rotate the shaft. Turn the adjuster plug in while rotating the wormshaft with the in.lb. torque wrench until a reading of between three and six in.lb. is observed on the torque wrench. Tighten the adjuster plug locknut without changing the preload setting.

4. Pack nine oz. of automotive chassis type lubricant in the case. Rotate the wormshaft from one stop to the other to lubricate the bearings and ball nut assembly.

5. Rotate the wormshaft until the ball nut is at mid-point between the two stops. Lubricate Pitman shaft bearing.

6. Cover the splined end of the Pitman shaft with Scotch tape and lubricate with Lubriplate® to protect the oil seal. Install the Pitman shaft in the case so that the center tooth of the sector gear enters the center groove of the ball nut.

7. Coat the side cover gasket with a thin coat of Permatex®, lubricate the needle bearing in the side cover and install the side cover on the case. Rotate the lash adjuster screw counterclockwise until the Pitman shaft contacts the cover assembly. Turn the adjusting screw clockwise ½ turn for starting lash adjustment. Install the locknut and torque while holding adjusting screw in predetermined position. Install the self locking cover capscrews and torque.

8. Rotate the wormshaft to position the ball nut in the center of the sector. To locate the center rotate the wormshaft clockwise until it stops then count the shaft revolutions as you rotate the wormshaft counterclockwise to the opposite stop. Turn the wormshaft one-half this number of revolutions clockwise. This will center the ball nut assembly on the wormshaft.

9. Loosen the Pitman shaft lash adjuster locknut and turn the Pitman lash adjusting screw clockwise to remove all the ball nut to sector lash without binding. Torque the locknut. The ball nut to worm lash should be "0" when the ball nut is in the center of the wormshaft and will require about ten in.lb. torque to rotate the wormshaft through the center Pitman shaft arc area.

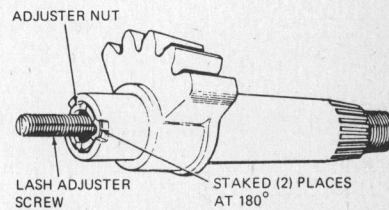

Pitman shaft, showing adjuster nut

10. Check the torque required to rotate the wormshaft with an $^{11}/_{16}$ in. socket and in.lb. torque wrench. The torque should not exceed eleven in.lb. to rotate the wormshaft when the ball nut is in the center area but should not be less than three in.lb. Recheck wormshaft bearing adjustment or Pitman shaft lash if necessary.

11. Install the bearings in the support tube. Use a suitable puller and press the end of the needle bearings that have the manufacturers name or identification.

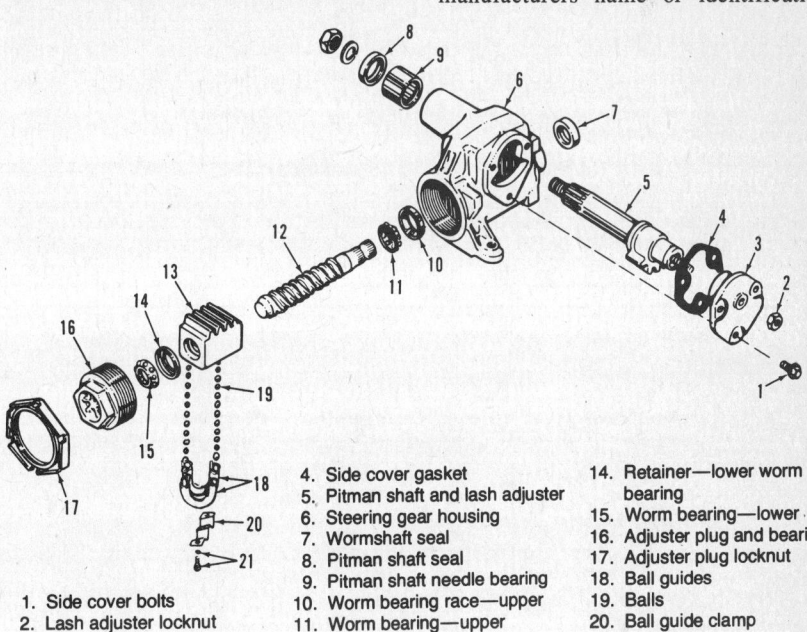

1. Side cover bolts
2. Lash adjuster locknut
3. Side cover and needle bearing
4. Side cover gasket
5. Pitman shaft and lash adjuster
6. Steering gear housing
7. Wormshaft seal
8. Pitman shaft seal
9. Pitman shaft needle bearing
10. Worm bearing race—upper
11. Worm bearing—upper
12. Wormshaft
13. Ball nut
14. Retainer—lower worm bearing
15. Worm bearing—lower
16. Adjuster plug and bearing
17. Adjuster plug locknut
18. Ball guides
19. Balls
20. Ball guide clamp
21. Clamp screw and washer assemblies

Steering gear components

Allis-Chalmers

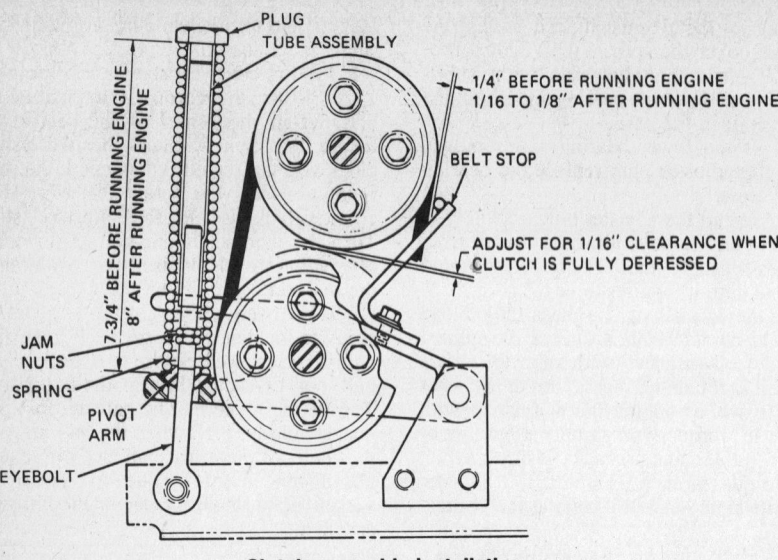

Clutch assembly installation

1. Spring
2. Roll pin
3. Lock rod
4. Spring
5. Spring
6. Pin, cotter
7. Stop
8. Setscrew
9. Anchor assembly
10. Key
11. Lever assembly, right
12. Sleeve bearing
13. Stop
14. Spring
15. Lever assembly, left
16. Retaining ring
17. Cam assembly
18. Knob
19. Cotter pin
20. Pin
21. Rear rod assembly
22. Jam nut
23. Turnbuckle
24. Jam nut
25. Front rod
26. Front rod
27. Cotter pin
28. Pin
29. Link plate
30. Locknut
31. Flat washer
32. Stud
33. Brake lever

Brake components

Lubricate the bearings with wheel bearing lubricant. Install the roller bearing in the bottom section of the support tube.

12. Install the retaining ring and thrust washer on the steering shaft.

13. Place the shaft in the support tube bearings. Install the universal joint, align the holes in the U-joint with the hole in the steering shaft and drive in the roll pin.

14. Install the gear case with the splined end of the wormshaft in the universal joint and secure to the tractor frame with three lockwashers and capscrews. Torque capscrews to 70-90 ft.lb.

15. Secure the steering arm to the Pitman shaft with lockwasher and nut. Torque nut 162-198 ft.lb.

16. Move the clutch assembly to its original position. Install the shims that were removed during disassembly in their original location. Secure the clutch assembly to the frame with four capscrews and lockwashers.

17. Install the thrust washer on the U-joint, move the shaft through the thrust washer and in the universal joint. Align the hole in the U-joint with the hole in the shaft and drive in the roll pin, retaining ring, thrust washer, etc.

18. Install the following:
• Instrument panel
• Choke and throttle cables
• Rubber washer
• Steering wheel

19. Install the clutch V-belts. Install the clutch eyebolt spacer on the pivot arm.

20. Install the jam nuts on the clutch eyebolt and adjust so that $1/16$ in. clearance is maintained between the drive and driven sheaves when the pivot arm is moved up as far as possible to release the clutch.

21. Install the spring and tube assembly. Adjust spring height to dimension in the operators manual.

22. Attach clutch pedal lever to the clutch and install washer and cotter pin.

23. Adjust belt stop.

24. Replace oil cooler and shrouds. Replace fuel tank and other items that were removed.

BRAKES

Brake Linkage

REMOVAL AND INSTALLATION

1. Raise the hood and remove the fuel tank.

2. Remove the six washer headed capscrews that attach the top cover to the tractor frame and remove the cover.

3. Unhook the brake return springs (4-5) and cotter pins to disconnect the front brake rods.

4. Remove setscrew (8) and move brake rod anchor assembly (9) left to permit removal of Woodruff key (10).

5. Remove Woodruff key (10) and re-

Allis-Chalmers

move righthand brake lever assembly from the tractor.

6. Remove lefthand brake lever assembly (15) and anchor assembly (9).

7. The front and rear rod assemblies can be removed by removing the pins (20) at the brake housing.

8. Inspect the bearings (12) in the lefthand brake lever assembly. If replacement is necessary remove the bearings with a suitable press or driver and install new bearings.

9. Inspect the righthand pivot bearings and replace if worn.

NOTE: These bearings are in the tractor frame but not in the illustration.

10. Check the brake rods and levers at the pivot areas and replace if worn. The link plate (29) and rear brake levers can be inspected after the drop housing and axle extensions are removed and the brake assembly is exposed.

11. Replace the brake levers, rods and springs in the reverse order of removal.

Brakes

REMOVAL AND INSTALLATION

1. Position the tractor safety on suitable blocks or stands and remove the tires and wheels.

2. Remove the four capscrews attaching the wheel guard to the drop housing.

3. Remove the four capscrews attaching the drop housing to the axle extension.

4. Remove the drop housing.

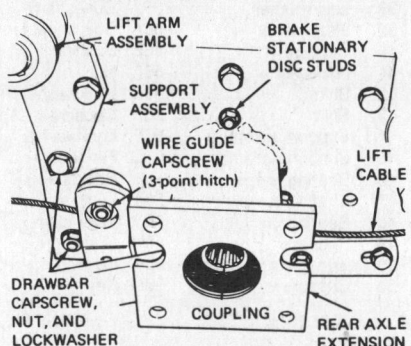

Rear axle extension housing

5. Disconnect the hydraulic lift cylinder from the left axle extension and lift shaft and move out of the way.

6. Remove the capscrews and pins and remove the lift shaft assembly.

7. Remove capscrews and nuts attaching the drawbar to the axle extensions and remove the drawbar assembly.

8. Remove the transmission oil filter.

9. Remove the four capscrews attaching each axle extension to the transmission case.

10. Remove the hex nuts and lockwashers from the stationary brake disc studs.

11. Remove the rear axle extension and coupling.

12. Remove the outer stationary disc.

13. Remove differential output shaft with outer rotating disc assembly and retaining ring.

14. Remove the two flat edge spacers and one round edge spacer from the stationary studs.

15. Remove pin attaching link plate to yoke link assembly and remove disc actuating assembly.

16. Clean all the brake parts and housing with suitable solvent to remove all grease and dirt from brake parts.

17. Inspect all linkage and pins for wear at the pivot areas and replace items that are worn excessively.

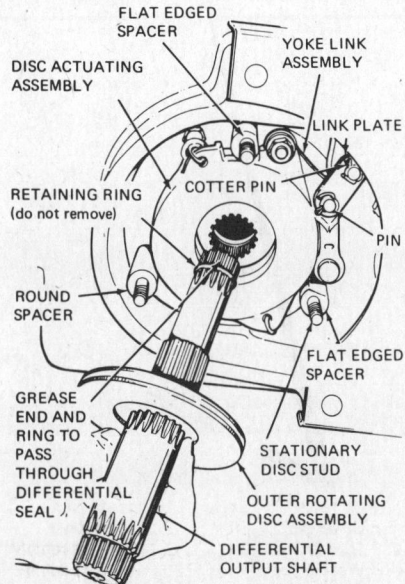

Removing differential output shaft

18. Check rotating disc assemblies for wear. Replace if necessary.

19. Check the outer stationary discs, actuating disc assembly and differential supports for scoring and wear. If new rotating discs are installed replace the other brake parts that are worn or scored.

20. Remove the oil seals from the differential supports.

21. Remove the springs from the actuating disc assemblies and replace if stretched or worn in the hook area. Check the three ¾ in. steel balls and ramps for wear or roughness. Replace worn or rough parts.

INSTALLATION

1. Install new oil seals in the differential supports and install the supports in the transmission case. Torque the capscrews.

2. Install the inner rotating disc assembly.

3. Install the two flat edge spacers on the stationary disc studs with the flat edge toward the yoke link assembly. Install the round spacer on the stationary disc stud.

4. Install the actuating disc assembly in the brake housing. Attach the yoke link assembly and actuating link to the link plate with a pin and cotter pin.

5. Install the retaining ring and outer rotating disc assembly on the differential output shaft. Lubricate the inner splines and retaining ring with Lubriplate® to prevent oil seal damage.

6. Install the outer stationary disc.

7. Install the axle extension and fasten with lockwashers, through hex nuts and four capscrews.

NOTE: If the tractor does not have a hole in the axle extensions, drill a ⅜ in. hole in the bottom area to allow water and oil to drain from the brake housing.

8. Install the coupling on the differential shaft, spring on the pinion shaft and install the drop housing.

9. Install the lift shaft assembly hydraulic ram, drawbar, wheel guard and accessories that were removed.

BRAKE ADJUSTMENT

1. Using a ⅜ in. wrench, remove six hex headed screws from the frame cover and remove the cover from the tractor.

2. Be sure the brake pedal locking pin is moved to the left, so the brakes can be applied individually.

3. Adjust the left brake first. You will need two 9/16 in. open end wrenches. Use one wrench to hold the turnbuckle, the other to loosen the locknuts at either end of the turnbuckle. Turn the turnbuckle counterclockwise (as you stand behind it looking toward the front of the tractor) to tighten the brake. Alternately turn the

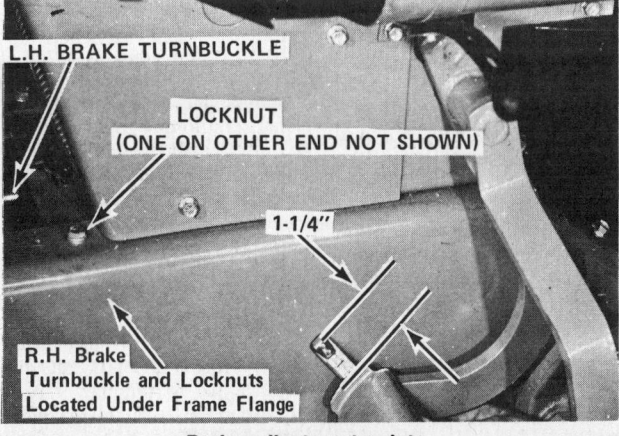

Brake adjustment points

Allis-Chalmers

turnbuckle and check the brake travel until the brake travel as measured in the accompanying illustration is 1¼ in. While holding the turnbuckle with one wrench, tighten the two nuts at either end of the turnbuckle against the turnbuckle to lock it in place.

4. Depress the left brake pedal and lock it in place with the parking brake lock. If the parking brake will not lock in place you may have to increase the amount of brake pedal free travel by turning the turnbuckle clockwise. With the parking brake lock set to hold the left brake pedal down, follow the procedure described in step (3) to adjust the right brake pedal. Alternately depress and tighten it until it is even with the left brake pedal when depressed. Checking the adjustment of the right brake pedal in this manner will not only give it the required 1¼ in. of free travel, but also insure that the two brakes will be activated at the same time when they are locked together. Tighten the locking nuts in place on either side of the turnbuckle.

5. Replace the cover over the frame and tighten the six hex headed screws securely in place.

CLUTCH AND DRIVESHAFT

REMOVAL

1. To remove the clutch and driveshaft remove the oil cooler, shrouds and fuel tank from the tractor.
2. Remove the bottom cover from the tractor frame.
3. Hold the clutch pedal forward and remove the clutch belts from the engine crankshaft pulley.
4. Remove two capscrews holding support and pivot shaft from the tractor frame.
5. Remove the shoulder bolts attaching rear driveshaft yoke to coupling disc.
6. Remove the driveshaft assembly from the tractor frame.
7. Inspect the shaft assembly bearings and the pivot bearings. Replace if necessary.
8. Check the splines on shaft assembly (47) and clutch pulley (36) and replace if they have excessive clearance.
9. Check the clutch belts and replace if cracked.
10. Check the rear driveshaft (55) and yoke splines. There will be a neoprene and felt washer on the driveshaft to form a grease seal at the coupling yoke. Clean the parts in a suitable solvent.
11. Check clutch spring (20), free height 9.75 in., compressed height 6.5 in., 62 to 68 lb. at compressed height. Clutch and spring (4) free length 1.375 in., compressed 57 lb. pressure.

INSTALLATION

1. Lubricate and install pivot bearings (46) in the pivot arm.

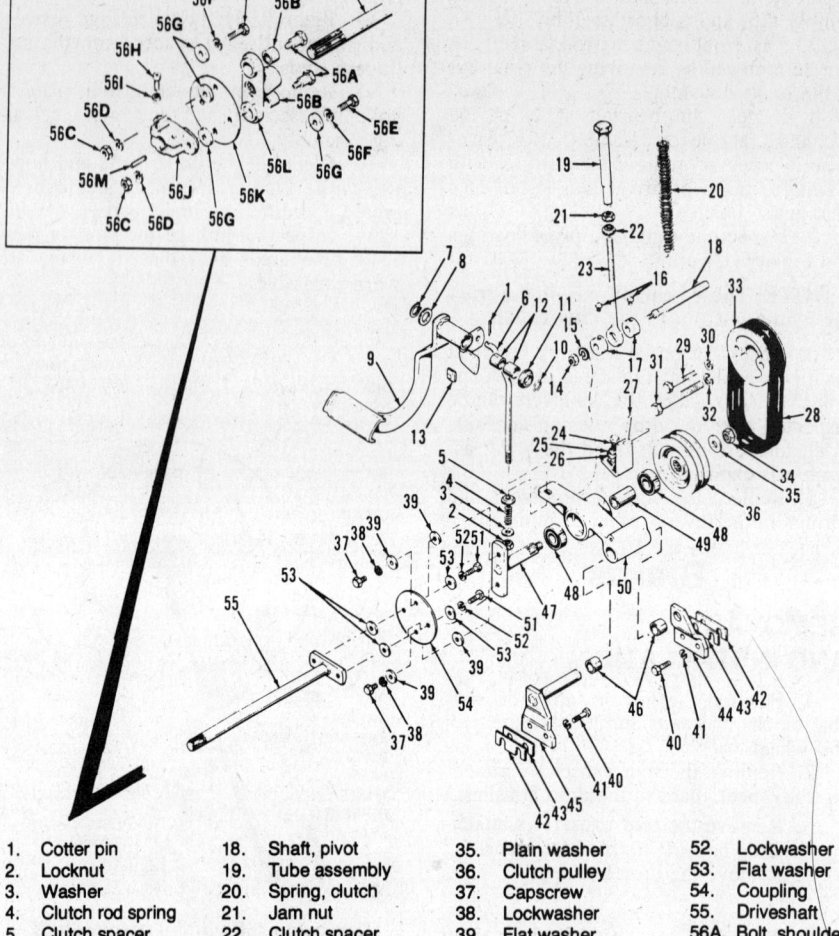

1. Cotter pin
2. Locknut
3. Washer
4. Clutch rod spring
5. Clutch spacer
6. Clutch rod
7. Retaining ring
8. Washer
9. Lever assembly
10. Retaining ring
11. Washer
12. Sleeve bearing
13. Bumper
14. Nut
15. Lockwasher
16. Setscrew
17. Collar
18. Shaft, pivot
19. Tube assembly
20. Spring, clutch
21. Jam nut
22. Clutch spacer
23. Eye bolt
24. Capscrew
25. Lockwasher
26. Flat washer
27. Belt stop
28. Matched "V" belts
29. Capscrew (2¾ lg.)
30. Lockwasher
31. Capscrew (4½ lg.)
32. Lockwasher
33. Engine pulley
34. Flange nut
35. Plain washer
36. Clutch pulley
37. Capscrew
38. Lockwasher
39. Flat washer
40. Capscrew
41. Lockwasher
42. Shim
43. Shim
44. Support assembly
45. Support assembly
46. Bearing, sleeve
47. Shaft assembly
48. Bearing
49. Spacer
50. Arm, pivot
51. Capscrew
52. Lockwasher
53. Flat washer
54. Coupling
55. Driveshaft
56A. Bolt, shoulder
56B. Bushing
56C. Nut
56D. Lockwasher
56E. Capscrew
56F. Lockwasher
56G. Flat washer
56H. Socket screw
56I. Lockwasher
56J. Yoke clamp
56K. Coupling
56L. Yoke
56M. Key

Clutch and driveshaft components

2. Install drive bearings (48) and spacer (49) in the pivot arm.
3. Install shaft assembly in the pivot assembly and pulley (36) on the shaft. Use a flange nut 2028644 to secure pulley to the shaft and torque to 75 ft.lb.
4. Install the pivot assembly, pivot shaft (18), collars (17), eyebolt (23), pivot supports (44-45), shims (42-43) and clutch belts. Install the shims (42-43) in their original locations.
5. Install front driveshaft coupler (54) to the driveshaft (47-55) with hardware shown.
6. Install rod spacer on the eyebolt.
7. Install the jam nuts on the eyebolt and adjust to maintain 1/16 in. clearance between the drive and driven sheaves when the pivot arm is raised as far as possible.

8. Install the spring, tube assembly, clutch rod and belts on engine pulley.
9. Adjust spring tension and belt stop as shown.
10. Assemble the rear driveshaft coupling with hardware and parts as shown.
11. Replace bottom cover, engine shrouds, oil cooler and fuel tank.

Traction Clutch Belt Tension

Every 100 hours of operating time—or whenever you suspect the clutch (drive belts) may be slipping the belt tension should be checked as follows:

1. Raise the tractor hood.
2. Remove capscrews and washers at

Allis-Chalmers

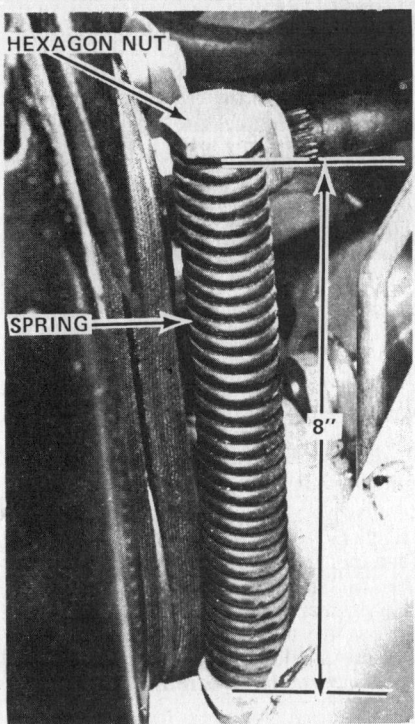

Clutch belt tension spring height adjustment

each end of the transmission oil cooler and raise the left end of it to expose the main drive belt tension adjustment.

3. Be sure the clutch is released and has fully returned to the engaged (pedal out) position.
4. Adjust hexagon nut to obtain a spring height of 8 in.
5. The belts top should be adjusted whenever the traction clutch belt tension is adjusted or if the clutch does not disengage completely when the clutch pedal is depressed. Using a ½ in. wrench, loosen the capscrew and adjust the belt stop so there is 1/16 in. clearance between the belt stop and each belt when the clutch pedal is released. Tighten the capscrew securely.

6. Replace the oil cooler and secure it in place with the washers and screws removed earlier.
7. The clutch free travel should also be checked each time the traction clutch belt tension is adjusted (See below).

Clutch Pedal Free Travel

Clutch pedal free travel is the distance which the clutch pedal can be easily pushed downward with finger pressure. This distance should be 1½ to 1¾ in. when measured as shown. If the distance becomes less than 1 in. or more than 1¾ in., readjust the clutch pedal free travel as follows:

1. Using a ⅜ in. wrench, remove the 10 capscrews holding the dust shield to the underside of the tractor frame.
2. Using a 9/16 in. wrench turn the locknut counterclockwise to increase the amount of free travel or clockwise to decrease the free travel. Alternately check the amount of free travel and turn the locknut until the amount of free travel measures 1½ to 1¾ in.
3. Replace the dust shield and 10 capscrews on the bottom of the tractor and tighten the capscrews securely.

Clutch pedal free travel adjusting nut

TRANSMISSION AND PTO

Front PTO Electric Clutch

REMOVAL AND INSTALLATION

If an implement is attached to the tractor it will be necessary to remove the drive belt.

1. Remove the four self threading capscrews to the armature assembly.
2. Remove the capscrew, lockwasher and washer attaching the clutch assembly to the engine crankshaft.
3. Remove the clutch armature assembly and key from the crankshaft.
4. Remove the four capscrews attaching the field to the engine block and remove the field assembly.
5. Check the bearing, clutch parts and key. Replace worn or damaged parts.
6. The field can be checked with an ohmmeter. Connect one ohmmeter lead to either female connection from the armature field and the other female lead to the field assembly for a ground. The normal resistance should be 2.75 to 3.60 ohms.
A reading outside these valves indicates a faulty field assembly. Replace the field.
7. Clean rust and burrs from the engine crankshaft and apply a coat of Lubriplate® to the shaft to aid in assembly of the clutch.
Install the components in the following order: Key, field assembly, four capscrews, clutch armature assembly, large washer, lockwasher and capscrew. Install the V-belt pulley with the four self threading capscrews.

PTO Shaft

REMOVAL

1. Raise the tractor hood and disconnect the negative battery terminal. Fasten

Clutch belt stop adjustment

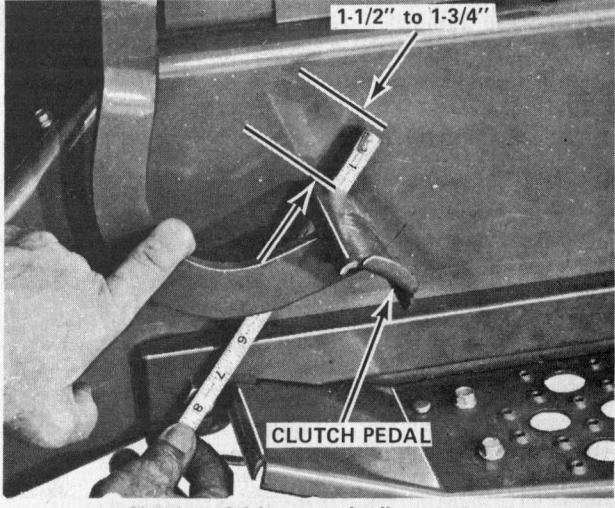

Clutch pedal free travel adjustment

77

Allis-Chalmers

Left side components   Right side components

battery cable to prevent it from swinging back and touching the negative.

2. Remove the four capscrews that hold the seat support to the top of the transmission housing and remove the seat and support.

NOTE: Be sure that electric wires to the lights in the righthand and lefthand fenders are disconnected from the terminal if the wires are strung over the seat support structure.

3. Remove the four ¼ in. capscrews that fasten the frame cover to the frame and remove the cover.

4. Remove the two capscrews that hold the lefthand end of the fuel tank straps to the frame. Disconnect the fuel lines and remove the fuel tank.

5. Remove the upper rear oil cooler shroud.

6. Remove the righthand and lefthand oil cooler capscrews and retaining washers. Swing the oil cooler out of the way to righthand side and remove the righthand and lefthand oil cooler shrouds.

7. Remove the capscrews that fasten the righthand and lefthand side of the console to the tractor frame to permit raising the rear side of console enough to provide clearance to remove rear PTO shaft assembly.

8. Remove the four capscrews that attach the electric clutch to the hub (34). Disconnect electric connector.

9. Remove the two capscrews, spacers and washers that attach the coupling disc (6) to the flywheel.

10. Allow the front section of the shaft with yoke to drop down and remove the electric clutch and PTO shaft from the rear.

DISASSEMBLY, OVERHAUL AND INSTALLATION

1. Remove yoke (10) and coupling (6) from the shaft (15).

2. Remove the felt (13) and rubber washer (14) from the shaft.

3. Remove two shoulder bolts (12) flat washers, lockwashers attaching coupling (6) to yoke (10).

4. Check the yoke bushings (8) and replace if worn or loose in yoke.

5. Remove capscrew (30), lockwasher and clutch collar (28).

6. Remove the socket head screw (22), lockwasher (21) from bearing retainer (19).

7. Remove the field and clutch parts from the shaft.

8. Remove Woodruff key (16), washer (20).

9. Remove the bearing (18) and bearing retainer (19) from the shaft.

10. To remove bearing (18) from retainer (19) remove retaining ring (17).

11. Clean parts in suitable solvent. Do not wash lubricant from the sealed bearings. Check all parts and replace worn, damaged or parts that will not function properly.

12. Install the bearing in the bearing retainer and secure with retaining ring.

13. Install the bearing and washer on the shaft.

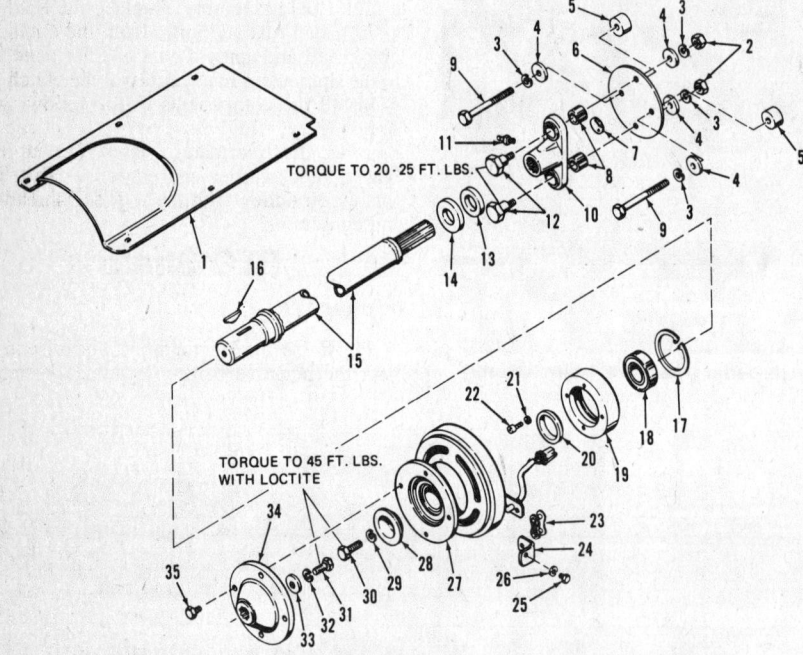

1. Top cover
2. Hex nut
3. Lockwasher
4. Flat washer
5. Spacer
6. Coupling
7. Plug
8. Bushing
9. Hex capscrew
10. Yoke
11. Grease fitting
12. Shoulder bolt
13. Felt washer
14. Rubber washer
15. Shaft assembly
16. Key
17. Retaining ring
18. Bearing
19. Bearing retainer
20. Washer
21. Lockwasher
22. Socket-head screw
23. Connector link
24. Bracket
25. Hex capscrew
26. Lockwasher
27. Electric clutch
28. Collar
29. Lockwasher
30. Hex capscrew
31. Hex capscrew
32. Lockwasher
33. Flat washer
34. Hub
35. Hex capscrew

PTO clutch components

78

Allis-Chalmers

14. Install the Woodruff key in shaft.
15. Attach armature to the bearing retainer with four socket head capscrews.
16. Install the remaining clutch parts on the shaft.
17. Use Loctite® on the capscrew threads and torque to 45 ft.lb. to hold the collar and clutch assembly in place.
18. Install the rubber washer (14) and felt washer (13) on the shaft as shown. Check yoke (10) and install plug (7) if plug was removed.
19. Assemble the coupling disc (6) and yoke (10) with the shoulder bolts, washers, lockwashers and nuts.
20. Place the shaft and clutch under the console between the tractor frame. Install the assembled front yoke on the shaft.
21. Align bracket (24) with the capscrew attaching hole in the frame.
22. Apply Loctite® to the four capscrews and attach the clutch to the hub (34). Torque capscrews to 30 ft.lb.
23. Secure the clutch bracket (24) to the frame with capscrew. Connect electric connection.
24. Apply Loctite® to the two capscrews (9) and attach front disc coupling (6) to the flywheel. Torque capscrews to prevent the yoke from moving freely on the flywheel.
25. Use a dial indicator to check the PTO shaft runout. Remove the spark plugs from the engine and rotate the engine by hand to check PTO shaft runout.
26. Torque the capscrews to 30 ft.lb. Lubricate the yoke splines through lube fitting.
27. Move the felt washer (13) and rubber washer (14) against the yoke (10). Lubricate yoke splines.
28. Replace console capscrews, top cover, bottom cover, oil cooler and other items that were removed for shaft removal.

PTO Top Transmission Shaft

REMOVAL

1. Remove the electric clutch and shaft assembly as outlined.
2. Remove the upper link supports and upper link from rear of transmission.
3. Remove the capscrews that fasten the transmission cover to the top of transmission.

NOTE: To remove the two rear capscrews first remove the two top bolts holding the 3 point hitch lift tube supports to the tractor frame and loosen the two lower bolts as shown. Rotate the lift tube out of way of the rear capscrews. Remove the transmission cover.

4. If the gears, bearings or other items are worn and damaged drain the oil from the transmission case. To drain the transmission case remove the ten washer head capscrews that attach the bottom cover to the tractor frame and remove the cover.
5. Place a pan under the front of the transmission case, remove the electrical wire and temperature sending unit to drain oil from case. Clean all abrasive and foreign material from the transmission case with a suitable solvent.
6. Remove oil seal (22) retaining ring (19) from the front bearing housing and remove the bearing, spacer and O-rings, then remove the shaft from the rear.

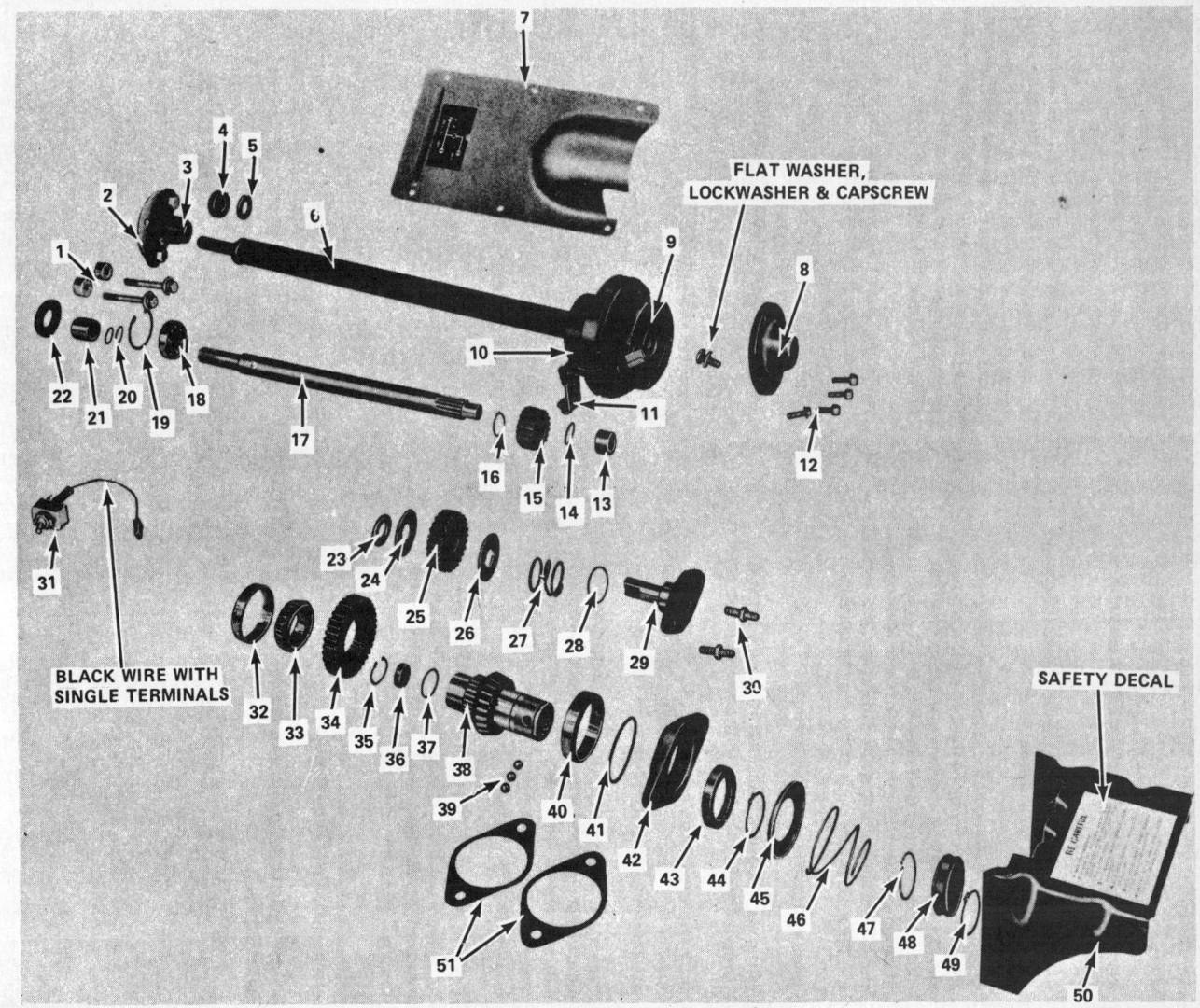

PTO top transmission components

79

Allis-Chalmers

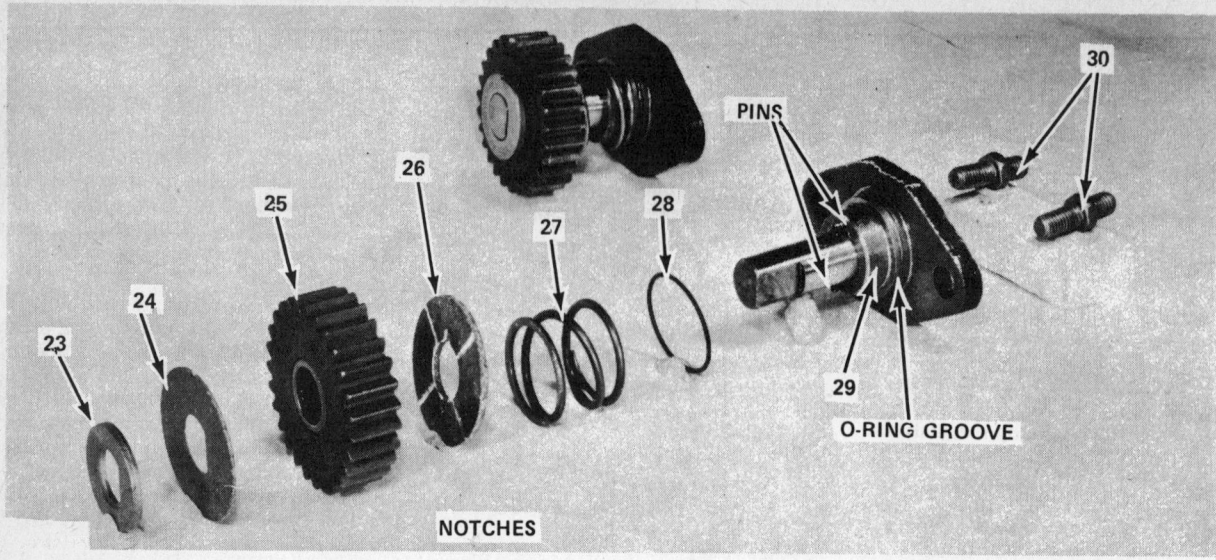

PTO center idler gear

7. Move the shaft forward as shown and remove the shaft from the case.

8. Check roller bearing (13) and remove if not satisfactory.

INSTALLATION

1. Replace roller bearing (13) if it was removed. Press against the end of the bearing with the manufacturers identification. Use a suitable fixture so that the bearing case is not damaged.

2. Install the shaft through the front bearing housing from the rear. Install the bearing (18) on the shaft (17). Install the two O-rings (20) in the recess in the spacer and install spacer on the shaft with the O-rings toward the bearing (18). Press the bearing in the housing and install the retaining ring (19) in the groove to hold the bearing in place.

3. Align the shaft so it will enter the roller bearing (13) and press shaft (17) with bearing (18) in front bearing housing.

4. Install the retaining ring (19) in the groove in bearing housing to hold the bearing and shaft in place.

5. Lubricate spacer (21) and oil seal (22) with Lubriplate® and press oil seal in the housing.

6. Install the front hub. Wedge a clean rag between the gears to prevent the shaft from turning while torquing hub retaining capscrew.

7. Apply Loctite® to capscrew threads and torque to 45 ft.lb.

PTO Center Idler Gear

The cap assembly (29) and thrust washers (24-26) can be removed without removing the top transmission PTO shaft (17), however, if the gear or bearing need to be removed from the case the top shaft will have to be removed.

REMOVAL AND INSTALLATION

1. Remove the two capscrews (30) and cap assembly (29). The steel washer and fiber thrust washer will drop to the bottom of the case.

2. If the roller bearing is worn or damaged press from the gear.

3. Clean all parts in a suitable solvent.

4. Check the two thrust washer retaining roll pins in the cap assembly and replace if distorted.

5. Replace the thrust washers whenever the center idler is removed. Check the spring and replace if broken or worn.

6. The accompanying illustration shows an enlarged view of the parts that make up the center idler gear. To assemble: Coat one side of washer (23) with grease and stick washer against the front support surface with the flat side of hole toward the left. Using grease, stick the grooved side of thrust washer (24) against the forward face of gear (25) and stick the grooved side of thrust washer (26) to the rear face of gear with the notches in washer in vertical position.

7. Place O-ring (28) in O-ring groove in cap assembly (29). Place spring (27) over shaft of cap assembly. While holding the gear and thrust washers down inside the transmission, insert the cap assembly through the rear of transmission with the flat side of shaft to left. Pick up the thrust washers and gear with the shaft. Make certain that thrust washer (24) and washer (23) align with the flat side of shaft as shown. When washers are in place on shaft, install long end of capscrews (30) through cap assembly and into tapped holes in transmission. As cap is tightened down make certain that notches in thrust washer (26) align and slip over the 2 pins in cap assembly. Tighten capscrews securely.

8. Install shield assembly with decal (50) on the exposed threaded end of the cap assembly capscrews.

PTO Lower Shaft

REMOVAL

The output shaft (38) and rear bearing can be removed by removing the locking collar and items 44 to 49 and cap (42) without removing the center idler gear (25) and top PTO transmission shaft.

To remove the gear (34) front bearing

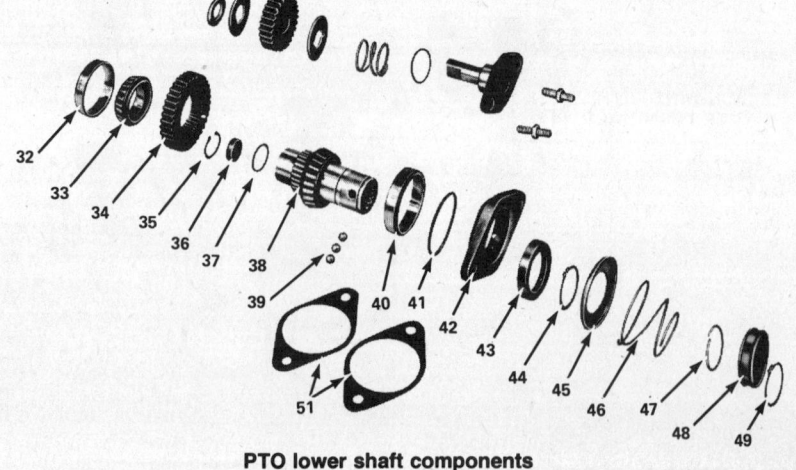

PTO lower shaft components

80

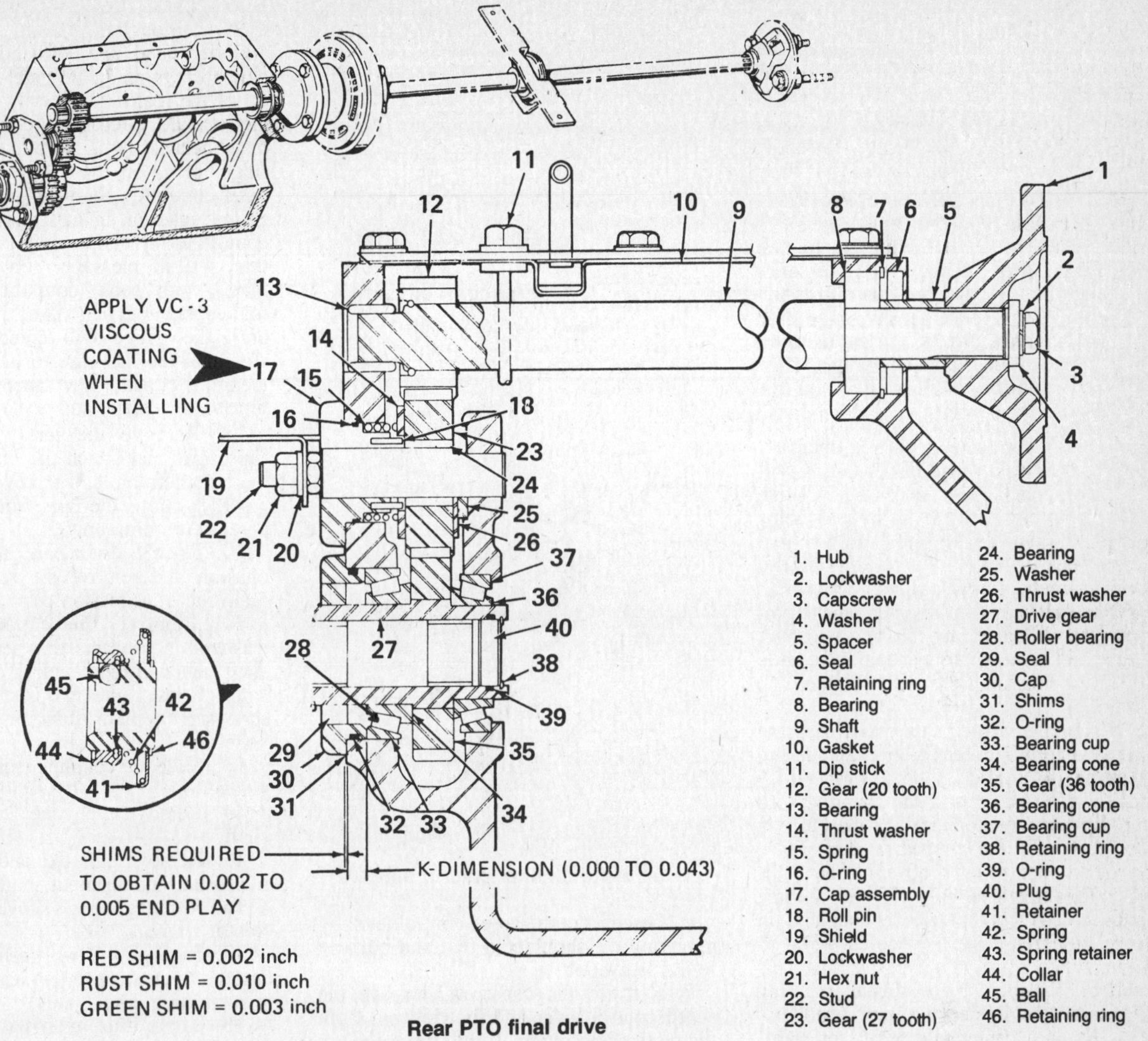

APPLY VC-3
VISCOUS
COATING
WHEN
INSTALLING
SHIMS REQUIRED
TO OBTAIN 0.002 TO
0.005 END PLAY
K-DIMENSION (0.000 TO 0.043)
RED SHIM = 0.002 inch
RUST SHIM = 0.010 inch
GREEN SHIM = 0.003 inch
1. Hub
2. Lockwasher
3. Capscrew
4. Washer
5. Spacer
6. Seal
7. Retaining ring
8. Bearing
9. Shaft
10. Gasket
11. Dip stick
12. Gear (20 tooth)
13. Bearing
14. Thrust washer
15. Spring
16. O-ring
17. Cap assembly
18. Roll pin
19. Shield
20. Lockwasher
21. Hex nut
22. Stud
23. Gear (27 tooth)
24. Bearing
25. Washer
26. Thrust washer
27. Drive gear
28. Roller bearing
29. Seal
30. Cap
31. Shims
32. O-ring
33. Bearing cup
34. Bearing cone
35. Gear (36 tooth)
36. Bearing cone
37. Bearing cup
38. Retaining ring
39. O-ring
40. Plug
41. Retainer
42. Spring
43. Spring retainer
44. Collar
45. Ball
46. Retaining ring

Rear PTO final drive

cone (33) and race it will be necessary to remove the top PTO shaft and center idler gear and shaft.

1. Remove the top shaft and center idler gear and shaft.
2. Remove retaining ring (49), collar (48), spring (46), washer (45), retaining ring (44) and (3 locking balls).
3. Remove the bearing cap (42) and shims (51).

RETAINER
RETAINING RING
FINAL DRIVE GEAR SHAFT
BALL
COLLAR AND SPRING RETAINER
SPRING

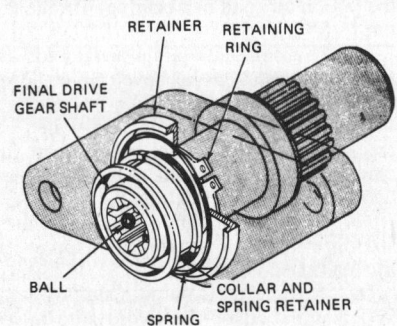

Locking collar

4. Pull the output shaft (38) with bearing cone and cup (40) out.
5. Remove the gear (34) and bearing cone (33) from the case.
6. Clean all parts in suitable solvent and inspect for wear or damage. Check the bearing cups and cones for wear or damage. Replace all parts that are needed.

NOTE: If bearing cone (33) requires replacement pull bearing cup (32) from the transmission case. Remove bearing cone from the output shaft with a suitable puller to prevent damage to the oil seal area of the output shaft. The plug and O-ring are held in the hollow shaft with retaining ring (35). If oil leaks through the center of the shaft replace O-ring.

INSTALLATION

1. Place bearing race (32) in front bore of lower hole in transmission with large end of cup facing the rear. Press race firmly in place against shoulder in casting with suitable adapter tool.
2. Place O-ring (37) in groove in front end of PTO output sleeve and carefully work plug (36) into place under O-ring. Install retaining ring (35) on outside of plug. Press rear bearing cone on output sleeve.
3. Coat bearing cone (33) with grease and place in bearing race (32). Lower gear (34) from top of transmission housing as shown and while holding it at the rear of bearing cone (33), place front end of PTO output sleeve (38) through the center of gear and into bearing cone (33). Align splines in gear and sleeve and carefully press sleeve fully into bearing race (33).
4. Install bearing race (40) over rear of PTO output sleeve and press it into rear of housing. Place shims under bearing cap (42) and bolt cap in place tightly to finally seat the bearing races. Tap end of shaft with plastic hammer to seat bearing cones and cups.
5. Set up a dial indicator against the end

0.002 TO 0.005 INCH END PLAY
BEARING CONE
CAP
O-RING
BEARING CUP

Measuring bearing cap to case clearance

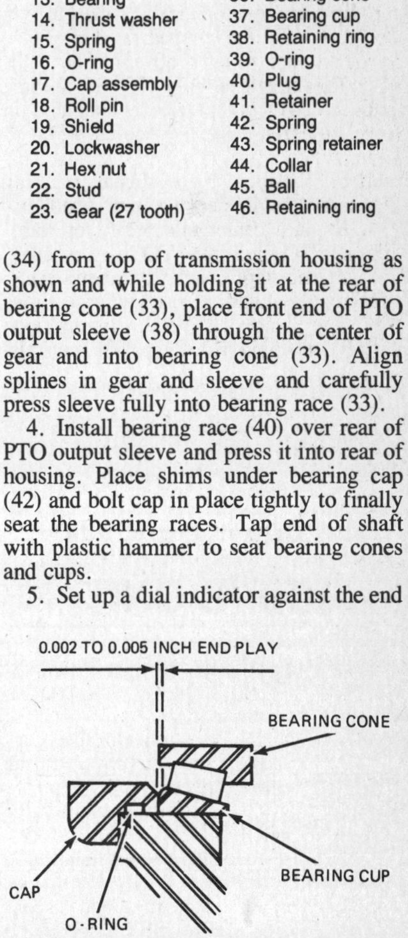

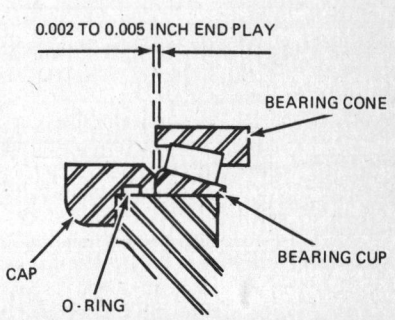

of the PTO output sleeve and measure the end play of sleeve. Correct end play is .002 in. to .005 in. The dark red gaskets are .010 in. thick, the green ones .003 in. and the light red gaskets are .002 in. thick. Remove gaskets as required to reduce end play to the .002 in. to .005 in. specifications or add to increase bearing clearance.

NOTE: To determine an approximate shim thickness hold the cap to the bearing cup with a slight torque on the two capscrews, then measure the clearance between the cap and the machined surface of the case with feeler gauge then add approximately .005 to the distance determined with the feeler gauge. Then select the quantity of shims to equal this figure.

6. When correct end play is obtained, remove bearing cap and install O-ring (41) in groove in cap. Press the oil seal (43) into the bearing cap with the lip of seal facing inward and outside the seal flush with rear face of bearing cap. Lubricate O-ring and seal lip freely. Cover snapring grooves in PTO output sleeve with Scotch tape to protect the oil seal. Place bearing cap over sleeve and work it into place taking care to protect O-ring and lip of seal. Tighten cap securely.

7. Remove Scotch tape from snapring groove in PTO output sleeve and install snapring (44). Slide cupped washer (45) onto sleeve and against the snapring with cup lip onto rear. Place spring (46) on sleeve against the cupped washer. Place spring retainer (47) in groove in collar (48). Place the three steel balls (39) in the three holes in PTO output sleeve and hold in place with grease, push the collar (48) with spring retainer (47) in it over the three balls until they snap in place. Install snapring (49) in groove in sleeve to rear of collar.

8. Replace center idler gear, top transmission PTO shaft rear PTO shaft, oil sending unit and wire bottom cover, transmission cover and gasket, seat and accessories that were removed. Fill the transmission with Power Fluid 401 (Dexron®) transmission fluid.

HYDROSTATIC DRIVE UNIT

REMOVAL

1. Remove the seat.
2. Remove the six washer headed capscrews that fasten the top cover to the frame and remove cover.
3. Disconnect the pump to spool valve oil tube (Steel).
4. Remove hose clamp to disconnect the cooler oil return line.
5. Remove the hose clamp and disconnect oil filter hose at the charge pump elbow.

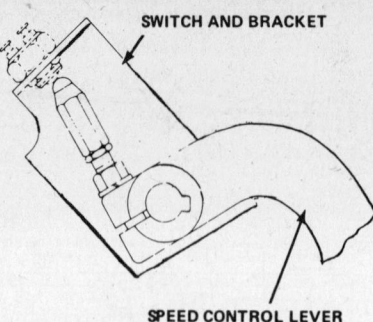

Speed control switch

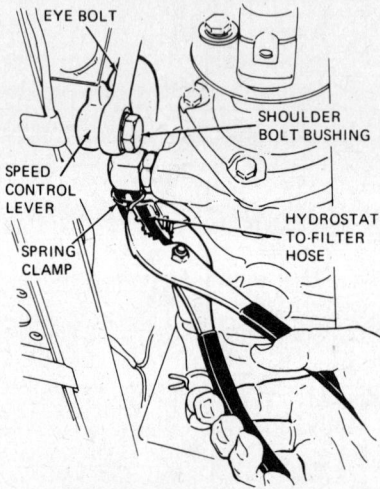

Removing oil filter control hose

6. Disconnect the speed control lever by removing the shouldered bolt and bushing from the eyebolt.
7. Remove the capscrew, remove the speed control lever and locating pin from the hydrostatic pump. It will be necessary to remove the switch and bracket from pump case.

FRONT PTO CLUTCH ADJUSTMENT

1. Turn each of the four locknuts on the front end of the four clutch support studs exactly 1/3 of a turn clockwise when standing in front of tractor facing the locknut. This will move the clutch brake plate and the driven clutch plate slightly closer to the drive plate of the clutch.
2. After adjusting the locknut check with the tip of a .010 in. feeler gauge that there is at least .010 in. running clearance between the driving surface of the electric clutch and the rear surface of the driven

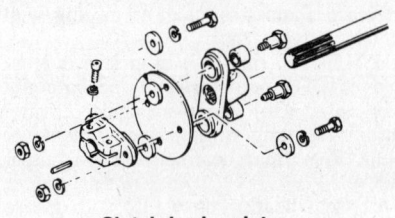

Clutch brake plate

clutch plate. The feeler gauge should be inserted through the oblong holes in the top and righthand and lefthand sides of the clutch brake plate. Be careful to insert the tip of the feeler gauge only 1/16 in. below the outside circumference of the clutch plate because with normal clutch wear a ridge forms on the clutch plate near the outer circumference which will give a false gauge reading if gauge is inserted far enough to lay on the ridge. If there is less than .010 in. clearance between the clutch plates, turn back (counterclockwise) the four locknuts about 1/12 of a turn at a time until the gauge will enter between the clutch plates. Be sure to turn each of the locknuts exactly the same amount each time to keep plates and brake plate parallel.

8. Remove the ten washer headed capscrews that attach the bottom cover to the tractor frame and remove the cover.
9. Remove the wire from the oil temperature sending unit.
10. Place a drain pan under the transmission and remove the sending unit to drain the transmission oil.
11. Remove the capscrews, nuts, washers and spacers that secure the drive shaft yokes to the coupling disc.
12. Loosen the two socket head capscrews in the hydrostatic pump input shaft yoke.
13. Slide the coupling from between the yokes.
14. Remove the yoke from the driveshaft.
15. Remove the yoke and key from the hydrostatic pump input shaft.
16. Remove access cover in lefthand tractor frame.
17. Remove the two washer nuts attaching the hydrostatic pump and motor to the transmission case.
18. Move the hydrostatic pump and motor forward to remove it through the opening in the bottom of the tractor frame.
19. Refer to the 616/620 tractor section for service on the components.

INSTALLATION

1. If a new bearing is installed in the hydrostatic motor housing or a new bevel pinion is installed on the pump shaft always install the correct number of shims between the pinion hub and ball bearing. Torque the nut to 55 ft.lb.
2. If a complete pump and motor assembly is installed the bevel pinion location must be correct.

Before the bevel pinion is installed on the shaft refer to the following chart for correct shim stock thickness to be between the pinion and ball bearing.

3. A four digit pinion locating number starting with 30— is stamped on top of the tractor transmission.
4. A three digit number starting with 900 is stamped on the hydrostatic motor mounting flange. The numerals in steps 3 and 4 are used to determine shim thickness for pinion location. Refer to the following chart.
5. One of the three digit numerals that will be stamped on the hydrostatic pump

Allis-Chalmers

flange will be in the row of numerals across the top of the chart. The four digit numeral on the top of the transmission case will be one of the numerals in the left column of the chart.

6. **Example:** If the four digit dimension "C" stamped on the transmission case is 3011 and the three digit dimension stamped on the pump flange is 904 locate the numeral 3011 in the left column of the chart and follow this row of figures to the right of the column directly under the pump dimension 904.

  C Transmission Dimension   3011
  D Motor Mounting Flange
     Dimension   904
     Shim Thickness is   .045

Shims are available in .002, .003 and .005 thickness.

7. Install .045 thickness of shims between pinion hub and ball bearing. Install pinion and torque nut to 55 ft.lb.

8. Position the hydrostatic pump and motor on a bench with the charge pump inlet port upward. Rotate the pump shaft counterclockwise while pouring Dexron® Power Fluid 401 transmission lubricant in the charge pump inlet port. This should fill the charge pump and oil cavities in the hydrostatic pump and motor to provide lubrication for the units initial start.

9. Install a new O-ring on the pump case and install the assembly in the tractor frame. Torque the two flange nuts that fasten the hydrostatic unit to the transmission case.

10. Install the driveshaft coupling assembly, pump shaft key and yoke. To use proper capscrews, etc. refer to diagram, for correct material to install coupling.

11. Move the pump shaft yoke to obtain ⅛ to ¼ in. clearance between the oil seal and yoke. Torque the two Allen head capscrews.

12. Install the transmission oil temperature sending unit.

13. Install the alignment pin in the pump trunnion shaft with the head of the pin toward the front of the tractor. Install the switch bracket. Install the speed control lever on the shaft and torque capscrew to hold the lever in place. The lever should be located so that a clearance of ⅛ in. is between the lever and oil seal.

NOTE: The safety switch actuating detent is attached to the speed control capscrew and will need to be adjusted to actuate the safety switch for starting. Adjust detent so switch closes the circuit when the hydrostatic lever is in neutral.

14. Install access hole cover in the tractor frame.

15. Connect the speed control lever to the eyebolt with the shouldered bolt and bushing.

16. Connect the transmission to oil filter hose.

17. Connect oil cooler to hydrostatic pump oil return line. Connect charge pump to spool valve oil line (Steel).

18. Remove the transmission filler plug and fill the transmission to the proper level with Dexron® Power Fluid 401 transmission fluid.

19. Remove the spark plugs from the engine.

20. Remove one pipe Allen type pipe plug from the front section of the hydrostatic pump and pour some Dexron® Power Fluid 401 transmission oil in this port.

21. Rotate the charge pump with the engine starter until oil flows from the open port.

22. Install a pressure gauge in the open port so that charge pump pressure and implement relief valve pressure can be checked.

23. Move the transmission gear selector lever to have the transmission in neutral.

24. Replace the spark plugs in the engine and operate the engine at idle speed to fill the hydraulic system with oil.

25. Move the spool valve lever to extend and retract the ram to eliminate air pockets.

26. Check the charge pump pressure. Correct charge pump pressure is 70-150 psi.

27. Move the spool valve to extend the ram and activate implement lift relief valve. The pressure should be 550-700 psi. If the pressures are incorrect refer to the 616/620 tractor section information.

28. Remove the pressure gauge and install the pipe plug. Use a sealer on the pipe plug threads.

29. Replace the top cover, bottom cover, seat, etc.

Hydrostatic Transmission Neutral Adjustment

If the tractor has a tendency to move forward or rearward when the hydrostatic transmission control lever is in the NEUTRAL position, the Neutral adjustment should be made. All clockwise and counterclockwise directions given are as you look up to the turnbuckle from below the tractor frame. Proceed as follows:

1. Remove 10 capscrews and dust cover.

TRANSMISSION NUMBER (On Case)	900	901	902	903	904	905	906	907	908	909	910	911	912	913	914	915
3000	.038	.037	.036	.035	.034	.033	.032	.031	.030	.029	.028	.027	.026	.025	.024	.023
3001	.039	.038	.037	.036	.035	.034	.033	.032	.031	.030	.029	.028	.027	.026	.025	.024
3002	.040	.039	.038	.037	.036	.035	.034	.033	.032	.031	.030	.029	.028	.027	.026	.025
3003	.041	.040	.039	.038	.037	.036	.035	.034	.033	.032	.031	.030	.029	.028	.027	.026
3004	.042	.041	.040	.039	.038	.037	.036	.035	.034	.033	.032	.031	.030	.029	.028	.027
3005	.043	.042	.041	.040	.039	.038	.037	.036	.035	.034	.033	.032	.031	.030	.029	.028
3006	.044	.043	.042	.041	.040	.039	.038	.037	.036	.035	.034	.033	.032	.031	.030	.029
3007	.045	.044	.043	.042	.041	.040	.039	.038	.037	.036	.035	.034	.033	.032	.031	.030
3008	.046	.045	.044	.043	.042	.041	.040	.039	.038	.037	.036	.035	.034	.033	.032	.031
3009	.047	.046	.045	.044	.043	.042	.041	.040	.039	.038	.037	.036	.035	.034	.033	.032
3010	.048	.047	.046	.045	.044	.043	.042	.041	.040	.039	.038	.037	.036	.035	.034	.033
3011	.049	.048	.047	.046	.045	.044	.043	.042	.041	.040	.039	.038	.037	.036	.035	.034
3012	.050	.049	.048	.047	.046	.045	.044	.043	.042	.041	.040	.039	.038	.037	.036	.035
3013	.051	.050	.049	.048	.047	.046	.045	.044	.043	.042	.041	.040	.039	.038	.037	.036
3014	.052	.051	.050	.049	.048	.047	.046	.045	.044	.043	.042	.041	.040	.039	.038	.037
3015	.053	.052	.051	.050	.049	.048	.047	.046	.045	.044	.043	.042	.041	.040	.039	.038
3016	.054	.053	.052	.051	.050	.049	.048	.047	.046	.045	.044	.043	.042	.041	.040	.039
3017	.055	.054	.053	.052	.051	.050	.049	.048	.047	.046	.045	.044	.043	.042	.041	.040
3018	.056	.055	.054	.053	.052	.051	.050	.049	.048	.047	.046	.045	.044	.043	.042	.041
3019	.057	.056	.055	.054	.053	.052	.051	.050	.049	.048	.047	.046	.045	.044	.043	.042
3020	.058	.057	.056	.055	.054	.053	.052	.051	.050	.049	.048	.047	.046	.045	.044	.043
3021	.059	.058	.057	.056	.055	.054	.053	.052	.051	.050	.049	.048	.047	.046	.045	.044
3022	.060	.059	.058	.057	.056	.055	.054	.053	.052	.051	.050	.049	.048	.047	.046	.045
3023	.061	.060	.059	.058	.057	.056	.055	.054	.053	.052	.051	.050	.049	.048	.047	.046
3024	.062	.061	.060	.059	.058	.057	.056	.055	.054	.053	.052	.051	.050	.049	.048	.047

Pinion shim thickness chart

Allis-Chalmers

2. Using two open end wrenches, loosen the locking nuts on either side of turnbuckle. To loosen the locknuts hold the turnbuckle stationary with one wrench and use the other to turn the locking nuts clockwise to loosen them.

3. While sitting on the tractor seat start the engine and let it run at a slow idle. Release parking brake.

4. Place the gear transmission in first gear.

5. Place the hydrostatic transmission control lever against the NEUTRAL stop.

6. If the tractor does not move, increase the engine speed until it does.

7. Stop the engine. Shift transmission to neutral, apply parking brake, remove key.

---- **CAUTION** ----
Do not attempt to make adjustment of the turnbuckle while the engine is running. The rear power take off shaft and main driveshaft are turning at high speed near the turnbuckle.

8. If the tractor moved forward, with the hydrostatic control lever in the notched neutral position, turn the turnbuckle clockwise (as viewed from below) about ½ turn. If the tractor moved rearward, turn the turnbuckle counterclockwise.

9. Alternately run the engine and adjust the turnbuckle until the tractor does not move at all when the engine is running at full speed. Be sure that the engine is stopped, transmission in neutral and parking brakes are on every time you get off the tractor seat.

10. Use one wrench to prevent the turnbuckle from turning while tightening the two locking nuts at either end against it with the other wrench.

11. Check your adjustment once again by running the engine to make sure you didn't change the adjustment while tightening the locking nuts.

12. Replace the dust shield and ten capscrews.

Neutral Safety Starting Switch

If the engine starter will not actuate when the hydrostatic control lever is in neutral, or will actuate when the hydrostatic transmission control lever is not in the NEUTRAL position, the neutral safety starting switch should be adjusted.

---- **CAUTION** ----
While making adjustments to the neutral safety starting switch make sure that the tractor engine is stopped, throttle lever is fully back to slow position, hydrostatic control lever is in neutral, transmission gears are in neutral, both PTO clutches are off and parking brakes are set.

1. Make sure that the hydrostatic control lever is exactly in the notched neutral position and that the neutral adjustment explained in 12 steps immediately above has been properly made.

2. Use a screwdriver to pry the snap-out cover from the inspection hole in the righthand tractor side frame near the operators position. Detent (A), locknut (B) and threaded stud (C) seen through the inspection hole are attached to the hydrostatic control arm (D) and make contact with the neutral start safety switch. Adjust as follows.

3. Tighten threaded stud (C) securely into control arm (D) (turn stud clockwise when viewed from above).

4. Loosen locknut (B) from detent (A), back off detent (A) by turning it clockwise until it does not touch the detent button of the neutral start safety switch. Check by turning switch key to START for a second or two, starter motor should not run.

5. Turn detent (A) counterclockwise slowly toward the neutral safety switch while holding the switch key to START. As soon as switch makes contact and starter motor turns over release the key, and turn the detent only $1/6$ turn more counterclockwise. Hold the detent with one wrench and tighten locknut (B) securely against lower end of detent with another wrench.

6. Recheck the adjustment by turning switch key to START. Starter motor should run, if not, repeat step 5, until neutral start switch is actuated.

7. Replace snap-out cover in inspection hole.

DROP HOUSING

DISASSEMBLY

1. Position the tractor safely on the suitable stands or blocks.
2. Remove the tire and wheel.
3. If the drop housing requires disassembly drain the lubricant from the housing before removing it from the tractor.
4. Remove the capscrews, nuts and lockwashers that attach the wheel guard to the drop housing.

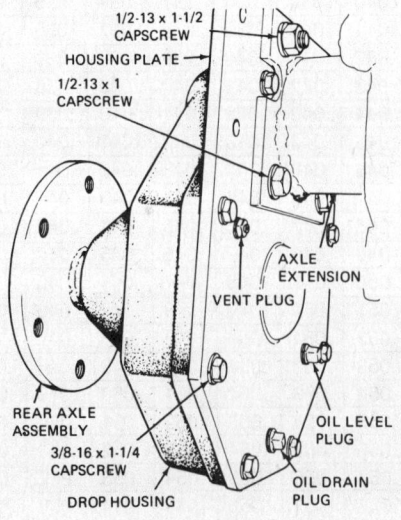

Drop housing

5. Remove the lockwashers and capscrews attaching the drop housing to the axle extension.

6. Slide the drop housing away from the axle extension. When the pinion shaft clears the coupling the assembly can be moved to a suitable work area.

7. Remove drop housing plate by removing the ⅜ x 1¼ capscrews and lockwashers. Tap plate to remove from housing.

8. Remove the spring from the bull pinion shaft and press the shaft and bearing from the plate.

9. Remove the bull pinion thrust washer from the drop housing.

10. Check the vent and remove if damaged.

11. Press the pinion shaft from the bearing and remove the slinger.

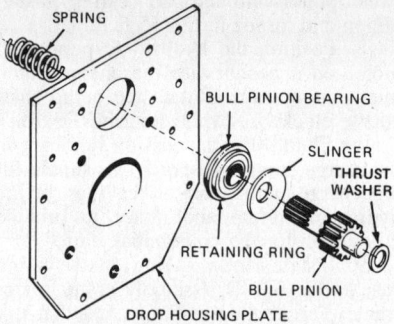

Bull pinion removal

12. To remove bull gear, bearings or axle shaft from the housing, remove locknut and flat washer from the rear axle assembly and remove bull gear.

13. Remove the rear axle assembly from the gear case.

14. Remove the inner and outer bearing cones and oil seal.

15. Clean the gear case and all parts in a suitable solvent and inspect the gear and bearings. Replace the gear if worn or chipped and bearings if they are rough and worn. Check bull pinion roller bearing and remove if replacement is necessary. Remove the two bearing cups if the axle bearing cones are to be replaced.

16. Use a brush to remove dirt or oil from the bull pinion shaft sealed bearings. Do not allow the sealed bearings to remain in the solvent. If EP90 oil enters the brake housing from the drop housing replace the sealed bull pinion shaft bearings.

17. Rotate the sealed bearings to check for damage or wear.

ASSEMBLY

1. Press the bull pinion roller bearing, inner and outer bearing cone in the drop housing. Replace the roll pins if they have been removed.

2. Lubricate the outer bearing cone with SAE90EP and place in the housing.

3. Press a new oil seal in the housing. Lubricate the shaft contact area of the seal with Lubriplate®.

4. Install the rear axle assembly through the oil seal and outer bearing cone. Install the inner bearing cone.

5. Install the bull gear on the axle shaft.

6. Install the flat washer and locknut.

7. Torque the locknut to slightly preload the bearings. Tap each end of the axle to seat the cones and cups.

8. Place a socket wrench on the nut and use an in.lb. torque wrench to check the effort required to rotate the axle assembly.

9. Carefully continue to torque the nut until it requires twelve (12) in.lb. to rotate the axle in the bearings.

10. Install the slinger on the bull pinion shaft and press the sealed bearing on the pinion shaft.

NOTE: The bearing must be installed on the shaft so that the narrowest area between the retaining ring and bearing edge is toward the pinion.

11. Press the sealed bearing in the drop housing plate.

12. Lubricate the bull pinion roller bearing and install the thrust washer on the pinion shaft.

13. Coat the gasket area of the plate and housing with a sealer and install a new gasket. Install new vent plug if necessary.

14. Install the assembled bull pinion and plate on the drop housing. Torque the eight ⅜ x 1¼ capscrews.

15. Place the coupling on the differential output shaft, spring on the bull pinion shaft and install the drop housing assembly on the axle extension. Torque the nuts and capscrews that attach the drop housing to the axle extension and to the wheel guards. Install the wheel and tire.

16. Fill the drop housing with SAE90EP oil to the correct level.

BEVEL GEAR, INPUT SHAFT AND GEARS

DISASSEMBLY

1. Remove two capscrews (1) lockwashers and cap (3) from the transmission case. Remove O-ring and discard.

2. Remove bearing cone (6) from the sliding gear shaft (8) and store with bearing cap (3). If bearings are to be reused always reinstall with the original cup and in same location.

3. Remove 20T gear (7) from the shaft.

4. Move the shaft to the left and remove the bearing cone (17) and gear (9) from the shaft and case.

5. Remove sliding gear shaft (8) from the case.

6. Remove two capscrews (10) and bearing cap (12) with shims from the transmission case. Store cap, shims, cup (16) with righthand bearing cone (17) removed from shaft (8).

7. Remove two capscrews (18) and cap (20) from the transmission case.

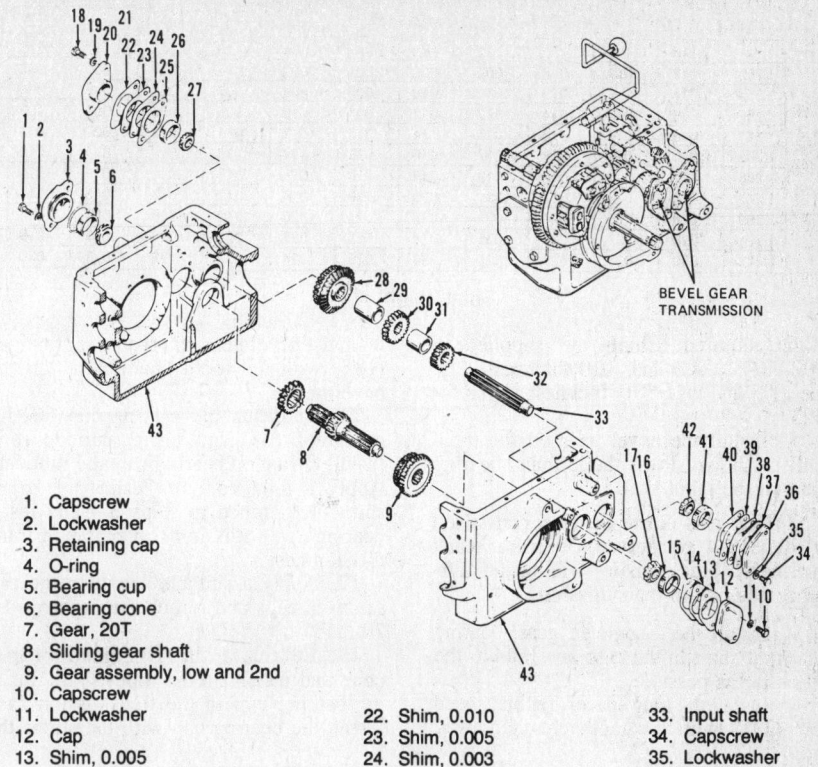

1. Capscrew
2. Lockwasher
3. Retaining cap
4. O-ring
5. Bearing cup
6. Bearing cone
7. Gear, 20T
8. Sliding gear shaft
9. Gear assembly, low and 2nd
10. Capscrew
11. Lockwasher
12. Cap
13. Shim, 0.005
14. Shim, 0.003
15. Shim, 0.002
16. Bearing cup
17. Bearing cone
18. Capscrew
19. Lockwasher
20. Cap
21. O-ring
22. Shim, 0.010
23. Shim, 0.005
24. Shim, 0.003
25. Shim, 0.002
26. Bearing cup
27. Bearing cone
28. Bevel gear assembly
29. Long spacer
30. Gear, 23T
31. Short spacer
32. Gear 18T
33. Input shaft
34. Capscrew
35. Lockwasher
36. Cap
37. Shim, 0.010
38. Shim, 0.005
39. Shim, 0.003
40. Shim, 0.002
41. Bearing cup
42. Bearing cone
43. Differential case

Bevel gear transmission components

8. Remove shims (22-25) and cup (26) from the transmission case and store with cap (20). Remove R-ring (21) and discard.

9. Move the input shaft (33) left as far as possible and remove bearing cone (42), gears (30-32) and two spacers (29-31). Remove input shaft, bevel gear (28) and bearing cone (27) from the top of the case.

10. Remove two capscrews (34), bearing cap (36), shims (37-40) and bearing cup (41). Store bearing cone (42) with cap (36) bearing cone (42) and shims.

11. Clean all parts and case with suitable cleaning solvent. Check all bearing cones, cups and other parts for wear or damage. If a bearing cone is replaced the original cup must be replaced also.

12. The lefthand cone is a press fit to the input shaft and should not be removed unless the bearing cone, bevel gear, or cup is to be replaced or the shim thickness (22-25) is to be changed. Use a suitable bearing puller to remove the cone from the shaft.

NOTE: DO NOT press cone (27) on the input shaft at this time.

12. Check the internal splines and gear teeth for wear. Check the shafts for wear or chipped splines.

ASSEMBLY

1. If the lefthand input shaft bearing cone (27), cup (26), cap (20) or transmission case is to be replaced the shim stack height will need to be corrected.

The bevel gear to pinion lash is determined by the location of the bevel gear in the transmission case with these shims.

2. To determine the height of the shim stack refer step 4.

3. Place the lefthand bearing cap with the hub up. Place the bearing cup on the cap hub and bearing cone in the cup. Rotate the cone to be sure cone is seated in the cup.

4. Measure the distance from the top surface of the cap to the top surface of the bearing cone inner race and record. This dimension will be 1.25+ in.

5. A case dimension is stamped on the transmission case to establish the bevel gear location. This dimension will be a four digit number 3.11+.

6. Refer to the chart to select correct shim stack height. The top line numerals are stamped on the transmission case. The lefthand column of numerals is bearing and cap dimensions determined in steps 3 and 4.

7. **Example:** Case dimension is 3.119 and bearing and cap dimension is 1.267. Locate the numeral 1.267 in the lefthand column and move toward the right on the same line to the column directly under the case dimension 3.119. The numeral 0.023 is in this column and is the correct shim

Allis-Chalmers

Shim height chart

DIAL INDICATOR READ OUT	DIMENSION "A"										
DIMENSION "B"		3.118	3.119	3.120	3.121	3.122	3.123	3.124	3.125	3.126	3.127	3.128
	1.255	.012	.011	.010	.009	.008	.007	.006	.005	.004	.003	.002
	1.257	.014	.013	.012	.011	.010	.009	.008	.007	.006	.005	.004
	1.259	.016	.015	.014	.013	.012	.011	.010	.009	.008	.007	.006
	1.261	.018	.017	.016	.015	.014	.013	.012	.011	.010	.009	.008
	1.263	.020	.019	.018	.017	.016	.015	.014	.013	.012	.011	.010
	1.265	.022	.021	.020	.019	.018	.017	.016	.015	.014	.013	.012
	1.267	.024	.023	.022	.021	.020	.019	.018	.017	.016	.015	.014
	1.269	.026	.025	.024	.023	.022	.021	.020	.019	.018	.017	.016
	1.271	.028	.027	.026	.025	.024	.023	.022	.021	.020	.019	.018
	1.272	.030	.029	.028	.027	.026	.025	.024	.023	.022	.021	.020

height required. Shims are supplied in .002, .003, .005 and .010 thickness. Use one .003 and two .010 thickness shims for .023.

8. Install the bevel gear on the input shaft as shown. Press the lefthand bearing cone on the input shaft.

NOTE: This is the bearing cone used in steps 3-4 to determine shim stock thickness. This bearing will be on lefthand side of the transmission.

9. Install the assembled gear, bearing and input shaft in the case and hold to the left as far as possible.

10. Install the long spacer, second speed gear (23T) with the tooth chamfer away from the bevel gear, short spacer, first gear (18T) with the tooth chamfer toward the bevel gear.

11. Lubricate the bearing cup (used in paragraph 3-4) and install cup on cone. Install O-ring on bearing cap and lubricate. Apply a light coat of Permatex® to the shims determined to steps 3-4 and install bearing cap with shims in place on transmission case.

12. Apply a suitable sealer to the capscrew threads and torque the capscrews to 20 ft.lb.

13. Lubricate the righthand bearing cone and install on the input shaft. Install the bearing cup in the transmission case. Install the bearing cap with the shims that were removed during disassembly. If the shims were destroyed or lost, install approximately .040 shims of various thickness to start. Shims are available in .002, .003, .005 and .010 thickness.

14. Tap the caps to seat the bearing cones and cups.

15. Check the input shaft for end play and bearing adjustment.

16. The shaft should rotate freely without end play. If a vertical or lateral movement of the shaft or gears is apparent remove shims from between the righthand bearing cap and transmission case to obtain correct adjustment. Do not preload the bearings.

17. When the adjustment is correct the spacers can be rotated without the gears and shaft rotating.

18. After the correct bearing adjustment is made remove the cap and shims. Apply a light coat of Permaxtex® to the shims and a suitable sealer to the capscrew threads. Torque the capscrews to 20 ft.lb.

19. Install the sliding gear shaft in the transmission case with the first and second speed gear installed on the shaft as shown.

20. Install the third speed gear on the sliding gear shaft with the hub toward left as shown. Lubricate and install the lefthand bearing cone on the shaft. Press the left-

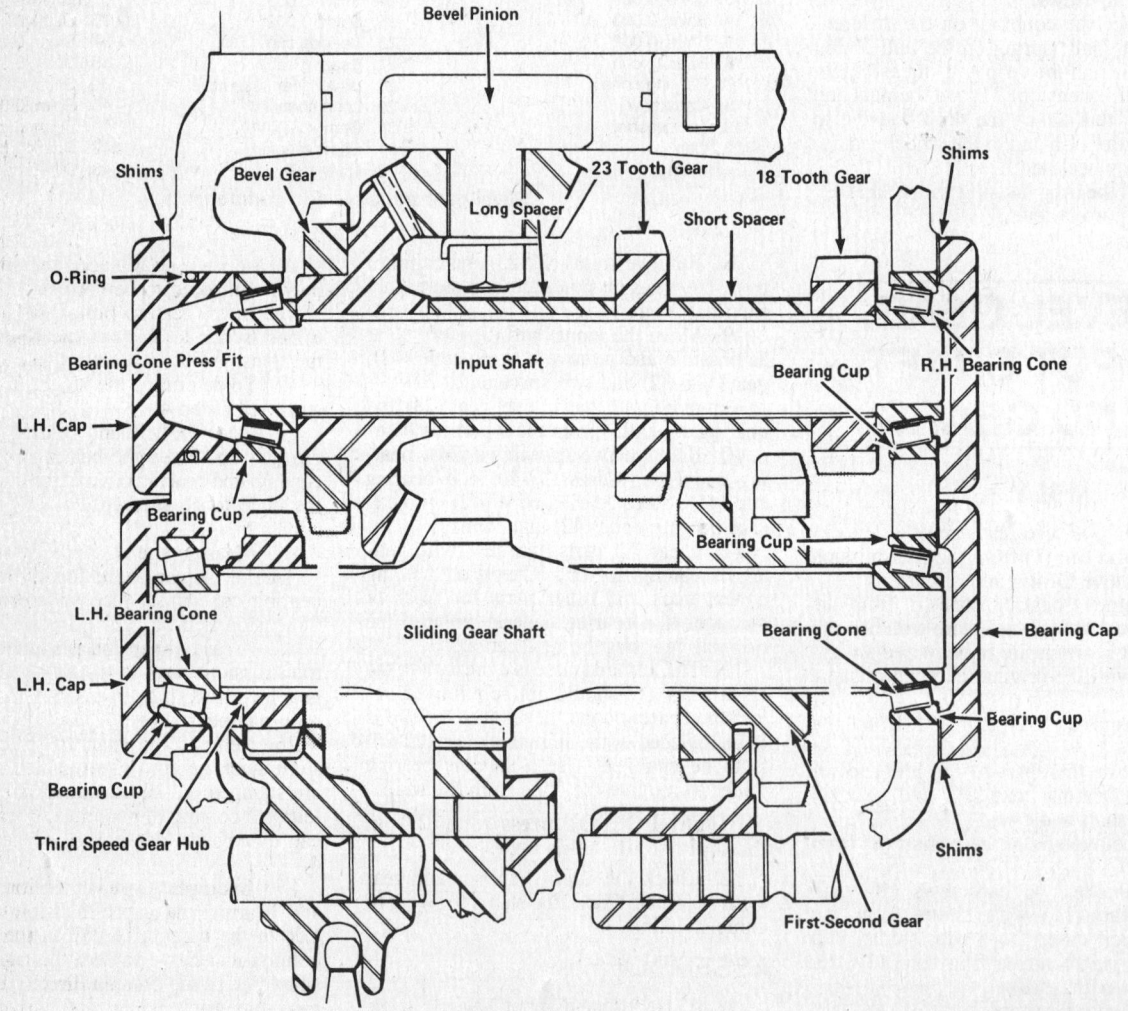

Bevel gear installation

hand bearing cup in the cap. Lubricate a new O-ring and install on the cap. Install the cap on the transmission case.

21. Apply a sealant to the capscrew threads and torque to 20 ft.lb. Lubricate the righthand torque cone on the sliding gear shaft. Install the bearing cup in the transmission case. Install the shims that were removed during disassembly on the cap and install the cap on the case. If the shims are destroyed and require new ones a shim stack of .040 of shims of various thickness could be used to start. Shims are available in .002, .003, .005 and .010 thickness.

22. Torque capscrews to 20 ft.lb. Tap the caps to seat the bearings.

23. The shaft should rotate freely without end play.

24. Check the shaft for vertical or lateral movement and remove shims to remove end play. If the bearings are preloaded add shims to adjust bearings to a free rolling fit.

25. After the bearings are adjusted correct remove cap and apply a light coat of Permatex® to the shims and sealer to the capscrew threads. Torque capscrews to 20 ft.lb.

26. Install the differential assembly in the transmission case.

27. Install new shifter shaft lever oil seal in transmission case and install shift shaft lever in the transmission case.

28. Move the shift forks (4-7) so that sliding gears are in the neutral position. Place shift stem (2) in the recesses of the shift forks and secure with capscrew and lockwasher.

29. Install top PTO shaft in the transmission case. Use a new oil seal on the shaft. Use Loctite® on the clutch hub attaching capscrews.

30. Install the hydrostatic pump and motor assembly on the transmission. Refer to Hydrostatic Pump and Motor Installation section.

31. Install transmission case in the tractor.

32. Install brakes (Refer to Brake Installation section for information).

33. Install axle extensions.

34. Lift shaft and lift ram.

35. Install transmission oil filter.

36. Install the tractor drawbar.

37. Install drop housing.

38. Install wheels.

39. Connect front PTO shaft to top transmission PTO shaft.

40. Install hydrostatic transmission if it was not done in step 15.

41. Install oil temperature sending unit and wire.

42. Connect oil filter to charge pump oil line.

43. Install the transmission safety switch and hydrostatic control linkage.

44. Refer to the charts and connect clutch shaft to pump input shaft as shown.

45. Install top hydraulic oil lines and items removed during disassembly.

46. Install transmission cover.

47. Fill transmission with Dexron® Power Fluid 401. Adjust the brakes.

48. Start the tractor and check hydraulic system and hydrostatic pump and motor and adjust. Adjust neutral start switch. Install seat, top cover and bottom cover.

DIFFERENTIAL

REMOVAL

1. Place the tractor on jack stands.
2. To drain transmission lubricant remove the ten washer head capscrews attaching the bottom cover to the frame, place a pan under the sending unit. remove the electrical wire from the transmission oil temperature sending unit and remove the sending unit.
3. Remove the wheels, drop housing, axle extensions, brake assemblies. Seat with seat support and transmission cover.
4. If the tractor is equipped with a rear PTO remove the top power take off shaft. Refer to PTO section for shaft removal.

NOTE: If the transmission case is to be removed from the tractor frame it is not necessary to remove the front PTO shaft. It is not necessary to remove transmission case from the tractor to remove the differential.

5. Remove the upper brake stationary disc studs.
6. Remove the two capscrews and lockwashers attaching each support (5) to the transmission case. Remove the supports from the case.

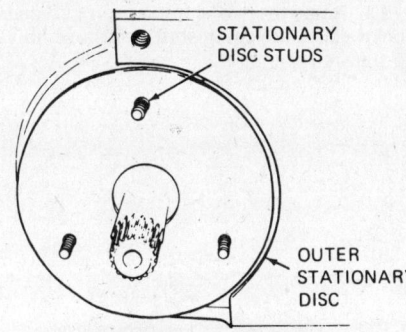

Stationary disc studs

7. Remove the discard oil seals and O-rings from the differential supports. Examine the bearing cups and replace if they are rough or pitted or if new bearing cones are to be used on the differential assembly.
8. Remove the differential assembly from the case.

DISASSEMBLY

1. Inspect bearing cones (9) on the differential gear assembly and gear carrier. If the cones require replacement remove with a suitable bearing puller.
2. Remove the pinion shaft retaining rings (12) and lock plates (13).
3. Remove six capscrews (10) and lockwashers to separate the gear carrier (24) from the gear assembly (15).
4. Dowel pins (14) are pressed into the gear assembly. Unless replacement is necessary do not remove.

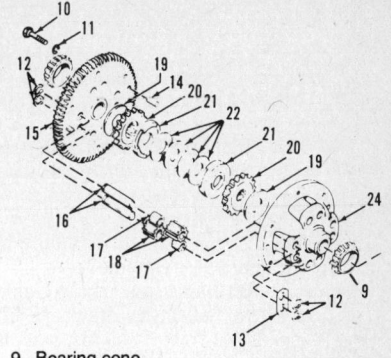

9. Bearing cone
10. Capscrew
11. Lockwasher
12. Retaining ring
13. Lockplate
14. Pin
15. Gear assembly
16. Shaft pinion
17. Spacer
18. Pinion gear
19. Thrust washer
20. Differential gear
21. Differential clutch
22. Spring washer
24. Gear carrier

Differential components

5. Remove the pinions, spacers and shafts from the gear carrier.
6. Remove two thrust washers, gears, differential clutches and Belleville washers from the gear carrier.
7. Remove all the oil from the transmission case. Clean the transmission case and all parts in a suitable solvent. Inspect all parts and replace any that are worn excessively or damaged.

ASSEMBLY

1. If bearing cones were removed from gear assembly hub (15) and gear carrier hub (24) press new bearing cones on the hubs.
2. Install pinion shafts (16) in the gear carrier and secure with lock plates (13) and retaining rings (12).
3. Install in the gear carrier in the following order, one thrust washer (19), one gear (20), one clutch plate (21) with the flat side toward carrier, one Belleville washer (22) with the concave side toward the clutch plate, another Belleville washer with the convex side toward the first Belleville washer, install the third Belleville washer with the concave side toward the second Belleville washer, install the clutch plate with the flat side away from the Belleville washers, install the differential gear and thrust washer as shown.
4. Install one pinion gear (18) on four alternate shafts, then place one spacer on each shaft.

Install one pinion gear on each of the four remaining shafts that are without pinions.

5. Align the gear assembly (15) on the carrier assembly so that the dowel pins in the gear align with the holes in the carrier.
6. Secure the gear assembly to the carrier with the six capscrews.
7. Install the retaining rings (12) on the gear assembly ends of the pinion shafts.
8. Install the differential assembly in the transmission case.
9. Press bearing cup in the differential supports.
10. Install new oil seals in the differen-

Allis-Chalmers

tial supports. Lubricate the O-rings and oil seal with Lubriplate®. Install differential supports in case.

11. Replace the brake stationary disc studs and secure each carrier with two capscrews and lockwashers.

12. Replace the top PTO shaft with new oil seal. Install the electric clutch hub. Use Loctite® on the hub attaching capscrew. Install the front PTO shaft.

13. Install the transmission assembly in the tractor frame if it was removed.

14. Install the brake assemblies, axle extensions, drop housings, drawbar, lift shaft, ram, oil filter, wheel shields, wheels and temperature sending unit.

15. Replace the transmission cover.

16. Fill transmission to correct level with Dexron® power fluid 401. Approximately 6 quarts of 401 power fluid will be required to fill the transmission.

17. Adjust the hydrostatic transmission control lever so that the tractor does not move when the lever is is neutral. Refer to the operators manual for correct adjustment.

18. Adjust the brakes for proper pedal free travel.

19. Replace the top cover, bottom cover, seat with support, connect clutch, electrical connections, light wire connectors and any items removed.

Shift Assemblies

DISASSEMBLY

If the complete transmission case is to be removed from the tractor frame the hydrostatic pump and motor assembly can be removed from the transmission case after the complete assembly is removed from the tractor frame.

1. Disconnect the hydrostatic oil lines, control linkage and the temperature sending unit wire.

2. Remove the wheels, drop housing, axle extensions and brakes.

3. Remove the ten washer headed capscrews that attach the bottom cover to the tractor frame and remove cover. Disconnect clutch shaft at U-joint.

4. Place a pan under the front of the transmission case, remove the temperature sending unit to drain the transmission oil.

5. Disconnect the front PTO shaft by removing the four hub attaching capscrews.

6. Remove the transmission assembly from the tractor and place in a suitable working area.

7. Remove the hydrostatic pump and motor from the transmission case. For hydrostatic pump and motor information refer to 616/620 tractor section.

8. Remove the top PTO shaft and differential assembly (Refer to Differential and Transmission Section) from the transmission case.

9. Remove capscrew (21) and lockwasher attaching shift stem (2) to shift shaft assembly and remove shift stem.

10. Remove shift lever assembly (22) and seal (19). Discard oil seal.

11. Remove two capscrews (13) and washer seals (14) attaching the shift guide (3) to the differential case.

12. Remove the shift guide and two spacers (18).

13. Remove two capscrews (11) and lockwashers attaching shift shaft assembly to transmission case.

14. Pull the shift shaft assembly (10) from the case and the two shifter forks (4-7). Remove the shaft from the forks carefully so that the detent springs (5) and balls (6) encased in the forks are not lost.

15. Remove the O-rings and gasket from the shift shaft assembly and discard.

ASSEMBLY

1. Install new O-rings (8-9) on shift shaft assembly (10) and lubricate. Apply a light coat of Permatex® to gasket (15) and install on shifter shaft plate. Start the shifter shaft assembly through the bore in lefthand side of the transmission case.

2. Install the detent spring (5) and ball (6) in the third speed shifter fork (7) and place the groove in the shifter fork over the 20T gear. Hold the detent ball and spring far enough in the drilling in the shifter fork to allow the shifter shaft to pass through the bore in the fork (7). Move the shaft through the fork a sufficient distance to retain the detent ball and spring.

3. Install the detent spring (5) and ball (6) in the drilling in first and second shifter fork (4) and install the shifter fork in the groove around the first and second speed gear.

4. Hold the detent ball and spring in the shifter fork so that the shifter shaft can move through the fork until the plate and gasket contact the transmission case. Coat the capscrew threads with a sealer, install lockwashers and secure. Secure shifter shaft assembly to the transmission case with the two capscrews.

5. Install seal washers on two capscrews (13) and start them through the differential case. Install one spacer (18) on each capscrew. Install the shift guide in the case and secure with the two capscrews (13).

6. Install the differential assembly in the tractor.

7. Install top PTO shaft in transmission case. Install new PTO shaft oil seal. Use Loctite® on clutch hub capscrew.

8. Install hydrostatic pump and motor assembly on the transmission case.

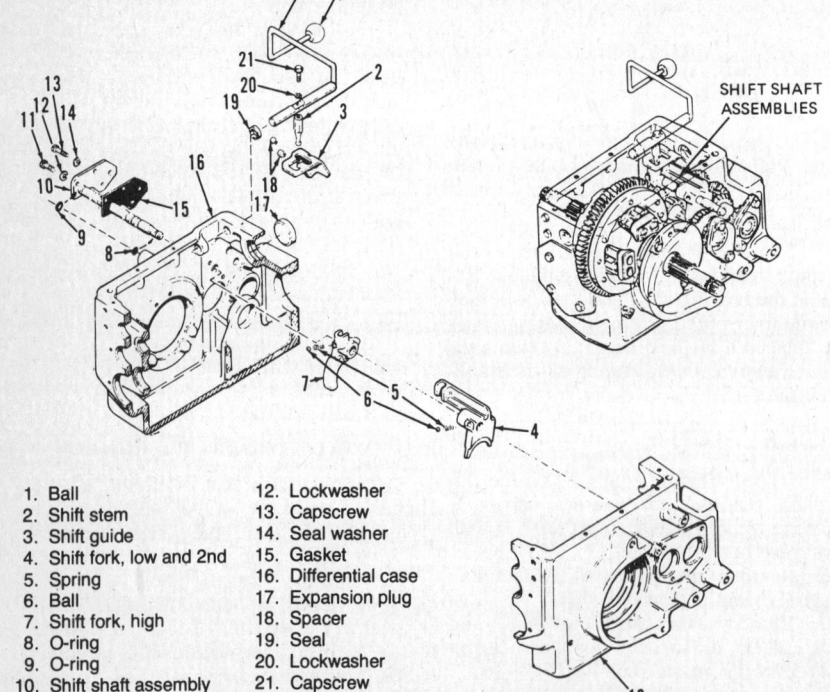

1. Ball
2. Shift stem
3. Shift guide
4. Shift fork, low and 2nd
5. Spring
6. Ball
7. Shift fork, high
8. O-ring
9. O-ring
10. Shift shaft assembly
11. Capscrew
12. Lockwasher
13. Capscrew
14. Seal washer
15. Gasket
16. Differential case
17. Expansion plug
18. Spacer
19. Seal
20. Lockwasher
21. Capscrew
22. Shift lever assembly

Shift assembly components

ENGINE

This tractor uses the Onan CCKB engine.

NOTE: For engine service, refer to the Onan Engine part of the Engine Unit Repair section.

REMOVAL

1. Remove the hood by removing the two shoulder capscrews from the hinge assembly. Remove the hood.

2. Remove the battery cover, disconnect the battery ground cable and the positive cable. Remove the battery clamp to remove the battery.

3. Disconnect the throttle and choke cable from the engine. Disconnect the

electrical connections at the front of the engine. Disconnect the fuel lines at the fuel tank. The clamps should be moved off of the end of the hose to prevent over-stretching the clamps.

4. Remove the capscrews and washers that fasten the oil cooler to the shrouds and move the oil cooler away from the shrouds. Remove the six capscrews that attach the righthand, lefthand and center shrouds to the engine and remove the shrouds.

5. Remove the two capscrews, washers, lockwashers and spacers that fasten the rear PTO yoke to the engine flywheel if the tractor is equipped with a PTO.

6. Hold the clutch pedal forward and remove the two clutch belts from the engine pulley.

7. Remove the cotter pin from the clutch rod and allow the clutch assembly to drop to the frame. Remove the bottom cover from the tractor frame.

8. Remove the engine mounting capscrews from the tractor frame. Four capscrews are used on the 720.

9. Attach a chain or web to the engine as shown, and raise the engine with a hoist to move to a suitable work area.

10. Check the condition of the clutch belts and sheaves and replace if necessary before installing the engine.

INSTALLATION

1. Position the electrical wires, throttle cable and choke cable so they will not be damaged during engine installation. Install the engine in the tractor frame and replace the engine mounting capscrews, lockwashers and nuts. Move the engine as far forward as possible and torque the mounting capscrews to 30 ft.lb.

2. Attach the front PTO yoke to the flywheel with the spacers, washers, lockwashers and capscrews. Apply Loctite® to the capscrews. Torque the capscrews sufficient to prevent the yoke from moving freely on the flywheel.

3. Rotate the engine and tap the coupling to align the shaft with the engine flywheel. A dial indicator should be used to align the shaft with the engine flywheel. Torque the capscrews to 25-30 ft.lb. Do not over torque.

4. If the yoke has a setscrew, torque the setscrew. Torque the center bearing carriage bolts after the shaft is aligned and replace the top cover. If the PTO yoke has a lube fitting lubricate with automotive type lube.

5. Attach the clutch rod to the clutch pedal arm. Hold the clutch pedal forward and install the two V-belts. Refer to the operators manual for clutch spring adjustment and pedal free travel. Replace the bottom cover

6. Install the righthand, lefthand and upper engine shrouds.

7. Install the oil cooler. A capscrew and special washer is used to fasten the righthand end of the oil cooler.

For the lefthand end of the cooler use two neoprene washers and one special steel washer. Place one neoprene washer between the cooler tube and the engine shroud. Place a steel washer on the capscrew and one neoprene washer between the steel washer and the cooler tube. Torque the two capscrews.

8. Attach the fuel lines to the fuel tank and secure with clamps.

9. Connect the electrical wires at the right front of the engine.

10. Install the battery and fasten with clamp at the front of the tractor frame. The long battery cable is attached to the starter solenoid and must be attached to the positive (+) battery terminal. The short cable is attached to the negative (−) battery terminal and tractor frames. Install the battery cover and secure with two springs.

11. Install the choke and throttle cables. Refer to the proper engine service section for correct cable installation and adjustments.

808GT, 810GT

NOTE: These models are essentially similar to the 608GT, 610GT series. The procedures given below cover those aspects which differ among the models. For all procedures not covered here, see the 608GT, 610GT tractor section.

WHEELS

Front Wheels

REMOVAL

1. Jack up the front of the tractor and safely block it before removing the wheels.
2. Remove the retaining ring and washer from the axle and remove the wheel.
3. Inspect the wheel bushings for unusual or excessive wear and replace the bushings if necessary. There are two bushings in each wheel.

INSTALLATION

1. Slide the wheel onto the axle shaft, then the flat washer and secure with the retaining ring.
2. Always grease the wheel assemblies when they have been removed.

Rear Wheels

REMOVAL

1. Jack up the rear of the tractor and safely block it before removing the wheels.
2. Remove the cotter pin from the pin and remove the pin.
3. Slide the wheel assembly off of the axle.

INSTALLATION

1. Slide the wheel assembly onto the axle.
2. Install the pin through the hub and axle.
3. Install the cotter pin through the pin and bend over.
4. There are no bushings or grease needed for the rear wheels.

SHEET METAL

Hood

The hood hinges at the front of the tractor's frame and rests on the upper dash. If the hood fits too loosely around the edge of the upper dash, the edges can be bent inward to give it a tighter fit over the upper dash.

If the seam between the upper dash and the hood overlaps or has too large a gap the hood can be moved forward or back on the side panel by loosening the four (4) carriage bolts attaching the hood to the bumper. After the hood is properly positioned in the slotted holes, tighten the four (4) carriage bolts that hold the hood to the side panel.

Seat Assembly

To remove the complete seat deck assembly remove bolts (A). When installing seat deck assembly tighten locknuts just enough so that deck does not bind. Remove nuts (B) to remove seat deck from the supports. Remove bolts (C) and the two handles to remove the seat from the seat deck. To adjust the seat, loosen the two handles and slide the seat forward or backward and tighten handles.

Grille Assembly

The grille assembly can be removed or replaced by removing the six (6) screws from the hood. These screws should be torqued to 30 in.lb. The wires to the lights will have to be disconnected before remov-

Allis-Chalmers

ing the grille assembly. The grille screen is removed by removing the seven (7) capscrews from the grille assembly.

Light Assembly

Remove the two (2) screws from the grille to remove the light assembly. Remove the light assembly to replace the bulb or lens.

Upper Dash and Fuel Tank Removal

1. Disconnect the battery ground cable from the battery.
2. Remove the fuel line at the carburetor and drain the fuel tank completely.

CAUTION

Do not drain fuel from the tank when engine is hot, while smoking, near an open flame, near anything that is hot, or in a closed building.

---

3. Remove the throttle and choke cable and ground wire from lefthand side of the engine.
4. Remove the steering wheel by removing the steering wheel cap, locknut and concaved washer.
5. Remove the throttle and choke knob. Heating the knob with a light bulb for 5-10 minutes will aid in removing. Do not let the light bulb rest on the dash.
6. Remove the two (2) screws that hold the upper dash to the dashboard brace, and the two (2) screws that hold the upper dash to the lower dash.
7. Carefully raise the upper dash a few inches to gain access to the wires. Pull the socket off of the ignition switch or remove the ignition switch from the dash.
8. Pull the clips from the light switch terminals, ammeter terminals and the two (2) neutral indicator light sockets.
9. Raise the dash further while feeding the fuel line and ground wire through the lower dash. Remove the upper dash and fuel tank.
10. The dashboard brace or bearing assembly can now be replaced, by removing the two (2) capscrews.
11. When installing the dashboard brace and bearing assembly put the flat washer next to the lower dash.

FUEL TANK INSPECTION

1. If fuel tank is leaking or damaged, do not repair it. Replace it with a new tank.
2. Replace the fuel line if it is hard and inflexible.
3. Clean all dirt and sediment from the tank.
4. Replace the fuel cap if it leaks or is damaged.

Fuel Tank and Upper Dash Installation

1. Install the fuel tank in the reverse order of removal and make sure the two (2) pads are in place and that the fuel line is securely attached to the tank fitting.
2. Feed the fuel line through the hole in the lower dash and the ground wire and throttle cable through the hole in the upper dash.
3. Install the ignition connector onto the ignition switch, the orange wire indicator light socket in the top hole and the yellow wire indicator light socket in the bottom hole. The black wire to the righthand side of the light switch and the red wire to the lefthand side. The red wire from the ignition connector connects to the righthand side of the ammeter and the red wire from the main harness to the lefthand side.

NOTE: If the throttle and choke handle was removed from the dashboard brace, install it onto the two (2) studs and put the black ground wire on the lower stud. The connector on this ground wire will have a larger I.D. hole than the black ground wire going to the engine.

4. Secure the upper dash to the dashboard brace with the two (2) screws, and bolt the upper dash to the lower dash in front with the two (2) bolts.
5. Install the throttle and choke knob onto the handle.

NOTE: Knob should be heated in hot water before installation.

6. Install the steering wheel onto the steering shaft. Put the concave washer, with the concave side down, over the shaft and tighten the locknut.
7. Install the steering wheel cap. This is a press fit.
8. Connect the battery ground cable to the battery.

Lower Dash Removal

1. Remove the upper dash as outlined in UPPER DASH AND FUEL TANK REMOVAL.
2. Disconnect the main wiring harness at the connectors and feed the main wiring harness connector down through the hole in the lower dash.
3. Remove the shift panel. There are two capscrews on each side of the panel.

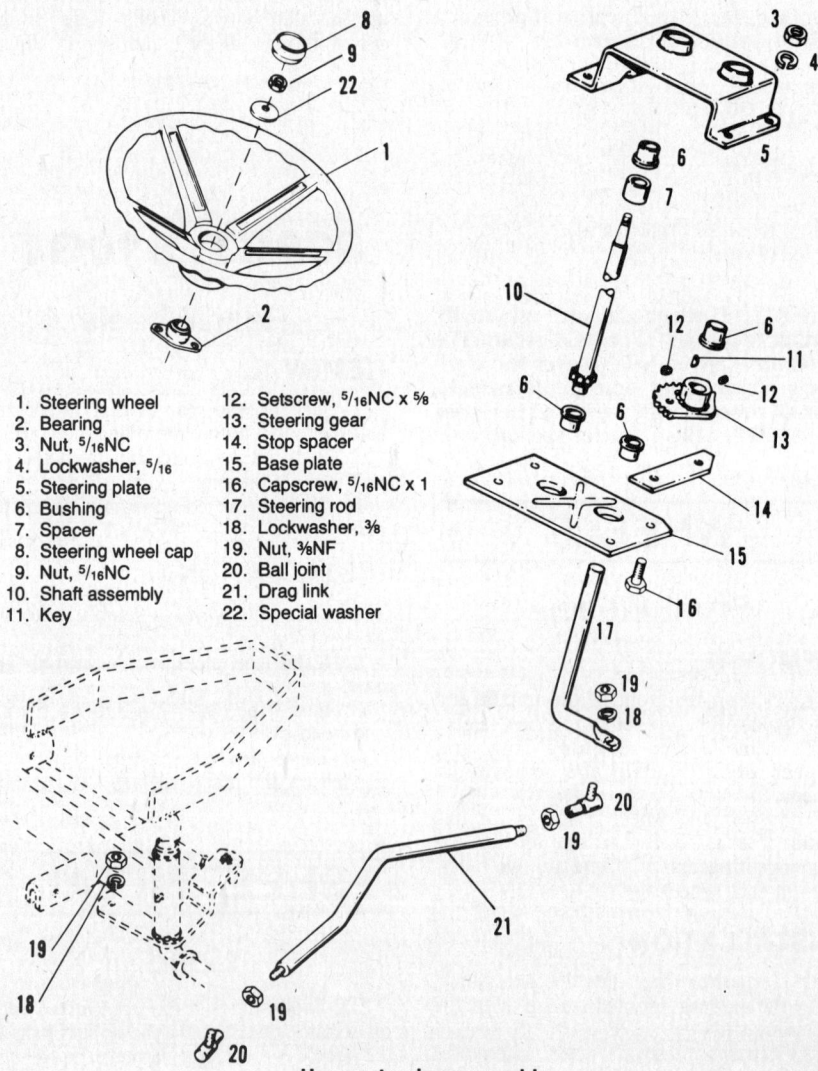

1. Steering wheel
2. Bearing
3. Nut, 5/16NC
4. Lockwasher, 5/16
5. Steering plate
6. Bushing
7. Spacer
8. Steering wheel cap
9. Nut, 5/16NC
10. Shaft assembly
11. Key
12. Setscrew, 5/16NC x 5/8
13. Steering gear
14. Stop spacer
15. Base plate
16. Capscrew, 5/16NC x 1
17. Steering rod
18. Lockwasher, 3/8
19. Nut, 3/8NF
20. Ball joint
21. Drag link
22. Special washer

Upper steering assembly

4. Remove the dashboard brace if repair is needed to brace or bearing assembly and ground wire from the throttle and choke handle assembly.

5. Remove the two (2) rear capscrews that hold down the lower dash and the two (2) front capscrews.

6. Remove the lower dash, while feeding the wires through the holes in the dash.

LOWER DASH INSTALLATION

1. Install dashboard, brace and bearings assembly to lower dash if it was removed.

2. Install the lower dash while putting the wires through the hole.

3. Tighten the four (4) capscrews that hold the lower dash, in the front and rear.

4. Install the ground wire to the throttle and choke handle assembly.

5. Install the upper dash in the reverse order of removal. See Fuel Tank and Upper Dash Installation.

6. Connect all electrical wires, fuel line and choke cable.

7. Connect the battery ground cable to the battery.

STEERING AND FRONT AXLE

Upper Steering Assembly

REMOVAL AND INSTALLATION

1. Remove the upper dash, fuel tank and lower dash as outlined in Upper Dash and Fuel Tank Removal and Lower Dash Removal.

2. Disconnect the drag link from the steering rod.

3. Remove the four (4) capscrews that hold the steering gear assembly to the tractor frame.

4. Check the two (2) bushings in the steering plate for wear or damage and replace if needed.

5. Inspect the steering shaft gear and the steering gear for wear, cracks or broken teeth and replace if needed.

6. Inspect the two (2) bushings in the base plate for wear or damage and replace if needed.

7. When removing the steering gear from the steering rod, use a press or puller. Put a brace between the gear and base plate or the bushing will be damaged when pressing gear out.

8. When reassembling the steering gear to the steering rod, install the key into the key way with the rounded end down towards the bent end of the steering rod.

9. Press gear on until the gear is flush or slightly above the key. If the key is above the gear, it will cut into the upper bushing.

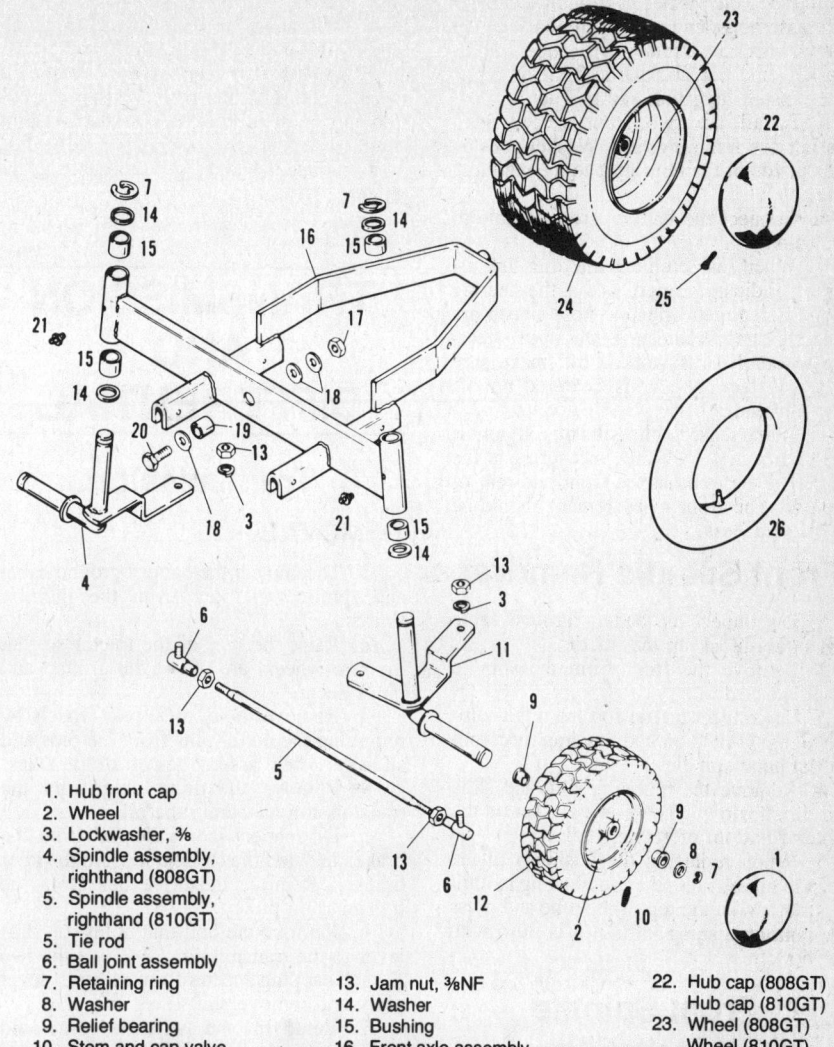

1. Hub front cap
2. Wheel
3. Lockwasher, ⅜
4. Spindle assembly, righthand (808GT)
5. Spindle assembly, righthand (810GT)
5. Tie rod
6. Ball joint assembly
7. Retaining ring
8. Washer
9. Relief bearing
10. Stem and cap valve
11. Spindle assembly, lefthand (808GT)
    Spindle assembly, lefthand (810GT)
12. Tire
13. Jam nut, ⅜NF
14. Washer
15. Bushing
16. Front axle assembly
17. Locknut, ½NC
18. Washer
19. Spacer
20. Capscrew, ½NC x 1½
21. Fitting
22. Hub cap (808GT)
    Hub cap (810GT)
23. Wheel (808GT)
    Wheel (810GT)
24. Tire (808GT)
    Tire (810GT)
25. Valve
26. Tube (808GT)
    Tube (810GT)

Lower steering assembly

10. Torque the two (2) setscrews to 25 ft.lb.

11. Install base plate with steering gear and rod on tractor frame. Install stop spacer, steering shaft and spacer and steering plate. Tighten to tractor frame.

NOTE: Bevel on the steering shaft spacer goes towards the welder gear.

12. Install upper and lower dash as outlined in Fuel Tank and Upper Dash Installation and Lower Dash Installation.

Lower Steering Assembly

1. Examine all parts of the lower steering assembly for damage or wear.

2. Check the drag link and tie rod ball joints for damage or excessive play.

3. Check the spindles and spindle bushings for damage and wear.

Front Axle Assembly Removal

1. Disconnect the battery ground cable and safely block up the tractor.

2. Remove the two (2) front wheels.

3. Disconnect the drag link at the steering rod or lefthand spindle.

4. Remove the bolt from the front axle that secures it to the tractor frame. There is a replaceable spacer inside of the axle. Replace if damaged or badly worn.

5. Pull the axle assembly out of the center pivot and remove.

6. Inspect drag link and tie rod ball joints. Replace where needed.

Front Axle Assembly Installation

1. Lubricate axle pivot stud and stick it into center pivot.

Allis-Chalmers

2. Place on washer in front and back of axle assembly, and one on the back of the tractor frame. Secure with bolt, lockwasher, and nut. Torque to 25 ft.lb.

3. Install drag link and tie rod.

4. Install the two front wheels with washer and retaining ring. Washer goes on the outside of the hub, next to the retaining ring.

5. Connect the battery ground cable to the battery.

6. When reassembling the drag link assembly, adjust its length so that the steering wheel is centered when the front wheels are pointing straight ahead. If the upper steering assembly was worked on, make sure that the steering gear is centered on the steering shaft.

7. When reassembling the tie rod, adjust the toe-in to 1/8 to 1/4 in., measuring at the center of the tread at the front and rear of the tire. The front measurement should be 1/8 to 1/4 in. less.

Front Spindle Removal

1. Disconnect the battery ground cable and safely block up the tractor.

2. Remove the front from the spindle being worked on.

3. Disconnect the tie rod from the spindle being worked on and the drag link from the lefthand spindle.

4. Remove the retaining ring and slide the spindle down. There is a washer on the top and bottom of each spindle.

5. When replacing the bushings in the spindle tubes, press the top bushing in until it is flush with the top of the tube and press the bottom bushing in until it is flush with the bottom.

Front Spindle Installation

1. Install the spindle with a washer on the top and bottom of the spindle tube. Lock spindle in place with the retaining ring.

2. Install drag link and tie rod. See steps 6 and 7 under Front Axle Assembly Installation for proper adjustments.

3. Install the front wheel with the washer and retaining ring. Washer goes on the outside of the hub, next to the retaining ring.

4. Connect battery ground cable to the battery.

TRANSMISSION AND DIFFERENTIAL

Transmission

REMOVAL

1. Disconnect the battery ground cable and remove the key from the ignition switch.

2. Raise the rear of the tractor so that the rear wheels are off of the ground and block securely.

3. Remove the two (2) rear wheels by removing the cotter pins from the pins and slide the wheel assemblies off of the axles.

4. Remove the drain plug from the transmission and drain the oil.

5. Disconnect the brake rod from the brake band and the clutch rod from the idler bracket. Remove the drive belt from the transmission pulley.

6. Remove the bolt that holds the shift lever to the transmission. Remove the two (2) self-tapping screws from the shift lever pivot and remove shift lever and pivot.

7. Removing the switch lever will aid in removing the two (2) transmission support bolts. Remove the switch lever bolt and switch lever.

8. Remove the two (2) nuts from the axle bolt that secure the righthand axle to the tractor frame.

9. Remove the two (2) self-tapping bolts from the transmission support plate. The support plate should not be removed from the transmission.

10. Remove the four (4) frame support bolts or the two (2) transmission support bolts and lower the transmission sliding it to the left and remove it from the tractor. If the two (2) transmission support bolts are removed, they will have to be sealed when reinstalling, because the holes are tapped into the transmission case.

DISASSEMBLY

1. Remove the brake drum by loosening the two setscrews and sliding the brake drum off the shaft and away from the transmission. Remove the transmission pulley by loosening the setscrew and sliding off the shaft. Remove the keys from the shafts.

2. If the transmission support plate is removed, put a screwdriver in the slotted end of the shaft to keep it from turning when removing the nut. If the shaft rotates, the lockout for the spring will be wrong.

3. Using fine emery cloth, remove any burrs and paint from the shafts and axles, so that the bearings will not be damaged.

4. Drive out the two roll pins located at the top and bottom of the transmission. Be sure the pins are driven completely out of the transmission cover. Remove the eight capscrews and lockwashers securing the transmission case and separate cover from case. Place a pan beneath case to catch residue.

Differential

REMOVAL AND DISASSEMBLY

1. Remove the righthand axle shaft and the driven gear assembly from the gear case.

2. DO NOT take the differential apart unless it is damaged.

3. Remove the low gear assembly from the case, and the two recessed washers.

4. Inspect the shaft and gear for wear. Replace any damaged parts.

5. Remove the hex nut securing the reverse pinion shaft from the outside of the transmission case. Grasp the reverse pinion shaft and pull the reverse pinion shaft out of the transmission case.

6. Inspect the shaft and gear for wear. Replace any damaged parts.

7. Remove the hex nut securing the reverse-low shaft from outside the righthand side of transmission case. Remove the shaft with reverse shift fork from transmission case.

8. Inspect the shaft and fork assembly for wear. Replace any damaged parts.

9. Remove the hex nut securing the high-low shaft from outside the righthand side of transmission case. This is done so that the high-low pinion can be removed with the shaft and fork assembly.

10. Remove the shaft, fork and PTO shaft from the case together.

Transmission mounting points

Allis-Chalmers

11. Remove three capscrews and lockwashers securing the axle housing to the transmission case. Remove the axle housing assembly from transmission.

12. Remove the two oil seals and O-ring and, using a bearing puller, remove both roller bearings from the axle housing. Replace old oil seals with new. Refer to Bearings and Oil Seals for bearing Cleaning and Inspection.

Bearings and Oil Seals

1. Remove and replace with new oil seals.

NOTE: Seals are to be replaced after shafts and axles are installed.

2. Remove all roller bearings from transmission cover, case, and if not already removed, from axle housing.

3. Thoroughly clean by soaking in an approved solvent all bearings. Bearing must soak long enough to loosen all grease and dirt. After all visible dirt is removed, rinse bearing in a clean container of clean solvent, and immediately dip in oil or light grease.

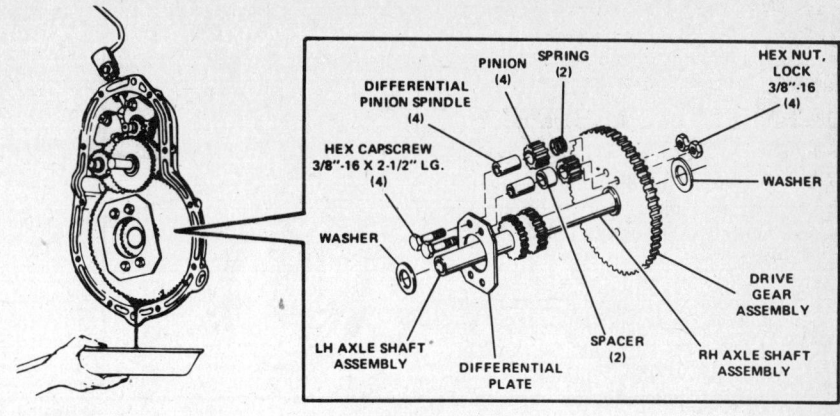
Differential removal

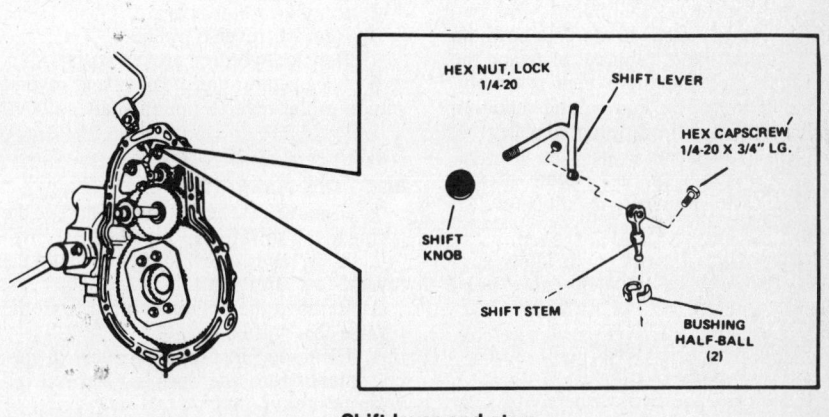
Shift lever and stem

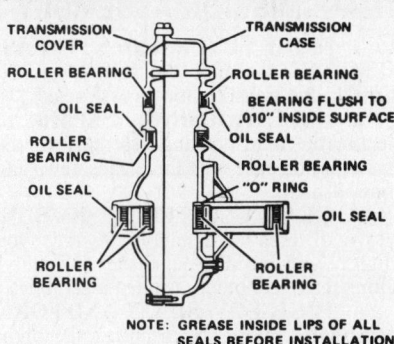
Bearings and seal

4. Inspect bearings for broken or cracked races, dented seals or shields, cracked or broken separators, balls or rollers. If bearing has been overheated, it will be brownish blue or blue-black color. If bearing is found to have any of the above conditions, replace it with a new one.

5. Bearings are to be recessed 0.010 in. from the machined surface inside of the case and cover.

Shift Lever and Stem Assembly

1. Remove capscrew and locknut securing the shift lever to the shift stem.
2. Shift knob already removed.
3. Remove the shift stem assembly and the two half-ball bushings.
4. Inspect and replace worn or broken parts.
5. Install the two half-ball bushings and the shift stem in the transmission case.
6. Secure the shift lever to the shift stem with a capscrew and new locknut.

Roll Pins

1. Check two roll pins for damage. One is located in the transmission case and the other in the transmission cover.
2. If replacement is necessary the proper installation is with the groove up and a distance of 0.808 in. (+0.000-0.010) from the face of the transmission cover to the end of the roll pin subassembly. The roll pin subassembly and the transmission case is properly installed with the groove up and a distance of 1.180 in. (+0.010) between the end of the roll pin subassembly and the face of the transmission case.

Cleaning

Clean all transmission assemblies, cover, and case with an approved solvent. Dry thoroughly. Do not use high pressure air on bearings.

2nd/3rd gear assembly

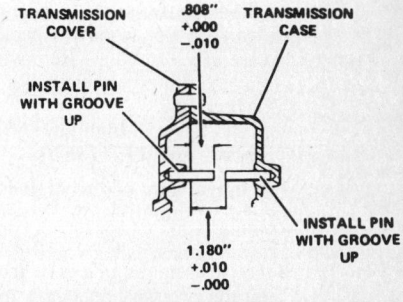
Roll pin installation

Allis-Chalmers

Pipe Plugs

Check, clean, and replace with new, if necessary, the three pipe plugs in the transmission case.

DIFFERENTIAL ASSEMBLY

1. Remove the four capscrews and locknuts securing the differential spindles, four differential pinions, two differential spacers, and two springs to the drive gear assembly.

2. Replace worn or broken parts.

3. Assemble the righthand axle assembly through the drive gear assembly and butt the gear of the lefthand axle assembly against the gear of the righthand axle assembly before mounting the differential pinions, spacers, springs and capscrews.

4. Assemble differential assembly with the righthand top and lefthand bottom capscrews holding two differential spindles, two differential pinions and two springs with the heads of the capscrews on the outside of the differential plate and the springs against the drive gear assembly. Assemble the left top and righthand bottom capscrew through the differential plate, two spindles, two spacers against the differential plate and two pinions. Secure all four capscrews with four new locknuts torqued to 20 ft.lb.

NOTE: Under torque will allow slippage and over torque will cause hard steering due to lack of differential action.

2nd - 3rd Gear Assembly

1. Remove the two washers.
2. The brake shaft and gear cluster assembly is best replaced as a complete assembly.
3. Replace broken or worn parts.
4. Slide gear cluster assembly over the roll pin and on the brake shaft.
5. Replace the two washers, one on each side of the gear cluster assembly.

Low Gear Assembly

1. Remove the two washers.
2. Remove the two retaining rings.
3. Replace broken or worn parts.
4. Secure the two retaining rings and washers to the gear and shaft assembly with the washers to the outside of the retaining rings.

High-Low Pinion Assembly

1. Remove the two washers.
2. Remove the two retaining rings.
3. Slide the high-low pinion assembly over the key and off the shaft.
4. Remove the key.
5. Replace worn or broken parts. Check the key way carefully for burrs. Remove any burrs with a fine grit emery cloth. Replace broken or bent key.
6. Assemble the key and high-low pinion assembly on the pulley PTO shaft.

NOTE: The longer end of the shaft is facing outward (transmission cover side).

7. Secure the two retaining rings to the shaft and mount the two washers outboard of the retaining rings.

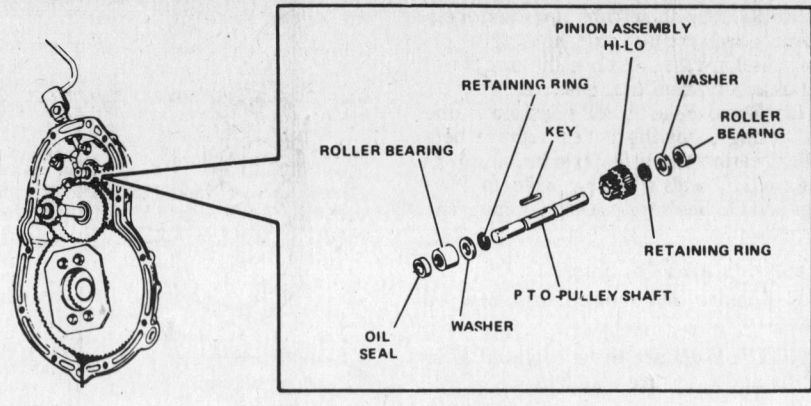

High-low pinion assembly

Reverse Pinion Assembly

1. Remove O-rings. Replace with new.
2. Remove roller bearing.
3. Remove reverse pinion.
4. Replace worn or broken parts.
5. Secure the roller bearing and reverse pinion to the reverse pinion shaft.

Reverse-Low Shaft and Fork Assembly

1. Remove the capscrew holding the spring to the shift fork. Remove the spring.
2. Slide reverse-low shifter fork off the reverse-low shift shaft.
3. Remove the ball and spring from the reverse-low shifter fork.
4. Clean and inspect all parts with special attention to the springs. Replace all worn or broken parts.
5. Install the spring and the ball into the reverse-low shift fork. Slide the shift fork onto the reverse-low shifter shaft.
6. Install the spring and secure with capscrew.
7. Install the retaining ring on the reverse-low shifter shaft.

High-Low Shaft and Fork Assembly

1. Remove the capscrew holding the spring to the high-low shifter fork. Remove spring.
2. Slide the high-low shifter fork off the high-low shift shaft.
3. Clean and inspect all parts with special attention to the springs. Replace all worn or broken parts.
4. Install the spring, then install the ball in the high-low shift shaft.
5. Install spring and secure with capscrew.

TRANSMISSION ASSEMBLY

1. **SHIFT LEVER AND STEM ASSEMBLY:** Assemble the shift lever and stem on the transmission as instructed.

2. **AXLE HOUSING ASSEMBLY:** Secure the axle housing assembly to the transmission case with three capscrews and lockwashers.

3. **GEAR CLUSTER ASSEMBLY:** Move differential assembly towards you slightly and install gear cluster assembly, tilting it to clear other installed assemblies.

4. **HIGH-LOW SHAFT AND FORK ASSEMBLY:** Before installing the high-low shift fork be sure the flat spot on the shift shaft is turned so the spring rests in it. When securing the shaft to the transmission case and support plate it is very important that this shaft not be allowed to turn. Use a screwdriver on the end of the shaft to hold it from turning. Use a new hex nut torqued to 50 ft.lb. to secure the high-low shaft and fork assembly to the transmission case and righthand side plate.

5. **REVERSE-LOW SHAFT AND FORK ASSEMBLY:** Before installing the reverse-low shift fork be sure the flat spot on the shift shaft is turned so the spring

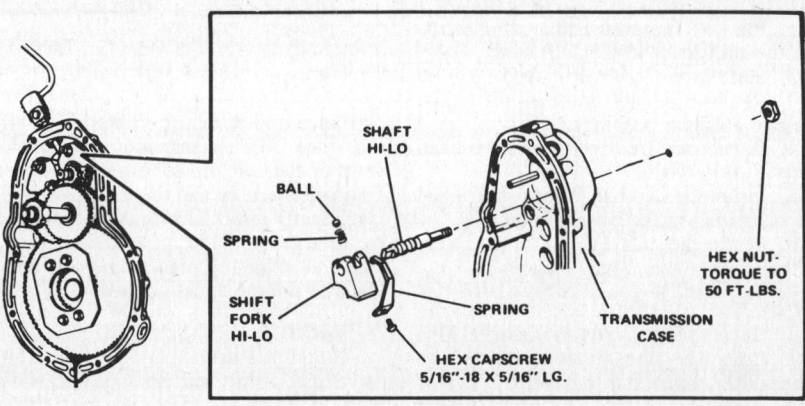

High-low shift fork assembly

rests in it. When securing the shaft to the transmission case and support plate it is very important that this shaft is not allowed to turn. Use a screwdriver on the end of the shaft to hold it from turning. Use a new hex nut torqued to 50 ft.lb. to secure the reverse-low shaft and fork assembly to the transmission case.

NOTE: After each fork is installed, check to see that there is clearance between it and the transmission case to allow easy movement in all positions. It may be necessary to grind away part of the case in some instances.

6. **REVERSE PINION ASSEMBLY:** Install the reverse pinion assembly in the transmission case and secure it with a hex nut, torqued to 50 ft.lb. on outside of case.

7. **HIGH-LOW PINION ASSEMBLY:** Mount the PTO pulley shaft through the transmission case carefully to prevent damage to bearing and oil seal. Tilt the high-low shifter fork out of the way for ease of removal.

8. **GEAR AND SHAFT ASSEMBLY:** Carefully insert gear and shaft assembly into transmission case bearing meshing its gears with the already installed gear assemblies. Be sure smaller gear faces out or away from case.

9. **DIFFERENTIAL ASSEMBLY:** Mount the differential assembly with the righthand axle assembly through the transmission case and axle housing assembly.

10. Carefully slide lefthand axle into transmission cover and install washer on axle.

11. Install a new gasket on transmission case. Drive two roll pins in transmission cover.

12. Secure the transmission cover on the case with eight capscrews and lockwashers.

13. Install the brake drum with key in place and secure with two setscrews.

14. Install the transmission pulley with the key and tighten.

TRANSMISSION INSTALLATION

1. Stick the righthand axle through the hole in the tractor frame and start the two (2) self-tapping bolts into the transmission support plate.

2. Install the four (4) frame support bolts or two (2) transmission support bolts and tighten. If the two (2) transmission support bolts were removed, use a good sealer on the bolts before installing.

3. Tighten the two (2) self-tapping bolts the rest of the way and tighten the axle bolt on the righthand axle.

4. Install the switch lever, shift lever pivot, and shift.

5. Secure the shift lever to the transmission with the bolt and locknut.

6. Install the brake rod to the brake band and the clutch rod to the idler lever.

7. Install the drive bolt and two rear wheel assemblies and lower the tractor to the ground.

8. Remove the fill plug and oil level screw and fill with Gear Lube 715 or equivalent, and install the plug and screw. Remove the draw bar to expose the fill plug.

9. Connect the battery ground cable and test run the tractor. Readjust the brake and clutch rods.

ENGINE

For detailed engine overhaul, see the Briggs & Stratton engine part of the Engine Unit Repair Section. These tractors are equipped with the models 191707 and 251707 engines.

ENGINE REMOVAL

The tractors have a larger hole in the frame for the crankshaft, allowing the engine to be removed from the tractor before removing the drive pulleys.

1. Set the parking brake and block the tractor securely so that it does not roll.

2. Disconnect the battery ground cable from the battery.

3. Remove the fuel line at the carburetor and drain the fuel tank completely.

—— **CAUTION** ——
Do not drain fuel from the tank when engine is hot, while smoking, near an open flame, near anything that is hot, or in a closed building.

4. Remove the throttle cable, engine ground wire, charging circuit wire and starter wire from the engine, and the drive belt from the main drive pulley.

5. Remove the four (4) capscrews that hold the engine to the frame and front brace and remove the engine. On 810GT tractors, removing the muffler will aid in removing the engine.

6. Remove the capscrew and lockwasher attaching the pulley assembly to the engine crankshaft.

7. Pull the engine pulley off of the crankshaft with a puller. Tap the pulley with a soft hammer just below the tractor drive pulley if needed.

ENGINE INSTALLATION

1. Install the engine in place on the frame. The drive pulley assembly can be installed on the crankshaft before engine installation. Torque the drive pulley capscrew to 55 ft.lb.

2. Install the four (4) capscrews and four (4) special washers that hold the engine to the frame. Make sure that the washers and belt stops are installed properly. On the two rear capscrews the belt stops go between the special washer and flat washer. The special washers are installed next to the tractor frame. Torque the four (4) capscrews to 25 ft.lb.

3. Attach the throttle cable, engine ground wire, charging circuit wire, starter wire and fuel line.

4. Connect the battery ground cable to the battery.

Allis-Chalmers

900 Series
Models 910 6-speed, 912 Hydro, 914 Shuttle, 916 Hydro, 917 Hydro

FRONT WHEELS AND AXLE

Front Wheel and Tire Assembly

REMOVAL

1. To avoid getting dirt in the bearings, the wheels and axle should be cleaned first. Use a forceful stream of water directed against the tire treads and sidewalls and both sides of the wheels and hubs, as well as the axle assembly.

2. Park the tractor on a level surface. Set the parking brake, and place all other controls in the "off" or "neutral" position. Remove the ignition switch key.

3. Jack or hoist the front of the tractor up so the front axle assembly and wheels are not supporting the tractor.

NOTE: Do not raise the front wheels more than two feet unless the battery has been removed.

4. Throughout the following procedures, the terms right, left, up, down, front, and rear are used viewed from normal driving position.

5. The dust caps are retained by friction and must be pryed out (if a dust cap pliers is available, removal is made easy). Start separation by carefully driving a large, blunt chisel into the groove (A) between the end of the hub (B) and the flange of the dust cap (C). Hold the chisel against the wheel rim so the angle of entry will be as shallow as possible. Avoid cocking the

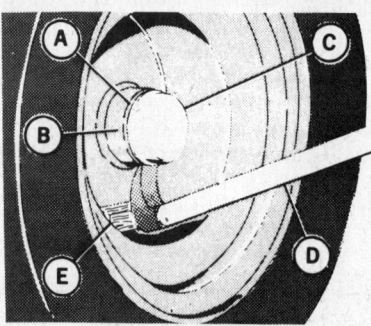

Removing dust cap

cap. Turn the wheel between successive blows to distribute the pressure uniformly around the groove.

6. When the groove has been opened enough to admit the nib of a wedge bar (D), substitute the latter for the chisel. Reset the heel of the wedge bar on a block of wood (E) about 1¼ in. high, laid against the wheel web. Rotate the wheel between successive applications of the wedge bar so the cap will not become cocked.

7. Set dust cap aside with open end up. With an Allen wrench, loosen the setscrew in the set collar. Slip off the collar and washer or washers. Pull the wheel out a half in. or so, and then push it back to initial position, leaving the outside, roller bearing cone accessible. Remove the cone, and place the collar, washers, and cone in the up-ended dust cap.

8. Pull off the wheel. Lay it aside with the dust cap and contents. Identify these parts as belonging to the right or left side of the tractor for reference at assembly.

9. With a rag, wipe the grease off of the outer end of the spindle. Wrap a clean cloth around the inner bearing cone and seal. With a fine file, remove any burrs raised on the spindle around the score mark left by the collar setscrew. Wipe the filings off of the spindle and remove the cloth from the bearing and seal.

10. Using finger pressure only, try to remove both inner bearing cones and seals. If they respond, lay them aside with the other parts for their respective spindles. If not, remove them as described in steps 11 through 13.

11. Insert a screwdriver blade between the spacer collar and the adjacent structure of the spindle assembly. Pry the collar outward so it forces the seal and bearing cone to move with it. Do not pry against the seal or bearing cone.

— CAUTION —
Do not touch either the seal or the cone with either the wooden block or the hammer.

12. When the spacer has been well started by prying drive it off, using a wooden block and a hammer. Distribute the blows as uniformly as possible around the spacer collar, so it will not tend to cock and pinch.

13. If the movement becomes sufficiently free to respond to finger pressure, remove them manually. Lay the bearing cone, seal, and spacer aside with the other parts of the same wheel and tire assembly.

14. If required, repeat steps 5 through 13 for the other wheel.

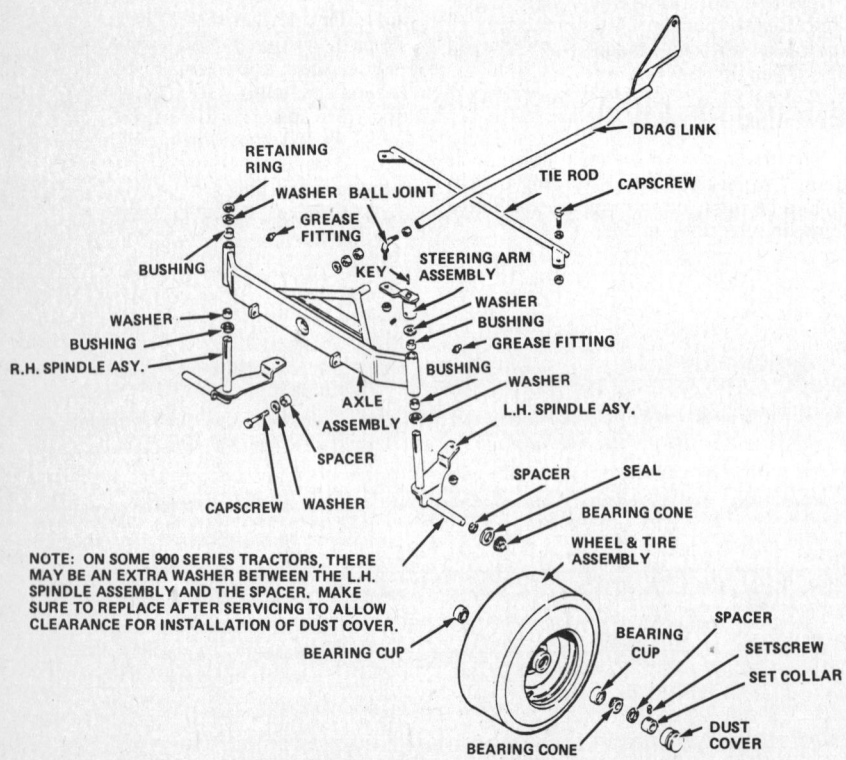

Front axle, wheels and tie rod

Allis-Chalmers

CLEANING

1. Using a grease solvent, clean all of the parts individually. Remove any caked deposits of grease and dirt. Flush all of the grease out from between the rollers in the bearing cones. Flush all of the grease out of the interior of the wheel hubs, particularly the bearing cups.
2. When the parts are clean and dry, examine them carefully for wear and damage.

CAUTION
Do not spin bearings with compressed air.

INSPECTION AND REPAIR

1. Examine the rollers of the roller bearings and the mating surfaces of the cones and cups for discoloration due to overheating, scoring, spalling, cracking, or chipping. If any defects are present on one part of a bearing, discard the entire cone and cup assembly and replace them with a new set.
2. If a bearing cup (in a wheel hub) is to be removed, use a bearing puller. Do not try to remove the cups with a rod and hammer.
3. Because of the importance of the seals in retaining grease and excluding dirt, and the difficulty of detecting minor wear or damage, they should be replaced with new seals.
4. Inspect tire for wear and tire valve for leakage. Replace as required. Inflate tire to specified pressure.

INSTALLATION

1. If a wheel bearing cap is to be replaced, press or drive the replacements into the wheel hub using a rod slightly smaller than the seat bore. Do not continue forcing a cup if it becomes cocked. Remove it, and start over. The smaller diameter of the cup must be placed inward. Seat the cup solidly against the inner shoulder of the bore.
2. If not already installed, assemble the seals and inner bearing cones on both spindles. Place flanged side of seal next to the spacer collar and the largest end of the bearing cone next to the seals.

NOTE: If force is required, use a hammer and a brass or aluminum tube against the outer face of the cone. The tube should have an inner diameter slightly larger than the spindle and a wall thickness no greater than 1/8 in. Drive the cone in until the seal is firmly confined between the spacer collar and the cone.

3. Lubricate the two inner bearing cones with prime quality wheel bearing grease. Work the grease thoroughly into the spaces between the rollers. Extend the grease a distance out on the spindles and spread a layer of grease over the portion of the seal that the wheel hub runs on.
4. Replace the wheels. The tire air valves should be on the inside nearest the tractor frame.

5. After placing each of the outer bearing cones partly on its spindle, lubricate it with grease as explained in step 3.
6. Assemble the spacer washers and the set collars. Spin each wheel while urging the set collar inward, and then, with an Allen wrench, tighten the setscrew in each collar. Spin the wheels again, and then test the fit by urging the tire tread back and forth sideways in a wobbling motion. If side play is more than barely perceptible, reposition the set collar.
7. Replace the dust caps, with a rubber mallet applied near rim of the caps while spinning the wheels. If a cap becomes cocked, do not force it; remove it and start over.
8. Lower the front of the tractor.

Right Front Spindle Assembly

REMOVAL

1. Remove right front wheel and tire assembly.
2. Disconnect the tie rod (B) from the right spindle assembly by removing the hex head capscrew (A) with its bushing, two washers, and nuts. (Mark end of the rod and mating spindle assembly for replacement.)
3. Remove the retaining ring (C) and the plain washer immediately under it. Pull the pintle (D) of the spindle assembly out of its bushings in the axle assembly. Remove the lower plain washer from the pintle.
4. Place all of the parts removed from the right end of the axle assembly together and identify them for future reference.

CLEANING

Using a grease solvent, clean all parts. Flush all grease out of pintle bearings in axle assembly.

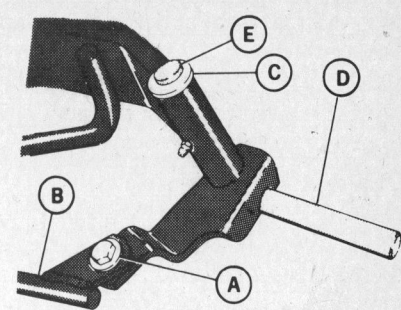
Right front spindle

INSPECTION AND REPAIR

1. Insert bushing into tie rod end lug and shake from side to side. If excessive play exists, discard bushing and tie rod.
2. Push the pintle of each spindle assembly temporarily back into its bushings in the end of the axle assembly. Test for wear by shaking the pintle sideways, noting the extent of the side play. The running fit allowance in these bushings is normally rather generous. The amount of shake, even with no wear, will be quite perceptible. If there is appreciable rattle, however, the bronze bushings should be replaced. If they are to be removed, press or drive them out using a rod slightly smaller than the seat bore.
3. Examine the pintle areas that run in the wheel bearings. If there is appreciable wear, or scoring indicative of seizing or galling, discard the spindle assembly.

INSTALLATION

1. Put a plain washer on the pintle (D) of the righthand spindle assembly, and insert the pintle upward into the bushings of the axle assembly. Put another plain washer on the pintle and then install the retaining ring (C).
2. Slip a plain washer and a spacer bush-

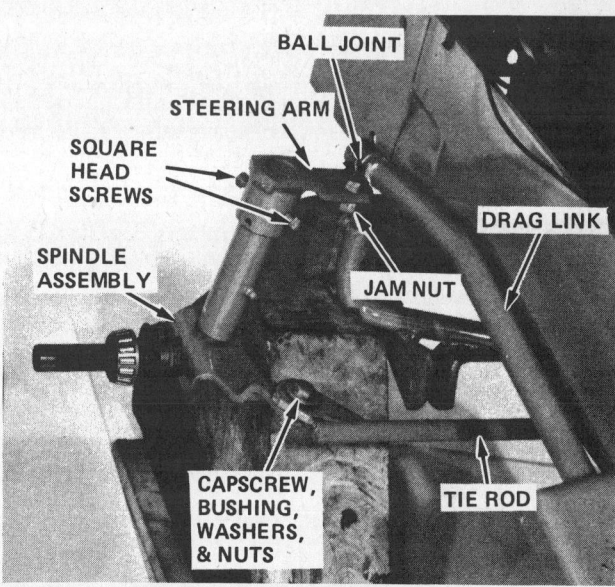
Lift front spindle

97

Allis-Chalmers

ing on hex head capscrew (A). Insert capscrew with bushing and washer downward through the hole in the tie rod lug.

3. Lay the end lug of the tie rod (B) on top of the bent lever arm and push the capscrew assembly through its hole.

4. Place a lockwasher on the capscrew and secure assembly with nut.

5. Install right front wheel and tire assembly.

LUBRICATION

1. Wipe grease fitting (A) clean. Using multi-purpose lithium base gun grease, lubricate the fitting until the grease is forced from the bushing.

2. Lubricate pivot points (B) with a film of grease.

Left Front Spindle Assembly

REMOVAL

1. Remove left front wheel and tire assembly.

2. Disconnect tie rod end lug from spindle assembly by removing hex head capscrew with its bushing, two washers, and nut.

3. Remove jam nut attaching ball joint to lefthand steering arm assembly and move drag link away from steering arm.

4. Loosen the two square head setscrews two turns each. Using a hammer and a piece of brass rod applied against the lower end of the steering arm hub, drive the steering arm up and off the pintle. Do not try to drive the pintle down, out of the steering arm. Lay the steering arm aside.

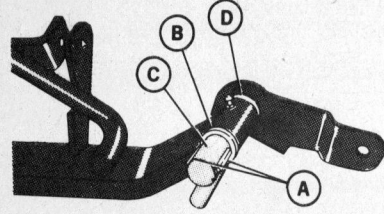

Pintle removal

5. Remove the two keys (A) and washer (B) from pintle (C).

6. With a cloth, wipe all of the dirt and grease off of the exposed end of the pintle. With a fine file, remove any burrs that may exist around the two key ways. Wipe off all of the fillings.

7. Pull the pintle out of its bushing in the axle assembly and remove the bottom plain washer (D).

8. Place all of the parts removed together and identify them for future reference.

CLEANING, INSPECTION AND REPAIR

1. See right front spindle assembly.

2. Examine keys for swelling caused by pressure of the setscrews when previously installed. Use a fine file to remove swollen material, but do not reduce size of keys. Replace keys that are worn, bent, or broken. Replace the bronze bushings if needed in the same manner as for the right front spindle.

INSTALLATION

1. Retrieve the spindle assembly having the longer pintle (C) with key ways.

2. Place a plain washer (D) on the pintle (C). Then insert the pintle upward through the bronze bushings in the left end of the axle. Place another plain washer (B) on the end of the pintle.

3. Assemble the keys, if necessary, driving them home in their key ways. Use a brass hammer, and back up the pintle with a lead mallet.

4. Place the steering arm over the pintle with its hub toward, and its arm opposite, the spindle.

5. Engage the keys in the key ways, and drive the arm onto the pintle with a plastic mallet while backing up the lower end of the pintle with a lead mallet. Drive the arm down on the pintle until there is barely perceptible end play. Then permanently tighten the two square head setscrews.

6. Secure the drag link ball joint to the steering arm with a jam nut and torque nut to 40-60 ft.lb.

7. Retrieve the tie rod and identify the left end lug (marked or tagged during disassembly). Slip a plain washer, and then a steel spacer bushing, on the hex head capscrew. Insert the capscrew with bushing and washer downward through the hole in the tie rod lug. Place the lug on top of the bent lever arm of the spindle assembly, and push the capscrew through its hole. Place a lockwasher on the capscrew, and then install the nut, tightening it permanently.

8. Install front wheel and tire assemblies.

Tie Rod

DESCRIPTION

The tie rod functions to transmit the same degree of turn to each wheel assembly by connecting or tying the left, and right spindle assemblies. It differs from the drag link, in that the drag link transmits turn commands from the steering gear to the left spindle assembly.

REPAIR

The tie rod can be removed and installed from the right and left spindle arms by removing the attaching nut, flat and lockwashers, capscrews, and bushings. Bushings should be cleaned, inspected, and replaced if worn. The tie rod should be replaced if distorted beyond repair or if bushings do not fit securely in rod end lugs.

Drag Link

DESCRIPTION

The drag link connects and transmits turn commands from the steering gear to the left spindle assembly. The tie rod connects both left and right spindle assemblies, thus providing uniform turning of the wheel assemblies.

REMOVAL

1. Disconnect the front ball joint (A) of the tie rod from the steering arm (B) by removing the nut (C). Prevent rotation of the stem by placing an open end wrench to the hex flange of the stem located just above the arm.

2. Disconnect the drag link from the steering arm assembly, by removing nut, lockwasher and flatwasher from capscrew. Spacer may remain on capscrew for reassembly.

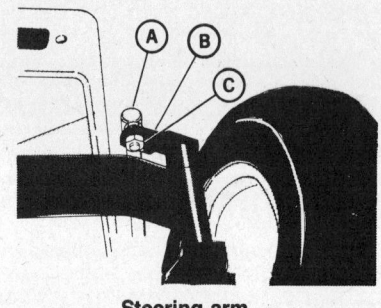

Steering arm

INSPECTION AND REPLACEMENT

Examine drag link for unusual bends or distortion and ball joint threads for damage or wear. Replace parts as required.

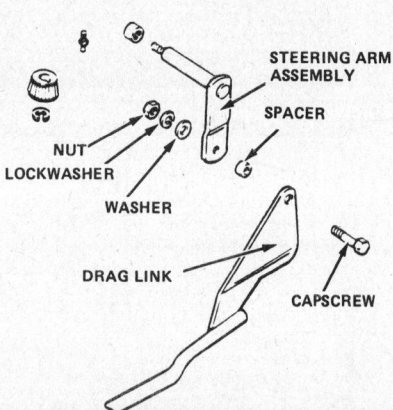

Drag link and steering arm components

INSTALLATION

1. If necessary install ball joint, at front of the drag link.

2. Connect rear of the drag link to steering arm assembly with capscrew, spacer, washer and nut, torque 40-60 ft.lb.

3. Turn the front wheels and/or the steering shaft as necessary and insert the stem of the front tie rod ball joint through the steering arm. Install and torque nut (C) to 40-60 ft.lb. preventing rotation with an open end wrench on hex flange of the ball joint stem.

Allis-Chalmers

Front Axle Assembly

REMOVAL

1. Remove both wheel and tire assemblies.
2. Remove both spindle assemblies, tie rod and axle assembly and drag link.
3. Remove the nuts from the back end of the hex head bolt.
4. While supporting the axle assembly withdraw the bolt, notice location of washers.
5. Still supporting the axle assembly, pull the axle toward the front of the tractor until the rear trunnion is free of the bearing in the frame.
6. Remove and lay aside the bushing from the center hole in the axle assembly that receives the bolt removed in step 4.
7. Take the now separated axle assembly to the bench and grasp the main bar in a vise.

CLEANING

Using a grease solvent, clean entire axle assembly. Flush all of the grease out from between the pintle bushings in the axle pivot tubes.

INSPECTION AND REPAIR

1. Push the pintle of each spindle assembly temporarily back into the axle pivot tube bushings and shake the assembly noting the extent of side play. If there is appreciable rattle, the bushings must be replaced. If bushings are to be replaced, use a rod slightly smaller than the seat bore.
2. Make sure grease fittings are clean, with unobstructed hole and are secure in axle pivot tube.
3. Check the front axle and hitch bar for damage, distortion, or broken or weak welds. Replace entire assembly if damaged excessively.
4. Check axle bushing for wear and replace if excessively worn.

INSTALLATION

1. Retrieve one of the two plain washers. Spread a layer of thick grease on its flat side. Using the grease as a temporary adhesive to hold the washer in place, center the washer over the bolt hole in the front face of the frame cross member.
2. Slip the second plain washer on the bolt rounded face next to the bolt head. Lay the bolt, with washer assembled, at the front of the tractor where it can be easily reached.
3. Bring the axle assembly to the front of the tractor. Spread some thick grease on the steel bushing for bolt, and slip the bushing into its hole at the center of the axle main member.
4. Supporting the axle assembly, insert the trunnion into its bushing in the frame, simultaneously aligning the holes for bolt. Insert bolt with washer through the axle main member, the second washer pasted to the frame, and then through the frame cross member. Finally, install and torque nut to 80-100 ft.lb. A jam nut is also required.
5. Install both spindle assemblies, tie rod, and axle assembly end of drag link.
6. Install both wheel and tire assemblies.

STEERING

Steering Shaft and Gear

REMOVAL

1. Park tractor on a hard, level surface. Place park brake in "off" position and all other controls in "off" or "neutral" positions. Remove ignition switch key.
2. Remove steering wheel and battery.
3. Block the rear wheels at front and back. Use a hoist or jack and raise tractor so that the steering gear assembly can be removed.
4. Remove drag link.

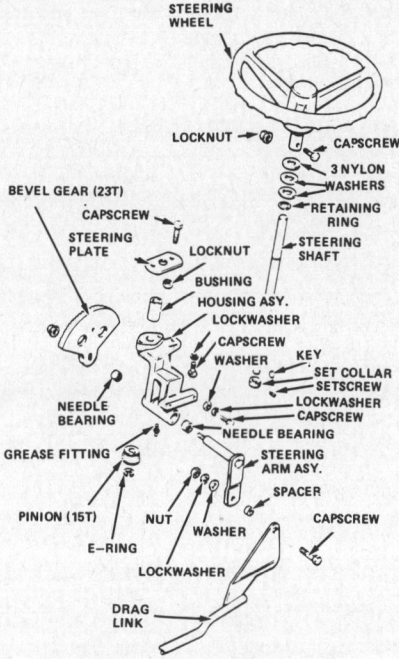

Steering gear components

5. Turn steering arm to allow access to capscrew. Remove capscrew and lockwasher.
6. Remove the two capscrews with lockwashers that attach steering gear casting to side of frame.
7. Move entire steering gear and shaft assembly forward until casting lug clears edge of frame opening. Lower assembly until shaft has passed completely out of opening.

CLEANING

Clean entire assembly with stream or with a grease solvent and stiff brush.

DISASSEMBLY

1. Secure casting in vise and remove locknut (B).
2. Using a plastic mallet, drive bevel gear (C) from steering arm shaft and casting. Remove gear by sliding it sideways and out of engagement with pinion gear (D).
3. Using bearing puller, remove both needle bearings from casting.
4. Remove retaining ring (E).
5. Secure steering shaft (G) in vise. Using a hammer and a piece of hardwood, drive pinion gear (D) off shaft.
6. Remove Woodruff key.

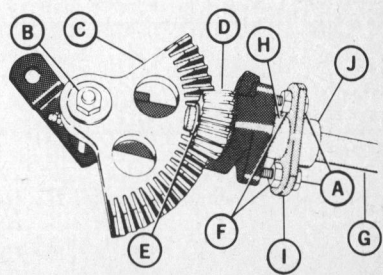

Steering gear assembly

7. Pull out shaft (G) from bearing bushing (H).
8. Loosen locknuts (F) and back out capscrews (A) to separate the plate (I) from the casting.
9. Remove bushing (H) from casting.
10. Remove collar (J) with setscrew from shaft (G).

CLEANING

1. Flush out casting bores and remove all grease from bearings.
2. Clean all teeth on both gears using a wire brush and solvent.

INSPECTION AND REPAIR

1. Examine steering arm shaft in areas that serve as journals for needle bearings. If there is excessive wear, or scoremarks showing evidence of seizing or galling, replace both the arm assembly and the needle bearings.
2. If no such defects are apparent, insert steering arm shaft into both needle bearings. Rotate shaft, at the same time urging it from side to side. If rotation is not perfectly free and smooth, or if there is more than a barely perceptible side shake, replace both the arm assembly and the needle bearings.
3. If needle bearings are to be replaced, press them out of the casting bore with a rod slightly smaller than the bore.
4. Inspect all teeth on both gears for damage. If damaged, replace gears.
5. Examine area of steering shaft that runs in casting bushing. If there is excessive wear or score marks showing evidence of seizing or galling, replace both shaft and bushing. Check shaft for bends or distortion.
6. Slip bushing on steering shaft to its normal location. Rotate shaft, at the same time urging it from side to side. If rotation

99

Allis-Chalmers

is not smooth, or if there is a side shake, replace both shaft and bushing.

7. If old shaft is to be reassembled, file away any burrs around retaining ring groove and key way. Clean the shaft and remove the filings.

ASSEMBLY

1. Press needle bearings, one from each end into the bore. Be sure the end of the bearing marked "Torrington" is facing outward—installation pressure must be applied on the marked end. The bearing at the gear end of the bore must be seated 1/8 in. below the face of the casting on which the gear runs. Press the bearing in using a rod slightly smaller than the bore. The bearing at the arm end of the bore must be seated flush with the face of the casting against which the arm runs. Press this bearing in, using a rod at least 1/4 in. larger than the bore.

2. Slip the collar (J) on the shaft, leaving the setscrew untightened. Spread a film of oil over the shaft from the key way back a distance of a couple of inches. Be sure that there is a retaining ring installed in the groove near the steering wheel end.

3. Slip the bushing (H) into the casting.

4. Assemble the plate (I) by running the two capscrews (A) into the casting until the end of the bushing opposite the plate protrudes about 1/64 in. from the adjacent face of the casting. Be sure that the plate is not cocked from one screw being run in farther than the other. Tightly run the locknuts back against the plate finger.

5. Insert the end of the shaft into the bushing until the shoulder emerges.

6. Assemble the Woodruff key by driving it in with a brass hammer while backing up the shaft with a lead hammer.

7. Drive the shaft into the pinion gear (D) up to the shoulder using a plastic or wooden mallet against its free end while supporting the pinion on a wooden block.

8. Install the retaining ring (E).

9. With the casting gripped between lead jaws in a vise, use a plastic mallet to drive the end of the shaft protruding from the pinion backward, so the retaining ring bears snugly against the face of the pinion.

10. Check that the plate is perpendicular to (square with) the shaft. Then move the collar (J) up against the plate and tentatively tighten its setscrew.

11. Test the fit by rotating the shaft. Any binding is probably due to the collar rubbing on the plate. Check that the locknuts are against the plate and then tap the free end of the shaft with a plastic mallet, shifting the shaft slightly through the collar.

12. Repeat the above procedure, resetting the collar if necessary, until the shaft rotates freely with only a barely perceptible end play.

13. Examine the face of the pinion at the small end. Note the rectangular depression embossed in the end of one tooth. This is the witness mark (A). Examine the top face of the sector gear (C). Note the circular dimple-like depression embossed at the

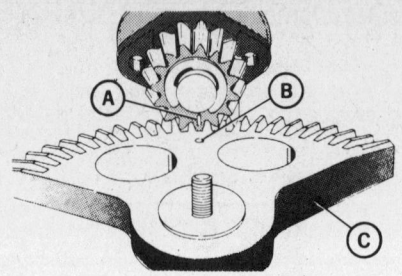

Witness marks on gears

center of the tooth arc, between the two large holes. This is the witness mark (B). In assembly, the two witness marks must coincide for proper timing.

14. Turn the steering shaft so the witness mark (A) is down; that is, nearest the space to be occupied by the sector gear. Then, aligning the witness mark (B) with (A), slide the sector gear into engagement with the pinion, centering its double-flat hole over the arm shaft bore.

15. Oil the arm shaft, and insert it through its needle bearings toward the sector gear. Rotate the arm shaft until its flats are aligned with those in the gear. Then, urge the shaft into the gear, tapping the shaft, if necessary, with a plastic mallet.

16. Run on the nut (B) and tighten it permanently.

ADJUSTMENT

The steering gear mechanism must be adjusted for minimum play in the steering before the assembly is installed in the tractor. Minimum play is obtained when the gears are adjusted for the least amount of backlash that can be obtained without creating binding of the gears. This adjustment is accomplished by loosening or tightening the capscrews (A) as follows:

NOTE: To make an adjustment, first loosen the locknuts (F). Then turn the capscrews small, equal amounts in the same direction to keep the plate (I) perpendicular to the steering shaft. After completing the above, retighten the locknuts and rotate the gears to check for binding or excessive backlash.

1. Tighten capscrews (A) until gears begin to bind. Due to size variations, the initial binding may occur at one specific position of the gears. If so, concentrate all adjustments on this specific position.

2. When binding occurs, loosen the capscrews slightly and tap the end of the steering shaft in the same direction with a plastic mallet. Retest and repeat this step until the binding is just eliminated.

3. Check the backlash of the gears by urging the sector gear back and forth while holding the steering shaft stationary. Note the amount of sector gear movement that is possible without moving the pinion gear. This play or backlash should be no more than a few thousandths of an inch or the thickness of a sheet of paper as seen at the tip of the sector gear. If the backlash is excessive, repeat steps 1 and 2 above.

4. When the backlash is properly adjusted, tighten both locknuts (F) and the

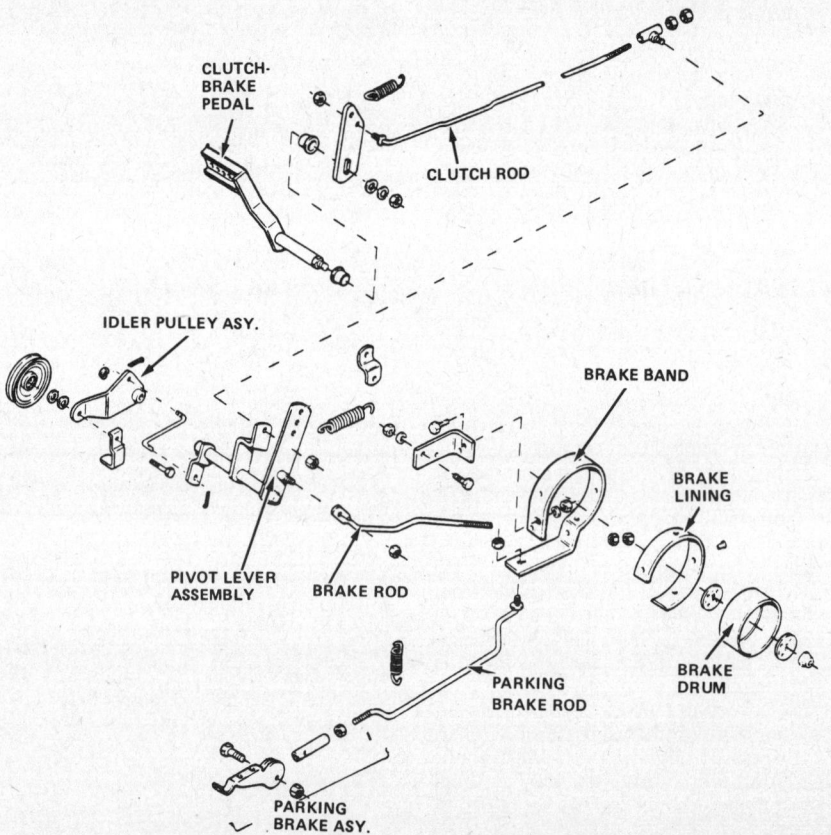

Brake and clutch group with shuttle drive

setscrew for the steering shaft set collar (J). Then repeat step 3 to determine if tightening of the locknuts and set collar affected the adjustment. If it did, loosen these items and repeat the adjustment steps.

5. Urge the steering shaft strongly toward the sector gear and retest the fit throughout the entire range of the gears. If binding occurs, repeat entire adjustment.

LUBRICATION

Using a grease gun, fill steering casting bore through grease fitting until grease seeps out at both arm and gear ends of the bore. Oil the gears and the shaft, the latter at three points: between the gear and the casting, between the bushing and the plate, and between the collar and the plate.

INSTALLATION

Prior to installation in the tractor, turn the steering shaft clockwise, as viewed from the steering wheel end, as far as it will go. Then install the steering gear mechanism into the tractor in the reverse order of removal. Replace the steering wheel and the battery.

CLUTCH AND BRAKE

Description

The brake and clutch group includes a foot pedal, an idler pulley, a brake drum and band, and a series of rods, links, and pivots that connect these items together. The connecting items change frequently.

The basic operation of the clutch and brake group is the same for all models. Pressing the foot pedal pulls the clutch rod forward to turn a pivot. As the pivot turns, it moves a linkage to pull the idler pulley away from the drive belt between the bevel gear box and the transmission. Thus, the belt tension is reduced to disengage the transmission from the bevel gear box and engine.

The parking brake is a lever connected by a separate rod to the brake end. This lever can be operated independently of the foot pedal to lock the brake.

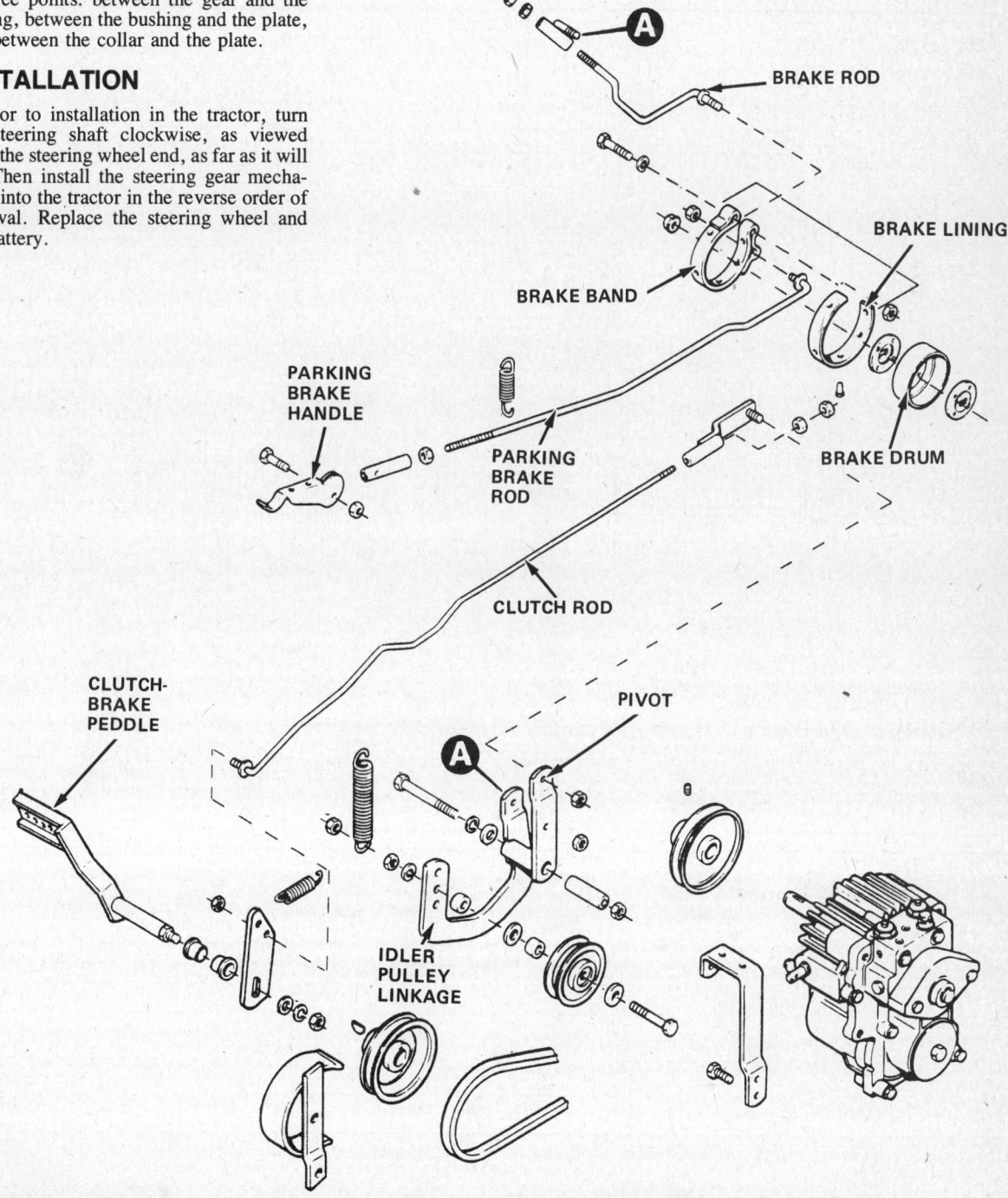

Brake and clutch group with hydrostatic drive

Allis-Chalmers

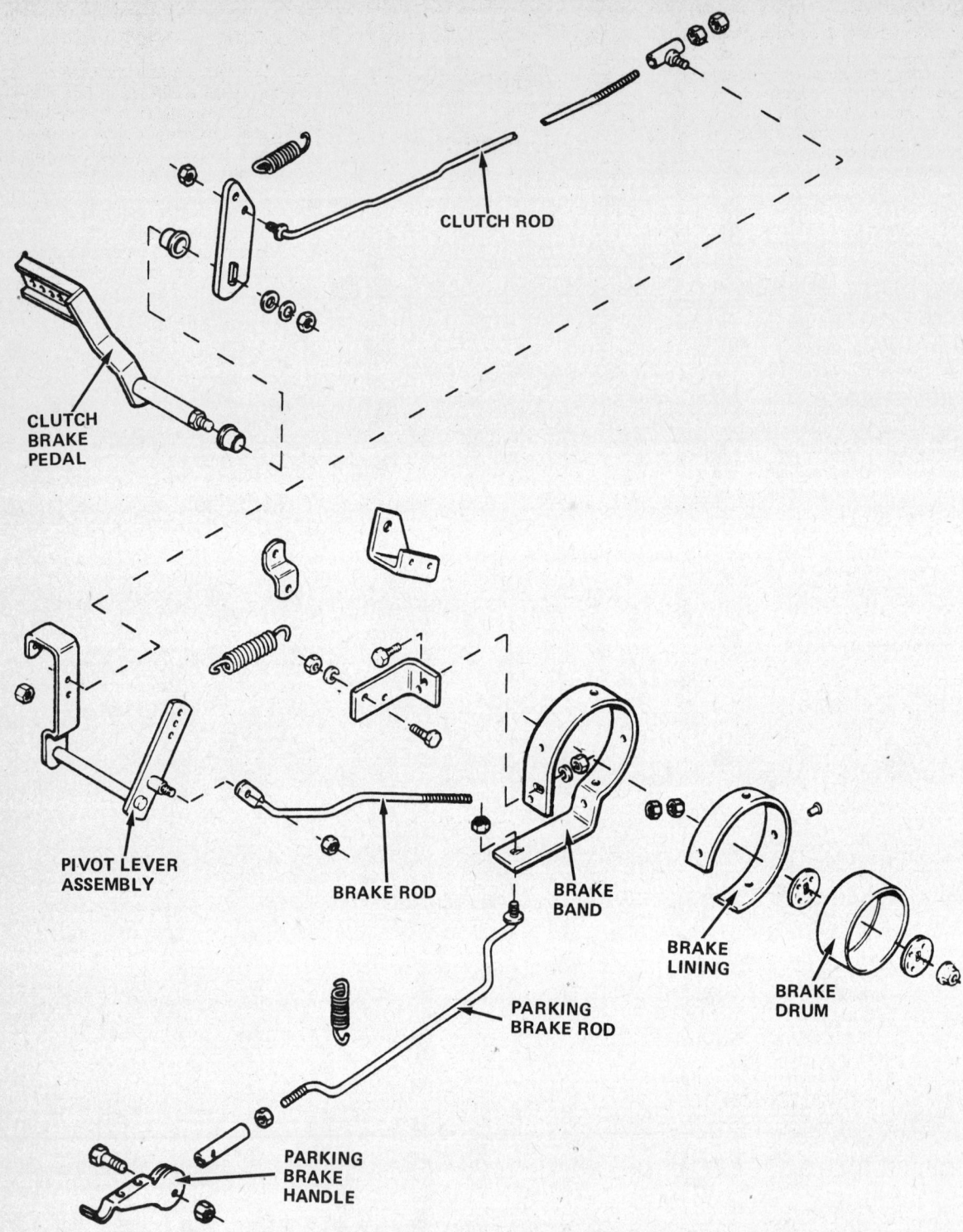

Brake and clutch group with dual range 6-speed drive

Pivot motion also moves a rod that passes through the brake band. If the pivot moves enough, a nut on this rod contacts the brake band and tightens it around the brake drum. The friction that results slows or locks a transmission gear shaft to brake the tractor.

Inspection

Before replacing items of the brake and clutch group, check the overall operation of both the clutch and brake. Be sure that the problem is not one that can be corrected simply by performing one of the adjustments described in the Adjustment Section.

If belt slippage is a problem, check both the belt and the pulleys. Be sure that both are clean of oil and grease. Use a clean rag soaked in solvent to remove any grease or oil.

If belt breakage or excessive wear is a problem, check the pulleys. The pulleys should be free of burrs and sharp edges that can cause belt breakage.

Clutch and Brake Adjustments

1. Adjust jam nuts on rear end of brake rod so that when foot clutch brake pedal is pushed firmly forward clutch arm will stop

Allis-Chalmers

with its forward edge ⅝ in. to rear of the rear lift cable guide near the bevel gear box when the brake is locked tight.

2. To adjust the parking brake, loosen jam nut on rod and turn lever and rod end so that brake is tight when parking brake lever is pulled fully upward against the finder opposite the position shown. Tighten jam nut against rod end.

3. To adjust tractor clutch, with clutch pedal up, in engaged position, adjust jam nuts on clutch rod ¼ in. away from clutch rod guide.

4. Slide the front belt guard forward or back in slot at bolt C until guard has ⅛ in. clearance to the outside of the front pulley at the front side. Make sure that the belt guard does not touch the pulley or the belt in any point while clutch is fully engaged.

5. Check that the upper 90° formed leg of the idler pulley belt retainer is resting on and is parallel to the top edge of the idler arm as shown to properly locate the belt retainer.

6. The transmission safety interlock switch is actuated by a special flat top carriage bolt screwed into the gear shift rod and held by a jam nut. The carriage bolt should be adjusted in or out of the rod to the correct height to actuate the switch when the shift rod is in neutral but not high enough to drag tightly against the switch. The bolt must not touch the switch when shift is in any of the 4 gears.

7. With the shuttle clutch control lever in neutral position make sure center of notch in brake detent is centered on brake pin. If necessary loosen setscrew in detent, move it, and retighten setscrew.

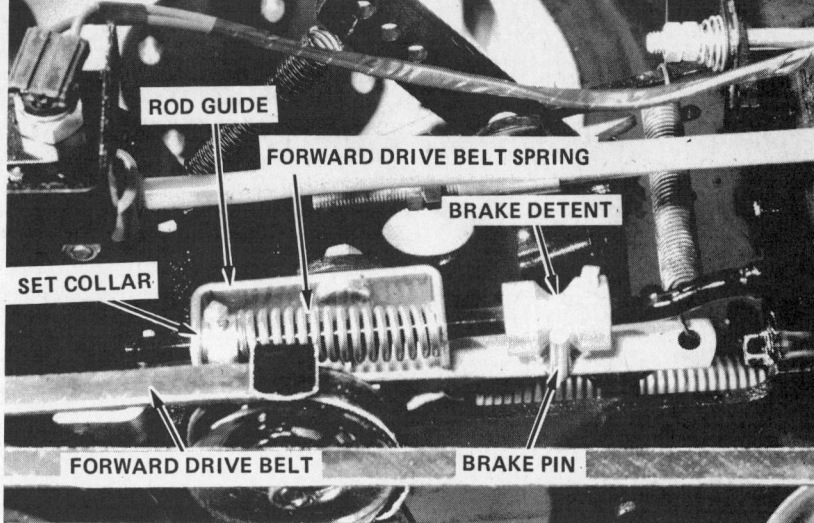

Control lever in neutral

8. With control lever in full forward position adjust brake pad assembly, to have 1/16 in. clearance between pad and surface of pulley. Check with control lever in full rear position, clearance between brake pad and pulley must be at least 1/16 in. or more. If less, readjust to a full /16 in. clearance. With the control lever in neutral there must be at least 1/16 in. clearance between brake pin and notch of brake detent.

9. With control lever in neutral position loosen setscrew in set collar behind forward drive belt spring. Move rod guide assembly and spring forward until slack is taken out of forward drive belt. Tighten setscrew. Move control lever to the full forward position. There should now be 3/16 in. clearance between rear surface of set collar and rear leg of rod guide assembly. If necessary reset set collar to obtain the 3/16 in. dimension when control lever is fully forward.

10. With control lever in neutral position rotate pivot on threads on brake rod so that when pin is reinstalled in brake lever hole, all of the slack will be taken out of brake band. Fasten pivot pin in place in brake lever with lock pin.

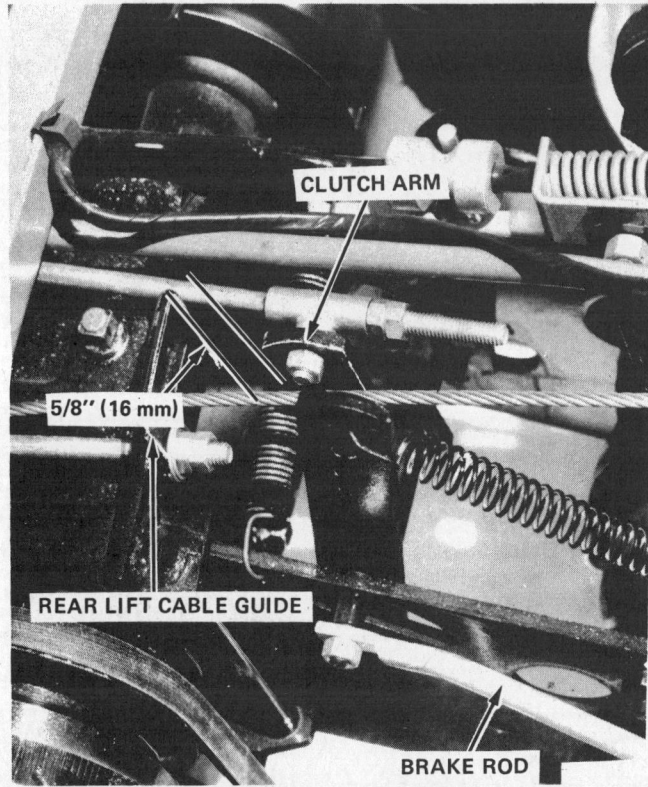

Foot brake adjustment, shuttle drive

ELECTRIC LIFT

Specifications

1. Rated load: 750 pounds.
2. Travel: 4″ ± .200 in.
3. Travel rate: 1.2 in. per second
4. 12 volt
5. 18 amp at rated load. 7 amp under no load.
6. Overload clutch set at 1050 pounds maximum.
7. Automatic brake.
8. Gear ratio, 25:1

Motor Inspection

Remove the long screws that attach the motor to the gear box.

1. The motor can now be removed from the gear box. Use care to prevent damage to the two housing gaskets during disassembly.
2. Remove the end cap. Check the condition of the brushes, springs and wire connections.
3. The armature can be checked and tested.
4. Clean the commutator with fine sandpaper.

103

Allis-Chalmers

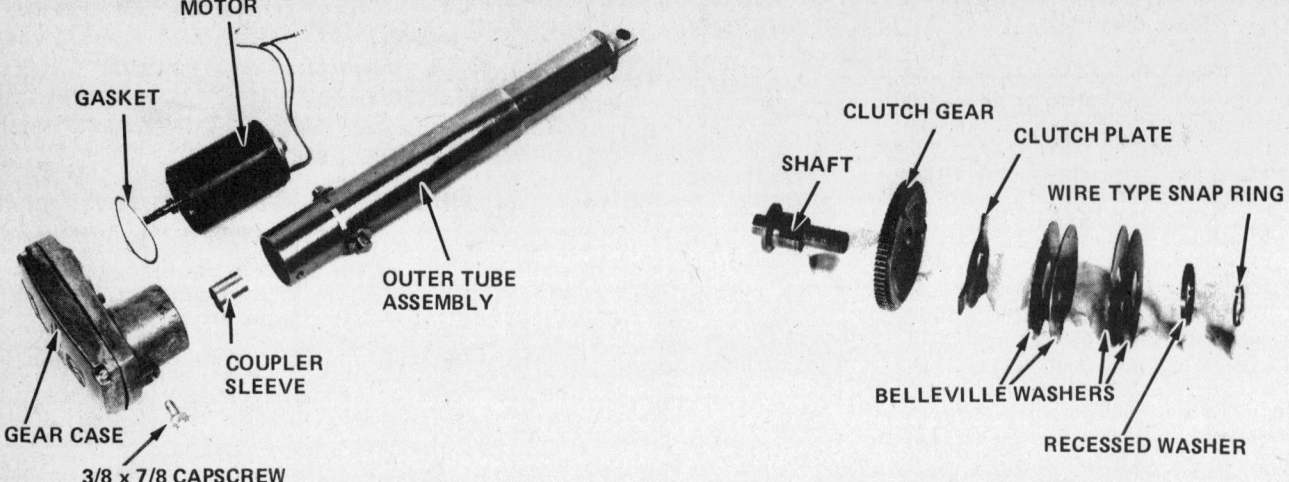

Outer tube components

Overload clutch components

MOTOR ASSEMBLY

1. Push the brushes back in the brush holders and hold them in so that the brushes will not be damaged when the shaft is installed in the end cap bushing. Place the gasket on the end cap.
2. Hold the end cap on the armature in the motor case. Note the permanent magnets are not centered in the case. If the case is properly installed the permanent magnets will be flush with the end of the armature at the gear case end.
3. Rotate the end cap so that the lug in the case is in the slot in the end cap.
4. Install the gasket on the motor case and rotate the motor assembly to align the lug in the motor case so it enters the slot in the gear case.
5. Install the two (2) long screws and tighten.

OUTER TUBE REMOVAL

Remove the ⅜ in. x ⅞ in. capscrew from the gear case. Apply some penetrating oil around the base of the tube and with a rotating movement the tube can be removed. Note it is not necessary to remove the gear case cover or gears to remove the outer tube.

INSTALLATION OF OUTER TUBE

Install coupling sleeve over coupling. Align the hole in the tube with the hole in the gear case for the ⅜ in. x ⅞ in. capscrew; press the tube in the gear case and replace the ⅜ in. capscrew.

GEAR CASE INSPECTION

1. Remove the four (4) machine screws at the corners of the gear case cover and remove the cover.
2. Remove the intermediate gear assembly, and note the location of wave washer.
3. To remove the gear and overload clutch assembly:
   a. Remove the ⅜ in. x ⅞ in. capscrew from gear case.
   b. Remove outer tube assembly.
   c. With a suitable punch, drive the ball bearing screw and brake drive pin from the shaft. It will be necessary to rotate the clutch and gear assembly to align the pin with the hole in the gear case. The punch must be small enough to pass through the ⅜ in. threaded capscrew hole in the gear case.
4. Check the condition of gear overload clutch and shaft before installation in the gear case.

INSTALLATION OF CLUTCH AND SHAFT ASSEMBLY IN GEAR CASE

1. Lubricate the shaft and insert end of shaft in the gear case.
2. Rotate the clutch and shaft assembly so that the hole in the end of the shaft aligns with the ⅜ in. threaded capscrew hole in the gear case. Insert the drive pin in the hole in the shaft and drive the pin into the shaft, so the coupler sleeve can be centered around the pin.
3. Install the outer tube assembly. (See outer tube and shaft installation instructions.) **NOTE:** Be sure the coupling sleeve is in place.
4. Install the intermediate gear assembly.
5. Put sufficient pressure gun grease in the gear case to lubricate the gears and bushings. Install the gear case cover.

OVERLOAD CLUTCH AND GEAR ASSEMBLY INSPECTION

1. Remove the overload clutch gear assembly.
   a. Place C clamps or vise grips on the edge of the recessed washer and compress the Belleville washers sufficient to permit the removal of the wire snapring from the groove in the shaft.
   b. Remove the recessed washer, Belleville washers and clutch plate from the shaft.

NOTE: When removing the Belleville washers, watch the direction of curve (concave or convex) in each washer and reassemble washers in the same way.

   c. Inspect the clutch plate, gear driving lugs and gear for wear or damage. Replace the complete assembly if any of the components are not in satisfactory condition.
2. Lubricate before placing the components on the shaft in the following order:
   a. The clutch gear, the clutch plate, and Belleville washers. Replace the washers, with the curves in the same direction as when they were removed. In some overload clutches, there may be only three Belleville washers.
   b. Place recessed washer over the shaft with the recess up. Compress the Belleville washers with C clamps or vise grips sufficient to install the wire type snaprings in the groove in the shaft.
   c. Replace the assembly in gear case.
   d. Check motor shaft end clearance.
   e. It should be .0051/.015 in.
   f. Install intermediate gear. Put in sufficient pressure gun grease to lubricate the bushings and gears.
   g. Install cover plate.

PTO SYSTEM

REMOVING PIVOT ARM AND PULLEY ASSEMBLY

1. Push the power take-off tensioning lever (A) down until it springs past center, and moves freely forward.
2. Disconnect tension spring.
3. Remove attachment drive belt.
4. Using pliers, straighten the legs of and remove, the cotter pin.
5. If you are removing the PTO to repair the bevel gear box or transmission, remove the nut. Restrain rotation of the shaft by applying a pipe wrench to the hub on the backside of the pulley. Do not re-

move the pulley. Leave it temporarily in place, so the pipe wrench can be applied to its hub when removing a nut on the other end of the shaft during a later step.

6. In order to remove the power take-off pulley assembly (A), by pulling its pivot shaft (B) out of the side plates, the capscrew (C) must first temporarily be removed. This can be done directly if you have a socket wrench small enough or by using a box end wrench. With the capscrew (C) out of the way, the entire pulley assembly can be separated from the side plates. After separation, replace the capscrew (C). Do not tighten it excessively.

7. Remove the nut (F), restraining rotation of the shaft with the pipe wrench applied to the pulley hub as in step 5.

8. Remove the clutch plate (G) by tapping the rim lightly with a brass hammer.

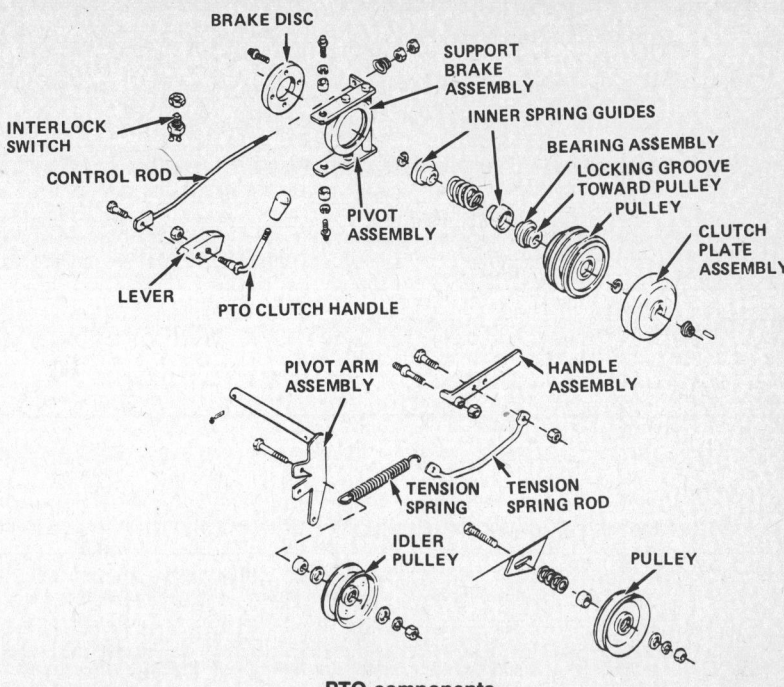

PTO components

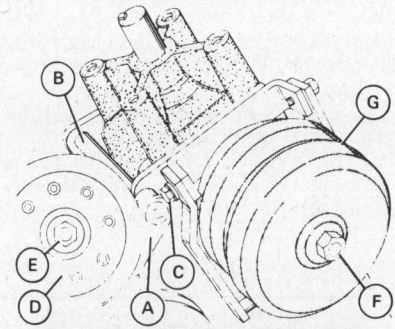

PTO and clutch plate assembly

9. Wrap two layers of thick rubber electrical tape around the threads to protect them. Insert a small screwdriver blade between the rubber tape and the overhanging end of the key and pry the key out of its key way.

―――――― **CAUTION** ――――――
Follow instructions in remaining steps carefully. Injury can occur if steps are not taken to prevent the pulley from being propelled off the shaft by an internal spring.

10. Apply a C-clamp as shown to compress the spring and hold the pulley. Be careful not to raise any burrs or gouge surfaces.

11. Tighten clamp until the gap between the pulley and the retaining ring is ⅛ in. or more.

12. Remove retaining ring. To avoid damage, do not stretch ring more than necessary.

13. Inspect edges of key way and retaining ring slot on shaft. Reduce any burrs to shaft level using a fine file and/or abrasive cloth.

14. Clean shaft to remove all dirt, grit and grease and remove electrical tape from threads.

15. Remove two pivot screws with related bushings.

16. Slowly release clamp pressure until spring ceases to push pulley off shaft.

17. Remove clamp, pulley, bushing or bearing, spring, and spring retainer(s) from pulley shaft.

18. To prevent loss, temporarily replace retaining ring and nut on shaft and the two pivot screws with bushings in the bracket. Also secure key to shaft with electrical tape.

19. If PTO is being removed to repair bevel gear box or transmission, complete pulley removal begun in step 5. If pulley does not slip off easily, use a wheel puller. To prevent loss of pulley key, secure it to the shaft with electrical tape.

INSPECTION, CLEANING AND REPAIR

1. Inspect the bevel gear box cross shaft for damage or excessive runout. The runout should be checked at the outer retaining ring groove on the PTO shaft with a dial indicator. If the runout is greater than .008 in.-.010 in. the shaft should be removed from the bevel gear box and straightened or replaced. If the shaft is replaced, replace it with shaft (Part No. 1657178) and nut (Part No. 922133). This shaft has ¾ in. fine threads on both ends, and the clutch plate nut should be torqued to 70 ft.lb. If the shaft with ½ in. threads is used, torqued to 50 ft.lb.

2. If the spring breaks, replace the spring and the inner and outer spring guide. Be sure that the inner retaining ring is seated properly in the groove before reassembling.

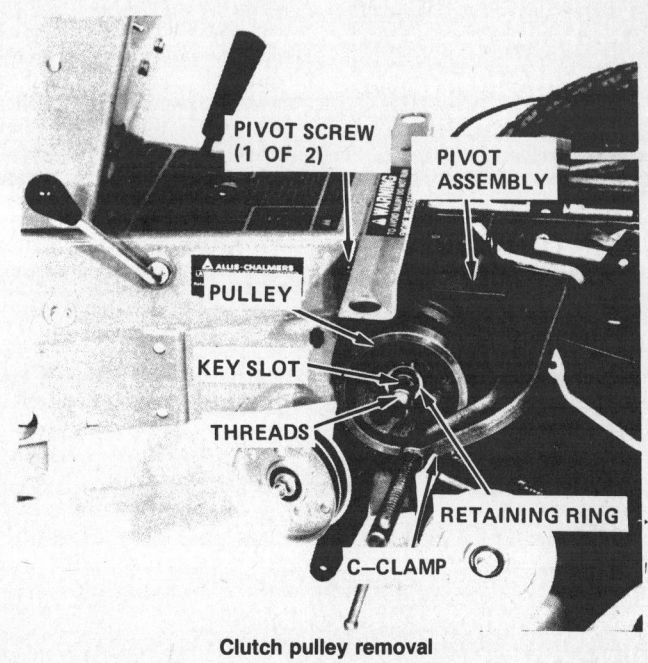

Clutch pulley removal

Allis-Chalmers

3. When the bearing fails, replace the bearing and only parts that are damaged. Be sure that the locking groove in the bearing is toward the pulley.

4. Check the clutch plate assembly for proper balance. This is a visual inspection to see if there is more material on one side than the other. Measure at the thickest point between the outside edge and inside edge of the clutch plate and then 180° from that point. If there is more than $3/16$ in. difference between the two edges, replace the clutch plate. Do not include the clutch lining in the measurement.

5. Inspect belts for wear and hardness. Old belts will be brittle or hard and must be replaced. Pulley misalignment often causes belt wear and therefore this alignment should be checked when replacing any worn belt.

6. Inspect pulleys for wear or damage both on the grooves and the shaft diameter. Clean and dry thoroughly before reassembly.

7. Inspect bearing and bushing assemblies for broken or cracked races, dented seals or shields, cracked or broken separators, balls or rollers. Inspect for overheating. Overheating can be detected by a brownish-blue or blue-black color. Any of the above are cause for replacement. Clean all bearings with a suitable solvent and dry before reassembly.

8. Inspect keys for distortion, bends or breaks. Check key ways for cleanliness and sharp edges which allow the key to seat securely.

9. Clean springs and inspect for weakness or broken ends. Replace if damaged.

INSTALLING PIVOT ARM AND PULLEY ASSEMBLY

1. Slip an inner spring guide (small I.D.), flanged side first onto the shaft and push it back against the retaining ring.

2. Slip pivot assembly, spring, spring guide (large I.D.), flanged side toward pulley and bearing, and pulley onto shaft. Temporarily install nut on end of shaft to hold other parts.

3. Insure that spring is properly seated against flange on spring guides. Then install clamp as shown.

4. Align the pivot assembly and then insert the two pivot arm capscrews with their bushings and lockwashers assembled. Tighten the top capscrew permanently, but leave the lower capscrew untightened.

5. Tighten clamp until gap between pulley and slot for retaining ring is $1/8$ in. or more.

6. Install the retaining ring and key. Position the latter with the outboard end overhanging the shaft shoulder a fraction of an inch so that it can be readily removed if ever necessary. Drive the key home in its key way with a brass hammer while backing up the shaft with a lead hammer.

7. Remove the C-clamp and the capscrew.

8. Install the clutch plate (G) and the nut (F), leaving final tightening of the nut until later.

9. Install the pivot arm assembly (A) by inserting its shaft (B) into the bearing holes in the side plates.

10. Replace the capscrew (C), with lockwasher and bushing assembled, tightening it permanently. If the space available is not sufficient for manipulation of the socket wrench, temporarily remove the pulley (D) by removing its spindle bolt (E). Replace pulley as soon as capscrew has been firmly tightened.

11. Permanently install the cotter pin bending its legs in opposite directions.

12. Install the spacer sleeve on the short end of the driveshaft with its painted end outward and seat it against the inner race of the ball bearing.

13. Install the Woodruff key, driving it into its key way with a brass hammer while backing up the shaft with a lead hammer.

14. If previously removed, install pulley with the hub inward. Secure by installing and tightening nut while holding pulley hub with pipe wrench.

15. Hold hub of pulley with pipe wrench and tighten nut that holds PTO clutch on other end of shaft.

16. Attach tension spring to tensioning rod, pull back and up on the tensioning lever until it is in a locked position.

PTO CLUTCH ADJUSTMENT

To check PTO clutch adjustment, observe the movement of the pulleys in relation to the clutch plate as the PTO clutch lever is moved from the engaged to the disengaged position. The pulley movement should be $1/16$ in. If not adjust the PTO clutch as follows:

1. Set PTO clutch lever to engaged position.
2. Loosen the rear jam nut.
3. Turn the front nut slightly clockwise to increase pulley travel or slightly counterclockwise to decrease pulley travel.
4. Tighten the rear nut against the front nut and repeat the check.
5. When proper adjustment is obtained, lower and lock the seat deck in place.

PTO BELT ADJUSTMENT

1. Pull up and back on pulley tensioning lever.
2. Measure the clearance between the front idler bracket and the stop.

The clearance should be $3/4$ in. to $7/8$ in. or in the green of the decal (if so equipped).

3. If clearance is set wrong, push the belt tensioning lever down.
4. The rear idler V-pulley is mounted in a slotted hole. The pulley can be moved forward or back to increase or decrease belt tension.

Change position of the rear idler pulley, move V-pulley forward in slot to bring idler pulley rearward, closer to stop. Move V-pulley rearward in slot to bring idler pulley forward away from stop.

5. Recheck the measurement with belt tensioning lever pulled up and back.
6. Drive mower for 15 minutes and recheck the measurement. Change the position of the rear idler pulley, if necessary.

REAR WHEELS, DIFFERENTIAL AND AXLE

Rear Wheel and Tire Assembly

REMOVAL

1. Raise rear wheels and support the tractor on jack stands.
2. Remove five hub bolts. Then remove wheel and tire assembly from hub.
3. Repeat step 2 on other side of tractor.

INSPECTION AND REPAIR

1. Check tire for excessive wear on tread, weak sidewalls, or cracked beads.

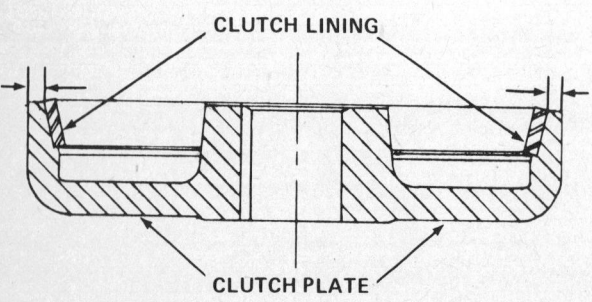
Clutch plate balance measurement

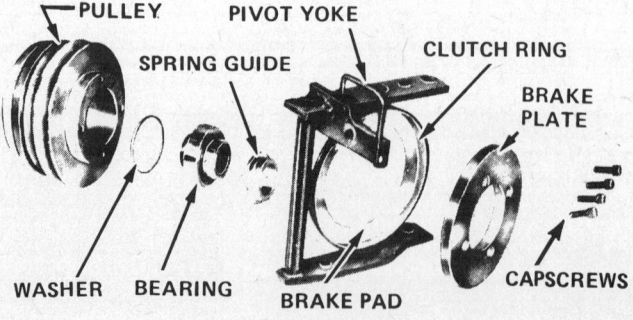
Pulley assembly

Allis-Chalmers

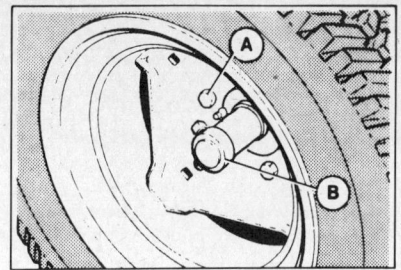

Right rear wheel and hub collar

2. Check tire valve for leakage.
3. Replace or repair defective tire or valve and inflate tire to specified pressure.

INSTALLATION
1. Place wheel and tire assembly against hub. Align holes in wheel and hub and loosely install the five hub bolts.
2. Tighten hub bolts evenly and snugly to secure wheel and tire assembly to hub.

Right Wheel Hub
REMOVAL
1. Prepare tractor and remove right wheel and tire assembly as described in paragraphs above.
2. Loosen the two setscrews and remove axle collar from right end of axle.
3. Use a fine file to remove any burrs left in axle by axle collar setscrews. Then wipe exposed end of axle to remove dirt and fillings.
4. Slide hub off right end of axle.
5. Wipe exposed end of axle to remove all dirt and grease.

CLEANING AND INSPECTION
1. Using a thin grease solvent, clean entire right hub assembly. Be sure to flush all grease from differential gear and bearings within center bore of hub.
2. Check differential gear on hub for signs of excessive wear and broken or chipped teeth.
3. Check bearings within bore of hub for discoloration and other signs of overheating. Insert a finger into the bore and check bearings for grooving or cracks. If any bearing defects are noted, also check portion of axle that runs in the hub. Axle defects, other than minor surface discoloration or shallow scratches and pitting that are easily removed by polishing, are cause for axle replacement.

DISASSEMBLY
The right wheel hub assembly includes a pair of bearings and a grease fitting. Disassembly consists only of removing these items from the combination hub and differential gear.
1. Remove grease fitting from hub.
2. Remove the bushing type bearings from the bore of the hub. The removal is best done with an internal bearing puller. If a puller is not available, the bearings can also be driven out of the hub. Begin by inserting a long rod about half the size of the axle into the bore of the hub. Tilt this rod slightly so that its end bears against the inner face of the bearing at the far end of the bore. Then drive the far bearing out, being careful not to scratch or mar the bore. Drive the second bearing out in the normal manner, using a rod just slightly smaller than the bore of the hub.

ASSEMBLY
1. Be sure that hub and bearings are clean.
2. Wipe a light coat of oil over the outer diameter of the bearings. Then press or drive the bearings into the hub until the outer faces are flush with the ends of the hub bore. When driving the bearings into place, use a driving rod slightly smaller in diameter than the hub bore. Never hammer directly on the bearings.
3. Install the grease fitting in the proper hole of the hub.

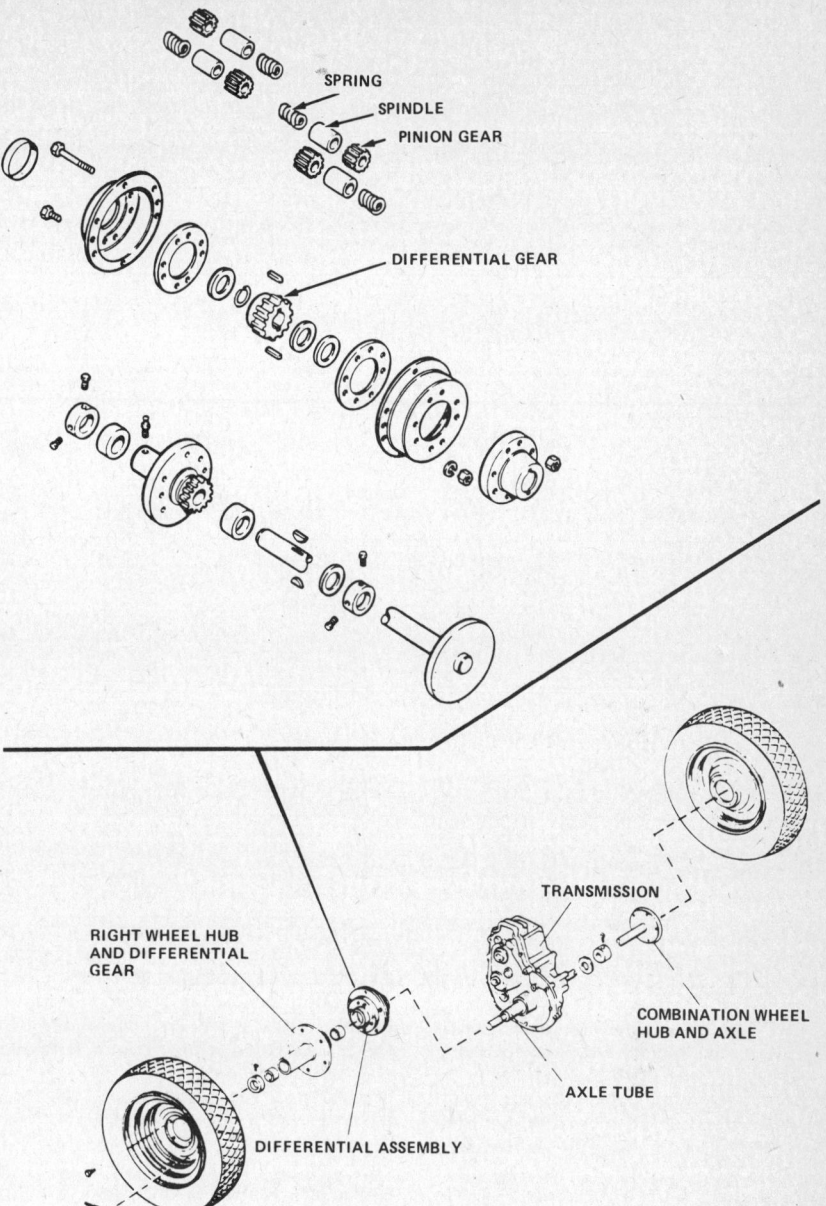

Rear wheels and axle

INSTALLATION
1. Check rubber seal in outer rim of differential assembly. Replace seal if it is worn or cracked.
2. Check axle collar. One face must be smooth and free of paint. The mating surface on the wheel hub must be in the same condition.
3. Spread a thin coat of grease over the exposed portion of the axle. Apply grease liberally to the differential gear on the wheel hub.
4. Slide wheel hub over end of axle with differential gear facing differential assembly. Rotate wheel hub as necessary to align teeth of its differential gear with teeth of pinion gears within differential assembly.
5. Grease paint-free surface of axle collar. Slide axle collar onto axle with greased surface nearest hub. If axle has not been replaced, axle collar setscrews should be aligned with old scoremarks in axle.

107

Allis-Chalmers

6. Press axle collar inward to firmly seat hub in differential assembly while tightening axle collar setscrews to secure hub and collar in place.

7. Check for end play of hub. If necessary, loosen axle collar setscrews and repeat step 6 until end play is elminated.

8. Adjust and lubricate wheel hub.

9. Install right wheel on hub.

DIFFERENTIAL

REMOVAL

1. Prepare tractor by raising and blocking up rear end.

2. Remove both rear wheel and tire assemblies from wheel hubs.

3. Remove right wheel hub.

4. Remove the paint and burrs from the lefthand axle shaft.

5. Loosen the lefthand set collar and move the axle all the way to the right, so that the set collar is against the lefthand hub.

6. Remove the recessed washer, retaining ring and the two keys from the right end of the axle shaft.

7. Pull the differential assembly off and remove the axle shaft. Remove all burrs and sharp edges from the axle shaft, before removing.

8. Remove the axle tube keys and retaining ring if replacement is needed or if the gear case is to be disassembled.

DISASSEMBLY

1. Remove the six screws with lockwashers and nuts around the outer rim of the differential covers.

2. Remove the eight locknuts from the pinion gear axle bolts. The locknuts should be loosened slowly in succession until the pressure of the internal springs is released.

3. Remove the inner differential cover and all internal parts.

4. Clean all parts using a thin grease solvent.

5. Check pinion gear teeth for excessive wear, breaks or chips. Check gear faces for scoring. Discard or replace any damaged gears.

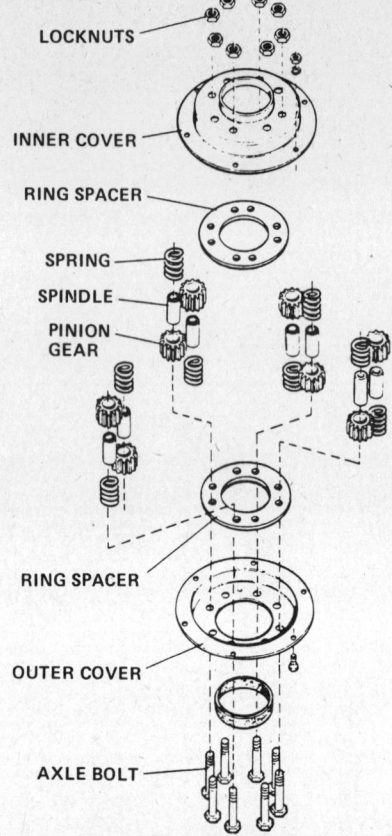

Differential components

6. Check the pinion gear spindles for scoring and signs of overheating. Discard and replace any bad spindles.

7. Springs should be checked for broken or brittle ends as well as general strength. Discard and replace any springs with even questionable defects.

8. Replace the rubber seal in the lip of the cover and replace any broken, stripped or scored pinion gear axle bolts.

9. Scrape clean mating surfaces of the two covers. Replace both covers if either is bent or cracked.

ASSEMBLY

1. Insert eight bolts that serve as pinion gear axles through holes from outside to inside of outer differential cover. The outer differential cover has the lip for the rubber seal.

2. Set outer differential cover with lip and bolt heads down on a flat surface. Coat portion of bolts that will be within covers with grease. (Use a good grade of lithium base automotive chassis lubricant).

3. Inspect one ring spacer and grease face that shows the last wear. Install this ring spacer over the eight pinion gear axle bolts with the greased side facing up.

4. Coat all eight spindles with grease and install one on each pinion gear axle bolt.

5. Coat one spring and one pinion gear with grease. Install these items, spring first, on a spindles. Repeat this process three additional times, installing a spring and pinion gear on every other spindle.

6. Place a differential gear between the four pinion gears that are already installed and rotate the latter until all four mesh with the differential gear at the same time. Then remove the differential gear. From this point on, exercise care to prevent rotation of pinion gears once they are installed.

7. Pinion gears and springs must still be installed on the four remaining spindles. Coat a pinion gear and spring with grease. Place pinion gear over an empty spindle, rotate it until its teeth are centered in gaps between teeth on adjacent previously installed pinion gear. Then allow the pinion gear to drop straight down the spindle and place spring over spindle on top of pinion gear. Repeat this process three more times until a pinion gear and spring are installed on all eight spindles.

8. Inspect remaining ring spacer and grease face with least wear. Then install ring spacer over eight pinion gear bolts with greased side facing spindles.

9. There are six larger bolt hole and one smaller alignment hole around the outer rim of the inner cover. Matching this alignment hole to a similar hole in the outer cover, install inner cover over the eight pinion gear axle bolts.

10. Obtain six ¼-20 by 2 in. long capscrews and mating nuts. Install these items in the six holes around the rims of the cover halves. Do not attempt to draw the cover halves together with these capscrews; the capscrews are being used only to keep the covers aligned during the next step.

11. Install locknuts on the eight pinion gear axle bolts. Tighten locknuts successively in small increments to draw covers together. To avoid cover damage, stop tightening of locknuts if resistance suddenly increases as pinion gears are drawn together. In such cases, it will be necessary to return to conditions of step 7 and realign the pinion gears.

12. When covers are drawn together, remove capscrews and nuts installed in step 10 and replace them, one at a time, with the normal ⅝ in. long capscrews, lockwashers and nuts.

13. Pack extra grease around all pinion gears within differential assembly.

INSTALLATION

The installation procedure given in the

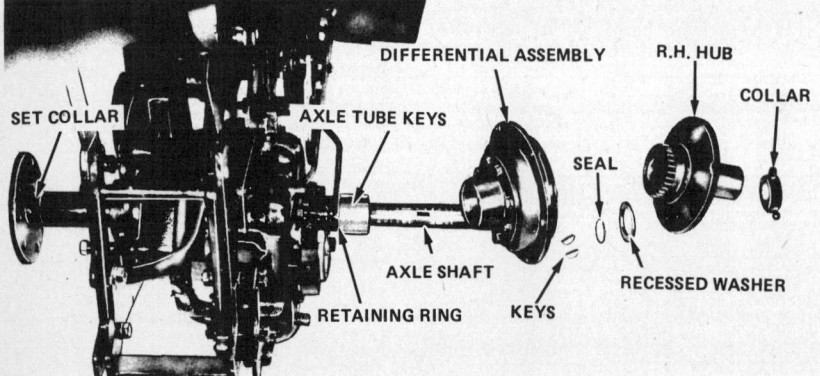

Differential assembly removal

steps below may be used whenever the axle is already installed in the tractor. If the axle, right hub and differential are to be installed as a unit, the differential carrier should be removed from the differential and installed on the transmission axle tube as detailed in the first four steps below. After cleaning and oiling the axle, the other parts can be installed by simply inserting and slipping the axle through the axle tube. To complete the installation, rotate unit until pinion gear axle bolts pass through, differential carrier, install and tighten locknuts to secure carrier to differential, and then install the left wheel hub and the two wheel and tire assemblies.

1. Insure that the retaining ring is installed on transmission axle tube.
2. Obtain keys (A) for differential carrier. File small chamfer along side and top edges at one end of keys. Then place keys in key ways with plain end against retaining ring (G). Tap keys home in key way with a brass hammer, being sure that keys remain butted against retaining ring.

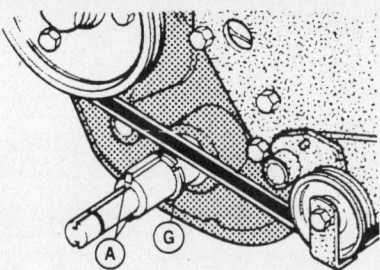

Differential gear installation

3. If axle is installed, put it from left side of transmission to right.
4. Coat keys (A) and axle tube outboard of retaining ring (G) with a film of oil. Place differential over keys and tap it with a brass hammer until it seats against the retaining ring.
5. Install the two Woodruff keys into the axle shaft key way and install the retaining ring onto the axle shaft.
6. Align the two Woodruff keys in the axle shaft with the key ways in the differential gear and push the axle shaft toward the left so that the gear encases the keys and the retaining ring contacts the gear. Move the lefthand set collar against the gear case and remove all the end play and tighten the setscrews.
7. Install the recessed thrust washer on the shaft with the recess toward the retaining ring, and install the hub seal.
8. Install the righthand hub and set collar, and remove all the end play from the axle shaft and tighten the setscrews.
9. Replace the drawbar and lift assembly.
10. Refer to the hydrostatic drive section for the following instructions:
    a. Installation of the hydrostatic drive unit.
    b. Adjustment of the hydrostatic drive unit, brake, clutch and other linkage.
    c. Filling of the gear case with oil. (Three quarts of Dexron® Hydraulic Fluid).
11. Install the wheels, seat, springs and other items that were removed and adjust as outlined in the operator's manual.

Axle

REMOVAL

1. Prepare tractor by raising and blocking up rear end.
2. Remove right wheel hub.
3. Remove differential assembly including differential gear and its keys from axle.
4. Remove any rough edges and burrs from edges of differential gear keyways in axle using a file. Use polishing cloth to smooth out any scratches in the axle. Then use a clean cloth to remove grease, dirt, fillings and abrasive particles from exposed righthand portion of axle. Polishing and cleaning axle in this manner is necessary to prevent damage to bearings within axle tube of transmission.
5. Grasp left wheel hub and pull axle out of axle tube from left side of tractor. Care must be taken to maintain axle alignment, particularly after axle slips through first bearing in transmission, to avoid binding within the bearings.

INSPECTION

1. Check axle for signs of overheating and excessive wear in the areas that run in the bearings of the axle tube and right wheel hub.
2. Check keyways and retaining ring slot to be sure that edges are sharp and able to perform retaining functions.
3. If damage is evident in checks above, axle must be replaced. In such cases, inspect and replace as necessary parts that mate with damaged area on axle.

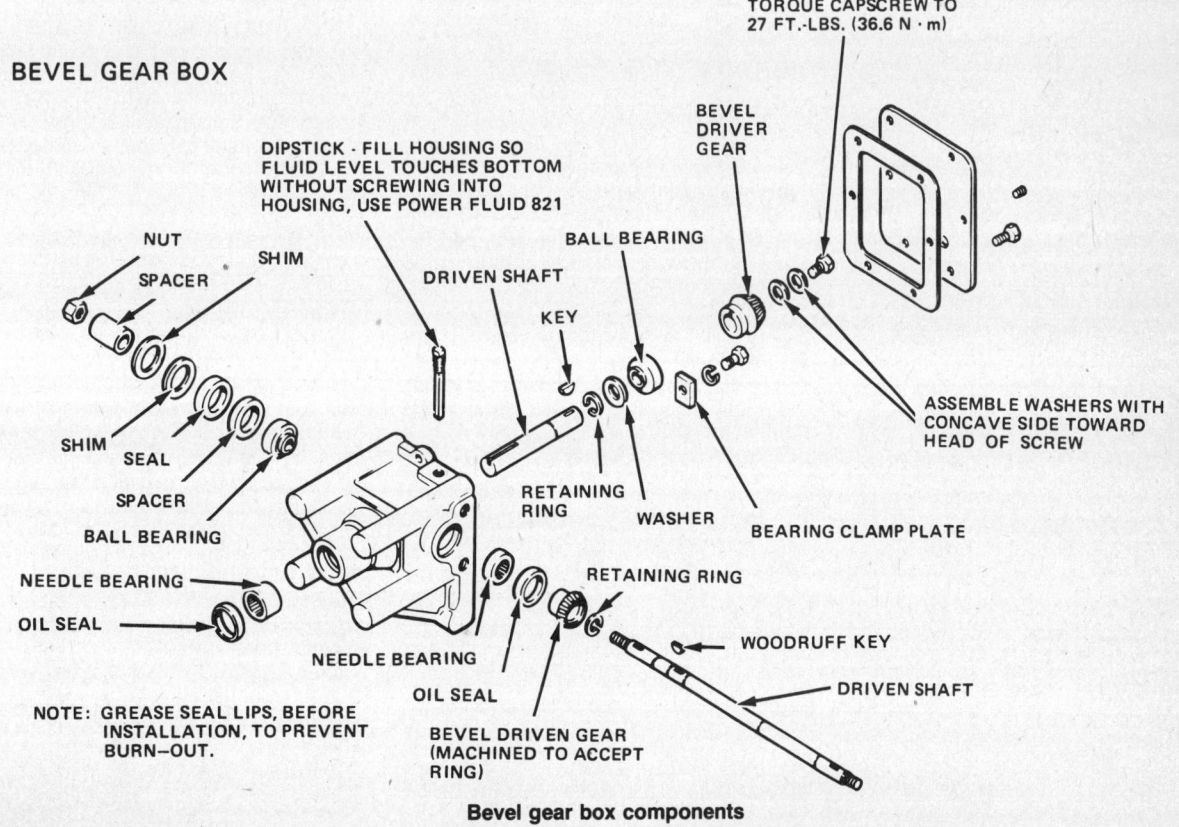

Bevel gear box components

Allis-Chalmers

INSTALLATION

1. Use a clean rag to give the axle a final cleaning. Then spread a light film of oil over the axle.
2. Carefully insert right end of axle into left end of axle tube. The axle is a running fit in the axle tube bearings. Any binding as the axle is installed will be due to axle misalignment. With this in mind, carefully continue insertion of axle through axle tube. Be especially careful when axle reaches near end of bearing at right side of axle tube. Once axle is protruding from right side, simply push axle through the rest of the way.
3. Install differential assembly.
4. Install right wheel hub.
5. Install wheel and tire assemblies.

BEVEL GEAR DRIVE

Description

The bevel drive houses a drive shaft, driven shaft and their respective bevel gear, bearings and seals. This assembly transfers the engine drive to the transmission through a drive pulley and V-belt arrangement.

Preparation

If the bevel gear drive requires servicing, the complete drive unit must be removed from the tractor in order to gain access to the bevel gear drive.

REMOVAL

1. Remove drive unit, drive belt, and pulleys.
   a. Refer to the shuttle drive section if tractor is equipped with shuttle transmission.
   b. Refer to the hydrostatic drive section if tractor is equipped with a hydrostatic transmission.
2. Refer to the PTO system section for removal of power take-off, which has to be removed before the side plates can be removed from the bevel gear housing. Remove drive pulley and key.
3. Remove the three capscrews from each side plate.
4. Pull the side plates apart, separating them from the bevel gear housing, but leaving the linkage parts still assembled to each.

INSPECTION AND CLEANING

1. Remove the large, thin, spacer washers from the short end of the driven shaft.
2. Examine the housing in the areas of the cover gasket and oil seals for signs of oil seepage. If so, note the location for future attention.
3. Remove all dirt and grease from the exterior of the housing and shafts. Steam cleaning is preferable, but a water soluble grease solvent may be used. Brush solvent on thickly, let the solvent act for a time, and then hose it off with hot water.
4. Spin the driver shaft several revolutions in both directions while applying light braking pressure to the driven shaft. Repeat this procedure, spinning the driveshaft. Note for future attention any noticeable binding, roughness, unevenness in the motion, or excessive backlash (lost motion) between the gears.

DISASSEMBLY

1. Remove drain plug (J) and drain fluid from housing. Then replace and tighten drain plug (J). If housing has filler plug with dipstick on top of housing, insert, remove filler plug during disassembly and assembly procedures.
2. Block the housing up on short pieces of wood on a bench.
3. Remove the nut (E) from the short end of the driven shaft.
4. Remove the Woodruff key (F).
5. Using a fine file, reduce any burrs around the keyway to shaft surface level. Do not remove any surface material.

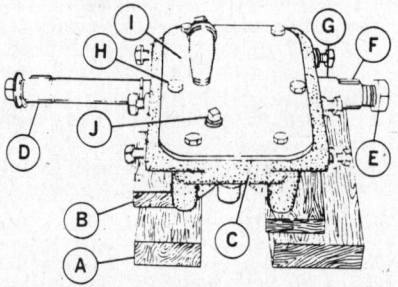

Blocking the bevel gear drive assembly

6. Remove the spacer sleeve (G).
7. With a strip of fine abrasive cloth, polish the exposed short end of the shaft to remove any file marks.
8. Using a clean rag, remove all abrasive dust from the shaft, the seal, and the keyway. Spread a film of light oil over the entire short end of the shaft.
9. Remove the six capscrews (H).
10. Using a screwdriver, carefully pry off the cover (I).
11. Applying a lead hammer as a buck to back up the gear (A), drive the shaft (B) through the gear to expose the key (C).
12. Remove the key (C).

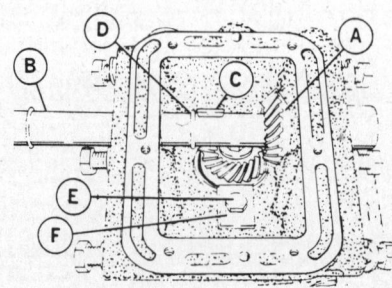

Bevel gear drive housing with cover removed

13. If it is present, disengage the retaining ring (D) from its groove and slide it over near the gear (A).
14. Stuff a clean rag into the housing under the driven shaft. Then use a fine file to remove any burrs from around the keyway and ring groove. Reduce all protrusions to shaft surface level, but do not remove any surface material. Do not use any abrasives.
15. Remove the rag, being careful not to spill any filings into the gears or bearings.
16. Pull the shaft (B) out of the housing with one hand, while holding the gear (A) with the other hand.
17. Remove the capscrew (E) and lift out the clamp plate (F).
18. Turn the housing over as shown.

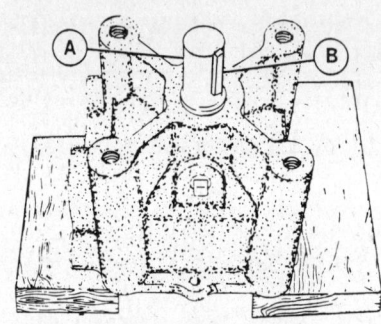

Removing driver shaft

19. Place a piece of brass rod on the overhanging end of the key (B) and drive the key out of its keyway.
20. Using a fine file, remove any burrs from around the keyway. Using a strip of abrasive cloth, polish the shaft to smooth the file marks.
21. With a clean cloth, clean grit from the shaft, the keyway, and around the seal. Spread a film of light oil over the entire exposed area of the shaft.
22. With a plastic mallet, drive the shaft downward until it begins to move freely. Be careful not to hit the seal with the mallet.
23. Turn the housing over and lift the shaft, gear, and ball bearing assembly out of the housing.

INSPECTION, CLEANING AND REPAIR

1. Hold the driver shaft with one hand and spin the outer race of the ball bearing with the other hand. Any binding or unevenness in the motion or perceptable side play of the outer race indicates the need for bearing replacement.
2. Examine the area of the driver shaft that serves as a journal of the needle bearing. Any scoring, galling, excessive wear, or discoloration due to overheating indicates the need for shaft and needle bearing replacement.
3. Examine the area of the driver shaft that turns in the oil seal. Any scratches or other surface defects indicate the need for shaft and oil seal replacement.
4. To disassemble the shaft from the gear and ball bearing for replacement of any of these items, proceed as follows:

a. Lay the shaft on a wooden block. Drive the key at the end opposite the gear back into the keyway. An end of the key should hang over the end of the shaft to permit easy removal.
b. With gear facing upward, grasp portion of driver shaft below bearing between thick lead jaws of a vise. The shaft key should be against a jaw to prevent shaft rotation in the next step.
c. With a socket wrench, back out capscrew (A) about ¼ in.
d. Remove shaft from vise. Grasp the gear only with a leather gloved hand and strike head of capscrew (A) with a plastic mallet until capscrew bottoms out on gear.
e. Remove capscrew (A) and washer(s). Then repeat step d using a piece of brass or plastic rod to drive shaft out of gear.

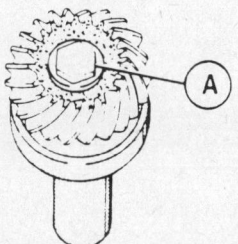

Driver shaft capscrew

f. Remove both keys from the driver shaft. Holding the shaft with the bearing above to keep filings out of bearing, remove any burrs from keyway with a fine file.
g. Use a clean rag to remove the filings and then apply a light coat of oil over the short end of the shaft.
h. Reinstall the capscrew (A). Hold the bearing with a gloved hand and gently tap the capscrew with a plastic mallet to drive the shaft out of the bearing.
5. If initial inspection indicated fluid leaks around shafts, replace related seals. Pry seals out carefully; do not scratch or score the bore that the seal fits.
6. Examine the needle bearings. To remove a damaged needle bearing, it is first necessary to remove the adjacent seal. Press the bearing inward, using a smooth faced plug with a diameter slightly smaller than the bore.
7. Spin the inner race of the ball bearing in the housing. Any binding, unevenness in motion, or sideplay indicate the need for replacement of the ball bearing and adjacent seal. To remove the bearing, push it outward with a smooth faced plug slightly smaller than the bore.
8. Examine the driven shaft. Any evidence of scoring, galling, seizing, excessive wear, or discoloration due to overheating indicates the need for shaft, seal, and needle bearing replacement. If the shaft is all right except for scratches in the area that runs in a seal, replace the shaft and the seal.
9. Examine both gear. If gears are worn or damaged or if initial spin tests indicate problems and bearing problems are not evident, replace both gears.
10. Remove all foreign material from housing gasket flange and the mating surfaces of the cover. Scrape surfaces until clean, but do not remove any surface material.
11. Thoroughly clean inside of housing with a grease solvent.

ASSEMBLY

Replacing Bearings and Seats

1. Install the ball bearing for the driven shaft into the housing. Use a press and press the bearing inward until it seats against the shoulder of the bore. Do not press against the inner race.
2. Install the related ball bearing onto the driver shaft. Use a press and press the bearing until it seats against the shoulder of the shaft. The pressing plug should be hollow with a rim that bears only against the inner race.
3. Install the needle bearings into the housing with the end marked "Torrington" facing outward. Press the bearing inward until it is flush with the bottom of the counterbore. Use a smooth faced plug about ⅛ larger in diameter than the bearing.
4. Install the seals into the housing with the spring element facing inward. Use a smooth faced plug with an outer diameter slightly smaller than the counterbore for the seal. When properly seated, the seals other than the one for the driven shaft ball bearing will be about flush with the housing surface; the seal for the ball bearing will be slightly below the flush point.

Assembly of Gears and Shafts

1. Before installing the shafts, grease the sealing lip of each seal.
2. Install the Woodruff key (A) in the gear end of the driver shaft (B) and, resting the gear (C), face down, tap the shaft downward with a plastic mallet until its hub is within a quarter of an inch of the ball bearing.

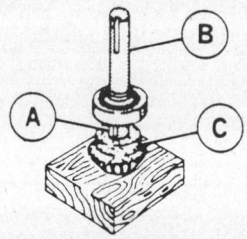

Assembling driveshaft components

3. With the key in the outboard end of the shaft temporarily installed to prevent rotation, grasp the driver shaft between thick lead jaws in a vise. Replace the washers. Tighten capscrew with concave side of washers facing the capscrew to 27 ft.lbs. drawing the gear down against the inner race of the ball bearing.
4. Remove the outboard key, clean the shaft, and cover shaft with a light film of oil.
5. Block the housing up.
6. Install the driver shaft assembly (B) in the housing, pushing it gently downward from the inside until the ball bearing enters

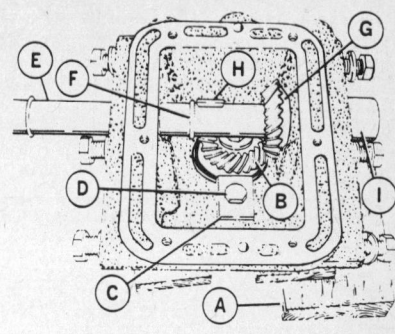

Installing shafts in housing

its seat. Do not strike the gear. Use a plastic rod held at an angle against the outer race of the bearing.
7. Install the clamp plate (C) with the arch upward. Then install and tighten the capscrew (D) until the clamp plate bends downward and is tightly clamped to the housing wall.
8. Clean the driven shaft (E) and cover it with a light coat of oil. Then insert the large threaded end of the driven shaft into the housing through the bore at the left side when the housing is viewed as shown.
9. Clean the driven shaft (E) and cover it with a light coat of oil. Then insert the large threaded end of the shaft into the housing through the bore at your left when viewing the housing as shown.
10. Slip retaining ring (F) onto the shaft. Then hold the gear (G) inside the housing while turning the shaft into the gear. When the slot appears, snap the retaining ring (F) into place.
11. Place Woodruff key (H) in its keyway on the shaft. Double electrical tape around both the shaft and the key to protect them and squeeze the key home with a pair of Channel-Lock® pliers. Do not use a hammer to drive the key into place.
12. Remove the electrical tape from the key and turn the shaft to align the keyway slot in the gear (G).
13. Install the spacer (I) with the painted end outward on the short end of the driven shaft. Turn the spacer gently through the oil seal until it seats against the ball bearing.
14. Bore a $^{15}/_{16}$ in. hole through a block of soft lumber, stand the housing with driven shaft (E) vertical and the outer end of sleeve (I) centered over the bored hole. Drive the shaft downward with a plastic mallet until the retaining ring contacts the face of the gear.
15. Test operation of the assembly by spinning first the driver shaft (B) and then the driven shaft (E). If the motion is not completely free, isolate the cause of the difficulty and correct it before proceeding further.
16. Lay a new gasket on the flange of the housing. Place the cover over the gasket and align the six screw holes in the housing, gasket and cover. Then install and tighten the six capscrews (H).
17. use a lead hammer to back up the driver shaft (A) and a brass hammer to drive the key (J) into its keyway. For easy future removal, the key should overhang the end of the shaft by about ⅛ in.

Allis-Chalmers

INSTALLATION

Install bevel gear box in reverse order of removal. Refer to removal procedure at beginning of this section.

TRANSMISSION

Description

This section provides repair instructions for the shuttle, and dual range 6-speed transmissions. Refer to Dual Range 6-speed Tractors for pulley repair.

Preparation

Removal is necessary for most transmission repairs. To prepare for this removal, it is necessary to raise the rear of the tractor and remove the rear axle. After transmission removal, the tractor must normally remain raised while the repairs are being made.

1. Select a good site for transmission removal. The site should have a level and hard floor. Block wheels to prevent the tractor from rolling forward.
2. Shift transmission to first or second gear and set the parking brake.
3. Remove seat deck. Seat deck is secured with nuts, lockwashers and bolts.
4. Drain fluid from transmission. Place large pan under transmission and remove drain plug. Then remove vent plug (A) to speed draining.
5. While fluid is draining from transmission, remove capscrew with its lockwasher and hex nut and disconnect the shifter rod from the transmission. To prevent loss, reinstall capscrew with its lockwasher and hex nut on the shifter rod.
6. To remove the transmission safety switch (B), disconnect electrical plug from this switch at this time.
7. When fluid has finished draining from transmission, reinstall both the vent and drain plugs and remove the drain pan from under the tractor.

CAUTION
Use the jack in the following step only for initial raising of the tractor. Use the pair of jack (safety) stands to support the tractor before beginning any work.

NOTE: The use of jack stands is a good practice. This practice becomes a necessity when working on the raised rear end of the tractor. Because of the pivot between the front wheels and frame of the tractor, support must be provided at two additional points to assure side-to-side stability of the raised tractor.

8. Place a portable or floor jack under the rear hitch and raise rear of tractor off floor. Adjust jack (safety) stands to keep rear wheels off floor and place one stand under each end of seat deck holddown bar. Then release and lower jack to lower tractor weight onto jack stands.
9. Remove rear wheels, differential, and rear axle. See Differential and Axle.

Planetary Gear and Sheave Removal, Shuttle Tractors

1. Remove hair pin from pivot at lower end of shuttle brake rod. After disconnecting pivot from lever, reinstall hair pin in pivot.
2. Using a screwdriver, pry grease cover off planetary gear assembly.
3. Insert a long screwdriver between the webs of the pulleys. Use this screwdriver to hold pulleys while using a wrench (B) to remove nut from end of pulley shaft. Then reinstall grease cover.

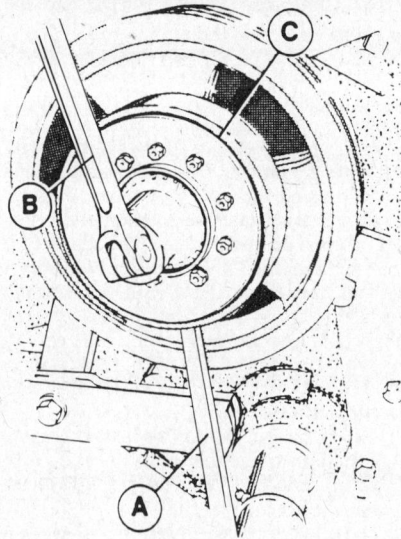

Planetary assembly with cover removed

4. Slip planetary gear assembly (C) off end of pulley shaft.
5. Remove key from pulley shaft.
6. Use a fine file and and polishing cloth to remove any burrs from edges of keyway in pulley shaft. Then use a clean cloth to remove any grit or filings.

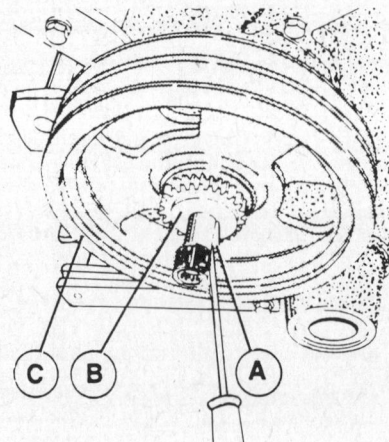

Right side view of sheave assembly

7. Remove brass washer (B) from pulley shaft.
8. Depress clutch-brake pedal while pulling main drive belt from sheave and gear assembly (C).
9. Remove sheave and gear assembly (C) from end of pulley shaft.
10. Remove spacer, second brass washer (B), and half-sheave (C) from pulley shaft.

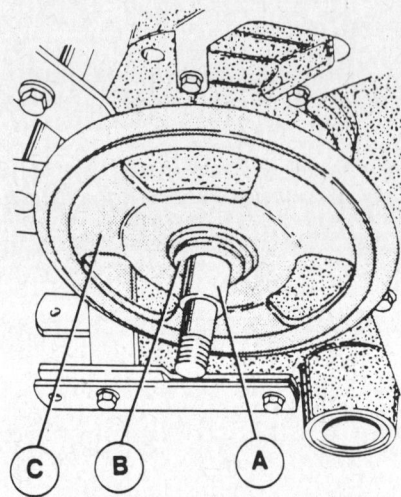

Right side view of half sheave

Transmission Removal

1. Prepare tractor for transmission removal.
2. Remove main drive belt and drive pulley or planetary gear assembly from driveshaft on transmission.
3. Remove three capscrews that secure hitch assembly to right side of transmission.
4. If tractor is equipped with a rear lift kit, remove cotter pin (C) and disconnect rod (B) from lift kit (D). To prevent loss, reinstall cotter pin in rod.

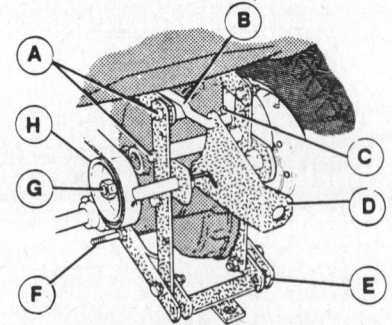

Hitch, lift and brake assembly

5. Remove two capscrews (A) and stud (F) to separate hitch assembly (E) from transmission. Then reinstall all hitch assembly fasteners in transmission holes.
6. With parking brake engaged to help prevent turning of brake drum, loosen and remove nut (G). Then disengage parking brake and slide brake drum (H) off its shaft. To prevent loss, reinstall nut (G) and any spacers on the brake drum shaft.

112

Allis-Chalmers

CAUTION

The transmission, which weighs approximately 140 lbs., will be separated from the tractor in the next step. Some form of support is required to prevent dropping of the transmission. Such support can be provided in many ways including:

7. Support the transmission from underneath using a floor jack.
8. Support the transmission from above using a sling attached to a chain hoist or to a portable crane.
9. Remove three capscrews on each side to separate the transmission from the sideplates. Use a crane, hoist or jack to hold the transmission up and to pull it clear from the tractor.
10. To prevent loss, temporarily reinstall the six capscrews in the transmission. The longer capscrew goes into the lower hole on the left side.

EXTERNAL CLEANING

To prevent the gears from being contaminated with dirt, it is necessary to clean the exterior of the transmission. While the need for cleanliness is at its greatest during assembly, the cleaning is best done prior to disassembly while the transmission is still on the floor.

Before beginning the cleaning, examine the case for signs of oil leakage. Note areas of suspected leaks so the related seal or gasket can be checked after disassembly.

Steam cleaning of the transmission case is preferred. If facilities for steam cleaning are not available, cap both ends of the axle tube with plastic bags held in place by rubber bands. Then brush a strong water soluble grease solvent thickly over the entire case. Allow the solvent to stand for a time to act. Then hose the casing off with hot water and dry it using compressed air or towels.

Transmission Disassembly

Disassembly of the transmission is best done on a workbench. The bench used should be small enough to permit access to the transmission from at least three sides.

Before hoisting the transmission to the bench, check the brake shaft and the portion of the axle tube on the same side of the transmission as the brake shaft. Burrs or paint on either must be removed before removing the gears and shafts to prevent damage to the bearings. To clean the shafts, stuff a clean rag into the open end of the axle tube. Then smooth and clean the surfaces using abrasive cloth and, if necessary, a file. To complete this process, remove the rag from the axle tube. Whether or not you polished the shafts, wipe both clean with a rag before applying a light coat of oil to each.

Hoist the transmission to the bench top using a chain hoist or a portable crane. Then lay the transmission on wooden blocking with cover facing upward.

COVER REMOVAL

1. Remove twin keys from axle tube.
2. Remove retaining ring (B) from axle tube.

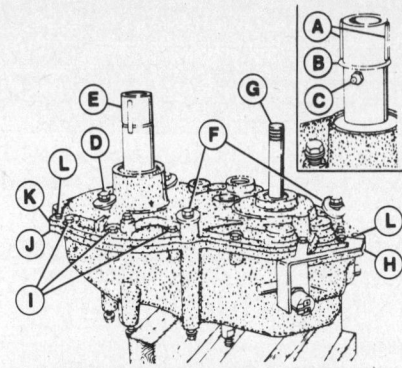

Removing transmission cover

3. Check axle tube (E) and pulley shaft (G) to be sure that both are free of paint and burrs. If so, omit steps 4 and 5.
4. Stuff a clean rag into the exposed open end of the axle tube (E).
5. Using abrasive cloth and, if necessary, a file, smooth and clean surfaces of axle tube (E) and pulley shaft (G) until all traces of paint and burrs that would otherwise damage bearings during cover removal are eliminated.
6. Use a clean cloth to remove grease, dirt, grit, and filings from axle tube and pulley shaft.
7. Remove grease fitting (C) from axle tube.
8. Wipe a light coat of oil over surfaces of axle tube and pulley shaft.
9. Remove 14 capscrews (I) from around outer edge of cover. If one is present, safety switch bracket (H) will come free and can be removed.
10. Remove two capscrews (D) on opposite sides of axle tube and two capscrews (F) on opposite sides of pulley shaft.
11. Drive the two dowel pins (L) downward out of the holes in the cover.
12. Use a screwdriver or other wedge-shaped item as a prying tool. Insert this tool at various points between the case (J) and its cover (K) and pry upward to separate the cover. Be sure to raise the cover straight up. Otherwise the cover bearings will bind on the axle tube or pulley shaft. If either the axle tube or pulley shaft rises with the cover, use a small rawhide hammer to pound the shafts back down.
13. Continue raising the cover straight up until it clears the axle tube. Then set it aside with the inner side facing up. Check the inside of the cover for any shaft washers that may have stuck to and been removed with the cover. Slip these washers back over the related shaft in the case.

GEAR REMOVAL

Gear removal can begin as soon as the cover is removed. To speed the process, gears are removed from the transmission still installed on the gear shafts for later disassembly. The shafts must be removed in a specific order as follows:

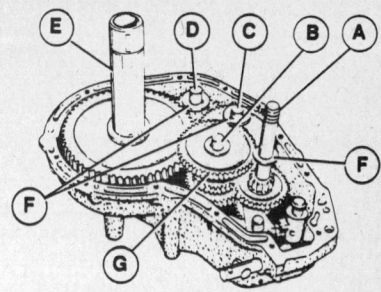

Transmission with cover removed

1. Remove washers above top gear on the pulley, second intermediate and third intermediate shafts (A, C, and D).
2. Remove top gear (G) from first intermediate shaft (B).
3. Lift axle tube (E) and its gear straight up and out of the transmission. Check bottom of case around axle tube hole for washers or spacers. If any are found, reinstall them on the axle tube before setting the latter aside.
4. Pull second intermediate shaft with its gear upward out of the transmission. Repeat check at bottom of case and reinstall any loose washers or spacers on shaft.

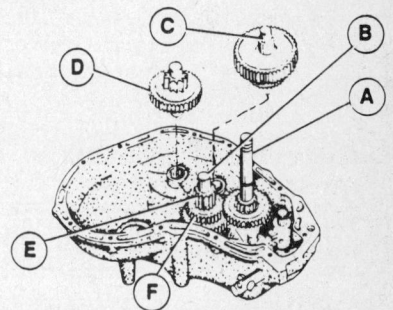

Removing second and third intermediate shafts

5. Lift up and remove third intermediate shaft (D) with its gears. Again, check for and reinstall any loose washers or spacers on the shaft.
6. Remove two upper gears (E and F) with any spacers from second intermediate shaft (B).
7. Push shaft rod fully into transmission. Insert something that can be used as a lever through the hole in the end of the shift rod and rotate the shift fully clockwise (as viewed) from lever end of shaft rod.

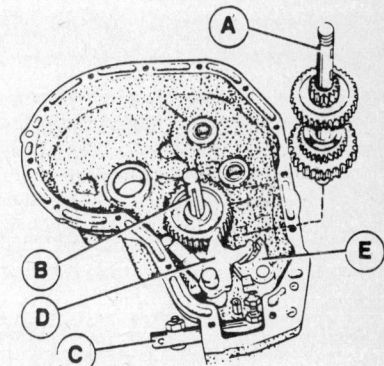

Removing first intermediate and pulley shafts

113

Allis-Chalmers

8. Pull shift fork (D) upward on its shaft until the fork's detent ball fill hole is just below the end of the shaft. Avoid lifting fork any higher to keep detent ball and spring in fork.

9. Keeping shift forks in positions established in steps 7 and 8, pull pulley shaft (A) up until its lower end comes clear of the bearing hole. Then slide pulley shaft sideways until gears clear shift forks before completing removal of shaft from transmission. Be sure that any washers or spacers left in bottom of transmission are reinstalled on the shaft.

10. Remove first intermediate shaft (B) by pulling it straight up and out of the transmission.

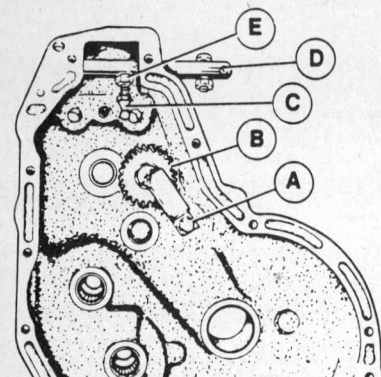

Removing reverse gear

11. On three-speed transmission only, remove locknut on opposite side of case from reverse gear. Then remove capscrew (A) and pull reverse gear (B) with its bracket out of the transmission.

SHIFT FORK AND ROD REMOVAL

Remove a shift fork or rod as described in the steps below. Avoid removing the shift rod except when necessary to replace the rod, its bearings, or the bearing seals.

1. Pull shift fork and its shaft out after loosening related setscrew (A).

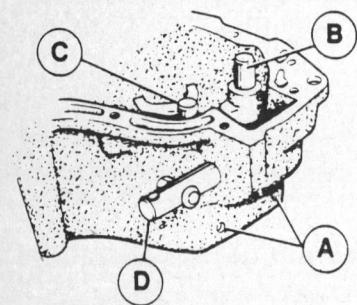

Removing shifter fork shafts

2. Once the shift fork and shaft are removed, separate them by sliding the work along the shaft away from the retaining ring. Cup a hand over the fork to prevent the loss of the denten ball or spring.

3. Before removing shift rod, loosen locknut and remove shift rod stem (C).

4. Remove shift rod by pulling it forward, out of the transmission.

CLEANING AND INSPECTION

1. Clean the inside of both the case and its cover. Use a thin grease solvent. Be sure to also remove any metal chips or filings.

2. Use a putty knife to scrape any remaining gasket material off flanges of case and cover.

3. Clean all internal gears, shafts, and spacers. Use a thin grease solvent, drying all parts thoroughly before use.

4. Inspect all gears. Replace any that have broken or chipped teeth or any that show signs of excessive wear.

5. Check all spacers. Replace any that show galling or deep scratches on the faces.

6. Check shafts for signs of excessive wear, overheating, galling, scratching in the areas that run within bearings. Replace the shaft and the related bearing when such defects are found.

7. Check and replace any damaged keys on the gear shafts.

BEARING REMOVAL

There are a total of 12 bearings and bushings in the transmission case and cover and two more bushings in the axle tube. Remove a bearing only when it is necessary to replace that bearing. Note that there are oil seals on the outer side of some bearings. These oil seals must be removed before removing the related bearing or when the initial exterior inspection indicated a possible leak from the seal. To remove an oil seal, pry it out using an awl or similar tool. Take care not to damage or scratch the seal seat in the case or cover. Once the seal is removed, discard and replace it. Never attempt to reuse an oil seal.

1. A ball bearing is used in the pulley shaft hole of the cover. On later models, a retaining ring against the inside surface of the bearing must be removed first. Use special retaining ring pliers. Then drive or press the bearing inward from the outside of the cover using a rod slightly smaller in diameter than the bearing.

2. To remove a needle bearing or the bearing which can be either a bronze or needle type depending upon the tractor model, drive or press the bearing inward from the outside of the case or cover. Use a driving rod slightly smaller than the outer diameter of the bearing. Be careful to avoid damage to the bearing housing.

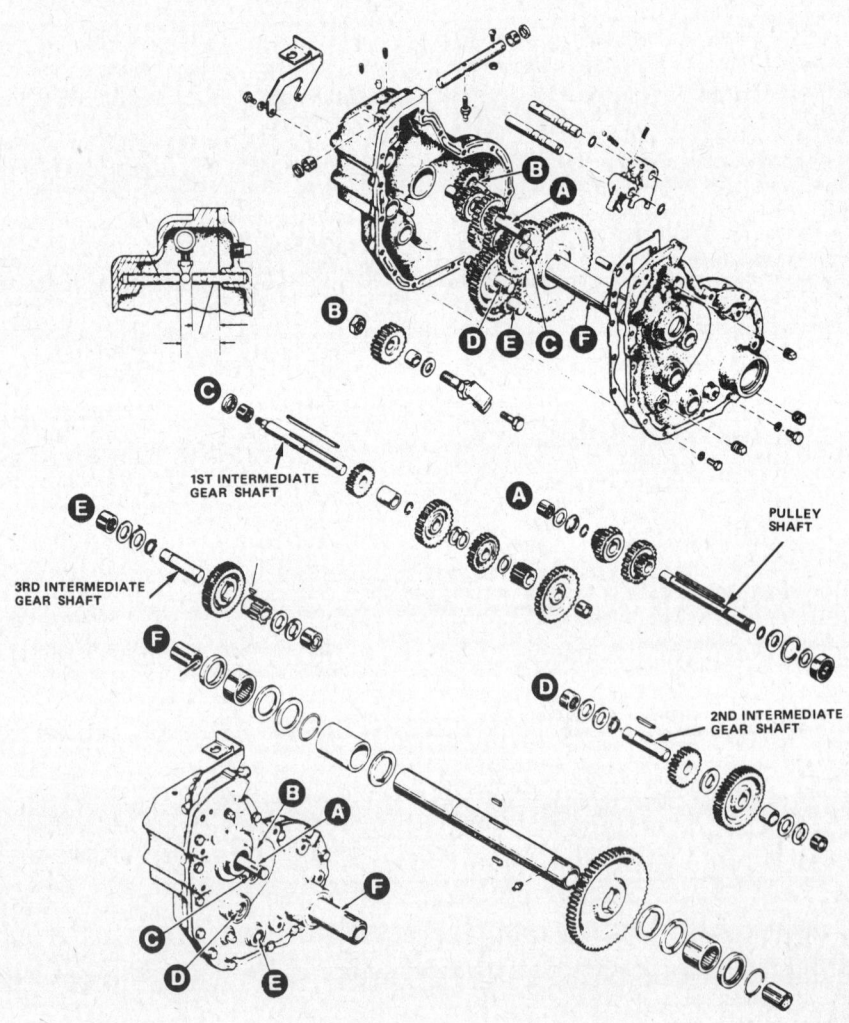

6-speed transmission components

3. Shift rod bushings can also be removed by driving or pressing them inward. Remember to remove the oil seals first.

4. Should it be necessary to remove a bushing inside the end of the axle tube, the bushing at the opposite end should also be removed. Use a bearing puller to remove these bushings.

Transmission Reassembly

GEAR REPLACEMENT

Once the gear shaft assemblies are removed from the transmission during disassembly, it is easy to replace individual gears. Simply disassemble one gear shaft assembly at a time, replacing parts as necessary. Use the exploded views to identify and properly locate parts on the shafts.

BEARING INSTALLATION

Bearing installation is done as the first step of transmission assembly. The accompanying chart lists the descriptions of each of the bearings. It also lists the step below that should be used to install the particular bearing.

Normally, it is best to press the bearings into place. However, the bearings can be driven into the housing when a press is not available. When driving a bearing into a housing, do not hammer directly on the bearing. Use a smooth, flat faced driving rod slightly smaller than the diameter of the bearing housing. For the ball bearing, the driving rod should be hollow or have its end machined away so that the rod bears only against the outer race of the bearing.

1. The ball bearing is installed from the inside of the cover. Drive or press the bearing into the housing until it bottoms out on the shoulder within the housing. Then install the retaining ring in the groove within the housing to hold the bearing in place.

2. The needle bearing for the pulley shaft and the four needle bearings for second and third intermediate gear shafts are all sealed on one end and are installed the same way. Start these bearings, unsealed end first, into the bearing housing from the outside of the case or cover. Drive the bearings inward until the unsealed end is flush with the machined surface of the housing on the inside of the case or cover.

3. The needle bearing in the cover for the first intermediate gear shaft is also sealed on one end. Start this bearing, unsealed end first, into the bearing housing from the outside of the cover. Drive the bearing inward until its unsealed end is positioned as shown.

4. The needle bearing in the case for the first intermediate gear shaft is not sealed on either end. Start this bearing with the end marked "Torrington" facing outward into the bearing housing from the outside of the case. Drive the bearing inward until its unmarked end is positioned as shown.

5. Bearings for the axle tube may be bronze bushings or needle bearings depending upon the tractor model. These bearings are always replaced in pairs. Drive the bearings into the housings from the outside to the inside of the case or cover. If needle bearings are used, the end marked "Torrington" must face outward. The inner end of the bearing in the case must be flush with the machined surface on the bearing housing inside the case. The outer end of the bearing in the cover must be flush with the second (smaller) counterbore visible within the bearing housing from the outside of the cover. If bronze bushings are used, install the one in the case first. Then place the cover on the case using the dowel pins in the flanges for alignment and insert the axle tube through the case and cover. Now slide the other bushing over the axle tube

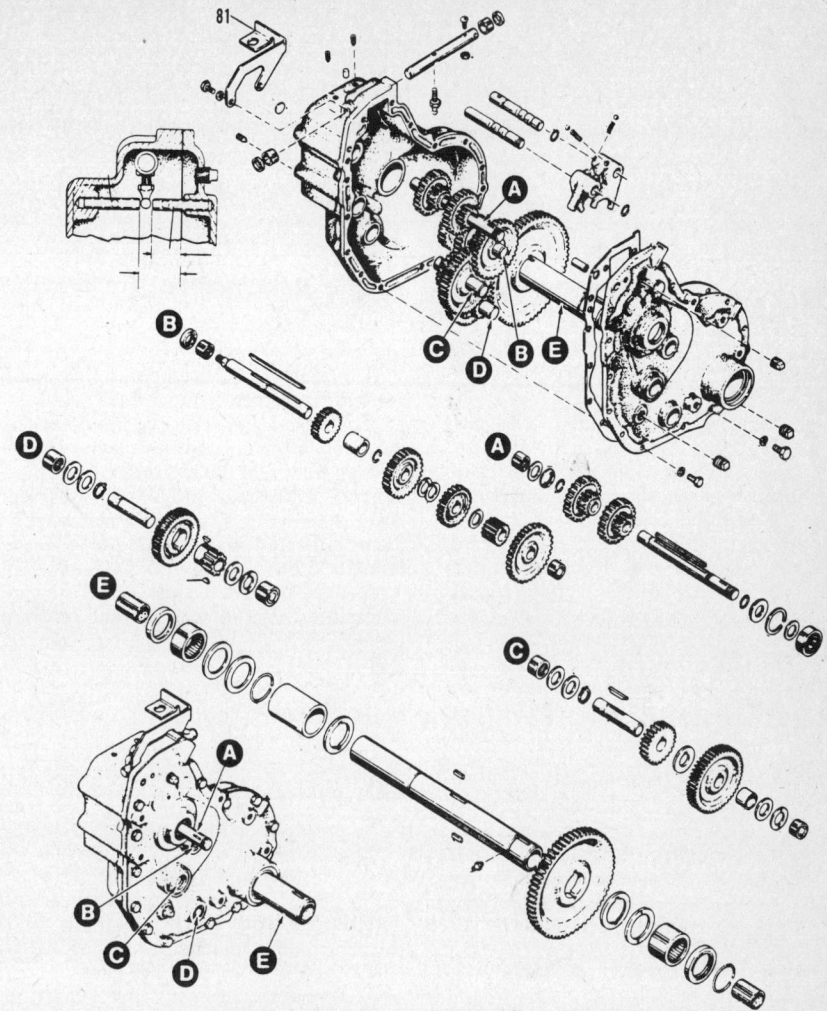

Shuttle drive transmission components

Bearing Location	Type	Installation Step
A	Ball Bearing	1
B	Needle Bearing	2
C	Needle Bearing	3
D	Needle Bearing	4
E	Needle Bearing	2
F	Needle Bearing	2
G	Bronze Bushing or Needle Bearing	5
H	Bronze Bushing	6
Inside Axle Tube	Bronze Bushing	7

Transmission bearing data

Allis-Chalmers

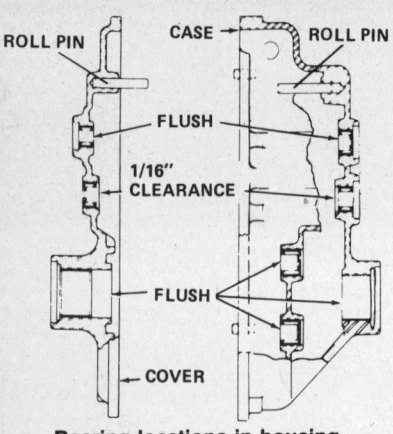

Bearing locations in housing

and press or drive it into place in the cover. Using the axle tube as an alignment mandrel in this manner prevents misalignment and binding of the bearings.

6. The bronze bushings for the shift rod are driven into the case from the outside until the inner end is flush with the end of the bearing housing on the inside of the case.

7. The bronze sleeve bushings in the axle tube are to be driven in flush with the ends of the axle tube. Use a rod at least a quarter of an inch larger than the bore of the tube. Drive in the first bushing, and then insert the axle shaft through it and out of the other end of the axle tube. Place the second bearing on the axle shaft and then center the bearing with the bore and drive it while guided by the axle shaft. This procedure insures exact alignment of the two bushings so that binding will not occur.

ROLL PIN INSTALLATION

There are two roll pins, one in the case

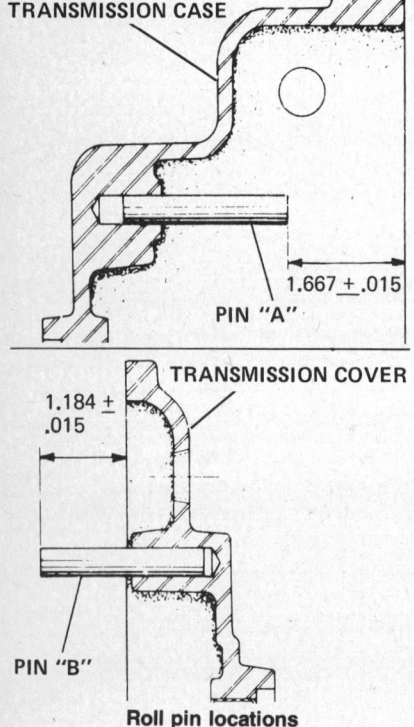

Roll pin locations

and the other in the cover. These pins limit movement of the shift mechanism and must be properly installed if the shift is to work. Check these pins even if replacement was not necessary. Be sure that the grooves in the pins face the bottom of the case or cover and that the pins protrude from the casting to the dimensions shown. Make measurements from the top of the pins to the machined surfaces on the flanges of the case or cover.

SHIFT ROD INSTALLATION

The shift rod is installed in the transmission case as follows:

1. If seals (C and E) are installed, run a greased finger gently around the seal lips.

2. Coat rod with a film of light machine oil. Then insert end opposite clevis pin hole through front seal (C) and bearing and push the rod gently rearward through the rear bearing and seal (E). If extra resistance is felt as rod enters rear seal, do not force rod through. Instead, pull rod forward again and pry out and discard rear seal. Take care to avoid damage to seal seat and bearing.

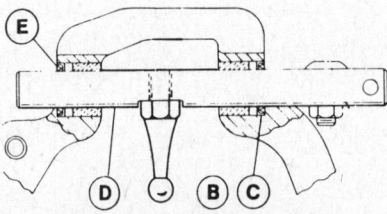

Shift rod installation

Then push rod rearward again until threaded hole is centered in housing. Seals, if removed, will be installed later.

3. Spread a coat of Loctite®, Grade A, on thread of shift stem (B) and its hole in the shift rod.

4. Install shift stem (B) into its hole in the shift rod. Tighten new style stem against flat of shift rod.

ROLL PIN ALIGNMENT TO CORRECT LOCKING IN GEAR, TESTING SHIFT ROD (1665525) FUNCTION

1. Remove shift rod assembly 2087693.

2. Apply clockwise (as viewed from rear) pressure to shift rod (1665525) while moving it forward and back. It should move in straight line, if not, the roll pin (which guides movement of shift rod) in the case needs adjusting.

3. Apply counterclockwise (as viewed from rear) pressure to shift rod (1665525) while moving it forward and back. It should move in straight line. If it deflects, the roll pin in the cover needs adjusting.

ADJUSTING ROLL PINS

1. If the results from the testing indicated that the roll pin(s) need to be moved in, follow the procedure(s) outlined below.

2. Using the template positioned on the case as shown, mark the casting, using a center punch.

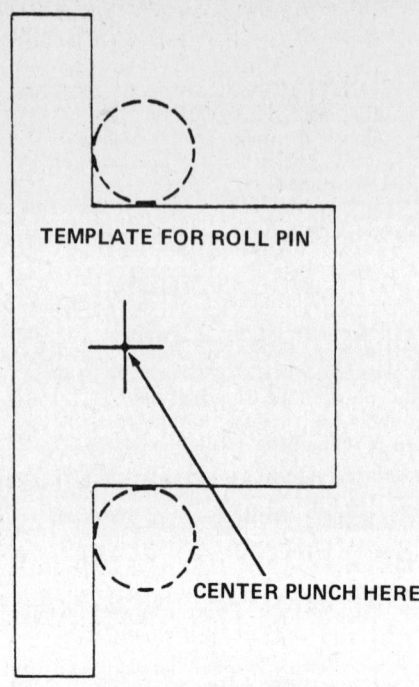

Roll pin template

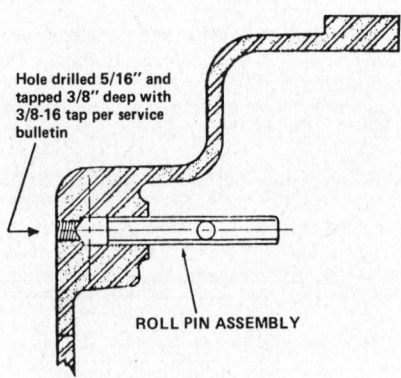

Drilling into the roll pin bore

3. Drill through the casting into the roll pin bore, using a 5/16 in. drill.

4. Thread the hole 3/8 in. deep, using a 3/8 in. 16 N.C. tap.

5. Remove top oil fill plug of case assembly and blow compressed air through the plug hole to pressurize case and to clear chips out of newly drilled and threaded hole.

NOTE: Use extreme care in operation 6, not to move roll pin in too far and block space in which shift stem moves.

6. Roll pin may be moved in, using a drift punch and hammer. Tap the roll pin in, to guide shift rod in a straight path, without deflection.

7. Remove right rear wheel and pulleys.

8. Mark casting in the center of the roll pin boss, using a center punch.

9. Drill through the casting into the roll pin bore, using a 5/16 in. drill.

10. Thread the hole 3/8 in. deep, using a 3/8 in. 16 N.C. tap.

11. Use extreme care in operation 12, not to move the roll pin too far in.

Allis-Chalmers

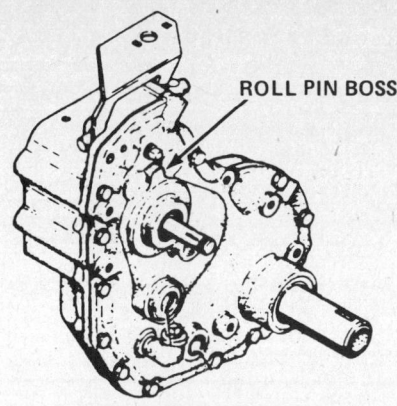

Marking the roll pin boss

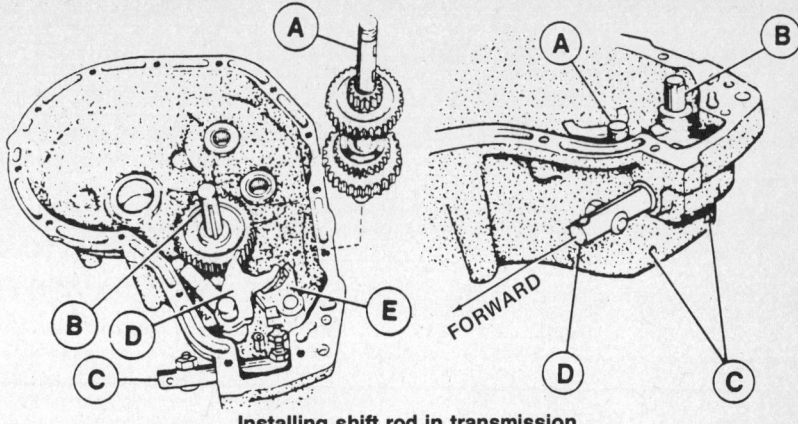

Installing shift rod in transmission

12. Roll pin may be moved in, using a drift punch and hammer.

SHIFT FORK POSITIONING

Remove the input pulleys. A shift fork that is not in the neutral position on the shift rail may be pushed or pulled into the center detent, through the top oil fill plug hole, using a heavy piece of brazing rod or wire 6-8 in. long. Bend the rod into an "L" shape with a ⅜ in. long leg and put through the plug hole to move the fork. Oil fill plug and input pulleys may be replaced once the shift forks are in the neutral position.

FINAL STEPS

The shift rod (2027693) may be attached, and the drilled and threaded holes can be plugged with a ⅜ in.-16 x ⅜ socket head screw with sealant applied to threads to prevent leakage.

SHIFT FORK INSTALLATION

The shift forks are identical. But, the shafts (A and B) for the forks are different. One shaft (B) is slightly longer than the other and has its retaining ring groove near one end. The other shaft (A) has its retaining ring groove near the setscrew hole (E).

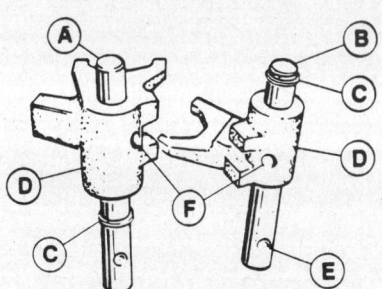

Pulling shift rod forward

Note that both forks are installed on the shafts with the long portion of the hub facing the retaining ring. To assemble and install the shift forks, proceed as follows:

1. Install retaining ring (C) on one shift fork shaft (A or B).
2. The retaining ring divides the shaft into two parts. Start longer part of the shaft into hub side of fork (D). Push shaft into fork only as far as near side of ball loading hole (F).
3. Install a detent spring and then a ball into ball loading hole. Insert a narrow flat screwdriver into the hole and press ball and spring down below shaft hole. Then push shaft forward through fork to trap the ball and spring in permanent operating position.
4. Repeat steps 1 through 3 to assemble other shift fork and shaft.
5. Pull shift rod forward as far as it will go.
6. Select shift fork that has its retaining ring (C) located on end of shaft furthest from setscrew hole (E).
7. Install selected shift fork in case with the setscrew hole down and facing the setscrew.
8. Tighten setscrew (C) while feeling for setscrew hole. When the two are in apparent alignment, loosen setscrew one turn and rotate shift fork shaft slightly. Then retighten setscrew and note whether or not setscrew forces shaft to return to prior position. If it does, setscrew is properly seated and can be tightened. If not, the entire step should be repeated.
9. Push shift rod as far rearward as it will go.
10. Install other shift fork assembly (A) into case as described in steps 7 and 8.

GEAR INSTALLATION

Once the shift rod and forks are installed in the case, gear installation can begin. To speed the process, all gears should be assembled on their shafts before installation.

1. On three-speed transmissions only, (used in 6-speed tractors), begin gear installation by installing reverse gear. Assemble gear with its spacer and washer on the bracket. Then install this assembly in case as shown. Torque capscrew (C) to 20 ft.lbs. Complete assembly by installing locknut on pin at backside of case and torque it to 55 ft.lbs.
2. Push shift rod fully into transmission, insert something that can be used as lever through hole in the end of the shift rod and rotate the shift rod fully clockwise (as viewed from lever end of shift rod).
3. Pull shift fork (D) upward on its shaft until the fork's detent ball fill hole is just below the end of the shaft. Avoid lifting fork any higher to keep detent ball and spring in fork.
4. Remove the two washers from lower end of pulley shaft (A). Coat these washers with grease and place them over pulley shaft hole in bottom of case. The composition washer must be on the bottom with its grooved side down.
5. Install first intermediate gear shaft assembly (B) into its needle bearing at bottom gear case. Then remove the three top gears from this shaft and set them aside for later installation.
6. Lower pulley shaft assembly (A) with long end up into bottom of gear case away from shift forks (D and E). While holding large gear on first intermediate gear shaft up for clearance and tilting pulley shaft (A) as necessary, slide pulley shaft sideways toward forks. Engage rear fork (E) in groove above lower gear on pulley shaft and front fork (D) in groove below upper gear on pulley shaft.
7. With gears properly engaged in forks, insert lower end of pulley shaft through washers installed in step 4 and into bearing at bottom of case.
8. Carefully tap front shift fork (D) back down to its neutral position, where detent ball will snap into groove in shaft.
9. Turn shift rod (C) back to neutral position to lower rear shift fork (E).
10. Pull shift rod (C) back and forth several times and be sure that shift stem alternately engages both forks.
11. Be sure that two spacers are placed above the upper gear on first intermediate gear shaft. Then install in order the larger gear with its beveled side up, another spacer, and the smaller gear.
12. Remove two washers from lower ends of both the second and third intermediate gear shaft assemblies. Coat these washers with grease and center them over respective bearings in bottom of case. The composition washers must be on the bottom with the groove sides down.
13. Install third intermediate gear shaft assembly into its bearing in bottom of case. The larger gear on this shaft should be on the bottom.
14. Install second intermediate gear shaft assembly into its bearing in bottom of case. The larger gear on this shaft should be at the top and a spacer should be above the gear.
15. Remove two washers from lower end of axle tube assembly. After coating

Allis-Chalmers

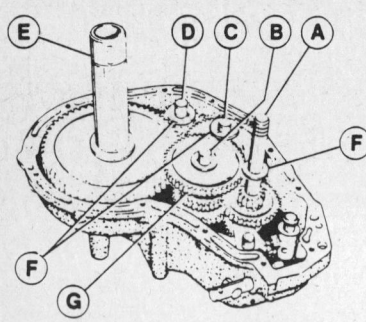

Finishing gear installation

these washers with grease, center them over the axle tube bearing in bottom of case. The composition washer must be on the bottom with the grooved side down.

16. Install axle tube assembly (E) into its bearing in bottom of case. Be sure that the two spacers above the gear on this shaft are installed.

17. Install top gear (G) with its beveled side down on first intermediate gear shaft (B).

18. Grease the two pairs of washers for the second and third intermediate gear shafts. Then install one pair of washers on each shaft (C and D). The composition washers must be on the top with the grooved side up.

19. Grease the two washers for the upper end of pulley shaft (A). Then install them on the shaft. The thicker washer must be on the bottom with its large flat face up.

COVER INSTALLATION

When gear installation is completed, the cover can be installed on the transmission case. It is then advisable to check the gears for proper shifting and lack of binding before completing assembly of the transmission.

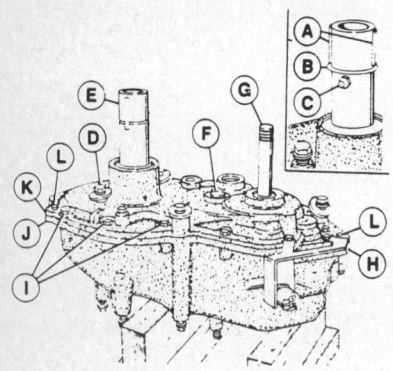

Installing transmission cover

1. Drive the two dowel pins (L) in flange of case upward. These pins must protrude above flange a distance equal to the thickness of the cover flanges.

2. Place cover gasket on flange of case. Be sure that gasket is flat and properly centered over dowel pins and holes in flange.

3. Coat the upper ends of the gear shafts in the case with a light film of oil.

4. Install the cover (K). Begin by centering the proper bearings in the cover over the axle tube (E) and pulley shaft (G). Then hold the cover level while lowering into place. Do not force the cover. If binding occurs, raise the cover slightly to relevel it before lowering it again. Continue until the dowel pins in the case flange enter the holes in the cover.

5. When the dowel pins begin entering the holes in the cover, it will be necessary to begin driving the cover into place. Be sure that the pins are centered in the cover holes. Then use a wooden mallet and strike the cover alternately over the two dowel pin holes.

6. When the cover is seated on the gasket, place lockwashers over the fourteen capscrews (I) and loosely install the capscrews through the cover and into the case. On later models, two of these capscrews also pass through the safety switch bracket (H) which should be installed at this time.

7. Select four capscrews evenly spaced around and near the top, bottom and two sides of the cover. Alternately tighten these four capscrews snugly, but not completely, to draw the cover evenly down to the case. Then proceed around the cover to snug up the other ten capscrews. Finish by tightening all fourteen capscrews a little at a time until the cover is firmly secured.

8. Check the gears for binding. Temporarily install a lathe dog or pulley on the pulley shaft (G). Insert a rod through the shift rod clevis and shift the gears into high. Then turn the pulley shaft while observing the axle tube. The tube should turn freely without binding in all gears. If binding occurs, try striking ends of pulley and first intermediate shafts (G and F) moderately hard blows with a wooden hammer. If this treatment does not relieve binding immediately, it will be necessary to disassemble the transmission to locate and correct the trouble.

OIL SEAL INSTALLATION

There are five oil seals over the bearings in the transmission case and cover. If any of these seals were removed, new seals should be installed only after the transmission is reassembled and checked. Installation should be done with the pressing tool and thimble described in the accompanying illustration. If a thimble is not available, heavy paper lubricated with grease or oil can be used as a substitute. Wrap this paper in a spiral fashion over the shaft beginning near the seal seat. Be sure that the paper or thimble covers any keyways in the shaft. Specific steps for seal installation are as follows:

1. Check shaft in area of seal seat. If shaft shows nicks, scratches, or burrs in this area it must be replaced. Attempts to file or remove flaws will only cause flat spots and oil leakage.

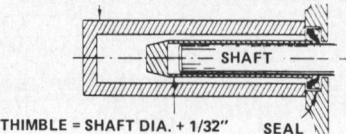

Seal installation tools

2. Check the seal seat or bore. Be sure to remove any nicks, scratches, burrs and foreign material.

3. Slide thimble over shaft and coat it lightly with clean grease.

4. Remove seal from package and gently lubricate it with clean grease. Do not run finger roughly around seal as it is easily damaged.

5. Place seal onto thimble with spring element facing case or cover.

6. Drive seal into seat in cover or case using the pressing tool. Avoid tilting the seal as it starts into seat and do not hammer directly on seal. When properly seated, outer surface of seal should be flush with shoulder of bore or seat in cover or case.

Final Assembly

All that remains to complete assembly of the transmission is to install parts on the axle tube. Do this as follows:

1. Install grease fitting in its hole in end of axle tube.

2. Before installing retaining ring (B) check it for damage. Discard it if it has bent out of round or if the ends are more than 1/8 in. apart.

3. Use retaining ring pliers to expand ring (B) and slide it over axle tube into its groove.

4. Install twin keys (A). Flat end of keys must butt up against retaining ring. Use a brass hammer to tap keys down into keyways. Be sure keys are fully bottomed along entire length.

Transmission Installation

1. Hoist transmission into place between side plates at rear of tractor.

2. Align holes and install capscrews. Installation is easiest if upper capscrew on each side is installed first. Note that one of the six capscrews is longer and must be installed into lower hole on left side.

3. Install brake drum on shaft.

4. Engage parking brake to help prevent turning of brake drum. Then install and tighten nut to secure brake drum in place.

5. Lift rear hitch into place on transmission. Secure hitch by installing two capscrews and stud on left side and three capscrews on right side.

6. If tractor is equipped with a rear lift kit connect to lift kit and install cotter pin cable clevis assembly with pivot pin.

Planetary Gear and Sheave Installation Shuttle Tractors

1. Use a clean rag to wipe all grease and oil off the sheave and half-sheave.

2. Install half-sheave on pulley shaft with belt groove facing outward. If necessary, use a plastic mallet to fully seat half-sheave.

3. Stretch shorter V-belt over pulley

shaft and into groove of half-sheave. Belt must be seated properly for good drive.

4. Install brass washer and spacer on pulley shaft.

5. Install sheave and gear assembly on pulley shaft.

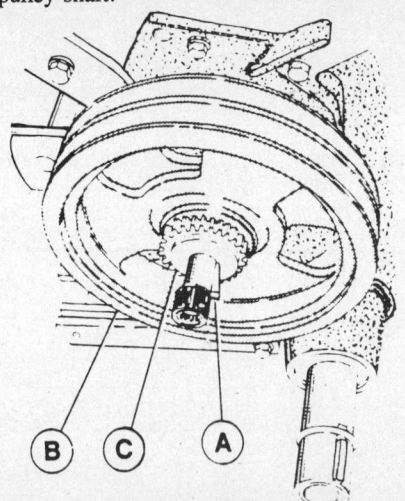

Installing sheave

6. Depress and hold clutch pedal of tractor while stretching main drive V-belt over pulley shaft and into groove of sheave.

7. Install composition washer (C) on pulley shaft.

8. Install key (A) using a plastic hammer to seat it in keyway.

9. Install planetary gear assembly aligning its keyway with key on pulley shaft. Bull gear (D) must be on outboard side. Use a plastic hammer to fully seat gear assembly on key.

10. Insert a long screwdriver (A) through webs of sheave and half-sheave. Use this screwdriver to hold sheaves while using a wrench (B) to install and tighten nut on end of pulley shaft. Torque nut to 50-55 ft.lbs.

11. Install grease cap. Wipe off any grease from planetary assembly brake drum.

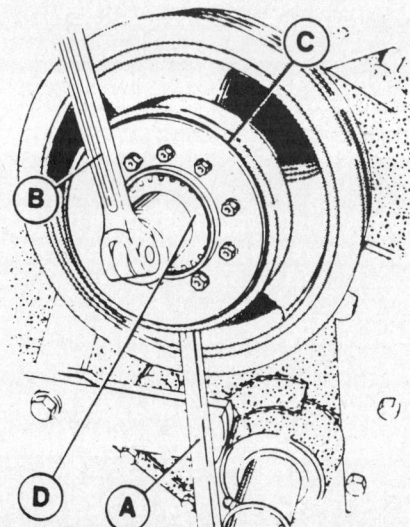

Installing planetary assembly

12. Install brake band. Hold hex shaft while tightening capscrew.

13. Connect pivot at lower end of brake rod to lever. Then install cotter pin through hole in pivot to secure pivot to lever.

Final Installation Steps

Complete installation of the transmission in the tractor as follows:

1. Install drive belt and pulley or planetary gear assembly on transmission driveshaft.

2. Install rear axle, differential, and wheels.

3. Connect shift rod to transmission using a capscrew, lockwasher, and nut.

4. If tractor is equipped with transmission safety switch, connect plug to switch.

5. Complete lubrication of transmission, axle, and on shuttle, tractors only, the planetary assembly.

6. Install seat deck. Secure it with nuts, lockwashers, and capscrews. On later models, be sure to install seat deck stop.

7. It is recommended that the tractor be operated at this time to check the repair as well as to identify any pre-existing conditions that might warrant adjustments. If adjustments of the clutch, brake or electrical parts are required, refer to the applicable section of this manual.

Allis-Chalmers

SHUTTLE CLUTCH UNIT

Overhaul

REMOVING SHUTTLE CLUTCH UNIT FROM TRACTOR

1. Jack up the rear of the tractor so that the right rear wheel may be removed. Remove the wheel.

2. Raise the seat deck to allow ample work room.

3. Remove the spring clip "A" from the brake band assembly. Loosen the capscrew "B" two turns. The brake band may now be removed from the bracket "C" and rotated downwards.

4. Using a large screwdriver, pry the dust cap from the planetary.

5. Place a larger screwdriver through the slots "A" in the two pulley sheaves to prevent them from turning. Remove the locknut "B" securing the planetary assembly.

6. Remove the planetary assembly from shaft.

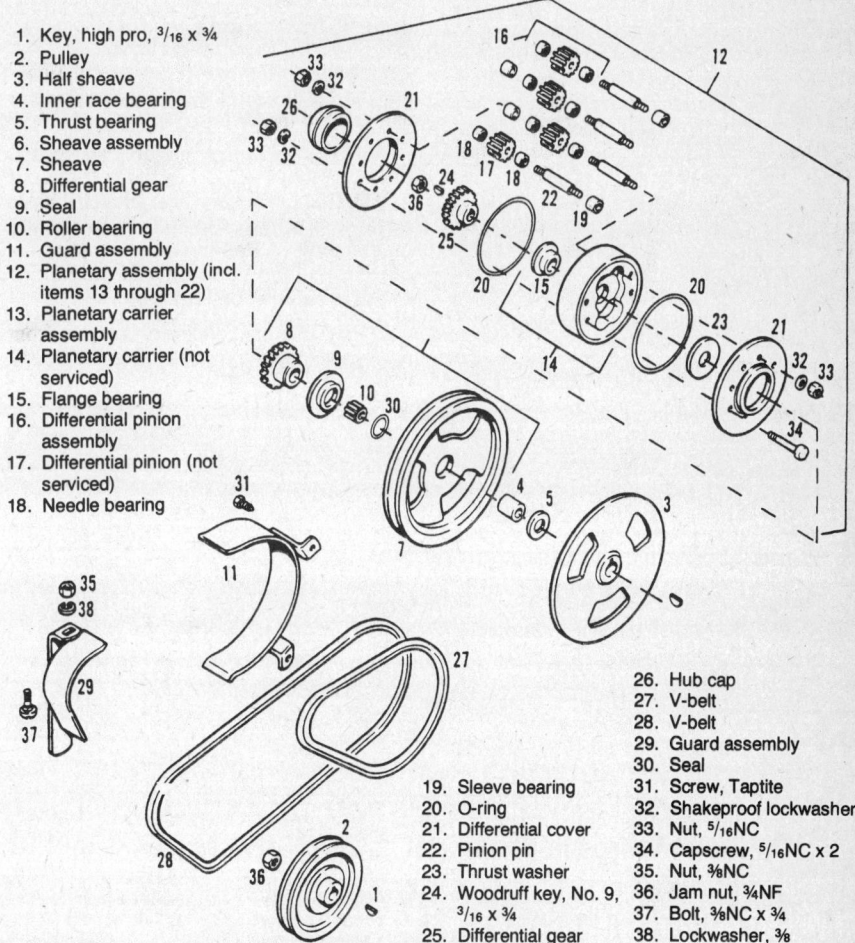

1. Key, high pro, 3/16 x 3/4
2. Pulley
3. Half sheave
4. Inner race bearing
5. Thrust bearing
6. Sheave assembly
7. Sheave
8. Differential gear
9. Seal
10. Roller bearing
11. Guard assembly
12. Planetary assembly (incl. items 13 through 22)
13. Planetary carrier assembly
14. Planetary carrier (not serviced)
15. Flange bearing
16. Differential pinion assembly
17. Differential pinion (not serviced)
18. Needle bearing
19. Sleeve bearing
20. O-ring
21. Differential cover
22. Pinion pin
23. Thrust washer
24. Woodruff key, No. 9, 3/16 x 3/4
25. Differential gear
26. Hub cap
27. V-belt
28. V-belt
29. Guard assembly
30. Seal
31. Screw, Taptite
32. Shakeproof lockwasher
33. Nut, 5/16NC
34. Capscrew, 5/16NC x 2
35. Nut, 3/8NC
36. Jam nut, 3/4NF
37. Bolt, 3/8NC x 3/4
38. Lockwasher, 3/8

Forward/reverse shuttle drive unit components

Allis-Chalmers

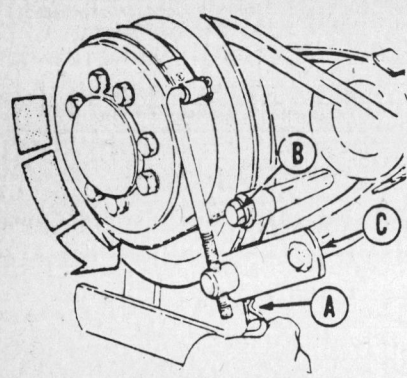

Removing brake band

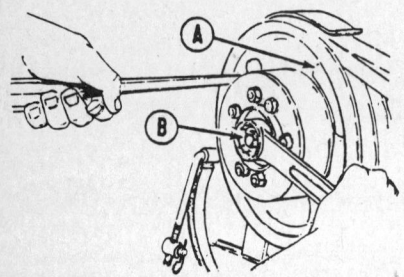

Removing planetary locknut

7. Remove the key "A" from the transmission pulley shaft. Be sure the key used is of the regular Woodruff design—NOT a high profile design.

8. Remove the thrust washer from the transmission pulley shaft. The pulley sheave may now be removed. This is accomplished by depressing the foot clutch and sliding the pulley sheave from the transmission input shaft. The sheave can then be removed from the belt.

9. Place the forward clutch belt down, away from the split sheave. Remove the split sheave which is keyed to the transmission shaft.

10. Disassembly and assembly of planetary unit

a. To disassemble the unit, remove the nuts from bolts (34) and the planetary shafts.

b. Remove cover (21), seal ring (20), and bearing (15). The needle bearings (18), planetary pinions (17), sleeves (19), and planetary carrier (14) can now be removed.

c. Clean all parts and replace any parts that show excessive wear or other damage.

d. Use new seal rings, and reassemble by reversing the disassembly procedure.

e. Tighten the nuts on capscrews (34) and planetary shafts (22) to a torque of 12-15 ft.lbs.

11. The assembly process is the reverse of the disassembly.

a. There are two sides to the planetary, one with eight hex nuts showing, and the other with four hex nuts and four capscrews showing. The side with the eight hex nuts faces outward from the tractor.

b. The locknut securing the planetary is torqued to 75 ft.lbs.

c. The capscrew securing the brake band assembly is tightened to a torque of 30 ft.lbs. When tightening the capscrew, use a wrench on the inside capscrew to prevent the assembly from turning.

d. Pack the planetary unit with No. 2 lithium grease. Do not get any grease on the planetary brake band.

e. Replace cover (26) and adjust shuttle linkage.

DUAL RANGE 6-SPEED TRACTORS

Description

The dual range 6-speed tractor has 3 forward gears and one reverse gear with a dual range pulley system. To select the High or Low range, tractor movement must be stopped and the clutch-brake pedal must be fully depressed. Push the dual range lever forward for High range or pull it back for Low range, and then select the proper gear.

PULLEY AND BELT REMOVAL

1. Remove the two capscrews that hold the front belt guard to the tractor frame, and remove the belt guard.

NOTE: The idler pulley can be replaced without removing the belt guards and belts. Depress the clutch-brake pedal and remove the belt from the top of the pulley. Remove the locknut, replace the pulley, and tighten the nut securely. Place the belt back on the pulley.

2. Remove the cotter pin from the front end of the pulley brake rod. Do not remove the brake rod from the rear belt guard, unless it has to be replaced.

3. Remove the two capscrews that hold the rear belt guard to the tractor frame, and remove the belt guard and pulley brake rod together.

4. Replace the needed belt, and install the belt guards and pulley brake rod in the reverse order that they were removed. If either the drive pulley or driven pulleys have to be replaced, remove both belts.

5. To replace the drive pulleys, remove the nut in the center of the pulleys, from the bevel gear box cross shaft.

6. Replace drive pulley and tighten nut. Replace belt guards, belts, and pulley brake rod.

7. To replace the driven pulleys, remove the three capscrews and pulleys. There are three spacers between the two pulleys. The driven pulley hub can be removed, by removing the nut on the transmission input shaft. This hub is keyed to the input shaft. When installing a new hub, replace the key.

8. Install the key, hub, and tighten the nut. Then install the smaller pulley, three spacers, the larger pulley and tighten the three capscrews. Install the two belts, belt guards, and pulley brake rod.

DUAL RANGE CONTROLS

1. To replace the idler pivot assembly (18), depress the clutch-brake pedal and lift the belt off of the idler pulley (17).

2. Disconnect the control rod (20) from the cam lever (21), and remove the setscrew from the cam lever (21).

3. Slide the pivot assembly (18) to the right, remove the key (19) and remove it from the tractor frame.

4. Replace the needed parts, and assemble in the reverse order.

5. If the bushings in the dual range lever need replacing, remove nut (7), washers (6 and 5), and control link (4). Slide the control lever (1) to the right and replace one bushing (3) from the outside and one from the inside of the tractor frame.

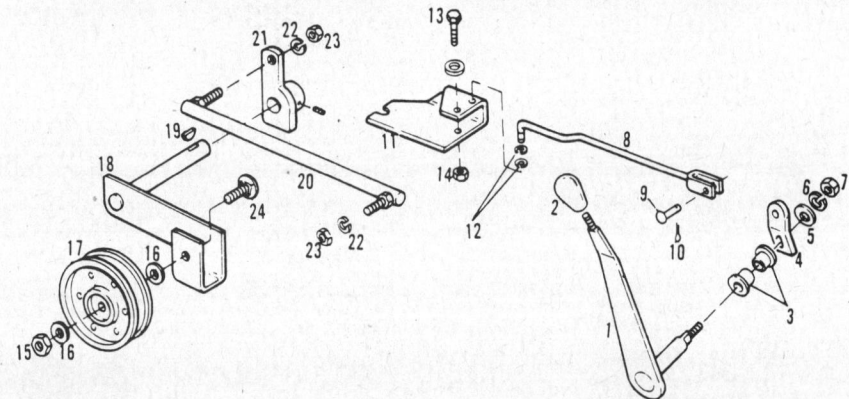

1. Control lever assembly
2. Knob
3. Flanged bushing
4. Link
5. Washer, ⅜
6. Lockwasher
7. Nut
8. Control rod assembly
9. Pin
10. Cotter pin
11. Shift fork
12. E-ring
13. Capscrew, ⅜NC x 2
14. Locknut, ⅜NC
15. Nut, ⅜NC, lock
16. Washer, 5/16
17. Idler pulley
18. Idler arm assembly
19. Woodruff key, no. 5, ⅛ x ⅝
20. Ball joint
21. Lever assembly
22. Lockwasher, 5/16
23. Nut, 5/16NF
24. Bolt, ⅜NC x 1¼

Dual range control components

Allis-Chalmers

6. Install the control link (4), washers (5 and 6), and secure with nut (7).

ADJUSTMENTS —DUAL RANGE

Rear Belt Guard Adjustment

1. Loosen, but do not remove the two capscrews which secure the rear belt guard to the side of the transmission.

2. Adjust the belt guard to obtain 1/8 in. clearance between the outside diameter of the large pulley and the inside diameter of the retainer.

3. Tighten the two capscrews and recheck for 1/8 in. clearance. It may be necessary to bend the belt guard to obtain proper clearance.

Cam Lever Adjustment

1. Depress the clutch-brake pedal and shift the dual range lever to high range (forward). The sliding idler arm should be tight against the right side plate bushing. This will slide the shifting cam lever into contact with the lefthand side of the gate finger.

2. Check to see if binding occurs between the cam lever and gate finger. To eliminate binding, loosen the setscrew in the hub on the shift cam lever and reposition the cam lever on the idler arm shaft, so that it is against the gate finger by sliding the cam lever right or left on the idler arm shaft. Tighten the setscrew securely.

Idler Height Adjustment

1. Depress the clutch-brake pedal and move the dual range lever until the cam lever is under the gate finger. Release the pedal and let the spring tension hold the cam lever in position.

2. With the outside belt resting on both the front and rear belt guards, loosen the idler adjusting bolt and position the idler pulley so there is a minimum of 1/8 in. clearance between the belt and outside diameter of the idler pulley. Tighten the adjustment bolt securely.

NOTE: If pulley is adjusted to end of slot and 1/8 in. clearance isn't held, bend rear belt guard up to support belt and gain clearance.

Clutch Rod Adjustment

1. Depress the clutch-brake pedal and move the dual range lever in High range.

2. With the pedal up in engaged position, there should be a clearance of 3/8 in. between the front jam nut and the rod guide. Pull back on the pedal to make sure it is completely in engaged position.

3. To make the proper adjustment, loosen the two jam nuts and turn the front jam nut in or out to obtain the 3/8 in. measurement. Hold the front jam nut in position and tighten the rear jam nut securely.

Transmission Brake and Parking Brake Adjustment

1. Depress the clutch-brake pedal and move the dual range lever unit the cam lever is under the gate finger.

2. Place a .030 in. feeler gauge between the cam lever and the end of the gate finger. Release the pedal and let the spring tension hold the feeler gauge in place.

3. Loosen the jam nut behind the parking brake lever and turn the lever clockwise to increase tension or counterclockwise to decrease tension, until tension is felt when putting the lever into engaged position. Tighten locknut when properly adjusted. Clockwise and counterclockwise direction given is standing in front of the tractor looking at the parking brake lever.

4. With the parking brake engaged, loosen the brake rod jam nuts, and turn the front jam nut in or out until it just comes in contact with the brake band plus 1/3 turn. Hold the front jam nut in position and tighten the rear adjusting nut securely.

Pulley Brake

1. With the clutch-brake pedal engaged and the dual range lever in High or Low range, set the coil length of the return spring (the rear spring) to 1 3/8 in. Loosen the jam nuts and turn the front jam nut in or out to obtain this measurement.

2. Depress the clutch-brake pedal and move the dual range lever until the cam lever is under the gate finger.

3. Loosen the jam nuts behind the front spring and turn the front jam nut in or out until the spring is compressed to 1 3/4 in. between the rod guide and the front nut. Tighten the jam nuts securely.

NOTE: The idler spring should be in the bottom hole of the idler arm. If belt slippage occurs with a heavy drawbar load, move the spring into the next hole. When the spring is moved into a higher hole, there will be an increase in tension on the clutch-brake pedal, due to the increased tension on the idler arm. This will require more pressure to depress the clutch-brake pedal.

HYDROSTATIC TRANSMISSION

Overhaul

Removal is necessary for most gear case repairs. To prepare for this removal, it is necessary to raise the rear of the tractor and remove the rear axle. After gear case removal, the tractor must normally remain raised while the repairs are being made.

1. Select a good site for gear case removal. The site should have a level and hard floor.

2. Park tractor with front wheels blocked to prevent forward motion of tractor.

3. Place wooden blocks or bricks behind front wheels to keep tractor from rolling backwards.

4. Remove seat deck. Seat deck is secured with nuts, lockwashers, and bolts.

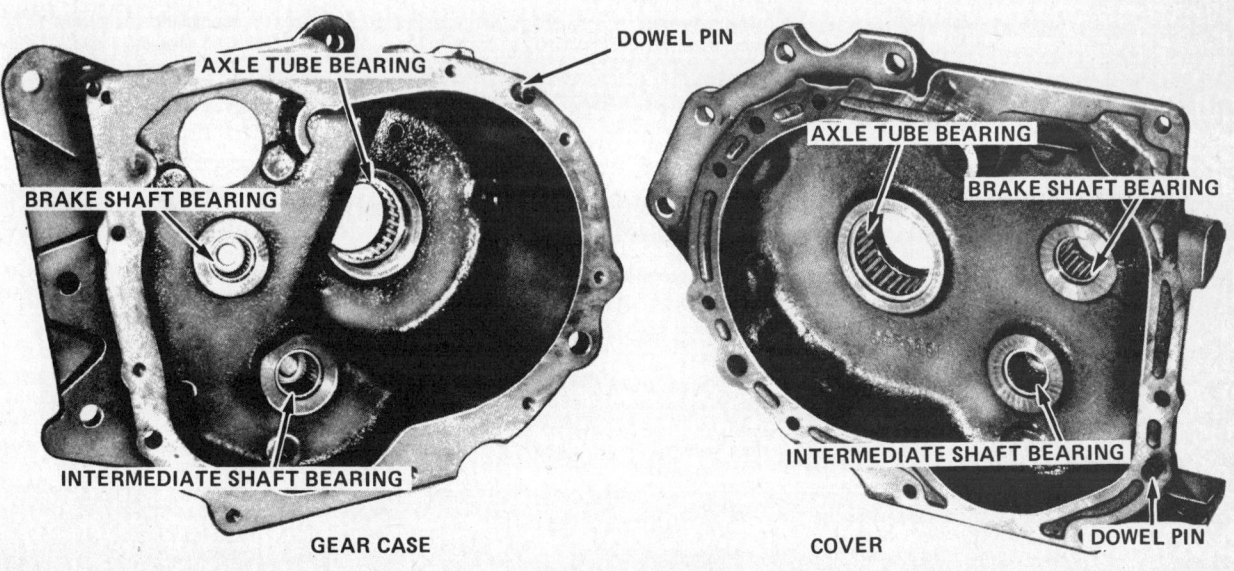

Bearing positions

Allis-Chalmers

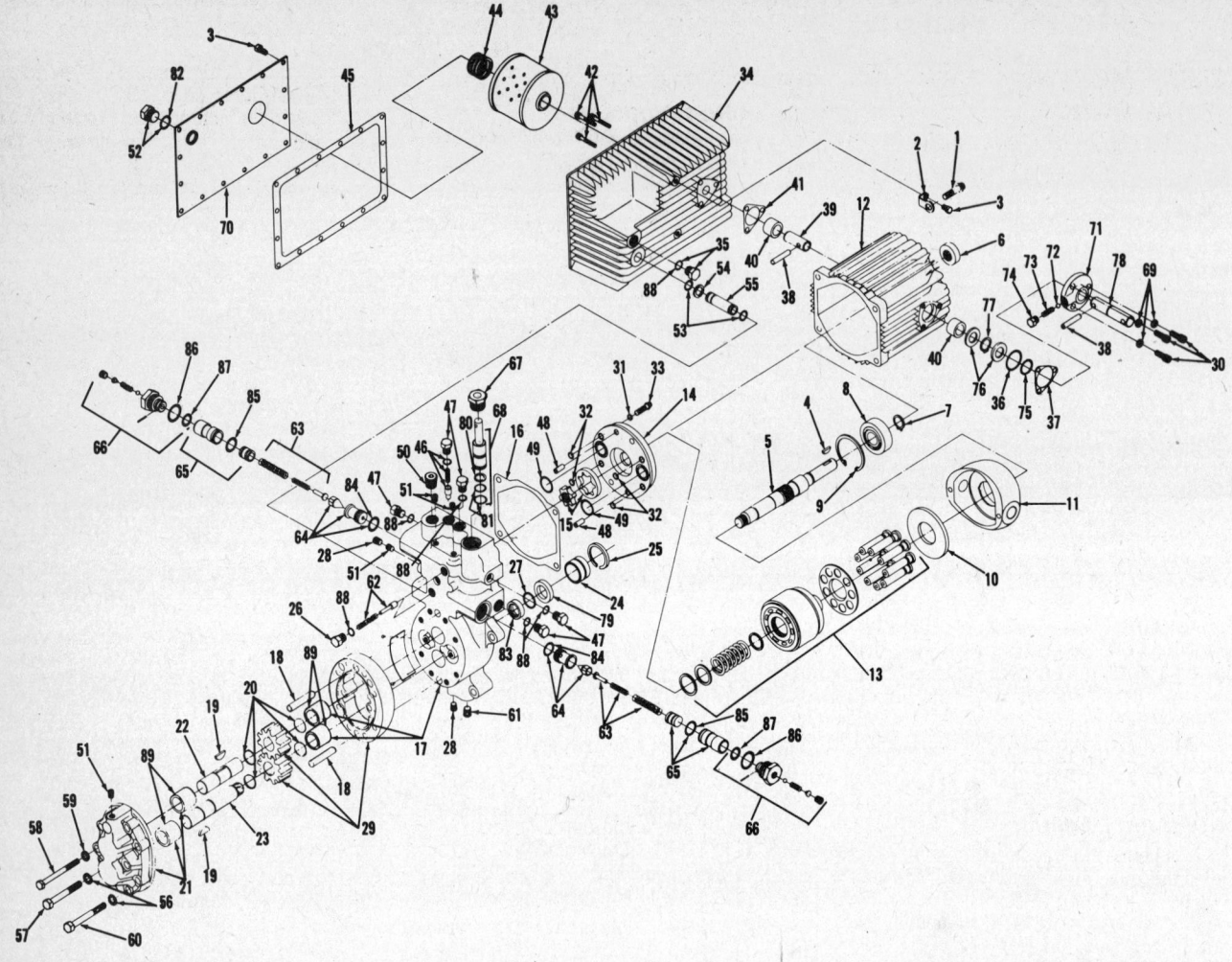

1. 12 point screw
2. Bracket
3. Self tap screw
4. Woodruff key
5. Pump driveshaft
6. Lip seal
7. Retaining ring
8. Ball bearing
9. Retaining ring
10. Thrust plate
11. Swashplate
12. Pump housing
13. Cylinder block assembly
14. Valve plate assembly
15. Charge pump
16. Pump gasket .010"
    Gasket .015"
    Gasket .021"
    Gasket .031"
17. Center section assembly
18. Hollow pin
19. Woodruff key
20. Retaining ring
21. End cap assembly
22. Motor idler shaft
23. Motor driveshaft
24. Pilot
25. Retaining ring
26. Plug
27. O-ring
28. Pipe plug
29. Gear spacer plate assembly
30. Screw
31. Copper washer
32. Rectangular O-ring
33. 12 point screw
34. Reservoir
35. Plug
36. O-ring
37. Trunnion cover shim
38. Spring pin
39. Hollow trunnion shaft
40. Needle bearing
41. Gasket
42. Screw
43. Filter
44. Filter spring
45. Gasket reservoir
46. Charge relief valve assembly
47. Plug
48. Spring pin
49. Rectangular O-ring
50. Pipe plug
51. Pipe plug
52. Plug
53. O-ring
54. Retaining ring
55. Transfer tube
56. Toothlock washer
57. Screw
58. Screw
59. Toothlock washer
60. Hex head screw
61. Pipe plug
62. Check valve assembly
63. Relief spring kit
64. Relief valve assembly
65. Sleeve assembly piston
66. Metering plug assembly
67. Drive-no drive valve plug
68. Drive-no drive assembly
69. Plain washer
70. Reservoir cover
71. Control trunnion plate
72. Ball
73. Spring
74. Threaded plug
75. O-ring
76. Trunnion washer
77. Retaining ring
78. Control trunnion shaft
79. Lip seal
80. Back-up ring
81. O-ring
82. O-ring
83. Insert
84. O-ring
85. O-ring
86. O-ring
87. O-ring
88. O-ring
89. Bearing

Hydrostatic gear case components

Allis-Chalmers

5. Drain fluid from transmission. Place large pan under transmission and remove drain plug. Then remove fill cap to speed draining.
6. Remove the hydrostatic drive unit from the gear case.

GEAR CASE REMOVAL

1. Remove the brake linkage and brake drum from the gear case. Disconnect the oil suction line from the bottom of the gear case.

CAUTION

Use the jack in the following step only for initial raising of the tractor. Use the pair of jack (safety) stands to support the tractor before beginning any work. If jack stands are not available, make sure blocking will support tractor under severe movement and working conditions.

The use of jack stands is a good practice. This practice becomes a necessity when working on the raised rear end of the tractor. Because of the pivot between the front wheels and frame of the tractor, support must be provided at two additional points to assure side-to-side stability of the raised tractor.

---

2. Place a portable or floor jack under drawbar and raise rear of tractor off floor. Adjust jack (safety) stands to keep rear wheels off floor and place one stand under each end of the seat deck holddown bar. Then release and lower jack to lower tractor weight onto jack stands.
3. Remove rear wheels, differential, and rear axle.
4. Remove the drawbar and rear lift assembly.
5. Remove the righthand capscrews and the lefthand capscrews (not shown) at the front of the gear case.
6. Move the gear case to a clean suitable work area.

GEAR CASE DISASSEMBLY

1. Remove any paint, rust, or burrs from the axle tube and brake drum shaft to prevent bearing damage when components are removed.
2. Drive the two (2) dowel pins from the cover into the case.
3. Remove the capscrews from around the edge of the cover and remove the cover. Be sure that the keys and grease fitting have been removed from the axle tube.
4. Remove the intermediate gear and shaft and two (2) thrust washers.
5. Remove the drive gear, axle tube and thrust washers.
6. Remove the brake shaft seal from the cover and axle tube seals from the case and cover.
7. Clean the case, cover and components that were removed with a suitable solvent and compressed air. Do not wash dirt or any foreign substance in the bearings.

BEARING INSTALLATION

1. Check the condition of the roller bearings. Do not remove bearing unless replacement is necessary. If the bearings are removed use a suitable fixture so that the bore in the gear case or cover is not scuffed.
2. When new bearings are installed always press against the end with the manufacturers name or number.
3. The bearings for the intermediate shaft and brake drum shaft in the gear case are open at one end only. The brake drum shaft bearing in the cover is open at both ends. The intermediate shaft bearing in the cover is open at one end only.
4. To install new bearings use a fixture that will not distort the bearing case. The internal edge of the bearing should be 1/8 in. below the inside machined surface of the case or cover.
5. Press the axle tube bearing in the cover and gear case so that the outside end of the bearing in flush with the outside end of the small bore in the case or cover.
6. Lubricate the roller bearings in the cover and case with Lubriplate®.
7. Inspect the bushings in the axle tube and replace if worn or damaged.

GEAR CASE REASSEMBLY

1. Install two keys in the axle tube keyway.
2. Install the short spacer and wire type retaining ring on the axle tube as shown.
3. Install the axle tube into the drive gear, with the two keys engaging the keyways in the drive gear.
4. Install the long spacer on the long end of the axle tube next to the gear.
5. Install two thrust washers next to the long spacer, and push the axle tube assembly through the large bearing in the case.

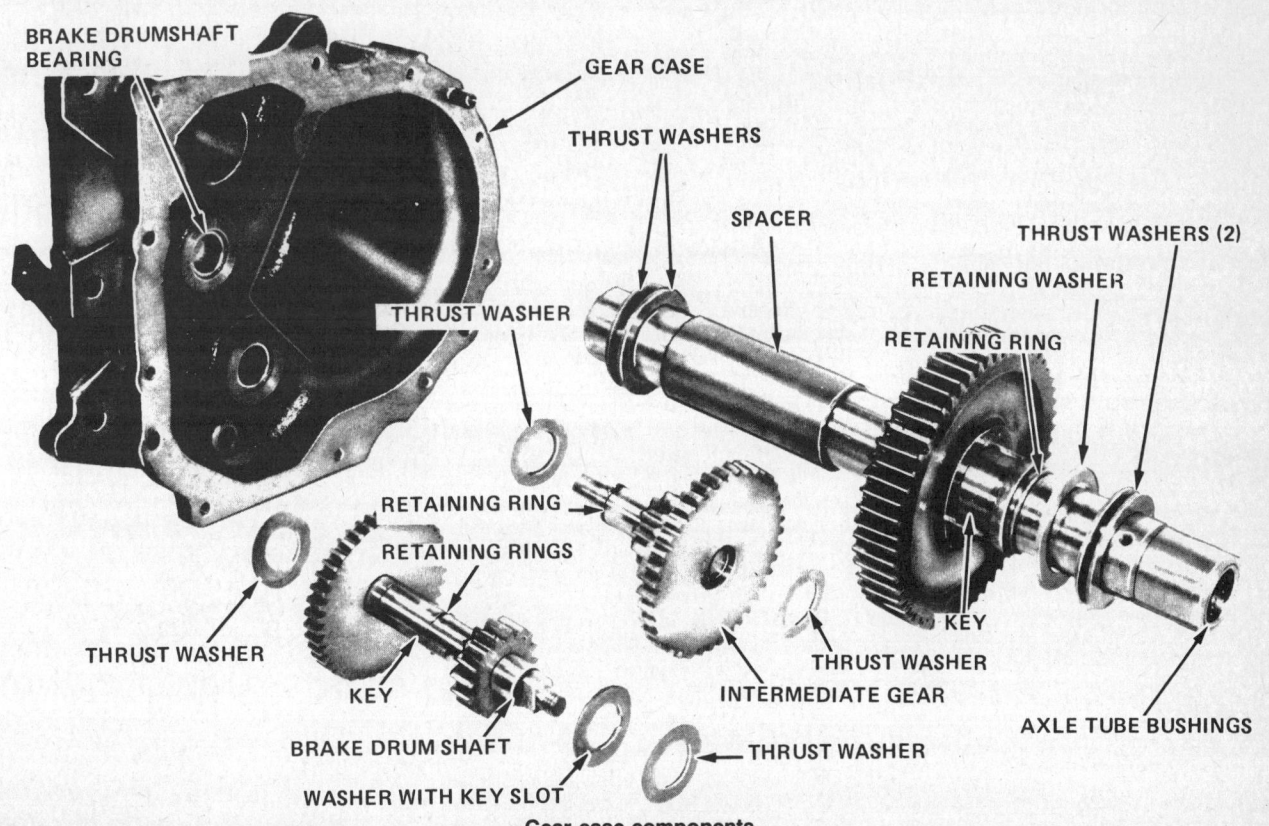

Gear case components

Allis-Chalmers

6. Install one large ID washer, that will encase the retaining ring on the axle tube end and install two thrust washers on the shaft against the large ID washer.

7. Install the true arc snapring on the intermediate shaft, and install the pinion and gear assembly on the shaft with the small pinion toward the retaining ring.

8. Install one thrust washer on the shaft next to the retaining ring and one thrust washer against the large intermediate gear.

9. Install the assembled gear and shaft in the gear case bearing so that the small pinion will engage the large drive gear.

10. Install two wire type retaining rings on the brake drum shaft, and one long key in the shaft keyway, between the retaining rings.

11. Install the large gear on the shaft end without the threads and a thrust washer as shown.

12. Install the small pinion on the shaft end, with the threads, and align the shaft key with the keyway in the pinion and push pinion against the retaining ring.

13. Install the thrust washer with the key slot in the ID over the shaft so that the shaft key is in the slot and the washer is against the pinion, and install the second thrust washer.

14. Install the assembled brake drum shaft in the gear case, with the small pinion engaging the intermediate gear.

15. Drive the two roll pins into the case, leaving enough stick up to engage the cover. Install a new gasket on the case and install the cover.

16. Install the capscrews and torque them to 17 ft.lbs.

17. Install new axle tube and brake drum shaft oil seals in the gear case and cover. Protect the oil seals from damage by covering the keyway, retaining ring grooves and other uneven or sharp areas on the axle tube or brake drum shaft. Use a suitable fixture when installing oil seals to prevent any distortion of the oil seals.

18. Install the gear case in the reverse order that it was removed from the tractor.

Hydraulic Motor

OVERHAUL

Removal

1. Remove covers and other tractor components to gain easy access to top and lefthand side of hydrostatic unit.

---- **CAUTION** ----

If rear wheels are removed, be certain tractor is properly supported.

2. Remove 3 capscrews attaching fan shroud and deflector to transmission and remove shroud and deflector.

3. Loosen and remove drive belt between bevel gear case and input shaft pulley. Remove pulley and fan.

NOTE: Belt guide capscrew is removed but guard was left in position to show location.

4. Disconnect control shaft lever spring from bracket. Remove bolt, cam roller, lockwasher, and nut connecting control shaft and control cam.

5. Remove oil filter and drain hydrostatic unit by removing ⅜ in. pipe plug from lower right side of gear case. Be certain oil drains into a proper container. Discard this oil. Disconnect 2 hydraulic hoses at oil filter assembly.

6. Remove 3 bolts attaching hydrostatic unit to gear case. Slide unit out, to clear gear case, and free of tractor.

7. Remove 2 hydraulic hoses and fittings, control shaft lever, and motor shaft pinion (on output shaft) from unit.

Disassembly

NOTE: The accompanying figure shows a plastic plug in the housing section. This is the way a new U Type hydrostatic transmission is shipped. This plastic plug MUST be replaced by O-ring, part no. 1651704, and plug, part no. 1651705, before unit is installed.

CHARGE PUMP

1. The charge pump with housing must be removed before you start the complete disassembly. Note the orientation of the charge pump housing to end section and either scribe a line or make punch marks to insure proper relocation. Clean the shaft extension to remove all sharp edges, burrs and abrasive residue to prevent shaft seal damage.

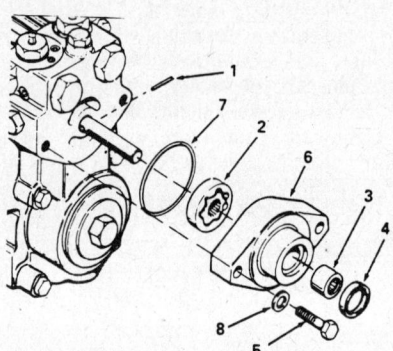

Removing charge pump

2. Remove 2 hex head screws (Item 5) and slide the housing assembly, (Items 3, 4, 6 and 7) over the shaft. Remove drive pin (Item 1), and charge pump (Item 2), from shaft. Only if replacement is necessary remove shaft, seal and bearing (Items 4 and 3) from housing.

3. Examine the wear surfaces of the gerotor assembly for excessive scratching or heavy wear patterns. Replace both parts of the gerotor, if necessary. Do not replace or interchange individual parts within the gerotor. The drive pin should always be replaced. Visually inspect bearing (Item 3), O-ring (Item 7), and shaft seal, (Item 4) and replace as required. Torque screws 50-55 ft.lbs.

CHARGE RELIEF VALVE

Remove plug and O-ring (Item 2) then slide the spring (Item 4) and poppet (Item 5) out of the housing. Do not alter the

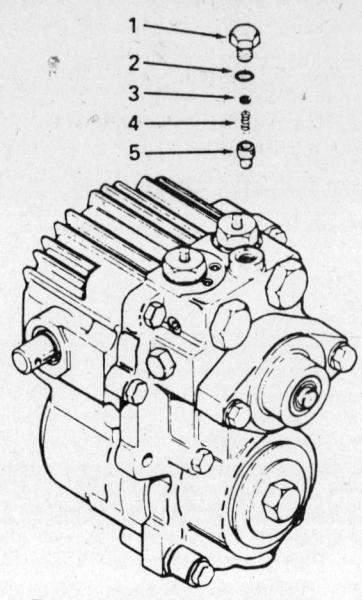

Removing charge relief valve

shims (Item 3) if used or interchange parts with another valve. Inspect the poppet and seat in housing for damage and remove any foreign material in the valve area. Replace parts as required and reinstall in housing bore.

CHARGE CHECK VALVES

Remove the check valve cartridges complete with seals (1, 2, 3, and 4) from the housing. Press on poppet and insure there is a slight spring load. Load the valves with solvent and allow them to sit for fifteen minutes during which time there should be no leakage. Inspect the valve seat in housing for damage and remove any foreign material in the valve area. Replace parts as required and reinstall in the housing bore.

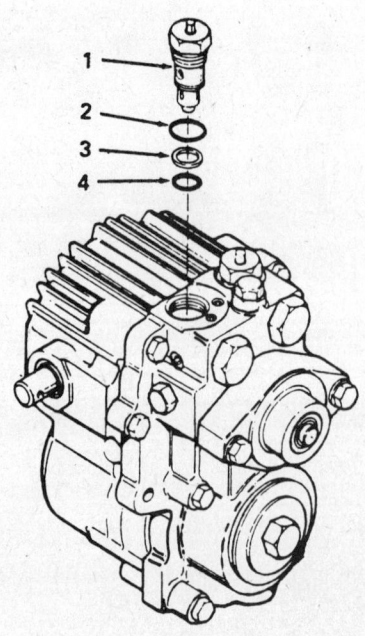

Removing charge check valve

Allis-Chalmers

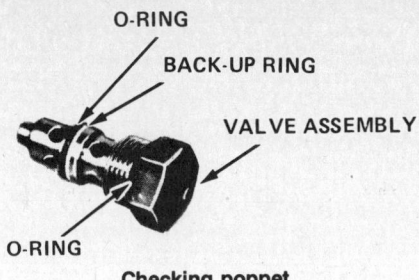

Checking poppet

SHAFT AND TRUNNION SEALS

Lip type seals are used throughout the transmission. These seals can be replaced without disassembly of the transmission; however, replacement of either the input or output seal requires removal of the transmission from the tractor. Pry the seal carefully out of the housing bore, using care not to distort the housing or damage the bore of shaft. Once removed, the seal is not reusable.

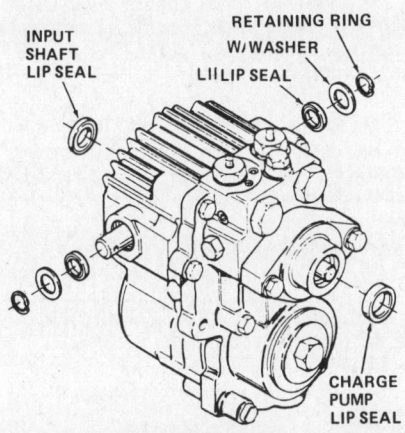

Seal replacement

Prior to installing the new seal, polish the shaft extension, wrap it in thin plastic and lubricate with hydraulic oil to insure that the seal is not damaged during assembly. Slide the seal over the shaft and press it into the housing bore.

In the case of trunnion shaft seals it is necessary that the retaining rings and washers be removed before removing the seals. The washer should be replaced if it is noticeably bent or distorted.

Major Disassembly and Repair

NOTE: Prior to disassembly, scribe a mark on the control side covering of the housing, end section, and charge pump. This will verify reassembly and indicate correct side for control trunnion shaft. Support transmission with shafts in a horizontal position. Remove 8 capscrews and lift end section off housing.

NOTE: Bolts attaching end section to housing are of 4 different sizes. Do not interchange these bolts. Do not let internal parts fall when removing end section. If valve plates remain on end section, remove them.

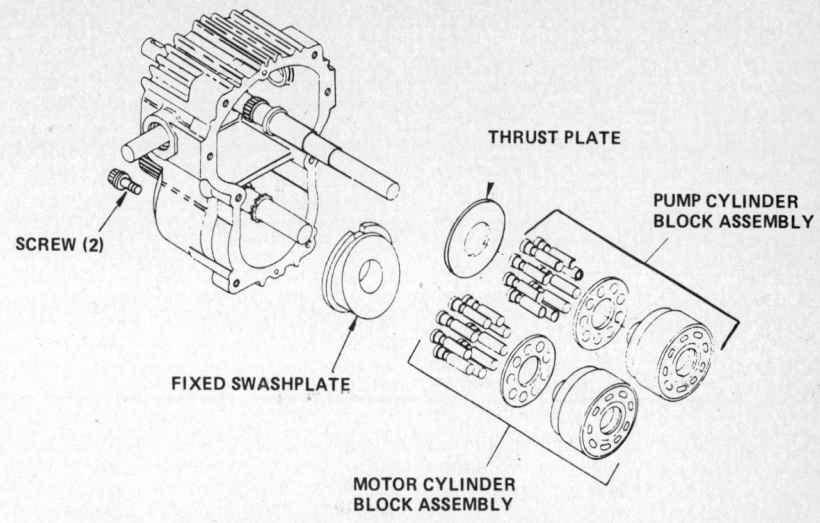

Removing cylinder block assembly

NOTE: Keep the motor valve plate separate from the pump valve plates as they are not interchangeable.

1. If the valve plates remain on the cylinder block, remove them at this time. Remove gasket, valve plate pins, and locating pins.

2. Lift out both the pump (upper) cylinder block assembly and the motor (lower) cylinder block assembly. This is the same for both pump and motor section. The pistons may come out of cylinder block bores. There is no special orientation of piston to bore that needs to be maintained.

3. Do not attempt to disassemble the spirng and other parts from the center bore of the cylinder block. The entire cylinder block assembly should be replaced if any of its components are damaged. Parts are not interchangeable between cylinder block assemblies.

4. Visually inspect wear surfaces of valve plate, cylinder block and slippers for damage. Check to be sure pistons are free in bores. Remove pump thrust plate from counterbore in housing. Visually inspect both sides for damage and flatness.

5. Remove 2 screws attaching fixed (motor) swashplate to housing. Remove swashplate from counterbore in housing and inspect wear surface for damage.

6. The pump shaft can be removed by pressing it out through the open end of housing. After the pump shaft has been removed, the pump shaft bearing may keep the swashplate from full, 15° movement. Press the bearing back into the housing if this should happen.

7. The motor shaft can also be removed by pressing it out through the open end of the housing. The motor shaft bearings can now be removed by pressing it out the other end of the housing.

8. Visually inspect 2 needle bearings in end section. If necessary, remove and replace needle bearings by pulling out of end section.

9. When replace needle bearings, press into end section leaving 1/16 in. to 1/8 in. of bearing protruding beyond counterbore. The valve plates pilot on these bearings.

10. Support housing with open end up. Use care not to damage housing. Place a 3/16 in. diameter punch in the control shaft and tilt movable (pump) swashplate to full angle of 15° (full forward), and hold in this position while removing pins.

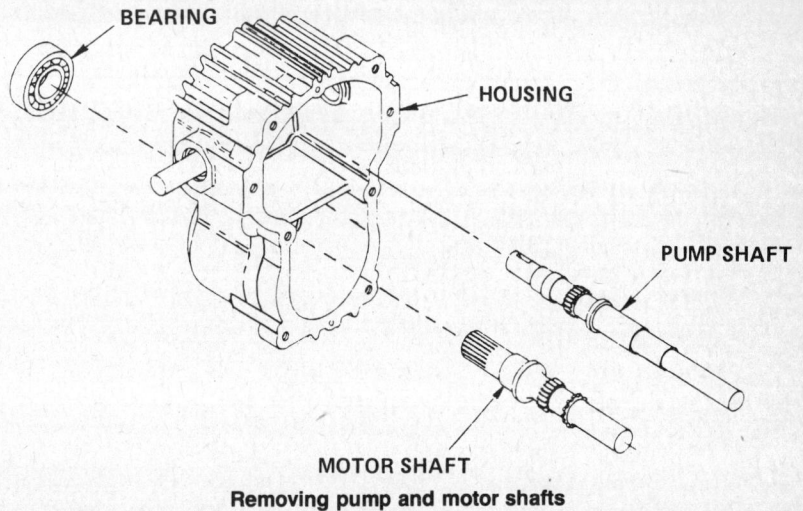

Removing pump and motor shafts

125

11. Using a second 3/16 in. diameter punch, tap the single pin out of trunnion shaft until it hits the housing. Tap both pins out of the control shaft until the first pin hits the housing.

CAUTION
Do not continue to drive pins or housing will be damaged.

12. Twist the swashplate (control shaft) back toward neutral and first control shaft pin and the trunnion shaft pin should fall into housing.
13. Tilt the swashplate back to its full angle (15°) and remove the second control shaft pin using the same procedure as for removing the first.

NOTE: The orientation of swashplate to housing and mark parts accordingly to insure proper reassembly.

14. Remove control shaft and trunnion shaft from swashplate toward outside of housing. The swashplate can now be removed from housing.
15. Visually inspect 2 needle bearings. If necessary, remove and replace bearings by pressing them to the outside of the housing. The pump shaft bearing can be removed by pressing it toward open end of housing.

Reassembly

1. Press motor shaft and pump shaft bearings into housing. Press needle bearings into each side of housing until flush to 1/64 in. below counterbore (for lip seals).
2. Place swashplate into housing with counterbore for thrust plate up. Install control and trunnion shafts being certain shafts are on proper side. Align holes in swashplate and shafts.
3. Install new roll pins through the swashplate and the two shafts. Use two roll pins on the control shaft and one roll pin on the trunnion shaft. Install one roll pin in each shaft and then install the second roll pin in the control shaft. Drive the top of the roll pins 1/4 in. below the swashplate. The swashplate should swing freely in the pump housing to 15° each side of center. Press pump and motor shafts into bearings.
4. Replace needle bearings in end section. Press bearings into end section leaving 1/16 in. to 1/8 in. of bearing protruding beyond counterbore. The valve plates pilot on these bearings.
5. Install fixed (motor) swashplate into counterbore of housing. Orient swashplate so that notch is at top and high point of cam angle is toward the bottom.
6. Assemble motor and pump cylinder block parts if necessary and lubricate with clean hydraulic fluid. There is no special orientation of piston to bore that needs to be maintained, however, parts are not interchangeable between cylinder block assemblies.
7. Hold the pump cylinder block in the palm of your hand with pistons up. Lubricate thrust plate and lay it on top of pistons and block. Lower housing assembly over cylinder block until shaft spline engages cylinder block assembly. Be certain that pistons and thrust plate remain in place. When properly installed a slight spring tension can be felt when pushing on cylinder block. Lubricate exposed surface of cylinder block with clean hydraulic oil.
8. Install motor cylinder block assembly using same installation procedure as used for pump cylinder block.

NOTE: Pump cylinder block must be held in place while installing motor cylinder block. A second set of hands is helpful for this procedure.

9. Insert 2 case guide pins into main housing. Insert 1 locating pin into pump (upper) portion of end section. Lubricate the slotted side of the pump valve plate and slip it over locating pin and protruding needle bearing.

NOTE: The pump valve plate has two vee notches. The motor plate has four vee notches. Valve plates are not interchangeable.

10. Insert 1 locating pin into the motor (lower) portion of end section. Lubricate the slotted side of the motor valve plate and slip over locating pin and protruding needle bearing.
11. Place gasket on center section using a small amount of hydraulic fluid to hold it in place.
12. Place the 8 screws that hold the end section to the housing in their respective holes. Tighten these screws equally and pull the two sections together. Torque screws to 20-30 ft.lbs.

NOTE: After screws are tightened, input shaft MUST be able to be turned by hand. If shaft cannot be turned by hand, disassemble unit and check pump and motor valve plates to see if locating pins are properly engaged in groove.

13. Install new lip seals. Install remaining components, charge pump and valves as described at the beginning of the manual.

Priming Procedure

1. Raise seat deck.
2. Remove filler cap from transmission filler tube and lift relief valve for a short period.
3. While holding relief valve up, add A-C Power Fluid 821, or its equivalent, until fluid is visible in the filler tube.

NOTE: The relief valve must be held in the up position while transmission fluid is added.

4. Remove spark plugs.
5. Raise the rear wheels of the ground, and block the tractor securely.
6. With the speed control lever about half-way forward, depress the neutral-start switch manually and crank the engine. When the wheels begin to move, stop cranking engine.
7. Replace spark plug and lower tractor wheels to ground, place control lever in neutral.
8. Start the tractor, and run for 1-2 minutes.
9. Stop the engine.
10. With the engine off and the relief valve held in the up position, add more transmission fluid until fluid is within 1/8 in. of the top of the filler tube.
11. Replace filler cap.
12. Check oil level after 5 hours of operation.

Testing After Repair

After repairing and reinstalling the hydrostatic unit, test for proper operation. Check for leaks, especially at lip seals and where sections join together. If necessary, refer to servicing section of this manual and correct malfunction.

Installation

Installation procedures are in reverse order of removal with the exceptions listed below. Adjust hydrostatic unit per applicable tractor unit.

1. Clean mounting surfaces between hydrostatic unit and gearcase. Clean gasket eliminator from 2 cavities on gear case.
2. Apply Loctite® Gasket Eliminator 504 to gear case mounting surface as shown. Reapply Gasket Eliminator to 2 cavities.

NOTE: Be certain motor shaft pinion is properly meshed with gear case gear during installation.

3. After installation, check that fan does not hit shroud, or that the belt hits the gear case.

Adjustments

1. To adjust parking brake, loosen nut from rod end at front of parking brake rod.

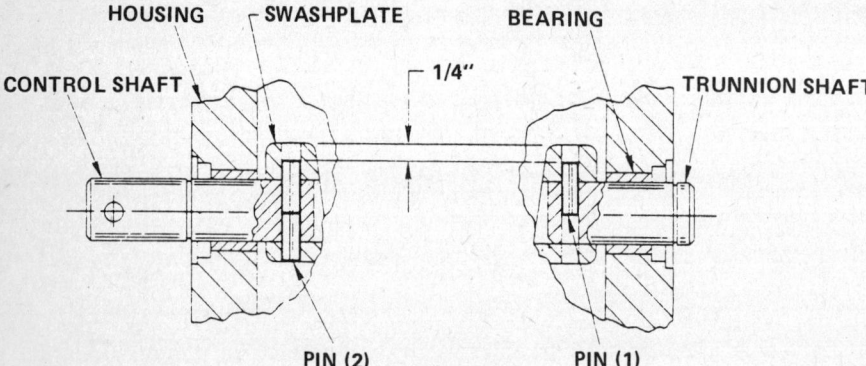

Attaching swash plate to trunnion shaft

Allis-Chalmers

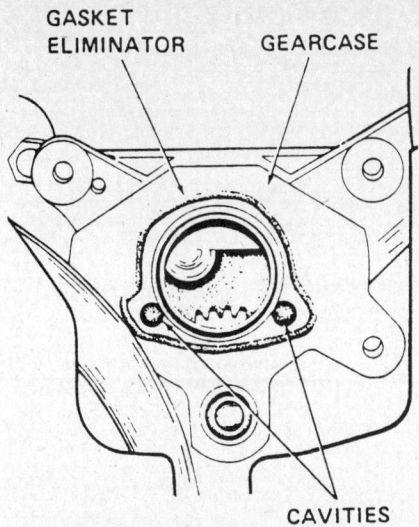

Applying gasket eliminator to gearcase mounting surfaces

Turn parking brake handle and rod end until parking brake is fully tight when parking brake handle is pulled up against fender as shown.

2. With parking brake tight, adjust jam nuts on end of foot brake rod to provide ¾ in. clearance to rod guide.

3. Adjust jam nuts on clutch rod to have ¼ in. clearance from clutch rod guide.

4. **HYDROSTATIC NEUTRAL ADJUSTMENT:** If tractor moves forward or backward when the hydrostatic control lever is in the neutral notch of the quadrant, proceed as follows to adjust the control linkage.

a. Park tractor on level ground, make sure hydrostatic control lever is firmly seated in the neutral notch of quadrant, stop engine and set parking brake.

b. Raise the seat deck and check if the pump control arm roller is exactly centered with the centering mark on the hydrostatic control cam. If it is not centered, loosen bolt and move the hydrostatic control cam assembly forward or backward in the slotted hole in the hydrostatic control strap until centering mark is centered on roller. Tighten bolt. Lower seat deck, get in the operator's seat, start engine and release parking brake. If tractor still creeps with hydrostatic control lever in neutral, note which direction it creeps and proceed with next step.

c. Again stop tractor engine, set parking brake and raise seat deck. Loosen jam nut "B" on end of the cam pivot shaft. If tractor creep had been in reverse turn adjusting nut ⅛ to ¼ turn clockwise when viewed from the righthand side of tractor. If tractor creep had been forward turn nut ⅛ to ¼ turn counterclockwise. Lock jam nut, lower seat deck, get on tractor seat, start engine and release parking brake. If tractor still creeps in neutral, repeat step C moving nut "C" a small amount at a time until no more creep occurs.

--- **CAUTION** ---
ALWAYS stop tractor engine before leaving tractor seat. NEVER attempt to make adjustments while engine is running.

NOTE: If drive belt is too long or too short, three holes have been provided on the idler lever assembly (2027445). The center hole is normally used, but with a longer belt, the top hole may be used or with a shorter belt, the bottom hole may be used.

ENGINE

For detailed engine overhaul and repair, see the Kohler Engine part of the Engine Unit Repair Section. The engines used in the 900 series tractors are as follows: 910-K241; 912-K301; 914-K321; 916-K341; 917-KT17.

Engine Removal

1. Disconnect the fuel line from the fuel pump and plug the hose or drain the fuel tank of all gasoline.

--- **CAUTION** ---
DO NOT drain fuel from the tank when the engine is hot, while smoking, near an open flame, near anything that is hot, or in a closed building

2. Disconnect the battery ground (−) cable, all electrical wires, choke cable, and throttle cable from the engine.

3. Remove the two capscrews with the thick washers and spacers that fasten the drive shaft to the engine flywheel.

4. Remove the four capscrews that hold the engine in the tractor frame, and remove the engine.

NOTE: On 917 Twin the left rear engine mounting bolt goes in from the bottom of the tractor because the starting motor interferes with installation from the top. Make sure flatwashers are in place to prevent capscrews from mislocating in slotted holes or from interfering with starter motor.

Engine Installation

1. Place the engine on the tractor frame and align the mounting holes in the engine with the slotted holes in the tractor frame.

2. Install the capscrews into the holes, with the flatwashers, lockwashers and nuts on the bottom side of the frame (except for 917 Twin, See Note in Engine Removal). Use a flatwasher on each slotted hole or the engine will not hold properly. Do not tighten the capscrews at this time.

3. Fasten the drive coupling to the engine flywheel in the following way. On each capscrew place one lockwasher and one thick washer and place the capscrew thru the coupling disc with a spacer between the thick washer and the engine flywheel. Install the second capscrew in the same order, and torque both capscrews.

4. Move the engine in the slotted holes so that the engine flywheel and the drive shaft are properly lined up. The drive coupling should be kept flat and not bowed or bent. The drive shaft has to be at right angles to the flywheel. Torque the four engine mounting capscrews.

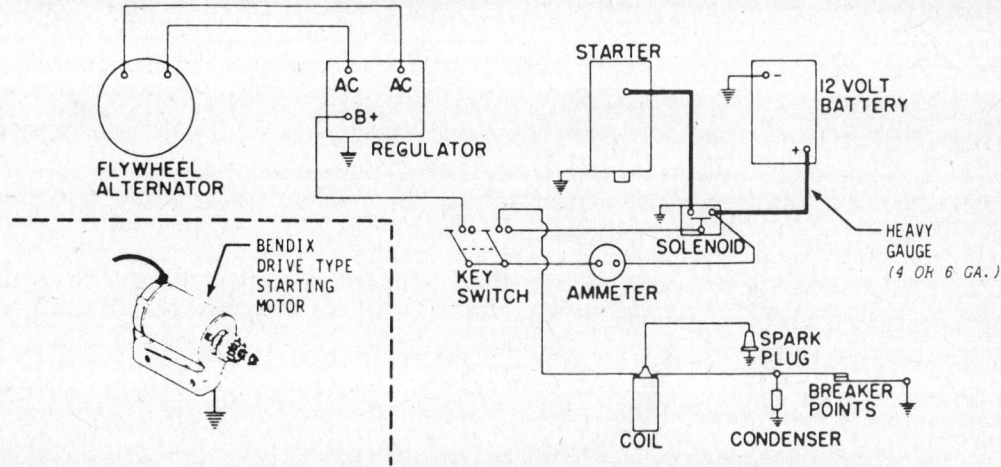

Ignition system wiring diagram with battery ignition

Allis-Chalmers

B Series

MODELS: B-1, B-10 SERIAL #15001-31227 AND UP, BIG TEN, B-10 SERIAL #50001 AND UP, B-12, B-110, B-112, HB-112, B-206 AND 206E, B-207 AND 207E, B-208 AND 208S, B-210, B-212, HB-212.

FRONT AXLE AND WHEELS

Axle Main Frame

REMOVAL AND INSTALLATION

All models except B-207 and 208
1. Raise the tractor front end, and support it on jack stands.
2. Disconnect the tie rod ball joint.
3. Remove the capscrew and spacer from the center of axle.
4. Lower the front of axle and pull forward to slide the stabilizer out of the frame angle. The frame angle is replaceable if it is excessively worn.
5. Installation is the reverse of removal.

B-207 and 208
1. Front axle main member on the B-207 and the B-208 is welded to the frame. New frame assembly is necessary if axle cannot be repaired.
2. Refer to the accompanying figure for disassembly and assembly of components.
3. Inspect the bushings on the spindle assemblies and the bearings in the front wheels.
4. Replace where needed.
5. Clean dirt and grease off of all parts.
6. Lubricate both of the spindle end and assembly unit.

Drag Link

REMOVAL AND INSTALLATION
1. Remove bolts, washers and spacers that hold tie rod to spindles.
2. Installation is the reverse of removal.

Steering Spindles

REMOVAL AND INSTALLATION
1. Raise tractor front end, remove wheels.
2. Remove drag link.
3. Remove steering arm and key from left spindle and remove spindle.
4. Remove cotter key from right spindle and remove spindle.
5. There are 4 spindle bearings. Two in each end of axle frame.
6. Installation is the reverse of removal.

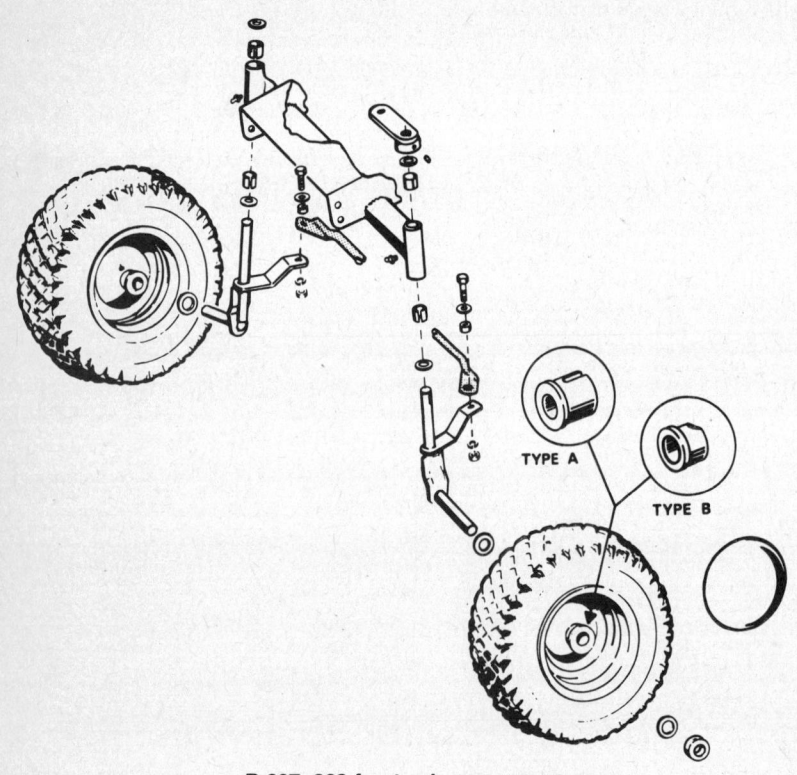

B-207, 208 front axle components

STEERING

Steering Wheel and Gear

REMOVAL AND INSTALLATION

B-1
1. Tilt seat assembly rearward, remove hood.
2. Loosen setscrew, remove steering wheel.
3. Remove keys and washers from top of steering shaft.
4. Remove battery.
5. Disconnect fuel line at tank.
6. Remove capscrews holding frame cover to frame.
7. Remove gear shift lever ball.
8. Remove capscrew from gear shift rod guide.
9. Lift off frame cover and tank assembly.
10. Disconnect rear tie rod ball joint.
11. Loosen setscrew and remove steering arm, Woodruff key and washer from steering gear.
12. Remove steering driven gear.

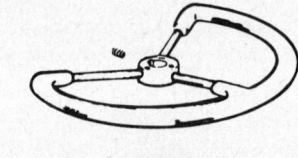

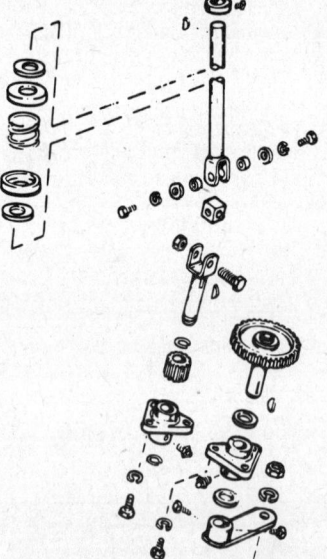

B-1 and B-10 steering components

13. Removal of shaft and pinion gear—remove snapring and washer from lower end of shaft.
14. Lift shaft out of bearing.
15. Remove capscrews to remove bearing.

Allis-Chalmers

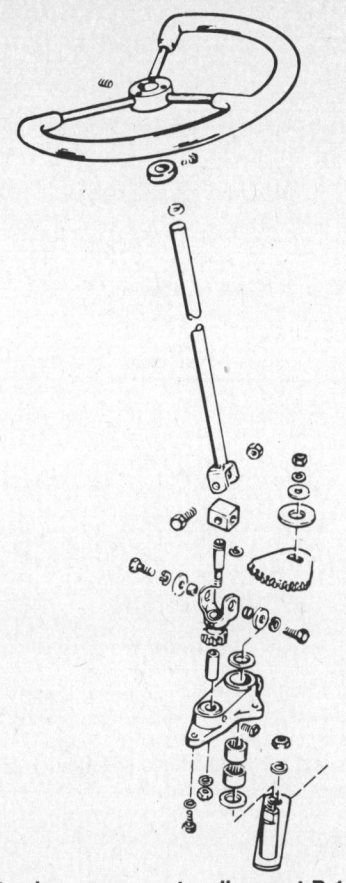

Steering components, all except B-1 and B-10

16. Installation is the reverse of removal.

B-10
1. Remove hood, side panels and battery.
2. Loosen setscrew and remove steering wheel and key.
3. Lift off dash.
4. Remove tank support.
5. Disconnect rear tie rod ball joint.
6. Loosen setscrews, remove steering arm.
7. Remove steering driven gear.
8. Remove snapring and washer from lower end of shaft.
9. Lift out shaft assembly.
10. Remove capscrews from main frame to remove bearings.
11. Reassembly is the reverse of removal.

Big Ten, B-10 S/N 50001 & Up, B-12, B-110, B-112, HB-112, B-210, B-212, HB-212
1. Remove hood and side panels.
2. Remove steering wheel and key.
3. Remove battery and fuel tank.
4. Remove dash assembly.
5. Remove locking collar or steering shaft.
6. Disconnect universal joint in steering shaft, remove shaft.
7. Disconnect tie rod.
8. Remove steering gear.
9. Remove steering bracket.

10. Installation is the reverse of removal.

B-207, 208
1. Remove hood and steering wheel.
2. Disconnect ignition wire and choke and throttle cables.
3. Unbolt and lift off dash assembly and upper steering shaft support.
4. Raise front of tractor and disconnect drag link from steering gear.
5. Unbolt and remove steering gear.
6. Remove nut securing steering pinion to frame and remove upper steering shaft, "U" joint and steering pinion.
7. When reassembling the steering units, move steering gear closer to steering pinion to remove excessive steering wheel play.
8. To reassemble, reverse disassembly procedure.

BRAKES AND CLUTCH

NOTE: Detailed repair procedures for brakes and clutch will be found in the transaxle section. Adjustments only, will be found here.

B-1

PARKING BRAKE

Depress foot pedal until brake holds securely; flip capscrew toward rear for park. Adjust length of screw to permit screw head to wedge on bottom of frame and hold lever in depressed position with brake applied.

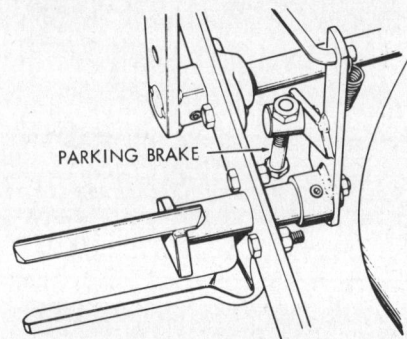

Parking brake adjustment points

CLUTCH AND BRAKE

1. With clutch and brake pedal in normal position (i.e., clutch engaged and brake released), adjust hex nuts (A) to give ¾ in. clearance between rod guide assembly (B) and nuts.
2. Pull brake band up by hand so that it is tight around brake drum. Adjust hex screw (C) to have a clearance of 1" between brake band and screw as shown.
3. Adjust nut (D) to permit clutch link (E) to pivot freely without excessive play.

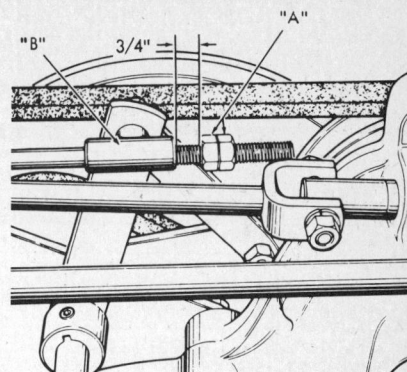

Rod guide adjustment points

Brake band adjustment

4. Adjust nut (F) to permit clutch link (E) to pivot freely without excessive play and check to see that nut has at least 1½ in. of travel before touching transmission case.
5. When clutch is disengaged and brake is applied, the clutch and brake lever assembly should have at least 1 in. of travel before touching bevel gear housing.
6. To compensate for belt stretch or other variances it may be necessary to move idler pulley (G) into the alternate hole provided in its lever arm.

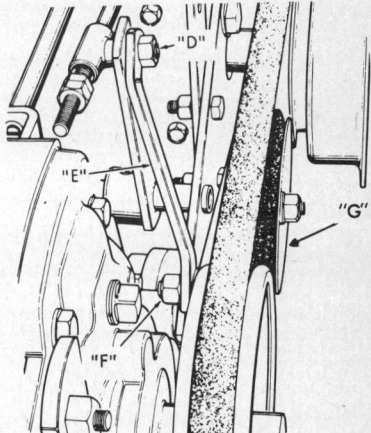

Clutch link play adjustment

B-10 (9 hp.)

BRAKE ADJUSTMENT

Pull the brake band up by hand so that it is tight around the brake drum. Adjust the

Allis-Chalmers

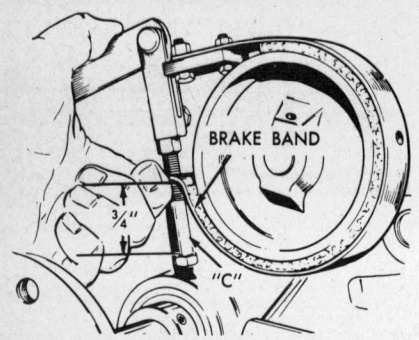

B-10 brake adjustment

hex screw (C) to give a clearance of about ¾ in. between the brake band and the screw head. Then check to see that the idle pulley releases the belt properly before the brake is applied. If the brake does not hold properly when the pedal is pushed all the way forward, reduce slightly the spacing between the head of the hex bolt and the brake band. Then recheck the clutch rod adjustment for proper idler release.

CLUTCH

With clutch and brake pedal in normal position (i.e., clutch engaged and brake released), adjust hex nuts (A) to give ¾ in. clearance between rod guide assembly (B) and nuts.

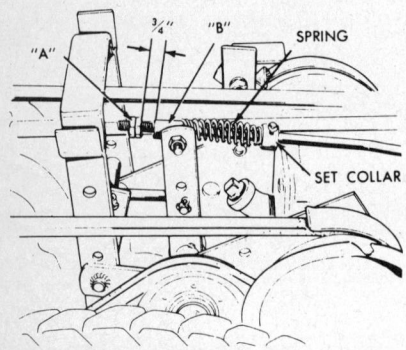

B-10 clutch adjustment

Position the set collar on the clutch rod to compress the spring about ⅝ in. Then recheck and position the locknuts "A" to leave a spacing of about ¾ in. between them and the end of the rod guide "B". Check to see that when the pedal is operated, the spring is completely decompressed as the locknuts engage the end of the rod guide.

B-10 (10 hp.), Big 10 and all 12 hp.

CLUTCH ADJUSTMENT

Adjust locknuts on clutch rod to give 11/16 in. space between them and idler pulley pivot arm.

BRAKE ADJUSTMENT

Pull the brake band up by hand so that it is tight around the brake drum. Adjust the locknuts to give a clearance of about 11/16 in. between the brake band and the locknut. Then check to see that the idler pulley releases the belt properly before the brake is applied. If the brake does not hold properly when the pedal is pushed all the way forward, reduce slightly the spacing between the locknuts and the brake band. Now recheck the clutch rod adjustment for proper idler release.

TRANSAXLE

Bevel Drive— All Models

REMOVAL AND INSTALLATION

1. Remove seat assembly.
2. Remove dash assembly.
3. Remove top frame cover.
4. Support tractor under frame just ahead of bevel gear housing.
5. Disconnect brake linkage, clutch-brake rod and transmission shift rod.
6. Remove transmission drive belt and capscrews securing transmission to side plates.
7. Roll transmission rearward from tractor.
8. Disconnect driveshaft.
9. Remove gear shaft flange from bevel gear shaft.
10. Remove capscrews holding bevel gear housing to frame, lift off housing.
11. Remove rope starter pulley.
12. Remove transmission drive pulley.
13. Remove PTO drive pulley.
14. Remove side plates and rear cover.
15. Back up the driven bevel gear and carefully drive the driven shaft to the left until the key is free of gear.
16. Remove key and driveshaft out left side of housing.
17. Remove bearing clamp plate.
18. Drive front shaft, bearing and bevel gear assembly out of housing.
19. Remove bevel gear retaining capscrew and washer.
20. Remove bevel gear and bearing from shaft.
21. Inspect bearings and seal, renew if necessary.
22. Installation is reverse of removal.

Differential—All Models

REMOVAL AND INSTALLATION

1. Block up tractor and remove wheels.
2. Remove left wheel hub and key.
3. Loosen setscrews and remove collar and washers on left side of transmission.
4. Remove right hub, differential and axle assembly.
5. Remove set collar from right end of axle shaft.
6. Remove bolts from outer edge of case.
7. Remove nuts from inner row of capscrews.

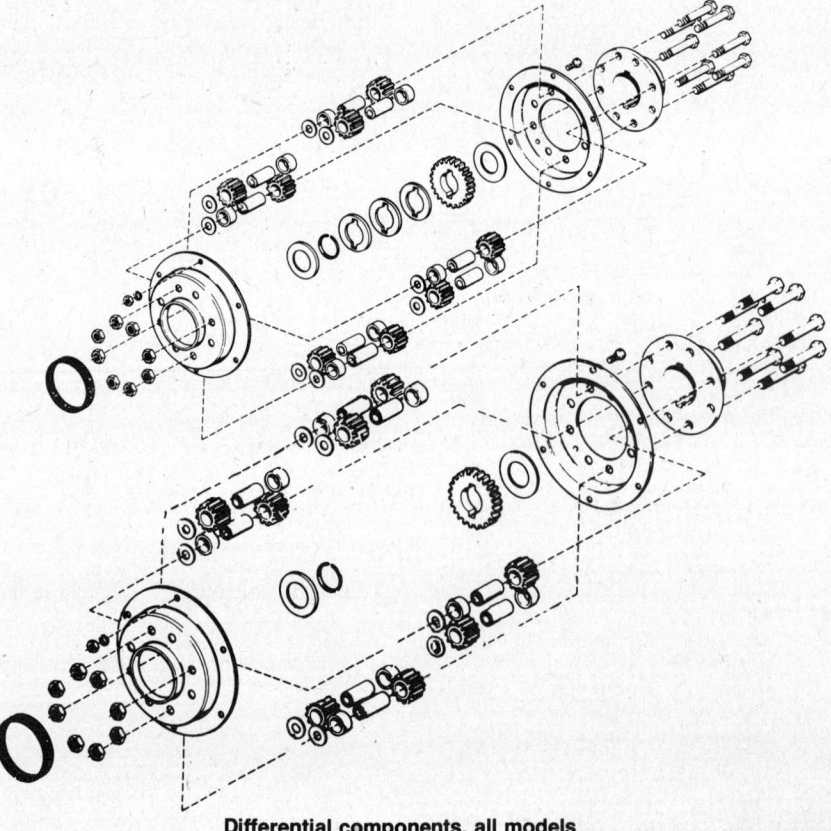

Differential components, all models

Allis-Chalmers

8. Separate case halves. Leave capscrews in position to hold parts in place.

9. When removing parts identify to aid in reassembly.

10. Reference to accompanying illustration will aid in disassembly and assembly.

11. When installing the assembled differential, the axle and differential are properly seated so the seal between them is compressed. The axle is held in place by the collar on the left side of the transmission.

3-Speed Transmission

ALL EXCEPT 206, 207, 208 AND HB SERIES

Removal

NOTE: Before attempting removal of the transmission from the tractor, place the tractor on a level surface and drain all the lubricating oil from the transmission. For fastest draining, remove the upper pipe plug from the transmission cover to allow air to enter transmission case.

1. It is necessary to lift or jack up the tractor to a point where the rear wheels will be free of the ground. Place a support (strong enough to bear the weight of tractor) at a point under the frame and ahead of PTO assembly.

2. If the tractor is equipped with a rear light, raise the tractor hood and disconnect the ground cable from the negative terminal of the battery.

3. Remove the rear light from the tractor seat back by removing the hex capscrew holding the light mounting bracket to the seat back. Remove the cable clip from the righthand arm assembly. This clip is held in place by a hex capscrew, flat washer, lockwasher and hex nut.

4. Remove the hex capscrew and locknut from each arm assembly. The seat assembly may now be lifted free of the tractor.

5. To remove the left rear wheel and hub complete; loosen the locknuts and setscrews "A" and "B" shown. Setscrew "A" locks against the key located in the axle shaft and setscrew "B" locks into a hole in the axle shaft. It will be necessary to loosen "B" until the screw is free of the hole in the axle shaft. If necessary, tap the edge of the wheel hub with a lead mallet to loosen from the axle shaft.

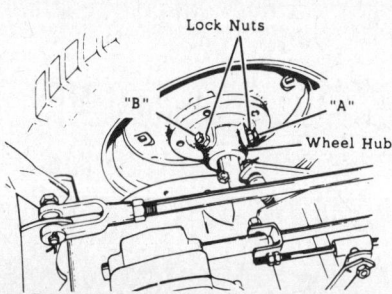

Removal of left rear wheel and hub

6. Loosen the 2 setscrews holding the set collar on the axle shaft, and remove the collar. Should the collar stick or bind on the paint on the axle shaft, remove the paint and sandpaper shaft. After the collar is removed, clean the axle shaft of any remaining paint and remove any burrs from edge of keyway or setscrew lock points by using a fine file. Burrs or paint, etc., will damage bearings when the axle shaft is removed, so be certain the axle shaft is smooth and clean.

7. From the righthand side of the tractor, remove the axle shaft, righthand wheel, hub and differential in one piece by tapping the edge of the differential hub with a lead mallet. When the differential hub is free of the 2 keys on the transmission axle tube as shown, pull the axle shaft, etc., straight out of the transmission.

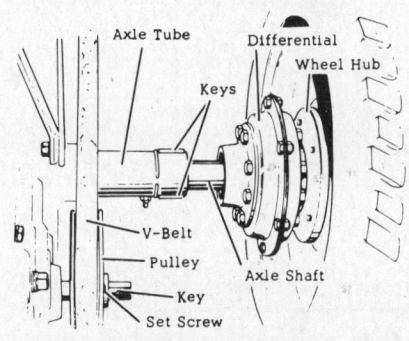

Axle shaft removal

8. Remove the V-Belt and pulley from the transmission pulley shaft, taking care not to use excessive thrust force on shaft. Use a wheel puller if necessary. The pulley is held in place by a setscrew and key. Ko not hammer or the snaprings on the shaft inside transmission may be damaged.

9. Disconnect the lift rod clevis from the rear lift bracket by removing the cotter pin, spacer, and pin as shown. Remove hex capscrews "A", "B", and "C" from both sides of the arm assembly and lift off the drawbar, rear lift bracket, and arm assembly as one piece.

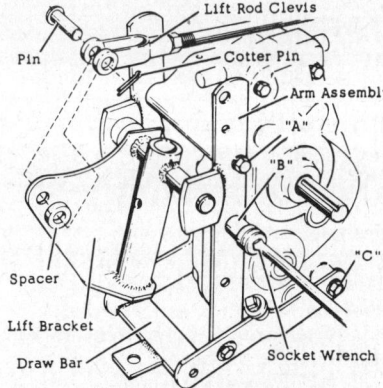

Lift rod clevis removal

10. Disconnect the shifter rod assembly from the transmission shift rod as shown.

11. Disconnect the brake band linkage as shown by removing the cotter pin and clevis pin and spring.

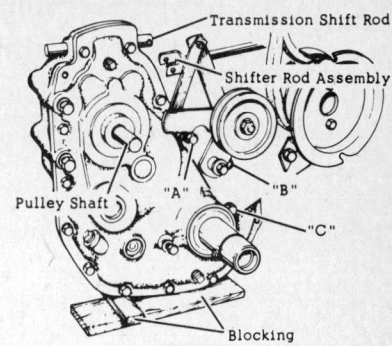

Shifter rod removal

12. Add blocking to support the weight of the transmission at points shown.

13. Remove the hex capscrews "A", "B", and "C" from the lefthand side of the transmission and "A" and "B" from the righthand side. Now loosen "C" on the righthand side and steady the transmission by gripping the pulley shaft with one hand while removing "C" with the other hand.

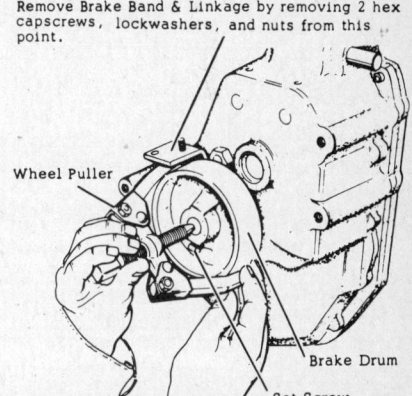

Brake band and linkage removal

NOTE: Before starting to disassemble the transmission, it will be necessary to remove the brake band and linkage as shown. To do this, remove the 2 hex capscrews, lockwashers, and hex nuts. Remove the brake drum by loosening the setscrew, and pull the drum from the shaft with a wheel puller as shown. Do not hammer or the snaprings on the shaft inside the transmission may be damaged.

Disassembly

Before beginning to disassemble the transmission be certain to file on the edges of the keyway on the brake drum shaft to avoid cutting the oil seal if any burrs have been raised there. Remove the 2 keys from the axle tube and check for burrs. File if necessary. Remove the grease fitting from the axle tube, and also the snapring. Check for burrs. Check the drive pulley shaft for burrs and file if necessary.

1. Remove the oil drain plug and allow any remaining oil to drain from the transmission. Draining will be speeded by removing the upper pipe plug from the transmission case, and setting the transmission in an upright position as if in place on the tractor.

131

Allis-Chalmers

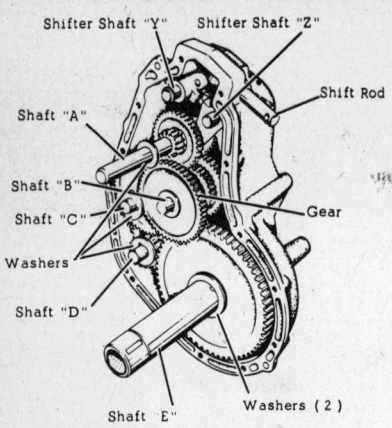

Disassembling transmission case

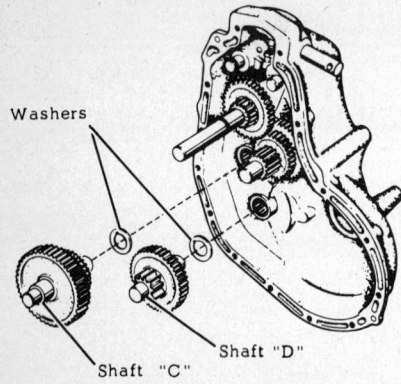

Gear shaft removal

2. Remove the 14 hex head capscrews and lockwashers from around the edge of the transmission cover. Drive the 2 dowel pins down into the transmission case holes and insert a screwdriver at several points between the cover and the case and pry upwards to break the cover loose from the case. When cover is free, lift off.

3. Remove the flat washer from shaft "A", shaft "C", and shaft "D". Now remove the gear from the end of shaft "B". Remove the axle tube and gear assembly "E" from the transmission case.

4. Lift out shaft "C" with gear assembly and then remove shaft "D" with its gear assembly.

5. To remove the pulley shaft "A" and its gear assembly it will be necessary to position the shift forks as follows:
   a. Place shift fork "T" in Neutral so that the lower edge of shift stem slot is even with the end of roll pin in the transmission case.
   b. Place the shift fork "Z" in top-most position.

---- **CAUTION** ----

Do not raise too far, or it may come off of the shaft and the lock ball or spring may be lost.

Now, lift the pulley shaft "A" with one hand to clear the lower bearing, and with the other hand, raise the cluster of gears on shaft "B" slightly to allow pulley shaft "A" and its gears to be moved away from the shift forks. When the yokes of the shift forks disengage from the shift rings on the gear clusters on shaft "A", the shaft and gears may be lifted free of the transmission. Now, lift out shaft "B" and its gear.

6. The reverse gear "F" may be removed now, by undoing the locknut on the back side of the transmission case, and removing the hex head capscrew from the bracket on the inside of the case.

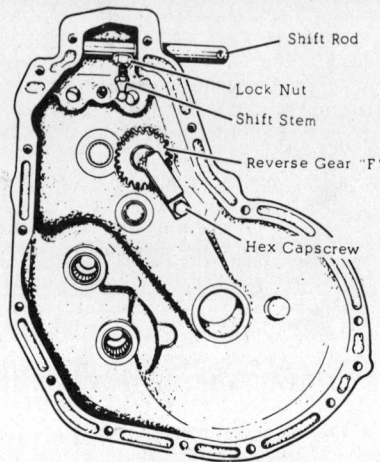

Reverse gear removal

7. Remove the shifter forks and shafts from the transmission case by loosening the setscrews located as shown. Loosen these screws sufficiently to clear the locating holes in the shafts.

8. To remove the shift rod, first loosen the locknut on the shift stem, and unscrew the shift stem from the shift rod. The shift rod may then be pulled from the case.

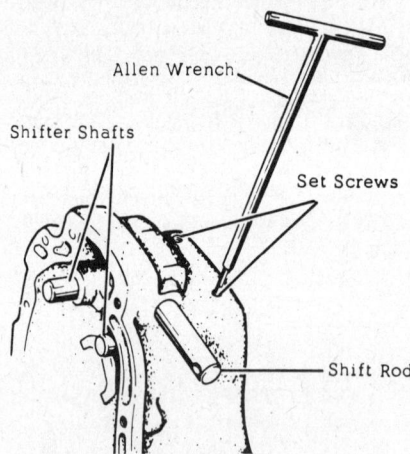

Shifter fork and shaft removal

Shifter Forks
DISASSEMBLY

1. To remove the shifter forks from the shifter shafts, slide the fork towards the end of the shaft without the retaining ring. Cup one hand over the lock ball loading hole while withdrawing the shaft to prevent the loss of the shift lock ball or spring.

2. The lock ball and spring are under tension when the fork is in place on the shaft, and unless care is taken the ball may fly out as the shaft is removed and be lost.

ASSEMBLY

The shifter shafts are of unequal length and the position of the fork is reversed from one shaft to the other. To properly assemble the shafts and forks, proceed as follows:

1. Identify the long shaft, (it has a retaining ring on the end of the shaft farthest from the setscrew hole) and insert the setscrew hole end into the hub of a shift fork. Before pushing the shaft through the hub of fork, insert the shift lock spring and ball through the loading hole and depress with a ⅜ in. rod so that the shaft may slide through.

2. Identify the short shaft, (it has a retaining ring located near the setscrew hole) and insert the end without the retaining ring into the hub of the shifter fork. Before pushing the shaft through the hub of the fork, insert the shift lock spring and ball through the loading hole and depress with a ⅜ in. rod so that the shaft may slide through.

3. To obtain a clear idea of the position of the shift forks when properly installed on the shifter shafts. This shows the general appearance of the shift forks and shafts and the shift rings that the forks engage. In this particular view, the parts are shown in their relative positions as they would be seen if the end of the transmission were to be cut off. Note that the gear cluster nearest the pulley end of the shaft is engaged by the shift fork on the short shaft, and the other gear cluster is engaged by the shift fork on the long shaft. Also note the relative position of the shift forks to each other.

Bearings

The transmission contains a total of 12 bearings. 4 needle bearings and 1 bronze bearing are located in the transmission case, 4 needle bearings and 1 bronze bear-

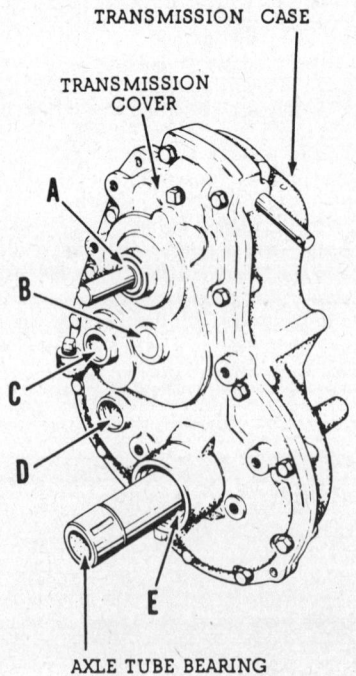

Bearing locations in case

ing are located in the transmission cover and 2 bronze bearings are located in the axle tube.

TRANSMISSION CASE

Location	Description
"A"	Needle brg., sealed on one end
"B"	Needle brg., open on both ends
"C"	Needle brg., sealed on one end
"D"	Needle brg., sealed on one end
"E"	Bronze brg.

TRANSMISSION COVER

Location	Description
"A"	Needle brg., open on both ends
"B"	Needle brg., sealed on one end
"C"	Needle brg., sealed on one end
"D"	Needle brg., sealed on one end
"E"	Bronze brg.

AXLE TUBE

Location	Description
Each end of tube	Bronze

REMOVAL

To remove the bearings, disassemble the transmission as outlined in transmission disassembly instructions. For best results, use a rod slightly smaller than the outside diameter of the bearing and press the bearing from its housing. Use caution to avoid damaging the bearing housing.

INSTALLATION

Needle Bearings

Be sure to use the correct size bearing for each location as listed in the chart above. Press bearings "A", "C", and "D" into the cover and case until the end of the bearing case is flush with the machined face of the bearing housing. This machined face of bearing housing is on inside of transmission. Bearing "B" is to be installed $1/16$ in. below surface of bearing housing.

NOTE: The needle bearings all have a number stamped on one end of bearing case. This is the end of bearing to press against, when installing. DO NOT press against the un-numbered end of the bearing or damage may result.

The bearings with the sealed end, are to be installed with the sealed end facing the outside of the transmission.

Bronze Bearings

Press one bronze bearing into the "E" bearing housing of the transmission case until it is flush with the inside machined face of bearing housing. Now insert the axle tube into the bearing and place the transmission cover over the dowel pins of case and slide the other bronze bearing into place in the housing of the cover. Press the bearing into the housing until the outer end of bearing is flush with bottom face of the smaller of two counterbores in housing.

NOTE: The axle tube acts as an aligning mandrel for the bearings, and must be used to prevent cocking the bearings when they are being installed.

Axle Tube

Press a bronze bearing into one end of the tube until it is flush with the end of the tube. Now insert the axle shaft through the axle tube and slide the other bronze bearing over the axle tube until it starts to enter the opposite end of axle tube. Press the bearing into position flush with the end of axle tube.

NOTE: The axle shaft acts as an aligning mandrel for the bearings and must be used to prevent cocking the bearings when they are being installed.

Oil Seals

The transmission contains a total of 4 oil seals; 2 seals are in the transmission cover, and 2 seals are in the transmission case. The following indicates the correct order for seal replacement:

Transmission cover:
 a. One seal for pulley shaft
 b. One seal for axle tube

Transmission case:
 c. One seal for brake drum shaft
 d. One seal for axle tube

REMOVAL

To remove old seals, carefully pry the seals out of their positions in the case and cover. Use caution to avoid damage to either the bearings or the cast iron seats or bores that the seals rest in.

NOTE: Do not attempt to reuse or salvage seals. When seals are removed from their positions in the transmission, they are not fit for reuse, and must be discarded.

INSTALLATION

The importance of properly installing oil seals cannot be minimized if they are expected to do their job and do it well. Failure to observe correct installation procedure will account for more seal failure than any other cause.

Inspect the surface of each shaft to be certain that no nicks, burrs, scratches, or sharp edges will be able to damage the seals during installation. Be particularly critical of the area of the shaft that the seal

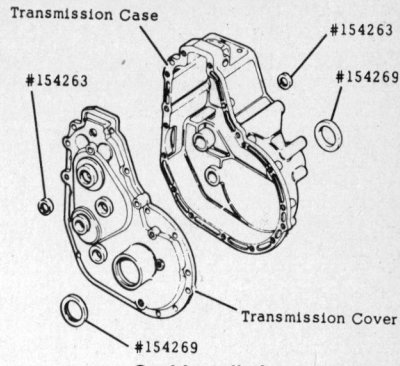

Seal installation

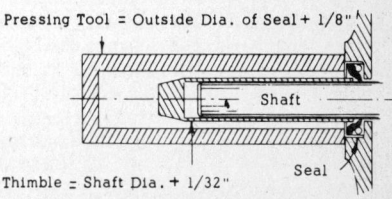

Seal installation tools

covers when in position. If a shaft shows any nicks, scratches, or burrs at this area, discard the shaft and replace with a new one. Any attempt to file or remove flaws at this point will only result in a flat spot and oil leakage.

1. Make sure that the seal is correct size.
2. Check the cast iron seat or bore that the seal will rest in. Remove all nicks, burrs, scratches, or foreign material.
3. When installing the seal, it is advisable to use a thimble that will fit over the shaft as shown. This will aid in stretching the spring element in the seal to allow it to slide over the shaft, and at the same time will protect the seal from damage or cutting by the edges of keyways.

NOTE: The thimble should be long enough to protect the seal until it is completely past all holes or keyways. The maximum diameter of the thimble should equal the shaft diameter plus $1/32$ in. Lubricate the surface of the thimble with clean grease to aid in sliding the seal in place.

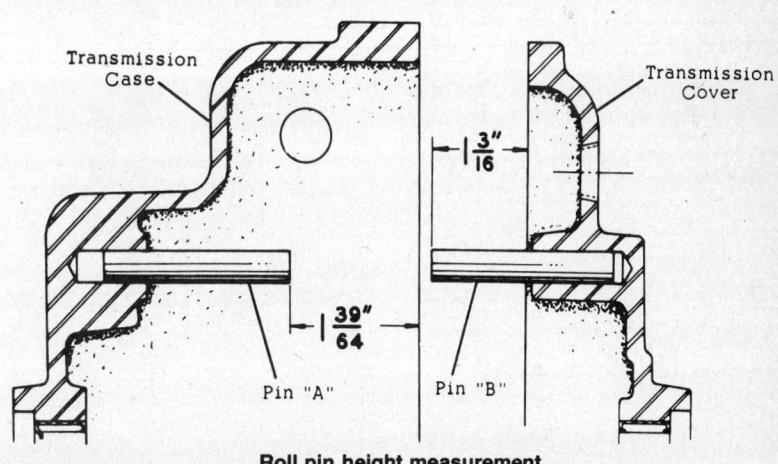

Roll pin height measurement

Allis-Chalmers

NOTE: If a thimble is unavailable, wrap the shaft with a heavy grade of paper lubricated with oil or grease. When wrapping the shaft, start at the seal end of the shaft and wrap in an overlapping spiral fashion, being sure to cover the keyways.

4. Remove the seal from its wrapper or package, and gently lubricate the sealing element with a light coating of absolutely clean grease. Do not run your finger roughly around the sealing element, as it is easily deformed and ruined.

5. As these seals are to be installed flush with the outer surface of the bore or seat, use a pressing tool at least 1/8 in. larger in diameter than the outside diameter of seal. Place the seal on the thimble with the spring element of the seal facing the liquid to be retained, and with the pressing tool, gently slide the seal towards its seat. Avoid cocking the seal as it starts into its seat, and gently tap the pressing tool with a hammer until the seal is in place. NEVER HAMMER ON THE SEAL ITSELF.

ASSEMBLY

Before beginning to assemble the transmission, make certain that the transmission case, cover, and all of the parts that go into the transmission have all been completely cleaned. Scrape the mating surfaces of the case and cover to remove any pieces of gasket material that may have stuck. As you put the various gear shafts into their bearings, apply a light coating of clean transmission oil to the bearing surfaces of each shaft.

1. Before beginning assembly, check the 2 roll pins that limit the movement of the shift stem. It is imperative that both of these pins be checked for proper height and adjusted if necessary. Note the dimensions shown: pin "A" in the transmission case is set to give a dimension of $1^{39}/_{64}$ in. from the end of the pin to the face of transmission case, and pin "B" in the transmission cover is set to give a dimension of $1^{3}/_{16}$ in. from the end of the pin to the face of cover. When replacing pins, the groove of the pin should face the top of the transmission case.

2. Insert the shift rod into the transmission case and position it to allow the shift stem with locknut to be screwed into the shift rod. Screw the shift stem into the shift rod until a distance of 5/8 in. from the round surface of the rod to the shoulder of shift stem is obtained. This setting is important to insure proper shifting, so check and adjust until correct. Be sure the locknut is tight enough to hold the shifter stem at this setting.

3. Assemble the shifter fork assembly with the longer shaft into the "Y" shaft hole of the transmission. Take care to be sure that the setscrew, which holds the shaft in place, is actually locked into the setscrew hole of the shifter shaft. Now assemble the shorter shaft into the "Z" shaft hole of the transmission and lock in place in same manner as other shifter fork assembly.

4. Assemble the reverse gear "F" as shown. When properly assembled, insert the gear assembly into the transmission case and position the bracket over the mounting hole in the case. Fasten the bracket to the case with a hex capscrew and tighten to 20 ft.lbs. Add a locknut to the pin protruding through the transmission case and tighten to 55 ft.lbs.

5. Place shaft "B" and gears into bearing "B" in the transmission case.

6. Place a greased flat washer over face of bearing "A" in transmission case. To install shaft "A" and gears, first move the "Y" shift fork into Neutral (see disassembly instructions) and move the "Z" fork into raised position (see disassembly

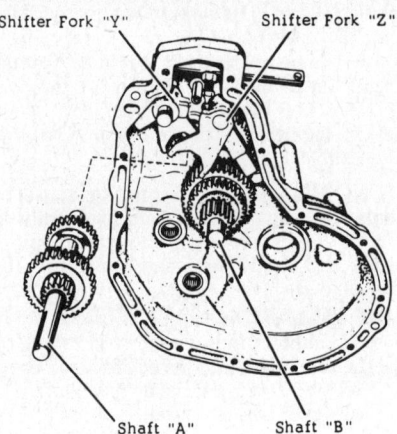
Preliminary parts installation

instructions). Raise the gear cluster on shaft "B" slightly and move shaft "A" toward the shift forks. When you slide the shift rings on the gear assemblies into position against the shift forks, it should be possible to lower shaft "A" into place in its bearing.

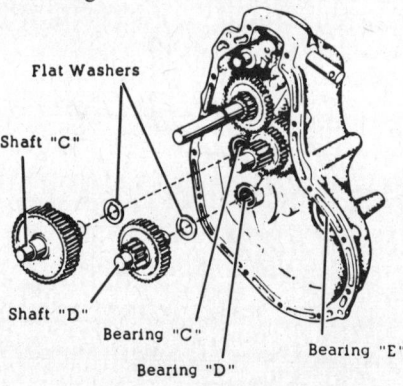
Secondary parts installation

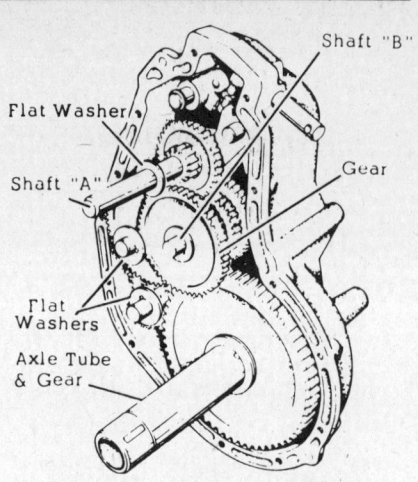
Axle tube and gear installation

7. Place a greased flat washer over face of bearing "D" and place shaft "D" and gears in place in the bearing.

8. Place a greased flat washer over face of bearing "C" and place shaft "C" and gears in place in the bearing.

9. Place 2 greased flat washers over face of bearing "E" and place axle tube and gear in place in the bearing.

10. Assemble the gear to end of shaft "B", with the beveled edge facing gear cluster on shaft "A". Place a greased flat washer on shaft "A" and on end of shaft "C" and shaft "D". Add 2 flat washers to axle tube.

11. Before putting gasket in position, drive the 2 dowel pins up until they protrude approximately 1/4 in. to 5/16 in. above the machined face of the transmission case. Now position the new gasket in place and seat the cover over the 2 dowel pins before inserting the capscrews. When cover seats properly, insert and tighten the capscrews.

12. Attach a pulley to the pulley shaft and rotate by hand to check gears for binding. Check all gear ranges to see that gears rotate freely. If a slight bind is noticed, tap the end of the pulley shaft and brake drum shaft with a rawhide mallet. It may be that one of the bearings is not seated far enough into the cover or case, and the impact of the mallet will drive it into position and remove the binding. If a severe binding is noticed, it will be necessary to disassemble the transmission and locate the cause.

13. Assemble the grease fitting and snapring to the axle tube.

INSTALLATION

1. Assemble the brake drum to the shaft with the key and setscrew. The setscrew side of the brake drum faces away from the transmission. Add the brake band and adjusting linkage and secure in place with 2 hex capscrews, lockwashers, and hex nuts.

2. Position the transmission case on supports so that hole "C" on righthand side of transmission lines up with the mounting hole of side plate. Insert the capscrew and tighten partially. Hold the pulley shaft with one hand and by using your knee for a brace, position the trans-

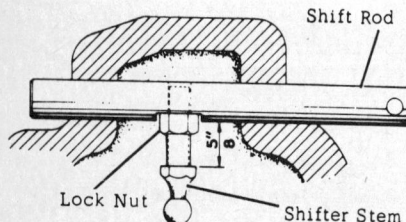
Shift rod and stem installation

Allis-Chalmers

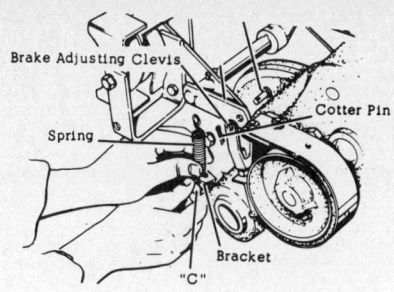

Bolt tightening sequence

mission so that bolt "A" may be inserted and partially tightened. Add bolt "B". Now add bolts "A" and "B" to the lefthand side of the transmission. With the spring bracket in place on bolt "C" and positioned as shown tighten the bolt. Now tighten all 6 mounting bolts securely.

3. Insert the clevis pin through the hole of brake adjusting clevis, through the hole in lever of clutch and brake shaft assembly, and through the end loop of spring. After the pin is in place, secure with cotter pin.

4. Use Vise-Grip® pliers and attach the large spring from the power take-off to the mounting hole on the spring retainer.

5. Attach the clevis of the shifter rod to the transmission shift rod, using hex capscrew, lockwasher, and hex nut as shown.

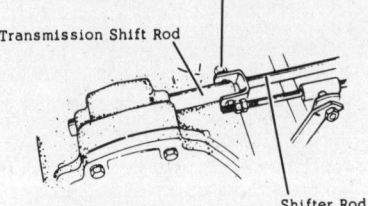

Shift rod clevis installation

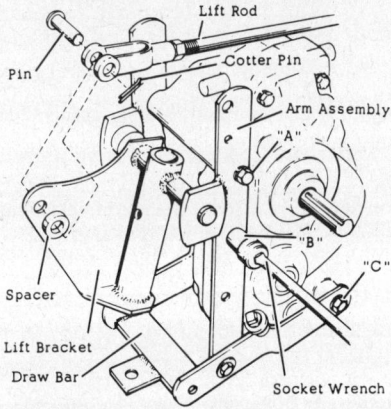

Draw bar, rear lift bracket and arm installation

6. Mount the drawbar, rear lift bracket, and arm assembly to the transmission as shown, and fasten with hex capscrews (3 on each side) "A", "B", and "C". Connect the lift rod clevis to rear lift bracket with pin, spacer and cotter pin. The spacer is to be placed on righthand side of the lift bracket.

7. Place the key in the keyway of pulley shaft and mount the pulley with the hub facing away from the transmission. Do not tighten the setscrew until after the V-belt is in place and the pulley is aligned with the pulley on the bevel gear box.

8. Insert the axle shaft into the axle tube on the righthand side of the transmission and place the 2 keys in the slots on the axle tube. Align the keyways in the differential hub with the keys and while holding the keys in place, push the axle shaft through the transmission. The differential hub is to seat against the snapring on the axle tube.

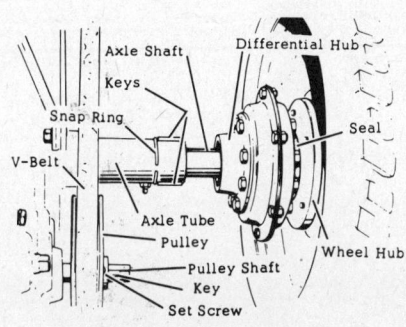

Mounting the pulley

9. Make sure that the axle and differential are properly seated so that the seal between the differential and the wheel hub is compressed. The axle is then held in this position by placing the set collar over the lefthand end of the axle shaft and locking it securely against the axle tube by means of the 2 setscrews. It is very important that this set collar be securely locked at all times to eliminate any end play of the axle shaft.

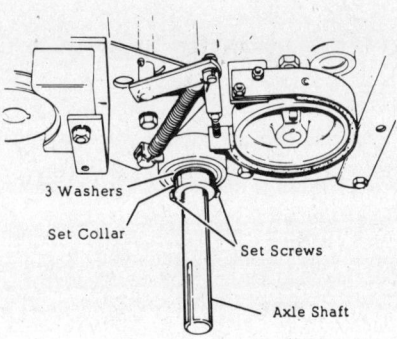

Shimming the axle shaft

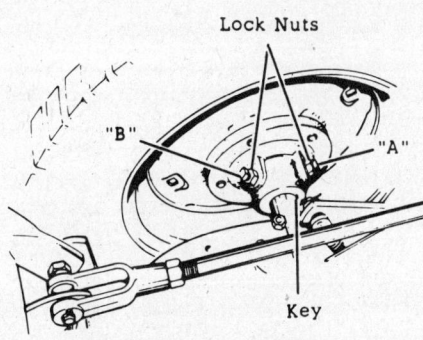

Locking the axle shaft

NOTE: Later production models have 3 washers between the set collar and the axle tube for the purpose of moving the set collar farther out on the axle shaft. Use washers No. 105050.

10. Place the key in the slot of the axle shaft and mount the lefthand wheel and hub in place over the key. Tighten the setscrew "B" into the setscrew hole on the axle shaft. Tighten the setscrew "A" and lock both setscrews with the locknuts. Tighten securely.

11. Mount the seat assembly to the tractor, placing the arms of the seat pivot assembly between the vertical arm assembly fastened to transmission. Fasten with hex capscrews and locknuts. Tighten enough to take out excessive "play" but leave loose enough to pivot properly.

12. Mount the rear light support bracket to the seat back as shown, using a hex capscrew and lockwasher. Fasten the cable clip to the arm assembly as shown, with a hex capscrew, flat washer, lockwasher, and hex nut.

13. Attach the ground cable to the negative terminal of the battery.

14. Refill transmission with 1½ quarts of SAE #90 Oil, and check drain plug, filler plug, and vent plug for tightness.

HYDROSTATIC TRANSMISSION

This unit is the Vickers T66 or TA6 Series 10 design.

HB-112 and HB-212

REMOVAL AND INSTALLATION

1. Raise and support the tractor with jack stands under the main frame just ahead of the bevel gear housing.

2. Remove the seat deck and fender housing.

3. Remove the rear wheels and hubs.

4. Remove the differential assembly and axle shaft.

5. Remove the bevel gear PTO belt pulley.

6. Disconnect the PTO tension spring.

7. Support the gear reduction housing and remove the left side plate.

8. Drain the transmission, disconnect the oil lines and control rod, then unbolt and remove the oil reservoir, oil cooler, shroud and cooler fan.

9. Unbolt and remove the transmission and brake assembly.

10. Remove the capscrews securing the gear reduction unit to the right side of the tractor and lift the unit clear.

11. Installation is the reverse of removal. Fill the reduction unit to the level plug with SAE90EP oil. Fill the hydrostatic reservoir with Dexron® type A automatic transmission fluid. Adjust the brake and idler gear as required.

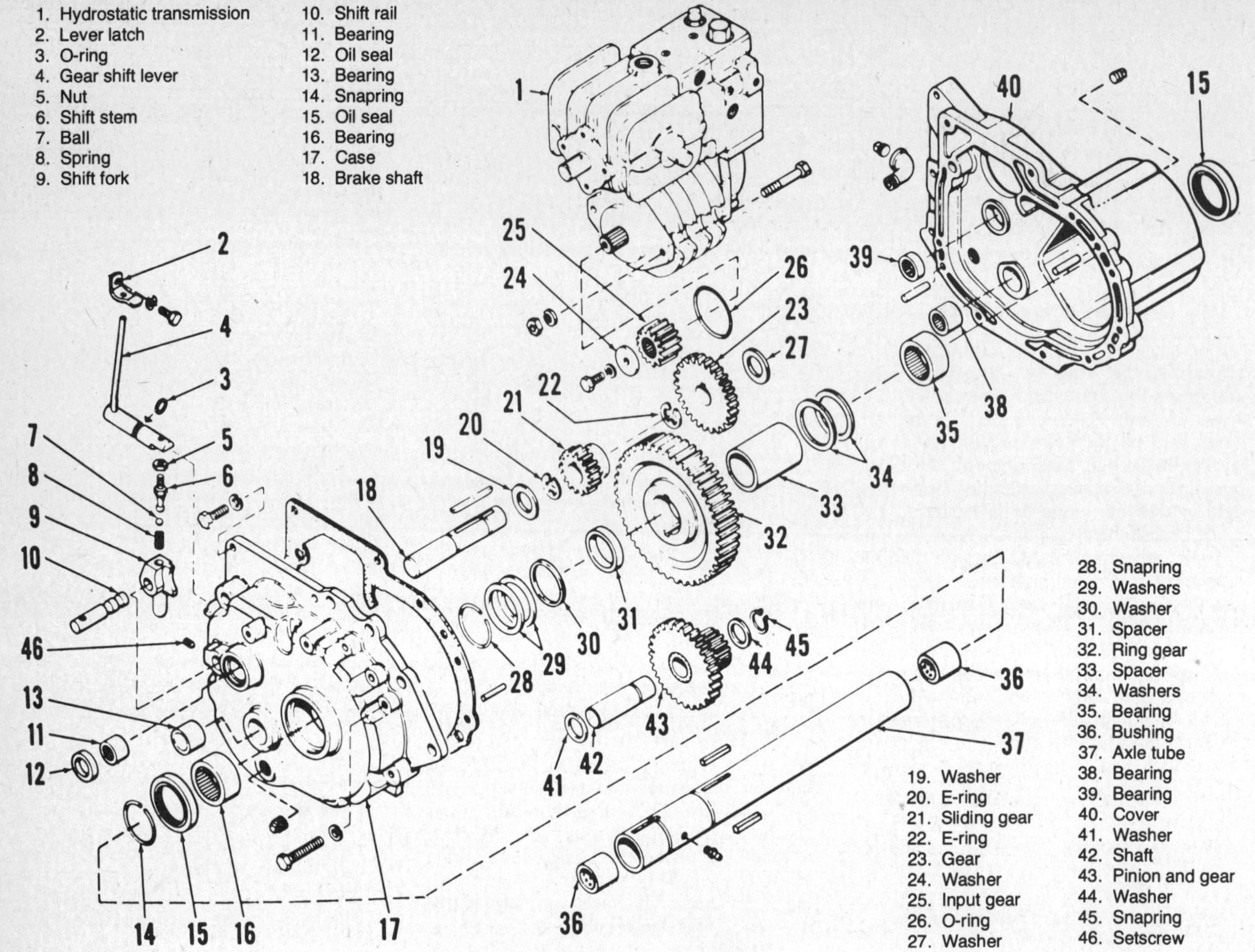

Allis-Chalmers

1. Hydrostatic transmission
2. Lever latch
3. O-ring
4. Gear shift lever
5. Nut
6. Shift stem
7. Ball
8. Spring
9. Shift fork
10. Shift rail
11. Bearing
12. Oil seal
13. Bearing
14. Snapring
15. Oil seal
16. Bearing
17. Case
18. Brake shaft
19. Washer
20. E-ring
21. Sliding gear
22. E-ring
23. Gear
24. Washer
25. Input gear
26. O-ring
27. Washer
28. Snapring
29. Washers
30. Washer
31. Spacer
32. Ring gear
33. Spacer
34. Washers
35. Bearing
36. Bushing
37. Axle tube
38. Bearing
39. Bearing
40. Cover
41. Washer
42. Shaft
43. Pinion and gear
44. Washer
45. Snapring
46. Setscrew

Gear reduction unit components

OVERHAUL

Transmission Case

With the unit removed from the tractor:
1. Remove the brake drum.
2. Thoroughly clean all paint, burrs and rust from the keyed end of the axle tube.
3. Unbolt and remove the cover from the case.
4. Remove the washer and first reduction gear from the case.
5. Remove the snapring, then withdraw the output gear and axle tube assembly.
6. Remove the second reduction gear and shaft assembly.
7. Loosen the setscrew and remove as a unit the shift fork and rail assembly and the brake shaft assembly. Use caution when removing the shift rail from the shift fork as the poppet ball and spring will be released.
8. Loosen the locknut, remove the shifter stem and withdraw the shift lever.
9. The oil seals and needle bearings may now be removed from the case as required.
10. Clean and inspect all parts. Replace as required. Installation is the reverse of removal.

Hydraulic Motor Overhaul

Special tools are required for this procedure, as well as an absolutely clean work area. The use of lint-free rags is also important.

1. Separate the motor assembly from the transfer block by removing four hex head screws. Discard O-rings and replace with new ones.

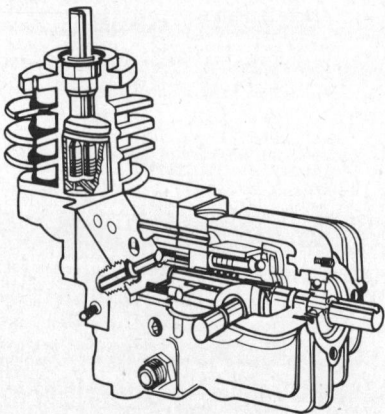

Pump cross-section

2. Separate motor valve plate from the motor housing by removing four screws.
3. If valve plate doesn't separate easily from motor housing, tap corner of valve plate with plastic mallet.
4. With one hand under rotating group end, tilt housing until rotating group slides into your hand.
5. If the rotating group does not need to be disassembled, place it on a clean surface and proceed to step 13. To disassemble this group proceed to step 7.
6. Remove swash plate from shoe plate.
7. Remove assembled parts as shown. Be careful not to scratch the pistons or cylinder running surfaces.
8. Generally, no further disassembly is required. However, if the cylinder block is to be disassembled, proceed to step 9.
9. To relieve cylinder block spring tension, refer to the accompanying illustration.

---- **WARNING** ----
Exercise extreme caution. Spring is under a great deal of tension.

10. To remove the motor shaft, first remove the large snapring with the 90° Truarc pliers.

Allis-Chalmers

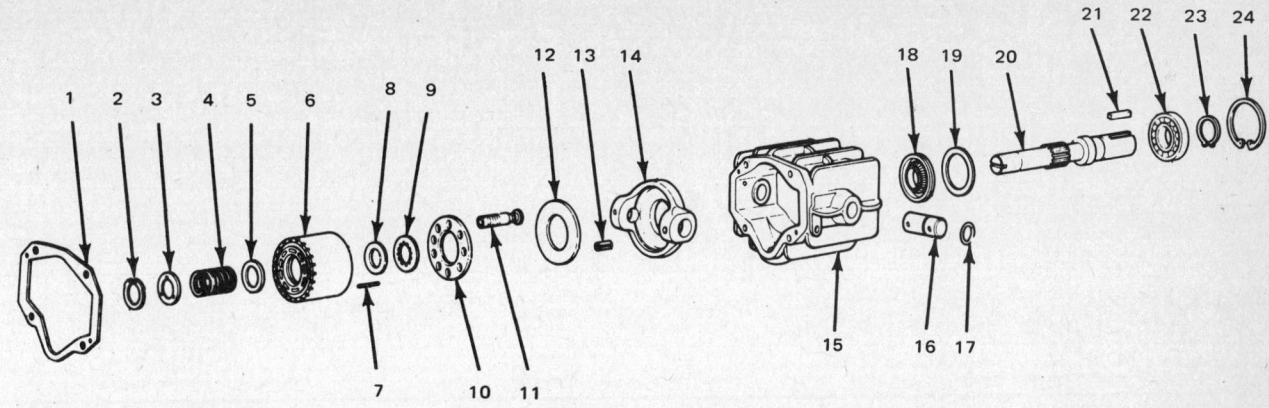

1. Gasket	4. Spring	7. Pin (3 required)	10. Shoe plate	13. Pin (2 required)	16. Pintle	19. Spacer	22. Bearing
2. Snapring	5. Thrust washer	8. Washer	11. Piston	14. Yoke	17. O-ring	20. Shaft	23. Snapring
3. Lift limiter	6. Cylinder block	9. Spherical washer	12. Swash plate	15. Pump housing	18. Shaft seal	21. Key	24. Snapring

T66 and TA6 pump components

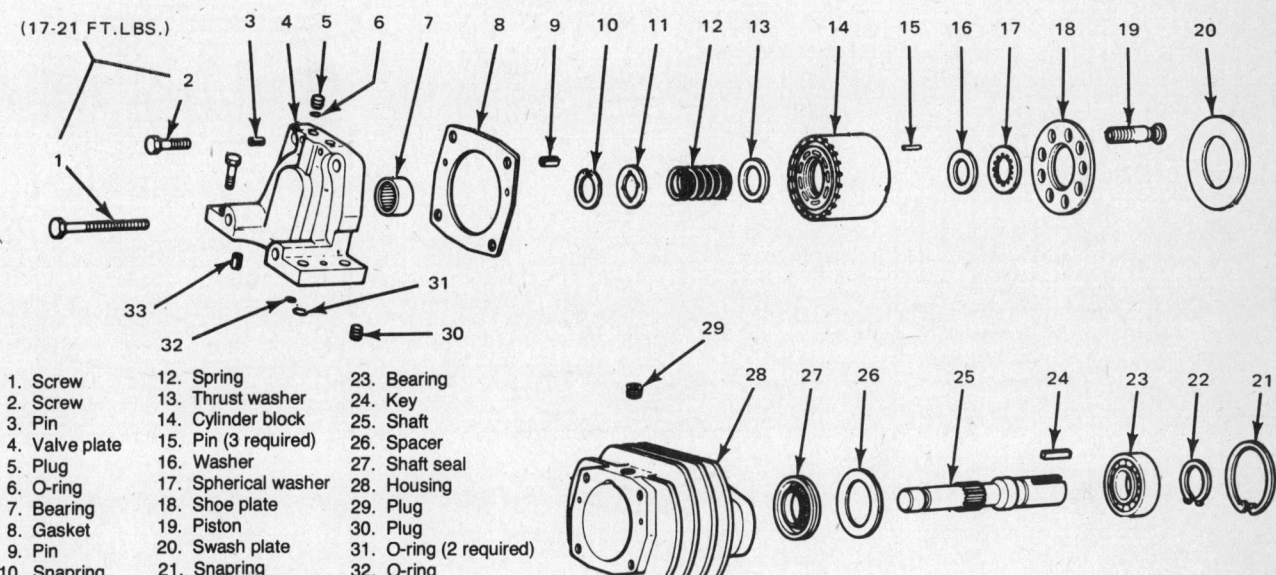

1. Screw	12. Spring	23. Bearing
2. Screw	13. Thrust washer	24. Key
3. Pin	14. Cylinder block	25. Shaft
4. Valve plate	15. Pin (3 required)	26. Spacer
5. Plug	16. Washer	27. Shaft seal
6. O-ring	17. Spherical washer	28. Housing
7. Bearing	18. Shoe plate	29. Plug
8. Gasket	19. Piston	30. Plug
9. Pin	20. Swash plate	31. O-ring (2 required)
10. Snapring	21. Snapring	32. O-ring
11. Lift limiter	22. Snapring	33. Pin

T66 motor components

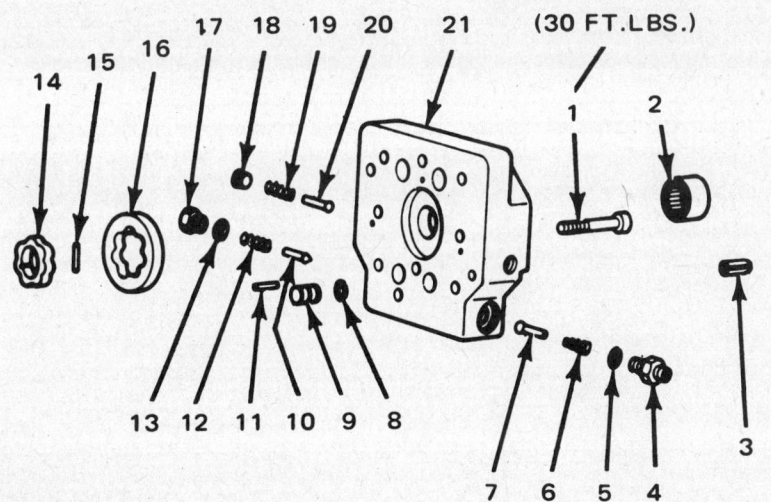

1. Screw
2. Bearing
3. Pin
4. Plug
5. O-ring
6. Spring
7. Valve poppet
8. O-ring
9. Plug
10. Valve poppet (deleted in non "S" unit)
11. Pin
12. Spring (deleted in non "S" unit)
13. O-ring
14. Inner rotor
15. Key
16. Outer rotor
17. Plug
18. Plug
19. Spring
20. Valve
21. Valve plate

T66 and TA6 valve plate components

137

Allis-Chalmers

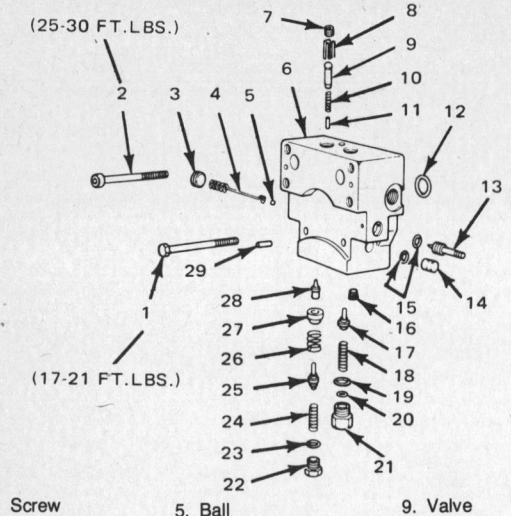

- 1. Screw
- 2. Screw
- 3. Plug
- 4. Spring
- 5. Ball
- 6. Transfer block
- 7. Plug
- 8. Guide
- 9. Valve
- 10. Spring
- 11. Pin
- 12. O-ring
- 13. Plunger
- 14. Plug
- 15. O-ring
- 16. Plug
- 17. Poppet
- 18. Spring
- 19. Shim
- 20. O-ring
- 21. Cap
- 22. Cap
- 23. O-ring
- 24. Spring
- 25. Pin
- 26. Spring
- 27. Seat
- 28. Poppet
- 29. Pin

T66 and TA6 transfer block components

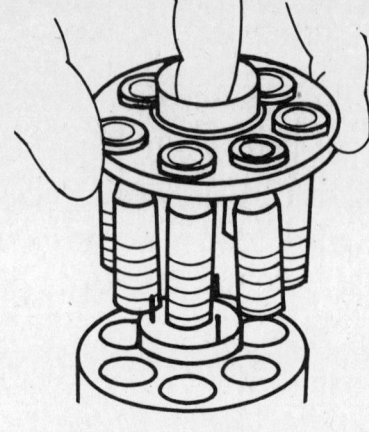

Removing pistons and shoe plate

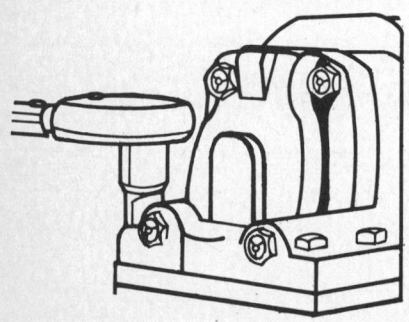

Removing transfer block-to-motor housing bolts

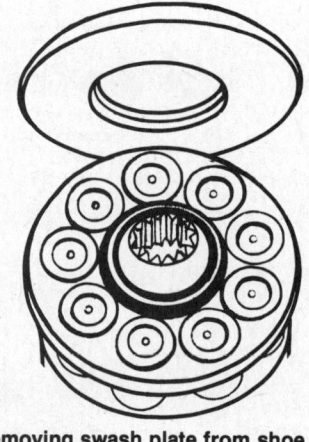

Removing swash plate from shoe plate

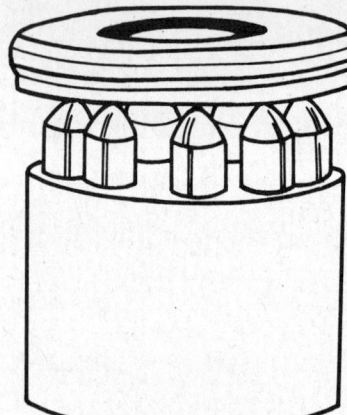

Rotating group

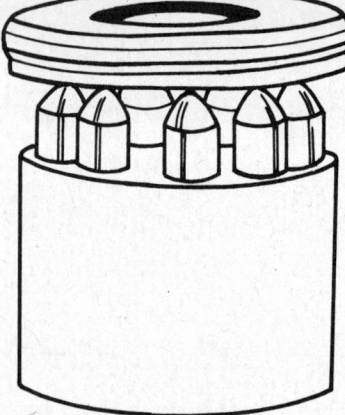

Motor shaft bearing removal

Lifting valve plate from motor

Disassembling shoe plate and piston

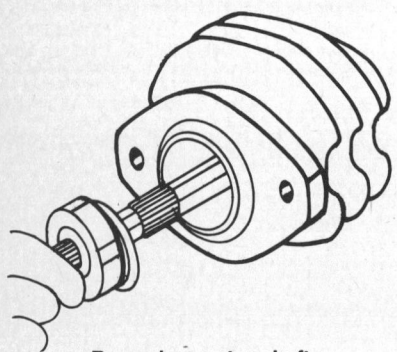

Removing motor shaft

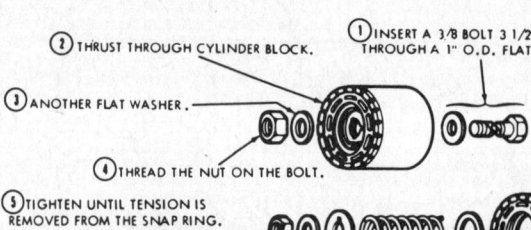
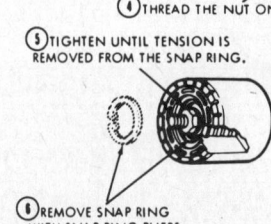
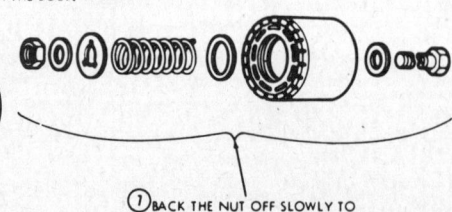

Relieving cylinder block spring tension

11. Remove the shaft by tapping on the small end with a soft tipped hammer or mallet.

12. Remove shaft. The spacer and press fit bearing should come out with it. Replace shaft seal.

13. Remove snapring and key from motor shaft before you remove the bearing.

14. If it is necessary to remove the bearing, first remove the key; then use an Owatanna 10-11 bearing puller, or equivalent puller, or an arbor press. Any other method of removal may damage bearing.

15. To disassemble the pump, remove the valve plate and transfer blocks as a unit by removing two recessed Allen head screws, and then the two hex head screws.

16. Pull valve plate and transfer block straight up from pump housing. Set it down on its painted side.

Removing valve plate and transfer block from pump housing

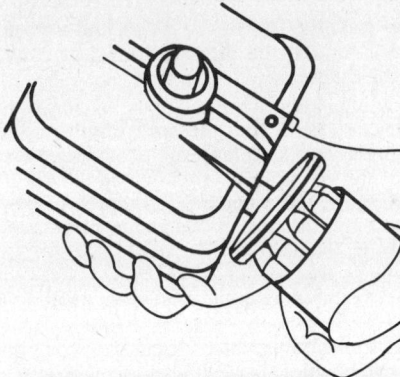

Removing rotating group from pump housing

Note: Line up pins with holes in valve plate and gerotor key with drive shaft slot.

17. Pick up pump housing with one hand and slowly tilt it forward to remove group as assembled unit. To disassemble rotating group perform steps 6 through 9.

18. Now to remove the pump shaft. First remove the snapring with 90° snapring pliers.

19. Remove the pump shaft by tapping the small end of the shaft with a plastic tip

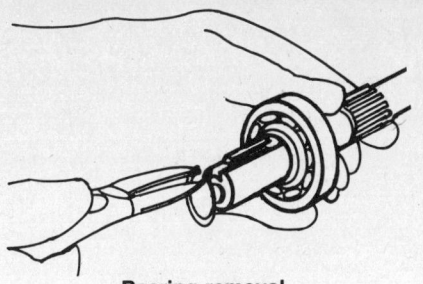

Bearing removal

hammer. Remove the shaft with the loose spacer and the press fit bearing installed on it. Replace shaft seal.

20. To remove the bearing, remove the key, and then the snapring. Refer to step 14 for bearing removal.

21. To remove both pintles and yoke from the housing, set a 3/16 in. punch on the roll pin. Tap punch with a hammer until roll pin is disengaged from yoke.

22. Now place a ¼ in. brass rod on the pintle, and tap the pintle out of the yoke.

23. Repeat this procedure on the other pintle. Remove yoke from housing. Pintles must not be installed backward.

24. To disassemble the valve plate, remove the two recessed Allen head screws.

25. Separate valve plate from transfer blocks by pulling them apart. If required, tap valve plate with a plastic mallet to separate them.

26. Remove replenishment pump from valve plate.

NOTE: Dots not to be visible when replenishing pump is in pocket.

27. Remove the two replenishing system check valves by removing the Allen-head plugs. Don't interchange valve parts.

Replenishment pump showing dots

28. Remove replenishing pump relief valve. (Some models have only one valve.)

29. To remove bearing, place valve plate on protective surface. Put brass shim stock or other protective stock under puller. Use Owatanna MD956-B-1, an equivalent puller, or an arbor press.

30. Set transfer block with finished surface facing up. Remove caps and take out both the soft-ride valve and high-pressure relief valve.

31. To remove high pressure check valves, first remove three O-rings.

NOTE: Be sure open ends of guide point outward.

32. Remove high-pressure check valve seats with an Allen head wrench.

NOTE: During assembly, torque valve seats to 30-35 ft.lbs.

MANUAL TRANSAXLE

B-207, B-208

REMOVAL AND INSTALLATION

1. Remove the drive belt from the transaxle pulley.
2. Remove the brake band.
3. Raise and support the tractor with jack stands under the frame. Remove the shift lever knob.
4. Remove the axle U bolts.
5. Roll the transaxle away from the frame.
6. Installation is the reverse of removal.

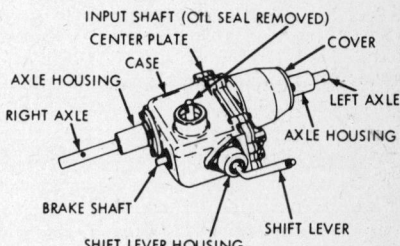

Transaxle unit

OVERHAUL

With the unit removed from the tractor:
1. Drain the oil and remove the input pulley.
2. Remove the brake drum, wheel and hub assemblies.
3. Place the shift lever in neutral.
4. Unscrew the shift housing capscrews, and pull the shift assembly from the case.
5. Remove all paint, burrs and rust from the axle shafts and place the unit in a vise with the right (longer) axle shaft pointing down.
6. Unscrew the capscrews from the case and drive out the dowel pins.
7. Separate the cover from the case and

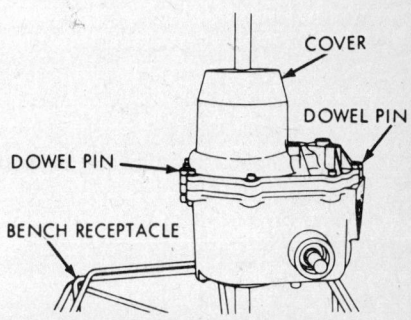

Cover showing dowel pin location

Allis-Chalmers

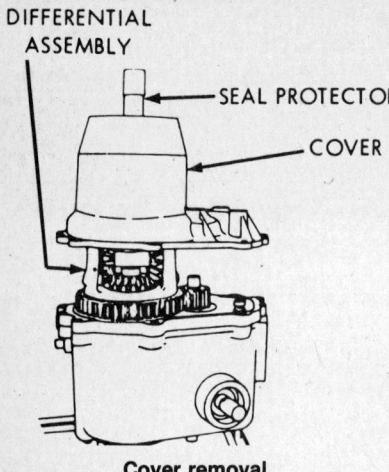
Cover removal

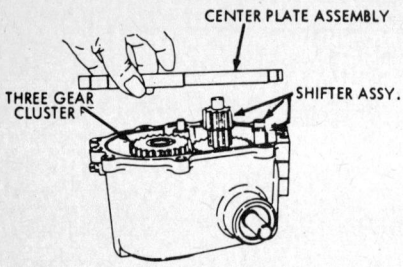
Center plate removal

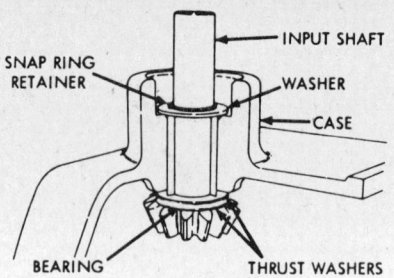
Removing input shaft

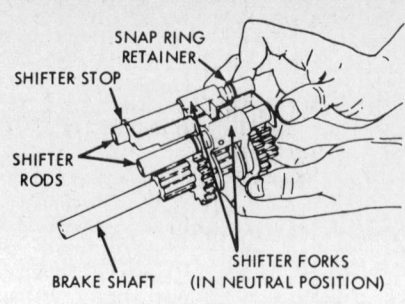
Correct shifter assembly

B-206

REMOVAL AND INSTALLATION

lift the cover up off the axle. The brake shaft and idler gear will be removed with the cover.

8. Remove the output shaft with the output gear spacer and washer.

9. Pull out the differential and axle shaft assembly and lay aside for further disassembly.

10. Hold the upper ends of the shifter rods together and lift out the shifter rods, forks, shifter stop, shaft and sliding gears as an assembly.

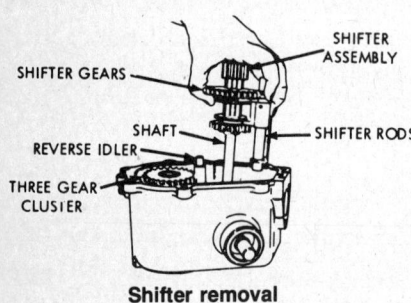
Shifter removal

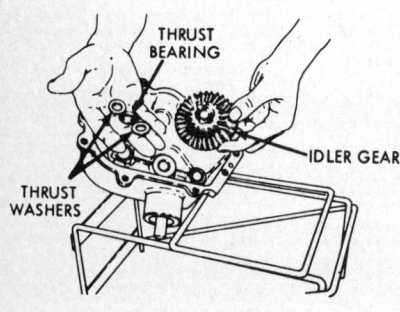
Removing idler

11. Remove the reverse idler gear, idler shaft, and spacer, then remove the idler shaft along with the idler gears and spacers.

12. Remove the input shaft and gear from the case.

13. Block up under the brake shaft gear and remove the shaft out of the gear with a press or puller, being careful that the pressure is not applied to the cover.

14. Replace the seals and bushings in the axle housings as required.

15. Unscrew the four capscrews and separate the axle shaft and carriage assemblies from the ring gear.

16. The drive blocks, bevel pinion gears and drove pin can now be removed from the ring gear.

17. Remove the snaprings and slide the axle shafts from the axle gears and carriages.

18. Clean and inspect all parts for damage or excessive wear. Replace as required.

19. Assembly is the reverse of disassembly. Note the following points:

a. When installing the needle bearings, press the bearings into the case and cover from the inside until the bearings are 0.015–0.020 in. below the thrust surfaces.

b. Use all new seals and gaskets.

c. Install the reverse idle gear in the case so that the rounded edge of the gear teeth and spacer are toward the cover.

d. When installing the idler shaft, place the shorter spacer between the gears and the longer spacer between the gears in front of it.

e. Bevels on all gear teeth must be on the side of the gear nearest the large gear.

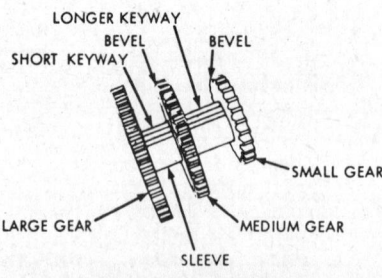
Correct gear installation

f. When installing the shifter assembly, position the shifter rods in neutral. Observe the following torque values:
Differential capscrew: 7 ft.lb.
Case-to-cover capscrews: 10 ft.lb.
Axle housing capscrews: 13 ft.lb.
Shift lever housing capscrews: 10 ft.lb.

1. Lift or jack up the tractor to a point where the rear wheels are free of the ground. Place a support strong enough to bear the weight of the tractor at a point under the frame ahead of the PTO pulley.

2. Disengage and remove transmission drive belt.

3. Remove hub caps from wheels.

4. Remove (two) snaprings from grooves (one at each end of transaxle). Slide wheels off axle. Remove axle spacers and washer from each end of axle.

5. Remove shifter lever knob.

6. Loosen and remove three capscrews connecting 1602064 bracket to frame and transaxle on left side—remove bracket.

7. Remove four washers and nuts and two "U" brackets holding transaxle to frame.

8. Standing at rear of tractor, lower the left side of the transaxle until the shift lever has passed through the frame and then slide the transaxle to the left until free of the brake band and frame.

NOTE: The brake band and rod do not have to be disassembled to remove the transaxle.

9. Standing at the rear of the tractor: Hold the transaxle with the brake drum on the left and the shift lever up, tilt the right end of the transaxle up slightly and slide it to the right allowing the shift lever to pass through the hole in the frame. The brake drum should fit inside the brake band.

10. Reassemble 2 "U" brackets (one on each end of axle shaft) to frame and fasten to frame with 4 washers and nuts.

11. Fasten 1602064 bracket to frame and transaxle with 3 capscrews, washers, and nuts the long side of the bracket should extend down and rearward from the bottom of the frame.

12. Slide a wheel spacer on each end of the transaxle. Assemble wheels on axle and assemble snapring in groove outside of each wheel. Replace hub caps.

NOTE: Wheel width of rear wheels can be changed by reversing the face of the wheel hub on the axle.

If both wheels are assembled with deep dish side of hub toward the transaxle a narrow stance can be obtained.

13. Install transmission drive belts.
14. Adjust brake and clutch.

Allis-Chalmers

OVERHAUL

1. Clean the outside surface of the transaxle, away from the area where disassembly will take place. (Position shift lever in neutral position to help disassembly. Remove screws (3) holding shift lever housing. Remove shift lever housing. Drain oil through the shift lever opening. Remove all keys from keyways, remove all burrs and dirt from shafts. On hardened shafts, use a stone to remove burrs. All seals should be replaced whenever a shaft is pulled through a seal. Always use a new gasket whenever the gasket surfaces have been separated.

2. After removing axle housings, place the unit in a receptacle, bench or clamp the transaxle in a soft jaw vise. Position the transaxle so that the socket head capscrews are facing up.

3. Remove the socket head capscrews holding the case and cover together. Drive out the dowel pins used for alignment of the case and cover.

4. Lift off the cover assembly. Use a seal protector on axle shaft and lift off transaxle cover assembly. Because this seal is a single lip type, it may be reused, if care is taken to see that it isn't scratched or cut. Discard gasket.

5. To remove differential assembly, it may be necessary to replace two or three screws to hold center plate assembly down. Pull assembly straight up. If tight, tap on lower axle with soft mallet.

CAUTION
Do Not Use Steel Hammer.

6. Remove temporary holding screws, is used, and lift off center plate assembly. Discard gasket.

7. Remove complete shifter assembly by grasping shifter gears, shaft and both shifter rods as a unit.

NOTE: Examine assembly carefully; if no service is required, retain assembly as a unit for easy reassembly.

8. Remove reverse idler shaft and spacer, cluster gear assembly and thrust washer. For removal and replacement of gears or cluster, see step 11.

9. Lift idler gear assembly out of case.

NOTE: For sequence of thrust washers and bearings, see the accompanying illustration.

NOTE: Caution required as needles from shifter and brake shaft bearing may fall out.

10. Remove input shaft oil seal to allow access to snapring and input shaft will slide out. A removed seal must be replaced by a new seal.

11. Cluster gear sub-assembly
   a. The cluster gear can be disassembled. All gears are replaceable if damaged or worn. Preferably use a press to drive the gears squarely.
   b. The small and middle gear bevel faces down, there is no beveled edge on large gear. Shorter section between middle and large gear.

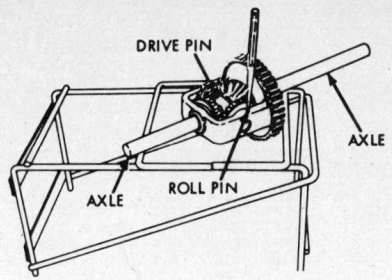

Roll pin removal

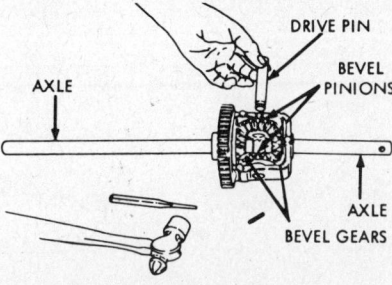

Simultaneous bevel pin removal

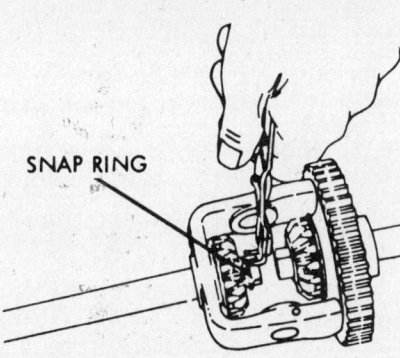

Removing snap ring and bevel gear from double D models

   c. Key edge ends must align with shaft ends.

12. Differential
   a. Drive out roll pin that secures drive pin with suitable driver.
   b. Remove drive pin.
   c. Thrust washers must be removed before attempting to remove the pinions. Remove bevel pinions simultaneously by rotating the gears in opposite directions; gears will move out of position.
   d. Drive out double roll pin and slide axle out. On roll pin drive types, drive the bevel gears from the axle.
   e. On double "D" type drives, remove snapring, bevel gear and thrust washer. Slide axle out.
   f. Inspect bushings and gears for wear and replace when necessary.

13. Shifting assembly
The shifting assembly is usually removed from and installed into the transaxle as a unit. The assembly is removed and replaced by grasping the shifting rods firmly. This will cause the binding necessary to hold the assembly together. Before removal or installation of the shifting asembly, notches in the shifter forks should be aligned with notches in the shifter stop. This indicates that shifting assembly is in a neutral position. The shifter stop must be so positioned that the notch aligns with notches in shifter forks.

14. Shift lever assembly
   a. Prior to removing a shift lever assembly from a transaxle, make note of the position of the shift lever so that it may be assembled correctly to the shift lever housing.
   b. Move the shifter lever to Neutral, if possible, before removing it from the transaxle. Clean around the lever housing to prevent dirt from falling into the transaxle. Cover this opening, if possible.
   c. Place the shift lever in a vise so that the shift lever housing is at least one inch from the top of the vise jaws.

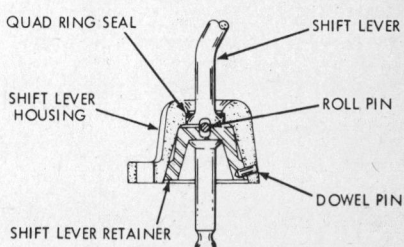

Dowel pin type shifter assembly

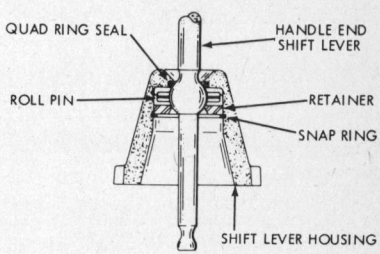

Snap ring type shifter

   d. Dowel Pin Type: Locate the dowel pin holding the retainer in the housing from the outside. Place a ¼ in. flat face punch on the gasket surface directly over the dowel pin. Strike the punch sharply but lightly with a hammer to dislodge the retainer from the shift lever housing. Always use a new dowel pin for reassembly.
Snapring Type: Use the proper compressing type tool for removing the snapring. Loosen the vise and disassemble the pieces.
   e. Remove the shift lever from the shift lever housing. Examine the roll pin in the ball of the shift lever, if bent or worn, replace. When inserting a new roll pin in the ball, position so that equal lengths protrude from both sides of the ball.
   f. Oil leakage past the point where the shift lever enters the shift lever housing will require replacement of the quad ring seal in the shift lever housing.
   g. Prior to reassembly, be sure that bends in the shift lever correspond to the mounting on the vehicle.

15. Shifting assembly
   a. Shifting assemblies are removed from and installed into transaxles by squeezing the top end of the shifter rods.

Allis-Chalmers

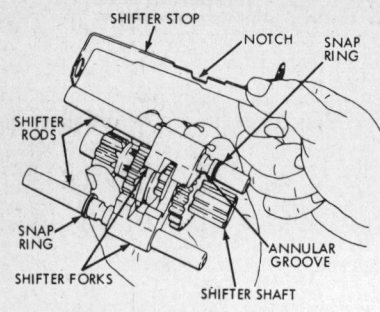
Shifter disassembly, step 1

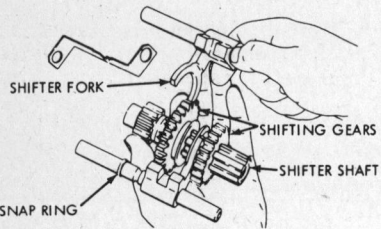
Shifter disassembly, step 2

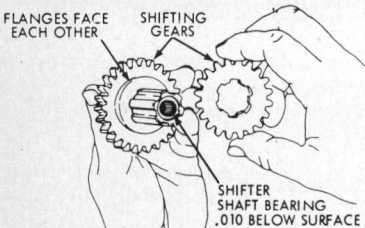
Shifter disassembly, step 3

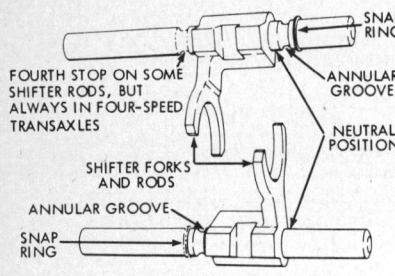
Shifter disassembly, step 4

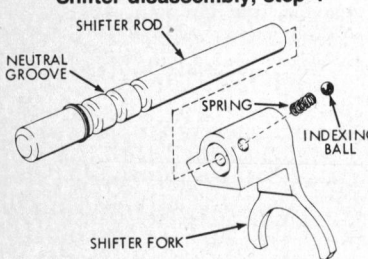
Shifter disassembly, step 5

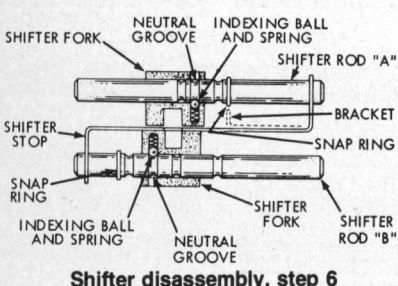
Shifter disassembly, step 6

This causes a binding that retains all parts during removal or installation.

b. Follow the accompanying illustrations in order. Prior to disassembly compare the assembly with the illustrations. This will aid during the reassembly.

c. Replace the shifter stop if worn or damaged.

d. Examine the teeth and internal splines of the two shifter gears. Replace damaged gears. The gears must slide freely on the shifter shaft. Excessive wear of the internal spline in the gears will create cocking and difficult shifting. Replace the gear if this condition is present.

e. Replace the shifter shaft needle bearing if wear is evident. Replace if the bearing surface of this shaft should it be scuffed, pitted or worn to a diameter less than .750 in.

f. Replace other parts showing wear, looseness, cracks, etc.

16. Reassemble the shifting assembly by following the accompanying illustrations. Lay the parts on the bench in the same manner as illustrated on a clean paper or shop cloth. Pay particular attention to the annular grooves in the shifter rods and the snapring.

17. Assemble the shifter forks to the shifter rods. The shifter forks are interchangeable.

18. Slide the shifter fork onto the shifter rod until it comes to the hole with the indexing ball and spring. With a flat blade screwdriver press the indexing ball into the hole and move the shifting fork completely onto the shifter rod.

19. Move the shifting fork to the Neutral position. The neutral groove is the center groove. If the shifter rod has four grooves, the neutral groove is the second groove from the shortest end. This neutral groove can be seen through the hole in the shifter fork.

20. When the shifter forks are properly assembled to the shifter rods and positioned in neutral, the ends of the notches in the shifter forks are in alignment.

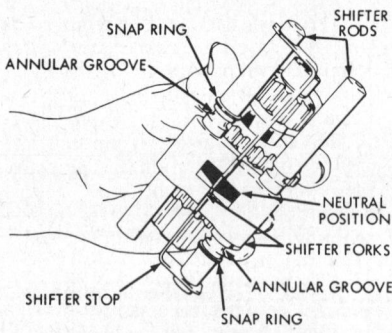
Final shifter assembly

21. Assemble the two flanged gears onto the shifter shaft. Note that the large gear is placed on the shaft first with the flange side toward the needle bearing in the end of the shifter shaft. Slide on the smaller gear with the flange toward that of the larger gear.

22. When assembling the shifter fork and rod to the flanged gears on the shifter shaft, make sure that shifter fork which is on shifter rod "A" always engages in flange in the larger gear. To determine which is shifter rod "A" compare the parts to illustrations. Hold the shifter shaft in the hand as illustrated during assembly.

23. After the shifter fork and rod assemblies have been engaged with the flanged gears allow the shifter rods to lay open in the hand and position the shifter stop. The notch in the shifter stop is the guide for correct positioning. Align this notch with the corresponding notches in the shifter forks and insert the shifter stop. Move the shifter rods together and insert into the transaxle. Remember to squeeze the ends of the shifter rods to cause the assembly to bind and stay together.

24. In three speed transaxles the needle bearing end is inserted first into the case to engage the end of input shaft.

25. When placing the shifting assembly into the four-speed transaxle be sure the thrust washer is on the bearing. Place the assembly into the transaxle with the needle bearing end of the shifter shaft up. Allow the end of the shifter shaft to protrude below the ends of the shifter rods, this will ease the alignment of the assembly.

26. The shifter assembly is correctly installed in the transaxle if the notches in the shifter forks are just about in the center of the opening in the case or cover of the transaxle.

27. Dowel Pin Type: Secure with a new dowel pin. A second dowel pin is used in some assemblies for alignment. This dowel pin is located in the gasket surface of the shift lever housing and fits into a mating hole in the transaxle.

28. Snapring Type: Secure parts with the snapring. Before installing the shift lever and housing to the transaxle housing, check the shifting forks for Neutral position.

29. Always use new gaskets between the shift lever housing and the transaxle.

30. Place axles (left and right) into differential gear assembly. Install thrust washers.

NOTE: The axles differ in length so select the proper axle.

31. On roll pin drive models, install double roll pins into holes in each shaft. Place bevel gears on shaft. Roll pins fit into the recess in back of the gears, bevel gears must be seated tightly on the roll pins or binding will occur.

32. On double "D" type drives, place bevel gears on the shaft and install snapring in groove on the shaft.

33. Install bevel pinions SIMULTANEOUSLY FROM OPPOSITE SIDES by rotating pinions in opposite directions while sliding into position in gear assembly. Check alignment by inserting fingers into drive pin holes. If not aligned, drive pin cannot be inserted. Remove and replace bevel pinions as only one tooth out of position will cause misalignment.

34. After aligning, insert thrust washers behind each pinion. Insert drive pin and secure with roll pin.

35. Install thrust washers and bearing on input shaft. Note sequence.

36. Install input shaft into case assembly. Lock on with snapring retainer. Install oil seal.

37. Set case assembly open side up. Insert the idler shaft gear assembly, thrust washers and bearing. Note sequence of washers and bearings.

NOTE: Place reverse idler shaft into bearing to aid in holding washers, thrust bearing, idler shaft and gear assembly prior to installing shifter assembly.

38. Insert the washer and then the three gear cluster assembly.

39. Insert shifter assembly. Check that rods are seated properly.

NOTE: Reverse idler shaft will be pushed out at this time.

40. Install reverse idler. Make sure beveled edge is up. Spacer on top of gear.

41. Place new gasket on case and install center plate.

42. Place new gasket on center plate and install differential assembly, longer axle in down position. Be sure gear on shifter shaft is on shaft.

43. Install gear case dowel pins. Leave dowel pins slightly exposed on top to locate cover assembly.

44. Install transaxle cover assembly, and secure with eight (8) capscrews.

45. Install bearings and/or bushings, if necessary. Install seal.

46. Install axle housing assembly. Fill with 1½ pints SAE EP 90 oil.

47. Inspection Note: For a neutral position, shift notches in forks and notch in shifter stop must be aligned and centrally located.

TWO SPEED PULLEY

All Models

REMOVAL

1. Block up the rear of the tractor and remove the right rear wheel.

2. Remove the shift fork assembly by removing the locknut and long pivot bolt.

3. Next remove the capscrews holding the support to the drawbar and seat support. The support, pivot bracket and stop can now be removed from the tractor in one piece.

4. Disconnect the spring from the belt idler arm and remove the belt from the pulley.

5. Shift the pulley assembly into low range by grasping the shift ring with both hands and pulling out.

6. Rotate the shift ring and cover assemblies until the setscrew over the shaft keyway is aligned with one of the holes in the pulley hub. Loosen setscrew. The other setscrew in the collar need not be loosened at this point as it merely holds the collar in position on the spider assembly.

7. Slide the Pulley Kit off the transmission shaft.

DISASSEMBLY

1. Place pulley kit in a vise with pulley side up. Clamp lightly on outer cover assembly so as not to deform the cover.

2. Before attempting to remove the set collar, back the long setscrew out a few more turns to make sure it is not engaging the hole in the shaft. Loosen second setscrew and remove the set collar and pulley assembly by lifting upward.

3. Remove the six flange nuts and capscrews holding the two cover assemblies, ring gear and shift ring together. Carefully lift the lefthand (inner) cover assembly and shift ring so as not to tear the gaskets.

4. To remove the right (outer) cover assembly from the spider assembly, grasp the spider shaft in the lefthand and hold the cover a couple inches above a bench. Insert a bar into the spider shaft bore and drive the cover out of spider assembly.

5. The ring gear and pinions can be disassembled by removing the three locknuts that hold the ring assembly and pinions to the spider.

6. Thoroughly clean and inspect all parts. Replace bearings, gears and other parts that are excessively worn.

ASSEMBLY AND INSTALLATION

1. Place ring assembly on a flat surface with studs pointing up.

2. Install three spacers and three pinions on studs. Apply a small quantity of Shell Durina EP #1 grease to bore and sides of pinions before placing over the spacers.

3. Place ring gear on ring assembly and engage with pinion gears.

4. Attach spider assembly to ring assembly with three locknuts and torque to 15 ft.lbs.

NOTE: If bronze bearings in the spider are to be replaced follow this procedure:

a. Place the spider in press and install the long 1601460 bearing. Bearing should be pressed into the bore .598 ± .001 in. from the end of the hub. A special pressing tool must be made for this operation in order to obtain the required dimension.

b. Insert the 1601461 snapring and press the short 1601459 bearing flush with the spider hub. The snapring must be free to expand and not pressed tight between the two bearings.

NOTE: If pressing equipment is not available, it is recommended that the

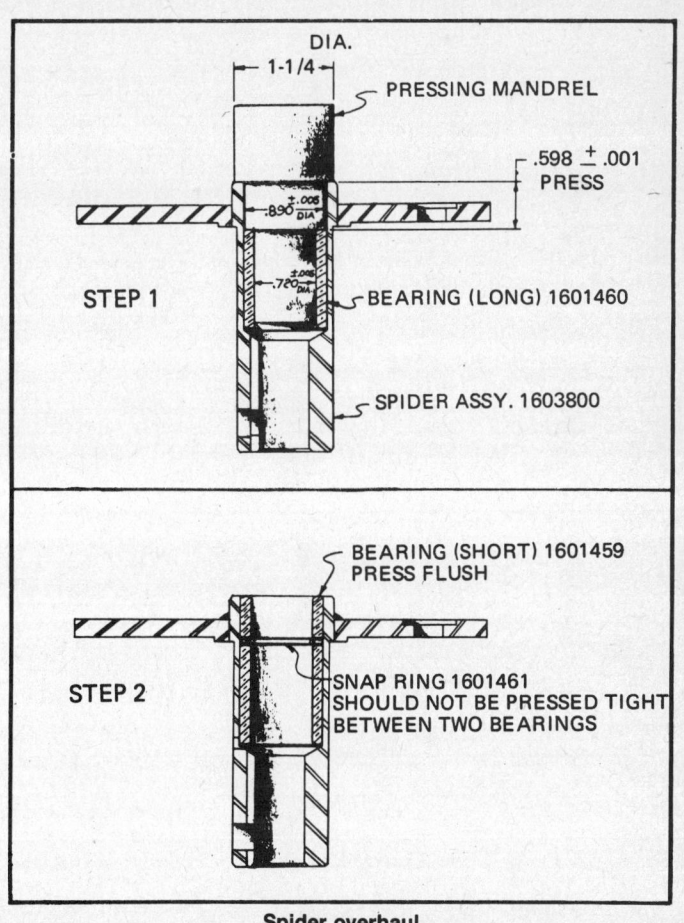

Spider overhaul

Allis-Chalmers

entire Spider and Bearing Assembly be replaced rather than attempting to replace the bearings.

5. Apply Shell Durina EP #1 grease liberally to the area around the gears and in the bore of the spider.

6. Place one gasket on the righthand cover assembly and press the spider assembly on to the cover assembly until the snapring engages the first detent groove.

7. Clamp the cover assembly in a vise and align the holes. Place the second gasket on the ring gear and align holes.

8. Place shift ring on the ring gear and align holes.

9. Place pulley on a solid surface or in a press and install one needle bearing with marked end out. Press bearing flush with surface of pulley hub. Turn pulley over and press the second bearng flush with hub. Marked end of bearing should also be out.

10. Work grease into needles of both bearings and fill the area between the bearings with grease to the approximate thickness of the bearings.

11. With pulley assembly resting on bench, place the lefthand cover assembly on the pulley and position the third gasket on the cover.

12. Holding the pulley and lefthand cover assemblies together, place on top of shift ring. Carefully align pulley gears with pinions in the spider assembly. Align holes in cover before engaging gear teeth.

13. Install six capscrews with heads toward pulley. Assemble nuts to screws and turn finger tight but do not tighten. Holding cover assemblies, turn the pulley in both directions to center gears. Tighten nuts to 75 inch pounds in 180° sequence.

NOTE: If pulley seems locked and will not turn, the spider assembly may have slipped out of low detent. If this has occurred, the unit will have to be disassembled and step six repeated.

14. If pulley turns freely and no binding exists, the set collar may be installed. Turn pulley until one hole in the hub aligns with the keyway in the spider shaft. Place the set collar over the shaft with the setscrews angled toward the pulley hole. Turn the long setscrew until end just protrudes into hole in shaft. Turn pulley until hole in pulley hub lines up with the short setscrew in collar. Tighten this setscrew securely to the shaft.

15. Install on tractor, reversing steps 1-6 in Disassembly.

ENGINE, DRIVESHAFT AND COUPLING

The following Briggs & Stratton Engines are used:

Tractor Model	Engine Model	Engine H.P.
B-1	19D	7¼
B-10	243431	10
Big Ten	242431	10
B-12	300401	12
B-110	243431	10
B-112	300401	12
HB-112	300401	12
B-206	146700	6
B-207	170700	7
B-210	243431	10
B-212	300401	12
HB-212	300401	12

For detailed engine service, see the Briggs & Stratton part of the Engine Unit Repair Section.

Removal and Installation

B-1

1. Remove hood.
2. Remove grille and grille support.
3. If equipped with electric starter, remove battery ground clamp.
4. Disconnect wire running from starter switch to starter.
5. Disconnect fuel line from tank.
6. Remove ignition wire from switch.
7. Disconnect choke and throttle cables from tractor when engine is being removed.
8. Disconnect front drive shaft coupling from engine.
9. Disconnect complete drive shaft when repairs are needed on the shaft or couplings.
10. Remove capscrews holding engine to frame.
11. Remove oil drain pipe.
12. Remove engine.
13. Installation is the reverse of removal.

B-10 (9 HP), B-10 (10 HP), BIG TEN, AND B-12 PRIOR TO SERIAL #50001

1. Remove hood, grille and grille support.
2. Disconnect battery ground clamp.
3. Disconnect ignition wire.
4. Remove fuel line.
5. Remove starter generator wires.
6. Remove choke and throttle cables when engine is being removed.
7. Remove oil drain pipe.
8. Disconnect front coupling of drive shaft from engine.
9. Disconnect complete drive shaft when repairs are needed on the shaft or couplings.
10. Remove capscrews holding engine to frame.
11. Slide engine forward to remove.
12. Installation is the reverse of removal.

B-10 SERIAL #50001 AND UP, B-12, B-110, B-112, HB-112, B-210, B-212, HB-212

1. Remove hood, side panels.
2. Remove battery, fuel tank.
3. Remove dash assembly.
4. Remove shift lever ball and brake lock.
5. Remove frame cover assembly.
6. Remove choke and throttle cables, when engine is being removed.
7. Disconnect front coupling of drive shaft from engine.
8. Disconnect complete drive shaft when repairs are needed on the shaft or couplings.
9. Remove oil drain plug.
10. Remove capscrews holding engine to frame.
11. Remove engine.
12. Installation is the reverse of removal.

NOTE: On units equipped with hydraulic system, removal of quadrant and lever assembly will simplify removal of drive shaft from engine.

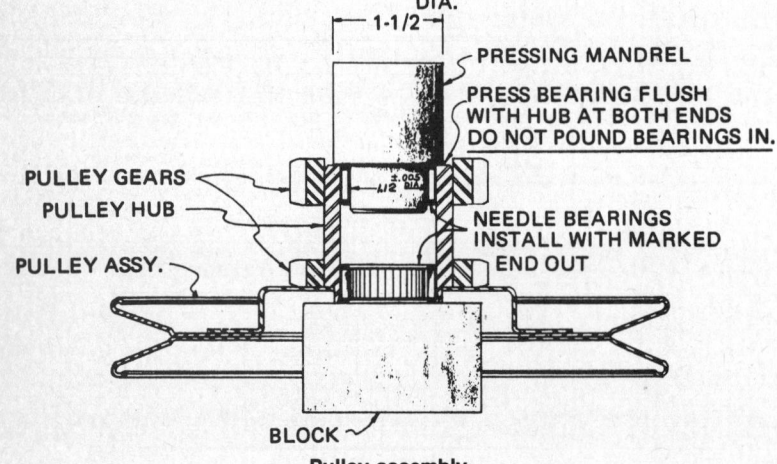
Pulley assembly

Allis-Chalmers

Models 5020 and 5030

NOTE: These models are identical to the Simplicity 9523 and 9528 tractors. These models use Hinomoto S-126 and S-148 diesel engines. Complete overhaul procedures for these engines are found in this section.

FRONT AXLE ASSEMBLY

REMOVAL FROM TRACTOR

1. Set the brakes on the tractor.
2. Remove the bolts through the end of the steering arms. Pull the steering arms off the spindles.
3. Jack the tractor up by using the front frame. Remove the bolts on the front of the pivot pin located on the frame. The pivot pin can now be pulled from under the tractor.

DISASSEMBLY

1. Remove the capscrews holding the cap to the hub, remove cap.
2. Remove the cotter key and castle nut on the end of the spindle.
3. Tap the hub lightly with a hammer and remove. The spindle can now be removed from the axle: this completes the disassembly.

INSPECTION

All parts (bearings, machined surfaces, etc.) must be inspected for any defects (stress cracks, nicks, abnormal rotating noise, etc.). If any defects are found the necessary parts must be replaced.

ASSEMBLY

Mount the axle to the tractor and then install spindles. Remove, clean and repack the front wheel bearings with grease every 500 hours of operation. In extremely wet or muddy conditions service the bearings more often. Always replace the seals when repacking the bearings.

WHEEL BEARING ADJUSTMENT

To adjust the wheel bearings proceed as follows:
1. Tighten the castle nut until the wheel has a definite drag.
2. Check the alignment of the cotter pin hole in the spindle with notches in the castle nut.
3. If necessary, back up castle nut just enough to make the cotter pin hole align with a notch in the castle nut.
4. Install new cotter pin.
5. Install the hub cap.

Check and adjust the wheel bearings periodically.

STEERING

REMOVAL OF STEERING MECHANISMS

1. Remove the two rings holding the pins in place on the universal joint. Tap out the two pins. The steering shaft and universal joint can now be removed.
2. Remove the seven capscrews holding the box joint to the torque housing. The box joint can now be lifted from the torque housing.
3. Remove the nut on the bent steering shaft. The steering gear and pinion gear can now be removed.
4. Remove the snapring located on the bottom of the bent steering shaft and remove the bearing. The bent shaft can now be pulled through the top hole.

INSPECTION

Inspect all parts for any defects; if any are found, replace the necessary parts.

INSTALLATION

1. Installation is a reversal procedure of the removal procedure. During installation be sure the bend in the bent steering shaft is on the left side of the input shaft. When installing the pinion and segment gears, be sure to install them simultaneously and position the gears such that the pinion gear is in the middle of the gear segment.
2. Check that the bend in the bent shaft is pointing away from the input shaft at a 90° angle.

NOTE: If the bent shaft is improperly positioned, it may hit the input shaft or the pitman arm will be in the wrong position.

CLUTCH

REMOVAL AND INSTALLATION

1. Remove the transmission as described later.
2. Install the clutch aligning tool to support the unit.
3. Remove the six bolts attaching the pressure plate to the flywheel.
4. Remove the return spring and release bearing.
5. Remove the three bolts to dismantle the release bearing guide. With a plastic

Spindle components

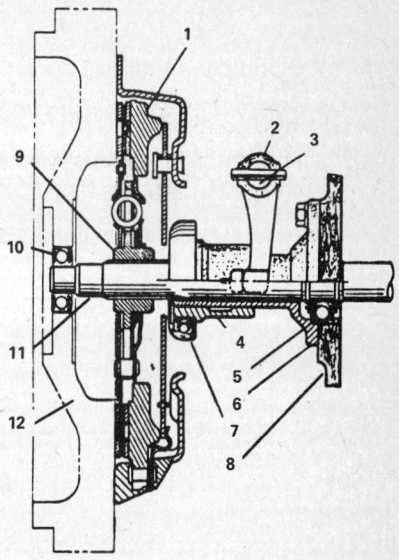

1. Pressure plate
2. Shifter
3. Shifter rod
4. Release bearing sleeve
5. Ball bearing
6. Release bearing guide
7. Release bearing
8. Clutch housing
9. Clutch disc
10. Bearing
11. Input shaft
12. Flywheel

Clutch components

145

Allis-Chalmers

hammer knock the input shaft out toward the engine side.

6. Remove the snapring, thrust washer, and ball bearing from the shaft.
7. Using a centering tool, position the clutch disc and pressure plate on the flywheel.
8. Tighten the six retaining bolts alternately and evenly to 11-14 ft.lbs.
9. Insert the input shaft into the clutch housing.
10. Insert the bearing guide into the shaft and tighten the three bolts.
11. Press the bearing into the bearing sleeve, insert the sleeve into the bearing guide and attach it to the shifter and return spring.
12. Connect the clutch housing to the clutch assembly and attach the transmission to the clutch.

TORQUE HOUSING

REMOVAL

1. Remove the negative cable at the battery terminal. Disconnect all electrical connectors under the instrument panel. Also disconnect the rear lights at the connectors under the seat and loosen the clamps holding the wire to the transmission.
2. Turn the fuel cock off and disconnect the fuel line at the injection pump. Disconnect the hydraulic lines at the engine (plug lines, to avoid oil spills).
3. Remove the water temperature sensor and the tachometer cable. Also disconnect the clutch, throttle and fuel shut-off linkages.
4. Remove the four capscrews holding the instrument panel to the torque housing, and the three capscrews holding the fuel tank to the torque housing. The fuel tank/instrument panel assembly can now be removed.
5. Remove cotter pin and castle nut on end of steering shaft. Tap steering arm of shaft.
6. Place blocks under the transmission and position a hoist over the engine as in "engine removal." Remove the nuts bolting the torque housing to the transmission and roll the engine and torque housing assembly forward.
7. Remove the bolts and nuts securing the torque housing to the engine. The torque housing may now be removed from the engine.

NOTE: BE CAREFUL of the torque housing, do not apply excess force to the torque housing as it is constructed of aluminum.

DISASSEMBLY AND ASSEMBLY

1. The input shaft is very easily removed from the torque housing and can be used as an aligning tool for installing the clutch disc.
2. The release bearing can be removed by pulling the two return springs. The release bearing can now be slid off the shaft. The bearing can now be pressed off the sleeve and replaced if necessary.
3. Remove the three capscrews holding the release bearing guide in the housing. The shaft can now be pulled forward out of the torque housing.
4. The release bearing guide can be slid off the end of the shaft. Remove the snapring on the flywheel end of the shaft and the bearing should slide off without too much difficulty.
5. When pressing the release bearing onto it's sleeve, be sure the rounded chamfered edge is facing away from the sleeve.
6. Place the bearing and sleeve assembly onto the input shaft (with bearing facing outwards) and slide into position. Replace the return springs.

INSTALLATION

Installation is the reverse of removal. The transmission to torque housing nuts and bolts should be torqued to 46-54 ft. lbs. The torque for mounting the torque housing to the engine is 46-54 ft.lbs. and the two bolts for mounting the starter motor should be torqued to 29-33 ft.lbs.

Brake Pedal Adjustment

1. Adjust right pedal first.
2. Loosen locknut and rotate turnbuckle so that pedal play is .8 to 1 in. (20-30 mm). (Play: Moving distance of pedal when gently pushed by hand).
3. Adjust left pedal in the same way so that pedal play is equal.
4. When pedal play is equal, set connecting rod. Pull side brake lever, and confirm that more than 2 notches remain.

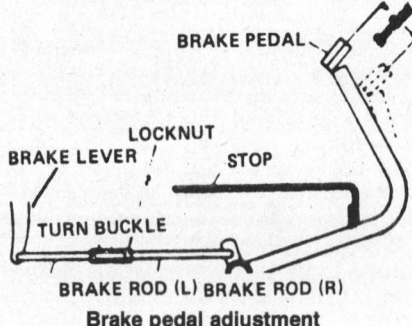
Brake pedal adjustment

Clutch Pedal Adjustment

To adjust the clutch pedal linkage, disconnect the clutch rod from the clutch lever and adjust yoke on rod until there is .8 to 1 in. (20-30 mm) of free travel at pedal pad, with rod attached to clutch lever and pedal against stop. When correct adjustment is obtained, attach rod to clutch lever with pin and insert the retaining cotter pin.

TRANSMISSION, 2-WHEEL DRIVE

Shifter Assembly

REMOVAL AND INSTALLATION

1. Refer to the Torque Housing Assembly paragraph, and remove the input shaft.
2. Using a punch, drive out the pins from the transmission side of the torque housing.
3. Remove the pin holding the clutch-lever to the shifter shaft. The shifter may now be removed.
4. Installation is the reverse of removal.

Transmission

REMOVAL

1. Disconnect the battery cables at the battery terminals.
2. Disconnect the two safety start switches and the rear lights at their respective connectors. Loosen the clamps on the transmission to allow freeing of the rear light wire.
3. Drain the fluid from the transmission. There is approximately six (6) gallons of fluid in the transmission rear housing sump.
4. Remove the four capscrews holding the brake cross shaft to the transmission and unfasten the springs hooked to the platform.
5. Remove the bolts holding the platforms to the transmission and remove the platforms. Disconnect the hydraulic lines at the rear housing.
6. Place blocks under the engine, and put one 2 in. x 2 in. x ½ in. piece of wood under each side of the engine, between the engine and front frame. Place a hoist over the transmission and wrap a chain around the middle of the transmission, so to form a sling.
7. Remove the bolts holding the torque housing to the transmission. Roll the rear housing and transmission rearward.
8. Place blocks under the rear housing and remove the bolts holding the transmission and rear housing together. Roll the transmission forward.

DISASSEMBLY

1. Remove ten capscrews in the transmission housing, and remove the shift cover with the control cover upward. Separate the control cover from the shift cover by removing six capscrews.
2. Turn the shift cover upside down, remove three detent springs and three detent balls.
3. Using a jig to drive out the lock pin connecting the fork and the rod. Drive the rod forward and, at the same time, the plug is removed.

Allis-Chalmers

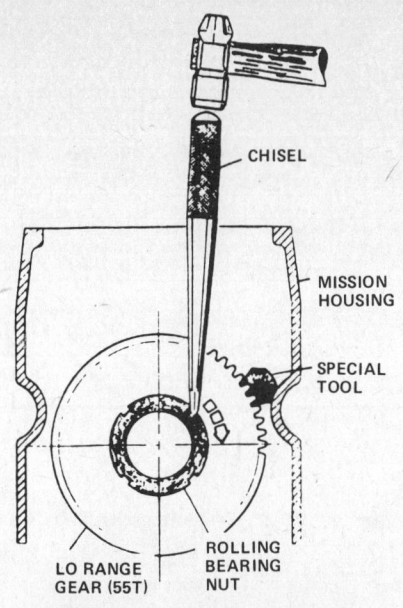

Loosening the pinion shaft nut

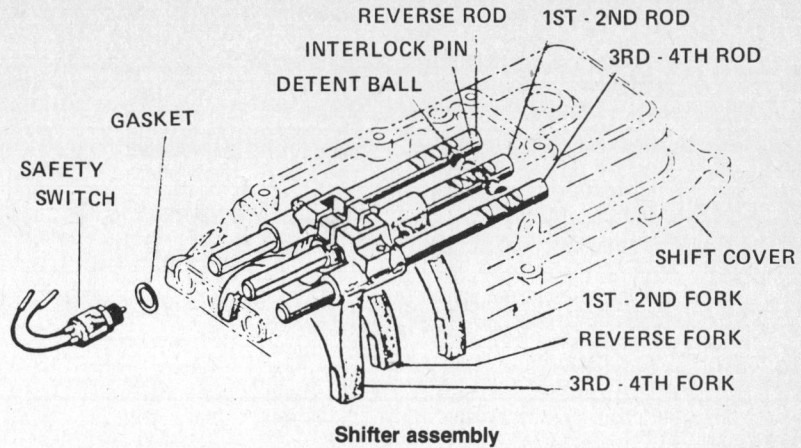

Shifter assembly

NOTE: Avoid removing the lock pin without turning the shift cover upside down.

4. Since the interlock pin system is employed, rods cannot be removed unless the three rods are set in the neutral position.

5. Remove the lockscrew (+) which prevents revolution of the shift lever and remove the spring pin (for spring) which holds the lever, then pull the lever downward. Remove the lockscrew (M16) and pull out the spring and control pin.

6. Remove eight nuts on the front cover. Strike the end of the main shaft with a plastic hammer to remove the cover, then pull the main shaft forward.

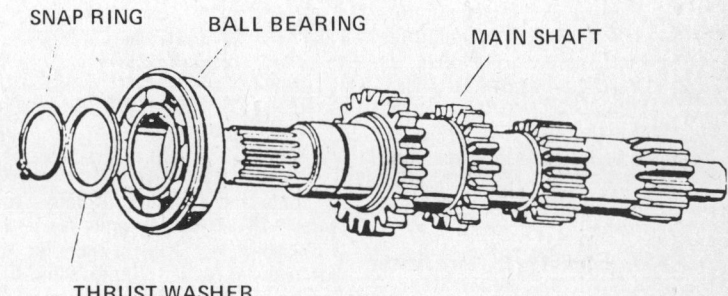

Mainshaft

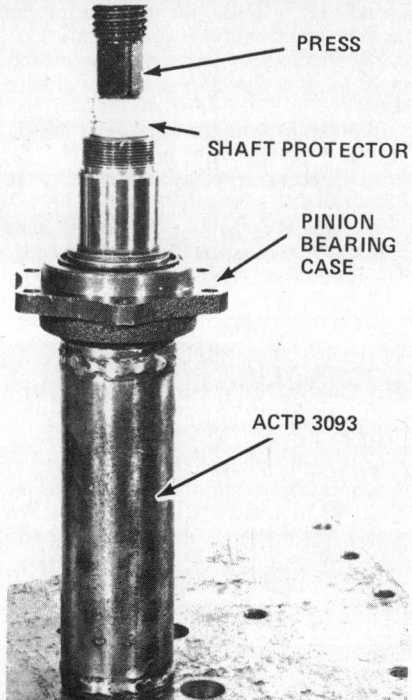

Removing pinion bearing

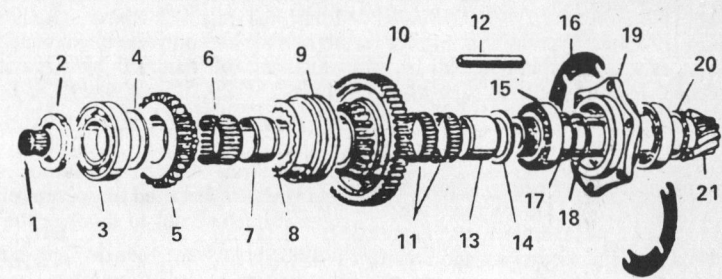

1. Needle bearing
2. Rolling bearing nut
3. Ball bearing
4. Thrust washer
5. High-range gear
6. Needle bearing
7. Inner ring
8. High-low range collar
9. High-low range coupling
10. Low-range gear
11. Needle bearing
12. Key
13. Inner ring
14. Thrust washer
15. Bearing cone
16. Adjusting shim
17. Drive pinion spacer
18. Adjusting washer
19. Drive pinion bearing case
20. Thrust washer
21. Drive pinion shaft

Model 5030 drive pinion shaft and gears

NOTE: Oil seal is included in the cover, so cover the main shaft spline so as not to scratch the lip when removing.

7. Remove the snapring fixing the ball bearing at the rear end of the shaft, then remove the snapring at the rear end of the gear (25T), and the shaft can be pushed out forward.

8. Remove the gear, inner ring, needle bearings, and thrust washer in this sequence.

9. Remove the lockscrew fixing the shaft, turn the shaft drive 180° in reverse direction, and use a small punch to drive out the lock pin fixing the distance collar. While pulling the shaft forward, remove the gear, its needle bearing, distance collar, and thrust washer (which eliminates axial clearance).

NOTE: Do not remove the lock pin unless the shaft is turned 180° from its installation position.

10. Remove the capscrew attaching the shifter rod stopper to the rear of the transmission housing, and the lockscrew locking the fork. While pulling the shifter rod backward, pull the fork (with detent spring) upward. In this case, the detent ball may be dropped inside the transmission so take great care.

11. In order to remove the pinion shaft, it is necessary to remove the reverse idler shaft. A large blunt edged chisel must be used to loosen the nut on the pinion shaft.

Allis-Chalmers

After the nut is loose a spanner wrench (ACTP-3091) can be used to remove the nut. When the nut becomes too close to the PTO driven gear, gradually loosen the capscrews on the pinion bearing case.

NOTE: The rolling bearing nut has left hand threads. Also the nut is loctited. Be sure to remove the needle bearing located in the pinion shaft.

12. After the nut and pinion bearing case have been removed, the shaft may be pulled rearward and the components may be removed in the following order. (5) Hi-range gear, (6) needle bearing, (7) inner ring, (8) High-low range collar, (9) High-low Coupling, (10) Low-range gear, (11) needle bearing, (12) key, (13) inner ring, and (14) thrust washer.

13. The two bearing cones, and bearing case will be removed with the shaft. Removal of the pinion bearing assembly requires pressing off, using ACTP-3093, as shown. To remove the bearing assembly, place the special tool ACTP-3093 over the pinion gear and then press.

14. Remove the detent screw retaining the shifter rod, and remove the lockscrew locking the fork. While pulling the rod forward, remove the fork upward.

NOTE: When removing the fork, take care not to drop the detent ball.

15. Remove the PTO control lever, remove the switch cover with the clutch-housing clamping nut, and the safety switch can be disassembled.

16. Adjust contact condition of the switch by means of the clamping nut.

17. Adjust washers are available as follows:
   a. .02 in. (0.5 mm)
   b. .04 in. (1.0 mm)

18. Remove the ball bearing at the rear end of the shaft, remove the snapring, push the shaft forward, and remove the gear, thrust washers, needle bearings and inner ring. The ball bearings at the front end of the shaft are pressed in, the snapring and the thrust washer must be removed before removing the ball bearing.

19. Remove the capscrew (M16) at the center of the tire, remove the tire, and

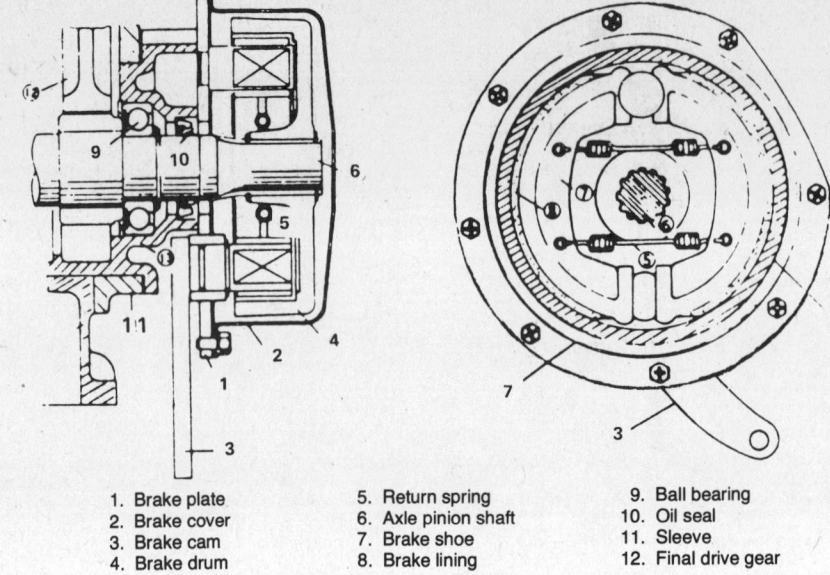

1. Brake plate
2. Brake cover
3. Brake cam
4. Brake drum
5. Return spring
6. Axle pinion shaft
7. Brake shoe
8. Brake lining
9. Ball bearing
10. Oil seal
11. Sleeve
12. Final drive gear

Brake components

disconnect the flasher. Remove four capscrews (M12) retaining the fender, and the two capscrews (M8) in the side step; and remove the fender. Remove the link from the brake pedal at the brake cam. Remove eight screws (+) to remove the brake cover. Remove the snapring at the end of the axle pinion shaft and pull out the brake drum, the result is shown. Remove the brake shoe and only the brake plate is left, which can be removed by removing the screws (M12 x 55), capscrews and nut.

NOTE: The screws (M12 x 55) are mounted to the curved portion of the sleeve. Since the portion is perforated, seal washers are used to prevent oil leakage. Take care not to damage the seal.

20. Remove the four capscrews mounting the seat to the lift housing.

21. Drain the fluid from the transmission—rear main sump. There is approximately 6 gallons of fluid in the sump.

22. Disconnect the hydraulic lines from the housing.

23. Disconnect the rear lights at their respective connectors. Place one 2 in. x 2 in. x ½ in. piece of wood under each side of the engine, between the engine and frame.

24. Remove the wheels and fenders. After the wheels and fenders are removed, be sure to support the rear main with a hoist.

25. Remove the draft arms and lift links. Also remove the upper link.

26. Remove the capscrews and nuts holding the lift housing to the rear main. The lift housing can now be removed.

27. Remove the nuts bolting the transmission and rear housing together. After the nuts are removed, the rear housing can be separated from the transmission.

28. Remove the four capscrews mounting the seat to the lift housing and remove seat.

29. Remove the lift links from the lift arms and the upper line bracket.

30. Disconnect the hydraulic line to the lift housing.

31. Remove the capscrews and nuts bolting the lift housing to the rear housing. The lift housing can now be removed.

ASSEMBLY

1. Assembly of the lift housing is the reverse of disassembly. Note the following specific procedures and torque valve. When installing the lift housing, torque the six long capscrews to 29-33 ft.lbs. The short capscrew should be torqued to 13-15 ft.lbs. When tightening the pipe coupling tighten until the turning force becomes heavy, then 1/16 turn more.

2. Install the brake plate, install the brake shoes in the direction marked by the cam. Fit the brake drum and retain with a snapring. Then install the brake cover with eight screws (+). Connect the link to the

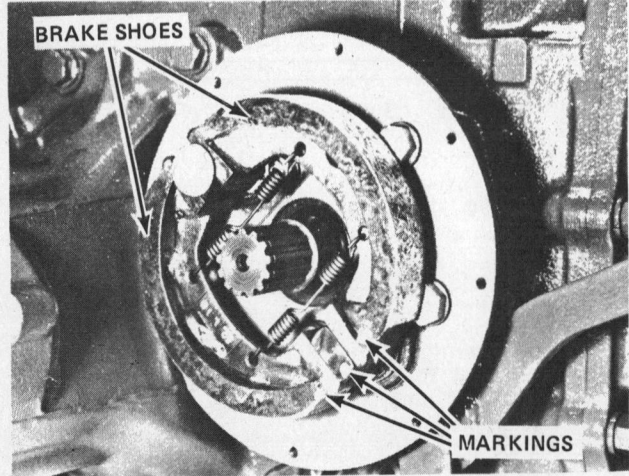

Removing link from the brake pedal at the cam

brake cover. The link has a turn buckle for adjustment. Install the fender and tire to complete installation.

NOTE: When installing two rear capscrews on brake plate, be sure to install seal washers on these capscrews.

3. With the transmission for the AC 5030 a one way clutch is fitted to the front of the shaft (in front of the PTO coupling collar).

4. The clutch should be assembled, disassembled, or replaced as a unit of the entire clutch assembly.

5. Press the ball bearings into the front end of the shaft, install the thrust washer and snapring. Place a snapring on the spline. Install a snapring to the PTO coupling collar.

6. Insert the shaft into the housing, fit the PTO coupling to the PTO coupling collar, and install the PTO driven gear, inner ring, and needle bearing.

7. Temporarily assemble the shaft, snapring, PTO coupling collar, and inner ring as shown, measure clearance with a thickness gauge.

Standard clearance value; .004 in. to .012 in. (0.1 to 0.3 mm)

8. If the measured clearance value is not within the above range, choose the suitable thrust washer among the following:
 a. .12 in. (3.0 mm)
 b. .13 in. (3.3 mm)
 c. .14 in. (3.6 mm)
 d. .15 in. (3.9 mm)

9. Fit the fork to the PTO coupling, insert the shift rod and attach with a lockscrew. Insert the detent ball and spring for positioning the fork and rod onto the fork, and tighten them with a detent screw.

NOTE: When inserting the rod into the transmission housing take care not to cut the O-ring located at the top of the rod.

10. Install the shaft assembly to the transmission housing. During the mounting process, insert ball bearing into the center wall of the housing. Install gears from inside the housing to the shaft. Install the bearing case with the chosen adjusting shim to the housing. Apply Loctite® No. 277 to the rolling bearing nut to prevent looseness, and tighten with standard torque, and clamp the bearing case with six capscrews.

NOTE: Be sure to insert pilot needle bearing during assembling process because the PTO driven gear prevents insertion after completion of assembly, (Only 5030). When installing the shaft, the reverse idler shaft should be removed in advance.

11. After installing the pinion shaft, check, by hand, that the shaft rotates smoothly and that the high-range gear and the low-range gear rotates smoothly.

12. Mesh the high-low range coupling with the low-range gear (24T) and measure preload with a spring balancer. Indication of the spring balancer (gear, 24T) 3.3 to 6.6 lb.

13. Press the outer ring of the bearing cones into the pinion bearing case from both sides of the case back seating both back faces towards the center. Slip the (20) thrust washer .110 in. (15) a bearing cone, (19) the bearing case, (18) adjusting washer, (17) pinion spacer, (15) bearing cone and (14) thrust washer onto the dummy pinion shaft (ACTP-3090). Tighten the pinion nut to 43-58 ft.lbs. or 52-69 lbs. when using the spanner wrench and a spring scale. This may be accomplished by clamping the assembly in a vise, using the spanner wrench (ACTP-3092) and holding onto the hex head with a wrench.

NOTE: The dummy pinion has left-hand threads.

14. After torquing the pinion nut to the proper torque, test the preload of the bearing. Two methods can be used to check the preload; (1) a spring scale and string, (2) an inch-pound torque wrench. To check the preload using method one, simply wrap a string around the dummy pinion, attach the spring scale and pull. The scale should read 5-15 in.lbs. If a torque wrench is used, simply attach the wrench to the hex head and read the rolling torque 5-15 in.lbs.

15. When an improper preload reading is obtained, shims are to be added or removed. If too great a preload is obtained, it is necesary to increase the shim stack. If too little a preload is obtained the shim stack is to be reduced. Shims are available in the following thicknesses:
 a. 0.004 in. (0.10 mm)
 b. 0.006 in. (0.15 mm)
 c. 0.008 in. (0.20 mm)
 d. 0.040 in. (1.00 mm)

NOTE: Shims are to be placed between the two bearing cones.

16. After the preload adjustment is completed, the proper shim for adjusting the pinion shaft and ring gear contact is to be selected. The selection is accomplished by using a micrometer to measure the distance from the bearing case to the non-threaded end of the dummy shaft. When this dimension is obtained, the dimension stamped on the end of the dummy pinion is subtracted from the measured dimension. The difference of the two dimensions should be equal to the following standard dimension.

Model	Difference
5020	1.20 in. (30.5 mm) (standard)
5030	1.22 in. (31.0 mm) (standard)

If the difference is incorrect, shims are to be added or removed to obtain the standard dimension. To increase the difference, add shims; to decrease the difference remove shims. Shims are available in the following thicknesses:
 a. 0.004 in. (0.1 mm)
 b. 0.008 in. (0.2 mm)
 c. 0.016 in. (0.4 mm)

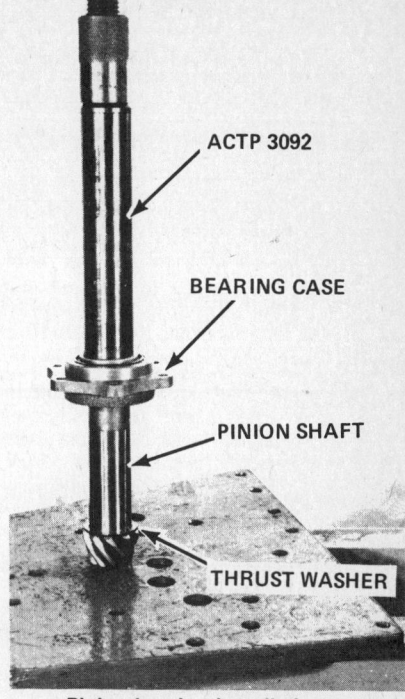

Pinion bearing installation

After the preload and pinion bearing has been measured, the pinion shaft may be assembled. Slip the thrust washer .110 in. (2.8 mm) onto the pinion shaft. The pre-selected bearing assembly should now be pressed onto the pinion shaft using ACTP-3092 as shown. After pressing the bearing assembly onto the shaft, (14) thrust washer, (13) inner ring, (11) needle bearings, and (12) key can be installed on the shaft.

17. Fix the shift fork to the gear, insert the rod into the fork, and fix with a shifter rod stopper. Then insert detent ball and detent spring and tighten the detent screw.

18. Fit O-ring to the shaft, and insert the distance collar, the chosen thrust washer, needle bearing, and gear, distance collar in this sequence. Tighten the lockscrew fixing the shaft, then tap the lock pin into the distance collar to fix it to the shaft.

NOTE: When inserting the shaft into the transmission housing, take care not to cut the O-ring. When striking the lock pin, set the shaft so that chamfered hole is in the upper side.

19. Temporarily install the reverse idler shaft, use the special tool instead of lock pin, and measure clearance with a thickness gauge. Standard clearance .0039 in. to .0118 in.. If not in the above range, choose the suitable thrust washer from the following.
 a. .114 in. (2.9 mm)
 b. .126 in. (3.2 mm)
 c. .138 in. (3.5 mm)

20. Press ball bearing into the front end of the shaft, mount the thrust washer and snapring, and insert the 3rd-4th gear and other gears.

NOTE: Before installation, press shell-shaped needle bearing into the upper center wall of the housing.

Allis-Chalmers

21. After installation, check to insure that the 3rd-4th gear, reverse idler gear, and second gear can slide smoothly and that the 1st gear rotates easily.

22. Press ball bearing onto the main shaft, mount the snapring and thrust washer. Press the new oil seal into the front cover with the lip facing toward the ball bearing until the seal reaches the cover shoulder.

NOTE: In the same manner as main shaft removal, cover the main shaft spline to protect the oil seal lip. Apply 3MEC-847 cement (Part No. 921164) to the gasket.

23. Insert the shift lever into the hole in the cover and fix it with the lockscrew. Insert the spring and washer into the lever from upper side and fix them with the lock pin. Mount the lockscrew, pin, and spring which permits reverse operating force to be greater than forward operating force.

24. Set the forks in neutral position, insert detent balls and three detent springs into the shift cover, replace the control cover assembly, and clamp it with six capscrews. Fit the whole assembly to the gears (in neutral position) included in the transmission housing and clamp with ten capscrews.

25. Start assembling with the reverse rod and fork, place two detent balls between rods. Drive down the lock pin which fixes the rod and fork from upper side (assembled state).

26. Insert the interlock pin into the hole at the rear of the 1st-2nd rod.

27. Applying pressure to the plug and insert it, mount the safety switch to the reverse rod hole of the insertion.

NOTE: When rod is moved after insertion of the interlock pin, the pin may be dropped out.

INSTALLATION

Installation is a reversal of the removal procedure. Transmission mounting nuts should be torqued to 46-54 ft.lbs.

REAR AXLE, 2-WHEEL DRIVE

REMOVAL

1. Remove the brake assembly, remove two capscrews (axle pinion housing), a nut, six capscrews (sleeve), both right and left side in same manner, and pull out the axle, then separate them from each other. First, remove the oil seal gradually with a screwdriver by using the dent in the front of the sleeve, and remove the snapring.

2. Remove the snapring on the final drive gear side, strike the top of the axle with a plastic hammer to remove the axle.

3. Remove the axle pinion housing from the sleeve.

4. Remove the snapring from the axle pinion housing, and take out the axle pinion shaft. Remove the snapring of the axle pinion shaft, and remove the ball bearing.

Differential Lock Controls

REMOVAL AND INSTALLATION

1. Remove the sleeve assembly. The differential lock coupling, thrust washers, and differential lock spring are fitted to the axle.

2. Remove the differential assembly, taking care not to damage the bearing. Remove the PTO extension shaft.

3. Remove the double lock pin fixing the foot lever, and remove the lever. Remove the snapring, thrust washer, cotter pin in the shift lever side, and washer. Insert the control lever into the housing to be removed. Remove the control guide 1.6 in. from the opposite side. (An O-ring is located at the lower neck of the guide). Remove the plug at the lower side of the shift lever. Remove two snaprings, the shaft and shift lever.

4. Install the shift rod to the control lever. Mount an O-ring to the control guide, and install the control guide to the housing. Insert the control lever from inside the housing. Insert a shaft into the perforated hole in the housing, mount a snapring, insert the shift lever, and mount another snapring.

5. Connect the shift rod with the shift lever. Press the plug into the bottom of the shaft. Mount the thrust washer and snapring which position the control lever, and attach the foot lever with a double lock pin.

REAR AXLE INSTALLATION

1. Press the oil seal into the axle pinion housing up to the shoulder of the housing.

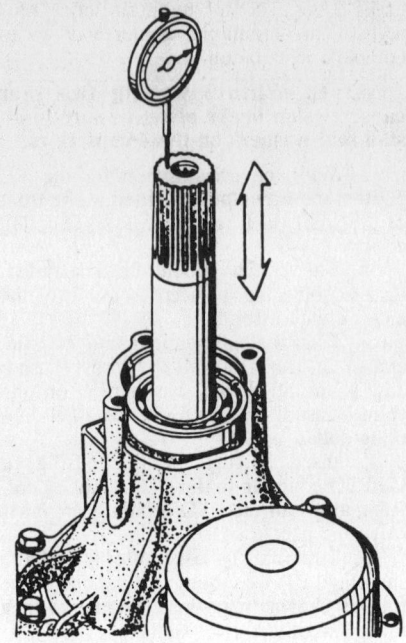

Measuring axle axial play

Press bearing onto the axle pinion shaft, mount the thrust washer and snapring, insert the shaft into the pinion housing, and retain the shaft with a snapring. Apply adhesive to the adjusted shim and temporarily install with the shaft to the sleeve.

2. Put the bearing and final drive gear inside the install sleeve. Insert the axle outside the sleeve, and the snapring positioning the final drive gear.

3. Press the bearing outside into the shaft and retain with the snapring. Insert the differential assembly, differential lock coupling and spring, and washer, then install the sleeve assembly.

NOTE: Mount the axle seal only after clearance adjustment. The ball bearing should be pressed into the axle pinion

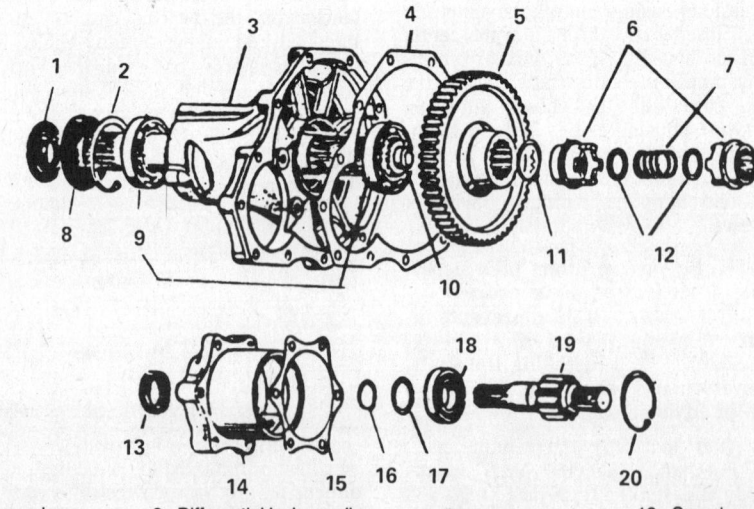

1. Inner ring
2. Snapring
3. Sleeve
4. Gasket
5. Final drive gear
6. Differential lock coupling
7. Differential lock spring
8. Oil seal
9. Ball bearing
10. Axle
11. Snapring
12. Thrust washer
13. Oil seal
14. Axle pinion housing
15. Shims
16. Snapring
17. Thrust washer
18. Ball bearing
19. Axle pinion shaft
20. Snapring

Final drive gear components

Allis-Chalmers

shaft so that the sealed side comes to the gear side.

NOTE: For AC 5030, when installing the left sleeve assembly to the rear housing, position the left differential lock coupling to the shift lever at first, after that install the sleeve assembly to the rear housing and coupling.

4. Double row bearings are used on the AC 5030 axles for supporting the final drive gear. Be sure the round notched side of the bearing faces the outside bearing.

5. Set up a dial indicator as shown to measure axle play in the axial direction. The standard clearance range is .004 in. to .012 in. (0.1 mm to 0.3 mm). If the measured value is not within the range, choose the suitable adjusting washer to adjust.
   a. .016 in. (0.4 mm)
   b. .024 in. (0.6 mm)

NOTE: When only one shim is required, the shim may be inserted to either side. When two shims are required, insert one shim to the right and the other to the left.

Differential Disassembly

Remove the sleeve, and the differential assembly can be removed without any outer ring. Remove the ring gear from housing if replacement of gear or housing is necessary. The ring gear is attached to housing flange with 12 lockscrews. Remove needle bearing from housing if replacement is necessary. Remove bearing cones from differential housing. Pull out lock pin 2.0 in. (50 mm) and the pinion shaft. Remove the pinion and thrust washer. Remove the side gear and thrust washer from housing.

Differential Assembly

1. Apply oil to the side gear, pinion, pinion shaft, and thrust washer .031 in. (0.8 mm) thick, install them in the housing. First, insert the two side gears. Set the pinion so that it engages with the gear. Insert the pinion shaft into the housing and the pinion, then attach the shaft with a lock pin.

2. Tighten the ring gear with twelve lockscrews at a torque of 36-43 ft.lbs.

3. Bend the lockwashers toward the screws to prevent looseness. Press the needle bearing into the housing so that it locates in the left side and comes to the outer surface of the housing. Press two bearing cones into both ends so that the two cones face towards each other and the outer ring comes off.

4. The differential assembly must be installed in the rear housing to enable check of the end play. Only one final drive and axle need be installed, but both axle pinion housings and all gaskets will have to be in place.

5. With one final drive, axle, pinion housing, and sleeve installed, install the differential assembly in the rear housing. After turning the rear housing so the installed axle is downward, install the other axle housing (be sure gasket is in place). Insert ACTP-3089 into the housing and rest it upon the differential bearing. Install extra shims and place the rear pinion housing in its proper position. Torque all bolts. Set a dial indicator on the end of ACTP-3089 and check the end clearance by lifting the differential assembly as shown.

6. The correct setting for the differential bearings is from 7 in.lbs. of rolling torque to .004 in. of end play. Check the end play as outlined below.
   a. Place housing so differential bearings can be adjusted vertically.
   b. Spin differential so bearings on bottom are fully seated (this is very important).
   c. Install dial indicator and ACTP-3089 tool as shown.
   d. Raise straight up on differential assembly and read end play.

NOTE: If assembly is not exactly vertical it will cause a reading less than actual end play.

   e. Make the measurement a number of times. Average out the readings.
   f. Remove the amount of shims required to get you into the range stated above the ideal setting is near zero end play.

7. Adjusting shims are available in the following thicknesses:
   a. 0.004 in. (0.10 mm)
   b. 0.006 in. (0.15 mm)
   c. 0.008 in. (0.20 mm)
   d. 0.012 in. (0.30 mm)
   e. 0.020 in. (0.50 mm)

8. After assembling all adjusted components, measure backlash as shown.

9. Standard backlash: .004 to .008 in. (0.1 to 0.2 mm).

10. If the measured backlash valve is not within the above range, adjust with the adjusting shims. When the left side shims are reduced, the right side shims are increased by the same quantity, and vice versa.

11. Apply red lead or blueing to several teeth of the ring gear, turn the drive pinion, and check minimum pattern on the teeth surface to judge tooth contact.
   a. Proper contact: Backlash of the drive pinion and ring gear is properly adjusted.
   b. Heel contact: The heel of the ring gear makes contact, due to insufficient mesh of the drive pinion. Move the ring gear slightly away from the pinion, and increase the number of thrust washers on the pinion side to adjust contact condition.
   c. Toe contact: Toe contact is caused by excessive mesh of the pinion. Move the ring gear toward the pinion and reduce the number of washers on the pinion side.

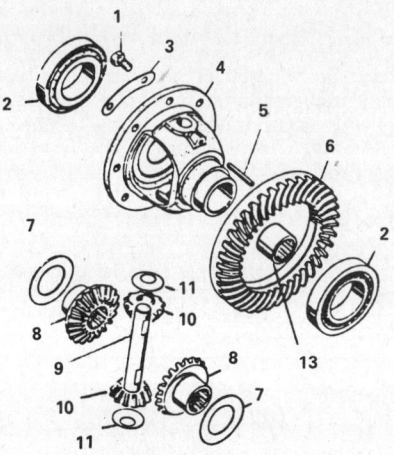

1. Lockscrew
2. Bearing cup and cone
3. Lockwasher
4. Differential housing
5. Lock pin
6. Ring gear
7. Differential—side gear thrust washer
8. Differential side gear
9. Differential pinion shaft
10. Differential pinion
11. Differential pinion thrust washer
12. Shell-type needle bearing

Differential components

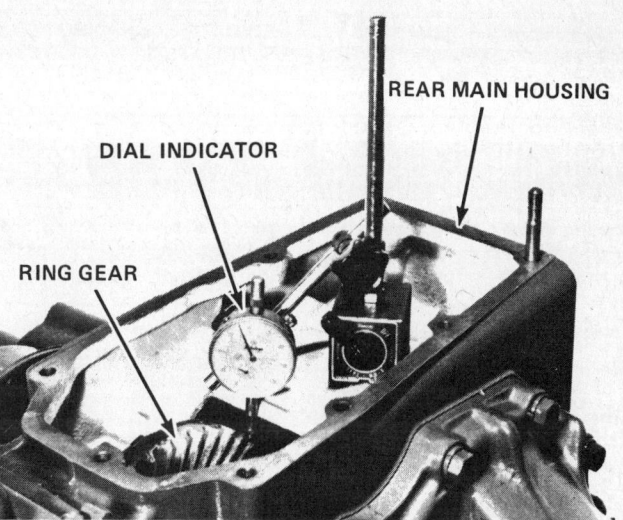

Measuring backlash

Allis-Chalmers

d. **Face contact:** Face contact means narrow contact at the tip of the ring gear. Increase the number of washers on the ring gear side and the drive pinion side.

e. **Flank contact:** Flank contact means deep mesh. Reduce the number of washers on the pinion side.

TRANSMISSION, 4-WHEEL DRIVE

REMOVAL AND INSTALLATION

Removal and Installation Procedure is the same as for 2-wheel drive tractors.

OVERHAUL

1. Remove six (6) nuts holding bearing retainer for main shaft and counter shaft. Remove snapring and thrust washer at rear of counter shaft. Remove snapring at rear of range transmission drive gear. Drive the shaft out.

2. Remove the snapring from center shaft and drive shaft out rearward.

3. Remove four (4) screws retaining the bearing cover. Remove the snapring on the front side of the rear main shaft. Drive the rear main shaft out.

NOTE: An impact wrench maybe required to remove the four (4) main shaft retainer capscrews due to increased bearing surface of tapered capscrews.

4. Remove four (4) nuts from bearing retainer at front of transmission. Remove snapring behind the last gear at rear of shaft. Remove the next snapring forward from that one. Remove the shaft forward.

5. Install the roller bearing in the transmission center wall from the rear with trade mark to rear. Install the front snapring on the shaft splined end. Install the snapring on shifter hub, enter the shaft from the front. Install hub with long end to the rear. Install shifter collar with shifter fork groove to the front. Install two (2) roller bearings with the wide bearing to the rear and inner race inside the 56T gear. Install 56T gear with engagement teeth toward the front.

6. Install the thrust washer next to the gear at the rear.

7. Push the shaft through the center wall of the transmission and install the roller bearing inner sleeve. Install the two (2) roller bearings with the spacer between them in the cluster gear 17 and 33T. Install a thrust washer next to the bearing and a snapring next to the thrust washer; but not in the groove. Install the snapring next to the bearing in the center wall. Install the thrust washer next to the shoulder with the chamfer to the front.

8. Install the thrust washer with the flat on I.D. in the cluster gear. Install the snapring next to the thrust washer. Install the ball bearing with snapring in the front bore. Install the thrust washer and snapring on the front of the shaft. Install the bearing retainer with a new gasket. Install the ball bearing with snapring in rear bore of housing.

NOTE: Do not drive too hard on this bearing as the front bearing retainer could be damaged.

NOTE: The reverse idler is installed and adjusted.

9. Check and replace if necessary the needle bearing in rear of shaft also check and install the roller bearing in front of shaft. Install the bearing retainer on the center shaft with recess toward front of shaft and cut away portion facing up.

10. Press the ball bearing with snapring toward rear on to the shaft against the shoulder. Install the thrust washer, inner bearing race and two (2) roller bearings on rear shaft and also the key. Install the ball bearing in the center wall of the housing from the rear driving on the trade name.

11. Install the 54T gear in the housing with the engaging teeth to the front. Install splined collar with the 40 tooth gear on it with the shifter fork groove toward the rear, engaged with the fork. Slide the shaft through until it is flush with the front of the splined collar. Lower the 23T gear with the thrust washer in front of the gear and with the engaging teeth toward the rear.

12. Slide the shaft through the gear, thrust washer and bearing. Install the four (4) screws in the bearing retainer with the relieved area at the top.

13. Select an adjusting washer by pressing together the adjusting washer and snapring with your fingers lightly to adjust the axial clearance.

Standard Clearance .004-.012 in. (0.1-0.3 mm)

14. If clearance is not in the above range, choose a suitable adjusting washer from the following:
 a. .087 in. (2.2 mm)
 b. .098 in. (2.5 mm)
 c. .110 in. (2.8 mm)
 d. .122 in. (3.1 mm)

15. Press the ball bearing with snapring to the front on the front end of the shaft.

16. Install the shaft in the transmission housing.

17. This shaft installs from the front. Install the front bearing with outer snapring to the front, on the front end of the shaft, this end has the long splines. Install the thrust washer and snapring on the shaft. Install the center roller bearing in the center wall of the transmission.

18. Place the 25T gear in the rear section of the housing with the hub toward the front. Place the 34 and 39T gear in the housing with the shifter fork groove to the rear.

19. Place the 37T gear in the housing with the shifter fork groove to the rear. Place the 51T gear in the housing with the engaging teeth to the front, this gear has a roller bearing inside of it and a thrust washer in a counterbore at the front an inner race and another thrust washer at the

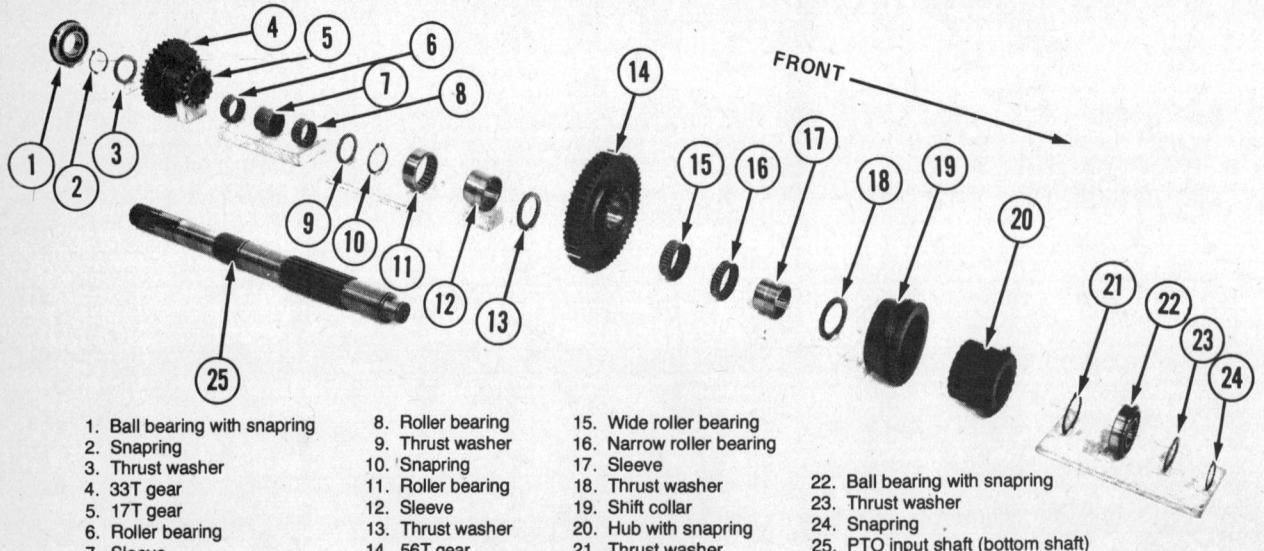

1. Ball bearing with snapring
2. Snapring
3. Thrust washer
4. 33T gear
5. 17T gear
6. Roller bearing
7. Sleeve
8. Roller bearing
9. Thrust washer
10. Snapring
11. Roller bearing
12. Sleeve
13. Thrust washer
14. 56T gear
15. Wide roller bearing
16. Narrow roller bearing
17. Sleeve
18. Thrust washer
19. Shift collar
20. Hub with snapring
21. Thrust washer
22. Ball bearing with snapring
23. Thrust washer
24. Snapring
25. PTO input shaft (bottom shaft)

Mainshaft components, 4-wheel drive 5020

Allis-Chalmers

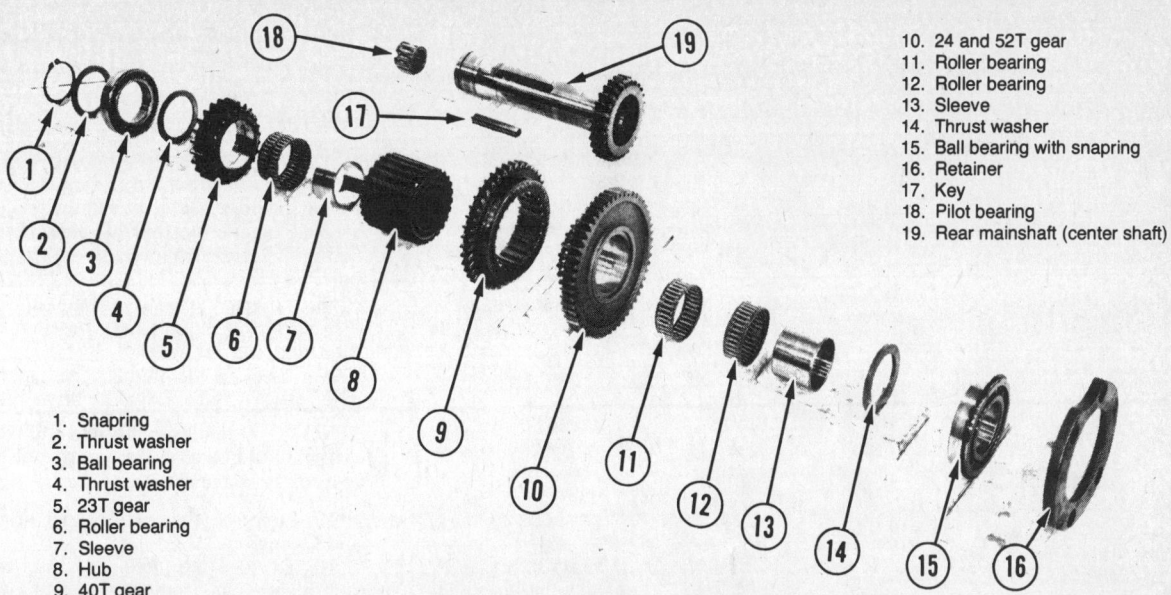

1. Snapring
2. Thrust washer
3. Ball bearing
4. Thrust washer
5. 23T gear
6. Roller bearing
7. Sleeve
8. Hub
9. 40T gear
10. 24 and 52T gear
11. Roller bearing
12. Roller bearing
13. Sleeve
14. Thrust washer
15. Ball bearing with snapring
16. Retainer
17. Key
18. Pilot bearing
19. Rear mainshaft (center shaft)

Countershaft components, 4-wheel drive 5020

rear. This bearing has an inner race that must be in place on the shaft.

20. Slide the shaft through these gears and thrust washers and place on the shaft the inner race for the center wall bearings the 25T gear and the snapring retaining this gear. Install this snapring after the shaft is in place.

21. Install the rear ball bearing, thrust washer and snapring on the shaft. Install the front bearing retainer cover after replacing the seal at the center shaft. Install the shift cover.

Differential Lock Controls

REMOVAL

1. Remove the righthand axle sleeve assembly. The differential lock coupling, thrust washers and differential lock spring are attached to the sleeve.
2. Remove the double roll pin attaching the foot lever, and remove the lever.
3. Remove the snapring and thrust washer, cotter pin in the shift lever side and washer. Insert the control lever into the housing to remove.
4. Remove the plug at the lower side of the shaft. Remove two cotter pins, shaft and shift lever.

INSTALLATION

1. Install the O-ring on the control lever. Install the lever in the housing from inside.
2. Install the shift rod to the lever and a shaft to the shift lever.
3. Attach the shaft to the shift lever with the cotter pins.
4. Connect the shift lever and the shift rod.
5. Press the plug into the bottom of the shaft. Install the thrust washer and the snapring which position the control lever, and attach the pedal with a double roll pin.

Pinion Shaft and Gears

REMOVAL

Remove the axle sleeve assemblies and the differential assembly. Lock the rotation of the front drive gear and unscrew the adjusting nut which has a LEFTHAND THREAD (spanner size—$1^{13}/_{16}$ in.). Remove a hub, spacer and gear. Drive the pinion shaft rearward with a plastic hammer.

PRE-LOAD ADJUSTMENT

Install a dummy pinion (ACTP-3096-1) with a tapered bearing assembled to the housing. Insert the adjust washers and spacer to the dummy pinion. Install another tapered bearing and a roller bearing nut, then tighten the nut at the following torque:
   Standard tightening torque 58-72 ft.lbs.
   Pre-load normal torque 5-10 in.lbs.
   3-5 lbs. at a notch of B portion.
If not in the above range, choose a suitable shim from the following:
1. .004 in. (0.10 mm)
2. .006 in. (0.15 mm)
3. .008 in. (0.20 mm)
4. .039 in. (1.00 mm)

PINION HEIGHT ADJUSTMENT

Install the pinion height gauge (ACTP-3096-2) in the housing. Measure the pinion height after adjusting. Measure the dimension "A".

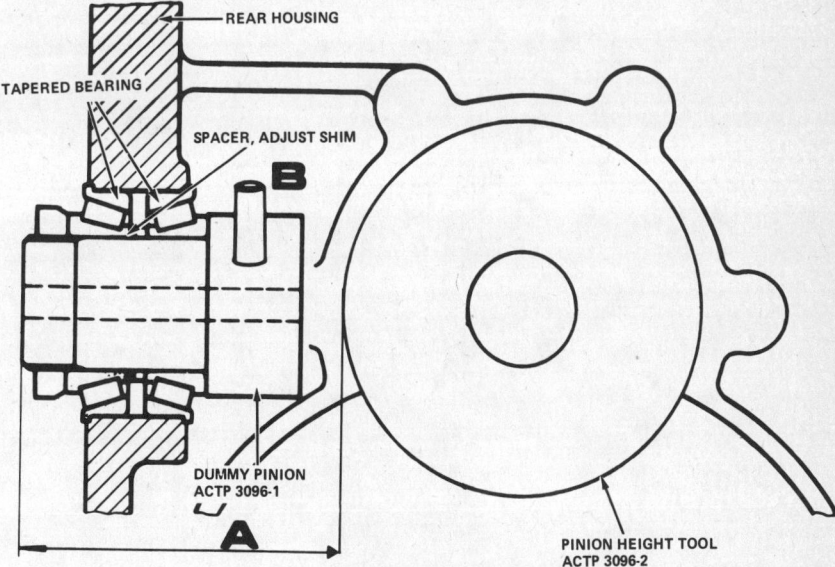

Preload adjustment

Allis-Chalmers

Tightening the pinion shaft locknut

The adjusting washer thickness = "A"—3.858 in. (97.99 mm)

Select an adjusting shim from the following:

.106 in. (2.7 mm) .122 in. (3.1 mm)
.110 in. (2.8 mm) .125 in. (3.2 mm)
.114 in. (2.9 mm) .130 in. (3.3 mm)
.118 in. (3.0 mm)

INSTALLING PINION SHAFT AND GEARS

1. Assemble the drive pinion shaft with the adjusting washers selected for pinion height adjustment and pre-load adjustment, tapered bearings used in determining the adjustment gear with hub toward the front, spacer, hub and nut.
2. Apply Loctite® No. 277 Part No. 930904 to the shaft.
3. Tighten the nut to a torque of 58-72 ft.lbs.

FRONT DRIVE AXLE

Axle Assembly

REMOVAL AND INSTALLATION

1. Loosen the front wheel installing capscrews.
2. Lift the tractor front end with a jack.
3. Remove the capscrews, lockwashers, flat washers and front wheels.
4. Remove the tie rod front ends on the steering arms.
5. Support the front axle on both sides with a hoist and slings.

NOTE: Support the axle level with the floor.

6. Loosen the three (3) capscrews securing the drive shaft cover at the rear end.
7. Remove the four (4) capscrews securing the carrier brackets to the front support.
8. Lower the axle slowly until the diff-carrier upper face becomes just lower than the carrier-bracket installing face (lower face) of the frame. Then, pull the axle forward and release the drive shaft.

NOTE: Over lowering and quick lowering should be avoided to prevent damage to the drive shaft.

9. Remove the two carrier-brackets and O-rings.
10. Remove the three (3) capscrews in step 6 above and remove the shaft cover.
11. Remove the drive shaft and coupling forward away from the transmission.
12. Insert the coupling on the transmission output shaft and insert the drive shaft with shaft cover and gasket into it.
13. Replace the rear carrier O-ring.
14. Install the O-rings, and the carrier-brackets (front and rear) on the diff-carrier as follows:
   a. Install the O-rings on the carrier-brackets as illustrated.
   b. Install the carrier-brackets on the diff-carrier and fasten tightly with wire.
   c. Then slide and install the O-rings correctly on the V-groove between the carrier and brackets. This protects the O-rings from damage.
14. Lift the front axle and install it under the front support in the reverse sequence of removing. At same time, insert the in-put shaft (spiral bevel pinion) into the drive shaft and the shaft cover front end into the carrier-bracket simultaneously.

Removing driveshaft end cover

Removing driveshaft

Allis-Chalmers

15. Mount the carrier-brackets on the front support with capscrews.
16. Retighten the rear shaft cover capscrews.
17. Install the front wheels making sure the flat washers contact the wheel disc. Tighten the capscrews to 57-67 ft.lbs.
18. Reinstall the tie rod front ends to the yokes (steering arms).
19. Lower the tractor front to the floor.

Front Wheel Hub and Gear Case Cover

REMOVAL AND DISASSEMBLY

1. Jack up the front of the tractor and remove the front wheels.
2. Drain the lubricating oil with the drain and filler plugs removed.
3. Six (6) capscrews and two (2) nuts should be removed to remove the gear case cover. Remove the gear case cover with hub.
4. If tight, place a screwdriver on the border-line between the gear case and the gear case cover, and hit the screwdriver head lightly with a hammer. Avoid damaging the gasket surfaces.
5. Remove a cotter pin and nut (castle nut) on the hub shaft.
6. Tap the gear case cover on a wood block to remove a ball bearing and gear on the hub shaft.
7. Remove the two ball bearings in the gear case cover.
8. Remove the inner-sleeve for the oil-seal on the shaft.

ASSEMBLY AND INSTALLATION

1. Remove the gasket and sealant on the joint surfaces of the case and cover and clean.
2. Reverse procedures 4 to 8 above.

NOTE: Install gear with hub in:

a. Tighten castle nut tight and install cotter pin.

NOTE: Do not back off nut.

Removing hub shaft ball bearing and gear by tapping assembly on wood block

b. Apply sealant on both surfaces of the gasket.
c. Apply Loctite® on the gear case cover securing capscrews.

Front Gear Case and Front Axle

REMOVAL AND DISASSEMBLY

1. Jack up the tractor front end then disconnect the tie rod front end and remove the wheel of the side to be disassembled.
2. Drain the diff-carrier oil by removing the drain plug at the bottom of diff-carrier.

CAUTION
Lower the tractor front end slightly to slant the axle assembly being removed downward to avoid pinching fingers between frame and axle.

---

3. Unscrew the eight (8) capscrews which secure the front axle to the diff-carrier.

4. Remove the front axle sleeve with O-ring and drive shaft.

NOTE: If the king pin assembly does not need to be overhauled be careful to remove shaft and axle sleeve together.

5. Disassembling the front axle, remove the four (4) capscrews securing the case-support of the front axle housing. Separate the front axle housing and front gear case.

NOTE: There are 105 balls in this bearing—don't lose them.

6. Remove the front gear cover and hub.
7. Remove a snapring on the joint shaft end. Remove a pinion and bearing, and pull out the universal joint.
8. Remove the front drive shaft and bearing from the front axle case. (Strike the joint side with a plastic hammer).
9. Remove the king pin thrust bearing outer ring and balls from the front axle.
10. Remove the king pin thrust bearing inner ring after removing the king pin seal and O-ring.
11. Drive the plug on the front axle case downward, if necessary to replace it.

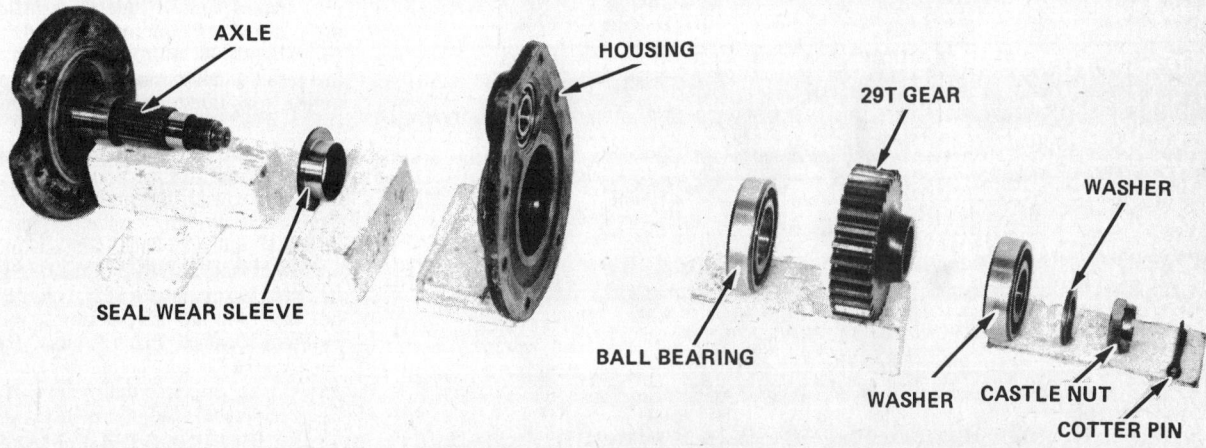

Front axle and gear case components

Allis-Chalmers

12. Remove the oil seal. (Use a new oil seal at assembly.)

13. Remove the bearing. (Pack the bearing at assembly.)

ASSEMBLY AND INSTALLATION

1. Install a plug, bearing and oil seal on the front axle. Use new ones.

2. Reassemble the king pin thrust bearing inner ring on the front axle. Use a soft hammer and tap the ring circumference lightly.

3. Install the king pin with the king pin thrust bearing outer ring as an installing jig.

4. Install the king pin thrust bearing outer ring and 105 small balls as shown. Be careful not to fold seal under outer ring.

5. Reassemble the universal joint to the front gear case and install a bearing, (13T) gear and snapring.

6. Install the bearing to the front drive shaft.

NOTE: This bearing is sealed on one side. Install with seal to outside.

7. Install the drive shaft in the axle housing.

8. Install an O-ring in the edge of the front gear case thrust bearing groove. Connect the front gear case and the front axle with case support and at the same time align the drive shaft to the universal joint. Do not damage the O-ring.

Reassembling torques are as follows:
Lift bolt 2.5 ft.lbs.
Locknut of lift bolt 7.5 ft.lbs.
Rolling torque of gear case is to be 2.5 ft.lbs. with this adjustment.

9. Install the gear case cover (assembled with hub, gear, bearing, and oil seal) to the front gear case with gasket.

10. Install new O-rings on the differential carrier sides.

11. Be sure thrust washer is in place. Install the front axle to the differential carrier.

NOTE: Do not damage the O-rings on the differential carrier sides.

12. Install the front wheels and tie rods making sure the flatwashers contact the wheel discs.

13. Adjusting the turning stop screws. Adjust the stop screws 2 in. (50 mm) from the center of the case support to the end of the capscrews head.

Differential Carrier

INSPECTION

1. Inspect tooth contact between the spiral bevel and the spiral bevel gear. Before disassembly, apply red lead or blueing to several teeth of the ring gear, turn the drive pinion, and check minimum pattern on the tooth surface to judge tooth contact.

a. Proper contact: Backlash of the drive pinion and ring gear is properly adjusted.

b. Heel contact: The heel of the ring gear makes contact, due to insufficient mesh of the drive pinion. Move the ring gear slightly away from the pinion, and increase the number of shims next to the pinion gear side to adjust contact conditions.

c. Toe contact: Toe contact is caused by excessive mesh of the pinion. Move the ring gear toward the pinion and reduce the number of shims on the pinion side.

d. Face contact: Face contact means narrow contact at the tip of the ring gear. Increase the number of shims on the ring gear side and the drive pinion.

e. Flank contact: Flank contact means deep mesh. Reduce the number of shims on the ring gear side and pinion. It will be necessary to readjust the backlash.

2. Ring gear inspection

a. It is very important that the tooth con-

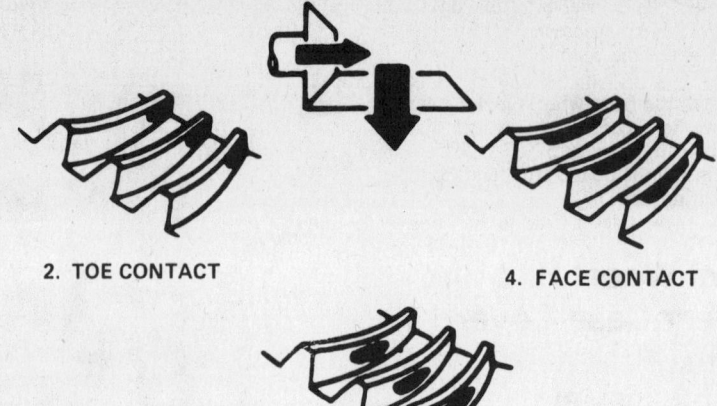

Gear tooth contact patterns

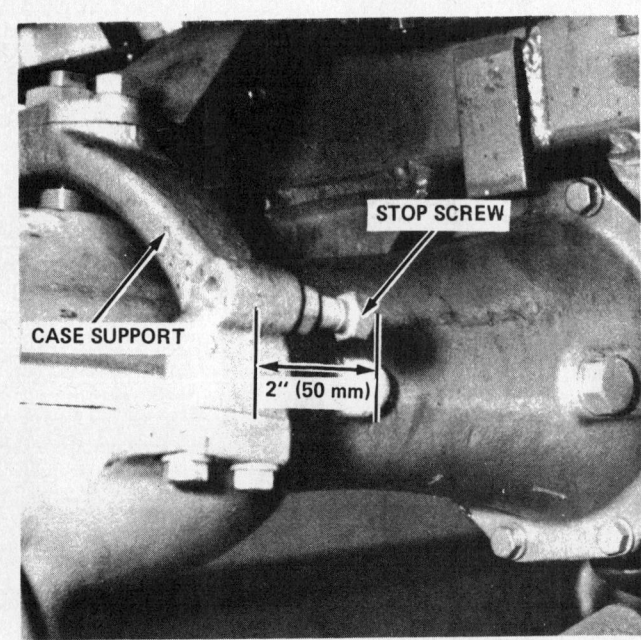

Adjusting stop screws

Allis-Chalmers

tact face is smooth, on the spiral bevel gears. So, if any burr or nick is found on the tooth, the gear kit—a ring gear and drive pinion—should be replaced with a new one. The small burrs or nicks can be polished with an oil-stone.

b. A deformed, bad tooth contact or a worn ring gear should be replaced by a new gear kit with a pinion.

DISASSEMBLY

1. Remove the front axles and carrier-brackets.
2. Remove the O-rings on the both sides of differential carrier.
3. Strike the lefthand side bearing with a plastic hammer, then the spacer and differential sub assembly can be removed.

NOTE: Do not lose and mix the adjusting shims installed on the bearing-backs. Count the number of adjusting shims on each side.

4. Place the spiral bevel pinion spline shaft (input shaft) in a vice and unscrew the nut.
5. Remove the spiral bevel pinion by striking its spline end with a soft hammer. A bearing inner ring, a distance collar and a seal collar can be removed simultaneously.

NOTE: Do not lose the adjusting shims.

6. Remove a bearing (tapered roller) from the spiral bevel pinion.

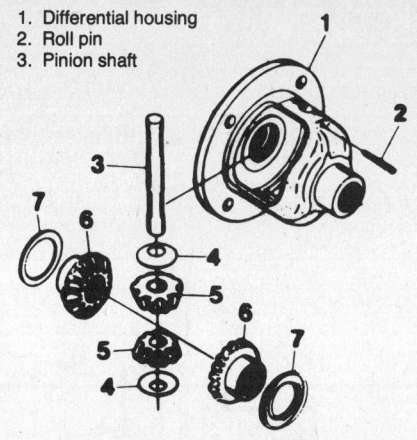

1. Differential housing
2. Roll pin
3. Pinion shaft
4. Pinion thrust washer
5. Differential pinion
6. Differential side gear
7. Thrust washer gear

Side gear disassembly

NOTE: Do not lose the adjusting shims between bevel pinion neck and bearing.

7. Remove the oil seal.
8. Remove two bearing outer rings and two snaprings in the differential carrier sleeve.

a. Remove the ring gear from differential by removing the four (4) securing capscrews.
b. Remove two bearings on both sides of differential case.
c. Remove a roll pin, then the pinion mainshaft, pinion thrust washers, differential pinions and differential side gears can be disassembled.

ASSEMBLY AND ADJUSTMENT

1. Reassemble the differential and install the ring gear.
2. Reassemble the drive pinion and the differential sub assembly in reverse of disassembly.

a. When no parts are replaced, use the shims that were removed.
b. When installing the spacer to the housing install it with the test hole at the bottom.
c. When a gear kit or any part is replaced, check the preload, clearance and measure the shims. The checking method is as follows:
d. The drive pinion pre-load adjustment. Assemble the dummy pinion, bearings and spacer and tighten the nut. Measure the pre-load with a spanner wrench and spring scale or torque wrench. The pre-load should be 5.3 to 7 in.lbs. If the pre-load is not in the above range, adjust with shims. Measure the clearance between the front dummy pinion shaft (ACTP 3096-3) and the pinion height gauge (ACTP 3096-4) with a feeler gauge.
e. The clearance "A" should be .007-.032 in. without shims.

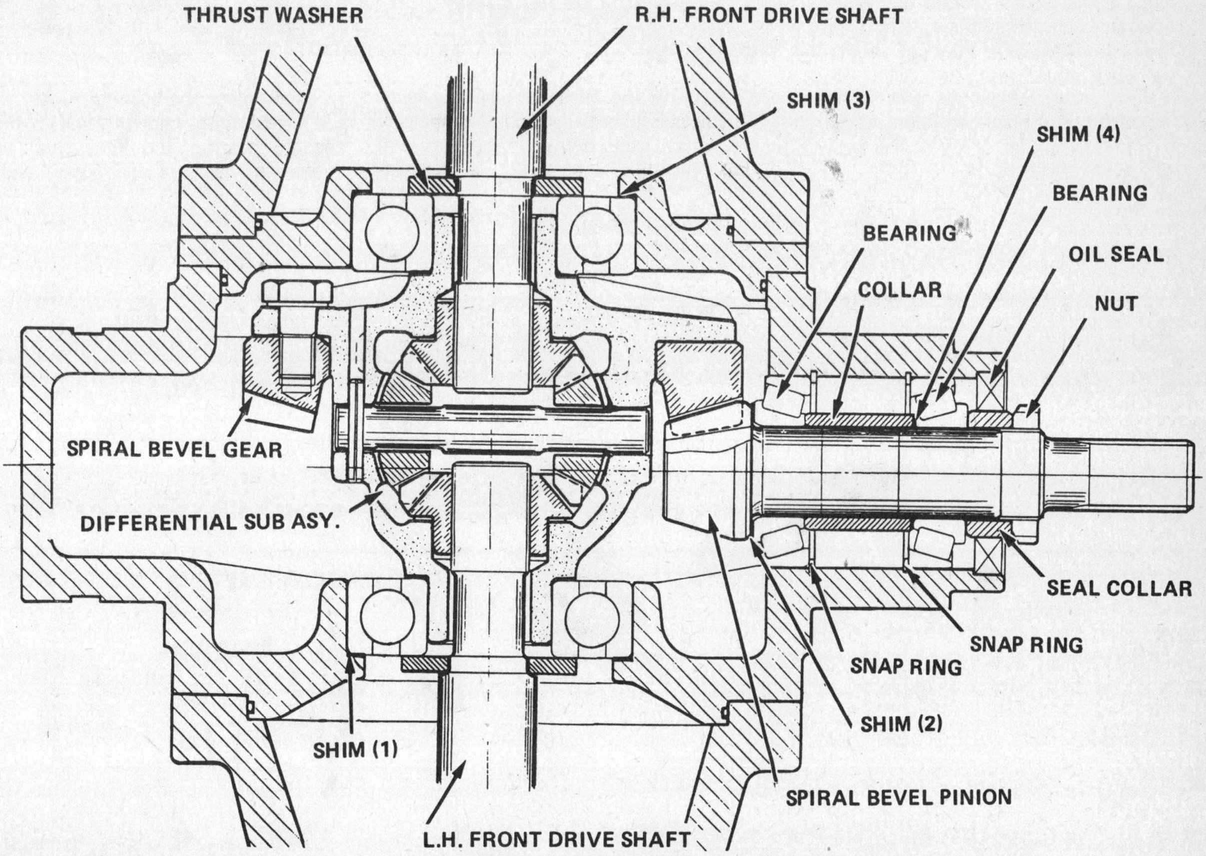

Differential components

157

Allis-Chalmers

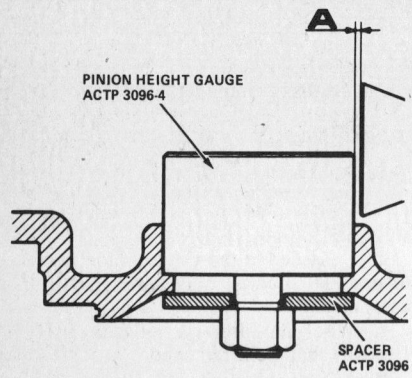

Measuring clearance between front dummy shaft and pinion height gauge

NOTE: The dummy pinion nut should be tightened to 43-57 ft.lbs. torque before measuring. Select shims for minimum clearance. Clearance should be zero at "A".

f. The pre-load of the differential bearings should be adjusted with shims item 1 and 3 in the accompanying figure. The clearance can be measured with feeler gauge. Select shims to allow as little clearance as possible without pre-loading the bearings.

g. Backlash adjustment of the ring gear and pinion.

Insert the dial indicator sensor into the test hole of spacer. Insert a front drive shaft to the differential. Rotate the differential slowly until a ring gear retaining capscrew head contacts the sensor of the dial indicator. Rotate the differential slightly to measure the backlash. The backlash should be .005 to .007 in. when the backlash is not in the above range adjust the shims. At final assembly of the spiral pinion install the seal, torque nut to 43-57 ft.lbs. and stake the nut in 3 places.

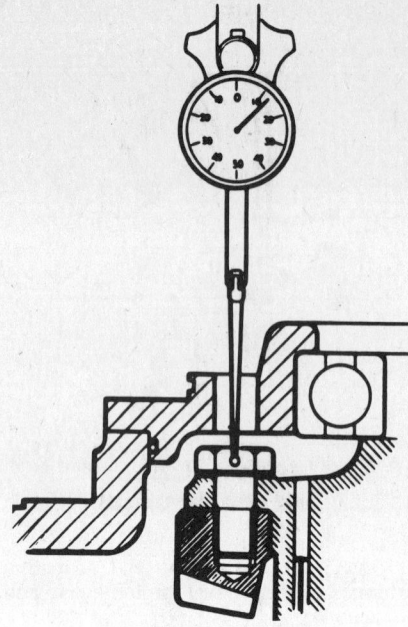

Backlash measurement

NOTE: When assembling differential housing to axle sleeves the spacer needs to be on the right side of the tractor.

NOTE: When reinstalling the front axle assembly install a new gasket on the drive shaft cover at the rear and a new O-ring at the front.

Front Drive Unit

REMOVAL

1. Remove the bushing, thrust washer, 21T gear, needle bearing, and idler shaft by unscrewing a capscrew on the center wall of the housing. Remove the oil seal installed on the front driveshaft with a screwdriver.

NOTE: Avoid damaging the shaft during seal removal.

2. Remove the snapring retaining the bearing in the housing at the front. Remove the snapring and thrust washer at rear of center wall. Remove the bearing and shaft forward. Remove the gear. Drive the roll pin out of the control lever and remove the lever.

3. Unscrew a lockscrew and remove the spring and detent ball, a rubber and metal washer seals this screw.

4. Remove a snapring and thrust washer and push the shift lever into the housing. The shaft has two O-rings on it. There is a bushing in the housing for this shaft that can be replaced.

INSTALLATION

1. Using new O-rings reassemble the shift lever, control lever and detent system. Install the shaft, gear with the shifter groove to the front, bearings thrust washers and snaprings. Replace the seal with a new one.

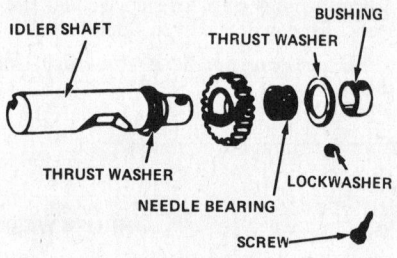

Idler shaft

2. Replace the bushing, thrust washers, 21T gear with engaging end to the front, needle bearing and idler shaft with the notch in the front end down, and secure

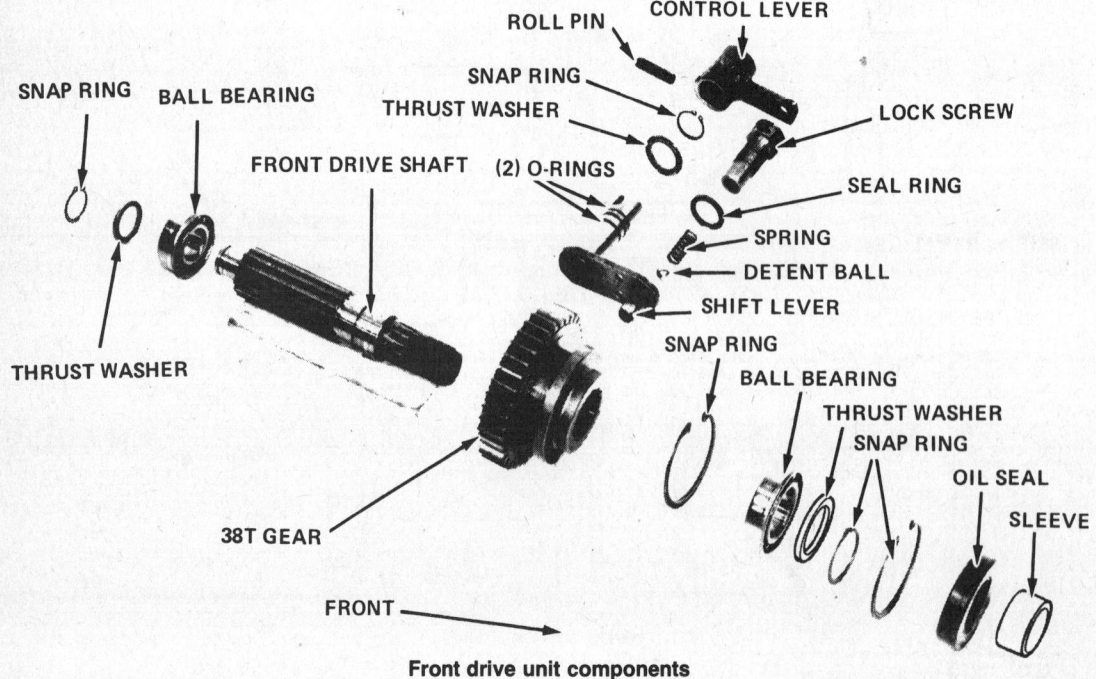

Front drive unit components

Allis-Chalmers

with the capscrew in the center wall of the housing.

3. If the idler shaft is not installed correctly the hydraulic pump suction screen will interfere with this shaft when installed.

PTO Extension

REMOVAL

Remove the snapring retaining the gear at front end of the shaft. Remove the six nuts retaining the rear bearing retainer. Then the shaft can be pushed out backward.

INSTALLATION

Use the same procedure as the 5020 two wheel drive to assemble the bearing retainer assembly on the PTO extension shaft. Insert the shaft assembly into the rear housing from rear side. When the shaft front end reaches the center wall of the housing, install the bearing, thrust washer, needle bearing, 28T gear, with hub to rear thrust washer and snapring. The coupler can be installed at the time the rear housing is attached to the transmission.

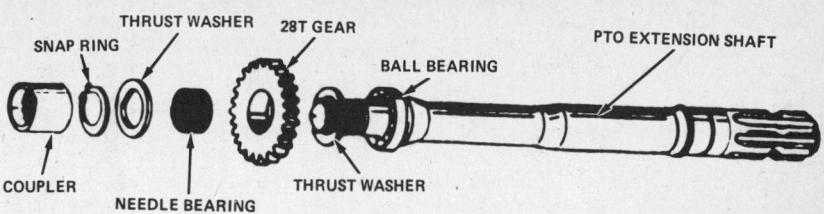

PTO extension shaft

HYDRAULIC PUMP

REMOVAL AND DISASSEMBLY

1. Disconnect the oil lines (suction and delivery). Cap all flanges of oil lines and the two outlets of the pump to prevent entrance of dirt.

2. Remove the four mounting nuts and remove pump from engine. Clean outside of pump thoroughly before disassembling. Scribe an angular mark across the back side of the pump so that all parts can be assembled in their original position.

3. Remove the six socket head capscrews (11) that hold the pump end cover (10) to the body (1). Separate the pump end cover and remove it.

NOTE: Do not pry between sections, as this will mark the mating surfaces and result in damage or leakage of the pump.

4. Remove the composite sealing element consisting of two sealing rings and two support rings. (7 and 8). Press the drive shaft (4) into the body so that a finger hold on the first pair of bushings (6) can be obtained. Remove the bushings and slide out the drive shaft (4) and gear, driven gear (5), second pair of bushings (3) with O-rings (2).

NOTE: A drift must not be used when extracting bushings. If they are tight, tap the pump body on a resilient surface (e.g., wood) to dislodge them.

5. Lay out the parts in the same order as they were withdrawn to maintain mating shafts and bushings in the same relative positions. Do not remove the shaft seal, it can be removed from inside the case, taking care not to score the counter bore in which the seal is housed.

6. Inspect the parts for wear or damage and, using discretion, renew worn or defective items.

7. The design of the pump is such that high efficiency is partly obtained by allowing the gear tips to wipe the wall of the pump body, a wear mark should be visible on the low pressure side of the pump. If excessive wear of the bushing bores or gear journals takes place due to long service and/or working with a dirty fluid system, the wear mark will become a wear track of some depth. If this track exceeds 0.002 in. in depth it would be unwise to fit new bushes in the body due to the fact that there would now be clearance between gear tips and body which would lead to low efficiency and loss of delivery.

8. It is not advisable to fit worn bushings in a new body unless careful running at low pressure can be undertaken.

9. Generally, the rate of wear on all moving parts is fairly uniform and it may be found more economic to renew a pump that has had long service in preference to fitting several new parts.

10. Wash all parts thoroughly in clean solvent (petrol or thinners) and blow out all passage ways with an airhose. All O-ring seals should be renewed. Examine the gears, reject if the shaft diameters are scored or if the teeth are damaged. The teeth will show a mating mark where contact takes place which should be evenly distributed over the tooth flank. Measure the shafts by a micrometer, reject if less than the dimensions indicated in the following: 0.496 in. (12.598 mm).

11. Examine the bushing bores for surface break-down or scoring. This fault, if present, in conjunction with discoloration of the gears, indicates over-heating and the hydraulic circuit should be checked for malfunction, i.e., a faulty relief valve. Bushings or gears with these defects must be rejected. Measure the overall bushing lengths and reject if less than the dimensions specified in the following: 0,810 in. (20.574 mm).

12. Measure the clearance between the shafts and bushing bores. Renew the bushings if the clearance is more than specified in the following: 0.007 in. (0.177 mm).

NOTE: Gears and bushings are mated for width to within .0002 in. (.005 mm) and must be renewed in pairs.

ASSEMBLY AND INSTALLATION

1. Ensure that all parts are perfectly clean. If the shaft seal (13) has been removed place the new seal (sealing lip fac-

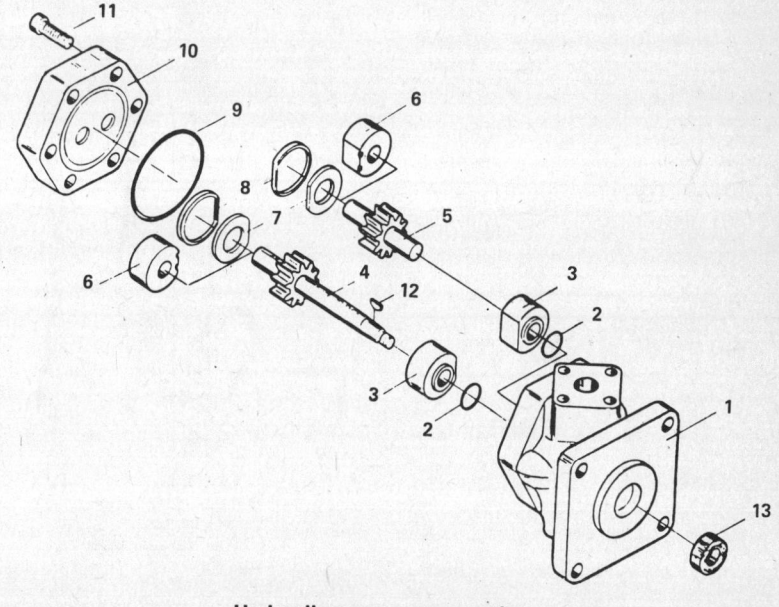

Hydraulic pump components

159

Allis-Chalmers

ing inwards) in the counterbore and press in flush with the spigot on the motor body. Place O-rings (2) on the bushings (3) and retain them with a smear of good quality mineral grease. Insert the bushings into the body bore (1) and slide them into place. The bushings must be kept square both to the bore and to each other during insertion to prevent damage.

2. Manufacturing tolerances are such that no force should be necessary to effect their entry. If a bushing has been damaged by incautious handling, it will jam. In this event the bushing should be extracted and the high spots removed by stoning lightly with a fine grit hand stone. The bushing faces should be rubbed on a smooth surface to ensure that there are no high spots. Clean thoroughly after stoning. An oiled bore will facilitate assembly and the bushings will slide in smoothly.

3. Check that the O-rings have not slipped during the installation of the bushings and that they are located between the bushings and the bottom of the bores. The pressure balancing system is dependent upon this assembly.

4. Ensure that the recess in the bushing faces are not staggered. Misalignment can cause the bushings to jam in the bores, rendering the pressure balancing system inoperative.

5. An assembly sleeve must be used to prevent damage to the shaft seal. The sleeve should be lubricated and placed over the drive shaft (4) which is then inserted into the body and followed by the driven gear (5). The assembly sleeve must be held firmly against the shoulder on the shaft when passing through the shaft seal.

6. If used gears are reassembled in the pump, check that the driven gear takes up its bedded position by mating the tooth contact marks with those on the drive gear. Freely lubricate and insert the second pair of bushings (5). Mount the sealing rings (8) on the support rings (7) and place in position on the bushings.

7. Place the O-ring seal (9) in its housing and hold the cover (10) down, then insert the socket headed screws (11) and tighten the screws evenly. Lock by center punching around the heads. Pour a small quantity of oil into the ports.

8. The pump is now ready for running, but it should be turned over to ensure that it is correctly assembled. It should rotate smoothly by hand from a radius of approximately 3 to 4 in. (100 mm). Undue stiffness must be investigated and remedied.

ENGINE

Cooling System

RADIATOR
Removal and Installation

1. Remove safety covers from both sides, right and left.
2. Drain cooling water from radiator.

1. Radiator
2. Radiator cap, 12.8 psi
3. Upper radiator hose
4. Lower radiator hose
5. Drain cock
6. Protector net

Radiator

3. Loosen air pipe band and remove air pipe from air cleaner.
4. Remove two support bolts (with split pins) located at bottom of radiator.
5. Loosen engine side hose bands of the upper and lower radiator hoses.
6. Installation is the reverse of removal.

THERMOSTAT
Removal and Installation

1. Drain coolant.
2. Disconnect the radiator hose from the thermostat upper case.
3. Remove thermostat upper case and the thermostat.
4. Installation is the reverse of removal.

WATER PUMP
Removal

1. Drain the coolant.

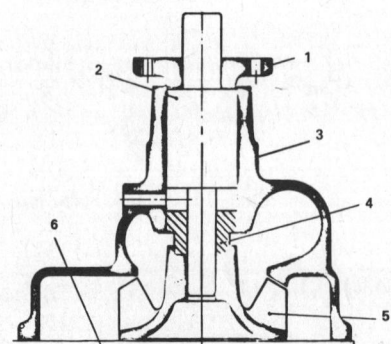

1. Fan pulley boss
2. Water pump bearing
3. Water pump body
4. Mechanical seal assembly
5. Impeller
6. Front plate

Water pump

2. Remove the water hose, and the by-pass hose, then remove the water pump assembly.
3. Installation is the reverse of removal.

Engine Overhaul

CYLINDER HEAD AND VALVES
Removal

1. Remove the injection pipe.
2. Remove the oil pipe connecting the cylinder block main gallery and cylinder head.
3. Loosen the rocker cover nut to remove the rocker cover.
4. Loosen the rocker bracket tightening nut and cylinder head bolt (long) to remove the valve rocker assembly (rocker, rocker shaft and rocker bracket).
5. Remove the nozzle holder assembly and pull the push rod.
6. Loosen the 10 cylinder head bolts to remove the cylinder head assembly.
7. Remove the cylinder head gasket.

Cylinder Head Inspection

1. Clean cylinder head and all its component parts; remove all carbon deposits. Check flatness of cylinder head gasket surface with a feeler gauge and a straight edge. The tolerance for flatness of cylinder head gasket surfaces is .002 in. (0.05 mm) total indicator reading.
2. Some discretion will have to be observed when checking the flatness of a cylinder head removed from an engine after a period of operation.
3. Side to side warpage of .002 in. (0.05 mm) can be tolerated if the low or high spots are not concentrated in small areas or around gasket grommets surrounding combustion chambers, water or oil passages.

Allis-Chalmers

4. Inspect cylinder head and component parts for wear or damage. Repair or replace any worn or damaged parts. If cylinder head is to be replaced, thoroughly inspect the parts removed from the old head before installing them in the new head. Make certain that cored water passages and drilled oil holes in cylinder head are clean.

SWIRL CHAMBER INSERT

Examine the swirl chamber inserts and retaining pins for looseness of fit, damage, and misalignment. When insert is loose, make certain retaining pin is present. Absence of retaining pin will allow the insert to move in the head, causing poor engine performance. If the insert is damaged, replace it with a new one.

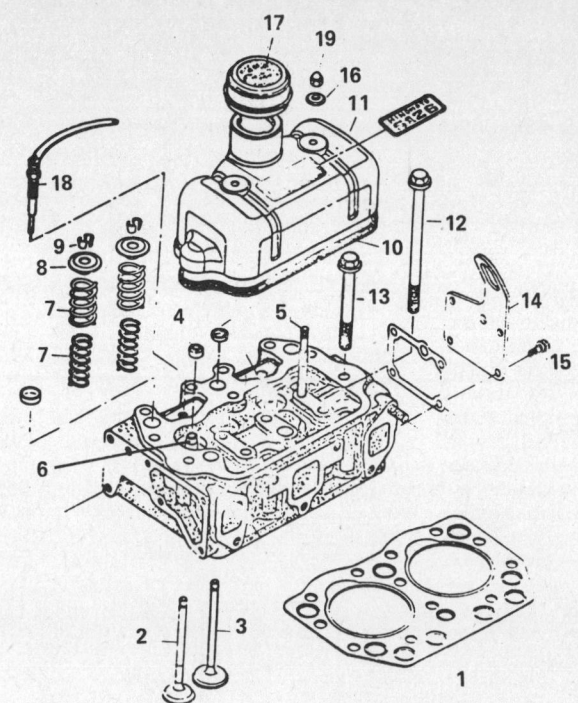

1. Gasket
2. Exhaust valve
3. Intake valve
4. Valve stem seal
5. Stud
6. Intake and exhaust valve guide
7. Intake and exhaust valve spring
8. Valve spring retainer
9. Valve cotter
10. Rocker cover gasket
11. Rocker cover
12. Head bolt long
13. Head bolt short
14. Head rear cover
15. Cover screws
16. Cover washers
17. Oil filler cap
18. Glow plug
19. Cover nuts

Cylinder head components

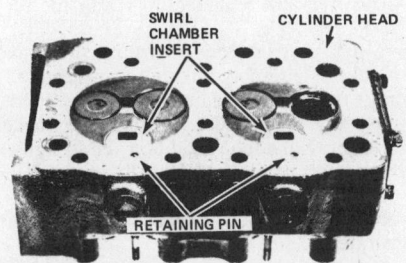

Removing swirl chamber inserts

Removal

A press should be used to remove the insert from the head. The inserts location requires pressing the insert out, via the glow plug hole.

NOTE: Be careful not to damage head during removal.

Installation

Due to the interference fit of the insert and head, freezing of the insert is required to allow installation. After freezing the insert, install it in the head and position the insert with the retaining pin. Allow the insert to rise to room temperature and the insert should be tight in the head.

VALVE SPRING REMOVAL

To remove the spring from the cylinder head, use a valve spring lifter as shown. Compress the valve spring and remove the valve cotter and spring retainer. To loosen the compressed valve spring, gradually release the valve spring lifter.

Valve Spring and Retainer Inspection

Inspect the valve springs and retainer for cracks. Check the valve spring tension with a spring tester. If the outer spring tension is less than 55.2 lb. and the inner spring tension is less than 19.9 lb. replace the valve spring.

VALVE SPRING INSTALLATION

With a spring lifter, compress the valve spring with the retainer, place the valve cotter and slowly return the valve lifter.

	Outer	Inner
Free length	223 in. (56.6 mm)	2.17 in. (55 mm)
Installed tension	26.6 lb. (118 N)	11 lb. (49 N)
Installed length	1.97 in. (50.0 mm)	1.89 in. (48 mm)

NOTE: When compressing the valve spring with the spring retainer, take care neither to damage the valve stem with the retainer nor to damage the valve stem seal due to overcompression.

NOTE: The valve end surface should have a recession of .016-020 in. (0.4 to 0.5 mm) as measured from the installation surface of the cylinder block.

VALVE, VALVE GUIDE AND VALVE SEAT RING

Remove the valve spring, retainer, exhaust valve and intake valve with a valve spring lifter and clean the valve guide bore with a cleaning tool.

NOTE: Do not use a wire brush for cleaning.

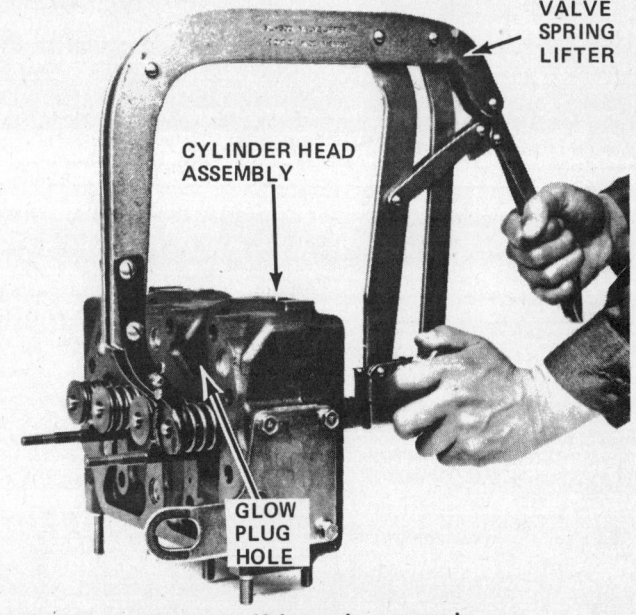

Valve spring removal

Allis-Chalmers

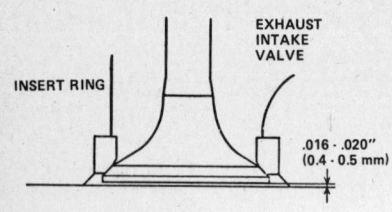

Measuring valve face recess

Valve Guide Inspection

The valve guide is the same for the exhaust and intake. When there is a scratch, scaffing, etc., replacement is necessary.

As shown, use a press to remove the valve guide with a removing tool, taking care not to cause a scratch in the cylinder head valve guide hole. Measure the dimension of the guide hole with a cylinder gauge, check for the press-in allowance (typically .004 to .0016 in.) (0.01 to 0.04 mm) and press a new valve guide in.

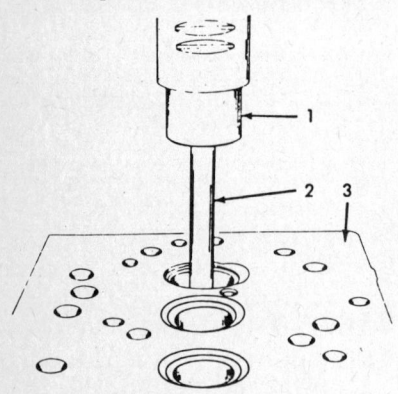

1. Press ram
2. Valve guide removing tool ACTP 2079
3. Cylinder head

Valve guide removal

NOTE: Before pressing the valve guide in, be sure to clean the cylinder head valve guide hole and measure the dimension.

Installing Valve Guide

With the cylinder block surface downward, place the cylinder head horizontal and press it in with a press. The press-in height of the valve guide should be such that the rocker cover surface and valve guide end are flush.

Reaming Valve Guide

Reaming the valve guide is difficult outside a factory because it requires concentricity with the valve seat. Before reaming, the valve guide has a bore diameter of .303 to .305 in. (7.70 to 7.75 mm). For reaming, position the cylinder head vertically with the cylinder block surface upward. After reaming, the dimension is .315 to .316 in. (8.000 to 8.018 mm).

Valve Seat Grinding

Grinding the valve seat is required in the following cases:

1. Considerable marring occurs on the seat surfaces.
2. There is a poor contact surface, or
3. Replacement is performed.

The valve seat ring, press fit to the cylinder head, cannot be replaced.

Before grinding the valve seat, clean the valve guide bore and check the inside diameter.

1. Recondition the valve seats with a valve seat grinder of 30°, 45° and 60°.

NOTE: When grinding the valve seat, the 45° seat should be ground last. The preferred order of grinding is 60°, 30° and 45°. This enables obtaining the proper seat as shown.

2. After grinding the valve seat, the valve should contact the seat exactly at the center, and finish the valve seat by lapping the seat to obtain the proper width of .055 in. (1.4 mm) and depth of .040 in. (1.0 mm).

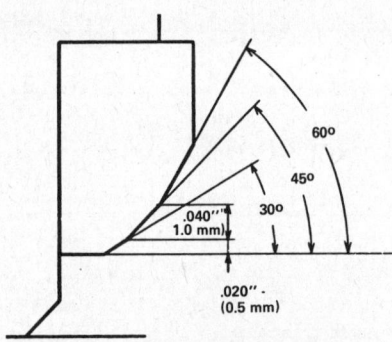

Proper valve seat dimensions

After contact production is completed, check for the proper sealing by supplying fuel at the port with the valve in close contact with the seat surface by hand. If leakage occurs, further adjustment is necessary.

Valve Grinding

Inspect the valves for wear and warpage. If the valves are excessively worn or bent, and the valve head edge is less than .040 in. (1.0 mm) replace the valves.
Overall length of valve
  Intake & exhaust: 4.921 in. (125 mm)
Stem diameter:
  Intake: .3128-.3134 in. (7.945-7.96 mm)
  Exhaust: .3188-.3124 in. (7.920-7.935 mm)
Valve seat angle: 45°
Grinding limit of stem end: .0197 in. (0.5 mm)

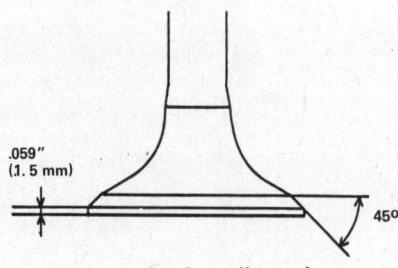

Proper valve face dimensions

CYLINDER HEAD INSTALLATION

1. Before installing the cylinder head asembly, clean the cylinder block top face and installing surface and remove carbon, oil, etc.

NOTE: Before installing head be sure swirl chamber insert and swirl chamber insert retaining pin are present.

2. Use a new cylinder gasket. The gasket is coated with sealing and separating material on both sides.
3. To tighten the cylinder head bolt, follow the sequence shown in the drawing. First, tighten with a torque of 29-36 ft.lbs. The head bolt (with seat) requires no lockwasher.
4. Install the push rods. Make certain push rods are seated properly in the tappets.
5. Position rocker arms, shaft and brackets assembly on cylinder head and align rocker arm adjusting screws in the push rod cap ends. Tighten the rocker bracket with a long head bolt. First, tighten with a torque of 29 to 36 ft.lbs. as with the method performed on the other long head bolts. Install nuts and lockwashers. The nut torque is 22 to 29 ft.lbs.

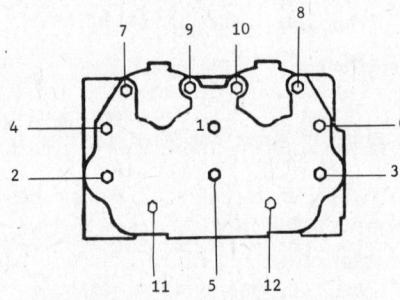

Cylinder head bolt torque sequence

6. Tighten all of the cylinder head bolts in the numerical sequence shown in the accompanying illustrations.
7. Adjust the valve lash (at low temperatures), .012 in. (0.3 mm) for intake and .014 in. (0.35 mm) for exhaust. Tightening torque of the adjusting nut is 29 to 33 ft.lbs.

NOTE: Apply a slight amount of oil to the push rod cap end.

8. Install the rocker cover containing the rocker cover gasket.
9. Install the oil tube which connects the cylinder block main gallery and connect the injection pipe from the injection pump with the nozzle holder.
10. Check the coolant and oil before starting the engine. Inspect the engine for fuel, water or oil leaks and correct any leaks found.

Timing Gear Cover
GENERAL

The timing gear cover is an aluminum casting, attached to the front side of the

Allis-Chalmers

engine. The timing gear cover contains a press fit front oil seal and timing point, and is provided with a hydraulic pump assembly.

REMOVAL AND INSTALLATION

Drain the cooling system and remove the air cleaner radiator, generator, generator bracket and crankshaft pulley. Remove the hydraulic tubes and hydraulic pump. Remove all capscrews and nuts on front and back side securing timing gear cover to cylinder block and front plate. Tap the timing gear cover and dowel pin with a soft wooden of plastic hammer to remove the gear cover from the front plate.

NOTE: Before reinstalling cover, install a new crankshaft front oil seal.

CRANKSHAFT FRONT OIL SEAL REPLACEMENT

1. Clean the timing gear cover, oil seal bore, and check for scratches, etc.
2. Apply a minute amount of clean grease or oil to the circumference of the front oil seal.
3. Place cover on a flat surface with the front side to the top as illustrated.
4. Position seal in cover with open side of seal facing down and position securely in the bore.
5. Drive or press seal into cover bore until it bottoms.

NOTE: Do not press on open face of seal or seal damage will occur.

6. After seal is installed in the cover, insert your fingers into the inner part of the seal and check for proper installation. If seal was installed properly, the inner part will turn when a firm force is applied through the fingers.

Valve Lash Adjustment

The valve lash (cold) is .012 in. (0.3 mm) for intake and .014 in. (0.35 mm) for exhaust. Valve lash adjustment is performed when the pistons are at the top of their compression stroke.

The firing order of the engine is (1) fires—180° of coast (2) fires—540° of coast—(1) fires.

To adjust the valve lash, use a feeler gauge in conjunction with the rocker adjusting bolt and locknut. The torque of the locknut is 29 to 33 ft.lbs. After adjusting the valve lash rotate the engine several turns at full compression and again check the valve lash.

Valve Timing

Check for valve timing with exhaust valve at .014 in. (0.35 mm) intake valve at .012 in. (0-30 mm).

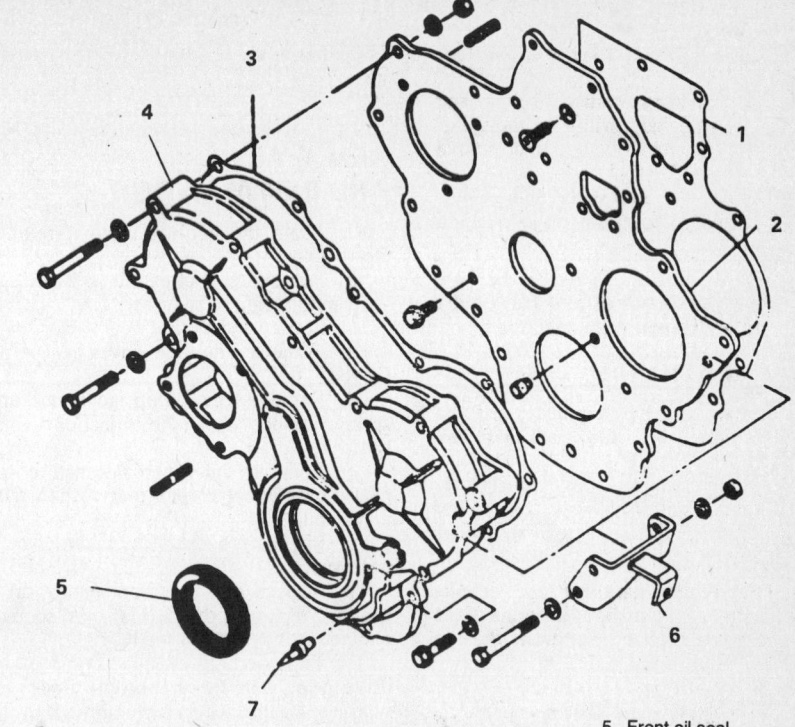

1. Front plate gasket
2. Front plate
3. Timing gear cover gasket
4. Timing gear cover
5. Front oil seal
6. Generator bracket
7. Timing point

Front cover components

	Intake Valve	Exhaust Valve
Opens	B.T.D.C. 20°	B.B.D.C. 56°
Closes	A.B.D.C. 62°	A.T.D.C. 16°
Valve is open	262°	252°

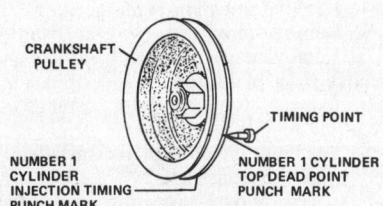

Timing marks

Valve lash adjustment

163

Allis-Chalmers

Crankshaft Pulley

REMOVAL

1. Drain cooling system and battery and radiator. Release the tension on the water pump and alternator drive belt and remove the belt.
2. Remove the crankshaft pulley tightening nut and remove the pulley with a puller.

NOTE: When using a puller, be sure to place the two jaws of tool OTC 1002 on the pulley re-enforcement tabs.

INSTALLATION

1. Remove any burrs if necessary and install Woodruff key into crankshaft key way.
2. Position crankshaft pulley on Woodruff key and install with water and pulley retaining nut. Tighten nut to a torque of 72-85 ft.lbs.

NOTE: When installing the crankshaft pulley, apply a slight amount of oil to the oil seal socket to prevent the tip from becoming damaged.

Injection Pump Drive Gear

The injection pump gear is positioned with a Woodruff and taper fastened on the injection pump camshaft. To remove the injection pump from the engine, remove the timing gear and then the injection pump gear.

REMOVAL AND INSTALLATION

1. Drain the cooling system; remove the radiator.
2. Remove the pulley nut so the crankshaft pulley can be removed.
3. Loosen the timing gear cover, capscrew so the timing gear cover can be removed.
4. Remove the pump gear nut and spring washer from the injection pump camshaft.
5. Remove the taper fastened pump gear on the injection pump driveshaft with a gear puller.
6. Inspect Woodruff key and replace if necessary.
7. Inspect the injection pump drive gear and replace if the gear is worn, scored, chipped or has broken teeth.
8. To assemble the injection pump drive gear, turn the crankshaft clockwise with a spanner until the timing mark of the crankshaft gear and idler gear and that of the idler gear and camshaft gear meet. With everything in said position, install injection pump gear on the pump shaft noting that Woodruff key is aligned. Also make sure that the timing mark on the idler gear and the pump gear is aligned.
9. Tighten the pump drive gear on the injection pump driveshaft with a lockwasher and nut. The torque is 43 to 51 ft.lbs.
10. Recheck injection pump timing marks. Adjust if necessary.
11. Inspect and clean the front plate gasket surface and the timing gear cover gasket surface. Use a new gasket. Install the timing gear cover with capscrews and lockwashers.
12. Complete the rest of the installation by a reversal of the removal procedure and fill the cooling system to the proper level.

Idler Gear and Shaft

The idler gear includes a press fit needle bearing. The idler gear shaft is pressed in the cylinder block and cannot be removed. The side clearance of the idler gear is .0020 to .0079 in. (0.05 to 0.2 mm). The thrust plate is quench hardened.

To install the idler gear, proceed as follows:

1. Inspect idler gear and replace if gear is worn, scored, chipped or has broken teeth.
2. Replacement is necessary if needle bearing is scratched or worn, or thrust plate is damaged.
3. Apply a slight amount of engine oil to the internal surface of the needle bearing and timing marks for assembling the idler gear.
4. Install both, the needle bearing and idler gear, with washer and bolt. Tighten to a torque of 33 to 36 ft.lbs.

Front Plate

GENERAL

The front plate is made of steel plate and is used to support the timing gear cover. It also provides a mounting surface for the fuel injection pump. The front plate is secured to the cylinder block with capscrews and lockwashers and is sealed to the cylinder block with the front plate gasket.

REMOVAL, INSPECTION AND INSTALLATION

1. Remove alternator and alternator bracket.
2. Remove water pump assembly.
3. Remove timing gear cover.
4. Remove injection pump gear.
5. Remove idler gear.
6. Remove capscrews, lockwashers and remove front plate from cylinder block.
7. Inspect front plate for damage or wear and replace if necessary.
8. Use new gaskets for the installation and install the front plate by a direct reversal of the removal procedure. After the installation is completed, fill the cooling system.

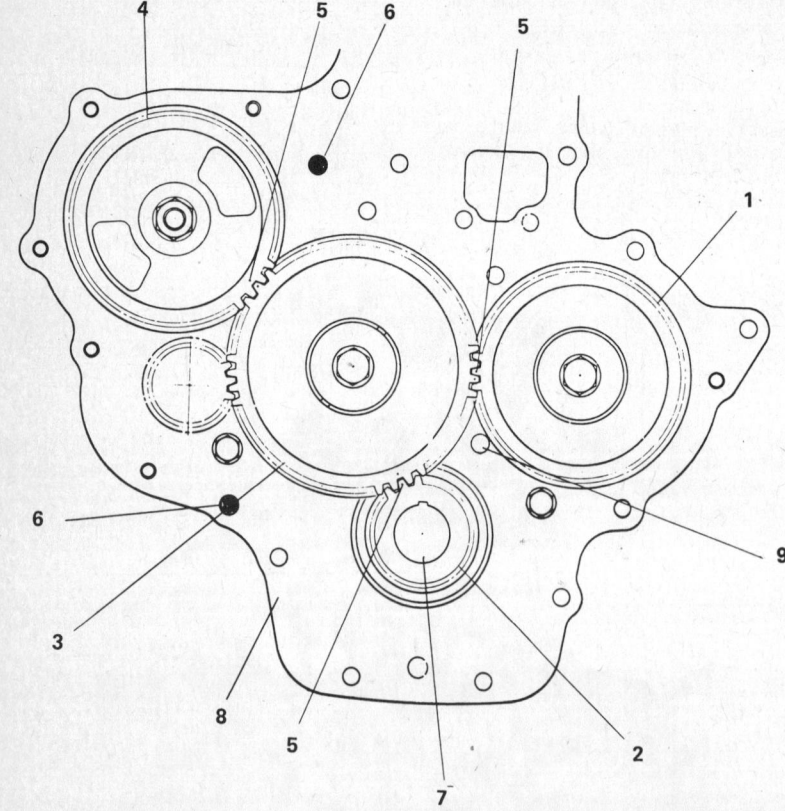

1. Cam gear
2. Pinion gear (crankshaft gear)
3. Idler gear
4. Injection pump gear
5. Timing mark
6. Dowel pins
7. Crankshaft
8. Front plate
9. Oil injector

Engine gear train

Allis-Chalmers

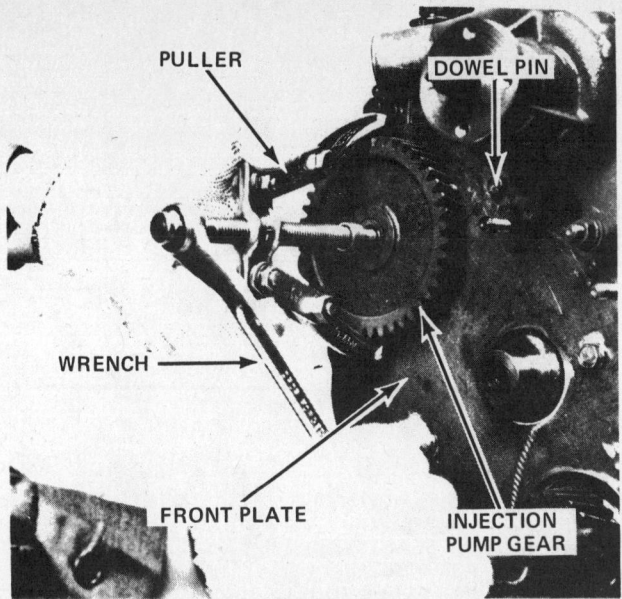

Injection pump drive gear removal

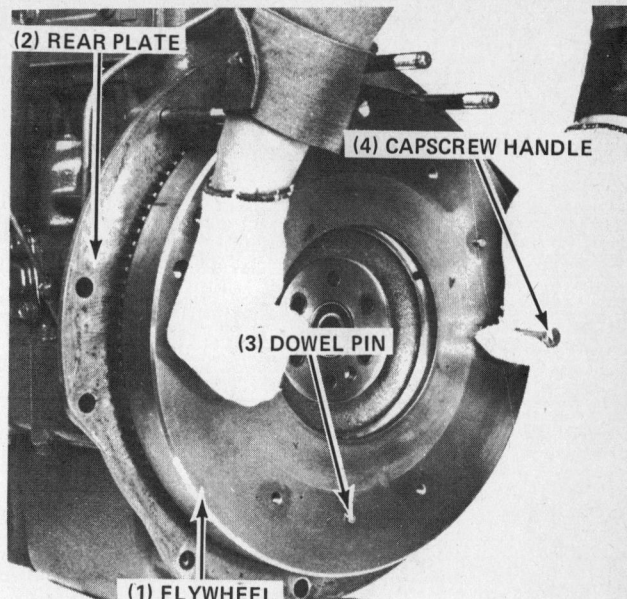

Flywheel removal

Flywheel

REMOVAL AND INSTALLATION

1. Remove engine clutch assembly from flywheel.
2. Remove flywheel lock bolts attaching flywheel to crankshaft flange.
3. Install two long capscrews to serve as handles in opposite holes of the flywheel face. Pull on handles and remove flywheel.

NOTE: If flywheel cannot be removed with a direct pull on the handles, it may be necessary to tap flywheel loose with a suitable brass bar. Turn flywheel and tap exposed part at intervals until flywheel can be removed.

4. Inspect the engine clutch shaft pilot bearing located in bore of flywheel. The bearing is lubricated when assembled.

When installing a new bearing, use a driver or tube that will provide for driving against the bearing outer race; then drive the bearing into position in the flywheel.

INSPECTION

1. Check to see if flywheel surface is scored or heat-checked.
2. Flywheel should be machined smooth or replaced if either condition exists. It is very important that all burrs and nicks be removed from the flywheel surface that fits against the crankshaft flange. If this surface is not smooth and true, the flywheel may have a slight wobble which will result in improper clutch operation and engine vibration.
3. If flywheel surface is scored or heat-checked it may be machined smooth; replace the flywheel if more than .157 in. (4 mm) of stock must be removed.

Rear Plate

REMOVAL AND INSTALLATION

1. Remove engine clutch from flywheel.
2. Remove flywheel from the engine crankshaft flange.
3. Remove lock bolts securing rear plate to the cylinder block.
4. Inspect rear plate for cracks and other damage; repair or replace if damaged.

Crankshaft Rear Oil Seal

REMOVAL

The crankshaft rear oil seal which is a press fit in the oil seal cover, is a spring-loaded, single-lip type seal.

1. Loosen the three rear oil seal cover nuts and three capscrews. Tap the circumference of the oil seal cover with a plastic hammer and remove the oil seal cover by pulling horizontally.
2. Remove oil seal cover gasket from block.

INSTALLATION

1. Clean oil seal bore in the oil seal cover.
2. Remove and discard crankshaft rear oil seal and install a new seal.
3. To press the oil seal on the rear oil cover, use the furnished tool taking care that it is not inclined.
4. Place the oil seal installing tool on the crankshaft rear end and install the oil seal cover gasket to the oil seal cover assembly.
5. To secure the oil seal cover, tighten nut and capscrews.

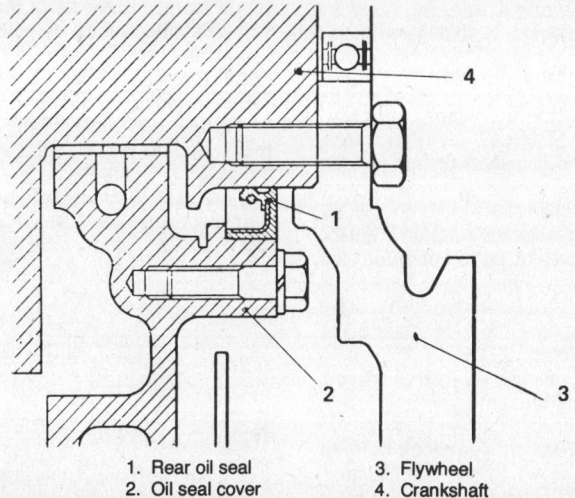

1. Rear oil seal
2. Oil seal cover
3. Flywheel
4. Crankshaft

Rear oil seal components

Allis-Chalmers

Installing oil seal in cover

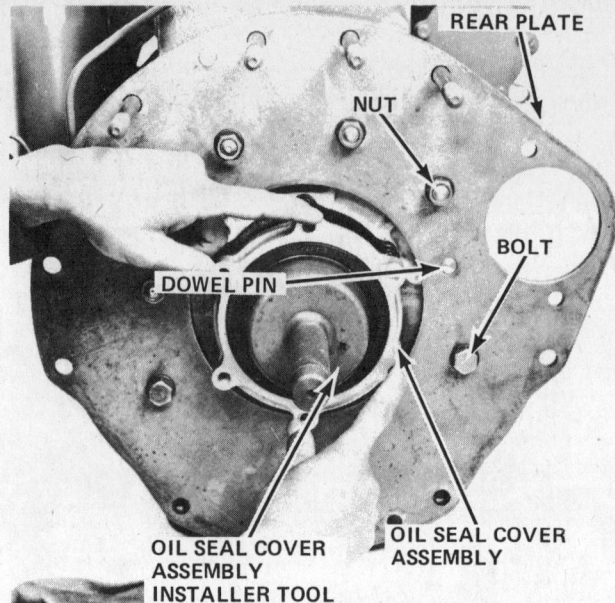

Installing oil seal assembly in block

NOTE: The oil seal cover has a socket only in the cylinder block (not in the crankcase); therefore, to tighten the capscrew and nut, press the oil seal cover to the block side.

Camshaft and Camshaft Gear

The removal of the camshaft with the camshaft gear is essentially the same regardless if the engine is mounted in a tractor, or has been removed for overhaul. If the engine is in a tractor it is necessary to remove the oil pan, oil pump and timing gear cover. To remove camshaft, proceed as follows:

GEAR REMOVAL

1. Remove the crankshaft pulley and timing gear cover.
2. Check the camshaft gear backlash before removing the camshaft from the cylinder block. The specified backlash between the mating gears of the idler and the camshaft is .0020-.0098 in. (0.05-0.2 mm). New parts must be installed when the backlash between any two mating gears exceeds .020 in. (0.5 mm).
3. Loosen the lock bolt on the camshaft, remove the thrust washer and remove the camshaft gear by pulling horizontally.

CAMSHAFT REMOVAL

1. Remove the camshaft pulley and timing gear cover.
2. Remove idler gear and camshaft gear.
3. Remove front plate.
4. Remove flywheel and rear plate.
5. Remove the cylinder head cover, rocker arms, shaft and brackets assembly and push rods.
6. Remove the oil pipe lubricating the valve rocker and the injection extending from the injection pump to remove the cylinder head assembly.
7. Tilt the engine for convenience of operation.

NOTE: The camshaft thrust plate cannot be removed without removing the camshaft gear.

8. Remove the crankcase and oil pump.
9. Loosen the two capscrews (fastening the thrust plate).
10. Remove the camshaft taking care that the camshaft bearing (pressed fit in to the cylinder block) is not damaged. Observe the following clearances:
Camshaft, camshaft gear and thrust plate inspection
Camshaft bearing journal a normal diameter.
1.9665 - 1.9675 in. (49.950 - 49.975 mm)
Camshaft bearing oil clearance
.0008 - .0040 in. (0.021 - 0.102 mm)

INSPECTION AND INSTALLATION

For a camshaft with a bearing oil clearance of .0047 in. (0.12 mm) or more, either replace the camshaft or use bearing of smaller size.

1. Inspect the camshaft lobes for roughness, scoring or excessive wear. Replace the camshaft if any of these conditions are found. These conditions can be caused by inferior quality of lubricating oil, excessive valve lash adjustment, contaminated oil or the tappets not rotating.
2. Replace camshaft gear if teeth are nicked, scored or broken.

NOTE: Tighten cam gear lock bolts to a torque of 36 to 40 ft.lbs. Check camshaft end play. The specified end play is .0020 to .0079 in. (0.05 - 0.2 mm).

Valve Lifters

REMOVAL

1. Remove the engine from the tractor.
2. Disassemble the front of the engine and remove the timing gear cover.
3. Remove flywheel and rear plate.
4. Remove crankcase and oil pump.
5. Remove the rocker arms shaft, bracket assembly, and push rods.
6. Remove the camshaft.

NOTE: Before camshaft removal, lay the cylinder block on its side so the valve lifters will not fall out of the block.

7. Remove the valve lifters and identify their location in the cylinder block so they can be reinstalled in their original location.

INSPECTION

Very little wear takes place on the valve lifter stems or bores in the block. The specific O.D. of the valve lifter stem is .5488 to .5496 in. (13.94 to 13.96 mm) and the I.D. of the bore in the cylinder block for the valve lifters is .5512 in. (14.00 mm). The fit of the valve lifter in the bore of the cylinder block is .0016 in. to .0031 in. (0.04 to 0.078 mm). If the clearance between the valve lifter stem and the bore in the cylinder block exceeds .0047 in. (0.12 mm) replace the valve lifter.

Inspect the valve lifter face which contacts the camshaft lobe for roughness, scuffing or excessive wear. These conditions can be caused by inferior quality of lubrication oil or contaminated oil which will make the valve lifters stick in the bores and not allow them to rotate. Replace any worn valve lifters.

INSTALLATION

1. Lubricate the valve lifters with clean engine oil and install them in their original location in the cylinder block.

Allis-Chalmers

2. Install the camshaft and the rest of the components by a direct reversal of the removal procedure.

Piston, Connecting Rod and Connecting Rod Bearing

REMOVAL

1. When overhauling the engine, removing the piston, or connecting rod, remove all components located on the front and rear sides of the cylinder block, and then remove the rocker arm assembly and cylinder head assembly.

2. Remove lock bolts securing connecting rod bearing caps. Remove bearing caps and free lower end of connecting rods from crankshaft. Remove bearing shells from bearing caps and connecting rods.

3. Carefully remove each piston and connecting rod assembly, pushing assembly out through top of cylinder.

REMOVAL OF CONNECTING ROD AND PISTON RINGS FROM PISTON

1. Remove piston rings from the piston using a ring remover and installer tool similar to the one shown.

2. Remove piston pin retainers from grooves in the piston at each end of the piston pin.

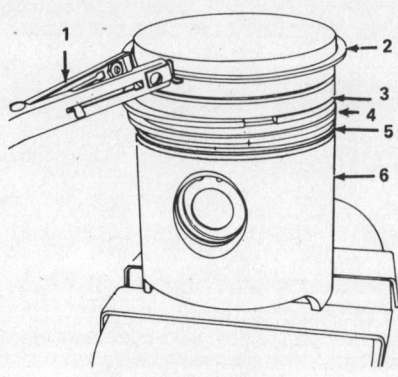

1. Piston ring remover and installer ACTP 2018
2. Top compression ring (chrome plated)
3. 2nd compression ring (chrome plated)
4. 3rd compression ring
5. Oil control ring
6. Piston

Removing piston rings

NOTE: When removing the piston ring and piston pin retainer, take care not to grip the circumference of the piston with a vise, etc. To prevent the piston circumference from becoming damaged, hold the connecting rod stem during piston ring removal.

3. Heat the piston to 158-176° F (70-80° C) and push out the piston pin with thumb.

PISTON AND PISTON RINGS INSPECTION

Piston

1. Clean the piston with light oil and remove the carbon present on the piston top surface and in the ring groove.

2. Be careful to remove only carbon or foreign material; do not scrape away any metal from the sides or bottom of the ring grooves.

NOTE: Some types of solvent contain chemicals harmful to aluminum alloy. Do not use this type of cleaning agent.

3. The piston skirt should be carefully examined for score marks or other indications of improper piston clearance. Inspect the inside of pistons for cracks; scored or cracked pistons should be replaced. Check pistons for wear.

4. The skirt diameter of a new piston (measured at right angles to piston pin bottom edge of the skirt):
5020: 3.6173-3.6185 in. (91.88-91.91 mm)
5030: 3.8169-3.8130 in. (96.95-96.85 mm)

5. The inside diameter of a new cylinder bore:
5020: 3.6220-3.6229 in (92.-92.022 mm)
5030: 3.8189-3.8198 in. (97.-97.022 mm)

6. Piston running clearance:
5020: .0035-.0056 in. (0.09-0.142 mm)
5030: .0028-.0059 in. (0.072-0.150 mm)

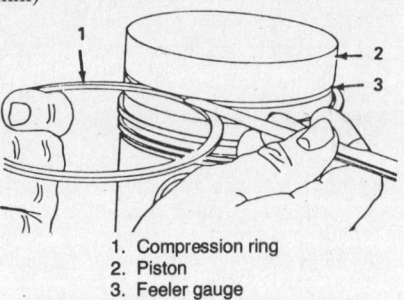

1. Compression ring
2. Piston
3. Feeler gauge

Measuring ring-to-groove clearance

7. The typical clearance between the piston and piston pin is such that the piston pin can be inserted with thumb when the piston is heated to 104-122° F (40-50° C).

8. Checking piston ring-to-groove clearance.

9. Top compression ring: .0016-.0031 in. (0.04-0.08 mm)
2nd and 3rd rings: .0012-.0028 in. (0.03-0.07 mm)
Oil control ring: .0012-.0035 in. (0.03-0.07 mm)
(new piston and new rings)

Piston Ring

The gap between ends of piston rings should be measured before rings are installed on pistons. Insufficient end gap will cause scored rings, scored cylinder walls and other damage. Check the ring gap by inserting each ring into the cylinder in which it is to be used.

Use a piston to push the ring squarely down the bore of the cylinder sleeve, far enough to be on the ring travel area. Check ring gap with a feeler gauge as shown. The specified ring end gaps, using standard cylinder bore of 3.6220-3.6229 in. (92.-92.022 mm) (5030: cylinder bore of 3.8189-3.8198 in. (97.-97.022 mm).
1st ring: .0079-.0157 in. (.2-.4 mm)
2nd ring: .0079-.0157 in. (.2-.4 mm)
3rd ring: .0079-.0157 in. (.2-.4 mm)
4th ring: .0079-.0157 in. (.2-.4 mm)

NOTE: The top compression and oil control ring should never be filed to open the gap because the chrome plating might be loosened by the filing and later distributed through the engine, causing damage or scoring of the piston and the cylinder walls.

2. Measure ring-to-groove clearance (top of ring to top of groove in piston) as shown. The specified ring to groove clearances, using a new piston and new rings, are as follows:

ASSEMBLY OF CONNECTING ROD TO PISTON

1. Before assembling connecting rod to piston, inspect the connecting rod.

2. Install one of the piston pin retainers in one end of the piston pin hole in the piston.

Top compression ring .0016-.0032 in. (.04-.08 mm)
2nd compression ring .0012-.0028 in. (.03-.07 mm)
3rd compression ring .0012-.0028 in. (.03-.07 mm)
Oil control ring .0012-.0028 in. (.03-.07 mm)
Coil expander

Ring-to-groove clearances

3. Insert upper end of connecting rod into the piston. Heat the piston in oil 158-176° F (70-80° C).

4. Insert the piston pin into the pin holes in the piston and connecting rod.

5. Install the other piston pin retainer at the opposite end of the piston pin bore.

NOTE: To install the connecting rod and piston, follow the arrangement shown in the accompanying illustration.

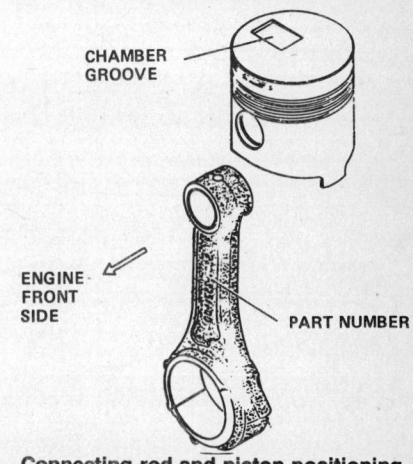

Connecting rod and piston positioning

Allis-Chalmers

PISTON
RINGS INSTALLATION

1. Install piston rings using a ring remover and installer tool. The rings must be installed in their respective grooves with the proper side to the top and bottom of the piston.

NOTE: Whenever a connecting rod and piston assembly is secured in a vise, use load jaw protectors to prevent damage to the piston skirt and connecting rod.

2. Install coil expander in bottom groove of piston. Then install oil control ring. Assemble tefron tube of coil expander so that it reaches control end.
3. Install third compression ring with point or dot mark toward top of piston.
4. Install second compression ring with point or dot mark toward top of piston.
5. Install chrome barrel-faced top compression ring with either side to top.
6. Recheck the rings to make certain they turn freely in their grooves.
7. Prior to installing the piston and connecting rod assembly coat lightly with engine oil and position ring gaps 180° apart and in line with the piston pin holes.

Crankshaft
REMOVAL

1. Drain the cooking system and engine lubrication oil. Disconnect wiring and remove engine assembly from tractor.
2. Remove starter, flywheel, and rear plate.
3. Remove oil pan, oil pump and oil pump discharge tube.
4. Remove components on front end of engine necessary to facilitate removal of timing gear cover. Remove timing gear cover.
5. Remove connecting rod bearing caps and connecting rod bearing shells.
6. Remove main bearing caps and lower main bearing shells.
7. Remove the crankshaft.
8. Remove upper main bearing shells from the cylinder block.

1. Crankshaft gear
2. Woodruff key
3. Crankshaft
4. Thrust bearing
5. Main bearing upper
6. Main bearing lower

Crankshaft and bearing components

CRANKSHAFT INSPECTION

1. Clean crankshaft thoroughly and inspect the journals for scoring, chipping, cracking or signs of over-heating. If crankshaft has been over-heated (usually indicated by discolored or blue bearing journal surfaces), is scored or excessively worn, reconditioning or replacement will be required.
2. Crankshaft magnetic particle inspection must be made before attempting to regrind the crankshaft to determine if it is in satisfactory condition for grinding. After regrinding, the inspection must be made again to make certain that no sub-surface casting flaws have been uncovered in the grinding operation.
3. If out-of-round or taper of journals exceeds .0012 in. (.0.3 mm) crankshaft must be reground to a standard undersize or replaced.

NOTE: Chrome plating or metallizing the bearing journals is not recommended.

4. Measure the crankshaft main bearings and connecting rod journals at several places on their diameter to check for roundness. The specified diameter of main bearing journals is 2.7543 to 2.7551 in. (69.96 to 69.98 mm), connecting rod journals is 2.2819 to 2.2827 in. (57.96 to 57.98 mm).

All main and connecting rod bearing journal surfaces of crankshaft are hardened to a minimum depth of approximately .0787 in. (2 mm). If regrinding of crankshaft journals becomes necessary, the work should be done by a reputible machine shop possessing suitable equipment to handle precision work of this type. Main bearing and connecting rod shells .0098 in. (0.25 mm) and .0197 in. (0.5 mm) undersize are available. If crankshaft is ground, the diameter of main bearing and connecting rod journals must be reduced in steps of .0098 in. (0.25 mm) or .020 in. (0.05 mm) below the specified journal diameters.

NOTE: When regrinding a crankshaft in the field it is important that the original journal fillet radii .0787-.0984 in. (2.00-2.5 mm) be maintained and not decreased. Decreasing these fillet radii will weaken the crankshaft to the extent that breakage can be expected.

NOTE: Camshaft grinding diameter when undersize connecting rod bearing and main bearing are used.

Connecting rod journal:
.0098 in. (0.25 undersize 2.7445-2.7453 in.; (69.710-69.730 mm)
.0197 in. (0.50 undersize 2.7346-2.7354 in.; (69.460 - 69.480 mm)

Main bearing journal
.0098 in. (0.25) undersize 2.2720-2.2728 in.; (57.510-57.730 mm)
.0197 in. (0.50) undersize 2.2622-2.2630 in.; (57.460-57.480 mm)

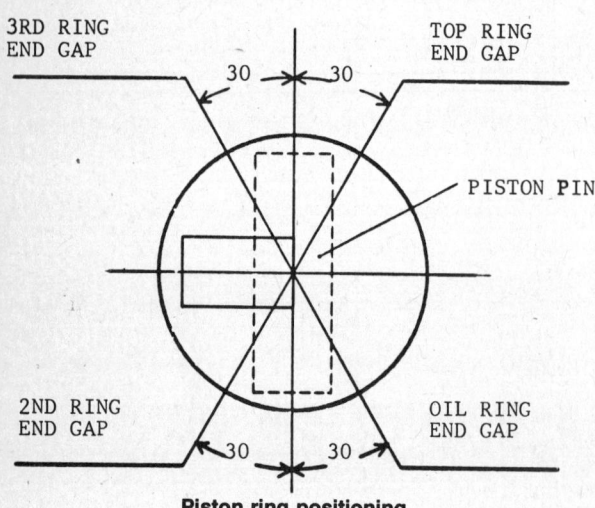

Piston ring positioning

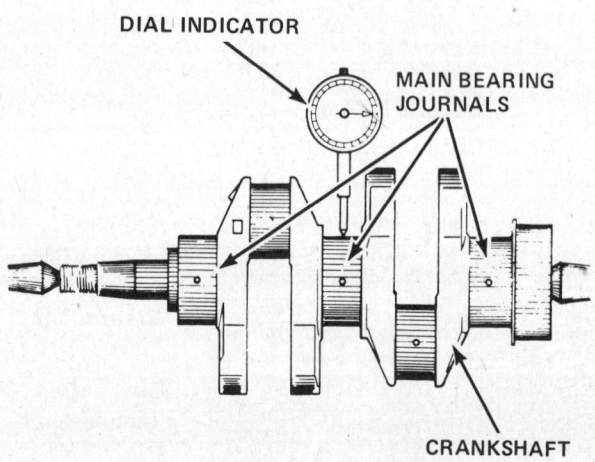

Measuring crankshaft journals

Main Bearing Removal

1. Remove the crankshaft.
2. Remove the main bearing upper shells from their seats in the cylinder block.

NOTE: Identify bearing shells as to their original location in the cylinder block and main bearing caps in the event inspection proves they can be reused.

3. Remove the main bearing lower shells from their seats in the main bearing caps.

Main Bearing Inspection

1. Any bearing shells that are scored, chipped, pitted or worn beyond the specified limits given below must be replaced. Inspect backs of the shells for bright spots. Bright spots on backs of the shells indicate shells have shifted in their supports; and are unfit for further use.
2. Main bearing: crankshaft main journal oil clearance .0008-.0033 in. (0.02-0.085 mm). New bearing shells must be installed when clearance exceeds .0059 in. (0.15 mm).

Main Bearing Installation

1. Install main bearing shell in each of the bearing seats in the cylinder block.

NOTE: Make certain the backs of the bearing shell are free from dirt and grit particles.

2. Lubricate all crankshaft main bearing journals, install crankshaft and main bearing caps, check crankshaft end play and reassemble engine.

PISTON, CONNECTING ROD AND CONNECTING ROD BEARING INSTALLATION

1. Before installation, inspect and clean the cylinder block bore, and crankshaft connecting rod bearing journal.
2. Clean the connecting rod big end bore.
3. Apply a slight amount of clean engine oil to the cylinder block bore, piston circumference and piston ring. Install the piston with a piston ring compressor and wooden hammer.

NOTE: Align all rings so the gaps are 180° apart. To install, position the connecting rod part number to point toward the engine front.

4. Lubricate and install a bearing shell in the connecting rod, with tang of bearing shell in the corresponding slot in connecting rod and position rod on crankshaft journal.

NOTE: Make certain the backs of the bearing shells are free from dirt and grit particles.

5. Lubricate and install bearing shell in the connecting rod bearing cap, with tang of bearing shell in corresponding slot in bearing cap. Install bearing cap and shell, making certain identification number stamped in the bearing cap is located on the same side as corresponding number stamped in the connecting rod.
6. The connecting rod lock bolt has a seat and requires no spring washer. Torque 36-40 ft.lbs.
7. Check to see that there is sufficient side clearance between connecting rods and crankshaft journals. Side clearance .0039-.0118 in. (0.1-0.3 mm).

MAIN BEARING, BEARING CAP AND CRANKSHAFT INSTALLATION

Installation at the Time of Overhaul

1. Before installation, inspect and clean the cylinder block main bearing bore and crankshaft connecting rod bearing journal and main journal.
2. Put the bearing shell tang into the cylinder block slot and apply a coat of clean oil.
3. Apply a slight amount of clean oil to the crankshaft main journal.
4. Install the crankshaft.
5. Install the main bearing shell to the bearing cap. Apply a slight amount of clean oil. When installing, do not confuse the numbers of the cylinder block and bearing cap. The center bearing cap contains four pressed-fit dowel pins, which includes a thrust bearing on each side.
6. The bearing cap lock bolt has a seat and requires no spring washer. Torque 43-58 ft.lbs.
7. Check crankshaft end play. End play .0039-.0118 in. (0.1-0.3 mm).
8. Check the crankshaft for smooth rotation.

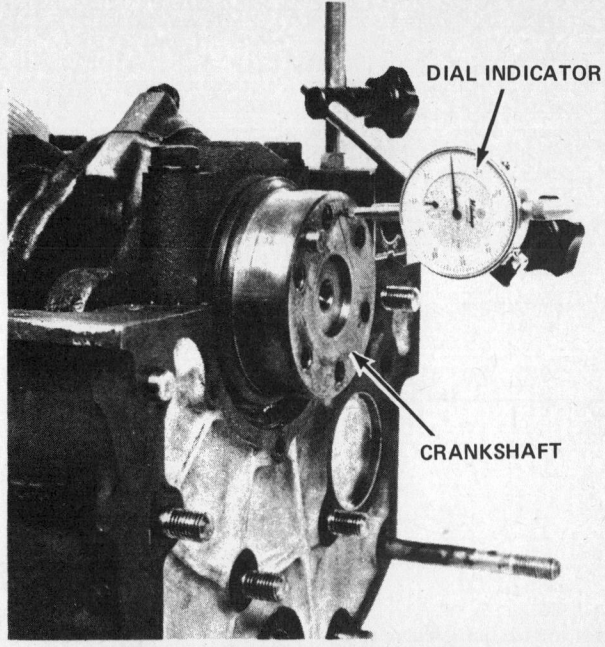

Measuring crankshaft end play

CRANKCASE

General

The crankcase, the frame of tractor, is a one piece casting made of an alloy cast iron. The crankcase is provided with the timing cover and front axle frame on the front end, and the rear plate and rear oil seal cover on the rear end. Oil gauge adaptor and stud are included in the crankcase assembly.

REMOVAL

The crankcase cannot be removed without removing the front plate and timing gear case (connected to the front end) and the rear oil seal cover, rear plate and flywheel (connected to the rear end).

To remove the crankcase from the cylinder block, loosen the capscrew and tap the four corners with a wooden hammer to release.

INSPECTION

Inspect crankcase for evidence of cracks or other dmage. Inspect drain plug boss for evidence of leakage. Make necessary repairs or replace crankcase if necessary.

INSTALLATION

1. Before installation, clean the crankcase (top, front and bottom faces) with solvent. Clean the dowel pin hole connected to the cylinder block.
2. Uniformly apply RTV Sealant to the crankcase (cylinder block installing surface). When installing the crankcase on the cylinder block, take care not to cause damage to the dowel pin installed in the block. (Put a drain plug in the crankcase.)

Allis-Chalmers

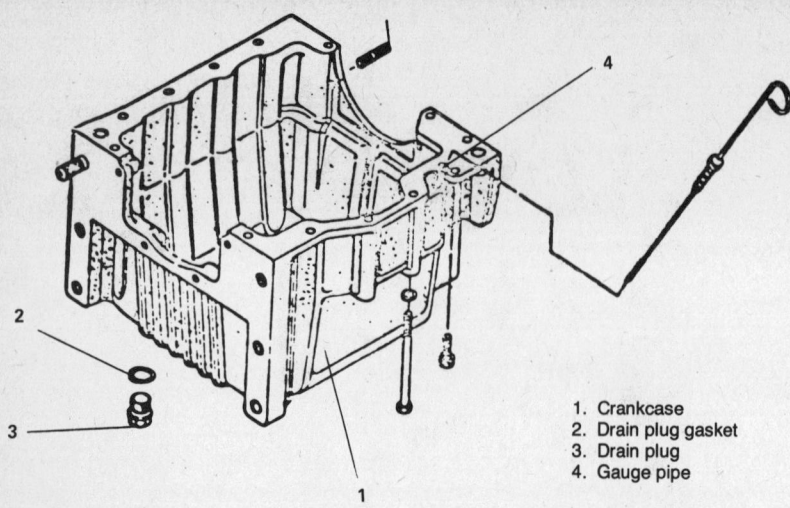

1. Crankcase
2. Drain plug gasket
3. Drain plug
4. Gauge pipe

Crankcase removal

3. Tighten with thirteen capscrews and lockwashers. Torque 22-25 ft.lbs.

4. Assemble rear plate, rear oil seal cover, flywheel, front plate, timing gear case and water pump.

NOTE: Check if the difference between the crankcase rear/front face and cylinder block is within .0039 in. (0.1 mm).

5. Run engine and check for oil leaks.

Oil Pump

REMOVAL AND DISASSEMBLY

1. Drain oil and remove the crankcase. Remove the oil screen assembly by removing two screws.

2. Remove the two pump attaching screws, and remove the oil pump assembly from engine by pulling out.

3. Disassemble engine oil pump.

4. Remove the pin, and pull the driven gear from the rotor shaft.

5. Remove the two capscrews retaining the cover to the pump housing, and remove the cover.

6. Remove the inner and outer rotor.

ASSEMBLY

1. Install the outer rotor into the pump-housing. Place a small quantity of sealing compound on the flange of the housing, and install the pump cover.

2. Torque the 6 mm capscrews alternately and evenly to 3.3-4.3 ft.lbs. The rotor must turn freely after pump is assembled, and the air never comes in through the retaining part of pump cover.

INSTALLATION

1. Install the engine oil pump in the cylinder block, using a new gasket.

2. Torque the retaining capscrews (8 mm) to 9-11 ft.lbs. Install the oil screen assembly using the new gasket. Torque the capscrews (8 mm) to 9-11 ft.lbs.

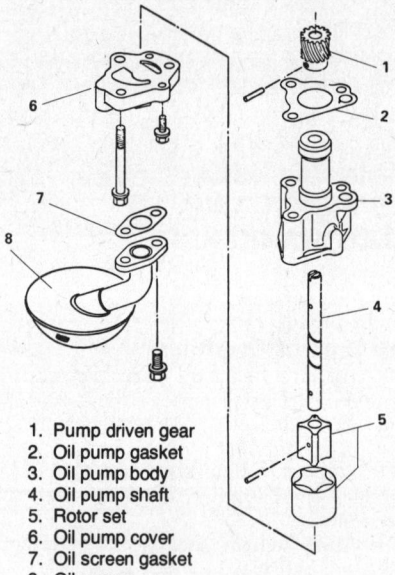

1. Pump driven gear
2. Oil pump gasket
3. Oil pump body
4. Oil pump shaft
5. Rotor set
6. Oil pump cover
7. Oil screen gasket
8. Oil screen

Oil pump components

ENGINE REMOVAL AND INSTALLATION

1. Remove the hood, then disconnect the battery cables and remove battery.

2. Drain all fluids and close the fuel valve. Drain the crankcase oil through the bottom plug and radiator water through the radiator drain cock and the block drain cock.

3. Disconnect the hydraulic oil lines attached to the hydraulic pump. Be sure to plug holes to prevent oil spillage.

4. Disconnect the water hoses from the engine. Remove the radiator support brackets and the two capscrews holding the radiator to the front frame. The radiator can now be lifted from the frame.

5. Disconnect the wires to the alternator and the starter motor. The two bolts mounting the starter motor to the back plate can now be removed and the starter should fall off.

6. Remove the water temperature sensor located in the thermostat, the tach drive located behind the oil filter, the wire connected to the oil pressure sensor and the fuel shut-off and throttle linkage.

7. Be sure the fuel valve cock is shut-off. Disconnect the fuel line at the injection pump.

8. Remove the cotter pin in the end of the steering shaft. After the cotter pin is removed, remove the castle nut and tap the steering arm with a hammer until it comes off the shaft.

9. Position a hoist over the engine, place a chain in the hoist's hook and in the engine lift supports. Draw the chain tight.

10. Remove the six (6) nuts holding the front frame to the engine. Place blocks under the transmission housing. The front frame assembly can now be rolled away from the engine.

11. Remove the nuts and capscrews bolting the torque housing to the engine. The engine can now be separated from the torque tube. Be careful of the torque tube and do not pry too hard or hammer too hard upon it as it is made of aluminum.

12. Installation is a reversal of the removal process. The torque for the bolts mounting the engine to front frame is 61-76 ft.lbs., and for the torque housing is 46-54 ft.lbs. The two bolts for mounting the starter motor should be torqued to 29-33 ft.lbs.

Bolens
INDEX

MEDIUM TUBE FRAME MODELS
Front Wheels and Axle 172
Steering 172
   Steering Knuckle R&R 172
   Steering Gear R&R 172
Brakes 173
Clutch 174
Transaxle 175
   R&R 175
   6-Speed Overhaul 175
   3-Speed Overhaul 175
   Hydrostatic Overhaul 178
Engine 180

XL SERIES MODELS
Front Axle 181
Steering 181
   Steering Gear R&R 181
   Steering Knuckle R&R 181
Brakes 182
Clutch 182
Manual Transaxle 182
Hydrostatic Transaxle 182
Power Take-Off 183
Engine 183

QS (1660) AND QT (1666) MODELS
Front Axle 185
   R&R 185
   Wheels and Bearings 185
Steering 185
Brakes 186
Transaxle 187
   R&R 187
   Overhaul 189
Power Take-Off 193
Engine 195

LARGE FRAME MODELS
Front Axle 196
Steering 197
Brakes 197
Transmission 198
Power Take-Off 198
Engine 198

BOLENS
Medium Tube Frame Models

600 (180/1), 650 (184/5),
750 (171/2), 853, 855 (G-9),
1055 (G-10), 1155 (G-11),
1220 (193), 1253, 1255 (G-12),
1257, 1435 (G-14), 1556, 1656
(H-16), 800(186/7), 850 (190)
900 (188/9), 1000 (190),
1050 (192), 1225 (194), 1053

FRONT WHEELS AND AXLE

REMOVAL AND INSTALLATION

1. Unbolt and remove the PTO support from the front axle support.
2. Disconnect the drag link ball joint end from the steering knuckle arm.
3. Jack up and support the front end on jack stands.
4. Remove the roll pin and drive the pivot pin forward out of the axle support.
5. Roll the front axle out from under the tractor.
6. Installation is the reverse of removal.

STEERING

Steering Knuckles

REMOVAL AND INSTALLATION

1. Raise and support the front end with jack stands under the axle main member.
2. Remove the wheels.
3. Disconnect the drag link at the knuckle.
4. Drive out the pins securing the tie rod end to the knuckles.
5. Using a soft mallet, drive the knuckles down and out of the steering arms and axle.
6. Installation is the reverse of removal.

Steering Gear

REMOVAL AND INSTALLATION

1. Disconnect the rear drag link ball joint end from the quadrant arm.
2. Remove the foot rest from the left side and remove the pins from the cross shaft.
3. Using a long, thin punch, drive the cross shaft toward the left side and remove the steering arm and quadrant gear.
4. Drive out the pinion gear roll pin and pull the steering shaft from the support.
5. Installation is the reverse of removal.

Make sure the front wheels are straight ahead when centering the quadrant gear on the pinion. The drag link can be lengthened or shortened as required to adjust the turning angle.

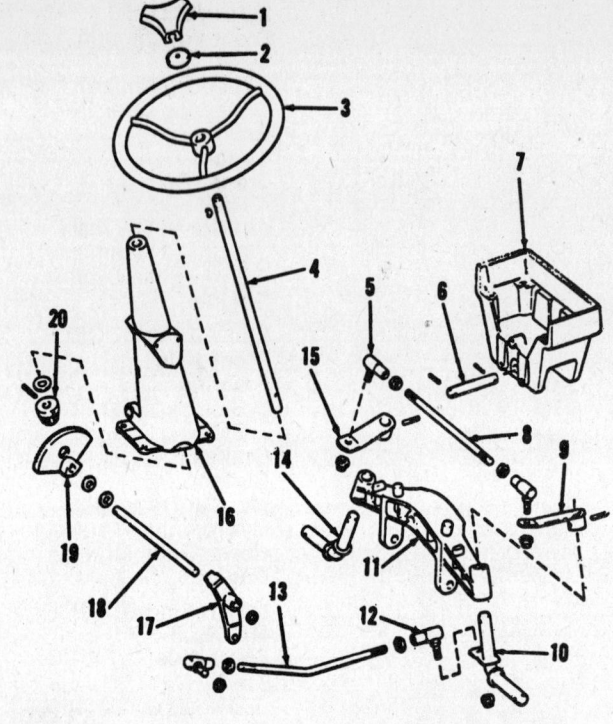

1. Cover
2. Washer
3. Steering wheel
4. Steering shaft
5. Tie rod end
6. Pivot pin
7. Axle support
8. Tie rod
9. Steering arm righthand
10. Steering knuckle righthand
11. Axle main member
12. Drag link end
13. Drag link
14. Steering knuckle lefthand
15. Steering arm lefthand
16. Steering support
17. Quadrant arm
18. Cross shaft
19. Quadrant gear
20. Pinion gear

Typical steering system components

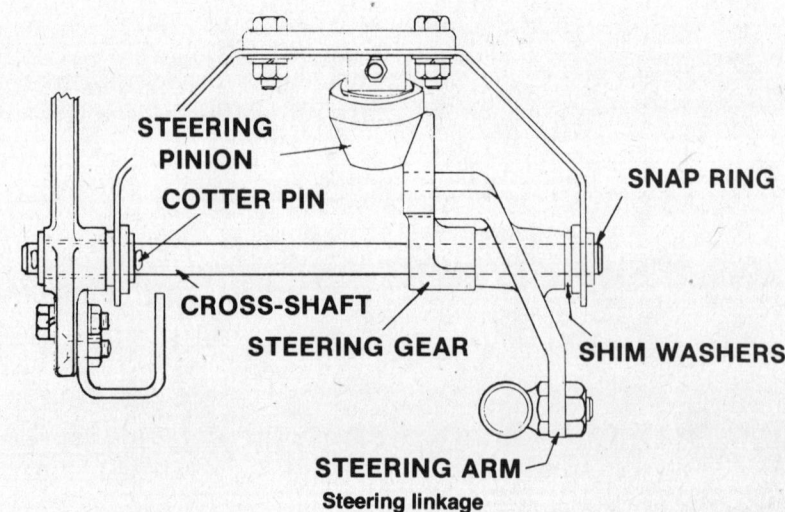

Steering linkage

Bolens

BRAKES

Adjustment

1255 AND ALL MODELS WITH 3-SPEED TRANSMISSION

Brake-to-Transmission Shaft Adjustment

1. Turn the adjusting nuts, located on either side of the brake rod pivot block to obtain the correct adjustment.
2. 600 series only: Adjust the brake by turning the two nuts, both of which are located on top of the pivot block.

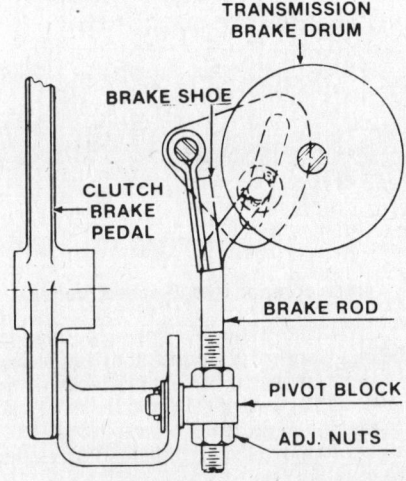

Brake on transmission shaft adjustment

Brake-to-Driveshaft Brake Drum

1. Loosen the adjusting nut on the brake rod.
2. Turn the nuts at both points a little at a time each until proper adjustment balance is obtained.
3. If the brake is set too closely, there won't be enough clutch pedal free travel.

1000 MODELS AND ALL MODELS WITH A 6-SPEED TRANSMISSION

Make sure that the main brake rod is positioned as shown in the activating lever.

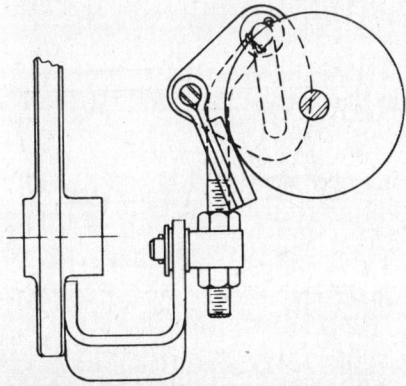

Brake adjustment for 6-speed models

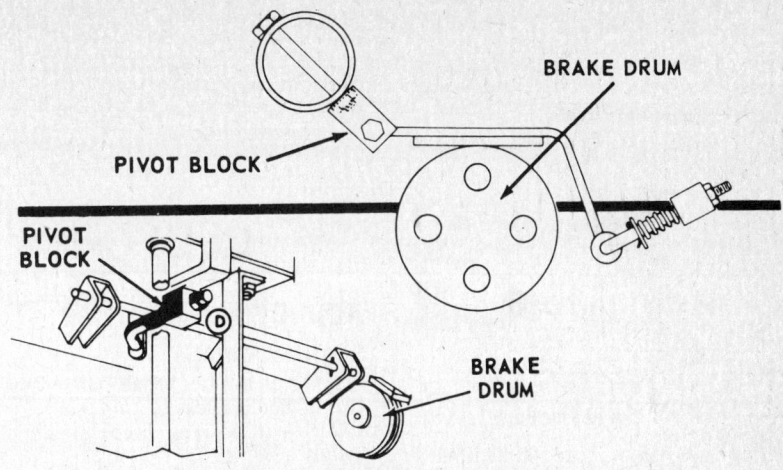

Brake on driveshaft adjustment

1. Depress the clutch/brake pedal about ⅔, or until the rear clutch flange has moved back enough to make contact with the stationary brake plate.
2. Adjust the brake rod up to the top of the activating lever slot.
3. Push down further on the pedal so that the brake shoe contacts the MAIN brake drum on the transmission wormshaft.
4. Measure the distance between the pedal stop and pedal. A gap of about 1 in., with the brake fully applied is required.
5. Adjust the brake rod to obtain the 1 in. distance.

1155 AND 1453 PTO BRAKE ADJUSTMENT

1. Place the PTO control handle in the disengaged position.
2. Adjust the lower nut until the bowed washer "B" is fully compressed.
3. Adjust the upper nut until there is a 1/16 in. gap as illustrated.
4. The sheave must stop within 5 seconds. If not, readjust the lower nut "A", to give more pressure on the sheave.

PARKING BRAKE ADJUSTMENT

With the parking lock in the PARK position, a small amount of resistance should be felt as the linkage passes over-center. If

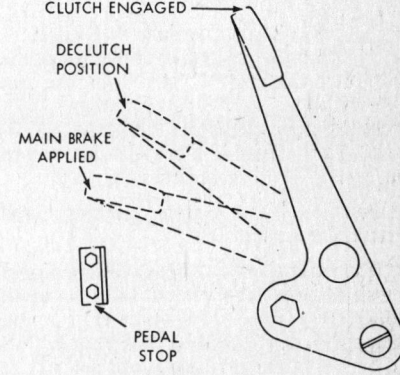

Clutch/brake adjustment

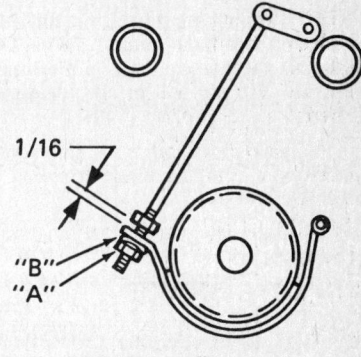

PTO brake adjustment

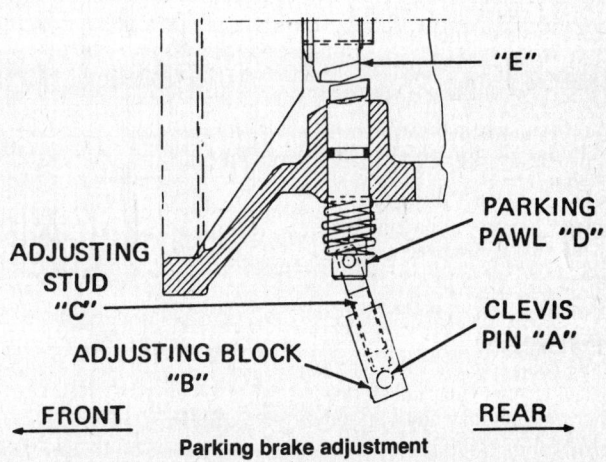

Parking brake adjustment

173

Bolens

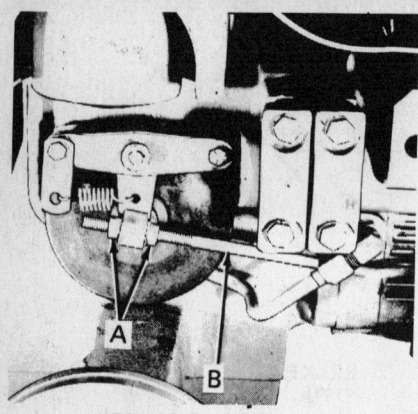

Disc brake adjustment

the differential is not locked in the PARK position, adjust as follows:
1. Raise and support the rear of the tractor on jack stands.
2. Place the parking lock in DRIVE.
3. Clamp a vise grip over the rear of the parking brake arm "F" and swing the arm rearward.
4. Remove clevis pin "A".
5. Turn adjusting block "B" in ½ turn increments to lengthen stud "C."

NOTE: Do not turn parking pawl "D" in transmission.

The parking pawl is beveled for a proper mesh with the bevel pinion. Incorrect mesh of the bevels will result in pawl or pinion damage.

6. When slight resistance is felt as the transmission is put into "Park" the pawl is adjusted properly. SEE NOTE.

NOTE: If parking pawl is not adjusted far enough into bevel pinion the parking pawl will ratchet over the pinion teeth when on an incline or if accidently pushed.

7. Replace clevis pin "A" into adjusting block and parking brake arm then remove vise grip.
8. Reinstall left rear wheel, and remove blocking.

1257 DISC BRAKE

To check brake adjustment, push brake pedal down by hand until brake linings contact brake disc. If there is less than 1 in. of travel left on the pedal the brake needs adjustment. To adjust the foot brake proceed as follows:
1. Loosen the two hex nuts "A."
2. Shorten brake rod "B" until there is 1½ in. of free pedal when brake linings touch the brake disc as the pedal is depressed by hand.
3. Tighten hex nuts "A" securely.
4. Check brake for proper braking and operation and readjust if necessary.

Brake Pads 1257

REPLACEMENT

1. Remove the clevis pin, unbolt the brake from the transmission housing and lift the brake assembly from the tractor.

2. Remove the snapring, washer and spring. Separate the remaining parts.
3. The pads are riveted to the carriers. Check the cam and actuator, and replace if worn.
4. Installation is the reverse of removal.

CLUTCH

ADJUSTMENT

3-Speed Models
1. Depress the pedal half way. The clutch should release.
2. If not, position the pedal at approximately a 60° angle.
3. Adjust the nut "C" until there is about 1/16 in. clearance between the clutch flange and the clutch facing with the pedal half way.

6-Speed Models
1. Depress the pedal about ⅔ of the way. At this point the clutch should disengage. If not:
2. Loosen nuts "A" and "C" and adjust the pedal to an angle of about 60° with a free play of ½ in.
3. Tighten the nuts.

REMOVAL AND INSTALLATION

3-Speed Models
Refer to the accompanying illustration.
1. Remove the belt guide (1) and drive belts.
2. Disconnect the clutch rod (2) from the pedal (3).
3. Remove the yoke (4) from the yoke support (5).
4. Remove the driveshaft (6) from the coupling (7).

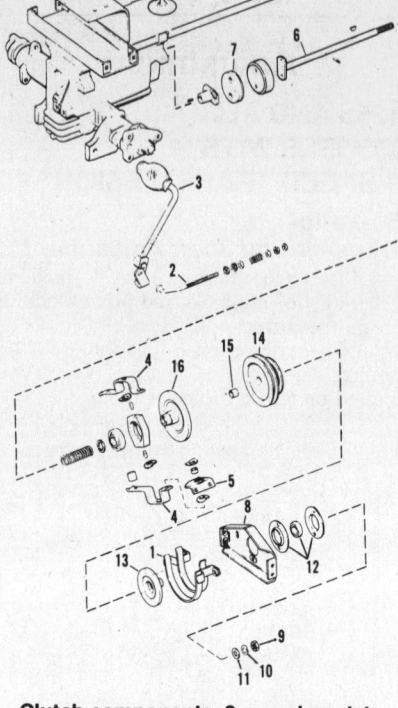

Clutch components, 3-speed models

5. Remove the four bolts securing the bearing support (8) to the mounting plate.
6. Remove the hex nut (9), lockwasher (10) and flat washer (11) from the front end of the driveshaft (6).
7. Remove the driveshaft (6) from the bearing (12).
8. Remove the clutch flange (13), pulley (14), sleeve bearing (15) and clutch flange (16) from the driveshaft.
9. Remove the old friction discs. The new discs must be cemented to the sides of the pulley (14).

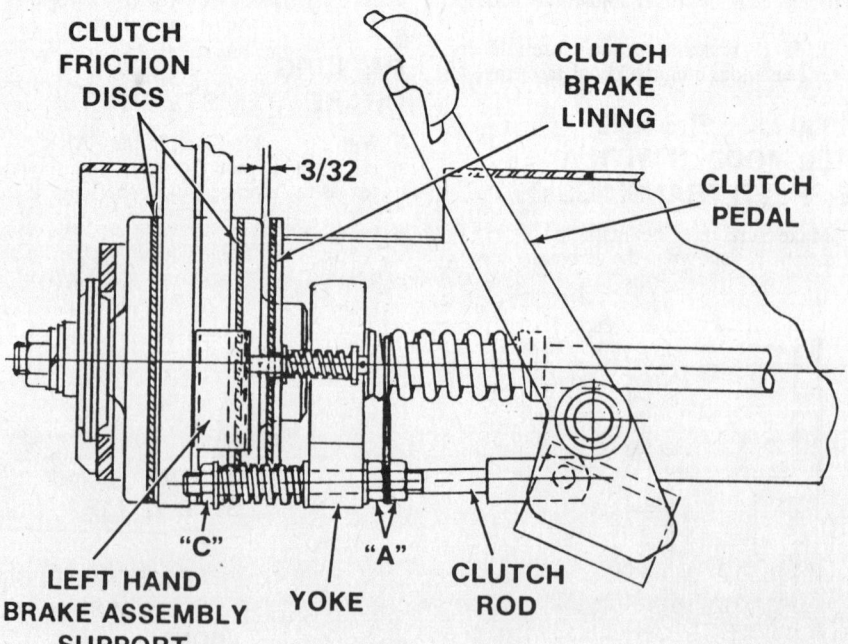

Clutch adjustment

174

Bolens

NOTE: To cement the discs to the pulley, thoroughly clean each side of the pulley. Place the new discs, coated with friction disc cement onto the pulley and clamp them securely. At least 50 psi of force is needed to properly secure the discs. Place the pulley in an oven which has been preheated to 400°F and heat the pulley for at least 15 minutes.

10. Installation is the reverse of removal.

6-Speed Models

Refer to the accompanying illustration.

1. Remove the drive belts (1) from the driveshaft pulley (2).
2. Remove the clevis pin (3) from the clutch rod (4).
3. Remove the clutch rod.
4. Remove the yoke (5) from the support (6).
5. Remove the driveshaft (7) from the bonded coupling (8).
6. Remove the cotter pins (9), flat washers (10), and springs (11) from the brackets (12).
7. Remove the four bolts (13), and bearing support (14).
8. Remove the hex nut (16), lockwasher (17) and flat washer (18) from the end of the driveshaft.
9. Remove the driveshaft from the bearing (19).

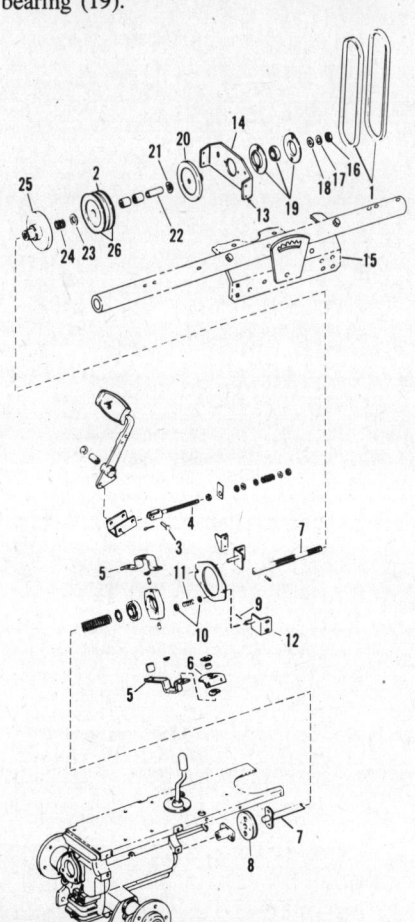

Clutch components, 6-speed models

10. Remove the clutch flange (20), bowed washer (21), pulley (2), spacer (22), thrust race (23), spring (24), and clutch flange (25) from the driveshaft.
11. Remove the old friction discs from the pulley. For friction disc installation, see the 3-speed procedure above.
12. Installation is the reverse of removal.

TRANSAXLE

All Models

REMOVAL AND INSTALLATION

1. Remove the seat and fenders.
2. On models with front disc brakes, remove the brake assemblies.
3. Remove the parking brake lever on models so equipped.
4. Raise and support the rear of the tractor on jack stands.
5. Disconnect the driveshaft.
6. Remove retaining bolts and roll the transaxle from under the tractor.

Overhaul

6-Speed

1. Remove the rear wheels and unbolt and remove the top cover.
2. On G10, 600, 650, 750 and 853 models, remove the cotter pin and castellated nuts from the axle shafts.
3. On 850, 800 and 900 models, remove the snapring, adjusting handwheel (Positraction), cotter pin, castellated nut, four spring washers, Positraction drive cone, cone drive hub and keys.
4. On all models, drive out the 3/16 in. shearproof pin next to the right gear hub.
5. Slide the axle shaft to the right until the side gear drive hub key can be removed.
6. Remove the left wheel hub, shims and thrust washer.
7. Unbolt and remove the transaxle bottom cover and pull the axle shaft from the right side while holding the differential and wormwheel assembly.
8. Remove the differential side gears through the bottom of the case.
9. Drive out the pins holding the shift forks to the rails.
10. Remove the plug from the interlock bore on the left side of the case. Pull the left shifter rail forward out of the case.

NOTE: Be prepared to catch the interlock balls, spring and pin.

11. Remove the right shift rail and lift out the forks.
12. Drive out the pin and remove the brake drum from the wormshaft.
13. Remove the capscrews from the bearing cap and tap the wormshaft out of the case.
14. Remove the bearing retainer cap, shims and rear bearing cup from the rear of the case.
15. The sliding gears can be removed as the wormshaft is removed.
16. Remove the driveshaft coupling flange, key and snapring. Pry out the front seal and remove the bearing retainer ring.
17. Bump the input shaft forward until the ball bearing is free of the case.
18. Remove the snapring from the rear of the input shaft, pull out the input shaft and lift the cluster gear from the case.
19. Drive out the pin securing the reverse idler shaft in the case and remove the shaft, reverse idler gear and thrust washers.
20. Assembly is the reverse of disassembly. When installing the wormshaft, add or delete shims to obtain a .004-.008 in. end play. Tighten the left axle nut just tight enough to avoid binding, then tighten the right axle nut to obtain a minimum end play. Apply a light coat of Prussian blue to the teeth on the wormwheel and check the wear pattern. The pattern should be centered on the wormwheel. If not, add or delete shims at the left hub as required.

3-SPEED

1. Remove the wheels.
2. Unbolt and remove the bottom cover, then the top cover.
3. Remove the snapring, adjusting handwheel, cotter pin, castellated nut, spring washers, Positraction drive cone, cone drive hub and keys, from the left end of the axle shaft.
4. Remove the setscrew from the side gear drive hub. Move the axle shaft to the right until the drive hub key can be removed.
5. Remove the left wheel hub, and, while holding the wormwheel and differential assembly, slide the axle shaft from the right side of the case.
6. Remove the differential assembly, side gears, thrust bearings and shims through the bottom of the case.
7. Drive out the pin securing the shifter forks to the rails. Remove the interlock pin bore plug from the right side of the case.
8. Pull the left shifter rail forward, out of the case and catch the detent balls, springs and pins.
9. Remove the right shifter rail and lift the shifter forks from the wormshaft sliding gears.
10. Drive out the pin and remove the brake drum from the wormshaft. Unbolt the bearing retainer at the rear of the case.
11. Tap the wormshaft forward while removing the bearing retainer cap, shims, rear bearing cup, and sliding gears.
12. Drive out the pin from the range shifter fork and rail and remove the plug, spring and detent ball from the right side of the case.
13. Pull out the range shifter rail and lift out the fork.
14. Remove the expansion plug and bearing retainer ring from the front of the case and the snapring and gear from the splined shaft.
15. Tap the splined shaft forward until the bearing is free from the case. Remove

175

Bolens

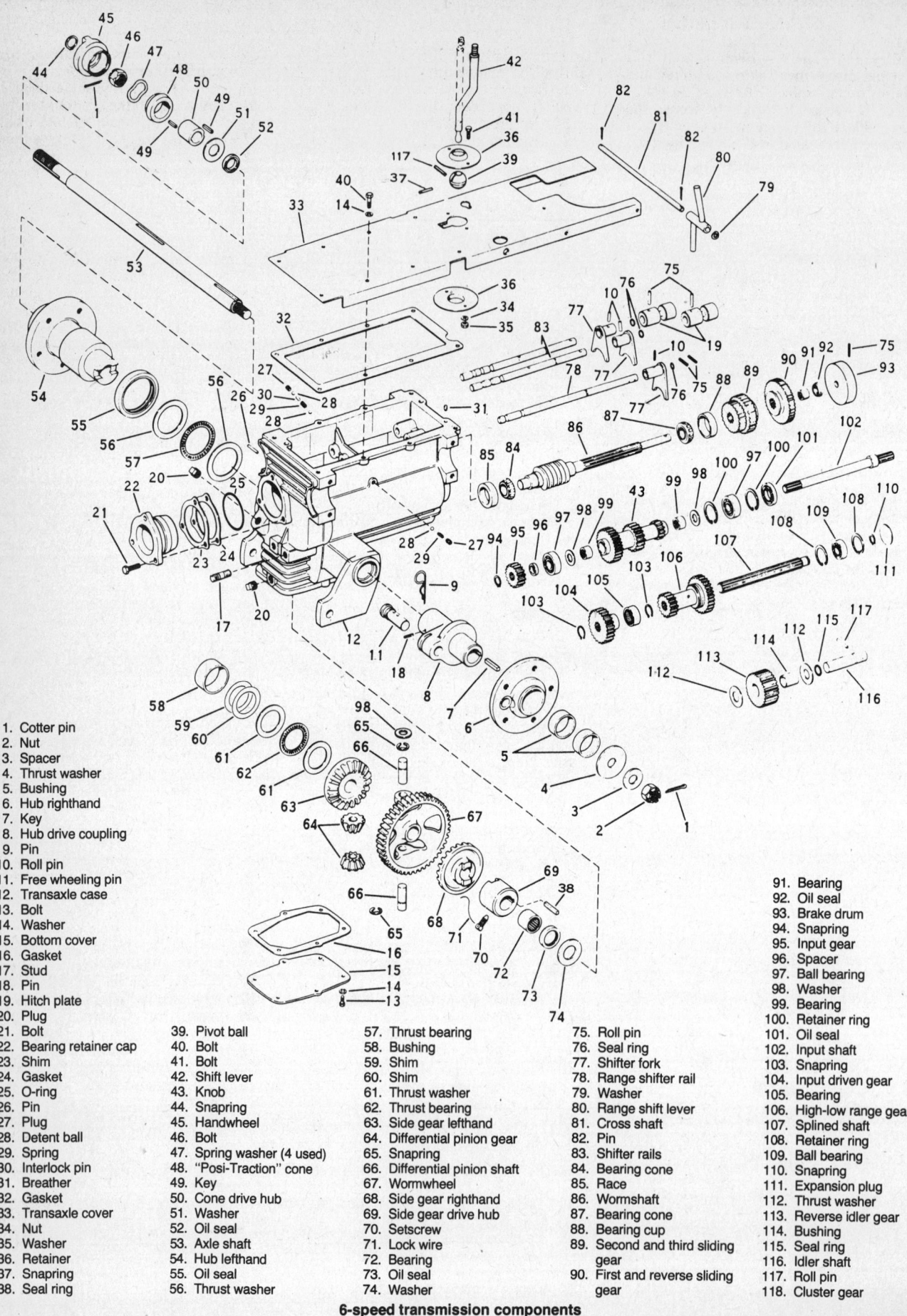

1. Cotter pin
2. Nut
3. Spacer
4. Thrust washer
5. Bushing
6. Hub righthand
7. Key
8. Hub drive coupling
9. Pin
10. Roll pin
11. Free wheeling pin
12. Transaxle case
13. Bolt
14. Washer
15. Bottom cover
16. Gasket
17. Stud
18. Pin
19. Hitch plate
20. Plug
21. Bolt
22. Bearing retainer cap
23. Shim
24. Gasket
25. O-ring
26. Pin
27. Plug
28. Detent ball
29. Spring
30. Interlock pin
31. Breather
32. Gasket
33. Transaxle cover
34. Nut
35. Washer
36. Retainer
37. Snapring
38. Seal ring
39. Pivot ball
40. Bolt
41. Bolt
42. Shift lever
43. Knob
44. Snapring
45. Handwheel
46. Bolt
47. Spring washer (4 used)
48. "Posi-Traction" cone
49. Key
50. Cone drive hub
51. Washer
52. Oil seal
53. Axle shaft
54. Hub lefthand
55. Oil seal
56. Thrust washer
57. Thrust bearing
58. Bushing
59. Shim
60. Shim
61. Thrust washer
62. Thrust bearing
63. Side gear lefthand
64. Differential pinion gear
65. Snapring
66. Differential pinion shaft
67. Wormwheel
68. Side gear righthand
69. Side gear drive hub
70. Setscrew
71. Lock wire
72. Bearing
73. Oil seal
74. Washer
75. Roll pin
76. Seal ring
77. Shifter fork
78. Range shifter rail
79. Washer
80. Range shift lever
81. Cross shaft
82. Pin
83. Shifter rails
84. Bearing cone
85. Race
86. Wormshaft
87. Bearing cone
88. Bearing cup
89. Second and third sliding gear
90. First and reverse sliding gear
91. Bearing
92. Oil seal
93. Brake drum
94. Snapring
95. Input gear
96. Spacer
97. Ball bearing
98. Washer
99. Bearing
100. Retainer ring
101. Oil seal
102. Input shaft
103. Snapring
104. Input driven gear
105. Bearing
106. High-low range gear
107. Splined shaft
108. Retainer ring
109. Ball bearing
110. Snapring
111. Expansion plug
112. Thrust washer
113. Reverse idler gear
114. Bushing
115. Seal ring
116. Idler shaft
117. Roll pin
118. Cluster gear

6-speed transmission components

Bolens

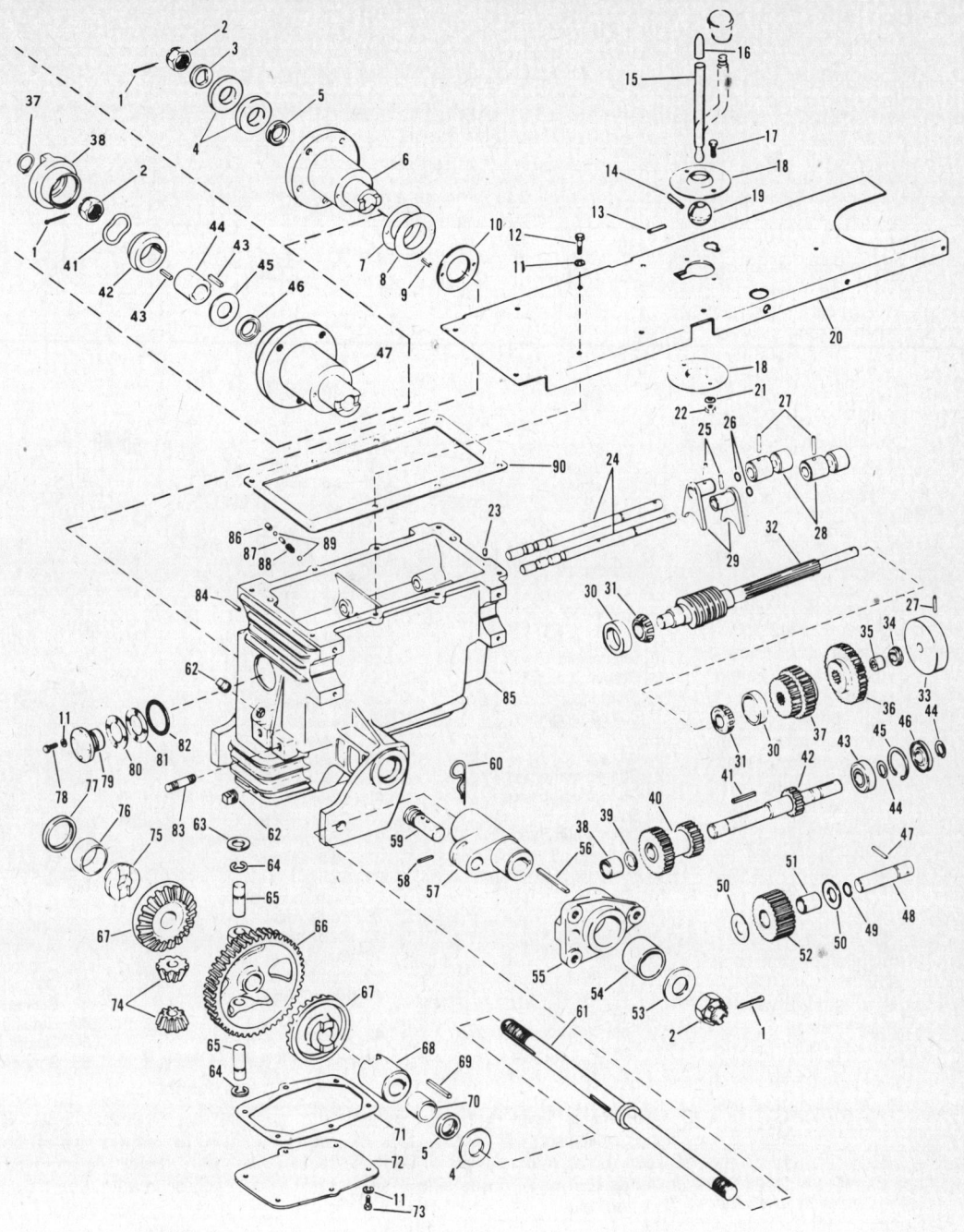

1. Cotter pin	21. Washer	41. Key	61. Axle shaft	81. Shim
2. Nut	22. Nut	42. Input shaft	62. Plug	82. O-ring
3. Spacer	23. Dowel	43. Ball bearing	63. Washer	83. Stud
4. Thrust washers	24. Shift rails	44. Snapring	64. E-clip	84. Pin
5. Seal	25. Pin	45. Retainer ring	65. Differential pinion shaft	85. Case
6. Hub	26. Seal ring	46. Oil seal	66. Wormwheel	86. Plug
7. Shim	27. Pin	47. Pin	67. Side gear	87. Spacer
8. Shim	28. Connector	48. Idler shaft	68. Side gear drive	88. Spring
9. Key	29. Shift fork	49. Seal ring	69. Key	89. Detent balls
10. Washer	30. Race	50. Thrust washer	70. Bushing	90. Gasket
11. Washer	31. Bearing	51. Bushing	71. Gasket	
12. Bolt	32. Wormshaft	52. Reverse idler gear	72. Bottom cover	Posi-Traction Units (Inserts)
13. Pin	33. Brake drum	53. Thurst washer	73. Bolt	37. Snapring
14. Pin	34. Oil seal	54. Bushing	74. Differential pinion gear	38. Handwheel
15. Shift lever	35. Bearing	55. Right hub	75. Side gear coupling	41. Spring washer
16. Cone	36. First/reverse gear	56. Key	76. Bushing	42. Posi-Traction® cone
17. Bolt	37. Second/third gear	57. Hub drive coupling	77. Oil seal	43. Key
18. Retainer	38. Bearing	58. Pin	78. Bolt	44. Cone drive hub
19. Pivot ball	39. Snapring	59. Free wheeling pin	79. Bearing retainer clip	46. Oil seal
20. Cover	40. Cluster gear	60. Pin	80. Shim	47. Hub

3-speed transmission components

177

Bolens

the snapring from the rear of the shaft and slide the shaft and lift the high-low range sliding gear from the case.

16. Pry the oil seal from the case and remove the bearing retainer ring.

17. Remove the snapring, input gear and spacer from the input shaft, and then remove the cluster gear and washers as the input shaft is pulled from the case.

18. Drive the pin from the idler shaft and remove the idler shaft, reverse idler gear and thrust washers.

19. Assembly is the reverse of disassembly. When installing the wormshaft, add or delete shims at the bearing retainer cap to obtain a wormshaft end play of .004-.008 in.

Apply a light coat of Prussian blue to the teeth on the wormwheel. Rotate the wormwheel and check the contact pattern. If the pattern is not centered on the teeth, add or delete shims as required.

HYDROSTATIC TRANSMISSION

Overhaul

1. Disconnect the hydraulic reservoir hose and remove the reservoir. Plug openings to prevent entrance of dirt into system. Unbolt and remove seat, fenders and frame top cover. Remove disc brake assembly.

2. On all models, disconnect control linkage and auxiliary hydraulic hoses from drive unit. Identify hoses to insure correct reassembly. Plug or cap all openings immediately. Support tractor frame and place scribe marks on frame tubes and reduction gear and differential housing for aid in reassembly. Unbolt frame clamps and roll the assembly rearward from tractor.

3. Remove filter tube and plug openings.

4. Remove control linkage from control shaft. Block up reduction and differential housing so that hydraulic drive unit input shaft is pointing upward.

5. Unbolt and remove hydrostatic drive unit.

NOTE: Before disassembling the hydrostatic drive unit, thoroughly clean the exterior of unit.

6. Remove venting plug and reservoir adaptor, invert the assembly and drain the fluid from the unit. Place the unit in a holding fixture so that the input shaft is pointing upward. Remove dust shield and snapring.

7. Remove the five capscrews from charge pump body. One capscrew is ½ in. longer than the others and must be installed in the original location (heavy section of pump body).

8. Remove charge pump body with ball bearing. Ball bearing and oil seal can be removed from body after first removing retaining ring. Remove the six pump rollers, snaprings and charge pump rotor. Remove O-rings and pump plate. Invert the drive unit in the holding fixture so that output shaft is pointing upward. Remove snapring and output gear.

9. Unscrew the two capscrews, then turn them in until two threads are engaged.

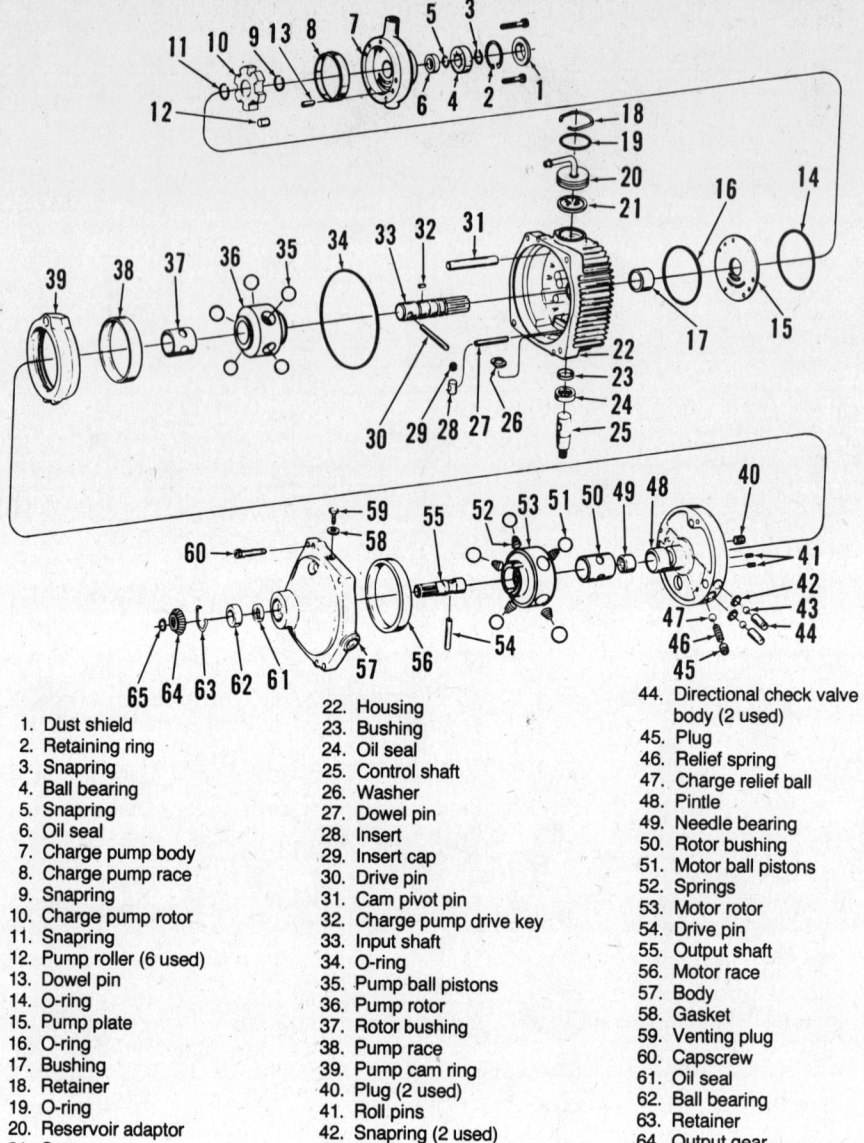

1. Dust shield
2. Retaining ring
3. Snapring
4. Ball bearing
5. Snapring
6. Oil seal
7. Charge pump body
8. Charge pump race
9. Snapring
10. Charge pump rotor
11. Snapring
12. Pump roller (6 used)
13. Dowel pin
14. O-ring
15. Pump plate
16. O-ring
17. Bushing
18. Retainer
19. O-ring
20. Reservoir adaptor
21. Screen
22. Housing
23. Bushing
24. Oil seal
25. Control shaft
26. Washer
27. Dowel pin
28. Insert
29. Insert cap
30. Drive pin
31. Cam pivot pin
32. Charge pump drive key
33. Input shaft
34. O-ring
35. Pump ball pistons
36. Pump rotor
37. Rotor bushing
38. Pump race
39. Pump cam ring
40. Plug (2 used)
41. Roll pins
42. Snapring (2 used)
43. Check valve ball (2 used)
44. Directional check valve body (2 used)
45. Plug
46. Relief spring
47. Charge relief ball
48. Pintle
49. Needle bearing
50. Rotor bushing
51. Motor ball pistons
52. Springs
53. Motor rotor
54. Drive pin
55. Output shaft
56. Motor race
57. Body
58. Gasket
59. Venting plug
60. Capscrew
61. Oil seal
62. Ball bearing
63. Retainer
64. Output gear
65. Snapring

Eaton model 10 components

10. Raise body until it contacts the heads of capscrews. Insert a fork tool between motor rotor and pintle until the tool extends beyond opposite side. The special fork tool can be fabricated from a piece of ⅛ in. flat stock approximately 3 in. wide and 12 in. long. Cut a slot 1 9/16 in. wide and 8 in. long. Taper the ends of the prongs.

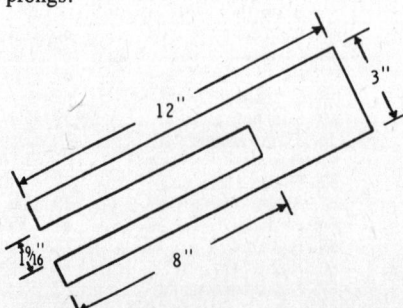

Special tool fabricated from a piece of ⅛ in. flat stock, used in disassembly and assembly of Eaton model 10

11. Remove capscrews and by raising ends of forked tool, lift off body and motor assembly. Place the removed assembly on a bench or in a holding fixture with output shaft pointing downward. Remove special fork tool and place a wide rubber band around motor rotor to hold ball pistons and springs in their bores. Carefully remove the motor rotor assembly and lay aside for later disassembly. Remove motor race and output shaft. Remove retainer, bearing and oil seal.

12. With housing assembly resting in holding fixture (input shaft pointing downward), remove pintle assembly.

CAUTION

Do not allow pump to raise with the pintle as the ball pistons may fall out of rotor. Hold pump in position by inserting a finger through hole in pintle. Remove plug, spring and charge pump relief ball. To remove the directional check valves from pintle, drill through the pintle with a

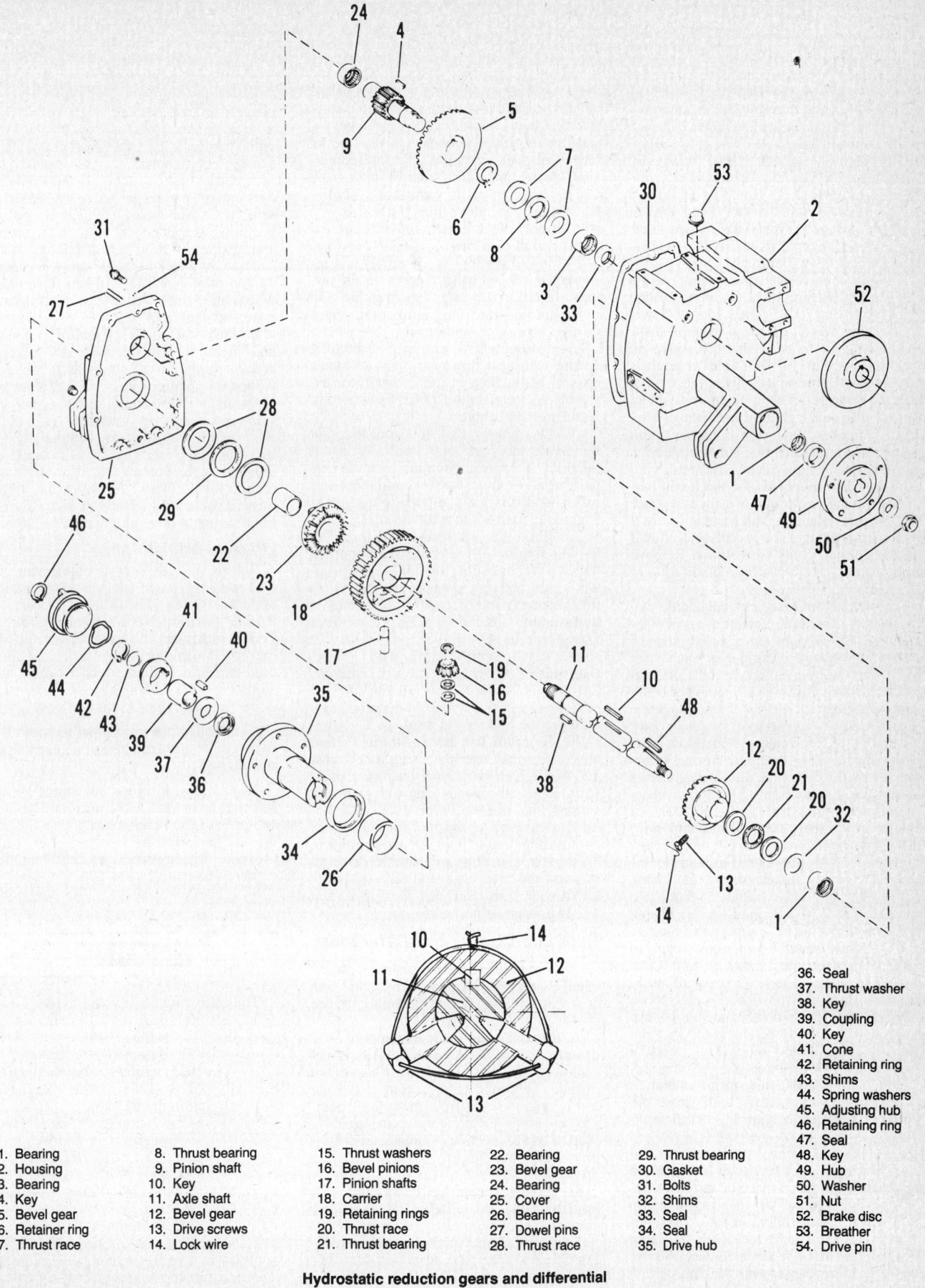

Hydrostatic reduction gears and differential

Bolens

drill bit that will pass freely through roll pins.

13. Redrill the holes from the opposite side with a ¼ in. drill bit. Drive or press roll pins from pintle. Using a 5/16-18 tap, thread the inside of valve bodies, then remove valve bodies using a draw bolt or a slide hammer puller.

14. Remove check valve balls and snaprings. Do not remove plugs.

15. Remove pump cam ring and pump race. Place a wide rubber band around pump rotor to prevent ball pistons from falling out. Carefully remove pump assembly and input shaft.

16. To remove the control shaft, drill a 11/32 in. hole through aluminum housing directly in line with center line of dowel pin. Press dowel pin from control shaft, then withdraw control shaft. Remove oil seal. Thread the drilled hole in housing with a ⅛ in. pipe tap. Apply a light coat of Loctite® grade 35 to a ⅛ in. pipe plug, install plug and tighten it until snug.

17. Number the piston bores on pump rotor and on motor rotor. Use a plastic ice cube tray or equivalent and mark the cavities 1P through 5P for the pump ball pistons and 1M through 5M for the motor ball pistons. Remove ball pistons one at a time, from pump rotor and place each ball in the correct cavity in the tray. Remove ball pistons and springs from motor rotor in the same manner.

18. Clean and inspect all parts and renew any showing excessive wear or other damage. Ball pistons are selective fitted to 0.0002-0.0006 in. clearance in rotor bores and must be reinstalled in their original bores. If rotor bushings are scored or badly worn (0.002 in. or more clearance on pintle journals) renew pump rotor or motor rotor assemblies. Check clearance between input shaft and housing bushing. Normal clearance is 0.0013-0.0033 in. If clearance is excessive, renew input shaft and/or housing assembly.

19. Install ball pistons in pump rotor and ball pistons and springs in motor rotor, then use wide rubber bands to hold pistons in their bores. Install snaprings, check valve balls and valve bodies in pintle and secure with new roll pins. Install charge pump relief valve ball, spring and plug.

20. When installing oil seals, apply a light coat of Loctite® grade 35 to the seal outer diameter. Renew oil seal and install control shaft in housing. Install special washer, then press dowel pin into control shaft until end of dowel pin is 1¼ in. from control shaft. Renew oil seal and reinstall output shaft with drive pin, bearing, retainer, output gear and snapring in body.

21. Insert input shaft with drive pin through bushing in housing. Install snapring in its groove on input shaft.

22. Place O-ring, pump plate and O-ring in housing, then install charge pump drive key, charge pump rotor and snapring. Apply light grease or vaseline to pump rollers and place rollers in rotor slots. Install oil seal and pump race in charge pump body, then install body assembly. Secure with the five capscrews, making certain that long capscrew is installed in its original location (in heavy section of pump body). Tighten capscrews to a torque of 28-30 ft.lb. Install snapring, bearing, retaining ring, snapring and dust shield.

23. Place charge pump and housing assembly in a holding fixture with input shaft pointing downward. Install pump race, insert cap and insert in cam ring, then install cam ring over cam pivot pin and control shaft dowel pin. Turn control shaft back and forth and check movement of cam ring. Cam ring must move freely from stop to stop. If not, check installation of insert and insert cap in cam ring.

24. Install pump rotor assembly and remove the rubber band used to retain pistons. Install pintle assembly over the cam pivot pin and into the pump rotor. Place O-ring in position on housing.

25. Place body assembly in a holding fixture with gear downward. Install motor race in body, then install motor rotor assembly and remove the rubber band used to retain pistons in rotor.

26. Using the special fork tool to retain motor assembly in body, carefully install the body and motor assembly over the pintle journal. Remove the fork tool, align bolt holes and install the two capscrews. Tighten capscrews to a torque of 15 ft.lb.

27. Place the hydrostatic unit on the holding fixture with reservoir adaptor opening and venting plug opening facing upward. Fill the unit with Bolens No. 171-9650 hydraulic oil or Type "A" automatic transmission fluid until fluid flows from fitting hole in body.

28. Install venting plug with gasket, then install reservoir adaptor, screen, O-ring and retainer. Plug all openings to prevent dirt or other foreign material from entering the hydrostatic unit.

29. Reinstall the unit on the reduction and differential housing, using a new gasket, and tighten capscrews to a torque of 20 ft.lb. Reinstall assembly on tractor by reversing the removal procedure. Fill hydraulic reservoir to dipstick full mark with approved hydraulic fluid. Operate engine at ⅓ throttle and shift hydrostatic drive to forward and reverse several times to purge air from system. Recheck fluid level and adjust linkage as necessary.

Reduction Gears and Differential

1. To remove reduction gear and differential assembly, disconnect hydraulic line at reservoir. Plug line to prevent entrance of dirt in hydraulic drive system.

2. Remove seat and fenders and on models 1256 and 1257 remove disc brake assembly and disconnect parking brake linkage. Disconnect speed control linkage and hydraulic hoses from drive unit. Identify hoses to insure correct reassembly. Plug or cap all openings immediately. Support rear of tractor and disconnect driveshaft.

3. Unbolt retaining bolts and roll transmission and differential assembly from tractor. Hydrostatic transmission can be separated from differential after removing parking brake pawl from side of differential case.

4. To overhaul gear reduction and differential unit, remove outer snapring, adjusting handwheel, spring washers, inner snapring, shim, Positraction drive cone, cone drive hub and keys from left end of axle. Remove left wheel hub, then remove cover from case. Remove right hub nut and right wheel hub. Withdraw axle and differential components. Input shaft components can now be removed.

5. Clean and inspect all components for damage or excessive wear. When reassembling differential, shims are used to position right side gear with drive pinions and also are used to adjust axle end play. Change number of size of shims to obtain an end play of 0.002-0.014 in. Install left wheel hub, seal, thrust washer, coupling, cone and snapring.

6. Push inward on left end of axle shaft and measure distance between coupling and snapring. Install shims as required to reduce the gap to 0.002-0.016. To assembly gear reduction and differential unit, reverse disassembly procedure.

7. After assembly is completed, fill transmission reservoir to full mark on dipstick with Bolens No. 171-9650 hydraulic oil or Type "A" automatic transmission fluid. Fill reduction and differential housing to level plug with SAE 90 oil.

Linkage Adjustment

Before making neutral adjustment, be sure parking pawl engages transmission drive gear when selector lever is placed in PARK position. With transmission in PARK position, differential should be locked. If differential is not locked, proceed as follows: Remove clevis pin. Turn adjusting block to lengthen stud until pawl correctly engages transmission gear.

NOTE: Do not turn pawl as mesh between pawl and transmission gear will be misaligned.

To accomplish neutral adjustment, place selector lever in PARK position. Loosen capscrews. Start engine and slide adjusting bracket back and forth until noise in hydrostatic transmission ceases. Retighten capscrews.

ENGINE

The following engines are used:
Models 600, 650: B&S 142302
750: B&S 170401
855: B&S 191707
1055, 1155: Tecumseh HH100
1220, 1253, 1255, 1257: Tecumseh HH120
1453, 1456: Tecumseh HH140
1556: Tecumseh HH150
1656: Tecumseh HH160
800, 850, 900: Wisconsin S-8D
1000, 1050, 1053: Wisconsin TR-10D
1225: Wisconsin TRA-12D

For engine service procedures, see the appropriate engine overhaul section in the Engine Unit Repair section.

Bolens

REMOVAL AND INSTALLATION

1. Disconnect the hood stop rod from the bracket on the cylinder head.
2. Unbolt and remove the hood and grille.
3. Disconnect all wiring and control cable from the engine.
4. Remove the drive belt or belts.
5. Remove the PTO belt guide. With the PTO lever in neutral, remove the belts.
6. Unbolt the engine and slide it forward. Lift it from the frame.
7. Installation is the reverse of removal.

XL Series Models

1058, 1060, 1258, 1261, 1458, 1461, 1658, 1661, 1858

FRONT AXLE

REMOVAL AND INSTALLATION

1. Remove the PTO drive belts.
2. Disconnect the ball joint at the forward end of the drag link, from the right side steering arm.
3. Jack up and support the tractor on jack stands under the frame rails. Remove the pivot pin.
4. Roll the axle from under the tractor.
5. Installation is the reverse of removal. Replace the pivot pin if excessive wear is evident.

STEERING

Steering Gear

REMOVAL AND INSTALLATION

1. Remove retaining ring and slide the steering arm shaft out.

NOTE: Keep track of all shims and their locations when disassembling.

2. Disassemble the steering wheel by removing roll pin (E) and washer.
3. Remove the steering shaft assembly from the steering column.

NOTE: It may be necessary to tap the steering shaft down until flange bushing (F) is pushed out of the steering support casting.

4. With flange bushing (F) slid onto the pinion shaft assembly, route it through the casting and through the steering column bearing. Slide on any needed shims at point (G) and secure the steering wheel with the roll pin removed earlier.
5. With the steering wheel emblem right side up position steering hear (H) in steering casting.
6. With the steering arm pointed about 10° to the rear, slide it through the steering casting, any shims removed earlier and steering gear (H).
7. If for some reason the two gears do not have maximum engagement, shim at point (I) and secure with retaining ring removed earlier.
8. Turn steering wheel to check for binding.

Steering Knuckles

REMOVAL AND INSTALLATION

1058, 1060, 1258, 1261, 1458, 1461, 1858

1. Raise and support the front axle with jack stands.

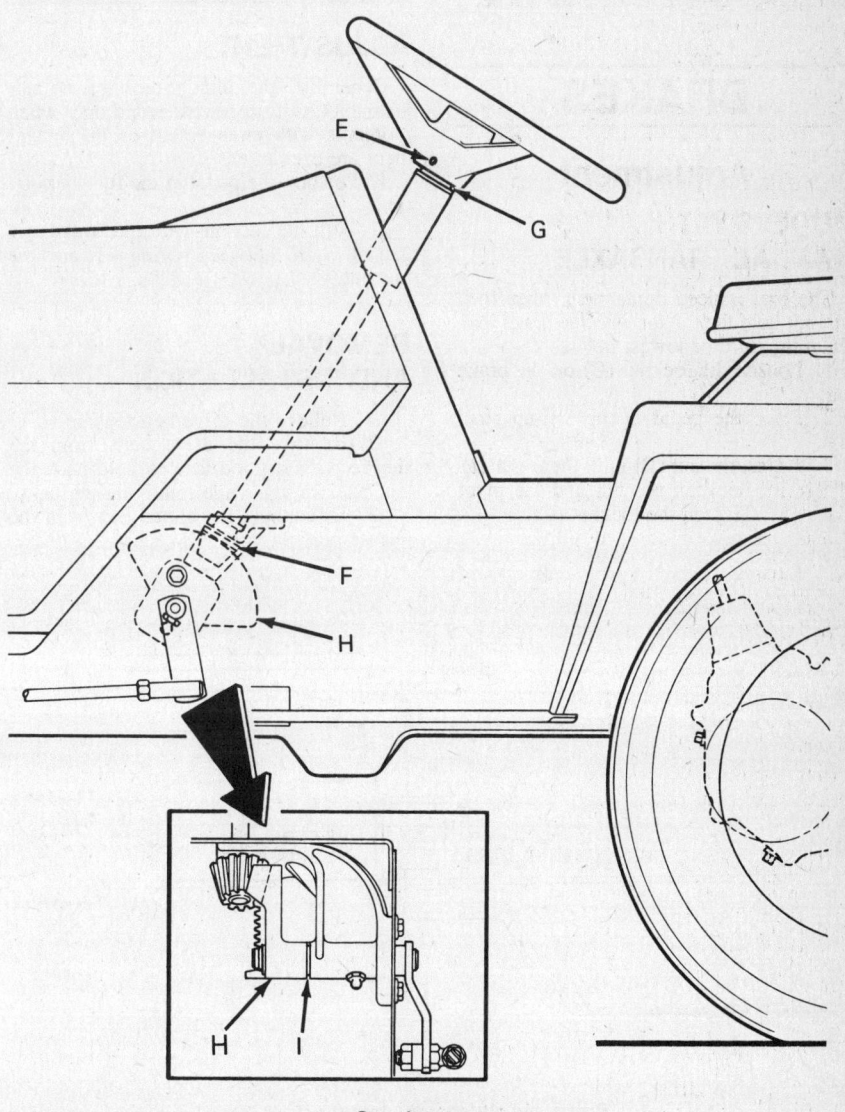

Steering system

Bolens

2. Remove the wheels.
3. Disconnect the tie rod and drag link.
4. Remove the clamp bolt from the steering arm and pull the arm from the knuckle.
5. Remove the key and lower knuckle from the axle end.
6. Clean and inspect all parts. Replace any parts showing excessive wear.
7. Installation is the reverse of removal. Lubricate the knuckle bearings and wheel bearings with multi-purpose chassis lube.

All other models

1. Raise and support the front axle on jack stands.
2. Remove the front wheels and bearings.
3. Disconnect the drag link at the ball joint and remove the tie rod from the steering arm.
4. Carefully drive out the roll pin securing the steering arm to the knuckle and lower the knuckle from the axle.
5. Check all parts and replace any that show excessive wear.
6. Installation is the reverse of removal. Lubricate the knuckle bearings and wheel bearing with multi-purpose chassis lube.

BRAKES

Adjustment

MODELS WITH MANUAL TRANSAXLE

The brakes need adjustment when they will not hold the tractor with the brake lock set in the third or lowest notch.
1. Loosen the locknut (C) on the brake rod.
2. Place the pedal in the full-up position.
3. Adjust the nut (D) until there is a gap of $1/16$ in. between the nut and the pivot block (E). Tighten the locknut (C), until it touches the pivot block.

MODELS WITH HYDROSTATIC TRANSMISSION

The brake needs adjustment when only light foot pressure is needed to latch the parking brake, or when the parking brake does not hold the tractor when latched.
1. Release the parking brake.
2. Turn the brake rod (A) until a gap of .010 in. exists between the brake pad and disc.

NOTE: The disc must turn freely with the brake adjusted. The tractor must not move with the brake in the first position.

Removal and Installation

For service on the brake system, see the clutch or Hydrostatic transmission section.

CLUTCH

ADJUSTMENT

Generally, the clutch requires no adjustment. Adjustment is needed only when problems with engagement or disengagement are encountered.
1. Position the pedal in the full-up position.
2. Turn the locknut (A) until there is a gap of $1/8$ in. between spring (B) and the locknut.

REMOVAL AND INSTALLATION

1. Relieve the drive belt tension.
2. Remove the drive belts from the sheaves. It's a good idea to hold back the tensioner spring while removing the belts.
3. Disconnect the clutch rod from the release bearing yoke.

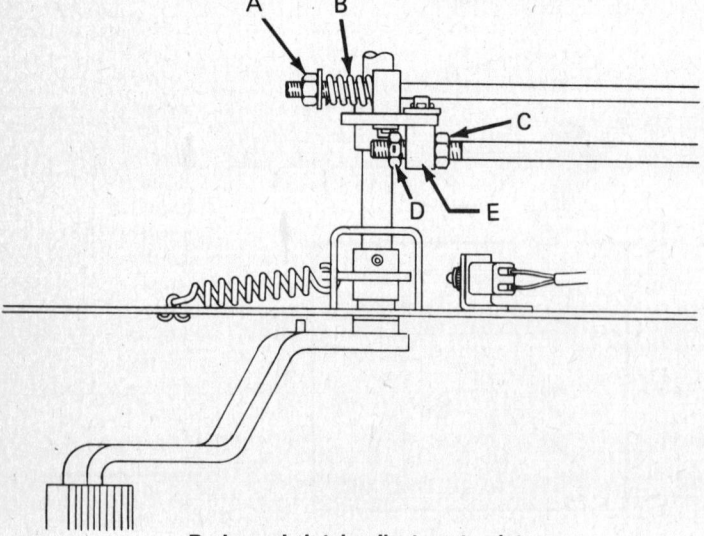

Brake and clutch adjustment points

4. Uncouple the driveshaft at the transaxle.
5. Unbolt the bearing support and lower the clutch and driveshaft assembly out from under the tractor.
6. Clamp the driveshaft securely in a holding fixture and back off the nut from the end of the shaft.
7. Lift the support off along with the bearing, retainers and spacers.
8. The forward clutch plate is threaded on the shaft. Take great care in backing it off the threaded section since the entire assembly is under spring tension.
9. Lift the remaining parts from the driveshaft.
10. Installation is the reverse of removal.

MANUAL TRANSAXLE

REMOVAL AND INSTALLATION

1. Remove the seat and fenders.
2. Raise and support the rear of the tractor on jack stands.
3. Remove the parking brake lever.
4. Unbolt the driveshaft at the rear coupling.
5. Unbolt the tranxaxle from the tractor frame rails so that it can be rolled rearward from under the tractor.
6. Installation is the reverse of removal.

OVERHAUL

For overhaul procedures turn to the Medium Tube Frame tractor section and follow the instructions under 3-speed transaxle overhaul.

HYDROSTATIC TRANSAXLE

REMOVAL AND INSTALLATION

1. Remove the rear seat and fenders.
2. Disconnect and remove the hydraulic fluid reservoir hose and reservoir along with all other hoses. Plug all openings to prevent fluid loss and contamination.
3. Disconnect the control rod from the linkage but don't remove the control levers from the control shaft.
4. Remove the disc brake assembly.
5. Raise the tractor rear and place jack stands under the frame.
6. Scribe marks on the frame tubes and reduction gear housings for reassembly.
7. Unbolt the frame rail clamps and roll the assembly from under the tractor.
8. Installation is the reverse of removal.

Bolens

OVERHAUL

For overhaul see the Medium Tube Frame tractor section under Hydrostatic Transaxle Overhaul.

POWER TAKE-OFF (PTO)

Electric Clutch

REMOVAL

NOTE: Do not use wheel puller to remove clutch/brake.

1. Remove the hood and front frame cross plate.
2. Disconnect polarized plug from clutch.
3. Remove four locknuts.
4. Remove mounting bolt from center of crankshaft.
5. On the 1458 serial no. 0100101 to 0199999 the engine must be removed to remove the electric clutch. On all others models the electric clutch can be removed while in the frame.
6. Remove sheave and armature/brake assembly.

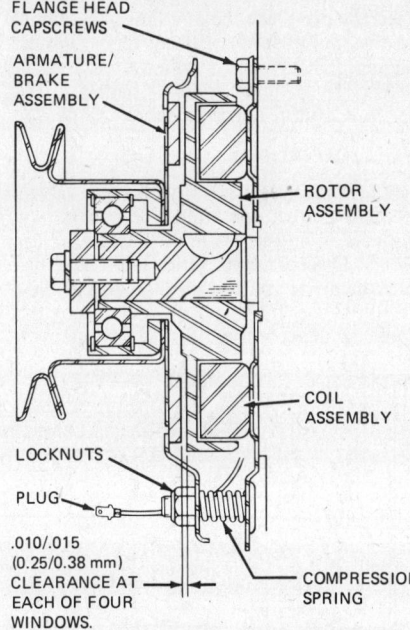

PTO sheave and armature brake assembly

7. Remove rotor assembly.
8. Remove four ⅜-16 x ¾ in. flange lock head capscrews securing coil assembly to crankcase.
9. Remove coil assembly from engine.

INSTALLATION

1. Position coil assembly on engine piloting in crankshaft seal bore. Secure with four ⅜-16 x ¾ flange lock capscrews.

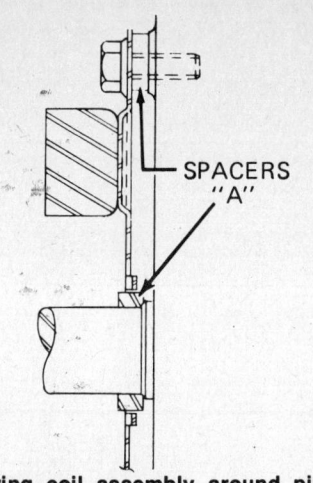

Centering coil assembly around pilot spacer on B&S engines

2. On Briggs and Stratton engines the coil assembly must be centered around pilot spacer (A). Center by feel. When mounting the assembly a spacer must be used at all four capscrew locations.
3. Place compression springs on brake mounting studs.
4. Remove rust, burrs and dirt from engine shaft with sandpaper.
5. Apply light coat of grease or oil to shaft.
6. Insert key into key way in shaft.
7. Push rotor and armature assembly with pulley onto engine shaft until hub bottoms on crankshaft shoulder.
8. Place large washer and lockwasher on clutch mounting capscrew.
9. Install capscrew into tapped hole in center of crankshaft. Hold clutch from rotating, while tightening capscrew. Torque to 23-27 ft.lb.
10. Adjust brake.
11. Reconnect plug lead to lead from switch making certain connection is secure and wire is not rubbing against rotating parts.

ADJUSTMENTS

NOTE: Once the clutch/brake has been installed, the clutch portion requires no further maintenance. The unit is self-adjusting for wear and never requires lubrication. The brake portion may require readjustment periodically depending upon unit usage. This can be accomplished as follows:

1. Position a .015 in. thick shim in each slot in brake flange and turn on clutch/brake. Tighten the (4) locknuts until they just contact brake flange.

CAUTION
Do not over torque locknuts as damage to brake flange may result.

2. Turn clutch/brake off and remove (4) shims. Recheck gap through the slots provided. A minimum of .010 in. and a maximum of .015 in. should be maintained.
3. If oil or grease contaminate clutch working surfaces, remove with a cleaning fluid such as barcothene alcohol or ammonia are acceptable substitutes. With engine off, pour a generous quantity of cleaning fluid between working surfaces.
4. If the clutch/brake has not been used over a long period of time, the following procedure is recommended prior to its use.
   a. Position tractor in neutral position.
   b. Start tractor engine, and put throttle in fast position.
   c. Turn clutch/brake switch on and off six times, engaging and disengaging driven attachments.

NOTE: Allow engine driven attachments to come to a complete stop between on-off cycles.

ENGINE

Removal and Installation

MODELS 1058, 1060

Remove all attachments and PTO kit.
1. Remove hood and side panels.

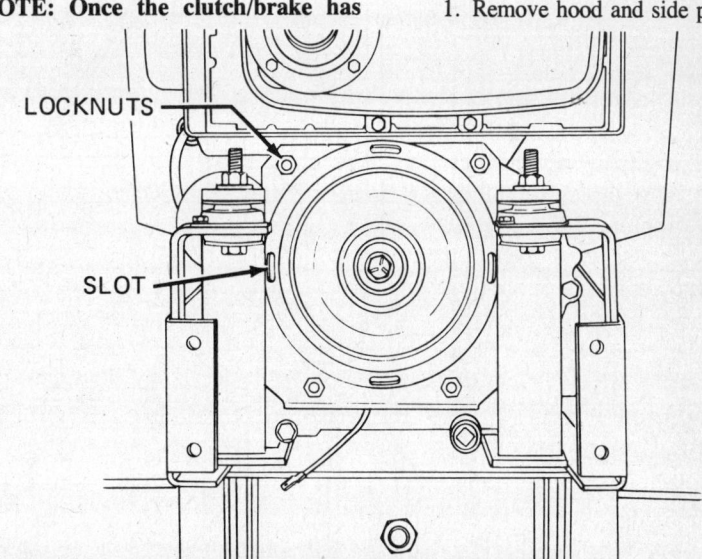

Positioning shims in slots

Bolens

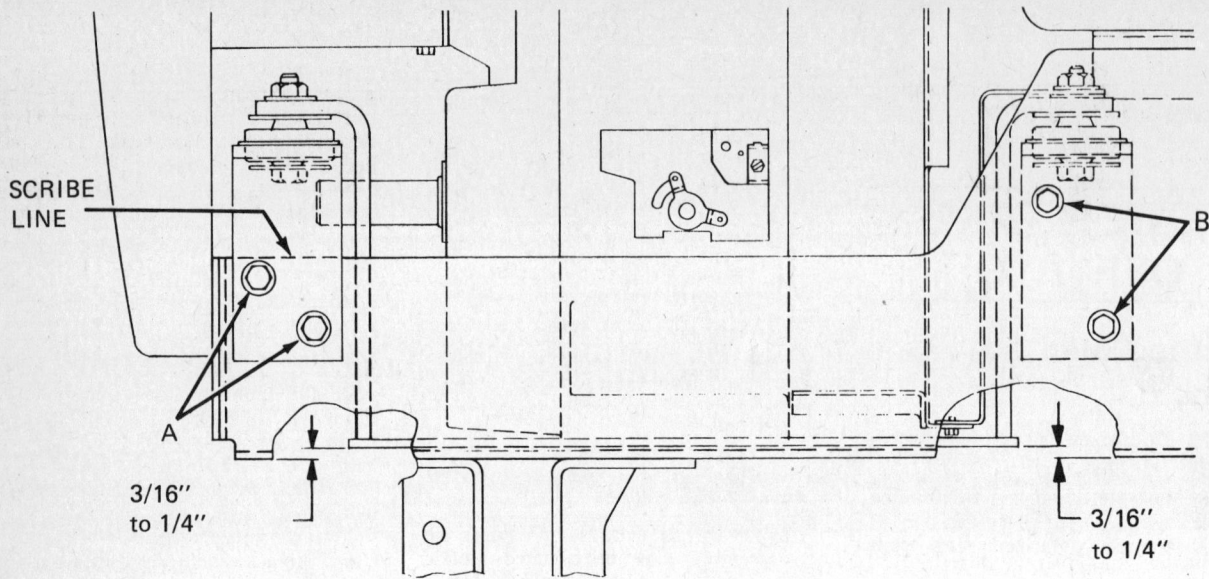

Model 1058, 1060 engine mounting

2. Disconnect negative battery cable.
3. Disconnect starter (red), magneto (blue), 12 volt lead (red), and electric clutch (purple) wires and pull through heat shield.
4. Disconnect engine ground wire (black) from battery support mounting hole on side of frame and pull through heat shield.
5. Close fuel tank petcock and disconnect fuel line at filter and pull through heat shield.
6. Disconnect choke and throttle cables at engine and pull through heat shield.
7. Remove the three rubber mounted driveshaft bolts at engine and disconnect the driveshaft.
8. Scribe line on the front engine supports along frame for reinstallation. Loosen two bolts (A) on each side of the engine supports.
9. Remove two rear engine mount bolts (B).

10. When ready to pull engine, remove four bolts from front engine supports.
11. Engine serial number is located behind heat shield on rear of engine.
12. Set engine on rubber mounts in rear and line up scribed line on front engine supports with frame, insert front four bolts (A). Insert proper engine mount spacers for engine height until a gap of 3/16 to 1/4 in. is reached between the bottom edge of the engine support and bottom edge of the side frame. This dimension should be held at the front and rear of the engine support. Secure engine supports to frame holding dimension given above.
13. Bring starter, magneto, rectifier, electric clutch and ground wire through heat shield and connect to proper terminals.
14. Pull fuel line through heat shield, connect to fuel filter and open petcock.
15. Pull choke and throttle through heat shield and connect (adjustment may be needed) to proper locations.

16. Connect driveshaft and tighten the three bolts on the driveshaft coupling. On hydrostatic models make sure the hub on the hydrostat is 5/16 in. away from the front face of the hydrostat. On gear drive models make sure the hub on the right angle gear box is slid all the way to the end of the spline.
17. With the hubs of the driveshaft in there proper place the couplings (2) should be flat, not pushed in. If not reposition the engine.
18. Attach hood and side panels.
19. Connect negative battery cable.

ALL OTHERS

1. Disconnect negative battery cable.
2. Remove hood and side panels (disconnect light harness at connector).
3. Disconnect throttle and choke cables at engine and pull through heat shield.
4. Close fuel petcock and disconnect fuel line at fuel filter.

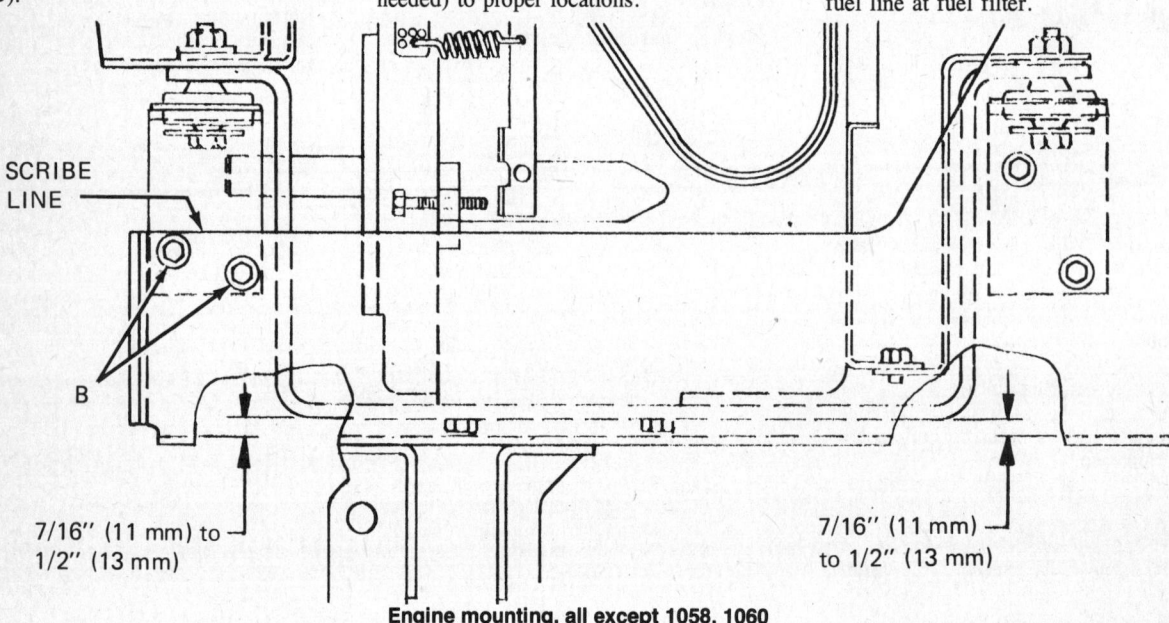

Engine mounting, all except 1058, 1060

Bolens

5. Disconnect electric clutch wire (purple).
6. Disconnect wire harness at connector (red and blue).
7. Disconnect starter wire (red) and ground wire (black) and pull all wires through heat shield.
8. Disconnect and remove three driveshaft bolts at flywheel.
9. Scribe line on front engine supports along frame. Loosen two bolts (B) on each side of the engine supports.
10. Remove two rear engine mount bolts.
11. When ready to pull engine, remove four bolts from front engine supports.
12. Set engine on rubber mounts in rear and line up scribed line on front engine supports with frame, insert front two bolts (B) on each side. Insert proper engine mount spacers for engine height until a gap of 7/16 in. to ½ in. is reached between the bottom surface of the engine support and inside surface of the frame. This dimension should be held at the front and rear of the engine support. Secure supports to frame holding dimension given above.
13. Bring starter (red), ground (black), electric clutch (purple) wires, and engine wire harness (red and green wires) through the heat shield and connect to proper terminals.
14. Connect fuel line to fuel filter and open fuel petcock.
15. Pull choke and throttle cable through heat shield and connect to proper locations (adjust later).
16. Connect driveshaft and tighten the three bolts on the driveshaft coupling. On hydrostatic models make sure the hub on the hydrostat is 5/16 in. away from the front face of the hydrostat. On gear drive models make sure the hub on the right angle gear box is slid all the way to the end of the spline.
17. With the hubs of the driveshaft in their proper place the couplings (2) should be flat, not pushed in. If not reposition the engine.
18. Attach hood and side panels.
19. Connect negative battery cable.

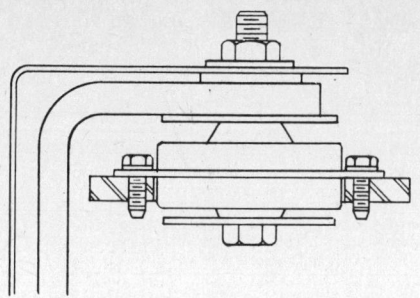

Engine mount replacement

Rubber Mount Replacement

When replacing the rubber engine mounts, make sure that the correct mount is replaced. To check, push down on the center of the rubber mount. The softer mount should always go to the rear.

Overhaul

Engines used in these tractors are:
Models 1058, 1060: B&S 253417-013101
      1258, 1261: Tecumseh OH120
      1458, 1461: Tecumseh OH140
      1658, 1661, 1858: Tecumseh OH160

For engine service, see the Engine Unit Repair section.

Models QS (1660) and QT (1666)

FRONT AXLE

REMOVAL AND INSTALLATION

1. Disconnect the forward end of the drag link from the left steering arm.
2. Raise and support the tractor on jack stands.
3. Remove the roll pin from the axle pivot pin.
4. Push the axle rearward and lower it from the tractor.
5. Inspect all parts and replace those showing excessive wear.
6. Installation is the reverse of removal.

Wheels and Wheel Bearings

REMOVAL AND INSTALLATION

1. Remove the dust cap (1).
2. Remove cotter pin (2) and unscrew slotted nut (3).
3. Remove outer bearing (4).
4. (Model QS and QT S/N 0100101 through 0599999): Remove wheel (5) and hub assembly (6) from spindle.
5. Remove seal (7) from hub. If this seal is damaged in any way replace it.

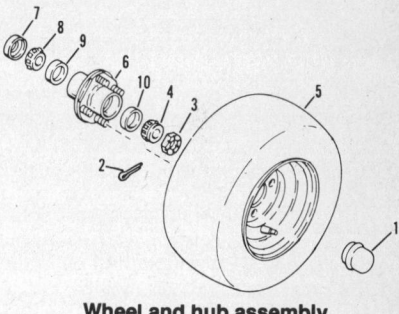

Wheel and hub assembly

1. Front axle
2. Tie rod
3. Lefthand ball joint
4. Righthand ball joint
5. Lefthand nut—½-20
6. Flange bearing
7. Righthand nut—1-20
8. Cotter pin
9. Woodruff key
10. Grease fitting
11. Spindle assembly
12. Righthand knuckle arm
13. Lefthand knuckle arm
14. Flat washer
15. Castellated nut—⅜-24
16. Flat washer
17. Lockscrew

Front axle components

STEERING

Steering Gear

REMOVAL AND INSTALLATION

1. Crank engine to move lift into down position.
2. Disconnect negative battery cable.
3. Close petcock, disconnect fuel line at filter and remove fuel tank.
4. Remove left footrest on earlier units.
5. Remove stop nut (A).
6. Remove front tie rod ball joint (B).
7. Rotate pinion gear (C) to left until sector (D) drops toward engine.

Bolens

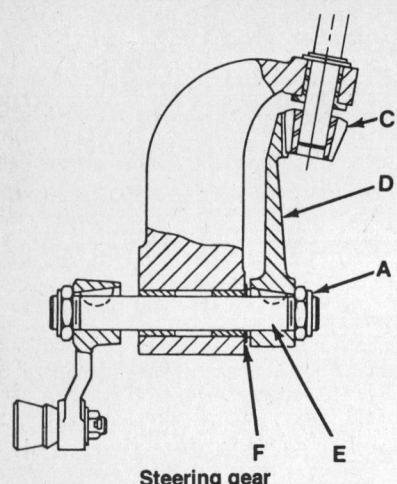

Steering gear

8. Slide sector gear (D) off of steering shaft (E). Some force may be needed to slide gear off.

9. Remove key from steering shaft.

NOTE: A punch can be used to partially remove key.

10. Slide steering shaft (E) out.
11. Remove washers (F) from steering support and save.
12. Disassemble steering support from frame if needed.
13. Position lift in down position.
14. Fasten steering support to frame.
15. Slide steering shaft (E) through support.
16. Slide washers (F) onto steering shaft.
17. Insert key into shaft (E) and slide steering sector (D) on shaft.
18. Run stop nut (A) onto shaft and secure.
19. With the Bolens steering cap logo in its normal level reading position, place a chalk mark on pinion shaft facing you.
20. Rotate steering wheel clockwise till chalk mark is straight forward.
21. Rotate sector gear into mesh with pinion gear. Turn steering wheel to left and engage pinion. Keep rotating till pinion is in the center of sector gear (D).
22. Reinstall front ball joint (B). Ball joints on tie rod should be 19¼ in. apart.
23. Remove inner bearing (8) from hub.
24. Clean both inner race (9) and outer race (10). Check both for wear or pitting. Replace if necessary.
25. Clean bearings in a commercial solvent. Dry with compressed air or a clean cloth. Do not spin the bearing with compressed air. Damage to the bearing could result.
26. Pack bearings with a good grade of wheel bearing grease.
27. Install inner bearing and seal into hub.
28. Install wheel and hub assembly to spindle.
29. Install outer bearing.
30. Tighten nut until it snugs up against the outer bearing. Spin wheel to align bearings, then back nut off to nearest slot in line with hole in spindle and install new cotter pin.
31. Install dust cap.

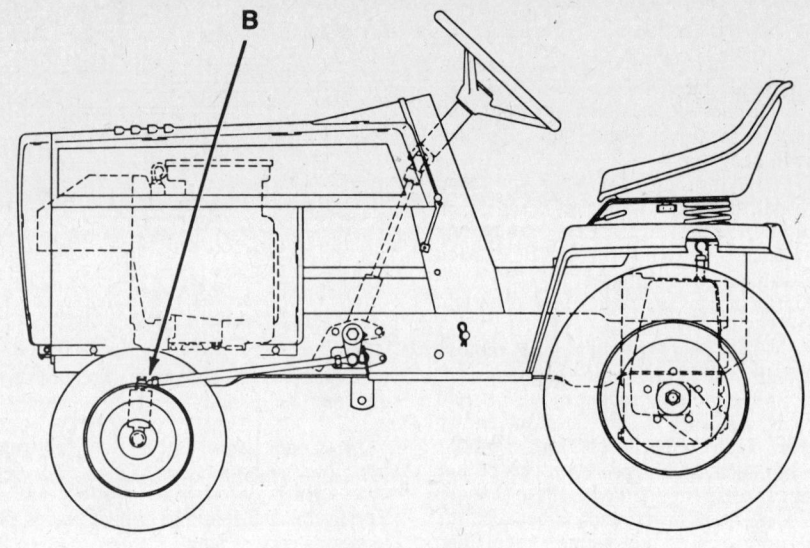

Ball joint location

BRAKES

Adjustments

PEDAL ADJUSTMENT

1. Loosen nut A.
2. Remove cotter pin from pin. Remove pin from control arm.
3. To bring toe of pedal closer to operator, shorten rod C.
4. To move toe of pedal farther away from operator, lengthen rod C.
5. After desired adjustment is reached, align hole in pedal end of rod C with hole in control arm and secure with pin and cotter pin. Lock hex nut A.

LINKAGE ADJUSTMENT

If the brake linkages have been disassembled for any reason readjust as follows. The brake system on these tractors has a "Neutral" interlock built into it. When the brake pedal is depressed, pin A moves into slot B of lock plate C moving the transmission controls into the "Neutral" position. If the parking brakes are locked the transmission is also locked in "Neutral."

1. With the brakes fully released, and arms D to the rear of the slot in the tractor frame, adjust the length of arms D to clear center section casting of transaxle.
2. Depress brake pedal until pin A of control just enters slot B of lock plate. Lock the brake pedal in this position.
3. Move brake arms D forward to remove brake pad/disc clearance. Adjust nuts E on brake rods to contact arm extension assemblies.

NOTE: Be sure to adjust the brakes on both wheels equally to avoid uneven braking. Check for proper operation.

Brake Pads

REPLACEMENT

1. Remove brake assembly F from tractor.

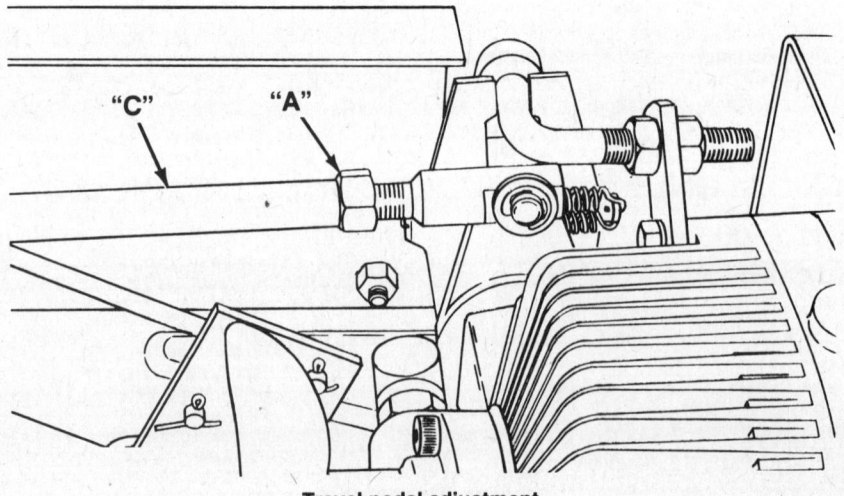

Travel pedal adjustment

Bolens

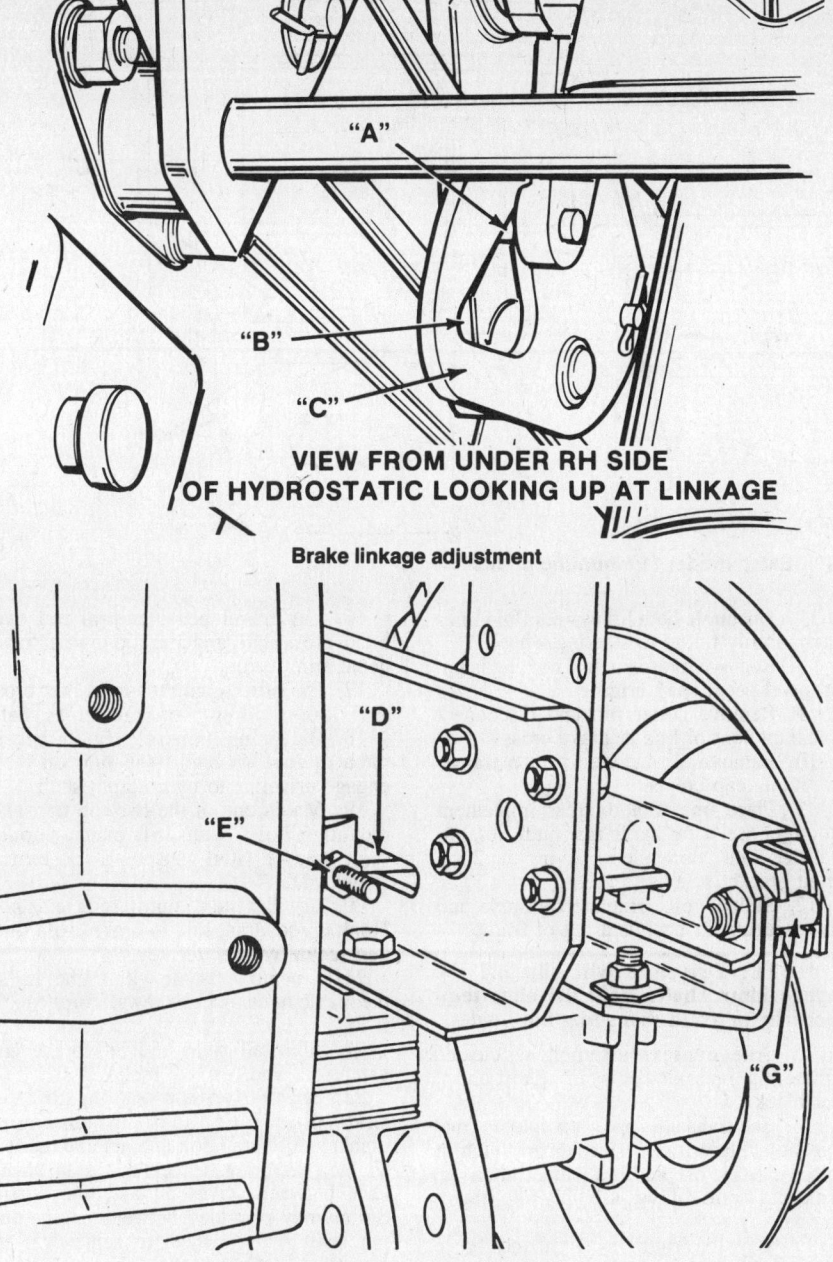

Brake linkage adjustment

Brake arm and nut location

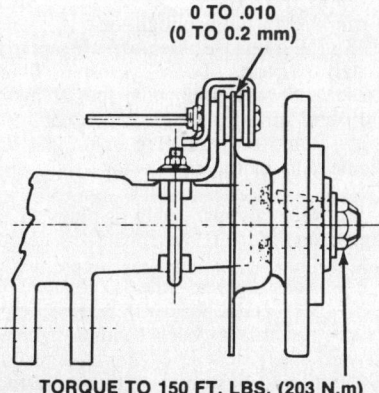

Brake pad installation

2. Remove one of the bolts G which hold the brake pads in the bracket.
3. The brake pads can now be removed.
4. Reverse the above procedure to reassemble.

TRANSAXLE

Both models come equipped with a hydrostatic transmission. The QS (1660) and early QT (1666) tractors use the Sundstrand in-line transmission. Later QT models (from serial #0500101 and up) use the Eaton model 11 transmission.

Removal and Installation

EATON MODEL 11

1. Remove seat and fender assemblies. Be careful to disconnect the seat switch wires before removing.
2. Remove tunnel (A).
3. If transaxle is to be worked on, drain fluid and dispose of.
4. Block tractor at midframe.
5. Disconnect two brake arms (B) from brake cross shaft.
6. Remove righthand and lefthand brake arm extensions (C) (2 bolts each).
7. Locate hydraulic line between hydro charge pump and lift control valve and disconnect at hydro.
8. Disconnect oil line between control valve and oil filter manifold at manifold.
9. Remove cotter pin and washer and disassemble control rod (D) from the foot pedal shaft.
10. Disconnect rod (E) from lock plate to brake cross shaft at cross shaft.
11. Remove the four screws securing the transaxle to the frame.
12. Remove the two top mounting screws (F).
13. Place one foot on the rear hitch or use a bar in rear hitch hole to balance the transaxle while removing.
14. Slowly roll the complete transaxle and hydro back and out of the frame.
15. After the transaxle assembly has cleared the frame, tip the transaxle backward to rest on a 6 in. block.
16. Position right and lefthand brake rods on inside of frame rails.
17. Slowly roll transaxle between the two side frames. At the same time position the following:
   a. Lay hydro control rod (D) over brake across shaft (G).
   b. Direct hydro control rod (D) toward arm (H).
   c. Lay lock plate rod (E) on top of cross shaft (G).
   d. Slide driveshaft onto hydro shaft.
18. Install two top transaxle mounting screws (F).
19. Connect hydro control rod (D) to foot pedal arm with clevis pin and cotter pin.
20. Connect lock plate arm (E) to cross shaft arm.
21. Connect two hydraulic lines to hydro and filter.
22. Line up transaxle to frame and install the four axle mounting screws (I). Torque to 60 ft.lb.
23. Torque two top transaxle bolts to 23 ft.lb.
24. Install new hydraulic fluid, 10 qts.

NOTE: A new oil filter is recommended.

25. Tighten all hydraulic fittings.
26. Check for interference around driveshaft and fan.
27. Remove spark plug or plugs from engine and motor engine for 15-20 seconds.

187

Bolens

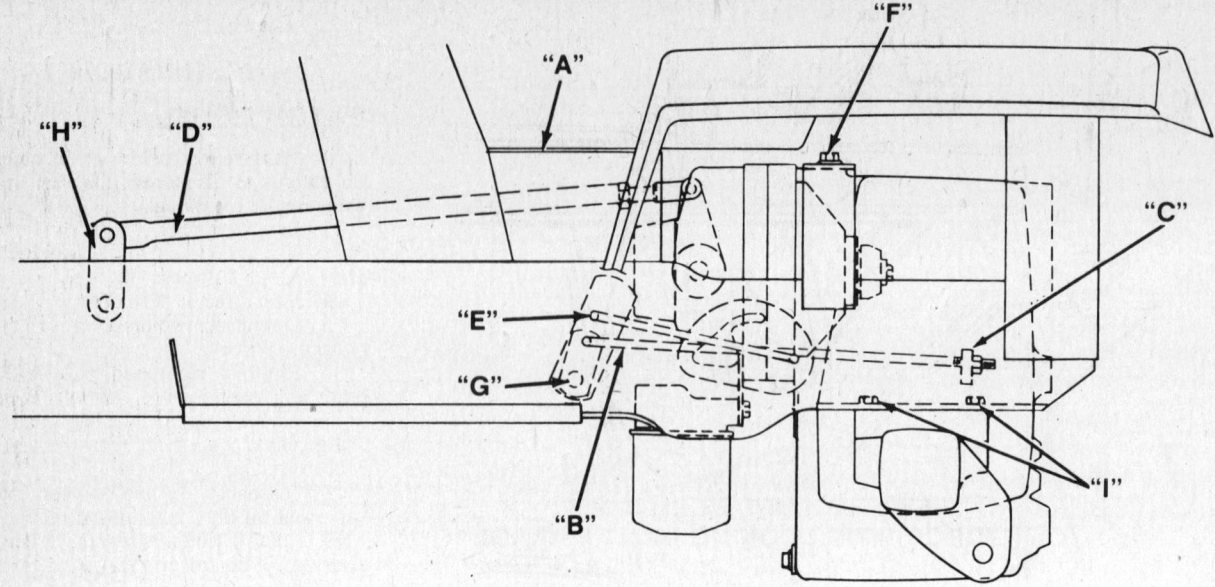

Eaton model 11 mounting points

28. Reinstall spark plug or plugs. Start engine and run at low rpm. Actuate lift system.

29. Block up rear wheels. Restart engine; press pedal forward, reverse, then back to Neutral. If wheels continue to creep forward when in Neutral, shorten adjusting rod (B) until wheels stop. Lengthen adjusting rod if wheels creep backwards.

30. Recheck transaxle fluid level. Add if necessary.

31. Secure tunnel (A) in place.

32. Position the fender in place and connect the two wires to the seat switch and secure fender.

33. Reinstall the seat.

SUNDSTRAND

1. Remove seat and fender assembly.
2. Drain transaxle fluid and discard.
3. Block tractor securely at mid-frame.
4. Temporarily, block up under filter until all subsequent removal steps have been performed.
5. Remove right and left brake rod adjusting nuts.
6. Remove right and left brake arm extensions (2 bolts each).
7. Uncouple both lift system fluid lines (temporarily tie up to steering wheel).
8. Remove hydro control rod clevis pin at travel pedal shaft arm.
9. Remove cotter pin and disconnect quadrant control link at brake cross shaft.
10. Remove all 4 (axle to main frame) mounting capscrews.
11. Place one foot on rear implement hitch (or use ¾ in. bar in rear hitch hole) to balance transaxle while removing the 2 top front bracket screws.
12. Slowly roll complete transaxle and hydro assembly back and out of frame.

NOTE: Driveline will slip off the hydro input shaft. Hold driveline temporarily to avoid damaging fan blades.

13. After transaxle assembly has cleared frame, tip transaxle backward to rest on a 6 in. block.
14. Set transaxle back up into normal position. Place foot on implement hitch or place a ¾ in. bar into implement hitch-pin hole and begin rolling assembly into tractor frame.
15. Lay the right and lefthand brake rods on inside frame ledge.
16. Lay travel pedal control rod over brake cross shaft and aim rod toward travel pedal arm clevis.
17. Be sure quadrant-to-brake cross shaft link is led up over brake cross shaft.
18. Balancing transaxle (with bar in hitch or foot on hitch) roll unit on in to engage driveline to hydro input shaft.
19. Match one of the two top transaxle mounting holes with drift punch through front bracket. Start capscrews and turn in fingertight.
20. Install treadle control rod clevis pin. Hook up quadrant link to brake cross shaft and secure with cotter key.
21. Line up transaxle with frame, install the 4 axle housing capscrews. Torque to 60 ft.lb.
22. Reinstall right and left brake arm extensions and readjust brakes.
23. Slide driveshaft onto hydro input shaft until hole lines up on engine crankshaft. Install bolt and nut and tighten. Tighten setscrew. Check for ⅛ in. clearance between driveshaft and hydrostatic. Shims may be added between engine output shaft and flywheel to achieve ⅛ in. clearance. Adjust engine mounts if necessary.
22. Fill a new Bolens oil filter with hydrostatic fluid (type "F"). Grease filter O-ring and install handtight only.
23. Fill transaxle assembly with approximately 10 quarts of new hydrostatic fluid.
24. Loosen pick-up tube line at hydro end, until fluid leaks out. Retighten.
25. Remove spark plug or plugs from engine and motor engine for 15-20 seconds.
26. Reinstall spark plug or plugs. Start engine and run at low rpm. Actuate lift system.
27. Block up rear wheels. Restart engine; press pedal forward, reverse, then back to Neutral. If wheels continue to creep forward when in Neutral, shorten adjusting rod until wheels stop. Lengthen adjusting rod if wheels creep backward.

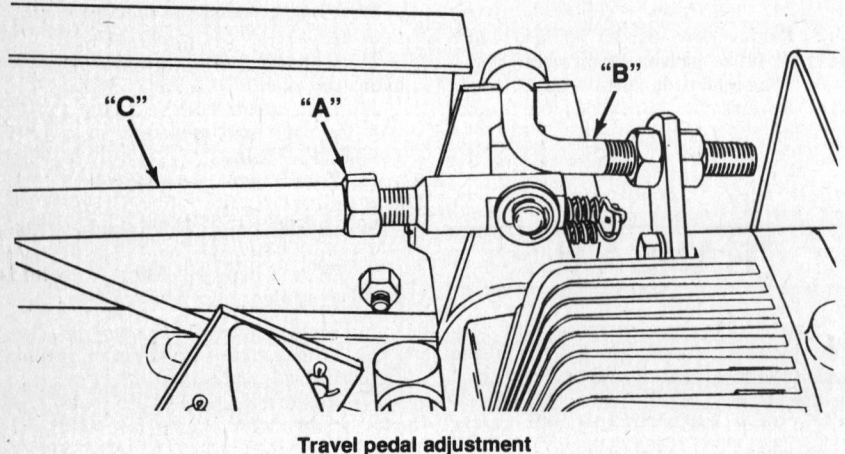

Travel pedal adjustment

28. Recheck transaxle fluid level. Add if necessary.

Overhaul—Eaton 11

HYDROSTATIC UNIT

Removal

1. Remove the transaxle from the frame. See transaxle removal.
2. Tip the transaxle back and rest on a 6 in. block.
3. Remove any dirt or grease from hydraulic suction line (A) fittings.
4. Disconnect hydraulic suction line (A) and save.
5. Remove nut and washer on control shaft and slide control arm assembly (B) off and save.

NOTE: A puller may have to be used when removing the control arm.

6. Remove remaining 2 oil filter lines at hydro.
7. Remove the four (4) mounting screws (C) securing the hydro and oil filter.
8. If replacing complete hydro, remove the three hydraulic tube fittings and install them into the new hydrostatic. Do not tighten.
9. If replacing complete hydro, remove bevel gear (D) on output shaft. To disassemble, hold spacer (E) in place when loosening nut. Position bevel gear onto new hydrostatic and fasten. Torque nut to 24 ft.lb.

Disassembly

NOTE: a. Before any repair work clean the area thoroughly.
b. O-ring (1) is replaceable as is control shaft seal, item (2).
c. Seal may be replaced without removing the transmission from vehicle, however, loss of fluid will occur.

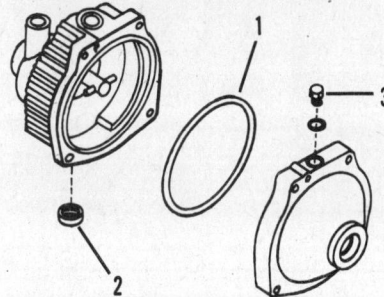

Hydrostatic unit separation

1. Remove control arm with approved tool (standard puller).

NOTE: Do not attempt to pry or drive the control arm off the shaft, as internal damage can occur.

2. Remove control shaft seal, item (2).

NOTE: A hook puller or screwdriver may be used.

3. Wipe seal counterbore clean and examine for damaged surfaces. Install new double lip seal, item (2), with the steel retainer to the outside. Press or tap lightly until seal is bottomed.

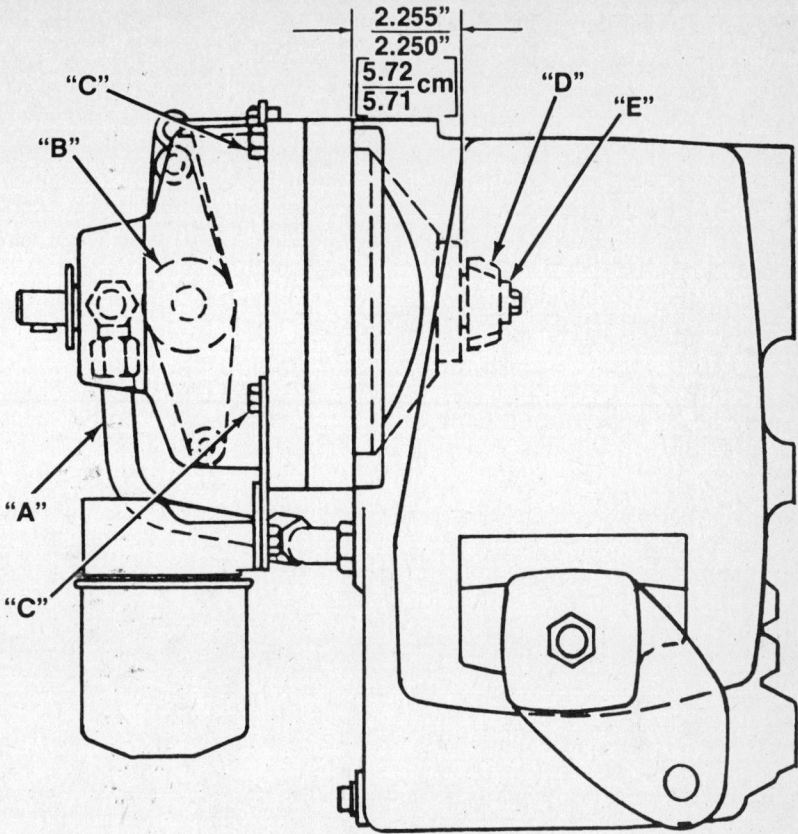

Hydrostatic unit removal

CAUTION
Over-travel of the seal will cut the rubber nose.

---

4. Use a square-faced tube with the O.D. slightly smaller than the O.D. of the seal and the I.D. a slip fit over the shaft. Grease the seal lips before installation. Install control arm.

NOTE: Lubricate the tapered portion of the shaft.

5. Remove capscrew, item (3), and add hydrostatic fluid until unit is full.
6. Replace and tighten capscrew, item (3).

Installation

1. Tip the transaxle back and rest on a 6 in. block.
2. Install spacer, nut (A), and bevel gear (B) on output shaft. Torque nut to 24 ft.lb.
3. Install control arm assembly (C) to control shaft. Secure with nut and washer.
4. Install hydro to transaxle housing.
5. Install the four mounting screws (D) to secure the hydro and oil filter. Torque nut to 24 ft.lb.
6. Install the three hydraulic tube fittings.
7. Connect hydraulic suction tube (E).

Overhaul—Sundstrand

HYDROSTATIC UNIT

Removal
1. Drain fluid from Transaxle.

NOTE: Do not reuse oil fluid.

2. Loosen setscrew and remove bolt holding driveshaft on engine crankshaft. Slide shaft forward until it clears hydro input shaft.
3. Disconnect hydraulic lines.
4. Disconnect pick-up tube line.
5. Remove hydro oil filter.
6. Remove bolt holding neutral adjusting rod.
7. Drive out roll pin holding linkage to control shaft. Remove linkage as an assembly.
8. Remove the remaining three bolts holding hydro.

NOTE: Hold hydro while removing last bolt to prevent it from dropping. Remove hydro from tractor.

9. Remove the two oil line fittings and the pickup tube fitting from old hydro.
10. Remove roll pin from free wheeling valve if replacing hydro.

Disassembly

NOTE: Before disassembly of the various sections, a line should be scribed across the top of the transmission so the proper positions of the sections will be maintained during reassembly.

REMOVAL OF CHARGE PUMP

1. Before removing housing, inspect the input shaft especially the key way, for burrs or sharp edges that could damage the lip seal.
2. Remove the four capscrews and pull housing off the input shaft.

Bolens

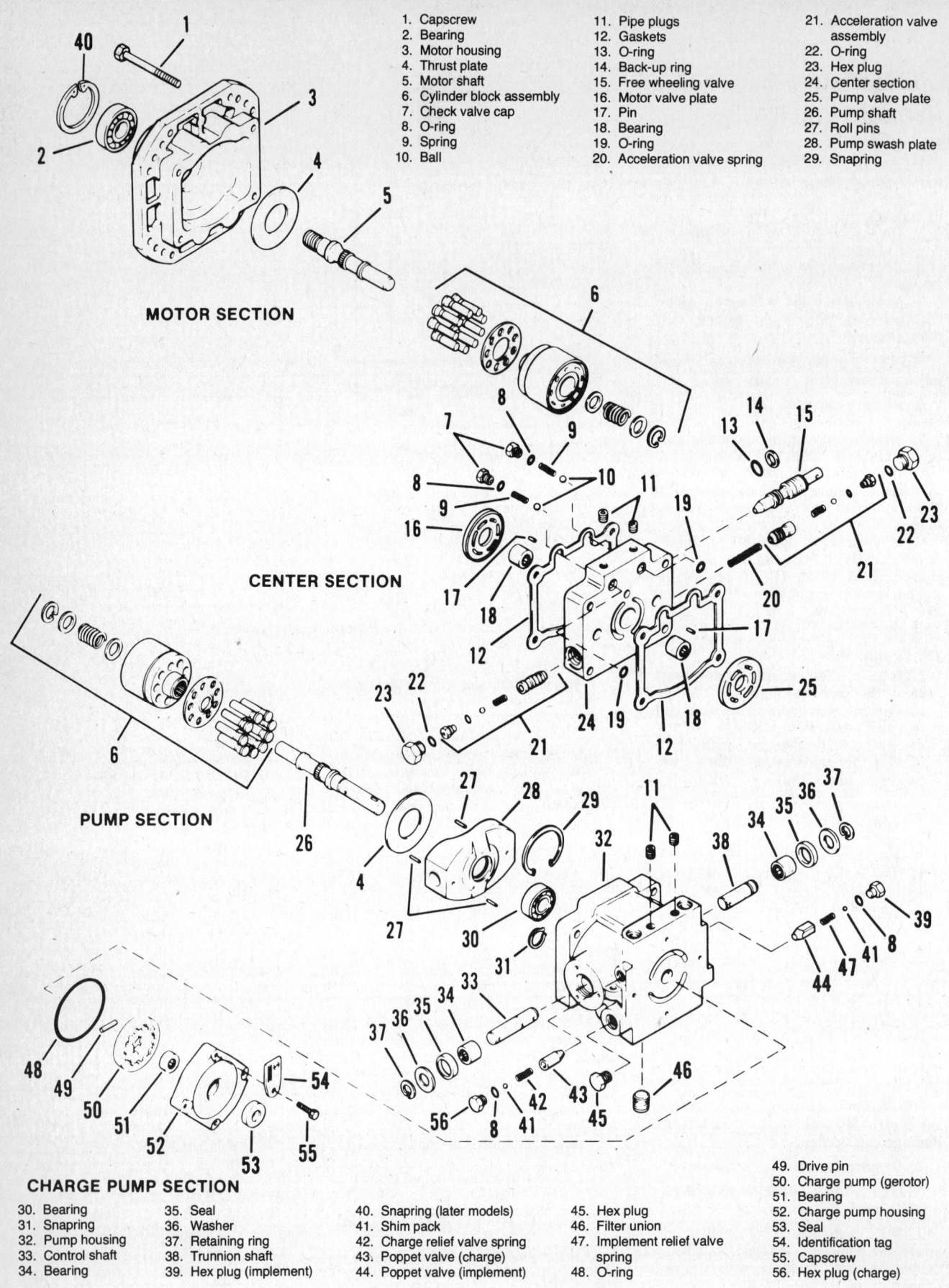

1. Capscrew
2. Bearing
3. Motor housing
4. Thrust plate
5. Motor shaft
6. Cylinder block assembly
7. Check valve cap
8. O-ring
9. Spring
10. Ball
11. Pipe plugs
12. Gaskets
13. O-ring
14. Back-up ring
15. Free wheeling valve
16. Motor valve plate
17. Pin
18. Bearing
19. O-ring
20. Acceleration valve spring
21. Acceleration valve assembly
22. O-ring
23. Hex plug
24. Center section
25. Pump valve plate
26. Pump shaft
27. Roll pins
28. Pump swash plate
29. Snapring
30. Bearing
31. Snapring
32. Pump housing
33. Control shaft
34. Bearing
35. Seal
36. Washer
37. Retaining ring
38. Trunnion shaft
39. Hex plug (implement)
40. Snapring (later models)
41. Shim pack
42. Charge relief valve spring
43. Poppet valve (charge)
44. Poppet valve (implement)
45. Hex plug
46. Filter union
47. Implement relief valve spring
48. O-ring
49. Drive pin
50. Charge pump (gerotor)
51. Bearing
52. Charge pump housing
53. Seal
54. Identification tag
55. Capscrew
56. Hex plug (charge)

Sundstrand hydrostatic unit components

Bolens

Charge pump components

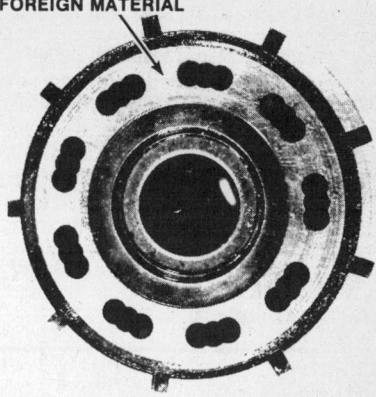

Inspecting cylinder block

3. The charge pump O-ring, drive pin, gerotor assembly, bearing and seal can now be inspected and replaced if necessary.

CHARGE PUMP INSPECTION

NOTE: The charge pump uses a needle bearing, small diameter seal, and a hardened input shaft. The hardened input shaft is needed to provide the inner race for the needle bearing.

1. The gerotor surface should be inspected on both sides for grooves or pickup of foreign material.
2. The drive pin should be inspected for cracks of fatigue points.
3. The charge pump oil cavity lead-ins and inner housing surface should be inspected for damage or excessive wear. Damaged or worn parts should be replaced.

SEPARATING PUMP, MOTOR AND CENTER SECTIONS

1. Set the hydrostatic pump end down on a soft wooden block with a center hole for the input shaft to extend into.
2. Remove the four capscrews located in the motor section which hold the 3 sections together. Remove all screws equally to prevent damage to threads or surfaces. The three sections are somewhat spring loaded so care must be taken to prevent the transmission from falling apart.
3. When the four capscrews have been removed, the transmission should separate into the three basic sections.

DISASSEMBLY OF PUMP SECTION

1. Remove cylinder block assembly from pump housing.
2. Remove thrust plate from swashplate.
3. Remove pump variable swashplate.
   a. Set pump housing, charge pump end down on a soft wooden block with a center hole for input shaft to extend into.
   b. With a punch and light hammer tap spring roll pins through variable swashplate.

NOTE: Excessive heavy pounding may damage bearings.

   c. Push shafts from swashplate and housing and remove the swashplate.
   d. Press the needle bearings from housing.
4. Remove pump driveshaft.
   a. Remove snaprings.
   b. Tap lightly on input end of pump shaft with a soft hammer to remove the bearing and shaft.
   c. Press bearing from driveshaft.

INSPECTION OF PUMP SECTION

1. Inspect control and trunnion shaft bearings and seals.
2. Inspect thrust plate for scratches or wear. Replace all worn or damaged parts.

INSPECTING CYLINDER BLOCK ASSEMBLIES

The pump cylinder block and the motor cylinder block are identical.

1. Inspect the cylinder blocks for wear or pickup of foreign material.

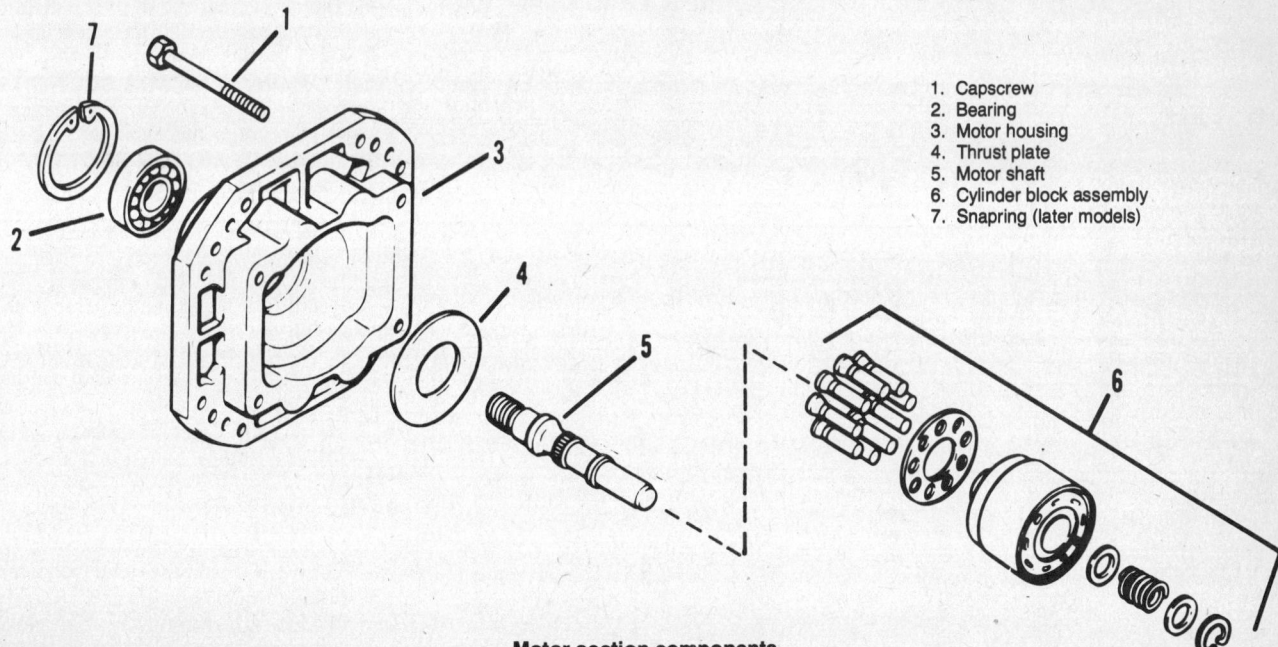

1. Capscrew
2. Bearing
3. Motor housing
4. Thrust plate
5. Motor shaft
6. Cylinder block assembly
7. Snapring (later models)

Motor section components

191

Bolens

2. Inspect the piston retainer for straightness. The retainer must be straight so the piston slipper surfaces will not wear uneven.

3. Inspect the piston assembly. The lubricant hole must be free of obstruction and the slipper surface even. If the slipper surface has more than $1/32$ in. taper, the cylinder block assembly must be replaced.

Individual parts are not available for the cylinder blocks because of the close tolerances. If any problem is evident the complete cylinder block must be replaced.

DISASSEMBLY
OF CENTER SECTION
1. Remove valve plates and valve plate locating pins.
2. Remove gaskets.
3. Remove bearings.
4. Remove check valve O-rings, pump side and check valves, motor side.

INSPECTION OF CENTER SECTION
1. Inspect bearings.
2. Inspect check valves and check valve seats. Make sure check valves operate freely.
3. Inspect back side of motor and pump valve plates for dirt or wear in the locating pin slots.

PUMP AND MOTOR VALVE PLATE IDENTIFICATION
Since the pump cylinder block assembly only turns in the lefthand or clockwise rotation, the lead-ins in the pump valve plate are located for lefthand rotation only. The motor cylinder block assembly turns in both directions thus the motor valve plate has lead-ins for both right and lefthand rotation.

DISASSEMBLY
OF MOTOR SECTION
1. Remove cylinder block assembly.
2. Remove thrust plate.

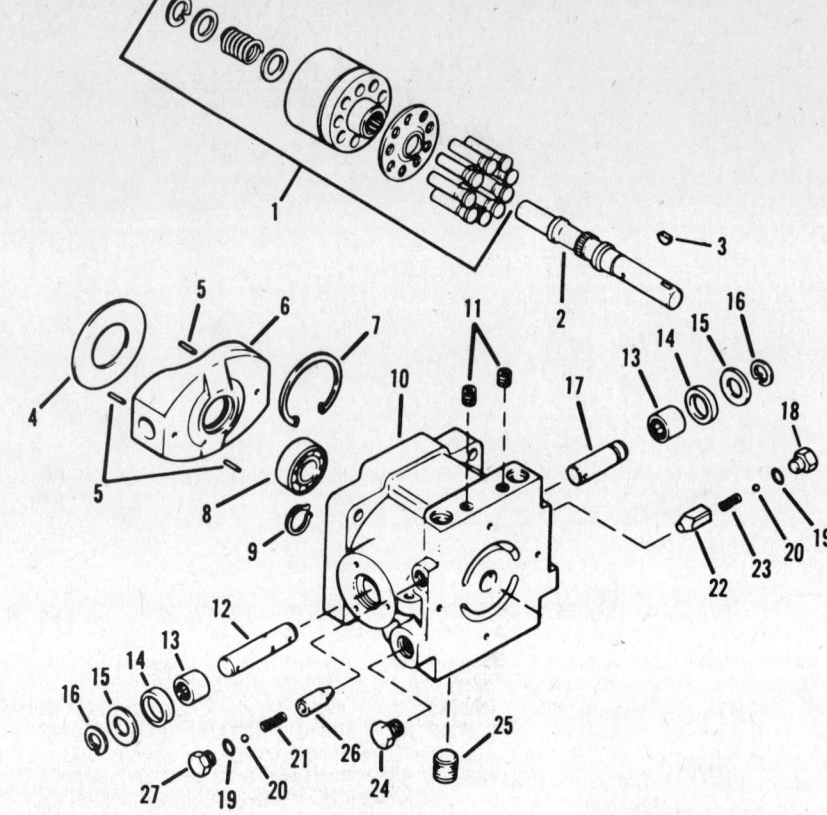

1. Cylinder block assembly
2. Pump shaft
3. Key
4. Thrust plate
5. Pin
6. Pump swash plate
7. Snapring
8. Bearing
9. Snapring
10. Pump housing
11. Pipe plugs
12. Control shaft
13. Bearing
14. Seal
15. Washer
16. Retaining ring
17. Trunnion shaft
18. Hex plug (implement)
19. O-ring
20. Shim pack
21. Charge relief valve spring
22. Poppet valve
23. Implement relief valve spring
24. Hex plug
25. Filter union
26. Poppet valve
27. Hex plug (charge)

Pump section components

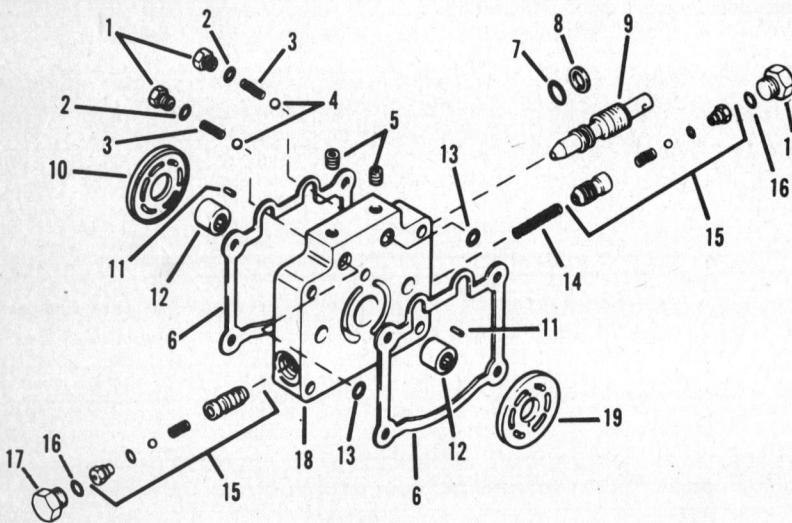

1. Check valve cap
2. O-ring
3. Spring
4. Ball
5. Pipe plugs
6. Gaskets
7. O-ring
8. Back-up ring
9. Free wheeling valve
10. Motor valve plate
11. Pin
12. Bearing
13. O-ring
14. Acceleration valve spring
15. Acceleration valve assembly
16. O-ring
17. Hex plug
18. Center section
19. Pump valve plate

Center section components

3. Remove snapring (later models).
4. Tap on internal end of motor shaft to remove shaft and bearing from housing.
5. Press bearing from motor shaft.

INSPECTION OF MOTOR SECTION
Inspect thrust plate for scratches or wear. Replace all worn or damaged parts.

ASSEMBLY OF MOTOR SECTION
1. Press bearing to shoulder on motor shaft.
2. Install motor shaft and bearing into housing.
3. Install snapring (later models).
4. Install thrust plate onto motor housing swashplate.
5. Align splines of cylinder block assembly with splines of motor shaft and install into housing.

ASSEMBLY OF CENTER SECTION
1. Install bearings into center section. Bearings must protrude .100 in. from polished surface, printed end out.
2. Install check valves on motor side of center section. Torque check valves to 10 ft.lb.
3. Install O-rings to pump side of center section.
4. Install valve plate pins and valve

plates. The pump valve plate has two lead-ins and goes on the pump side of center section and the motor valve plate has four lead-ins and goes to the motor side of center section.

5. Use new gaskets on both sides of center section.

ASSEMBLY OF PUMP SECTION

1. Press bearings into housing, lettered edge out, until flush to .005 in. below lip seal recess of housing.
2. Press bearing on pump shaft to shoulder.
3. Install input shaft and bearing into housing.
4. Position pump housing with check valve passageways up.
5. Place swashplate over driveshaft into housing.
6. Insert shorter trunnion shaft into housing and swashplate from left side and longer control shaft (small hole end) into housing and swashplate from right side.
7. Install the spring roll pins. Two on control shaft side, one on trunnion shaft side. Proper installation is when pins are ¼ in. into swashplate.
8. Install thrust plate to swashplate.
9. Align splines of cylinder block kit with input shaft splines and install into housing.

ASSEMBLY PROCEDURE OF CHARGE PUMP

Clean all parts and lubricate with new hydrostatic transmission fluid before assembly.

1. Install the input shaft bearing into the charge pump housing. Bearing must be installed from the inside and be flush to .005 in. below its bore, with the lettered side out.
2. Install the seal into the housing until seated. Press only on the outer edge.
3. Install new O-ring into housing groove.
4. Install drive pin into hole in input shaft.
5. Install gerotor assembly over drive pin.
6. Install charge pump housing. Be careful not to damage lip seal. Torque capscrews to 20 ft.lb.

ASSEMBLY OF HYDROSTATIC UNIT

1. Align pump, center, and motor sections properly.
2. Install the 4 capscrews and torque to 35 ft.lb.
3. Check the torque needed to turn the input shaft, output shaft, and control shaft. Torque reading should not exceed 25 in.lb.

NOTE: If torque exceeds this amount, hydro will have to be disassembled to locate binding.

Installation

1. Remove protection cap from oil filter stud and stake stud to prevent it from turning further into hydro during filter installation.
2. Remove protective tape from oil filter pad on hydro.

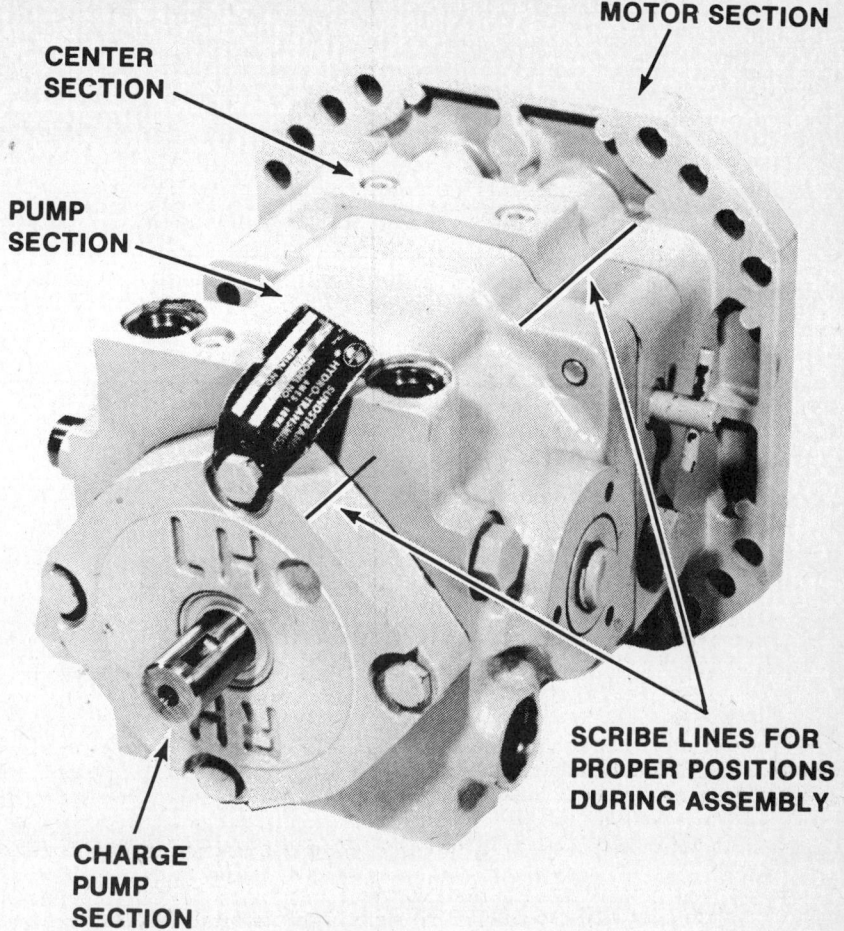

Hydrostatic unit assembly

NOTE: If filter is installed over protective tape, oil passage will be blocked and damage to the hydro will occur.

3. When replacing the hydro package the bevel output gear must be removed from the old hydro and installed on the new one. The bevel gear is held on with either an external snapring or a nut. Make sure snapring is properly installed in the groove. The snapring should be replaced when it shows signs of weakness or distortion. On hydro package, with nut holding bevel gear, bend the thin washer up around one or two of the flats on the nut to secure it.
4. Install roll pin into free wheeling valve.

NOTE: Be careful not to damage valve.

5. Install the two oil line fittings and pick-up tube fitting. Do not tighten packing nuts so fittings can be adjusted to lines.
6. Install gasket onto hydro using grease to hold it in place.

NOTE: Refer to old hydro for bolt pattern to locate gasket.

7. Install hydro and the three bolts. (Save longer bolt to attach neutral adjusting rod.)
8. Install linkage on control shaft and drive in roll pin to hold in place.
9. Install netural adjusting rod and torque all four bolts to 31 ft.lb.
10. Install pick-up tube and tighten packing nut.
11. Install the two hydraulic oil lines and tighten packing nut.

POWER TAKE-OFF (PTO)

DRIVE BELT REPLACEMENT

1660 (QS)

1. Remove hood.
2. Place attachment drive lever into the "On" position.
3. Remove cotter pin (A) and pin in lower brake shoe pivot and drop shoe (B).
4. Place attachment drive lever into the "Off" position.
5. Remove spring (C) and cable for attachment drive control.
6. Remove idler spring (D).
7. Remove old belts.
8. Reverse above procedure to install new belts.

NOTE: If the PTO brake adjustment has been disturbed while replacing the belts readjust as shown.

Bolens

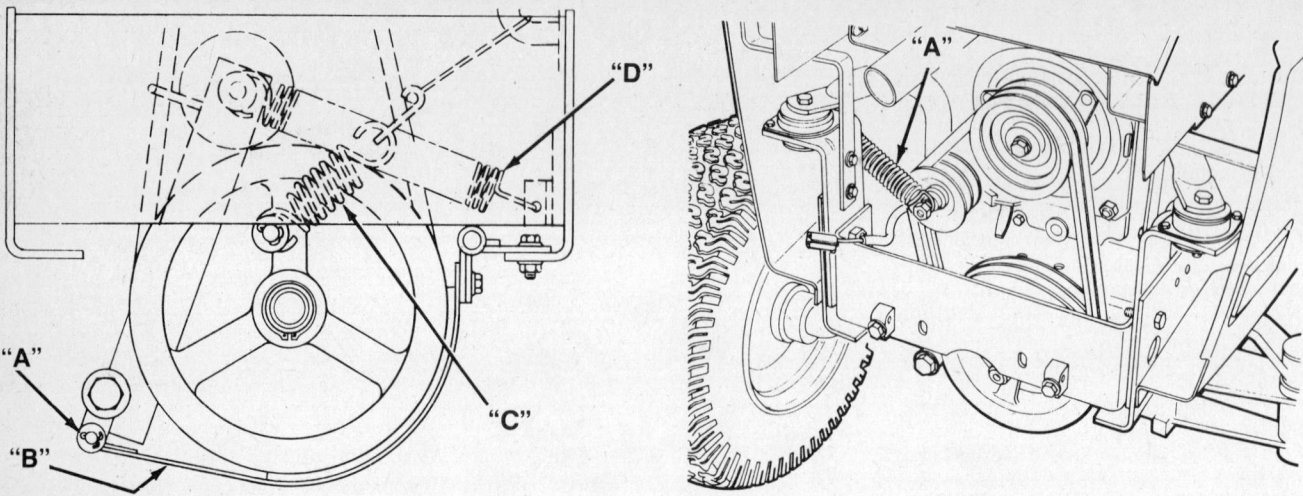

1660 PTO attachment points

1666 PTO attachment points

1666 (QT)
1. Place switch in OFF position.
2. Remove hood.
3. Unhook spring (A) from idler shaft.
4. Remove old belts.
5. Reverse the above procedure to install the new belts.

NOTE: Shredding of belts is generally caused by misaligned idler pulley. If this problem exists, make sure idler is aligned before new belts are installed.

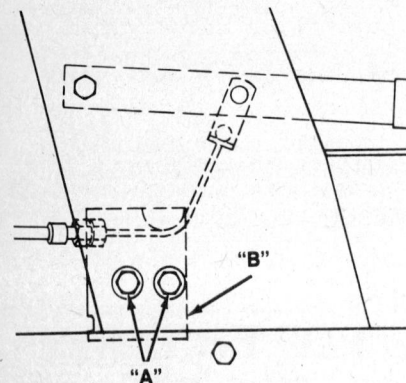

PTO cable adjustment

CABLE ADJUSTMENT

Should the belts on the attachment drive slip adjust as follows:
1. Loosen the two capscrews (A) on the left side of the tractor console.
2. Move bracket (B) down in slots.
3. Test control for proper belt tension.

BRAKE SHOE ADJUSTMENT

Model 1660 (QS)

Should the attachment drive brake need adjustment proceed as follows:
1. Loosen two hex nuts and capscrews (A).
2. Move brake shoe up in slots until there is .012 in. clearance between the brake shoe and the rim of the attachment drive sheave.

REMOVAL

NOTE: Do not use wheel puller to remove clutch/brake.

1. Disconnect polarized plug (A) from clutch.
2. Remove four locknuts (B) and remove brake plate mounting.
3. Remove mounting bolt from center of crankshaft.
4. Remove sheave and armature assembly.
5. Remove rotor assembly.
6. Remove four 10-32 x ½ in. socket head capscrews and lockwashers securing coil assembly to crankcase.
7. Remove coil assembly from engine.

INSTALLATION

1. Position coil assembly on engine piloting in crankshaft seal bore. Secure with four 10-32 x ½ in. socket capscrews and lockwashers.
2. Install the four brake mounting studs into engine block.

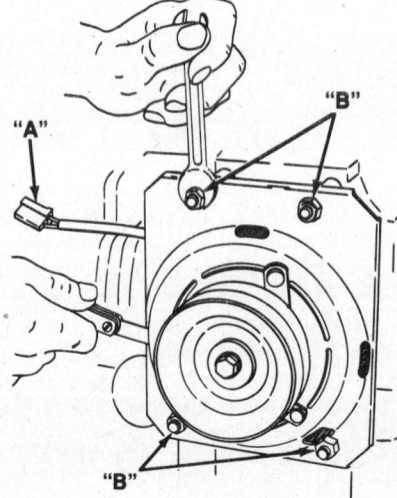

PTO unit removal

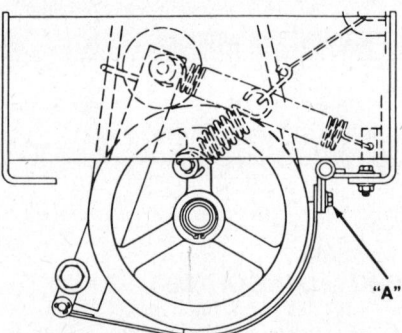

1660 PTO brake shoe adjustment

3. Place compression springs on brake mounting studs.
4. Remove rust, burrs and dirt from engine shaft with sandpaper.
5. Apply light coat of grease or oil to shaft.
6. Insert key into key way in shaft.
7. Push rotor and armature assembly with pulley onto engine shaft until hub bottoms on crankshaft shoulder.
8. Place plain washer and lockwasher on clutch mounting capscrew.
9. Install capscrew into tapped hole in center of crankshaft. Hold clutch from rotating, while tightening capscrew.
10. Adjust brake as described under maintenance.
11. Reconnect coil assembly lead to lead from switch making certain connection is secure and wire is not rubbing against rotating parts.

BRAKE/CLUTCH ADJUSTMENT

1. The brake portion may require readjustment periodically depending upon unit usage. This can be accomplished as follows:
2. Position a .015 in. thick shim in each slot provided (4) in brake flange and turn on clutch/brake.
3. Push on the brake flange until it bottoms out. Tighten the (4) locknuts until they just contact brake flange.

Bolens

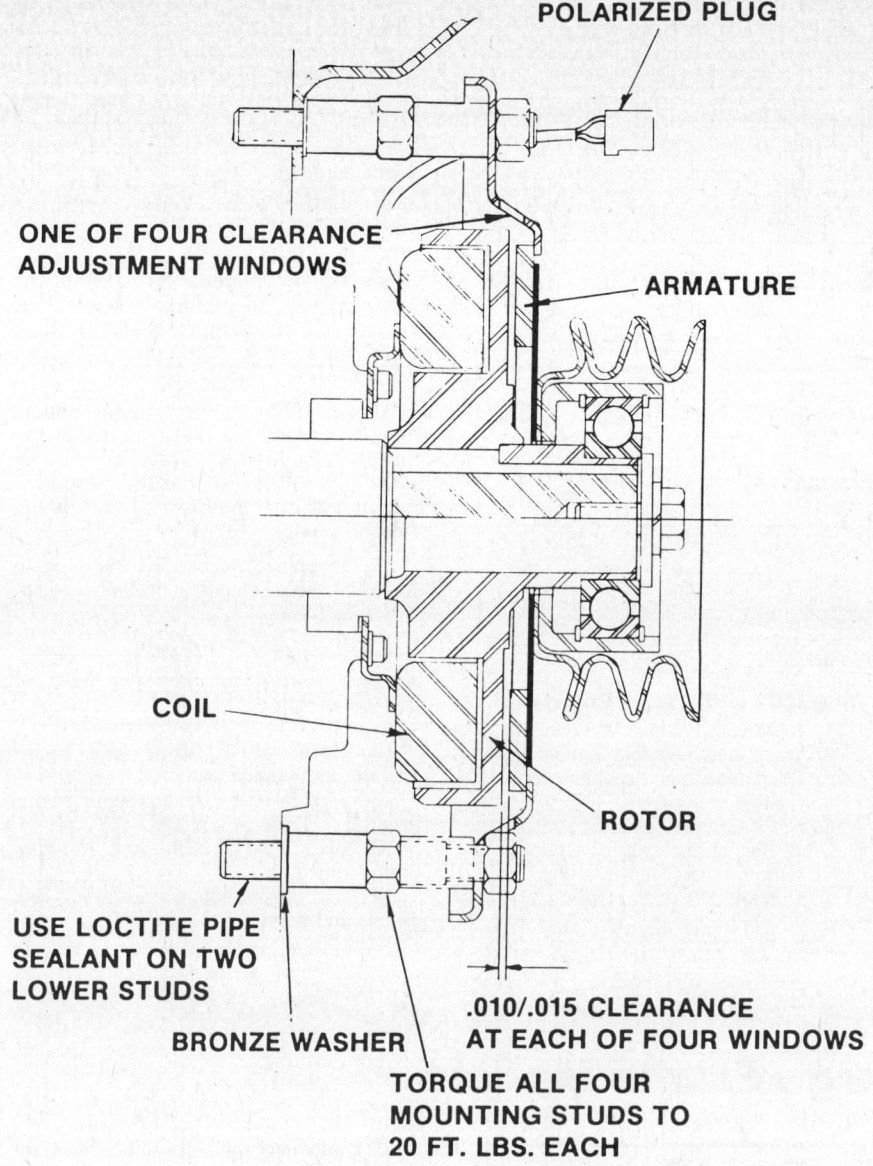

PTO brake/clutch adjustment

CAUTION

Do not over torque locknuts as damage to brake flange may result.

---

4. Turn clutch/brake off and remove (4) shims. Recheck gap through the slots provided. A minimum of .010 in. and a maximum of .020 in. should be maintained.

ENGINE

The following engines are used:
Model 1660 (QS): Kohler K-341
Model 1666 (QT) serial number 0600101 and later: Kohler KT-17
Model 1666 (QT) serial number 0100101 through 0599999: Onan BF/MS

For engine service, see the Engine Unit Repair section of this book.

REMOVAL AND INSTALLATION

1660 (QS)

1. Remove hood and side panels.
2. Disconnect negative battery cable.
3. Disconnect headlight, alternator-ignition connectors and starter cable.
4. Disconnect choke and throttle cables.
5. Close fuel valve and unhook fuel tank hold down spring.
6. Disconnect fuel line at fuel pump inlet. Remove fuel tank, fuel filter and complete fuel line from tractor.
7. Disconnect PTO clutch return spring. With clutch lever in "ON" position, remove cotter pin and washer at lower righthand brake anchor stud and remove brake band from stud.
8. With clutch lever in "OFF" position, remove belts from pulleys.
9. Remove belt idler pivot bolt from axle bolster and remove complete idler arm assembly.
10. Remove the 4 rubber mount bolts and driveshaft staybolt. Hold driveshaft at hydro to prevent damage to fan as rear joint clears hydro input shaft. Slide driveline forward on engine shaft. Remove driveshaft.
11. Raise engine and cradle assembly up and out of tractor frame.

NOTE: Before reinstalling engine, check rubber mounts for damage or fatigue cracks. Inspect hydro input shaft and key. Inspect engine shaft and key. Coat both with Never-Seez® before reinstalling.

NOTE: If the rubber engine mounts have been removed for any reason, the more rigid mount with the lower center is the "front" mount and the more flexible mount with the higher center is the "rear" mount.

12. Set engine and cradle assembly into frame on the 4 mounting brackets and insert bolts loosely.
13. Reinstall driveline and insert staybolt into engine shaft.
14. Tighten four engine mounting bolts securely. Right rear carries extra halfnut to attach electrical ground.
15. Reinstall PTO belts, brake band and linkages.
16. Run engine and observe belt guide behavior before reinstalling side panels and hood.

1666 (QT) with Kohler Engine

1. Disconnect light harness (A).
2. Remove hood and side panels.
3. Disconnect electric clutch (B), alternator-ignition connectors and starter cable (C). Pull through engine heat shield.
4. Disconnect battery cables and remove battery.
5. Remove air cleaner from engine.
6. Disconnect choke and throttle cables. Pull through heat shield.
7. Close fuel valve and disconnect fuel line from filter at engine side (D). Pull through shield.
8. Remove fuel tank.
9. Loosen side panel supports (E) and pivot away from engine.
10. Disconnect PTO clutch idler spring and remove PTO belts.
11. Disconnect right rear engine support (F). Take note the position of the ground wires.
12. Remove the three remaining engine support bolts (G).
13. Remove driveshaft bolt. Slide driveshaft off of engine stub shaft.
14. Raise engine and support assembly up and out of tractor frame.

NOTE: Before reinstalling engine, check rubber mounts for damage or fatigue cracks. Inspect engine shaft and key. Coat both with Never-Seez® before installing.

NOTE: If the rubber engine mounts have been removed for any reason, the

195

Bolens

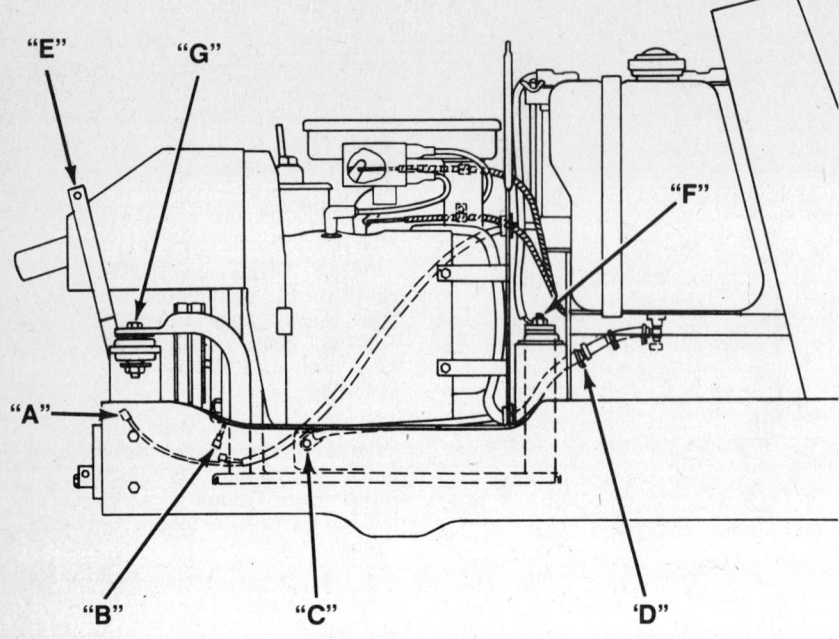

Kohler engine removal

more rigid mount with the lower center is the "front" mount and the more flexible mount with the higher center is the "rear" mount.

15. Reverse the above procedure when reinstalling engine.
16. Torque engine driveshaft bolt to 49 in.lb.
17. Torque four engine support bolts to 23 ft.lb.

1666 (QT) with Onan Engine

1. Remove hood and side panels.
2. Disconnect negative battery cable.
3. Disconnect electric clutch, coil, alternator and starter cable.
4. Close fuel tank petcock and disconnect fuel line at fuel pump. Unhook gas line at bottom blower housing clip.
5. Unhook PTO idler spring and remove PTO belts.
6. Remove air cleaner assembly, choke and throttle cables.
7. Remove the 4 rubber mount bolts and driveline staybolt. Remove driveshaft.
8. Remove the top screws and loosen bottom screws from the top panel right and left support brackets. Pivot brackets forward and down.
9. Raise engine and cradle assembly up and out of tractor frame.

NOTE: Before reinstalling engine, check rubber mounts for damage or fatigue cracks. Inspect hydro input shaft and key. Inspect engine shaft and key. Coat both with Never-Seez® before reinstalling.

NOTE: If the rubber engine mounts have been removed for any reason, the more rigid mount with the lower center is the "front" mount and the more flexible mount with the higher center is the "rear" mount.

10. Set engine and cradle assembly into frame on the 4 mounting brackets and insert bolts loosely.

---- CAUTION ----
Recheck routing of alternator leads to prevent abrasion.

11. Reinstall driveline and insert staybolt into engine shaft.
12. Tighten four engine mounting bolts securely.
13. Install PTO belts and position spring.
14. Run engine before reinstalling side panels and hood.

Large Frame Models

1886s-05, 1886s-06 (HT-18), 2086 (HT-20), 2087 (HT-20), 2389 (HT-23)

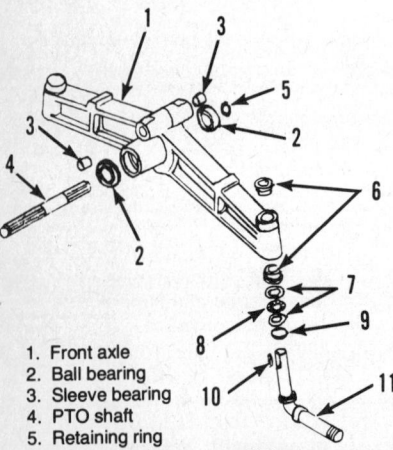

1. Front axle
2. Ball bearing
3. Sleeve bearing
4. PTO shaft
5. Retaining ring
6. Flange bearing
7. Thrust race
8. Thrust bearing
9. O-ring
10. Key ¼ x 1
11. Spindle assembly

Front axle components

FRONT AXLE

REMOVAL AND INSTALLATION

1. Remove the PTO drive belts.
2. Disconnect the ball joint at the right side steering arm.
3. Raise and support the front end on jack stands.
4. Remove the pivot pin from the axle and roll the axle from under the tractor.
5. Inspect all parts and replace any showing excessive wear.
6. Installation is the reverse of removal.

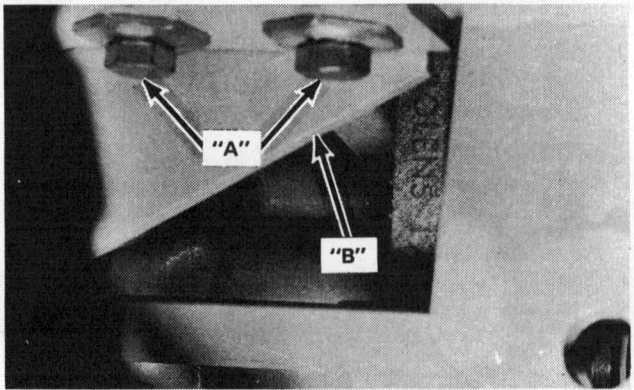

Front axle stop adjustment screws

Bolens

FRONT AXLE STOP ADJUSTMENT

1. Loosen the axle stop screws.
2. Move the axle stop assemblies up against the axle and tighten the screws.

Front Wheels and Bearings

REMOVAL, INSTALLATION AND REPACKING

1. Remove the dust cap (1).
2. Remove cotter pin (2) and unscrew slotted nut (3).
3. Remove outer bearing (4).
4. Remove wheel (5) and hub assembly (6) from spindle.
5. Remove seal (7) from hub. If this seal is damaged in any way replace it.
6. Remove inner bearing (8) from hub.
7. Clean both inner race (9) and outer race (10). Check both for wear or pitting. Replace if necessary.

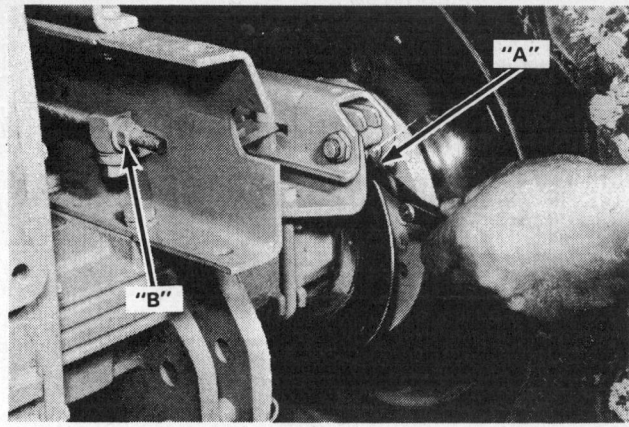

Brake adjustment points

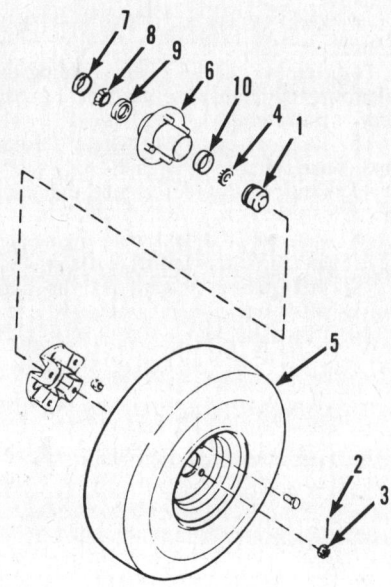

Front wheel and bearing

8. Clean bearings in a commercial solvent. Dry with compressed air or a clean cloth.
9. Pack bearings with a good grade of wheel bearing grease.
10. Install inner bearing and seal into hub.
11. Install wheel and hub assembly to spindle.
12. Install outer bearing.
13. Tighten nut until it snugs up against the outer bearing. Spin wheel to align bearings, then back nut off to nearest slot in line with hole in spindle and install new cotter pin.
14. Install dust cap.

STEERING

Steering Knuckles

REMOVAL AND INSTALLATION

1. Raise the tractor and support it with jack stands under the front axle.
2. Remove the wheels and bearings.
3. Disconnect the tie rod and drag link.
4. Remove the clamp bolt from the steering arm and pull the steering arm from the knuckle.
5. Remove the key and lower steering knuckle from the end of the axle.
6. Inspect all parts and replace those that are excessively worn.
7. Lubricate the bearings with multi-purpose chassis lube and install all parts in reverse order of removal.

Steering Gear

REMOVAL AND INSTALLATION

1. Disconnect the drag link rear ball joint from the lever portion of the quadrant gear.
2. Remove the steering wheel using a puller.
3. Remove the cotter pin from the lower end of the steering shaft. Lift the shaft, disengage the pinion from the quadrant and lift the shaft clear of the support plate. Take care to avoid damage to the self-aligning bearing.
4. Back out the pivot bolt and remove the quadrant gear.
5. Inspect all parts and replace any that show excessive wear.
6. Installation is the reverse of removal. The steering wheel should be straight ahead and the quadrant gear properly centered on the steering pinion. Lubricate all moving parts with clean engine oil. Adjust the steering stops. When the drag link is installed, measure the overall length. Adjust as needed to obtain a length of 27 1/8 in.

BRAKES

ADJUSTMENT

After every 50 operating hours check clearance of brake pads. If there is more than .010 in. clearance between the brake pads and brake disc (A), the brakes need adjustment. To adjust place brake pedal in the OFF position, then turn nut (B) clockwise to bring the brake pads closer to the brake disc. The correct clearance is 0 to 0.10 in.

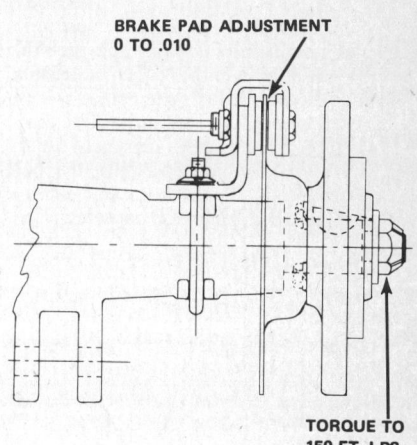

Measure pad clearance

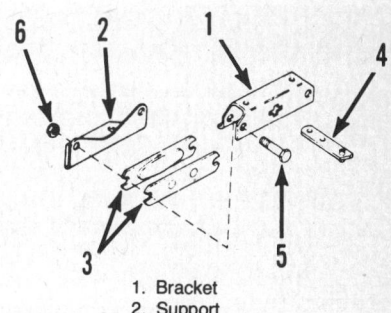

1. Bracket
2. Support
3. Brake pads
4. Lever
5. Bolt
6. Locknut

Brake components

Bolens

NOTE: Be sure to adjust the brakes on both wheels equally to avoid uneven braking. Check for proper operation.

PAD REPLACEMENT

1. Remove brake assembly from tractor.
2. Remove one of the bolts which hold the brake pads in the bracket.
3. The brake pads can now be removed.
4. Reverse the above procedure to reassemble.

TRANSMISSION

Models 1886, 2086 and 2087 use the Sundstrand in-line hydrostatic unit. Model 2389 uses the Eaton Model 11. For removal, installation and overhaul of these units, see the QS and QT tractor section, immediately preceding this section.

POWER TAKE-OFF (PTO)

PTO Lever

ADJUSTMENT

1. Place the lever in the OFF position.
2. Remove the cotter pin and turn the PTO control rod in until the desired tension is achieved. Install the cotter pin.
3. With the PTO lever in the ON position, loosen the hex head capscrew securing the upper belt guide. Adjust the belt guide for 1/8 to 1/4 in. clearance between the belt and guide. Tighten the capscrew.

PTO Belts

REMOVAL AND INSTALLATION

1. Place PTO lever in the OFF position.
2. Do not disturb upper belt guide, if 1/8 to 1/4 in. gap is evident with PTO lever in the "ON" position.
3. Remove lower belt guide and old belts.
4. With hood open place the three new belts over both engine and idler pulley with your right hand while feeding belts up from below with left hand, then feed into grooves of PTO pulley.

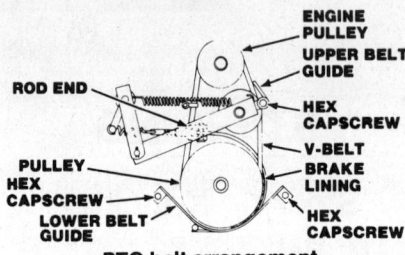

PTO belt arrangement

5. Reinstall lower belt guide and adjust for a 3/32 to 1/8 in. clearance between belt guide and belts with PTO lever in the ON position. Tighten the two hex capscrews securely. Check upper belt guide and adjust if necessary. Place PTO lever in OFF position.

PTO Brake (Model 2087 S/N 0400101 and up and 2389 and up)

ADJUSTMENT

With PTO handle in the engaged position or ON, adjust rod end to obtain a gap, not more than 1/16 in. between the PTO pulley and the brake lining. PTO pulley must stop within 5 If not, adjust rod end till stopping time is reached.

ENGINE

The following engines are used:
Models 1886s-05 & 06: Kohler K482S 35028B & 35113B
2086: Kohler K532S 53100A
2087: Kohler K532S 553102A & 53138A

For all engine service, see the Engine Unit Repair section of this book.

REMOVAL AND INSTALLATION

1. Disconnect battery, negative side first.
2. Remove bolt and setscrew from driveshaft.
3. Disconnect head light wires, remove hood strap, and front mounting bolts. Remove the hood.
4. Shut off gas at the fuel tank. Disconnect the fuel line.
5. Remove the four bolts holding the platform and remove the gas tank and battery as an assembly.
6. Remove electrical wires, throttle and choke cable from engine.
7. Remove the four engine mounting bolts.
8. Remove PTO belts from engine pulley. (Tip engine forward if necessary.)
9. Pull engine forward and up until driveshaft is free of crankshaft.
10. Lift engine out.

J.I. Case

INDEX

MODELS 130 & 160
- Front Axle 200
- Steering 200
- Transmission 201
- Engine 204

MODELS 155 & 195, 150-190, T90, 2310, 2510, 2712
- Brakes 204
- Front Axle and Steering 206
  - Axle 206
  - Enclosed Steering 207
  - Open Steering 208
  - Adjustments 208
- Transmission 208
- Hydraulic Pump 211
- Control Valve 212
- Orbital Motor and Brake Valve .. 213
- Engine 214

MODELS 107, 108, 117, 118, 200, 400, 644, 646
- Steering and Front Axle 215
  - Axle 215
  - Steering Gear 215
- Manual Transaxle 216
  - Model 107 216
  - Model 108 220
  - Model 210 223
  - Models 117 & 188 225
- Attachment Drive Clutch 227
- Hydrostatic Transmission 231
  - Models 117 & 118 231
  - Hydraulic Motor 233
- Hydraulic System 236
- Engine 244

J.I. CASE
Models 130 and 180

FRONT AXLE AND STEERING

130 Series

REMOVAL AND DISASSEMBLY

1. Raise and support the front end with jack stands.
2. Disconnect the drag link (4) from the spindle (13).
3. Remove wheel cover (if so equipped) cotter pin (8), washer (9) and wheel assembly.
4. Remove cotter pin (7), pin (6) and tie rod.
5. Remove cotter pin (15), washer (16) spindle and king pin (13 or 14).
6. Remove axle mounting bolt nut (11), bolt (10) and axle (12).

ASSEMBLY AND INSTALLATION

NOTE: When installing new front wheel bushings, make sure that the groove in the bushings lines up with the index mark of the wheel hub.

1. Install the axle (12) using the mounting bolt (10) and nut (11). Tighten nut securely so there is a slight drag when the axle is oscillated.
2. Install the spindle and king pin (13) in the axle (12) and secure with washer (16) and cotter pin (15).
3. Install tie rod (15) using pin (6) and cotter pin (7).
4. Install front wheel and secure with washer (9) and cotter pin (8). Install hub caps (if so equipped.)
5. Connect the drag link (4) using pin (6) and cotter pin (7).
6. Refer to operator's manual for proper lubrication.

180 Series

REMOVAL AND DISASSEMBLY

1. Raise and support the front end on jackstands.
2. Disconnect the drag link from the spindle.
3. Remove the cotter pins, washers and front wheels.
4. Remove the nut, lockwasher and tie rod assemblies.
5. Remove the cotter pins, washers and king pins from the spindles.
6. Remove the axle mounting bolt nut, drive out the bolt with a soft mallet and roll the axle from under the frame.

ASSEMBLY AND INSTALLATION

1. Install the axle using the mounting bolt and nut. Tighten the nut securely so there is a slight drag when the axle is turned.
2. Install the spindle and king pin in the axle and secure with the washers and cotter pins.
3. Install the tie rod, nut and lockwasher.
4. nstall the front wheels and secure them with the washers and cotter pins.
5. Connect the drag link.
6. Lubricate all moving parts with chassis lube or SAE 90EP oil as required.

STEERING

REMOVAL AND DISASSEMBLY

1. Remove the steering wheel medallion (1), nut (2), steering wheel (3), seals

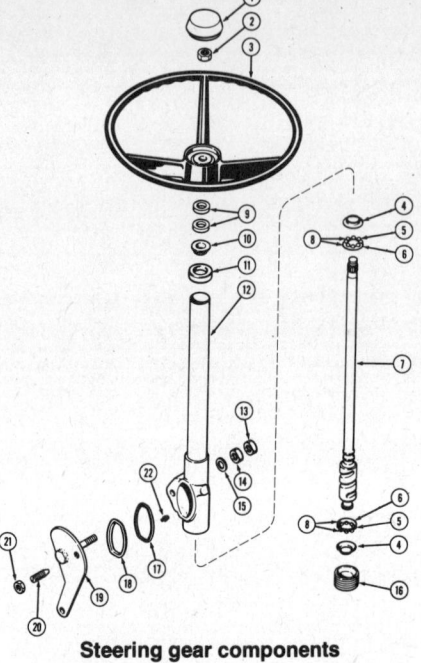

Steering gear components

(9), bearing spacer (10) and retainer (11).
2. Disconnect the steering gear linkage at the lever assembly.
3. Remove mounting capscrews and U-bolt which mount the steering gear assembly to the tractor main frame, then maneuver the steering gear assembly out the bottom of the tractor.
4. Remove locknut (13), adjusting nut (14), lever assembly (19), retainer (18) and gasket (17) from the housing (12).
5. Remove adjustment plug (16).
6. Install cam-tube (7) and bearing assemblies (8) out the bottom of the housing.
7. Remove bearing cups (4) and bearing assemblies (8).
8. Remove jam nut (21) and adjusting screw (20) from the lever assembly (19).
9. Thoroughly clean all parts before inspection. Replace all damaged or badly worn parts.
10. Inspect bearing assemblies (8) for worn or pitted balls; damaged retainer or cups. Replace as necessary.
11. Inspect cam and tube assembly (7) for misalignment, cracks or damaged threads.
12. Replace seals, bearing spacer and gasket each time the steering gear is disassembled.

ASSEMBLY AND INSTALLATION

1. If necessary, install new balls (5) in retainers (6). Then install ball and retainer assemblies (8), cup (4) onto cam and tube assembly (7).
2. Install cam-tube (7) and bearing assemblies (8) into housing assembly (12),

130 steering linkage

and secure with adjustment plug (16). Tighten adjustment plug until it is tight against the balls. Through a hole in the housing assembly, stake the adjusting plug with a center punch.

3. Install gasket (17), retainer (18) lever assembly (19) to housing and secure with washer (15), adjusting nut (14) and locknut (13).

4. Install adjusting screw (20) and locknut (21).

5. If necessary, replace fitting (22).

6. Fill the gear box with approximately ¼ pound of No. 1 Lithium base grease.

7. Mount the steering gear assembly to the tractor main frame. Tighten capscrews and nuts.

8. Connect the steering gear linkage to the lever assembly (19).

9. Install retainer (11), bearing (10), seals (9), steering wheel (3), nut (2) and medallion (1).

ADJUSTMENTS

Worm Shaft End Play Adjustment

1. Raise and support the front end on jack stands.

2. Loosen or tighten the adjusting plug located at the bottom end of the steering gear assembly until there is no noticeable play at the steering wheel.

Worm Gear Adjustment

1. Raise and support the front end on jack stands.

2. Loosen the jam nut on the slotted adjusting screw located on the right side of the gear box on the drop arm assembly lever shaft.

3. Turn the screw in just enough to eliminate end play.

4. Turn the steering wheel all the way right and left to check for binding.

5. Tighten the locknut.

TRANSMISSION

REMOVAL AND DISASSEMBLY

1. Drain the transmission oil, then remove the transmission cover capscrews (2) lockwashers (3), plain washers (4), cover (1) (with fenders, seat, tool box and gasket (6).

2. Remove the orbital motor and input shaft coupling (7) from the transmision housing (22).

3. Slide input shaft (9) out through the orbital motor hole in the gear housing (22). Then remove the input gear cluster (10) and thrust washer (11).

4. Drive out the roll pin (13) and remove the yoke lever (14) with yoke (12).

5. Drive out roll pin (13) to remove yoke (12) and pin (15) from the yoke lever (14).

6. Pull out the shift lever (28) being careful not to lose the ball (32) and spring (31).

7. Block up rear of tractor and remove the wheels.

8. Loosen the hub retainer screws (44) and remove hubs (42) and keys (43).

9. Remove axle housing retainer capscrews (40), lockwashers (41) axle housing (36 and 39) and gaskets (37).

10. Remove oil seals (34) from the axle housings (36 and 39).

11. Drive out axle retaining roll pins (25) and pull out the axles (23 and 24).

12. Lift the differential gear housing (21) from the transmission housing (22).

13. Remove the roll pins (20), pinion shafts (19), pinion gears (16), spacers (17) and differential gears (18) from the differential housing (21).

14. Thoroughly clean all parts before inspection. Replace all worn or badly damaged parts.

15. Inspect all gears and shafts for burrs, nicks, or excessive wear, both to the teeth and to the splines. Light burrs and nicks can be removed with a hone or crocus cloth.

16. Inspect bearing (26) for worn or pitted rollers. There should be not pitts or other visible damage. If damaged, drive out from inside of transmission case.

17. Inspect differential gear housing for cracks and for burrs, nicks, or excessive wear on teeth. Light burrs and nicks can be removed with a hone or crocus cloth.

18. Inspect axle housings (36 and 39) for cracks, and hubs (42) for cracks, stripped or damaged threads.

19. Inspect bushings (35 and 38) for score marks, nicks, or other damage. Remove if necessary using a suitable puller.

20. Replace oil seals (34) each time the rear axles are disassembled.

21. Inspect detent spring (31).
Free length .88"
Total no. coils 10
Active no. coils 8
Wire diameter .034"
7.5 to 8 lbs. at compressed height of .46"

22. Remove and inspect the filter screen

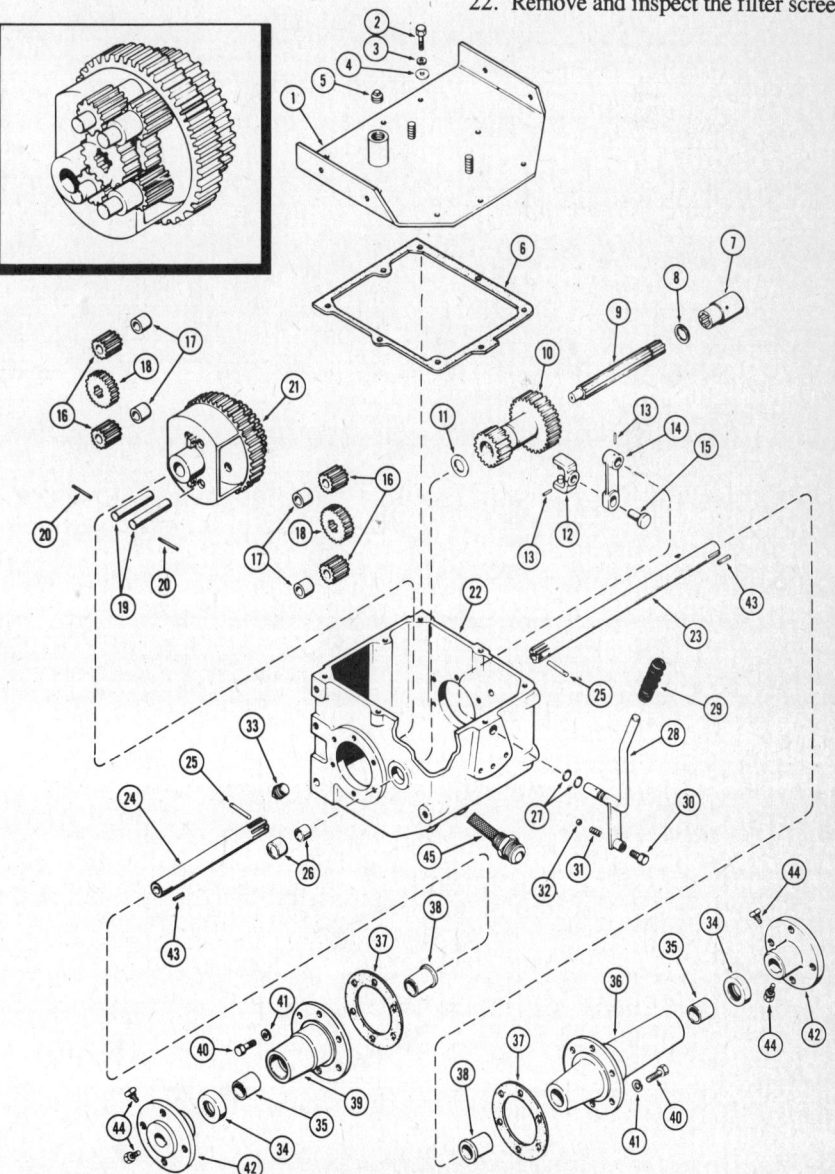

2-speed transmission components

J.I. Case

(45). Thoroughly clean and install. If damaged replace.

Hydraulic Pump

DISASSEMBLY

NOTE: Before disassembly, scribe a line across the pump covers and body so alignment can be made upon reassembly.

1. Remove key (8) from the drive gear shaft (6).
2. Remove bolts (1), washers (2) and cover (3) from body (9). Remove seal (4) from cover (3).
3. Remove adapter cover (12) and remove seal (13), brass seal (10), spacer (11) and seal (4).
4. From body (9), remove drive and driven gears (6 and 7) and bearings (5).
5. Clean all parts thoroughly before inspection. Replace all worn or damaged parts.
6. Inspect all metal parts for score marks, burrs, or other damage. Slight nicks and burrs can be removed with a hone or crocus cloth.
7. Inspect gears (6 and 7) for chipped or damaged teeth.

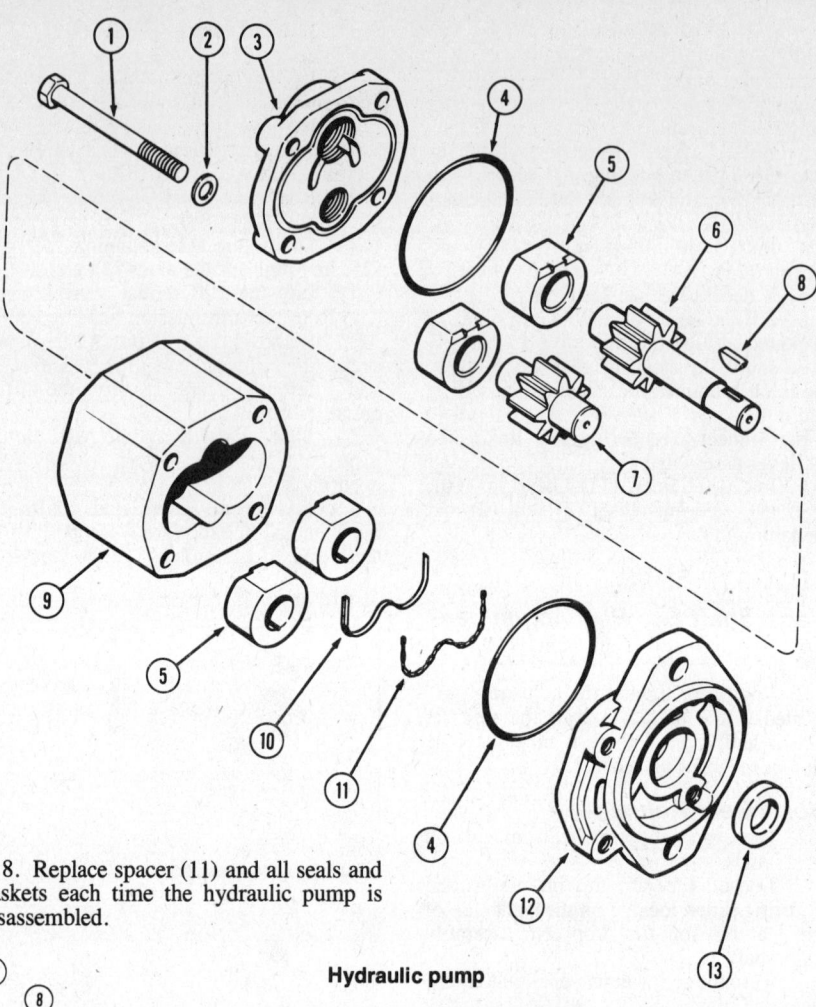

8. Replace spacer (11) and all seals and gaskets each time the hydraulic pump is disassembled.

Hydraulic pump

ASSEMBLY

1. Lubricate each part—except seal (13) with clean, fresh, SAE No. 10 oil at reassembly.
2. Apply a light coat of Permatex® No. 3, or equivalent, to the bore of cover (12), then press a new oil seal (13) into the cover until it bottoms with the oil seal lip to the inside of the pump.
3. Place bearings (5) on gears (6 and 7) with chamfered edge next to the gear. Then place gears in body (9).
4. Place rubber spacer (11) in cover, then also install brass seal (10) and seal (4) in cover (12).
5. Place seal (4) in cover (3).
6. Install both covers onto body, aligning the scribe marks during installation.

CAUTION
Exercise care when installing cover (12) onto gear shaft (6) so as not to damage oil seal (13).

---

7. Install bolts (1) and washers (2). Torque bolts (1) to 28-32 ft.lbs.

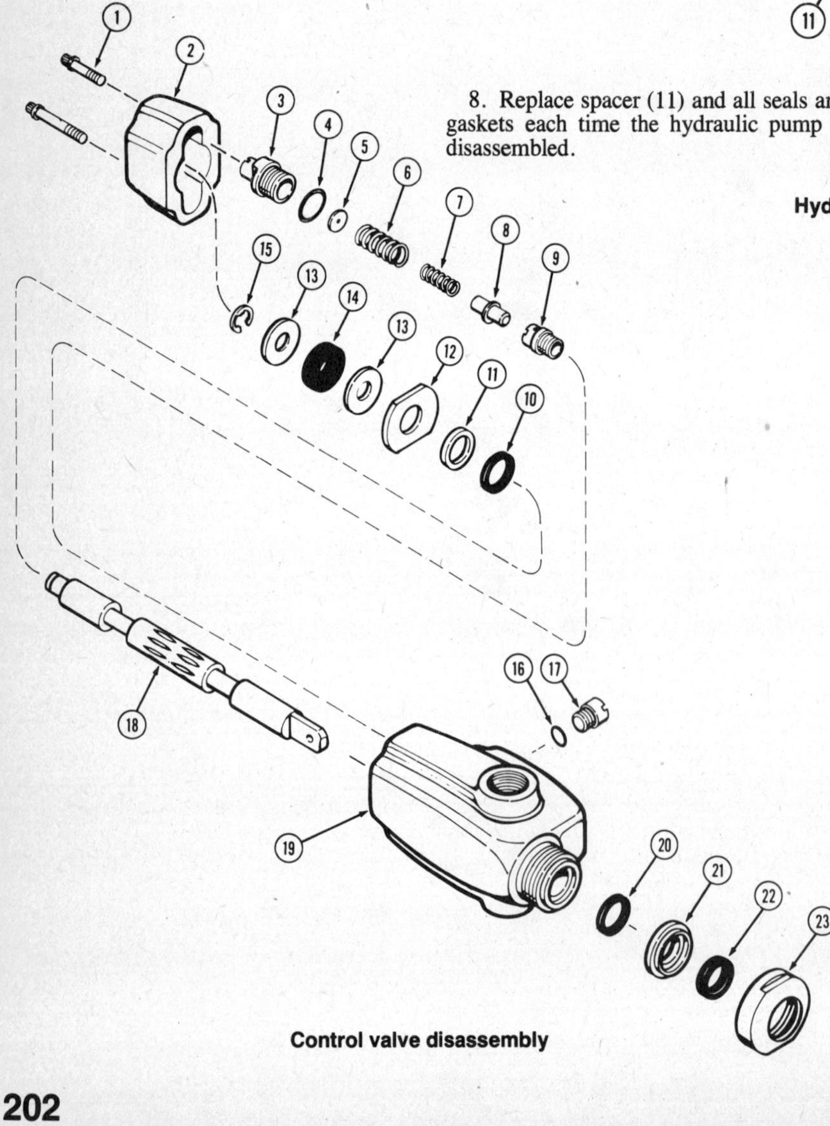

Control Valve

DISASSEMBLY

1. Remove the valve spool end cap (23) scraper seal (22), seal retainer (21) and

Control valve disassembly

quad ring (20) from the front end of the valve assembly.

2. Remove the ferry head capscrews (1) and end cap (2).

3. Grasp the end of the spool (16) and pull out of the valve body (19).

4. Remove the snapring (15), retainer (13), spacer (14), retainer (13), relief valve plug retainer (12), spacer (11) and quad ring (10) from the spool (18).

5. Remove the relief valve plug (3), O-ring (4), shim (5), outer spring (6), inner spring (7), poppet (8) and poppet seat (9).

6. Remove the plug (16) with O-ring (17).

Outer Spring (6) Inspection: (in.)
Free length   1.65
Total coils   11.5
Active coils   9.5
Wire diameter   .1055
250 lbs. per inch

Inner Spring (7) Inspection:
Free length   1.23
Total coils   15
Active coils   13
Wire diameter   .059
95 lbs. per inch

ASSEMBLY

1. Install the relief valve poppet set (9), poppet (8), inner spring (7), outer spring (6), shim (5) and plug (3) with a new O-ring (4) in the valve body (19).

2. Install the quad ring (10), spacer (11), relief valve plug retainer (12), retainer (13), spacer (14), retainer (13) and snapring (15) on the rear of the valve spool (19) into the valve body (19).

3. Install a new seal (22) in the retainer (21) with the lip towards the larger diameter opening.

4. Install the quad ring (20), retainer (21) with seal (22) and valve spool end cap (23).

5. Install the plug (16) with O-ring (17).

6. Install the end cap (2) using the ferry head capscrews (1).

Orbital Motor

NOTE: Before removing the orbital motor, clean the area around the motor thoroughly. If a vise is used, clamp across the port area, not the housing.

REMOVAL AND DISASSEMBLY

1. Remove the seven capscrews and washers (20) and the end cap (19).

2. Using an indelible pencil mark a line across the stator, rotor and motor body in line with the keyway in the shaft for proper assembly.

3. Remove the spacer (18) stator and rotor assembly (17) and spacer (16).

4. Remove the thrust bearing (15) and coupling shaft (14).

5. Remove key (9) and mounting flange capscrews (13) and washers.

6. Tap lightly on the mounting flange (2) and remove from motor shaft (8).

7. Remove O-ring (5) and quad ring (4) from inside the mounting flange (2).

8. With a small screwdriver, or knife remove the oil seal (1) from the outside of the mounting flange (2).

9. Remove bearing race (6) thrust bearing (7) and motor shaft (8).

10. Remove plug (11) with O-ring (12).

ASSEMBLY

1. Install the plug (11) with a new O-ring (12).

2. Install the motor shaft (8) thrust bearing (7) and bearing race (6).

3. Install a new oil seal (1) lip out, quad ring (4) and O-ring (5) in the mounting flange (2).

4. Install the mounting flange on the motor shaft and secure to motor body using capscrews and washers (13). Torque capscrews evenly to 225-275 in.lbs.

5. Install the coupling shaft (14) in the motor shaft.

NOTE: Make sure the keyway in the shaft is lined up with the mark made during disassembly. If no mark was made turn the shaft so the keyway is directly between the ports.

6. Install the thrust bearing (15) and spacer plate (16). Make sure the spacer plate is lined up with the mark made during disassembly.

7. Install the stator-rotor assembly (17) in alignment with the mark made during disassembly. If no mark was made or new assembly used the rotor must be installed 15° off the center line of the shaft keyway.

8. Install spacer (18) end cover (19) and secure with capscrews and washers (20). Torque capscrews evenly to 175-200 in.lbs.

Transmission Assembly and Installation

1. Install pinion gears (16), spacers (17), differential gears (18), pinion shafts (19) in the differential housing (21). Secure the pinion shafts (19) with roll pins (20).

2. Place the differential housing assembly in the transmission case with the hub to the right.

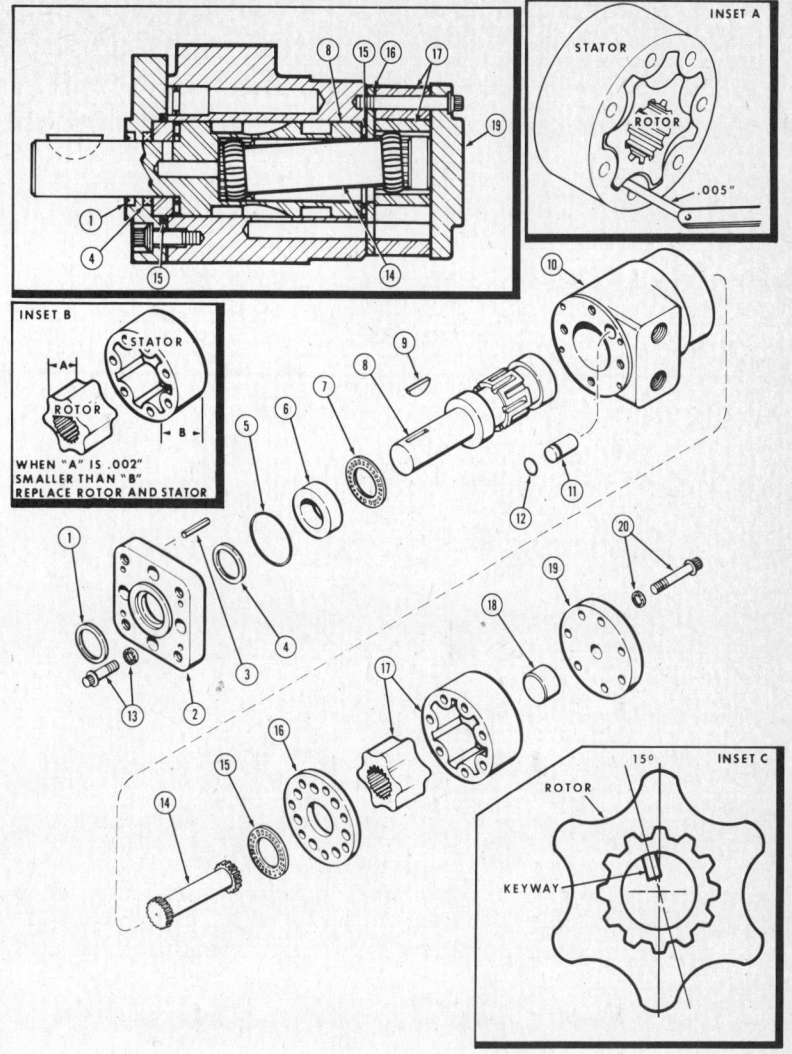

Orbital motor disassembly

J.I. Case

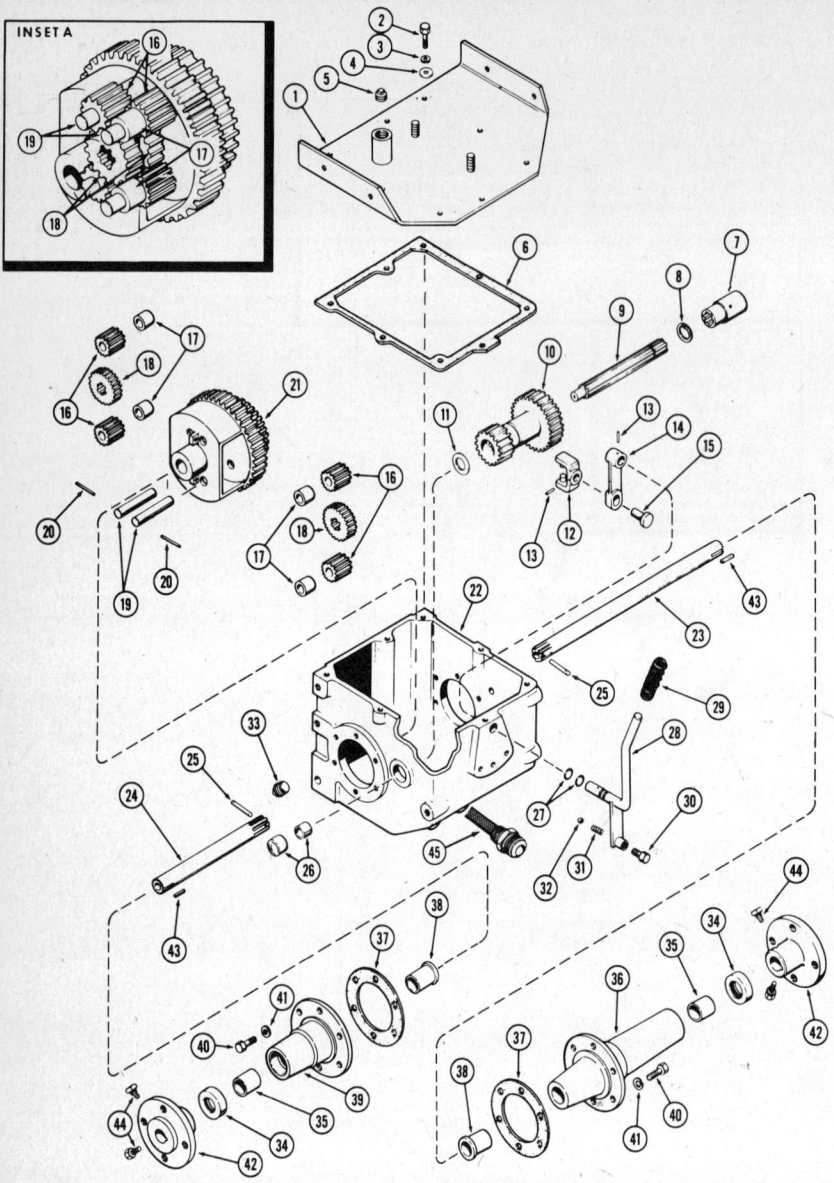

Transmission assembly

NOTE: Bushings must be reamed after assembly to 1.001 in.-1.002 in. inside diameter.

5. Press in new oil seals (34), lip in until flush with housing.
6. Using new gaskets (37), install the axle housing over the axles and fasten to transmission case using lockwashers (41) and capscrews.
7. Install keys (43), hubs (42) and secure with capscrews.
8. Install rear wheels and remove blocking from under tractor.
9. Install shift lever (28) using two new O-rings (27). Make sure ball (32) and spring (31) are in place.
10. Install yoke (12) to the yoke lever (14) using pin (15) and roll pin (13).
11. Install yoke lever (14) on the shift lever (28) and secure with roll pin (13).
12. If the bearing (26) was removed, press the bearing in until flush with outside transmission case.
13. Hold the input gear cluster (10) then install the input shaft (9) thru the gear cluster. Install the thrust washer (1) and slide the input shaft into the bearing (26).
14. Install the retainer ring (8) and input shaft coupling (7). Install orbital motor.
15. Install gasket (6), cover (1) with fenders seat and tool box, plain washers (4), lockwashers (3) and capsrews (2). Install drain plug (33).
16. Fill the transmission with clean oil, 1 inch from the top.

3. Install the axle shafts (23 & 24) into the differential housing and gears and secure with roll pins (25).
4. If the axle housing bushings were removed, press in new bushings (35 & 38). Press shoulder bushings (38) in until seated in the housing. Press straight bushing (35) in until flush with bore of housing.

ENGINE

These tractors use the Kohler K241S and K301S engines. For detailed engine repair procedures, see the Kohler Engine part of the Engine Unit Repair section.

REMOVAL AND INSTALLATION

For removal and installation procedures see the 155 and 195 Section immediately following this section.

Models 155 and 195, 150-190, T90, 2310, 2510, 2712

BRAKES

REMOVAL AND DISASSEMBLY

1. Loosen the rear wheel lug nuts.
2. Raise and support the tractor on jack stands.
3. Remove cotter pin (27), with washer (26) from brake rod (25).
4. Remove cotter pin (22), clevis pin (24), brake rod (25) and clevis (23).
5. Remove cotter pin (17), clevis pin (18), cotter pin with washer (9) and pull rod (11).
6. Remove the right hand brake assembly cotter pins and washers (31) from point A and also from the righthand brake assembly cam (30). Remove pull strap (29).
7. Remove roll pin (20), lever (21) and brake shaft (19).
8. Remove four bolts (2) securing the brake assemblies to the axle housings and lift off the brake assemblies.

NOTE: Both the right and left brake assemblies are identical and are disassembled in the same manner.

9. Remove nuts (14), bolts (1) and lift off the carrier (3), shims (4) and spacers (5).

NOTE: The brake linings are riveted to the carriers and should be replaced as an assembly when required.

10. Remove snapring (16), spring (15), cam actuator (13), cam (10), shims (8), brake bracket (7) and carrier (6).

J.I. Case

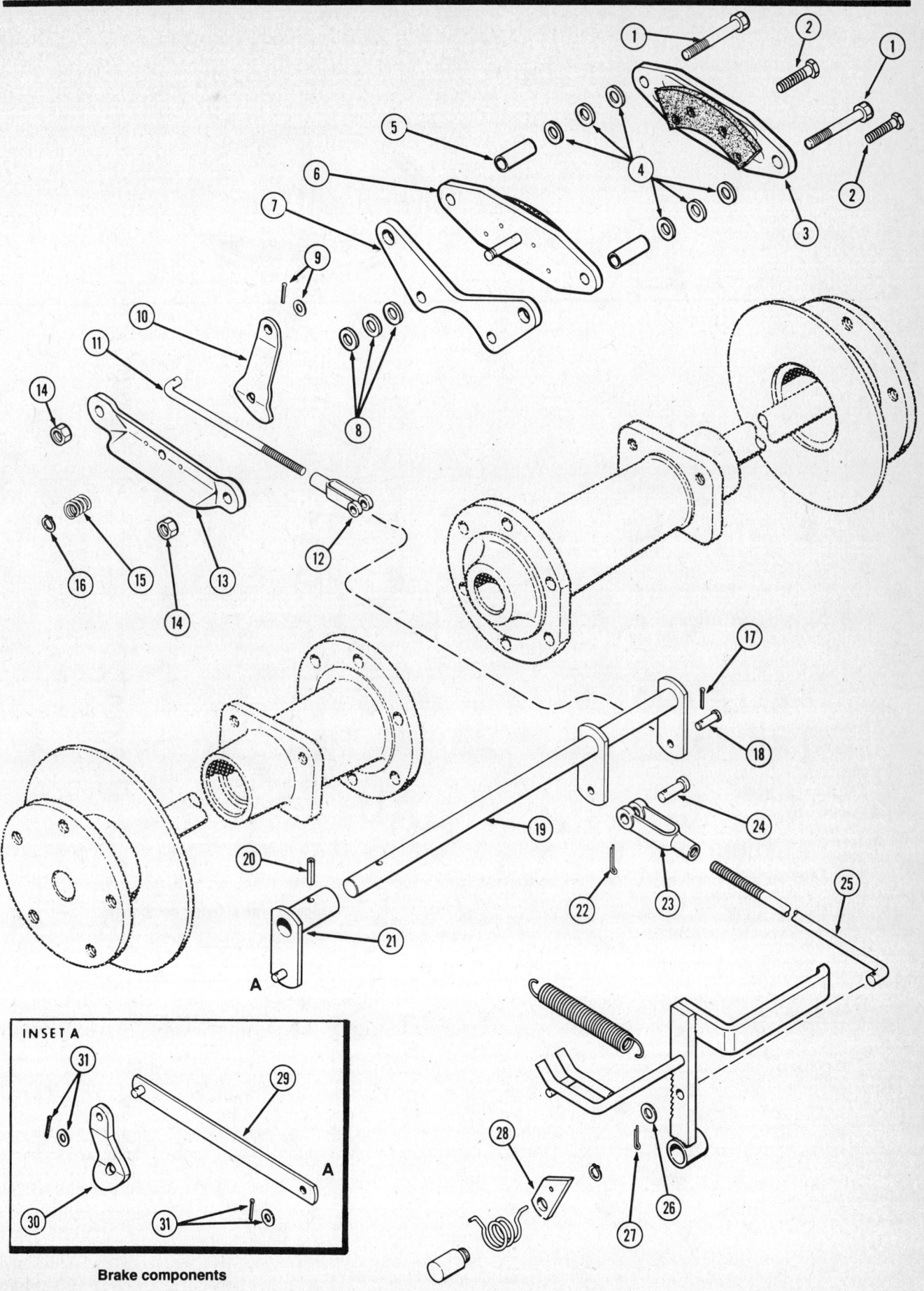

Brake components

205

J.I. Case

NOTE: Shims (4 and 8) may or may not be used. It is important that an equal amount be used at each location.

ASSEMBLY AND INSTALLATION

1. Assemble the brake assemblies (1 thru 16) being sure that shims (4 and 8) are equal and secure to the axle housings with four bolts (2).
2. Install the brake shaft (19), lever (21) and roll pin (20).
3. Install right brake assembly pull strap (29) with cotter pins and washers (31) at point A and the right brake assembly cam (30).
4. Position pull rod (11) and install washer and cotter pin (9), clevis (12), clevis pin (18) and cotter pin (17).
5. Position brake rod (25), clevis (23), clevis pin (24) and cotter pin (22).
6. Install washer (26) and cotter pin (27).
7. Check the brake adjustment and turn clevis (23) on or off the rod as required. The brake lock (28) should position no lower than the sixth tooth on the brake pedal nor higher than the fourth.
8. Turn the clevis (12) on or off the pull rod (11) until both the right and left brake linings contact the disc simultaneously.
9. Position the rear wheels and install the lug bolts.
10. Remove the blocks and lower the tractor. Torque the lug bolts to 90 ft.lbs.

FRONT AXLE AND STEERING

Axle

REMOVAL AND INSTALLATION

10 HP Models

1. Remove cotter pin (18), clevis pin (19) nut (10) and lift off the drag link as-

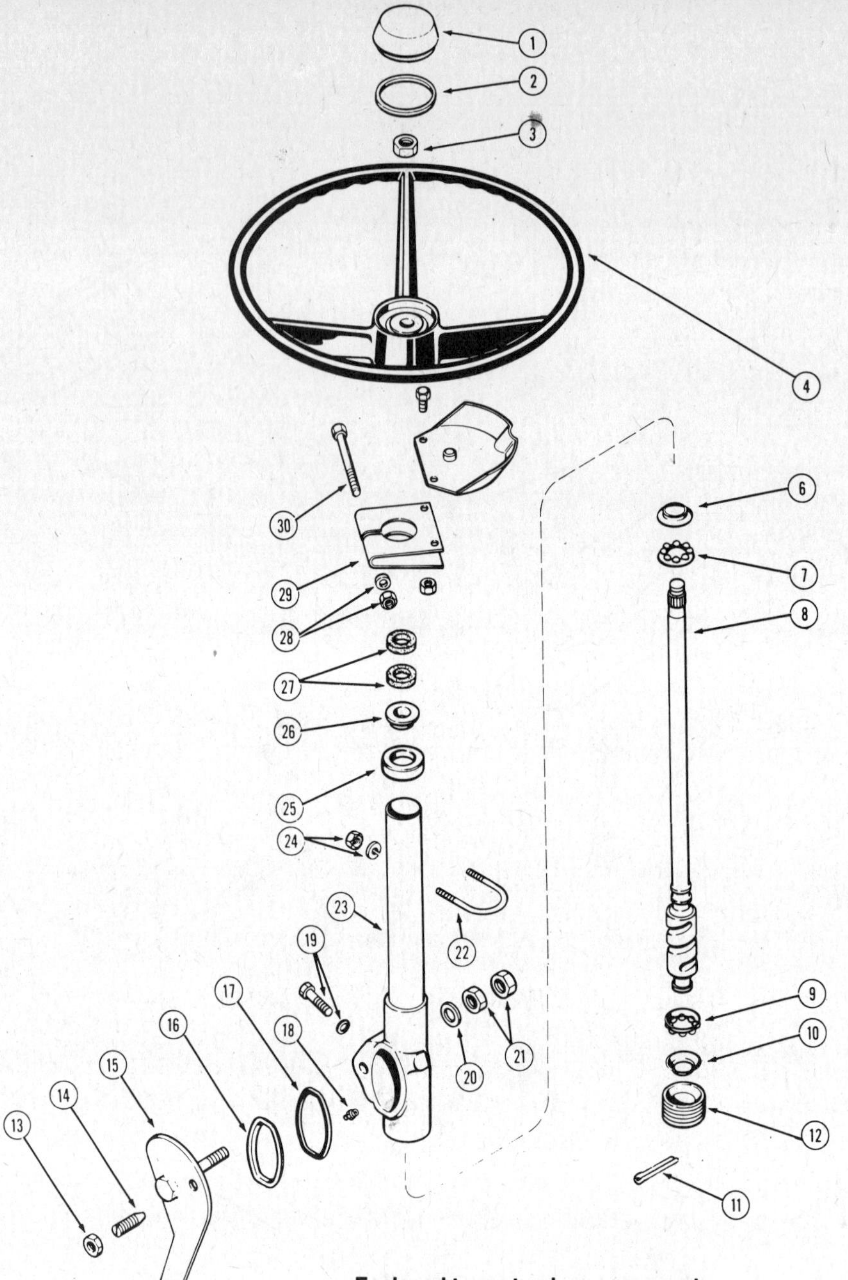

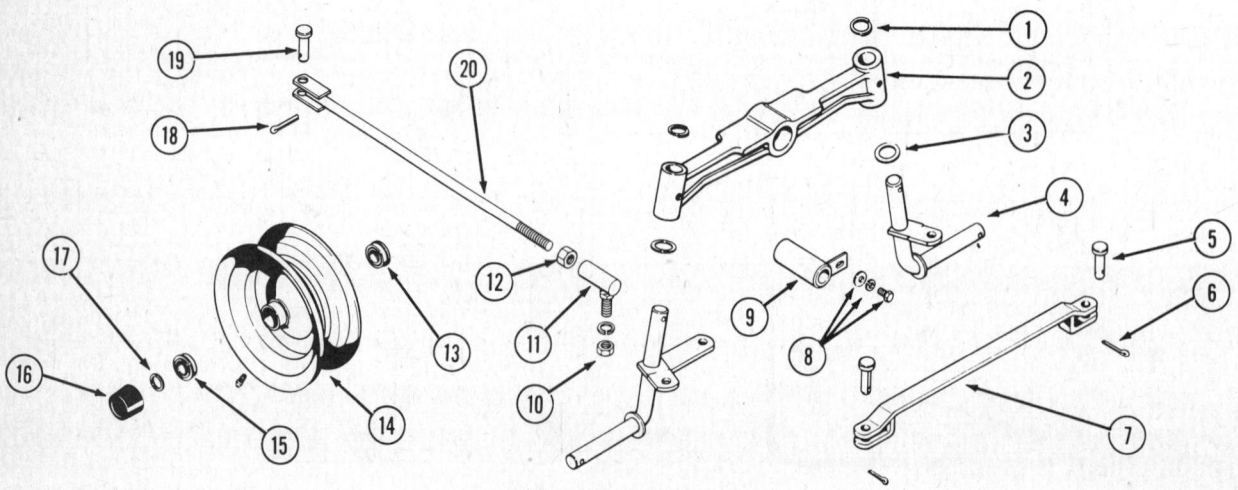

Enclosed type steering components

10 hp tractor axle components

J.I. Case

sembly. If necessary, turn off the ball joint (11) counting the number of turns or marking the drag link rod (20) to facilitate accurate assembly.

2. Block up the front of the tractor.

3. Remove the dust cover (16), retaining ring (17) outer bearing (15), wheel (14) and inner bearing (13).

4. Remove cotter pins (6), clevis pins (5) and lift off the tie rod (7).

5. Remove spindle retaining rings (1), spindles (4) and washers (3).

6. Remove bolt, lockwasher and plain washer (8) and tap out the axle pivot pin (9). Lower the front axle (2).

7. With the axle completely removed and disassembled, the component parts can easily be cleaned and inspected for wear, cracks or burrs. All parts that are worn or damaged must be replaced.

8. The wheel bearings must be inspected for cracks and excessive wear. Insert the spindle in the wheel and check for excessive side clearance. Check the fit of the king pin in the front axle. Replace all worn parts.

9. Position the axle (2) and install the axle pivot pin (9) using bolt (8) with lockwasher and plain washer to secure the axle.

10. Position washers (3) and install spindles (4) securing with spindle retaining rings (1).

11. Position the tie rod (7) and secure it with clevis pins (5) and cotter pins (6).

12. Install wheel bearings (13 and 15) and position the wheels (14) securing them with retaining rings (17).

13. Install the dust cover (16).

14. Assembly the drag link rod (20) and secure it with clevis pin (19), cotter pin (18) and nut (10).

12 HP Models

1. Remove cotter pin (16), clevis pin (17), nut (21) and lift out the drag linkage. If necessary, remove the ball joint (20) with locking nut (19). Count the number of turns or mark the drag link rod (18) to insure proper reassembly.

2. Block up the front of the tractor.

3. Remove dust cover (13), retaining ring (14), outer bearing (15), wheel (12) and inner bearing (11) with washer.

4. Remove nuts (7) with washers (6) and lift off the tie rod (5). If necessary, remove ball joints following the same procedures outlined in step 1 above.

5. Remove spindle retaining rings (1), spindles (4) and washers (3).

6. Remove nut, lockwasher and plain washer (9) and tap out the axle pivot pin (10). Lower the front axle (2).

7. With the axle completely removed and disassembled, the component parts can easily be cleaned and inspected for wear, cracks, or burrs. All parts that are worn or damaged must be replaced.

8. The wheel bearings must be inspected for cracks and excessive wear. Insert the spindle in the wheel and check for excessive side clearance. Check the fit of the king pin in the front axle. Replace all worn parts.

9. Position the axle (2) and install the axle pivot pin (10) and securing bolt (9) with lockwasher and plain washer (9).

10. Position washers (3) and install spindles (4) securing with spindle retaining rings (1).

11. Reassemble the tie rod (5) and position and secure it with lockwashers (6) and nuts (7).

12. Install wheel bearings (15 and 11) and position the wheels (12) securing them with retaining rings (14).

13. Install the dust cover (13).

14. Reassemble the drag link rod (18) and secure with clevis pin (17), cotter pin (16), nut (21) and lockwasher.

Enclosed Steering Gear Shaft System

REMOVAL AND DISASSEMBLY

1. Remove the steering wheel medallion (1) rubber retaining ring (2), nut (3), steering wheel (4), seals (27), bearing spacer (26) and retainer (25).

2. Loosen the clamping nut (28) on the shifter bracket (29) and open the clamp enough to allow the column housing to pass thru.

3. Disconnect the drag link rod from the lever assembly (15).

4. Remove U-bolt nuts (24), U-bolt (22) and bolts (19). Manuever the steering assembly out through the bottom of the tractor.

5. Remove locknut and adjusting nut (21), washer (20) and pull the entire lever assembly (13 thru 17).

6. Remove cotter pin (11) and adjustment plug (12).

7. Remove worm gear (8), bearing cups (6 and 10) and bearing assemblies (7 and 9).

8. Thoroughly clean all parts before inspection. Replace all damaged or badly worn parts.

9. Inspect bearing assemblies (7 and 9) for worn or pitted balls or damaged bearing cups (6 and 10) and replace if necessary.

10. Inspect the worm gear (8) and housing (23) for cracks or damaged threads.

11. Replace seals (27), bearing spacer (26) and gasket (17) each time the steering gear is disassembled.

ASSEMBLY AND INSTALLATION

1. If necessary, install new balls in bearing retainers (7 and 9).

2. Install bearing assemblies (7 and 9) with bearing cups (6 and 10) and position the worm gear (8) in the column housing (23).

3. Secure the worm gear (8) in the column housing (23) with adjustment plug (12). Tighten the plug until it is tight against the bearings and secure it with cotter pin (11).

4. Install gasket (17), retainer (16), lever assembly (15) to housing (23) and secure with washer (20), adjusting nut and locknut (21).

5. Install adjusting screw (14) and locknut (13).

6. If necessary, replace grease fitting (18).

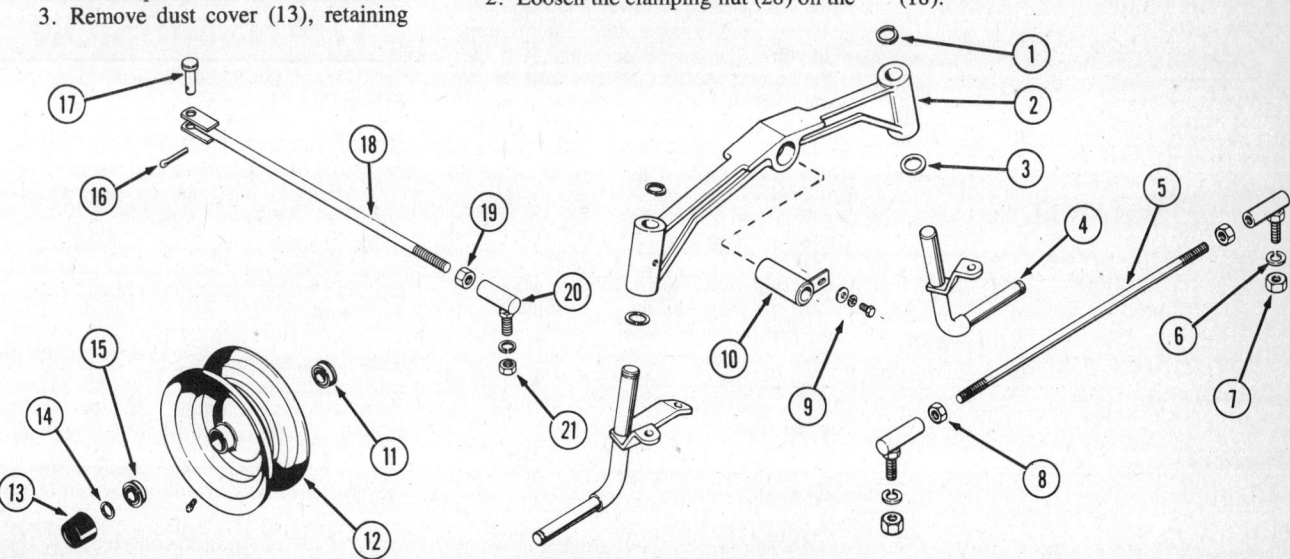

12 hp tractor axle components

J.I. Case

7. Fill with approximately ¼ pound of No. 1 Lithium base grease.

8. Secure the steering assembly to the tractor frame with U-bolt (22), nuts and washers (24) and bolts (19). Tighten the clamping nut (28) on the shifter bracket (29).

9. Connect the drag link rod to the lever assembly.

10. Install retainer (25), bearing (26), seals (27), steering wheel (4), nut (3), rubber retainer ring (2) and medallion (1).

Open Gear Shaft Steering System

REMOVAL AND DISASSEMBLY

1. Remove the steering wheel medallion (1) and retainer ring (2).

2. Remove nut (3), lift off steering wheel (4) and tap out spacer (5).

3. Disconnect the drag linkage from the steering sector (13).

4. Block up the tractor, remove retaining ring (10), washer (11) and lower the steering sector (13).

5. Remove bushing (12).

6. Remove retaining ring (6), washer (7) and steering gear (9) through the bottom of the tractor and tap out bushing (8).

7. Thoroughly clean all parts before inspection.

8. Inspect bushings for cracks or excessive wear. Replace if necessary.

9. Inspect steering gear and sector for broken or chipped teeth and replace if necessary.

10. Install bushing (8) in mounting bracket and position the steering gear (9).

11. Install new washer (7) and retaining ring (6) over the top of the steering gear and slide down the steering gear.

12. Push the steering gear into position and install bushing (5), steering wheel (4), nut (3), retaining ring (2) and medallion (1).

13. Install bushing (12) in the steering sector mounting bracket, position the steering sector (13) and install new washer (11) and snapring (10).

14. Position the steering sector (13) arm at a right angle to the drag linkage with the front wheels in the straight ahead position and connect the drag linkage.

Steering Gear Adjustments

CLOSED GEAR SHAFT

Worm Shaft End Play Adjustment

1. Raise and support the front end with jack stands.

2. Loosen or tighten the adjusting plug on the bottom end of the gear case until there is no play at the steering wheel.

Worm Gear Adjustment

1. Raise and support the front end.

2. Loosen the locknut and tighten the adjusting nut on the lower left side of the gear case, until a slight drag is felt midway in the turn. Torque the locknut to 50 ft.lbs.

3. Loosen the locknut and turn the adjusting screw located on the right side of the gear case on the drop arm lever shaft, in until a slightly heavier drag is felt midway in the turn. Tighten the locknut to 40-50 ft.lbs.

OPEN GEAR SHAFT

Steering gear adjustments are necessary only if the drag link has been replaced.

1. Position the front wheels straight ahead.

2. Turn the steering wheel lock to lock, counting the number of turns. Turn it back one half way.

3. The sector arm should be at a right angle to the drag link. Lengthen or shorten the drag link as necessary in order to insert into the sector arm.

4. When the correct alignment is attained, tighten the drag link adjusting nut.

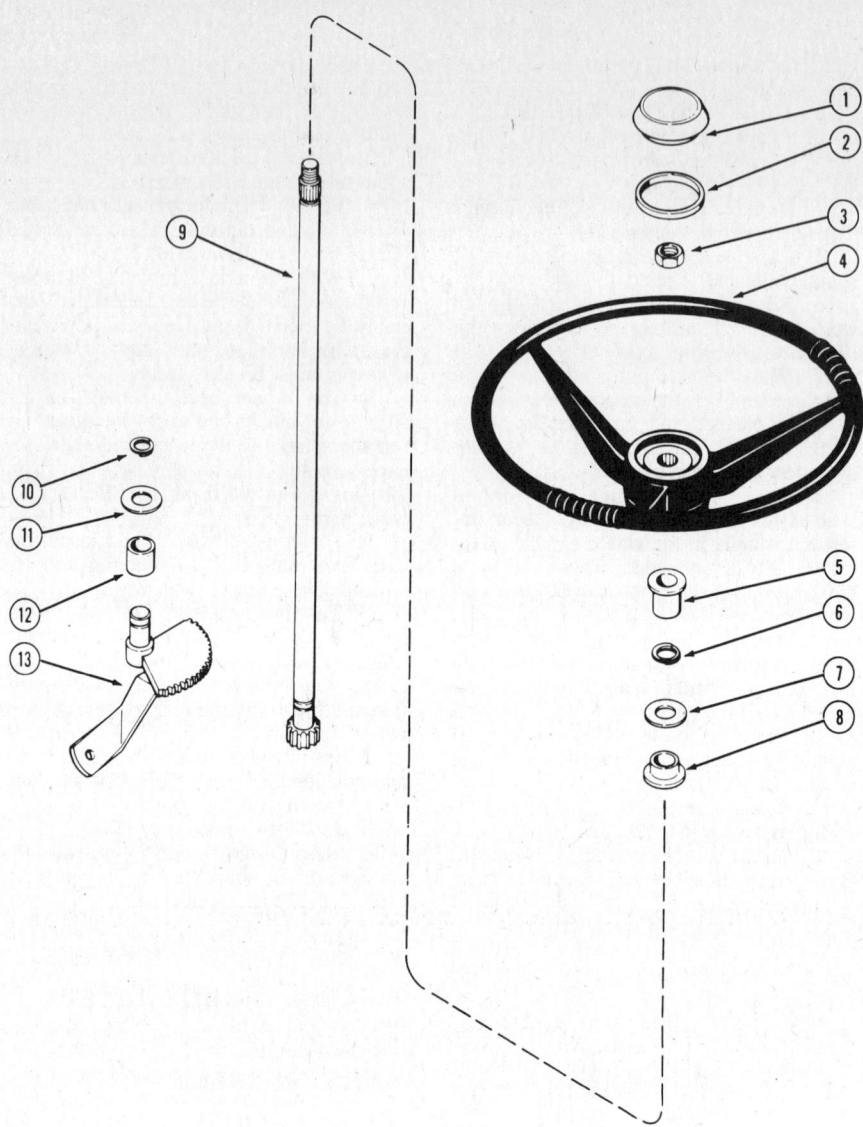

Open gear shaft steering system

TRANSMISSION

REMOVAL

Utility Transporter

1. Drain transmission oil.

2. Loosen lub bolts on rear wheels, raise and securely block up the tractor frame and remove the rear wheels.

3. Disconnect implement light wire (16) and brake light wires (1) and tag for reassembly.

4. Remove bolts (9), pull off shift lever knob and bushing (12), and lift off shifter indicator plate (11).

5. Remove cotter pin (10) and unscrew shift linkage (8) from yoke (3) and pull out.

NOTE: Shift linkage tension spring (15) will slide off as linkage is removed.

6. Remove cotter pin (19) from brake linkage (7) at brake control vlave (4) and disconnect brake tension spring (15).

J.I. Case

7. Remove securing bolts from frame, platform, and transmission case and lift off fender and seat assembly.

8. Disconnect the brake control lines (5 and 6) from the brake control valve (4).

NOTE: Cap all lines and openings to insure against entry of dirt or other foreign matter.

9. Disconnect transmission pump line (13) and heat exchanger line (14) at transmission case.

10. Place suitable supports under the transmission, remove bolts (17) and lift out the transmission.

— CAUTION —
The transmission assembly is heavy. Avoid injury by securing help when removing or installing the transmission.

Compact Tractors

1. Drain transmission oil.
2. Loosen lug bolts on rear wheels, raise and securely block up the tractor frame and remove the rear wheels.
3. Remove the transmission cover plate bolts (1) and lift off the fender and seat assembly and cover gasket.
4. Remove cotter pin (6), tension spring (2) and disconnect brake linkage (7) from brake control valve (3).
5. Disconnect upper and lower brake control lines (4 and 5) from the brake control valve (3).

NOTE: Cap all lines and openings to insure against entry of dirt or other foreign matter.

6. Disconnect transmission pump line (8) and heat exchanger line (9) from transmission.
7. Place suitable supports under the transmission, remove bolts (10), and lift out the transmission assembly.

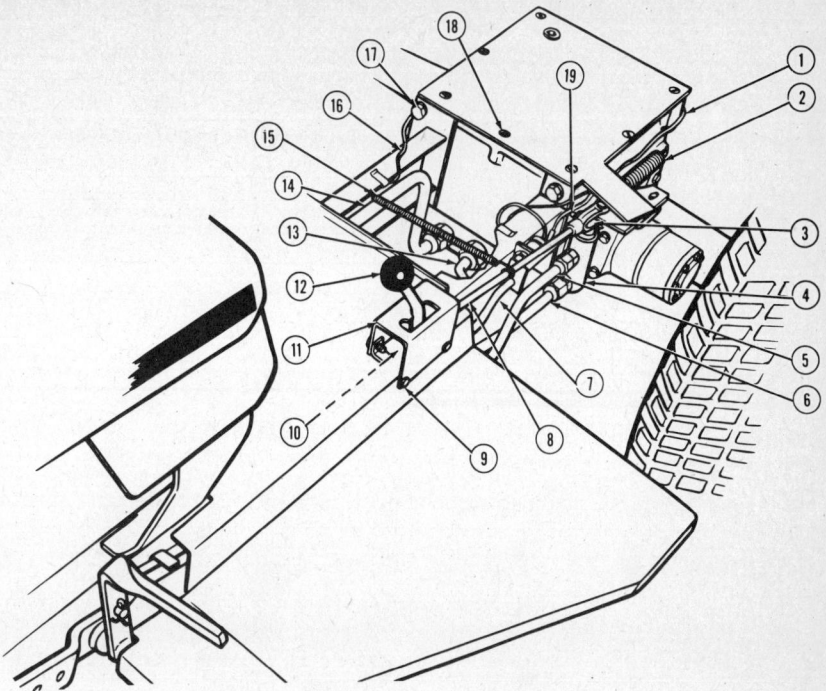
Transmission removal, utility tractor

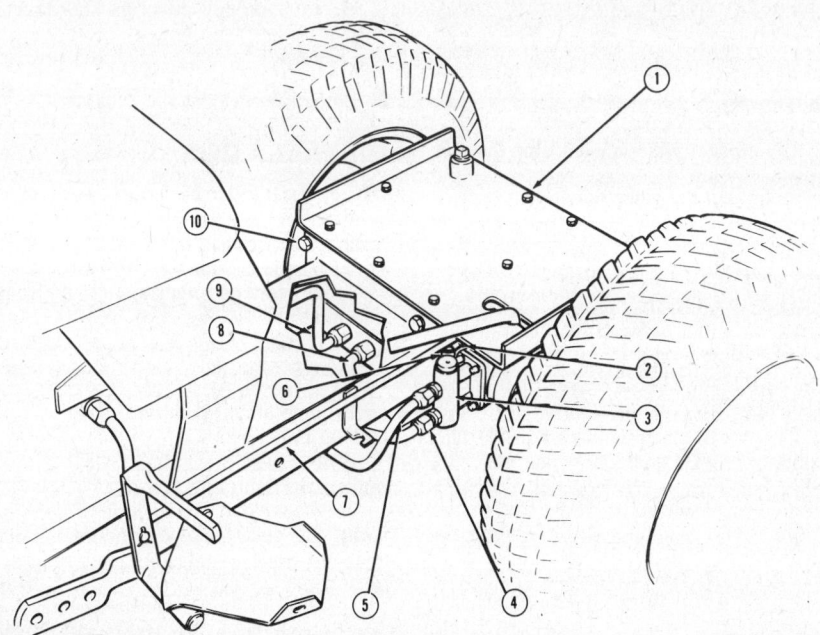
Transmission removal, compact tractor

— CAUTION —
The transmission assembly is heavy. Avoid injury by securing help when removing or installing the transmission.

DISASSEMBLY

NOTE: The following disassembly procedures apply to both the Utility Transporter and the Compact Tractor. Disassembly steps begin with the removal of the transmission cover plate although it is generally removed with the seat and fender assembly when the transmission is removed from the vehicle. The transmission cover plate used on the utility transporter differs slightly from that used on the compact tractors as stated in step one. The transmission may be equipped with a six tooth or twenty tooth input shaft. Differences are called out in steps 2, 5 and 6. Thoroughly clean the transmission before disassembly.

1. Remove transmission cover plate screws (2), shifter bracket securing nut (utility transporter only), cover plate and gasket (3).
2. Remove roll pins (5), shifting rod (4) and shifter fork (9). Press out nylon bushing (8) (20 tooth input shaft only).
3. Remove orbital motor attaching bolts, orbital motor, gasket and input shaft coupling (18).
4. Remove snapring (17) and coupling retaining ring (16).
5. Lightly tap the inner bearing race (31) off of the input shaft (14); remove the wire retainer ring (13) (six tooth input shaft only). Input shaft (14) and gear clutter (10) from the transmission housing.
6. Lightly tap the middle bearing race (11) off the input shaft (14). Remove the wire retainer ring (15) (6 tooth input shaft only). If necessary, remove bearing (12).
7. Remove roll pins (26 and 28) and pull out rear axles (25 and 27).
8. Remove bolts (22 and 33), axle housings (21 and 36) and gaskets (24 and 35).
9. Remove seals (19 and 29) and bushings (20-23-30-34).
10. Lift out the differential housing assembly (38) and remove transmission washers (37).
11. Remove cotter pins (45), roll pins (43), pinion shafts (44), pinion gears (40),

209

J.I. Case

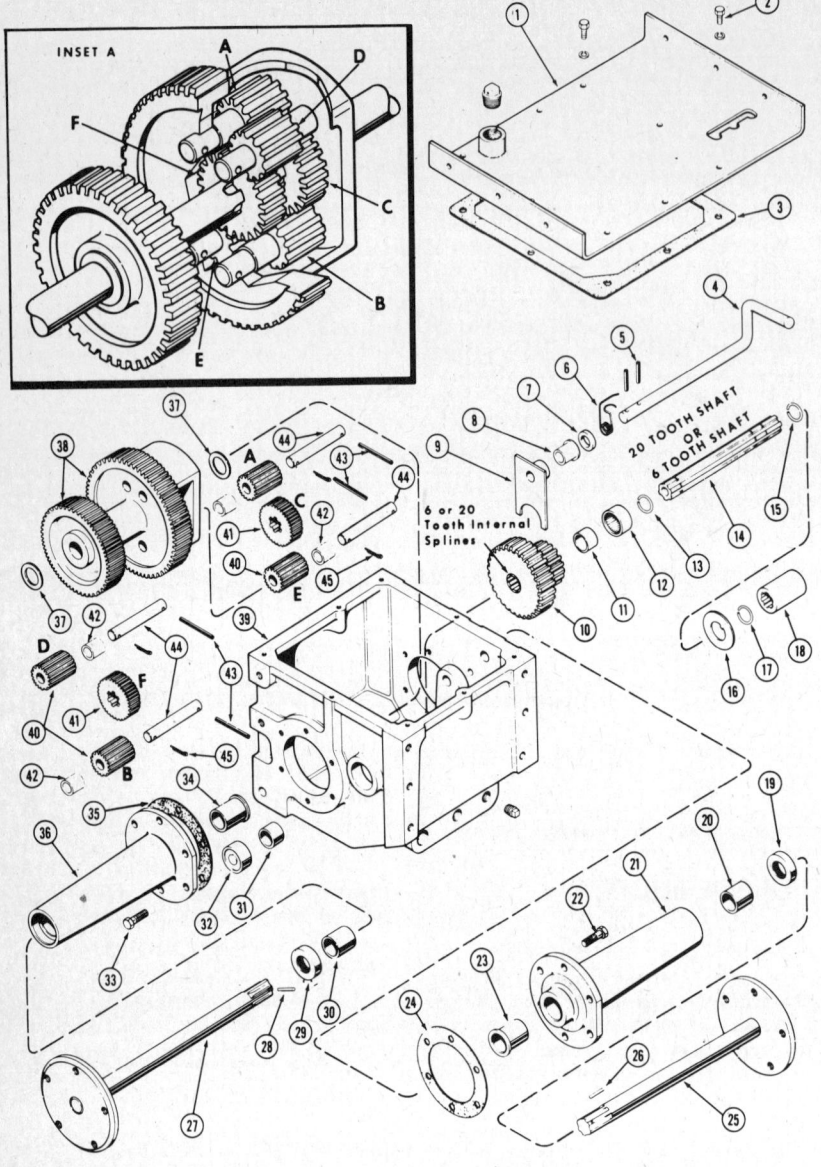

Transmission components

bushing (20 and 30) in until flush with the housing. Bushings must be reamed after assembly to 1.001-1.002 in. inside diameter. Install oil seals (19 and 29) lip side first.

4. Using new gaskets (24 and 35), position the axle housings (21 and 36) and secure with bolts (22 and 33).

5. Install the axle shafts (25 and 27) thru the axle housings (21 and 36) and into the differential assembly, being sure that transmission washers (37) are in place. Secure the axle shafts with locking pins (26 and 28).

6. Place the input shaft end bearing (32) into the transmission case until it is flush to the surface of the case.

NOTE: Use caution when installing the axles to insure a good oil seal. The lips on seals (19 and 29) must not be nicked or "rolled" in any way.

7. Press the input shaft middle bearing (12) in until flush with the inner wall.

8. If equipped with 6 tooth input shaft. Install wire retaining ring (15), inner race (11) and place input shaft in the transmission.

NOTE: If the transmission is equipped with a twenty tooth input shaft, wire retaining rings (13 and 15) will be omitted.

9. Install the cluster gear (10), retaining ring (13) (6 tooth shaft only), inner race (31) and position the input shaft.

10. Install coupling retaining ring (16), snapring (17) and coupling (18).

11. Position orbital motor and gasket and secure with ferry head capscrews.

12. Install new shifter shaft oil seal (7) flush with the transmission case.

13. Slide the shifter shaft tension spring (6) onto the shaft, position the shaft in the transmission and install the shifter fork (9) and roll pins (5).

14. Position new cover gasket (3) cover (1) and secure with bolts (2).

15. Install shifter bracket securing nuts (utility transporter only).

INSTALLATION
Utility Transport

1. Position transmission to frame and install securing bolts and lockwashers (17).

2. Connect transmission pump line (14) and heat exchanger line (13) to transmission.

3. Connect lower brake control line (6) to brake control valve.

4. Position seat, fender assembly, cover plate and gasket and secure with attaching bolts.

5. Connect brake light wires (1) and implement light wire (16).

6. Slide shift lever linkage tension spring (15) onto linkage (8) and screw linkage into yoke (3). Install new cotter pin (10).

7. Position shift indicator plate (11), secure with bolts (9), and push on shift lever knob and bushing (12).

8. Position brake pedal linkage (7), at-

spacers (42), and differential gears (41) from the differential housing (38).

12. Thoroughly clean and wipe dry all parts before inspection.

13. Inspect all gears and shafts for burrs, nicks or excessive wear, both to the teeth and to the splines. Light burrs and nicks may be removed with a hone or crocus cloth.

14. Inspect bearings for worn or pitted rollers. There should be no pits or other visible damage. If damaged, replace with new parts.

15. Inspect differential gear housing for cracks, burrs or nicks. Examine gear teeth for excessive wear. Light burrs and nicks may be removed with a hone or crocus cloth.

16. Inspect axle housings and hubs for cracks and stripped or damaged threads.

17. Inspect bushings for score marks, nicks, or other damage and replace if necessary.

18. Replace oil seals each time the unit is disassembled.

19. Inspect the filter screen and replace if necessary.

ASSEMBLY

NOTE: All bolts should have Permatex® applied before installation.

1. Install pinion gears (40), spacers (42), differential gears (41), and pinion shafts (44), into the differential gear housing (38). Secure the pinion shafts (44) with locking pins (43) and cotter pins (45).

2. Position the differential assembly with the gear side to the right side of the transmission case and set the transmission washers (37) in place.

3. If the axle bushings (23 and 34) were removed, press in new bushings.

NOTE: Press the flange bushing (23 and 34) in until seated and the straight

210

tach brake tension spring (2) and install new cotter pin (19).
9. Connect upper brake control line (5) to brake control valve (4).
10. Install rear wheels and remove supports.
11. Fill transmission to one inch below cover, using SAE 5W20 Motor Oil for winter operation (below 32° F.) or SAE 20W40 Motor Oil for summer operation. Use only MS or DS Service Classification Oil that has passed AMA Test Sequences I, II, and III.

Compact Tractors

1. Position transmission to frame and install bolts and lockwashers (10).
2. Connect transmission pump line (9) and heat exchanger line (8) to transmission.
3. Connect upper and lower brake control lines (5 and 4) to the brake control valve (3).
4. Position brake linkage (7) in brake control valve retainer eye, attach brake tension spring (2) and install new cotter pin (6).
5. Position new transmission cover plate gasket and install transmission cover plate and fender and seat assembly with bolts (1).
6. Install rear wheels and remove supports.
7. File transmission to one inch below cover plate using SAE5W20 Motor Oil for winter operation (Below 32° F.) or SAE 20W40 Motor Oil for summer operation. Use only MS or DM Service Classification Oil that has passed AMA Test Sequences I, II and III.

HYDRAULIC PUMP MOTOR

REMOVAL

1. Disconnect inlet and outlet tubes.
2. Remove capscrews and lockwashers from the hydraulic pump support.
3. Remove pump and support assembly out the right side of the tractor.

INSTALLATION

1. Hold pump and support in place and secure with capscrews and lockwashers.
2. Connect inlet and outlet tubes.

DISASSEMBLY

NOTE: Before disassembly, scribe a line across the pump covers and body so alignment can be made upon assembly.

1. Remove 90° fitting (8) with O-ring (9) and straight fitting (21) with O-ring (22) from the pump body (10).
2. Remove capscrews (13), lockwashers (14) and pump support.
3. Loosen setscrew (17) and remove spider hub (18) and key (20).
4. Remove bolts (1), washers (2) and cover (3) from the body (10). Remove seal (4) from cover (3).
5. Remove adapter cover (15) and remove seal (4), brass seal (11), rubber spacer (12) and seal (16).
6. From body (10), remove drive and driven gears (6 and 7) and bearings (5).
7. Clean all parts thoroughly before inspection. Replace all worn or damaged parts.
8. Inspect all metal parts for score marks, burrs, or other damage. Slight nicks and burrs can be removed with a hone or crocus cloth.
9. Inspect gears for chipped or damaged teeth.
10. Replace spacer and all seals and gaskets each time the hydraulic pump is disassembled.

INSTALLATION

1. Lubricate each part-except seal with clean, fresh, SAE No. 10 oil at assembly.
2. Apply a light coat of Permatex® No. 3, or equivalent, to the bore of cover (15), then press a new oil seal (16) in the cover until it bottoms with the oil seal lip to the inside of the pump.
3. Place bearings (5) on gears (6 and 7) and with chamfered edge next to the gear. Then place gears in body (10).
4. Place rubber spacer (12) in cover (15), then install brass seal (11) and seal (4).
5. Place seal (4) in cover (3).
6. Install both covers on body, aligning the scribe marks made during assembly.

CAUTION
Exercise care when installing cover on gear shaft so as not to damage oil seal.

7. Install bolts (1) and washers (2). Torque bolts to 28-32 ft.lbs.
8. Install key (20) and spider hub (18). Tighten setscrew (17) in hub securely.
9. Install pump support (19) using lockwashers (14) and capscrews (13).
10. Install 90° fitting (8) with O-ring (9) and straight fitting (21) with O-ring (22).

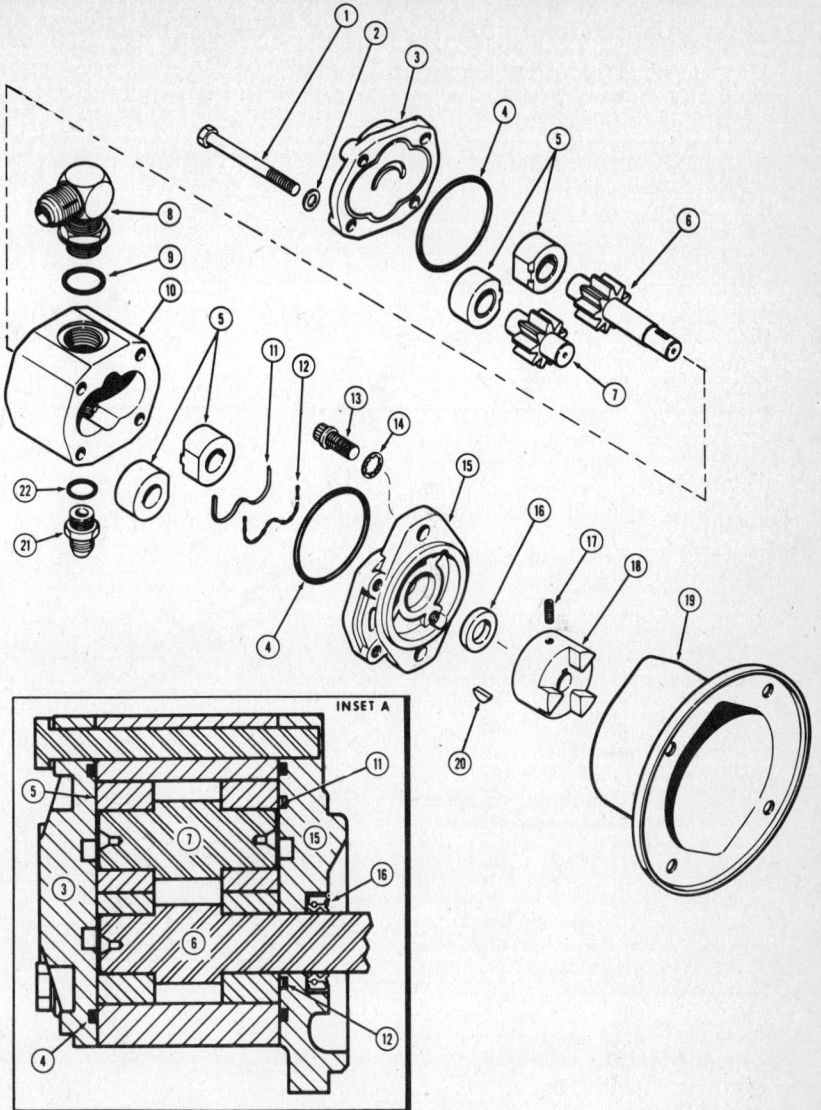

Hydraulic pump motor components

J.I. Case

CONTROL VALVE

REMOVAL

1. Drain the transmission.
2. Disconnect the four hydraulic tubes at the valve, (pump-to-valve tube, valve-to-heat exchanger tube, front valve port-to-top brake port tube and rear valve port-to-bottom brake port tube).
3. Disconnect the valve spool linkage at the valve spool.
4. Remove the valve mounting nuts, lockwashers and valve assembly.

DISASSEMBLY

1. Remove the valve spool end cap (26), spacer (25), O-ring (24), mounting bracket (23) scraper seal retainer (21) with seal (22) and quad ring (20) from the front end of the valve assembly.
2. Remove the ferry head capscrews (1) using a ¼ in. 12 pt. wrench and end cap (2).
3. Grasp the end of the spool (18) and pull out of the valve body (19).
4. Remove the snapring (15), retainer (13), spacer (14), retainer (13), relief valve plug retainer (12), spacer (11) and quad ring (10) from the spool (18).
5. Remove the relief valve plug (3), O-ring (4), shim (5), outer spring (6), inner spring (7), poppet (8) and poppet seat (9).
6. Remove the plug (16) with O-ring (17).
7. The valve body must be inspected for grooves, deep scratches and excessive wear. If the valve body has damaged threads, cracks or groove marks the body and spool must be replaced.
8. The valve spool must be inspected for grooves, deep scratches and excessive wear.

Outer Spring (6) Inspection:
Free length   1.65 in.
Total coils   11.5
Active coils   9.5
Wire diameter   .1055 in.
250 lbs. per inch

9. Check to see if the spool fits it's respective body bore with hand pressure and without excessive side clearance. If it is loose, scored or damaged the spool and body must be replaced.
10. Replace all O-rings and oil seals during assembly.

Inner Spring (7) Inspection:
Free length   1.23 in.
Total coils   15
Active coils   13
Wire diameter   .059 in.
95 lbs. per inch

ASSEMBLY

1. Install the relief valve poppet seat (9), poppet (8), inner spring (7), outer spring (6), shim (5), and plug (3) with a new O-ring (4) in the valve body (19).
2. Install the quad ring (10), spacer (11), relief valve plug retainer (12), retainer (13), spacer (14), retainer (13) and snapring (15) on the rear of the valve spool (18) and then into the valve body (19).

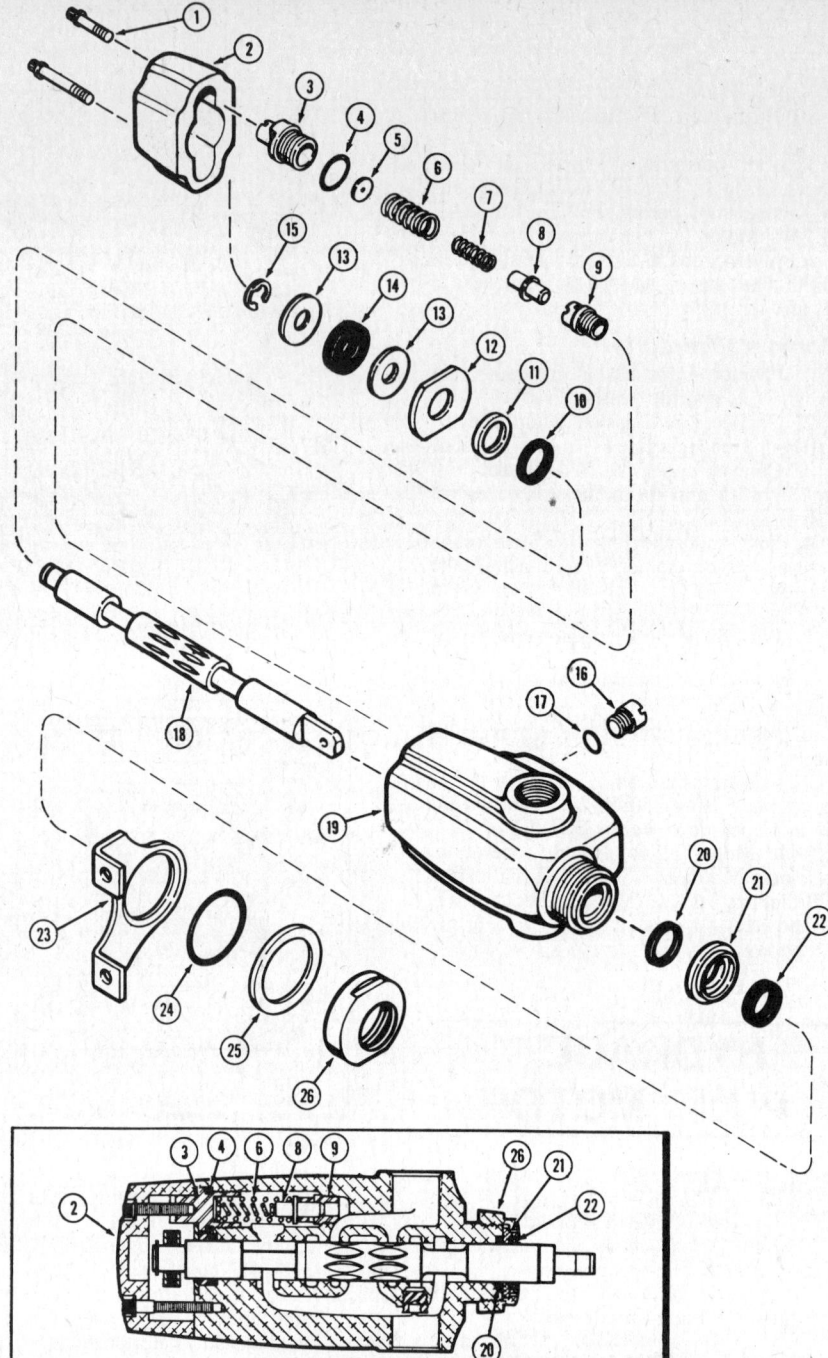

Control valve components

3. Install a new seal (22) in the retainer (21) with the lip towards the larger diameter opening.
4. Install the quad ring (20), retainer (21) with seal (22), mounting bracket (23), O-ring (24), spacer (25) and valve spool end cap (26).
5. Install the plug (16) with O-ring (17).
6. Install the end cap (2) using the ferry head capscrews (1). Torque to 225-275 in.lbs.

INSTALLATION

1. Install valve assembly using lockwashers and nuts.
2. Connect the valve spool linkage at the valve spool.
3. Connect the four hydraulic tubes at the valve, (pump to valve tube, valve to heat exchanger tube, front valve port to top brake valve port tube and rear port to bottom brake valve port tube).
4. Fill transmission with clean oil to one inch below top cover using SAE5W20 Motor Oil in winter (Below 32° F) or SAE20W40 Motor Oil in summer. Use only MS or DM Service Classification Oil that has passed AMA Test Sequences I, II, III.

ORBITAL MOTOR AND BRAKE VALVE

REMOVAL

1. Drain transmission.
2. Block up transmission and remove rear wheel.
3. Disconnect brake rod and hydraulic tubes. Cap hydraulic tubes to prevent the entry of foreign particles.
4. Remove transmission top cover including seat, fenders and seat support.
5. Remove mounting bolts, brake-motor assembly and gasket.

DISASSEMBLY

NOTE: Before removing the orbit motor, thoroughly clean the area around the orbit motor to prevent the entry of foreign material. Also make sure the disassembly area is clean.

1. Remove brake valve mounting capscrews (24), lockwashers (23), valve assembly and O-rings (21).
2. To disassemble brake valve, remove snapring (25), spool (19) with spool eye (18) and O-rings (20) from the body (22).

CAUTION
If a vise is used, avoid excess pressure which will distort the housing. Clamp across the port area not the housing.

3. Remove the seven capscrews (1) and the end cap (2).

IMPORTANT! Mark a line across the stator, rotor, motor body and shaft for proper assembly.

4. Remove the spacer (3) stator and rotor assembly (4) and (5) and spacer plate (6).
5. Remove the thrust bearing (7) and coupling shaft (8).
6. Remove mounting flange capscrews (29), washers (28) and flange (10).
7. Remove O-ring (13), quad ring (12) and oil seal (9) from inside the mounting flange (10).
8. Remove bearing race (14) thrust bearing (15) and motor shaft (16).
9. Remove plug (26) with O-ring (27).
10. Clean all parts before inspection being careful not to damage any machined surfaces.
11. Check the thrust bearing for excess wear, scratches and scoring. A polished pattern on the spacer plate and end plate due to rotor action is normal.
12. Check rotor to stator clearance using narrow feeler guage. The clearance should not exceed .005 in.
13. Check rotor to stator thickness with a micrometer. If rotor thickness is more than .002 in. less than the thickness of the stator, replace stator and rotor.
14. Replace all O-rings and oil seals during assembly.

ASSEMBLY

1. Install the plug (26) with a new O-ring (27).
2. Install the motor shaft (16) thrust bearing (15) and bearing race (14).
3. Install a new oil seal (9) lip out, quad ring (12) and O-ring (13) in the mounting flange (10).
4. Install the mounting flange on the motor shaft and secure using capscrews (29) and washers (28). Torque capscrews evenly to 225-275 in.lbs.
5. Install the coupling shaft (8) in the motor shaft.

NOTE: Be sure that the stator, rotor, motor body and shaft are lined up with the mark made during disassembly. If no mark was made, continue these assembly procedures and install the orbital motor. Follow the steps in the installation procedures to obtain correct motor alignment.

6. Install the thrust bearing (7) and spacer plate (6). Make sure the spacer plate is lined up with the mark made during disassembly.
7. Install the stator-rotor assembly (4) and (5) in alignment with the mark made during disassembly.
8. Install spacer (3), end cover (2) and secure with capscrews (1). Torque capscrews evenly to 175-200 in.lbs.
9. Install O-rings (20), spool (19) eye end up, spool eye (18), in the body (22) and secure with snapring (25).

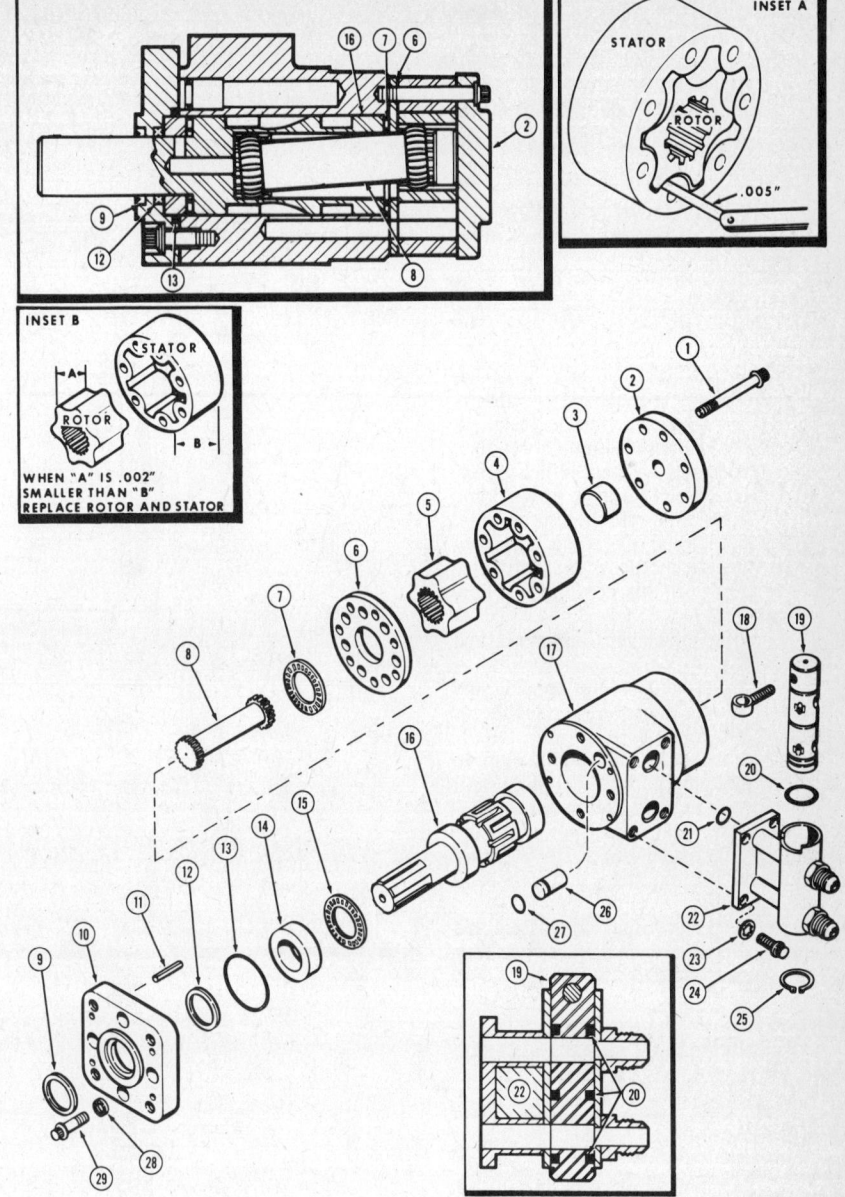

Orbital motor components

10. Install brake valve assembly O-rings (21), capscrews (24) and washers (23).

INSTALLATION

1. Mount the brake-motor assembly to the transmission using a new gasket and mounting bolts.

2. Install transmission top cover including seat, fenders and seat support.

3. Connect hydraulic tubes and brake lines.

4. Fill transmission with clean oil to one inch below top cover using SAE 5W20 Motor Oil in winter (Below 32° F) or SAE 20W40 Motor Oil in summer. Use only MS or DM Service Classification Oil that has passed AMA Test Sequence I, II and III.

5. With the orbital motor installed, check for proper motor alignment by placing the transmission in gear. When in forward, the tractor should drive forward and when in reverse, it should drive in reverse. If the results of these operations are opposite, the orbital motor is not properly aligned. To correct this condition, follow steps 6 thru 10.

6. Remove the seven capscrews and the end cap.

7. Mark the stator to the housing and the rotor to one of the splines on the motor shaft.

8. Move the stator and rotor enough to take the spacer out; mark the coupling shaft in line with the mark made on the rotor.

9. Slide the stator and rotor off the coupling shaft, rotate one tooth in either direction and slide the stator and rotor back onto the coupling.

10. Now the motor may be reassembled and the cap and spacer moved as necessary to align the bolt holes.

10. Install the fuel tank supports and fuel tank.

11. Install generator, generator belt and oil cooler fan.

12. Install exhaust system and hydraulic pump drive housing.

13. Install the engine from left side of tractor using four mounting bolts and nuts.

14. Install the hydraulic pump (10) and lines.

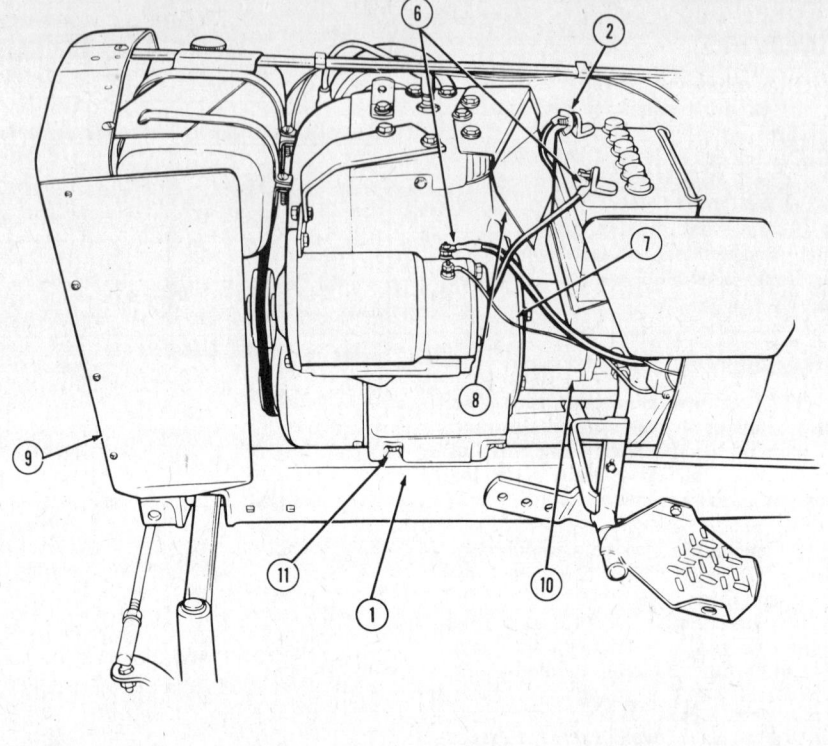

ENGINE

For detailed engine services, see the Kohler Engine part of the Engine Unit Repair section.

ENGINE REMOVAL AND INSTALLATION

1. Drain the engine crankcase oil (1).
2. Disconnect positive battery cable (2) from battery.
3. Disconnect fuel line (3) and drain fuel.
4. Disconnect choke control (4), throttle control (5), cables (6) and wires (7) from engine.
5. Remove the engine block air deflector (8) and lefthand grille side sheet (9).
6. Disconnect and remove the hydraulic pump (10). Cap the hydraulic lines.
7. Remove the four engine mounting bolts (11).
8. Remove the engine from the left side of the tractor.
9. Remove the fuel tank, supports, generator belt, generator, oil cooler fan, exhaust system and pump drive housing from the engine.

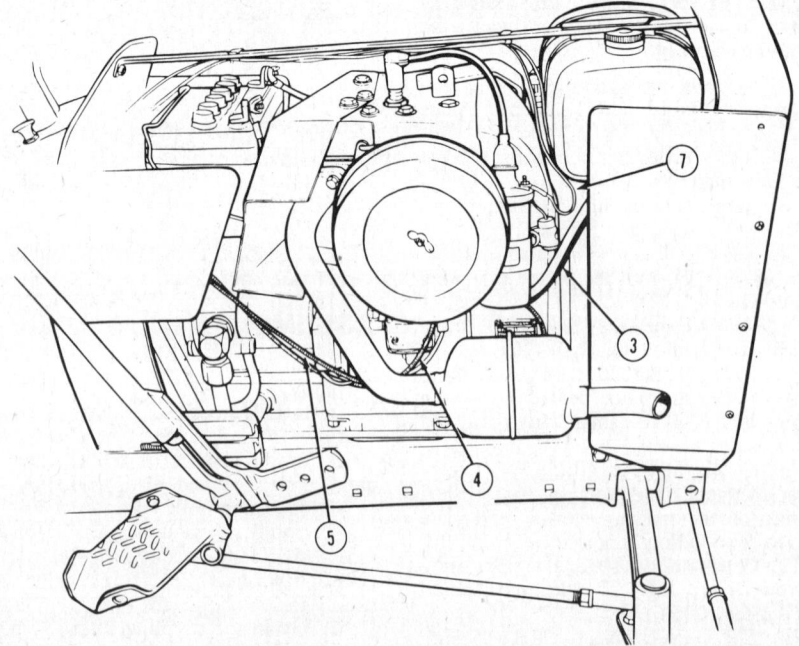

Engine removal and installation

J.I. Case

Models 107, 108, 117, 118, 200, 400, 644, 646

STEERING AND FRONT AXLE

Front Axle

REMOVAL

1. Support axle on blocks.
2. 107, 117, 108, and 118
   a. Remove roll pin from axle pin (107 and 117 has no roll pin in hole) put drift pin in axle pin hole to keep pin from turning, and remove nut from pin.
   b. Tap pin out of axle from front with drift.
3. 200, 400 series and loaders
   a. Remove retaining bolt from tab weld on axle pin.
   b. Pull pin out or tap from behind with drift.
   c. Remove axle.

INSTALLATION

1. On 107, 117, 108 and 118, position axle install pin and secure with nut. On 108 and 118 install roll pin in rear end of pin.
2. On 200, 400 series and loaders, position axle, install the axle pivot pin and secure with bolt.

Steering Shaft

REMOVAL

1. Remove Woodruff key from top of shaft.
2. Remove steering shaft from bottom of tractor except on loaders where it is necessary to loosen the two setscrews which hold the shaft to the U-joint, and then pull the shaft out of the top. Then remove snapring from U-joint mounting pin and remove washers and U-joint.

INSTALLATION

1. Push down on steering wheel shaft to be sure it is seated on bearing between pinion gear and mounting bracket.
2. Install Woodruff key in steering wheel shaft.
3. Place steering wheel on shaft, line up keyway with key, and push steering wheel down until it contacts nylon bushing on top of steering column.
4. Install washers over shaft and key to top of key or higher.
5. Install locknut and tighten until it just contacts washer.
6. Install cap.

Spindles

REMOVAL

1. Remove the rod.
2. Remove drag link ball joint from spindle steering arm.
3. Remove steering arm retaining bolt and remove steering arm. (On 200 and 400 series the steering arm is part of the spindle and is removed with the spindle.)
4. Remove key from spindle.
5. After removing the steering arm, that spindle is ready to remove. Remove roll pin from remaining spindle (snapring on loader) and remove spindles.

INSTALLATION

1. Place one shim washer on each spindle then place spindle into axle and secure.
2. Spindles should be shimmed to remove end play. This is done by adding shims between the axle and roll pin or snapring, and between the axle and the steering arm on models with removable steering arm. Add washers until you have a minimum amount of end play.
3. Install tie rod between spindles.

Steering Gear

REMOVAL

1. Remove drag link ball joint from sector gear.
2. Remove sector gear:
   a. On the 107, 117, 108 and 118 remove the two bolts which hold the sector gear pivot pin to the frame and remove gear.
   b. On the 200 and 400 series remove the gear retaining bolt from the end of the gear mounting pin, and remove gear.
   c. On the loaders remove the snapring which holds the gear to the mounting pin and remove the gear.
3. Steering gear mounting brackets can be removed from all models by removing the respective retaining bolts. The sector gear mounting pin is threaded into the mounting bracket on the 200 and 400 series models and can be removed.

INSTALLATION

1. Install mounting bracket in reverse of removal procedures.
2. Install respective sector gears in reverse of removal procedure.
3. Install steering shafts and steering wheels in reverse of removal procedure.
4. Connect drag link.

Steering Gear Adjustments

MODELS 107, 108, 117, 118

1. **STEERING WHEEL NUT:** The steering wheel nut is properly adjusted when there is no end play on the steering shaft and the pinion gear is in full alignment with the sector gear. Overtightening the nut can cause the pinion gear to bind against and distort the support bracket causing hard steering and travel lever operation.

2. **SECTOR GEAR VERTICAL END PLAY.** Vertical end play between the sector gear and the frame cutout should not exceed $1/16$ in. If adjustment is necessary, shims can be added between the sector gear and the frame cutout. Check the alignment between the sector and pinion gears to determine whether the shims should be placed above or below the sector gear.

3. **STEERING WHEEL FREE PLAY.** Steering wheel free play should not exceed 2½ in. Excessive free play can be caused by loose or worn ball joints on the drag link and tie rod or excessive clearance between the sector and pinion gears. The sector and pinion gears should mesh snugly without binding. To tighten the mesh between these gears, either equally remove shim washers between the support bracket and frame or equally add shim washers between the pivot pin and the frame.

MODELS 200 AND 400

The tractor is designed with two or more shim washers between the steering gear and support bracket as illustrated. As the gear teeth wear in, additional steering wheel free play may occur. If the free play becomes excessive, one (more if necessary) of the shims can be relocated to the bottom side of the steering gear.

1. Make certain there is not excessive end play on the steering shaft. Tighten steering wheel locknut to remove excessive end play without causing binding. If nut contacts key before play is out of shaft add shims between steering wheel and nut.
2. Disconnect the drag link from the steering gear. Remove the mounting bolt, lockwasher and plain washer (shims also if present) from the base of the pivot shaft. Slip the steering gear and one of the shim washers off the pivot shaft. Place the gear back on the pivot shaft and secure with the original mounting bolt, lockwasher, plain washer, shim(s) plus the shim removed from the upper side. The total number of shim washers must remain the same.
3. Make certain that some free play remains since a tight fit with no clearance between the two gears may cause binding and possible tooth failure.
4. Always coat all gear teeth with grease each time the two steering fittings are lubricated or at least each 50 hours operation.
5. Excessive steering wheel free play may not require gear adjustment as covered above. First check to make certain all ball joints on the drag link and tie rods are tight.

MODELS 644, 646

Steering Shaft

Check steering wheel and shaft for proper installation according to parts diagram. There should be a bearing, a washer and a snapring holding the U-joint to its mounting pin. The steering wheel should be adjusted so that when the pinion gear is resting on the bearing the lower side of the

215

J.I. Case

steering wheel hub just barely contacts the nylon bushing on the top of steering column. Adjust as follows:

Steering Wheel Free Play

Free play should not exceed two inches at the outside diameter of the steering wheel. If free play is excessive, first check to make certain all ball joints are properly tightened to the king pins, tie rod, steering arm, sector gear and drag link. Visually check each pivot point in the steering system to determine the source of free play. Thrust washers and bushings are located at the front axle, pinion joint and sector gear as shown in illustration.

NOTE: If the bushings in the front axle are replaced, make certain the spacers are reinstalled with the "split" in line with the lubrication holes.

TRANSAXLE

Model 107

TRANSAXLE UNIT

Removal and Installation

1. Jack up tractor so that transaxle is accessible. Use wood blocks to prevent equipment movement. Do not use bricks, cement, or cinder blocks. Visually inspect transaxle for oil leaks, cracked housing, binding or rubbing of parts, or other symptoms of malfunction.
2. Disconnect brake linkage.
3. Remove wheels and drive belt. Be aware of positioning of parts. Scribe mark, if in doubt, as to ability to re-assemble parts quickly.
4. If shifter lever will interfere with unit in any way, remove it before unit is removed.
5. Remove U-bolts and bracket holding transaxle to tractor frame. With transaxle free and supported, remove it from the area of the tractor to the work bench.
6. Reverse removal procedure to install.

Disassembly

1. Clean unit thoroughly of dirt, oil, debris. Remove shift housing and drain oil from unit.
2. Position the shifter forks in neutral.

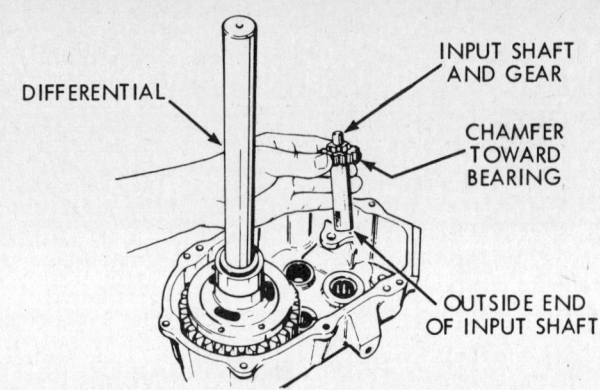

Input shaft removal

3. When disassembling the rest of the unit, it should be held so that:
   a. It lies on the case, properly blocked so that no weight rests on the input shaft or differential, yet the case is rigid.
   b. It can be worked on without the chance of falling, or causing injury.
4. Oil seals have a double lip so seal sleeves do not offer much protection during removal. Upon replacement, new seals should be used.
5. Tap dowel pins into the case and remove socket head capscrews. Lift the cover off from case. Discard gasket.
6. Remove output gear and shaft.
7. Remove the shifting assembly as one unit.
8. Remove the reverse idler shaft, spacer and gear.
9. Lift out the three gear cluster.
10. Remove the differential.
11. Tap the input shaft out of the case.
12. Check for cracks, stripped threads, metal chips, flat sealing surfaces, and rust. Clean out any rust. Replace parts if any damage is found that cannot be repaired.
13. Inspection of the case and cover may indicate the need for replacement of the axle housings.
14. Use an arbor press to drive out the housing, and a protective piece of bar stock between the housing and press when replacing the housing.
15. Press the housing in squarely until the flange seats against the case and cover.
16. Bearings, bushings and bearing surfaces should be thoroughly cleaned prior to examination. Examine closely for scuffing, wear, pitting and abnormal conditions. Replace if any conditions mentioned appear.
17. Use a good grade of clean solvent to clean bearings. After cleaning, always use clean lint-free cloth to dry and wipe bearings. Immediately coat cleaned bearing with lubricant to prevent rusting or corrosion.
18. Take care of bearings in the case and cover. Cover them to keep out foreign matter. Place gasket surface down on clean paper and cover with clean cloth.
19. Never clean the lubricant from new bearings. This lubricant prevents damage before the transaxle lubricant enters the bearing.
20. Use a bearing tool to press out the needle bearing. Insert the proper tool in the bearing and with an arbor press, press out the bearing from the inside.
21. When installing open end needle bearings, always apply pressure to the stamped side.
22. Use only the recommended tools to insert bearings. The opposite end of the same tool used for removal is used for replacement.
23. The inside face of the bearing housing should be below the thrust face on the case or cover. This distance is controlled by the design of the inserting tool. By using the proper tool, bearing life will be extended. Bearings should be pressed into the case or cover .015 to .020 in. below the thrust surface.
24. When removing bushings use the combined bushing remover and installation tool. Position the piece to be serviced on

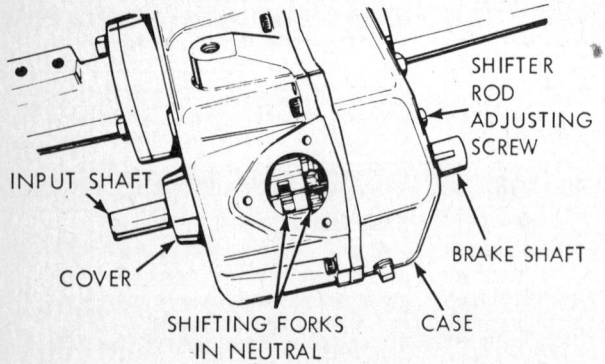

Positioning the shift forks in neutral

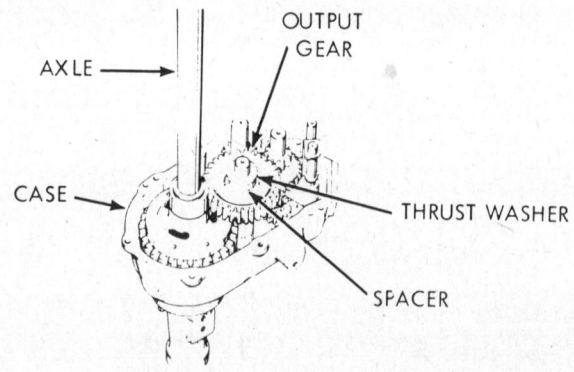

Removing output gear and shaft

J.I. Case

the table of an arbor press with an opening to allow the bushing to pass through.

25. After new bushings are pressed into the piece they must be sized. See the tool list for the proper sizing ball and driver. Use an arbor press and push the steel ball through the new bushing to expand it to the required size.

Assembly

1. Install input shaft in case. Use a soft mallet to seat shaft and gear completely. Binding can occur if the shaft is driven in only part way.
2. Install the differential assembly. The four capscrew heads should go down into the case.
3. Install the three gear cluster, with the smallest gear up.

NOTE: Bevels of small and middle gear go down toward large gear. Large gear bevel is up. The short spacer goes between the large and middle gears.

4. Position the reverse idler shaft in the unit, then install gear and spacer.
5. Install the shifter assembly as a unit into the case. When installed correctly, the neutral square formed by the shifting forks should appear through the case opening for attaching the shift housing. Both shift gears should be out of mesh.
6. Install the output shaft, gear, spacer and thrust washer.
7. Install brake shaft in the unit cover.
8. Position a new gasket on the cover mounting surface, then install cover.
9. Align cover with the dowel pin and secure with the socket head capscrew. Torque to 10 ft.lbs.
10. Install axle seals using sleeve and driver.
11. Install a new gasket and shift lever housing. Torque screws to 10 ft.lbs. Be sure the shift lever is in the proper position to allow shifting.

SHIFT LEVER UNIT

Disassembly

1. Prior to removing the shift lever assembly from the transaxle, make note of the position of the shift lever so that it may be assembled correctly to the shift lever housing.
2. Move the shift lever to Neutral, if possible, before removing it from the transaxle. Clean around the lever housing to prevent dirt from falling into the transaxle. Cover this opening, if possible.
3. Place the shift lever in a vise so that the shift lever housing is at least one inch from the top of the vise jaws.
4. Use the proper compressing type of tool for removing the snapring. Loosen the vise and disassemble the pieces.
5. Remove the shift lever from the shift lever housing. Examine the roll pin in the ball of the shift lever, if bent or worn, replace.
6. When inserting a new roll pin in the ball, position so that equal lengths protrude from both sides of the ball.
7. Oil leakage past the point where the shift lever enters the shift lever housing will require replacement of the quad ring seal in the shift lever housing.

Assembly

1. Prior to reassembly, be sure that bends in the shift lever correspond to the mounting on the vehicle.
2. Snapring type: Secure parts with the snapring. Before installing the shifter lever and housing to the transaxle housing, check the shifting forks for Neutral position.
3. Always use new gaskets between the shift lever housing and the transaxle.

SHIFTING ASSEMBLY

Removal

Shifting assemblies are removed from and installed into transaxles by squeezing the top end of the shifter rods. This causes a binding that retains all parts during removal or installation.

Disassembly

1. Follow the accompanying illustrations in order. Prior to disassembly compare the assembly with the illustration. This will aid during the reassembly.
2. Replace the shifter stop if worn or damaged.
3. Examine the teeth and internal splines of the two shifter gears. Replace damaged gears. The gears must slide freely on the shifter shaft. Excessive wear of the internal spline in the gears will create cocking and difficult shifting. Replace the gear if this condition is present.
4. Replace the shifter shaft needle bearing as follows if wear is evident. Replace the shaft if the bearing surface is scuffed, pitted or worn to a diameter less than .750 in.
5. To remove the needle bearing in the splined shifter shaft proceed as follows:

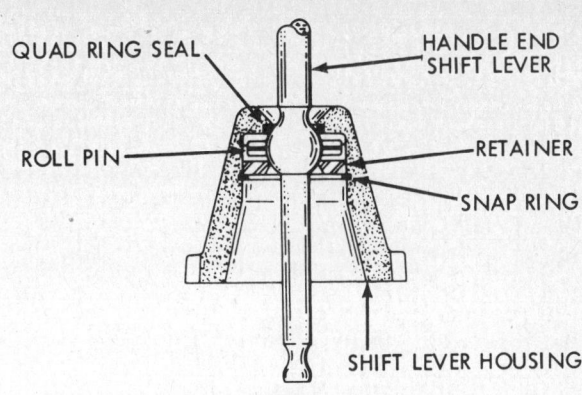

Shift lever

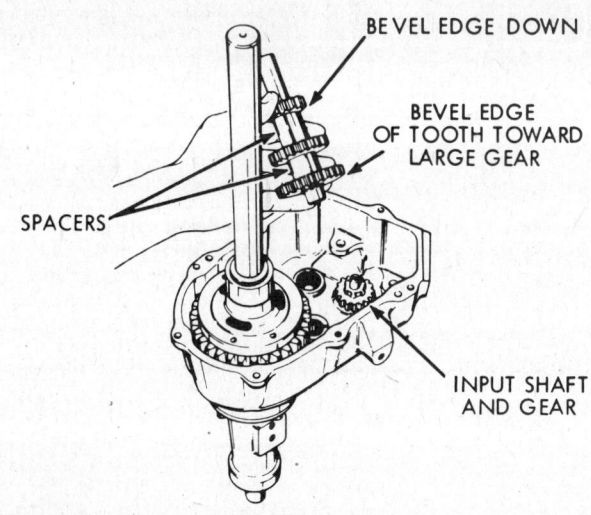

Input shaft installation

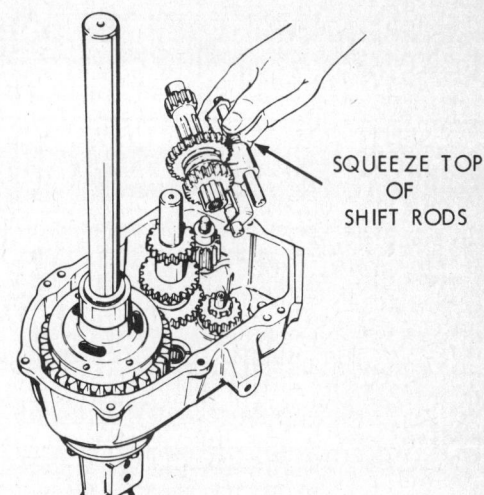

Shifter installation

217

J.I. Case

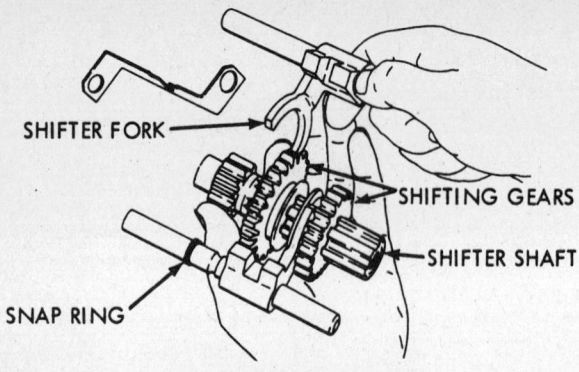

Disassembly sequence-A

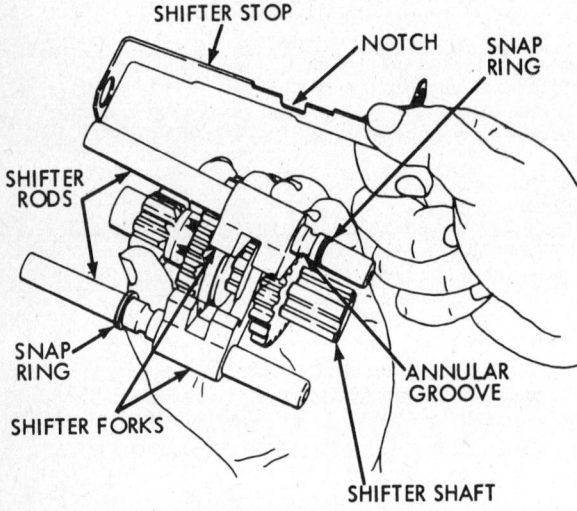

Disassembly sequence-B

Disassembly sequence-C

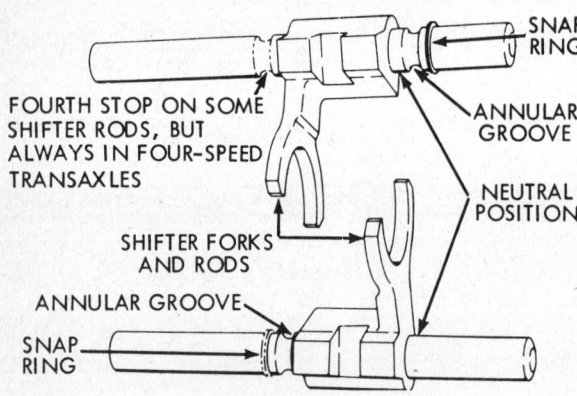

Disassembly sequence-D

NOTE: Blind bearing pullers are available to remove this bearing. There is a space between the bottom of the drilled hole and the inside end of the bearing to accommodate the ridges of the bearing puller. If no puller is available, however, proceed as follows.

6. With the needle bearing up, clamp the splined shaft vertically in a soft jaw vise so that the lower end of the shaft rests on a block of wood.
7. Prepare some pieces of paper toweling, newspaper, etc. by soaking in water.
8. Tear paper into pieces, approximately one to two inches square. Stuff these wet pieces of paper into the needle bearing until full.
9. Insert a $7/16$ in. metal rod into this bearing. With a mallet strike the rod sharply. This will compress the wet paper. Continue to add more wet paper, this will hydraulically lift the bearing out of the shaft.
10. Use the authorized tool to install the new bearing. Needle bearings in shifter shafts should be installed .010 in. below flush.
11. Replace other parts showing wear, looseness, cracks, etc.

Assembly

1. Reassemble the shifting assembly by following the accompanying illustrations. Lay the parts on the bench in the same manner as illustrated on a clean paper or shop cloth. Pay particular attention to the annular grooves in the shifter rods and the snapring.
2. Assemble the shifter forks to the shifter rods as illustrated. The shifter forks are interchangeable.
3. Slide the shifter fork onto the shifter rod until it comes to the hole with the indexing ball and spring. With a flat blade screwdriver press the indexing ball into the hole and move the shifting fork completely onto the shifter rod.
4. Move the shifting fork to the Neutral position. The neutral groove is the center groove. This neutral groove can be seen through the hole in the shifter fork. The arrow in the accompanying figure from the words "Neutral Groove" is passing through the hole for viewing.
5. When the shifter forks are properly assembled to the shifter rods and positioned in neutral, the ends of the notches in the shifter forks are in alignment.
6. Assemble the two flanged gears onto the shifter shaft. Note that the large gear is placed on the shaft first with the flange side toward the needle bearing in the end of the shifter shaft. Slide on the smaller gear with the flange toward that of the larger gear.
7. When assembling the shifter fork and rod to the flanged gears on the shifter shaft, that shifter fork which is on shifter rod "A" always engages in flange in the larger gear.
8. To determine which is shifter rod "A" compare the parts to illustrations. Hold the shifter shaft in the hands as illustrated during assembly.
9. After the shifter fork and rod as-

J.I. Case

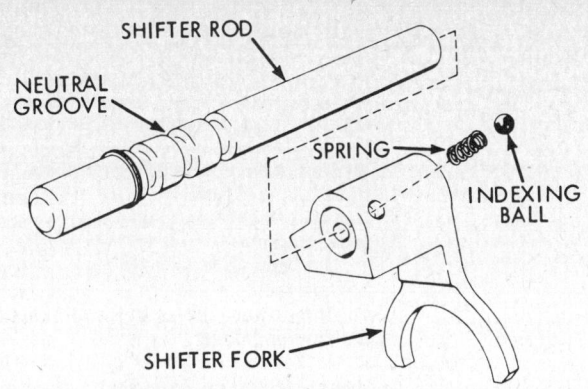

Disassembly sequence-E

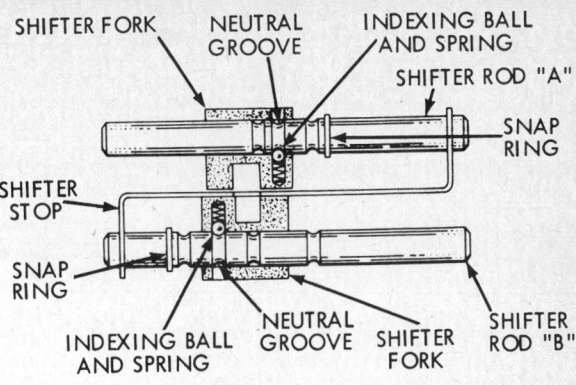

Disassembly sequence-F

semblies have been engaged with the flanged gears allow the shifter rods to lay open in the hand and position the shifter stop. The notch in the shifter shop is the guide to correct positioning. Align this notch with the corresponding notches in the shifter forks and insert the shifter stop. Move the shifter rods together, and insert into the transaxle. Remember to squeeze the ends of the shifter rods to cause the assembly to bind and stay together.

The needle bearing end is inserted first into the case to engage the end of input shaft.

The shifter assembly is correctly installed in the transaxle if the notches in the shifter forks are just about in the center of the opening in the case or cover of the transaxle.

DIFFERENTIAL UNIT
Disassembly

1. Clean the differential assembly, then check and note the axle lengths and their relation to the heads of the four hex head bolts.

2. If the unit will not turn freely, note where the unit binds. Check and replace those parts.

3. Place the differential in a large vise with soft jaws (hex head bolts up). Do not clamp the vise on the bearing race of a differential carrier.

3. Remove the four hex head bolts and the upper axle and differential carrier. Re-

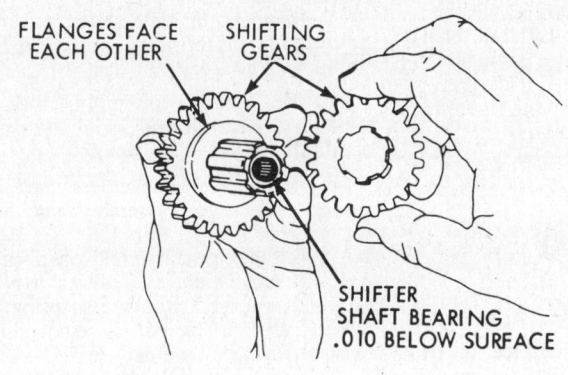

Shifting gear assembly

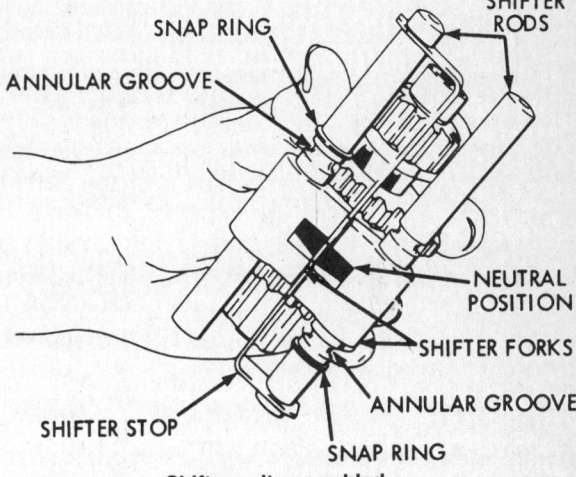

Shifter unit assembled

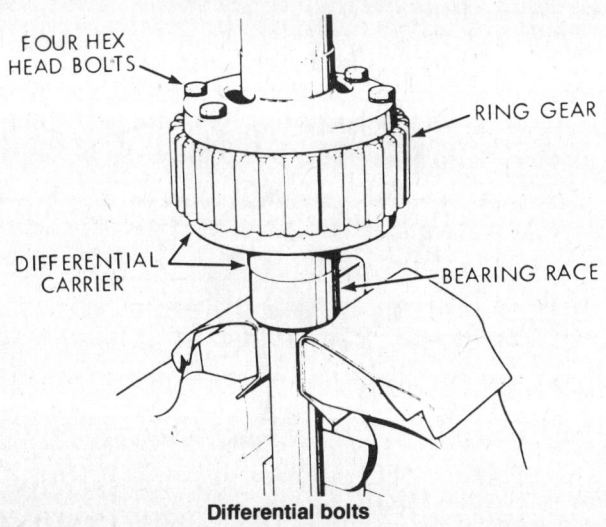

Differential bolts

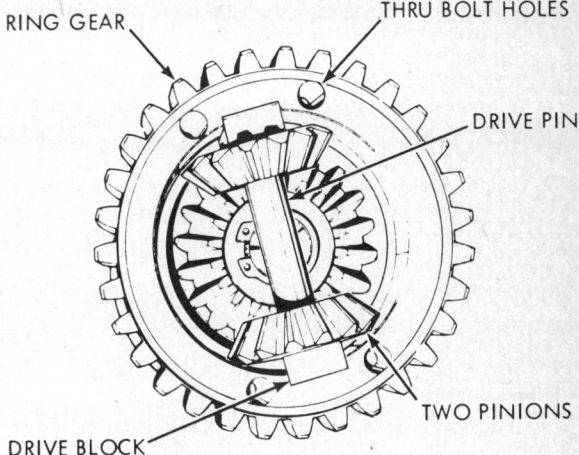

Differential with one carrier removed

219

J.I. Case

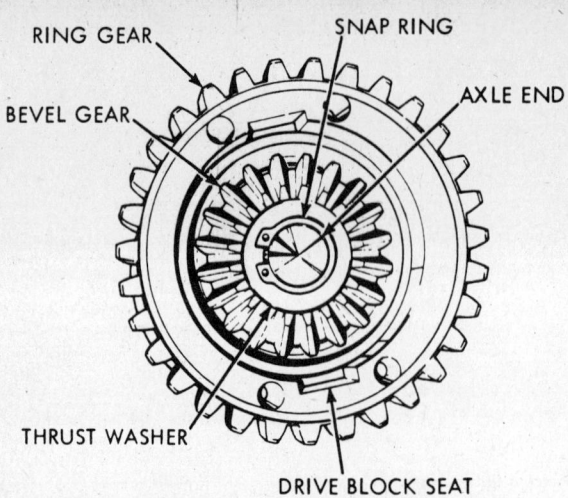

Drive block, pin and pinion gear removed

move the drive blocks, pinions, drive pin and thrust spacer if used, by lifting out of the ring gear. Tap the ring gear lightly with a mallet to loosen from the differential carrier.

4. If a snapring is used, the axle assembly may be disassembled. If the axle end has been rolled, do not attempt to break the rolled retaining edge. The parts are to be replaced as an assembly.

5. Remove the snapring and the thrust washer, if used. Separate the bevel gear and differential carrier from the axle.

6. Examine gears for wear, cracked or chipped teeth. Check the internal splines of the gears and the axle if the gear is removeable. If excess play is noted, it may be necessary to replace the individual parts or both the gear and axle.

7. Examine drive pinions, drive pins and drive blocks for wear and damage. Replace excessively worn pinion or the drive pin.

8. Examine the differential carriers. One has threaded holes and the other has larger holes so that the bolts will pass through.

9. Examine the internal bearing diameter of the differential carriers. If dimension is in excess of 1.004 in. at point A, replace the differential carrier or bushing, is used.

10. When assembling thrust bearings, always place a hardened thrust washer on each side of the caged thrust rollers. Never use the caged thrust rollers without the thrust washers.

Assembly

Oil all parts during reassembly.

1. Select the correct axle for the side of the differential opposite the hex head bolts. If the wrong axle is used, it will require complete tear down of the differential, or possibly the entire transaxle if the error is not detected until later.

2. Clamp the axle, in a soft jaw vise (not bearing or oil seal surfaces). The differential carrier with threaded holes is assembled to this axle.

3. Refer to the accompanying figure for the proper arrangement of parts.

4. Torque the four hex head bolts to 7 ft.lbs.

Model 108

TRANSAXLE UNIT

Disassembly

1. Clean the outside surface of the transaxle, away from the area where the disassembly will take place. (Position shift lever in neutral position to help disassembly.) Remove screws (3) holding shift lever and shift lever housing. Remove shift lever housing. Drain oil through the shift lever opening. Remove all keys from keyways, remove all burrs and dirt from shafts. On hardened shafts, use a stone to remove burrs. All seals should be replaced whenever a shaft is pulled through a seal. Always use a new gasket whenever the gasket surfaces have been separated.

2. After removing axle housings, place the unit in a receptacle, bench or clamp the transaxle in a soft jaw vise. Position the transaxle so that the socket head cap screws are facing up.

3. Remove the socket head capscrews holding the case and cover together. Drive out the dowel pins used for alignment of the case and cover.

4. Lift off the cover assembly. Use a seal protector on axle shaft and lift off transaxle cover assembly. Because this seal is a single lip type, it may be reused, if care is taken to see that it isn't scratched or cut. Discard gasket.

5. To remove differential assembly, it may be necessary to replace two or three screws to hold center plate assembly down. Pull assembly straight up. If tight, tap on lower axle with soft mallet.

CAUTION
Do not use steel hammer.

Remove gear on top of shifter shaft.

6. Remove temporary holding screws, if used, and lift off center plate assembly. Discard gasket.

7. Remove complete shifter assembly by grasping shifter gears, shaft and both shifter rods as a unit.

NOTE: Examine assembly carefully; if no service is required, retain assembly as a unit for easy reassembly.

8. Remove reverse idler shaft and spacer, cluster gear assembly and thrust washer.

9. Lift idler gear assembly out of case.

10. Remove input shaft oil seal to allow access to snapring. Remove snapring and input shaft will slide out. A removed seal must be replaced by a new seal.

11. **CLUSTER GEAR SUB-ASSEMBLY:**

a. The cluster gear can be disassembled. All gears are replaceable if damaged or worn. Preferably use a press to drive the gears squarely.

b. The small and middle gear bevel faces down, there is no beveled edge on

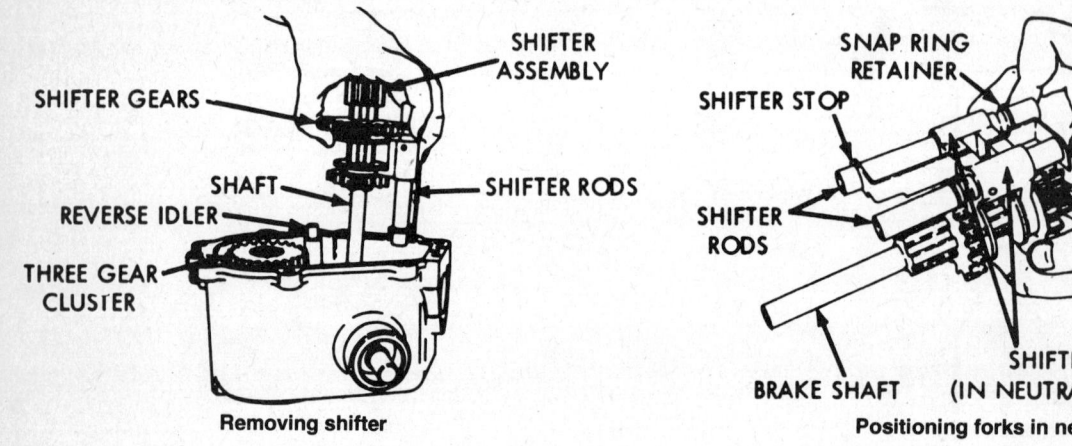

Removing shifter — **Positioning forks in neutral**

J.I. Case

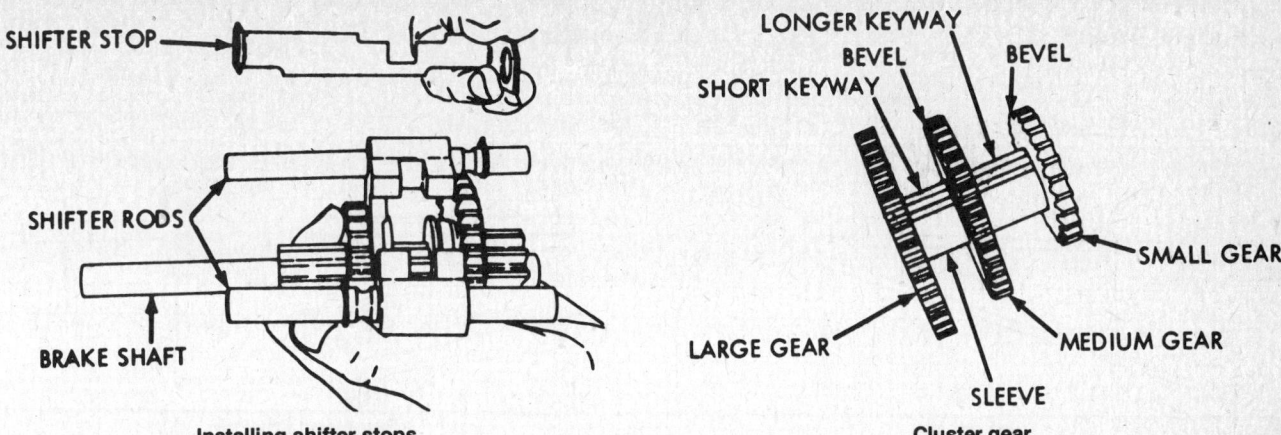

Installing shifter stops

Cluster gear

large gear. Shorter section between middle and large gear.

c. Key edge ends must align with shaft ends.

12. SHIFTING ASSEMBLY: The shifting assembly is usually removed from and installed into the transaxle as a unit. The assembly is removed and replaced by grasping the shifting rods firmly. This will cause the binding necessary to hold the assembly together. Before removal or installation of the shifting assembly, notches in the shifter forks should be aligned with notches in the shifter stop. This indicates that shifting assembly is in a neutral position. The shifter stop must be so positioned that the notch aligns with notches in shifter forks.

Assembly

1. Install thrust washers and bearing on input shaft.

2. Install input shaft into case assembly. Lock on with snapring retainer. Install oil seal.

3. Set case assembly open side up. Insert the idler shaft gear assembly, thrust washers and bearing. Note sequence of washers and bearings.

NOTE: Place reverse idler shaft into bearing to aid in holding washers, thrust bearing, idler shaft and gear assembly prior to installing shifter assembly.

4. Insert the washer and then the three gear cluster assembly.

5. Insert shifter assembly. Check that rods are seated properly.

NOTE: Reverse idler shaft will be pushed out at this time.

6. Install reverse idler. Make sure beveled edge is up. Spacer on top of gear.

7. Place new gasket on case and install center plate.

8. Place new gasket on center plate and install differential assembly, longer axle in down position. Be sure gear on shifter shaft is on shaft.

9. Install gear case dowel pins. Leave dowel pins slightly exposed on top to locate cover assembly.

10. Install transaxle cover assembly, and secure with eight (8) capscrews.

11. Install bearings and/or bushings, if necessary, using bearing driver and bushing tool. See bearing chart below.

12. Install axle housing assembly. Fill with 1½ pints SAE EP90 oil.

13. Inspection Note: For a neutral position, shift notches in forks and notch in shifter stop must be aligned and centrally located.

SHIFTER UNIT

Shift Lever
DISASSEMBLY

1. Place the shift lever in a vise so that the shift lever housing is at least one inch from the top of the vise jaws.

2. Use the proper compressing type tool for removing the snapring. Loosen the vise and disassemble the pieces.

3. Remove the shift lever from the shift lever housing. Examine the roll pin in the ball of the shift lever, if bent or worn,

replace. When inserting a new roll pin in the ball, position so that equal lengths protrude from both sides of the ball.

4. Oil leakage past the point where the shift lever enters the shift lever housing will require replacement of the quad ring seal in the shift lever housing.

5. Prior to reassembly, be sure that bends in the shift lever correspond to the mounting on the vehicle.

ASSEMBLY

1. Secure parts with the snapring. Before installing the shift lever and housing to the transaxle housing, check the shifting forks for Neutral position.

2. Always use new gaskets between the shift lever housing and the transaxle.

DIFFERENTIAL

Disassembly

1. Drive out roll pin that secures drive pin with suitable driver.

2. Thrust washers must be removed before attempting to remove the pinions. Remove bevel pinions simultaneously by rotating the gears in opposite directions; gears will move out of position.

3. Remove snapring, bevel gear and thrust washer. Slide axle out.

4. Inspect bushings and gears for wear and replace when necessary.

Assembly

1. Place axles (left and right) into differential gear assembly. Install thrust washers.

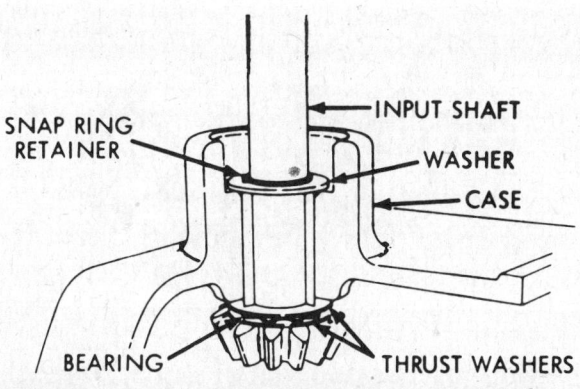

Installing thrust washers and bearing on input shaft

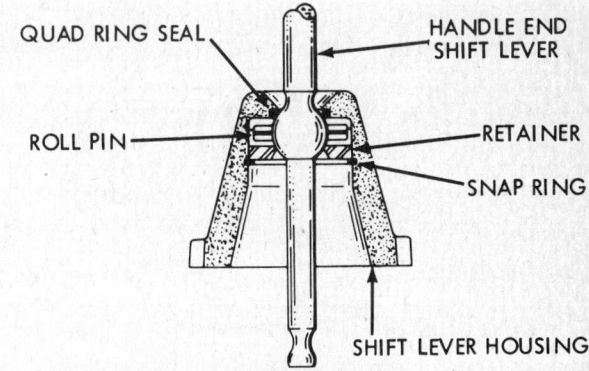

Shift lever

J.I. Case

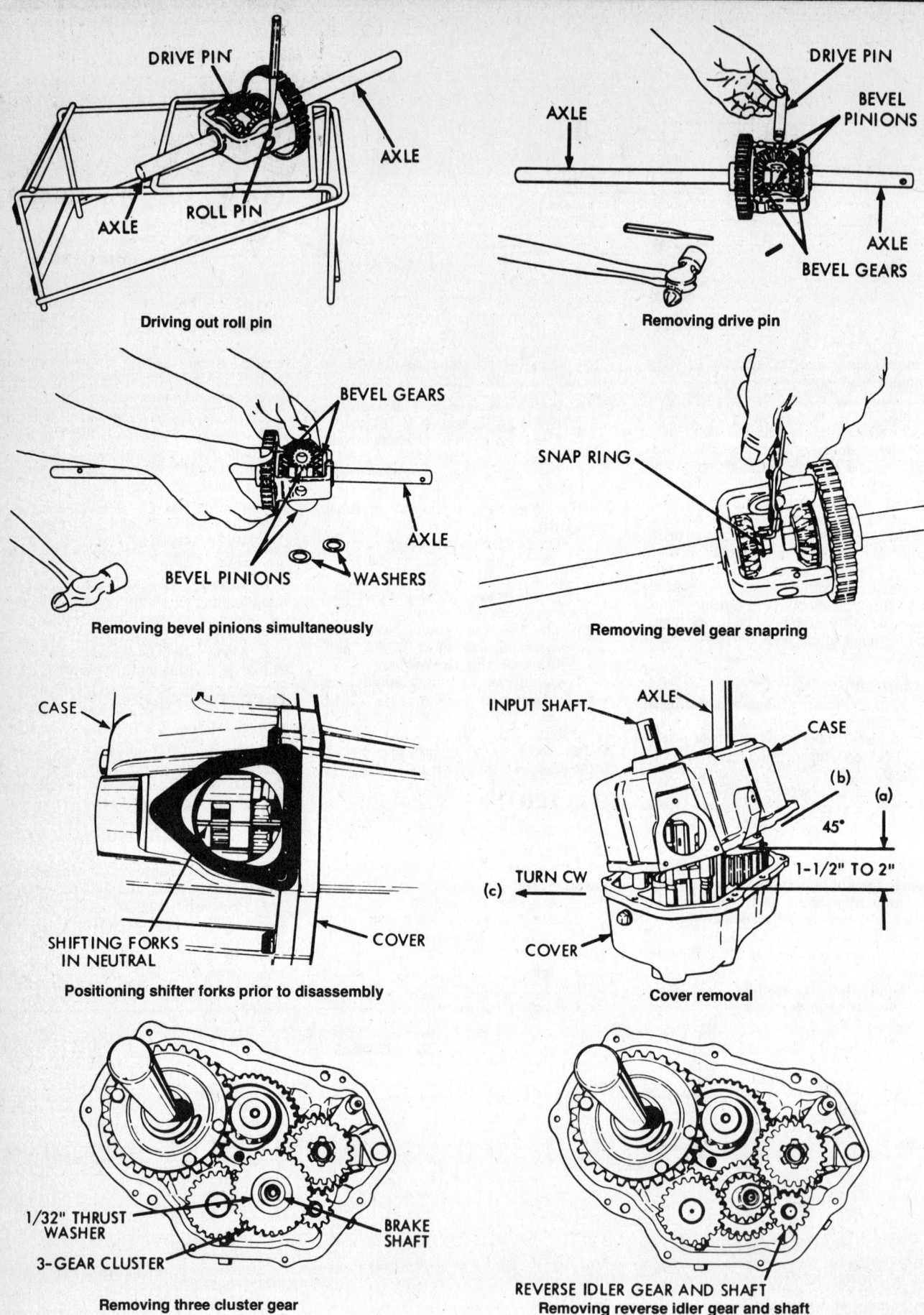

J.I. Case

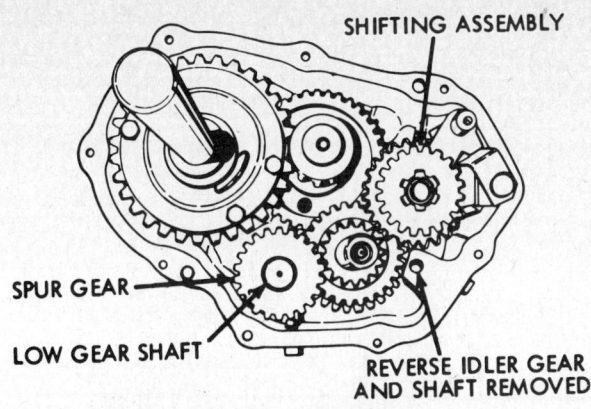

Removing shifting assembly

Brake shaft

NOTE: The axles differ in length so select the proper axle.

2. Place bevel gears on the shaft and install snapring in groove on the shaft.

3. Install bevel pinions SIMULTANEOUSLY FROM OPPOSITE SIDES by rotating pinions in opposite directions while sliding into position in gear assembly. Check alignment by inserting fingers into drive pin holes. If not aligned, drive pin cannot be inserted. Remove and replace bevel pinions as only one tooth out of position will cause misalignment.

4. After aligning, insert thrust washers behind each pinion. Insert drive pin and secure with roll pin.

Model 210

Disassembly

1. Position the shifter forks in neutral before disassembly.

2. Remove both axle housings and use the exposed axle as a ram to separate the seal retainers from the case and cover.

3. When disassembling the rest of the unit, it should be held so that:

a. It lies on the cover, properly blocked up, so that no weight rests on the brake shaft.

b. The cover should sit rigidly so that removal of parts can be done in a systematic step by step procedure.

c. It will not fall causing an accident or injury.

4. Oil seals are of the double lip type so sleeve protectors do not offer much protection when removing them. Upon replacement, new seals should be used.

5. Tap dowel pins into the cover and remove eight socket head capscrews.

6. To separate the case from the cover:
a. Lift the case 1½ to 2 in. (38.1–50.8 mm) above the cover.
b. Tilt the case so that shift rods will clear edge.
c. Rotate the case so that boss hidden inside will clear gears, then lift free of the differential.

7. Remove thrust washer and three gear cluster from brake shaft, noting whether the cluster has a sloppy fit.

8. Inspect gear teeth for wearing, chipping or breaks. Wear or chipping on the bevel area only, indicates shifting while the equipment is in motion.

9. Remove the reverse idler gear, spacer, and shaft from boss in cover.

NOTE: The spacer goes between the gear and the gear bevels go down.

10. Excessive wear on teeth bevels indicates improper shifting technique.

11. Lift out the shifter assembly.

12. If it is evident that the shifter assembly needs no further teardown, place it aside, in a clean place, intact, for easy re-assembly.

13. Remove the low gear and shaft, and splined spur gear. Separate gear and shaft. Note that NO thrust washer is between the gear and case.

14. Remove the two gear cluster and spacer from the brake shaft.

15. Lift the differential unit out of the cover.

16. Remove the output gear and shaft and thrust washer from each end of shaft.

17. Remove the brake shaft.

NOTE: The brake shaft idler separates from the shaft. If separated, be sure that when re-assembled, the idler gear chamfers are away from the cover.

18. Remove input shaft from case by tapping with a non-metallic hammer.

Assembly

1. Install input shaft in case. Use a soft mallet to seat shaft and gear completely. Often binding in the assembled unit can be traced to a partially installed input shaft.

2. Center on 1/32 in. (.79 mm) thick by 1 in. (25.4 mm) I.D. thrust washer on the cover brake shaft needle bearing, then install the brake shaft and gear (chamfer side away from cover).

3. Install the output shaft and gear after centering a 1/16 in. (1.59 mm) thick by 15/16 in. (23.81 mm) I.D. thrust washer on each end of the shaft.

4. Insert the differential assembly in the cover. Note that the four bolt heads should be out away from the output gear.

5. Install the two gear cluster and spacer on the brake shaft.

6. Install a 1/16 in. (1.59 mm) thick by ¾ in. (19.05 mm) I.D. thrust washer, gear, and low gear idler shaft in cover. Do not put a thrust washer on the exposed end of this shaft. Be sure the small gear meshes with the larger gear of the two gear cluster.

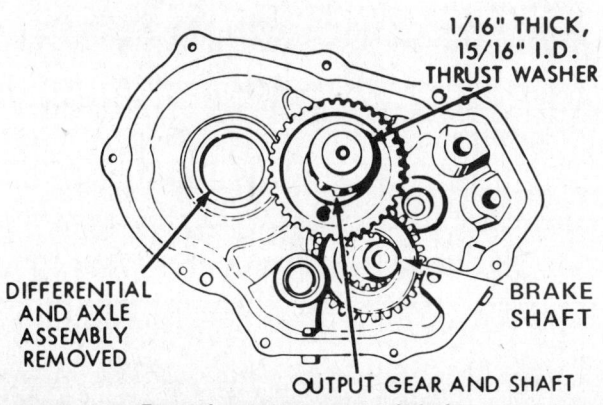

Removing output gear and shaft

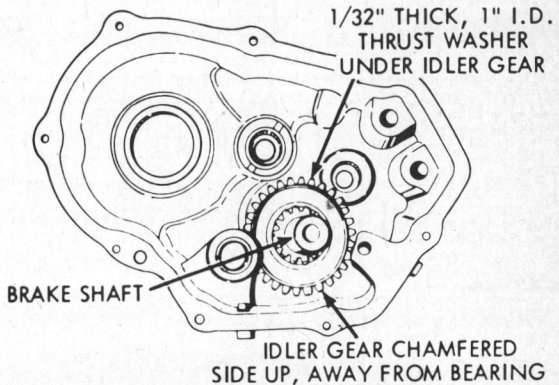

Removing idler gear and brake shaft

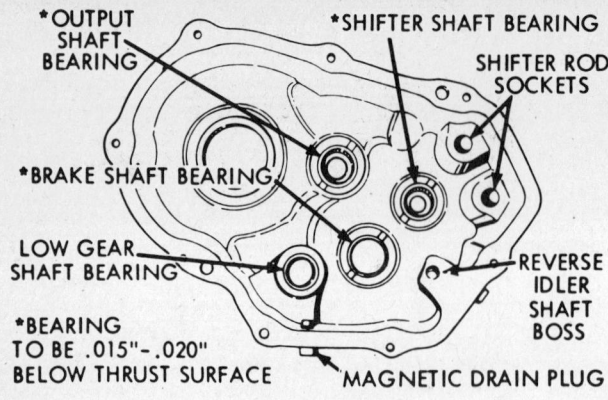
Cover assembly

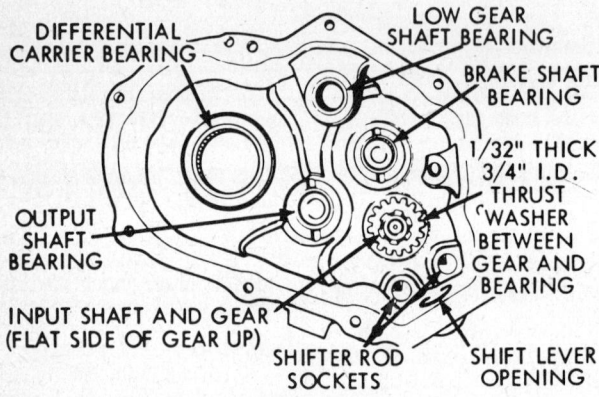
Case assembly

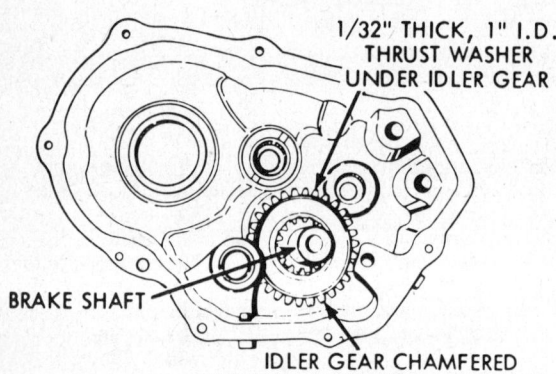
Installing idler gear and brake shaft

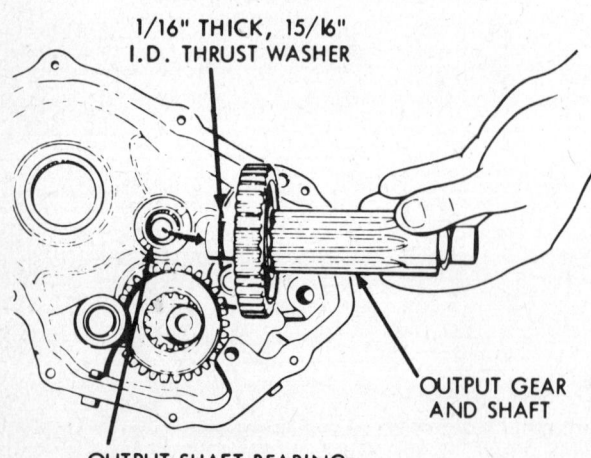
Installing output shaft bearing

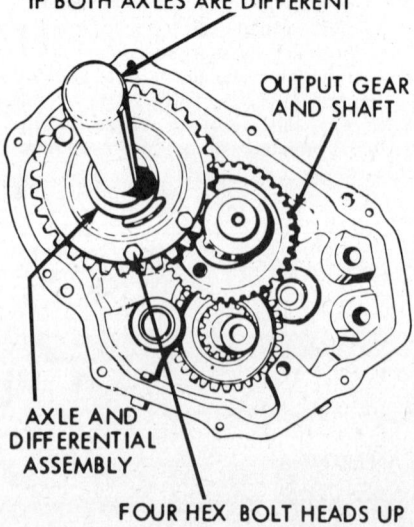
Installing differential

7. Center one 1/32 in. (.79 mm) thick by 7/8 in. (22.22 mm) I.D. thrust washer on cover shifter shaft bearing.

8. Install shifter assembly as a unit into the cover.

9. With the small gear of the three gear cluster toward the spacer, install the three gear cluster and other 1/32 in. (.79 mm) thick by 7/8 in. (22.22 mm) I.D. thrust washer on the brake shaft.

10. Install the reverse idler shaft, spacer, and gear into the cover. The beveled side of the idler gear should be down into the cover.

11. Position the gasket on the cover sealing surface, then install case over the differential shaft. Be sure the boss goes under gears and that edge of the case goes over the shaft rods in the opposite manner from which it was removed.

12. Once in position, if case hangs ½ (12.7 mm) to 1 in. (25.4 mm) high, turn the input shaft to get gears to mesh. The case should drop to about ¼ in. (6.35 mm) from closing.

13. Use a pair of needle nose pliers on the shifter stop on each shifter fork to agitate the shifter rod ends into their machined recesses in the case.

14. Align the case and cover with the two dowels, then install and tighten the eight socket head capscrews. Torque screws to 10 ft.lbs. (13Nm). Unit can now be placed flat on the work bench. Position seal retainers and new seals in position.

NOTE: Sleeves must be used to protect seals, especially axle ends or where wheels attach.

15. Install new O-rings on seal retainers and position axle supports to case and cover. Be sure mounting pads face in same position as when removed. Install capscrews and torque to 13 ft.lbs.

16. Install shift lever housing and new gasket.

J.I. Case

SHIFTER UNIT

Shift Lever Assembly
OVERHAUL

1. Prior to removing a shift lever assembly from a transaxle, make note of the position of the shift lever so that it may be assembled correctly to the shift lever housing.

2. Move the shift lever to Neutral, if possible, before removing it from the transaxle. Clean around the lever housing to prevent dirt from falling into the transaxle. Cover this opening, if possible.

3. Place the shift lever in a vise so that the shift lever housing is at least one inch from the top of the vise jaws.

4. Use the proper compressing type tool for removing the snapring. Loosen the vise and disassemble the pieces.

5. Remove the shift lever from the shift lever housing. Examine the roll pin in the ball of the shift lever, if bent or worn, replace. When inserting a new roll pin in the ball, position so that equal lengths protrude from both sides of the ball.

6. Oil leakage past the point where the shift lever housing will require replacement of the quad ring seal in the shift lever housing.

7. Prior to reassembly, be sure that bends in the shift lever correspond to the mounting on the vehicle.

8. Secure parts with the snapring. Before installing the shift lever and housing to the transaxle housing, check the shifting forks for Neutral position.

9. Always use new gaskets between the shift lever housing and the transaxle.

DIFFERENTIAL

Disassembly

1. Drive out roll pin that secures drive pin with suitable driver.
2. Remove drive pin.
3. Thrust washers must be removed before attempting to remove the pinions. Remove bevel pinions simultaneously by rotating the gears in opposite directions; gears will move out of position.
4. Remove snapring, bevel gear and thrust washer. Slide axle out.
5. Inspect bushings and gears for wear and replace when necessary.

Assembly

1. Place axles (left and right) into differential gear assembly. Install thrust washers.
2. Place bevel gears on the shaft and install snapring in groove on the shaft.
3. Install bevel pinions SIMULTANEOUSLY FROM OPPOSITE SIDES by rotating pinions in opposite directions while sliding into position in gear assembly. Check alignment by inserting fingers into drive pin holes. If not aligned, drive pin cannot be inserted. Remove and replace bevel pinions as only one tooth out of position will cause misalignment.
4. After aligning, insert thrust washers behind each pinion. Insert drive pin and secure with roll pin.

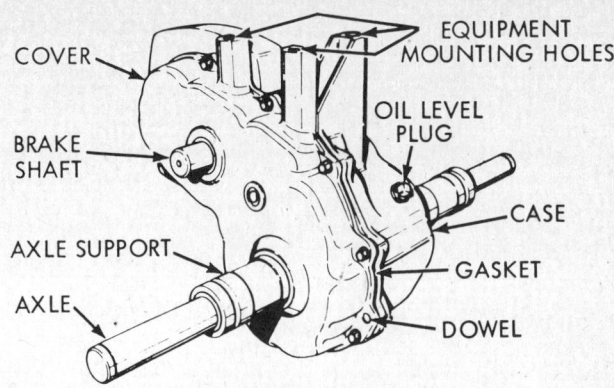

Differential exterior details

Models 117 and 118

TRANSAXLE UNIT

Removal and Installation

1. Jack up tractor so that transaxle is accessible. Use wood blocks to prevent equipment movement. Do not use bricks, cement, or cinder blocks. Visually inspect transaxle for oil leak, cracked housing, binding or rubbing of parts, or other symptoms of malfunction.
2. Disconnect brake linkage and hydrostatic transmission linkage.
3. Remove drive belt.
4. Remove rear frame plate.
5. Remove U-bolts at axle housing and two capscrews at frame crossmember in front of transaxle housing.
6. Pull transaxle and hydrostatic transmission as a unit, out the rear of tractor.
7. With transaxle free and supported, remove it from the area of the tractor to the work bench.
8. Reverse removal procedure to install.

Disassembly

1. Remove lockscrews and tap dowel pins out of cover. Lift off cover and discard gasket.
2. Lift out brake shaft, gear, and thrust washers on each side of gears.
3. Lift output shaft, gear, spacer, and thrust washer from case. At the same time, lift out the differential assembly.

NOTE: No thrust washer is located between the output shaft and case.

4. Bearings, bushings and bearing surfaces should be thoroughly cleaned prior to examination. Examine closely for scuffing, wear, pitting and abnormal conditions. Replace if any conditions mentioned appear.
5. Use a good grade of clean solvent to clean bearings. After cleaning, always use clean lint-free cloth to dry and wipe bearings. Immediately coat cleaned bearing with lubricant to prevent rusting or corrosion.
6. Take care of bearings in the case and cover. Cover them to keep out foreign matter. Place gasket surface down on clean paper and cover with clean cloth.
7. Never clean the lubricant from new bearings. This lubricant prevents damage before the transaxle lubricant enters the bearing.
8. Use a bearing tool to press out the bearing. Insert the proper tool in the bearing and with an arbor press, press out the bearing from the inside.
9. When installing open end needle bearings, always apply pressure to the stamped side.
10. Use only the recommended tools to insert bearings. The opposite end of the same tool used for removal is used for replacement.
11. The inside face of the bearing housing should be below the thrust face on the case or cover. This distance is controlled by the design of the inserting tool. By using the proper tool, bearing life will be extended. Bearings should be pressed into the case or cover .015 to .020 in. below the thrust surface.
12. To separate axle supports from the case and cover, use an arbor or hydraulic press. A piece of bar stock should be used to protect the support from the press ram.

Assembly

1. Inspect case and cover for cracks, stripped threads, marred sealing surfaces, and bearing condition. Cause of any oil leakage should be corrected. If parts can't be repaired, replace them.
2. Check shafts and gears for worn or chipped teeth. Check bearing surfaces for scratches which might affect oil seal performance. Check for wear.
3. Check differential for rigidity. Wobble indicates wear.
4. Check needle bearing for presence, seal, and smoothness. Also, be sure bearings are not corroded or rusty. Replace bearings of doubtful condition.
5. When installing axle support, be sure case and cover alignment is true with the press. Press supports in until flanged surfaces contact case and cover.
6. Install differential and output shaft simultaneously. Position gear ¾ in. I.D. spacer, and thrust washer on shaft.
7. Center one ¾ in. I.D. thrust washer over case needle bearing then install brake shaft gear, and other 1⅛ in. I.D. thrust washer.
8. Position a new gasket on the mounting surface of the case, then install cover. Align cover and case by tapping dowel pins

J.I. Case

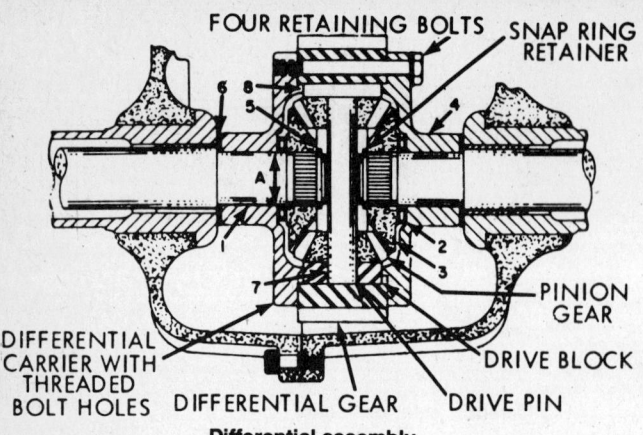
Differential assembly

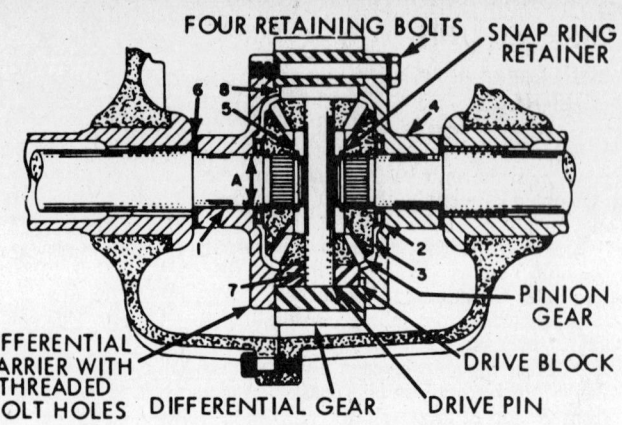
Proper arrangement of parts

into cover and secure with lockscrews torqued to 10 ft.lbs.

9. Install new brake shaft oil seal using sleeve Peerless number 670179 and drive number 670180.

10. Install new axle support oil seals using sleeve Peerless number 670179 and drive number 670180.

11. Add 2¾ (44 oz.) pts. oil (SAE EP90) before securing hydrostatic drive to the unit. Clean mounting surfaces and use a new gasket between the units. Torque 4 mounting bolts to standard torque for bolt used.

DIFFERENTIAL UNIT

Carrier Disassembly

1. The differential carrier is supported directly on the axle (1). Roller thrust bearings (2) are used between the bevel gear (3) and the differential carrier (4). This illustration shows axles with snapring (5) retainers, some earlier production had rolled over axle ends to secure the assembly. Thrust washers (6) are used at the ends of the differential carriers and case/cover thrust face. Replace the differential carrier if worn in excess of .878 in. at point A.

2. Clean the differential assembly, then check and note the axle lengths and their relation to the heads of the four hex head bolts.

3. If the unit will not turn freely, note where the unit binds. Check and replace those parts.

4. Place the differential in a large vise with soft jaws (hex head bolts up). Do not clamp the vise on the bearing race of a differential carrier.

5. Remove the four hex head bolts and the upper axle and differential carrier. Remove the drive blocks, pinions, drive pin and thrust spacer if used, by lifting out of the ring gear. Tap the ring gear lightly with a mallet to loosen from the differential carrier.

6. If a snapring is used, the axle assembly may be disassembled. If the axle end has been rolled, do not attempt to break the rolled retaining edge. The parts are to be replaced as an assembly.

7. Remove the snapring and the thrust washer, if used. Separate the bevel gear and differential carrier from the axle.

8. Examine gears for wear, cracked or chipped teeth. Check the internal splines of the gears and the axle if the gear is removeable. If excess play is noted, it may be necessary to replace the individual parts or both the gear and axle.

9. Examine drive pinions, drive pins and drive blocks for wear and damage. Replace excessively worn pinion or the drive pin.

10. Examine the differential carriers. One has threaded holes and the other has larger holes so that the bolts will pass through. Be sure to order the correct replacement pieces.

11. Examine the internal bearing diameter of the differential carriers. If wear is in excess of the tolerance noted at point A, replace the differential carrier or bushing, if used.

12. When assembling thrust bearings, always place a hardened thrust washer on each side of the caged thrust rollers. Never use the caged thrust rollers without the thrust washers.

Assembly

Oil all parts during reassembly.

1. Select the correct axle for the side of the differential opposite the hex head bolts. If the wrong axle is used, it will require complete tear down of the differential, or possibly the entire transaxle if the error is not detected until later.

2. Clamp the axle, in a soft jaw vise (not bearing or oil seal surfaces). The differential carrier with threaded holes is assembled to this axle.

3. Refer to the accompanying figure for the proper arrangement of parts.

4. Torque the four hex head bolts to 7 ft.lbs.

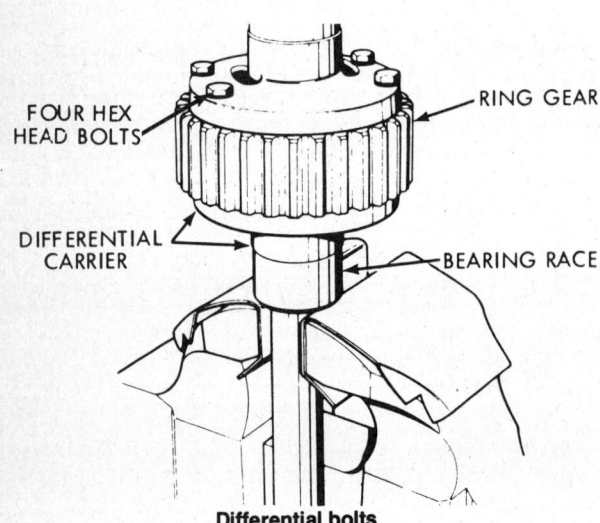
Differential bolts

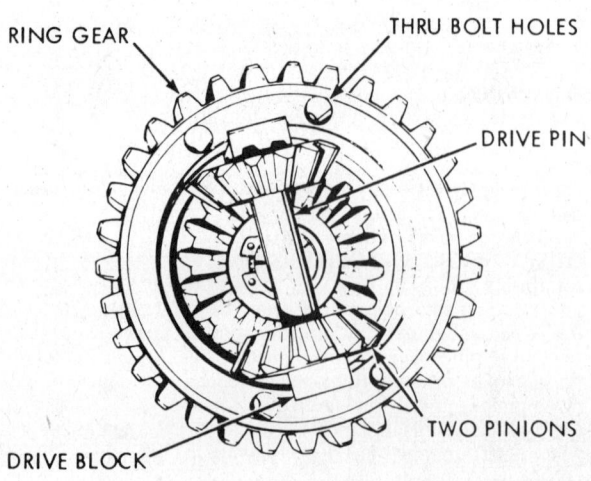
Differential with one carrier removed

J.I. Case

ATTACHMENT DRIVE CLUTCH

Models with a One Piece Hub Shaft and Snapring Located Between the Cams

CLUTCH DISC CLEARANCE ADJUSTMENT

1. Remove hood.
2. For tractors with muffler under the hood, remove three screws securing oil cooler to front support and swing cooler out.
   It is not necessary to drain hydraulic oil.
3. For tractors with muffler not under the hood, remove the four bolts holding the front support to the tractor frame and swing support with cooler and oil reservoir out.
4. It is not necessary to drain hydraulic oil.
5. Place attachment drive clutch lever in the OFF position.
6. Make a measurement using two blade type feeler gauges inserted 180 degrees apart between the clutch disc and backing plate. Record this measurement.
7. Shims are available in three thicknesses, .005 in., .010 in., and .050 in.. Consult your parts catalog for correct part numbers. Minimum clearance should be maintained at .002 in. to .007 in. to provide longest wear life before requiring readjustment. Shims should be added to reduce clearance. Shims should be removed to increase clearance.

Example: If the measurement made in step 6 was .018 in. clearance, the desired clearance is .002 in. to .007 in. Therefore, between .011 in. and .016 in. of shims should be added to reduce clearance.
Solution: Add one .010 in. and one .005 in. shim for total of .015 in. Clearance is then .003 in. which is within the .002 in. to .007 in. specification.

SHIM INSTALLATION PROCEDURE

1. Place attachment drive clutch in the OFF position.
2. Note the position of cam notches before disassembly. The front cam is installed with the shorter side up and the rear cam with shorter side down. The cams must be reassembled in this manner to insure proper operation.
3. Grip spacer (15) with suitable slip joint pliers and remove bolt (17). (Right hand thread on 646 Loaders, left hand thread on all other models.) Remove fan (19), spacer (15), springs (13), shim (8) and front cam and bearing assembly (10). Swing engaging arm (18) down.
4. Remove snapring (14), rear cam and bearing assembly (12). Check for shims that might have stuck to rear cam and bearing assembly (12).

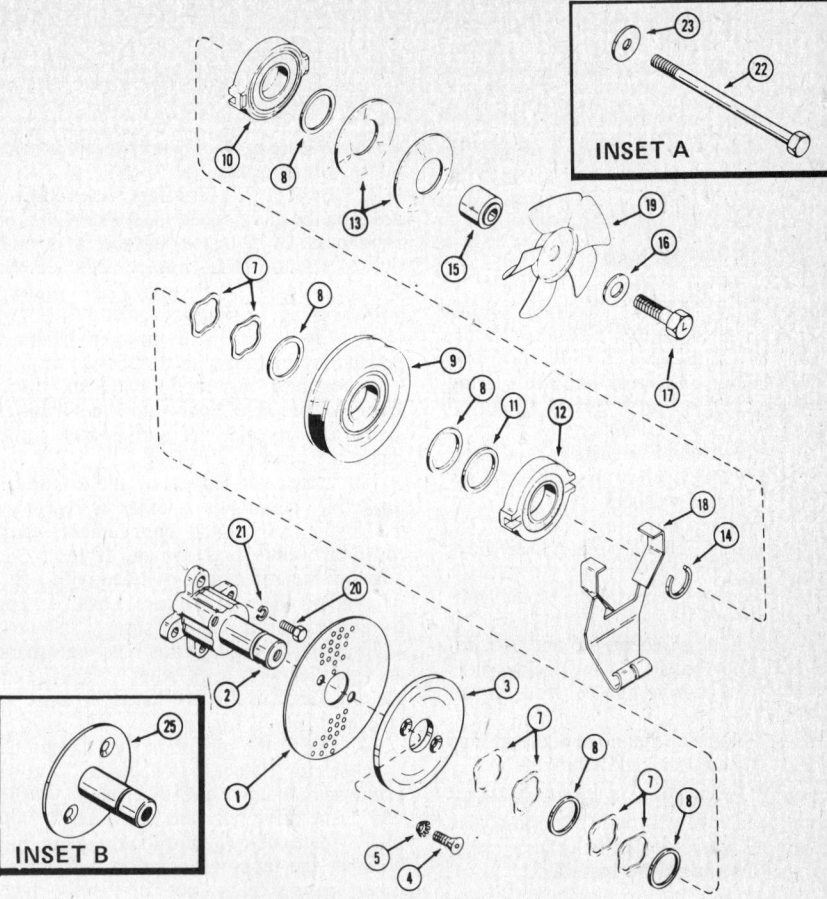

Attachment drive clutch components

Shim installation

J.I. Case

5. Add proper amount of shims as previously determined. Install shims so there is one .050 in. shim on each end of shim stack.

6. Reinstall components in reverse order of disassembly. Be sure to position cams—rear cam, shorter side down, front cam, shorter side up.

7. Torque bolt (17) to 45 ft.lbs.

CLUTCH REMOVAL AND INSTALLATION

1. Follow steps for CLUTCH DISC CLEARANCE ADJUSTMENT up to and including removal of rear cam and bearing assembly (12). (Step 4, SHIM INSTALLATION PROCEDURE).

2. Remove shim washers (8) and (11), pulley, disc and bearing assembly (9) six springs (7) and 3 washers (8).

3. Remove machine screws (4), lockwashers (5), clutch plate (3) and grass screen (1).

This completes removal on Models 446 and 646.

4. All models except 446 and 646, remove the four bolts (20) and lockwashers (21) and detach drive hub (2) from engine flywheel.

5. Check to see that none of the springs (7) and (13) are cracked or broken.

NOTE: If clutch was slipping under heavy load even though friction disc clearance was within tolerance, new springs (13) should be installed.

6. Inspect the cam notches (10) and (12) and engaging arm (18) for excessive wear. Also check to make certain the bearing flanges in the cams are not cracked or damaged.

7. Rotate the three sealed bearings to check for freeness and quiet operation. Replace bearings which are rough and noisy or do not turn freely.

8. Inspect the friction disc for glaze and wear. Replace if thickness measures less than ⅛ in. or if it is glazed.

9. The disc should be bonded to the pulley and must be heated under pressure to insure proper adhesion per the following procedure.

a. The pulley surface must be clean and free of rust, oil or grease. Wear patterns on the pulley surface will not effect the bonding unless they are severe.

Bonding material from the previous friction disc may be removed by heating (such as with a propane torch) and scraping.

b. A clamping device such as the one shown below should be fabricated:

c. Position friction disc, with coated side facing pulley, on pulley shoulder.

d. Bolt clamping device as illustrated and torque bolt to 80 ft./lbs.

e. Preheat oven to 400° F.

f. Place assembly in 400° F oven for 30 minutes.

g. Allow to cool, disassemble clamping device and install new bonded disc and pulley assembly in tractor.

NOTE: Do not use microwave oven.

CAUTION
If it is elected not to bond the disc to the pulley, remove all bonding material from the disc before installation.

h. Failure to do this may result in erratic clutch operation.

10. Inspect the friction disc contact areas on the clutch plate (3) for scratches or roughness. A polished surface is normal due to friction disc contact. Replace the clutch plate if score marks are present which cannot be polished out.

11. Check the fit of the cam bearings and pulley bearings on the drive hub (2). The bearings must slide back and forth freely for proper clutching and declutching. Polish off any nicks or burrs which could cause the bearings to bind.

12. Check the edges of the engaging arm which contact the notches in the cams. To insure full clutch engagement, they must be rounded as shown. If required, edges of the arm can be rounded with a file or grinding wheel. Take care, however, not to reduce overall width of arm.

13. Both engaging arms must be square and parallel with each other as shown. If the arms are misaligned, the lever must be replaced.

14. Check the mounting face of engine flywheel for flatness. If necessary file or grind smooth to insure that runout at the end of the drive hub does not exceed .006 in. Pay particular attention to the mounting holes as the tapping operation will sometimes leave a high spot on the flywheel face.

15. All clutches except 646 and 446 models:

a. Connect the drive hub (1) to the engine flywheel with original four capscrews (20) and new ⅜ in. lockwashers (21).

b. Secure the screen and clutch plate to the drive hub with original machine screws (4) and new lockwashers (5). Tighten the machine screws securely.

16. All 646 and 446 clutches only: Secure the drive hub to the flywheel with original machine screws and new lockwashers. Tighten the machine screws securely.

17. Fit two springs (7) together and Place on drive hub. Install .050 in. shim (8) then match two more springs and install. Place second .050 in. shim on hub and install the last two springs and .050 in. shim for a total of six springs and three shims.

18. Place disc pulley and bearing assembly on clutch hub with disc facing backing plate (3). If disc is not bonded to pulley, make sure disc remains on pulley shoulder during assembly procedure.

19. Install correct amount of shims.

a. If the reassembly contains no new parts, install shims as explained in CLUTCH DISC CLEARANCE ADJUSTMENT section.

b. If the reassembly contains some new parts, assemble clutch with .180 in. shims and then follow procedure in the clutch disc adjustment section of this manual and adjust if necessary.

20. Place rear cam (12) on drive hub (2) or (25) so that notch faces out, then install retaining ring (14) and front cam with notches facing in.

21. Rotate the cams until the lever notches are misaligned as shown. The rear cam on drive hub must be positioned so the shorter side is downward and the front cam must have the shorter side upward.

22. Separate the cams enough to insert the engaging lever. The engaging lever must be installed so the bend for the control rod is toward the front of the tractor.

23. Install .050 in. shim (8) and Belleville spring washers (13), dished out as pictured.

24. Assemble washer (16), fan if equipped (19) and spacer (15) on special bolt (17) or (22) and install on hub (2) or (25). Torque bolt to 45 ft.lbs.

25. With the clutch disengaged, check the friction disc. Clearance should measure between .002 in. and .007 in.. Use two "blade-type" feeler gauges 180 degrees apart when measuring. To increase clearance, add shims (8) or (11) as required between the pulley (9) and rear cam (12).

NOTE: If clutch does not engage or disengage properly, check for correct cam assembly according to steps 21 and 22 above. When correctly installed, the facing notches on the front and rear cams are "out of alignment" in the manner shown with the clutch in both the engaged or disengaged positions. Repeat steps 21, 22 and 23 to correct either friction clearance or cam assembly.

26. Reassemble oil cooler and the hood, or grille and headlight panel.

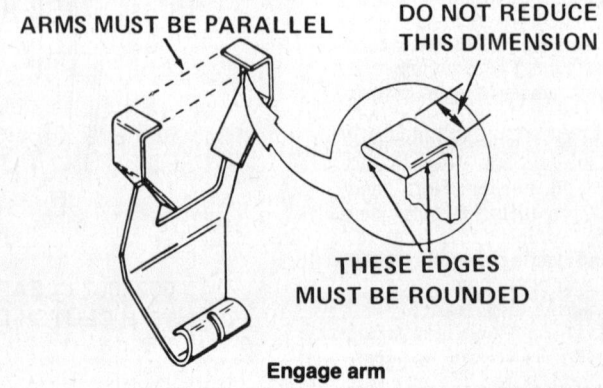

Engage arm

J.I. Case

Models with Two Piece Hub Shaft and Hub Assembly Retained by a Bolt (No Snapring)

NOTE: This unit is quite similar to the one covered immediately prior. Use the service procedures above with the following exceptions.

CLUTCH DISC CLEARANCE ADJUSTMENT

No change except that there is no snapring.

REMOVAL AND INSTALLATION

1. Check fit of front cam bearing on the spacer hub (15). Place the two load springs, one .050 in. spacer and bearing on the spacer hub as shown. The spacing measurement from end of the hub to extreme edge of the bearing inner race should be .075 in.-.090 in.. Use .010 in. or .050 in. shim washers, positioned between the load spring and the bearing inner race, as required to obtain the .075 in.-.090 in. dimension.

2. Place rear cam on drive hub so that notches face out. Assemble washer (16) or (23), fan (19) if equipped, spacer hub (15) or (24), Belleville spring washers (13) dished outward, .050 in. shim washer (8), cam (12) and spacer (14) or (23) in that order on special bolt (17) or (22).

3. Place assembly from step 2, above, on hub with engaging lever properly positioned between cams. The bend for the control rod on the engaging lever must face forward.

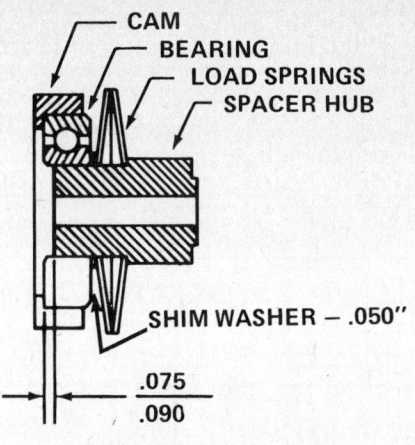

Adjustment measurements

Early Models with One Piece Hub Shaft and Front Cam Retained by a Snapring

REMOVAL AND INSTALLATION

1. Remove the tractor hood.
2. Remove the four bolts which hold the head exchanger support to the frame but leave the hose connected to the valve return tube.
3. Rotate the heat exchanger ahead for access to the clutch.
4. Disconnect the PTO control rod from the engaging arm (1).
5. Remove the RIGHT HAND THREAD bolt. Insert "A", (20) with lockwasher and plain washer and remove the fan (17) and spacer (16).
6. Carefully remove the clutch with a "puller" using the original ⅜x3½ in. long bolt as a guide on the engine shaft.
7. Before disassembling check and record the total clearance between the pulley and friction liner. Use two "blade-type" feeler gauges of the same thickness—insert 180° apart. The proper total clearance is between .002 in. and .012 in. with clutch disengaged. This will determine the number of shims to be added during assembly providing a new friction liner is not installed.
8. Carefully clamp the clutch assembly in a vise and compress the load springs (4) enough to remove the snapring (5).
9. Remove the clutch from the vise. Clean off paint and remove any burrs from the snapring end of the hub assembly (10).
10. Remove the springs (4), outer cam (2) and engaging arm (1).
11. Remove the inner cam (2). Check for and remove any shim washers (9) or (11) which may have come off attached to the cam.
12. Remove the shim washers (9) and (11), pulley (7), friction disc (6), and shim washers (11) and springs (12) from the drive hub (10).
13. Check to see that none of the springs (4) and (12) are cracked or broken.

NOTE: If clutch was slipping under heavy load even though friction disc clearance was within tolerance, new springs (4), part number C16786, should

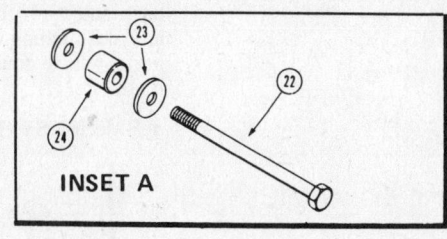

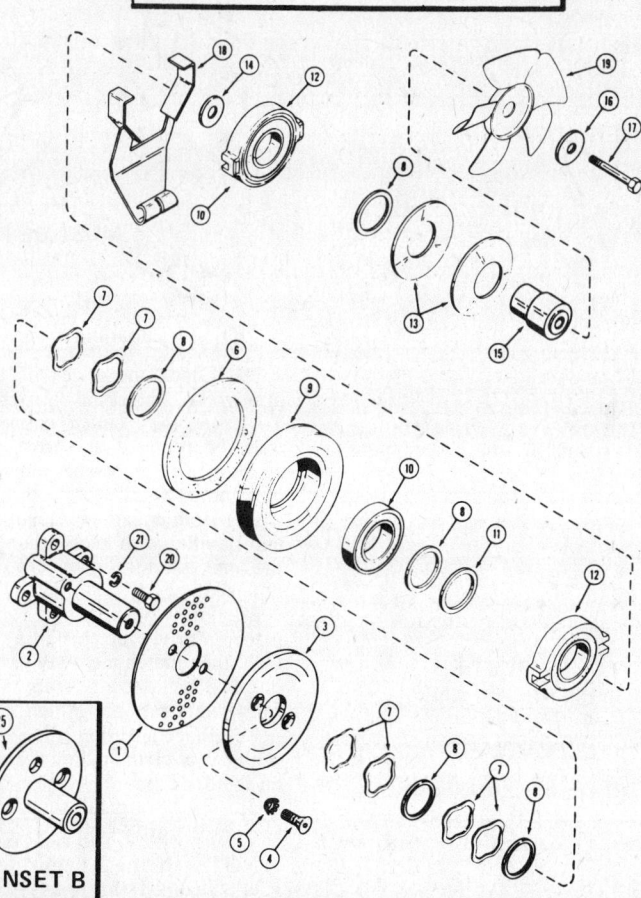

Two-piece hub type intermediate drive clutch

229

J.I. Case

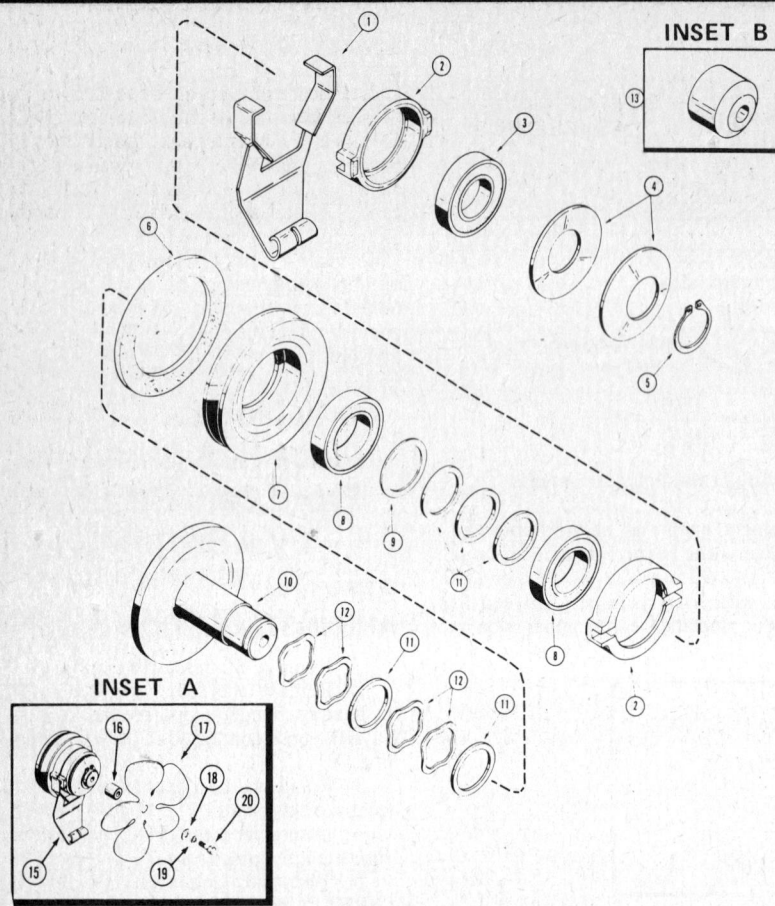

Early style attachment drive clutch

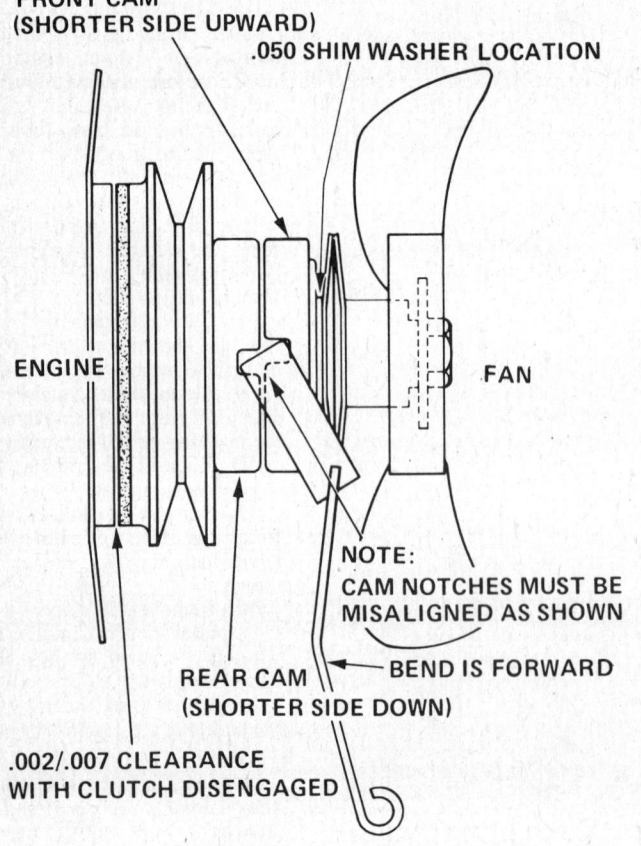

Shimming procedure

be installed. Also, if new springs are installed, the clutch must be reassembled with the spacer kit (13), part number C16483, instead of the snapring.

14. Inspect the cam notches (2) and engaging arm (1) for excessive wear. Also check to make certain the bearing flanges in the cams are not cracked or damaged.

15. Rotate the three sealed bearings to check for freeness and quiet operation. Replace bearings which are rough and noisy or do not turn freely.

16. Inspect the friction disc (6) for glaze and wear. Replace if thickness measures less than 1/8 in. or if it is glazed.

17. Inspect the friction disc contact areas on the drive hub (10) and pulley (7) for scratches or roughness. A polished surface is normal due to friction disc contact. Replace the clutch plate or pulley if score marks are present which cannot be polished out.

18. Check the fit of the outer cam bearing (3) on the drive hub (10) and the fit of the inner cam and pulley bearings (8) on the drive hub (10). The bearings must slide back and forth freely for proper clutching and declutching. Polish off any nicks or burrs which could cause the bearings to bind.

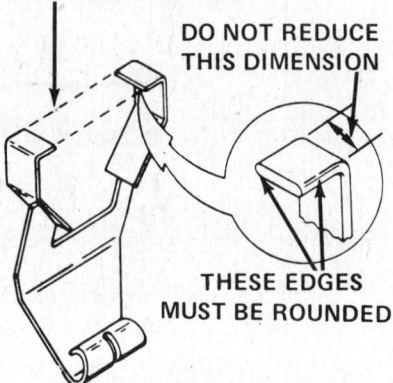

Engaging arms

19. Check the edges of the engaging arm which contact the notches in the cams. To insure full clutch engagement, they must be rounded as shown. If required, edges of the arm can be rounded with a file or grinding wheel.

20. Both engaging arms must be square and parallel with each other as shown. If the arms are misaligned, the lever must be replaced.

21. Fit two springs (12) together and place on drive hub. Install a .050 in. shim (11), then match the outer two springs and install. Place a second shim (11) on the hub for a total of 4 springs and 2 shims.

22. Place friction disc (6) over the shoulder of clutch pulley (7) and place both on the drive hub.

23. Place original shim stock (9) and (11) on the drive hub.

NOTE: If the friction disc clearance was measured and noted prior to clutch disassembly and no new parts were installed, adjust the shim stacks at this point to correct specification (.002 in. to .007 in.) with clutch disengaged.

24. Place the two cam assemblies (2) on the drive hub.
25. Rotate the cams until the lever notches are misaligned as shown. The inner cam must be positioned so the shorter side is downward and the outer cam must have the shorter side upward.
26. Separate the cams enough to insert the engaging lever (1). The engaging lever must be installed so the bend for the control rod is toward the front of the tractor.
27. Place the two springs (4) dished outward on the drive hub.
28. If the snapring (5) is to be used rather than the spacer kit (13), compress the load springs enough in a vise to expose the groove.

NOTE: Make certain the friction disc is flush against the face of the pulley so it is over the shoulder before compressing the springs and installing the snapring.

HYDROSTATIC TRANSMISSION

Models 117 and 118

NOTE: The manufacturer states that if any part of these units is disassembled by other than a qualified, authorized technician, the warranty will be considered void.

INPUT SHAFT REPLACEMENT

1. Remove the screen and fan.
2. Loosen the setscrews on the input pulley and remove the pulley.
3. Steam clean the hydrostatic transmission and transaxle areas.
4. Lift the expansion tank off its mounting bracket and remove the four bolts which connect the hydrostatic transmission to the transaxle. Note the oil level in the expansion tank and replace the vent cap with a "seal" cap (or install the original cap with plastic film to seal the tank.)
5. Carefully move the hydrostatic transmission away from the transaxle and lift it out of the chassis.
6. Thoroughly steam clean the input shaft area and blow dry with compressed air. Do the same to the new replacement input shaft and bearing assembly.

NOTE: Do not use cloth or paper towels to clean this area since lint particles deposited will cause the transmission to seize if they get inside.

7. Place the transmission onto blocks or a frame so the input shaft faces straight upward. Also, if possible keep the expansion tank anchored higher than the transmission.
8. Remove the snapring, with an internal snapring pliers and carefully lift the

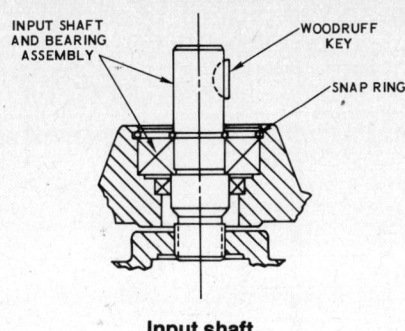

Input shaft

input shaft assembly straight upward and out of the housing.

--- CAUTION ---
Keep in mind that even a minute particle of dirt or paint can cause this transmission to seize or fail.

9. Carefully insert the new input shaft and bearing assembly into the housing and secure with the new snapring provided. Also use the new Woodruff key provided in the kit.
10. Install a new gasket (and O-ring on tractors above S/N 9646800) on the transmission mounting flange.
11. Place the transmission back on the transaxle and engage the transmission output shaft with the transaxle idler gear. Secure the transmission to the transaxle with the four original bolts and lockwashers.
12. Install the original vent cap, place the expansion tank back on its mounting bracket and check oil level. If the oil level is higher than when transmission was removed it will be necessary to "bleed" the system following the procedure outlined in the Oil Level Check section.

NOTE: If the pulley has ¼ in. setscrews, drill and tap the two holes for 5/16 in. setscrews. Also check to make certain the setscrews can be turned into the bore of the pulley before installing.

13. Install a new input pulley, with 5/16 in. setscrews, using the new Woodruff key provided in the kit. Make certain the pulley fits snugly to the shaft. Install the setscrews with Loctite®.
14. Install the fan by piloting it on the roll pin and secure with original plain washer, lockwasher and bolt.

NOTE: The roll pin must not extend beyond the outer face of the fan. Tap it further into the pulley if necessary. Also be certain, the I.D. of the plain washer is only slightly larger than the bolt. If necessary, use a new plain washer, part number 195-2012. Do not use a plain washer between the fan and the pulley.

15. Install the screen and check operation of tractor.

NOTE: If tractor operation is erratic or jerky, check for air in the system. Refer to oil level check section.

Transaxle Overhaul

1. Remove transmission cover plate screws (46), lockwashers (45), cover plate and gasket.
2. Place a large drift pin against the inside end of the brake drum shaft (40) and rap sharply with a hammer. This will dislodge the retainer which holds the brake drum shaft to the brake gear (38). Remove the brake drum (40), key (41), gear (38), retainer (39), seal (4), and bushings (5 and 6).
3. Remove snaprings (15), rear axles (11 and 13), differential assembly, washers (36), seals (4), and bushings (2 and 3).
4. Disassemble the complete differential assembly by removing the four special bolts (34) and locknuts (35).
5. Remove the four capscrews which attach the hydraulic motor to the transmission case and remove the motor and input shaft (16) as an assembly. If necessary, remove bearing (9).
6a. Model 220 Prior to S/N 9649062
    Model 222 Prior to S/N 9647203
    Model 442 Prior to S/N 9632450
When the input shaft is part way removed, slide off shims (48 and 49), thrust

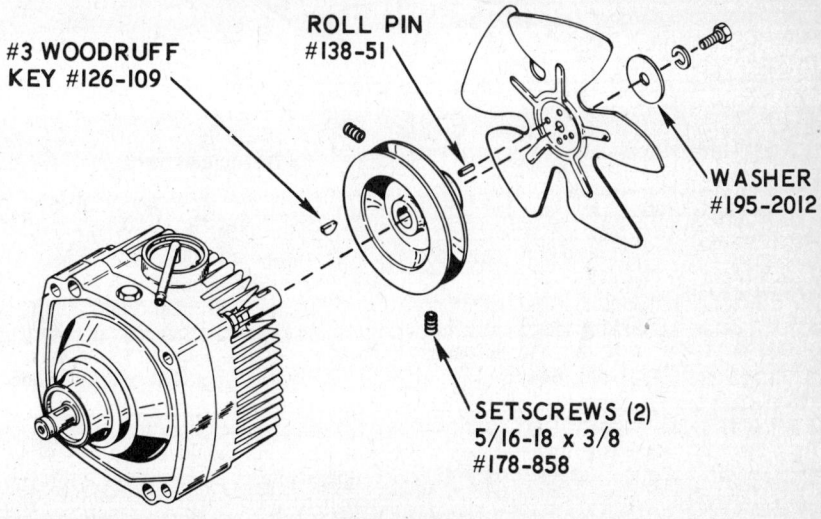

Pulley and fan

J.I. Case

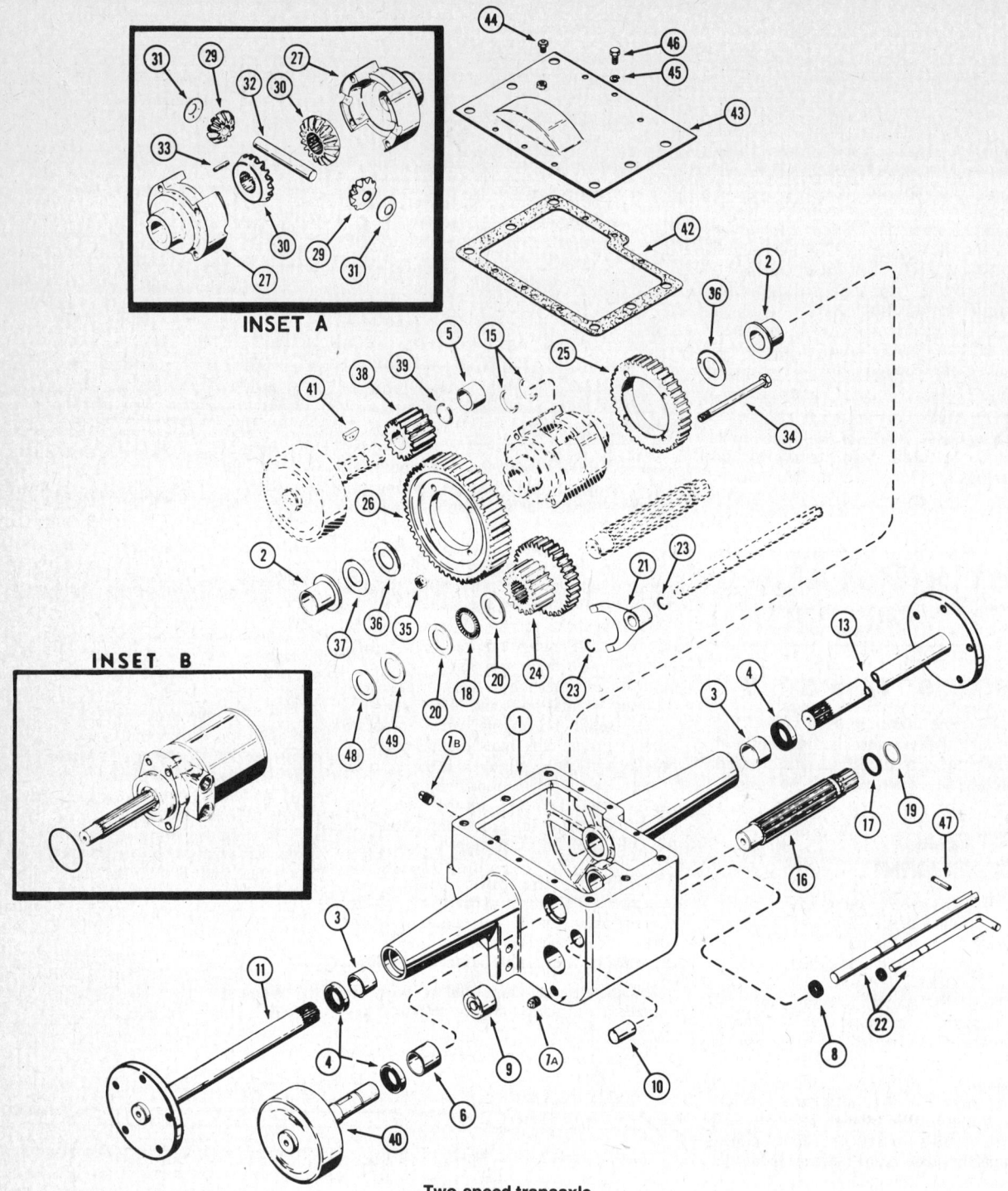

Two-speed transaxle

washers (20), bearing (18), and the drive gear (24) to prevent them from dropping. Pull out or gently tap the input shaft (16) from the hydraulic motor.

6b. Model 220 S/N 9649062 and after
Model 222 S/N 9647203 and after
Model 442 S/N 9632450 and after
All 224, 444, 446, 644, 646

Refer to inset B. Shims (48 and 49), thrust washers (20), bearing (18) are not used. The input shaft is an integral part of the hydraulic motor and cannot be removed unless the motor is disassembled.

7. Remove snaprings (23), yoke (21), shifting shaft (22) and seal (8). If necessary remove plug (10).

8. Thoroughly clean and wipe dry all parts before inspection.

9. Inspect all gears and shafts for burrs, nicks or excessive wear both to the teeth and to the splines. Light burrs and nicks may be removed with a hone or crocus cloth.

10. Inspect bearings for worn or pitted rollers. There should be no pits or other visible damage. If damaged, replace with new parts.

11. Inspect the differential housings and transmission case for cracks, burrs or nicks.

12. Inspect axles for nicks, cracks and stripped or damaged threads.

13. Inspect bushings for score marks, nicks or excessive wear and replace if necessary.

14. Replace all oil seals each time the unit is disassembled.

15. Install new seal (8) and reassemble shifting shaft (22) and yoke (21). Secure in place with snaprings (23). Replace plug (10) if previously removed.

16. Install needle bearing (9) flush with the inside of the transmission case.

J.I. Case

17. Model 220 Prior to S/N 9649062
    Model 222 Prior to S/N 9647203
    Model 442 Prior to S/N 9632450

Carefully install a new O-ring (17) and back up washer (19) on the input shaft (16). The O-ring must be installed first and be located to the inside of the back up washer. Check the O-ring for installation damage and carefully insert the shaft into the hydraulic motor until it bottoms. If necessary, tap the end of the shaft lightly with a plastic hammer. Install a new O-ring on the face of the hydraulic motor flange.

18. Start the assembled input shaft (16) into the case. Slide the drive gear (24), thrust bearing and washers (18 and 20), and shims (48 and 49) on the input shaft and place the yoke (21) between the gears. The thicker shim should be used next to the bearing (9).

19. Slide the input shaft (16) into the needle bearing (9) until the hydraulic motor flange is flush and tight against the machined face of the case. Check the clearance between thrust washer (20) and shim (48). Clearance must not be less than .001 in. (.02 mm) and not more than .015 in. (.38 mm). If necessary, adjust the shim stack keeping a .030 in. (.76 mm) shim next to the needle bearing.

20. Install the four hydraulic motor capscrews using Loctite® and torque to 110-125 ft.lbs. Install new O-rings on the hydraulic motor ports.

21. Model 220 S/N 9649062 and after
    Model 222 S/N 9647203 and after
    Model 442 S/N 9632450 and after
    All 224, 444, 446, 644, 646

a. Start the hydraulic motor and input shaft assembly into the case. Slide the drive gear (24) on the input shaft and place the yoke (21) between the gears.
b. Install the four hydraulic motor cap screws using Loctite® and torque to 110-125 ft.lbs.

22. Assemble the differential according to inset "A" and install the ring gears (25 and 26). Use new locknuts on the special bolts (34) and torque to 50 ft.lbs.

23. If the axle bushings (2 and 3) were removed, press in new bushings.

CAUTION

Install the flange bushings (2) with the oil groove downward. Press the flange bushings in until seated and the straight bushings (3) (if so equipped) in until flush with the oil seal shoulder. Ream bushings if necessary after assembly to 1.1876 in.-1.1877 in. (30.165 mm-30.167 mm) inside diameter.

---

24. Install new oil seals lip side first being sure the seal lip is not in contact with the outer axle bushing.

25. Grease the axle shafts (11 and 13) thoroughly and carefully install them through the seals, bearings and into the differential assembly, being sure that stop washers (36), and shim washers (37), if required, are in place.

NOTE: Maximum end play on the differential assembly is .030 in. (.76 mm). Minimum end play is .005 in. (.13 mm).

Use .015 in. (.38 mm) shim washers (37) as required. A coating of grease on the shim washers will help to hold them in position while inserting the axle. Secure the axle shafts with snaprings (15).

26. If the brake shaft bushings (5 and 6) were removed, press in new bushings until flush with their seats. Ream the inner bushing if necessary to 1.004 in.-1.005 in. (25.50 mm-25.52 mm). Ream the outer bushing if necessary to 1.192 in.-1.193 in. (30.28 mm-30.30 mm). Install a new seal (4).

27. Install a new wire retainer (39) to the brake gear (38). With the Woodruff key (41) in place, insert the brake shaft (40) through the case and into the brake gear. When the inside end of the shaft is in contact with the wire retainer, rap the outside end sharply with a leather mallet to seat the retainer.

28. Fill the transmission to level plug (7B) using approximately 3 quarts of clean 20W-40 motor oil or No. 80 or 90 EP gear lubricant. Install a new cover gasket (42) and replace cover (43) with six bolts (46) and lockwashers (45). Check to see that the breather plug (44) is functional.

29. Position the transmission under the tractor frame and place the brake drum into the brake band.

30. Secure the transmission to the frame with the four bolts and lockwashers.

NOTE: Check and make sure the two right frame channel spacers are in place between the mounting holes.

31. Model 220 Prior to S/N 9649062
    Model 222 Prior to S/N 9647203
    Model 442 Prior to S/N 9632450

Make sure the new O-rings are in place on the hydraulic motor ports. Secure the oil line manifold to the hydraulic motor with the original four capscrews and tighten to 24-28 ft.lbs.

32. Model 220 S/N 9649062 and after
    Model 222 S/N 9647203 and after
    Model 442 S/N 9632450 and after
    All 224, 444, 446, 644, 646

Connect the two valve to motor tubes.

33. Return the fuel tank to its normal location and install the seat support with the original bolts.

NOTE: Before installing the seat support mounting bolts, place the slotted end of the shift lever over the roll pin on the transmission shift rod.

34. Install the rear wheels and remove the blocks from the tractor frame.

HYDRAULIC MOTOR

REPLACEMENT

MODEL 220 S/N 9649062 AND AFTER
MODEL 222 S/N 9647203 AND AFTER
MODEL 442 S/N 9632450 AND AFTER
ALL 224, 444, 446, 644, 646

NOTE: For all models prior to those listed above, the transmission must be opened to remove the hydraulic motor. Refer to the "Removal" and "Disassembly" sections of this section.

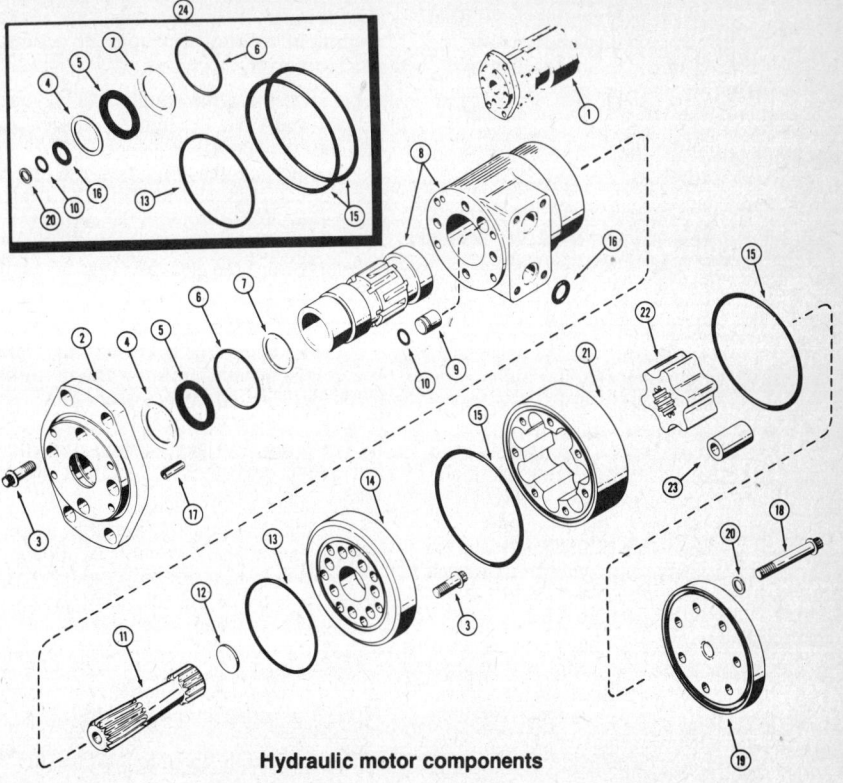

Hydraulic motor components

J.I. Case

1. Remove the drain plug (7A) to drain the transmission oil. Replace the plug when the oil has completed draining.

2. Remove the two valve to motor port hydraulic tubes.

3. Place the transmission in low range.

4. Place a jack stand under the left axle housing and remove the left rear tire and wheel.

5. Remove the four capscrews that hold the hydraulic motor.

6. Slide the hydraulic motor and input shaft assembly from the transmission case. The high-low cluster gear (24) will remain inside the case.

7. Reassemble in the reverse order of disassembly.

DISASSEMBLY

PART NUMBERS C-13047 AND C-13498

Important, before removing the hydraulic motor, thoroughly clean it and the attached parts to prevent the entry of foreign material. Also make sure the disassembly area is clean.

1. Mark a line across the geroler housing (21), spacer plate (14) and main housing (8). Also make a mark on the housing port side of the mounting flange (2) and dye mark a line across the front end of the spool bore and the mounting flange bore.

2. Remove the seven capscrews (18) and remove the end cap (19), O-ring (15) and coupling spacer (12).

3. Carefully remove the geroler and housing assembly (21) so it remains intact. Remove O-ring (15).

4. Remove the seven capscrews (3) from the spacer plate (14) and remove the plate (14), O-ring (13) and drive coupling (11).

5. Remove the four capscrews (3) from the mounting flange (2) and tap off the flange with a lead or leather mallet.

6. Remove back-up washer (4), quad ring (5), O-ring (6) and thrust washer (7) from the flange.

7. Remove the plug (9) from the main housing (8) by inserting a small pin or screwdriver through holes in the opposite end of the housing. Remove the O-ring (10) from the plug.

8. Discard all original seals and washers which are included in the C 14826 kit and the thrust washer (7).

9. Clean all parts in solvent and air dry before inspection being careful not to damage any machined surfaces or to remove the scribe and dye marks.

10. Inspect all parts for scratches, scoring and excessive wear giving particular attention to the spool and housing and geroler assemblies. The components of these assemblies should fit snugly and be free of wear spots. Also refer to paragraph "d" under introduction. A polished pattern on the spacer plate and end cap due to geroler action is normal but they should not be grooved. Check the housing, geroler and roller thickness with a micrometer. If the geroler and roller thickness is more than .002 in. less than the thickness of the housing, install a new geroler kit.

11. Use all new seals from the C 14826 kit plus a new thrust washer (7) during assembly.

ASSEMBLY

1. Install the plug (9) with a new O-ring (10). The O-ring end of the plug is toward the outside.

2. Install a new thrust washer (7), back up washer (4), quad ring (5), and O-ring (6) in the mounting flange (2) in that order. Position the mounting flange on the main housing so the side marked is on the same side as the motor parts. Roll pins (17) must be drawn back through mounting flange (2) until they are flush with the inner side.

3. Position the mounting flange on the main housing so the side marked is on the same side as the motor ports. Tap the mounting flange (2) onto the main housing with a lead or leather mallet and secure with four capscrews (3). Torque capscrews evenly to 215 in.lbs. Make certain the roll pins (17) do not protrude beyond the face of the mounting flange.

4. Slide the spool into the housing (8) until it bottoms on the thrust washer (7) in the flange. Align the dye marks on the spool (8) and mounting flange (2).

5. Place new O-rings (13) and (15) on the spacer plate (14). Align the scribe marks on the spacer plate (14) and main housing (8) and secure the plate with O-ring (13) toward housing with seven capscrews (3). Torque the capscrews evenly to 175 in.lbs.

NOTE: If for any reason the alignment marks were not made or were removed while handling or cleaning, continue these assembly procedures and install the motor. Refer to the "Important" paragraph at the end of this instruction if it is necessary to correct alignment to obtain proper direction of tractor travel.

6. Slide the drive coupling (11), longer splined end first, through the spacer plate and into the spool.

7. Recheck the alignment of the dye marks on the spool and flange and install the geroler assembly with the scribe marks in alignment and with the splined end of the geroler toward the end cap (19). Be careful not to allow the geroler and rollers to fall out of the housing.

NOTE: It may be necessary to wobble the drive coupling to align the mounting holes between the geroler assembly and spacer plate (14). This can be facilitated by inserting a finger through the flange bore and moving the coupling slightly back.

8. With the geroler assembly properly aligned, stand the motor on its mounging flange and install the end cap (19) with seven new washers (20) and original capscrews. Torque the capscrew evenly to 175 in.lbs.

CAUTION
With the hydraulic motor installed, check for proper geroler alignment by placing the transmission in gear. When in forward, the tractor should drive forward and when in reverse, it should drive in reverse. If the results of these operations are opposite, the geroler assembly is not properly aligned. To correct this condition, proceed as follows:

a. Remove the end cap (19) and coupling spacer (12) wihtout removing the hydraulic motor from the transmission.

b. Scribe mark the geroler housing (21) to the spacer plate (14) and ink mark the geroler (22) to one of the splines on the drive coupling (11).

c. Slide the geroler assembly off the drive coupling. Move it one tooth in either direction and slide it back on the drive coupling.

d. Install the end cap (19) using new washers (20) and original capscrews (18). Torque capscrews evenly to 175 in.lbs.

DISASSEMBLY

PART NUMBERS C-14655 AND C-16697

NOTE: Before removing the hydraulic motor, thoroughly clean the motor and the attached parts to prevent the entry of foreign meaterial. Also make sure the disassembly area is clean (a piece of clean wrapping paper makes an excellent and disposable work bench top).

CAUTION
Use care in handling close-fitting parts as nicks and dents result in serious damage. Do not force or abuse these precision parts. Avoid wiping parts with a cloth because lint and foreign particles may adhere to the working parts of the motor.

1. Remove retaining ring (1).

2. If only shaft seal (3) and spacer (2) are required, complete disassembly of the hydraulic motor is not necessary.

a. With retaining ring (1) removed, set the motor, shaft downward in a suitable clean can.

b. Plug motor "in" port and connect "out" port to 120 psi air hose or use a piston type hydraulic hand pump. Charge the motor with air or oil pressure to remove the shaft and seal spacer.

NOTE: The air connection should have provisions for quick shut-off after the seal spacer (2) and seal (3) have been ejected.

c. Repack the seal spacer (2) with Aero-Shell #14 grease or E.P. equivalent each time the shaft seal (3) is replaced.

NOTE: If the hydraulic motor is being completely disassembled, the seal can be removed without air or oil pressure after the shaft is removed.

3. Clamp the motor housing port boss in a padded jaw vise with the shaft pointing downward.

4. Using a 15 in. wrench with a 12 point 9/16 in. socket, remove the seven capscrews (16).

J.I. Case

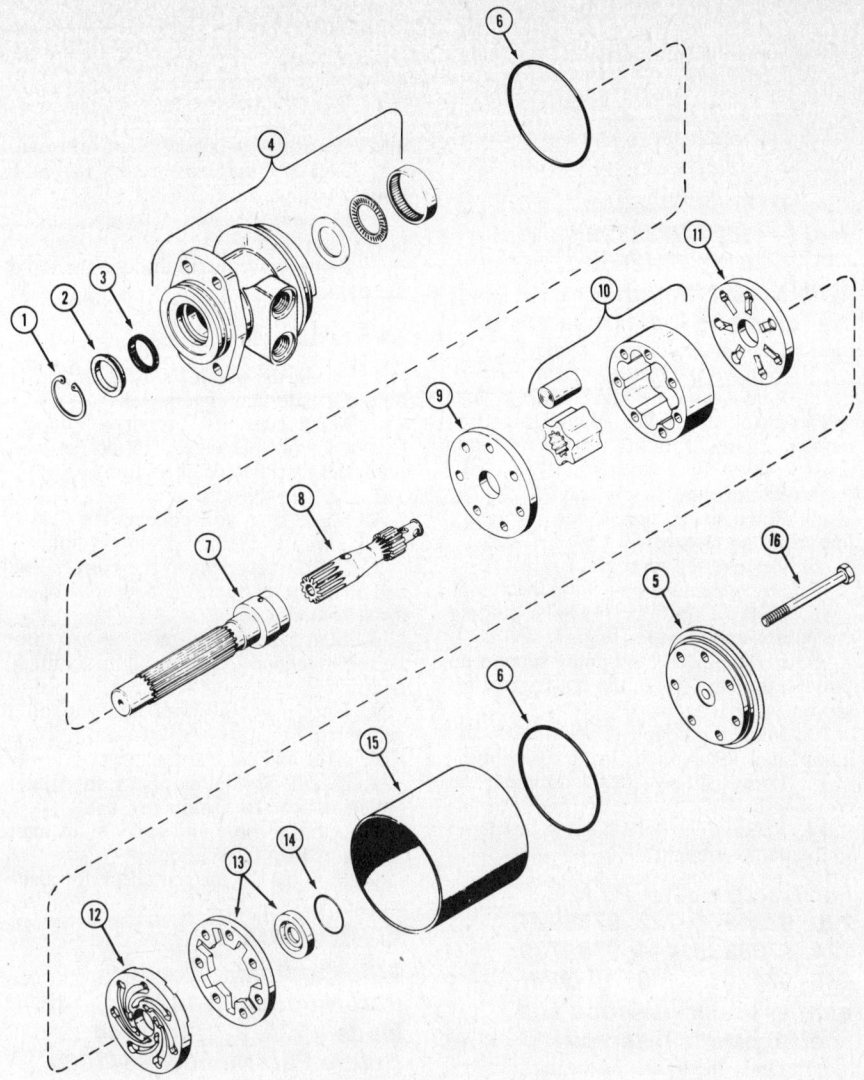

Hydraulic motor components

ASSEMBLY

1. Carefully clamp the housing (4) in a padded vise with the seven tapped holes upward.
2. Insert the shaft (7) and drive link (8).
3. Lubricate the O-ring (6) and slide into the groove on the housing.
4. If the rotor assembly was disassembled place the stator on the wear plate (9) and carefully insert the rotor and rollers.
5. Place the rotor assembly (10) and wear plate (9) over the drive link (8) and onto the housing, with the rotor counter bore facing upward.

NOTE: Two capscrews, ⅜ in. x 4½ in., with head removed, can be used to align the seven holes in the housing with the holes in the wear plate (9), rotor assembly (10), manifolds (11) and (12) and the commutator (13).

6. Install the manifold plate (11) with the slots toward the rotor.
7. Install the manifold (12) over the plate (11) with the swirl grooves toward the rotor and the diamond shaped holes upward.
8. Place the commutator and commutator ring (13) onto the manifold with the bronze ring groove facing upward. Place the bronze ring (14) into the groove with the rubber side downward.
9. Place the sleeve (15) over the assembled components and carefully force down over the lubricated O-ring (6) with arbor press or similar tool. Observe the O-ring closely while the sleeve is being pressed over to make certain it is not nicked or rolled.
10. Lubricate the new O-ring (6) and slide into the groove on the end cap (16). Carefully force the end cap into the sleeve with an arbor press or similar tool. Observe the O-ring closely while pressing in the end cap to make certain it is not nicked or rolled.
11. Remove the line up bolts and insert the seven capscrews (16). Using a tightening sequence similar to the cylinder head on an internal combustion engine, torque the seven capscrews to 50 ft.lbs.
12. Remove the motor from the vise and place it on the bench shaft end up. Lubricate the seal (3) and spacer (2) with Aero-Shell #14 grease and place onto the shaft. Seat the spacer and seal squarely into the housing using a tubular tool having an I.D. slightly larger than the shaft and O.D. slightly smaller then the housing bore. Install the retainer ring (1).
13. Lubricate the motor by introducing oil into the ports and turn the shaft several revolutions. Plug the ports to keep out foreign material unless motor is being connected back to the tractor immediately.
14. Install a new O-ring on the motor flange before installing to the transaxle. Operate the motor under load in both forward and reverse for several minutes and check the end cap and housing closely for leakage.

5. Pry off the end cap (5) using two large screwdrivers 180° apart. Discard the O-ring (6).
6. Remove the commutator and commutator ring (13) and place them flat on the end cap (5).
7. Pry the sleeve (15) off the housing (4) using two screwdrivers 180° apart. Discard the O-ring (6).
8. Remove mainfolds (11) and (12) by lifting them off the rotor section (10) and place them on top of the commutator assembly (13).
9. Remove the drive link (8) by lifting straight up with the wear plate (9) and rotor assembly (10), then place all three parts carefully on the work bench.

NOTE: Be careful when handling the rotor assembly (10) to keep the rollers and rotor intact with the housing.

10. Push the coupling shaft (7) up and out of the bearing housing (4).
11. Remove retainer ring (1), spacer (2) and oil seal (3).

NOTE: Since factory tooling and gauges are required to properly install the thrust and needle bearings in the housing (4), these items are not serviced.

12. Wash all parts thoroughly in clean petroleum base solvent and blow dry with clean, dry air. Do not use cloth to wipe off parts as lint and foreign material may cause binding and sticking of closely fitted components.
13. Inspect the bearing housing (4) for nicks at the O.D. O-ring (6) sealing surface. Check the rotor assembly (10) for finish.
14. The rotor and vanes (rollers) must be smooth and not show signs of galling or have nicks. Check the wear plate (9) for excessive wear, or poor sealing surface for the rotor. Check the wearing surfaces of the manifolds (11) and (12) for wear. Inspect the commutator (13) at the sealing areas for excess wear. Check the O-ring (6) sealing areas of the sleeve (15) closely for smoothness.

235

J.I. Case

HYDRAULIC SYSTEM

Hydraulic Pump

REMOVAL

For Tractors Before P.I.N. 9646800 (with panels that can be removed on the side of the instrument tower)

TRACTORS WITH MECHANICAL LIFT

1. Drain the hydraulic system.
2. Remove the panel from the righthand side of the control tower.
3. Disconnect the inlet and outlet tubes from the pump.
4. Fasten the choke and throttle cables up to clear the work area.
5. Remove the pump support from the engine. Pull the pump and support through the righthand side of the control tower.
6. Mark the position of the coupling on the pump shaft for a guide during assembly.
7. Loosen the setscrew and remove the coupling.
8. Remove the pump from the support.

TRACTORS WITH HYDRAULIC LIFT

1. Drain the hydraulic system.
2. Remove the panel from the lefthand side of the control tower.
3. Disconnect the inlet and outlet tubes from the pump. Remove the outlet tube from the tractor.
4. Remove the lefthand foot rest.
5. Disconnect the brake rod and spring from the brake pedal.
6. Slide the brake pedal off the shaft by holding the travel control lever of the tractor in the down position.
7. Remove the pump support from the engine. Pull the pump and support through the lefthand side of the control tower.
8. Mark the position of the coupling on the pump shaft for a guide during assembly.
9. Loosen the setscrew and remove the coupling.
10. Remove the pump from the support.

For Tractors Between P.I.N. 9646800 and 220: 970246, 222: 9706651, 224: 9708665, 444: 9711027 and 446: 9728158 (with external muffler but no side panels that can be removed)

1. Drain the hydraulic system.
2. Remove the battery and battery mounting plate.
3. Disconnect the lines to the pump and put a cap on them to prevent the entry of dirt.
4. Remove the four (4) bolts holding the pump support to the engine.
5. Pull the pump and pump support up through the area where the battery was.
6. Mark the coupling position on the pump shaft for a guide during assembly.
7. Loosen the setscrew and remove the coupling half.
8. Remove the bolts holding the pump to the pump support.

For Tractors Between P.I.N. 220: 970294-9734879, 222: 9706651-9766998, 224: 9708665-9738335, 444: 9711027-9739739, 446: 9728158-9742953 (with the muffler under the hood and metal reservoir)

1. Drain the hydraulic system.
2. Remove the battery.
3. Remove the voltage regulator and solenoid from the battery tray. Do not remove the wires from the voltage regulator and solenoid.
4. Remove the battery tray.
5. Remove the suction line and return line from the reservoir.
6. Remove the reservoir.
7. Disconnect the lines from the pump.
8. Remove the four (4) bolts holding the pump support to the engine.
9. Pull the pump and pump support up through the area where the battery and reservoir were.
10. Mark the coupling position on the pump shaft for a guide during assembly.
11. Loosen the setscrew and remove the coupling half.
12. Remove the bolts holding the pump to the pump support.

For Tractors After P.I.N. 220: 9734870, 222: 9766998, 224: 9738335, 444: 9739739, 446: 9742953, 448: All (with muffler under the hood and special plastic reservoir)

1. Drain the hydraulic system.
2. Remove the battery.
3. Remove the access cover from the control tower.
4. Disconnect the lines from the hydraulic pump.
5. Remove the right and left heat exchanger brackets.
6. Remove the four engine mounting bolts.
7. Slide the engine ahead a small amount.
8. Disconnect the return line from the reservoir.
9. Remove the reservoir.
10. Remove the four (4) bolts holding the pump support to the engine.
11. Remove the pump and pump support through the area that the reservoir and battery were.
12. Mark the coupling position on the pump shaft for a guide during assembly.
13. Loosen the setscrew and remove the coupling half.
14. Remove the bolts holding the pump to the pump support.

For Loader Tractors 644 and 646 (All)

1. Remove the battery and the mounting plate for the battery.
2. Remove the access panel from the control tower.
3. Disconnect the hydraulic lines at the pump. Put a cap on the lines to keep dirt out.
4. Disconnect the inlet line at the loader valve. Loosen this line at the travel valve.
5. Remove the two pump mounting bolts.
6. Loosen the two setscrews in the pump coupling.
7. Remove the pump through the top of the tractor.

INSTALLATION

1. Install the pump by using the procedure opposite that of removal.
2. Make sure that the drive coupling halves are fully engaged, but do not compress the flexible center section.
3. Tighten the setscrews.
4. Make sure you connect the battery ground wire to the pump support bolt.
5. Fill the reservoir to the correct level and with the correct oil. See your operator's manual.
6. Start and run the tractor at half throttle for 30 seconds. Operate all the hydraulic controls.
7. Check the oil level and add oil if necessary.
8. Start and run the tractor:
   a. at half throttle and apply short intervals of load for 3 minutes, and
   b. at full throttle and apply short intervals of load for 3 minutes.
9. Stop the tractor and check for leaks.

OVERHAUL

Wooster Pump
Part Numbers C-14243, C-14244
Models 220, 222, 442, 444
Before Part Number 9641000

DISASSEMBLY

NOTE: Before disassembly, make a mark across the pump cover and body. Use this mark for alignment during assembly.

1. Remove the 90 degree fitting (8) and O-ring (9). Remove the straight fitting (21) and the O-ring (22).
2. Remove bolts (1), washers (2), and cover (3) from the body (10). Remove the seal (4) from the cover (3).
3. Remove the adapter cover (15) and remove the seal (4), brass seal (11), rubber spacer (12) and seal (16).
4. Remove the drive gear (6), driven gear (7) and bearings (5) from the body (10).

INSPECTION

1. Completely clean all the parts before inspection. Replace all worn or damaged parts.
2. Check all metal parts for scoring or other damage. Small scratches can be removed with a hone or crocus cloth.
3. Check the gears for damaged teeth.
4. Replace the spacer and all seals and gaskets each time the hydraulic pump is disassembled.

J.I. Case

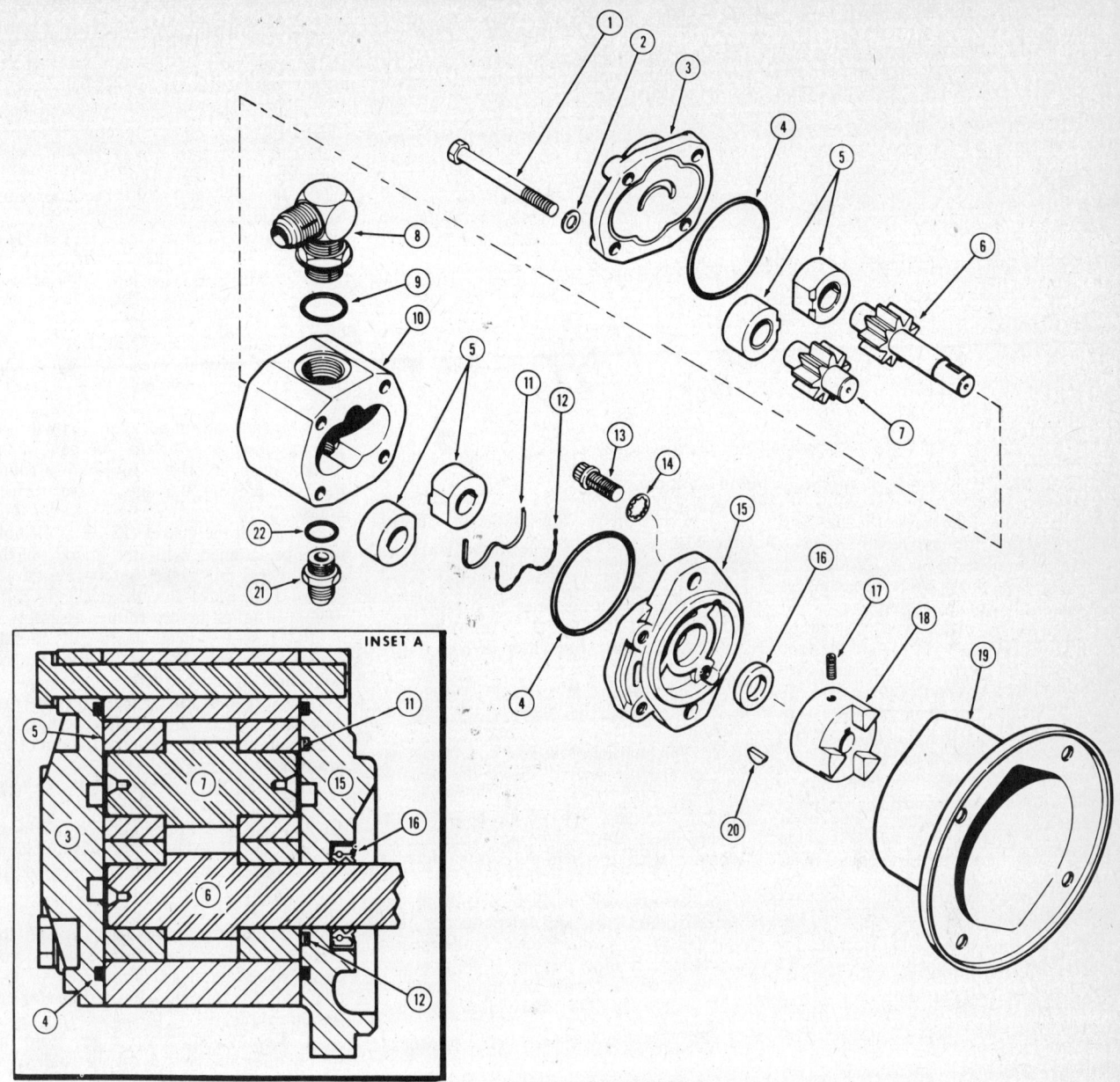

Wooster pump components

ASSEMBLY

1. Lubricate all the parts, except the seal, with a clean SAE No. 10 oil during assembly.

2. Apply a light coat of Permatex® No. 3, or equivalent, to the bore of the cover (15). With the seal lip to the inside, press a new oil seal (16) fully into the cover.

3. With the chamfer toward the gear, assemble the bearings (5) on the gears (6 and 7). Put this assembly into the pump body (10).

4. Put the rubber spacer (12) in the front cover (15), then install the brass seal (11) and the seal (4).

5. Install the seal (4) in the rear cover (3).

6. Install both the covers on the body. Align the marks made during disassembly.

NOTE: Be careful while installing the cover on the gear shaft so the oil seal is not damaged.

7. Install the bolts (1) and washers (2). Tighten to a torque of 28 to 32 ft.lbs.

8. Install the key (20) and drive coupling (18). Tighten the setscrew (17) in the coupling hub.

9. Install the pump support (19) using the washers (14) and capscrews (13).

10. Install the 90 degree fitting (8) and O-ring (9). Install the straight fitting (21) and O-ring (22).

Parker Hannifin Pump Part Numbers C15481 and C19045 For Models
220 P.I.N. 9641001 to P.I.N. 9674235,
222 P.I.N. 9641001 to P.I.N. 9681170,
224 P.I.N. 9667000 to P.I.N. 9676354,
442 All,
444 P.I.N. 9641001 to 967749,
446 Before P.I.N. 9728158
644 Before P.I.N. 9698284

DISASSEMBLY

1. Remove the key (15) from the drive shaft. If the key cannot be removed with a pliers:
   a. Tighten the key in a vise with a square jaw.
   b. Lift the rear of the pump.

2. Completely clean all external areas of the pump with a solvent.

3. Put a mark across all three sections of the pump. This mark will align the sections during assembly.

J.I. Case

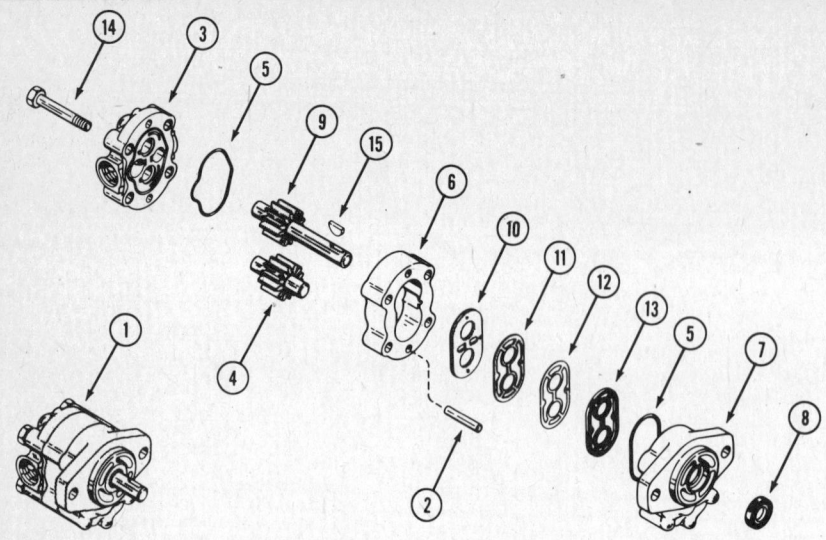

Parker Hannifin pump components

4. Remove the four bolts (14) that hold the three pump sections together.

5. Separate the three pump sections. Do not use a prybar as this will cause damage. If necessary use this procedure:

a. Hold the pump in a vertical position with the front cover (7) toward the top.

b. Lightly hit the sides of the back cover (3) with a soft hammer.

c. As the sections separate make sure the sections are parallel. This will prevent damage of the dowel pins (2).

6. Remove the wear plate (10).

7. Put a mark on the side of the front cover next to the small pressure hole in the heat shield (11). This mark will be a guide for correct alignment of the pressure hole. Remove the heat shield (11), gasket (12) and seal (13) by putting a small wire into the pressure hole.

8. Remove the O-rings (5) from the inside face of both front and back covers (3 and 7).

9. Do not remove the shaft seal (8) unless:

a. the seal was leaking or

b. wear or damage can be seen.

If replacement is necessary, carefully remove the seal to prevent damage to the cover bore. Use the following procedure:

a. Heat the cover to approximately 250° F (120° C).

b. Use a small puller that will make contact with the steel retainer section of the seal.

c. Pull the seal straight and even to prevent damage to the cover bore.

INSPECTION

1. Completely clean and air dry all parts before inspection. Replace all worn or damaged parts.

NOTE: The thickness of the center section (6) must be between .0002 in. and .0018 in. (0.005 and 0.045 mm) wider than the width of the gears.

2. Check the gear shafts and tooth faces and sides for damage or sharp edges. Small defects can be removed with emery cloth. Do not use emergy cloth on the edges of the gear O.D.

3. Check the positions of the bearings in the covers. These bearings are pressed fit and are not replaceable. The correct position for these bearings are as follows:

a. The split in the bearings must not be more than $1/16$ in. (1.58 mm) from a vertical center line through both bearings.

b. The bearing in the front cover must be even with the face of the bearing bore.

c. The bearing in the rear cover must be recessed below the face of the bearing bore.

4. Check the ends of the bearings for wear. There must not be any wear due to contact with the retaining rings of the gears.

5. Remove any small defects from the faces of the covers and center section. Do not damage any square edges while repairing defects. Follow this procedure to repair the faces of the covers and center section:

a. Put a No. 320 emery cloth on a flat surface.

b. Slide the face on this surface to repair the damage.

Also use this procedure to remove the wear ridge from the surface of the back cover.

6. Replace the seals, gaskets and wear

WEAR TOLERANCE CHART

Gear shaft diameter	.4993" (12.682 mm) minimum
Gear diameter	1.2390" (26.639 mm) minimum
Gear width	1.0557" (26.456 mm) minimum
Cover bearing I.D.	.5025" (12.764 mm) maximum
Gear wear ridge in back cover and wear plate	.0005" (0.013 mm) maximum
Gear bore diameters in center section	1.243" (31.572 mm) maximum

plate each time the pump is disassembled. Do not replace the shaft seal unless this seal is damaged. See step 9 under Disassembly.

ASSEMBLY

1. Make sure all parts are completely clean. Select a clean area for assembly. Lubricate each part, except the shaft seal, with clean SAE No. 10 oil.

2. If a new shaft seal is needed, follow this procedure for installation:

a. Put the cover (7) on a flat clean surface, with the face down.

b. Put the seal in position with the sealing lip toward the inside of the cover.

c. Use a steel rod with a flat end that is larger than the seal O.D. Hold the rod exactly vertical. Press the seal evenly into the cover bore until the seal is level with the rim.

3. Install the rubber seal (13) with the lip side down and the oil hole next to the mark you made. Use a tool with a round head to prevent damage to the surface areas.

4. Install the gasket (12). The oil hole must be aligned with the mark on the cover. Press this gasket tightly into the recessed area. Make sure the gasket is completely installed before further assembly.

5. Press the heat shield (11) tightly over the gasket (12). Make sure the oil holes are aligned correctly.

6. Apply oil to the faces of the front and back covers. Install the O-rings (5). Make sure the O-rings cannot fall out of the grooves.

7. Install the wear plate (10) over the heat shield. Make sure the oil holes are aligned and the bronze surface is toward the outside. Press until the wear plate is approximately even with the face of the cover. Be careful that you do not bend this plate.

8. Install the drive gear (9) and the driven gear (4) into the front cover. If you do not have a ½ in. (12.7 mm) seal protector, use the following procedure for installation:

a. Remove all sharp edges from the chamfer and keyway of the shaft.

b. Apply a layer of oil on the shaft.

c. With a rotating motion, push the shaft through the seal.

9. Put the center section (6) over the gears and on the front cover. Make sure the guide marks are aligned and the wear plate and O-ring are seated.

10. Install the dowel pins (2). Add a large amount of oil to the gear cavities and rotate the gears.

11. With the guide marks aligned install the back cover. Make sure the wear plate is still in place. Install the capscrews and tighten to a torque 24 to 26 ft.lbs.

12. Add a large amount of oil to both pump ports. Manually rotate the shaft for oil distribution and to make sure the shaft has free movement.

Borg-Warner Pump Part Numbers C20757, C19743 and C22771 For Models 200 P.I.N. 9674235 to P.I.N. 9734870,

J.I. Case

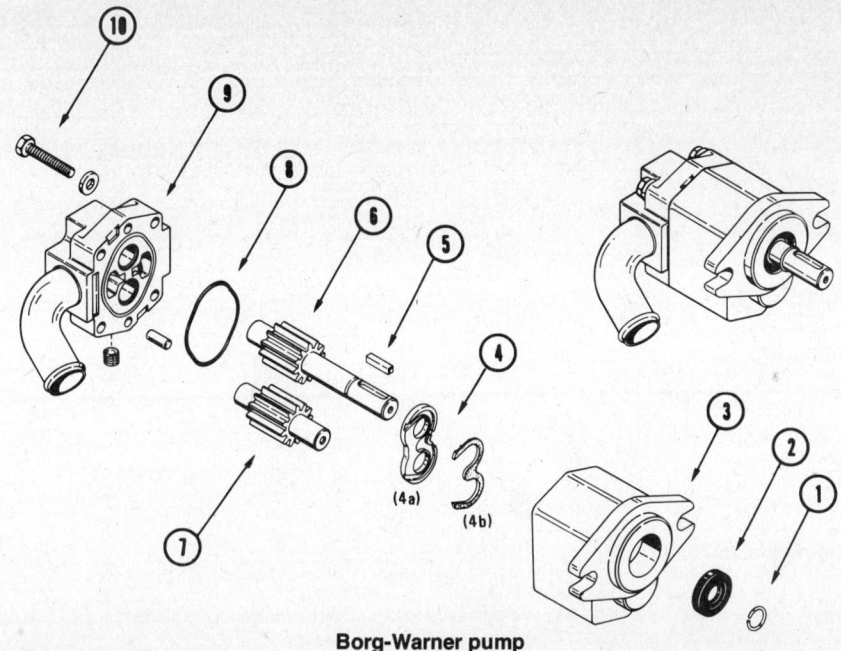

Borg-Warner pump

222 P.I.N. 9681170 to
P.I.N. 9736998,
224 P.I.N. 9676354 to
P.I.N. 9738335,
444 P.I.N. 9677449 to
P.I.N. 9739739,
446 P.I.N. 9728158 to
P.I.N. 9742953,
644 P.I.N. 9698284 to
P.I.N. 9757937

DISASSEMBLY

1. Remove the key (5) from the drive shaft. If the key cannot be removed with a pliers:
   a. Tighten the key in a vise with square jaws.
   b. Lift the rear of the pump.
2. Completely clean all external areas of the pump with a solvent similar to kerosene.
3. Put a mark across the pump sections for alignment during assembly.
4. Remove the four bolts (10) that hold the two pump sections together.
5. Put a screwdriver in each slot on opposite sides of the pump body. Carefully separate the two sections.
6. Remove the O-ring (8) from the pump cover (9).
7. Remove the driven gear (7) from the pump body (3).
8. Remove the snapring (1) from the drive gear shaft (6). Remove the drive gear from the pump body.
9. Remove the wear plate assembly (4) from the pump body (3).
10. Remove the pressure load seal (4b) from the wear plate (4a).
11. Do not remove the shaft seal (2) unless:
    a. the seal was leaking or
    b. wear or damage can be seen.
12. If replacement is necessary, carefully remove the seal to prevent damage to the cover bore. Use the following procedure:
    a. Heat the cover to approximately 250° F (120° C).
    b. Use a small internal puller that will make contact with the steel retainer section of the seal.
    c. Pull the seal straight and even to prevent damage to the cover bore.

INSPECTION

1. Completely clean and air dry all parts before inspection. Replace all worn or damaged parts.
2. Check the gear shafts and tooth faces and sides for damage or sharp edges. Small defects can be removed with emery cloth. Do not use emery cloth on the edges of the gear O.D.
3. Check the positions of the bearings in the covers. These bearings are pressed fit and are not replaceable. The correct position for these bearings is as follows:
    a. The split in the bearings must not be more than 1/16 in. (1.58 mm) from a vertical center line through both bearings.
    b. The bearing in the front cover must be even with the face of the bearing bore.
    c. The bearing in the rear cover must be recessed below the face of the bearing bore.
4. Remove any small defects from the faces of the covers and center section. Do not damage any square edges while repairing defects.
5. Follow this procedure to repair the faces of the covers and center section:
    a. Put a No. 320 emery cloth on a flat surface.
    b. Slide the face on this surface to repair the damage.
    Also use this procedure to remove the wear ridge from the surface of the back cover.
6. Replace the seals, gaskets and wear plate each time the pump is disassembled.

Do not replace the shaft seal unless this seal is damaged. See step 11 under Disassembly.

ASSEMBLY

1. Make sure all parts are completely clean. Select a clean area for assembly. Lubricate each part, except the shaft seal, with a clean SAE No. 10 oil.
2. If a new shaft seal is needed, follow this procedure for installation:
   a. Put the pump body (3) on a flat clean surface with the face down.
   b. Put the seal in position with the sealing lip toward the inside of the body.
   c. Use a steel rod with a flat end that is larger than the seal O.D. Hold the rod exactly vertical. Press the seal evenly into the body bore until the seal is .188" (4.78 mm) below the rim. Make sure the seal does not prevent the flow of oil through the oil passage.
3. Install the pressure load seal (4b) in the wear plate (4a).
4. Install the wear plate assembly (4) into the pump body (3). Make sure the seal (4b) is toward the casting.
5. Install the drive gear (6) into the body (3). If you do not have a seal protector, use the following procedure for installation:
   a. Remove all sharp edges from the chamfer, keyway and groove for the snapring.
   b. Apply a layer of oil on the shaft.
   c. With a rotating motion, push the shaft through the seal. Install the snapring on the shaft.
6. Install the driven gear (7) in the body (3). The long end of the shaft goes into the body.
7. Fit the O-ring (8) into the groove in the pump cover (9).
8. Align the guide marks on the pump body and cover. Install the cover.
9. Install the four retaining bolts. Tighten to a torque of 28 to 32 ft.lbs.

Parker Hannifin Pump Part Numbers C24828 and C24868 (Service Pump), and C25053 (Service Pump) for Some Models
220: P.I.N. 9734870 and after,
222: P.I.N. 9736998 and after,
224: P.I.N. 9738335 and after,
444: P.I.N. 9739789 and after,
446: P.I.N. 9742935 and after,
448: All,
644: P.I.N. 9757937 and after

DISASSEMBLY

1. Remove the key (15) from the drive shaft. If the key cannot be removed with a pliers:
   a. Tighten the key in a vise with a square jaw.
   b. Lift the rear of the pump.
2. Completely clean all external areas of the pump with a solvent.
3. Put a mark across all three sections of the pump. This mark will align the sections during assembly.
4. Remove the four bolts (14) that hold the three pump sections together.

J.I. Case

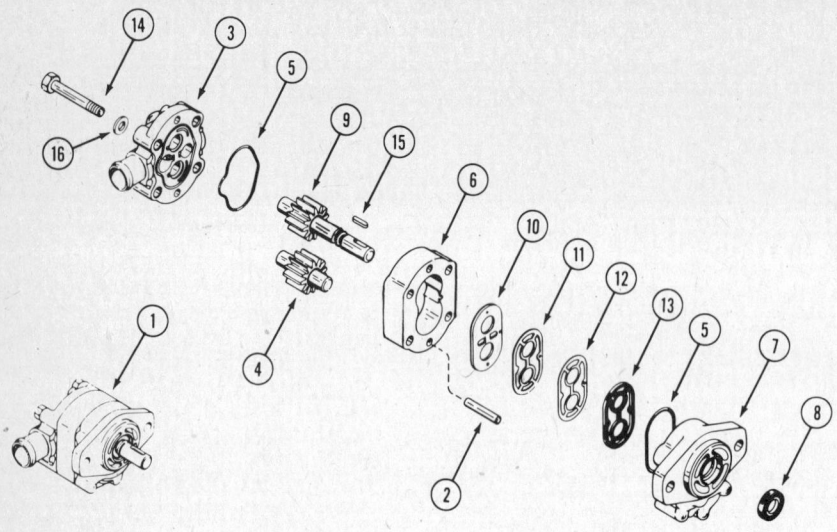

Parker Hannifin pump components

5. Separate the three pump sections. Do not use a prybar as this will cause damage. If necessary use this procedure:
   a. Hold the pump in a vertical position with the front cover (7) toward the top.
   b. Lightly hit the sides of the back cover (3) with a soft hammer.
   c. As the sections separate make sure the sections are parallel. This will prevent damage of the dowel pins (2).

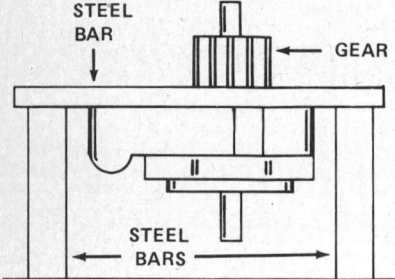

Steel bar placement

6. Remove the spiral retaining ring from the upper gear shaft.

7. Put a thin hard steel bar between the gear and the front cover and under each side of the gear.

8. Put this assembly in a press. Support the assembly on the ends of the hard steel bars.

9. Use the press to remove the gear from the shaft.

10. Remove the key and second spiral retaining ring.

11. Pull the shaft through the front cover bushing and seal.

12. Remove the wear plate (10).

13. Put a mark on the side of the front cover next to the small pressure hole in the heat shield (11). This mark will be a guide for correct alignment of the pressure hole. Remove the heat shield (11), gasket (12) and seal (13) by putting a small wire into the pressure hole.

14. Remove the O-rings (5) from the inside face of both front and back covers (3 and 7).

15. Do not remove the shaft seal (8) unless:
   a. the seal was leaking or
   b. wear or damage can be seen.
If replacement is necessary, carefully remove the seal to prevent damage to the cover bore. Use the following procedure:
   a. Heat the cover to approximately 250° F (120° C).
   b. Use a small internal puller that will make contact with the steel retainer section of the seal.
   c. Pull the seal straight and even to prevent damage to the cover bore.

INSPECTION
1. Completely clean and air dry all parts before inspection. Replace all worn or damaged parts.

NOTE: The thickness of the center section (6) must be between .0002 in. and .0018 in. (0.005 and 0.045 mm) wider than the width of the gears.

2. Check the gear shafts and tooth faces and sides for damage or sharp edges. Small defects can be removed with emery cloth. Do not use emery cloth on the edges of the gear O.D.

WEAR TOLERANCE CHART

Gear shaft diameter	.4993" (12.682 mm) minimum
Gear diameter	1.2390" (26.639 mm) minimum
Gear width	1.0557" (26.456 mm) minimum
Cover bearing I.D.	.5025" (12.764 mm) maximum
Gear wear ridge in back cover and wear plate	.0005" (0.013 mm) maximum
Gear bore diameters in center section	1.243" (31.572 mm) maximum

3. Check the positions of the bearings in the covers. These bearings are pressed fit and are not replaceable. The correct position for these bearings are as follows:
   a. The split in the bearings must not be more than $1/16$ in. (1.58 mm) from a vertical center line through both bearings.
   b. The bearing in the front cover must be even with the face of the bearing bore.
   c. The bearing in the rear cover must be recessed below the face of the bearing bore.

4. Check the ends of the bearings for wear. There must not be any wear due to contact with the retaining rings of the gears.

5. Remove any small defects from the faces of the covers and center section. Do not damage any square edges while repairing defects. Follow this procedure to repair the faces of the covers and center section:
   a. Put a No. 320 emery cloth on a flat surface.
   b. Slide the face on this surface to repair the damage.
Also use this procedure to remove the wear ridge from the surface of the back cover.

6. Replace the seals, gaskets and wear plate each time the pump is disassembled. Do not replace the shaft seal unless this seal is damaged. See step 15 under Disassembly.

ASSEMBLY
1. Make sure all parts are completely clean. Select a clean area for assembly. Lubricate each part, except the shaft seal, with clean SAE No. 10 oil.

2. If a new shaft seal is needed, follow this procedure for installation:
   a. Put the cover (7) on a flat clean surfae, with the face down.
   b. Put the seal in position with the sealing lip toward the inside of the cover.
   c. Use a steel rod with a flat end that is larger than the seal O.D. Hold the rod exactly vertical. Press the seal evenly into the cover bore until the seal is level with the rim.

3. Install the rubber seal (13) with the lip side down and the oil hole next to the mark you made. Use a tool with a round head to prevent damage to the surface areas.

4. Install the gasket (12). The oil hole must be aligned with the mark on the cover. Press this gasket tightly into the recessed area. Make sure the gasket is completely installed before further assembly.

5. Press the heat shield (11) tightly over the gasket (12). Make sure the oil holes are aligned correctly.

6. Apply oil to the faces of the front and back covers. Install the O-rings (5). Make sure the O-rings cannot fall out of the grooves.

7. Install the wear plate (10) over the heat shield. Make sure the oil holes are aligned and the bronze surface is toward the outside. Press until the wear plate is approximately even with the face of the cover. Be careful that you do not bend this plate.

J.I. Case

8. Install the gear shaft in the front cover. If you do not have a ½ in. (12.7 mm) seal protector, use the following procedure for installation:

a. Remove all sharp edges from the chamfer and keyway of the shaft.
b. Apply a layer of oil on the shaft.
c. With a rotating motion, push the shaft through the seal.

9. Install the spiral retaining ring in the groove nearest to the corner.
10. Install the key in the keyway.
11. Heat the gear to 275° F (135° C) and install on the shaft and key.
12. Install the second spiral retaining ring.
13. Install the bottom shaft and gear.
14. Put the center section (6) over the gears and on the front cover. Make sure the guide marks are aligned and the wear plate and O-ring are seated.
15. Install the dowel pins (2) if needed. Add a large amount of oil to the gear cavities and rotate the gears.
16. Align the guide marks and install the back cover. Make sure the wear plate is still in place. Install the capscrews and tighten to a torque of 200 in.lbs.
17. Add a large amount of oil to both pump ports. Manually rotate the shaft for oil distribution and to make sure the shaft has free movement.

Cessna Pump Part Number C25179 For Some Models
220: P.I.N. 9734870 and after,
222: P.I.N. 9736998 and after,
224: P.I.N. 9738335 and after,
444: P.I.N. 9739739 and after,
446: P.I.N. 9742953 and after,
448: All

DISASSEMBLY

1. Remove the key (9) from the shaft.
2. Completely clean the external areas of the pump with a solvent.
3. With the shaft down, put the pump in a vise.
4. Remove the four retainer bolts (15).
5. Put a mark across the pump sections for alignment during assembly.
6. Remove the pump from the vise. Lightly hit the pump shaft against a block of wood. This will separate the front cover (3) or the rear cover (4).
7. To separate the body (6) from the cover, put the drive gear (11) into the bearing. Lightly hit the shaft with a soft hammer.
8. Use a small knife to remove the diaphragm (2) from the front cover.
9. Remove the two springs (1) and two steel balls (13) from the front cover.
10. Lift the backup gasket (8) and the protector gasket (10) from the front cover.
11. Lift the diaphragm seal (7) from the front cover.
12. Remove the shaft seal (5) from the front cover.

GENERAL INSPECTION

1. Completely clean and dry all parts.
2. Use emery cloth to remove small defects from the parts.

GEAR ASSEMBLIES INSPECTION

1. Check the shaft of the drive gear (11) for a damaged keyway.
2. Check the gear shafts at the bearing and seal contact areas for rough surfaces and wear.
3. Replace the gear assembly if the shaft is worn too much in the bearing contact area. See the Wear Tolerance Chart.
4. Check the gear face for scoring and wear.
5. Check the gear width. Replace the gear assembly if worn too much. See the Wear Tolerance Chart.
6. Make sure the snaprings are fully installed in the grooves on either side of the gears.
7. Use emery cloth to remove any sharp edge from the gear teeth.

FRONT AND REAR COVERS INSPECTION

1. The oil grooves in the bearings in both covers must be aligned with the dowel pin holes. The grooves must also be 180 degrees apart. Replace the cover if the grooves are not aligned.
2. Check the I.D. of the bearing in both covers. Replace the cover if the wear is too much. See the Wear Chart.
3. The bearings in the front cover must be level with the inside rim of the cover bore.
4. Check for scoring on the face of the back cover. If the wear is too much, replace the cover. See the Wear Tolerance Chart.

BODY INSPECTION

1. Check the gear bore for scoring or wear.
2. Replace the body if the gear bore is worn too much. See the Wear Tolerance Chart.

WEAR TOLERANCE CHART

Gear shaft diameter	.5605" (14.237 mm) minimum	
Gear width	.803" (20.40 mm) minimum	
Cover bearing I.D.	.5655" (14.364 mm) maximum	
Wear ridge in rear cover	.0015" (0.038 mm) maximum	
Gear bore I.D. in pump body	1.404" (35.661 mm) maximum	

ASSEMBLY

1. Replace the diaphragm, backup gasket, diaphragm seal, protector gasket, shaft seal, steel balls and springs before assembly.
2. Push the diaphgram seal (7) into the grooves in the front cover. The open part of the "V" section must be to the inside. Use a dull tool.

NOTE: Make sure the inner lip of the diaphragm seal does not turn out during assembly.

3. Push the protector gasket (10) and the backup gasket (8) into the diaphragm seal.
4. Put the steel balls (13) into the seats and put the springs (1) over the balls.
5. With the bronze face up, put the diaphragm on the backup gasket. The complete diaphragm must fit inside the rim of the diaphragm seal.
6. Apply oil to the gear assemblies and push into the bearings in the front cover.
7. Install the dowel pins (14) into the body (6).
8. Apply a thin layer of heavy grease to both faces of the body (6). Align the guide marks and install the body on the front cover.
9. Slide the back cover over the gear shafts until the dowel pins fit into the cover. Make sure the guide marks are aligned.
10. Install the retaining bolts (15). Tighten to a torque of 23 to 25 ft.lbs.
11. Make sure all sharp edges are removed from the drive gear shaft. Apply a large amount of oil to the shaft and oil seal (5).
12. With a rotating motion, carefully push the seal (5) over the shaft. Lightly hit the seal with a soft hammer to fit the seal fully into the seat.
13. Put a large amount of oil into each pump port. Manually rotate the pump shaft until the shaft turns freely.

Borg Warner Pump Part Numbers C19176, C20025, C20122 and C23951 For Some Model 646

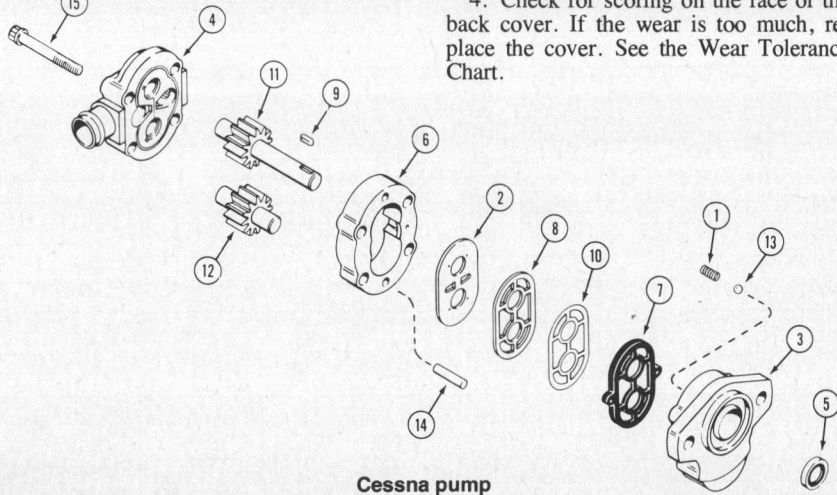

Cessna pump

241

J.I. Case

DISASSEMBLY

1. Remove the key (15) from the drive shaft.
2. Completely clean all external areas of the pump with a solvent.
3. Put a mark across the pump sections for alignment during assembly.
4. Remove the four bolts (14) that hold the two pump sections together.
5. Put a screwdriver in each slot on opposite sides of the pump body. Carefully separate the two sections.
6. Remove the thrust plate (11) from the pump cover (2).
7. Remove the quad ring (12) from the pump cover (2).
8. Remove the driven gear (10) from the pump body (3).
9. Remove the snapring (5) from the pump body (3).
10. Remove the wear plate (8) from the pump body (3).
11. Remove the pressure load seals (7) from the wear plate (8).
12. Do not remove the shaft seal (6) unless:
    a. the seal was leaking
    b. wear or damage can be seen.
13. If replacement is necessary, carefully remove the seal to prevent damage to the body bore. Use the following procedure:
    a. Remove the seal retaining ring (4).
    b. Heat the cover to approximately 250° F (120° C).
14. Use a small internal puller that will make contact with the steel retainer section of the seal.
15. Pull the seal straight and even to prevent damage to the body bore.

INSPECTION

1. Completely clean and air dry all parts before inspection. Replace all worn or damaged parts.
2. Check the gear shafts and tooth faces and sides for damage or sharp edges. Small defects can be removed with emery cloth. Do not use emery cloth on the edges of the gear O.D.
3. Check the positions of the bearings in the covers. These bearings are pressed fit and are not replaceable. The correct position for these bearings are as follows:
    a. The split in the bearings must not be more than $1/16$ in. (1.58 mm) from a vertical centerline through both bearings.
    b. The bearing in the front cover must be even with the face of the bearing bore.
    c. The bearing in the rear cover must be recessed below the face of the bearing bore.
4. Remove any small defects from the faces of the covers and center section. Do not damage any square edges while repairing defects.
    a. Put a No. 320 emery cloth on a flat surface.
    b. Slide the face on this surface to repair the damage.
    Also use this procedure to remove the wear ridge from the surface of the back cover.
5. Replace the seals, gaskets and wear plate each time the pump is disassembled.

Do not replace the shaft seal unless this seal is damaged. See step 12 under disassembly.

ASSEMBLY

1. Make sure all parts are completely clean. Select a clean area to assemble the pump. Lubricate each part, except the shaft seal, with clean SAE No. 10 before assembly.
2. If a new shaft seal is needed, follow this procedure for installation:
    a. Put the pump body (3) on a flat clean surface, with the face down.
    b. Put the seal (6) in position with the sealing lip toward the inside of the cover.
    c. Use a steel rod with a flat end that is larger than the seal O.D. Hold the rod exactly vertical. Press the seal evenly into the cover bore until the seal is level with the rim.
3. Install the seal retaining ring (4).
4. Install the pressure load seals (7) into the wear plate (8). The softer seal is installed first.
5. Install the wear plate assembly (8) into the pump body (3). The pressure load seals must be toward the body.
6. Install the drive gear (9) into the pump body (3). If you do not have a seal protector, use the following procedure for installation:
    a. Remove all sharp edges from the chamfer, keyway and groove for the snapring.
    b. Apply a layer of oil on the shaft.
    c. With a rotating motion, push the shaft through the seal.
    d. Install the snapring (5) on the pump shaft.
7. Install the driven gear (10) in the body (3). The long end of the shaft fits into the body.
8. Install the thrust plate (11) on the gear shafts.
9. Put the quad ring (12) into the groove in the pump cover (2).
10. Align the guide marks and install the cover (2) on the pump body.
11. Install the four retainer bolts (14).

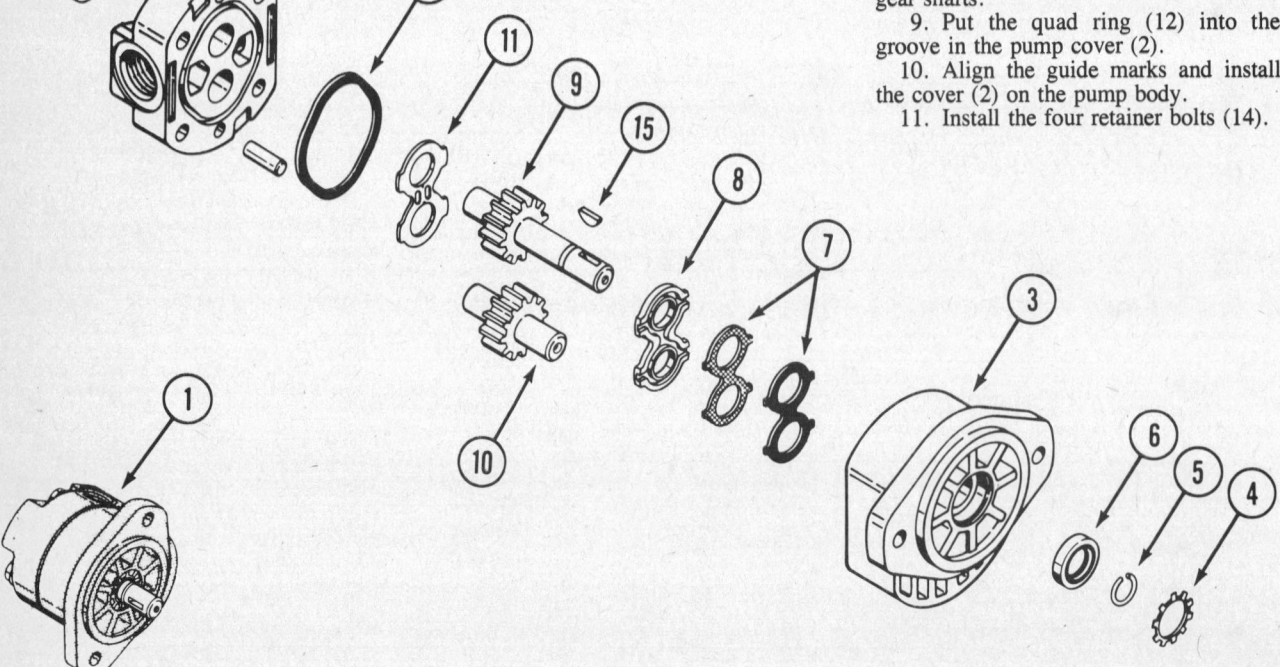

Borg-Warner pump components

J.I. Case

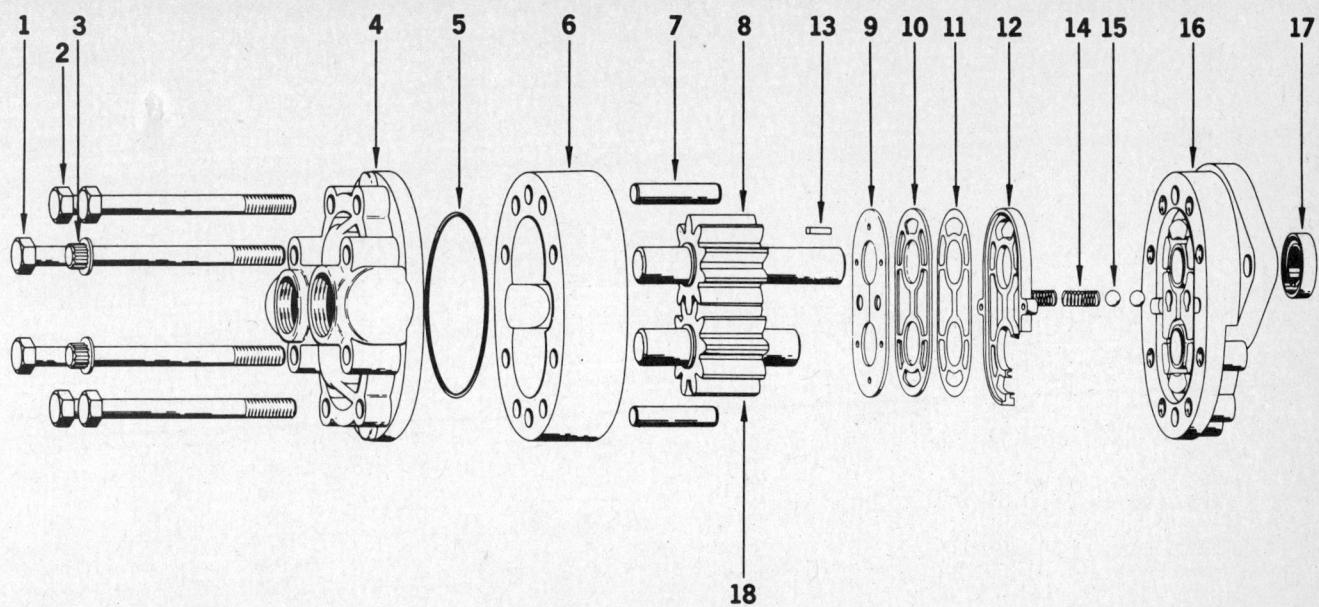

Cessna pump components

Cessna Pump Part Number C25287 for Some Model 646

DISASSEMBLY
1. Remove the key (13) from the shaft.
2. Completely clean all external areas of the pump with a solvent.
3. Put a mark across the pump sections for alignment during assembly.
4. With the shaft down, put the pump in a vise.
5. Remove the retainer bolts (1) (2) (3).
6. Remove the pump from the vise. Lightly hit the pump shaft against a block of wood. This will separate the front cover (16) or the rear cover.
7. To separate the body (6) from to cover, put the drive gear into the bearing. Lightly hit the shaft with a soft hammer.
8. Remove the O-ring (5) from the rear cover.
9. Use a small knife to remove the diaphragm (9) from the front cover.
10. Remove the two springs (14) and the two steel balls (15) from the front cover.
11. Lift the backup gasket (10) and the protector gasket (11) from the front cover.
12. Lift the diaphragm seal (12) from the front cover.
13. Remove the shaft seal (17) from the front cover.

GENERAL INSPECTION
1. Clean and dry all parts.
2. Use emery cloth to remove small defects from the parts.

GEAR ASSEMBLIES INSPECTION
1. Check the shaft of the drive gear (8) for a damaged keyway.
2. Check the gear shafts at the bearing and seal contact areas for rough surfaces and wear.
3. Replace the gear assembly if the shaft is worn too much in the bearing contact area. See the Wear Chart.
4. Check the gear face for scoring and wear.
5. Check the gear width. Replace the gear assembly if worn too much. See the Wear Chart.

WEAR TOLERANCE CHART

Gear shaft diameter	.6850" (17.399 mm) minimum
Gear width	.441" (11.20 mm) minimum
Cover bearing I.D.	.691" (17.55 mm) maximum
Wear ridge in the rear cover	.0015" (0.038 mm) maximum
Gear bore I.D. in pump body	1.719" (43.66 mm) maximum

6. Make sure the snaprings are fully installed in the grooves on either side of the gears.
7. Use emery cloth to remove any sharp edges from the gear teeth.

FRONT AND REAR COVERS INSPECTION
1. The oil grooves in the bearings in both covers must be aligned with the dowel pin holes. The grooves must also be 180 degrees apart. Replace the cover if the grooves are not aligned.
2. Check the I.D. of the bearing in both covers. Replace the cover if the wear is too much. See the Wear Tolerance Chart.
3. The bearings in the front cover must be level with the inside rim of the cover bore.
4. Check for scoring on the face of the back cover. If the wear is too much, replace the cover. See the Wear Tolerance Chart.

BODY INSPECTION
1. Check the gear bore for scoring or wear.
2. Replace the body if the gear bore is worn too much. See the Wear Tolerance Chart.

ASSEMBLY
1. Always replace the diaphragm, backup gasket, diaphragm seal, protector gasket and shaft seal before assembly.
2. Push the diaphragm seal (12) into the grooves in the front cover. The open part of the "V" section must be to the inside. Use a dull tool.
3. Push the protector gasket (11) and the backup gasket (10) into the diaphragm seal.
4. Put the steel balls (15) into the seats and put the springs (14) over the balls.
5. With the bronze face up, put the diaphragm (9) on the backup gasket. The complete diaphragm must fit inside the rim of the diaphragm seal.
6. Apply oil to the gear assemblies and push into the bearings in the front cover.
7. Install the dowel pins (7) into the body (6).
8. Apply a thin layer of heavy grease to both faces of the body (6). Align the guide marks and install the body on the front cover.
9. Install the O-ring (5) in the groove in the back cover (4).
10. Slide the back cover over the gear shafts until the dowel pins fit into the cover. Make sure the guide marks are aligned.
11. Install the retaining bolts (1) (2) (3). Tighten to a torque of 23 ft.lbs.
12. Make sure all sharp edges are removed from the drive gear shaft. Apply a large amount of oil to the shaft and the oil seal (17).
13. With a rotating motion, carefully push the seal (17) over the shaft. Lightly hit the seal with a soft hammer to fit the seal completely into the seat.
14. Put a large amount of oil into each pump port. Manually rotate the pump shaft until the shaft turns freely.

# J.I. Case

ENGINE

For detailed engine overhaul and repair, see the Engine Unit Repair section of this book for the engine in your tractor.

REMOVAL AND INSTALLATION

Kohler Engines

See the procedure outlined in the 155 and 195 tractor section.

Tecumseh Engines

1. Remove the attachment drive and drive belts.
2. Remove the two hood attaching screws, disconnect the lights and remove the hood.
3. Disconnect the carburetor linkage.
4. Remove and plug the fuel line at the carburetor.
5. Disconnect the battery ground and the starter leads.
6. Disconnect the studs at the engine front mounting plate.
7. Remove the top nuts and lockwashers from the engine rear mounts.
8. Remove the botton nut, lockwasher and flat washer from the front mount.
9. Lift the front of the engine until the front mount is free, then pull the engine forward and up to remove it.
10. Reverse the above procedure for installation.

Briggs and Stratton Engines

1. Remove the drive and attachment drive belts.
2. Remove the hood lanyard, disconnect the headlights and remove the hood.
3. Remove the carburetor linkage at the carburetor.
4. Disconnect the fuel line at the carburetor and plug it.
5. Disconnect the battery ground cable and disconnect the starter cables.
6. Unbolt the engine at the mounting plate and remove the engine.
7. Installation is the reverse of removal.

Onan Engines

For tractors with Onan engines, follow the procedure outlined under Briggs and Stratton engines above.

John Deere

INDEX

MODELS 850 & 950
- Front Axle 246
  - 2-wheel drive 246
  - 4-wheel drive 247
- Front Hubs and Bearings 248
- Brakes 248
- Steering 251
- Clutch 253
- Manual Transmission 254
- Rear Axle 264
- Final Drive 264
- Differential 270
- Power Take-Off 273
- Mechanical Front Wheel Drive 277
- Front Drive Axle Unit 280
  - Differential Driveshaft 282
  - Axle Pivots 283
- Hydraulic Pump 284
- Separation 284
- Engine 284
  - Overhaul 284
  - Cooling System 291
  - Fuel System 292
  - Electrical System 298

MODELS 60 & 70
- Front Wheels and Axle 298
- Steering 298
- Brakes 300
- Transaxle 301
- Power Take-Off Clutch 305
- Drive Belt Transmission 307

MODELS 110 & 112
- Front Wheels and Axle 308
- Steering System 311
- Hydraulic System 313
- Clutch, Brake, and Variable Speed Drive 316
  - Primary V-Belt 316
  - Secondary Belt 317
  - Variator 317
  - Brake Bands 318
- Three-Speed Transaxle 325
- Four-Speed Transaxle 329
- Power Take-Off Clutch 335

MODELS 108 & 111
- Front Wheels and Axle 339
- Steering System 339
- Traction Drive Clutch 340
- Brakes 341
- Power Take-Off 342
- Electrical Power Take-Off 344
- Five-Speed Transaxle 344
- Engine 347
- Fuel Tank 348

MODEL 100
- Front Wheels and Axle 348
- Steering 348
- Brakes 350
- Transaxle 350
- Clutch 353
- Power Take-Off Clutch 354
- Engine 355

MODEL 120
- Front Wheels and Axle 355
- Steering 356
- Brakes 358
- Clutch 359
- Transmission 360
- Control Cam 364
- Differential and Axle 364
- Engine 367

MODEL 140
- Front Wheels and Axle 368
- Steering 369
- Brakes 371
- Clutch 372
- Transmission 374
- Power Take-Off 377
- Engine 378

MODELS 200, 208, 210, 212, 214, 216
- Front Wheels and Axle 378
- Steering 379
- Brakes 381
- Clutch and Variable Speed Drive 383
- Transaxle 385
- Power Take-Off 389
- Hydraulic System 393
  - Control Valve 393
  - Pump 394
  - Cylinder 395
- Engine 396

JOHN DEERE
Models 850 and 950

FRONT AXLE

2-Wheel Drive

REMOVAL AND INSTALLATION

1. Set parking brake and block rear wheels.
2. Attach a hoist as shown.
3. Raise front of tractor until weight is off of wheels.
4. Disconnect drag link from steering arm.
5. Remove snapring and front pivot pin nut.
6. Remove cotter pin, slotted nut, and washer from rear of pivot pin.
7. Place floor jack under center of axle.
8. Install nut (A), washer (B), and spacer (C).
9. Tighten nut until pivit pin is loose, then remove pin.

NOTE: If pin cannot be removed, remove nut and install a 24H1540 washer over pin and repeat above procedure. Use additional washers as necessary to remove pin.

10. Raise front of tractor with the hoist and roll front axle away.

──────── CAUTION ────────
Lower front end of tractor onto splitting stand, while repairs are being made.

11. Inspect pivot pin at bushing surfaces for wear and scoring. The diameter of the pin at bushing surfaces should be 1.3760 to 1.3770 inch (34.95 to 34.98 mm).
12. Inspect pivot pin bushings in axle housing for wear and scoring. They should measure 1.3779 to 1.3795 inch (35.0 to 35.04 mm) I.D.
13. The clearance between pivot pin and pivot pin bushings should be 0.001 to 0.004 inch (0.03 to 0.10 mm), but no more than 0.016 inch (0.41 mm).
14. Drive out old bushings with a long drift. Drive new bushings in flush with end of tube using 27507 Disc and Driver.

Reverse the removal steps given on the preceding pages and note the installation instructions that follow.

NOTE: The 2-wheel drive tractor is equipped with an adjustable tread front axle. The axle can be adjusted to five different widths. These adjustments are made in 3.9 inch (100 mm) steps.

15. Install pivot pin by tapping it into front support from the front.
16. After thrust washer and front nut are installed, tighten nut to 239 to 297 ft.lbs.
17. Install washer and rear nut. Tighten nut securely. Do not install cotter pin at this time.
18. Check axle end play by pulling axle fully forward on pivot pin.
19. Inserting a blade-type feeler gauge between rear of axle housing and support.
20. Pivot pin end play should be 0.000 to 0.005 inch (0.00-0.13 mm). If there is excessive end play, tighten rear nut and repeat end play check. Install cotter pin.

NOTE: After tightening rear nut and registering correct end play, axle housing must be free to pivot.

21. Lubricate pivot pin with multi-purpose grease.

DISASSEMBLY

1. Position a floor jack under the front axle. Set parking brake and block rear wheels. Raise tractor until the front wheels clear the ground. Place a splitting stand under tractor.
2. Remove wheels.
3. Remove capscrew on steering arm and remove arm. It may be necessary to tap the steering arm with a hammer to loosen it from spindle shaft.
4. Remove key (B) from spindle shaft.
5. Remove spindle and knuckle assembly from knee. Knee does not have to be removed from axle.

NOTE: Do not allow spindle to fall on floor.

INSPECTION AND REPAIR

1. Inspect knee bushings (E) for wear and scoring. Bushing I.D. should be 1.1811 to 1.1824 inch (30.0 to 30.03 mm).
2. Inspect spindle shaft for wear and scoring. Shaft O.D. should be 1.1795 to 1.1803 inch (29.96 to 29.98 mm).

NOTE: The shaft-to-bushing clearance should be 0.001 to 0.003 inch (0.03 to 0.08 mm), but no more than 0.010 inch (0.25 mm).

3. Inspect thrust bearing seal (C) and O-ring (I) for wear. Replace if necessary.
4. If bushings must be replaced, drive old bushings out with a long drift.
5. Drive new bushings in flush with end of tube using 27504 Disc and Driver. Measure new bushings after installation. If necessary, hone new bushings to the new part specification.

NOTE: Since the shaft and spindle is an integrated unit, it should be carefully inspected for cracks. If there is any question about the condition of the unit, it should be inspected by Magnaflux or a similar crack-detection process.

ASSEMBLY

1. If knee was removed from axle, install knee. Install knee-to-axle bolts and tighten nuts to 123 to 152 ft.lbs. torque.

Attaching hoist

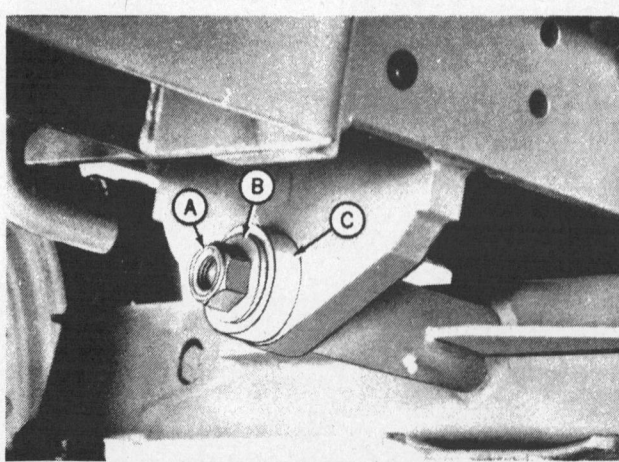

Installing front axle pivot pin

John Deere

2. Lightly lubricate bushings and install spindle assembly in knee.

3. Place O-ring and washer over spindle shaft. Install steering arm and key on shaft. Tighten capscrew to 58 to 72 ft.lbs.

4. Install a dial indicator as shown. Move the spindle up and down as shown while observing the dial indicator. Play should be 0.001 to 0.024 inch (0.03 to 0.61 mm).

5. Install wheel and tire assembly. Tighten six capscrews to 87 to 109 ft.lbs. torque.

6. Remove stand and lower tractor.

7. Using a multi-purpose grease lubricate spindle shafts. To lubricate, remove plug. Inject grease until it flows from plug opening. Replace plug and repeat for other spindle shaft.

4-Wheel Drive

REMOVAL AND INSTALLATION

1. Set parking brake and block rear wheels.
2. Disconnect headlight wire.
3. Disconnect hood holding bracket.
4. Remove hood mount bracket capscrews.
5. Remove hood.
6. Disconnect battery cables.
7. Remove battery clamps.
8. Remove battery.
9. Remove front and rear locknuts.
10. Loosen clamps.
11. Remove capscrews.
12. Slide drive shaft cover rearward.
13. Disconnect spring loaded collar from front axle drive pinion.
14. Set drive shaft and cover assembly aside.

NOTE: Be careful not to lose metal balls at front and rear of driveshaft collars.

15. Disconnect drag link from steering arm.
16. Attach chain hoist to weight support bracket.
17. Lift tractor and remove wheels.
18. Place jack under axle and remove capscrews from axle mounting bracket.

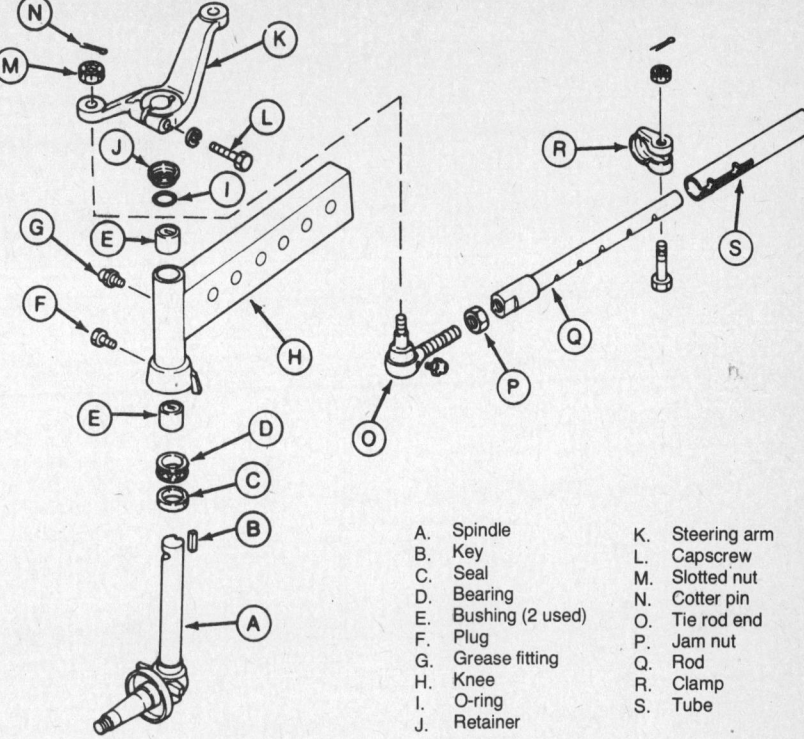

A. Spindle
B. Key
C. Seal
D. Bearing
E. Bushing (2 used)
F. Plug
G. Grease fitting
H. Knee
I. O-ring
J. Retainer
K. Steering arm
L. Capscrew
M. Slotted nut
N. Cotter pin
O. Tie rod end
P. Jam nut
Q. Rod
R. Clamp
S. Tube

Front axle knee components, 2-wheel drive

19. Lower jack and remove axle assembly.

- CAUTION -
Lower front end of tractor onto splitting stand, while repairs are being made.

Reverse the removal steps given on the preceding pages and note the installation instructions that follow.

20. While lifting axle assembly into place be sure dowel pins in lower mounting bracket are properly aligned with holes in upper bracket.
21. Mounting bolts and nuts must be coated with Loctite®.
22. Mounting bolts and nuts are tightened to 109 ft.lbs.
23. Remove cap from the front axle support.
24. Drive front axle forward by tapping with hammer.

25. Measure end play between plate and liner with a feeler gauge.

NOTE: End play should read between 0.002-0.012 in. (0.05-0.30 mm).

26. If out of specification remove plate capscrews and plate.
27. Either add or deduct shims to get proper end play.
28. Install plate and tighten capscrews to 109 ft.lbs.
29. Measure end play and repeat steps 4 thru 9 until correct end play is maintained.
30. Install front axle support cap.

DISASSEMBLY

For overhaul of this unit, see the Mechanical Front Wheel Drive part of this section.

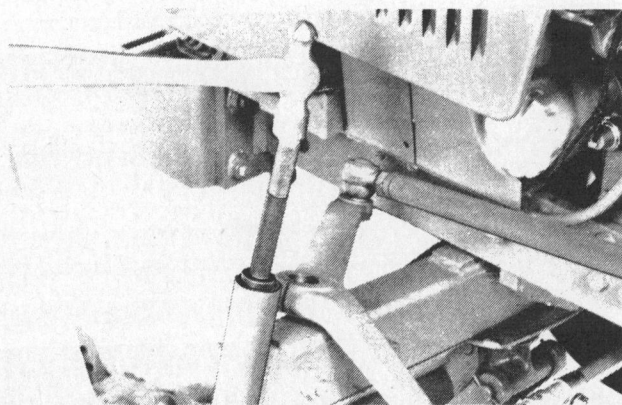

Installing bushings

Measuring end play

John Deere

Adjusting end play with shims

Drive out bearing cup, only at point indicated by the arrow

FRONT HUBS AND BEARINGS

2-Wheel Drive

REMOVAL

1. Set parking brake and block rear wheels.
2. Position a floor jack under front of tractor. Raise tractor.
3. Remove wheel.
4. Remove four screws (L) and remove hub cap (K).
5. Remove cotter pin (J) and nut (I).
6. Pull hub (D) forward slightly to unseat other bearing (G). Remove washer (H) and outer bearing.
7. Remove hub from spindle.
8. Remove oil seal and bearing (B) by driving out bearing cup (C) with a drift. Drive only at cut-outs in hub.

INSPECTION AND REPAIR

1. Inspect inner and outer bearings for worn rollers. Replace as necessary.
2. Inspect inner and outer bearing cups for wear. If cups are worn, drive them out using a drift.
3. Install inner bearing cup using 27522 Disc and Driver. Install outer bearing cup using 27515 Disc and Driver. Be sure cups are fully seated in hub.

INSTALLATION

1. Pack inner and outer bearings with wheel bearing grease.
2. Insert inner bearing in hub.
3. Install new oil seal in hub using 27522 Disc and Driver.
4. Place hub on spindle. Install outer bearing and washer.
5. Install slotted nut and tighten to 8.7 to 10.2 ft.lbs torque.
6. If slots in nut do not line up with cotter pin hole, remove nut and use different size washer, so that torque remains the same and pin hole lines up.
7. Install hub cap using four screws. Tighten screws securely.
8. Install wheel on hub. Tighten six capscrews to 87-109 ft.lbs. torque.

BRAKES

Pedal and Linkage

REMOVAL AND DISASSEMBLY

1. Remove pedal return spring from arm at each end of pedal shaft.
2. Remove cotter pin and headed pin from yoke and arm at each end of pedal shaft.
3. Disconnect clutch pedal linkage from clutch pedal.
4. Remove retaining ring from right-hand end of shaft.
5. Remove grease fitting from pedal.
6. Remove brake pedals from shaft.
7. Remove key from pedal shaft and remove shaft to the left through the clutch housing. Use care not to damage O-rings in clutch housing.

OVERHAUL AND INSTALLATION

1. Inspect pedal shaft for wear or damage and replace as necessary. O.D. at brake pedal is 1.257 to 1.260 in. (31.94-32.00 mm).
2. Inspect and assemble clutch pedal.
3. Inspect shaft bushings (B). Bushing I.D. is 1.263 to 1.266 in. (32.07-32.15 mm). If bushings are replaced, install 0.394 in. (10 mm below outer surface of bore for installation of O-rings.
4. Install bushings with grease.
5. Coat shaft (D) with grease and install through lefthand side of clutch housing. Position O-rings in bore in clutch housing.
6. Inspect brake pedals. Pedal I.D. is 1.260 to 1.262 in. (32.05-32.10 mm).
7. Install a spacer washer (C) over shaft and install pedals with key (G).
8. Install as many spacers as required on the outer end of the shaft to provide a maximum 0.040 in. (1 mm) shaft side play.
9. Install shaft retaining ring.

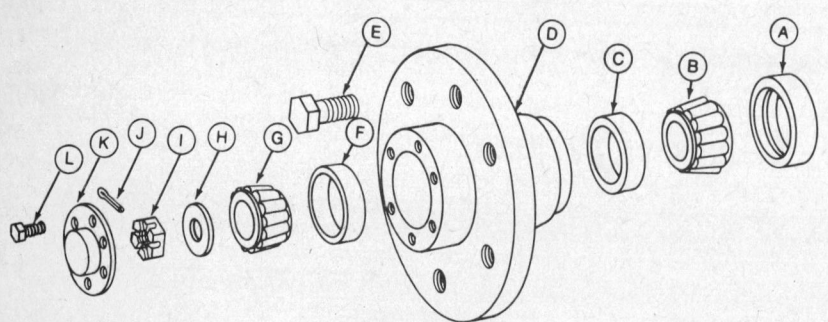

A. Oil seal
B. Inner bearing cone
C. Inner bearing cup
D. Hub
E. Capscrew (6 used)
F. Outer bearing cup
G. Outer bearing cone
H. Washer
I. Slotted nut
J. Cotter pin
K. Hub cap
L. Screw (4 used)

Two-wheel drive front hub assembly

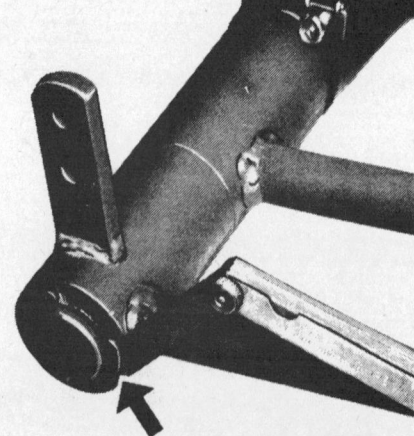

Measuring shaft side play

John Deere

- A. Brake pedals
- B. Clutch shaft
- C. Brake rod
- D. Return spring
- E. Retaining ring
- F. Brake housing

Brake pedal assembly

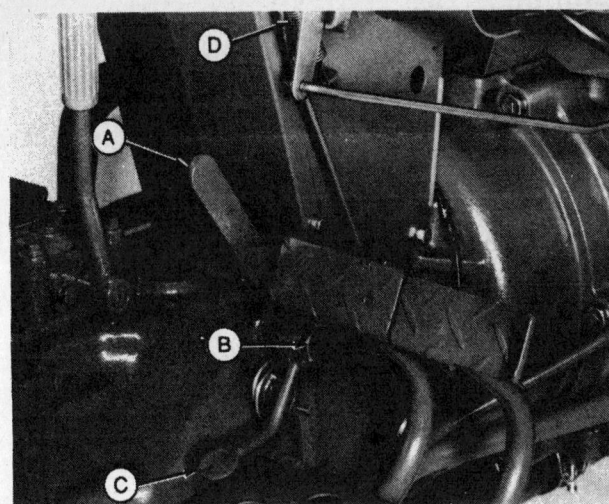

- A. Park lever
- B. Special screw
- C. Throttle pedal
- D. Park spring

Parking brake assembly

10. Inspect return springs for 9.13 in. length at 78.9 lbs.

11. Connect brake linkage, clutch linkage and install return springs.

12. Assemble parking brake as shown. Be sure lever engages properly after pedal adjustment. Spring must engage lever so it engages over center.

13. Install throttle pedal (C) and tighten special screw (B) securely.

ADJUSTMENT

1. Loosen jam nuts on each brake rod to unlock adjusting nut.

2. Depress each brake pedal individually to measure free travel of pedals before engagement.

3. Maximum free travel is 1⅜ in. (35 mm). Adjust so pedals are matched to 1 in. (25 mm) free travel.

4. Depress brake pedals and engage parking brake. Check operation of lever for proper engagement and disengagement.

Brake Housing Assembly

REMOVAL

1. Disconnect brake rod yoke from operating arm.

2. Remove cover-to-final drive housing capscrews.

3. Remove cover assembly with shoes.

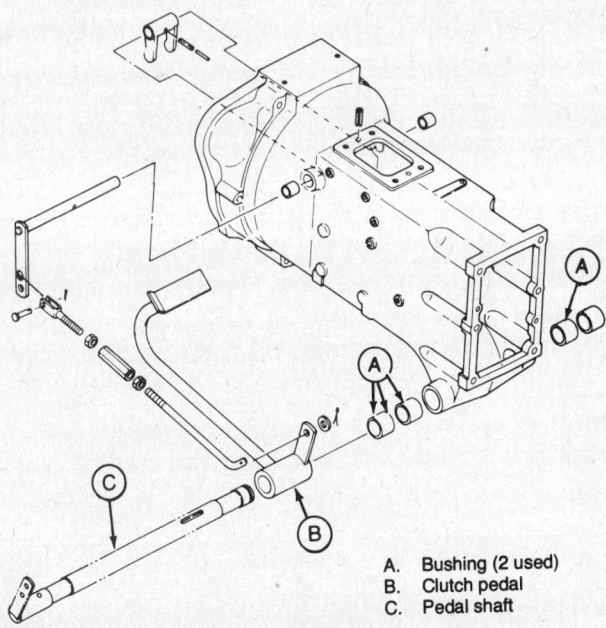

- A. Bushing (2 used)
- B. Clutch pedal
- C. Pedal shaft

Pedal shaft bushings

DISASSEMBLY

1. Remove return springs and remove shoes (E).

2. Remove operating arm with pin (D).

3. Remove nut from outside of cover and remove anchor pin (A).

4. If oil contamination is present in the brake housing or on the shoes, pinion shaft seals must be replaced.

ASSEMBLY AND INSTALLATION

1. If pinion shaft was removed, replace seals.

2. Install seal in brake housing-to-pinion shaft bore. Use a $3^1/_{16}$ in. driver disc and a 2⅛ in. pilot disc (OTC No. 27532 and 27517).

Inspect operating arm cam. 850 cam pin O.D. is 0.864-0.865 in. (21.95-21.98 mm), 950 cam pin O.D. is 1.100-1.102 in. (27.95-27.98 mm).

3. Check that retaining washer on end of cam and anchor pin is not loose.

4. Install O-ring (A) and coat with grease.

5. Inspect pin and cam pin bores in cover for wear or damage.

6. Install operating arm cam (B) so arm faces downward.

7. Inspect anchor pin (D). O.D. is 0.550-0.551 in. (13.97-14.0 mm) on 850, and 0.865-0.866 in. (21.98-22.00 mm) on 950. Width across flats at end of pin is 0.6654 to 0.6732 in. (16.9-17.1 mm). Thickness at end of pin is 0.5472 to 0.5551 in. (13.9-14.1 mm).

8. Install anchor pin (D).

9. Install washer (D) and lockwasher (E) and tighten nut (C).

10. Install vent tube (B) so open end faces down.

11. Check brake shoe return spring for damage or wear.

John Deere

- A. Anchor pin
- B. Shoe return springs
- C. Cover
- D. Operating arm with cam pin
- E. Brake shoes

Brake shoe disassembly

- A. O-ring
- B. Operating arm cam
- C. Cover
- D. Anchor pin

Left side cover assembly

Model	Working Load
850	2.05 in. at 56 lbs.
950	2.44 in. at 62 lbs.

12. Inspect brake linings for wear or damage.

Model	New Part Dimension	Replacement Limit
850	0.198 in. (4.80 mm)	0.118 in. (3 mm)
950	0.187 in. (4.75 mm)	0.118 in. (3 mm)

If linings are worn but still within specifications, adjust anchor pin.

13. Assemble brake shoes onto cover and securely attach shoe return springs.
14. Inspect I.D. of brake drum for wear or damage.

Model	New Part Dimension	Replacement Limit
850	5.512-5.516 in. (140-141 mm)	5.591 in. (142 mm)
950	6.693-6.697 in. (170-171 mm)	6.772 in. (172 mm)

NOTE: Put tractor in gear and engage differential lock.

15. Install brake drum on shaft. On 850 Tractor use tongued washer and torque capscrew to 36-44 ft.lbs. On 950 Tractors use spacer and washer. Torque retaining nut to 123 to 181 ft.lbs.
16. Coat cover gasket with Permatex® No. 2 and install cover on final drive housing.
17. Engage brake while torquing cover capscrews. Torque 850 to 17-22 ft.lbs. Torque 950 to 36-43 ft.lbs.
18. Connect brake rod to operating arm and adjust pedal free play.

ADJUSTMENTS

1. With brake pedals operating independently, measure free play in each pedal. Free play should be 1.0-1.375 in. (25-35 mm).
2. Loosen jam nut and adjust turnbuckle until correct amount of free play is obtained and retighten jam nut.

NOTE: Rotating the brake shoes will give added life and adjustment after the shoes are worn. Adjust as instructed above after shoes are rotated.

3. To compensate for brake drum wear, the anchor pin may be rotated to allow brake shoes to be positioned closer to the drum.
4. Loosen jam nut on anchor pin at the outside of the brake cover. Remove brake shoes and using a wrench, rotate the anchor pin 90°.
5. Seat the outside bore of the anchor pin using Permatex® #2 beneath the flat washer and tighten jam nut.
6. Install brake shoes and brake cover

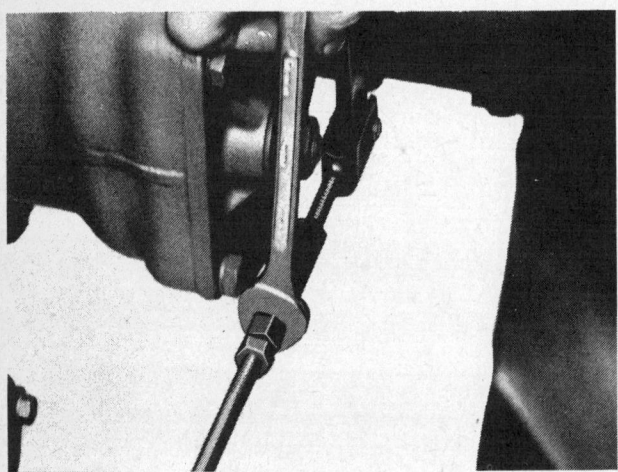

Adjusting brake pedal free play at turnbuckle

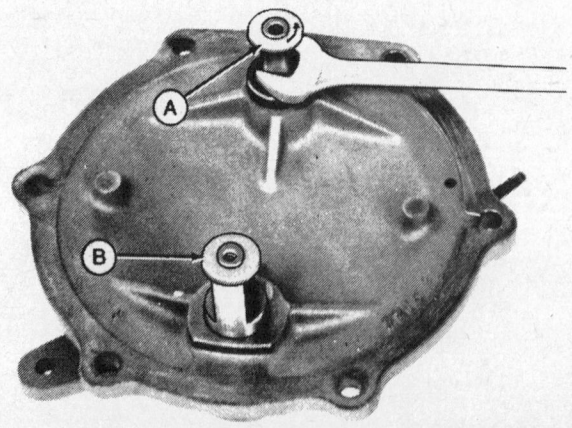

- A. Anchor pin
- B. Brake operating cam

Rotating anchor pin

John Deere

A. D-01204AA puller
B. D-01218AA attachment
Steering wheel removal

A. Steering column C. Steering arm
B. Locknut D. Cross shaft
Steering column and arm removal

on tractor. Adjust brakes as previously instructed. A fourth adjustment may be obtained by rotating the brake shoes again.

NOTE: Before rotating anchor pin be sure the brake shoes and brake drum meet specifications. This also applies before rotating the shoes after the anchor pin has been rotated.

STEERING

Steering Mechanism
REMOVAL

1. Remove steering wheel cap and locknut.
2. Disconnect battery ground cable.
3. Remove access panel at rear of control support and remove wiring harness from connectors.
4. Remove tachometer drive cable from tachometer.
5. Disconnect throttle linkage from righthand side.
6. Disconnect fuel lines from tank.
7. Remove two bottom capscrews from each side of the control console.
8. Install puller (A) with attachment (B) and remove steering wheel.
9. Remove Woodruff key from end of shaft.
10. Remove two capscrews at the front bottom of the support to clutch housing.
11. Lift control support off steering column.

DISASSEMBLY

1. Remove four side cover-to-housing capscrews (B) and remove jam nut from adjusting screw (D).
2. Turn adjusting screw to remove side cover (A).

NOTE: Housing (C) need not be removed from clutch housing unless there is damage to the housing casting or to its bushings.

3. If housing is removed, remove four retaining capscrews and lift from clutch housing. Housing is located on two dowel pins.
4. Loosen steering arm retaining nut from cross shaft (D) until it is flush with end of shaft.

A. Cross shaft C. Adjusting screw
B. Shim D. Shaft nut
Cross shaft and ball nut

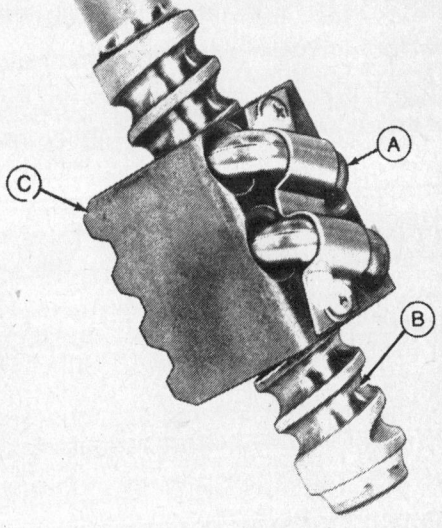

A. Guide with retainer
B. Steering shaft
C. Ball nut
Steering shaft with ball nut installed

251

John Deere

A. Driver discs
B. Seal

Installing cross shaft seal

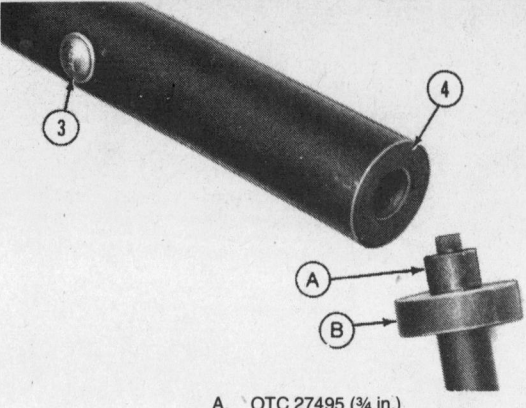

A. OTC 27495 (¾ in.)
B. OTC 27511 (1¼ in.)

Steering column seal installation

5. Remove four capscrews holding column to housing.
6. Remove adjusting screw (C) with shim (B).
7. Turn steering arm so cross shaft gear teeth (A) are aligned with wider opening in housing as shown.
8. Drive cross shaft on nut end (D) through cover side of steering housing. Remove nut and cross shaft.
9. Remove steering shaft with ball nut (D) (if column has been removed).

ASSEMBLY

1. Inspect housing bushings for wear or damage. If bushings are excessively worn or damaged, housing with bushings must be replaced. I.D. of bushings is 1.126 to 1.127 in. (28.60-28.62 mm).
2. If seal (B) is damaged or leaking, replace using OTC 27503 and 27518 driver discs (A) (1¼ in. and 2³/₁₆ in.). Coat seal with grease. Seal should be flush with end of housing.
3. Install seal in end of column using discs (A and B) flush with end of column. Coat lips of seal with clean grease.
4. Inspect bearing race in bottom of steering column and replace if worn or damaged. Coat mating surface with grease.
5. Replace packing on end of column and coat with grease.
6. If lower bearing race is damaged, install new race tight in bottom of housing.
7. Coat lower bearing and race with grease and place bearing in race.
8. Thoroughly clean steering shaft (B), and ball nut (C). Turn the nut on the shaft, feeling for roughness, binding or sticking. Examine worm path for damage.

NOTE: Do not remove ball nut from shaft. The two parts are replaceable as a unit only.

9. Generously coat shaft and nut mechanism with grease.
10. Coat upper bearing with grease and install on shaft as shown.

NOTE: Ball bearing (A) should have already been installed in housing race.

11. Carefully insert shaft with ball nut in steering column and through upper seal. Install column assembly in housing.
12. Install shaft key (A) and turn shaft while pushing downward to seat the bearing against lower race.
13. Install a dial indicator so it is zeroed on the end of the shaft.
14. Pry up on bottom of ball nut while reading shaft end play.
15. End play should be 0.001 to 0.004 in. (0.025-0.102 mm). To adjust add shims between column attaching plate and housing.
16. Examine cross shaft (A). O.D. at bushing is 1.125 to 1.126 in. (28.57-28.59 mm).
17. Turn steering shaft so ball nut (D) is centered in its travel.
18. Coat cross shaft with grease and install shaft (A) so center tooth engages the center of ball nut (D).
19. Install adjusting screw (C) with shim (B).
20. Fill housing with John Deere Multi-Purpose Lubricant. Capacity is 0.06 gal. (0.03 L).
21. If steering housing was removed, coat threads of mounting screws (B) with thread sealant and torque to 72 to 88 ft.lbs.
22. Install steering arm by lining up indexing marks on cross shaft and steering arm.
23. Install washer and nut (A). Torque nut to 109-145 ft.lbs.

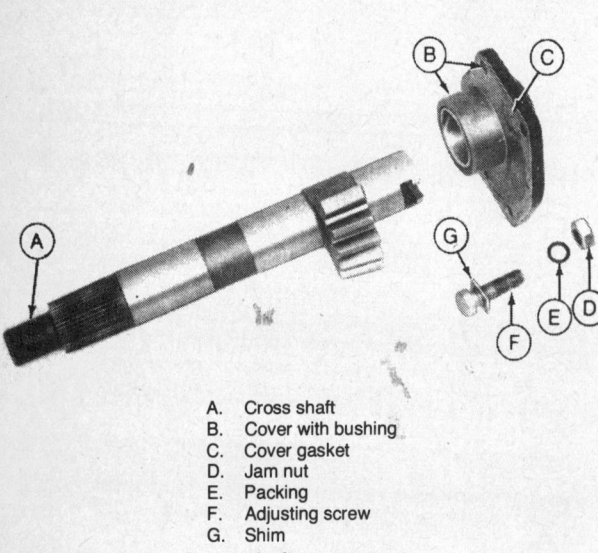

A. Cross shaft
B. Cover with bushing
C. Cover gasket
D. Jam nut
E. Packing
F. Adjusting screw
G. Shim

Cross shaft components

A. Cross shaft
B. Shim
C. Adjusting screw
D. Ball nut

Installing cross shaft

John Deere

24. Thread side cover to adjusting screw. Be sure gasket is in place. Use gasket adhesive.
25. Torque side cover capscrews to 17.22 ft.lbs. Be sure to keep adjusting screw loose when tightening.
26. Install jam nut over adjusting screw.
27. Check steering arm for smooth and free movement.
28. Install control support over steering column.
29. Install the two support-to-clutch housing capscrews at the bottom front of the support. Tighten screws securely.
30. Install steering wheel and tighten shaft nut.
31. Tighten the two capscrews at bottom of each side of control support.
32. Connect fuel lines and throttle linkage.
33. Connect tachometer drive cable.
34. Connect wiring harness.
35. Install access panel at rear of console support.
36. Connect battery ground cable.

ADJUSTMENT

Adjustment of steering play is made by adjusting backlash of the gears. Turn the adjusting screw in the adjusting cover to obtain 1-2 in. (25-50 mm) of free play measured at outer rim of steering wheel.

NOTE: By turning the screw to the right, the amount of play can be reduced. The standard position for the screw is a quarter-turn back when wheel becomes hard to turn.

CLUTCH

Separation

1. Remove seat.
2. Remove brake linkage and spring.
3. Remove steps.
4. Disconnect fender and Roll-Gard connectors.
5. Drain transmission oil.
6. Disconnect hydraulic lines.

NOTE: On tractors equipped with power steering, remove steering pressure and return lines to separate tractor.

7. Support both ends of tractor.
8. Block under side frame.
9. Remove connecting bolts.

NOTE: This will prevent clutch shaft from moving rearward and letting clutch thrust washer fall out of place.

10. Roll rear of tractor back.

NOTE: When separating transmission from clutch housing make sure clutch is engaged. If clutch is disengaged and PTO clutch shaft comes out, a clutch disk will fall, which would require a clutch housing to engine separation.

Installation

Reverse the removal steps and note the following installation instructions.

NOTE: If there is trouble removing old gaskets, spray on John Deere Paint and Decal Remover. Let it set for a few minutes; then scrape off gaskets.

Tighten clutch housing-to-transmission case capscrews to 87 ft.lbs.

Clutch Assembly

REMOVAL

1. Separate the clutch housing from the engine as instructed above.

NOTE: Before clutch removal, release plate height could be checked.

2. Remove six clutch cover-to-flywheel capscrews.

IMPORTANT: Do not remove cotter pins and three slotted nuts (D). Disturbing the slotted nuts affects initial factory release lever plate adjustment.

If the clutch cover assembly is damaged or worn, the entire assembly must be replaced.

3. Remove clutch noting two installation pins and using care not to drop drive disk which will be loose on flywheel. Use clutch shaft as a handle.
4. Remove two bearing release springs from bearing sleeve. Remove springs from pins in clutch housing.
5. Remove clutch shaft (A) and release bearing assembly (B).

NOTE: If clutch shaft-to-transmission drive shaft splined coupler remain on transmission driveshaft, remove it for inspection. Coupler may have fallen to bottom of clutch housing.

6. To remove clutch release yoke and linkage; drive out spring pins (D), disconnect rod (on outside of clutch housing)

A. Clutch shaft
B. Release bearing
C. Yoke and release shaft
D. Spring pins
E. Release spring

Clutch shaft and release mechanism

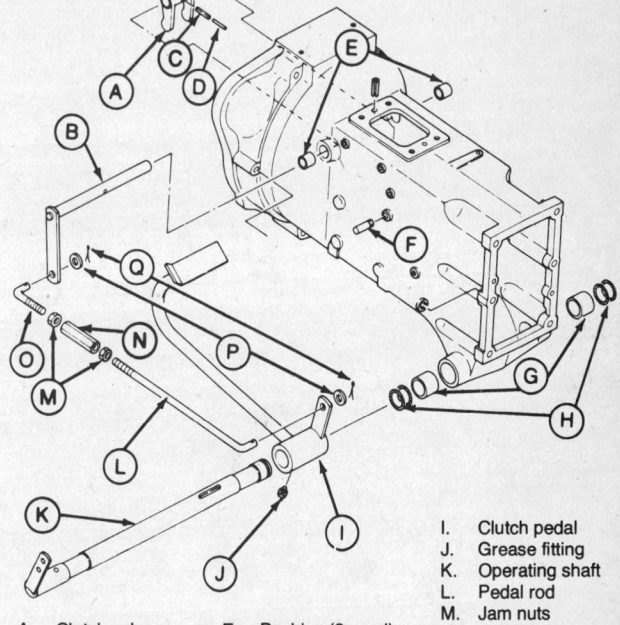

A. Clutch yoke
B. Yoke arm shaft
C. Spring pin
D. Spring pin
E. Bushing (2 used)
F. Pin (2 used)
G. Bushing (2 used)
H. O-ring (4 used)
I. Clutch pedal
J. Grease fitting
K. Operating shaft
L. Pedal rod
M. Jam nuts
N. Adjusting nut
O. Clutch rod
P. Washer
Q. Cotter pin

Clutch housing and pedal linkage, 2-wheel drive

253

John Deere

A. Clutch pedal
B. Grease fittings
C. Rod links
D. Yoke shaft arm

Clutch pedal linkage

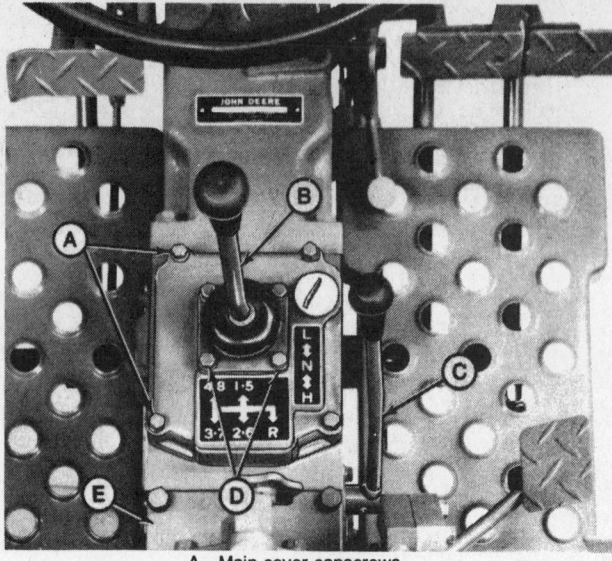

A. Main cover capscrews
B. Change shift lever
C. Range shift lever
D. Lever retainer capscrews
E. Differential lock cover

Transmission cover and shift levers

from yoke shaft arm, and remove arm shaft and yoke (C).

NOTE: On tractors without front wheel drive, spring pins (D) consist of two pins, one inside the other. Note position of splits in pins for reassembly. They should face 180° from one another in a vertical plane. Tractors with front wheel drive have one spring pin with a retaining wire through pin and over top of clutch yoke.

INSTALLATION

1. Install clutch disc in the flywheel with the long end of the hub (C) rearward.
2. Install clutch pressure plate assembly over pilot end of clutch shaft and install shaft in pilot bushing in flywheel.
3. Align clutch pressure plate cover aligning pins on flywheel and install clutch cover-to-flywheel capscrews (A).
4. Torque capscrews (A) to 20 ft.lbs.
5. Remove clutch shaft and install in clutch housing as described in the following sequence.
6. Install clutch release fork arm shaft bushings using a ¾ in. pilot disc and 1 in. driver disc (OTC No. 27495 and 27499).
7. Immerse clutch release bearing sleeve in transmission fluid.
8. Install clutch yoke arm shaft through bushings in clutch housing and through release shift yoke (C). Note that flat side of yoke faces to rear.
9. Rotate yoke shaft outer arm down so hole in shaft aligns with hole through yoke.
10. On tractors without front wheel drive, install the two spring pins (D) through hole in yoke and arm shaft (C). Note that one spring fits inside the other and be sure splits in pins are positioned 180° (opposite) one another. On tractors with front wheel drive, install spring pin into yoke and armshaft. Then install retaining wire through spring pin and twist ends together at top of yoke.
11. Install clutch driveshaft (A) through clutch housing into splined coupler at front of transmission drive shaft.
12. Wipe all excess oil from release bearing sleeve and install over clutch shaft (A) against yoke.
13. Install two release springs in release sleeve and over pins inside clutch housing. Open ends of springs should face down.
14. Join the clutch housing to the engine as instructed above.

NOTE: Place transmission shift lever in Neutral to ease clutch disk alignment with the clutch drive shaft.

15. Install clutch and brake pedals and operating shaft (K).
16. Install and connect pedal linkage as shown. Note that the longer rod has righthand thread while the shorter rod has lefthand thread.
17. Grease fittings (B).
18. Install brake pedal linkage.
19. Adjust pedal.

PEDAL FREE PLAY ADJUSTMENT

1. Loosen rod jam nuts. Note that front nut has lefthand thread while rear nut has righthand thread.

NOTE: Torque for lefthand footrest capscrews is 38 ft.lbs. The footrest is the clutch pedal stop and loosening of the screws affects free travel.

2. With a straight scale held in the line of clutch pedal travel, measure free play of clutch pedal at the pedal footrest. Free travel should be $9/16$ to 1 inch (15-25 mm).
3. If free play is excessive, turn adjusting nut counterclockwise (when viewed from front of tractor). If there is too little free play, turn adjusting nut clockwise (as viewed from front of tractor).
4. While holding adjusting nut, lock jam nuts.

MANUAL TRANSMISSION

Disassembly

SHIFT LEVER

1. If only change shift lever (B) is to be removed, remove four retainer-to-case cover cap screws (D). Remove change shift lever (B) with retainer.
2. If shift detents, interlock pins, shift forks, or shifter shafts are to be removed or inspected; the lever retainer need not be removed.
   a. Place change shift lever in neutral.
   b. Remove dip stick from case main cover.
   c. Remove four main cover-to-transmission case capscrews.
   d. Remove change shift lever (B) with main cover.

NOTE: As cover is removed, do not drop change detent springs which are located between case cover and transmission case at the rear of the case main cover.

RANGE SHIFT LEVER

1. Use a spring pin punch to remove spring pin from range shift lever (C) and from the lever arm.
2. Remove range shift lever.

John Deere

CHANGE AND REVERSE SHIFTER SHAFTS AND FORKS

1. Remove transmission main cover as described earlier.
2. Remove four differential lock cover-to-transmission case capscrews and remove lock cover (E).
3. Separate the tractor between the transmission and the clutch housing.
4. Remove detent springs (A) from transmission case.
5. Use a small magnetic pickup tool to remove detent balls (B) from detent bores.
6. Remove two nuts which secure seal cap retainer plate to case and remove retainer plate (G).
7. Remove reverse shift stop plate (D).

NOTE: **Remove shifters from left to right. Removing the reverse shifter (F) last is the easiest sequence of removal.**

8. Bend a piece of light wire to fit through the bottom of spring pin (J) and use a a spring pin punch to drive the "3-4" fork spring pin down through the shaft. Catch the spring pin using the wire.
9. Drive shaft (K) from rear of transmission case wall using a brass drift. Use care not to damage shaft bore in case wall.

NOTE: **It may be necessary to use a pliers to remove shaft through fork since the fork fits tightly on the shaft. If this is necessary, cut a one inch (25 mm) length of rubber hose lengthways to fit over shaft and protecting the shaft with the hose, remove the shaft.**

10. As the shaft is moved forward remove seal cap at front of case which will be displaced by the shaft.
11. Remove the shifter fork.
12. Repeat steps 8 through 11 for the remaining two shifters. Note that the reverse shifter assembly has two spring pins (J) which must be removed.
13. To remove the two shaft interlock pins, remove pipe plug (C) using a metric Allen wrench and use a small magnetic pickup tool to remove interlock pins through plug bore.

REVERSE COUNTERSHAFT AND GEAR

NOTE: **Remove reverse countershaft and gear only if it is necessary to replace shaft O-ring, shaft, gear or bushings. Excessive shaft removal can damage the shaft.**

1. Remove transmission drain plug from the oil filter housing at the bottom of the transmission case. Drain approximately one gallon (4 L) of transmission fluid from case and reinstall plug.
2. Remove transmission case side cover from the lefthand side of the transmission case.
3. Separate the tractor between the transmission case and the clutch housing.
4. Place change shift lever in speed "4"

A. Detent springs
B. Detent ball (3 used)
C. Pipe plug
D. Stop plate
E. Seal cap (3 used)
F. Reverse shifter fork
G. Retainer plate
H. "1-2" ("5-6") fork
I. "3-4" ("7-8") fork
J. Spring pins
K. "3-4" shaft

Shifter shafts and forks

and drive countershaft retaining spring pin through hole in side of transmission case and shaft (C). Catch spring pin using a piece of light wire bent to fit through inner end of spring pin.
5. Using a small punch as a lever, pry through hole in shaft pushing shaft forward enough to remove O-ring (A).
6. Slide reverse gear (B) forward against case.
7. Use a pliers with brass or rubber around the shaft to push shaft forward through front of transmission case. Avoid damaging the finish on the gear area of the shaft.
8. Remove reverse gear from case.

A. O-ring
B. Reverse idler gear
C. Spring pin holes
D. Reverse countershaft

Reverse idler gear and countershaft

John Deere

- A. Axle housing
- B. Rockshaft housing
- C. Change shift lever
- D. Differential driveshaft cover
- E. Transmission driveshaft
- F. Disassembly stand

Case and axle housings

NOTE: Although the accompanying figure shows reverse gear removal with the reverse shifter fork removed, the gear can be removed with the shifter forks in place.

TRANSMISSION DRIVESHAFT REMOVAL

NOTE: If transmission case is to be completely disassembled, or if one or more final drive housings are to be removed, the use of a disassembly stand is desirable.

1. Drain transmission fluid at oil filter housing drain plug.
2. If disassembly stand (F) is used, install as instructed below.
3. Separate the tractor between the transmission case and clutch housing.
4. Remove the rockshaft housing (B).
5. If there is a possibility that the differential driveshaft is to be removed, the differential driveshaft retaining nut should be loosened at this time.
   a. Remove the differential driveshaft front cover (D). Avoid losing shims located behind bearing retainer.
   b. Place the change shift lever (C) in "R".
   c. Loosen the differential driveshaft retaining nut.
6. Remove transmission case main cover with shift lever.
7. Remove differential lock cover.
8. Remove shifter seal cap retainer plate nuts and remove change and reverse shafts and forks.
9. Remove four front oil seal housing nuts.
10. Install JDT-40 Transmission Puller Body (B) with JDE-114-1 Forcing Screw (A) over transmission driveshaft.

NOTE: Small diameter of forcing screw fits in tapped hole in front end of transmission driveshaft. Do not bottom forcing screw in shaft.

11. Install JDT-39 Gear Spacer between front (1-2) gear set (C) and center (3-4) gear set (D). Install spacer with its open side to the rear.
12. Remove snapring from groove at rear of rear bearing on transmission drive shaft. Snapring is accessible through opening for differential lock shaft and spring.
13. Install 1⅝ in. open-end wrench between center (3-4) gear set (D) and reduction gear set (E). Be sure wrench (or spacer) clears snapring in front of reduction gear thrust washer.
14. Tighten nut on forcing screw (A) pulling transmission driveshaft through rear bearing. After shaft has been pulled about ¼ inch (6 mm) (or until the gear bearing snapring groove is no longer visible), remove wrench spacer tool and remove snapring from groove at front of reduction gear set thrust washer. Be sure to move the snapring to the rear of the groove.
15. Reinstall wrench spacer tool between the two rear gear sets.
16. Continue tightening forcing screw rear nut disassembling the transmission drive shaft parts as they are displaced.
   a. First remove splined coupler between rear of transmission driveshaft and front of PTO driveshaft.
   b. Second, remove snapring and washer from rear of rear bearing on transmission drive shaft.
   c. Next, remove thrust washer which is located between low reduction gear and rear bearing.
   d. Then, thrust washer and snapring be-

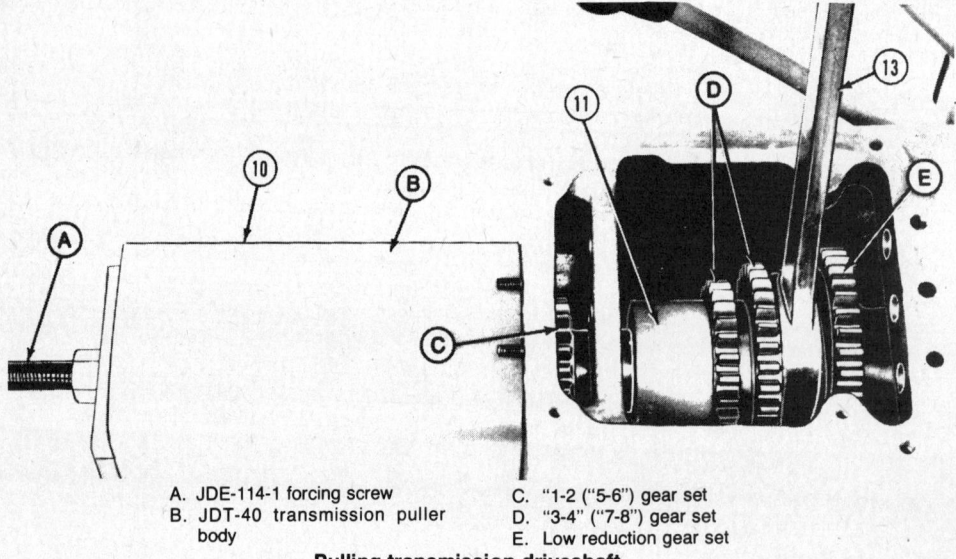

- A. JDE-114-1 forcing screw
- B. JDT-40 transmission puller body
- C. "1-2" ("5-6") gear set
- D. "3-4" ("7-8") gear set
- E. Low reduction gear set

Pulling transmission driveshaft

John Deere

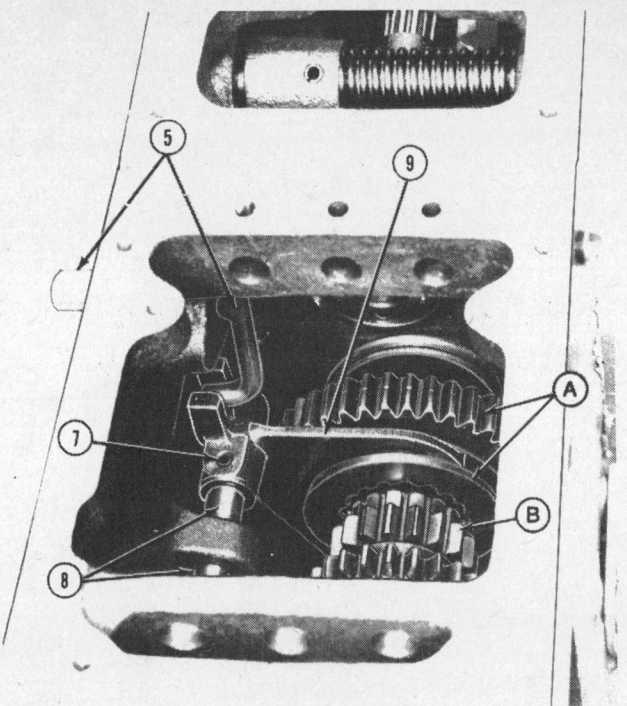

A. High-low shifter gear
B. Cluster gear
Removing range shifter assembly

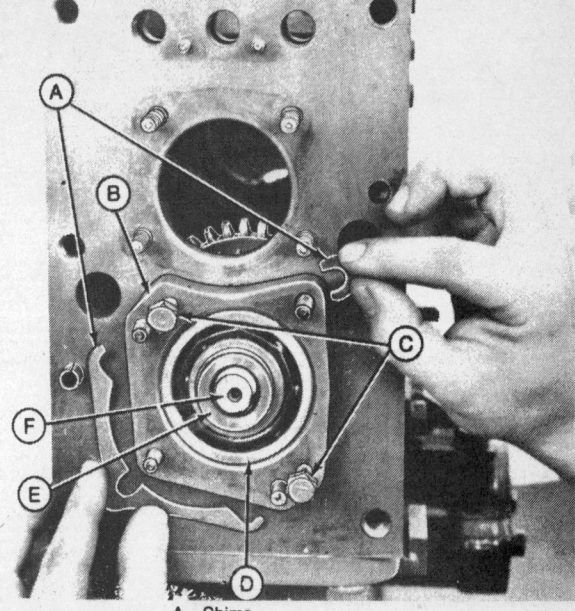

A. Shims
B. Front bearing retainer
C. Jack screws
D. Front bearing
E. Sleeve
F. Differential driveshaft
Removing differential driveshaft shims

tween reduction gear and 3-4 gear set (D) can be removed.

e. After shaft is removed, 3-4 gear set (D), 1-2 gear set, and the low reduction gear set (E) can be removed from the case.

17. Remove transmission driveshaft rear bearing using a $2^{5}/_{16}$ in. driver disc and a $^{11}/_{16}$ in. pilot disc (OTC No. 27520 and 27498). Use handle No. JDC-400-7.

RANGE SHIFTER ASSEMBLY REMOVAL

NOTE: Remove range shifter fork and shaft only if they must be removed for inspection or replacement, or if high-low gear (A) needs to be replaced. High-low shift detent is very difficult to install and final drive must be removed to install the detent assembly. Differential driveshaft parts (other than the high-low gear) can all be removed and reassembled with the high-low gear in place.

1. Remove spring pin from range shifter lever and arm and remove shifter lever.
2. Remove transmission driveshaft as instructed previously.
3. Remove two capscrews and keeper plate at lever arm shaft at outside of transmission case.
4. Drain transmission fluid at PTO drain plug and remove righthand final drive.
5. Remove range shifter arm shaft from transmission case and from shifter fork.
6. Slide the high-low gear (A) forward to engage the differential driveshaft cluster gear assembly (B).
7. Remove spring pin from shifter fork arm shaft using a spring pin punch. Catch spring pin using a piece of wire bent through the bottom of the spring pin.
8. Drive shifter shaft through fork and front wall of case using a brass drift. A pliers can also be used if the shaft is protected from scratches by using rubber hose or soft metal jaws on the pliers.
9. Remove shifter fork as shaft is driven through it. Drive shaft through the front of the transmission case displacing the seal cap at the front of the case.

NOTE: As shifter shaft is moved out of its detent bore, detent spring and ball is free to fall out the side of the transmission case or rearward into the differential case. Use a magnetic pickup tool in the detent bore to avoid losing detent ball.

DIFFERENTIAL DRIVESHAFT

1. Remove transmission drain plug from oil filter housing on bottom of transmission case and drain transmission.
2. Remove drain plug at rear bottom of transmission case and drain PTO and final drive case.
3. Separate the tractor between the clutch housing and the transmission case.
4. Remove the transmission driveshaft.

NOTE: Be sure to loosen differential drive shaft front retaining nut prior to removing transmission driveshaft.

5. Remove both final drive nuts.
6. Remove differential and differential lock.
7. Install jack screws (C) in front bearing retainer (B) tapped holes.
8. Tighten jack screws pulling retainer enough to remove shims (A).

NOTE: Do not lose shims. If differential drive shaft or differential ring gear are not replaced, the same shims must be reinstalled behind front bearing retainer (B).

9. Use a brass drift to drive differential driveshaft (F) rearward from transmission case.

a. If reverse gear was not removed for inspection or repair, be sure to align it with the differential driveshaft front cluster gear to prevent it binding.
b. If range shift assembly and high-low gear do not require inspection, do not remove. Place the range shift lever in "Low" to prevent damaging the shifter fork while driving the shaft.
c. Continue driving differential driveshaft rearward. As differential driveshaft parts are displaced, remove them.

10. Remove differential driveshaft through lefthand side of differential compartment.

Inspection, Repair and Subassembly

DIFFERENTIAL DRIVESHAFT

NOTE: If the differential driveshaft is replaced, the differential ring gear must also be replaced since they are supplied as a matched set.

1. Inspect all gears for uneven wear, chips, scoring and cracks. Replace any gear that is in questionable physical condition.
2. Inspect needle bearing (J) by removing outer retaining ring (L). Replace gear

John Deere

and bearing if rollers are chipped or worn excessively. Be careful not to damage the bore of cluster gear.

3. Check for excessive clearance between gears (H and O) and the cluster assembly and its key (I). If key is worn excessively, remove and drive new key into cluster (M).

4. Inspect splines on I.D. of high-low sliding gear (K) for wear. Check for wear on collar and internal gear.

5. Inspect bearing sleeve (C). I.D. is 0.9051 to 0.9059 in. (22.99-23.01 mm). O.D. is 1.1808 to 1.1814 in. (29.99-30.01 mm).

6. Install retaining ring (E) in smaller end of cluster (D). Be sure ring is seated in its groove.

7. Install needle bearing (F) and install outer retaining ring (E).

8. Install flat end of washer (G) inside larger diameter end of cluster (D).

9. Install bearing (H) against thrust washer and seat snapring (I) in its groove to retain the bearing.

10. Drive key (C) into groove on O.D. of cluster (D).

11. Align keyway in first speed gear (B) with key and drive gear onto cluster assembly using a plastic hammer. Be sure flat side of gear faces outward.

12. Install retaining ring (J) in groove in second speed gear (A) and install second speed gear on cluster assembly against first speed gear (B).

13. If differential driveshaft rear bearing must be replaced, press off using care not to damage shaft. Press with retaining nut (A) protecting shaft threads.

14. Inspect differential driveshaft diameters as shown. Diameter at rear bearing inner race is 1.3780 to 1.3786 in. (35.00-35.02 mm).

15. When installing differential driveshaft rear roller bearing, be sure flared end of roller cage (A) is toward the bevel pinion. Also be sure to press only on the inner bearing race (B).

16. Coat all mating parts with John Deere Hy-Gard Transmission and Hydraulic Oil or its equivalent to ease assembly and to minimize wear.

TRANSMISSION DRIVESHAFT

1. Install new oil seal (B) (if replaced) in seal housing (A) using 1¾ in. driver disc and 1⅛ in. pilot disc (OTC No. 27511 and 27501).

   a. Coat sealing lip with grease and be sure not to lose seal spring.

   b. Bottom seal in seal housing. Sealing lip should face transmission case.

2. Inspect transmission driveshaft (J) and its splines for excessive wear.

   a. Diameter at front ball bearing is 1.1812 to 1.1817 in. (30.002-30.015 mm).

   b. Diameter at reduction gear needle bearings is 1.1806 to 1.1811 in. (29.987-30.000 mm).

   c. Diameter at rear ball bearing is 0.9844 to 0.9849 in. (25.002-25.015 mm).

3. Replace any bearings that are excessively loose, chipped, or noisy.

4. Examine gears for excessive wear, cracks and chips. Replace any damaged snaprings.

5. Inspect reduction gear needle bearings. I.D. of bearings is 1.1814 to 1.1821 in. (30.009-30.025 mm). Replace when clearance between bearings and shaft is 0.004 in. (0.1 mm) or greater.

6. Install snaprings and needle bearings in reduction gear set.

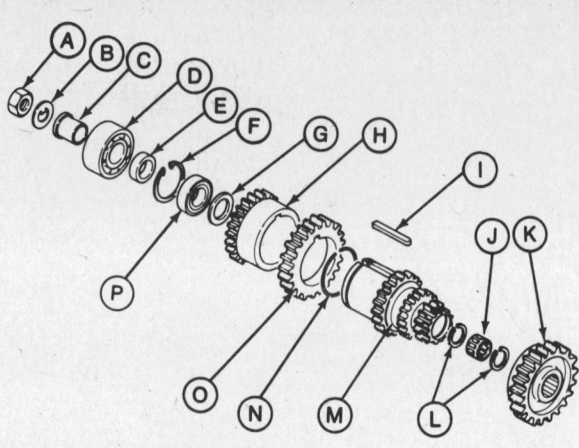

A. Retaining nut
B. Tanged washer
C. Sleeve
D. Ball bearing
E. Spacer
F. Snapring
G. Thrust washer
H. "2-6" speed gear
I. Key
J. Needle bearing
K. High-low sliding gear
L. Retaining rings
M. "3-4" ("7-8") cluster
N. Retaining ring
O. "1-5" speed gear
P. Ball bearing

Differential driveshaft components

A. Seal housing   B. Oil seal

Installing front oil seal

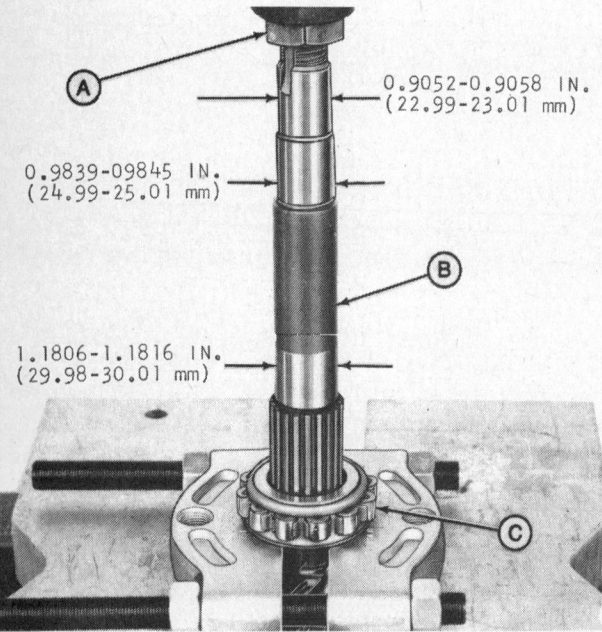

A. Retaining nut   B. Differential driveshaft   C. Rear roller bearing

Removing differential driveshaft rear bearing

A. Bearing cage
B. Inner race

Differential rear bearing installation

John Deere

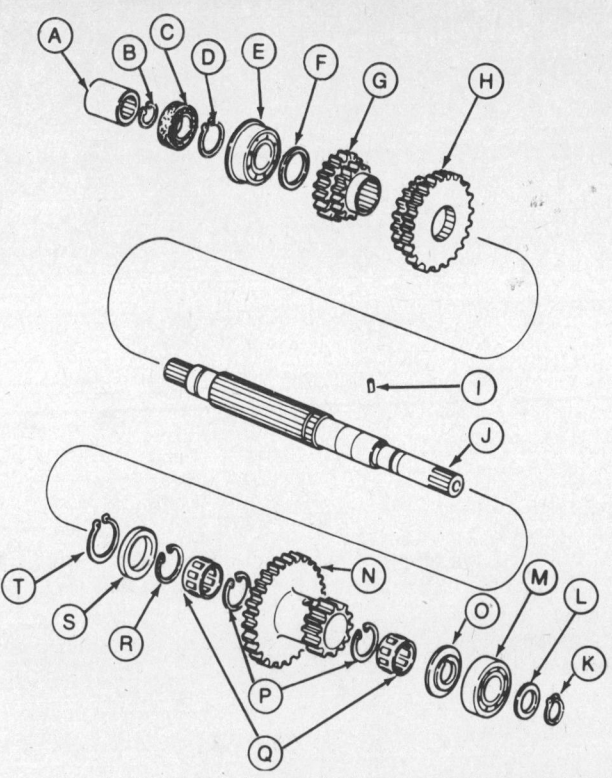

A. Splined coupler
B. Coupler snapring
C. Oil seal
D. Snapring
E. Ball bearing
F. Spacer washer
G. "1-2" gear
H. "3-4" gear
I. Pin
J. Transmission driveshaft
K. Snapring
L. Washer
M. Ball bearing
N. Reduction gear set
O. Thrust washer
P. Snaprings
Q. Needle bearings
R. Snapring
S. Thrust washer
T. Snapring

Transmission driveshaft components

Assembly and Installation

1. With rear roller bearing (F) installed on shaft as previously instructed, install outer race (E). Be sure shoulder on race is forward.
2. Install thrust washer (D). The flat side of the washer should face forward.

NOTE: Thrust washers with one flat side and the other side having a recess such as this should always be installed with the flat side away from the bearing. This prevents drag on the outer bearing race.

3. Install snapring (C) in shaft groove retaining washer and race against pinion.
4. Install cluster gear assembly (A) in transmission case with the larger end of the assembly facing forward.
5. Install high-low sliding gear (B) (if removed) into transmission case so its collar end is forward.
6. Install differential driveshaft through differential compartment and through high-low sliding gear.
7. Slide shaft through I.D. of high-low sliding gear (D) and through cluster gear assembly bearings.
8. Drive differential driveshaft forward through front bearing in cluster gear assembly. Drive on rear end of pinion shaft using a soft-head hammer.

NOTE: When installing shaft, use care not to damage shaft threads or cluster gear assembly needle bearing and snaprings.

9. Check to be sure high-low gear (D) slides freely on the shaft splines (A).
10. Check to be sure that cluster gear bearings roll freely and quietly on the shaft.
11. Coat O-ring (K) with oil and install on O.D. of bearing retainer (L).
12. Install bearing retainer (A) with O-ring over studs and into front of transmission case.
13. Install spacer (C) over end of drive shaft.
14. Press sleeve (E) into I.D. of bearing inner race. Be sure to support bearing at inner race when installing.
15. Install bearing (D) with sleeve part way into retainer (A) and over driveshaft end (B).

NOTE: Coat all mating parts with John Deere Transmission and Hydraulic Oil or its equivalent to ease assembly and to minimize wear.

16. Install a pry behind differential

A. Front bearing retainer
B. Differential driveshaft
C. Spacer
D. Ball bearing
E. Sleeve

Installing front bearing

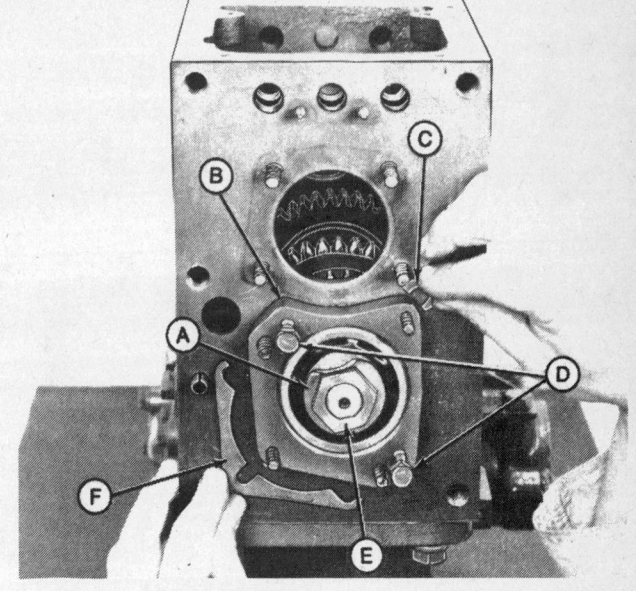

A. Tanged washer
B. Bearing retainer
C. 1-corner shim
D. Jack screws
E. Retaining nut
F. 3-corner shim

Installing front shims

259

John Deere

drive shaft bevel pinion (or outer race) to keep shaft from being driven rearward.

17. While prying forward on drive shaft, drive bearing sleeve onto shaft using JDT-38 Bearing Installer. Bearing will bottom against retainer shoulder and against spacer (C).

18. Install tanged lockwasher (A) over end of differential driveshaft.

NOTE: Replace the tanged lockwasher (A) whenever nut (E) is removed.

19. Install retaining nut (E). Torque nut to 58 to 72 ft.lbs.

NOTE: If a suitable spanner-type wrench is not available to hold the differential driveshaft from turning, install transmission driveshaft, range shifter and reverse countershaft, at this time.

The differential driveshaft can be locked by engaging the high-low gear and the 3-4 sliding gear while the reverse gear is engaged.

Damage to the differential driveshaft gears could result if the shaft is not locked up adequately while tightening retaining nut. Instructions for installing the transmission driveshaft and reverse gear assembly are given on the following pages.

20. Bend the lockwasher (A) over the retaining nut as shown.

21. Install jack screws (D) in retainer (B) tapped holes to install shims (C and F).

NOTE: If differential driveshaft and differential ring gear were replaced, differential cone point and backlash must be checked and adjusted accordingly. Therefore the same shims should NOT be reinstalled. If the number of shims that were on the case prior to disassembly is in question, also readjust cone point. Whenever installing the original driveshaft, the same number of shims that were removed must be installed.

22. Install cone point adjusting shims (C and F) by tightening jack screws (D) enough to insert the shims. Be sure the same number of one-corner shims (C) are installed as three-corner shims (F).

NOTE: Shims (C and F) come in two thicknesses—0.008 in. and 0.012 in. (0.2 mm and 0.3 mm). Be sure the same thickness of each type shim is used.

23. While holding the shims in place behind the bearing retainer (B) (using grease or clay), remove jack screws (D).

24. Install gasket on front bearing cover and install cover over bearing retainer (B). Be sure to coat gasket with sealant.

25. Torque front bearing cover-to-transmission case stud nuts to 20 ft.lbs. Note that stud nuts have integral lockwashers.

26. Install differential and differential lock.

27. Install range shifter assembly (if removed) and install final drive units.

28. Install transmission driveshaft as instructed.

29. Install shifters and covers as instructed.

30. Join the tractor.

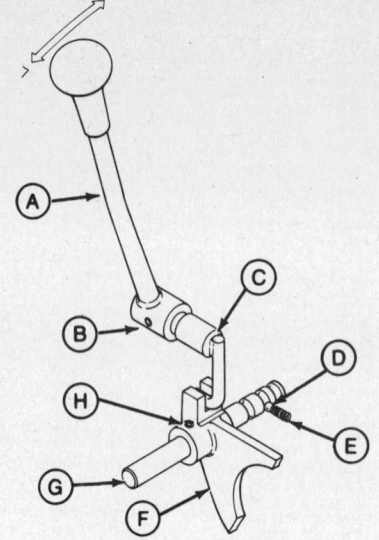

A. Range shift lever
B. Spring pin
C. Shifter armshaft
D. Detent ball
E. Detent spring
F. Shifter fork
G. Shifter shaft
H. Spring pin

Range shifter components

RANGE SHIFTER ASSEMBLY

1. Inspect shift fork (F). End of fork thickness is 0.264 to 0.272 in. (6.7-6.9 mm). Width of high-low gear collar groove is 0.280 to 0.287 in. (7.1-7.3 mm). Replace when parts are worn to 0.040 in. (1 mm) clearance or more.

2. Replace parts when clearance between shaft (G) and fork (F) is 0.008 in. (0.2 mm). O.D. of shaft is 0.5892 to 0.5899 in. (14.96-14.98 mm). I.D. of fork is 0.5906 to 0.5922 in. (15.00-15.04 mm).

3. Inspect detent spring (E). Note that all transmission shift detent springs are common.

Spring Free Length	Working Load
1.33 in. (33.8 mm)	1.01 in. at 16.1-17.8 lbs. (26.4 mm at 72-79 N)

4. Clearance between shift armshaft (A) and transmission case bore should not be 0.02 in. (0.5 mm) or greater.
  a. Diameter of bore in case is 0.7882 to 0.7895 in. (20.02-20.05 mm).
  b. O.D. of shifter armshaft (A) is 0.7854 to 0.7874 in. (19.95-20.00 mm).

5. Start spring pin (E) through top of shifter fork (B).

NOTE: Spring pin is one inch long (25 mm). Be sure all spring pins are installed with split facing the direction of force (forward).

6. Install fork in high-low gear collar groove (C).

7. Coat shaft (D) with oil and install through front of transmission case and shift fork. Be sure spring pin is aligned with hole in shaft.

8. Insert the shifter detent spring through the bore in the transmission case (E).

9. Apply grease to the 5/16 in. (8 mm) steel detent ball and place the ball on the end of a length of 1/4 in. (6 mm) copper tubing.

NOTE: Shifter shaft bore is drilled through the transmission case rear wall. Tubing (or similar tool) helps ball stay in place while compressing detent spring.

A. Keeper plate
B. Spring pin
C. Shifter armshaft
D. Shift lever
E. Detent bore

Installing range shifter

260

John Deere

10. While compressing detent spring with ball and tubing, insert shaft (D) through rear wall in transmission case until the end of the fork shaft is past the detent ball. Remove tubing.

11. Install spring pin (E) through shifter shaft (D) and through bottom part of fork. Spring pin should not protrude through either side of the fork.

12. Install the seal cap on the front of the transmission case (in the shaft removal bore).

13. Install the shifter armshaft (A) in the fork and through the side of the transmission case.

14. Install O-ring over armshaft (C) into transmission case. Be sure to coat O-ring with oil.

15. Install keeper plate (A) against case and torque capscrews with lockwashers to 20 ft.lbs.

16. Install range shift lever (D) over armshaft (C). Be sure lever is away from case on armshaft and that spring pin holes align.

17. Install spring pin (B) flush with either side of lever. Split in spring pin should face direction of lever travel.

18. Install transmission drive shaft as instructed on the following pages.

19. Install righthand final drive.

20. Join the tractor.

TRANSMISSION DRIVESHAFT

1. Remove PTO pinion shaft and PTO driveshaft.

NOTE: To assure proper assembly of rear bearing thrust washer (K) over pin (F), the rear bearing (J) must be installed from the rear of the transmission driveshaft.

2. Install spacer washer (L) over front of transmission driveshaft (G) against splined shoulder.

3. Install front bearing (M) against spacer washer (L) using a press. Be sure retaining ring on ball bearing is toward the front end of the shaft.

4. Seat snapring (N) in groove at front of front ball bearing (M) and install pin (F). (Hold in place with grease.)

NOTE: Coat all parts with John Deere Hy-Gard Transmission and Hydraulic Oil prior to assembly. Also replace any damaged snaprings.

5. Place low reduction gear set (E) in case with larger gear forward. Be sure needle bearings stay in place inside gear set.

6. Start transmission driveshaft through front of transmission case.

7. Install "1-2" sliding gear set over shaft. Be sure larger gear (second speed) is forward in the case.

8. Install "3-4" sliding gear set on shaft. Be sure smaller (third speed) gear is forward in the case.

9. Continue sliding shaft rearward through the gears.

10. Install snapring and thrust washer over end of shaft. Flat side of thrust washer must face the front. Note the tanged I.D. of the washer which locates it on the shaft splines.

11. Insert shaft through reduction gear needle bearings and through rear transmission case wall.

12. Seat snapring in groove on shaft splines.

13. Install gasket on the front bearing housing and install bearing housing (A) on front of transmission case. Coat gasket with sealant prior to assembly.

14. Torque stud nuts (with lockwashers) (B) to 20 ft.lbs.

15. Install thrust washer with pin slot over pin in driveshaft.

NOTE: The thrust washer should be installed with its chamferred side forward. Grease the chamferred side so it adheres to the rear of the reduction gear set over the locating pin.

16. Insert a feeler gauge between the top of the transmission case and in front of the rear wall (G). Flex gauge to fit the rear of the thrust washer.

NOTE: This holds the thrust washer over its locating pin while driving on the rear bearing. As bearing tightens against the thrust washer, pull the feeler gauge out so it just contacts the recessed diameter of the thrust washer.

17. Install rear ball bearing (E) over rear end of shaft.

18. Install washer (D) behind bearing.

19. Using a 1 inch pipe (approximately

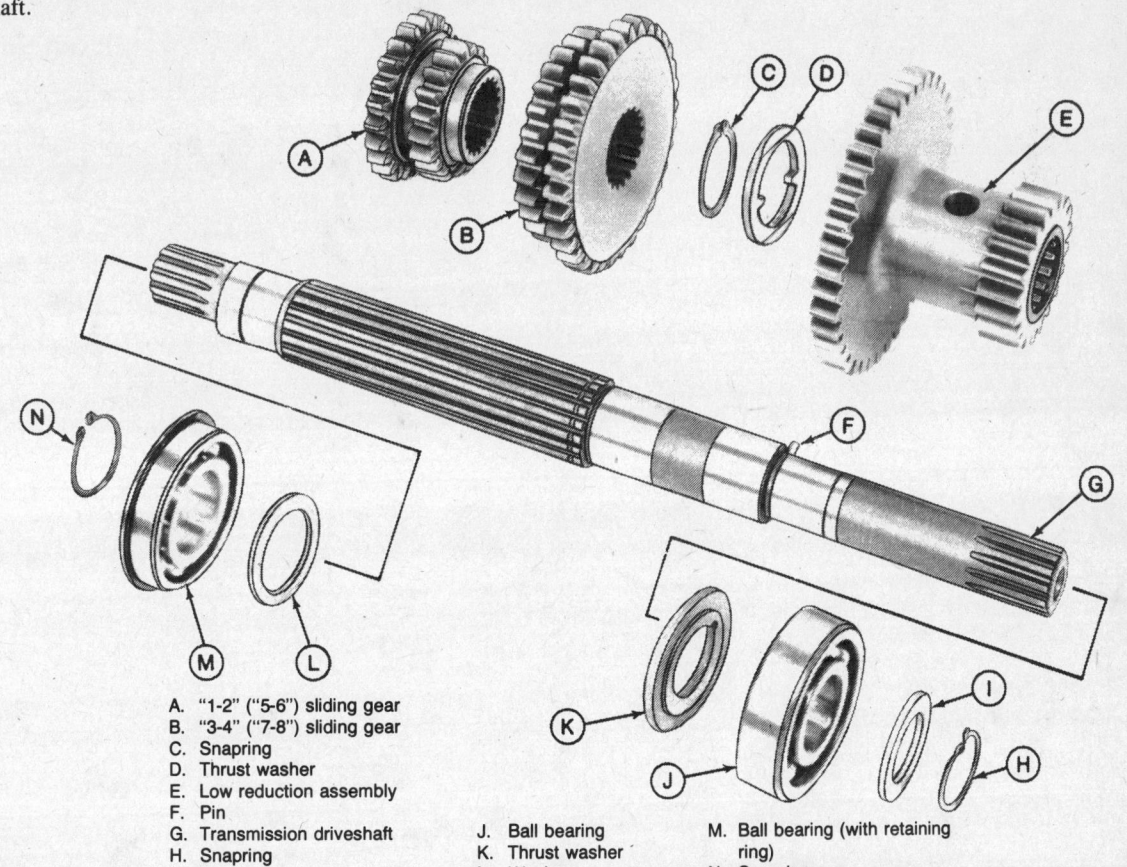

A. "1-2" ("5-6") sliding gear
B. "3-4" ("7-8") sliding gear
C. Snapring
D. Thrust washer
E. Low reduction assembly
F. Pin
G. Transmission driveshaft
H. Snapring
I. Washer
J. Ball bearing
K. Thrust washer
L. Washer
M. Ball bearing (with retaining ring)
N. Snapring

Transmission driveshaft components

John Deere

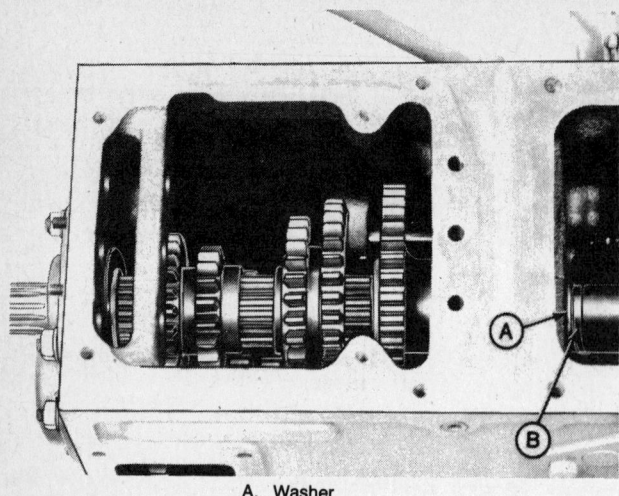

A. Washer
B. Rear snapring

Installing rear bearing rear snapring

A. O-ring
B. Reverse gear
C. Spring pin hole
D. Reverse countershaft

Installing reverse countershaft

20 in. long), drive against washer (D) and rear bearing inner race until feeler gauge is snug between thrust washer and bearing outer race.

NOTE: Be sure to hold feeler gauge in place while driving bearing. Thrust washer should still be in place over pin.

20. Remove feeler gauge and continue driving rear bearing forward until snapring groove is visible behind washer (D).
21. Remove driver pipe (F).
22. Install rear bearing rear snapring (B) over rear end of transmission driveshaft.
23. Seat snapring (B) in groove behind washer (A).
24. Check sliding gears for freedom of operation.
25. Install speed change shifters as instructed on the following pages.
26. Join tractor.
27. Install PTO driveshaft and pinion shaft.
28. Install the rockshaft as instructed.

REVERSE COUNTERSHAFT AND GEAR

1. Inspect gear for chips and wear and replace as necessary.
2. Inspect the two bushings inside the gear and check shaft for wear.
   a. I.D. of bushings is 0.7882 to 0.7910 in. (20.02-20.09 mm).
   b. O.D. of reverse countershaft is 0.7869 to 0.7874 in. (19.99-20.00 mm).
   c. Clearance between shaft and gear bushings should not be 0.008 in. (0.2 mm) or greater.
3. If bushings in reverse gear are replaced, use a ¾ in. pilot disc and a 1 in. driver disc (OTC No. 27495 and 27499). Install bushings flush with ends of gear.
4. Install O-ring (A) on end of shaft. Be sure to coat O-ring with oil before installation.

5. Insert shaft (D) through front of case and reverse gear (B) aligning spring pin holes (C). Be sure collar end of gear is rearward.
6. Install 1.4 in. (36 mm) long spring pin through hole in shaft and transmission case (C). Spring pin split should face the front of the transmission case.
7. Install change shifters (if removed) as instructed on the following pages.
8. Install transmission side cover and gasket. Gasket should be coated with sealant.
9. Torque side cover-to-transmission case capscrews and nut with lockwashers to 20 ft.lbs.
10. Fill transmission case with oil and join tractor.

CHANGE AND REVERSE SHIFTER ASSEMBLY

1. Inspect seal caps (A). Replace any cap that may leak.
2. Replace any distorted spring pins.
3. Inspect detent balls and springs for damage and wear.

Free Length	Working Load
1.33 in.	1.01 in. at 16.1-17.8 lbs.
(33.8 mm)	(26.4 mm at 72-79 N)

4. Inspect shifter forks. Thickness at end of fork is 0.264 to 0.272 in. (6.7-6.9 mm). Width of collar groove on sliding gears is 0.280 to 0.287 in. (7.1-7.3 mm). Replace parts when worn to 0.040 in. (1 mm) or more clearance.
5. Clearance between shafts and forks should be less than 0.008 in. (0.2 mm). Shaft O.D. is 0.5892 to 0.5899 in. (14.96-14.98 mm). Fork I.D. is 0.5906 to 0.5922 in. (15.00-15.04 mm).

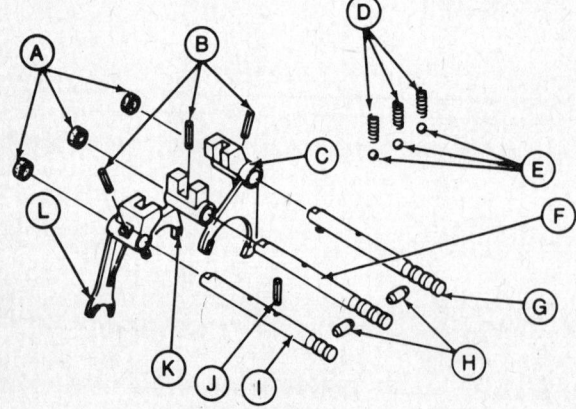

A. Seal caps
B. Spring pins (1 in. [25 mm] long)
C. "3-4" shifter fork
D. Detent springs
E. Detent balls
F. "1-2" shifter shaft
G. "3-4" shifter shaft
H. Interlock pins
I. Reverse shifter shaft
J. Reverse stop pin (1 in. [25 mm] long)
K. "1-2" shifter fork
L. Reverse shifter fork

Change and reverse shifter lever

John Deere

6. Coat shafts with John Deere Hy-Gard Transmission and Hydraulic Oil or its equivalent prior to installation.

7. Remove pipe plug (C).

8. Place righthand shifter fork (L) on "3-4" sliding gear collar.

NOTE: Center and righthand shifter shafts are interchangeable. Reverse shaft has one less detent groove and an extra spring pin hole.

9. Install righthand shifter shaft (K) through front of transmission case and its fork (I). Be sure spring pin holes align.

10. Install spring pin (J) through righthand shifter shaft and fork.

NOTE: Install all spring pins so split faces the direction of force toward the front of the tractor). Also, spring pins (J) should be installed using a spring pin punch approximately 0.002 in. (0.5 mm) below flush in the fork.

11. Install the shift interlock pin through pipe plug (C) bore against shifter shaft detent groove.

12. Install center "1-2" shifter fork on front "1-2" sliding gear.

13. Install center shaft through fork and into rear transmission case wall. Be sure spring pin holes align.

14. Install spring pin through fork and shaft.

15. Install second shift interlock pin through pipe plug bore against shaft detent groove.

16. Install lefthand (reverse) shifter fork (F) on reverse shifter gear collar.

17. Install reverse shaft through front of case and through its fork (F). Be sure spring pin holes align.

18. Install the reverse fork spring pin.

19. Install the reverse stop spring pin in rear hole in shaft. Spring pin should be installed to protrude approximately ⅜ in. (10 mm) from shaft.

20. Install fork stopper plate (D) as shown. Be sure it is positioned properly.

21. Torque fork stopper plate-to-transmission case capscrews with lockwashers to 20 ft.lbs.

22. Check fork shifter operation and install pipe plug (C).

23. Install the three seal caps at the front of the transmission case (E).

24. Install retainer plate (G) and torque stud nuts with lockwashers to 8 ft.lbs.

25. Place all three forks in neutral.

26. Install the three detent balls (B) and springs (A).

27. Install the transmission case cover and lever assembly as instructed on the following pages.

28. Join the tractor.

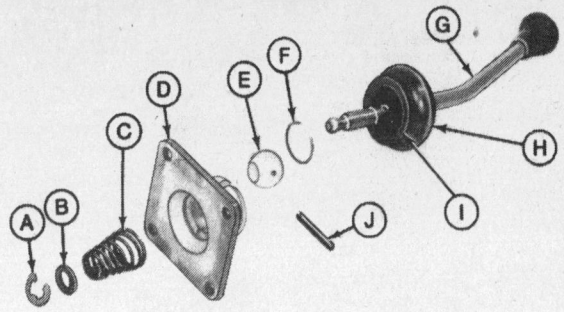

A. Retaining ring
B. Washer
C. Spring
D. Retainer
E. Lever ball
F. Retaining ring
G. Lever
H. Boot
I. O-ring
J. Spring pin

Shift lever components

SPEED CHANGE SHIFT LEVER

1. Replace any damaged retaining rings and spring pin.

2. Examine finished I.D. of retainer (D). Replace if scored or excessively worn.

3. Check plastic ball (E) for cracks, chips or wear and replace if necessary.

4. Lever arm (G) should be replaced if finished area at lower end (which contacts shifter forks) is damaged or worn.

5. Replace boot (H) and O-ring (I) if they show any sign of stress or cracking, or if they do not fit snugly over retainer and lever.

6. Inspect spring (C) for damage or wear.

Spring Free Length	Working Load
1.3 in. (33 mm)	1 in. at 10 lbs. (25 mm)

7. Install O-ring (I) over lip of rubber boot (H).

8. Install boot and O-ring over lever (G).

9. Insert plastic ball (E) over lever and install spring pin through ball hole and lever hole.

a. Install spring pin (J) so its split side faces up on the lever.

b. The spring pin is 1 9/16 in. (40 mm)

A. Detent springs
B. Detent ball (3 used)
C. Pipe plug
D. Stop plate
E. Seal cap (3 used)
F. Reverse shifter fork
G. Retainer plate
H. "1-2" ("5-6") fork
I. "3-4" ("7-8") fork
J. Spring pins
K. "3-4" shaft

Change shifter assembly

263

John Deere

A. Main cover capscrews
B. Change shift lever
C. Range shift lever
D. Retainer capscrews
E. Differential lock cover

Installing cover and lever retainer

A. Hinge bracket
B. Lock cover
C. Keeper plate
D. Hinge pin
E. Lock pedal

Lock pedal removal

long and should protrude equally from each side of the ball.

10. Install the lever ball into the lever retainer (D) and install retaining ring in the groove.

11. Install spring (C) with its larger end facing up on the lever.

12. Retain the spring with washer (B) and retaining ring (A). Be sure retaining ring is seated properly in the lever groove.

13. Install gasket on bottom of lever retainer (D). Be sure to coat the gasket with sealant.

NOTE: Be sure detent springs are in place in top of transmission case (if removed with cover).

14. Torque capscrews (A and D) to 20 ft.lbs.

15. Secure boot over lever groove and check the lever shifts for ease of operation.

MOUNTING TRANSMISSION STAND

If work is going to be done on the final drives, drain transmission and mount the transmission stand before splitting.

1. Remove drawbar and drawbar holding brackets.
2. Drain transmission oil.
3. Position mounting stand under transmission and bolt it to the transmission case.

Once the stand is mounted, it can be used in splitting and to hold the back of the tractor for final drive removal.

REAR AXLE

Axle Housing

REMOVAL

1. Remove hitch lift links and center link.
2. Remove spring between draft links.
3. Remove draft links and sway chain.
4. Remove drawbar (if stand is to be used).
5. Drain transmission.
6. Remove seat.
7. Disconnect fender and Roll-Gard connectors (if equipped).
8. Remove fender.
9. Remove brake linkage and spring.
10. Remove steps.
11. Remove Roll-Gard (if equipped).
a. Remove capscrews from bottom plate to Roll-Gard.
b. Remove capscrews from top plate to axle housing.
12. Remove fender mounting brakets.

NOTE: On tractors equipped with power steering, remove steering pressure and return lines to allow removal of left axle.

13. Support rear of tractor.
14. Remove tire.
15. Remove rubber plugs from jack-screw holes.
16. Remove all but two axle housing-to-transmission case capscrews.
17. Thread one long capscrew into the transmission at the front of axle housing to keep it from falling. Remove two retaining capscrews.
18. Thread two long capscrews into the holes that the rubber plugs were removed from. Turn clockwise to separate axle housing from transmission case.

19. Support axle housing using chain hoist.
20. Remove axle housing.

INSTALLATION

Reverse the removal steps on the preceding pages. Also follow the special installation instructions which follow.

NOTE: If there is trouble removing old gaskets, spray on John Deere Paint and Decal Remover. Let it set for a few minutes; then scrape off gaskets.

1. When installing axle housing, make sure dowel pins in housing line up with holes in transmission case.
2. Tighten axle housing-to-transmission case capscrews to 36 ft.lbs.

FINAL DRIVE

Final Drive Housing and Axle

REMOVAL

NOTE: Vent hose (B) is used on the early model tractors only. On later tractors, the elbow fitting points downward to prevent collection of moisture in the brake drum.

1. Drain transmission oil at the PTO (rear) drain plug in the bottom of the transmission case.
2. Remove rear wheel and fender.
3. Remove final drive housing-to-transmission case capscrews (C).

NOTE: As the capscrews (C) are removed, install at least three of the longer

John Deere

capscrews where the shorter capscrews were removed.

4. Remove the two closure plugs (A). These plugs are located opposite one-another on the housing flange.

NOTE: See the section on Mounting Transmission Stand which can be used to support the transmission case during removal of one or both of the final drive housings.

5. Install one of the longer capscrews in each of the two closure plug holes (B).

NOTE: Jack screws are used to separate the axle flange from the transmission case. Dowel pins and the gasket sealant cause the housing to "hang up" on the case.

6. Tighten the jack screws (B) until the final drive housing is free. Do not over-tighten jack screws beyond threaded diameter.

7. Remove brake cover (C) and brake shoe as instructed.

8. Wrap a chain around housing near flange and while balacing at outer end of axle remove the housing using an overhead hoist.

NOTE: Housing will balance for installation and easy removal if the chain is wrapped around housing at flanged area toward inner side of housing and around axle at outer end just inside wheel flange.

DISASSEMBLY

850
AXLE SHAFT

NOTE: If final drive pinion shaft (A) is to be removed, loosen its screw by inserting a drift to lock the final drive gear.

1. Remove snapring (D) and remove large final drive gear (B).
2. If outer bearing or seal are not to be removed, remove capscrews retaining axle outer cover (E) to axle housing.
3. Remove spacer shims (C).
4. Drive axle shaft (D) from the housing. If the outer bearing cover (E) was left on, the outer bearing and seal will press off during axle removal.
5. To remove the inner ball bearing (E), use a long drift or punch to drive from the outer end of the housing. Drive only on the outer bearing race.

A. Seal cap
B. Rockshaft housing
C. Housing retainer
D. Retainer plates
E. Final drive housing
F. Jack screw
G. Roller bearing
H. Bearing retainer

Removing final drive

FINAL DRIVE PINION

1. Remove the two Phillips head screws (F) which retain the bearing (B).
2. Remove the brake drum and remove the final drive gear (not the axle) as instructed earlier.
3. Reinstall the brake drum-to-pinion shaft retaining capscrew and drive the shaft and bearing inward through the housing.
4. Remove all gasket material from final drive housing, outer bearing cover and transmission case.
5. To remove drive pinion-to-brake oil seal, remove seal retaining ring and drive seal from housing.
6. Use a press to remove ball bearing

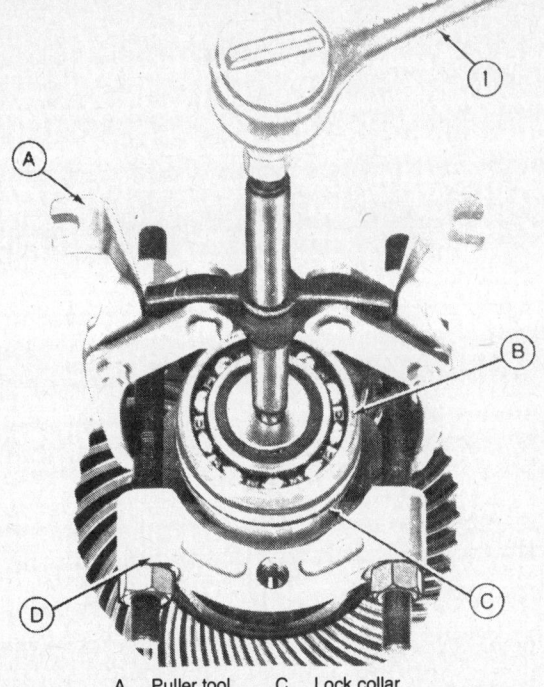

A. Puller tool
B. Ball bearing
C. Lock collar
D. Knife-edge puller

Removing right bearing and lock collar

A. Lock plate
B. Differential housing
C. Lefthand ball bearing
D. Capscrews
E. Lock collar
F. Righthand ball bearing
G. Bevel lock gear
H. Ring gear

Differential housing and ring gear

John Deere

- A. Differential housing cover
- B. Feeler gauge
- C. Spiral bevel pinion
- D. Gauge surface
- E. JDT-41 cone point adjustment mandrel

Measuring cone point distance

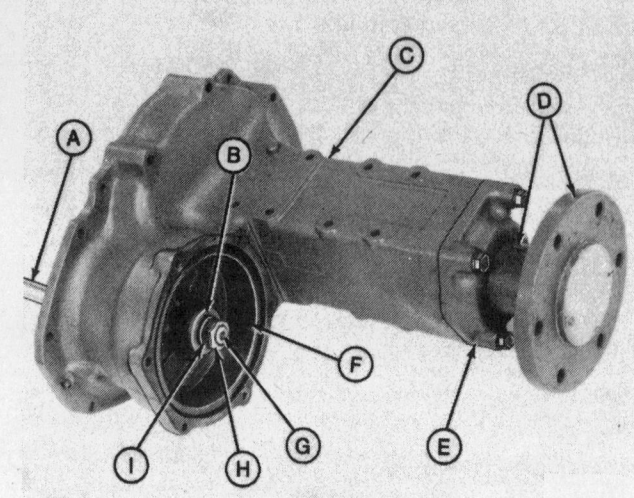

- A. Bar
- B. Washer
- C. Housing
- D. Axle shaft
- E. Bearing cover
- F. Brake drum
- G. Pinion shaft
- H. Nut
- I. Lockwasher

Removing final drive pinion nut

from differential output shaft. Be sure to press only on the inner bearing race.

950
AXLE SHAFT

NOTE: If final drive pinion shaft (G) is to be removed, loosen its nut by inserting a bar (A) to lock the final drive gear. Also, if seal and bearing in outer bearing cover (E) is not to be removed, remove cover-to-axle housing capscrews at this time.

1. Remove the three capscrews (B) and remove washer (C) and final drive gear (F).
2. Drive axle shaft from housing. If the outer bearing cover (E) was left on housing, the outer bearing and seal will press off during axle removal.
3. To remove the inner ball beaing, use a long drift or punch to drive from the outer end of the housing. Drive only on the outer bearing race.

FINAL DRIVE PINION

1. Remove the final drive gear (F) as instructed earlier.
2. Remove the brake drum.
3. Remove the five capscrews (G) which retain bearing retainer (H).
4. Reinstall the pinon shaft nut and

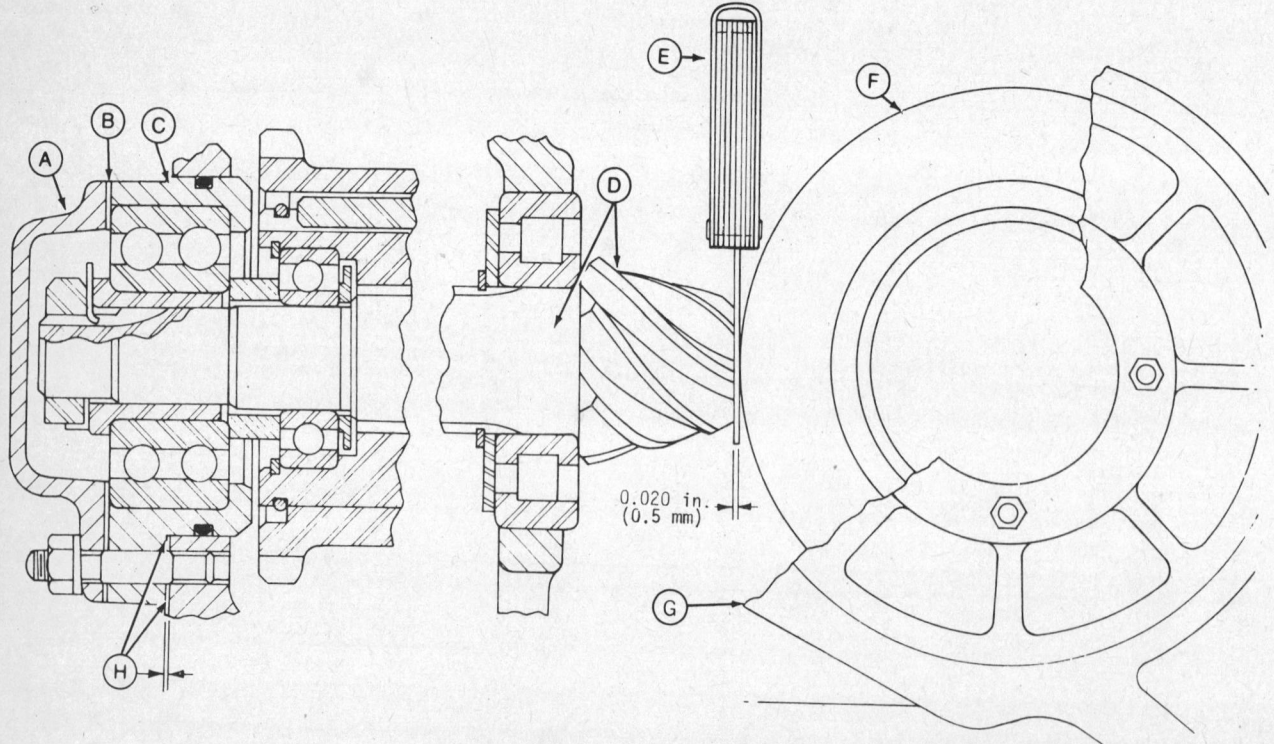

- A. Front bearing cover
- B. Gasket
- C. Front bearing retainer
- D. Differential driveshaft
- E. Feeler gauge
- F. Adjusting mandrel
- G. Transmission case
- H. Adjusting shims required

Adjusting differential driveshaft cone point

John Deere

- A. Final drive pinion
- B. Capscrews
- C. Spacer washer
- D. Lock plate
- E. Outer bearing cover
- F. Final drive gear
- G. Capscrews
- H. Retainer

Removing final drive gear

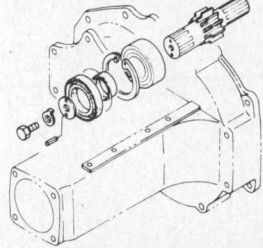

- A. Final drive housing
- B. Seal collar
- C. Oil seal
- D. Lockwasher
- E. Capscrew
- F. Spring pin
- G. Washer
- H. Retaining ring
- I. Sealed bearing
- J. Pinion shaft

Axle pinion components

drive the pinion shaft and bearing inward through the housing.

5. Remove the oil seal spacer ring.

6. If drive pinion-to-brake drum oil seal is faulty, remove it.

7. Use a press to remove ball bearing from differential final pinion shaft. Press only on the inner bearing race.

8. Remove all gasket material from final drive housing, outer bearing cover and transmission case.

ASSEMBLY

850

NOTE: Whenever the final drive housing is removed, check and adjust final drive gear end play. This is especially important after replacing axle shaft parts.

FINAL DRIVE PINION

1. Inspect teeth on pinion shaft (J). Replace shaft if teeth are chipped or worn excessively.

2. Diameter of pinion shaft at inner bearing is 1.1787 to 1.1795 in. (29.94-29.96 mm).

3. Diameter of pinion shaft at outer bearing is 1.3780 to 1.3787 in. (35.00-35.02 mm).

4. Inspect pinion shaft splines for wear or damage and replace as necessary.

5. Replace oil seal (C) if it has been leaking or if it is worn or damaged.

6. If I.D. or O.D. of seal collar (B) is damaged or worn, replace it.

7. Inspect sealed bearing (I).

a. If the bearing seal is damaged, replace it.

b. Check the fit of the bearing on the shaft and in the housing. It should not be loose.

c. Bearing should not be noisy when rotated.

8. Always replace lockwasher (D) if it is removed.

9. Replace retaining ring (H) if it is damaged or bent.

- A. Outer bearing cover
- B. Gasket
- C. Seal driver
- D. Oil seal

Installing outer cover seal

- A. O-ring
- B. O-ring spacer
- C. Axle shaft
- D. Inner oil seal
- E. Seal collar
- F. Cover gasket
- G. Outer bearing cover

Installing outer bearing seal collar

John Deere

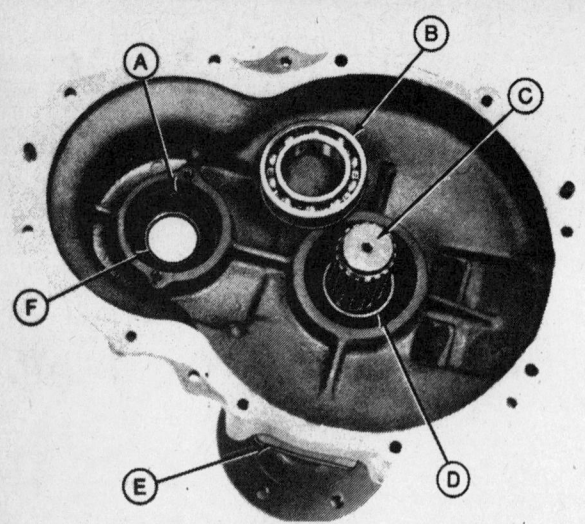

A. Retaining ring
B. Inner axle bearing
C. Axle shaft
D. Spacer sleeve
E. Outer bearing cover
F. Oil seal

Installing axle shaft

A. Outer bearing cover
B. Gasket
C. Seal driver
D. Inner oil seal

Installing outer bearing cover oil seals

AXLE SHAFT

1. Inspect outer bearing and seal for damage or wear and replace as required.
2. Replace axle seal collar and O-ring if damaged.
3. Inspect axle shaft. O.D. at both inner and outer bearing is 1.9685 to 1.9692 in. (50.00-50.02 mm).
4. Replace axle if splines are worn excessively.
5. If outer oil seal (D) is replaced, install new seal using a 3⅛ in. driver disc and 2¼ in. pilot disc (OTC No. 27533 and 27519).
   a. Coat seal with grease before installing.
   b. Install seal so it bottoms in the bearing cover.
   c. Be sure spring in seal is in place and the lip faces inward.
6. Install new gasket (B) on cover. Be sure to coat the cover with sealant.
7. Install outer cover (C) with seal over axle shaft (D).
8. Press the seal collar (E) against the axle shoulder. Be sure the larger I.D. of the collar faces up.
   a. Drive collar on using spacer sleeve (A) as a driver.
   b. Remove driver sleeve (A).

c. Install O-ring and spacer ring as shown.
9. Heat the outer ball bearing to 300° F (150° C).

----------- **CAUTION** -----------
Do not exceed 300° F (150° C). Plan a safe handling procedure to avoid burns.

10. Drop bearing (B) with the retaining ring (F) toward the axle flange onto the axle shaft.
11. Bottom bearing (B) against seal collar (E) using the spacer sleeve (A) as a driver.
12. After bearing has cooled, repeat step 11.
13. Coat final drive side of cover gasket with sealant.
14. Install axle assembly (C) into housing.
15. Torque outer bearing cover-to-final drive housing capscrews (E) to 20 ft.lbs. Do not forget lockwashers and be sure outer bearing starts squarely into the housing.
16. Install spacer sleeve (D) over axle shaft (C).
17. Install bearing (B) over axle shaft (C). Be sure bearing starts squarely into

housing bore. Drive onto axle shaft until inner race of bearing bottoms on spacer sleeve (D).

NOTE: Use the large final drive gear to press the bearing onto the shaft. After the gear has bottomed, check to be sure the bearing has bottomed on the sleeve by driving on bearing inner race and outer race alternately.

18. Install oil seal (F) in axle housing pinion bore.
   a. Coat seal with grease before installation.
   b. Drive seal using 3 in. driver disc and 2⅛ in. pilot disc (OTC No. 27531 and 27517).
   c. Drive seal until snapring groove can be seen, then install snapring (A).
   d. Drive seal from the other side of the housing so the seal bottoms against the snapring (A).

DRIVE PINION AND FINAL DRIVE GEAR

1. Press ball bearing (B) (if removed) on pinion shaft against pinion shoulder. Press on inner bearing race only.
2. Coat I.D. and O.D. of seal collar (C) with grease and install over shaft tight against the inner bearing race.
3. Install pinion shaft assembly in bore in housing. Bearing should bottom against snapring.
4. Torque bearing retaining screws to 10.5 ft.lbs. Be sure to coat threads with Loctite®.
5. Install final drive gear on axle shaft so the pinion teeth mesh with the gear teeth.
6. Install snapring in the groove in the axle shaft.
7. Measure gap between final drive gear and snapring using a feeler gauge. This thickness is the total thickness of shims to use behind the final drive gear.

NOTE: If a variety of shims is available, simply insert the maximum amount of shim pack that will fit be-

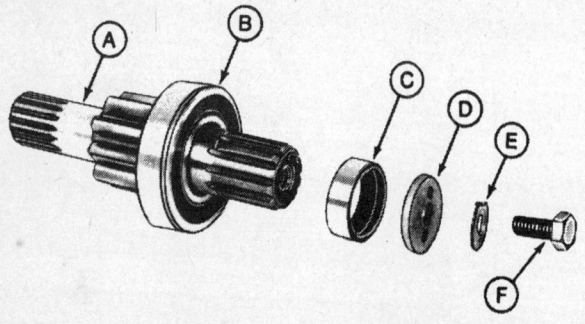

A. Pinion shaft
B. Sealed bearing
C. Seal collar
D. Spacer washer
E. Lockwasher
F. Capscrew

Final drive pinion components

tween the snapring and the gear. Shims are available in sizes of 0.004 in. (0.1 mm), 0.012 in. (0.3 mm) and 0.020 in. (0.5 mm).

8. Remove snapring and remove final drive gear.
9. Install shims (as determined in Step 7) over axle inner bearing.
10. Install the final drive gear and install the snapring.
11. Install brake drum and shoe. Be sure to replace tanged lockwasher. Torque drum-to-pinion shaft capscrew to 40 ft.lbs. To tighten, insert a drift in the hole in the final drive gear.
12. Reverse the steps of Removal to install the final drive housing. Use the guide pins in the housing to help align the housing during installation.
13. When installing axle, be sure to install gasket between housing and transmission case. Gasket should be coated with sealant.

950
FINAL DRIVE PINION
1. Inspect pinion teeth and splines. Replace if they are chipped or excessively worn.
2. Diameter of pinion shaft at inner bearing is 1.2976 to 1.2982 in. (32.96-32.98 mm).
3. Diameter at outer bearing is 1.5748 to 1.5756 in. (40.00-40.02 mm).
4. Replace oil seal and O-ring if damaged or leaking.
5. If bearing is noisy, loose or damaged, replace.

AXLE
1. Inspect outer bearing and seal for damage or wear and replace as required.
2. Inspect axle shaft splines. O.D. at both inner and outer bearing is 2.3623 to 2.3630 in. (60.00-60.02 mm).
3. If axle outer cover is removed, replace seal collar inner O-ring.
4. If the two oil seals in the outer bearing cover (A) are replaced, install using $3^{15}/_{16}$ in. driver disk and $2^{15}/_{16}$ in. pilot disc (OTC No. 27546 and 27530).
  a. Install outer seal first with its sealing lip facing out. Note that the outer seal is noticeably thinner than the inner seal. Bottom the seal in the cover.

NOTE: Be sure to coat the I.D. and O.D. of both seals with grease.

  b. Install the inner oil seal (D) with the sealing lip facing inward. Do not allow seal spring to come out of place.
  c. Bottom the back of the inner oil seal against the back of the outer seal. The inner seal should be just to the bottom of the cover chamfer.
5. Install the gasket (B) on the bearing cover. Coat only the cover side of the gasket with sealant at this time.
6. Install the outer cover (G) with seals over the axle (C).
7. Press the seal collar (E) against the axle shoulder. Be sure to install with recessed I.D. of collar toward the driver.
  a. Use the bearing spacer sleeve as a driver.
  b. Coat the I.D. and O.D. of the collar with grease prior to assembly.
8. Coat the O-ring with oil and install over shaft and inside the seal collar (E).
9. Install the O-ring spacer ring (B) against the O-ring inside the seal collar.
10. Heat the outer ball bearing to 300° F (150° C).

CAUTION
Do not exceed 300° F (150° C). Plan a safe handling procedure to avoid burns.

11. Drop heated bearing (H) over axle shaft (D).
12. Drive bearing against seal collar (C) using bearing spacer sleeve (I) as a driver.
13. After bearing has cooled, repeat Step 12 and remove spacer sleeve (I).
14. Coat the final drive side of gasket (E) with sealant.
15. Install the axle assembly into the axle housing (J).
16. Torque the outer bearing cover-to-final drive housing capscrews to 40 ft.lbs. Be sure bearing (H) starts squarely into the housing (J) and do not forget lockwashers.
17. With axle housing setting on outer cover, install axle inner bearing over the shaft. Be sure bearing starts squarely in housing bore. Bottom the bearing inner race against the bearing spacer sleeve.

NOTE: Use the large final drive gear to press the bearing over the shaft. After the gear has bottomed, check to be sure the bearing has bottomed on the spacer sleeve by driving on the inner race of the ball bearing.

18. Remove final drive gear.
19. Install oil seal (B) in brake housing-to-pinion shaft bore.
  a. Coat I.D. and O.D. of seal with grease prior to installation.
  b. Drive using a $3^{1}/_{16}$ in. driver disc and a $2^{1}/_{8}$ in. pilot disc (OTC No. 27532 and 27517). Install to bottom of chamfer in bore.
  c. Be sure seal spring stays in place.

DRIVE PINION
AND FINAL DRIVE GEAR
1. Coat shaft and I.D. of its parts with oil.
2. Coat O-ring with oil and install it in its groove.
3. Press ball bearing and seal collar on shaft against pinion shoulder. Press on the face of the collar only.
4. Install spacer washer (C) on top of

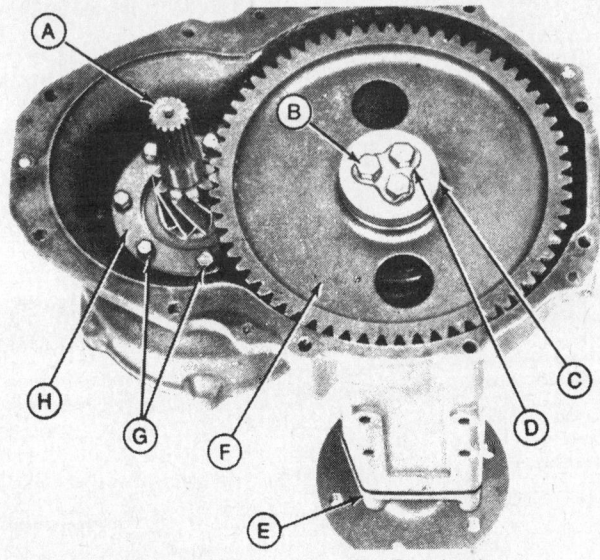

A. Final drive pinion
B. Capscrew
C. Spacer washer
D. Lock plate
E. Outer bearing cover
F. Final drive gear
G. Capscrews
H. Bearing retainer

Installing final pinion and gear

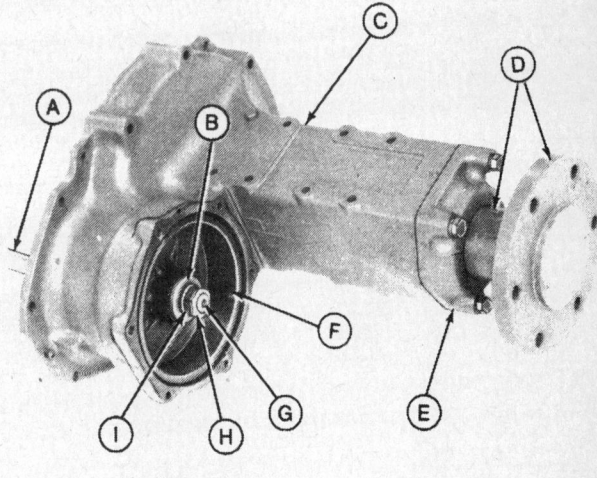

A. Bar
B. Spacer washer
C. Axle housing
D. Axle shaft
E. Outer cover
F. Brake drum
G. Pinion shaft
H. Nut
I. Lockwasher

Tightening pinion shaft and nut

John Deere

seal. Seal spacer should be flush with shoulder.

5. Install pinion shaft and driveshaft in so the outer bearing race bottoms on the shoulder.
6. Install bearing retainer (H) and draw bearing and retainer in place with the five capscrews (G) with lockwashers. Torque the capscrews to 20 ft.lbs.
7. Install the final drive gear (F).
8. Install spacer washer (C) aligning holes in washer with holes in axle shaft.
9. Install the lock plate (D) and three capscrews (B). Torque the three capscrews to 40 ft.lbs.
10. Bend lock plate (D) to lock the capscrews.
11. Check to be sure the axle and gear turns freely.
12. Install brake drum (F), washer (B) and lockwasher (I).
13. Torque pinion shaft retaining nut to 123-181 ft.lbs.

NOTE: Torque with a bar (A) in the cast hole in the final drive gear. If nut (H) is torqued after housing installation, put the tractor in first speed, compress the differential lock pedal and torque.

Both the righthand and the lefthand pinion nuts may be torqued simultaneously to the same specification if both axle housings were disassembled.

14. Reverse the steps of removal to install the final drive housings.

DIFFERENTIAL

Disassembly

DIFFERENTIAL LOCK SHIFTER

1. Remove lock hinge bracket-to-transmission case capscrews and remove lock pedal bracket (A) with pedal.
2. To disassemble pedal hinge, remove cotter pin from hinge pin (D) and remove pin and bushing from pedal assembly.
3. Remove four differential lock cover-to-transmission case capscrews and remove lock cover (B).
4. Remove two keeper plate-to-case capscrews and remove keeper plate (E).
5. Use a spring pin punch to remove spring pin (C) from lock fork (B) and shaft (A).

NOTE: Use a length of wire bent under and through spring pin (C) to prevent it from falling to the bottom of the differential compartment.

6. Drive lock shifter shaft (A) to the left through the side of the transmission case displacing the seal cap on the lefthand side of the shaft. Use a brass drift on the end of the shaft to avoid damage while driving on pedal contact area.
7. Use a screwdriver or large pliers to pry release spring (D) from transmission case.
8. If lock shifter fork (B) is removed, the rockshaft housing must be removed. Remove rockshaft housing and rotate fork (B) to the rear on its shift collar.

DIFFERENTIAL UNIT

1. Drain transmission oil at PTO drain plug and remove rockshaft housing (B).
2. Remove both final drive housings (E).
3. Remove differential lock shifter fork.
4. Remove the retainer plate-to-transmission case capscrew from each keeper plate (D).

NOTE: If bearing retainer (H) is removed, do not lose shims located between housing retainer (C) and bearing retainer (H).

5. Drive differential housing assembly from transmission case. Drive housing from righthand side of transmission case to lefthand side displacing retainer (C). Use a 2 in. driver disc and a 1¼ in. pilot disc (OTC No. 27515 and 27503).

NOTE: Do not allow differential to fall to floor as the bearings are driven from the transmission case. A long drift through the housing will ease removal.

Inspection

NOTE: If differential ring gear is replaced, differential drive shaft must also be replaced since they are a matched set.

DIFFERENTIAL LOCK SHIFTER

1. Inspect fork lock spring and replace as necessary.

Free Length	Working Load
3.8 in. (96.5 mm)	2.5 in. at 93 lbs. (64.5 mm at 412 N)

2. Replace any parts that may be damaged or worn.
3. Replace fork spring pin if it is loose in its bore or was damaged during removal.
4. Replace shaft O-ring prior to assembly. Be sure to coat shaft and O-ring with John Deere Hy-Gard Transmission and Hydraulic Oil prior to assembly.

DIFFERENTIAL

NOTE: If differential ring gear is replaced, differential drive shaft must also be replaced since they are a matched set.

NOTE: Remove differential housing ball bearings only to replace the bearings, lock collar (C) or differential housing.

1. Inspect righthand ball bearing. If it is loose or noisy, remove using a knife-edge bearing puller. On 950 Tractor, install puller under shift collar (C) to avoid pulling on bearing outer race. On 850 Tractor pull between collar and bearing. (A snapring separates the collar and bearing.)
2. If lefthand ball bearing is replaced, use the same puller as in step 1. Bearing may be damaged during removal since outer bearing race must be pulled.
3. Bend down lock plate tabs (A) and remove the six differential housing-to-ring gear capscrews (D).

NOTE: If the differential ring gear is not to be replaced, mark the ring gear and differential housing so they will be reassembled in the same position.

4. Use a soft-headed mallet to drive differential ring gear (H) from housing (B). Drive around the circumference of the ring gear as evenly as possible.

NOTE: The differential ring gear must be removed to obtain clearance for pinion shaft removal.

5. Drive spring pin (B) from differential housing and pinion shaft (A) using a spring pin punch.
6. Remove pinion shaft (A) from housing.
7. Remove gears and pinions by walking pinions (G) around gears. Remove thrust washers (F).

Assembly

1. Inspect bushing (D) and replace if worn excessively.

Tractor	Bushing I.D.	Bushing O.D.
850	1.1835-1.1847 in. (30.06-30.09 mm)	1.3012-1.3020 in. (33.05-33.07 mm)
950	1.2992-1.3022 in. (33.00-33.08 mm)	1.4587-1.4600 in. (37.05-37.09 mm)

2. Inspect housing (C) for damage or wear. I.D. for pinion shaft is 0.7874 to 0.7882 in. (20.00-20.02 mm).
3. Inspect pinion shaft. O.D. is 0.7858 to 0.7866 in. (19.96-19.98 mm).
4. Inspect differential pinions (G). I.D. of pinion is 0.7890-0.7898 in. (20.04-20.06 mm).
5. Maximum clearance between pinion and pinion shaft is 0.016 in. (0.4 mm).
6. Inspect thrust washers (A and F). New thickness is 0.037-0.041 in. (0.95-1.05 mm). Replace if washer is damaged or grooved or if they are less than 0.024 in. (0.6 mm) thick.
7. Grease thrust washers (F) to pinions (G).
8. Grease thrust washers (A) to the bevel lock gear (E) and the free bevel gear (B).
9. Install bevel gear (H) in differential housing. Be sure its thrust washer stays in place.
10. Install the differential lock bevel gear (E) in housing. Note that it has holes on its side for lock collar engagement.
11. Install bevel pinons (G) with their

John Deere

thrust washers in place between the two bevel gears.

NOTE: To properly align the bevel pinions walk them around the bevel gears to align with the shaft bores.

12. Insert the pinion shaft (A) through the bevel pinions (G) and their thrust washers (F). Be sure to align the spring pin hole in the shaft with the spring pin hole in the housing. The shaft should be free of burrs.

13. Install the shaft spring pin (B). Be sure the split in the pin is facing 90° from the shaft centerline.

14. Check to see that the gears and pinions turn freely in the housing.

15. Heat the differential ring gear and any ball bearings that were removed from the differential housing to 300° F (150° C).

NOTE: Do not press cold bearings onto differential housing. The housing is easily distorted.

---- **CAUTION** ----
Heat parts in a bearing heater. Use a thermometer and do not exceed 300° F (150° C). Plan a safe handling procedure to avoid burns.

16. Note the number etched on the end of the differential drive shaft and be sure it matches the number on the ring gear.

17. Install differential ring gear (D) so ring gear holes align with housing holes. If the original ring gear is reinstalled, be sure the marks on the ring gear and housing are in the same position as when removed.

18. Install three equally spaced capscrews (B) to pull ring gear (D) tight against housing face.

19. Remove capscrews (B).

20. Install new lock plates (A) and torque differential housing-to-ring gear capscrews (D) to 40 ft.lbs.

21. Bend lock plates (A) over capscrews (D) to lock them.

22. Install heated lefthand ball bearing (C) onto housing tight against housing.

23. Invert differential housing assembly.

NOTE: Do not use a press to install bearings on housing unless the bearings have been heated. Pressing on cold bearings will distort the differential housing.

24. If righthand ball bearing was removed, install lock collar (E) and drop heated bearing onto housing.

NOTE: The 850 Tractor has a snapring on both sides of the ball bearing. Be sure snapring is installed just after lock collar.

25. With bearing just clearing the housing snapring groove, seat the snapring (C) on the housing.

26. If the differential ring gear was not replaced, install the housing retainer with the same bearing shims as were removed behind the bearing retainer.

NOTE: If differential ring gear was replaced, differential drive shaft must also be replaced.

If the ring gear was replaced or if the number of cone point or backlash adjusting shims are in question, be sure to adjust differential bevel pinion cone point and differential backlash.

If the original ring gear and pinion shaft are to be installed and neither the cone point nor backlash shim pack are in question, disregard the following cone point and backlash adjustments.

DIFFERENTIAL CONE POINT ADJUSTMENT

NOTE: This adjustment places the differential drive shaft spiral bevel pinion in proper fore and aft mesh relation with the differential ring gear.

NOTE: Be sure the differential drive shaft has the same number etched on the end of its pinion as on the differential ring gear.

1. Install differential driveshaft.

2. Install 0.036 inch (0.9 mm) shim pack behind front bearing retainer (B). Use jack screws (D) to pull retainer forward for installation of shims.

NOTE: Shims (C and F) are available in two thicknesses. These thicknesses are 0.012 in. (0.3 mm) and 0.008 in. (0.2 mm). Install three of the thicker shims to assure clearance between driveshaft pinion and adjusting mandrel.

Be sure to install the same amount of one-corner shims (C) as three-corner shims (F).

3. Remove jack screws (D).

4. Install gasket on front bearing cover and install cover over front bearing retainer. Do not coat gasket with sealant at this time.

5. Torque front cover stud nuts to 20 ft./lbs.

6. Install JDT-41 Cone Point Adjustment Mandrel (E) in differential housing side cover (A) and install Mandrel in differential compartment and through righthand bearing bore.

NOTE: Do not damage any of the finished surfaces on the adjustment mandrel. If the tool will not clear the differential drive shaft spiral bevel pinion (C), add more shims behind front bearing retainer until it will.

7. Use a feeler gauge (B) to measure distance between finished end of spiral bevel pinion (C) and adjustment mandrel gauge surface (D).

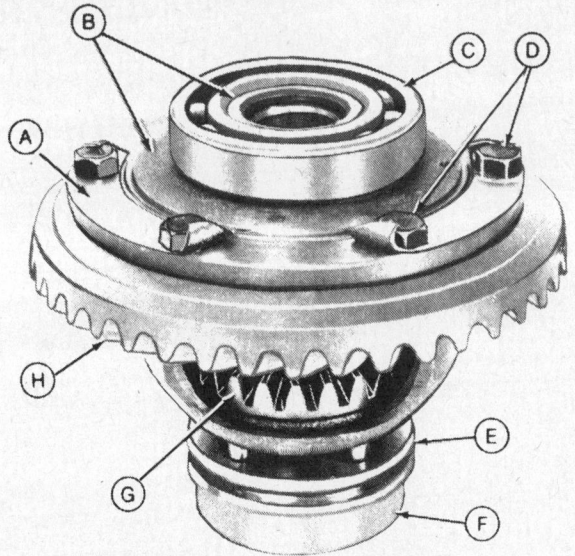

A. Lock plate
B. Differential housing
C. Lefthand ball bearing
D. Capscrews
E. Lock collar
F. Righthand ball bearing
G. Bevel lock gear
H. Ring gear

Differential housing assembly

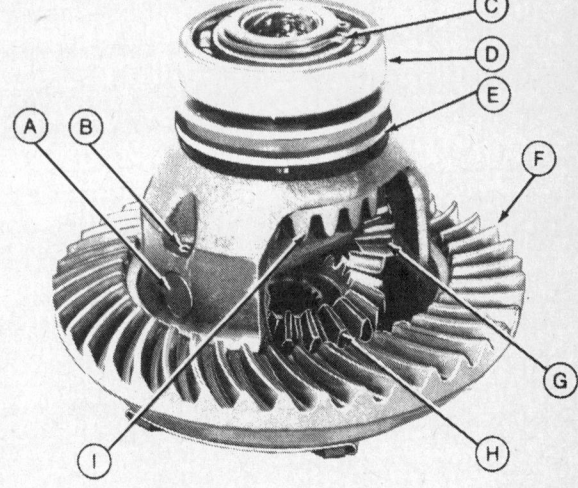

A. Pinion shaft
B. Spring pin
C. Snapring
D. Ball bearing
E. Lock collar
F. Ring gear
G. Bevel pinion
H. Bevel gear
I. Bevel lock gear

Installing right side ball bearing

John Deere

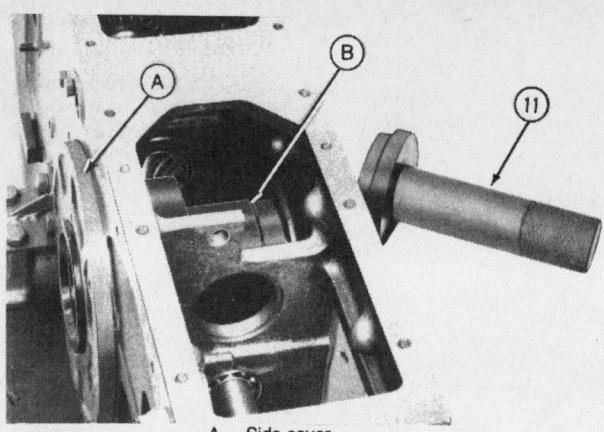

A. Side cover
B. Adjusting mandrel
Removing adjusting mandrel

a. If distance measured at feeler gauge (B) is 0.018 to 0.022 in. (0.45 to 0.55 mm), cone point shim pack is correct. Install the differential and adjust backlash.

b. If distance measured at feeler gauge (B) is less than 0.018 in. (0.45 mm), add shims to move the differential drive shaft pinion away from the mandrel. The correct shim pack to add is the feeler gauge measurement minus 0.020 in. (0.5 mm).

NOTE: Shim pack available does not consistently allow the ideal cone point setting of 0.020 in. (0.50 mm). Set cone point using various combinations of the 0.012 in. (0.3 mm) and 0.008 in. (0.2 mm) shims to obtain 0.018 to 0.022 in. (0.45 to 0.55 mm) clearance between pinion and mandrel.

c. If distance measured at feeler gauge (B) is more than 0.022 in. (0.55 mm), remove shims to move the differential drive shaft toward the mandrel. The correct shim pack to remove is the feeler gauge measurement minus 0.020 in. (0.5 mm).

8. Recheck cone point clearance after the shim pack (H) and the bearing cover (A) with its gasket (B) have been installed.

9. Remove cover (A) and apply sealant to the gasket.

10. Install the cover with gasket and torque front cover to transmission case stud nuts to 20 ft.lbs.

11. Remove adjusting mandrel (B) by driving it with the side cover from the side of the case (G).

a. Use OTC No. 27489 Handle with 3½ in. drive disc and 3 1/16 in. pilot (OTC No. 27539 and 27532).

b. Use care not to damage tool during removal.

12. Install the differential and adjust backlash as instructed on the following pages.

DIFFERENTIAL UNIT PLACEMENT

1. Install differential side cover (B) on lefthand ball bearing (F) of differential assembly.

2. Install the same shims (D) on cover (B) as were removed (if they were removed).

3. Place bearing retainer (C) over shims and install retainer-to-cover capscrews. Torque capscrews to 20 ft.lbs.

4. Install differential assembly with side cover (B) into differential compartment of the transmission case.

NOTE: If ring gear was replaced, more shims may have to be added under bearing retainer (C) to draw the side cover (B) flush with the side of the transmission case.

5. Install the two retainer plates (E) and tighten the four capscrews with lockwashers to 20 ft.lbs.

6. If backlash is to be checked, install a metal strap (A) to retain the side cover flush with the finished side of the transmission case. Construct the strap using the accompanying figure.

7. Install the differential lock shifter assembly and pedal.

8. If the differential ring gear was replaced or if backlash adjusting shim pack is in question, adjust backlash as follows.

DIFFERENTIAL BACKLASH ADJUSTMENT

NOTE: This adjustment places the differential ring gear in proper relation with the spiral bevel pinion from side to side.

NOTE: Always check and adjust backlash after cone point adjustment has been made. Do not adjust backlash unless the ring gear was replaced or if the amount of side shims are in question.

1. Install a dial indicator with extension (A).

a. Be sure base of indicator is firmly attached.

b. Extension of indicator should be toward outer part of ring gear tooth and as close to perpendicular to the tooth as possible.

2. Compress lock pedal to force differ-

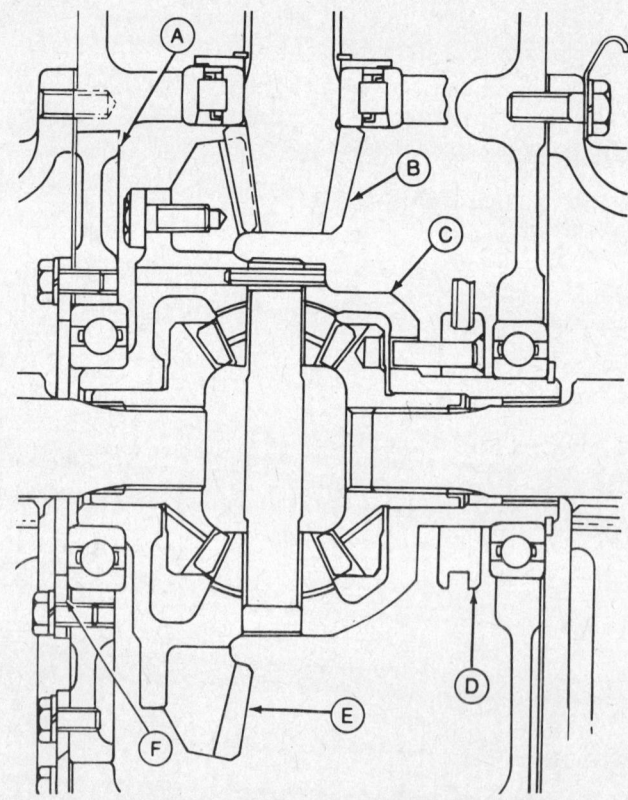

A. Side cover
B. Spiral bevel pinion
C. Differential housing
D. Lock collar
E. Ring gear
F. Bearing retainer shims

Top view of the differential

John Deere

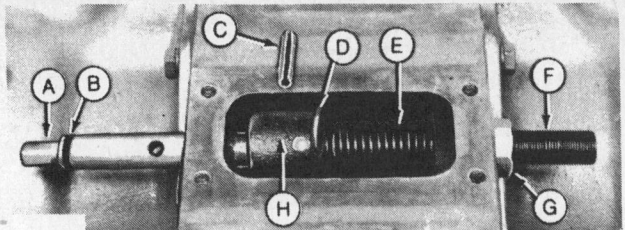

A. Lock shaft
B. O-ring
C. Spring pin
D. Spring locking pin
E. Lock release spring
F. JDT-36 lock shaft tool
G. Nut
H. Lock fork

Installing differential lock shaft

A. Lock shaft
B. Lock fork
C. Spring pin
D. Release spring
E. Keeper plate

Differential lock shifter installed

ential housing (C) and lefthand bearing against bearing retainer. Turn the differential drive shaft around at least three times with the pedal compressed to be sure housing is against bearing and bearing is against the retainer.

NOTE: Turn the transmission drive shaft at its front coupler with the range and change shifters engaged.

3. Release lock pedal and check backlash between the differential drive shaft spiral bevel pinion and the differential ring gear (E).
   a. Hold the differential drive shaft so differential shaft gear backlash is not involved. Shaft can be held at the spiral bevel pinion.
   b. While holding the pinion stationary rotate the ring gear noting the dial indicator readings.
   c. Backlash should be 0.005 to 0.007 in. (0.13-0.18 mm).
4. If backlash is excessive, remove side shims (F) to move the ring gear closer to the spiral bevel pinion.
5. If there is too little backlash, add side shims (F) to move the ring gear away from the spiral bevel pinion.

NOTE: Be sure lefthand bearing is tight against its retainer whenever backlash is checked.

NOTE: Side shims are available in three sizes. These sizes are 0.004 in. (0.1 mm), 0.012 in. (0.3 mm) and 0.039 in. (1 mm).

6. Recheck backlash after bearing retainer has been reinstalled and differential has been forced to the left.

FINAL INSTALLATION

1. Install both final drive assemblies.
2. Install the transmission drive shaft if it is not already installed.
3. Join the tractor (if it was separated).

--- **CAUTION** ---
Be sure to replace safety shield and guard after removing the PTO.

4. Install all covers.
5. Fill the transmission case with oil and install the rockshaft.

DIFFERENTIAL LOCK

1. Place differential lock release spring (E) in line with shaft bore.
2. Insert JDT-36 Lock Shaft Tool (F) through lefthand side of transmission case and through spring.
3. Insert spring locking pin (D) through end of Tool (F).
4. Tighten nut (G) until lock shifter fork (H) will fit between spring (E) and righthand side of transmission case.
5. Insert lock shaft (A) through righthand side of case and through fork (H). Be sure spring pin hole in shaft aligns with spring pin hole in fork.
6. Install O-ring (B) on shaft. Be sure to coat it with oil before installing.
7. Remove pin (D) and slide the lock shaft through the spring displacing the tool (F).
8. Install spring pin (C) through fork and shaft. Be sure split in pin faces the side of the transmission case.
9. Install keeper plate (E) with lockwashers and capscrews. Torque capscrews to 20 ft.lbs.
10. Install pedal bracket (A). Torque capscrews to 20 ft.lbs. Do not forget lockwashers.
11. Install hinge pin (D) through pedal and bracket and install cotter pins in each end of hinge pin.
12. Check differential lock linkage for proper operation.
13. Install gasket with sealant on lock cover (B) and install cover.
14. Torque lock cover-to-transmission case capscrews to 33 ft.lbs.
15. Install seal cap on lefthand side of lock shaft bore in transmission case. Coat with grease before installing.

POWER TAKE-OFF UNIT

Disassembly

PTO PINION SHAFT

1. Remove rockshaft housing.
2. Remove four PTO master shield retaining capscrews and remove master shield.

A. JDT-40 puller body
B. Rear ball bearing
C. PTO shaft
D. JDE-114-1 forcing screw
E. PTO pinion shaft

Removing PTO pinion shaft

273

John Deere

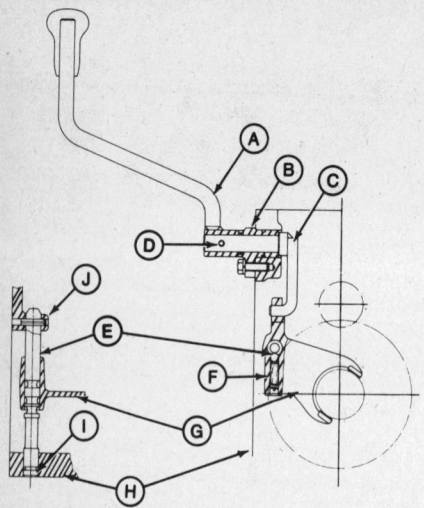

A. Shifter lever
B. Sleeve
C. PTO armshaft
D. Spring pin
E. Shifter shaft
F. Detent
G. Shifter fork
H. Transmission case
I. O-ring
J. Spring pin

PTO shifter

3. Remove four PTO pinion shaft rear bearing retainer capscrews and remove rear bearing retainer.

NOTE: As PTO pinion shaft (H) is removed, be sure to catch all parts that are loose on the shaft. A thrust washer (located between the two clutch drums), the rear clutch drum (J), spring (I) and washer (F) are all free to fall to the bottom of the PTO compartment.

NOTE: Snapring (G) is not used on later models. Replacement shaft and later models have a shoulder to support washer (F).

4. Install JDE-114-1 forcing screw in end of PTO pinion shaft (E). Small diameter end of screw (D) is installed in tapped end of shaft. Do not bottom screw in shaft bore.
5. Install puller body (A) as shown.
6. Tighten nut pulling PTO pinion shaft and rear ball bearing (B).
7. As shaft is removed, be sure to catch all loose parts.

PTO DRIVESHAFT

1. Remove PTO pinion shaft as instructed on the preceding page.
2. If transmission driveshaft has been removed, drive driveshaft from front of transmission case using a long brass drift.
3. If transmission driveshaft has not been removed, remove shaft by driving on front of rear bearing or using a slidehammer and claw-type puller around the front clutch drum.

NOTE: If shaft or its bushing are damaged, a blind hole puller may be used to pull PTO driveshaft from rear of transmission case.

4. As PTO driveshaft is removed, do not allow splined coupler (B) to fall to bottom of differential compartment.

PTO SHIFTER

1. Remove spring pin (D) and remove shifter lever (A).
2. Remove rockshaft.
3. Drain transmission oil at PTO drain plug.
4. Pull shifter armshaft (C) toward center of case and remove.
5. Drive spring pin (J) from case and shifter shaft (E).
6. Push shifter shaft (E) through shifter fork (G) through the rear of the transmission case. Plug (I) will be displaced by the shaft.
7. Remove shifter fork (G).

A. Spring pin hole
B. Shifter shaft bore
C. PTO drive gear
D. JDT-40 puller body
E. JDT-37 removal tool
F. Pin
G. Seal housing
H. Snapring
I. PTO shaft
J. Ball bearing

Removing PTO shaft

PTO SHAFT

1. Remove PTO pinion shaft.
2. Remove PTO shifter as instructed previously.
3. Remove the four seal housing capscrews.
4. Remove snapring (H) from its groove and slide snapring and shifter collar (not shown) forward.
5. Remove snapring behind shifter collar (in front of PTO gear (C)) and push it toward front bearing (J).
6. Install JDT-37 Removal Tool (E) over end of PTO shaft. Insert pin (F) through shaft and tool.
7. Install JDT-40 Puller Body (D) and JDE-114-1 Forcing Screw into tool (E). Larger diameter end of screw fits into Removal Tool (E). Do not bottom screw against shaft.

NOTE: Remove snapring in front of front bearing (J) when shaft has been removed part way.

8. Pull PTO shaft (I). Note the position of parts as they are removed. Bearing (J) will press off as shaft is pulled from transmission case.
9. Remove the Removal Tool.

Inspection
PTO DRIVESHAFT

1. Inspect PTO driveshaft for wear or damage and replace as necessary.
2. Inspect driveshaft bearing. Bearing balls should not be loose or noisy when turning. Be sure retaining ring on outer diameter of bearing is properly seated in its groove.
3. Be sure shaft oil passages are free and inspect shaft bushing. If replaced, drive new bushing flush with bottom of chamfer in I.D. of shaft.
   a. I.D. of bushing is 0.7086 to 0.7094 in. (18.00-18.018 mm).
   b. O.D. of bushing is 0.8833 to 0.8841 in. (22.435-22.456 mm).
   c. Drive new bushing into shaft using a 7/8 in. driver disc and 11/16 in. pilot (OTC No. 27497 and 27494).
4. Inspect front clutch drum. Replace if drive surface is excessively worn or if splines on I.D. are worn.
5. Replace any snaprings or retaining rings that are damaged or worn.
6. Inspect splined coupler on end of transmission driveshaft. Splines should not be worn excessively. Be sure coupler snapring is seated properly in its groove.

PTO PINION SHAFT

1. Inspect rear clutch drum (B). If drive surface or splines are worn, replace.
2. Install anti-rotation pin (E) into shaft. Coat with grease so it will not slip out during installation.
3. Inspect clutch spring (C).

Free Length	Working Load
2.9 in.	1 1/8 in. at 15 lbs.
(74 mm)	(29 mm at 66 N)

John Deere

4. Inspect pinion and shaft for wear or damage.
5. Check that double-row ball bearing (H) is not loose or noisy and replace as necessary.
6. Replace snapring (F) if damaged.

PTO SHAFT

1. Inspect shaft and parts for wear or damage.
2. Check I.D. of PTO gear (G). New part diameter is 1.7778 to 1.7753 in. (45.07-45.09 mm).
3. Inspect bushing (O). New part I.D. is 1.5743 to 1.5748 in. (39.99-40.00 mm). O.D. is 1.7712 to 1.7717 in. (44.99-45.00 mm).
4. Replace any damaged snaprings or retaining rings.
5. If seal collar (I) is removed, replace it. Be sure to coat with oil prior to reassembly.

NOTE: Coat all mating parts with John Deere Hy-Gard Transmission and Hydraulic Oil or its equivalent prior to assembly.

PTO SHIFTER

1. Replace O-ring.
2. Inspect detent spring (B).

Free Length	Working Load
1.33 in.	1.01 in. at 16.1-17.8 lbs.
(33.8 mm)	(26.4 mm at 72-79 N)

3. Install steel ball (C) in detent bore in fork (G). Install spring (B).
4. Install spring pin (A) so split faces the ball. Compress spring against ball using a small punch.
5. Inspect shifter fork (G). Fork thickness is 0.264 to 0.272 in. (6.7-6.9 mm). Width of collar groove is 0.280 to 0.287 in. (7.1-7.3 mm). Replace when parts are worn to 0.040 in. (1 mm) or more clearance.
6. Replace parts when clearance between shaft (E) and fork (G) is 0.008 in. (0.2 mm). O.D. of shaft is 0.5892 to 0.5899 in. (14.96-14.98 mm). I.D. of fork is 0.5906 to 0.5922 in. (15.00-15.04 mm).

Assembly and Installation

PTO SHAFT

1. Install thrust washer (F) against shoulder on shaft. Be sure grooves on washer are toward front of PTO shaft (I).
2. Press rear ball bearing (D) on shaft tight against washer (F). Ball bearing retaining ring must face rear of shaft (I).
3. Install spacer washer (L) and snapring (K). Be sure snapring is seated in its groove.
4. Coat I.D. of seal collar (H) with oil and install it against snapring (K).
5. Install retaining ring (J) in groove to rear of seal collar (H).
6. Insert PTO shaft (I) through rear of transmission case.
7. Install PTO gear (E) with bushing over shaft so spur cut I.D. is toward the front.
8. Install splined thrust washer in front of PTO gear (E).
9. Install snapring in front of thrust washer.
10. Install shifter collar (D). Be sure collar end is forward.
11. Install the collar front snapring (C).
12. Install front ball bearing (B) on end of shaft and insert a ⅛ in. piece of steel plate (A) between bearing and front PTO compartment wall.
13. Drive PTO shaft (I) forward using a mallet until the shaft bottoms against plate (A). Be sure bearing is aligned squarely on shaft.
14. Remove plate (A) and align bearing (B) with its bore in the transmission case.
15. Insert a 1⅛ inch open-end wrench in front of front bearing (A) so it clears the PTO shaft O.D.

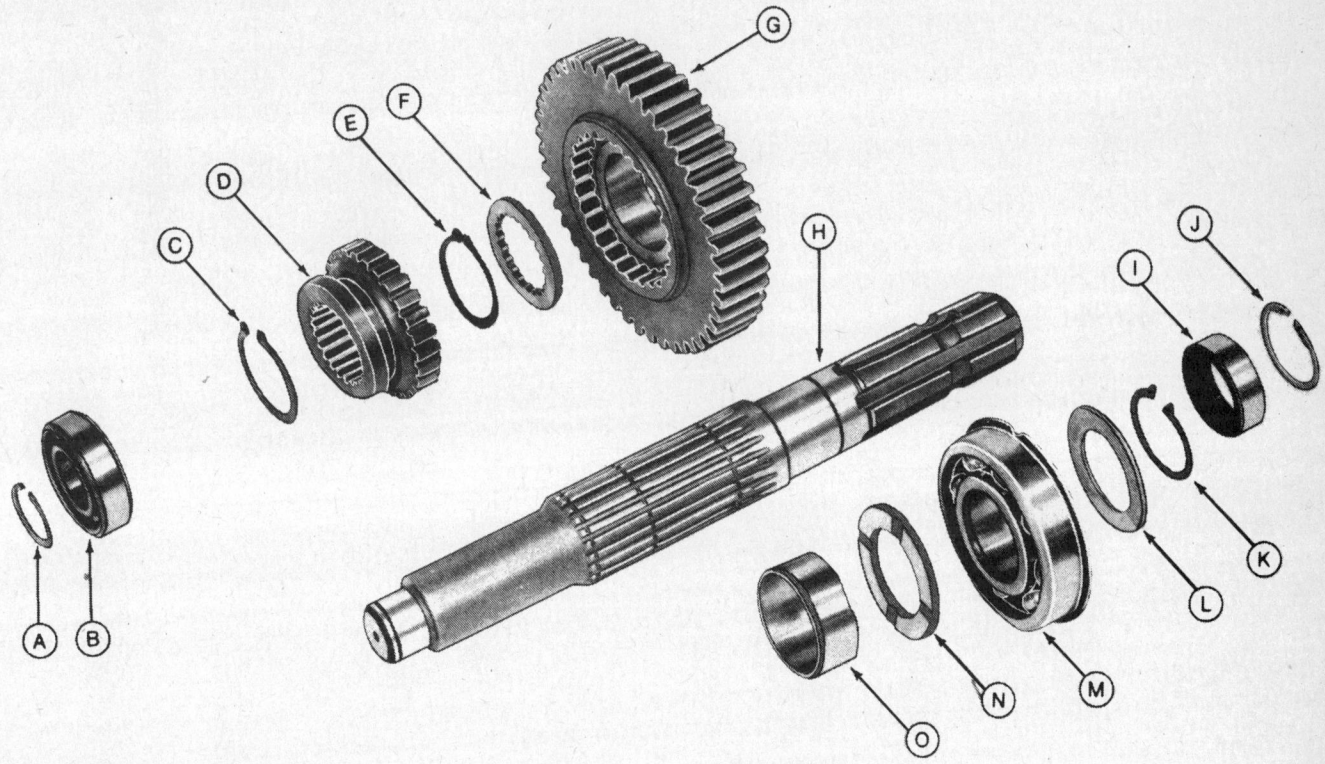

A. Snapring
B. Ball bearing
C. Snapring
D. Shift collar
E. Snapring
F. Thrust washer
G. PTO gear
H. PTO shaft
I. Seal collar
J. Retaining ring
K. Snapring
L. Spacer washer
M. Ball bearing with retaining ring
N. Thrust washer
O. Bushing

PTO shaft components

John Deere

A. Front bearing	D. Ball bearing		A. Lever collar	D. Spring pin	G. Shifter collar
B. PTO gear	E. Spacer washer		B. O-ring	E. PTO pinion shaft	H. Shifter fork
C. Shifter collar	F. PTO shaft		C. Shifter shaft	F. PTO shaft	I. Shifter armshaft

Driving on PTO front bearing   **Installing PTO shifter**

16. Drive shaft (F) against wrench until bearing bottoms against shoulder on shaft.
17. Remove wrench and install snapring in front of front bearing (A). Be sure snapring seats properly in its groove.
18. Continue driving shaft forward until bearing (D) retaining ring is seated against shoulder at rear of transmission case.
19. Slide PTO gear (B) with its bushing and front thrust washer to the rear until the snapring to the front of it can be seated. Seat the snapring.
20. Slide the shifter collar to the rear to engage the PTO gear.
21. Seat the shifter collar stop snapring in the second groove from the front. Note that the front groove on the shaft splines contains no snapring.
22. Install rear oil seal (B) in rear housing (A). Use a 1 5/16 in. driver disc and a 1 9/16 in. pilot disc (OTC No. 27520 and 27508). Drive until seal bottoms in seal housing.
23. Install gasket on finished side of housing and O-ring (C) on its O.D. Coat gasket with sealant.
24. Torque seal housing-to-transmission case capscrews to 20 ft.lbs. Do not forget lockwashers.
25. Install PTO shifter assembly as instructed following.
26. Install PTO pinion shaft.
27. Fill the transmission case with oil and install guard and shield.
28. Install rockshaft housing.

PTO SHIFTER ASSEMBLY

NOTE: Replace all O-rings. Coat them with oil prior to installation. All parts should be clean prior to installation.

1. Install PTO shifter fork (H) so it is in place in the collar groove.
2. Install shifter shaft (C) through rear of transmission case and through shifter fork (H). Be sure hole in shaft aligns with hole in case (D).
3. Install spring pin (D) to retain shifter shaft (C) in transmission case. Spring pin should face front of transmission case. Spring pin should be driven 1/8 in. (3 mm) below flush into casting.
4. Install lever collar (A) into transmission case tight against the side of the case. Be sure O-ring (B) is in place on collar O.D.
5. Insert shifter armshaft (I) through side of case and through collar (A). Be sure arm is positioned in the fork (H).
6. Install O-ring from outside of armshaft (I) into collar (A) I.D.
7. Install spacer washer on outside of O-ring.
8. Install shifter lever over armshaft aligning the spring pin holes in the lever and the armshaft (I).
9. Install the 1 in. (25 mm) spring pin through the lever and armshaft so the spring pin is forward.
10. Torque the lever collar-to-transmission case capscrew with lockwasher to 20 ft.lbs.
11. Fill the transmission case with oil and install the rockshaft housing.

---- **CAUTION** ----
Be sure to replace safety shield and guard after servicing the PTO.

PTO DRIVESHAFT

1. Install splined coupler (A) on rear end of transmission driveshaft.
2. Press ball bearing (C) onto shaft so it bottoms against splined area on shaft. Install front snapring.
3. Install front clutch drum (D) and retain it in place on shaft splines by properly seating rear snapring in its groove.
4. Drive assembly into place in transmission case using JDC-400-7 Driving Handle with 1 3/8 in. driver disc and 1 1/16 in. pilot disc (OTC No. 27505 and 27500).

NOTE: Be sure PTO driveshaft front splines align with splined I.D. of coupler (A). Bottom the bearing retaining ring in the case wall.

5. Install PTO pinion shaft as instructed following.

PTO PINION SHAFT

1. Install the rear ball bearing on the pinion shaft tight against the rear pinion shoulder.
2. Install snapring (H) over front of shaft.

NOTE: Later shafts have a shoulder in place of snapring (H) and its groove.

3. Install spring support washer (G) and clutch spring (E).
4. Install PTO pinion shaft (F) with spring (E) into transmission case through rear of case.
5. Install rear clutch drum (M) over shaft splines against spring (E).

NOTE: Be sure anti-rotational pin is located properly in the shaft O.D.

6. As the two shafts are brought together, install the thrust washer (N). Be sure washer is located over pin in shaft and that grooved side of washer faces the front of the transmission case.
7. Install PTO pinion shaft rear bearing retainer with gasket. Be sure to coat gasket with sealant.
8. Torque rear bearing retainer-to-transmission case capscrews to 20 ft.lbs.
9. Compress spring support washer (F) against clutch spring (G) and drum and install snapring (E) in the forward of the two grooves (if there are two grooves) on the shaft splines.

NOTE: On early models, the snapring (E) must be installed in the front groove

John Deere

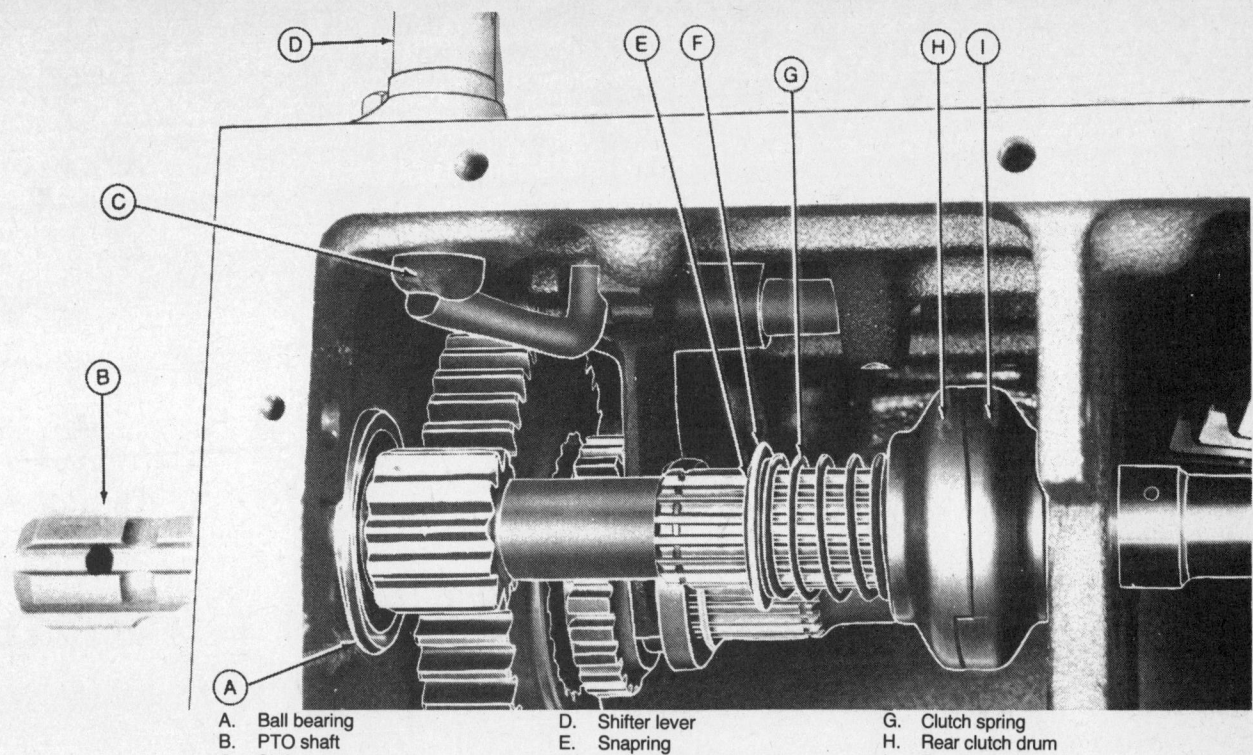

A. Ball bearing
B. PTO shaft
C. Shifter armshaft
D. Shifter lever
E. Snapring
F. Support washer
G. Clutch spring
H. Rear clutch drum
I. Front clutch drum

PTO gear train components

or the clutch will slip. Note that the rear groove in the shaft is not used and later shafts have no snapring grooves.

10. Check that the shaft turns freely and that the PTO shift lever (D) operates properly.
11. Install the rockshaft housing (A).

CAUTION
Be sure to replace safety shield and guard after servicing the PTO.

---

12. Install PTO guard (C) over PTO shaft.
13. Install PTO master shield (E). Be sure to tighten all four capscrews which secure the shield to the transmission case.
14. Fill transmission case to the proper level with John Deere Hy-Gard Transmission and Hydraulic Oil or its equivalent.
15. Operate the PTO to be sure it is operating properly.

MECHANICAL FRONT WHEEL DRIVE

Removal and Disassembly

SHIFTER SHAFT AND FORK ASSEMBLY

1. Separate the tractor between the clutch housing and transmission case.

2. Separate the clutch housing from the engine.
3. Remove transmission clutch driveshaft.
4. Remove plug from threaded hole in clutch housing and drive out spring pin (A) securing shifter shaft.

5. Pull shifter shaft (B) from clutch housing, and remove shifter fork (C) and engaging collar.

TOP FRONT WHEEL DRIVESHAFT

1. Remove front wheel drive shifter

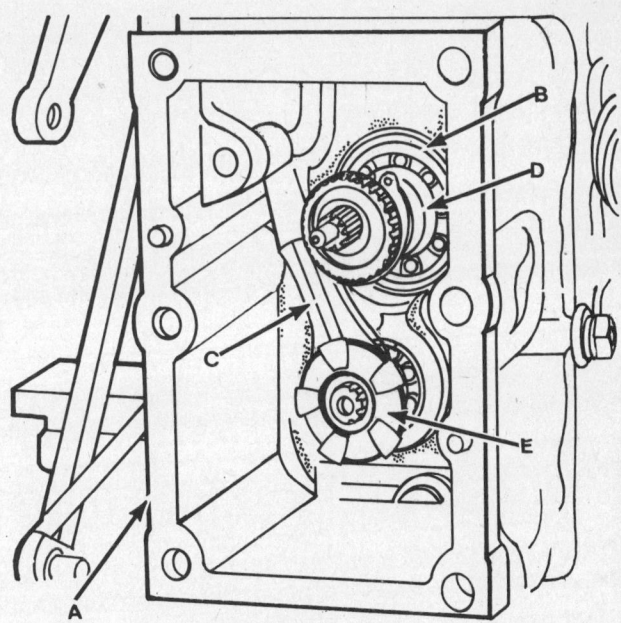

A. Rockshaft housing
B. Rear bearing retainer
C. PTO guard
D. Rear seal housing
E. PTO master shield

Removing shifter shaft and fork

John Deere

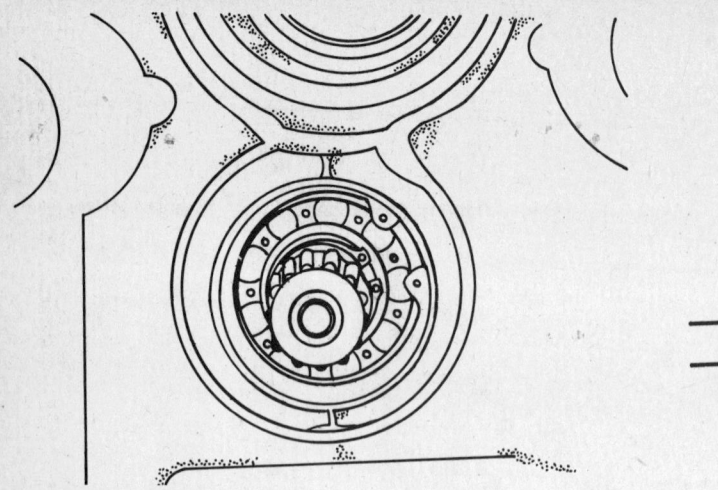

Front wheel driveshaft snapring

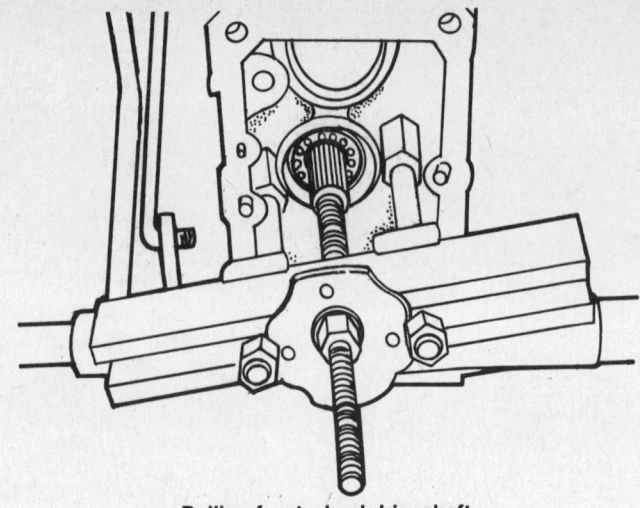

Pulling front wheel driveshaft

shaft and fork assembly as previously instructed.

2. Remove four capscrews securing drop housing and remove drop housing.

3. Remove snapring from rear bearing on top front wheel driveshaft.

4. Use D-1200AA Push Puller (A) and JDE-114-1 special forcing screw (B) to pull front wheel driveshaft until rear bearing comes out of its bore in transmission case.

5. Reach up through opening for drop housing in bottom of transmission case, and use a snapring pliers to move snapring toward front of driveshaft.

6. Continue pulling shaft. Remove front bearing, snapring, gear and spacer as they are displaced from shaft.

7. Press shaft from rear bearing.

DROP HOUSING

1. Drive roll pin (A) from shaft and drop housing.

2. Using a piece of pipe with a ⅞ in. (22 mm) I.D. (B) support drop housing in a press and push idler gear shaft from bearings and gear.

3. Use a slide hammer puller to remove bearings.

4. Remove snapring (A) and rubber dust shield (B) from output shaft.

5. Pry oil seal out of its bore.

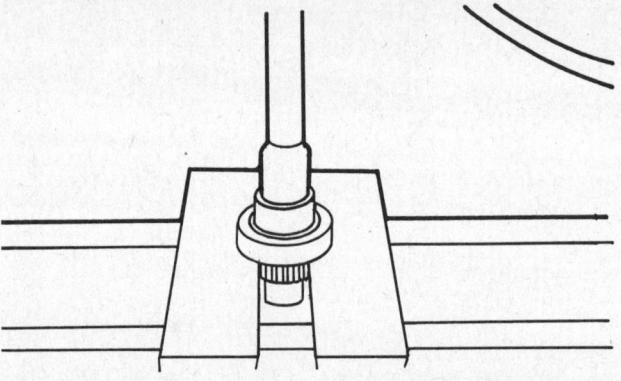

Removing front wheel driveshaft from the bearing

6. Remove snapring.

7. Use D-01200AA Push Puller (A) and JDE-114-1 special forcing screw (B) to pull output shaft and bearings from housing.

8. Press shaft from wear sleeve and bearing.

A. Roll pin
B. Pipe
C. Idler gear

Removing idler shaft

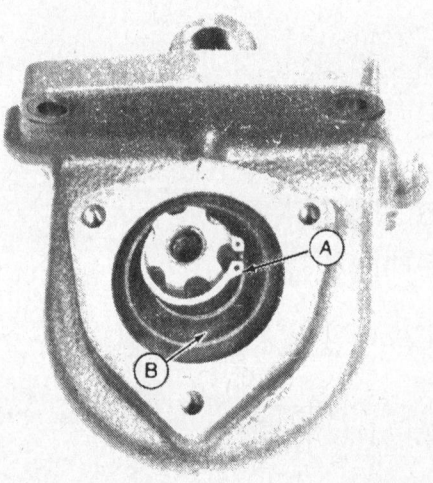

A. Snapring
B. Dust shield

Removing snapring and dust shield

John Deere

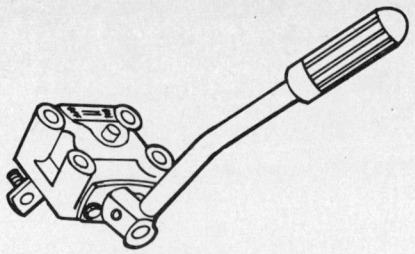
Front wheel drive shifter

FRONT WHEEL DRIVE SHIFTER

1. Remove the four capscrews securing shifter cover, and remove cover.
2. Drive out roll pin (A) and remove shifter arm (B).
3. Remove hex plug (C) in top of cover and drive out roll pin.
4. Remove shaft retainer plate capscrews (D) and remove plate (E).
5. Remove shaft (F).

Inspection

FRONT WHEEEL DRIVE SHIFTER

1. Drive out roll pin (A) in shifter fork (B) to inspect detent spring.

Free Length	Working Load
1.33 in.	1.04 in. at 17 lbs.
(33.8 mm)	(26.4 mm at 75 N)

2. Inspect shift fork (B). End of fork thickness is 0.264 to 0.272 in. (6.7 to 6.9 mm). Width of coupler collar groove is 0.280 to 0.287 in. (7.1 to 7.3 mm). Replace when parts are worn to 0.04 in. (1.0 mm) clearance or more.
3. Replace parts when clearance between shaft (C) and fork (B) is 0.02 in. (0.5 mm). O.D. of shaft is 0.589 to 0.590 in. (14.97 to 1498 mm). I.D. of fork is 0.5906 to 0.5922 in. (15.00 to 15.04 mm).
4. Inspect shifter arm shaft (A) and housing (B) for wear. Replace parts if clearance between shaft and housing is 0.02 in. (0.5 mm). O.D. of shaft is 0.7854 to 0.7874 in. (19.95 to 20.00 mm). I.D. of bore in housing is 0.788 to 0.789 in. (20.02 to 20.04 mm).

IDLER GEAR AND SHAFT

1. Inspect gear for worn or broken teeth.
2. Inspect remainder of shaft components.
   a. Shaft O.D. should be 0.7870 to 0.7874 in. (19.99 to 20.00 mm).
   b. Gear I.D. should be 2.0461 to 2.0472 in. (51.97 to 52.00).

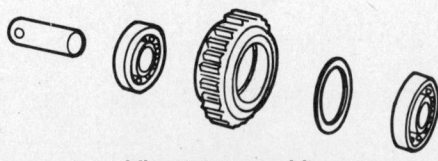
Idler gear assembly

OUTPUT GEAR AND SHAFT

1. Inspect output gear (C) for worn or broken teeth.
2. Inspect remainder of output shaft components.
   a. O.D. of output shaft (E) at front bearing is 1.1807 to 1.1815 in. (29.99 to 30.01 mm).
   b. O.D. of output shaft at rear bearing is 0.9839 to 0.9846 in. (24.99 to 25.01 mm).

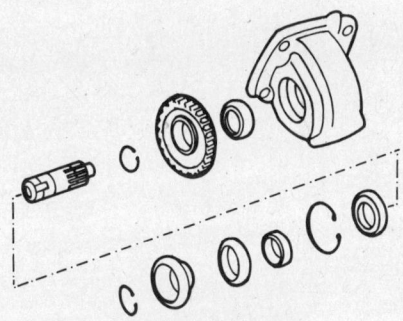
Output shaft components

Assembly

DROP HOUSING

1. Install output shaft rear bearing in drop housing. Press in bearing until it bottoms in housing.
2. Install snapring (D) on output shaft.
3. Place gear (C) in housing.
4. Place shaft (A) in housing making sure splines on shaft enter splines in gear.
5. Use a press to push shaft into rear bearing.
6. Place front bearing (B) over shaft and press on until bearing inner race contacts snapring.
7. Install snapring (C).

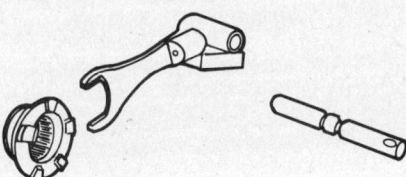
Shifter fork components

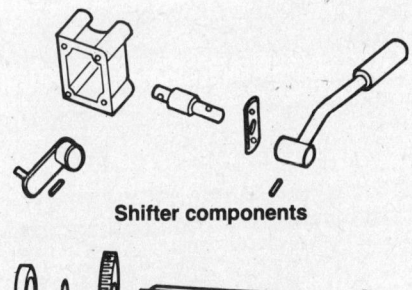
Shifter components

Top driveshaft components

8. Install oil seal wear sleeve. Press on wear sleeve until it bottoms on bearing.
9. Install oil seal with open side facing drop housing.
10. Install dust shield (B) and snapring (A).

IDLER SHAFT ASSEMBLY

1. Install spacer ring (D) in gear (C).
2. Using a calipers, measure from side of spacer ring to outside of gear. Record measurement.
3. Measure again on opposite side of gear.

NOTE: This measurement is to determine which side of gear the spacer washer (E) is to be inserted from. Spacer ring (D) is not centered in gear.

4. Insert spacer ring from side of gear with largest measurement.
5. Press in bearing (F) from spacer washer side of gear until it bottoms on spacer washer.
6. Press in other bearing (B) until it bottoms against spacer ring (D).
7. Use a press to install idler shaft into housing and gear assembly. Make sure roll pin holes in shaft and housing (B) are aligned.

TOP FRONT WHEEL DRIVESHAFT

1. If plug with O-ring was removed, install in clutch housing. Press in plug until it bottoms in bore.
2. Unstall bearing in clutch housing. Start bearing squarely in bore and use a brass drift to seat bearing against plug.
3. While holding shaft (A) in place insert gear (B) through hole for drop housing. Slide gear onto shaft.
4. Again reach up through hole for drop housing, and install front snapring on shaft.
5. Use a soft lead hammer to drive shaft into front bearing. Drive in shaft until it bottoms against bearing.
6. Install spacer over shaft.
7. Place rear bearing onto shaft, and using a piece of pipe with a 1 in. (25.4 mm)

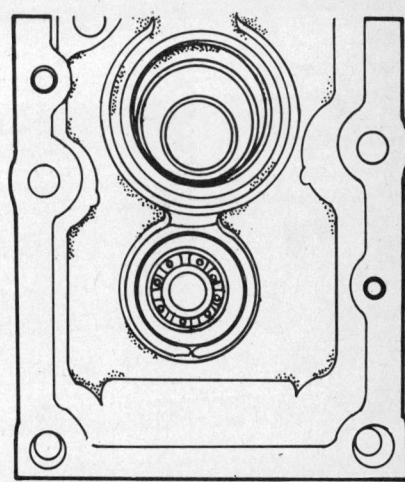
Rear bearing installed

279

John Deere

I.D. drive bearing into clutch housing until it clears snapring groove.

8. Install snapring.
9. Make sure shaft turns freely.
10. Install drop housing under clutch housing. Make sure idler gear and top drive gear mesh. Torque drop housing capscrews to 33-44 ft.lbs.

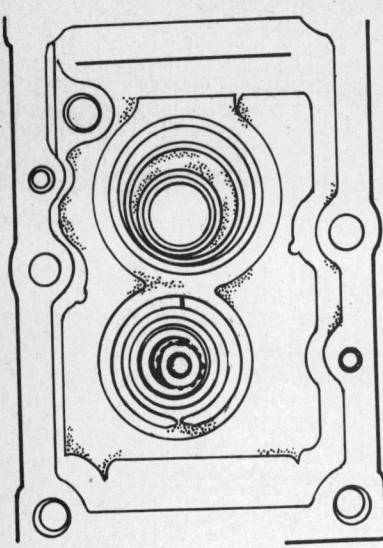

Shaft and gear installed

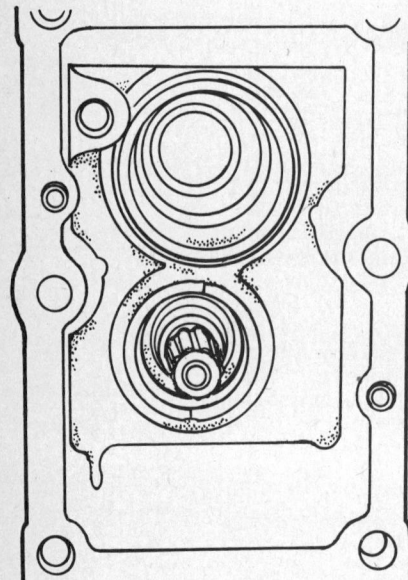

Spacer installed

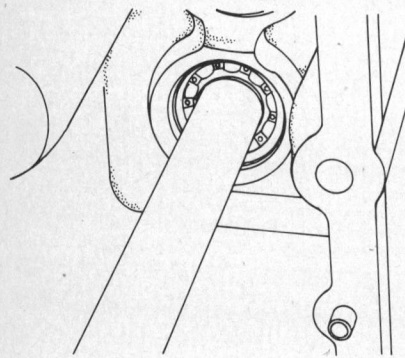

Rear bearing installed

FRONT AXLE UNIT

Removal

1. Remove front axle assembly.
2. Remove front wheels.
3. Remove drag link from king pin support.

Disassembly

OUTER DRIVE HOUSING

─── CAUTION ───
Be sure to support outer drive housing (A) while removing steering arm-to-housing capscrews (B). When cap screws are removed, drive housing may drop from spindle.

1. Remove steering arm-to-outer drive housing capscrews (B) and remove outer drive housing (A).

NOTE: Do not loose shims between steering arm and outer drive housing.

2. Remove the eight outer cover-to-housing capscrews (C).

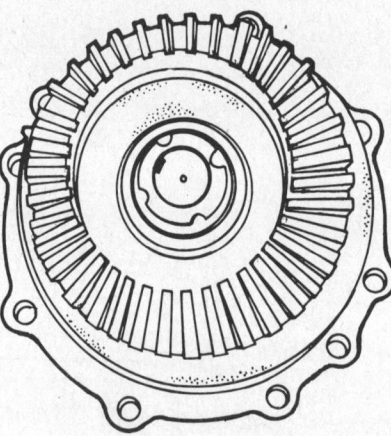

Outer drive gear

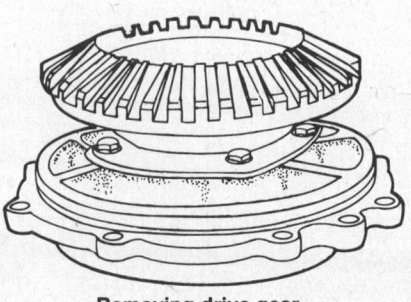

Removing drive gear

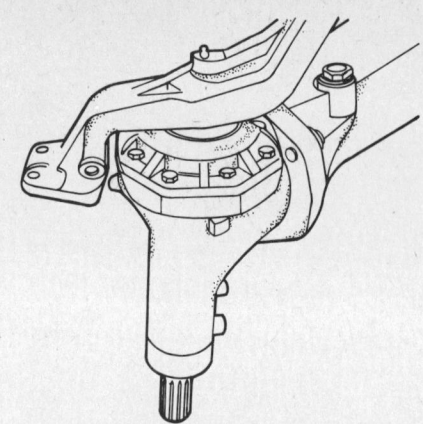

Removing steering arm

3. Insert two of the capscrews removed into jack screw holes (D).
4. While removing cover, note number of shims and placement.
5. Use a dull punch and hammer to remove retaining nut (A). Remove washer (B).
6. Use a brass drift to drive outer drive hub (A) from gear (B).
7. Bearing (C) can now be removed from hub.
8. Inspect splines on hub (A) for wear or damage.
9. Inspect oil seal wear sleeve (B) for scratches.
10. Inspect teeth on drive gear (A) for wear.
11. To remove drive gear from housing, first remove bearing retainer plate capscrews (B).
12. Use a brass drift to drive gear with bearing from housing.
13. Inspect oil seal (C) in housing for wear.
14. Make sure bearing turns freely.
15. Remove steering arm (A) from spindle (B).
16. Inspect bushing (A) in I.D. of arm, and O-ring (B) for wear. I.D. of bushing should be 1.3787 to 1.3807 (35.02 to 35.07 mm).
17. Remove the eight capscrews (C) securing top cover and remove cover.
18. Use a lead hammer to tap on bottom of spindle (B) and remove spindle and gear teeth for wear.
19. Inspect splines on spindle and gear teeth for wear.
20. Inspect needle bearing wear surface on shaft.
21. To remove shaft from bearing and gear, place in a press and press shaft from gear and bearing.
22. Use a knift edge puller and support under bearing to press gear from bearing.
23. Inspect needle bearings (A) in bottom of spindle housing for wear.
24. Remove snapring (B) and bearings.
25. Inspect lower bevel gear teeth for wear. Make sure gear turns freely in bearing.
26. Use a brass drift to drive gear with bearing from housing.
27. Inspect bevel gear (A) and bearing (B). Make sure bearing turns freely. Re-

John Deere

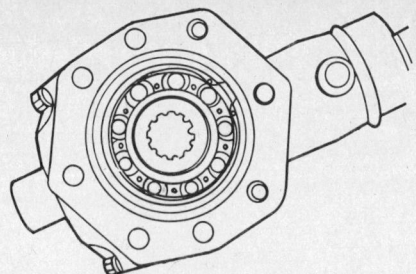

Outer axle bevel gear

move snapring (C) and remove gear with bearing. Inspect gear teeth for wear.

NOTE: Do not lose shims (D) if bearing is removed.

Assembly and Adjustment
OUTER DRIVE HOUSING

1. Install roller bearings (A) in spindle housing.
2. Install snapring (B).
3. Install bearing (A) on bevel gear (B).
4. Install gear with bearing on shaft.
5. Install spindle shaft assembly into spindle housing.
6. Install top cover and torque capscrews to 17 to 22 ft.lbs.
7. If either the spindle shaft top bevel gear (B) or outer axle bevel gear (A) were replaced, the clearance between the two gears should be checked.
8. With outer axle bevel gear (A), bearing (B) and snapring (C) in place, push bevel gear toward spindle shaft bevel gear. This is to obtain maximum gear mesh.
9. Insert a feeler gauge between snapring (C) and gear (A). Clearance should be 0.003 to 0.007 in. (0.07 to 0.17 mm). Three shim thicknesses are available: 0.008 in (0.2 mm), 0.012 in. (0.3 mm), and 0.039 in. (1.0 mm).

NOTE: Before assembling spindle housing to axle housing make sure the gap between eyes of snapring (C) and notch in housing (D) is aligned. This is to insure lubricant flow to outer drive housing.

10. Install the outer seal (B) with the open side of seal facing outward. Outer seal is thinner than inner seal. Drive in seal using a $3^{7}/_{16}$ in. driver disc and $2^{5}/_{8}$ in. pilot disc (OTC No. 27538 and 27525) with handle No. 27488.
11. Install the inner seal (A) with the open side of seal facing inward. Drive in seal using a 3½ in. driver disc and a $2^{5}/_{8}$ in. pilot disc (OTC No. 27539 and 27525) with handle No. 27488.
12. Install oil seal wear sleeve (B) with O-ring onto wheel hub shaft (A).
13. With bearing retainer plate (C) and bearing on ring gear (A), install ring gear in housing (D).
14. Install wheel hub shaft from oil seal side of housing.
15. Use a soft lead hammer to drive shaft until wear ring bottoms on gear.
16. Install bearing inner race with shoulder toward gear. Drive on race with a hammer and brass drift until it bottoms on gear.
17. Install bearing (C) over inner race.
18. Install washer (B) and retainer (A). Use a dull chisel and hammer to tighten retainer.
19. Install oil seals in lower part of outer drive housing.
 a. Use a $3^{7}/_{16}$ in. driver disc and a $2^{5}/_{8}$ in. pilot disc (OTC No. 27538 and 27525) to install both upper (A) and lower (B) seals.
 b. Make sure upper seal is installed with open side up.
 c. Make sure lower seal is installed with open side down.
20. Install lower bevel gear and cover (E).
21. Install same number of shims (C) as were removed on outer drive housing (B).
22. Install ring gear (A) into outer drive housing (B). Be sure to lubricate O-ring (D) before assembly.
23. Install the eight housing capscrews and tighten to 40 ft.lbs.
24. Install a dial indicator as shown. Check backlash between bevel pinion and ring gear. Backlash should be 0.004 to 0.006 in. (0.10 to 0.15 mm). There are two shim thicknesses available: 0.008 in. (0.2 mm) and 0.012 in. (0.3 mm).
25. Install spindle housing onto axle housing as shown.
26. Install arm (A) on spindle housing.
27. Install outer drive housing (A) and tighten capscrews (B) to 40 ft.lbs.
28. Insert a feeler gauge between top

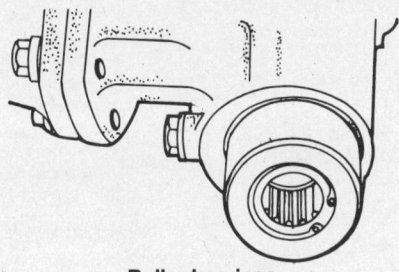

Roller bearings

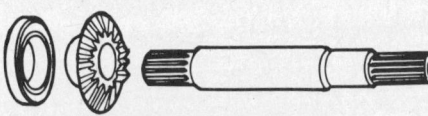

Spindle shaft components

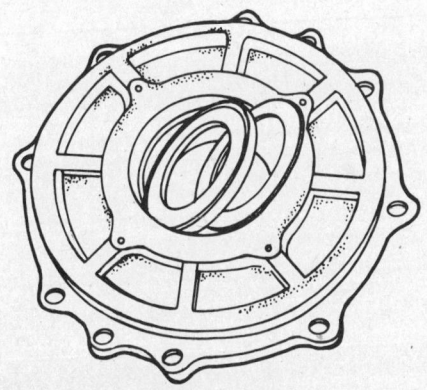

Installing oil seals

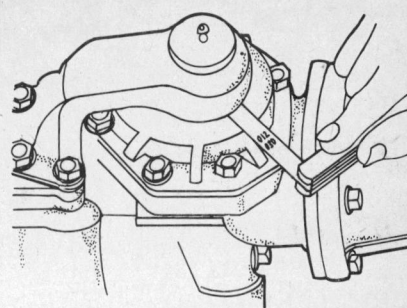

Checking support clearance

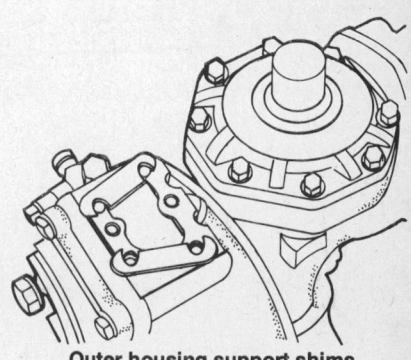

Outer housing support shims

cover (A) and arm (B). Clearance should be 0.001 to 0.006 in. (0.02 to 0.14 mm).

29. To adjust clearance, remove arm (B).
30. Add or subtract shims to obtain correct clearance. Shims are available in 0.012 in. (0.3 mm).
31. Install supprt arm and capscrews and tighten to 47 to 61 ft.lbs.

Front Axle Differential
DISASSEMBLY

1. Remove right axle housing with axle.

NOTE: Use care when removing left axle housing, as differential carrier will be loose in center housing when axle housing is removed.

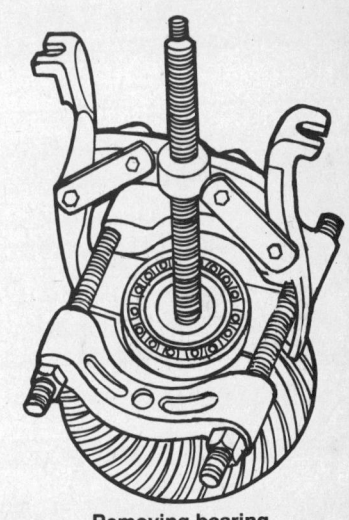

Removing bearing

John Deere

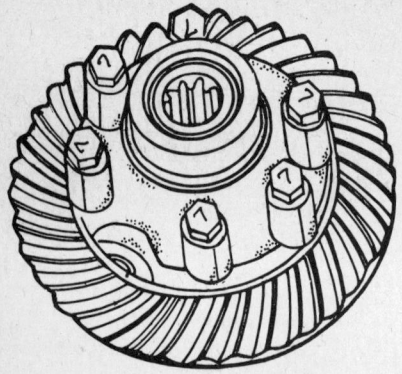

Differential carrier assembly

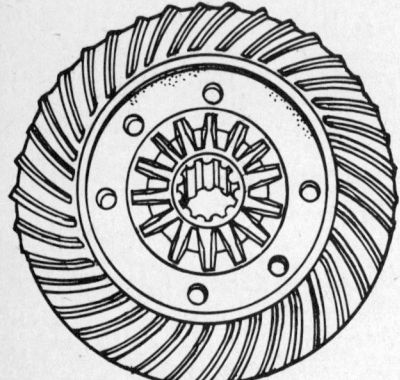

Ring gear removed

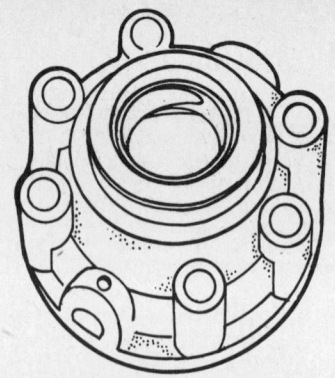

Bevel gear bushing

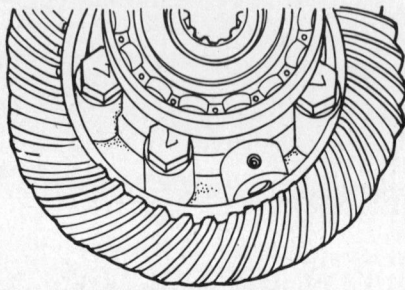

Roll pins installed

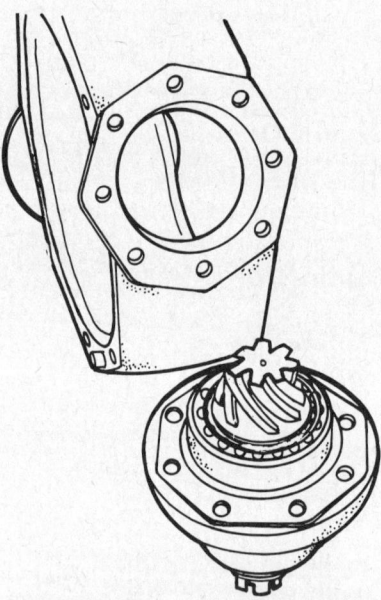

Cone point adjusting shims

Differential Driveshaft

DISASSEMBLY

1. Remove the eight Allen head differential driveshaft housing capscrews and remove housing.
2. Note cone point adjusting shims between differential driveshaft housing and center housing when removing differential driveshaft housing.
3. Remove snapring (A).
4. Remove cotter pin (B).
5. Hold differential driveshaft (D) carefully with pliers to protect splines, and remove differential driveshaft nut (C).
6. Use a brass drift to drive differential driveshaft from splined end out of housing.
7. Use a brass drift to remove bearings and seals.
8. Inspect differential driveshaft parts for loose bearings, worn races, and worn splines on shaft.

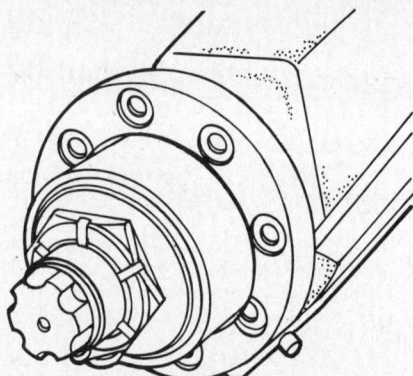

Removing differential driveshaft

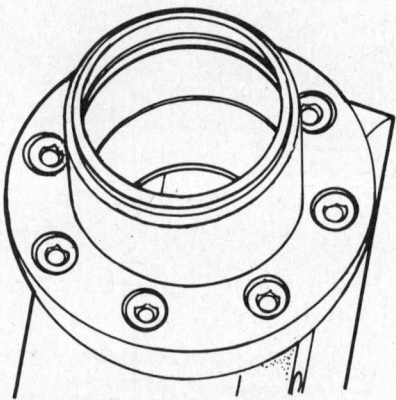

Bearing races installed

ASSEMBLY AND ADJUSTMENT

1. Use a 2 in. pilot and 2⅜ in. driver disc (OTC No. 27515 and 27521) to install both front (A) and rear (B) bearing races. Seat races in housing.
2. Install spacer ring (C). Seat against bearing race (B).
3. Install differential driveshaft with front bearing into housing.
4. Install rear bearing while holding differential driveshaft in place.
5. Install O-ring in groove in shaft.

2. Remove left axle housing with axle.
3. Remove differential carrier from center housing.
4. Use a knife edge puller (A) and gear puller as shown to right side bearing (B).

NOTE: The ends of the differential carrier-to-ring gear capscrews are peened over to prevent loosening. These capscrews if removed, must be replaced with new.

5. Remove the differential carrier-to-ring gear capscrews (A).
6. Use a lead hammer to gently tap ring gear (C) from differential carrier (B).
7. Inspect teeth on both ring gear (A) and bevel gear (B) for wear.
8. Remove bevel gear (B).
9. Inspect bevel gear bushing in ring gear and in carrier for wear. Bushing I.D. should be 1.5748 to 1.5764 in. (40.00 to 40.04 mm). O.D. of gear should be 1.5709 to 1.5717 in. (39.90 to 39.92 mm). Clearance should not exceed 0.01 in (0.3 mm).
10. Drive out double roll pins (A) from differential carrier and pinion shaft (B).
11. Remove shaft and bevel pinions.
12. Inspect O.D. of shaft and I.D. of bevel pinions. Shaft O.D. should be 0.6287 to 0.6293 in. (15.97 to 15.98 mm). Bevel pinion I.D. should be 0.6307 to 0.6311 in. (16.02 to 16.03 mm). Clearance should not exceed 0.02 in. (0.4 mm).

ASSEMBLY

1. Reverse disassembly steps and note the following instructions when assembling differential carrier assembly.

2. If bevel gear bushings in differential carrier and ring gear were removed, install using a 1⅞ in. driver disc and 1⅝ in. pilot disc (OTC No. 27513 and 27509). Bushing in differential carrier should be installed flush with inside of carrier. Bushing in ring gear should be installed flush with counter-bored recess in inside of gear.
3. When installing roll pins into differential carrier and pinion shaft, make sure splits in pins are opposite (180°) from one another.
4. When installing differential carrier-to-ring gear capscrews (A) torque to 25 to 38 ft.lbs. Do not use old capscrews.

John Deere

6. Install oil seal (B) in housing. Seat oil seal against spacer ring.

7. Install oil seal wear sleeve. Lubricate O-ring, lip of oil seal and O.D. of wear sleeve before installation.

8. Install same number of shims as removed on top wear sleeve.

NOTE: These shims are for adjusting position of castellated nut so that when the proper starting drag torque is obtained the castellations (slots) in nut will align with cotter pin hole (B) in shaft.

9. Using an inch-pound torque wrench and a $2^{5}/_{16}$ in. socket, check starting drag torque of differential driveshaft.

   a. Place socket with torque wrench on differential driveshaft nut.

   b. Rotate differential driveshaft with torque wrench noting torque at which differential driveshaft begins to turn. Starting drag torque should be 3-14 in. lbs.

   c. If the starting drag torque is less than 3-14 in.lbs. hold differential driveshaft and tighten differential driveshaft nut. If less than specified, loosen differential driveshaft nut.

10. When the correct specification is obtained, check alignment of slot in nut and hole in differential driveshaft for inserting cotter pin.

11. If slot in nut and hole in differential driveshaft do not align, remove differential driveshaft nut. Insert or add shims (A) to adjust nut position. Shim thickness is 0.004 in. (0.1 mm).

12. Reinstall nut (C) and recheck starting drag torque.

13. Install cotter pin (B).

14. Install differential driveshaft housing into center housing and install the eight Allen head capscrews.

15. Install differential carrier and left side axle housing.

Adjusting Ring Gear-to-Pinion Backlash

1. With left side axle housing attached to center housing, attach a dial indicator as shown.

2. Hold differential driveshaft stationary while moving ring gear to determine backlash. Backlash should be 0.007 to 0.009 in. (0.18 to 0.23 mm). Shims are available in 0.004 in. (0.10 mm) and 0.006 in. (0.15 mm).

3. To add or remove shims, remove left side axle housing.

4. Add shims to decrease backlash and subtract to increase.

Adjusting Differential Driveshaft Cone Point

1. With left side axle housing installed, attach a dial indicator as shown.

2. While turning differential driveshaft, observe run-out of differential carrier on dial indicator.

3. Use a lead hammer to tap on carrier and bring as close as possible to "O" run-out.

4. Rotate differential carrier (A) until one of the machined surfaces on the O.D. of carrier is in line with pinion (B).

5. Insert JDG-48-2 Gauge (C) between pinion (B) and machined surface of carrier (A) to check differential driveshaft cone point.

NOTE: The thinner end of JDG-48-2 Gauge (C) indicates a "Go" thickness and the thicker end indicates a "No-Go" thickness. If Gauge cannot be inserted between pinion and carrier, shims will have to be added. If Gauge can be inserted to "No-Go" thickness shims will have to be removed.

6. To add or subtract shims, remove the eight Allen head capscrews securing differential driveshaft housing to center housing.

7. Add shims to increase clearance and subtract shims to decrease clearance. Three shim thicknesses are available: 0.004 in. (0.1 mm), 0.008 in. (0.2 mm) and 0.012 in. (0.3 mm).

Axle Pivot Assemblies

REMOVAL

Front Pivot

1. Remove capscrews securing front cover and remove cover.

2. Remove capscrews (A) securing plate (B) and remove plate.

3. Remove shims (A) and thrust washer (B).

4. Remove pivot (C).

5. Remove thrust washer.

INSPECTION

1. Inspect thrust washers for wear. Thickness should be 0.08 in. (2.0 mm).

2. Inspect front and rear pivot bushings and pivot shaft for wear. Bushing I.D. should be 2.954 to 2.957 in. (75.03 to 75.12 mm). Shaft O.D. should be 2.952 to 2.953 in. (74.97 to 75.00 mm).

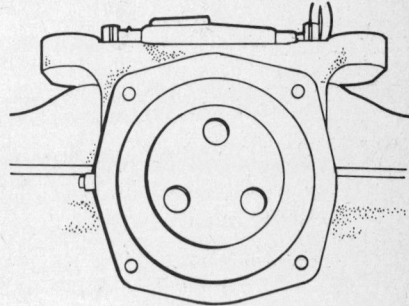

Removing shims

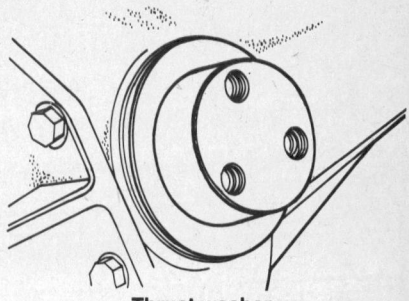

Thrust washer

INSTALLATION AND ADJUSTMENT

1. Reverse removal steps for assembly and note the following instructions.

2. Install thrust washers with chamfers toward center housing.

3. Lubricate O-ring (B) on O.D. of front pivot shaft and O-ring in I.D. of rear pivot before assembly.

4. Torque the three retainer plate capscrews to 110 ft.lbs.

5. Install axle assembly.

6. Using a hammer, tap axle forward taking up end play.

7. Measure clearance between plate (A)

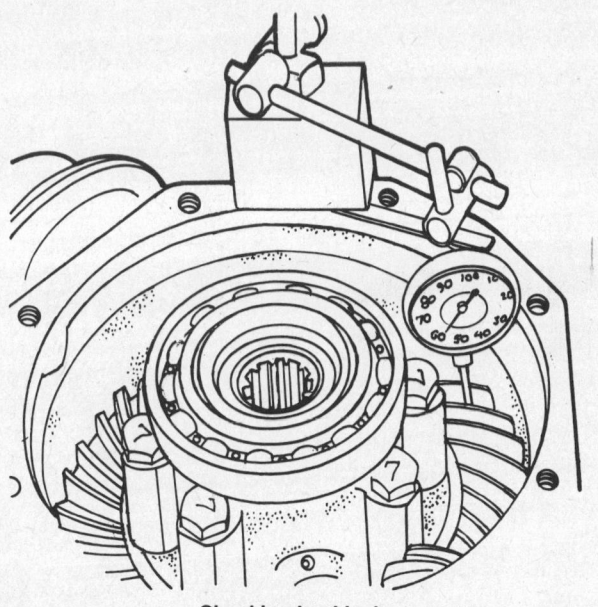

Checking backlash

John Deere

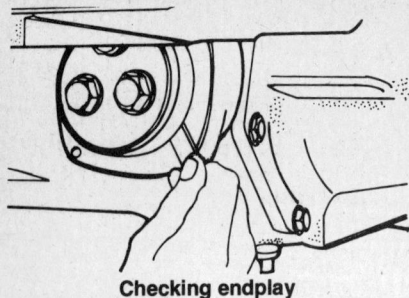

Checking endplay

and center housing (C). Clearance should be 0.002 to 0.012 in. (0.05 to 0.30 mm).

8. To add or subtract shims, first remove plate (A).

9. Shims (A) are available in 0.020 in. (0.5 mm) and 0.040 in. (1.0 mm) thicknesses. Add shims to increase end play, and remove shims to decrease end play.

Hydraulic Pump

NOTE: The 9cc and optional 11cc pumps are not repairable. Service is limited to replacement.

REMOVAL

---- **CAUTION** ----

Before attempting to remove or disassemble any hydraulic components, lower rockshaft and cycle all hydraulic controls with the engine off to relieve hydraulic oil pressure.

---

1. Remove capscrews from both inlet and outlet pipes (A, B) at the hydraulic pump (D).
2. Remove the nuts (C) holding the pump to the engine timing cover and remove pump. Observe the direction of the arrow on the cover plate as the pump is removed.
3. Remove the nut from the pump shaft and remove drive (B).

INSTALLATION

1. Torque nuts mounting pump to 19.5 ft.lbs.
2. Connect pipes to hydraulic pump and torque capscrews to 7 ft.lbs.

SEPARATION

Front End Separation

FRONT END REMOVAL

1. Disconnect ground cable from frame or battery.
2. Disconnect headlight wire.
3. Disconnect hood holding bracket.
4. Remove hood mount bracket capscrews.
5. Remove hood.
6. Remove fanbelt guard.
7. Disconnect cable from starter.
8. Drain cooling system.
9. Disconnect air intake hose from intake manifold.
10. Disconnect radiator overflow hose from expansion tank.
11. Disconnect radiator upper and lower hose from engine.
12. Disconnect drain line from engine block.
13. Disconnect radiator bracket.
14. Disconnect drag link from pitman arm.
15. Support tractor under clutch housing.
16. Insert blocks between side frames and front axle.
17. Remove side frame bolts.
18. Roll front end away from tractor.

INSTALLATION

Reverse the removal steps given on the preceding page and note the installation instructions that follow.

1. Tighten side frame to engine capscrews to 65 ft.lbs.
2. If coolant was lost, add anti-freeze to the radiator.

FRONT END FROM CLUTCH HOUSING SEPARATION

1. Disconnect ground.
2. Disconnect fuel line to filter housing and drain fuel tank.
3. Disconnect hydraulic lines from pump.
4. Disconnect throttle linkage from pump throttle lever.
5. Disconnect tachometer.
6. Disconnect line from thermostart bowl.
7. Disconnect drag link from Pitman arm.
8. Support under clutch housing.
9. Insert blocks under side frames.
10. Attach lifting eyes to engine.
11. Disconnect wires from:
   a. Starter
   b. Temperature sensor
   c. Thermostart
   d. Oil pressure sensor
   e. Alternator
12. Remove clutch housing to engine capscrews.
13. Slowly roll front of tractor away. (Driveshaft stays with rear of tractor.)

INSTALLATION

Reverse the removal steps and note the installation instructions that follow.

NOTE: If there is trouble removing old gaskets, spray on John Deere Paint and Decal Remover. Let it set for a few minutes; then scrape off gaskets.

1. Insure that both clutch springs are properly attached.
2. Engage PTO and turn shaft while pushing tractor together.
3. Tighten hydraulic lines-to-pump capscrews to 7.5 ft.lbs.
4. Tighten clutch housing-to-engine capscrews to 65 ft. lbs.

ENGINE

Removal

1. Separate front end from engine as previously instructed.
2. Disconnect fuel line from filter housing and plug line or drain tank.
3. Disconnect hydraulic lines from pump.

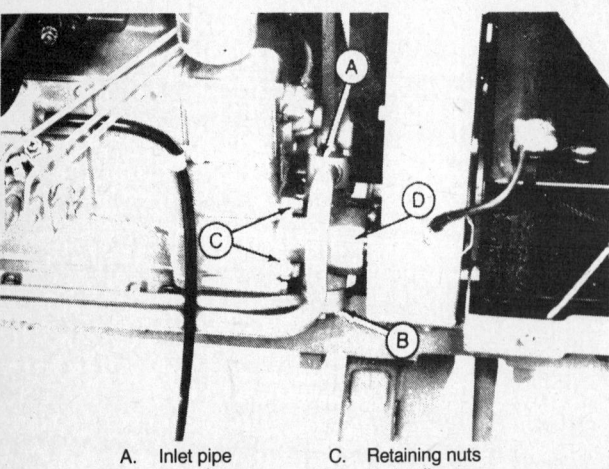

A. Inlet pipe
B. Outlet pipe
C. Retaining nuts
D. Hydraulic pump

Hydraulic pump removal

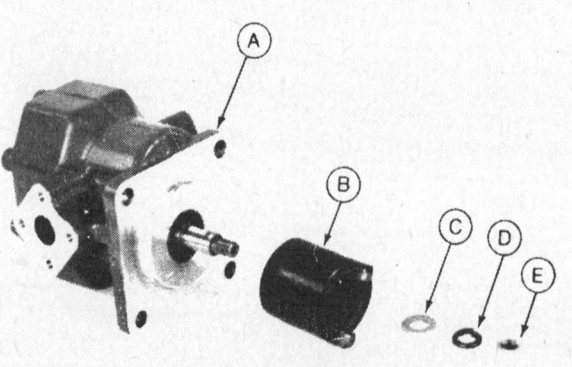

A. Hydraulic pump
B. Pump drive
C. Flat washer
D. Lockwasher
E. Nut

Hydraulic pump and drive

John Deere

4. Disconnect throttle linkage from pump throttle lever.
5. Disconnect tachometer cable.
6. Disconnect line from thermostart bowl.
7. Disconnect wires from:
   a. Starter
   b. Temperature sensor
   c. Thermostart
   d. Oil pressure sensor
   e. Alternator
8. Connect lifting eyes to engine.
9. Attach an overhead hoist to the lift eyes.
10. Remove clutch housing-to-engine capscrews.
11. Slowly pull engine straight out from clutch housing.

NOTE: When setting the engine down, support it at the edges of the oil pan. Be sure that no weight is put on the center of the pan.

Installation

Reverse the removal steps on the preceding pages and note the installation instructions that follow.
1. Insure that both clutch springs are properly attached.
2. Tighten hydraulic lines-to-pump capscrews to 7.5 ft.lbs.
3. Tighten clutch housing-to-engine capscrews to 65 ft.lbs.
4. Tighten side frame-to-engine capscrews to 65 ft. lbs.

Intake Manifold

REMOVAL

1. Thoroughly clean area around intake manifold before removing parts.
2. Disconnect the air intake hose (A), thermostart fuel supply hose (B), and wiring lead (C).
3. Remove fuel pipes (D).

A. Intake hose
B. Supply hose
C. Wiring lead
D. Fuel pipes
E. Stud nuts

Intake manifold removal

4. Remove three nuts (E) which secure manifold to cylinder head, and carefully remove manifold.

REPAIR

1. Inspect the intake manifold to see that it is not cracked, or has any other defects which would permit unfiltered air to enter the combustion chamber.
2. Replace the intake manifold, should it be defective.

INSTALLATION

1. If gasket is not in good condition, remove all the old gasket material from cylinder head and intake manifold, and replace with a new one.
2. Install manifold using a new gasket. Torque stud nuts to 18 ft.lbs.
3. Install fuel pipes and air intake hose.
4. Connect thermostart hose and wiring lead.
5. Bleed the fuel system.
6. Start engine and check operation.

Valves

ADJUSTMENT

1. On engine numbers -009000 (850) or -012000 (950) and up, disconnect decompression linkage.
2. Remove thermostart reservoir support capscrews.
3. Remove expansion tank bracket capscrews.
4. Remove rocker arm cover capscrews, and remove rocker arm cover.
5. Crank engine until No. 1 cylinder is at TDC of its compression stroke. Both valves should be in the up position (rocker arms loose).
6. Check the intake and exhaust valve clearances of the No. 1 cylinder. No. 1 CYLINDER IS AT REAR OF ENGINE. Adjust to the following specifications.

INTAKE AND EXHAUST VALVE CLEARANCE SPECIFICATIONS

850	0.2 mm (0.008 in.)
950	0.15 mm (0.006 in.)

7. Turn the cranksahft clockwise 240° to align the TDC mark of the No. 3 cylinder.
8. Check the valve clearance of the intake and exhaust valves of the No. 3 cylinder, and adjust to proper specification.
9. Turn the crankshaft another 240° clockwise to align the TDC mark of the No. 2 cylinder.
10. Check the valve clearance of the intake and exhaust valve of the No. 2 cylinder, and adjust to proper specification.

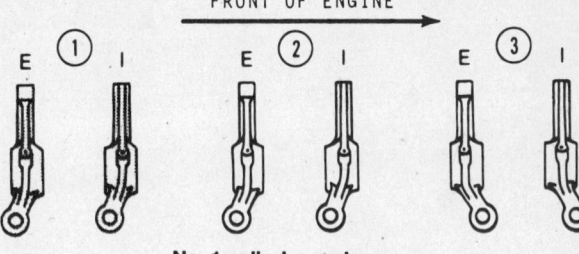

No. 1 cylinder rocker arms

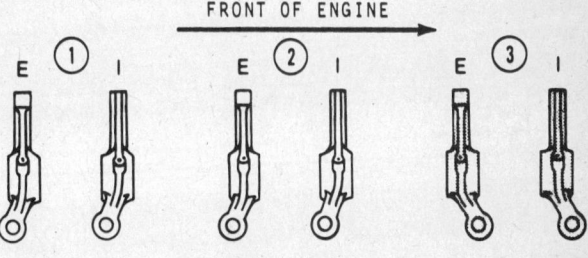

No. 3 cylinder rocker arms

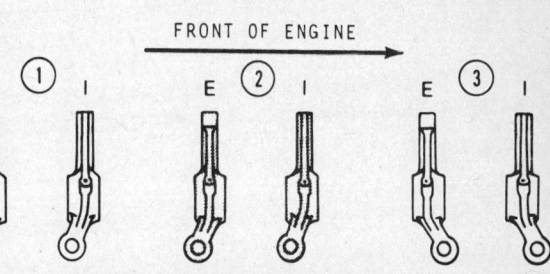

No. 2 cylinder rocker arms

John Deere

A. Decompression lever (early models only)

Decompression service windows

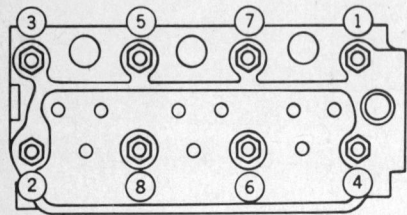

Cylinder head bolt loosening sequence, all 850 and early 950

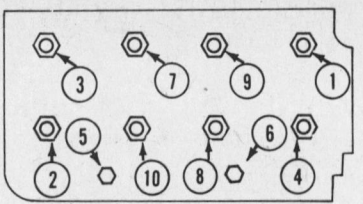

Cylinder head bolt loosening sequence, 950 from serial #012001 and up

ADJUSTING DECOMPRESSION DEVICE

Engine nos. -009000 (850) -012000 (950) and up

1. Crank engine until No. 1 cylinder is at TDC of its compression stroke. Both valves should be in the up position (rocker arms loose).
2. Remove the decompression service windows from rocker arm cover.
3. Push decompression lever (A) rearward into decompression position.
4. Loosen locknut (A).
5. Turn in adjusting screw (B) to the 0 in. valve clearance position, and screw in adjusting screw one additional turn.

NOTE: One turn adjusts the decompression clearance 0.03 in. (0.8 mm). This means that when the decompression device is used, the exhaust valve is held 0.03 in. (0.8 mm) from its sealed position.

6. Tighten locknut.
7. Adjust the remaining decompression levers, one at a time, by rotating crankshaft to bring each piston to TDC.

NOTE: After adjusting, manually turn the crankshaft to check whether the valves come in contact with the pistons due to excessive decompression lift.

Rocker Arm Assembly

REMOVAL AND INSTALLATION

1. Remove rocker arm cover.

NOTE: When removing arm components, identify for reassembly into original positions.

2. On 850 Tractor remove set plate securing rocker arm to shaft. On 950 remove capscrew securing rocker arm to shaft.
3. Measure O.D. of rocker arm shafts and I.D. of bushings in rocker arm and compare measurements taken with the following specifications:

ROCKER ARM ASSEMBLY SPECIFICATIONS

	New Part	Wear Limit
Shaft O.D.		
850	15.98 to 16.00 mm (0.6291 to 0.6299 in.)	15.90 mm (0.6260 in.)
950	16.98 to 17.00 mm (0.6685 to 0.6693 in.)	16.90 mm (0.6653 in.)
Bushing I.D.		
850	16.02 to 16.03 mm (0.6306 to 0.6313 in.)	16.10 mm (0.6338 in.)
950	17.02 to 17.03 mm (0.6699 to 0.6706 in.)	17.10 mm (0.6732 in.)
Bushing-to-Shaft Clearance		
	0.016 to 0.052 mm (0.0006 to 0.002 in.)	0.15 mm (0.006 in.)

Cylinder Head, Valves, Valve Springs

REMOVAL

1. Remove rocker arm cover as previously described.
2. Remove fan belt and radiator hoses.
3. Remove air intake pipe.
4. Remove fuel leak-off and injection lines.
5. Remove injection nozzles (see Fuel System).
6. Remove intake manifold.
7. Remove cylinder head lubrication oil line capscrews.
8. Remove exhaust manifold with muffler.
9. Remove cylinder head capscrews by following loosening sequence.

NOTE: A new cylinder head was adopted for 950 Tractors with Serial No. (012001 and up). It is equipped with 2 additional capscrews. Remove according to the sequence illustrated.

10. Use a valve spring compressor to remove valves from head.

INSTALLATION

1. Apply AR44402 Lubricant, or equivalent, to valve stems and guides, and install valves in guides from which they were removed.

NOTE: Valves must move freely and seat properly.

2. Install valve springs and rotators.
3. Compress valve springs and install retainer locks.
4. Strike the end of each valve three or our times, with a soft mallet, to ensure proper positioning of the retainer locks.
5. Install new head gasket on cylinder block. Install dry—Do not use bonding materials.
6. Install head on block.
7. On 850 Tractors, and 950 Tractors with Serial No. prior to -012000, coat cyl-head nuts with clean engine oil and torque to 43 ft./lbs. following the sequence illustrated.
8. Retighten capscrews in sequence to 87 ft./lbs. and finally tighten to 116 to 130 ft./lbs.

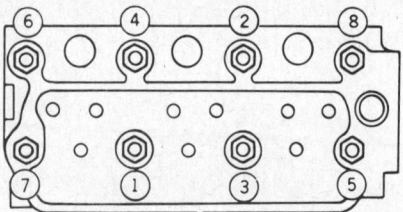

Cylinder head tightening sequence, all 850 and early 950

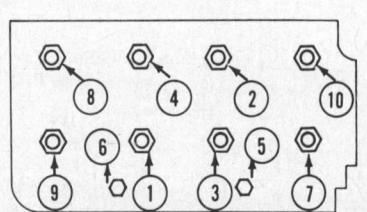

Cylinder head bolt tightening sequence, 950 from serial #012001 and up

John Deere

9. On 950 Tractors with Serial No. 012001-and up, coat cylinder head nuts and capscrews with clean engine oil and torque to 43 ft.lbs. following sequence illustrated.

NOTE: On late model 950 Tractors only, torque numbers 5 and 6 capscrews to 20 ft./lbs. DO NOT OVERTIGHTEN.

10. Retorque capscrews in sequence to 87 ft./lbs. and finally to 116 to 130 ft./lbs.

NOTE: On late model 950 Tractors only, retorque numbers 5 and 6 capscrews to 40 ft./lbs. and finally to 55 ft./lbs. DO NOT OVERTIGHTEN.

11. Install wear caps on valves.
12. Install rocker arm assemblies on head.
13. Torque rocker arm support nuts to 41 ft./lbs. on 850 Tractor and 48 ft./lbs on 950 Tractor.
14. Adjust the valve clearance.
15. Install rocker arm cover.
16. On earlier model tractors, adjust decompression device.

Camshaft

REMOVAL

1. Remove tractor front end and timing gear cover.
2. Remove rocker arm cover.
3. Remove cylinder head.

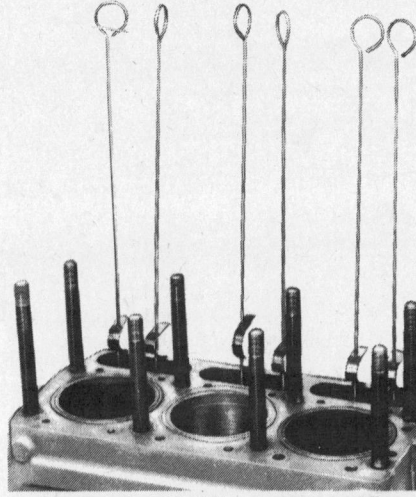
Using the D-15001NU magnetic holding tool set

Weight support retaining nut

4. Remove camshaft bearing retainer screw.
5. Use D-15001NU magnetic holding tool set to hold cam followers away from camshaft lobes during removal.
6. Carefully remove camshaft from cylinder block so that camshaft lobes do not drag in bores.

Removing Camshaft Bearing and Gear

1. Inspect camshaft ball bearings, making sure they rotate smoothly, and are not loose or worn.
2. Slide governor sleeve from camshaft.
3. Remove weight support retaining nut from camshaft.

NOTE: Weight support retaining nut has lefthand threads.

4. Remove camshaft gear from camshaft.
5. Support camshaft bearings in a press.

NOTE: Prevent camshaft from striking floor when pressing camshaft from gear.

6. Press camshaft from bearings.

ASSEMBLY

1. Support camshaft in a press, and press on bearing with JDE-121 Installation Tool. Bearing should be pressed on until it bottoms on shoulder of camshaft.
2. Make sure key for cam gear is in place before installing gear. Install gear on camshaft.
3. Install governor weight support as shown.
4. When installing retaining nut, make sure shoulder on nut faces toward camshaft gear. Retaining nut has lefthand threads.

INSTALLATION

1. If cam followers have been removed, reinstall using the D-15001NU magnetic holding tool set to hold cam followers away from camshaft bore until camshaft is installed.
2. Coat camshaft journals and bearing bores with clean engine oil.
3. When installing camshaft, make sure lobes do not drag in bores.
4. Align timing marks on camshaft and crankshaft gears during installation.
5. Install bearing retainer screw with retainer.
6. Install governor sleeve. Make sure governor sleeve is installed on two governor weights.

Pistons

REMOVAL

NOTE: It is not necessary to remove the engine to service pistons, rods and liners.

1. Remove cylinder head.
2. Remove oil pan.

NOTE: Keep bearing inserts with their respective rods and caps, and mark rods, pistons and caps to insure correct reassembly.

3. Remove rod capscrews and rod caps.
4. Gently tap piston through top of cylinder.

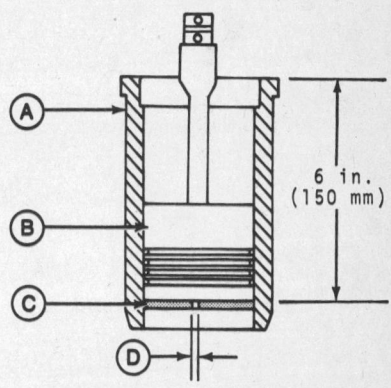

A. Liner
B. Piston head
C. Piston ring
D. Ring gap

Checking ring gap

Bearing retainer

Camshaft timing marks

John Deere

INSTALLATION

1. Install pin bushing using JDE-117 removal and installation tool.
2. Install one pin retaining snapring in each piston.
3. Heat pistons at 80° C (176° F) for 10 to 15 minutes.
4. Coat piston pins with engine oil.
5. Remove pistons from heat and install piston connecting rod and piston pin.
6. Install remaining snaprings.

NOTE: Make sure connecting rod moves smoothly with piston.

7. Push each piston ring (one at a time) into the cylinder using the piston head. Remove piston.
8. Measure the ring gap with a feeler gauge and compare measurement taken with the following specifications.

NOTE: Push ring into liner approximately 6 in. (150 mm) before measuring ring gap.

PISTON RING GAP SPECIFICATIONS

	New Part	Wear Tolerance
Top, Second, Third and Oil Ring	0.01 to 0.02 in. (0.3 to 0.5 mm)	0.06 in. (1.5 mm)

9. If there is excessive gap, replace rings. If gap is too small, file to meet specifications.

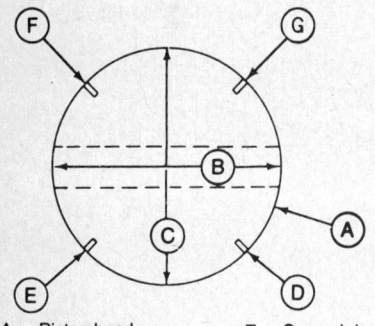

- A. Piston head
- B. Piston pin direction
- C. Piston thrust direction
- D. Top ring
- E. Second ring
- F. Third ring
- G. Oil ring

Staggering ring gaps

NOTE: Make sure manufacturers identification on ring is facing toward top of piston during assembly.

10. Remove rings from cylinder and install on piston using tool from JDE-120 ring expander kit.
11. Before installing pistons in cylinders, make sure ring gaps are between piston pin direction (A) and piston thrust direction (B).
12. Coat pistons, liners and inside of ring compressor with clean engine oil. Use JDE-115 ring compressor on 850 engine and JDE-116 ring compressor on 950 engine.
13. Carefully place piston in ring compressor making sure identification marks on rod face flywheel side of engine.
14. Carefully place ring compressor with piston and rod over liner.

NOTE: Be sure crankshaft journals are not damaged when pushing piston into cylinder.

15. With piston centered in installing tool and rings staggered correctly, push piston into liner.
16. Install connecting rod caps, and tighten to 40 ft.lbs. on 850 tractor and 47 ft.lbs. on 950 tractors.

Cylinder Liners

REMOVAL

1. Use a dial indicator to measure the height of liners before removal from block.

LINER HEIGHT SPECIFICATION

850	0.05 to 0.13 mm (0.002 to 0.005 in.)
950	0.03 to 0.10 mm (0.001 to 0.004 in.)

NOTE: When using cylinder liner puller (A) to remove liners, insure that jaw of puller is correctly positioned before attempting to remove liner.

2. Use D-01062AA cylinder liner puller to remove cylinder liner.

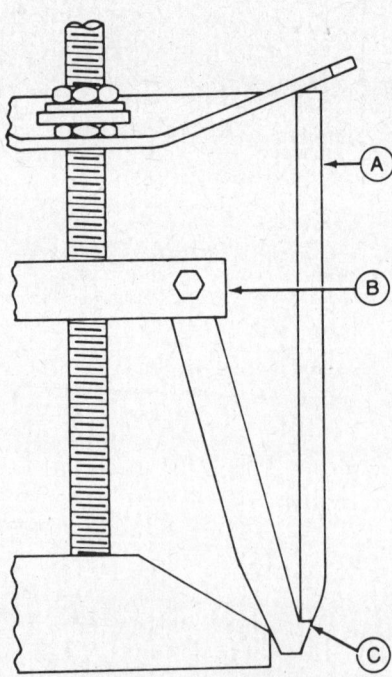

- A. Cylinder liner
- B. D-01062AA (no. 974) liner puller
- C. Proper jaw position

Installing cylinder liner puller

3. Measure cylinder liner inside diameter and compare reading taken with the following specifications:

CYLINDER LINER I.D. SPECIFICATIONS

	New Part	Wear Tolerance
850	80.00 to 80.03 mm (3.1496 to 3.1510 in.)	80.20 mm (3.157 in.)
950	90.00 to 90.03 mm (3.5433 to 3.5445 in.)	90.20 mm (3.551 in.)

4. Compare liner I.D. measurements to piston skirt measurements. See specifications.

LINER-TO-PISTON CLEARANCE

	New Part	Wear Tolerance
850	0.08 to 0.15 mm (0.003 to 0.006 in.)	0.30 mm (0.012 in.)
950	0.13 to 0.20 mm (0.0051 to 0.0079 in.)	0.45 mm (0.018 in.)

LINER INSTALLATION

1. Be sure liner bore is clean.
2. Install O-rings (gray O-ring on top, red on bottom).
3. Lubricate O-rings and bottom of liner with AR54749 lubricant soap.
4. Carefully push cylinder in place. Seat liner with a wood block and hammer.
5. Check liner height. Excessive height may be a result of rust or scale not removed from shoulder in block.

Crankshaft

REMOVAL

1. Remove crankshaft pulley-to-crankshaft capscrew.
2. Use D-01207AA puller to remove pulley.
3. Remove timing gear cover.
4. Use No. 515 puller to remove crankshaft gear.
5. Remove oil pan.

NOTE: Removal of crankshaft will be easier if cylinder block is inverted so that flywheel end is up.

6. Remove the rear main bearing housing capscrews. If housing cannot be removed by hand, insert capscrews in the two jack screw holes and turn in screws evenly to remove housing.
7. Secure a chain hoist to crankshaft.

John Deere

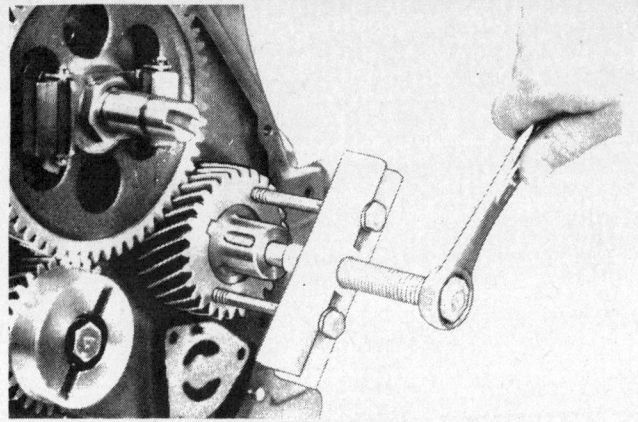

Removing crankshaft gear

A. Driver
B. No. 27532 disc
C. No. 27520 disc
D. Oil seal

Installing front seal

8. Slightly lift crankshaft so weight is taken off of center main bearing housings.
9. Remove the two set bolts from center main bearing housings.
10. Carefully lift crankshaft from block.

NOTE: Be careful not to damage the outer surfaces of the intermediate main bearing bodies during crankshaft removal.

INSPECTION

1. Check crankshaft for cracks or signs of load stress.
2. Measure each main journal O.D. at points shown and compare measurement taken with the following specifications. Also, compare bearing I.D. with journal O.D. to determine oil clearance.

CRANKSHAFT JOURNAL SPECIFICATIONS 850

New Part	Wear Tolerance
2.754 to 2.755 in. (69.95 to 69.96 mm)	2.753 in. (69.92 mm)

CRANKSHAFT JOURNAL SPECIFICATIONS 950

	New Part	Wear Tolerance
Rear	3.541 to 3.542 in. (89.95 to 89.96 mm)	3.540 in. (89.92 mm)
Front and Intermediate	2.754 to 2.755 in. (69.95 to 69.96 mm)	2.753 in. (69.92 mm)

INSTALLATION

1. Coat intermediate bearings with engine oil and install in bearing housing halves.

NOTE: Make sure tangs on bearings fit in recesses in bearing housing.

2. Assemble bearing housings with bearings on crankshaft with "F" mark toward flywheel side of crankshaft.

NOTE: Make sure thrust bearing is installed in bearing housing closest to flywheel.

3. Tighten main bearing housing capscrews to 33 to 36 ft.lbs.
4. Stand cylinder block upright, and lower crankshaft into block.
5. Before completely installing, align setscrew holes between intermediate main bearing housing and block.
6. Lightly oil setscrews.
7. Completely lower crankshaft into position and install intermediate bearing housing setscrews. Do not tighten.
8. Tighten setscrew in bearing housing nearest flywheel first (thrust bearing housing) to 58 ft.lbs.
9. Tighten the remaining setscrew to 58 ft.lbs.
10. After tightening setscrews, make sure the crankshaft rotates smoothly.
11. Apply a light coat of oil to the rear main bearing and oil seal.
12. When installing rear main bearing housing on 850 tractor, make sure mark on housing is facing downward (toward oil pan). This so oil hole in block and oil hole in bearing housing are aligned. The bolt hole pattern in 950 tractor bearing housings prevents incorrect installation.
13. Carefully place rear main bearing housing on block and tighten capscrews to 32 to 36 ft.lbs.
14. Install flywheel and tighten capscrews to 54 ft.lbs.

Front Oil Seal

REMOVAL

1. Remove timing gear cover.
2. Check oil seal for wear or deterioration.
3. Use a punch to remove oil seal from timing gear cover.

INSTALLATION

The front oil seal is a "pumping" seal. A rib design in the seal prevents oil from leaking during crankshaft operation. DO NOT apply grease to the lip of seal. This would eliminate seals pumping ability.

1. Place seal over bore as shown.
2. Use No. 27489 Driver with No. 27532 ($3^{1}/_{16}$ in. Disc and No. 27520 ($2^{5}/_{16}$ in.) Disc to install oil seal.

Rear Oil Seal

REMOVAL

1. Separate the engine from clutch housing.
2. Remove rear main bearing housing from cylinder block.
3. Use a pry bar to remove oil seal.

INSTALLATION

The rear oil seal is a "pumping" seal. A rib design in the seal prevents oil from leaking during crankshaft operation. DO

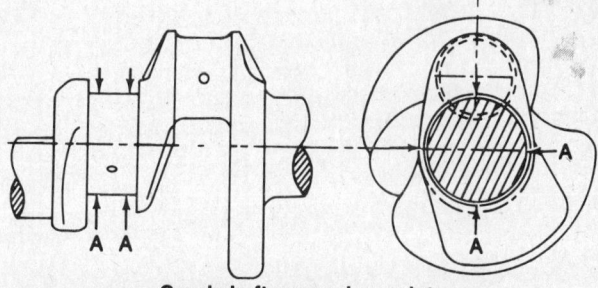

Crankshaft measuring points

John Deere

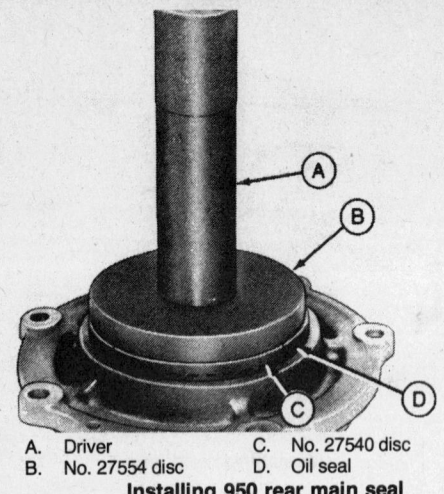

A. Driver
B. No. 27554 disc
C. No. 27540 disc
D. Oil seal

Installing 950 rear main seal

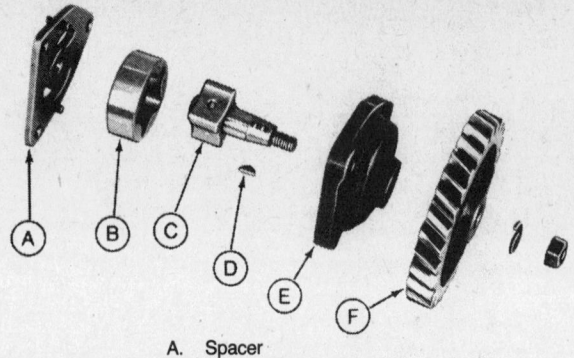

A. Spacer
B. Outer rotor
C. Inner rotor-shaft assembly
D. Key
E. Pump body
F. Gear

Oil pump components

NOT apply grease to the lip of seal. This would eliminate seal's pumping ability.

1. When installing oil seal, make sure open side of seal faces cylinder block side of housing.
2. Use No. 27489 Driver with No. 27554 (4$^{7}/_{16}$ in.) Disc and No. 27540 (3$^{9}/_{16}$ in.) Disc to install oil seal on 950 tractor. Use No. 27489 Drive with No 27532 (3$^{1}/_{16}$ in.) Disc and No. 27520 Disc to install rear oil seal in 850 tractor.

Flywheel

REMOVAL AND INSTALLATION

1. Separate the engine from the clutch housing.
2. Remove clutch-to-flywheel capscrews and remove clutch assembly.

NOTE: Always use new flywheel bolts when they are loosened or removed.

3. Installation is the reverse of removal.

Oil Pump

REMOVAL

1. Remove timing gear cover.
2. Remove the three capscrews securing pump body to block and remove pump.

INSPECTION

1. Place a straightedge across machined surface on pump body.
2. Use a feeler gauge to check recess of pump rotors in pump body. Compare measurement taken with the following specifications.

PUMP ROTOR RECESS SPECIFICATIONS

New Part	Wear Tolerance
0.0004 to 0.002 in. (0.01 to 0.05 mm)	0.004 in. (0.10 mm)

If measurement exceeds wear tolerance, replace pump.

3. Use a feeler guage to check clearances between outer rotor and pump body. Compare measurement taken with the following specifications:

OUTER ROTOR-TO-PUMP BODY CLEARANCE

New Part	Wear Tolerance
0.002 to 0.004 in. (0.05 to 0.10 mm)	0.006 in. (0.15 mm)

If measurement exceeds wear tolerance, replace pump.

4. Check inner rotor-to-outer rotor clearance between a high point on inner rotor, and high point on outer rotor. Compare measurement taken with the following specifications:

INNER ROTOR-TO-OUTER ROTOR SPECIFICATIONS

New Part	Wear Tolerance
0.002 to 0.004 in. (0.05 to 0.10 mm)	0.006 in. (0.15 mm)

If measurement exceeds wear tolerance, replace pump.

ASSEMBLY AND INSTALLATION

1. Install inner rotor-shaft assembly (C) in pump body (E).
2. Install outer rotor (B) in pump body.
3. Install key in inner rotor-shaft assembly.
4. Install capscrews in pump body.
5. Install gear on shaft.
6. Install lockwasher and nut on shaft.
7. Tighten nut to 12 to 15 ft.lbs. DO NOT OVERTIGHTEN NUT.
8. Install spacer plate (A) on pump body.
9. Install pump so spring pin (B) enters hole in block.

A. Spacer plate
B. Spring pin

Oil pump installation

Oil pressure regulating valve

John Deere

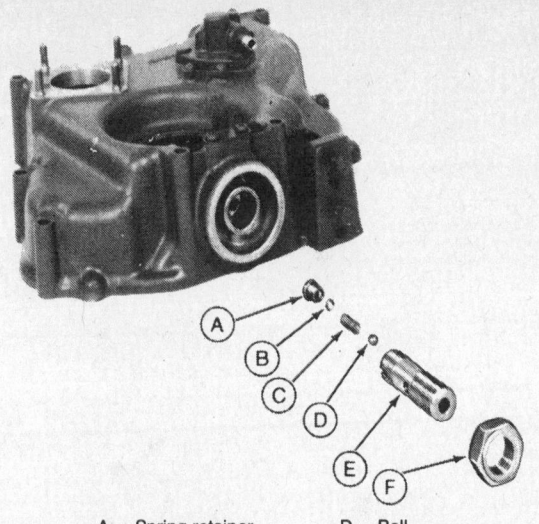

A. Spring retainer
B. Shim
C. Spring
D. Ball
E. Valve body
F. Locking nut

Oil pressure regulating valve components

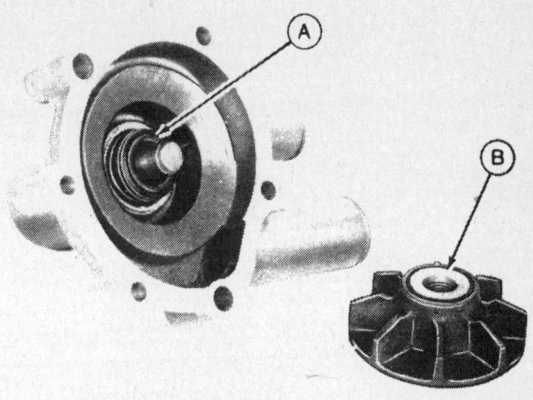

A. Seal
B. Ceramic insert

Impeller, seal and housing

Oil Pressure Regulating Valve

REMOVAL, DISASSEMBLY AND INSPECTION

1. Remove oil filter.
2. Remove oil pressure regulating valve body.
3. Remove spring retainer from valve body.
4. Remove shim(s), spring and check ball from valve body.
5. Add shim(s) to increase oil pressure and subtract shims to decrease oil pressure. Addition of one shim increases oil pressure 2.6 psi.

INSTALLATION

1. Install ball (D), spring (C), shim(s) (B), and spring retainer (A) in valve body (E).
2. Install valve body in timing gear cover and screw completely in by hand.
3. Install locking nut (F) on valve body.
4. Install oil filter.

COOLING SYSTEM

Tractors Prior to Serial Number 009000 (850) and 012000 (950)

WATER PUMP

Removal

1. Remove expansion tank.
2. Remove coolant hoses from water pump.
3. Remove radiator shroud capscrews.
4. Remove fan belt from alternator pulley.
5. Remove water pump-to-thermostat housing capscrews and remove water pump with radiator shroud from right side of tractor.

Disassembly

1. Use a press to push bearing shaft from fan hub.
2. Remove bearing retainer snapring from water pump housing.
3. Support water pump in a press. Press bearing from impeller and water pump housing.

Inspection, Assembly and Installation

1. Inspect seal (A) and ceramic insert in impeller (B) for wear.
2. If seal was removed, use a one-inch socket (A) and hydraulic press to reinstall.
3. Use JD-243 Installation Tool to press

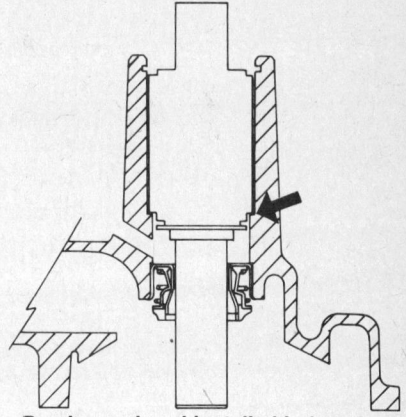

Bearing and seal installed in housing

Spring retainer

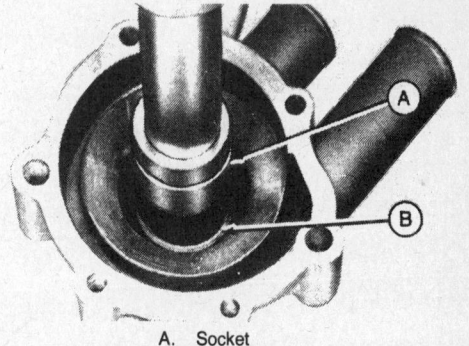

A. Socket
B. Seal

Installing seal

John Deere

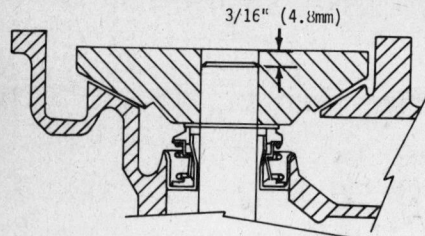

Measuring impeller installation depth

in water pump bearing. Press in bearing until it bottoms on shoulder in water pump housing.

4. Support water pump in a press as shown. Press in impeller until impeller comes within 3/16 in. (4.8 mm) of being flush with bearing shaft.

5. Place a spacer between water pump shaft and support on press. This is to prevent disturbing impeller position on shaft when installing hub. Press on hub until hub is flush with end of bearing shaft.

THERMOSTAT

Removal and Installation

1. Visually check area around thermostat housing for leaks.
2. Remove radiator hose from thermostat cover.
3. Remove thermostat cover.
4. Remove thermostat.
5. Check thermostat in D-05103ST Thermostat Tester. Thermostat should open at 160° F (71° C).

Tractors from Serial Number 009001 (850) and 012001 (950)

WATER PUMP

Removal

1. Drain cooling system.
2. Remove fan belt from alternator pulley.
3. Remove water pump hoses from water pump.
4. Remove radiator shroud-to-radiator capscrews.
5. Remove water pump-to-cylinder head capscrews and remove water pump with radiator shroud.

Disassembly

1. Use a press to push fan hub from bearing shaft.
2. Support housing in press.

NOTE: Be sure there is enough clearance to allow impeller to pass through supports.

3. Using JD-243 Bearing Driver press bearing shaft with impeller from pump housing.
4. Support impeller by hub (A) on press. Use a capscrew (B) to press impeller from shaft.

Assembly

1. Install new seal (A) and ceramic insert (B).
2. Use a 1 in. socket (A) and hydraulic press to install new seal (B) in pump housing.

NOTE: Heat pump body to 80 to 90° C (176 to 194° F) in hot water before installing bearing shaft.

3. Using JD-243 Bearing Driver (A) press bearing shaft into housing until bearing surface (B) is flush with housing surface (C).
4. Support pump in press as shown. To keep shaft from moving during installation, place a spacer (A) under it.
5. Use a 1 in. socket and hydraulic press to push fan hub (8) onto bearing shaft until the distance from the end of the bearing shaft to the top surface of the fan hub is 8.5 mm (11/32 in).
6. To prevent damage to impeller during installation place a spacer (A) on impeller hub surface (B).
7. Support pump in press as shown. Press impeller in until it comes within 9.5 mm (3/8 in.) of being flush with end of shaft.

Thermostat

REMOVAL AND INSTALLATION

1. Visually check area around thermostat housing for leaks.
2. Remove radiator hose from thermostat cover.
3. Remove thermostat cover.
4. Remove thermostat.
5. Check thermostat in D-05103ST Thermostat Tester. Thermostat should open at 71° C (160° F).

FUEL SYSTEM

Fuel Tank

REMOVAL

1. Disconnect the fuel tank-to-fuel filter line (rubber hose) at the fuel filter. Drain the fuel tank.
2. Disconnect the battery ground cable.
3. Remove the steering wheel.
4. Remove cowl and instrument panel.
a. Remove access panel (A).
b. Disconnect wiring harness leads from all electrical components (switches, lights, etc.).
c. Disconnect tachometer drive cable from tachometer (B) and decompression device* control cable (C), at engine end.
d. Remove instrument panel and cowl (D) from tractor.
e. Disconnect the fuel return hose (E) from tank (G).
f. Disconnect and remove tank mounting straps (F).
5. Remove fuel tank.

INSTALLATION

1. Make sure that the fuel tank is thoroughly clean and dry inside before installing on tractor.
2. If fuel outlet fitting (2) was removed, or new packing (3) is required, install packing on fitting. Use a long-handled magnet

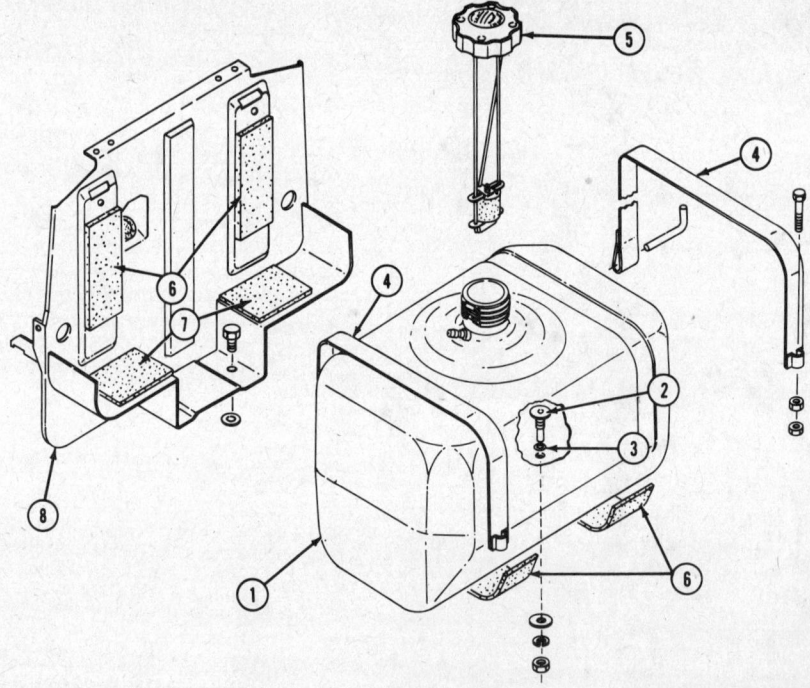

1. Fuel tank
2. Outlet fitting
3. Packing
4. Mounting strap
5. Fuel cap/gauge
6. Pads (front and rear)
7. Mounting pads (bottom)
8. Firewall

Fuel tank components

John Deere

A. Inlet hose
B. Outlet hose
C. Air vent screws
D. Filter element
E. Shut-off lever
F. Attaching screw

Fuel filter assembly

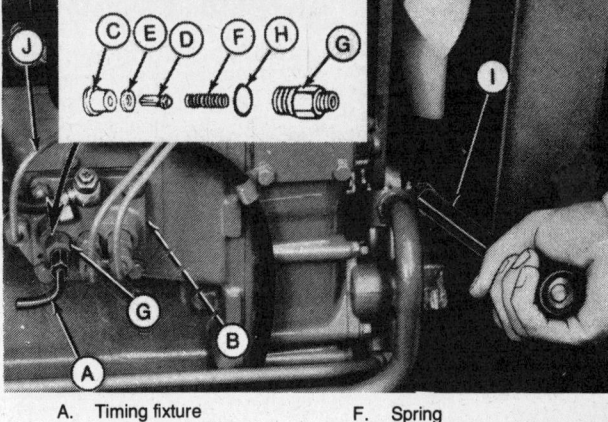

A. Timing fixture
B. Shims
C. Delivery valve housing
D. Delivery valve
E. Copper washer
F. Spring
G. Delivery valve holder
H. O-ring
I. Rachet handle
J. No. 1 fuel pipe

Checking pump timing

to insert outlet fitting through its hole from inside the tank. Secure fitting with a flat washer, lockwasher, and nut.

3. Check mounting pads (6 and 7) to see that they are in good condition and positioned as shown in the above illustration.

4. Position fuel tank on firewall (8) and locate the rear pads under end of both mounting straps (4). Tighten mounting strap screws.

5. Reverse the removal procedures given on the preceding page.

6. Fill fuel tank with the correct grade of clean diesel fuel.

7. Bleed the fuel system.

Fuel Filter

REMOVAL

1. Disconnect inlet hose (A) and plug the disconnected end to prevent the loss of fuel.

NOTE: Since no fuel tank shut-off valve is used, it is necessary to perform step 1, or else drain the tank completely.

2. Disconnect the outlet hose (B) at filter.

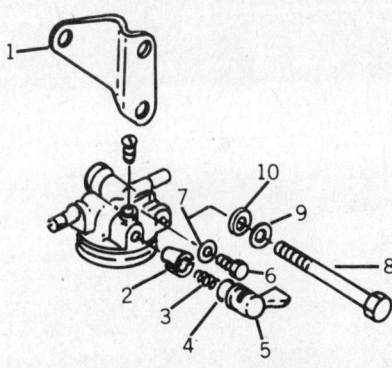

1. Mounting bracket
2. Taper valve
3. Spring
4. O-ring
5. Lever
6. Vent screw
7. Packing
8. Attaching screw
9. Lockwasher
10. Flat washer
11. Setscrew

Exploded view of filter housing and mounting parts

3. Remove the filter housing attaching screw (F).

NOTE: It may be necessary at some time to repair the filter housing because of fuel leakage from around the shut-off lever or bleed screws. Use the above illustration as a guide for replacing parts.

4. Installation is the reverse of removal.

REPLACING FILTER ELEMENT

Before replacing the filter element, shut-off the fuel supply to the filter by turning the fuel shut-off lever straight up; then, do the following:

1. Unscrew large nut (D) which retains the sediment bowl from filter housing.

2. Remove nut and sediment bowl (C) from filter housing.

3. Remove filter element (B), and install a new element in its place.

4. Thoroughly clean sediment bowl. Be sure that O-ring seal (A) is in good condition and install on sediment bowl. Also, be sure spring (C) is positioned in bowl.

5. Install sediment bowl, being careful not to overtighten nut.

6. Turn on fuel supply.

7. Open both air vent screws (E) on filter and vent screw on injection pump to bleed air from fuel. Tighten screws when fuel flows free of bubbles.

Fuel Injection Pump

REMOVAL

Clean the injection pump, pipes, and area around the pump with cleaning solvent or a steam cleaner.

NOTE: Never steam clean or pour cold water on an injection pump while the pump is running or while it is warm. To do so may cause seizure of pump parts.

1. Close the fuel shut-off valve at fuel filter.

2. Remove attaching screw from fuel inlet banjo fitting.

3. Disconnect the speed control rod and swivel from pump throttle lever.

4. Disconnect and remove the fuel pipes. Be sure to plug or cap both ends of each pipe to keep contaminants out.

5. Remove cover plates from pump chamber and governor chamber.

6. Disconnect and remove governor spring. Use care when unhooking spring, not to deform it.

7. Remove spring clip and spacer (use needle nose pliers with 90° bend as shown) from vertical pin on pump control rack. Disconnect governor link from injection pump.

NOTE: Do not drop spring clip and spacer during removal. These parts may fall down into crankcase.

8. Remove the injection pump-to-engine block screws (two nuts and two capscrews). Be careful not to damage shims as pump is separated from engine block. After pump flange is free from dowel (on lefthand side), position pump control rack to clear opening in engine block as pump is removed.

9. Note number and thickness of shims under mounting flange of pump to facilitate pump installation.

INSTALLATION

To install the injection pump:

1. Place the same number and thickness of shims on pump mounting flange as when removed. Shims should be clean and not deformed.

2. Install injection pump using new sealing washers under mounting screws.

NOTE: It may be necessary to add or remove shims from under pump mounting flange to obtain the correct timing (beginning of injection must take place at the proper time for optimum performance). Check pump timing following instructions given in next column.

3. Place the governor link over vertical pin of pump control rack. Install spacer (A) and spring clip (B). A pair of needle nose pliers having a 90-degree bend (C, Utica

John Deere

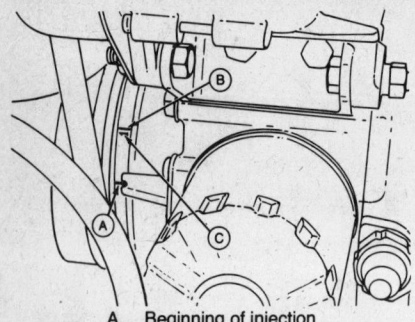

A. Beginning of injection
B. Top dead center
C. Cylinder number

Timing marks for no. 1 cylinder

No. 888-6 or equivalent) will be helpful when installing these parts.

4. Install governor spring.
5. Install cover plates.
6. Install fuel pipes and connect fuel inlet hose to pump.
7. Connect speed control rod to pump throttle lever.
8. Open fuel shut-off valve on filter.
9. Open both bleed screws on filter and one bleed screw on pump. When fuel flows without air, close bleed screws.

Pump Timing

To check pump timing:

1. Shut off fuel at filter and disconnect No. 1 fuel pipe (J).

NOTE: Remember that No. 1 engine cylinder or injection pump plunger is the one closest to the engine flywheel.

2. Unscrew delivery valve holder (G) from pump. Remove delivery valve (D) and spring (F). Reinstall holder and copper washer (E) back on pump.
3. Install JDF-14 Timing Fixture (A) on No. 1 outlet, with end of pipe pointing toward rear of tractor (horizontal).
4. Assemble a 19 mm (or ¾ in.) socket, a 2-in. extension, and a ratchet handle together and place socket on crankshaft pulley capscrew. Ratchet handle should be long enough to come out between radiator and engine as shown in the illustration. A longer handle permits the engine to be rotated more smoothly.
5. Pull out the engine decompression knob on dash.
6. Rotate engine in normal clockwise rotation (as viewed from front of tractor) until No. 1 piston is coming up on the compression stroke.
7. Have a small container close at hand, and turn on fuel.

NOTE: Since the delivery valve was removed from the pump, fuel will flow from timing fixture in a continuous stream, but will stop at a point of crankshaft rotation. The point at which the fuel just stops flowing is known as the "beginning of injection."

If the fuel does not stop flowing, even when the TDC mark (B) has passed the pointer, it is an indication that No. 1 piston was on the exhaust stroke instead of the compression stroke.

8. Rotate engine until fuel flow just stops. During this step, the fuel flow will change from a steady stream to droplets, and then stop when the beginning of injection point is reached. Disregard the last drop of accumulated fuel from the end of pipe.
9. Observe beginning of injection mark (A) on crankshaft pulley. Mark will be aligned with pointer when correct thickness of shims (B) is under pump flange. If mark has gone past pointer, remove shims. If mark has not reached pointer, add shims.

NOTE: A 0.1 mm (0.004 in.) shim change corresponds to a 1 degree difference in crankshaft position.

NOTE: The beginning of injection mark is at 26 ± 2 degrees before TDC on 850 Tractor and 25 ± 2 degrees before TDC on 950 Tractor.

10. When the pump timing is correct, remove JDF-14 timing fixture from pump.
11. Install delivery valve in valve housing while wet with clean diesel fuel. Place a NEW rubber O-ring (H) on delivery valve holder. Insert spring (F), and a NEW copper washer (E) in holder, then install holder on pump. Tighten holder to 29-33 ft.lbs. torque. Loosen holder by backing off ¼ turn, and then finalize installation by bringing up to full torque again.
12. Connect fuel pipe to pump. Tighten connector to 20 ft.lbs. torque.
13. Bleed air from fuel system. Start engine and check for leaks.

Fuel Injection Nozzles

REMOVAL

1. Disconnect leak-off hoses from injection nozzle and remove banjo fitting from top of each nozzle.
2. Disconnect thermostart system fuel hoses.
3. Disconnect fuel pipes from injection nozzles.
4. Remove retainer from top of each injection nozzle.
5. Remove injection nozzle from cylinder head. Note cylinder location of each nozzle to permit nozzle to be installed back in the same bore from which it was removed.

NOTE: If the injection nozzle can not be removed, install JDE-122 Adapter (A) in nozzle leak-off hole, and use JDE-38 Nozzle Puller (B) to remove nozzle.

If necessary to remove the precombustion chambers from cylinder head, withdraw the following parts: Outer half of precombustion chamber, copper gasket washer, inner half of precombustion chamber and a copper gasket washer.

TESTING

Test the injection nozzle before disassembling to determine its condition. Test for: Opening Pressure, Leakage, Chatter and Spray Pattern.

--- CAUTION ---
The nozzle tip should always be directed away from the operator. Fuel spray can penetrate clothing and skin, causing serious personal injury. It is recommended that the spray be collected in a container as shown.

Before applying pressure to the nozzle tester, be sure that all connections are tight, and that the fittings are not damaged. Fuel escaping from a very small hole can be almost invisible. Use a piece of cardboard or wood, rather than hands, to search for suspected leaks.

If injured by escaping fuel, see a doctor at once. Serious infection or reaction can develop if proper medical treatment is not administered immediately.

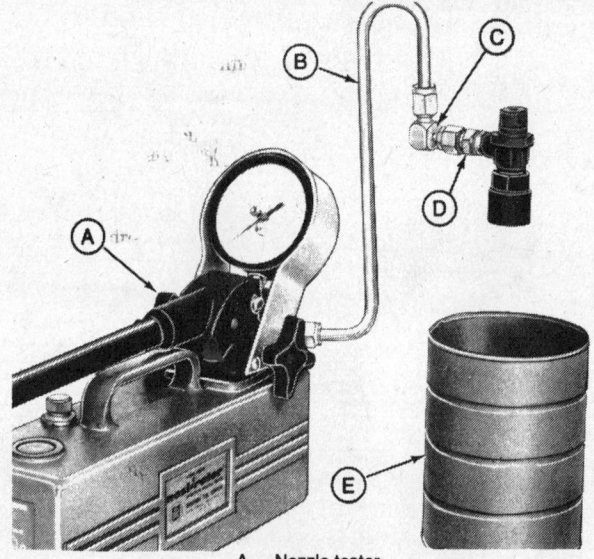

A. Nozzle tester
B. Fuel line
C. Adaptor (90°)
D. Adaptor (straight)
E. Container

Testing nozzle opening pressure

John Deere

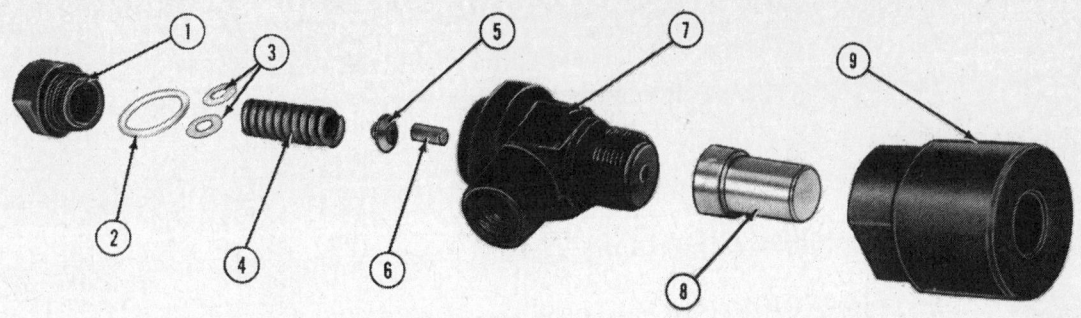

1. Upper holder nut
2. Copper washer
3. Shims
4. Spring
5. Lower spring seat
6. Inter spindle
7. Nozzle holder
8. Nozzle assembly
9. Nozzle nut

Exploded view of YDN injection nozzle

NOTE: Always use clean, filtered fuel when testing injection nozzle performance. Testing nozzles with dirty fuel will severely damage the precision parts of the nozzle.

Opening Pressure Test

1. To make the opening pressure test, connect the YDN injection nozzle to the No. Y-900 nozzle tester (A), using No. Y-900-2 fuel line (B), Y-900-3 90° adaptor (C), and JDF-17 straight adaptor (D). Place a container (E) under injection nozzle to catch fuel.
2. Pump the handle several times to flush out nozzle fittings. Tighten the fittings.
3. Expel air from the nozzle by operating the pump handle for several strokes. Then raise the pressure slowly and steadily. Observe the gauge pressure at which the valve opens. Recheck by completely releasing the pressure, and gradually building pressure until the valve opens.
4. The nozzle should open at approximately 2275 psi. If the opening pressure is not correct, disassemble the injection nozzle and change shims until nozzle opens at the proper pressure.
5. Each 0.004 in. (0.1 mm) shim changes the opening pressure approximately 100-142 psi.

NOTE: Always use John Deere nozzle adjusting shims, which are specially hardened. Other shims will not be satisfactory.

Leakage Test

1. To check for a leaking nozzle, wipe the nozzle dry. Bring the pressure up slowly to 1990 psi and watch for an accumulation of fuel from the spray orifice, indicating a bad seat. If the nozzle drips within 10 seconds, replace the nozzle assembly.
2. Check for leakage around the nozzle retaining nut (9) thread connection with nozzle holder (7). Leakage indicates a bad seat between nozzle assembly (8) and nozzle holder.
3. Be sure that the nozzle retaining nut is tight on the nozzle holder before disassembling or replacing parts.
4. Lightly lap nozzle body-nozzle holder seating surfaces to insure a good seal. If leakage persists, replace nozzle body and holder.

Chatter and Spray Pattern Test

1. The injection nozzle should chatter very softly, and only when the hand lever movement is slow (1o2 downward movements per second). Failure to chatter may be caused by a binding or bent nozzle valve.
2. Until the chattering range is reached, the test oil emerges as non-atomized streams. When the lever movement is accelerated, (4-6 times per second) the spray should be very broad and finely atomized.
3. A partially clogged or eroded throttling valve will usually cause the spray to deviate from the correct angle. The spray will also be streaky rather than finely atomized.
4. Disassemble the nozzle for cleaning or reconditioning if it fails to chatter or spray properly.

DISASSEMBLY

1. Clamp handle of open-end wrench (use correct size to fit flats on nozzle holder) in a vise.
2. Insert nozzle holder (B) into wrench with leak-off end facing up. Remove upper holder nut (A) and copper washer.
3. Remove nozzle holder from wrench and withdraw shims (3), spring (4), lower spring seat (5), and inter spindle (6) from holder.
4. Insert nozzle holder upside down in wrench. Loosen and remove the nozzle retaining nut (C). Remove nozzle from holder.
5. Withdraw nozzle valve from nozzle. If valve is stuck, it may be necessary to soak the nozzle assembly in Bendix cleaner, acetone, or other commercial cleaners sold especially for freeing stuck valves.

--- **CAUTION** ---

Use these nozzle cleaning fluids in accordance with the manufacturer's instructions.

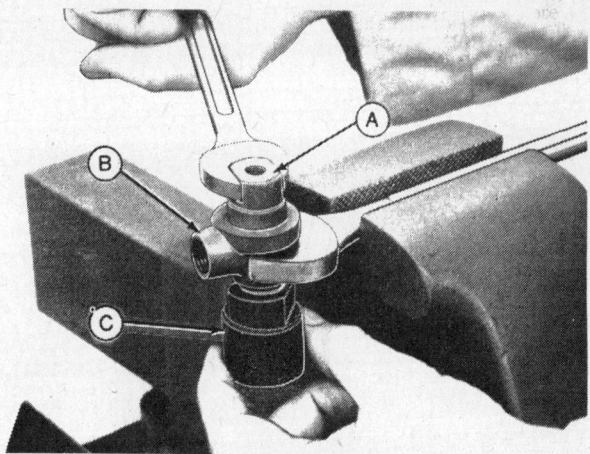

A. Upper holder nut
B. Nozzle holder
C. Nozzle nut

Disassembling YDN injection nozzle

John Deere

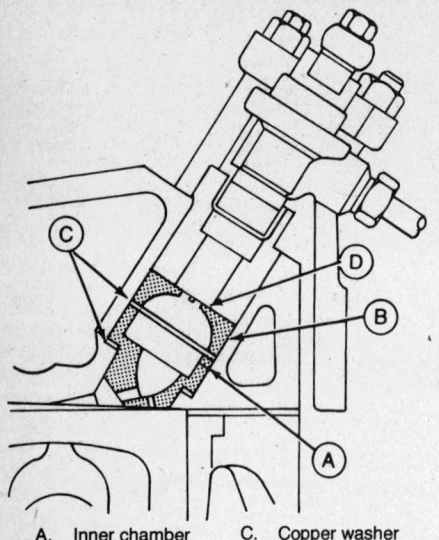

A. Inner chamber
B. Outer chamber
C. Copper washer
D. Packing

Sectional view of precombustion chamber

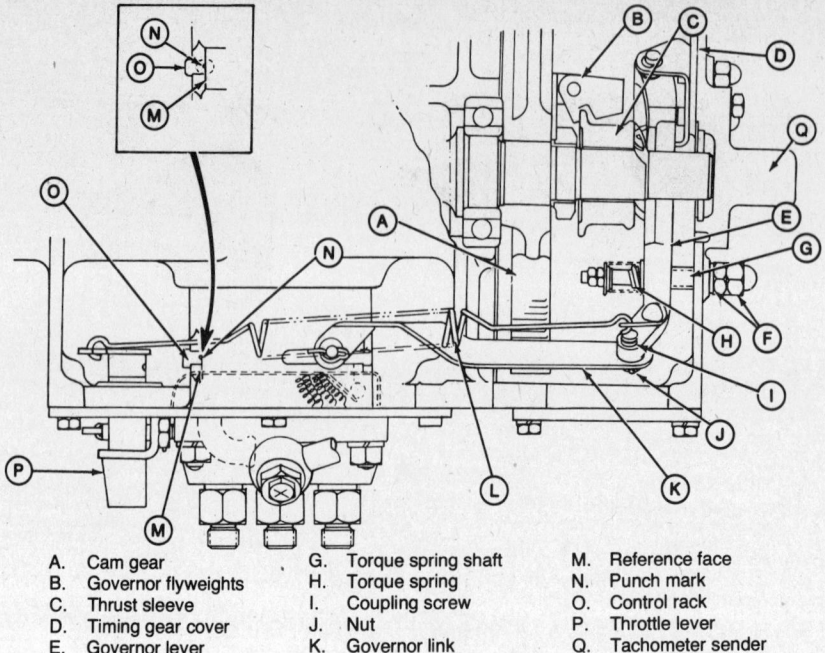

A. Cam gear
B. Governor flyweights
C. Thrust sleeve
D. Timing gear cover
E. Governor lever
F. Locknut and cap nut
G. Torque spring shaft
H. Torque spring
I. Coupling screw
J. Nut
K. Governor link
L. Governor spring
M. Reference face
N. Punch mark
O. Control rack
P. Throttle lever
Q. Tachometer sender

Sectional view of governor

6. The valve and nozzle are individually fitted and hand lapped. Keep these mated parts together, and do not permit the lapped surfaces to come in contact with any hard substance. Do not touch the valve unless hands are wet with fuel.

INSTALLATION

1. Before installing the YDN injection nozzle, check outer part of precombustion chamber for condition of the heat insulating packing. The packing must be in good condition in order to reduce heat transfer from the precombustion chamber to the injection nozzle, and thereby extending life of the injection nozzle.
2. Insert the injection nozzles into cylinder head. To insure correct positioning of nozzle for each cylinder, connect fuel pipes to injection nozzles before installing retainers.
3. Install retainer on injection nozzle with side of retainer having two "V"-type protrusions resting on injection nozzle.
4. Tighten the injection nozzle retainer studs nuts to 15 ft.lbs. torque. Be sure to keep retainer even as nuts are tightened.
5. Install the leak-off banjo connectors and washers on injection nozzles. Connect rubber hoses.
6. Tighten the fuel delivery pipe connectors to 20 ft.lbs. torque.
7. Connect thermostart system fuel hoses.
8. Bleed the fuel system.
9. Start the engine and check for leaks.

Precombustion Chamber

REMOVAL AND INSTALLATION

1. To remove the precombustion chamber, first remove the injection nozzle. Then, using any suitable tool with a hook on one end, withdraw outer chamber (B), outer copper washer (C), inner chamber (A), and inner copper washer (C).

NOTE: The inner chamber has a tang which fits into a slot in cylinder head. Its purpose is to maintain alignment of the chamber with the cylinder to insure that the fuel (and gaseous mixture) is injected into the cylinder at the correct angle. If the precombustion chamber can not be removed as described above, it will be necessary to remove the cylinder head and drive out the chambers using a suitable soft driver or sleeve driver.

2. The heat insulating packing improves the durability of the injection nozzle by preventing combustion heat from being transferred to the nozzle valve.
3. When the heat insulating packing becomes defective, pry out the old packing and replace with new.
4. Reverse the removal instructions given on the previous page.
5. Install the injection nozzles as described on the previous page.

Governor

NOTE: For service information on the governor flyweights (B) and governor thrust sleeve (C).

GOVERNOR LINK AND SPRING

Removal

1. Remove cover plates from pump chamber and governor chamber.
2. Disconnect and remove the governor spring (L).
3. Straighten ear on lockplate for nut (J). Remove nut and lockplate; then, remove link and cone washer from coupling screw (I).

NOTE: Do not allow parts to drop during their removal, as they can fall into crankcase. It is suggested that a clean shop towel be placed in opening to catch any parts that fall, and that a

Heat insulation packing

John Deere

magnet be at hand to retrieve them. Don't forget to remove shop towel when parts have been removed.

4. Disconnect rear end of link by removing spring clip and spacer.

Installation and Adjustment

1. Reverse the removal steps to install the governor link. Leave governor spring disconnected until adjustment is completed.

2. Before tightening the nut on coupling screw, the link must be correctly positioned.

NOTE: Use the following instructions only when the torque spring and shaft was NOT disassembled or its adjustment altered.

To adjust the governor link:

1. Move the governor lever (E) rearward until it touches the torque spring assembly (H), but does not compress the spring.

2. Move the governor link either right or left, until the punch mark (N) on control rack (O) is centered on vertical reference face (M) of pump housing.

3. Hold coupling screw with a screwdriver, and tighten nut. Make sure adjustment is correct; then, bend ear of lockplate against nut.

4. Connect governor spring, and install cover plates on engine.

TORQUE SPRING AND SHAFT

Removal

1. Disconnect the governor spring (L) and governor link (K) from the governor lever (E).

2. Remove the timing gear cover (D) from the engine.

3. Remove cap nut and locknut (F) from torque spring shaft (G). Unscrew shaft out of timing gear cover.

4. Inspect assembly. If any parts are worn or damaged, discard entire assembly and replace.

NOTE: Torque spring and shaft are available as complete factory adjusted assembly only. All necessary adjustments are preset. Do not attempt to service or adjust.

Installation and Adjustment

NOTE: This procedure will be accurate only if new assembly is installed, or old one has NOT been disturbed or changed in any way. If there is doubt about condition of old assembly, discard it and replace with new one.

1. Thread torque spring shaft assembly into timing gear cover, and install timing gear cover on engine.

2. Connect governor link to governor lever. Leave link retaining nut loose and adjust following the procedure below. Do not connect governor spring at this time.

3. To adjust the torque spring shaft:

a. Screw the torque spring shaft (G) inward (toward camshaft gear, (A)) far enough that the governor lever (E) can not contact the torque spring retainer, even when forced.

b. Move the governor link (K) all the way to the left (rearward). Position governor lever toward camshaft gear as far as it will go, and tighten nut (J) on governor link screw. Bend ear of lockplate against nut.

c. Move governor link to the right (frontward) until the punch mark (N) on control rack (O) is centered on reference face (M) of pump housing.

d. Screw torque spring shaft out of cover (frontward) until front torque spring retainer just contacts the governor lever. Secure adjustment with locknut and install cap nut (F).

e. Connect governor spring. Install timing gear cover plate and injection pump chamber cover plate on engine.

f. Start engine and check operation.

Themostart System

REMOVAL

1. Disconnect wiring lead (A) from thermostart plug (C).

2. Disconnect fuel supply hose (B) from plug and drain fuel from reservoir (D).

3. Remove thermostart plug from intake manifold.

INSPECTION

1. Inspect the thermostart plug, fuel reservoir, and hoses for evidence of flaws, cracks, or other deficiencies which would cause the system to not work properly.

2. Replace the parts as required when any of the above deficiencies are present.

DIAGNOSIS AND TESTING

If a visual inspection does not reveal the cause for the thermostart plug not working, the plug should be tested for ball check

A. Wiring lead
B. Supply hose
C. Thermostart plug
D. Fuel reservoir
Thermostart system

valve operation and heater coil/igniter continuity.

Check Valve Test

Perform the check valve test to make sure that the ball check valve works correctly.

Use the following procedure:

1. Connect shop air supply (A) to inlet line (B) and to thermostart plug (C). Regulate pressure to 20 psi.

NOTE: Use any suitable line as a means to connect air hose to the thermostart plug. A piece of copper tubing bent in the shape shown works well.

2. Immerse the thermostart plug in a container of diesel fuel (D) for 10 seconds.

3. Watch for air bubbles coming out from body of plug as a result of the ball check valve unseating. If bubbles appear, replace the thermostart plug.

INSTALLATION

Install thermostart plug in the intake manifold. Connect wiring lead to terminal, and connect supply hose to plug.

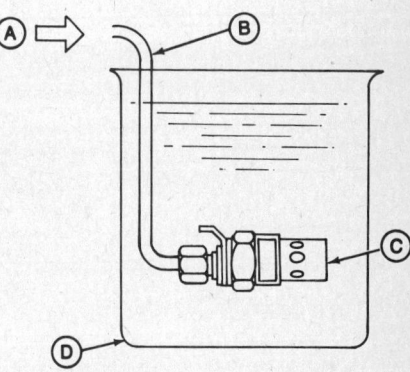

A. Shop air supply
B. Inlet line
C. Thermostart plug
D. Container
Check valve test

John Deere

NOTE: If it is necessary to use the thermostart system just prior to starting the engine, be sure to first fill the reservoir with clean diesel fuel. The reservoir can be filled by disconnecting the front rubber hose (hose from No. 1 injection nozzle to reservoir) at the injection nozzle and pouring diesel fuel through hose into reservoir.

ELECTRICAL SYSTEM

Alternator

REMOVAL

1. Disconnect battery ground cable at negative (−) battery terminal.
2. Remove three screws and remove lower shield.
3. Disconnect harness connector and terminal wire and ground wire.
4. Support alternator and remove two mounting bolts and adjusting bolt. Lift off alternator and belt guard.

INSTALLATION

1. Position alternator and belt guard and install mounting bolts and adjusting bolt. Do not tighten bolts at this time.
2. Attach harness connector. Connect terminal wire to "A" terminal and black ground wire to "E" terminal.
3. Install lower shield with three screws.
4. Attach battery ground cable to negative (−) terminal on battery.
5. Adjust belt to have ⅜ to ⅝ inch (10 to 16 mm) deflection when a 20 lb. force is applied midway between alternator and engine pulley.

Voltage Regulator

NOTE: On 850 Tractors (before 4738) and on 950 Tractors (before 7142) a CH13808 Wiring Harness Extension with diode is available to prevent the possibility of the "OIL" or "CHG" indicator lamps from burning out when the key switch is turned off. Install the wiring harness extension with diode between the voltage regulator and the main wiring harness. On 850 Tractors (4739 and up) and on 950 Tractors (7143 and up) a diode was added to the voltage regulator making wiring harness extension with diode no longer necessary.

REMOVAL AND INSTALLATION

1. Remove instrument access panel by removing five screws.
2. Remove voltage regulator by removing two screws. Disconnect wiring harness and lift out voltage regulator.
3. Install new regulator by reversing the above procedure.

Starter

INSPECTION, REMOVAL AND INSTALLATION

1. Inspect starting motor for external defects before removal. Check for loose mounting bolts and pole shoe retaining screws. Operate motor and listen for rattling, squealing or grinding noises.
2. Disconnect battery ground cable.
3. Disconnect battery cable and switch wiring.
4. Remove two nuts and separate starter from engine.
5. Install starter using two nuts. Tighten securely.
6. Connect switch wiring and battery cable.
7. Connect negative (−) battery cable to battery.

Models 60 and 70

FRONT WHEELS AND AXLE

Front Wheels

REMOVAL AND INSTALLATION

1. Raise and support the front of the tractor.
2. Remove the bearing grease cap.
3. Using snapring pliers, remove the snapring and retaining washer.
4. Installation is the reverse of removal.

Spindle

REMOVAL AND INSTALLATION

1. Disconnect tie rod end.
2. Remove the snapring and retaining washer.
3. Slip spindle out of axle.
4. Apply a light coat of grease on the upper spindle shaft.
5. Install the spindles into the axle.

Wheel Bearings

REMOVAL AND INSTALLATION

1. Using a driver or tool fabricated for the purpose, drive the bearings from the wheel hub.
2. Using a driver or socket, drive the new bearing onto the wheel hub.

NOTE: When installing the bearing, the notch in the bearing should be aligned with the grease fitting in the wheel.

STEERING

Steering Wheel

REMOVAL AND INSTALLATION

Model 60

1. Remove the upper capscrew from the coupling.
2. Pull the wheel and shaft up and out of the coupling and the shaft bearing.
3. Installation is the reverse of removal.

Model 70

1. Remove the steering shaft nut and install a wheel puller.
2. Screw an open nut onto the steering shaft to protect the threads.
3. Pull the steering wheel.
4. To install the wheel, place it on the shaft and tighten the nut.

Steering Gear

REMOVAL AND INSTALLATION

1. Disconnect the drag link.
2. Remove the three capscrews holding the gear housing to the frame.
3. Slip the gear assembly out from below.
4. Coat the threads of the retaining bolts with Loctite,® position the gear on the frame, install and tighten the bolts.
5. Connect the drag link to the lever arm and tighten the nuts.
6. Adjust the steering gear.

OVERHAUL

1. Remove the locknut from the cross bolt.

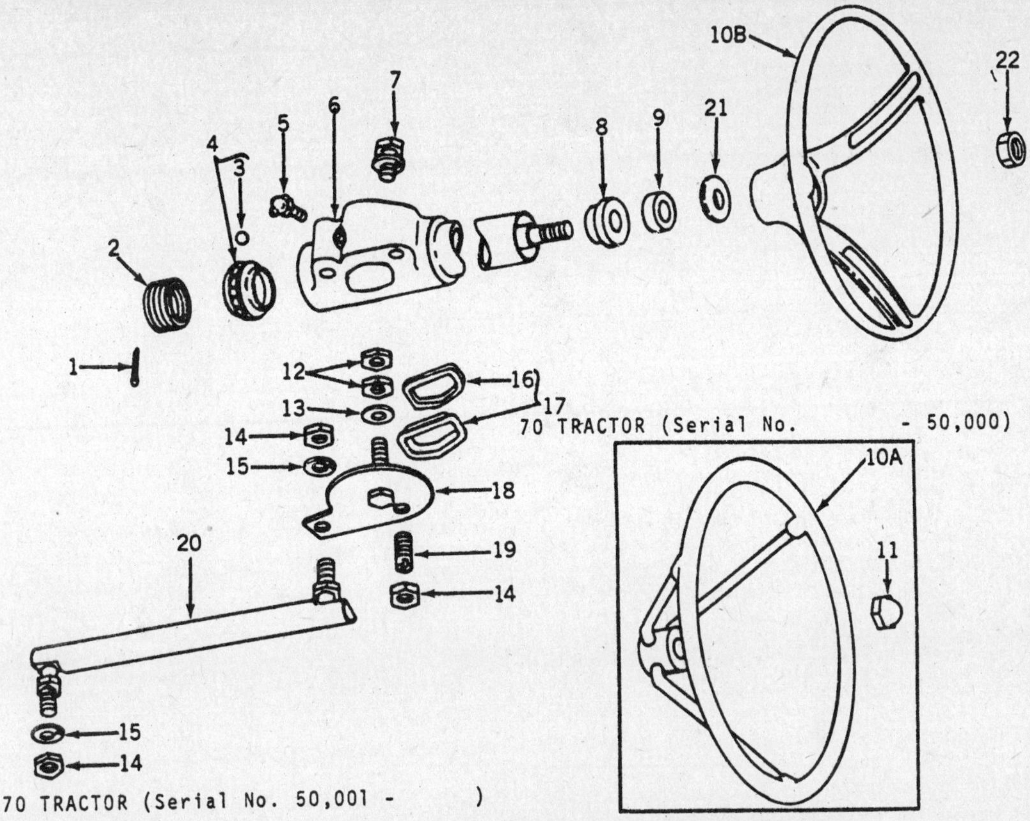

John Deere

1. Cotter pin
2. Adjusting plug
3. Steel ball (16 used)
4. Bearing kit
5. Grease fitting
6. Steering gear housing
7. Capscrew washer
8. Bearing
9. Seal
10A. Steering wheel (through 50,000)
10B. Steering wheel (50,001 and up)
11. Acorn nut (through 50,000)
12. Jam nut (2 used)
13. Flat washer
14. Jam nut (3 used)
15. Lockwasher
16. Seal
17. Seal Kit
18. Lever arm and cross bolt
19. Stud
20. Drag link
21. Grommet
22. Jam nut (50,001 and up)
23. Steering wheel cap (50,001 and up)
24. Decal (50,001 only)

70 TRACTOR (Serial No. - 50,000)

70 TRACTOR (Serial No. 50,001 -)

70 series steering linkage components

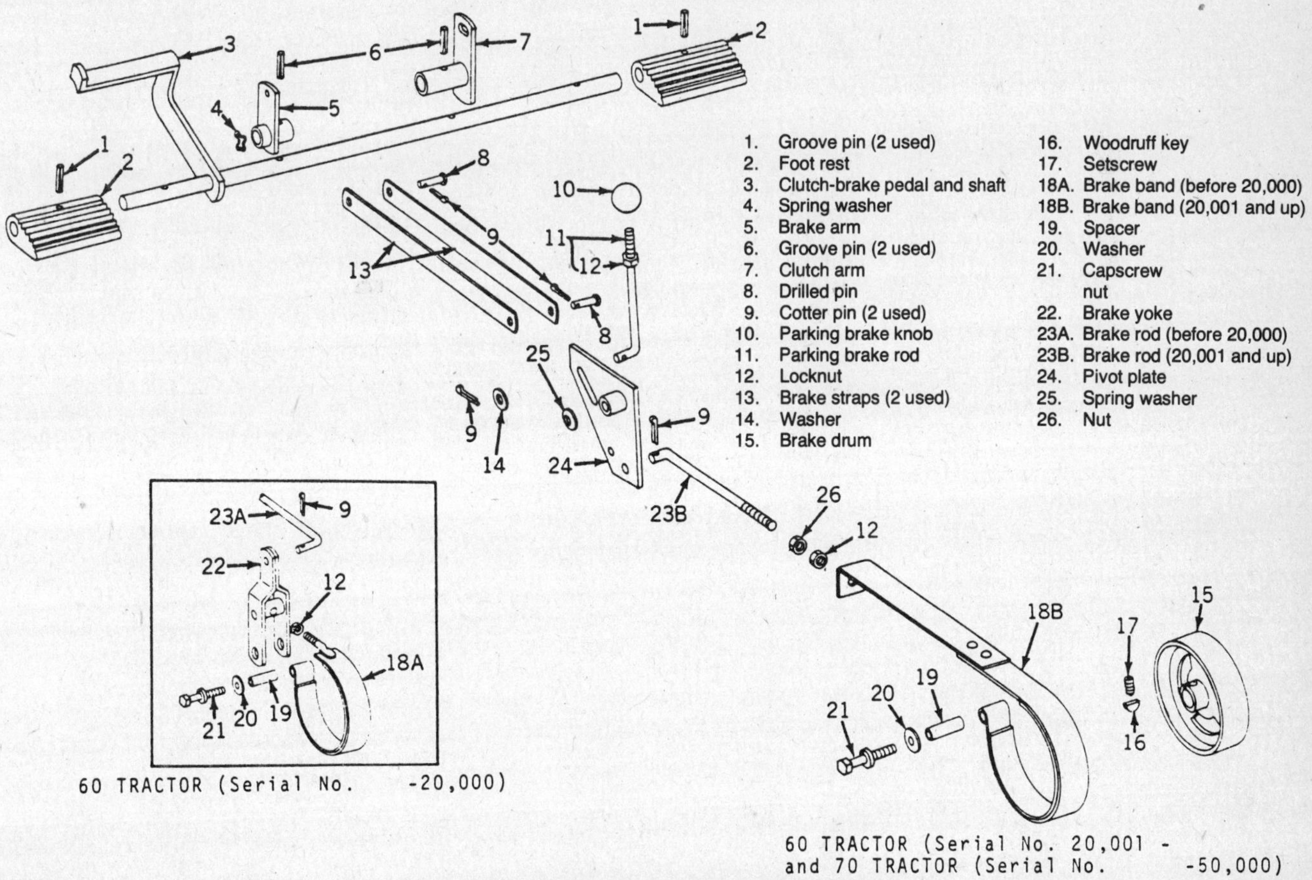

1. Groove pin (2 used)
2. Foot rest
3. Clutch-brake pedal and shaft
4. Spring washer
5. Brake arm
6. Groove pin (2 used)
7. Clutch arm
8. Drilled pin
9. Cotter pin (2 used)
10. Parking brake knob
11. Parking brake rod
12. Locknut
13. Brake straps (2 used)
14. Washer
15. Brake drum
16. Woodruff key
17. Setscrew
18A. Brake band (before 20,000)
18B. Brake band (20,001 and up)
19. Spacer
20. Washer
21. Capscrew nut
22. Brake yoke
23A. Brake rod (before 20,000)
23B. Brake rod (20,001 and up)
24. Pivot plate
25. Spring washer
26. Nut

60 TRACTOR (Serial No. -20,000)

60 TRACTOR (Serial No. 20,001 -) and 70 TRACTOR (Serial No. -50,000)

Brake components for tractors up to serial #50,000

John Deere

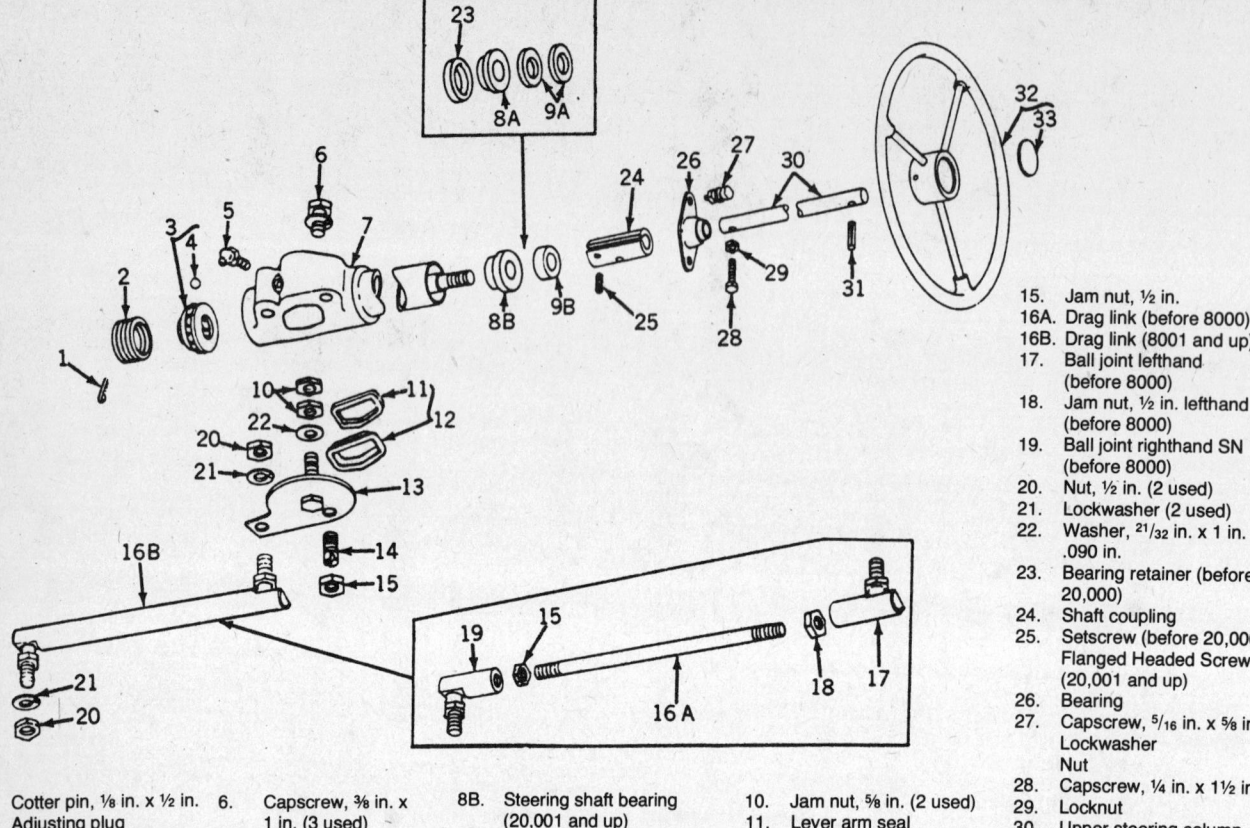

15. Jam nut, ½ in.
16A. Drag link (before 8000)
16B. Drag link (8001 and up)
17. Ball joint lefthand (before 8000)
18. Jam nut, ½ in. lefthand (before 8000)
19. Ball joint righthand SN (before 8000)
20. Nut, ½ in. (2 used)
21. Lockwasher (2 used)
22. Washer, 21/32 in. x 1 in. x .090 in.
23. Bearing retainer (before 20,000)
24. Shaft coupling
25. Setscrew (before 20,000) Flanged Headed Screw (20,001 and up)
26. Bearing
27. Capscrew, 5/16 in. x 5/8 in. Lockwasher Nut
28. Capscrew, ¼ in. x 1½ in.
29. Locknut
30. Upper steering column
31. Groove pin, ¼ in. x 1¾ in.
32. Steering wheel
33. Emblem

1. Cotter pin, ⅛ in. x ½ in.
2. Adjusting plug
3. Bearing kit
4. Steel ball (16 used)
5. Grease fitting
6. Capscrew, ⅜ in. x 1 in. (3 used)
7. Steering gear
8A. Steering shaft bearing (before 20,000)
8B. Steering shaft bearing (20,001 and up)
9A. Felt washer (before 20,000)
9B. Bearing seal (20,001 and up)
10. Jam nut, ⅝ in. (2 used)
11. Lever arm seal
12. Seal kit
13. Lever arm and cross bolt
14. Tapered stud

60 series steering linkage components

2. Remove the lever arm and cross bolt from the housing.

3. Remove the steering shaft coupling on models so equipped.

4. Remove the adjusting plug from the housing and slide the shaft and bearings from the column.

5. Wash all parts in a safe solvent and dry with compressed air. Inspect all parts for wear or damage.

6. Apply grease to all moving parts and place the ball bearings, ball cups and retaining rings on both ends of the cam.

7. Grease the cam lightly with chassis lube.

8. Slide the cam and tube assembly into the housing and jacket. Install the plug and tighten it to 7 to 12 ft.lbs.

9. Lock the adjusting plug with a cotter pin.

10. Before installing the lever arm, back out the stud.

11. Install a new seal and retainer.

12. Turn the stud out until contact with the steering cam is felt. Rotate the steering shaft all the way to the right and left, adjusting the stud for light resistance at the center point. Tighten the stud locknut.

ADJUSTMENTS

1. Disconnect ball joint from the lever arm.

2. Loosen the jam nut and stud two or three turns.

3. Remove the cotter pin retaining the adjusting plug in the gear housing. Use a screwdriver to turn the adjusting plug in until it is tight. Back the plug out until the steering wheel turns freely with no drag. Lift up on the wheel to check end play.

4. Turn the plug in only far enough to allow the cotter pin to drop in. Insert and secure the cotter pin.

5. Loosen the jam nut on the pivot bolt and tighten only the inside nut using a long thin open end wrench until all end play is removed, or until the distance between the steering arm and the gear housing is between 1/16 and 3/32 inch. Tighten the jam nut.

6. Find the center of the steering range by turning the steering arm until a line between the centers of the pivot bolt and the ball joint is vertical.

7. Turn the stud in until snug to remove all backlash. Move the lever arm through its full steering range in both directions. The steering wheel will turn as this check is made. When properly adjusted, a slight drag can be felt in the mid-point of the range. Tighten the jam nut to 40 ft.lbs.

8. Connect the ball joint to the lever arm.

BRAKES

Model 60 and Model 70 to Serial No. 50,000

The position of the brake adjusting nut, determines the point at which braking pressure is applied when depressing the clutch-brake pedal.

The brake may require adjustment after a period of time due to normal wear and seating of parts. Adjustment is required when the secondary belt idler drops low enough to strike the lower belt guide when the clutch-brake pedal is fully depressed.

To adjust brake, turn adjusting nut, clockwise until idler no longer strikes guide.

NOTE: Turning the brake adjusting nut in too far can cause simultaneous braking and driving, which can seriously damage transaxle.

Lower seat section and install thumb screws on 60 Tractors below Serial No. 20,001; capscrews on all other 60 and 70 Tractors.

CAUTION
Failure to replace thumb screws can cause injury by tipping the operator backward during operation.

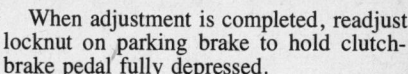

When adjustment is completed, readjust locknut on parking brake to hold clutch-brake pedal fully depressed.

Model 70 After Serial No. 50,000

Remove spring pin and washer. Turn brake rod clockwise into brake strap to tighten brakes.

NOTE: Turning brake rod in too far can cause simultaneous braking and driving, which can seriously damage transaxle.

TRANSAXLE

Removal and Installation

1. In preparation for removal of the transaxle assembly, drain lubricant by removing drain plug from bottom of transaxle case.
2. Remove fender-deck, disconnect brake linkage, unhook clutch spring, loosen rear belt guide and remove drive belt.
3. Support tractor with stands or blocks.
4. Remove four transaxle and four hitch plate capscrews and roll transaxle rearward.
5. Remove three capscrews from shift lever housing and lift shift lever from transaxle.
6. Remove drive sheave and brake pulley and their respective keys.
7. Install wheels, brake pulley, brake band, and drive sheave. On 60 Tractors prior to Serial No. 20,001 install small side of 2-speed drive sheave to the outside.
8. Roll transaxle into place. Install four transaxle-to-frame capscrews and install hitch plate.

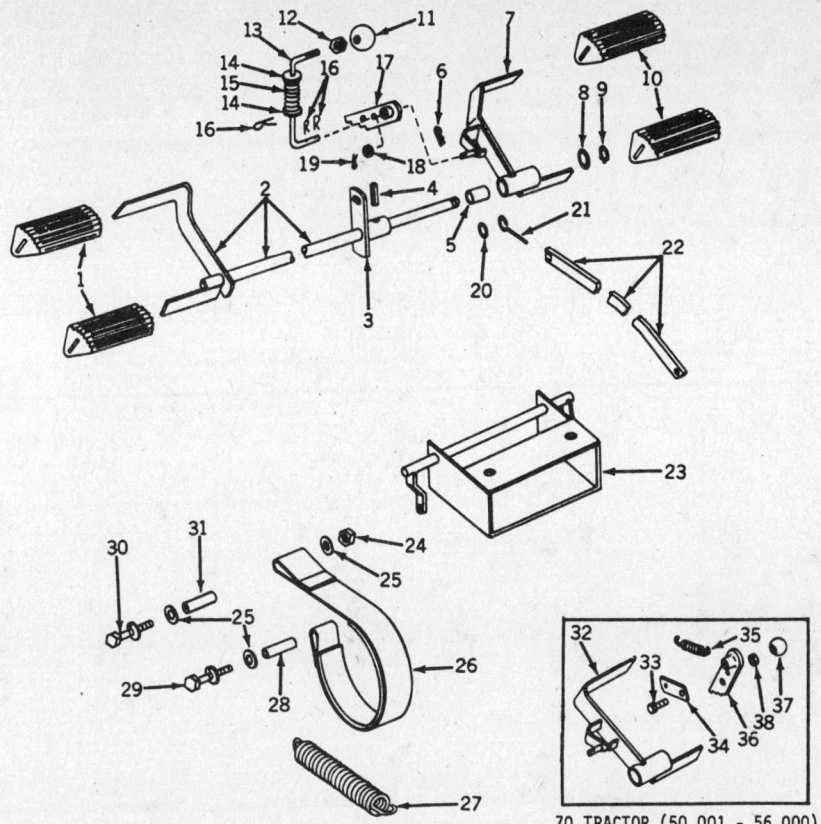

70 TRACTOR (56,001 -

1. Pad
2. Clutch pedal
3. Clutch arm
4. Groove pin
5. Bushing
6. Spring pin
7. Brake pedal (56,001 and up)
8. Washer
9. Retaining ring
10. Pad
11. Parking brake knob (56,001 and up)
12. Nut (56,001 and up)
13. Brake lock rod (56.001 and up)
14. Washer (2 used) (56,001 and up)
15. Spring (56.001 and up)
16. Cotter pin (3 used) (56,001 and up)
17. Brake stop assembly (56.001 and up)
18. Washer (56,001 and up)
19. Cotter pin (56.001 and up)
20. Washer
21. Brake rod
22. Brake strap
23. Pedestal
24. Nut
25. Washer (3 used)
26. Brake band
27. Spacer
28. Spacer
29. Screw and washer
30. Screw and washer
31. Spacer
32. Brake pedal (before 56,000)
33. Screw (before 56,000)
34. Strap (before 56,000)
35. Spring (before 56,000)
36. Brake stop assembly (before 56,000)
37. Parking brake knob (before 56,000)
38. Nut (before 56,000)

Brake components for 70 series tractors, serial #50,001 and up

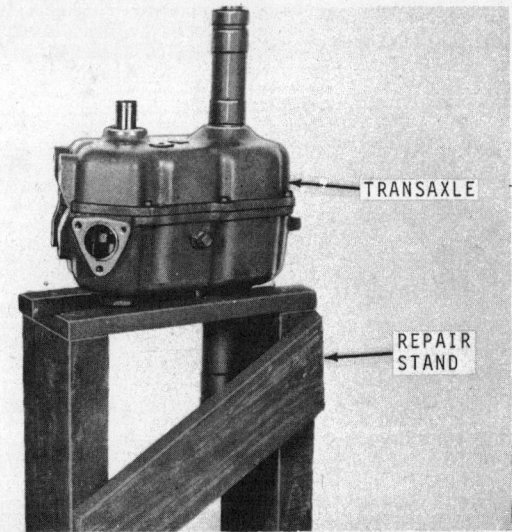

Construct a work stand for the transaxle

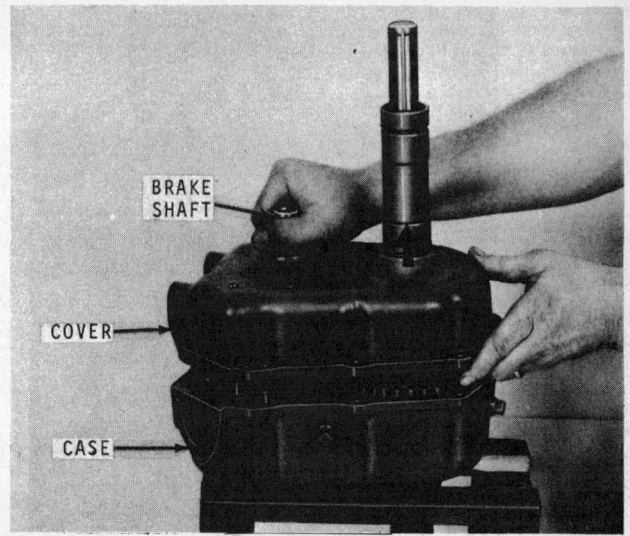

Removing cover

301

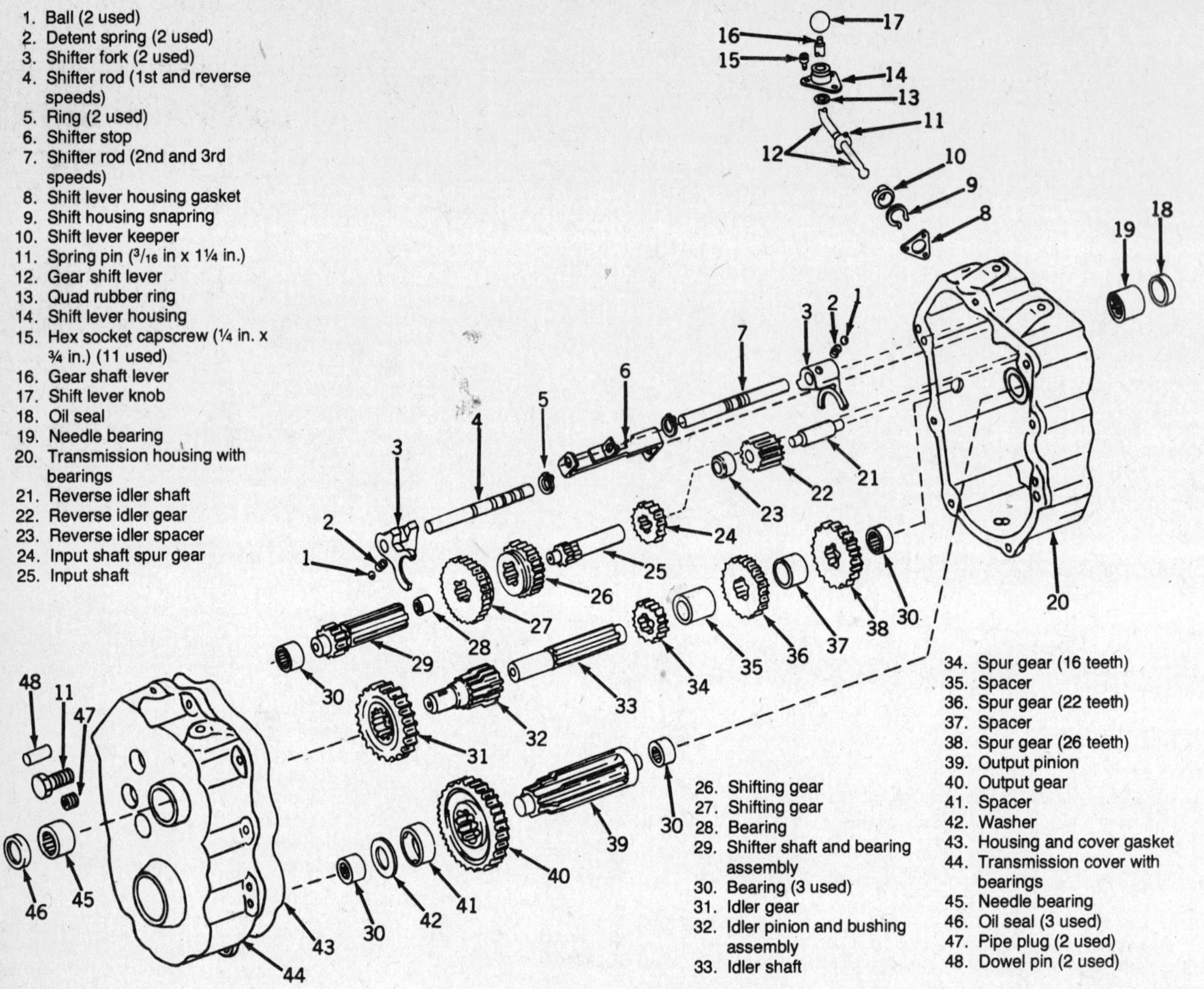

John Deere

1. Ball (2 used)
2. Detent spring (2 used)
3. Shifter fork (2 used)
4. Shifter rod (1st and reverse speeds)
5. Ring (2 used)
6. Shifter stop
7. Shifter rod (2nd and 3rd speeds)
8. Shift lever housing gasket
9. Shift housing snapring
10. Shift lever keeper
11. Spring pin (3/16 in x 1¼ in.)
12. Gear shift lever
13. Quad rubber ring
14. Shift lever housing
15. Hex socket capscrew (¼ in. x ¾ in.) (11 used)
16. Gear shaft lever
17. Shift lever knob
18. Oil seal
19. Needle bearing
20. Transmission housing with bearings
21. Reverse idler shaft
22. Reverse idler gear
23. Reverse idler spacer
24. Input shaft spur gear
25. Input shaft
26. Shifting gear
27. Shifting gear
28. Bearing
29. Shifter shaft and bearing assembly
30. Bearing (3 used)
31. Idler gear
32. Idler pinion and bushing assembly
33. Idler shaft
34. Spur gear (16 teeth)
35. Spacer
36. Spur gear (22 teeth)
37. Spacer
38. Spur gear (26 teeth)
39. Output pinion
40. Output gear
41. Spacer
42. Washer
43. Housing and cover gasket
44. Transmission cover with bearings
45. Needle bearing
46. Oil seal (3 used)
47. Pipe plug (2 used)
48. Dowel pin (2 used)

3-speed transaxle components

9. Connect brake linkage, and clutch spring. Install drive belt, position belt guide and tighten securely. Install fender-deck assembly.

10. Check brake and clutch linkage and adjust if required.

Overhaul Transaxle
Case Disassembly

1. For ease of transaxle disassembly construct a stand as shown. As an option two, 2-inch holes cut into a work bench top will work satisfactorily.

2. Place transaxle in stand vertically with capscrews up. Remove eight screws. Leave dowel pins in place.

3. While holding case halves together, invert entire transaxle and reposition in stand.

NOTE: Transaxle cover must be removed first.

4. Tap cover to loosen, then grasp brake shaft and cover as shown. Lift cover slowly, shaking gently, so all parts remain in the lower case.

5. To remove brake shaft from case cover, support pinion gear from bottom side and tap end of brake shaft with a mallet. The brake shaft is a light press fit into pinion gear and will slide out with minimum of force.

6. Remove the gear group sub-assemblies in the following order: output shaft, differential and axle assembly, idler shaft, reverse idler, shifter shaft and shifter rods and forks.

Removing brake shaft

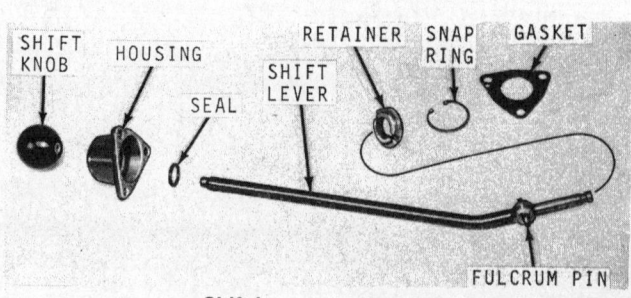

Shift lever components

7. Input shaft is a press fit installation. If close inspection reveals that gears and bearings are in satisfactory condition, DO NOT remove from case.

8. Wash all internal parts in a safe cleaning solvent. Brush and scrape foreign matter from all parts and dry thoroughly. After all parts are thoroughly cleaned, oil to prevent rust formation on gears and bearings.

SHIFT LEVER DISASSEMBLY

1. To disassemble gear shift assembly, remove snapring and knob, and slip housing and retainer off shift lever.

2. When servicing transaxle equipped with other than the snapring style shift lever assemblies, a complete new style snapring shift lever assembly will be required.

REMOVING AXLE HOUSINGS

To remove axle housing from transaxle case, position case halves in press as shown and press in the direction indicated.

REPLACING SEALS

1. Install new axle and brake shaft seals, whenever transaxle is disassembled for repair.

2. Remove seals with a seal hook or driver.

3. Install seals with drivers of the proper size. Be sure seal lips face interior of transaxle.

REPLACING BEARINGS AND BUSHINGS

1. To remove case needle bearings, press bearings from case with a driver of the proper size.

2. To install a new case bearing, press bearing into case until open end of bearing is within .020-inch of interior of case bearing boss.

3. To remove axle shaft bushings cut bushings out of housing with a bushing cutter.

4. Install new pre-sized bushings with a driver of correct size.

NOTE: A driver of incorrect size will damage bushing. Drive bushings into their original depth.

5. Disassemble, clean and inspect parts shown. Look for gear wear and tooth damage. Replace worn parts and reassemble.

6. Apply Loctite® to carrier capscrew threads when assembling differential. Use lockwashers with carrier capscrews and torque to 20 ft.lbs.

7. The axles should rotate freely in opposite directions when assembled. Lay the differential assembly aside for later installation.

8. Broken detent springs, can cause gear damage. When the springs are broken, the shifter fork is free to move, thus allowing gear pressure to slide the gears out of mesh.

9. When the gears slide out of gear, especially under load, gear chipping or cracking will result.

10. Prolonged heavy drawbar loads and wheel slippage are the most common cause of bevel pinion gear failure, in the differential section of the transaxle.

11. Damage to the input shaft spline, is caused by improper coupling of the shifter shaft and input shaft when transaxle is shifted.

12. A broken detent spring or a loose shift lever and housing can cause improper coupling.

1. Output shaft
2. Differential and axle assembly
3. Idler shaft
4. Reverse idler
5. Shifter shaft and shifter rods and forks

Gear removal sequence

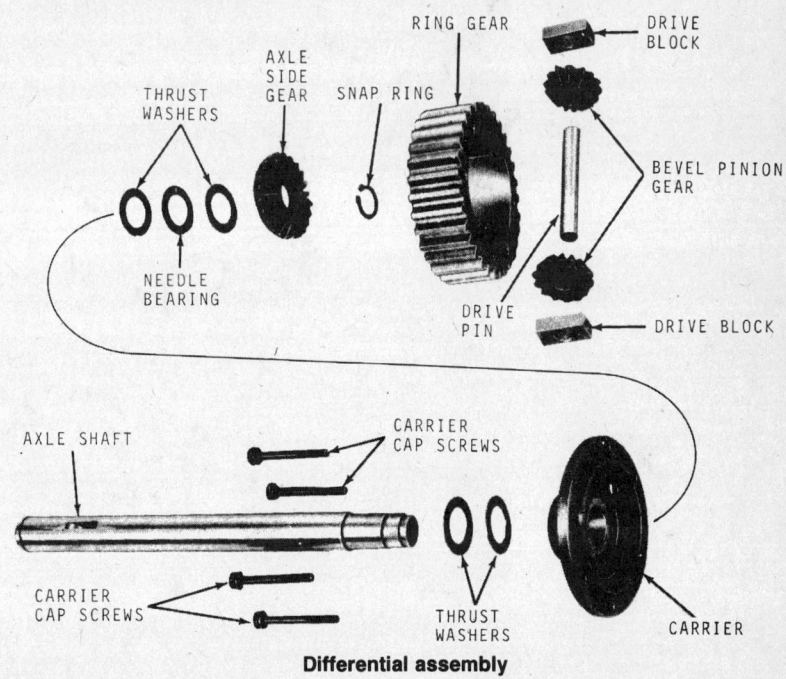

Differential assembly

John Deere

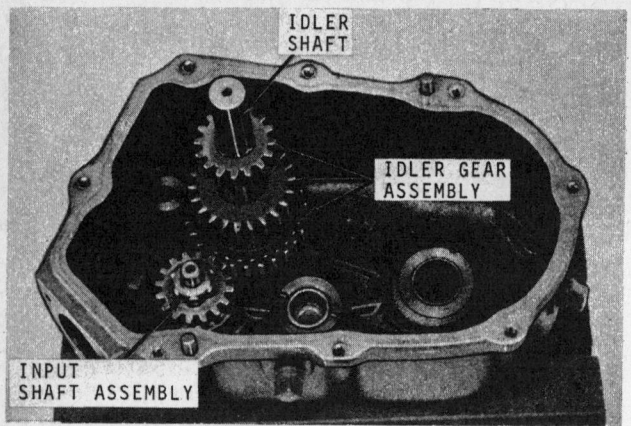
Idler shaft installation

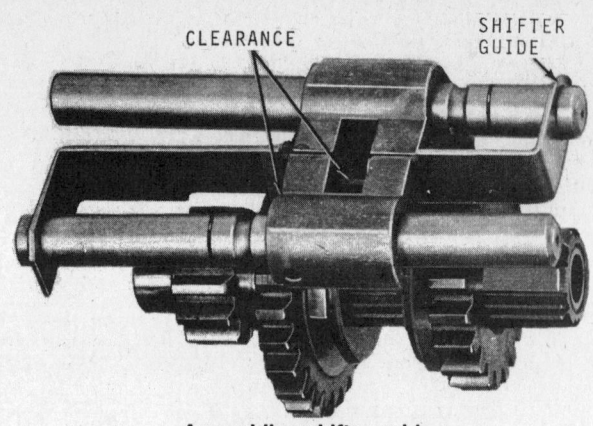
Assembling shifter guide

Assembly

INPUT SHAFT AND GEAR

1. Assemble input shaft and gear. Counterbored gear spline must face to right as shown. Gear is a press fit onto shaft.
2. Install input shaft and gear into case as shown. Flat side of gear should now face upward.
3. Assemble idler shaft and gear assembly. These gears are a slip fit on the spline. Notice that raised hub of large gear faces short spacer. The teeth on the medium and small gear have round engagement edges that must face the large gear.

INSTALLING AXLE HOUSINGS

1. To install axle housings into transaxle case halves, support case and press axle housing until firmly seated against flange.
2. Bushings and seals within axle housings can be replaced without removing housings from case half. Note that the housings are of different lengths. Assemble as shown.
3. The long round end of the idler shaft turns in the bushing on the brake shaft. Be sure end of shaft is not battered.

INSTALLING IDLER SHAFT AND GEARS

Install idler shaft and gears in case. Long end of shaft faces upward. Large gear engages the input gear.

REVERSE IDLER SHAFT AND GEAR

1. Assemble reverse idler shaft assembly as illustrated. Round edge of teeth faces spacer.
2. Install reverse idler gear as shown. Position spacer above gear.

SHIFTER RODS AND FORKS

1. Because of heavy detent pressure, the assembly of these rods can be difficult.
2. Assemble forks as shown. Both forks should face to the right for assembly. The 2nd and 3rd rod must have the unequally spaced grooves at the top and away from the fork as shown. The 1st and reverse rod must have the shortest ungrooved end face the fork as shown. Start the rod into the fork. Depress detents and complete the assembly. Slide forks along rod. A good snap should be felt in each detent.
3. Place forks in center of neutral detent positions at this time.

SHIFTER SHAFT ASSEMBLY

1. To assemble shifter, lay out parts as shown. Be sure forks are in center grooves. Note that the exposed groove on the unequally spaced 2nd and 3rd shifter faces the gear on the shifter shaft. The exposed groove of the 1st and reverse equally spaced shifter faces away from the gear on the shifter shaft.
2. The shifter shaft assembly should appear as shown. The slot in the forks should line up when the large gear is slipped as far as possible on the spline. Note the position of exposed grooves on shifter shafts.
3. Assemble shifter guide over shifter rods. Slot in guide should match rectangular opening between the forks. The long notch in underside of guide should clear the large 1st and reverse shifter gear.
4. Grasp shifter assembly firmly in hand and lower it into case. The input shaft stud should enter needle bearing in end of shifter shaft. The shifter rods should now enter the two machined sockets in bottom of the case.
5. All parts assembled thus far should appear as shown.

Reverse idler shaft and gear

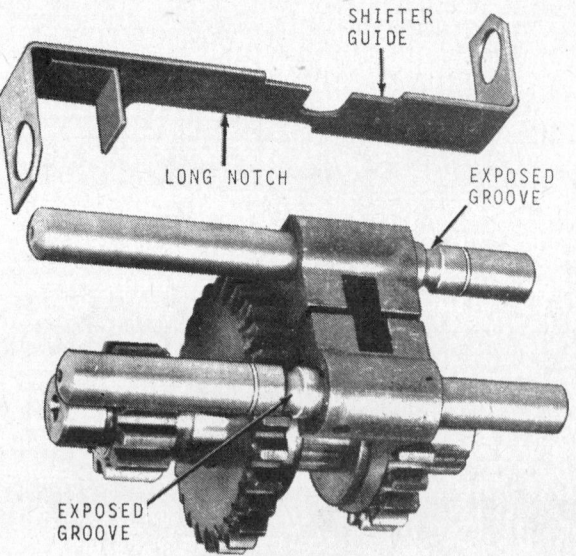
Shifter assembly

John Deere

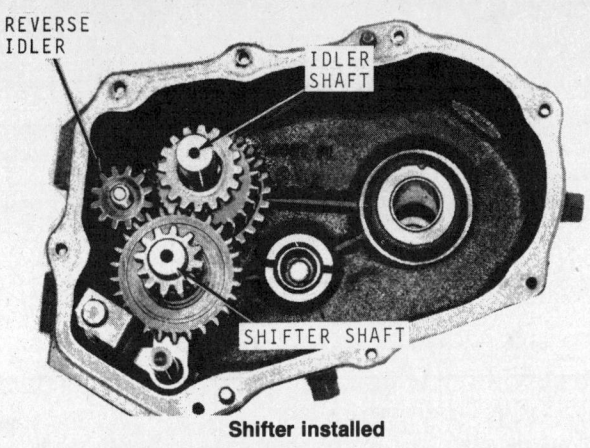
Shifter installed

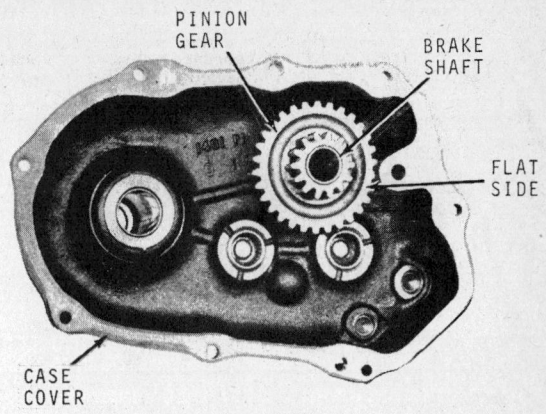
Brake shaft installed

INSERTING DIFFERENTIAL ASSEMBLY IN CASE

1. Install differential and axle assembly into the case with the bolt heads on the righthand carrier downward.
2. The assembly should now appear as shown.

OUTPUT SHAFT

1. The output gear shaft is assembled on the pinion with a press fit. A slight looseness can be tolerated because the spacer and washer will hold the gear in place.
2. Install the output shaft assembly in the case as shown. It may be necessary to lift the differential about ½ inch to mesh gears during assembly.
3. Position the gasket on the lower case at this time.

BRAKE SHAFT AND COVER

1. To assemble brake shaft and pinion gear to case cover, position bevel gear in cover with flat side facing out. Insert brake shaft into gear and case bearing, mate splines and press gear and shaft together until firmly seated.
2. When pressing or tapping brake shaft into place, properly support case and bearing to prevent damage.
3. Brake shaft and pinion gear will appear as shown, when properly installed.

PLACING COVER ON CASE

1. Loosen or remove setscrew on transaxles so equipped.
2. Install the case cover as illustrated. Shake the case lightly, and all shafts and bearings will align themselves. To close the last ½ inch, tap the cover horizontally at the corner indicated.
3. Invert transaxle and tighten case screws securely to 120 in.lbs. torque. Put Loctite® on threads, then tighten setscrew against the shifter shaft, on transaxles so equipped.

POSITIONING SHIFTER FORKS

Inspect the shifter forks to be sure they are aligned and in neutral. Failure to do this will cause damage to the transmission when engaged under power.

ASSEMBLING SHIFT LEVER

1. Assemble shift lever components in the order shown. Insert seal carefully into lever housing, use a small amount of gasket cement, if necessary, to hold in place during assembly.
2. Position shift lever assembly onto transaxle case. Check angle of lever in relation to case. Lever angle can be changed by removing snapring and rotating housing and retainer 180°. A correctly installed shift lever should appear as shown.
3. When lever is in proper position, install three screws and torque to 120 in.lbs.
4. With the transaxle assembled and positioned as shown, remove filler plug and add 3 pints of SAE 90 Gear Lubricant.

POWER TAKE-OFF CLUTCH

Disassembly

1. To remove the clutch cone, remove

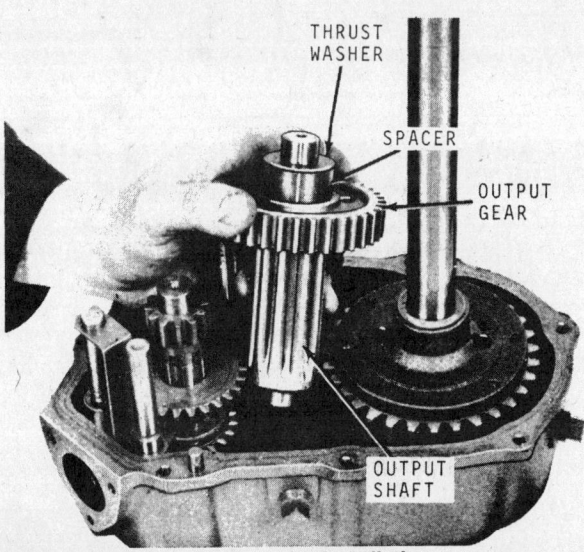
Output shaft installation

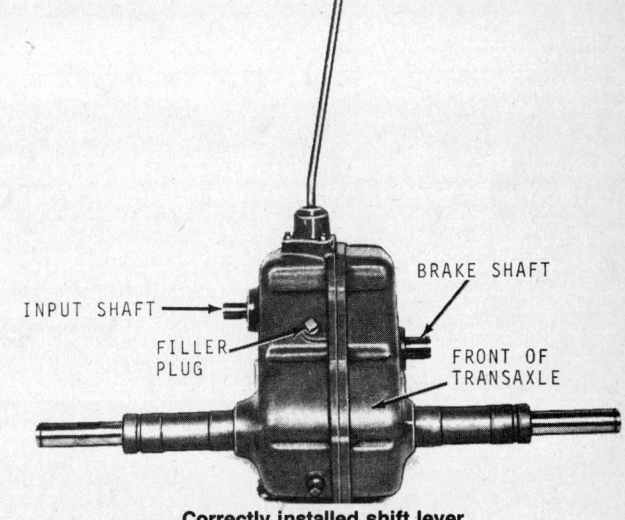
Correctly installed shift lever

John Deere

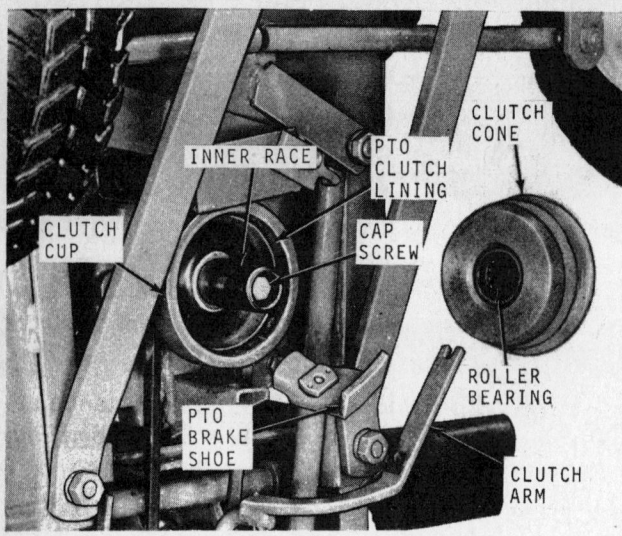

Removing clutch cone assembly

Clutch cone components

arm retaining pin and PTO brake shoe. The entire clutch cone will then slide off easily.

2. Remove capscrew from end of crankshaft and slide roller bearing inner race and clutch cup from engine crankshaft.

3. The clutch lining in the clutch cup can not be replaced. Replace entire clutch cup if lining is worn excessively.

4. To replace clutch cone bearings, remove large snapring from bottom end of clutch cone. Press clutch pivot and ball bearing out bottom of clutch cone. Remove small snapring and press ball bearing off clutch pivot.

5. Press roller bearing and seals out of clutch cone. If clutch cone mating surface is badly worn, entire clutch assembly must be replaced.

6. The roller bearing and inner race are not sold as a matched set, but if either is damaged replacement of both parts is recommended.

Assembly

1. Slide clutch cup and inner race on end of crankshaft, and secure with retaining washer and capscrew. Torque capscrew to 39 ft.lbs.

2. Press roller bearing and seals onto clutch cone, following dimensions given in the accompanying illustration. Lips on seals must point toward roller bearing.

3. Press ball bearing onto clutch pivot by pressing on inner race. Insert small snapring in pivot groove. Press ball bearing into clutch cone by pressing on outer race and install large snapring.

4. Grease roller bearing liberally with John Deere High Temperature Grease (AT17659T), being certain to pack groove above roller bearing with grease.

5. Slide cone assembly onto roller bearing inner race, connect the clutch arm, and attach the PTO brake shoe.

6. Replace cone brake shoe if lining is worn excessively.

Adjustment

Check clutch adjustment periodically to be certain clutch is not slipping and power driven equipment is engaging and disengaging properly. To check adjustment:

1. Engage PTO clutch.

2. Observe that clutch arm is parallel with tractor frame and dimensions "A" and "B" are equal. If not, remove locking pin and lower clutch arm. Loosen jam nut and turn fulcrum either up or down until arm is approximately level when PTO clutch is engaged and locking pin is in place.

3. Engage and disengage the clutch slowly with the PTO lever. Observe at

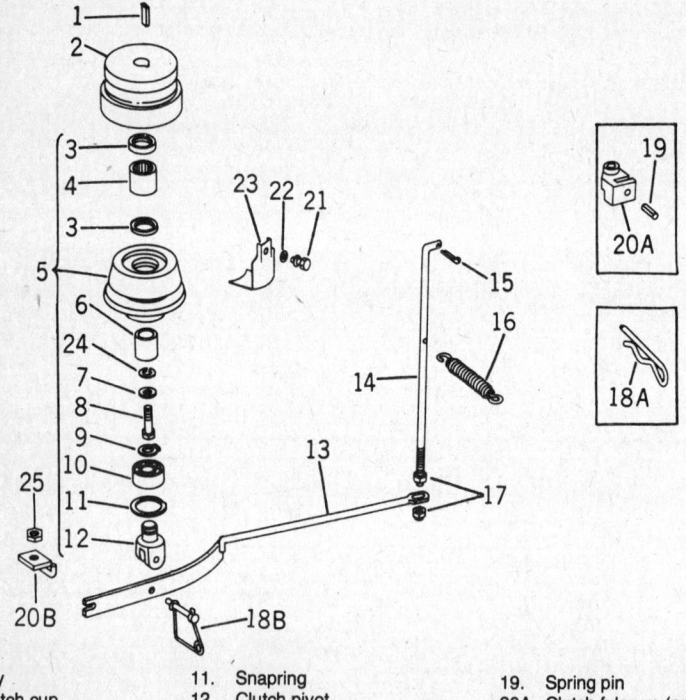

1. Key
2. Clutch cup
3. Oil seal
4. Roller bearing
5. Clutch cone assembly
6. Bearing race
7. Retaining washer
8. Capscrew
9. Snapring
10. Ball bearing
11. Snapring
12. Clutch pivot
13. Clutch arm
14. Clutch rod
15. Cotter pin
16. Spring
17. Elastic stop nut
18A. Locking pin (prior to 13,107)
18B. Locking pin (13,108 and up)
19. Spring pin
20A. Clutch fulcrum (prior to 20,000)
20B. Clutch Fulcrum (20,001 and up)
21. Capscrew
22. Washer
23. PTO brake shoe
24. Snapring
25. Jam nut

PTO clutch components

John Deere

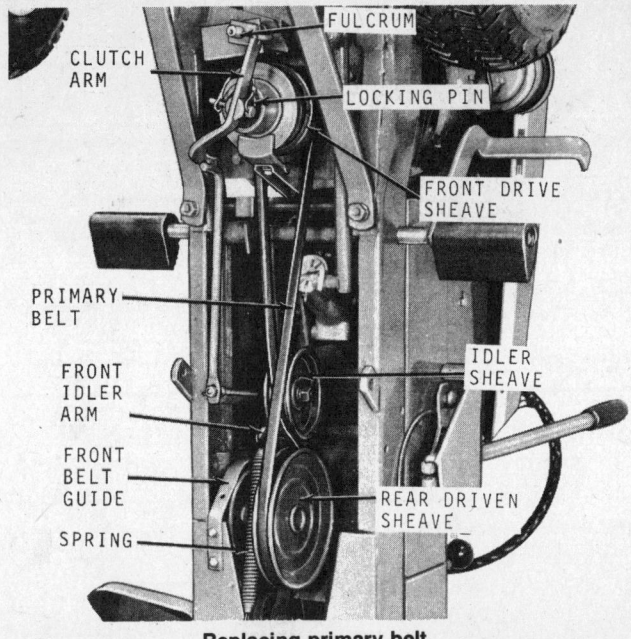
Replacing primary belt

what point the clutch cones begin to contact each other. From this point to full engagement of the PTO lever, the clutch arm should deflect ½ inch. This will properly tension clutch cones for positive engagement.

4. Move nuts on end of clutch rod as required until proper arm deflection is obtained.

Adjusting Power Take-Off Clutch Brake

1. The PTO clutch brake stops the mower blades or snow thrower rotor quickly after disengaging the drive with the PTO lever.
2. Engage the PTO lever and check the distance between the brake and clutch cone. When properly adjusted, this distance should be approximately $1/16$ inch.
3. If adjustment is required, loosen capscrew, and slide brake shoe in slotted hole until $1/16$ inch clearance is obtained. Tighten capscrew firmly.

DRIVE BELT TRANSMISSION

Replacing Primary V-Belt

CAUTION
Remove ignition key and disconnect spark plug wire when replacing belts.

1. Remove spring locking pin. Disconnect clutch arm from fulcrum and clutch cone. Unhook spring from front idler arm. Slip belt off both the front and rear sheaves.
2. Place a new primary belt on sheaves with bottom strand running from rear driven sheave to lefthand side of front sheave. Reassemble spring, clutch arm, and spring locking pin.

NOTE: Installing belt with a reverse twist will result in reverse tractor travel.

Replacing Secondary V-Belt

60 TRACTORS (BEFORE SERIAL NO. 20,000)

1. Pivot tractor seat and fender section rearward. Loosen the pair of wing nuts on tractor frame, and slide wing nuts and belt guide to the rear position of slotted hole. Also loosen and move similar pair of wing nuts found below tractor frame on righthand side.
2. Remove thumb nut and guide fork.
3. With parking brake set and idler depressed, slip belt out of sheaves and toward righthand side of tractor. Lift belt out.
4. Reverse procedure for belt installation. Be sure to install and position all belt guides properly. Install thumb screws in seat and fender section.

60 TRACTORS (SERIAL NO. 20,001- AND UP) AND 70 TRACTORS

1. Depress clutch-brake pedal and set parking brake. Remove four capscrews (one under each fender and two under seat) holding fender-deck in place.
2. With fender-deck removed, loosen capscrews and slide rear belt guide out of the way. Remove front belt guide. Unhook spring from front idler arm. Remove primary belt from rear sheave.
3. Remove spring hooked to rear idler arm. Loosen capscrews on rear idler mount. Remove secondary belt.
4. Install new secondary belt on sheaves. Tighten capscrews on rear idler mount. Replace primary belt and hook spring to front idler arm. Install front belt guide. Hook spring to rear idler arm.
5. Reposition rear belt guide and tighten capscrews. Replace fender-deck and tighten capscrews.

NOTE: Adjust all belt guides to within $1/16$ inch of belt to avoid chafing.

Changing Axle Speed Range

60 TRACTORS (BEFORE SERIAL NO. 20,000)

1. To change speed ranges, remove thumb screws from point "A", on both sides of tractor frame and tilt seat section rearward.
2. Loosen pair of wing nuts on tractor frame, and slide wing nuts and attached belt guide to the rear position in the slot. Also loosen and move similar pair of wing nuts found below tractor frame on righthand side.
3. Check to be sure parking brake is set. Parking brake in the locked position will lower idler and give needed belt slack.
4. Remove secondary belt from largest drive sheave first. Then slip belt into alternate position in both belt guide forks.
5. Slip belt into V-grooves of both front and rear sheaves.
6. Move both pairs of wing nuts and belt guides to the original position (nearest the drive sheave) and tighten nuts finger tight.
7. Close seat section and start tractor momentarily.
8. Shut off engine, move seat section rearward again and check to be sure belt does not rub either belt guide fork or the belt guides near front and rear drive sheaves. Bend belt guides if necessary to eliminate rubbing.
9. Lower seat section and replace thumb screws.

Counter Sheave Repair

60 TRACTOR (BEFORE SERIAL NO. 20,000)

A counter sheave assembly, in the 60 and 70 Lawn Tractor performs three basic functions.
1. It reduces engine rpm to a usable input speed for the transaxle input shaft.
2. The counter sheave arrangement permits the use of two short belts in place of one long belt.
3. By using two belts, a more efficient clutch arrangement can be used.

NOTE: 60 Lawn tractors (Before Serial No. 20,000) are equipped with a counter sheave assembly as shown.

307

John Deere

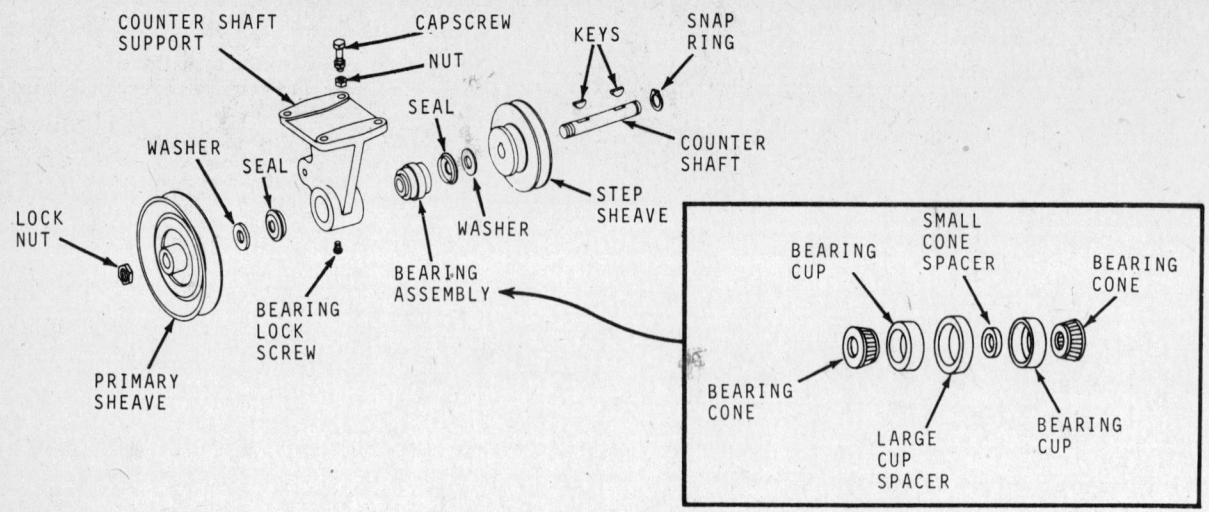

Counter sheave assembly

4. To service the counter sheave assembly, remove the locknut retaining primary sheave and countershaft and slide countershaft out of support. With shaft removed, cone bearings, inner spacer, seals, and washers can be removed.

5. The bearings of the countershaft assembly are serviced only as an assembly, which consists of following: (2) bearing cones, (2) bearing cups, (1) large bearing cup spacer, (1) small bearing cone spacer.

6. To assemble the bearing unit, press new bearing cup spacer into support bore until it is seated in center groove.

7. Press bearing cups into support until they bottom against spacer. Assemble bearing cones, countershaft, and small bearing cone spacer to support. Place seals, washers, sheaves onto shaft. Install snapring on end of countershaft and install locking nut onto opposite end of shaft and torque to 45-50 ft.lbs.

8. Install assembly to tractor and install belts and belt guides.

60 TRACTORS (SERIAL NO. 20,001 AND UP) ALL 70 TRACTORS

1. 60 lawn tractors above Serial No. 20,000 and 70 lawn tractors utilize a counter sheave assembly, which cannot be disassembled.

2. In the event of a bearing or sheave failure, replace the complete counter sheave assembly.

Models 110 and 112

FRONT WHEELS AND AXLE

Front Wheels

REMOVAL AND INSTALLATION

1. Raise and support the front end on jack stands.
2. Remove the dust cap, cotter pin, slotted nut, wheel and bearings.
3. Install all parts in the reverse order of removal.
Adjust the bearings as follows:
a. While rotating the wheel, torque the slotted nut to 60-120 in.lbs. to seat the bearings.
b. Back off the slotted nut until the wheel turns freely.
c. Align the cotter pin hole by backing off the nut, never by tightening it.

Spindle

REMOVAL AND INSTALLATION

NOTE: Tractors prior to serial #15000 have a keeper ring at the top of the spindle.

1. Disconnect the tie rod.
2. Remove the spring pin with a blunt punch and slip the spindle out of the axle.
3. On tractors from serial number 15000-100,000, and on all 112 series, use retaining ring pliers to remove the retaining ring and washer.
4. Installation is the reverse of removal.

NOTE: Coat all moving parts with chassis grease prior to assembly.

John Deere

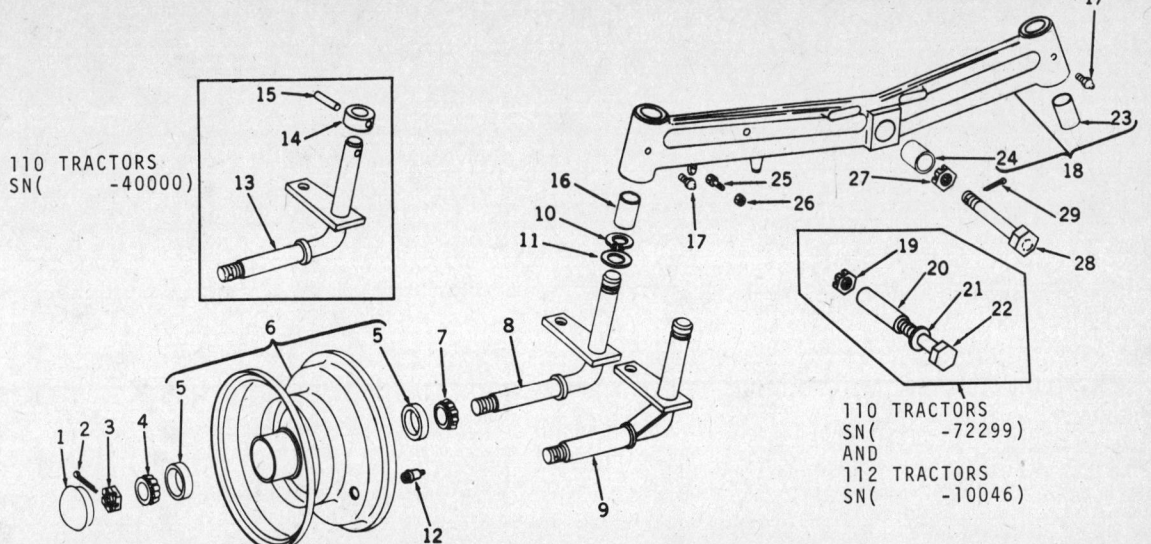

1. Spindle cap (2 used)
2. Cotter pin (2 used)
3. 5/8 in. slotted hex nut
4. Outer cone (2 used)
5. Cup (4 used)
6. Front wheel (2 used)
7. Inner cone with seal (2 used)
8. Front wheel spindle (110 tractors) (2 used)
9. Front wheel spindle (112 tractors and 110 tractors with loaders) (2 used)
10. Snapring
11. Spindle washer
12. Front tire valve (high flotation tires only)
13. Front wheel spingle (110 tractors)
14. Spindle keeper (110 tractors)
15. Spring pin (110 tractors)
16. Bronze bushing (4 used)
17. Grease fitting (2 used)
18. Front axle with bushings
19. 1/2 in. hex slotted nut
20. King pin bearing
21. Washer
22. 1/2 in. UNF x 3 1/2 in. king pin bolt
23. Bronze bushing (4 used)
24. King pin bushing
25. Steering stop bolt
26. 1/2 in. hex jam nut
27. 3/4 in. UNF hex slotted nut
28. 3/4 in. UNF x 3 5/8 in. king pin bolt
29. Cotter pin

Front wheels and axle tractors through serial #100,000

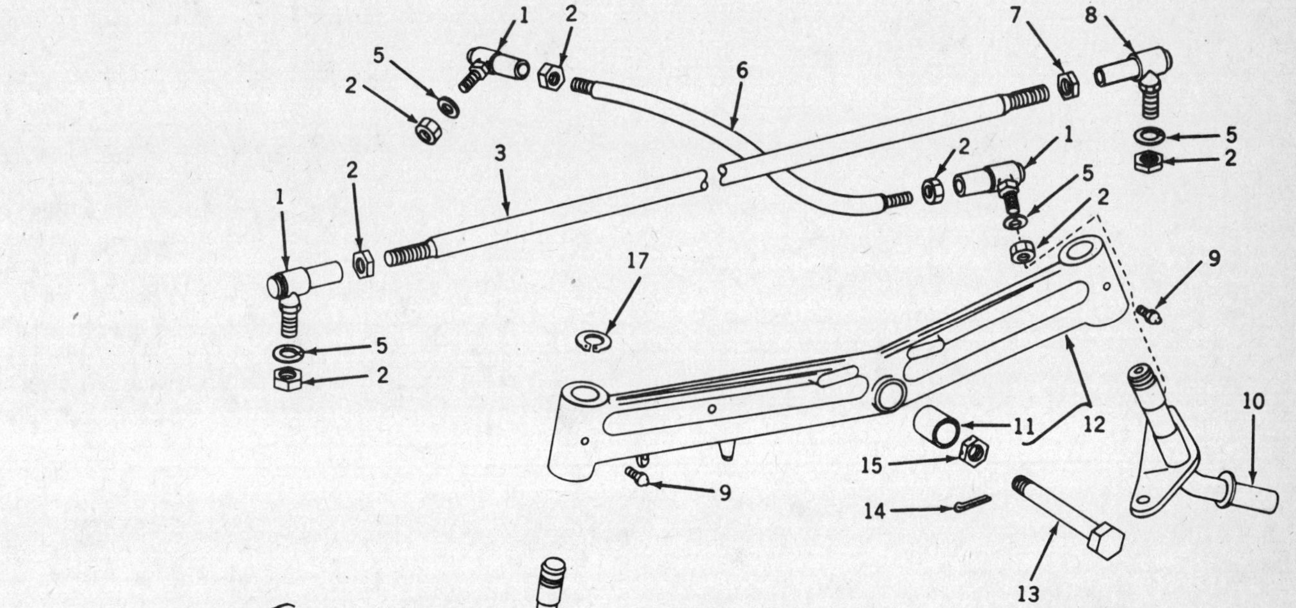

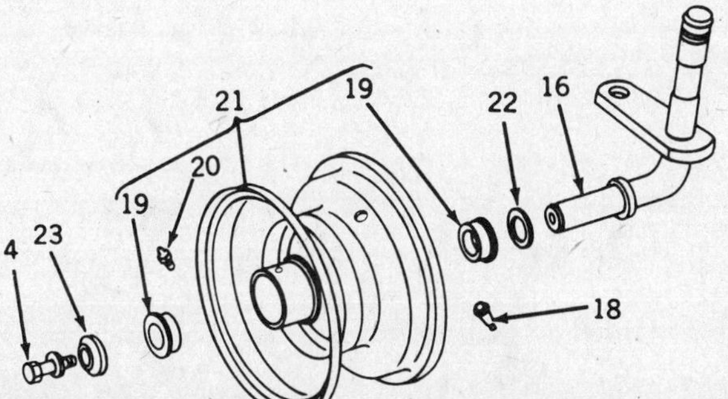

1. Ball joint assembly
2. Jam nut
3. Tie rod
4. Capscrew
5. Lockwasher
6. Drag link
7. Jam nut
8. Ball joint assembly
9. Grease fitting
10. Lefthand front wheel spindle
11. King pin bushing
12. Axle with bushing
13. King pin bolt
14. Cotter pin
15. Slotted nut
16. Righthand front wheel spindle
17. Snapring
18. Valve stem
19. Bearing
20. Grease fitting
21. Front wheel disc
22. Spring washer
23. Spindle cap

Front wheels and axle, 110 tractors serial #250,001 and up

309

John Deere

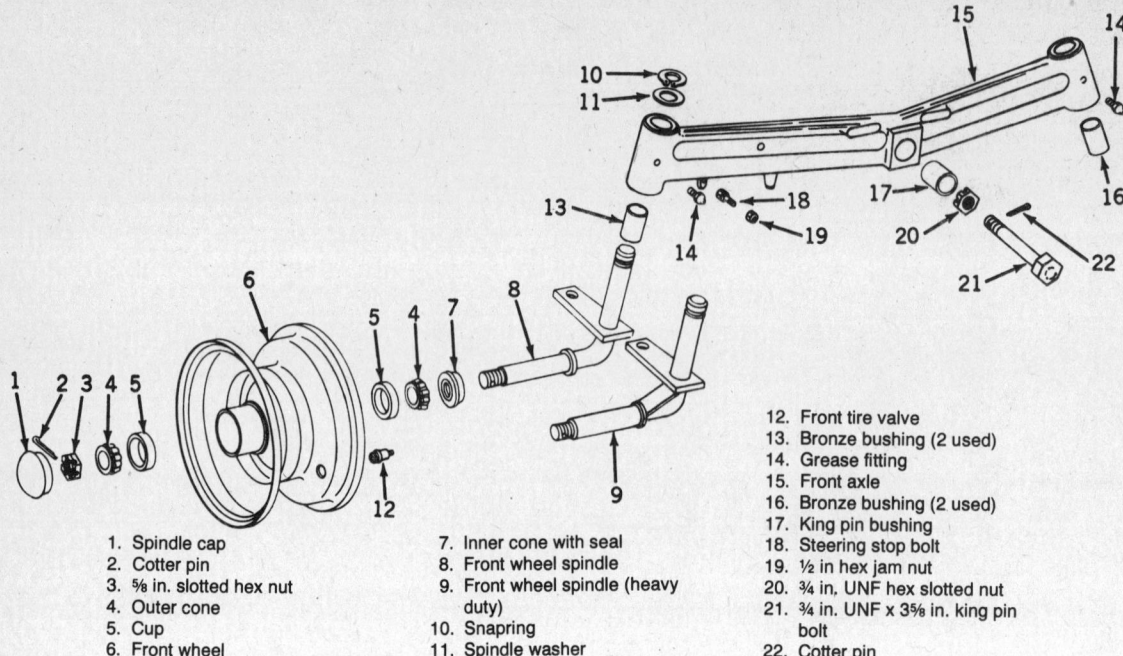

12. Front tire valve
13. Bronze bushing (2 used)
14. Grease fitting
15. Front axle
16. Bronze bushing (2 used)
17. King pin bushing
18. Steering stop bolt
19. ½ in hex jam nut
20. ¾ in. UNF hex slotted nut
21. ¾ in. UNF x 3⅝ in. king pin bolt
22. Cotter pin

1. Spindle cap
2. Cotter pin
3. ⅝ in. slotted hex nut
4. Outer cone
5. Cup
6. Front wheel
7. Inner cone with seal
8. Front wheel spindle
9. Front wheel spindle (heavy duty)
10. Snapring
11. Spindle washer

Front wheels and axle, tractors serial #100,001-250,000

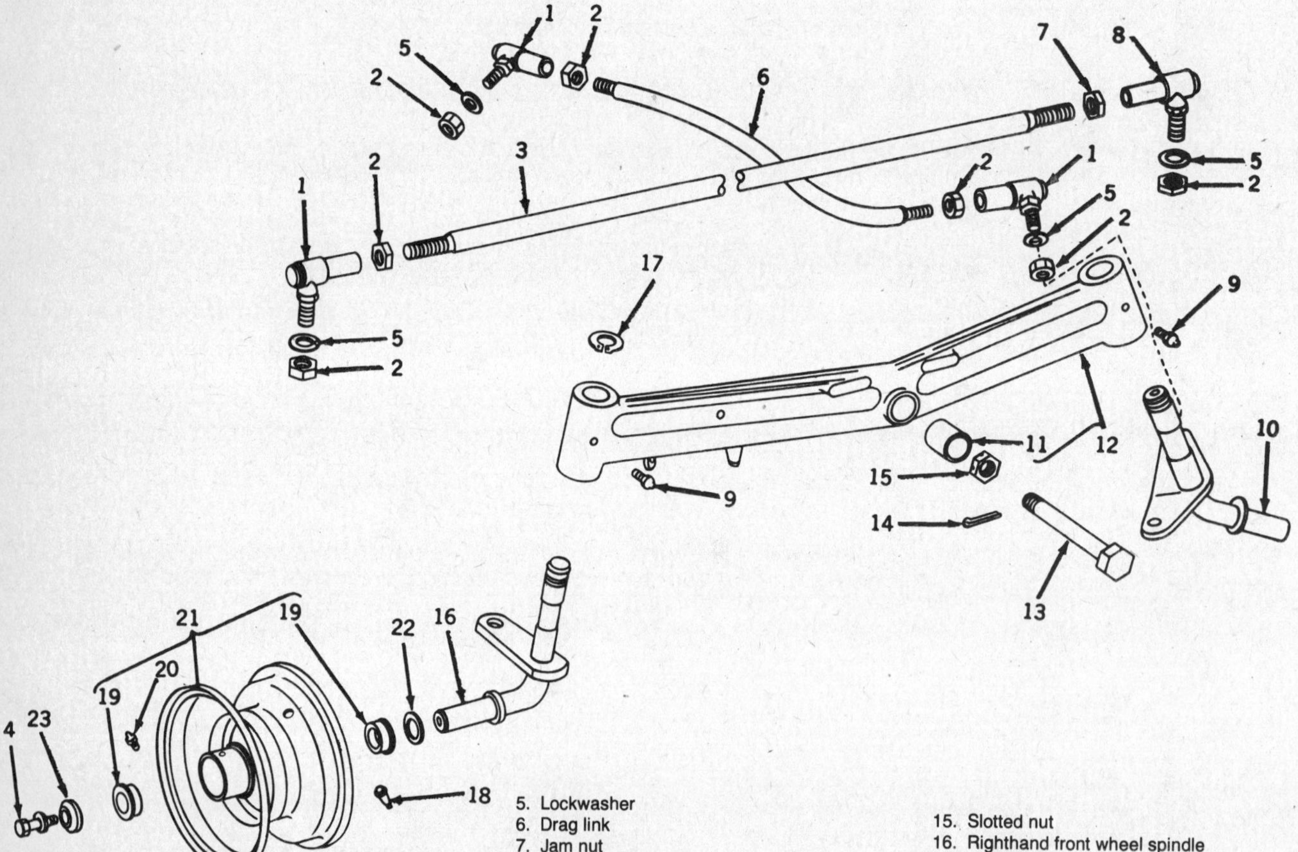

1. Ball joint assembly
2. Jam nut
3. Tie rod
4. Capscrew
5. Lockwasher
6. Drag link
7. Jam nut
8. Ball joint assembly
9. Grease fitting
10. Lefthand front wheel spindle
11. King pin bushing
12. Axle with bushing
13. King pin bolt
14. Cotter pin
15. Slotted nut
16. Righthand front wheel spindle
17. Snapring
18. Valve stem
19. Bearing
20. Grease fitting
21. Front wheel disc
22. Spring washer
23. Spindle cap

Front wheels and axle, 112 tractors serial #250,001 and up

John Deere

STEERING SYSTEM

These tractors are equipped with a Ross cam and lever type steering system.

Steering Wheel

REMOVAL AND INSTALLATION

1. Remove the cap, nut and retaining washer.
2. Using a puller, such as the one illustrated, remove the steering wheel.
3. Installation is the reverse of removal. Torque the retaining nut to 10-12 ft.lbs.

Steering Gear

REMOVAL AND INSTALLATION

1. Remove the battery.
2. Remove the steering jacket clamp.
3. Disconnect the drag link.
4. Remove the gear housing-to-frame bolts and lower the gear assembly from the frame.
5. Installation is the reverse of removal. Apply Loctite® to the gear retaining bolts. Torque the bolts to 20-25 ft.lbs.

NOTE: It is important that the drag link is positioned with the bend facing the center of the tractor.

DISASSEMBLY

1. Loosen the jam nut on the tapered stud in the lever arm.
2. Turn the stud counterclockwise until resistance is felt.
3. Remove the nuts from the lever arm cross bolt. Remove the housing.
4. Remove the plug in the steering gear housing and slide the shaft with the cam and bearing from the column.
5. Wash all parts in a safe solvent and blow them dry with compressed air or wipe them with a clean, lint free cloth.

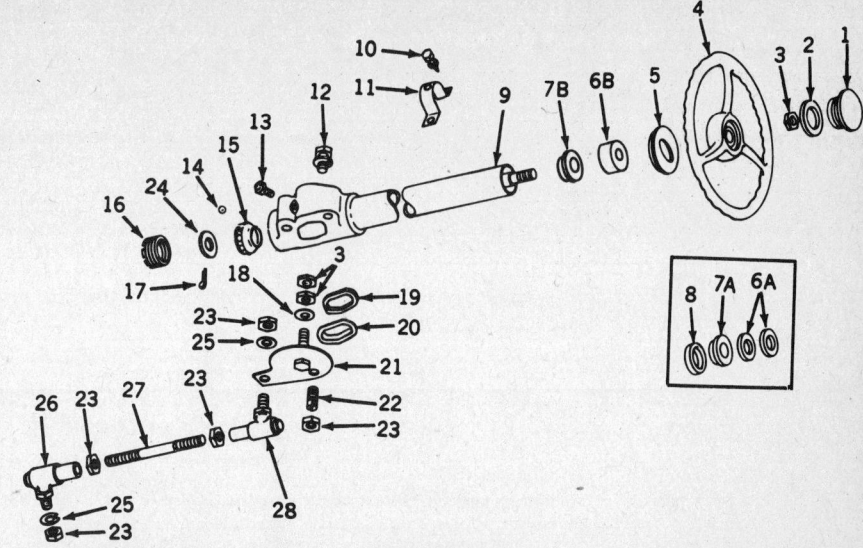

1. Steering wheel emblem
2. O-ring
3. ⅝ in. hex nut
4. Steering wheel
5. Grommet
6A. Felt washer (2 used) (before 130,000)
6B. Bearing seal
7A. Bearing spacer (before 130,000)
7B. Bearing
8. Bearing retainer
9. Steering shaft and column
10. Thread cutting screw (2 used)
11. Clamp
12. Capscrew (3 used)
13. Grease fitting
14. Steel balls (16 used)
15. Bearing cups and retainers (2 used)
16. Adjusting plug
17. Cotter pin
18. Washer
19. Lever arm seal
20. Retainer
21. Lever arm and cross bolt
22. Tapered stud
23. ½ in. hex jam nut (fine thread) (3 used)
24. Spring washer (130,000 and up)
25. ½ in. lockwasher (6 used)
26. Ball joint (righthand threads) (2 used)
27. Drag link
28. Ball joint (lefthand threads)

Steering system components, tractors serial #100,001-250,000

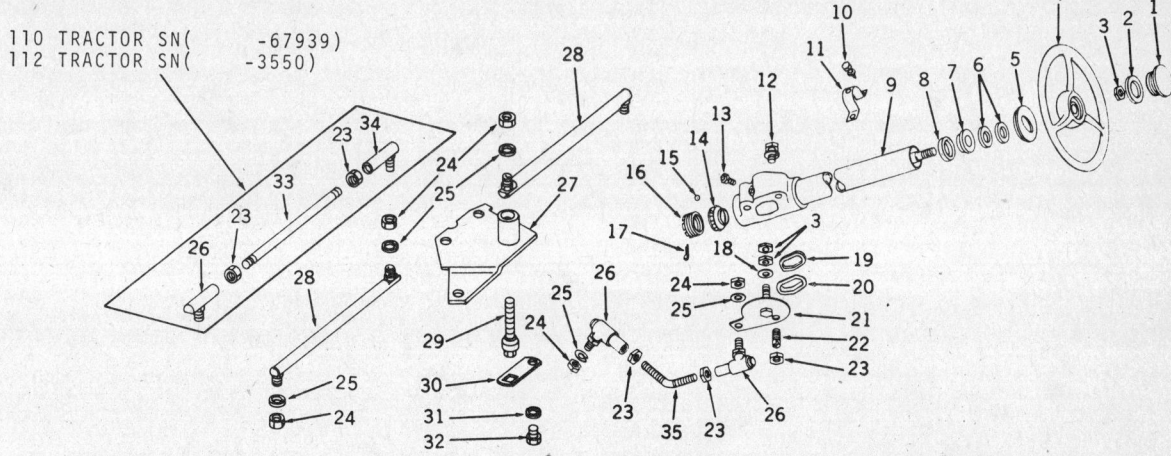

1. Steering wheel emblem
2. O-ring
3. ⅝ in. hex nut (3 used)
4. Steering wheel
5. Grommet
6. Felt washer (2 used)
7. Bearing spacer
8. Bearing retainer
9. Steering shaft and column
10. Thread cutting screw (2 used)
11. Clamp
12. Capscrew (3 used)
13. Grease fitting
14. Bearing cups and retainers (2 used)
15. Steel balls (16 used)
16. Adjusting plug
17. Cotter pin
18. Washer
19. Lever arm seal
20. Retainer
21. Lever arm and cross bolt
22. Tapered stud
23. ½ in. hex jam nut (fine thread) (3 used)
24. ½ in. hex nut (6 used)
25. ½ in. lockwasher (6 used)
26. Ball joint (righthand threads) (2 used)
27. Steering arm
28. One-piece tie rod assembly (2 used)
29. Bolt and cone assembly
30. Locking strap
31. Washer
32. Self-tapping screw
33. Adjustable tie rod
34. Ball joint (lefthand threads)
35. Drag link

Steering system components, tractors through serial #100,000

John Deere

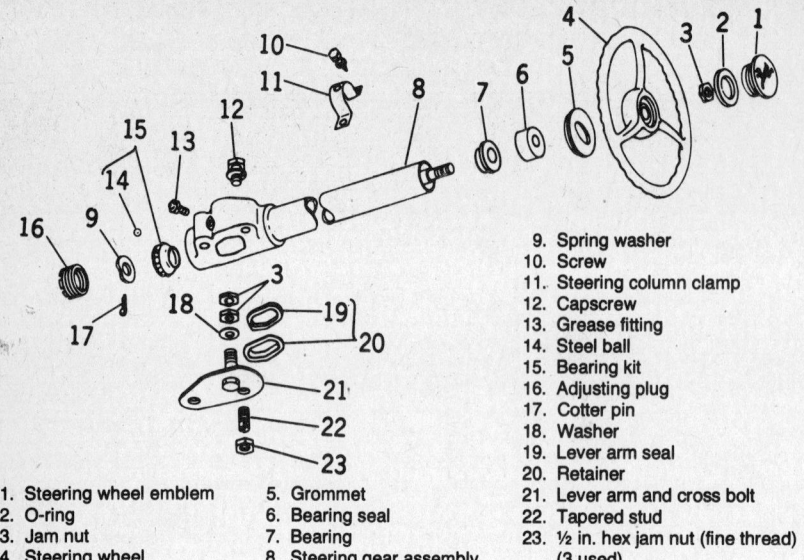

1. Steering wheel emblem
2. O-ring
3. Jam nut
4. Steering wheel
5. Grommet
6. Bearing seal
7. Bearing
8. Steering gear assembly
9. Spring washer
10. Screw
11. Steering column clamp
12. Capscrew
13. Grease fitting
14. Steel ball
15. Bearing kit
16. Adjusting plug
17. Cotter pin
18. Washer
19. Lever arm seal
20. Retainer
21. Lever arm and cross bolt
22. Tapered stud
23. ½ in. hex jam nut (fine thread) (3 used)

Steering system components, 110 tractors serial #250,001 and up

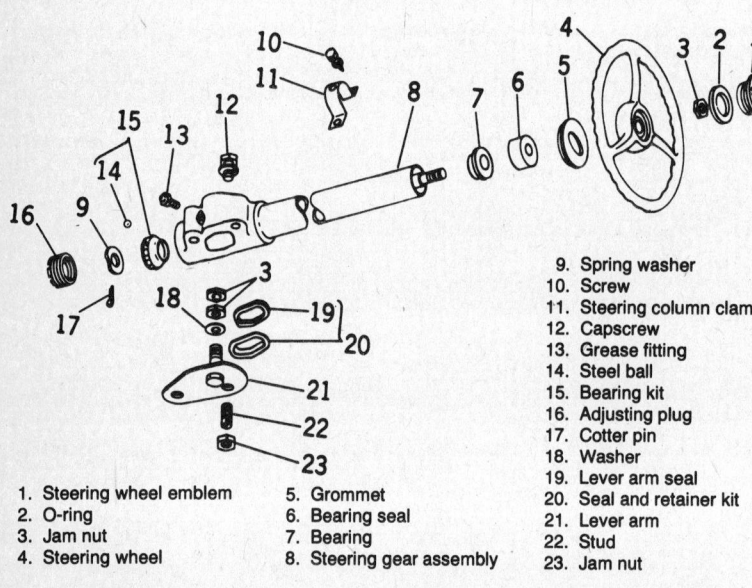

1. Steering wheel emblem
2. O-ring
3. Jam nut
4. Steering wheel
5. Grommet
6. Bearing seal
7. Bearing
8. Steering gear assembly
9. Spring washer
10. Screw
11. Steering column clamp
12. Capscrew
13. Grease fitting
14. Steel ball
15. Bearing kit
16. Adjusting plug
17. Cotter pin
18. Washer
19. Lever arm seal
20. Seal and retainer kit
21. Lever arm
22. Stud
23. Jam nut

Steering system components, 112 tractors serial #250,001 and up

Adjusting arm lever

ASSEMBLY

1. Apply grease to all moving parts prior to assembly.
2. Place the baring balls, ball cups and retaining rings on both ends of the cam.
3. Slide the cam and tube assembly into the housing and jacket tube. Install the plug and torque to 7-12 ft.lbs. After torquing, lock the plug with a center punch.
4. Insert the cotter pin and make sure that the steering column turns freely.

ADJUSTMENTS

NOTE: The steering mechanism must be adjusted in the sequence described below.

1. Disconnect ball joint from lever arm.
2. Loosen jam nut and turn stud counterclockwise two or three turns.
3. Remove cotter pin holding adjusting plug in gear housing. Steering columns on tractors prior to serial #51052 do not have a cotter pin. Turn adjusting plug into housing until 7-12 ft.lbs. torque is obtained. Back plug out until steering wheel turns freely with no drag.
4. Lock plug after adjustment is obtained. On tractors prior to serial #51052, lock plug by upsetting plug threads with a punch. On tractors after serial #51053, lock plug by turning plug only far enough to insert cotter pin through housing and closest slot in plug.
5. Loosen jam nut on cross bolt and tighten only the inside nut using a thin open-end wrench, until all end play is removed or until the distance between the steering arm and gear housing is between $1/16$ and $3/32$ inch. After adjustment is completed, torque the lever arm cross bolt to 22-25 ft.lbs.
6. Turn steering arm until the arm is parallel with steering gear body.
7. Turn stud in (clockwise) until snug to remove all backlash. Then move steering arm through its full steering range in both directions (front to rear). Steering wheel will turn as this check is made. When properly adjusted, a slight drag can be detected in the midpoint of the range (when line between the cross bolt and ball joint is vertical). After adjustment is completed, torque the jam nut to 40 ft.lbs. Make final test by turning steering arm through full range.
8. Set front wheels straight forward and turn steering wheel so that lever arm is parallel with steering gear housing (center of lever arm travel). Connect drag link.

NOTE: It may be necessary to lengthen or shorten drag link by turning drag link end. It is important that drag link is positioned with bend facing the center of the tractor before tightening nuts.

9. Check steering for equal turn in both directions.
10. Readjust ball joint if necessary.

John Deere

Adjusting steering arm on tractors prior to serial #100,000

Steering arm cone adjustment, tractors serial #100,001 and up

Adjusting Steering Arm on 110 Tractors Serial Numbers 40001-100,000 and 112 Tractors Prior to Serial Number 100,000

1. Disconnect drag link and tie rod at "A".
2. Loosen lock retaining screw and remove lock from bolt head.
3. Remove steering bolt, cone and arm assembly. Apply grease to both inner and outer cones and reassemble.
4. Tighten bolt only until a slight amount of drag can be felt when turning the steering arm through its range and all end play has been removed.
5. Position lock plate over bolt head and tighten lock plate capscrew. Be sure plain washer is used with lock plate capscrew. Reassemble tie rods and drag link to steering arm and tighten nuts firmly.

Adjusting Steering Arm Cone
TRACTORS
SERIAL NO. 100,000 AND UP

1. Block up front end of tractor so that front tires are off the ground.
2. Disconnect ball joints at points "A".
3. Turn steering arm by hand and notice freedom of movement. When properly adjusted, the steering arm will pivot freely through the entire steering range with only a slight amount of drag.
4. If steering arm turns hard or has worn and loosened so that you can feel end play in the steering arm bearing, remove lock plate.
5. Remove steering bolt with lower cone and steering arm. Apply grease to lower cone and upper cone (in tractor frame) and reassemble.

6. Tighten bolt only until a slight amount of drag can be felt when turning the steering arm and all end play has been removed.
7. Position lock plate over bolt head and tighten lock plate capscrew. Be sure plain washer is used with lock plate capscrew.
8. Reassemble ball joints in position "A".

Toe-In Adjustment

Measure distances "A" and "B" above. The tractor has proper toe-in or alignment when dimension "A" is 3/16 inch less than dimension "B". When required, loosen jam nuts and turn both righthand and lefthand tie rods "C" equally until proper toe-in is obtained. Tighten jam nuts firmly.

HYDRAULIC SYSTEM

Control Valve
REMOVAL

1. Lower equipment to ground and with

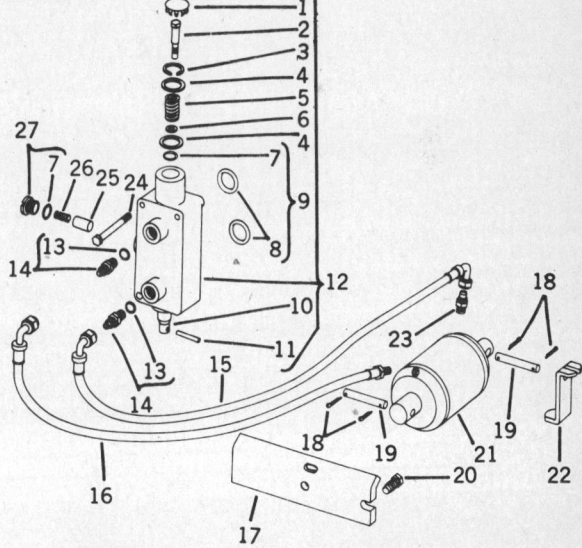

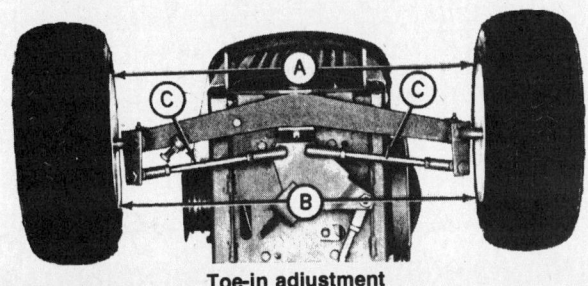

Toe-in adjustment

1. Button plug
2. Spool screw
3. Snapring
4. Washer (4 used)
5. Inner and outer springs
6. Spool spacer
7. O-ring (3 used)
8. Control valve O-ring (2 used)
9. Control valve O-ring kit
10. Spool
11. Pin
12. Control valve assembly
13. O-ring for connector (2 used)
14. Control valve connectors (2 used)
15. Lower hose, 30 in. long
16. Upper hose, 27¼ in. long
17. Cylinder bracket
18. Cotter pin (4 used)
19. Cylinder pin (2 used)
20. Tapping screw (2 used)
21. Hydraulic cylinder
22. Locking clip
23. Cylinder connector
24. Capscrew (3 used)
25. Lift check plunger
26. Lift check spring
27. Lift check plug

Control valve components

313

John Deere

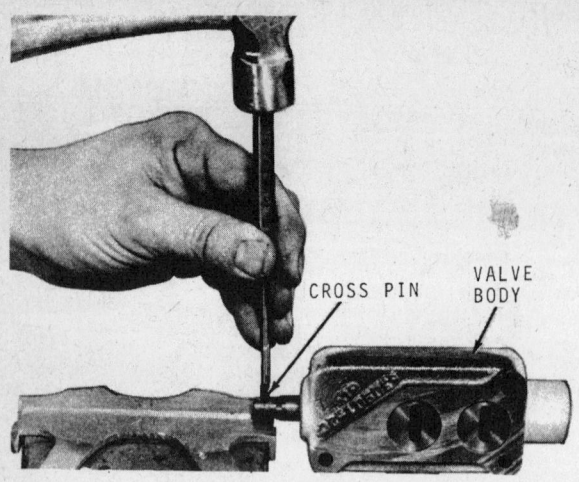

Removing crosspin from spool

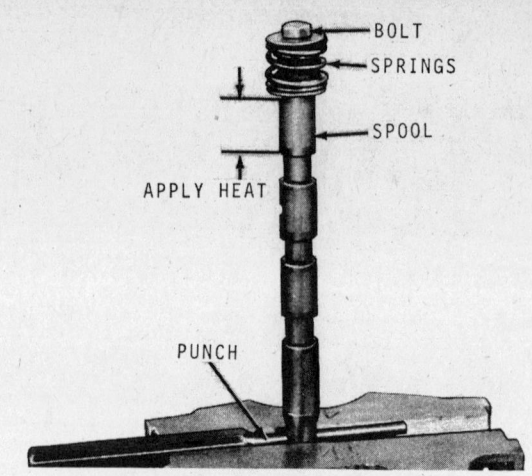

Removing spool centering rings

engine stopped, move control lever up and down to release all pressure in system.

2. Wipe all dirt from connections on valve body. Disconnect hoses at valve body. Cap connections on valve body and plug hoses.

3. Loosen idler, remove drive belt, drive sheave and key. Remove two mounting bolts.

4. Twist unit to disconnect end of spool from control bracket arms and remove hydraulic unit.

5. Thoroughly wash outside of assembly with clean, safe cleaning solvent.

6. Drain reservoir before removing valve assembly from pump body.

DISASSEMBLY

Removing Valve Assembly from Pump

1. Remove valve assembly from pump by removing three capscrews which hold valve assembly to pump back plate.

2. Discard O-rings between the valve body and pump back plate.

Disassembling Valve

1. Rest small diameter of spool end on a partially closed vise and very carefully remove crosspin.

— **CAUTION** —
Use special care to prevent marring or bending spool.

2. Remove cap and snapring from valve body. Pull valve spool out spring end of valve body.

3. Remove and discard O-rings from inside diameter of each end of spool bore.

4. With a large screwdriver or impact tool, loosen and remove plug, lift check spring and plunger. Discard O-ring from slotted plug.

5. Do not remove bolt from spool unless springs are broken.

6. Insert punch through hole in spool and clamp spool in a vise with soft jaws.

NOTE: Apply heat to threaded end of spool before attempting to remove shoulder bolt from spool.

7. Clean and dry all parts thoroughly and inspect parts for wear and damage. Clean O-ring grooves in valve body of all foreign matter.

8. Check valve housing for cracks or damaged threads. Inspect inside diameter of valve for scratches or excessive wear.

9. The lift check seat is machined into the valve body. Inspect lift check seat in body for damage. It is important that the lift check seat be smooth.

Spool Plunger and Springs

1. Remove burrs from spool with fine emery cloth. Inspect spool for wear, scratches or other damage. The housing and spool must always be replaced as a matched assembly.

2. Inspect lift check plunger for scratches or unevenness of seating surface.

3. Whenever lift check seat is scratched or pitted, dress seat surface until plunger seating area is smooth and even.

4. Inspect inner and outer spool centering springs for breakage or excessive weakness. Replace weak or broken springs.

ASSEMBLY

NOTE: Replace all control valve O-rings with new O-rings whenever the valve is disassembled for service.

Installing O-Rings in Valve Body

Apply oil to new O-rings and install in valve body. Always use new O-rings.

Installing Lift Check Plug

1. Install new O-ring on lift check plug. If lift check plunger or spring is damaged, replace them. Install lift check plunger and lift check spring in valve body and secure with lift check plug. Tighten plug firmly.

2. If valve housing is to be replaced, a new spool must be used because the valve body and spool are a matched assembly. If spool centering springs are broken or show signs of cracking, use new springs.

3. If spool has been disassembled, place spool in vise with soft jaws and secure inner and outer springs to spool with washer and shoulder bolt. Apply Loctite® or equivalent to threads of shoulder bolt.

4. Torque spool centering spring bolt to 60-65 ft.lb.

5. Apply grease to O-rings in spool bore and insert spool assembly from spring end of valve body. Insert spool slowly while rotating spool so as not to cut O-ring as spool lands pass through O-ring.

6. Secure spool assembly in valve body with snapring.

7. Place cap on spring end of valve body.

Rest small end of spool on partially closed vise and install cross pin.

8. Wipe a light film of clean grease on O-rings and place O-rings on valve body.

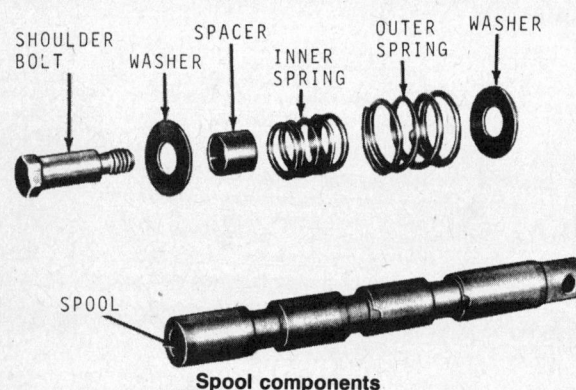

Spool components

John Deere

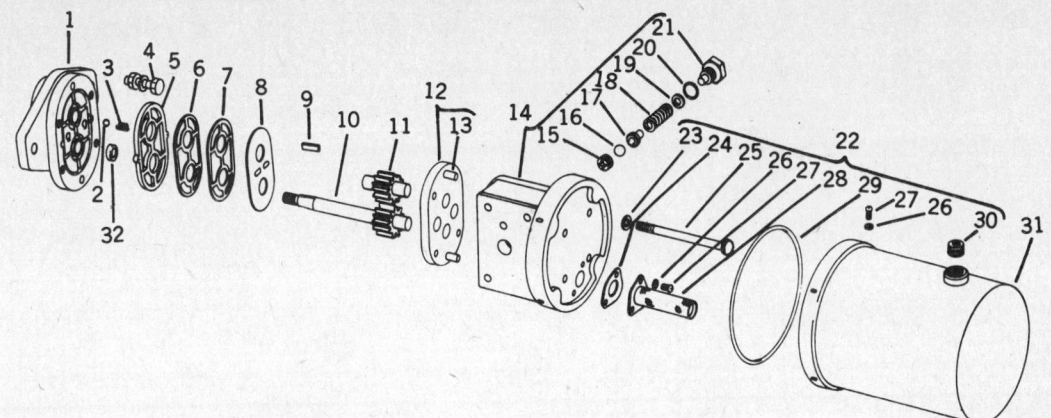

1. Front plate
2. Steel ball (2 used)
3. Check spring (2 used)
4. Capscrew (2 used)
5. Diaphragm seal
6. Protector gasket
7. Backup gasket
8. Diaphragm
9. Key
10. Driveshaft and gear
11. Idler shaft and gear
12. Body and dowel
13. Dowel (2 used)
14. Back plate assembly
15. Relief valve seat
16. Steel ball
17. Relief valve adaptor
18. Relief valve spring
19. Relief valve shim
20. O-ring
21. Hex hd. plug
22. Filter replacement kit
23. Aluminum washers (4 used)
24. Filter gasket
25. 12 point capscrews (4 used)
26. Internal tooth washer (6 used)
27. Machine screw (6 used)
28. Filter
29. O-ring
30. Filler plug
31. Reservoir
32. Driveshaft seal

Hydraulic pump components

Installing Valve Assembly on Pump

1. With new O-rings between valve body and back plate, secure valve assembly to pump back plate with three capscrews.
2. Torque the valve body capscrews to 7-10 ft.lb.

Place new O-rings on connectors and screw connectors into valve body. Tighten connectors firmly.

INSTALLING ON TRACTOR

1. Position hydraulic base on engine making sure washers and fiber washers are positioned as shown in inset.
2. Position end of spool in lever arms and secure front plate to bracket with two bolts, lockwashers and nuts.
3. Install key in shaft, install sheave on shaft and secure with elastic stop nut. Install drive belt.
4. Adjust drive belt tension.
5. Connect hoses to valve assembly. Fill reservoir with fluid.

Hydraulic Pump

DISASSEMBLY

1. Before separating pump assembly, scribe a clear line across outside of pump assembly. This will assure proper reassembly.
2. Remove reservoir and four 12-point capscrews.
3. Tap against front plate to separate front plate, body and back plate. Do not use sharp tools or screwdriver to separate parts.
4. Place a screwdriver under the diaphragm seal being careful not to damage front plate. Lift diaphragm seal and gaskets from plate. Discard diaphragm seal and gaskets.
5. The relief valve seat is locked in place. Do not attempt to remove seat unless repair is necessary. Apply heat to back plate and use screwdriver to remove seat.
6. When replacing seat, apply Loctite® or equivalent and turn in to 1.776-1.786 in.

INSPECTION

1. Wash all parts in a clean safe cleaning solvent and dry them with compressed air.
2. Inspect all parts for wear, and remove all scratches, nicks, burrs and rough spots with emery cloth. Check condition of springs.
3. Inspect the drive gear and idler gear shafts at bearing points and seal areas for rough surfaces and excessive wear. Use a micrometer to measure the shafts. Wear tolerance is 0.4359 in. Inspect driveshaft for broken keyway. Shafts and gears are available as assemblies only.
4. Inspect the face of the gear for scoring and excessive wear. Use a micrometer to measure gear width. Snaprings should be in groove in drive and idler shaft gears. If gears require replacing, replace gear and shaft as an assembly. If edges of teeth are sharp, break edges with emery cloth.
5. Use a telescope gauge to measure bearing wear in the front and back plate. Bearing thickness is 0.4386-0.4389 in.

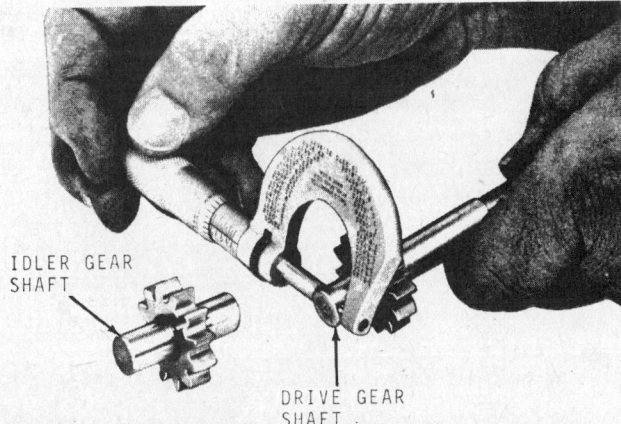

Measuring gear shafts

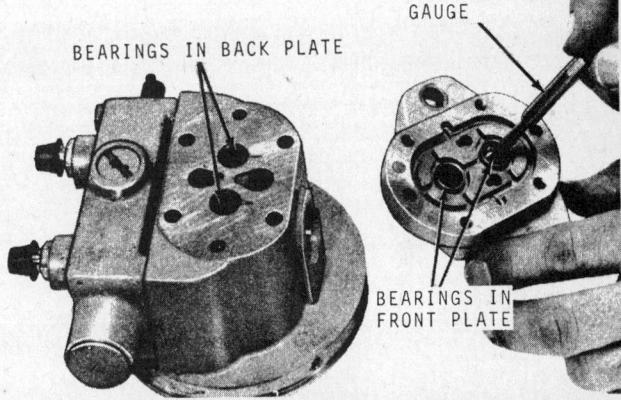

Measuring inside diameter of bearings

John Deere

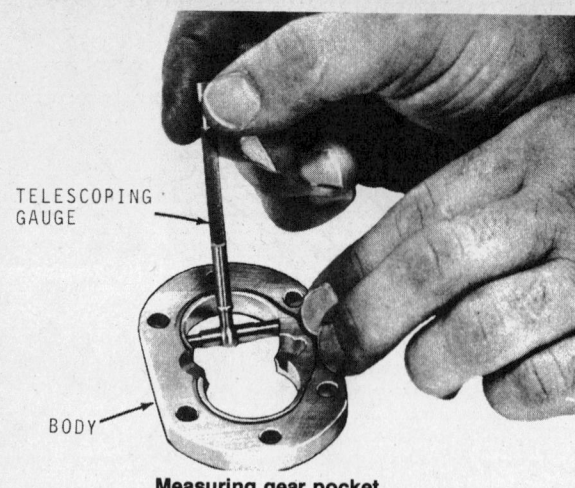
Measuring gear pocket

Bearings in front plate should be flush with islands in groove pattern. Bearings are available for service only as a plate and bearing assembly. Replace front or back plate if scored or if bearings are worn beyond specifications.

*6. Small scratches and some wear pattern should be considered normal and will not affect pump operation. Check plate wear. Back plate wear is 0.0015 in. Replace back plate if worn beyond specification.

7. Inspect the gear pockets for scoring or wear. Gear pocket diameter is 1.675-1.681 in. If gear pockets are scored or worn, beyond specifications, replace body.

8. Inspect condition of relief valve seat, ball, adapter and spring. Replace parts showing abnormal wear.

9. If relief valve seat removal is necessary.

ASSEMBLY

1. Install new diaphragm seal, protector gasket, backup gasket and diaphragm when reassembling pump. Install diaphragm seal in grooves of front plate with seal "V" groove down. Use small blunt screwdriver to position seal in grooves. Press protector gasket and backup gasket into diaphragm seal. Drop steel balls into respective seats and place springs over balls.

2. Place diaphragm on top of gaskets with bronze face up and coined indents on suction side. The entire diaphragm must fit inside the raised rim of the diaphragm seal. Insert dowel pins in front plate.

3. Dip gear assemblies in light, clean oil and slip into front plate bearings.

4. Apply a thin layer of "Copper Coat" or equivalent to both milled surfaces of body. Slip body over gears onto front plate. Half moon port cavities in body must face away from front plate and scribe lines should be aligned. The cavity with the small hole drilled in it must be on the pressure side of pump.

5. Place new aluminum washers over the four 12-point capscrews. Install capscrews through back plate and secure front plate to back plate. Torque the bolts to 7-12 ft.lb.

6. Place scotch tape over keyway in shaft. Oil seal liberally and work shaft seal over drive shaft.

7. Tap seal in place with a deep well socket and hammer.

8. The outer face of the seal should be flush with outer edge of front plate when seal is in place.

9. Rotate the driveshaft to make sure there is no interference with rotating parts. A smooth, heavy drag indicates a good pump. A jerky drag or frozen shaft indicates an improperly assembled pump. (Pump rotation is counterclockwise from end of shaft.)

10. Whenever relief valve seat has been removed, install seat in back plate as shown. Seat depth is 1.776-1.786 in.

NOTE: Seat must be held in place with Loctite® or equivalent. Clean threads and seat thoroughly before applying Loctite®. Wipe off excess Loctite® after positioning seat.

11. After relief valve seat is properly located, install ball adapter and spring in back plate. Place new O-ring on plug and secure parts with plug. See the Pressure Adjustment procedure.

12. Refer to exploded view, and install new filter gasket and filter to back plate with two washers and two machine screws.

13. Install new O-ring over reservoir mounting shoulder and carefully slide reservoir onto pump. Be sure port in reservoir is in correct location. Secure reservoir to back plate with four washers and machine screws. Turn filler plug loosely into reservoir port.

14. Install the assembly on the tractor. Connect the hydraulic hoses to the valve assembly.

15. Fill the reservoir with fluid.

16. Adjust drive belt tension and relief valve pressure as explained on this page.

ADJUSTMENTS

Drive Belt Tension

1. Loosen the idler bolt and move idler against belt until a 3 to 4 pound pressure midway between the sheaves deflects the belt ½ inch.

2. Tighten the idler nut firmly to maintain proper belt tension.

Relief Valve Pressure

1. A pressure gauge having sufficient capacity must be used to obtain proper relief valve pressure. Excessive pressure can do severe damage to various components, thus voiding warranty. Add or remove shims as necessary until 800 (-0 +100) psi is obtained.

2. Always follow instructions supplied by test equipment manufacturer.

CLUTCH, BRAKE AND VARIABLE SPEED DRIVE

PRIMARY V-BELT REPLACEMENT

Tractors without PTO

--- CAUTION ---
To prevent possibility of injury, always remove spark plug cable before removing belts.

1. Remove muffler and disconnect safety switch leads. Remove belt guards, belt guide, hydraulic drive belt and mower drive if tractor is so equipped. Move vari-

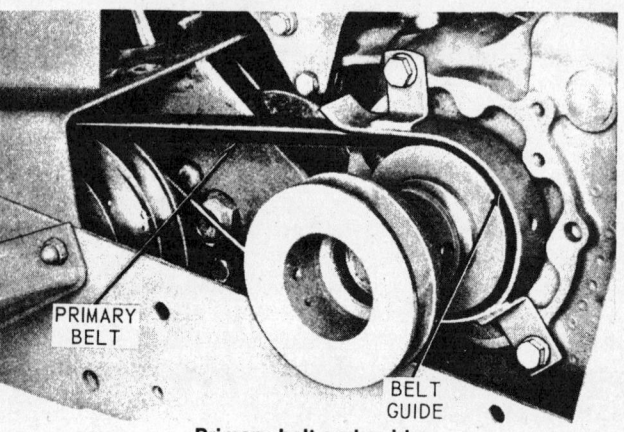
Primary belt and guide

John Deere

able speed control lever forward (fast speed position). Turn engine over slightly until variator is fully back. Then raise secondary idler and slip secondary belt off variator sheave.

2. After secondary belt is removed from variator, depress clutch-brake pedal and lock parking brake to hold variator forward. Remove primary belt guide and loosen bolt holding belt guide. Remove belt from variator and engine sheave.

3. Reverse above procedure to install new primary belt. Adjust primary belt guide.

Tractors with PTO
SERIAL NO. 100,000-249,000

1. Remove PTO shield by lifting up with fingers and pulling out.
2. Remove PTO brake capscrew.
3. Remove PTO brake.

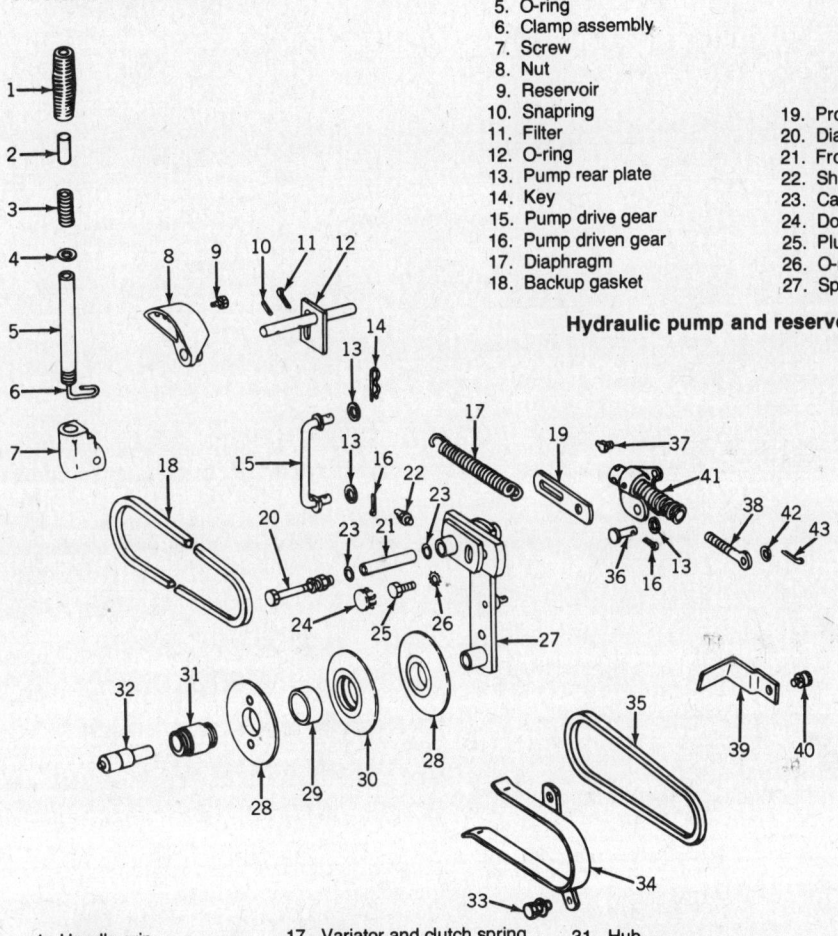

1. Pump body assembly
2. O-ring
3. O-ring
4. Plug assembly
5. O-ring
6. Clamp assembly
7. Screw
8. Nut
9. Reservoir
10. Snapring
11. Filter
12. O-ring
13. Pump rear plate
14. Key
15. Pump drive gear
16. Pump driven gear
17. Diaphragm
18. Backup gasket
19. Protector gasket
20. Diaphragm seal
21. Front plate assembly
22. Shaft seal
23. Capscrew
24. Dowel pin
25. Plug assembly
26. O-ring
27. Spring
28. Check ball
29. Spring
30. Check ball
31. Spring seat

Hydraulic pump and reservoir components—90 degree unit

1. Handle grip
2. Thumb release
3. Spring
4. Washer
5. Speed control lever
6. Speed control rod
7. Lever hub
8. Lever quadrant
9. Screw
10. Spring pin
11. Cotter pin
12. Speed control shaft
13. Washer
14. Spring locking pin
15. Speed control shaft link
16. Cotter pin
17. Variator and clutch spring
18. Secondary belt
19. Spring link
20. Capscrew
    Nut
    Lockwasher
21. Pivot ferrule
22. Grease fitting
23. O-ring
24. Button plug
25. Capscrew
26. Lockwasher
27. Variator arm
28. Outer variator half sheave
29. Center sheave bearing
30. Center sheave
31. Hub
32. Bearing and shaft assembly
33. Capscrew
    Lockwasher
34. Belt guide
35. Primary belt
36. Drilled pin
37. Grease fitting
38. Clutch override link
39. Belt guide
40. Capscrew
    Lockwasher
41. Clutch override
42. Washer
43. Cotter pin

Variator assembly, tractors under serial #272,000

4. Unsnap clutch arm clip and slide clutch arm and PTO sheave off shaft. Be careful not to get any dirt or foreign material into sheave bearings.

NOTE: Secondary belt and hydraulic pump drive belt must be removed to replace primary belt.

5. Depress clutch-brake pedal and lock parking brake. Remove belt guides and shield mounting bracket. Loosen 5/16 inch capscrew on variator belt guide and slide guide up far enough to remove belt from variator sheave and engine sheave.

6. Replace belt and reassemble parts removed.

SERIAL NO. 250,000 AND UP

Remove PTO shield by pushing down and unhook bottom of shield from lower attaching pin. Then lift outward and upward on bottom of shield to unhook top of shield from two attaching pins.

8 hp Tractors (Serial No. 250,001-272,000) and 10 hp Tractors (Serial No. 250,001-285,000)

1. Remove PTO brake capscrew.
2. Lift and pivot clip to release clutch arm. Lower clutch arm and remove PTO drive sheave. Be careful not to get dirt or foreign material into sheave bearings.

NOTE: Secondary belt and hydraulic pump drive belt 10 hp tractors (Serial No. 260,001-272,000) must be removed to replace primary belt.

3. Depress clutch-brake pedal and lock pocking brake. Remove belt guides and

317

John Deere

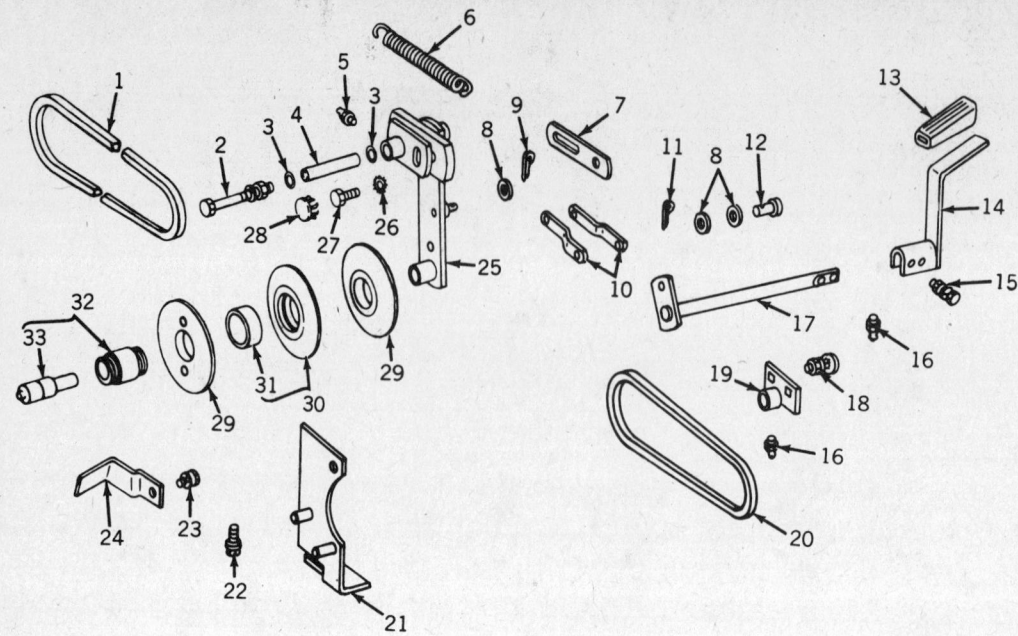

1. Secondary belt
2. Capscrew with lockwasher and nut
3. O-ring
4. Pivot ferrule
5. Grease fitting
6. Variator spring
7. Spring link
8. Washer
9. Cotter pin
10. Clutch link
11. Cotter pin
12. Drilled pin
13. Pedal pad
14. Clutch pedal
15. Capscrew with lockwasher and nut
16. Grease fitting
17. Clutch shaft
18. Carriage bolt with lockwasher and nut
19. Clutch shaft bearing
20. Primary belt
21. Belt guide
22. Capscrew with lockwasher
23. Capscrew with lockwasher
24. Belt guide
25. Variator arm
26. Lockwasher
27. Capscrew
28. Button plug
29. Outer half variator sheave
30. Center variator sheave with bearing
31. Center sheave bearing
32. Variator bearing and hub assembly
33. Bearing and shaft assembly

Variator assembly, tractors serial #272,001 and up

shield mounting bracket. Loosen 5/16 inch capscrew on variator belt guide and slip guide up far enough to remove belt from variator sheave and engine sheave.

4. Replace belt and reassemble parts removed.

NOTE: After replacing primary belt, readjust variator.

8 hp Tractors (Serial No. 272,001-285,000 and 310,001-320,000)
10 hp Tractors (Serial No. 285,001-310,000)

Remove spring locking pin from pivot pin and remove pivot pin from clutch arm pivot. Remove clutch arm and proceed as outlined previously.

10 hp Tractors (Serial No. 320,001 and up)

Lift and pivot clip to release clutch and proceed as outlined previously.

SECONDARY BELT REPLACEMENT

1. To replace worn or broken secondary belt, move variable speed control lever forward (fast speed position). Turn engine over momentarily to allow variator to move to fast speed position. Then raise secondary idler and slip secondary belt off variator. Remove three screws from input sheave and slide sheave off hub far enough to remove belt.

2. Install new belt around variator sheave. Block up secondary idler to release belt tension and install belt and input sheave.

NOTE: If transmission has been moved rearward, to take up secondary belt slack prior to belt replacement, loosen bolts and move transmission forward before installing new secondary belt. Tighten bolts holding transmission.

3. Readjust variator and brake linkage after moving transmission.

Variator

REMOVAL

Tractors without PTO

1. Remove primary and secondary belt from variator.

NOTE: Do not pry belts over sides of variator.

Disassembling center variator sheave

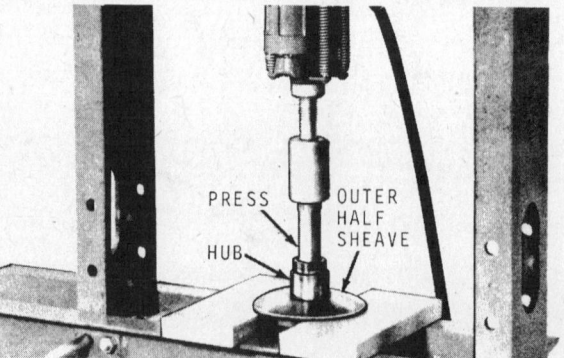

Pressing variator bearing from hub

2. Remove two capscrews from variator arm to remove variator.

3. Place variator half sheave (next to bearing support) in a vise with soft jaws. Insert ends of two large punches in holes of sheave and a bar between punches. Then turn counterclockwise to remove sheave. Lift center sheave from variator hub.

Tractors with PTO

1. Remove primary and secondary belt from variator.

NOTE: Do not pry belts over sides of variator.

2. Loosen 5/16 inch capscrew on primary belt guide and slide guide up far enough to remove belt from variator sheave and engine sheave.

3. Disconnect variator spring. Remove battery and battery base. Disconnect clutch override at variator arm and clutch shaft arm. Disconnect speed control shaft link and remove variator pivot capscrew. Guide pivot end of variator arm through notch in tractor frame. Variator must be removed from underside of tractor.

DISASSEMBLY

1. Place variator bearing and hub assembly under press and press bearing from hub. Be sure to press against outer race only.

2. Place hub in a vise and remove half sheave with two punches and a bar.

INSPECTION

1. Measure I.D. of center sheave bearing and O.D. of variator hub after cleaning parts thoroughly. Sheave bearing I.D. is 2.0015-2.0025 in. Variator hub O.D. is 1.999-2.001 in. Replace center sheave or hub if wear limits are exceeded. Do not attempt to service center sheave bearing. Bearing and center sheave are available only as a factory assembly.

2. Check center sheave and sheave halves for wear on the sheave faces or for evidence of damage or nicks. Replace parts which may cause excessive belt wear or which would upset the delicate balance of the variator assembly.

3. Measure press fit between bearing and hub. Hub I.D. is 1.17945-1.800 in.; bearing O.D. is 1.1806-1.1811 in. Also check press fit of bearing shaft in bearing support. Bearing shaft O.D. is 0.6262-0.6267 in.; bearing support I.D. is 0.6240-0.6255 in. Replace parts necessary to obtain proper fit.

NOTE: The center sheave is lubricated with a special grease at the factory and will last for the lifetime of the sheave. Do not attempt to lubricate center sheave.

ASSEMBLY

1. Coat bearing case with light film of oil. Place hub with sheave on press bed and press bearing into hub until distance between end of bearing shaft and hub face is 0.031-0.047 inch beyond hub face.

2. Wipe light film of oil on bearing shaft. Place bearing support on bearing shaft with weld down or under cut-up, depending on type of support. Press bearing support on shaft until distance between bearing support and sheave is 0.13 inch.

3. Clamp assembly in vise having soft jaws. Place center sheave assembly on hub and thread half sheave on hub. Using two large punches and a bar, tighten sheaves firmly by turning sheave in opposite direction.

4. Spike threads three or four places on both sides of variator. After spiking threads, recheck distance between bearing support and sheave. If distance is greater than 0.13 inch, press bearing support further on shaft until proper distance is obtained.

5. Attach variator and primary belt guide on variator arm with two capscrews. Install belts and adjust primary belt guide. Tighten capscrews firmly.

6. After installation, readjust variator linkage.

Connecting Variator Spring

1. Connect variator spring to spring link with an automotive brake spring pliers.

2. Install primary belt. To ease primary belt installation, move variator lever rearward and depress and lock parking brake. This will hold variator sheave in a forward position.

3. Next, install secondary belt. To facilitate this belt installation, release parking brake and move the variator lever to the forward position before placing belt over sheave. In most cases it will be helpful to move secondary belt tightener upward to gain the additional belt length required for installation.

4. Position primary belt guide approximately 1/16 inch from sheave and tighten retaining capscrew firmly.

5. After installation is completed, make final adjustments to variator.

Brake Bands

REPLACEMENT

NOTE: A brake band with bonded lining is used on all 110 and 112 Tractors. Whenever brake band servicing is required due to worn or oily lining or other damage, the following procedure should be used depending on the tractor serial number.

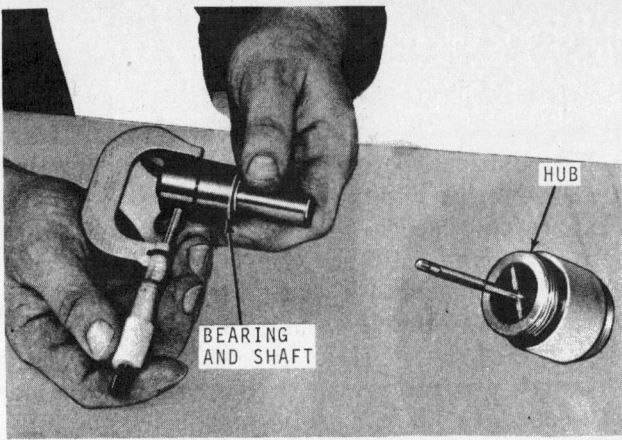

Checking variator bearing and shaft

110 Tractors Prior to Serial Number 3550

1. The frame must be separated from the transaxle to replace the brake band or brake pulley.

2. To separate, remove three capscrews from transaxle sheave on input shaft, disconnect brake clevis, idler spring and neutral start wires.

3. Remove eight capscrews securing tractor base to hitch assembly. Raise transaxle assembly and place blocks between transaxle and tractor base.

4. Remove brake pulley with puller.

5. Remove brake band pivot bolt and raise assembly. Slip brake pin out of brake arms and lever.

6. Drive spring pins from arms and band.

7. Lubricate lever pivot before reassembly.

8. Reverse disassembly procedure to assure correct installation.

9. After installing transaxle on tractor, adjust linkage accordingly.

110 Tractors Prior to Serial Number 9082

1. Loosen pulley setscrew and remove brake pulley with a puller.

2. Disconnect brake rod from brake arm on clutch shaft. Bend end of lever stop far enough to clear brake lever. Remove brake pivot bolt and lower assembly from brake bracket. Remove brake pin.

3. Use light grease to lubricate lever pivot before reassembling pivot in lever.

4. Apply Loctite® to threads before tightening setscrew in brake pulley.

5. Adjust lever stop against lever to prevent brake arms from dragging on brake pulley.

6. To check lever stop adjustment, place shifter lever in neutral position. If brake adjustment is correct, the brake pulley should be free enough to rotate by hand.

7. After assembling brake, adjust linkage accordingly.

John Deere

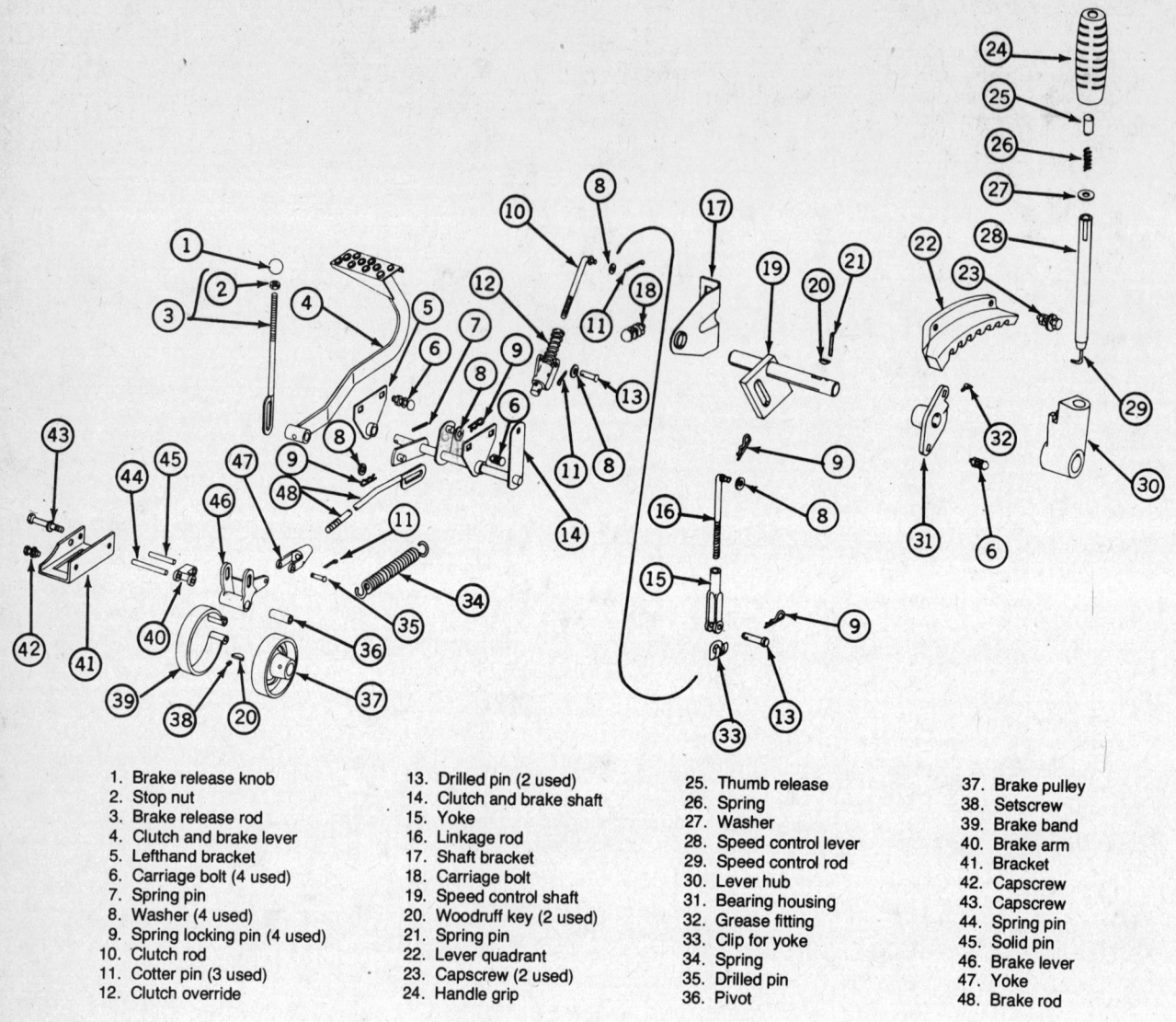

1. Brake release knob
2. Stop nut
3. Brake release rod
4. Clutch and brake lever
5. Lefthand bracket
6. Carriage bolt (4 used)
7. Spring pin
8. Washer (4 used)
9. Spring locking pin (4 used)
10. Clutch rod
11. Cotter pin (3 used)
12. Clutch override
13. Drilled pin (2 used)
14. Clutch and brake shaft
15. Yoke
16. Linkage rod
17. Shaft bracket
18. Carriage bolt
19. Speed control shaft
20. Woodruff key (2 used)
21. Spring pin
22. Lever quadrant
23. Capscrew (2 used)
24. Handle grip
25. Thumb release
26. Spring
27. Washer
28. Speed control lever
29. Speed control rod
30. Lever hub
31. Bearing housing
32. Grease fitting
33. Clip for yoke
34. Spring
35. Drilled pin
36. Pivot
37. Brake pulley
38. Setscrew
39. Brake band
40. Brake arm
41. Bracket
42. Capscrew
43. Capscrew
44. Spring pin
45. Solid pin
46. Brake lever
47. Yoke
48. Brake rod

Brake system components, tractors through serial #100,000

Variator Brake Shoe

REPLACEMENT

110 Tractors Prior to Serial Number 9082

1. A brake shoe was provided on these tractors to stop the flywheel action of the heavy cast iron input sheave on the transaxle when clutching drive train.
2. The brake shoe is not required on 110 tractors having the lighter sheet metal input sheave. The shoe may be removed on these tractors.
3. If the brake shoe causes undesirable squealing on tractors having the brake shoe and cast iron sheave, remove the brake shoe and replace the cast iron input sheave with the lighter sheet metal sheave. Remove the variator brake assembly only on tractors equipped with sheet metal sheave.

Brake Band

REPLACEMENT

110 and 112 Tractors Serial Numbers 15001-100,000

1. Remove lefthand fender by removing three capscrews. Refer to brake band replacement for 110 tractors prior to serial number 3550 to remove brake band having brake bracket with one hole.
2. To remove brake band on 110 tractors with two holes in brake bracket, and all 112 tractors, remove lefthand fender by removing three capscrews. Loosen brake pulley setscrew and pull brake pulley from shaft with a puller.
3. Remove brake band pivot bolt through slotted hole in tractor frame.
4. Then lift brake band until brake pin is aligned with hole in brake bracket. Using a needle nose pliers, pull brake pin through hole.
5. Lubricate lever pivot before reassembly.
6. Apply Loctite® to threads before tightening setscrew in brake pulley.
7. After assembling brake, adjust linkage accordingly.

Tractors from Serial No. 100,000-249,000

1. Remove spring locking pin and drilled pin from brake arm.
2. There are two access holes on the left frame through which a socket and extension can be inserted for removal of the brake band retaining capscrews. Remove these two capscrews and slip band off bottom of brake pulley. Lift band assembly upward and to the right to remove.
3. Remove brake pin and separate brake band assembly from arm and bracket.

John Deere

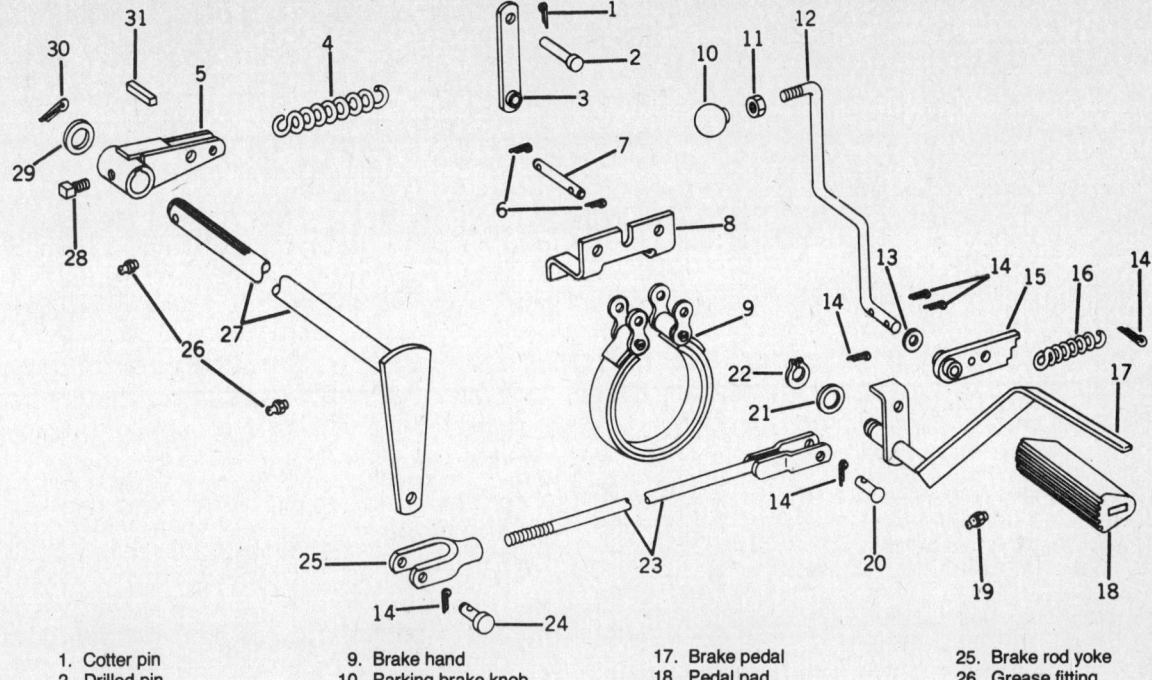

1. Cotter pin	9. Brake hand	17. Brake pedal	25. Brake rod yoke
2. Drilled pin	10. Parking brake knob	18. Pedal pad	26. Grease fitting
3. Brake link	11. Nut	19. Grease fitting	27. Brake shaft
4. Return spring	12. Parking brake lever	20. Drilled pin	28. Screw
5. Brake arm	13. Washer	21. Washer	29. Washer
6. Cotter pin	14. Cotter pin	22. Snapring	30. Cotter pin
7. Drilled pin	15. Parking brake ratchet	23. Brake rod	31. Key
8. Bracket	16. Parking brake spring	24. Drilled pin	

Brake system components, 110 tractors serial #272,001 and up and 112 tractors serial #260,001 and up

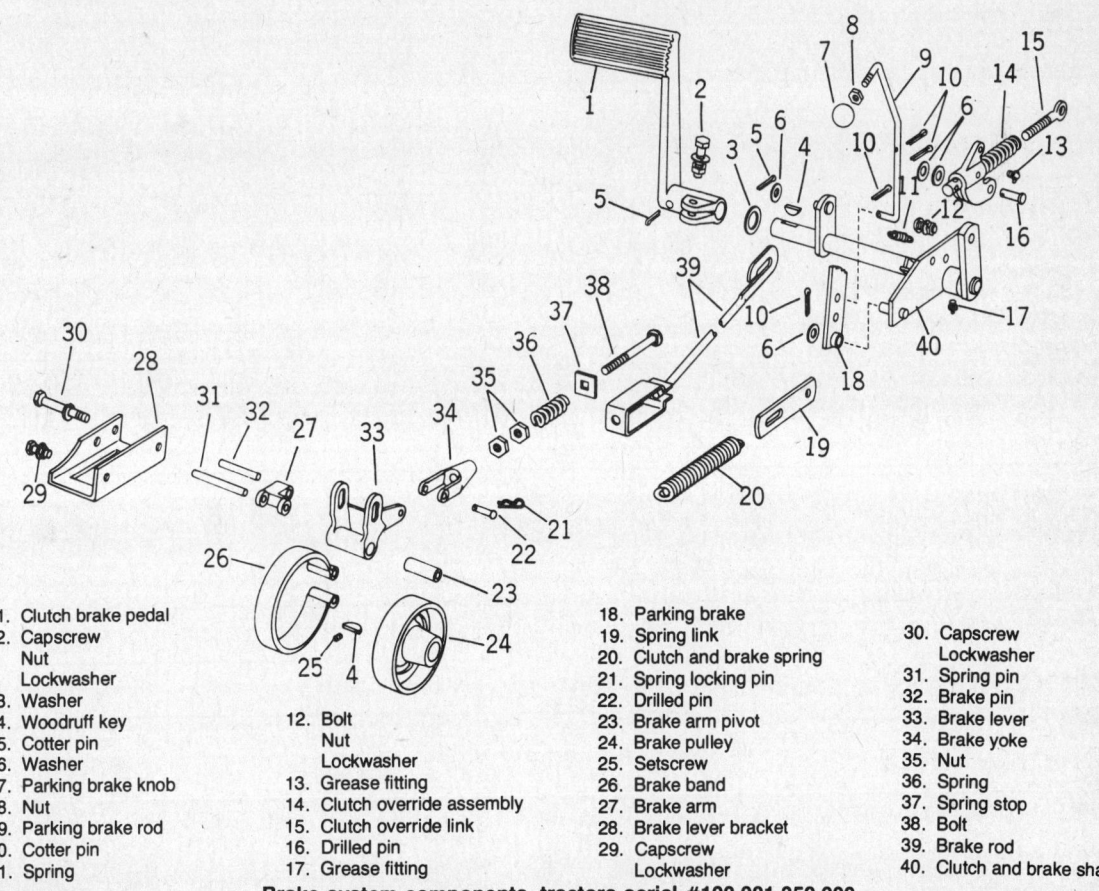

1. Clutch brake pedal	12. Bolt	18. Parking brake	30. Capscrew
2. Capscrew	Nut	19. Spring link	Lockwasher
Nut	Lockwasher	20. Clutch and brake spring	31. Spring pin
Lockwasher	13. Grease fitting	21. Spring locking pin	32. Brake pin
3. Washer	14. Clutch override assembly	22. Drilled pin	33. Brake lever
4. Woodruff key	15. Clutch override link	23. Brake arm pivot	34. Brake yoke
5. Cotter pin	16. Drilled pin	24. Brake pulley	35. Nut
6. Washer	17. Grease fitting	25. Setscrew	36. Spring
7. Parking brake knob		26. Brake band	37. Spring stop
8. Nut		27. Brake arm	38. Bolt
9. Parking brake rod		28. Brake lever bracket	39. Brake rod
10. Cotter pin		29. Capscrew	40. Clutch and brake shaft
11. Spring		Lockwasher	

Brake system components, tractors serial #100,001-250,000

John Deere

4. Drive spring pins from arm and band.
5. Lubricate lever pivot before reassembly.
6. Reverse disassembly procedure to assure correct installation.
7. After installing brake band assembly on tractor, adjust brake linkage accordingly.

Tractors from Serial No. 250,000 and up

1. Remove spring locking pin and drilled pin from brake arm.
2. On tractors (Serial No. 272,001 and up), remove cotter pin (14) and drilled pin (24) from brake arm.
3. There are two access holes on the left frame through which a socket and extension can be inserted for removal of the brake band retaining capscrews. Remove these two capscrews and slip band off bottom of the brake pulley. Lift band assembly upward and to the right to remove.
4. Remove brake pin and separate brake band assembly from bracket.
5. Drive spring pins from arm band. Tractors (Serial No. 272,001 and up) do not require removal of pins or brake arms.
6. Lubricate lever pivot before reassembly.
7. Reverse disassembly procedure to assure correct installation.
8. After installing brake band assembly on tractor, adjust brake linkage accordingly.

1. Clutch brake pedal
2. Capscrew
3. Pedal pad
4. Woodruff key
5. Cotter pin
6. Washer
7. Parking brake knob
8. Nut
9. Parking brake rod
10. Cotter pin
11. Parking brake spring
12. Bolt
13. Grease fitting
14. Clutch override assembly
15. Clutch override link
16. Drilled pin
17. Grease fitting
18. Parking brake stop arm
19. Spring link
20. Clutch and brake spring
21. Cotter pin
22. Drilled pin
23. Brake arm pivot
24. Brake pulley
25. Setscrew
26. Brake band
27. Brake arm
28. Brake lever bracket
29. Capscrew
30. Capscrew
31. Spring pin
32. Brake pin
33. Brake lever
34. Brake yoke
35. Spring
36. Washer
37. Brake adjusting bolt
38. Brake rod
39. Pivot pin
40. Clutch and brake shaft

Brake system components, tractors serial #250,001-272,000 and 112 tractors serial #250,001-260,000

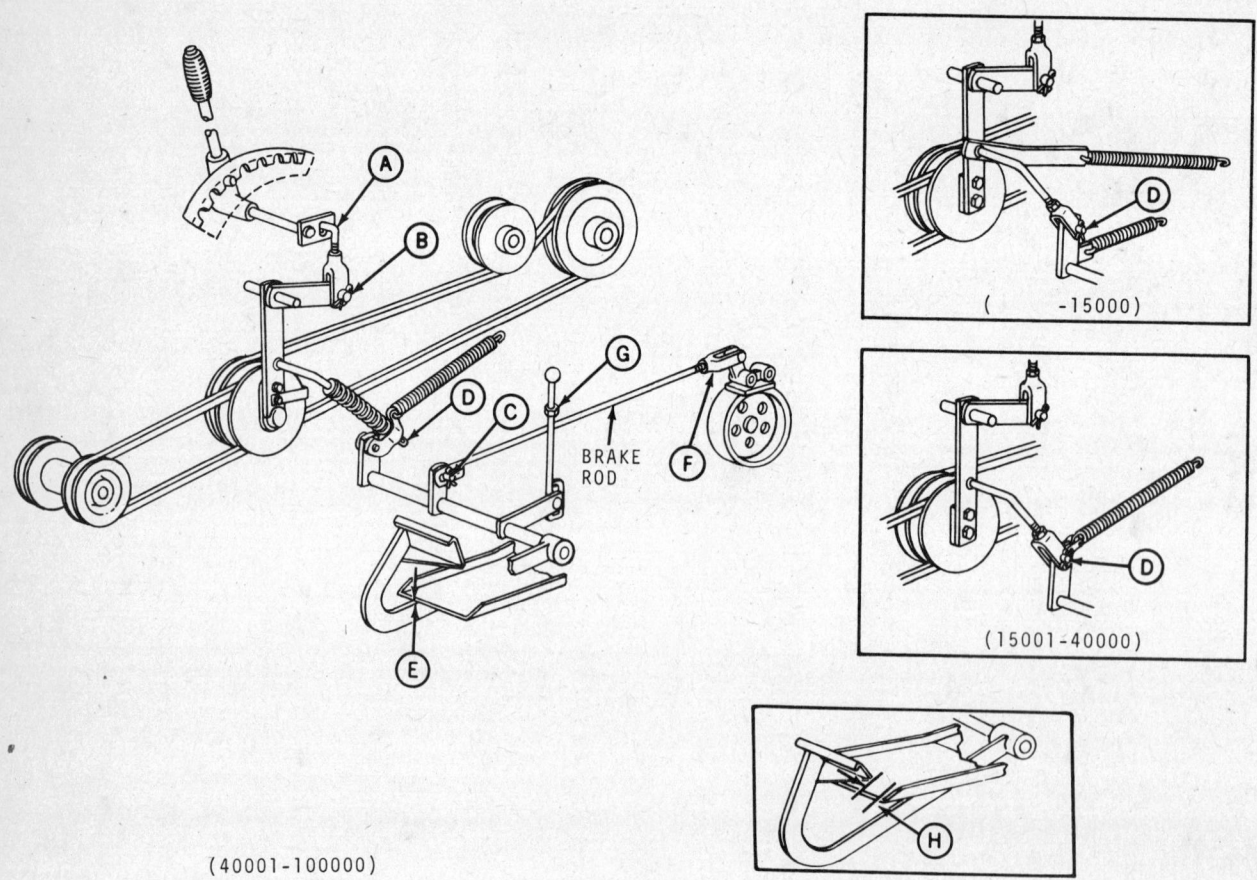

(40001-100000)
Clutch, brake and variable speed drive linkages, tractors through serial #100,000

322

John Deere

Adjustments

LINKAGE ADJUSTMENT

Brake, clutch and variator adjustments should not be made individually because each adjustment affects the other. Always adjust the entire linkage as explained in the following procedures, when adjustment is required.

When tractor linkage is properly adjusted, the variable speed control lever will increase tractor speed when moved forward from quadrant notch 7 through notch 3.

Linkage adjustment is necessary when either of the following occurs:

A. Tractor is inoperative when variable speed lever is in notch 7 (thru serial no. 100,000) or notch 1 (serial no. 100,000 and up) on the quadrant (slow speed position).

B. Clutch-brake pedal strikes bottom of footrest during normal operation.

Adjust tractor linkage as follows:

Through Serial No. 100,000

1. Remove inspection plate from pedestal to open adjusting hole and disconnect linkage "B."
2. Disconnect brake linkage at "C."
3. Place variable speed lever in notch 3 on the quadrant which is the third notch from the front of the tractor.
4. Disconnect spark plug cable and turn engine with key starter several revolutions.
5. Measure distance at "E" which is the distance between the bottom of the footrest and the top of the clutch-brake arm. This distance should be ½ inch. If dimension "E" is not ½ inch, adjust according to tractor Serial No. as follows:

110 Tractors Prior to Serial Number 4000

Disconnect "D," and turn threaded clevis either up or down until dimension "E" is ½ inch.

110 Tractors Serial Number 40001 and up and All 112 Tractors

Insert tapered punch or screwdriver at "D," and turn rod either up or down until dimension "E" is ½ inch.

6. Hold link "A" to top of slot and turn threaded clevis up or down as required until pin can easily be inserted at "B." Insert spring locking pin.
7. Connect pin "C" temporarily.
8. Turn engine several times with key starter while moving ground speed control lever to notch 7 (slow position).
9. Depress clutch-brake pedal as far as possible. The top of the clutch-brake pedal should now be ¾ inch above the top of the footrest (dimension "H"). If not, turn brake rod into clevis "F" until the ¾ inch dimension can be obtained. Insert spring locking pin into pin "C."
10. Turn nut "G" on parking brake rod either up or down until the clutch-brake pedal can be held in the lowered position.

NOTE: If, after adjusting linkage, tractor still will not move when ground speed control lever is in first notch on the quadrant (slow speed position), remove inspection plate and turn threaded clevis up one or two turns on link "A." If necessary, install a new primary belt.

From Serial No. 100,000 and up

1. Place variable speed lever in notch 5 on the quadrant, which is the third notch from the front of the tractor.
2. Pry plug button from adjusting hole in tractor pedestal and loosen capscrew (one or two turns) with a ¾ inch socket wrench.
3. Disconnect spark plug cable and turn engine several revolutions with key starter until the clutch-brake pedal rises as high as it will go.
4. Center capscrew in adjusting hole. Tighten capscrew firmly and replace plug button in adjusting hole.

NOTE: If, after adjusting variator linkage, tractor still will not move when variable speed control lever is in first notch on the quadrant (slow speed position), and the clutch-brake pedal is released, a new primary belt must be installed.

V-BELT TENSION ADJUSTMENT

1. V-belt tension should be adjusted if:
   a. Clutch-brake pedal strikes the bottom

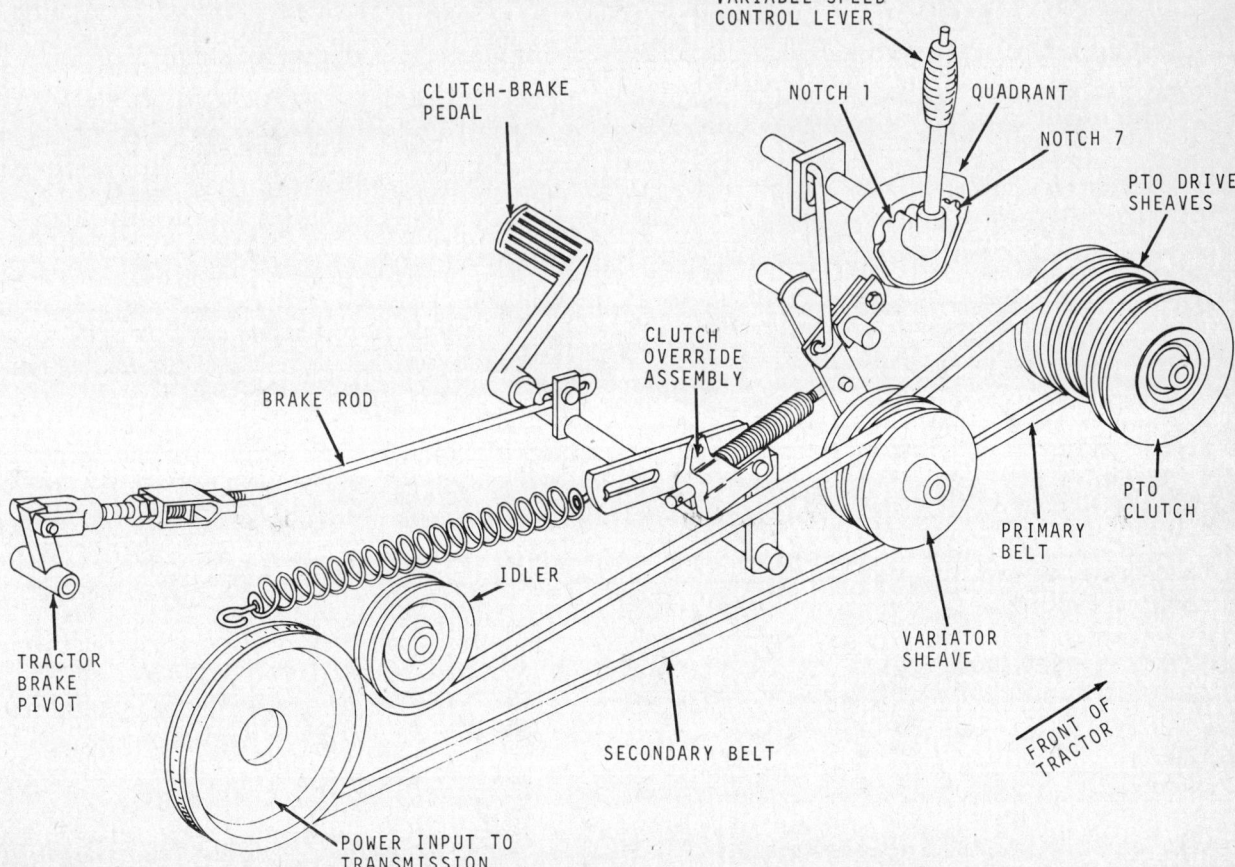

Clutch, brake and variator linkages, tractors serial #100,001-250,000

John Deere

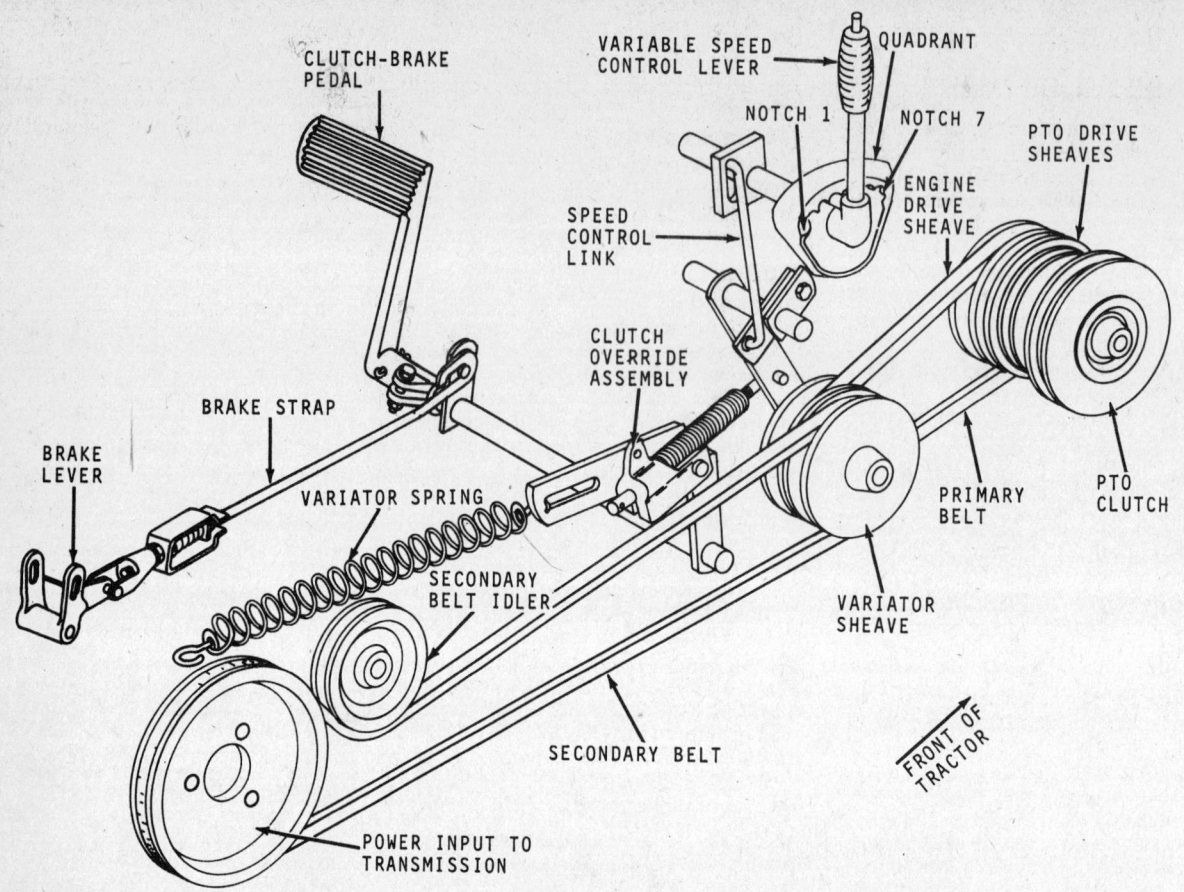

Clutch, brake and variator linkages, 110 tractors serial #250,001 and up

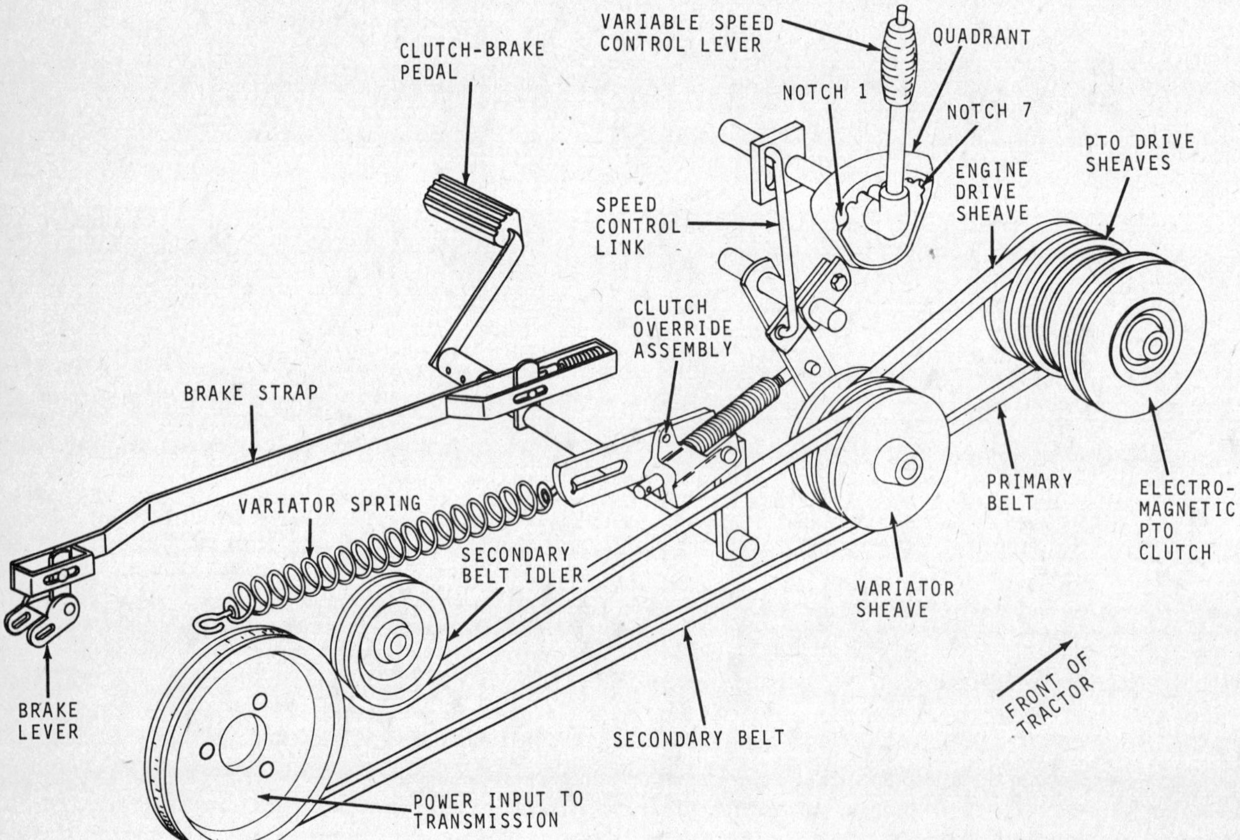

Clutch, brake and variator linkages 112 tractors serial #250,000-260,000

John Deere

Moving transmission

Moving transaxle, tractors serial #15,000-100,000

of footrest when variable speed control lever is in the forward position.

b. Tractor does not move when variable speed control lever is in the rearmost notch in quadrant.

c. Secondary belt strands operate less than ¾-inch apart.

2. Adjust V-belts as follows for each condition A through C.

A-B Primary Belt Tension

1. If at any time the clutch-brake pedal strikes the bottom of the footrest or if the tractor does not move with the variable speed control lever in the rearmost notch in the quadrant, the tractor linkage will require adjustment.

2. If, after making the adjustment, the tractor is still inoperative with the variable speed control lever in the rearmost notch in the quadrant, install a new primary belt.

C Secondary Belt Tension

If excessive belt stretching allows the idler to rub on the lower belt strand or operates less than ¾ inch at the closest point, additional belt tension can be obtained by moving transmission rearward as follows:

110 TRACTORS PRIOR TO SERIAL NUMBER 15000

1. Loosen capscrew "A," and move transmission in slotted holes toward rear of tractor until desired tension is obtained. Tighten nuts firmly.

CAUTION

Do not allow transmission and wheels to angle to one side in slotted holes. This causes "dog tracking" of the rear wheels and secondary belt wear.

---

2. After moving transmission, be sure to readjust variator and brake rod linkage.

110 AND 112 TRACTORS SERIAL NUMBERS 15001-100,000

1. Remove capscrews "A," and move transmission rearward into second set of holes. Insert capscrews through rear holes and tighten firmly.

2. After moving transmission, be sure to readjust variator and brake rod linkage.

BELT GUIDE ADJUSTMENT

1. If the primary belt jumps the variator sheave when the clutch-brake pedal is depressed, the distance between the variator and primary belt guide should be checked. Distance between guide and sheave should not exceed ⅛ inch.

2. Under certain conditions, the secondary belt of the 110 tractor may jump out of the variator groove. This usually happens when the tractor is driven down a steep incline in second or third gear, with the variator in the forward (fast speed) position, while at the same time the engine acts as a brake.

3. A secondary belt guide was not installed at the factory on 110 tractors serial numbers 3551-10076. Install belt guide to prevent belt jumping out of variator groove.

CLUTCH-BRAKE PEDAL ADJUSTMENT

Serial No. 100,000 and up

1. The clutch-brake pedal is properly adjusted when the lowest point on the pedal is not less than ¾ inch from the top of the footrest when fully depressed. If less than ¾ inch above the footrest when fully depressed, adjust the brake as follows:

2. Remove spring locking pin and drilled pin from clevis.

3. Turn clevis onto brake rod until a ¾ inch dimension can be obtained. Replace drilled pin and insert spring locking pin.

NOTE: There is no separate clutch adjustment on tractors beginning with Serial No. 272,001 and above, because clutch-variator functions have been separated from the braking process.

ADJUSTING CLUTCH OVERRIDE

Serial Number 100,000 and up

1. The clutch-brake pedal height should not exceed 7 to 8 inches from the footrest when clutch is released. Place the variable speed control lever in notch 7 (fast drive position) when this measurement is taken.

2. If the clutch-brake pedal rises higher than 7 to 8 inches when released it will be necessary to adjust the clutch override.

3. To adjust the clutch override insert a punch or narrow screwdriver into the hole in the adjusting screw. Turn screw counterclockwise until the 7 to 8-inch dimension is obtained.

BRAKE PEDAL ADJUSTMENT

Serial No. 272,001 and up

1. When brake pedal pushes down to fender deck or no longer gives braking pressure it is necessary to adjust brake. Lack of parking brake pressure also indicates need for adjustment.

2. Remove pin and turn clevis on brake rod as far as necessary to put brake in first notch of park lock.

3-SPEED TRANSAXLE

REMOVAL

NOTE: Tractor repair stands are available for mounting and inverting the tractor. These stands greatly facilitate transaxle removal. If a stand is not available, jack up the tractor and place stands under the frame members.

NOTE: Steps 1-5 apply only if the tractor is to be mounted on a repair stand.

1. Shut off fuel at sediment bowl.
2. Remove gas tank.
3. Remove battery.
4. Drain engine crankcase.
5. Replace vented filler cap on hydraulic reservoir with pipe plug to prevent leakage.
6. Disconnect brake clevis pin, idler spring and neutral-start wires from switch.

John Deere

1. Shifter fork (2 used)
2. Spring (2 used)
3. Ball (2 used)
4. Shifter rod (1st and reverse speeds)
5. Shifter stop
6. Shifter rod (2nd and 3rd speeds)
7. Needle bearing for shifter shaft
8. Shifter shaft and gear
9. Needle bearing for input shaft
10. 26 tooth shifter gear (1st and reverse speeds)
11. 20 tooth shifter gear (2nd and 3rd speeds)
12. Input shaft and pinion
13. 16 tooth input shaft gear
14. Pin (2 used)
15. Shifter lever knob
16. Shifter lever
17. Socket head capscrew (11 used)
18. Lever housing
19. Rubber seal
20. Spring pin
21. Keeper
22. Gasket
23. Case
24. Needle bearing
25. Oil seal (1¼ in. O.D.)
26. Reverse idler gear
27. Spacer (7/16 in. long)
28. Reverse idler shaft
29. 26 tooth idler shaft gear
30. Spacer (¾ in. long)
31. 22 tooth idler shaft gear
32. Spacer (1 3/16 in. long)
33. Idler shaft
34. 16 tooth idler shaft gear
35. Bronze bushing (1¼ in. long)
36. 30 tooth idler gear
37. Idler pinion and brake shaft
38. Case and cover gasket
39. Cover
40. Dowel pin (4 used)
41. Oil seal (1½ in. O.D.)
42. Needle bearing for brake shaft
43. Needle bearing for axle (2 used)
44. Washer
45. Spacer (⅝ in. long)
46. 36 tooth output gear
47. Output shaft
48. Pipe plug (2 used)
49. Capscrew (4 used)
50. Lockwasher (4 used)
51. Spacer (3/16 in. thick)
52. Righthand carriage
53. Thrust bearing (2 used)
54. Thrust washer (4 used)
55. Axle (2 used)
56. Bevel pinion (2 used)
57. Drive block (2 used)
58. Drive pin
59. Spacer (1⅝ in. long)
60. Ring gear
61. Lefthand carriage
62. Oil seal for axle housing (2 used)
63. Lefthand axle housing
64. Bearing for axle housing (2 used)
65. Righthand axle housing
66. Capscrew (8 used)
67. Lockwasher (8 used)
68. Setscrew
69. Rear tire valve (2 used)
70. Setscrew (2 used)
71. Hex jam nut (2 used)
72. Woodruff key (2 used)
73. Rear wheel hub (2 used)
74. Wheel bolt (6 used)
75. Rear wheel (2 used)
76. Needle bearing

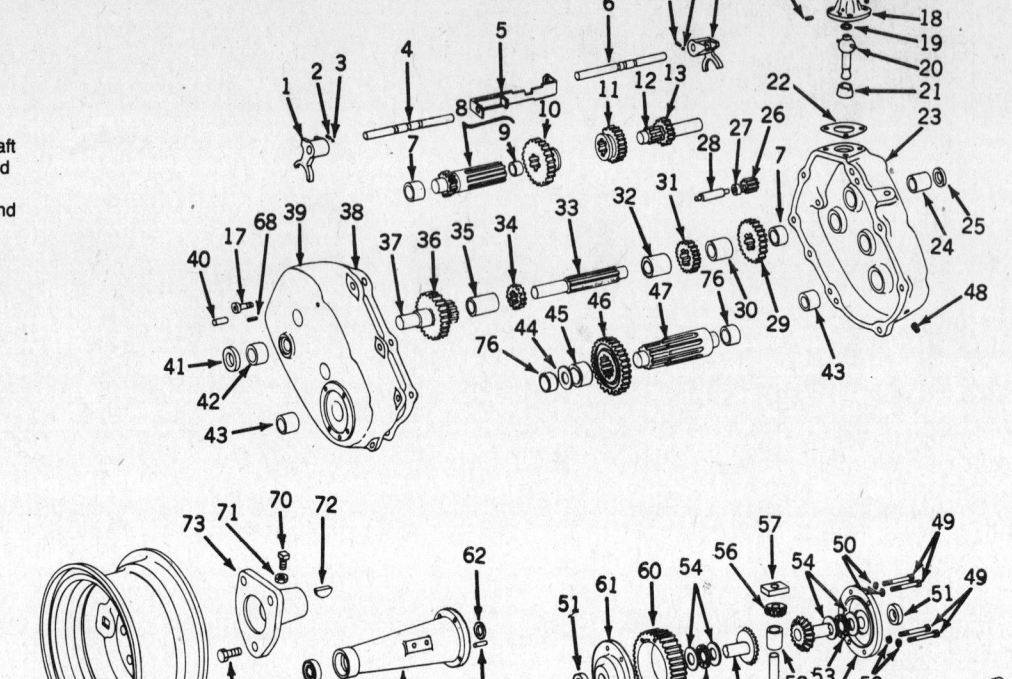

3-speed transaxle components, 110 tractors serial number through 3,571

7. Remove three capscrews from driven pulley on input shaft.

8. Remove wheels, then remove the remaining screws that hold the transaxle support and hitch plate to tractor base. Lift transaxle away from tractor.

DISASSEMBLY

1. Remove brake, idler arm, hitch, input hub and wheel hub assemblies from transaxle. Use a puller to prevent hub breakage or internal differential damage.

--- CAUTION ---
Never use hammer on end of axles or driveshafts toward transmission.

2. Position shift lever in neutral. Remove neutral start bracket with switch, shifter assembly, axle supports and retainers with seal. Use extreme care when removing axle supports since they are machined to a light press fit.

NOTE: Mark locations of right-and-lefthand axle housings on transaxles in 110 tractors prior to serial number 3571.

3. Clean and polish axles as necessary to permit easy removal of axle support.

4. Drill two holes in a sturdy work bench about 8 in. from the front of the bench. A wooden stand may be used instead.

5. Place transaxle in bench or stand vertically with socket head capscrews up. Remove eight screws. Leave dowel pins in place.

6. While holding case halves together, invert entire transaxle and reposition in bench. THIS IS IMPORTANT. Transaxle cover must be removed first.

7. Drive out dowel pins. Grasp the brake shaft with the lefthand and transaxle cover with the righthand. Lift case slowly and shake lightly so all loose parts remain in lower case.

8. Removing the following parts in order:
   a. Output shaft.
   b. Differential and axle assembly.
   c. Idler shaft.
   d. Reverse idler.
   e. Shifter shaft and forks assembly.
   f. Input shaft.

NOTE: Input shaft is installed with a press fit. If close inspection reveals that gears and bearings are satisfactory, do not remove input shaft.

Shift Lever

To disassemble the shifter, it will be necessary to self shear the cross pin be-

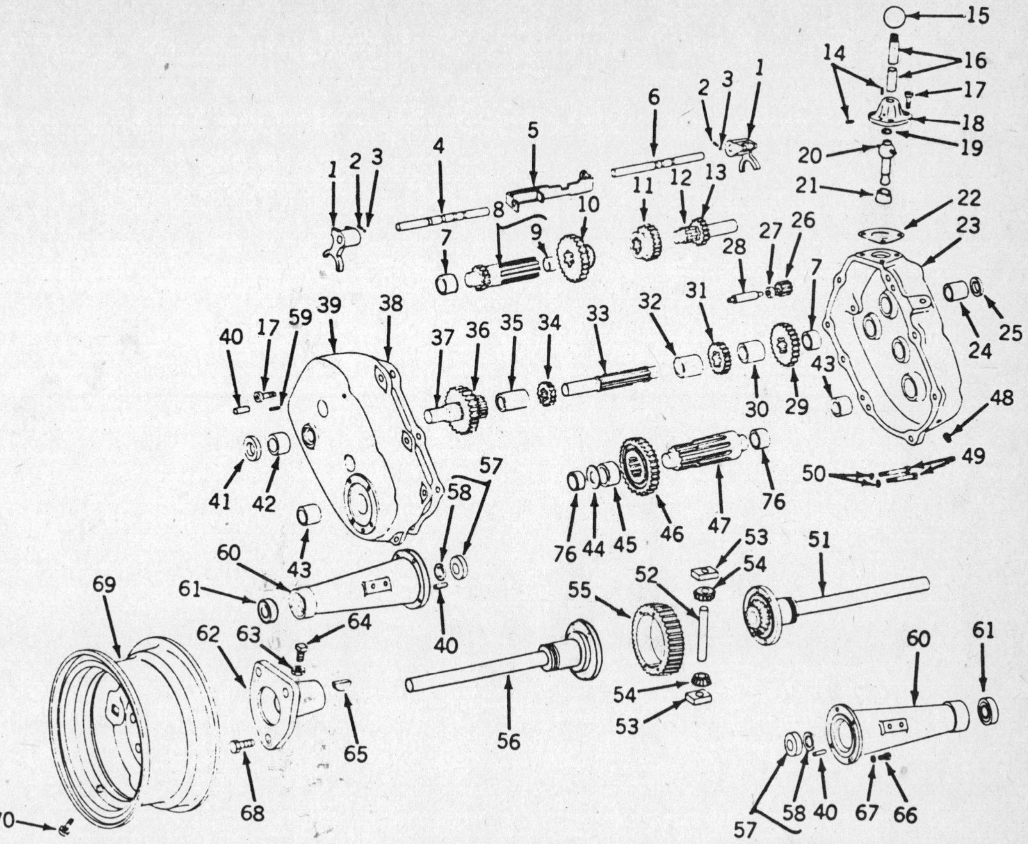

1. Shifter fork (2 used)
2. Spring (2 used)
3. Ball (2 used)
4. Shifter rod (1st and reverse speeds)
5. Shifter stop
6. Shifter rod (2nd and 3rd speeds)
7. Needle bearing for shifter shaft
8. Shifter shaft and gear
9. Needle bearing for input shaft
10. 26 tooth shifter gear (1st and reverse speeds)
11. 20 tooth shifter gear (2nd and 3rd speeds)
12. Input shaft and pinion
13. 16 tooth input shaft gear
14. Pin (2 used)
15. Shifter lever knob
16. Shifter lever
17. Socket head capscrew (11 used)
18. Lever housing
19. Rubber seal
20. Spring pin
21. Keeper
22. Gasket
23. Case
24. Needle bearing
25. Oil seal (1¼ in. O.D.)
26. Reverse idler gear
27. Spacer (7/16 in. long)
28. Reverse idler shaft
29. 26 tooth idler shaft gear
30. Spacer (¾ in. long)
31. 22 tooth idler shaft gear
32. Spacer (1 3/16 in. long)
33. Idler shaft
34. 16 tooth idler shaft gear
35. Bronze bushing (1¾ in. long)
36. 30 tooth idler gear
37. Idler pinion and brake shaft
38. Case and cover gasket
39. Cover
40. Dowel pin (4 used)
41. Oil seal (1⅜ in. O.D.)
42. Needle bearing for brake shaft
43. Needle bearing for axle (2 used)
44. Washer
45. Spacer (⅝ in. long)
46. 36 tooth output gear
47. Output shaft
48. Pipe plug (2 used)
49. Capscrew (4 used)
50. Lockwasher (4 used)
51. Righthand axle and carriage assembly
52. Drive pin
53. Drive block (2 used)
54. Bevel pinion (2 used)
55. Ring gear
56. Lefthand axle and carriage (tapped carrier)
57. Axle retainer with seal
58. Oil seal
59. Setscrew
60. Righthand or lefthand axle housing (2 used)
61. Axle housing bearing (2 used)
62. Rear wheel hub (2 used)
63. Hex jam nut (2 used)
64. Setscrew (2 used)
65. Woodruff key
66. Capscrew (8 used)
67. Lockwasher (8 used)
68. Capscrew (6 used)
69. Rear wheel (2 used)
70. Rear tire valve (2 used)

3-speed transaxle components, 110 tractors serial #3,572-15,000

tween the housing and keeper. Use a vise and blunt shaft or punch.

Axle Shafts and Pinion Gear

110 tractors prior to serial number 3571

The axle shafts and bevel gears for transaxles in this serial number range are factory assembled and can only be serviced by replacing the shaft and gear assembly.

110 tractors serial numbers 3572-15000

1. The axle shaft and bevel gear (1 and 7) for transaxles in this serial number range are factory assembled with the bevel gear rolled or peened on the splined shaft. A loose bevel gear indicates trouble and should be serviced.

2. When either the axle shaft or bevel gear must be serviced, the AM30744 gear kit consisting of two axle bevel gears (1) and two pinion gears (2) must be used.

NOTE: Gears in this kit are a matched set. Do not mix old and new gears.

3. New axles (7), thrust washers (3) and snaprings (4) must also be used.

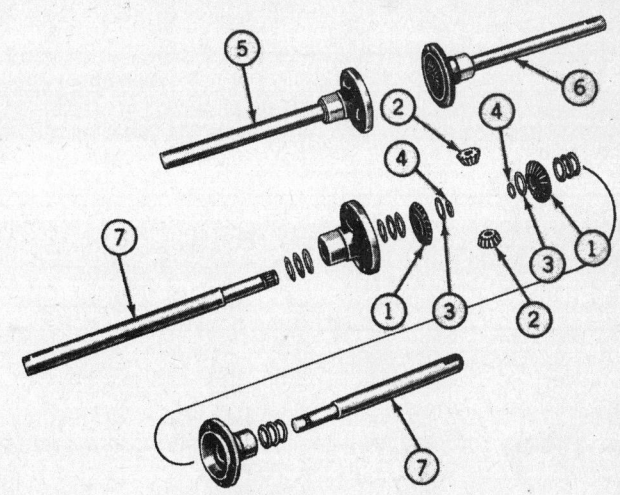

1. Axle bevel gear
2. Bevel pinion gear
3. Thrust washer
4. Inner axle snapring
5. Lefthand axle and carriage (prior to 15000)
6. Righthand axle and carriage (prior to 15000)
7. Lefthand axle shaft (3572-15000) (replacement axle)

Old and new style differential parts

John Deere

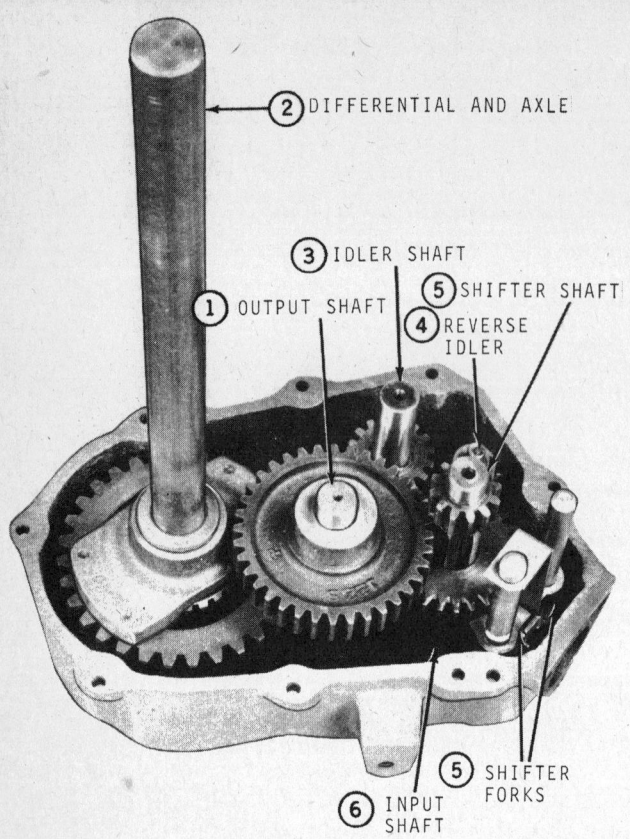

Gear removal sequence

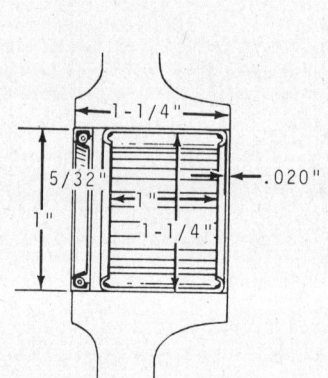

Installing bearings

Shifter assembled

ASSEMBLY

1. Assemble all parts shown above depending upon the tractor serial number.
2. Apply Loctite® or equivalent to ends of threads and assemble capscrews through carrier into tapped carrier. Be sure lockwasher is under head of screw. Torque capscrews to 10-12 ft.lb.
3. The axles should rotate freely in opposite directions when assembled. Lay the differential assembly aside for later installation.
4. All bearings are pressed into the bearing bores from the inside of the case interior. Be sure seal is installed with lip inward.
5. Bearing drivers are necessary to install bearings properly. As a general rule, all bearings should be pressed into the bearing bore to a depth of 0.020 in. beyond flush with case interior.
6. Assemble input shaft and gear. Counterbored gear spline must face to right. Gear is a press fit onto shaft.
7. Install input shaft and gear into case as shown. Flat side of gear should now face upward.

Idler Shaft and Gears

1. Assemble idler shaft. These gears are a slip fit on the spline. Notice that raised hub of large gear faces short spacer. The teeth on the medium and small gear have round engagement edges that must face the large gear. Spacers are of different length.
2. The long round end of the idler shaft turns in the bushing on the brake shaft. Be sure end of shaft is not battered.
3. Install idler shaft and gears in case. Long end of shaft faces upward. Large gear engages the input gear.

Reverse Idler Shaft and Gear

1. Assemble reverse idler shaft assembly as illustrated. Round edge of teeth faces spacer.
2. Install reverse idler and shaft in case with round edge of teeth and spacer upward.

Shifters

1. Because of heavy detent pressure, the assembly of these shafts can be difficult.
2. Assemble forks. Both forks should face to the right for assembly. The 2nd and 3rd shaft must have the unequally spaced grooves at the top and away from the fork as shown. The 1st and reverse shaft must have the shortest ungrooved end face the fork as shown. Start the shaft into the fork. Depress detents and complete the assembly. Slide forks along shaft. A good snap should be felt in each detent.
3. Place forks in center or neutral detent positions at this time.
4. To assemble shifter, lay out parts. Be sure forks are in center grooves. Note that the exposed groove on the unequally spaced 2nd and 3rd shifter faces away from the gear on the shifter shaft. The exposed groove on the 1st and reverse equally spaced shifter faces away from the gear on the shifter shaft.
5. The slot in the forks should line up when the large gear is slipped as far as possible on the spline. Note the position of exposed grooves on shifter shafts.
6. Assemble shifter guide over shifter shafts. Slot in guide should match rectangular opening between the forks. The long notch in underside of guide should clear the large 1st and reverse shifter gear.
7. Grasp shifter assembly firmly in left hand and lower it into case. The input shaft stud should enter needle bearing in end of shifter shaft. The shifters should now enter the two machined sockets in bottom of the case.

Inserting Differential Assembly in Case

Install differential and axle assembly into the case with the bolt heads on the

Input shaft and gear assembly

Brake shaft components

righthand carrier downward. A thick, hardened spacer or thrust bearing assembly must be on each side of the differential as shown for 110 tractors prior to serial no. 3571. For 110 tractors serial nos. 3572-15000, be sure needle bearing is flush with inside of case and cover and no lower than 0.020 in. below flush.

Output Shaft

1. The output gear shaft is assembled to the pinion with a press fit. A slight looseness can be tolerated because the spacer and washer will hold the gear in place.
2. Install the output shaft assembly in the case. It may be necessary to lift the differential about ½ in. to mesh gears during assembly.
3. Position the gasket on the lower case at this time.

Brake Shaft and Cover

1. The brake shaft and large pinion are a press fit to each other within the case. This is necessary because of the overhang of the reverse idler support bearing. Do not use the case itself to support any part of the pressure required to install the pinion and shaft. To assemble, support the case on a sleeve slightly larger than the needle bearing. Start the shaft through the gear with flat side of teeth up and press from the side.
2. New brake bushing tolerances are 0.749 to 0.751 in. Replace if worn beyond wear tolerance limit of 0.756 in.
3. If bushing is replaced, check I.D. It may require reaming. Bushing I.D. dimension is 0.749-0.751 in.

Placing Cover on Case

1. Loosen or remove setscrew on transaxles so equipped.
2. Install the lefthand case half. Shake the case lightly, and all shafts and bearings will align themselves. To close the last ½ in., tap the case horizontally at the corner indicated.
3. Align and insert dowel pins. Start socket head capscrews from bottom.
4. Invert transaxle and tighten case screws securely to 120 in.lb. Put Loctite® on threads, then tighten setscrew against the shifter-shaft, on transaxles so equipped.

Installing Seals

1. Use seal driver to install seals or seal with retainer. Install seal after shaft has been installed.
2. Be sure seal is installed with lip inward.

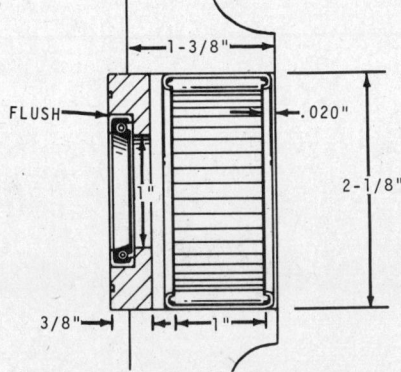

Seal retainer and bearing installation

3. Install axle supports. Torque bolts to 10-12 ft.lb.

Positioning Shifter Forks

Inspect the shifter forks to be sure they are aligned and in neutral. Failure to do this will cause damage to the transmission when engaged under power.

Assembling Shifter Lever

The shifter is assembled in the order shown. To prevent incorrect positioning of the quad ring in the housing, a little shellac or gasket cement will be helpful. Align the small cross pin holes between housing and keeper and drive in new pin. Then install the second pin as a retainer and locater in the housing. Position neutral start bracket lever and gasket on the transaxle. When the locating pin is positioned on the housing, the slight bend in the lever should point to the left. If it does not, reassemble the shift lever and shift lever housing. Tighten three screws to 120 in.lb.

Installing Transaxle

1. Install brake, input hub, secondary idler, hitch, neutral start bracket with switch, and wheel hub assemblies on transaxle. Apply Loctite® or equivalent to threads on all bolts and setscrews used in assembling components to transaxle.
2. Adjust neutral start switch and bracket.
3. Before installing transaxle in tractor base, check transaxle by turning input hub and shifting transaxle in each gear.
4. Place transaxle in tractor base. Install capscrews holding transaxle support and hitch plate to tractor base.
5. Place secondary belt on transaxle sheave and install sheave on hub with three capscrews. Connect brake clevis and secondary idler spring.
6. Bolt wheels to hubs with wheel bolts.
7. Connect neutral start switch leads.
8. Readjust brake and variator linkage.
9. Add lubricant after turning tractor upright.

4-SPEED TRANSAXLE

REMOVAL

NOTE: A tractor repair stand is available for mounting and inverting the tractor to facilitate transaxle removal. If you do not have such a stand, jack up the tractor and mount it on jack stands.

If a repair stand is being used, observe steps 1-5.

1. Shut off fuel at sediment bowl.
2. Remove gas tank.
3. Remove battery.
4. Drain engine crankcase.
5. Replace vented filler cap on hydraulic reservoir with pipe plug to prevent leakage.
6. With tractor inverted, disconnect brake clevis pin, idler spring and neutral start wires from switch.
7. Slip secondary belt off variator and remove shift quadrant from deck.
8. Remove wheels. Then remove the remaining screws that hold the transaxle support and hitch plate to tractor base. Lift transaxle away from tractor.

DISASSEMBLY

NOTE: All 112 tractors with hydraulic lifts above serial no. 130,000 are equipped with limited slip differentials, identified by the number 2317 stamped on the serial number plate fixed to the differential.

1. Remove brake, idler arm, hitch plate, input hub and wheel hub assemblies from transaxle. Use a puller to prevent hub breakage or internal differential damage.

--- **CAUTION** ---
Never use hammer on end of axles. Never drive shafts toward transmission.

2. Position shift lever in neutral. Remove neutral start bracket with switch, shifter assembly, axle housings, O-rings and retainers with seal. Use extreme care when removing axle supports since they are machined to a light press fit.
3. Clean and polish axles as necessary to permit easy removal of axle housing.
4. Drill two holes in a sturdy work bench about 8 in. from the front of the bench. A wooden stand may be used instead.
5. Place transaxle in bench or stand vertically with socket head capscrews and input shaft upward. Remove eight screws.
6. Drive out dowel pins. Grasp the input shaft with the righthand and the transaxle case with the lefthand. Lift case slowly and

John Deere

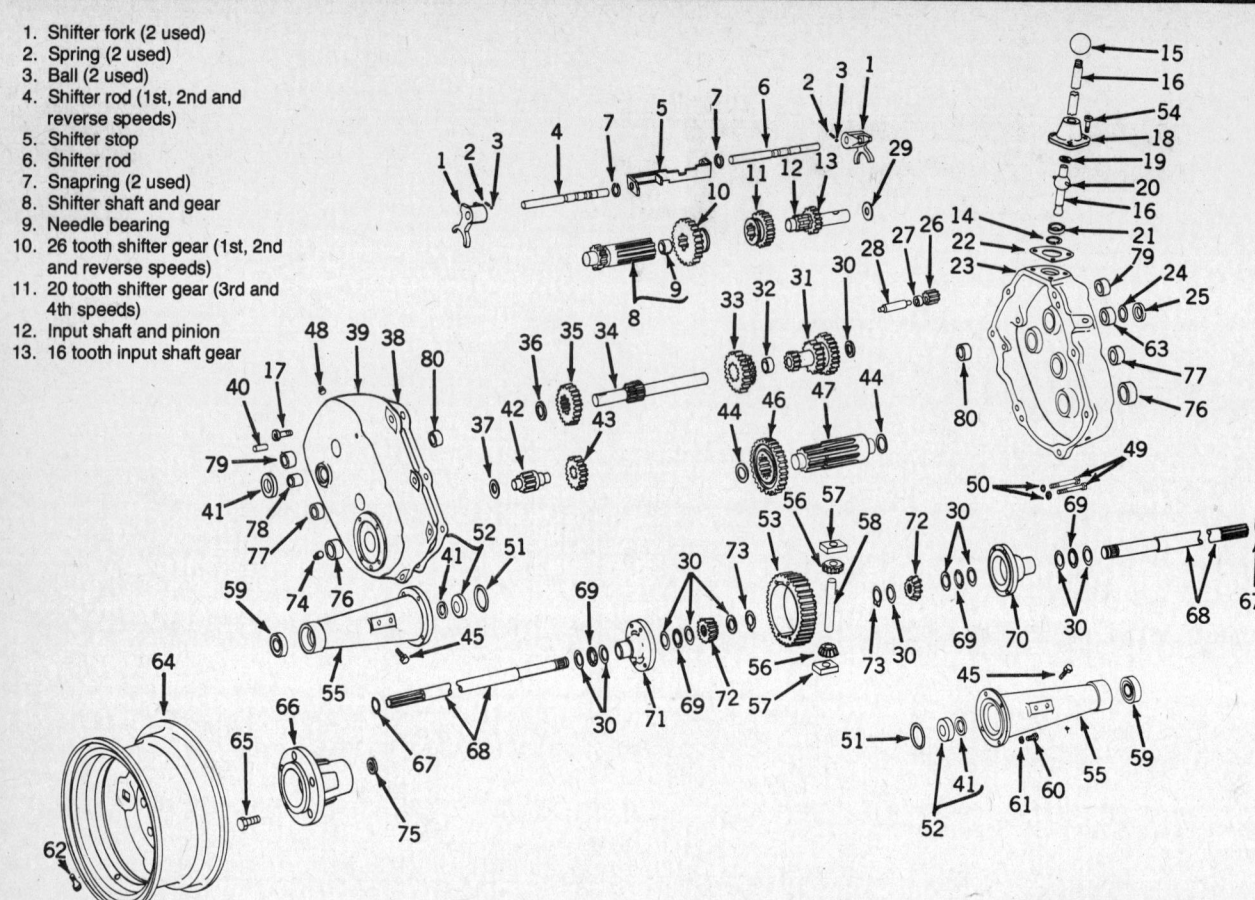

1. Shifter fork (2 used)
2. Spring (2 used)
3. Ball (2 used)
4. Shifter rod (1st, 2nd and reverse speeds)
5. Shifter stop
6. Shifter rod
7. Snapring (2 used)
8. Shifter shaft and gear
9. Needle bearing
10. 26 tooth shifter gear (1st, 2nd and reverse speeds)
11. 20 tooth shifter gear (3rd and 4th speeds)
12. Input shaft and pinion
13. 16 tooth input shaft gear
14. Snapring
15. Knob
16. Shifter lever
17. Capscrew (8 used)
18. Shifter lever housing
19. Rubber seal
20. Spring pin
21. Keeper
22. Gasket
23. Case
24. Input shaft bearing
25. Oil seal (1⅝ in. O.D.)
26. Reverse idler gear
27. Spacer (1¹¹⁄₁₆ in. long)
28. Reverse idler shaft
29. Thrust washer (¾ in. I.D. x 1¼ in. O.D.)
30. Thrust washer (⅞ in. I.D. x 1⁷⁄₁₆ in. O.D.) (13 used)
31. 3 cluster gear
32. Spacer ¹⁵⁄₃₂ in. long)
33. 2 cluster gear
34. Brake shaft and pinion
35. 30 tooth idler gear
36. Thrust washer (1 in. I.D. x 1½ in. O.D.)
37. Idler shaft washer
38. Gasket
39. Cover
40. Dowel pin (2 used)
41. Oil seal (1⅜ in. O.D.)
42. Idler shaft and pinion
43. 22 tooth idler gear
44. Thrust washer (1⁵⁄₁₆ in. I.D. x 1½ in. O.D.) (2 used)
45. Capscrew (5 used)
46. 36 tooth output gear
47. Output shaft
48. Pipe plug (2 used)
49. Capscrew (4 used)
50. Lockwasher (4 used)
51. O-ring (2 used)
52. Axle retainer (2 used)
53. Ring gear
54. Capscrew (3 used)
55. Righthand or lefthand axle housing (2 used)
56. Bevel pinion (2 used)
57. Drive block (2 used)
58. Drive pin
59. Axle housing bearing (2 used)
60. Capscrew (8 used)
61. Lockwasher (8 used)
62. Rear tire valve
63. Needle bearing (1 in. O.D.) (3 used)
64. Rear wheel (2 used)
65. Wheel bolt (10 used)
66. Rear wheel hub (2 used)
67. Snapring (2 used)
68. Righthand or lefthand axle (2 used)
69. Thrust bearing (⅞ in. I.D. x 1⁷⁄₁₆ in. O.D.) (4 used)
70. Righthand carriage
71. Lefthand carriage (tapped)
72. Bevel gear (2 used)
73. Snapring
74. Drain plug
75. Take-up washer (4 used)
76. Needle bearing for axle shaft (2 used)
77. Needle bearing for output shaft (2 used)
78. Needle bearing for brake shaft
79. Needle bearing for shifter shaft

4-speed transaxle components

shake lightly so all loose parts remain in cover.

7. Disassemble the case half in the following order:
   a. Gasket.
   b. Differential and axle assembly.
   c. Washer, 3-cluster gear and spacer from shaft and pinion brake.
   d. Gear pinion and washer.
   e. Reverse idler assembly.
   f. Shifter rod and shaft assembly.
   g. 2-cluster gear.
   h. Output shaft and washers (one at each end of shaft).
   i. Shaft and pinion, idler gear and washer.
   j. Input shaft.

NOTE: Input shaft is installed with a press fit. If close inspection reveals that gears and bearing are satisfactory, do not remove input shaft.

NOTE: If it is necessary to remove the input shaft, do not use the case itself to support any of the pressure required to separate the input assembly or brake shaft assembly from the case halves. Use a large pipe to support the pinion and press the shaft from the opposite side.

8. To disassemble shift lever, remove snapring in shifter housing and slide assembly apart.

INSPECTION

Wash all internal parts in a safe cleaning solvent. Brush and scrape foreign matter from all parts and dry thoroughly.

NOTE: Oil the bearings immediately after cleaning to prevent rusting.

Inspecting Gears and Shafts

1. Replace all gears having chipped, broken or worn teeth. Badly scored gears must be replaced.
2. Replace any shaft that is bent, scored or worn. Replace any shaft showing side wear or if any of the splines are damaged.
3. Chipped, broken or excessive wear on gear teeth ends is usually caused by

John Deere

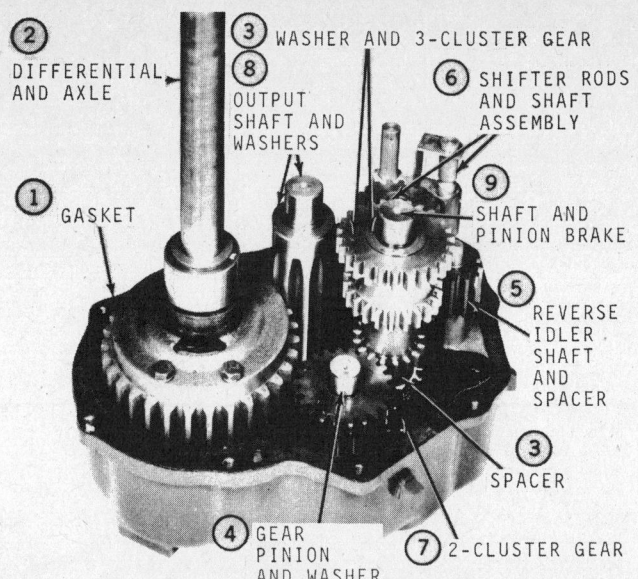

Gear removal sequence

1. GASKET
2. DIFFERENTIAL AND AXLE
3. WASHER AND 3-CLUSTER GEAR
4. GEAR PINION AND WASHER
5. REVERSE IDLER SHAFT AND SPACER
6. SHIFTER RODS AND SHAFT ASSEMBLY
7. 2-CLUSTER GEAR
8. OUTPUT SHAFT AND WASHERS
9. SHAFT AND PINION BRAKE
3. SPACER

shifting transaxle while tractor is still moving or by gears not being properly meshed when tractor is under load. Gear wear can cause gears to jump out of position.

4. Broken detent springs can cause gear damage. When the springs are broken, the shifter fork is free to move, thus allowing gear pressure to slide the gears out of mesh.

5. When the gears slide out of gear, especially under load, gear chipping or cracking will result.

6. Prolonged heavy drawbar loads and wheel slippage are the most common cause of bevel pinion gear failure in the differential section of the transaxle.

7. Damage to the input shaft spline is caused by improper coupling of the shifter shaft and input shaft when transaxle is shifted into high range. A broken detent spring or an improperly adjusted quadrant are normally the cause of improper coupling.

8. A damaged shifter gear spline is caused by improper coupling of the shifter and input shaft. A worn or damaged shifter gear will cause gear jump-out when the tractor is operated in high range or under heavy drawbar loads.

Inspecting Oil Seals and O-rings

Always replace oil seals in axle housings whenever transaxle is disassembled. Always use new O-rings on axle housings.

Inspecting Transmission Case

Inspect the transmission case halves for cracks, worn or damaged bearing bores, damaged threads and case mating surfaces.

Inspecting Shifter Assembly

Check condition of the shifter forks, shift rods and detent springs. Slide forks along the shaft to inspect grooves. If a good snap is felt in each detent position, disassembly is not necessary.

Inspecting Drive Blocks

Check condition of differential drive blocks. Replace if cracked or broken.

ASSEMBLY

Replacing Axle Shafts and Pinion Gear

110 TRACTORS SERIAL NUMBERS 15001-42035

1. When either of the axle bevel gears or pinions are to be replaced on transaxles in this tractor serial number range, the AM30744 gear kit consisting of two axle bevel gears and two pinion gears must be used.

NOTE: Gears in this kit are a matched set. Do not mix old and new gears.

2. The axles and thrust washers do not have to be replaced unless servicing is required.

110 AND 112 TRACTORS SERIAL NUMBERS 42036-100,000

1. All differential parts can be replaced as individual parts items for all transaxles in this tractor serial number range.
2. Assemble all parts.
3. Apply Loctite® or equivalent to ends of threads and assemble capscrews through

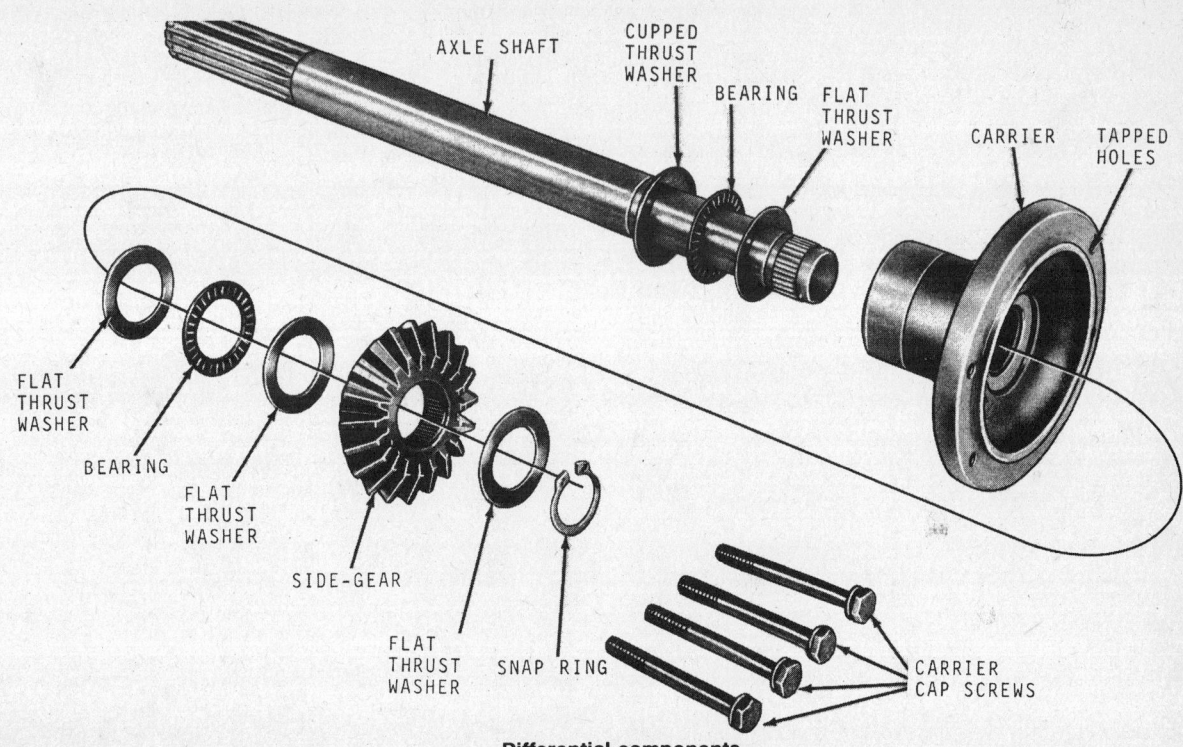

Differential components

John Deere

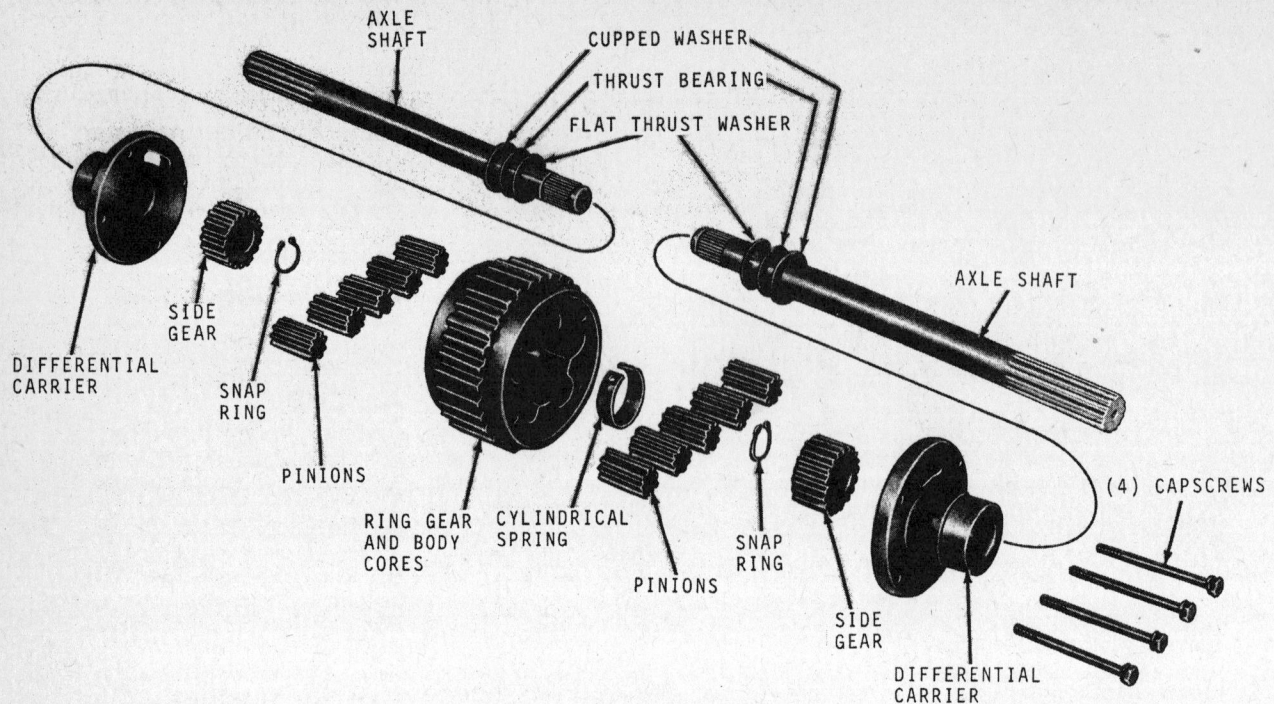

Limited slip differential components

Input shaft and gears

carriers. Be sure lockwasher is under head of screw.

4. The axles should rotate freely in opposite directions when assembled. Lay the differential assembly aside for later installation.

5. All bearings are pressed into the bearing bores from the inside of the case interior.

6. Bearing drivers are necessary to install bearings properly. As a general rule, all bearings should be pressed into the case to a depth of 0.020 in. beyond flush with case interior.

Limited Slip Differential Assembly

1. Install body cores to ring gear so that pockets in one core are out of alignment with pockets in the other core.

2. Assemble thrust washers, differential carriers and side gears to axles and secure with the snaprings.

3. Install pinion gears on one side, then use the differential carrier and axle to hold them from falling out when unit is turned over. The side gear must mesh with the five pinions.

4. Insert the remaining five pinions to mesh with those previously installed.

5. Insert the cylindrical spring with a pair of large 90° tip snapring pliers. It should bottom on the side gear and contact most of the ten pinions when properly positioned.

6. Install other axle, insert four capscrews and tighten securely.

Input Shaft and Gear

1. Assemble input shaft, gear and thrust washer. Counterbored gear spline must face to left. Gear is a light press fit onto shaft.

2. Install washer, input shaft and gear into case as shown. Use special tool to protect seal when slipping shaft through seal. Flat side gear should now face upward.

Idler Gear and Pinion Shaft

1. Use a seal sleeve tool and assemble thrust washer, idler gear and pinion shaft. Beveled edge of teeth must face away from pinion shaft. Pinion shaft is a light press fit through idler gear.

2. When thrust washer, idler gear and pinion shaft are properly assembled and installed, the flat edge of the idler gear should now face upward.

Output Shaft and Gear

1. The output gear is assembled to the output pinion shaft with a press fit. A thrust washer is used on both ends of output shaft.

2. Install output gear, pinion shaft and thrust washers into lefthand case.

3. Install compound gear with bushing into lefthand case.

Shifter Shaft

1. Because of heavy detent pressure, the assembly of these shafts can be difficult. Assemble forks as shown. 1st, 2nd and

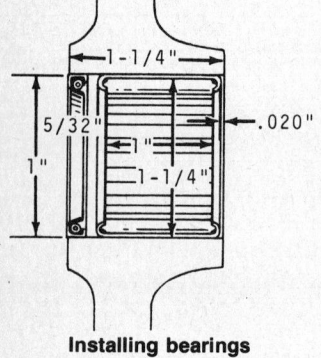

Installing bearings

Pinion gear arrangement

John Deere

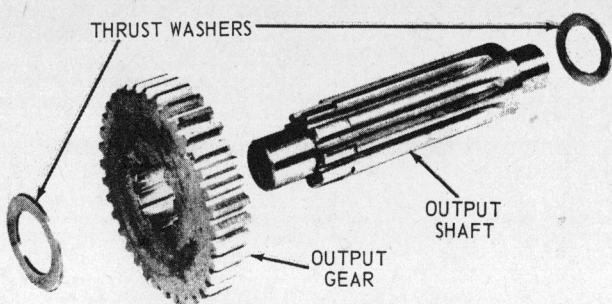

Output shaft and gear assembly

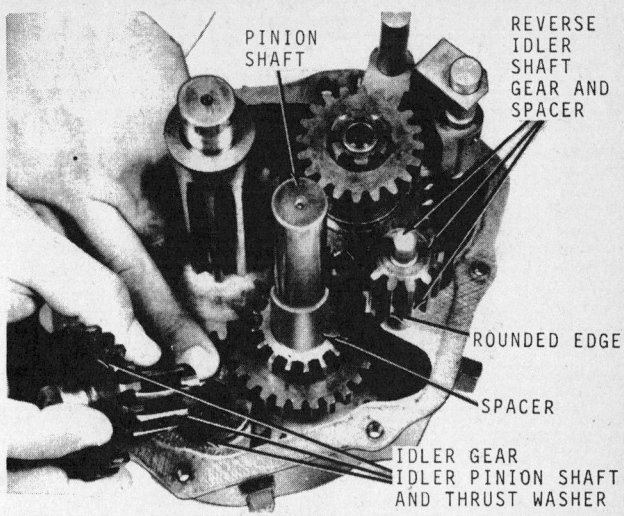

Installing idler and reverse idler

reverse fork will face to the left and 3rd and 4th fork will face to the right or away from shaft. The 1st, 2nd and reverse shaft must have the short end of shifter shaft toward fork. The 3rd and 4th shifter fork must have end opposite stop snapring toward fork as shown. Start the shaft into the fork. Depress detent and complete the assembly. Slide forks along shaft. A good snap should be felt in each detent. Place forks in neutral positions at this time.

2. To assemble shifter, be sure forks are in neutral detent. 1st, 2nd and reverse will have one detent showing on either side of fork. 3rd and 4th will have one detent showing on side of shifter fork or one detent showing between fork and snapring. Be sure shifter rod with one detent showing on either side of fork is used with 1st, 2nd and reverse shifter gear and that shifter rod with one detent between fork and snapring is used with 3rd and 4th shifter gear.

3. The shifter shaft assembly should appear as shown. The slot in the forks should line up when the large gear is slipped as far as possible on the spline. Note the position of exposed grooves on shifter rods.

4. Assemble shifter guide over shifter rods. Slot in guide should match rectangular opening between the forks. The long notch in underside of guide should clear the large 1st, 2nd and reverse shifter gear.

5. Place thrust washer over needle bearing. Grasp shifter assembly firmly in lefthand and lower it into case. When lowered and positioned, shifter shaft should be through thrust washer and in shifter shaft bearing case. The shifter rods should now enter the two machined sockets in lefthand case.

Idler Gear, Pinion and Thrust Washer

The inside of the idler gear is splined to slip freely onto splined end of idler pinion.

Reverse Idler Shaft and Gear

Assemble reverse idler shaft assembly as illustrated. Round edge of teeth faces spacer.

NOTE: Shaft is the same on both ends.

Installing Reverse Idler, Idler Gear Assembly and Spacer

1. Install reverse idler assembly.
2. Install thrust washer, idler pinion shaft and idler gear. The accompanying figure shows proper assembly before lowering into lefthand case.
3. Place spacer on pinion shaft.

Installing Cluster Gear and Thrust Washer

Install gear cluster and thrust washer on pinion shaft.

Installing Differential

1. Install differential assembly into lefthand case with bolt heads facing upward.
2. The internal components should now appear as shown.
3. Position the gasket on the lower (lefthand) case at this time. Use new gasket.

Placing Cover On Case

1. Assemble the righthand case half. Shake the case lightly and all shafts and bearings will align themselves. Also, a short turn in both directions on the input shaft will help align gears.

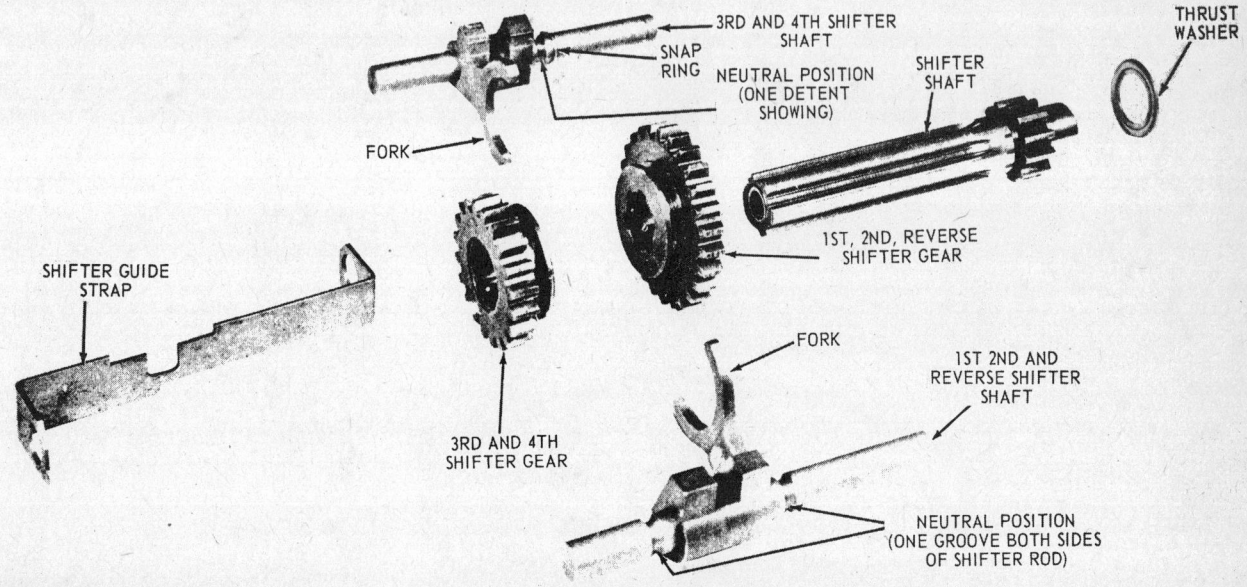

Shifter shaft and gear components

John Deere

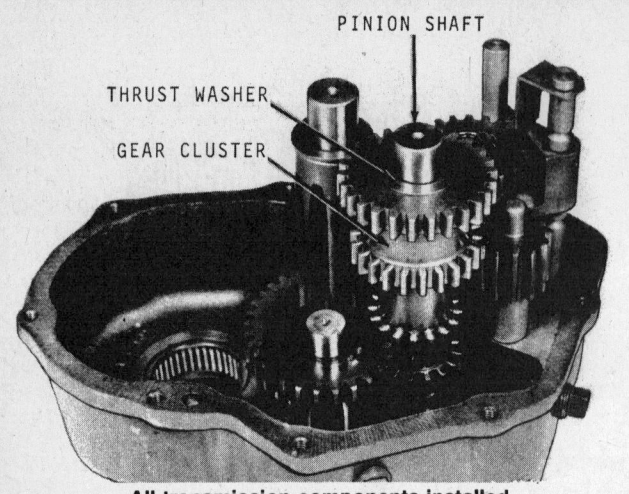
All transmission components installed

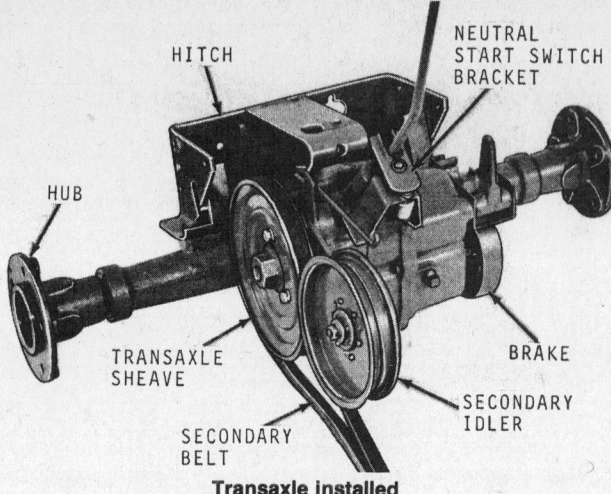
Transaxle installed

2. To close the last one-half inch, tap the righthand case horizontally. If case will not close, reach through round hole in righthand case with a screwdriver and move shifter rods. This will help align shifter rods so they will fall into shifter rod sockets in righthand case.

Installing Seals
1. Insert dowel pins and bolt case halves together with eight socket head capscrews. Torque capscrews to 120 in.lb.

2. Install retainer and new seal or shim stock to prevent cutting seal when sliding it over splined end of axle. Oil seal lip must face inward.

Installing Axle Supports
Install O-rings and axle supports with bearings. Always use new O-rings.

Positioning Shifter Forks
Inspect the shifter forks to be sure they are aligned and in neutral position. Failure to do this will cause damage to the transmission when engaged under power.

Assembling Shifter Lever
The shifter is assembled in the order shown. When assembling shifter, be sure rubber seal is positioned properly in shifter housing. A little shellac or gasket cement will be helpful to prevent incorrect positioning of the rubber seal in the housing. Align housing, keeper and spring pin in shift lever and place snapring in groove in shifter housing. Torque capscrews to 120 in.lb.

INSTALLING TRANSAXLE

1. Position neutral start bracket with switch, shift lever and gasket on transaxle. Secure with three screws.
2. Install brake, input hub, driven sheave with belt, secondary idler and hitch assembly to transaxle.
3. Before installing transaxle in tractor base, check transaxle by turning driven sheave and shifting transaxle in each gear.
4. Apply Loctite® to threads on all bolts and setscrews used in assembling components to transaxle.
5. Adjust neutral start switch and bracket.
6. Place transaxle in tractor base. Install capscrews holding transaxle support and hitch plate to tractor base.
7. Connect brake clevis and secondary idler spring. Then slip the secondary belt on variator.
8. Install wheel hubs with washers and snaprings. Bolt wheels to hubs with wheel bolts.
9. Connect neutral start switch leads.

Adjusting Quadrant
1. Turn tractor upright and install shift quadrant. Apply Loctite® to shift lever threads and tighten knob on lever. Position quadrant before tightening screws. Secure seat spring to tractor base.
2. Adjust brake and variator linkage.
3. Fill the unit with lubricant.

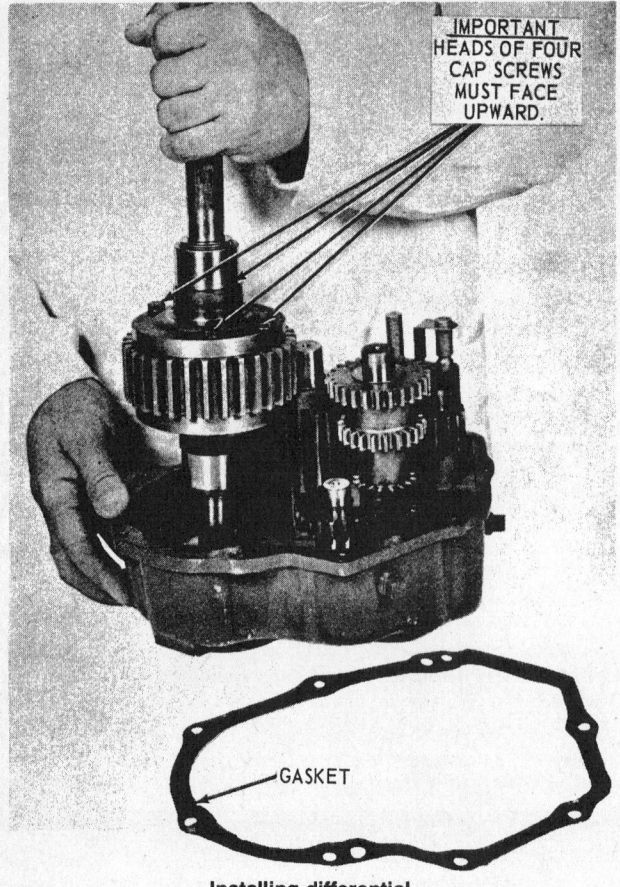
Installing differential

John Deere

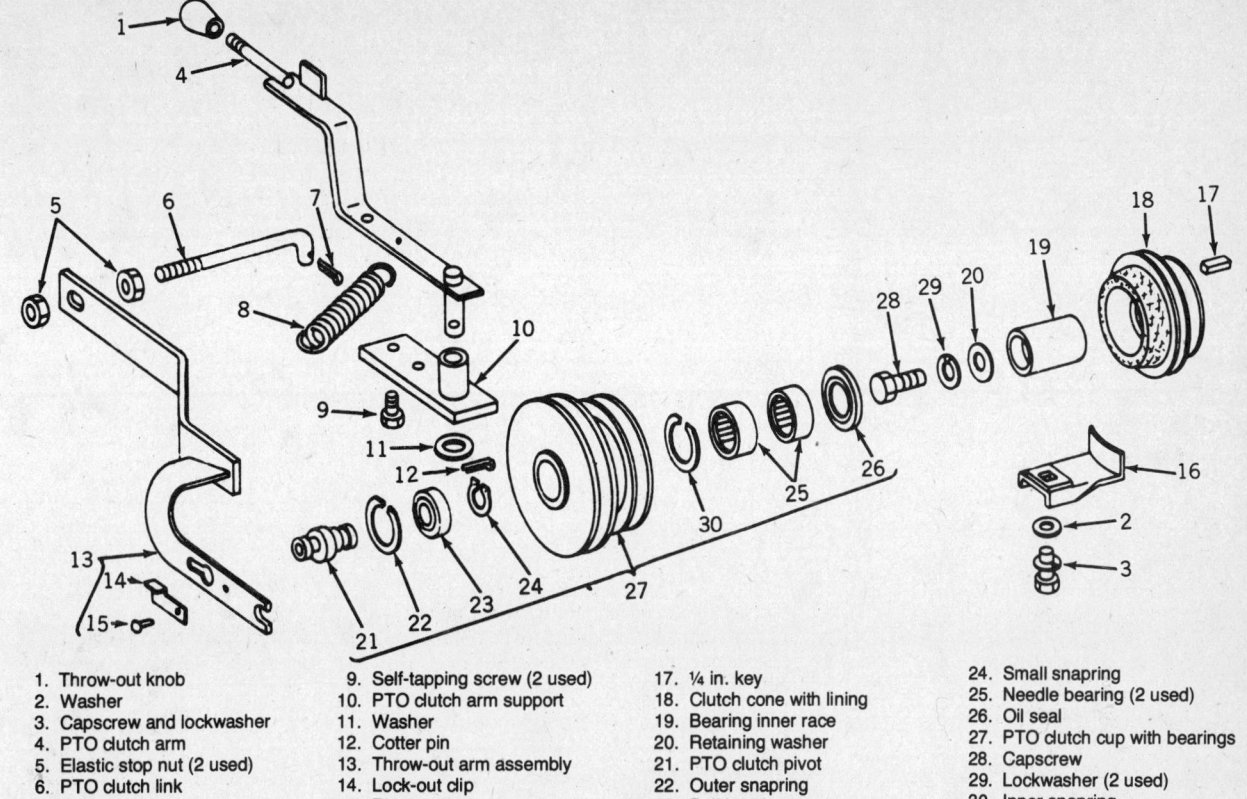

1. Throw-out knob
2. Washer
3. Capscrew and lockwasher
4. PTO clutch arm
5. Elastic stop nut (2 used)
6. PTO clutch link
7. Cotter pin
8. Spring
9. Self-tapping screw (2 used)
10. PTO clutch arm support
11. Washer
12. Cotter pin
13. Throw-out arm assembly
14. Lock-out clip
15. Rivet
16. Clutch brake shoe
17. ¼ in. key
18. Clutch cone with lining
19. Bearing inner race
20. Retaining washer
21. PTO clutch pivot
22. Outer snapring
23. Ball bearing
24. Small snapring
25. Needle bearing (2 used)
26. Oil seal
27. PTO clutch cup with bearings
28. Capscrew
29. Lockwasher (2 used)
30. Inner snapring

Power take-off clutch assembly

POWER TAKE-OFF CLUTCH

Tractors Through Serial No. 272,000

DISASSEMBLY

1. Remove clutch throw out arm assembly (13) from PTO clutch pivot and fulcrum bolt. Loosen clutch brake shoe sufficiently to allow PTO clutch cup to be removed. Remove PTO clutch cup. Remove capscrew and washers from end of crankshaft and remove bearing inner race. Remove clutch cone and key. If tractor has hydraulic lift, hydraulic pump drive belt must be removed prior to removing cone.

2. Remove outer snapring and press PTO clutch pivot out of PTO drive sheave. Remove inner snapring. Press two needle bearings and oil seal out of PTO drive sheave. Remove small snapring from PTO clutch pivot and press ball bearing off PTO clutch pivot.

INSPECTION

1. Inspect clutch linings and mating surfaces for excessive wear. Inspect bearings, bearing inner race and seal. Inspect PTO brake shoe for excessive wear. Replace parts as necessary.

2. Inspect clutch linkage and linkage return spring to be certain nothing is bent, broken, or stretched.

ASSEMBLY

1. Install key in crankshaft and slide clutch cone onto crankshaft. Install belt for hydraulic pump if tractor is so equipped.

2. Install bearing inner race and secure with retaining washer, lockwasher and capscrew. Tighten securely.

3. Press bearings and seal into PTO drive sheave following dimensions given in the accompanying figure. Pack bearings with John Deere High Temperature Grease.

4. Press clutch pivot into ball bearing and secure with small snapring. Install inner snapring into PTO drive sheave, press clutch pivot assembly into PTO drive sheave, and secure with outer snapring.

5. Slide complete PTO clutch sheave assembly onto bearing inner race.

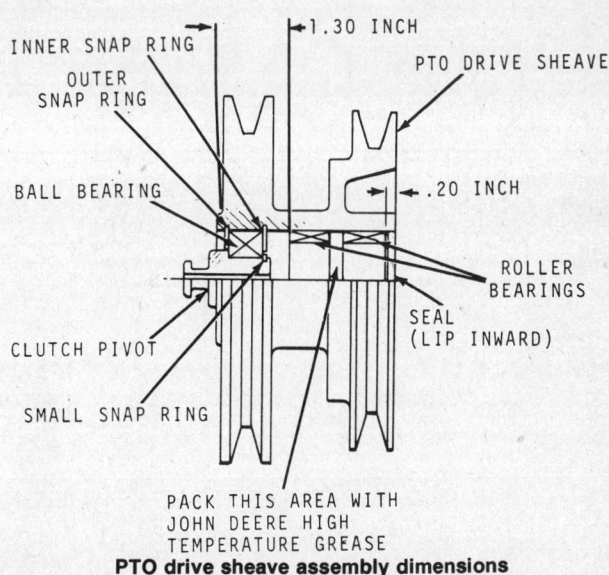

PTO drive sheave assembly dimensions

335

John Deere

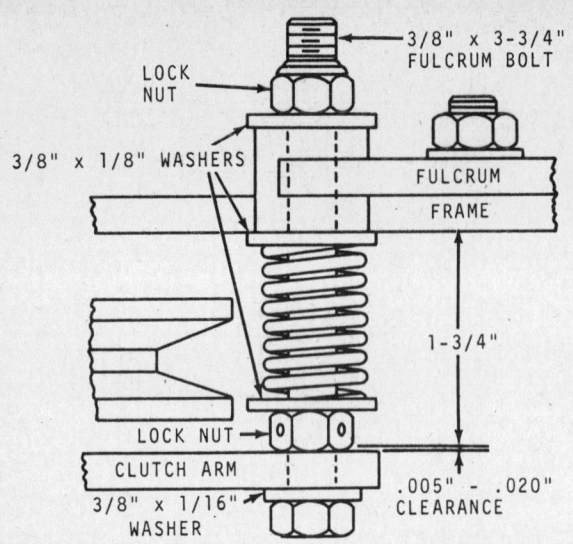

Spring loaded fulcrum bolt dimensions, serial #250,001-272,000

6. Replace clutch fulcrum with new spring-loaded type if tractor is not already equipped with this fulcrum. This is available as a complete kit. Adjust to dimensions shown.

7. Connect clutch throw-out arm to PTO clutch pivot and fulcrum.

8. Check adjustment of clutch components.

ADJUSTMENT

1. The inner elastic stop nut on the PTO clutch link should be screwed on the rod so that $1^{5}/_{16}$ in. of thread remain between the nut and the righthand end of link. The outer elastic stop nut should be screwed on the link until ⅝ in. of thread remains.

NOTE: Do not thread the nuts on the rod beyond these dimensions because there must be $^{7}/_{16}$ in. clearance between nuts.

NOTE: On tractors below Serial No. 130,000, it may be necessary to lower the fulcrum until there is ⅜ to ¾-in. clearance between the rear end of the clutch arm and the edge of the channel reinforcement. If fulcrum strikes tractor frame before clearance is achieved, notch the frame and slot the attaching holes so the fulcrum can be lowered further. Lower fulcrum until the ⅜ to ¾ in. measurement is obtained.

2. Make final adjustment to PTO clutch as follows:

3. The PTO clutch must start to engage when the PTO clutch lever (located on the instrument panel pedestal) is halfway between the engaged and the disengaged position, thus giving ½ slot of free travel.

4. If less than ½ slot of free travel is present, lengthen fulcrum bolt slightly.

5. If more than ½ slot of free travel is present, shorten fulcrum bolt slightly.

6. With PTO clutch engaged, adjust the brake shoe so there is approximately $^{1}/_{16}$ in. clearance between the brake shoe and clutch cup sheave.

8hp Tractors, Serial No. 272,001-285,000 and 310,001-320,000
10hp Tractors, Serial No. 285,001-310,000

DISASSEMBLY

1. Remove clutch lever from pivot bolt and drilled pin.

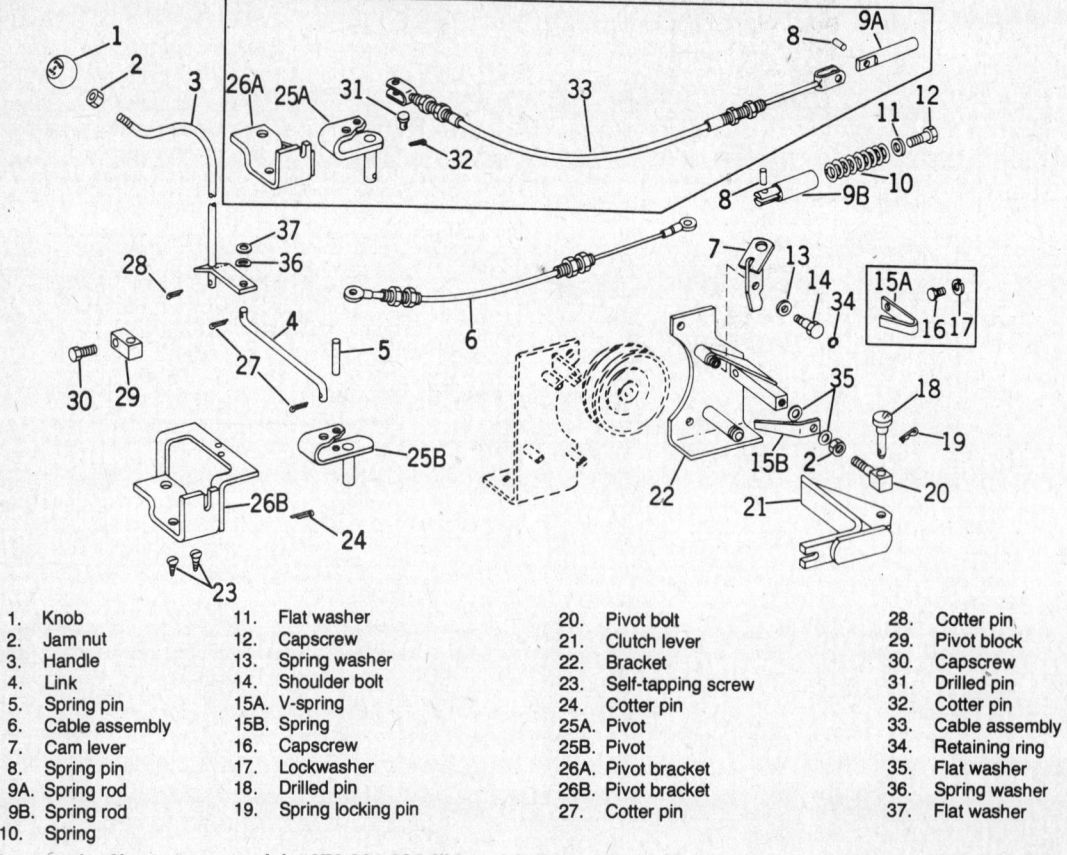

1. Knob	11. Flat washer	20. Pivot bolt	28. Cotter pin
2. Jam nut	12. Capscrew	21. Clutch lever	29. Pivot block
3. Handle	13. Spring washer	22. Bracket	30. Capscrew
4. Link	14. Shoulder bolt	23. Self-tapping screw	31. Drilled pin
5. Spring pin	15A. V-spring	24. Cotter pin	32. Cotter pin
6. Cable assembly	15B. Spring	25A. Pivot	33. Cable assembly
7. Cam lever	16. Capscrew	25B. Pivot	34. Retaining ring
8. Spring pin	17. Lockwasher	26A. Pivot bracket	35. Flat washer
9A. Spring rod	18. Drilled pin	26B. Pivot bracket	36. Spring washer
9B. Spring rod	19. Spring locking pin	27. Cotter pin	37. Flat washer
10. Spring			

PTO controls, 8hp tractors serial #272,001-285,000 and 310,001-320,000; 10hp tractors serial #285,001-310,000

John Deere

1. PTO clutch pivot
2. Outer snapring
3. Ball bearing
4. Small snapring
5. Inner snapring
6. Needle bearing (2 used)
7. Oil seal
8. PTO clutch cup with bearings
9. Capscrew
10. Lockwasher (2 used)
11. Retaining washer
12. Bearing inner race
13. Clutch cone with lining
14. ¼ in. key
15. Clutch brake shoe
16. Washer
17. Capscrew and lockwasher

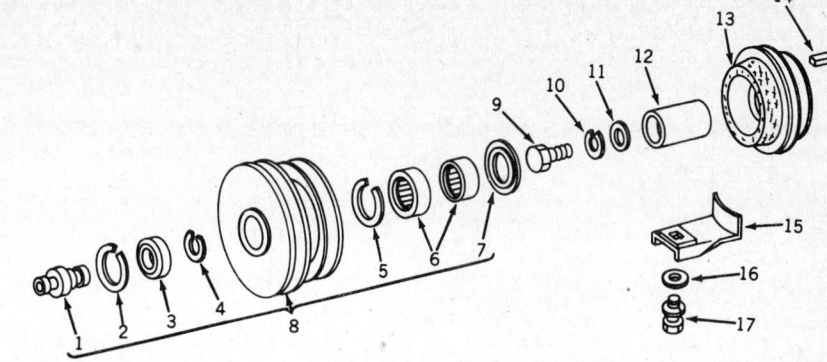

PTO clutch drive components 10hp tractors serial #310,001 and up

2. Loosen clutch brake shoe sufficiently to allow PTO clutch cup to be removed. Remove PTO clutch cup. Remove capscrew and washers from end of crankshaft and remove bearing inner race. Remove clutch cone and key.

3. Remove outer snapring and press PTO clutch pivot out of PTO clutch cup. Remove inner snapring. Press two needle bearings and oil seal out of PTO clutch cup. Remove small snapring from PTO clutch pivot and press ball bearing off PTO clutch pivot.

INSPECTION

1. Inspect clutch linings and mating surfaces for excessive wear. Inspect bearings, bearing inner race and seal. Inspect PTO brake shoe for excessive wear. Replace parts as necessary.

2. Inspect clutch linkage and linkage return spring to be certain nothing is bent, broken, or stretched.

ASSEMBLY

1. Install key in engine crankshaft and slide clutch cone onto crankshaft.

2. Install bearing inner race and secure with retaining washer, lockwasher and capscrew.

3. Press bearings and seal into PTO clutch cup following dimensions give in the accompanying illustration. Pack bearings with John Deere High Temperature Grease (AT30408).

4. Press clutch pivot into ball bearing and secure with small snapring. Install inner snapring into PTO clutch cup. Press clutch pivot assembly into PTO clutch cup and secure with outer snapring.

5. Slide complete PTO clutch cup assembly onto bearing inner race.

Replace clutch fulcrum with new spring-loaded type if tractor is not already equipped with this fulcrum. Adjust dimensions as shown.

ADJUSTMENT

Adjusting PTO Clutch

1. Move PTO clutch lever on the control panel to the disengaged position (down).

2. Loosen front jam nut and tighten rear jam nut on cable conduit until ⅛ in. clearance exists between the spring pin and the cam lever when the PTO clutch lever is moved to the engaged position (up).

3. Check clutch engagement and disengagement by engaging and disengaging PTO clutch lever on the control panel.

Adjusting PTO Brake

1. The PTO clutch brake stops the mower blades, snow thrower rotor, rotary tiller tines and any of the PTO-operated equipment soon after disengaging the drive with the PTO clutch lever.

2. With the engine shut off, engage the PTO clutch lever (up position) and check the distance between the brake and clutch cup sheave using a blade-type feeler gauge. When properly adjusted, the distance should be .030 in. or 1/32 in. when the clutch is engaged.

3. If adjustment is required, use a ½ in. socket with extension to loosen capscrew. Slide brake shoe in slotted hole until proper adjustment is obtained. Tighten capscrew firmly.

10hp Tractors, Serial No. 320,001 and Up

DISASSEMBLY

1. Remove clutch arm assembly from PTO clutch pivot and fulcrum bolt. Loosen clutch brake shoe sufficiently to allow PTO clutch cup to be removed. Remove PTO clutch cup. Remove capscrew and washers from end of crankshaft and remove bearing inner race. Remove clutch cone and key.

2. Remove outer snapring (17) and press PTO clutch pivot (16) out of PTO clutch cup. Remove inner snapring (25). Press two needle bearings (25) and oil seal (27) out of PTO drive sheave. Remove small snapring (19) from PTO clutch pivot and press ball bearing (18) off PTO clutch pivot.

INSPECTION

1. Inspect clutch linings and mating surfaces for excessive wear. Inspect bearings, bearing inner race and seal. Inspect PTO brake shoe for excessive wear. Replace parts as necessary.

2. Inspect clutch linkage and linkage return spring to be certain nothing is bent, broken or stretched.

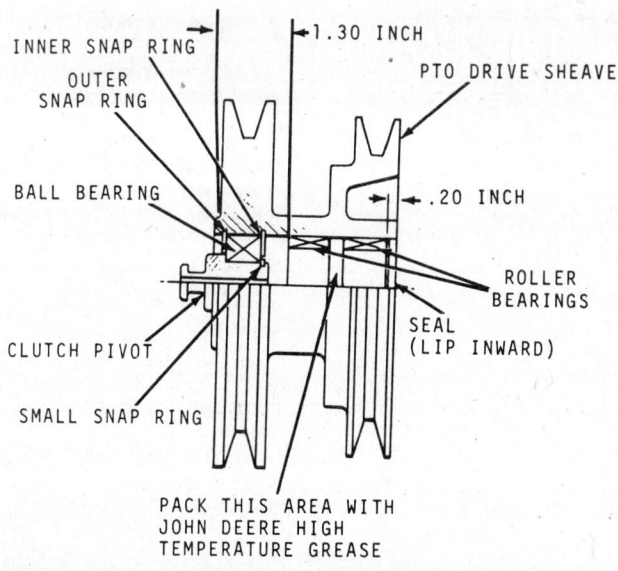

PTO drive cup dimensions

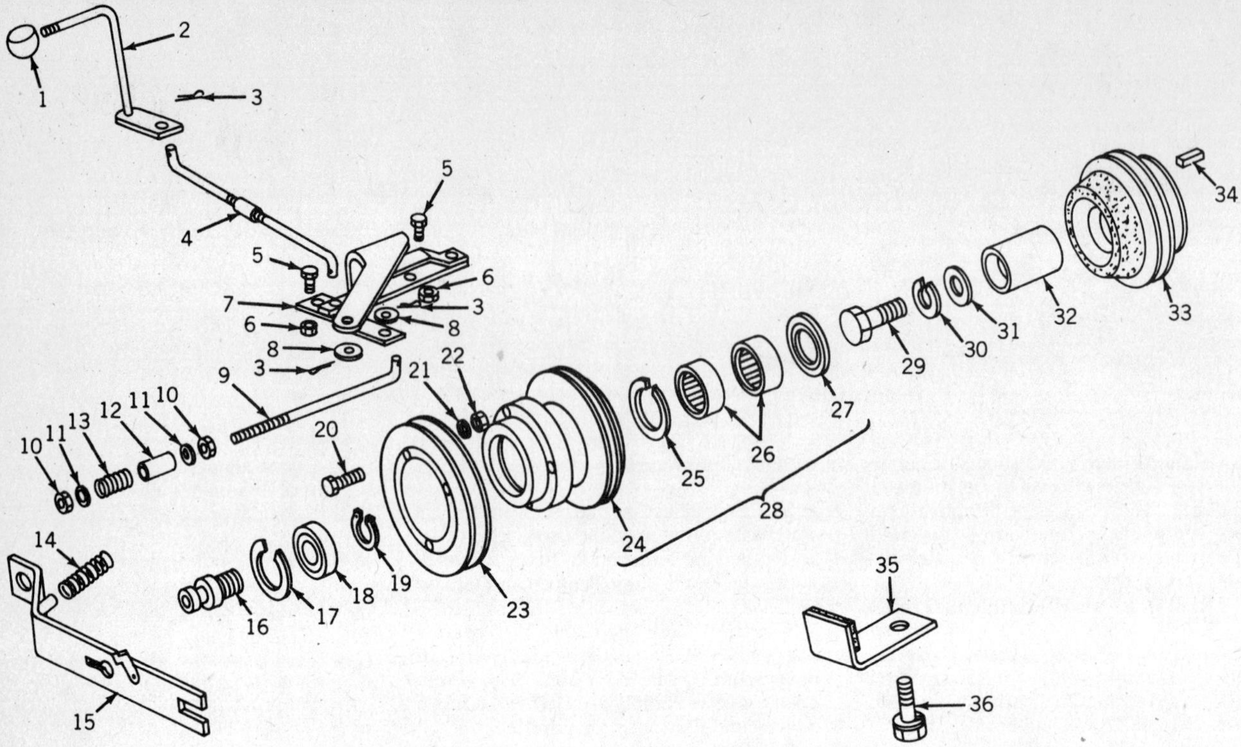

John Deere

1. Knob
2. Handle
3. Cotter pin
4. Rod
5. Bolt
6. Nut
7. Pivot
8. Washer
9. Adjusting link
10. Adjusting locknut
11. Washer
12. Spacer
13. Spring
14. Brake spring
15. Clutch arm
16. PTO clutch pivot
17. Outer snapring
18. Ball bearing
19. Small snapring
20. Sheave bolt
21. Lockwasher
22. Nut
23. Sheave
24. PTO clutch cup
25. Inner snapring
26. Needle bearing (2 used)
27. Oil seal
28. PTO clutch cup with bearings
29. Capscrew
30. Lockwasher
31. Retaining washer
32. Bearing inner race
33. Clutch cone with lining
34. Key
35. Clutch brake shoe
36. Capscrew and lockwasher

PTO clutch components 10hp tractors serial #320,000 and up

ASSEMBLY

1. Install key in crankshaft and slide clutch cone onto crankshaft.

2. Install bearing inner race and secure with retaining washer, lockwasher and capscrew and tighten securely.

3. Press bearings and seal into PTO drive sheave following dimensions given in the accompanying illustration. Pack bearing with John Deere High-Temperature Grease or equivalent. Press clutch pivot into ball bearing and secure with small snapring. Install inner snapring into PTO drive sheave, and secure clutch pivot and ball bearing with outer snapring.

4. Slide complete PTO clutch sheave assembly onto bearing inner race.

ADJUSTMENT

1. Loosen locknut on fulcrum bolt, and adjust fulcrum bolt to preliminary setting of 2½ in.

2. Adjust clutch spring position by drawing up locknuts and washers tight against spacer.

3. With PTO in the engaged position, adjust for $11/64$ in. gap between the clutch arm and washer. The clutch arm must be kept approximately parallel with the clutch sheave.

4. If necessary, readjust fulcrum bolt to keep clutch arm parallel with the clutch sheave, and still retain correct gap between clutch arm and washer. Tighten locknut on fulcrum bolt.

5. With clutch engaged, loosen capscrew on PTO brake. Adjust to $1/32$ in. gap between PTO sheave and the brake shoe, then tighten capscrew.

NOTE: The brake must be readjusted every time the PTO clutch is adjusted.

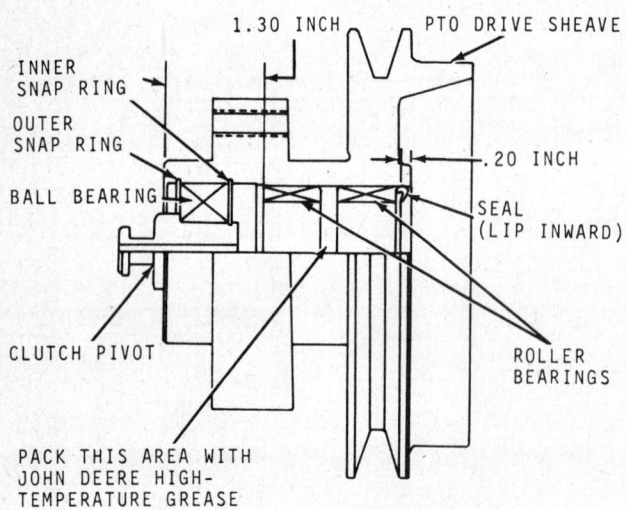
Drive sheave assembly dimensions

John Deere

Models 108 and 111

FRONT WHEELS AND AXLES

Wheels

REMOVAL AND INSTALLATION

Securely support front of tractor.
1. Remove spindle cap.
2. Remove snapring and pull wheel off spindle.
3. Installation is the reverse of removal.

Spindles, Tie Rod and Drag Link

REMOVAL AND INSTALLATION

1. Remove jam nut from tie rod.
2. Remove snapring and washer.
3. Remove spindle from axle.
4. Remove jam nut from tie rod.
5. Remove jam nut from drag link.
6. Remove snapring and washer.
7. Remove spindle from axle.
8. Installation is the reverse of removal.

Axle

REMOVAL AND INSTALLATION

1. Remove hood and muffler.

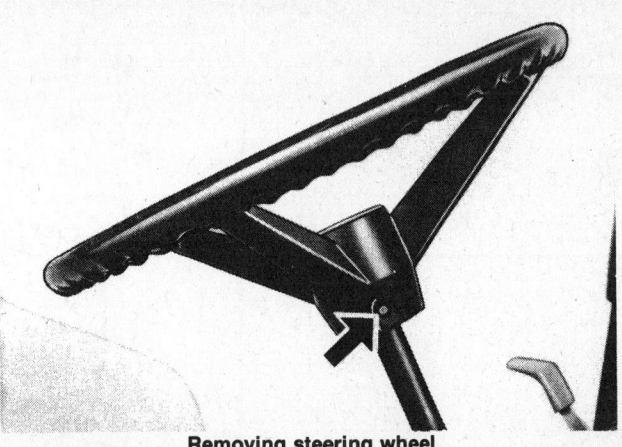
Removing steering wheel

2. Remove four capscrews from pivot anchors.
3. Remove axle.
4. Installation is the reverse of removal.

STEERING SYSTEM

Steering Wheel

REMOVAL AND INSTALLATION

1. Remove pin (arrow) and lift steering wheel off shaft.
2. Installation is the reverse of removal.

Steering Wheel Shaft and Bearing

REMOVAL AND INSTALLATION

1. Raise hood and remove battery and battery base.
2. Remove pin and steering wheel.
3. Remove two bearing support nuts.
4. Remove three instrument panel bolts.
5. Disconnect throttle control and switch from instrument panel. Slide instrument panel and bearing off shaft.
6. Remove two nuts, bolts, spacers, clamp, washers and steering wheel shaft.

Steering Gear Case

REMOVAL AND INSTALLATION

1. Disconnect steering wheel shaft from gear case assembly.
2. Disconnect drag link from arm on sector gear shaft.

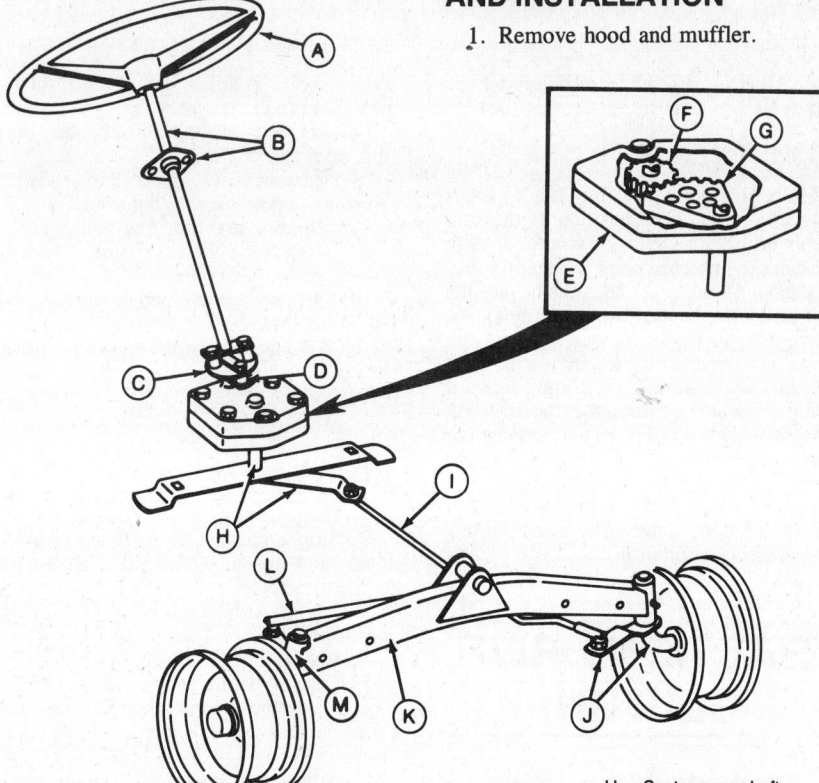

A. Steering wheel
B. Steering wheel shaft and bearing
C. Leather disc
D. Pinion gear shaft
E. Gear case
F. Pinion gear
G. Sector gear
H. Sector gear shaft
I. Drag link
J. Lefthand spindle
K. Axle
L. Tie rod
M. Righthand spindle

Wheels, front axle and steering system

339

John Deere

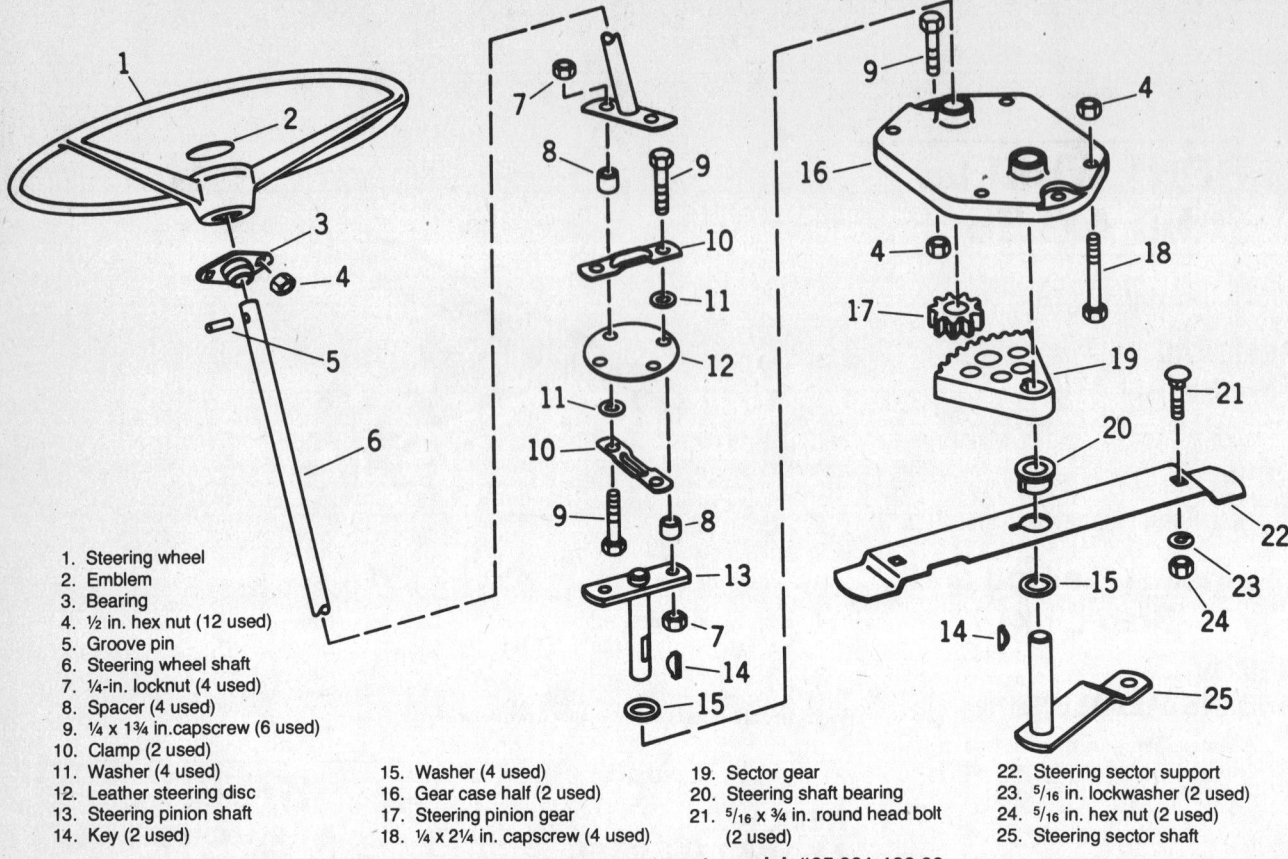

1. Steering wheel
2. Emblem
3. Bearing
4. ½ in. hex nut (12 used)
5. Groove pin
6. Steering wheel shaft
7. ¼-in. locknut (4 used)
8. Spacer (4 used)
9. ¼ x 1¾ in. capscrew (6 used)
10. Clamp (2 used)
11. Washer (4 used)
12. Leather steering disc
13. Steering pinion shaft
14. Key (2 used)
15. Washer (4 used)
16. Gear case half (2 used)
17. Steering pinion gear
18. ¼ x 2¼ in. capscrew (4 used)
19. Sector gear
20. Steering shaft bearing
21. 5/16 x ¾ in. round head bolt (2 used)
22. Steering sector support
23. 5/16 in. lockwasher (2 used)
24. 5/16 in. hex nut (2 used)
25. Steering sector shaft

Steering wheel and gear components, serial #95,001-120,00

3. Remove two support bolts.
4. Remove four nuts from bolts securing gear case assembly to tractor frame.
5. Remove gear case by lowering it out of frame between traction drive belt.

Steering Gear Case

OVERHAUL

1. Remove two bolts, clamp, washers, leather disc and spacers.
2. Remove two gear case bolts. Separate gear case halves by tapping sides of case with a plastic hammer.
3. Remove washers from top of gear shafts and gears. Remove keys and washers from under gears.
4. Slip gear shafts out of case halves.
5. Remove support and bearing from sector shaft.
6. Remove grease from case and parts. Wash all parts in solvent and then blow them dry with compressed air.
7. Assemble in reverse of disassembly.

NOTE: Apply a liberal amount of multi-purpose-type grease on bearings, steering pinion gear and sector gear. Grease should completely cover gear teeth.

IMPORTANT: Apply a liberal amount of multi-purpose-type grease on bearings, steering pinion gear and sector gear. Grease should completely cover gear teeth.

ADJUSTMENTS

Before connecting the drag link to steering sector shaft arm position both front wheels on tractor straight forward. If drag link can then be connected to steering sector shaft arm easily without disturbing the straight forward position of wheels, the steering is properly adjusted. If drag link cannot be connected to steering sector shaft arm without disturbing the straight forward position of wheels, disconnect the drag link from the wheel spindle arm and lengthen or shorten it as necessary so it can be connected to steering sector shaft arm without disturbing the straight forward position of wheels. Turn ball joint on end of drag link to lengthen or shorten it.

TRACTION DRIVE CLUTCH

Traction Drive Belt

REPLACEMENT

1. Remove mower from tractor.
2. Engage PTO lever on tractor.
3. Loosen belt tightener bolt and sheave bolt. Position guard and sheave away from belt.
4. Remove nut and disconnect drag link from steering arm.
5. Remove two bolts and position steering support parallel between belt.
6. Remove two nuts and belt guide.
7. Remove clutch support to frame bolts.
8. Depress clutch pedal and remove belt from sheaves.
9. Replace belt in reverse order of removal.
10. Adjust belt.

ADJUSTMENT

1. Remove mower from tractor.
2. Stop engine. DO NOT set parking brake. Put transmission shift lever in gear to hold tractor.
3. Foot clutch must be "UP" in engaged position.
4. Loosen nut on idler.
5. Turn adjusting bolt (in or out) until a 3.7 in. dimension is obtained between inner surface of flat idler and inside of frame.
6. Adjust belt guide for 3/16 in. clearance between guide and belt on the engine side of guide.
7. Tighten nut on idler.
8. Loosen bolt and move guide until belt is centered and tighten bolt.
9. Reinstall mower on tractor.

John Deere

Traction Drive Clutch

REMOVAL

1. Disconnect clutch spring from tractor frame.
2. Disconnect two links from bell crank and clutch shaft.
3. Remove hex nut from round head bolt.
4. Remove bell crank and clutch sheave assembly.
5. Disassemble sheaves, bushings and spacer from clutch assembly.

INSTALLATION

1. Apply Never-Seez® lubricant on two bushings and ends of bell crank sleeve.
2. Reinstall traction drive clutch assembly in reverse order of disassembly.
3. Reinstall and adjust traction drive belt.
4. Reinstall mower on tractor.

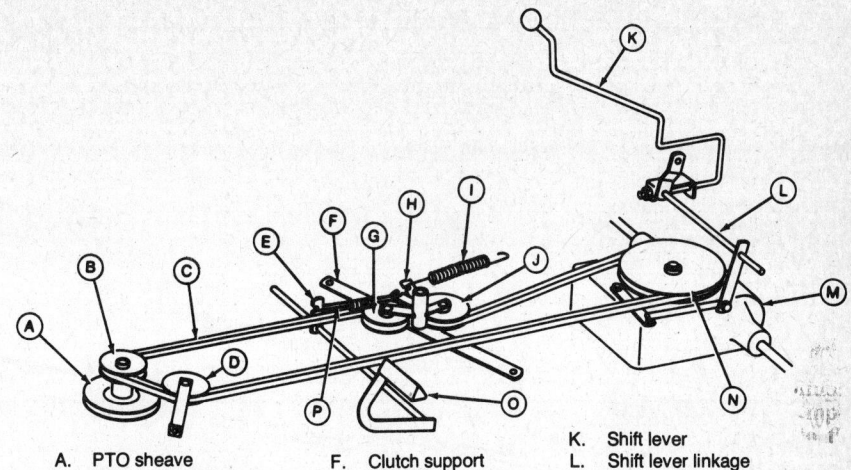

A. PTO sheave
B. Engine drive sheave
C. Traction drive belt
D. Adjustable belt tightener
E. Clutch shaft arm
F. Clutch support
G. Clutch sheave
H. Bell crank assembly
I. Clutch spring
J. Clutch sheave
K. Shift lever
L. Shift lever linkage
M. Transaxle
N. Transaxle driven sheave
O. Clutch pedal shaft
P. Return spring assembly

Traction drive clutch components

BRAKES

Brake Assembly

REMOVAL

1. Block up tractor frame and remove righthand rear wheel.
2. Disconnect brake rod from lever.

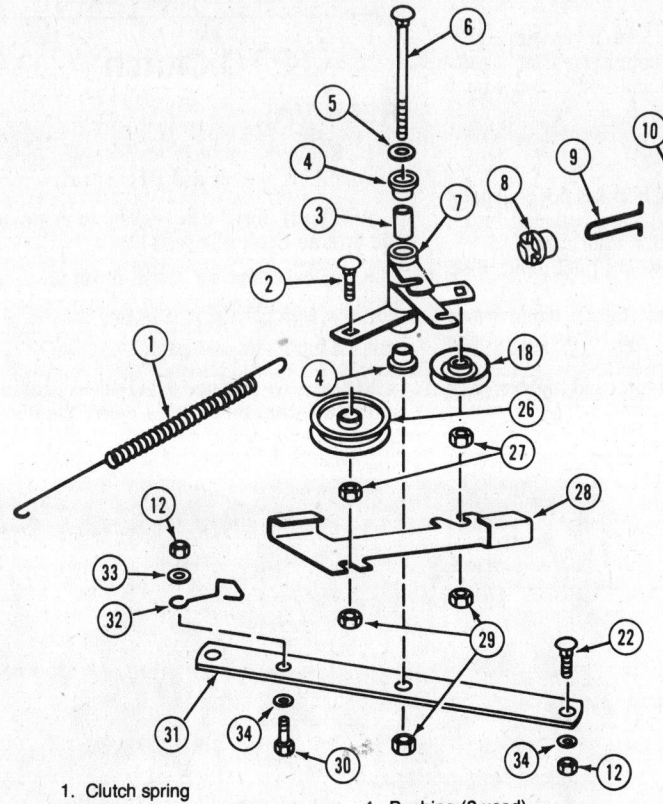

1. Clutch spring
2. ⅜ in. x 1¾ in. round head bolt (2 used)
3. Bell crank sleeve
4. Bushing (2 used)
5. 17/32 in. x 15/16 in. x 0.060 in. washer
6. ⅜ in. x 4½ in. round head bolt
7. Bell crank
8. End cap (2 used)
9. Link (2 used)
10. Return spring
11. Clutch shaft
12. 5/16 in. hex nut
13. Pad
14. 5/16 in. x 3½ in. capscrew
15. Belt tightener link
16. ⅜ in. x 2¼ in. round head bolt
17. Cam roller
18. Sheave (2 used)
19. Belt guard
20. 13/32 in. x 13/16 in. x 0.060 in. washer
21. ⅜ in. locknut
22. 5/16 in. x ¾ in. round head bolt
23. Clutch shaft mount
24. 41/64 in. x 13/16 in. x 0.060 in. washer (2 used)
25. Snapring
26. Sheave
27. ⅜ in. jam nut (2 used)
28. Belt guard
29. ⅜ in. hex nut (3 used)
30. 5/16 in. x 1 in. capscrew
31. Support
32. Belt guide
33. 11/32 in. x 1 in. x 0.194 in. washer
34. 5/16 in. lockwasher (2 used)

Drive exploded view

341

John Deere

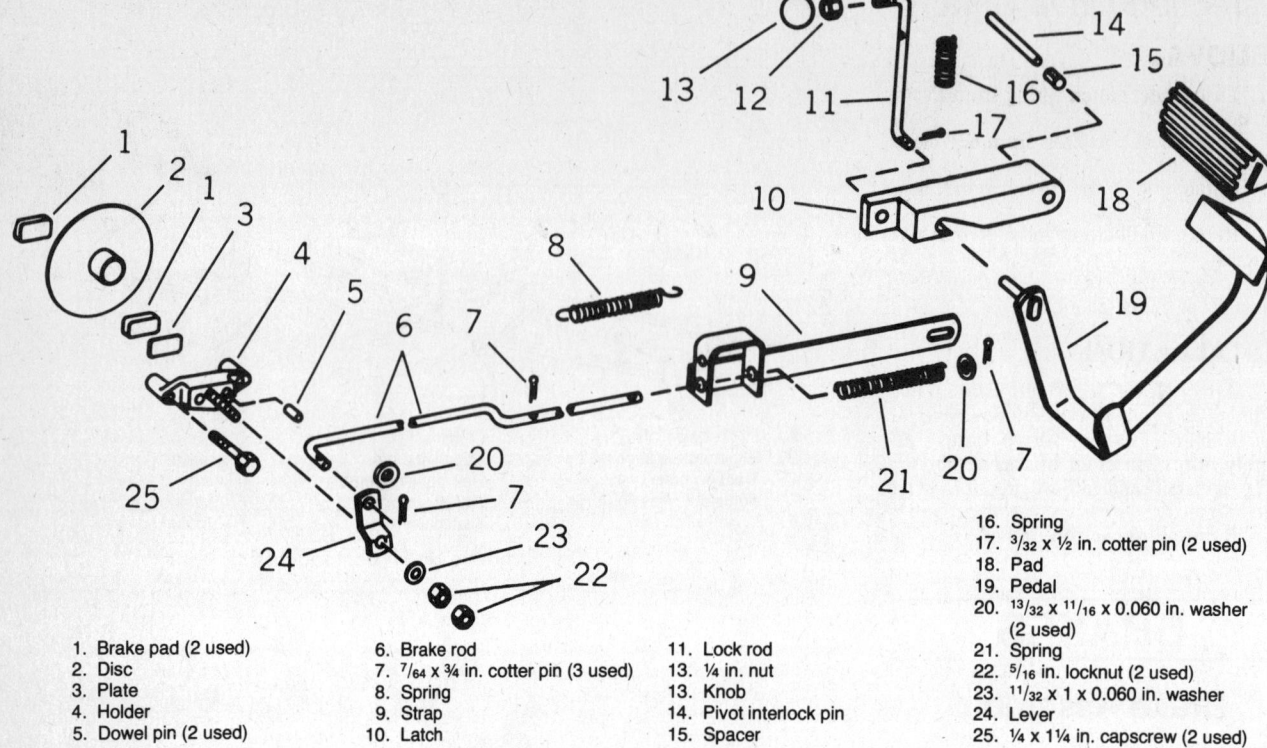

1. Brake pad (2 used)
2. Disc
3. Plate
4. Holder
5. Dowel pin (2 used)
6. Brake rod
7. 7/64 x 3/4 in. cotter pin (3 used)
8. Spring
9. Strap
10. Latch
11. Lock rod
13. 1/4 in. nut
13. Knob
14. Pivot interlock pin
15. Spacer
16. Spring
17. 3/32 x 1/2 in. cotter pin (2 used)
18. Pad
19. Pedal
20. 13/32 x 11/16 x 0.060 in. washer (2 used)
21. Spring
22. 5/16 in. locknut (2 used)
23. 11/32 x 1 x 0.060 in. washer
24. Lever
25. 1/4 x 1 1/4 in. capscrew (2 used)

Brake system components

3. Remove two capscrews and brake assembly.
4. Remove disc from shaft.
5. Remove pad from transaxle case.

INSPECTION AND REPAIR

Inspect pads and plate for wear or damage. Inspect disc for wear or cracks. Replace parts as necessary.

INSTALLATION

1. Install one of pads in transaxle case.
2. Place disc on splined shaft with hub out.
3. Install plate and other pad in holder.
4. Attach holder to transaxle case with two capscrews.
5. Install two dowel pins in holder.
6. Attach lever to holder with two locknuts.
7. Connect brake rod to lever with cotter pin. Spread ends of cotter pin around rod.
8. Reinstall righthand wheel and adjust brake.

ADJUSTMENTS

1. Be sure parking brake is not engaged.
2. Place a feeler gauge between brake puck and disc to check adjustment.
3. Clearance between puck and disc should be 0.010 in.
4. To adjust brakes, loosen jam nut and turn inside adjusting nut to obtain proper clearance.
5. Hold adjusting nut and tighten jam nut.

PTO SYSTEM

PTO Clutch

REMOVAL

1. Remove mower, capscrew, lockwasher, flat washer and PTO sheave.

NOTE: It may be necessary to remove the engine drive sheave (12).

2. Release spring tension on clutch spring, by lifting spring arm off cam roller.
3. If further disassembly is required, remove hardware and parts.

**NOTE: To remove PTO lever, pivot assembly, cam follower or cam, it is nec-

Measuring brake gap

Adjusting brake gap

John Deere

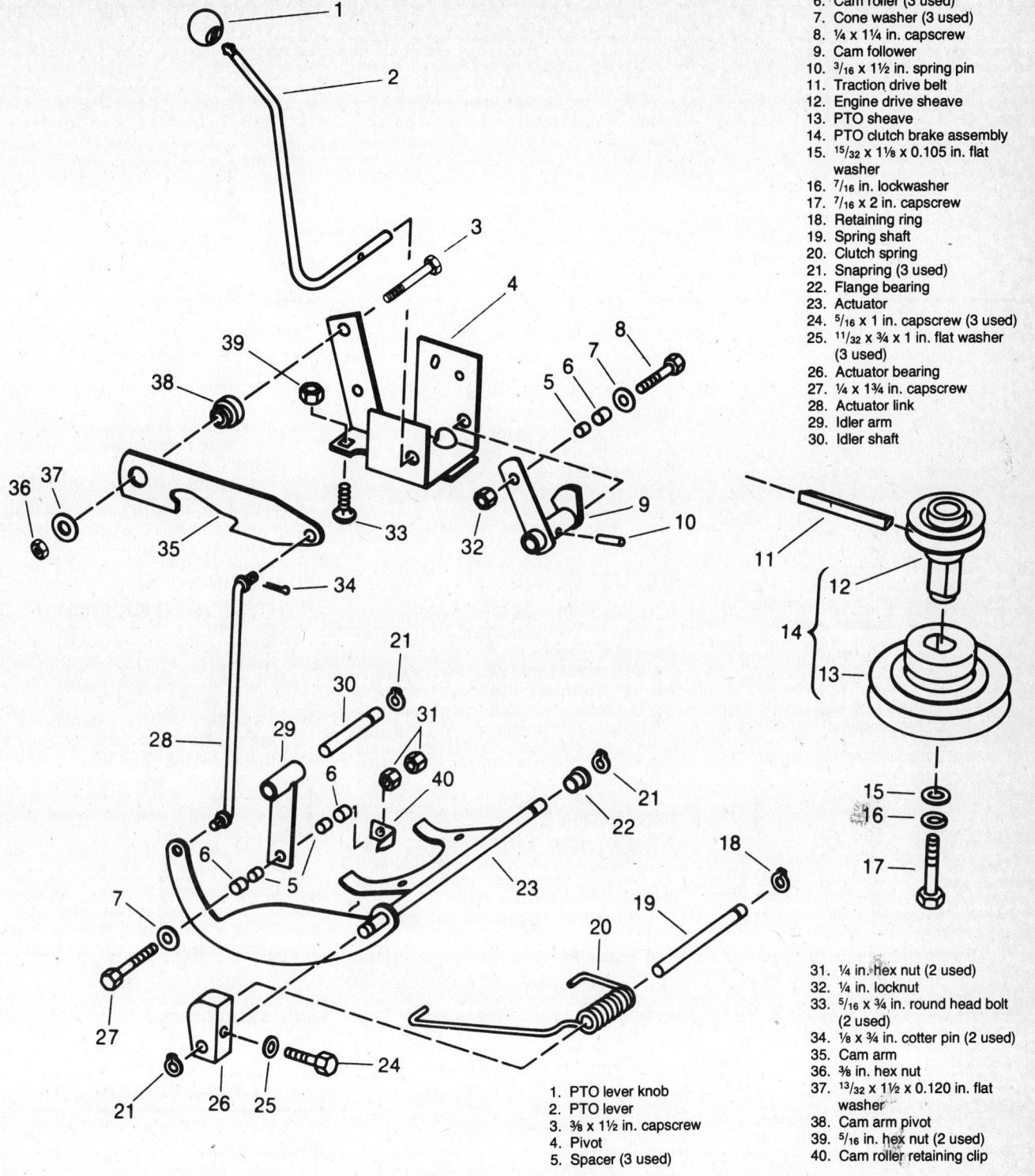

PTO clutch components

1. PTO lever knob
2. PTO lever
3. 3/8 x 1 1/2 in. capscrew
4. Pivot
5. Spacer (3 used)
6. Cam roller (3 used)
7. Cone washer (3 used)
8. 1/4 x 1 1/4 in. capscrew
9. Cam follower
10. 3/16 x 1 1/2 in. spring pin
11. Traction drive belt
12. Engine drive sheave
13. PTO sheave
14. PTO clutch brake assembly
15. 15/32 x 1 1/8 x 0.105 in. flat washer
16. 7/16 in. lockwasher
17. 7/16 x 2 in. capscrew
18. Retaining ring
19. Spring shaft
20. Clutch spring
21. Snapring (3 used)
22. Flange bearing
23. Actuator
24. 5/16 x 1 in. capscrew (3 used)
25. 11/32 x 3/4 x 1 in. flat washer (3 used)
26. Actuator bearing
27. 1/4 x 1 3/4 in. capscrew
28. Actuator link
29. Idler arm
30. Idler shaft
31. 1/4 in. hex nut (2 used)
32. 1/4 in. locknut
33. 5/16 x 3/4 in. round head bolt (2 used)
34. 1/8 x 3/4 in. cotter pin (2 used)
35. Cam arm
36. 3/8 in. hex nut
37. 13/32 x 1 1/2 x 0.120 in. flat washer
38. Cam arm pivot
39. 5/16 in. hex nut (2 used)
40. Cam roller retaining clip

essary to remove the battery and battery base.

INSPECTION AND REPAIR

Inspect cam arm, cam follower, and cam arm pivots for wear. Inspect linkage for looseness. Loose linkage can cause erratic operation. Inspect PTO sheave for damage or wear. Inspect fork on actuator to be sure it is not bent.

INSTALLATION AND ADJUSTMENT

1. Reinstall parts in reverse order of removal.
2. Move PTO lever forward (engaged position).
3. Loosen capscrew (A), and slide block (B) back and forth to center clutch cone (C) in fork. Clearance should be approximately 1/16 in. on each side of cone between cone and fork.
4. Tighten capscrew (A).
5. Reinstall mower on tractor.

John Deere

ELECTRIC PTO SYSTEM

PTO Clutch

REMOVAL

1. Remove capscrew and washer (A).
2. Disconnect wiring harness at connector (B).

NOTE: If clutch is stuck to crankshaft, tap it with a plastic mallet to free it from crankshaft.

DISASSEMBLY

1. Remove three locknuts.
2. Remove bearing retaining ring.
3. Place three 3/16 in. spacer washers between pulley and armature housing for support.
4. Support pulley and press bearing collar with bearing out of rotor and pulley assembly.

NOTE: Be careful not to drop bearing and bearing collar on floor when they are pressed out.

5. Remove spacer from bearing collar.
6. Separate field coil and rotor assembly from pulley and armature assembly.
7. Remove three springs from stud bolts.
8. Press bearing collar out of bearing.

NOTE: Discard bearing. Install a new bearing whenever clutch is disassembled.

INSPECTING ROTOR AND PULLEY

1. Test field coil.
2. Inspect field coil and rotor assembly (A) for defective bearing. If bearing is defective, the complete field coil and rotor assembly must be replaced.

NOTE: The pulley and armature bearing (F) must be replaced whenever the clutch is disassembled.

3. Inspect for broken springs in pulley and armature assembly.
4. If rotor or armature is scored or grooved excessively, consider replacing the complete clutch rather than attempting to repair it.

INSTALLATION

1. Install two rear drive wheels and move transaxle under tractor.
2. Attach transaxle to tractor frame with two U-bolts.
3. Connect wire to neutral start switch.
4. Bolt shift arm to transaxle shifter shaft.
5. Install two support bolts.
6. Depress clutch pedal and slip drive belt on transaxle sheave.
7. Connect brake rod to brake arm with washer and cotter pin. Spread ends of cotter pin around rod.
8. Adjust drive belt.
9. Adjust brake.
10. Adjust transmission shift lever as follows.

ADJUSTMENTS

1. When shift lever is properly adjusted, it should be in center of slot and 0.039 in. from edge of slot.
2. To adjust shift lever, first turn nuts, one on each side, to move shift lever forward or rearward until it centers in slot.
3. Then, loosen jam nut and turn stop screw until lever is held 0.039 in. from edge of slot. Tighten jam nut after adjustment.

INSPECTING ARMATURE AND ROTOR

1. Inspect armature for weak or broken contact springs (A).
2. Inspect between rotor and armature contacts (B) for paint. Be sure surfaces are clean.

TESTING FIELD COIL

1. Attach test lead wires to field coil leads and battery terminals.
2. Place a chisel across rotor. Electromagnetic action will hold the chisel to rotor if PTO field is in good condition. If chisel will not stick to rotor, field coil is defective and the complete rotor and field coil assembly must be replaced.

NOTE: Battery must be in good condition and fully charged for proper test.

ASSEMBLY

1. Support inboard end of rotor and press bearing collar into rotor.
2. Place spacer in bottom of pulley hub.
3. Assemble three springs on rotor studs.
4. Install pulley assembly on rotor.
5. Press bearing into pulley over collar.
6. Install retaining ring.
7. Install three locknuts.

NOTE: Adjust clutch before installing clutch on tractor. After clutch is adjusted properly, install it.

ADJUSTING ELECTRIC PTO CLUTCH

1. Before adjusting PTO clutch, disconnect spark plug wire and remove key switch from ignition to prevent engine from accidentally starting. Also, be sure PTO switch is in the "OFF" (down) position.
2. Insert a flat feeler gauge in each of three slots (A) in armature housing. Clearance between armature plate and rotor should be 0.012 to 0.014 in. when clutch is properly adjusted. Loosen or tighten three locknuts (B) to adjust clutch.
3. After checking PTO clutch adjustment, connect spark plug wire to spark plug and start engine. Move PTO switch to "ON" position. If clutch still does not engage, remove and disassemble clutch for inspection and repair.

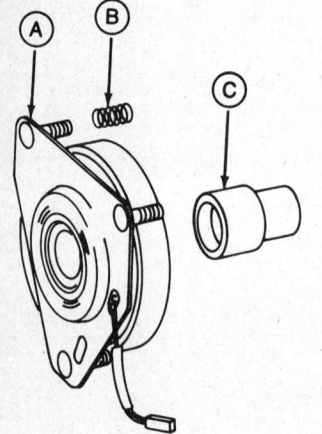

A. Field coil and rotor assembly
B. Spring (3 used)
C. Bearing collar
D. Pulley and armature assembly
E. Spacer
F. Bearing
G. Retaining ring
H. Elastic stop nut (3 used)

Exploded view of the Ogura electric PTO clutch

5-SPEED TRANSAXLE

REMOVAL

1. Remove mower from tractor.
2. Securely block up rear of tractor for access to parts.

NOTE: Place transmission shift lever on tractor in neutral position before removing transaxle.

3. Remove cotter pin and washer. Disconnect brake rod from brake arm.
4. Depress clutch pedal and remove traction drive belt from transaxle driven sheave.
5. Remove two support bolts.

John Deere

6. Remove shift arm capscrew and disconnect shift arm from transaxle.
7. Disconnect wire from transaxle neutral start switch.
8. Place a floor jack under transaxle case.
9. Remove hex nuts from two U-bolts. Slowly, lower transaxle to the surface.
10. Remove two rear drive wheels and place transaxle assembly on a bench.

DISASSEMBLY

1. Remove snapring, washer, sheave and key.
2. Remove snapring and washer. Slide hubs off axles.
3. Remove two self-tapping screws and support.
4. Remove neutral start switch.
5. Remove two capscrews and brake assembly.
6. Remove Allen head screw. Remove spring and detent ball with a magnet.
7. Remove seventeen capscrews securing case halves.
8. Tap sides of case with a soft plastic hammer. Grasp input shaft and lift top half of case off bottom half.
9. Slip input shaft and gear out of case.
10. Remove snapring, gear, washer and seal from input shaft.
11. Remove old grease from case by washing it in solvent.

NOTE: Observe how parts are assembled before removing them from case.

12. Remove shifter fan (B) from case.
13. Remove transmission gear and shaft assemblies as a unit by lifting them out of case.
14. Lift differential assembly out of case.

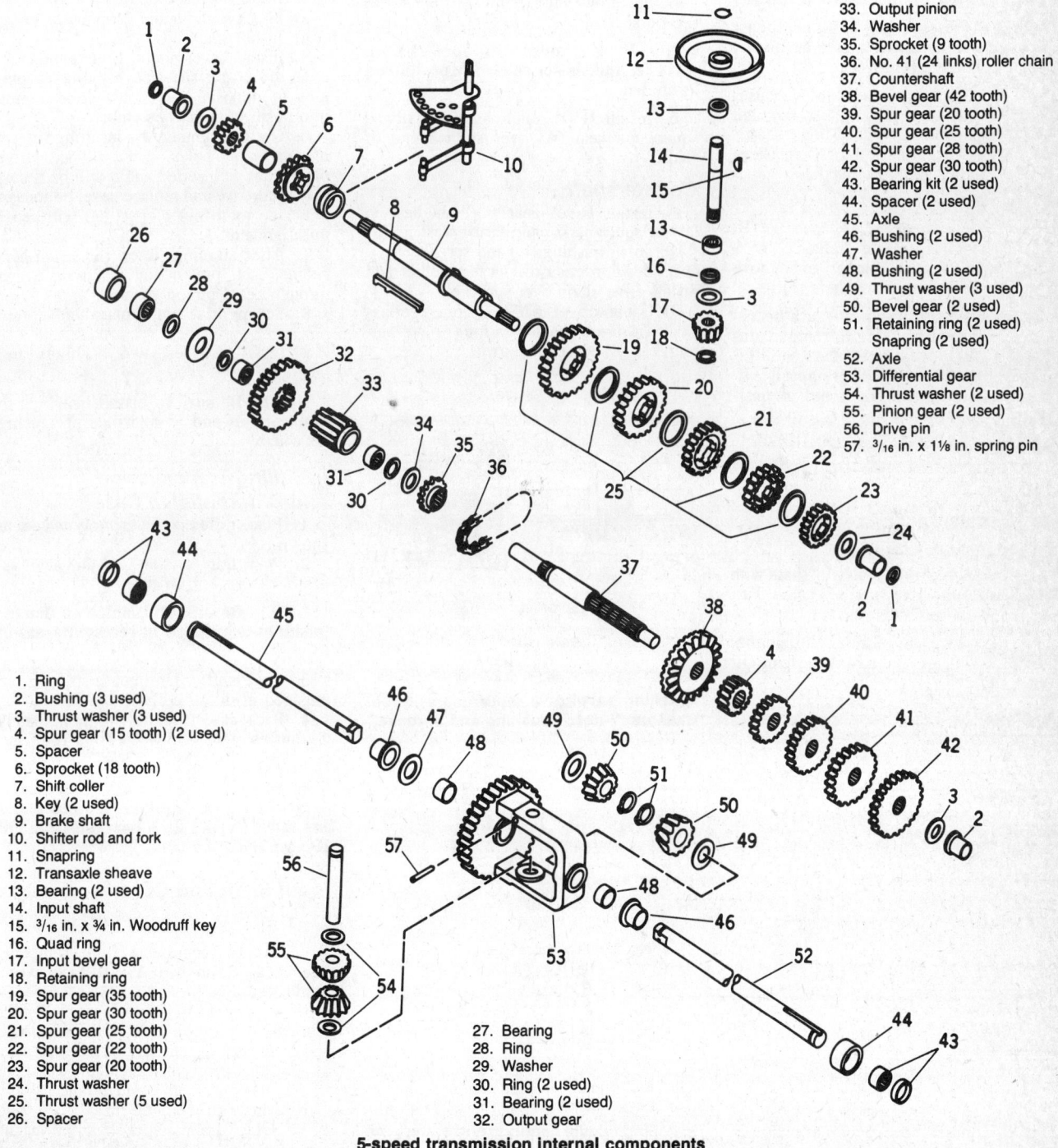

1. Ring
2. Bushing (3 used)
3. Thrust washer (3 used)
4. Spur gear (15 tooth) (2 used)
5. Spacer
6. Sprocket (18 tooth)
7. Shift collar
8. Key (2 used)
9. Brake shaft
10. Shifter rod and fork
11. Snapring
12. Transaxle sheave
13. Bearing (2 used)
14. Input shaft
15. 3/16 in. x 3/4 in. Woodruff key
16. Quad ring
17. Input bevel gear
18. Retaining ring
19. Spur gear (35 tooth)
20. Spur gear (30 tooth)
21. Spur gear (25 tooth)
22. Spur gear (22 tooth)
23. Spur gear (20 tooth)
24. Thrust washer
25. Thrust washer (5 used)
26. Spacer
27. Bearing
28. Ring
29. Washer
30. Ring (2 used)
31. Bearing (2 used)
32. Output gear
33. Output pinion
34. Washer
35. Sprocket (9 tooth)
36. No. 41 (24 links) roller chain
37. Countershaft
38. Bevel gear (42 tooth)
39. Spur gear (20 tooth)
40. Spur gear (25 tooth)
41. Spur gear (28 tooth)
42. Spur gear (30 tooth)
43. Bearing kit (2 used)
44. Spacer (2 used)
45. Axle
46. Bushing (2 used)
47. Washer
48. Bushing (2 used)
49. Thrust washer (3 used)
50. Bevel gear (2 used)
51. Retaining ring (2 used)
    Snapring (2 used)
52. Axle
53. Differential gear
54. Thrust washer (2 used)
55. Pinion gear (2 used)
56. Drive pin
57. 3/16 in. x 1 1/8 in. spring pin

5-speed transmission internal components

John Deere

Remove old grease from case and then wash off case and parts with solvent.

15. Slip bearing (A), flat washer (B), output gear (C), output pinion (D) and flat washer (E) off countershaft.

NOTE: Remove square cut seal from each end of output pinion.

16. Position ends of shafts toward each other and carefully remove drive chain (arrow) from sprockets.

17. Remove sprocket (A), flanged bushing (E), thrust washer (D), spur gears (C) and bevel gear (B) from countershaft.

18. Remove square cut seal (A), flanged bushing (B) and spur gears (D) with thrust washers (C) from shifter and brake shaft.

NOTE: Observe position of thrust washers (on ends and between each spur gear) as they are removed.

19. Remove square cut seal (A), bushing (B), spur gear (C), spacer (D) and sprocket (E) from shifter and brake shaft.

20. Remove shifter collar (A) and keys (B) from shifter and brake shaft.

Differential

21. Remove bearings (A), bushings (B) and washer (C) from axle shafts.

22. Drive out spring pin (D) and remove drive pin (E) and thrust washers (F).

NOTE: Thrust washers must be removed before attempting to remove pinion gears. Remove pinion gears simultaneously by rotating them in opposite directions so they will move out of position.

23. Remove pinion gears (G).

24. Remove snaprings (A), bevel gears (B), thrust washers (C) and axles (D).

ASSEMBLY

Shifter and Brake Shaft

1. Grease both key ways in shaft with John Deere High Temperature Grease (Part No. AT30408) or equivalent.

NOTE: Two equivalent lubricants are, Shell Darina EP2 and American Oil Rykon 2EP.

2. Assemble shifter collar (A) over keys (B) and slide on shaft.

NOTE: The thick side of shifter collar must face the shoulder (C) on shaft as shown.

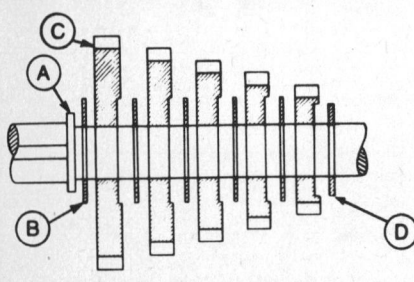

A. Shoulder
B. Chamfered thrust washers
C. Shifter spur gears
D. Plain thrust washer

Installing shifter spur gears and thrust washers

3. Install five chamfered thrust washers (B) and shifter spur gears (C) on shaft.

NOTE: The 45 degree chamfer on inside of thrust washers must face the shoulder (A) on shaft. Also, the flat side of gears must face the shoulder as shown.

4. Install plain thrust washer (D) on shaft next to outside gear.

NOTE: The thrust washer on end of shaft does not have a chamfer on inside diameter.

5. Install sprocket (E), spacer (D), small spur gear (C), plain thrust washer (B), "V" notch bushing (I) and square cut seal (A) on shaft.

NOTE: Be sure shoulders (F) on sprocket and shifter collar are positioned as shown.

6. Install "V" notch bushing (I) and square cut seal (A) over splined end of shaft.

Countershaft

1. Install bevel gear (C), smallest to largest spur gears, plain thrust washer (E), "V" notch bushing (D) and sprocket (A).

2. Position ends of shafts together and install drive chain over sprockets.

3. Using driver 670252 from Peerless Tool Kit, press needle bearing (F) into pinion gear (B) to depth shown.

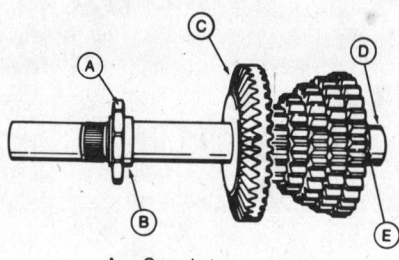

A. Sprocket
B. Collar
C. Bevel gear
D. "V" notch bushing
E. Plain thrust washer

Installing bevel gear, spur gears, thrust washers, V-notch bushing and sprocket

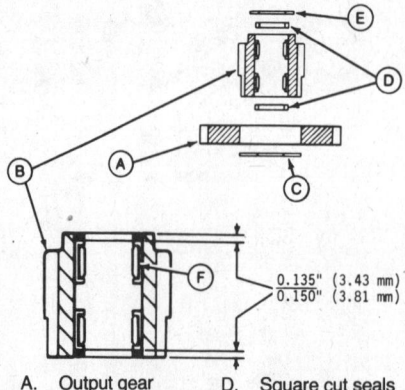

A. Output gear
B. Pinion gear
C. Large washer
D. Square cut seals
E. Small washer
F. Needle bearing

Installing needle bearings and square cut seals in pinion gear

4. Install square cut seals (D) in each end of pinion gear.

NOTE: The accompanying figure shows the order of assembly for output gear (A) and output pinion (B).

5. Install small washer (D), output pinion (E), output gear (F) and large washer (G).

6. Install a square cut seal in recessed end of closed end needle bearing (H). Install closed end needle bearing on end of shaft.

Differential

1. Insert shortest axle through gear hub.
2. Insert longest axle through yoke.
3. Place washers and gears on axles.
4. Install snaprings.
5. Install bevel pinion gears simultaneously from opposite sides by rotating pinions in opposite directions while sliding into position in gear assembly.
6. Check alignment by inserting fingers into drive pin holes. If not aligned, drive pin cannot be inserted and it will be necessary to remove and replace bevel pinions as only one tooth out of position will cause misalignment.
7. After aligning, insert thrust washers behind each pinion and insert drive pin through bevel pinion gears.
8. Secure drive pin in place with spring pin.
9. Install washer, two bushings and bearings.

NOTE: Be sure to install bearings so seal cups in end of bearings are facing outward.

Installing Transmission and Differential in Case

1. Place differential assembly in case as shown.
2. Place transmission gear and shaft assemblies as a unit in case.

NOTE: Be sure "V" notch on flanged bushings are seated in recessed areas of case.

3. Place shifter fan in case so forks are engaged in shifter collar.
4. Pack case level full (approximately 24 ounces) with John Deere High Temperature Grease (Part No. AT30408) or equivalent.

NOTE: Two equivalent lubricants are Shell Darina EP2 and American Oil Rykon 2EP.

Input Shaft and Cover

1. Using a driver press needle bearing flush with top of cover.
2. Apply John Deere High Temperature Grease (Part No. AT30408) between needle bearings.
3. Secure bevel gear to input shaft with retaining ring. Place thrust washer on shaft and install square cut seal in recessed area of case against needle bearing.
4. Insert shaft up through hole in case and secure with retaining ring.
5. Grasp end of input shaft and install cover on case. Secure cover in place with

John Deere

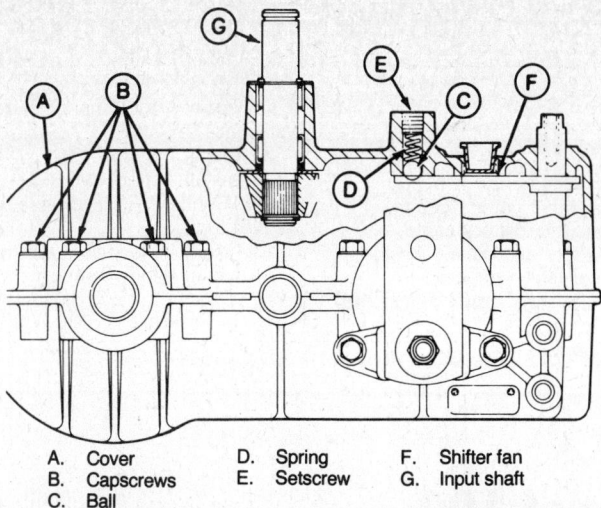

A. Cover
B. Capscrews
C. Ball
D. Spring
E. Setscrew
F. Shifter fan
G. Input shaft

Installing cover, shifter fan index ball, spring and setscrew

seventeen capscrews. Tighten capscrews to 90-100 in.lb. torque.

NOTE: Snug capscrews up slightly and then tighten each capscrew alternately from left to right of cover. Be sure square cut seals are fitting in recessed areas of cover and case properly when tightening screws.

6. Install index ball, spring and setscrew. Tighten setscrew one turn below flush with case cover.

NOTE: Prior to installing brake assembly, apply a light film of Lubriplate® to the inside of the brake lever portion which contacts the pins and to the outside of the lever which contacts the flat washer. Also, apply lubricant between bore of brake disc and splined end of shifter and brake shaft.

NOTE: Brake pads and disc must be free of grease and oil. Be sure these parts are clean.

7. Install brake assembly on transaxle opposite order of disassembly.
8. Install neutral start switch.
9. Install support with two self-tapping screws.
10. Slide hubs on axles. Install flat washer and snapring.

11. Place key in slot of input shaft and slip sheave over shaft. Secure in place with washer and snapring.

ADJUSTMENTS

Adjust Transmission Drive Belt

1. Stop engine. DO NOT set parking brake. Put transmission shift lever in gear to hold tractor.
2. Foot clutch must be "UP" in engaged position.
3. Loosen nut on idler.
4. Turn adjusting bolt (in or out) until a 3.7 in. dimension is obtained between inner surface of flat idler and inside of frame.
5. Adjust belt guide for 3/16 in. clearance between guide and belt on the engine side of guide.
6. Tighten nut on idler.
7. Loosen bolt and move guide until belt is centered and tighten bolt.

Adjust Transmission Shift Lever

If transmission shift lever does not go in "NEUTRAL" slot when in neutral position, adjust as follows:

1. Stop engine. Set parking brake.
2. Turn nuts (one on each side) to move shift lever forward or rearward until it centers in slot.
3. Loosen jam nut and turn stop screw until lever is 0.039 in. from edge of slot. Tighten jam nut.

ENGINE

REMOVAL AND INSTALLATION

1. Tilt hood forward.
2. Remove springs from rod in hood.
3. Lift hood up to remove.
4. Remove fuel line from carburetor and drain fuel tank.

CAUTION

Gasoline is dangerous. Avoid fires due to smoking or careless maintenance practices.

5. Disconnect throttle cable and white wire from throttle bracket.
6. Remove starter wire from starter and disconnect diode wire. Remove wires from clip on engine.
7. Remove front bumper.

NOTE: Chalk mark the front bumper as shown to aid in reassembly. Bumper position determines the hood fit to the panel.

8. Remove muffler and exhaust pipe.
9. Hold flywheel with a screwdriver on 108 and 111 tractors serial no. 120,000 and below.

NOTE: 108 and 111 tractors serial no. 120,001 and above, are equipped with an electric PTO clutch.

10. Remove PTO sheave and loosen traction drive idler.
11. Pull PTO lever to the rear (disengaged position).
12. Push clutch and brake pedals down and engage parking brake. This locks the clutch pedal down and relieves tension on the drive belt.
13. Remove drive belt from engine drive sheave.
14. Remove engine drive sheave and key.
15. Remove engine mounting bolts.
16. Rotate engine to free oil drain from hole in frame and lift engine out.
17. Place engine in position with driveshaft in the slotted hole in frame.
18. Rotate engine until oil drain goes through hole in frame and mounting holes line up.
19. Install mounting bolts, lockwashers and nuts. DO NOT tighten nuts at this time.
20. Raise engine slightly and place screwdriver between engine and frame.

NOTE: This provides more clearance between clutch fork and engine to provide easier installation of the engine drive sheave.

NOTE: On tractors serial no. 120,001 and above, install spacer on crankshaft first and then install drive sheave as explained in step 21 below.

21. Apply Never-Seez® to the crankshaft and then start engine drive sheave on shaft. Place key in shaft key way. Rotate sheave until key way in sheave aligns with key.
22. Install drive belt around sheave and push sheave all the way up on shaft.

NOTE: 108 and 111 tractors are equipped with an electric PTO clutch.

23. Move PTO lever forward (engaged position). This moves the clutch fork up and holds the engine drive sheave in place.
24. Remove screwdriver and tighten engine mounting bolts.
25. Install PTO sheave. Hold flywheel

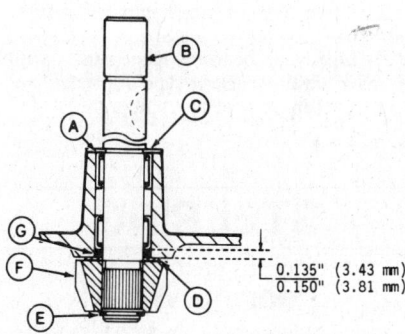

A. Needle bearing
B. Input shaft
C. Retaining ring
D. Thrust washer
E. Retaining ring
F. Bevel gear
G. Square cut seal

0.135" (3.43 mm)
0.150" (3.81 mm)

Installing needle bearing and input shaft

347

John Deere

with a screwdriver and tighten PTO sheave retaining bolt to 45-50 ft.lb. torque.

26. Tighten traction drive idler sheave. Belt guide should be square with belt.
27. Install muffler and exhaust pipe.
28. Align front bumper with chalk mark made during removal and install bolts and flat nuts.
29. Connect starter wire to starter. Connect diode wire and place wires in clip on engine.
30. Attach throttle cable and white wire to the throttle bracket. Adjust throttle cable.
31. Connect fuel line to the carburetor.

FUEL TANK

REMOVAL

1. Remove battery hold down.
2. Remove negative battery cable.
3. Remove positive battery cable.
4. Remove battery and battery base.
5. Remove fuel tank vent line.
6. Remove fuel line from filter and drain fuel tank.
7. Remove fuel tank.

INSTALLATION

1. Attach fuel tank to support with four self-tapping screws.
2. Connect fuel line to filter and replace vent line.
3. Install battery base.
4. Install battery unless it is to be serviced. Connect positive cable first and then negative.
5. Install battery hold down.
6. Install hood. Secure with springs.

Model 100

FRONT WHEELS AND AXLE

Wheels

REMOVAL AND INSTALLATION

1. Remove spindle cap.
2. Remove snapring and washer.
3. Remove wheel from spindle.
4. Lubricate wheel hub generously with an SAE multi-purpose-type lubricant.
5. Install wheel on spindle.
6. Secure wheel with washer and snapring. Install snapring with flat side toward end of axle.
7. Install spindle cap.

NOTE: Wheels should be installed with tire valve toward tractor.

Spindles

REMOVAL AND INSTALLATION

1. Remove jam nut from tie rod.
2. Remove snapring and washer.
3. Remove spindle from front axle.
4. Apply a light coat of an SAE multi-purpose-type lubricant to upper spindle shaft.
5. Install spindle on axle.
6. Secure spindle with washer and snapring.

Axle

REMOVAL AND INSTALLATION

1. Remove jam nuts.
2. Remove tie rods.
3. Remove cotter pin.
4. Remove slotted nut and king pin bolt.
5. Remove axle from tractor.
6. Position axle in tractor frame. Spindle grease fittings should be toward front of tractor.
7. Lubricate king pin bolt generously with an SAE multi-purpose-type lubricant.
8. Install king pin bolt. Secure king pin bolt with slotted nut and cotter pin.
9. Attach tie rods to steering arm with jam nuts.

Replacing Wheel Bearings

1. Drive old bearings out of wheel hub.

NOTE: When installing bearing, be sure notch in bearing is aligned with grease fitting in wheel.

2. Drive new bearings into hub with a driver of the appropriate size.
3. Lubricate bearings generously with an SAE multi-purpose-type lubricant before installing wheel on spindle.

STEERING

Steering Wheel

REMOVAL AND INSTALLATION

1. Remove steering wheel cap.
2. Loosen jam nut. Leave jam nut par-

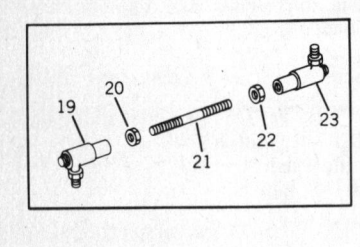

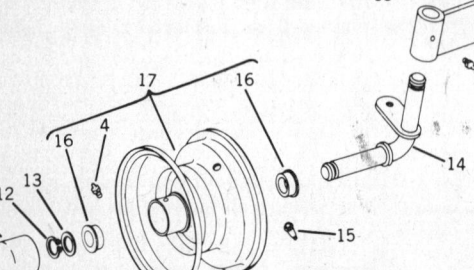

11. Front axle
12. Snapring (4 used)
13. Spindle washer (4 used)
14. Wheel spindle (2 used)
15. Tire valve
16. Wheel bearing (4 used)
17. Front wheel (2 used)
18. Spindle cap
19. Ball joint (heavy-duty)
20. Jam nut (heavy-duty)
21. Tie rod (heavy-duty)
22. Jam nut (heavy-duty)
23. Ball joint (heavy-duty)

1. ½ in. x 4 in. capscrew
2. ½ in. jam nut
3. Bearing
4. Grease fitting (5 used)
5. Steering arm
6. ⅜ in. UNF jam nut (6 used)
7. Tie rod (2 used)
8. ½ in. x 1¾ in. drilled bolt
9. ½ in. slotted nut
10. ⅛ in x 1 in. cotter pin

Front axle and wheels

John Deere

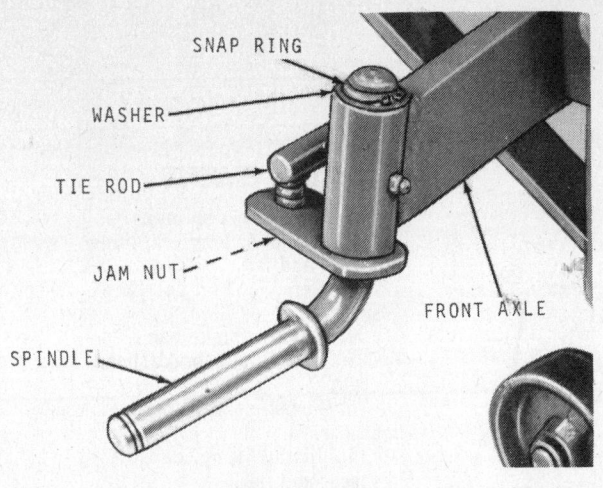

Installing spindle

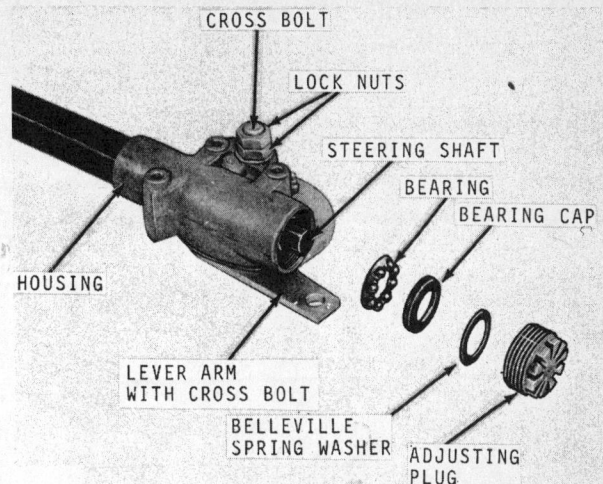

Steering gear components

tially on steering shaft to protect shaft from being damaged by wheel puller.

3. Install a wheel puller to loosen steering wheel.
4. Remove jam nut.
5. Remove steering wheel.
6. Place steering wheel on steering shaft so that center leg of wheel points down when front wheels are perfectly straight.
7. Install nut. Torque nut to 10-12 ft.lb.
8. Install steering wheel cap so that leaping deer is horizontal when front wheels are positioned straight ahead.

Steering Gear

REMOVAL AND INSTALLATION

1. Remove jam nut.
2. Remove three capscrews.
3. Remove steering gear from bottom of tractor.
4. Attach steering gear to frame with capscrews. Apply Loctite® or its equivalent to threads of screws.
5. Connect drag link to lever arm with jam nut. Tighten jam nut.

DISASSEMBLY

1. Remove locknuts from cross bolt. Remove lever arm and cross bolt from housing.
2. Remove adjusting plug from housing and slide shaft and bearings from column.
3. Apply grease and place bearing balls, ball cups and retaining rings on both ends of cam.
4. Grease cam lightly with multi-purpose-type grease.
5. Slide cam and tube assembly into housing, and jacket tube. Install bearing cap and Belleville spring washer over bearing. Install plug and tighten to 7-12 ft.lb. torque.
6. After torquing, lock adjusting plug with a cotter pin. Be sure steering shaft turns freely after torquing.
7. Before installing lever arm, back out stud. This will prevent damage to steering shaft cam when assembling.
8. Install new seal and retainer from repair kit.
9. Assemble lever arm, retainer, seal washer and jam nuts on housing.
10. Turn stud until contact with steering shaft cam is felt. Rotate steering shaft completely to the right and left, adjusting stud for light resistance at center point. Tighten stud locknut.

ADJUSTMENTS

1. Disconnect steering linkage from lever arm.
2. Loosen jam nuts on cross bolt.
3. Loosen locknut and tapered stud several turns.
4. Remove cotter pin (see inset) from adjusting plug.
5. Turn adjusting plug into housing to 7-12 ft.lb. torque.
6. Install cotter pin.
7. Connect steering linkage to lever arm.
8. To properly adjust backlash, turn steering wheel until lever arm is positioned with ball joint directly below pivot bolt.
9. Place a 0.100 in. thick shim between lever arm and housing face.
10. Tighten inside jam nut. Loosen inside jam nut just enough to remove shim. Tighten outside jam nut to 40 ft.lb. torque.
11. Turn tapered stud into housing until it is snug.
12. Tighten locknut to 40 ft.lb. torque.
13. Test adjustment by turning steering wheel through its full steering range in both directions. When properly adjusted, a slight drag can be detected at mid-point of range (when line between the pivot bolt and ball joint is vertical).

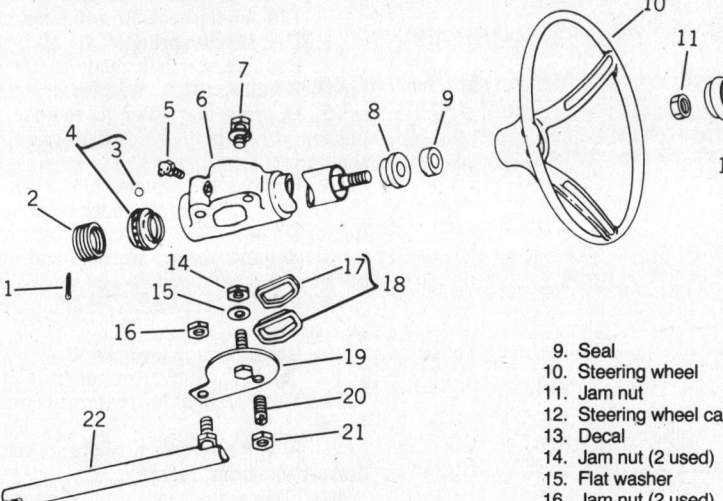

1. Cotter pin
2. Adjusting plug
3. Steel ball (6 used)
4. Bearing kit
5. Grease fitting
6. Steering gear housing
7. Capscrew
8. Bearing
9. Seal
10. Steering wheel
11. Jam nut
12. Steering wheel cap
13. Decal
14. Jam nut (2 used)
15. Flat washer
16. Jam nut (2 used)
17. Seal
18. Seal kit
19. Lever arm and cross bolt
20. Stud
21. Jam nut
22. Drag link

Steering linkage components

John Deere

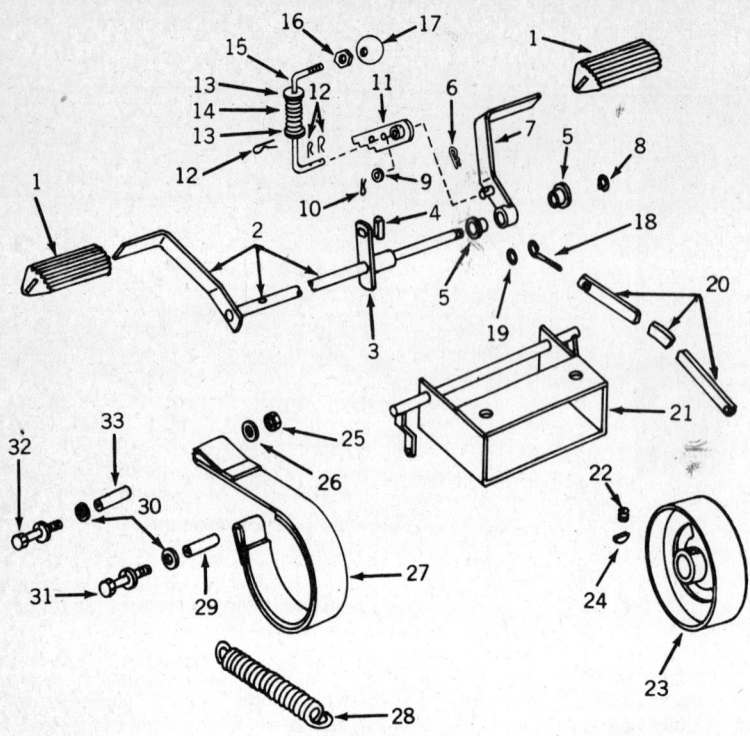

1. Pad (2 used)
2. Clutch pedal
3. Clutch arm
4. Groove pin
5. Bushing
6. Cotter pin
7. Brake pedal
8. Retaining ring
9. Washer
10. Cotter pin
11. Parking brake stop
12. Cotter pin (3 used)
13. Washer (2 used)
14. Spring
15. Brake lock rod
16. Nut
17. Parking brake knob
18. Brake rod
19. Washer
20. Brake strap
21. Pedestal
22. Setscrew
23. Brake drum
24. Key
25. Nut
26. Washer
27. Brake band
28. Spring
29. Spacer
30. Washer (2 used)
31. Capscrew
32. Capscrew
33. Spacer

Brake components

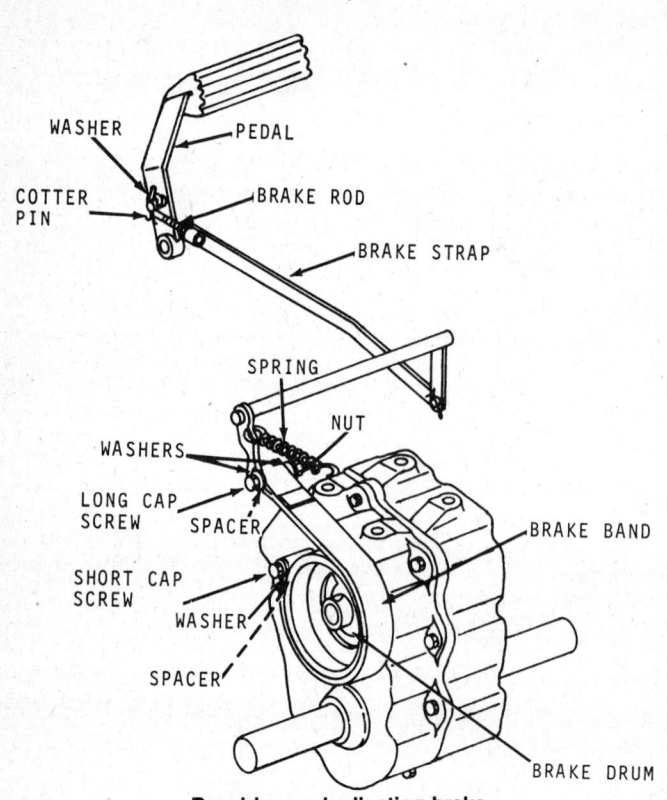

Repairing and adjusting brake

BRAKES

Brake Band

REPLACEMENT

1. Disconnect spring.
2. Remove long capscrew, washers, spacer and nut.
3. Remove short capscrew, washer and spacer.
4. Remove brake band.
5. Install new brake band over brake drum.
6. Install short capscrew, washer and spacer.
7. Install long capscrew, washers, spacer and nut.
8. Connect spring to brake linkage.
9. Readjust brake.

ADJUSTMENTS

1. Remove cotter pin and washer from pedal stud.
2. Remove brake rod from pedal.
3. Turn brake rod into brake strap to tighten, or out of brake strap to loosen brake adjustment.
4. Install brake rod on pedal stud.
5. Install washer and cotter pin.

NOTE: A brake that is adjusted too tightly causes simultaneous braking and driving which can seriously damage the transaxle.

TRANSAXLE

REMOVAL AND INSTALLATION

1. Remove fender-deck screws.
2. Lift fender-deck from tractor.
3. Disconnect spring.
4. Remove rear belt guide screws and rear belt guide.
5. Depress rear idler to remove secondary drive belt from transaxle drive sheave.
6. Remove hitch plate.
7. Place jack stands under each side of tractor frame.
8. Remove screw, washers and nut to disconnect brake band.
9. Remove capscrews holding transaxle to frame.
10. Remove shift lever knob.
11. Guide shift lever out of cam.
12. Roll transaxle rearward out of frame.
13. Remove screw, washer, spacer and brake band from transaxle.
14. Loosen setscrew.
15. Remove brake drum.
16. Remove key.
17. Loosen setscrews.
18. Remove drive sheave.
19. Remove key.
20. Remove wheels before disassembling transaxle.

John Deere

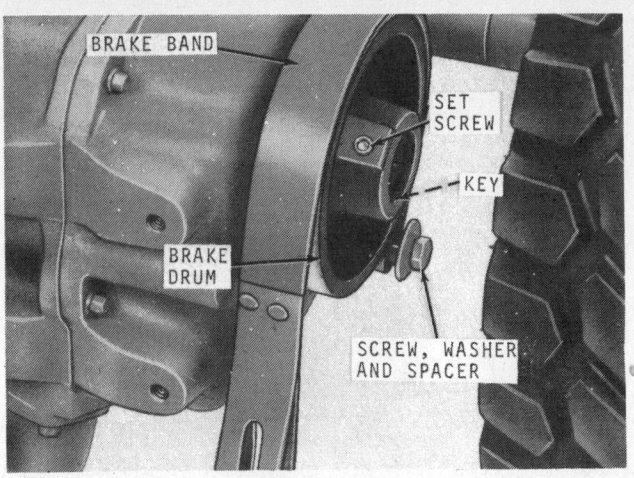

Brake removal

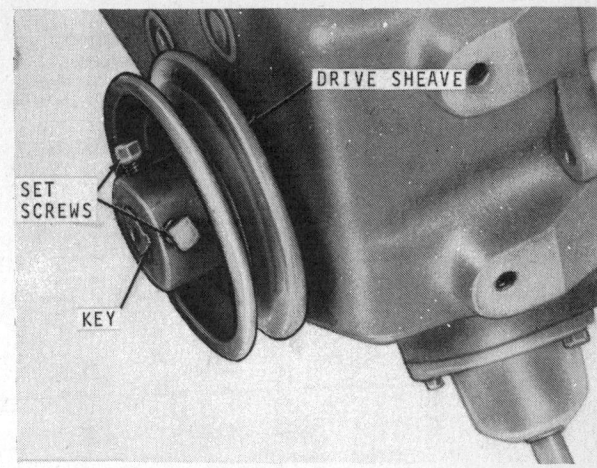

Removing drive sheave

DISASSEMBLY

1. For ease of transaxle disassembly construct a stand. As an option, two 2 in. holes cut into a work bench top will work satisfactorily.
2. Place transaxle in stand vertically with capscrews up. Remove eight capscrews. Leave dowel pins in place.
3. Tap cover to loosen, then grasp brake shaft and cover. Lift cover slowly, shaking gently, so all parts remain in the lower case.
4. To remove brake shaft from cover support pinion gear from bottom side and tap end of brake shaft with a mallet. The brake shaft is a light press fit into pinion gear and will slide out with minimum of force.
5. Lift gear group assemblies from case in the following order:
   a. Output shaft.
   b. Differential and axle assembly.
   c. Idler shaft.
   d. Reverse idler.
   e. Shifter shaft, shifter rods and forks.
6. To remove axle housing from transaxle case, position case halves in press.
7. To disassemble gear shift assembly, remove snapring and knob. Slip housing and retainer off shift lever.

Differential

1. Disassemble, clean and inspect parts shown. Look for gear wear and tooth damage. Replace worn parts and reassemble.
2. Apply Loctite® to carrier capscrew threads when assembling differential. Use lockwashers with carrier capscrew. Torque capscrews to 20 ft.lb.
3. The axles should rotate freely in opposite directions when assembled. Lay the differential assembly aside for later installation.

INSPECTION

1. Replace all gears having chipped, broken or worn teeth. Badly scored gears must be replaced.
2. Replace any shaft that is bent, scored or worn. Replace any shaft showing side wear or spline damage.
3. Chipped, broken or excessive wear on gear teeth ends is usually caused by shifting transaxle while tractor is still moving or by gears not being properly meshed when tractor is under load. Gear wear can cause gears to jump out of position.
4. Input shaft is a press fit installation. If close inspection reveals that gears and bearing are in satisfactory condition, DO NOT remove them from case.
5. Wash all internal parts in a safe cleaning solvent. Brush and scrape foreign matter from all parts and dry thoroughly. After all parts are thoroughly cleaned, oil them to prevent rust.
6. Broken detent springs can cause gear damage. When the springs are broken, the shifter fork is free to move, thus allowing gear pressure to slide the gears out of mesh.
7. When the gears slide out of gear, especially under load gear chipping or cracking will result.
8. Prolonged heavy drawbar loads and wheel slippage are the most common cause of bevel pinion gear failure in the differential section of the transaxle.
9. Damage to the input shaft spline is caused by improper coupling of the shifter shaft and input shaft when transaxle is shifted.
10. A broken detent spring or a loose shift lever and housing can cause improper coupling.

ASSEMBLY

1. Install new axle and brake shaft seals whenever transaxle is disassembled for repair.
2. Remove seals with a seal hook or driver.
3. Install seals with drivers of the proper size. Be sure seal lips face interior of transaxle.
4. To remove case needle bearings press bearings from case with a driver of the proper size.
5. To install a new case bearing, press bearing into case until open end of bearing is within 0.020 in. of interior of case bearing boss.
6. To remove axle shaft bushings cut bushings out of housing with a bushing cutter.
7. Install new pre-sized bushings with a driver of correct size.

NOTE: A driver of incorrect size will damage bushing. Drive bushings into their original depth.

8. Assemble input shaft and gear. Counterbored gear spline must face to right as shown. Gear is a press fit onto shaft.
9. Install input shaft and gear into case. Flat side of gear should now face upward.
10. Assemble idler shaft and gear assembly. These gears are a slip fit on the spline. Notice that raised hub of large gear faces short spacer. The teeth on the medium and small gear have round engagement edges that must face the large gear.
11. Spacers are of different length. Assemble them.
12. The long round end of the idler shaft turns in the bushing on the brake shaft. Be sure end of shaft is not battered.
13. To install axle housings into transaxle case halves, support case and press axle housing until firmly seated against flange.
14. Bushings and seals within axle housings can be replaced without removing housings from case half.
15. Install idler shaft and gears in case. Long end of shaft faces upward. Large gear engages the input gear.
16. Assemble reverse idler shaft assembly. Round edge of teeth faces spacer.
17. Install reverse idler as shown. Position spacer above gear.

NOTE: Because of heavy detent pressure, the assembly of these rods can be difficult.

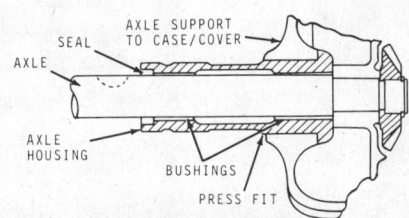

Replacing axle bushings

351

John Deere

1. Ball (2 used)
2. Detent spring (2 used)
3. Shifter fork (2 used)
4. Shifter rod (1st and reverse speeds)
5. Ring (2 used)
6. Shifter stop
7. Shifter rod (2nd and 3rd speeds)
8. Shift lever housing gasket
9. Shift housing snapring
10. Shift lever keeper
11. 3/16 in. x 1¼ in. spring pin
12. Gear shift lever
13. Quad rubber ring
14. Shift lever housing
15. ¼ in. x ¾ in. hex socket capscrew (11 used)
16. Gear shift lever
17. Shift lever knob
18. Oil seal
19. Needle bearing
20. Transmission housing with bearings
21. Reverse idler shaft
22. Reverse idler gear
23. Reverse idler spacer
24. Input shaft spur gear
25. Input shaft
26. Shifting gear
27. Shifting gear
28. Bearing
29. Shifter shaft and bearing assembly
30. Bearing (3 used)
31. Idler gear
32. Idler pinion and bushing assembly
33. Idler shaft
34. Spur gear (16 teeth)
35. Spacer
36. Spur gear (22 teeth)
37. Spacer 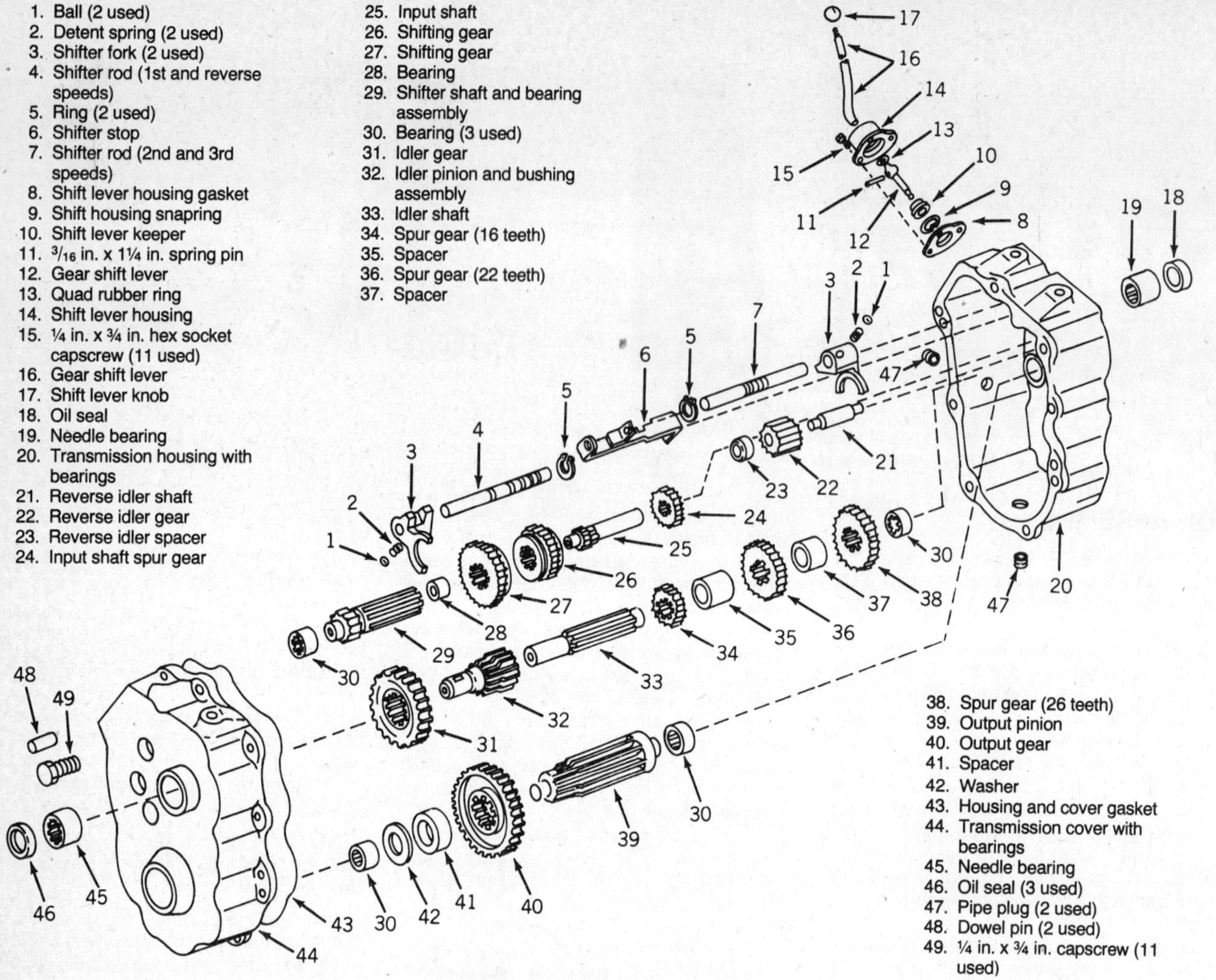
38. Spur gear (26 teeth)
39. Output pinion
40. Output gear
41. Spacer
42. Washer
43. Housing and cover gasket
44. Transmission cover with bearings
45. Needle bearing
46. Oil seal (3 used)
47. Pipe plug (2 used)
48. Dowel pin (2 used)
49. ¼ in. x ¾ in. capscrew (11 used)

3-speed transaxle components

18. Assemble forks as shown. Both forks should face to the right for assembly. The 2nd and 3rd rod must have the unequally spaced grooves at the top and away from the fork. The 1st and reverse rod must have the shortest ungrooved end face the fork. Start the rod into the fork. Depress detents and complete the assembly. Slide forks along rod. A good snap should be felt in each detent.

19. Place forks in center or neutral detent positions at this time.
20. Install output shaft assembly in case.

NOTE: If necessary, lift differential approximately ½ in. to mesh gears with output shaft.

21. Place gasket on the lower case.

Brake Shaft and Cover

1. Identify brake shaft and pinion gear.
2. Install brake shaft.

NOTE: If tapping brake shaft into place is necessary, properly support case and bearing to prevent damage.

3. Place pinion gear in case cover flat side up.
4. Insert brake shaft into pinion gear and

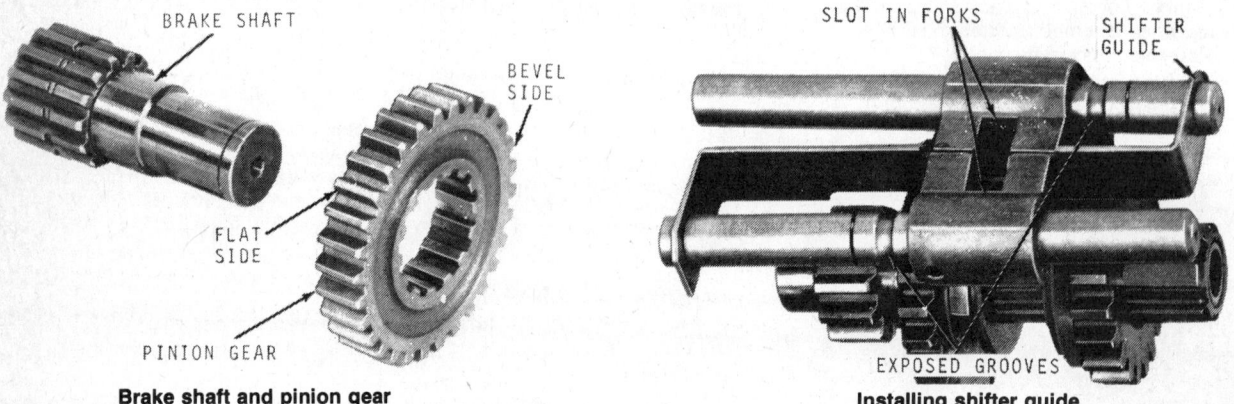

Brake shaft and pinion gear

Installing shifter guide

John Deere

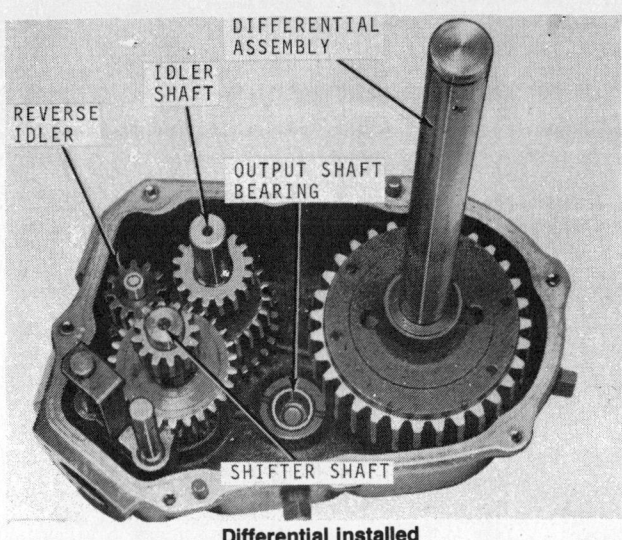

Differential installed

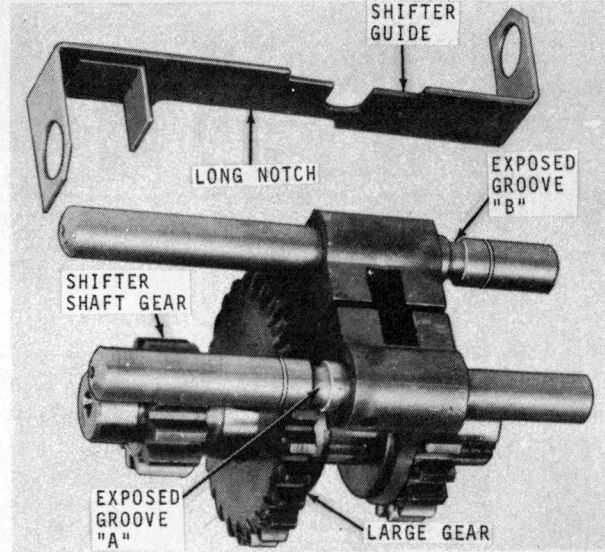

Assembling gears

case bearing, mate splines and press gear and shaft together until firmly seated.

Placing Cover on Case

1. Hold brake shaft and cover as shown.
2. Shake case lightly during assembly to align all shafts and bearings.
3. To close the last ½ in., tap cover horizontally at the corner indicated.
4. Lay parts on a bench exactly.

NOTE: Forks are assembled in the center grooves of shifters.

5. Assemble gears.

NOTE: Exposed groove "A" faces shifter shaft gear. Exposed groove "B" faces away from shifter shaft gear.

6. Move large gear toward shifter shaft gear as far as possible.
7. Hold gears in this position to install shifter guide.
8. Check that slots are aligned and exposed grooves are positioned.
9. Install shifter guide so that slot in guide matches slot in forks.

NOTE: The long notch in shifter guide should allow clearance for large (1st and reverse) gear.

10. Hold shifter assembly firmly.
11. Lower shifter assembly into case. The input shaft should enter the needle bearing. The shifter rods should rest in the two shifter rod sockets.
12. When installed, the shifter assembly should look exactly as shown.

Inserting Differential Assembly in Case

1. Install differential and axle assembly into the case with the bolt heads on the righthand carrier downward.
2. The assembly should now appear as shown.

Output Shaft

The output gear shaft is assembled on the pinion with a press fit. A slight looseness can be tolerated because the spacer and washer will hold the gear in place.

INSTALLATION

Installing Drive Sheave

1. Place drive sheave on input shaft of transaxle.
2. Align key ways to install key.
3. Tighten setscrews.

Installing Brake

1. Install brake drum on brake shaft.
2. Align key ways to install key.
3. Tighten setscrew.
4. Install brake band, spacer, washer, and screw. Tighten screw.

Installing Transaxle

1. Roll transaxle into position under frame.
2. Guide shift lever upward through cam.
3. Install capscrews with spring clip located with capscrew shown.
4. Torque screws to 35 ft.lb.
5. Install brake band, screw, washers and nut on brake arm. Tighten nut.
6. Install shift lever knob.
7. Install secondary drive belt on idler and transaxle drive sheave.
8. Install spring.
9. Install rear belt guide and rear belt guide screws. Adjust for 1/16 in. clearance. Tighten screws.
10. Install hitch plate.

Installing Fender-Deck

1. Place fender-deck on tractor frame.
2. Install fender-deck screws. Tighten screws.
3. Check brake adjustment.
4. Invert transaxle.
5. Install capscrews. Torque capscrews to 120 in.lb. torque.

Positioning Shifter Forks

Make sure shifter forks are in neutral position and aligned.

NOTE: If shifter forks are not in neutral and aligned before installing shift lever, damage will occur when the transmission is engaged under power.

Assembling Shift Lever

1. Assemble shift lever components in the order shown. Insert seal carefully into lever housing, use a small amount of gasket cement, if necessary, to hold in place during assembly.
2. Position shift lever assembly onto transaxle case. Check angle of lever in relation to case. Lever angle can be changed by removing snapring and rotating housing and retainer 180°.
3. When lever is in proper position, install three screws and torque to 120 in.lb.
4. Remove filler plug.
5. Add 3 U.S. pints of SAE 90 Gear Lubricant.
6. Install filler plug. Tighten plug.

CLUTCH

Removal

REMOVING PRIMARY DRIVE BELT

1. Depress idler sheave.
2. Remove primary drive belt from rear driven sheave.
3. Remove locking pin.
4. Disconnect clutch arm from fulcrum and clutch cone.

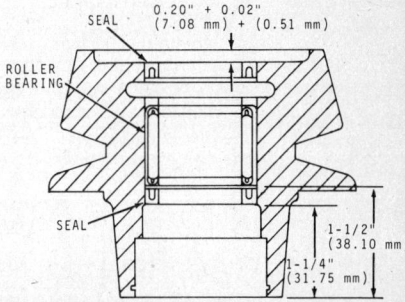

Installing roller bearing and seals

353

John Deere

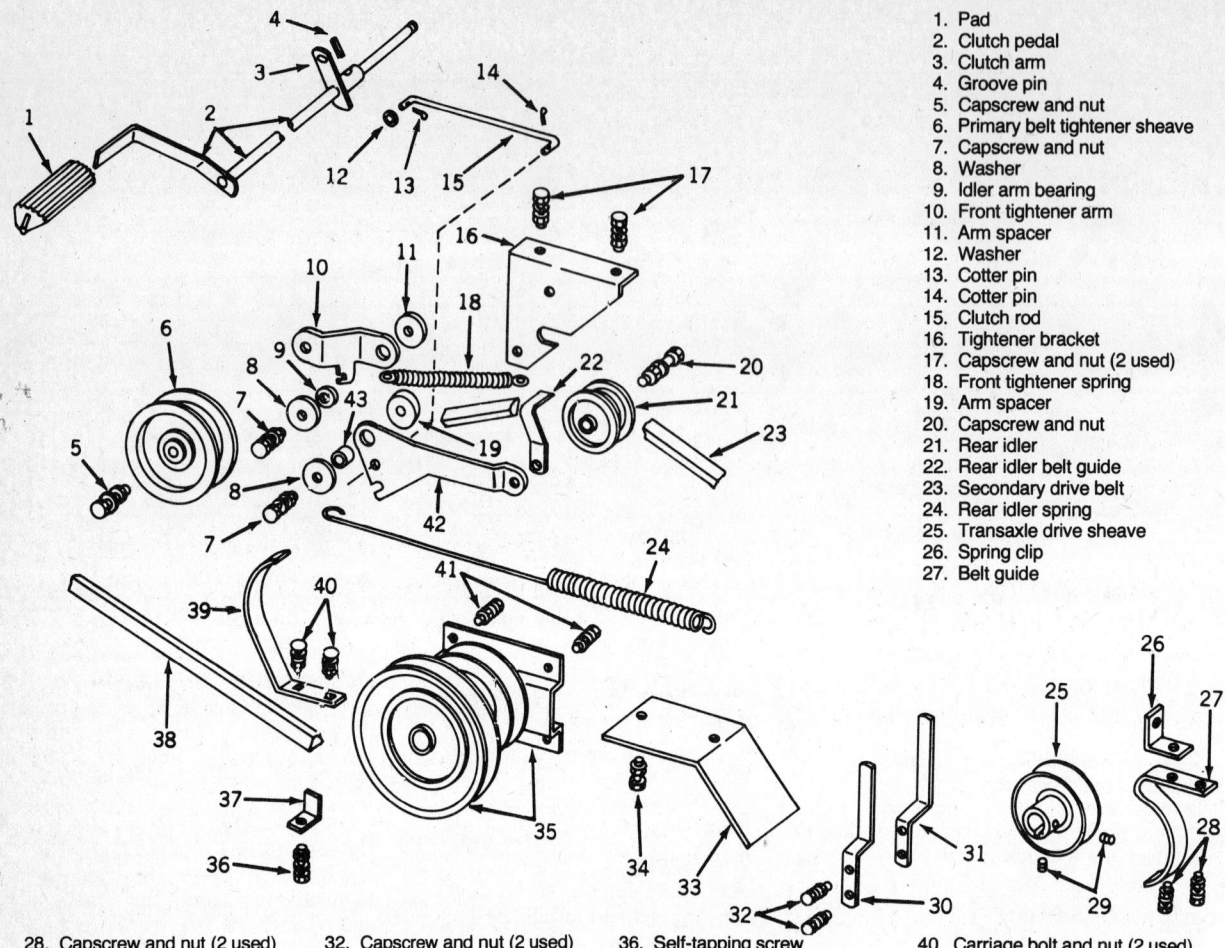

1. Pad
2. Clutch pedal
3. Clutch arm
4. Groove pin
5. Capscrew and nut
6. Primary belt tightener sheave
7. Capscrew and nut
8. Washer
9. Idler arm bearing
10. Front tightener arm
11. Arm spacer
12. Washer
13. Cotter pin
14. Cotter pin
15. Clutch rod
16. Tightener bracket
17. Capscrew and nut (2 used)
18. Front tightener spring
19. Arm spacer
20. Capscrew and nut
21. Rear idler
22. Rear idler belt guide
23. Secondary drive belt
24. Rear idler spring
25. Transaxle drive sheave
26. Spring clip
27. Belt guide

28. Capscrew and nut (2 used)
29. Setscrew (2 used)
30. Inner belt guide
31. Outer belt guide
32. Capscrew and nut (2 used)
33. Upper belt guide
34. Capscrew and nut (2 used)
35. Counter sheave assembly.
36. Self-tapping screw
37. Lower primary belt guide
38. Primary drive belt
39. Lower secondary belt guide
40. Carriage bolt and nut (2 used)
41. Capscrew and nut (4 used)
42. Rear tightener arm
43. Idler arm bearing

Clutch components

5. Remove capscrew and brake shoe.
6. Remove clutch cone.
7. Remove primary drive belt.

REMOVING
SECONDARY DRIVE BELT

1. Remove fender-deck.
2. Loosen capscrews and slide rear belt guide rearward. Remove belt guide at front of counter sheave. Depress rear idler.
3. Remove secondary drive belt from counter sheave and transaxle drive sheave.

Installation

INSTALLING
SECONDARY DRIVE BELT

1. Install secondary drive belt on transaxle drive sheave and counter sheave.
2. Depress rear idler to install secondary drive belt on idler. Move belt idlers within 1/16 in. of belt and tighten capscrews.
3. Install fender-deck.

INSTALLING
PRIMARY DRIVE BELT

1. Install primary drive belt in drive sheave.

2. Install clutch cone, brake shoe and capscrew.
3. Install clutch arm on clutch cone and fulcrum.
4. Install locking pin as shown.
5. Depress idler sheave to install primary drive belt on rear driven sheave. Adjust belt guide within 1/16 in. of belt. Tighten capscrews.

NOTE: Make sure twist in belt is exactly as shown.

6. Adjust PTO clutch and PTO clutch brake if necessary.

PTO CLUTCH

Removal

REMOVING
CLUTCH CONE

1. Depress idler sheave.
2. Remove primary drive belt from rear driven sheave.
3. Remove U-shaped locking pin.

4. Disconnect clutch arm from fulcrum and clutch cone.
5. Remove capscrew and brake shoe.
6. Remove clutch cone.
7. Remove primary drive belt.

REMOVING
CLUTCH CUP

1. Remove capscrew, spacer and clutch cup.
2. Remove Woodruff key from engine crankshaft.

Installation

INSTALLING
CLUTCH CUP

1. Install clutch cup, spacer and capscrew on engine PTO shaft.
2. Torque capscrew to 45-50 ft.lb.

INSTALLING
CLUTCH CONE

1. Install primary drive belt in drive sheave.

John Deere

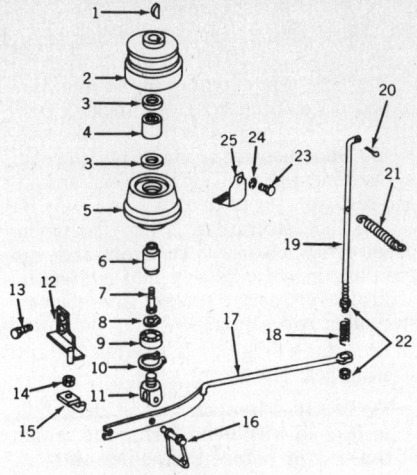

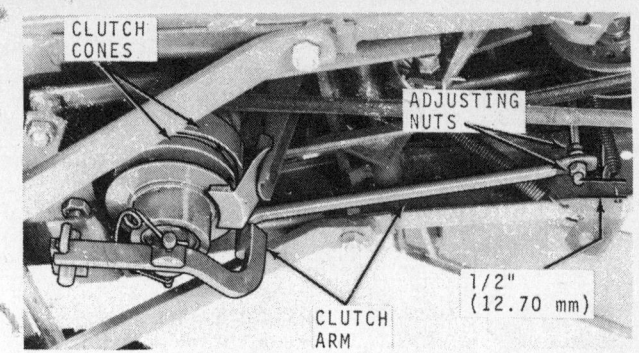

Adjusting clutch engagement

1. 5/32 in. x 5/8 in. key
2. Clutch cup
3. Oil seal (2 used)
4. Bearing
5. Clutch cone
6. Bearing race
7. 3/8 in. x 3 in. capscrew
8. Snapring
9. Bearing
10. Snapring
11. Clutch pivot
12. Fulcrum bracket
13. 5/16 in. x 5/8 in. capscrew (2 used)
14. 1/2 in. jam nut
15. Fulcrum
16. Locking pin
17. Clutch arm
18. Spring
19. Rod
20. 1/8 in. x 3/4 in. cotter pin
21. Spring
22. Nut, 5/16 in. (2 used)
23. 5/16 in. x 1/2 in. capscrew
24. Flat washer
25. Brake shoe

PTO clutch exploded view

2. Install clutch cone, brake shoe and capscrew. Tighten capscrew.
3. Install clutch arm on clutch cone and fulcrum.
4. Install U-shaped locking pin.
5. Depress idler sheave.

6. Install primary drive belt on rear driven sheave.

IMPORTANT: Make sure belt is installed exactly as shown.

Adjustments

ADJUSTING PTO CLUTCH

1. Engage PTO clutch and check that clutch arm is parallel with tractor frame and dimensions "A" and "B" are equal.
2. If not, remove U-shaped locking pin and lower clutch arm. Loosen jam nut, and turn fulcrum either up or down. Clutch arm should be approximately level when PTO clutch is engaged with clutch arm in place and locking pin installed.
3. Engage and disengage the clutch slowly. Observe at what point the clutch cone contacts the clutch cup. At this point the clutch arm should have raised 1/2 in. as shown.
4. This will provide proper tension of the clutch for positive engagement.
5. Turn the adjusting nuts clockwise to increase or counterclockwise to decrease the amount of clutch arm movement.
6. Engage PTO clutch and check that approximately 1/16 in. exists between clutch cone and brake shoe.
7. If not, loosen capscrew, raise or lower brake shoe until 1/16 in. clearance exists between brake shoe and clutch cone. Tighten capscrew.

ENGINE REMOVAL AND INSTALLATION

1. Remove secondary belt and then primary belt from the variator sheave. Loosen and unhook idler spring.
2. Remove hood, grille, and cowls.
3. Disconnect battery ground cable.
4. Shut off gas, remove gas tank, and fuel line to fuel pump.
5. Remove air cleaner assembly and disconnect choke and throttle cables from engine.
6. Disconnect two ground wires from rear cylinder head capscrew.
7. Disconnect starter cable at starter.
8. Remove mechanical PTO clutch arm and belt guide.

NOTE: On tractors with electric PTO, disconnect couplers on lefthand side of engine and from PTO clutch. Remove PTO wire from clips and tie strap by PTO clutch.

9. Remove capscrews securing engine to engine base.
10. Attach hoist to engine head bolt and lift engine, with primary belt, out of tractor.
11. Drain crankcase oil and clean engine exterior prior to disassembly.
12. Installation is the reverse of removal.

Model 120

FRONT WHEELS AND AXLE

Wheel, Spindle and Axle

REMOVAL AND INSTALLATION

1. Block up or hoist front of tractor until wheel clears the ground. Remove cap from wheel. Remove cotter pin, slotted nut, wheel and bearings from spindle inside cap.
2. To remove spindle from axle disconnect tie rod. Use snapring pliers and remove snapring and washer. Slip spindle out of axle.
3. To remove axle, block up or hoist front of tractor. Remove cotter pin from slotted nut. Remove nut and king pin bolt. Ease axle to floor. Remove tie rods from either steering arm or spindle arm, as desired.
4. Place axle end on press bed and press bushings out of axle.
5. Wipe axle bushing bore clean. Coat bushing with oil. Place axle on press and press bushing in axle until bushing is flush with axle face.
6. Place axle in a vise and turn reamer through axle bushings. Axle bushing diameter, new, is .751-.755 in.
7. Check king pin bushing and other king pin components for wear or other damage. Replace parts as necessary.
8. Grease king pin assembly and install axle on tractor frame. Secure king pin bolt with slotted nut and cotter pin.
9. Apply light coat of grease on upper

355

John Deere

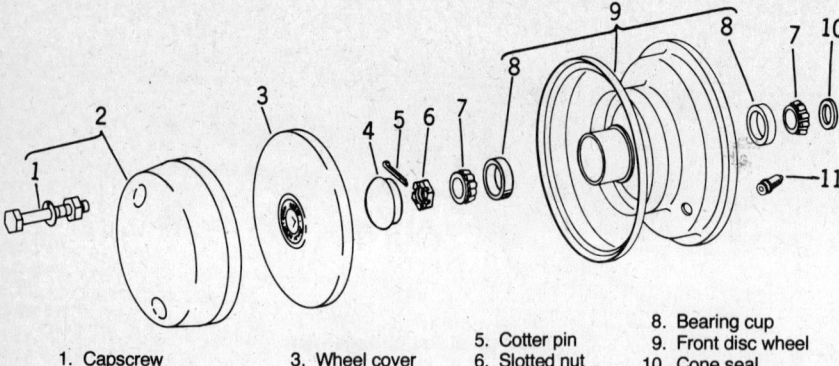

1. Capscrew
2. Front wheel weight
3. Wheel cover
4. Spindle cap
5. Cotter pin
6. Slotted nut
7. Bearing cone
8. Bearing cup
9. Front disc wheel
10. Cone seal
11. Tire valve

Front wheel components

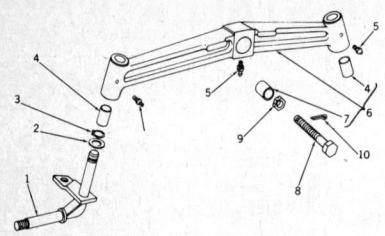

1. Front wheel spindle
2. Spindle washer
3. Snapring
4. Spindle bushing
5. Grease fitting
6. Front axle
7. King pin bushing
8. King pin bolt
9. Slotted nut
10. Cotter pin

Front axle components

spindle shaft. Install spindles into axle bushing.

10. Pack wheels with SAE multi-purpose-type grease. Install bearing with seal, wheel, outer bearing and slotted nut on axle. Adjust wheel bearings according to the following instructions. Place grease cap on wheel.

Front Wheel Bearing Adjustment

Adjust the front wheel bearings if the wheel is loose on the spindle or if the wheel does not rotate freely.

1. Raise the tractor until the front tires clear the floor.
2. Remove the grease cap from wheel.
3. Wipe the excess grease from the end of the spindle and remove cotter pin and slotted nut.
4. While rotating the wheel and tire, tighten the slotted nut to 60-120 in.lb. torque to seat the bearings. Back off slotted nut until wheel turns freely.
5. Using a $^{15}/_{16}$ in. open end wrench, back off the nut until the slot in nut aligns with cotter pin hole in spindle.
6. Install a new cotter pin and bend the long end of the cotter pin around the end of the axle.
7. Install cap.

STEERING

Steering Wheel and Gear

REMOVAL AND INSTALLATION

1. Remove steering wheel with puller. Remove battery, battery base and fuel tank.
2. Check instrument panel under the "John Deere" decal for two dimples, resulting from spot welding panel support bracket to panel. Center punch these dimples and drill a $^5/_{16}$ in. hole to eliminate weld.
3. Swing panel support bracket away from steering column. Remove clamp holding column to pedestal. Disconnect drag link from lever arm.
4. Remove three capscrews holding steering column to frame. Remove capscrew from hydrostatic control lever shaft bearing support last, using a ratchet and screwdriver. For ease of removal, an access hole can be drilled in the right side of pedestal to permit use of a socket extension.
5. Slide steering gear assembly upward in instrument panel. Lift lower end upward and remove over engine.
6. Slide steering gear into instrument panel grommet and lower front end into position. Install capscrew into hydrostatic control lever bearing support first, then install other two capscrews.
7. Attach drag link to steering arm and steering gear lever arm.

NOTE: It is important that drag link is positioned with bend facing the center of the tractor before tightening nuts.

8. Install clamp holding steering column to pedestal.
9. Adjust the steering gear mechanism.
10. Install instrument panel support bracket to instrument panel using two chrome-plated ¼ in. stove bolts. Paint slotted heads black to match instrument panel.
11. Install battery base, battery and fuel tank. Install steering wheel and tighten nut to 10-12 ft.lb. torque. Replace O-ring and steering wheel emblem.

Steering Gear

DISASSEMBLY

1. Loosen and remove both jam nuts from cross bolt. Slide cross bolt out of housing. Remove adjusting plug, Belleville washer, and bearing. Pull steering shaft out of gear housing.
2. Wash parts in a clean, safe solvent and dry with compressed air and clean cloth.
3. Check bearing condition. Inspect cam, housing and plug for cracks, scoring and other damage especially in the bearing

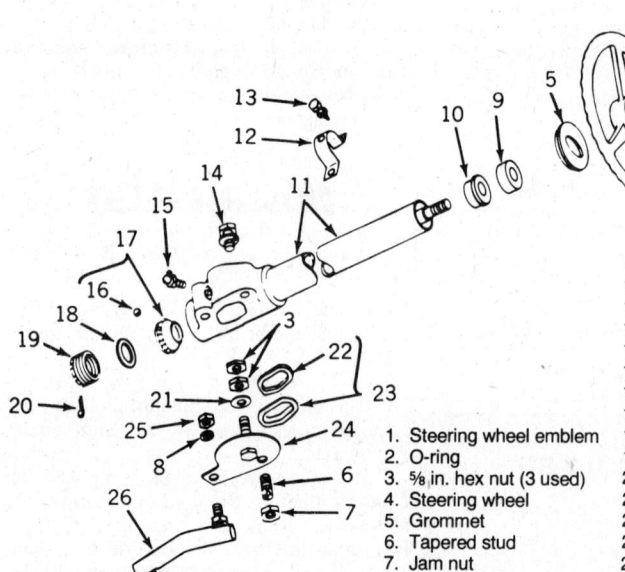

1. Steering wheel emblem
2. O-ring
3. ⅝ in. hex nut (3 used)
4. Steering wheel
5. Grommet
6. Tapered stud
7. Jam nut
8. Lockwasher
9. Bearing seal
10. Shaft bearing
11. Steering gear assembly
12. Steering column clamp
13. Screw
14. Capscrew and lockwasher
15. Grease fitting
16. Steel ball
17. Bearing kit
18. Belleville washer
19. Adjusting plug
20. Cotter pin
21. Washer
22. Lever arm seal
23. Retainer with seal
24. Lever arm with pivot bolt
25. Jam nut
26. Drag link assembly

Steering linkage components

356

John Deere

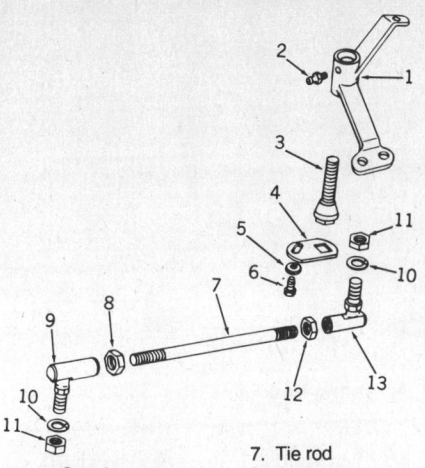

1. Steering arm
2. Grease fitting
3. Bolt with cone
4. Locking strap
5. Washer
6. Self-tapping screw
7. Tie rod
8. Jam nut
9. Ball joint
10. Lockwasher
11. Nut
12. Jam nut
13. Ball joint

Steering arm and tie rods

Adjusting steering cone

area. Replace parts showing excessive wear or damage.

4. Apply grease and place ball bearing, ball cups and retaining rings on both ends of cam.

5. Grease cam lightly with multi-purpose-type grease.

6. Slide cam and shaft assembly into housing and steering column. Install plug and tighten to 10-14 ft.lb. torque.

7. After tightening, lock plug with cotter pin. Be sure steering shaft turns freely.

8. Install new seal and retainer from repair kit. Attach lever arm to steering gear housing with washer and two jam nuts.

ADJUSTMENT

NOTE: It is important that the steering gear be adjusted in the sequence described below!

Cone Adjustment

1. Block up front end of tractor so that front tires are off the ground.
2. Disconnect ball joints at points "A".
3. Turn steering arm by hand and notice freedom of movement. When properly adjusted, the steering arm will pivot freely through the entire steering range with only a slight amount of drag.
4. If steering arm turns hard or has worn and loosened so that you can feel end play in the steering arm bearing, remove lock plate, cone bearing and pivot bolt. Lubricate bearing cone and install pivot bolt.
5. Tighten bolt only until a slight amount of drag can be felt when turning the steering arm and all end play has been removed.
6. Position lock plate over bolt head and tighten lock plate capscrew. Be sure plain washer is used with lock plate capscrew.

NOTE: If lock plate cannot be installed without turning bolt out of adjusted position, turn plate over.

7. Connect ball joints at points A.

Steering Gear Alignment

1. Steering effort is minimized when the steering gear is properly aligned. Always align steering gear as detailed below before adjusting the steering gear.
2. Visually check alignment of lever arm with steering gear housing. When properly aligned, lever arm will be parallel with the steering gear housing when the front wheels are pointed straight forward. Install drag link and tighten nuts securely.

NOTE: Be sure drag link is positioned with the bend pointing toward the center of the tractor, before tightening jam nuts.

3. Steering effort is minimized when the steering gear is properly aligned. The drag link is a one-piece, non-adjustable rod. If no parts have been bent or damaged, the steering gear will remain properly aligned as long as tie rods are adjusted to equal lengths.

Steering Gear Adjustment

Adjust steering gear as explained below to correct loose steering and steering wheel end play. Remove battery and battery base from tractor.

1. Disconnect ball joint from lever arm.
2. Loosen jam nut and stud two or three turns.
3. Remove cotter pin holding adjusting plug in steering gear housing. Use screwdriver socket and torque wrench to turn

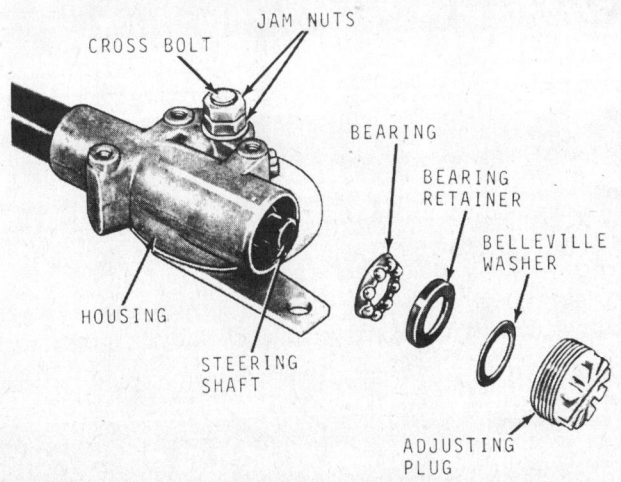

Steering gear components

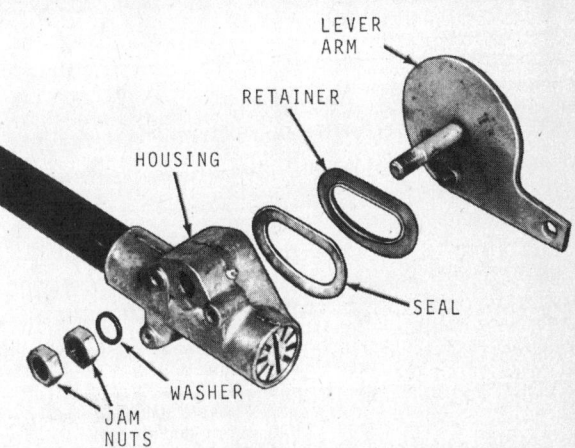

Attaching lever arm to steering gear

357

John Deere

Toe-in adjustment

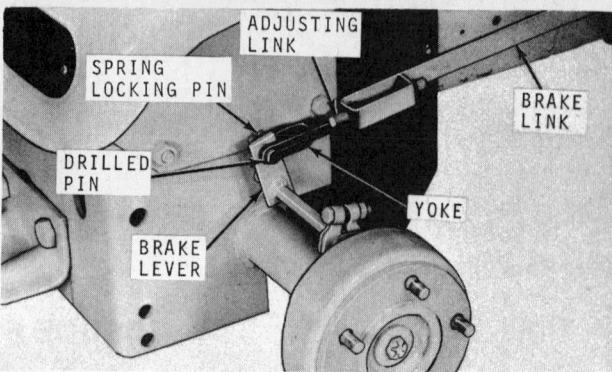

Adjusting brake link

adjusting plug into steering gear housing until end play is removed. Torque adjusting plug 10 to 14 ft.lb.

4. Turn plug only far enough after adjustment to insert cotter pin through housing and closest slot in plug. Spread cotter pin.

5. Loosen jam nut on pivot bolt and tighten only the inside nut using a thin open end wrench, until all end play (see arrow) is removed or until the distance between the lever arm and steering gear housing is between $1/16$ and $3/32$ in.

6. Turn steering wheel until the lever arm is parallel with steering gear housing.

7. Turn stud in (clockwise) until snug to remove all looseness. Then move steering wheel through its full steering range in both directions (right to left). When properly adjusted, a slight drag can be detected in the midpoint of the range (when line between the pivot bolt and ball joint is vertical). Tighten jam nut firmly. Make final test by turning steering arm through full range.

8. Connect ball joint to lever arm.

Front Wheel Toe-In Adjustment

Measure distances A and B. The tractor has proper toe-in or alignment when dimension A is $3/16$ in. less than dimension B. If adjustment is needed, loosen jam nuts and turn both right and lefthand tie rods C equally until proper toe-in is obtained. Tighten jam nuts firmly.

BRAKE SYSTEM

Brake Shoes and Drum
REPLACEMENT

1. Remove wheel hub retaining capscrews and washers and wheel hub. Install a puller similar to the one shown. If the puller being used does not have a broad push point, be sure to use a spacer of some sort to protect the wheel stud threads. With puller properly attached, remove drum.

2. Examine lining on brake shoes for wear and oil contamination. Replace shoes and linings if either condition is found.

3. To replace brake shoes, disconnect shoe pull-back springs and unhook hold down springs.

4. Replace brake drum if it is found to be worn or damaged.

5. To replace brake drum, press wheel studs out of drive hub. Press wheel studs through new brake drum and drive hub.

6. Assemble back plate to axle flange with four offset capscrews, lockwashers and nuts.

7. Before tightening capscrews, align

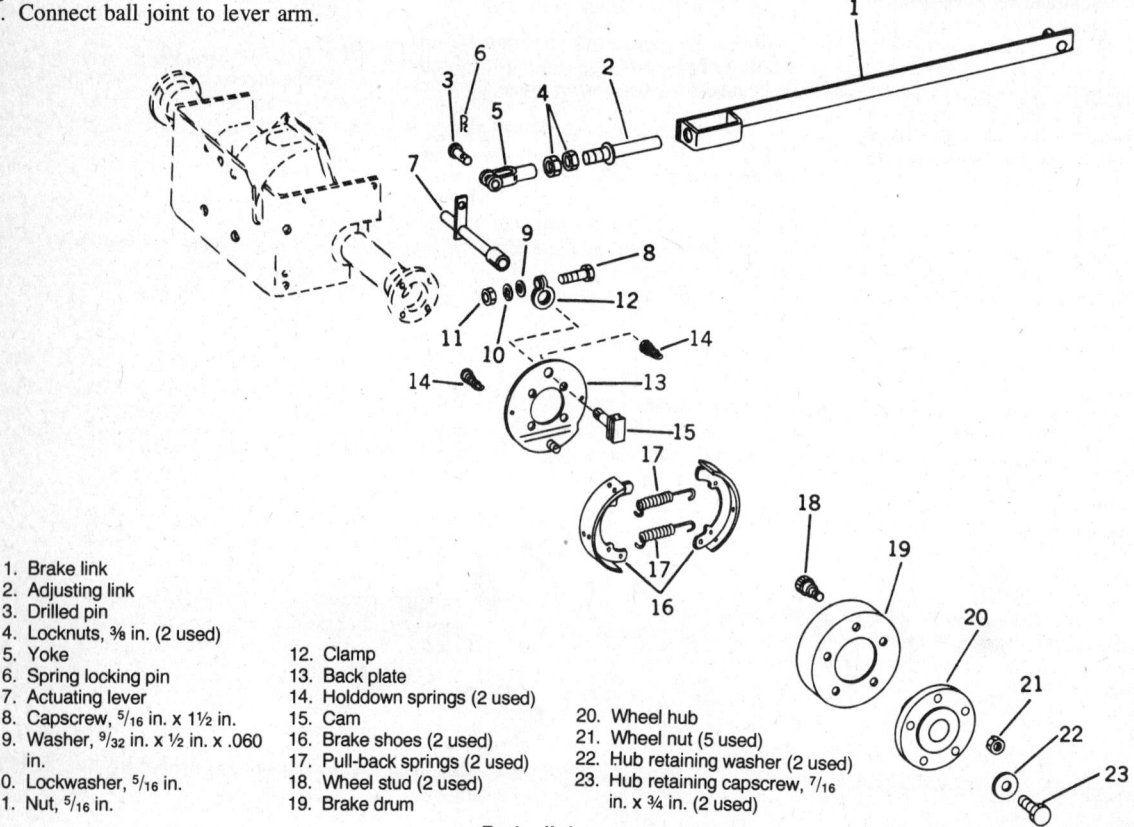

1. Brake link
2. Adjusting link
3. Drilled pin
4. Locknuts, 3/8 in. (2 used)
5. Yoke
6. Spring locking pin
7. Actuating lever
8. Capscrew, 5/16 in. x 1½ in.
9. Washer, 9/32 in. x ½ in. x .060 in.
10. Lockwasher, 5/16 in.
11. Nut, 5/16 in.
12. Clamp
13. Back plate
14. Holddown springs (2 used)
15. Cam
16. Brake shoes (2 used)
17. Pull-back springs (2 used)
18. Wheel stud (2 used)
19. Brake drum
20. Wheel hub
21. Wheel nut (5 used)
22. Hub retaining washer (2 used)
23. Hub retaining capscrew, 7/16 in. x ¾ in. (2 used)

Brake linkage components

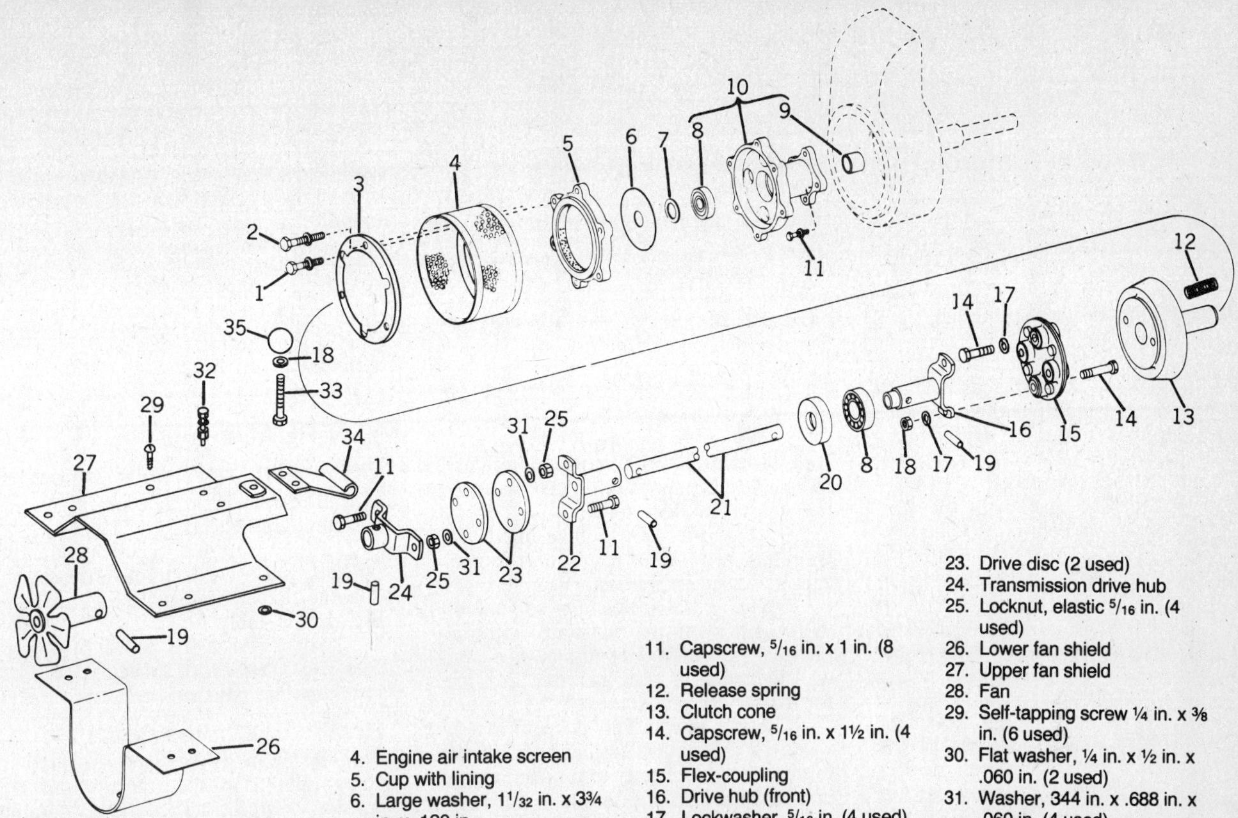

John Deere

Clutch components

1. Capscrew, 5/16 in. x 1¼ in. (3 used)
2. Capscrew, 5/16 in. x 1¾ in. (3 used)
3. Screen end cap
4. Engine air intake screen
5. Cup with lining
6. Large washer, 1 1/32 in. x 3¾ in. x .120 in.
7. Small washer, 1 1/32 in. x 1 3/8 in. x .060 in.
8. Throw-out bearing
9. Bushing
10. Drive hub assembly
11. Capscrew, 5/16 in. x 1 in. (8 used)
12. Release spring
13. Clutch cone
14. Capscrew, 5/16 in. x 1½ in. (4 used)
15. Flex-coupling
16. Drive hub (front)
17. Lockwasher, 5/16 in. (4 used)
18. Nut, 5/16 in. (2 used)
19. Spring pin
20. Bearing retainer
21. Driveshaft
22. Drive hub (rear)
23. Drive disc (2 used)
24. Transmission drive hub
25. Locknut, elastic 5/16 in. (4 used)
26. Lower fan shield
27. Upper fan shield
28. Fan
29. Self-tapping screw ¼ in. x 3/8 in. (6 used)
30. Flat washer, ¼ in. x ½ in. x .060 in. (2 used)
31. Washer, 344 in. x .688 in. x .060 in. (4 used)
32. Machine screw 3/16 in. x ½ in. (2 used)
33. Free-wheeling valve screw
34. Valve actuating plate
35. Knob

hub seal and back plate with JDST-9 seal tool.

NOTE: The seal tool is required to prevent installing the seal off-center.

8. Tighten 5/16 in. back plate retaining nuts to 15 ft.lb. torque.

ADJUSTMENT

1. Adjust brakes by shortening the brake link.
2. To shorten brake link, remove spring locking pin and drilled pin. Turn yoke onto adjusting link until drilled pin can be inserted without the brakes being applied. Install spring locking pin and test the brake by parking tractor on an incline.

CLUTCH

REMOVAL

1. Remove spring pins which secure the driveshaft hubs to driveshaft. Remove capscrews that attach rear driveshaft hub to flexible discs. Slide hub forward, deflect driveshaft to left side, and remove hub. With rear hub removed, the driveshaft can be pulled rearward and out.
2. After the driveshaft has been removed, the front hub and flex-coupling can be removed from the clutch cone. Remove two 5/16 in. retaining capscrews and move flex-coupling, hub and throw-out bearing assembly up and out.
3. To service the clutch assembly, it is recommended that the engine and clutch be removed as a unit from the tractor. To do so, first remove the engine mounting bolts. Disconnect fuel lines, electrical wiring, and control cables. Disconnect flex-coupling from clutch cone. Attach hoist and lift complete engine and clutch assembly forward and up.
4. Remove clutch assembly from engine by removing the four capscrews. Disassemble and inspect internal components.

ASSEMBLY AND INSTALLATION

1. With the proper size arbor or round shaft, press the bronze pilot bushing into drive hub flush to 1/8 in. from inside edge of drive hub. Invert the hub and press the ball bearing into the hub until it bottoms on the outer ring.

NOTE: When pressing in bronze bushing from the front of clutch drive hub, be sure to adequately support casting around ball bearing bore to prevent cracking the casting. When pressing ball bearing into place, press on outer ring only.

Cone

2. Install four clutch springs on cone studs. Install a 3¾ in. washer, then a 1 3/8 in. washer on pilot shaft. Install cone and pilot shaft assembly into drive hub, being sure springs are in their proper position.
3. Align drive hub on flywheel. Install four 5/16 x 1 in. capscrews in drive hub and flywheel. Tighten capscrews to 20 ft.lb. torque.

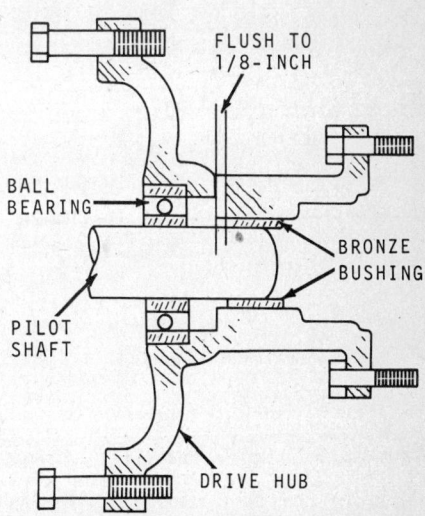

Installing bearings in drive hub

John Deere

4. Install cup on drive hub with three 5/16 x 1¾ in. capscrews and lockwashers equally spaced. The long capscrews are to pull the cup and drive hub together because the short capscrews would not reach the drive hub. Assemble the screen and end cap with three 5/16 x 1¼ in. capscrews and lockwashers. Tighten all 5/16 in. capscrews to 20 ft.lb. torque.

5. To install driveshaft assembly with the engine removed, install throw-out bearing on front drive hub. Insert driveshaft into hub and secure with spring pin.

6. Bolt flex-coupling to clutch cone and place driveshaft assembly in tractor.

7. Install engine and clutch assembly in tractor. Connect clutch linkage and attach rear driveshaft hub to flexible disc coupling on transmission.

8. Adjust clutch linkage.

9. Fill the engine crankcase, transmission, and gas tank. Replace the battery.

ADJUSTMENT

The clutch linkage is properly adjusted when the distance between the throw-out arm and the throw-out bearing is between 1/32 to 1/16 in., with clutch pedal in released (up) position. Adjust clutch when the above dimension exceeds 1/16 in. as follows:

1. Remove battery and battery base.
2. Loosen nut and move adjusting bolt in slotted hole until correct adjustment is obtained.
3. Tighten nut securely.

TRANSMISSION

REMOVAL AND INSTALLATION

1. Remove fender-deck assembly. Wipe dirt and grass from the transmission and the area around it.
2. Disconnect driveshaft at flexible coupling and PTO shaft extension, if so equipped. Disconnect control rod at cam. Remove brake link from lever.
3. Disconnect hydraulic oil lines.
4. Remove hitch plate capscrews and transmission mounting bracket capscrews.
5. With tractor properly supported and with the wheels still in place, roll axle, transmission, and hitch plate assembly rearward away from tractor.
6. Disconnect oil intake pipe from transmission and remove the four capscrews securing transmission to axle.
7. Use a safe cleaning solvent and thoroughly clean the outside of the transmission. Move the transmission to a clean, well lighted work area and use clean tools for the service procedure.
8. Use a new gasket between front and rear axle housings. Install housing capscrews, lockwashers, and nuts. Tighten nuts to 18 to 23 ft.lb. torque.
9. Apply a light coat of gasket sealer to axle mounting flange and position a new axle-to-hydrostatic unit gasket.
10. Place hydrostatic unit into position and install four capscrews and spacers. Tighten capscrews to 25 to 30 ft.lb. torque.
11. Install transmission mounting brackets on top mounting capscrews and secure with lockwashers and nuts.
12. Install oil intake tube, and oil filter (if filter had not been installed previously).
13. Move axle and transmission assembly into frame and install the following:
    a. Driveshaft assembly.
    b. Wheels
    c. Brake linkage
    d. Fender-Deck
    e. All control knobs
    f. Hydraulic lines
    g. Lubricant in axle housing

OVERHAUL

Disassembling Transmission

1. The outside of the transmission must be thoroughly clean.
2. Many internal parts have highly polished surfaces. Extreme care must be taken to prevent damage during disassembly and assembly.
3. To minimize the possibility of rust, coat your hands with transmission oil before handling polished parts.

Removing Charge Pump

1. Remove capscrews securing the charge pump to the transmission front housing. Carefully slide pump assembly off input shaft.
2. Remove pump drive pin from input shaft.
3. Remove oil filter.

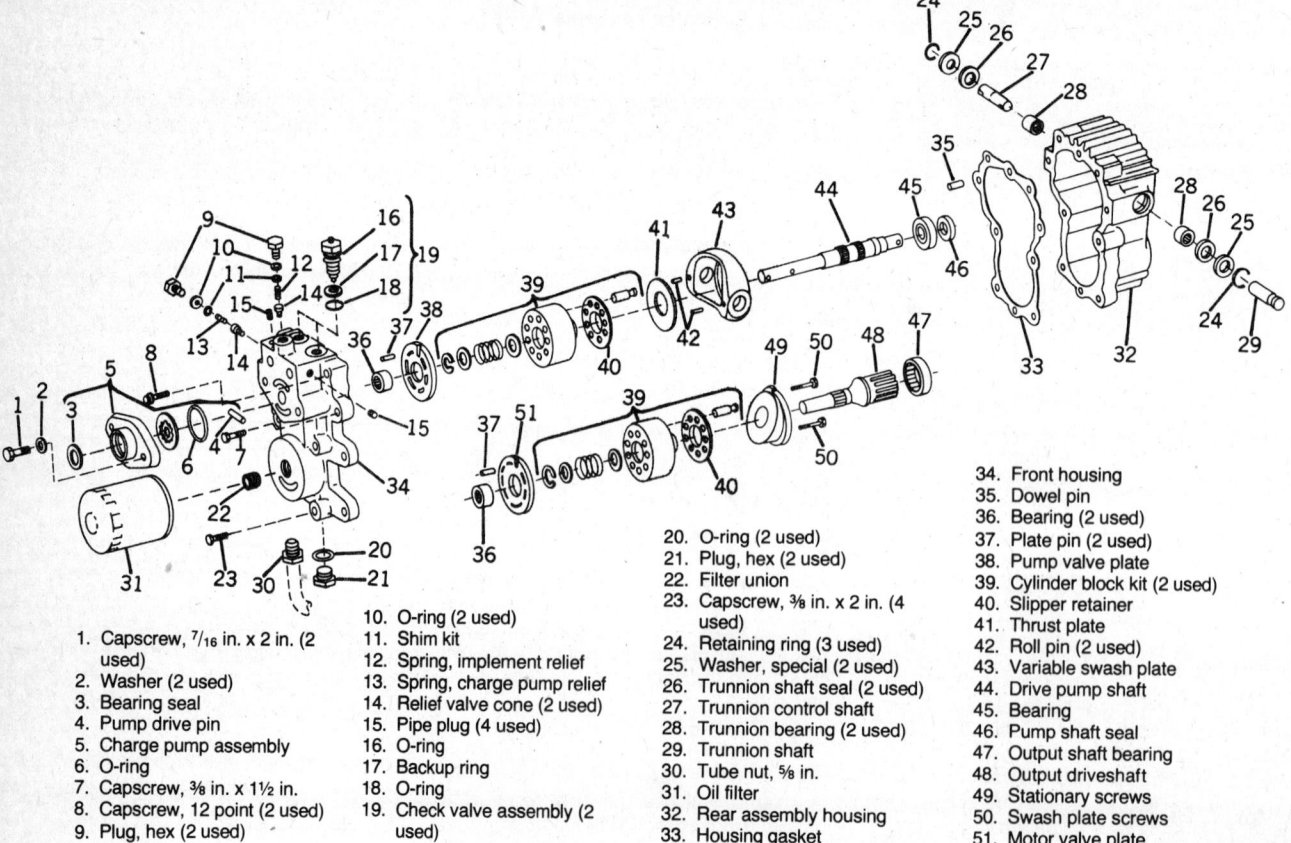

1. Capscrew, 7/16 in. x 2 in. (2 used)
2. Washer (2 used)
3. Bearing seal
4. Pump drive pin
5. Charge pump assembly
6. O-ring
7. Capscrew, 3/8 in. x 1½ in.
8. Capscrew, 12 point (2 used)
9. Plug, hex (2 used)
10. O-ring (2 used)
11. Shim kit
12. Spring, implement relief
13. Spring, charge pump relief
14. Relief valve cone (2 used)
15. Pipe plug (4 used)
16. O-ring
17. Backup ring
18. O-ring
19. Check valve assembly (2 used)
20. O-ring (2 used)
21. Plug, hex (2 used)
22. Filter union
23. Capscrew, 3/8 in. x 2 in. (4 used)
24. Retaining ring (3 used)
25. Washer, special (2 used)
26. Trunnion shaft seal (2 used)
27. Trunnion control shaft
28. Trunnion bearing (2 used)
29. Trunnion shaft
30. Tube nut, 5/8 in.
31. Oil filter
32. Rear assembly housing
33. Housing gasket
34. Front housing
35. Dowel pin
36. Bearing (2 used)
37. Plate pin (2 used)
38. Pump valve plate
39. Cylinder block kit (2 used)
40. Slipper retainer
41. Thrust plate
42. Roll pin (2 used)
43. Variable swash plate
44. Drive pump shaft
45. Bearing
46. Pump shaft seal
47. Output shaft bearing
48. Output driveshaft
49. Stationary screws
50. Swash plate screws
51. Motor valve plate

Hydrostatic transmission components

John Deere

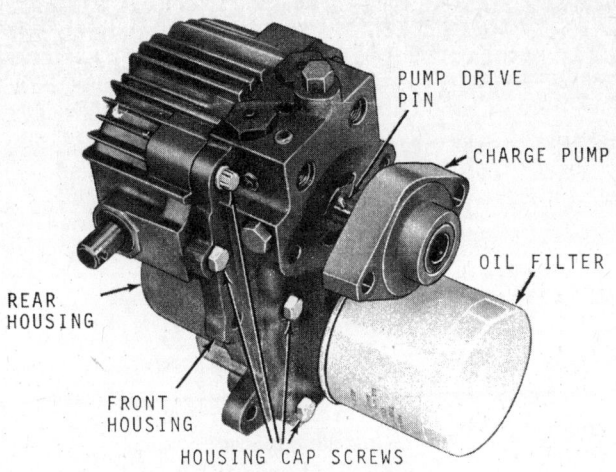

Removing charge pump

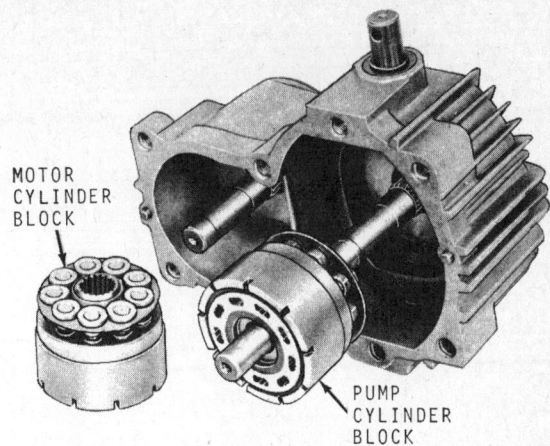

Removing cylinder blocks

Removing Front Housing
1. Remove the eight capscrews which retain the front housing to the rear housing.
2. Lift front housing from rear housing.

Removing Valve Plates
NOTE: The valve plates may stick to the front housing. Be extremely careful not to drop them.

1. Remove the pump and motor valve plates, noting the location of each plate.
2. The plate with two relief notches is used on the piston pump assembly and the plate with four notches is used on the motor assembly.
3. Remove plate anchor pins to avoid losing or dropping them during service.

Removing Cylinder Block
1. Tip the rear transmission housing when the blocks are being removed to prevent piston assemblies from falling out of the cylinder blocks.
2. Grasp the cylinder blocks and gently slide them off the shafts. Handle them carefully to avoid dropping and damaging the units.

Removing Motor Swashplate
1. Remove the two socket head capscrews retaining the motor swashplate.
2. Tip housing on its side and remove swashplate and motor shaft.

Removing Pump Swashplate
To remove the pump swashplate, drive spring pins into trunnion shaft until they are flush with the shaft surface.

NOTE: DO NOT drive spring pins more than $1/16$ in. below surface of trunnion shaft. If pins are driven too deep, further disassembly will be difficult.

INSPECTION

Pump Shaft
1. Remove and inspect pump shaft and bearing assembly. Replace bearing if rough or worn.
2. Inspect pump shaft bearing and seal surfaces. If shaft is found to be rough or grooved, replace shaft.

Cylinder Block Bore and Pistons
1. Gently lift all pistons. Check for free movement of pistons in cylinder block bores.
2. A scored piston or cylinder block bore will require replacement of the complete assembly.

NOTE: Do not interchange pistons between cylinder blocks. Pistons and cylinder blocks are matched.

Slipper Retainer
1. Check slipper retainer for flatness. Replace retainer if bent. Reusing a bent retainer can cause pistons to bind, scoring the cylinder block.
2. If slipper retainer is bent, check for the following possible causes:
   a. Charge pump inoperative.
   b. Wrong viscosity oil.
   c. Oil filter plugged.
   d. Air leaking into intake line.

Pistons and Slippers
1. If pistons have scored barrels or slippers have edges rounded more than $1/32$ in., replace the entire cylinder block assembly.
2. Inspect lubricant hole for blockage. If blocked, open with compressed air.
3. Individual pistons and cylinder blocks are not available for service because of the close tolerance-match fitting required for proper operation. A new cylinder block assembly complete with pistons must be installed.

Cylinder Block Face
1. Inspect polished face of cylinder

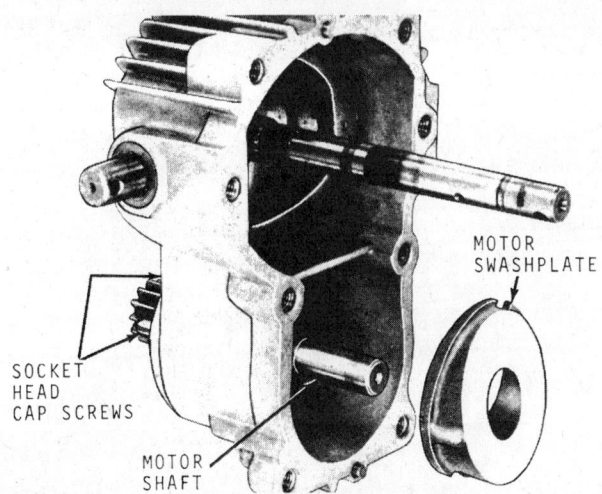

Removing motor swash plate

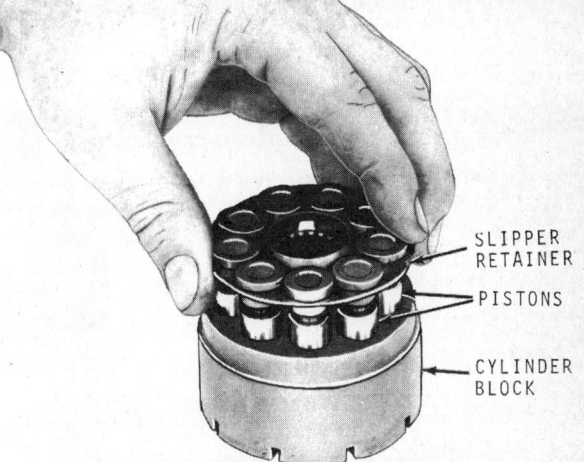

Lifting out pistons to inspect block

361

John Deere

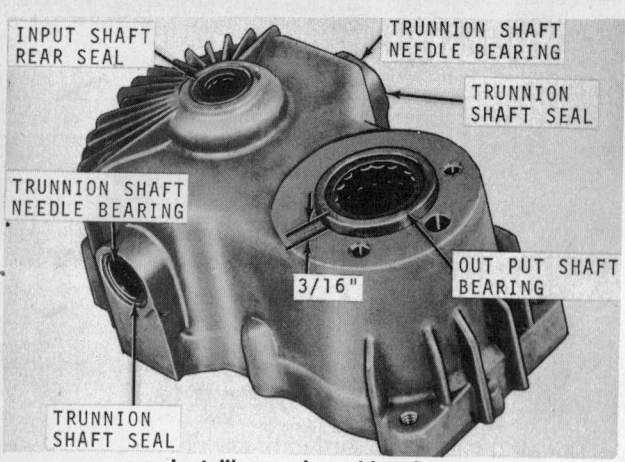

Installing seals and bearings

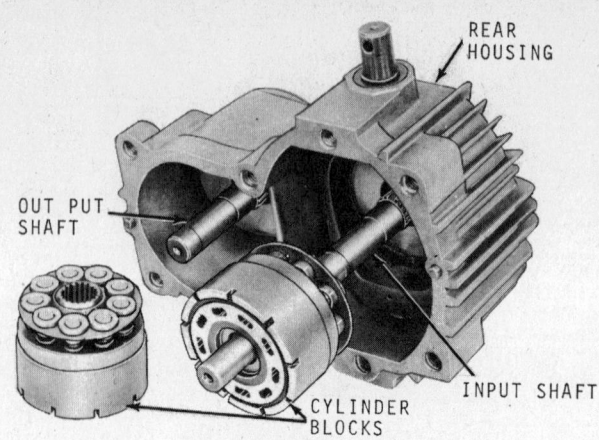

Installing blocks

block for scoring. Replace cylinder block assembly if scored.

2. Check spring for breakage. If broken, replace cylinder block assembly.

Valve Plates

1. Thoroughly clean valve plates and dry with compressed air.

2. Inspect valve plates for scratches, excessive wear or erosion. A worn or scored valve plate reduces pump efficiency.

NOTE: To check valve plates for wear, run your finger nail across face of plate. If wear is felt, replace the plate.

3. Inspect the valve plate pin slot and grooves. Remove any foreign material and deburr the surface, if necessary.

Swashplates

Inspect pump and motor swashplates for wear or scoring of the polished bearing surfaces. Replace plates if worn or scored.

Charge Pump

1. Inspect gerotor set, pump housing, and needle bearing for wear or damage. Replace entire charge pump if wear or damage is evident.

2. Always replace O-ring and pump seal before reassembling charge pump to transmission.

Check Valves

1. Remove and inspect both check valves for free movement of check ball.

2. Use new O-rings when reinstalling check valves. Note position of white back-up ring in lower O-ring groove. Replace this ring if damaged.

Relief Valves

1. Remove and inspect implement and charge pump relief valve cones and springs.

2. Check for rust and pitting of springs. Inspect cone points for wear. Replace springs and cones if worn, pitted, or rusty.

Housing Bearings

Inspect housing bearings for wear and loose bearing rollers. Replace any questionable bearings.

Housing Seals

To prevent costly rework, install new seals when assembling the transmission.

ASSEMBLY

After all parts have been cleaned and inspected, oil them lightly with John Deere Type 303 Special-Purpose-Oil or Automatic Transmission Fluid—Type "A".

Seals and Bearings

1. Install trunnion shaft needle bearings by pressing the bearings into the housing from the outside until they bottom in the bore.

2. Start trunnion shaft seals into housing bore by hand. Complete installation with a seal driver or pipe of the same diameter.

Drive seals in until they bottom against needle bearings.

3. When installing a new output shaft bearing press bearing into housing until 3/16 in. of outer race remains extended above mounting flange.

4. Install front housing bearings so that 7/64 in. of the bearing remains above the machined surface. This is important because these two bearings pilot the valve plates when unit is assembled.

5. Install input shaft bearing onto shaft by pressing on inner bearing race only. Lubricate bearing with oil and place shaft and bearing assembly into rear housing.

Variable Swashplate

1. Place pump swashplate into rear housing with the thin pad toward top of housing.

2. Position trunnion shafts. Tap shafts into swashplate using a soft metal hammer. Be careful to align spring pin holes in swashplate and shaft to facilitate installation of spring pins. Use a small drift punch to align the two parts before installing spring pins. Drive spring pins flush with the top of swashplate casting.

3. Place thrust plate into machined recess of swashplate with the highly polished surface of plate facing outward.

Motor Swashplate

Place output shaft into housing and insert motor swashplate with the thin edge toward center of housing. Secure plate with two

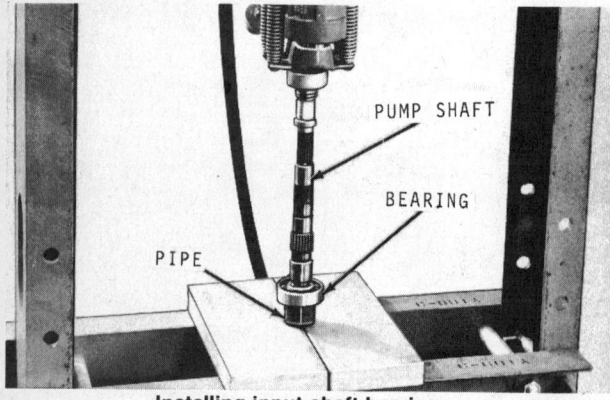

Installing input shaft bearing

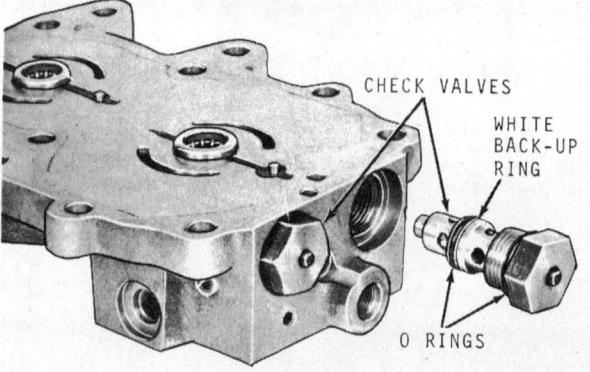

Installing check valves

John Deere

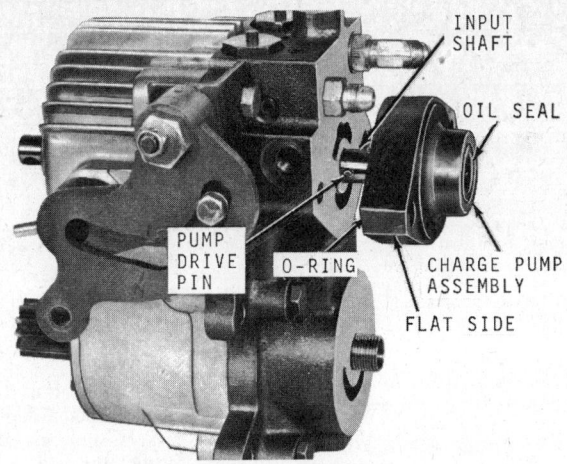

Installing charge pump

socket head screws. Tighten screws to 100 in.lb. torque.

Cylinder Block and Pistons

Lubricate the pistons. Put the slipper retainer and pistons into the cylinder block. Be sure each piston moves freely in its bore.

NOTE: DO NOT interchange pistons between cylinder blocks. Pistons and cylinder blocks are matched.

Installing Cylinder Blocks in Housing

Place rear housing on its side as shown. Install cylinder blocks. Pistons will then remain in their proper position until slippers are in contact with the swashplates.

Check Valves

1. Place new O-rings on check valves. Note position of white nylon back-up ring on lower end of valve. Be sure this ring is positioned correctly.

2. Dip check valves into clean transmission oil and install in housing. Both valves are alike and can be interchanged.

Relief Valves

1. Assemble relief valves in front housing by inserting cones, springs, and caps. Notice the difference in spring wire diameter. Place the spring with the heavier or larger diameter wire in the implement circuit, located at the top of the front housing.

2. Insert the lighter or smaller diameter spring in the charge pump relief circuit, located at the side of the front housing.

3. Place new O-rings on caps. Install and tighten caps firmly.

4. Shim kits are available to increase relief spring tension if testing reveals pressure readings below 75 to 110 psi for the charge pump and 500 psi for the implement circuit.

Valve Plates

Position valve plates on front housing. Apply a small amount of petroleum jelly between plate and front housing to hold plates in place during assembly.

Installing Front Housing

Place gasket on rear housing and carefully lower front housing until valve plates and cylinder blocks meet. Observe valve plates before bolting housings together to be sure valve plates are positioned properly.

Installing Control Cam Assembly

1. Install swashplate arm and cam spring on trunnion shaft and insert spring pin.

2. Position control cam and bracket assembly on housing and secure with housing capscrews. After tightening all housing capscrews to 35 ft.lb. torque, check for free movement of control cam. If binding is noted, loosen bracket capscrews and reposition bracket to relieve binding.

Installing Charge Pump

1. Install a new pump shaft oil seal prior

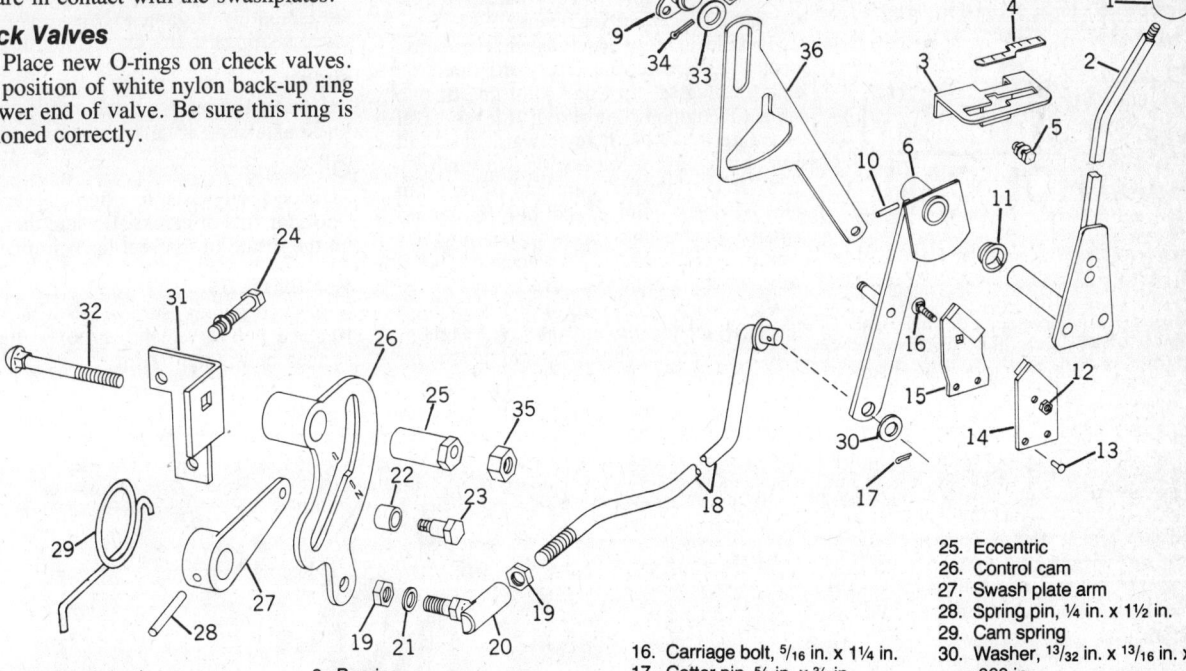

1. Plastic knob
2. Hydrostatic control lever
3. Control lever quadrant
4. Quadrant decal
5. Capscrew, ¼ in. x ½ in.
6. Control arm
7. Hex nut
8. Lockwasher
9. Bearing
10. Groove pin, ¼ in. x 1¼ in.
11. Nylon bearing
12. Elastic stop nut size
13. Rivet, btn. hd., 3/16 in. x 3/8 in.
14. Brake plate with shoe
15. Adjustable brake plate with shoe
16. Carriage bolt, 5/16 in. x 1¼ in.
17. Cotter pin, 5/8 in. x ¾ in.
18. Control rod
19. Jam nut, UNF, 3/8 in.
20. Ball joint
21. Lockwasher, 3/8 in.
22. Cam roller
23. Shoulder bolt
24. Capscrew, 12 point
25. Eccentric
26. Control cam
27. Swash plate arm
28. Spring pin, ¼ in. x 1½ in.
29. Cam spring
30. Washer, 13/32 in. x 13/16 in. x .060 in.
31. Cam bracket
32. Round bolt, 5/16 in. x 2½ in.
33. Washer, ¾ in. x 1 5/16 in. x .060 in.
34. Cotter pin, 1/8 in. x 1½ in.
35. Locknut 5/16 in.
36. Neutral cam

Control cam components

363

John Deere

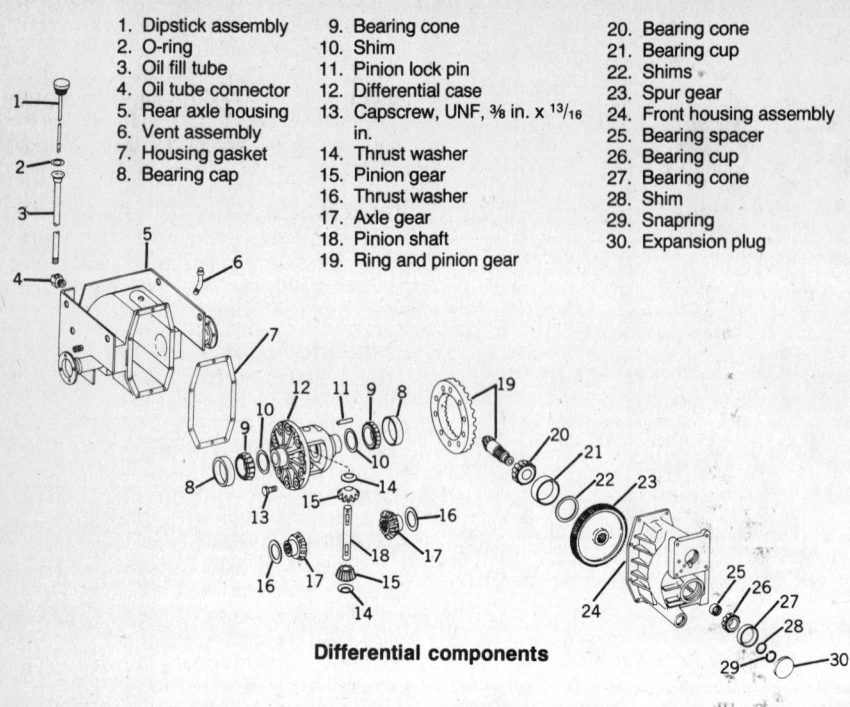

1. Dipstick assembly
2. O-ring
3. Oil fill tube
4. Oil tube connector
5. Rear axle housing
6. Vent assembly
7. Housing gasket
8. Bearing cap
9. Bearing cone
10. Shim
11. Pinion lock pin
12. Differential case
13. Capscrew, UNF, 3/8 in. x 13/16 in.
14. Thrust washer
15. Pinion gear
16. Thrust washer
17. Axle gear
18. Pinion shaft
19. Ring and pinion gear
20. Bearing cone
21. Bearing cup
22. Shims
23. Spur gear
24. Front housing assembly
25. Bearing spacer
26. Bearing cup
27. Bearing cone
28. Shim
29. Snapring
30. Expansion plug

Differential components

to installing pump on the front housing. Position O-ring on pump face and insert pump drive pin in shaft.

2. Slide charge pump assembly with gerotor set in place on input shaft. Be sure flat side of casting is on relief valve side.

3. Carefully position charge pump on front housing being careful not to dislodge O-ring. Install two 7/16 x 2 in. capscrews and tighten to 50 to 52 ft.lb. torque.

CONTROL CAM

ADJUSTMENT

Whenever the control cam or linkage has been removed or disconnected, make the following adjustments to eliminate forward or reverse creep. With tractor on level surface, place the hydrostatic control lever in neutral:

1. Block one rear wheel off the ground. Block under the frame only, not under the transmission.

2. Close the free wheeling valve, release parking brake, and place hydrostatic control lever in neutral.

3. Remove control rod ball joint and align cam roller with neutral indicator mark (N) on cam.

4. Holding the cam in this position, adjust ball joint to drop into its hole in cam. Install lockwasher and nut and tighten jam nut. Start the engine.

5. Run engine at idle. With 7/8 in. open end wrench, turn eccentric bearing nut forward until wheel creeps, then rearward until wheel creeps. Adjust for center position of wrench swing.

NOTE: This should place the cam roller in the center of the neutral range to provide equal forward and reverse travel of the hydrostatic control lever before wheel rotation occurs.

6. With the tractor still blocked up and engine running, move control lever forward out of neutral. Check the distance that lever travels before wheel rotates. Follow the same procedure for reverse.

7. If lever travel (before wheel rotation occurs) is greater in a forward direction than in reverse, turn ball joint one or two turns to lengthen control rod. If lever travel is greater rearward than forward, turn ball joint onto control rod one or two turns.

8. Reconnect ball joint and tighten jam nut. Shut off engine and tighten nut on small bolt through eccentric bearing nut. Repeat steps 6 and 7, if necessary, until lever travel is equal.

9. Remove blocks and test tractor on a level surface for forward and reverse creep. Check at various engine operating speeds.

DIFFERENTIAL AND AXLE

Axle

DISASSEMBLY

1. To remove the righthand axle hub, remove 7/16 x 3/4 in. capscrew and washer from center of hub. Install a universal hub and drum puller, and pull hub from axle shaft.

2. Remove the four 5/16 in. nuts, lockwashers and special capscrews. Remove seal, gaskets, and bearing retainer.

3. Service the lefthand axle in the same manner described above except that the hub can be removed with either a puller or a press.

4. Grasp end of axle shaft and pull outward. Axle bearing and races are cemented together with an epoxy adhesive, and in most cases will remain together. In case bearing and race part remove bearing race from housing with your fingers.

5. Press axle bearing from axle shaft by supporting inner bearing race on press bed and applying pressure to splined end of shaft.

INSPECTING AXLE PARTS

Roller Bearings

Clean bearings in a safe solvent. The bearings should be free from rust and other foreign material. The bearings should rotate smoothly without excessive play. Replace bearings if any defects are detected.

Axles

Inspect splines for wear or breaks. Replace axle shaft if defects are evident.

Oil Seals

Inspect oil seals for signs of leaking. Look for cuts or cracks. Be sure the spring on the inside of the seal lip is in place.

Wheel Hubs

Inspect sealing surface of wheel hub. Replace hub if seal has grooved the hub surface more than 1/64 in.

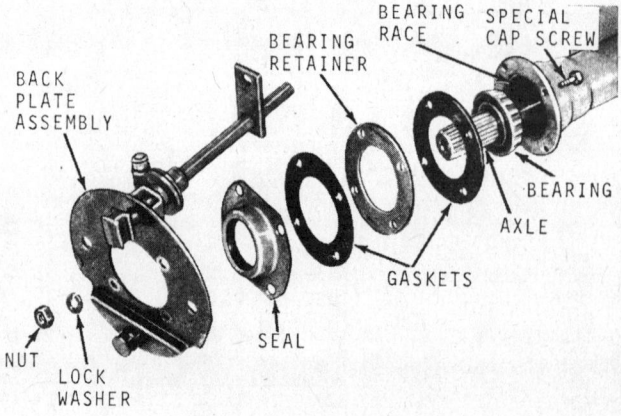

Removing axle and seal

John Deere

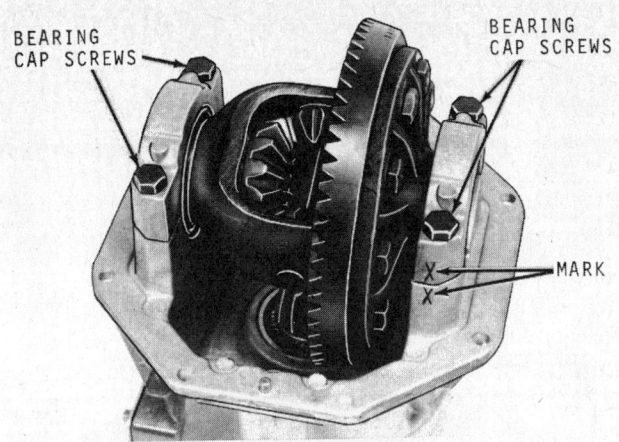
Removing bearing caps

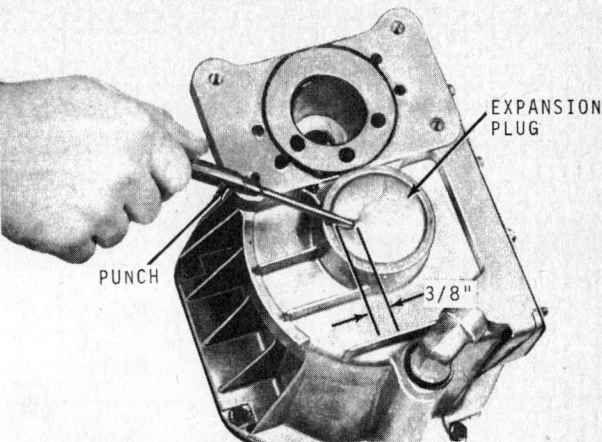
Removing pinion expansion plug

ASSEMBLY

1. Place bearing on axle shaft with locking ring to the outside. Position bearing and axle shaft in press. Press bearing onto shaft until firmly seated against shaft shoulder.
2. Slide O-ring on axle shaft until it seats against axle bearing inner race.
3. Assemble gaskets, bearing retainer, seal, and back plate.
4. Install a felt seal on wheel hub.
5. Position hub and drum on axle shaft and push on by hand as far as possible.
6. Using a $7/16 \times 2$ in. NF capscrew and nut, pull hub and drum onto axle shaft.

NOTE: Do not hammer hub and drum on axle shaft because it will damage bearing.

Differential

REMOVAL

1. Place blocks under side frame rails and remove fender-deck assembly. Disconnect brake linkage at brake arm and transmission linkage at control cam ball joint.
2. Disconnect hydraulic oil lines at transmission.
3. Remove six self-tapping screws from top fan shield and remove shield. Drive spring pin from fan and remove fan and bottom fan shield.
4. Remove transmission guard (if tractor is so equipped) and transmission mounting brackets. Drive both spring pins from rear flexible coupling hubs and slide coupling assembly forward.
5. Remove the four $½ \times 1$ in. capscrews securing rear axle and hitch plate assembly to frame side rails.
6. Tip transmission and axle assembly slightly forward to clean control cam attaching bolt. Slowly and carefully roll entire assembly rearward.

NOTE: Support axle adequately so that it doesn't fall forward from frame when bolts are removed.

7. Drain oil from axle and transmission by removing oil cooling tube and oil filter.
8. After units are drained, remove four transmission mounting capscrews, and lift transmission from front axle housing.

NOTE: Protect transmission from contamination by installing oil filter. Cover oil outlet fittings.

DISASSEMBLY

1. Remove right and lefthand axle assemblies.
2. Remove eight $5/16 \times 1¼$ in. housing capscrews, and separate front and rear axle housings.
3. Mark one bearing cap prior to cap removal.
4. Remove four bearing capscrews and remove caps. Place caps in a safe place to avoid damaging their machined surfaces.
5. To remove differential assembly, place two wooden handles under differential case and pry sharply upward.
6. Prying is required because of the built-in bearing preload of housing.
7. Using a long thin drift punch, drive pinion pin out of pinion shaft.
8. Work on a clean surface and protect bearings from contamination.
9. Support differential properly to avoid bearing or ring gear damage during disassembly.
10. Drive pinion shaft from differential case with a long drift punch.
11. Avoid ring gear damage. Be careful not to slip off drift punch with hammer.
12. Remove pinion gears and thrust washers by rotating both gears 90 degrees to the openings in differential case.
13. Remove axle drive gears and thrust washers from case.
14. Do not remove bearings from differential case unless bearing failure is evident because bearings are easily damaged in removal.
15. Remove case side bearings with a narrow jaw puller. Be sure to insert jaws into indentations provided in the differential case.
16. Remove 10 ring retaining capscrews. Using a hardwood block and ham-

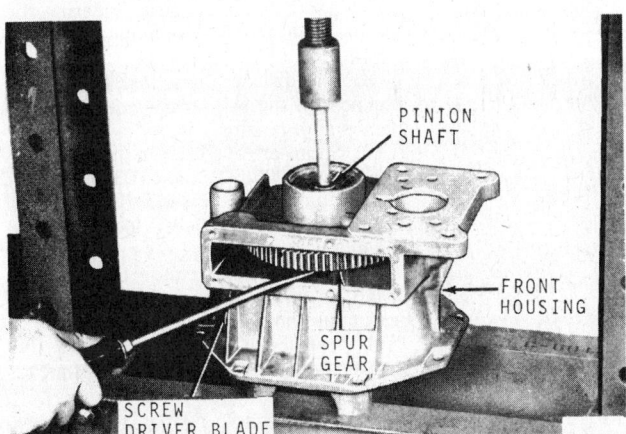
Removing differential pinion drive gear

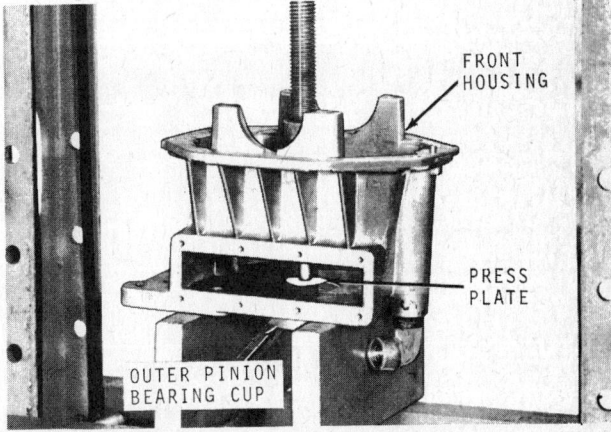
Removing outer pinion bearing cup

365

John Deere

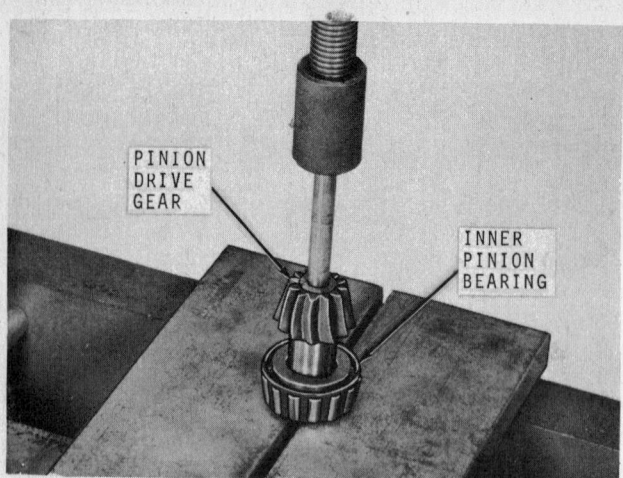
Installing inner pinion bearing

Installing outer pinion bearing cup

mer, drive ring gear off differential case.

17. Remove pinion expansion plug by driving a pointed punch through plug about ⅜ in. from outer edge. When hole is large enough, insert a large screwdriver and pry plug outward.

18. To remove pinion drive gear, remove snapring and shim from end of pinion. Position housing and pinion drive gear assembly on hydraulic press bed.

19. Before starting to press gear, remove side cover and place a ⅛ in. piece of steel or a screwdriver blade under the edge of spur gear. This will prevent spur gear from cocking and possibly cracking housing.

20. When pinion gear is close to being pressed completely out of bearing, reach under housing and catch gear in your hand to prevent damage to gear.

21. Removing the pinion drive gear releases the spur gear, spacer, and outer pinion bearing for removal.

22. Clamp inner pinion bearing in a universal gear remover with a thin edge.

23. Position unit in press and carefully push pinion drive gear out of bearing. DO NOT allow pinion drive gear to drop on the floor. Damage will result.

24. To remove outer pinion bearing cup, position housing in press. Place a press plate of the proper size against cup. Press cup out of housing.

25. Position front housing on press bed with bearing saddles resting on press bed. Protect bearing saddles with a strip of wood if press bed is rough.

26. Insert a press plate of proper size and press bearing cup toward inside of housing. Retain shims located under bearing cup.

INSPECTION

Bearings

1. Inspect all bearing rollers and cups for galling, rust or flaking.
2. Replace any bearing that is discolored or looks questionable.

Gears

Check ring, pinion, and pinion drive gears for abnormal wear and damage. Replace if worn.

Inspect spur gear for spline wear and tooth wear. Replace if worn.

Axle Housings

Inspect housings for cracks and external damage that could affect the operation of axle assembly.

Differential Case

Inspect differential case for wear in the axle gear and pinion gear area. Replace case if machined areas are scored or if pinion shaft fits loosely in bore.

ASSEMBLY

1. Press inner pinion bearing on pinion drive gear. Support bearing on inner cup only when installing.

2. Position front housing on press, and using a press plate, push outer pinion bearing cup into housing until it bottoms in housing.

3. When installing a new inner pinion bearing and cup in the original housing, reuse the original shim pack.

4. Use a press plate to push bearing cup into housing until it bottoms against housing flange.

NOTE: If a new housing is installed, measure for proper shim pack thickness as explained below.

5. When installing a new front housing, press inner pinion bearing cup into housing without shims. Place pinion drive gear (with bearing installed) into cup.

6. Position JDST-10 pinion depth gauge into differential bearing saddles.

7. Use a feeler gauge to measure the distance from tool depth pin to pinion gear button face.

8. The distance measured will be the thickness of the shim pack required under inner pinion bearing cup.

9. Shims are available in the following sizes: .003, .005, .010, and .030 in. Select and combine shims as needed to equal the measured distance between tool depth pin and pinion gear.

10. Remove inner pinion bearing cup. Install required shim pack and reinstall bearing cup into housing.

11. Insert spur gear into front housing with chamfered area of center spline toward pinion drive gear.

12. Insert pinion drive gear into spur gear. Place spacer over pinion drive gear shaft.

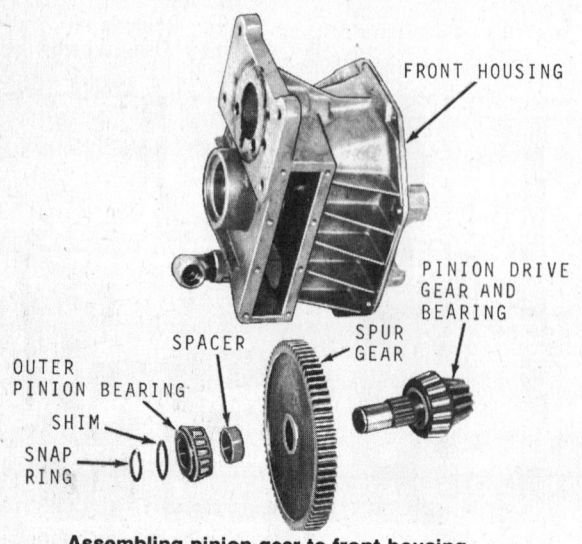
Assembling pinion gear to front housing

13. Position housing and pinion drive gear assembly into press. With pinion drive gear supported, place outer pinion bearing over shaft. With a press sleeve of proper diameter, push bearing onto pinion shaft until a slight drag is noticed when gear is turned by hand. If drag is too severe, tap pinion shaft with a soft metal hammer until drag is reduced.

14. Install a shim and snapring on end of pinion shaft. Use the thickest shim possible which will permit installation of the snapring.

15. After pinion drive gear is positioned satisfactorily, install expansion plug and spur gear cover.

16. When installing new differential bearings, reuse original shims or use new shims of the same thickness. Press bearings into case.

17. If a new differential case is being installed, start with a .020 in. pack of shims under each differential bearing.

18. Shims are available in .003, .005, .010 and .030 in. sizes.

19. Position ring gear on differential case and start capscrews into gear with fingers. Proceed to tighten screws, alternating back and forth across gear to allow gear to be pulled evenly into place. Torque capscrews to 50 to 55 ft.lb.

20. Install thrust washers behind axle gears and place gears in differential case. Install differential pinion gears and thrust washers. Rotate both pinion gears at the same time until pinion shaft can be inserted. Install pinion shaft and secure with pinion pin.

21. To assemble differential to front housing, position differential bearing caps on bearings and insert assembly into bearing cradles. Position assembly with ring gear facing same side as spur gear cover.

22. The bearing cradles are designed to apply a slight preload to bearings, therefore, it is important to push both bearing assemblies simultaneously into their saddles.

23. Install bearing caps in their original position as previously marked. Torque capscrews to 40 to 45 ft.lb.

ADJUSTING RING GEAR AND PINION

1. Using a dial indicator, check ring gear backlash. Ring gear backlash should be .003 to .007 in.

2. If backlash is not in this range, move shims which are located beneath differential bearings from one side to the other until correct backlash is attained.

3. Shims are available in .003, .005, .010, and .030 in. sizes.

4. To check ring gear and pinion pattern, paint teeth of ring and pinion gear with gear pattern compound. Rotate pinion gear until ring gear has made one complete revolution.

5. Study the patterns illustrated and correct if necessary.

6. This is the preferred pattern on both sides of ring gear tooth.

7. To move toe pattern toward heel, increase backlash within .003 to .007 in. limits by shimming ring gear away from pinion gear.

8. To move heel pattern toward toe, decrease backlash within .003 to .007 in. limits by shimming ring gear toward pinion gear.

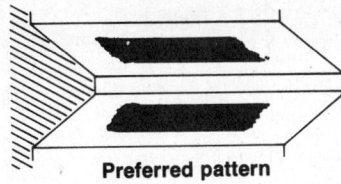
Preferred pattern

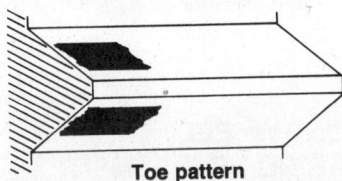
Toe pattern

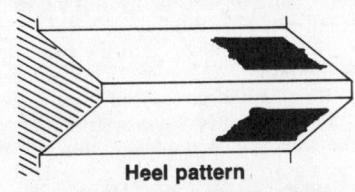
Heel pattern

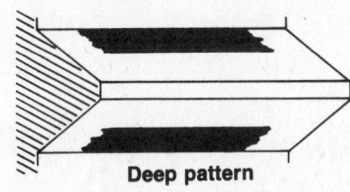
Deep pattern

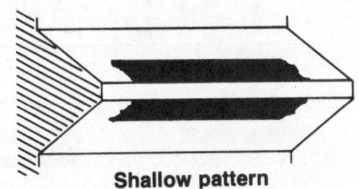
Shallow pattern

9. To correct a deep pattern on the ring gear, reduce shim pack thickness under inner pinion bearing cup.

10. To correct a shallow pattern on the ring gear, increase shim pack thickness under inner pinion bearing cup.

INSTALLATION

1. Position a new housing gasket between front and rear axle housings and install eight housing capscrews. Tighten capscrews to 18 to 23 ft.lb. torque.

2. Install a new gasket between hydrostatic transmission and axle housing. Install four mounting capscrews and spacers. The ⅜ x 5½ in. capscrews are used in the top mounting holes and the ⅜ x 4¾ in. capscrews in the bottom mounting holes. Tighten capscrews to 25 to 30 ft.lb. torque.

3. Place transmission mounting brackets on the top mounting capscrews. Secure brackets with lockwashers and nuts.

4. Install oil intake tube.

5. Attach wheels to axle and roll assembly into tractor frame. Insert hitch plate capscrews and tighten securely. Position mounting brackets and attach to frame sides.

6. Install fan shields, fan, and driveshaft to transmission. Fill transmission with Automatic Transmission Fluid-Type "A" or John Deere Type 303 Special-Purpose Oil. With hydrostatic control lever in neutral and engine running, check fluid level. Fill to "full" mark on dipstick.

7. Attach control cam linkage and adjust if necessary.

8. After making necessary adjustments, install fender-deck and all control knobs.

ENGINE

The engine used is the Kohler K301AS. For engine service, see the Engine Unit Repair section.

REMOVAL AND INSTALLATION

1. For some service applications, it may be desirable to mount the tractor on an

Engine support stand

John Deere

OTC Motor-Rotor Stand prior to engine removal. This is particularly true when engine removal is necessary to service the underneath side of the tractor or parts below the battery.

NOTE: The OTC Motor-Rotor Stand is not necessary for engine removal, but provides easy service access to all components from a convenient standing position.

2. If the stand is used, first shut off the fuel below the gas tank, disconnect the fuel lines and remove the tank. Also remove the battery and drain crankcase and transmission.

3. Loosen four bolts from tractor frame and remove hood and grille components as a unit. Mount the tractor on the stand. The tractor can now be rotated.

4. Disconnect all electrical wires at engine. Remove choke and throttle cables from engine.

5. Remove two capscrews to free rear drive disc hub from drive disc. Remove rear hub roll pin and remove hub from the driveshaft.

6. Loosen four bolts securing engine to engine base. Pivot tractor on stand, so engine once more is on top. Remove bolts from engine base.

7. Bolt a hoist chain to engine head bolts. Move engine forward and upward over front bumper until clear of tractor.

8. Installation is the reverse of removal.

Model 140

FRONT WHEELS AND AXLE

Wheels, Spindle and Axle

REMOVAL AND INSTALLATION

1. Block up or hoist front of tractor until wheel clears the ground. Remove cap from wheel. Remove cotter pin, slotted nut, wheel and bearings from spindle inside cap.

2. To remove spindle from axle disconnect tie rod. Use snapring pliers and remove snapring and washer. Slip spindle out of axle.

3. To remove axle, block up or hoist front of tractor. Remove cotter pin from slotted nut. Remove nut and king pin bolt. Ease axle to floor. Remove tie rods from either steering arm or spindle arm, as desired.

4. Place axle end on press bed and press bushings out of axle.

5. Wipe axle bushing bore clean. Coat bushing with oil. Place axle on press and press bushing in axle until bushing is flush with axle face.

6. Place axle in a vise and turn reamer through axle bushings.

7. Check king pin bushing and other king pin components for wear or other damage. Replace parts as necessary.

8. Grease king pin assembly and install axle on tractor frame. Secure king pin bolt with slotted nut and cotter pin.

9. Apply light coat of grease on upper spindle shaft. Install spindles into axle bushing.

10. Pack wheels with SAE multipurpose-type grease. Install bearing with seal, wheel, outer bearing and slotted nut on axle. Adjust wheel bearings according to the following instructions. Place grease cap on wheel.

FRONT WHEEL BEARING ADJUSTMENT

Adjust the front wheel bearings if the wheel is loose on the spindle or if the wheel does not rotate freely.

1. Raise the tractor until the front tires clear the floor.
2. Remove the grease cap from wheel.
3. Wipe the excess grease from the end of the spindle and remove cotter pin and slotted nut.
4. While rotating the wheel and tire, tighten the slotted nut to 60 to 120 in.lb. torque to seat the bearings. Back off slotted nut until wheel turns freely.
5. Using a $^{15}/_{16}$ in. open end wrench, back off the nut until the slot in nut aligns with cotter pin hole in spindle.
6. Install a new cotter pin and bend the long end of the cotter pin around the end of the axle.
7. Install cap.

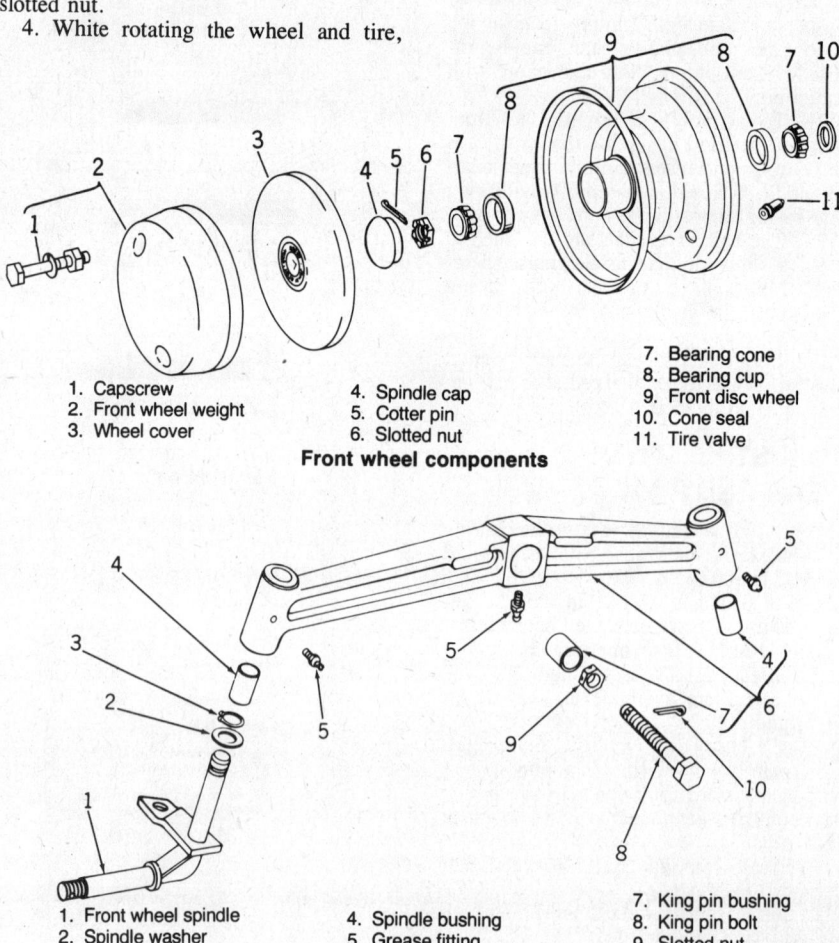

1. Capscrew
2. Front wheel weight
3. Wheel cover
4. Spindle cap
5. Cotter pin
6. Slotted nut
7. Bearing cone
8. Bearing cup
9. Front disc wheel
10. Cone seal
11. Tire valve

Front wheel components

1. Front wheel spindle
2. Spindle washer
3. Snapring
4. Spindle bushing
5. Grease fitting
6. Front axle
7. King pin bushing
8. King pin bolt
9. Slotted nut
10. Cotter pin

Front axle components

John Deere

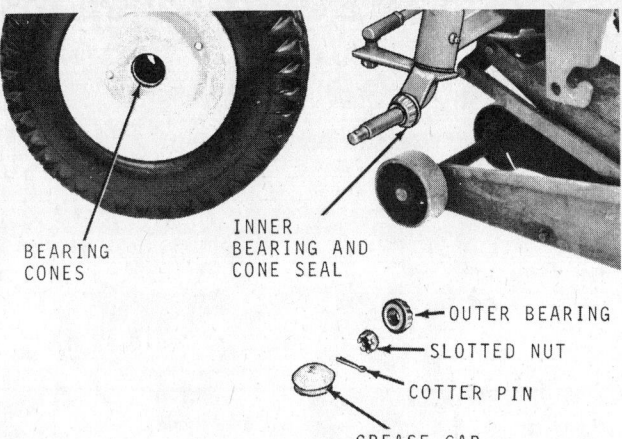

Installing bearings and wheels

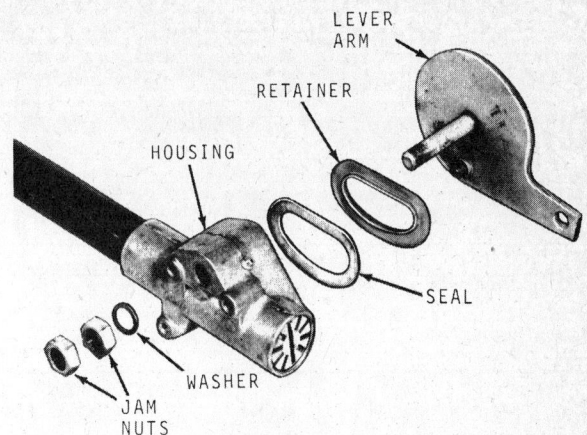

Attaching lever arm to gear

STEERING

Steering Wheel and Gear

REMOVAL

1. Remove steering wheel with puller. Remove battery, battery base and fuel tank.
2. Check instrument panel under the "John Deere" decal for two dimples, resulting from spot welding panel support bracket to panel. Center punch these dimples and drill a 5/16 in. hole to eliminate weld.
3. Swing panel support bracket away from steering column. Remove clamp holding column to pedestal. Disconnect drag link from lever arm.
4. Remove three capscrews holding steering column to frame. Remove capscrews from hydrostatic control lever shaft bearing support last, using a ratchet and screwdriver. For ease of removal, an access hole can be drilled in the right side of pedestal to permit use of a socket extension.

NOTE: On H-3 tractors below serial number 10,000 it may be necessary to remove hydraulic lever linkage to obtain adequate clearance to remove steering gear.

DISASSEMBLING STEERING GEAR

Loosen jam nut on tapered stud in lever arm. Turn stud counterclockwise until resistance is felt. Remove nuts from lever arm pivot bolt and remove from housing. Remove adjusting plug in steering gear housing and slide shaft with cam and bearings from column.

INSPECTING STEERING GEAR PARTS

1. Wash parts in a clean, safe solvent and dry with compressed air and clean cloth.
2. Check bearing condition. Inspect cam, housing and plug for cracks, scoring and other damage especially in the bearing area. Replace parts showing excessive wear or damage.

ASSEMBLING STEERING GEAR

1. Apply grease and place bearing balls, ball cups and retaining rings on both ends of cam. On tractors above serial number

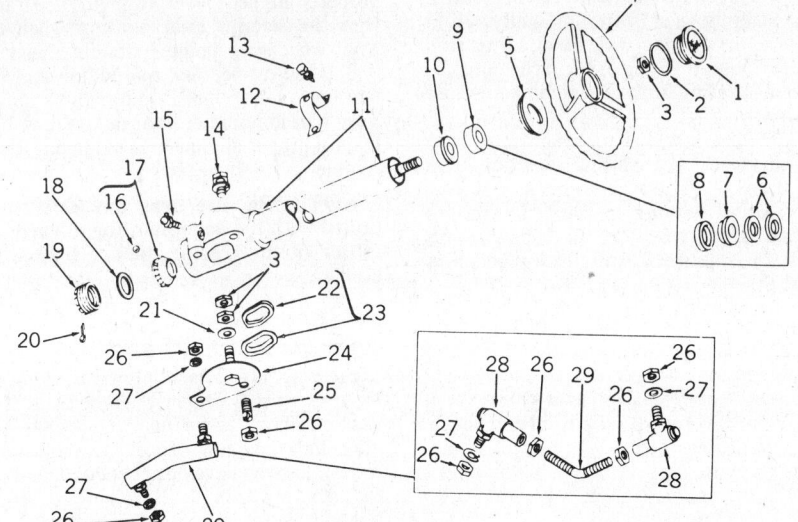

1. Steering wheel emblem
2. O-ring
3. 5/8 in. hex nut (3 used)
4. Steering wheel
5. Grommet
6. Felt washer (2 used)
7. Bearing spacer
8. Bearing retainer
9. Bearing seal
10. Shaft bearing
11. Steering gear assembly
12. Steering column clamp
13. Screw
14. Capscrew and lockwasher
15. Grease fitting
16. Steel ball
17. Bearing kit
18. Belleville washer (10,684 and up)
19. Adjusting plug
20. Cotter pin
21. Washer
22. Lever arm seal
23. Retainer with seal
24. Lever arm with pivot bolt
25. Tapered stud
26. Jam nut
27. Lockwasher
28. Ball joint
29. Drag link assembly

Steering system components

1. Steering arm
2. Grease fitting
3. Bolt with cone
4. Locking strap
5. Washer
6. Self-tapping screw
7. Tie rod
8. Jam nut
9. Ball joint
10. Lockwasher
11. Nut
12. Jam nut
13. Ball joint

Steering linkage components

369

John Deere

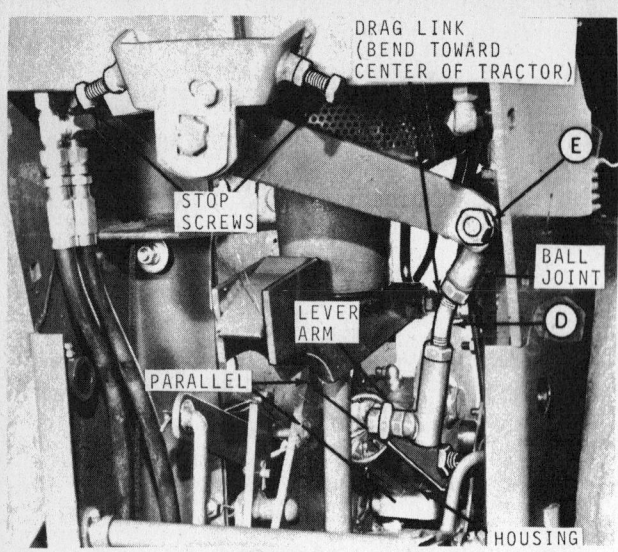

Aligning steering gear

Adjusting steering cone

10,684 install Belleville washer on lower end of cam.
2. Grease cam lightly with multi-purpose-type grease.
3. Slide cam and tube assembly into housing and jacket tube. Install plug and tighten 7-12 ft.lb. for tractors through #10,683; 10-14 ft.lb. for tractors 10,684 and up.
4. After tightening, lock plug with cotter pin. Be sure steering column turns freely.
5. Install new seal and retainer from AM30980 kit. Attach lever arm to steering gear housing with washer and two jam nuts.

INSTALLING STEERING GEAR

Slide steering gear into instrument panel grommet and lower front end into position. Install capscrew into hydrostatic control lever bearing support first. Install other two capscrews.

INSTALLING STEERING LINKAGE

Install drag link into lever arm. Install bracket holding steering column to pedestal.

NOTE: It is important that drag link is positioned with bend facing the center of the tractor before tightening nuts.

COMPLETING INSTALLATION

1. Install instrument panel support bracket to instrument panel using two chrome-plated ¼ in. stove bolts.

NOTE: If hydraulic lever linkage has been removed in the case of H-3 tractors below serial number 10,000 connect linkage.

2. Install battery base, battery and fuel tank. Install steering wheel and tighten nut. Replace O-ring and steering wheel emblem.
3. Adjust the steering gear mechanism.

ADJUSTMENTS

Steering Arm Stop Screws

1. Steering lever pivot bolt breakage can occur if the steering gear is not allowed to move through its full range. Incorrectly adjusted steering arm stop screws will restrict this full-range movement.
2. To adjust the steering arm stop screws, loosen both jam nuts and turn screws out until they no longer contact the steering arm.
3. Have someone turn steering wheel until steering gear stops internally. Hold pressure on steering wheel and adjust stop screw closest to steering arm until it touches steering arm; then tighten one-half turn further.
4. Hold screw and lock jam nut. Turn steering wheel in opposite directions and repeat procedure to adjust other stop screw.
5. Adjust steering mechanism in the sequence shown when excessive play (loose steering) is noticed, or if steering becomes difficult.

Steering Cone Adjustment

1. Block up front end of tractor so that front tires are off the ground.
2. Disconnect ball joints at points A.
3. Turn steering arm by hand and notice freedom of movement. When properly adjusted, the steering arm will pivot freely through the entire steering range with only a slight amount of drag.
4. If steering arm turns hard or has worn and loosened so that you can feel end play in the steering arm bearing, remove lock plate.
5. Tighten bolt only until a slight amount of drag can be felt when turning the steering arm and all end play has been removed.
6. Position lock plate over bolt head and tighten lock plate capscrew. Be sure plain washer is used with lock plate capscrew.

NOTE: If lock plate cannot be installed without turning bolt out of adjusted position, turn plate over.

7. Connect ball joints at points A.
8. Set steering arm stop screw adjustment.

Steering Gear Alignment (Before 10,000)

1. Steering effort is minimized when the steering gear is properly aligned. Always align steering gear as detailed below before adjusting the steering gear.
2. Visually check alignment of lever arm with steering gear housing. When properly aligned, lever arm will be parallel with the steering gear housing when the front wheels are pointed straight ahead.
3. If adjustment is required, loosen front jam nut on drag link D and remove nut E. Turn ball joint either in or out on drag link D until proper alignment is obtained. Reassemble and tighten nuts.

NOTE: Be sure drag link D is positioned with the bend pointing toward the center of the tractor, before tightening jam nuts.

Steering Gear Alignment (10,001 and Up)

Steering effort is minimized when the steering gear is properly aligned. On 140 tractors above serial number 10,000, the drag link is a one-piece, non-adjustable rod. If no parts have been bent or damaged, the steering gear will remain properly aligned as long as tie rods are adjusted to an equal length.

Stop Screws

1. The stop screws limit the amount of turn in both directions and should be adjusted frequently.
2. Check steering wheel for equal turn in both directions and reset stop screws if necessary. *Be sure setscrews are turned in far enough to prevent wheels from turning past linkage center.*

John Deere

Steering Gear Adjustment (Before 10,683)

Adjust steering gear as explained below to correct loose steering and steering wheel end play.

Remove battery and battery base from tractor.

1. Disconnect ball joint from lever arm.
2. Loosen jam nut and stud two or three turns.
3. Remove cotter pin holding adjusting plug in steering gear housing. Use screwdriver to turn adjusting plug into steering gear housing until end play is removed. Lift up on steering wheel to check end play. Back plug out until steering wheel turns freely with no drag.
4. Turn plug only far enough after adjustment to insert cotter pin through housing and closest slot in plug. Spread cotter pin.
5. Loosen jam nut on pivot bolt and tighten only the inside nut using a thin open end wrench, until all end play (see arrow) is removed or until the distance between the lever arm and steering gear housing is between $1/16$ and $3/32$ in.
6. Turn steering wheel until the lever arm is parallel with steering gear housing.
7. Turn stud in (clockwise) until snug to remove all looseness. Then move steering wheel through its full steering range in both directions (right to left). When properly adjusted, a slight drag can be detected in the midpoint of the range (when line between the pivot bolt and ball joint is vertical). Tighten jam nut firmly. Make final test by turning steering arm through full range.
8. Connect ball joint to lever arm.

Steering Gear Adjustment (10,684 and Up)

To adjust steering gear for tractors above serial number 10,683 follow the instructions given above, disregarding steps 3 and 4. Adjusting the plug is not necessary due to the Belleville washer which should maintain the proper torque setting.

BRAKES

Brake Shoe and Disc

REPLACEMENT (BEFORE 10,001)

1. Jack up tractor and remove rear wheels. Remove drilled pin holding clevis to brake lever. Remove two nuts and bolts holding brake shoes and brake plate to axle. If the spring in the bracket needs replacing, compress the spring and remove the cotter pin. Slide the brake rod back through the bracket and remove the spring from the bracket.
2. Remove cap, cotter pin and nut from rear axle. Install a puller on the hub and remove from axle. The disc is held on the hub with four capscrews.
3. Position disc on hub and install bolts. Torque bolts to 10 ft.lb. Be sure key is in place and install hub and disc assembly on

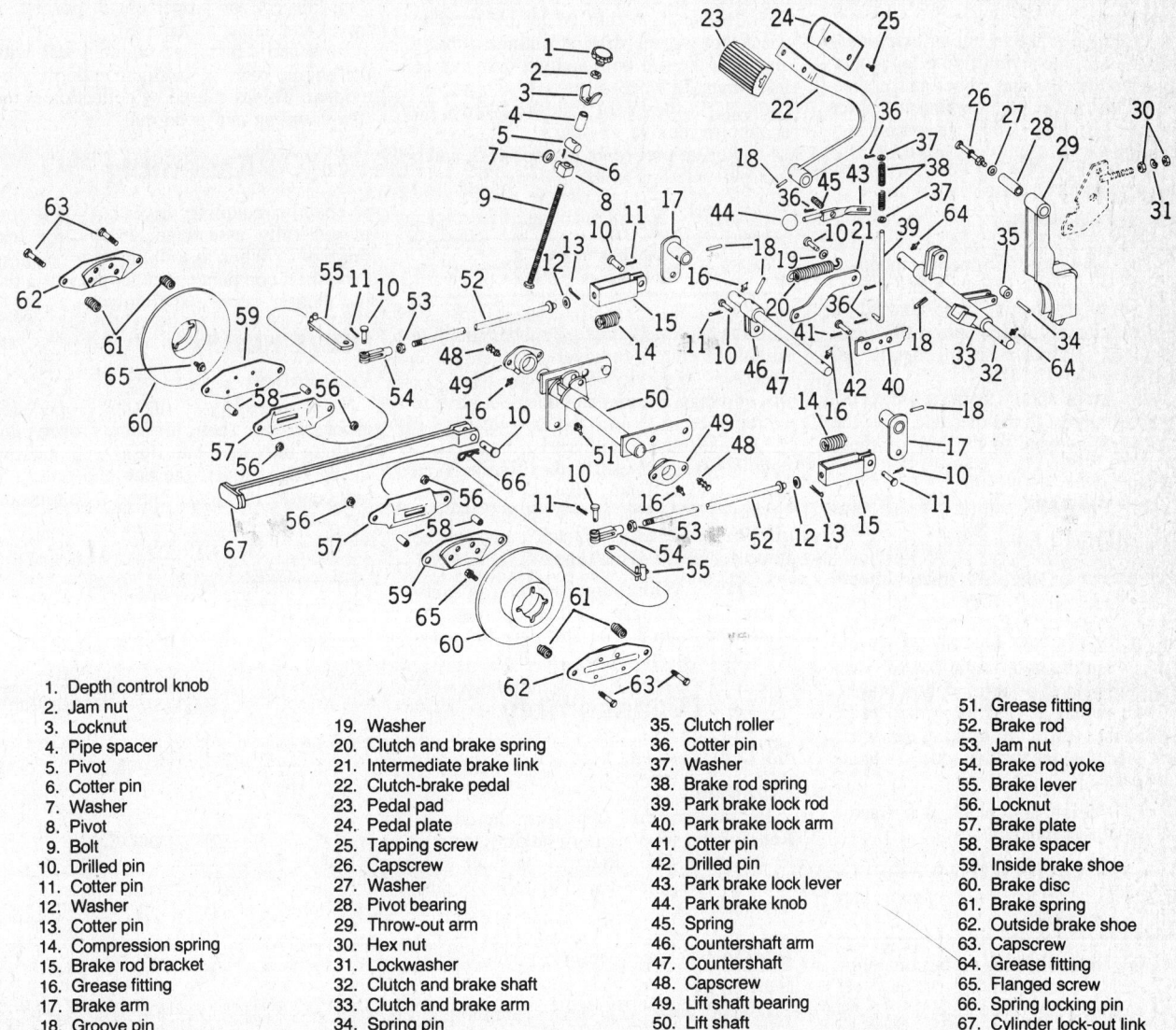

1. Depth control knob
2. Jam nut
3. Locknut
4. Pipe spacer
5. Pivot
6. Cotter pin
7. Washer
8. Pivot
9. Bolt
10. Drilled pin
11. Cotter pin
12. Washer
13. Cotter pin
14. Compression spring
15. Brake rod bracket
16. Grease fitting
17. Brake arm
18. Groove pin
19. Washer
20. Clutch and brake spring
21. Intermediate brake link
22. Clutch-brake pedal
23. Pedal pad
24. Pedal plate
25. Tapping screw
26. Capscrew
27. Washer
28. Pivot bearing
29. Throw-out arm
30. Hex nut
31. Lockwasher
32. Clutch and brake shaft
33. Clutch and brake arm
34. Spring pin
35. Clutch roller
36. Cotter pin
37. Washer
38. Brake rod spring
39. Park brake lock rod
40. Park brake lock arm
41. Cotter pin
42. Drilled pin
43. Park brake lock lever
44. Park brake knob
45. Spring
46. Countershaft arm
47. Countershaft
48. Capscrew
49. Lift shaft bearing
50. Lift shaft
51. Grease fitting
52. Brake rod
53. Jam nut
54. Brake rod yoke
55. Brake lever
56. Locknut
57. Brake plate
58. Brake spacer
59. Inside brake shoe
60. Brake disc
61. Brake spring
62. Outside brake shoe
63. Capscrew
64. Grease fitting
65. Flanged screw
66. Spring locking pin
67. Cylinder lock-out link

Brake-clutch components, tractors through serial #10,000

371

John Deere

Adjusting brakes

axle. Position washer on axle and tighten nut to proper torque. Install new cotter pin and old cap.

4. Assemble brake to tractor axle referring to the accompanying figure for proper part placement. Be sure ribbed brake shoe is toward the outside and the flat shoe is on the inside. Install bolts through spacers and tighten nuts.

BRAKE REPAIR (AFTER 10,001)

1. The brake pedals operate independently on a split shaft. Each shaft is connected to its respective axle through a spring-loaded strap which returns the brake pedal to its original position after its release.

2. The straps should be attached to the lever holes closest to the disc and hub. The brake disc is welded to the hub on these tractors.

ADJUSTMENT (ALL MODELS)

The brakes on the 140 tractor should wear very little because they are used mainly for parking and "panic" stops. The hydrostatic transmission will do all normal braking when the hydrostatic control lever is placed in neutral position. Adjust brakes, however, when excessive brake pedal travel causes difficult braking or when brakes become ineffective due to brake band wear.

1. Stop engine and block rear tractor wheels off the ground.

---CAUTION---
Block securely under tractor frame; not under transmission.

2. Open free-wheeling valve on transmission and release parking brake. Rear wheels can then be turned freely.

3. Remove cotter and drilled pins from brake clevis on both rear brakes.

4. Turn righthand brake clevis onto the rod several turns and reconnect drilled pin.

5. Rotate righthand wheel by hand to check for wheel drag. Continue turning clevis on the rod until a slight drag can be felt when the wheel is turned.

6. Back clevis off the rod only far enough to remove all drag.

7. Repeat procedure in steps 4, 5, and 6 to adjust lefthand brake.

CLUTCH

If the 140 tractor has the rubber coupling inside the clutch, removal of the engine is necessary to service the clutch, driveshaft or couplings. The engine does not have to be removed to replace the drive disc.

On 140 tractors equipped with the flex-coupling-type driveshaft, the driveshaft can be serviced without removing the engine. However, it is recommended that the engine be removed, so clutch parts can be inspected at the same time.

NOTE: Whenever service is performed on the driveshaft components, it is recommended that the new flex-coupling driveshaft kit be installed.

Disassembly

CLUTCH
Removal

Remove the driveshaft from the clutch. Remove the six capscrews holding the cup to the drive hub. Remove the four capscrews holding the drive hub to the engine flywheel.

CONE
Removal

Discard cone with rubber coupling inside. Install new flex-coupling driveshaft kit.

DRIVE HUB
Removal

NOTE: Do not remove the bearings or bushing unless necessary. Removing these bearings may cause them to be damaged and require replacement.

1. Using a puller or slide hammer, pull the bronze bushing out. Invert the hub and press the ball bearing out, using a 1⅛ in. shaft.

2. Whenever driveshaft problems occur, it is recommended that the rubber coupling-type cone, the driveshaft and related parts be replaced with the flex-coupling driveshaft kit.

3. To install the new driveshaft assembly, remove old clutch cone and driveshaft assembly, clutch cup with lining, and drive disc.

4. Slip previously removed clutch cup with lining over the new driveshaft assembly and attach complete unit to the engine.

5. Complete assembly by attaching drive disc to driveshaft and transmission. Install flat washers between disc and lockwashers.

6. Inspect for proper clutch pedal free travel and adjust as required.

7. When depressing clutch pedal with the engine running, a slight rattle may be evident. This is caused by deflection of the flex-coupling and is normal.

Assembly

The flex-coupling driveshaft kit is furnished fully assembled, except for rear drive disc. When assembling flex-coupling driveshaft components, other than at initial installation, service as follows:

DRIVE HUB
Installation

1. If necessary, using the proper size arbor or round shaft, press the bronze pilot bushing into drive hub flush to ⅛ in. from inside edge. Invert the hub and press the ball bearing into the hub until it bottoms on the outer ring.

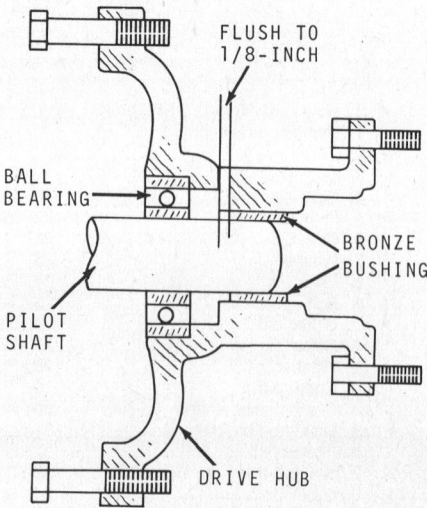

Installing bearings in drive hub

John Deere

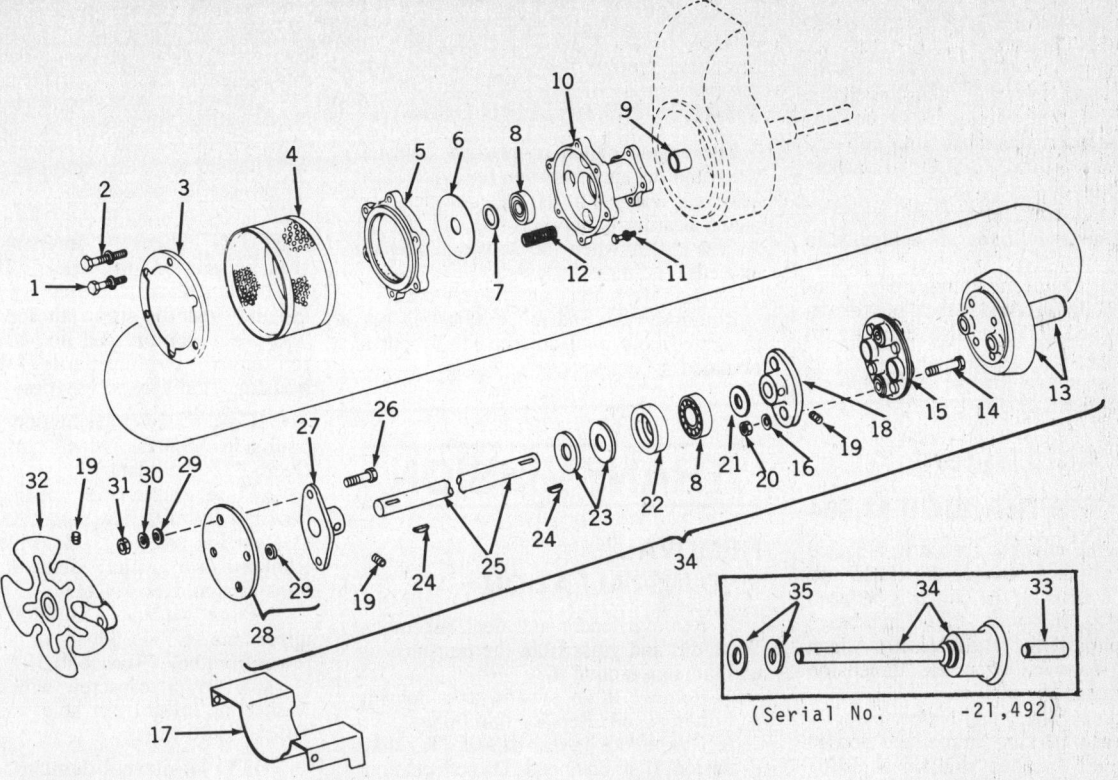

1. Capscrew with lockwasher
2. Capscrew with lockwasher
3. Screen end cap
4. Intake screen
5. Clutch cup with lining
6. Flat washer
7. Flat washer
8. Ball bearing
9. Bushing
10. Drive hub
11. Capscrew with lockwasher
12. Clutch spring (4 used)
13. Clutch cone and pilot shaft
14. Capscrew
15. Vibration coupling
16. Lockwasher
17. Fan guard
18. Driveshaft hub
19. Setscrew
20. Nut
21. Flat washer
22. Bronze thrust bearing
23. Thrust washer (2 used)
24. Woodruff key
25. Driveshaft
26. Capscrew
27. Drive disc hub
28. Drive disc
29. Flat washer
30. Lockwasher
31. Nut
32. Fan and hub
33. Pilot shaft
34. Clutch cone and driveshaft
35. Thrust bearing (2 used)

Clutch, driveshaft and couplings

NOTE: When pressing in bronze bushing from the front of clutch drive hub, be sure to adequately support casting around ball bearing bore to prevent cracking the casting. When pressing ball bearing into place, press on outer ring only.

2. Fill cavity between bronze pilot bushing and ball bearing with multipurpose-type grease. Align drive hub on flywheel.

Install four 5/16 x 1 in. capscrews in drive hub and flywheel and torque to 20 ft.lb.

CONE

Installation

Install four clutch springs on studs. install a 3¾ in. washer, then a 1⅜ in. washer on pilot shaft. Install cone and pilot shaft assembly into drive hub being sure springs are in their proper position.

CUP, SCREEN AND DRIVESHAFT

Installation

1. Install cup on drive hub with three 5/16 x 1¾ in. capscrews and lockwashers equally spaced. The long capscrews are to pull the cup and drive hub together because the short capscrews would not reach the drive hub. Assemble the screen and end

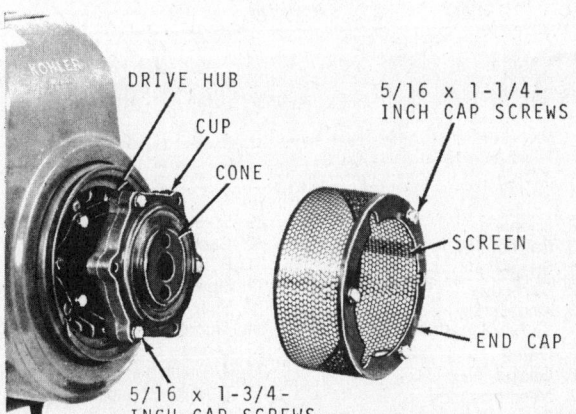

Installing cup and screen

Measuring clutch adjustment

John Deere

cap with three 5/16 x 1¼ in. capscrews and lockwashers. Torque all 5/16 in. capscrews to 20 ft.lb.

2. To install driveshaft with the engine removed, install throw-out bearing on front drive hub. Insert driveshaft into hub and secure with setscrew. Place thrust washers on driveshaft.

3. Bolt flex-coupling to clutch cone and place engine and driveshaft assembly in tractor.

4. Install engine mounting bolts. Connect clutch linkage and attach rear driveshaft hub and flexible disc coupling to transmission.

5. Adjust clutch linkage and complete the engine installation.

Adjustment

TRACTORS THROUGH 11,904

The engine clutch is properly adjusted when the distance between the clutch arm and wear washer on the clutch is between 1/32 and 1/16 in. when the clutch-brake pedal is up (parking brake released). Adjust clutch as follows when the dimension shown exceeds 1/16 in.:

1. Remove battery and battery base.
2. Release parking brake and loosen both nuts on threaded stud on adjusting plate.
3. Move adjusting plate forward (towards front of tractor) and tighten both nuts firmly after correct adjustment is obtained.

TRACTORS AFTER 11,904

The engine clutch is properly adjusted when the distance between the clutch arm and wear washers or the wear washers and clutch bearing is between 1/32 and 1/16 in. Adjust clutch when the above dimension exceeds 1/16 in. as follows:

1. Remove battery and battery base.
2. Loosen nut and move adjusting bolt in slotted hole until correct adjustment is obtained.

TRANSMISSION

REMOVAL AND INSTALLATION

1. Remove fender and deck assembly. Wipe dirt and grass from the transmission and the area around it.
2. Remove drain plug in axle housing and drain all oil. Replace drain plug.
3. Disconnect driveshaft and PTO shaft extension, if so equipped. Disconnect control rod at cam.
4. Disconnect hydraulic lines. Remove four bolts and lift transmission from tractor. Plug all external ports.
5. Use a safe cleaning solvent and thoroughly clean the outside of the transmission. Move the transmission to a clean, well lighted work area and use clean tools for the service procedure.

NOTE: Whenever internal damage, chip generation, or severe oil contamination is present, completely disassemble and flush the pump, motor and valve body. Use compressed air, if available, to remove contaminants from valve housings and system cavities.

6. Coat a new transmission gasket on both sides with Permatex® No. 3 or John Deere PT502 Sealant.
7. Place gasket over openings in axle housing. Install a new square O-ring over axle suction tube. Oil axle drive gear and transmission output gear and mount transmission over gasket.
8. Place two 7/16 x 4 in. through-bolts in upper hole, a 7/16 x 3¼ in. through-bolt in lower front hole. Torque to 55 ft.lb. Install a 7/16 x 1½ in. capscrew with aluminum washer in lower rear hole and tighten. Torque to 36 ft.lb.

NOTE: If internal damage, chip generation or severe oil contamination was

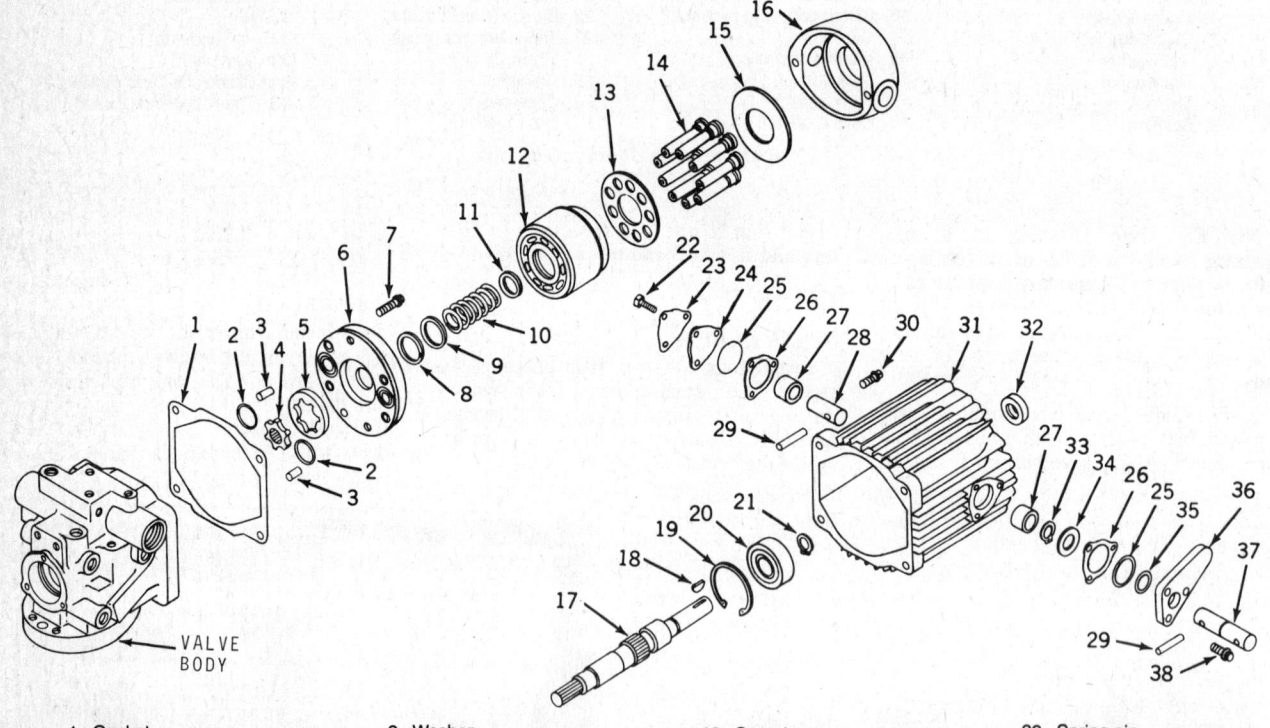

1. Gasket
2. O-ring
3. Spring pin
4. Charge pump rotor (Gerotor)
5. Charge pump stator (Gerotor)
6. Valve plate
7. Capscrew
8. Snapring
9. Washer
10. Spring
11. Washer
12. Cylinder block
13. Slipper retainer
14. Slippers
15. Thrust plate
16. Swash plate
17. Driveshaft
18. Key
19. Snapring
20. Ball bearing
21. Snapring
22. Capscrew
23. Model plate
24. Trunnion cap
25. O-ring
26. Gasket
27. Needle bearing
28. Trunnion shaft
29. Spring pin
30. Capscrew
31. Pump housing
32. Lip seal
33. Snapring
34. Trunnion washer
35. O-ring
36. Trunnion plate
37. Trunnion control shaft
38. Capscrew

Variable displacement piston pump components

John Deere

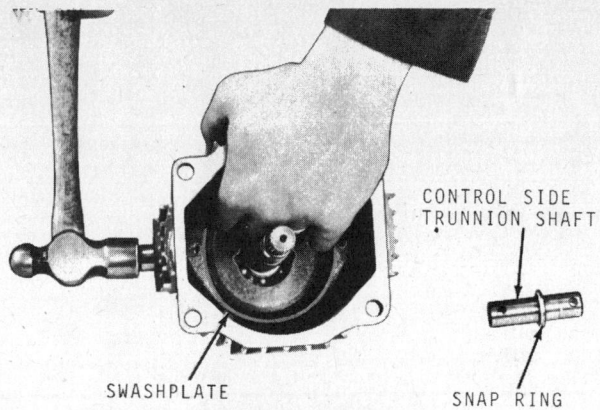

Installing swash plate

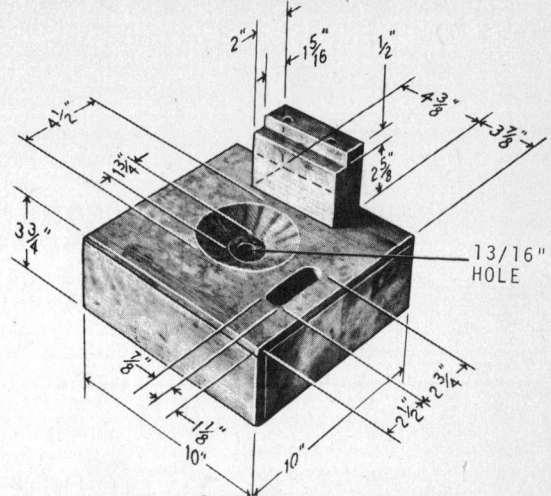

Wood mounting block for pump repairs

found when servicing either the pump, motor or valve body, be sure to flush the axle housing with clean oil before installing the transmission. Clean the intake screen in the axle housing and replace the oil filter. Clean the magnetic drain plug.

9. Fill the system with oil.

Overhaul

MAKING TRANSMISSION REPAIR STAND

Make the wooden repair stand shown to facilitate servicing the pump or motor assemblies.

DISASSEMBLING PUMP

1. Remove fan assembly. Loosen four capscrews holding finned aluminum housing to valve body. Hold the pump in a horizontal position and tap the finned housing on either side with a plastic or rubber hammer to break the gasket seal. Let oil drain from the pump case into a pan.

2. After the oil has drained, remove the four capscrews and slowly pull the pump housing from the valve body.

NOTE: Retain oil. Check for metal chips or other forms of contamination which might indicate cause of problem.

NOTE: Insert fingers between housing and valve plate to prevent the cylinder block and piston assembly from falling out. Be careful during disassembly to prevent damage to these critical parts.

3. Carefully remove the cylinder block and piston assembly from the driveshaft. Use a stiff wire hook to remove the thrust plate located in the swashplate.

Removing Swashplate

1. Support the pump housing in the wooden mounting block repair stand. This removes strain from the main pump bearing and places it evenly on the pump case.

2. If it is necessary to remove the cylinder block spring use two washers, a bolt and a nut to compress the spring. Remove the retaining ring, back off the bolt and the spring.

NOTE: Do not allow bolt and washers to distort any surface or part of the cylinder block assembly.

3. Remove trunnion cap (24), trunnion plate (36), and related parts.

4. Drive spring pins on both sides of swashplate down until they bottom in the pump housing. Insert a punch through the hole in the outer end of the trunnion control shaft and with a turning, pulling motion, remove the shaft.

5. Remove the trunnion shaft by tapping it into the housing. Lift the swashplate out of the case. Remove spring pins from swashplate.

6. Remove the snapring and lift the driveshaft and bearing from the pump housing. To remove the bearing from the driveshaft, remove the snapring and press the bearing off the input end of the shaft.

Removing Lip Seal

1. Replace the lip seal in the pump housing whenever the driveshaft is removed.

2. To remove the seal, rest the pump housing vertically on a piece of wood. Use a socket or piece of pipe about the same outside diameter as the seal to press the seal into the housing until it drops from its bore.

Removing Trunnion Bearings

The trunnion bearings can be removed the same way as the lip seal. Press the bearings into the pump housing until they drop out.

Removing Valve Plate

1. Remove the four bolts holding the valve plate to the valve body. Remove the valve plate by lifting it upward with a screwdriver. Pry on both sides of the valve plate in the valve body reliefs.

NOTE: Do not try to insert screwdriver between valve plate and valve body. These are lapped surfaces and any nicks or scratches will cause oil leakage which will have to be corrected before the unit can be reassembled.

2. When removing the valve plate, be careful that the charge pump rotor and stator (4 and 5) do not fall out and scratch the lapped surface on the valve body.

ASSEMBLING PUMP

Seals and Bearings

1. After all parts have been reworked to specifications and are clean and free from lapping dust and other contaminants, oil them lightly with Automatic Transmission Fluid—Type "F".

2. Use a piece of pipe or a socket of about the same diameter, and press the pump shaft seal (numbers up) into the pump housing until it is flush with the outside of seal bore.

3. Press the trunnion bearings into the pump housing with the numbers facing to the outside. Be certain the bearings are pressed below the recess in the pump housing.

4. Install pump shaft bearing on pump shaft by pressing on inner ring only. Install snapring on pump shaft.

5. Wrap cellophane tape around key way or install an oil seal tool in oil seal prior to inserting shaft into pump housing. Seat the front bearing in the housing with a light tap on the splined end of the shaft. Install the bearing retaining snapring.

Swashplate

1. Holding the swashplate in position with one hand, tap the short trunnion shaft into the housing and swashplate, lining up holes for spring pin.

2. Install the control side trunnion shaft into the swashplate in a similar manner with the snapring toward the housing (trunnion washer out).

3. Drive spring pins into swashplate and trunnion shafts until pins are flush with swashplate. Install trunnion shims, O-rings, trunnion cover and trunnion plate. Torque trunnion plate bolts to 7-10 ft.lb.

Thrust Plate

Apply a light coat of grease to the underside of the thrust plate to hold it in place when installing cylinder block and pistons.

John Deere

Install thrust plate with lapped (shiny) side up.

Checking Stack-Up

Before assembling the pump housing to the valve body, check for stack-up in order to install gaskets of the proper thickness between the pump housing and the valve. Refer to the numbered steps below and install the following parts before making the stack-up measurement: driveshaft, swashplate, thrust plate and cylinder block without cylinder block spring.

1. Remove the cylinder block spring from the cylinder block.
2. Put the slipper retainer and nine pistons into the cylinder block. Be sure the slipper retainer is positioned properly on the ball socket of the cylinder block.
3. Hold the pump housing at about a 30 degree angle and slide the cylinder block and piston assembly into the housing.
4. Be sure that thrust plate is seated properly in the swashplate, the cylinder block is all the way down in the pump housing and the slippers are in contact with the thrust plate.
5. Use a depth micrometer to measure the distance from the top of the pump housing to the uppermost part of the cylinder block (balance land). This should be measured in two or three places around the housing to get an average dimension.
6. Use a 1 to 2 in. micrometer and measure the valve plate thickness.
7. The valve plate thickness should be greater than the cylinder block depth. If the cylinder block depth is more than 0.004 in. greater than the valve plate thickness, replace the part from which the most surface was removed during the lapping process. If more than one part required lapping, more than one part *might* have to be replaced.
8. Subtract the cylinder block depth from the valve plate thickness. This figure, Dimension "A" in the chart below, is the amount the valve plate stands above or below (minus dimension) the pump housing. Refer to this chart for the proper gasket thickness when assembling the pump housing to the valve body.

Dimension "A"	Gasket Thickness
0.004" to 0.001"	0.010"
0.002" to 0.004"	0.015"
0.005" to 0.007"	0.021"
0.008" to 0.012"	0.031"
0.013" to 0.020"	0.047"
0.021" to 0.027"	0.060"

9. Use only one gasket of the correct thickness. At no time should a gasket thicker than 0.060 in. be used.

NOTE: Gasket thickness is more than twice the difference in measurements to permit compression of the gasket and to compensate for slipper-to-retainer clearance.

10. After determining the gasket required, remove the cylinder block from the pump housing and install the block spring. Be sure the small washer goes into the block first, then the spring, large washer and retaining ring.
11. After the spring is in place, the cylinder block and piston assembly can be installed into the pump housing for final assembly.

NOTE: Be sure the thrust plate is seated properly in the swashplate.

Valve Plate

1. Install new O-rings and place gerotor set in valve plate. Apply Automatic Transmission Fluid—Type "F" to gerotor set and position plate on valve gody.

NOTE: Position notch toward gear motor end of unit. Be sure spring pin holes and spring pins are aligned before forcing plate into position. Tighten capscrews to 20-22 ft.lb.

2. Install pump housing gasket of proper thickness selected from stack-up chart.
3. Apply a liberal amount of Automatic Transmission Fluid—Type "F" to lapped surfaces. Holding the cylinder block with fingers, install pump housing. Slowly turn the pump shaft until the spline mates with the gerotor set and PTO coupling, if unit is so equipped.
4. Install the four capscrews into the housing. Torque to 36-38 ft.lb.
5. Use the key and an adjustable wrench and rotate the input shaft. If the torque required to turn the shaft is greater than 36 in.lb. recheck the unit for proper assembly and gasket selection. If the thrust plate and other parts are positioned correctly and the measurements are correct, use the next largest size gasket and reassemble.
6. Assemble pump to valve body so that trunnion cap is on top and trunnion shaft is on the bottom.
7. Install cam and roller assembly and tighten long capscrew securely.

DISASSEMBLING GEAR MOTOR

1. Before disassembly, scribe a line across the gear motor sections. This will assure proper reassembly.
2. Remove the ten capscrews which hold the motor together. Tap the motor firmly with a plastic hammer to separate sections. Do not pry the motor apart with sharp tools or a screwdriver.
3. The motor sections are aligned with two dowel pins—one at each end of the end cap. You will feel resistance when separating the sections.
4. After removing the end cap and spacer plate from the valve body, the idler gear and idler shaft may be removed as a unit. To remove the output shaft, remove in order: the snapring, gear, Woodruff key, and another snapring. The output shaft can be pulled from the end cap.
5. New snaprings and O-rings must be installed in reassembly. Output shaft seals can be removed with a small screwdriver.
6. Remove the snaprings to separate the idler gear from the idler shaft.

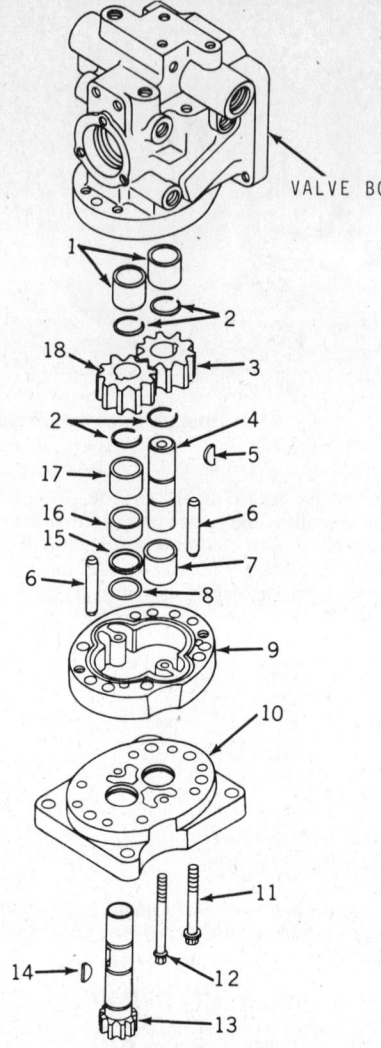

1. Needle bearing
2. Snapring
3. Idler gear
4. Idler shaft
5. Key
6. Dowel pin
7. Needle bearing
8. O-ring
9. Spacer plate
10. End cap
11. Capscrew
12. Capscrew
13. Output shaft with drive gear
14. Key
15. O-ring
16. Needle bearing
17. Needle bearing
18. Output gear

Gear motor components

7. Use a punch to drive the output shaft bearings toward the inside of the end cap.

ASSEMBLING GEAR MOTOR

Bearings

1. Before assembling, oil all parts lightly with Automatic Transmission Fluid—Type "F". install needle bearings with a press or tap them into place with a

John Deere

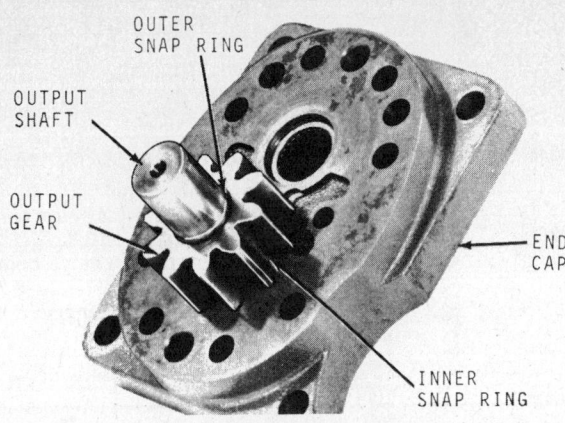

Assembling output shaft

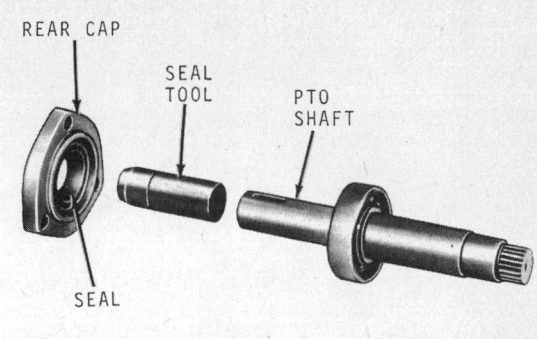

Installing rear cap with seal tool

hammer. Use a round shaft or a socket that is approximately the same size and press on the outside of the bearing only. Install the bearings with the writing facing out. Press the bearings in to the proper depth (inner .095-.140 in.) (outer .901-.911 in.) below lapped face.

2. After installing the bearings, use a surface plate and check all lapped surfaces for flatness.

3. Install new gear motor output shaft O-ring seals in end cap. The split nylon O-ring goes toward the outside and the rubber O-ring goes toward the gear motor.

Output Gear and Shaft

Oil the output shaft and insert it into the end cap. Install a new inner snapring. Place the drive gear key into the key way and slip the gear over the shaft and key. Install a new outer snapring.

Idler Gear and Shaft

Assemble the idler gear in similar manner and install it in valve body.

Spacer Plate

1. Apply a light coat of thin shellac and alcohol mixture to the outer edge (¼ in.) of the spacer plate mating surfaces. Do not allow the mixture to enter the drain grooves in the mating surfaces.

2. Align scribe lines and install spacer plate. Drive dowel pins into valve body just enough to hold spacer plate in place.

3. Oil gears and mating parts liberally with Automatic Transmission Fluid—Type "F". Rotate the gears to distribute oil over all surfaces.

4. Turn the output shaft until the gears mesh and install the end cap. The dowel pins should be ⅜ in. above the end cap. Install the two tie bolts first and torque to 17 ft.lb. Torque the remaining capscrews to 32 ft.lb. in a cross pattern.

5. Put a cloth over the output gear. Use a pliers to rotate the output gear and shaft to check the torque. A turning torque of greater than 36 in.lb. (with free-wheeling valve open) indicates a lack of proper clearance or that excess shellac was extruded into the gears. Disassemble the motor and correct the problem before installing the unit in the tractor.

POWER TAKE-OFF (PTO)

DISASSEMBLY

1. Remove the three capscrews holding the PTO rear cap on to the valve body and pull the power take-off shaft out of the valve body. The coupling on the inside of the valve body can be removed with a magnet or a wire hook. Slide rear cap from shaft.

2. Remove the snapring on the PTO shaft. Remove the ball bearing by pressing on the inner bearing ring only.

3. If it is necessary to remove the needle bearing in the valve body, remove the pump housing and valve plate. Punch the bearing out toward the PTO end. Be careful not to damage the highly lapped surface of the valve body.

4. Press the seal from the rear cap by providing proper backing for the cap.

ASSEMBLY

1. Press the needle bearing into the housing (numbers out) until it is 2.76 to 2.86 in. from the outside of the machined surface.

2. Press the ball bearing on the PTO shaft using a piece of pipe to back up the inner ring.

3. Press the seal into the rear cap until it is flush with the outside of the rear cap. Be sure to install the seal so that the numbers will be facing out when the rear cap is installed.

4. When installing rear cap and seal, tape key slot or install seal tool on shaft to protect oil seal from being damaged by key slot.

5. Install a new gasket on the rear cap. Insert spined end of PTO shaft into the splined coupling and install the PTO assembly into the valve body.

6. Install three capscrews with internal tooth washers in rear cap. Torque to 10-12 ft.lb.

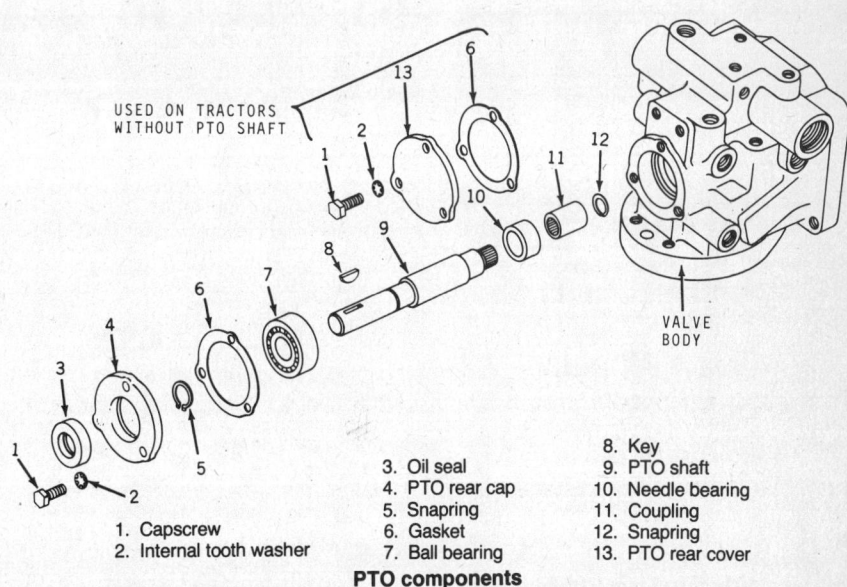

1. Capscrew
2. Internal tooth washer
3. Oil seal
4. PTO rear cap
5. Snapring
6. Gasket
7. Ball bearing
8. Key
9. PTO shaft
10. Needle bearing
11. Coupling
12. Snapring
13. PTO rear cover

PTO components

377

John Deere

Engine service support stand

ENGINE

The engines used are the Kohler K301AS in tractors through serial number 10,000 and the Kohler K321AS used in tractors serial number 10,001 and up. For complete engine service see the Engine Unit Repair section.

REMOVAL AND INSTALLATION

For some service applications, it may be desirable to mount the tractor on an OTC Motor-Rotor Stand prior to engine removal. This is particularly true when engine removal is necessary to service the underneath side of the tractor or parts below the battery.

NOTE: The OTC Motor-Rotor Stand is not necessary for engine removal, but provides easy service access to all components from a convenient standing position.

1. If the stand is used, first shut off the fuel at the shut-off valve, disconnect the fuel lines and remove the fuel tank. Also remove the battery and drain crankcase and transmission.
2. Loosen four bolts from tractor frame and remove hood and grille components as a unit. Mount the tractor on the stand. The tractor can now be rotated for convenient disassembly.
3. Disconnect all electrical wires at engine. Remove choke and throttle cables from engine.
4. Pivot tractor on stand so fan guard can be removed from bottom of tractor. Remove two capscrews to free drive disc hub from drive disc. Loosen setscrew and remove hub from the driveshaft. If tractor has flex-coupling on front of driveshaft; remove two capscrews to disconnect coupling from clutch cone, leaving flex-coupling and driveshaft in tractor.
5. Loosen four bolts securing engine to engine base. Pivot tractor on stand, so engine once more is on top. Remove bolts from engine base.
6. Bolt a hoist chain to the engine lifting strap (on tractors so equipped). Move engine forward and upward over front bumper until clear of tractor.
7. Insert keyed driveshaft into clutch drive coupling. Swing hoist over tractor and lower engine into position on engine base.
8. Install drive disc hub on driveshaft. Position setscrew directly over center of key on shaft. Apply Loctite® and tighten setscrew. Attach drive disc hub to drive disc. Install fan guard.
9. Fasten engine to engine base with four bolts.
10. Install battery and gas tank.

NOTE: When installing gas tank, attach mounting straps with bolts in upper position. This will simplify future tank removals.

11. Connect fuel line to carburetor. Attach choke cable. Insert throttle cable in lower hole of adjusting screw lever.

NOTE: To properly adjust throttle cable, hold the adjusting screw lever rearward against stud while installing throttle cable clip on choke bracket.

12. Connect all electrical wires. Remove tractor from OTC Motor-Rotor Stand, if used, and install hood and grille components.

Models 200, 208, 210, 212, 214, 216

FRONT WHEELS AND AXLE

Front Wheels

REMOVAL AND INSTALLATION

1. Jack up tractor until wheel clears the ground. Remove capscrew. Remove spindle cap, outer bearing, wheel, inner bearing and spring washer.
2. Pack wheel bearings with John Deere Multipurpose lubricant or an equivalent SAE multipurpose-type grease.
3. Install spring washer, inner bearing, wheel, outer bearing, washer, spindle cap and capscrew.

NOTE: There is no adjustment necessary on the front wheel bearings.

Front Axle

REMOVAL AND INSTALLATION

1. Raise and support the front end on jack stands.

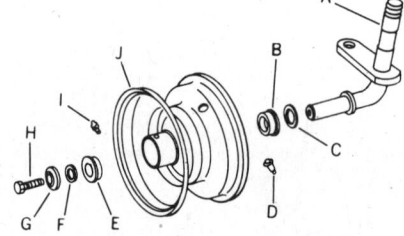

A. Spindle
B. Inner bearing
C. Spring washer
D. Valve stem
E. Outer bearing
F. Washer
G. Spindle cap
H. Capscrew
I. Grease fitting
J. Wheel

Bearings and wheels

John Deere

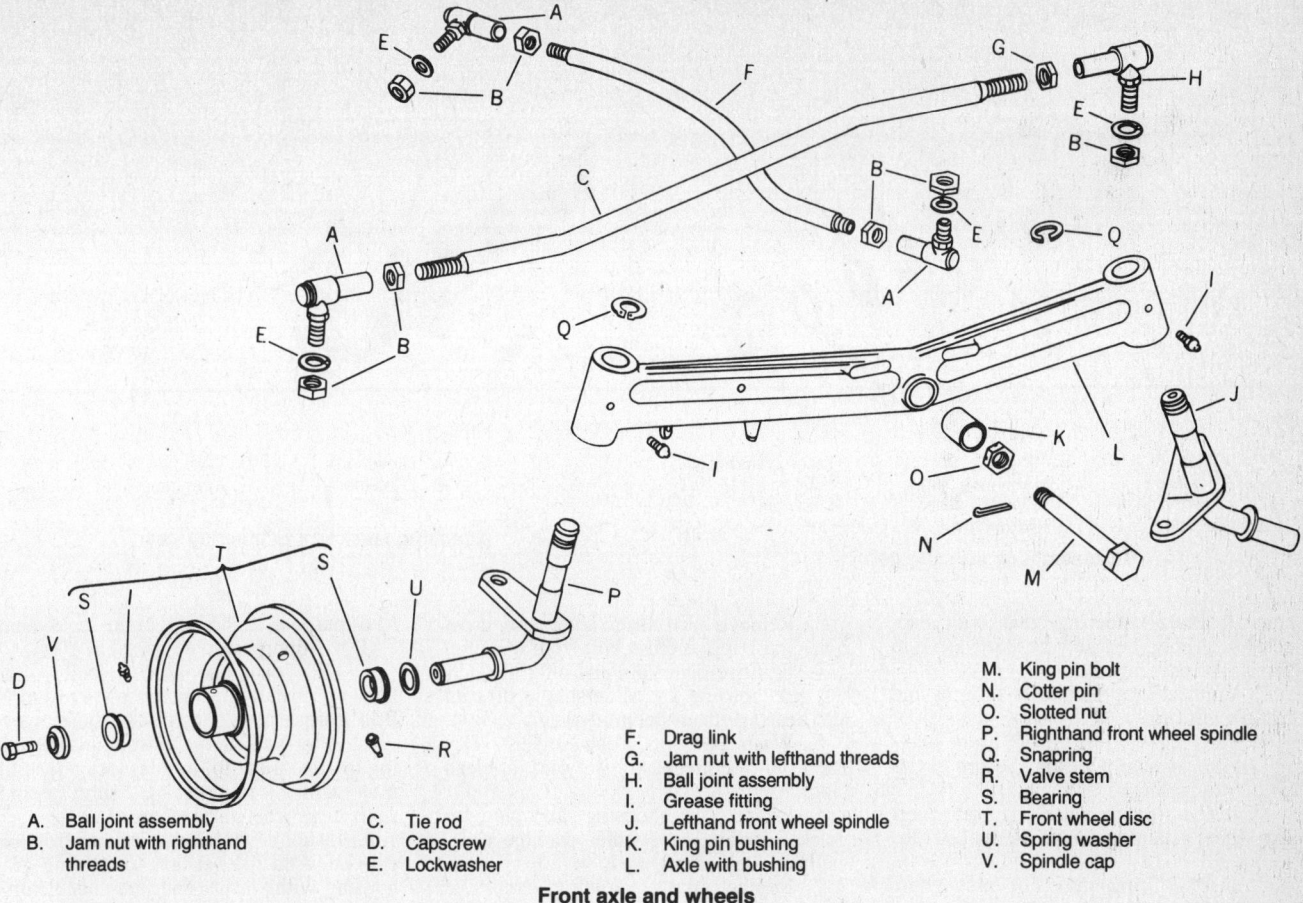

A. Ball joint assembly
B. Jam nut with righthand threads
C. Tie rod
D. Capscrew
E. Lockwasher
F. Drag link
G. Jam nut with lefthand threads
H. Ball joint assembly
I. Grease fitting
J. Lefthand front wheel spindle
K. King pin bushing
L. Axle with bushing
M. King pin bolt
N. Cotter pin
O. Slotted nut
P. Righthand front wheel spindle
Q. Snapring
R. Valve stem
S. Bearing
T. Front wheel disc
U. Spring washer
V. Spindle cap

Front axle and wheels

2. Remove the cotter pin from the slotted nut and remove the nut.
3. Remove the king pin bolt and roll the axle from under the tractor.
4. Grease king pin assembly and install axle. Tighten king pin bolt until it takes 25 to 75 lb. force at the end of the axle to rotate it up or down. Secure slotted nut with cotter pin.

Spindles

REMOVAL AND INSTALLATION

1. To remove spindle, disconnect tie rod end, and using a snapring pliers, remove snapring from spindle. Slip spindle out of axle.
2. Apply a light coat of grease on upper spindle shaft. Install spindle into axle. Secure with snapring.

STEERING

Steering Wheel and Gear

REMOVAL

1. Remove steering wheel with a puller.

NOTE: The wrong puller will damage the steering wheel.

2. Remove battery and battery box.
3. Remove clamp securing steering column to pedestal.
4. Disconnect drag link from lever arm.
5. Remove capscrews holding housing to frame and slip steering gear out from below tractor.

INSTALLATION

1. Position steering gear assembly in

A. Adjusting plug
B. Spring washer
C. Ball
D. Bearing kit
E. Grease fitting
F. Capscrew
G. Clamp
H. Capscrew
I. Steering gear assembly
J. Bearing
K. Steering wheel
L. Jam nut
M. O-ring
N. Cap
O. Jam nut
P. Seal
Q. Seal and retainer kit
R. Lever arm
S. Stud
T. Jam nut
U. Washer
V. Cotter pin

Steering system components

John Deere

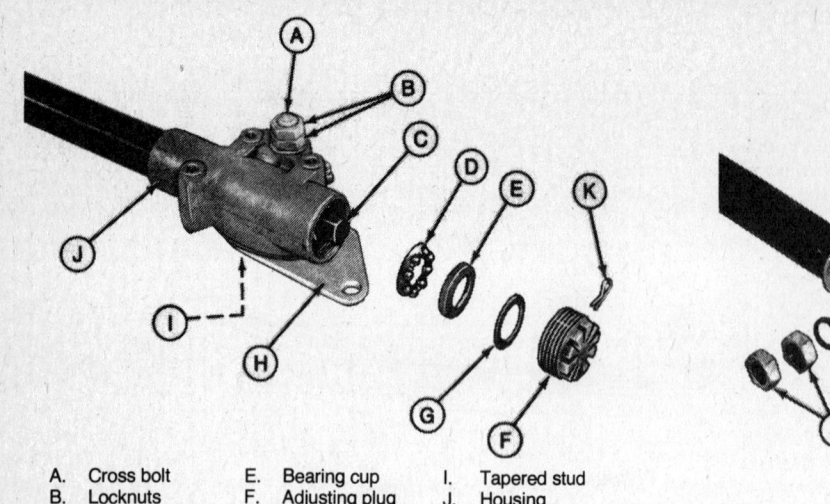
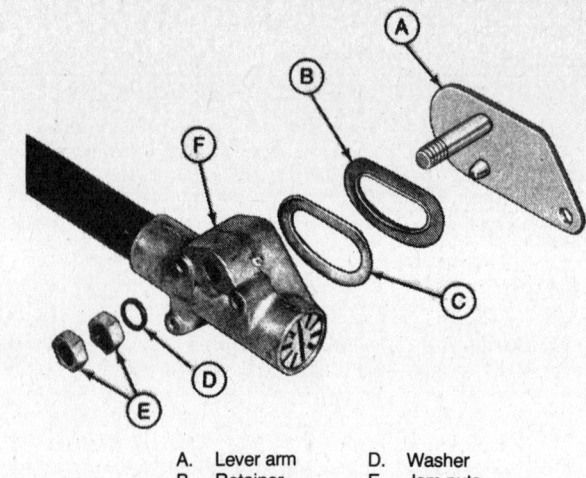

A.	Cross bolt	E.	Bearing cup	I.	Tapered stud
B.	Locknuts	F.	Adjusting plug	J.	Housing
C.	Steering shaft	G.	Spring washer	K.	Cotter pin
D.	Bearing	H.	Lever arm		

Disassembling steering gear

A.	Lever arm	D.	Washer
B.	Retainer	E.	Jam nuts
C.	Seal	F.	Housing

Attaching lever arm to steering gear

tractor. Install steering gear with capscrews. Apply Loctite® or equivalent to threads of capscrews.

2. Connect drag link to lever arm, and tighten nut firmly.

3. Place clamp over upper part of steering column inside pedestal. Secure clamp with two capscrews and nuts.

4. Install steering wheel. Tighten steering wheel retaining nut to 10 to 12 ft.lb. torque.

5. Insert O-ring into slot in steering wheel cap and press cap into steering wheel.

6. Adjust the steering gear.

DISASSEMBLY

1. Loosen jam nut on tapered stud in lever arm.

2. Turn stud in until resistance is felt.

3. Remove nuts from lever arm cross bolt, and remove cross bolt from housing.

4. Remove cotter pin and plug in steering gear housing and slide shaft with cam and bearings from column.

5. Wash parts in clean solvent. Dry them with compressed air and a clean cloth.

6. Inspect cam, housing, and plug for cracks, scoring, and other damage especially in the bearing area.

7. Replace parts showing excessive wear or damage.

ASSEMBLY

1. Apply grease (John Deere Multipurpose Lubricant or equivalent) and place bearing balls, ball cups and retaining rings on both ends of cam.

2. Grease cam lightly with John Deere Multipurpose Lubricant or an equivalent SAE multipurpose-type grease.

3. Slide cam and shaft assembly into housing. Install plug and tighten to 10 to 14 ft.lb. torque.

4. After tightening adjusting plug torquing to 10 to 14 ft.lb. torque, lock adjusting plug with a cotter pin. Be sure steering shaft turns freely after torquing.

5. Install new seal and retainer from repair kit. Attach lever arm to steering gear housing with washer and jam nuts.

ADJUSTMENTS

Steering Gear

1. Disconnect drag link (not illustrated) from lever arm (C).

2. Loosen jam nuts (A) on cross bolt (D).

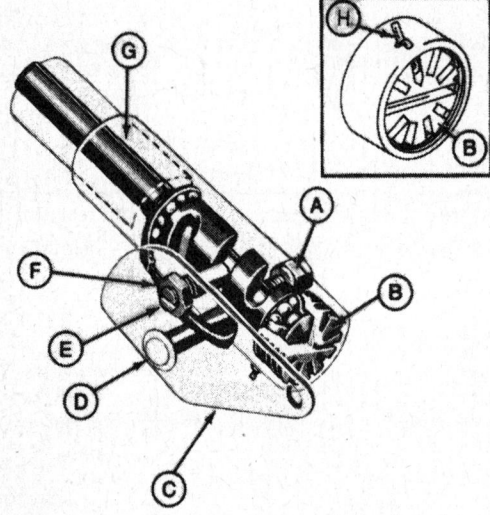
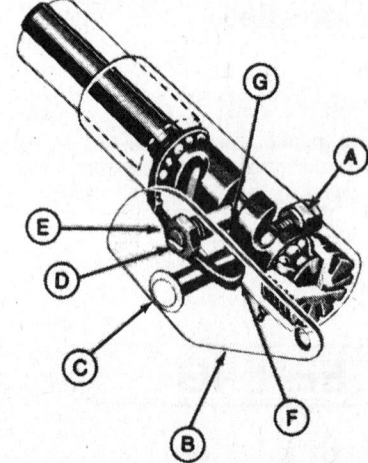

A.	Jam nuts	E.	Tapered stud
B.	Adjusting plug	F.	Jam nut
C.	Lever arm	G.	Housing
D.	Cross bolt	H.	Cotter pin

Adjusting bearings

A.	Jam nuts	E.	Jam nut
B.	Lever arm	F.	Housing face
C.	Cross bolt	G.	Mid-point on gear
D.	Tapered stud		

Adjusting backlash

John Deere

3. Loosen jam nut (F) and tapered stud (E) several turns.
4. Remove cotter pin (H) from adjusting plug.
5. Tighten adjusting plug (B) into housing (G) to 10 to 14 ft.lb. torque.
6. Install cotter pin (H).
7. Connect drag link to lever arm.
8. Adjust the steering gear so there is a slight drag at mid-point (G) as the steering wheel is turned through its full range of travel.
9. To find mid-point (G), count the number of turns from one extreme to the other.
10. Turn steering wheel one-half the total number of turns. Adjust at this point.
11. Loosen tapered stud (D) two or three turns.
12. To adjust backlash, place a 0.100 in. thick spacer between lever arm (B) and housing face (F).
13. Tighten inside jam nut (A) until spacer is held tight.
14. Then loosen inside jam nut just enough to remove spacer.
15. Tighten outside jam nut (A) to 40 ft.lb. torque.
16. Turn tapered stud (D) into housing until it is snug.
17. Tighten jam nut (E) to 40 ft.lb. torque.
18. Turn steering wheel through full range of travel, for final test.

Linkage

1. The overall length of the drag link determines left and right turning radius.
2. To equalize turning radius, disconnect the drag link from lever arm. Point wheels straight ahead.
3. Turn steering wheel through its full range, counting the number of turns. Turn wheel half-way back. Loosen jam nuts at the drag link. Then lengthen or shorten drag link until it can be attached to lever arm.
4. If drag link interferes with left front tire, loosen drag link ball joint jam nuts. Rotate drag link inward (toward center of the tractor). Lock ball joint jam nuts and re-test for equal turning.

BRAKES: 200 SERIES (SER. No. 30,001-95,000)

Brake Band
REPLACEMENT

1. Remove cotter pin and drilled pin from brake arm.
2. There are two access holes (A) in the lefthand frame through which a socket and extension can be inserted to remove the brake band retaining capscrews.
3. Remove the lefthand rear wheel. Remove the two capscrews and slip brake band (B) off bottom of brake pulley (C). Lift band assembly upward and to the right to remove.
4. Remove brake pin (A), and separate brake assembly from bracket (B). Reverse disassembly procedure to install new brake band. Adjust brake linkage.

Brake Pulley
REPLACEMENT

1. Loosen pulley retaining setscrew with hex socket wrench (A). Remove brake pulley (B) using a puller (C).
2. When replacing pulley, tap pulley onto shaft with a soft mallet and install setscrew. Use a thread lock compound to secure setscrew. Tighten setscrew to 280 in.lb. torque.

BRAKES: MODELS 210, 212, 214, 216

Brake Assembly
REMOVAL AND INSTALLATION

1. Block up tractor and remove lefthand rear wheel.
2. Remove two capscrews and gas line retaining clip
3. Remove cotter pin and drilled pin.
4. Slide brake band off pulley and remove assembly.
5. Installation is the reverse of removal.

Clevis
REPLACEMENT

1. Remove cotter pin and drilled pin.

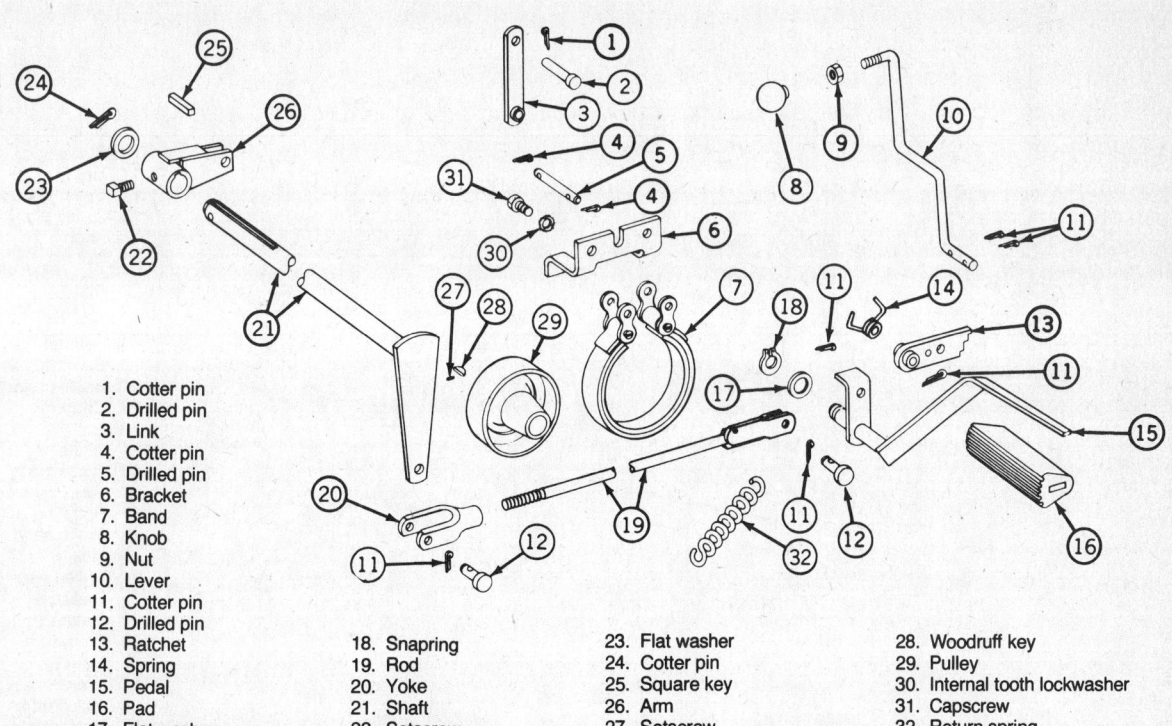

1. Cotter pin
2. Drilled pin
3. Link
4. Cotter pin
5. Drilled pin
6. Bracket
7. Band
8. Knob
9. Nut
10. Lever
11. Cotter pin
12. Drilled pin
13. Ratchet
14. Spring
15. Pedal
16. Pad
17. Flat washer
18. Snapring
19. Rod
20. Yoke
21. Shaft
22. Setscrew
23. Flat washer
24. Cotter pin
25. Square key
26. Arm
27. Setscrew
28. Woodruff key
29. Pulley
30. Internal tooth lockwasher
31. Capscrew
32. Return spring

Brake components, tractors models 200, 210, 212, 214 serial #30,001-55,000

John Deere

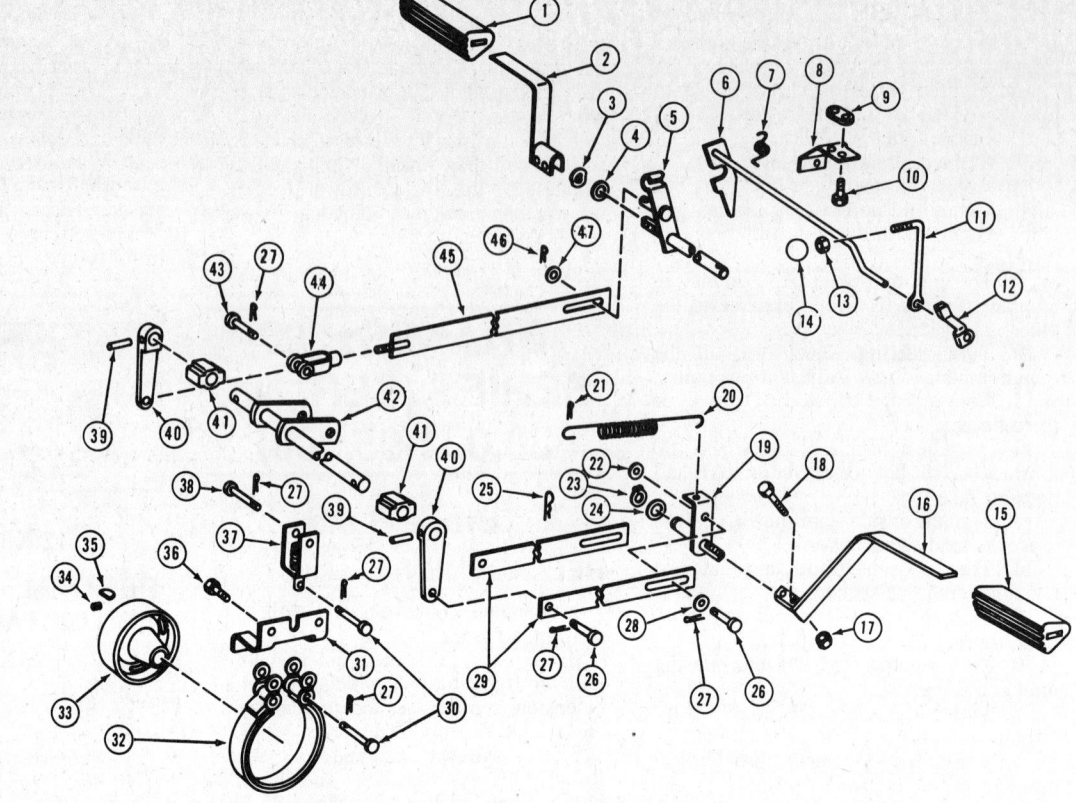

1. Pad
2. Clutch-brake pedal
3. Wave washer
4. Washer
5. Clutch-brake shaft
6. Parking brake shaft
7. Spring
8. Bracket
9. Strap
10. Capscrew
11. Parking brake lever
12. Guide
13. Hex nut
14. Parking brake knob
15. Pad
16. Brake pedal
17. Hex nut
18. Capscrew
19. Brake pedal shaft
20. Spring
21. Cotter pin
22. Washer
23. Snapring
24. Washer
25. Spring locking pin
26. Drilled pin
27. Cotter pin
28. Washer
29. Brake pedal straps
30. Drilled pin
31. Bracket
32. Brake band
33. Brake pulley
34. Setscrew
35. Key
36. Capscrew
37. Brake spring clevis
38. Drilled pin
39. Pin

40. Brake arm
41. Bearing
42. Brake arm shaft
43. Drilled pin
44. Clevis
45. Clutch-brake rod strap
46. Cotter pin
47. Washer

Brake components, tractors models 210, 212, 214, 216 serial #95,001 and up

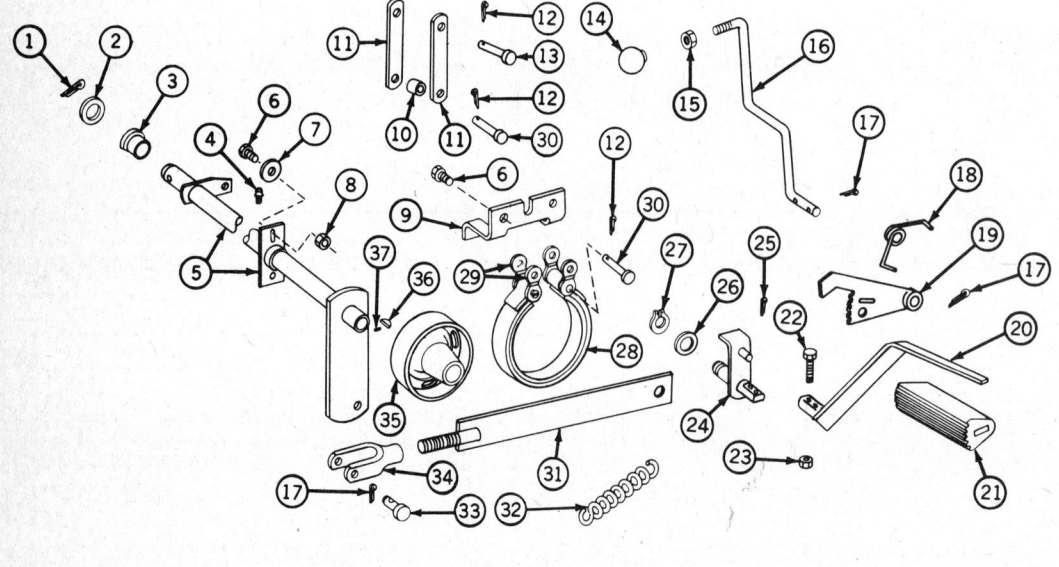

9. Bracket
10. Spacer
11. Link
12. Cotter pin
13. Drilled pin
14. Knob
15. Nut
16. Brake rod
17. Cotter pin
18. Springs
19. Ratchet
20. Brake pedal
21. Brake pedal pad
22. Capscrew
23. Nut
24. Brake pedal pad
25. Cotter pin
26. Flat washer
27. Snapring
28. Brake band
29. Link
30. Drilled pin
31. Brake rod
32. Return spring
33. Drilled pin
34. Yoke
35. Brake pulley
36. Woodruff key
37. Setscrew

1. Cotter pin
2. Flat washer
3. Bearing
4. Grease fitting
5. Brake shaft
6. Capscrew
7. Flat washer
8. Nut

Brake components, tractors models 200, 210, 212, 214 serial #55,001-95,000

John Deere

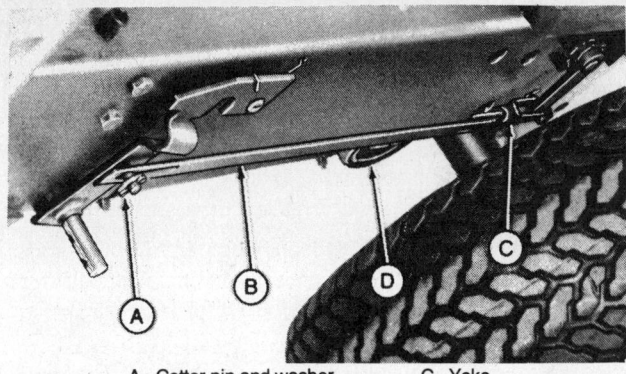

A. Cotter pin and washer
B. Strap
C. Yoke
D. Brake

Brake adjustment, tractor models 210, 212, 214, 216 serial #95,001 and up

2. Remove spring pin, drilled pin, spring and clevis.
3. Replace parts in reverse order.

Brake Band

REPLACEMENT

1. Remove cotter pins, drilled pins and brake band.
2. Replace parts in reverse order.

Brake Pulley

REPLACEMENT

1. Remove brake pulley setscrew.
2. Remove pulley from shaft. If pulley is stuck to shaft, use a puller to remove it.
3. Slide new pulley on shaft over key.
4. Apply thread lock compound on setscrew. Install setscrew and tighten to 280 in.lb. torque.

NOTE: When reinstalling brake assembly, be sure to install gas line retaining clip vertically so it does not pinch gas line.

Brake Adjustments

MODELS 200, 208, 210, 212, 214 (SER. NO. 30,001-95,000)

1. When brake pedal pushes down to fender-deck or no longer gives braking pressure, adjust brake clevis on righthand side of tractor. Lack of parking brake pressure also indicates need for adjustment.
2. To adjust, remove pin and turn clevis on brake rod as far as necessary to put brake in first notch of park lock.

MODELS 210, 212, 214, 216 (SERIAL NO. 95,001 AND UP)

To adjust brake, remove cotter pin and washer and disconnect strap from clutch-brake pedal stud on lefthand side of tractor. Turn strap into yoke two or three turns to tighten brake. Reconnect strap with washer and cotter pin and test brake by operating tractor on a level surface at a slow travel speed. The brake is properly adjusted when clutch-brake pedal is depressed and strong resistance is encountered when the pedal is approximately 1 in. from deck.

CLUTCH AND VARIABLE SPEED DRIVE

Primary Belt

REPLACEMENT

1. Remove righthand side panel. Unclip PTO arm (A) and push it to the rear away from the machine.
2. Loosen capscrew securing PTO brake shoe (B) enough so PTO clutch sheave (C) can be pulled off the engine crankshaft.
3. On 200, 210, 212, 214 and 216 tractors, move the variable speed control lever

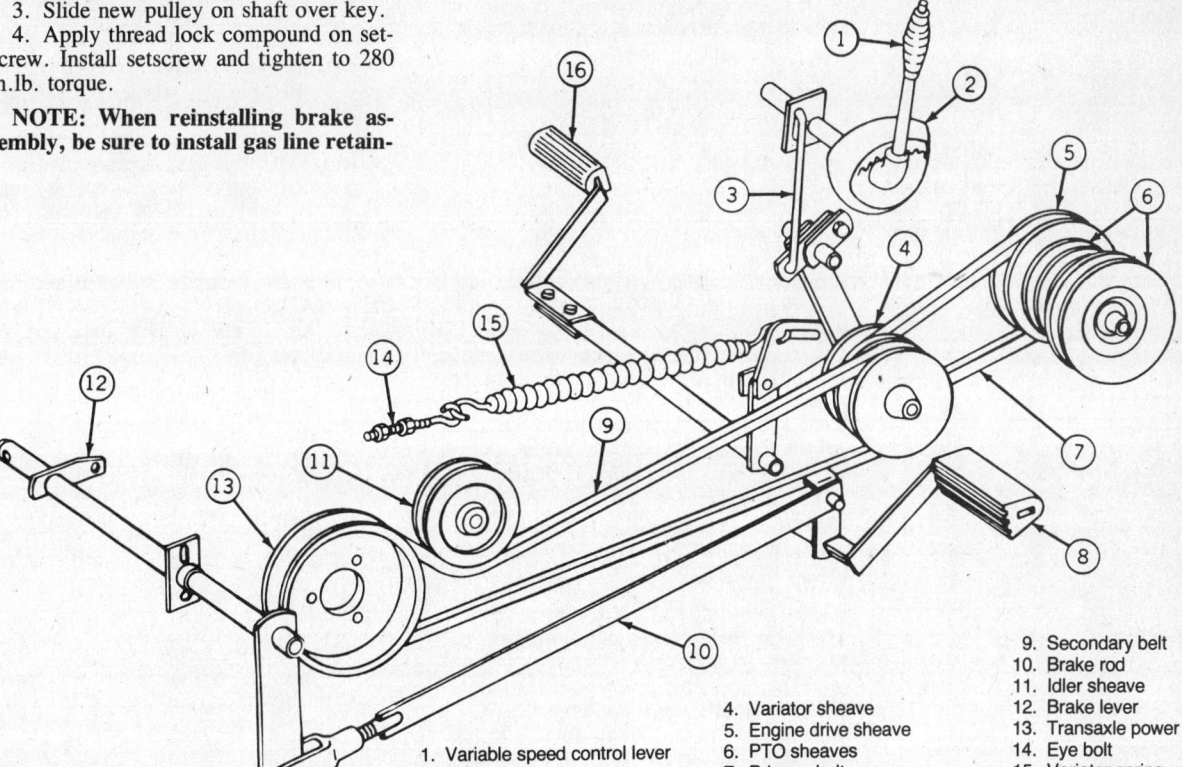

1. Variable speed control lever
2. Quadrant
3. Speed control link
4. Variator sheave
5. Engine drive sheave
6. PTO sheaves
7. Primary belt
8. Brake pedal
9. Secondary belt
10. Brake rod
11. Idler sheave
12. Brake lever
13. Transaxle power input sheave
14. Eye bolt
15. Variator spring
16. Clutch pedal

Power train components for models 200, 210, 212, 214 serial #30,001-95,000

John Deere

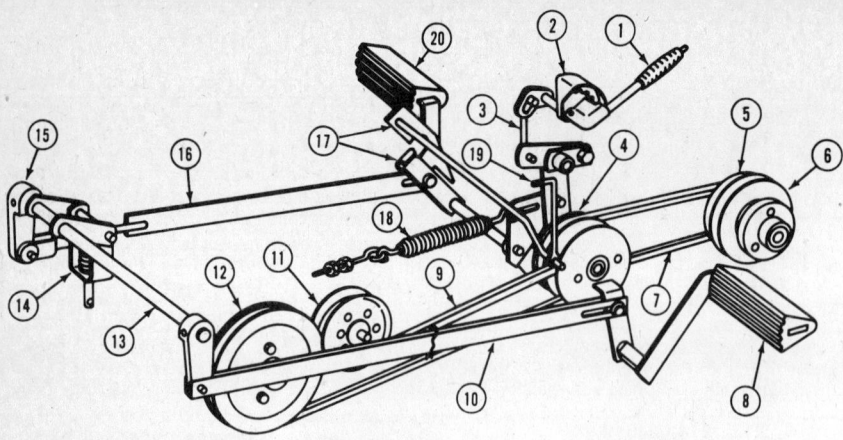

1. Variable speed control lever
2. Quadrant
3. Speed control link
4. Variator sheave
5. Engine drive sheave
6. PTO sheave
7. Primary belt
8. Brake pedal
9. Secondary belt
10. Brake pedal straps
11. Idler sheave
12. Transaxle power input sheave
13. Brake shaft
14. Brake shock absorber
15. Brake arm
16. Clutch-brake strap rod
17. Parking brake assembly
18. Variator spring
19. Parking brake rod
20. Clutch brake pedal

Power train components for models 210, 212, 214, 216 serial #95,001 and up

forward. Push up on secondary belt idler (A) and remove the secondary belt (B) from variator or secondary sheave.

4. Depress clutch pedal so secondary belt (B) may be pulled past brake pedal shaft (C).

5. Loosen the primary belt guide (D) at the variator or secondary sheave and remove the primary belt (E).

6. Install new belt in opposite order.

NOTE: After replacing primary belt, readjust variator on 200, 210, 212, 214 and 216 tractors.

Secondary Belt
REPLACEMENT

1. To replace a worn or broken secondary belt (B), move variable speed control lever forward on 200, 210, 212, 214 and 216 tractors. Raise the secondary belt idler (A) and slip secondary belt (B) off secondary or variator sheave.

2. Depress clutch pedal to allow sec-

ondary belt (B) to be pulled past the pedal shaft (C).

3. Loosen three capscrews on the transaxle driven sheave (A) and slide sheave off hub far enough to remove secondary belt (B).

4. Install new belt on variator or secondary sheave.

5. Raise secondary belt idler (D) and install belt on transaxle driven sheave (A). Tighten capscrews (C) onto driven sheave (A).

6. After belt replacement, check variator and brake adjustments on 200, 210, 212, 214 and 216 tractors.

Variator
REMOVAL

1. Remove secondary belt (A) from the variator sheave.

2. Loosen 5/16 in. capscrew securing primary belt guide (B) to variator arm. Pivot guide away to allow room for primary belt (C) to be removed.

3. Depress clutch pedal to pivot variator forward. Remove primay belt from variator sheave.

4. Remove battery and battery base to provide access. Disconnect speed control shaft link from speed control shaft.

5. Move the variable speed control lever forward.

6. Disconnect variator spring by removing nut from eyebolt.

--- **CAUTION** ---
Variator spring is under slight tension.

7. Remove plug from pedestal and remove variator pivot shoulder bolt.

8. Guide variator assembly out through bottom of tractor.

INSTALLATION

1. Install variator assembly from under

tractor. Guide pivot through notch in tractor frame.

2. Install shoulder bolt and tighten securely.

3. Connect speed control link with spring locking pin.

4. Attach variator spring to variator and eyebolt and tighten eyebolt for desired tension.

5. Reinstall battery base and battery.

6. Reinstall belts on variator.

7. After installation is completed, make final adjustments to variator.

DISASSEMBLY

1. Place variator half sheave in a vise with soft jaws. Insert ends of two large punches in holes of sheave. Place a bar between punches. Then turn counterclockwise to remove half sheave. Lift center sheave from variator hub.

2. Using the same procedure as described above, a special tool can be used. This tool is not available from a supplier, but can be made by drilling a bar of steel and inserting 3/8 in. round stock, 2¾ in. apart at the centers and 1 in. long. After tool has been tried and checked, weld round stock in place.

3. Place variator bearing and hub assembly under press with sheave down. Press bearing from hub. Be sure to press against outer race only.

4. Place hub in a vise with soft jaws and remove other half sheave using punches or the special tool. Press bearing out of variator arm.

5. After cleaning parts thoroughly, measure I.D. of center sheave bearing and O.D. of variator hub. Replace center sheave or hub if wear limits are exceeded. Do not attempt to service center sheave bearing. Bearing and center sheave are available as an assembly only.

6. Check center sheave and half sheaves for wear on the sheave faces or for evidence of damage or nicks. Replace parts that may cause excessive belt wear or would upset the balance of the variator assembly.

7. Measure press fit between bearing and hub. Check press fit of bearing shaft in variator arm. Replace parts necessary to obtain proper fit.

NOTE: The center sheave bearing is lubricated with a special grease at the factory that will last for the lifetime of the sheave. Do not attempt to lubricate this bearing.

8. Inspect brass bushings and shoulder bolt for wear or damage. The variator arm must pivot freely.

ASSEMBLY

1. Thread one sheave onto hub. Coat bearing case with a light film of oil. Place hub with sheave on press bed. Press bearing into hub shaft first until bearing end is 1/8 in. below hub face.

NOTE: Press on outer race of bearing only.

2. Wipe a light film of oil on bearing

A. Driven sheave
B. Eyebolt
C. Spring
D. Half sheave
E. Drive sheave
F. Primary belt
G. Center sheave
H. Secondary belt

Variator components

John Deere

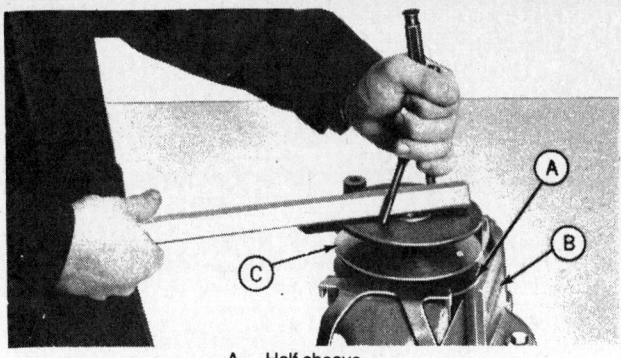

A. Half sheave
B. Vise
C. Center sheave

Removing variator sheaves

A. Hub assembly
B. Press

Pressing hub bearing

shaft. Place variator arm on bearing shaft with weld down. Press variator arm on bearing shaft until end of bearing shaft is flush with outside of variator arm.

3. Clamp assembly in vise having soft jaws. Place center sheave on hub and thread half sheave on hub. Using two large punches and bar, or special tool, tighten sheaves firmly by turning sheave in clockwise direction.

4. Spike threads three or four places on both sides of variator.

ADJUSTMENTS

Variator Linkage

1. To adjust the variator linkage, place the variable speed control lever (A) in notch 5 on the quadrant, which is the third notch from the front of the tractor.

2. Pry button plug from adjusting hole in right side of the tractor pedestal and loosen capscrew (D) one to two turns with a ¾ in. socket wrench (B).

3. Disconnect spark plug cable and ground. Turn ignition key (C) to crank engine several revolutions with starter until the clutch pedal raises as high as it will go.

4. Take up slack in linkage by pushing down on capscrew (D). Tighten capscrew (D) and replace button plug in adjusting hole.

NOTE: If, after adjusting variator linkage, tractor will not move when the variable speed control lever is in first notch on the quadrant (slow speed position) and the clutch pedal is released, install a new primary belt.

Spring

1. To obtain desired torque and load sensing characteristics, adjust the variator spring as follows:

2. For greater load sensitivity (variator increases torque earlier under load) loosen the spring tension by lengthening the eyebolt. For less load sensitivity, tighten the spring tension by shortening the eyebolt.

Belt Guide

If the primary belt jumps the variator sheave when the clutch pedal is depressed, the distance between the variator and primary belt guide should be checked. Distance between guide and sheave should not exceed ⅛ in.

TRANSAXLE

REMOVAL AND INSTALLATION

1. Remove fender-deck.
2. Remove fuel tank.
3. Remove secondary belt.
4. Block up tractor frame and remove rear wheels.
5. Disconnect rods or straps from brake arm.

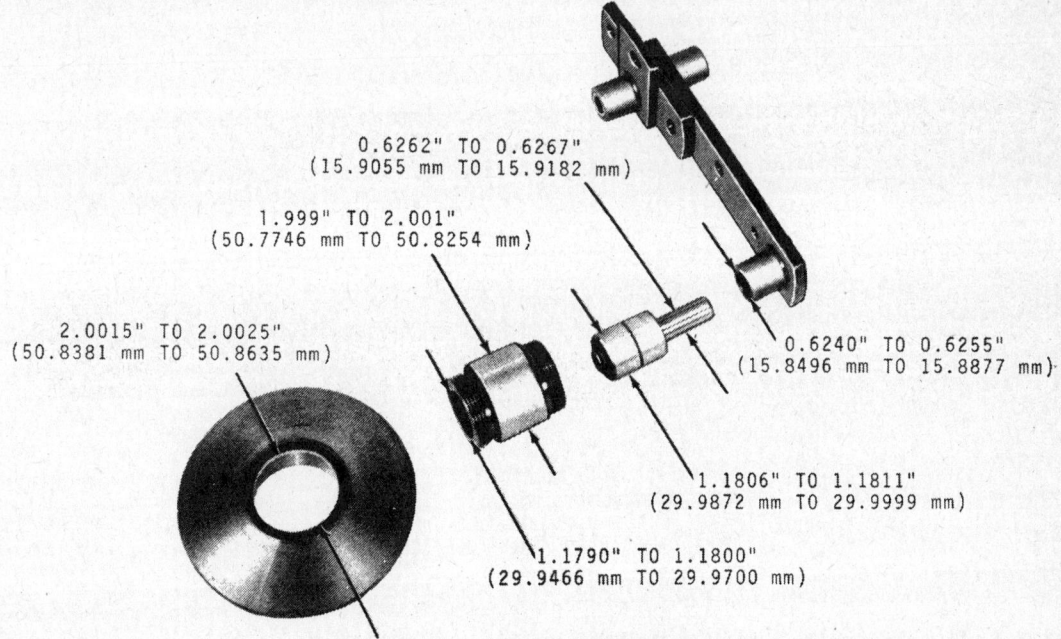

Variator components dimensions

385

John Deere

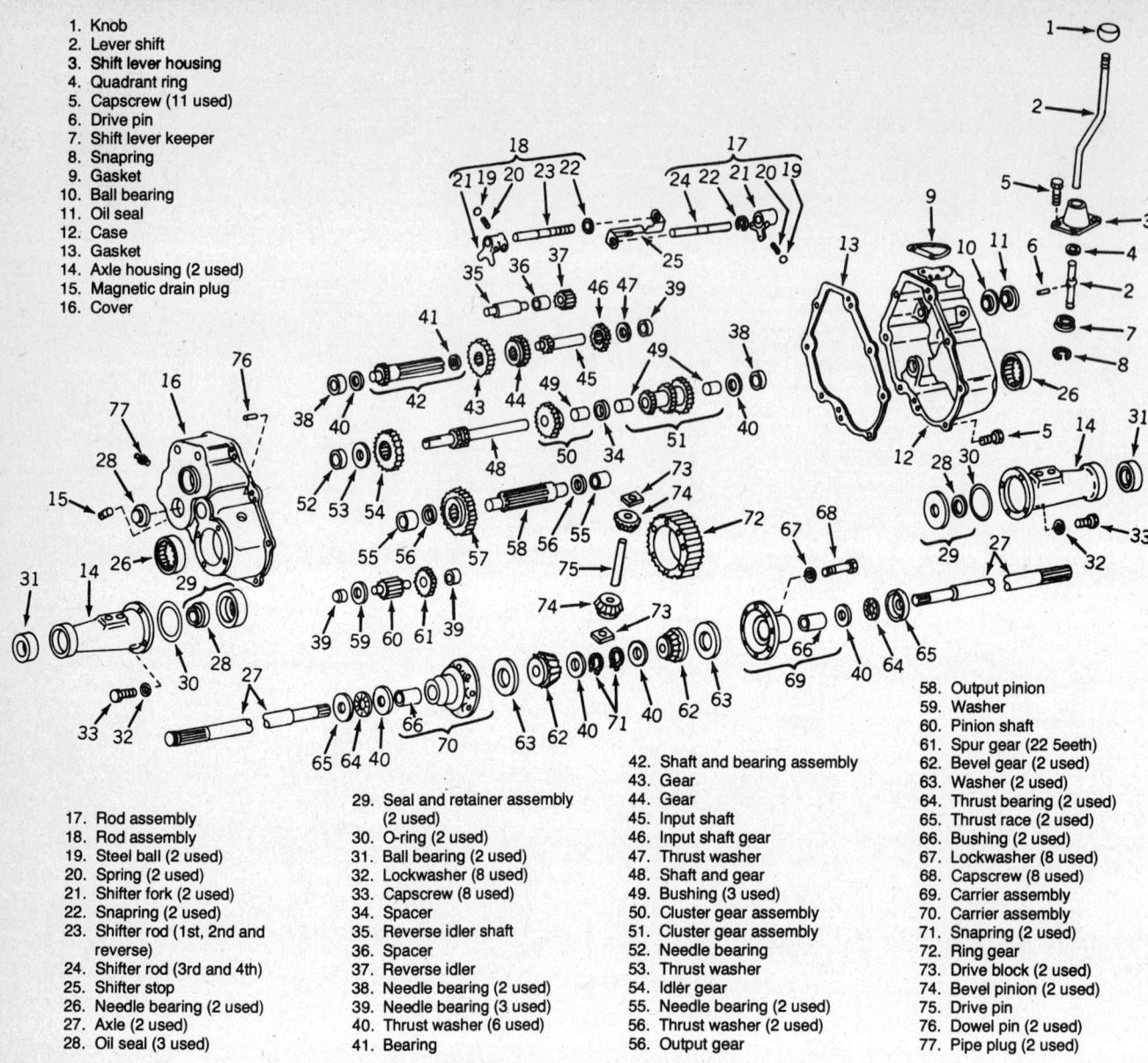

1. Knob
2. Lever shift
3. Shift lever housing
4. Quadrant ring
5. Capscrew (11 used)
6. Drive pin
7. Shift lever keeper
8. Snapring
9. Gasket
10. Ball bearing
11. Oil seal
12. Case
13. Gasket
14. Axle housing (2 used)
15. Magnetic drain plug
16. Cover
17. Rod assembly
18. Rod assembly
19. Steel ball (2 used)
20. Spring (2 used)
21. Shifter fork (2 used)
22. Snapring (2 used)
23. Shifter rod (1st, 2nd and reverse)
24. Shifter rod (3rd and 4th)
25. Shifter stop
26. Needle bearing (2 used)
27. Axle (2 used)
28. Oil seal (3 used)
29. Seal and retainer assembly (2 used)
30. O-ring (2 used)
31. Ball bearing (2 used)
32. Lockwasher (8 used)
33. Capscrew (8 used)
34. Spacer
35. Reverse idler shaft
36. Spacer
37. Reverse idler
38. Needle bearing (2 used)
39. Needle bearing (2 used)
40. Thrust washer (6 used)
41. Bearing
42. Shaft and bearing assembly
43. Gear
44. Gear
45. Input shaft
46. Input shaft gear
47. Thrust washer
48. Shaft and gear
49. Bushing (3 used)
50. Cluster gear assembly
51. Cluster gear assembly
52. Needle bearing
53. Thrust washer
54. Idler gear
55. Needle bearing (2 used)
56. Thrust washer (2 used)
56. Output gear
58. Output pinion
59. Washer
60. Pinion shaft
61. Spur gear (22 5eeth)
62. Bevel gear (2 used)
63. Washer (2 used)
64. Thrust bearing (2 used)
65. Thrust race (2 used)
66. Bushing (2 used)
67. Lockwasher (8 used)
68. Capscrew (8 used)
69. Carrier assembly
70. Carrier assembly
71. Snapring (2 used)
72. Ring gear
73. Drive block (2 used)
74. Bevel pinion (2 used)
75. Drive pin
76. Dowel pin (2 used)
77. Pipe plug (2 used)

4-speed transaxle components

6. Remove idler spring on 200, 210, 212, 214 and 216 tractors.

CAUTION

On 200, 210, 212, 214 and 216 tractors idler spring tension is severe. Place variator control in "FAST" position to reduce tension. Then, remove spring.

7. Place jack under transaxle and remove six capscrews holding hitch plate to tractor frame. Roll transaxle and hitch plate to the rear.
8. To facilitate transaxle disassembly, remove the components illustrated.
9. When removing the shift lever assembly, place lever in neutral before removing retaining screws.
10. When removing wheel hubs, use a wheel or gear puller to prevent damage.
11. Installation is the reverse of removal.

DISASSEMBLY

1. Drill two 2 in. holes 8 in. apart in a sturdy work bench or stand.
2. Place transaxle in bench or stand with socket head capscrews and input shaft upward. Remove eight screws.
3. Grasp the input shaft and the transaxle case. Lift case slowly and shake lightly so all loose parts remain in cover.
4. The accompanying illustrations will identify the group assemblies for the 4-speed transaxle. Lift them from the case in the following order:
   a. Gasket.
   b. Differential and axle assembly.
   c. Washer, 3-cluster gear, and spacer from shaft and pinion brake.
   d. Gear pinion and washer.
   e. Reverse idler assembly.
   f. Shifter rod and shaft assembly.
   g. 2-cluster gear.
   h. Output shaft and washers (one at each end of shaft).
   i. Shaft and pinion, idler gear and washer.
   j. Input shaft.

NOTE: Input shaft is installed with a press fit. If close inspection reveals that gears and bearing are satisfactory, do not remove input shaft.

5. If it is necessary to remove the input shaft, do not use the case itself to support any of the pressure required to separate the input assembly or brake shaft assembly from the case halves. Use a large pipe to support the pinion and press the shaft from the opposite side.
6. To disassemble shift lever remove snapring in shifter housing and slide assembly apart.

John Deere

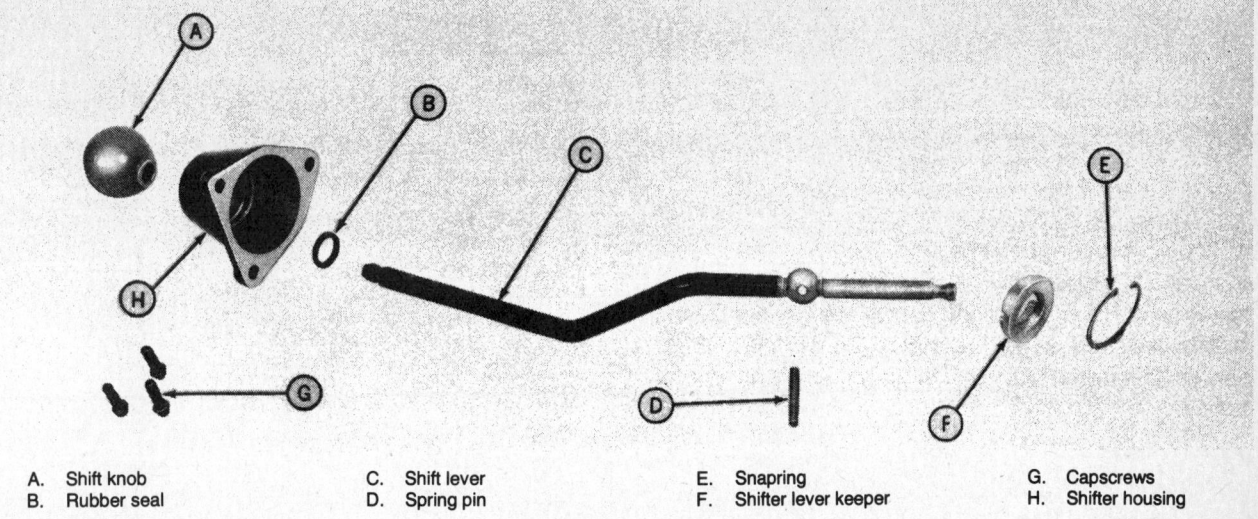

A. Shift knob
B. Rubber seal
C. Shift lever
D. Spring pin
E. Snapring
F. Shifter lever keeper
G. Capscrews
H. Shifter housing

Shift lever components

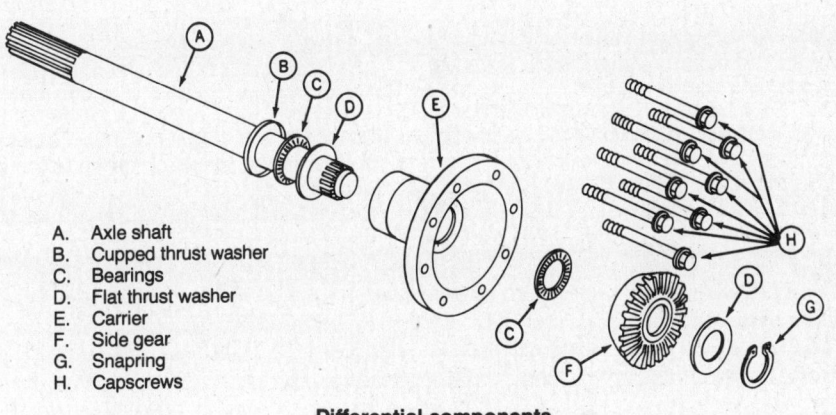

A. Axle shaft
B. Cupped thrust washer
C. Bearings
D. Flat thrust washer
E. Carrier
F. Side gear
G. Snapring
H. Capscrews

Differential components

INSPECTION AND ASSEMBLY

1. Wash all internal parts in a safe cleaning solvent. Brush and scrape foreign matter from all parts and dry thoroughly.

NOTE: Oil the bearings immediately after cleaning to prevent rusting.

2. Replace all gears having chipped, broken, or worn teeth. Badly scored gears must be replaced.

3. Replace any shaft that is bent, scored, or worn. Replace any shaft showing side wear or if any of the splines are damaged.

4. Chipped, broken, or excessive wear on gear teeth ends is usually caused by shifting transaxle while tractor is still moving or by gears not being properly meshed when tractor is under load. Gear wear can cause gears to jump out of position.

5. Broken detent springs can cause gear damage. When the springs are broken, the shifter fork is free to move, thus allowing gear pressure to slide the gears out of mesh.

6. When the gears slide out of gear, especially under load, gear chipping or cracking will result.

7. Prolonged heavy drawbar loads and wheel slippage are the most common cause of bevel pinion gear failure in the differential section of the transaxle.

8. Damage to the input shaft spline is caused by improper coupling of the shifter shaft and input shaft when transaxle is shifted into high range. A broken detent spring or an improperly adjusted quadrant are normally the cause of improper coupling.

9. A damaged shifter gear spline is caused by improper coupling of the shifter and input shaft. A worn or damaged shifter gear will cause gear jump-out when the tractor is operated in high range or under heavy drawbar loads.

10. Inspect the transmission case halves for cracks, worn or damaged bearing bores, damaged threads, and case mating surfaces.

11. Check condition of the shifter forks, shift rods, and detent springs. Slide forks along the shaft to inspect grooves. If a good snap is felt in each detent position, disassembly is not necessary.

12. Always replace oil seals in axle housings whenever transaxle is disassembled. Always use new O-rings on axle housings.

13. Check condition of differential drive blocks. Replace if cracked or broken.

14. To assemble the bevel gear differential, install the components on the axle shaft (A) in the order shown in the accompanying illustration. Secure with snapring (G).

15. Position the thrust washers (B and

A. Axle shaft
B. Pinion gears
C. Drive blocks
D. Pinion shaft
E. Ring gear
F. Carrier

Assembling ring and pinion gears

387

John Deere

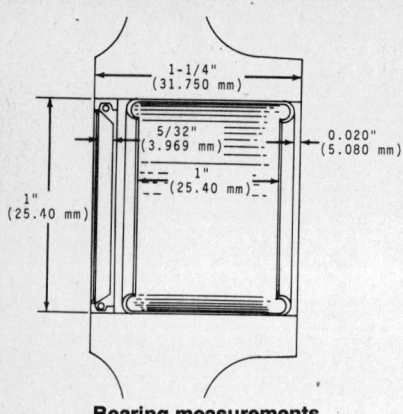

Bearing measurements

D) exactly as shown in the accompanying illustration. It is important that the cupped thrust washer (B) be placed onto axle first.

16. Place ring gear (E) onto one of the carriers (F) and install the pinion gears (B), pinion shaft (D), and drive blocks (C).

17. Position the other carrier and install the eight capscrews and lockwashers. Tighten capscrews to 8 to 10 ft.lb. torque.

18. The axles (A) should rotate freely in opposite directions when assembled. Place the differential aside for later installation.

19. All bearings are pressed into the bearing bores from the inside of the axle housing interior.

20. Bearing drivers to install bearings properly are available in most good supply houses. As a general rule, all bearings should be pressed into the housing to a depth of 0.020 in. beyond flush with housing interior.

21. Assemble input shaft (C), gear, and thrust washer (A). Chamfered gear spline (B) must be toward outer end of shaft (C). Gear is a light press fit onto shaft.

22. Install washer, input shaft and gear assembly (B) into case as shown. Use special tool to protect seal when slipping shaft through seal. Flat side of gear (C) should now face upward.

23. Assemble thrust washer (A), idler gear (B), and pinion shaft (C). Beveled edge on teeth (E) must face away from pinion shaft (C) as shown. Pinion shaft is a light press fit through idler gear.

24. When thrust washer, idler gear, and pinion shaft (A) are properly assembled and installed, they will appear as shown in the accompanying illustration. The flat edge of the idler gear (B) should now face upward.

25. The output gear (C) is assembled on the output pinion shaft (B) with a press fit. Thrust washer (A) is used on both ends of output shaft.

26. Install output gear (A), pinion shaft (B), and thrust washers (C) into lefthand case.

27. Install compound gear with bushing (D) into lefthand case.

28. Because of heavy detent pressure, the assembly of these rods can be difficult. Assemble forks as shown in the accompanying illustration.

29. 1st, 2nd and reverse fork (D) is assembled onto the shifter rod (H) so the fork will face away from the short end of the rod (G).

30. The 3rd and 4th shifter rod (B) must have the end opposite snapring (A) toward fork when assembled.

31. To assemble rod into forks, depress detents and slide forks along rod. A good snap should be felt in each detent. Place forks in neutral positions at this time.

32. To assemble shifter, lay out parts as shown in the accompanying illustration. Place forks (H) and (J) in neutral position. First, second, and reverse shifter rod (F) will have one detent showing on each side of fork (G).

Third and fourth shifter rod (A) will have one detent showing between fork and snapring (B). Be sure shifter rod (F) with detent showing on each side of fork is used with 1st, 2nd and reverse shifter gear (E), and shifter rod (A) is used with 3rd and 4th shifter gear (K).

33. The shifter shaft assembly should appear as shown in the accompanying illustration. The slot (E) in the forks should line up when the large gear (F) is slipped as far as possible on the spline. Note the position of the exposed grooves on shifter rods (C) and (D).

34. Assemble shifter guide (B) over shifter rods (C) and (D). Slot in guide (B) should match rectangular opening between the forks (E). The long notch in underside of guide (B) should clear the large 1st, 2nd and reverse shifter gear (A).

35. Place thrust washer (D) over needle bearing (E). Grasp shifter assembly (C) in left hand and lower it into case.

36. The shifter rods (B) should now enter the two machined sockets (A) in lefthand case.

37. The inside of the idler gear (A) is splined to slip freely onto splined end of idler pinion shaft (B).

38. Assemble reverse idler shaft with rounded edge of teeth facing spacer.

39. Install reverse idler assembly.

40. Install thrust washer (C), idler pinion shaft (E), and idler gear (D).

41. The accompanying illustration shows proper assembly before lowering into lefthand case.

42. Place spacer (B) on pinion shaft (F).

43. Install gear cluster (C) and thrust washer (B) on pinion shaft (A).

44. Install differential assembly (B) into lefthand case with capscrews (A) facing upward.

45. Position a new gasket (K) on lower (lefthand) case at this time.

46. Install righthand case half (B) over axle and input shaft (A).

47. Shake case slightly to align shafts and shifter rods (E). Also a short turn of the input shaft (A) will help align shafts and gears.

48. To close the last one-half in., tap the righthand case horizontally at point (F).

49. If case will not close, reach through round hole in righthand case (B) with a screwdriver and move shifter rods (E).

50. This will help align shifter rods (E) so they will fall into shifter rod sockets in righthand case (B).

A. Pinion shaft, idler and washer
B. Flat edge of gear up

Idler gear and shaft assembly

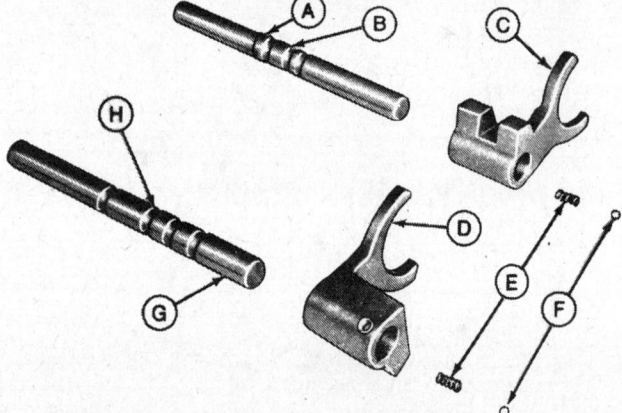

A. Snapring
B. 3rd and 4th shifter rod (3 detents)
C. 3rd and 4th shifter fork
D. 1st, 2nd and reverse shifter fork
E. Detent springs
F. Detent balls
G. Short end of rod
H. 1st, 2nd and reverse shifter rod (4 detents)

Shifter components

John Deere

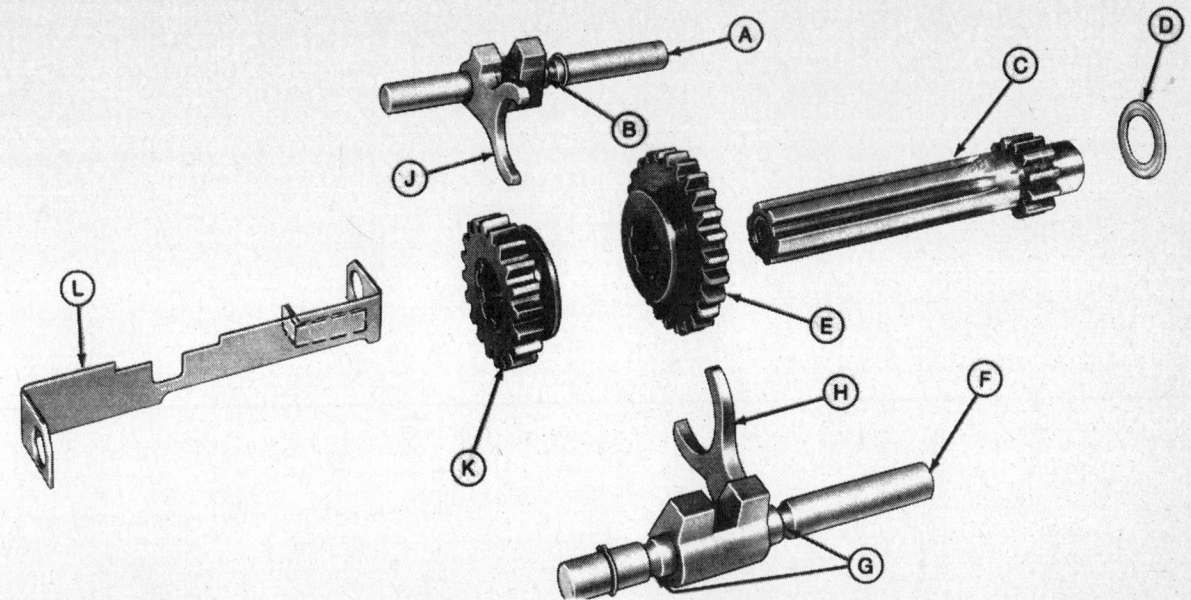

A. 3rd and 4th shifter rod
B. Neutral position (one detent showing between fork and snapring)
C. Shifter shaft
D. Thrust washer
E. 1st, 2nd and reverse shifter gear
F. 1st, 2nd and reverse shifter rod
G. Neutral position (two detents showing)
H. Fork 1st, 2nd and reverse
J. Fork 3rd and 4th
K. 3rd and 4th shifter gear
L. Shifter guide

Shifter shaft and gear components

51. Insert eight socket head screws (C) and tighten to 120 in.lb. torque.
52. Install retainer and new seal (B) with special oil seal cone tool (A) to prevent cutting seal during installation.
53. The seal is properly installed when the open face is facing inward toward the gears.
54. Install O-rings (B) and axle supports (A) with bearings.
55. Always use new O-rings. Tighten support bolts.
56. Inspect the shifter forks to be sure they are aligned and in neutral position.

57. Failure to do this will cause damage to the transmission when engaged under power.
58. The shifter is assembled in the order shown in the accompanying illustration.
59. When assembling shifter, be sure rubber seal is positioned properly in shifter housing. Shellac or gasket cement will hold seal in position during assembly.
60. Align housing, keeper and spring pin in shift lever and place snapring in groove in shifter housing.
61. Tighten screws to 120 in.lb. torque.

POWER TAKE-OFF (PTO)

PTO Clutch

DISASSEMBLY

1. Remove clutch arm (X) from PTO clutch pivot (A) and fulcrum bolt (W). Loosen clutch brake shoe (T) to allow PTO clutch cup (M) to be removed. Remove

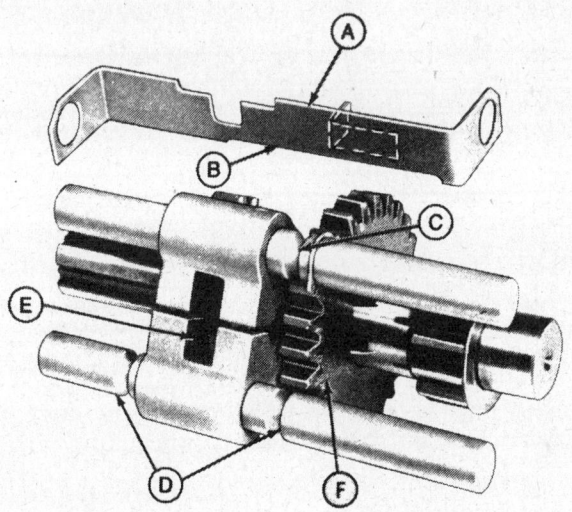

A. Shifter guide
B. Long notch
C. Shifter rod neutral position
D. Shifter rod neutral position
E. Slot (forks)
F. Large gear

Shifter in the neutral position

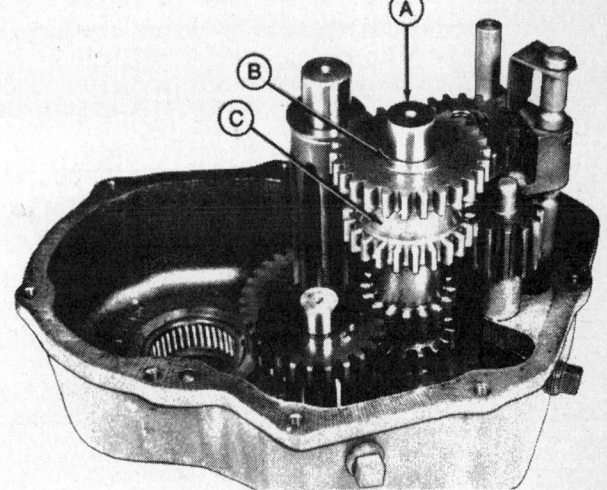

A. Pinion shaft
B. Thrust washer
C. Gear cluster

Transmission assembled

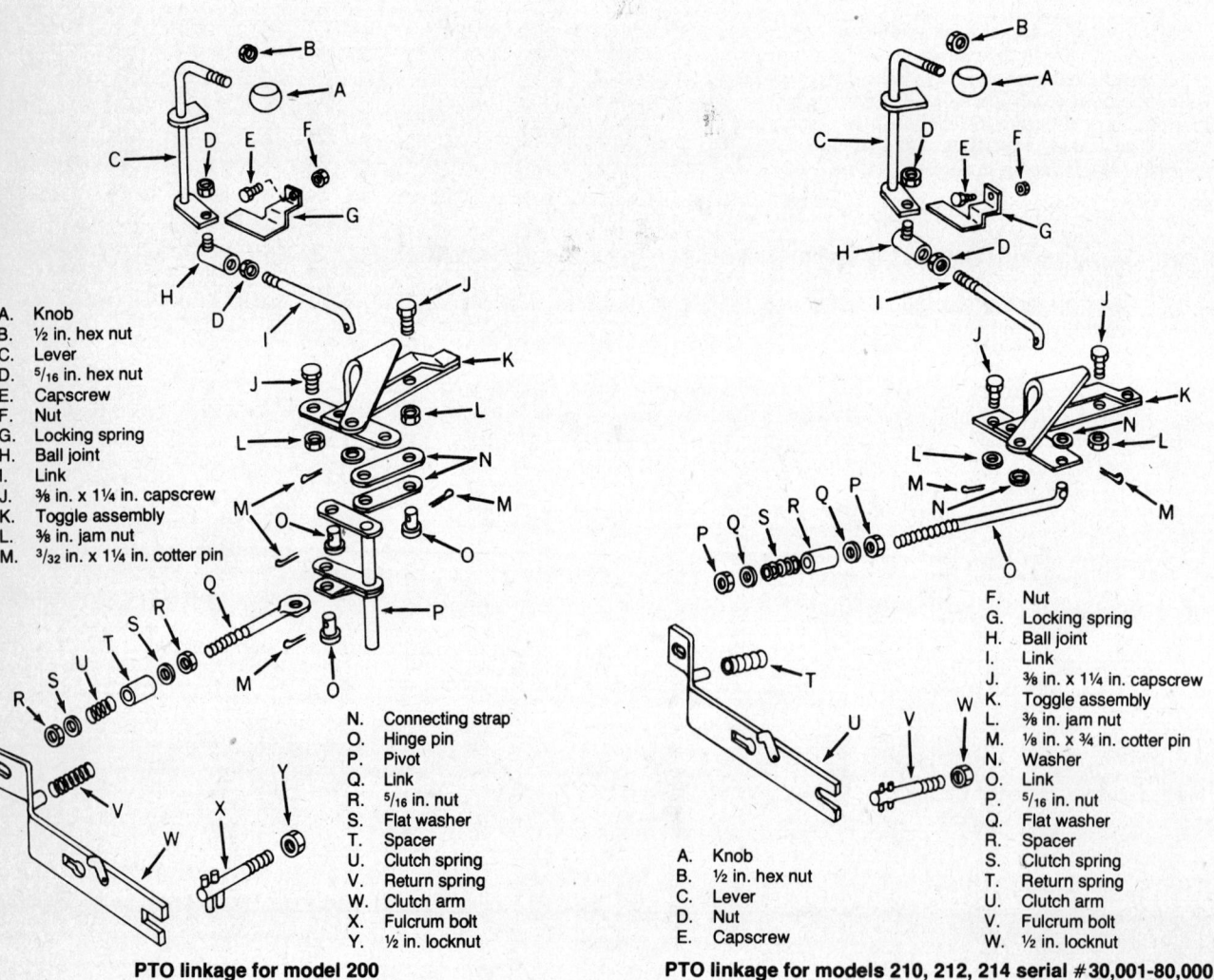

John Deere

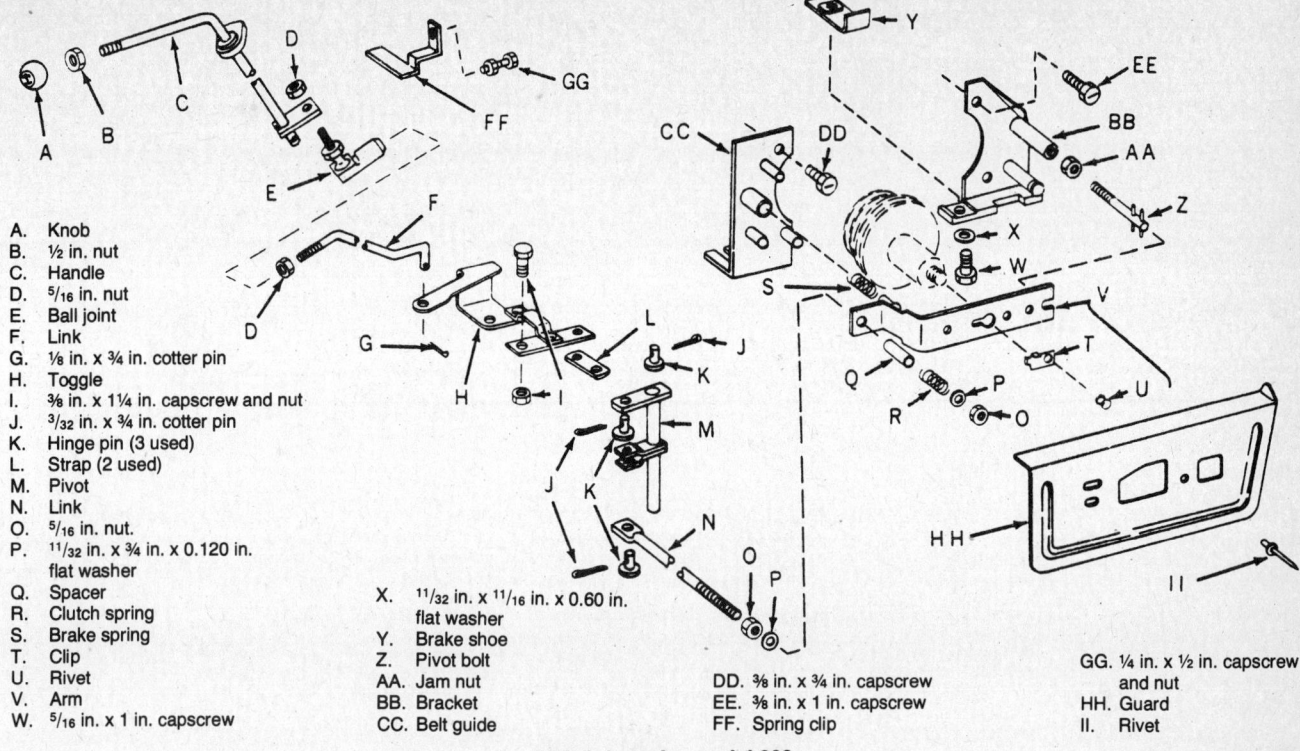

A. Knob
B. ½ in. nut
C. Handle
D. 5/16 in. nut
E. Ball joint
F. Link
G. ⅛ in. x ¾ in. cotter pin
H. Toggle
I. ⅜ in. x 1¼ in. capscrew and nut
J. 3/32 in. x ¾ in. cotter pin
K. Hinge pin (3 used)
L. Strap (2 used)
M. Pivot
N. Link
O. 5/16 in. nut.
P. 11/32 in. x ¾ in. x 0.120 in. flat washer
Q. Spacer
R. Clutch spring
S. Brake spring
T. Clip
U. Rivet
V. Arm
W. 5/16 in. x 1 in. capscrew

X. 11/32 in. x 11/16 in. x 0.60 in. flat washer
Y. Brake shoe
Z. Pivot bolt
AA. Jam nut
BB. Bracket
CC. Belt guide

DD. ⅜ in. x ¾ in. capscrew
EE. ⅜ in. x 1 in. capscrew
FF. Spring clip

GG. ¼ in. x ½ in. capscrew and nut
HH. Guard
II. Rivet

PTO linkage for model 208

capscrew (N) and washers (O and P) from end of crankshaft and remove bearing inner race (Q). Remove clutch cone (R) and key (S).

2. Remove outer snapring (B). Press PTO clutch pivot (A) out of PTO clutch cup (I). Remove inner snapring (J). Press two needle bearings (K) and oil seal (L) out of PTO drive sheave. Remove small snapring (D) from PTO clutch pivot (A) and press bearing (C) off clutch pivot (A).

INSPECTION

Inspect clutch linings and mating surfaces for excessive wear. Inspect bearings, bearing inner race and seal. Inspect PTO brake shoe for excessive wear. Replace parts as necessary.

ASSEMBLY

1. Install key in crankshaft and slide clutch cone (A) onto crankshaft.

2. Install bearing inner race (B), retaining washer, lockwasher and capscrew (C). Tighten capscrew securely.

3. Press bearings (C) and seal (B) into PTO drive sheave (A) to dimension given in the accompanying illustration. Pack area (D) with John Deere High-Temperature Grease or equivalent. Press clutch pivot (E) into ball bearing (F) and secure with small snapring (not illustrated). Install inner snapring (H) into PTO drive sheave (A).

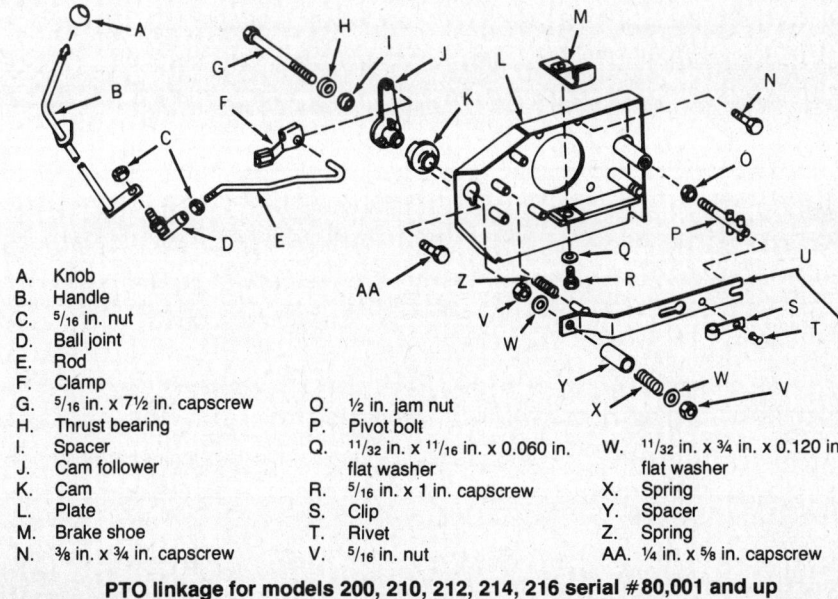

A. Knob
B. Handle
C. 5/16 in. nut
D. Ball joint
E. Rod
F. Clamp
G. 5/16 in. x 7½ in. capscrew
H. Thrust bearing
I. Spacer
J. Cam follower
K. Cam
L. Plate
M. Brake shoe
N. ⅜ in. x ¾ in. capscrew

O. ½ in. jam nut
P. Pivot bolt
Q. 11/32 in. x 11/16 in. x 0.060 in. flat washer
R. 5/16 in. x 1 in. capscrew
S. Clip
T. Rivet
V. 5/16 in. nut

W. 11/32 in. x ¾ in. x 0.120 in. flat washer
X. Spring
Y. Spacer
Z. Spring
AA. ¼ in. x ⅝ in. capscrew

PTO linkage for models 200, 210, 212, 214, 216 serial #80,001 and up

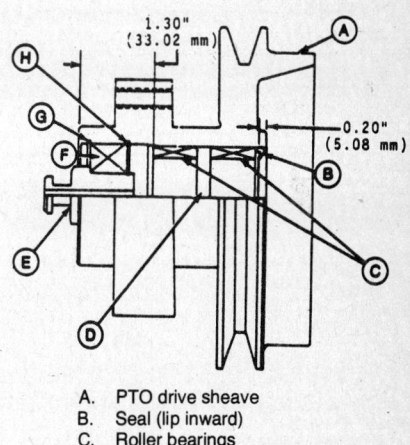

A. PTO drive sheave
B. Seal (lip inward)
C. Roller bearings
D. Pack this area with grease
E. Clutch pivot
F. Ball bearing
G. Outer-snapring
H. Inner snapring

PTO drive sheave components

John Deere

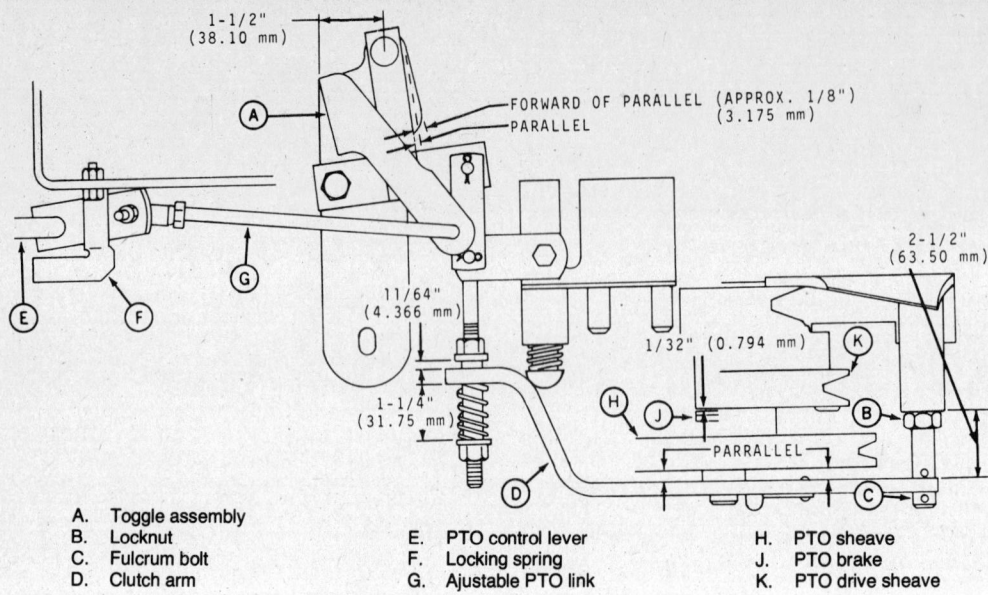

- A. Toggle assembly
- B. Locknut
- C. Fulcrum bolt
- D. Clutch arm
- E. PTO control lever
- F. Locking spring
- G. Ajustable PTO link
- H. PTO sheave
- J. PTO brake
- K. PTO drive sheave

Model 200 and 208 linkage adjustments

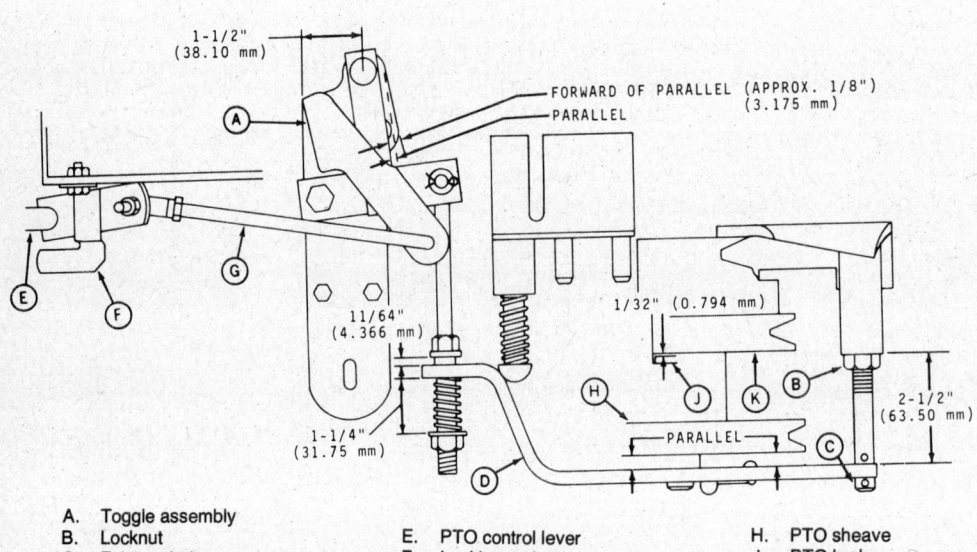

- A. Toggle assembly
- B. Locknut
- C. Fulcrum bolt
- D. Clutch arm
- E. PTO control lever
- F. Locking spring
- G. Adjustable PTO link
- H. PTO sheave
- J. PTO brake
- K. PTO drive sheave

Model 210, 212, 214 serial #30,001-80,000 linkage adjustments

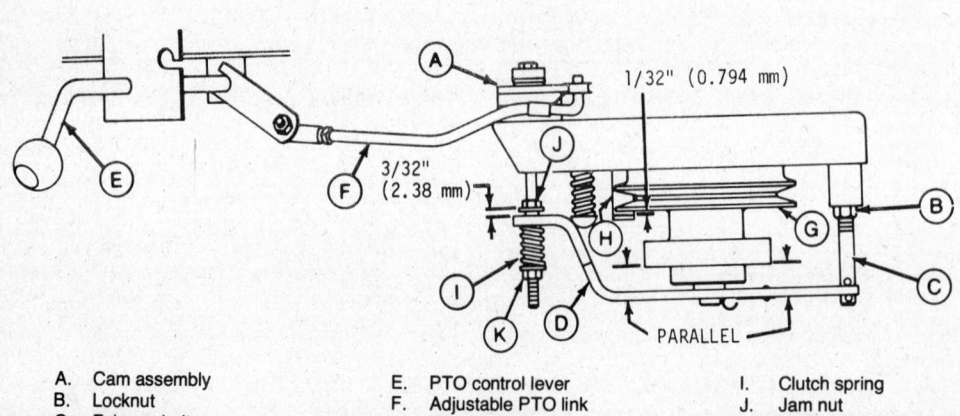

- A. Cam assembly
- B. Locknut
- C. Fulcrum bolt
- D. Clutch arm
- E. PTO control lever
- F. Adjustable PTO link
- G. PTO sheave
- H. PTO brake
- I. Clutch spring
- J. Jam nut
- K. Adjusting nut

Model 210, 212, 214, 216 serial #80,001 and up adjustments in the engaged position

John Deere

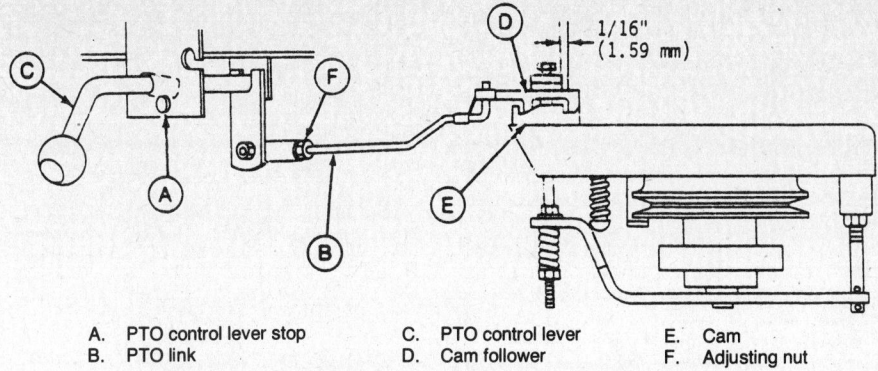

- A. PTO control lever stop
- B. PTO link
- C. PTO control lever
- D. Cam follower
- E. Cam
- F. Adjusting nut

Model 210, 212, 214, 216 serial #80,001 and up linkage adjustments

Secure clutch pivot and ball bearing assembly in place with outer snapring (G).

4. Slide complete PTO clutch sheave assembly onto bearing inner race.

ADJUSTMENTS

Before Serial No. 80,000

1. Adjust length of PTO link so the PTO lever points forward and the toggle is over-center approximately 1½ in. or ⅛ in. forward of parallel.
2. Adjust the fulcrum bolt to a preliminary setting of 2½ in. from the back of the locknut to the inside of clutch arm.
3. Adjust the clutch spring so when the PTO is in the engaged position the distance from the clutch arm to the outside washer is 1¼ in.
4. Readjust the fulcrum bolt and clutch spring if necessary to maintain a parallel condition between the clutch arm and PTO sheave with the PTO engaged.
5. After the clutch adjustments have been made, adjust the PTO brake for a $1/32$ in. gap between the brake pad and the PTO drive sheave with the PTO engaged.

After Serial No. 80,000

1. Place PTO control lever in disengaged position and check clearance between lug of cam follower and notch of cam. Clearance should not exceed $1/16$ in. If necessary, disconnect PTO link and turn adjusting nut on or off link until clearance of $1/16$ in. exists when link is connected and PTO clutch control lever is disengaged.
2. Engage PTO control lever. Adjust clutch arm parallel with PTO sheave. Loosen locknut and turn fulcrum bolt until clutch arm is parallel with face of PTO sheave. Tighten locknut after adjustment.
3. Spring tension gap should be $3/32$ in. between washer and clutch arm with PTO control lever engaged. To adjust, loosen jam nut and turn adjusting nut on or off link until spring tension gap is correct. Tighten jam nut after adjustment.
4. With PTO control lever engaged, place a $1/32$ in. feeler gauge between PTO brake and PTO sheave. When properly adjusted, clearance between PTO brake and PTO sheave should not exceed $1/32$ in.

NOTE: The PTO brake must stop the PTO sheave in 4 seconds at an engine speed of 3600 rpm under no load when PTO lever on dash of tractor is disengaged. If the PTO will not stay engaged, replace the cam and readjust PTO.

HYDRAULIC SYSTEM

Control Valve

REMOVAL

1. Lower equipment to the ground. With the engine stopped, move the control lever forward and back to relieve all pressure in the system.
2. Remove the battery and battery box.
3. Wipe all hoses and connections clean. Disconnect hose from reservoir to pump. Drain oil into can.
4. Disconnect hoses to cylinder. Disconnect hose from pump.
5. Remove the two bolts and nuts holding the lefthand pedestal side. Remove the pedestal side with reservoir and valve.
6. Disconnect link between lift lever arm and control valve.

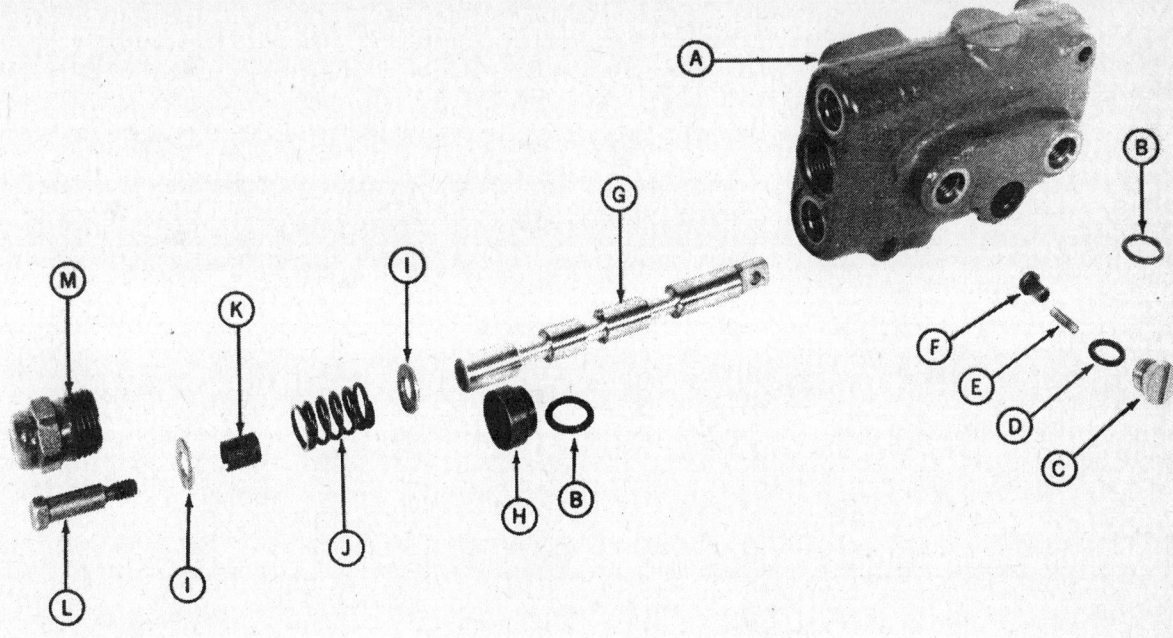

- A. Valve body
- B. O-ring (2 used)
- C. Lift check plug
- D. O-ring
- E. Lift check spring
- F. Lift check plunger
- G. Spool
- H. Sleeve
- I. Washer (2 used)
- J. Spring
- K. Spacer
- L. Shoulder bolt
- M. Cap

Control valve components

John Deere

7. Disconnect line from reservoir at valve.
8. Remove mounting bolts from the control valve and remove valve from side pedestal.
9. Plug hydraulic lines and hoses to prevent dirt from entering.

DISASSEMBLY

1. Plug all ports and clean outside of valve thoroughly.
2. Remove the spool cap and spool. Do not remove the capscrew unless the spring needs to be replaced. The spring need not be replaced unless it is broken.
3. Remove the spool from the valve body.

NOTE: Spool and valve body are matched sets. If one is damaged, both must be replaced.

4. Remove the sleeve and O-ring from the spool.
5. Remove lift check plug, spring and plunger from the valve body. Remove the O-ring from the plug.
6. Remove O-ring from control valve body, using a small wire or paper clip. Discard all O-rings.

INSPECTION

1. Remove all nicks and burrs from ports and inspect the spool and valve body for excessive wear.
2. Inspect the spool spring and lift check spring for breakage.
3. Do not inspect O-rings. Replace them.

ASSEMBLY

1. Thoroughly clean and dry all parts. Metal parts should be lightly oiled prior to assembly.

NOTE: Install all new O-rings.

2. Position new O-ring in spool bore.
3. Install O-ring on lift check plunger. Position spring on plunger and place in valve body. Install plug.
4. If spring on spool was removed for replacement, capscrew should be installed with Loctite® and tightened to 5 to 8 ft.lb. torque.
5. Slide spool sleeve over spool. Place O-ring over spool and position it on sleeve. Dip spool in clean oil and insert spool and spring assembly into valve body.
6. Install the spool cap to secure spool assembly in valve.
7. Reverse the removal procedure.

Pump

REMOVAL

1. Disconnect inlet line from pump and drain oil from reservoir into a container.
2. Disconnect outlet line from pump.

NOTE: Plug hydraulic lines to keep out dirt.

3. Loosen pump mounting capscrews to loosen drive belt. Remove pump drive sheave. Remove key. Do not lose key.
4. Remove pump mounting capscrews, nuts and lockwashers and remove pump.

DISASSEMBLY

1. Scribe a line across the three sections of the pump as a guide for reassembly.
2. Remove the four capscrews and the front cover. The center section will remain attached to either the front or back cover.
3. Place the drive gear into the unseparated sections; then, remove the center section being careful to avoid cocking it on the dowel pins.
4. Remove the gears.
5. Mark the front cover island next to the pressure vent hole in the wear plate to act as a guide for reassembly. The location of this vent hole determines pump rotation.
6. Use a small diameter wire (a paper clip will do) to remove the wear plate, heat shield, gasket and V-seal. Discard these parts. Wear plate wear ridges of more than 0.0005 in. indicates a worn pump in need of repair.
7. Remove and discard both O-rings.
8. Do not remove shaft seal in front cover unless it is damaged or leaking. If the seal is to be replaced be careful not to damage the seal recess. Heat the cover in an oven to 250°F to reduce the press fit.
9. If the relief valve is defective, replace it as a complete unit.

INSPECTION

1. Inspect shafts for roughness in bearing and sealing areas. Minimum shaft diameter is 0.4998 in.
2. Inspect bearing bore for wear. Maximum bore diameter should not exceed 0.5015 in.
3. Inspect key way and key for damage or excessive wear.
4. Inspect gear end faces, outside diameter and teeth for roughness and score marks. Minimum gear width is 0.2770 in. Minimum gear outside diameter is 1.2395 in.
5. Inspect the pump center section at the wall of gear bore diameters for excessive wear or score marks. The center section will show wear at the inlet side of the pump. This wear ridge should not exceed $1/32$ in.

ASSEMBLY AND INSTALLATION

1. Clean all parts thoroughly before assembly.
2. If it was removed, install shaft seal into front cover with spring-loaded lip facing inward. Place the front cover on a smooth, flat surface. Use a flat steel rod slightly smaller in diameter than the outside diameter of the seal to force the seal into the front cover.
3. In this order, install the V-seal, gasket, heat shield and wear plate into the front cover as follows:
4. The small vent hole through all these parts must be in line and positioned next to the scribe mark made during disassembly. This locates the vent holes on the outlet side of the pump.
5. The lips of the V-seal must face

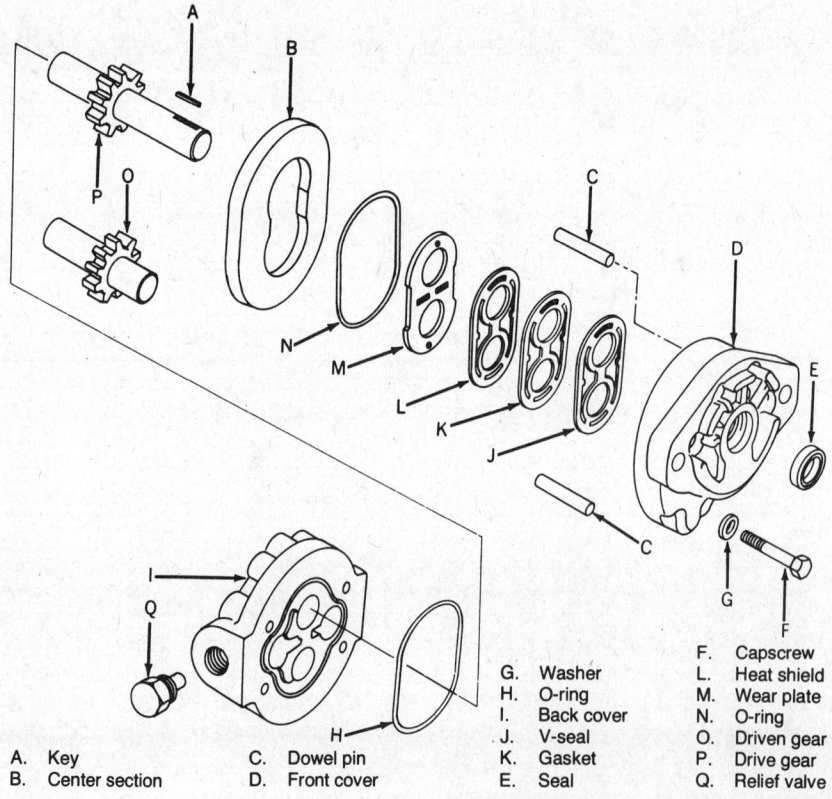

A. Key
B. Center section
C. Dowel pin
D. Front cover
E. Seal
F. Capscrew
G. Washer
H. O-ring
I. Back cover
J. V-seal
K. Gasket
L. Heat shield
M. Wear plate
N. O-ring
O. Driven gear
P. Drive gear
Q. Relief valve

Pump motor components

toward the cavity. Use a small, dull screwdriver or similar tool to carefully tuck the seal into the groove.

6. Press the gasket and heat shield firmly toward the bottom of the cavity with your thumbs to provide sufficient space for the wear plate.

7. Install the wear plate with the bronze surface up (toward the gears) and the small vent hole in line with the vent hole in the heat shield.

8. Install O-ring into front cover. If necessary, oil and stretch O-ring slightly so it will remain in its groove.

9. Install drive gear and driven gear into the front cover. Apply oil to the shaft at the drive end to prevent damage to the shaft seal. Make sure that the shaft seal lip and spring is not pushed out by the shaft.

10. Install dowel pins into front cover.

11. Make sure the wear plate is still seated in its cavity.

12. Align the scribe mark on the outside of the center section with the scribe mark on the front cover and install the center section over the gears. Also, the small slot must align with the small vent hole in the wear plate.

13. Add a generous amount of oil into the gear cavities. Rotate the gears to distribute the oil.

14. Install the O-ring in the back cover.

15. Oil the face of the back cover.

16. Align the scribe mark on the back cover with the scribe marks on the center section and front cover.

17. Install the back cover with the four capscrews and washers. Tighten the capscrews to 190 to 210 in.lb. torque.

18. Add a generous amount of oil to both pump ports to insure adequate lubrication.

19. Install the pump in reverse order of removal.

Engine service stand

Cylinder

REMOVAL

1. Lower attachment to the ground.
2. With the engine shut off, move the hydraulic control lever back and forth to relieve all pressure in the system.
3. Remove the battery and battery box to provide access.
4. Disconnect the hydraulic cylinder hoses from the control valve. Plug the ends of the hoses to keep out dirt.
5. Loosen the hose guide nut and pull the hoses down below the frame.
6. Remove the capscrew, flat washer and hex nut holding the offset strap and pin.
7. Remove the offset strap and pin.
8. Remove the spacer from the end of the lift cylinder ram.
9. Remove the cotter pin and flat washer from the pivot pin. Slide the rear part of the cylinder off the pivot pin. Remove the cylinder with hoses.
10. Disconnect hose lines from cylinder. Remove fittings and discard defective cylinder.

INSTALLATION

1. Install the fittings in the new cylinder and connect the longer hose and line to the cylinder rear port. Connect the shorter hose and line to the cylinder front port.
2. Pull the spring locking pin and remove the drilled pin to disconnect the lift link.

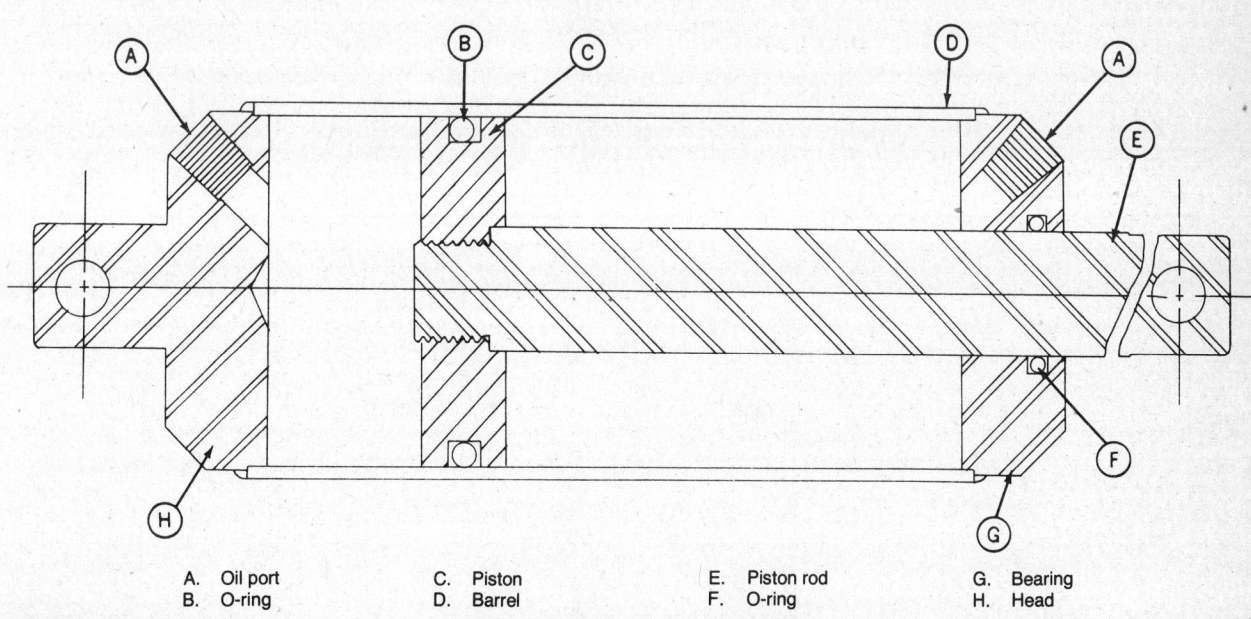

| A. Oil port | C. Piston | E. Piston rod | G. Bearing |
| B. O-ring | D. Barrel | F. O-ring | H. Head |

Sectional view of the control cylinder

John Deere

3. Position cylinder and hoses in tractor. Slide rear of cylinder on pivot pin. Secure cylinder to pivot pin with a flat washer and cotter pin.

4. Slide spacer through hole in cylinder ram.

5. Manually rotate primary lift arm rearward.

NOTE: Remove plugs from ends of hoses. The cylinder ram can then be moved in and out to make installing the cylinder easier. However, be sure to keep dirt from entering hoses.

6. Attach ram to quadrant with offset strap and pin through spacer. Bolt offset strap and pin to primary lift arm with a 5/16 x 7/8 in. capscrew, flat washer and hex nut.

7. Connect lift link with drilled pin and spring locking pin.

8. Connect the hose from the cylinder rear port to the control valve front port.

9. Connect the hose from the cylinder front port to the control valve rear port.

10. Position hoses in hose guide slot. Push down on hose guide and tighten nut.

NOTE: Make sure hydraulic hoses clear all linkage.

11. Install battery box and battery.

ENGINE

The engines used in these tractors are:
Model 200: Kohler K181QS
208: Kohler K181S
210: Kohler K241AQS
212: Kohler K301AQS
214: Kohler K321AQS
216: Kohler K341AQS

For complete engine service, see the Engine Unit Repair Section.

REMOVAL AND INSTALLATION

1. On 200, 210, 212, 214 and 216 tractors remove grille first; then, raise hood and remove side panels. Disconnect lead to headlight wiring harness and remove hood and cowl assembly. On 208 tractors remove grille first; then remove hood and cowl assembly.

2. Drain crankcase oil.

3. Disconnect coupler from engine to main wiring harness.

4. Disconnect battery cables (negative cable first).

5. Disconnect lead to starter.

6. Close fuel shut off at tank. Remove air cleaner and air cleaner base. Disconnect fuel line at fuel pump.

7. Disconnect throttle and choke cables.

8. Loosen idler assembly spring by turning knob on tractor front bumper. Disconnect spring and remove idler assembly from engine.

9. Remove PTO clutch.

10. Remove engine drive sheave bolt. Remove drive sheave with primary belt by depressing clutch pedal and sliding sheave with belt off end of crankshaft.

11. Remove engine base bolts and remove engine.

12. Installation is the reverse of removal.

Ford

INDEX

MODELS 70, 75, 85, 90, 100, 120, 125, 145, 165, 195LGT
- **Front Axle** 398
- **Steering** 399
  - Steering Gear 399
  - Power Steering Valve 401
- **Brakes** 402
- **Manual Transaxle** 404
  - Models 70, 75, 85, 90, 100, 120 404
- **Hydrostatic Transaxle** 412
  - Models 125, 145, 165
- **Hydrostatic Transmission** 415
- **Hydraulic System** 420
  - Testing and Adjustment 420
  - Control Valve 421

MODELS 1000 & 1600
- **Front Axle** 423
- **Steering** 424
- **Brakes** 425
- **Clutch** 425
- **Transmission** 426
- **Differential** 428
- **Rear Axle** 428
- **Power Take-Off** 430
- **Hydraulic System** 430
  - Testing and Adjustment 430
  - Lift Cover 430
  - Pump 431
- **Engine** 432
  - R&R 432
  - Cylinder Head 434
  - Crankshaft and Bearings 436
  - Camshaft 437
  - Flywheel 437
  - Timing Gears 437
- **Engine Lubrication** 437
- **Engine Cooling** 438
- **Fuel System** 439
- **Engine Electrical** 440

FORD
70, 75, 85, 90, 100, 120, 125, 145, 165, 195 LGT Series

FRONT AXLE

All Except 195

REMOVAL

1. Park tractor on a flat surface, set parking brake, and block rear wheels.
2. Remove linkage mounting bolt, bushing, two (2) washers, and locknut connecting front hanger frame on both sides.

NOTE: Left front wheel removed for picture clarity.

3. Disconnect drag link from steering arm.
4. Raise front end of tractor 6 in. and block frame just behind the tie rod assembly.
5. Place a jack under center of axle and raise jack until it just touches axle.
6. Remove roll pin from axle mounting pin with a punch and remove axle mounting pin.
7. Lower jack slowly and allow axle to come to rest on tires.
8. Roll axle assembly from under tractor.
9. Disassemble axle as follows:
  a. Remove front wheels.
  b. Remove tie rod.
  c. Remove spindles by loosening capscrews and tapping spindle assemblies free.

INSPECTION AND REPAIR

1. Check axle for straightness.
2. Check spindle hole for excessive wear, replace if necessary.
3. Check axle pivot bushings for wear, replace if necessary.

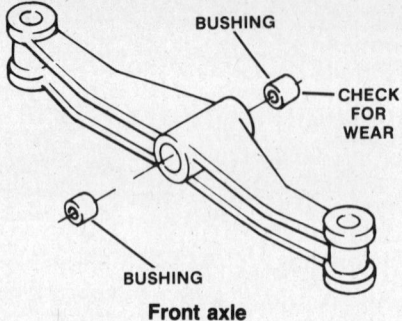
Front axle

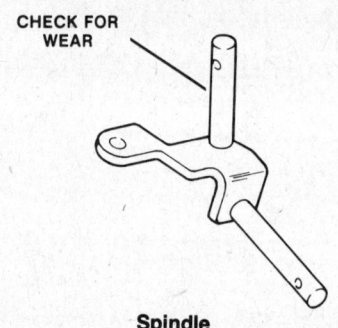
Spindle

4. Replace axle pivot bushings as follows:
  a. Drive bushings from axle with a punch.
  b. Press new bushings in flush with surface.
5. Check axle mounting pin for wear.
6. Check spindles for wear.
7. Check ball joints for wear.

INSTALLATION

1. Install spindles in axle, place washer over righthand spindle end and steering arm over lefthand, secure with roll pin. Grease spindles.

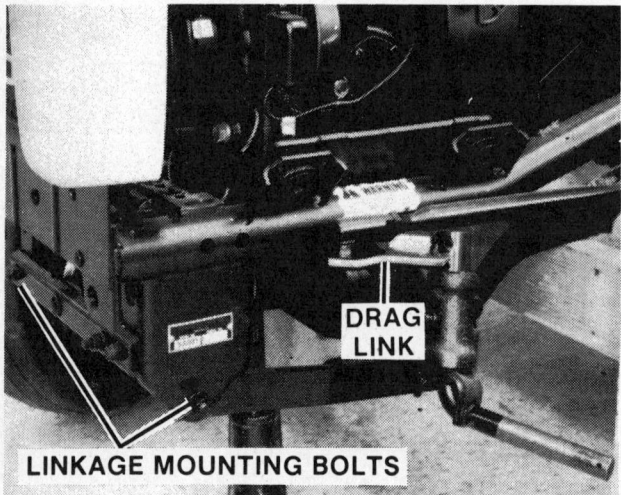
Steering linkage mounting bolt removal

2. Install tie rod assembly to axle, secure with lockwasher and nut.
3. Place axle assembly on a jack and position beneath tractor. Raise jack until mounting hole lines up.
4. Place anti-rotation pin in mounting pin. Drive in mounting pin and secure with roll pin. Grease pivot pin.
5. Install drag link ball joint to lefthand steering arm using internal tooth lockwasher. Secure.
6. Remove blocks from behind tie rod assembly.
7. Secure front hangers to frame with bolt, bushing, two (2) washers and locknut.
8. Lower jack and proceed to steering adjustment.

195

REMOVAL

1. Park tractor on flat surface, set brake, and block rear wheels.
2. Remove capscrew and locknut connecting power steering cylinder to righthand steering arm assembly.
3. Remove drag link ball joint from lefthand steering arm.
4. Raise front end of tractor 6 in. and block frame just behind tie rod assembly.
5. Place a jack under center axle, raise jack until it just touches axle.
6. Remove cotter pin from slotted hex nut; remove nut and washer from axle mounting pin; and drive axle mounting pin from frame.
7. Lower jack and remove axle.
8. Disassemble axle as follows:
  a. Remove front wheels.
  b. Remove tie rod.
  c. Remove spindles by loosening capscrews securing steering arm assemblies, and tapping spindle assemblies free.

INSPECTION AND REPAIR

1. Inspect all axle bushings for wear.
  a. Check inside diameter of center pivot bushings for wear, replace if necessary.
  b. Check inside diameter of spindle bushings for wear, replace if necessary.
2. Check axle for straightness.
3. Pump grease through grease fittings.

NOTE: If grease will not pump through fittings, check for hardened grease or obstruction inside.

4. Check ball joints for wear.
5. Check spindles and axle mounting pin for wear.

INSTALLATION

1. Position axle beneath tractor. Raise axle with jack until mounting pin holes line up.

Ford

2. Install mounting pin with its "tab" up.

3. Install cotter pin in slotted hex nut.

4. Install spindle assemblies.

5. Install tie rod and adjust ball joints as follows:

   a. Make sure cut out in ball joint is parallel to mounting bracket after assembly and before securing.

   b. Maintain ½ in. minimum thread engagement when adjusting ball joints.

6. Connect drag link and lefthand steering arm assembly with drag link ball joint.

7. Connect power steering cylinder to righthand steering arm assembly, with capscrew and locknut.

8. Install front wheels.

STEERING

Steering Gear

REMOVAL AND DISASSEMBLY

1. Remove steering wheel hub cap and loosen steering wheel retaining nut.

2. With the use of a three-armed wheel puller, remove steering wheel.

---- CAUTION ----
Under no circumstances hammer on steering gear shaft. Hammering WILL cause internal damage to steering gear boxes.

3. Disconnect battery and all electrical plugs and wires from back of dash panel.

---- CAUTION ----
Do not allow red lead to contact tractor frame.

4. Mark position of choke and throttle control cables at their respective clips on the engine, with a short piece of tape or wire. Remove choke and throttle cables from carburetor.

5. Remove four (4) bolts (2 on each side) retaining dash panel to console support arms. Lift dash panel off tractor.

6. Remove hex head capscrew on steering arm on steering gear output shaft. Slide steering arm off shaft.

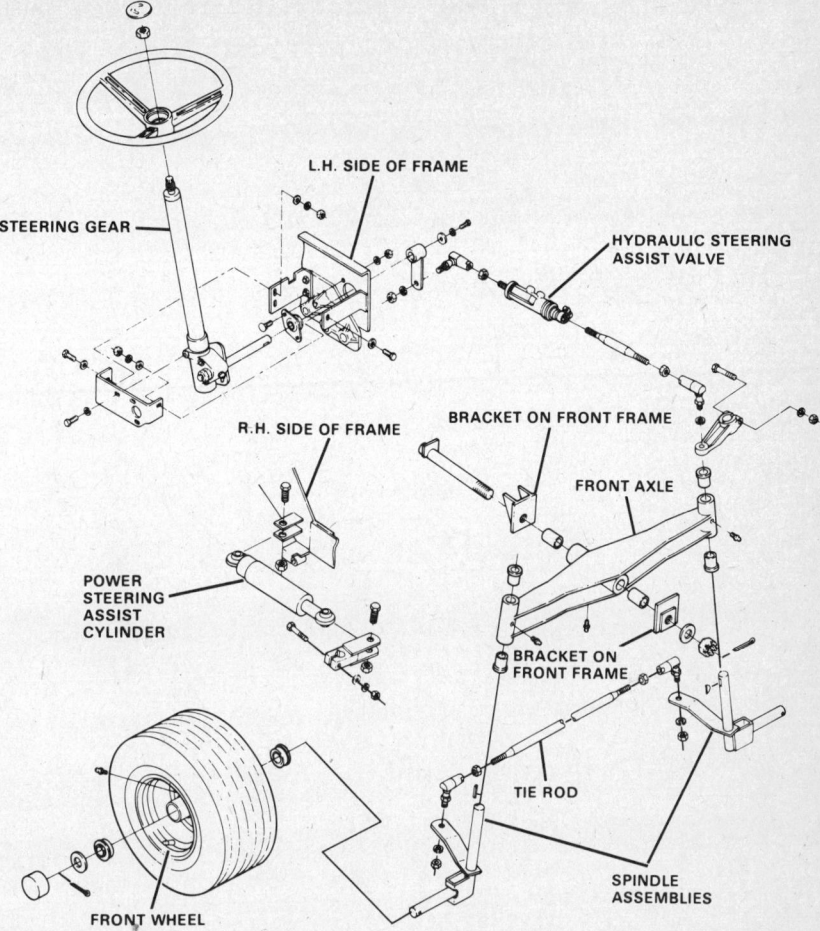

195 front axle and steering components

7. Remove steering gear box brackets as follows:

   a. All except 195. Remove bearing retaining hardware on left side of steering gear box and, on the right side, remove two bolts securing gear box support bracket to mounting frame on tractor.

   b. 195. Remove two (2) capscrews re-

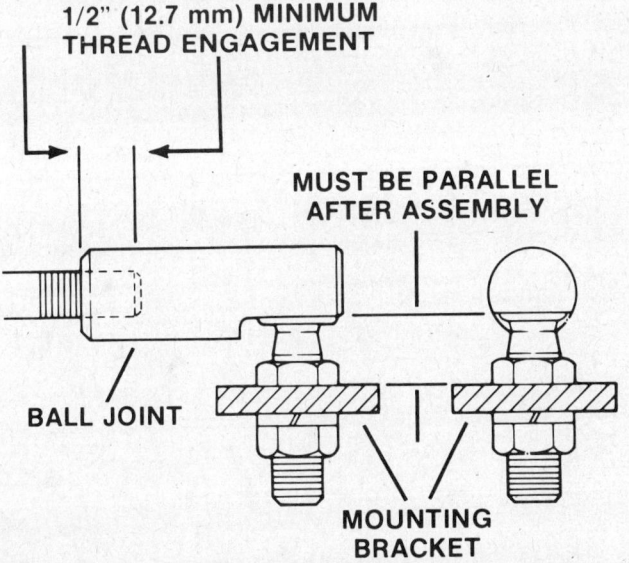

Ball joint adjustment

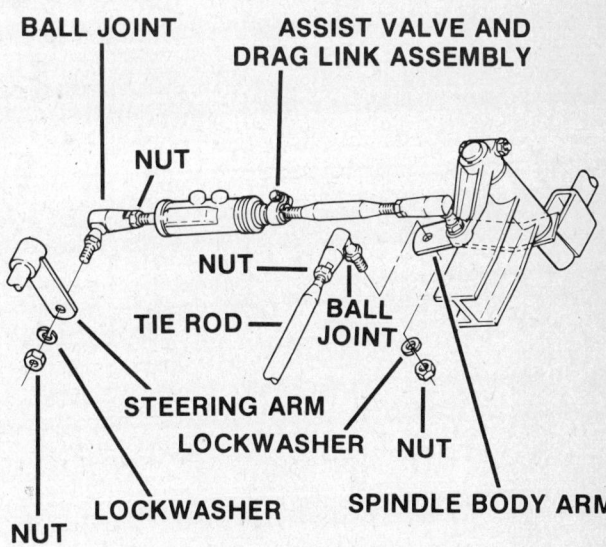

Installing drag link

399

Ford

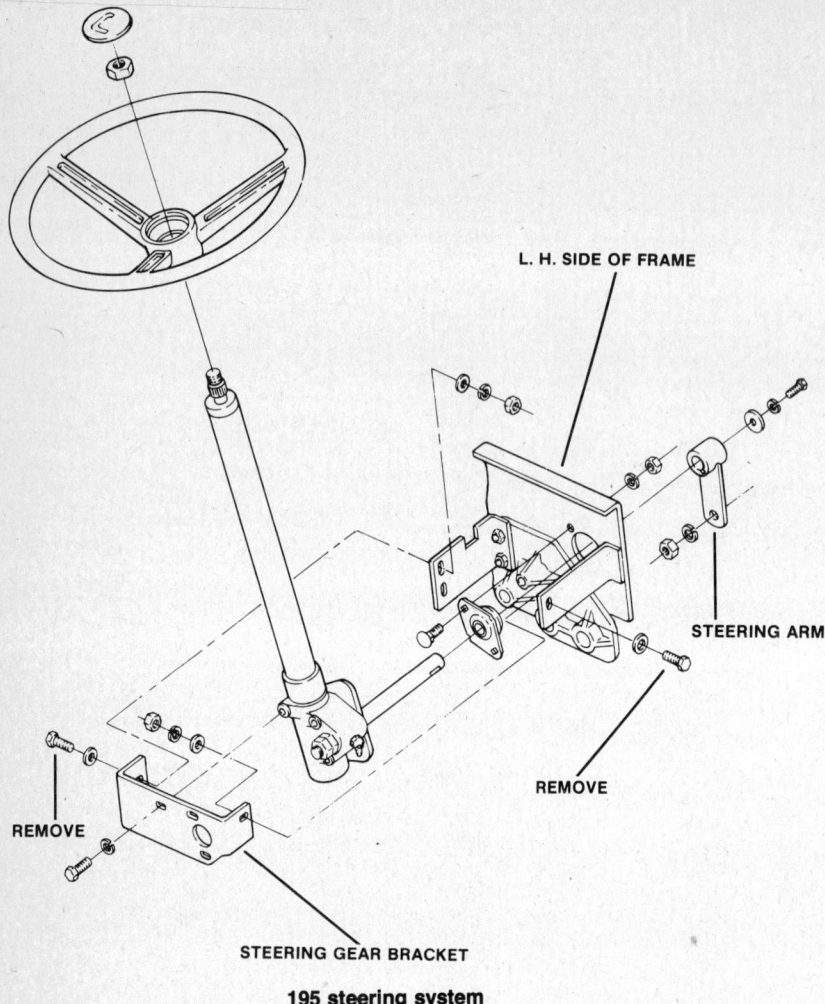

195 steering system

taining steering gear bracket to frame. Slide steering gear assembly out from bearing and remove.

8. Remove two bolts securing support bracket to steering gear box on right side. Carefully turn and lift steering gear box to remove from tractor.

9. Remove cotter key and end plug from steering gear boxes. Note position of Belleville washer and bearing assembly. Remove the two (2) jam nuts from steering lever assembly.

10. Slip lever assembly from steering gear housing. Remove oil seal and retainer. Inspect seal for damage and replace if necessary (Part #546071).

11. Note upper bearing on tube assembly and remove the cam and tube assembly from the steering gear housing.

INSPECTION AND REPAIR

1. Thoroughly clean all parts in a suitable solvent.
2. Inspect bearings for excessive wear (side play in ID of bearings) and replace as necessary.
3. Liberally lubricate all internal components with lithium grease-general purpose lubricant.

REASSEMBLY

1. Slide upper bearing and cup over cam and tube assembly and insert bearing, cam and tube assembly into steering gear housing.
2. Install lower bearing and cup assembly and Belleville washer on cam and tube assembly shaft. Install steering gear adjustment plug fingertight.
3. Install oil seal and retainer. Slide lever assembly into steering gear housing assembly using hex jam nuts to tighten in position.

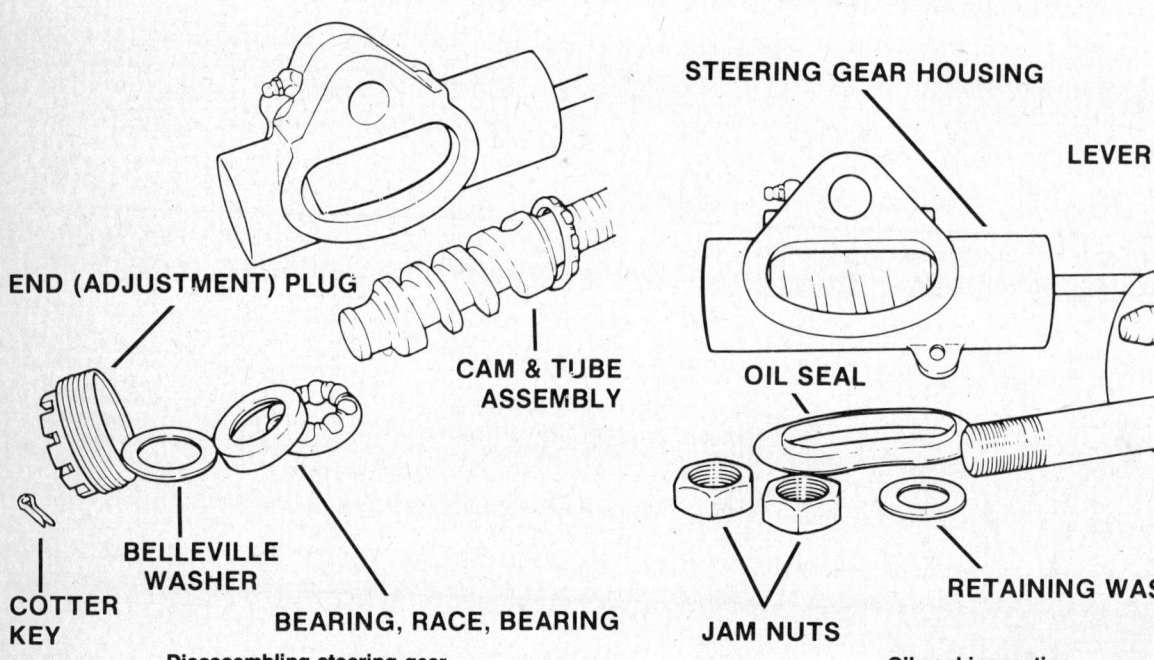

Disassembling steering gear

Oil seal inspection

Ford

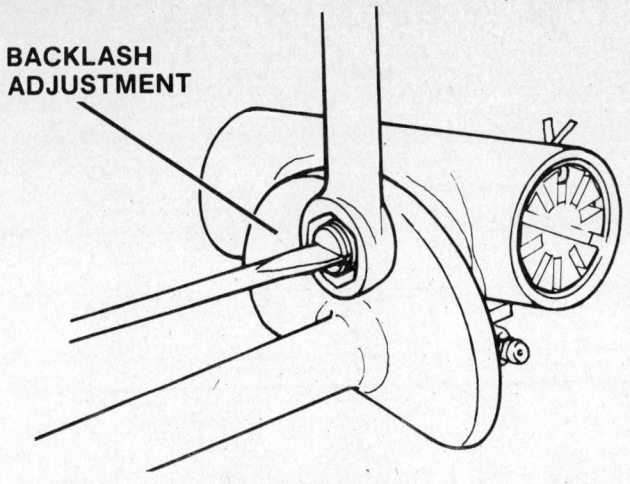

Backlash adjustment

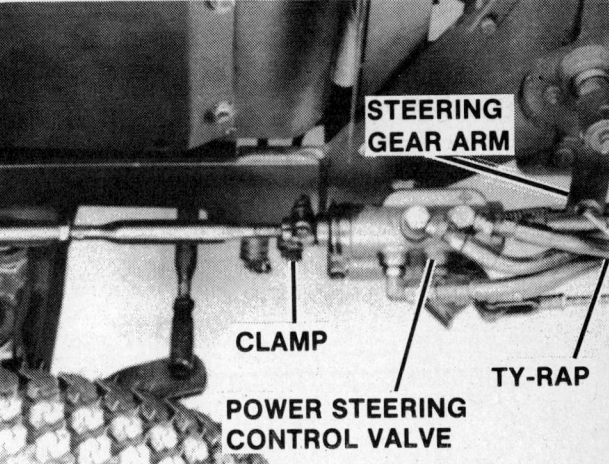

Power steering control valve removal

ADJUSTMENT

NOTE: Check free travel (backlash) with steering gear in neutral (center) position.

1. Adjust as follows:
a. Loosen locknut on adjusting stud and turn stud clockwise to reduce free travel.

NOTE: Maximum torque on the adjusting stud is 5 in.lb.

b. Hold stud stationary with a wide blade screwdriver and retighten locknut.
2. Check drag link ball joint to make sure there is no interference between ball joint and both the tractor frame and brake lock pawl pin. Do this at both full right and left steering lock. If no interference, proceed to step C. If interference is noted, proceed as follows:
  a. Remove inner bolt on bracket.
  b. Remove outer bolt on bracket.
  c. Loosen steering gear mounting screws.
  d. Loosen bearing mounting nuts on bearing mounting screws.
  e. Position steering gear so that at full right steering lock the drag link rear ball joint passes below brake lock pawl pin.
  f. Tighten steering mounting screws to a maximum torque of 6-7 ft.lb.
  g. Tighten bearing mounting nuts to a maximum torque of 8-11 ft.lb.
  h. Reinstall inner bolt.
  i. Tighten both inner and outer bolts firmly.
3. Recheck for interference between drag link ball joint and both the tractor frame and brake lock pawl pin. If any interfernce is still noted, a new steering arm assembly must be installed.

INSTALLATION

1. Install steering gear box in tractor and secure support bracket to box with two (2) bolts.
2. Install bracket to frame with two (2) bolts; install hardware and bearing on left side of steering gear box.
3. Slide steering arm onto shaft and install capscrew on steering arm.
4. Place dash panel in place and secure with four (4) bolts.
5. Connect choke and throttle control cables.
6. Connect electrical plugs, and wires at back of dash panel. Connect battery.

— **CAUTION** —
Be sure to connect red lead to positive (+) terminal of ammeter and yellow lead to negative (−) terminal. If leads are reversed, ammeter will show "discharge" when it is actually charging.

7. Coat the I.D. of steering wheel hub and steering shaft with grease. Install steering wheel. Secure with retaining nut and hub cap.

Power Steering Valve—195

REMOVAL AND INSTALLATION

1. Clean around fittings.
2. Remove Ty-rap® securing hoses to steering arm.
3. Disconnect hoses from valve.

NOTE: Tag hoses to insure proper reassembly.

4. Cap or plug hoses and valve ports preventing dirt from entering system.
5. Remove nut retaining ball joint to steering gear arm.
6. Loosen jam nut and remove ball joint.
7. Loosen nut and capscrew holding clamp and remove steering control valve.
8. Position front wheels so they are heading straight ahead.
9. Position front steering wheel at center of its travel, 1½ turns from either right or left stop.
10. Thread valve on to drag link.
11. Install jam nut and ball joint to spool end.
12. Adjust the length and valve and drag link assembly as necessary to reassemble it to steering arm without changing position of the steering wheel or front wheels.
13. When length is properly adjusted, install ball joint to steering arm, tighten jam nut against ball joint and clamp on drag link.
14. Remove all caps and plugs from hoses and fittings.
15. Connect hoses and tighten fittings.
16. Start engine and slowly turn steering wheel from lock to lock 6-8 times, purging air from system. Check for leaks.
17. Stop engine.

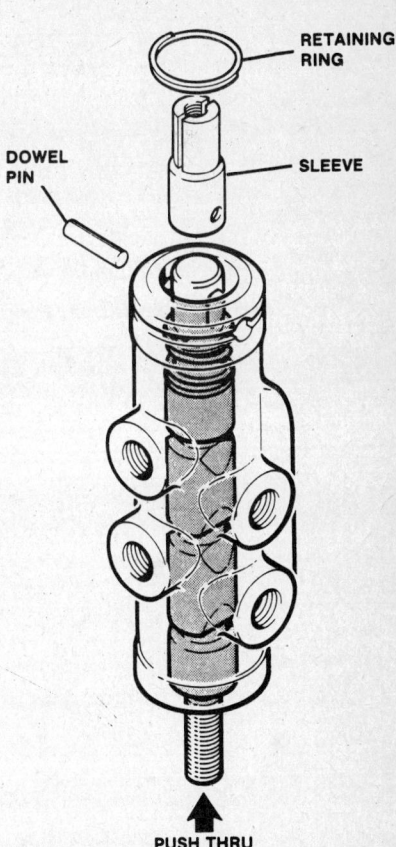

Power steering valve disassembly

401

Ford

18. Install Ty-rap® securing hoses to steering arm.
19. Check fluid level in transaxle.

DISASSEMBLY

1. Clean valve assembly in a suitable solvent. Wipe dry.
2. Remove retaining ring.
3. Drive dowel pin from valve.
4. Remove sleeve.
5. Remove spool from body.

NOTE: It is not necessary to remove spring for seal replacement.

INSPECTION

After disassembly, inspect as follows:
1. Visually inspect parts. Service parts are available for all components except spool and body. These are replaced by a complete valve assembly only.
2. Remove all burrs and small scratches from spools and spool bores in body with very fine emery cloth.

--- CAUTION ---
Use extreme care, do not remove enough metal to cause internal leaking. Do not alter O-ring grooves.

Replace Spool Return Spring

To replace spool return spring, proceed as follows:
1. Compress spring and remove snap-ring.
2. Remove washer and spring.
3. Replace spring, place washer over end of spool and compress spring.
4. Replace snapring.

O-Rings and Back-Up Ring

Replace all O-rings and back-up rings with new parts, whenever valve is disassembled. O-ring (A) is located inside body. Use extreme care when removing.

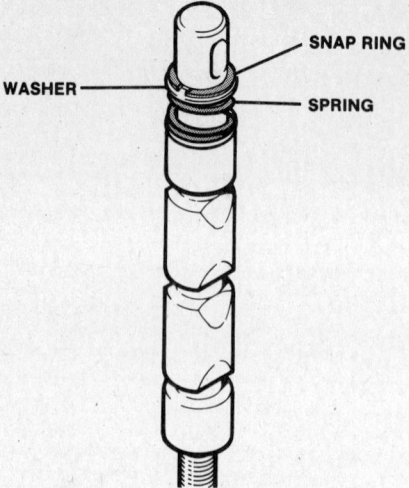

Spool return spring installation

--- CAUTION ---
Do not use hardened tool or scratches in bore will result.

Install O-ring (B) and back-up ring in groove at opposite end. Note position of back-up ring.

REASSEMBLY

Use clean hydraulic fluid to lubricate all parts during reassembly. To reassemble the power steering control valve, proceed as follows:
1. Carefully guide spool assembly into body. Align slotted hole in spool with hole in body.
2. Install sleeve over end of spool. Align hole in sleeve with holes in body and spool.
3. Install dowel pin. Ends of dowel pin must be flush with retaining ring groove, both sides.
4. Install retaining ring in groove.
5. Install clamp over sleeve.

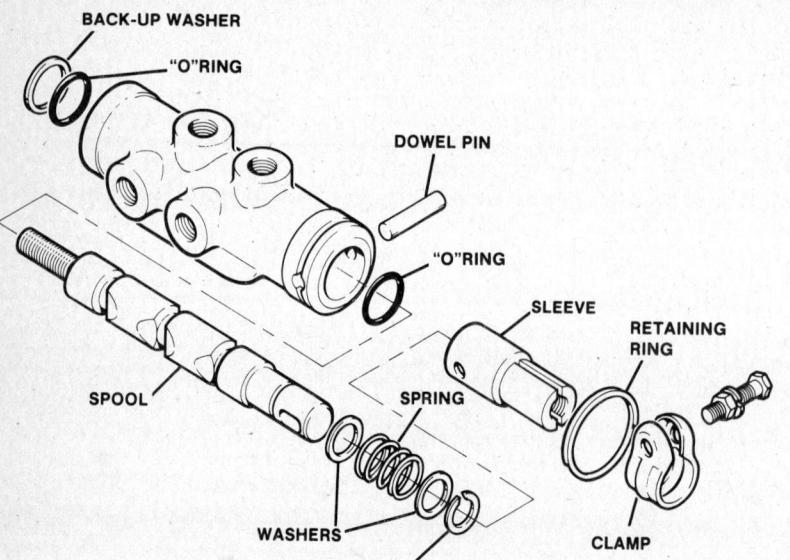

Power steering valve assembly

BRAKES

All Except 195

REMOVAL

1. Remove two (2) hex locknuts, washer and compression spring from brake rod.
2. Remove capscrew from band pivot assembly. Pull band pivot assembly from frame.

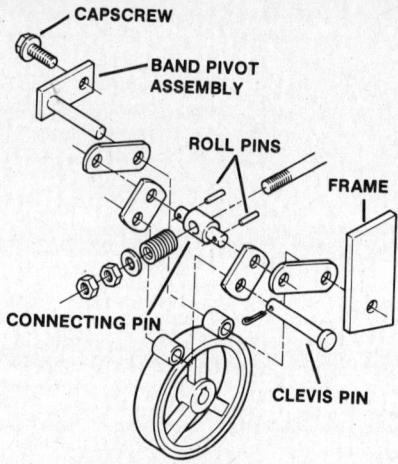

Brake components

3. From rear of tractor, turn brake band and lining assembly clockwise (seen from outside) and remove cotter pin and clevis pin.
4. From rear of tractor, continue to turn and pull brake band and lining assembly clockwise until it is removed from tractor.
5. Remove two (2) roll pins from connecting pin. Remove pin and links from brake band.

INSPECTION AND REPAIR

1. Inspect brake drum for wear. Surface may be remachined.
2. Inspect links for wear. Replace if holes appear out of round.
3. Inspect band pivot assembly pin. It must be free of grooves and be at right angles to plate. Straighten or replace.
4. Check brake band lining. It must be free of breaks and extend above rivets a minimum of $1/16$ in. Brake band assembly must be replaced if not within these specifications. Lining is not replaceable.

INSTALLATION

1. Assemble connecting pin and links. Secure with roll pins.
2. Position brake band assembly by pulling it clockwise around and on to brake drum.
3. Place clevis pin through free ends of short links and through rear end of band assembly. Secure with cotter pin.
4. Slide brake rod through hole in connecting pin.

Ford

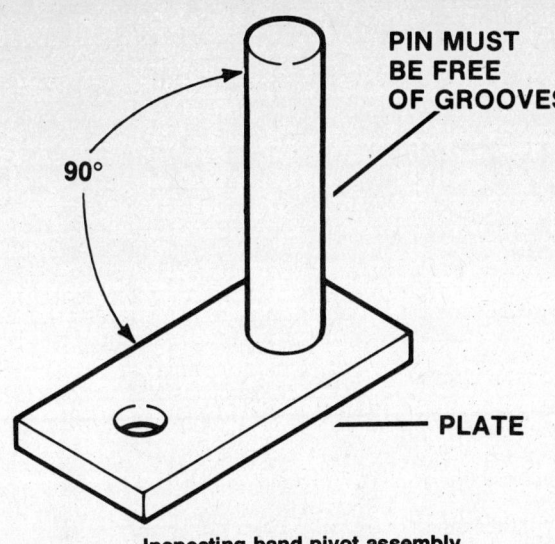

Inspecting band pivot assembly

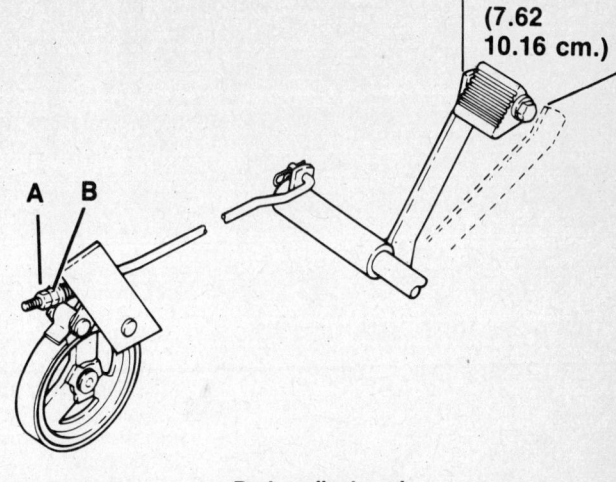

Brake adjustment

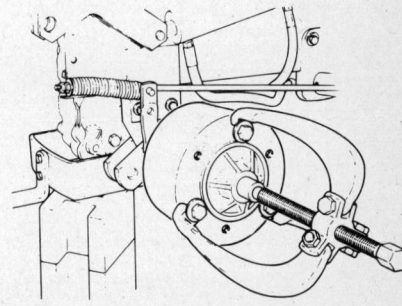

Brake drum removal

5. Line up free end of brake band assembly with hole in frame; and slid band pivot assembly pin through hole in frame, through free ends of large links and through end of band assembly. Secure with capscrew.

6. Place compression spring and washer on end of brake rod. Secure with locknuts.

ADJUSTMENT

NOTE: Brake pedal travel is adjusted by changing the "working length" of the brake actuator rod. The brake pedal should travel about 3 to 4 in. before brake applies.

Adjust brake pedal travel as follows:
1. Back off locknut "A" to free adjusting nut "B".
2. Turn nut "B" counterclockwise (facing front of tractor) to lengthen rod for more pedal travel; or turn "B" clockwise to shorten rod for less pedal travel.
3. When pedal is adjusted, tighten locknut "A" while holding nut "B" still.

CAUTION
Do not overtighten, or brake life will be shortened.

CAUTION
Parking brake must lock in fully depressed position to prevent tractor movement when parked.

195

REMOVAL

1. Park tractor on a level surface, set parking brake and block front wheels.
2. Raise rear axle.
3. Remove rear wheel lugs and wheels.

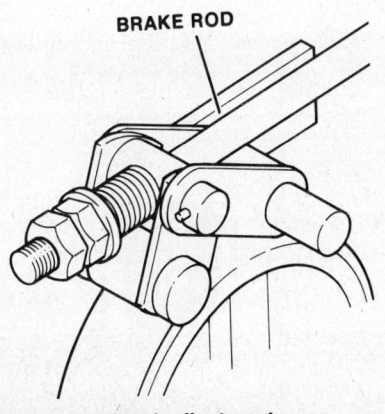

Band adjustment

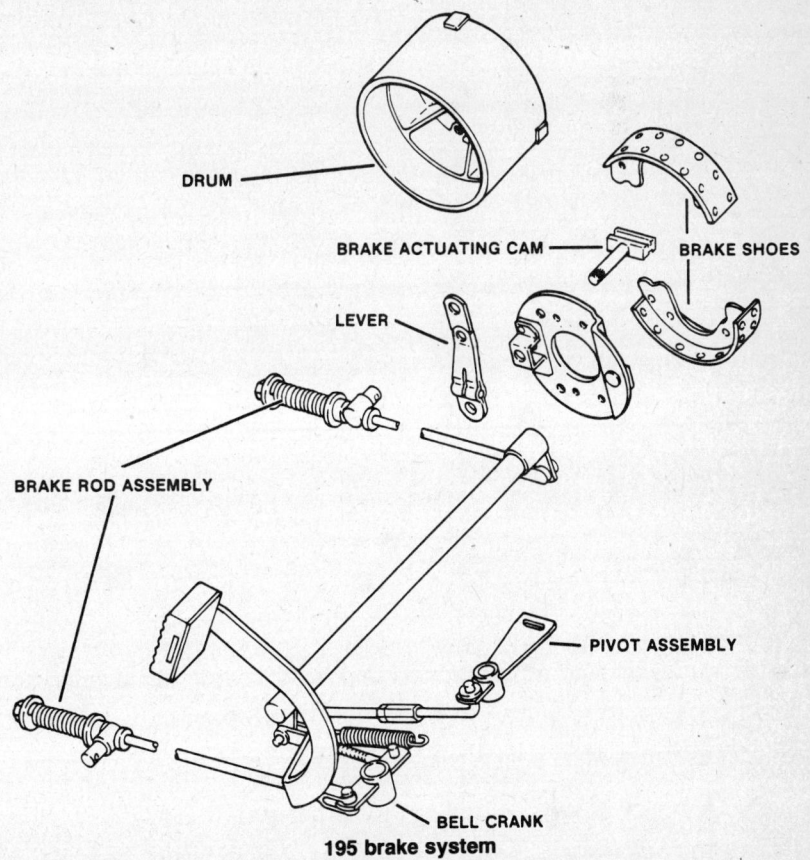

195 brake system

403

Ford

4. Remove nuts securing brake drums to axles.
5. Disengage parking brake. Place washers behind lug nuts and remove brake drums, using a three claw wheel puller.
6. Remove two (2) holddown springs (brake shoe retaining springs), brake shoe return spring, and brake shoes from backing plate.

INSPECTION AND REPAIR

1. Inspect springs for wear or damage.
2. Inspect brake shoes for wear. Measure distance from lining plate to rivets. Replace if less than 1/16 in. above rivets.
3. Inspect drums for wear or scoring. Oversize brake shoes are not available. Drums must be replaced if worn.

INSTALLATION

1. Install brakes as follows:
a. Install brake shoes. Connect holddown springs and brake shoe return spring.
b. Install brake hub.
c. Engage parking brake and secure brake hub with jam nut.
d. Install wheels.

ADJUSTMENT

NOTE: Adjust linkage so brakes are applied when pedal is pressed approximately 3 in. Adjust one brake rod at a time.

1. Remove cotter pin and clevis pin holding brake rod and yoke to bell crank.
2. Back off jam nut on brake rod.
3. Turn yoke to lengthen rod, increase pedal travel. Shorten rod to decrease travel.
4. Reassemble brake rod and yoke to bell crank. Depress pedal to check adjustment. Remove and readjust, if needed.
5. Tighten jam nut up against yoke to lock in adjustment.

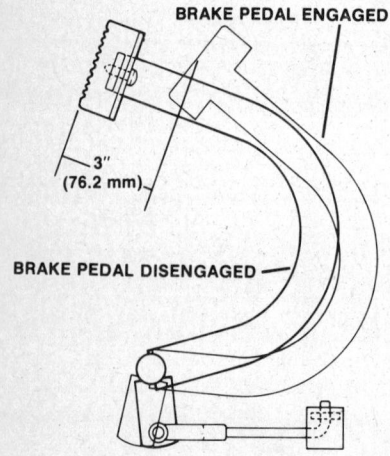

Brake adjustment

MANUAL TRANSAXLE

70, 75, 85, 90, 100, 120

REMOVAL AND INSTALLATION

1. Park tractor on flat surface and block front wheels.
2. Raise hood and remove side panels.
3. Disconnect battery.
4. Remove body insert.
5. Disconnect and remove brake band.
6. Remove clutch spring.
7. Remove capscrews securing transaxle to frame.
8. Raise and support rear of tractor, roll transaxle out and away.
9. Replace rear wheels.
10. Roll transaxle into position under frame.
11. Lower rear of tractor onto transaxle. Replace capscrews.
12. Install drive belt.
13. Replace clutch spring at front of transmission housing.
14. Install brake band and reconnect linkage.
15. Replace body insert.
16. Connect battery.
17. Replace side sheets.

OVERHAUL

1. Preparation for disassembly.
a. Visually inspect for evidence of oil seepage, tampering, misalignment, freedom of rotating shafts, etc.
b. Clean unit thoroughly of dirt, oil, debris.
c. Remove shift housing and drain oil from unit. Observe oil to see if metal particles are present.
d. Check axle shafts carefully for smoothness. Use a stone or suitable hard abrasive to rub down high spots and eliminate rust or paint.
e. Check model number. It is advisable to have the exploded parts view handy.
f. Have seal sleeves, driver, tools, shop clothes and information material at hand.
2. Remove wheel from axles.
3. Remove shift handle assembly.
a. Prior to removing a shift lever assembly from a transaxle, make note of the position of the shift lever so that it

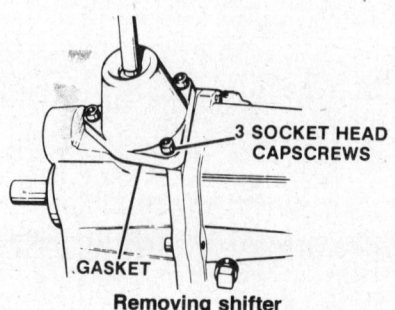

Removing shifter

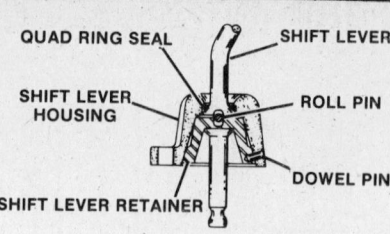

Dowel pin type shift lever

may be assembled correctly to the shift lever housing.
b. Move the lever to Neutral, if possible, before removing it from the transaxle. Clean around the lever housing to prevent dirt from falling into the transaxle. Cover this opening, if possible.
4. Disassemble shift handle assembly.
a. Place the shift lever in a vise so that the shift lever housing is at least one inch from the top of the vise jaws.
b. **DOWEL PIN TYPE:** Locate the dowel pin holding the retainer in the housing from the outside. Place a 1/4 in. flat face punch on the gasket surface directly over the dowel pin. Strike the punch sharply but lightly with a hammer to dislodge the retainer from the shift lever housing. Always use a new dowel pin for reassembly.

SNAPRING TYPE: Use the proper compressing tool for removing the snapring. Loosen the vise and disassemble the pieces.

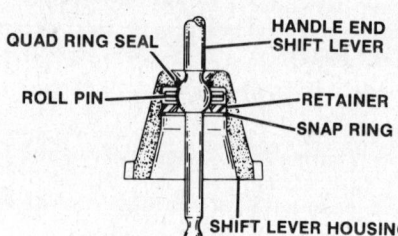

Snapring type shift lever

c. Remove the shift lever from the shift lever housing. Examine the roll pin in the ball of the shift lever, if bent or worn, replace. When inserting a new roll pin in the ball, position so that equal lengths protrude from both sides of the ball.
d. Oil leakage past the point where the shift lever enters the shift lever housing will require replacement of the quad ring seal in the shift lever housing.

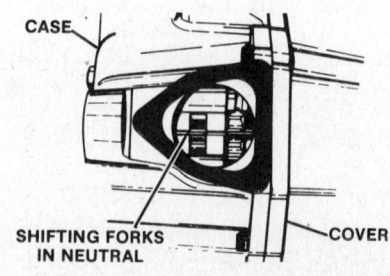

Positioning the shift forks in neutral

Ford

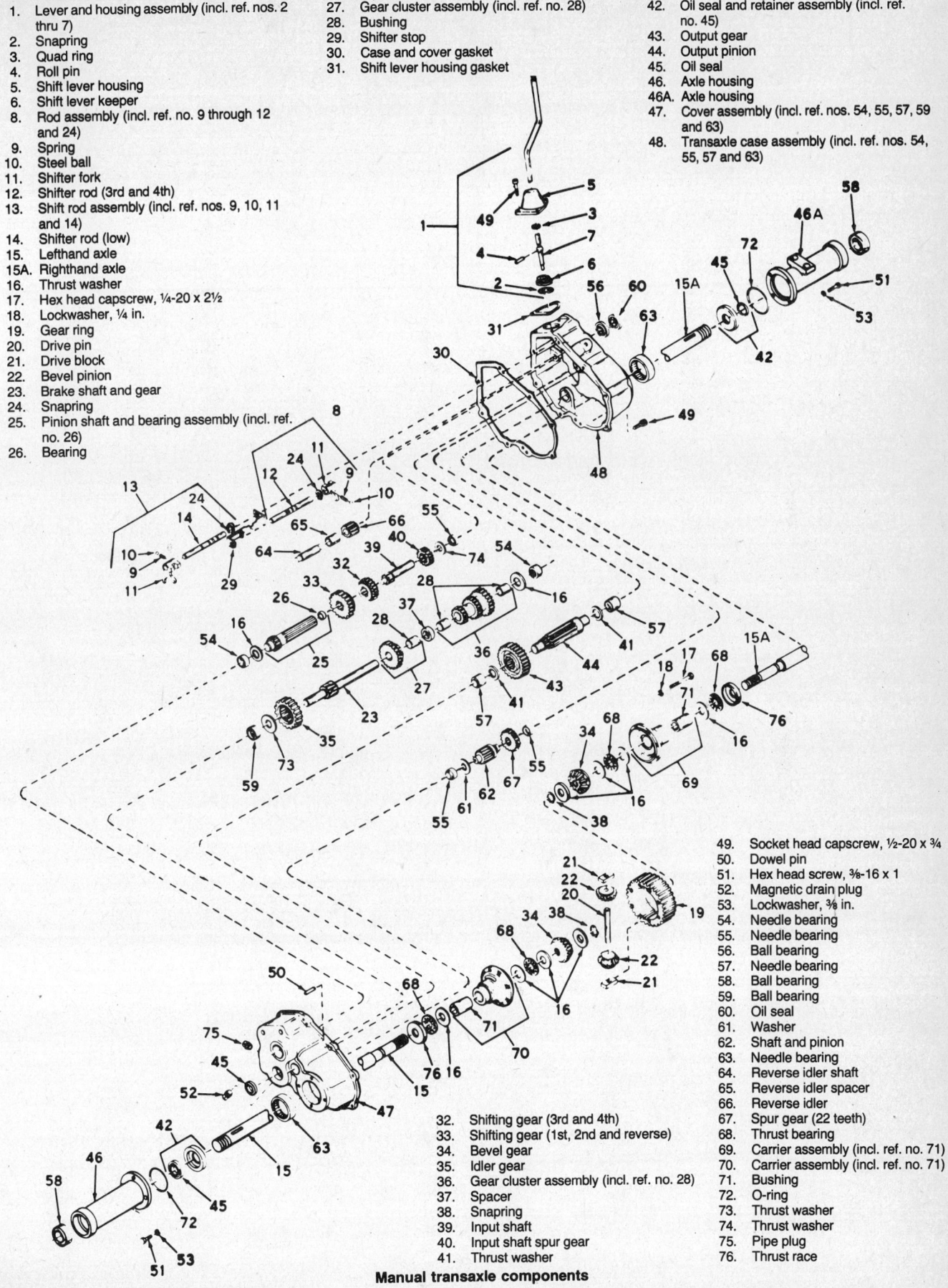

1. Lever and housing assembly (incl. ref. nos. 2 thru 7)
2. Snapring
3. Quad ring
4. Roll pin
5. Shift lever housing
6. Shift lever keeper
8. Rod assembly (incl. ref. no. 9 through 12 and 24)
9. Spring
10. Steel ball
11. Shifter fork
12. Shifter rod (3rd and 4th)
13. Shift rod assembly (incl. ref. nos. 9, 10, 11 and 14)
14. Shifter rod (low)
15. Lefthand axle
15A. Righthand axle
16. Thrust washer
17. Hex head capscrew, 1/4-20 x 2 1/2
18. Lockwasher, 1/4 in.
19. Gear ring
20. Drive pin
21. Drive block
22. Bevel pinion
23. Brake shaft and gear
24. Snapring
25. Pinion shaft and bearing assembly (incl. ref. no. 26)
26. Bearing
27. Gear cluster assembly (incl. ref. no. 28)
28. Bushing
29. Shifter stop
30. Case and cover gasket
31. Shift lever housing gasket
32. Shifting gear (3rd and 4th)
33. Shifting gear (1st, 2nd and reverse)
34. Bevel gear
35. Idler gear
36. Gear cluster assembly (incl. ref. no. 28)
37. Spacer
38. Snapring
39. Input shaft
40. Input shaft spur gear
41. Thrust washer
42. Oil seal and retainer assembly (incl. ref. no. 45)
43. Output gear
44. Output pinion
45. Oil seal
46. Axle housing
46A. Axle housing
47. Cover assembly (incl. ref. nos. 54, 55, 57, 59 and 63)
48. Transaxle case assembly (incl. ref. nos. 54, 55, 57 and 63)
49. Socket head capscrew, 1/2-20 x 3/4
50. Dowel pin
51. Hex head screw, 3/8-16 x 1
52. Magnetic drain plug
53. Lockwasher, 3/8 in.
54. Needle bearing
55. Needle bearing
56. Ball bearing
57. Needle bearing
58. Ball bearing
59. Ball bearing
60. Oil seal
61. Washer
62. Shaft and pinion
63. Needle bearing
64. Reverse idler shaft
65. Reverse idler spacer
66. Reverse idler
67. Spur gear (22 teeth)
68. Thrust bearing
69. Carrier assembly (incl. ref. no. 71)
70. Carrier assembly (incl. ref. no. 71)
71. Bushing
72. O-ring
73. Thrust washer
74. Thrust washer
75. Pipe plug
76. Thrust race

Manual transaxle components

Ford

e. Prior to reassembly, be sure that bends in the shift lever correspond to the mounting on the vehicle.
5. Remove shifting assembly.
a. Place shifter forks in neutral.
b. Shifting assemblies are removed from and installed into transaxles by squeezing the top end of the shifter rods. This causes a binding that retains all parts during removal or installation.
6. Disassemble shifting assembly.
a. Follow the next 5 illustrations (a-e) in order. Prior to disassembly compare the assembly with the illustrations. This will aid during the reassembly.
7. Remove axle housings, use exposed axle to separate seal retainers from case and cover.
8. When disassembling the rest of unit:
a. Place cover side down. Support on blocks so no weight rests on brake shaft.
b. Cover should sit rigidly on blocks so removal of parts can be done in systematic step by step procedure.
9. Tap dowel pins into the cover, remove (8) eight socket head capscrews.
10. Separate case from cover.
a. Lift the case 1½ to 2 in. above the cover.

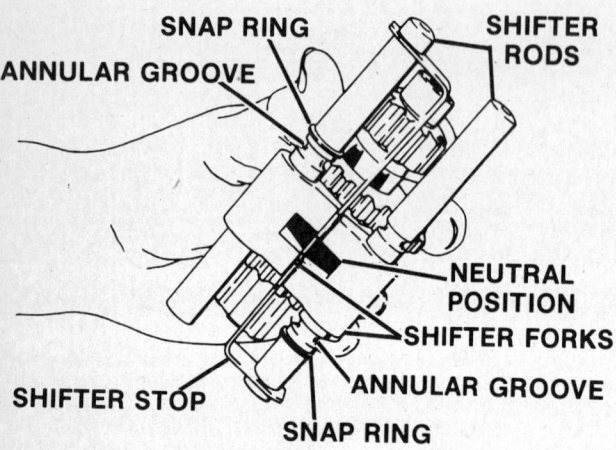

Figure A: shifting assembly

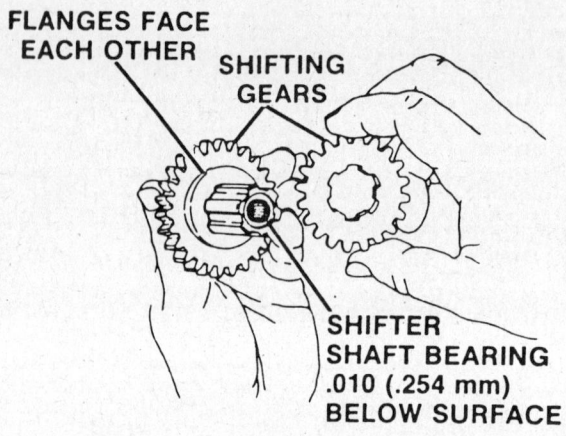

Figure D: two flanged gears on shifter shaft

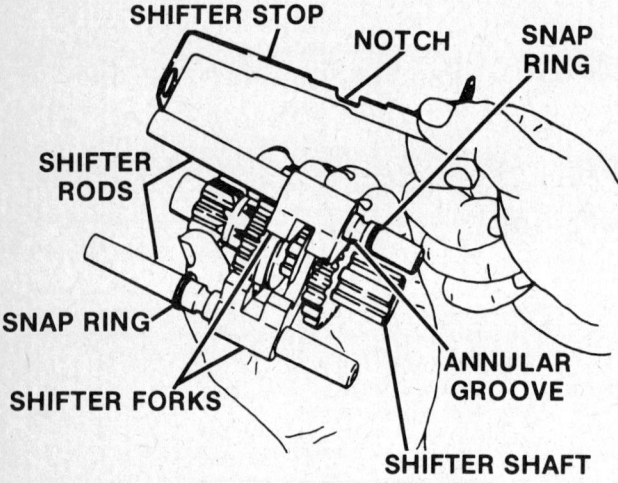

Figure B: removing shifter stop

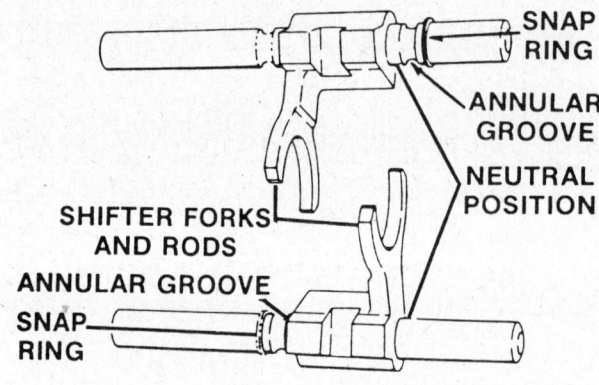

Figure E: shifter forks and rods in the neutral position

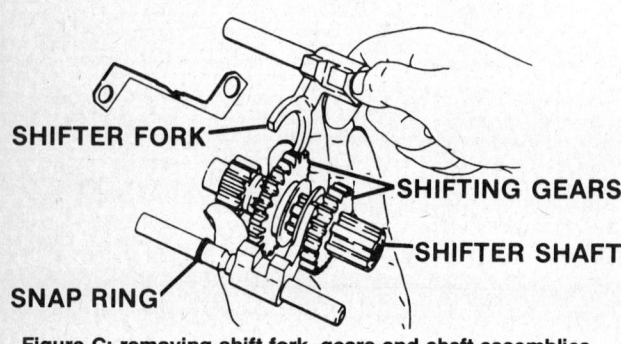

Figure C: removing shift fork, gears and shaft assemblies

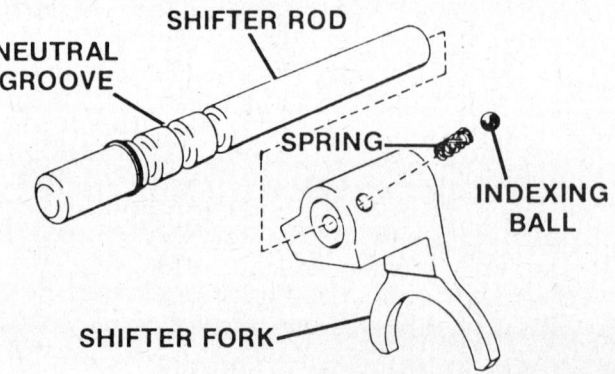

Removing shifter forks from rods

Ford

b. Tilt the case so that shift rods will clear edge.

c. Rotate the case so that boss hidden inside will clear gears, then lift free of the differential.

11. Remove thrust washer and three (3) gear cluster from brake shaft.

NOTE: Inspect gear teeth for wear, chips or breaks. Wear or chipping on bevel area only indicates shifting while equipment is in motion.

12. Remove the reverse idler gear, spacer, and shaft from boss in cover.

Note that the spacer goes between the gear and that the gear bevels go down. Excessive wear on teeth bevels indicates improper shifting technique.

13. Lift out shifter assembly.

NOTE: If it is evident that shifter assembly needs no further teardown, place it aside, in a clean place, intact, for easy assembly.

14. Remove the low gear and shaft, and splined spur gear. Separate gear and shaft. Note that NO thrust washer is between the gear and case.

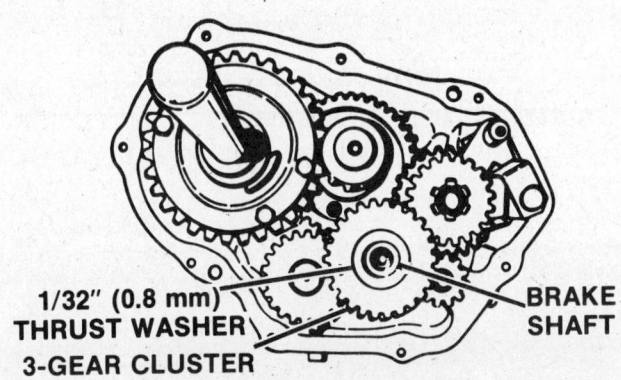

Removing the 3-gear cluster

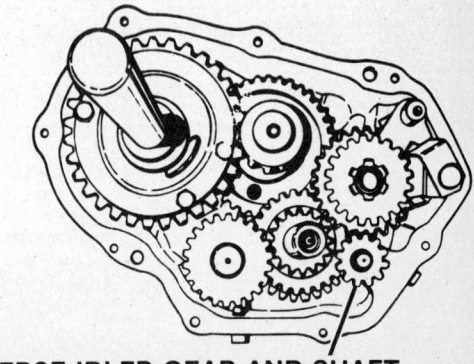

Reverse idler gear and shaft

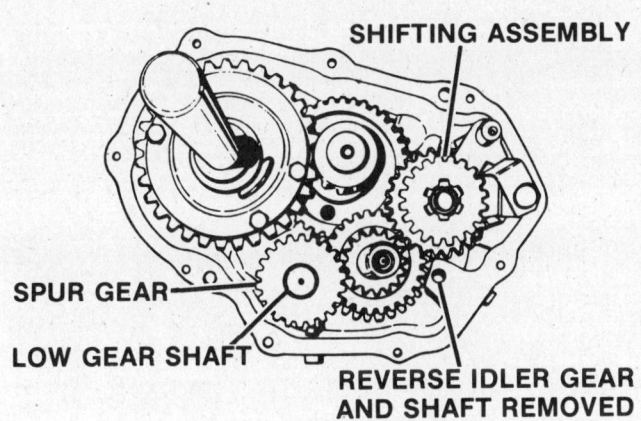

Removing shifting assembly

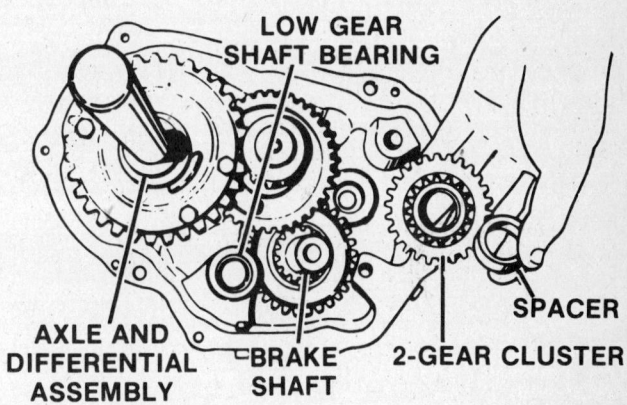

Removing 2-gear cluster from brake shaft

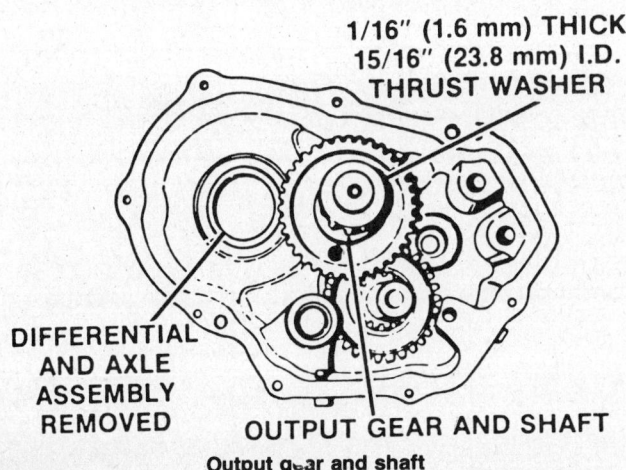

Output gear and shaft

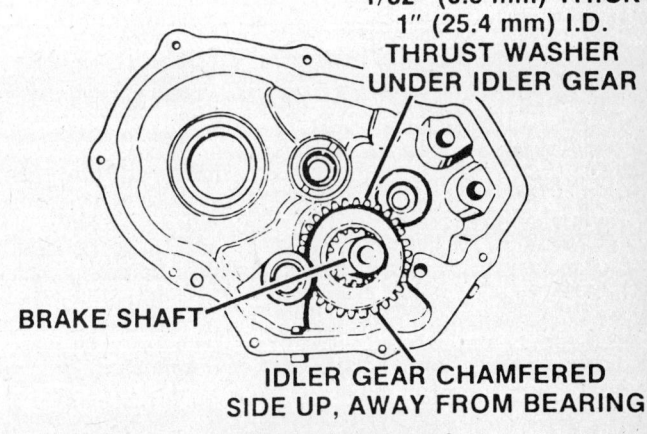

Brake shaft removal

407

Ford

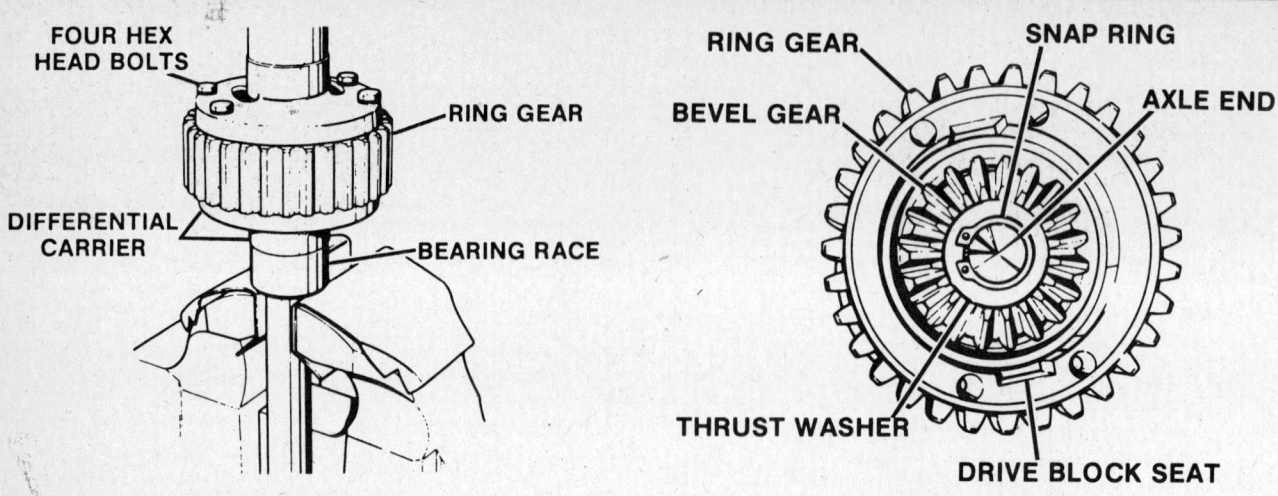

Differential bolt removal

Differential components

15. Remove two (2) gear cluster from brake shaft.
16. Lift differential unit out of cover, set aside.
17. Remove the output shaft and gear and thrust washer from each end of shaft.
18. Remove the brake shaft.

NOTE: Brake shaft idler separates from the shaft. If separated, be sure that when re-assembled, idler gear chamfers are away from the cover.

19. Remove input shaft from case by tapping with a non-metallic hammer.
20. Disassemble differential.
   a. Clean the differential assembly, then check and note the axle lengths and their relation to the heads of the four hex head bolts.
   b. If the unit will not turn freely, note where the unit binds. Check and replace those parts.
   c. Place the differential in a large vise with soft jaws (hex head bolts up). Do not clamp the vise on the bearing race of a differential carrier.
21. Remove the four hex head bolts and the upper axle and differential carrier. Remove the drive blocks, pinions, drive pin and thrust spacer if used, by lifting out of the ring gear. Tap the ring gear lightly with a mallet to loosen from the differential carrier.
22. If a snapring is used, the axle assembly may be disassembled. If the axle end has been rolled, do not attempt to break the rolled retaining edge. The parts are to be replaced as an assembly.
23. Remove the snapring and the thrust washer, if used. Separate the bevel gear and differential carrier from the axle.

INSPECTION AND REPAIR

Gears

1. Check bevels for evidence of galling due to improper shifting.

NOTE: Transaxles must be stopped for shifting.

2. Check face of teeth for wear, large shiny areas indicate much tooth contact and possible excessive wear. Replace gears indicating damage or excessive wear.

Shafts and Axles

1. Check surface for rust, pitting, scratches or wear.
2. Check key ways, splines, threads, and grooves for wear. Replace parts if worn or damaged beyond a refinishable state.

Case and Cover

Check for cracks, stripped threads, metal chips, flat sealing surfaces, and rust. Clean out any rust. Replace parts if any damage is found that cannot be repaired.

Thrust Washers and Spacers

Replace if wear is evident. Try to determine cause of thrust washer wear such as: lack of end play due to re-use of gasket or use of wrong thrust washer.

Shifting Assembly

1. Replace the shifter stop if worn or damaged.
2. Examine the teeth and internal splines of the two shifter gears. Replace damaged gears. The gears must slide freely on the shifter shaft. Excessive wear of the internal spline in the gears will create cocking and difficult shifting. Replace the gear if this condition is present.
3. Replace the shifter shaft needle bearing if wear is evident. Replace if the bearing surface of this shaft should it be scuffed, pitted or worn to a diameter less than .750 in. If replacing needle bearing, see Needle Bearing Service.
4. Replace other parts showing wear, looseness, cracks, etc.

Differential Assembly

Examine the external bearing race on the differential carriers (1) for wear or pitting, and replace if evident. The differential carriers in this assembly have replaceable bushings (2). Replace if worn in excess of .878 in. point A. These differentials have been built with rolled axle ends and also snaprings (3) as illustrated.

1. Examine gears for worn, cracked or chipped teeth. Check the internal splines of the gears and the axle if the gear is removeable. If excess play is noted, it may be necessary to replace the individual parts or both the gear and axle.
2. Examine drive pinions, drive pins and drive blocks for wear and damage. Replace excessively worn pinion or the drive pin.
3. Examine the differential carriers. One has threaded holes and the other has larger holes so that the bolts will pass through. Be sure to order the correct replacement piece.
4. Examine the internal bearing diameter of the differential carriers. If wear is in excess of the tolerance noted at point A, replace the differential carrier or bushing, if used.

Gaskets

Replace all gaskets.

Oil Seals

Replace all oil seals.

Bearings and Bushings

Bearings, bushings and bearing surfaces should be thoroughly cleaned prior to examination. Examine closely for scuffing, wear, pitting and abnormal conditions. Replace if any conditions mentioned appear. Use a good grade of cleaning solvent to clean bearings. After cleaning, always use clean lint-free cloth to dry and wipe bearings. Immediately coat cleaned bearing with lubricant to prevent rusting or corrosion. If the bearing is to be stored, wrap in oil proof paper until needed.

--- CAUTION ---
Never use compressed air to spin a race. The cage will explode causing serious injury.

Take care of bearings in the case and cover. Cover them to keep out foreign matter. Place gasket surface down on clean paper and cover with clean cloth. Never clean the lubricant from new bearings. This lubricant prevents damage before the transaxle lubricant enters the bearing.

Ball Bearing Service

The ball bearings used in the outer ends of the axle supports are sealed. Without

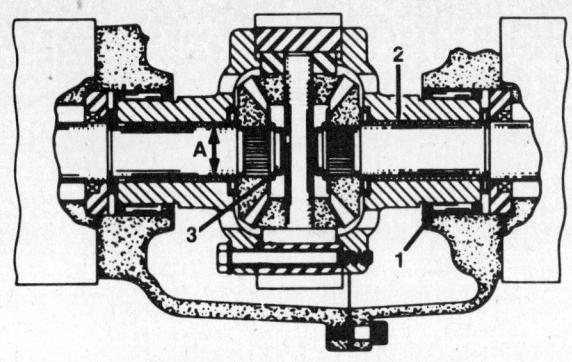
Differential measurement points

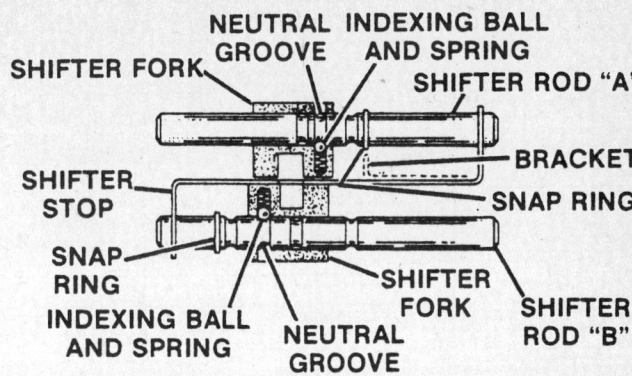
Four-stop, transaxle shifter, and fork assembly

removing, but with the axle out, rotate the inner race with the fingers. If any roughness is noted replace the ball bearing assembly. These ball bearings are factory lubricated and additional lubricants cannot be added. When driving in these ball bearings, use a proper tool that drives on the outer race.

1. Install the needle and ball bearing combination for the input shaft into the cover prior to installation of the input shaft.

2. When installing ball bearings use a tool to drive on the race which is encountering the restricted fit. For example, install the input shaft ball bearing into the case by driving on the outer race. After the input shaft bearings are installed assemble the input shaft. Press the input shaft into the bearing combination while supporting the inner race of the ball bearing on a hollow tube.

Needle Bearing Service

It is advisable to use an arbor press to remove and install needle bearings.

1. Use a bearing tool to press out the bearing. Insert the proper tool in the bearing and with an arbor press, press out the bearing from the inside.

2. When installing open end needle bearings, always apply pressure to the stamped side.

3. The inside face of the bearing housing should be below the thrust face on the case of cover. This distance is controlled by the design of the inserting tool. By using the proper tool, bearing life will be extended. Bearings should be pressed into the case or cover .015 to .020 in. below the thrust surface. The open end bearing in the low speed shaft ear is to be .010 in. below the thrust surface. The open end bearing in the shifter shaft should be .010 in. below the end.

4. To remove the needle bearing in the splined shifter shaft proceed as follows:

  a. With the needle bearing up, clamp the splined shifter shaft vertically in a soft jaw vise so that the lower end of the shaft rests on a block of wood.

  b. Prepare some pieces of paper toweling, newspaper, etc. by soaking in water.

  c. Tear paper into pieces, approximately one to two in. square. Stuff these wet pieces of paper into the needle bearing until full.

  d. Insert a $7/16$ in. metal rod into this bearing. With a mallet strike the rod sharply. This will compress the wet paper. Continue to add more wet paper, this will hydraulically lift the bearing out of the shaft.

  e. Install the new bearing. Needle bearings in shifter shafts should be installed .010 in. below flush.

Bushing Service

When removing bushings position the piece to be serviced on the table of an arbor press with an opening to allow the bushing to pass through.

1. The bushings in the three gear cluster are both removed at the same time. The bushing from one end will contact the bushing in the opposite end and both may be pushed out.

2. After new bushings are pressed into the piece they must be sized.

REASSEMBLY

To reassemble the transaxle proceed as follows:

Differential

1. Oil all parts during reassembly.

  a. Select the correct axle for the side of the differential opposite the hex head bolts. If the wrong axle is used, it will require complete teardown of the differential, or possibly the entire transaxle if the error is not detected until later.

  b. Clamp the axle, in a softjaw vise (not bearing or oil seal surfaces). The differential carrier with threaded holes is assembled to this axle.

  c. Refer to the accompanying figure, titled "differential measurement points" for the proper arrangement of parts.

  d. Torque the four hex head bolts to 7 ft.lb.

2. When assembling thrust bearings, always place a hardened thrust washer on each side of the caged thrust rollers. Never use the caged thrust rollers without the thrust washers.

3. Test differential action by holding the upper axle vertically, and spinning the differential. The unit should spin and rotate freely. Place the assembly on the bench and rotate both axles in different directions. If any binding is noted in either test check retaining bolt torque, gear meshing, or bearing surfaces in the differential carriers. Little or no end play should be apparent between the axles and carriers.

Shifter

1. Reassemble the shifting assembly by following the accompanying illustrations. Lay the parts on the bench in the same manner as illustrated on a clean paper or shop cloth. Pay particular attention to the annular grooves in the shifter rods and the snapring.

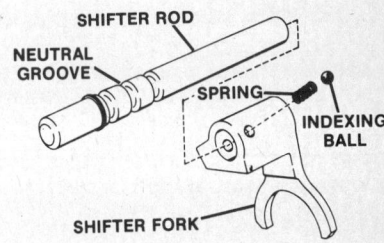
Assembling shifter forks and rods

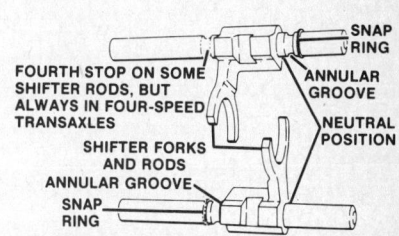
Forks and rods in the neutral position

Removing blind bearing

Ford

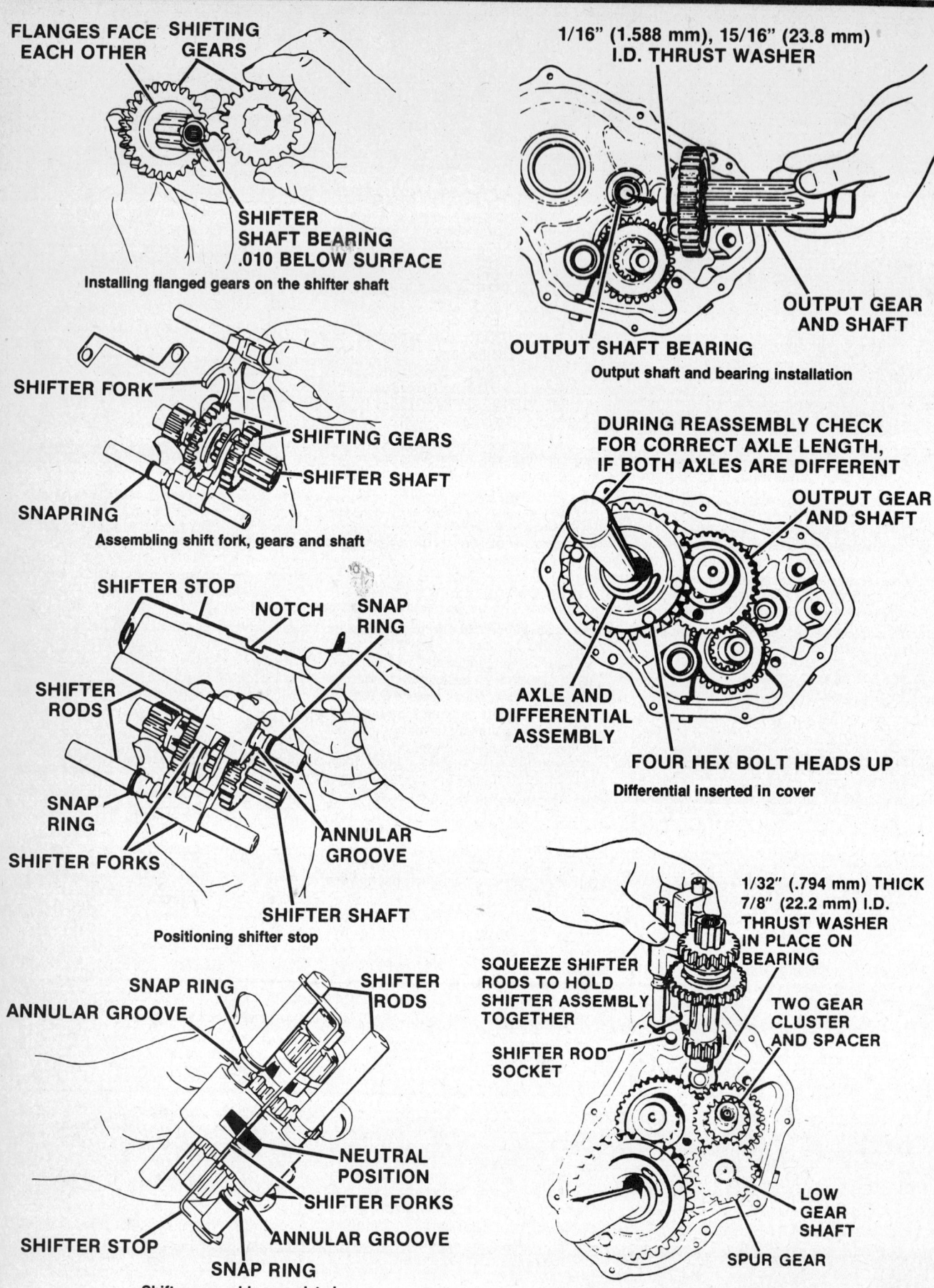

a. Assemble the shifter forks to the shifter rods as illustrated. The shifter forks are interchangeable.

b. Slide the shifter fork onto the shifter rod until it comes to the hole with the indexing ball and spring. With a flat blade screwdriver press the indexing ball into the hole and move the shifting fork completely onto the shifter rod.

c. Move the shifting fork to the Neutral position. The neutral groove is the center groove. If the shifter rod has four grooves, the neutral groove is the second groove from the shortest end. This neutral groove can be seen through the hole in the shifter fork. The arrow from the words "Neutral Groove" is passing through the hole for viewing.

d. When the shifter forks are properly assembled to the shifter rods and positioned neutral, the ends of the notches in the shifter forks are in alignment.

2. Assemble the two flanged gears onto the shift shaft. Note that the larger gear is placed on the shaft first with the flange side toward the needle bearing in the end of the shifter shaft. Slide on the smaller gear with the flange toward that of the larger gear.

3. When assembling the shifter fork and rod to the flanged gears on the shifter shaft, the shifter fork which is on shifter rod "A" always engages in flange in the larger gear. To determine which is shifter rod "A" compare the parts to illustrations. Hold the shifter shaft in the hand as illustrated during assembly.

4. After the shifter fork and rod assemblies have been engaged with the flanged gears allow the shifter rods to lay open and position the shifter stop. The notch in the shifter stop is the guide for correct positioning. Align this notch with the corresponding notches in the shifter forks and insert the shifter stop. Move the shifter rods together.

5. Install input shaft in case. Use a soft mallet to seat shaft and gear completely. Often, binding in the assembled unit can be traced to a partially installed input shaft.

6. Center one $1/32$ in. thick by 1 in. I.D. thrust washer on the cover brake shaft

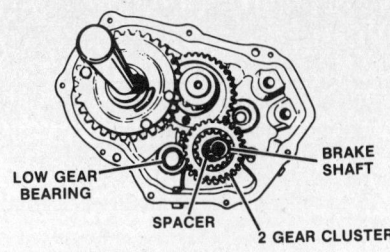

2-cluster gear installation

needle bearing, then install the brake shaft and gear (chamfer side away from cover).

7. Install the output shaft and gear after centering a $1/16$ in. thick by $15/16$ in. I.D. thrust washer on each end of the shaft.

8. Insert the differential assembly in the cover. Note that the four bolt heads should be out away from the output gear.

9. Install the two gear cluster and spacer on the brake shaft.

10. Install a $1/16$ in. thick by $3/4$ in. I.D. thrust washer, gear, and low gear idler shaft in cover. Do not put a thrust washer on the exposed end of this shaft. Be sure the small gear meshes with the larger gear of the gear cluster.

11. Center one $1/32$ in. thick by $7/8$ in. I.D. thrust washer on cover shifter shaft bearing.

12. Install shifter assembly as a unit into the cover.

a. Place the shifting assembly into the transaxle, be sure the thrust washer is on the bearing. Place the assembly into the transaxle with the needle bearing end of the shifter shaft up. Allow the end of the shifter shaft to protrude below the ends of the shifter rods, this will ease the alignment of the assembly.

b. The shifter assembly is correctly installed in the transaxle if the notches in the shifter forks are just about in the center of the opening in the case or cover of the transaxle.

13. With the small gear of the three gear cluster toward the spacer, install the three gear cluster and other $1/32$ in. thick by $7/8$ in. I.D. thrust washer on the brake shaft.

14. Install the reverse shaft, spacer, and gear into the cover. The beveled side of the idler gear should be down into the cover. Position the gasket on the cover sealing surface, then install case over the differential shaft. Be sure the boss goes under gears and that edge of the case goes over the shaft rods in the opposite manner from which it was removed. Position the gasket on the cover sealing surface, then install case over the differential shaft. Be sure the boss goes under gears and that edge of the case goes over the shaft rods in the opposite manner from which it was removed.

15. Once in position, if case hangs ½ to 1 in. high, turn the input shaft to get gears to mesh. The case should drop to about ¼ in. from closing.

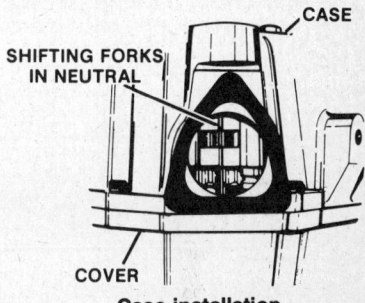

Case installation

16. Use a pair of needle nose pliers on the shifter stop on each shifter fork to agitate the shifter rod ends into their machined recesses in the case.

17. Align the case and cover with the two dowels, then install and tighten the eight socket head capscrews. Torque screws to 10 ft.lb. Unit can now be placed flat on the work bench.

--- **CAUTION** ---
Sleeves must be used to protect seals; especially axle ends or where wheels attach.

18. Position seal retainers and new seals in position. Install new O-rings on seal retainers and position axle supports to case and cover. Be sure mounting pads face in same position as when removed. Install capscrews and torque to 13 ft.lb.

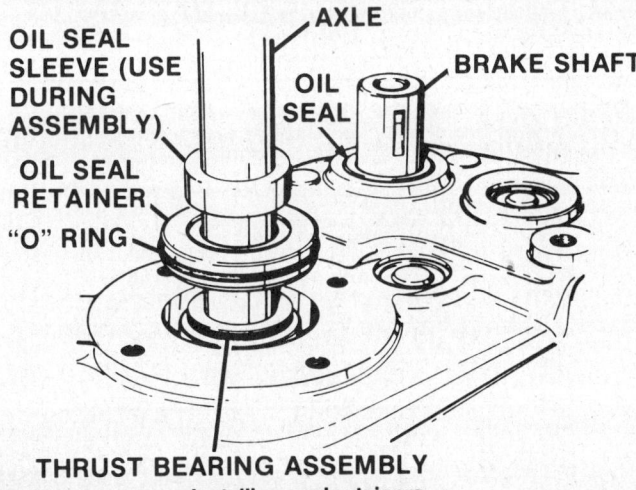

Installing seal retainers

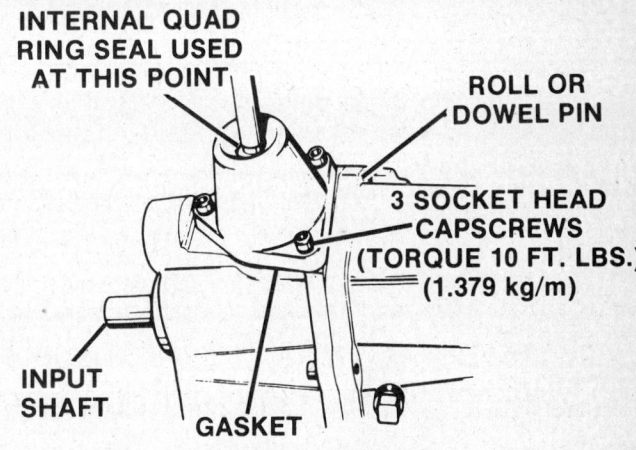

Shift lever housing installation

Ford

19. Install shift lever housing and new gasket.
20. Refill transaxle.
21. Testing. The absence of binding and oil leakage are the best indications that the unit has been properly reassembled.
22. With the shift forks in neutral, rotate both axle ends in the same direction. They should turn smoothly although a little effort may be necessary. The brake shaft should rotate whenever the axles turn together, but in neutral, the input shaft should not turn.
23. By moving any shifter gear into mesh, a greater drag should be felt on the axles on both the input and brakeshaft should turn.
24. To ease in turning of the various shafts, insert a tool (such as a punch or a socket head screw key) into the key way, however, do not force if the shaft is binding.

HYDROSTATIC TRANSAXLE

REMOVAL AND INSTALLATION

125, 145, 165

1. Park tractor on a flat surface, block front wheels and drain transmission.
2. Raise hood and remove side panels.
3. Disconnect battery.
4. Loosen instrument panel by removing four hex head screws on console arms.
5. Remove four truss head screws on deck and four hex nuts from fender, then remove fender-deck assembly.
6. Clean dirt and grass from transmission area.
7. Remove three hex head screws, nuts and washers from coupling assembly.
8. Disconnect tube assembly at joint from valve to filter. Plug openings.
8. Disconnect tube assembly at joint from hydro to valve. Plug openings.
10. Disconnect hydro-actuating arm and hydro control springs located under hydro unit.
11. Disconnect brake rod by removing locknuts and compression spring, then remove pivot assembly band.

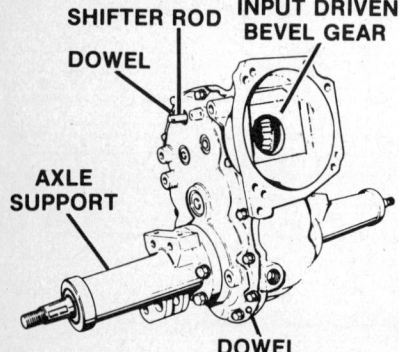

Front view of the transaxle

12. Disconnect hydro actuating arm assembly at ball joint coupling and remove centering springs.
13. Remove six hex head screws, washers, and two spacers from frame assembly and transaxle assembly.
14. Raise rear of tractor and guide transaxle from beneath it.
15. Lower tractor to transaxle and secure.
16. Connect hydro-actuating arm at ball joint and install centering springs.
17. Connect brake linkage.
18. Install shift bell crank mounting bracket and install bell crank.
19. Remove plugs from opening of hydraulic line, hydro to valve, (14, 16 hp) hydro to filter (12 hp), and then connect line.
20. Remove plugs from openings of hydraulic line, hydro to filter, connect line.
21. Install driveshaft coupling.
22. Replace fender-deck assembly.
23. Replace hardware on instrument panel.
24. Connect battery.
25. Start engine and purge hydraulic system.
26. Check fluid level.

195

1. Park tractor on a flat surface, block front wheels and drain transmission.
2. Raise hood, remove slide panels and disconnect battery.
3. Remove control knobs and panels.
4. Shut off gas at tank. Disconnect gasline hose at gas tank.
5. Disconnect wiring harness.
6. Remove fender and deck assembly as follows:
   a. Remove four truss head screws from deck.
   b. Remove four hex screws, nuts and washers from fender.
   c. Fender and deck assembly may now be removed (gas tank is removed with fender-deck assembly).
7. Disconnect hydro lines.
   a. Disconnect tube assembly (hydro to valve) at hydro.
   b. Disconnect tube assembly (hydro to filter) at hydro.
   c. Disconnect tube assembly (filter to tee) at filter and tee.
   d. Remove filter and tube assembly.
8. Remove three hex screws, nuts and washers from coupling assembly and disconnect coupling assembly.
9. Disconnect right and left brake rod arms.
10. Disconnect hydro actuating arm and hydro control springs located under hydro unit.
11. Disconnect bell crank from transaxle shift rod and clevis pin. Remove bell crank.
12. Remove four hex screws, nuts and washers from transaxle mounting bracket on righthand of frame and four from lefthand of frame.
13. Remove left and righthand hydro mounting brackets (two hex head screws).
14. With frame supported, roll transaxle rearward away from tractor.

15. Lower tractor to transaxle and secure.
16. Install shift rod bell crank.
17. Connect hydro actuating arm and install centering springs.
18. Connect brake rods.
19. Install driveshaft coupling.
20. Remove plugs and connect hydraulic lines.
   a. Install filter and line.
   b. Install tube assembly, filter to tee.
   c. Install tube assembly, hydro to filter.
   d. Install tube assembly, hydro to valve.
21. Install fender-deck assembly.
22. Connect lighting wiring harness.
23. Connect fuel line.
24. Replace body insert and knobs on controls.
25. Connect battery.
26. Start engine and purge hydraulic system.
27. Check fluid level.

OVERHAUL

Disassembly

1. Remove the axle supports.
2. Remove and discard square O-ring seal.
3. If the tapered roller bearings are loose, remove them.
4. Position the unit on the "cover up" side, then remove the dowel and screws. Lift off the cover and discard the gasket.
5. Remove, in order.

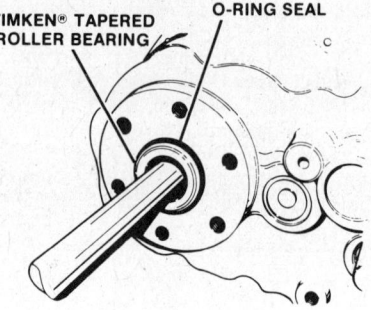

Axle support removed

a. Output shaft, thrust washer and gears.
b. Lift out the differential assembly.
c. Lift out the input bevel, gear and shaft as an assembly.
d. Work the shifter shaft and gears out of mesh with the shifter fork and rod and remove.
e. To remove the shifter rod. The setscrew, spring and ball should be removed at the outside of the case.

Inspection and Repair
AXLE SUPPORTS

1. Check ball bearings and bearing races for wear, rust and ease of rotation. Clean interior of the support as necessary.
2. Replace oil seals in axle supports.
3. Check for cracks.

CASE AND COVER

1. Check for leaks and cracks.
2. As necessary, replace needle bearings.
3. Do not replace brake shaft seal until the unit is reassembled.

Ford

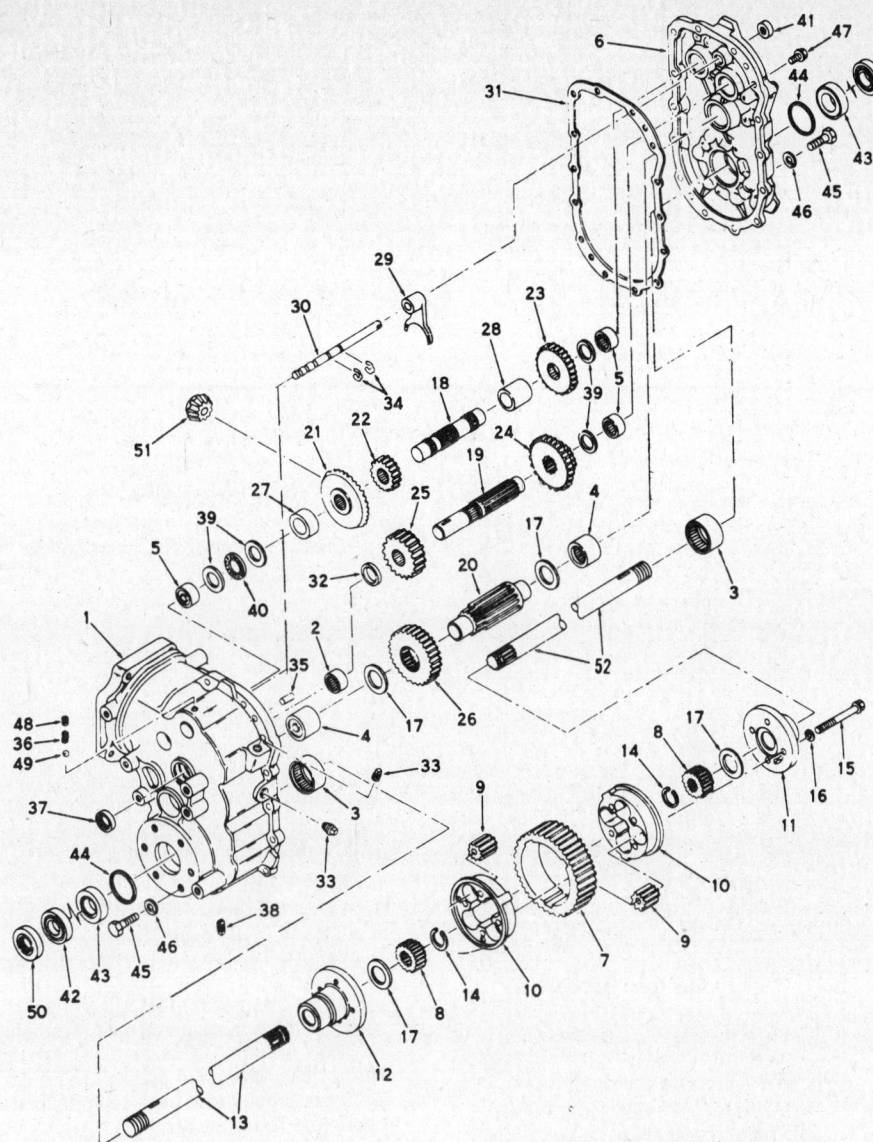

1. Transaxle case assembly (incl. nos. 2 through 5)
2. Needle bearing
3. Needle bearing
4. Needle bearing
5. Needle bearing
6. Transaxle case assembly (incl. nos. 3, 4 and 5)
7. Ring gear
8. Side gear
9. Pinion gear
10. Body core
11. Differential carrier
12. Differential carrier
13. Lefthand axle shaft
14. Snapring
15. Screw, 3/8-16 x 3 3/8 hex head
16. Lockwasher, 3/8
17. Thrust washer
18. Countershaft
19. Brake shaft
20. Output pinion
21. Bevel gear (30 teeth)
22. Spur gear (16 teeth)
23. Spur gear (23 teeth)
24. Cluster gear (20 and 27 teeth)
25. Spur gear (17 teeth)
26. Output gear (33 teeth)
27. Spacer
28. Spacer
29. Shift fork
30. Shift rod
31. Case and cover gasket
32. Spacer
33. Pipe plug
34. Snapring
35. Dowel pin
36. Spring
37. Oil seal
38. Magnetic drain plug
39. Thrust washer
40. Thrust bearing
41. Oil seal
42. Ball bearing
43. Thrust bearing
44. Square cut seal
45. Screw, 1/2-13 x 1 1/2 hex head
46. Lockwasher, 1/2
47. Screw, 5/16-18 x 1 thrd. forming hex head
48. Screw, 1/4-20 x 1/2 set
49. Steel ball
50. Oil seal
51. Bevel pinion
52. Righthand axle shaft

Hydrostatic transaxle components

SHAFTS AND GEARS
1. Check teeth for wear, pitting or breakage.
2. Inspect bearing surfaces for smoothness.
3. Inspect gears for concentricity and out of round.
4. Splines should allow a smooth fit. Rotate meshing parts for a better fit if binding seems excessive.

SHIFTER MECHANISM
1. Check spring tension and ball for wear.
2. Check shifter rod grooves for wear. Be sure the sharp edge of snapring goes away from shifter fork.
3. Inspect shifter fork for straightness and wear.

DIFFERENTIAL
1. Check and smooth axle hub ends.
2. Check security of parts.
3. Check snaprings for condition and presence.

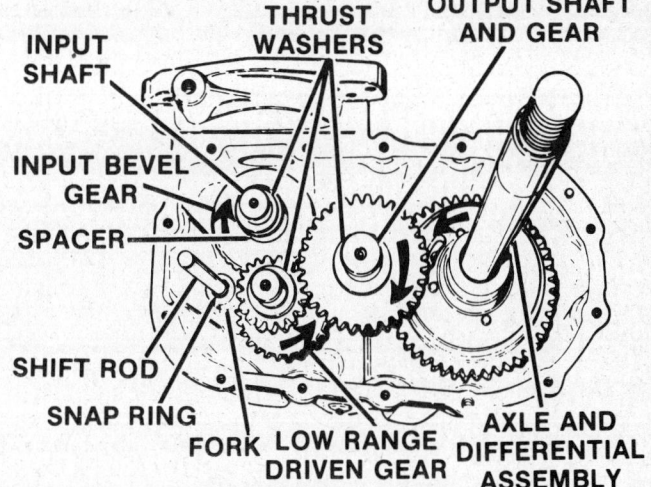

Cover removed

413

Ford

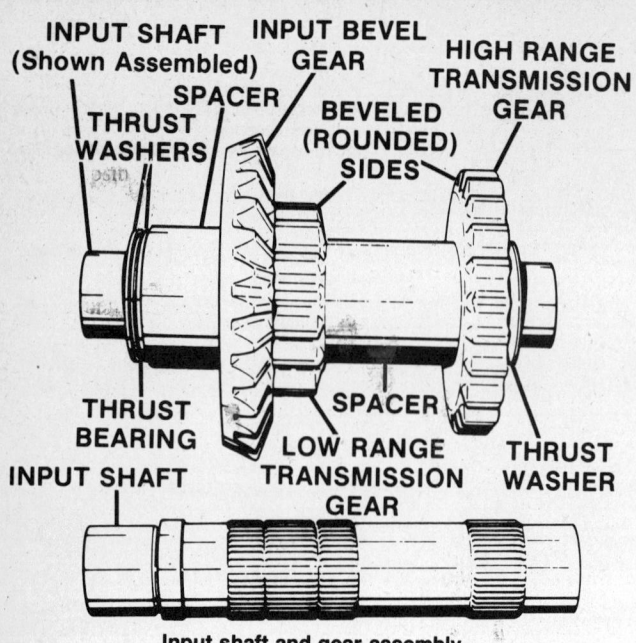
Input shaft and gear assembly

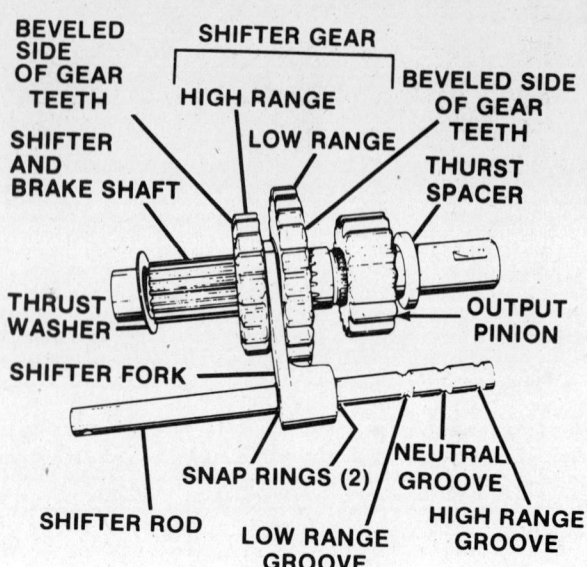
Shifter assembly

OTHER
1. Check thrust washers for wear.
2. Check tapered roller bearings for wear and ease of rotation.
3. Replace shifter rod oil seal after assembly of all parts.
4. Check thrust bearing for wear and presence of rollers.
5. Replace any parts necessary.

Assembly
1. Install parts reversing the disassembly procedure, watching out for the following:
a. After shifter rod is positioned, install ball, then spring and setscrew. Turn setscrew in slowly while raising and lowering rod until ball stops rod movement.
b. Be sure thrust washers and spacers are between every shaft and case and cover.

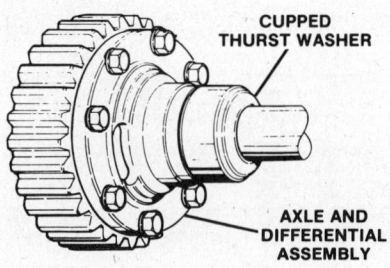
Differential gear assembly

c. Install new gasket. It may be helpful to dampen the gasket to get it to lie flat.
d. Be sure differential bolt heads go opposite output gear (large gear).
e. Use seal sleeve 670196 to protect axle support oil seals during installation.

NOTE: 670196 is the 8-12 hp crankshaft seal sleeve.

f. Install axle supports correctly. Do not rotate the support to a new position when pressed tight against the square O-ring seal or the seal may be cut.
2. After assembly:
a. To install brake shaft oil seal, use seal sleeve 670179 and driver 670180.
b. To install shifter rod oil seal, use seal sleeve 670211.
c. When filling with fluid, allow fluid to settle behind the tapered bearings into the axle supports. This may necessitate filling, checking, and adding. Fluid capacity is about 7 to 8 pints. After filling, check the fluid level.
3. Install hydrostatic unit to transaxle.
4. Install wheels on axles and roll unit to position under tractor frame.

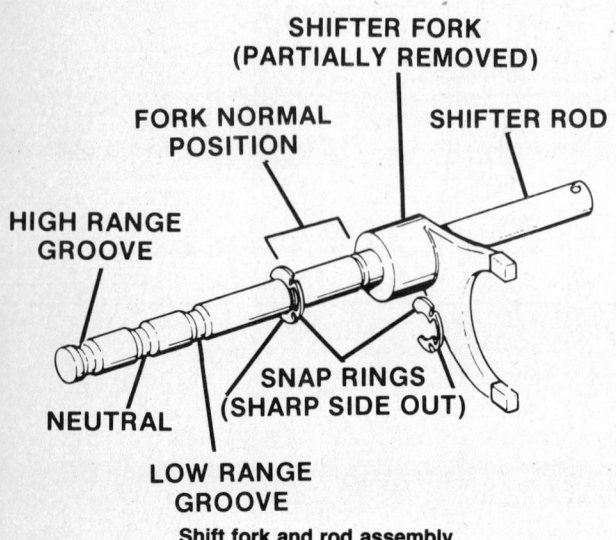
Shift fork and rod assembly

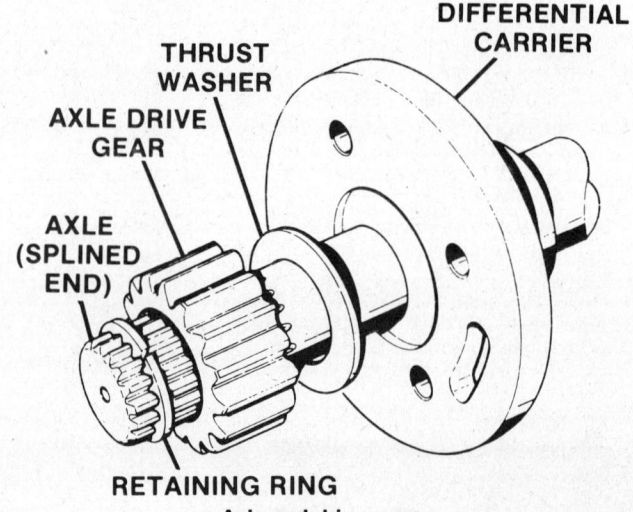
Axle and drive gear

Ford

HYDROSTATIC TRANSMISSION

REMOVAL AND INSTALLATION

1. Clean dirt from exterior of transmission.
2. Drain transmission fluid.
3. Remove screws securing coupling assembly.
4. Disconnect hydraulic tube assembly, hydro to filter at hydro.
5. Disconnect hydraulic tube assembly, valve to hydro at hydro.
6. Disconnect hydraulic tube assembly, hydro to transaxle at hydro.
7. Unhook springs from hydro actuating arm.
8. Remove nut securing speed control rod to hydro actuating arm, remove speed control rod from hydro actuating arm.
9. Remove four (4) capscrews securing hydro unit to transaxle. Carefully guide hydro down and out of tractor.
10. Place the clean transmission on a bench fixture with the output shaft in an UP position.

NOTE: A block of wood 2 in. x 6 in. x 10 in. with ¾ in. hole in the center is recommended as a stand.

NOTE: Replacement of the control shaft seal (36) is not necessary unless leakage is observed.

CAUTION
Do not pry or drive the control arm off the control shaft (37) as internal damage can occur. A puller must be used.

11. Clean gasket surface of transaxle.
12. Position new gasket on face of hydro and lift in place, secure with capscrews and lockwashers previously removed.
13. Connect drive coupling. Use caution not to damage cooling fan.
14. Connect hydraulic lines.
15. Connect speed control rod to hydro actuating lever, install front and rear centering springs.
16. Refill transaxle. Refer to checking Hydrostatic Transmission Fluid Level.

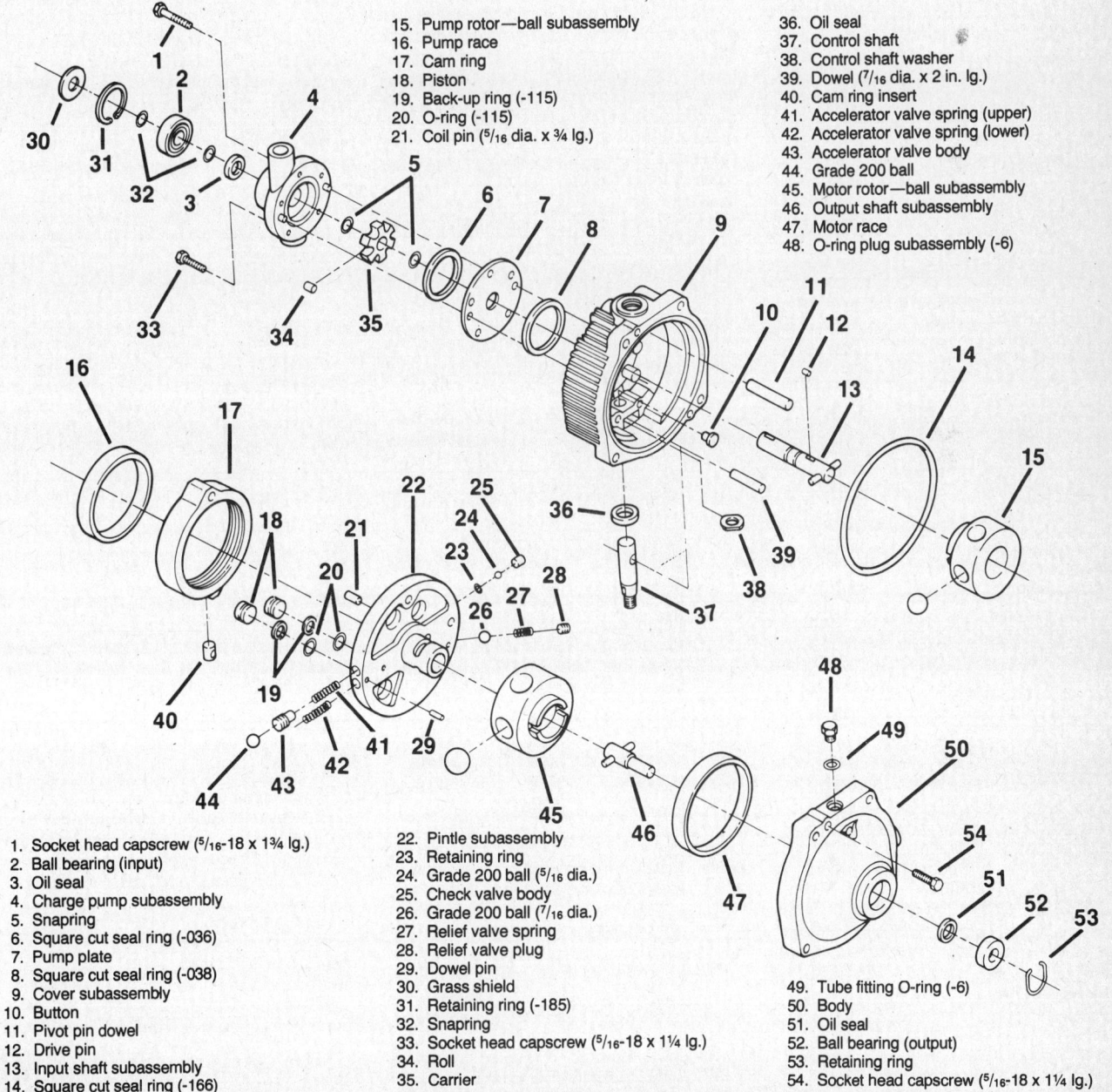

1. Socket head capscrew (5/16-18 x 1¾ lg.)
2. Ball bearing (input)
3. Oil seal
4. Charge pump subassembly
5. Snapring
6. Square cut seal ring (-036)
7. Pump plate
8. Square cut seal ring (-038)
9. Cover subassembly
10. Button
11. Pivot pin dowel
12. Drive pin
13. Input shaft subassembly
14. Square cut seal ring (-166)
15. Pump rotor—ball subassembly
16. Pump race
17. Cam ring
18. Piston
19. Back-up ring (-115)
20. O-ring (-115)
21. Coil pin (5/16 dia. x ¾ lg.)
22. Pintle subassembly
23. Retaining ring
24. Grade 200 ball (5/16 dia.)
25. Check valve body
26. Grade 200 ball (7/16 dia.)
27. Relief valve spring
28. Relief valve plug
29. Dowel pin
30. Grass shield
31. Retaining ring (-185)
32. Snapring
33. Socket head capscrew (5/16-18 x 1¼ lg.)
34. Roll
35. Carrier
36. Oil seal
37. Control shaft
38. Control shaft washer
39. Dowel (7/16 dia. x 2 in. lg.)
40. Cam ring insert
41. Accelerator valve spring (upper)
42. Accelerator valve spring (lower)
43. Accelerator valve body
44. Grade 200 ball
45. Motor rotor—ball subassembly
46. Output shaft subassembly
47. Motor race
48. O-ring plug subassembly (-6)
49. Tube fitting O-ring (-6)
50. Body
51. Oil seal
52. Ball bearing (output)
53. Retaining ring
54. Socket head capscrew (5/16-18 x 1¼ lg.)

Hydrostatic transmission components

Ford

OVERHAUL

1. Place the clean transmission on a work bench with the control shaft (37) UP in a vertical position.
2. Pierce the top of the metal portion of the control shaft seal (36) with a narrow sharp edged tool. Pry the control shaft seal (36) upward to remove.

--- CAUTION ---
Do not scratch the control shaft (37) or distort the control shaft seal counterbore in the cover (9) with the tool.

3. Clean the control shaft seal counterbore in the cover (9) with fast drying fluid. Blow dry. Inspect for tool marks caused by removal of the control shaft seal (36). Repair as required.
4. Inspect control shaft (37) for tool marks caused by removal of the control shaft seal (36). Repair as required.
5. Apply clean grease to the control shaft seal lip and a thin coating of Loctite® to the control shaft seal O.D. and the seal counterbore in the cover (9). Loctite® Grade No. 271 is recommended.
6. Guide the new control shaft seal (36) over the control shaft (37) with the rubber coated lip on the O.D. toward the counterbore in the cover (9) and press or tap the control shaft seal (36) to a bottomed position in the counterbore.

--- CAUTION ---
Do not overdrive as this may cut the rubber seal on the open end.

NOTE: A tube with the I.D. slightly larger than the control shaft (37) and the O.D. slightly smaller than the O.D. of the control shaft seal (36) with the open end faced square or slightly concave, is recommended as a tool for driving the seal into position.

Replacing the Input Shaft Seal

1. When applicable, see: Disassemble, Inspect, and Reassemble Auxiliary Charge Pump with Input Shaft Ball Bearing (2).

NOTE: A block of wood 2 in. x 6 in. x 10 in. with ¾ in. hole in the center is recommended as a stand.

NOTE: It is not necessary to remove the input shaft seal (3) unless leakage is observed.

2. Remove the grass shield (30).
3. Inspect for burrs where coupling contacted shaft and repair as necessary.
4. Remove the input shaft seal (3) following the same procedure as outlined for removal of the control shaft seal (36).
5. Inspect the input shaft (13) and input shaft seal counterbore in the charge pump body (4) for damage caused when removing the input shaft seal (3). Repair as required.
6. Apply clean grease to the input shaft seal lip and a thin coating of Loctite® to the input shaft seal O.D. and seal counterbore in the charge pump body (4). Loctite® Grade No. 271 is recommended.
7. Guide the new input shaft seal (3) over the input shaft (13) with the rubber coated O.D. toward the input shaft seal counterbore in the charge pump body (4) and press or tap the input shaft seal (3) to a bottomed position in the counterbore.

--- CAUTION ---
Do not overdrive as this may cut the rubber seal on the open end.

NOTE: A tube with the I.D. slightly larger than the input shaft (13), approximately $^{25}/_{32}$ in., and the O.D. slightly smaller than the input shaft seal (3), approximately $1^{7}/_{32}$ in., is recommended for placing seal into position.

8. Replace the grass shield (30).

Disassembly of Auxiliary Charge Pump Without Input Shaft Ball Bearing (2)

NOTE: It is rarely ever necessary to disassemble the auxiliary charge pump. The pump may be checked for performance. The specification is: at 3600 rpm with the relief valve set at 500 psi, the flow should be 1½ gpm, fluid temperature is 120°F. If there is an oil leak between the charge pump body (4) and the cover (9), the O-rings (6 and 8) are either damaged or missing and the auxiliary charge pump must be removed to replace either O-ring (6 and 8).

1. Place the clean transmission in a stand on a level work bench with the input shaft (13) UP in a vertical position.

NOTE: A block of wood 2 in. x 6 in. x 10 in. with ¾ in. hole in the center makes a suitable stand.

2. Remove the grass shield (30). Polish input shaft (13) to remove any raised surface.
3. Remove the four $^{5}/_{16}$ - 18 x 1¼ in. screws (33) and the one $^{5}/_{16}$ -18 x 1¾ in. screw (1).
4. Remove the charge pump body subassembly (4). Pull the charge pump body (4) carefully off shaft (13). Do not damage seal (3).
5. Remove the six rolls (34).
6. Remove the first snapring (5).
7. Mark the carrier (35) indicating "up" side and remove.

NOTE: Do not mark face in such a manner that the marked surface is raised. Remove any raised metal with a fine grade stone.

8. Remove carrier drive pin.

--- CAUTION ---
Do not drop pin in open ports as complete transmission disassembly may be necessary.

9. Remove pump plate (7).
10. Inspect input shaft (13) for worn key way and check for excessive clearance between the shaft and the bushing; specified clearance is .0013 in. to .0033 in. If replacement is required, it will be necessary to follow the procedure for complete disassembly and inspection.

Inspection and Reassembly

NOTE: Clean all parts in preparation for inspection and reassembly. Keep hands clean.

1. Inspect the two O-rings (6 and 8). Replacement is recommended but not required, if intact.
2. Apply clean light grease to O-ring (8) and install in the machined groove in the face of the cover (9).
3. Inspect pump plate (7) and if the face is scored, replace. If replacement is necessary, body (4), carrier (35), and rolls (34) must also be replaced as a set.

NOTE: The pump plate surface stamped "A" should be down on clockwise rotation and up on counterclockwise rotation. Nonported plates (sump cooled) must be installed with same side up as they were removed.

4. Install one snapring (5) to shaft (13) against pump plate (7).
5. Inspect pump drive pin. Replace if worn and reinstall.

NOTE: Clean light grease in the key way will retain the pin during assembly.

6. Inspect the contact surfaces of the carrier (35) for measurable wear. Place on shaft with marked face up. If replacement is necessary, body (4), rolls (34), and plate (7) must also be replaced as a set.
7. Install new snapring (5).
8. Inspect the six rolls (34), for wear on the end radius and O.D. Replace worn rolls. Apply clean light grease to the rolls and assembly in position in carrier. If replacement is necessary, body (4), carrier (35), and plate (7) must also be replaced as a set.

NOTE: The grease will hold the rolls in position.

9. Inspect the pocket of the charge pump body (4) for end milling.

NOTE: If measurable wear is observed, replace charge pump body. If replacement is necessary, carrier (33), rolls (34), and plate (7) must be replaced as a set. Inspect the cam insert (40) for any unusual wear pattern.

NOTE: If badly worn or scored, replace charge pump body.

10. Inspect input shaft seal (3), if seal lip is damaged or hard, replace.
11. Apply clean light grease to O-ring (6) and install in machine groove in the face of the charge pump body (4).
12. Apply clean light grease to the input shaft seal (3) lip.
13. Guide the charge pump body subassembly (4) with O-ring in place over the input shaft (13) and guide into position over the rolls and into the dowel pin holes.
14. Install four $^{5}/_{16}$-18 x 1¼ in. long hex socket capscrews (1).

NOTE: If the capscrews are not hex socket screws, it is recommended that they be replaced.

Ford

CAUTION

The 5/16-18x1¾ in. long capscrew (1) must be installed in the heavy section of the body. If installed in any of the other four holes, internal damage to the die casting will unknowingly occur.

15. Torque the 5/16 in.-18 screws (1) and (33) to 15 ft.lb.

NOTE: The input shaft should turn by hand.

16. Inspect the grass shield (30). If bent, replace and install in position.

Disassembly of Auxiliary Charge Pump with Input Shaft Ball Bearing

1. Place the CLEAN transmission in a stand on a level work area with the input shaft (13) UP in vertical position.

NOTE: A block of wood 2 x 6 x 10 in. with ¾ in. hole in the center makes a suitable stand.

2. Polish shaft (13) to remove any raised surface.

3. Remove the four 5/16-18 x 1¼ in. long socket head capscrews (33) and the one 5/16-18 x 1¾ in. long socket head capscrew (1).

NOTE: The capscrews (1) and (33) may be 5/16-18 twelve-point capscrews. If so, a thin wall twelve-point socket wrench is required.

4. Remove the retaining rings (31) and (32).

5. Remove the charge pump body subassembly (4) with input shaft ball bearing (2) in position. Then remove bearing from body.

NOTE: To remove: Looking at transmission with the control shaft down, tap the top hole and the two bottom holes in the auxiliary pump body with a ⅜-16 NC tap. Be careful not to run tap too deep into the aluminum cover (only tap about ½ in. deep). Now use a puller.

CAUTION

Do not pound on puller—apply a steady pull away from the transmission on puller while removing auxiliary body (4). On units equipped with puller slots on auxiliary pump body, use a 2-fingered puller.

6. Remove the six rolls (34).
7. Remove snapring (32).
8. Mark the carrier (35) indicating the "up" side and remove.

NOTE: Do not mark face in such a manner that the marked surface is raised. Remove any raised metal with a fine grade stone.

9. Remove pump drive pin.

CAUTION

When applicable, cover the open ports in the plate to prevent pin from dropping into the unit.

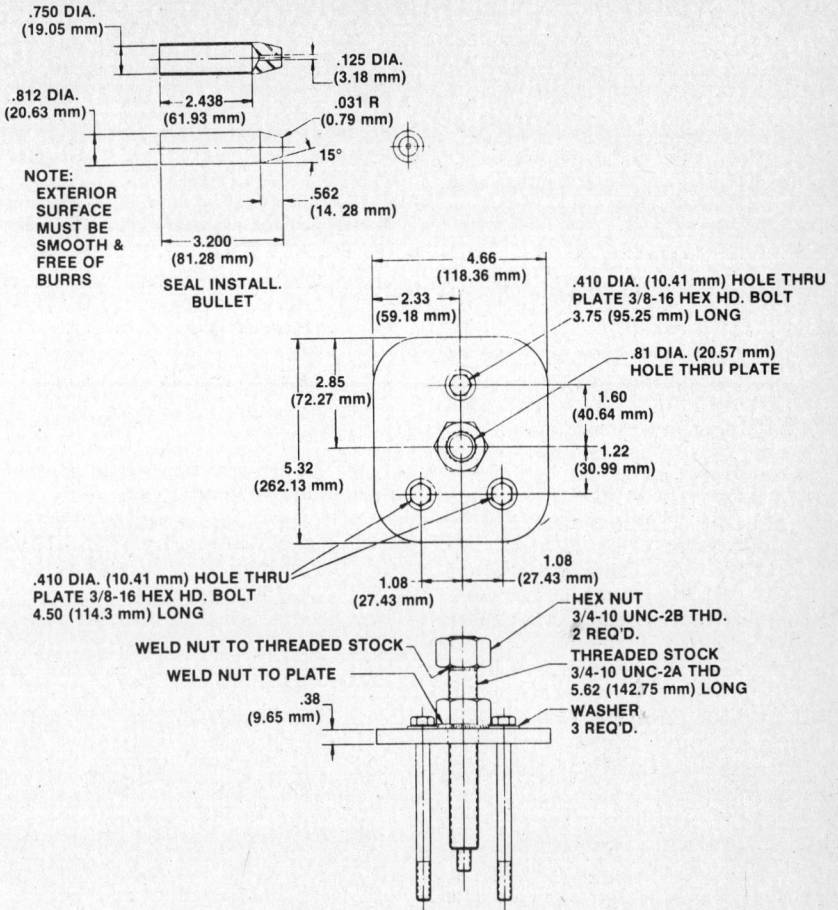

Puller required for auxiliary pumps with ball bearings

10. Remove the pump plate (7).
11. Inspect input shaft (13) for worn key way and excessive wear in the seal or bearing area; specified clearance .0013 in. to .0033 in. If replacement is required, it will be necessary to follow the procedure for complete disassembly and inspection. CLEAN ALL parts before inspection.

Inspection and Reassembly

NOTE: Complete transmission must be disassembled before ball bearing charge pump can be reassembled. Proceed to disassemble the transmission. It is not necessary to break down all subassemblies if the auxiliary pump is the only portion that requires service. The pump rotor subassembly (15) and cam ring (17) must be removed from the aluminum cover. The aluminum cover, with input shaft installed, must be supported properly for reassembly of the auxiliary pump. The cover should be placed on a clean bench with the shaft pointing up and with the shaft supported from underneath the cover so that drive cross pin is tight against the cover. A solid block (steel bar stock) 2 in. in diameter by at least 2.5 in. long may be used to support the shaft. It is important to support the shaft when installing the body and bearing because if the shaft slides down the cover more than 1/32 in. (0.794 mm), damage to the seal will occur.

NOTE: Clean all parts before inspection. Keep hands clean.

1. Inspect O-rings (6 and 8). Replacement is recommended but not required if in good condition.

2. Apply clean light grease to O-ring (8) and install in the machined groove in the face of the cover (9).

3. Inspect pump plate (7), and if face is scored, replace and install with surface stamped "A" UP on counterclockwise rotation and DOWN on clockwise rotation. If replacement is necessary carrier (35), body (4), and rolls (34) must also be replaced as a set.

NOTE: Non-ported plates (sump cooled) must be installed in same position as removed.

4. Inspect pump drive pin. Replace if worn and install.

NOTE: Clean light grease in the shaft key way will retain the pin for assembly of the carrier (35).

5. Inspect the contact surfaces of the carrier (35) for measurable wear. Replace if worn. Install with marked face up. If replacement is necessary, pump plate (7), body (4), and rolls (34) must also be replaced as a set.

Ford

6. Inspect the six rolls (7) for wear on the end radius and O.D. Replace worn rolls. Apply light grease to the rolls and assemble in the carrier (6). If replacement is necessary, carrier (35), body (4), and pump plate (7) must also be replaced as a set.

NOTE: The grease will hold the rolls in position for assembly of the body subassembly (4).

7. Inspect the pocket surface of the charge pump body (4) for end milling or scoring.

NOTE: If measurable wear is observed, replacement of the charge pump body subassembly (4) is recommended. If replacement is necessary, pump plate (7), rolls (34), and carrier (35) must also be replaced as a set.

8. Inspect the cam insert for any unusual wear pattern.

NOTE: If badly worn or scored, replacement of the charge pump subassembly (4) is recommended. If replacement is necessary, pump plate (7), rolls (34), and carrier (35) must also be replaced as a set.

9. Inspect ball bearing (2). Roll check; if noisy or binds, replace.
10. Inspect input shaft seal (3). Replace if the seal lip is cut or hard.
11. Apply clean light grease to O-ring (6) and install in the machined groove in the face of the charge pump body (4).
12. Apply clean light grease to the input shaft seal lip (3) and guide the charge pump body subassembly (4) with O-ring (6) in place over the input shaft (13), and then guide into position over the rolls (34) and into the two dowel pin holes.

NOTE: A bullet nose tool over the input shaft (13) must be used to protect the shaft seal. The bullet must be long enough to cover both snapring grooves.

――――――― CAUTION ―――――――
Before bottoming the charge pump body subassembly (4) to the cover subassembly (10), check to be certain the O-ring (6) is in position.

13. Install the four $5/16$-18 x $1\frac{1}{4}$ in. long hex socket capscrews (33) and the one $5/16$-18 x $1\frac{3}{4}$ in. long hex socket capscrew (1).

――――――― CAUTION ―――――――
The $5/16$-18 x $1\frac{3}{4}$ in. long capscrew (1) MUST be installed in the heavy section of the body (4). If installed in any of the other four holes, internal damage to the die casting will unknowingly occur.

14. Torque the $5/16$ in.-18 screws to 15 ft.lb.
15. Install the snapring (32) in the bottom groove.
16. Install input shaft bearing (2). Position the input shaft ball bearing (2) to the input shaft (13). Draw NOT PRESS the bearing (2) to a bottomed position on the shaft (13).

NOTE: A sleeve $1\frac{5}{8}$ in. long with a .760 in. I.D. and $1\frac{1}{8}$ in. O.D.; a washer $1/8$ in. thick and O.D. $1\frac{1}{4}$ in. with a $5/32$ in. hole; a Class 5 hex head $1/4$-20 bolt 1 in. long are recommended for drawing the bearing in position.

――――――― CAUTION ―――――――
Do not draw bearing in position with draw tool contacting the outer race.

17. Install the retaining rings (32) and (31).

Disassembly of Transmission

It is absolutely essential that the transmission be thoroughly cleaned before proceeding with disassembly. Cap all fittings before cleaning.

NOTE: Steam cleaning is an approved method for cleaning.

――――――― CAUTION ―――――――
Do not expose seal lips to high air pressure when cleaning.

1. Remove the fluid from the CLEAN transmission by removing the reservoir fitting (if any) in the aluminum cover and/or the bleed plug (48) from the body (50). Allow fluid to drain from these openings.
2. Place the clean transmission, less fluid, in the stand on a level work area with the output shaft UP in a vertical position.

NOTE: A block of wood 2 in. x 6 in. x 10 in. with $3/4$ in. hole in the center, or front coupling, makes a suitable stand.

Separating Aluminum Cover and Cast Iron Housing Body

1. Remove the two $5/16$-18 x $1\frac{1}{4}$ in. long (54) twelve-point capscrews or socket head capscrews.

NOTE: A thin wall $5/16$ in. twelve-point socket wrench is required.

2. Reinstall the two $5/16$-18 x $1\frac{3}{4}$ in. long twelve-point capscrews (54) with two threads only engaged.
3. Lift the body subassembly (50) to separate the cover subassembly (9). Lift to the bolt stops.

NOTE: The bleed plug hole should be toward the disassembler.

NOTE: If difficulty is experienced in breaking the fluid seal, tap body and/or cover with a plastic hammer.

4. Insert a fork tool ($1/8$ in. flat stock approximately 3 in. wide and 12 in. long with a $1\frac{9}{16}$ in. wide slot 8 in. long, taper the ends of the prongs) from the right side into the opening. Guide the tip of the prongs between the motor rotor subassembly (45) and the pintle subassembly (22). Lift the motor rotor subassembly (45) and slide the fork tool through until it extends beyond the opposite side. This procedure is to prevent loss of motor balls.
5. Remove the two $5/16$-18 x $1\frac{3}{4}$ in. long twelve-point capscrews (54) and rotate the fork tool 90 degrees.
6. Remove the body subassembly (50) with the rotor subassembly (45) held in a fixed position by the fork tool. Place on a work bench with the output shaft in a DOWN position.

Disassembly of Motor

1. Place a split thin wall sleeve, approximately $4\frac{1}{4}$ in. I.D. x $4\frac{1}{2}$ in. O.D. x $2\frac{1}{2}$ in. long over the motor rotor subassembly (45); a large rubber band may be used in place of this sleeve. Remove the motor rotor subassembly (45) from the body subassembly (50) and place on work bench.
2. Remove the five motor ball pistons from the motor rotor by working clockwise from the letter stamped in the face of the rotor.

NOTE: Remove one motor ball piston at a time and place in a marked receptacle. A plastic ice cube tray is suggested.

3. Inspect the motor race (47). The contact line for the motor ball pistons must be smooth, complete, and free of raised material.

NOTE: If raised metal is observed, the race (47) in the body must be replaced. It is also reasonable to assume that one or more motor rotor ball pistons and bores will be damaged. Therefore, the motor rotor subassembly (45) should also be replaced.

4. Remove the output shaft (46). If broken, replace the output shaft subassembly (46).
   a. Remove output gear snapring and gear, if supplied.
   b. Remove snapring.
   c. Remove output shaft subassembly (46) by pressing or tapping inward.

NOTE: The shaft is a slip to light press fit to bearing (52).

5. Inspect output shaft bearing (52) located in body. Replace if necessary.
6. Inspect output shaft seal (51). Replace if necessary.
7. Inspect for a broken or loose output shaft dowel pin. If broken, replace with output shaft subassembly (46).

Disassembly of Charge Relief Valve Located in Pintle

NOTE: It usually is not necessary to remove the charge relief valve.

Disassembly of Pintle Subassembly

1. Place aluminum cover assembly input shaft DOWN in wooden stand and pintle UP.
2. Remove the pintle subassembly (22) by lifting away from cover (9), light tapping on cover may be necessary. During this process, hold the pump rotor subassembly (15) to a bottomed position through the 1 in. hole in the pintle. Hold in position and complete removal of the pintle subassembly.

NOTE: Do not remove pintle pipe plugs, they are solid body plugs.

1. Locate plug (28) in outer circumference of pintle (22) and remove.
2. Remove spring (27) and ball (26).

Ford

3. Inspect spring for cracks or bends. Inspect ball and its seat in pintle for pitting or grooves.

Removal of Check Valves Two Used

NOTE: It usually is not necessary to remove the check valves. The check valve balls (23) must roll free, in the pintle cavity, by their own weight. If this is the case, removal is generally not necessary.

1. Remove spiral pin (21).

NOTE: To remove pin, select drill that fits through spiral pin and drill through pintle, redrill pintle from side away from pin with ¼ in. drill. Press or drive pin (21) from pintle. Newer units are through drilled from factory.

2. Tap the hole in valve body(s) (25), insert puller or bolt and pull valve body(s) from pintle.

3. Remove ball (24) and inspect for pitting or grooves.

4. Replace retaining ring (23) as required.

5. Clean pintle passageways thoroughly.

Removal of Acceleration Valves—Some Models

Located in pintle out circumference, 180° from check valves.

NOTE: To permit removal of acceleration valves, check valves must first be removed.

1. Refer to removal of check valves section.

2. Remove pin (29) from pintle.

3. Obtain drive rod ³⁄₁₆ in. diameter by approximately 8 in. long. Place rod through vacated check valve passageways in pintle, locating end of rod into acceleration valve (43). Tap lightly on valve(s) to remove ball(s) (44), valve (43), and springs (42) from each bore.

NOTE: Do not damage any wall or seat in passageways or replacement of pintle subassembly (22) is required.

Removal of Damping Pistons (18)

1. Hold pintle with the damping pistons away from you and facing down.

2. With a down stroke, firmly tap outside edge of pintle on work bench.

---- **CAUTION** ----

Do not hit journal on anything or the pintle will be ruined.

3. If this does not dislodge pistons from bores, they more than likely do not need service. But, if they are scored on the surface and need replacement, they may be removed by cementing a bolt or other object with a smooth surface to the face of the piston. Use a "super glue" such as Loctite® 404 and use sparingly; do not get any on the pintle. Use the glue according to instructions and then pull straight on the piston. A side load on the bolt will usually break the glue.

Reassembly of Components into Pintle

1. Make sure all parts are clean as well as hands.

2. Install charge relief valve.

a. Place ball (26) then spring (27) into proper passage.

b. Screw plug (28) into threads until just below pintle O.D. surface.

3. Install acceleration valves (43).

a. Place springs (41 and 42) in proper bores, then valve bodies (43).

b. With clean tool or rod, push the valves down by hand to check for freeness. The valves hould seat all the way down and return with no binding.

c. Press the ⅝ in. diameter balls (44) in bores just until pin (29) will go into place. Tap pin in place with light hammer.

4. Install check valves.

a. Place retaining rings (23) in bores. It may be necessary to push the rings down flat with a long punch by hand.

b. Drop balls (24) in bores.

c. Put new check valve bodies (25) in bores and press into position.

---- **CAUTION** ----

A hand arbor press is recommended so valves will not be overpressed. If too much pressure is used, the retainer rings can be forced into pintle passage.

d. Install roll pin (21). The roll pin must be flush or slightly below pintle surface.

5. Install damping piston (18).

a. Replace O-ring (20) on piston.

b. Determine which face of piston is the smoothest.

c. Install new back-up ring (19) nearest smooth surface.

d. Grease bore and piston O.D. with clean light grease.

e. Push into bore with smooth side out-taking care not to cut O-ring or back-up ring.

Disassembly of Pump— Pintle Subassembly Removed

NOTE: Place aluminum cover assembly in wooden block (2 in. x 6 in. x 10 in. with hole in center) input shaft DOWN.

NOTE: Each ball must be reassembled in the same bore from which it was removed. Before removing any balls, prepare an egg carton or ice cube tray, etc. for ball storage.

1. Remove pump rotor subassembly (15).

a. Place a thin wall sleeve or large rubber band over pump rotor assembly (15) to retain ball pistons in rotor. DO NOT permit balls to fall from rotor or become mixed.

b. Remove pump rotor assembly (15) from input shaft subassembly.

c. Remove the five pump ball pistons from pump rotor by working clockwise from letter stamped on face of rotor. Place balls in marked receptacle. An ice cube tray is suggested.

d. If scoring or seizure has occurred, or clearance of balls to rotor seems excessive, the pump rotor assembly must be replaced.

2. Remove the cam ring subassembly (17) from control shaft pin (39) and pivot pin (11).

3. Remove two buttons (10).

4. Remove cam ring insert (40) from cam ring (17).

5. Inspect pump race (16). If scored, replace. Reinspect piston bores in rotor (15) if pump race (16) is replaced.

NOTE: It is reasonable to assume if the pump race (16) is scored that one or more pump rotor ball pistons and bores will be damaged. Therefore, the pump rotor subassembly (15) should also be replaced.

6. Remove input shaft subassembly (13). (Charge pump subassembly removed.)

a. Remove snapring (5) from input shaft (13) located on charge pump side of aluminum cover (9).

b. Pull input shaft subassembly (13) from cover (9).

c. Inspect shaft (13), its key way or spline, and input shaft dowel pin.

d. Inspect for excessive clearance between input shaft and bushing. If excessive, the cover subassembly (9) must be replaced.

NOTE: Specified clearance .0013 in. to .0033 in.

7. Inspect for excessive clearance between the control shaft (37) and cover (9). If excessive, the cover subassembly (9) must be replaced.

8. Inspect for loose or broken control shaft dowel pin (39). If loose to shaft, replace both parts (37 and 39). If control shaft washer (38) is bent or dowel pin (11) is broken, a new cover subassembly (10) is required. Drill a ¹¹⁄₃₂ in. hole in the cover directly in line with the control shaft dowel pin (39) and thread hole with ⅛ in. pipe tap. Press broken pin out. Apply Loctite® Grade No. 271 to ⅛ in. pipe plug and install to cover. Remove shaft and replace with new shaft. If washer (38) is bent, replace. Place new shaft in cover and washer, press new dowel pin (39) through shaft until 1¼ in. of dowel extends from shaft.

---- **CAUTION** ----

The key way in the control shaft (37) must be to be assembler's left with the threaded portion of the shaft pointing to the assembler.

9. If the cover (9) is broken or cracked, especially around the control shaft area, a new cover is required.

Reassembly

All parts have passed inspection or otherwise have been replaced with new parts. They must be clean and dry.

Ford

CAUTION
The work area and workers hands should be clean.

1. Install output shaft subassembly (46).
  a. Guide output shaft subassembly (46) through oil seal (51) using a bullet nose tool to position seal lip spring and protect the seal lip from damage. Remove tool and visually determine if seal spring is in position.
  b. Support shaft with drive pin flush to body and press ball bearing (52) to bottomed position.

CAUTION
Do not permit shaft to slip out of seal during assembly.

2. Install snapring (53).
3. Install input shaft subassembly (13). Guide input shaft through the cover bushing.
4. Install snapring (5), when applicable.
5. When applicable, install charge pump following the instructions provided.
6. Position the cover subassembly with the auxiliary charge pump assembled to the cover on the work stand with the input shaft down in a vertical position.
7. Install the cam ring insert (40) in the cam ring subassembly (17).

CAUTION
The hole in the cam ring insert (40) must be to the outside of the cam ring.

8. Install the cam ring subassembly (17).
  a. Guide the cam ring over the dowel pins (11) and (39).

CAUTION
The side with the cam ring (17) flush with the race (16) must be down facing the cover. The cam ring insert (40) must be in position.

  b. Press firmly until cam has bottomed.

CAUTION
The cam ring must move freely from one stop to the other. If there is a bind at either end, rotate cam ring insert (40) 180°. Recheck travel.

9. Install the pump ball pistons (40) in the pump rotor bores. Slip the thin wall sleeve over the rotor to retain the ball pistons.

CAUTION
The pump ball pistons must be reassembled in the same bore in the pump rotor from which they were removed. They are selectively fitted electronically to .0002 in. to .0006 in. clearance.

10. Install the pump rotor subassembly (15). Turn to insure proper alignment with the input drive pin.

NOTE: Remove thin wall sleeve.

11. Install pintle subassembly (22). Guide pintle into the pump rotor and over the dowel pin (11). Press to bottomed position in the cover (9). **Do not force** pintle through rotor.
12. Install slug in reservoir fitting in the top of the aluminum cover where applicable, and fill cover cavity with new fluid.
13. Install O-ring (14).
14. Install the motor ball pistons in the motor rotor bores. Slip the thin wall sleeve over the motor rotor subassembly receptive (45) to retain pistons.

CAUTION
The motor ball pistons must be reassembled in the same bore in the motor rotor from which they were removed. They are selectively fitted electronically to .0002 in. to .0006 in. clearance.

15. Install the motor rotor subassembly (45) in the body subassembly (50) and remove thin wall sleeve.
  a. Retain the motor rotor subassembly (45) in the body subassembly (45) in the body subassembly (50); with the fork tool, guide the motor rotor subassembly (20) over the journal on the pintle subassembly (22). When bottomed, remove the fork tool.

NOTE: Do not force motor rotor subassembly on pintle journal.

16. Install the two ⁵/₁₆-18 capscrews (54) and torque to 15 ft.lb.

NOTE: The output shaft (46) should rotate finger free.

17. Place the transmission with the reservoir fitting, if any, in the cover and/or ⅜-16 tapped hole in the body in an **UP** position. Fill the transmission through either of these openings with proper fluid. Continue filling until fluid flows through overflow hole in body or, when applicable, the bleed hole.
18. Install bleed plug with O-ring (48).

HYDRAULIC SYSTEM

Testing and Adjustment

ALL EXCEPT 195

Preparation for Test

The following preparations must be made before testing:
1. Check, and adjust if necessary, engine high rpm (3200 rpm). This speed must meet the engine manufacturer's recommendations.

NOTE: Overspeeding of the engine can cause shortened charge pump life and premature pump failure.

2. Obtain an accurate 0 to 1000 psi pressure gauge with a hose 36 in. long.
3. Bring hydraulic system to operating temperature; minimum of 70°F.
4. Set parking brake, retract cylinder completely and stop engine.
5. Move hydraulic control lever back and forth to relieve pressure in system.
6. Raise hood and remove side panels.
7. Remove cotter pins retaining cylinder to mounting pins and slide cylinder off.
8. Remove dirt from cylinder and disconnect hose fitting from rod end of cylinder.
9. Install a tee in cylinder port, reconnect hose to tee.

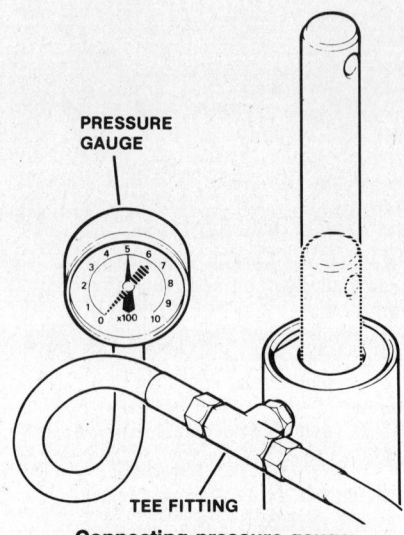

Connecting pressure gauge

10. Install pressure gauge in tee.
11. Start engine and place throttle in high idle position.

Hydraulic Pressure Test

To test the hydraulic pressure, actuate hydraulic control lever back and forth several times to purge air from lines and cylinder. Hold hydraulic control lever in the RAISE position. Gauge should indicate 500 psi.

Adjustment

Hydraulic system pressure can be adjusted at the relief valve located on the hydraulic lift valve.

To adjust the relief on the AICO lift valve proceed as follows:

1. If pressure reading is higher than 500 psi, remove plug and one (1) shim. Replace plug and take another reading.

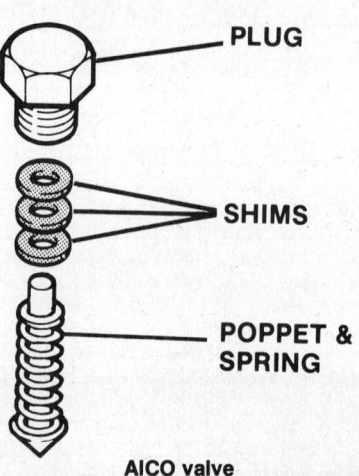

AICO valve

420

If pressure drops below 500 psi with shim removed, replace shim after reducing thickness by rubbing on emery cloth. Repeat until correct pressure is reached.

2. If pressure reading is lower than 500 psi, add shim(s) until pressure is normal.

3. If readings are consistently low, check relief valve for damage.

To adjust relief on the VFP lift valve proceed as follows:

1. If pressure reading is higher than 500 psi, remove acorn nut and aluminum crush washer from control valve. Back-out relief valve adjusting screw (counterclockwise) until reading is normal. Tighten jam nut, install aluminum crush washer and acorn nut.

2. If reading is lower than 500 psi, turn in relief valve adjusting screw (clockwise) until reading is normal.

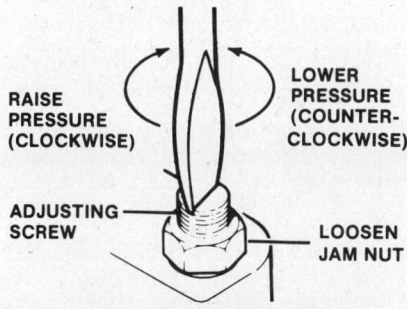

VFP pressure adjustment

3. If readings are consistently low, replace relief valve.

4. If readings are still low with new relief valve replace charge pump.

NOTE: Charge pump or relief valve should be replaced only after all troubleshooting checks have been made.

195

Preparation for Test

The following preparation must be made before testing:

1. Check and, if necessary, adjust engine high rpm (3300 rpm). This speed must meet the engine manufacturer's recommendations.

---- **CAUTION** ----

Overspeeding of the engine can cause shortened charge pump life and premature pump failure.

---

2. Obtain an accurate 0 to 1000 psi pressure gauge with a hose 36 in. long.

3. Bring hydraulic system to operating temperature; minimum of 70° F.

4. Set parking brake, retract cylinder completely and stop engine.

5. Move hydraulic control lever back and forth to relieve pressure in system.

6. Plug gauge into coupling of front hydraulic outlet.

7. Start engine and place throttle in high idle position.

Hydraulic Pressure Test

To test the hydraulic pressure, actuate hydraulic control lever back and forth several times to purge air from lines and cylinder. Hold hydraulic control lever in the RAISE position. Gauge should indicate 800 psi.

Adjustment

To adjust relief on the lift valve, follow procedure used on the VFP lift valve.

Hydraulic Control Valves

ALL EXCEPT 195

Removal and Installation

1. Raise hood and remove side panels.
2. Remove battery.
3. Clean dirt from around fittings.
4. Disconnect all lines and linkage at valve.
5. Plug line openings and valve ports to eliminate dirt from entering hydraulic system.
6. Remove (2) capscrews, nuts and lockwashers retaining valve to frame.
7. Remove valve.
8. Installation is the reverse of removal.

Disassembly

To disassemble the VFP valve, see procedure for the two spool valve.

To disassemble the AICO valve, proceed as follows:

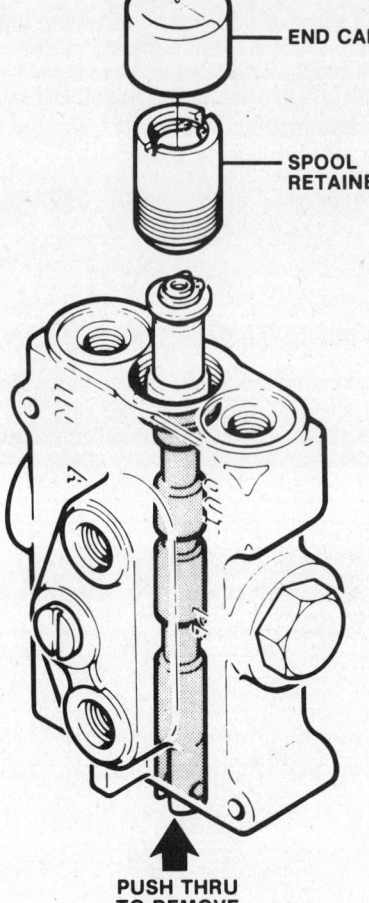

AICO valve disassembly

NOTE: Service parts available for AICO and VFP valves are limited to seal kits. Clean outside of valve with a suitable solvent.

1. Remove spool as follows:
a. Remove end cap.
b. Unscrew and remove spool retainer.
c. Slide spool back a little to expose the O-ring.

NOTE: Extreme care must be used to prevent damaging machined spool surfaces; even a tiny nick will ruin the valve.

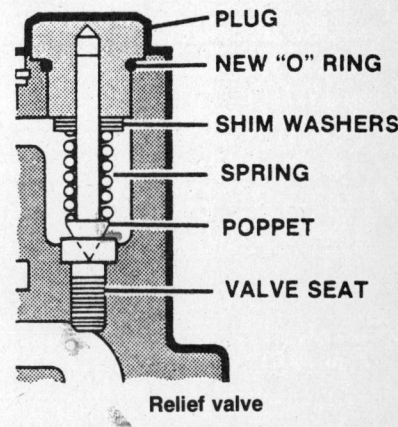

Relief valve

d. Remove retaining ring, back-up washer, and quad-ring from valve body.
e. Remove quad-ring and washer.

2. Remove relief valve (AICO valve) as follows:
a. Remove plug and washers, spring, and poppet will drop out.
b. Use a broad bladed screwdriver to remove seat.

Inspection and Repair (AICO Valve)

See procedure for VFP valve.

Replacing O-Rings and Quad Rings (AICO Valve)

See procedure for VFP valve.

Assembly (AICO Valve)

Clean all parts in a suitable solvent and visually inspect all bores and ports for dirt.

NOTE: All components must be free of dirt, oil, rust, debris.

To assemble the lift valve, proceed as follows:

1. Install relief valve as follows:
a. Install relief valve seat in body.
b. Place spring on poppet and drop in cavity.
c. Install new O-ring on plug.
d. Place shim washers on end of poppet and install plug.

2. Install spool as follows:
a. Install new quad-ring, nylon back-up washer and metal washer in spool bore and install retaining ring.
b. Install new quad-ring in other end and carefully guide spool into valve body.
c. Install cap over spool end.

Ford

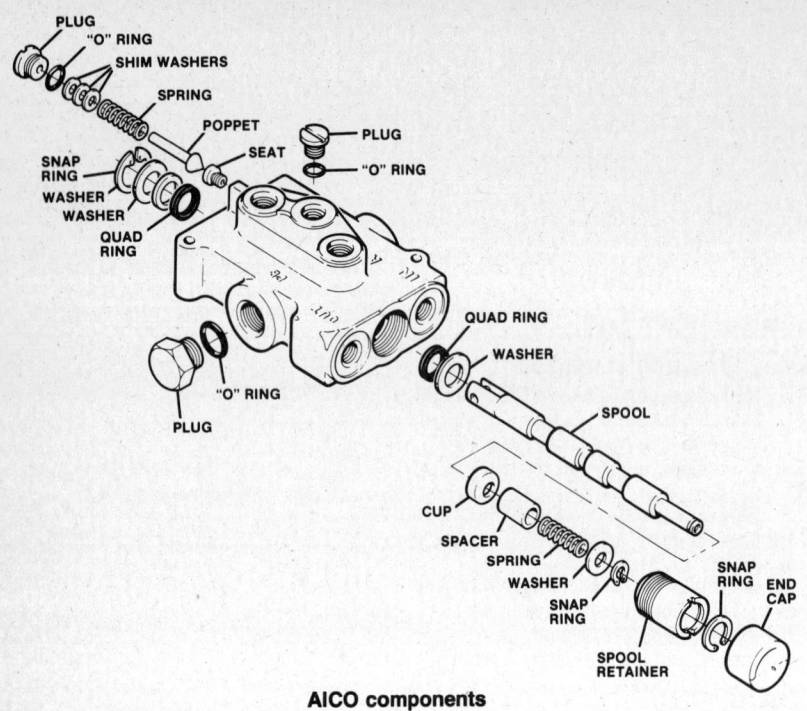

AICO components

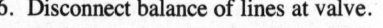

195 hydraulic system

195

Removal and Installation

1. Remove body insert.
2. Remove hydraulic lever links from levers.
3. Shut off fuel supply at the fuel tank, remove fuel line.
4. Clean valve and fittings thoroughly.
5. Disconnect line, charge pump to valve, at charge pump.
6. Disconnect balance of lines at valve.

NOTE: Plug all ports and tube or fitting openings preventing dirt from entering.

7. Remove (3) three capscrews, nuts and lockwashers securing valve to tractor.
8. Remove valve from tractor.
9. Installation is the reverse of removal.

Disassembly

1. Remove spools as follows:
   a. Remove (2) two screws retaining end cap.
   b. Remove end cap.
   c. Push spool out of valve body far enough to remove O-rings from spool and discard.
2. Remove relief valve as follows:
   a. Remove acorn nut and aluminum crush washer.
   b. Loosen jam nut and remove it along with adjusting screw.
   c. Remove spring, ball retainer, ball, and valve seat from valve body. Valve seat is threaded—remove with a 3/16 in. Allen wrench.
3. Removing power beyond bushing as follows:
   a. Remove bushing.
   b. Remove O-rings and discard.

Inspection and Repair

Visually inspect all parts. Service kits are available for all components except

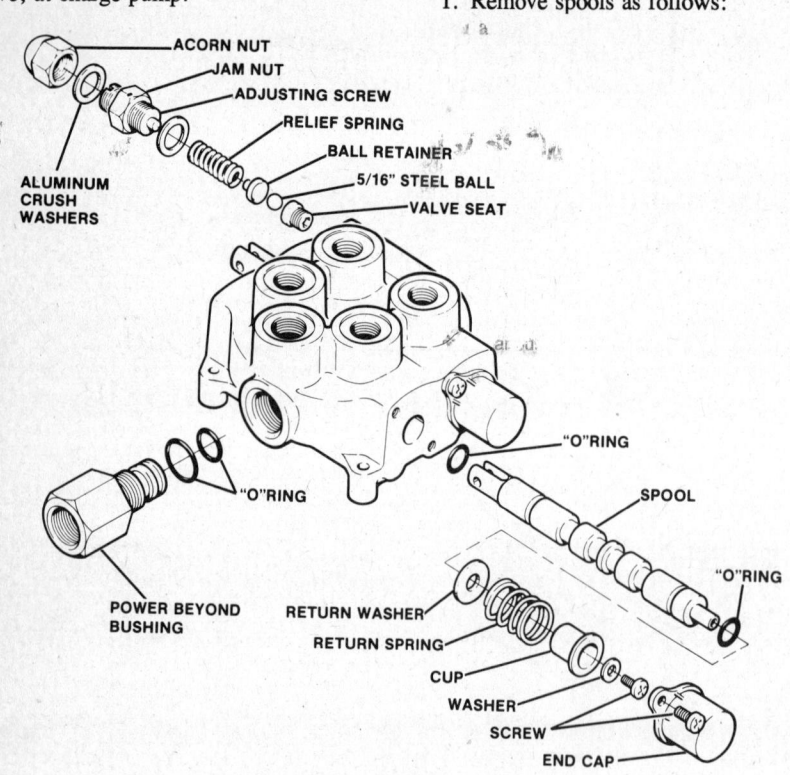

Two-spool valve components

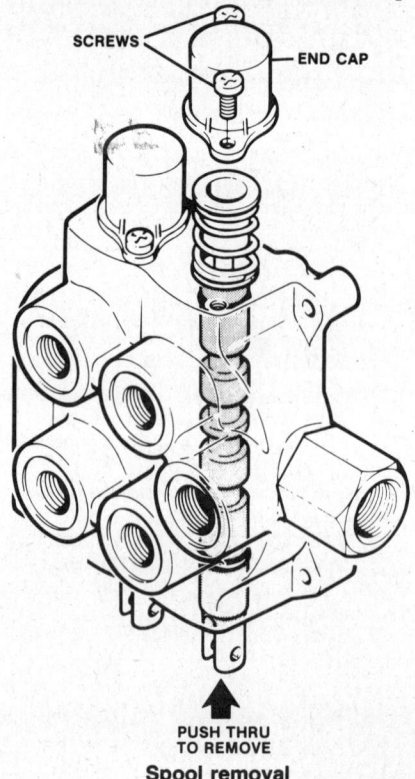

Spool removal

Ford

spools and the body itself. These are replaced by a complete valve assembly only.

CAUTION
Do not use hardened tools to remove O-rings. Do not over stretch, cut or nick O-rings when installing.

Replace Spool Return Spring
To replace spool return spring, proceed as follows:
1. Remove screw, spacer, spring and washer.
2. Place washer, spring and cup on spool.
3. Insert screw and tighten.

Assembly
Clean all parts in a suitable solvent. Visually inspect all bores and ports for dirt.

NOTE: Dirt is the biggest enemy of a hydraulic system. Prior to assembly, coat all parts with clean hydraulic oil.

To assemble lift valve, proceed as follows:
1. Install relief valve as follows:
 a. Install valve seat in body.
 b. Drop in ball, take care that ball is resting on seat face.
 c. Place ball retainer in end of spring and insert in bored hole.
 d. Insert adjusting screw pilot into end of spring and thread into body approximately ¼ in.

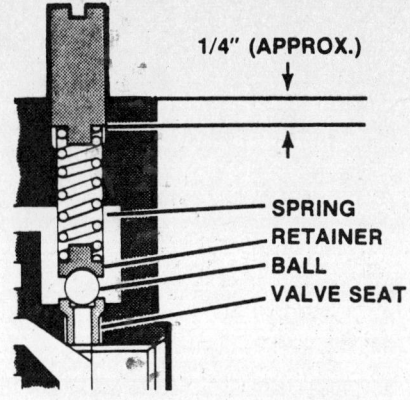

Adjusting lift valve pilot screw

 e. Place (1) one aluminum crush washer over adjusting screw, thread on jam nut fingertight.
 f. Place aluminum crush washer over end of adjusting screw and install acorn nut on screw, fingertight.

NOTE: When lift valve is reassembled to tractor, set relief valve at 500 psi.

2. Install spools as follows:
 a. Install new O-rings on spool.
 b. Immerse spool assembly in oil and install into correct location.

NOTE: Follow same procedure for other spool.

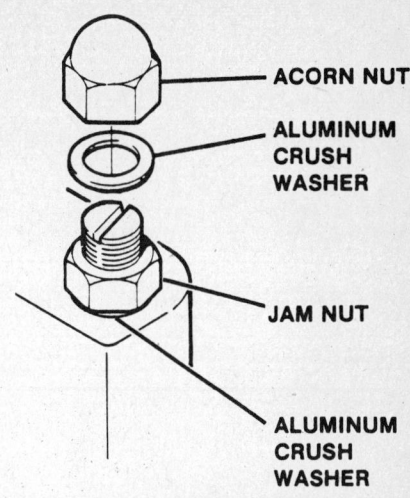

Relief valve assembly

 c. Place end cap over spool, install and tighten two screws.
3. Install power beyond bushing as follows:
 a. Install new O-rings on bushings.
 b. Dip bushing in oil and install in valve body.
 c. Tighten securely. Keep wrench as close to valve body as possible when tightening to avoid distortion of fitting.

Models 1000 and 1600

FRONT AXLE

Front Axle and Steering Linkage

REMOVAL AND DISASSEMBLY

1. Remove the hood panel and wiring.
2. Drain the cooling system and remove the radiator.
3. Place a suitable jack under the clutch housing and raise the front of the tractor.
4. Remove the six retaining bolts from each wheel and remove the wheel and tire assembly.
5. Remove the retaining nut and drag link from the Pitman arm.
6. Remove the six retaining bolts on each side of the cylinder block and remove the front axle support from the block.
7. Remove the set bolt and pivot shaft cover and pull out the pivot shaft.
8. Remove the retaining nuts and bolts from the spindle arms and remove the spindle arms from the spindles.
9. Remove the key and shims from the spindle.
10. Extract the spindle and thrust bearing and washer from the axle extension.
11. Remove the retaining bolts, cap, and gasket from the front hub.
12. Remove the cotter pin locating the castellated nut retaining the wheel hub.
13. Remove the nut and thrust washer, the outer bearing, and then the wheel hub.
14. Remove the snapring, thrust washer, grease retainer, and the inner bearing from the rear of the wheel hub.

INSPECTION AND REPAIR

1. Clean components with a suitable solvent and air dry. Clean the machined surfaces and lightly lubricate.
2. Inspect the roller bearing cones, rollers, and cups for signs of excessive wear or damage. Renew if necessary.
3. Inspect the spindle bushings in the axle extension housings for wear or scor-

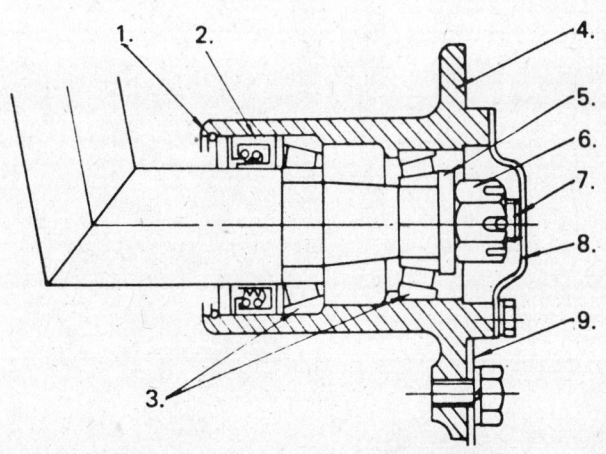

1. Snapring
2. Oil seal
3. Bearing
4. Front wheel hub
5. Spacer
6. Castle nut
7. Split pin
8. Wheel cap
9. Front wheel

Wheel and hub

423

Ford

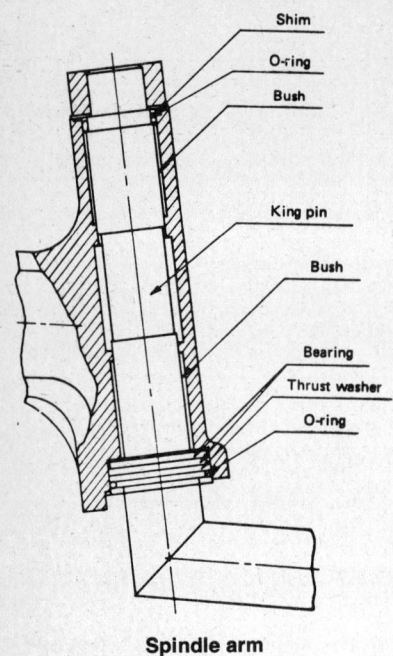

Spindle arm

ing. If necessary to renew, remove the axle extension from the center axle by removing the retaining bolts and their associated nuts and lockwashers. Remove the bushings making sure that the bores are not damaged, and install new bushings in the bores.

4. Inspect the spindle thrust bearing for correct operation and replace if necessary.
5. Inspect the wheel spindle bearing surfaces for scoring or excessive wear. If severe damage is apparent, then replace the spindle.
6. Inspect the pivot shaft bushings for wear or scoring. Replace if necessary.
7. Inspect the pivot shaft for extensive wear or damage. Replace the shaft if necessary.
8. Inspect all O-rings, prior to assembly, to make sure that they are not cracked, cut, or excessively worn. Replace if necessary.

ASSEMBLY AND INSTALLATION

1. Pack the cone and roller assembly with clean, short-fiber grease. Pack approximately .25 in. of grease in the space between the bearing cups in the hub, but do not pack the hub completely. Apply a film of grease on the surface of the spindle.
2. Install the inner bearing, the grease retainer, thrust washer, and snapring in the rear of the wheel hub.
3. Place the wheel hub on the spindle and install the outer bearing, the thrust washer, and castellated nut.
4. Tighten the nut, at the same time turning the wheel, until a slight drag is felt.
5. Back off the nut until the nearest slot in the nut lines up with the hole in the spindle. Install a new cotter pin.
6. Install the front hub gasket and cap and secure with the retaining bolts.
7. Pack the wheel spindle thrust bearing and install the thrust washer and bearing on the spindle.
8. Install the wheel spindle into the axle section housing making sure that it rotates freely.
9. Install a new spindle shaft O-ring and install the spacer and shims over the spindle.
10. Install the key and spindle arms over the spindles and secure in place with the bolts and nuts.
11. Install the pivot shaft through the front axle shaft and front axle support. Install the set bolt and pivot shaft cover and retaining bolts.
12. Position the front axle support over the cylinder block and secure in place with the six retaining bolts on each side.
13. Install the drag link and nut on the Pitman arm.
14. Install the wheel and tire assembly using the six retaining bolts on each side.
15. Lower the tractor and remove the jack.
16. Install the radiator.
17. Fill the cooling system.
18. Install the hood panel and wiring harness.
19. Check the toe-in.

STEERING GEAR

REMOVAL

1. Remove the nut, washer, and Pitman arm from the steering gear sector shaft.
2. Remove the steering wheel cap, retainer, washer, and spacer from the steering shaft and remove the steering wheel.
3. Remove the instrument panel.
4. Remove the four retaining bolts and remove the steering gear assembly from the clutch case.

DISASSEMBLY AND INSPECTION

1. Remove the sector shaft cover and adjuster screw assembly and pull out the sector shaft.
2. Remove the four retaining bolts and remove the steering shaft housing and shims. Pull the steering shaft out of the housing.
3. Inspect the steering shaft and sector shaft for broken or severely worn teeth. Replace those parts which are severely damaged.
4. Replace any bushings or oil seals that are worn or damaged.

ASSEMBLY

1. When a new steering shaft or bearing is installed the number of shims must be changed to obtain the proper shaft thrust bearing pre-load.
2. Install the sector shaft.
3. Install the adjuster screw assembly and the sector shaft cover and secure with the four retaining bolts.
4. Adjust the steering wheel play.
5. Fill the steering gear box.

INSTALLATION

1. Position the steering gear assembly on the clutch case and secure with the retaining bolts.
2. Install the instrument panel.
3. Position the steering wheel on the shaft and secure it with the spacer, washer, retainer, and cap.
4. Install the Pitman arm, washer, and nut onto the sector shaft.

ADJUSTMENTS

Steering Wheel Play

Steering wheel play in the direction of rotation should be between .78-1.38 in. If

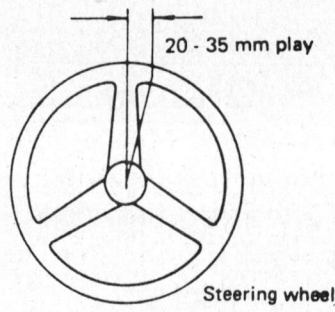

Measuring steering wheel free play

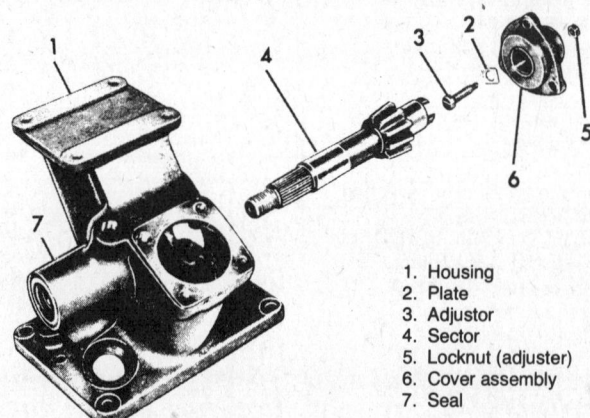

1. Housing
2. Plate
3. Adjustor
4. Sector
5. Locknut (adjuster)
6. Cover assembly
7. Seal

Steering gear housing and sector

the play exceeds 1.97 in., then adjustment is necessary.

1. Make sure that all link bolts are tightened properly. If severe wear is apparent, install new parts.
2. Loosen the adjuster locknut on the right side of the steering gear box and turn the adjuster screw. Turning the screw clockwise will decrease the free play while turning the adjuster screw counterclockwise will increase the steering wheel free play.
3. Once the adjustment is made, tighten the adjuster locknut securely.

Thrust Play

1. Tighten the steering shaft housing bolts if loose.
2. If retightening the steering shaft housing bolts is not effective, then remove the four bolts and the steering shaft housing.

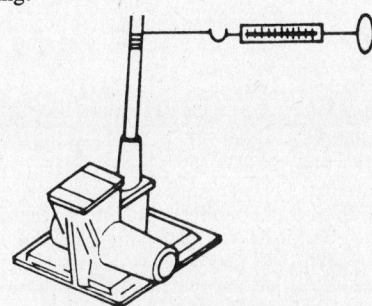

Measuring steering shaft preload

3. Add or remove shims from between the shaft housing and the steering gear housing unit until the thrust bearing preload meets the specified limit of .072-.210 ft.lb. The preload is measured with a pull scale and rope on the steering shaft with the sector shaft removed.

Steering Wheel Rotation

If the steering wheel is difficult to turn, the trouble may be in the linkage, kingpins, or steering gear box. To adjust:
1. Grease all linkage and kingpin bushings.
2. If steering is still difficult then increase the backlash of the steering gear box as outlined in this section under "Steering Wheel Play."

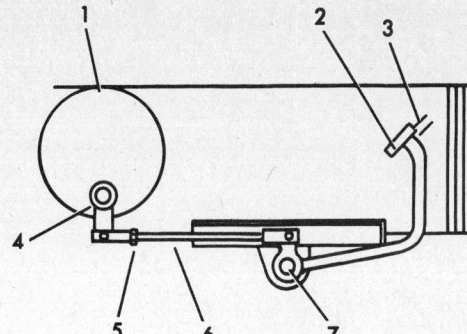

1. Brake
2. Pedal
3. Free play
4. Brake lever
5. Locknut
6. Adjustable rod assembly
7. Cross shaft

Brake adjustment points

washer from the pivot pin and remove the brake shoes.

INSPECTION AND REPAIR

1. Clean and inspect all parts and install new parts where worn or damaged.
2. Inspect the brake shoe linings. If they are deeply scored or badly worn new shoes should be fitted.
3. Check the brake drum for streaks, discoloration or other damage. Minor imperfections can be removed with a lathe. If the inside diameter of the brake drum is less than the specified limit stamped on the drum, replace the drum.

ASSEMBLY

1. Position the brake shoes over the pivot pin and retain with the flat washer, lockwasher, and nut.
2. Place the brake shoes against the brake cam and install the return spring.

INSTALLATION

1. Position the brake drum onto the brake drum shaft and retain in place with the spacer, lockwasher and nut.
2. Position the brake cover assembly over the dowel pins and install the retaining bolts.
3. Connect the brake rod to the brake lever.

ADJUSTMENT

1. Loosen the locknut on the brake rod assembly.
2. Rotate the brake rod until the .7874-1.180 in. free play is obtained on both brake pedals.
3. Retighten the locknut and check to see that the brake pedal free play is within the specified limit.

BRAKES

REMOVAL

1. Disconnect the brake rod assembly at the brake cover lever.
2. Remove the retaining bolts and the brake cover assembly.
3. Remove the locknut, locknut washer, and spacer and then remove the brake drum from the brake drum shaft.

DISASSEMBLY

1. Using brake spring pliers, remove the return spring from the brake shoes.
2. Remove the nut, lockwasher, and flat

CLUTCH

Adjustments

CLUTCH PEDAL FREE TRAVEL

To obtain maximum clutch life, it is essential that the clutch pedal free travel be checked every 50 hours to maintain a free travel of .75-1.20 in. as measured along the path of travel of the pedal. Free travel is the distance the pedal can be moved before resistance is met. To adjust the clutch pedal free travel:
1. Remove the clutch rod cotter pin and clevis pin, and rotate the clevis as necessary. Increasing the rod length will increase the pedal free travel and decreasing the rod length will decrease the clutch pedal free travel.
2. After the proper free travel is obtained, install the clevis pin and secure with a new cotter pin.

SAFETY START SWITCH

The tractor cannot be started until the clutch pedal is fully depressed. The pedal shank, when depressed, is rotated to the point where the plunger of the safety start switch protrudes and activates the switch. To adjust the safety start switch, rotate the adjustable stop which contacts the plunger of the switch until correct operation is obtained.

CLUTCH RELEASE LEVER HEIGHT

Uneven height of the release levers can cause improper clutch operation and premature clutch disc wear. Therefore, it is important that all three release levers are adjusted to the same height. It is also important that new pressure plate and cover assemblies be checked, and if necessary adjusted, prior to installation in the tractor. To make this adjustment easier, remove the flywheel from the engine.
1. Prior to adjusting the clutch, the clutch disc and pressure plate should be securely mounted to the flywheel.
2. Remove the retainer clips which are located over each of the adjuster screws.
3. Set the outer legs of the Special Tool SJ-101 so that they rest against the flywheel surface.
4. Loosen the locknut and turn the adjuster screw. Turning the adjuster screw clockwise will raise the release lever tip and turning the screw counterclockwise will lower the lever tip.
5. The clutch is properly adjusted when the center leg of the special tool contacts each of the three release levers.

Ford

6. After the adjustment, the locknuts must be securely tightened and the retainer clips installed.

7. Depress each release lever several times and re-check to be sure the levers are properly adjusted.

Clutch Plate and Disc

REMOVAL

1. Separate the engine from the clutch housing.

2. Once the engine is separated from the clutch housing inspect the clutch assembly for any oil or water which may have leaked into the housing or for damaged release levers on the pressure plate and cover assembly.

3. Loosen the bolts retaining the pressure plate to the flywheel and remove the pressure plate and clutch disc.

INSTALLATION

It is important that all new pressure plates and clutch assembies be checked and, if necessary, adjusted prior to installation in the tractor. Adjust the clutch release lever height.

When installing a new pressure plate and cover assembly make sure that the friction face of the pressure plate is free from dirt or oil film. Lightly lubricate the hub splines of the input shaft with silicon grease prior to installing the clutch disc.

1. If the flywheel has been removed, install the flywheel onto the crankshaft and tighten the nut to 723-868 ft.lb.

2. Position the clutch disc on the flywheel using the input shaft splines as an alignment tool or a pilot shaft.

3. Locate the pressure plate assembly on the flywheel and install the locating bolts and lockwashers. Tighten the bolts evenly to 13-18 ft.lb., then remove the pilot shaft.

4. Install the transmission input shaft if removed.

5. Install the release beraing on the bearing release hub and move the hub to the rear of the clutch case. Connect the return spring to the hub and spring hanger.

6. Reconnect the engine and clutch housing.

7. Check and adjust the clutch free travel play.

TRANSMISSION

REMOVAL AND DISASSEMBLY

1. Remove the rear wheels, fenders, and roll bar (if equipped) from the tractor.

2. Remove the pin at the brake rod joint and remove the bolts and brake cover.

3. Straighten the lockwasher, loosen the nuts, and remove the brake drums. Use push bolts to remove the brake drum if it cannot be removed by hand.

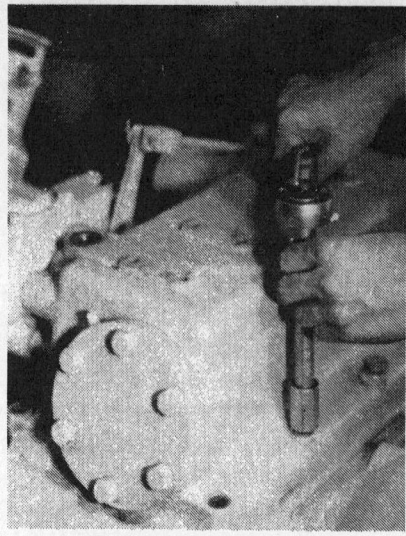

Hydraulic lift cover removal

4. If oil is leaking into the brake chamber, remove the differential seal cover by using two push bolts, and replace the seals.

5. Remove the seat assembly and the hydraulic lift cover assembly.

6. Remove the bolts and the PTO cover and remove the PTO shaft and bearing holder from the rear of the housing.

7. Remove the retaining bolts and the shift cover and case cover.

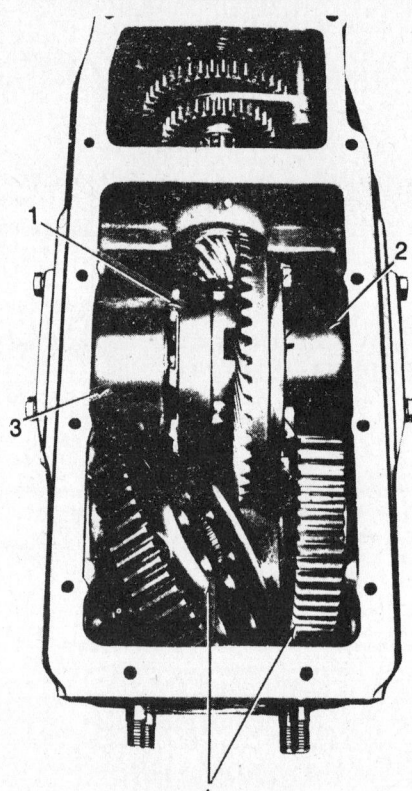

1. Differential assembly
2. Righthand carrier
3. Lefthand carrier
4. Final drive gears

Differential removal

8. Pull out the spring pin on the left side of the differential lock pedal shaft.

9. Shift the differential lock clutch to the right and remove the snapring securing the final gears.

10. Remove the bolts from the axle housing and remove the rear axle housings leaving the final gears in the transmission case.

11. Remove the final drive pinion cover and seal assemblies and final drive pinions.

12. Lay the righthand final drive gear as far down in the center housing as possible so as to clear the bottom of the righthand carrier.

13. Using two jack screws, withdraw the righthand carrier from center housing, at the same time keeping the differential assembly in the lefthand carrier.

14. Lift the righthand final drive gear between the differential and the center housing and withdraw it from the housing.

15. Pull the differential assembly to the right, out of the lefthand carrier, and lift it from the center housing.

16. Lay the lefthand final drive gear as far down in the center housing as possible to clear the bottom of the lefthand carrier.

17. Using two jack screws, withdraw the lefthand carrier from the center housing and remove the lefthand final drive gear.

18. Remove the drive pinion assembly.

19. The unit may now be disassembled.

20. Remove the clutch rod and the retaining bolts which secure the clutch case to the transmission case and separate the transmission and clutch case.

21. Pull out the spring pin from the shifter fork.

22. Remove the shifter rods one at a time, then remove the detent balls and springs. Be careful not to drop the balk pin into the transmission case.

23. Remove the shifter fork.

24. Pull the PTO counter shaft forward leaving the PTO sliding gear in the transmission case.

25. Drive the main shaft forward and out of the case using a brass hammer.

26. Remove the snapring from the counter shaft groove by pulling the shaft toward the front and then remove the snapring from the rear bearing. Pull the counter shaft out from the rear of the transmission case.

27. Remove the spring pin from the reverse idler gear shaft and remove the gear.

NOTE: The gears, collars, and other parts remaining in the case should be kept in an orderly manner to aid in the assembly procedure.

INSPECTION

1. Clean all old lubricant and dirt from the parts with a suitable cleaning solvent.

2. To clean the bearings, rotate the bearings in cleaning solvent until all of the old lubricant has been removed. Dry the bearings with compressed air, being careful not to spin them.

3. Inspect all bearings for wear, scores, discoloration from overheating or for missed rollers. Check the bearing cups for wear, cracks, or scores. Check bearings for

bent or worn retainers. Discard all defective bearings and cups.

4. After inspecting the bearings, lubricate them thoroughly and keep clean until ready to use.

5. Inspect the transmission case for cracks, worn bearing bores, damaged threads, or other damage. Install a new case if any of these conditions exist.

6. Discard all gears that are worn, chipped, or broken or otherwise damaged. Small nicks or burrs should be removed with a fine grinding stone.

7. Inspect all shift lever forks and rails for wear or damage. Discard all defective parts as required and replace with new parts.

8. Install new oil seals and gaskets at time of assembly.

9. Lubricate all parts thoroughly with clean oil before the assembly procedure.

ASSEMBLY AND INSTALLATION

1. Install the mid range gear and the rear bearing on the countershaft, and install the countershaft from the rear of the transmission case. Insert the countershaft until the front of the shaft is in the rear compartment of the transmission case.

NOTE: Prior to installing the countershaft, be certain that the upper bearing is installed in the web at the rear of the middle compartment.

2. Place the 39 tooth high range gear and bearing onto the countershaft and insert the front of the shaft into the middle compartment.

3. Place the 43 tooth 2nd gear, the 39 tooth 3rd gear, the spacer and snapring, and the 38 tooth and 46 tooth—1st and reverse gear on the countershaft and insert the countershaft through the web into the front compartment.

4. Install the bearing and snapring into the web at the rear of the front compartment. Lay the 47 tooth sliding PTO gear in the compartment but do not install the PTO driveshaft.

NOTE: The mainshaft must be installed prior to installing the sliding PTO gear on the PTO driveshaft.

5. Install the snapring in the groove of the web at the rear of the middle compartment.

6. Install the double bearing on the mainshaft. Then install the snapring, the reverse gear, and the other snapring on the mainshaft.

7. Install the 2nd-3rd sliding gear on the shaft and carefully insert the mainshaft through the webs until the bearing seats against the snapring in the second web.

8. Position the input shaft coupler so that the larger diameter is facing inward, and engage it with the splines of the mainshaft.

9. Install the rear tapered bearing on the pinion shaft and carefully insert the shaft through the rear web. Install the front tapered bearing, locknut, lockwasher,

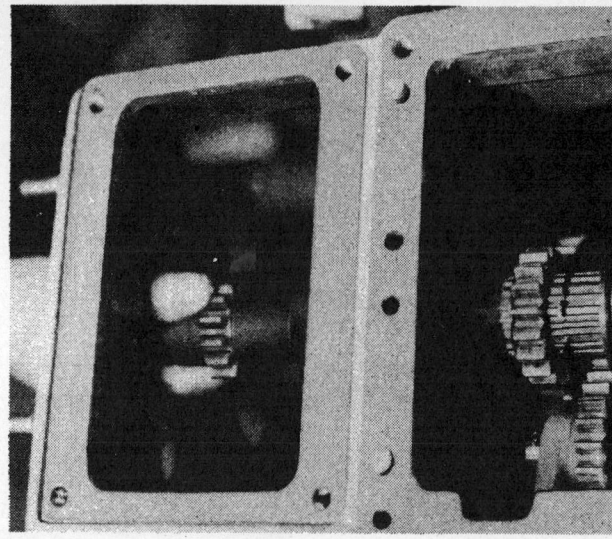
Mainshaft removal

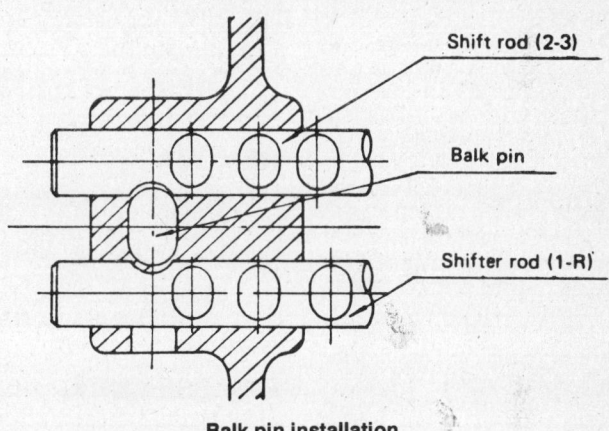
Balk pin installation

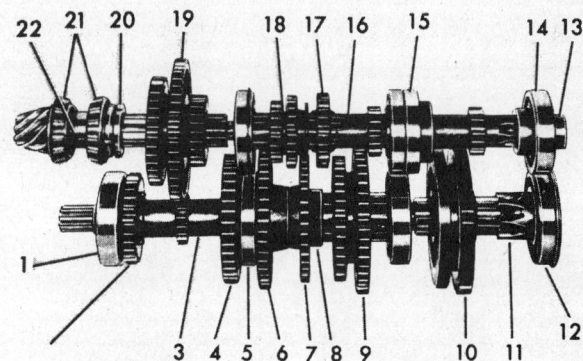

1. Bearing
2. Mid range gear
3. Countershaft
4. High range gear
5. Bearing
6. 2nd gear
7. 3rd gear
8. Spacer
9. 1st and reverse gear
10. PTO gear
11. PTO driveshaft
12. Bearing
13. Input shaft coupler
14. Bearing
15. Double bearing
16. Mainshaft
17. Reverse gear
18. 2nd-3rd gear
19. Sliding range gear
20. Locknuts and washer
21. Tapered bearings
22. Pinion shaft

Gear assemblies

locknut and the sliding range gear. Tighten the locknut to 14 ft.lb.

10. Install the bearing on the end of the PTO driveshaft and install the inner and outer snaprings.

11. Insert the shaft through the first web of the front compartment and place the sliding PTO gear onto the shaft.

12. Install the countersunk spacer, the thrust bearing, and the square cut spacer on the shaft.

NOTE: The concave side of the first spacer is positioned towards the thrust bearing.

13. Install the PTO driveshaft through the countershaft until the bearing outer snapring seats in the front web.

14. Install the reverse gear and shaft. Install the roll pins at each end of the shaft.

15. Insert the shifter forks in their respective grooves and insert the shifter rods. Make sure the balk pin is between the shifter rods 1-R and 2-3.

16. Install the shifter boss on the shifter fork and install the spring pins into the shifter fork bosses. Secure the assembly with safety wire. Install the detent balls, springs, and pins.

17. Install the shift cover and the case cover and tighten the bolts to 10 ft.lb.

18. Position the lefthand final drive gear in the transmission case and install the lefthand carrier in the center housing.

19. Place the differential assembly into the center housing and position it into the lefthand carrier.

20. Lower the righthand final drive gear between the differential assembly and the center housing and lay the gear as far down in the center housing as possible.

21. Install the righthand carrier assembly into the center housing and position the final drive gear in the carrier.

22. Install the final drive pinions and bearings and install the seal and cover assembly.

23. Align the transmission case with the clutch case and secure with the retaining bolts. Install the clutch rod assembly.

24. Install the axle housings.

25. Install the differential lock pedal, fork, and spring and secure in position with the spring pin.

26. Install the PTO connector and bearing on the PTO output shaft and insert through the bearing retainer and gasket assembly. Insert the shaft through the rear of the transmission case and position the PTO connector over the PTO driveshaft.

27. Install the bearing, oil seal, retainer gasket and outer PTO cover over the PTO output shaft and tighten the bolts.

28. Install the hydraulic life cover assembly and gasket and tighten the bolts.

29. Install the seat assembly.

30. Install the brake drum and secure with the spacer, locknut washer, and locknut.

31. Install the brake cover and retaining bolts. Connect the brake rod joint with the pin.

32. Install the fenders and rear wheels, (and roll bar if equipped) and tighten the wheel bolts.

DIFFERENTIAL

REMOVAL

1. Remove the differential assembly from the center housing by following the procedure outlined in "Transmission Removal and Disassembly."

2. Clean the outer surface of the differential assembly with a suitable solvent and allow to dry.

DISASSEMBLY

1. Straighten the ring gear retaining bolt lock straps and remove the bolts and ring gear from the differential assembly.

2. Straighten the differential case set bolt lock straps, and remove the bolts and the differential case.

3. Remove the spider assembly from the righthand case by pulling out the pinion shafts. Remove the pinions, joint and thrust washers from the case. Remove the side gear and thrust washer from the case.

4. Remove the side gear and thrust washer from the lefthand case.

INSPECTION

1. Clean all parts with a suitable solvent and blow dry with compressed air.

2. Inspect all gears for damage or excessive wear. Minor nicks can be removed with a grinding stone. Replace those parts which are severely damaged.

3. Check the differential thrust washer and differential pinion thrust washers for wear or damage. If wear is beyond the specified limit of .0078 in. and .0472 in. respectively, replace the thrust washers.

4. Inspect the pinion shafts for wear or damage. If the pinion shaft wear exceeds the specified limit of .670 in diameter, replace the pinion.

ASSEMBLY

1. Insert the thrust washer and side gear into the lefthand case, and install the thrust washer and side gear into the righthand case. Insert the longer pinion shaft through one of the four shaft holes in the righthand case. A thrust washer and pinion are assembled on the shaft and the shaft is passed through the joint. Another pinion and thrust washer are added to the shaft and the shaft is then inserted through the opposite hole in the case.

2. Install the two shorter shafts, pinions, and thrust washers in the same manner as previously described. The shorter shafts will stop against the longer shaft at the center of the joint. Turn the flats on the outer ends of the shafts so that they are parallel with the inner face of the case.

3. Install the lefthand case over the righthand case and against the flats preventing the shafts from rotating with the pinions. Install the lock straps and retaining bolts and tighten the bolts. Bend the lock straps over the bolt heads.

4. Install the ring gear on the righthand case and secure with the lock straps and retaining bolts.

INSTALLATION

1. Install the differential assembly as outlined in "Transmission Assembly and Installation."

2. Tighten all bolts.

REAR AXLE

REMOVAL

1. Drain the oil from the center housing.

2. Remove the rear wheels, fenders (and rollbar if equipped) from the tractor.

8. Pinion shafts
9. Thrust washers
10. Pinion gear bushings
11. Bushings
12. Differential pinion joint
13. Bearings
14. Righthand lock straps
15. Bolt
16. Bolt
17. Lefthand lock straps

1. Ring gear
2. Righthand differential housing
3. Lefthand differential housing
4. Side gears
5. Thrust washers
6. Pinion gears
7. Pinion shaft

Differential components

3. Disconnect the brake linkage and remove the brake cover assembly and brake drum.
4. Remove the seat assembly and the hydraulic lift cover assembly.
5. Remove the final drive gear retaining snapring and the axle housing to center housing bolts.
6. Remove the complete housing and axle assembly from the tractor.

DISASSEMBLY

1. Remove the axle housing cover retaining bolts.
2. Using a suitable press, apply pressure to the inner end of the axle to push the axle through the inner bearing which at the same time will push the outer bearing out of the housing bore.
3. Withdraw the axle assembly from the housing.
4. Remove the lock collar and lockwasher from the axle shaft.
5. Using a suitable press, remove the outer bearing from the axle shaft and then remove the spacers, oil seal and axle housing cover.

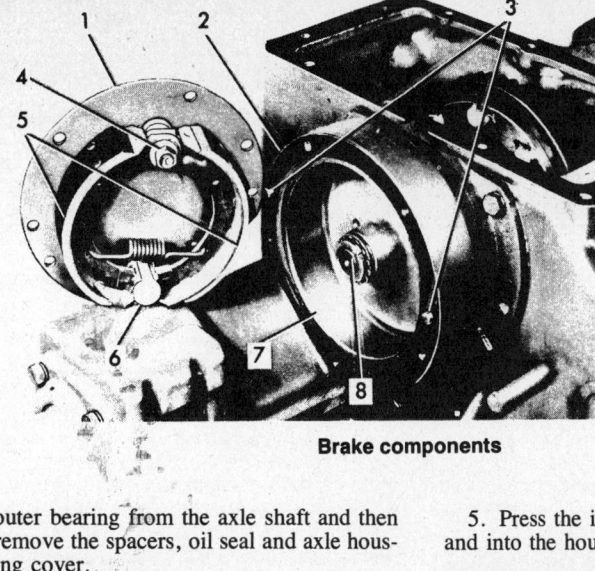

1. Cover
2. Axle housing
3. Dowel pins
4. Pivot pin
5. Brake shoes
6. Actuating cam
7. Brake drum
8. Final drive pinion gear shaft

Brake components

INSPECTION

1. Clean all parts in a suitable solvent and allow them to dry.
2. Inspect the bearings for wear, discoloration, looseness, or any other damage. Replace if necessary.
3. Check the oil seal for wear, cracks, or other damage, and replace if necessary.
4. Check the axle to make sure it is not twisted or bent. Inspect the splines at the end of the axle shaft for wear or damage. Replace the axle shaft if necessary.

ASSEMBLY

1. Place the axle housing cover, gasket, spacer, oil seal, and spacer over the end of the axle shaft. Make sure the oil seal is positioned toward the axle housing cover.
2. Using a suitable press, install the outer bearing onto the axle shaft and secure in place with the lockwasher and locknut.
3. Insert the axle shaft into the housing, and press the outer bearing into the housing outer bore.
4. Install the axle cover to the axle housing with the retaining bolts.

5. Press the inner bearing onto the axle and into the housing counter bore.

INSTALLATION

1. Install the righthand housing and axle assembly first. Insert the splines on the axle shaft into the final drive gear. At this time position the snapring over the axle shaft and then hold the differential lock coupling in the fork, with the lugs on the coupling facing toward the lefthand final drive gear. Continue to push the right axle shaft through the final drive gear and differential coupling until the axle housing contacts the center housing. Install the axle housing-to-center housing retaining bolts and the final drive gear retaining snapring.
2. Install the lefthand housing and axle assembly to the center housing. Make sure that the splines on the axle shaft line up with the final drive gear and that the bushed hole in the end of the lefthand axle pilots over the short, stub shaft, on the end of the righthand axle. Install the axle housing-to-center housing retaining bolts and the final drive gear retaining snapring.
3. Install the hydraulic lift cover and the seat assembly.
4. Install the brake drum, cover, and connect the brake linkage.

1. Oil seal
2. Outer bearing
3. Press

Outer bearing installation

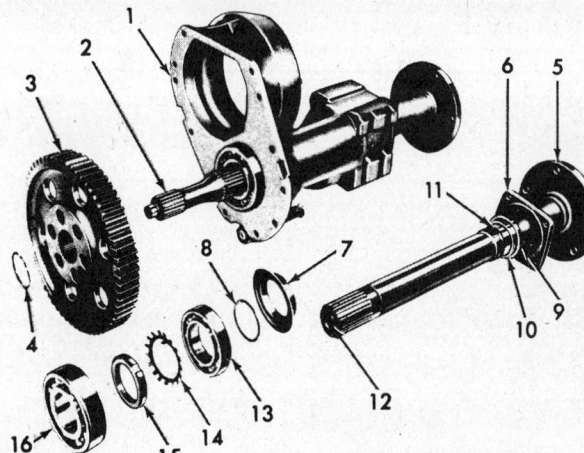

1. Righthand housing
2. Righthand axle
3. Final drive gear
4. Snapring
5. Lefthand axle
6. Cover
7. Oil seal
8. Plate
9. Flange
10. Hub
11. Oil seal retainer
12. Pilot bushing
13. Outer bearing assembly
14. Lockwasher
15. Nut
16. Inner bearing

Rear axle components

Ford

5. Install the fenders, wheels, and rollbar if equipped.
6. Fill the center housing with oil.

POWER TAKE-OFF (PTO)

REMOVAL

1. Remove the four bolts and the PTO cover from the rear of the center housing.
2. Pull the PTO shaft out from the rear of the housing.
3. Pry the bearing retainer from the center housing using a screwdriver.

DISASSEMBLY

1. Using a suitable press, remove the two bearings from the PTO shaft.
2. Remove the oil seal from the rear cover.

INSPECTION

1. Inspect the PTO shaft splines for wear or damage. Minor burrs can be removed with an oil stone. If severe damage is apparent, replace the PTO shaft.
2. Inspect the bearings for wear or discoloration from heat. Replace the bearings if necessary.
3. Check the oil seal for wear, cracks, or other damage, replace if necessary.
4. Check the bearing retainer for severe wear or damage, replace if necessary.

ASSEMBLY

1. Using a suitable press, install the bearings on the PTO shaft.
2. If a new oil seal is installed, press it into the PTO cover.

INSTALLATION

1. Position the gasket on the bearing retainer and insert the retainer into the center housing.
2. Install the coupling on the end of the PTO shaft and insert the shaft assembly through the bearing retainer. Make sure the coupling lines up with the PTO driveshaft.
3. Install the PTO cover and the retaining bolts.

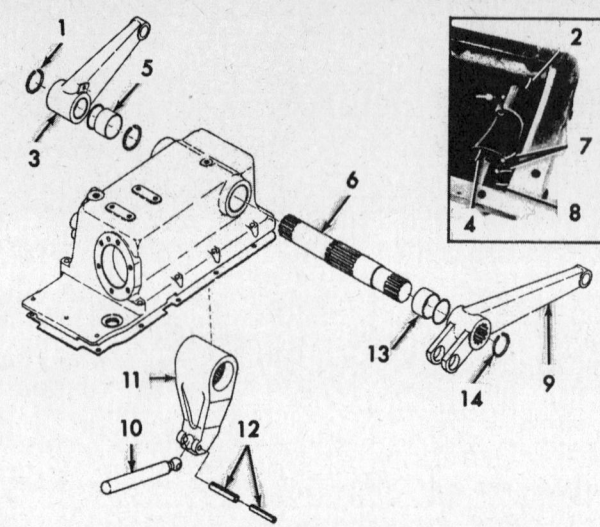

1. Snapring
2. Ram arm
3. Lift arm
4. Snapring
5. Bushing
6. Lift shaft
7. Alignment marks
8. Lift shaft
9. Lift arm
10. Piston rod
11. Ram arm
12. Roll pin
13. Bushing
14. Snapring

Lift shaft components

HYDRAULIC SYSTEM

Testing and Adjustments

POSITION CONTROL LINK ADJUSTMENT

The length of the position control link is critical. If it is too short, the control spool will be held forward in a raised position. In that case, the lift arms will fully raise without the spool returning to neutral. The system relief valve must then open to relieve the pressure build up. If the link is too long, full lift action cannot be attained. The pivot will be too far rearward to allow the actuating lever to push the spool forward when the lift arms are near maximum lift. To properly adjust the length of the position control link:

1. Start the tractor engine.
2. Raise the lift control lever to the top of the quadrant.
3. When the lift arms raise to full lift, adjust the length of the position control link to the shortest length which will not cause the system relief valve to blow.

1. Lift arm
2. Position control link
3. Position control arm
4. Lift control lever
5. Position control cam
6. Acuating lever
7. Lift control cam
8. Control valve spool
9. Return spring

Position control linkage

HYDRAULIC SYSTEM RELIEF VALVE—PRESSURE TEST

The hydraulic system operates at a system pressure of 1375-1425 psi. To check the system relief pressure, the lift arms must be chained down to obtain system relief pressure at the cylinder. To perform the pressure test:

1. Remove the test port plug from the cylinder head.
2. Install the adaptor into the cylinder head.
3. Install the hose onto the adaptor.
4. Install the pressure gauge onto the hose coupling.

NOTE: The adaptor NPD threads do not exactly match the British pipe threads of the test port. However, the amount of leakage that could result is very minimal and can be virtually eliminated by the application of a thread sealer. This installation will not damage the threads of the test port, and the plug will provide a leak free fit when reinstalled using a thread sealing compound.

5. With the test fittings installed, and the lift arms chained down, start the tractor engine and set the throttle to obtain an engine speed of 1,000 rpm. Raise the hydraulic lift control lever until the system relief valve blows. The guage reading should stabilize at the specified system pressure. If the pressure isn't 1425 psi, add shims in the relief valve to increase pressure or remove shims from the relief valve to decrease pressure, to bring the reading to within specification.

Hydraulic Lift Cover

REMOVAL

1. Remove the seat assembly from the hydraulic lift cover.
2. Disconnect the hydraulic oil tubes from the hydraulic lift cover and control valve.

3. Remove the lift rods from the hydraulic lift arms.
4. Remove the retaining bolts and the hydraulic lift cover and gasket from the transmission case.

DISASSEMBLY
Piston and Cylinder
1. Remove the retaining bolts and cylinder head cover from the hydraulic lift cover.
2. Push the piston out through the head end of the cylinder using a wooden handle.
3. If the sleeve must be replaced, it can be removed using a sleeve puller.

Lift Shaft
1. Prior to removing the lift arms, mark both the lift arms and lift shaft to facilitate alignment during the installation procedure.
2. Remove the lift arm retaining snapring and lift arms from the lift shaft.
3. Remove the ram arm snapring.

NOTE: The ram arm and lift shaft have alignment marks to assure proper alignment during installation.

4. Withdraw the lift shaft from the ram arm and hydraulic lift cover.

INSPECTION AND REPAIR
Piston and Cylinder
1. Check the piston and O-ring for wear or damage and replace if necessary.
2. Check the lift cylinder sleeve for wear or damage and replace if necessary.
3. Check all O-rings for wear, cracks, or other damage and replace if necessary.

Lift Shaft
1. Check the lift shaft bushings and O-rings for wear or other damage and replace if necessary.
2. Check the splines on the lift shaft, lift arms, and ram arm for wear or damage. Minor nicks or burrs can be removed with an oil stone. If severe damage is evident, replace the damaged parts.
3. Check the piston rod and roll pins for wear or damage and replace if necessary.

ASSEMBLY
Piston and Cylinder
1. If the cylinder sleeve was removed, install the sleeve into the hydraulic lift cover and install the snapring in the sleeve.
2. Install the O-ring on the piston and insert the piston into the sleeve from the head end. Make sure that the concave side of the piston faces away from the head.
3. Install the O-ring on the head and install the head and retaining bolts. Tighten the bolts to 40-50 ft.lb.

Lift Shaft
1. Insert the lift shaft through the hydraulic lift cover and through the ram arm. Install the snapring on the lift shaft to secure the ram arm in position.
2. Install the O-rings and lift arms onto the ends of the lift shaft and install the retaining snaprings.

NOTE: When reassembling these components, make sure that the ram arm aligns with the mark on the lift shaft, and the lift arms align with the marks on the lift shaft. Misalignment of these components may result in reduction of the range of lift, or the inability to fully lower the lift arms.

INSTALLATION
1. Position the gasket and hydraulic lift cover on the transmission case and install the retaining bolts.
2. Install the lift rods onto the hydraulic lift arms.
3. Connect the hydraulic oil tubes to the hydraulic lift cover and control valve.
4. Install the seat assembly onto the hydraulic lift cover.
5. Start the engine, operate the hydraulic control valve and check for leaks.

Hydraulic Pump
REMOVAL
1. Disconnect the hydraulic oil tubes from the hydraulic pump.

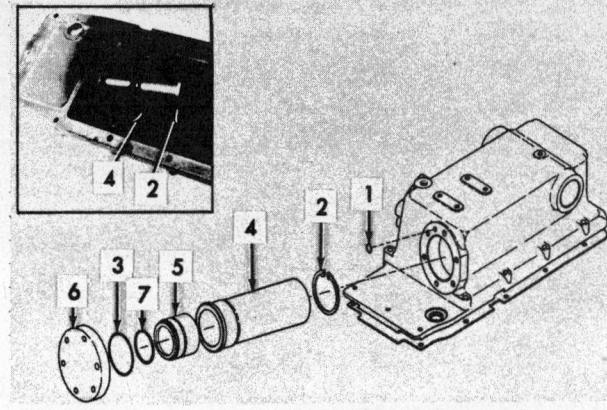

Piston components
1. Cylinder port O-ring
2. Piston snapring
3. Sleeve O-ring
4. Cylinder sleeve
5. Piston
6. Cylinder head cover
7. Piston O-ring

2. Remove the retaining nuts and the hydraulic pump and drive gear from the engine front cover.

DISASSEMBLY AND INSPECTION
The hydraulic pump is only serviced as an assembly. It is not practical to attempt to rebuild the pump because of the difficulty in obtaining proper gear and bearing clearances with the pump body. If the hydraulic pump is to be disassembled, the following procedure must be followed:
1. Remove the retaining nut and drive gear from the hydraulic pump shaft.
2. Remove the retaining bolts and cover from the pump body.
3. Remove the front seals, front bearings, driveshaft and gear, and the rear bearings and O-rings from the pump body.
4. Check the O-rings and seals for wear or damage and replace if necessary.
5. Check the bearings and pump driveshaft and gear for excessive wear or damage. Replace the complete pump if either conditions exist.

ASSEMBLY
1. Install a new O-ring in the hydraulic pump body.

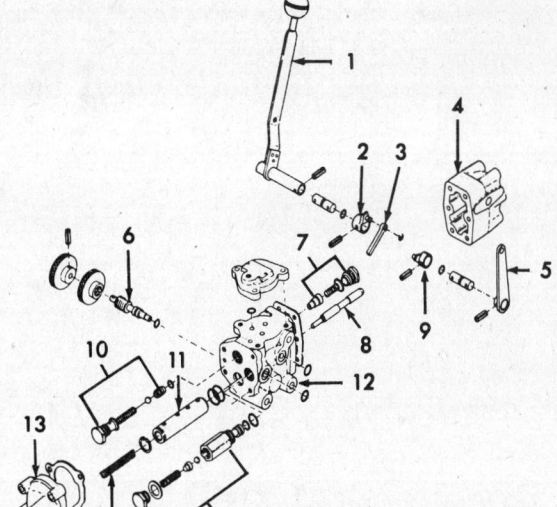

Hydraulic control valve components
1. Lift control lever
2. Lift control cam
3. Actuating lever
4. Rear cover
5. Position control arm
6. Flow control valve
7. Unload valve
8. Control valve spool
9. Position control cam
10. Check valve
11. Control valve spool bushing
12. Control valve body
13. Front cover
14. Spool return spring
15. System relief valve

Ford

2. Install the two O-rings, the rear bearings, the pump driveshaft and gear, the front bearings and front seals in the pump body.

NOTE: Make sure that the bearings are reinstalled with the notches on the bearings facing the gears.

3. Insert a new O-ring in the hydraulic pump cover and install the cover and retaining bolts.

4. Install the pump drive gear and retaining nut.

INSTALLATION

1. Position the hydraulic oil pump and gasket on the engine front cover and install the retaining nuts.
2. Connect the hydraulic oil tubes to the pump.
3. Start the engine and check for leaks.

ENGINE

The engine used in both models is the Japanese made Shibaura LE892A, 2-cylinder diesel with a displacement of 77.6 cubic in. Complete overhaul procedures follow.

Engine Unit

REMOVAL

1. Disconnect and remove the battery cables, the starter relay terminals, the headlight terminals, the oil pressure sensor terminals, glow plug terminals and the water temperature gauge, then remove the battery.
2. Remove the nuts, the wiring harness, the air cleaner cap, and unlatch the back of the hood panel and remove the panel.
3. Remove the cotter pin, washer, and accelerator rod.
4. Close the fuel tank valve, loosen the clamps, and remove the fuel pipe and return pipe. Then remove the fuel tank and base.
5. Loosen the hose clamps, nuts and bolts, and remove the radiator and hose.
6. Remove the hydraulic pump suction and delivery tubes.
7. Remove the cotter pin, nut and steering drag link from the Pitman arm.
8. Support the tractor by placing a suitable jack under the clutch housing.
9. Install a chain hoist to the engine and raise the hoist until the chain is taut.
10. Remove the six bolts on each side of the front axle support and remove the entire front axle assembly.
11. Loosen the bolts and clamps and remove the muffler and air cleaner.
12. Remove the bolts retaining the cylinder block to the clutch housing and remove the engine.
13. Loosen the bolts and remove the pressure plate and clutch disc from the flywheel.

DISASSEMBLY

1. To remove the flywheel nut, raise the lockwasher and remove the nut.
2. Remove the oil pressure sensor.
3. Pull out the oil level dipstick.
4. Remove the three bolts and the air cleaner flange.
5. Remove the injection pipes from the injectors.
6. Loosen the four nuts and remove the fan.
7. To remove the alternator assembly, remove the adjusting plate holder nuts and take out the alternator assembly and V-belt.
8. Remove the return pipe.
9. Remove the nozzle holder.
10. Disconnect and remove the glow plug assembly from the cylinder head.
11. Remove the starting motor.
12. Remove the six bolts and the water pump.
13. Remove the thermostat cover.
14. Remove the nuts and the rocker arm cover assembly.
15. Loosen the center nut and remove the rocker arm assembly and push rods.
16. Loosen the nuts evenly and remove the cylinder head and gasket from the cylinder block.
17. Remove the oil filter.
18. Remove the bolts and the crankshaft pulley and key.
19. Loosen the nuts and lockwashers and remove the hydraulic pump.
20. Remove the bolts and timing gear cover.
21. Remove the nut and the oil pump gear and key.
22. Loosen the nut and remove the injector coupling.
23. Remove the three bolts and the injection pump gear.
24. Remove the nuts and camshaft gear.
25. After removing the camshaft gear remove the bolt and tachometer assembly.

NOTE: The relief valve should be removed only when servicing of the valve is necessary.

26. Remove the bolts and idler gear, idler gear shaft and oil pipe.
27. Remove the bolts and injection pump from the front plate.
28. Loosen the bolts and remove the front plate.
29. To remove the flywheel, place a block of wood on the end of the crankshaft and tap with a hammer.
30. Turn the cylinder block upside down and remove the bolts and oil pan.
31. Remove the capscrews and oil suction filter.
32. Remove the two bolts and oil pump assembly.
33. Remove the bolts and bearing caps from the connecting rods. Then remove the piston and connecting rod assembly by

1. Center nut
2. Rocker arm shaft support
3. Rocker arm
4. Push rods
5. Cylinder head nuts

Rocker arm and shaft

1. Spring pin
2. Thrust bearing

Installing thrust bearing

432

Ford

1. Idler gear
2. Camshaft gear
3. Matching marks
4. Crankshaft gear

Timing gear installation

1. Dial gauge
2. Idler gear

Checking timing gear backlash

tapping the assembly out towards the top of the cylinder block with a hammer handle.

NOTE: Before removing the piston assembly, it may be necessary to use a cylinder ridge reamer to remove any ridge or carbon from the top of each cylinder.

34. Remove the bolts and flywheel cover. Take care not to damage the thrust bearing or oil seal.
35. Remove the crankshaft from the rear of the block.
36. Remove the rear camshaft bearing and the camshaft.
37. Pull out the tappets from the bottom of the cylinder block.

ASSEMBLY

1. Prior to assembly, apply clean engine oil to allow moving and sliding components. Replace all old gaskets and packings with new ones. If necessary, use liquid packing to prevent oil leaks.
2. Insert the tappets into the cylinder block. Make sure taat the tappets move up and down easily.
3. Insert the camshaft into the front camshaft bearing, and install the rear camshaft bearing. Make sure the camshaft rotates easily within the bearings.
4. Insert the front thrust bearing into the inside front wall of the cylinder block. Insert the crankshaft from the rear of the block into the front bearing. Take care not to damage the bearing.
5. If a new thrust bearing is being installed use a spring pin in the procedure. Make sure that the oil groove is facing the crankshaft thrust area and that the spring pin sinks .020 in. below the thrust bearing.
6. To install the flywheel cover, fit the greased O-ring to the camshaft bearing, apply liquid packing to both sides of the packing, and secure the flywheel cover with bolts tightened diagonally to 32-36 ft.lb. Take care not to damage the thrust bearing during the installation of the flywheel cover.
7. Before inserting the pistons make sure each piston ring end gap is offset by 90°.
8. Unsert the piston into the cylinder bore using a piston ring compressor. Tap the piston into the cylinder bore using a hammer handle. Make sure that the piston head front mark notch is facing the injection pump.

NOTE: When installing the piston make sure that the top ring gap is not at a right angle to the piston pin. Also, make sure to install the connecting rod with the alignment mark of the smaller figure in the No. 1 cylinder.

9. Install the connecting rod caps and tighten to 50-55 ft.lb. Measure crankshaft end float in the axial direction to make sure that it is within .0040-.0170 in. After tightening the connecting rod caps check to see if the crankshaft can be rotated.
10. Place the oil pump assembly into position and secure with the two washers and bolts. Make sure that the oil pump gear moves freely.
11. Place the oil suction filter into position and secure it with the two capscrews.
12. Position the gasket and oil pan on the cylinder block and tighten the bolts to 10-14 ft.lb.
13. Position the flywheel key on the crankshaft.
14. When installing the flywheel, make sure that there is clearance between the key head and the groove in the flywheel. Install the lockwasher and nut and tighten the nut until it is snug. The nut is torqued later in the assembly procedure.
15. Place the front plate into position and secure with the bolts. Tighten bolts to 10-14 ft.lb.
16. Position the injection pump on the front plate and secure into position with the bolts.
17. Install the idler gear shaft and oil pipe. Do not overtighten the bolt.
18. To install the idler gear, coat the bearing with oil and align the idler gear mark with the crankshaft gear mark. Tighten to 14 ft.lb.

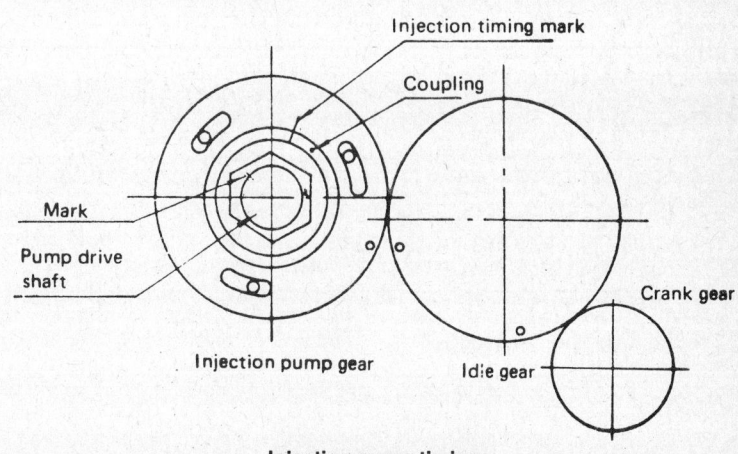

Injection pump timing

Ford

19. If the relief valve has been serviced then install the relief valve and tighten to 65-72 ft.lb.
20. Install the tachometer assembly and secure it with the two bolts.
21. Engage the two tachometer gears together and then install the camshaft gear onto the end of the camshaft. Align the mark on the camshaft gear with the ''O'' mark on the idler gear, and tighten the camshaft gear nut to 108-116 ft.lb.
22. At this time it is necessary to check the timing gear backlash.
   a. Measure the backlash of the timing gear with a dial gauge.
   b. If the backlash exceeds .004 in., then the timing gear unit should be replaced.
23. Position the injector pump coupling over the injection pump shaft and key and secure with the washer and nut. Tighten the nut to 29-36 ft.lb.
24. Align the injection pump gear mark with the idler gear mark by turning the crankshaft counterclockwise. Align the injection pump coupling with the slash (/) mark on the injection pump gear, and tighten the bolts.

NOTE: When the injection pump gear, or injection pump coupling is replaced, the injection timing must be set through the following procedure.

25. Injection timing is set at 21°BTDC and the following procedure must be followed:
   a. Align the injection pump gear mark with the idler gear mark as previously stated.
   b. Remove the delivery valve from the No. 1 cylinder pumping element on the injection pump.
   c. Align the injection pump driveshaft mark with the elongated hole in the injection pump gear which is opposite the injection pump-to-idler gear timing mark.
   d. Install the three pump gear to coupling retaining bolts but do not tighten.
   e. Rotate the injection pump coupling driveshaft until fuel ceases to flow out of the delivery valve holder. This is the spill-timing location.
   f. Tighten the injection pump gear-to-coupling retaining bolts to 29-36 ft.lb.
   g. Using a chisel, inscribe a injection timing mark on the injection pump gear and coupling.
   h. Replace the delivery valve in the No. 1 cylinder pumping element.
26. Install the oil pump gear and key over the oil pump shaft and secure with the washer and nut. Tighten the nut to 21-25 ft.lb.
27. Position the timing gear case and gasket on the cylinder block and secure with the bolts. Be careful not to damage the oil seal in the timing gear case.
28. Place the hydraulic pump into position and secure it with the nuts and lockwashers.
29. Install the crankshaft pulley and key over the end of the crankshaft and secure it with the washer and nut. Tighten the nut to 29-36 ft.lb.
30. Install the oil filter and hand tighten only.
31. Place a new head gasket on the cylinder block, then carefully position the cylinder head on the gasket. Coat the cylinder head bolts and nuts with engine oil and install the nuts finger tight.
32. Tighten the cylinder head nuts in the proper sequence, which is from the two center bolts, alternating to the four corner bolts. Tighten the head nuts in three steps progressively. Finally, tighten the head nuts to 108-112 ft.lb.
33. Insert the push rods into the holes in the cylinder head from which they were removed. Make sure that the oil holes in the push rods are free from obstructions.
34. Position the rocker arm assembly on the cylinder head. Make sure that the holes in the rocker arm support are aligned with the roll pins in the cylinder head, and that the ends of the rocker arm adjusting screws are seated in the push rods.
35. Install the center nuts over the rocker arm assembly support bracket and tighten to 35-47 ft.lb.
36. Rotate the engine and set the valve lash. The clearance for both the intake and exhaust valves is .012 in. The adjustment should be made while the engine is cool.
37. Position the gasket and rocker cover over the cylinder head and tighten the cap nuts to 36-60 in.lb. Take care not to damage the oil seal washer.
38. Place a light coat of liquid packing on the area of the flywheel cover where the starting motor and cover meet. Then position the starting motor in the flywheel cover and secure with the bolts.
39. Install the thermostat, cover, and gasket onto the cylinder head and secure with the bolts. Make sure that the thermostat spring is installed inside the cylinder head.
40. Position the gasket and water pump onto the cylinder block and secure with the bolts.
41. Fit the alternator to the holder and secure it to the gear case with the bolts. (Models up to No. 1375 have no holder.)
42. Fit the V-belt over the water pump pulley, the crankshaft pulley, and the alternator pulley and adjust the alternator so that the V-belt can be pressed down approximately $7/16$ to $9/16$ in. and then tighten the adjusting plate bolt.
43. Install the fan and secure with the four bolts.
44. Install the nozzle holder and tighten to 43-50 ft.lb.
45. Install the return pipe.
46. Install the glow plug assembly into the cylinder head.
47. Position the air cleaner flange and gasket onto the cylinder head and secure with the bolts.
48. Install the injection pipes onto the injectors and tighten. Do not over tighten the injection pipes, as damage may occur.
49. Insert the oil level dipstick.
50. Apply liquid packing to the threads of the oil pressure sensor and install the sensor into the cylinder block.
51. Tighten the flywheel nut to 723-868 ft.lb. After tightening, bend the lockwasher tab over the flywheel nut.

INSTALLATION

1. Center the clutch disc assembly to the flywheel using a tool such as a pilot shaft or the transmission input shaft, and then position the pressure plate over the clutch disc and tighten the pressure plate bolts evenly to the flywheel.
2. Using a chain hoist, position the engine in line with the clutch housing. Move the engine slowly towards the clutch housing making sure that the transmission input shaft is aligned with the clutch disc. Once the clutch disc and the transmission input shaft have been aligned secure the engine block to the clutch housing with the bolts. Tighten the bolts to 32-36 ft.lb.
3. Install the muffler and air cleaner and tighten the bolts and clamps.
4. Position the front axle assembly under the engine cylinder block and secure the front axle support to the cylinder block using the six bolts on each side of the front axle support. Tighten the bolts.
5. At this time the floor jacks can be removed from under the clutch housing and the chain hoist can also be removed.
6. Position the steering drag link on the Pitman arm, tighten the nut and install the cotter pin.
7. Install the hydraulic pump suction and delivery tubes.
8. Position the radiator, radiator support, and hoses, and tighten all nuts, bolts, and clamps.
9. Place the fuel tank base into position and tighten the bolts. Place the fuel tank on the base and tighten the fuel tank bands to secure the tank.
10. Place the fuel pipe and return pipe into position and install the clamps. Then open the fuel tank valve.
11. Install the accelerator rod, washer, and cotter pin.
12. Move the hood panel into position over the engine. Install the nuts under the hood support assembly, the wiring harness, the air cleaner cap, and latch the back of the hood panel.
13. Install the battery, the water temperature gauge, glow plug terminals, oil pressure sensor terminal, headlight terminals, starter relay terminals and connect the battery cables.

Engine Component Overhaul

CYLINDER HEAD

Disassembly

1. Position a valve spring compressor over the valve and spring and compress the spring. Remove the valve keeper, the valve spring retainer, the valve spring and valve.

NOTE: Mark the valves with their appropriate cylinder to aid in reassembly.

2. Remove the valve guide oil seal.

Ford

INSPECTION AND REPAIR

1. Inspect the cylinder head for cracks, nicks or burrs. Install a new head if necessary. Minor nicks or burrs can be removed with an oil stone. Make sure that the gasket contact area is clean.

2. Place the cylinder head on a surface plate. Measure for distortion of the cylinder head by inserting a feeler gauge at four points. If distortion is more than .0047 in. it may be skimmed with a surface grinder to within .002 in. flatness.

Valve Guide and Valve Stem
INSPECTION AND REPAIR

1. Measure the valve stem diameter with a micrometer at three points I, II, and III. Valve stem size and allowable wear limits are listed in the specification chart. If wear exceeds these limits, replace the valve.

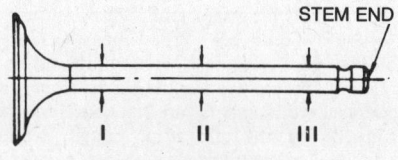

Measuring valve stem

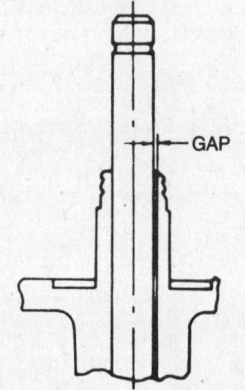

Measuring stem-to-guide clearance

2. Measure the gap between the valve guide and valve stem. If the gap exceeds specified limits replace the cylinder head and valves.

Valve Seat
INSPECTION AND REPAIR

1. Valve guide wear should be measured first to determine if valve seat repair is necessary. Refer to "Valve Guide and Valve Stem" inspection, above.

2. Seat cutters of 15°, 45°, and 75° are used to correct the valve seat so that the contacting width becomes equivalent to the standard. Contacting width of both intake and exhaust valve seats is .0472-.0591 in.

3. If the depth of the seat exceeds .0787 in. regrind the valve seats or replace the cylinder head.

4. Check the head and stem of the intake and exhaust valves for any seizure, wear or deformation. If any of these conditions exist, replace the valves.

5. Check the thickness of the valve head. If the thickness is less than .0512 in., replace the valve.

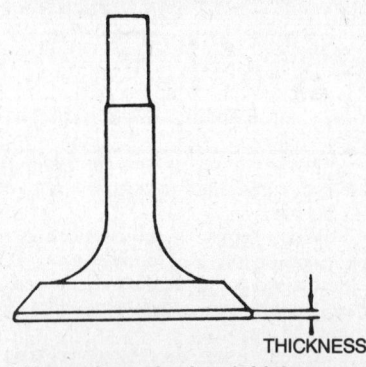

Measuring valve seat depth

Measuring valve head thickness

6. Check the valve seat contact by applying compound on the valve seat and then rotate the valve. Check that the valve contacting width is within the specified limit and that the contact is even.

NOTE: If the contacting width is too wide, carbon will accumulate on the valve. If the contact is too narrow, rapid wear will result.

Valve Springs
INSPECTION

1. Discard any valve springs that show signs of erosion or rust.

2. Check each valve spring for squareness, and free length. Measure the spring vertically with a square on a surface plate. Discard any valve springs that do not meet the specified limits.

3. Measure each valve spring with a spring tester. Weak valve springs cause poor engine performance, therefore; if the springs do not meet specified limits, replace the spring.

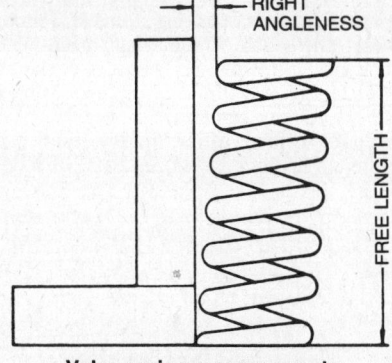

Valve spring measurements

Push Rods and Tappets
INSPECTION

1. Check the ends of the push rods for nicks, grooves, roughness or excessive wear. If the push rods are not straight, or if any of the above wear conditions exist, replace the rods. Do not attempt to straighten push rods.

2. Check the tappets for nicks, grooves, roughness, or excessive wear. If any of these conditions exist, replace the tappets.

Rocker Arm Assembly
DISASSEMBLY

1. Identify the rocker arms with their appropriate cylinders to assure reassembly in the same location.

2. Remove the snaprings at both ends of the rocker arm shaft and slide the rocker arms off the shaft.

INSPECTION AND REPAIR

1. Measure the diameter of the rocker arm shaft with a micrometer. If the shaft diameter is not within .5334-.5394 in., replace the shaft.

2. Measure the inside diameter of the rocker arm. Replace the rocker arm if wear exceeds .0016 in.

3. Check for any uneven wear or damage on the valve end of the rocker arm. Remove imperfections with an oil stone or grinder. If wear or damage is severe, replace the rocker arm.

4. If rocker arm assembly parts are within specifications, clean them thoroughly in solvent and make sure oil passages are clean of obstructions.

ASSEMBLY

1. Coat the rocker arm shaft with engine oil prior to assembly. Lubricate the valve pads on all rocker arms.

2. Coat the inside bore of the rocker arms with engine oil prior to assembly.

3. Slide the rocker arm shaft through the rocker arm support.

4. Install the rocker arms in their original position on each end of the rocker arm shaft and retain in place by installing the snaprings on each end of the rocker arm shaft.

CYLINDER HEAD ASSEMBLY

1. Insert each valve into the guide bore from which it was removed and lap it into position to give an even seat around the valve. On completion of this operation remove the valve and carefully clean the valve seat and seat insert of any lapping compound.

2. Lubricate all moving parts with engine oil prior to installation.

3. Insert each valve in the guide bore from which it was removed or to which a new valve was fitted. Position a new valve seal over each intake valve and guide.

4. Install the valve spring and retainer over the valve guide.

5. Compress the spring and spring retainer and install the valve keeper.

Piston, Piston Pin and Rings
DISASSEMBLY

1. Using a ring expander, remove the piston rings.

2. Remove the piston pin snapring with snapring pliers.

3. Heat the piston to 122-140°F and remove the piston pin.

INSPECTION AND REPAIR

1. Inspect pistons for cracks, streaks, seizure, damage at the ring lands, skirts, and pin bosses. Replace any piston that has these characteristics.

2. Measure the gap between the longer diameter of the piston skirt and the cylinder body. If the gap exceeds the specified limit a new piston should be installed.

3. Replace the piston rings if they are worn or damaged, or if the engine is being overhauled.

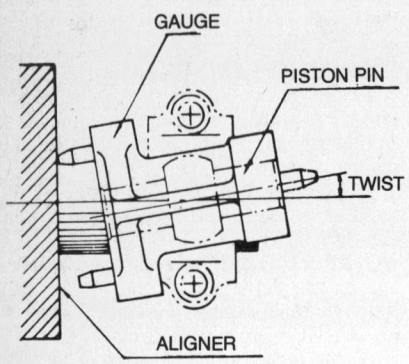

Measuring rod torsion

4. Place the ring at a right angle to the cylinder bore and measure the end gap with a thickness gauge. If the ring end gap exceeds the specified limit install new rings.

5. Measure the gap between the piston ring groove and the ring. If the gap exceeds specified limits, install new rings.

NOTE: Piston rings should be installed with the R mark upward and with each ring gap being offset by 90°.

6. Measure the diameter of the piston pin. If the wear exceeds 1.258 in. install a new pin.

7. The piston pin fit with the connecting rod small end bushing is proper if the piston pin, with oil on its surface, can be pushed in under slight pressure when the piston is at normal room temperature.

Connecting Rods and Bearings
INSPECTION AND REPAIR

Measure the connecting rod for twist, warpage or other signs of damage.

2. Measure the large-end and small-end twist and straightness by using a connecting rod alignment fixture. If the measured value exceeds .008 in., install a new connecting rod.

3. Check the connecting rod bolts. Any part that shows signs of wear or damage should be replaced.

4. Inspect the connecting rod bearings for signs of wear, uneven contact, fatigue failure, scratches, seizure, or improper tension. Replace the bearings if any of these conditions exist.

5. Replace the bearing or crankshaft when oil clearance is excessive resulting from wear in the crankpin and bearing. Oil clearance limits are specified in the crankshaft chart.

6. Insert the bearings into the connecting rod cap.

Rod bearing cap installation

Piston and Connecting Rod
ASSEMBLY

1. Heat the piston up to 158°-212°F and install the rod and pin. Make sure that the piston head mark and the connecting rod "F" mark are set as shown. Alignment marks of figures are inscribed on the connecting rod.

2. When replacing the connecting rod, piston or piston pin, choose one of the nearest in weight to the old part. Difference in weight between cylinders should be kept within .35 oz.

3. When fitting the connecting rod to the crafkshaft, install the connecting rod so that the piston mark "*" faces toward the combustion chamber jet, and measure the play in axial direction. If the play exceeds .004-.012 in., replace the connecting rod.

Cylinder Block
Inspection and Repair

1. Inspect the expansion plugs for evidence of rust. If rust is present this indicates leakage and new plugs should be installed. Remove the defective plugs. Apply sealer to the new plugs and install them securely.

2. Inspect the cylinder block for cracks, nicks, or burrs. Minor nicks and burrs may be removed from the top of the cylinder block with a surface grinder. Replace the cylinder block if severe damage has occured.

3. Check for distortion in the cylinder block in the same manner as for the "Cylinder Head." If distortion exceeds .0047 in. resurfacing of the cylinder block must be performed.

4. Check the cylinder bores for waviness, scratches, scuffing, out of round, wear, and taper. These irregularities and scratches, although in most cases too small to be measured with the naked eye, usually can be felt by running a finger over the cylinder surface. Minor imperfections can be removed with a cylinder hone. If more severe damage is apparent the cylinder block should be rebored and new oversize pistons should be fitted.

5. Check the cylinder bore wear at the top, middle, and bottom of the bore with a cylinder bore gauge. Bore top is at the top piston ring position with the piston at TDC or about .433 in. below the top of the cylinder block and bore bottom is at the piston skirt position at BDC. Wear is usually more severe at the top than at the bottom of the bore. Therefore, wear can be calculated by deducting the minimum diameter at the skirt from the maximum measured bore diameter. If the wear in the cylinder bore exceeds 3.543 in., rebore the cylinder and fit oversize pistons to the block.

CRANKSHAFT
Inspection and Repair

1. Clean the cranksahft with a suitable solvent. Clean all drilled passages and blow out the passages with compressed air.

2. Place the crankshaft on V-blocks to check the crankshaft run-out. Measure the run-out by fitting a dial gauge to the crankshaft pulley and flywheel side oil seal surface area. Read the T.I.R. value after rotating the crankshaft one full turn. If the T.I.R. exceeds .0024 in., replace the crankshaft.

3. Inspect the main and connecting rod journals and oil seal contact area of the crankshaft for cracks, scratches, grooves or scores. Minor imperfections may be dressed with an oil stone or crocus cloth. A severely damaged crankshaft should be replaced.

4. Measure the roundness and taper of the journals and pin with a micrometer. If the measured valve exceeds .002 in., replace the crankshaft.

5. Measure the end float of the crankshaft with a feeler gauge at the crankshaft rear bearing position as shown. If the end float exceeds .0040-.0170 in., replace the thrust bearing.

NOTE: Install the thrust bearing with the oil groove facing the crankshaft thrust surface.

MAIN BEARINGS
Inspection and Repair

1. Clean the bearing liners and caps thoroughly. Inspect each bearing carefully. Bearings that have signs of wear, fatigue

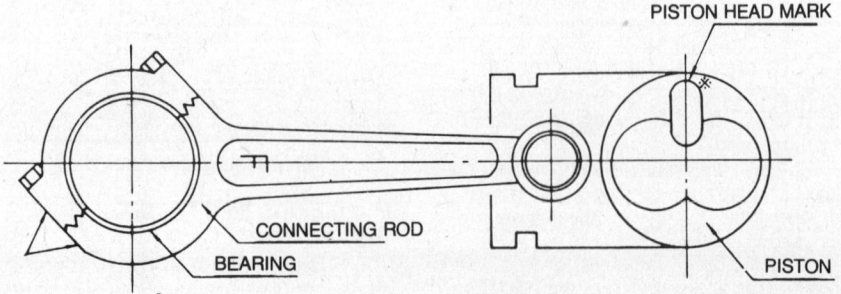

Piston and rod alignment

failure, scratches, chipped or scored surfaces should be replaced.

2. Replace the bearing when oil clearance is excessive resulting from wear between the bearing and crankshaft. Oil clearance limits are specified in the crankshaft chart.

3. When replacing the main bearings, the oil hole in the bearing must align with the oil passages in the block and flywheel cover.

CAMSHAFT

Inspection and Repair

1. Inspect the camshaft journals and lobes for roughness, scores, nicks, pits, or discoloration from heat. Minor imperfections can be removed with an oil stone or crocus cloth.

2. Place the camshaft on V-blocks to check the camshaft run-out. Measure the camshaft run-out by placing a dial guage at the center of the camshaft. Read the maximum value after rotating the camshaft one full turn. If the T.I.R. exceeds .004 in., replace the camshaft.

3. Measure the height of the camshaft lobes with a micrometer. If the height of the camshaft lobes does not meet the specified limit of 1.26 in., replace the camshaft.

4. Check the camshaft roller bearings for wear. Ball bearings can be checked by rotating the outer race with one hand while holding the inner race with the other. Good bearings produce smooth movement while worn bearings produce vibration in the inner race. If any type of roughness, discoloration, or looseness of the balls is apparent, replace the bearing.

FLYWHEEL AND RING GEAR

Inspection and Repair

1. Inspect the flywheel for cracks, scores, or excessive roughness. Minor imperfections can be removed by resurfacing the flywheel. If excessive warpage or damage is apparent, the flywheel should be replaced.

2. Measure the flywheel run-out with a dial indicator. Measure the run-out in the area where the clutch disc contacts the flywheel. If the run-out exceeds .008 in., the flywheel should be machined to within specifications.

3. Check the ring gear for rough edges and for missing teeth which could scuff or gouge the teeth on the drive gear. If necessary, dress the teeth with a wire wheel to smooth up the edges. If the ring gear has signs of minor wear it can still be used by changing the position of the gear by 90°.

4. When installing a new ring gear, heat the gear to approximately 248-302°F (120-150°C). Quickly place the hot gear on the flywheel with the flat gear face against the shoulder of the flywheel. Be sure the ring gear face is flush with the flywheel then quench the gear with water to cool it rapidly.

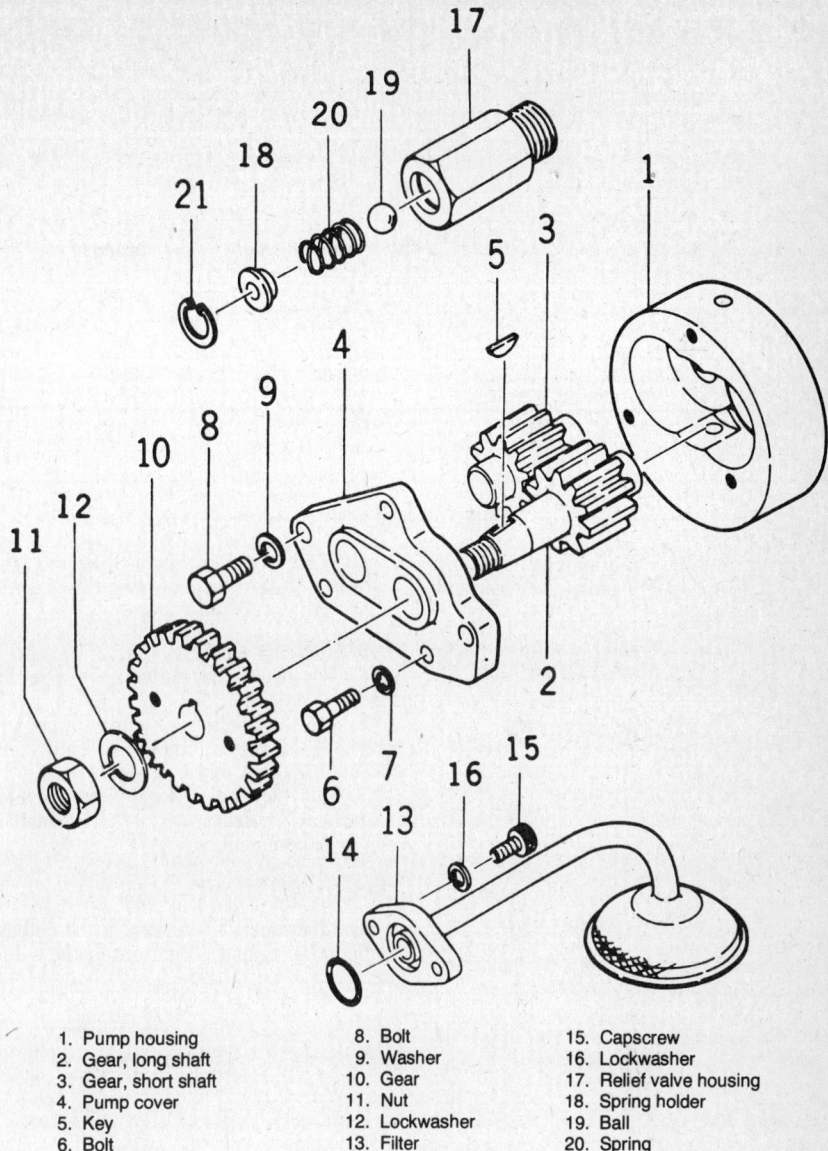

1. Pump housing
2. Gear, long shaft
3. Gear, short shaft
4. Pump cover
5. Key
6. Bolt
7. Washer
8. Bolt
9. Washer
10. Gear
11. Nut
12. Lockwasher
13. Filter
14. O-ring
15. Capscrew
16. Lockwasher
17. Relief valve housing
18. Spring holder
19. Ball
20. Spring
21. Snapring

Oil pump components

TIMING GEAR

Inspection and Repair

1. Inspect the timing gears for missing teeth, scores, nicks, burrs, and the condition of the teeth wear pattern. Minor imperfections can be removed with a wire wheel. More severely damaged gears should be replaced.

2. To check timing gear backlash refer to "Engine Assembly."

Engine Lubrication

OIL PUMP

Disassembly and Inspection

1. Remove the cover from the oil pump.
2. Pull out the oil pump gears.
3. Inspect the oil pump housing cover, and gears for excessively worn or damaged parts. Replace any parts that are in poor condition.
4. Measure the clearance between the oil pump gear and oil pump case. If the measured value exceeds .0028-.0056 in., replace the gear.
5. Measure the clearance between the oil pump gear teeth and the oil pump case. If the clearance exceeds .0004-.0012 in., replace the gear.

Assembly

1. Coat all moving parts with clean engine oil.
2. Insert the oil pump gears into the oil pump housing.
3. Place the oil pump cover onto the oil pump housing.

FILTER

Removal and Installation

To remove the oil filter, turn the cartridge body counterclockwise and then discard the filter.

Ford

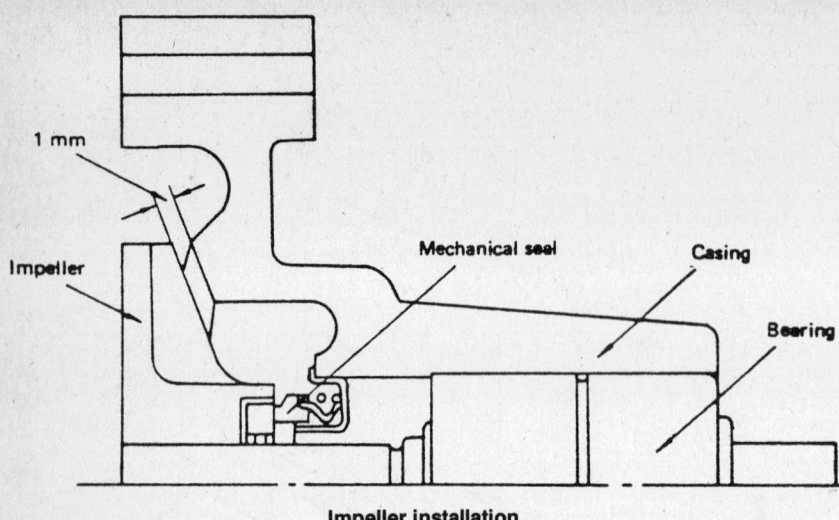

Impeller installation

2. Place a thin coat of engine oil on the seal of the new filter. Install the filter and hand tighten only.

NOTE: The oil filter should be replaced every 200 hours of operation.

Engine Cooling

RADIATOR

Removal
1. Drain the cooling system as previously outlined.
2. For easy accessibility, remove the hood panel as outlined under "Engine Removal."
3. Loosen the hose clamps and slide the clamps toward the middle of the hose.
4. Remove the nuts from the rubber bumpers located at the bottom of the radiator.
5. Disconnect the radiator support bracket and remove the radiator from the tractor.

Inspection and Repair
1. Check the upper tank for leaks.
2. Check the fins to be sure they are not bent or clogged.
3. Check the lower tank for leaks.

NOTE: Any repairs done on the radiator should be performed by a qualified radiator repair shop.

Installation
1. To install the radiator, reverse the removal procedure.
2. Fill the cooling system with coolant and add the proper amount of antifreeze, depending upon the season and locality.
3. Run the engine for several minutes and check for leaks in the radiator and at the hose connections.

THERMOSTAT

The thermostat is located in the coolant outlet connection on the front of the cylinder head.

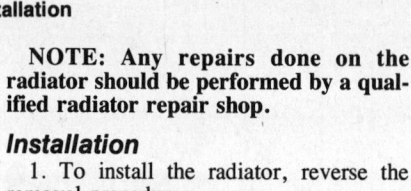

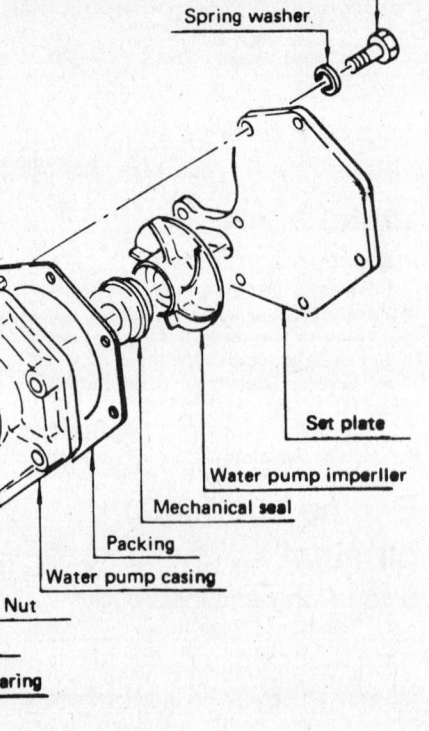

Water pump components

Removal
1. Drain the cooling system to below the level of the coolant outlet connection.
2. Remove the thermostat cover retaining bolts and slide the cover with the hose attached to one side.
3. Remove the thermostat and gasket.

Inspection
Place the thermostat in a container of water and heat the water. If the thermostat valve does not open at or near the opening temperature (160°F) or if it fails to close when the water temperature decreases, install a new thermostat.

Installation
1. Clean the thermostat cover and the cylinder head surface of any foreign material.
2. Coat the new gasket with a sealer and position the gasket on the cylinder head.
3. Position the thermostat so that the heat element will be in the cylinder head.
4. Position the thermostat cover over the thermostat and install the retaining bolts.
5. Fill the radiator and operate the engine. Check for leaks around the thermostat cover.

WATER PUMP

Removal
1. Drain the cooling system as previously outlined.
2. Remove the radiator.
3. Loosen the alternator adjusting arm bolt and the two pivot bolts and relax the tension on the belt.
4. Remove the six water pump attaching bolts and loosen the hose clamp and remove the water pump and gasket.

Disassembly
1. Remove the four attaching bolts and remove the fan from the pump pulley.
2. Using a gear puller, remove the pump pulley from the shaft.
3. Remove the three attaching bolts and remove the set plate and gasket from the casing. Loosen the water pump bearing set bolt.
4. Using a press, remove the impeller and shaft from the pump casing.

NOTE: Avoid using a hammer on the impeller. It is made of cast iron and can be easily broken if hit with a hammer.

Inspection and Repair
1. Check each component of the water pump for any cracks, wear or damage. Components which are damaged should be replaced.
2. Check the impeller for worn or damaged vanes and check the seal to be sure it is in good condition. Install a new impeller if the seat or vanes are damaged.
3. Check the bearing shaft for nicks, scores, or other damage. If the shaft is damaged a new shaft should be installed.
4. Check the water pump bearing for looseness in the radius direction and if it exceeds .008 in., then replace the bearing.
5. Check the pump casing for cracks, fractures, or signs of leakage.

Ford

Assembly

1. Use the exploded view for reference during reassembly. Using a press, install the bearing into the fan pulley.
2. Place the bearing into the pump casing using a press. Align the water pump casing bolt hole with the bearing outer race set hole.
3. Coat the casing side of the mechanical seal with a sealer and insert the seal into the casing.
4. Coat the mechanical seal impeller side with oil and install the impeller over the shaft using a press.
5. Coat a new washer pump gasket with a sealer and install the gasket and set plate onto the pump. Tighten the bolts.
6. Install the fan on the pulley. Tighten the bolts.
7. Rotate the pulley to make sure that the water pump operates smoothly.

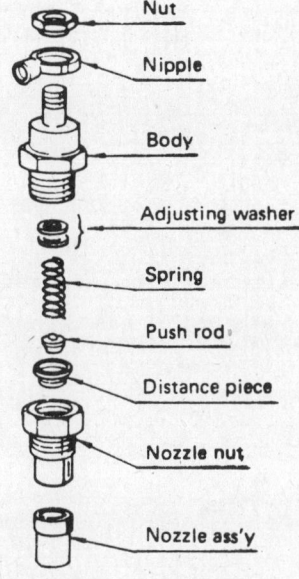

Injector components

Installation

1. Coat a new water pump gasket with a sealer and position the gasket and water pump onto the front of the cylinder block and install the six attaching bolts. Tighten the bolts. Position the hose onto the pump and tighten the clamp.
2. Position the V-belt and adjust the belt tension to obtain a ½ in. deflection. Tighten the adjusting arm bolt and the two alternator pivot bolts.
3. Install the radiator.
4. Fill the cooling system. Run the engine for several minutes and check for leaks.

FUEL SYSTEM

Injection Nozzles

REMOVAL

1. Clean all loose dirt from around the injectors and lines. Disconnect the leak-off lines from the injectors.
2. Disconnect the injection pump lines at the pump and injectors. Cover the ends of the lines and the injector inlet and leak-off ports to prevent the entry of foreign material or dirt.
3. Remove the injectors from the cylinder head and discard the dust sealing washers.
4. Remove and discard the copper injector sealing washers from the injector locating bores. Cover the bores to prevent entry of dirt.

DISASSEMBLY AND INSPECTION

1. Loosen and remove the nozzle nut being careful not to drop the nozzle.
2. Remove the nozzle and needle valve.
3. Clean the nozzle body and needle valve and check the nozzle for any burn or score marks. Also check the seat for any fuel leakage. Fuel leakage from the seat can be corrected by lightly polishing the seat.
4. Inspect the upper and lower contact surfaces of the nozzle holder distance piece for clean contact.
5. Check the contacting area between the push rod and nozzle needle valve for wear. Check the spring seat for any cracks. Replace worn or damaged parts.
6. Clean all of the injector components in a good solvent.

ASSEMBLY

1. Reassembly should be done in the reverse procedure of disassembly.
2. During reassembly be careful not to damage the needle valve. Tighten all connections securely to prevent leakage.

INSTALLATION

1. Place new dust sealing washers around the injector bodies.
2. Install a new copper sealing washer in each injector locating bore. Install the injectors and tighten to 43-50 ft.lb.
3. Install the leak off lines, using new copper sealing washers above and below each connection.

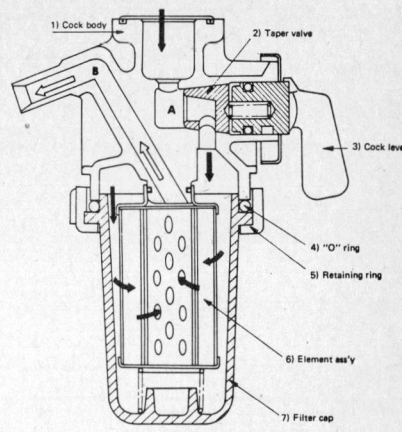

Fuel filter

4. Install the injector lines. Finger tighten the fittings at the injectors until after bleeding the fuel system. Tighten the fittings at the pump to 18-22 ft.lb.
5. Bleed the fuel system of all air then tighten the injector lines at the injectors to 18-22 ft.lb.

TESTING

Nozzle Injection Pressure

1. Attach a nozzle tester to the injection assembly, and operate the nozzle tester handle to clean the nozzle assembly.
2. Pump the tester to approximately 1425 psi of oil pressure and check the top of the needle valve and the threaded portion of the nozzle nut. If an oil leak is apparent, the seat and nozzle head can be repaired by polishing. Refer to "Injection Nozzle—Disassembly and Inspection." If leakage still occurs after servicing the seat and nozzle head, replace the nozzle assembly.
3. Adjust the adjusting washers so that the injection will start at 1700 psi of pressure. A washer width of .004 in. can increase the pressure by approximately 140 psi.
4. During the testing there should be no drops in the mist pattern coming from the nozzle.
5. Check to make sure that the mist is being injected to an area of about 12° from the nozzle center line.

1. Air bleed screw 2. Drain plug

Injection pump

439

Ford

6. Place a piece of white paper under the nozzle and repeat the test. A good spray pattern will leave a completely round circle on the paper.

Fuel Filter

REMOVAL

1. Turn the fuel shut off valve to the "OFF" position.
2. Loosen the retaining ring and remove the sediment bowl and element.

REPAIR

Clean the element in a suitable cleaner, keeping the element ends closed off with your fingers.

NOTE: The fuel filter element should be cleaned every 100 hours of operation and replaced at every 200 hours.

Installation

1. Position the element in the shut off valve body.
2. Install the O-ring filter bowl over the element and tighten securely. Make sure the O-ring is seated properly.
3. Turn the fuel shut off valve to the "ON" position.
4. Bleed the air from the filter assembly by loosening the air bleed screw, until the air bubbles are forced from the sediment bowl, then retighten the screw.

Fuel Tank

REMOVAL

1. Remove the air cleaner inlet cap, unlatch the hood panel and tilt the panel forward.
2. Turn the fuel shut off valve to the "OFF" position.
3. Disconnect the fuel pipe.
4. Remove the tensioner screws from the fuel tank bands and remove the fuel tank.

INSTALLATION

1. Place the fuel tank in the bands and install and tighten the band tensioner screws.
2. Connect the fuel pipe.
3. Turn the fuel shut off valve to the "ON" position.
4. Position the hood panel over the fuel tank and secure the hood panel latches.
5. Install the air cleaner cap over the air cleaner assembly.

Injection Pump

REMOVAL

1. Clean all dirt from the injection pump and surrounding tractor parts.
2. Turn the fuel shut off valve to the "OFF" position.
3. Drain the coolant from the radiator and remove the radiator.
4. Disconnect the injection pump inlet line and the fuel injector lines.
5. Disconnect the throttle control rod.
6. Remove the engine front cover.
7. Remove the nut and the injector coupling from the injection pump driveshaft.
8. Remove the three retaining bolts and the injection pump from the engine front mounting plate.

INSTALLATION

1. Place a new pump-to-front mounting plate gasket on the injection pump and install the injection pump and retaining bolts. Tighten the bolts to 32-36 ft.lb.
2. Install the injection pump coupling over the pump driveshaft and secure with the retaining nut.
3. Install the engine front cover and components.
4. Connect the throttle control rod.
5. Connect the injection pump inlet lines and the fuel injector lines.
6. Install the radiator and fill it with coolant.
7. Turn the fuel shut off valve to the "ON" position.
8. Loosen the fuel injection pump air vent screw, and bleed the air out of the pump, then tighten the screw.

ENGINE ELECTRICAL

Alternator

REMOVAL AND INSTALLATION

1. Disconnect the battery ground (negative) cable from the battery.
2. Disconnect all wires from the alternator.
3. Remove the alternator from the tractor by removing the attaching bolts.
4. Installation is the reverse of removal.

Starter

REMOVAL AND INSTALLATION

1. Disconnect the battery ground (negative) cable from the battery.
2. Disconnect the positive battery cable from the solenoid or starting motor terminal.
3. Remove the starting motor mounting bolts and remove the starting motor.
4. Reinstall the starting motor on the tractor in the reverse order of steps 1 through 3.

DISASSEMBLY AND ASSEMBLY

1. Disconnect the lead wire on the solenoid.
2. Remove the retaining bolts and the solenoid assembly from the starting motor.
3. Remove the retaining bolts and rear cover from the starting motor.
4. Remove the brush holder and brushes from the starting motor.
5. Remove the pin from the shift lever.
6. Separate the armature and yoke from the case by tapping with a plastic hammer.
7. Remove the clips and the pinion stopper.
8. Remove the pinion and center bearing.
9. Assemble the starting motor in the reverse order of steps 1 through 8.

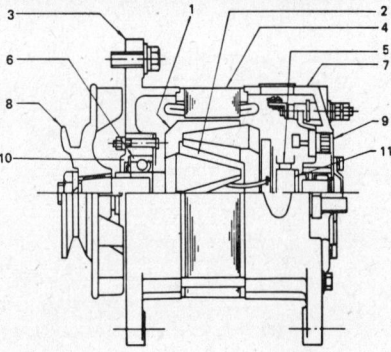

1. Rotor
2. Rotor coil
3. Front cover
4. Stator
5. Brush
6. Ball bearing
7. Bracket
8. Pulley
9. Silicon diodes
10. Bearing retainer
11. Slip ring

Alternator

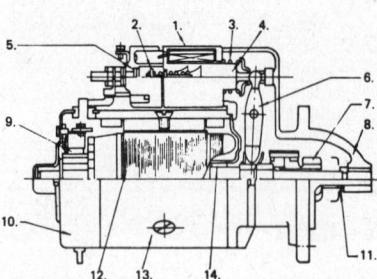

1. Solenoid
2. Moving contactor
3. Return spring
4. Plunger
5. Stationary contactor
6. Shift lever
7. Pinion
8. Gear case
9. Brush
10. Rear cover
11. Pinion stopper
12. Armature
13. Yoke
14. Center bearing

Starter

Gilson

INDEX

- Front Axle 442
- Steering 442
- Brakes 443
- Clutch 443
- Drive Belts 444
- Drive Chain 444
- Driveshaft 445
- Manual Transaxle 446
- Reduction Gears and Differential 449
- Hydrostatic Transaxle 450
- Engine 456

GILSON
Models 52037, 53024, 53025, 53026, 53027, 53028, 53030, 53031, 53032, 53033, 53034

FRONT AXLE

REMOVAL AND INSTALLATION

1. Raise and support the front end on jack stands.
2. Disconnect all steering linkage points.
3. Remove the pivot pin.
4. Roll the axle from under the tractor.
5. Installation is the reverse of removal.

STEERING

Bellcrank

REMOVAL AND INSTALLATION

Single Cylinder Models

1. Remove attachment from tractor (if any).
2. Disconnect and remove battery from tractor.
3. Thoroughly clean area where engine sump meets engine block, and around dipstick.
4. Drain approximately ½ of the oil in crankcase.
5. Remove completely the four (4) ¼-20 x ½ hex head capscrews holding the PTO engagement lever cup and lever support plate to the gas tank support assembly.
6. Position front wheels so they are turned to the right as far as possible and unbolt tie rod end from left spindle.
7. Loosen jam nut on bellcrank end and unscrew short steering link from bellcrank. Note or mark number of threads showing to aid in setting during reassembly.
8. Remove the hex nut of the tie rod end going through the opposite side of steering bellcrank under tractor.
9. Remove cotter pin from bellcrank pivot pin.
10. Remove all four capscrews securing engine to base/sump assembly and replace them with ⅜-16 x 3 or 3½ in. long bolts in the accessible hole location(s). (These will act as temporary "guides" when engine is lifted to obtain additional clearance for removal of the bellcrank pivot pin.)
11. By an appropriate means, raise engine up to the point where removal of the pin can be accomplished.

NOTE: Care must be taken to prevent damage to the sump gasket. If its condition is questionable or damage is done to it, replacement must be made.

12. Move bellcrank out of tractor frame and while doing so, separate the tie rod end of drag link from bellcrank arm (loosened in step 8.)
13. Slip tie rod end of drag link into new bellcrank, position bellcrank in tractor frame and insert pivot pin from the top down. Reinstall hex nut on tie rod end.
14. Reverse balance of procedure to install.
15. After engine is retightened, balance of engine oil should be drained and fresh oil installed.

Two Cylinder Models

1. Remove attachment from tractor, if any.
2. Disconnect battery and remove from tractor.
3. Position front wheels so they are turned to the right as far as possible and remove tie rod end from left spindle.
4. Loosen jam nut on bellcrank end and unscrew short steering link from bellcrank. Note or mark number of threads showing to aid in setting during reassembly.
5. Remove the hex nut of the tie rod end going through opposite side of steering bellcrank.
6. Remove the two front engine mount bolts, and the two side mounting bolts which run through the frame rails into the rear engine mount.
7. Remove cotter pin from bellcrank pivot pin.
8. Gently lift engine up on the left side—just enough to allow complete removal of the pivot pin.
9. Move bellcrank out of tractor frame and while doing so, separate the tie rod end of drag link from bellcrank arm (loosened in step 5.)
10. Reverse procedure to install.

NOTE: Make certain the pivot pin is inserted from top down.

Steering Knuckles

REMOVAL AND INSTALLATION

1. Raise and support the front end on jack stands.
2. Remove steering linkage at all attachment points.
3. Remove clamp bolt or cotter pin and, where used, keys.
4. Drive out roll pins and remove the knuckles.
5. Installation is the reverse of removal.

Steering Gear

REMOVAL AND INSTALLATION

1. Disconnect the steering linkage at all attaching points.
2. Drive out the roll pins and remove the steering wheel and tube.
3. Unbolt and remove the steering gear unit.
4. Installation is the reverse of removal.

ADJUSTMENTS

Model 52037

If excessive play develops in the steering wheel the steering gear can be adjusted by repositioning the flange bearing.

1. Loosen the two (2) capscrews that secure the flange bearing to the frame.
2. Carefully slide the flange bearing towards the gear sector assembly.
3. When the steering gear and gear sector mesh together evenly and smoothly tighten both capscrews.

All Other Models

If excessive play develops in steering wheel, the gear mesh can be adjusted by loosening the locking nut and turning the

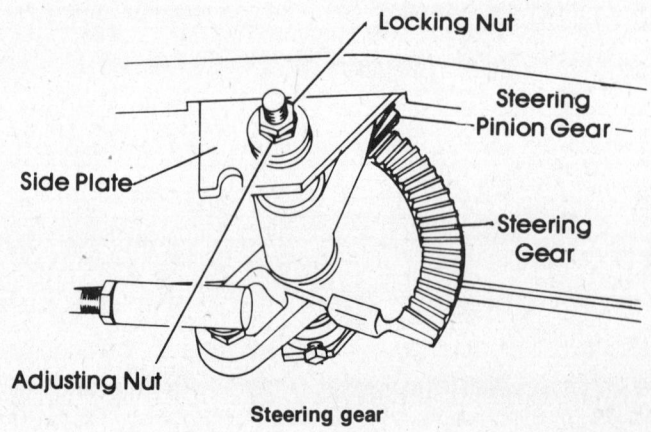

Steering gear

Gilson

adjustment nut clockwise until play is removed. If steering wheel turns freely, tighten the locknut. If not, the gear mesh is too tight and must be loosened. Recheck steering wheel so it is not too loose.

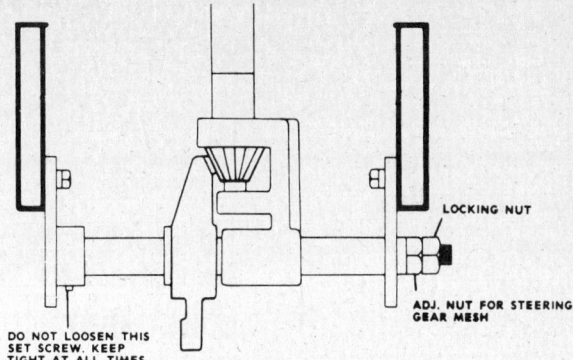
Steering gear adjustment

BRAKES

Models With Disc Brakes

BRAKE PAD ADJUSTMENT

The brake is a disc type brake and is on the lefthand side of transaxle. The brake is factory adjusted and will not require adjustment until the brake begins to lose its braking ability. At that time adjust as described below.

1. Remove cotter pin from castellated nut.
2. Turn castellated nut clockwise to tighten brake pad against disc to compensate for brake pad wear. Slight drag permitted. Overtightening will lock brake. Roll tractor back and forth to test drag.
3. Depress foot pedal. Brake should engage.
4. Replace cotter pin. To replace cotter pin, turn nut until groove in nut lines up with hole.

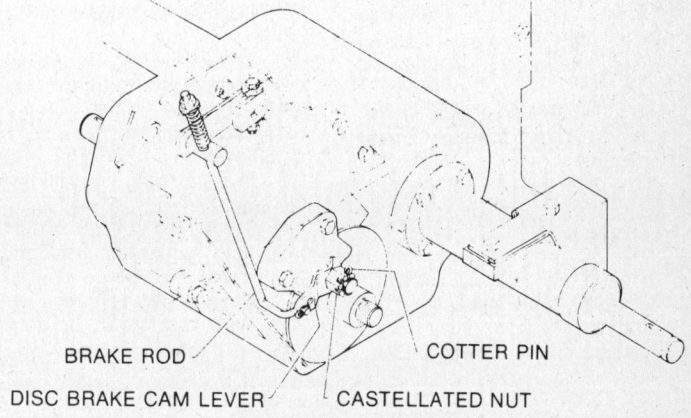
Brake adjustment

BRAKE PAD REPLACEMENT

When the brake can no longer be adjusted to hold the unit the brake pads will have to be replaced.

1. Remove brake rod from the disc brake cam lever.
2. Remove the two (2) capscrews that secure the disc brake to the transaxle.

NOTE: When removing disc brake be careful that the brake assembly is held firmly together.

3. Remove worn brake pad from brake.
4. Remove metal brake disc from transaxle.
5. Remove worn brake pad from transaxle.
6. Reassemble carefully, using new brake pads, by reversing the above procedure.
7. Adjust disc brake as outlined in the Brake Pad Adjustment section.

NOTE: All other models use a clutch/brake system which is covered in the following section.

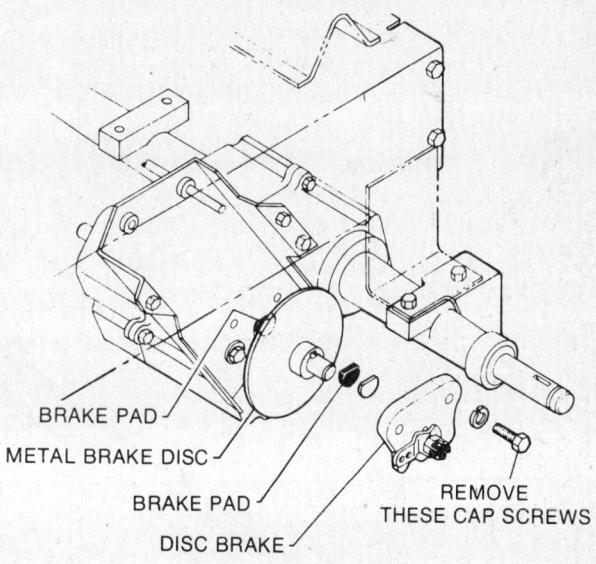
Brake pad replacement

CLUTCH AND BRAKE

8 hp tractors use a spring loaded friction drive clutch. All other manual transmission models use a belt idler type clutch system.

ADJUSTMENT

8 hp Models

Adjustment is required only if the unit has been disassembled.

1. Disconnect the brake rod from the arm.
2. Adjust the clevis on the clutch rod so that, when the clutch/brake pedal is pushed halfway, the carrier moves the friction wheel away from the drive disc.
3. Reconnect the brake rod and adjust the nut so that the caliper brake is applied after the friction wheel is moved away from the drive disc.

All Other Models

The idler pulley is spring loaded and the clutch is not adjustable. Adjust the brake by turning the adjusting nut on the brake rod until proper braking action is obtained. The brakes should not bind, but should hold the tractor on an incline.

Gilson

DRIVE BELTS

Traction Belt—Engine-to-Inner Jackshaft Pulley

ADJUSTMENT

1. Lift up the fender and seat support.
2. Loosen nut on underside of frame.
3. Turn nut on top of frame clockwise ½ a turn to tighten belt. If belt needs to be loosened turn nut counterclockwise ½ a turn.
4. Retighten nut on underside of frame.
5. Check belt tension. At approximately the midpoint of belt, place your finger on belt and press down. The belt should move approximately 1 in.
6. If belt can be depressed more or less than 1 in., repeat steps 1 through 5 until belt tension is correct.

REPLACEMENT

1. Lift up the fender and seat support.
2. Loosen nut on top righthand side of frame. **Do not remove.**
3. Remove the engine pulley belt cover.
4. Remove belt from inner jackshaft pulley.
5. Remove belt from engine pulley.
6. Remove old belt and replace with new one by reversing the preceding procedure.
7. Adjust belt for proper tension. See Adjustment procedure.

Traction Belt—Outer Jackshaft-to-Transaxle

ADJUSTMENT

This belt is self-adjusting.

REPLACEMENT

1. Depress foot pedal and set parking brake.
2. Loosen the "V" groove idler pulley.
3. Loosen the setscrew in the outer jackshaft pulley.
4. Slide pulley out on shaft so belt can be removed.
5. Remove old belt and replace with new one.
6. Reassemble by reversing the above procedure. Adjust the belt retainer. With the belt engaged there must be $1/16$ in.-$1/8$ in. clearance between retainer and belt. Make sure setscrew in the outer jackshaft pulley is tightened securely.

DRIVE CHAIN

REPLACEMENT

1. Loosen the adjustment sleeve located under the left foot rest.
2. Turn the adjusting rod counterclockwise until the chain can be lifted from the front sprocket.
3. Remove the chain guard and lift off the chain.
4. Installation is the reverse of removal.

ADJUSTMENT

Depress clutch-brake pedal completely down and hold. Loosen the adjustment sleeve (located under left running board in front of left rear tire) about 5 complete

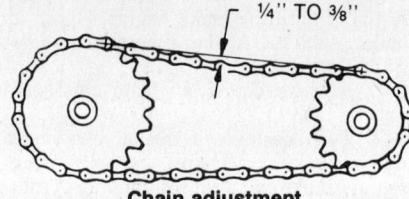

Chain adjustment

turns. Insert a screwdriver into slot on bottom end of adjustment rod assembly and turn clockwise until the chain is properly adjusted. Retighten adjustment sleeve.

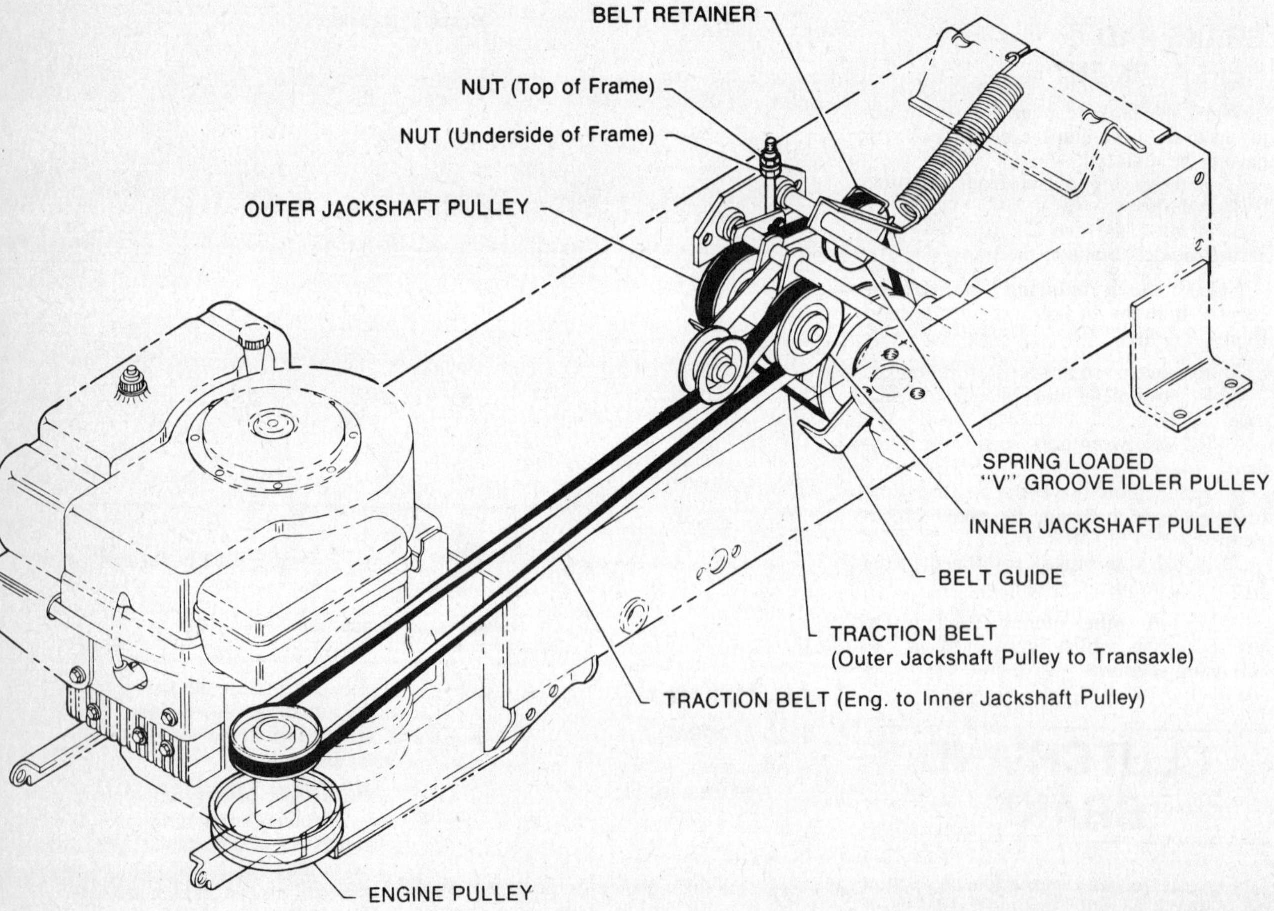

Drive belts

Gilson

DRIVESHAFT

REPLACEMENT

1. Disconnect spark splug wire and remove attachments from tractor.
2. Disconnect battery cables.
3. Remove shift plate between seat and instrument panel (4 bolts).
4. Remove roll pin from steering gear pinion. Loosen steering lash adjustment.
5. Remove bolt mounting steering wheel, remove steering wheel from shaft, then remove key and snapring from steering shaft.
6. Remove four bolts securing lower steering assembly to frame rails.
7. Pull lower assembly down and position out of the way.
8. Remove 2 bolts securing front coupling to flywheel.
9. Unbolt the 2 bolts securing rear flex coupling to driveshaft. Shaft may be removed by pulling it out underside of unit.
10. Loosen (2) setscrews in hub on transmission or gear box. Check for wear in bore of hub and on transmission of gear box input shaft. Replace hub on shaft and leave setscrews loose.
11. Assemble new driveshaft properly.
12. Position driveshaft in line with engine flywheel and rear coupling assembly. (Do not lose the 2 spacers on disc to flywheel mount bolts.) Tighten two bolts 16 to 24 ft.lb.
13. Line up transmission hub bolt holes with disc holes. Insert bolts with heads toward engine. Tighten bolts 16 to 24 ft.lb.
14. Reassemble steering gear assembly, install roll pin and reattach steering wheel. Make certain unit turns equally in both directions.
15. Grease steering gears.
16. Readjust steering gear as follows:
If excessive play develops in steering gears, the gear mesh can be adjusted by loosening the locking nut and turning the adjustment nut clockwise until play is removed. If steering wheel turns freely, tighten the locknut. If not, the gear mesh is too tight and must be loosened. Recheck steering wheel so it is not too loose.

NOTE: Steering wheel should have maximum of 1 in. free movement.

17. Check clearance between driveshaft and steering shaft upper roll pin and washer above steering gear bracket (minimum 3/16 in. clearance). If the clearance is less than 3/16 in., loosen square head setscrew in righthand steering support bracket and move steering shaft and gear assembly towards left to obtain more clearance. Retighten setscrew and readjust steering gear as outlined in step 16 above.
18. Reinstall battery, reconnect cables (attach positive cable first) and reconnect spark plug wire.
19. Start up tractor and allow hub mounted to transmission or gear box input shaft to seek its proper position. Run tractor approximately 30 seconds to 1 minute.
20. Turn ignition key off, torque the 2 setscrews in hub mounted on transmission shaft 9 to 14 ft.lb.
21. Reinstall shift plate.
22. Restart engine and check engine rpm at full throttle (should be 3400-3600 rpm).

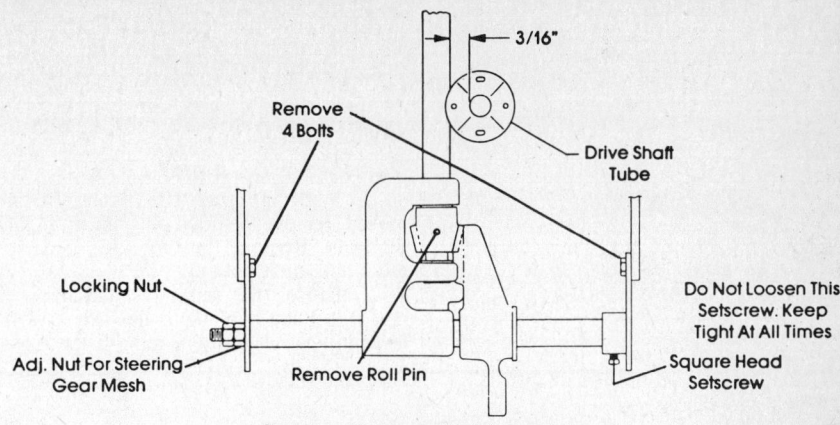

Steering linkage attaching points

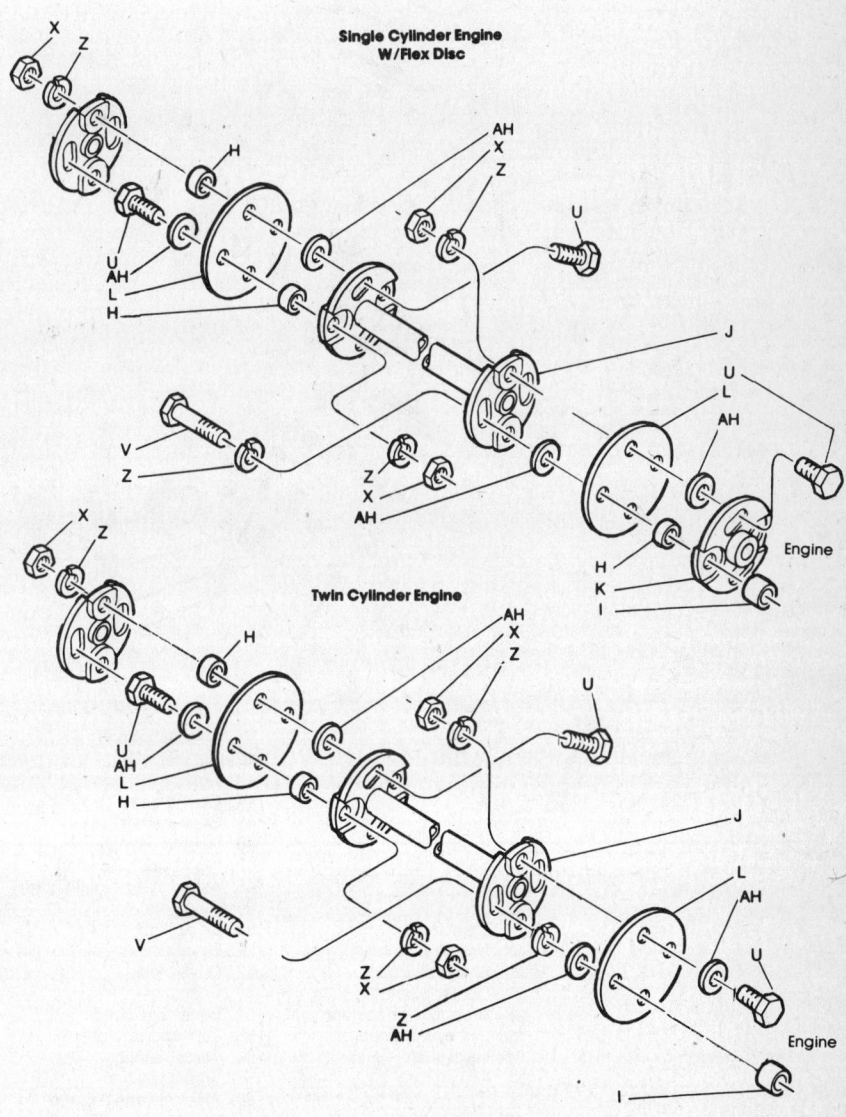

Single cylinder (top) and twin cylinder driveshafts. The reference letters in the illustration refer to codes in the parts book

445

Gilson

MANUAL TRANSAXLE

REMOVAL AND INSTALLATION

1. Support the tractor frame on jack stands.
2. Remove all drive belts.
3. Disconnect brake and/or clutch brake linkage.
4. Unbolt the transaxle from the frame and roll it away.
5. Installation is the reverse of removal.

OVERHAUL

Peerless 1708-A and 1709

1. With the transaxle assembly removed, remove the drain plug and drain the lubricant. Remove the brake band, brake drum and input pulley.
2. Loosen the setscrews and remove rear wheel and hub assemblies. Place shift lever in neutral position, remove the three capscrews from shift lever housing, then withdraw shift lever assembly. Remove the capscrews securing transaxle cover (59) to case (14). Place the unit on the edge of a bench or in a vise with the right axle pointing downward. Drive the dowel pins down out of cover. Remove all rust, paint and burrs from outer ends of axle shafts. Separate cover from case and lift cover upward off the axle shaft. Brake shaft (33) and idler gear (32) will be removed with cover. Remove output shaft (45) with output gear (44), spacer (43) and washer (42).
3. Withdraw the differential and axle shaft assembly and lay aside for later disassembly.
4. Hold the upper ends of shifter rods together and lift out shifter rods, forks, shifter stop, shaft (26) and sliding gears (28 and 29) as an assembly. Remove reverse idler gear (17), shaft (15) and spacer (18), then remove idler shaft (34) with idler gears (35, 37 and 39) and spacers (36 and 38). Input shaft (30) and input gear (31) can now be removed from case.
5. To remove the brake shaft (33) and gear (32) from cover, block up under gear (32) and press shaft out of gear.

CAUTION
Do not allow the cover to support any part of the pressure required to press brake shaft from gear.

6. To disassemble the differential, remove the four capscrews, then separate axle shaft and carriage assemblies from ring gear (49). Drive blocks (52), bevel pinion gears (51) and drive pin (50) can now be removed from ring gear.
7. Remove snaprings (53) and slide axle shafts (47 and 55) from axle gears (54) and carriages (48 and 58).
8. Unbolt axle housings (12 and 64) and renew bearings (13 and 65) and seals (11 and 63) as required.
9. Clean and inspect all parts and renew any showing excessive wear or other damage. When installing needle bearings, press bearings in from inside of case and cover until bearings are 0.015-0.020 in. below thrust surfaces.
10. Renew all seals and gaskets and reassemble by reversing the removal procedure, keeping the following points in mind: When installing the reverse idler (17) in the case, rounded edge of gear teeth and spacer (18) will be to the top. When installing idler shaft (34) and idler gears, position gears and spacers as follows: Gear (39) with raised hub up, short spacer (38), gear (37) with rounded teeth edge down, long spacer (36) and gear (35) with rounded teeth edge down.
11. Tighten transaxle capscrews to the following torque:

Differential capscrews	7 ft.lb.
Case to cover capscrews	10 ft.lb.
Axle housing capscrews	13 ft.lb.
Shift lever housing capscrews	10 ft.lb.

12. After the assembled transaxle is installed on tractor, fill the unit to level plug opening with SAE 90 EP gear oil.

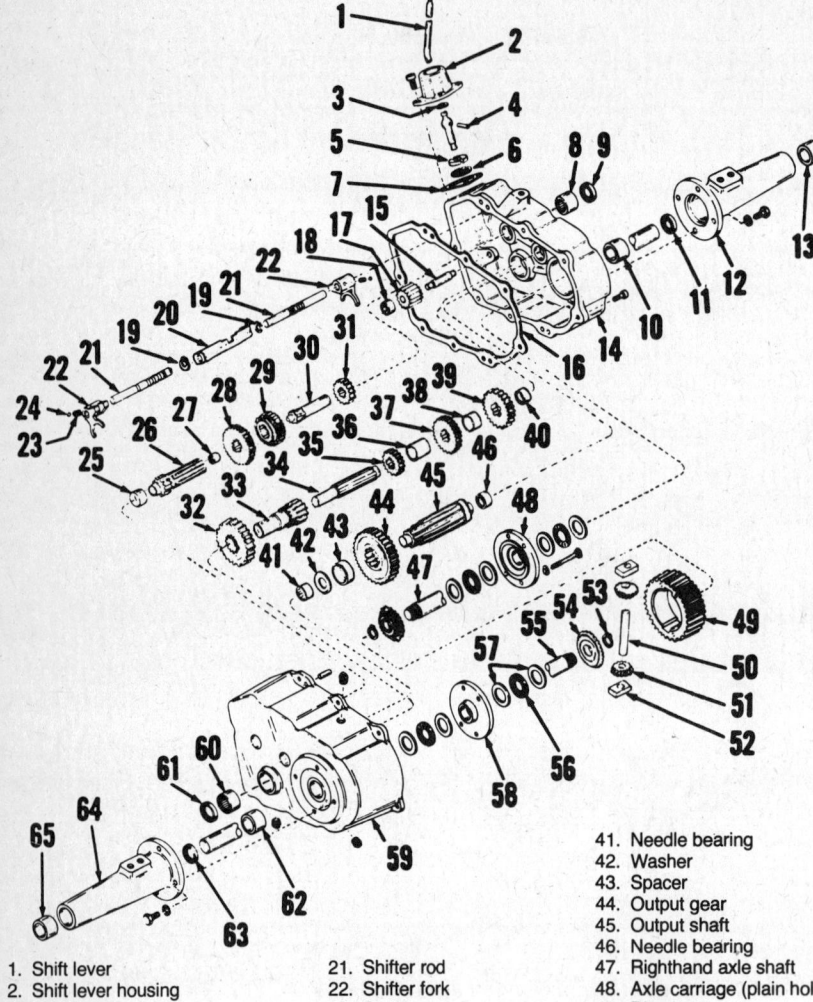

1. Shift lever
2. Shift lever housing
3. Seal ring
4. Roll pin
5. Retainer
6. Snapring
7. Gasket
8. Needle bearing
9. Oil seal
10. Carrier bearing
11. Oil seal
12. Righthand axle housing
13. Axle outer bearing
14. Transaxle case
15. Reverse idler shaft
16. Gasket
17. Reverse idler gear
18. Spacer
19. Snapring
20. Shifter stop
21. Shifter rod
22. Shifter fork
23. Spring
24. Detent ball
25. Needle bearing
26. Shifter shaft
27. Center bearing
28. 1st and reverse gear
29. 2nd and 3rd gear
30. Input shaft
31. Input gear
32. Idler gear
33. Brake shaft
34. Idler shaft
35. Idler gear
36. Long spacer
37. Idler gear
38. Short spacer
39. Idler gear
40. Needle bearing
41. Needle bearing
42. Washer
43. Spacer
44. Output gear
45. Output shaft
46. Needle bearing
47. Righthand axle shaft
48. Axle carriage (plain holes)
49. Ring gear
50. Drive pin
51. Bevel pinion gear
52. Drive block
53. Snapring
54. Bevel axle gear
55. Lefthand axle shaft
56. Thrust bearing
57. Thrust washer
58. Axle carriage (tapped holes)
59. Transaxle cover
60. Needle bearing
61. Oil seal
62. Carrier bearing
63. Oil seal
64. Lefthand axle housing
65. Axle outer bearing

Peerless Model 1708-A and 1709

Gilson

Peerless 616

1. Unscrew drain plug and drain lubricant.

2. Remove brake assembly and input pulley. Loosen setscrews and remove wheel and hub assemblies. Place shift lever in neutral position, unscrew retaining capscrews and remove shift lever assembly. Remove axle housings (52 & 16). Place unit in a vise so that socket head capscrews are pointing up. Remove all rust, paint and burrs from axle shafts.

3. Drive dowel pins out of case and cover. Unscrew socket head capscrews and lift cover from case. Screw two or three socket head screws into case to hold center plate (76) down while removing differential assembly.

4. Pull differential assembly straight up out of case. It may be necessary to gently bump lower axle shaft to loosen differential assembly.

5. Remove center plate. Hold shifter rods and lift out shifter rods, forks, shifter stop, shaft (27), sliding gears (25 and 26) and spur gear (24). Be careful when removing shaft (27) as rollers in bearing (13) may be loose and fall out.

6. Remove idler shaft (29) and gear (30).

7. Remove reverse idler shaft (79), spacer (80), gear (81), cluster gears (35, 36, and 37) on shaft (41) and thrust washer (42).

8. Remove bevel gear (31), washers (32 and 34) and thrust bearing (33).

9. Remove input shaft oil seal (9), snapring (10), input shaft (48) and gear (49). Washers (45 and 47) and thrust bearing (46) are removed with input shaft and gear. Remove bearing (11) and bushing (12).

10. To disassemble differential assembly, drive roll pin (71) out of drive pin (74).

11. Remove drive pin, thrust washers (63 and 66) and pinion gears (64 and 65).

12. Remove snaprings (56 and 70) and withdraw axle shafts from side gears (57 and 69). Remove side gears.

13. To disassemble cluster gear assembly, press gears from shaft. Note that beveled edge of gears (35 and 36) is on side closest to large gear (37). Remove gears from key (39). Note that short raised portion of key (39) is between middle gear (36) and large gear (37). Long raised section of key (39) is between small gear (35) and middle gear (36).

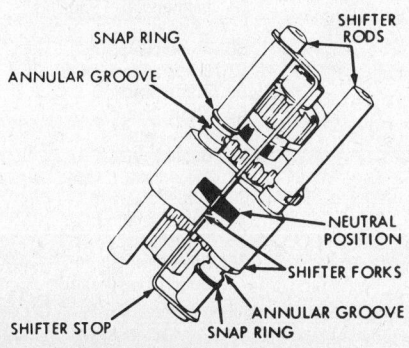

Positioning shifters in neutral

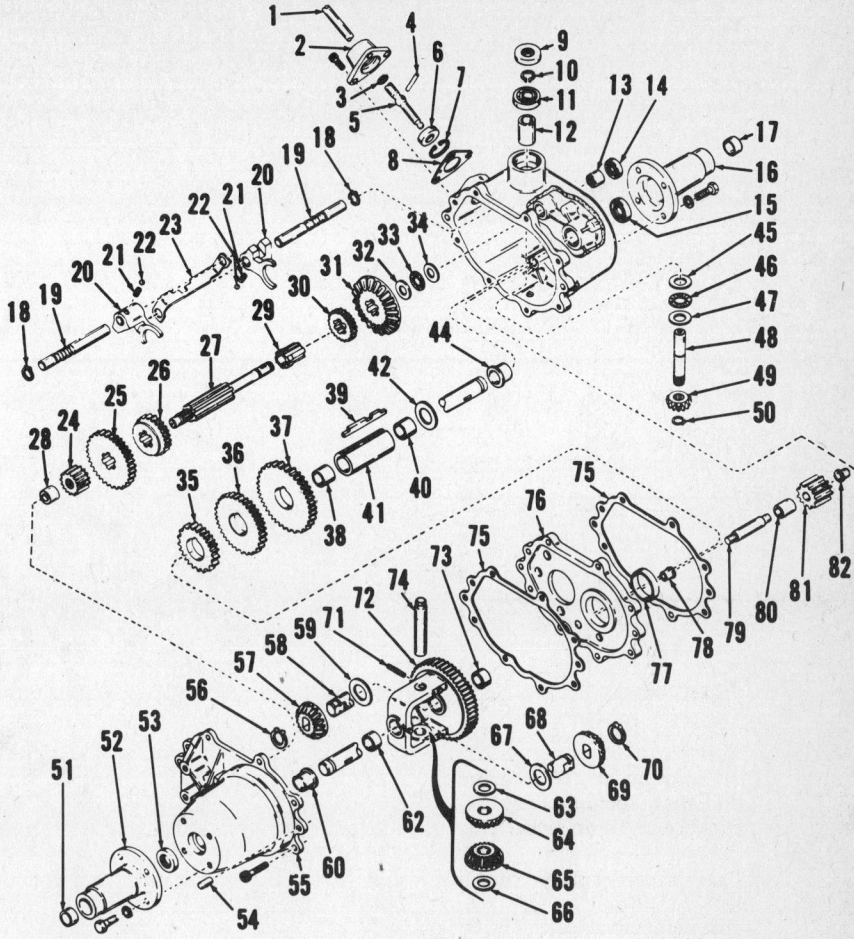

1. Shift lever
2. Lever housing
3. Quad ring
4. Roll pin
5. Shift lever
6. Retainer
7. Snapring
8. Gasket
9. Oil seal
10. Snapring
11. Ball bearing
12. Bushing
13. Roller bearing
14. Oil seal
15. Oil seal
16. Axle housing
17. Bushing
18. Snapring
19. Shift rod
20. Shift fork
21. Spring
22. Steel ball
23. Shifter stop
24. Spur gear
25. Sliding gear (1st and rev.)
26. Sliding gear (2nd and 3rd)
27. Shift and brake shaft
28. Needle bearing
29. Idler shaft
30. Gear
31. Bevel gear
32. Washer
33. Thrust bearing
34. Washer
35. Gear (25 tooth)
36. Gear (34 tooth)
37. Gear (39 tooth)
38. Bushing
39. Key
40. Bushing
41. Sleeve
42. Thrust washer
44. Bushing
45. Washer
46. Thrust bearing
47. Washer
48. Shaft
49. Pinion gear
50. Snapring
51. Bushing
52. Axle housing
53. Oil seal
54. Dowel pin
55. Cover
56. Snapring
57. Side gear
58. Axle shaft
59. Thrust washer
60. Bushing
62. Bushing
63. Thrust washer
64. Pinion gears
65. Pinion gears
66. Thrust washer
67. Thrust washer
68. Axle shaft
69. Side gear
70. Snapring
71. Roll pin
72. Differential carrier and gear
73. Bushing
74. Drive pin
75. Gasket
76. Center plate
77. Bushing
78. Bushing
79. Reverse idler shaft
80. Spacer
81. Reverse idler gear
82. Bushing

Peerless Model 616

14. Clean and inspect components for excessive wear or damage. Renew all seals and gaskets. Check for binding of shift forks on shift rods. Position shift forks in neutral position by aligning notches on shift forks with notch in shifter stop (23).

15. To assemble transaxle, install input shaft assembly by reversing disassembly procedure. Position case so that open side is up. Install bearing (13) and seal (14) if removed during disassembly. Install idler shaft (29), gear (30), bevel gear (31), washers (32 & 34) and thrust bearing (33). Be sure thrust bearing is positioned between washers.

16. Reverse idler shaft (79) may be used

Gilson

to temporarily hold idler gear assembly in position. Position cluster gears (35, 36 & 37) on key (39) so that bevel on gears (35 & 36) is toward large gear (37) and short section of key (39) is between middle gear (36) and larger gear (37). Press gears and key on shaft (41). Install shifter assembly components (18 through 27) in case, being sure that the shifter rods are properly seated. Install reverse idler shaft (79), gear (81) and spacer (80). Beveled edge of gear should be up.

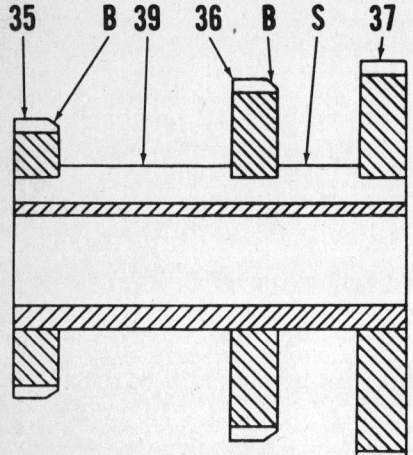

Note the position of bevels (B) on gears (35 and 36) and short section of key (39) between gears (36 and 37) on Model 616.

17. Install gasket, center plate (76) and gasket on case.

18. Assemble differential assembly by reversing disassembly procedure. Install differential assembly in case with longer axle pointing down. Be sure shift shaft gear (24) mates correctly with ring gear (72). Install locating dowel pins and secure cover (55) to case.

19. Install seals (15 and 53), axle housings (52 and 16) and shifter assembly (1 through 8).

20. Fill transaxle with 1½ pints of SAE 90 EP oil.

Peerless 2312 and 2313

1. Remove the drain plug and drain the lubricant.

2. Loosen the setscrews, remove snaprings, then remove rear wheel and hub assemblies. Remove brake drum and transaxle input pulley assemblies. Place shift lever (1) in neutral position, then unbolt and remove the shift lever assembly.

3. Remove axle housings (14 and 64) and remove seal retainers (11) with oil seals (12) and O-rings (13) by pulling each axle shaft out of case and cover as far as possible.

4. Place transaxle unit on edge of a bench with left axle shaft pointing downward.

5. Remove capscrews securing case (16) to cover (66) and drive aligning dowel pins out of case.

6. Lift case (16) up 1½ to 2 in., tilt case about 45 degrees, rotate case clockwise and remove it from the assembly.

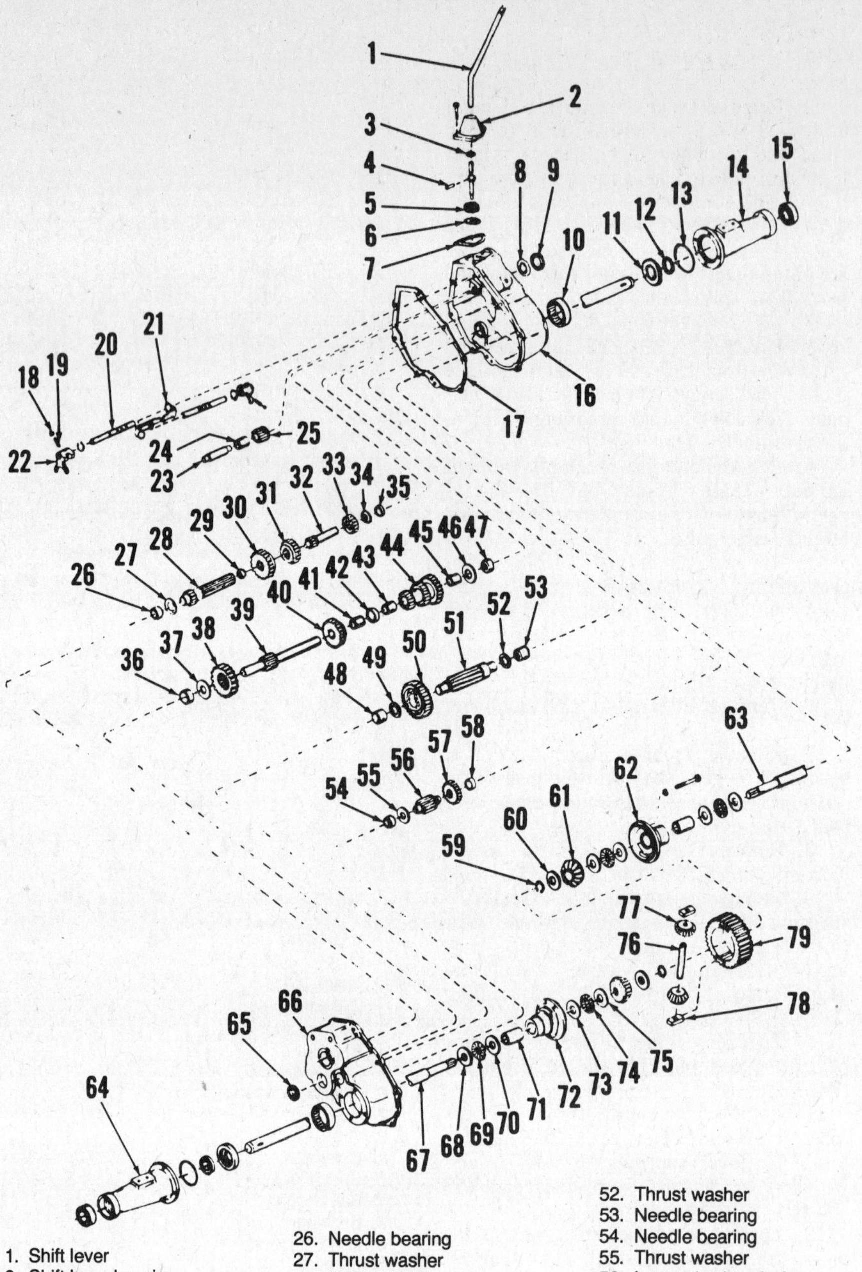

1. Shift lever
2. Shift lever housing
3. Seal ring
4. Roll pin
5. Retainer
6. Snapring
7. Gasket
8. Ball bearing
9. Oil seal
10. Carrier bearing
11. Seal retainer
12. Oil seal
13. O-ring
14. Righthand axle housing
15. Axle outer bearing
16. Transaxle case
17. Gasket
18. Detent ball
19. Spring
20. Shifter rod
21. Shifter stop
22. Shifter fork
23. Reverse idler shaft
24. Spacer
25. Reverse idler gear
26. Needle bearing
27. Thrust washer
28. Shifter shaft
29. Needle bearing
30. 1st, 2nd and reverse gear
31. 3rd and 4th gear
32. Input shaft
33. Input gear
34. Thrust washer
35. Needle bearing
36. Needle bearing
37. Thrust washer
38. Idler gear
39. Brake and cluster shaft
40. 2-cluster gear
41. Bushing
42. Spacer
43. Bushing
44. 3-cluster gear
45. Bushing
46. Thrust washer
47. Needle bearing
48. Needle bearing
49. Thrust washer
50. Output gear
51. Output shaft
52. Thrust washer
53. Needle bearing
54. Needle bearing
55. Thrust washer
56. Low reduction shaft
57. Low reduction gear
58. Needle bearing
59. Snapring
60. Thrust washer
61. Axle gear
62. Axle carrier (plain holes)
63. Righthand axle shaft
64. Lefthand axle housing
65. Oil seal
66. Transaxle cover
67. Lefthand axle shaft
68. Thrust washer
69. Thrust bearing
70. Thrust washer
71. Bushing
72. Axle carrier (tapped holes)
73. Thrust washer
74. Thrust bearing
75. Thrust washer
76. Drive pin
77. Bevel pinion gears
78. Drive block
79. Ring gear

Peerless Model 2312 and 2313

Gilson

Input shaft (32) and input gear (33) will be removed with case.

7. Withdraw differential and axle shaft assembly and lay aside for later disassembly.

8. Remove the 3-cluster gear (44) with its thrust washer (46) and spacer (42).

9. Lift out reverse idler gear (25), spacer (24) and shaft (23). Hold upper ends of shifter rods together and lift out shifter rods, forks, shifter stop (21), sliding gears (30 and 31) and shaft (28) as an assembly.

10. Remove low reduction gear (57), reduction shaft (56) and thrust washer (55), then remove 2-cluster gear (40) from brake shaft.

11. Lift out the output gear (50), shaft (51) and thrust washers (49 and 52). To remove brake shaft (39) and gear (38) from cover (66), block up under gear (38) and press shaft out of gear.

CAUTION
Do not allow cover or low reduction gear bearing boss to support any part of the pressure required to press brake shaft from gear.

12. Remove input shaft (32) with input gear (33) and thrust washer (34) from case (16).

13. To disassemble the differential, remove the four capscrews and separate axle shaft and carrier assemblies from ring gear (79).

14. Drive blocks (78), bevel pinion gears (77) and drive pin (76) can now be removed from ring gear.

15. Remove snaprings (59) and withdraw axle shafts (63 and 67) from axle gears (61) and carriers (62 and 72).

16. Clean and inspect all parts and renew any showing excessive wear or other damage.

17. When installing new needle bearings, press bearing (29) in spline shaft (28) to a depth of 0.010 in. below end of shaft and low reduction shaft bearings (54 and 58) 0.010 in. below thrust surfaces of bearing bosses.

18. Carrier bearings (10) should be pressed in from inside of case and cover until bearings are 0.290 in. below face of axle housing mounting surface. All other needle bearings are to be pressed in from inside of case and cover to a depth of 0.015-0.020 in. below the thrust surfaces.

19. Renew all seals and gaskets and reassemble by reversing the disassembly procedure, keeping the following points in mind: When installing brake shaft (30) and idler gear (38), beveled edge of gear teeth must be up away from cover.

20. Install reverse idler shaft (23), spacer (24) and reverse idler gear (25) with rounded end of gear teeth facing spacer.

21. Install input gear (33) and shaft (32) so that chamfered side of input gear is facing case (16).

22. Tighten transaxle capscrews to the following torque:
Differential capscrews 7 ft.lb.
Case to cover capscrews 10 ft.lb.
Axle housing capscrews 13 ft.lb.
Shift lever housing capscrews 10 ft.lb.

23. Fill transaxle, after unit is installed on tractor, to the level plug opening with SAE 90EP gear oil. Capacity is approximately 4 pints.

REDUCTION GEARS AND DIFFERENTIAL

8 hp Tractors

REMOVAL AND INSTALLATION

1. To remove the reduction gear and differential assembly, first loosen adjustment sleeve (53) and turn adjusting rod (52) counterclockwise until drive chain can be removed from front of sprocket.

2. Remove swivel pin from end of torque arm (54). Disconnect brake linkage.

1. Wheel hub
2. Washer
3. Axle bracket
4. Collar
5. Bushing
6. Grease retainer
7. Righthand axle housing
8. Oil seal
9. Gear cover
10. Gasket
11. Bearing plate
12. Carrier bearing
13. Differential case half
14. Righthand axle shaft
15. Thrust washer
16. Bevel axle gear
17. Thrust washer
18. Bevel pinion gear
19. Drive pin
20. Snapring
21. Bevel axle gear
22. Thrust washer
23. Lefthand axle shaft
24. Differential case half
25. Two piece ring gear
26. Carrier bearing
27. Bearing plate
28. Gear case
29. Lefthand axle housing
30. Oil seal
31. Thru bolts
32. Chain sprocket
33. Spacer
34. Thrust washer
35. Needle bearing
36. Second reduction gear
37. Spacer
38. Shaft
39. Shaft
40. Spacer
41. Thrust washer
42. Needle bearing
43. First reduction gear
44. Input shaft
45. Ball bearing
46. Snapring
47. Spacer
48. Input gear
49. Thrust washer
50. Snapring
51. Ball bearing
52. Chain adjusting rod
53. Adjustment sleeve
54. Torque arm
55. Brake disc
56. Snaprings

Reduction gears and differential used in 8hp tractors

Gilson

3. Support tractor frame and unbolt axle brackets (3) from frame.
4. Raise rear of tractor and roll reduction gear and differential assembly rearward from tractor.
5. Reinstall by reversing the removal procedure and adjust drive chain tension and brake linkage as required.

OVERHAUL

1. To disassemble the reduction gear and differential assembly, unbolt and remove caliper brake assembly and the torque arm (54). Remove snaprings (56) and brake disc (55).
2. Unbolt and remove chain guard and drive chain, then remove nut, washers and sprocket (32). Remove rear wheel and hub assemblies and axle brackets (3).
3. Unbolt and remove axle housings (7 and 29).
4. Clamp heads of bolts (31) in a vise so that left axle shaft is pointing downward.
5. Unbolt and remove gear cover (9) and bearing plate (11). Withdraw differential and axle assembly (12 through 26).
6. Reduction gears, shafts, bearings, gear, spacer and thrust washer (44 through 51) can now be removed.
7. To disassemble the differential, unbolt the two piece ring gear (25) from differential case, then separate the case halves (13 and 24).
8. Remove thrust washers (17), bevel pinions (18) and drive pin (19). Remove snaprings (20) and separate bevel axle gears (16 and 21), thrust washers (15 and 22) and axle shafts (14 and 23) from differential case halves (13 and 24).
9. If necessary, carrier bearings (12 and 26) can now be removed from the case halves.
10. Remove oil seals (8 and 30) from gear cover (9) and gear case (28).
11. Clean and inspect all parts and renew any showing excessive wear or other damage.
12. Renew gasket and oil seals and reassemble by reversing the disassembly procedure. Fill assembly with 1 pound of Shell Alvania EPRO 71030 grease or equivalent.

HYDROSTATIC TRANSAXLE

REMOVAL AND INSTALLATION

Eaton Model 10

1. Remove seat/fender assembly and shift plate.
2. Remove brake return spring from brake rod clevis pin. Disconnect rear brake link. Both of these items are located on the parking brake ratchet assembly.
3. Disconnect clevis from transmission control lever assembly.
4. Disassemble driveshaft coupling assembly at transmission end. To gain more room for puck assembly removal, loosen setscrews and slide fan hub assembly towards transmission.
5. Disconnect hydraulic hose from auxiliary pump. Fitting is located at top front of transmission. Plug or cap end of hose to avoid contamination and loss of oil.
6. Disconnect valve-to-filter hose at lift valve end.
7. To eliminate possible damage to the transmission from falling to the floor, place sturdy box or brace directly under transmission oil filter. Loosen the two large side mount bolts but do not remove at this time.
8. Remove cone locknuts from four axle bolts and pull bolts out of axle mount assemblies.
9. Remove the two large side mount bolts.

NOTE: Do not allow transmission to fall to floor.

10. Raise rear end of frame away from transmission assembly. Be careful not to bend fan blades. Place frame away from transmission assembly.
11. Block rear wheels and tilt transmission assembly backwards until fan faces upward. Block in this position.
12. Remove fan assembly. Disconnect large intake hose and filter hose from transmission. Remove four mounting bolts and lift from differential assembly.
13. Place transmission on bench and transfer fittings, control arm and pinion gear to new transmission.

NOTE: Use small gear puller to remove control arm. DO NOT strike end of control shaft or any part of control arm. Internal damage could occur. Pinion gear should slide onto shaft quite easily. DO NOT use a hammer as internal damage could occur.

14. Reverse procedure to install new hydrostatic. Change filter before starting engine. Adjust speed control lever.

1. Dust shield
2. Retaining ring
3. Snapring
4. Ball bearing
5. Snapring
6. Oil seal
7. Charge pump body
8. Charge pump race
9. Snapring
10. Charge pump rotor
11. Snapring
12. Pump roller (6)
13. Dowel pin
14. O-ring
15. Pump plate
16. O-ring
17. Bushing
18. Plug
19. Gasket
20. Neutral spring cap (2)
21. Neutral spring (2)
22. Housing
23. Bushing
24. Oil seal
25. Control shaft
26. Washer
27. Dowel pin
28. Insert
29. Insert cap
30. Drive pin
31. Cam pivot pin
32. Charge pump drive key
33. Input shaft
34. O-ring
35. Pump ball pistons
36. Pump rotor
37. Rotor bushing
38. Pump race
39. Pump cam ring
40. Plug (2)
41. Roll pins
42. Snapring (2)
43. Check valve ball (2)
44. Directional check valve body (2)
45. Plug
46. Relief spring
47. Charge relief ball
48. Pintle
49. Needle bearing
50. Rotor bushing
51. Motor ball pistons
52. Springs
53. Motor rotor
54. Drive pin
55. Output shaft
56. Motor race
57. Body
58. Gasket
59. Venting plug
60. Capscrew
61. Oil seal
62. Ball bearing
63. Retainer
64. Output gear
65. Snapring

Eaton Model 10 components

Gilson

Vickers Model

1. Remove seat pan and shift plate.
2. Remove the two transmission to valve hoses from lift control valve.
3. Remove bottom suction hose from transmission at swivel nut connection. (Keep hose pointing up so you won't lose oil from unit.)
4. Remove pin from neutral rod.
5. Loosen setscrews in coupling hub.
6. Remove four bolts holding transmission to axle. Take out the lower front bolt first and top bolt last.
7. Slide transmission over to right side away from transaxle.
8. Remove return to sump hose from transmission at swivel nut.
9. Pry transmission shaft out of coupling hub. Be careful not to damage fan.
10. Tip front end of transmission down and lift out in this tilted position.

NOTE: Keep the ends of hoses that you have removed from the transmission, pointing up to prevent oil loss. Oil should not have to be added to the system, if none is lost out of these hoses.

11. Tilt the front end of transmission down and set in position within the frame. Line up the key and shaft with the coupling hub and pry transmission forward within the hub. Use your other hand on the driveshaft to level and line up the transmission.
12. Tighten return to sump hose fitting on side of transmission.
13. Secure return to sump hose to fitting.
14. Lift transmission into position and tighten the four bolts securing transmission to axle. Be sure to replace the gasket.
15. Secure bottom hose to bottom fitting. Install pin into neutral control rod and secure with spring clip.
16. Tighten setscrews in coupling hub. Secure the two hoses onto the hydraulic valve (hydraulic lift units).
17. Check to be sure fan blades are not bent—straighten if necessary.
18. Assemble the seat pan and shift plate.

OVERHAUL— EATON MODEL 10

Input Shaft Seal Replacement

1. Disassemble the driveshaft coupling assemblies. This will allow access to the input shaft.
2. Remove fan and hub assembly. Remove grass shield from transmission input shaft.
3. Pierce the metal portion of the seal with a narrow sharp edged tool. Pry the seal outwards using caution not to scratch the input shaft or distort the seal counterbore in the charge pump body with the prying tool. Repair if necessary.
4. Apply a coating of grease to the seal inner lip and a light coating of Loctite® (Grade #35) to the charge pump counterbore and the O.D. of the seal.
5. With the lip or spring side of seal facing the pump, slide seal onto shaft and tap into a bottoming position in the pump counterbore.

— **CAUTION** —
Do not over drive as this may cut the rubber seal on the open end.

NOTE: A tube with the open end face square or slightly concave and I.D. slightly larger than the input shaft (approximately $25/32$ in.) and the O.D. slightly smaller than the O.D. of the input shaft seal (approximately $17/32$ in.) is recommended as a tool for driving seal into position.

6. Replace grass shield, coupling and driveshaft. Test run unit and check for leaks. Check oil level and add if necessary.

Control Shaft Seal Replacement

1. Clean hydrostatic transmission area thoroughly, especially the large hose fitting at bottom of transmission.
2. Disconnect the large hose at the transmission. Remove the control arm from the control shaft.

— **CAUTION** —
Do not pry or drive the control arm off the control shaft. Do not strike end of control shaft. Internal damage to the transmission could occur. A screw type puller is recommended.

3. Pierce the metal portion of the seal with a narrow sharp edged tool. Pry the seal outward using caution not to scratch the control shaft or distort the seal counterbore in cover with the prying tool.
4. Inspect the steel control shaft bushing in the cover. It should have a press fit and be positioned flush or slightly below the control shaft seal counterbore in cover. If out of position, push bushing flush and, with a punch, stake the casting to hold bushing in place. Three or four punch marks equally spaced is recommended.
5. Inspect for excessive wear between the control shaft and the steel bushing. If excessive wear is observed, replace transmission.

NOTE: Wear is rarely ever observed if the steel bushing has a press fit. Inspect control shaft for scratches, nicks, etc. Repair as required. Be sure shaft does not have sharp edges.

6. Apply a coating of grease to the seal inner lip and a light coating of Loctite® (Grade #35) to the counterbore in cover and the O.D. of the seal.
7. With the lip or spring side of seal facing the case, slide seal onto shaft and tap into a bottomed posiiton in the counterbore. Seal should be flush with or slightly below the cover surface. Do not overdrive as this may cut the rubber seal on the open end.

NOTE: A tube with the open end face square or slightly concave and I.D. slightly larger than the control shaft (approximately $25/32$ in.) and the O.D. slightly smaller than the O.D. of the control shaft seal (approximately $1 3/32$ in.) is recommended as a tool for driving seal into position.

Auxiliary Charge Pump

1. Clean the transmission external area thoroughly. (A steam cleaner or good degreaser works the best.)
2. Disassemble the driveshaft coupling assemblies and push driveshaft out of the way. Remove fan and hub assembly.
3. Remove the grass shield (B). Polish input shaft to remove any raised surface.
4. Remove the four $5/16$-18 x $1 1/4$ in. screws (A) and the one $5/16$-18 x $1 3/4$ in. screws (C).
5. Remove the charge pump body subassembly (D). Pull the charge pump body (D) carefully off shaft. Do not damage seal (E).
6. Remove the six rolls (G).
7. Remove the first snapring (F).
8. Mark the carrier (H) indicating "up" side and remove. (Use a marking pen.)

NOTE: Do not mark face in such a manner that the marked surface is raised.

Remove any raised metal with a fine grade stone.

9. Remove carrier drive pin (I).

— **CAUTION** —
Do not drop pin in open ports as complete transmission disassembly may be necessary.

10. Remove pump plate (L).
11. Inspect input shaft for worn key way and check for excessive clearance between the shaft and the bushing; specified clearance is .0013 in. to .0033 in. If replacement is required, it will be necessary to follow the procedure for complete disassembly and inspection.
1. Inspect the two O-rings (J & K). Replacement is recommended but not required, if intact.
2. Apply clean light grease to O-ring (K) and install in the machined groove in the face of the cover.
3. Inspect pump plate (L) and if the face is scored, replace.

NOTE: The pump plate surface stamped "A" should be installed face up. Non-ported plates (sump cooled) are reversible.

4. Install one snapring (F) to shaft against pump plate (L).
5. Inspect pump drive pin (I). Replace if worn and reinstall.

NOTE: Clean light grease in the key way will retain the pin during assembly.

6. Inspect the contact surfaces of the carrier (H) for measurable wear. Place on shaft with marked face up. Replace if worn.
7. Install new snapring (F).
8. Inspect the six rolls (G) for wear on the end radius and O.D. Replace worn rolls. Apply clean light grease to the rolls and assemble (E) in position in carrier (H).

NOTE: The grease will hold the rolls in position.

Gilson

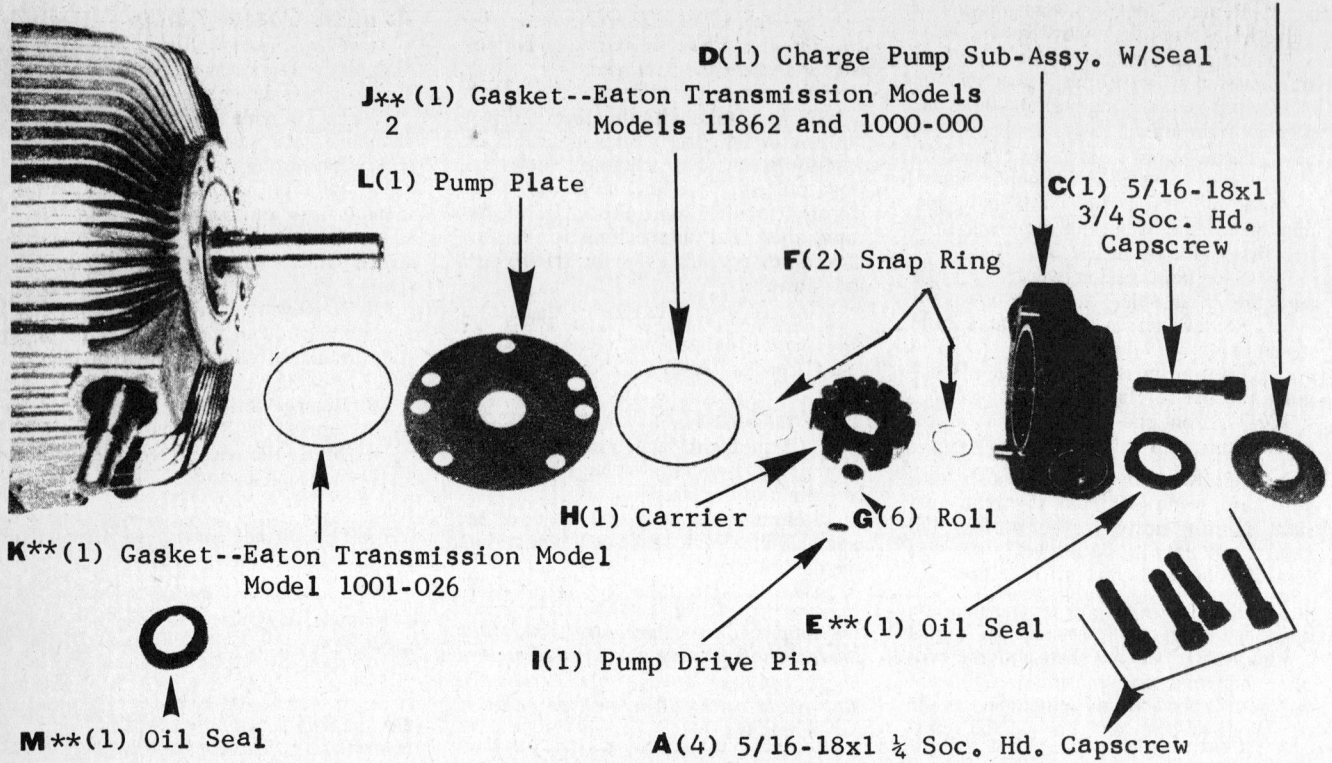

Auxiliary charge pump components

9. A. Inspect the pocket of the charge pump body (D) for end milling.

NOTE: If measurable wear is observed, replacement of the charge pump subassembly (D) is recommended.

B. Inspect the cam insert for any unusual wear pattern.

NOTE: If badly worn or scored, replacement of the charge pump subassembly (D) is recommended.

10. Inspect input shaft seal (E), if seal lip is damaged or hard, replace.
11. Apply clean light grease to O-ring (J) and install in machine groove in the face of the charge pump body (D).
12. Apply clean light grease to the input shaft seal (E) lip.
13. Guide the charge pump body subassembly (D) with O-ring in place over the input shaft and guide into position over the rolls and into the dowel pin holes.
14. Install four $^5/_{16}$-18 x 1¼ in. long hex socket capscrews (A) and one $^5/_{16}$-18 x 1¾ in. long hex socket capscrew (C).

NOTE: If the capscrews are not hex socket screws, it is recommended that they be replaced.

— **CAUTION** —
The $^5/_{16}$-18 x 1¾ in. long capscrew (C) must be installed in the heavy section of the body. If installed in any of the other four holes, internal damage to the die casting will unknowingly occur.

15. Torque the $^5/_{16}$-18 in. screws (A) and (C) to 28-30 ft.lb.

NOTE: The input shaft should turn by hand.

16. Inspect the grass shield (B). If bent, replace and install in position. Reassemble driveshaft coupling, check oil level. Inspect for leaks.
17. Start engine. Purge system by moving lift lever back and forth.

OVERHAUL— VICKERS MODEL

1. Place the transmission on a clean work bench. Have a supply of clean, lint-free rags, shop paper, or craft paper handy to lay parts on and to cover parts from dirt and foreign particles.
2. Separate the motor assembly from the transfer block by removing four hex head screws. Discard O-rings and replace with new one.
3. Separate motor valve plate from the motor housing by removing four screws.
4. If valve plate doesn't separate easily from motor housing, tap corner of valve plate with plastic mallet.
5. With one hand under rotating group end, tilt housing until rotating group slides into your hand.
6. If this group does not need to be disassembled, place it on a clean surface and proceed to step 14. To disassemble this group proceed to step 7.
7. Remove swash plate from shoe plate.
8. Remove assembled parts as shown.

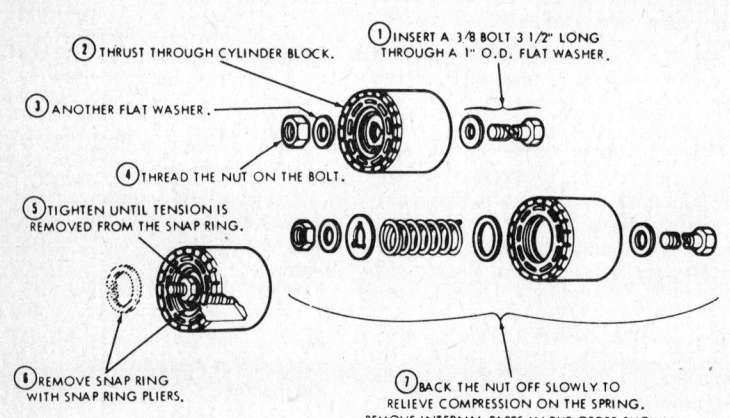

Relieving cylinder block spring tension

Gilson

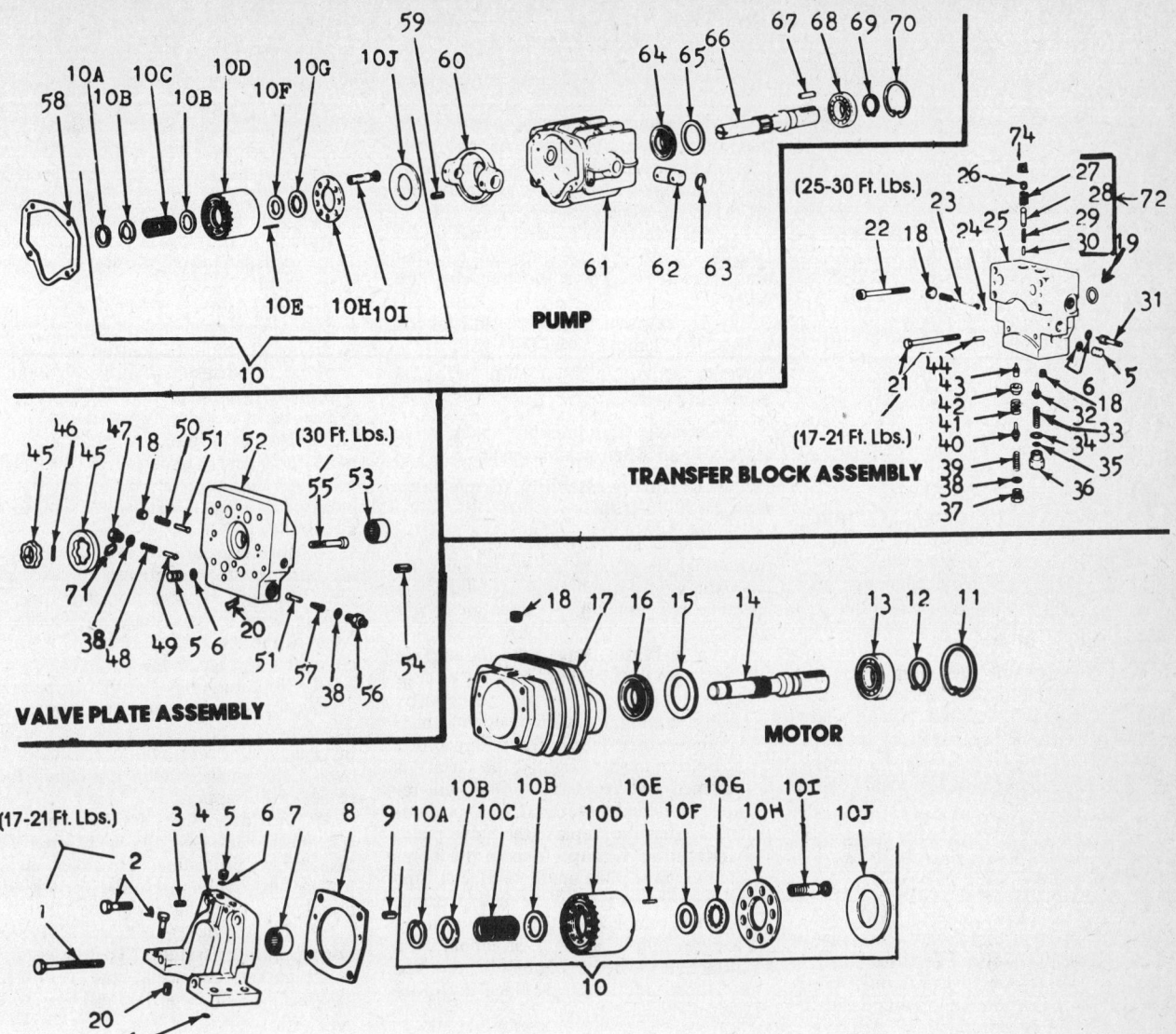

1. Hex head capscrew, 5/16-18 x 2¾
2. Hex head capscrew, 5/16-18 x 1¼
3. Roll pin, 3/16 x 1
4. Motor valve plate w/bearing
5. Plug
6. O-ring
7. Needle bearing
8. Motor gasket
9. Dowel pin
10. Rotating group
10A. Snapring
10B. Washer
10C. Spring
10D. Cylinder block—use #89282
10E. Pin
  Pin—use (3) #89260 and (1) #89259
10F. Washer—use (3) #89260 and (1) #89259
10G. Spherical washer
  Spherical washer—use (3) #89260 and (1) #89259
10H. Shoe plate
10I. Piston sub assembly— use #89282
10J. Swash plate
11. Retainer ring
12. Retainer ring
13. Bearing (ONLY FOR 23229 and 23082)
  Bearing (ONLY FOR 16461, 28031, and 14647)
14. Splined output shaft
15. Spacer
16. Oil seal (ONLY FOR 14647 and 16461)
17. Motor housing
18. Hydraulic plug, ¼-18
19. O-ring
20. Roll pin
21. Hex head capscrew, 5/16-18 x 3¾
22. Socket head screw, 5/16-18 x 3½
23. Spring
24. Ball
25. Transfer block
26. Plug
27. Guide
28. Check valve
29. Spring
30. Pin
31. Tow valve
32. Poppet relief valve H.P.
33. Spring
34. Shim
35. O-ring
36. Cap H.P.
37. Cap (soft ride valve)
38. O-ring
39. Spring
40. Pin
41. Spring
42. Seat
43. Soft ride poppet and pin
44. Roll pin, tow valve
45. Rotor—inner and outer w/key
46. Key
47. Plug
48. Spring
49. Poppet
50. Spring
51. Valve ball check
52. Pump valve plate
53. Bearing
54. Roll pin
55. Socket head screw, 5/16-18 x 1½
56. Connector
57. Spring
58. Gasket
59. Spiral pin
60. Yoke
61. Housing (ONLY FOR 23229 and 28032)
  Housing (ONLY FOR 16461, 28031 and 14647)
62. Pintle
63. O-ring
64. Oil seal
65. Spacer
66. Input shaft
67. Key, 3/16 x 3/16 x 1
68. Bearing
69. Retaining ring
70. Retaining ring
71. Special washer
72. Valve repair kit
73. Seal kit
74. Hydraulic plug, 3/8-18

Vickers hydrostatic components

Gilson

Be careful not to scratch the pistons or cylinder running surfaces.

9. Generally, no further disassembly is required. However, if the cylinder block is to be disassembled, proceed to step 10.

10. To relieve cylinder block spring tension, refer to the accompanying illustration.

CAUTION
Exercise extreme caution. Spring is under a great deal of tension.

---

11. To remove the motor shaft, first remove the large snapring with the 90° tru-arc pliers.

12. Remove the shaft by tapping on the small end with a soft tipped hammer or mallet.

13. Remove shaft. The spacer and press fit bearing should come out with it. Replace shaft seal.

14. Remove snapring and key from motor shaft before you remove the bearing.

15. If it is necessary to remove the bearing, first remove the key; then use an Owatanna 10-11 bearing puller, or equivalent puller, or an arbor press. Any other method of removal may damage bearing.

16. To disassemble the pump, remove the valve plate and transfer blocks as a unit by removing two recessed Allen head screws, and then the two hex head screws.

17. Pull valve plate and transfer block straight up from pump housing. Set it down on its painted side.

NOTE: Line up pins with holes in valve plate and gerotor key with driveshaft slot.

18. Pick up pump housing with one hand and slowly tilt it forward to remove group as assembled unit. To disassemble rotating group perform steps 7 through 10.

19. Now to remove the pump shaft. First remove the snapring with 90° snapring pliers.

20. Remove the pump shaft by tapping the small end of the shaft with a plastic tip hammer. Remove the shaft with the loose spacer and the press fit bearing installed on it. Replace shaft seal.

21. To remove the bearing, remove the key, and then the snapring. Refer to step 15 for bearing removal.

22. To remove both pintles and yoke from the housing, set a $3/16$ in. punch on the roll pin. Tap punch with a hammer until roll pin is disengaged from yoke.

23. Now place a $1/4$ in. brass rod on the pintle, and tap the pintle out of the yoke.

24. Repeat this procedure on the other pintle. Remove yoke from housing. Pintles must not be installed backward.

25. To disassemble the valve plate, remove the two recessed Allen head screws.

26. Separate valve plate from transfer blocks by pulling them apart. If required, tap valve plate with a plastic mallet to separate them.

27. Remove replenishment pump from valve plate.

NOTE: Dots not to be visible when replenishing pump is in pocket.

28. Remove the two replenishing system check valves by removing the Allen head plugs. Don't interchange valve parts.

29. Remove replenishing pump relief valve. (Some models have only one valve.)

30. To remove bearing, place valve plate on protective surface. Put brass shim stock or other protective stock under puller. Use Owatanna MD956-B-1, an equivalent puller, or an arbor press.

31. Set transfer block with finished surface facing up. Remove caps and take out both the soft-ride valve and high-pressure relief valve.

32. To remove high pressure check valves, first remove three O-rings.

NOTE: Be sure open ends of guide point outward.

33. Remove high-pressure check valve seats with an Allen head wrench.

NOTE: During assembly, torque valve seats to 30-35 ft.lb.

34. Clean all parts thoroughly with mineral spirits prior to inspection and after any stoning or machining operation. Inspection and repair procedures are as follows:

a. **Valve Plate:** Inspect the flat surface mates with the cylinder block for wear or scoring. Remove minor defects by lightly stoning the surface with a hard Arkansas stone that is flat within 0.001 in. Be sure to stone lightly; the surface is hardened and excessive stoning will remove hardened surface. If wear or damage is extensive, replace the valve plate.

b. **Rotating Group:** Inspect the bores and the valve plate mating surface of the cylinder block for wear and scoring. Remove minor defects on the running face by lightly stoning or lapping the surface. If the defects cannot be removed by these methods, replace cylinder block.

c. If one or more piston and shoe subassemblies need to be replaced, check that all piston and shoe subassemblies in the unit ride properly on the swash plate. In a set of nine pistons, variations in thickness greater than 0.001 of an in. from one shoe to another, will result in excessive internal leakage and shoe wear. The replacement of all nine piston and shoe subassemblies in the pump and motor, as well as the cylinder block, is recommended for maximum service between overhauls.

d. If necessary, hand-lap the shoes with 500-A emery paper (Tuff-Bak Durite Silicon Carbide) backed-up by a lapping plate. Good results may be obtained by dipping the emery paper in kerosene and keeping it wet during polishing.

35. **Swash Plate:** Inspect the swash plate for wear and scoring. If the defects are minor, lightly stone the swash plate. If wear or damage is extensive, replace the swash plate.

36. **Bearings and Driveshaft:** Inspect all bearings for roughness or excessive play; replace if necessary. Examine the sealing area of the shaft for scoring or wear. If the driveshaft is bent or worn excessively, replace it.

37. **Replenishing Pump:** Inspect the surface of all parts subject to wear. Remove light scoring from the face of the inner and outer rotor with crocus cloth laid over a flat surface, with a medium India stone, or by lapping.

38. The procedures for assembling the transmission are basically the reverse of the disassembly procedures. However, the following instructions describe certain additional procedures that should be adhered to:

39. Install new gaskets, seals, and O-rings during assembly. To ease assembly of the gaskets and seals, apply a thin film of Vaseline® or clean hydraulic oil to the

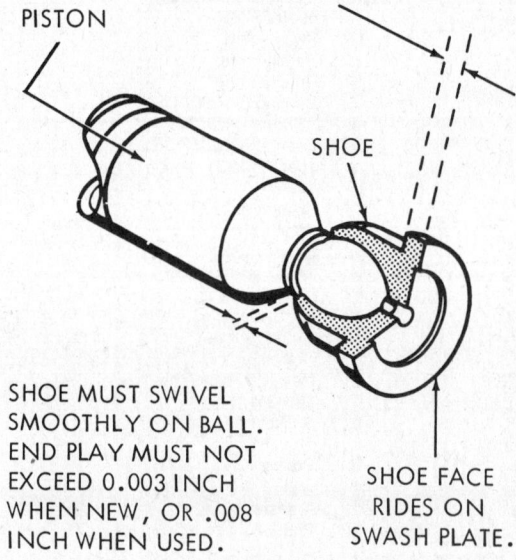

Inspecting piston and shoe assemblies

Gilson

O-rings. If a new rotating group is being used, squirt clean oil on it.

Differential

REMOVAL AND INSTALLATION

1. Remove rear wheels and wheel hubs.
2. With chisel, cut or drive the bearing retainer (V) off the end of axle housing. The axle and bearing assembly can now be removed from axle housing.
3. To remove the oil seal (S) first remove the snapring (T) on inside of the axle tube. Pry out the oil seal with a screwdriver.

NOTE: Do not remove oil seal unless it is to be replaced.

4. To replace the bearing (U), remove the snapring (T) holding the bearing on the axle. Press the bearing from the axle. Press the new bearing onto axle shaft (numbers facing out) until it bottoms on shaft. Install snapring on axle. Pack bearing with good multipurpose grease. Press only on inside race of bearing.
5. To replace the seal apply grease and install seal in axle tube (numbers facing out). The seal must be far enough into tube to start snapring and bearing axle assembly into tube. Tap on the outer ring at the bearing with a piece of round shaft or with a seal driver until the bearing pushes the snapring into place. This can be felt by solid backing on the bearing.
6. After axle is in place, pack bearing retainer with a good grade of multipurpose type grease and install a **NEW** retainer over axle tube. Crimp retainer into the slot in the tube using a dull chisel. Install felt seal, wheel hub, and wheel.

Removing bearing retainer

Installing bearing retainer

DISASSEMBLY

Remove seven capscrews holding housing together. Drive out the two spring pins to separate housing. Gears normally will be held in the lefthand housing by the oil suction tube. Remove the two retaining clips and lift the suction tube from the housing. The differential assembly and countershaft gear can now be removed.

Remove the spring pin from differential case with a long punch. Press out the pinion shaft (AG). To remove the pinion and bevel gears rotate the pinion gears 90 degrees in the differential case.

If the differential drive gear is damaged, remove it from the differential case. To do so, remove the ten screws (AA) holding the gear to the case and drive off the gear (AT) with a wooden block and hammer.

Countershaft gear needle bearing cages can be removed by punching them in or out of the case. Axle bearing cages must be split with a chisel and removed with a pliers. Use care when chiseling so as not to damage the area where the cage fits in the case.

Installing bearing on axle

ASSEMBLY

Needle Bearings

1. Thoroughly clean all parts and make sure gasket surface is free of all foreign material. Lightly oil all moving parts. Grease bearing retainers. Be sure the needles in all needle bearings are in place; eliminating one or more of the needles will cause axle to lock up.
2. If differential bearings are being replaced, press new bearings in (numbers towards inside of the housing) until flush with bottom of oil grooves in axle housing edge. Countershaft needle bearings are pressed in from outside of housing.

Drive Gear

3. Install drive gear on differential case. Align holes and install three oppositely positioned capscrews until gear fits snugly against case. Install rest of capscrews.

455

4. Install thrust washers on bevel gear hubs and insert geart into differential case. Align pinion gears **directly opposite** eacb other. Assemble pinion thrust washer on outside of each pinion gear and roll pinions into place by turning bevel gears until the holes line up. Install the pinion shaft into case and pinions, being careful to line up spring pin holes in the shaft and differential case. Drive spring into case and shaft.

NOTE: Pinion shaft is a light press fit in differential case and can often be driven in with a punch.

5. Install suction tube in lefthand housing, using new retaining clips. Assemble thrust washer on countershaft and install countershaft gear into lefthand housing. Assemble thrust washers on each side of differential assembly and install into lefthand housing. Assemble outside thrust washer on countershaft. Be sure all needle bearings are in place.

6. Coat both sides of the gasket with a good gasket cement. Position gasket on lefthand housing. Assemble righthand housing to lefthand housing making sure needle bearings are in place. Drive spring pins into place and install seven capscrews with lockwashers and nuts.

7. Attach elbow for oil filter adaptor and drain plug. Be sure breather screen is free of dirt.

ENGINE

Gilson tractors use a variety of engines by such manufacturers as Briggs & Stratton, Kohler, Onan and Tecumseh. Check the engine identification plate on your tractor and see the appropriate engine chapter in the Engine Unit Repair section of this book.

REMOVAL AND INSTALLATION

1. Disconnect all linkage points attached to the engine.
2. Disconnect all hoses and wires attached to the engine.
3. Remove the gas tank where necessary.
4. Remove the grille and hood.
5. Unbolt and remove the engine.
6. Installation is the reverse of removal.

Gravely

INDEX

MODELS 10, 10A, 12, 400, 424, 430, 432, 450, & 500/5000 SERIES
Front Axle 458
Steering 458
Brakes 458
Clutch 458
Transmission 458
Differential 459
Power Take-Off 462
Engine 463

MODELS 810, 812, 816S, 817
Front Axle 464
Steering 464
Brakes 464
Clutch 465
Transmission 466
Engine 469

GRAVELY

Models 10, 10A, 12
400 Series, Models 424, 430, 432, 450
500/5000 Series

FRONT AXLE

REMOVAL AND INSTALLATION

1. Raise and support the front end on jack stands.
2. Disconnect the tie rod ends at the steering arms.
3. Remove the axle pivot pin.
4. Roll the axle from under the tractor.
5. Installation is the reverse of removal.

STEERING

King Pin

REMOVAL AND INSTALLATION

1. Raise and support the front end on jack stands.
2. Remove the wheels.
3. Disconnect the tie rod ends at the steering arm.
4. Remove the cotter key from the king pin and slide the pin down, out of the axle.
5. Installation is the reverse of removal. Grease the king pin bushings with EP chassis lube prior to pin installation.

Steering Gear

Early production models have an exposed type gear. Late production models have an enclosed steering box.

REMOVAL AND INSTALLATION

Exposed Type

1. Disconnect the tie rods from the gear.
2. Remove the cotter pin securing the pivot pin in position.
3. Remove the pivot pin from the gear.
4. Slide the gear from the tractor.
5. Installation is the reverse of removal. Liberally grease the moving parts of the gear with EP chassis lube.

Enclosed Type

1. Remove the hood.
2. Remove the fuel tank.
3. Remove the battery.
4. Remove the steering wheel and the snapring from the steering column weldment.
5. Disconnect the tie rods from the steering arms.
6. Unbolt the steering gear box from the frame and lower it from the tractor.
7. Installation is the reverse of removal. Pack the gear box with EP chassis lube prior to installation.

BRAKES

Brake Lining

REPLACEMENT

1. Remove the clevis and retainers.
2. Remove the lining.
3. Install the new lining and adjust as described below.

ADJUSTMENT

1. Tighten the brake adjusting nut until the brake bands bind slightly.
2. Back off the nut ½ turn.

CLUTCH

The clutch is integral with the transmission. For service, refer to the transmission section following.

TRANSMISSION

ADJUSTMENTS

High Adjustment

1. Move the jam nuts (G) on the lower end of the clutch rod (F) closer to the clutch lever for a more sensitive adjustment and closer to the end of the rod for a less sensitive adjustment.
2. Adjust the jam nut (B) on the lower end of the clutch lever so that when the lever is in the high position there is .010 in. gap between the coils of the spring.

Low Adjustment

1. Move the jam nuts (D) on the upper end of the clutch lever closer to the end of the lever (C) for a more sensitive adjustment or closer to the clutch rod for a less sensitive adjustment.
2. Adjust the jam nuts (E) on the upper end of the clutch rod when the lever is in the low position, to give a .010 in. gap between the coils of the spring.

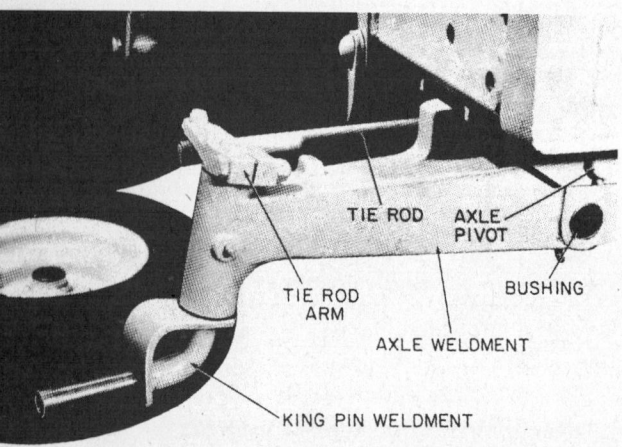

Steering axle and king pin

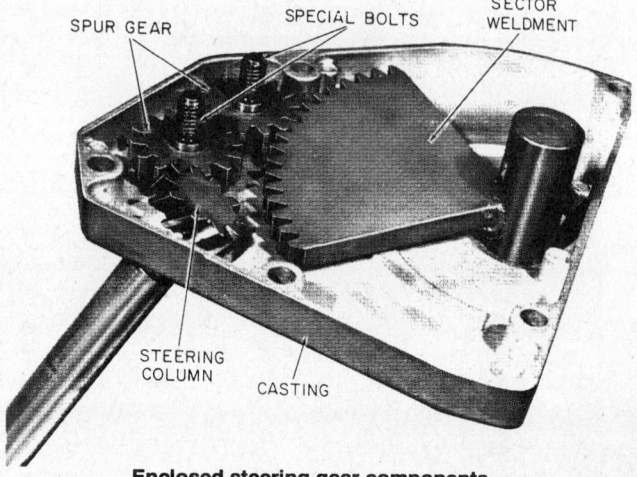

Enclosed steering gear components

Gravely

Planetary

REMOVAL

1. Remove engine as described later.
2. Remove the clutch lever from the actuating shaft.
3. Turn the actuating shaft to release clutch slide rods.
4. Slide rear space, ring gear, clutch up, slide rods, high-low planetary and ring gear and bushing assembly out of chassis.
5. Carefully examine all parts for damage and unusual wear. Check orbit gears for wear both on teeth and wear on the pins. Carefully check for gears that may be tight or starting to seize to the pin.
6. Check the surface of the clutch cup for wear or damage. Also, check the friction surfaces of the ring gears for scoring.
7. To reassemble for installation, place the planetary in the ring gear and bushing assembly, fit the clutch cup on the ring gear and bushing assembly, fit the ring gear in the clutch cup.
8. The notches of the slide rods fit on the outside of the clutch cup with the teeth upwards to mesh in the teeth of the actuating shaft.
9. Install the rear spacer over the slide rods to fit on the ring gear.
10. Install the unit in the tractor. Reinstall engine.

DIFFERENTIAL

500/5000 SERIES REMOVAL AND DISASSEMBLY

1. Drain oil from transmission.
2. Raise tractor and remove wheels.
3. Remove righthand wheel hub.
4. Disconnect swiftamatic shift linkage.
5. Remove righthand axle housing with axle.
6. Carefully examine the shifting yoke, shifting clutch, stationary clutch and

Enclosed steering gear box

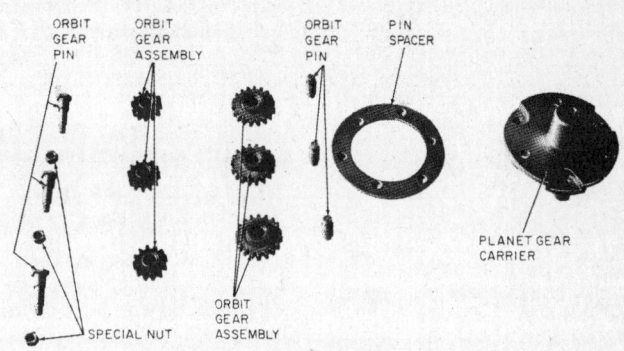

Planetary components

Brakes and controls, 400 series shown, others are similar

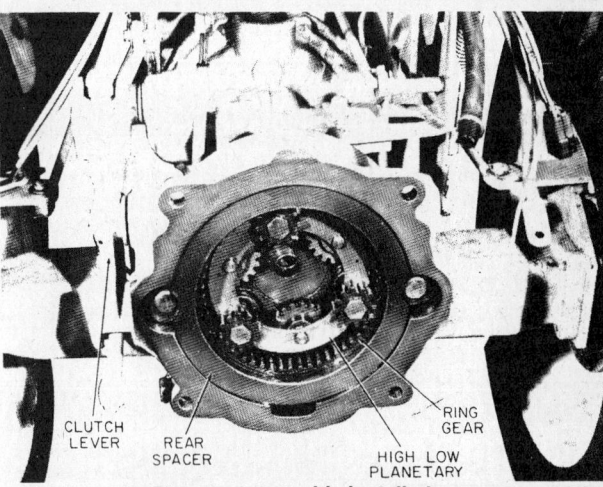

Planetary assembly installed

Gravely

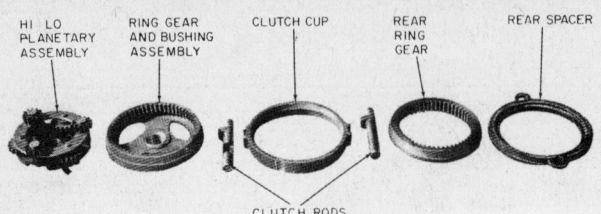

Transmission gear assemblies

clutch housing. Particular attention should be given to see that the stationary clutch is properly secured against rotation. Also make sure the shifting clutch slides freely on the shifting gear.

7. If no discrepancies have been found at this point or if the failure is obviously a part of the differential assembly, remove the differential assembly.

8. Remove lefthand wheel hub.

9. Remove lefthand axle housing with axle.

10. Remove the differential assembly from the tractor.

11. Using wire cutters, remove wire securing bolts on each side of worm gear. Secure differential assembly upright in vise.

12. Remove bolts securing clutch housing to worm gear. Remove clutch housing, gears, and shims.

13. Turn worm gear over and remove pinned housing and gears.

14. Tilt differential mechanism and remove from worm gear. The mechanism does not ride on the inside surface of the gear.

15. Disassemble the differential mechanism—carefully check the bore of the gears and the spider arms for scoring, check the backs of the shifting train pinions.

ASSEMBLY AND INSTALLATION

1. Reassemble the differential mechanism using new thrust needle bearings if any other parts were replaced. On the two spider arms with snapring grooves at the end, assemble in this sequence: spider gear, shifting train pinion, bearing, spacer and snapring. On the other two arms assemble in this sequence: spider gear, race, bearing, spacer and snapring.

2. Tilt the differential mechanism and install it in the worm gear.

3. Install the pinned gear and differential gear on the pinned housing and install it on the worm gear. Use new drilled bolts; clean bolts and apply a stud locking preparation to the bolt threads. Torque bolts to 17-20 ft.lb.

4. Turn worm gear over and install the clutch housing, with differential gear, shifting gear and shims on the worm gear. Torque bolts to 20 ft.lb. Replace with new bolts as in step 3.

5. Check end play of shifting gear .00-.010 in. maximum. Remove clutch housing and adjust shims as necessary. Torque bolts to 20 ft.lb.

6. Insert wire through one bolt head, cross and insert through other, than twist ends together securely. Trim twist to ½ in. length. Repeat, securing other bolts on other side of gear.

7. Install differential on lefthand axle in lefthand axle housing. (Pinned housing side of differential assembly in axle housing.)

8. Be sure thrust pin is in place in the differential assembly so it will be between the axles.

9. Place equal amount of shims and a shellacked gasket on each axle housing. Install lefthand axle housing and differential assembly in tractor. Secure axle housing to tractor.

10. Install righthand axle housing on tractor, being sure the shifting clutch fits on the shifting gear. Secure axle housing to tractor. Torque bolts to 45 ft.lb.

11. Remove the engine and high-low planetary or the advance casting and the forward-reverse planetary.

12. Using a screwdriver or similar tool, check the differential assembly for play. Running clearance should be .008-.012 in. Adjust by adding or removing an equal amount of shims from each axle housing.

13. Adjust axles for .008-.012 in. end play. Add or remove shims behind the bearing cap.

MODELS 10, 10A, 12 OVERHAUL

1. Drain oil from transmission.
2. Raise tractor and remove wheels.
3. Remove righthand wheel hub.
4. Disconnect Swiftamatic shift linkage.
5. Remove righthand axle housing with axle.
6. Carefully examine the shifting yoke, shifting clutch, stationary clutch and clutch housing. Particular attention should be given to see that the stationary clutch is properly secured against rotation. Also make sure the shifting clutch slides freely on the shifting gear.
7. If no discrepancies have been found at this point or if the failure is obviously a part of the differential assembly, remove the differentail assembly.
8. Remove lefthand wheel hub.
9. Remove lefthand axle housing with axle.
10. Remove the differential assembly from the tractor.

NOTE: The following steps (11 through 21) refer to tractor prior to serial number J-65504. Refer to step 22 through 32 for tractor serial number above J-65504. Steps 33 through 38 apply to all tractors.

11. Secure the differential assembly upright in a vise.

12. Bend down locking plates and remove six (6) bolts securing clutch housing to worm gear.

13. Lift off clutch housing. Remove shims.

14. Remove internal parts of differential. Carefully check all parts for damage or abnormal wear.

15. Examine internal surfaces of the worm gear. An even wear pattern is normal, however it should not exceed .010 in. If the gear is pitted in spots or has small gouges, polish these areas.

16. Reassemble the differential mechanism. Use new needle thrust bearings if any other parts have been replaced. Assemble on two opposite spider arms in this sequence: spider gear, shifting train pinion, needle thrust bearing, needle thrust race, shims, spider thrust washer. On the two remaining spider arms, assemble in this sequence, spider gear and thrust spacer. Install in bronze gear.

17. Using a differential gear, check shimming of spider gears. Adjust shims so that gear sits solid on all four spider gears.

18. Install shims, differential gear, shifting gear and clutch housing on worm gear. Use new drilled bolts; clean bolts and apply a stud locking preparation to the bolt threads. Torque bolts to 17-20 ft.lb.

19. Check end play of shifting gear—.0-.010 maximum. If necessary, remove clutch housing and adjust shims accordingly. Reinstall clutch housing and bolts. Torque to 20 ft.lb.

20. Insert wire through one bolt head, cross and insert through another, then twist ends securely. Trim twist to ½ in. length. Repeat, securing other bolts.

21. Turn differential assembly over and remove bolts. Replace with new bolts as in steps 18 and 20.

22. Using wire cutters, remove wire securing bolts on each side of worm gear. Secure differential assembly upright in vise.

23. Remove bolts securing clutch housing to worm gear. Remove clutch housing, gears, and shims.

24. Turn worm gear over and remove pinned housing and gears.

25. Tilt differential mechanism and remove from worm gear. The mechanism does not ride on the inside surface of the gear.

26. Disassemble the differential mechanism—carefully check the bore of the gears and the spider arms for scoring, check the backs of the shifting train pinions.

27. Reassemble the differential mechanism using new thrust needle bearings if any other parts were replaced. On the two spider arms with snapring grooves at the end, assemble in this sequence: spider gear, shifting train pinion, bearing, spacer and snapring. On the other two arms assemble in this sequence: spider gear, race, bearing, spacer and snapring.

28. Tilt the differential mechanism and install it in the worm gear.

29. Install the pinned gear and differen-

Gravely

tial gear on the pinned housing and install it on the worm gear. Use new drilled bolts; clean bolts and apply a stud locking preparation to the bolt threads. Torque bolts to 17-20 ft.lb.

30. Turn worm gear over and install the clutch housing, with differential gear, shifting gear and shims on the worm gear. Torque bolts to 20 ft.lb. Replace with new bolts as in step 29.

31. Check end play of shifting gear .00-.010 in. maximum. Remove clutch housing and adjust shims as necessary. Torque bolts to 20 ft.lb.

32. Insert wire through one bolt head, cross and insert through other, then twist end together securely. Trim twist to ½ in. length. Repeat, securing other bolts on other side of gear.

33. Install differential on lefthand axle in lefthand axle housing. (Pinned housing side of differential assembly in axle housing.)

Be sure thrust pin is in place in the differential assembly so it will be between the axles.

34. Place equal amount of shims and a shellacked gasket on each axle housing. Install lefthand axle housing and differential assembly in tractor. Secure axle housing to tractor.

35. Install righthand axle housing on tractor, being sure the shifting clutch fits on the shifting gear. Secure axle housing to tractor. Torque bolts to 45 ft.lb.

36. Remove the engine and high-low planetary or the advance casting and the forward-reverse planetary.

37. Using a screwdriver or similar tool, check the differential assembly for play. Running clearance should be .008-.012 in. Adjust by adding or removing an equal amount of shims from each axle housing.

38. Adjust axles for .008-.012 in. end play. Add or remove shims behind the bearing cap.

400 SERIES REMOVAL AND DISASSEMBLY

1. Drain transmission oil.
2. While transmission is draining, raise rear of tractor and remove rear wheels. Remove axle mount rear hitch or frame assembly axle connector (if so equipped).
3. Loosen the jam nut and special setscrew and remove wheel hub from lefthand axle. Remove Woodruff key.
4. Remove bolts ecuring brake support to the axle housing. Slide brake support and band off axle and out of way.
5. Remove high-low clutch link and the high-low clutch arm. Remove bolts securing axle housing to chassis casting. Remove axle housing from axle.
6. Remove righthand wheel hub and brake assembly as in steps 3 and 4.
7. Disconnect two-speed shifter link.
8. Remove bolts securing axle housing to chassis casting.
9. Remove axle housing and differential assembly from tractor.
10. Hold the axle housing in vise, and remove differential assembly from axle housing.

Carefully examine the bearings on the differential assembly. Check for roughness and scoring. Be sure the bearings are fully pressed in position on the housings.

11. Secure the differential assembly upright in a vise shifting gear side up.

NOTE: The following steps (12 through 18) refer to the differential in tractors serial numbers prior to 7928.

Refer to steps 20 through 30 for tractor serial number 7928 and up.

12. Bend down locking plates and remove six (6) bolts securing clutch housing to the worm gear.
13. Lift off clutch housing with axle. Remove shims.
14. Remove internal parts of differential. Carefully check all parts for damage or abnormal wear. Abnormal conditions include:
   a. Scoring of thrust washers.
   b. Scoring of spider pin arms.
   c. Scoring of bore of gears.
   d. Scoring on the back of gears.
   e. Chipped teeth.
   f. Heavy wear on gear teeth.
15. Examine the internal surfaces of the worm gear. An even wear pattern is normal, however it should not exceed .010 in. If the gear is pitted in spots or has small gouges, polish these areas.
16. Reassemble the differential mechanism. Use new needle thrust bearings if thrust washers or gears have been replaced. Assemble in this sequence on **all four spider arms:** spider gear, shifting train pinion, needle thrust bearing, needle thrust race, shims, spider thrust washer. Install in bronze gear.

Earlier production tractors had shifting train pinions on two spider arms and two thick thrust washers on the arms with only spider gears. These should be converted to the style described above.

17. Adjust shimming using a differential gear. The gear should be solid on all four spider gears. If the differential gear wobbles, decrease shimming on gears which gear is sitting solid on. This allows these gears to move away from center and in effect lowers them.
18. Reinstall clutch housing and shims

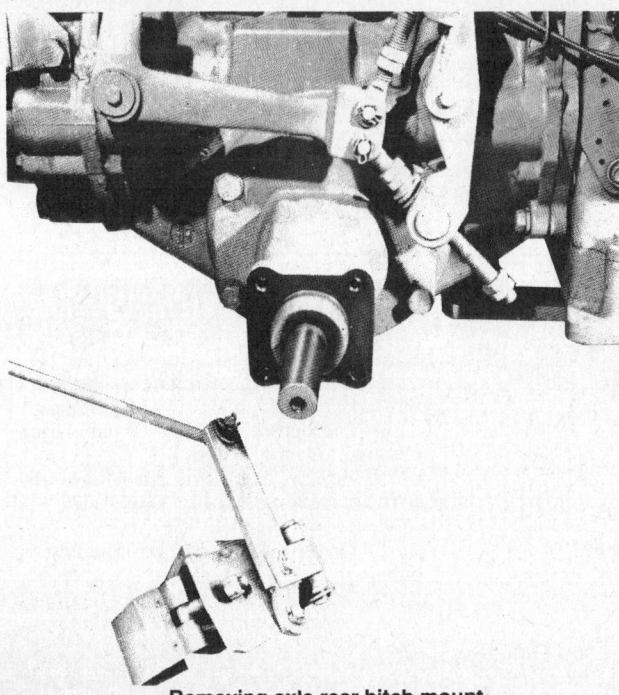
Removing axle rear hitch mount

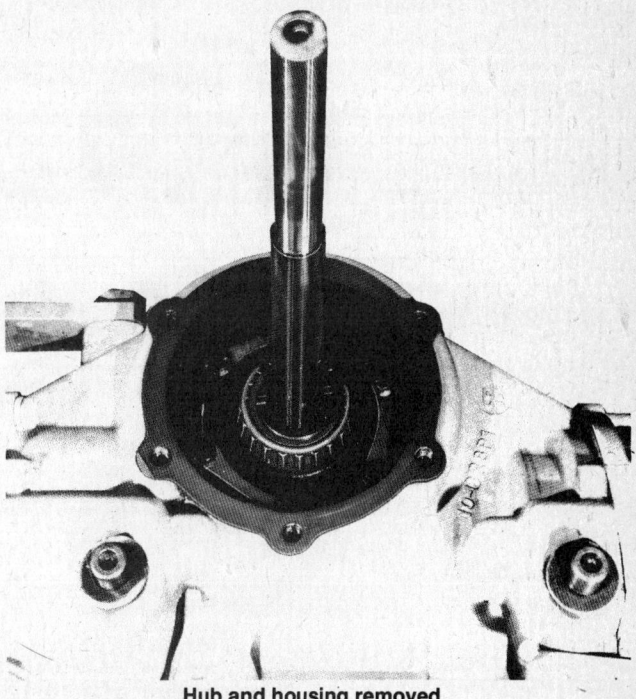
Hub and housing removed

Gravely

Differential components

back on worm gear. Install bolts using new locking plates. Torque to 20 ft.lb.

19. Check end play of shifting gear .0-.010 in. maximum.

Remove clutch plate and adjust shims accordingly if necessary. Reinstall bolts. Torque to 20 ft.lb. Secure locking plates to bolts.

20. Bend down locking plates and remove bolts securing clutch housing to worm gear.

21. Lift off clutch housing with axle. Remove shims.

22. Turn differential over and bend down locking tabs and remove bolts securing pinned housing to worm gear.

23. Lift off pinned housing with axle.

24. Remove internal parts of differential, carefully check all parts for damage or abnormal wear. Abnormal wear includes:
 a. Scoring of spider pin arms.
 b. Scoring of bore of gears.
 c. Scoring on back of gears.
 d. Chipped teeth.
 e. Heavy wear on teeth.

25. The internal differential assembly does **not** ride on the inside surface of the bronze gear.

26. Reassemble the differential mechanism. Use new needle thrust bearings of gears have been replaced. Assemble in this sequence on **all four spider arms**: spider gear, shifting train pinion, bearing, spacer, snapring.

27. Install differential mechanism in bronze gear. It will be necessary to tilt the differential mechanism to fit it inside the gear.

28. Install pinned housing with differential gear, pinned gear and axle back on bronze gear. Install bolts using new locking plates. Torque to 20 ft.lb. Secure locking plates to bolts.

29. Reinstall clutch housing and shims back on worm gear. Install bolts using new locking plates. Torque to 20 ft.lb.

30. Check end play of shifting gear 0-.010 in. maximum. Remove clutch plate and adjust shims accordingly, if necessary. Reinstall bolts and torque to 20 ft.lb. Secure locking plates to bolts.

400 SERIES ASSEMBLY AND INSTALLATION

1. Wrap key area of righthand axle with ultra-thin shim stock to protect oil seal and install differential in righthand axle housing, being sure the shifting clutch lines up properly.

2. Be sure the machined surface of the axle housings, chassis casting and shims are clean of any oil or debris.

3. Put an equal amount of .020 in. and .005 in. shim and a new gasket on each axle housing.

4. Install the righthand axle housing—differential assembly in place. Secure with 2-3 bolts.

5. Wrap axle with ultra-thin shim stock and install lefthand axle housing. Secure with 2-3 bolts.

6. Check end play of each axle. End play should be approximately .020 in. Adjust accordingly.

7. After adjusting end play, install all bolts and torque to 45 ft.lb.

NOTE: The bottom bolt on the lefthand axle housing (oil drain bolt) cannot be torqued. Tighten securely.

8. Clean axle with crocus cloth.
9. Reinstall brake band supports on each axle housing.
10. Install Woodruff key in axles and wheel hubs. Tighten special setscrews and jam nuts.
11. Connect two-speed shifter link.
12. Install rear wheels.
13. Add 5 quarts of SAE 90W EP gear lubricant to chassis.
14. Test operation.

POWER TAKE-OFF (PTO)

REMOVAL AND DISASSEMBLY

1. Disconnect both brade rods at brake levers.
2. Disconnect two speed shifter rod from shifter weldment.
3. Disconnect attachment throwout link from attachment throw out shifter arm.
4. Disconnect the high-low link from the high-low clutch link.
5. Remove the clevis pin holding the forward-reverse clutch rods in the forward-reverse clutch link.
6. Remove cable from positive post of battery.
7. Turn off fuel and separate fuel line.
8. Separate wiring harness at connector.
9. Separate tail light wire (if so equipped).

Differential spider

Gravely

Differential removal

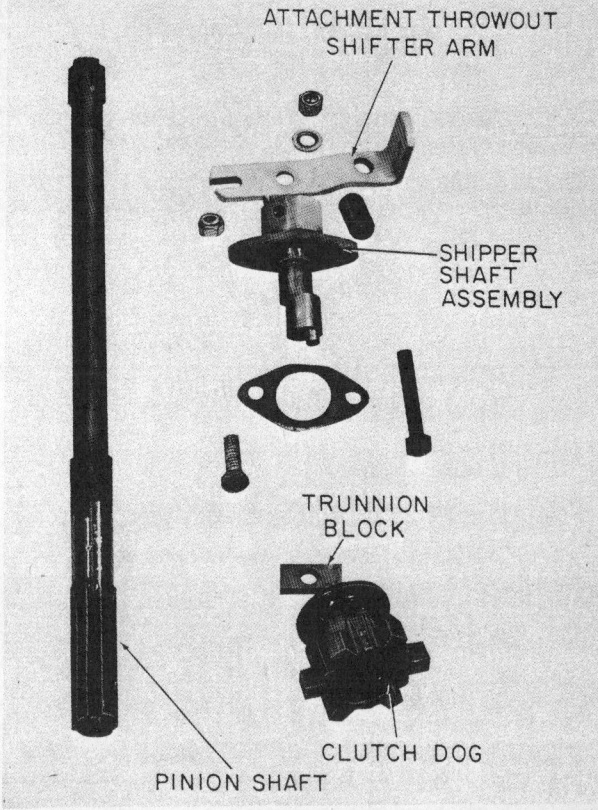

PTO components

10. Remove wire from starter motor.
11. Disconnect throttle and choke cables.
12. Support tractor frame just in front of the PTO and remove the two bolts and two nuts securing chassis to frame.
13. Roll chassis back from frame.
14. Lower engine to floor.
15. Remove advance casting cover plate/dog clutch assembly.
16. Remove the attachment throw out shifter arm/shifter shaft assembly.
17. Remove clutch dog from pinion shaft.
18. Remove pinion shaft.
19. Carefully check all parts for wear or binding. Sight down splines of pinion shaft to see if the shaft is twisted.
20. Carefully check the bearings of the advance casting cover plate for wear.
21. If there was no gasket under the shipper shaft guide on the advance casting, be sure to reassemble using a gasket.
22. If the PTO seemed not to be engaging fully, reassemble using two attachment gaskets between the advance casting and the cover plate assembly.

ASSEMBLY AND INSTALLATION

1. Install pinion shaft.
2. Install clutch dog on pinion shaft.
3. Install attachment throw out shifter arm/shifter shaft assembly.
4. Install advance casting cover plate/dog clutch assembly.
5. Roll chassis in frame, use care to be sure controls are in the proper locations to be connected.
6. Secure the frame in place with the two bolts and two nuts. Torque to 65 ft.lb.
7. Connect the throttle and choke controls.
8. Attach wire to starter motor.
9. Connect wiring harness at connector.
10. Connect tail light wire (if so equipped).
11. Connect cable to positive post of battery.
12. Connect fuel line, turn on fuel.
13. Put in forward-reverse clutch rods in place in the forward-reverse clutch link. Secure with clevis pin.
14. Connect the attachment throw out link to the attachment throw out shifter arm.
15. Connect the two-speed shifter rod to the shifter weldment.
16. Connect both brake rods to brake levers.

ENGINE

The following engines are used:
Models 10, 10A: Kohler K241; 12: Kohler K301
424: Kohler K241 or Onan NB
430: Kohler K301 or Onan NB
432: Kohler K321
450: Onan CCKA
500/5000: various Kohler engines

For complete engine service, see the appropriate engine part of the Engine Unit Repair section.

REMOVAL

NOTE: Make stubs out of two 7/16 in. bolts by removing heads. Grind wrench flats or cut a screwdriver slot in the unthreaded end of stud. This will aid removal and installation of the engine.

1. Drain oil from transmission.
2. Disconnect choke control from carburetor.
3. Disconnect governor control from engine.
4. Shut off fuel at tank and separate fuel line.

--- CAUTION ---
A small amount of fuel will be spilled as the line is separated.

5. Remove cable from positive (+) battery post.
6. Remove cable from starter motor.
7. Separate wiring harness at connector(s).
8. Remove the bolts securing the rear hitch to the engine oil base. Swing rear hitch to one side.
9. Remove the two top bolts securing the engine to the chassis. Replace with two studs.
10. Remove the lower lower bolts securing the engine to the chassis.
11. Slide engine back off chassis.

463

Gravely

Engine removal

INSTALLATION

1. Install special studs in top holes of engine adaptor plate. Be sure gasket is in place.
2. Slide engine in place on transmission, slide studs through top holes of the chassis casting.
3. Install two lower bolts securing engine to chassis. Do not tighten.
4. Swing rear hitch in place and install bolts securing hitch to engine oil base.
5. Remove studs from two top holes and install bolts. Tighten all bolts securely. Torque to 35 ft.lb.
6. Connect wiring harness at connector.
7. Install cable on starter motor.
8. Connect fuel line. Turn on fuel.
9. Connect governor control to engine.
10. Connect choke control to carburetor.
11. Replace + cable on battery.
12. Add five quarts SAE 90W EP gear lubricant to transmission.

800 Series, Model 810, 812, 816S, 817

FRONT AXLE

REMOVAL AND INSTALLATION

1. Raise and support the front of the tractor on jack stands.
2. Disconnect the tie rods from the steering arms.
3. Remove the axle pivot pin.
4. Roll the axle from under the tractor.
5. Installation is the reverse of removal.

STEERING

King Pins and Knuckles

REMOVAL AND INSTALLATION

1. Raise and support the front end on jack stands.
2. Remove the front wheels.
3. Disconnect the tie rods at the steering arms.
4. Slide the kingpins down out of the axle.
5. Loosen the clamp bolt and remove the steering arm.
6. Remove the knuckles.
7. Installation is the reverse of removal.

Steering Gear

REMOVAL AND INSTALLATION

1. Remove the hood.
2. Remove the fuel tank.
3. Remove the battery.
4. Remove the steering wheel snapring.
5. Disconnect the tie rod ends from the steering arm.
6. Unbolt the steering gear box from the frame.
7. Lower the steering gear down and out of the frame.
8. Installation is the reverse of removal. Pack the gear box with multipurpose EP chassis lube.

ADJUSTMENTS

1. Turn the wheels completely right and tighten the left side adjusting nut.
2. Turn the wheels all the way left and tighten the right side adjusting nut.

NOTE: The hood must be on securely while making adjustments.

BRAKES

Brake Lining

REPLACEMENT

1. Remove the brake band clevis and mounting bolt.
2. Remove the brake band.
3. Install the new band, clevis and mounting bolt. Adjust as described below.

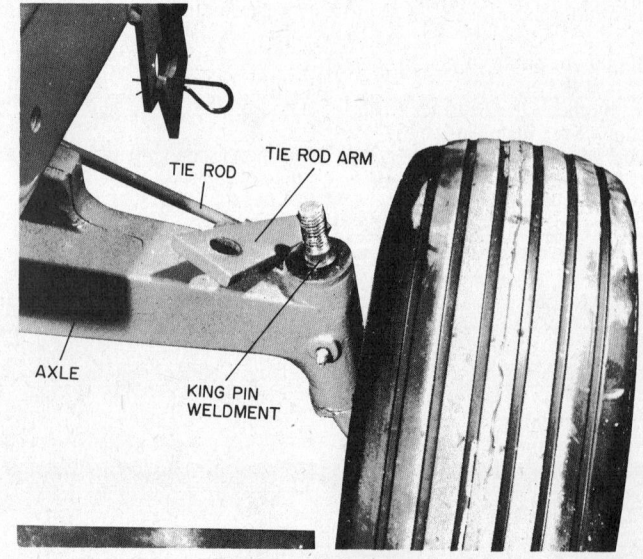
Front axle and king pins

Gravely

Steering system mounting points

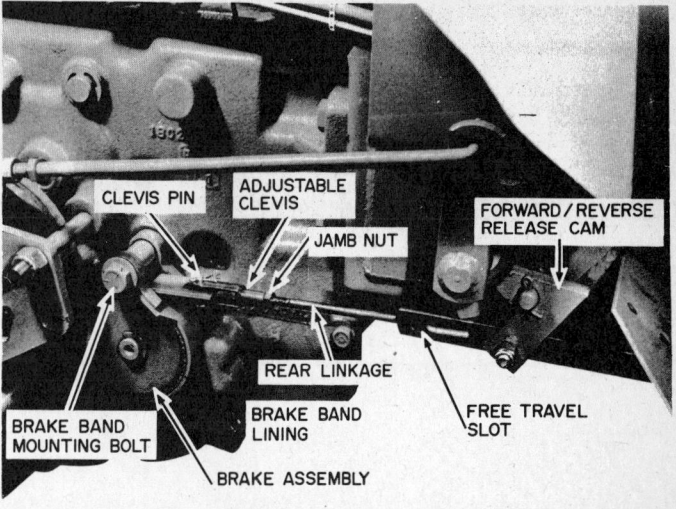

Brake end clutch

ADJUSTMENT

1. Park the tractor on level ground and block the wheels.
2. Place the direction control in FORWARD.
3. Slowly depress brake pedal until control lever returns to neutral. Engage stop rod as explained in Operation section.
4. Observe free travel of brake linkage rod in slot. (Braking action should start at this point).
5. Remove clevis pin, loosen jam nut and turn clevis until the free movement is taken up in slot. Reconnect clevis and tighten jam nut.
6. Repeat steps 2, 3 and 4. Braking action should not start until direction control is returned to neutral.

NOTE: Optional individual rear wheel brakes are available for 800 series tractors. Contact your Gravely dealer for more information.

CLUTCH

REMOVAL

1. Remove the clutch rod from the clutch cam.
2. Remove the two nuts holding clutch plate. Remove the clutch plate, carefully check the clutch rollers and clutch pin. The rollers should turn free on the pins.
3. Remove the clutch assembly from the clutch shaft.
4. Check shaft for rust or debris buildup.

DISASSEMBLY

1. Remove the spring and snapring from the bore of clutch disc assembly.
2. Compress the Belleville spring washers and remove the snapring.
3. Disassemble the clutch assembly.
4. Check all parts for wear, damage or debris buildup.

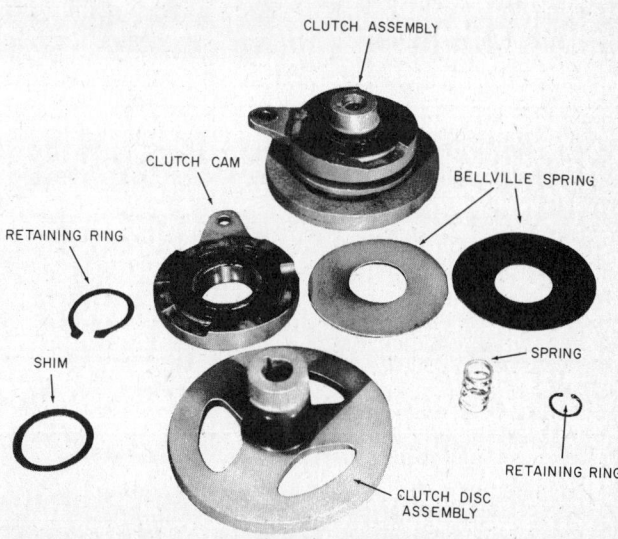

Clutch components

465

Gravely

ASSEMBLY

1. Install parts in the following sequence on clutch disc assembly: Belleville spring washer (cup up); Belleville spring washer (cup down); shim, clutch cam with bearing (notches up).
2. Compress Belleville spring washer and install snapring.
3. Install spring and snapring in bore of clutch disc assembly.

INSTALLATION

1. Be sure shaft is clean and key is in place, slide the clutch assembly on shaft.
2. Place the clutch plate over the studs being sure the clutch pin and rollers remain in place.
3. Tighten the two nuts down until they are on or about .050 in. gap all around.

TRANSMISSION

REMOVAL

NOTE: Remove engine as covered later.

1. Remove hydraulic lines from the pump and drain hydraulic reservoir, if so equipped.
2. Remove the forward clutch rod from the right clutch cam.
3. Remove the reverse clutch rod from the left clutch cam.
4. Remove the two-speed rod from the shifter arm.
5. Remove the shifting rods from the 1-3 and 2-4 shifter arms.
6. Remove the PTO rod from the PTO lever.
7. Remove the lift rod from the cross shaft weldment.
8. Remove the brake rod from the brake band assembly.
9. Remove the bolts securing the cross shaft weldment to the transmission case.
10. Raise the cross shaft weldment up to clear the transmission.
11. Block tires from rolling and support transmission case in front of wheels.
12. Remove the bolts and nuts securing the transmission case to the frame.
13. Lift and roll frame forward and out of the way.

DISASSEMBLY

1. When the transmission has been disconnected and clear of frame, turn the transmission up on the left wheel.
2. Remove right wheel from hub.
3. Remove the E-ring securing the wheel hub to the axle. Lift hub and Woodruff key from the axle.
4. Remove the four bolts securing the axle bearing retainer and lift retainer from axle.
5. Remove the nuts from the clutch support studs. Lift off the clutch plate. There are three pins and rollers on the clutch plate. Be sure these are not lost.

Hub removal

Removing bearing retainer

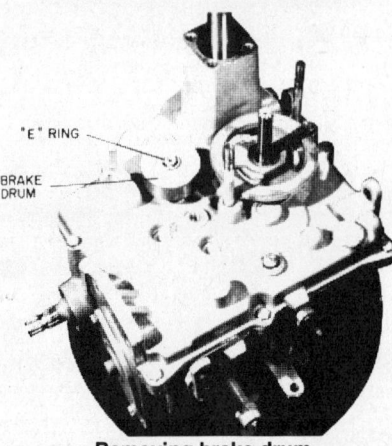

Removing brake drum

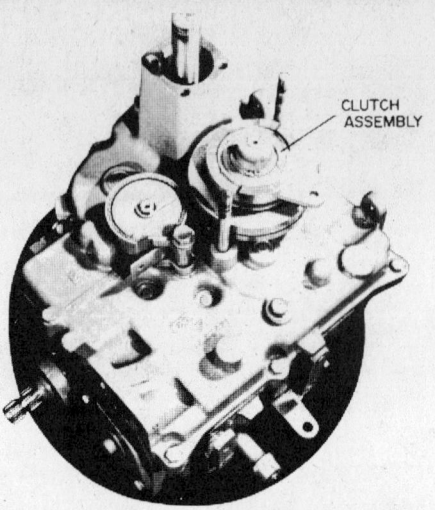

Clutch assembly removal

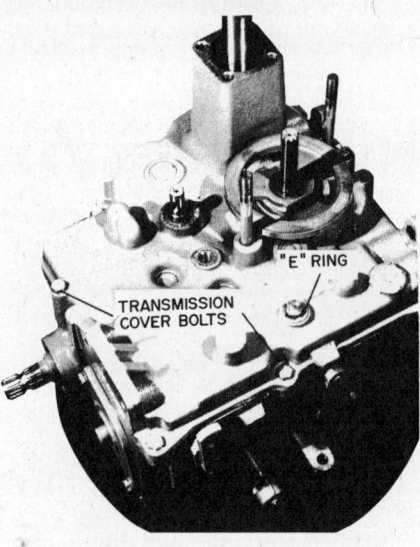

Cover and E-ring removal

6. Lift clutch assembly, spring and washer from shaft.
7. Remove the retaining ring from the splined shaft.
8. Remove the brake band mounting bolt, and lift off brake band.
9. Remove the E-ring securing brake drum to shaft. Remove drum and Woodruff key from shaft.
10. Remove the E-ring shown in the accompanying illustration. Remove the bolts securing the transmission cover to transmission case and lift the cover from the case. If difficulty is experienced in removing the cover, tap the protruding shafts with a soft hammer.
11. When cover is removed, carefully examine the transmission for damage or abnormalities. When removing the shafts, examine the splined shafts for burring and chipped or broken gears. Shifting ranges with the tractor in motion will cause the gears to burr and not mesh. If this happens the tractor will have no power in forward, but power in reverse.
12. When removing the gear assemblies, it is good practice to keep the gears, shaft, race and bearings on the shaft

being removed where possible. When removing the shafts note the beveled gears. (Bevel meshes with bevel.) Carefully examine the assembly for burring, chipped gears, worn bearings and race.

13. The shafts in the accompanying illustration are numbered for identification only. The shafts are pictured and described in removal sequence to give the least amount of difficulty in removing from the transmission case. The shafts may be removed in any sequence as required for repair of the transmission.

14. To remove shaft #6, first remove the splined shaft. Swing the shifting fork to a position where the remaining gears can be removed. Place these back on the shaft in order. There is a .0900 in. race against the transmission case. Put this on the shaft.

15. Before removal of the high-low (2 Speed Gear) from shaft #1, the high-low shifting fork will have to be removed. Remove the plug using caution: the plug compresses a spring. Remove the spring and ball from the transmission case. Remove the shifting fork and high-low gear from shaft #1.

16. To remove the #2 shaft, the reverse clutch assembly will have to be removed, using the same procedure as described for removal of the forward clutch assembly.

17. Remove the E-ring securing the PTO lever to the shaft. Remove PTO lever and Woodruff key from shaft.

18. Remove the shaft and yoke from the transmission case.

19. Remove the retaining ring from the PTO shaft. Using a punch and hammer, gently tap around the bearing and remove from the transmission case. The bearing is removed to the outside of the case.

20. Remove the two bolts securing the bearing cap assembly to the transmission case. Remove the bearing cap assembly.

21. Remove the PTO assembly from transmission case.

22. To remove the four-speed shifting forks, remove the two plugs. Use caution, the plugs compress two springs. Remove the top plug first (as shown) and the two springs and two balls behind plug.

23. Lift out the top shifting fork. Before removing the bottom plug and shifting fork, remove the interlock pin from the transmission case located between the two shifting forks.

24. Remove the bottom plug, two springs and two balls. Lift out bottom shifting fork.

25. To remove the shifter weldment, remove the roll pin from the outside shifter arm. Remove the shifter arms. The inside arm has a Woodruff key. With a punch and hammer, tap around the brass bushing from inside to out and remove the bushing from the transmission case. Pull the shifter weldment through the transmission case.

26. Remove the transmission case prior to the removal of the PTO spur gear and PTO shaft.

27. To remove the PTO spur gear and PTO shaft, block the spur gear from turning and remove the locknut from the PTO shaft. Remove the spur gear from the shaft

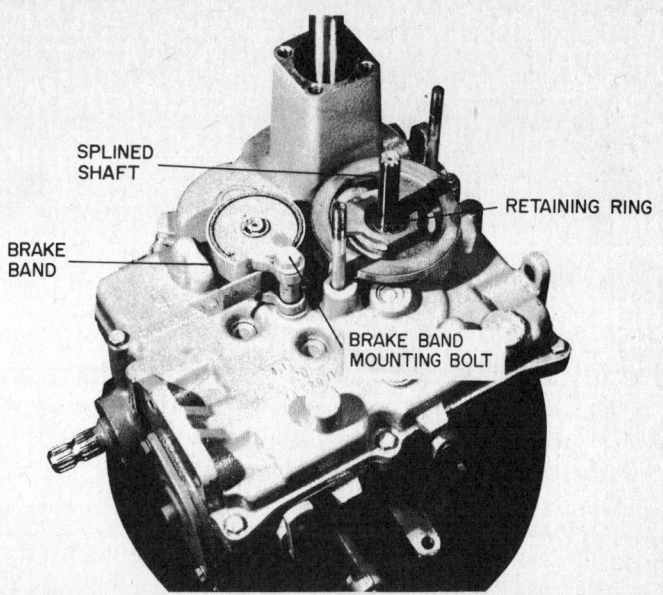

Removing brake band

and remove the PTO shaft from the transmission case.

Differential Removal

1. Lay the transmission case over, and block so left wheel is clear.
2. Remove the left wheel, wheel hub and axle bearing retainer from the axle as described in removal of the right side.
3. Remove the differential assembly from the transmission case.
4. Carefully examine the differential gears and bearings for wear or damage.
5. Remove the four nuts from the four bolts through the differential.
6. Separate the differential cap from the final drive gear.
7. Carefully examine the assembly for wear or damage. Also, check the axle for splitting.
8. Replace any damaged or worn parts and reassemble.

PTO Disassembly

1. Holding the shifting collar assembly, press the exposed end of the PTO shaft and compress the spring. Remove the snapring on the PTO shaft inside the clutch cup.
2. Remove the clutch cups, clutch cone assembly and keys.
3. Exercising extreme caution, release the tension on the spring.
4. Remove the spring and other components from the shaft.
5. Carefully check part for wear or damage. Replace any part showing wear or damage.

ASSEMBLY AND INSTALLATION

1. Before reassembling the transmission carefully check the bearings and case for damage. Replace any bearings that are worn or damaged.

2. Coat the machined areas and bearings in the case with grease.
3. Install the differential assembly in the transmission case.
4. Install the left axle bearing retainer, wheel hub, Woodruff key and wheel. Turn the transmission up on the left wheel.
5. Install the PTO assembly in the transmission case.
6. Install the bearing cap assembly.
7. Install the retaining ring on the PTO clutch cup in the bearing assembly.
8. Install the shifting yoke and shaft, making sure the heavier prongs on the yoke are toward the PTO cups.
9. Install PTO lever, Woodruff key and E-ring.
10. Install the gear and shaft assemblies. Install high-low shifting fork when installing #1 shaft.
11. Install detent ball, spring and plug securing the high-low shifting fork.
12. Install transmission case cover.
13. Install transmission case cover bolts and E-ring.
14. Install Woodruff key, brake drum and E-ring securing brake drum.
15. Install brake band and mounting bolt.
16. Install retaining ring on splined clutch shaft.
17. Install clutch assembly with cam toward top of transmission.
18. Install clutch plate, making sure the three (3) rollers and pins are in place. Install clutch plate with the roller on the long edge toward the top of the transmission.
19. Install the bearing retainer and four (4) bolts.
20. Install Woodruff key, wheel hub, E-ring, wheel and bring transmission down on both wheels.
21. When transmission has been connected to frame and engine installed, adjust both clutch assemblies to .040 in. all around. Fill transmission with oil.

Gravely

Shafts numbered for identification purposes

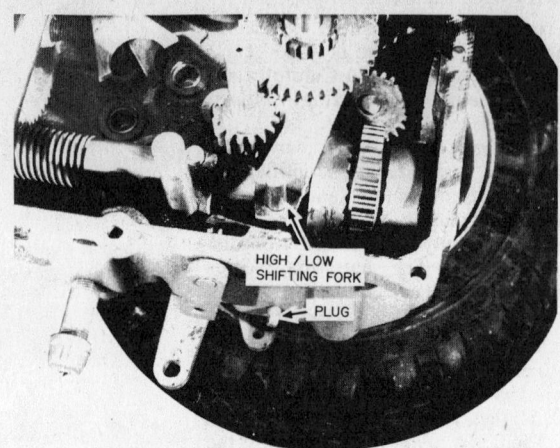

Removing high-low shift fork

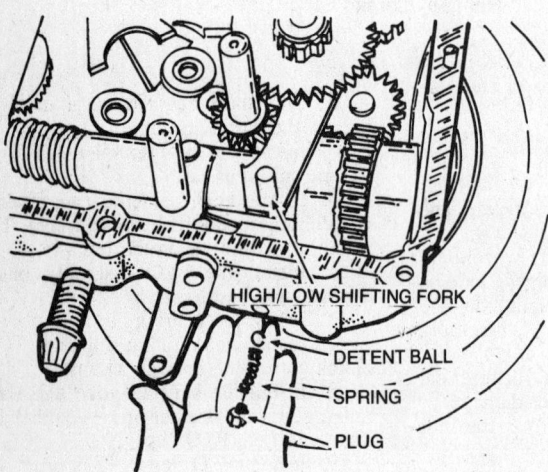

Removing high-low detent spring and ball

PTO shaft bearing

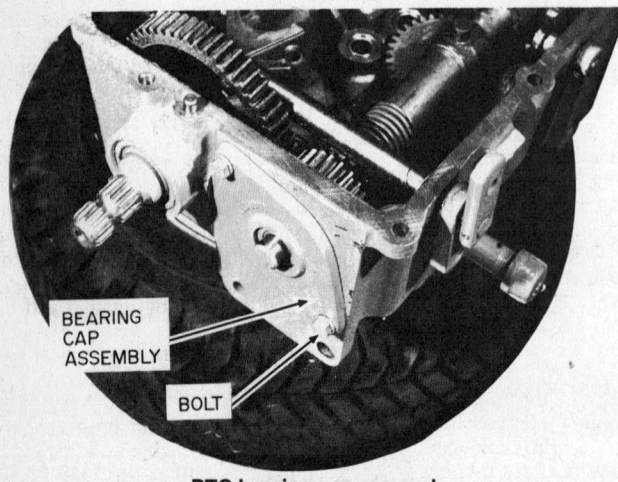

PTO bearing cap removal

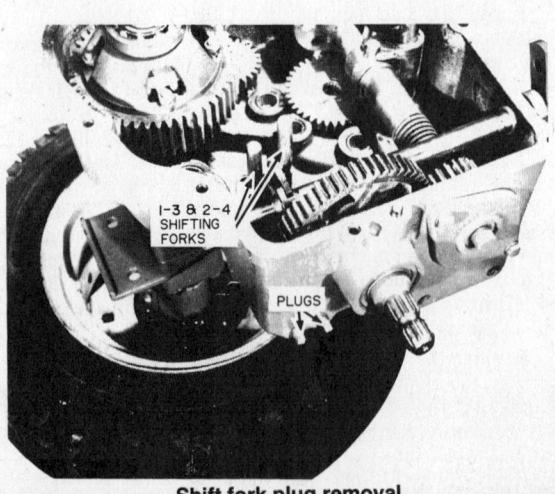

Shift fork plug removal

Gravely

Engine adaptor and input gear

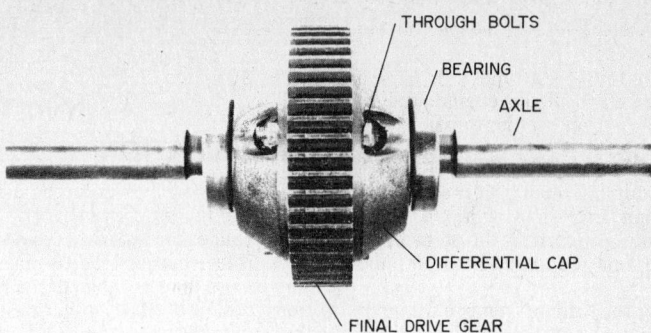

Differential unit

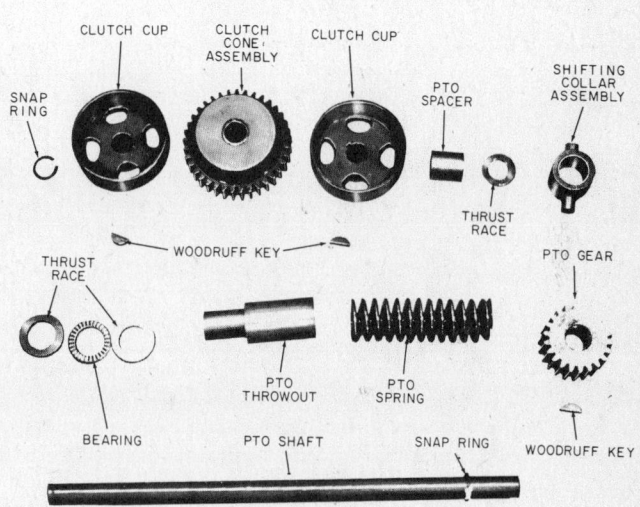

PTO components

PTO Clutch Adjustment

1. It is necessary that the PTO lever have free travel. If the free travel is too great, the clutch will not disengage and if the free travel is too little, the clutch will activate too hard and operation will be affected. With the PTO lever engaged (IN POSITION) there would be 1⅜ in. to 1½ in. of free travel between the lever and the back of the slot in the cover plate.
2. When the free travel reaches 1 in., adjust the clevis on the PTO rod until there is 1⅜ in. to 1½ in. of free travel of the lever again.

NOTE: When operating the tractor, without a powered attachment, engage the PTO to increase oil circulation within the transmission.

ENGINE

Engines used in these models are:
Model 810: Kohler K241
812: Kohler K301
816: Briggs and Stratton 16 hp
817: Onan CCKA

For complete engine service, see the appropriate engine part of the Engine Unit Repair section.

REMOVAL

1. Shut off fuel at tank.
2. Drain oil from transmission.
3. Remove ground cable from battery.
4. Remove the choke and throttle controls from linkage at engine.
5. Separate fuel line at connection.
6. Remove wire from starter motor.
7. Separate wiring harness at connector.
8. Remove rear hitch from tractor.
9. Remove 2 bottom bolts securing the engine and engine adaptor to the transmission case.

Engine adaptor thrust bearing components

Gravely

10. Next remove the 2 top bolts securing the engine and engine adaptor to the transmission case.

11. Using a floor jack under the engine oil pan, carefully roll the engine straight back and away from the transmission.

Engine Adaptor and Input Gear

The transmission input gear is secured to the crankshaft extension with a key. A thrust bearing between 2 thrust races is located between the input gear and the adaptor.

To remove the adaptor, remove the input gear, thrust bearing and races. Remove the 3 bolts and socket head screw securing the adaptor plate to the engine block. The O-rings and gasket should be replaced prior to reinstallation.

INSTALLATION

1. Support the engine on a floor jack, raise it to the proper level.
2. Slide the engine straight into the back of the transmission. It may be necessary to turn the PTO shaft at the front of the transmission by hand to get the gear teeth lined up so that the input gear will slide in.
3. Secure in place with the two upper bolts.
4. Install the two lower bolts.
5. Connect wiring harness at connector.
6. Connect cable to starter motor.
7. Connect fuel line.
8. Connect choke and throttle controls to linkage at engine.
9. Install rear hitch.
10. Add oil transmission to full lever. Capacity is 6 quarts.
11. Install ground cable on battery.
12. Turn on fuel tank.

International Harvester

INDEX

CUB CADET MODELS 70 & 100
Front Wheels and Axle 472
Steering 472
Brakes 474
Splitting and Recoupling 475
Clutch 475
Transmission and Differential .. 476
Creeper Drive 477
Reduction Drive 479
Rear Axle 480
Engine 480

CUB CADET MODELS 71, 102, 122, 123
Front Wheels and Axle 482
Steering 482
   Steering Knuckle 482
   Steering Gear 483
Brakes 484
Clutch 484
Splitting and Recoupling 485
Transmission and Differential .. 488
Hydrostatic Transmission 489
Reduction Drive 493
   Overhaul 494
Creeper Drive 498
Rear Axle 499
Power Take-Off 500
Engine 502

CUB CADET MODELS 72, 104, 105, 124, 125
Front Wheels and Axle 503
Steering 504
Brakes 505
Splitting and Recoupling 506
Clutch 508
Rear Power Take-Off 508
Reduction Drive 508
Creeper Drive 508
Transaxle 508
Differential 508
Hydrostatic Transmission 508
Rear Axle 508
Front PTO 508
Engine 509

CUB CADET MODELS 73, 106, 107, 126, 127, 147
Front Wheels and Axle 509
Steering 510
Brakes 511
Clutch 512
Splitting and Recoupling 514
Transmission and Differential .. 517
   Models 73, 106, 126 517

Reduction Drive 519
Hydrostatic Transmission 521
Rear Axle 521
Front Power Take-Off 521
Rear Power Take-Off 521
Electric Power Lift 521
Engine 523

CUB CADET MODELS 86, 108, 109, 128, 129, 149, 169, 800, 1000, 1200, 1250, 1450, 1650
Front Wheels and Axle 524
Steering 524
Brakes 525
   External System 525
   Internal System 526
Clutch 526
Splitting and Recoupling
   Hydrostatic Drive 527
   Gear Drive 529
Reduction Drive 529
Standard Transmission and Differential 530
Hydrostatic Transmission and Differential 531
Rear Axle and Housing 532
Creeper Drive 532
Mechanical Front Power Take-Off 532
Electric Front Power Take-Off .. 534
Rear Power Take-Off 536
Electric Power Lift 536
Engine 536

CUB 154, 184, AND 185 LO-BOY
Front Axle and Wheels 536
Steering 537
Brakes 539
Clutch 540
Splitting and Recoupling 543
Transmission and Differential .. 545
Creeper Drive 547
Final Drive 548
Power Take-Off 550
Electric Power Take-Off 552
Engine 553
   R&R 553
   Cylinder Head 554
   Valves 554
   Pistons, Rings, Connecting Rods 555
   Timing Cover and Gears 558
   Camshaft 559
   Crankshaft 559
Engine Lubrication 560

Fuel System 561

MODELS 482 & 1100
Front Wheels and Axle 563
Steering 564
Brakes 565
Drive Belt and Clutch Spring ... 565
Right Angle Drive 566
Transaxle 567
Differential 569
Electric Front Power Take-Off .. 570
Engine 571

MODELS 582, 682, 782, 982
Front Wheels and Axle 571
Steering 572
   Models 582, 582, 782 572
   Model 982 574
Internal Brakes 575
External Brakes 577
Clutch 578
Splitting and Recoupling 579
Reduction Drive 579
Transmission and Differential .. 580
Creeper Drive 581
Hydrostatic Transmission 582
Rear Axle 582
Front Power Take-Off 583
Rear Power Take-Off 583
Engine 584

MODEL 284
Front Axle 584
Steering 585
Brakes 586
Clutch 587
Splitting and Recoupling 587
Transmission 589
Differential 594
Rear Axle 595
Power Take-Off 595
Hydraulic System 597
Engine 598
   R&R 598
   Manifolds 598
   Cylinder Head 598
   Valves 598
   Pistons and Connecting Rods .. 598
   Timing Chain and Sprockets .. 599
   Camshaft 599
   Crankshaft 600
Engine Lubrication 600
Cooling System 600
Fuel System 601

INTERNATIONAL HARVESTER
Cub Cadet Models 70 and 100

FRONT AXLE AND WHEELS

Front Axle

REMOVAL AND INSTALLATION

1. Disconnect the drag link ball joint from the drag link arm.
2. With the front of the tractor frame supported by a suitable stand, drive out the retaining pin from the front of the axle pivot pin (1).
3. Remove the pivot pin (1). The front axle (2) is now free of its mounting and can be removed.
4. Apply chassis lubricant liberally to the axle pivot pin and its bore in the axle.
5. Position the axle in its support bracket channel, align the pivot pin holes and insert the pin.
6. Align the retaining pin holes (through the front of the pivot pin and through the front collar of the support bracket) then drive the retaining pin through both parts.

Wheels

REMOVAL AND INSTALLATION

1. Lock the brake and block the rear wheels. Jack up the front axle.
2. Remove the capscrew and flat washer from the outer end of the front spindle.
3. Slide the wheel and bearings from the spindle.

NOTE: The bearings are a press fit in the wheel and a slip fit on the spindle.

4. Wheel bearings can be driven from the wheel hub with a hammer and long drift punch. Drive from the inside toward the outside.
5. Inspect the entire wheel and hub for weld separation, split hub tube and rim bending.
6. Bearings should be inspected for wear, seizure and seal condition.
7. If the bearings were removed, press in new ones. Be sure force is directed to the outer race only when being pressed in.

NOTE: These wheels may be used on older model cub cadet tractors by pressing in service bushings instead of ball bearings.

8. Slide wheel and bearing assembly over the spindle and secure with capscrew and flat washer.
9. If excessive end play exists, place a sufficient thickness of shim washers (¾ in. ID) over the outer end of the spindle and between the retaining washer and wheel bearing.

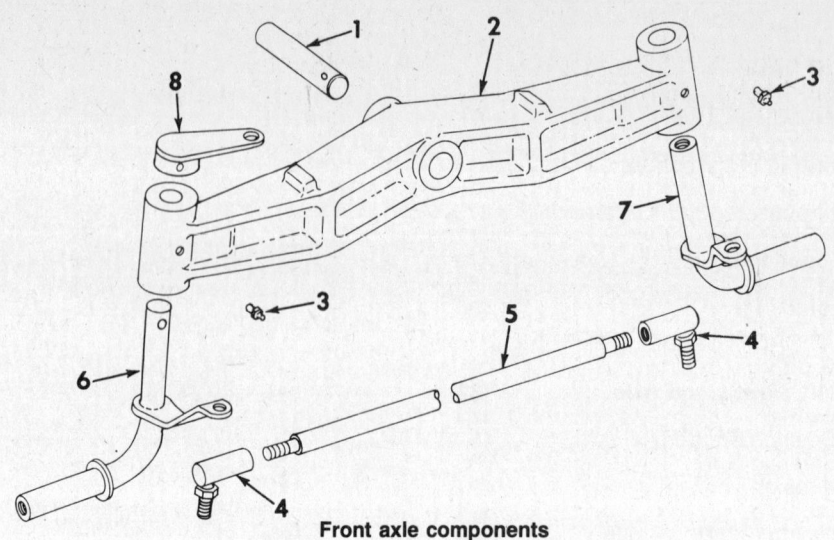

Front axle components

STEERING

Steering Knuckle

REMOVAL AND INSTALLATION

1. Lock the brake, jack up the front of the tractor and support it on a suitable stand.
2. Remove the front wheels.
3. Disconnect the tie rod ball joints (4) from left steering knuckle (6) and right steering knuckle (7).
4. Using a pin punch and hammer, drive out the coiled spring pin from the drag link arm (8) and steering knuckle (6).
5. Remove the steering knuckle (6) from the axle (2).
6. Remove the capscrew and flat washer from the upper end of the steering knuckle (7).

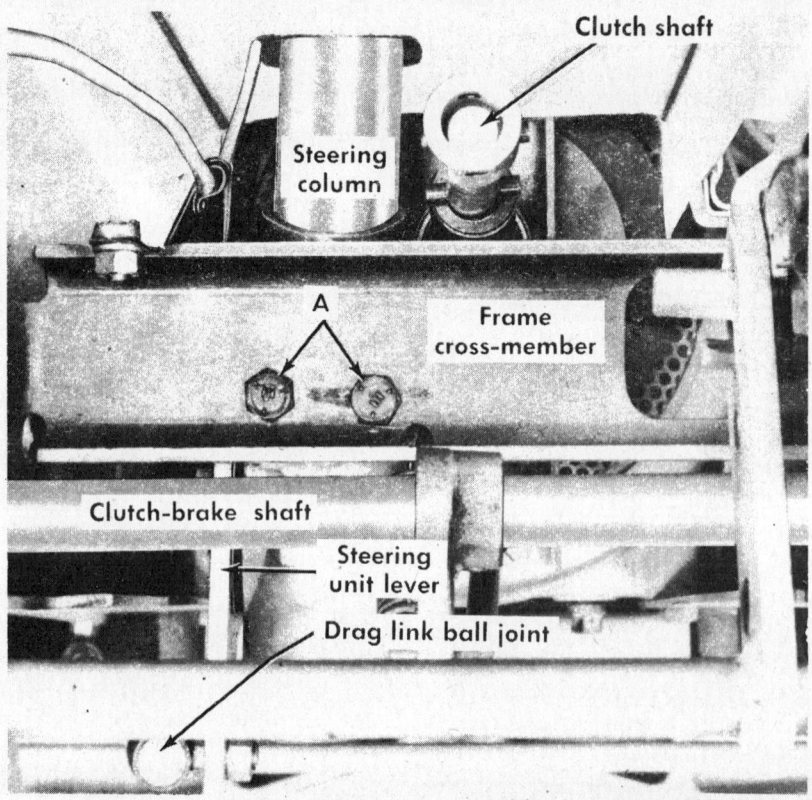

Underside view of the steering support and frame cross member

International Harvester

7. Remove the steering knuckle (7) from the axle (2).

8. Apply chassis lubricant liberally to the steering knuckle arm thrust surface and the bottom of the axle ends.

9. Insert the righthand knuckle pin (7) in its respective bore in the axle and secure with the capscrew and flat washer.

10. Insert the lefthand knuckle pin (6) in its bore in the axle and position the drag link arm (8) on the pin.

11. Secure the drag link arm (8) and knuckle (6) with the coiled spring pin.

NOTE: Spring pin must extend through the drag link arm an equal amount on each side.

12. Install the tie rod (5).
13. Install the front wheels and check toe-in adjustment.

Steering Gear

REMOVAL

1. Remove the steering wheel.
2. Remove the felt seal, retainer, bearing and bearing retainer from the upper end of the steering column.
3. Remove the drag link rear ball joint from the steering unit lever.
4. Remove capscrews "A" from the frame cross member and steering unit.
5. Lower the steering column assembly through the instrument panel pedestal and grommet.

DISASSEMBLY

1. Secure the steering lever and bolt in a vise.
2. Remove the adjusting plug.
3. Remove the lever bolt jam nut, adjusting nut, and bronze washer.
4. Slide the column and housing assembly away from the lever, bolt and cam follower.
5. Remove the steering cam and bearings from the housing.
6. Remove the bearing race retainer snaprings.

INSPECTION AND REPAIR

1. Wash all parts in cleaning solvent, then dry thoroughly.
2. Inspect the cam follower for wear (flat spots).
3. Inspect the cam ends, bearings and races for wear, roughness and pitting.
4. Inspect the cam grooves for wear, roughness and galling.
5. Inspect the housing for cracks and stripped threads.
6. Inspect the upper bearing (nylon bushing) for wear or damage.

REASSEMBLY AND ADJUSTMENT

1. Thoroughly coat the cam ends, balls and races with chassis lubricant.
2. Install the balls and races on the cam ends with their retaining snaprings.
3. Thoroughly coat the cam and bearings with chassis lubricant then install them into the housing and column assembly.

NOTE: Be sure the races enter the housing squarely and are not cocked.

4. Install the adjusting plug. Screw the plug inward until end play (of the cam) is removed and the cam turns free.
5. Stake the plug by inserting a small center punch through the housing holes and spread the plug threads to keep the plug from loosening.
6. Fill the housing with chassis lubricant.

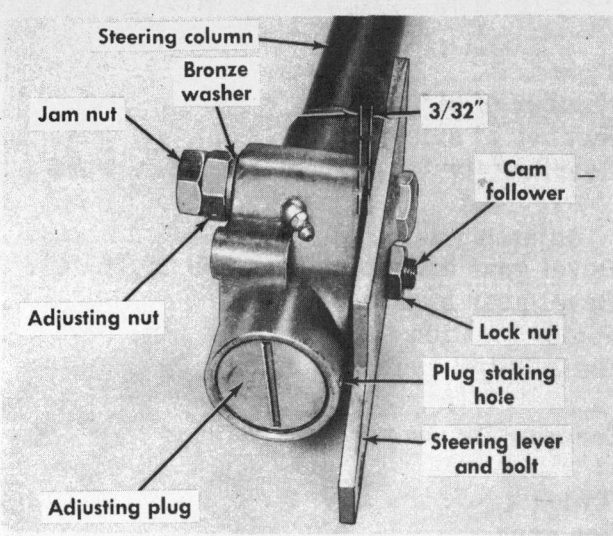
Steering gear

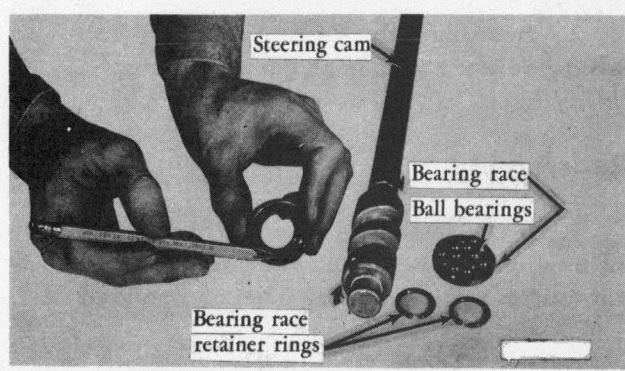
Inspecting steering gear components

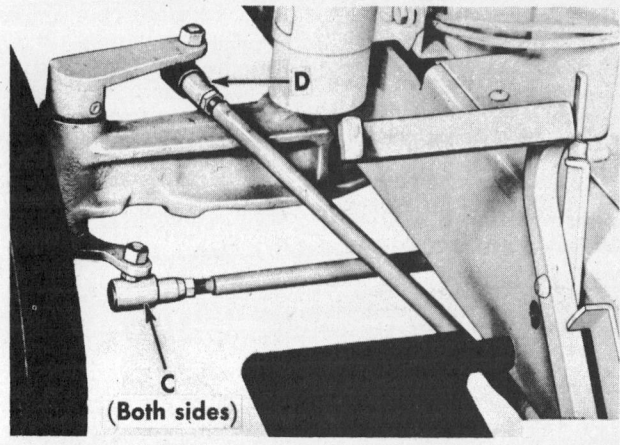
Tie rod and drag link ball joints

7. Loosen the cam follower locknut, then back out the cam follower one turn.
8. Install the seal, retainer and lever-bolt assembly to the housing.
9. Install the bronze washer and adjusting nut. Tighten the adjusting nut sufficiently to provide good seal compression. Secure with the jam nut. Tighten jam nut to 40 ft.lb. Lubricate the lever-bolt at the fitting.
10. Center the steering cam by rotating the steering shaft half way between full right and full left turn.
11. Adjust the cam follower inward to

473

International Harvester

eliminate backlash, then tighten locknut to 60 ft.lb. Turn steering shaft full right and left to check for binding.

12. Replace the steering assembly in the tractor chassis. Secure with two capscrews through the frame cross member.
13. Connect the drag link.
14. Install the upper column bearing, retainer, felt seal and retainer.
15. Replace the steering wheel and secure with nut.
16. Adjust the drag link "D" to proper length to place front wheels in the straight ahead position when the steering assembly is centered.
17. Adjust tie rods "C" to provide 1/32 in. to 1/8 in. toe-in.

BRAKES

Brake Unit

REMOVAL

Tractors equipped with a "creeper" attachment will require "splitting", and the creeper removed before complete brake service can be performed.

1. Drain the transmission lubricant.
2. Remove the brake adjusting screw "C" and jam nut "B" from its lever.
3. Remove the brake lever, pivot pin and push rod.
4. Remove the reduction housing front cover plate and slide it forward on the clutch shaft.
5. Remove the reduction gear from the front of the transmission mainshaft.
6. Move the gear upward and the bottom of the gear forward to clear the cover screw bosses as the gear is lifted from the housing.
7. Slide the brake disc forward on the countershaft as the front lining and retainer are moved forward in their bore.

NOTE: Both linings and the disc can be removed without removing the front lining retainer; however, removal of the retainer is recommended for inspection and replacement of the retainer O-ring.

INSPECTION

1. Inspect the control rods and levers for wear at their connecting pivot points.
2. Inspect the linings and disc for wear.
3. Inspect the disc hub splines for wear.
4. Check the splines on the countershaft for wear.
5. Check the pedal return spring ends for wear.

ASSEMBLY AND ADJUSTMENT

1. Clean the brake cavity and lining recess in the reduction housing.
2. Place a small quantity of grease in the rear brake lining recess in the reduction housing then insert the lining.
3. Install the disc on the countershaft and slide it rearward against the rear lining.

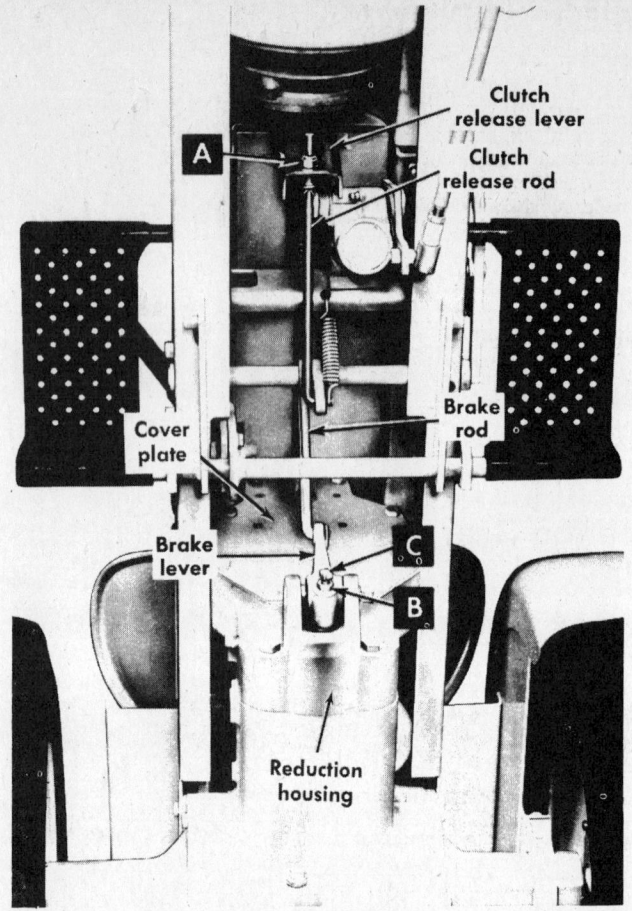

Brake and clutch linkage

4. Install a new O-ring on the front lining retainer, lubricate the retainer and O-ring then push them into the retainer bore in the reduction housing.
5. Install the front lining in the retainer lining recess and push the retainer and lining rearward against the disc.
6. Install the reduction gear on the transmission mainshaft and secure with flat washer, lockwasher and capscrew. Tighten capscrew to 55 ft.lb.
7. Install a new cover gasket, then replace the cover plate.
8. Be sure the ball is in place in the front lining retainer then replace the push rod, lever, pivot pin, adjusting screw and locknut.
9. Fill transmission to proper level with Hy-Tran fluid or SAE 30 engine oil.

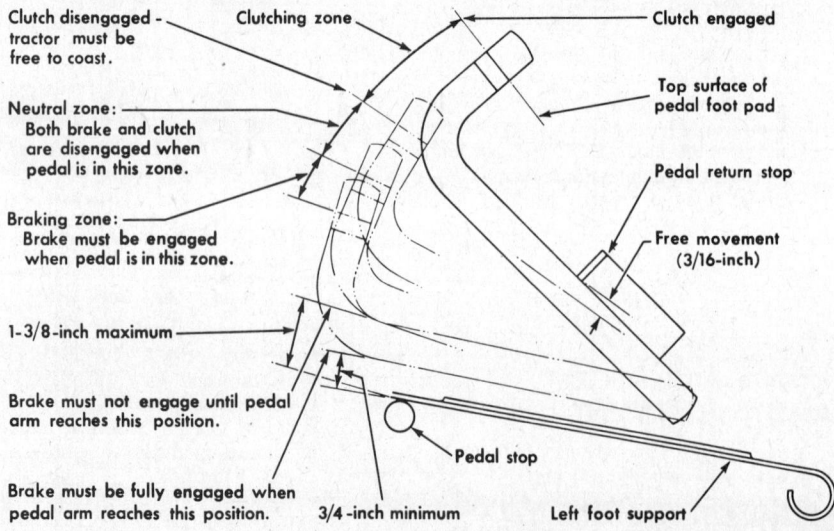

Brake and clutch adjustment

International Harvester

SPLITTING THE TRACTOR

1. Remove the fenders and their support brackets (if tractor is so equipped).
2. Remove the seat and its support bracket.

NOTE: Disconnect tail light wire at junction under seat support (if tractor is so equipped).

3. Remove the clutch shaft coupling rear pin. (Connects coupling to creeper or reduction driveshaft.)
4. Drive a small wooden wedge between the front axle and frame on each side.
5. Block both front wheels.
6. If tractor is equipped with creeper attachment, shift the lever forward then remove the lever knob and the breather.
7. Disconnect the brake rod from the brake lever.
8. If tractor is equipped with three-point hitch, remove the lift bar and its attaching plate.
9. Place a jack under the implement lift handle cross shaft to support the frame.
10. Remove three capscrews from each side of frame.

NOTE: Capscrews hold frame to the rear axle, transmission and reduction drive housings.

11. Push down on the drawbar and pull rearward on the transmission shifter lever.
12. Move the transmission-differential-rear end assembly rearward far enough to disengage from the clutch shaft coupling.
13. Lower the front end (creeper) to clear creeper shift lever through frame slot, then assembly can be moved rearward and away from the frame.
14. Support the transmission housing on a stand or block and drain the lubricant if internal service is to be performed.
15. Recoupling of the tractor is basically the reverse of splitting; however, precautions should always be taken to safeguard against damage to shafts, bearings, seals etc. when aligning and securing components which work together.

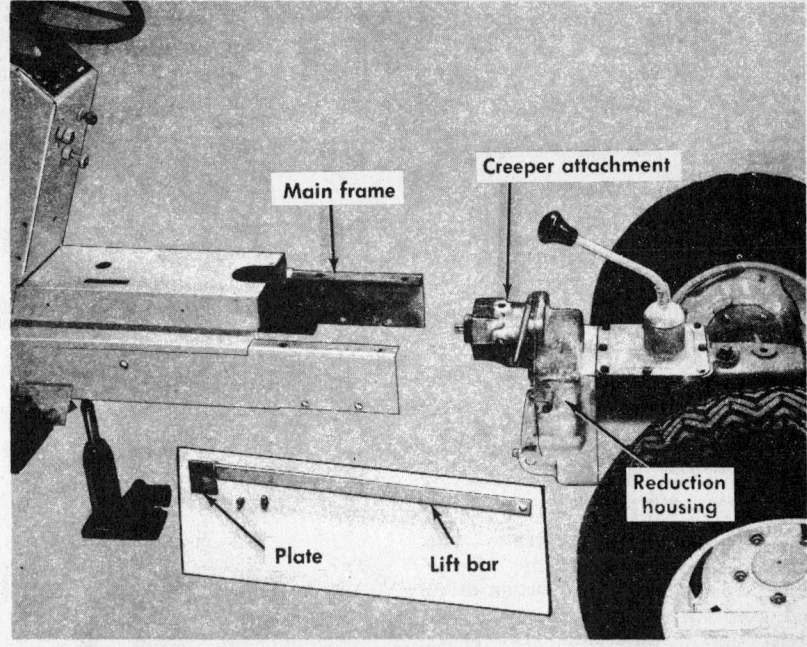

Splitting the tractor

6. Replace the pressure plates, driving disc or driving plate as necessary.

NOTE: Wiring cable clips (on tractors equipped with electric starting) will need spreading so that wires can follow engine as it is moved forward.

7. Slide the engine rearward while aligning the clutch shaft and driving plate pins.
8. Depress and lock the clutch-brake pedal.
9. Replace the pressure plate locating pin.
10. Align the engine to the frame and replace the securing capscrews.
11. Secure the wires and cables in their clips.
12. Connect the ground cable (if it was disconnected).
13. Replace the clutch shield.
14. Place the clutch shaft over the frame cross member and move it rearward to engage the coupling to the reduction unit or creeper driveshaft.
15. Align the release lever to its bracket, install the pin and secure with cotter.
16. Apply chassis lubricant or Lubriplate® liberally to the clutch shaft pilot bushing.
17. Slide the engine rearward while aligning the clutch shaft into its pilot bushing and the drive plate pins into their holes in the driven disc.
18. Align the engine mounting capscrew holes and install the capscrews.
19. Align the clutch shaft coupling rear pin hole with pin hole in reduction drive (or creeper drive) shaft and install the pin.

CLUTCH

REMOVAL AND INSTALLATION

1. Depress the clutch and brake pedal and lock it.
2. Remove the clutch shield.
3. Using a hammer and punch, drive out the pressure plate locating pin.
4. Remove the four capscrews (two on each side) which hold the engine to the tractor frame.
5. Release the clutch and brake pedal, then slide the engine forward in the frame.

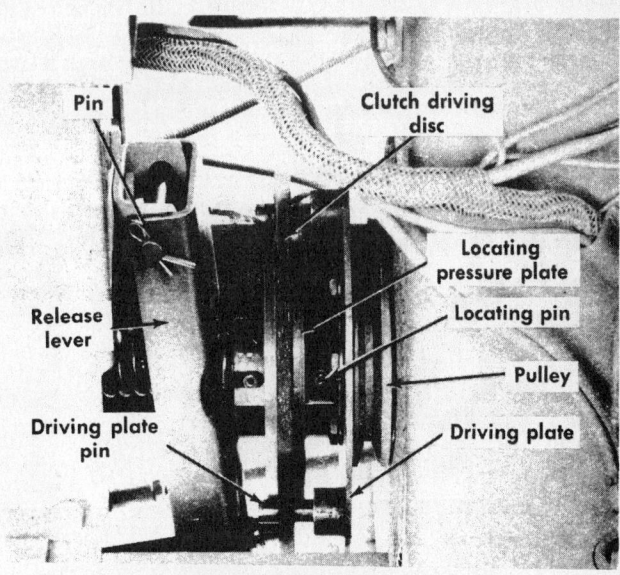

Clutch drive

475

International Harvester

20. Connect the clutch release rod to its pedal shaft lever and secure with the return spring.
21. Depress the pedal and release the clutch to allow components to move into final alignment.
22. Tighten the engine mounting capscrews.
23. Position the wires and cables in their clips and secure.
24. Connect the battery ground cable.
25. Replace the clutch shield.
26. Check the clutch pedal and linkage for proper adjustment.

TRANSMISSION AND DIFFERENTIAL

NOTE: Complete service of the transmission requires splitting of the tractor, removal of the reduction drive (and creeper if so equipped), rear axles, carriers and the differential. The differential can be removed and replaced without disassembling the transmission, however, the transmission countershaft should be removed when checking preload of the differential carrier bearings. The transmission and differential are therefore covered together.

REMOVAL AND DISASSEMBLY (DIFFERENTIAL)

1. Drain the lubricant.
2. Split the tractor.
3. Remove the reduction drive.
4. Remove the rear axles and their carriers.
5. Remove the differential carrier bearing cage and shims from each side. Keep the shims with each cage and identified for each side.
6. Remove the differential from the transmission case.

NOTE: The differential must be turned into position before it can be removed.

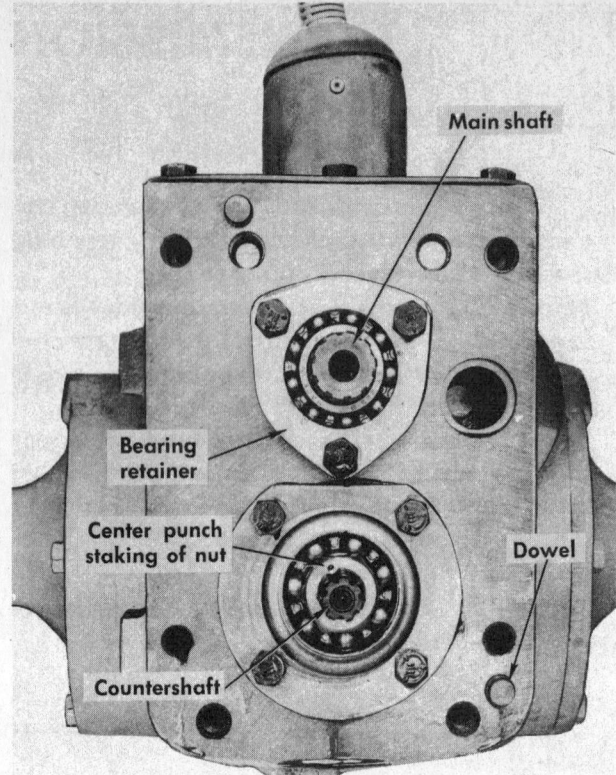

Transmission front view

7. Drive out the pinion shaft lock pin.
8. Remove the pinion shaft.
9. Remove the pinion gears and side gears.
10. If the differential drive gear requires separate replacement, press out the eight retaining rivets.
11. Remove the bearing cones from the differential carrier if they are to be replaced.
12. Remove the bearing cups from each cage if replacement is necessary.

DISASSEMBLY (TRANSMISSION)

1. Remove the differential as previously described.
2. Remove the gear shift lever and cover assembly.

Shift the transmission into two gear speeds to lock the transmission then remove the nut from the countershaft.

3. Remove the shifter fork setscrews then drive the shifter rods forward and out of the transmission.

— CAUTION —
Cover the gear shift poppet ball hole to prevent the ball and spring from flying out as the rods are removed.

4. Remove the capscrews from the mainshaft front bearing retainer.
5. Pull the mainshaft forward and out of the transmission as the gears are removed.
6. Push the countershaft rearward and out of the transmission as the gears and spacers are removed. Note the sequence of spacers and gears for reassembly.

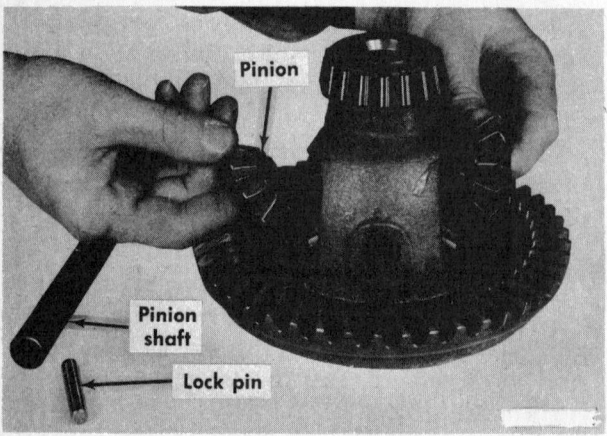

Removing pinion gear

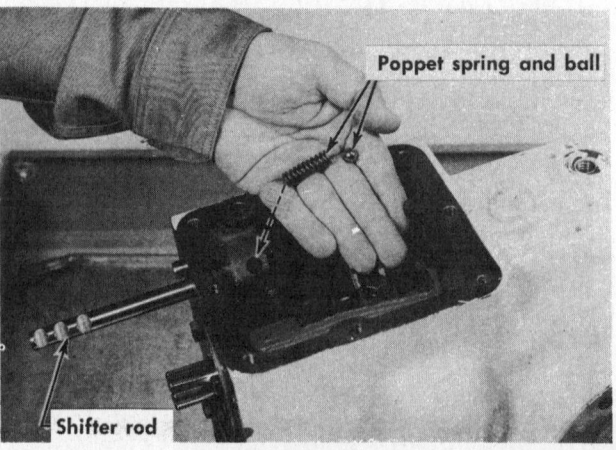

Removing shifter rods

International Harvester

7. Pull the mainshaft needle bearing from the housing.
8. Remove the reverse idler shaft and gear.
9. Remove the countershaft front bearing, retainer and shims.

INSPECTION AND REPAIR

1. Wash all parts in cleaning solvent and dry with compressed air. Do not spin bearings.
2. Check all bearings for looseness, wear, roughness, pitting and scoring.
3. Check gears and shafts for wear and burrs. Remove any burrs with a fine stone.
4. Inspect the housing for cracks, restricted oil passages and dents or raised places on its machines faces. Smooth off raised places with a file.

REASSEMBLY

Reassembly is basically the reverse of disassembly; however, particular attention should be given to the following steps.

1. Be sure all bearings are thoroughly lubricated.
2. The transmission mainshaft needle bearing must be installed with its oil hole aligned with the oil hole in the housing.
3. Assemble the differential, carrier bearings, cages and shims. Check bearing pre-load and adjust as necessary before replacing the transmission countershaft. Install or remove shims as necessary. Preload is correct when a steady pull of one to three pounds is necessary to rotate the differential assembly.
4. Remove the differential assembly, keep the shims with the cages then install the transmission countershaft, bearings, gears, spacers, front bearing retainer, shims and nut. Tighten the nut to 85 ft.lb. Tighten retainer capscrews to 20 ft.lb.
5. Install the differential assembly, keeping the pre-load shim pack correct as previously established. Drive gear must be on the right with teeth facing left.
6. Check the backlash between the drive gear and pinion and the gear teeth bearing pattern as follows:
7. Apply a thin coat of red lead or Prussian blue to the bevel pinion teeth faces, then rotate the gears by hand and observe the bearing pattern.

NOTE: Some deflection will occur under load. Allowance is made in gear design to prevent concentration of load on teeth edges.

8. Hand testing and very light loads should provide a pattern as shown. When load and deflection increases the pattern will progress.
9. The desirable (no load) pattern is the result of adjusting the bevel gear lateral position to the specified range of .003 in. to .005 in. backlash.
10. Tooth bearing position from the root to the crown of the tooth is controlled by lateral position of the pinion. If low tooth bearing on bevel pinion is indicated the pinion must be adjusted toward the bevel gear. If high tooth bearing on the bevel pinion is indicated the pinion must be adjusted away from the bevel gear.

NOTE: If it is necessary to move the pinion in or out to correct "Root-to-crown" bearing, the bevel gear must also be moved laterally to maintain the specified backlash.

11. Stake the countershaft nut by center-punching the face of the nut over a spline groove.
12. Continue the assembly in reverse order of disassembly.
13. Fill housing to proper level with specified lubricant.

NOTE: Creeper attachment has its own lubricant separate of the transmission. Fill creeper at breather and check at side plug in creeper housing.

Checking differential bearing preload

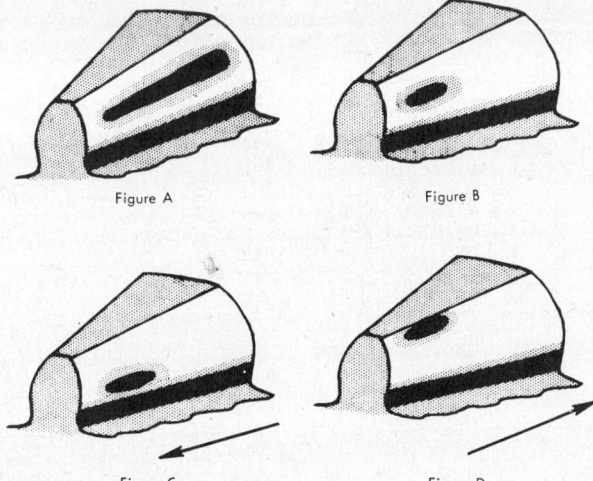

Tooth patterns

CREEPER DRIVE

REMOVAL

1. Split the tractor.
2. Support the transmission and reduction housing on a suitable block, place an

International Harvester

oil drip pan under the creeper unit and drain the creeper lubricant.

3. Four capscrews hold the creeper housing to the reduction housing cover plate. Remove the capscrews.

4. Bump the creeper to the side to loosen the housing from its gasket and dowels. Pull the creeper forward from the reduction cover and splined coupling.

5. If the driven coupling or the pilot bushing needs replacing, support the coupling and drive out the coiled spring pin. Remove the coupling.

DISASSEMBLY

1. Remove the snapring which holds the input shaft bearing cage in the housing.

2. Pull the shaft, bearing, retainer, planetary assembly and the direct drive coupling from the housing.

NOTE: The spline grooves of the direct drive coupling must align with the splines of the shifter collar.

3. Support the direct drive coupling and drive the coiled spring pin out. Remove the coupling from the shaft.

4. Slide the planet carrier off the input shaft and the planet gears off the carrier pins.

5. Remove the flat thrust washer.

6. A snapring holds the ball bearing outer race in the retainer. Remove the snapring then push the bearing and shaft from the retainer.

7. The ball bearing is held on the shaft by a snapring. Remove the snapring. The shaft can now be pressed from the bearing.

8. Press the oil seal from the bearing retainer.

9. Drive the shift poppet pin from the shaft and remove the poppet.

10. Shift the lever and shifter collar toward the rear of the case and at the same time lift the shifter collar up to disengage it from the shift yoke.

11. Drive the pin out of the shift yoke and lever shaft.

12. Slide the lever shaft from the yoke and housing.

13. Remove the O-rings from the shaft, housing and bearing retainer.

14. Wash all parts in cleaning solvent then dry thoroughly.

INSPECTION AND REPAIR

1. Check the input driveshaft for oil seal groove wear, worn or chipped teeth on the integral gear and pilot bushing wear on the rear end.

2. Check the splines of the direct drive coupling, planet carrier and the shifter collar for wear and chipping.

3. Check the housing for cracks and the integral sun gear for wear and broken teeth.

4. Inspect the ball bearing for pitting, scoring, wear and rough operation.

REASSEMBLY

Reassembly is basically the reverse of disassembly however, particular attention should be given the following:

1. Always use new O-rings, gaskets and oil seals. O-rings and oil seals should be coated with Lubriplate® or chassis lubricant to assist in installation and provide initial lubrication.

2. Install the oil seal after completing the drive assembly in the housing.

3. The pins which secure the direct drive coupling and the driven coupling to their respective shafts must be flush or below the spline groove so as not to interfere with shifting.

4. The long internal splines of the shifter collar go toward the rear.

5. The machined shoulder of the direct drive coupling goes toward the planet carrier.

6. Lubricate the components and rotate the driveshaft several turns with the shifter in each speed selection to insure freedom of movement and rotation.

INSTALLATION

1. Place a new gasket on the mounting face of the creeper housing. The dowels will hold it in place.

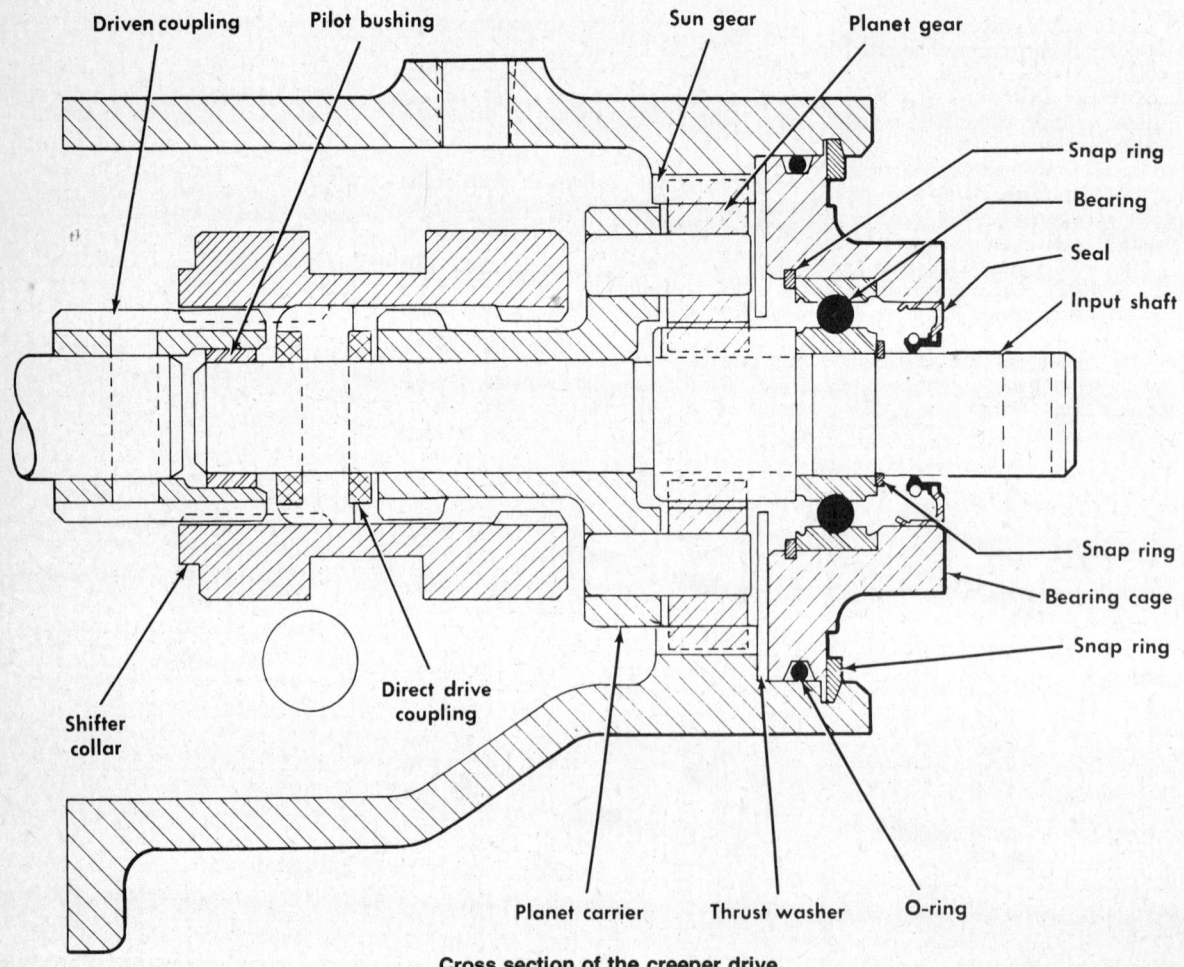

Cross section of the creeper drive

International Harvester

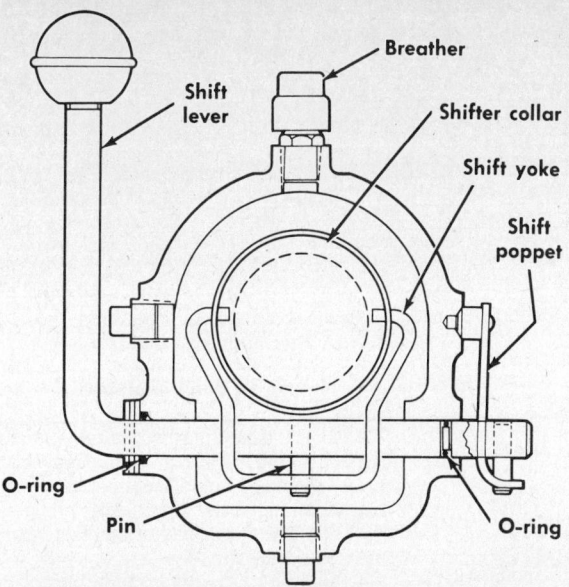

End view showing creeper drive shift components

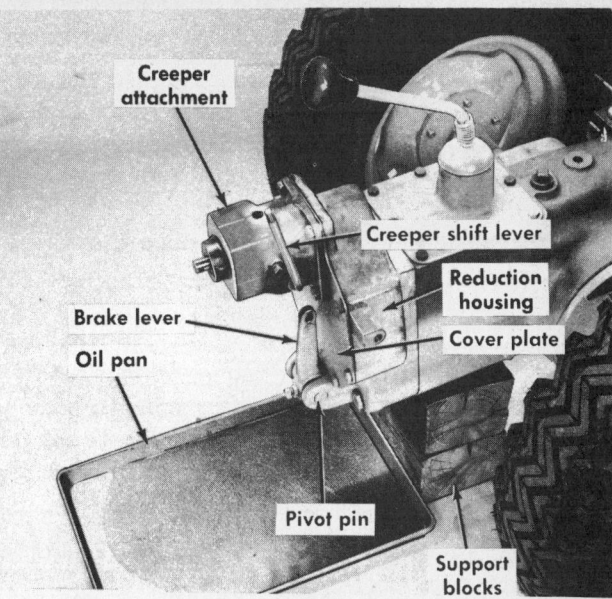

Supporting power train

2. With the shifter collar in its rear position, rotate the driveshaft so the collar will engage the driven coupling as the dowels engage the reduction cover plate and the driveshaft enters its pilot bushing in the driven coupling.

3. Secure the creeper to the reduction housing cover plate.

4. Fill the creeper housing to proper level (check plug on side of housing) with the same lubricant as specified for the transmission-differential. Hy-tran fluid or SAE 30 engine oil.

5. Shift the unit several times to insure freedom of movement. Leave the shifter lever in its forward position for re-coupling the split tractor.

6. Reassemble the split tractor.

REDUCTION DRIVE

REMOVAL AND DISASSEMBLY

1. Split the tractor.
2. Place an oil pan under the reduction housing and remove the creeper assembly (if tractor is so equipped).
3. Remove the brake lever, pivot pin and push rod.
4. Remove the reduction housing front cover plate.
5. Hold the drive coupling and shaft from turning and remove the reduction gear retaining capscrew and washers. Remove the gear spacer.
6. Remove the reduction gear from the transmission shaft and from the housing.

NOTE: It may be more convenient to pull the reduction driveshaft, seal and bearing before removing the reduction gear from the housing. Clearance between the gear and the capscrew bosses is restricted on some tractors.

7. Remove capscrews from holes "A" and "B".

NOTE: Soft copper sealing washers are used under the "B" capscrew heads.

8. Move the housing forward and away from the transmission housing as the brake disc slides off the transmission countershaft.

9. Pull the reduction driveshaft, seal and bearing from the reduction housing if it was not removed in step 6 NOTE.

10. Support the driveshaft splined coupling and drive out the coiled spring pin.

NOTE: The splined coupling is used only on tractors equipped with creeper attachment.

11. Press the driveshaft from the ball bearing.
12. Press the needle bearing rearward from the housing.
13. Remove the brake components.

INSPECTION AND REPAIR

1. Inspect the driveshaft for wear on the gear teeth, needle bearing area, oil seal contact area and drive pin hole.
2. Inspect the reduction gear teeth for

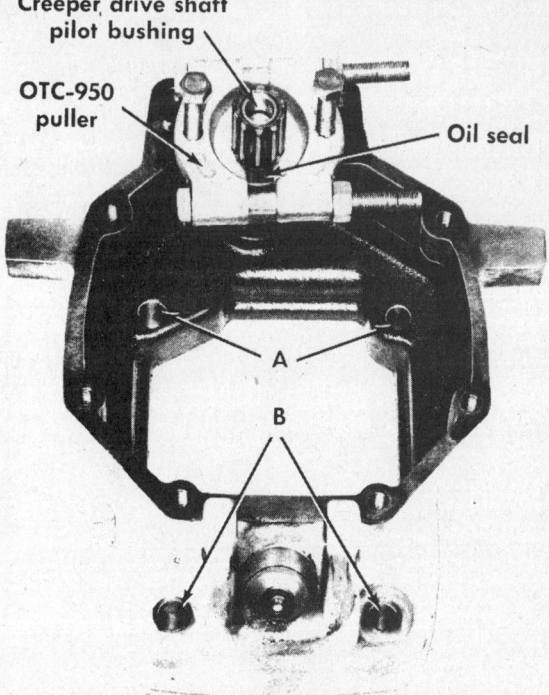

Pulling the reduction driveshaft, seal and bearings

International Harvester

wear or chipping and the fit of the gear on the transmission shaft.

3. On tractors with creeper, inspect wear of pilot bushing for creeper driveshaft.

4. Inspect needle and ball bearings for wear, pitting and roughness of operation.

5. Thoroughly clean the reduction housing.

REASSEMBLY AND INSTALLATION

1. Install a new gasket to the front of the transmission case. Dowels in the case will hold the gasket in place.

2. Press the needle bearing into the reduction housing from the rear. Rear end of bearing should be flush with housing.

3. Press the ball bearing onto the driveshaft.

4. Lubricate the lip of a new oil seal and install the seal onto the shaft. Be careful when sliding the seal lip over the pin hole in the shaft. Flat face of seal case must be forward.

5. Install the splined coupling and coiled pin (if tractor is equipped with a creeper). Coiled pin ends must be even with or below the spline root to avoid interference when shifting the creeper unit.

6. Install a new O-ring onto the brake lining retainer and install in the reduction housing.

7. Install the reduction housing to the transmission case. Be sure the gasket and dowels are in place.

8. Install new copper sealing washers on the two lower capscrews. Tighten capscrews to 80 ft.lb.

9. Install the brake linings and disc then push the front lining retainer rearward to hold disc and linings in place.

10. Install the reduction gear and spacer to the transmission mainshaft and secure with the capscrew, flat and lockwasher. Tighten capscrew after driveshaft is installed.

11. Install the driveshaft with ball bearing, seal and splined coupling (on tractors with creeper). Seal case should be flush with housing.

NOTE: Seal holds ball bearing in place and front cover holds seal in place.

12. Tighten reduction gear retaining capscrew to 55 ft.lb.

13. Install new gasket and housing front cover.

14. Install new gasket and creeper unit (on tractors so equipped).

15. Replace the brake push rod, ball, lever and pivot pin.

16. Recouple the tractor by reversing the splitting procedure.

17. Fill transmission and creeper to proper level with specified lubricant.

REAR AXLES

REMOVAL

1. Drain the transmission lubricant.

2. Stabilize the tractor by driving wooden wedges between the front axle and frame on each side and block the front wheels.

3. Remove the drawbar and differential housing rear cover.

4. Place a jack under either rear axle carrier and raise the rear wheel off the floor. Remove the "C" type snapring from the axle shaft inner end.

5. Slide the axle out of the differential side gear and axle carrier.

6. Support the transmission-differential housing on a block and remove the axle carrier.

7. Press the carrier bushing from the carrier.

8. Drive the oil seal from the carrier.

9. Clean the gasket surface of the axle carrier and differential housing.

INSPECTION

1. Inspect the axle shaft for wear at the oil seal area, bushing location and splines on the inner end.

2. Roll the axle shaft along a flat surface to detect any warping or bending.

3. Check the axle carriers for cracks or breaks. Remove any high spots from the gasket surface with a flat file.

REASSEMBLY

1. Press a new bushing into the axle carrier. The oil groove must be at the bottom.

2. Press a new oil seal into the axle carrier.

3. Using a new gasket, install the axle carrier to the differential housing. Capscrew threads should be coated with a non-hardening sealer (Permatex®) to avoid oil leaks. The frame pad of the axle carrier must be on the top.

4. Fill the cavity between the lips of the oil seal with chassis lube or heavy oil.

5. Lubricate the axle shaft and bushing then slide the shaft through the seal, bushing, carrier and differential side gear. Rotate the axle as it is pushed through to avoid damage to the seal. Wipe off excess lubricant.

6. Install a new "C" type snapring to the inner end of the axle shaft.

7. Replace the rear cover and drawbar.

8. Fill differential housing to proper level with specified lubricant.

ENGINE

Engine used in these models are:
Model 70 with electric starter:
Kohler K161S with recoil starter:
Kohler K161T 100:
Kohler 241AS

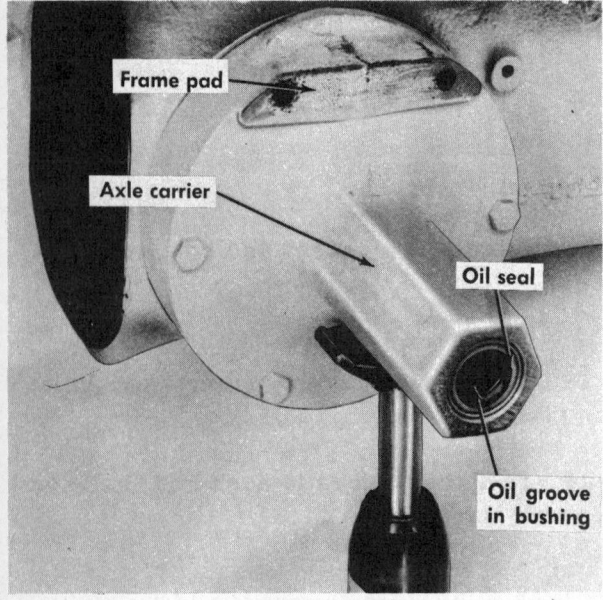

Rear axle removed

Rear axle carrier removed

International Harvester

For complete engine service, see the Kohler Engine section of the Engine Unit Repair section of this book.

REMOVAL AND INSTALLATION

Manual Starter

1. Remove the grille and hood.

NOTE: Four capscrews (2 on each side) hold the grille to the tractor main frame.

2. Remove the clutch shield.
3. Disconnect the throttle wire from the governor speed control disc.
4. Disconnect the choke wire from the carburetor.
5. Disconnect the engine from the main frame.

NOTE: Four capscrews (two on each side) go through the frame into tapped holes in the engine oil pan base.

6. Slide the engine forward in the frame to disengage the clutch drive plate pins from the driven disc and disengage the clutch shaft from its pilot bushing.
7. The engine can now be lifted from the tractor frame and chassis.

NOTE: One cylinder head capscrew can be removed and a ⅜ in. N.C. eyebolt installed to lift the engine if desired.

8. Installation is the reverse of removal.

Electric Starter

1. Disconnect the battery cables and remove the battery.
2. Remove the grille and hood. Lay the grille and hood (grille face down) in front of the tractor.
3. Remove the clutch shield.
4. Remove the air cleaner.
5. Remove capscrew "A" and loosen screw "B". Remove the choke wire from the carburetor, and the throttle wire from the speed control bracket.
6. Disconnect the positive (+) coil wire from the coil. Spread the wire clip.
7. If the ground cable is connected to the starter-generator pivot capscrew, disconnect it.
8. Disconnect the two wires from the Generator "A" terminal and one wire from the "F" terminal. Tie the two wires removed from the "A" terminal together and identify them for reassembly.
9. Disconnect the engine from the main frame. Two engine mounting capscrews on each side hold the engine to the frame.
10. Slide the engine forward to disengage the drive pins from the clutch driven disc.
11. The engine can now be lifted from the frame. An eyebolt (⅜ in. N.C.) can be installed in place of one cylinder head bolt.

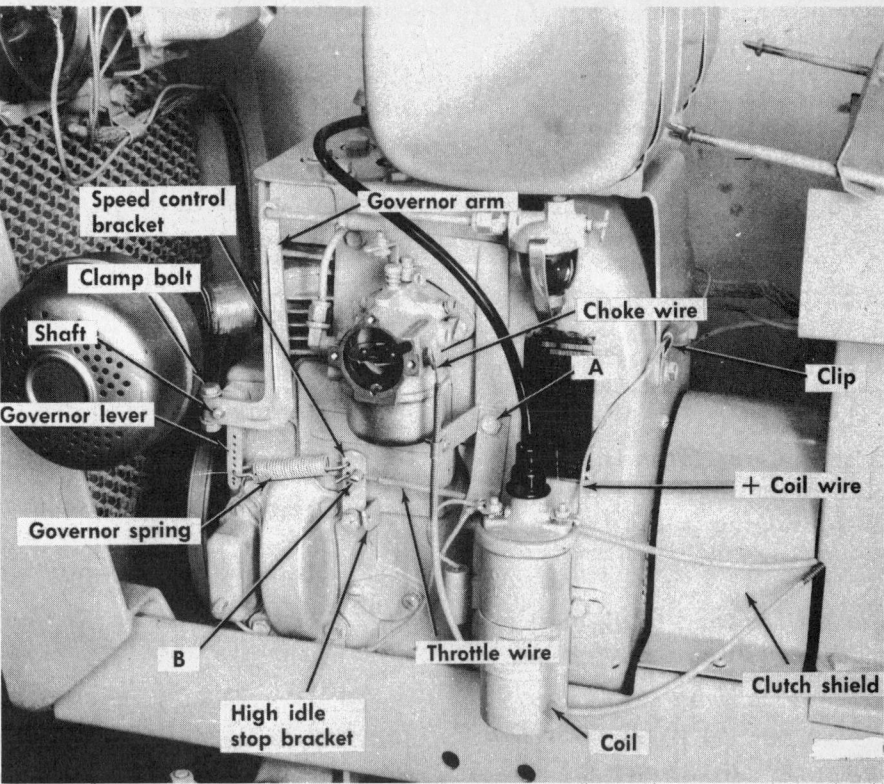

Left side view

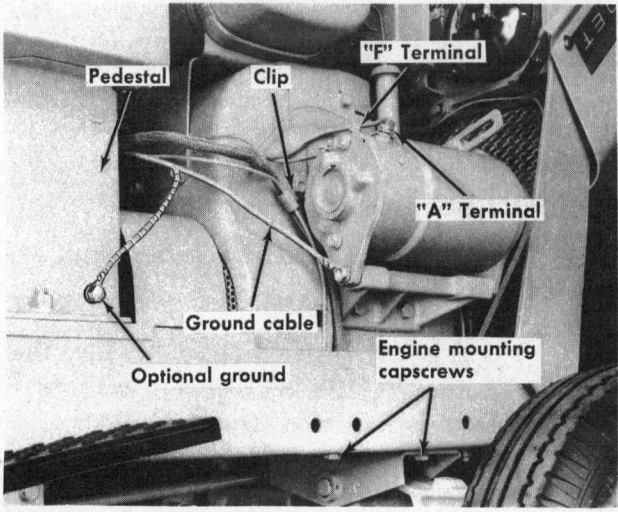

Right side view

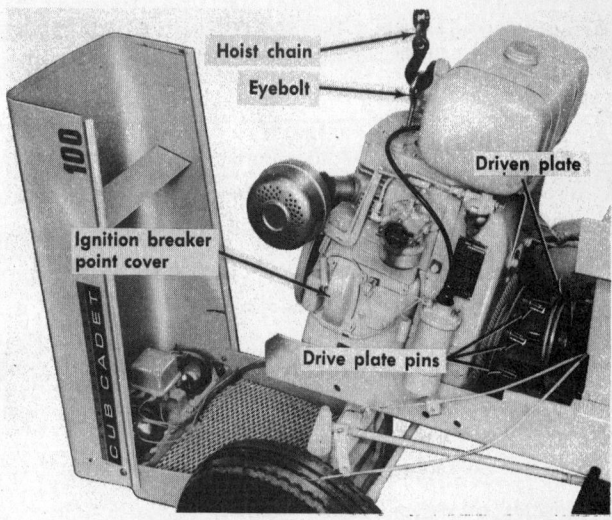

Lifting the engine from the frame

International Harvester

Cub Cadet Models 71, 102, 122, 123

FRONT WHEELS AND AXLE

Front Axle

REMOVAL AND INSTALLATION

1. Disconnect the drag link ball joint "D" from the drag link arm (8).
2. With the front of the tractor frame supported by a suitable stand, drive out the retaining pin from the front of the axle pivot pin (1).
3. Remove the pivot pin (1). The front axle (2) is now free of its mounting and can be removed.
4. Apply chassis lubricant liberally to the axle pivot pin and its bore in the axle.
5. Position the axle in its support bracket channel, align the pivot pin holes and insert the pin.
6. Align the retaining pin holes (through the front of the pivot pin and through the front collar of the support bracket) then drive the retaining pin through both parts.

Front Wheels and Bearings

REMOVAL

1. Lock the brake and block the rear wheels. Jack up the front axle.
2. Remove the capscrew and flat washer from the outer end of the front spindle.
3. Slide the wheel and bearings from the spindle.

NOTE: The bearings are a press fit in the wheel and a slip fit on the spindle.

DISASSEMBLY

1. Wheel bearings can be driven from the wheel hub with a hammer and long drift punch. Drive from the inside toward the outside.

INSPECTION AND REPAIR

1. Inspect the entire wheel and hub for weld separation, split hub tube and rim bending.
2. Bearings should be inspected for wear, seizure and seal condition.

REASSEMBLY

If the bearings were removed, lubricate and press in new ones. Be sure force is directed to the outer race only when being pressed in.

INSTALLATION

1. Slide the wheel and bearing assembly over the spindle and secure with capscrew and flat washer.

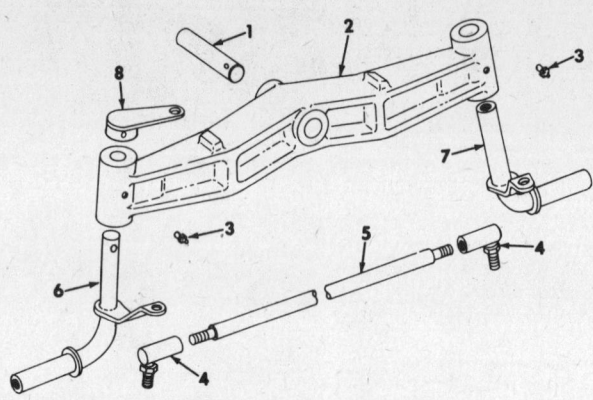

Front axle components

2. If excessive end play exists, place a sufficient thickness of shim washers (¾ in. ID) over the outer end of the spindle and between the retaining washer and wheel bearing.

STEERING

Steering Knuckle

REMOVAL

1. Lock the brake, jack up the front of the tractor and support it on a suitable stand.
2. Remove the front wheels.
3. Disconnect the tie rod ball joints from left steering knuckle and right steering knuckle.
4. Using a pin punch and hammer, drive out the coiled spring pin from the drag link arm and steering knuckle.
5. Remove the steering knuckle from the axle.
6. Remove the capscrew and flat washer from the upper end of the steering knuckle.
7. Remove the steering knuckle from the axle.

INSTALLATION

1. Apply chassis lubricant liberally to the steering knuckle arm thrust surface and the bottom of the axle ends.
2. Insert the righthand knuckle pin in its respective bore in the axle and secure with the capscrew and flat washer.
3. Insert the lefthand knuckle pin in its bore in the axle and position the drag link arm on the pin.
4. Secure the drag link arm and knuckle with the coiled spring pin.

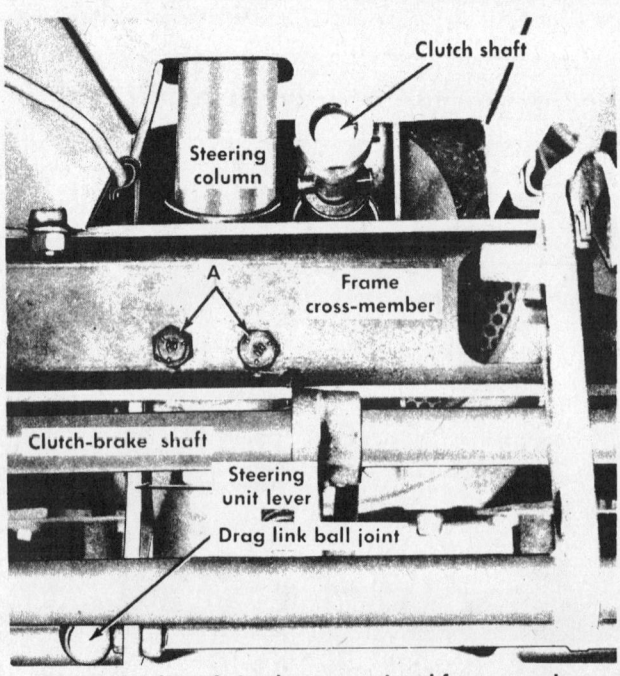

Underside view of steering support and frame member

International Harvester

NOTE: Spring pin must extend through the drag link arm an equal amount on each side.

5. Install the tie rod.
6. Install the front wheels and check toe-in adjustment.

Steering Gear

REMOVAL

1. Hold the steering wheel with front wheels in straight position. Remove the steering wheel.
2. Remove the felt seal, retainer, bearing and bearing retainer from the upper end of the steering column.
3. Remove the drag link rear ball joint from the steering unit lever.
4. Remove capscrews "A" from the frame cross member and steering unit. Remove stabilizer bolt from right side of frame if tractor is so equipped.
5. Lower the steering column assembly through the instrument panel pedestal and grommet.

DISASSEMBLY

1. Secure the steering lever and bolt in a vise.
2. Remove the adjusting plug.
3. Remove the lever bolt jam nut, adjusting nut, and bronze washer.
4. Slide the column and housing assembly away from the lever, bolt and cam follower.
5. Remove the steering cam and bearings from the housing.
6. Remove the bearing race retainer snaprings.

INSPECTION AND REPAIR

1. Wash all parts in cleaning solvent, then dry thoroughly.
2. Inspect the cam follower for wear (flat spots).
3. Inspect the cam ends, bearings and races for wear, roughness and pitting.
4. Inspect the cam grooves for wear, roughness and galling.
5. Inspect the housing for cracks and stripped threads.
6. Inspect the upper bearing (nylon bushing) for wear or damage.

REASSEMBLY AND ADJUSTMENT

1. Thoroughly coat the cam ends, balls and races with chassis lubricant.
2. Install the balls and races on the cam ends and secure with their retaining snaprings.
3. Thoroughly coat the cam and bearings with chassis lubricant then install them into the housing and column assembly.

NOTE: Be sure the races enter the housing squarely and are not "cocked."

4. Install the adjusting plug. Screw the plug inward until end play (of the cam) is removed and the cam turns free.
5. If the plug has only one slot (old style), adjust and stake the plug by inserting a small center punch through the housing hole and spread the plug threads to keep the plug from loosening. If the plug has several slots (new style), adjust and insert cotter pin in the nearest hole.
6. Fill the housing with chassis lubricant.
7. Loosen the cam follower locknut, then back out the cam follower one turn.
8. Install the seal, retainer and lever-bolt assembly to the housing.
9. Install the bronze washer and adjusting nut. Tighten the adjusting nut sufficiently to provide good seal compression. Secure with the jam nut. Tighten jam nut to 40 ft.lb. Lubricate the lever-bolt at the fitting.
10. Center the steering cam by rotating the steering shaft half way between full right and full left turn.
11. Adjust the cam follower inward to eliminate backlash, then tighten locknut to 40 ft.lb. Turn steering full right and left to check for binding.
12. Replace the steering assembly in the tractor chassis. Secure with two capscrews through the frame cross member.
13. Connect the drag link.

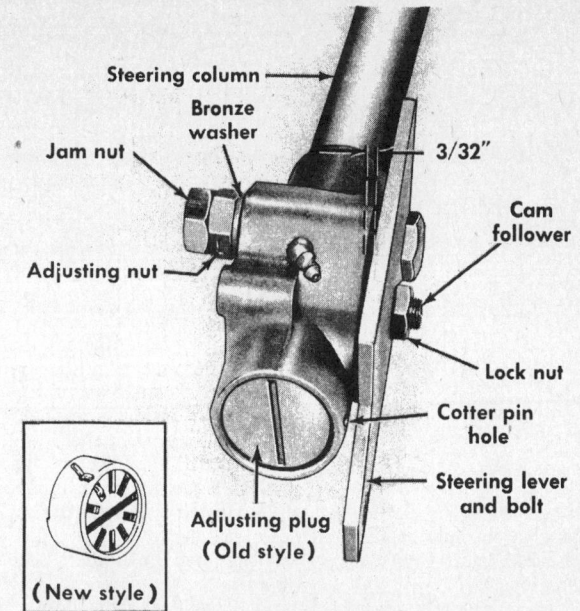

Steering gear

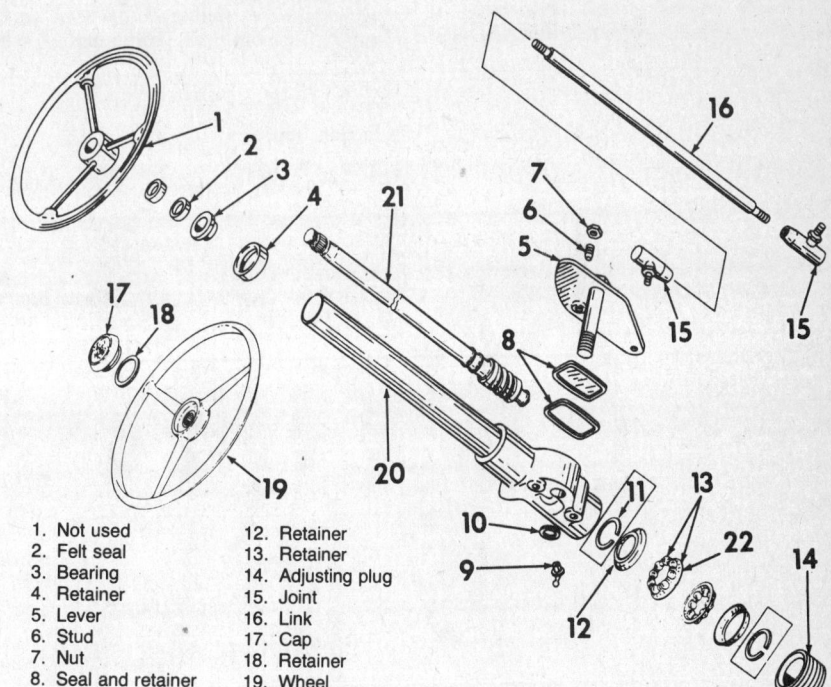

1. Not used
2. Felt seal
3. Bearing
4. Retainer
5. Lever
6. Stud
7. Nut
8. Seal and retainer
9. Fitting
10. Bronze washer
11. Not used
12. Retainer
13. Retainer
14. Adjusting plug
15. Joint
16. Link
17. Cap
18. Retainer
19. Wheel
20. Tube
21. Cam and tube
22. Retainer

Steering system components

International Harvester

14. Install the uper column bearing, retainer, felt seal and retainer.
15. Replace the steering wheel and secure with nut.
16. Adjust the drag link to proper length to place front wheels in the straight ahead position when the steering assembly is centered.
17. Adjust tie rods to provide $1/32$ in. to $1/8$ in. toe-in.

BRAKES

NOTE: Tractors equipped with a creeper attachment will require splitting, and the creeper removed before complete brake service can be performed.

REMOVAL

1. Drain the transmission lubricant.
2. Remove the brake adjusting screw and locknut from its lever.
3. Remove the brake lever, pivot pin and push rod.
4. Remove the reduction housing front cover plate and slide it forward on the clutch shaft.
5. Remove the reduction gear from the front of the transmission mainshaft.
6. Move the gear upward and the bottom of the gear forward to clean the cover screw bosses as the gear is lifted from the housing.
7. Slide the brake disc forward on the countershaft as the front lining and retainer are moved forward in their bore.

NOTE: Both linings and the disc can be removed without removing the front lining retainer; however, removal of the retainer is recommended for inspection and replacement of the retainer O-ring.

INSPECTION

1. Inspect the control rods and levers for wear at their connecting pivot points.
2. Inspect the linings and disc for wear.
3. Inspect the disc hub splines for wear.
4. Check the splines on the countershaft for wear.
5. Check the pedal return spring ends for wear.

ASSEMBLY AND ADJUSTMENT

1. Clean the brake cavity and lining recess in the reduction housing.
2. Place a small quantity of grease in the rear brake lining recess in the reduction housing then insert the lining.
3. Install the disc on the countershaft and slide it rearward against the rear lining.
4. Install a new O-ring on the front lining retainer, lubricate the retainer and O-ring then push them into the retainer bore in the reduction housing.
5. Install the front lining in the retainer lining recess and push the retainer and lining rearward against the disc.
6. Install the reduction gear on the transmission mainshaft and secure with flat washer, lockwasher and capscrew. Tighten capscrew to 55 ft.lb.
7. Install a new cover gasket, then the cover plate.
8. Be sure the ball is in place in the front lining retainer than replace the push rod, lever, pivot pin, adjusting screw and locknut.
9. Fill transmission to proper level with Hy-Tran fluid of SAE 30 engine oil.

Adjusting the Brake

1. The brake should engage when the pedal arm is pressed down to within a maximum of $1^5/16$ in. and a minimum of $3/4$ in. distance above the top of the left foot support, which serves as the pedal stop.
2. It may be possible to push the pedal all the way down to the pedal stop, but this is of no concern as long as the brake is engaged when the pedal arm is at least $3/4$ in. above the pedal stop.
3. To adjust the brake, loosen the locknut and turn the brake lever adjusting screw in or out as required to get this measurement. The brake must not engage before the pedal arm is within the maximum distance of $1^5/16$ in. above the pedal stop.

CLUTCH

REMOVAL

1. Depress the clutch and brake pedal and lock it.
2. Remove the clutch shield.
3. Using a hammer and punch, drive out the pressure plate locating pin.
4. Remove the four capscrews (two on each side) which hold the engine to the tractor frame.
5. Release the clutch and brake pedal, then slide the engine forward in the frame.
6. Replace the pressure plates, driving disc or driving plate as necessary.

NOTE: Wiring cable clips (on tractors equipped with electric starting) will need spreading so that wires can follow engine as it is moved forward.

7. Slide the engine rearward while aligning the clutch shaft and driving plate pins.
8. Depress and lock the clutch-brake pedal.
9. Replace the pressure plate locating pin.

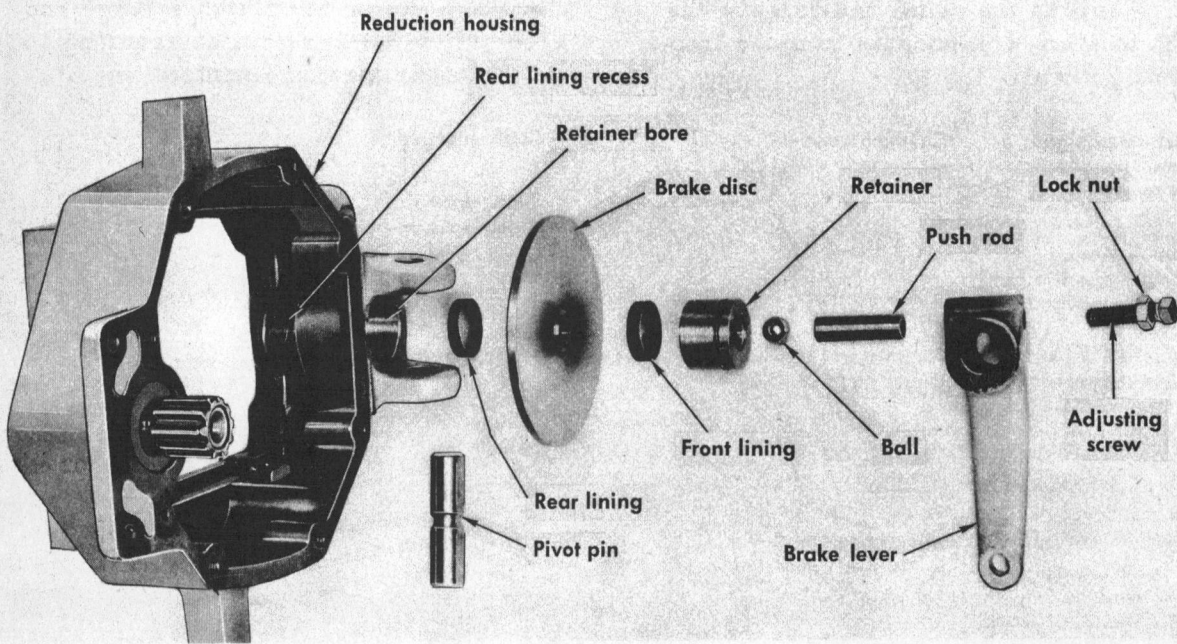

Brake system components

International Harvester

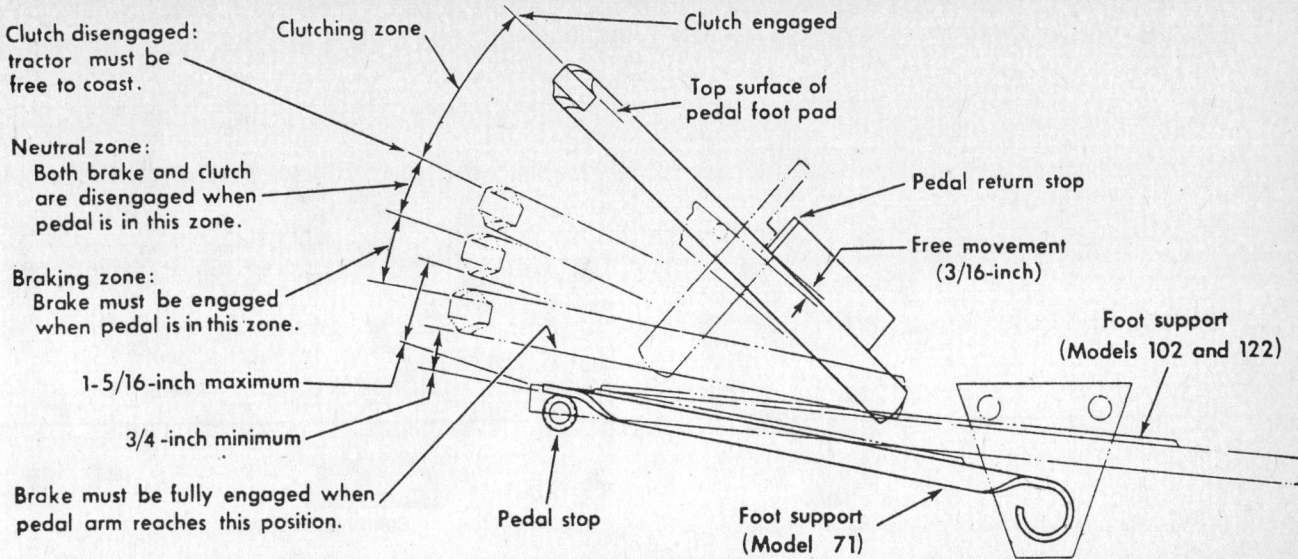

10. Align the engine to the frame and replace the securing capscrews.
11. Secure the wires and cables in their clips.
12. Connect the ground cable (if it was disconnected).
13. Replace the clutch shield.

INSTALLATION

1. Place the clutch shaft over the frame cross member and move it rearward to engage the coupling to the reduction unit or creeper driveshaft.
2. Align the release lever to its bracket, install the pin and secure with cotter.
3. Apply chassis lubricant or Lubriplate® liberally to the clutch shaft pilot bushing.
4. Slide the engine rearward while aligning the clutch shaft into its pilot bushing and the drive plate pins into their holes in the driven disc.
5. Align the engine mounting capscrew holes and install the capscrews.
6. Align the clutch shaft coupling rear pin hole in reduction drive (or creeper drive) shaft and install the pin.

Brake and clutch adjustments

7. Connect the clutch release rod to its pedal shaft lever and secure with the return spring.
8. Depress the pedal and release the clutch to allow components to move into final alignment.
9. Tighten the engine mounting capscrews.
10. Position the wires and cables in their clips and secure.
11. Connect the battery ground cable.
12. Replace the clutch shield.
13. Check the clutch pedal and linkage for proper adjustment.

ADJUSTMENT

It is important that a clearance of .050 in. be maintained between the clutch release lever and the clutch release bearing. In order to maintain this clearance, the pedal should have a free movement of approximately 3/16 in. This measurement is taken at the point of contact of the pedal arm with the front edge of the pedal return stop.

The clutch pedal adjustments are set at the factory and should not require frequent attention unless the linkage has been disturbed or when the pedal movement becomes less than 3/16 in. When it is necessary to adjust the clutch, turn the adjusting nut "A" on the clutch release rod in or out as required to get the proper measurements.

SPLITTING AND RECOUPLING THE TRACTOR
Models 71, 102, 122

SPLITTING THE TRACTOR

Remove the fenders and their support assemblies (if tractor is so equipped) or the seat support and fender assembly.
2. Remove the seat and its support bracket.

NOTE: Disconnect tail light wire at junction under seat support (if tractor is so equipped).

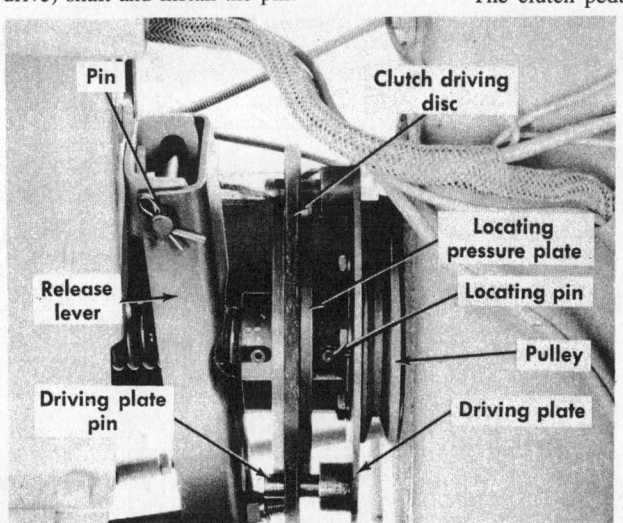

Clutch drive visible with clutch shield removed

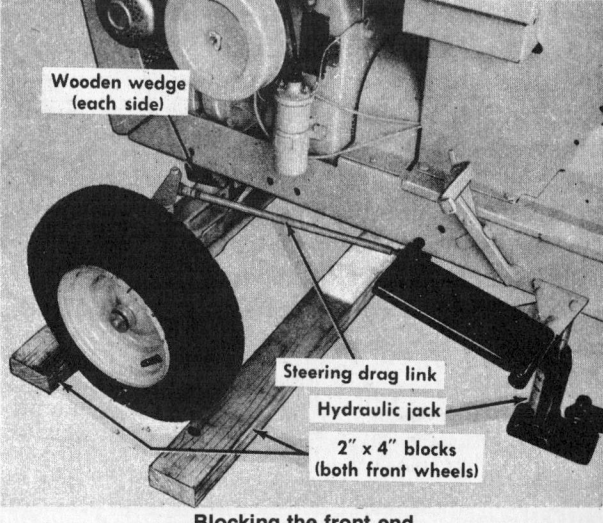

Blocking the front end

485

International Harvester

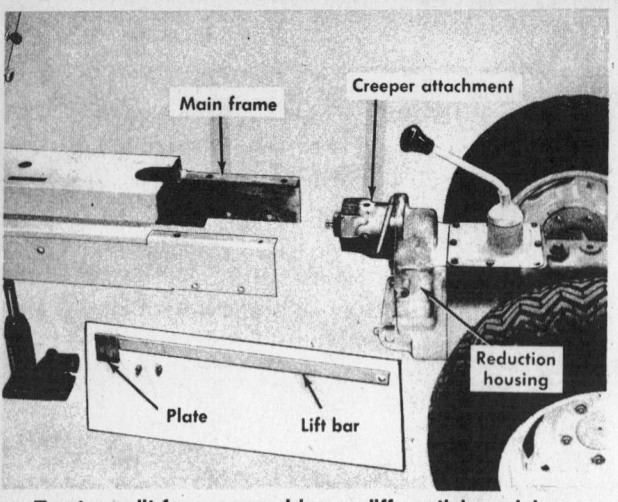

Tractor split for creeper drive or differential servicing

Splitting the tractor

3. Remove the clutch shaft coupling rear pin. (Connects coupling to creeper or reduction driveshaft.)

4. Drive a small wooden wedge between the front axle and frame on each side.

5. Block both front wheels.

6. If tractor is equipped with creeper attachment, shift the lever forward then remove the lever knob and the breather.

7. Disconnect the brake rod from the brake lever.

8. If tractor is equipped with three-point hitch, remove the lift bar and its attaching plate.

9. Place a jack under the implement lift handle cross shaft to support the frame.

10. Remove three capscrews from each side of frame.

NOTE: Capscrews hold frame to the rear axle, transmission and reduction drive housings.

11. Push down on the draw bar and pull rearward on the transmission shifter lever.

12. Move the transmission-differential-rear end assembly rearward far enough to disengage from the clutch shaft coupling. Lower the front end (creeper) to clear creeper shift lever through frame slot, then assembly can be moved rearward and away from the frame.

13. Support the transmission housing on a stand or block and drain the lubricant if internal service is to be performed.

RECOUPLING THE TRACTOR

Recoupling of the tractor is basically the reverse of splitting; however, precautions should always be taken to safeguard against damage to shafts, bearings, seals etc. when aligning and securing components which work together.

Model 123

SPLITTING THE TRACTOR

1. Remove the seat, seat support and fenders.

NOTE: Disconnect the tail light wire (if so equipped) at the junction under the seat support.

2. Remove the air deflector.
3. Disconnect the ball joint (13) from the speed control cam (6).

Air deflector

4. Disconnect the brake rod (38) from the rod lever (37). Loosen the jam nut (40) and brake adjusting screw (39).

5. Drive small wooden wedges between the front axle and frame on each side to stabilize the front end of the tractor.

6. Block the front wheels in the front and back.

7. Place a jack on each side of the tractor main frame in front of the foot rests.

8. Support the transmission-differential rear end assembly with blocks. Be sure not to contact the oil tube.

9. Remove six capscrews (three each side) from the frame.

10. Position the tow lever in the "Tow" position. Move the transmission-differential rear end assembly rearward far enough to permit access to the hydrostatic unit. Be sure the driveshaft pin clears the hole in the separator.

RECOUPLING THE TRACTOR

1. Tip the unit into recoupling position and lubricate the cam slot with chassis lubricant.

2. Fill the transmission-differential case with seven (7) quarts of Hy-Tran fluid.

3. Recouple the split sections of the tractor being sure of the following:

   a. That the brake rod goes over the lift handle cross shaft. Be sure to guide the driveshaft through the fan shroud and air baffle plate.

   b. That the dowel pin on the front end of the driveshaft is positioned to engage the slot in the drive plate hub on the engine.

4. Install and tighten securely the three capscrews on each side of the frame.

5. Connect the speed control rod ball joint (13), to the speed control cam (6).

6. Depress the brake pedal and release it to be sure the control lever is in the "N" position.

7. Connect the brake rod (38), to the rod lever (37).

Adjusting the Speed Control Lever

NOTE: The brake pedal must be properly adjusted before beginning the speed control lever adjustment. If the tractor "creeps" in the "N" position or,

International Harvester

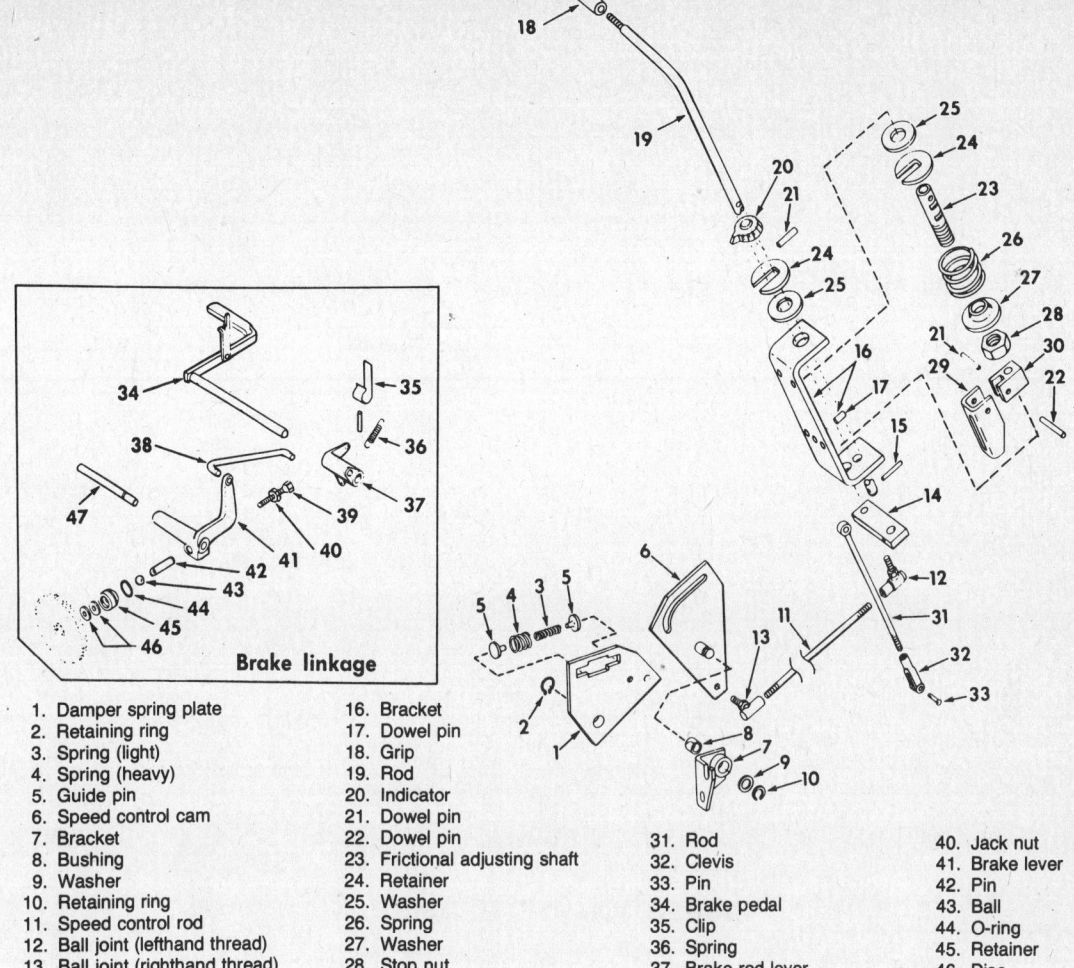

1. Damper spring plate
2. Retaining ring
3. Spring (light)
4. Spring (heavy)
5. Guide pin
6. Speed control cam
7. Bracket
8. Bushing
9. Washer
10. Retaining ring
11. Speed control rod
12. Ball joint (lefthand thread)
13. Ball joint (righthand thread)
14. Bar
15. Slotted pin
16. Bracket
17. Dowel pin
18. Grip
19. Rod
20. Indicator
21. Dowel pin
22. Dowel pin
23. Frictional adjusting shaft
24. Retainer
25. Washer
26. Spring
27. Washer
28. Stop nut
29. Cam channel
30. Swivel
31. Rod
32. Clevis
33. Pin
34. Brake pedal
35. Clip
36. Spring
37. Brake rod lever
38. Brake rod
39. Brake adjusting screw
40. Jack nut
41. Brake lever
42. Pin
43. Ball
44. O-ring
45. Retainer
46. Disc
47. Pin

Brake and clutch linkages for 123

If the speed control linkage has been disassembled or removed for any reason, the following adjustment must be made.

1. Block the tractor so the left rear wheel is off the ground.
2. Start the engine at half throttle or faster.
3. Move the speed control lever to the forward position. The rear wheel should rotate in the forward direction. Depress the brake pedal all the way down and release. The speed control lever should return to the "N" position and the rear wheel stop turning.
4. If the rear wheel turns in the forward direction, loosen jam nut "D" and turn the connecting rod "E" counterclockwise to lengthen it until the wheel stops turning.
5. If the wheel turns in the reverse direction turn the connecting rod "E" clockwise. Tighten the jam nut "D".
6. Proper friction adjustment is necessary on speed control lever for proper operation. The lever friction should be adjusted as follows:
   a. Remove the battery.
   b. Place a small wedge between the hand-control mounting bracket and the adjusting nut.

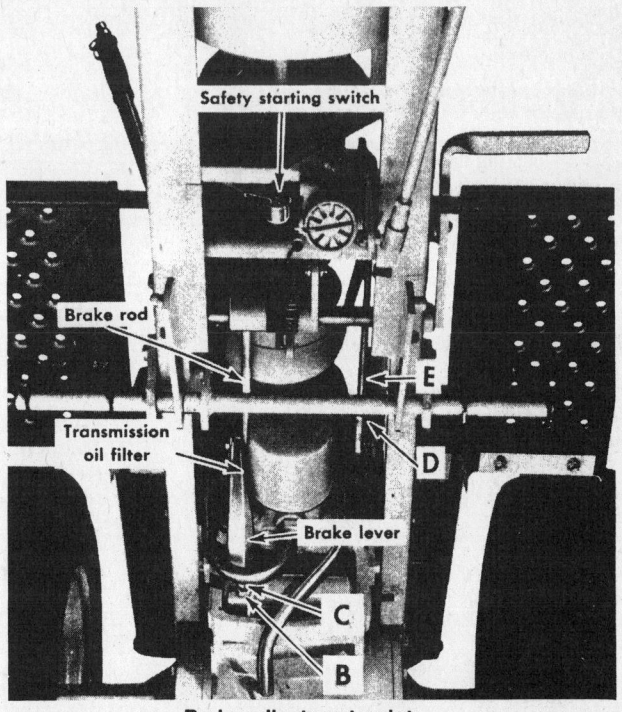

Brake adjustment points

487

International Harvester

c. Move the lever towards the "F" or forward position to tighten the nut.
d. Remove the wedge.
e. Check the friction adjustment with a scale. The reading should be 10 lb. when pulled in either direction from the offset in the lever.
f. Replace the battery.

TRANSMISSION AND DIFFERENTIAL

Models 71, 102, 122

REMOVAL AND DISASSEMBLY (DIFFERENTIAL)

1. Drain the lubricant.
2. Split the tractor.
3. Remove the reduction drive.
4. Remove the rear axles and their carriers.
5. Remove the differential carrier bearing cage and shims from each side. Keep the shims with each cage and identified for each side.
6. Remove the differential from the transmission case.

NOTE: The differential must be turned into position before it can be removed.

7. Drive out the pinion shaft lock pin.
8. Remove the pinion shaft.
9. Remove the pinion gears and side gears.
10. If the differential drive gear requires separate replacement, press out the eight retaining rivets.
11. Remove the bearing cones from the differential carrier if they are to be replaced.
12. Remove the bearing cups from each cage if replacement is necessary.

DISASSEMBLY (TRANSMISSION)

1. Remove the differential.
2. Remove the gearshift lever and cover assembly.

Shift the transmission into two gear speeds to lock the transmission then remove the nut from the countershaft.

3. Remove the shifter fork setscrews then drive the shifter rods forward and out of the transmission.

CAUTION

Cover the gearshift poppet ball hole to prevent the ball and spring from flying out as the rods are removed.

---

4. Remove the capscrews from the mainshaft front bearing retainer.

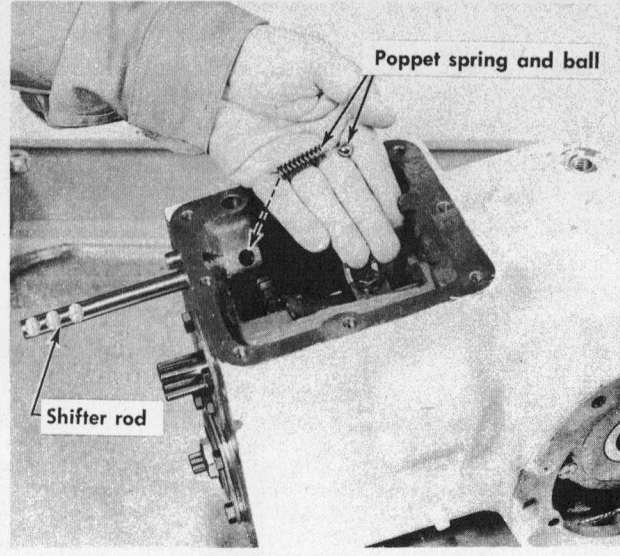
Removing shifter rods

5. Pull the mainshaft forward and out of the transmission as the gears are removed.
6. Push the countershaft rearward and out of the transmission as the gears and spacers are removed. Note the sequence of spacers and gears for reassembly.
7. Pull the mainshaft needle bearing from the housing.
8. Remove the reverse idler shaft and gear.
9. Remove the countershaft front bearing, retainer and shims.

INSPECTION AND REPAIR

1. Wash all parts in cleaning solvent and dry with compressed air. Do not spin bearings.
2. Check all bearings for looseness, wear, roughness, pitting and scoring.
3. Check gears and shafts for wear and burrs. Remove any burrs with a fine stone.
4. Inspect the housing for cracks, restricted oil passages and dents or raised places on its machines faces. Smooth off raised places with a file.

REASSEMBLY

Reassembly is basically the reverse of disassembly; however, particular attention should be given to the following steps.

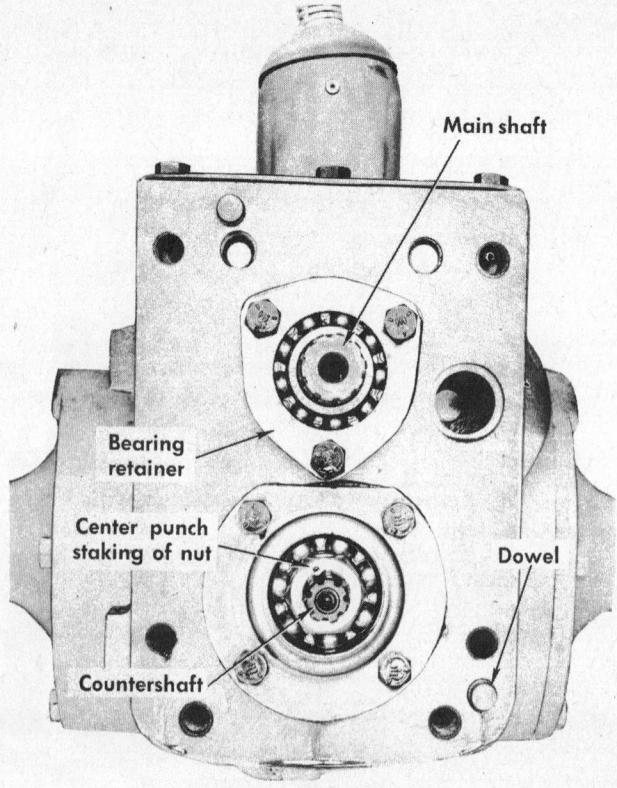
Transmission front view

1. Be sure all bearings are thoroughly lubricated.

2. The transmission mainshaft needle bearing must be installed with its oil hole aligned with the oil hole in the housing.

3. Assemble the differential, carrier bearings, cages and shims. Check bearing pre-load and adjust as necessary *before* replacing the transmission countershaft. Install or remove shims as necessary. Preload is correct when a steady pull of one to three lb. is necessary to rotate the differential assembly.

4. Remove the differential assembly, keep the shims with the cages then install the transmission countershaft, bearings, gears, spacers, front bearing retainer, shims and nut. Tighten the nut to 85 ft.lb. Tighten retainer capscrews to 20 ft.lb.

5. Install the differential assembly, keeping the preload shim pack correct as previously established. Drive gear must be on the right with teeth facing left.

6. Check the backlash between the drive gear and pinion and the gear teeth bearing pattern as follows.

7. Apply a thin coat of red lead or Prussian blue to the bevel pinion teeth faces, then rotate the gears by hand and observe the bearing pattern.

NOTE: Some deflection will occur under load. Allowance is made in gear design to prevent concentration of load on teeth edges.

8. Hand testing and very light loads should provide a pattern as shown in Figure "B". When load and deflection increases the pattern will progress as in Figure "A".

9. The desirable (no load) pattern in Figure "B" is the result of adjusting the bevel gear lateral position to the specified range of .003 in. to .005 in. backlash.

10. Tooth bearing position from the root to the crown of the tooth is controlled by lateral position of the pinion. If low tooth bearing on bevel pinion is indicated (as shown in Figure "C") the pinion must be adjusted toward the bevel gear. If high tooth bearing on the bevel pinion is indicated (as shown in Figure "D") the pinion must be adjusted away from the bevel gear.

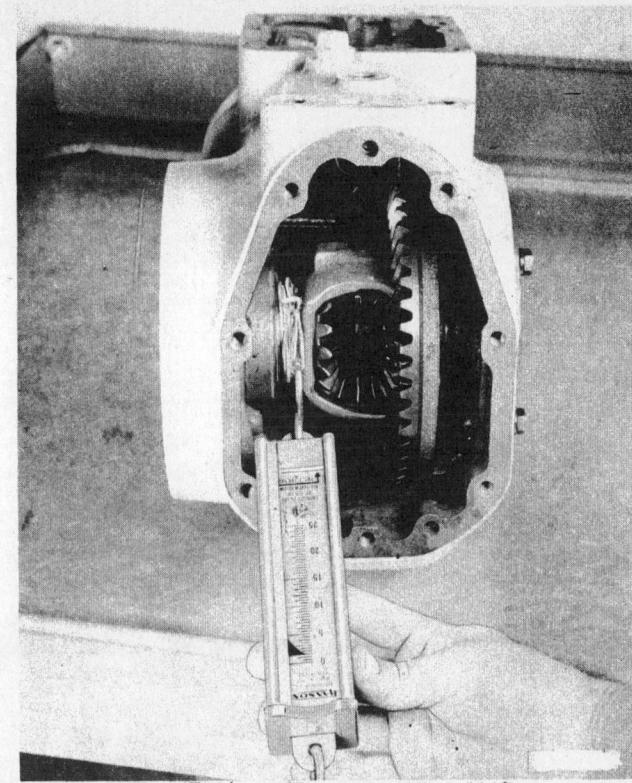

Checking differential bearing preload

NOTE: If it is necessary to move the pinion in or out to correct "Root to-crown" bearing, the bevel gear must also be moved laterally to maintain the specified backlash.

11. Stake the countershaft nut by center-punching the face of the nut over a spline groove.

12. Continue the assembly in reverse order of disassembly.

13. Fill housing to proper level with specified lubricant.

NOTE: Creeper attachment has its own lubricant separate of the transmission. Fill creeper at breather and check at side plug in creeper housing.

HYDROSTATIC TRANSMISSION AND DIFFERENTIAL

Model 123

REMOVAL (TRANSMISSION)

1. Split the tractor.

2. Loosen the transmission rear cover plate and drain the lubricant from the transmission-differential case.

3. Remove the oil filter. Plug the opening. Tape all openings.

4. Block the rear wheels and tip the unit up.

5. Be sure to support the collar and shaft firmly. Remove the rear pin securing the main driveshaft coupling to the hydrostatic unit shaft.

6. To remove the suction tube it will be necessary to loosen both connections. Lift up on the end of the tube that goes into the differential case to move it part way out. Then, remove the end of the tube that goes into the hydrostatic unit. Completely remove the tube. Plug the openings.

7. Remove the cam bracket capscrews

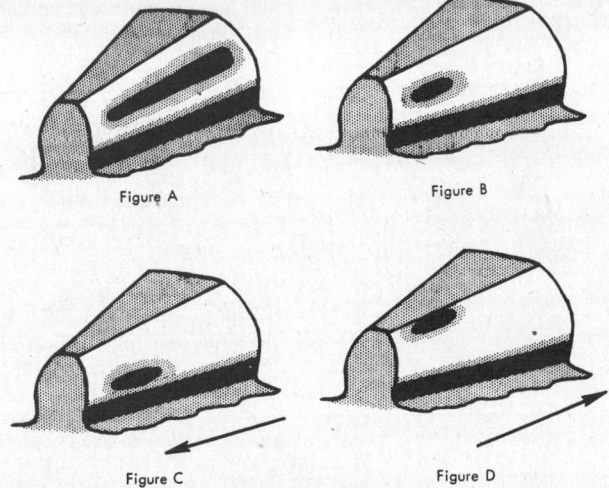

Wear patterns

International Harvester

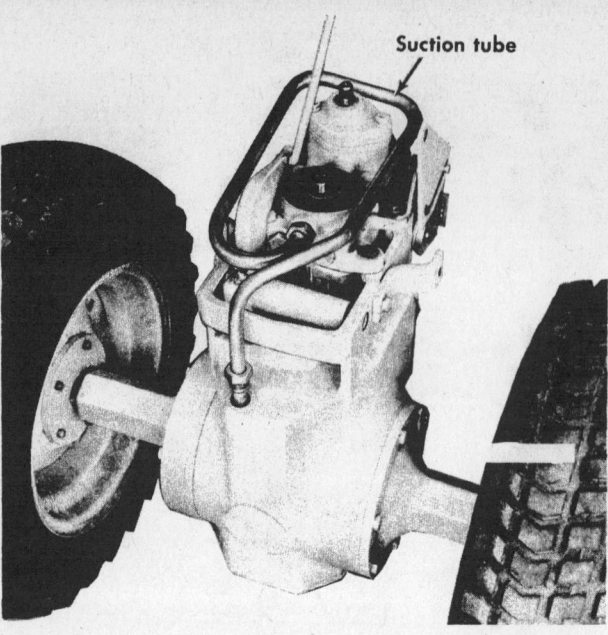

Suction tube partially removed

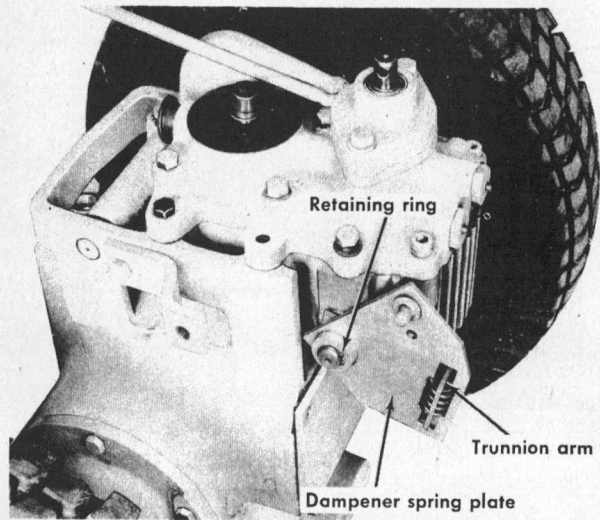

Damper spring plate retaining ring

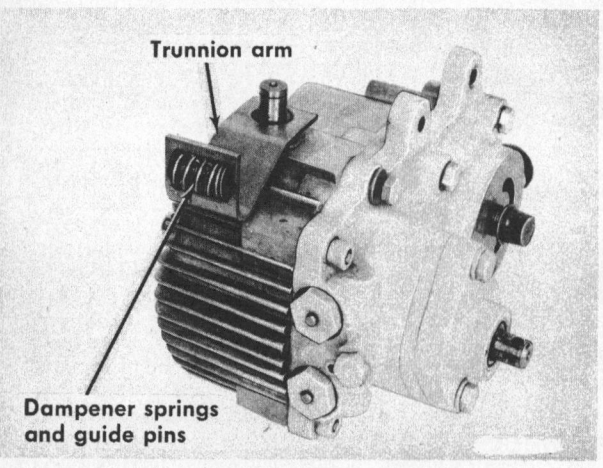

Damper springs and guide plate installed

and remove the bracket and speed control cam assembly. If it is necessary to service this assembly, remove the snapring that holds the speed control cam in the bracket.

8. Remove the retaining ring (2), that secures the dampener spring plate to the unit. Remove the plate.

9. Remove the springs and the guide pins from the trunnion arm.

10. Remove the three remaining mounting capscrews and remove the unit.

INSTALLATION

1. **IMPORTANT:** Prior to attaching a new hydrostatic drive replacement unit onto the transmission-differential case, squirt oil into the drain which is directly underneath the motor output shaft and the filter port. Turn the unit upside down to allow oil to flow into the passages. Rotate the input pump shaft and output motor shaft to insure they are free to turn.

2. Install the light and heavy dampener springs with guide pins into the trunnion arm.

3. Install the dampener spring plate and secure it to the shaft with the retaining ring.

4. If the speed control cam was removed from the bracket, reassemble it and install the retaining ring.

5. Install the cam and bracket assembly on the unit so it is at the bottom end of the slot and fasten it in place just snug using the slotted end of the bracket as a fastening point. Be sure there is a washer between the capscrew and the bracket.

NOTE: The elongated hole in the bracket is also used as a mounting bolt hole. Do not install this capscrew at this time.

6. Using a new gasket, install the unit on the transmission-differential case and fasten it securely with the three capscrews. Torque the capscrews to 30 ft.lb. The mounting capscrew that goes through the cam bracket should be tightened just snug.

7. Install the oil tube on the unit. Install the differential case end of the tube first, then the end that goes into the hydrostatic drive unit. Be sure the ferrules seat in their shoulders in the cases. Start the nuts and screw them in until they just touch the ferrules, then tighten the nuts ¼ turn only. Do not paint the oil tube.

8. Install a new oil filter as follows:
 a. Apply oil to gasket.
 b. Thread filter on, by hand until tight to seat gasket.
 c. Loosen filter.
 d. Turn again until gasket contacts base.
 e. Tighten filter an additional three quarter turn.
 f. Check for leaks.

9. Install the driveshaft and coupling to the drive unit input shaft. Firmly support the input shaft and install the pin to lock the shafts together.

10. If the cooling fan was removed from the driveshaft it will be necessary to position it on the shaft so that the rear edges of the blades are ¾ of an in. from the front edge of the collar.

International Harvester

11. Recouple the tractor.

REMOVAL AND DISASSEMBLY (DIFFERENTIAL)

1. Drain the lubricant.
2. Split the tractor.
3. Remove the rear axles and their carriers.
4. Remove the differential carrier bearing cage and shims from each side. Keep the shims with each cage and identified for each side.
5. Remove the differential from the transmission case.

NOTE: The differential must be turned before it can be removed.

6. Drive out the pinion shaft lock pin.
7. Remove the pinion shaft.
8. Remove the pinion gears and side gears.
9. If the differential drive gear requires separate replacement, press out the eight retaining rivets.
10. Remove the bearing cones from the differential carrier if they are to be replaced.
11. Remove the bearing cups from each cage if replacement is necessary.
12. Remove the hydrostatic drive.
13. Remove the bevel pinion shaft expansion plug.
14. Remove the snapring securing the bevel pinion shaft in the transmission case.
15. Using a brass drift and hammer, tap the bevel pinion shaft to the rear which will release it from the front bearing and the constant mesh gear.

INSPECTION AND REPAIR

1. Wash all parts in cleaning solvent and dry with compressed air. Do not spin bearings.
2. Check all bearings for looseness, wear, roughness, pitting and scoring.
3. Check gears and shafts for wear and burrs. Remove any burrs with a fine stone.
4. Inspect the housing for cracks, restricted oil passages and dents or raised places on its machines faces. Smooth off raised place with a file.

ASSEMBLY

1. Assemble the differential, carrier bearings, cages and shims. Check bearing preload and adjust as necessary before replacing the bevel pinion shaft. Install or remove shims as necessary. Preload is correct when a steady pull of one to three lb. is necessary to rotate the differential assembly.
2. Remove the differential assembly, keep the shims with the cages.
3. If the original bevel pinion shaft and transmission case is used, skip to step 4.
4. If a new bevel pinion shaft, transmission case or rear bearing cup and cone are used proceed as follows:
   a. Take the number stamped on the case and the number stamped on the

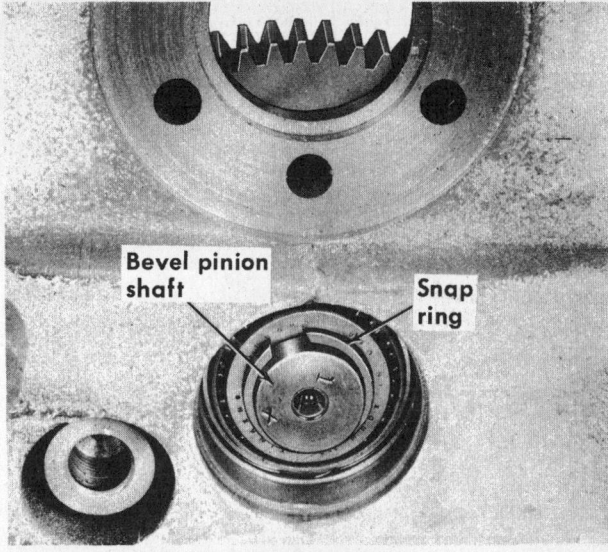

Bevel pinion shaft snapring

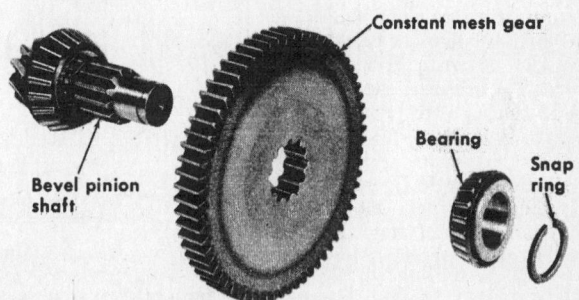

Bevel pinion shaft components

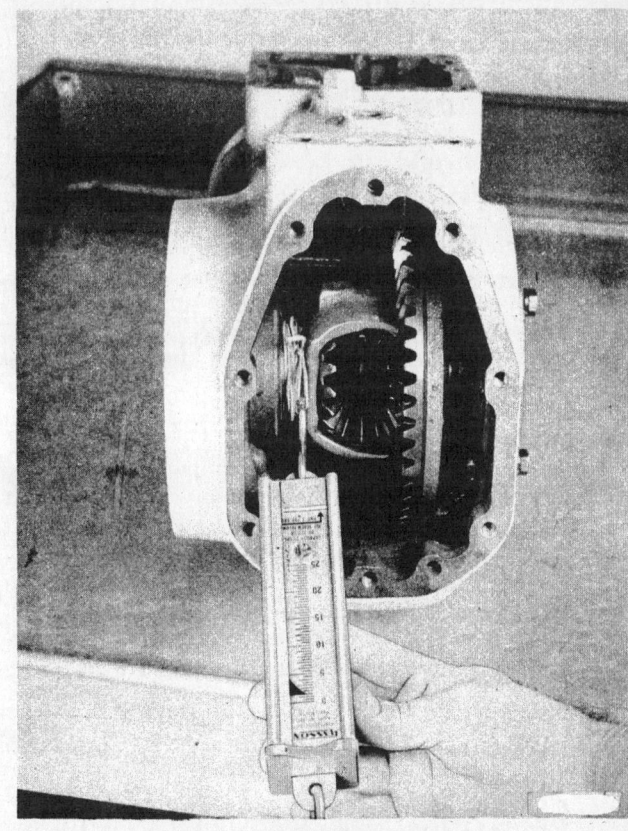

Measuring differential bearing preload

International Harvester

end of the bevel pinion shaft and add them together.

b. Add .015 in.

c. Install sufficient shims, to total the figure obtained previously, in the bore for the rear bearing cup.

d. Press the cup in its bore till it bottoms against the shims.

NOTE: Shims are available in .004, .007 and .015 in.

5. If it was removed, press the rear beraing cone on the bevel pinion shaft until it bottoms against the shoulder on the shaft.

6. Install the constant mesh gear retaining ring on the pinion shaft. Be sure the brake lining disc is installed and in position in transmission case. Hold with grease.

7. Start the bevel pinion shaft assembly in its bore in the rear of the transmission case. Install the constant mesh gear and complete the installation of the bevel pinion shaft assembly.

8. With the bevel pinion shaft supported at the gear end, gradually press or tap the front bearing cone onto the shaft. Rotate the shaft while installing the bearing to be sure the bearing does not get cocked or damaged. Press the bearing cone onto the shaft until the bearings are preloaded within the range of 5 in.lb. to 30 in.lb. rolling torque.

9. With the rolling torque figure obtained in step 8, refer to the table below to determine the amount of axial preload in the assembly at this time.

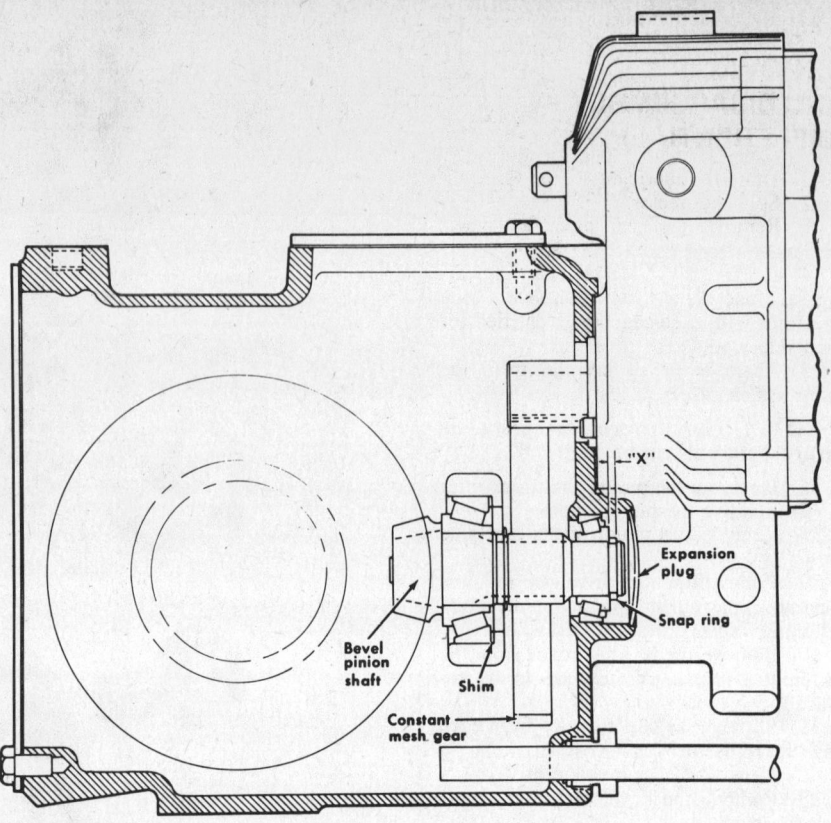

Bevel pinion shaft installed in case

Rolling Torque, in. lb.	Preload①
5	.0045
10	.0060
15	.0075
20	.0090
25	.0105
30	.0120

①The distance the bearings are telescoped beyond the desired .003 in. end play.

10. Using a feeler gauge, measure the gap between the cone surface of the front bearing to the furthest edge of the groove for the snapring. The measurement obtained is to be considered "X".

Class	Snapring Thickness Chart (In.)
A	.122 or up
B	.122-.116
C	.116-.110
D	.110 or under

11. Using the dimension obtained from the table in step 9, subtract the value of preload from the figure obtained for "X" in step 10. This value will be the correct size for the snapring to give an average of .003 in. end play.

Example:

a. The distance from the cone surface to the furthest side of the snapring groove is .117 in.

b. Rolling torque is 15 in.lb.

c. From the table, 15 in.lb. torque equals .0075 in. preload.

d. Exact snapring thickness should be .117-.0075 = .1095 in.

e. .1095 would use a class "D" snapring.

12. Install the snapring selected and be sure it bottoms in its groove. Tap the pinion shaft back to seat the front bearing against the snapring.

13. Install a new expansion plug.

14. Install the differential assembly, keeping the preload shim pack correct as previously established. Drive gear must be on the right with teeth facing left.

15. Check the backlash between the drive gear and pinion and the gear teeth bearing pattern as follows.

16. Apply a thin coat of red lead or Prussian blue to the bevel pinion teeth faces, then rotate the gears by hand and observe the bearing pattern.

NOTE: Some deflection will occur under load. Allowance is made in gear design to prevent concentration of load on teeth edges.

17. Hand testing and very light loads should provide a pattern as shown in figure

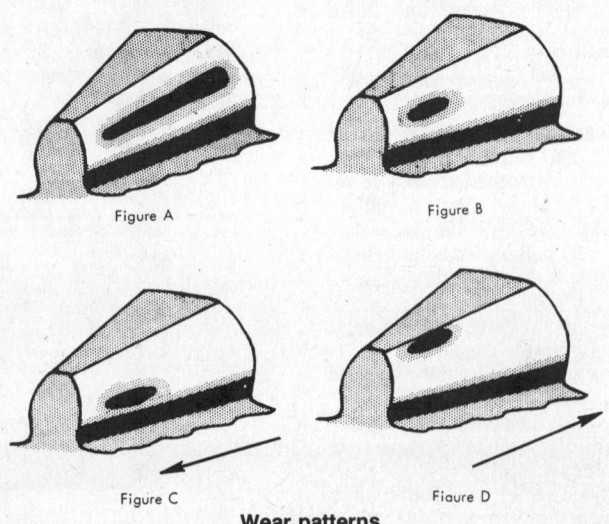

Wear patterns

International Harvester

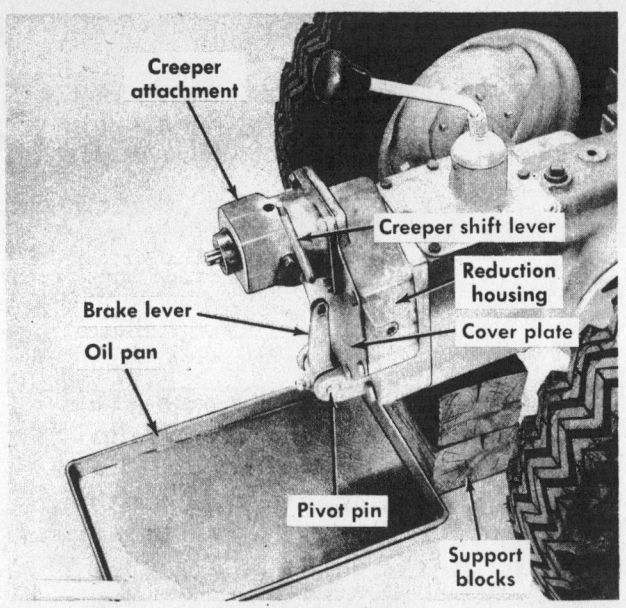

Supporting the tractor

Reduction drive

"B". When load and deflection increases the pattern will progress as in Figure "A".

18. The desirable (no load) pattern in figure "B" is the result of adjusting the bevel gear lateral position to the specified range of .003 in. to .005 in. backlash.

19. Tooth bearing position from the root to the crown of the tooth is controlled by lateral position of the pinion. If low tooth bearing on bevel pinion is indicated (as shown in Figure "C") the pinion must be adjusted toward the bevel gear. If high tooth bearing on the bevel pinion is indicated (as shown in figure "D") the pinion must be adjusted away from the bevel gear.

NOTE: If it is necessary to move the pinion in or out to correct "Root-to-crown" bearing, the bevel gear must also be moved laterally to maintain the specified backlash.

20. Install the rear axles and wheels.
21. Install the hydrostatic drive.
22. Recouple the tractor.

REDUCTION DRIVE

REMOVAL AND DISASSEMBLY

1. Split the tractor.
2. Place an oil pan under the reduction housing and remove the creeper assembly (if tractor is so equipped).
3. Remove the brake lever, pivot pin and push rod.
4. Remove the reduction housing front cover plate.
5. Hold the drive coupling and shaft from turning and remove the reduction gear retaining capscrew and washers. Remove the gear spacer.

6. Remove the reduction gear from the transmission shaft and from the housing.

NOTE: It may be more convenient to pull the reduction driveshaft, seal and bearing before removing the reduction gear from the housing. Clearance between the gear and the capscrew bosses is restricted on some tractors.

7. Remove capscrews from holes "A" and "B".

NOTE: Soft copper sealing washers are used under the "B" capscrew heads.

8. Move the housing forward and away from the transmission housing as the brake disc slides off the transmission countershaft.

9. Pull the reduction driveshaft, seal and bearing from the reduction housing if it was not removed in step 6 NOTE.

10. Support the driveshaft splined coupling and drive out the coiled spring pin.

NOTE: The splined coupling is used only on tractors equipped with creeper attachment.

11. Press the driveshaft from the ball bearing.
12. Press the needle bearing rearward from the housing.
13. Remove the brake components.

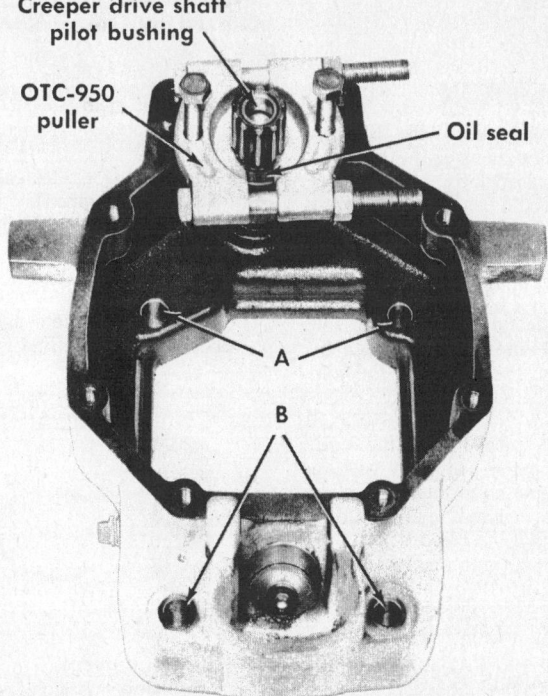

Removing reduction driveshaft, seal and bearing

493

International Harvester

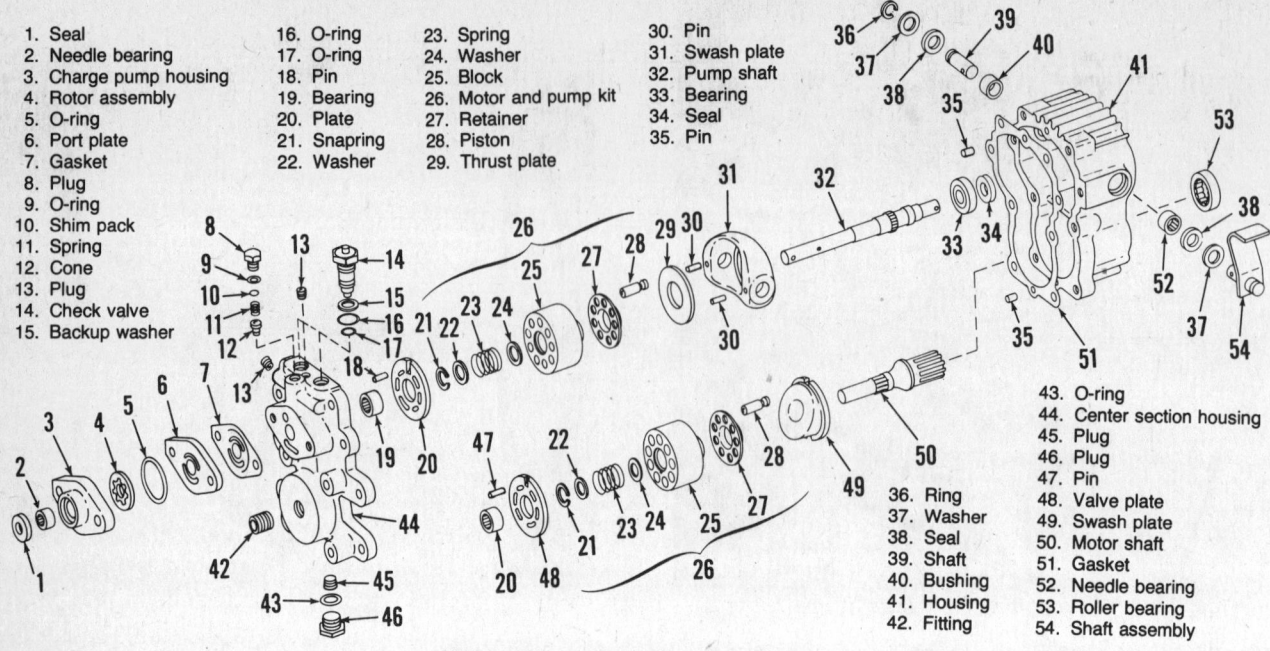

1. Seal
2. Needle bearing
3. Charge pump housing
4. Rotor assembly
5. O-ring
6. Port plate
7. Gasket
8. Plug
9. O-ring
10. Shim pack
11. Spring
12. Cone
13. Plug
14. Check valve
15. Backup washer
16. O-ring
17. O-ring
18. Pin
19. Bearing
20. Plate
21. Snapring
22. Washer
23. Spring
24. Washer
25. Block
26. Motor and pump kit
27. Retainer
28. Piston
29. Thrust plate
30. Pin
31. Swash plate
32. Pump shaft
33. Bearing
34. Seal
35. Pin
36. Ring
37. Washer
38. Seal
39. Shaft
40. Bushing
41. Housing
42. Fitting
43. O-ring
44. Center section housing
45. Plug
46. Plug
47. Pin
48. Valve plate
49. Swash plate
50. Motor shaft
51. Gasket
52. Needle bearing
53. Roller bearing
54. Shaft assembly

Hydrostatic transmission components

INSPECTION AND REPAIR

1. Inspect the driveshaft for wear on the gear teeth, needle bearing area, oil seal contact area and drive pin hole.
2. Inspect the reduction gear teeth for wear or chipping and the fit of the gear on the transmission shaft.
3. On tractors with creeper, inspect wear of pilot bushing for creeper driveshaft.
4. Inspect needle and ball bearings for wear, pitting and roughness of operation.
5. Thoroughly clean the reduction housing.

ASSEMBLY AND INSTALLATION

1. Install a new gasket to the front of the transmission case. Dowels in the case will hold the gasket in place.
2. Press the needle bearing into the reduction housing from the rear. Rear end of bearing should be flush with housing.
3. Press the ball bearing onto the driveshaft.
4. Lubricate the lip of a new oil seal and install the seal onto the shaft. Be careful when sliding the seal lip over the pin hole in the shaft. Flat face of seal case must be forward.
5. Install the splined coupling and coiled pin (if tractor is equipped with a creeper). Coiled pin ends must be even with or below the spline root to avoid interference when shifting the creeper unit.
6. Install a new O-ring onto the brake lining retainer and install in the reduction housing.
7. Install the reduction housing to the transmission case. Be sure the gasket and dowels are in place.
8. Install new copper sealing washers on the two lower capscrews. Refer to "B".

Tighten capscrews at "A" and "B" to 80 ft.lb.

9. Install the brake linings and disc then push the front lining retainer rearward to hold disc and linings in place.
10. Install the reduction gear and spacer to the transmission mainshaft and secure with the capscrew, flat and lockwasher. Tighten capscrew after driveshaft is installed.
11. Install the driveshaft with ball bearing, seal and splined coupling (on tractors with creeper). Seal case should be flush with housing.

NOTE: Seal holds ball bearing in place and front cover holds seal in place.

12. Tighten reduction gear retaining capscrew to 55 ft.lb.
13. Install new gasket and housing front cover.
14. Install new gasket and creeper unit (on tractors so equipped).
15. Replace the brake push rod, ball, lever and pivot pin.
16. Recouple the tractor by reversing the splitting procedure.
17. Fill transmission and creeper to proper level with specified lubricant.

Overhaul of Hydrostatic Transmission

CHARGE PUMP

Removal and Disassembly

1. Thoroughly clean and deburr the outside of the transmission before attempting any disassembly. Remove paint from shaft surfaces.
2. Remove the capscrews securing the charge pump housing to the center section housing. Carefully remove the pump housing. The rotor assembly may stick to the housing. Do not drop the assembly.
3. Remove the rotor assembly (if it was not removed in step 2). Because of the polished surface, be sure to protect the assembly against nicks, scratches and rust.
4. Remove the pump port plate.

NOTE: The position of the port plate gasket.

The new gasket must be installed with the circular groove to the top of the center section housing with the flat end toward the flat of the housing.

5. Using a screwdriver, pry the lip seal out of the pump housing.

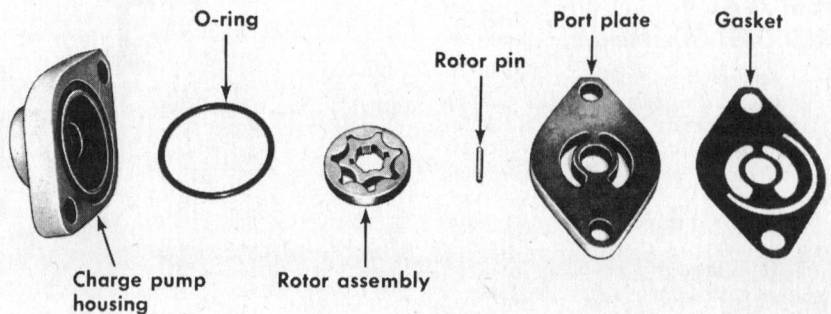

Charge pump components

494

International Harvester

6. To replace the needle bearing, press the bearing out of the pump housing.

Assembly and Installation

Reassembly and installation of the charge pump to the center section housing is the exact reverse of removal and disassembly with particular attention paid to the following:

1. Thoroughly lubricate all parts in clean Hy-Tran.
2. If removed, install the new needle bearing in the pump housing. Be sure the bearing is "bottomed" in the bore.
3. Install a new oil seal in the pump housing. Install a new O-ring.
4. Install the pump port plate gasket with the circular groove to the top of the center section housing and flat end toward the flat of the housing.
5. Torque the pump housing capscrews evenly to 52 ft.lb. Rotate the pump shaft while tightening the capscrews. Loosen and retighten the capscrews evenly as necessary to relieve any binding of the shaft.

RELIEF VALVE

Removal and Installation

1. Remove the relief valve plug, shim pack (if any), spring and cone from the center section housing.
2. Wash and dry the components.
3. Check the spring for pitting and rust.
4. Check the cone for wear or damage. Check the valve seat in the center section housing for dirt, nicks and scratches.
5. Install the relief valve in the reverse order of removal. Be sure to use a new O-ring on the plug.
6. Recouple the tractor and check the operation of the relief valve.

HYDROSTATIC UNIT

Disassembly

1. Be sure the outside surfaces of the transmission have been thoroughly cleaned. Place the transmission assembly in the holding fixture.
2. Remove the capscrews securing the center section housing to the transmission housing.
3. Lift the center section housing from the transmission housing.

IMPORTANT: The valve plates may stick to the center section housing surface. Be extremely careful not to drop them.

4. Remove the pump and motor valve plates (if not removed in step 3 above) noting the location of each plate. The valve plate with two notches is used on the pump assembly and the plate with four notches on the motor assembly. Remove the valve plate pins.
5. Tip the transmission housing so that the pump and motor cylinder block assemblies can be removed. Grasp the assemblies so that the pistons will not fall out and be damaged.
6. Remove the trunnion shaft assemblies from the hydraulic pump swash plate by driving on the spring pins suffi-

Lifting center section

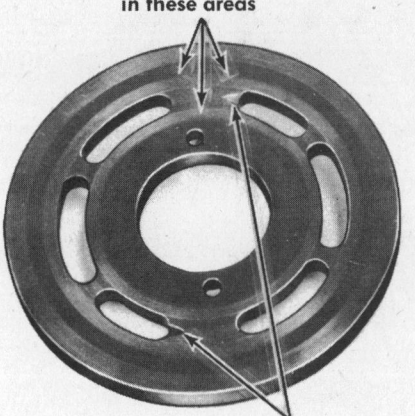
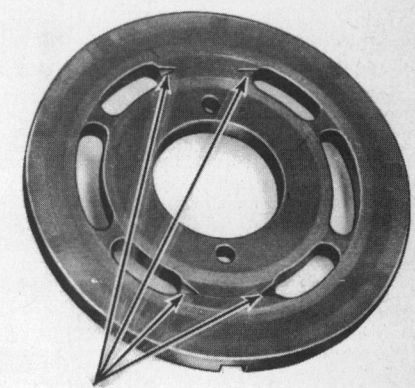

Pump valve plate (two notches)   Motor valve plate (four notches)

Check for wear in these areas

Pump and motor valve plates

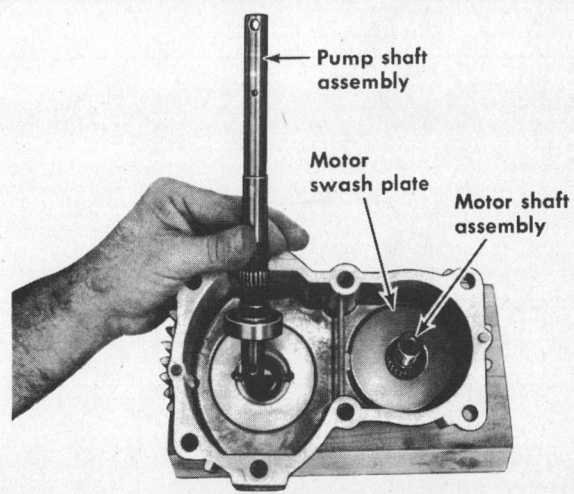

Removing the pump shaft

International Harvester

Removing the hydraulic motor swash plate capscrews

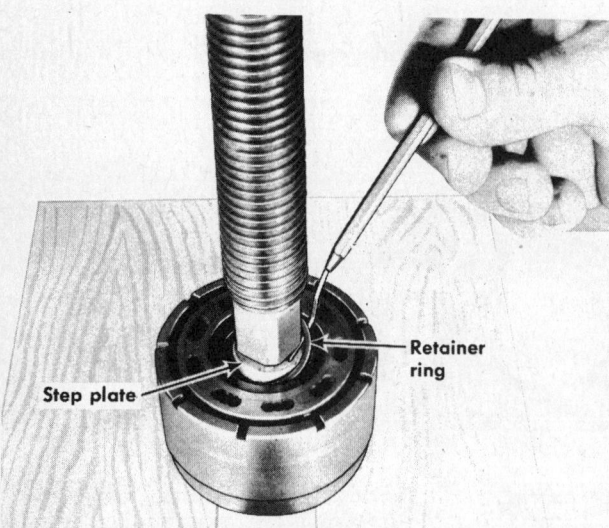

Removing retaining ring

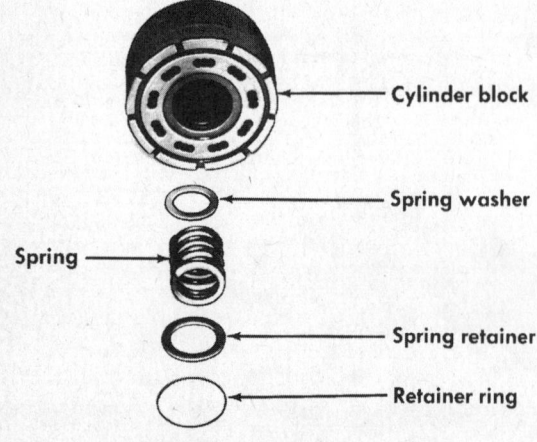

Cylinder block components

ciently to remove the shaft assemblies. Remove the swash plate.
7. Remove the pump shaft assembly.
8. Remove the socket head capscrews securing the hydraulic motor swash plate.
9. Remove the transmission motor shaft and swash plate from the housing.

Inspection and Repair

1. Remove piston and slipper assembly.
2. Place the cylinder block assembly in a press on wood blocks.
3. Press on the spring retainer, using a step plate to compress the cylinder block spring. Remove the retaining ring.
4. Carefully release the press. Remove the spring retainer, spring and spring seat. Remove the cylinder block from the press.
5. Thoroughly clean all parts and blow dry with compressed air.
6. Check the spring.
7. Check the cylinder block valve face for damage and the piston bushings for excessive wear. Any linear scratches along the length of the bore will reduce efficiency. Check piston fit in the bushings.
8. Install the spring washer (bevel side in), spring and spring retainer. Place the assembly in a press.
9. Compress the spring, using a step plate. Install the retaining ring.
10. Release the press and wrap the assembly in clean paper or lint free cloth before setting aside.
11. Remove the pistons from the slipper retainer. Thoroughly clean the pistons and blow dry with compressed air. Be certain center oil passage is open.
12. Carefully inspect each piston for scoring, wear or scratches.
13. Check the slippers for severe scratches or embedded material. Slippers may be lapped, but do not remove more than .005 in. All slippers must be within .002 in. thickness of each other.
14. Replace any pistons that are badly worn or scored.
15. Wrap the pistons in clean paper or lint free cloth or reinstall in the cylinder block and wrap the assembly.
16. Thoroughly clean the valve plate and blow dry with compressed air.
17. Inspect the valve plate for scratches, excessive wear or erosion. A worn or scored valve plate reduces pump efficiency.

NOTE: To check the plate for wear, run your finger nail or a sharp pencil across the face of the plate. If wear is felt, replace the plate.

18. Inspect the pin slot and grooves of the valve plate. Clean out any foreign matter and deburr the surface as necessary.
19. Inspect the slipper retainer for damage. A slight wear pattern where the slippers ride is normal. Replace if wear is excessive.
20. Inspect the thrust plate (for the hydraulic pump swash plate) for wear, embedded material, or scoring.
21. Inspect all the bearings and replace as necessary.

International Harvester

Assembly

Reassembly is basically the reverse of disassembly however, particular attention should be given the following:

1. Thoroughly lubricate all parts in clean Hy-Tran. Pipe plugs in the center section housing showing leakage must be removed, doped (Teflon® tape on the threads is ideal) and reinstalled.
2. Be sure to install the pump swash plate with the thin pad toward the top of the transmission housing.
3. Use all new O-rings, seals and gaskets.
4. Pistons are interchangeable between the pump and motor block assemblies. The thickness of the piston slippers in the block assembly must not vary more than .002 in. of each other.

NOTE: Some units have thick wall pistons. If it is necessary to replace one thick wall piston, the complete set (nine) in the block assembly must be replaced and then the remaining set (nine) of thick wall pistons must be used ONLY in pump block assembly.

NOTE: The thin wall pistons can be serviced separately however all the slippers in a block assembly must be within .002 in. of each other.

5. New center section needle bearings must be installed so that they extend .100 in. above the machined surface of the center section. The bearings "pilot" the valve plates when the unit is reassembled.

ADJUSTMENTS

Brake

1. The brake should engage when the pedal arm is pressed down to within a maximum of 1 5/16 in. and a minimum of 3/4 in. distance above the top of the left foot support, which serves as the pedal stop.
2. It may be possible to push the pedal all the way down to the pedal stop, but this is of no concern as long as the brake is engaged when the pedal arm is at least 3/4 in. above the pedal stop.
3. To adjust the brake, loosen the jam nut "B" and turn the brake lever adjusting screw "C" in or out as required to get this measurement. The brake must not engage before the pedal arm is within the maximum distance of 1 5/16 in. above the pedal stop.

Speed Control Lever

NOTE: The brake pedal must be properly adjusted before beginning the speed control lever adjustment. If the tractor "creeps" in the "N" position or, if the speed control linkage has been disassembled or removed for any reason, the following adjustment must be made.

1. Block the tractor so the left rear wheel is off the ground.
2. Start the engine at half throttle or faster.
3. Move the speed control lever to the forward position. The rear wheel should rotate in the forward direction. Depress the brake pedal all the way down and release. The speed control lever should return to the

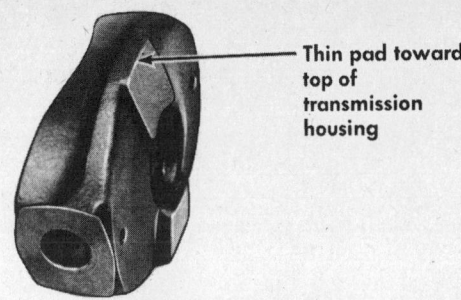

Pump swash plate pads

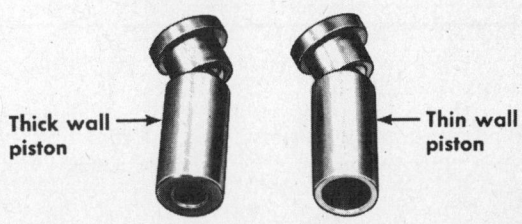

Piston identification

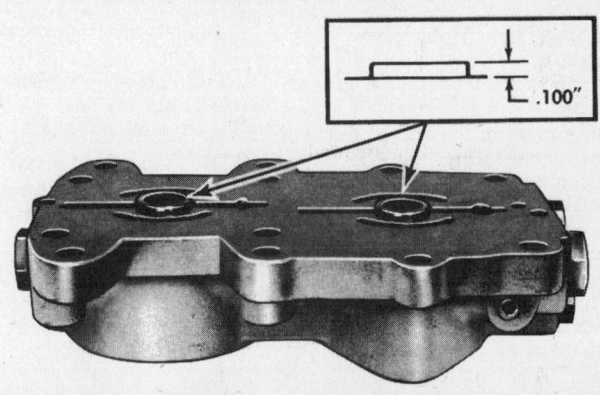

Center section needle bearings

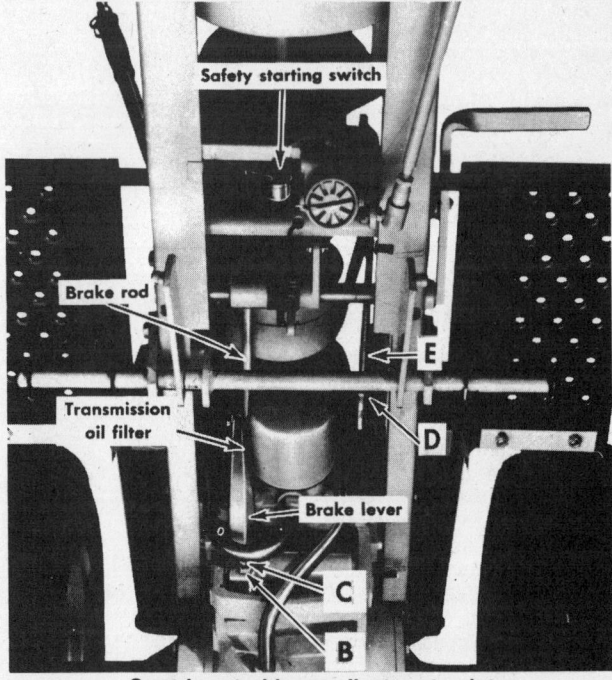

Speed control lever adjustment points

International Harvester

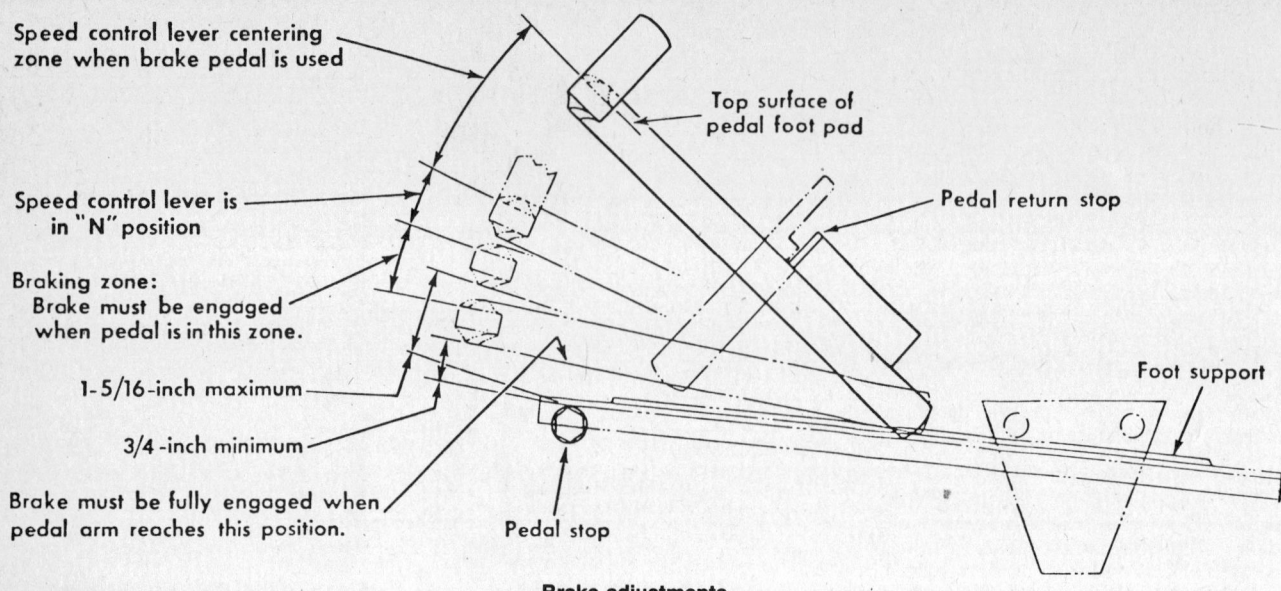

Brake adjustments

"N" position and the rear wheel stop turning.

4. If the rear wheel turns in the forward direction, loosen jam nut "D" and turn the connecting rod "E" counterclockwise to lengthen it until the wheel stops turning.

5. If the wheel turns in the reverse direction turn the connecting rod "E" clockwise. Tighten the jam nut "D".

6. Proper friction adjustment is necessary on speed control lever for proper operation. The lever friction should be adjusted as follows:
  a. Remove the battery.
  b. Place a small wedge between the hand-control mounting bracket and the adjusting nut.
  c. Move the lever towards the "F" or forward position to tighten the nut.
  d. Remove the wedge.
  e. Check the friction adjustment with a fish scale. The reading should be 10 lb. when pulled in either direction from the offset in the lever.
  f. Replace the battery.

CREEPER DRIVE

REMOVAL

1. Split the tractor.
2. Support the transmission and reduction housing on a suitable block, place an oil drip pan under the creeper unit and drain the creeper lubricant.
3. Four capscrews hold the creeper housing to the reduction housing cover plate. Remove the capscrews.
4. Bump the creeper to the side to loosen the housing from its gasket and dowels. Pull the creeper forward from the reduction cover and splined coupling.
5. If the driven coupling or the pilot bushing needs replacing, support the coupling and drive out the coiled spring pin. Remove the coupling.

DISASSEMBLY

1. Remove the snapring which holds the input shaft bearing cage in the housing.
2. Pull the shaft, bearing, retainer, planetary assembly and the direct drive coupling from the housing.

NOTE: The splined grooves of the direct drive coupling must align with the splines of the shifter collar.

3. Support the direct drive coupling and drive the coiled spring out. Remove the coupling from the shaft.
4. Slide the planet carrier off the input shaft and the planet gears off the carrier pins.
5. Remove the flat thrust washer.
6. A snapring holds the ball bearing outer race in the retainer. Remove the snapring then push the bearing and shaft from the retainer.
7. The ball bearing is held on the shaft by a snapring. Remove the snapring. The shaft can now be pressed from the bearing.
8. Press the oil seal from the bearing retainer.
9. Drive the shift poppet pin from the shaft and remove the poppet.
10. Shift the lever and shifter collar toward the rear of the case and at the same time lift the shifter collar up to disengage it from the shift yoke.
11. Drive the pin out of the shift yoke and lever shaft.
12. Slide the lever shaft from the yoke and housing.
13. Remove the O-rings from the shaft, housing and bearing retainer.

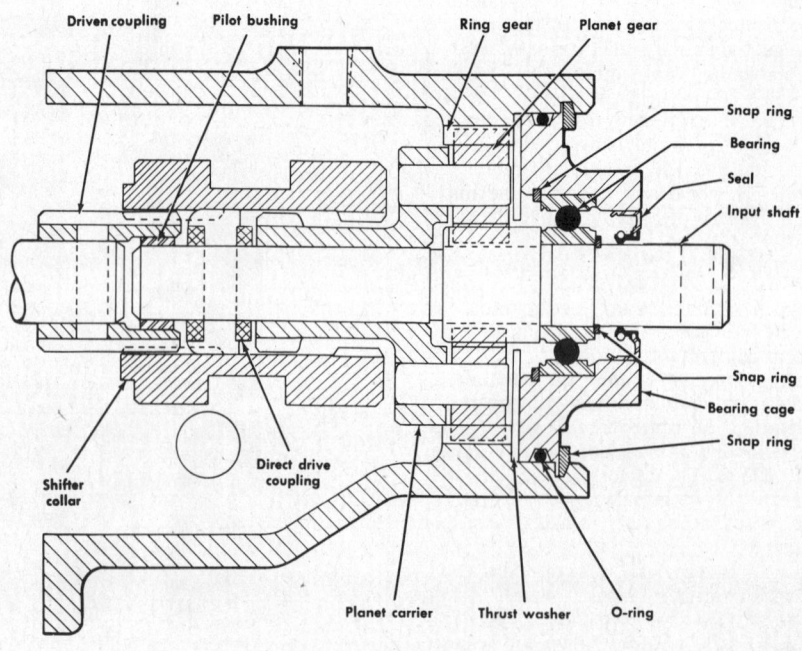

Side view cross section of the creeper drive

14. Wash all parts in cleaning solvent then dry thoroughly.

INSPECTION AND REPAIR

1. Check the input driveshaft for oil seal groove wear, worn or chipped teeth on the integral gear and pilot bushing wear on the rear end.
2. Check the splines of the direct drive coupling, planet carrier and the shifter collar for wear and chipping.
3. Check the housing for cracks and the integral sun gear for wear and broken teeth.
4. Inspect the ball bearing for pitting, scoring, wear and rough operation.

REASSEMBLY

Reassembly is basically the reverse of disassembly however, particular attention should be given the following:

1. Always use new O-rings, gaskets and oil seals. O-rings and oil seals should be coated with Lubriplate® or chassis lubricant to assist in installation and provide initial lubrication.
2. Install the oil seal after completing the drive assembly in the housing.
3. The pins which secure the direct drive coupling and the driven coupling to their respective shafts must be flush or below the spline groove so as not to interfere with shifting.
4. The long internal splines of the shifter collar go toward the rear.
5. The machined shoulder of the direct drive coupling goes toward the planet carrier.
6. Lubricate the components and rotate the driveshaft several turns with the shifter in each speed selection to insure freedom of movement and rotation.

INSTALLATION

1. Place a new gasket on the mounting face of the creeper housing. The dowels will hold it in place.
2. With the shifter collar in its rear position, rotate the driveshaft so the collar will engage the driven coupling as the dowels engage the reduction cover plate and the driveshaft enters its pilot bushing in the driven coupling.
3. Secure the creeper to the reduction housing cover plate.
4. Fill the creeper housing to proper level (check plug on side of housing) with the same lubricant as specified for the transmission-differential. Hy-tran fluid or SAE 30 Engine oil.
5. Shift the unit several times to insure freedom of movement. Leave the shifter lever in its forward position for recoupling the split tractor.
6. Reassemble the split tractor.

REAR AXLES

REMOVAL

1. Drain the transmission lubricant.
2. Stabilize the tractor by driving wooden wedges between the front axle and frame on each side and block the front wheels.
3. Remove the drawbar and differential housing rear cover.
4. Place a jack under either rear axle carrier and raise the rear wheel off the floor. Remove the "C" type snapring from the axle shaft inner end.
5. Slide the axle out of the differential side gear and axle carrier.
6. Support the transmission-differential housing on a block and remove the axle carrier.
7. Press the carrier bushing from the carrier.
8. Drive the oil seal from the carrier.
9. Clean the gasket surface of the axle carrier and differential housing.

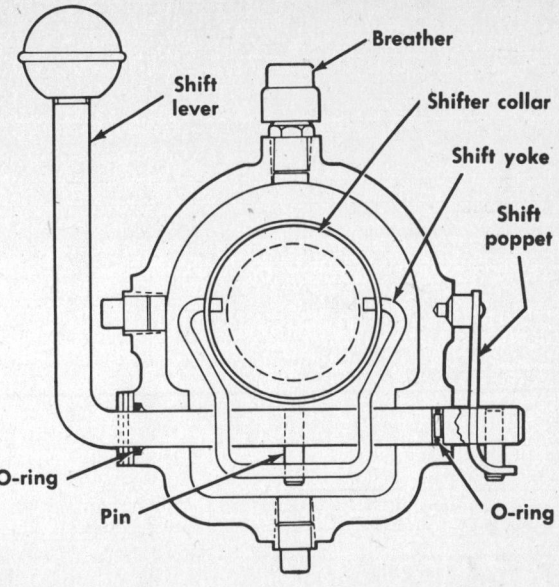

End view showing creeper drive shift mechanism

INSPECTION

1. Inspect the axle shaft for wear at the oil seal area, bushing location and splines on the inner end.
2. Roll the axle shaft along a flat surface to detect any warping or bending.
3. Check the axle carriers for cracks or breaks. Remove any high spots from the gasket surface with a flat file.

REASSEMBLY

1. Press a new bushing into the axle carrier. The oil groove must be at the bottom.
2. Press a new oil seal into the axle carrier.
3. Using a new gasket, install the axle carrier to the differential housing. Cap-

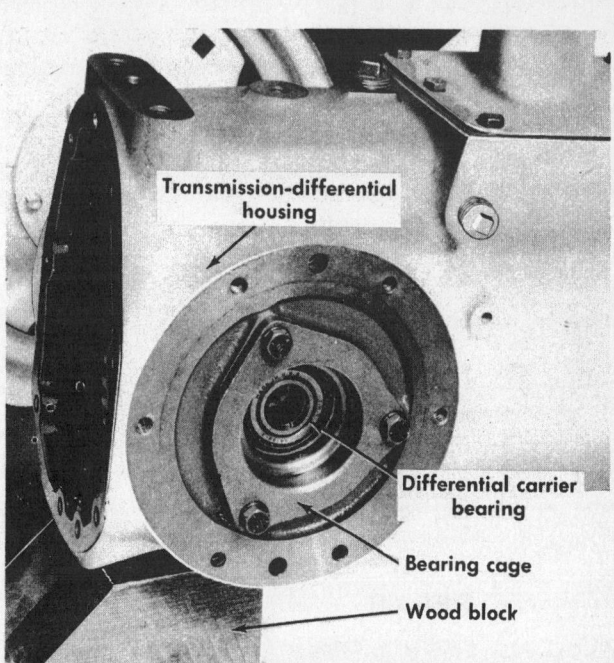

Rear axle carrier removed

International Harvester

screw threads should be coated with a non-hardening sealer (Permatex®) to avoid oil leaks. The frame pad of the axle carrier must be to the top.

4. Fill the cavity between the lips of the oil seal with chassis lube or heavy oil.

5. Lubricate the axle shaft and bushing then slide the shaft through the seal, bushing, carrier and differential side gear. Rotate the axle as it is pushed through to avoid damage to the seal. Wipe off excess lubricant.

6. Install a new "C" type snapring to the inner end of the axle shaft.

7. Replace the rear cover and drawbar.

8. Fill differential housing to proper level.

POWER TAKE-OFF

Front PTO Clutch

REMOVAL

1. Loosen the idler pulley adjusting nut.
2. Remove the nut, lockwasher and capscrew on each side of the grille guard that secures the mounting bracket to the tractor frame.
3. Disconnect the front clutch rod from the turnbuckle.
4. Remove the four nuts, lockwashers and capscrews securing the grille guard to the tractor frame. Remove the grille guard, hood and clutch shaft with rod from the tractor as an assembly.
5. Remove the jam setscrew and lock setscrew from each of the three holes in the clutch pulley housing.

NOTE: These setscrews lock the clutch to the bearing on the tractor crankshaft.

6. Remove the clutch from the tractor as an assembly.

DISASSEMBLY

1. Install a three jaw puller on the clutch assembly so the jaws hook into belt pulley groove and are located between the throw-out levers. Use a step plate between the thrust button and the puller bolt.
2. Tighten up on the puller bolt just until the friction disc disengages the pressure plate.
3. Remove the throw-out lever screw jam nuts and remove the pressure plate and friction disc.
4. Release the pressure applied to the clutch button by the puller and the remaining components of the clutch assembly can be removed.

INSPECTION AND REPAIR

1. Inspect the pressure plate for scoring or excessive warpage. If it is .010 in. or more out of flat it must be replaced.

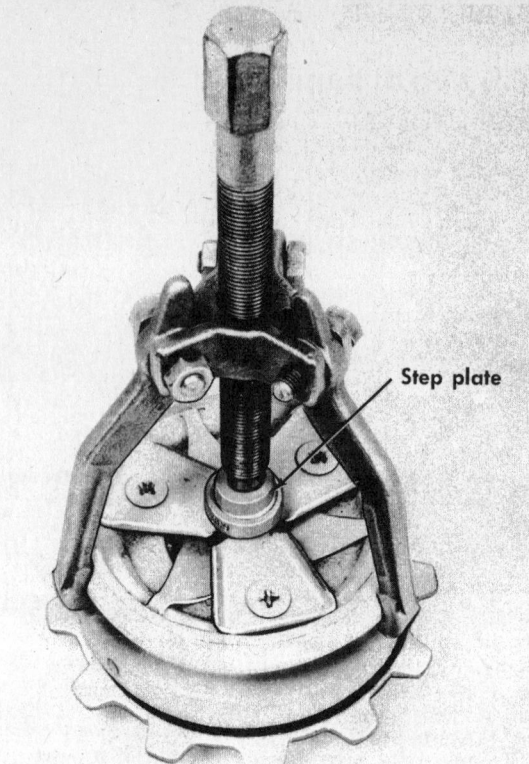

Puller installed on clutch

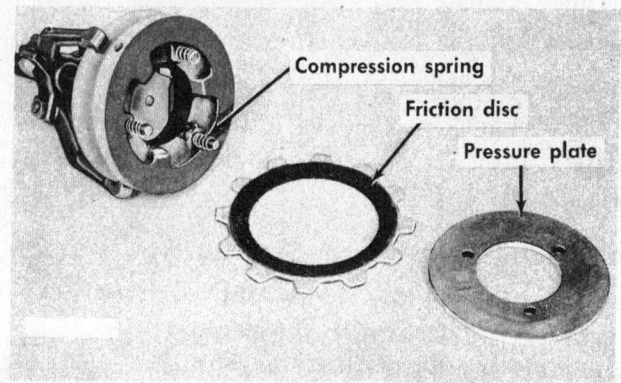

Pressure plate and friction disc removed

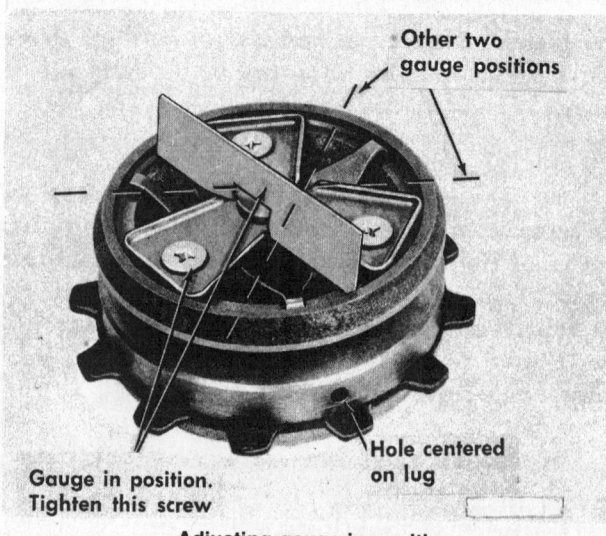

Adjusting gauge in position

International Harvester

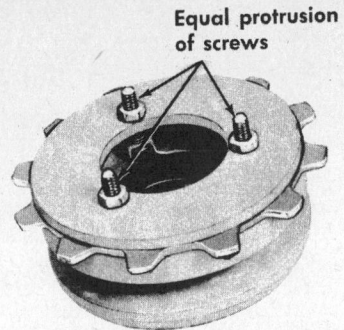

Screw protrusion approximately equal

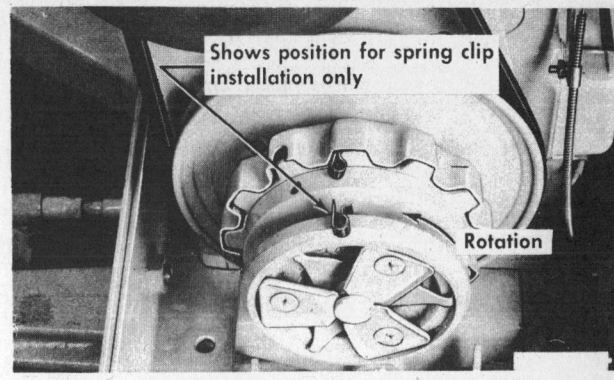

Installing clutch assembly

2. Inspect the friction disc for scoring, cracks or damaged lugs.

NOTE: The clutch will operate satisfactorily if four lugs are broken, providing the four broken lugs are not grouped together. If it is necessary to replace the friction disc, the pressure spring must be replaced at the same time.

3. Inspect the thrust button for damage or wear and replace if necessary.

4. Inspect the pulley groove for burrs, nicks or grooves that may cause damage to the belt. Repair or replace the pulley if necessary.

5. Inspect the bearing snapring for grooves or wear and replace if necessary.

6. Inspect the clutch bearing on the engine crankshaft to be sure it turns free. Replace the bearing if there is any doubt as to its serviceability.

7. Inspect the wear button on the clutch shaft assembly for wear and replace if necessary.

8. If new throw-out lever screws are to be used in reassembly, be sure they are those as listed in the parts catalog as these are special machine screws.

REASSEMBLY

1. Position the pressure spring on the actuating pulley so the tips are equally spaced between the screw holes.

2. Install the thrust button in the pressure spring.

3. Install the three throw-out levers so they engage the slot in the thrust button. Line up the screw holes and install the screws.

4. Holding the throw-out levers, screws, thrust button and pressure spring in place, turn the assembly over and install the compression springs (one to each screw), friction disc and pressure plate.

5. Install the nuts on the screws and tighten them finger tight. Be sure the friction disc is centered and that a lug on the disc is centered with a setscrew hole in the pulley housing.

ADJUSTMENT

1. Install the adjusting gauge in position shown.

2. Tighten the special machine screw (in line with the center of the gauge) until the gauge ends contact the recessed machined surface of the pulley. The gauge should not rock the tips.

3. Repeat step 2 for the remaining two gauge positions.

4. Recheck each of the three positions with the gauge a second time after all three screws have been adjusted to be sure all three adjustments are equal.

5. The specified pressure applied to the pressure spring is now set. The protruding ends of the screws should be approximately equal. If they are quite different, something is wrong and the clutch will have to be disassembled, checked, reassembled and adjusted.

6. Install the machine screw locknuts and tighten them to 6-7 ft.lb. torque.

INSTALLATION

1. If a new clutch bearing is to be used, install it on the crankshaft so it is flush with the end of the crankshafts. Lock it in place with the locking collar. Be sure to lock the collar to the bearing in the direction of crankshaft rotation. Lock the collar in place with the setscrew and nut.

2. Install the clutch assembly on the bearing part way. Be sure the setscrew holes in the clutch pulley housing line up with the slots in the crankshaft pulley.

3. Equally space and install 3 disc springs on the friction disc lugs on the non-drive side of the lugs. The non-drive side of the lugs is the lefthand side of the lug when looking at the front of the clutch. Place flat side of springs inside the cup. Push the clutch assembly the rest of the way on the bearing until the snapring in the clutch is flush with the bearing. Be sure the complete thickness of the disc is under the drive pulley cup.

4. Install the three ¼ x ½ in. cone point hex setscrews in the clutch pulley and torque them to 5-6 ft.lb.

5. Install the three ¼ x ¼ in. flat point hex setscrews and torque them 6-7 ft.lb.

6. Install the grille and hood being sure the clutch rod is inserted through the engine front plate on hand start units and between the engine and motor-generator on electric start units.

7. Install the capscrew, lockwashers and nut on each side of the grille guard that secures the mounting bracket to the tractor frame.

8. Connect the front clutch rod to the turnbuckle. With the hand lever in the forward position (clutch fully engaged), adjust the turnbuckle so there is 1/64 in. minimum clearance between the wear button and thrust button. Secure the turnbuckle with the jam nut.

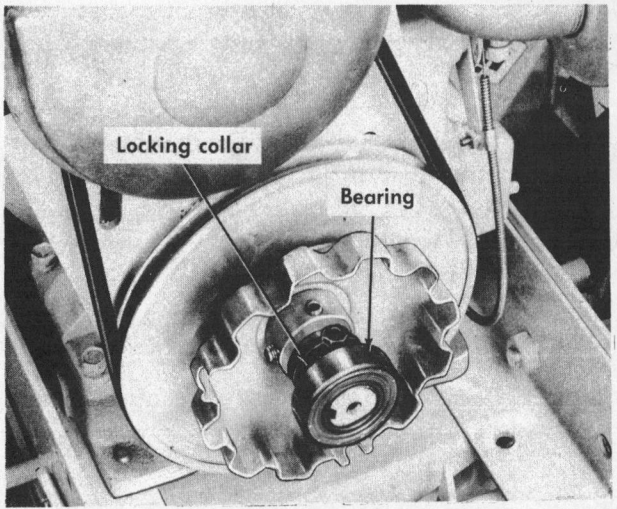

Clutch bearing and collar installed on shaft

501

International Harvester

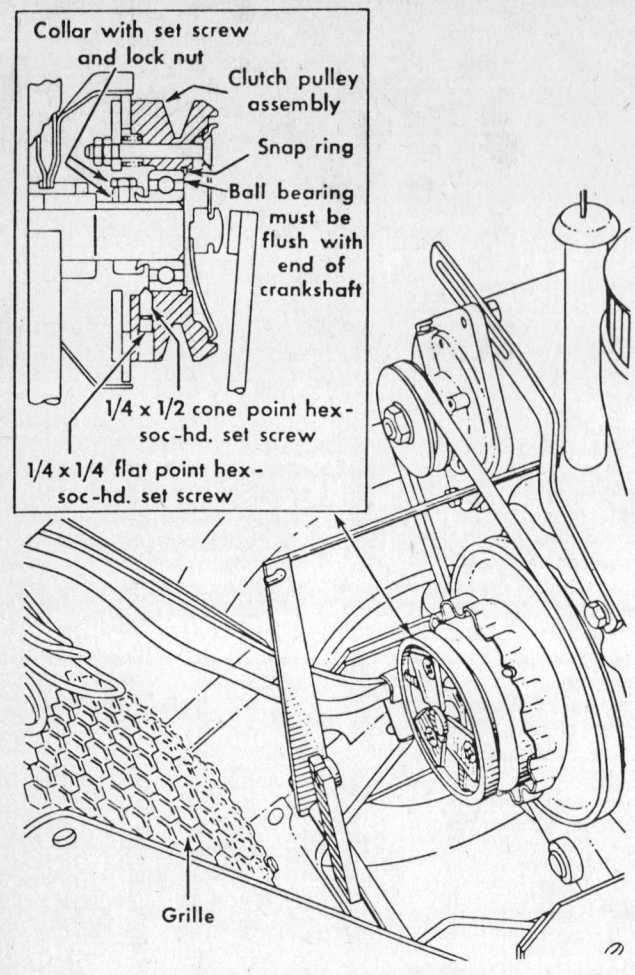

Clutch linkage adjustments

Rear PTO

The servicing of the rear PTO is basically easy as shown. It i removed by removing the seven capscrews securing it to the transmission case. The PTO shaft bearing (11) is held on the shaft with snapring (21) and in the retainer (5) with snapring (13). Always be sure to replace the oil seal (3) when service is performed to the assembly.

ENGINE

Engines used in these models are:
Model 71: Kohler K161S (with electric starter)
Kohler K161T (with manual starter)
102: Kohler K241AS
122 and 123: Kohler K301AS

For complete engine service, see the Kohler Engine part of the Engine Unit Repair section of this book.

REMOVAL

Engine with Manual Starters

1. Remove the grille and hood.

NOTE: Four capscrews (2 on each side) hold the grille to the tractor main frame.

2. Remove the clutch shield.
3. Disconnect the throttle wire from the governor speed control disc.
4. Disconnect the choke wire from the carburetor.
5. Disconnect the engine from the main frame.

NOTE: Four capscrews (two on each side) go through the frame into tapped holes in the engine oil pan base.

6. Slide the engine forward in the frame to disengage the clutch drive plate pins from the driven disc and disengage the clutch shaft from its pilot bushing.
7. The engine can now be lifted from the tractor frame and chassis.
8. Installation is the reverse of removal.

NOTE: One cylinder head capscrew can be removed and a ⅜ in. N.C. eyebolt installed to lift the engine if desired.

Engines with Electric Starters

1. Disconnect the battery cables and remove the battery.
2. Remove the grille and hood. Lay the

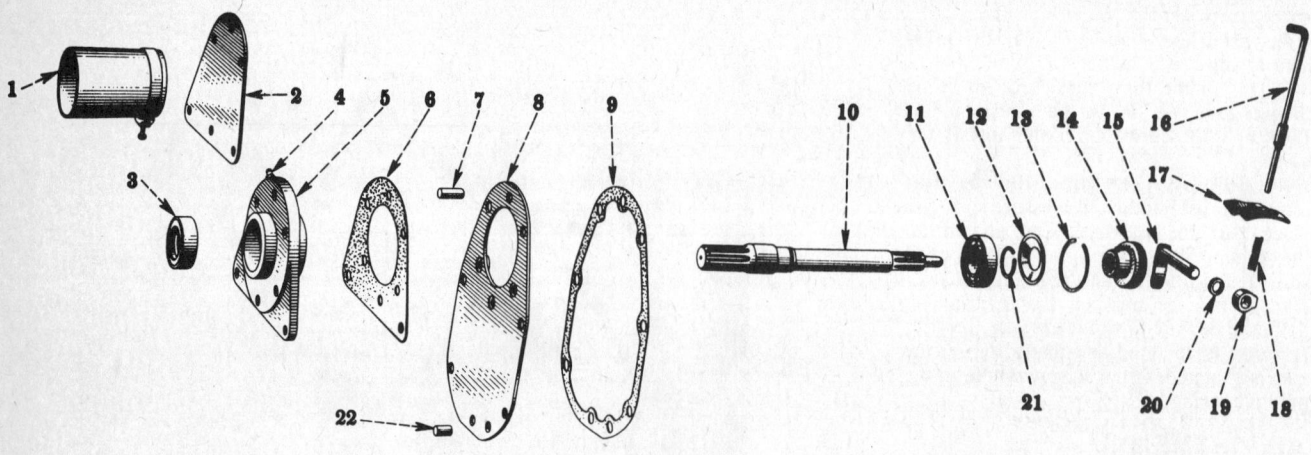

1. Guard	6. Gasket	11. Bearing	16. Shifter rod	21. Snapring
2. Adaptor cover	7. Dowel	12. Shield	17. Guide	22. Dowel
3. Oil Seal	8. Adaptor	13. Snapring	18. Spring	
4. Fitting	9. Gasket	14. Shifter clutch	19. Collar	
5. Retainer	10. Shaft	15. Lever and shaft	20. Seal	

Rear PTO components

International Harvester

grille and hood (grille face down) in front of the tractor.

3. Remove the clutch shield.
4. REmove the air cleaner.
5. Remove capscrew "A" and loosen screw "B". Remove the choke wire from the carburetor, and the throttle wire from the speed control bracket.
6. Disconnect the positive (+) coil wire from the coil. Spread the wire clip.
7. If the ground cable is connected to the starter-generator pivot capscrew, disconnect it.
8. Disconnect the two wires from the generator "A" terminal and one wire from the "F" terminal. Tie the two wires removed from the "A" terminal together and identify them for reassembly.
9. Disconnect the engine from the main frame. Two engine mounting capscrews on each side hold the engine to the frame.
10. Slide the engine forward to disengage the drive pins from the clutch driven disc.
11. The engine can now be lifted from the frame. An eyebolt (⅜ in. N.C.) can be installed in place of one cylinder head bolt.

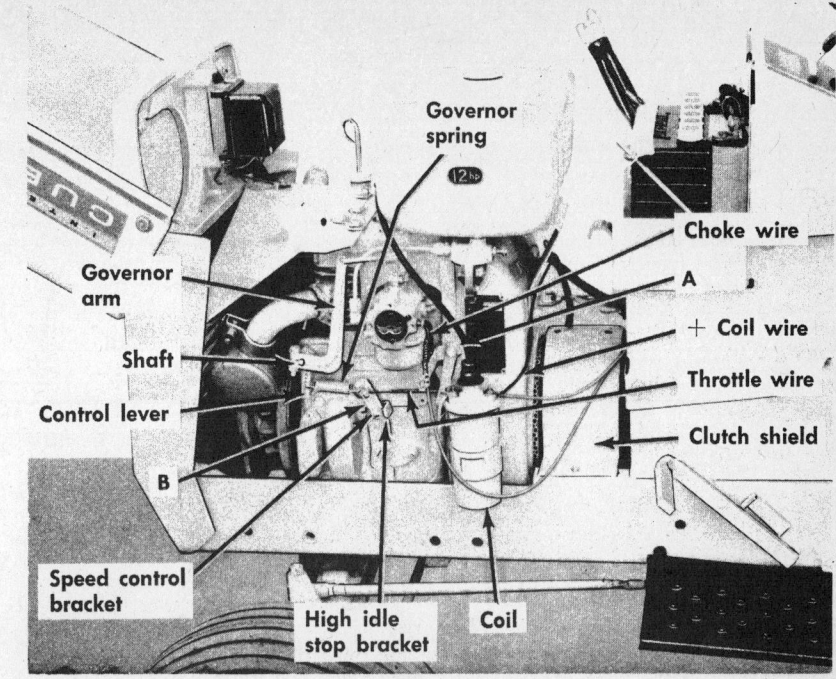

Left side view of a 122

Cub Cadet Models 72, 104, 105, 124, 125

FRONT WHEELS AND AXLE

Front Wheel and Bearings

REMOVAL

1. Lock the brake and block the rear wheels. Jack up the front axle.
2. Remove the capscrew and flat washer from the outer end of the front spindle.
3. Slide the wheel and bearings from the spindle.

NOTE: The bearings are a press fit in the wheel and a slip fit on the spindle.

DISASSEMBLY

Wheel bearings can be driven from the wheel hub with a hammer and long drift punch. Drive from the inside toward the outside.

INSPECTION AND REPAIR

1. Inspect the entire wheel and hub for wear or damage.
2. Bearings and seal should be inspected and replaced as necessary.
3. Bearing fit to wheel must be tight. If not, replace or repair wheel.

ASSEMBLY

1. If the bearings were removed, lubricate and press in new ones. Be sure force is directed to the outer race only.

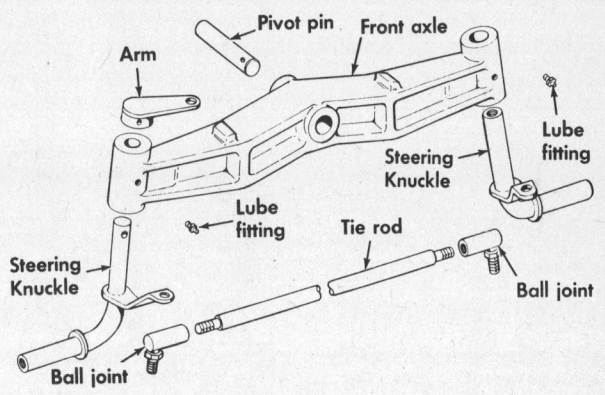

Front axle

2. Slide the wheel and bearing assembly over the spindle and secure with capscrew and flat washer.
3. If excessive end play exists (maximum 1/32 in.), place a sufficient thickness of shim washers (3-4 in. ID) over the outer end of the spindle and between the retaining washer and wheel bearing to take up excess end play.

Front Axle

REMOVAL AND INSTALLATION

1. Disconnect the drag link ball joint "D" from the drag link arm.
2. With the front of the tractor frame supported on a suitable stand, drive out the retaining pin from the front of the axle pivot pin.
3. Remove the pivot pin. The front axle is now free of its mounting and can be removed.
4. Apply chassis lubricant liberally to the axle pivot pin and its bore in the axle.
5. Position the axle in its support bracket channel, align the pivot pin holes and insert the pin.
6. Align the retaining pin holes (through the front of the pivot pin and through the front collar of the support bracket) then drive the retaining pin through both parts.

International Harvester

STEERING

Steering Knuckle

REMOVAL AND INSTALLATION

1. Lock the brake, jack up the front of the tractor and support it on a suitable stand.
2. Remove the front wheels.
3. Disconnect the tie rod ball joints from left and right steering knuckles.
4. Using a pin punch and hammer, drive out the coiled spring pin from the drag link arm on the left steering knuckle.
5. Remove the steering knuckle from the axle.
6. Remove the capscrew and flat washer from the upper end of the right steering knuckle.
7. Remove the right steering knuckle from the axle.
8. Apply chassis lubricant liberally to the steering knuckle arm thrust surface and the bottom of the axle ends.
9. Insert the righthand steering knuckle in its respective bore in the axle and secure with the capscrew and flat washer.
10. Insert the lefthand steering knuckle in its bore in the axle and position the drag link arm on it.
11. Secure the drag link arm and knuckle with the coiled spring pin.

NOTE: Spring pin must extend through the drag link arm an equal amount on each side.

12. Install the tie rod.
13. Install the front wheels and check toe-in adjustment.

Steering Gear

REMOVAL

1. Hold the steering wheel with front wheels in straight ahead position. Remove the steering wheel.
2. Remove the drag link rear ball joint from the steering unit lever.
3. Remove clutch shield bottom sheet.
4. Remove capscrews "A" from the frame cross member and steering unit.
5. Lower the steering column assembly through the instrument panel pedestal and grommet.

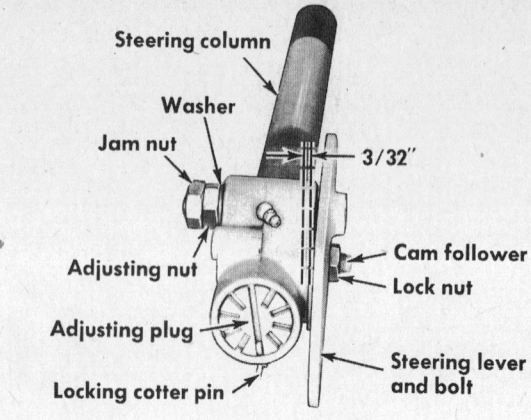
Steering gear

DISASSEMBLY

1. Secure the steering lever and bolt in a vise.
2. Remove the lever bolt jam nut, adjusting nut, and washer.
3. Slide the column and housing assembly away from the lever, bolt and cam follower.
4. Remove the adjusting plug.
5. Remove the steering cam and bearings from the housing.

INSPECTION AND REPAIR

1. Wash all parts in cleaning solvent, then dry thoroughly.
2. Inspect the cam follower for wear (flat spots).
3. Inspect the cam ends, bearings and races for wear, roughness and pitting.
4. Inspect the cam grooves for wear, roughness and galling.
5. Inspect the housing for cracks and stripped threads.
6. Inspect the upper bearing (nylon bushing) for wear or damage.

REASSEMBLY AND ADJUSTMENT

1. Thoroughly coat the cam ends, balls and races with chassis lubricant.
2. Install the balls and races on the cam ends.
3. Thoroughly coat the cam with chassis lubricant then install into the housing and column assembly.

NOTE: Be sure the races enter the housing squarely and are not cocked.

4. Install the adjusting plug. Screw the plug inward until end play of the cam is removed but turns freely. Insert the cotter pin in the nearest hole.
5. Fill the housing with chassis lubricant.
6. Loosen the cam follower locknut, then back out the cam follower two turns.
7. Install the seal, retainer and lever-bolt assembly to the housing.
8. Install the washer and adjusting nut. Tighten the adjusting nut sufficiently to provide good seal compression. Secure with the jam nut. Tighten jam nut to 40 ft.lb. Lubricate at the fitting in the housing slowly until lubricant begins to seep out.
9. Center the steering cam by rotating the steering shaft half way between full right and full left turn.
10. Adjust the cam follower inward to eliminate backlash, then tighten locknut to 40 ft.lb. Turn steering shaft full right and left to check for binding.
11. Replace the steering assembly in the tractor chassis. Secure with two capscrews through the frame crossmember.
12. Replace clutch shield bottom sheet.
13. Connect the drag link.
14. Replace the steering wheel and secure with nut.
15. Adjust the drag link "E" to proper length to place front wheels in the straight ahead position when the steering assembly is centered.

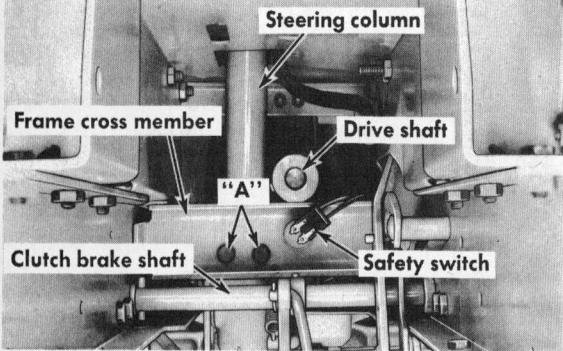
Underside view of steering support and frame

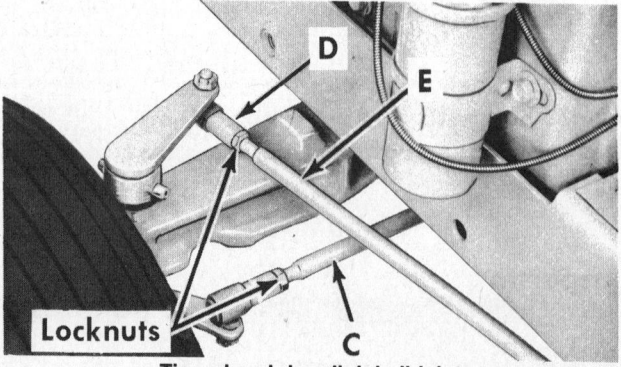
Tie rod and drag link ball joints

16. Adjust tie rod "C" to provide 1/32 in. to 1/8 in. toe-in.

NOTE: Do not remove ball joint ends. Loosen locknuts and turn rod clockwise or counterclockwise to adjust.

BRAKES

NOTE: Tractors equipped with a creeper attachment will require splitting, and the creeper removed before complete brake service can be performed.

REMOVAL

1. Drain the transmission lubricant.
2. Remove the brake adjusting screw and locknut from its lever.
3. Remove the brake lever, pivot pin and push rod.
4. Remove the reduction housing front cover plate and slide it forward on the clutch shaft. Also see "Note" above.
5. Remove the reduction gear from the front of the transmission mainshaft.
6. Move the gear upward and the bottom of the gear forward to clear the cover screw bosses as the gear is lifted from the housing.
7. Slide the brake disc forward on the countershaft as the front lining and retainer are moved forward in their bore.

NOTE: Both linings and the disc can be removed without removing the front lining retainer; however, removal of the retainer is recommended for inspection and replacement of the retainer O-ring.

INSPECTION

1. Inspect the control rods and levers for wear at their connecting pivot points.
2. Inspect the linings and disc for wear.
3. Inspect the disc hub splines for wear.

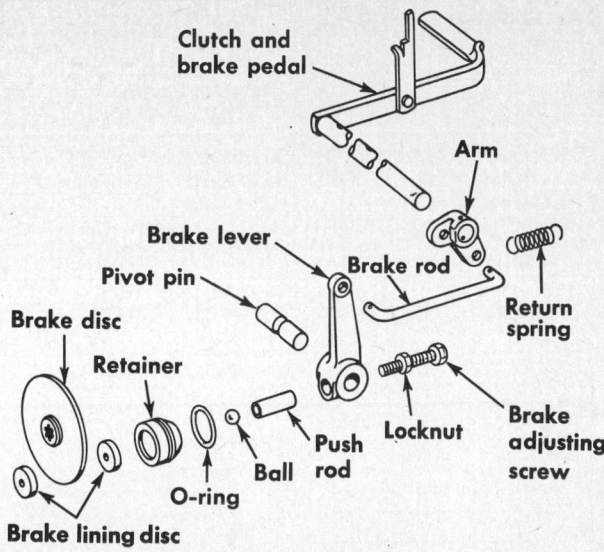

Brake components

4. Check the splines on the countershaft for wear.
5. Check the pedal return spring ends for wear.

REASSEMBLY AND ADJUSTMENT

1. Clean the brake cavity and lining recess in the reduction housing.
2. Place a small quantity of grease in the rear brake lining recess in the reduction housing then insert the lining.
3. Install the disc on the countershaft and slide it rearward against the rear lining.
4. Install the front lining in the retainer.
5. Install a new O-ring on the front lining retainer, lubricate the retainer and O-ring then push them into the retainer bore in the reduction housing.
6. Install the reduction gear on the transmission mainshaft and secure with flat washer, lockwasher and capscrew. Tighten capscrew to 55 ft.lb.
7. Install a new cover gasket, then replace the cover plate.
8. Be sure the ball is in place in the front lining retainer then replace the push rod, lever, pivot pin, adjusting screw and locknut.
9. Fill transmission to proper level with Hy-Tran fluid or SAE 30 engine oil.

ADJUSTING THE BRAKE

Models 72, 104, 124

NOTE: For models 105 & 125 see the Hydrostatic transmission section.

1. Loosen the locknut and turn the brake lever adjusting screw in or out as required.
2. The brake should engage when the pedal arm is pressed down to within a maximum of 1 5/16 in. and a minimum of 3/4 in. distance above the top of the left foot support, which serves as the pedal stop.
3. It may be possible to push the pedal all the way down to the pedal stop, but this

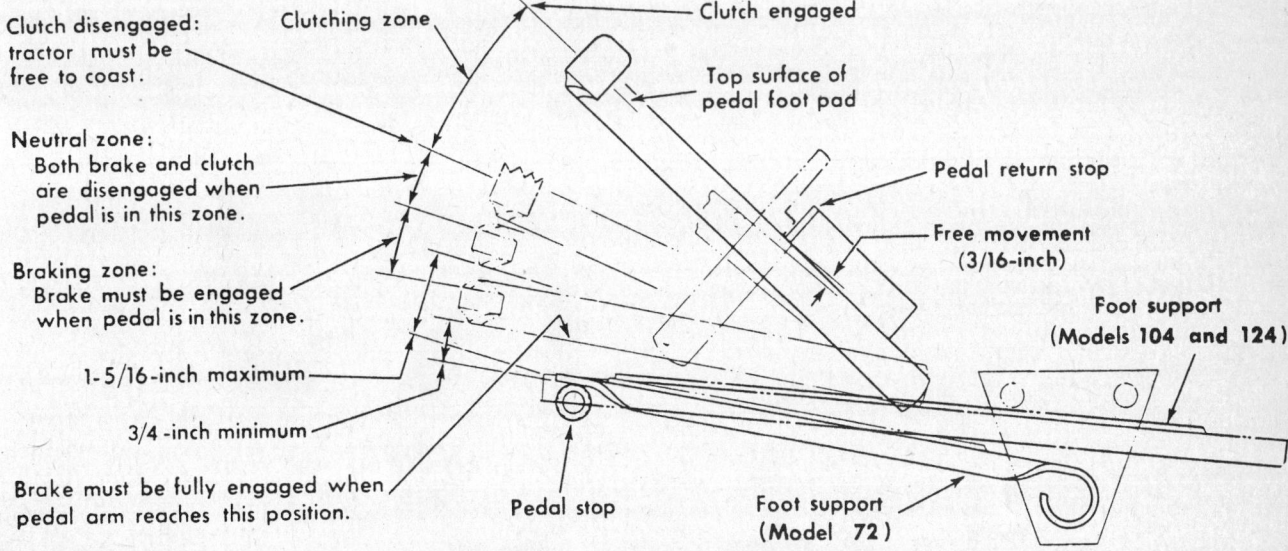

Clutch and brake adjustments

International Harvester

is of no concern as long as the brake is engaged within above limits.

SPLITTING AND RECOUPLING

Model 105 and 125

SPLITTING THE TRACTOR

NOTE: Disconnect battery ground cable.

1. Remove the seat, seat support and fenders.
2. Disconnect the tail light wire (if so equipped) at the junction under the seat support.
3. Disconnect the ball joint from the speed control cam.
4. Disconnect the brake rod.
5. Drive small wooden wedges between the front axle and frame on each side to stabilize the front end of the tractor.
6. Block the front wheels in the front and back.
7. Place a jack under the lift handle cross shaft.
8. Remove six capscrews (three each side) from the frame.
9. Move the transmission-differential rear end assembly rearward far enough to permit access to the hydrostatic unit.

RECOUPLING THE TRACTOR

1. Tip the unit into recoupling position and lubricate the cam slot with chassis lubricant.
2. Fill the transmission-differential case with seven quarts of Hy-Tran fluid.
3. Recouple the split sections of the tractor being sure of the following:
   a. That the brake rod goes over the lift handle cross shaft. Be sure to guide the driveshaft through the fan shroud and air baffle plate.
   b. That the dowel pin on the front end of the driveshaft is positioned to engage the slot in the drive plate hub on the engine.
4. Install and tighten securely the three capscrews on each side of the frame.
5. Connect the brake rod.

SPEED CONTROL LEVER LINKAGE AND CAM BRACKET ADJUSTMENT

NOTE: The brake pedal must be properly adjusted before beginning the speed control adjustment.

1. If the tractor creeps in the "N" position or, if the speed control linkage has been disassembled or removed for any reason, the following adjustment must be made.
2. Proper friction adjustment is necessary on speed control lever for proper operation. The lever friction should be adjusted as follows:
   a. Remove the battery.

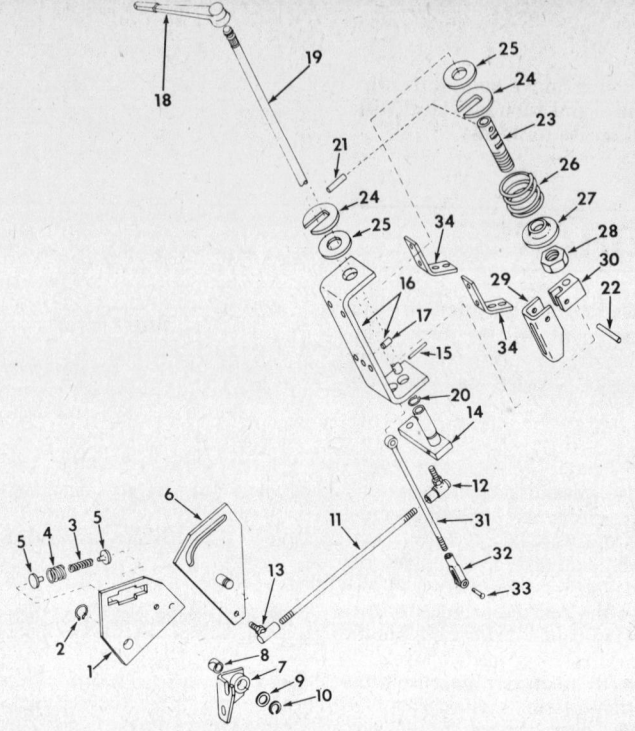

1. Damper spring plate
2. Retaining ring
3. Spring (light)
4. Spring (heavy)
5. Guide pin
6. Speed control cam
7. Bracket
8. Bushing
9. Washer
10. Retaining ring
11. Speed control rod
12. Ball joint (lefthand thread)
13. Ball joint (righthand thread)
14. Lever assembly
15. Slotted pin
16. Bracket
17. Dowel pin
18. Handle
19. Rod
20. Ring
21. Dowel pin
22. Dowel pin
23. Frictional adjusting shaft
24. Retainer
25. Washer
26. Spring
27. Washer
28. Stop nut
29. Cam channel
30. Swivel cam
31. Rod
32. Clevis
33. Pin
34. Brace

Clutch and speed control linkage

b. Place a small wedge between the hand-control mounting bracket and the adjusting nut. Refer to foldout at back of manual.
c. Move the lever towards the "F" or forward position to tighten the nut.
d. Remove the wedge.
e. Check the friction adjustment with a scale. The reading should be 10 lb. when pulled in either direction from the offset in the lever.
f. Replace the battery.

3. Block the tractor so the left rear wheel is off the ground and tractor is secured so it cannot move forward or backward.
4. Depress brake pedal and lock to set brake.
5. Disconnect connecting rod at the ball joint (13).
6. Loosen cam bracket mounting capscrews if not previously left loose. Move

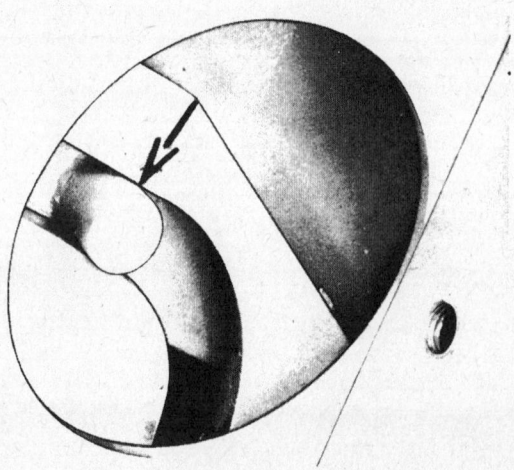

Cam plate with arrow and pin viewed through access hole in left frame

International Harvester

the cam bracket to its lowest position in the slotted holes and tighten slightly to hold in position.

7. Start the engine and run at half throttle or faster.

8. With the engine running align the arrow with the pin. At this position the hydrostatic unit is pulling against the brake. Adjust the cam bracket upward with a punch and hammer until a neutral condition is reached while the arrow and pin are kept aligned. Tighten the cam bracket capscrews. Connect the rod and ball joint to the cam plate.

Connect the rod and ball joint to the cam plate.

9. Release the brake pedal and move the speed control lever to the forward position. The wheel should rotate in the forward direction. Depress the brake pedal all the way down and release. The speed control lever should return to the "N" position and the wheel stops turning.

10. If the wheel continues to turn in the forward direction, loosen jam nut "D" and turn the connecting rod "E" to lengthen it until the wheel stops turning after depressing the pedal.

11. If the wheel turns in the reverse direction turn the connecting rod "E" to shorten it until wheel stops. Tighten the jam nut "D".

Models 72, 104, 124

SPLITTING THE TRACTOR

NOTE: Disconnect battery ground cable.

1. Remove the clutch shaft coupling rear pin. (Connects coupling to creeper or reduction driveshaft.)
2. Drive a small wooden wedge between the front axle and frame on each side.
3. Block both front wheels as shown.
4. If tractor is equipped with creeper attachment, shift the lever forward then remove the lever knob and the breather. Also remove gear shift knob.
5. Disconnect the brake rod from the brake lever.
6. If tractor is equipped with three-point hitch, remove the lift bar and its attaching plate.
7. Place a jack under the implement lift handle cross shaft to support the frame.
8. Remove three capscrews from each side of frame.
9. Raise frame and fender assembly high enough to clear gear shifter.
10. Push down on the drawbar and pull rearward. Guide transmission and rear end assembly from under frame.

RECOUPLING THE TRACTOR

Recoupling of the tractor is the reverse of splitting; however, precautions should always be taken to safeguard against damage to shafts, bearings, seals etc. when aligning and securing components which work together.

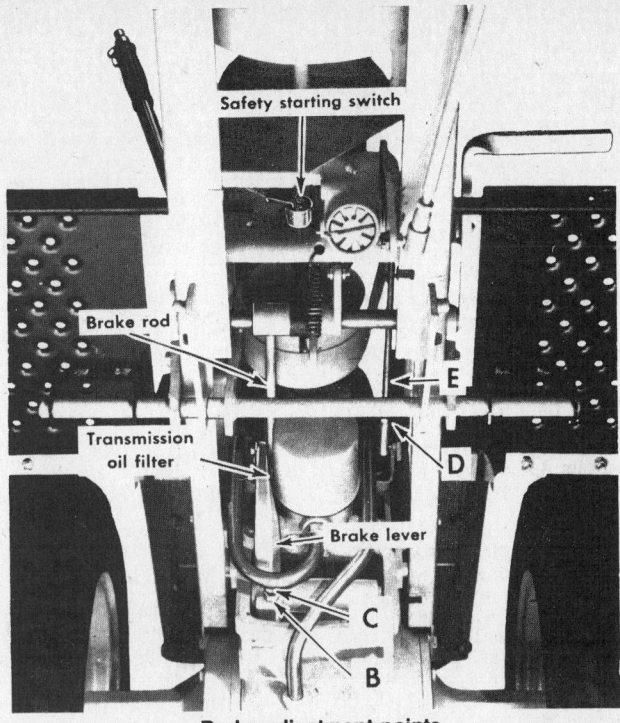

Brake adjustment points

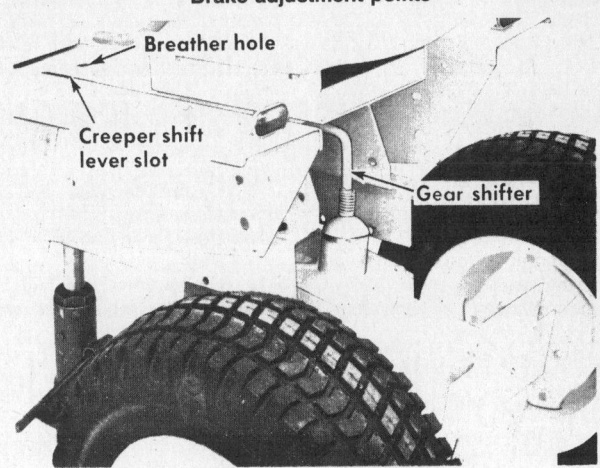

Splitting the tractor

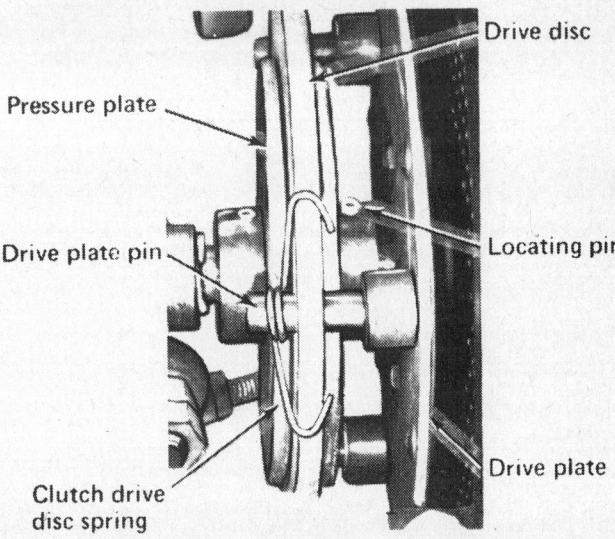

Clutch drive as viewed with shield removed

507

International Harvester

CLUTCH

REMOVAL

Complete service of the clutch shaft, loading spring, release bearing, and cushion spring will require removal of the tractor engine, then the clutch assembly.

NOTE: When minor service (replacing the driving disc, driving plate or pressure plates) only is required, perform the following steps 1 through 13.

1. Depress the clutch and brake pedal and lock it.
2. Remove the clutch shield.
3. Using a hammer and punch, drive out the pressure plate locating pin.
4. Remove the four capscrews (two on each side) which hold the engine to the tractor frame.

NOTE: Wiring cable clips (on tractors equipped with electric starting) will need spreading so that wires can follow engine as it is moved.

5. Release the clutch and brake pedal, then lift engine far enough to clear clutch parts.
6. Replace the pressure plate, driving disc or driving plate as necessary. Replace clutch driving disc spring(s) on driving plate pins.
7. Slide the engine rearward while aligning the clutch shaft and driving plate pins.
8. Depress and lock the clutch-brake pedal.
9. Replace the pressure plate locating pin.
10. Align the engine to the frame and replace the securing capscrews.
11. Secure the wires and cables in their clips.
12. Connect the ground cable (if it was disconnected).
13. Replace the clutch shield.

INSTALLATION AND ADJUSTMENT

1. Place the clutch shaft over the frame cross member and move it rearward to engage the coupling to the reduction unit or creeper driveshaft.

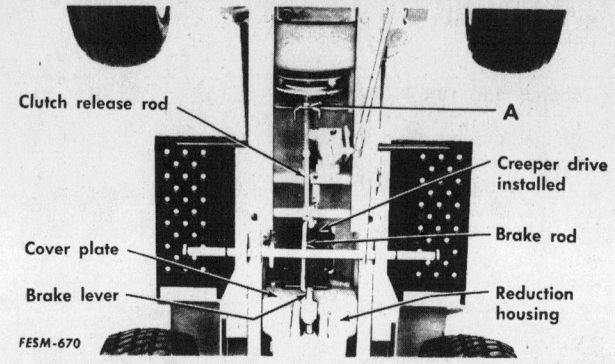

Underside view of clutch and brake linkage

2. Align the release lever to its bracket, install the pin and secure with cotter.
3. Apply chassis lubricant or Lubriplate® liberally to the clutch shaft pilot bushing.
4. Slide the engine rearward while aligning the clutch shaft into its pilot bushing and the drive plate pins into their holes in the driven disc.
5. Align the engine mounting capscrew holes and install the capscrews.
6. Align the clutch shaft coupling rear pin hole with pin hole in reduction drive (or creeper drive) shaft and install the pin.
7. Connect the clutch release rod to its pedal shaft lever and secure with the return spring.
8. Depress the pedal and release the clutch to allow components to move into final alignment.
9. Position the wires and cables in their clips and secure.
10. Connect the battery cables.
11. Replace the clutch shield.
12. Check the clutch pedal and linkage for proper adjustment.

Pedal Adjustment

1. It is important that a clearance of .050 in. be maintained between the clutch release lever and the clutch release bearing. In order to maintain this clearance, the pedal should have a free movement of approximately $3/16$ in. This measurement is taken at the point of contact of the pedal arm with the front edge of the pedal return stop.

2. The clutch pedal adjustments should not require frequent attention. When it is necessary to adjust the clutch, turn the adjusting nut "A" on the clutch release rod in or out as required to get the proper measurements.

NOTE: For complete service on the following items, see the 71, 102, 122, 123 section, immediately preceding this section:
REDUCTION DRIVE
CREEPER DRIVE
TRANSAXLE
DIFFERENTIAL
HYDROSTATIC TRANSMISSION
REAR AXLE
FRONT PTO

REAR PTO

REMOVAL

Remove 7 capscrews m and n around rear oil seal and bearing retainer. Pull shaft, retainer, shifter clutch to the rear letting the front clutch disengage from the shift lever shaft.

INSPECTION

Check shifter clutch for excessive wear. Check lockscrew and nut for damage. Inspect bearing and splines. Check lever shaft pin for flat sides.

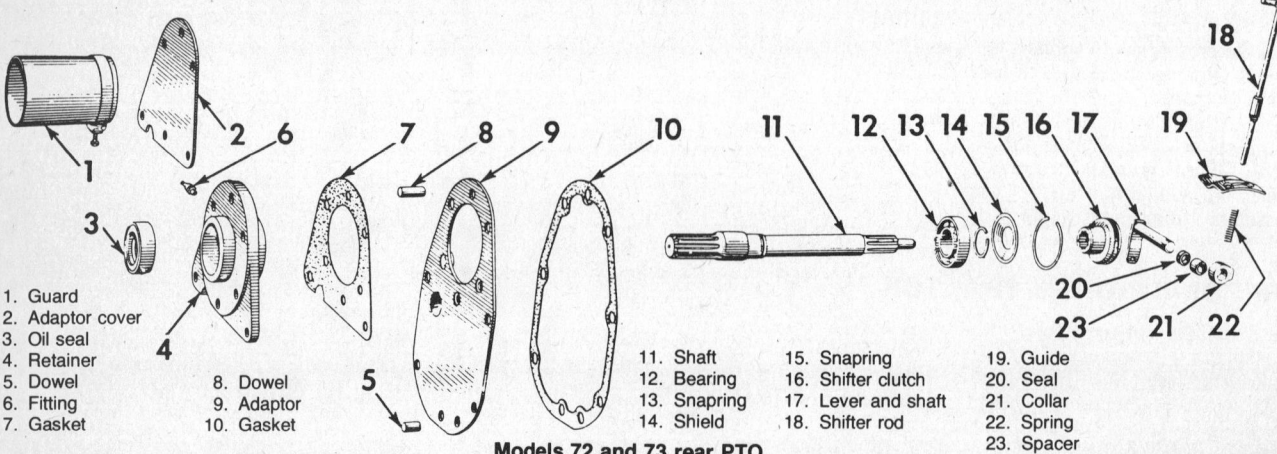

1. Guard
2. Adaptor cover
3. Oil seal
4. Retainer
5. Dowel
6. Fitting
7. Gasket
8. Dowel
9. Adaptor
10. Gasket
11. Shaft
12. Bearing
13. Snapring
14. Shield
15. Snapring
16. Shifter clutch
17. Lever and shaft
18. Shifter rod
19. Guide
20. Seal
21. Collar
22. Spring
23. Spacer

Models 72 and 73 rear PTO

International Harvester

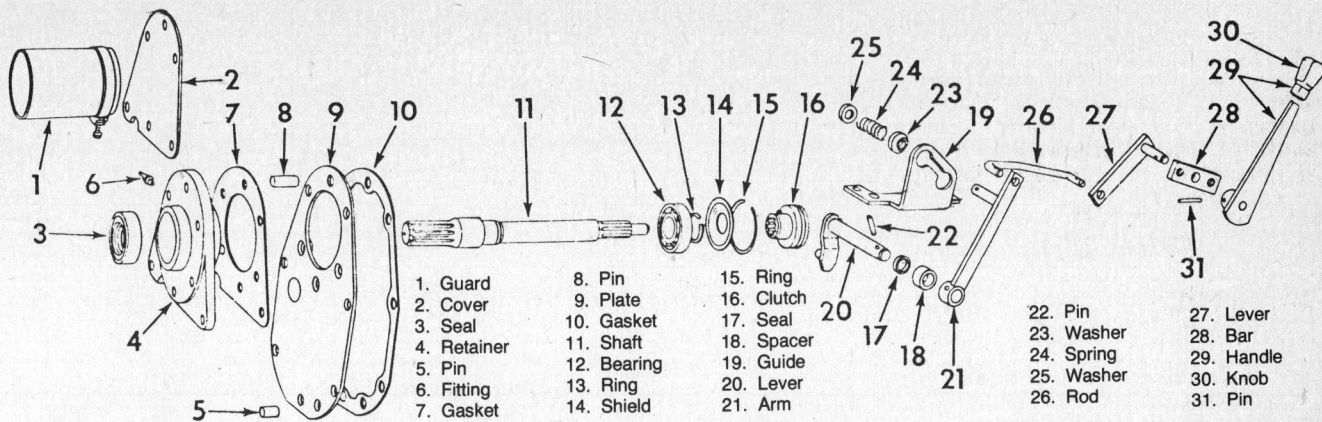

Models 104, 106, 124 and 126 rear PTO

1. Guard
2. Cover
3. Seal
4. Retainer
5. Pin
6. Fitting
7. Gasket
8. Pin
9. Plate
10. Gasket
11. Shaft
12. Bearing
13. Ring
14. Shield
15. Ring
16. Clutch
17. Seal
18. Spacer
19. Guide
20. Lever
21. Arm
22. Pin
23. Washer
24. Spring
25. Washer
26. Rod
27. Lever
28. Bar
29. Handle
30. Knob
31. Pin

ASSEMBLY

1. Slide shifter clutch on to the power take off shaft and secure with setscrew and nut.

2. Apply a coating of grease on the pilot end of the power take off shaft. Making sure oil seal is installed with lip towards inside and gaskets are in place insert power take off shaft assembly through the hole in the adaptor plate and engage the pin on the shifter shaft lever with the groove in the clutch.

3. Move the assembly forward and engage the pilot end of the shaft with the female end of the transmission input shaft.

4. Fasten the power take off assembly into place with the capscrews removed.

ENGINE

The following engines are used:
Models 72: Kohler K161S (with electric starter)
Kohler K161T (with manual starter)
104, 105: Kohler K241AS
124, 125: Kohler K301AS

For complete engine service, see the Kohler Engine part of the Engine Unit Repair section.

REMOVAL AND INSTALLATION

Manual Starter

1. Remove grille and hood:
   a. Remove four capscrews and nuts that hold the grille to the tractor main frame.
   b. Slide bottom of grille forward 3½ in. and remove clutch release pivot shaft from grille casting.
   c. Remove the grille and hood. Lay the grille and hood (grille face down) in front of the tractor.
2. Remove clutch shield.
3. Remove air cleaner.
4. Disconnect throttle wire.
5. Disconnect choke wire from the carburetor.
6. Disconnect switch wire.
7. Remove four capscrews holding engine to the frame.
8. Attach lifting equipment to motor lift bracket on head bolt.
9. Slide engine forward to disengage the drive pins from the clutch driven disc.
10. The engine can now be lifted from the tractor.
11. Installation is the reverse of removal.

Electric Starting

1. Disconnect the battery cables.
2. Remove four capscrews and nuts that hold the grille to the main frame. Slide the bottom of the grille forward 3½ in. and remove the clutch release pivot shaft from the grille casting. Remove the grille and hood. Lay the grille and hood (grille face down) in front of the tractor.
3. Remove the clutch shield.
4. Remove the air cleaner.
5. Remove the choke wire from the carburetor, and the throttle wire from the speed control lever.
6. Disconnect the positive (+) coil wire from the coil.
7. Disconnect the wires from the Generator "A" terminal and from the "F" terminal

NOTE: Large wire to "A" terminal. Small wire to "F" terminal.

8. Disconnect the ground cable connected to a starter-generator bracket capscrew and remove starter-generator and coil wires from the generator bracket.
9. Remove four capscrews holding the engine to frame. Two on each side.
10. Slide the engine forward to disengage the drive pins from the clutch driven disc.
11. The engine can now be lifted from the frame using the engine lift bracket on a head bolt.
12. Installation is the reverse of removal.

Cub Cadet Models 73, 106, 107, 126, 127, 147

FRONT WHEELS AND AXLE

Front Wheels and Bearings

REMOVAL

1. Lock the brake and block the rear wheels. Jack up the front axle.
2. Remove the capscrew and flat washer from the outer end of the front spindle.
3. Slide the wheel and bearings from the spindle.

NOTE: The bearings are a press fit in the wheel and a slip fit on the spindle.

DISASSEMBLY

Wheel bearings can be driven from the wheel hub with a hammer and long drift punch. Drive from the inside toward the outside.

INSPECTION AND REPAIR

1. Inspect the entire wheel and hub for wear or damage.
2. Bearings and seal should be inspected and replaced as necessary.
3. Bearing fit to wheel must be tight. If not, replace or repair wheel.

REASSEMBLY

1. If the bearings were removed, lubricate and press in new ones. Be sure force is directed to the outer race only.

509

International Harvester

2. Slide the wheel and bearing assembly over the spindle and secure with capscrew and flat washer.

3. If excessive end play exists (maximum 1/32 in.), place a sufficient thickness of shim washers (3-4 in. ID) over the outer end of the spindle and between the retaining washer and wheel bearing to take up excess end play.

Front Axle

REMOVAL AND INSTALLATION

1. Disconnect the drag link ball joint (4) from the drag link arm (No. 5).
2. With the front of the tractor frame supported on a suitable stand, drive out the retaining pin from the front of the axle pivot pin.
3. Remove the pivot pin. The front axle is now free of its mounting and can be removed.
4. Apply chassis lubricant liberally to the axle pivot pin and its bore in the axle.
5. Position the axle in its support bracket channel, align the pivot pin holes and insert the pin.
6. Align the retaining pin holes (through the front of the pivot pin and through the front collar of the support bracket) then drive the retaining pin through both parts.

STEERING

Steering Knuckles

REMOVAL

1. Lock the brake, jack up the front of the tractor and support it on a suitable stand.
2. Remove the front wheels.
3. Disconnect the tie rod ball joints from left and right steering knuckles.
4. Using a pin punch and hammer, drive out the coiled spring pin from the drag link arm on the left steering knuckle.
5. Remove the steering knuckle from the axle.
6. Remove the capscrew and flat washer from the upper end of the right steering knuckle.
7. Remove the right steering knuckle from the axle.

INSTALLATION

1. Apply chassis lubricant liberally to the steering knuckle arm thrust surface and the bottom of the axle ends.
2. Insert the righthand steering knuckle in its respective bore in the axle and secure with the capscrew and flat washer.
3. Insert the lefthand steering knuckle in its bore in the axle and position the drag link arm on it.
4. Secure the drag link arm and knuckle with the coiled spring pin.

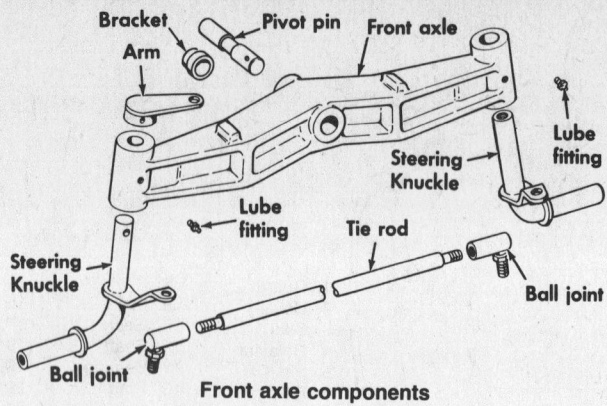

Front axle components

NOTE: Spring pin must extend through the drag link arm an equal amount on each side.

5. Install the tie rod.
6. Install the front wheels and check toe-in adjustment.

Steering Gear

REMOVAL

1. Hold the steering wheel with front wheels in straight ahead position. Remove the steering wheel.
2. Remove the drag link rear ball joint from the steering unit lever.
3. Remove clutch shield bottom sheet.
4. Remove capscrews from the frame cross member and steering unit.
5. Lower the steering column assembly through the instrument panel pedestal and grommet.

DISASSEMBLY

1. Secure the steering lever and bolt in a vise.
2. Remove the lever bolt jam nut, adjusting nut, and washer.
3. Slide the column and housing assembly away from the lever, bolt and cam follower.
4. Remove the adjusting plug.
5. Remove the steering cam and bearings from the housing.

INSPECTION AND REPAIR

1. Wash all parts in cleaning solvent, then dry thoroughly.
2. Inspect the cam follower for wear (flat spots).
3. Inspect the cam ends, bearings and races for wear, roughness and pitting.
4. Inspect the cam grooves for wear, roughness and galling.
5. Inspect the housing for cracks and stripped threads.
6. Inspect the upper bearing (nylon bushing) for wear or damage.

REASSEMBLY AND ADJUSTMENT

1. Thoroughly coat the cam ends, balls and races with chassis lubricant.
2. Install the balls and races on the cam ends.
3. Thoroughly coat the cam with chassis lubricant then install into the housing and column assembly.

NOTE: Be sure the races enter the housing squarely and are not cocked.

4. Install the adjusting plug. Screw the plug inward until end play of the cam is removed but turns freely. Insert the cotter pin in the nearest hole.
5. Fill the housing with chassis lubricant.
6. Loosen the cam follower locknut, then back out the cam follower two turns.
7. Install the seal, retainer and lever-bolt assembly to the housing.
8. Install the washer and adjusting nut. Tighten the adjusting nut sufficiently to provide good seal compression. Secure with the jam nut. Tighten jam nut to 40 ft.lb. Lubricate at the fitting in the housing slowly until lubricant begins to seep out.

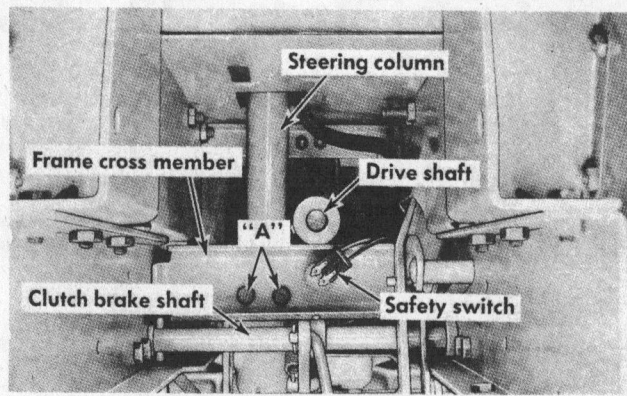

Underside view of steering supports and frame

International Harvester

9. Center the steering cam by rotating the steering shaft half-way between full right and full left turn.

10. Adjust the cam follower inward to eliminate backlash, then tighten locknut to 40 ft.lb. Turn steering shaft full right and left to check for binding.

11. Replace the steering assembly in the tractor chassis. Secure with two capscrews through the frame cross member.

12. Replace clutch shield bottom sheet.

13. Connect the drag link.

14. Replace the steering wheel and secure with nut.

15. Adjust the tie rod to provide 1/32 in. to 1/8 in. toe-in as follows:

   a. Remove one of the tie rod ball joints and loosen the locknut.

   b. Screw the ball joint in or out to obtain the specified toe-in of 1/32 in. to 1/8 in. and tighten the locknut.

   c. Connect the ball joint to the steering knuckle and be sure to install the cotter pin.

16. Adjust the drag link to proper length to place front wheels in the straight ahead position when the steering assembly is centered.

BRAKES

Model 73

NOTE: Tractors equipped with a creeper attachment will require splitting, and the creeper removed before complete brake service can be performed.

REMOVAL

1. Drain the transmission lubricant.

2. Remove the brake adjusting screw and locknut from its lever.

3. Remove the brake lever, pivot pin and push rod.

4. Remove the reduction housing front cover plate and slide it forward on the clutch shaft. Also see "Note" above.

5. Remove the reduction gear from the front of the transmission mainshaft.

6. Move the gear upward and the bottom of the gear forward to clear the cover screw bosses as the gear is lifted from the housing.

7. Slide the brake disc forward on the countershaft as the front lining and retainer are moved forward in their bore.

NOTE: Both linings and the disc can be removed without removing the front lining retainer; however, removal of the retainer is recommended for inspection and replacement of the retainer O-ring.

INSPECTION

1. Inspect the control rods and levers for wear at their connecting pivot points.

2. Inspect the linings and disc for wear.

3. Inspect the disc hub splines for wear.

4. Check the splines on the countershaft for wear.

5. Check the pedal return spring ends for wear.

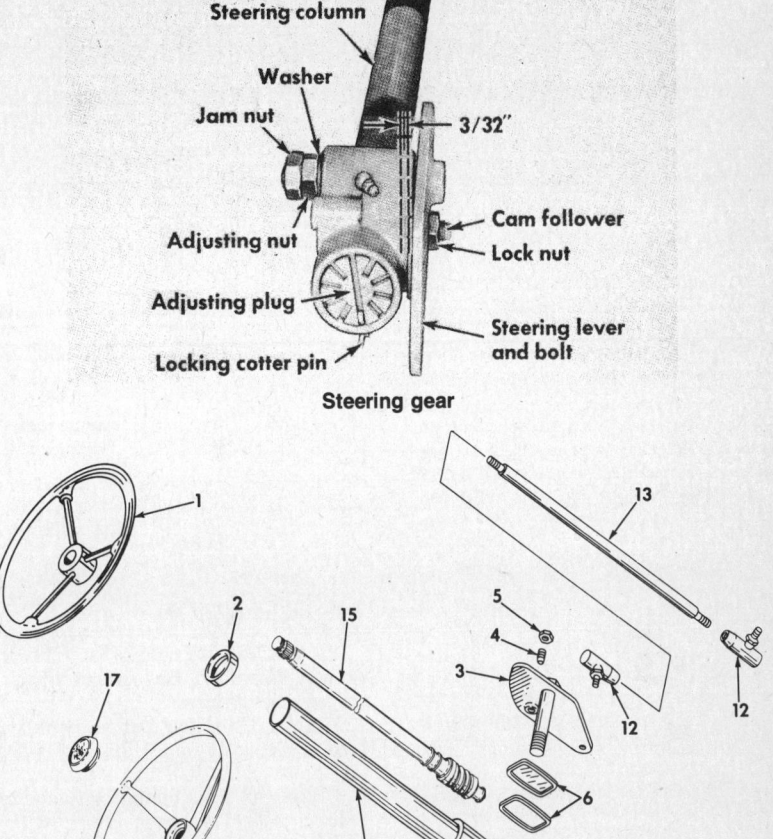

1. Wheel (Model 73)
2. Bearing
3. Lever
4. Stud
5. Nut
6. Seal and retainer
7. Fitting
8. Washer
9. Bearing
10. Retainer and ball assembly
11. Plug, adjusting
12. Joint
13. Link, steering
14. Tube assembly
15. Cam and tube
16. Retainer
17. Cap
18. Steering wheel

Steering system components

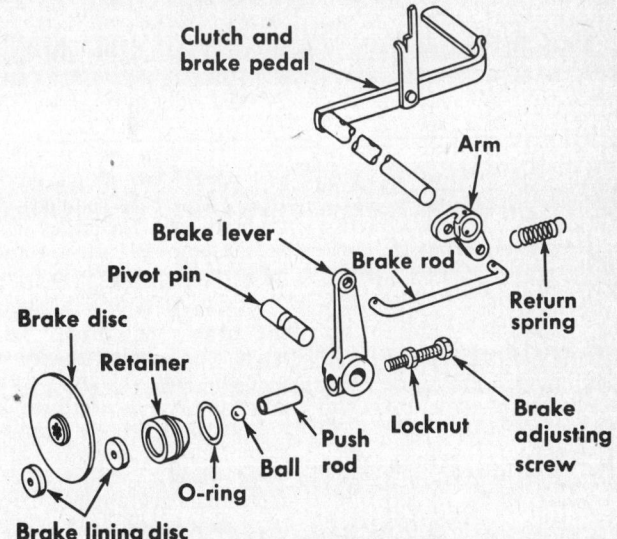

Model 73 brake system components

International Harvester

REASSEMBLY AND ADJUSTMENT

1. Clean the brake cavity and lining recess in the reduction housing.
2. Place a small quantity of grease in the rear brake lining recess in the reduction housing then insert the lining.
3. Install the disc on the countershaft and slide it rearward against the rear lining.
4. Install the front lining in the retainer.
5. Install a new O-ring on the front lining retainer, lubricate the retainer and O-ring then push them into the retainer bore in the reduction housing.
6. Install the reduction gear on the transmission mainshaft and secure with flat washer, lockwasher and capscrew. Tighten capscrew to 55 ft.lb.
7. Install a new cover gasket, then replace the cover plate.
8. Be sure the ball is in place in the front lining retainer then replace the push rod, lever, pivot pin, adjusting screw and locknut.
9. Fill transmission to proper level with Hy-Tran fluid.

ADJUSTMENT

1. The brake should engage when the pedal arm is pressed down to within a maximum of $1^5/_{16}$ in. and a minimum of ¾ in. distance above the top of the left foot support which serves as the pedal stop.
2. It may be possible to push the pedal all the way down to the pedal stop, but this is of no concern as long as the brake is engaged when the pedal arm is at least ¾ in. above the pedal stop.
3. To adjust the brake, loosen the locknut and turn the brake lever adjusting screw in or out as required to get the above measurements. The brake must not engage before the pedal arm is within the maximum distance of $1^5/_{16}$ in. above the pedal stop.

Models 106, 107, 126, 127, 147

REMOVAL

1. Remove the brake rod (10) from the brake arm (9).
2. Remove the capscrews (6) from the mounting flange (7) and remove the brake assembly from the axle carrier.
3. Remove the capscrews (4) securing the caliper assembly (1) to the bracket assembly (8).

INSPECTION AND REPAIR

1. Inspect the control rods and levers for wear at their connecting pivot points.
2. Inspect the brake pads for excessive wear. Replace if necessary.
3. Check the brake discs for excessive wear or damage.
4. Check the pedal return spring ends for wear.

1. Caliper assembly
2. Brake pads
3. Brake disc
4. Capscrew
5. Spacer and spring
6. Capscrew
7. Mounting flange
8. Bracket assembly
9. Brake arm
10. Brake rod

Model 106, 107, 126, 127, 147 brake system components

INSTALLATION

1. Assemble the caliper assembly to the bracket assembly. Be sure to install the spacers and springs (5).
2. Install the caliper assembly and bracket assembly on the disc and axle carrier.
3. Assemble the mounting flange to the bracket assembly and secure with the capscrews.
4. Install the brake rods in the brake arms.

ADJUSTMENT

1. The disc brakes should engage when the pedal arm is pressed down to within a maximum of $1^5/_{16}$ in. and a minimum of ¾ in. distance above the pedal stop.
2. It may be possible to push the pedal all the way down to the pedal stop, but this is of no concern as long as the brake is engaged when the pedal arm is at least ¾ in. above the pedal stop.

NOTE: At ¾ in. distance the brakes must withstand a torque of 100 ft.lb. per wheel.

3. To adjust the brakes block the front wheels securely and raise the tractor so the rear wheels are off the ground and turn freely. Adjust the jam nuts on the ends of the brake rods. The brakes must not engage before the pedal arm is within the maximum distance of $1^5/_{16}$ in.

NOTE: It is very important to have the brakes equalized. To check the equalization of the brakes start the engine and shift the gears to third speed. After the wheels are turning, apply the brakes. Both wheels should stop at the same time. If one wheel stops and the other wheel continues to revolve when the brakes are applied, adjust the jam nuts on the brake rod of the wheel that stops so both wheels stop simultaneously.

CLUTCH

REMOVAL

Complete service of the clutch shaft, loading spring, release bearing, and cushion spring will require moving the engine forward, then removing the clutch assembly.

NOTE: Models 106 and 126 will require the removal of the muffler and heat shield before moving the engine forward.

To remove the clutch assembly, remove the clutch shield, clutch shaft coupling rear pin, clutch release lever pin and disconnect the clutch release rod from the pedal arm.

When minor service (replacing the driving disc, driving plate or pressure plates) only is required, perform the following steps 1 through 13.

1. Depress the clutch and brake pedal and lock it.
2. Remove the clutch shield.
3. Using a hammer and punch, drive out the pressure plate locating pin.
4. Remove the four capscrews (two on each side) which hold the engine to the tractor frame.

NOTE: Wiring cable clips will need spreading so that wires can follow the engine as it is moved.

5. Release the clutch and brake pedal, then move the engine forward far enough to clear the clutch parts.
6. Replace the pressure plate, driving disc or driving plate as necessary. Replace clutch driving disc spring(s) on driving plate pins.
7. Slide the engine rearward while aligning the clutch shaft and driving plate pins.
8. Depress and lock the clutch-brake pedal.

International Harvester

9. Replace the pressure plate locating pin.
10. Align the engine to the frame and replace the securing capscrews.
11. Secure the wires and cables in their clips.
12. Connect the ground cable (if it was disconnected).
13. Replace the clutch shield.

INSTALLATION

1. Place the clutch shaft over the frame crossmember and move it rearward to engage the coupling to the reduction unit or creeper driveshaft.
2. Align the release lever to its bracket, install the pin and secure with cotter.
3. Apply chassis lubricant or Lubriplate® liberally to the clutch shaft pilot bushing.

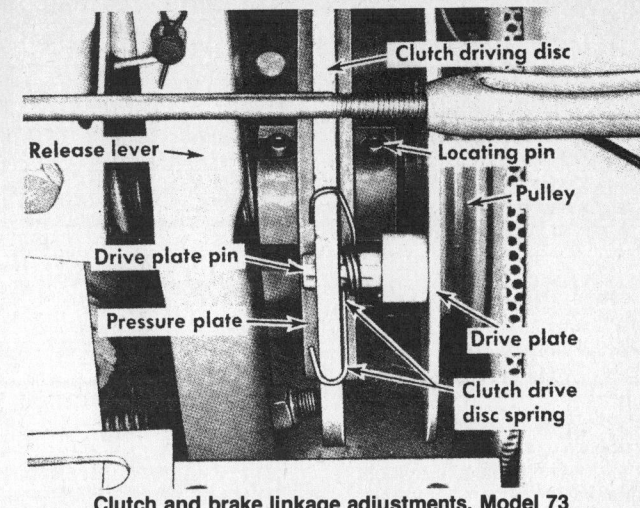

Clutch and brake linkage adjustments, Model 73

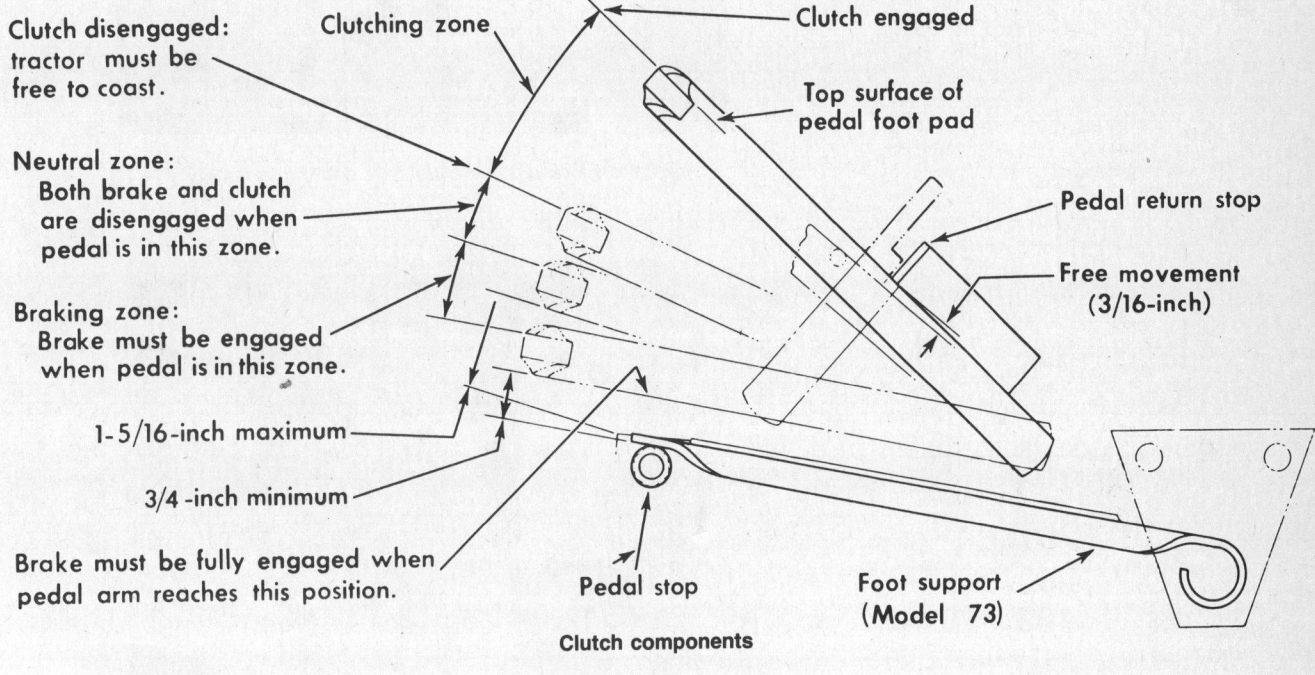

Clutch components

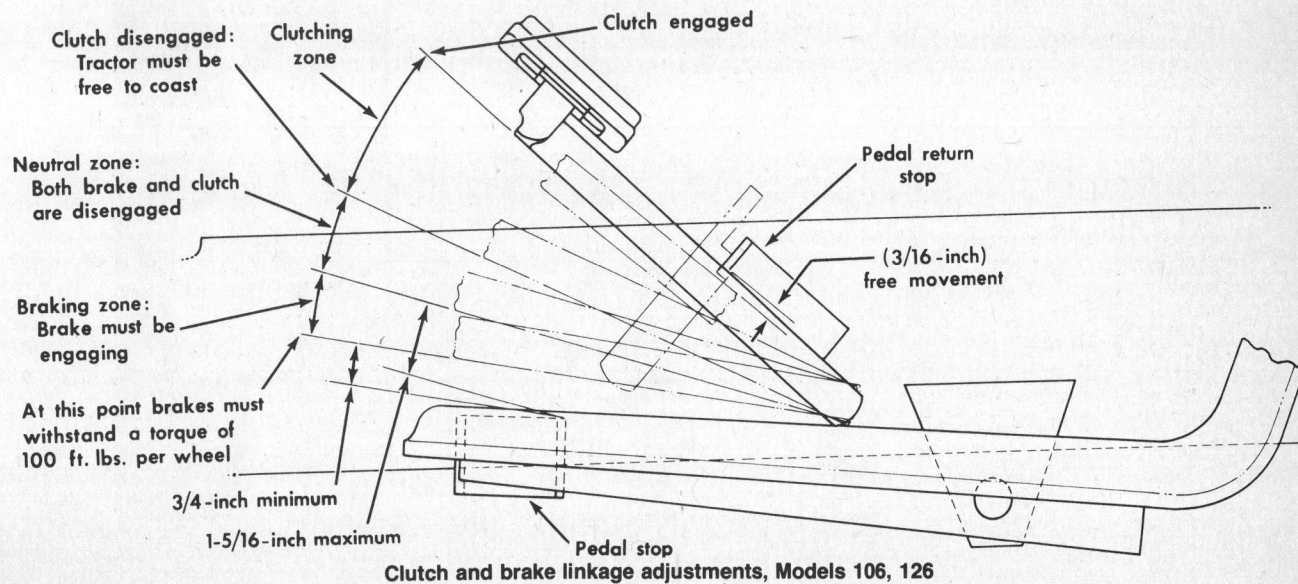

Clutch and brake linkage adjustments, Models 106, 126

513

International Harvester

4. Slide the engine rearward while aligning the clutch shaft into its pilot bushing and the drive plate pins into their holes in the driven disc.

5. Align the engine mounting capscrew holes and install the capscrews.

6. Align the clutch shaft coupling rear pin hole with pin hole in reduction drive (or creeper driveshaft) and install the pin.

7. Connect the clutch release rod to its pedal shaft lever and secure with the return spring.

8. Depress the pedal and release the clutch to allow components to move into final alignment.

9. Position the wires and cables in their clips and secure.

10. Connect the battery cables.

11. Replace the clutch shield.

12. Check the clutch pedal and linkage for proper adjustment.

ADJUSTMENT

It is important that a clearance of .050 in. be maintained between the clutch release lever and the clutch release bearing. In order to maintain this clearance, the pedal should have a free movement of approximately 3/16 in. This measurement is taken at the point of contact of the pedal arm with the front edge of the pedal return stop.

The clutch pedal adjustments should not require frequent attention. When it is necessary to adjust the clutch, turn the adjusting nut (3) on the clutch release rod in or out as required to get the proper measurements.

SPLITTING AND RECOUPLING

Models 73, 106, 126

SPLITTING THE TRACTOR

NOTE: Disconnect battery ground cable.

1. Disconnect the tail light wires (if so equipped).

2. Remove the electric lift foot guard (if so equipped).

3. Remove the fender and seat support assembly by removing the four capscrews in the bottom of the tool box under the seat and removing the six truss head machine screws, three each on the left and right foot supports.

NOTE: It is not necessary to remove the seat and fenders on the Model 73.

4. Drive small wooden wedges between the front axle and frame on each side to stabilize the front of the tractor.

5. Block both front wheels so the tractor is secure and can not move.

6. If the tractor is equipped with creeper attachment, shift the lever forward

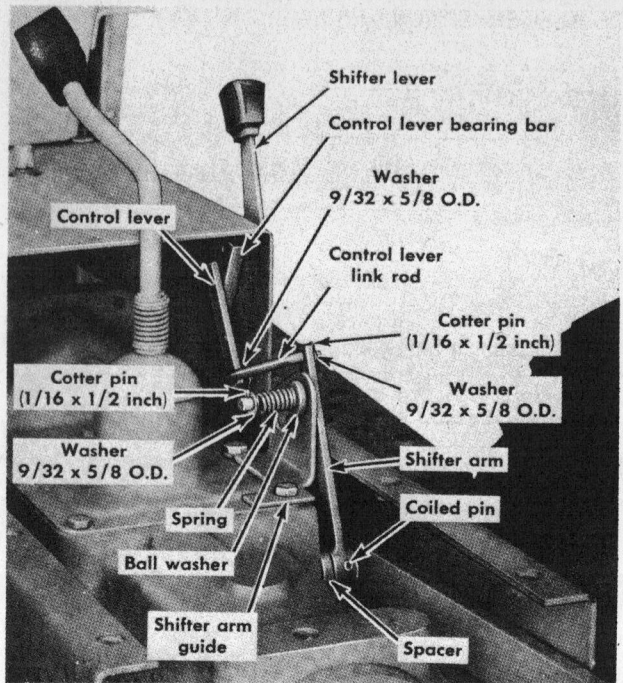

Models 106 and 126 PTO shifter control assembly

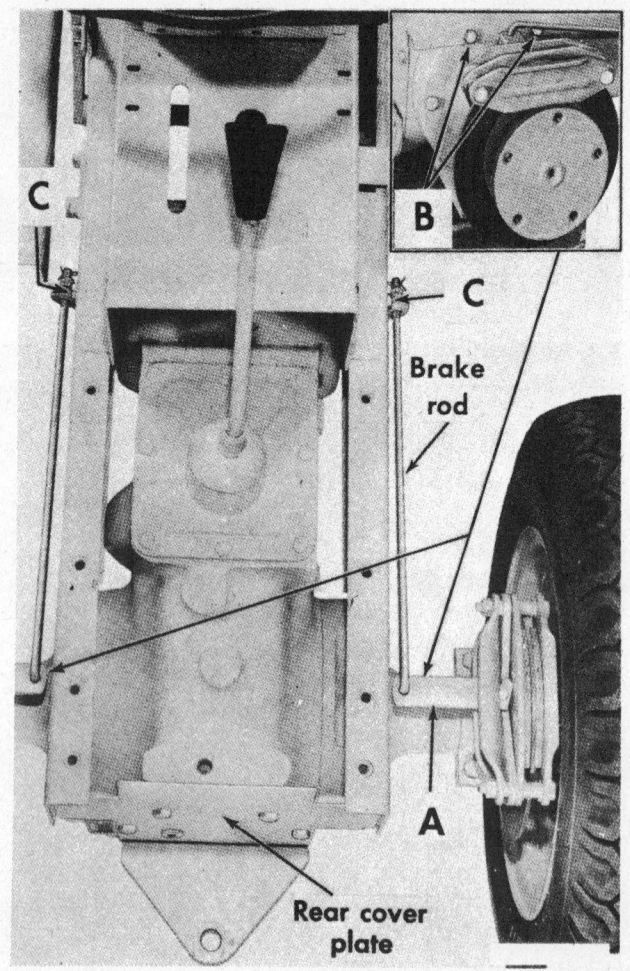

Location of frame capscrews

International Harvester

and remove the lever knob and the breather.

7. If the tractor is equipped with rear PTO, remove the control rod from the shifter arm.

8. If the tractor is equipped with three point hitch, remove the lift bar and its attaching plate.

9. Disconnect the brake rods from the brake arms adjacent to the disc brakes.

10. Remove the cotter pin and disconnect the brake rod from the brake lever.

11. Remove the coiled spring pin "D" from the rear hole in the clutch shaft coupling.

12. Place a jack under the implement lift cross shaft to support the frame.

13. Remove the four capscrews "B" holding the frame to the rear axle carriers and remove the two capscrews "C" holding the frame to the reduction gear housing.

14. Push down on the draw bar and pull the transmission-rear end assembly rearward. Guide the assembly from under the frame.

RECOUPLING THE TRACTOR

1. Tip the unit into recoupling position.

2. Recouple the split sections of the tractor being sure to align the clutch shaft and secure with the six capscrews.

3. Attach the coupling at the end of the clutch shaft using the coiled spring pin "D".

4. Connect the brake rod to the brake lever.

5. Connect the brake rods to the brake arms adjacent to the disc brakes.

6. If the tractor is equipped with three point hitch, install the lift bar and attaching plate.

7. Install the rear PTO control rod in the shifter arm (if so equipped).

8. Replace the creeper lever knob and breather (if so equipped).

9. Install the fenders and seat support assembly (except Model 73). Connect the tail light wires (if so equipped).

10. Install the electric lift foot guard (if so equipped).

11. Check oil level and fill to proper level with Hy-Tran fluid.

Models 107, 127, 147

SPLITTING THE TRACTOR

NOTE: **Disconnect battery ground cable.**

1. Disconnect the tail light wires (if so equipped).

2. Remove the electric lift foot guard (if so equipped).

3. Remove the fenders and seat support assembly by removing the four capscrews in the bottom of the tool box under the seat and the six truss head machine screws, three each on the left and right foot rests.

4. Drive small wooden wedges between the front axle and frame on each side to stabilize the front of the tractor.

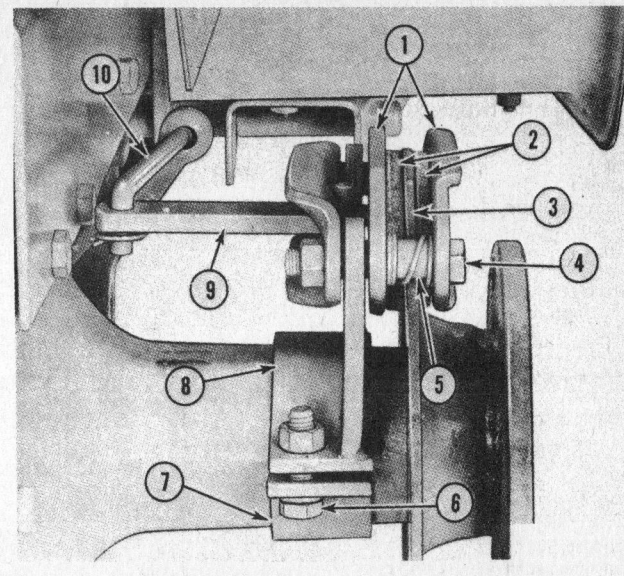

Brake assembly with wheel removed

5. Block both front wheels so the tractor is secure and cannot move.

6. If the tractor is equipped with three point hitch, remove the lift bar and its attaching plate.

7. Disconnect the brake rods (10) from the brake arms (9) adjacent to the disc brakes.

8. Depress the brake pedal and lock it in place.

9. Disconnect the brake rod (9) from the brake cross shaft (10).

10. Disconnect the ball joint (12) from the speed control cam (6).

11. Remove the pin from the "N" return rod and disconnect the rod from the brake cross shaft.

12. Remove the clutch shield. This will enable you to guide the dowel pin on the front end of the driveshaft through the fan shroud and air baffle plate.

13. Place a jack under the implement lift cross shaft to support the frame.

14. Remove the six capscrews (2), three on each side, holding the frame to the transmission assembly.

15. Push down on the draw bar and pull the transmission assembly rearward. Guide the assembly from under the frame.

RECOUPLING

NOTE: **The cam pivot bracket capscrews should be loosened slightly for ease of cam bracket adjustment.**

1. Lubricate the cam slot with chassis lubricant.

2. Recouple the split sections of the tractor being sure of the following:

a. The brake rod and "N" return rod go over the implement lift cross shaft and into position on the brake cross shaft.

b. Be sure to guide the driveshaft through the fan shroud and air baffle plate.

Underside view

International Harvester

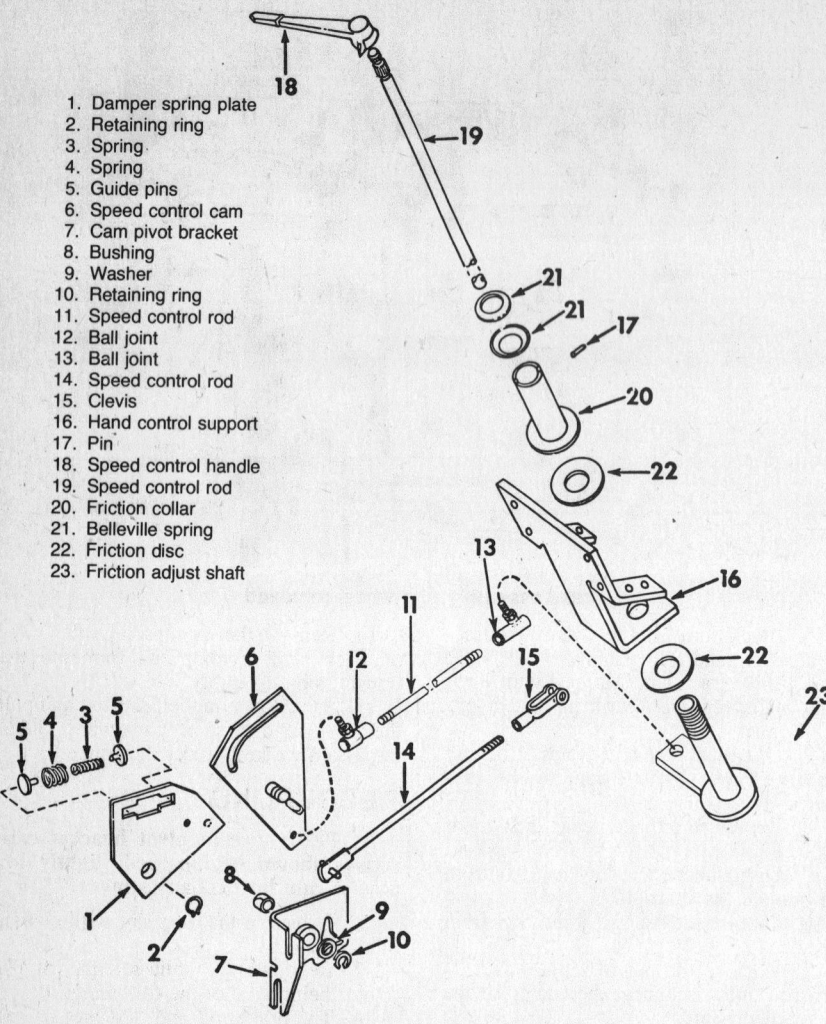

1. Damper spring plate
2. Retaining ring
3. Spring
4. Spring
5. Guide pins
6. Speed control cam
7. Cam pivot bracket
8. Bushing
9. Washer
10. Retaining ring
11. Speed control rod
12. Ball joint
13. Ball joint
14. Speed control rod
15. Clevis
16. Hand control support
17. Pin
18. Speed control handle
19. Speed control rod
20. Friction collar
21. Belleville spring
22. Friction disc
23. Friction adjust shaft

Speed control lever and linkage

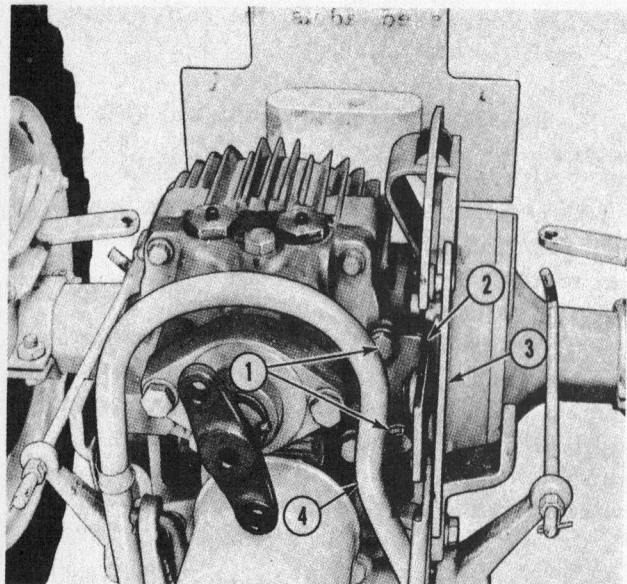

Location of cam bracket capscrews

c. The dowel pin on the front end of the driveshaft is positioned to engage the slot in the drive plate hub on the engine.

3. Install and tighten the three capscrews on each side of the frame.
4. Install the clutch shield.
5. Install the "N" return rod and pin on the brake cross shaft.
6. Install the ball joint on the speed control cam.
7. Depress the brake pedal and lock in place. Install the brake rod on the brake cross shaft.
8. Release the brake pedal and install the brake rods on the brake arms adjacent to the disc brakes.
9. Check for oil leaks. Check the oil level and fill to correct level with Hy-Tran fluid.
10. If the tractor is equipped with three point hitch, install the lift bar and its attaching plate.
11. Install the fender and seat support assembly. Connect the tail light wires (if so equipped).
12. Install the electric lift foot guard (if so equipped).
13. Adjust the linkage.

SPEED CONTROL LEVER LINKAGE AND CAM BRACKET ADJUSTMENT

NOTE: The brake pedal must be properly adjusted before beginning the speed control linkage adjustment.

1. If the tractor creeps in the "N" position or, if the speed control linkage has been disassembled or removed for any reason, the following adjustment must be made.
2. Proper friction adjustment is necessary on speed control lever for proper operation. The lever friction should be adjusted as follows:
   a. Remove the battery.
   b. Place a small wedge or key stock of proper size between the mounting bracket and the adjusting nut. Refer to foldout at back of manual.
   c. Move the lever towards the "R" or reverse position to increase the friction or toward the "F" or forward position to decrease the friction.
   d. Remove the wedge.
   e. Check the friction adjustment with a scale. The reading should be 10 lb. when pulled in either direction from the offset in the lever.
   f. Replace the battery.
3. Remove the seat and fender assembly.
4. Block the tractor so the rear wheels are off the ground and the tractor is secured so it can not move.
5. Move the speed control lever to the "F" or forward position.
6. Loosen the cam bracket mounting capscrews if not previously left loose.

NOTE: It may be necessary to deflect the suction tube slightly for clearance in order to loosen the cam bracket capscrews.

International Harvester

Cam slot clearance

7. Move the cam bracket to its highest position in the slotted holes and tighten the capscrews slightly to hold it in place.

8. Start the engine and with a punch and hammer adjust the cam bracket downward until the wheels stop turning.

9. Move the speed control lever to the forward position. Depress the brake pedal and lock in place.

10. If there is excessive vibration or noise in the transmission when the brake pedal is depressed, adjust the cam bracket to eliminate the noise.

11. Release the brake pedal and stop the engine.

12. Move the speed control lever to the "F" position and tighten the cam bracket capscrews.

13. With the engine running, move the speed control lever to the forward position. Depress the brake pedal all the way down and release it. The speed control lever should return to the "N" position and the wheels should stop turning.

12. If the speed control lever does not return to the "N" position, loosen the jam nut and turn the connecting rod to lengthen or shorten it until the speed control lever is in the "N" position when the brake pedal is depressed. Tighten the jam nut.

13. Check the rod in the speed control cam slot. The rod should not be touching the end of the slot when the brake pedal is fully depressed.

14. If the rod touches the end of the slot adjust as follows:
Remove the clevis end of the rod from the brake cross shaft. Loosen the jam nut and turn the clevis to lengthen the rod to prevent it from hitting the end of the slot.

15. Tighten the jam nut and install the clevis on the brake cross shaft.

16. Install the seat and fender assembly.

TRANSMISSION AND DIFFERENTIAL

Models 73, 106, 126

Complete service of the transmission requires splitting of the tractor, removal of the reduction drive (and creeper if so equipped), rear axles, carriers and the differential. The differential can be removed and replaced without disassembling the transmission, however, the transmission countershaft should be removed when checking preload of the differential carrier bearings. The transmission and differential are therefore covered together.

REMOVAL AND DISASSEMBLY (DIFFERENTIAL)

1. Drain the lubricant.
2. Split the tractor.
3. Remove the reduction drive.
4. Remove the rear axles and their carriers.
5. Remove the differential carrier bearing cage and shims from each side. Keep the shims with each cage and identified for each side.
6. Turn the differential and remove it from the transmission case. If the assembly will not clear the side of the transmission case, it will be necessary to remove one of the differential carrier bearings.
7. Drive out the pinion shaft lock pin.
8. Remove the pinion shaft.
9. Remove the pinion gears and side gears.
10. If the drive gear requires separate replacement, press out the eight retaining rivets.
11. Remove the carrier bearing cones from the differential carrier if they are to be replaced.
12. Remove the bearing cups from each cage if replacement is necessary.

DISASSEMBLY (TRANSMISSION)

1. Remove the differential.
2. Remove the gear shift lever and cover assembly.

Shift the transmission into two gear speeds to lock the transmission then remove the nuts from the countershaft.

3. Remove the shifter fork setscrews.

--- **CAUTION** ---
Cover the gear shift poppet ball hole to prevent the ball and spring from flying out as the rods are removed.

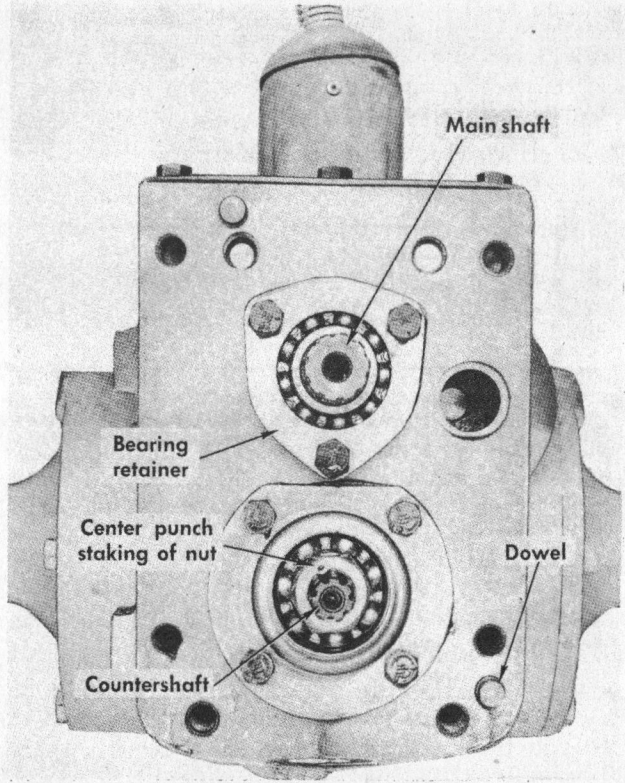

Front view of transmission

517

International Harvester

Drive the shifter rods forward and out of the transmission.

4. Remove the capscrews from the mainshaft front bearing retainer.

5. Pull the mainshaft forward and out of the transmission as the gears are removed.

6. Push the countershaft rearward and out of the transmission as the gears and spacers are removed. Note the sequence of spacers and gears for reassembly.

7. Pull the mainshaft needle bearing from the housing.

8. Remove the reverse idler shaft and gear.

9. Remove the countershaft front bearing, retainer and shims.

INSPECTION AND REPAIR

1. Wash all parts in cleaning solvent and dry with compressed air. Do not spin bearings.

2. Check all bearings for looseness, wear, roughness, pitting and scoring.

3. Check gears and shafts for wear and burrs. Remove any burrs with a fine stone.

4. Inspect the housing for cracks, restricted oil passages or raised places on its machine faces. Smooth off raised places with a file.

REASSEMBLY

Reassembly is basically the reverse of disassembly; however, particular attention should be given to the following steps.

1. Be sure all bearings are thoroughly lubricated.

2. The transmission mainshaft needle bearing must be installed with its oil hole aligned with the oil hole in the housing.

3. Assemble the differential, carrier bearings, cages and shims. Check bearing preload and adjust as necessary before replacing the transmission countershaft. Install or remove shims as necessary. Preload is correct when a steady pull of one to eight lb. is necessary to rotate the differential assembly.

4. Remove the differential assembly, keep the shims with the cages, then install the transmission countershaft, bearings, gears, spacers, front bearing retainer, shims and nut. Tighten the nut to 85 ft.lb. Tighten retainer capscrews to 20 ft.lb.

5. Install the differential assembly, keeping the preload shim pack correct as previously established. Drive gear must be on the right with teeth facing left.

6. Check the backlash between the drive gear and bevel pinion and the gear teeth bearing pattern as follows.

7. Apply a thin coat of red lead or Prussian blue to the bevel pinion teeth faces, then rotate the gears by hand and observe the bearing pattern.

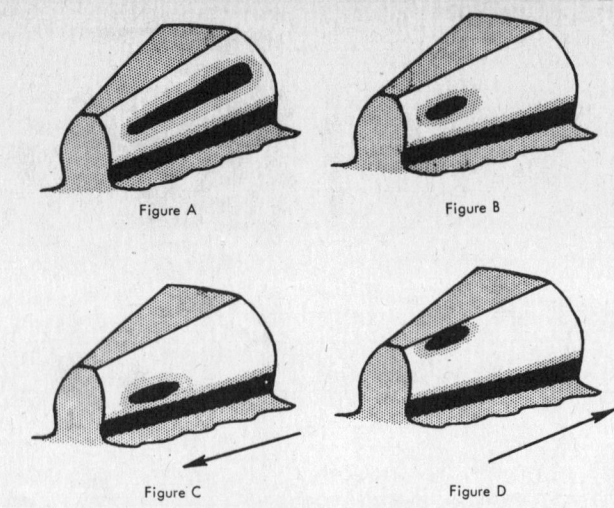
Gear tooth wear patterns

NOTE: Some deflection will occur under load. Allowance is made in gear design to prevent concentration of load on tooth edges.

8. Hand testing and very light loads should provide a pattern as shown in figure "B". When load and deflection increases the pattern will progress as in figure "A".

9. The desirable (no load) pattern in figure "B" is the result of adjusting the drive gear lateral position to the specified range of .003 in. to .005 in. backlash.

10. Tooth bearing position from the root to the crown of the tooth is controlled by lateral position of the bevel pinion. If low tooth bearing on bevel pinion is indicated (as shown in figure "C") the bevel pinion must be adjusted toward the drive gear. If high tooth bearing on the bevel pinion is indicated (as shown in figure "D") the bevel pinion must be adjusted away from the drive gear.

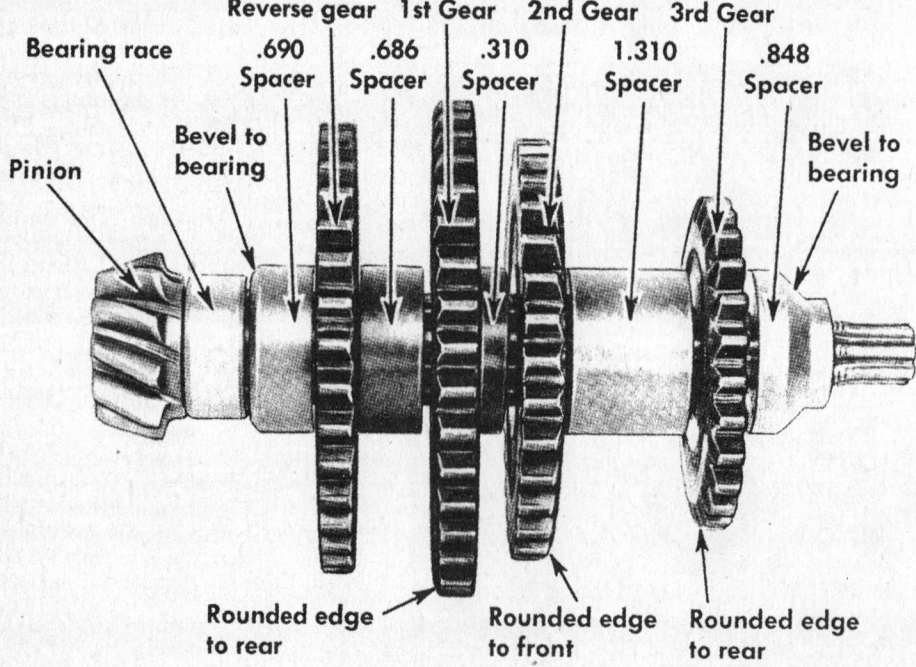
Countershaft

International Harvester

NOTE: If it is necessary to move the bevel pinion in or out to correct "Root-to-crown" bearing, the drive gear must also be moved laterally to maintain the specified backlash.

11. Stake the countershaft nut by center-punching the face of the nut over a spline groove.
12. Continue the assembly in reverse order of disassembly.
13. Fill housing to proper level.

NOTE: Creeper attachment has its own lubricant separate of the transmission. Fill creeper at breather and check at side plug in creeper housing.

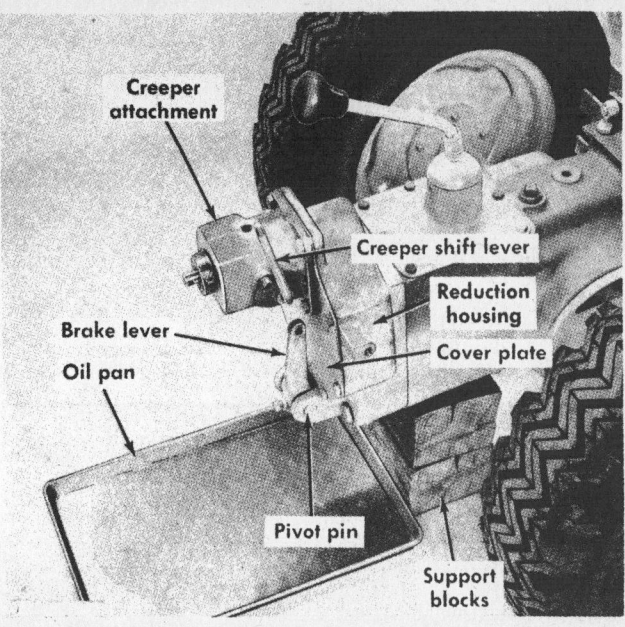

Power train supported for disassembly

REDUCTION DRIVE

Model 73

REMOVAL AND DISASSEMBLY

1. Split the tractor.
2. Place an oil pan under the reduction housing and remove the creeper assembly (if tractor is so equipped).
3. Remove the brake lever, pivot pin and push rod.
4. Remove the reduction housing front cover plate.
5. Hold the drive coupling and shaft from turning and remove the reduction gear retaining capscrew and washers. Remove the gear.
6. Remove the reduction gear spacer from the transmission shaft.

NOTE: It may be more convenient to pull the reduction driveshaft, seal and bearing before removing the reduction gear from the housing. Clearance between the gear and the capscrew bosses is restricted on some tractors.

7. Remove capscrews from holes "A" and "B".

NOTE: Soft copper sealing washers are used under the "B" capscrew heads.

8. Move the housing forward and away from the transmission housing as the brake disc slides off the transmission countershaft.
9. Pull the reduction driveshaft, seal and bearing from the reduction housing if it was not removed in step 6 NOTE.
10. Support the driveshaft splined coupling and drive out the coiled spring pin.

NOTE: The splined coupling is used only on tractors equipped with creeper attachment.

11. Press the driveshaft from the ball bearing.
12. Press the needle bearing rearward from the housing.
13. Remove the brake components.

INSPECTION AND REPAIR

1. Inspect the driveshaft for wear on the gear teeth, needle bearing area, oil seal contact area and drive pin hole.
2. Inspect the reduction gear teeth for wear or chipping and the fit of the gear on the transmission shaft.
3. On tractors with creeper, inspect wear of pilot bushing for creeper driveshaft.
4. Inspect needle and ball bearings for wear, pitting and roughness of operation.
5. Thoroughly clean the reduction housing, bearings, gears and brake parts.

REASSEMBLY AND INSTALLATION

1. Install a new gasket to the front of the transmission case. Dowels in the case will hold the gasket in place.
2. Press the needle bearing into the reduction housing from the rear. Rear end of bearing should be flush with housing.
3. Press the ball bearing onto the driveshaft.
4. Lubricate the lip of a new oil seal and install the seal onto the shaft. Be careful when sliding the seal lip over the pin

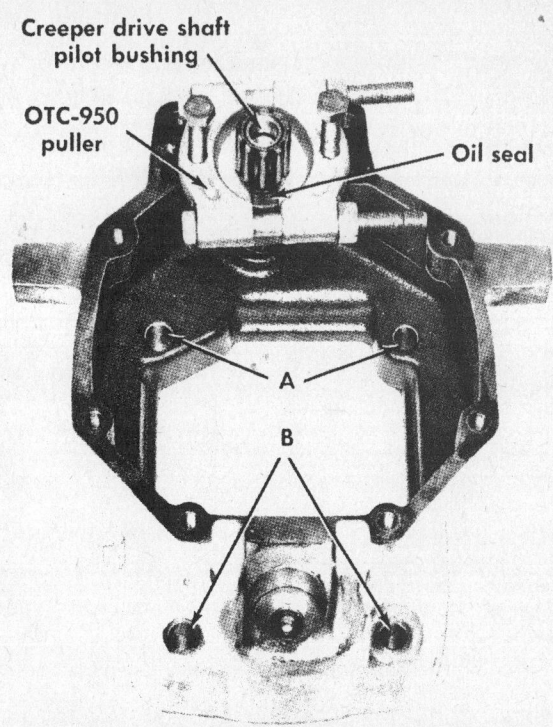

Pulling reduction driveshaft, seal and bearing

International Harvester

hole in the shaft. Flat face of seal case must be forward.

5. Install the splined coupling and coiled pin (if tractor is equipped with a creeper). Coiled pin ends must be even with or below the spline root to avoid interference when shifting the creeper unit.

6. Install a new O-ring onto the brake lining retainer and install in the reduction housing.

7. Install the reduction housing to the transmission case. Be sure the gasket and dowels are in place.

8. Install new copper sealing washers on the two lower capscrews. Refer to "B". Tighten capscrews at "A" and "B" to 80 ft.lb.

9. Install the brake linings and disc then push the front lining retainer rearward to hold disc and linings in place.

10. Install the reduction gear and spacer to the transmission mainshaft and secure with the capscrew, flat and lockwasher. Tighten capscrew after driveshaft is installed.

11. Install the driveshaft with ball bearing, seal and splined coupling (on tractors with creeper). Seal case should be flush with housing.

NOTE: Seal holds ball bearing in place and front cover holds seal in place.

12. Tighten reduction gear retaining capscrew to 55 ft.lb.

13. Install new gasket and housing front cover.

14. Install new gasket and creeper unit (on tractors so equipped).

15. Replace the brake push rod, brake lever and pivot pin.

16. Recouple the tractor by reversing the splitting procedure.

17. Fill transmission and creeper to proper level with specified lubricant.

Models 106, 126

REMOVAL AND DISASSEMBLY

1. Split the tractor.
2. Remove the transmission drain plug and allow the lubricant to drain completely; then replace the plug.
3. Remove the creeper assembly (if tractor is so equipped).
4. Remove the three coiled spring pins (E) to disconnect the brake arms and brake lever from the reduction housing.
5. Remove the reduction housing cover plate.
6. Remove the reduction driven gear by removing the capscrew, lockwasher and driven gear retainer.

NOTE: It may be more convenient to pull the reduction driveshaft, seal and bearing before removing the reduction gear from the housing. Clearance between the gear and capscrew bosses is restricted on some tractors.

7. Remove the spacer from the input shaft.

8. Remove the capscrews securing the transmission input shaft bearing retainer to the transmission and remove the retainer.

9. Remove the capscrews from holes "A" and "B".

NOTE: Soft copper sealing washers are used under the "B" capscrew heads.

10. Move the reduction housing forward and away from the transmission housing.

11. Pull the reduction driveshaft, seal and bearing from the reduction housing if it was not removed in step 6 NOTE.

12. Support the driveshaft splined coupling and drive out the coiled spring pin.

NOTE: The splined coupling is used only on tractors equipped with creeper attachment.

13. Press the reduction driveshaft from the ball bearing.

14. Press the needle bearing rearward from the housing.

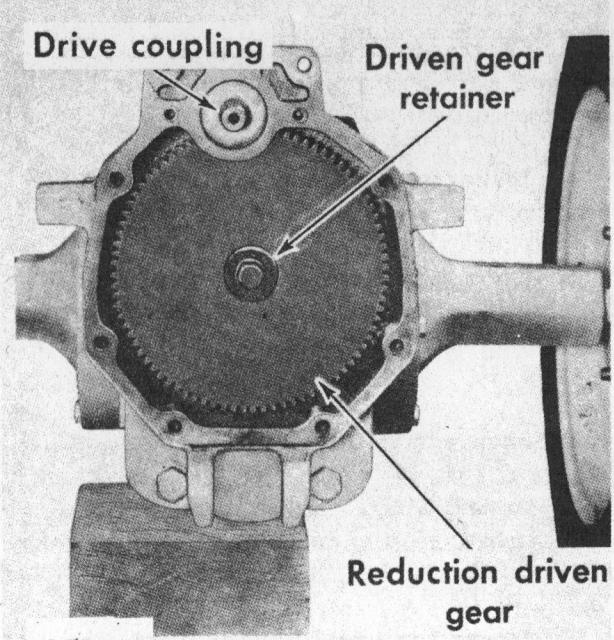

Reduction drive

INSPECTION AND REPAIR

1. Inspect the driveshaft for wear on the gear teeth, needle bearing area, oil seal contact area and drive pin hole.

2. Inspect the reduction gear teeth for wear or chipping and the fit of the gear on the transmission shaft.

3. On tractors with creeper, inspect wear of pilot bushing for creeper drive shaft.

4. Inspect needle bearings and ball bearings for wear, pitting and roughness of operation.

5. Thoroughly clean the reduction housings, bearings and gears.

REASSEMBLY AND INSTALLATION

1. Install a new gasket to the front of the transmission case.

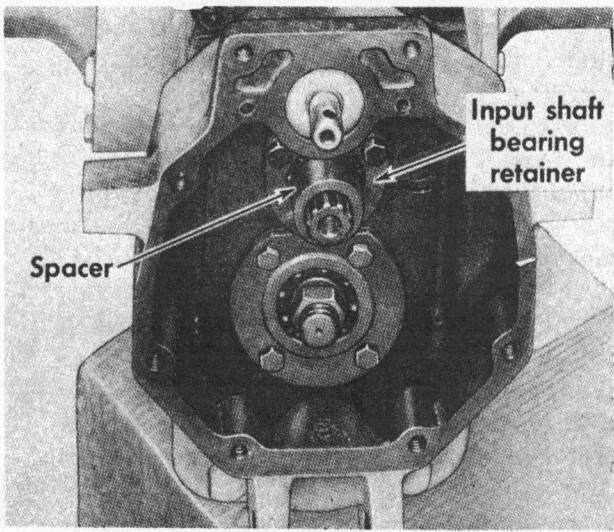

Transmission input shaft

2. Press the needle bearing into the reduction housing from the rear. The rear end of the bearing should be flush with the housing.

3. Press the ball bearing onto the reduction driveshaft.

4. Lubricate the lip of a new oil seal and install the seal onto the shaft. Be careful when sliding the seal lip over the pin hole in the shaft.

5. Install the splined coupling and coiled pin (if tractor is equipped with a creeper). Coiled pin ends must be even with or below the spline root to avoid interference when shifting the creeper unit.

6. Install the reduction housing to the transmission case. Be sure the gasket and dowels are in place.

7. Install new copper sealing washers on the two lower capscrews. Refer to "B". Tighten the capscrews at "A" and "B" to 80 ft.lb.

8. Install the reduction gear and spacer on the transmission mainshaft and secure with the retainer, capscrew and lockwasher.

9. Tighten the reduction gear retaining capscrew to 55 ft.lb. and lock in place.

10. Install the reduction driveshaft with ball bearings, seal and splined coupling (on tractors with creeper). Seal case should be flush with housing.

NOTE: Seal holds ball bearing in place and front cover holds seal in place.

11. Install a new gasket and install the housing front cover.

12. Install a new gasket and install the creeper unit (on tractors so equipped).

13. Install the brake lever and brake arms on the pivot shaft and reduction housing.

14. Recouple the tractor.

15. Fill the transmission and creeper to proper level with specified lubricant.

HYDROSTATIC TRANSMISSION AND DIFFERENTIAL

For complete service see the Hydrostatic transmission part of the Model 71, 102, 122, 123 section.

REAR AXLE

For complete rear axle service, see the Rear Axle part of the Model 71, 102, 122, 123 section.

FRONT PTO CLUTCH

For complete Front PTO Clutch service, see the Front PTO Clutch part of the Model 71, 102, 122, 123 section.

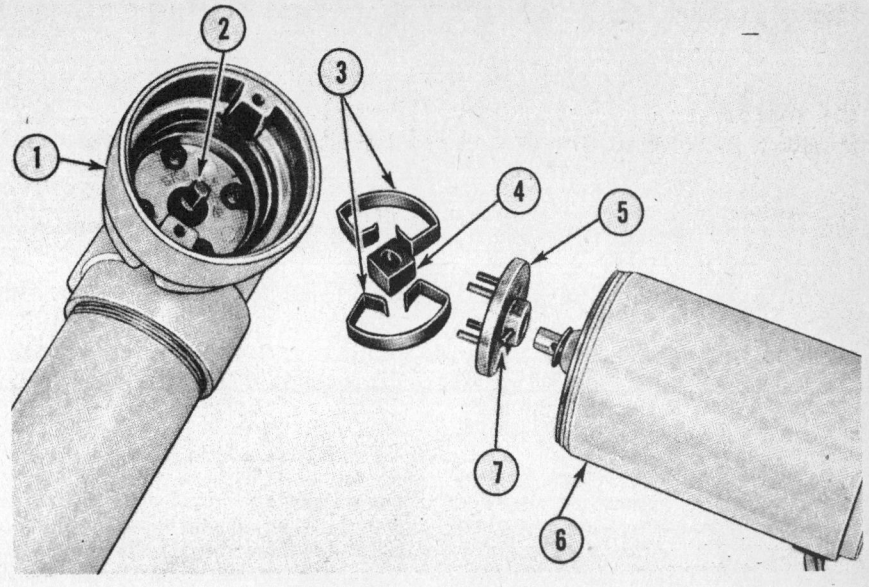

1. Motor adaptor
2. Pinion shaft
3. Brake springs
4. Brake cam
5. Drive coupling
6. Motor
7. Pin

Electric lift motor and brake components

REAR PTO

For complete Rear PTO service, see the Rear PTO part of the Model 72, 104, 105, 124, 125 section.

ELECTRIC POWER LIFT

REMOVAL

IMPORTANT: Disconnect battery ground cable.

1. Disconnect the wires to the motor at the multiple connector.
2. Remove the cotter pin and washer securing the lift assembly to the mounting bracket.
3. Remove the cotter pin, washer, locking clip pin and locking clip from the clevis and the lift arm.
4. Remove the electric lift assembly.

INSTALLATION

1. Remove two screws and pry the motor and drive coupling from the motor adaptor.

NOTE: Do not let motor end plates separate from motor body.

2. To remove the drive coupling from the motor, press the grooved pin (7) out of its bore. Do not remove the coupling unless the coupling or motor is to be replaced.
3. Remove the brake springs (3) and brake cam (4) from the pinion shaft (2).
4. Remove the four screws and remove the motor adaptor.
5. Pull the pinion shaft and bearing out of the housing.
6. Remove the retaining ring securing the pinion bearing to the pinion shaft and remove the bearing.
7. Loosen the setscrew in the housing and unscrew the housing from the outer tube.
8. Pull the screw, load bearing, and spiroid gear out of the housing if it was not removed in the above step.
9. To remove the spiroid gear and load bearing from the screw shaft, press the pin (7) out of the gear hub groove and screw shaft bore. Pull the gear, bearing and washer off the shaft.
10. **Early model:**
Press the pin (6) securing the clevis to the translating tube out and remove the clevis.
Late model:
Remove the screws securing the clevis

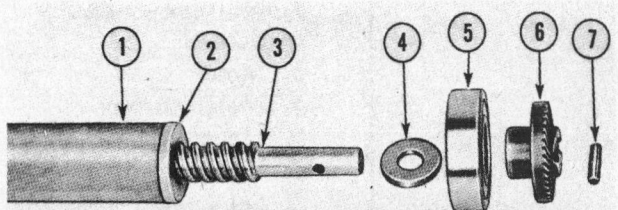

1. Translating tube
2. Lifting nut
3. Screw shaft
4. Washer
5. Load bearing
6. Spiroid gear
7. Pin

Screw with spiroid gear removed

International Harvester

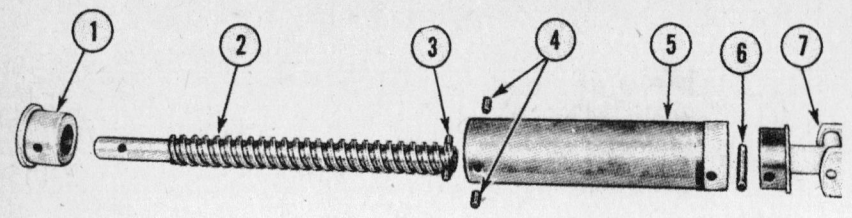

1. Lifting nut
2. Screw shaft
3. Stop pin
4. Pin (2)
5. Translating tube
6. Pin
7. Clevis assembly

Screw and nut removed from translating tube

and slip clutch assembly to the translating tube and remove the clevis.

11. Remove the screw and nut from the translating tube.
    a. Remove the screw from the nut and translating tube.
    b. Press the two pins (4), securing the nut to the translating tube, out through the tube and nut and press the nut (1) out of the tube.

NOTE: Do not attempt to press the two pins out of the translating tube while the screw is still in the nut.

12. Drive the stop pin (3) out of the screw.

INSPECTION AND REPAIR

1. Clean and inspect all bearings for damage or noticeable play indicating excessive wear. If bearings are not to be discarded, they should be oiled and wrapped in oil proof paper to keep them clean until reassembly.
2. Inspect the bushings for wear or damage and if there is any doubt of their serviceability replace the bushings.
3. Inspect all gear teeth for pitting, chipping and rounding off of teeth due to excessive wear.
4. Inspect the screw and nut for damage or excessive wear.
5. Inspect the snapring and snapring groove for wear or damage.
6. Inspect the brake springs, brake cam and brake liner insert for damage or noticeable wear.
7. Inspect the drive coupling for wear or damage.
8. Replace the oil seal in the outer tube.
9. The motor is serviced as a complete unit, and should be replaced if necessary.
10. Inspect all wiring for damage and loose or grounded connections.
11. Lubricate all parts before reassembly with grease, IH 251 HEP or equal.

ASSEMBLY

1. Drive the stop pin (23) into the end of the screw. The pin must be equally spaced with the same amount extending on either side of the screw. Liberally coat the screw with grease, IH 251 HEP or equivalent.
2. Mate the translating tube to the nut being sure the holes in the nut line up with the holes in the tube. Drive the two pins (18) into the translating tube and nut until they are flush with the translating tube outer diameter.
3. Be sure to grease the screw and assemble it in the translating tube and nut.
4. **Early model:** Align the pin bores and install the clevis in the translating tube. Drive the pin (22) into its bore until the ends are flush with the outer diameter of the translating tube.
   Late model: Attach the clevis with slip clutch to the tube with screws.
5. Install a new oil seal in the outer tube. Be sure the lip of the seal in pointing into the tube when installed. Peen the tube at three points to lock the seal in place.
6. Lubricate the seal and assemble the outer tube on the translating tube.
7. Assemble the washer, load bearing and spiroid gear on the end of the screw shaft. Drive the pin (7) into the gear hub groove and the bore in the screw. The pin must be equally spaced with the same amount extending on each side of the screw.
8. Grease the spiroid gear and load bearing.
9. Screw the housing into the outer tube until tight. Lock the tube in place with the setscrew.

NOTE: If a new outer tube is installed, tighten the tube and spot drill (No. 25) a hole through the tapped hole in the housing into the outer tube about $1/16$ in. deep. Lock the tube in place with the setscrew.

10. Assemble the pinion bearing onto the pinion shaft and secure with the snapring retainer. Grease the assembly with IH 251 HEP or equivalent.
11. Install the pinion and bearing assembly in the housing. Install the motor adaptor on the housing and secure it with the four screws. Be sure the two screw holes in the motor adaptor are in line with the outer tube.
12. Lightly grease the inside of the motor adaptor and install the brake cam on the pinion shaft. Install the brake springs on the cam with the outside of the springs against the steel insert brake lining.

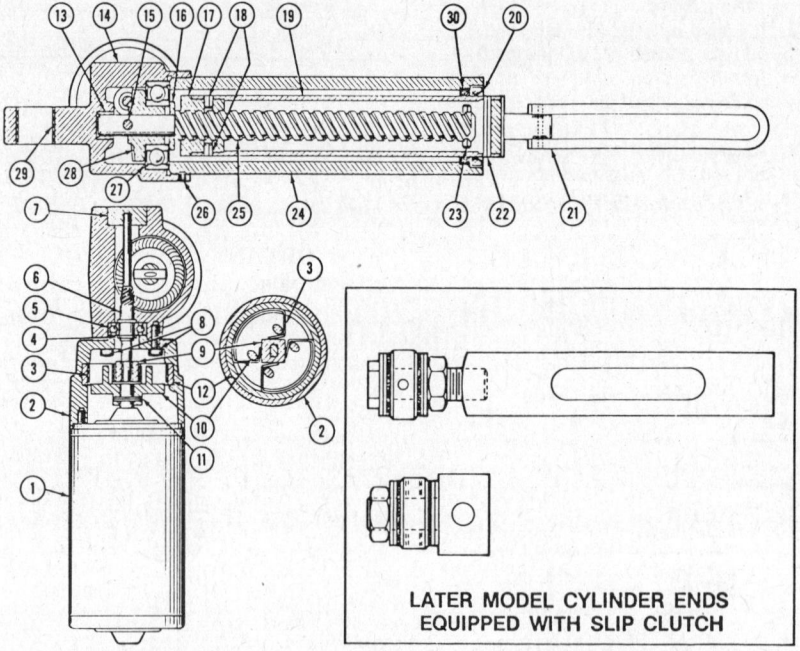

1. Motor
2. Motor adaptor
3. Brake springs
4. Retaining ring
5. Pinion bearing
6. Pinion shaft
7. Pinion bushing
8. Tap-Tite® screws
9. Brake cam
10. Drive coupling
11. Pin
12. Liner insert
13. Screw bushing
14. Housing
15. Pin
16. Washer
17. Lifting nut
18. Pins
19. Translating tube
20. Oil seal
21. Clevis assembly
22. Pin
23. Stop pin
24. Outer tube
25. Screw
26. Setscrew
27. Load bearing
28. Spiroid gear
29. Bushing
30. Guide bushing

Lift components

International Harvester

13. Pin the drive coupling to the motor if it was removed.
14. Install the motor and drive coupling on the motor adaptor. Be sure the drive pins are installed in the brake springs (3) correctly. Be sure the motor is positioned so the multiple connector and wires are pointing toward the tube.

INSTALLATION

1. Install the electric lift assembly on the mounting bracket and the lift arm.
2. Install the locking clip, locking clip pin, washer and cotter pin on the clevis and the lift arm.
3. Install the washer and cotter pin on the mounting bracket.
4. Connect the wires to the motor at the multiple connector.
5. Connect the battery ground cable.

ADJUSTMENT

1. The slip clutch is adjusted on the tractor by lifting an implement and observing the slip clutch action.
2. The clevis type end is adjusted externally.
3. The sand trap rake end must be removed from the translating tube to make the adjustment.
4. Adjustments must be made in small increments. 1/8 of a turn will change the tension quickly and considerably as you approach the correct setting. The lift should pick the load up without slipping but must rotate freely at the end of the stroke. A slight slip on engagement of load is permissible but once the load is moving slippage should not occur until maximum stroke is reached.

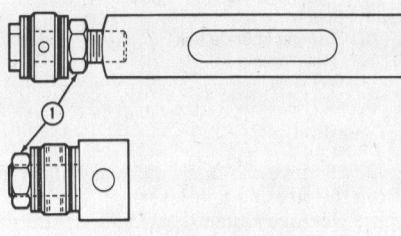

1. Adjust here

Lift cylinder adjustment

ENGINE

Engines used are: Models 73: Kohler K161S 106, 107: Kohler K241AS 126, 127: Kohler K301AS 147: Kohler K321A.
Complete engine service will be found in the Kohler Engine part of the Engine Unit Repair section of this book.

REMOVAL AND INSTALLATION

1. Disconnect the battery ground cable.
2. Remove six capscrews and nuts that hold the grille to the main frame. Slide the bottom of the grille forward 3½ in. and remove the clutch release pivot shaft from the grille casting. Remove the grille and hood. Lay the grille and hood (grille face down) in front of the tractor.
3. Remove the clutch shield.
4. Remove the air cleaner.
5. Remove the choke wire from the carburetor, and the throttle wire from the speed control lever.
6. Disconnect the positive (+) coil wire from the coil.
7. Disconnect the wires from the generator "A" terminal and from the "F" terminal.

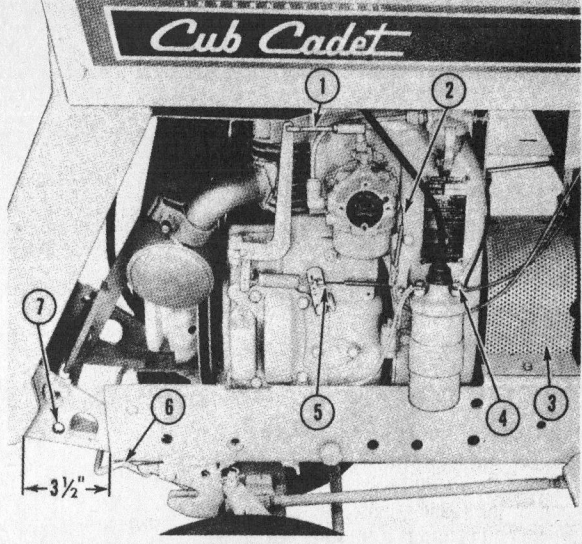

1. Throttle link
2. Choke wire
3. Clutch shield
4. Positive coil wire
5. Throttle wire
6. Quick attaching latch
7. PTO clutch release pivot shaft

Left side view

NOTE: Large wire to "A" terminal. Small wire to "F" terminal.

8. Disconnect the ground cable connected to a starter-generator bracket capscrew and remove starter-generator and coil wires from the generator bracket.
9. Remove the four capscrews two on each side, holding the engine to the frame.
10. Slide the engine forward to disengage the drive pins from the clutch driven disc.
11. The engine can now be lifted from the frame using the engine lift bracket on a head bolt.

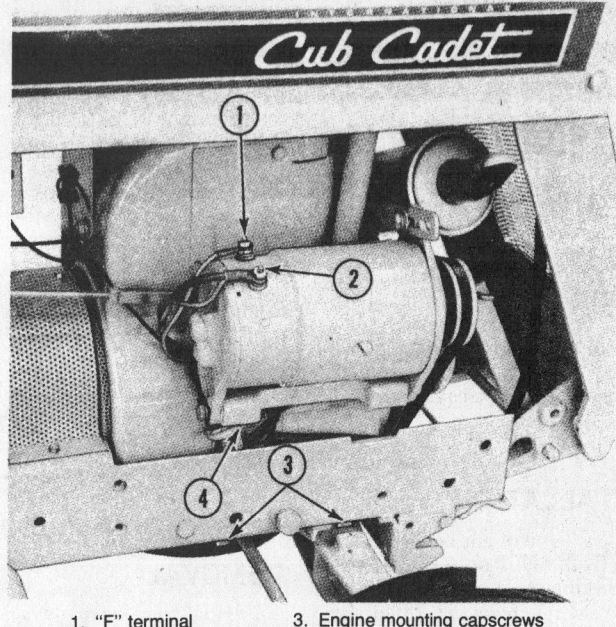

1. "F" terminal
2. "A" terminal
3. Engine mounting capscrews
4. Ground wire attached here

Right side view

523

International Harvester

Cub Cadet Models: 86, 108, 109, 128, 129, 149, 169, 800, 1000, 1200, 1250, 1450, 1650

FRONT WHEELS AND AXLE

Front Wheels and Bearings

REMOVAL

1. Lock the brake and block the rear wheels. Jack up the front axle.
2. Remove the capscrew and flat washer from the outer end of the front spindle.
3. Slide the wheel and bearings from the spindle.

NOTE: The bearings are a press fit in the wheel and a slip fit on the spindle.

4. Wheel bearings can be driven from the wheel hub with a hammer and long drift punch. Drive from the inside toward the outside.

INSTALLATION

1. If the bearings were removed, lubricate and press in new ones. Be sure force is directed to the outer race only.
2. Slide the wheel and bearing assembly over the spindle and secure with capscrew and flat washer.
3. If excessive end play exists (maximum 1/32 in.), place a sufficient thickness of shim washers (3-4 in. ID) over the outer end of the spindle and between the retaining washer and wheel bearing to take up excess end play.

Front Axle

REMOVAL AND DISASSEMBLY

1. Lock the brake, raise the front of the tractor and support it with suitable stands.
2. Remove the front wheels.
3. Disconnect the drag link from the left steering knuckle.
4. Remove the tie rod.
5. Remove the steering knuckle bolt, steering knuckle and spacer from both sides.
6. Drive out the retaining pin from the axle pivot pin.
7. Remove the axle pivot pin and front axle.

ASSEMBLY AND INSTALLATION

1. Coat the axle pivot pin and its bore in the axle with IH 251 HEP or its equivalent.
2. Position the axle in the support bracket and install the pivot pin. Install the retaining pin through the bracket and pivot pin.
3. Install the steering knuckles. Tighten the nut to 80 ft.lb. Install the cotter key. If the cotter key cannot be installed, tighten the nut only enough to line up the slot with the hole.
4. Install the tie rod and connect the drag link to the left steering knuckle.
6. Check and adjust toe-in. Adjust the tie rod to provide approximately 1/8 in. toe-in.
7. Adjust the drag link to proper length to place front wheels in the straight ahead position when the steering assembly is centered.

STEERING

Steering Gear

REMOVAL

1. Hold the steering wheel with front wheels in straight ahead position. Remove the steering wheel.
2. Remove the drag link rear ball joint from the steering unit lever.
3. Remove clutch shield bottom sheet.
4. Remove capscrews from the frame cross member and steering unit.
5. Lower the steering column assembly through the instrument panel pedestal and grommet.

DISASSEMBLY

1. Secure the steering lever and bolt in a vise.
2. Remove the lever bolt jam nut, adjusting nut, and washer.
3. Slide the column and housing assembly away from the lever, bolt and cam follower.
4. Remove the adjusting plug.
5. Remove the steering cam and bearings from the housing.

ASSEMBLY AND ADJUSTMENT

1. Thoroughly coat the cam ends, balls and races with chassis lubricant.

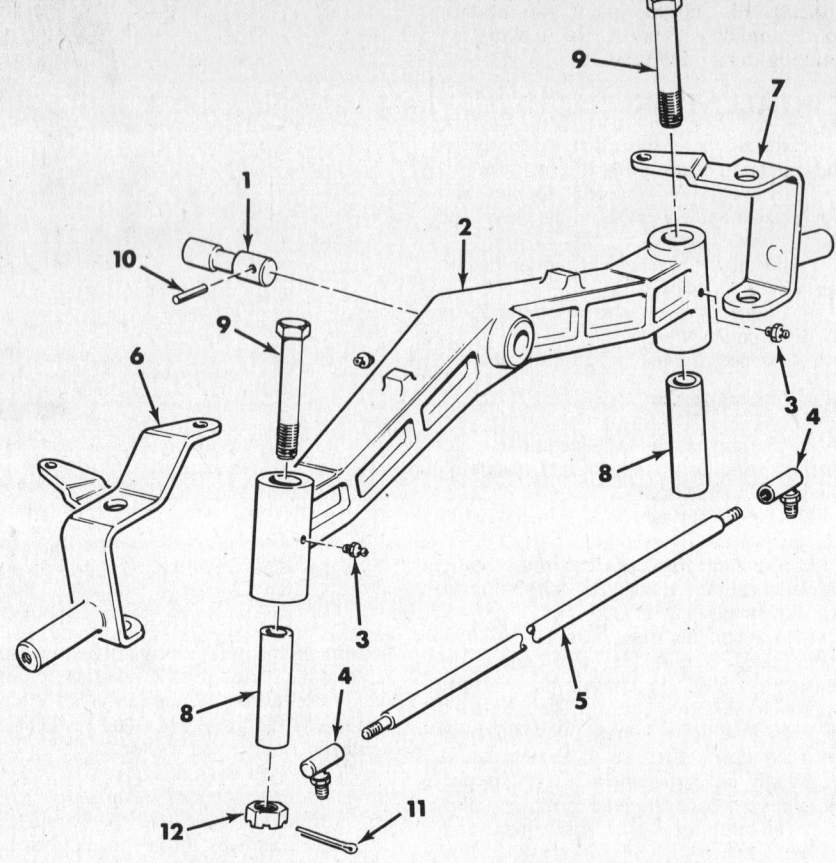

1. Axle pivot pin
2. Axle
3. Fitting
4. Ball joint
5. Tie rod
6. Left hand steering kuckle
7. Right hand steering knuckle
8. Spacer
9. Steering knuckle bolt
10. Retaining pin
11. Cotter key
12. Nut

Front axle components

International Harvester

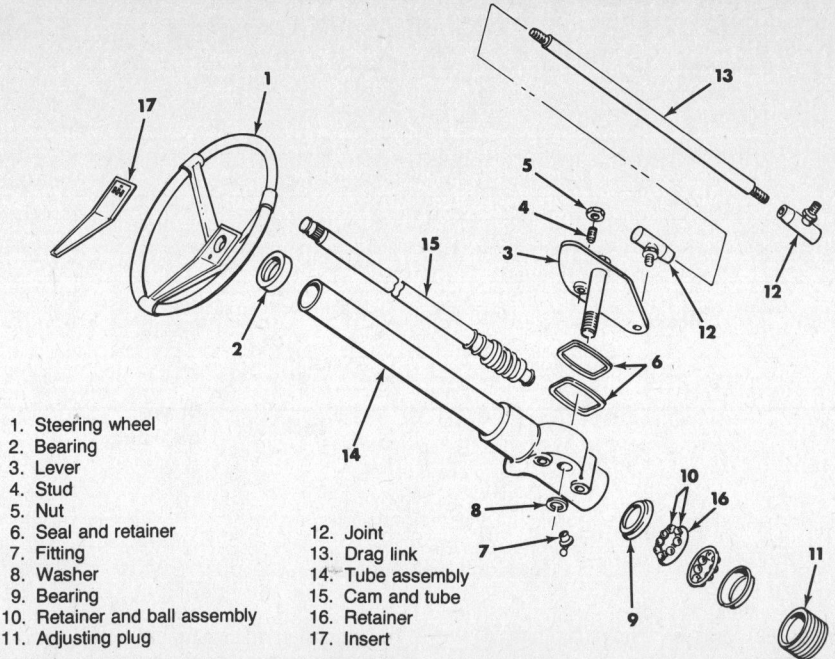

1. Steering wheel
2. Bearing
3. Lever
4. Stud
5. Nut
6. Seal and retainer
7. Fitting
8. Washer
9. Bearing
10. Retainer and ball assembly
11. Adjusting plug
12. Joint
13. Drag link
14. Tube assembly
15. Cam and tube
16. Retainer
17. Insert

Steering system components

2. Install the balls and races on the cam ends.

3. Thoroughly coat the cam with chassis lubricant then install into the housing and column assembly.

NOTE: Be sure the races enter the housing squarely and are not cocked.

4. Install the adjusting plug. Screw the plug inward until end play of the cam is removed but turns freely. Insert the cotter pin in the nearest hole.

5. Fill the housing with chassis lubricant.

6. Loosen the cam follower locknut, then "back out" the cam follower two turns.

7. Install the seal, retainer and lever-bolt assembly to the housing.

8. Install the washer and adjusting nut. Tighten the adjusting nut sufficiently to provide good seal compression. Refer to illustration for adjustment dimensions. Secure with the jam nut. Tighten jam nut to 40 ft.lb. Lubricate at the fitting in the housing slowly until lubricant begins to seep out.

9. Center the steering cam by rotating the steering shaft half-way between full right and full left turn.

10. Adjust the cam follower inward to eliminate backlash, then tighten locknut to 40 ft.lb. Turn steering shaft full right and left to check for binding.

11. Replace the steering assembly in the tractor chassis. Secure with two capscrews through the frame crossmember.

12. Replace clutch shield bottom sheet.

13. Connect the drag link.

14. Replace the steering wheel and secure with nut.

15. Adjust the tie rod to provide approximately 1/8 in. toe-in as follows:
   a. Remove one of the tie rod ball joints and loosen the locknut.
   b. Screw the ball joint in or out to obtain the specified toe-in of approximately 1/8 in. and tighten the locknut.
   c. Connect the ball joint to the steering knuckle and be sure to install the cotter pin.

16. Adjust the drag link to proper length to place front wheels in the straight ahead position when the steering assembly is "centered."

BRAKES

External Brake System

REMOVAL

1. Remove the brake rod (10) from the brake arm (9).

2. Remove the capscrews (6) from the mounting flange (7) and remove the brake assembly from the axle carrier.

3. Remove the capscrews (4) securing the caliper assembly (1) to the bracket assembly (8).

INSTALLATION

1. Assemble the caliper assembly to the bracket assembly. Be sure to install the spacers and springs (5).

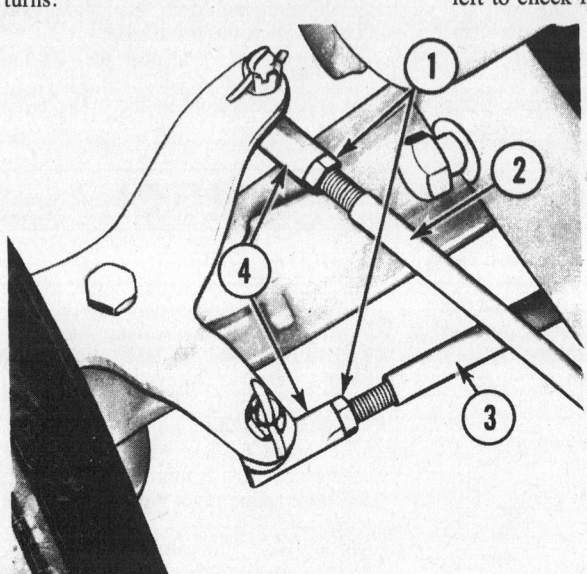

1. Locknuts
2. Drag link
3. Tie rod
4. Ball joints

Steering linkage adjustment points

1. Caliper assembly
2. Brake pads
3. Brake disc
4. Capscrew
5. Spacer and spring
6. Capscrew
7. Mounting flange
8. Bracket assembly
9. Brake arm
10. Brake rod

External brake system components

International Harvester

2. Install the caliper assembly and bracket assembly on the disc and axle carrier.

3. Assemble the mounting flange to the bracket assembly and secure with the capscrews.

4. Install the brake rods in the brake arms.

ADJUSTMENT

Standard Transmission

The disc brakes should engage when the pedal is pressed down to within a maximum of $1^{3}/_{16}$ in. and a minimum of ¾ in. above the platform.

It may be possible to push the pedal all the way down to the platform, but this is of no concern as long as the brake is fully engaged when the pedal arm is at least ¾ in. above the platform.

NOTE: The brakes must not engage before the pedal is within the maximum distance of $1^{3}/_{16}$ in. top of the braking zone.

To adjust the brakes, block the front wheels securely and raise the tractor so the rear wheels are off the ground and turn freely. Tractor must be in neutral.

Disconnect the left brake rod at the pinned end, rotate the right wheel by hand and adjust the jam nuts on the brake rod until wheel brakes firmly. Then, disconnect the right brake rod at the pinned end and reconnect the left brake rod. Turn the left wheel by hand and adjust the jam nuts until the wheel brakes firmly. Reconnect the right rod.

NOTE: To check the equalization of the brakes start the engine and shift the gears to third speed. After the wheels are turning apply the brakes. Both wheels should stop at the same time. If one wheel stops and the other wheel continues to revolve when the brakes are applied, stop the engine, adjust the jam nuts on the brake rod of the wheel that does not stop, enough so that both wheels stop simultaneously.

Hydrostatic Drive

The disc brakes should engage when the pedal is pressed down to within a maximum of $1^{3}/_{16}$ in. and a minimum of ¾ in. above the pedal stop.

The brake is engaged when the pedal arm is at least ¾ in. above the pedal stop.

To adjust the brakes block the front wheels securely and raise the tractor so the rear wheels are off the ground.

NOTE: The brakes must not engage before the pedal is within the maximum distance of $1^{3}/_{16}$ in.

With the rear wheels off the ground and the brake pedal in the locked position, the brake settings should be equalized as follows:

Disconnect left brake rod at the pinned end, rotate the right wheel by hand and adjust the jam nuts on the brake rod until the wheel brakes firmly. Then, disconnect the right brake rod at the pinned end and reconnect the left brake rod. Turn the left wheel by hand and adjust the jam nuts until the wheel brakes firmly. Reconnect the right rod.

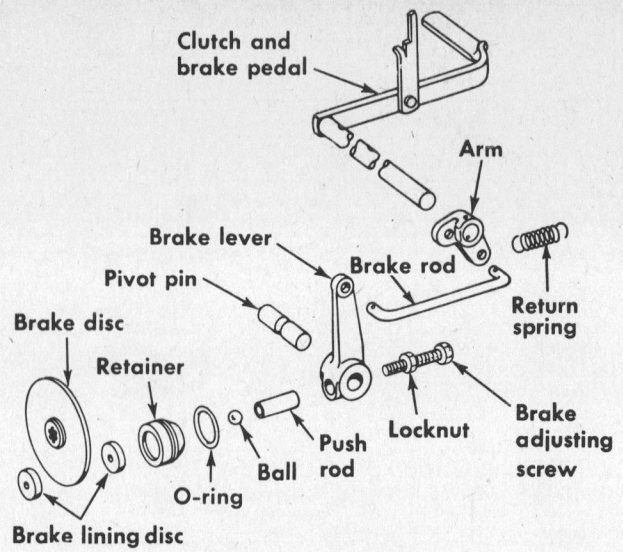

Internal brake components

Internal Brakes

REMOVAL

NOTE: Tractors equipped with a creeper attachment will require splitting, and the creeper removed before complete brake service can be performed.

1. Drain the transmission lubricant.
2. Remove the brake adjusting screw and locknut from its lever.
3. Remove the brake lever, pivot pin and push rod.
4. Remove the reduction housing front cover plate and slide it forward on the clutch shaft. Also see "Note" above.
5. Remove the reduction gear from the front of the transmission mainshaft.
6. Move the gear upward and the bottom of the gear forward to clean the cover screw bosses as the gear is lifted from the housing.
7. Slide the brake disc forward on the countershaft as the front lining and retainer are moved forward in their bore.

NOTE: Both linings and the disc can be removed without removing the front lining retainer; however, removal of the retainer is recommended for inspection and replacement of the retainer O-ring.

INSTALLATION

1. Clean the brake cavity and lining recess in the reduction housing.
2. Place a small quantity of grease in the rear brake lining recess in the reduction housing then insert the lining.
3. Install the disc on the countershaft and slide it rearward against the rear lining.
4. Install the front lining in the retainer.
5. Install a new O-ring on the front lining retainer, lubricate the retainer and O-ring then push them into the retainer bore in the reduction housing.

6. Install the reduction gear on the transmission mainshaft and secure with flat washer, lockwasher and capscrew. Tighten capscrew to 55 ft.lb.
7. Install a new cover gasket, then replace the cover plate.
8. Be sure the ball is in place in the front lining retainer; then install the push rod, lever, pivot pin, adjusting screw and locknut.
9. Fill transmission to proper level with Hy-Tran fluid or SAE 30 engine oil.

ADJUSTMENT

Standard Transmission

Loosen the locknut and turn the brake lever adjusting screw in or out as required.

The brake should engage when the pedal arm is pressed down to within a maximum of $1^{5}/_{16}$ in. and a minimum of ¾ in. distance above the top of the left foot support, which serves as the pedal stop.

It may be possible to push the pedal all the way down to the pedal stop, but this is of no concern as long as the brake is engaged within above limits.

CLUTCH

REMOVAL

1. Remove the extension panels.
2. Remove the frame cover.
3. Disconnect the battery ground cable from the battery. Remove the electric lift and bracket assembly if equipped.

Models 86, 108 and 128:

4. Remove the roll pins from the rear coupling and slide it forward on the shaft.
5. Remove the pivot pin and the hanger assembly.

Models 800, 1000 and 1200:

4a. Remove the flex coupling from the rear of the driveshaft.

NOTE: There is a steel ball spacer located in the flex coupling to properly locate the driveshaft. Be sure the ball is in place during reassembly.

International Harvester

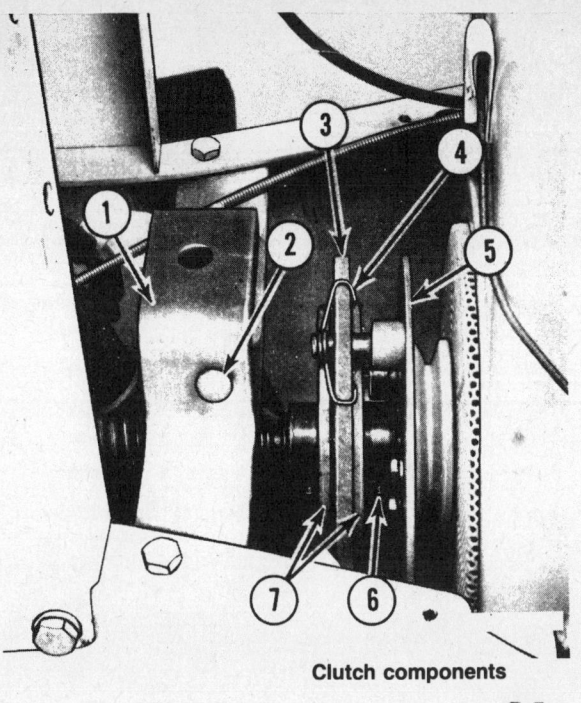

1. Hanger assembly
2. Pivot pin
3. Driving disc
4. Drive disc spring
5. Drive plate
6. Locating pin
7. Pressure plate assembly

Clutch components

1. Loading spring
2. Coiled pin

Pin positioning

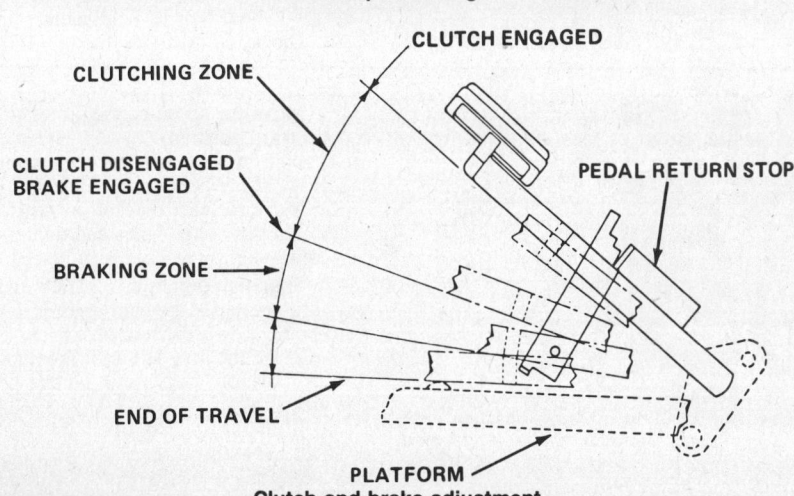

Clutch and brake adjustment

5a. Remove the pivot pin and the hanger assembly.

NOTE: On these models one of the drive pins is longer than the other two. This longer pin should have the drive disc spring installed on it.

6. Slide the clutch shaft to the side and out of the pressure plate assembly and driving disc.

7. Using a vise equipped with brass jaws, clamp the clutch shaft snug and tap the shaft down enough to compress the loading spring slightly. Remove the coiled spring pin.

--- CAUTION ---
Slowly release the vise allowing the spring to extend as the shaft slips through the vise jaws.

INSTALLATION

1. Assemble the front coiled spring pin, teaser spring, spring spacer, throwout bearing, bushing, loading spring and washer onto the clutch shaft.

NOTE: The bushing shown with the lubrication fitting is used on models with S/N 612808 and above.

2. Using a vise equipped with brass jaws, clamp the clutch shaft snug. Tap the shaft down enough to compress the loading spring and install the rear coiled spring pin.

3. Continue reassembling the clutch by reversing the disassembly procedure.

4. On Cub Cadet S/N 612808 and above use only powder bushing grease when packing the clutch drive plate bushing and when coating the clutch shaft.

5. Adjust the clutch.

ADJUSTMENT

Adjust the clutch linkage. It is important that a clearance of .050 in. be maintained between the clutch release lever and the clutch release bearing. In order to maintain this clearance, the pedal should have a free movement of approximately $9/32$ in. This measurement is taken at the point of contact of the pedal arm with the front edge of the pedal return stop. When it is necessary to adjust the clutch, turn the adjusting nut on the clutch release rod in or out as required to get the proper measurements.

SPLITTING AND RECOUPLING— HYDROSTATIC DRIVE

Splitting the Tractor

1. Remove the battery and battery holddown bracket.
2. Remove the rear fender to frame bolts and the battery ground wire.

527

International Harvester

3. Disconnect the wire harness from the voltage regulator and mounting clip, if equipped.
4. Remove the mounting screws from the foot platforms and the implement height adjustment knob.
5. Remove the frame cover.
6. **Models 1250, 1450 and 1650:**
Remove an extension panel and disconnect the front flex coupling.
7. Remove the cam bracket mounting bolts and move the cam bracket and linkage up out of the way.
8. Disconnect the rear brake rods from the caliper assemblies and the brake lever rod from the cross shaft.
9. Support the frame of the tractor, remove the frame mounting bolts and roll the rear end out. Tractors equipped with a three point hitch attachment will require the removal of the lift lever. Before rolling the rear end out, disconnect the lift bar from the lift lever and raise the frame high enough to remove the lift lever and shaft.

Recoupling the Tractor

1. Recouple the tractor by reversing the splitting procedure.
2. Check the oil level in the rear frame and fill to proper level with IH Hy-Tran or its equivalent.
3. Adjust the cam bracket.
4. Recheck the oil level of the rear frame.

SPEED CONTROL LEVER ADJUSTMENT

Models 109, 129, 149 and 160—Below Serial No. 503725

1. Remove the frame cover.
2. Loosen the friction adjusting screw.
3. Slide the shaft up to the limit of its travel.
4. Measure the gap between the control rod arm and the friction bushing flange.
5. If the gap measures more than .064 in., determine the number of washers (20 869 R1) required to reduce the gap to .064 in. or less.
6. Remove the bushing support. Clean

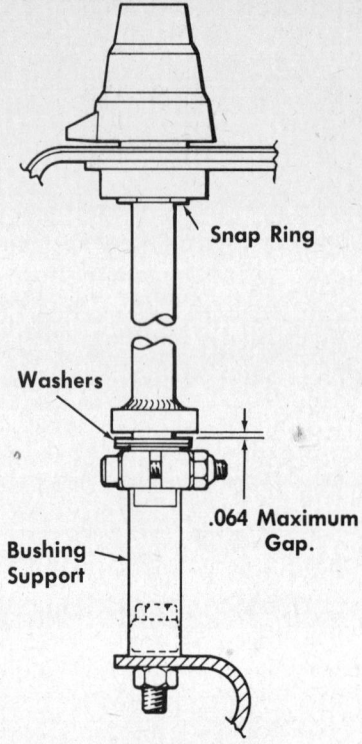

Upper snapring

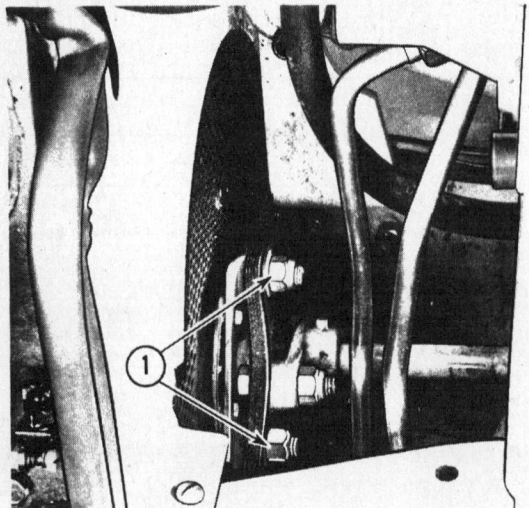

Flex coupling; disconnect at the point marked #1

the lower end of the shaft and apply a light coat of grease to the lower end.
7. Install the washers as determined by step 5. Install the bushing support and torque the mounting bolts to 9-10 ft.lb.
8. Recheck the clearance to be sure the gap is correct.
9. Tighten the adjusting bolt until a pull of 7-12 lb. on a scale keeps the handle in motion.

Models 109, 129, 149 and 160—Serial No. 503275 and Above

1. Tighten the friction adjustment until a pull of 7-12 lb. on a scale keeps the handle in motion.
2. Check the gap between the control handle and the instrument panel. The handle should not rub on the panel or exceed

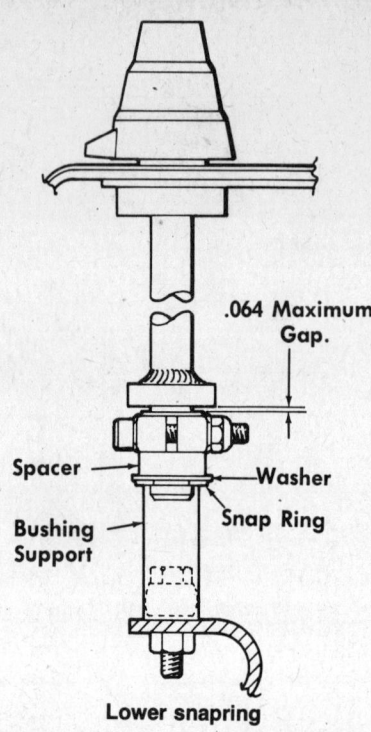

Lower snapring

.25 in. gap. Add or remove spacers as necessary.

Models 1250, 1450 and 1650

Adjust the friction control until a pull of 7-12 lb. on a scale keeps the handle in motion.

CAM BRACKET ADJUSTMENT

NOTE: The brake pedal must be properly adjusted before beginning the speed control linkage adjustment.

If the tractor creeps in the N position or, if the speed control linkasge has been disassembled or removed for any reason, the following adjustment must be made.

1. Proper friction adjustment on the speed control lever is necessary for proper operation. Check and adjust the lever friction as necessary.
2. Lubricate the T slot in the cam bracket and the cam control assembly slot with IH 251 HEP or its equivalent.
3. Block the tractor so the rear wheels are off the ground and the tractor is secured so it can not move.
4. Move the speed control lever to the F or forward position.
5. Loosen the cam bracket mounting capscrews if not previously left loose.
6. Move the cam bracket to its highest position in the slotted holes and tighten the capscrews slightly to hold it in place.
7. Start the engine and with a punch and hammer adjust the cam bracket downward until the wheels stop turning.
8. Move the speed control lever to the forward position. Depress the brake pedal and lock in place.
9. If there is excessive vibration or noise in the transmission when the brake pedal is depressed, adjust the cam bracket to eliminate the noise.

International Harvester

10. Release the brake pedal and stop the engine.

11. Move the speed control lever to the F position and tighten the cam bracket capscrews.

12. With the engine running, move the speed control lever to the forward position. Depress the brake pedal all the way down and release it. The speed control lever should return to the N position and the wheels should stop turning.

13. If the speed control lever does not return to the N position, loosen the jam nut and turn the connecting rod to lengthen or shorten it until the speed control lever is in the N position when the brake pedal is depressed. Tighten the jam nut.

14. Check the rod in the speed control cam slot. The rod should not be touching the end of the slot when the brake pedal is fully depressed.

15. If the rod touches the end of the slot adjust as follows:

Remove the clevis end of the rod from the brake cross shaft. Loosen the jam nut and turn the clevis to lengthen the rod to prevent it from hitting the end of the slot.

16. Tighten the jam nut and install the clevis on the brake cross shaft.

SPLITTING AND RECOUPLING— GEAR DRIVE

Splitting the Tractor

1. Remove the battery and battery hold-down bracket.
2. Remove the rear fender to frame bolts and the battery ground wire.
3. Disconnect the wire harness from the voltage regulator and mounting clip, if equipped.
4. Remove the mounting screws from the foot platforms and the implement height adjustment knob.
5. Remove the frame cover.
6. Remove the fender assembly.
7. Disconnect the rear brake rods from the caliper assemblies and the brake lever rod from the cross shaft.
8. Disconnect the rear coupling.

NOTE: Models with a flexible rear coupling have a steel ball spacer located in the flex drive coupling to properly locate the driveshaft. Be sure the ball is in place during reassembly.

9. Support the frame of the tractor, remove the frame mounting bolts and roll the rear end out of the frame. Tractors equipped with a three point hitch attachment will require the removal of the lift lever. Before rolling the rear end out, disconnect the lift bar from the lift lever and raise the frame high enough to remove the lift lever and shaft.

Recoupling the Tractor

1. Recouple the tractor by reversing the splitting operation.
2. Check oil level of the rear frame and fill to proper level with 1H Hy-Tran or equivalent.

REDUCTION DRIVE

REMOVAL AND DISASSEMBLY

1. Split the tractor.
2. Remove the transmission drain plug and allow the lubricant to drain completely; then replace the plug.
3. Remove the creeper assembly (if tractor is so equipped).
4. Remove the three coiled spring pins to disconnect the brake arms and brake lever from the reduction housing.
5. Remove the reduction housing cover plate.
6. Remove the reduction driven gear by removing the capscrew, lockwasher and driven gear retainer.

NOTE: It may be more convenient to pull the reduction driveshaft, seal and bearing (see step 11) before removing the reduction gear from the housing. Clearance between the gear and capscrew bosses is restricted on some tractors.

7. Remove the spacer from the input shaft.
8. Remove the capscrews securing the transmission input shaft bearing retainer to the transmission and remove the retainer.
9. Remove the capscrews.

NOTE: Soft copper sealing washers are used under one of the capscrew heads.

10. Move the reduction housing forward and away from the transmission housing.
11. Pull the reduction driveshaft, seal and bearing from the reduction housing if it was not removed in step 6.
12. Support the driveshaft splined coupling and drive out the coiled spring pin.

NOTE: The splined coupling is used only on tractors equipped with creeper attachment.

13. Press the reduction driveshaft from the ball bearing.
14. Press the needle bearing rearward from the housing.

ASSEMBLY AND INSTALLATION

1. Install a new gasket to the front of the transmission case.
2. Press the needle bearing into the reduction housing from the rear. The rear end of the bearing should be flush with the housing.
3. Press the ball bearing onto the reduction driveshaft.
4. Lubricate the lip of a new oil seal and install the seal onto the shaft. Be care-

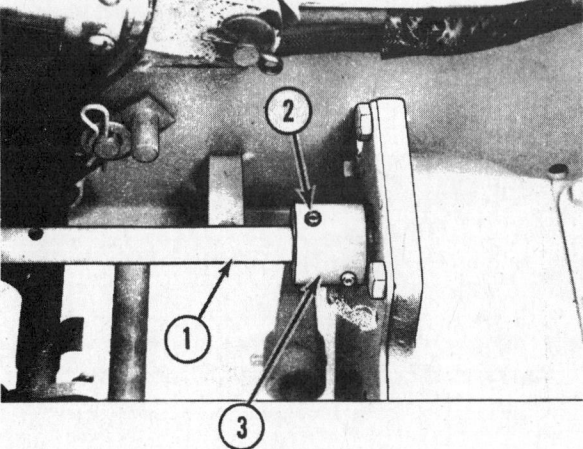

1. Driveshaft
2. Roll pin
3. Rear coupling

Driveshaft and rear coupling

1. Brake rod
2. Brake lever
3. Coiled pins
4. Brake arms

Reduction drive mounting points

529

International Harvester

ful when sliding the seal lip over the pin hole in the shaft.

5. Install the splined coupling and coiled pin (if tractor is equipped with a creeper). Coiled pin ends must be even with or below the spline root to avoid interference when shifting the creeper unit.

6. Install the reduction housing to the transmission case. Be sure the gasket and dowels are in place.

7. Install new copper sealing washers on the two lower capscrews. Tighten the capscrews to 80 ft.lb.

8. Install the reduction gear and spacer on the transmission mainshaft and secure with the retainer, capscrew and lockwasher.

9. Tighten the reduction gear retaining capscrew to 55 ft.lb.

10. Install the reduction driveshaft with ball bearings, seal and splined coupling (on tractors with creeper). Seal case should be flush with housing.

NOTE: Seal holds ball bearing in place and front cover holds seal in place.

11. Install a new gasket and install the housing front cover.

12. Install a new gasket and install the creeper unit (on tractors so equipped).

13. Install the brake lever and brake arms on the pivot shaft and reduction housing.

14. Recouple the tractor.

15. Fill the transmission and creeper to proper level with IH Hy-Tran or its equivalent.

STANDARD TRANSMISSION AND DIFFERENTIAL

Complete service of the transmission requires splitting of the tractor, removal of the reduction drive (and creeper if so equipped), rear axles, carriers and the differential. The differential can be removed and replaced without disassembling the transmission, however, the transmission countershaft should be removed when checking preload of the differential carrier bearings. The transmission and differential are therefore covered together.

REMOVAL AND DISASSEMBLY

Differential

1. Drain the lubricant.
2. Split the tractor.
3. Remove the reduction drive.
4. Remove the rear axles and their carriers.
5. Remove the differential carrier bearing cage and shims from each side. Keep the shims with each cage and identified for each side.
6. Turn the differential into the posi-

Positioning the differential gear for removal

tion shown and remove it from the transmission case. If the assembly will not clear the side of the transmission case, it will be necessary to remove one of the differential carrier bearings.

7. Drive out the pinion shaft lock pin.
8. Remove the pinion shaft.
9. Remove the pinion gears and side gears.
10. If the drive gear requires separate replacement, press out the eight retaining rivets.
11. Remove the carrier bearing cones from the differential carrier if they are to be replaced.
12. Remove the bearing cups from each cage if replacement is necessary.

DISASSEMBLY
Transmission

1. Remove the differential
2. Remove the gear shift lever and cover assembly.

Shift the transmission into two gear speeds to lock the transmission then remove the nut from the countershaft.

3. Remove the shifter fork setscrews.

— CAUTION —
Cover the gearshift poppet ball hole to prevent the ball and spring from flying out as the rods are removed.

Drive the shifter rods forward and out of the transmission.

4. Remove the capscrews from the mainshaft front bearing retainer.
5. Pull the mainshaft forward and out of the transmission as the gears are removed.
6. Push the countershaft rearward and out of the transmission as the gears and spacers are removed. Note the sequence of spacers and gears for reassembly.
7. Pull the mainshaft needle bearing from the housing.
8. Remove the reverse idler shaft and gear.

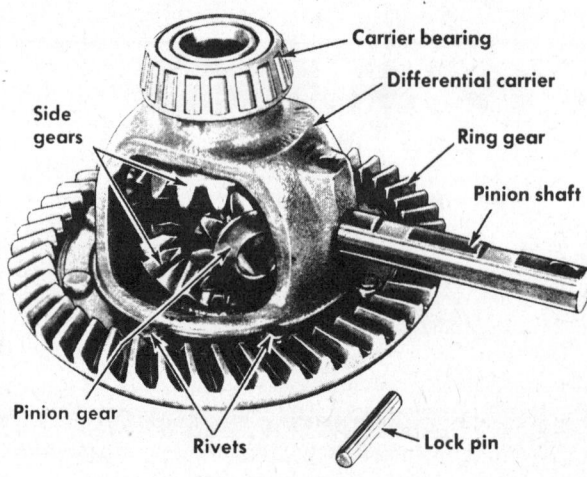

Disassembling differential

International Harvester

9. Remove the countershaft front bearing, retainer and shims.

ASSEMBLY

Reassembly is basically the reverse of disassembly; however, particular attention should be given to the following steps.

1. Be sure all bearings are thoroughly lubricated.
2. The transmission mainshaft needle bearing must be installed with its oil hole aligned with the oil hole in the housing.
3. Assemble the differential, carrier bearings, cages and shims. Check bearing preload and adjust as necessary before replacing the transmission countershaft. Install or remove shims as necessary. Preload is correct when a steady pull of one to eight lb. is necessary to rotate the differential assembly.
4. Remove the differential assembly, keep the shims with the cages then install the transmission countershaft, bearings, gears, spacers, front bearing retainer, shims and nut. Tighten the nut to 85 ft.lb. Tighten retainer capscrews to 20 ft.lb.
5. Install the differential assembly, keeping the preload shim pack correct as previously established. Drive gear must be on the right with teeth facing left.
6. Check the backlash between the drive gear and bevel pinion and the gear teeth bearing pattern as follows.
7. Apply a thin coat of red lead or Prussian blue to the bevel pinion teeth faces, then rotate the gears by hand and observe the bearing pattern.

NOTE: Some deflection will occur under load. Allowance is made in gear design to prevent concentration of load on tooth edges.

8. Hand testing and very light loads should provide a pattern as shown in Figure "B". When load and deflection increases the pattern will progress as in Figure "A".
9. The desirable (no load) pattern in Figure "B" is the result of adjusting the drive gear lateral position to the specified range of .003 in. to .005 in. backlash.
10. Tooth bearing position from the root to the crown of the tooth is controlled by lateral position of the bevel pinion. If low tooth bearing on bevel pinion is indicated (as shown in Figure "C") the bevel pinion must be adjusted toward the drive gear. If high tooth bearing on the bevel pinion is indicated (as shown in Figure "D") the bevel pinion must be adjusted away from the drive gear.
11. Stake the countershaft nut by center-punching the face of the nut over a spline groove.
12. Continue the assembly in reverse order of disassembly.

NOTE: The right side axle carrier mounting bolt hole located at the nine o'clock position is tapped through the differential case. The bolt installed in this hole should be coated with IH Gasketmaker to prevent oil seepage past the threads.

13. Fill housing to proper level with specified lubricant.

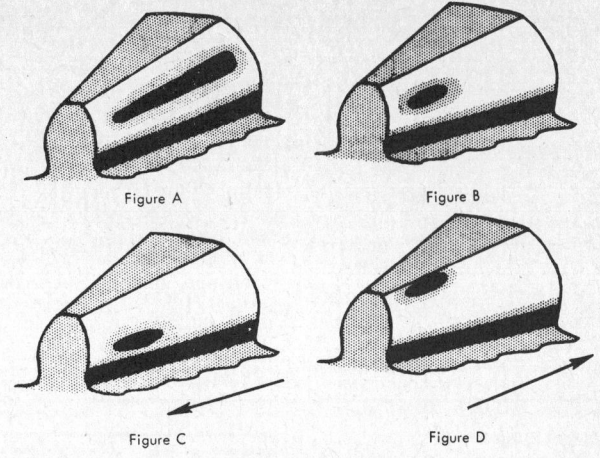
Wear patterns

NOTE: Creeper attachment has its own lubricant separate of the transmission. Fill creeper at breather and check at side plug in creeper housing.

HYDROSTATIC TRANSMISSION AND DIFFERENTIAL

For complete transmission service, see the Model 71, 102, 122, 123 tractor section.

REMOVAL

1. Drain the lubricant.
2. Split the tractor.
3. Remove the differential.
4. Remove the bevel pinion shaft and constant mesh gear.

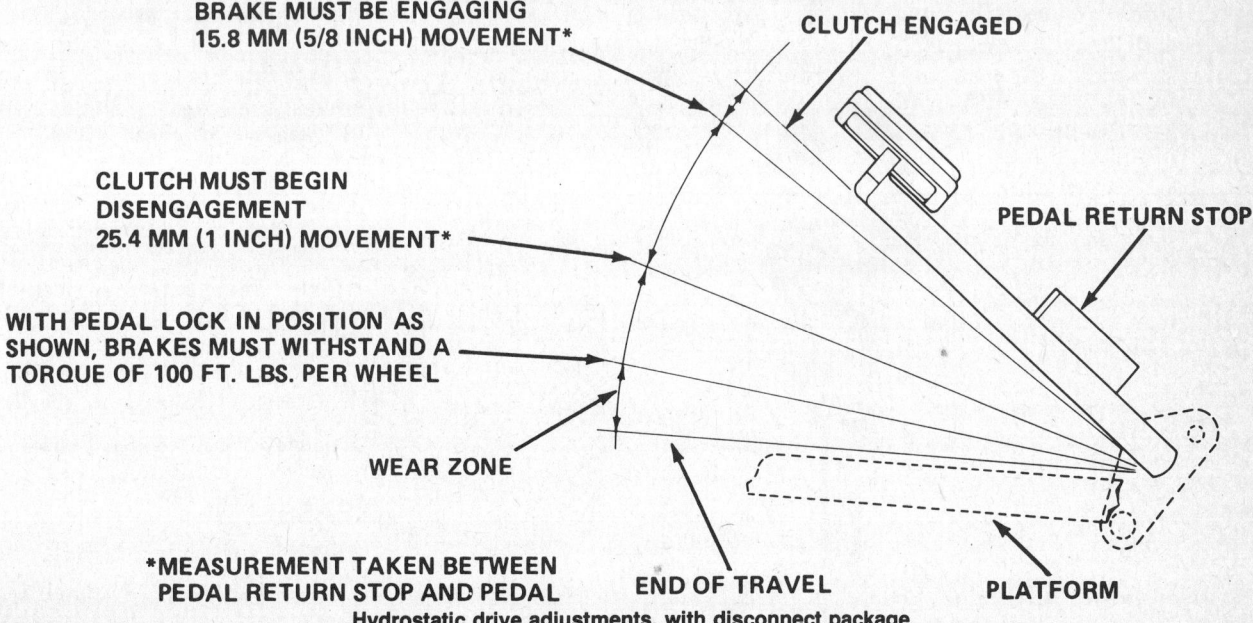
Hydrostatic drive adjustments, with disconnect package

531

International Harvester

5. Remove the rear brake lining disc.
6. Remove the brake retainer by pushing it out of the housing forward.
7. Remove the brake lining disc from the retainer.

INSTALLATION

1. Clean the brake cavity and lining recess in the transmission case.
2. Place a small quantity of grease in the rear brake lining recess in the transmission case; then insert the brake lining disc.
3. Install the front brake lining disc in the retainer.
4. Install a new O-ring on the front lining retainer, lubricate the retainer and the O-ring then push them into the retainer bore in the transmission case.
5. Reassemble the transmission case.
6. Fill transmission to proper level with Hy-Tran fluid or SAE 30 engine oil.

ADJUSTMENTS

1. Block the front tires and support the rear end so one rear tire is off the ground and free to turn.
2. Place the brake pedal in the up position.
3. Loosen the jam nut. Then tighten the brake lever adjusting screw until finger tight (8-10 in.lb.). Retighten the jam nut while holding the adjusting screw.
4. Actuate the brake pedal through a full stroke at lease one time.
5. Repeat step 3.
6. If the brake drags with the brake pedal in the up position, loosen the jam nut and back off the adjusting screw slightly and retighten the jam nut.

With Disconnect Package
CLUTCH

1. To adjust clutch free travel, turn the adjusting nut so the clutch just begins to release when the clutch pedal is depressed one in.
2. The one in. dimension is measured at the point of contact between the clutch pedal and the front of the pedal return stop.

EXTERNAL BRAKE

1. Adjust the clutch.
2. Block the front wheels securely and raise the tractor so the rear wheels are off the ground.
3. Disconnect the left brake rod at the pinned end, rotate the right rear wheel by hand and adjust the jam nut on the right brake rod until the wheel just begins to brake with the clutch pedal depressed to ⅝ in. between the pedal arm and the pedal return stop.
4. Disconnect the right brake rod at the pinned end and reconnect the left brake rod.
5. Adjust the left brake in the same manner as described for the right side.
6. Reconnect the right rod.

REAR AXLE AND HOUSING

REMOVAL AND DISASSEMBLY

1. Raise the rear of the tractor and block securely.
2. Remove the wheel and tire.
3. Remove the brake assembly from the axle housing.
4. Place a drip pan under the rear frame housing and remove the draw bar and differential housing cover.
5. Remove the "C" type snapring from the axle shaft and slide the axle out of the housing.
6. The axle housing outer oil seal may be removed with the axle housing on the tractor. Collapse the seal with a hammer and chisel and pry the seal from the bore. Be careful not to nick or damage the seal bore.
7. The needle bearing may be removed using a puller without removing the axle housing from the tractor.
8. To remove the axle housing, remove the capscrews attaching the frame to the axle housing and reduction housing. Raise the frame clear of the axle housing and remove the housing.

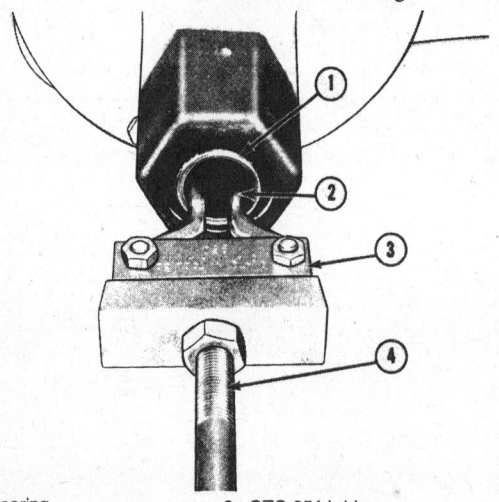

1. Bearing
2. OTC 960—8 legs
3. OTC 954 bridge
4. Slide hammer attachment

Pulling needle bearings

ASSEMBLY AND INSTALLATION

1. Using a new gasket, install the axle carrier to the differential housing. Capscrew threads should be coated with a non-hardening sealer (Permatex®) to avoid oil leaks. The frame pad of the axle carrier must be to the top.
2. Install the needle bearing into the housing with a suitable tool. One side of the bearing is chamfered to aid in installation. Install the bearing flush with the outer edge of its bore.
3. Install the seal with a suitable tool. Fill the cavity between the lips of the oil seal with IH 251 HEP or its equivalent before installing the seal.
4. Lubricate the axle shaft and bushing then slide the shaft through the seal, bushing, carrier and differential side gear. Rotate the axle as it is pushed through to avoid damage to the seal. Wipe off excess lubricant.
5. Install a new "C" type snapring to the inner end of the axle shaft.
6. Replace the rear cover and drawbar.
7. Install the brake assemblies on the outer ends of the axle carriers.
8. Install the wheel.
9. Fill differential housing to specified level with Hy-Tran or its equivalent.

CREEPER DRIVE

For complete creeper drive service, see the Model 72, 104, 105, 125 section.

MECHANICAL FRONT POWER TAKE-OFF

Without Brake
REMOVAL

1. Remove the grille.
2. Disengage the clutch and align the setscrew holes with the notches in the crankshaft pulley.
3. Remove the jam setscrew and lock setscrew from each of the three holes in the clutch pulley housing.

NOTE: These setscrews lock the clutch to the bearing on the tractor crankshaft.

4. Disconnect the PTO linkage and rotate the clutch release shaft forward out of the way.
5. Remove the clutch from the tractor as an assembly.

DISASSEMBLY

1. Remove the jam nuts from the throwout lever screws.
2. Loosen the throwout lever adjusting screws evenly and remove the screws.

International Harvester

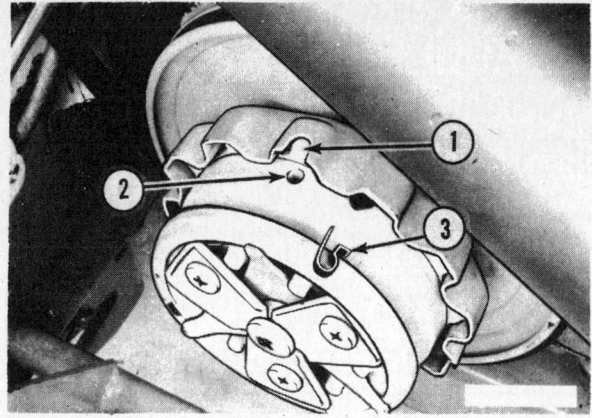

1. Notch
2. Setscrew hole
3. Friction disc spring installing position

PTO clutch, without brake

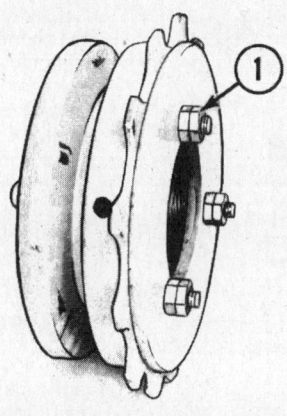

#1 is the jam nut

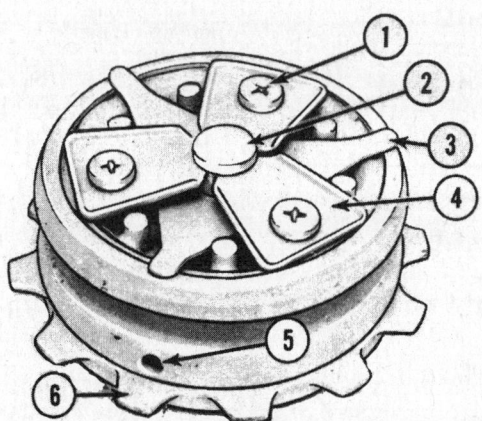

1. Screw
2. Thrust button
3. Pressure springs
4. Throwout lever
5. Setscrew hole
6. Friction disc lug

Clutch components

ASSEMBLY

1. Position the pressure springs on the actuating pulley so the tips are between the cast lugs.
2. Install the thrust button on the pressure spring.
3. Install the three throwout levers so they engage the slot in the thrust button. Line up the screw holes and install the screws.
4. Holding the throwout levers, screws thrust button and pressure spring in place, turn the assembly over and install the compression springs (one to each screw), friction disc and pressure plate.

NOTE: The 14 hp model is equipped with dual friction discs. A dual disc repair kit may be applied to the 10 or 12 hp model if additional reliability is required.

ADJUSTMENT

1. Install the adjusting gauge in position shown.

NOTE: Adjusting gauge is supplied with clutch disc repair kit.

2. Tighten the special machine screw (in line with the center of the gauge) until the gauge ends contacts the machined surface of the pulley. The gauge should not rock the tips.
3. Repeat step 2 for the remaining two gauge positions.
4. Recheck each of the three positions with the gauge a second time after all three screws have been adjusted to be sure all three adjustments are equal.
5. The specified pressure applied to the pressure spring is now set. The protruding ends of the screws should be approximately equal. If they are quite different, something is wrong and the clutch will have to be disassembled, checked, reassembled and adjusted.
6. Install the machine screw lock nuts and tighten them to 6-7 ft.lb. torque.

INSTALLATION

NOTE: The drive pulley hub must be 1½ in. from end of crankshaft on 8 hp model and 1¼ in. on 10, 12 and 14 hp models.

IMPORTANT: The setscrew that contacts the key must be tightened first; then tighten the setscrew that contacts the engine shaft. Tighten the setscrews to 12 to 14 ft.lb. torque.

1. If a new clutch bearing is to be used, install it on the crankshaft so it is flush with the end of the crankshaft. Lock it in place with the locking collar. Be sure to lock the collar to the bearing in the direction of crankshaft rotation. Lock the collar in place with the setscrew and nut equipped.
2. Install the clutch assembly on the bearing part way. Be sure the setscrew holes in the clutch pulley housing line up with the slots in the drive pulley cup.
3. Equally space and install 3 friction disc springs on the friction disc lugs on the non-drive side of the lugs. Then non-drive

International Harvester

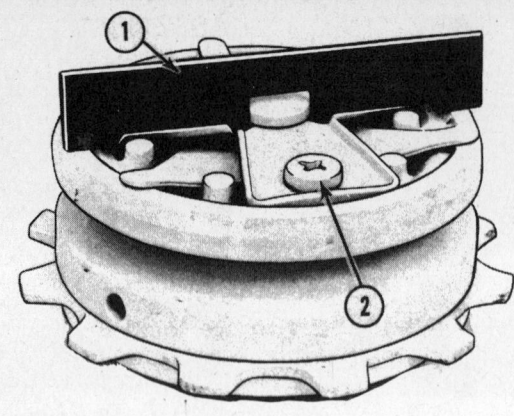

1. Gauge
2. Adjust here

Adjusting gauge positioned

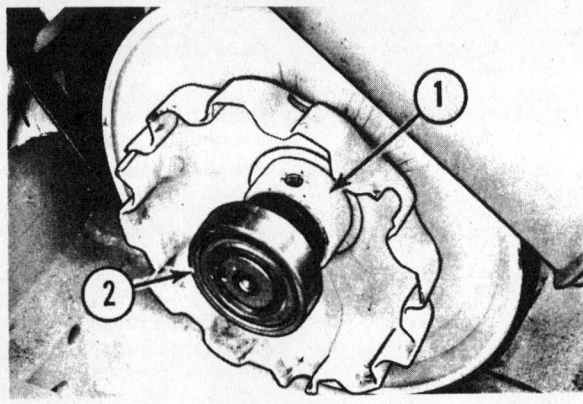

1. Drive pulley hub
2. Clutch bearing

Installing drive pulley hub

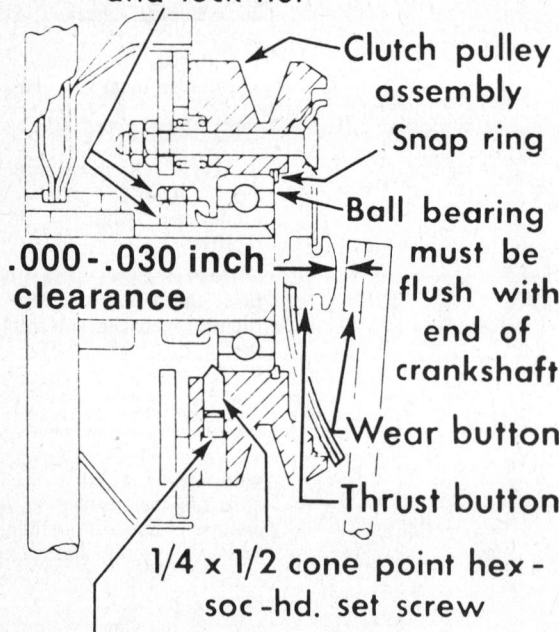

Clutch installation completed, showing proper measurements

side of the lugs is the lefthand side of the lug when looking at the front of the clutch. Place flat side of springs inside the cup. Push the clutch assembly the rest of the way on the bearing until the snapring in the clutch is flush with the bearing. Be sure the complete thickness of the disc is inside the drive pulley cup.

4. Install the three ¼ x ½ in. cone point hex socket setscrews in the clutch pulley and torque them to 60-72 in.lb.
5. Install the three ¼ x ¼ in. flat point hex setscrews and torque them to 72-84 in.lb.
6. With the hand lever in the forward position (clutch fully engaged), adjust the turnbuckle so there is 0 to .030 in. minimum clearance between the wear button and thrust button. Secure the turnbuckle with the jam nut.
7. Install the grille.

With Brake

A PTO clutch with brake is effective on tractors with serial nos. U482000-U5299999. The brake increases safety in the operation of the PTO driven equipment. This clutch along with the braking assembly may also be used on Cadets between serial nos. U400000-U481999 if rockshaft 59 978 C1 and brake shoe assembly 59 973 C1 are installed.

Removal, disassembly, assembly and installation are the same as for clutches without brakes, with the following exception: the brake disc must be removed from the clutch prior to disassembly.

ADJUSTMENT

1. Place the hand lever in the rear position (clutch fully disengaged) and tighten the turnbuckle until the clutch disengages.
2. Check for disengagement of the clutch by rotating the clutch pulley slightly. The brake will prevent complete rotation of the clutch pulley.
3. Then tighten the turnbuckle one full turn and lock.
4. Disengage the lever and check for running clearance between the brake and the clutch.

ELECTRIC FRONT PTO CLUTCH

REMOVAL

1. Remove the engine side hood panels and disconnect the clutch wire.
2. Remove the grille housing and hood as an assembly.
3. Remove the brake flange.
4. Remove the retaining bolt, washer and spacer.
5. Using a suitable puller, remove the clutch rotor assembly.
6. Remove the driving hub and key from the crankshaft.
7. If it is necessary to take off the field coil, remove the four stud bolts.

International Harvester

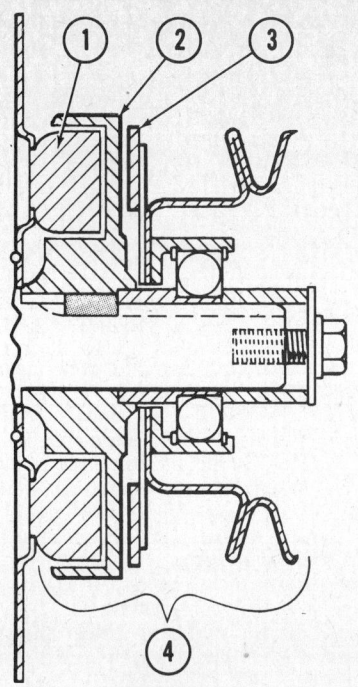

1. Field coil
2. Driving hub
3. Driven disc
4. Rotor assembly

Electric PTO clutch

1. Driving hub
2. Stud bolts (4)
3. Feeler gauges (.008-.010 in.)

Driving hub-to-field coil measurement

INSTALLATION AND ADJUSTMENT

NOTE: The field coil for the electric IPTO clutch is centered on the crankshaft by four tabs that fit into the bore for the main bearing oil seal. On some blocks the chamfer for the main bearing oil seal may be too deep, and the field coil may be off center. This would result in interference between the field coil and the driving hub. The field coil must be correctly located as follows. (Step 1 through 3).

1. Install the field coil and install the four stud bolts finger tight.
2. Slide the driving hub onto the crankshaft. Then using .008-.010 in. feeler gauges, determine that there is clearance at four locations between the field coil and the driving hub. The field coil may be shifted slightly if necessary.
3. Remove the feeler gauges and tighten the four stud bolts to securely hold the field coil in its proper location.
4. Place the driven disc onto the driving hub on a bench.
5. Determine the clearance between the driven disc and the driving hub using a feeler gauge at three evenly spaced locations. There should be a gap of .060-.090 in.
6. Add or remove shims as required onto the driving hub.

NOTE: Some late production tractors have an electric IPTO clutch which does not require shims. The free air gap is permanently set. This clutch is unpainted, where as the clutch which requires shims is painted black.

7. Install the driving hub along with the necessary shims onto the crankshaft.
8. Install the clutch rotor assembly.
9. Install the spacer, washer and the retaining bolt.
10. Install the brake flange.
11. Adjust the clutch as follows:
    a. Disengage the clutch.
    b. Check the clearance between the driven disc and the driving hub by inserting a feeler gauge in the four slots in the brake flange. There should be a .010 in. gap between the driven face of the rotor assembly and the driving hub.
    c. Tighten or loosen the brake flange mounting nuts to obtain the correct gap around the rotor assembly.
12. Connect the clutch wire.
13. Install the grille housing and hood as an assembly.
14. Install the engine side hood panels.

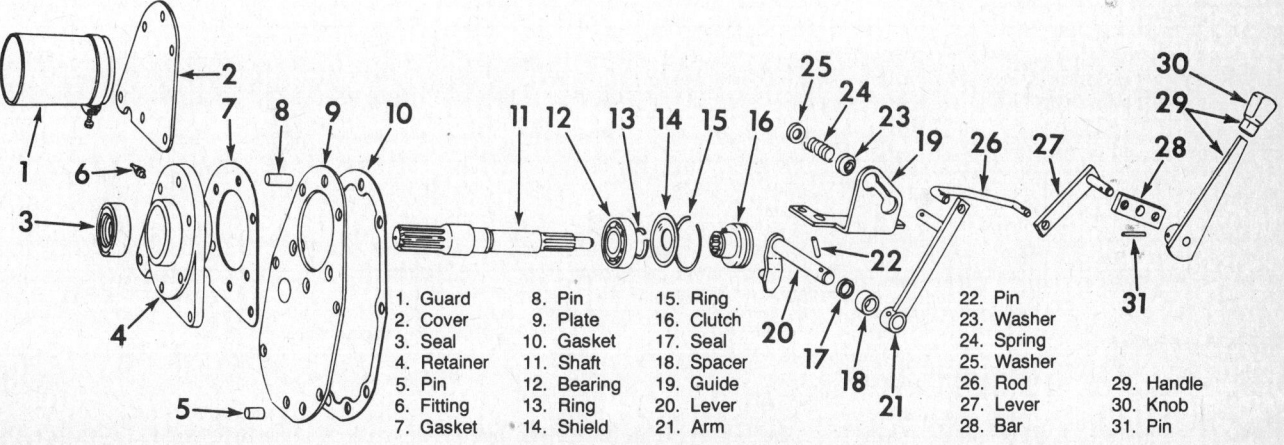

1. Guard	8. Pin	15. Ring	22. Pin	29. Handle
2. Cover	9. Plate	16. Clutch	23. Washer	30. Knob
3. Seal	10. Gasket	17. Seal	24. Spring	31. Pin
4. Retainer	11. Shaft	18. Spacer	25. Washer	
5. Pin	12. Bearing	19. Guide	26. Rod	
6. Fitting	13. Ring	20. Lever	27. Lever	
7. Gasket	14. Shield	21. Arm	28. Bar	

Rear PTO clutch components

International Harvester

REAR POWER TAKE-OFF

REMOVAL

Remove 7 capscrews around rear oil seal and bearing retainer. Pull shaft, retainer, shifter clutch to the rear letting the front clutch disengage from the shift lever shaft.

INSTALLATION

Slide shifter clutch on to the power take off shaft and secure with setscrew and nut.

Apply a coating of grease on the pilot end of the power take off shaft. Making sure oil seal is installed with lip towards inside and gaskets are in place, insert power take off shaft assembly through the hole in the adaptor plate and engage the pin on the shifter shaft lever with the groove in the clutch.

Move the assembly forward and engage the pilot end of the shaft with the female end of the transmission input shaft.

Fasten the power take off assembly into place with the capscrews removed.

ELECTRIC POWER LIFT

For complete service, see the Model 73, 106, 107, 126, 127, 147 section.

ENGINE

Engines used are various Kohler models. Check the engine identification plate on your tractor and see the appropriate engine in the Kohler Engine part of the Engine Unit Repair section of this book.

Models 86, 108, 109, 128, 129, 149, 169

REMOVAL AND INSTALLATION

1. Disconnect the battery ground cable.
2. Remove the fuel tank and panel extensions as an assembly.
3. Disconnect the wire harness from the coil and motor-generator.
4. Remove the motor-generator.
5. Disconnect and remove the PTO linkage.
6. Remove the motor-generator belt.
7. Disconnect the choke and throttle cables.
8. Remove the engine mounting bolts. The front of the tractor will have to be raised to remove the front bolts.
9. Slide the engine forward and lift it out.
10. Installation is the reverse of removal.

Models 800, 1000, 1200, 1250, 1450, 1650

REMOVAL AND INSTALLATION

1. Disconnect the battery ground cable.
2. Raise the hood. Disconnect the spring and remove the wing nut under the cowl to take off the side panels.
3. Disconnect the wire harness, starter cable and ground cable.
4. Disconnect head light wiring and remove front grille and hood as an assembly.
5. Disconnect the electric clutch wire.
6. Remove the air cleaner assembly.
7. Disconnect the choke and throttle cables.
8. Shut off fuel and disconnect the fuel line at the carburetor.
9. Raise the front of the tractor and remove the engine mounting bolts.
10. **Models 800 and 1000:** Slide the engine forward and lift it out.
Model 1200: Remove the gas tank, heat shield and mounting brackets.
Model 1200: Remove the gas tank, heat shield and mounting brackets. Lift the engine out.
Models 1250, 1450 and 1650: Remove the gas tank, heat shield and mounting brackets. Disconnect the driveshaft front flexible coupler from the engine. Lift the engine out.
11. Installation is the reverse of removal.

Cub 154, 184 and 185 Lo-Boy

FRONT AXLE AND WHEELS

Front Wheels and Bearings

REMOVAL

1. Lock the brake and block the rear wheels. Jack up the front axle.
2. Remove the hub cap.
3. Remove the capscrew and flat washer from the outer end of the spindle.
4. Slide the wheel and bearings off the spindle.

NOTE: The bearings are a press fit in the wheel or hub and a slip fit on the spindle.

DISASSEMBLY

Wheel bearings can be driven from the wheel hub with a hammer and long drift punch. Drive from the inside toward the outside.

1. Front axle assembly
2. Lefthand steering knuckle assembly
3. Righthand steering knuckle assembly
4. Front axle pivot pin
5. Tie rod
6. Hub cap
7. Ball joint
8. Bushing
9. Wheel ball bearing assembly

Front axle components

INSPECTION AND REPAIR

1. Inspect the entire wheel and hub for wear or damage.
2. Inspect the bearings and seals and replace as necessary.
3. Bearing fit to wheel or hub must be tight. If not, replace the wheel.

International Harvester

REASSEMBLY AND INSTALLATION

1. If the bearings were removed, lubricate and press in new ones. Be sure force is directed to the outer race only.
2. Slide the wheel and bearing assembly over the spindle and secure with the capscrew and flat washer. Tighten the capscrew to 80 ft.lb. torque.
3. Install the hub cap.

Front Axle

REMOVAL

1. Disconnect the tie rod ball joints from the steering lever.
2. With the front of the tractor frame supported on a suitable stand, drive out the retaining pin from the front of the axle pivot pin (4).
3. Remove the pivot pin and then remove the front axle (1) from the tractor.

INSPECTION

Thoroughly clean all parts. Inspect all parts closely for wear or damage and replace as necessary.

INSTALLATION

1. Apply chassis lubricant liberally to the axle pivot pin and its bore in the axle.
2. Position the axle in its support bracket channel. Align the pivot pin holes and insert the pin.
3. Align the retaining pin holes (through the pivot pin and front collar of the support bracket) and drive the retaining pin through both parts.
4. Connect the tie rod ball joints to the steering lever. Be sure to tighten the locknuts securely.

STEERING

Steering Knuckle

REMOVAL

1. Lock the brake, jack up the front of the tractor and support it on a suitable stand.
2. Remove the front wheels.
3. Disconnect the tie rod ball joints from the left and right steering knuckles.
4. Remove the capscrew and flat washer and remove the steering knuckle from the axle.

INSTALLATION

1. Thoroughly lubricate the steering knuckle shaft.
2. Install the righthand and lefthand steering knuckles in their respective bores in the axle and secure with the capscrews and flat washers. Tighten the capscrews securely.
3. Connect the tie rod ball joints to the steering knuckle arms. Be sure to tighten the locknut securely.
4. Install the front wheels and check the toe-in adjustment.

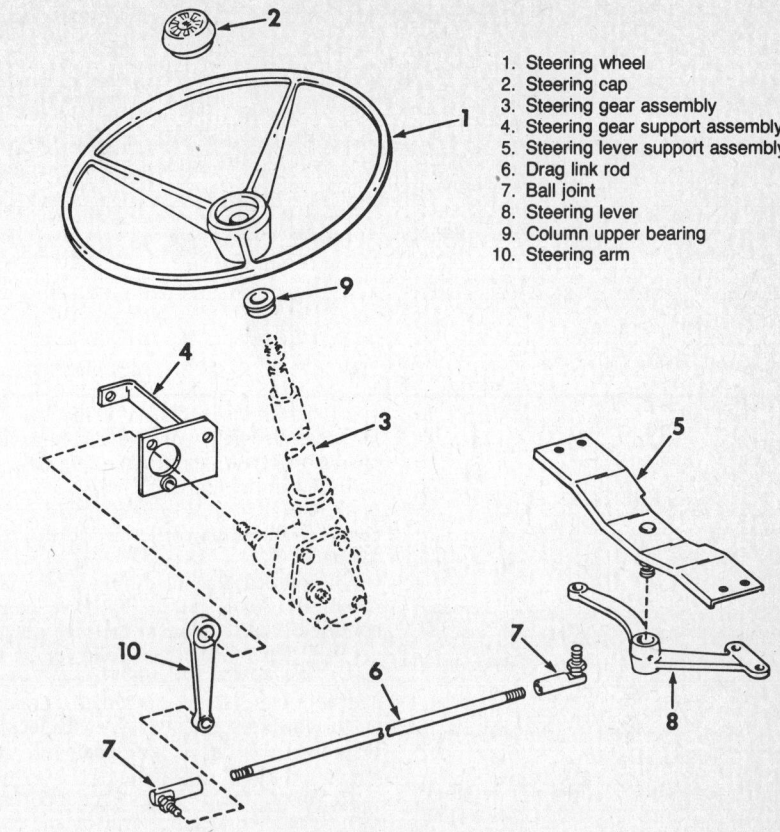

1. Steering wheel
2. Steering cap
3. Steering gear assembly
4. Steering gear support assembly
5. Steering lever support assembly
6. Drag link rod
7. Ball joint
8. Steering lever
9. Column upper bearing
10. Steering arm

Steering system components

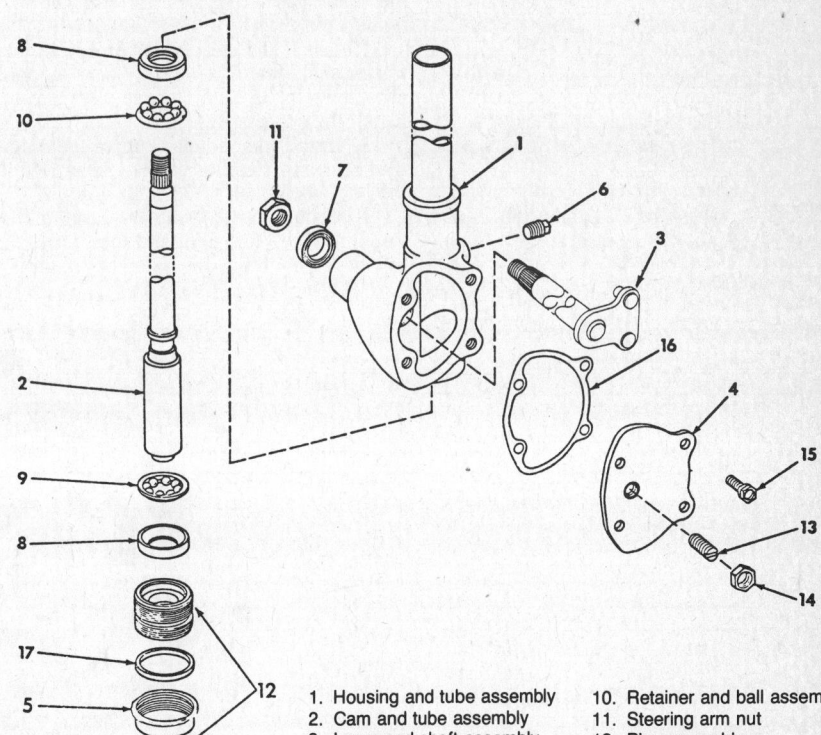

1. Housing and tube assembly
2. Cam and tube assembly
3. Lever and shaft assembly
4. Side cover
5. Locknut
6. Pipe plug
7. Oil seal
8. Ball cup
9. Ball retainer
10. Retainer and ball assembly
11. Steering arm nut
12. Plug assembly
13. Adjusting screw
14. Adjusting screw locknut
15. Capscrew
16. Side cover gasket
17. Seal

Steering gear components

537

International Harvester

STEERING UNIT

REMOVAL

1. Remove the steering cap and nut. Using a puller, remove the steering wheel. Remove the dust seal and the column upper bearing.
2. Disconnect the drag link ball joint from the steering arm.
3. Remove the capscrews in the steering gear support and remove the steering gear assembly from the tractor.

DISASSEMBLY

1. Remove the nut securing the steering arm to the lever and shaft assembly. Using a puller, remove the arm from the shaft.
2. Remove the side cover.
3. Pull the lever and shaft assembly out of the housing and tube assembly. Remove the oil seal from the housing.
4. Bend the locknut locking tab up and remove the locknut.
5. Remove the plug assembly, ball cup and retainer and ball assembly.
6. Remove the cam and tube assembly from the housing.
7. Remove the second retainer and ball assembly from the shaft. Remove the ball cup if it is to be replaced.

NOTE: Keep inner and outer retainer and ball assemblies identified for proper location as they are not interchangeable.

INSPECTION AND REPAIR

1. Wash all parts in cleaning solvent, then dry thoroughly.
2. Inspect the lever and shaft cam followers for wear and replace if necessary.
3. Inspect the bearings, ball cups and the cam and tube assembly for roughness and pitting.
4. Inspect the cam grooves for wear, roughness and galling. Replace the cam and tube assembly if necessary.
5. Inspect the housing for cracks and stripped threads.
6. Inspect the column upper bearing for wear or damage.
7. Be sure to install new gaskets and seals in reassembly.

REASSEMBLY AND ADJUSTMENTS

1. Thoroughly coat the ball bearings and ball cups with recommended chassis lubricant.
2. Install the retainer and ball assemblies and the ball and the ball cups on the cam and tube assembly.
3. Thoroughly coat the cam with recommended chassis lubricant and then install it in the housing and tube assembly.

NOTE: Be sure that ball cups are in the housing squarely and are not "cocked."

4. Install the plug assembly and tighten until end play of the cam is removed but turns freely. Be sure to use a new seal.
5. Install the locknut and tighten securely. Stake the nut into a housing slot.
6. Install a new oil seal in the housing. Pack the housing with the recommended chassis lubricant, and then install the lever and shaft in the housing.
7. Install a new gasket and the side cover. Tighten the capscrews to 20 ft.lb. torque.
8. Lubricate at the fitting in the plug assembly slowly until lubricant begins coming out of the hole at the pipe plug.
9. "Center" the cam follower on the cam. Loosen the adjusting screw locknut and adjust the screw inward to eliminate cam follower backlash. Tighten the locknut securely. Turn the steering shaft full right and left to check for binding.
10. Install the steering gear support and then install the steering arm on the lever and shaft assembly. Be sure the mark on the steering arm is in line with the mark on the shaft. Install the steering arm nut and tighten securely.
11. Install the steering assembly in the tractor and tighten the steering gear support capscrews to 35 ft.lb. torque. Be sure to install the spacers.
12. Install the column upper bearing, dust seal and steering wheel. Secure with the nut and tighten to 35 ft.lb. torque.
13. Center the steering by turning the steering wheel full right and then turn full left while counting the number of turns.

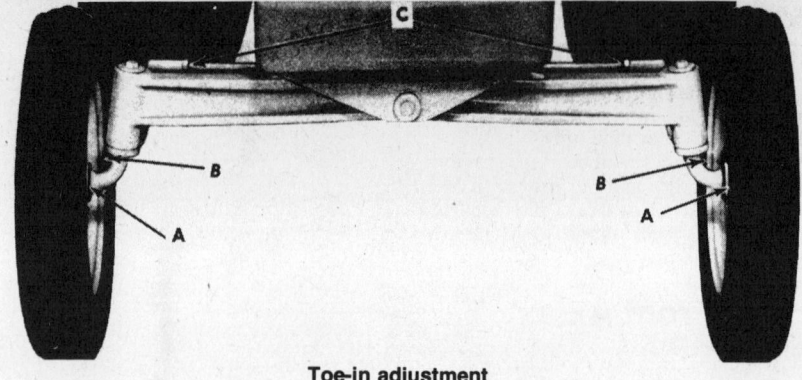

Toe-in adjustment

Turn the wheel to the right one-half of the above figure.
14. Adjust the drag link rod to place the front wheels in the straight ahead position and connect the rod to the steering arm. Install the locknut and tighten securely.
15. Install the steering cap.
16. Adjust the front wheel toe-in to the specified ¼ in. ± 1/16 in. (¼ in. closer in front than in the rear) as follows:

a. Place chalk marks at points "A" on each rim at hub height, and measure the distance between them.
b. Move the tractor forward a distance equal to one-half revolution of the front wheels. The chalk marks will now be at points "B".
c. Measure the distance between points "B". The distance between points "B" must be ¼ in. ± 1/16 in. greater than at "A".
d. To adjust, disconnect the tie rod ball joints from the steering knuckle arms. Loosen the locknuts "C" and turn the ball joints in or out as required. Be sure to make the tie rod adjustments equal, and be sure the steering knuckle arms stop on the axle.
e. Connect the tie rod ball joints to the steering knuckle arms. Tighten the locknuts "C" securely.

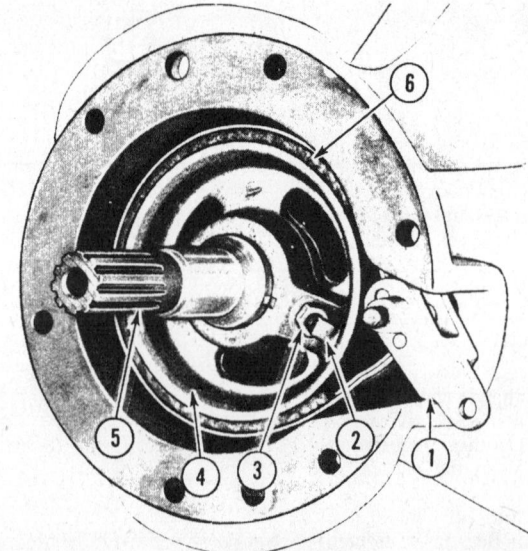

1. Actuating toggle
2. Setscrew
3. Locknut
4. Brake drum
5. Differential shaft
6. Brake band

Brake components

International Harvester

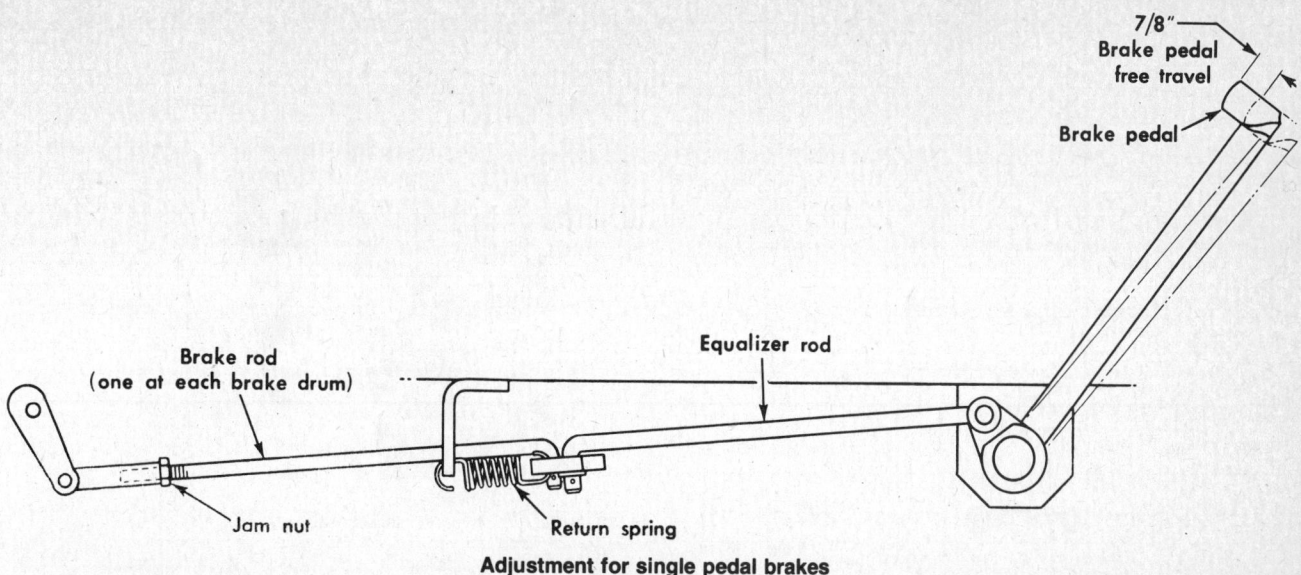

Adjustment for single pedal brakes

BRAKES

REMOVAL

1. Split the tractor.
2. Remove the final drive assemblies.
3. Remove the brake bands (6) from the brake drums (4).
4. Loosen the locknut (3) and setscrew (2) and remove the brake drum (4) from the differential shaft (5).

INSPECTION AND REPAIR

1. Inspect the brake band and drum for damage and excessive wear. Replace them if there is any doubt of their serviceability.
2. Inspect the control rods for wear at their connecting pivot points.
3. Check the pedal return spring ends for wear.

INSTALLATION

1. Align the key way in the brake drum with the key on the differential shaft and install the drum on the shaft. Tighten the setscrew and locknut securely.
2. Install the brake band on the drum.
3. Install the final drive assembly on the transmission case.
4. Recouple the tractor.

ADJUSTMENT

Single Pedal Brakes

To adjust the brakes, jack up the rear end of the tractor. Loosen the jam nuts at the end of the brake rods and remove the cotter pins at the brake equalizer rod. Unhook the brake rods and turn them in or out of the clevises. The brakes should not drag before they take hold. Adjust the brake linkage so there is brake pedal free travel, by hand, of approximately 7/8 in.

It is very important to have the brakes equalized. To check the equalization of the brakes, jack up both rear wheels so they will turn freely, block the tractor securely and then start the engine. Shift the gears to third speed and engage the clutch; while the wheels are turning, apply the brakes. Application of the brakes should slow down both wheels at the same time and also reduce the speed of the engine.

CAUTION
Avoid high speed throttle settings that might cause the tractor to slip off of the blocking.

If one wheel stops and the other wheel continues to revolve when the brakes are applied, adjust the brake rod on the wheel that stops so both wheels stop simultaneously when the brakes are applied.

Two-Pedal Brakes

To check the brakes for proper adjustment, latch the pedals together, engage the first or second notch of the brake pedal lock against the platform, place the transmission in first gear, and set the engine at low idle. Release of the clutch should stall the engine without moving the tractor.

To adjust the brakes, jack up the rear end of the tractor. Loosen the jam nuts at the end of the equalizer rods and turn the turnbuckles in or out. Brakes are properly adjusted when each wheel drags slightly when turned.

It is very important to have the brakes equalized. To check the equalization of the brakes, jack up both rear wheels so they will turn freely, block the tractor securely and then start the engine. Shift the gears to

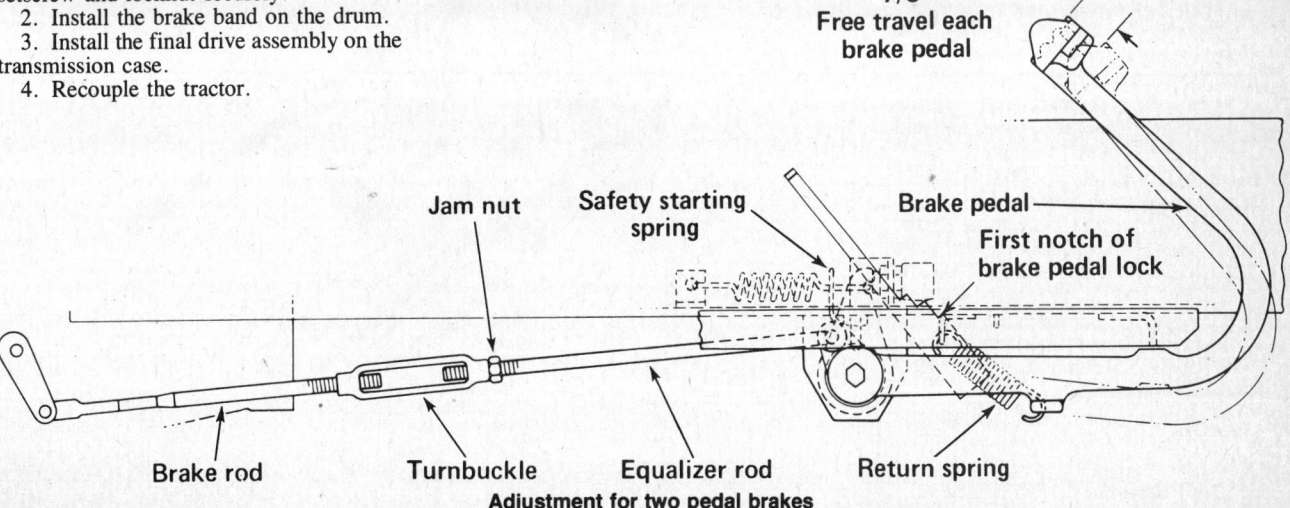

Adjustment for two pedal brakes

International Harvester

third speed and engage the clutch; while the wheels are turning, lock the brake pedals together so they operate in unison, and apply the brakes. Application of the brakes should slow down both wheels at the same time and reduce the speed of the engine.

If one wheel stops and the other wheel continues to revolve when the brakes are applied, adjust the turnbuckle on the wheel that stops just enough so both wheels stop simultaneously when the brakes are applied.

CLUTCH

154 Models Through Serial # U018709

REMOVAL

1. Remove the frame top cover and pedestal side sheet sections.
2. Drive out the pin or remove the socket head bolt and nut from the clutch coupling.
3. Loosen two capscrews on the shaft and pulley adjuster and loosen the PTO drive belt tension. Loosen four capscrews securing the front mainshaft bearing support to the frame. Remove the drive belts (1) from the top pulley.
4. Loosen the bearing locking collar (6).
5. Remove the nuts (7) from the bearing flange bolts.
6. Remove the clutch coupling capscrews.
7. Slide the mainshaft forward into the clutch coupling (3).
8. Remove six capscrews securing the engine to the frame (four in rear and two in front).
9. Move the engine forward approximately ½ in. and insert small wooden wedges between the frame and engine to hold the engine forward.
10. Remove the clutch brake yoke from the clutch release yoke.
11. Remove the pin securing the clutch release yoke to the clutch release bracket.
12. Remove the pin securing the clutch release yoke to the clutch adjusting rod. Remove the release yoke.
13. Remove the bolts and nuts securing the clutch release bracket to the frame and remove the bracket.
14. Remove the clutch brake assembly from the transmission case.
15. Remove the six capscrews in the clutch assembly. Move the pressure plate and clutch disc assembly forward on the shaft.
16. Remove the socket head bolt and nut or drive out the pin in the rear pressure plate hub. Move the pressure plate rearward on the transmission shaft.
17. Work the shaft out of position and remove the clutch driven disc assembly, pressure plate and release bearing. Remove the pressure plate from the transmission shaft.

1. PTO drive belts
2. Pin (or nut and socket head bolt)
3. Clutch coupling
4. Main clutch shaft
5. Pulley
6. Bearing locking collar
7. Nut (2)

154 clutch components

INSTALLATION

1. Install the rear pressure plate on the transmission shaft.
2. Install the clutch release bearing, pressure plate, and clutch disc assembly on the shaft.
3. Install the socket head bolt and nut in the pressure plate hub and transmission shaft. Tighten the nut securely.
4. Install the six capscrews holding the clutch assembly together and tighten to 25 ft.lb. torque.
5. Install the clutch brake retainer on the transmission case and tighten the capscrews to 80 ft.lb. torque.
6. Install the clutch release bracket on the frame and tighten the nuts and bolts securely.
7. Install the clutch release yoke to the clutch adjusting rod and secure with the pin.
8. Install the pin securing the clutch release yoke to the clutch release bracket.
9. Assemble the clutch brake yoke to the clutch release yoke and secure with the pin.
10. Install the belts on the lower pulley.
11. Move the engine rearward to its operating position. Install the six capscrews securing the engine to the frame and tighten them securely.
12. Slide shaft forward into the pilot hole in the coupling retainer, and install the four capscrews in the clutch coupling and tighten securely.

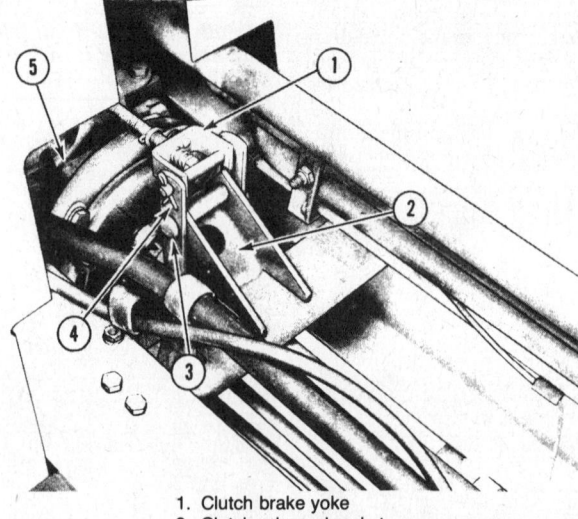

1. Clutch brake yoke
2. Clutch release bracket
3. Pin
4. Clutch release yoke
5. Clutch brake assembly

Clutch linkage

International Harvester

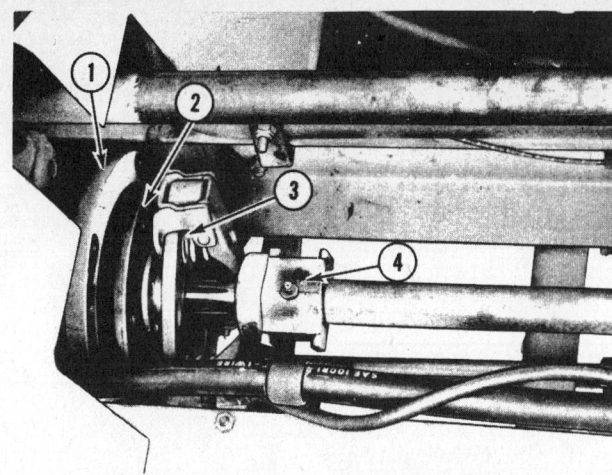

1. Pressure plate
2. Clutch disc assembly
3. Pressure plate assembly
4. Release bearing

Clutch plate units

13. Install the bearing flange bolts and nuts. Tighten only snug.
14. Slide the shaft rearward and install the socket head bolt and nut and tighten securely.
15. If the engine flywheel does not have a 1 in. pilot hole, proceed with the following steps:
 a. Install the capscrews in the clutch coupling and tighten until snug.
 b. Install the nut and bolt in the clutch coupling.
 c. Using a dial indicator, check the shaft run-out while cranking the engine. Be sure indicator reading is taken on smooth area of shaft.

NOTE: Removal of the spark plugs will make cranking of engine easier.

 d. Tap the coupling in the direction required to reduce run-out to minimum possible.

NOTE: If minimum run-out of .004 in. cannot be obtained, replace the clutch coupling and/or shaft.

 a. Securely tighten the capscrews in the clutch coupling.
 f. Install the nut and bolt in the clutch coupling hub.
16. Securely tighten the bolts and nuts (7) in the bearing flanges.
17. Tighten the 4 capscrews securely in the cross member slotted holes.
18. Tighten the bearing locking collar.
19. Install the PTO drive belts in the pulley grooves.
20. Tighten the drive belts enough to prevent slippage and then tighten the capscrews in the pulley adjuster.

NOTE: Do not over-tighten the belts.

21. Connect the hydraulic line (if equipped) and tighten the support clamp.
22. Adjust the clutch.

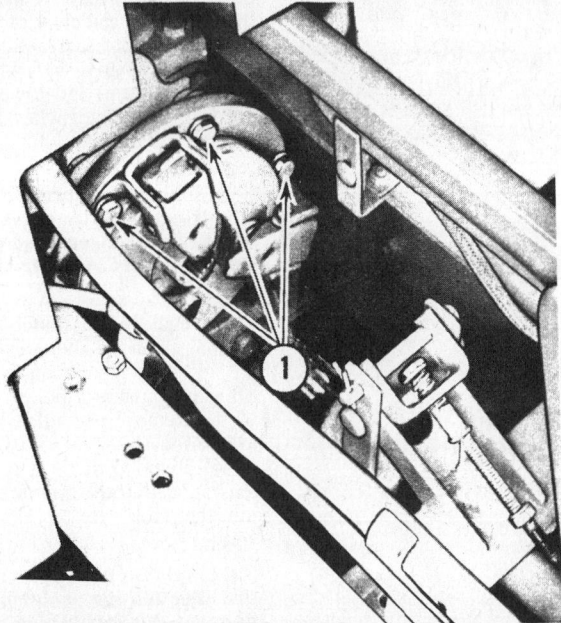

Clutch plate, the #1 shows the pressure plate bolts

23. Install the frame top cover and pedestal side sheet assemblies.

ADJUSTMENTS

1. Remove the frame top cover. The clutch linkage should not be adjusted to provide .090 in. clearance between the release fingers and the bearing. Adjust by loosening the two nuts on the clutch rod adjusting yoke and positioning the clutch adjusting rod to give the specified free travel. Retighten the nuts.
2. After adjusting the clutch pedal free travel, check the clearance between the clutch brake arm and the clutch brake lining. A clearance of .010 in. is specified. To adjust, loosen the self locking nuts on the clutch brake bolt. Adjust the clutch brake bolt to obtain the specified .010 in. clearance. Retighten the self locking nuts. Install the frame top cover.

154 and 185 Models, Serial #U018709 and Up

REMOVAL

1. Remove the frame top cover and pedestal side sheet sections.
2. Remove the Allen head bolt from the clutch shaft coupling and slide the shaft ahead.
3. Remove the pin securing the clutch rod to the clutch release yoke. Remove the nut from the end of the clutch brake bolt. Remove the four bolts securing the clutch release bracket to the frame and slide the clutch release bracket ahead.
4. Remove the six bolts securing the pressure plate assembly to the input flywheel. Remove the pressure plate assembly and the clutch disc.
5. To remove the driveshaft, remove the four bolts from the front clutch shaft coupling. Slide the shaft ahead and remove the clutch release bearing. Lower the rear end of the shaft and slide it to the rear.

INSTALLATION

1. Place the shaft in its approximate position inside the frame. Slide the front clutch shaft coupling onto the shaft.
2. Slide the shaft ahead and install the clutch release bearing and bracket assembly onto the shaft.
3. If the engine clutch was disassembled, assemble the release lever, pressure springs and lever brackets on the pressure plate and install the lever pivot pin.
4. Install the rear pressure plate on the transmission shaft.
5. Install the socket head bolt and nut in the pressure plate hub and transmission shaft. Tighten the nut securely.
6. Install the pressure plate and clutch disc assembly.
7. Install the six capscrews holding the clutch assembly together and tighten to 25 ft.lb. torque.
8. Install the clutch release bracket on the frame and tighten the nuts and bolts securely.

541

International Harvester

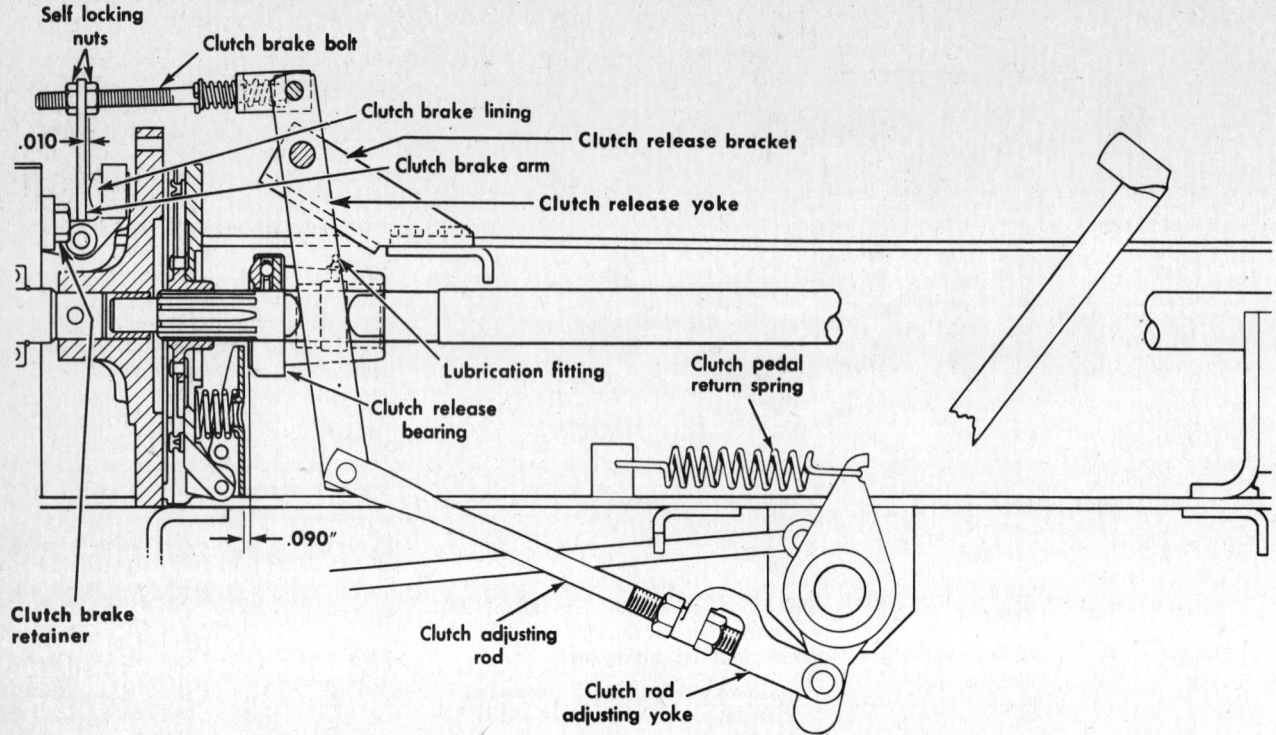

Clutch release adjustments, 154 models above serial #UO18709 and 185 tractors

9. Install the clutch release yoke to the clutch adjusting rod and secure with the pin.

ADJUSTMENTS

1. Remove the frame top cover. The clutch linkage should be adjusted to provide .090 in. clearance between the release fingers and the bearing. Adjust by loosening the two nuts on the clutch rod adjusitng yoke and positioning the clutch adjusting rod. Retighten the nuts.

2. After adjusting the clutch pedal free travel, check the clearance between the clutch brake arm and the clutch brake lining. A clearance of .010 in. is specified. To adjust, loosen the self locking nuts on the brake bolt. Adjust the clutch brake bolt to .010 in. clearance. Retighten the self locking nuts. Install the frame top cover.

Model 184 clutch adjustment

184 Model

REMOVAL

1. Disconnect the battery cables from the battery.

NOTE: Remove the ground cable first to reduce electrical hazards.

2. Remove the pedestal side sheets and the frame top cover.

3. Remove the setscrews from the driveshaft knuckle. Move the snapring forward from the groove. Slide the long end of the knuckle onto the shaft until the cross of the knuckle is against the end of the driveshaft and clear of the transmission or creeper input shaft.

Lift out the driveshaft when connected to the transmission and lower the driveshaft out from the tractor when connected to the creeper drive.

NOTE: If the tractor is equipped with a hydraulic selector control valve then it will have to be removed when the drive shaft connects to the transmission input shaft only.

4. Relieve the clutch belt tension by pushing in on the idler tension arm nut with a ¾ in. wrench and then slip the drive belts off the tension arm pulley.

5. Remove the idler pulley from the idler arm.

6. Remove the bolt securing the idler tension arm to the tractor frame and lower out the idler tension arm and the idler spring.

NOTE: Do not use any hand tools on the idler tension spring since this could damage the spring and result in early failure.

International Harvester

NOTE: Rotate the main drive clutch to obtain the necessary clearance for lowering the idler arm out of the frame.

7. Remove the rockshaft from under the tractor.

8. Disconnect the clutch release arm from the clutch release bearing. Remove the clutch release bearing.

9. Remove all the capscrews from both the clutch drive pulley and the pressure plate assembly. Then lower both the clutch pulley and pressure plate out of the tractor at the same time.

10. If necessary, remove the flywheel and press out the pilot bearing.

INSTALLATION

1. If the engine clutch was replaced assemble the release levers, pressure springs and lever brackets on the new pressure plate.

2. Position both the clutch drive pulley and the pressure plate assembly against the flywheel and position in place with the existing driveshaft. Secure the clutch drive pulley and pressure plate to the flywheel with the capscrews.

3. Carefully remove the driveshaft.

4. With clutch pulley on flywheel, measure from flywheel to clutch drive pulley groove (Dim. A). Then measure from flywheel to corresponding location of IPTO clutch pulley groove (Dim. B). This dimension should be the same as dimension A 1.3 mm (+.050 in.) If not, make the following adjustment.

5. Loosen the locking collar on rear bearing of power take-off shaft; then loosen the nuts mounting the front bearing to the support bracket and install shim washers until the required dimension is obtained. Tighten the bolts and recheck. Tighten locking collar on power take-off shaft rear bearing.

6. Install the clutch release bearing and release arm.

7. Apply Never Seez® compound to the key way end of the driveshaft before installing the knuckle. Assemble the key in the shaft and slide the long end of the knuckle onto the shaft until the cross of the knuckle is against the end of the shaft.

8. Install the shaft back into the tractor drive clutch assembly. Slide the knuckle onto the transmission or creeper input shaft until it seats against the shaft shoulder. Position the snapring in the shaft groove and slide the driveshaft back until the snapring shoulders against the knuckle. Tighten the setscrews over the shaft. Then tighten the setscrews over the Woodruff keys.

9. Install the rockshaft.

10. Install one end of the idler spring to the idler arm and the other end to the spring bracket under the tractor. Lift the idler arm up through the opening between the main drive clutch and the tractor frame and remount the idler arm.

11. Install the idler pulley on the idler arm and torque the pulley nut to standard specifications. Then reinstall the drive belts, making certain belts and pulleys are free of grease and oil.

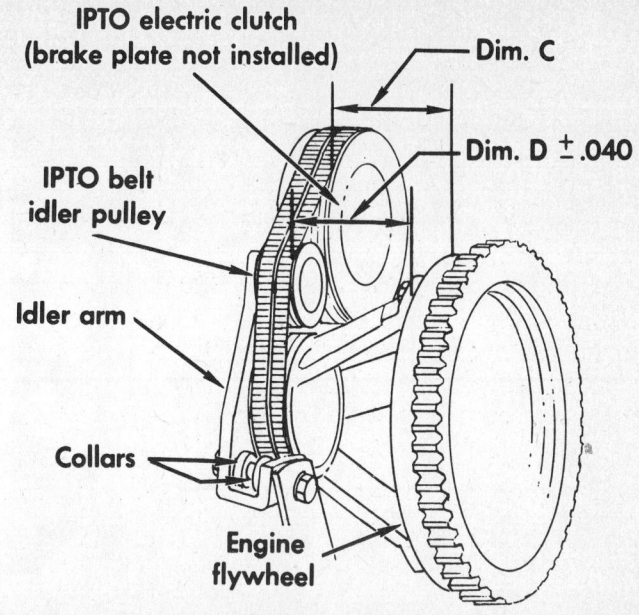

Model 184 IPTO adjustment

12. Measure from the forward edge of forward power take-off belt to face of flywheel at top of power take-off clutch pulley (Dim. C). Then measure from the forward edge of forward power take-off belt where the belt leaves the idler pulley to face of flywheel (Dim. D). Dimensions should be the same ± 1 mm (±.040 in.). If not, loosen the locking collars and slide the idler arm on the trunnion bushing until the above measurement is reached. After adjustment has been made, slide the locking collars against the inside surface of idler arm and tighten setscrews to 20 in.lb.

13. The clutch pedal must maintain a free travel distance of 1⅛-1¼ in.

ADJUSTMENT

With the clutch depressed, check clearance between clutch plate and pressure plate. If clearance is less than .015 in. or more than .020 in., adjustment is necessary. To adjust, remove cotter pin and header pin, shorten or lengthen clutch rod by turning clevis until correct clearance is reached.

SPLITTING AND RECOUPLING

SPLITTING

1. Block the front wheels of the tractor so it can not move. Drive wooden wedges between the front axle and frame on each side to stabilize the tractor.

2. Remove the two capscrews and spacers (no spacers w/creeper drive) securing the front of the transmission to the frame.

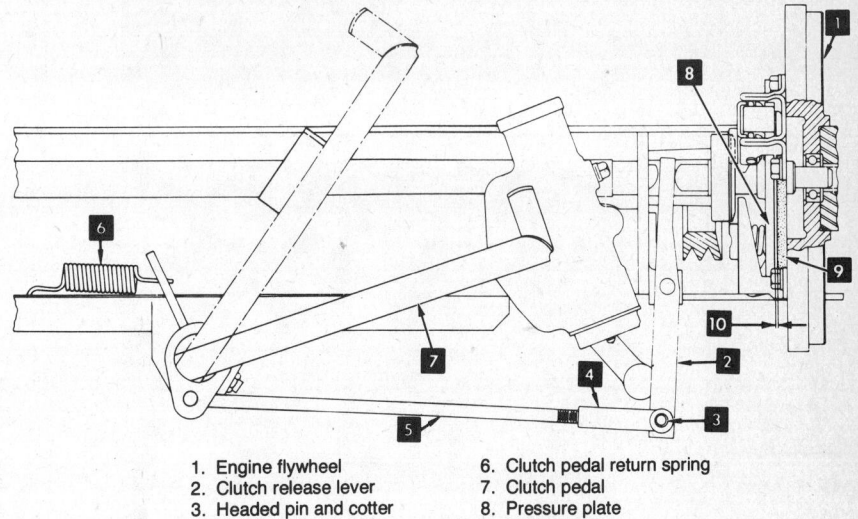

1. Engine flywheel
2. Clutch release lever
3. Headed pin and cotter
4. Clevis
5. Clutch rod
6. Clutch pedal return spring
7. Clutch pedal
8. Pressure plate
9. Clutch plate
10. Dimension is: .015 to .020 in.

Model 184 main engine clutch adjustment

International Harvester

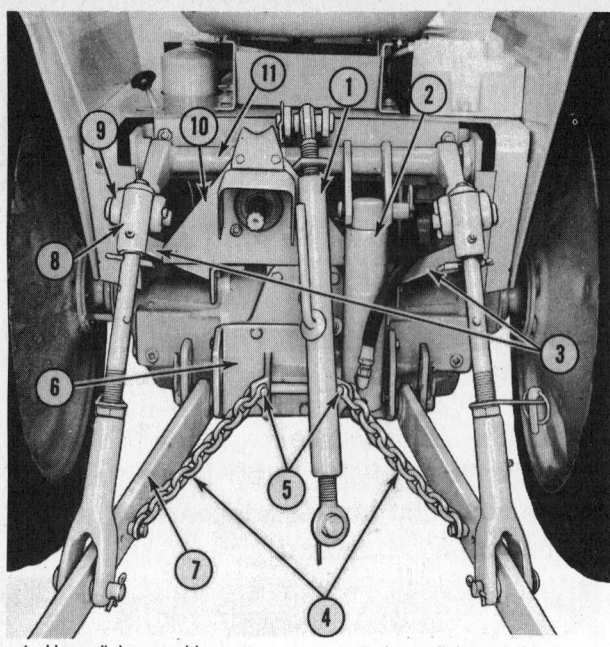

1. Upper link assembly
2. Hydraulic cylinder
3. Rear frame shield braces
4. Limiter chains
5. Chain pins
6. Link mounting plate
7. Lower link assembly
8. Lift link swivel
9. Rockshaft arm swivel
10. Shield support
11. Rockshaft support assembly

Splitting the tractor, showing main points of interest

3. Disconnect the brake rods from the actuating toggles.
4. Disconnect the oil lines to the transmission and drain the Hy-Tran fluid.
5. Remove the shift lever knob.
6. Remove the frame top cover and the pin securing the clutch brake yoke to the clutch release yoke.
7. Remove the seat, and seat support assembly as a unit.
8. Remove the battery.
9. Remove the fenders and rear frame cover assembly as a unit.
10. Remove the seat support sheet assemblies. If the tractor is equipped with three point hitch, remove the rockshaft support assembly as follows:
 a. Remove the pins from the upper and lower ends of the hydraulic cylinder.
 b. Remove the limiter chain shackle pins.
 c. Remove the lower link assemblies from the link mounting plate assembly.
 d. Remove the rear shield braces.
 e. Remove the rockshaft support assembly.
11. Remove the IPTO as follows:
 a. Remove the PTO rear shield and support and engage the PTO clutch.
 b. Remove the PTO brake support and the shaft bearing mounting bracket support. Loosen the clutch cleat bracket capscrews.
 c. Disconnect the clutch control arm.
 d. Remove the PTO clutch assembly from the tractor.
12. Support the frame of the tractor using one jack stand FES 142-4 with an adaptor, FES 142-13 and remove the capscrews from both sides that secure the transmission case to the frame. Raise the rear of the frame to clear the axle housing.
13. Fasten a crescent wrench to the drawbar or the link mounting plate assembly (three point hitch equipped) and move the assembly rearward.

RECOUPLING

1. Be sure to install the reinforcement block on the righthand final drive assembly if it was removed.
2. Recouple the tractor by reversing the splitting procedure.

NOTE: When securing the transmission case to the frame, the two 1⅜ in. capscrews go in the left side and the two 2⅞ capscrews are used on the right side.

3. After filling the transmission to correct fluid level with Hy-Tran fluid, start and run the tractor for several minutes. Recheck the fluid level. Check and adjust the main clutch, brakes and IPTO unit.
4. Recouple the split sections of the tractor being sure the driveshaft splines mate with the transmission input splines.
5. Install the capscrews securing the transmission case to the frame and tighten securely.

NOTE: The two 1⅜ in. capscrews go in the left side and the two 2⅞ in. capscrews are used on the right side.

NOTE: If Tractor is equipped with a creeper drive and PTO, install PTO clutch fenders, seat, and seat support assembly.

7. Install the rear shield braces.
8. If the tractor is equipped with three point hitch, proceed as follows:
 a. Install the upper link assembly.
 b. Install the hydraulic cylinder and secure the upper and lower ends with pins.
 c. Install the lower link assembly and the lower link pins.
 d. Assemble the lift link swivels to the rockshaft arm swivels.
 e. Install the limiter chains and chain shackle pins.
9. Install the rear PTO shield support (if so equipped).
10. Install the pin securing the clutch brake yoke to the clutch release yoke.
11. Install the shift handle knob.
12. Install the two capscrews and spacers securing the front of the transmission to the frame and tighten securely.
13. Connect the transmission oil lines.
14. Connect the brake rods to the actuating toggles.
15. Fill the transmission to the correct level with Hy-Tran fluid.
16. Connect the battery ground strap to the battery.
17. Adjust the clutch and brakes.

1. Brake support
2. Shaft bearing mounting bracket
3. Clutch cleat bracket
4. Clutch control arm

IPTO removal points

International Harvester

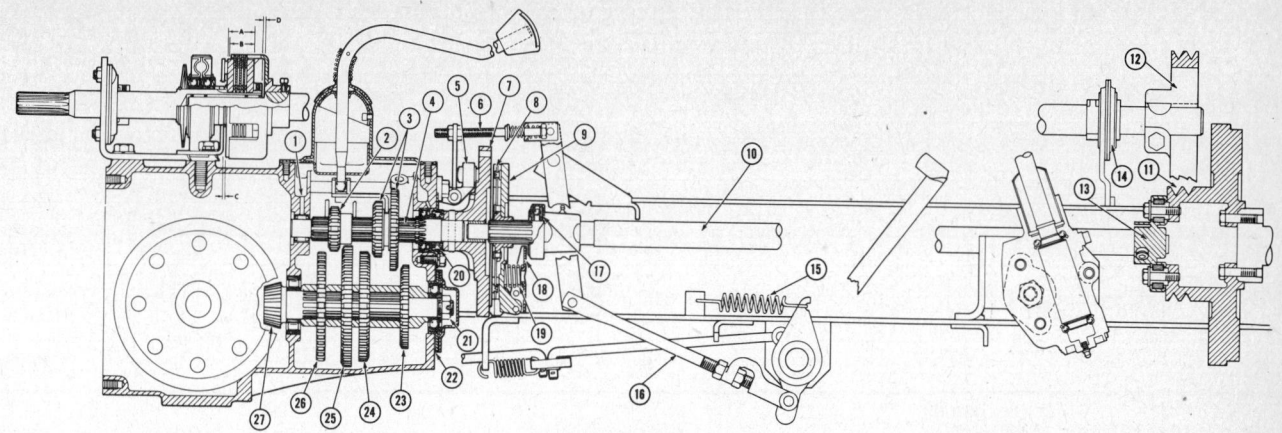

1. Oil passage
2. 1st and reverse speed sliding gear
3. 2nd and 3rd speed sliding gear
4. Spline and clutch shaft
5. Clutch brake assembly
6. Clutch brake adjusting bolt
7. Rear pressure plate assembly
8. Clutch driven disc assembly
9. Front pressure plate assembly
10. Main clutch shaft
11. PTO drive pulley
12. PTO driven pulley
13. Clutch coupling
14. Bearing
15. Clutch return spring
16. Clutch adjusting rod
17. Clutch release bearing
18. Release lever
19. Pressure spring
20. Shaft oil seal
21. Countershaft nut
22. Shims
23. 3rd speed gear
24. 2nd speed gear
25. 1st speed gear
26. Reverse speed gear
27. Countershaft and bevel pinion

Transmission and differential components, serial #UO18709 and above

TRANSMISSION AND DIFFERENTIAL

Differential

REMOVAL

1. Split the tractor.
2. Remove the final drive assemblies.
3. Remove the hitch link mounting plate (if so equipped) and the transmission rear cover plate.
4. Remove the bearing retainers and shims from each side of the differential. Be sure to keep the shims with each retainer and identified for each side.
5. Turn the differential into position as shown and remove it from the transmission case. If the assembly will not clear the side of the transmission case, it will be necessary to remove the righthand bearing cone.
6. Drive out the pinion shaft lock pin and remove the pinion shaft.
7. Remove the pinion gears and side gears.
8. If the differential drive gear requires separate replacement, press out the eight retaining rivets.
9. Remove the bearing cones from the differential carrier if they are to be replaced.
10. Remove the bearing cups from the bearing retainers if replacement is necessary.
11. Remove the oil seals from the bearing retainers.

Transmission

DISASSEMBLY

1. Remove the differential.
2. Remove the gear shift lever and cover assembly.
3. Remove the countershaft bearing retainer cap.

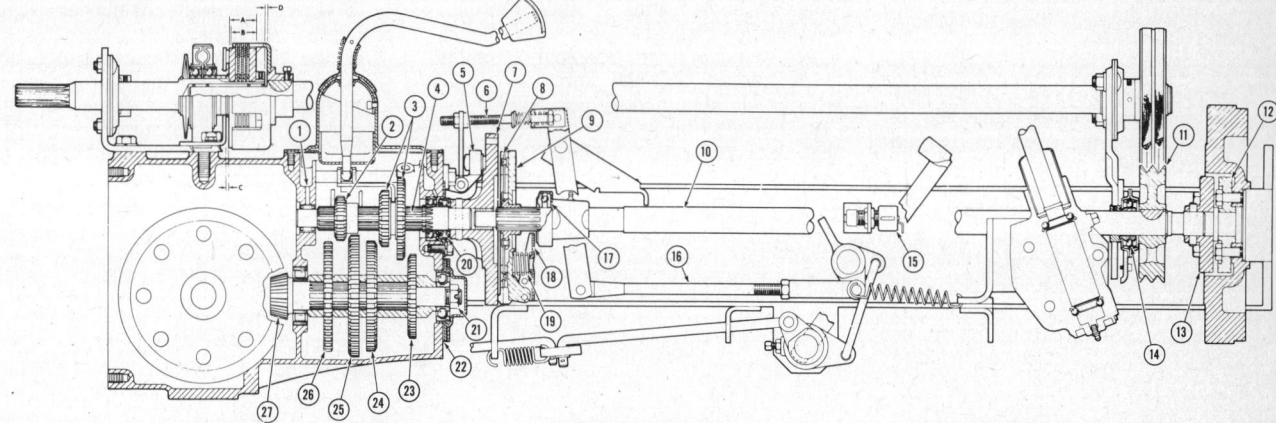

1. Oil passage
2. 1st and reverse speed sliding gear
3. 2nd and 3rd speed sliding gear
4. Spline and clutch shaft
5. Clutch brake assembly
6. Clutch brake adjusting bolt
7. Rear pressure plate assembly
8. Clutch driven disc assembly
9. Front pressure plate assembly
10. Main clutch shaft
11. PTO drive pulley
12. Coupler retainer
13. Clutch coupling
14. Bearing
15. Neutral start switch
16. Clutch adjusting rod
17. Clutch release bearing
18. Release lever
19. Pressure spring
20. Shaft oil seal
21. Countershaft nut
22. Shims
23. 3rd speed gear
24. 2nd speed gear
25. 1st speed gear
26. Reverse speed gear
27. Countershaft and bevel pinion

Transmission and differential components, through serial #UO18708

545

International Harvester

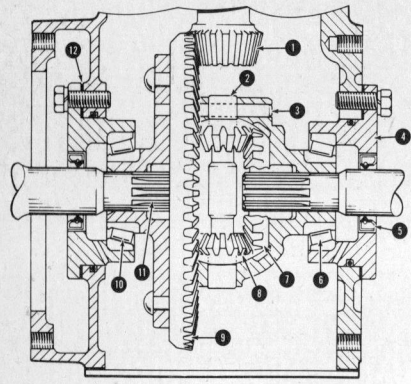

1. Bevel pinion
2. Pinion shaft
3. Lock pin
4. Bearing retainer
5. Oil seal
6. Righthand bearing
7. Side gear (2)
8. Pinion (2)
9. Drive gear
10. Lefthand bearing
11. Differential shaft
12. Shims

Differential gear components

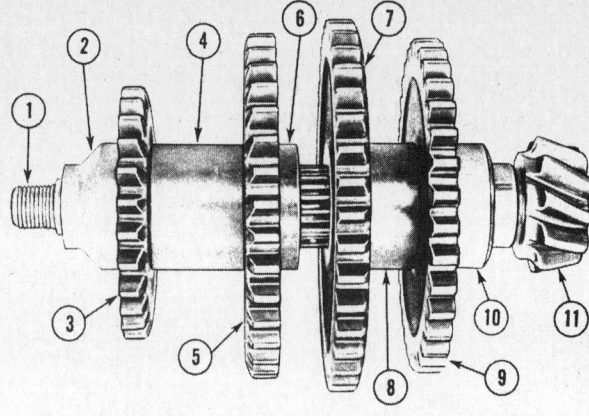

1. Countershaft
2. Spacer
3. 3rd speed gear
4. Spacer
5. 2nd speed gear
6. Spacer
7. 1st speed gear
8. Spacer
9. Reverse speed gear
10. Spacer
11. Bevel pinion

Countershaft components

4. Remove the cotter pin from the countershaft nut.

5. Shift the transmission into two gear speeds to lock the transmission, then remove the nut from the countershaft.

6. Shift the transmission into the neutral position and remove the shifter fork setscrews.

7. Using a brass rod, drive the shifter rods (2) forward and out of the case, and remove the shifter forks.

CAUTION
Cover the shifter poppet ball holes to prevent the balls and springs from flying out as the rods are removed.

8. Remove the countershaft bearing retainer with shims. Be sure to keep the shims with the retainer for use in reassembly.

9. Remove the spline and clutch shaft front bearing retainer and oil seal.

10. Move the spline and clutch shaft forward and out of the transmission case as the gears are removed. Keep the gears in correct order for proper reassembly.

11. Move the countershaft rearward and out of the transmission case as the gears and spacers are removed. Note the sequence of spacers and gears for reassembly.

12. To remove the reverse idler shaft and gear assembly, remove the setscrew and remove the shaft, reverse idler gear assembly and expansion plug. Press the bushing out of the reverse idler gear if the bushing is to be replaced.

13. Remove the bearings from the transmission case if they are to be replaced.

ASSEMBLY AND INSTALLATION

Transmission and Differential

1. Be sure all bearings are thoroughly lubricated before reassembly.
2. Be sure to replace all gaskets, O-rings and oil seals.
3. Install new bearings in the transmission case if they were removed. BE SURE the spline and clutch shaft rear bearing is installed with its oil passage aligned with the oil passage in the case.

4. Assemble the side gears and the pinion gears in the differential case assembly.

5. Install the pinion shaft and drive the pinion shaft lock pin into place.

6. Install the differential assembly in the transmission case. The drive gear must be on the left with the teeth facing right.

7. Press the righthand bearing on the differential carrier if it was removed during disassembly.

8. Be sure to install new O-rings and oil seals. Install the bearing retainers and shims. Install the capscrews and tighten to 45 ft.lb. torque.

9. Check bearing preload before installing the transmission countershaft. Preload is correct when a steady pull of one to eight lb. is necessary to rotate the differential assembly.

10. Add or remove an equal amount of shims on both bearing retainers to adjust for specified preload.

11. Remove the differential assembly being sure to keep the shims with each retainer and identified for each side.

12. Install the transmission countershaft, spacers, gears, front bearing, bearing retainer and shims and the countershaft nut. Do not torque nut at this time. Be sure to install a new O-ring and gasket.

13. If the reverse idler bushing is to be replaced, press the bushing into the gear until the edge is flush with the gear face. Ream the bushing to the specified I.D. of .612 to .613 in.

14. Install the reverse idler shaft and idler gear assembly in the case. Install the setscrew and tighten securely. Be sure to install a new expansion plug.

15. Install the spline and clutch shaft, bearing and gears in the transmission case.

16. Install a new gasket and shaft oil seal on the clutch and spline shaft. Use a double lip neoprene seal, do not use a leather seal.

17. Install the capscrews and tighten to 25 ft.lb. torque.

NOTE: Be sure the slot in the oil seal is aligned with the oil passage in the case.

18. Install the gear shift poppet springs and balls in their bores.

19. Depress the springs and balls and install the shifter forks and shifter rods.

20. Lock the forks in place with the setscrews and tighten securely.

21. Be sure to install new expansion plugs in the front shifter rod bores.

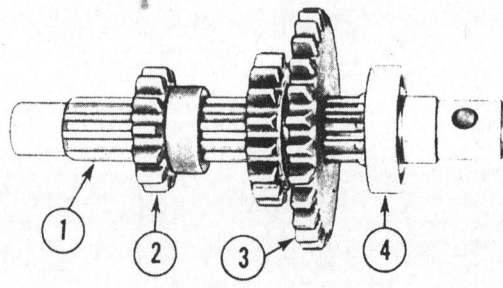

1. Spline and clutch shaft
2. 1st and reverse speed sliding gear
3. 2nd and 3rd speed sliding gear
4. Shaft front bearing

Spline and clutch shaft components

22. Shift the transmission into two speeds to lock the gears, and then tighten the countershaft nut to 85 to 100 ft.lb. torque.
23. Install the cotter pin in the nut and countershaft. Install the bearing retainer cap and gasket and tighten the capscrews to 25 ft.lb. torque.
24. Install the differential assembly in the case. The drive gear must be on the left with teeth facing right.
25. Install the righthand carrier bearing if it was removed.
26. Keeping preload shim pack correct as previously established, install the bearing retainers and capscrews and tighten to 45 ft.lb. torque.
27. Check the backlash between the drive gear and bevel pinion and the gear teeth bearing pattern as follows.

   a. Apply a thin coat of red lead or Prussian blue to the bevel pinion teeth faces, then rotate the gears by hand and observe the bearing pattern.

Some deflection will occur under load. Allowance is made in gear design to prevent concentration of load on tooth edges.

   b. Hand testing and very light loads should provide a pattern as shown in Figure B. When load and deflection increases the pattern will progress as in Figure A.

   c. The desirable (no load) pattern in Figure B is the result of adjusting the differential drive gear lateral position to the specified range of .003 in. to .005 in. backlash.

   d. Adjust the drive gear lateral position by removing shims from one side and installing the shims removed on the opposite side.

NOTE: Do not add or remove shims to change the total amount of shims in the previously established shim pack as this will change the bearing preload.

   e. Tooth bearing position from the root to the crown of the tooth is controlled by lateral position of the bevel pinion.

     (1) If low tooth bearing position on the bevel pinion is indicated (as shown in Figure C) the pinion must be adjusted towards the drive gear.

     (2) If high tooth bearing position on the bevel pinion is indicated (as shown in Figure D), the pinion must be adjusted away from the drive gear.

   f. Adjust the bevel pinion by adding or removing shims between the bearing retainer and the transmission case.

NOTE: If it is necessary to move the bevel pinion in or out to correct "Root-to-crown" bearing, the drive gear must also be moved laterally to maintain the specified backlash.

28. Install the final drive assemblies on the transmission case. Be sure the spacer is in place on the righthand final drive.
29. Install the gear shift lever and cover assembly.
30. Install the transmission rear cover plate and the link mounting plate assembly (if equipped).
31. Recouple the tractor.

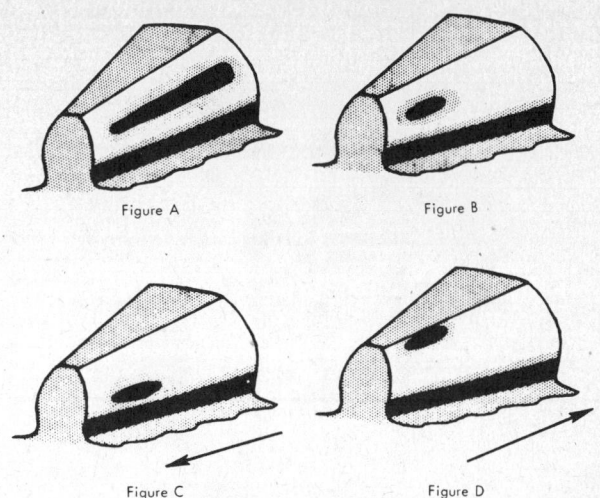

Contact patterns

CREEPER DRIVE

154 and 185 Models

REMOVAL

1. Remove as a unit, the fender and seat assembly, and if equipped with PTO, remove the PTO clutch assembly from the PTO driveshaft.
2. Split the tractor.
3. Support the transmission and place an oil drip pan under the creeper unit and drain the creeper lubricant.
4. Remove the engine clutch assembly from the shaft.
5. Remove the clutch brake assembly and the clutch brake support from the creeper housing.
6. Remove the capscrews securing the creeper unit to the transmission case.
7. Pull the creeper forward and remove it from the transmission.
8. Drive out the pin and remove the splined coupling from the transmission shaft if it is to be serviced.

DISASSEMBLY

1. Drive the coiled spring pin out of the gear shift plate and the control lever.
2. Remove the control lever and shift plate from the creeper assembly.
3. Remove the capscrew, poppet spring and ball from the case.
4. Remove the gear shift fork and fork spindle and the spindle drive sleeve.
5. Loosen the setscrew and remove the shift fork spindle from the shift fork if service is necessary.
6. Press the main gear housing out of the creeper housing.
7. Drive the coiled spring pin out and remove the direct drive spline gear from the sun gear driveshaft.
8. Remove the low gear drive plate and planet gears from the shaft. Remove the planet gears.
9. Remove the outer retaining ring and then remove the bearing retainer and wear plate from the main gear housing.
10. Remove the snapring holding the bearing in the bearing retainer and then remove the bearing and shaft from the bearing retainer.
11. Remove the oil seal from the bearing retainer.
12. Remove the bearing retaining ring and press the shaft out of the bearing.

INSPECTION AND REPAIR

1. Inspect the sun gear driveshaft for oil seal groove wear, worn or chipped teeth on

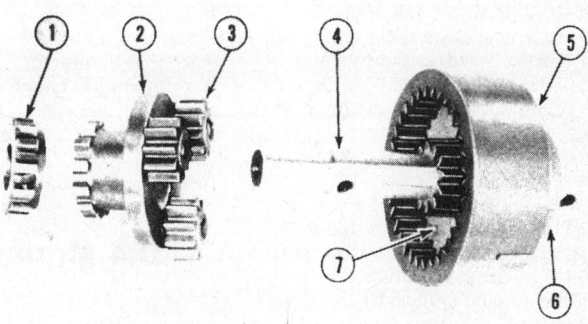

1. Direct drive spline gear
2. Low gear drive plate
3. Planet gear
4. Sun gear driveshaft
5. Main gear housing
6. Bearing retainer
7. Wear plate

Creeper drive gear components

International Harvester

1. Low gear drive plate
2. Planet gear
3. Planet gear pin

Creeper drive planetary components

the gear and spline drive bushing wear on the rear end of the shaft.

2. Check the splines and gear teeth for wear and chipping. Replace any damaged parts.

3. Inspect the ball bearing for pitting, scoring, wear and rough operation and replace if necessary.

4. Inspect the spline drive bushing for wear and damage. Install a new bushing in the spindle drive sleeve if replacement is necessary.

5. Check the housing for cracks.

REASSEMBLY

1. Press the ball bearing on the sun gear driveshaft. Be sure the bearing bottoms against the gear shoulder. Install the snapring retainer.

2. Install a new oil seal in the bearing retainer.

3. Press the shaft and bearing assembly in the bearing retainer. Be sure the assembly bottoms against the retaining shoulder. Install the snapring retainer.

4. Install the wear plate and then install the bearing retainer and shaft assembly in the main gear housing. Be sure to install a new O-ring on the bearing retainer. Install the outer retaining ring.

5. Install the planet gears on the pins.

6. Install the low gear drive plate and planet gears on the shaft and in the main gear housing.

7. Install the direct drive spline gear on the shaft and drive the coiled spring pin into its bore. Be sure the pin does not protrude above the gear hub.

8. Align the slot in the main gear housing with the slot in the creeper case and press the main gear housing into the case until it is flush with the case. Drive the coiled spring pin into the hole made by the two slots.

9. Install the shift fork spindle in the shift fork, if it was removed, and tighten the setscrew securely.

10. Position the shift fork in the spindle drive sleeve groove, and as a unit, slide the spindle drive sleeve and shift spindle into place on the direct drive spline gear and low gear drive plate gear.

11. Install the poppet ball and spring in the housing bore and then install the capscrew and tighten securely.

12. Position the shift plate on the shift fork. Install the control lever in the housing and shift plate.

13. Align the holes in the control lever and shift plate and drive the coiled spring pin into the hole.

INSTALLATION

1. Replace the transmission seal and the bearing retainer.

2. Install the splined coupling on the transmission shaft if it was removed. Drive the coiled spring pin into its bore.

3. Install a new gasket on the transmission case and then install the creeper on the case.

4. Install the capscrews securing the creeper unit to the transmission case and tighten to 80 ft.lb. torque.

5. Install the clutch brake support and clutch brake assembly on the creeper housing.

6. Install the engine clutch assembly on the creeper shaft.

7. Fill the creeper drive with 1½ pints of Hy-Tran fluid.

8. Recouple the tractor.

9. Install the PTO clutch assembly if it was removed.

10. Adjust the clutch brake and the PTO unit.

11. Install the fender and seat assembly.

184 Model

REMOVAL

1. Remove the driveshaft.
2. Remove the crossmember.
3. Disconnect the brake linkage for single pedal brake.
4. Remove the creeper drive.

5. Service the creeper drive as outlined in the steps for the 154 & 185 tractor.

INSTALLATION

Installation is the reverse of the removal procedure.

FINAL DRIVE UNIT

Final Drive

REMOVAL

1. Split the tractor.
2. Remove the drain plugs in each rear axle housing and drain the lubricant.
3. Support the transmission and remove the rear wheels.
4. Remove the capscrews securing the rear axle housings to the transmission case.
5. Support the transmission and final drive assemblies and remove the assemblies from the transmission.

Differential Shaft

DISASSEMBLY

1. Remove the bearing retainer cap.
2. Loosen the locknut and setscrew securing the brake drum to the differential shaft and remove the brake assembly from the shaft.
3. Remove the differential shaft from the rear axle housing.
4. Remove the retaining ring and pull the ball bearing assembly off the shaft.
5. Remove the oil seal from the housing.

ASSEMBLY

1. Install a new oil seal.
2. Press the ball bearing assembly on the shaft and install the bearing lock ring.
3. Install the differential shaft in the housing and be sure the outside snapring is against the housing.
4. Align the brake drum key way with the key on the differential shaft and install the brake drum on the shaft. Tighten the setscrew and locknut securely.
5. Install the brake band on the brake drum. Be sure the actuating toggles are positioned correctly.
6. Install the bearing retainer cap.

INSTALLATION OF FINAL DRIVE

1. Be sure the spacer is on the righthand final drive assembly and install the final drive assemblies on the transmission case.
2. Install the capscrews and tighten securely.

NOTE: The 2⅞ in. long capscrews are used on the righthand final drive assembly and the 1⅞ in. long capscrews are used on the lefthand assembly. Apply sealer to the bolt in the 9 o'clock position on right side before installing.

International Harvester

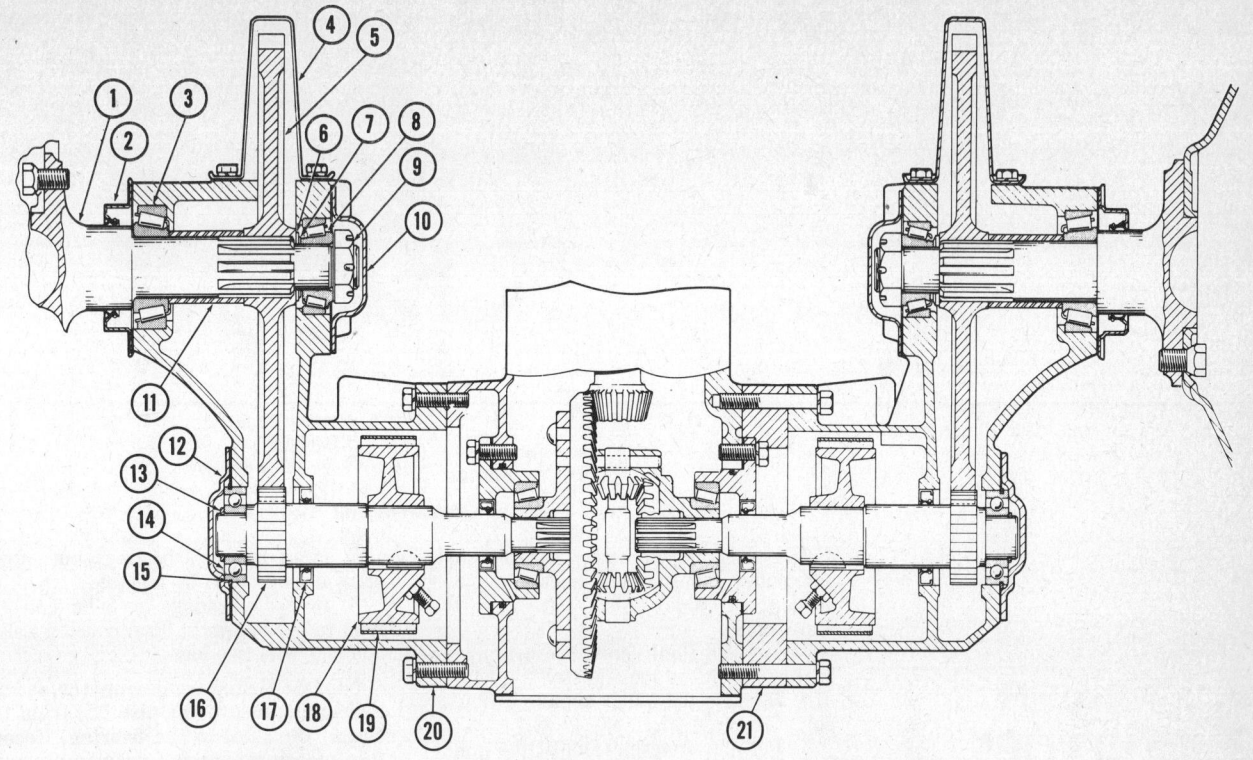

1. Lefthand axle
2. Oil seal
3. Outer bearing
4. Housing pan
5. Drive gear
6. Drive gear spacer
7. Shims
8. Inner bearing
9. Axle nut
10. Bearing cap
11. Bearing spacer
12. Shaft bearing retainer
13. Bearing lockring
14. Bearing
15. Bearing snapring
16. Differential shaft and gear
17. Oil seal
18. Brake drum
19. Brake band
20. Lefthand rear axle housing
21. Rear axle spacer

Final drive components

3. Support the transmission and final drive assemblies and install the rear wheels. Tighten the lub bolts securely.
4. Fill the housings to the correct level with Hy-Tran fluid.
5. Recouple the tractor.

Rear Axle

DISASSEMBLY

1. Remove the rear axle housing pan and gasket.
2. Remove the rear axle bearing cap.
3. Remove the cotter pin from the rear axle nut. Block the rear axle drive gear to prevent the shaft from turning and remove the nut.
4. Remove the capscrews securing the rear axle oil seal to the housing.
5. Press the rear axle out of the housing and remove the rear axle drive gear, spacers and shims. Be sure to retain the shims and spacers for use in reassembly.
6. Remove the outer bearing and remove the oil seal.

ASSEMBLY

NOTE: Install the rear axles before installing the differential shafts.

1. Be sure to replace the outer oil seals and gaskets.

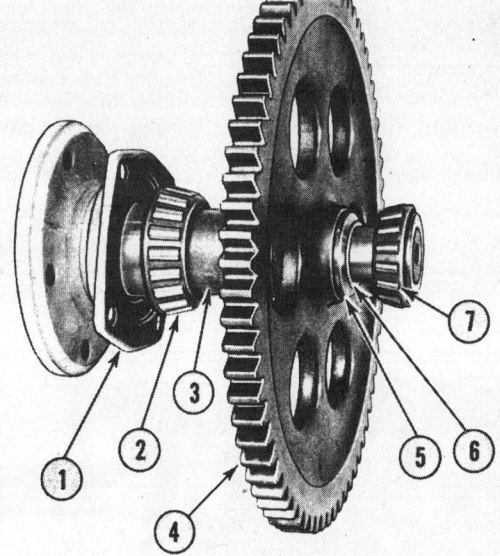

1. Outer oil seal
2. Outer bearing
3. Spacer
4. Rear axle drive gear
5. Spacer
6. Shim
7. Inner bearing

Rear axle drive gear components

549

International Harvester

2. Press the outer bearing on the rear axle. Be sure the bearing bottoms against the shoulder on the shaft.

3. Install the rear axle assembly in the housing. Be sure the spacers and shims are installed in the proper order.

4. Press the inner bearing on the shaft.

5. Install the rear axle nut and tighten securely.

6. Check for specified bearing preload as follows:

   a. A drag torque of 10 to 20 in. lb. is necessary to rotate the rear axle assembly before the outer oil seal is bolted in place. Read rolling torque, not starting torque.

   b. To adjust for specified preload, remove the rear axle assembly from the housing and add or remove shims.

7. Install the cotter pin in the rear axle nut.

8. Bolt the outer oil seal in place.

9. Install the bearing cap.

10. Install the axle housing pan and gasket.

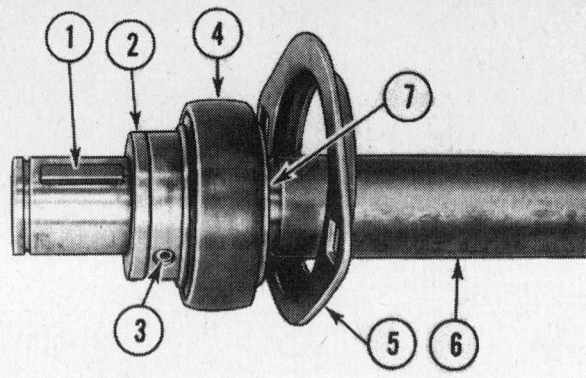

1. Key
2. Spacer
3. Setscrew
4. Front ball bearing
5. Bearing flange
6. PTO driveshaft
7. Snapring (not shown)

Driveshaft unit

POWER TAKE-OFF

154 and 185 Models

REMOVAL

IPTO Clutch Assembly

1. Remove the PTO rear shield and support.

2. Remove the PTO brake support and the shaft bearing mounting bracket support. Loosen the clutch cleat bracket capscrews.

3. Disconnect the clutch control arm.

4. Remove the PTO clutch assembly from the tractor.

NOTE: Tractors equipped with a 3 pt. hitch will require removal of the rockshaft to remove the clutch assembly.

IPTO Drive Shaft

1. Remove the frame top cover and pedestal side sheet sections.

2. Loosen the capscrews on the shaft and pulley adjuster and release the belt tension and remove the belts from the upper pulley.

3. Remove the bolts and nuts securing the bearing flanges to the adjuster.

4. Remove the pulley retaining ring or bolt.

5. Pull the driveshaft out of the pulley and remove the pulley.

6. Remove the shaft from the tractor.

DISASSEMBLY

IPTO Clutch Assembly

1. Loosen the bearing locking collar and remove the PTO shaft rear ball bearing, locking collar and bearing flanges.

2. Remove the setscrew in the locknut or snapring and shims from the shaft. Be sure to keep the shims for use in reassembly.

3. Remove the clutch assembly components from the PTO shaft being sure to note their position for reassembly.

4. Remove the cam cover and then remove the actuating cam and balls.

5. Compress the load spring and remove the snapring retainer.

6. Carefully release the load spring pressure and remove the stationary cam and the load spring from the shaft.

7. Press the cam bearings out of the cams if they are to be replaced.

8. If needle bearings are to be replaced, remove the outermost bearing using a slide hammer with two legs.

NOTE: Before removing the second bearing, measure the distance from the end of the shaft to the bearing. Record this measurement for reassembly purposes.

9. To remove the innermost bearing, use a slide hammer with only one appropriate length leg as shown.

IPTO Driveshaft

1. Remove the key from the shaft.

2. Remove the spacer from the shaft.

3. Loosen the setscrews in the driveshaft front bearing and remove the bearing.

4. Remove the setscrews from the drive cup assembly and remove it from the driveshaft.

ASSEMBLY AND INSTALLATION

IPTO Driveshaft

1. Install the key and the drive cup assembly on the driveshaft and tighten the setscrews securely.

2. Install the cam cover on the drive cup if it was removed. Use rubber cement to hold the cover in position.

3. Install the rear snapring, bearing flange, front ball bearing, spacer and Woodruff key on the shaft.

4. Position the bearing against the rear snapring and securely tighten the bearing setscrew.

5. Install the PTO driveshaft assembly in the tractor.

6. Install the second bearing flange on the shaft. Place the PTO driven pulley in position and move the driveshaft into it. Install the retaining snapring or bolt.

7. Install the capscrews and nuts securing the bearing flanges to the adjuster. Tighten nuts only finger tight at this time.

8. Install the belts on the pulley and tighten only enough to remove the slack. Do not tighten to operating tightness at this time.

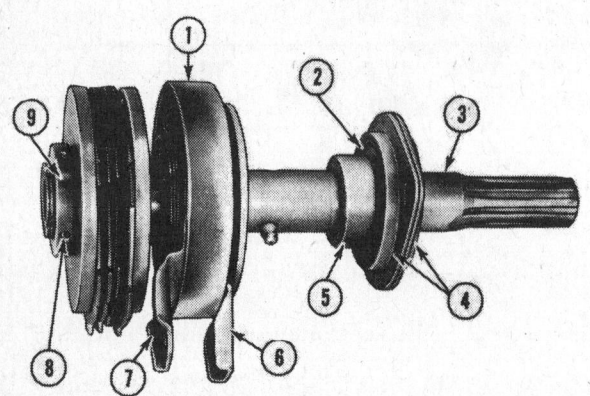

1. Cam cover
2. Rear ball bearing
3. PTO shaft
4. Bearing flanges
5. Locking collar
6. Stationary cam
7. Actuating cam
8. Setscrew
9. Locknut or snapring

PTO clutch unit

International Harvester

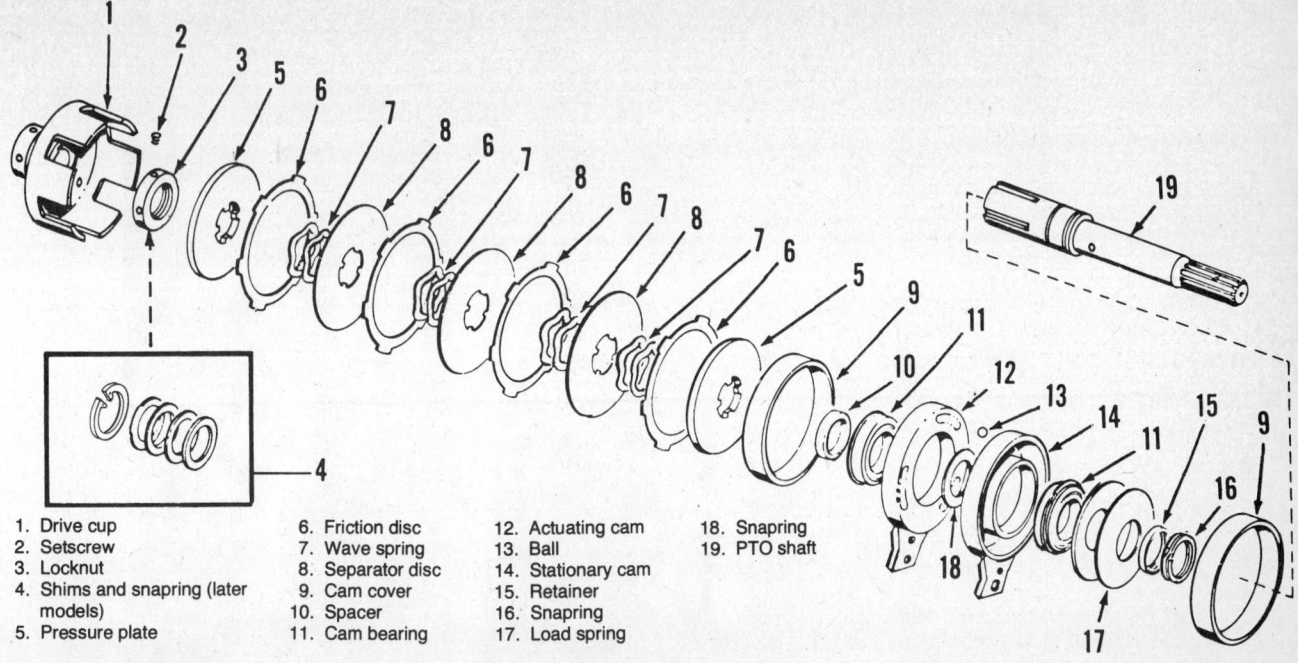

1. Drive cup
2. Setscrew
3. Locknut
4. Shims and snapring (later models)
5. Pressure plate
6. Friction disc
7. Wave spring
8. Separator disc
9. Cam cover
10. Spacer
11. Cam bearing
12. Actuating cam
13. Ball
14. Stationary cam
15. Retainer
16. Snapring
17. Load spring
18. Snapring
19. PTO shaft

IPTO clutch components

IPTO Clutch Assembly

1. If the needle bearings inside the PTO shaft have been removed, new bearings must be installed.

2. Press new cam bearings into the cams if they were removed. Be sure the snapring around the bearing bottoms against the cam body.

3. Install the load spring and stationary cam on the shaft.

4. Compress the load spring and install the snapring retainer.

5. Be sure to grease the cam ball slots and balls. Install the balls in the stationary cam ball slots and then install the actuating cam on the shaft.

6. Install the cam cover around the cams. Use rubber cement to hold the cover in place.

7. Install the spacer on the shaft and then install the pressure plate friction discs, wave spring, and separator discs.

8. Install the shims and snapring or the locknut. Do not tighten the setscrew at this time.

IMPORTANT: The following adjustment must be made before installing the PTO clutch assembly on the tractor.

9. Measure the clutch assembly in the disengaged position A. Move the acutating cam to the engaged position B and measure the clutch assembly. The differenece in the two measurements must be .050 in. Adjust by tightening or loosening the locknut or by installing shims between the snapring and the pressure plate. When proper adjustment has been made, install and tighten the setscrew.

NOTE: Use caution when tightening the locking setscrew. OVER TIGHTENING will result in distortion of the needle bearings.

10. Install the bearing flanges, locking collar and shaft ball bearing. Do not tighten the locking collar at this time.

11. Install the clutch unit in the drive cup. Be sure the clutch cleat pin mates with the stationary cam.

NOTE: If the clutch cleat pin fits tightly in the bottom of the slot in the stationary cam, shims must be installed under the clutch cleat bracket. Failure to do this will result in misalignment of the PTO shafts.

12. Connect the clutch control arm to the actuating cam.

13. Install the shaft bearing mounting bracket support and the brake support on the transmission case.

14. Tighten the front bearing flange nuts securely.

15. Install and securely tighten the rear bearing flange bolts and nuts.

ADJUSTMENTS

NOTE: Measurements "A" and "B" are taken with the PTO assembly removed from the tractor.

1. Measure the clutch assembly in the disengaged position A. Move the actuating cam to the engaged position B and measure the clutch assembly. The difference in the two measurements must be .050 in. Adjust by tightening or loosening the locknut or by installing shims between the snapring and the pressure plate.

NOTE: Use caution when tightening the locknut setscrew. OVER TIGHTENING will result in distortion of the idler shaft needle bearings.

2. Install the PTO clutch unit on the tractor.

3. Measurement D, shaft to drive cup clearance, must be ⅛ in. This is obtained by bottoming the PTO assembly against the drive cup and then moving the assembly rearward ⅛ in. Tighten the rear PTO shaft bearing locking collar and setscrew securely.

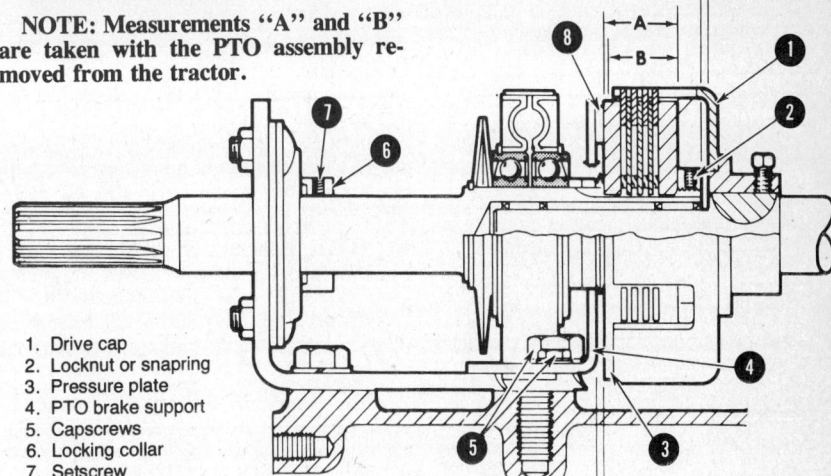

1. Drive cap
2. Locknut or snapring
3. Pressure plate
4. PTO brake support
5. Capscrews
6. Locking collar
7. Setscrew
8. Brake friction pad

PTO adjustments

551

International Harvester

4. Be sure to get full pin contact between the clutch cleat bracket and the stationary cam to prevent vibration. Position the clutch cleat bracket to get full pin contact.

NOTE: If the clutch cleat pin fits tightly in the bottom of the slot in the stationary cam, shims must be installed under the clutch cleat bracket. Failure to do this will result in misalignment of the PTO shaft. Tighten the capscrews securely.

5. Measurement C, clearance between the pressure plate (engaged position B) and the brake friction pad, must be .040 in. This is obtained by loosening the capscrews and properly locating the brake plate at the slotted holes.

6. Check the belt tension. The belts are properly adjusted when 50 ft.lb. of torque is required to turn the output shaft. To check this adjustment, lock the tractor brakes with the padal lock, shift the transmission into third gear (the creeper drive unit in high if tractor is so equipped) and engage the PTO clutch. Place two Allen wrenches in any splines of the shaft. Slide a $^{15}/_{16}$ in. twelve-point socket over the shaft and the Allen wrenches. Using a torque wrench, apply force until a reading of 50 ft.lb. is obtained.

7. If 50 ft.lb. torque reading cannot be obtained remove the left side panel and adjust the belts. If after adjustment the torque reading of 50 ft.lb. cannot be obtained, remove the PTO clutch and check for damage or excessive wear. Readjust the PTO clutch.

8. Remove the rear bearing flange nuts and install the rear PTO shields. Install and securely tighten the nuts.

1. Idler tension arm
2. Idler tension arm nut
3. Anchor strap
4. Clutch belts
5. Idler tension pulley

Electric PTO drive belts

ELECTRIC POWER TAKE-OFF

184 Models

REMOVAL AND DISASSEMBLY

1. Disconnect the battery cables from the battery.

NOTE: Remove the ground cable first to reduce electrical shorting hazards.

2. Remove the pedestal side sheets, frame top cover and rear frame cover.

NOTE: Tractors without a three point hitch will have the rear frame cover.

3. Relieve the clutch belt tension by pushing in on the idler tension arm nut with a ¾ in. wrench and then slip the drive belts off the tension arm pulley.

4. Disconnect the anchor strap at the PTO clutch assembly.

5. Remove the PTO clutch brake plate locknuts and remove the brake plate assembly and springs and slip off the drive belts.

6. Disconnect the electrical terminal.

7. Remove the $^{5}/_{16}$ x 1¼ in. capscrew from the pulley assembly and install in its place a ⅝ x 1 in. capscrew. Tighten down the ⅝ x 1 in. capscrew approximately one and a half turns to loosen the pulley assembly. Then remove the ⅝ x 1 in. capscrew and slide the pulley assembly off the key way end of the PTO shaft.

8. Install the breaker bar and 1⅝ in. socket to the lefthand threaded field bearing nut and secure to the tractor frame with twine or wire.

9. Install at the spline end of the PTO shaft a $^{15}/_{16}$ in. socket and a ⅛ in. Allen wrench between the socket points and shaft spline end.

10. Turn the PTO shaft clockwise and remove the field bearing jam nut, field assembly and clutch spacer spring.

11. Remove the two setscrews from the bearing collar at the key way end of the PTO shaft.

NOTE: Bearings on the PTO shaft are not interchangeable. The bearing on the key way end of the shaft is secured to the shaft with a collar and two setscrews. Tighten the setscrew in the counterbore of the shaft first.

12. Remove the PTO bracket at the rear end of the tractor and pull the PTO shaft out of the front bearing. If the rear bearing at the spline end of the PTO shaft requires replacement remove the lock collar and pull off bearing.

ASSEMBLY AND INSTALLATION

1. Before installing the PTO shaft bearings apply Never Seez® compound to both ends of the PTO shaft.

2. Install the bearing flanges and bearing on the threaded end of the PTO shaft. Make certain the bearing is against the shaft shoulder then tighten the first setscrew into the shaft counterbore and then tighten the second setscrew.

3. Install the PTO shaft in the tractor and loosely assemble the front bearing flanges to the front bearing bracket.

4. Install the lock collar and the rear bearing to the spline end of the PTO shaft and loosely assemble the bearing flanges to the rear bearing bracket. Lock the rear bearing collar in the direction of PTO shaft rotation.

IMPORTANT: Check alignment between IPTO idler pulley and IPTO clutch pulley, making certain specifications are met. (Refer to Main Clutch Shaft and Engine Clutch—184 Tractor —Assembly and Installation.)

5. After the PTO shaft has been preassembled into the tractor, complete tightening of the bearing flanges.

6. Install the clutch spacer spring and field assembly and torque the field jam nut to 35-40 ft.lb.

NOTE: Position the high points of the clutch spacer spring against the bearing flange and not on the carriage bolt heads.

7. Align the clutch pulley assembly to

International Harvester

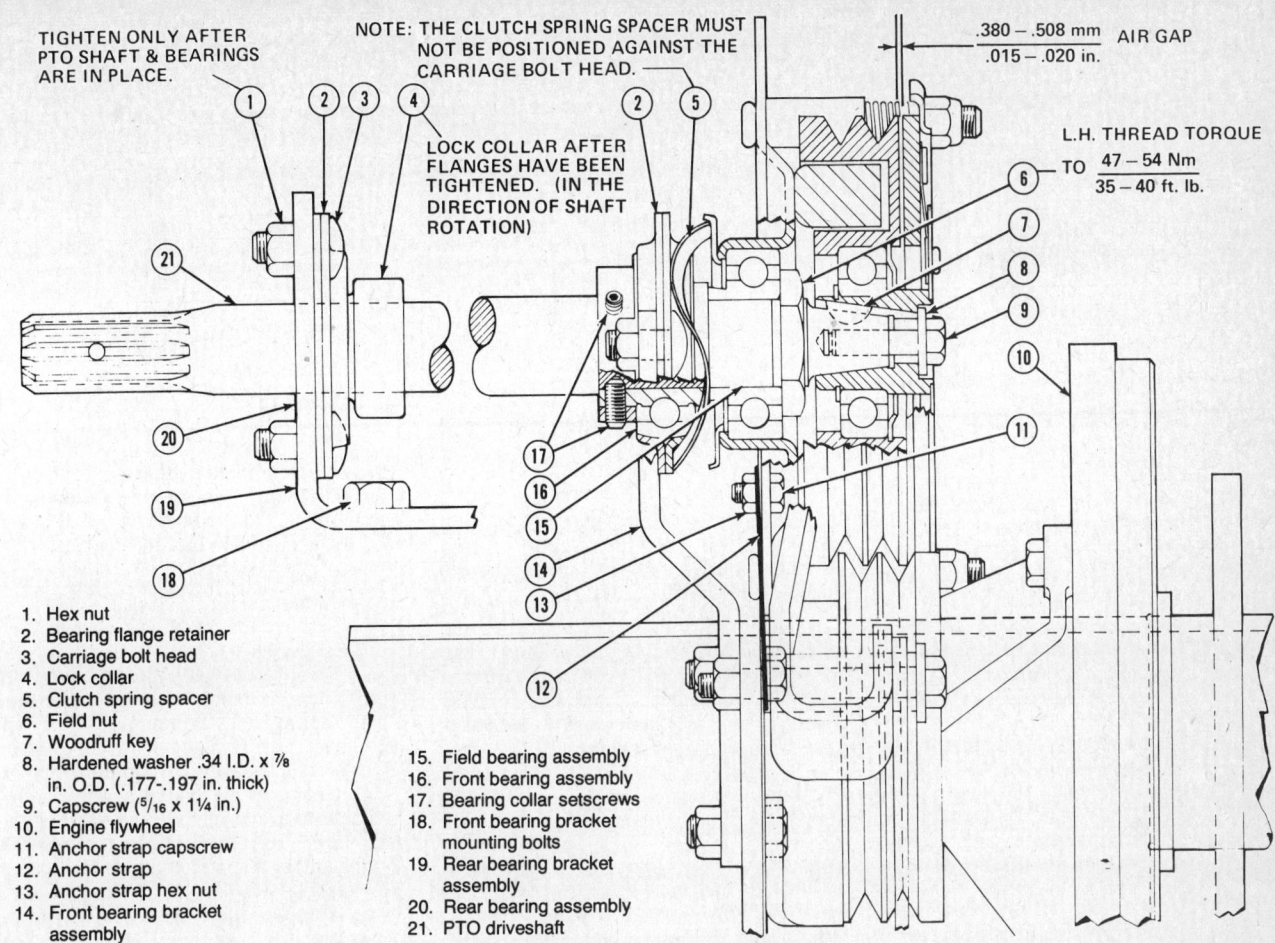

1. Hex nut
2. Bearing flange retainer
3. Carriage bolt head
4. Lock collar
5. Clutch spring spacer
6. Field nut
7. Woodruff key
8. Hardened washer .34 I.D. x 7/8 in. O.D. (.177-.197 in. thick)
9. Capscrew (5/16 x 1¼ in.)
10. Engine flywheel
11. Anchor strap capscrew
12. Anchor strap
13. Anchor strap hex nut
14. Front bearing bracket assembly
15. Field bearing assembly
16. Front bearing assembly
17. Bearing collar setscrews
18. Front bearing bracket mounting bolts
19. Rear bearing bracket assembly
20. Rear bearing assembly
21. PTO driveshaft

Electric PTO adjustment

the shaft key way and torque the capscrew to 19-21 ft.lb.

NOTE: Make certain belts and pulleys are free of grease and oil.

8. Air gap for the PTO clutch must be set between .015-.020 in.
9. Make certain that the idler tension arm pulley nut is torqued to standard specifications.
10. Make certain to reconnect the anchor strap.

ENGINE

REMOVAL

1. Disconnect the battery ground cable at the battery.
2. Drain the crankcase oil and remove the drain plug in the water inlet elbow and drain the coolant.
3. Remove the hood and side sheet sections.
4. Shut off the fuel at the fuel strainer and disconnect the fuel line from the fuel strainer and the carburetor.
5. Remove the fuel tank.
6. Disconnect the choke cable from the carburetor.
7. Remove the air cleaner and air cleaner pipe.
8. Disconnect the oil lines to the hydraulic pump if so equipped.

NOTE: Be sure to plug all openings in tubes and parts to prevent dirt from entering the system.

9. Disconnect and remove the radiator hoses.
10. Remove the cooling fan assembly and lay in shroud.
11. Disconnect the throttle cable from the governor.
12. Disconnect the wires to the generator, voltage regulator, coil positive terminal and headlights (alternator and cranking motor on 184).

NOTE: Be sure to tag and identify all wires so they can be reconnected correctly.

13. Disconnect the oil pressure switch assembly.
14. Remove the capscrews securing the pedestal front sheet to the engine and remove the front sheet.
15. Relieve the IPTO clutch belt tension by pushing in on the idler tension arm nut with a ¾ in. wrench and then remove the drive belts from the tension arm pulley.
16. Remove the frame top cover. Move the snapring forward from the groove on the driveshaft. Slide the driveshaft back until it clears the flywheel retainer.
17. Remove the capscrews securing the clutch coupling hub to the flywheel.
18. Remove the six capscrews, two in front and four in back, securing the engine to the frame.
19. Using attaching brackets FES 100, attach a chain hoist to the engine and lift it out of the tractor.

INSTALLATION

1. Using a chain hoist and attaching brackets FES 100, install the engine in the tractor. Install the six capscrews securing the engine to the frame and tighten securely.
2. Install the capscrews securing the clutch coupling hub to the fly wheel. Slide shaft forward into flywheel retainer to pilot coupling and tighten securely. Position shaft in operating position and install belts.
3. Install the pedestal front sheet and secure with the capscrews.
4. Connect the oil pressure switch assembly.
5. Connect the wires to the generator, voltage regulator, coil positive terminal and head lights (alternator and cranking motor on 184).
6. Connect the throttle cable to the governor.
7. Install the cooling fan assembly. Be sure the fan blades clear the radiator shroud. If clearance is not sufficient, reposition fan shroud.
8. Connect the radiator hoses.
9. Install the air cleaner and air cleaner

553

International Harvester

pipe. Be sure air cleaner pipe connections are good to prevent dirt from entering the engine.

10. Connect the oil lines to the hydraulic pump, (if equipped).
11. Connect the choke cable to the carburetor.
12. Install the fuel tank and connect the fuel line to the carburetor and the fuel strainer.
13. Install the frame top cover, side sheets and hood.
14. Fill the crankcase with oil and the radiator with coolant.
15. Connect the battery ground cable.

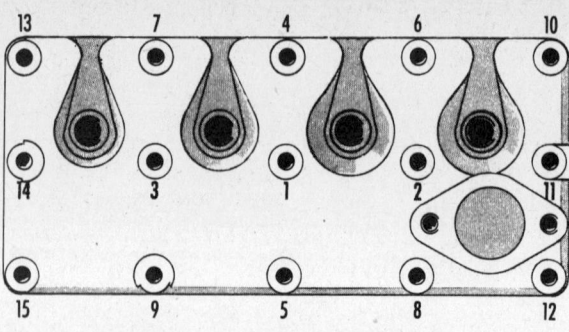

Cylinder head tightening sequence

Cylinder Head

REMOVAL

1. Remove the pipe plug in the water inlet elbow and drain the coolant.
2. Remove the hood, fuel tank, air cleaner and water outlet elbow.
3. Remove the spark plugs.
4. Remove the cylinder head capscrews and lift off the cylinder head and gasket.
5. Use a straight edge and inspect for warped head, particularly in any area which shows "blow-by."
6. Inspect water jacket in head for an accumulation of rust or lime deposit which would affect circulation of cooling water and cause hot spots. Clean if necessary.
7. Thoroughly clean the gasket surface to insure proper sealing of the new gasket.
8. Be sure to use a new gasket.

INSTALLATION

1. Using a new gasket, install the cylinder head on the engine.
2. Install the cylinder head capscrews. Using the sequence shown, tighten evenly in steps to 45 ft.lb. torque. Be sure to install all brackets and mounting clips under the capscrew heads before tightening.
3. Install the spark plugs and tighten to 30 ft.lb. torque.
4. Install the water outlet elbow, air cleaner, muffler, fuel tank and hood.
5. Refill the radiator with coolant.

Valves

LASH ADJUSTMENT

1. Remove the intake and exhaust manifold assembly. Remove the valve tappet cover. With the cover removed, inspect the entire valve assembly for rust and dirt. Clean the assembly with cleaning solvent. Inspect for looseness in the valve assembly and for worn or broken valve springs.
2. Remove the spark plugs from No. 1 cylinder (nearest the radiator) and No. 4 cylinder.
3. Place a thumb over the No. 1 spark plug opening and slowly hand crank the engine until an outward pressure can be felt. Pressure indicates the piston is moving toward top-dead-center of the compression stroke.
4. Continue cranking slowly until the O mark on the fan drive pulley is in line with the timing pointer on the crankcase front cover.

NOTE: Valve tappet have self-locking tappet screws. Adjustment requires two wrenches, one to hold the tappet and one to turn the tappet screw.

5. Insert the feeler gauge between the valve tappet and the valve stem. The specified clearance is .015 in. (engine cold). Turn the adjusting screw in or out as necessary to give a slight drag on the feeler gauge. Adjust valves number 1, 2, 3, 5.
6. Crank the engine until the No. 4 piston is on TDC (compression) and the O mark on the fan drive pulley is in line with the timing pointer. Adjust valves number 4, 6, 7, 8.
7. Install the valve cover being sure to use a new gasket. Check for any oil leaks.
8. Install the intake and exhaust manifold assembly.

REMOVAL

NOTE: When valve assemblies are removed, all parts should be kept in order. They may then be reinstalled in the same ports, from which removed, if they are to be used for further service.

1. Drain the cooling system and remove the cylinder head.
2. Remove the intake and exhaust manifold assembly.
3. Remove the valve tappet cover, and turn down the tappet screws several turns so the springs may be removed easily and to prevent interference with valve stems after seats and faces are reground.
4. Compress the valve springs with a suitable tool and remove the valve spring seat keys. Be careful not to compress the springs more than necessary as they can be distorted.
5. Remove the valves, valve spring seats and valve springs. Be sure to keep valves in order so they may be installed in the same port.

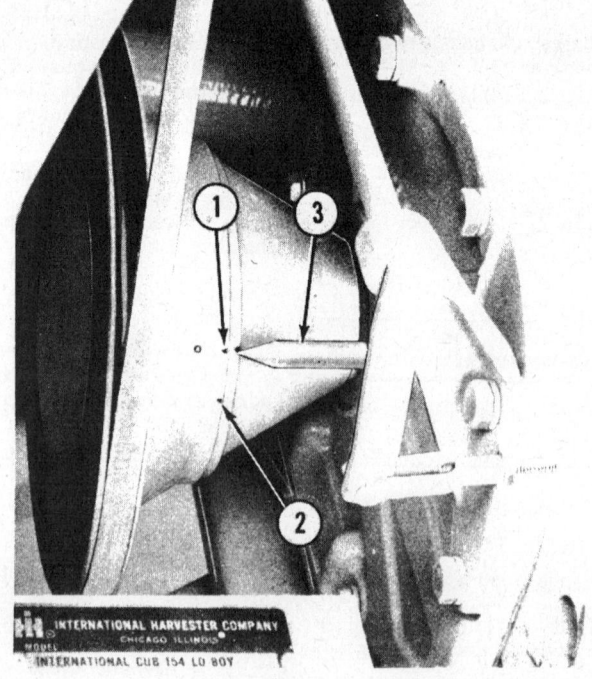

1. O mark
2. 16° mark
3. Timing pointer

Engine timing marks

International Harvester

NOTE: To remove the tappets the camshaft must be removed.

INSTALLATION

1. Coat the valve stems with engine oil and insert the intake and exhaust valves into their original positions.
2. Install the valve springs and valve spring seats. Compress the valve springs and install the valve seat retainer keys. Release the springs and remove the valve compressor.
3. Adjust the valve tappets.
4. Install the valve tappet cover using a new gasket.
5. Install the intake and exhaust manifold assembly using a new gasket. Tighten the nuts evenly in steps to 20 ft.lb. torque.
6. Install the cylinder head.
7. Refill the radiator with coolant.

Pistons, Rings and Connecting Rods

REMOVAL

1. Remove the cylinder head.
2. Remove the drain plug and drain the engine lubricating oil from the crankcase oil pan. Replace the drain plug.
3. Remove the capscrews securing the oil pan, and remove the oil pan and gasket.

---- CAUTION ----
Before proceeding with piston and connecting rod removal, the ridge, existing on the cylinder wall at the upper end of the ring travel, must be removed by using a ridge reamer. This prevents damage to the piston ring lands during removal of pistons, and prevents damage to new top piston rings after the installation of new rings.

4. Remove the oil pump screen and tube assembly.
5. Remove the connecting rod bearing cap nuts. Remove the bearing cap. Be sure that each bearing and cap can be identified with the connecting rod from which it was removed. Each connecting rod should be found numbered on the camshaft side of the rod, indicating its position in the engine.
6. Push the connecting rod and piston assembly to the top and lift out from the crankcase. Replace the cap on the connecting rod to avoid damage.

---- CAUTION ----
Pistons must be handled carefully to avoid damage and knocking out-of-round or alignment. When removing a piston from the crankcase, do not allow the skirt of the piston to strike the crankcase or connecting rod. Mark the pistons so they can be installed in the same position and cylinder from which they were removed. The dome of the piston is stamped with an arrow, indicating its position when properly installed.

7. Crank the engine by hand to make each rod and cap accessible and remove all pistons and connecting rods in the same manner.

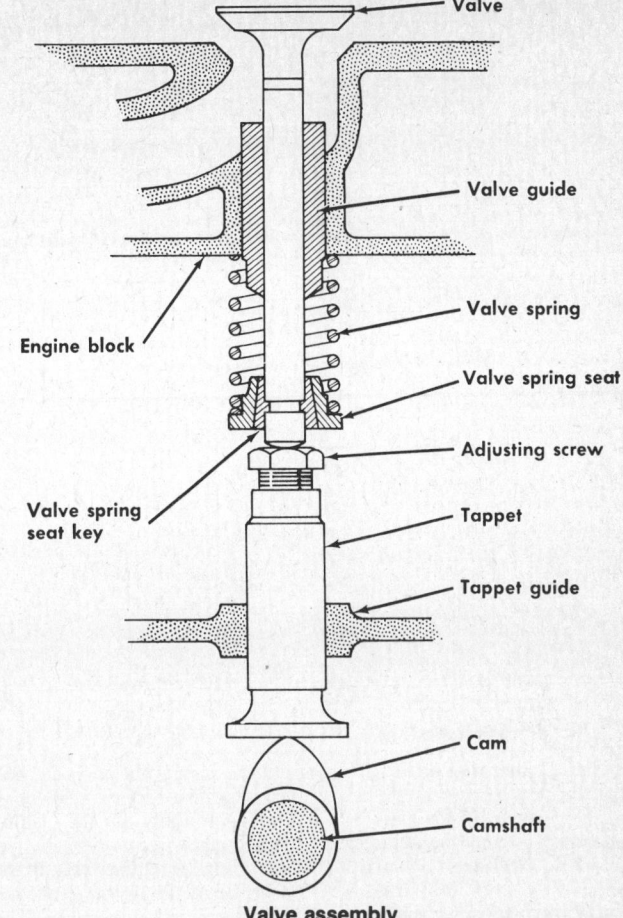

Valve assembly

DISASSEMBLY

1. Remove the piston rings with a piston ring expander. Remove the top ring first and the remaining rings in order.
2. Remove the piston pin retainer rings.
3. Remove the piston pins from the pistons. As an aid in removing the pin, heat the piston in a piston heater or dip the piston in hot-to-boiling water and then remove the pin, being careful not to damage the piston.

ASSEMBLY

1. Before assembling the piston and connecting rod, check the fit of the piston pin in the piston for proper end clearance as follows:
   a. Prepare the piston and the pin for assembly as outlined in step 2.
   b. Push the pin into the piston and install a retainer ring at each side of the piston.
   c. Push one end of the piston pin until it

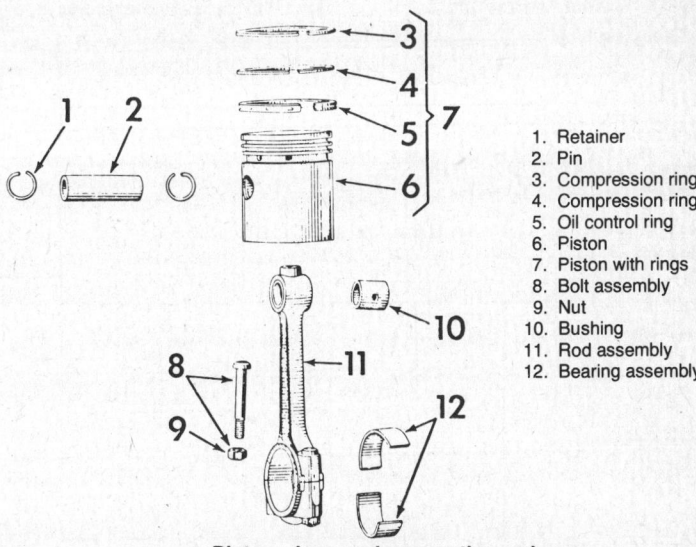

1. Retainer
2. Pin
3. Compression ring
4. Compression ring
5. Oil control ring
6. Piston
7. Piston with rings
8. Bolt assembly
9. Nut
10. Bushing
11. Rod assembly
12. Bearing assembly

Piston, rings and connecting rods

555

International Harvester

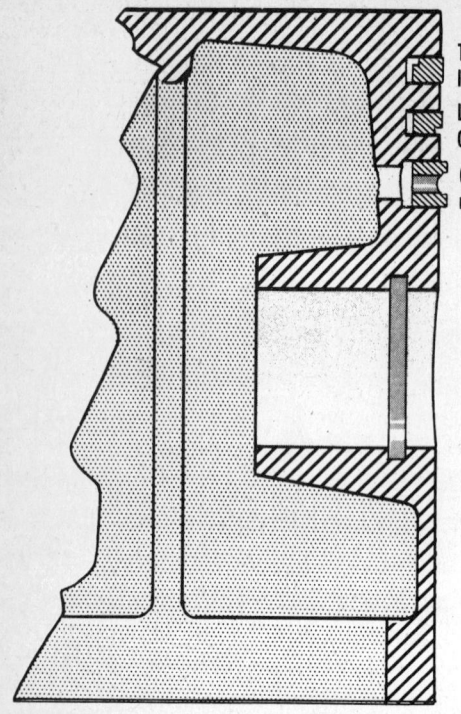

Ring installation

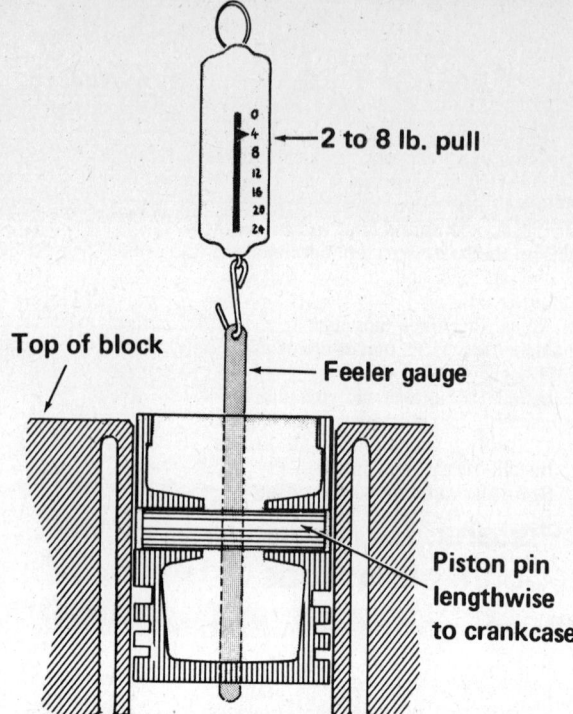

Piston pull in bore

stops against the retainer ring on the opposite side of the piston.

d. Using a feeler gauge, in the gap between the piston pin and the retainer ring, check for end clearance. Specified end clearance is .010 to .030 in.

e. Remove the retainer rings and proceed with the assembly as follows:

NOTE: When assembling the pistons to the rods, the front of the piston will be indicated by an arrow.

2. With the piston pin at room temperature (70°) and generously coated with clean engine lubricating oil, and the piston heated in hot water to approximately 150°F the piston pin can be entered into one boss of the piston by pushing with the hand. While the piston is hot, quickly and correctly position the connecting rod inside the piston, align the bushing in the rod bore with the piston pin holes in the piston and push the piston pin completely into position. Thoroughly dry the piston with compressed air.

3. Install a retainer ring in the groove at each side of the piston to secure the piston pin.

4. Using a piston ring expander, install the rings, oil control ring first, into the grooves of the pistons.

Position the ring gaps 90 degrees from the thrust side of the piston (in line with the piston pin bore) and 180 degrees from one gap to another.

Piston Fit in Bore

Specified piston-to-bore clearance is .0016 to .0024 in. and can be determined using a ½ in. wide feeler gauge and a spring-type tension scale (FES 108).

The thickness of the feeler gauge that can be removed with a 2 to 8 lb. pull represents the piston-to-bore clearance as outlined in the "Piston Clearance Chart." Clearances should conform to specifications.

The chart shows the relationship between the feeler guage thickness and lb. pull in measuring piston-to-bore clearance. Note that with a given feeler gauge thickness the actual clearance is less than the feeler guage used when the lb. pull is towards the high side of the lb. pull range. This is especially true with the thinner feeler gauges.

To determine piston-to-bore clearance proceed as follows:

1. Select a feeler gauge (free of dents or burrs) of one of the thicknesses listed in the chart. Position the feeler gauge in the cylinder bore so that it extends the entire length of the piston 90° from the piston pin location.

2. Invert the piston and install it in the bore so that the end of the piston is about 1½ in. below the top of the cylinder block and the piston pin is parallel to the crankshaft axis.

3. Hold the piston and slowly pull the scale in a straight line with the feeler gauge, noting the pull required to remove the feeler gauge. Check three times and record the average of the three readings obtained. Do not bend or kink the feeler guage.

4. Refer to the chart to determine the actual clearance. The clearance is shown where the horizontal column indicating lb. pull and the vertical column indicating the thickness of the feeler gauge used intersect.

Example: If a .003 in. feeler gauge is used and it takes 8 lb. pull to remove the feeler guage, the clearance is .0023 in.

5. Repeat step 3 with the piston at right

Pull in Lbs.	Feeler Gauge Thickness					
	.0015	.002	.003	.0035	.004	.0045
	Clearance in Inches					
2	.0016	.0022	.0033	.0039	.0044	.005
4	.0013	.0018	.0029	.0035	.004	.0046
6	.001	.0015	.0026	.0031	.0036	.0042
8	.0008	.0013	.0023	.0028	.0033	.0038

Piston clearance chart

International Harvester

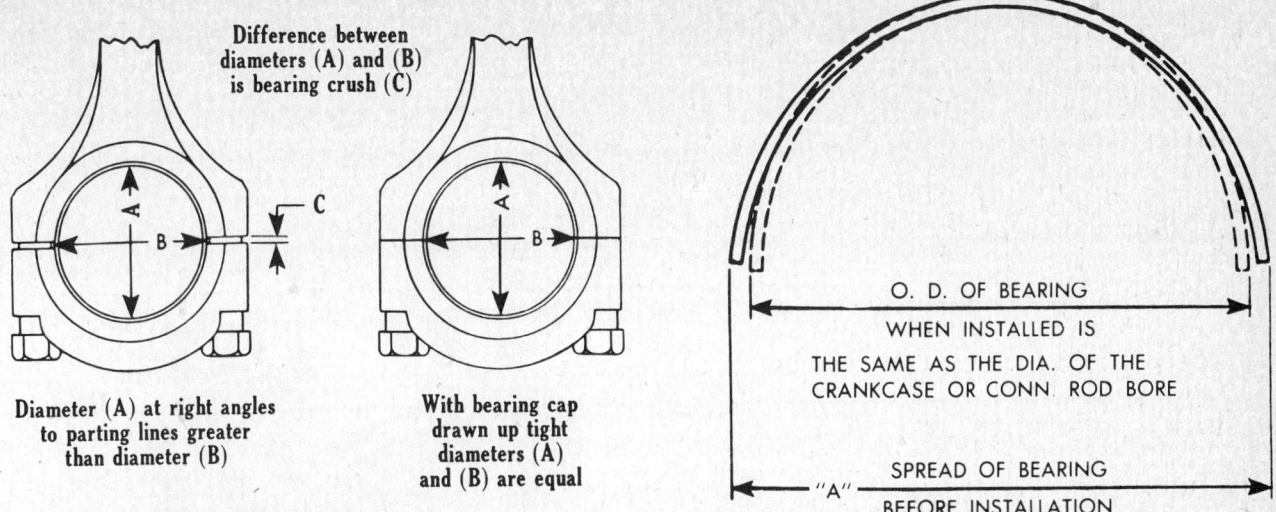

Diameter (A) at right angles to parting lines greater than diameter (B)

With bearing cap drawn up tight diameters (A) and (B) are equal

Difference between diameters (A) and (B) is bearing crush (C)

Fitting the bearings

Checking bearing speed

angles to the crankshaft axis. Determine the clearance as instructed in step 4.

6. Measuring piston-to-bore clearance with the piston pin parallel and at right angles to the crankshaft axis will reflect any "out of round" in the bore.

Bearing Fitting Procedure

CAUTION

Bearings or bearing caps must not be filed, lapped or modified in any manner to reduce journal-to-bearing clearance. Premature bearing failure will result from attempts to reduce journal-to-bearing running clearances. While such methods will make a tighter fit at the top and bottom of the bearing, it will result in an out-of-round bore and distortion of the bearing shell. New bearing shells will have to be installed eventually and additional problems will be encountered. Such modification alters the engineered fit of the bearing shells in their bores and destroys the desired "crush".

1. When installing precision type bearings, it is important that the bearing shells fit tightly in the rod or crankcase bore. To accomplish this, the diameter of the bearing at right angles to the parting line is slightly larger than the actual diameter of the bore onto which the bearing will be assembled. When the bearing cap is drawn up tight the bearing is compressed, assuring a positive contact between the bearing back and bore. The increased bearing diameter is called "bearing crush." Be certain the bearings are fully seated and the locking tangs (1) on the bearings fit into the recesses.

2. To assemble the bearings with the correct "bearing crush," tighten the clamping bolts alternately and evenly to the specified torque with a torque wrench.

3. Main and connecting rod bearings are designed with the "spread" (width across the open ends) slightly greater than the diameter of the crankcase bore or connecting rod bore into which they are to be assembled. For example, the width across the open ends of the connecting rod bearing not in place is approximately .025 in. more than when the bearing is in position in the rod. This condition causes the bearing to fit snugly in the rod bore and the bearing must be "snapped" or lightly forced into its seat.

NOTE: Rough handling in shipment, storage, or normal use in an engine, may cause the bearing spread to be increased or decreased from the specified width. Bearing spread should therefore be carefully measured and corrected as necessary before installation in an engine. Bearing spread can be safely adjusted as follows if care and judgment are exercised.

a. **EXCESSIVE SPREAD:** If measurement of bearing indicates that dimension A is excessive, place bearing on a wood block and strike the side lightly and squarely with a soft mallet. Recheck measurement and, if necessary, continue until correct width is obtained.

b. **INSUFFICIENT SPREAD:** If measurement of bearing indicates insufficient spread, place bearing on a wood block and strike the back of the bearing

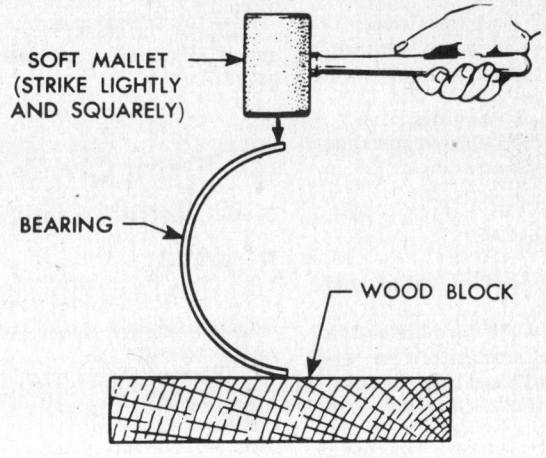

DECREASING SPREAD

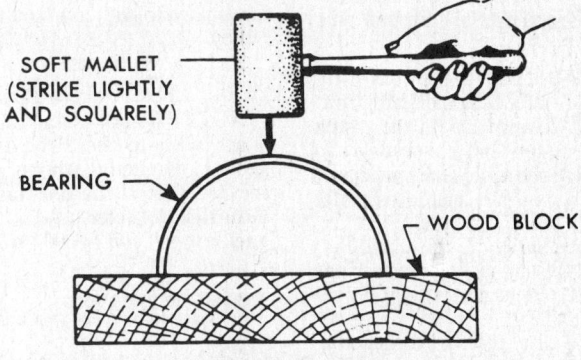

INCREASING SPREAD

Correcting excessive or insufficient bearing spread

lightly and squarely with a soft mallet. Recheck measurement and if necessary continue until correct width is obtained.

4. **BEARING CLEARANCE:** When installing bearings in an engine, the proper clearance between bearing surface should be checked closely. Specified bearing clearance is .002 to .003 in. To get an accurate measurement of this clearance, the Plastigage method, or virgin lead, can be used. The following instructions can be used when measuring with Plastigage:

a. Remove bearing cap and wipe bearing surface and exposed half of crankshaft journal free of oil.
b. Place a piece of Plastigage the full width of bearing insert.
c. Reinstall the bearing cap and tighten the self-locking capscrews to 16 ft.lb. torque.
d. Remove the bearing cap. The flattened plastic material will be found adhering to either the bearing shell or the crankshaft.
e. To determine the bearing clearance, compare the width of the flattened plastic material at its widest point with the graduations on the envelope. The number within the graduation on the envelope indicates the clearance in thousandths of an in.
f. If using virgin lead, carefully remove the flattened lead and measure its thickness with a micrometer.

NOTE: Do not turn crankshaft during the above procedure.

Should the readings not fall within the specified limits, and the torque wrench is known to be accurate in its measurement, remove the bearing from the connecting rod and replace it with a new one. However, with the precision bearings used, no difficulty should be encountered providing the crankshaft and/or connecting rod are in proper condition.

INSTALLATION

NOTE: When reinstalling a piston and connecting rod assembly, install the assembly in the same cylinder bore and in the same position from which it was removed. Connecting rods are stamped with the cylinder number on one side of the rod and on the same side of the bearing cap, No. 1 starting at the front end of the engine. Be sure to install the numbered side of both the rod and bearing cap so both are on the camshaft side of the engine.

1. Generously coat the piston ring compressor and bore with lubricating oil. Install the ring compressor on the piston and insert the piston and connecting rod assembly through the top of the crankcase.
2. Push down on the piston carefully until it is in the crankcase bore.
3. Wipe clean and oil the crankshaft journals and fit the connecting rod bearings as outlined in "Bearing Fitting Procedure."
4. Install all the pistons, connecting rods and bearings in the same manner.
5. Check the connecting rod side clearance by inserting a feeler gauge between the bearing cap and lobe of the crankshaft. The specified side clearance is .005 to .012 in.
6. Install the oil pump screen and tube assembly.
7. Install the crankcase oil pan and new gasket. Fill the crankcase to the level on the gauge.
8. Install the cylinder head and gasket.

Front Cover and Timing Gears

REMOVAL

To service timing gears only, the radiator assembly should be removed for ease of service.

1. Remove the cooling fan assembly.
2. Remove the distributor and distributor drive.
3. Remove the engine governor assembly.
4. Remove the capscrew securing the fan pulley to the crankshaft. Remove the pulley.
5. Remove the capscrews from the front cover and remove the cover.

NOTE: Before removing any gears it is advisable to check the backlash of the gears to determine which, if any, require service. Check the backlash with a dial indicator or feeler gauges. The specified backlash is .003 to .006 in.

6. Remove the idler gear shaft bolt and remove the idler gear.
7. Remove the crankshaft gear with a puller.
8. Using a puller, remove the camshaft gear.

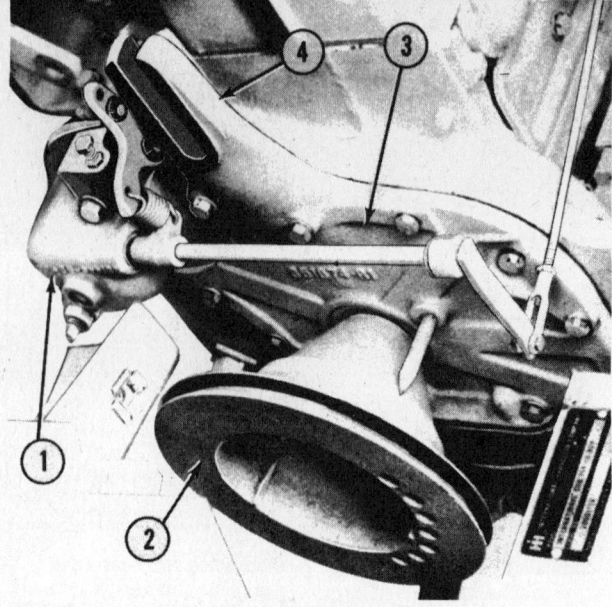

1. Governor assembly
2. Fan drive pulley
3. Front cover
4. Fan mounting location

Timing gear cover

9. Remove the ignition drive seal.

ASSEMBLY AND INSTALLATION

1. Install a new ignition drive oil seal in the crankcase. The seal must be square in its bore and positioned $23/32$ in. from the ignition mounting flange face.
2. Install the key in the camshaft if it was removed. Heat the camshaft gear in boiling water or a piston heater. Install the camshaft gear on the shaft with the timing marks facing out.
3. Install the key in the crankshaft if it was removed. Heat the crankshaft gear and install the gear, being sure the single timing mark of the camshaft gear and the single mark on the crankshaft gear are aligned.
4. Install the idler gear and shaft. Tighten the bolt to 90 ft.lb. torque. The idler gear is correctly timed by lining up the double punch mark on the idler gear with the double punch mark on the crankshaft gear.

NOTE: Before installing the crankcase front cover, mark the top surfaces of the two teeth on each side of the single punch mark on the idler gear with chalk.

5. Install the crankcase front cover with a new oil seal and gasket. Do not tighten the capscrews at this time.
6. Install the fan drive pulley on the crankshaft. Install the capscrew and tighten to 80 ft.lb. torque.
7. Tighten the crankcase front cover capscrew to 20 ft.lb. torque.

NOTE: Before installing the governor assembly, mark the front surface of the ignition drive gear having a single punch mark with chalk.

International Harvester

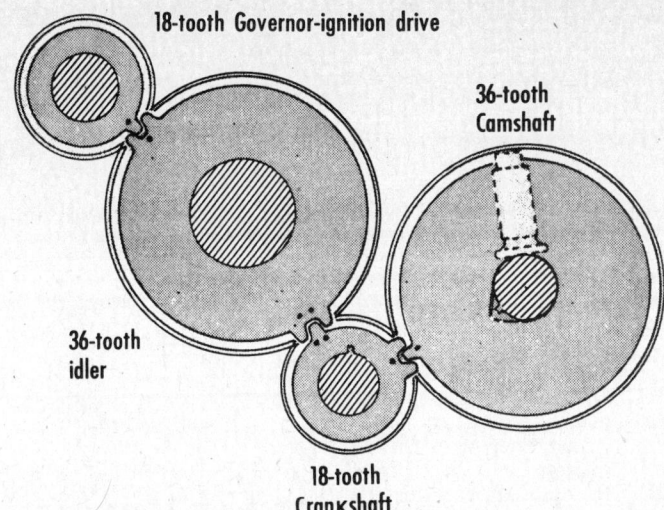

Timing gear alignment

8. Install the governor assembly with a new housing gasket. Time the ignition drive gear as follows:

With the engine on top dead center of number 1 firing stroke, mesh the marked tooth of the ignition drive gear with the two chalk marked teeth on the idler gear.

9. Position the ignition unit distributor arm and driveshaft lug (A) for firing number one cylinder. Install the ignition unit on the engine, meshing distributor lug and governor drive slots.

NOTE: Remove spark plug cables 2, 3 and 4 and ground them to prevent any change of accidentally starting the engine.

10. Place the number one spark plug cable in a position so that the spark to ground will be audible when hand cranking the engine. Then advance or retard the ignition distributor until spark occurs as the O mark on the fan drive pulley aligns with pointer while hand cranking engine.

NOTE: Final ignition timing must be made with a timing light. The specified ignition timing is 16 degrees before TDC at 2420 engine rpm.

11. Install the cooling fan assembly and the radiator assembly.

Camshaft

REMOVAL

Removal of the camshaft only, requires the removal of the engine from the tractor and removal of head, valves, oil pan, flywheel, oil pump, and crankcase front cover from the engine.

1. Remove the valve tappet cover, and remove the valve assemblies. Be sure to identify each assembly for proper reassembly.
2. Remove the oil pump body and gears. Remove the key in the rear of the camshaft.
3. Remove the crankcase front cover.
4. Remove the two capscrews in the camshaft retainer plate through openings in the cam drive gear.
5. Turn the crankcase upside down so the tappets will fall away from the camshaft to provide clearance for removal.
6. Withdraw the camshaft from the front carefully so the crankcase bores are not damaged by nicks from the edges of the cams.
7. If necessary to remove the camshaft gear, press the gear off.
8. Remove the thrust plate and key if the camshaft gear was removed.
9. Lift the valve tappets out of the crankcase. Be sure to identify the tappets so they can be installed in their original bores.

INSTALLATION

1. Install the valve tappets in their original bores.
2. Place the camshaft thrust plate on the shaft and install the key in the keyway. Heat the camshaft gear in boiling water and install the gear (with the timing mark facing out).
3. Check the end clearance with a feeler gauge between the camshaft front journal and the thrust plate. Be sure the drive gear is in place against the shoulder on camshaft. The specified end clearance is .003 to .012 in. If the end play is excessive, replace the thrust plate with a new one.
4. Coat the camshaft with engine oil and install the camshaft in the crankcase. Be sure the camshaft gear is correctly indexed with the timing mark on the crankshaft gear.
5. Secure the thrust plate to the crankcase and tighten the capscrews to 20 ft.lb. torque. Access to each of the two screws is through the holes in the camshaft gear.
6. Install the front cover and fan drive pulley.
7. Install the key in the rear of the camshaft, and install the oil pump gears and body. Install the flywheel.
8. Install the valve assemblies and head.
9. Install the oil pan.
10. Install the engine in the tractor.
11. Start the engine and bring up to operating temperature.
   a. Inspect for oil leaks and check for correct engine oil pressure.
   b. Check and adjust the ignition timing.
   c. Check and adjust the valve clearance if necessary.

Crankshaft

REMOVAL

To completely service the crankshaft and bearings, the crankcase oil, cooling and hydraulic systems must be drained and the engine removed from the tractor.

1. Remove the fan drive pulley.
2. Remove the front cover.
3. Remove the flywheel and the rear oil seal retainer.
4. Remove the crankcase oil pan, and remove the oil pump screen and tube.
5. Remove the crankshaft bearing cap

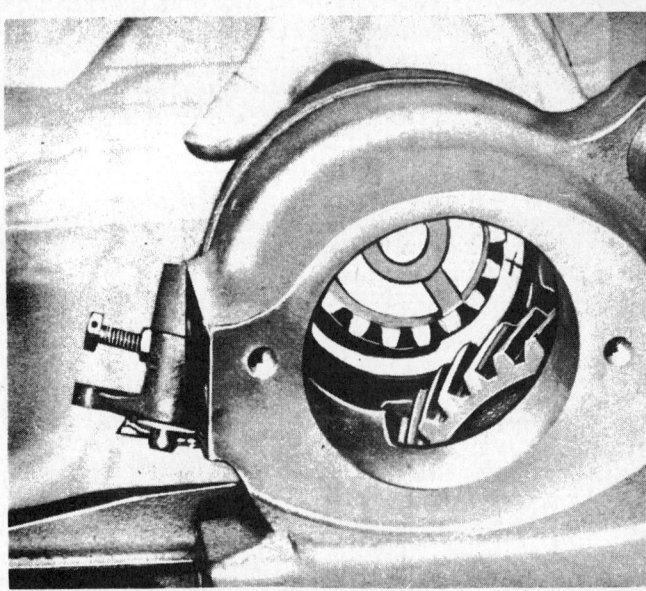

Marking idler gear

International Harvester

bolts. Tap the caps lightly with a lead hammer, if necessary, to dislodge them and remove squarely from position.

6. Remove the lower bearing from each cap. If they are to be reassembled, be certain that they are identified as to their original positions. Wrap them in clean cloths and store until reassembly.

NOTE: If the crankshaft is to be removed disregard step 7.

7. Remove the upper bearing halves from between the crankshaft and the crankcase with a thin piece of flexible soft sheet metal. Push against the end of the bearing without the positioning nib, while turning the crankshaft in the direction of rotation. The bearing will slide from position easily.

8. Remove the connecting rod bearing cap nuts and bolts and remove the caps. Push the piston and rod assemblies to the top of their travel.

9. Lift the crankshaft out of the crankcase.

INSPECTION AND REPAIR

1. Clean all parts with cleaning solvent and dry with compressed air.
2. Inspect the bearings for wear and evidence of uneven bearing support. Examine the bearing caps and supporting surfaces of the crankcase for high spots and burrs.
3. Inspect the crankshaft journals for scoring and measure the diameter of each journal with a micrometer. Specified journal diameter is 1.623 to 1.624 in. Measure each journal at two points, one at right angles to the other, in order to detect any out-of-roundness.

NOTE: Move the micrometer over the entire width of the journal.

4. Check the connecting rod journals for out-of-round condition. Use a micrometer and take measurements at least three places around the journals. The specified rod journal diameter is 1.498 to 1.499 in.
5. Inspect the crankshaft gear teeth for excessive wear and chipping. If necessary to replace it, pull the gear with a puller.
6. Inspect the crankcase for sludge and deposits and thoroughly clean it.
7. Replace all seals and gaskets with new ones.

INSTALLATION

1. Wipe all surfaces of the crankshaft bearing bores of crankcase and bearing caps free of oil, and place bearing halves in the bore of the crankcase and bearing caps. Be certain the bearings are fully seated, oil holes are in alignment, and locking tangs on the bearings fit into the recesses.
2. Apply a film of engine oil on the bearing surfaces and place the crankshaft into position.

NOTE: When installing the crankshaft, be certain to correctly index the timing marks on the crankshaft gear with the idler gear and camshaft gear.

3. Install the bearing caps over the crankshaft journals, being certain to install the caps in their correct positions and with the numbered side of the caps to the camshaft side of the engine.

4. Check the main bearing clearances as follows:
a. Remove the bearing cap and wipe the bearing surface and exposed half of the crankshaft journal free of oil.
b. Place a piece of Plastigage or virgin lead, the full width of the bearing insert, on the crankshaft journal.
c. Reinstall the bearing cap and tighten the capscrews to 55 ft.lb. torque.
d. Remove the bearing cap. The flattened section of the virgin lead or Plastigage represents the clearance present between the bearing surface and the crankshaft journal. Measure the thickness with a micrometer or match the flattened Plastigage at several points (on either the bearing insert or the crankshaft), with the corresponding graduation on the Plastigage envelope, which indicates the clearance in thousandths of an in. Running clearance must be .002 to .003 in.

NOTE: Do not turn the crankshaft during the bearing clearance check.

e. Should the readings not fall within the specified limits, and the torque wrench is known to be accurate, remove the bearing and replace it with a new one. However, with the precision bearings used, no difficulty should be encountered providing the crankshaft and/or crankcase and caps are in good order.

5. Install the bearing caps to their original position and tighten the capscrews to 55 ft.lb. torque.

NOTE: When installing center main bearing cap, hold crankshaft against the rear thrust face of the upper half of the bearing. Tighten center cap bolts lightly and tap cap toward the rear before final tightening of cap bolts. This lines up the upper and lower thrust surfaces of the bearing halves and prevents binding the shaft on the thrust surfaces.

6. Check the crankshaft thrust bearing side clearance with a feeler guage at the front side of the center bearing on both upper and lower thrust faces. Specified side clearance is .004 to .008 in.

While making this check, be sure the crankshaft is held against the rear thrust face of the bearing to show total clearance at front side.

7. Install the connecting rod bearings and caps.
8. Install the hydraulic oil pump (if so equipped).
9. Install new front and rear crankshaft oil seals in their retainers.
10. Install the rear oil seal retainer.
11. Install the flywheel and secure with the capscrews. Tighten the capscrews to 45 ft.lb. torque.
12. Install the oil pan.
13. Install the front cover and the fan drive pulley.
14. Install the engine in the tractor.

LUBRICATION SYSTEM

Oil Pump

REMOVAL AND DISASSEMBLY

1. Remove filter element and clean filter case. If filter case is exceptionally sludgy, install bolt without cover and flush thoroughly.
2. Remove hex head regulator valve retaining plug and remove the spring and valve.
3. Remove the crankcase oil pan. Remove the oil intake pipe and screen.
4. Remove the engine.
5. Remove the flywheel.
6. Remove the oil pump body and pump gears.

ASSEMBLY AND INSTALLATION

1. Install new filter element along with cover and drain plug.
2. Install regulator valve, spring and retainer, using new retainer gasket.
3. Install oil intake pipe and screen.
4. Install crankcase pan, using new gasket.
5. Install oil pump gears.

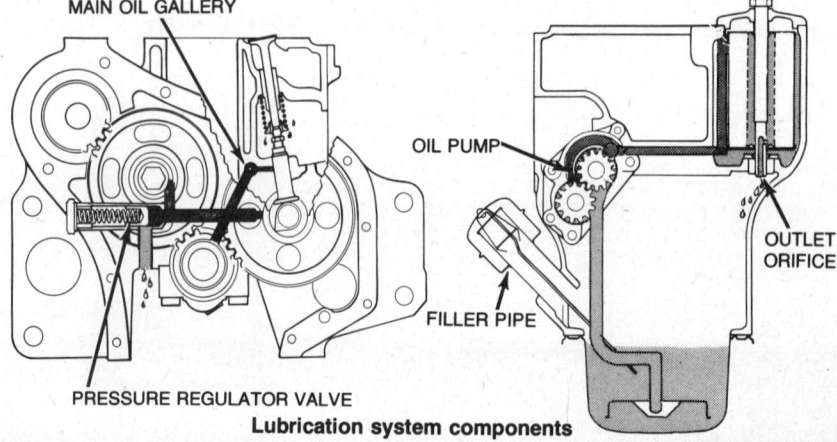

Lubrication system components

International Harvester

6. Install pump body, using new gasket.
7. Install the flywheel and secure with the four capscrews. Tighten capscrews to 45 ft.lb. torque.
8. Install the engine in the tractor.
9. Fill crankcase with proper amount of specified oil.
10. Prime oil pump, start engine and check oil pressure.
11. Recheck crankcase oil level.

FUEL SYSTEM

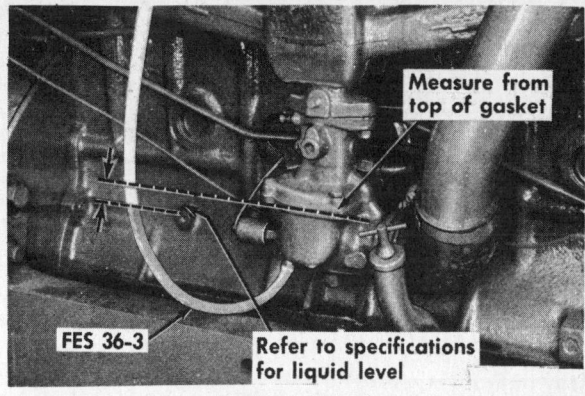
Fuel level check

FUEL LEVEL CHECK

1. Close the fuel shut-off at the fuel tank.
2. Remove the drain plug at the bottom of the carburetor. Attach the tool FES 36-3 as shown.
3. Open the fuel shut-off. Fuel will flow into the tube and seek the same level as the liquid level in the carburetor.
4. Measure the distance between fuel level in the tube to the top of the fuel bowl gasket. This will be the liquid level in the carburetor.

GOVERNOR ADJUSTMENT

1. With the engine positioned at top dead center of number one cylinder firing stroke, locate the single punch mark between teeth of idler timing gear. Use chalk to mark top surface of two teeth on each side of punch mark.
2. Chalk the rear end of the punch marked tooth on the governor drive gear.
3. Install the governor assembly, meshing the marked gear teeth.
4. Position the ignition unit distributor rotor arm and driveshaft lugs for firing number one cylinder. Install ignition unit on engine, meshing lugs and drive slots.

NOTE: Remove spark plug cables 2, 3 and 4 and ground them to prevent any chance of accidentally starting the engine.

5. Remove the number one spark plug cable from the number one spark plug and position the end of the cable so a spark discharge to "ground" will be audible while hand cranking the engine.
6. Advance or retard ignition distributor, until spark occurs as the O mark on fan drive pulley aligns with pointer while hand cranking engine.

SYNCHRONIZING THE GOVERNOR

1. Synchronizing the governor-to-carburetor throttle movement.
Because of possible change in center-to-center distance between governor and

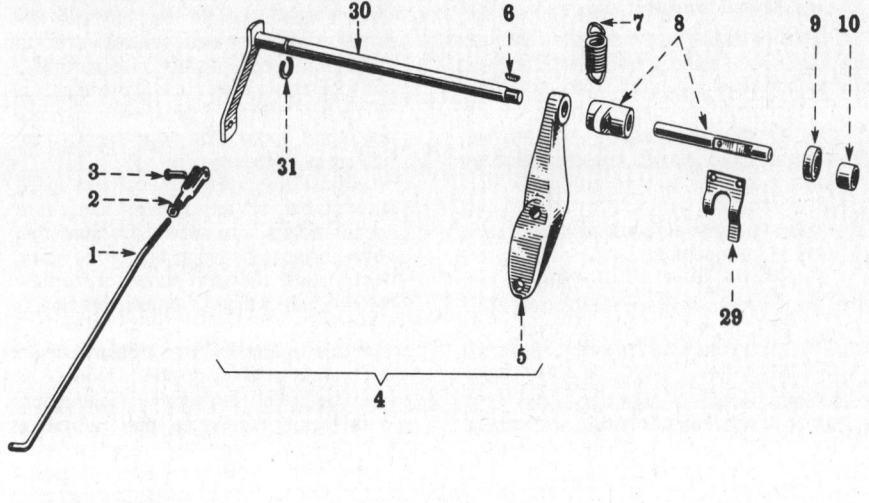
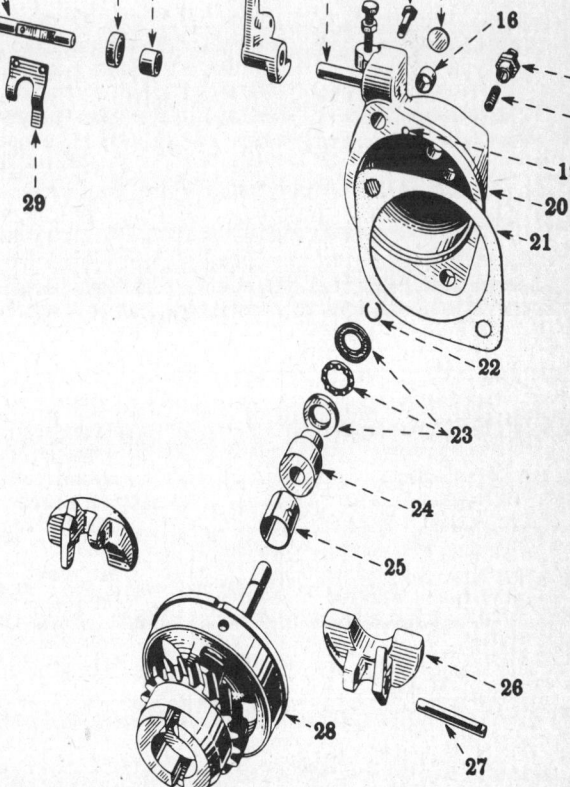

1. Governor connecting rod
2. Adjusting rod end clevis
3. Rod end pin
4. Rockshaft and bracket extension assembly
5. Rockshaft extension bracket
6. Woodruff key
7. Governor spring
8. Governor and spring rockshaft assembly
9. Rockshaft oil seal
10. Rockshaft bearing
11. Spring throttle lever
12. Throttle lever shaft
13. Speed change lever stop
14. Screw
15. Expansion plug
16. Governor shaft bushing
17. Bumper spring body
18. Bumper spring
19. Governor base dowel pin
20. Governor housing assembly
21. Governor housing gasket
22. Governor sleeve stop ring
23. Governor thrust ball bearing
24. Governor thrust bearing
25. Governor base bushing
26. Governor weight
27. Governor weight pin
28. Governor with carrier and pin shaft
29. Governor tension fork
30. Rockshaft extension assembly
31. Rockshaft extension stop ring

Governor components

International Harvester

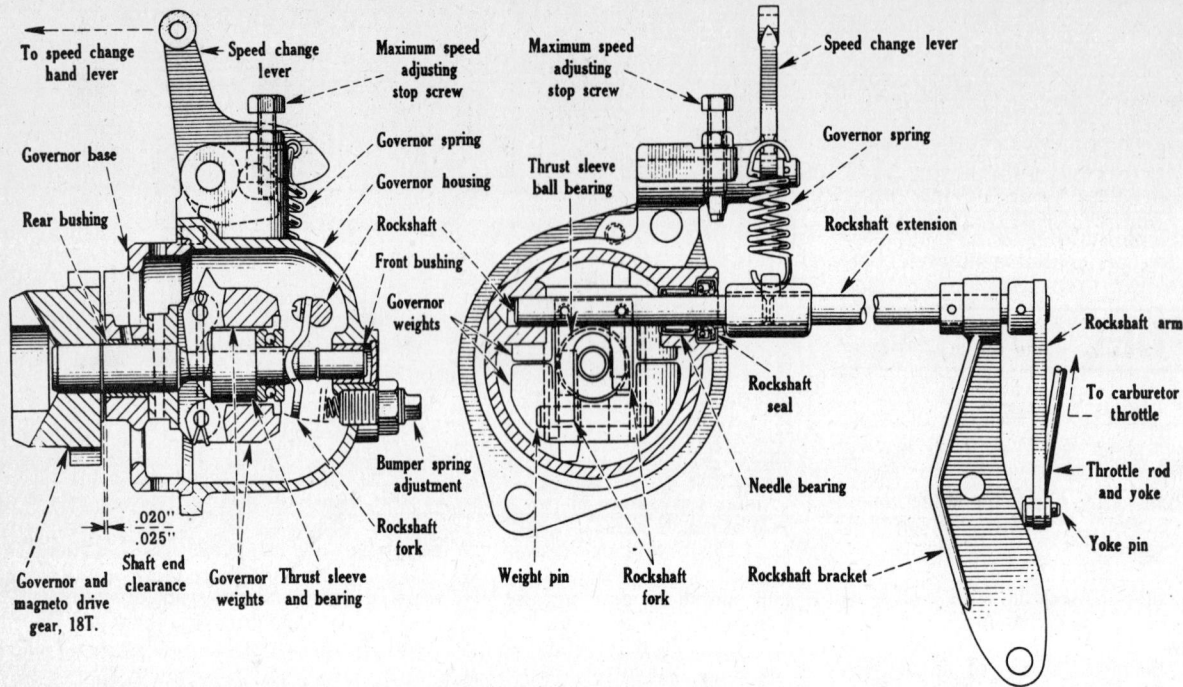

carburetor, due to removal and replacement of manifold, carburetor or governor assemblies, the linkage between the governor and carburetor must be adjusted to establish the throttle position in relation to governor weight position. This adjustment insures the full power response of a wide open throttle when the governor weights are collapsed by reduction in rpm by application of heavy load. This governor-to-carburetor linkage must be free from binding throughout its range of movements. Adjustment procedure for all engines follows:

  a. With engine stopped, advance the operator's engine speed control lever to about half speed position; sufficient to create tension on the governor spring.
  b. Disconnect governor-to-carburetor control rod (either end). Hold carburetor throttle against its stop in wide-open position and adjust length of governor-to-carburetor control rod so that it may be reconnected freely without moving throttle lever or governor lever.
  c. Shorten control rod one turn from the above condition, to compensate for wear, and reconnect.
  d. After tightening the control rod clevis locknut, check to be sure that both ends of the control rod are in the same plane, to eliminate possibility of binding on levers.
  e. Move operator's engine speed control lever a few times between half speed and low speed position, checking the governor-to-carburetor control rod in all positions for interference or binding.

2. Low idle speed adjustment.

Smooth low speed engine operation depends upon careful adjustment of carburetor idle air-fuel mixture at the specified engine low idle speed. Good governor performance also is dependent on this smooth engine operation and free throttle shaft movement near closed throttle positions.

Governor speed adjustments

Any tendency of the carburetor throttle to stick or bind in its low idle (closed) position will cause the governor to surge excessively. The governor is equipped with an adjustable bumper spring to counteract the effect of manifold vacuum on the closed position of the throttle. Causes for binding or sticking of the throttle shaft are misalignment due to wear or interference due to improper assembly. Excessive tension adjustment of bumper spring, in an attempt to overcome these ills, will prevent the throttle from closing against its stop, resulting in greater than specified low idle speed.

  a. Start engine and allow it to reach operating temperature.
  b. Place operator's speed change lever in the extreme low speed position. See that operator's speed change lever linkage will allow the throttle to close against its stop screw. Adjust speed change linkage if necessary. See also that governor bumper spring adjustment is not interfering with closing of throttle.
  c. Adjust carburetor throttle stop screw to secure the specified low idle speed and set idle fuel mixture screw for smoothest engine operation.
  d. Advance operators speed change lever for a few seconds and again idle the engine, rechecking adjustments for specified low idle speed and smoothest operation.

3. Adjusting governed fast idle speed.

To protect the engine from excessive speed, and also to provide sufficient speed to maintain the engine's rated load, the governed fast idle speed adjustment must be properly made. Be sure the service tachometer used is accurate. Do not expect the tractor tachometer to be sufficiently accurate for this operation. Adjustment procedure for all engines follows:

  a. Before adjustment is attempted, the engine must be brought up to operating temperature. Engine lubricant viscosity should be correct for the season of use and should be near operating temperature.
  b. With engine running and accurate service tachometer in use, advance operator's engine speed control lever to maximum speed position. Be sure also that operator's speed change linkage is being held firmly against the governor maximum speed stop adjustment; reset linkage is necessary.
  c. Adjust the governor maximum speed stop screw or adjustment to secure specified fast idle speed. Be sure that governor speed change linkage is being held against the stop screw in its new position when the tachometer reading is taken.
  d. Place operator's speed change lever in maximum speed position. Notice the fast idle speed on service tachometer. With thumb and finger, pull carburetor throttle lever toward open position, sufficient to gain 50 rpm fast idle speed. Release throttle lever instantly; the governor will react by closing the throttle and opening again, seeking its balance. Under this condition two surges of the governor are considered normal. Excessive surging would indicate binding in carburetor throttle assembly or governor rockshaft and linkage assembly as outlined previously under carburetor and governor headings. This may be corrected by adjusting the bumper spring.
  e. Bumper Spring Adjustment: The adjuster may be turned in one-half turn at a time, just sufficient to reduce surging to normal. Test, as in operation (d) above, after each slight adjustment. If screwed in too far, the bumper spring will prevent the throttle from closing to low idle stop screw. Where such extreme setting of

International Harvester

bumper spring is found necessary, it would indicate excessive friction or sticking is occurring in throttle assembly or governor rockshaft assembly. This should be corrected and the bumper spring readjusted. After the bumper spring has been adjusted properly, lock it in place with the jam nut. Where use of the bumper spring is not required to control surging, screw in unitl it just touches at low idle speed and then backed out ¼ turn and locked.

NOTE: Adjustment of the maximum speed stop, to allow increased tension to be placed on the governor spring by the operator's engine speed control lever, will result in increased engine speed. Adjustment to reduce tension which can be placed on the governor spring, will result in reduced engine speed.

THROTTLE ADJUSTMENT

1. Loosen the screw of the throttle control swivel pin (1) on the governor spring throttle lever (2).
2. Loosen the throttle control clamp (3) and position the throttle cable (4) in the lower part of the clamp. Leave the clamp loose.
3. Pull the cable wire through the control swivel pin about ½ in. and tighten screw on control swivel pin.
4. Move throttle control lever to maximum speed position.
5. Pull throttle cable back to bring governor spring throttle lever against high idle stop screw. Tighten throttle control clamp.

1. Governor control swivel pin
2. Governor spring throttle lever
3. Throttle control clamp
4. Throttle cable

Throttle adjustment

Models 482 and 1100

FRONT WHEELS AND AXLE

Front Wheels

REMOVAL

1. Lock the brake and block the rear wheels. Jack up the front axle.
2. Remove the capscrew and flat washer from the outer end of the front spindle.
3. Slide the wheel and bearings from the spindle.

NOTE: The bearings are a press fit in the wheel and a slip fit on the spindle.

4. Wheel bearings can be driven from the wheel hub with a hammer and long drift punch. Drive from the inside toward the outside.

INSTALLATION

1. If the bearings were removed, lubricate and press in new ones. Be sure force is directed to the outer race only.
2. Slide the wheel and bearing assembly over the spindle and secure with capscrew and flat washer.
3. If excessive end play exists (maximum) $1/32$ in. place a sufficient thickness of shim washers 3-4 in. ID over the outer end of the spindle and between the retaining washer and wheel bearing to take up excess end play.

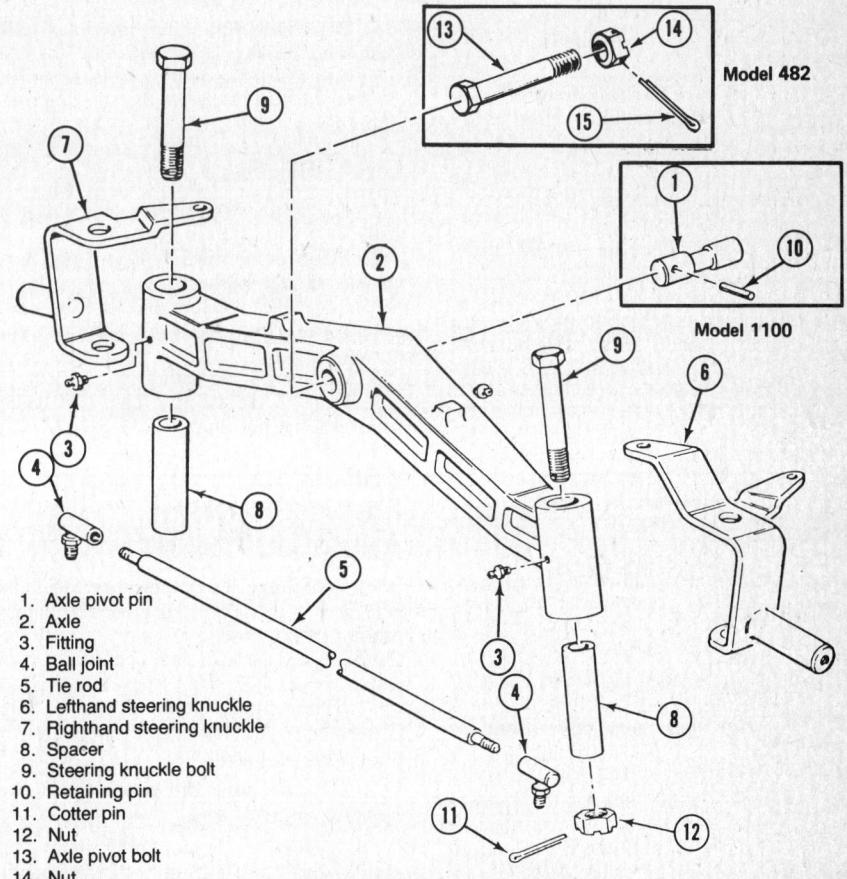

1. Axle pivot pin
2. Axle
3. Fitting
4. Ball joint
5. Tie rod
6. Lefthand steering knuckle
7. Righthand steering knuckle
8. Spacer
9. Steering knuckle bolt
10. Retaining pin
11. Cotter pin
12. Nut
13. Axle pivot bolt
14. Nut
15. Cotter pin

Front axle components

563

International Harvester

Front Axle

REMOVAL AND DISASSEMBLY

1. Lock the brake, raise the front of the tractor and support it with suitable stands.
2. Remove the front wheels.
3. Disconnect the drag link from the left steering knuckle.
4. Remove the tie rod.
5. Remove the steering knuckle bolt, steering knuckle and spacer from both sides.

Model 482:

6. Remove the cotter pin and nut from the axle pivot bolt.
7. Remove the axle pivot bolt and the front axle.

Model 1100:

6. Drive out the retaining pin from the axle pivot pin.
7. Remove the axle pivot pin and remove the front axle.

ASSEMBLY AND INSTALLATION

1. Coat the axle pivot pin and its bore in the axle with IH 251 HEP or its equivalent.
2. Position the axle in the support bracket and install the pivot pin. Install the retaining pin through the bracket and pivot pin.
3. Install the steering knuckles. Tighten the nut to 80 ft.lb. Install the cotter key. If the cotter key cannot be installed, tighten the nut only enough to line up the slot with the hole.

NOTE: Be sure to check for free rotation of the steering knuckles after securing the nut.

4. Install the tie rod and connect the drag link to the left steering knuckle.
5. Install the wheels.
6. Check and adjust toe-in. Adjust the tie rod to provide approximately ⅛ in. toe-in.
7. Adjust the drag link to proper length to place front wheels in the straight ahead position when the steering assembly is centered.

STEERING

Steering Gear

REMOVAL

Model 482:
1. Remove the steering wheel insert.

Model 1100:
2. Peel back the metal cover from the wheel hub.

Both Models:
2. Remove the nut securing the wheel to the column.
4. Place a ⅜ in. bolt into the column to protect the column end from damage.
5. Remove the steering wheel using a puller.

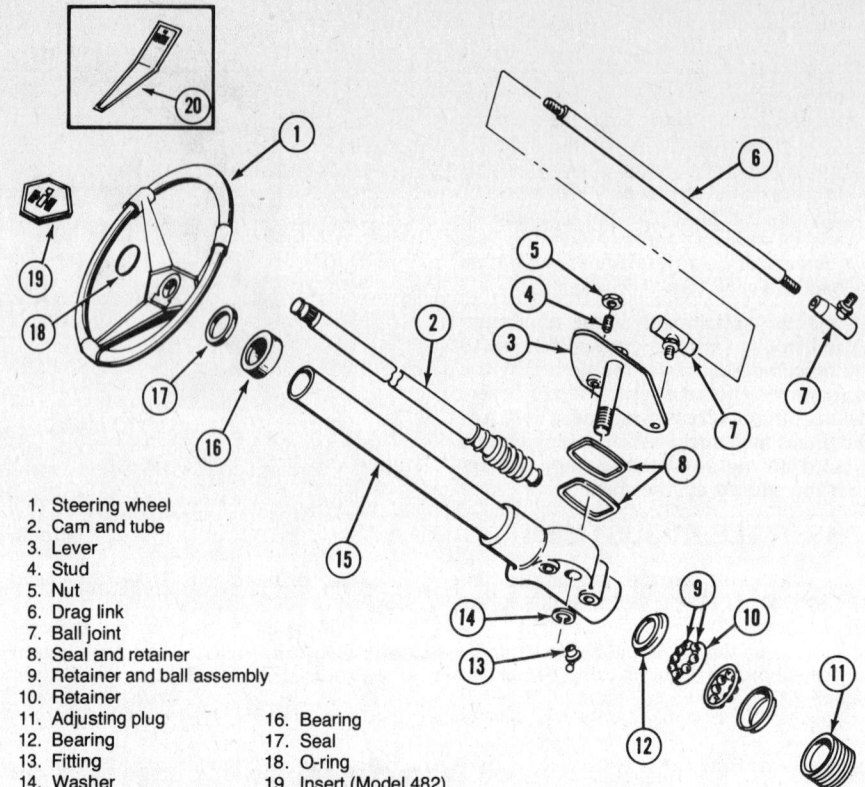

1. Steering wheel
2. Cam and tube
3. Lever
4. Stud
5. Nut
6. Drag link
7. Ball joint
8. Seal and retainer
9. Retainer and ball assembly
10. Retainer
11. Adjusting plug
12. Bearing
13. Fitting
14. Washer
15. Tube assembly
16. Bearing
17. Seal
18. O-ring
19. Insert (Model 482)
20. Insert (Model 1100)

Steering system components

6. Disconnect the drag link rear ball joint from the steering lever.
7. Remove the mounting bolts securing the steering unit. Remove the steering unit and column.

DISASSEMBLY

1. Secure the steering lever and bolt in a vise.
2. Remove the lever bolt jam nut, adjusting nut and washer.
3. Slide the column and housing assembly away from the lever, bolt and cam follower.
4. Remove the adjusting plug.
5. Remove the steering cam and bearings from the housing.

ASSEMBLY, INSTALLATION AND ADJUSTMENT

1. Thoroughly coat the cam ends, balls and races with IH 251 HEP (or equivalent lithium base grease).
2. Install the balls and races on the cam ends.
3. Thoroughly coat the cam with chassis lubricant then install into the housing and column assembly.

NOTE: Be sure the races enter the housing squarely and are not cocked.

4. Install the adjusting plug. Screw the plug inward until end play of the cam is removed but turns freely. Insert the cotter pin in the nearest hole.
5. Fill the housing with IH 251 HEP (or equivalent lithium base grease).
6. Loosen the cam follower locknut, then back out the cam follower two turns.
7. Install the seal, retainer and lever-bolt assembly to the housing.
8. Install the washer and adjusting nut. Tighten the adjusting nut sufficiently to provide good seal compression. Refer to illustration for adjustment dimensions. Secure with the jam nut. Tighten jam nut to 40 ft.lb. Lubricate at the fitting in the housing slowly until lubricant begins to seep out.
9. Center the steering cam by rotating the steering shaft half-way between full right and full left turn.
10. Adjust the cam follower inward to eliminate backlash, then tighten locknut to 40 ft.lb. Turn steering shaft full right and left to check for binding.
11. Install the steering assembly in the tractor chassis. Secure with two capscrews through the frame crossmember.
12. Connect the drag link.
13. Install the steering wheel and secure with nut.
14. Adjust the tie rod to provide approximately ⅛ in. toe-in as follows:
   a. Mark an "X" in the center of the front of the tires. Measure the distance across.
   b. Roll the tractor until the "X"s are in the center of the tires at the rear. Measure the distance across. The front measurement should be ⅛ in. less than the rear measurement (toe-in).
   c. If the measurement is off, discon-

International Harvester

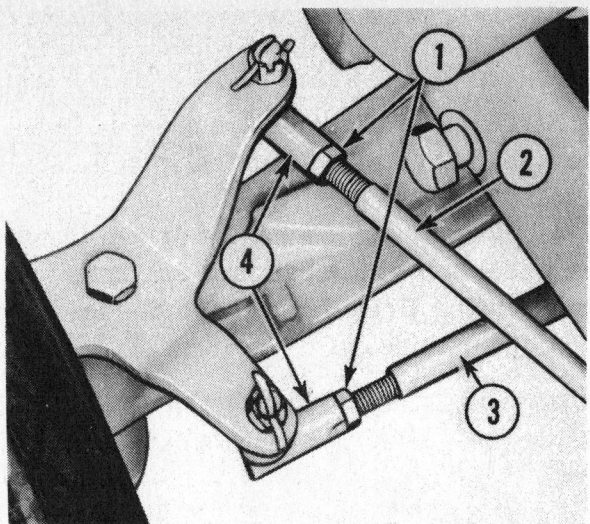

1. Locknuts
2. Drag link
3. Tie rod
4. Ball joints

Steering linkage

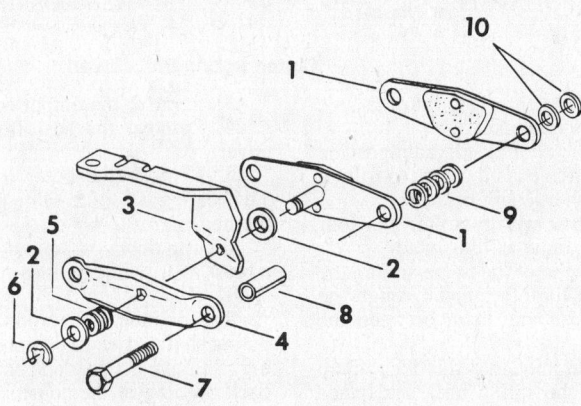

1. Brake plate
2. Thrust washer
3. Actuating cam
4. Actuating plate
5. Spring
6. Retaining ring
7. Bolt
8. Bushing
9. Spring
10. Spacer washer

Brake components

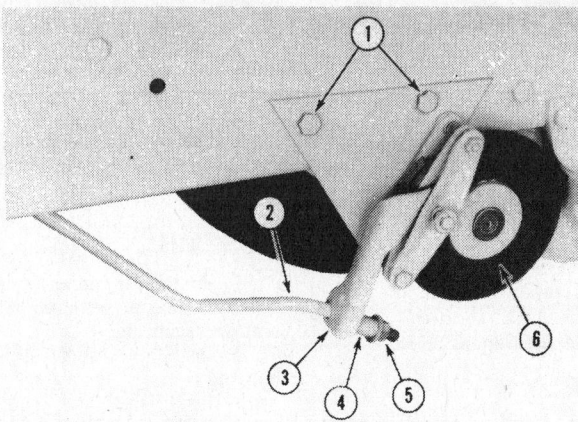

1. Mounting bolts
2. Control rod
3. Actuating cam
4. Sleeve spacer
5. Adjusting nut
6. Brake disc

Brake linkage

nect the ball joint from the tie rod on the lefthand side steering knuckle. Loosen the locknut and turn the ball joint in or out until specified toe-in is obtained.

BRAKES

REMOVAL AND DISASSEMBLY

1. Disconnect the brake control rod from the clutch shaft.
2. Remove the mounting bolts securing the brake assembly to the frame and remove the brake assembly.
3. Remove the retaining ring from the brake plate and disassemble the brake.

ASSEMBLY AND INSTALLATION

1. Reassemble and install in the reverse order of removal and disassembly.
2. Adjust the brake properly as follows:

ADJUSTMENT

1. The clutch-brake pedal should be in the raised (clutch engaged) position.
2. Move the actuating cam forward by hand until the brake pads contact the disc.
3. Turn the adjusting nut to obtain ¼ in. clearance between the spacer sleeve and the actuating cam.
4. Put the tractor in gear and push it while slowly depressing the clutch-brake pedal. The brake should start to apply as the clutch disengages. There should be no neutral or free zone between the braking and clutch action. Adjust the brake as necessary.

DRIVE BELT AND CLUTCH SPRING

Drive Belt

REMOVAL

1. Disconnect the battery.
2. Remove the drawbar assembly.
3. Remove the center frame cover.
4. Depress the clutch-brake pedal and lock into its lowest position.
5. Loosen the bolts securing the two drive belt guides.
6. Remove the idler pulley.
7. Remove the two mounting bolts securing the right angle drive to the cross support.
8. Rotate the right angle drive to bring the output pulley down. Remove the drive belt.

INSTALLATION

1. Slip the drive belt into position on the output and input pulleys.

565

International Harvester

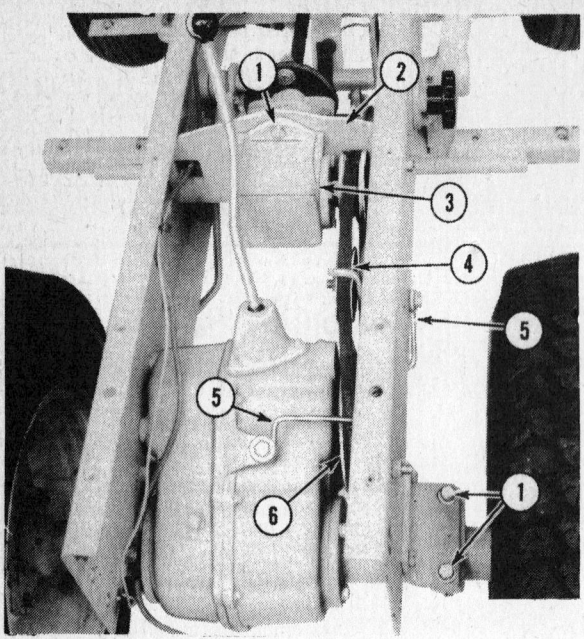

1. Mounting bolts
2. Cross support
3. Right angle drive
4. Idler pulley
5. Drive belt guides
6. Input pulley

Drive belt replacement

1. Clutch spring
2. ⅜ in. x 3½ in. bolt
3. Washer
4. Twine
5. Frame
6. Washer-nut (not shown)
7. Washer-nut-washer

Clutch spring installation

2. Rotate the right angle drive back into position and secure with the mounting bolts.
3. Install the idler pulley.
4. Release the clutch-brake pedal and adjust and secure the drive belt guides. There should be a gap of ⅛-3/16 in. between the belt and the guides.
5. Install the center frame cover.
6. Install the drawbar assembly and reconnect the battery.

Clutch Spring and Linkage

REMOVAL

NOTE: First perform steps 1 through 8 under DRIVE BELT REMOVAL. After performing these steps, proceed as follows:

1. Release the clutch-brake pedal to the raised position.
2. Remove the battery.
3. Insert a screwdriver through the hole in the frame to support the spring load.
4. With the spring load under control, remove the mounting bolt and slowly ease the spring forward.
5. Work the spring free from the belt tension bracket and remove the spring.
6. Disconnect the control rod from the clutch shaft.
7. Remove the mounting bolt securing the belt tension bracket to the frame and remove the bracket.

INSTALLATION

1. Install the belt tension bracket with the control rod to the frame.
2. Connect the control rod to the clutch shaft.

3. Insert the clutch spring into the hole in the belt tension bracket.
4. Using a piece of plastic baling twine 50 in. long, tie the ends together to form a closed loop with a length of 25 in.
5. Install a bolt with two nuts and four washers to the frame as illustrated.
6. Loop one end of the twine around the bolt head. Run the other end through the spring, then over and down between the washers.
7. Insert your foot into this loop. Step down to bring the spring back into place and secure with the mounting bolt.
8. Continue reassembly following steps 1 through 6, Drive Belt Installation.

RIGHT ANGLE DRIVE

REMOVAL

1. Disconnect the battery.
2. Remove the center frame cover.
3. Depress the clutch-brake pedal and lock into its lowest position.
4. Drive the roll pin from the rear half of the flexible coupler.
5. Remove the two mounting bolts securing the right angle drive to the cross support.
6. Remove the right angle drive.

DISASSEMBLY

1. Remove the rear cover and clean the lubricant from the inside of the housing.
2. Remove the snapring securing the output pulley. Back out the setscrew and remove the pulley and key.
3. Remove the output cover.
4. Remove the seal from the output cover.
5. Remove the outer output bearing. The bearing is a loose fit on the shaft and in the housing.
6. Tap down on the output gear, using a brass drift, while pulling up on the output shaft.
7. Continue this procedure until the output shaft is free of the inner output bearing. Remove the gear and shaft through the back opening of the housing.
8. Remove the inner output bearing.
9. Remove the input seal as follows:
   a. Punch two small holes in the seal at the outer edge using an awl. Do not drill the holes as the ball bearing under the seal could be damaged.
   b. Insert sheet metal screws into the holes and pull the seal out. The seal can be forced out if the screws are long enough by turning the screws in.
10. Remove the snapring (at the outer input bearing) securing the input shaft into the housing.
11. Remove the input shaft with the inner input bearing and input gear. Remove the snapring to remove the gear and bearing.

ASSEMBLY

1. Position the inner input bearing and the input gear onto the input shaft. Secure with the snapring.
2. Install the input assembly into the housing.
3. Install the outer input bearing and secure to the input shaft with the snapring.
4. Install a new input seal being sure to seal it fully in the housing.
5. Install the inner output bearing.
6. Insert the output shaft in to the out-

International Harvester

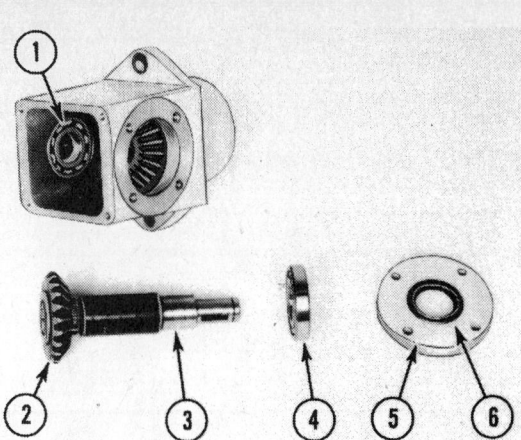

1. Inner output bearing
2. Output gear
3. Output shaft
4. Outer output bearing
5. Output cover
6. Seal

Output shaft components, right angle drive

1. Axle housing
2. Scribe mark
3. O-ring
4. Bearing retainer
5. Oil seal

Axle shaft removal

put gear until it is flush with the bottom face of the gear.

7. Position the shaft and gear into the housing.
8. Using a brass hammer, tap the output shaft to its seat in the gear.
9. Install the outer output bearing.
10. Install a new output seal .045 in. below flush into the output cover.
11. Install the output cover, torquing the capscrews to 7 ft.lb.
12. Install the output pulley with key and secure with the setscrew and snapring.
13. Fill the housing with the remainder of the Molykote lithium grease (four ounces required).
14. Install the back cover with gasket and secure with the four capscrews.
15. Installtion is in the reverse order of removal.

TRANSAXLE

REMOVAL

1. Disconnect and remove the battery.
2. Remove the four capscrews in the battery pit which secure the fender assembly to the rear frame.
3. Remove the drawbar assembly.
4. Remove the mounting screws from the foot platforms and remove the fender assembly.
5. Remove the center frame cover.
6. Block the front axle on both sides to prevent accidental tipping.
7. Depress the clutch-brake pedal to its lowest position.
8. Loosen the bolts securing the two drive belt guides.
9. Work the drive belt off of the transaxle input pulley.
10. Support the rear of the tractor frame with safety stands.
11. Remove the mounting bolts securing the transaxle assembly to the frame. Roll the assembly out of the frame.

DISASSEMBLY

1. Remove the rear wheels.
2. Place the assembly on a work bench and drain the lubricant.
3. Remove the snapring securing the wheel hub to the axle and remove the hub with Woodruff key.
4. Remove the brake disc from the brake shaft.
5. Remove the input pulley being sure to first loosen the Allen head screw.
6. Position the shift lever into neutral position. Remove the shift lever housing and gasket.

NOTE: If necessary, disassemble the shift lever assembly by removing the snapring. Be sure to match mark the shift lever to the housing before disassembly so the lever is not reassembled 180° out of line.

7. Scribe mark the axle housings to the case and cover for easy reassembly.
8. Remove the axle housings.
9. Remove the O-rings and oil seals with retainers. During reassembly install new oil seals.
10. If necessary, remove the axle support bearing by driving the bearing out of the housing from the inside. During reassembly install new bearings.

NOTE: To support the transaxle properly in a vise, make the support locally from 3/16 in. angle iron.

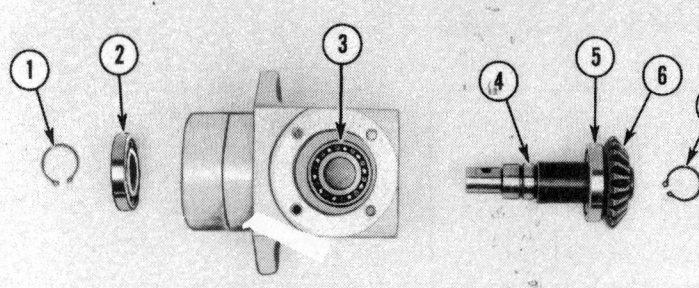

1. Snapring
2. Outer input bearing
3. Inner output bearing
4. Input shaft
5. Inner input bearing
6. Input gear

Input shaft components, right angle drive

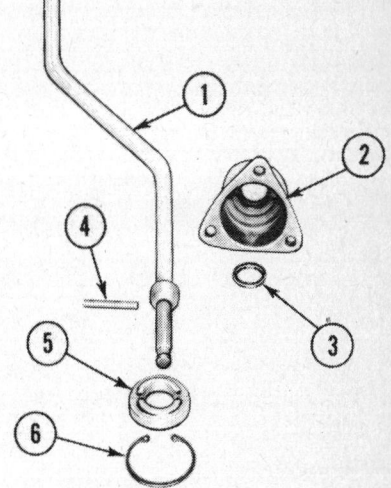

1. Shift lever
2. Housing
3. Quad ring
4. Pin
5. Keeper
6. Snapring

Shift lever components

567

International Harvester

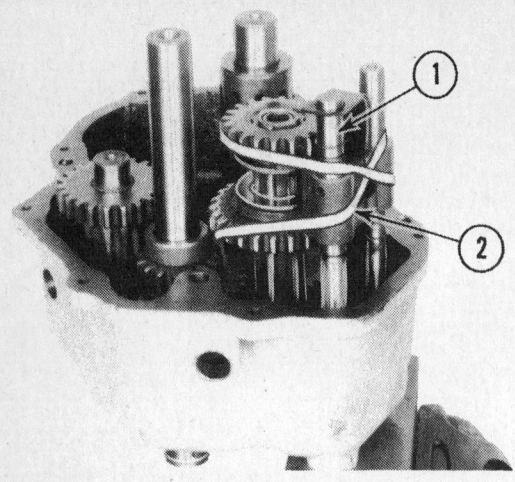

1. Shifter assembly
2. Rubber band

Securing the shifter assembly with a rubber band

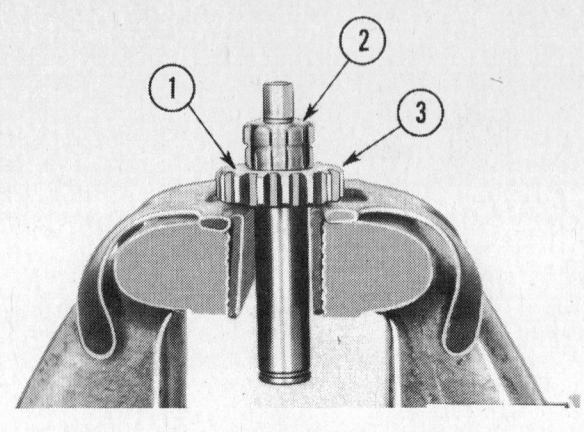

1. Spur gear
2. Input shaft
3. Beveled edge

Installing the spur gear onto the input shaft

11. Mount the transaxle in a vise with the case to the top.
12. Tap the two dowel pins into the cover and remove the eight capscrews.
13. Separate the case from the cover.
14. Remove the differential assembly.
15. Remove the thrust washer and three gear clusters from the brake shaft.
16. Remove the reverse idler gear, spacer and shaft from the cover.

NOTE: The beveled edge of the reverse idler gear goes down towards the cover.

17. Secure the shifter assembly with a rubber band. Remove the shifter assembly.
18. Remove the shifter shaft and slide off the gears.
19. Slide the forks off the rails. Be careful to catch the poppet ball as the fork comes off the rail.
20. Remove the low gear shaft and spur gear. Note that a thrust washer is used only between the low gear shaft and the cover.
21. Remove the two gear clusters and spacer from the brake shaft.
22. Remove the output shaft and gear. Remove the thrust washers from each end of the shaft.
23. Remove the brake shaft and idler gear.

NOTE: If separated, be sure that when reassembled the chamfer on the idler gear faces up.

24. Remove the input shaft with spur gear by tapping with a brass hammer.
25. If necessary, separate the spur gear from the input shaft by supporting the assembly in a vise with protectors. Using a brass hammer, tap the shaft from the gear.

ASSEMBLY

1. If necessary, install the spur gear onto the input shaft by supporting the assembly in a vise with protectors. Using a brass hammer seat the spur gear on the shaft.

NOTE: Be sure the beveled edge of the spur gear is in the "up" position.

2. Install the input shaft and spur gear with thrust washer into the case. Use a brass hammer to seat the shaft. Binding in the assembled unit can often be traced to a partially installed input shaft.
3. Install the brake shaft and idler gear with thrust washer.

NOTE: Be sure that the chamfer on the idler gear faces up.

4. Install the output shaft and gear with a thrust washer on each end of the shaft.
5. Install the two gear clusters (small gear up) and spacer onto the brake shaft.
6. Install the thrust washer, low gear shaft and the spur gear.
7. Assemble the forks to the shift rails.
8. Assemble the shift rails and stop as shown in the illustration. This will position the forks in neutral.
9. Set the shifter shaft and gears in place on the forks.
10. Position a thrust washer on the cover shifter shaft bearing.
11. Install the shifter shaft assembly.
12. Install the differential assembly. The differential through bolts should be up and away from the output gear.
13. Install the reverse idler shaft, spacer

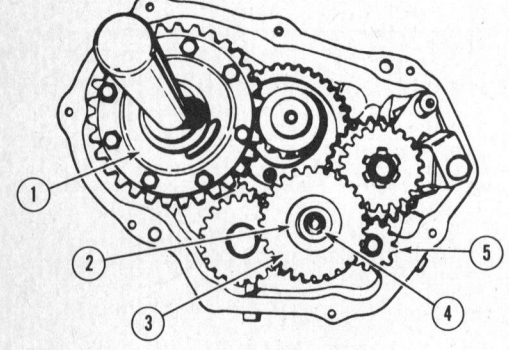

1. Differential assembly
2. Thrust washer
3. Three gear cluster
4. Brake shaft
5. Reverse idler gear

Differential components in case

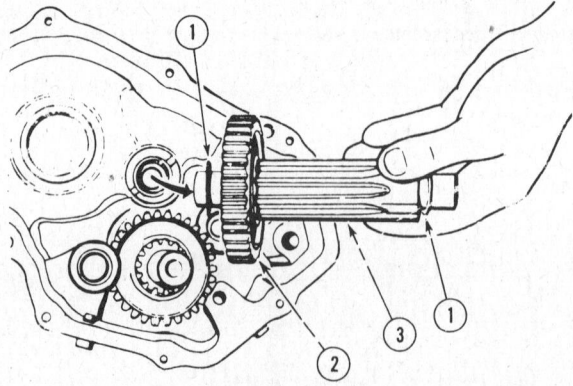

1. Thrust washer
2. Output gear
3. Output shaft

Removing output shaft and gear

International Harvester

and gear into the cover. Be sure the beveled edge of the gear is down.

14. Install the three gear clusters with the small gear down. Install the thrust washer.
15. Install the gasket onto the cover.
16. Lower the case into position on the cover. If the case hangs ½ to 1 in. high, turn the input shaft to get the gears to mesh. The case should now drop within ¼ in. of the cover.
17. Using a screwdriver, jiggle the shifter forks and rods into their machined recesses in the case. When the rods align, the case will drop into place.
18. Align the case and cover with the two dowels, Install the eight capscrews, torquing them to 10 ft.lb.
19. Install the bearing retainers with new oil seals and O-rings and the axle housings. Torque the capscrews to 13 ft.lb.

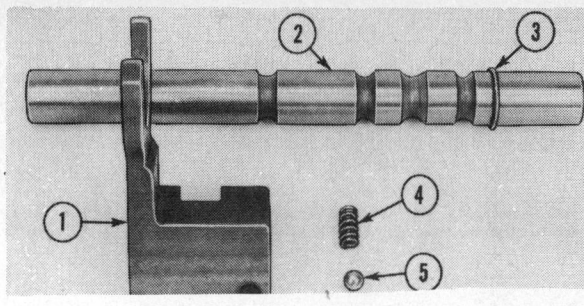

1. Fork
2. Rail
3. Snapring
4. Spring
5. Ball

Assembling the forks onto the shift rail

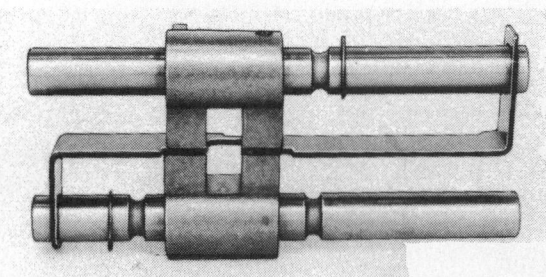

Shift forks in the neutral position

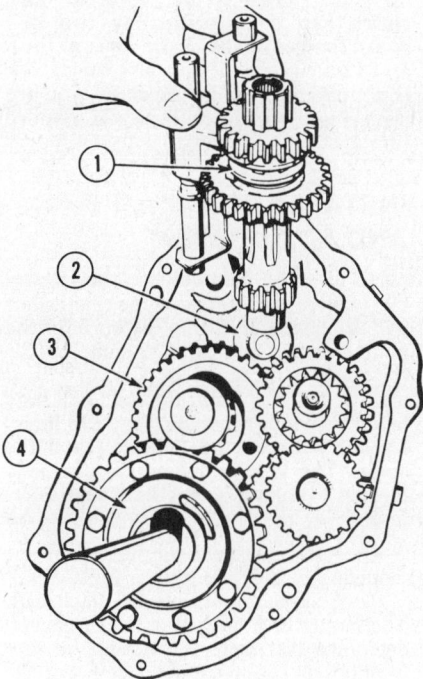

Installing the shifter shaft assembly

NOTE: Install the retainers so that the oil seals are towards the outside.

20. Reassemble and install the shift lever assembly to the case, torquing the capscrews to 10 ft.lb.

INSTALLATION

1. Fill the transaxle with 2 litres (4 pints) of IH gear lubricant (SAE 90 oil).
2. Install the rear wheels.
3. Install the brake disc.
4. Roll the assembly into position. Slip the drive belt onto the input pulley and secure the transaxle to the frame with mounting bolts.
5. Secure and adjust the drive belt guides being sure to obtain a gap of ⅛-³⁄₁₆ in. between the belt and the guides.
6. Install the center frame cover.
7. Install the fender assembly. Reconnect the battery and electrical leads.
8. Install the drawbar assembly.

DIFFERENTIAL

REMOVAL AND DISASSEMBLY

NOTE: For removal of the differential, see Transaxle Removal, steps 1-11 and Transaxle Disassembly, steps 1-14.

1. Secure the differential assembly in a vise with brass jaws. Clamp on the left axle to keep the through bolts facing upward.

NOTE: Do not clamp on the carrier bearing surface as damage could result to this surface.

2. Remove the through bolts and lift off the right axle and carrier.
3. Lift out the pinion shaft with the drive blocks and the pinion gears.
4. Remove the ring gear from the left carrier.
5. Remove the snapring securing the left axle into the carrier.
6. Disassemble the carrier assembly.

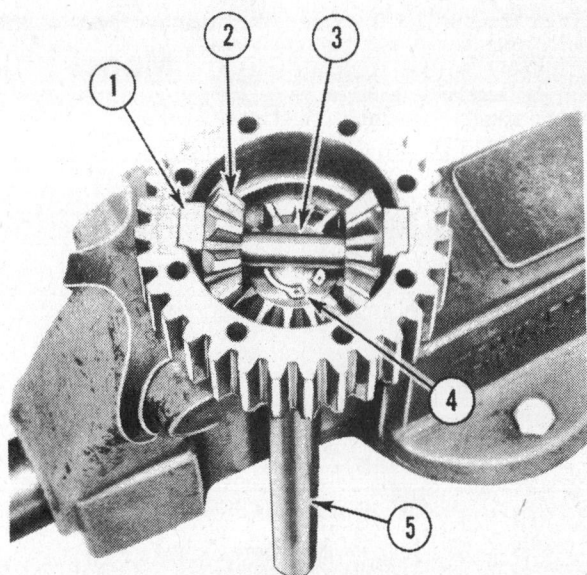

Removing pinions

569

International Harvester

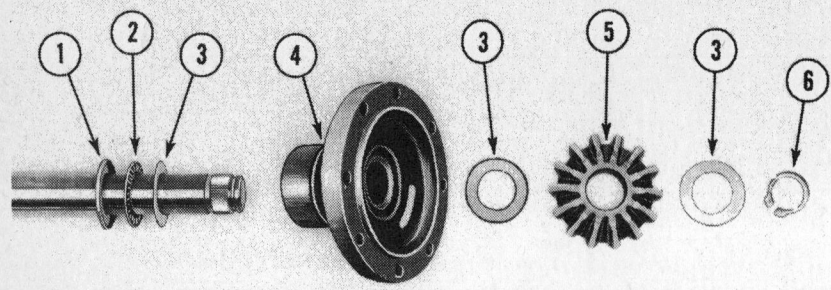

1. Cupped thrust washer
2. Thrust bearing
3. Thrust washer
4. Differential carrier
5. Bevel gear
6. Snapring

Differential carrier components

IMPORTANT: Note that the outer thrust washer is cupped in towards the thrust bearing. The inner race of the thrust bearing however, is cupped towards the outside.

7. Assembly and installation are the reverse of the above procedures. Coat all parts with 135 HEP or SAE 90 gear oil, prior to assembly. Torque the through bolts to 7 ft.lb.

ELECTRIC FRONT POWER TAKE-OFF CLUTCH

Model 482

REMOVAL

1. Disconnect the battery ground cable and the clutch wire.
2. Remove the grille housing and hood as an assembly.
3. Remove the brake flange.
4. Remove the retaining bolt, washer and armature assembly.

NOTE: It may be necessary to use a puller to remove the armature assembly.

5. Remove the rotor and key from the crankshaft.
6. To remove the field coil, remove the mounting bolts.

INSTALLATION AND ADJUSTMENT

1. Prior to installation, check the clearance between the armature assembly and the rotor. Using a feeler gauge, check the clearance at three evenly spaced locations. There should be a gap of .060 to .125 in. If the gap is exceeded, the armature assembly and rotor should be replaced.
2. Install the field coil.
3. Install the rotor and key on the crankshaft.
4. Install the armature assembly, washer and retaining bolt.
5. Install the brake flange.
6. Adjust the clutch as follows:
   a. With the clutch disengaged, check the clearance between the rotor and armature assembly. Insert a feeler gauge into both access slots between the ears of the brake flange. The air gap must be .010-.015 in.
   b. Tighten or loosen the brake flange mounting nuts to obtain the correct air gap.
7. Connect the clutch wire and the battery ground cable.
8. Install the hood and grille assembly.

Model 1100

REMOVAL

1. Disconnect the battery ground cable and the clutch wire.

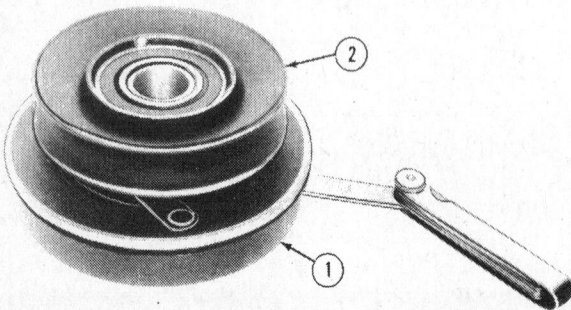

1. Armature assembly
2. Rotor

Armature-to-rotor clearance adjustment, 482 series

2. Remove the grille housing and hood as an assembly.
3. Remove the brake flange.
4. Remove the retaining bolt, washer and armature assembly. It may be necessary to use a puller to remove the armature assembly.

NOTE: There may be factory installed shims on the rotor driving hub. Be sure they are in place during reassembly to assure proper braking action of the PTO when disengaged.

5. Remove the rotor and key from the crankshaft.
6. To remove the field coil, remove the mounting bolts.

IMPORTANT: When ordering a replacement clutch it is possible to receive a clutch from a different manufacturer than the clutch to be replaced. This clutch can be identified by two belt guard flanges on the brake cover at the 6 and 12 o'clock positions, also this clutch is not painted. Unlike the clutch you are replacing, this clutch does not require shims.

INSTALLATION AND ADJUSTMENT

NOTE: The field coil for the electric PTO clutch is centered on the crankshaft by four tabs that fit into the bore for the main bearing oil seal. On some blocks the chamfer for the main bearing oil seal may be too deep, and the field coil may be off center. This would result in interference between the field coil and the rotor. The field coil must be correctly located as follows (steps 1 through 3).

1. Install the field coil and install the mounting bolts finger tight.
2. Slide the rotor onto the crankshaft. Then using .008 to .010 in. feeler gauges; determine that there is clearance at four locations between the field coil and the rotor. The field coil may be shifted slightly if necessary.
3. Remove the feeler gauges and tighten the four mounting bolts to securely hold the field coil in its proper location.
4. Place the armature assembly onto the rotor on a bench.
5. Determine the clearance between the armature assembly and the rotor using a feeler gauge at three evenly spaced locations. There should be a gap of .060 to .090 in.
6. Add or remove shims as required onto the rotor driving hub.
7. Install the rotor along with the necessary shims onto the crankshaft.
8. Install the armature assembly.
9. Install the washer and the retaining bolt.
10. Install the brake flange.
11. Adjust the clutch as follows:
   a. Disengage the clutch.
   b. Check the clearance between the armature assembly and the rotor by inserting a feeler gauge in the four slots in the

International Harvester

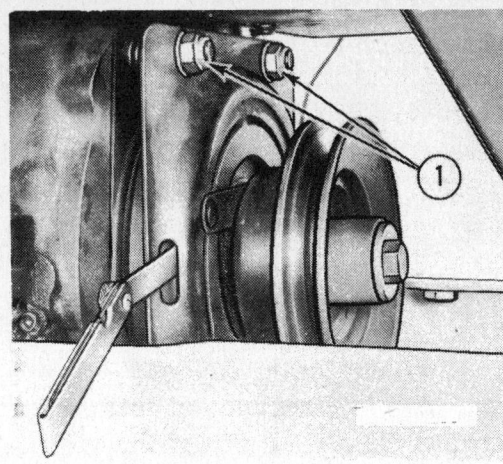

Armature-to-rotor clearance adjustment, 1100 series; #1 is the brake flange mounting nuts

brake flange. There should be a .010 in. gap between the armature assembly and rotor.

c. Tighten or loosen the brake flange mounting nuts to obtain the correct gap around the armature assembly.

12. Connect the clutch wire.
13. Install the grille housing and hood as an assembly.

ENGINE

These tractors use Kohler engines. For complete engine service, see the Kohler Engine part of the Engine Unit Repair section of this book.

Model 482

REMOVAL AND INSTALLATION

1. Disconnect the battery ground cable.
2. Disconnect the spark plug wire.
3. Disconnect the starter wires.
4. Disconnect the PTO clutch wire and the ground wire.
5. Remove the air filter shroud and disconnect the choke cable. Reinstall the air filter shroud.
6. Disconnect the diode charge wire, the engine shut-off ground wire and the throttle cable.
7. Close the fuel shut-off valve at the tank. Disconnect the fuel line at the carburetor and be sure to plug the line.
8. Remove the grille housing and hood as an assembly.
9. Disconnect the flex coupler from the engine (two capscrews).
10. Disconnect the engine mounting brackets from the mounting frames.
11. Using a suitable hoist and sling, remove the engine.
12. Installation is the reverse of removal.

Model 1100

REMOVAL AND INSTALLATION

1. Disconnect the battery ground cable.
2. Disconnect the spark plug wire.
3. Close the fuel shut-off at the fuel tank and disconnect and plug the fuel line.
4. Remove the required mounting bolts to remove the fuel tank and supports as an assembly.
5. Remove the air filter shroud and disconnect the choke control cable. Reinstall the air filter shroud.
6. Disconnect the wire to the starting motor.
7. Disconnect the diode charge wire (red wire), the engine shut-off ground wire (black wire) and the throttle control cable.
8. Disconnect the wire to the electric PTO clutch and the ground wire.
9. Remove the grille housing and hood as an assembly.
10. Disconnect the flex coupler from the engine (two capscrews).
11. Disconnect the isomounts at the front and back.
12. Using a suitable hoist and sling, remove the engine.
13. Installation is the reverse of removal.

Models 582, 682, 782, 982

FRONT AXLE AND WHEELS

Front Wheels

REMOVAL

1. Lock the brake and block the rear wheels. Jack up the front axle.
2. Remove the capscrew and flat washer from the outer end of the front spindle.
3. Slide the wheel and bearings from the spindle.

NOTE: The bearings are a press fit in the wheel and a slip fit on the spindle.

4. Wheel bearings can be driven from the wheel hub with a hammer and long drift punch. Drive from the inside toward the outside.

INSTALLATION

1. If the bearings were removed, lubri-

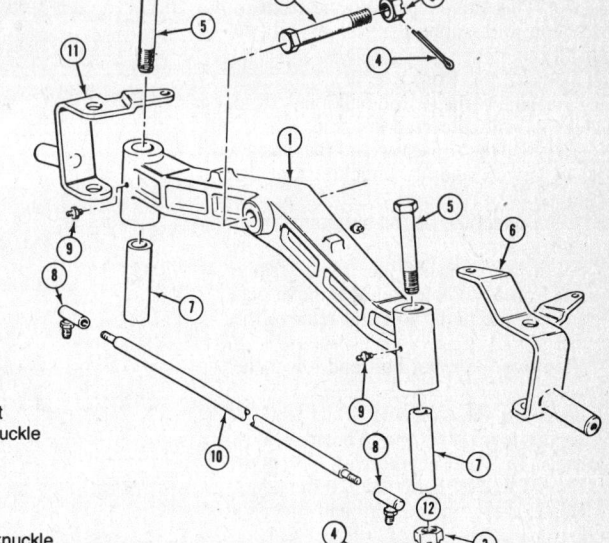

1. Axle
2. Axle pivot bolt
3. Nut
4. Cotter pin
5. Steering knuckle bolt
6. Lefthand steering knuckle
7. Spacer
8. Ball joint
9. Fitting
10. Tie rod
11. Righthand steering knuckle

Front axle, all except 982 models

571

International Harvester

cate and press in new ones. Be sure force is directed to the outer race only.

2. Slide the wheel and bearing assembly over the spindle and secure with the capscrew and flat washer.

Front Axle

MODELS 582, 682, 782

Removal

1. Lock the brake, raise the front of the tractor and support it with jack stands.
2. Remove the front wheels.
3. Disconnect the drag link from the lefthand steering knuckle.
4. Remove the tie rod.
5. Remove the steering knuckle bolt, steering knuckle and spacer from both sides.
6. Remove the axle pivot bolt and remove the axle.

Installation

1. Coat the axle pivot pin and its bore in the axle with IH 251H E.P. or its equivalent.
2. Position the axle in the support bracket and install the axle pivot bolt.
3. Install the steering knuckles. Tighten the nut to 80 ft.lb. Install the cotter key. If the cotter key cannot be installed, tighten the nut only enough to line up the slot with the hole.

NOTE: Be sure to check for free rotation of the steering knuckles after securing the nut.

4. Install the tie rod and connect the drag link to the left steering knuckle.
5. Install the wheels.
6. Check and adjust toe-in. Adjust the tie rod to provide $1/32$ to $1/8$ in. toe-in.
7. Adjust the drag link to proper length to place front wheels in the straight ahead position when the steering assembly is centered.

MODEL 982

Removal

1. Lock the brake, jack up the front of the tractor and support it on a suitable stand.
2. Remove the front wheels.
3. Disconnect the tie rod ball joints from the left and right steering knuckles.
4. Remove the capscrew and flat washer and remove the steering knuckle from the axle.
5. Disconnect the tie rod ball joints from the steering lever.
6. With the front of the tractor frame supported on a suitable stand, remove cotter pin from the pivot bolt and remove the nut.
7. Remove the pivot bolt and lower the front axle.
8. Remove the snapring which retains the steering lever pivot pin. Thread a $5/8$-18 Female $3/8$-16 Male adaptor into the pivot pin. Thread a slide hammer into the adaptor. Remove the pivot pin and steering lever.
9. Drive the steering knuckle bushings from the axle if necessary.

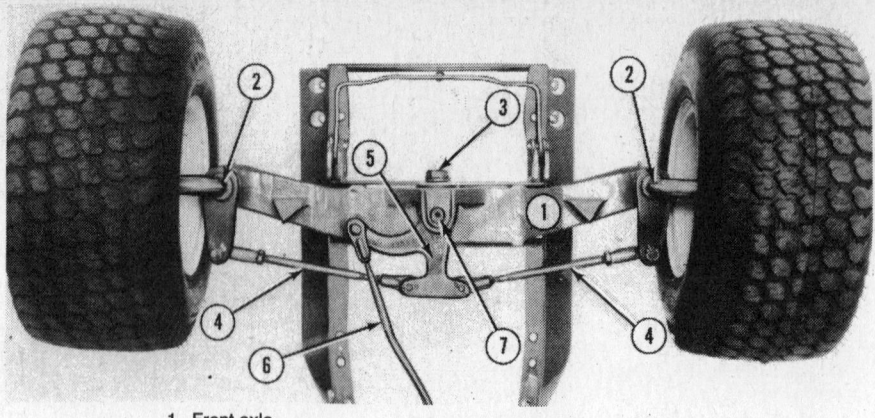

1. Front axle
2. Steering knuckle
3. Pivot bolt
4. Tie rod
5. Steering lever
6. Drag link
7. Steering lever pivot pin and snapring

Front axle, 982 models

Installation

1. Install the steering lever in the front axle. Install the steering lever pivot pin and the snapring.
2. Apply chassis lubricant liberally to the axle pivot bolt and its bore in the axle.
3. Position the axle in its support bracket channel.
4. Install the pivot bolt. Tighten the pivot bolt until the axle becomes snug in the support bracket channel. Tighten the nut additionally until the cotter pin can be installed. Secure the nut with the cotter pin.
5. Connect the tie rod ball joints to the steering lever. Be sure to tighten the locknuts securely.
6. Thoroughly lubricate the steering knuckle shaft.
7. Install the righthand and lefthand steering knuckles in their respective bores in the axle and secure with the capscrews and flat washers. Tighten the capscrews securely.
8. Connect the tie rod ball joints to the steering knuckle arms. Be sure to tighten the locknuts securely.
9. Install the front wheels and check the toe-in. Adjust the tie rods to provide $1/32$ to $1/8$ in. toe-in.

STEERING

Models 582, 682, 782

REMOVAL

1. Remove the steering wheel cover.
2. Remove the nut securing the wheel to the column.
3. Place a $3/8$ in. bolt into the column to protect the column end from damage. Remove the steering wheel using a puller.

Model 782:
Disconnect the hydraulic lines (from the

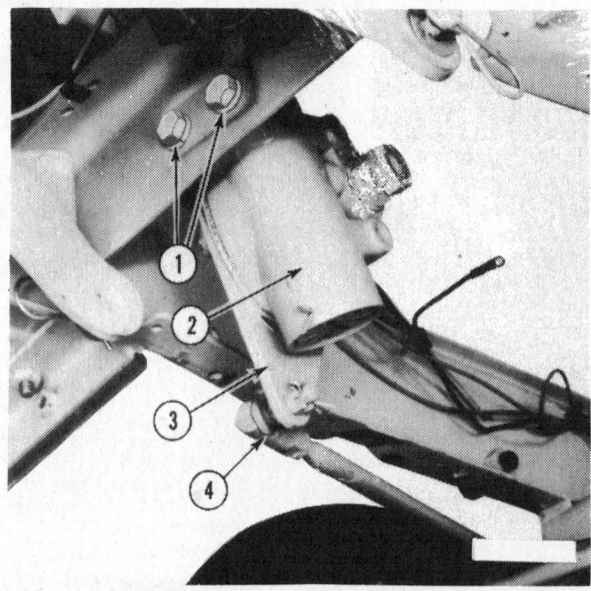

1. Mounting bolts
2. Steering unit
3. Steering lever
4. Drag link rear ball joint

Removing steering gear, all except 982 models

International Harvester

transmission) at the control valve. Remove the connecting link(s) from the control valve spools. Remove the control valve from the mounting plate. Scribe a line on the steering column just above and below the control valve mounting plate and remove the mounting plate.

4. Disconnect the drag link rear ball joint from the steering lever.

5. Remove the mounting bolts securing the unit. Remove the steering assembly by lowering it through the control panel.

DISASSEMBLY

1. Secure the steering lever and bolt in a vise.
2. Remove the lever bolt jam nut, adjusting nut and washer.
3. Slide the column and housing assembly away from the lever, bolt and cam follower.
4. Remove the adjusting plug.
5. Remove the steering cam and bearings from the housing.

ASSEMBLY AND ADJUSTMENT

1. Thoroughly coat the cam ends, balls and races with IH 251 H E.P. (or equivalent lithium base grease).
2. Install the balls and races on the cam ends.
3. Thoroughly coat the cam with chassis lubricant then install into the housing and column assembly.

NOTE: Be sure the races enter the housing squarely and are not cocked.

4. Install the adjusting plug. Screw the plug inward until end play of the cam is removed but turns freely. Insert the cotter pin in the nearest hole.
5. Fill the housing with IH 251 H E.P. (or equivalent lithium base grease).
6. Loosen the cam follower locknut, then back out the cam follower two turns.
7. Install the seal, retainer and leverbolt assembly to the housing.
8. Install the washer and adjusting nut. Tighten the adjusting nut sufficiently to provide good seal compression. Secure with the jam nut. Tighten jam nut to 40 ft.lb. Lubricate at the fitting in the housing slowly until lubricant begins to seep out.
9. Center the steering cam by rotating the steering shaft half-way between full right and full left turn.
10. Adjust the cam follower inward to eliminate backlash, then tighten locknut to 40 ft.lb. Turn steering shaft full right and left to check for binding.
11. Install the steering assembly in the tractor chassis. Secure with two capscrews through the frame cross member.
12. Connect the drag link.
13. Install the steering wheel and secure with nut.
14. **Model 782:**
 a. Install the control valve mounting plate aligning it with the scribe marks on the column.
 b. Install the control valve and connect the hydraulic lines. Install the connecting link(s).

1. Jam nut
2. Adjusting nut
3. Washer
4. Steering column
5. 3/32 in.
6. Cam follower with locknut
7. Steering lever
8. Cotter pin
9. Adjusting plug

Steering gear, all except 982 models

15. Adjust the tie rod to provide $1/32$ to $1/8$ in. toe-in as follows:
 a. With the wheels straight ahead place a chalk mark on each rim at points "A" (wheel hub height). Measure the distance between the two points.
 b. Move the tractor straight forward a distance equal to one-half revolution of the front wheels. The chalk marks will now be at point "B".
 c. Measure the distance between points "B". This distance must be .8 to $1/32$ to $1/8$ in. less than distance "A".
 d. To adjust, remove one of the tie rod ball joints and loosen the locknut.
 e. Screw the ball joint in or out to obtain the specified toe-in and tighten the locknut.
 f. Connect the ball joint to the steering knuckle and be sure to install the cotter pin.

16. Adjust the drag link to the proper length to place the front wheels in the straight ahead position when the steering assembly is centered.

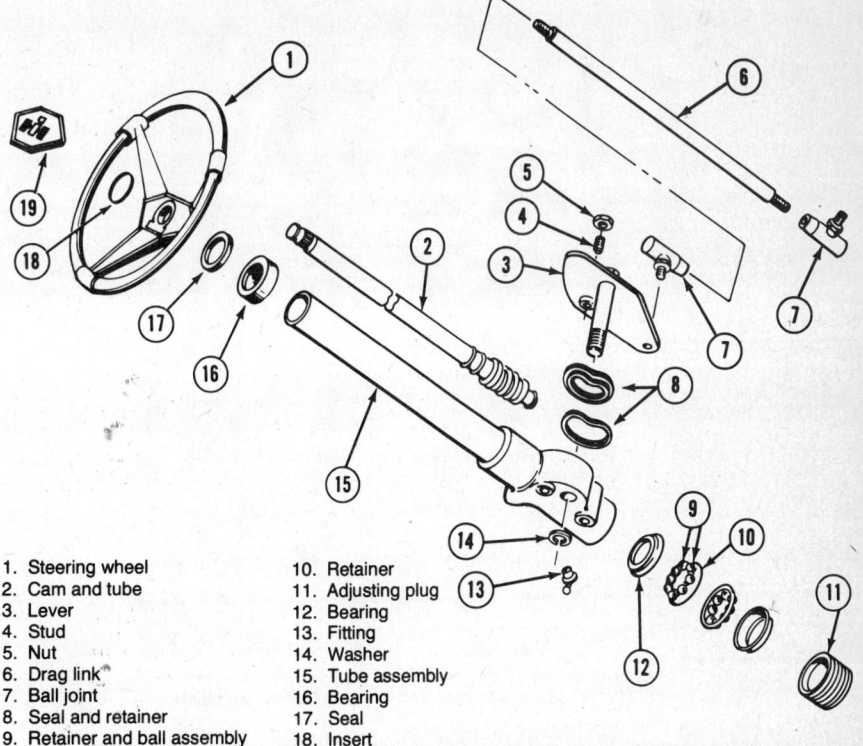

1. Steering wheel
2. Cam and tube
3. Lever
4. Stud
5. Nut
6. Drag link
7. Ball joint
8. Seal and retainer
9. Retainer and ball assembly
10. Retainer
11. Adjusting plug
12. Bearing
13. Fitting
14. Washer
15. Tube assembly
16. Bearing
17. Seal
18. Insert

Steering system components, all except 982 models

International Harvester

Model 982

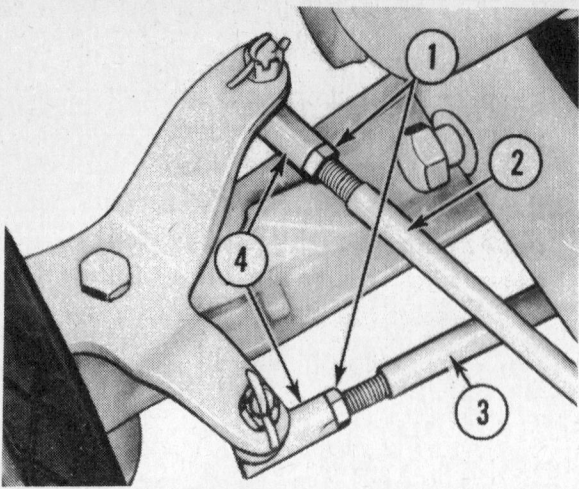

1. Locknuts
2. Drag link
3. Tie rod
4. Ball joints

Steering adjustment points, all except 982 models

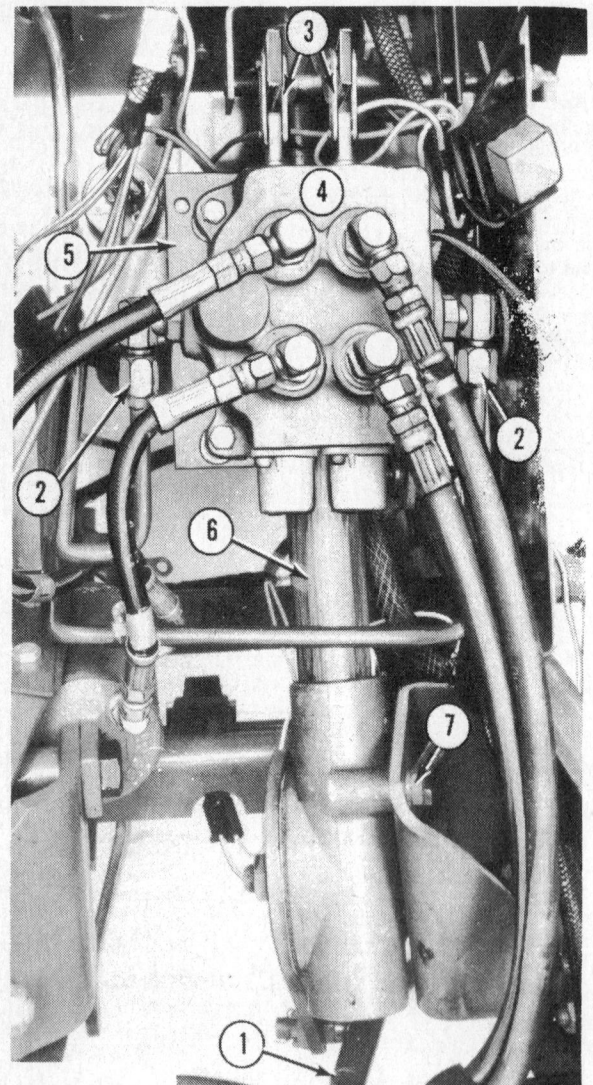

1. Drag link rear ball joint
2. Hydraulic lines
3. Connecting links
4. Control valve
5. Control valve mounting plate
6. Steering column
7. Steering column assembly mounting bolts

Hydraulic steering system installed, 982 models

REMOVAL

1. With the front wheels held in the straight ahead position, remove the steering wheel. The steering wheel is fitted on a tapered shaft and may require a puller for removal.
2. Shut off the fuel at the fuel tank. Disconnect the fuel line. Remove the fire wall and fuel tank as an assembly.
3. Remove the drag link rear ball joint from the steering lever.
4. Disconnect the hydraulic lines (from the transmission) at the control valve. Remove the connecting links from the control valve spools. Remove the control valve from the mounting plate. Scribe a line on the steering column just above and below the control valve mounting plate. Remove the mounting plate.
5. Remove the three steering column assembly mounting bolts. Lower the steering column assembly to remove it.

DISASSEMBLY

1. Secure the steering lever and bolt in a vise.
2. Remove the lever bolt jam nut, adjusting nut and washer.
3. Slide the column and housing assembly away from the lever, bolt and cam follower.
4. Remove the adjusting plug.
5. Remove the steering cam and bearings from the housing.

ASSEMBLY AND ADJUSTMENT

1. Thoroughly coat the cam ends, balls and races with chassis lubricant.
2. Install the balls and races on the cam ends.
3. Thoroughly coat the cam with chassis lubricant then install into the housing and column assembly.

NOTE: Be sure the races enter the housing squarely.

4. Install the adjusting plug. Screw the plug inward until end play of the cam is removed but turns freely. Insert the cotter pin in the nearest hole.
5. Fill the housing with chassis lubricant.
6. Loosen the cam follower locknut, then "back out" the cam follower two turns.
7. Install the seal, retainer and lever-bolt assembly to the housing.
8. Install the washer and adjusting nut. Tighten the adjusting nut sufficiently to provide good seal compression. Refer to illustration for adjustment dimensions. Secure with jam nut. Tighten jam nut to 40 ft.lb. Lubricate at the fitting in the housing slowly until lubricant begins to seep out.
9. "Center" the steering cam by rotating the steering shaft half-way between full right and full left turn.
10. Adjust the cam follower inward to eliminate backlash, then tighten locknut to 40 ft.lb. Turn steering shaft full right and left to check for binding.

International Harvester

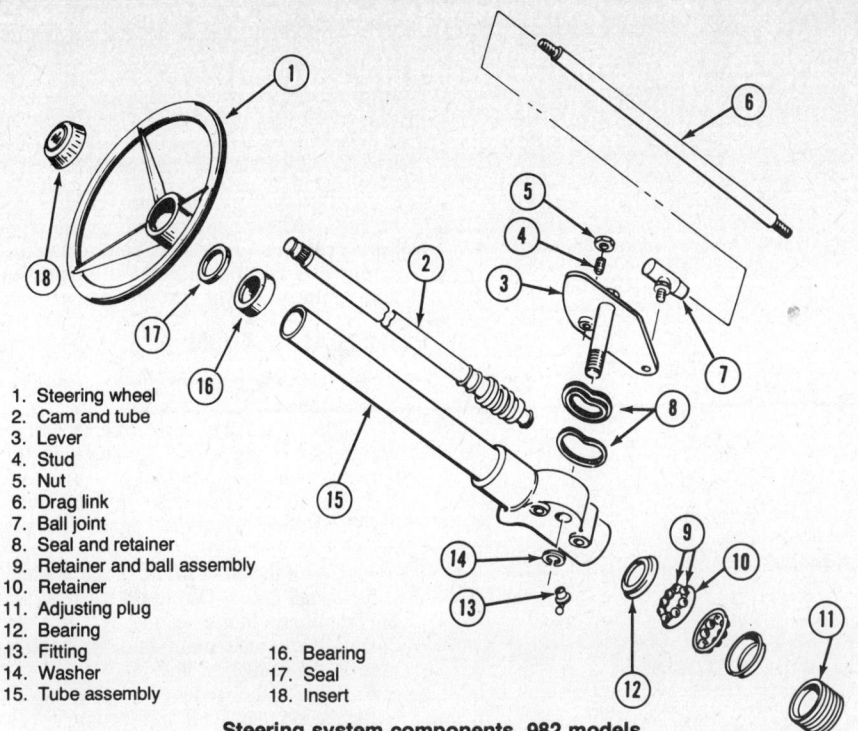

1. Steering wheel
2. Cam and tube
3. Lever
4. Stud
5. Nut
6. Drag link
7. Ball joint
8. Seal and retainer
9. Retainer and ball assembly
10. Retainer
11. Adjusting plug
12. Bearing
13. Fitting
14. Washer
15. Tube assembly
16. Bearing
17. Seal
18. Insert

Steering system components, 982 models

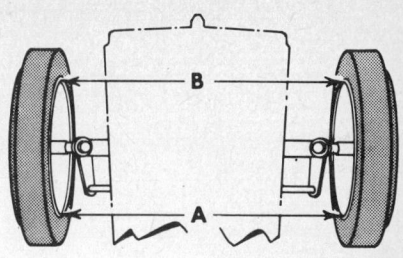

Steering adjustment

e. Reconnect the ball joints to the steering knuckle arms. Tighten the ball joint locknuts.

18. Adjust the drag link to the proper length to place the front wheels in the straight ahead position when the steering assembly is centered.

INTERNAL BRAKES

Model 582

REMOVAL

NOTE: Tractors equipped with a creeper drive attachment will require removal of the creeper unit before complete brake service can be performed. Refer to Standard Transmission—Creeper Drive.

1. Drain the transmission lubricant.
2. Remove the brake adjusting screw and locknut from its lever.
3. Remove the brake lever, pivot pin and push rod.
4. Remove the reduction housing front cover plate and slide it forward on the clutch shaft. Also see preceding "Note".
5. Remove the reduction gear from the front of the transmission mainshaft.

11. Install the steering column assembly in the tractor chassis.
12. Install the control valve mounting plate aligning it with the scribe marks on the column.
13. Install the control valve and connect the hydraulic lines. Install the connecting links.
  Connect the drag link.
15. Install the fire wall and fuel tank. Connect the fuel line.
16. Install the steering wheel and secure with nut.
17. Adjust the tie rods to provide 1/32 to 1/8 in. toe-in as follows:
  a. Place a chalk mark on the rim at points "A" at the hub height and measure the distance between them.
  b. Move the tractor straight forward a distance equal to one-half revolution of the front wheels. The chalk marks will now be at points "B".
  c. Measure the distance between points "B". This distance must be 1/32 to 1/8 in. less than distance "A".
  d. To adjust, disconnect the tie rod ball joints from the steering knuckle arms. Loosen the locknuts and turn the ball joints in or out as required. Be sure to make the tie rod adjustments equal on both sides, and be sure the knuckle arms stop on the axle.

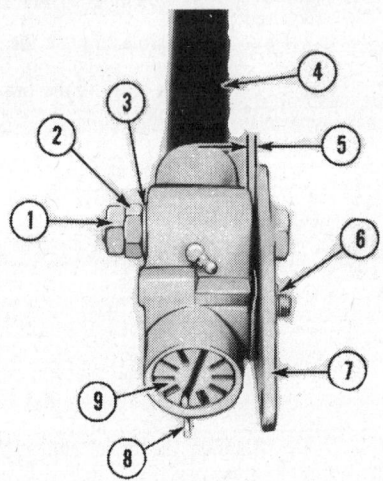

1. Jam nut
2. Adjusting nut
3. Washer
4. Steering column
5. 3/32 in.
6. Cam follower with locknut
7. Steering lever
8. Cotter pin
9. Adjusting plug

Steering gear, 982 models

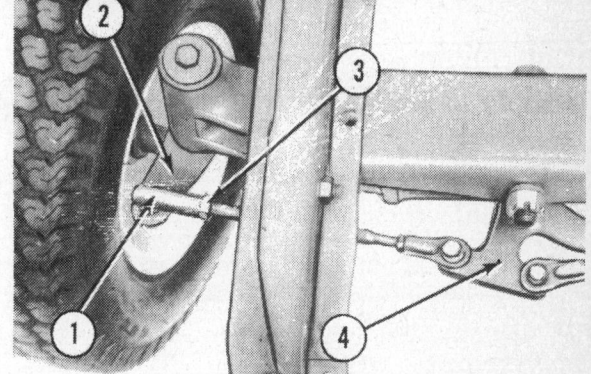

1. Tie rod ball joint
2. Steering knuckle arm
3. Locknut
4. Steering lever

Steering linkage

575

International Harvester

1. Clutch and brake pedal
2. Arm
3. Return spring
4. Adjusting screw
5. Locknut
6. Push rod
7. Ball
8. O-ring
9. Lining disc
10. Brake disc
11. Retainer
12. Pivot pin
13. Brake lever
14. Brake rod

Brake components, model 582

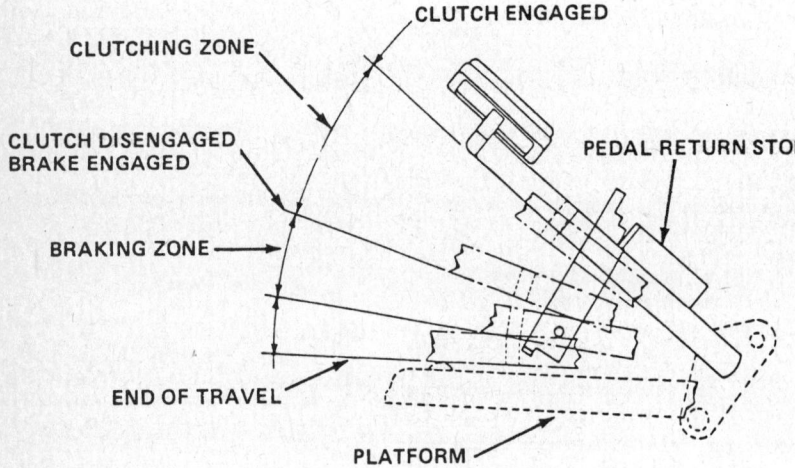

Brake adjustment, model 582

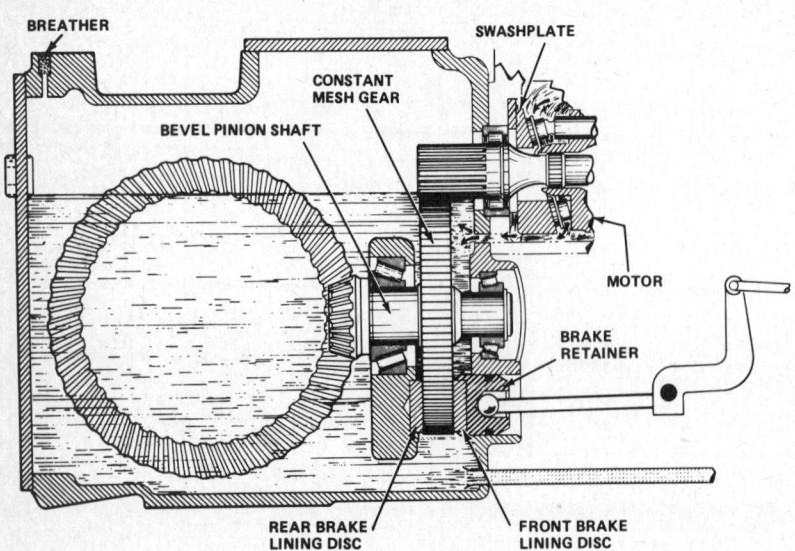

Brake system, models 682, 782

6. Move the gear upward and the bottom of the gear forward to clear the cover screw bosses as the gear is lifted from the housing.

7. Slide the brake disc forward on the countershaft as the front lining and retainer are moved forward in their bore.

NOTE: Both linings and the disc can be removed without removing the front lining retainer; however, removal of the retainer is recommended for inspection and replacement of the retainer O-ring.

INSTALLATION

1. Clean the brake cavity and lining recess in the reduction housing.
2. Place a small quantity of grease in the rear brake lining recess in the reduction housing then insert the lining.
3. Install the brake disc on the countershaft and slide it rearward against the rear lining.
4. Install the front lining in the retainer.
5. Install a new O-ring on the front lining retainer, lubricate the retainer and O-ring then push them into the retainer bore in the reduction housing.
6. Install the reduction gear on the transmission mainshaft and secure with flat washer, lockwasher and capscrew. Tighten capscrew to 55 ft.lb.
7. Install a new cover gasket; then install the cover plate.
8. Be sure the ball is in place in the front lining retainer; then install the push rod, lever, pivot pin, adjusting screw and locknut.
9. Fill transmission to proper level with Hy-Tran.

ADJUSTMENT

The brake should engage when the pedal arm is pressed down to within a maximum of $1^{5}/_{16}$ in. and a minimum of ¾ in. distance above the top of the left foot support, which serves as the pedal stop.

To adjust, loosen the locknut and turn the brake lever adjusting screw in or out as required.

If may be possible to push the pedal all the way down to the pedal stop, but this is of no concern as long as the brake is engaged within above limits.

Models 682, 782

REMOVAL

1. Drain the lubricant.
2. Split the tractor.
3. Remove the differential.
4. Remove the bevel pinion shaft and constant mesh gear.
5. Remove the rear brake lining disc.
6. Using pliers, remove the brake disc retainer.
7. Remove the brake lining disc from the retainer.

INSTALLATION

1. Clean the brake cavity and lining recess in the transmission case.
2. Place a small quantity of grease in the

International Harvester

rear brake lining recess in the transmission case; then install the brake lining.

3. Install the front brake lining disc in the retainer.

4. Install a new O-ring on the front lining retainer, lubricate the retainer and O-ring, then push them into the retainer bore in the transmission case.

5. Reassemble the transmission case. Refer to Hydrostatic Transmission—Differential—Reassembly.

6. Fill the transmission to the proper level with Hy-Tran.

ADJUSTMENT

1. Block the front tires and support the rear end so one rear tire is off the ground and free to turn.

2. Place the brake pedal in the up position.

3. Loosen the jam nut "B". Then tighten the brake lever adjusting screw "C" until finger tight (8 to 10 in.lb). Retighten the jam nut "B" while holding the adjusting screw "C".

4. Actuate the brake pedal through a full stroke at lease one time.

5. Repeat step 3.

6. If the brake drags with the brake pedal in the up position, loosen the jam nut and back off the adjusting screw slightly and retighten the jam nut.

EXTERNAL BRAKES

Model 982

REMOVAL

1. Remove the cotter pin and washer and disconnect the brake rod from the brake arm.

2. Remove the capscrews from the mounting flange and remove the brake assembly from the axle carrier.

3. Remove the capscrews securing the caliper assembly to the bracket assembly.

INSTALLATION

1. Assemble the caliper assembly to the bracket assembly. Be sure to install the spacers and springs.

2. Install the caliper assembly and bracket assembly on the disc and axle carrier.

3. Install the brake rods in the brake arms.

ADJUSTMENT

1. Raise the rear of the tractor and support with safety stands. Be sure both tires are off the ground.

NOTE: Adjust the lefthand side brake first and make all adjustments one turn at a time.

2. Rotate the wheel and notice the

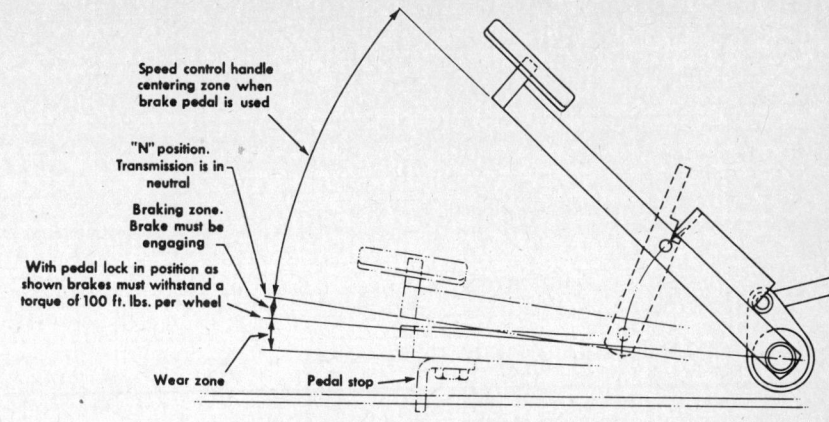

Brake adjustment, models 682, 782

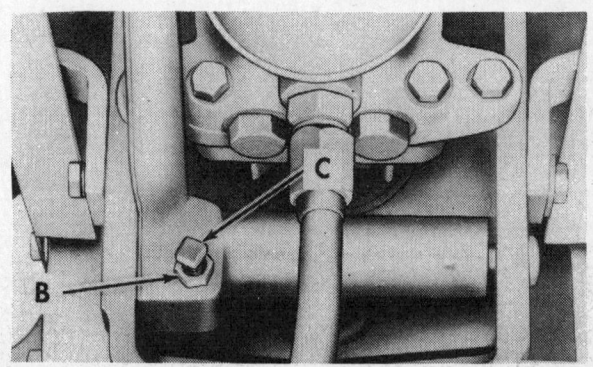

Adjusting nut and screw

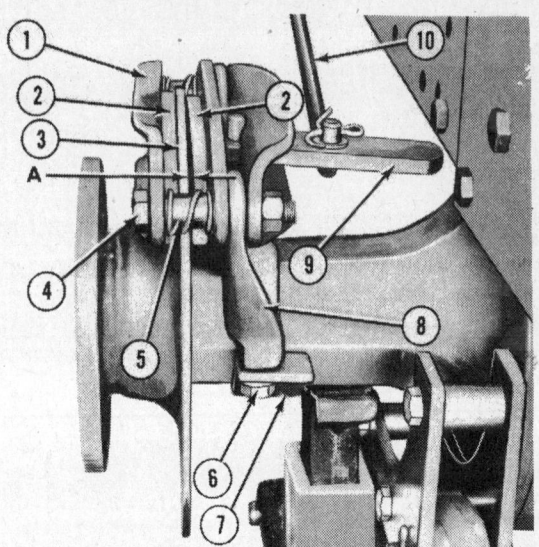

1. Caliper assembly
2. Brake pads
3. Brake disc
4. Capscrew
5. Spacer and spring
6. Capscrew
7. Mounting flange
8. Bracket assembly
9. Brake arm
10. Brake rod

Brake system, model 982

International Harvester

amount of drag put on the brake disc by the brake pads.

3. Disconnect the brake rod from the brake arm. Shorten the rod by turning it clockwise in the clevis until drag is noticeable on the brake disc when the wheel is rotated.

4. Once drag is obtained, turn the rod counterclockwise one turn. Drag should be eliminated.

5. Check the gap between the inner brake pad and brake disc at the rear of the caliper assembly, dimension "A". The gap must be .030 to .035 in.

If the gap is unobtainable by turning the brake rod, finer (½ turn) adjustments are possible by removing the clevis from the brake pivot shaft and rotating it ½ turn.

6. Repeat the above procedures for the righthand side brake.

7. Once both brake assemblies are adjusted and clearance gaps obtained, test the tractor for equal braking.

Driving forward at a moderate speed, apply the neutral return (clutch) pedal. If the tractor brakes unevenly (pulls left or right), adjust the righthand side brake rod until equal braking is obtained.

CLUTCH

REMOVAL

1. Remove the engine side panels.
2. Remove the frame cover.
3. Disconnect the battery ground cable.
4. Remove the pivot pin and hanger assembly. Remove the drive disc spring.
5. Disconnect the clutch release rod from the clutch release lever.
6. Remove the bolts from the flex coupling.
7. Drive out the driveshaft coupling roll pin.
8. Drive out the coupling arm roll pin.
9. Slide the couplings forward on the clutch shaft and move the shaft to the side of the transmission or creeper input shaft.
10. Remove the clutch shaft assembly including pressure plates, drive plate and clutch release lever.
11. Remove the drive plate, pressure plate and clutch release lever from the clutch shaft.
12. Using a vise equipped with brass jaws, clamp the clutch shaft snug and tap the shaft down enough to slightly compress the spring. Remove the coiled spring pin.

CAUTION
Slowly release the vise allowing the spring to expand as the shaft slips through the vise jaws.

INSTALLATION

1. Assemble the front coiled spring pin, teaser spring, spring spacer, throw-out bearing, bushing, loading spring and washer onto the clutch shaft.
2. Using a vise equipped with brass jaws, clamp the clutch shaft snug. Tap the

1. Driveshaft coupling
2. Flex coupling
3. Coupling arm

Clutch flex coupling

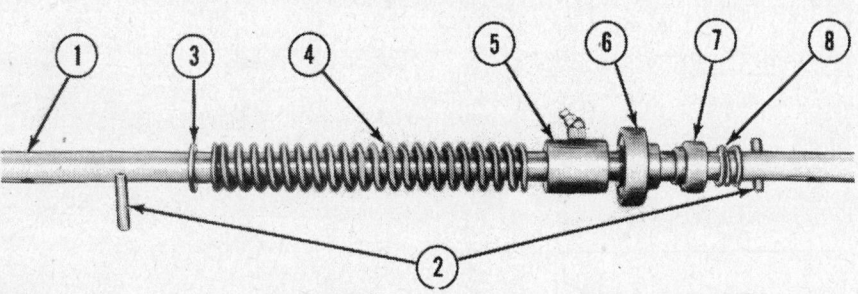

1. Clutch shaft
2. Coiled pin
3. Washer
4. Loading spring
5. Bushing
6. Throw out bearing
7. Spring spacer
8. Teaser spring

Front coiled spring

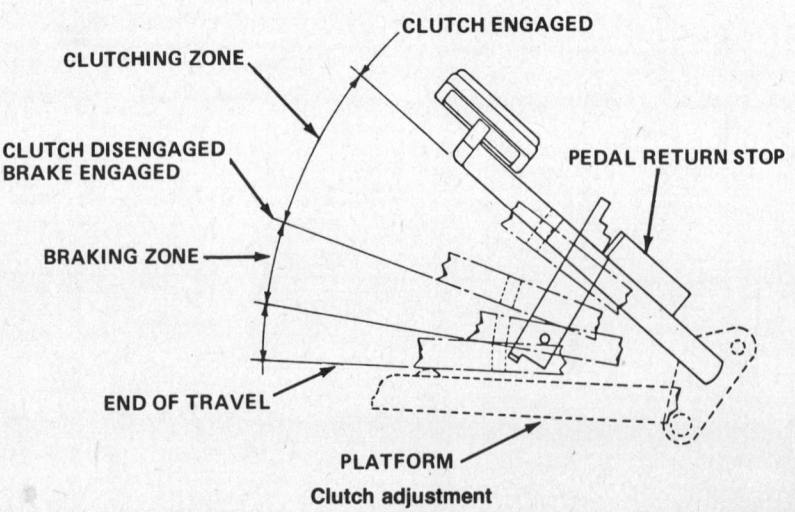

Clutch adjustment

International Harvester

shaft down enough to compress the loading spring and install the rear coiled spring pin.

3. Continue reassembling the clutch by reversing the disassembly procedure.

NOTE: Before installing the clutch shaft assembly, pack the clutch drive plate bushing and coat the clutch shaft with 139 933 C1 powered bushing grease.

4. Install the clutch shaft assembly in the tractor by reversing the removal procedure.

5. Adjust the clutch.

ADJUSTMENT

Adjust the clutch linkage. It is important that a clearance of .05 in. be maintained between the clutch release lever and the clutch release bearing. In order to maintain this clearance, the pedal should have a free movement of approximately 9/32 in. This measurement is taken at the point of contact of the pedal arm with the front edge of the pedal return stop. When it is necessary to adjust the clutch, turn the adjusting nut on the clutch release rod in or out as required to get the proper measurements.

SPLITTING AND RECOUPLING

Splitting

1. Remove the battery strap and the battery.
2. Disconnect the electrical leads from the solenoid.
3. Disconnect the tail light leads and the seat switch.
4. Remove the rear fender to frame bolts and the battery ground wire.
5. Remove the mounting screws from the foot platforms.
6. Remove the fender assembly.
7. Remove the frame cover.
8. Remove the brake rod.
9. Disconnect the rear flex coupling.

NOTE: Tractors equipped with a three point hitch attachment require removal of the lift lever before splitting.

10. Support the frame of the tractor, remove the frame mounting bolts and roll the rear end out of the frame as illustrated.

Recoupling

1. Recouple the tractor by reversing the splitting procedure.
2. Check the oil level of the rear frame and fill to proper level with IH Hy-Tran.

REDUCTION DRIVE

REMOVAL AND DISASSEMBLY

1. Split the tractor.

Removing reduction driveshaft

2. Remove the transmission drain plug and allow the lubricant to drain completely; then replace the plug.
3. Drain and remove the creeper assembly (if unit is so equipped).
4. Remove the brake adjusting screw, pivot pin, brake lever and push rod.
5. Remove the reduction housing cover plate.
6. Remove the reduction driven gear by removing the capscrew, lockwasher and driven gear retainer.
7. Using pliers, remove the brake disc retainer. Remove the brake disc and brake lining discs.

NOTE: Check the condition of the O-ring on the brake disc retainer; replace as necessary.

8. Remove the reduction housing mounting bolts and remove the housing.

NOTE: Soft copper sealing washers are used under the lower two bolts. Replace the washers during reassembly.

9. Using a split jawed puller, remove the reduction driveshaft, seal and bearing from the reduction housing.
10. To remove the splined coupling

1. Adjusting nut
2. Clutch release lever
3. Release rod

Clutch linkage

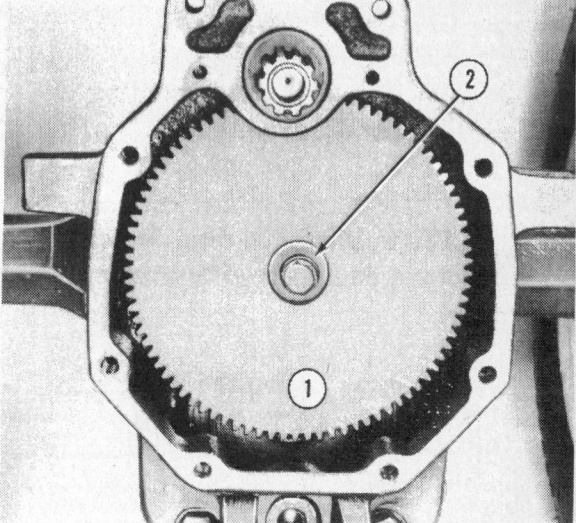

1. Reduction driven gear
2. Driven gear retainer

Reduction drive with cover plate removed

International Harvester

from the driveshaft, drive out the spring pin.

NOTE: The splined coupling is used only on tractors equipped with creeper attachment.

11. If necessary, press the reduction driveshaft from the ball bearing.

12. If necessary, press the needle bearing from the rear of the housing.

ASSEMBLY AND INSTALLATION

1. If removed, press the needle bearing into the reduction housing from the rear. The rear end of the bearing must be flush with the housing.

2. If removed, press the ball bearing onto the reduction driveshaft.

3. Lubricate the lip of a new oil seal and install the seal onto the shaft. Be careful when sliding the seal lip over the pin hole in the shaft.

4. Install the splined coupling on the driveshaft (if tractor is equipped with a creeper). Spring pin ends must be even with or below the spline root to avoid interference when shifting the creeper unit.

5. Install the reduction housing on the transmission case. Be sure to install a new gasket.

6. Install new copper sealing washers on the two lower mounting bolts. Torque the bolts to 80 ft.lb.

7. Clean the brake cavity and the brake lining recess in the housing.

8. Apply a small quantity of petroleum jelly in the rear brake lining recess in the reduction housing, then install the brake lining.

9. Install the brake disc on the countershaft and slide it rearward against the rear brake lining.

10. Install a new O-ring on the front lining retainer and install the front lining in the retainer. Lubricate the retainer and push it into the bore in the reduction housing.

11. Install the reduction gear on the transmission mainshaft and secure with the retainer, lockwasher and capscrew. Tighten the reduction gear capscrew to 55 ft.lb.

12. Install the reduction driveshaft with bearing, seal and splined coupling (on tractors with creeper). Seal case must be flush with housing.

13. Install a new gasket and install the housing cover plate.

14. With the ball in place in the front brake lining retainr, install the push rod, brake lever, pivot pin, adjusting screw and locknut.

15. Install a new gasket and install the creeper unit (on tractors so equipped).

16. Recouple the tractor.

17. Fill the transmission and creeper to proper level with IH Hy-Tran.

18. Adjust the brake. Refer to Internal Brakes-Adjustment.

TRANSMISSION AND DIFFERENTIAL

Gear Drive Units

REMOVAL AND DISASSEMBLY

Differential

1. Drain the lubricant.
2. Split the tractor.
3. Remove the reduction drive.
4. Remove the rear axles and their carriers.
5. Remove the differential carrier bearing cage and shims from each side. Keep the shims with each cage and identified for each side.
6. Turn the differential into the position shown and remove it from the transmission case. If the assembly will not clear the side of the transmission case, it will be necessary to remove one of the differential carrier bearings.
7. Drive out the pinion shaft lock pin.
8. Remove the pinion shaft.
9. Remove the pinion gears and side gears.
10. If the drive gear requires separate replacement, press out the retaining rivets.
11. Remove the carrier bearing cones from the differential carrier if they are to be replaced.
12. Remove the bearing cups from each cage if replacement is necessary.

Transmission

1. Remove the differential.
2. Remove the gear shift lever and cover assembly.

Shift the transmission into two gear speeds to lock the transmission then remove the nut from the countershaft.

3. Remove the shifter fork setscrews.

— CAUTION —
Cover the gearshift poppet ball hole to prevent the ball and spring from flying out as the rods are removed. Drive the shifter rods forward and out of the transmission.

4. Remove the capscrews from the mainshaft front bearing retainer.
5. Pull the mainshaft forward and out of the transmission as the gears are removed.
6. Push the countershaft rearward and out of the transmission as the gears and spacers are removed. Note the sequence of spacers and gears for reassembly.
7. Pull the mainshaft needle bearing from the housing.
8. Remove the reverse idler shaft and gear.
9. Remove the countershaft front bearing, retainer and shims.

ASSEMBLY AND INSTALLATION

Reassembly is basically the reverse of disassembly; however, particular attention should be given to the following steps.

1. Be sure all bearings are thoroughly lubricated.
2. The transmission mainshaft needle bearing must be installed with its oil hole aligned with the oil hole in the housing.
3. Assemble the differential, carrier bearings, cages and shims. Check bearing preload and adjust as necessary before replacing the transmission countershaft. Install or remove shims as necessary. Preload is correct when a steady pull of one to eight lb. is necessary to rotate the differential assembly.
4. Remove the differential assembly, keep the shims with the cages then install the transmission countershaft, bearings, gears, spacers, front bearing retainer, shims and the nut. Tighten the nut to 85 ft.lb. Tighten retainer capscrews to 20 ft.lb.
5. Install the differential assembly, keeping the preload shim pack as previously established. Drive gear must be on the right with teeth facing left.

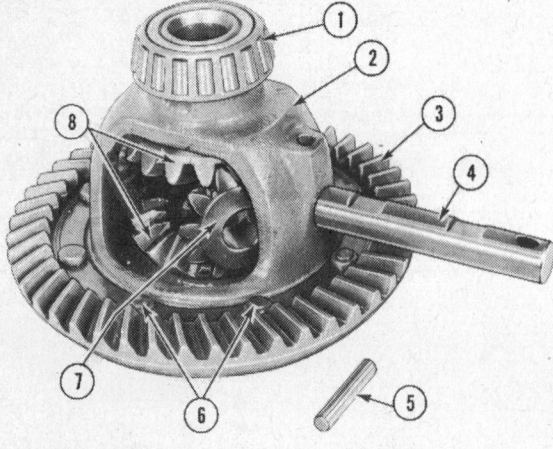

1. Carrier bearing
2. Differential carrier
3. Ring gear
4. Pinion shaft
5. Lock pin
6. Rivets
7. Pinion gear
8. Side gears

Disassembling the differential carrier

International Harvester

6. Check the backlash between the drive gear and bevel pinion and the gear teeth bearing pattern as follows.

7. Apply a thin coat of red lead or Prussian blue to the bevel pinion teeth faces, then rotate the gears by hand and observe the bearing pattern.

NOTE: Some deflection will occur under load. Allowance is made in gear design to prevent concentration of load on tooth edges.

8. Hand testing and very light loads should provide a pattern as shown in Figure "B". When load and deflection increases the pattern will progress as in Figure "A."

9. The desirable (no load) pattern in Figure "B" is the result of adjusting the drive gear lateral position to the specified range of .003 to .005 in. backlash.

10. Tooth bearing position from the root to the crown of the tooth is controlled by lateral position of the bevel pinion. If low tooth bearing on bevel pinion is indicated (as shown in Figure "C") the bevel pinion must be adjusted toward the drive gear. If high tooth bearing on the bevel pinion is indicated (as shown in Figure "D") the bevel pinion must be adjusted away from the drive gear.

11. Stake the countershaft nut by center-punching the face of the nut over a spline groove.

12. Continue the assembly in reverse order of disassembly.

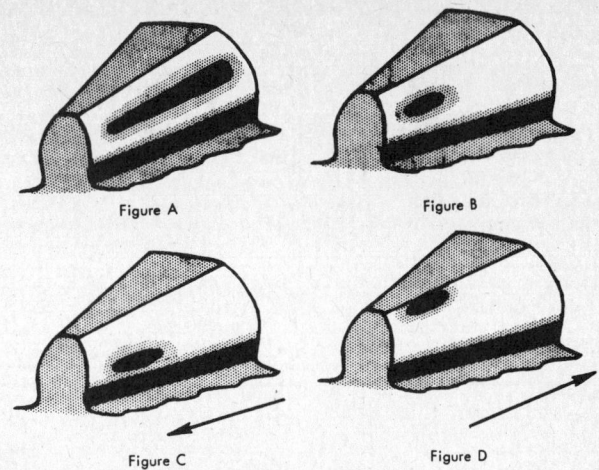

Gear wear patterns

NOTE: The right side axle carrier mounting bolt hole located at the nine o'clock position is tapped through the differential case. The bolt installed in this hole should be coated with IH Gasketmaker to prevent oil seepage past the threads.

13. Fill housing to proper level with specified lubricant.

Creeper attachment has its own lubricant separate of the transmission. Fill creeper at breather and check at side plug in creeper housing.

CREEPER DRIVE

REMOVAL

1. Remove the frame cover.
2. Drive out the roll pins from the flex coupling arms and slide the couplers forward on the driveshaft.
3. Place a drip pan under the tractor and remove the creeper drive unit. It may be necessary to bump the creeper unit to

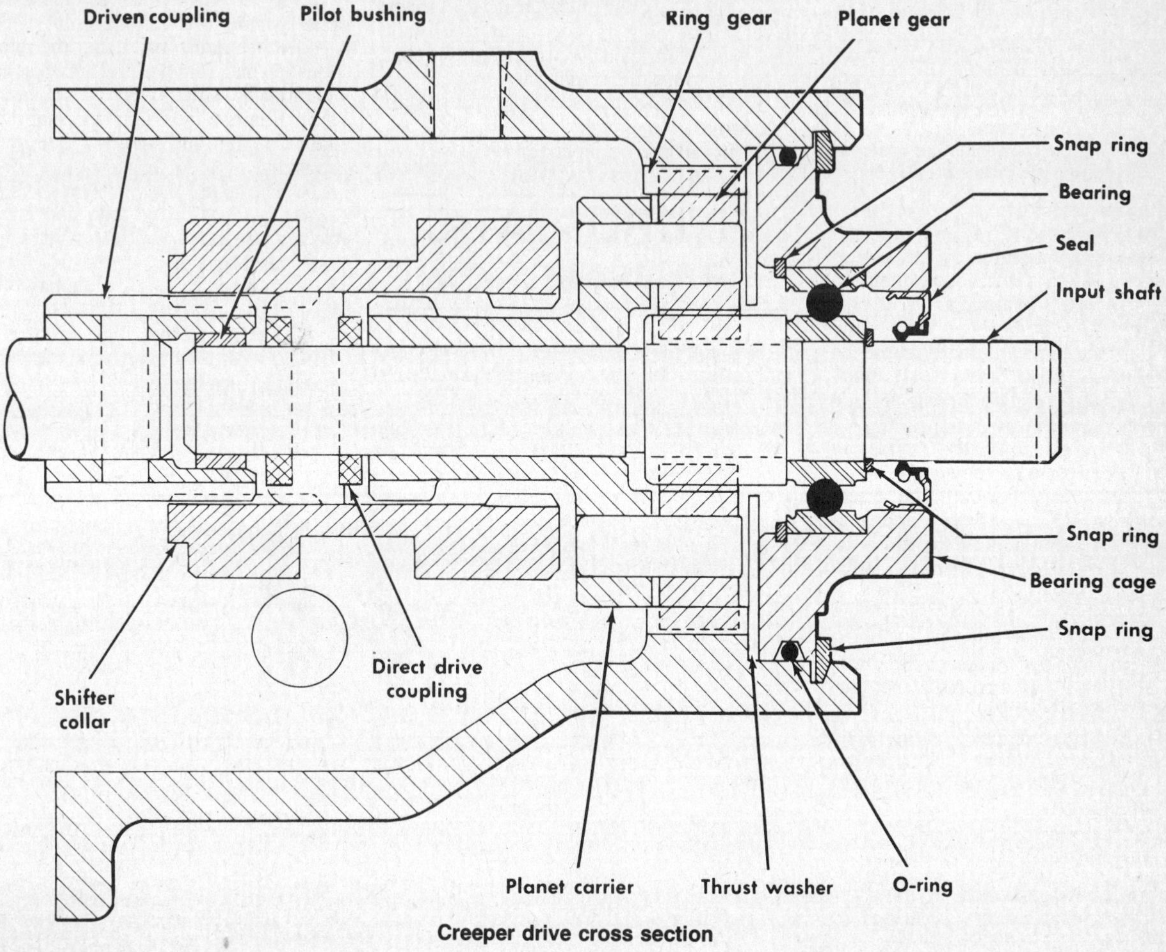

Creeper drive cross section

International Harvester

loosen the housing from its gasket and dowels.

DISASSEMBLY

1. Remove the snapring which holds the input shaft bearing cage in the housing.
2. Pull the shaft, bearing, retainer, planetary assembly and the direct drive coupling from the housing.

NOTE: The spline grooves of the direct drive coupling must align with the splines of the shifter collar.

3. Support the drive coupling and drive the coiled spring pin out. Remove the coupling from the shaft.
4. Slide the planet carrier off the input shaft and the planet gears off the carrier pins.
5. Remove the flat thrust washer.
6. A snapring holds the ball bearing outer race in the retainer. Remove the snapring then push the bearing and shaft from the retainer.
7. The ball bearing is held on the shaft by a snapring. Remove the snapring. The shaft can now be pressed from the bearing.
8. Press the oil seal from the bearing retainer.
9. Drive the shift poppet pin from the shaft and remove the poppet.
10. Shift the lever and shifter collar toward the rear of the case and at the same time lift the shifter collar up to disengage it from the shift yoke.
11. Drive the pin out of the shift yoke and lever shaft.
12. Slide the lever shaft from the yoke and housing.
13. Remove the O-rings from the shaft, housing and bearing retainer.
14. Wash all parts in cleaning solvent then dry thoroughly.

ASSEMBLY

Reassembly is basically the reverse of disassembly; however, particular attention should be given the following:

1. Always use new O-rings, gaskets and oil seals. O-rings and oil seals should be coated with Lubriplate® or chassis lubricant to assist in installation and provide initial lubrication.
2. Install the oil seal after completing the drive assembly in the housing.
3. The pins which secure the direct drive coupling and the driven coupling to their respective shafts must be flush or below the spline groove so as not to interfere with shifting.
4. The long internal splines of the shifter collar go toward the rear.
5. The machined shoulder of the direct drive coupling goes toward the planet carrier.
6. Lubricate the components and rotate the driveshaft several turns with the shifter in each speed selection to insure freedom of movement and rotation.

INSTALLATION

1. Place a new gasket on the mounting face of the creeper housing. The dowels will hold it in place.

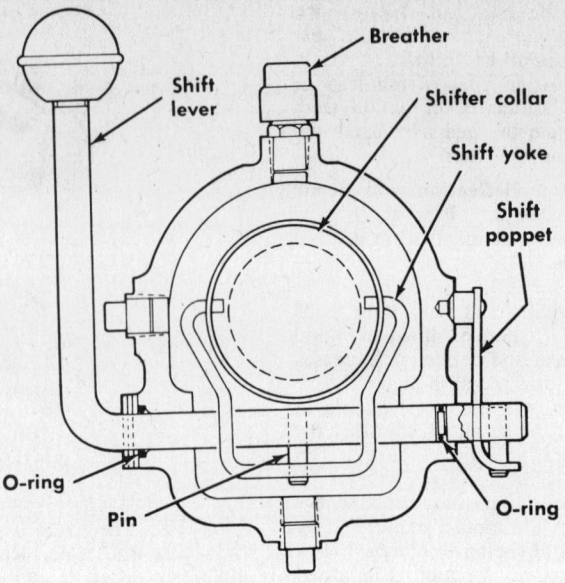

Creeper drive shift lever

2. With the shifter collar in its rear position, rotate the driveshaft so the collar will engage the driven coupling as the dowels engage the reduction cover plate and the driveshaft enters its pilot bushing in the driven coupling.
3. Secure the creeper to the reduction housing cover plate and reconnect the rear coupling.
4. Fill the creeper housing to proper level (check plug on side of housing) with the same lubricant as specified for the transmission-differential.
5. Shift the unit several times to insure freedom of movement.
6. Install the frame cover.

HYDROSTATIC TRANSMISSION

For complete hydrostatic transmission and differential service, see the Model 71, 102, 122, 123 tractor section.

REAR AXLE

DISASSEMBLY

1. Raise the rear of the tractor and support with a jack stand.
2. Remove the rear wheel.
3. Remove the brake assembly from the axle housing (Model 982 only).
4. Place a drip pan under the rear frame housing and remove the drawbar and differential housing cover.
5. Remove the "C" type snapring from the axle shaft and slide the axle out of the housing.
6. The axle housing outer oil seal may be removed with the axle housing on the tractor. Collapse the seal with a hammer and chisel and pry the seal from the bore. Be careful not to nick or damage the seal bore.
7. The needle bearing may be removed using a puller without removing the axle housing from the tractor.

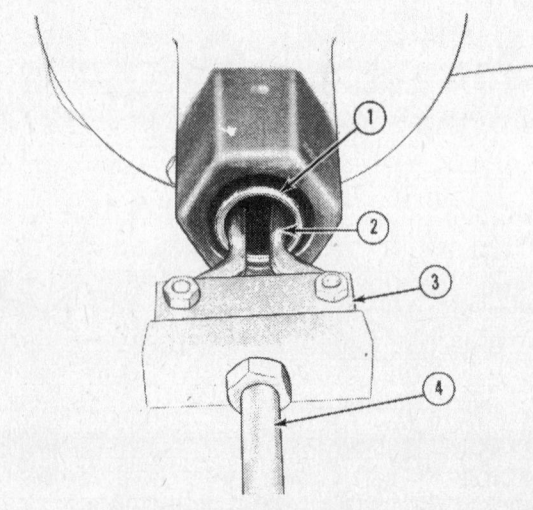

Removing needle bearing

International Harvester

8. To remove the axle housing, remove the capscrews attaching the frame to the axle housing and reduction housing. Raise the frame clear of the axle housing and remove the housing.

ASSEMBLY

1. Using a new gasket, install the axle carrier to the differential housing. Capscrew threads should be coated with a non-hardening sealer to avoid oil leaks. The frame pad of the axle carrier must be to the top.

2. Install the needle bearing into the housing with a suitable tool. One side of the bearing is chamfered to aid in installation. Install the bearing flush with the outer edge of its bore.

3. Install the seal with a suitable tool. Fill the cavity between the lips of the oil seal with IH 251H E.P. before installing the seal.

4. Lubricate the axle shaft and bushing then slide the shaft through the seal, bushing, carrier and differential side gear. Rotate the axle as it is pushed through to avoid damage to the seal. Wipe off excess lubricant.

5. Install a new "C" type snapring to the inner end of the axle shaft.

6. Replace the rear cover and drawbar.

7. Install the brake assemblies on the outer ends of the axle carriers (Model 982 only).

8. Install the wheel.

9. Fill differential housing to specified level with Hy-Tran.

FRONT POWER TAKE-OFF CLUTCH

REMOVAL

1. Disconnect the battery ground cable and the clutch wire.
2. Remove the grille housing and hood as an assembly.
3. Remove the brake flange.
4. Remove the retaining bolt, washer and armature assembly.
5. Remove the rotor and key from the crankshaft.
6. To remove the fiel coil, remove the mounting bolts.

INSTALLATION AND ADJUSTMENT

1. Prior to installation, check the clearance between the armature assembly and the rotor. Using a feeler gauge, check the clearance at three evenly spaced locations. There should be a gap of .060 to .125 in. If the gap is exceeded, the armature assembly and rotor should be replaced.

2. Install the field coil.

3. Install the rotor and key on the crankshaft.

1. Chamfered end
2. Bearing
3. 1 in. pilot
4. 1 7/16 in. plate
5. Installing tool

Installing needle bearing

4. Install the armature assembly, washer and retaining bolt.
5. Install the brake flange.
6. Adjust the clutch as follows:
a. With the clutch disengaged, check the clearance between the rotor and armature assembly. Using a feeler gauge, check the clearance at both access slots between the ears of the brake flange. The air gap must be .010 to .015 in.
b. Tighten or loosen the brake flange mounting nuts to obtain the proper air gap.
7. Connect the clutch wire and battery ground cable.
8. Install the hood and grille assembly.

REAR POWER TAKE-OFF CLUTCH

REMOVAL

To remove the PTO clutch assembly it is necessary to first remove the three point hitch. Use the following procedure:

1. Remove the drawbar support plate.
2. With the hitch at its highest point, remove the headed pin securing the implement lift bar to the lift bracket. Remove the pin through the access hole in the righthand side of the frame.
3. Remove the rockshaft support plate mounting bolts. Lift the rockshaft over the PTO clutch and rest the assembly on the ground.

DISASSEMBLY

1. Remove the brake flange, retaining bolt, washer and armature assembly.
2. Remove the rotor and key.
3. Remove the field coil and spacer.
4. To remove the driveshaft, remove the snapring from the shaft and press the shaft from the housing.
5. Inspect the bearings for smoothness of operation. If necessary to replace a bearing, remove the snapring and press the bearings and spacer from the housing.

ASSEMBLY AND ADJUSTMENT

Reassembly is the reverse of the disas-

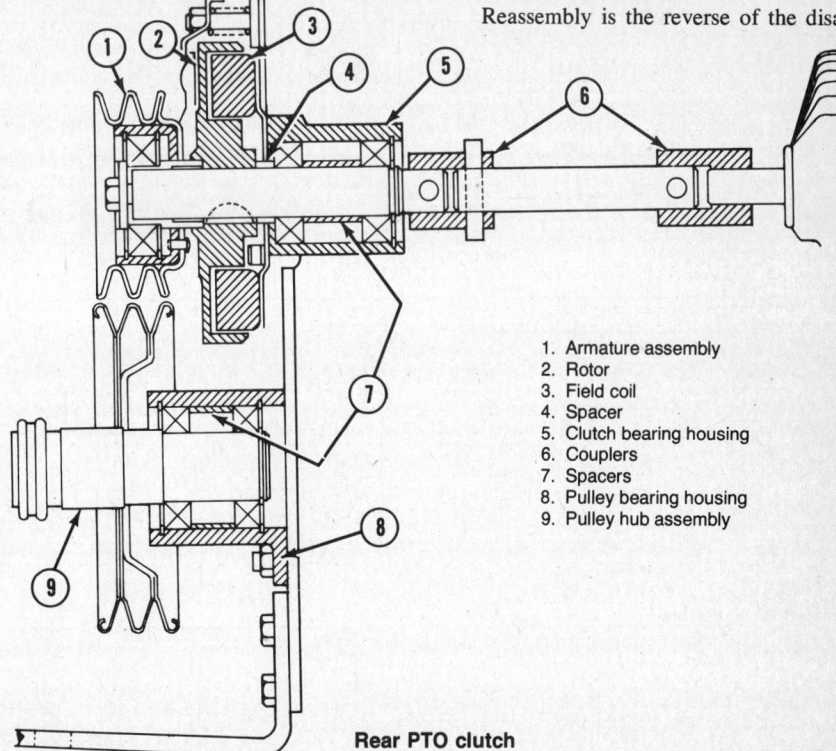

1. Armature assembly
2. Rotor
3. Field coil
4. Spacer
5. Clutch bearing housing
6. Couplers
7. Spacers
8. Pulley bearing housing
9. Pulley hub assembly

Rear PTO clutch

International Harvester

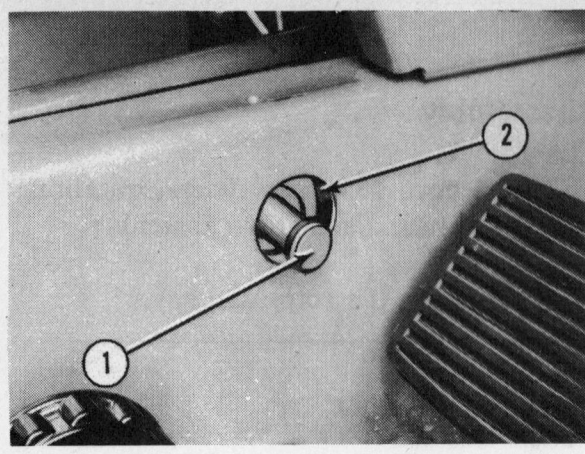

1. Headed pin
2. Access hole

Headed pin removal

sembly procedure, paying special attention to the following.

1. Prior to installation, check the clearance between the armature assembly and the rotor. Using a feeler gauge, check the clearance at three evenly spaced locations. There should be a gap of .054 to .1 in. If the gapis exceeded, the armature assembly and rotor should be replaced.
2. Once assembled, check the clearance between the armature assembly and the rotor by inserting a feeler gauge in the four slots in the brake flange. The gap should be .010 to .015 in.
3. Tighten or loosen the brake flange mounting nuts to obtain the correct gap.

ENGINE

The engines used in these tractors are Kohler Onan, and Briggs and Stratton engines. For complete engine service, see the Onan, Kohler or Briggs and Stratton engine part of the Engine Unit Repair section of this book.

REMOVAL AND INSTALLATION

Model 582

1. Disconnect the battery ground cable.
2. Raise the hood and remove the engine side panels secured by wing nuts and a spring.
3. Disconnect the headlight wiring and remove the hood and grills as an assembly.
4. Disconnect the alternator-regulator wire.
5. Disconnect the starter wire and the PTO clutch wire.
6. Remove the air cleaner assembly.
7. Disconnect the choke and throttle cables. Disconnect the engine shut-off wire.
8. Shut off the fuel and disconnect the fuel line at the carburetor. Be sure to plug the line.
9. Remove the engine mounting bolts and slide the engine forward.

10. Using a suitable hoist and sling, lift out the engine.
11. Installation is the reverse of removal.

Models 682, 782

1. Disconnect the battery ground cable.
2. Raise the hood and remove the engine side panels secured by wing nuts and a spring.
3. Disconnect the headlight wiring and remove the hood and grille as an assembly.
4. Remove the air cleaner assembly. Disconnect the choke cable, throttle cable and wire harness.
5. Disconnect the PTO clutch wire and the starter wire (lefthand side of engine).
6. Shut off the fuel and disconnect the fuel line at the tank. Be sure to plug the line.
7. Remove the nuts securing the front flex coupler to the flywheel flange.
8. Remove the engine mounting bolts. Using a suitable hoist and sling, remove the engine.
9. Installation is the reverse of removal.

Model 982

1. Disconnect the battery ground cable.
2. Raise the hood and remove the engine side panels secured by wing nuts and a spring.
3. Disconnect the headlight wiring and remove the hood and grille as an assembly.
4. Disconnect the PTO clutch wire and the starter wire.
5. Remove the air cleaner assembly and disconnect the choke and throttle cables.
6. Shut off the fuel and disconnect the fuel line from the fuel pump. Be sure to plug the line.
7. Disconnect the lead at the positive (+) terminal on the coil.
8. Disconnect the lead at the rectifier.
9. Remove the nuts securing the front flex coupler to the flywheel flange.
10. Remove the engine mounting bolts.
11. Install lifting brackets. Using a suitable hoist and sling, remove the engine.

Model 284

FRONT AXLE

REMOVAL AND DISASSEMBLY

1. Using a floor jack, raise the front of the tractor enough to remove the weight from the front axle.
2. Support the tractor using two FES 142-1B adaptor plates with safety stands.
3. Disconnect the drag link from the knuckle arm.
4. Loosen the locknut on the set bolt at the rear of the pivot pin. Remove the set bolt.
5. Remove the pivot pin using a slide hammer with FES 544-2 axle pivot pin pulling adaptor.

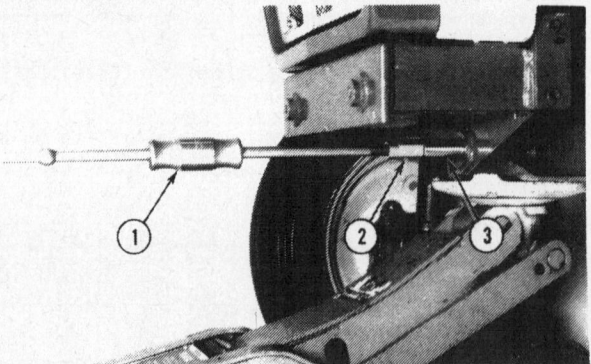

1. Slide hammer
2. FES 544-2 axle pivot pin pulling adaptor
3. Pivot pin

Removing pivot pin with a slide hammer

International Harvester

6. Raise the front of the tractor until the front axle clears the chassis.

7. Roll the front axle forward and support the axle with safety stands.

8. Disconnect the tie rod ball joint assemblies from the tie rod arms.

9. Remove the axle extension mounting bolts. Remove the axle extensions.

10. If necessary remove the center pin bushings using a brass drift.

ASSEMBLY AND INSTALLATION

1. If the clearance between the pivot pin and the bushings exceeds .012 in., replace the bushings using an OTC 27488 handle with OTC 27501 pilot and OTC 27505 driver disc.

2. Insert the axle extensions, torquing the bolts to 170-228 ft.lb.

3. Connect the tie rod ball joints to the tie rod arms. Torque the castellation nuts to 36-51 ft.lb. and secure with the cotter pins.

4. Position the front axle in place under the tractor.

5. Insert a 14 mm bolt into the pivot pin and tap the pivot pin into place.

6. Install the set bolt securing the pivot pin and tighten the locknut.

STEERING

Spindle and Hub

REMOVAL

1. Remove the load from the front axle and wheels using a floor jack.

2. Remove the front wheel weights if equipped. Then remove the front wheels.

3. Remove the hub cap.

4. Remove the cotter pin, castellated nut and washer.

5. Remove the outer bearing and the hub from the spindle using an OTC 951 puller and an OTC 927 bridge.

6. Disconnect the tie rod from the tie rod arm.

7. Remove the knuckle arm.

8. Tap the spindle down through the axle extension arm and remove.

9. If necessary remove the inner bearing and grease seals from the spindle. Replace the thrust washer if necessary.

NOTE: Inspect the inner and outer bearings for wear or damage. Replace if necessary.

10. Inspect the oil seal for wear. Remove the bushings if necessary using a slide hammer and puller.

INSTALLATION

1. If necessary, install new bushings into the axle extension arm using an OTC 27488 handle with an OTC 27501 pilot and OTC 27505 driver disc.

2. Install the thrust washer onto the spindle.

3. Install a new grease seal and bearing if they were removed.

NOTE: Repack the bearings and hub with IH 251 HEP grease or equivalent.

4. Insert the spindle into the axle extension arm. Secure with the knuckle arm and key. Torque the securing bolt to 47-54 ft.lb.

5. Connect the tie rod to the tie rod arm. Torque the castellation nut to 36-51 ft.lb. and secure with the cotter pin.

6. Install the hub with the outer bearing.

7. Install the castellation nut with washer, torquing the nut to 36-72 ft.lb. Then loosen the nut $1/6$ of a turn and secure with the cotter pin.

8. The rolling force of the hub should now measure 1.6-3.0 lb. force.

9. Install the hub cap.

10. Install the front wheels torquing the mounting bolts to 72-90 ft.lb. Install the front wheel weights.

Steering Gear

REMOVAL

1. Remove the muffler.
2. Remove the radiator grill.

NOTE: The radiator grill is removed so the engine hood mounting bolts can be removed easier.

3. Remove the engine hood with side sheets.

4. Shut off the fuel cock valve at the bottom of the fuel tank.

5. Disconnect the fuel line from the sediment bowl.

6. Disconnect the wiring at the fuel gauge.

7. Remove the fuel tank strap. Remove the fuel tank.

8. Disconnect the choke cable and the tachometer cable from the engine.

9. Disconnect the governor control rod, the throttle control rod and the accelerator rod.

10. Remove the bolt securing the governor control lever to the steering gear box.

Steering gear and column

Remove the governor control lever.

11. Remove the steering wheel using an OTC 951 puller with an OCT 927 bridge.

12. Remove the bolts securing the fuel tank support to the steering gear box and the center section.

13. Remove the instrument panel with the fuel tank support.

14. Disconnect the drag link from the Pitman arm.

NOTE: Before removing the Pitman arm from the steering gear box, note the timing marks on the sector shaft and the arm for proper reassembly.

15. Remove the nut securing the Pitman arm and remove the arm using a puller if necessary.

16. Remove the steering gear box from the center section.

DISASSEMBLY

1. Remove the bolts securing the side cover to the steering gear box.

2. Unlock the adjusting nut and screw, the adjusting screw in through the side

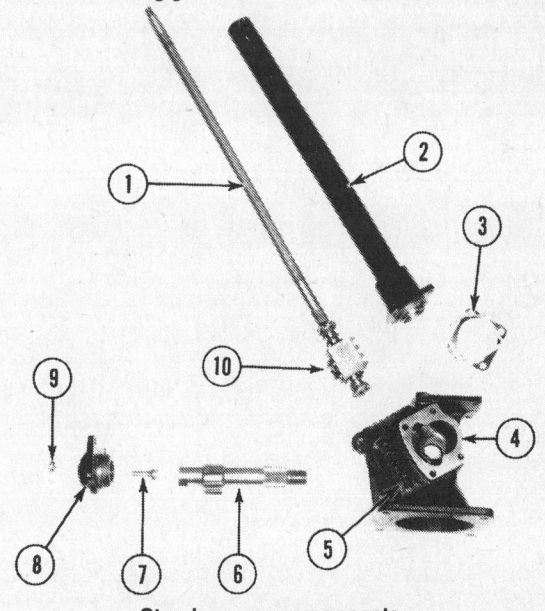

Steering gear components

International Harvester

cover. Remove the side cover and drain the steering gear box fluid.

3. Remove the sector shaft from the steering gear box.

4. Remove the bolts securing the steering column to the gear box.

5. Remove the steering column and shaft with the ball-nut attached.

ASSEMBLY AND INSTALLATION

NOTE: New gaskets and seals must be used during reassembly. Grease the bushings and oil seals using IH HEP 251 grease.

1. Position the lower steering shaft bearing into the steering gear box.

2. Install the steering shaft and the ball-nut assembly with the upper bearing in place into the gear box.

3. Install the steering column with the shim set and secure with capscrews.

4. Insert the sector shaft being sure to center it properly with the teeth on the ball-nut assembly.

5. Install the adjusting screw with retaining strap on the sector shaft.

6. Install the gear box side cover with gasket and secure the adjusting screw with the locknut.

7. Fill the steering gear box to the bottom of the filler plug, lefthand side, using IH gear lubricant (SAE 90 oil).

8. Install the steering box assembly onto the center section. Continue installation in the reverse order of removal.

NOTE: When installing the Pitman arm align the timing marks on the arm with the timing mark on the sector shaft. Torque the nut to 109-145 ft.lb.

ADJUSTMENTS

Steering Wheel Free Play

The specified rotating free play is $^{13}/_{32}$-$^{15}/_{32}$ in. measured at the outer circumference of the steering wheel. Adjust the free play as follows:

1. Loosen the locknut on the adjusitng screw at the steering gear box.

2. Turning the adjusting screw in clockwise decreases the free play. Backing the adjusting screw out counterclockwise increases the free play.

3. After adjustment, torque the locknut to 18-24 ft.lb.

1. Brake piston
2. O-ring
3. Bearing retainer
4. 8 mm capscrews
5. ⅜ in. bolts
6. FES 544-1 brake piston tool

Brake disc removal

BRAKES

Brake Discs and Pistons

REMOVAL

1. Remove the axle carrier assembly.
2. Remove the inner brake ring.
3. Disconnect the brake pressure line.
4. Install the FES 544-1 brake piston tool as shown. Insert two 8 bolts into the threaded holes in the brake piston.
5. Pull the piston out evenly using the two ⅜ in. bolts. Alternately tighten each bolt no more than 1½ turns.

INSTALLATION

1. Using petroleum jelly, lubricate the O-rings and the contact surfaces of the differential retainers and brake pistons. Position the piston into the bore.

2. Install the FES 554-1 brake piston tool as shown. Hand tighten the three 12mm bolts until the tool is in firm contact with the piston in three places.

3. Tighten the ⅜ in. bolts evenly, no more than 1½ turns at a time.

4. Push the brake piston in until it is just below the surface of the rear frame.

Brake Cylinders

REMOVAL AND INSTALLATION

1. Disconnect the brake fluid supply hose from the elbow on the brake cylinder. Plug the line quickly to prevent fluid loss.

2. Disconnect the brake linkage from the brake pedal.

3. Disconnect the brake pressure line from the brake cylinder.

4. Remove the necessary mounting bolts and remove the brake cylinder.

5. Installation is the reverse of removal.

Parking Brake

ADJUSTMENT

Adjust the parking brake lever by loosening or tightening the adjusting nuts on the parking brake rod. The specified pull at the

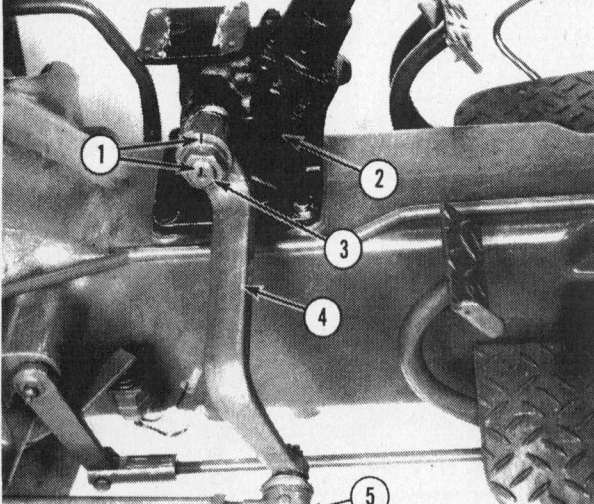

1. Timing marks
2. Steering gear box
3. Sector shaft
4. Pitman arm
5. Drag link

Steering arm installation

International Harvester

1. Brake pressure lines
2. Fluid supply line
3. Fluid supply hoses
4. Brake cylinders
5. Brake linkage

Brake cylinder removal

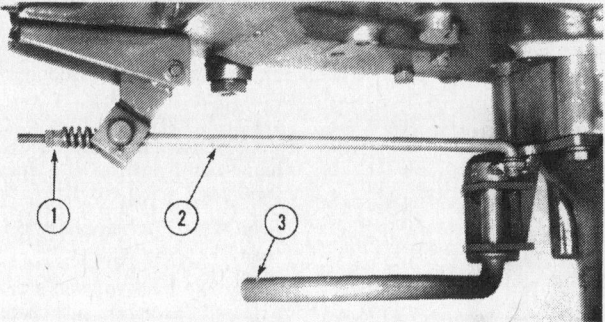

1. Adjusting nuts
2. Parking brake rod
3. Parking brake lever

Parking brake adjustment

end of the lever is 44 lb. as measured with a spring scale.

CLUTCH

REMOVAL

1. Perform a front section split.
2. Index mark the clutch cover assembly to the flywheel for proper reassembly.
3. Remove the mounting bolts securing the clutch assembly to the flywheel and remove the assembly.
4. Cut the lock wire on the release fork set bolt and loosen the bolt.
5. Pull out the clutch release shaft and remove the release fork with the release bearing.
6. Remove the flywheel.

INSTALLATION

1. Position the flywheel on the crankshaft and align the reamer bolt hole on the crankshaft with the "O" mark on the flywheel.
2. Install the longer reamer bolt into this hole, then secure with the other five bolts. Torque the bolts to 64 ft.lb.
3. Position the clutch assembly with the clutch disc to the flywheel. Secure the assembly loosely with the mounting bolts.
4. Insert the input shaft into the clutch assembly and pilot bearing to properly center and align the clutch disc.
5. With the input shaft still inserted, torque the mounting bolts to 17-20 ft.lb.
6. Install the release shaft and fork into the clutch housing. Torque the set bolt to 9-11 ft.lb. and secure with a wire.

ADJUSTMENTS

Pedal Adjustment

As a result of normal clutch facing wear, the free travel between the clutch release lever and the release bearing is reduced. The specified free travel of the clutch pedal is $1^{3}/_{16}$-$1\frac{1}{2}$ in. measured at the front edge of the pedal.

The amount of free travel will decrease as the clutch lining wears. Adjustment should be made before the free travel reaches $^{5}/_{8}$ in.

The free travel is adjusted at the yoke as shown. When screwing the yoke in the free travel increases. Screwing the yoke will decrease the pedal free travel.

NOTE: The limit of adjustment is less than $^{13}/_{16}$ in.

Neutral Start Switch Adjustment

When the clutch pedal is fully depressed the safety switch should be on. Adjust the safety starting switch by loosening the bracket mounting bolts.

The switch should be on at $^{3}/_{16}$ in. of extension tab travel. There should be a free travel of $^{1}/_{8}$ in.

SPLITTING AND RECOUPLING

Splitting

FRONT SECTION

1. Remove the muffler.
2. Remove the radiator grill.

NOTE: The radiator grill is removed so the engine hood mounting bolts can be removed easier.

3. Remove the engine hood with side sheets.
4. Disconnect the hydraulic return line from the hydraulic pump and the filter base.
5. Disconnect all electrical connections to the engine.
6. Disconnect the throttle control rod, the tachometer drive cable and the choke control cable from the carburetor.
7. Shut off the fuel cock valve at the bottom of the fuel tank. Disconnect the fuel hose at the inlet of the fuel pump.
8. Disconnect the drag link from the knuckle arm.
9. Disconnect and remove the high pressure hydraulic line from the hydraulic pump and the control valve.
10. Support the center section using a floor jack. Support the engine using an FES 138-1 sling with an overhead hoist.

International Harvester

1. 10 x 45 mm bolt (1.77 in. long) with lockwasher
2. 10 x 55 mm bolt (2.16 in. long) with nut
3. 10 x 45 mm bolt (1.77 in. long) with lockwasher
4. 10 x 55 mm bolt (2.16 in. long) with lockwasher and nut
5. 10 x 30 mm bolt (1.18 in. long) with lockwasher

Coupling bolt locations

1. Transmission input shaft
2. Coupler
3. Parking brake drum
4. Parking brake band
5. Aligning dowels (made locally)
6. Parking brake rod

Coupling rear section

NOTE: Wedge a wooden block between the front axle and the engine side channel to prevent tipping when splitting. Support the front bolster weights with a safety stand.

11. Remove the mounting bolts securing the center section to the engine. Split the tractor rolling the rear section rearward.

NOTE: **When making a front split, the transmission input shaft may remain in the clutch assembly. Loosen the clutch mounting bolts to remove the input shaft.**

REAR SECTION

1. Disconnect the battery.
2. Drain the Hy-Tran from the rear frame.
3. Remove the hydraulic pressure line cover. Then disconnect the pressure line from the control valve.
4. Place the tractor in neutral and remove the transmission speed lever and retainer.
5. Remove the righthand platform step and disconnect the brake valve linkage from the brake pedals. Remove the hydraulic return line.
   Disconnect and remove the brake supply line from the brake reservoir and the brake cylinders.
6. Disconnect the clutch pedal spring.
7. Disconnect the clutch yoke from the clutch shaft. Then remove the clutch pedal pin.
8. Remove the clutch pedal and rod from the tractor.
9. Disconnect the spring to the parking brake lever. Then disconnect the parking brake rod at the rear.
10. Remove the lefthand platform step.
11. Disconnect the necessary electrical harnesses from the rear frame and bring forward.

NOTE: **The wires are color-coded for ease in reconnecting.**

12. Wedge a wooden block between the axle and the engine side channel to prevent the tractor from tipping when splitting.
13. Support the tractor at the center section using two FES 142-1B adaptor plates and safety stands.
14. Install two aligning dowels (made locally from 12 mm coarse thread bolts 4 in. or longer) into the rear frame, one each side, at the bottom corners.
15. Position a hydraulic floor jack under the tractor to support the rear frame.
16. Remove the mounting bolts securing the center section to the rear frame and split the tractor.

Recoupling

FRONT SECTION

1. Loosen the clutch assembly mounting bolts and install the transmission input shaft into place. Then torque the bolts to 17-20 ft.lb.
2. Place the input shaft coupler onto the transmission input shaft.

International Harvester

3. Roll the rear section towards the engine until the transmission input shaft with the coupler contacts the input shaft in the rear frame.
4. The center section and the engine rear plate should now be within 1½ in. of each other.
5. Insert two mounting bolts (one each side) through the center section and hand start them into the engine rear plate. Tighten them two turns more to pull the center section to the engine.
6. Engage the PTO lever and hand turn the PTO output shaft until the input shaft in the rear frame properly engages with the coupler.
7. Push the center section to the engine and secure with mounting bolts.
8. Continue reassembly in the reverse order of disassembly.

REAR SECTION

1. Place the coupler onto the transmission input shaft in either the center section or the rear frame.
2. Roll the rear frame towards the center section until the coupler contacts the transmission input shaft.
3. With the PTO lever engaged, turn the PTO output shaft by hand until the transmission input shaft properly engages into the coupler.
4. Push the rear frame and center section together and secure them with the mounting bolts. Torque the bolts to 72-88 ft.lb.
5. Continue reassembly in the reverse order of disassembly.

TRANSMISSION

REMOVAL

1. Disconnect the battery.
2. Drain the Hy-Tran from the rear frame.
3. Remove the hydraulic pressure line cover. Then disconnect the pressure line from the control valve.
4. Remove the righthand platform step and disconnect the brake linkage from the brake pedals.
5. Disconnect the brake supply lines from the reservoir.
6. Remove the hydraulic return line.
7. Disconnect the clutch pedal spring.
8. Disconnect the clutch yoke from the clutch shaft. Then remove the clutch pedal pin.
9. Remove the clutch pedal and rod from the tractor.
10. Disconnect the spring to the parking brake lever. Then disconnect the parking brake rod at the rear.
11. Remove the lefthand platform step.
12. Disconnect the necessary electrical harnesses from the rear frame and bring them forward.

NOTE: The wires are color coded for ease in reconnecting.

13. Wedge a wood block between the axle and the engine side channel to prevent the tractor from tipping when splitting.
14. Support the tractor at the center section using two 17-142-1B adaptor plates and safety stands.
15. Remove the leveling link assemblies.
16. Disconnect the necessary electrical lines and remove the fenders.
17. Position a hydraulic floor jack under the rear frame.

1. Hydraulic return line
2. Brake cylinders
3. Brake linkage

Hydraulic line disconnections

18. Raise the rear of the tractor slightly and remove the tires and wheels.
19. Remove the battery.
20. Support the axle carriers using an 17-138-1 sling with an overhead hoist.
21. Remove the axle carrier assemblies. Then remove the inner brake ring.
22. Remove the mounting bolts securing the hitch cover to the rear frame. Remove the hitch cover.
23. Install two aligning dowels (made

1. 17-142-1B adaptor plate
2. Wood block
3. Safety stand

Supporting the tractor

589

International Harvester

from 12 mm coarse thread bolts 4 in. or longer) into the rear frame, one each side, at the bottom corners.

24. Support the rear frame using three 17-100-3 lifting brackets and two 17-138-1 slings with an overhead hoist.

25. Remove the mounting bolts securing the rear frame to the center section and split the tractor.

INSTALLATION

1. Support the rear frame using three 17-100-3 lifting brackets and two 17-138-1 slings with an overhead hoist.

2. Position the coupler onto the transmission input shaft.

3. Push the rear frame to the center section. Properly engage the coupler with the input shaft by hand turning the PTO output shaft when the PTO is engaged.

NOTE: Be sure the parking brake is in the unlock position so damage does not occur to the brake band.

4. Secure the rear frame to the center section torquing the mounting bolts to 72-88 ft.lb.

5. Support the rear frame with a floor jack, removing the hoists and slings.

6. Install the hitch cover torquing the mounting bolts to 43-51 ft.lb.

7. Install the inner brake ring.

8. Support the axle carrier with an 17-138-1 sling and an overhead hoist.

9. Install the axle carrier torquing the mounting bolts to 69-90 ft.lb.

10. Install the hydraulic return line.

11. Connect the brake supply line to the reservoir.

12. Install the leveling link assemblies.

13. Raise the rear slightly and install the fenders, tires and wheels.

NOTE: Torque the rear wheel mounting bolts to 180-230 ft.lb.

14. Connect the electrical harnesses.

15. Install the lefthand platform step.

16. Connect the parking brake rod to the lever. Connect the parking brake spring to the lever.

17. Install the clutch pedal and rod to the tractor.

18. Install the clutch pedal pin and secure with the bolt.

19. Connect the clutch yoke to the clutch shaft. Then connect the clutch spring.

NOTE: Adjust the parking brake lever by loosening or tightening the adjusting nuts on the parking brake rod. The specified pull at the end of the lever is 44 lb. as measured with a spring scale.

20. Connect the brake linkage to the brake pedals. Install the righthand platform.

21. Connect the hydraulic pressure line to the control valve. Install the hydraulic pressure line cover.

22. Refill the rear frame with IH Hy-Tran through the filler cap in the hitch cover; Approximately 3.4 gals.

23. Start the engine and operate the hydraulic systems.

24. Check the fluid level again, filling the rear frame until Hy-Tran flows from the side fluid level hole with the plug removed.

OVERHAUL

1. Support the rear frame with an 17-142-13 adaptor plate and safety stands as shown.

2. Remove the speed transmission top cover.

3. Disconnect the battery.

4. Using a magnet probe, remove the springs and detent balls.

5. Remove the washers, springs and the differential lock fork along with the differential lock rod.

6. Remove the roll pins securing the transmission shift rods. Then drive the shift rods out through the front of the rear frame.

7. Remove the PTO input shaft and retaining cover by inserting two capscrews into the threaded holes in the cover.

NOTE: If the input shaft remains in the inner bearing, remove the shaft by driving against the coupler using a brass drift.

1. PTO retaining cover
2. Capscrews
3. Shims
4. PTO input shaft
5. Coupler

PTO shaft components in transmission

8. Remove the transmission input shaft cover.

9. Remove the snapring securing the transmission input gear in place. Slide the snapring, input gear and the spacer rearward.

10. Using a brass drift, drive the transmission input shaft out of the housing.

11. Install the 12-544-1 brake piston tool. Insert two 8 mm bolts into the threaded holes in the brake piston.

12. Pull the piston out evenly using the two ⅜ in. bolts. Alternately tighten each bolt no more than 1½ turns.

13. Support the differential using an 17-138-1 sling with a suitable hoist.

14. Remove the bolts securing the differential bearing retainers in place.

15. Install two capscrews into the threaded holes in the retainers. Tighten the bolts alternately and evenly to force the bearing retainer out of the rear frame.

16. Remove the retainer and shims. Keep the shims with the side from which they were removed.

17. Remove the differential.

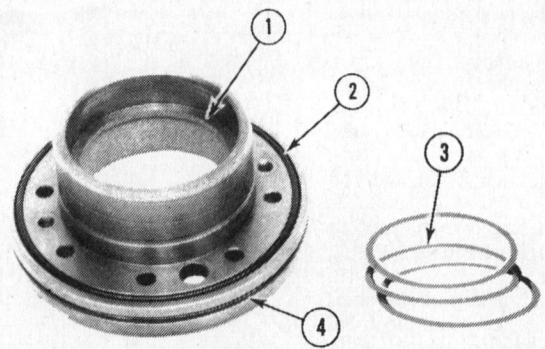

1. Differential bearing retainer
2. O-ring
3. Shims (as required)
4. O-ring

Differential bearing retainer

International Harvester

Mainshaft Removal

1. Release the nut lock and remove the nut securing the parking brake drum to the mainshaft.
2. Remove the brake drum, washer and key.
3. Release the lock straps and remove the capscrews securing the mainshaft bearing retainer to the rear frame.

NOTE: Install two capscrews into the threaded holes in the retainer. Tighten the bolts alternately and evenly to force the retainer from the housing.

4. Remove the bearing retainer using an OTC 1001 puller. Use a step plate to prevent damage to the mainshaft.

NOTE: Be sure to keep the shims together when the bearing retainer is removed.

5. Support the gears on the mainshaft by placing a wood block between the 2nd and 3rd gear.
6. Install the end nut onto the mainshaft. Using a brass hammer, drive the mainshaft rearward until the tapered inner roller bearing can be removed.
7. Remove the gears, spacers and needle bearings from the mainshaft.
8. If necessary, remove the mainshaft inner bearing by first removing the snapring securing the bearing in place.

Countershaft Removal

1. Remove the set bolts securing the reverse fork rod and the reverse idler shaft in place.
2. Remove the reverse fork and the reverse fork rod.
3. Disengage the snaprings on the reverse idler shaft and remove the reverse idler gear, the needle bearings and the shaft.
4. Remove the set bolt securing the high-low shift rod.
5. Remove the high-low shift rod and the shift fork being careful not to lose the detent ball and spring under the plug.
6. Using a brass hammer, tap the countershaft out towards the front until the outer bearing clears the frame.
7. Remove the bearing using an OTC 951 puller and OTC 927 bridge.
8. Remove the countershaft assembly.
9. Remove the bolt, locking washer, securing pin and flat washer (shims where applicable).

Countershaft Installation

1. Install the snapring or split ring, gears and spacers onto the countershaft.

NOTE: Should the snapring have to be replaced on tractors S/N J011052 and below, the countershaft, 4th gear, and snapring must be replaced due to production modifications.

2. Press the inner bearing onto the countershaft.

IMPORTANT: Tractors with S/N J011052 and below require shims between the flat washer and the end of the countershaft. To determine the number of shims necessary, follow steps 3 and 4.

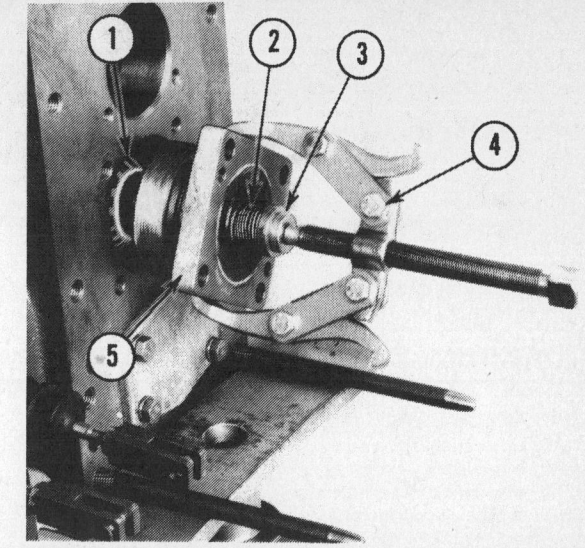

Pulling the mainshaft

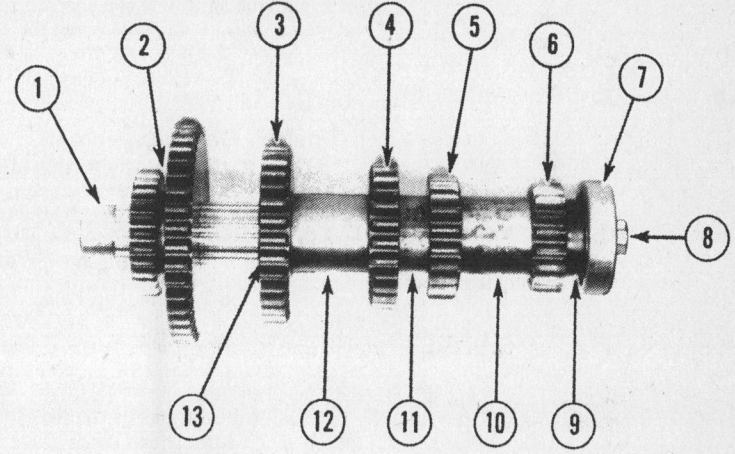

1. Countershaft
2. High-low gear
3. 4th gear
4. 3rd gear
5. 2nd gear
6. 1st gear
7. Inner bearing
8. Securing pin, washers, bolt
9. 11 mm (.433 in.) spacer
10. 40 mm (1.574 in.) spacer
11. 16.5 mm (.650 in.) spacer
12. 43 mm (1.693 in.) spacer
13. Snapring or split ring

Countershaft components

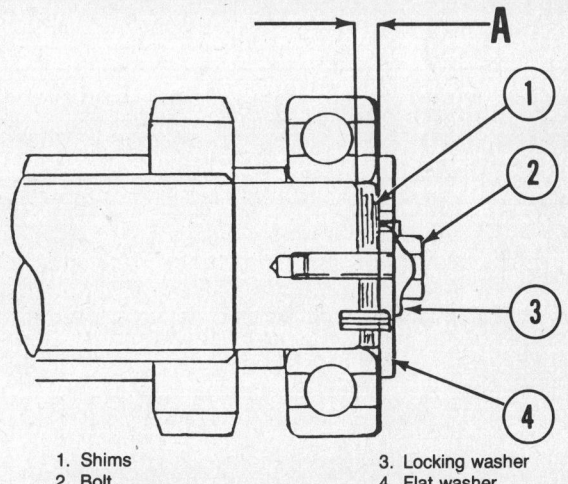

1. Shims
2. Bolt
3. Locking washer
4. Flat washer

Bearing-to-end of shaft measurement

International Harvester

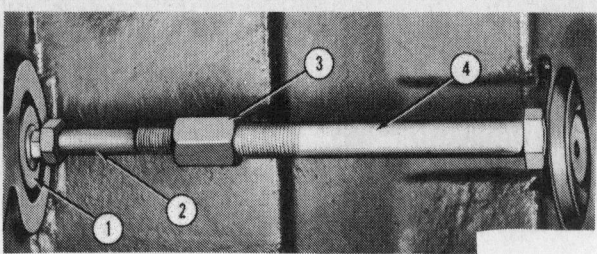

1. Countershaft assembly
2. ½ in. x 2¾ in. NF bolt
3. OTC 916 connector
4. ⅝ in. x 5½ in. NF bolt

Supporting the end of the countershaft

Tractors with S/N J011053 and above, proceed to step 5.

3. Measure the depth from the bearing to the end of the shaft, dimension "A".
4. Select shims so that the thickness is the same to .002 in. less than the depth.
5. Install the flat washer, securing pin, locking washer and bolt. For S/N J011053 and above, shims are not required. Torque the bolt to 11-14.5 ft.lb. Do not exceed recommended torque.
6. Position the countershaft in place in the rear frame. Using a brass drift, drive the shaft into the housing until the inner bearing is fully seated in the counterbore of the rear frame.
7. Support the rear end of the countershaft using an OTC 916 reducing connector, a ⅝ in. × 5½ in. NF bolt and a ½ in. × 2¾ in. NF bolt.
8. Using an OTC 27528 driver disc, start the outer bearing into the housing and continue to drive until the bearing is started onto the countershaft.
9. With a pipe of a suitable ID, drive the outer bearing to its seat on the countershaft.
10. Measure dimension "A" on the countershaft cover and dimension "B" on the housing. Subtract "A" from "B" to find the required shim pack "C".

11. Insert the required shim pack. Then install the countershaft cover, torquing the mounting bolts to 23-34 ft.lb.
12. Install the reverse idler shaft with the rear snapring. Then install the thrust washers, needle bearings and the reverse gear (tapered teeth to the rear). Secure with the front snapring.
13. Install the reverse fork and the reverse fork rod.
14. Insert the high-low shift rod into place with the shift fork. Be sure the spring and detent ball are in place under the plug.
15. Secure the high-low shift rod in position with the set bolt torqued to 18-25 ft.lb.

Mainshaft Bearing Preload

NOTE: If the mainshaft bearing retainer or bearings are replaced, the bearing preload must be checked. Be sure to lubricate the bearings using clean IH Hy-Tran before testing the preload.

1. Press the bearing cups into the retainer.
2. Position the outer bearing into the retainer with a ⅞ in. × 4 in. bolt through the outer bearing.
3. Using the original spacer, assemble the retainer assembly and support in a vise. Torque the ¾ in. bolt to 30-40 ft.lb. while rotating the bearing retainer.
4. Using a string and spring scale, measure the force required to rotate the retainer. A force of 3½-4½ lb. should be obtained.
5. If a proper force is not obtained, select a suitable spacer from the list. Going to a thinner spacer will increase the rotating force; a thicker spacer will decrease the force.

Part Number	Spacer Thickness	
	mm	inches
973 930 C1	8.56	0.3370
" 931 C1	8.58	0.3378
" 932 C1	8.60	0.3386
" 933 C1	8.62	0.3394
" 934 C1	8.64	0.3402
" 935 C1	8.66	0.3410
" 936 C1	8.68	0.3417
" 937 C1	8.70	0.3425
" 938 C1	8.72	0.3433
" 939 C1	8.74	0.3441
" 940 C1	8.76	0.3449
" 941 C1	8.78	0.3457
" 942 C1	8.80	0.3465
" 943 C1	8.82	0.3472
" 944 C1	8.84	0.3480
979 740 C1	8.86	0.3488
" 741 C1	8.88	0.3496
" 742 C1	8.90	0.3500
" 743 C1	8.92	0.3512
" 744 C1	8.94	0.3520
" 745 C1	8.96	0.3528
" 746 C1	8.98	0.3535
" 747 C1	9.00	0.3543

Mainshaft Installation

1. Install the inner bearing onto the mainshaft if it was removed. Secure the bearing into place with the snapring.
2. Place the end nut onto the mainshaft and drive the shaft through the inner wall of the rear frame.
3. Install the gears, thrust washers and needle bearings onto the mainshaft.
4. Using a brass hammer, drive the mainshaft rearward until the inner bearing is flush with the housing.
5. Support the mainshaft using a wood block with a nut and bolt.
6. Drive the tapered inner roller bear-

Measuring countershaft shim pack

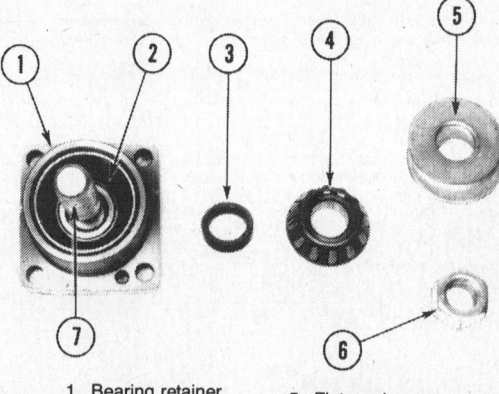

1. Bearing retainer
2. Outer bearing
3. Spacer
4. Inner bearing
5. Flat washers
6. Nut
7. ⅞ in. x 4 in. bolt

Mainshaft bearing components

ing to its seat on the mainshaft using a suitable pipe.

7. Install the bearing spacer.

8. Drive the outer tapered roller bearing to its seat using a suitable pipe.

NOTE: If the mainshaft was replaced with new one, proceed with steps 9 and 10. If the original shaft is being used, proceed to step 11.

9. Continue driving on the bearing race until the bearing retainer is seated against the rear frame.

10. Set a pinion gear (mainshaft) to the specified distance as follows:

a. Position the 89-143-2 bevel pinion gauge bar as shown. Use one of the existing threaded holes in the bar with a nut and bolt on one side. Brace the bar on the other side using a nut and bolt.

Example:

measured distance	1.088 inch
	−.015
actual distance	1.073

NOTE: Two examples for step "c" will be shown, for the case when the third figure is a two digit number, and for the case when the third figure is a single digit number.

978 ±0 (+12)	978 −0.35 (+3)
Move decimal point one place to the left.	Move decimal point two places to the left
0 (+1.2)	−0.35 (+0.03)
Then subtract −	
0.0	0.35
−1.2	−0.03
−1.2 mm	−0.38 mm
Convert to inches	
−1.2 mm = −.047 inch	
−0.38 mm = −.015 inch	
Add this result to figure in step "b"	
1.073	1.073
+(−.047)	+(−.015)
1.026 required mounting distance	1.058
Subtract required from specified	
1.062	1.062
−1.026	−1.058
.036 required shim pack	.004

b. Measure the distance between the bar and the mainshaft (pinion gear) using a telescoping gauge. Then subtract .015 in. from this reading to account for the gap between the housing bore and the bar.

IMPORTANT: When measuring with the gauge be sure it is as close to the center of both the pinion gear and the bar as possible to obtain a proper reading.

c. Three sets of numbers will be found etched on the face of the pinion gear. These numbers refer to measurements in the metric system. Subtract the third set from the second set. Add this result to the figure obtained in step "b" (actual distance). This will give you the required mounting distance.

d. The specified distance from the

Positioning the bevel pinion gauge bar

edge of the differential retainer bore to the pinion gear is 1.062 in.

e. Subtracting the required mounting distance in step "c" from the specified distance in step "d", will give you the required shim pack thickness ± .002 in.

NOTE: After the measurements are taken, tap the mainshaft forward slightly to exert the proper load when installing the bearing retainer.

11. Install the required shim pack and torque the mainshaft bearing retainer mounting bolts to 40-54 ft.lb. Secure with the locking straps.

12. Install the key and the parking brake drum. Torque the mainshaft end nut to 130 ft.lb. and secure with the nut lock.

13. Install new O-rings onto the differential bearing retainers, lubricating the O-rings with petroleum jelly. Install the bearing retainers and shims. Torque the bolts evenly to 20-24 ft.lb.

NOTE: Install a new O-ring onto the brake piston, being sure to lubricate the O-ring with petroleum jelly.

14. Install the 12-544-1 brake piston tool. Hand tighten the three 12 mm bolts until the tool is in firm contact with the piston in three places.

15. Tighten the ⅜ in. bolts evenly, no more than 1½ turns at a time. Install the piston until it is just below the surface of the housing.

16. Install the transmission input shaft into the housing.

17. Install the transmission input cover torquing the capscrews to 12-17 ft.lb.

18. Install the transmission shift rods as follows:

a. Install the "1st-2nd" gear shift rod with the fork and hub and secure with roll pins.

b. Remove the side plug on the housing and insert a drift.

c. Drop two detent balls into the center hole, and with the drift push them towards the "1st-2nd" gear shift rod.

d. Push the "3rd-4th" gear shift rod into place and install two more detent balls as in step "c", between the "3rd-4th" gear shift rod and the reverse gear shift rod.

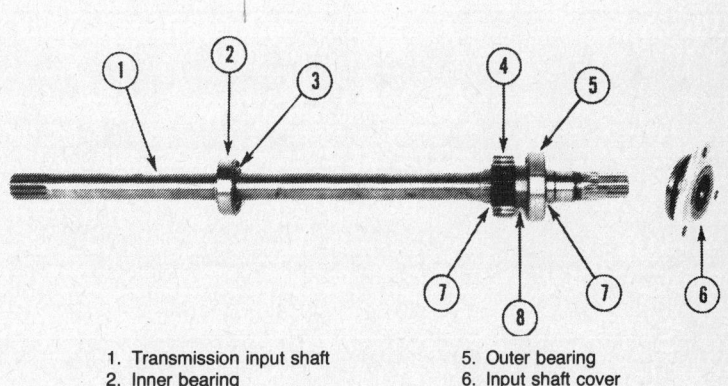

1. Transmission input shaft
2. Inner bearing
3. Snapring
4. Input gear
5. Outer bearing
6. Input shaft cover
7. Snapring
8. Spacer

Input shaft components

International Harvester

e. Install the reverse gear shift rod and secure with roll pins.

19. Install the three shift rod housing end caps.

20. Install the PTO input shaft with the inner bearing against the counterbore in the housing.

21. Install the input cover torquing the capscrews to 20-24 ft.lb.

22. Install the differential lock rod, fork, springs and washers. Secure with the roll pins.

23. Install the shift rod detent balls and springs.

24. Install the speed transmission top cover.

DIFFERENTIAL

REMOVAL

For removal procedures, see the first 23 steps under Transmission Removal.

OVERHAUL

1. Unlock the locking plates securing the ring gear capscrews. Remove the capscrews and the locking plates.

2. Pull the bearing from the ring gear using an OTC 951 puller with an OTC 927 bridge.

3. Insert two capscrews into the threaded holes in the ring gear and force the ring gear off.

4. Position the bevel gear so its lubrication holes are in line with the pinion shaft and setscrew.

5. Turn the bevel gear over and note that the lubrication holes in the pinion gears are in line with the lubrication holes in the bevel gear. The lubrication holes in the inner bevel gear are 90° out of line.

6. Remove the setscrew retaining the pinion shaft in place.

7. Remove the inner bevel gear, the pinion gears and the thrust washers.

8. Pull the lefthand bearing from the differential case using an OTC 951 puller with an OTC 927 bridge.

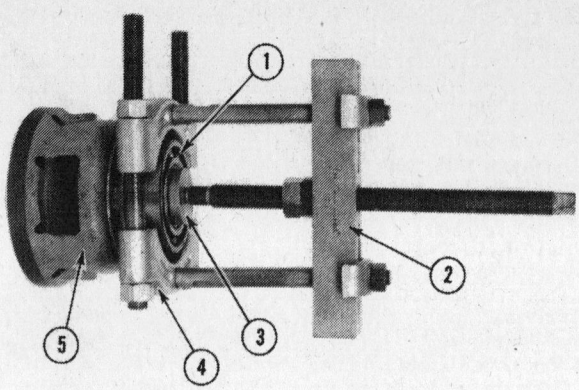

1. Lefthand bearing
2. OTC 927 bridge
3. Step plate
4. OTC 951 puller
5. Differential case

Pulling lefthand bearing from the differential case

ASSEMBLY

NOTE: During reassembly be sure to lubricate all moving parts with petroleum jelly.

1. Install the differential lock clutch and the spacer.

2. Install the lefthand bearing onto the differential case using an OTC 27488 handle with an OTC 27505 pilot and an OTC 27522 driver disc.

3. Install the inner bevel gear with its thrust washer into the differential case.

4. Install the pinion gears and thrust washers. Be sure to keep the pinion gear lubrication holes 90° out of line from the lubrication holes of the inner bevel gear.

5. Secure the pinion gears with the pinion shaft and install the setscrew to retain the shaft.

6. Install the outer bevel gear with its lubrication holes in line with the lubrication holes of the pinion gears.

7. Install a new thrust washer. Then install the ring gear to the differential case.

8. Secure the ring gear to the differential case. Torque the capscrews to 43-51 ft.lb. then bend the lock plates.

9. Install the bearing to the ring gear using an OTC 27488 handle with an OTC 27505 pilot and an OTC 27522 driver disc.

NOTE: After reassembly, install a sun shaft into the differential to confirm that the gears rotate smoothly.

DIFFERENTIAL BEARING ADJUSTMENT

1. Using a suitable hoist and an FES 138-1 sling support the differential in position in the rear frame.

2. Install the right and left differential bearing retainers without O-rings or shims. Secure the retainers with four capscrews.

NOTE: The lubrication holes should be positioned at the bottom.

3. Slide the differential assembly to the left so that the backlash becomes zero.

4. Using a feeler gauge, measure the gap between the bearing and the bearing retainer. Obtain a reading from the right and left sides.

5. Determine the required shims by subtracting .012 in. from the reading at the right side in step 4. Insert the shims into the right bearing retainer. Install the retainer, torquing the bolts to 20-24 ft.lb.

6. For the left retainer, determine the required shims by adding .012 in. to the reading obtained on the left side in step 4.

NOTE: In no case should the shims on the left side exceed the above thickness. Excessive shimming will result in bearing damage.

7. Insert the required shims into the left bearing retainer. Install the retainer torquing the bolts to 20-24 ft.lb.

8. Using an FES 67 dial indicator, measure the backlash between the ring gear and the pinion gear. The backlash should be .005-.008 in. Take readings at two points 180 degrees apart.

NOTE: When checking the backlash be sure to block the mainshaft to keep it from turning. Keep the dial indicator in line with the ring gear and at the very tip of the gear tooth.

9. Adjust the backlash by moving the

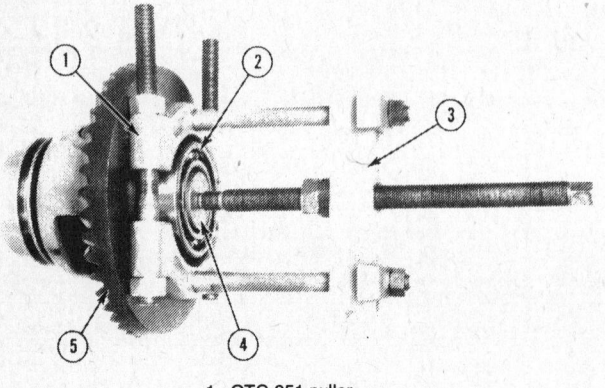

1. OTC 951 puller
2. Righthand bearing
3. OTC 927 bridge
4. Step plate
5. Ring gear

Pulling the bearing from the ring gear

bearing retainer shims from one side to the other. In no case should shims be added or removed, since this will change the preload.

10. Remove the bearing retainers and install the O-rings being sure to lubricate the O-rings with petroleum jelly.

11. Install the bearing retainers torquing the bolts to 20-24 ft.lb. Then secure with the locking straps.

12. Continue reassembly in reverse of disassembly.

DIFFERENTIAL LOCK ADJUSTMENT

1. Rotate the differential assembly so that the lock clutch will not engage with the inner bevel gear in the differential.

2. Depress the differential lock pedal until the lock clutch comes in contact with the inner bevel gear.

3. The gap between the differential lock bracket and the differential lock pedal should be between .004-.040 in.

4. Check the gap using a $1/32$ in. drill. If the gap is excessive, add washers as required between the lock pedal and the retaining ring.

REAR AXLE

REMOVAL AND DISASSEMBLY

1. Drain the Hy-Tran from the rear frame.

2. Disconnect the necessary electrical lines and remove the fender.

3. Remove the leveling link assembly from the side being worked on.

4. Support the axle carrier using an FES 138-1 sling with an overhead hoist.

5. Remove the mounting bolts securing the axle carrier to the rear frame. Then separate the axle carrier from the rear frame.

6. Remove the brake disc and the sun shaft.

7. Remove the outer brake ring by prying evenly with a screwdriver.

8. Remove the axle retaining ring. Then remove the planetary system from the axle carrier.

9. Using an OTC 943 puller, pull the ring gear out of the axle carrier.

10. Remove the axle cap mounting bolts.

11. Push the axle assembly out of the carrier using an OTC 938 bridge.

12. If necessary, remove the axle cap and outer bearing using an OTC 952 puller with an OTC 938 bridge and OTC 930-C-1 legs.

13. Inspect the outer bearing and oil seal for damage. Replace if necessary.

14. If necessary, remove the inner bearing, the spacer and the oil seal from the axle carrier. Use a pipe of suitable length to drive the parts out. Replace these parts with new during reassembly.

ASSEMBLY AND INSTALLATION

1. Install the axle cap with oil seal onto the axle.

2. Press the outer bearing to its seat on the axle using a hydraulic press.

3. Secure the outer bearing to the axle with the snapring.

4. Press the axle into the axle carrier using a hydraulic press. Install the axle cap bolts torquing them to 20-24 ft.lb.

5. If necessary, install the inner bearing using a brass drift. Tap lightly and evenly around the bearing until it is flush with the beveled edge of the carrier.

6. Tap the ring gear in flush with the axle carrier using a wooden block. Then install the retaining pins.

7. Using a brass drift, seat the ring gear fully into the axle carrier.

8. Install the planetary carrier and secure with the snapring.

9. Install the outer brake ring into the axle carrier.

10. Continue installation in the reverse order of removal.

Planet Gear

REMOVAL

For removal procedures, see steps 1-8 of Rear Axle Removal.

DISASSEMBLY

1. From the outside, tap the roll pin into the planet shaft.

2. Drive the planet shaft out of the carrier.

3. Remove the gear, roller bearings and the shims. Inspect these parts for excessive wear or damage. Replace as necessary.

ASSEMBLY AND INSTALLATION

1. Lubricate the bearings lightly and install them into the planet gear.

2. Using original shims, install the planet gear with bearings into the carrier. Secure with the planet shaft.

3. Check to see that the planet gear rotates lightly by hand and that excessive play is not present.

4. Remove or add shims as required and secure the planet shaft with the roll pin.

5. Install the planetary carrier into the axle carrier and secure with the snapring.

6. Continue installation in the reverse order of removal.

POWER TAKE-OFF

REMOVAL AND DISASSEMBLY

1. Disconnect the hydraulic pressure line from the control valve.

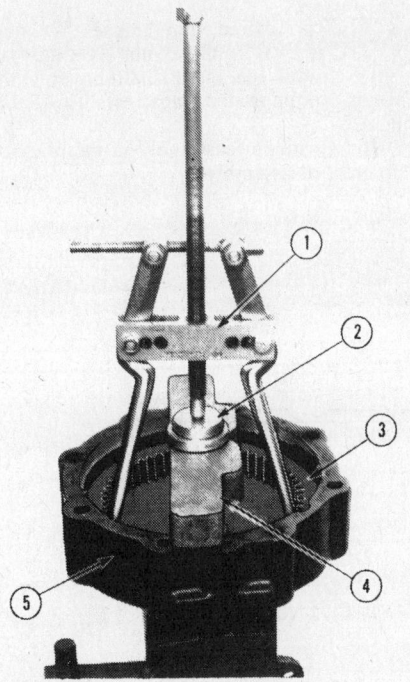

1. OTC 943 puller
2. Step plate
3. Ring gear
4. Cross member
5. Axle carrier

Pulling the ring gear from the axle carrier

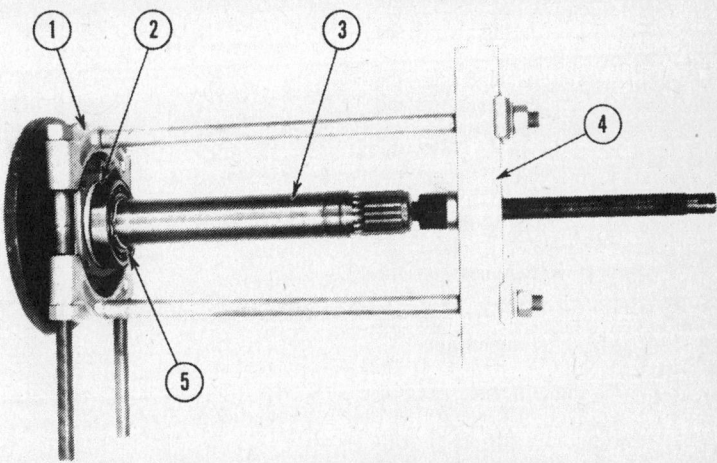

Pulling the axle cap and outer bearing

International Harvester

2. Move the necessary electrical lines out of the way.

3. Disconnect and then plug the brake supply line just below the reservoir.

4. Remove the necessary mounting bolts and remove the hitch cover.

5. Remove the PTO shield.

6. Remove the PTO input shaft and retaining cover by inserting two capscrews into the threaded holes in the cover.

NOTE: If the input shaft remains in the inner input bearing, drive against the coupler using a brass dift to remove the shaft and bearing.

7. Place the retaining cover in a vise with brass jaws.

8. Install an OTC 951 puller on the shaft behind the gear.

9. Tap the puller gently with a brass hammer to remove the shaft from the retaining cover.

10. If necessary, remove the outer input bearing from the shaft and the inner input bearing from the rear frame.

11. Remove the bolt retaining the shifter rod into the rear frame.

12. Pull the shifter rod out of the shift fork, being careful not to lose the ball bearing which is under a spring load in the shift fork.

13. Remove the capscrews securing the output shaft cover to the rear frame. Remove the cover and the shims.

14. Dislodge the snapring which holds the spacer to the outer bearing and slide the snapring towards the front on the PTO output shaft.

15. Pull the output shaft from the inner bearing using an OTC 951 puller.

16. Remove the output shaft, spacers and snapring.

17. Remove the inner and outer input shaft bearings if necessary.

ASSEMBLY AND INSTALLATION

1. If it was removed, install the outer driven bearing onto the driven shaft using a hydraulic press.

2. Install the inner output bearing into the rear frame using an OTC 27488 handle with OTC 27501 pilot disc and OTC 27523 driver disc. Bottom the bearing against the shoulder in the housing.

3. Position the output shaft into the rear frame with the spacers, snapring and slide gear in their related positions.

4. Using a brass hammer, drive the PTO output shaft into place. Secure the snapring into place.

5. Grease the oil seal in the output shaft cover using IH HEP 251 grease.

6. Adjust the shim thickness on the PTO output shaft cover as follows:

a. Measure dimension "A" on the output shaft cover and dimension "B" on the rear frame.

b. The required shim thickness is equal to "A"-"B" minus .004 in.

7. Install the output shaft cover with shims torquing the capscrews to 20-24 ft.lb.

1. PTO input shaft
2. OTC 951 puller
3. Retaining cover
4. Outer bearing

Pulling input shaft gear

8. Position the PTO shift fork and shifter rod in place. Insert the spring and then the ball bearing into the shift fork.

9. While holding the ball down into place with a punch, slide the shifter rod into place.

NOTE: The shifter rod is held in place by a retaining bolt. Using a screwdriver in the slotted end of the shifter rod, turn the rod until the retaining bolt can be properly installed. Install a new end cap.

10. Install the inner input bearing into the rear frame using an OTC 27488 handle with OTC 27501 pilot disc and OTC 27523 driver disc. Bottom the bearing against the shoulder in the housing.

11. Install the outer input bearing onto the input shaft.

12. Position the input shaft assembly into the rear frame.

13. Slide the coupler onto the input shaft for proper alignment with the transmission input shaft. Using a brass hammer, drive the input shaft assembly into place.

14. Adjust the shim thickness on the PTO input shaft cover as follows:

a. Measure dimension "A" and dimension "B" on the input cover, and measure dimension "C" on the rear frame.

b. The required shim thickness is equal to "A"-"B"-"C" minus .004 in.

15. Install the input shaft cover with shims, torquing the capscrews to 20-24 ft.lb.

16. Continue reassembly in the reverse order of disassembly.

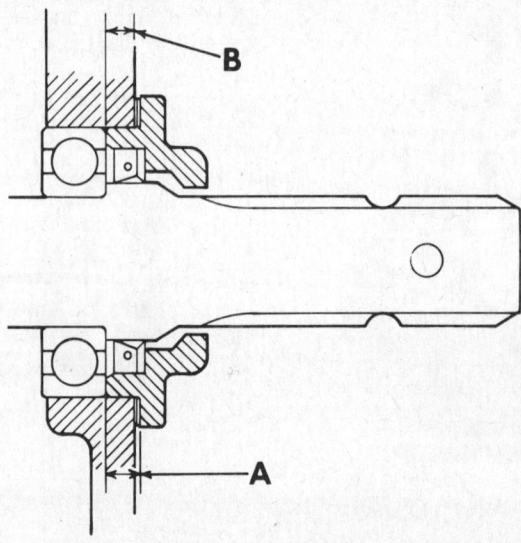

Measuring output shaft cover shim thickness

International Harvester

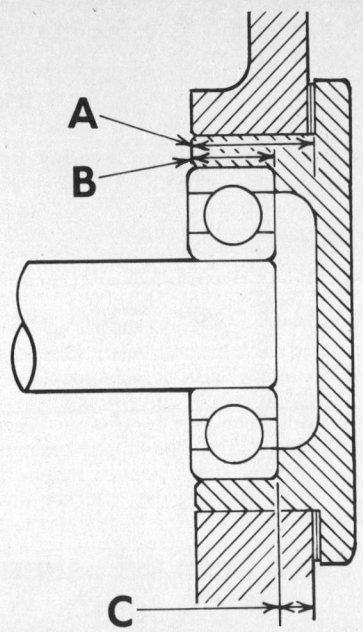

Measuring input shaft cover shim thickness

HYDRAULIC SYSTEM

Load and Position Control Hitch

REMOVAL AND DISASSEMBLY

1. Relieve the system pressure by lowering the hitch to its lowest position.
2. Remove the seat.
3. Remove the leveling link assemblies.
4. Disconnect the hydraulic pressure line from the auxiliary valve.
5. If equipped, disconnect and remove the pressure line from the auxiliary valve.
6. Disconnect the brake fluid supply line from the brake reservoir.
7. Remove the mounting bolts securing the hitch cover to the rear frame. Remove the hitch cover and place it on a work bench.
8. Disconnect the position control rod from the rockshaft arm.
9. Remove the necessary mounting bolts securing the control valve to the hitch cover. Remove the control valve.
10. Remove the capscrews securing the end plates to the rockshaft arms.
11. Remove the rockshaft arms.
12. Remove the setscrew from one side of the housing. Tap the rockshaft and bushing out of the housing using a brass hammer.

NOTE: When the center of the connecting pin is in line with the rib on the housing, a timing mark can be found on a rockshaft spline also in line with the rib.

13. Remove the setscrew from the other side. Drive the bushing out using an OTC 27488 handle with an OTC 27511 pilot and an OTC 27518 driver disc.
14. Remove the front cover from the hitch cover. Remove the piston.
15. Inspect the piston, the O-ring and the back-up ring for wear. Inspect the O-ring in the front cover.

ASSEMBLY AND INSTALLATION

NOTE: During reassembly replace all O-rings and be sure to dip all parts in Hy-Tran before installation

1. Install the piston into the hitch cover. Then install the front cover.
2. Install the lefthand bushing using an OTC 27488 handle with an OTC 27511 pilot and OTC 27518 driver disc. Secure with a setscrew.
3. Connect the piston rod and the inner arm with the connecting rod. Secure with the snaprings and position the assembly into the hitch cover.
4. Insert the rockshaft with the timing spline in line with the housing rib. When inserting the rockshaft, the center of the connecting pin should also be in line with the rib.

NOTE: This properly aligns the inner arm with the rockshaft. The timing marks on the rockshaft are also now in the proper position for the timing marks on the rockshaft arms.

5. Insert the righthand bushing and secure with the setscrew.
6. Install the rockshaft arms noting the position of the timing marks.
7. Install the end caps securing them with a lockwasher and capscrew.
8. Secure the hitch cover onto the rear frame, torquing the counting bolts to 43-51 ft.lb.
9. Continue installation in the reverse order of removal.

Control Valve

REMOVAL

NOTE: Relieve the system pressure by lowering the hitch to its lowest position before disassembly.

1. Thoroughly clean the outer surface around the auxiliary valve and control valve.
2. Disconnect the coupler pressure lines from the auxiliary valve.
3. Remove the necessary mounting bolts and remove the auxiliary valve.
4. Disconnect the position control rod from the control valve.
5. Disconnect the high pressure line to the control valve.
6. Remove the position control panel.
7. Remove the necessary mounting bolts and remove the control valve.

DISASSEMBLY

1. Remove the front cover.
2. Remove the relief valve.
3. Remove the spool and sleeve assembly from the control valve.
4. Remove the check valve assembly.
5. The ball seat can be removed using a short wire.
6. Back out the setscrew securing the adjusting shaft.
7. Remove the adjusting shaft.
8. Drive out the roll pins securing the position control lever and link. Remove the control lever and link.
9. Remove the rear cover.
10. Remove the unloading valve assembly.

ASSEMBLY

NOTE: Reassemble in the reverse order of disassembly paying attention to the following:

1. All O-rings and gaskets should be replaced with new.
2. A gasket sealer should be used on the gaskets between the end covers and the control valve body.
3. All moving parts should be dipped in Hy-Tran prior to reassembly.
4. If the position control lever and the position control link have been properly installed, moving the control lever back to the raise position should bring the control link forward.
5. If the lever and link do not operate properly, alternately reinstall the lever or the link, rotating them 180° on the shafts. Check for proper functioning as in step 4.

Auxiliary Valve

REMOVAL AND DISASSEMBLY

1. Remove the auxiliary valve following steps 1-3 of Control Valve Removal.
2. Disconnect the control lever linkage from the spool.
3. Remove the two Allen screws securing the lever and linkage to the auxiliary valve.
4. Remove the O-ring and wiper.
5. Remove the two Allen screws securing the rear end cap to the valve.
6. Remove the end cap and withdraw the spool assembly.
7. Remove the O-ring and wiper.
8. Remove the check poppet from the valve body.
9. Inspect the spool assembly, the wiper and O-ring, and the check poppet for excessive wear or damage. Replace as necessary.
10. Reassemble in the reverse order of disassembly.

Hydraulic Pump

REMOVAL AND DISASSEMBLY

1. Remove the front radiator grille.
2. Remove the engine side sheets.
3. Disconnect the hydraulic return and high pressure lines from the pump.
4. Remove the hydraulic pump.

597

International Harvester

5. Remove the six Allen head bolts and disassemble the pump.
6. The support ring, seal ring, O-rings and oil seal are presently being serviced.
7. Reassembly in the reverse order of disassembly.

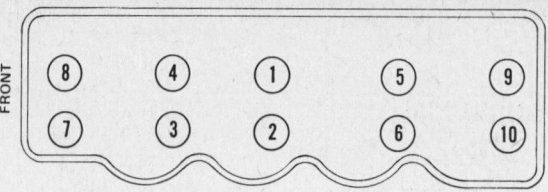
Cylinder head bolt tightening sequence

ENGINE

This engine is a Japanese made, Toyo-Kogyo model M471G.

REMOVAL AND INSTALLATION

1. Remove the muffler.
2. Remove the radiator grill.

NOTE: The radiator grill is removed so the engine hood mounting bolts can be removed easier.

3. Remove the engine hood with side sheets.
4. Disconnect the hydraulic return line from the hydraulic pump and the filter base.
5. Disconnect all electrical connections to the engine.
6. Disconnect the throttle control rod, the tachometer drive cable and the choke control cable from the carburetor.
7. Shut off the fuel cock valve at the bottom of the fuel tank. Disconnect the fuel hose at the inlet of the fuel pump.
8. Disconnect the drag link from the knuckle arm.
9. Disconnect and remove the high pressure hydraulic line from the hydraulic pump and the control valve.
10. Support the center section using a floor jack. Support the engine using an FES 138-1 sling with an overhead hoist.

NOTE: Wedge a wooden block between the front axle and the engine side channel to prevent tipping when splitting. Support the front bolster weight with a safety stand.

11. Remove the mounting bolts securing the center section to the engine. Split the tractor, rolling the rear section rearward.

NOTE: When making a front split, the transmission input shaft may remain in the clutch assembly. Loosen the clutch mounting bolts to remove the input shaft.

12. Open the drain cocks on the radiator and the cylinder block to drain the coolant from the engine.
13. Disconnect the upper and lower radiator hoses from the engine.
14. Remove the radiator mounting bolts and remove the radiator.
15. Remove the hydraulic pump from the crank pulley.
16. With the engine supported as in step 10, remove the engine mounting bolts and remove the engine.

NOTE: Be sure to keep the engine mounting shims on the side which they belong. All of the hydraulic pump shims should also be kept in place.

17. Installation is the reverse of removal.

Intake Manifold

REMOVAL AND INSTALLATION

NOTE: The manifold is made of aluminum.

1. Drain the coolant to a level just below the manifold.
2. Unbolt and remove the manifold.
3. Use a new gasket and install the manifold. Torque the bolts to 14-18 ft.lb.

Exhaust Manifold

REMOVAL AND INSTALLATION

1. Unbolt and remove the manifold.
2. Using a new gasket, install the manifold and torque the bolts to 12-17 ft.lb.

Cylinder Head

REMOVAL AND INSTALLATION

1. Remove the intake and exhaust manifolds.
2. Remove the valve cover.
3. Remove all cables, hoses and wires connected to the head.
4. Remove the rocker arm assembly, and push rods. Keep the push rods in order.
5. Unbolt and remove the head.
6. Install the cylinder head in the reverse order of removal paying attention to the following:

a. Be sure the longer push rod is installed on the exhaust valve side.
b. Move the exhaust side rocker arm supports so that the valve stem-to-rocker arm center is offset by .040 in. Then, temporarily tighten the support hex nuts.
c. Install the hex nuts and torque the nuts to 49 ft.lb. in the sequence illustrated.

Valve Lash Adjusting Procedure

Using the chart below and the simplified procedure outlined, all valves can be adjusted by cracking the engine only twice. The valve lash is adjusted with the engine warm.

Four valves are adjusted when the No. 1 piston is at TDC (Compression) and the remaining four are adjusted when the No. 4 piston is at TDC (Compression). The following chart shows the numbering sequence of the valves which correspond to the chart. Set the valve lash to the specified clearance of .010 in.

Pistons and Connecting Rods

REMOVAL

1. Remove the head.
2. Remove the oil pan.
3. Unbolt the connecting rod caps.
4. Using a ridge reamer, remove the ridge from the top of the cylinder liner.
5. Push the piston and rod up out of the cylinder.

WITH	ADJUST VALVES (Engine Warm)							
No.1 Piston at T.D.C. (Compression)	1	2	3			6		
No.4 Piston at T.D.C. (Compression)				4	5		7	8

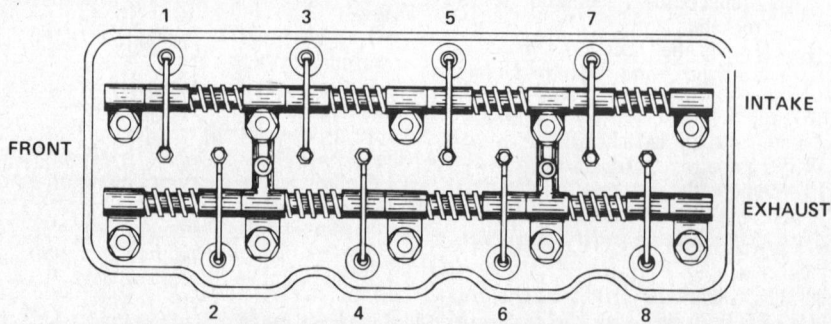
REPLACE HEAD GASKET IF ALL HEAD BOLTS ARE LOOSENED.
Valve adjusting sequence chart

International Harvester

6. Using a press, remove the piston pins.

ASSEMBLY AND INSTALLATION

1. If the piston pin bushing must be replaced, be sure to align the oil hole of the new bushing with the oil hole in the connecting rod.
2. When reassembling the piston and connecting rod, align the "F" mark on the piston with the oil hole on the connecting rod.
3. The oil ring must be assembled to the piston so that the upper and lower rail ends are 30 to 40 degrees to each side of the spacer ends.
4. Place a piece of Plastigage on the bearing surface the full width of the bearing about ¼ in. off center.
5. Install the cap and torque to 30 ft.lb.

NOTE: Do not turn the crankshaft while the Plastigage is in place.

6. Remove the bearing cap and use the Plastigage scale to measure the widest point of the Plastigage. This reading indicates the bearing clearance in millimeters (thousandths of an inch). Specified clearance is .0011-.003 in.
7. Check the connecting rod side clearance using a feeler gauge. Excessive clearance may require replacement of rods or shaft. The check should be made to make certain that the specified running clearance exists. Lack of clearance could indicate a damaged rod or perhaps a rod bearing out of position.
8. The crankpin journal of the crankshaft can be reground to .010 in., .020 in. and .030 in. undersize.
9. Be sure to finish grind the radius to .010-.016 in.

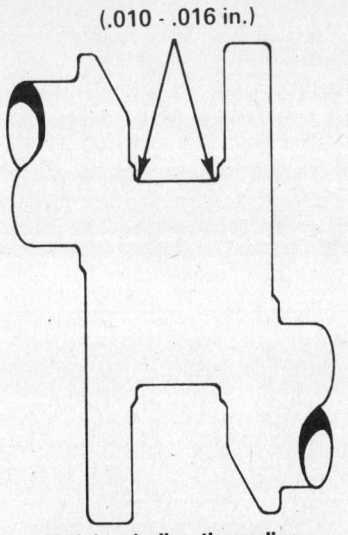

Finish grinding the radius

Front Cover Oil Seal

REPLACEMENT

NOTE: It is not necessary to remove the engine front cover in order to replace the front oil seal.

1. Drain the engine coolant out of the radiator.
2. Remove the hood and side sheets.
3. Remove the radiator.
4. Disconnect the hydraulic lines from the front mounted pump.
5. Remove the pump. Note and record the number of shims below the pump mounting.
6. Using a suitable puller, remove the crankshaft pulley.
7. Remove the oil seal from the front cover. Be careful not to mar the seal contact surface of the cover.
8. Using a suitable driver, install the new seal.

Timing Chain and Sprockets

REMOVAL

1. Remove the valve tappets. Be sure to keep the tappets in the sequence removed so that they may be reinstalled in their original locations.
2. Remove the oil pump driven gear and shaft assembly with the shims and thrust washer.
3. Remove the chain adjuster assembly.
4. Remove the vibration damper assembly.
5. Remove the oil slinger and key from the crankshaft.
6. Remove the capscrews securing the camshaft thrust plate.
7. Simultaneously remove the camshaft assembly, the timing chain and the crankshaft gear from the engine.
8. Check the condition of the camshaft.
9. Check the timing chain and sprockets for wear and damage. Replace as needed.

INSTALLATION

1. If it was removed, install a new front plate gasket and the front plate.
2. Position number one piston at top dead center on compression stroke.

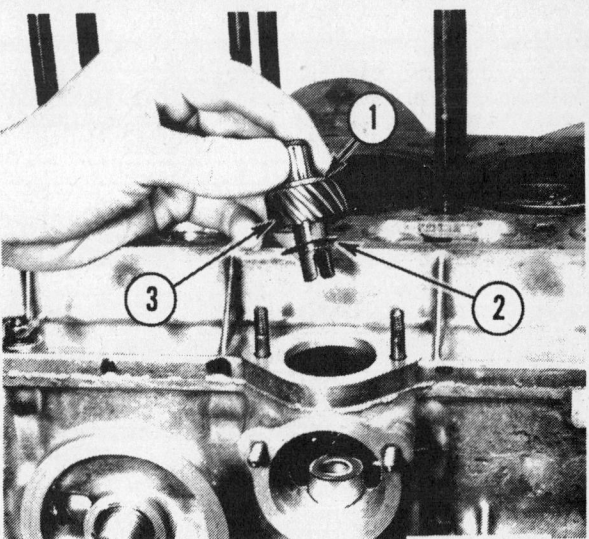

Removing oil pump driven gear

3. Thoroughly lubricate the camshaft journals with clean engine oil.
4. Aligning the timing marks of the chain and sprockets, position the sprockets in the chain.
5. Mount the assembly on the crankcase and camshaft.
6. Rotate the camshaft sprocket until the key way of the crankshaft sprocket is aligned with the crankshaft key way. Install the key.
7. Secure the camshaft thrust plate with hex nuts.
8. Install the oil slinger.
9. Secure the timing chain vibration damper to the crankcase.
10. Position a new gasket on the front plate. Assemble and install the chain adjuster.

Camshaft

REMOVAL AND INSTALLATION

1. Camshaft removal is covered under Removal of the Timing Chain and Sprocket.
2. Check camshaft runout. Standard runout is less than .0004 in. Maximum permissible runout is .001 in. Repair or replace the camshaft as needed.
3. Check the cam lobe height for wear. Minimum allowable lobe height is 1.432 in. for both intake and exhaust valves.
4. Measure the camshaft journal diameters. Replace the camshaft if the diameters are below the minimum allowable dimensions.
5. Minimum permissible journal diameters:

Front	1.885 in.
Second	1.807 in.
Third	1.757 in.
Fourth	1.728 in.
Rear	1.689 in.

6. Measure the camshaft journal bores in the crankcase and the camshaft journals

International Harvester

to determine running clearance. Maximum allowable running clearance is .006 in. Replace the camshaft or crankcase if clearance is found to be excessive.

7. Check camshaft end play with a feeler gauge. Specified end play is .0008-.007 in. Maximum allowable end play is .008 in.

8. Check the distributor drive gear and the oil pump drive gear for worn or damaged teeth. If the oil pump drive gear is in doubtful condition, replace the camshaft.

9. Install the camshaft as described in Installation: Timing Chain and Sprockets.

Crankshaft

REPLACING MAIN BEARINGS

1. Remove one bearing cap and bearing insert. The remaining caps are left tight while checking the fit of this bearing.

2. Wipe the oil from all contact surfaces such as crankshaft journal, bearing insert and bearing cap.

3. Place a piece of Plastigage the full width of the bearing surface on the cap insert (or the crankshaft journal). Install the cap and torque the cap bolts to 42 ft.lb.

NOTE: Do not turn the crankshaft while making the check with Plastigage.

4. Remove the bearing cap and insert.

5. Do not disturb the Plastigage. Using the Plastigage scale, measure the widest point of the Plastigage. This reading indicates bearing clearance. Specified clearance is .001-.002 in. Maximum allowable clearance is .003 in.

6. If bearing clearance is not within specifications, the crankshaft must be reground and undersize bearings installed.

7. Undersize bearings of .010, .020 and .030 in. are available.

8. Refer to the chart for crankshaft regrind finish dimensions.

9. Specified crankshaft runout is less than .0008 in. Maximum allowable runout is .001 in.

10. The rear bearing controls the crankshaft thrust. Use a dial indicator to check end play. Specified end play is .004-.011 in. Maximum allowable end play is .012 in.

11. Oversize thrust washers are available in .010, .020 and .030 in. sizes.

IMPORTANT: When installing the thrust washers, the oil grooves must face away from the bearing inserts (against the crankshaft sides).

LUBRICATING SYSTEM

Oil Pump

OVERHAUL

1. Wash all pump parts and the screen assembly in cleaning solvent.

2. With the outer rotor in the pump body, use a feeler gauge to measure the clearance between the rotor and body. Allowable clearance is .012 in.

NOTE: The inner and outer rotors are a matched set and must be replaced as a unit.

3. Using a straight edge, check the clearance between the rotor and the face of the pump body.

4. Using a straight edge, check for excessive wear of the pump cover face with a feeler guage.

If the total clearance of step 3 and step 4 is greater than .006 in., lap or replace the pump cover.

5. Check the fit of the pressure control plunger in its bore. The plunger should slide freely in the bore.

6. Check the free length of the pressure control spring. Specified length is 2.165 in. If less, replace the spring.

7. When reassembling the pump, be sure to align the pump marks of the inner and outer rotors.

8. Install the pump.

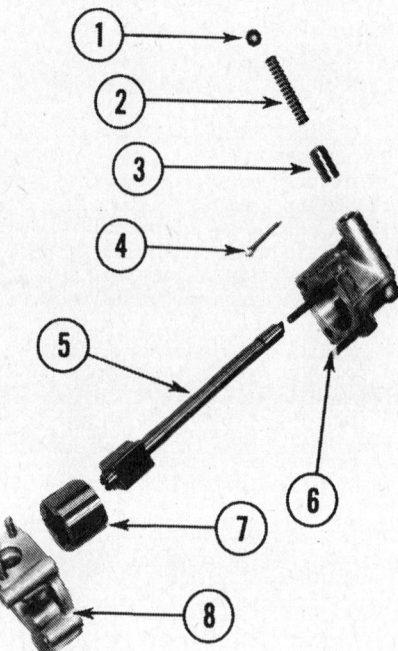

1. Spring seat
2. Spring
3. Plunger
4. Cotter pin
5. Driveshaft
6. Pump body
7. Outer rotor
8. Cover

Oil pump components

COOLING SYSTEM

Water Pump

OVERHAUL

1. Drain the fan, pulley and drive belt.
2. Drain the cooling system.
3. Remove the water pump.
4. Unbolt and remove the back cover.
5. Check the bearings for excessive wear. Replace the bearings if in doubtful condition.
6. Carefully check the contact surfaces of the seal and the impeller surface. If any wear, nicks or cracks are found, replace the parts.
7. Install the bearings with the sealed sides facing away from the bearing spacer.
8. Fill approximately ⅓ of the space between the bearings and the spacer with IH 251 HEP grease or equivalent lithium base grease.
9. Install the new water seal until the bent portion of the plate touches the body.

NOTE: Apply a small amount of engine oil on sliding face of the water seal. This will prevent water leakage while operating just after reassembly.

10. Supporting the opposite end of the pump shaft, press the impeller on the shaft until the impeller hub is flush with the shaft. Be sure the shaft rotates freely after impeller installation.

11. Supporting the impeller end of the pump shaft, press on the pulley hub. Check to see that the pump shaft rotates freely after hub installation.

CRANKSHAFT REGRIND FINISH DIMENSIONS

Undersize	Crankpin Diameters		Crankshaft Main Journal Diameters	
	mm	inches	mm	inches
0.25mm (.010 inch)	44.690-44.705	1.7591-1.7598	55.694-55.709	2.1923-2.1930
0.50mm (.020 inch)	44.440-44.455	1.7493-1.7499	55.444-55.459	2.1825-2.1831
0.75mm (.030 inch)	44.190-44.205	1.7394-1.7401	55.194-55.209	2.1726-2.1733

International Harvester

Radiator

REMOVAL AND INSTALLATION

1. Remove the muffler.
2. Remove the radiator grill.

NOTE: The radiator grill is removed so the engine hood mounting bolts can be removed easier.

3. Remove the engine hood with side sheets.
4. Drain the coolant from the radiator.
5. Disconnect the electrical wiring from the headlights.
6. Disconnect the upper and lower radiator hoses.
7. Remove the radiator mounting bolts and remove the radiator.
8. Installation is the reverse order of removal.

FUEL SYSTEM

Fuel Tank

REMOVAL AND INSTALLATION

1. Remove the muffler.
2. Remove the radiator grill.

NOTE: The radiator grill is removed so the engine hood mounting bolts can be removed easier.

3. Remove the engine hood with side sheets.
4. Shut off the fuel cock valve at the bottom of the fuel tank.
5. Disconnect the fuel line from the sediment bowl.
6. Disconnect the wiring at the fuel tank.
7. Remove the fuel tank strap. Remove the fuel tank.
8. Installation is the reverse order of removal.

Carburetor

DISASSEMBLY

Air Horn

1. Disconnect all links on the carburetor.
2. Remove the accelerator pump plunger spring.
3. Separate the air horn assembly from the body.
4. Remove the accelerator pump plunger assembly and gasket.
5. Remove the float pin and float.
6. Remove the float valve seat, valve and spring.

Body

1. Separate the body from the flange.
2. Remove the gasket.
3. Remove the following parts:
   - Slow air bleed.
   - Outlet weight and steel ball.
   - Inlet check valve.
   - Main jet.
   - Plug.
   - Main air bleed.
   - Slow jet.

ASSEMBLY

Assemble the carburetor in the reverse order of disassembly paying attention to the following:
1. Install all new gaskets.
2. Set float height to .157 in. by adding or removing shims from the float needle valve seat.
3. With all connecting links installed, check throttle valve position with the choke fully closed.
4. With the choke closed the throttle valve should be at the dimension illustrated at "A". To adjust the valve position, bend the connecting link "B".

ADJUSTMENT

1. Seat the mixture adjusting screw and back it off 3 turns.

NOTE: Do not overtighten the screw on its seat as damage to the seat will occur.

2. Turn in the throttle adjusting screw 2 to 3 turns and start the engine.
3. Slowly back out the throttle screw until the engine starts to run rough.
4. Slowy turn in (towards its seat) the mixture adjusting screw until the engine starts to run smoothly at a higher rpm.
5. Slowly back out the throttle screw to lower the engine speed to specified 650 ± 50 rpm.
6. The engine must run smoothly at the specified low idle. If not, repeat the above procedures.

Governor

REMOVAL

1. Disconnect the governor link from the carburetor.
2. Slacken the tension of the governor belt idler pulley.

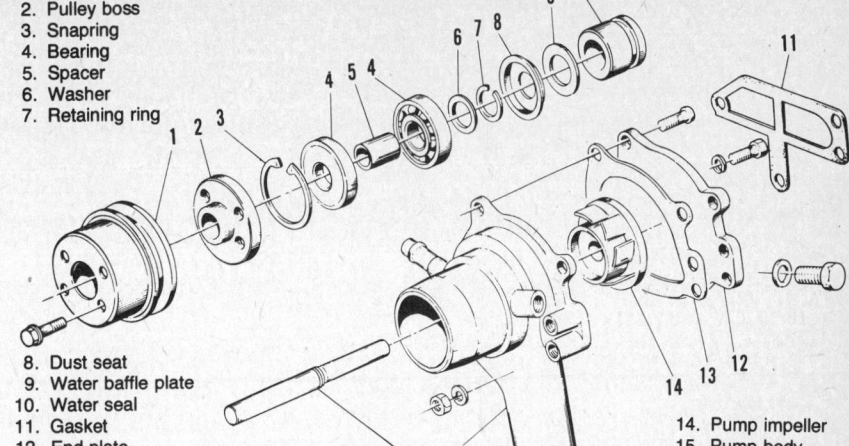

1. Pulley
2. Pulley boss
3. Snapring
4. Bearing
5. Spacer
6. Washer
7. Retaining ring
8. Dust seat
9. Water baffle plate
10. Water seal
11. Gasket
12. End plate
13. Gasket
14. Pump impeller
15. Pump body
16. Shaft

Water pump components

3. Remove the idler pulley bracket, governor and governor bracket as an assembly.
4. Thoroughly wash the outside of the governor before disassembly.

DISASSEMBLY

1. Remove the drain plug and drain the oil from the housing.
2. Remove the tachometer cover and gear from the governor.
3. Separate the governor housing from the cover.
4. Remove the screw from the fork and control shaft assembly. Remove the control shaft.
5. Remove the tension spring from speed lever assembly. Remove the spring and fork assembly.
6. Remove the taper pins from the speed shaft and lever assembly. Pull the speed shaft from the cover.

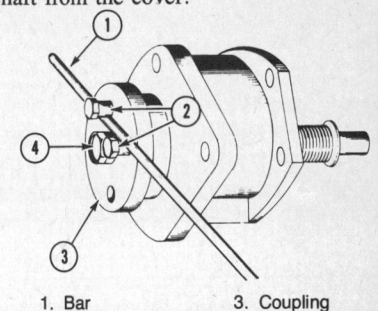

1. Bar
2. Bolts
3. Coupling
4. Capscrew

Removing coupling from governor shaft

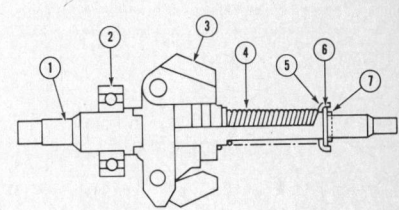

1. Governor shaft
2. Bearing
3. Flyweights
4. Spring
5. Spring seat
6. Shims
7. Pin

Governor shaft components

601

International Harvester

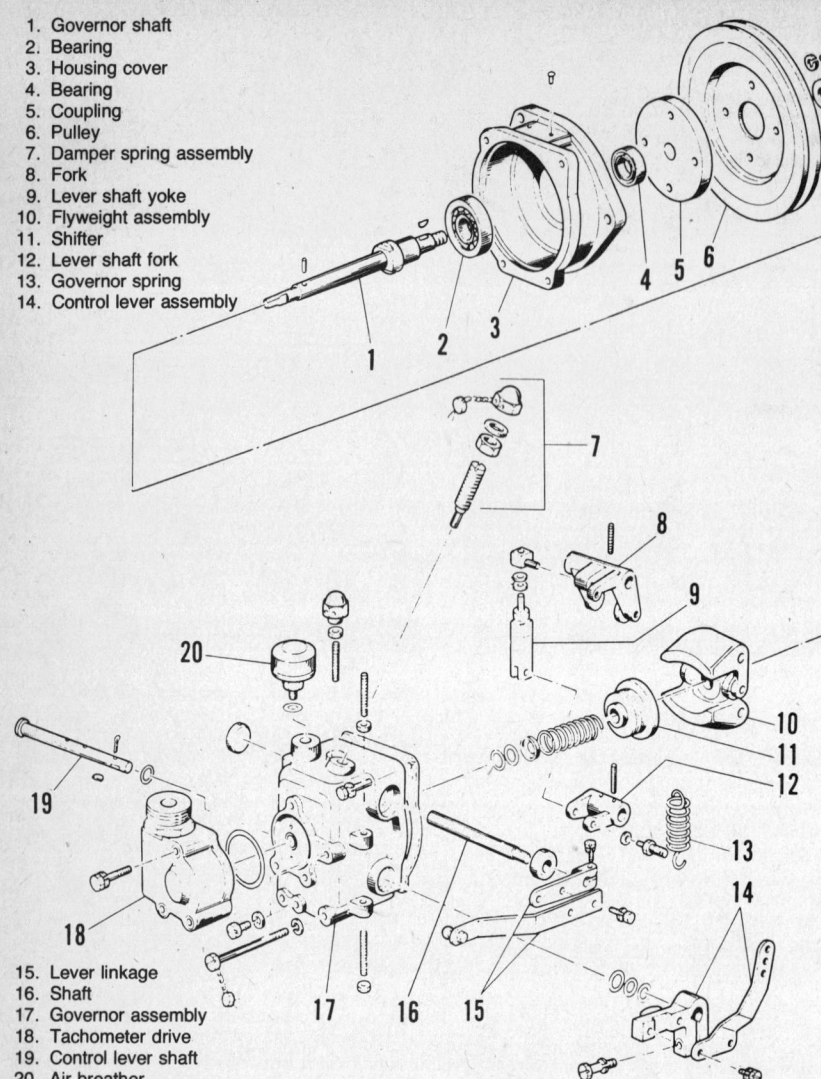

1. Governor shaft
2. Bearing
3. Housing cover
4. Bearing
5. Coupling
6. Pulley
7. Damper spring assembly
8. Fork
9. Lever shaft yoke
10. Flyweight assembly
11. Shifter
12. Lever shaft fork
13. Governor spring
14. Control lever assembly
15. Lever linkage
16. Shaft
17. Governor assembly
18. Tachometer drive
19. Control lever shaft
20. Air breather

Governor components

7. To remove the coupling from the governor shaft:
   a. Install two bolts in the coupling. Using a bar through the bolts as a stop, remove the capscrew securing the coupling.
   b. Tap on the coupling end of the governor shaft to free the coupling. Remove the coupling and governor housing from the shaft.

8. Remove the retaining pin to disassemble the governor shaft. Note and record any amount of shims behind the spring seat for reassembly purposes.

9. If necessary, press the bearing off of the governor shaft. Press on the inner race when installing a new bearing.

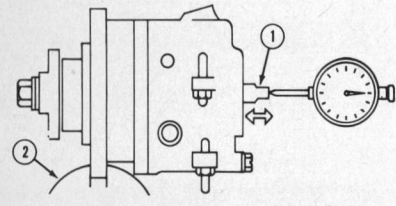

1. Governor shaft   2. Vise
Measuring shaft end play

ASSEMBLY

Reassemble the governor in the reverse order of disassembly paying attention to the following:

1. Use new O-rings, oil seals and gaskets. Coat the lips of the O-rings and oil seals with engine oil.

2. When assembling the central spring on the governor shaft, do not depress the spring more than $13/32$ in. past the need to install the seat (shims if any) and pin securing the spring assembly.

If the spring is compressed more than is needed to secure the assembly, the spring constant may be changed.

3. Install the governor shaft and flyweights vertically into the housing in order to prevent damage to the housing.

4. Install the coupling on the shaft. Torque the hex nut securing the coupling to 18-22 ft.lb.

5. Adjust end play of the speed shaft to .008 in. Shims are available in .004 in.; .008 in.; .012 in.; .020 in.; and .039 in. sizes.

6. If it was removed, install speed arm assembly (damper spring adjusting screw) 1.77 in.

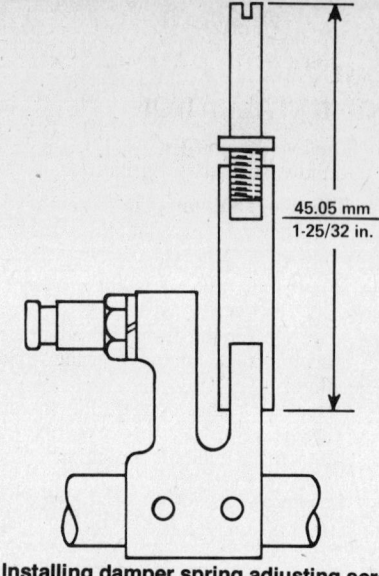

Installing damper spring adjusting screw

7. After reassembling the cover and housing, position the assembly in a vise. Using a dial indicator, check the end play of the governor shaft.

8. Specified end play is 0.1-0.3 mm (.004-.012 in). Shims are available in the following sizes:

   .004 in.
   .008 in.
   .012 in.
   .020 in.

9. To install the shims, the housings must be separated and the shims installed behind the spring seat.

10. Fill the governor to the level plug, through the breather hole with the same grade engine oil as in being used in the engine.

INSTALLATION

Install the governor in the reverse order of removal paying attention to the following:

- Adjust belt tension:
- Scale pull at 22 lb. force.
- Deflection:
- New belt—½-²¹⁄₃₂ in.
- Used belt—½-²⁵⁄₃₂ in.

ADJUSTMENT

Adjust the high idle with the full load stop bolt to 2950 ± 50 rpm.

Adjust low idle speed with the idle stop bolt to 650 ± 50 rpm.

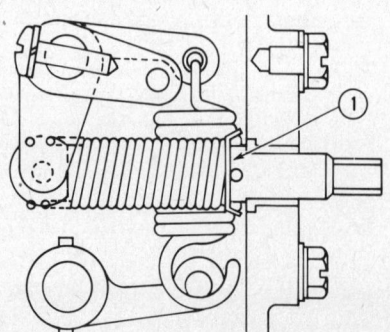

Shim location is noted by the #1

Jacobsen

INDEX

Front Axle 604
Steering 604
Brakes 609
Hydrostatic Transmission 610
Two-Speed Transaxle 611
Three-Speed Transaxle 613
Four-Speed Transaxle 615
Differential 617
Right Angle Drive 619
Engine 620

JACOBSEN
GT and LT Models

LAWN KING 1060, LT885, LT860, LT750, GT16, GT14, GT12, GT10

FRONT AXLE

REMOVAL AND INSTALLATION

1. Disconnect the implement lift bars.
2. Disconnect the drag link ball joint end from the steering arm.
3. Raise and support the front end on jack stands.
4. Remove the axle pivot pin.
5. Roll the axle from under the tractor.
6. Installation is the reverse of removal.

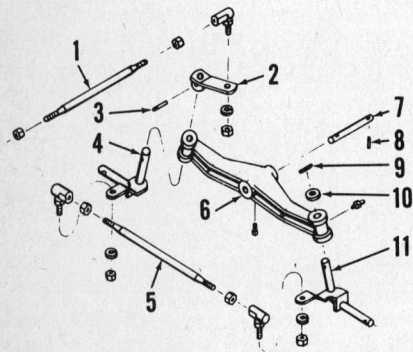

1. Drag link
2. Steering arm
3. Roll pin
4. Steering knuckle, lefthand
5. Tie rod
6. Axle main member
7. Pivot pin
8. Roll pin
9. Roll pin
10. Washer
11. Steering knuckle, righthand

GT series front axle

STEERING

Steering Knuckles

REMOVAL AND INSTALLATION

1. Raise and support the front with jack stands under the axle.
2. Remove the front wheels.
3. Disconnect the tie rod ball joint ends from the steering knuckle arms.
4. Disconnect the drag link ball joint end from the steering arm.
5. On the left knuckle, remove the snapring or pin retaining the knuckles.
7. Installation is the reverse of removal.

1. Ball joint assembly
2. Lockwasher, ½ med.
3. Hex jam nut, ½-20
4. Hex jam nut, ½-20
5. Steering drag link
6. Hex jam nut, ½-20
7. Tie rod
9. Tie rod arm and hub assembly
10. Bushing
11. Front axle assembly
12. Cap
13. Front axle pivot shaft
14. Lockwasher, ½ med.
15. Hex jam nut, ½-13
16. Bushing
17. Steering and tie rod arm and hub assembly
18. Lockwasher, ½ med.
19. Hex jam nut, ½-20
20. Wheel spindle assembly, front
21. Front tire
22. Front wheel
22A. Front wheel assembly (incl. Ref. 21 and 22)
23. Cotter pin, 5/32 x 1¼
24. Washer
25. Cap
26. Nut
27. Hub cap
28. Rear wheel assembly, including tire, rim, and tire valve
29. Bushing

LT series front axle

Steering Gear

REMOVAL AND INSTALLATION

LT Models

1. Remove the hood, disconnect the battery cables and remove the battery.
2. Disconnect the drag link from the steering arm.
3. Remove the steering wheel.
4. Drive out the pinion gear retaining pin and remove the gear.
5. Remove the steering shaft.
6. On early models, remove the snapring and pull out the steering arm and shaft. On later models, remove the steering arm and quadrant gear.
7. Installation is the reverse of removal.

GT Models

1. Remove the hood.
2. Remove the driveshaft.
3. Remove any implements obscuring the gear.
4. Drive out the steering lever retaining pin and remove the steering lever from the steering gear shaft.
5. Unscrew the steering wheel retaining nut and remove the wheel.
6. Unbolt and remove the steering gear.
7. Installation is the reverse of removal.

OVERHAUL

The steering gear assembly is serviced as an assembly only. However, if it is necessary to disassemble, clean, and adjust the unit, follow the procedure outlined below.

1. With the gear assembly removed from the tractor, remove the cross shaft and actuator plate assembly by unscrewing the two jam nuts and thrust washer.
2. Remove the worm gear and shaft assembly from the housing by pulling out the cotter pin from the catellated threaded plug, unscrewing the plug from the housing, and removing the lower bearing race, bearing assembly, and worm gear along with the upper bearing assembly and race.
3. Thoroughly clean all parts of the assembly.

Jacobsen

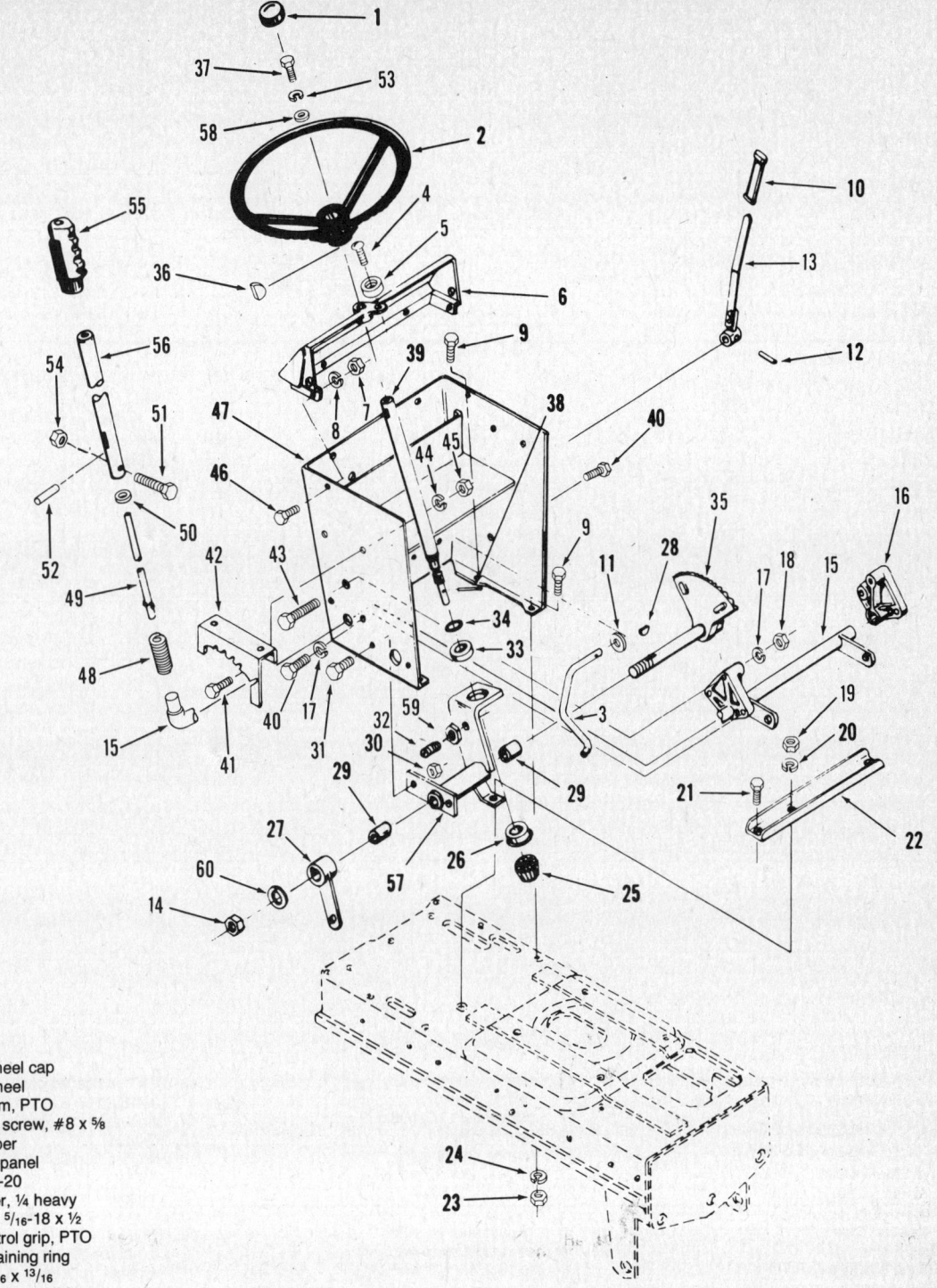

1. Steering wheel cap
2. Steering wheel
3. Acutator arm, PTO
4. Road head screw, #8 x 5/8
5. Hood bumper
6. Instrument panel
7. Hex nut, 1/4-20
8. Lockwasher, 1/4 heavy
9. Hex screw, 5/16-18 x 1/2
10. Clutch control grip, PTO
11. Tru-arc retaining ring
12. Roll pin, 3/16 x 13/16
13. Clutch lever assembly
14. Hex jam nut, 3/4-16
15. Lift handle shaft assembly
16. Bracket
17. Lockwasher, 5/16 heavy
18. Hex nut, 5/16-18
19. Nut, 7/16-14
20. Lockwasher, 7/16 heavy
21. Hex screw, 5/16-18 x 3/4
22. Channel frame
23. Hex nut, 5/16-18
24. Lockwasher, 5/16 heavy
25. Pinion steering
26. Steering shaft bearing
27. Steering arm assembly
28. Woodruff key, #9
29. Steering arm pivot bushing
30. Hex c/lock nut, 5/16-18
31. Hex screw, 5/16-18 x 1/2
32. Steering gear thrust spacer
33. Felt bearing protection seal
34. O-ring seal protector
35. Steering gear and shaft assembly
36. Key
37. Hex screw, 5/16-18 x 3/4
38. Roll pin, 1/8 x 3/4
39. Steering shaft
40. Hex screw, 5/16-18 x 3/4
41. Hex screw, 1/4-20 x 5/8
42. Lift quadrant
43. Hex screw, 5/16-18 x 1 3/4
44. Lockwasher, 1/4 heavy
45. Hex nut, 1/4-20
46. Hex screw, 1/4-20 x 1/2
47. Console assembly
48. Lift lever latch spring
49. Lift lever rod
50. Lift handle spacer
51. Hex screw, 1/4-20 x 1 1/4
52. Roll pin, 1/4 x 1 1/2
53. Lockwasher, 5/16 heavy
54. Hex c/lock nut, 1/4-20
55. Lift lever grip
56. Lift handle
57. Steering shaft support assembly
58. Washer
59. Hex jam nut, 5/8-11
60. Lockwasher, 3/4

1060 steering components

605

Jacobsen

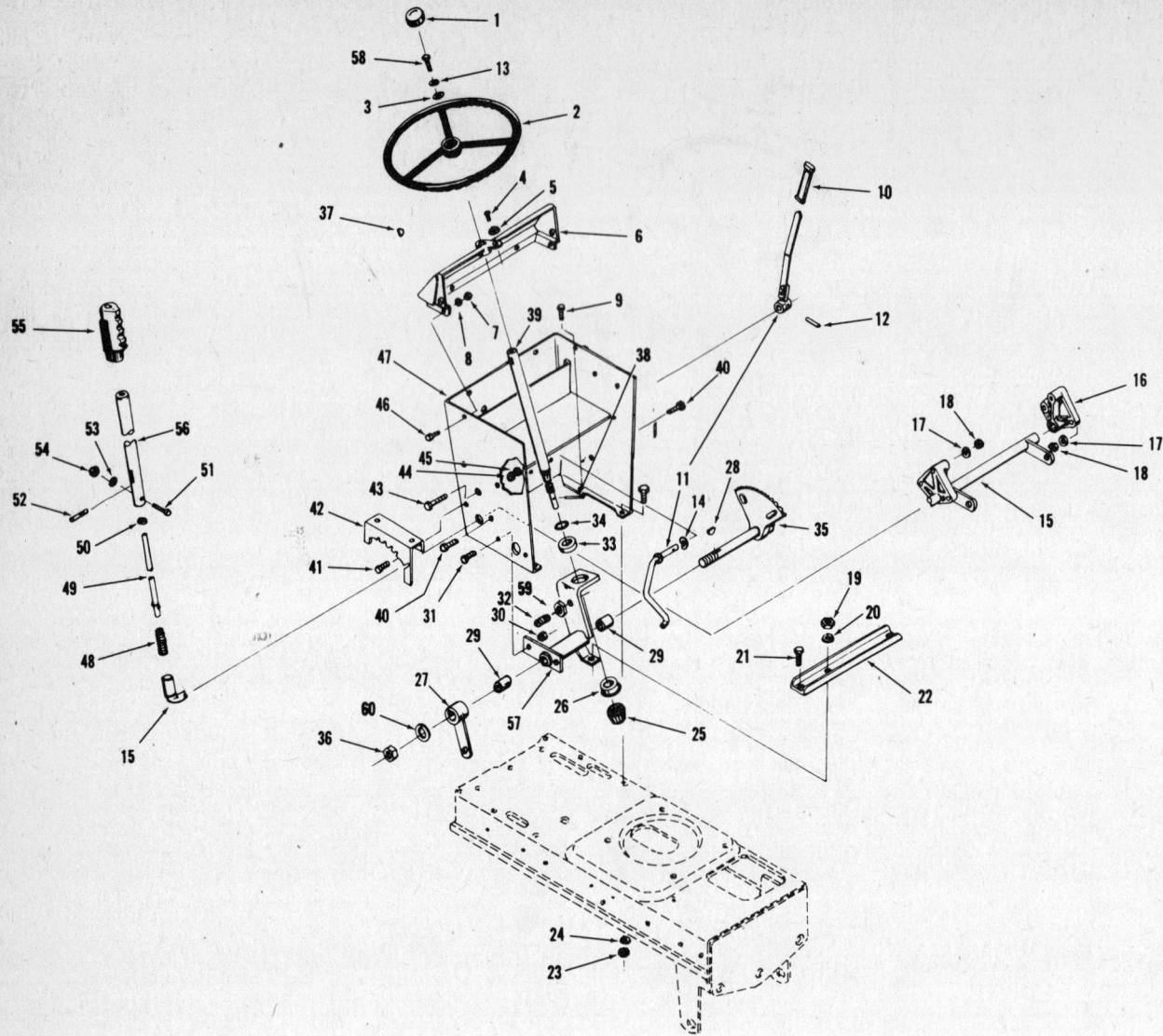

1. Steering wheel cap assembly
2. Steering wheel
3. Washer
4. Road head screw, #8 x 5/8
5. Hood bumper
6. Instrument panel
7. Hex nut, 1/4-20
8. Lockwasher, 1/4 med
9. Hex head capscrew, 5/16-18 x 1/2
10. Clutch control grip, PTO
11. Clutch arm and lever assembly
12. Roll pin, 3/16 x 13/16
13. Lockwasher, 5/16 heavy
14. Shakeproof washer (spring)
15. Lift handle shaft assembly
16. Bracket
17. Lockwasher, 5/16 medium
18. Hex Nut, 5/16-18
19. Hex Nut, 7/16-14
20. Lockwasher, 7/16 medium
21. Hex head capscrew, 5/16-18 x 3/4
22. Frame channel
23. Hex nut, 5/16-18
24. Lockwasher, 5/16 medium
25. Steering pinion
26. Steering shaft bearing
27. Steering arm assembly
28. #9 Woodruff key
29. Bushing
30. Hex jam nut, 5/16-18
31. Hex head capscrew, 5/16-18 x 1/2
32. Spacer
33. Felt seal
34. O-ring
35. Steering gear and shaft assembly
36. Hex jam nut, 3/4-16
37. Steering wheel mounting key
38. Roll pin, 1/8 x 3/8
39. Steering shaft
40. Hex screw, 5/16-18 x 3/4
41. Hex screw, 1/4-20 x 5/8
42. Lift quadrant
43. Hex head capscrew, 5/16-18 x 1 3/4
44. Lockwasher, 1/4 medium
45. Hex nut, 1/4-20
46. Hex head capscrew, 1/4-20 x 1/2
47. Console assembly
48. Lift lever latch spring
49. Lift lever rod
50. Lift handle spacer
51. Hex head capscrew, 1/4-20 x 1 1/4
52. Type A groove pin, 1/4 x 1 1/2 SAE
53. Lockwasher, 1/4 medium
54. Hex nut, 1/4-20
55. Lift lever grip
56. Lift handle
57. Steering shaft support assembly
58. Hex head capscrew 5/16-18 x 3/4
59. Hex jam nut, 5/8-11
60. Arm mounting lockwasher, 3/4

LT885 steering components

Jacobsen

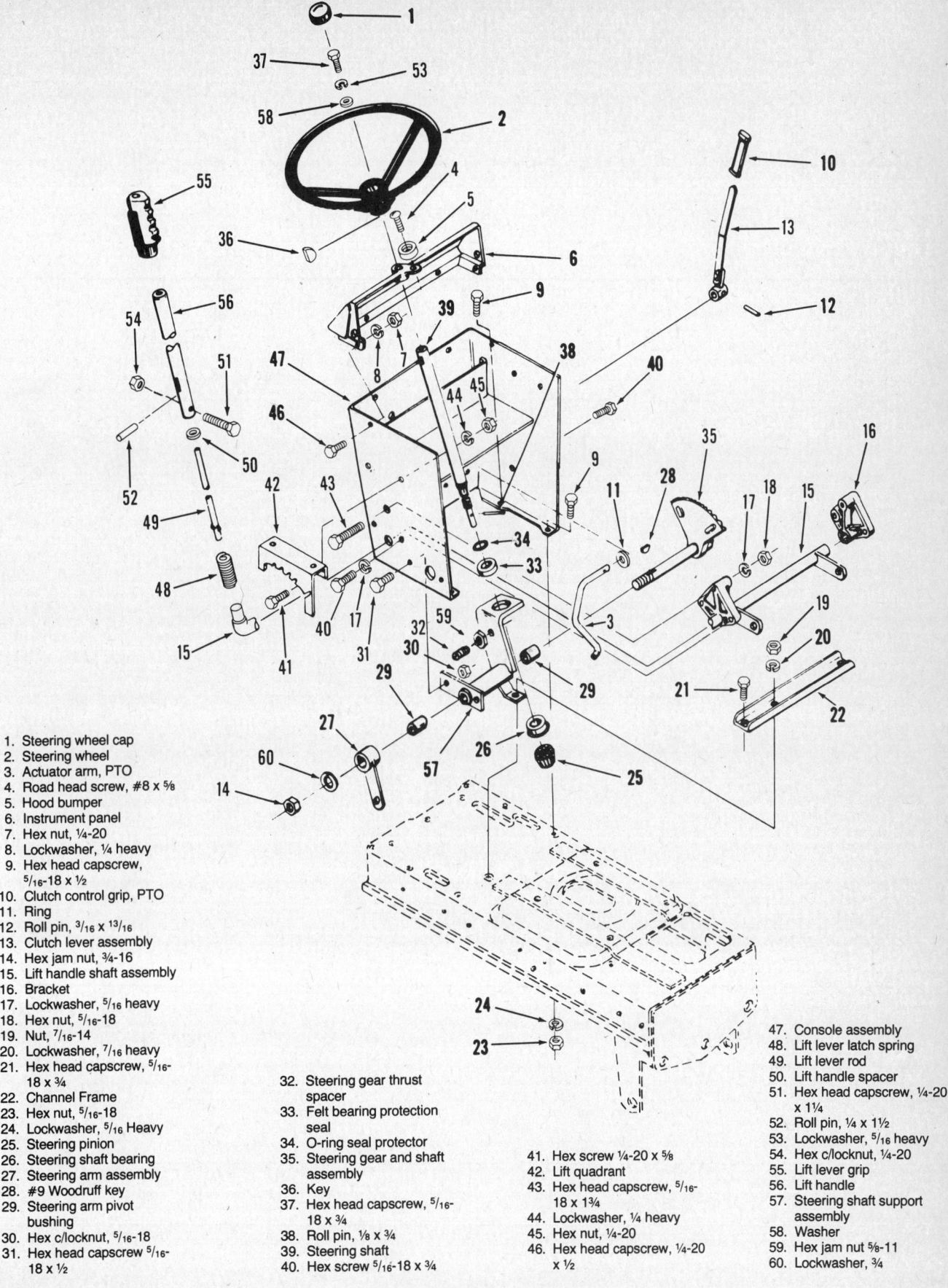

1. Steering wheel cap
2. Steering wheel
3. Actuator arm, PTO
4. Road head screw, #8 x ⅝
5. Hood bumper
6. Instrument panel
7. Hex nut, ¼-20
8. Lockwasher, ¼ heavy
9. Hex head capscrew, 5/16-18 x ½
10. Clutch control grip, PTO
11. Ring
12. Roll pin, 3/16 x 13/16
13. Clutch lever assembly
14. Hex jam nut, ¾-16
15. Lift handle shaft assembly
16. Bracket
17. Lockwasher, 5/16 heavy
18. Hex nut, 5/16-18
19. Nut, 7/16-14
20. Lockwasher, 7/16 heavy
21. Hex head capscrew, 5/16-18 x ¾
22. Channel Frame
23. Hex nut, 5/16-18
24. Lockwasher, 5/16 Heavy
25. Steering pinion
26. Steering shaft bearing
27. Steering arm assembly
28. #9 Woodruff key
29. Steering arm pivot bushing
30. Hex c/locknut, 5/16-18
31. Hex head capscrew 5/16-18 x ½
32. Steering gear thrust spacer
33. Felt bearing protection seal
34. O-ring seal protector
35. Steering gear and shaft assembly
36. Key
37. Hex head capscrew, 5/16-18 x ¾
38. Roll pin, ⅛ x ¾
39. Steering shaft
40. Hex screw 5/16-18 x ¾
41. Hex screw ¼-20 x ⅝
42. Lift quadrant
43. Hex head capscrew, 5/16-18 x 1¾
44. Lockwasher, ¼ heavy
45. Hex nut, ¼-20
46. Hex head capscrew, ¼-20 x ½
47. Console assembly
48. Lift lever latch spring
49. Lift lever rod
50. Lift handle spacer
51. Hex head capscrew, ¼-20 x 1¼
52. Roll pin, ¼ x 1½
53. Lockwasher, 5/16 heavy
54. Hex c/locknut, ¼-20
55. Lift lever grip
56. Lift handle
57. Steering shaft support assembly
58. Washer
59. Hex jam nut ⅝-11
60. Lockwasher, ¾

LT860, 750 steering components

Jacobsen

1. Front axle
2. Grease fitting
3. Righthand spindle assembly
4. Lefthand spindle assembly
5. Steering arm assembly
6. Roll pin, 5/16 x 1½
7. Washer
8. Tie rod
9. Hex jam nut, ½-20
10. Ball joint assembly
11. Lockwasher, ½ heavy
12. Hex jam nut, ½-20
13. Pin
14. Roll pin (safety), 3/16 x 1
15. Steering gear assembly
16. Steering gear brace
17. Hex nut, 3/8-16
18. Lockwasher, 3/8 heavy
19. Hex head capscrew, 3/8-16 x ¾
20. Bearing
22. Square neck bolt, 5/16-18 x ¾
23. Lockwasher, 5/16 heavy
24. Hex nut, 5/16-18
25. Steering arm assembly
26. Roll pin, 3/8 x 1½
27. Steering drag link
28. Hex jam nut, ½-20
29. Ball joint assembly
30. Lockwasher, ½ heavy
31. Hex jam nut, ½-20
32. Front wheel bearing
33. Front wheel grease fitting
34. Front wheel
35. Front tire
36. Washer
37. Cotter pin, 5/32 x 1¼
38. Front wheel hub cap
39. Steering wheel
40. Steering wheel cap
41. Hex jam nut, 5/8-18
42. Square head cup point setscrew, 3/8-16 x 1
43. Cotter pin

GT16 steering components

1. Front axle
2. Grease fitting
3. Righthand spindle assembly
4. Lefthand spindle assembly
5. Steering arm assembly
6. Roll pin, 5/16 x 1½
7. Righthand spindle retainer washer
8. Tie rod
9. Hex jam nut, ½-20
10. Ball joint assembly
11. Lockwasher, ½ heavy
12. Hex jam nut, ½-20
13. Pin
14. Pin retainer roll pin (safety), 3/16 x 1
15. Steering gear assembly
16. Steering gear brace
17. Hex nut, 3/8-16
18. Lockwasher, 3/8 heavy
19. Hex head capscrew, 3/8-16 x ¾
20. Bearing
22. Square neck bolt, 5/16-18 x ¾
23. Lockwasher, 5/16 heavy
24. Hex nut, 5/16-18
25. Steering arm assembly
26. Roll pin, 3/8 x 1½
27. Steering drag link
28. Hex jam nut, ½-20
29. Ball joint assembly
30. Lockwasher, ½ heavy
31. Hex jam nut, ½-20
32. Front wheel bearing
33. Front wheel grease fitting
34. Front wheel
35. Front tire
36. Washer
37. Cotter pin, 5/32 x 1¼
38. Front wheel hub cap
39. Steering wheel
40. Steering wheel cap
41. Hex jam c/locknut, 5/8-18
42. Square head cup point setscrew, 3/8-16 x 1
43. Cotter pin

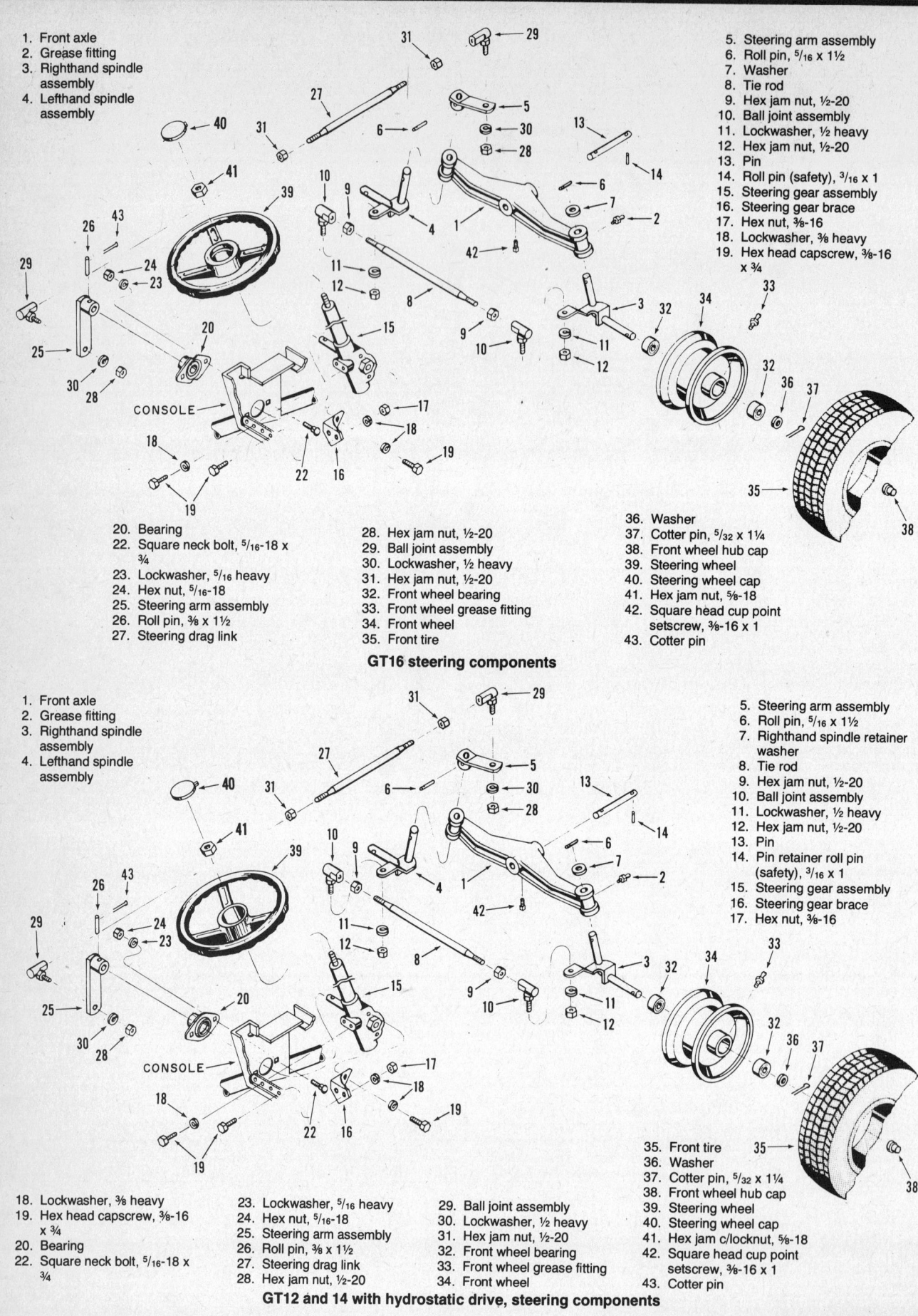

GT12 and 14 with hydrostatic drive, steering components

608

Jacobsen

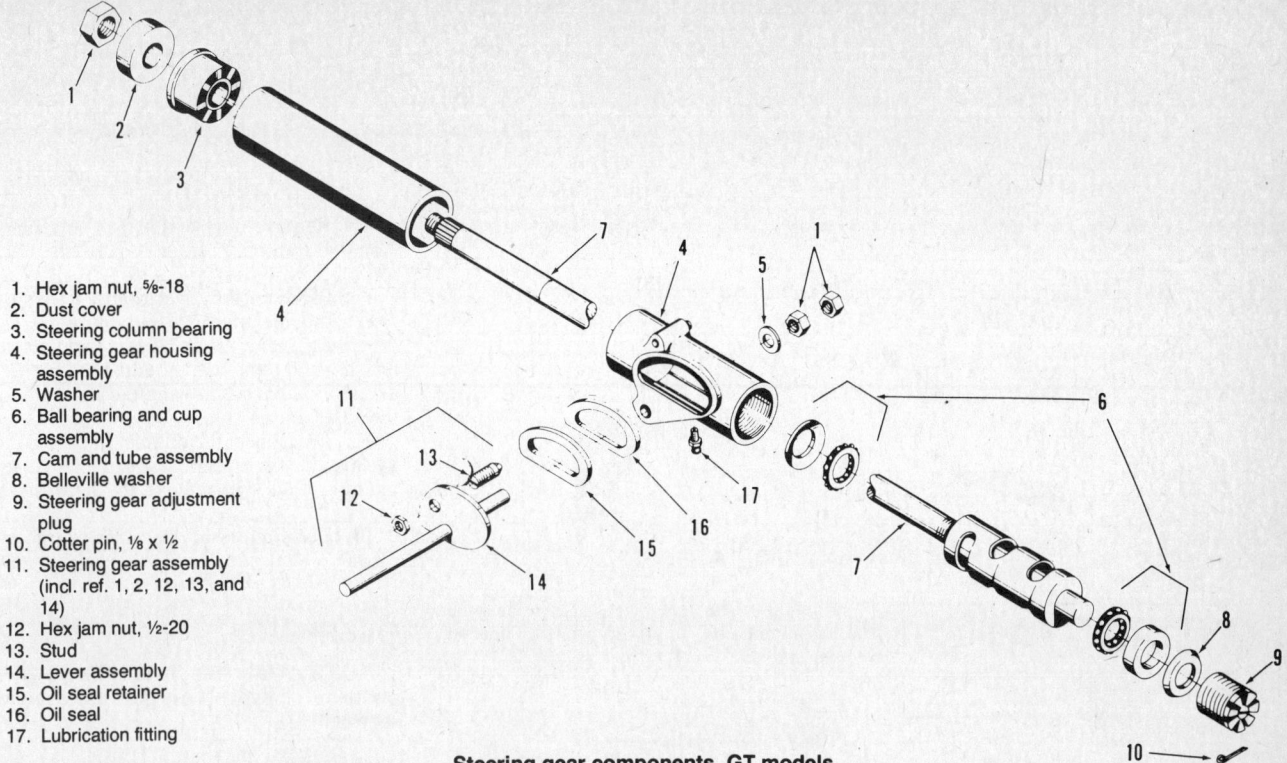

1. Hex jam nut, 5/8-18
2. Dust cover
3. Steering column bearing
4. Steering gear housing assembly
5. Washer
6. Ball bearing and cup assembly
7. Cam and tube assembly
8. Belleville washer
9. Steering gear adjustment plug
10. Cotter pin, 1/8 x 1/2
11. Steering gear assembly (incl. ref. 1, 2, 12, 13, and 14)
12. Hex jam nut, 1/2-20
13. Stud
14. Lever assembly
15. Oil seal retainer
16. Oil seal
17. Lubrication fitting

Steering gear components, GT models

4. Coat all parts with a light coat of grease.
5. Reassemble the bearings and worm gear into the housing, placing the lower race and Belleville washer in the threaded plug and installing the threaded plug into the housing.
6. Tighten the plug until all end play is out of the worm gear and shaft assembly, then tighten one more notch on the castellated plug to partially compress the Belleville washer.
7. Install the cotter pin to retain the plug.
8. Install the cross shaft and steering actuator plate.
9. Fill the worm gear cavity with clean grease and install the metal sealing plate over the opening.
10. Adjust the cross shaft and steering drive lug to the worm gear as follows:
    a. Loosen the jam nut on the drive lug.
    b. Back the drive lug out as far as possible.
    c. Install the thrust washer and two nuts on the cross shaft.
    d. Tighten the jam nuts in position so there is no end play in the cross shaft but so free movement remains in the actuator plate.
    e. Locate the drive lug in the worm gear so that a slight drag may be felt in one spot through the total travel of the worm, then lock the drive lug in position using the jam nut.
11. Grease the unit lightly through the grease fitting and install in the tractor.

BRAKES

ADJUSTMENT

1. Remove the rear cover plate.
2. Set the transmission shift lever to neutral.
3. Press the clutch-brake pedal and engage the parking brake lock in the upper notch on the brake arm.
4. Block the wheels securely, and (after making sure that the spark plug is connected, and the oil is in the crankcase) start the engine. Then, run the engine at medium throttle.
5. Look through the shift lever hole and check if the transmission pulley is stopped. If the pulley is not stopped, take off the brake adjustment nut on the brake actuator rod and move the parking lock to the second notch on the brake arm. If necessary, repeat the procedure using the next notch until the transmission pulley stops. Shut off the engine.
6. With the parking brake lock still engaged, turn the brake adjustment nut clockwise until the brake friction pads are tight against the brake disc.
7. Check if the brake is holding by:
   a. Attempting to rotate the brake disc manually; it should remain stationary.
   b. Removing any wheel obstruction and attempting to push the tractor; it should remain stationary or skid on the wheels.
8. Release the parking brake.
9. Check if the parking brake is off by attempting to rotate the disc brake manually; it should rotate.
10. Start the engine and check that the transmission pulley comes to a fast stop when the foot brake is applied while the engine is at full rpm.
11. Reinstall the rear cover plate.

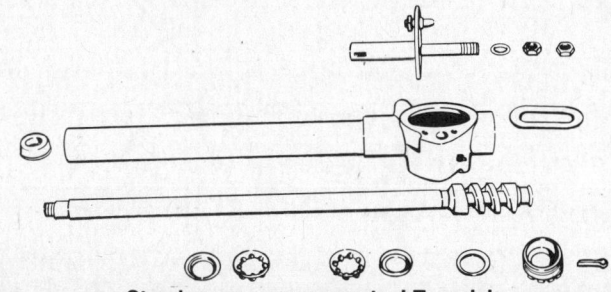

Steering gear components, LT models

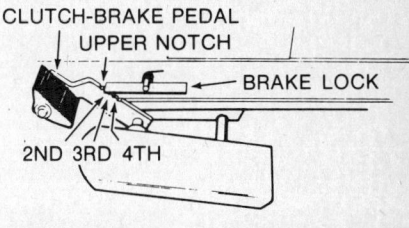

Parking brake lock engaged

609

Jacobsen

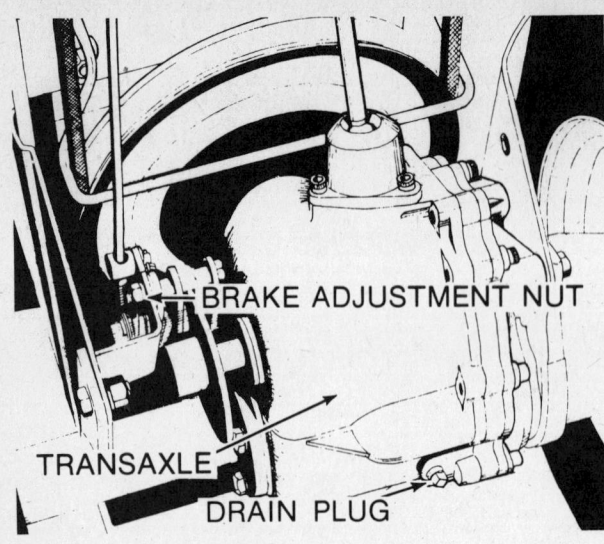

Brake adjustment points

HYDROSTATIC TRANSMISSION

LINKAGE ADJUSTMENTS

1. To adjust the hydrostatic transmission linkage, turn adjusting nuts on the speed control rod where the rod is attached to the speed control arm. Moving the adjusting nuts towards the front of the tractor increases forward speed, while turning the nuts in the opposite direction increases reverse speed.
2. Adjust the nuts so that the tractor does not creep in neutral. Turn the adjusting nuts to increase or decrease spring tension against the speed control arm until the tractor remains stationary in the neutral position.

REMOVAL AND INSTALLATION

1. Drain the oil from the transaxle bottom drain plug.
2. Remove the high-low shift lever, hydraulic lever, and brake lock knobs.
3. Remove the cover by removing the four Phillips head screws.
4. Disconnect the driveshaft coupler from the transmission flange.
5. Disconnect the hydraulic valve line from the top of the transmission hydraulic pump and disconnect the oil return line from the valve to transmission filter.
6. Remove the brake linkage, brake band, and brake anchor bracket from the transaxle by removing the two brake anchor brackets.
7. Remove the rod and spring from the hydroactuating arm.
8. Remove the transaxle high-low actuating linkage and pivot brackets from the transaxle.
9. Remove the frame-to-transaxle retaining bolts from both sides.
10. Remove the hydrostatic transmission and transaxle assembly from the tractor.
11. Remove the hydrostatic transmission from the transaxle by disconnecting the filter line from the hydrostatic transmission and removing the filter and bracket assembly after unscrewing the nut holding the filter on the stud.
12. Disconnect the oil inlet line from the hydrostatic transmission and the three remaining bolts and remove the hydrostatic unit.
13. Installation is the reverse of removal.

OVERHAUL

NOTE: Only the hydraulic booster pump of the hydrostatic gear assembly is serviced.

1. Remove the pump housing by unscrewing five capscrews which secure the pump housing to the hydrostatic housing and removing the Woodruff key from the shaft and the shaft seal shield from the shaft. Then, using a soft tip hammer, tap the housing and dowels away from the hydrostatic.
2. Before removing the pump rotor, note the direction it is installed so that the straight sides of the rotor are driving the rollers. Remove the outer snapring from the shaft, then the rotor, key, and pump backing plate.
3. If additional tear-down of the hydro

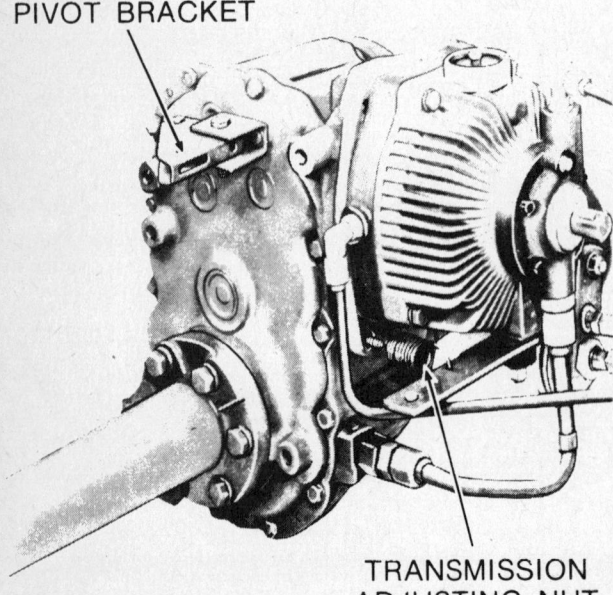

Transmission adjusting point

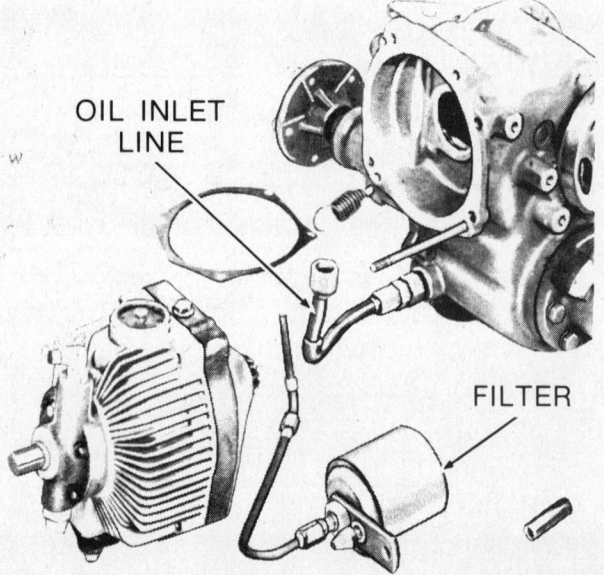

Removing the transmission from the transaxle

Jacobsen

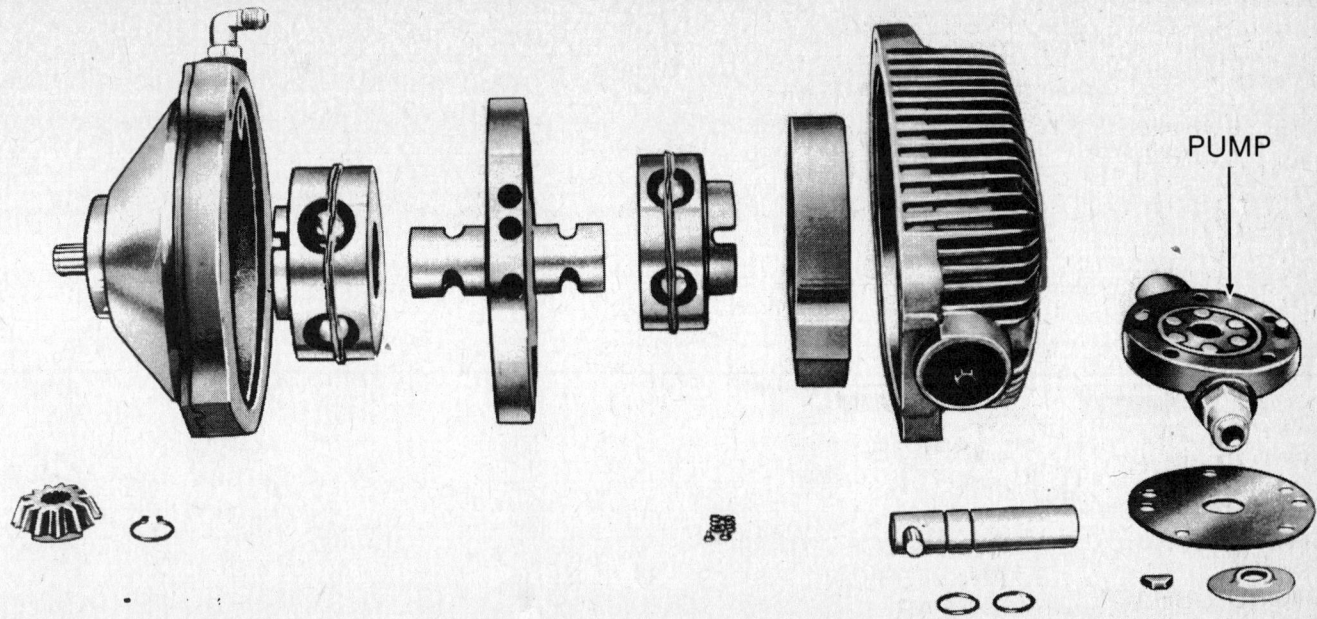

Hydrostatic transmission and pump components

gear is to be done, remove the inner snapring from the input shaft.

4. To separate the variable speed hydraulic pump housing from the hydraulic motor housing, remove the two retaining bolts, and set the assembly on the motor end.

5. Lift the variable pump section including the oil distributor assembly from the motor end.

— CAUTION —
The ball pistons are a select fit to each bore; they should be retained in their respective positions.

6. Remove distributor assembly from the variable pump housing.
7. Remove the pump components.

— CAUTION —
Do not mix the ball pistons; they must be reinstalled in the original bores.

8. Thoroughly clean, check, lubricate, and reassemble all parts, using reverse order of the disassembly procedure.

TWO-SPEED TRANSAXLE

The two-speed transaxle unit is powered by the hydrostatic unit and both share a common lubricating system.

DISASSEMBLY

1. Remove the axle supports and square O-ring seal.
2. Position the unit with the cover side up.
3. Remove the dowels and hex bolts and separate the cover from the case assembly.
4. Lift out the differential assembly and remove the output gear and shaft.
5. On the outside of the case, remove the setscrew, spring, and ball.
6. Remove both the input and shifter gear and shaft assemblies, along with the shifter rod and fork.

INSPECTION

Check the components for the following:
1. Axle support ball bearings and bearing races for wear.
2. Case and cover for leaks or cracks.
3. Gear teeth for wear, pitting or chips.
4. Gear for concentricity and roundness.

NOTE: Splines should fit smoothly. Meshing parts should be rotated for a better fit if binding seems excessive.

5. Shifter mechanism for tension and the ball for wear.
6. Shifter rod grooves for wear.
7. Snapring position, the sharp edge should face away from the shifter fork.
8. Shifter for straightness and wear.
9. Axle hub ends for burrs.

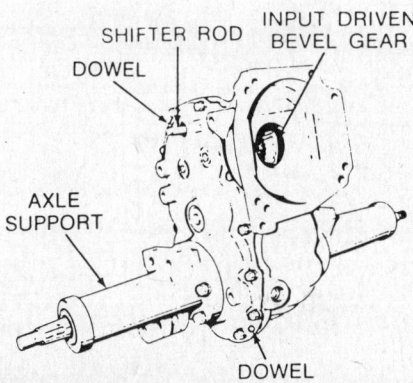

Two-speed transaxle

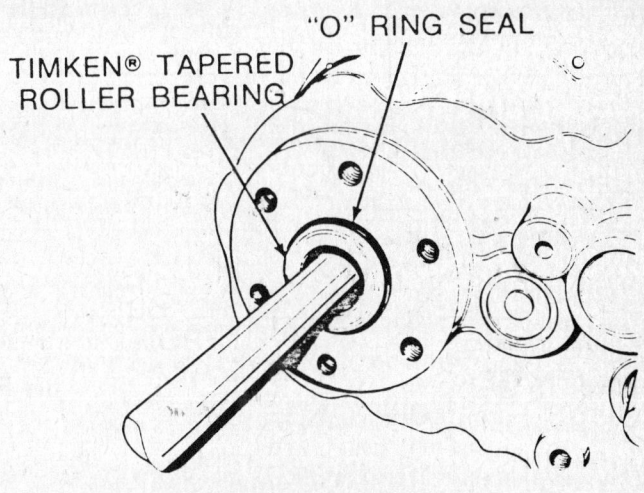

Axle support seal

Jacobsen

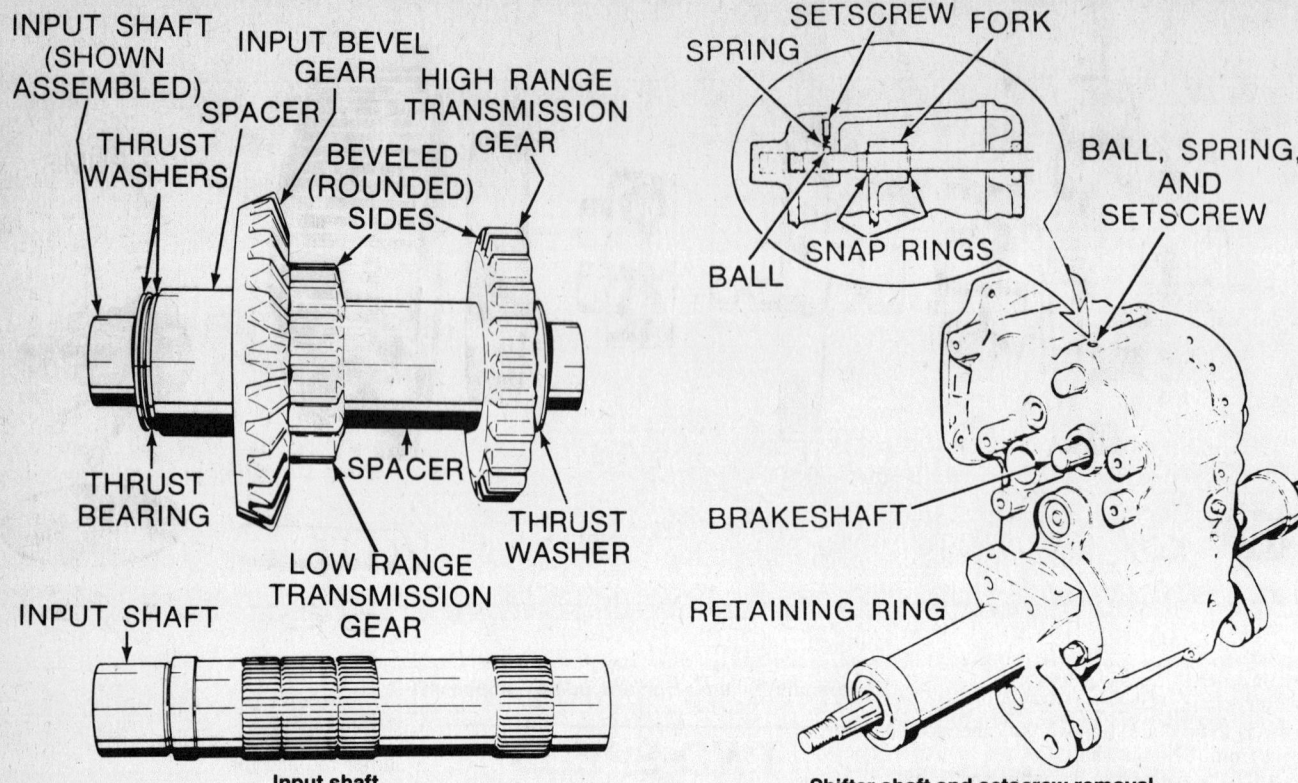

Input shaft

Shifter shaft and setscrew removal

10. Thrust washers for wear.
11. Tapered roller bearings for smooth rotation.
12. Thrust bearings for wear.
13. Rollers for proper placement.

ASSEMBLY

1. Install the input and shifter gear shaft assemblies.

2. Position the shifter rod and fork and install the ball, spring, and setscrew and turn the setscrew in slowly while raising and lowering the rod until the ball stops the rod movement. Install the output gear and shaft, and the differential assembly.

NOTE: Thrust washers and spacers must be between every shaft and the case and cover.

3. Install a new cover gasket and position the cover on the case.
4. Install the dowel pins and hex head cover bolts.
5. Reinstall the tapered roller bearings if they were removed from the case during disassembly.
6. Install new square O-ring seals and slide the axle supports on the axles.
7. Install the axle support hex head bolts.

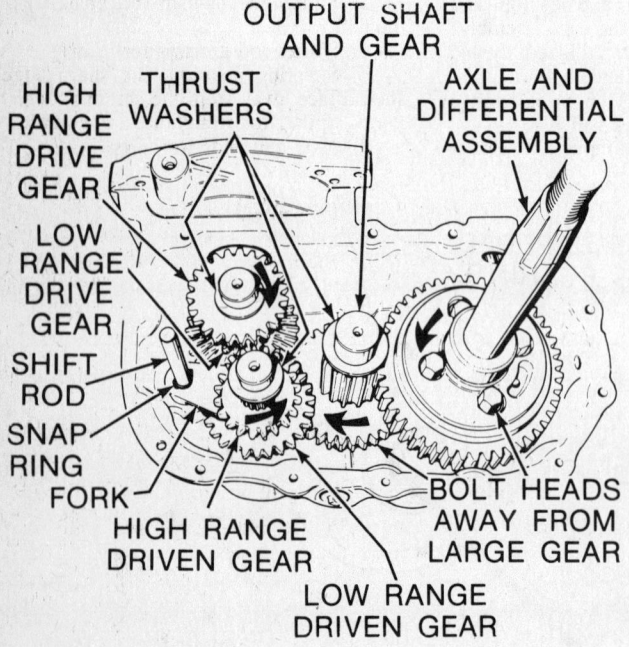

Output gear and shaft

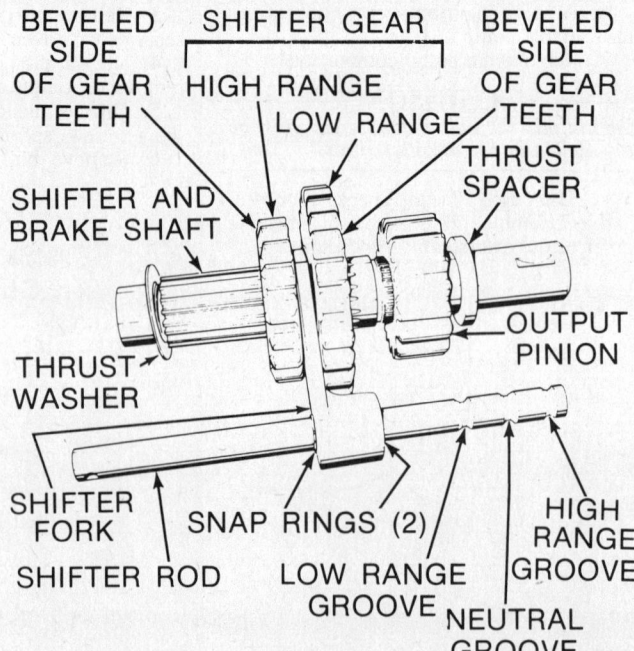

Shifter assembly

Jacobsen

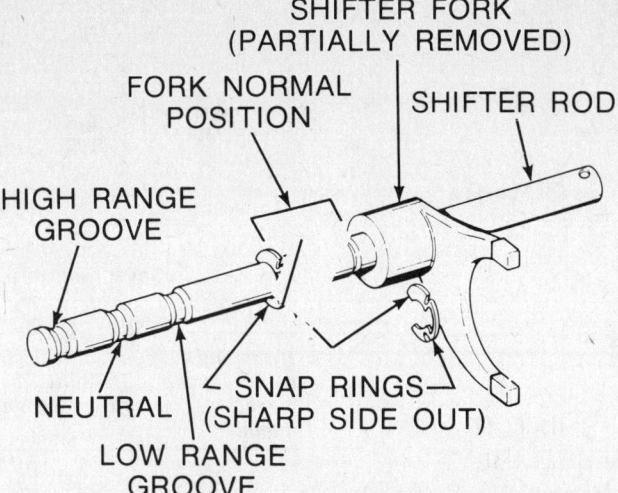

Shifter rod and fork

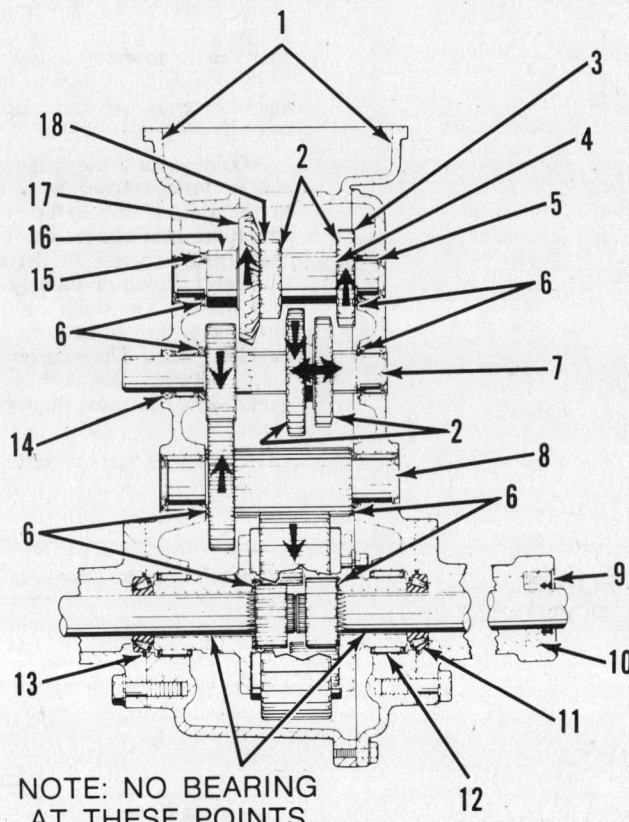

1. Hydrostatic drive mounting surface
2. Beveled sides of gears
3. High range drive gear
4. Spacer
5. Input shaft
6. Thrust washer
7. Shifter and brake shaft
8. Output shaft
9. Seal
10. Ball bearings (2)
11. Tapered bearings (2)
12. Needle bearings (2)
13. O-ring seal
14. Oil seal
15. Thrust bearing
16. Spacer
17. Driven input gear
18. Low range drive gear

Two-speed components

THREE-SPEED TRANSAXLE

DISASSEMBLY

NOTE: Position the shift lever in neutral to facilitate disassembly.

1. Clean the outside of the transaxle.
2. Remove the three screws securing the shift lever to the shift lever housing.
3. Remove the shift lever housing.
4. Drain the oil from the housing through the shift lever opening.
5. Remove all keys from key ways.

NOTE: A stone may be required to remove burrs from hardened shafts.

6. Remove burrs and dirt from shafts.
7. Remove the axle housings.
8. Place the transaxle assembly in a soft-jawed vise or clamp with the socket head capscrews up.
9. Remove the socket head capscrews that secure the case to the cover.
10. Drive out the dowel pins that align the case with the cover.

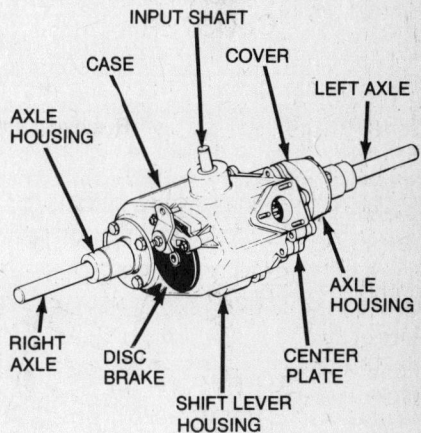

Three-speed transaxle

11. Remove the cover.

NOTE: The seal is a single lip type and may be reused if removed carefully. Discard the gasket.

12. Using a seal protector on the axle shaft, lift the transaxle cover free.

NOTE: Two or three screws may have to be reinstalled to hold down the center plate during removal of the differential assembly.

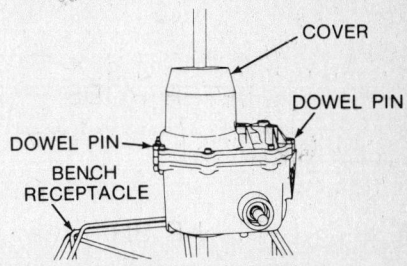

Transaxle mounted in a vise

613

Jacobsen

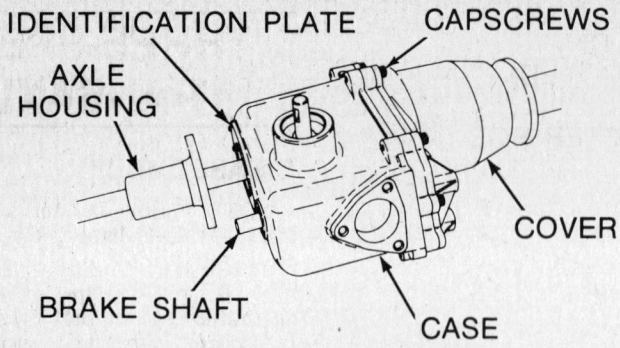

Case-to-cover attachment points

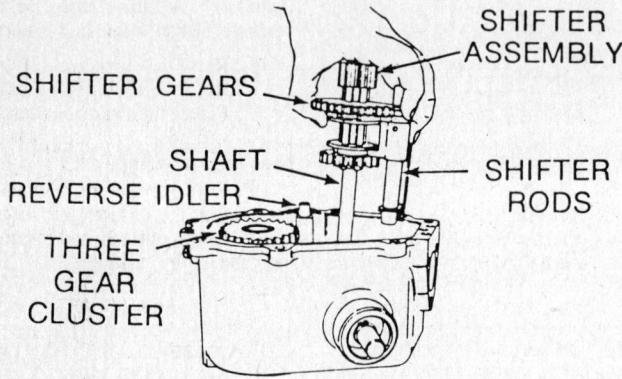

Removing shifter shaft gear

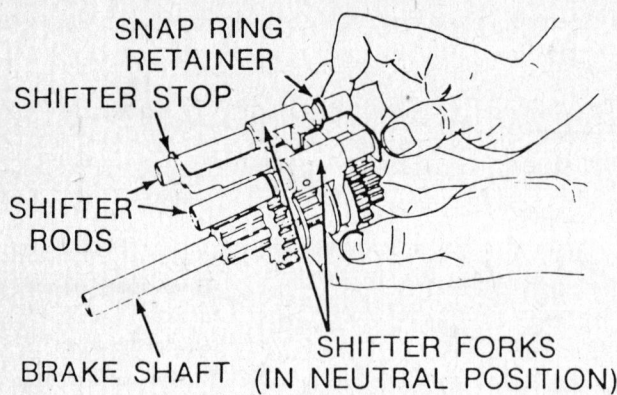

Shifter assembly removal

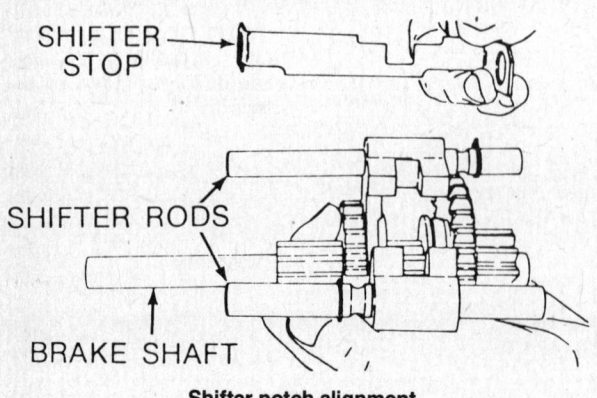

Shifter notch alignment

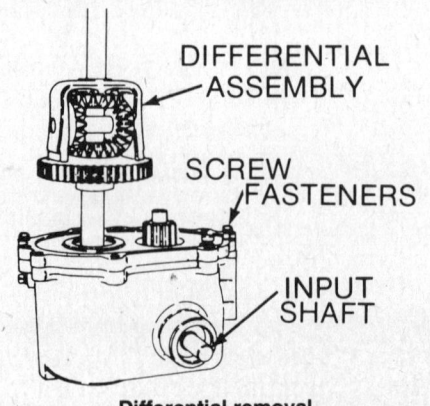

Cover removal

CAUTION
If tapping is required to separate components during differential assembly removal, use a soft mallet, not a metal hammer.

13. Remove the differential assembly.
14. Remove the gear from the shifter shaft.
15. Remove the screws used to temporarily secure the center plate.
16. Remove the center plate and discard the gasket.
17. Remove the shifter assembly by grasping and pulling the shifter gears, shaft, and both shifter rods.
18. Check the shifter assembly for damage or excessive wear. Replace a damaged or badly worn assembly.

NOTE: Mark the alignment of the shifter fork notches with the notches in the shifter stop before disassembly. The three gear cluster can be disassembled (with a press to drive the gears squarely). Worn or damaged gears can be replaced. The small and middle gear bevel faces down (the large gear has no beveled edge). The shorter section belongs between the middle and larger gear. Key edges must align with the shaft ends.

19. Remove the reverse idler shaft and spacer, three gear cluster and thrust washers.

CAUTION
Do not drop or lose the needle bearings.

20. Lift the idler gear assembly out of the case.

NOTE: Replace the shaft oil seal.

Differential removal

21. Remove the shaft oil seal and then the snapring and input shaft.

ASSEMBLY

1. Install the thrust washers on the input shaft.
2. Install the input shaft into the case assembly.
3. Set the case assembly open side up.

NOTE: To facilitate reassembly of the washers, thrust bearing, idler shaft, and gear assembly before installing the shifter assembly, insert the reverse idler shaft in place.

4. Install the idler shaft gear assembly, thrust washers, and bearings.
5. Install the washer and three gear cluster assembly.

NOTE: The reverse idler shaft will be pushed out when the shifter assembly is installed.

6. Install the shifter assembly and make sure the rods are seated properly.
7. Install the reverse idler, making sure the beveled edge is up and the spacer is on top of the gear.
8. Install a new gasket on the case.
9. Install the center plate.
10. Install a new gasket on the center plate.
11. Install the differential assembly with the longer axle down, making sure the gear is on the shifter shaft.
12. Install the gear case dowel pins so they are slightly exposed on top to facilitate aligning and installing the cover.
13. Install the cover and secure it with the eight capscrews.
14. Install the bearings and/or bushings (if necessary) with a bearing driver and bushing tool.
15. Install the axle housing and fill it with 1½ pints of SAE 90 EP oil.

NOTE: Make sure the shift notches in the forks and the notch in the shifter stop are aligned and centered so the unit is in neutral.

FOUR-SPEED TRANSAXLE

DISASSEMBLY

1. Remove both axle housings and separate the seal retainers from the case and cover.
2. Block the axle assembly so the assembly lies on its cover without placing weight on the brake shaft.
3. Tap the dowel pin into the cover and remove the eight socket head bolts.
4. Separate the case from the cover by lifting the case 1½ to 2 in. above the cover, tilting the case so the shift rods clear the edge, and rotating the case to allow the boss hidden inside to clear the gears.
5. Remove the thrust washer and three gear clusters from the brake shaft.

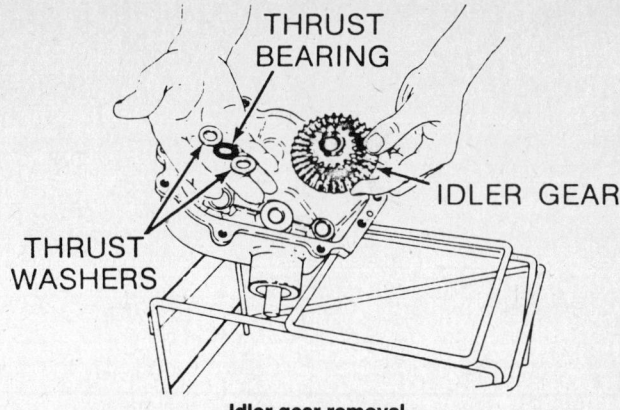

Idler gear removal

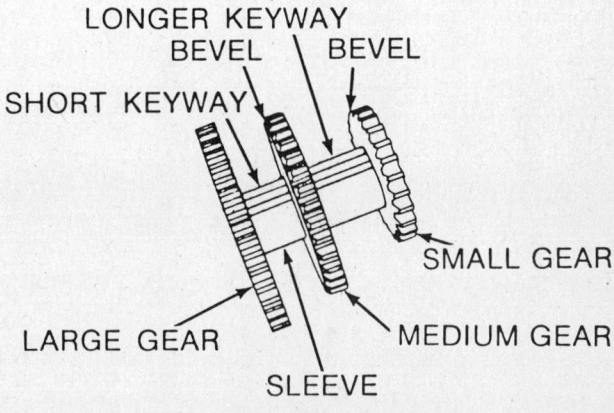

Three-cluster gear

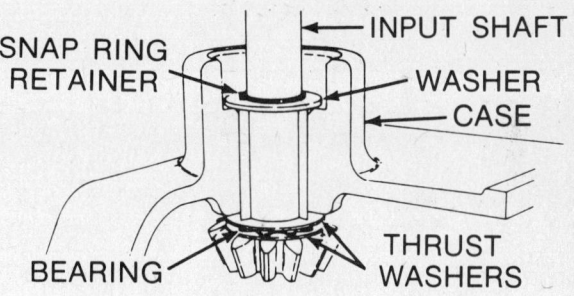

Snapring, input shaft and thrust washers

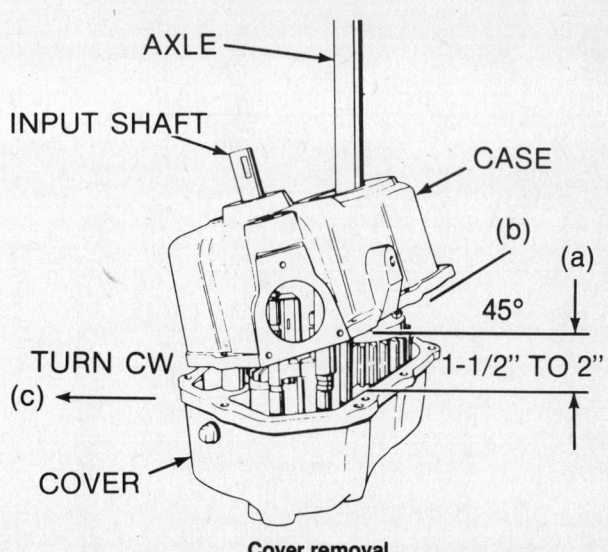

Cover removal

Jacobsen

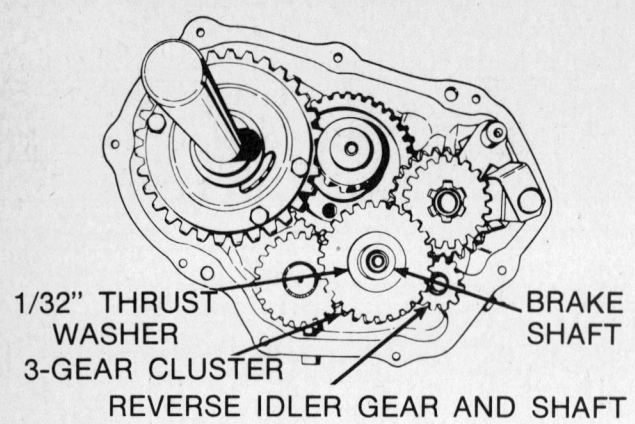

Three-cluster gear

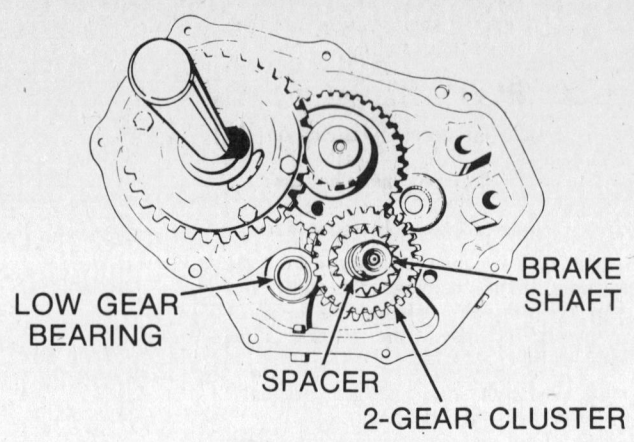

Two-gear cluster

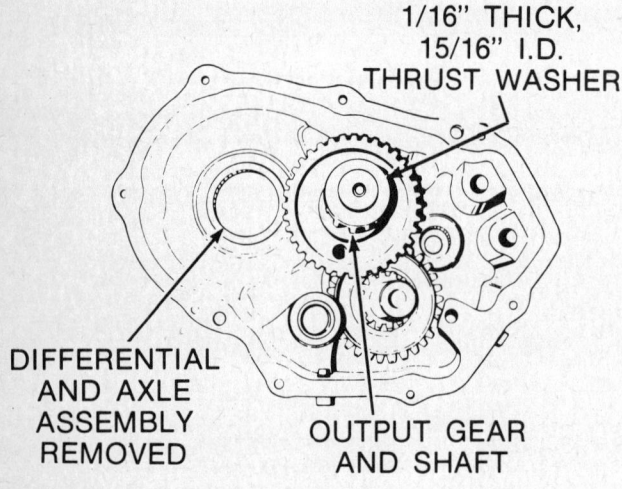

Output gear and shaft

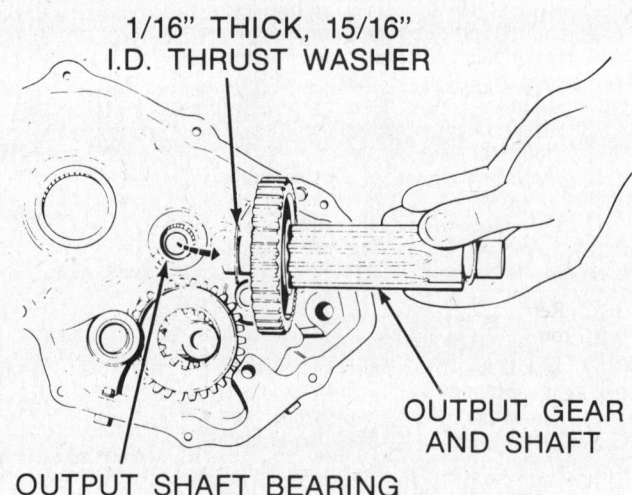

Removing output shaft

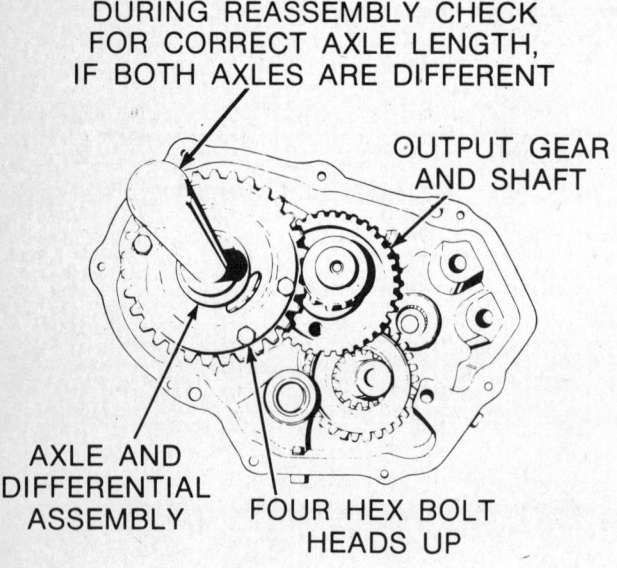

Differential assembly

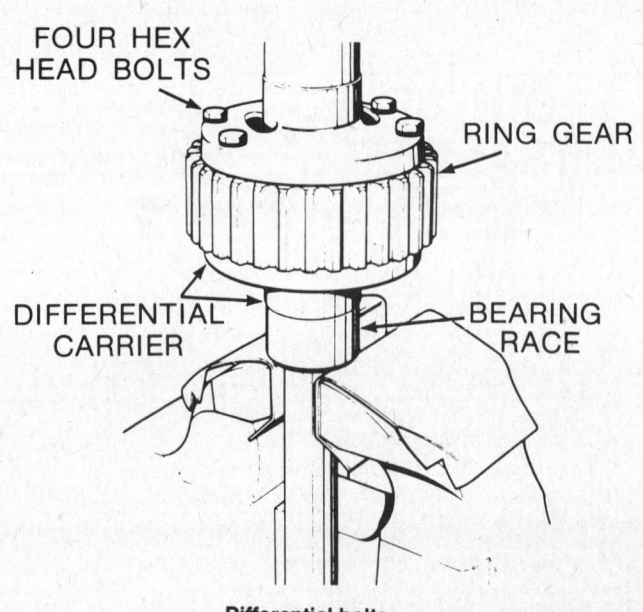

Differential bolts

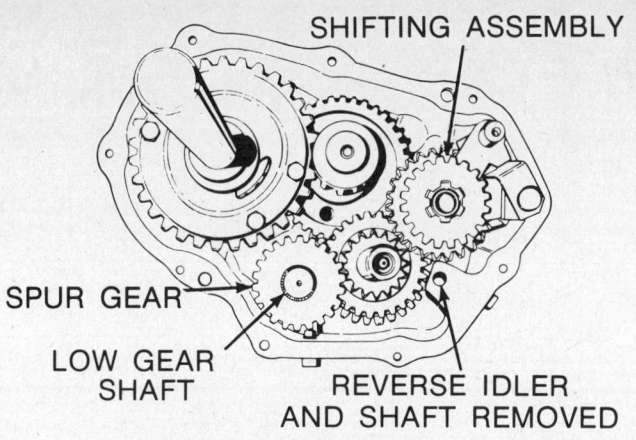

Shifting assembly removal

Brake shaft

6. Remove the reverse idler gear, spacer, and shaft from the boss in the cover.
7. Lift out the shifter assembly.
8. Remove the low gear and shaft, and splined spur gear.
9. Disassemble the two gear cluster and spacers from the brake shaft.
10. Lift the differential unit out of the cover.
11. Remove the output shaft, gear, and thrust washer.

NOTE: If necessary, tap the input shaft with a non-metallic hammer.

12. Remove the brake shaft and input shaft from the case.

INSPECTION

Check the components for the following:
1. Gear bevels for galling.
2. Faces of gear teeth for wear (indicated by large shiny areas).

NOTE: Excessively worn parts should be replaced.

3. Shafts and axles for rust, pitting, scratches, or wear.
4. Key ways, splines, threads, and grooves for wear.
5. Case and cover for cracks, stripped threads, chips, warped or damaged mating surfaces that are sealed, and rust.
6. Thrust washers and spacers for wear, indicated by shiny areas.

ASSEMBLY

NOTE: Use a soft mallet to seat shaft and gear completely.

1. Install the input shaft in the case.
2. Center the 1 in. x $1/32$ in. thrust washer on the cover brake shaft needle bearing, then install the brake shaft and gear.

NOTE: The chamfer side of the gear must be away from the cover.

3. Install the output shaft and gear after centering the $15/16$ in. I.D. x $1/16$ in. thrust washer on each end of the shaft.
4. Insert the differential assembly in the cover with the four hex bolt heads away from the output gear.

5. Install the two gear clusters and spacer on the brake shaft. Install the ¾ in. I.D. x $1/16$ in. thrust washer, gear, and low gear idler shaft in the cover.

NOTE: Do not put a thrust washer on the exposed end of the low gear idler shaft. Be sure the small gear teeth are in mesh with the larger gear teeth of the two gear cluster.

6. Center the ⅞ in. inside diameter x $1/32$ in. thrust washer on the cover shifter shaft bearing.
7. Install the shifter assembly as a unit in the cover.
8. With the small gear of the three gear cluster toward the spacer, install the three gear cluster and ⅞ in. inside diameter x $1/32$ in. thrust washer on the brake shaft.

NOTE: The bevel side of the idler gear should be facing toward the cover.

9. Install the reverse idler shaft, spacer and gear into the cover.
10. Position the gasket on the cover sealing surface, then install the case over the differential shaft.

NOTE: Be sure the boss goes under the gears and that the edge of the case goes over the shaft rods in the reverse order of removal. If the case hangs ½ in. to 1 in. above closing, turn the input shaft to get the gears to mesh. The case will then drop to about ¼ in. from closing.

11. Using a pair of needle nose pliers, move the shifter stop on each shifter fork to agitate the shifter rod ends in their machined recesses in the case.
12. Align the case and cover with the two dowels and install and tighten the eight hex head bolts. Torque the hex head bolts to 10 ft.lb.
13. Position the seal retainers and new seals in/on the axles.
14. Install new O-rings on the seal retainers.
15. Position the axle housing on the case and install the eight ⅜ in.—16 x 1 in. hex head bolts and torque the bolts to 13 ft.lb.

16. Install new shift lever gasket, position the shift lever assembly, and install the three Allen head bolts. Torque the bolts to 10 ft.lb.

DIFFERENTIAL

DISASSEMBLY

1. Clean the differential.
2. Check and record (for reassembly purposes) the axle lengths and their relation to the heads of the four hex head bolts.
3. Replace any parts causing binding.

--- **CAUTION** ---
Do not clamp the bearing race in the vise.

4. Clamp the differential, hex head bolts up, in a soft jawed vise.
5. Remove the four hex head bolts and upper axle and differential carrier.

NOTE: Tap the ring gear lightly with a mallet to separate it from the differential carrier.

6. Remove the drive blocks, pinions, drive pin, and thrust spacer by lifting out of the ring gear.
7. Replace defective axle assemblies that have a rolled axle end. Disassemble the assemblies with snapring retainers by removing the snapring and thrust washer and separating the bevel gear and differential carrier from the axle.

INSPECTION

Check the following parts and, if excessively worn or damaged, replace:
1. Gears: internal splines and, if the gear is removable, the axle.
2. Pinions, drive pins, and drive blocks.
3. Differential carriers: internal bearing diameter. If wear at point A is excessive, replace the carrier or bushing.

NOTE: One differential carrier has threaded holes and the other has larger (bolt) holes.

Differential cross section

Differential with one carrier removed

Drive block, pin and pinion gear removed

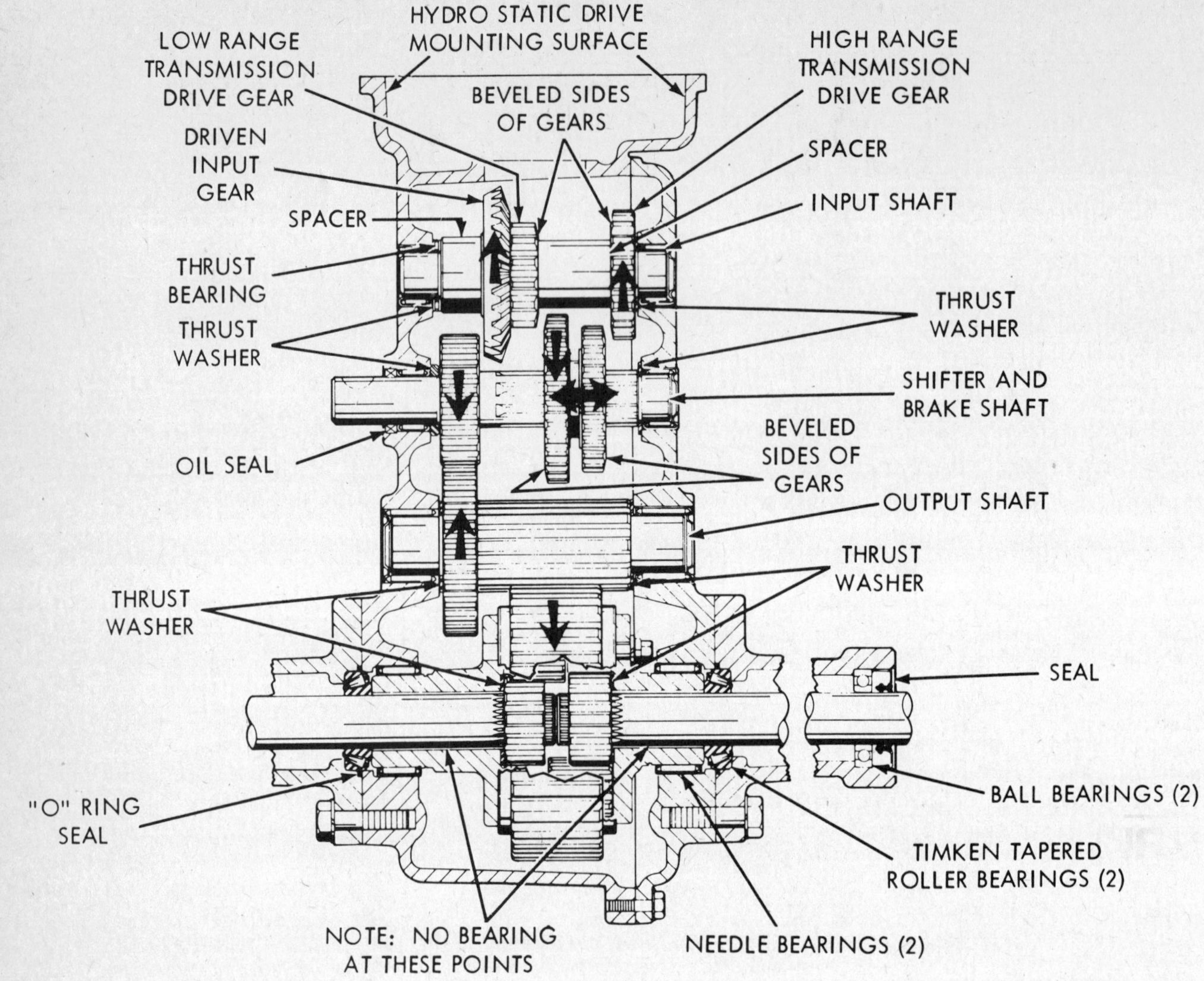

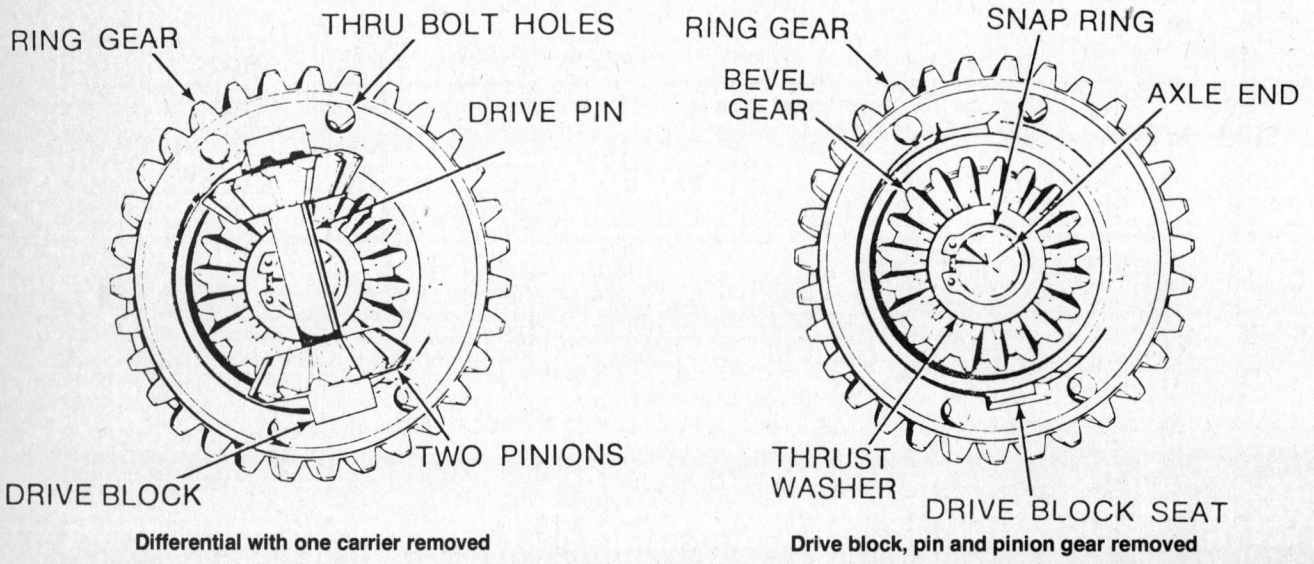

Jacobsen

ASSEMBLY

CAUTION

Select the correct axle for the side of the differential opposite the hex head bolts. If the wrong axle is used, it will require complete tear down of the differential, or possibly the entire transaxle if the error is not detected until later.

NOTE: Lightly oil all parts before reassembly.

CAUTION

Do not clamp the bearing or oil seal surfaces.

1. Clamp the axle in a soft jawed vise.
2. Assemble the differential carrier to the axle in the reverse order of disassembly.
3. Torque the four hex head bolts to 7 ft.lb.

TESTING

Test differential action by holding the upper axle vertically, and spinning the differential. The unit should spin and rotate freely. Place the assembly on the bench and rotate both axles in different directions. If any binding is noted in either test check retaining bolt torque, gear meshing, or bearing surfaces in the differential carriers. Little or no end-play should be apparent between the axles and carriers.

RIGHT ANGLE DRIVE

REMOVAL

1. Remove the drive belt from the right angle drive pulley and remove the lower drive pulley.
2. Unbolt and remove the right angle drive unit.

DISASSEMBLY AND REASSEMBLY

1. Drain the lubricant and remove the gasket (2).
2. Remove the cap (7), seal (8), and gasket (9).
3. Pull the input shaft (11) with the bearing (10) from the housing (18).
4. Remove the snapring (13) and gear (12).
5. Remove the oil seal (15) and snapring (16).
6. Drive the output shaft (6) with gear (4) and bearing.
7. Remove the gear (4) from the output shaft (6).
8. Remove bearings (5) and (17) from the case.
9. Clean and inspect all parts and replace any showing excessive wear or other damage.
10. Reassemble in the reverse order of disassembly and fill the unit with 4 ounces of Moly E.P. Lithium Grease.

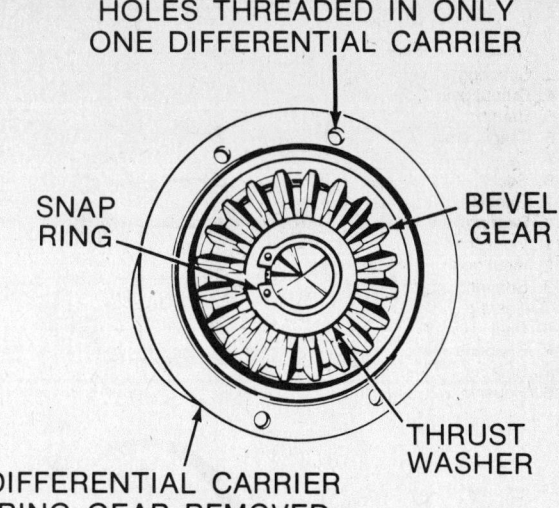

Ring gear removed

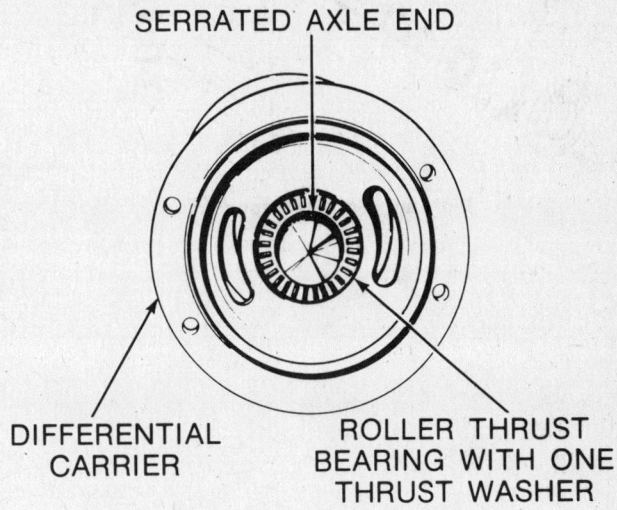

Snapring, washer and bevel gear removed

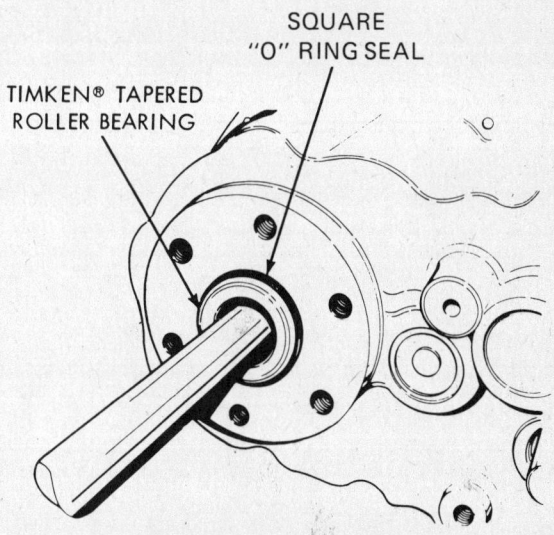

Tapered roller bearing location

Jacobsen

1. Cover
2. Gasket
3. Snapring
4. Output gear
5. Bearing
6. Output shaft
7. Cap
8. Seal
9. Gasket
10. Bearing
11. Input shaft
12. Input gear
13. Snapring
14. Bearing
15. Seal
16. Snapring
17. Bearing
18. Housing

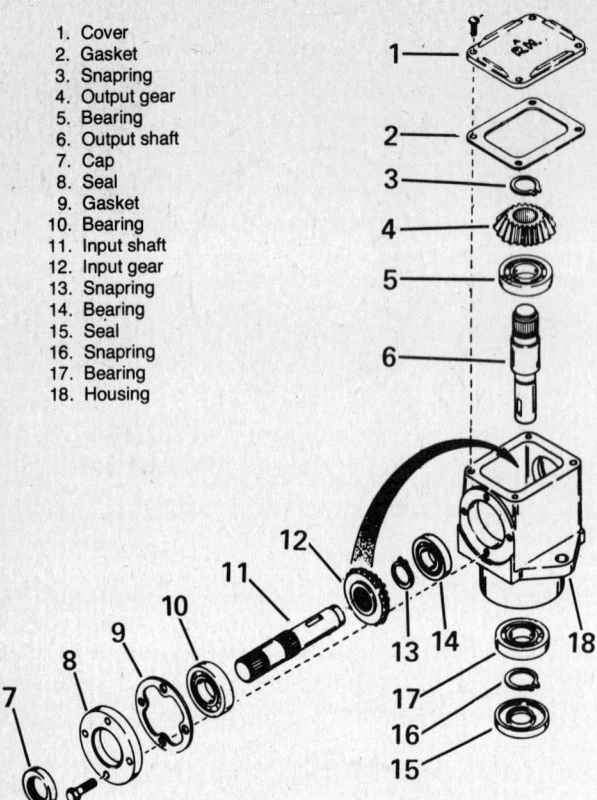

Right angle drive components

ENGINE

Engine used are as follows:
Model 1060: Briggs & Stratton 251707
LT885: Briggs & Stratton 191707
LT860: Briggs & Stratton 191707
LT750: Briggs & Stratton 170707
GT16: Kohler KS341AS
GT14: Kohler KS321AS
GT12: Kohler K301AS
GT10: Kohler K241AS

REMOVAL AND INSTALLATION

1. Remove hood.
2. Shut off fuel valve.
3. Disconnect all lines, cables and wires attached to the engine.
4. On LT series, disconnect PTO clutch actuator rod at the front end. On GT tractors, disconnect the PTO belt.
5. On LT tractors, remove the rear cover, belt guide and main drive belt, by depressing the clutch pedal so that tension is released on the shaft.
6. On GT tractors, remove the driveshaft shield and disconnect the driveshaft.
7. Drain the oil, unbolt and remove the engine.
8. Installation is the reverse of removal.

Kubota

INDEX

MODELS B5100, 6100, 7100
Front Wheels and Axle **622**
   Bracket and Gear Case 622
   Axle Differential 623
Steering **624**
Brakes **625**
Splitting and Recoupling **626**
Clutch **628**
Transmission, Differential, Rear Axle **629**
   R&R 629
   Differential Gear Case 632
   Axle Cases 632
Hydraulic System **634**
Engine **635**
Fuel System **636**

MODEL 175
Front Wheels and Axle **637**
Steering **638**
Brakes **639**
Clutch and Mainshaft **639**
Transmission **640**
Differential and Rear Axle **641**

Hydraulic System **643**
   Pump 643
   Control Valve 643
   Cylinder 644
Engine **644**
Fuel System **646**

MODEL L225
Front Wheels and Axle **646**
Steering **647**
Brakes **648**
Clutch and Mainshaft **648**
Transmission and Power Take-Off **649**
Differential and Rear Axle **650**
Hydraulic System **652**
Engine **652**

MODEL L285
Front Wheels and Axle **652**
Steering **653**
Brakes **654**
Clutch **654**
Transmission **654**

Differential **655**
Hydraulic System **657**
Engine **659**

MODEL L185, 245, 295
Two-Wheel Drive Front Axle **659**
Four-Wheel Drive Front Axle ... **661**
Steering **662**
Brakes **663**
Clutch **663**
Transmission **663**
   Clutch Housing Side 663
   Transmission Case Side 667
Differential **669**
Rear Axle **671**
Hydraulic System **671**
   Pump 671
   Position Control Valve 673
   Draft Control Valve 673
   Relief Valve 674
   Linkage 674
   Cylinder 675
Engine **677**
Fuel System **682**

KUBOTA
Models B5100, B6100, B7100

FRONT WHEELS AND AXLE

Front Wheel Bracket and Gear Case

REMOVAL AND DISASSEMBLY

1. Remove the snap pins holding the hood bolts with vibration-proof rubber rings on the front of the hood.
2. Remove the bolts.
3. Remove the snap pins holding the radiator mounting bolts.
4. Rmove the bolts.
5. Remove the right and left front wheel mounting nuts.
6. Detach wheels.
7. Remove the cotter pin holding the tie rod end mounting nut.
8. Remove the slotted nut.
9. Set a tie rod pin puller, and remove the tie rod by tightening the puller bolt.
10. Remove the seven bolts holding the periphery of the front wheel gear case cover.
11. Detach the front wheel gear case cover by lightly tapping the periphery with a plastic hammer.
12. Remove the knuckle arm mounting bolts.
13. Remove the kingpin together with the lock plate by lightly tapping the sides of the periphery of the knuckle arm.
14. Remove the two bolts holding the lock plate which is under the front wheel gear case.
15. Detach the lock plate.
16. Remove the kingpin from under the gear case.
17. Remove the 14 bolts holding the dust cover.
18. Push and release the dust cover.
19. Push and release the felt gasket.
20. Push and release the dust seal.
21. Push and release the dust seal holder.
22. Push and release the gasket.
23. Remove the front wheel gear case.
24. Remove all the seals which have been pushed and released.
25. Remove the joint shaft circlip with a set of snapring pliers.
26. Remove the bearing with a bearing puller.
27. Remove the 12T gear from the joint shaft.
28. Remove the joint shaft by lightly tapping it with brass rod and hammer and pushing it toward the transmission shaft.

Removing tie rod

Removing gear case

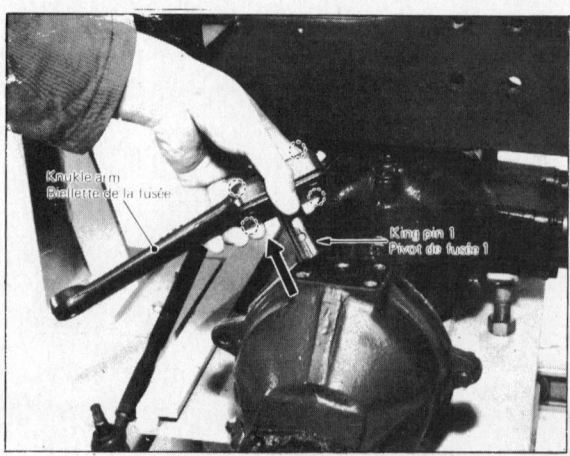

Removing knuckle pin

Kubota

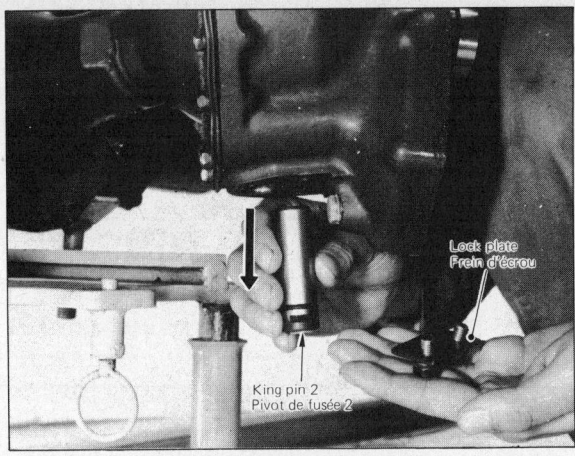

Removing lower king pin

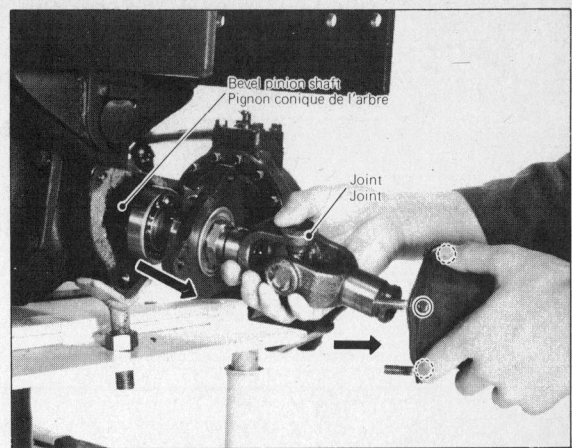

Removing bevel pinion shaft

ASSEMBLY AND INSTALLATION

Installation is basically the reverse of removal and disassembly. Observe the following points:

1. When installing the joint shaft, apply a thin film of gear oil to the oil seal inside the front wheel gear case.
2. Polish the spherical surfaces with sandpaper if they are rusted.
3. Be careful about the direction of installation of the dust seal.
4. Apply oil or grease to the felt gasket and turn it several times.
5. Apply 2 small amounts of grease to the spherical sliding parts. When tightening the dust cover, be careful not to allow the front wheel gear case to move.
6. Apply a thin film of gear oil to the O-ring, and carefully insert the king pin.
7. Tighten the knuckle arm bolts to 28.93 to 36.17 ft.lb.
8. Tighten the slotted nut to 21.70 to 28.93 ft.lb.
9. Tighten the locknut to 21.70 to 25.31 ft.lb.
10. Be sure to fit the M10 spring washer when reassembling.
11. Tighten the front wheel mounting nuts to 39.98 to 43.40 ft.lb.
12. Measure the tire pressure with a tire pressure gauge. Tire pressure: 17.01 psi.
13. The vibration-proof rubber rings must be 0.8268 in. thick (including upper and lower washers).
14. The vibration-proof rubber rings must be 0.2165 to .02362 in. thick.

Front Axle Differential
DISASSEMBLY

1. Remove the joint case mounting bolts.
2. Remove the joint case.
3. Detach the bevel pinion shaft together with the front axle case by lightly tapping it with a copper hammer.
4. Remove the eight bolts for the front axle case.
5. Remove the front axle case toward you by lightly tapping it with a wooden hammer.
6. Pull out the differential gear toward the front axle case (to the left). (The differential side shim is located inside the right and left front axle cases. Be careful not to confuse right and left.)
7. Remove the center pin set pin
8. Remove the center pin slotted nut.
9. Remove the center pin by tapping it backwards with a copper hammer.
10. Detach the front axle case by shifting it to the side.
11. Loosen the front axle case step bolts and nut on the engine cradle, and lower the front axle case step.
12. Remove the front axle case by pulling it to the side.

ASSEMBLY

Installation is the reverse of removal. Note the following:

1. Apply a thin film of gear oil to the center pin, and install the front axle case on the front wheel bracket.
2. Tighten the center pin slotted nut to 7.23 ft.lb. If the set pin and the pin hole do not align with each other, screw out the slotted nut a little and install the set pin.
3. Place the center pin set pin in the right position. At this point, insert the set pin from the left side.
4. Wrong amount of gear backlash may cause noise while the tractor is travelling.
5. After checking to see if the differential side shim is properly installed, set the differential gear.
6. Tighten the bolts to 17.36 to 20.25 ft.lb.
7. Apply a little grease to the universal joint, and install the joint case.

TOE-IN ADJUSTMENT

1. Equalize the pressure of right and left tires.

Removing front axle case

Removing differential gear

623

Kubota

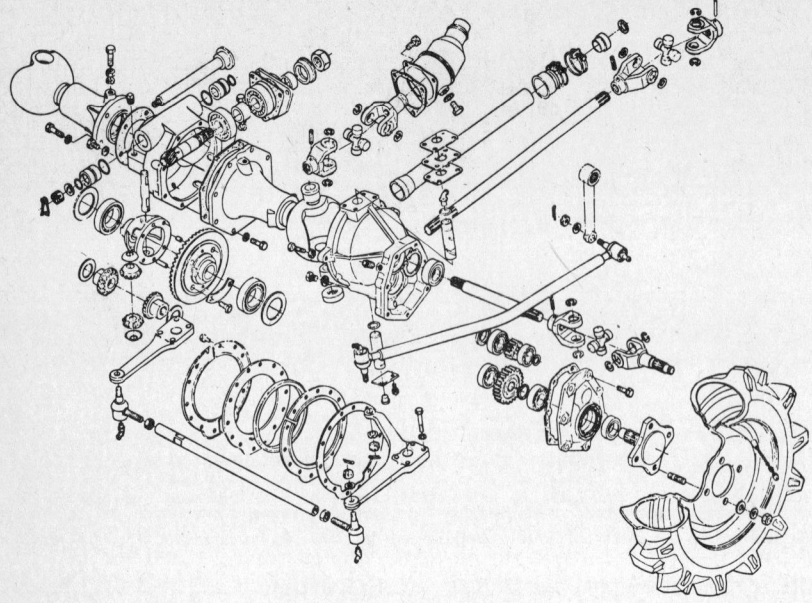

Front axle components

2. Place the tractor on a horizontal base with the front wheels straight forward.
3. Find the distances between the front two wheels at their front and rear, and the difference.
4. If the difference exceeds 0.1969 in., adjust by turning the tie rod.

STEERING

Steering Gear

REMOVAL AND DISASSEMBLY

1. Disconnect the negative battery cord from the negative terminal.
2. To release the instrument panel, detach the coupler under the panel.
3. Disconnect the decompression wire from the engine.
4. Remove the steering wheel cap by inserting a regular screwdriver into the notch of the cap.

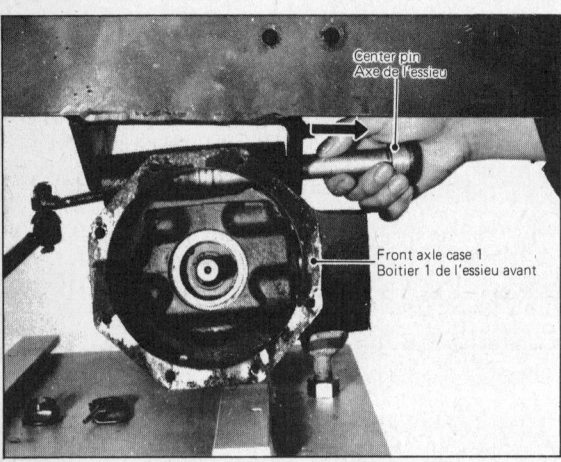

Removing center pin

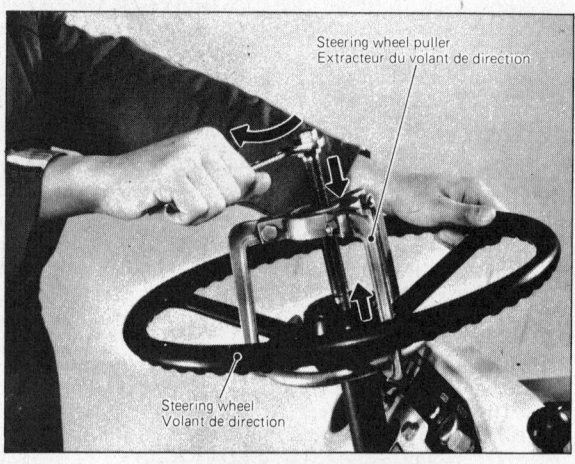

Pulling the steering wheel

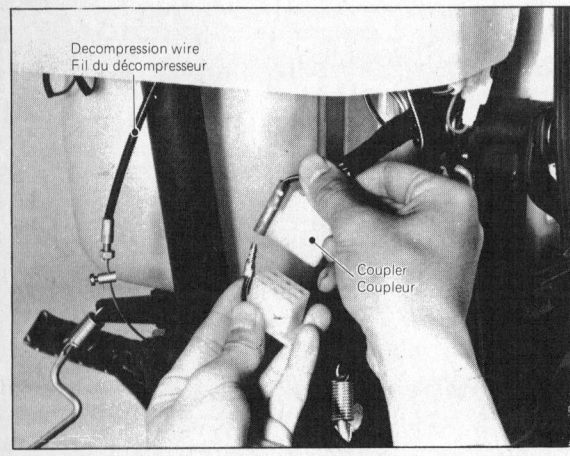

Coupler and decompression wires

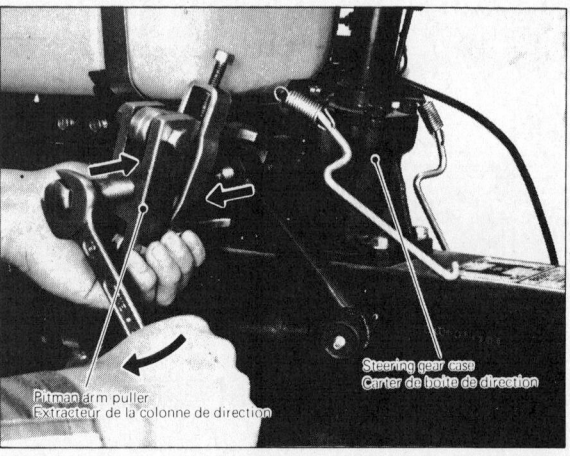

Removing Pitman arm

5. Loosen the nut for the steering shaft.
6. Set a steering wheel puller on the wheel, and pull it out by screwing in the puller bolt and tapping the wheel with a hammer.
7. Remove the four bolts holding the instrument panel.
8. Lift the panel to detach along the steering post.
9. Remove the sector shaft nut.
10. Set a Pitman arm puller on the Pitman arm which is attached to the sector shaft.
11. Remove the Pitman arm by screwing in the puller bolt and lightly tapping the arm with a hammer.
12. Remove the four bolts holding the steering gear case.
13. On the side of the steering gear case remove the bolt which is painted red, and drain oil. (Removal of the upper bolt may quicken oil drainage.)

ASSEMBLY AND INSTALLATION

Reverse the removal and disassembly steps, noting the following:
1. When pouring oil into the gear case, tilt the gear case such that it will hold the maximum possible amount of oil, and the prescribed amount of oil (0.05GA) will then be supplied.
2. Tighten the steering gear case mounting bolts (7T) to 14.47 to 21.70 ft.lb.
3. Attach the Pitman arm after checking to see if the alignment mark on the arm aligns with that on the sector shaft.
4. Tighten the Pitman arm to 43.40 to 57.86 ft.lb.
5. When installing the instrument panel and wire harness, be careful not to make any error in their mounted position.
6. Always check to see if the front wheels are facing straight forward before installing the steering wheel, and secure it properly.
7. Tighten the steering wheel mounting nut to 14.47 to 21.70 ft.lb.
8. When replacing the decompression wire, connector check to see if it is given enough play to allow the decompression wire to disengage in the off position.

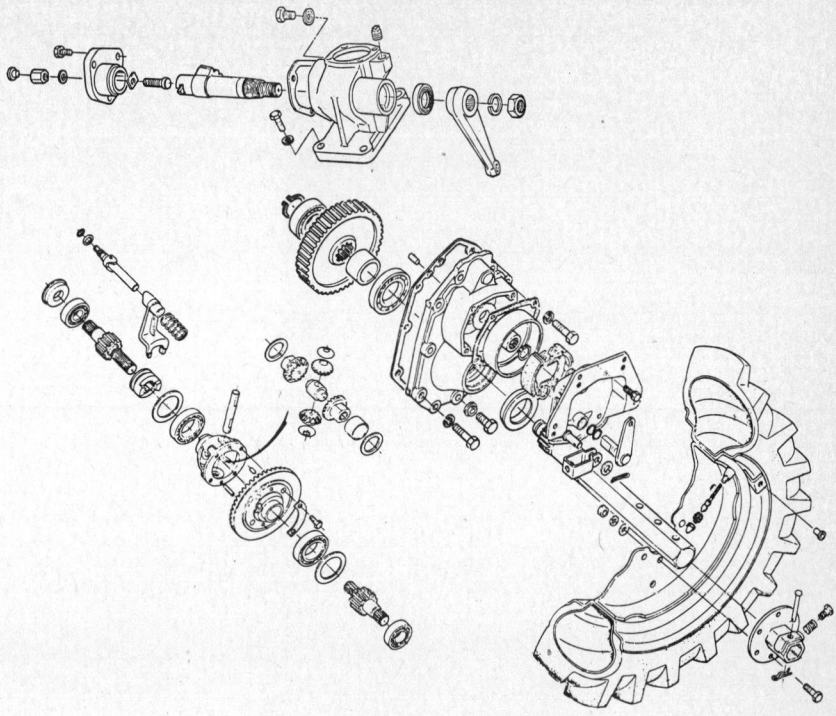

Steering system components

BRAKES

Brake Shoe and Drum

REMOVAL

1. Set differential transmission case cradle and rail under the tractor.
2. Set the fitting base to differential transmission case.
3. Jack up the rear wheels 0.3937 to 0.7874 in.
4. Loosen the wheel hub set bolt, and remove the axle pin.
5. Detach a rear wheel.
6. Remove the coupler from the flasher lamp cord.
7. Remove the step plate and fender mounting bolts.
8. Remove the brake return spring, and the differential lock return spring.
9. Loosen the turnbuckle of the brake rod, and detach the rod.
10. Remove the bolts holding the brake cover.
11. Remove the external circlip with a set of snapring pliers.
12. Remove the brake drum.

INSTALLATION AND ADJUSTMENT

Installation is the reverse of removal. Note the following points:
1. After installing the rod, measure and adjust the play of the brake pedal with its turnbuckle.
2. Tighten the locknut for the turnbuckle securely.

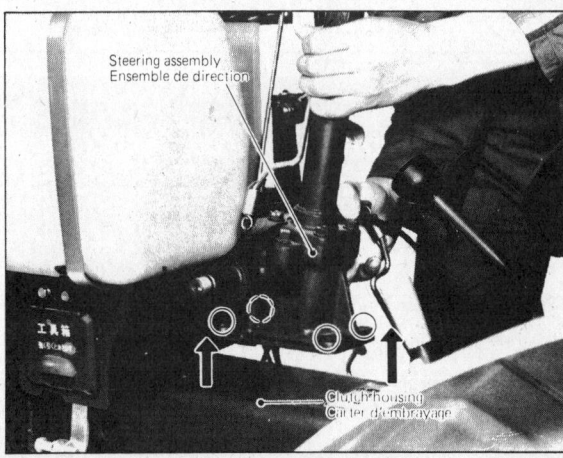

Removing steering gear box

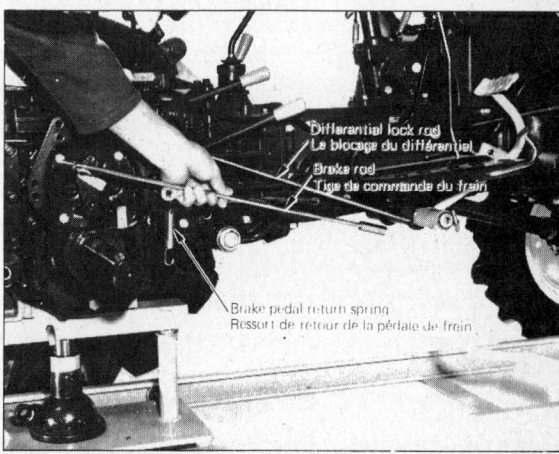

Removing brake rod and return spring

Kubota

Removing brake cover

Removing brake drum

3. Equalize the play of both brake pedals.

4. Install the turnbuckles so that the screws of the right and left brake rods are equal in length.

5. Tighten the hub set bolts to 65-72 ft.lb.

6. Tighten the hub and wheel set bolts to 50-58 ft.lb.

PEDAL ADJUSTMENT

1. Remove the brake lock so that the play of right and left brake pedals can be measured separately. Pedal play should be 0.0787-0.3937 in.

2. While pressing down the upper face of a pedal lightly by hand, measure the play of the pedal before braking begins to take effect.

3. If play is excessive, adjust by turning the turnbuckle of the brake rod.

4. Equalize right and left brake pedals' play.

SPLITTING AND RECOUPLING

Front End Separation

1. Drain coolant through the drain cock on the bottom of the radiator. Remove radiator cap for fast draining. If the engine is still hot, removing the cap should be done only after first loosening slightly to release pressure.

2. Remove the drain plug at the lower right of the front axle case, and drain oil. Drain oil from the right and left front wheel gear cases.

3. Disconnect the negative cord from the negative battery terminal.

4. Detach the air cleaner assembly.

5. Remove the muffler.

6. Remove the left side cover.

7. Remove the right side cover.

8. Loosen the bands with a regular screwdriver or a set of double-end wrenches, and draw out the water pipes gently.

9. Loosen the driveshaft band from the right side (viewed when facing forward). When the band becomes loose enough to turn, bring it backward.

10. Remove the nut connecting the knuckle arm to the drag link. Draw out the rod end with a tie rod pin puller.

11. Before setting the disassembly and assembly base, remove blind plugs on the housing case and front wheel step.

12. Place the engine cradle rail under the tractor, and set the clutch housing case cradle.

13. Attach the front wheel step to the engine cradle.

14. Detach the head lamp coupling at the back of the radiator.

15. Remove the twelve bolts fixing the front axle support.

16. Jack up the tractor body about 0.7874 in. (Make sure that oil filter is not in contact with the front axle support.)

17. Move the front axle support forward.

Front End Recoupling

Recoupling is the reverse of separation. Note the following:

1. Tighten the mounting bolts to 28.93 to 47.74 ft.lb. When installing the front axle support, leave the side cover bolts loose so that the covers can be easily attached later.

3. Tighten the drag link nut to 14.47 to 28.93 ft.lb.

Engine Separation

1. Loosen the drain plug on the lower left side of the engine, and drain oil.

2. Drain oil in the axle case on the rear side of the tractor, from the drain plugs.

3. Referring to Front End Separation, separate the engine from the front axle support.

4. Disconnect the starter wiring.

5. Disconnect the oil switch wiring.

6. Disconnect the glow plug wiring.

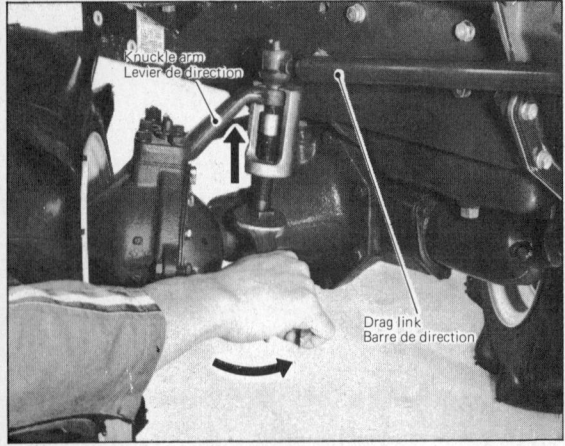

Removing drag link

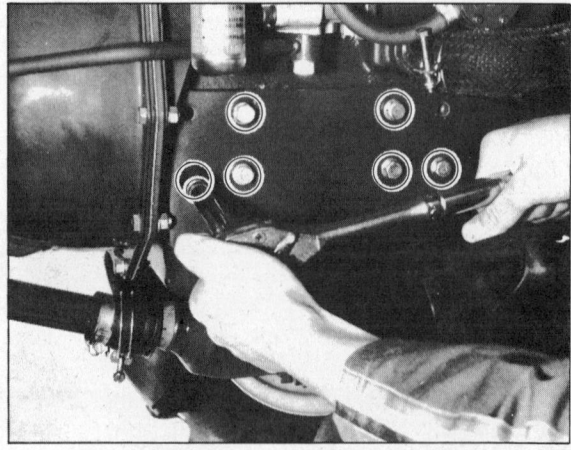

Removing front axle support bolts

Kubota

Starter removal

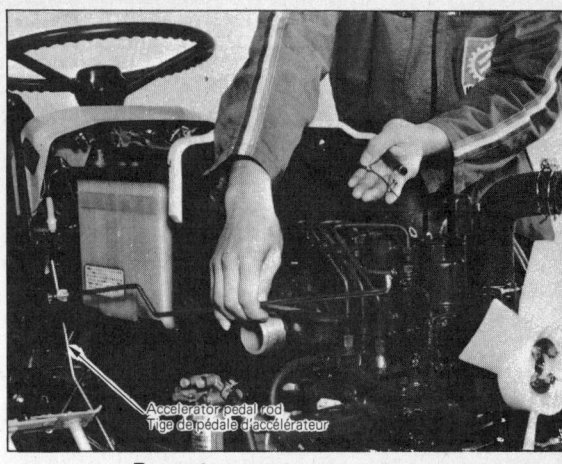

Removing accelerator pedal rod

7. Disconnect the alternator wiring coupling.
8. Remove the starter.
9. Release the release rod on the speed control lever side.
10. Detach the accelerator pedal rod.
11. Release the decompression wire on the decompression lever side.
12. Detach the overflow pipe, first with the nozzle near the fuel tank.
13. Remove the heat insulator.
14. Close the fuel filter cock.
15. Loosen the clip, near the fuel feed pump, for pipe and draw out the pipe.
16. Remove the fuel filter together with the filter stay.
17. Remove the pipe connecting the fuel filter and the tank, from the fuel filter, and plug it so that no fuel spills.
18. Loosen the two bolts holding the fuel tank band.
19. Open the band, and remove the tank.
20. Detach the hydraulic pipe clamp.
21. Detach the hydraulic pipe joint bolt.
22. Detach the pipe joint.
23. Detach the pipe joint.
24. Detach the oil filter.
25. Remove the hydraulic pipes.
26. Detach the engine cradle from the front axle support, and attach it to the bottom of the engine.
27. Install the fitting base to the side of the engine.
28. Set a jack under the engine and the clutch housing case so that it supports them equally.
29. Remove the eight connecting bolts.
30. Check to see if the engine and the clutch housing case are in horizontal position. If not, adjust with a jack.
31. Separate the engine from the case by slowly moving it forward.

Recoupling the Engine

Recoupling is the reverse of separation. Note the following points:
1. Tighten the engine set bolts to 14.47 to 21.70 ft.lb.
2. When placing the engine cradle, be careful not to let the front wheel driveshaft touch the cradle at the left side of the engine.

Clutch Housing Separation

1. Remove the negative battery cable from the negative terminal.
2. Remove the drain plug on the bottom of the rear axle case, and drain oil.
3. Remove the air cleaner assembly.
4. Detach the muffler.
5. Detach the left side cover.
6. Detach the right side cover.
7. Remove the cotter pin.
8. Remove the nut connecting knuckle arm and drag link.
9. Draw out the rod end with a tie rod pin puller.
10. Disconnect the lamp wiring coupling.
11. Disconnect the starter wiring.
12. Disconnect the oil switch wiring.
13. Disconnect the AC dynamo wiring.
14. Disconnect the glow plug wiring.
15. Remove the starter.
16. Loosen the driveshaft band from the tractor right side (viewed when facing forward.)
17. When the band becomes loose enough to turn, bring it backward.
18. Release the release rod on the speed control lever side.
19. Detach the accelerator pedal rod.

Detaching the release rod and accelerator rod

20. Release the decompression wire on the decompression lever side.
21. Detach the overflow pipe, first with the nozzle near the fuel tank.
22. Detach the heat insulator.
23. Close the fuel filter cock.
24. Loosen the pipe clip on the side of the fuel feed pump, and draw out the pipe.
25. Detach the fuel filter together with the filter stay.
26. Remove the pipe connecting the fuel filter and the tank, from the filter, and plug it to prevent oil from spilling.
27. Loosen the two bolts holding the fuel tank band.
28. Open the tank band, and detach the tank.
29. Detach the hydraulic pipe clamp.
30. Detach the hydraulic pipe joint bolt.
31. Detach the pipe joint.
32. Detach the pipe joint.
33. Detach the oil filter.
34. Remove the hydraulic pipes.
35. Remove the blind plugs on the housing case and front wheel step.
36. Place the cradle rail under the clutch housing.
37. Move the clutch housing cradle to the engine along the rail and together with the engine fitting base fix it to the side of the clutch housing.
38. Move the engine cradle to the en-

Kubota

gine along the rail, and together with the engine fitting base fix it to the side of the engine.

39. Move the whole of the tractor forward along the rail, and fix the engine cradle to the cradle rail.
40. Remove eight connecting bolts.
41. Check to see if the engine and the clutch housing case are in horizontal position. If not, adjust with a jack.
42. Separate the engine from the clutch housing by shifting the housing case gently backwards.

Recoupling the Clutch Housing

Recoupling is the reverse of separation. Observe the following points:
1. With the front wheel drive lever in neutral, join the engine and the housing case, align the splines, and tighten the connecting bolts.
2. Tighten the engine mounting bolts to 14.47 to 21.70 ft.lb.

Transmission Separation

1. Remove the drain plugs at the bottom of the rear axle cases, and drain oil from them.
2. Remove the negative battery cord from the negative terminal.
3. Detach the air cleaner assembly.
4. Detach the right side cover.
5. Remove the cotter pin on the fuel injection pump, and then the release rod.
6. Remove the upper and lower cotter pins holding the accelerator pedal rod, and detach the rod.
7. Close the fuel filter cock.
8. Loosen the pipe clip on the side of the fuel feed pump, and draw out the pipe.
9. Remove the fuel filter together with the filter stay.
10. Disconnect the pipe connecting fuel filter and tank, from the fuel filter, and plug it to prevent fuel from spilling.
11. Detach the hydraulic pipe clamp.
12. Detach the hydraulic pipe joint bolt.
13. Detach the pipe joint.
14. Detach the pipe joint.
15. Detach the oil filter.
16. Remove the hydraulic pipes.
17. Remove two bolts holding the stay which connects step and step support plate.
18. Detach the right and left step support plates.
19. Release the parking lock.
20. With a set of pliers, remove the cotter pin holding the brake rod. Detach the brake rod from under the step.
21. Detach the differential lock rod from the differential pedal in the same manner as above.
22. Disconnect the flasher lamp wirings from the cord holders (right and left), at the couplings.
23. Remove the blind plugs which may hinder setting of the disassembly and assembly base.
24. Move the differential transmission case cradle backward, and fix it to the two-point link bracket with the differential transmission case fitting base and pins.
25. Bring the clutch housing cradle to the housing along the rail, and fix it together with the housing fitting base.
26. Remove the bolts connecting the housing case and the transmission case.
27. Check to see if the transmission case and the housing case are in horizontal position.
28. Separate the housing case from the transmission case by moving the housing case gently forward.

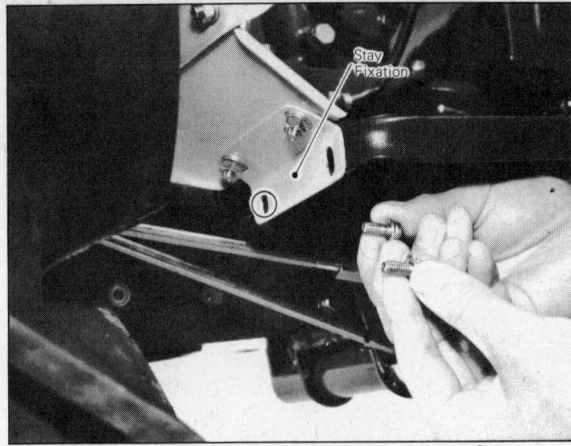

Removing the stay, step and support plate

Recoupling the Transmission

Recoupling is the reverse of separation. Not the following points:
1. With the speed change lever in neutral, check to see that both propeller and driveshafts are engaged with the hole on the clutch housing case.
2. Engage the brake lock lever with the third groove from the bottom.
3. Adjust the travel of the differential lock pedal to 0.9055 in. with no play.
4. Reassembly will be easier if the bolts on the step support plates are loosened.

CLUTCH

REMOVAL

1. Separate the clutch housing.
2. Insert the clutch aligning tool into the clutch disc.
3. Remove the bolts holding the pressure plate assembly, and detach the assembly.
4. Remove the clutch aligning tool and the clutch disc.
5. Remove the spring for clutch rod.
6. Remove the cotter pin and the head

Installing clutch aligning tool

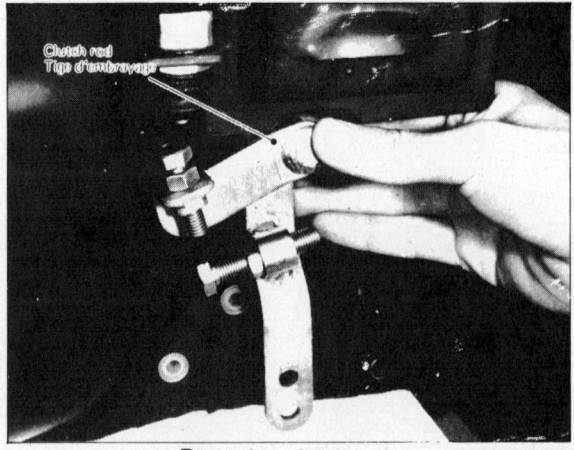

Removing clutch rod

Kubota

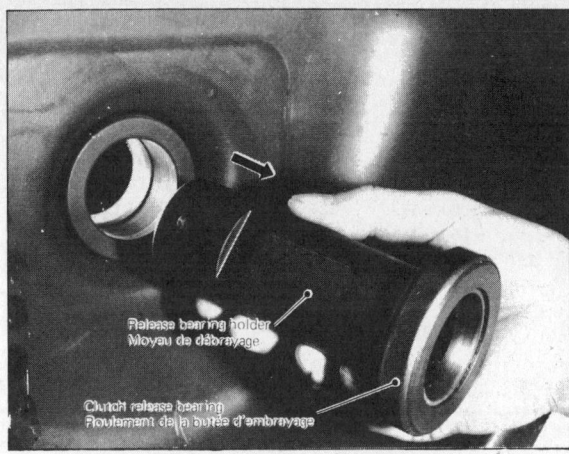
Removing the release bearing holder

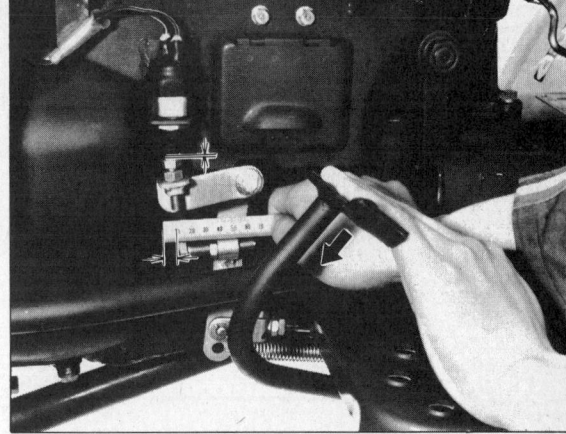
Clutch pedal adjustment

pin holding the U-bracket from the clutch intermediate rod.

7. Remove the clutch rod mounting nut.

8. While pulling out the clutch rod to the left side of the clutch housing, pull the clutch release fork, too.

9. Remove the release spring pulling the bearing holder backwards in the clutch housing case.

10. Remove the grease nipple.

11. Remove the release bearing holder by lightly tapping it with a hammer and a bearing drive guide 1.1811 to 1.3780 in. in outer diameter. At this point, be careful not to scratch the clutch release bearing.

INSTALLATION

Installation is the reverse of removal. Note the following points:

1. After installation, bend the release spring securely.

2. The bearing holder must be installed so that it can move lightly by the force of the spring.

3. Tighten the clutch rod nut to 11 to 13 ft.lb.

4. Install the fork so that the enlarged end faces backwards.

ADJUSTMENT

1. With a ruler at the bolt of the clutch rod, measure the travel and play of the pedal.

2. If the pedal travel or play exceeds 0.1181 ± 0.39 in., remove the cotter pin and head pin at the U-bracket in front of the clutch intermediate rod, and adjust the play by turning the U-bracket.

3. Adjust the safety switch by means of the adjusting bolt of the clutch rod.

TRANSMISSION

Transmission, Differential and Rear Axle

REMOVAL AND INSTALLATION

1. Tilt the seat forward, remove the head pin at the fulcrum, and detach the seat.

2. Loosen the right and left wheel hub set bolts (0.7087 in.) with a set of double-ended wrenches.

3. Remove the axle pins, and detach the both wheels.

4. Remove the five bolts inside each fender (both right and left).

5. Detach the rear seat support.

6. Remove the springs for brake and differential lock rods.

7. Detach the differential lock pedal by removing the head pin.

8. Remove a brake rod by loosening the locknut and turning the rod.

9. Remove the right and left mud guard covers.

10. Remove the cotter pin connecting the check rod to the control lever.

11. Detach the check rod.

12. Remove the three set bolts for the control valve.

13. Remove the control valve horizontally being careful not to drop the two O-rings.

14. Remove the nine set bolts on the rear case cover.

15. Remove the cover by lightly tapping the back of it upward with a wooden hammer.

NOTE: Because of the knock pin, the cover must by all means be tapped upward.

16. Remove the six front case cover set bolts.

17. Remove the cover by lightly tapping the back of it upward with a plastic hammer.

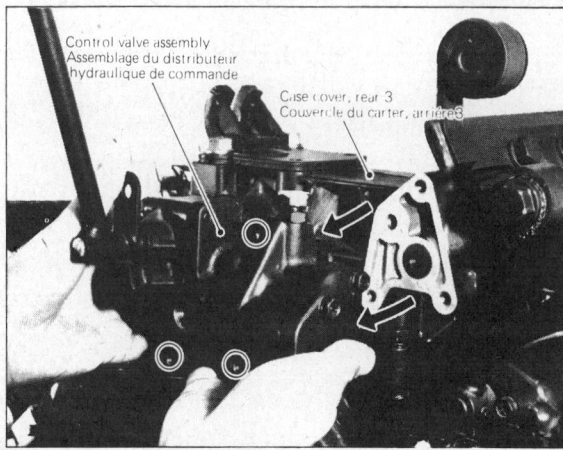
Control valve removal

Removing the rear case cover

Kubota

Removing front case cover

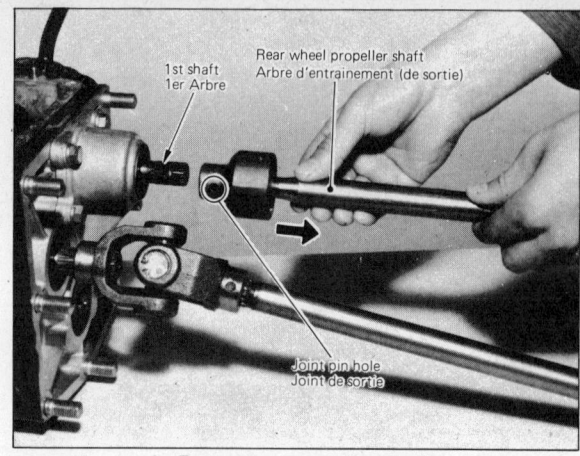

Removing propellor shaft

NOTE: Because of the knock pin, the cover must by all means be tapped upward.

18. Remove the ten bolts on the periphery of the axle case. When removing an axle case, prepare a drip tray so as to catch what little oil as may spill out.

19. Remove the axle case from the differential gear case by lightly tapping the axle with a wooden hammer. When a gap appears between axle and gear case, separate them carefully by hand.

20. Remove the two bolts and ten nuts connecting transmission case and differential gear case.

21. Remove the transmission case from the differential gear case by lightly tapping the grooves on both sides of the transmission case with a wooden hammer.

22. When the transmission case is separated, the collar at the end of the reverse shaft will fall. Do not lose it. Also be careful about the coupling (spline boss) connecting the 3rd and 5th shaft.

DISASSEMBLY

1. Detach the propeller shaft by removing the snap pin and the head pin on the ball joint which is located at the connection between propeller shaft and 1st shaft.

2. Remove the driveshaft by removing the spring pin on the universal joint which is located at the connection between the driveshaft and reverse shaft.

3. Remove the six bolts. Remove the 1st shaft cover by lightly tapping the periphery of the cover with a plastic hammer.

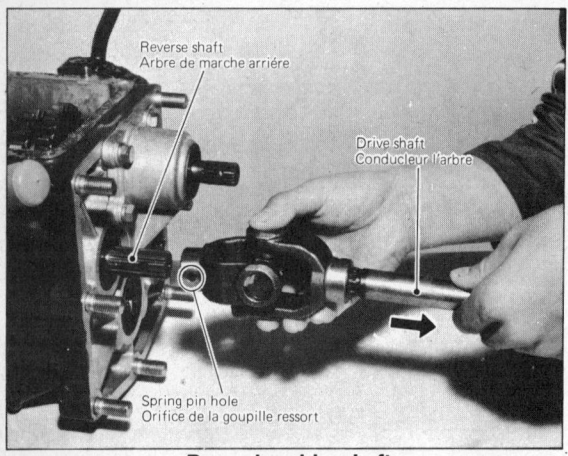

Removing driveshaft

Removing the first shaft cover

Removing the auxiliary speed change fork shaft

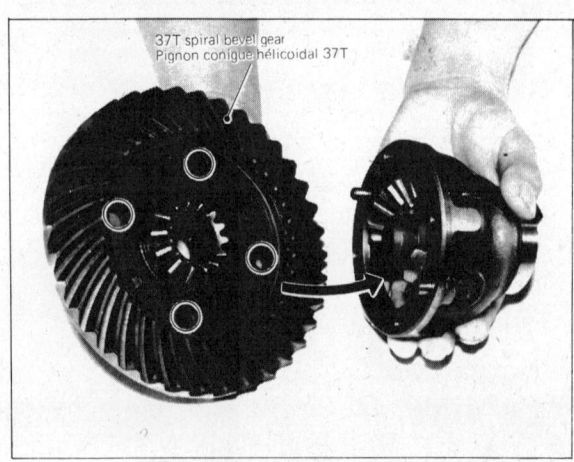

Removing spiral bevel gear

Kubota

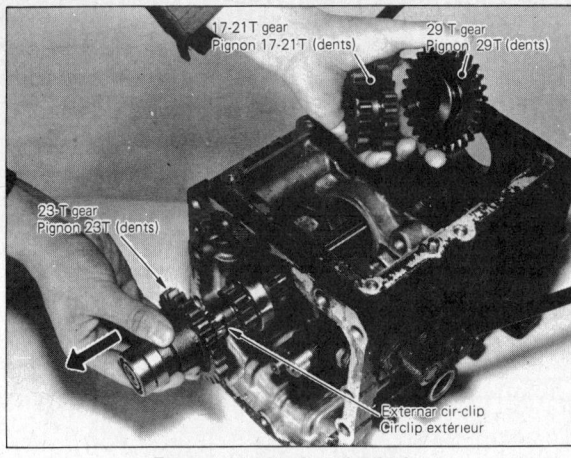

Removing the fourth shaft

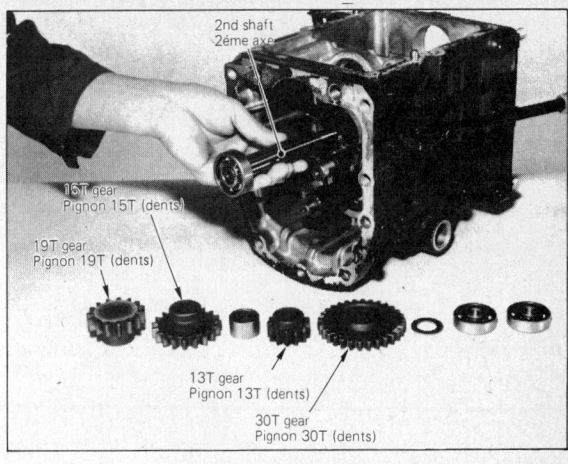

Removing the second shaft

4. Straighten the tang of the washer, and remove the M8 bolt.

5. Get an M8 bolt 1.9685 in. long, and screw it into the hole of the M8 bolt above.

6. Drive out the fork shaft backwards by lightly tapping the bolt with brass rod and hammer. Be careful not to lose the ball which will fall when the shaft is driven out.

7. The fork itself cannot be removed, and thus should be left as it is.

8. Draw out the pinion shaft by lightly shaking it back and forth. It can be removed only with the front wheel drive gear.

9. Remove the auxiliary speed change fork.

10. Remove the spacer and shim from the transmission case.

11. Spread the 4th shaft circlip with a set of snapring pliers, and shift it toward the low-speed gear by 0.3937 in.

12. Shift the 4th shaft high-speed gear toward the low-speed gear.

13. Detach the 4th shaft by shifting it backward by 0.3937 in. by inserting a brass rod from the front side.

14. Drive the 2nd shaft out of the back of the transmission case by lightly tapping it with brass rod and hammer.

15. As the shaft and its gears are separately removed, the order in disassembling four gears, two collars and three bearings must be carefully observed.

16. Remove the internal circlips holding the ball bearing 6204 which is on the back of the reverse shaft.

17. Remove the external circlip holding the reverse gear from the groove, and bring it forward by 0.3937 in.

18. Drive the reverse shaft out of the back of the transmission case by lightly tapping it with brass rod and hammer.

19. Remove the stopper.

20. Drive out the PTO speed change fork shaft forward by lightly tapping it with brass rod and hammer.

21. Take the fork out of the shaft. Be careful not to lose the ball.

22. Remove the internal circlip close to the auxiliary speed change chamber.

23. Drive the 3rd shaft out of the back of the transmission case by lightly tapping it with brass rod and hammer, being careful about the oil seal lip.

24. Take the gear out of the transmission case.

25. Straighten the tangs of the tongued washers.

26. Remove the fork shaft fixing bolts.

27. Detach the fork shafts.

ASSEMBLY

Assembly is the reverse of disassembly. Note the following points:

1. Fix the balls in the forks with a 0.5112 in. diameter ball guide, and install the forks.

2. Securely spread the cotter pins of the forks.

3. Securely bend the tangs of the tongued washers against one side of the bolts.

4. Check to see that the shim is in front of the 3rd shaft.

5. Two thrust collars are on the sides of the 29T claw gear. Check to see that the grooves of these thrust collars face the gear so that the bushing can be lubricated easily.

6. Before installing the 29T claw gear, apply oil to the bushing.

7. Before installing the oil seal, apply oil to the oil seal lip surface.

8. Be careful about the direction of installation of the fork shaft.

9. Install the stopper with the larger C surface backwards.

10. Fix the ball in the fork with a 0.5118 in. diameter ball guide, and install the fork.

11. Do not lose the fork ball.

12. Before installing the arm, apply oil to the arm and O-ring.

13. The reverse shaft is provided with the 18-22T gear on both sides of which there are thrust collars. Check to see that the grooves on the side of these thrust collars face the side of the 18-22T gear so that the needle bearings can be lubricated with ease. Install the 18-22T gear so that the

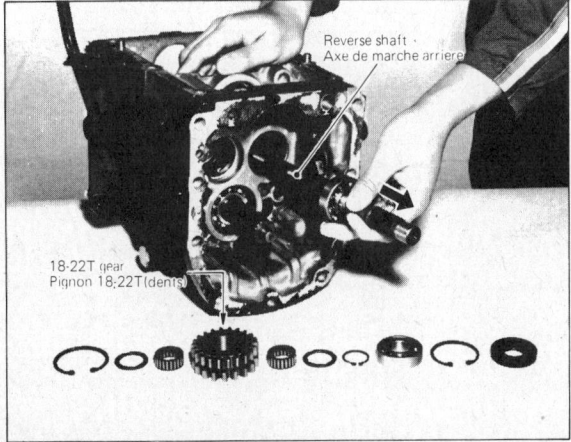

Removing the reverse shaft

Removing the PTO speed change fork

Kubota

Removing the rear cover and PTO shaft

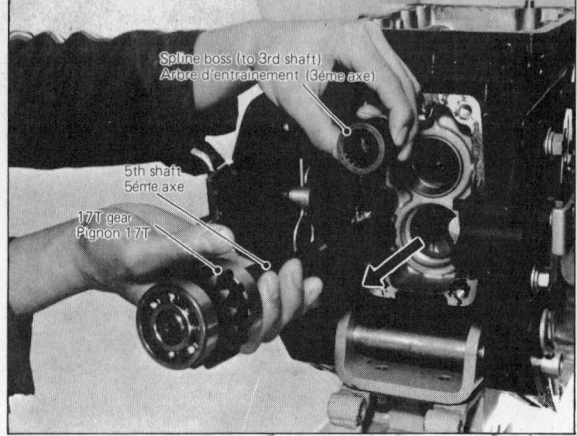

Removing the fifth shaft

18T gear is closer to the clutch housing case.

14. Install the auxiliary speed change fork shaft so that the end with longer distance between the ball groove and the edge faces backwards.

15. When installing the fork shaft, use a ball guide as long as 0.7874 in. Too long a ball guide cannot do the work well.

16. Before installing auxiliary speed change arm, apply a thin film of oil to the fork shaft and the O-ring.

17. Before driving in the oil seal, apply grease to the oil seal lip surface.

18. Before attaching the 1st shaft to the transmission case, lubricate the needle bearing which must be then incorporated with the shaft.

Differential Gear Case

DISASSEMBLY

1. Remove the six bolts on the rear cover.

2. Remove the rear cover and PTO shaft by lightly tapping the back of the cover with a plastic hammer.

3. Shift the 5th shaft backwards by lightly tapping it from the front side of the differential gear case.

4. Pull out the shaft from the back of the case.

5. Remove the spline boss which connects the 3rd and 5th shaft.

6. Straighten the bend in the lock plates which fix the differential bearing holder set bolts.

7. Remove the four set bolts for the right and left differential bearing holders.

8. Detach the differential bearing holders. Be careful not to confuse the right and left holders; they have a backlash adjusting shim.

9. Detach the differential gear assembly.

10. Put the differential gear assembly on a disassembly and assembly base.

11. Place a 1.1811 in. diameter doubling plate at the end of the differential case, set a special-purpose puller (bearing puller), and remove the right and left bearings.

12. Straighten the lock plates holding the bolts on the side of the spiral bevel gear.

13. Remove the four bolts on the spiral bevel gear.

14. Detach the spiral bevel gear from the differential case by lightly tapping the periphery of the bevel gear with a plastic hammer.

15. Remove from the differential case the straight pin holding the differential pinion shaft.

16. Remove the differential pinion shaft.

17. Remove the differential pinions, thrust collar and side gear.

ASSEMBLY

Assembly is the reverse of removal. Note the following points:

1. Check to see that the differential side gear turns smoothly.

2. If the backlash is excessive, adjust with three kinds of shims.

3. Tighten the spiral bevel gear set bolts to 21.70 to 25-32 ft.lb.

4. After tightening the differential bearing holder set bolts, securely bend the lock plates.

5. Check to see that the backlash adjusting shim is properly installed.

6. If there is excessive backlash between spiral bevel pinion and spiral bevel gear, adjust with a shim.

Axle Cases

DISASSEMBLY

1. Remove the differential lock spring.
2. Remove the differential lock shift fork.
3. Remove the differential lock clutch.
4. Remove the rear axle circlip.
5. Remove the 6306 ball bearing with a bearing puller.
6. Draw out the 55T gear.

Removing the differential bearing holder

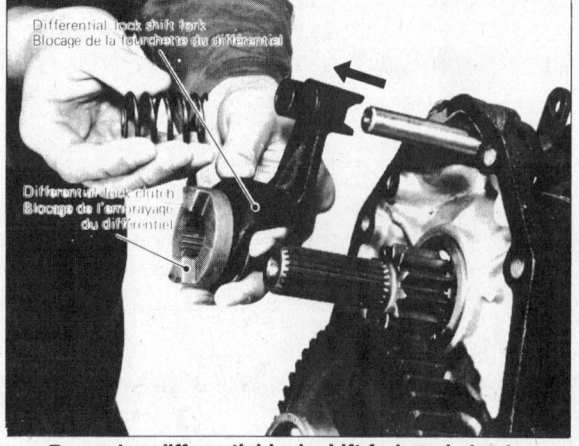

Removing differential lock shift fork and clutch

Kubota

Removing 55T gear

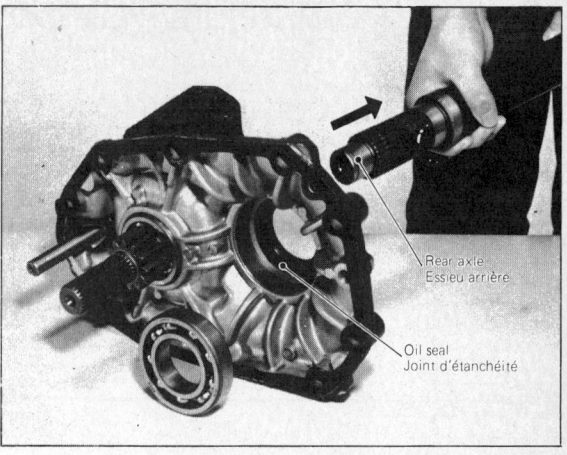

Removing rear axle

7. Drive the rear axle toward the inside of the axle case by lightly tapping it with a wooden hammer, and detach.

8. Remove the four bolts on the brake cover.

9. Detach the brake cover with the brake shoe on it.

10. Detach the external circlip with a set of snapring pliers.

11. Detach the brake drum from the differential gear shaft.

12. Drive the differential gear shaft toward the inside of the axle case by lightly tapping it with a plastic hammer.

ASSEMBLY

Assembly is the reverse of disassembly. Note the following:

1. Apply oil to the oil seal lip surface, and install the shaft with the seal straight.

2. Be careful not to allow too much oil to stick to the brake drum. Check to see that there is no paint or oil inside the brake chamber (brake drum, cover and shoe).

3. Apply grease to the oil seal lip surface, and install the axle with the seal straight.

4. Install the 55T gear so that the longer boss is closer to the differential gear case.

5. Securely engage the external circlip with the groove.

6. Before installing the differential lock fork shaft, apply grease to the O-ring.

Removing brake cover

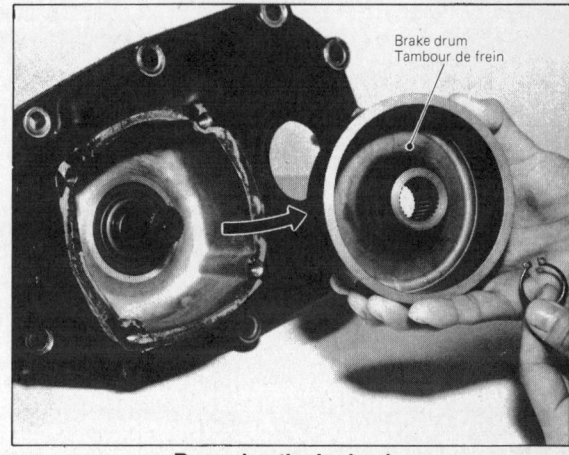

Removing the brake drum

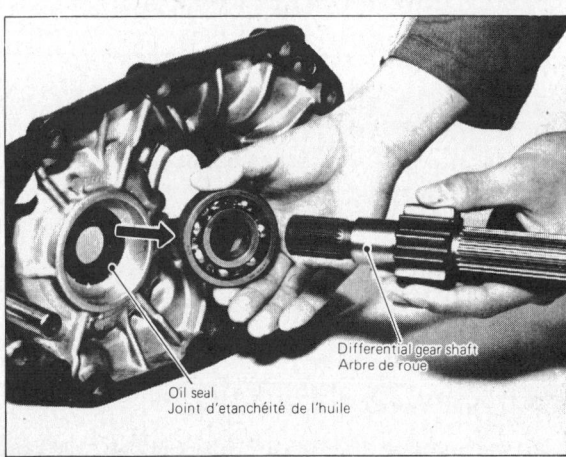

Removing differential gear shaft

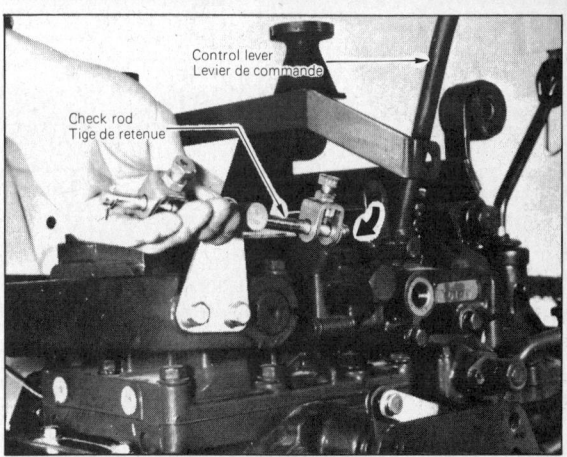

Removing the check rod

Kubota

HYDRAULIC SYSTEM

Control Valve

REMOVAL

1. Loosen the drain plug on the side of the transmission case, and drain oil.
2. Remove the air cleaner assembly.
3. Remove the right side cover.
4. Remove the cotter pin on the front of the release rod, and the rod itself.
5. Remove the cotter pins on the top and bottom of the accelerator pedal rod, and detach the rod.
6. Close the fuel filter cock.
7. Loosen the pipe clip on the side of the pipe fuel feed pump, and detach the pipe.
8. Remove the fuel filter and filter stay.
9. Remove the pipe connecting fuel filter and tank from the filter, and plug it to prevent fuel from overflowing.
10. Set the differential transmission case cradle and its rail under the tractor body.
11. Set the fitting base to differential transmission case.
12. Jack up the rear wheels by 0.3937 to 0.7874 in.
13. Loosen the wheel hub setscrew, and remove the axle pin.
14. Detach the rear wheel.
15. Remove the coupler for the flasher lamp cord.
16. Remove step support plate and fender mounting bolts.
17. Tilt the seat forward, and detach it by removing the head pin at the fulcrum.
18. Remove the hydraulic pipe clamp.
19. Remove the hydraulic pipe joint bolt.
20. Remove the pipe joint.
21. Remove the pipe joint.
22. Remove the oil filter.
23. Remove the hydraulic pipes.
24. Remove the cotter pin connecting check rod to control lever.
25. Detach the check rod.
26. Remove the control valve mounting bolts.
27. Remove the control valve, being careful about the two O-rings.
28. Remove the nine bolts holding the rear case cover.
29. Remove the cover upward by lightly tapping the back of it.

NOTE: Because of the knock pin, tap the back of the cover upward by all means.

30. Remove control lever and bolt.
31. When removing the bolt, be careful not to scratch or lose ball and spring.
32. Remove the four valve cover mounting bolts.
33. Remove the valve cover.
34. Detach the spool from the valve body.

Removing the control lever and ball stopper spring holder

35. After removal, be careful not to scratch the surface of the spool.
36. Remove the bolts holding the relief cover. Before removal, check to see how much the adjusting screw is projecting from the surface of the valve body.
37. Remove the adjusting screw and ball.
38. Remove the holder.

INSTALLATION

Installation is the reverse of removal. Note the following:

1. Install the O-ring and relief valve, being careful not to scratch the holder surface.
2. First place the ball or the spring holder, and install them in the collar.
3. Before tightening the screw, check to see that there is a flat washer and shim inside the adjusting screw.
4. Tighten the relief adjusting screw to the same tightness as before disassembly.
5. Apply a little gear oil to the spool, and install it. After reassembling, check to see that the spool moves smoothly.
6. Install the valve so that the spring pin of the spool is engaged with the groove of the valve guide arm.
7. Tighten the valve guide arm bolts to 7.233 to 13.02 ft.lb.
8. Keep the valve guide arm in the valve cover in a position where it can be seen through the hole by moving the control lever, and then install the ball.
9. Then, install spring and bolt.
10. When installing the case cover to the differential gear case, adjust and install the lift arm so that the hydraulic piston rod would press against the piston.
11. Check to see that the O-rings are properly fitted.
12. Tighten the bolts attaching the control valve assembly to the rear case cover, to 13.02 to 15.19 ft.lb.
13. Tighten the hub set bolts to 65.10 to 72.33 ft.lb.
14. Tighten the wheel and hub set bolts to 50.63 to 57.86 ft.lb.

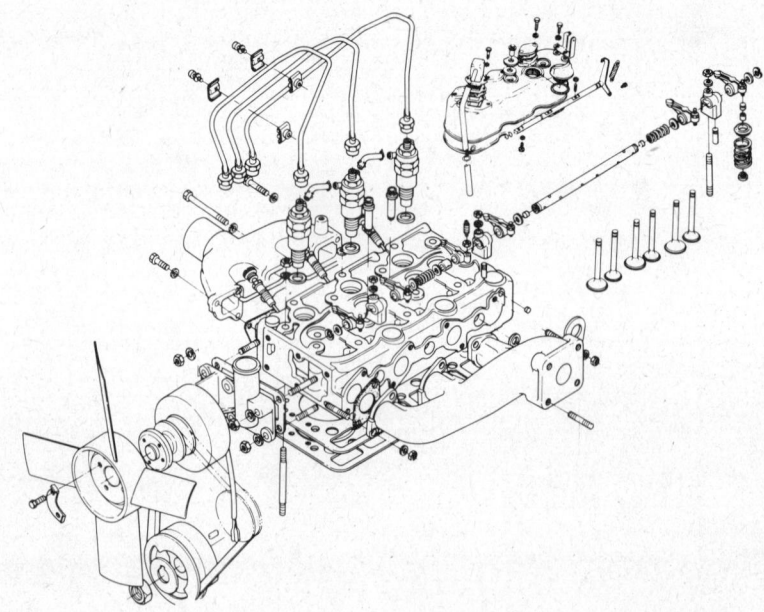

Cylinder head components

Kubota

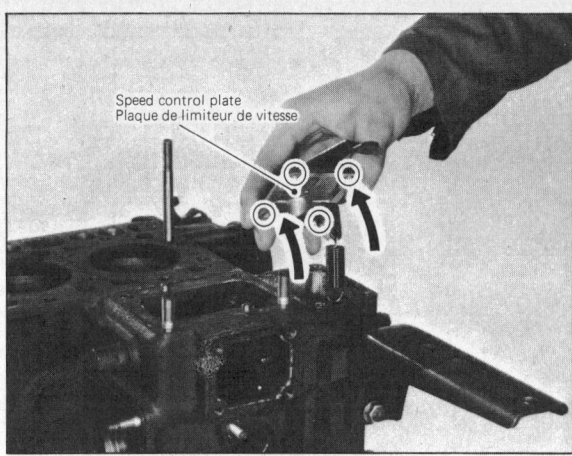

Removing speed control lever

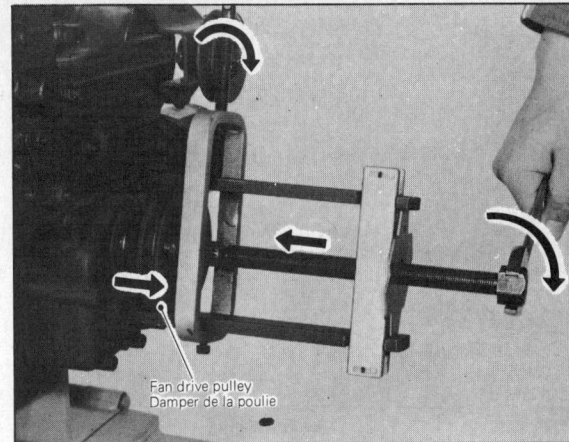

Removing fan drive pulley

ENGINE

Cylinder Head

REMOVAL AND INSTALLATION

1. Remove the head cover nuts.
2. Remove the head cover.
3. Loosen the washer-faced machine screws on the pipe clamp.
4. Detach the injection pipes 1, 2 and 3 in this order.

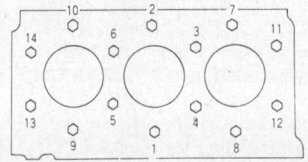

Cylinder head bolt tightening sequence

5. Remove the pipe clip holding the fuel overflow pipe, and detach the pipe.
6. Detach the nozzle holder using a 1.0630 in. nozzle holder socket wrench.
7. Loosen the tension pulley shaft set nut.
8. Loosen the tension bolt for the tension pulley and loosen the V-belt.
9. Remove the V-belt by hand.
10. Remove the set nuts for the rocker arm brackets.
11. Detach the whole rocker arm.
12. Remove the six push rods.
13. Remove the twelve cylinder head set bolts and two nuts.
14. Lift up the cylinder head to detach.
15. Installation is the reverse of removal. Torque the head bolts to 31-34 ft.lb. Torque the rocker arm shaft bolts to 12-16 ft.lb. Torque the injector nozzles to 22-36 ft.lb.

Timing Case Cover, Camshaft and Oil Pump

REMOVAL AND INSTALLATION

1. Remove the four bolts holding the speed control plate.
2. Lift the plate to detach.
3. Remove the starter spring near the gear case with a set of cutting pliers.
4. Straighten the crankshaft washer.
5. Loosen and remove the crankshaft nut. Remove the pulley, and then the key.
6. Set a puller to the pulley.
7. By tightening the puller center bolt, draw out the pulley.
8. Remove the fourteen mounting bolts on the periphery of the gear case cover.
9. Remove the case cover by lightly tapping the periphery of the cover with a plastic hammer.
10. Remove the crankshaft collar.
11. Remove the O-ring.
12. Detach the crankshaft oil slinger.
13. Remove the external circlip with a set of snapring pliers.
14. Detach the idle gear.
15. Align the round hole on the cam gear with the camshaft stopper mounting bolt position.
16. Remove the camshaft stopper mounting bolt.
17. Detach the camshaft stopper.
18. Detach the camshaft by drawing out the cam gear.
19. Remove the fuel camshaft stopper mounting bolts.
20. Detach the fuel camshaft stopper.
21. Remove the two fork lever holder mounting bolts.
22. Detach the fork lever holder and fork lever assembly.
23. Draw out the fuel camshaft and the injection pump gear.
24. Straighten the claw of the claw washer on the oil pump, and remove the nut.
25. Set a gear puller on the oil pump

Crankshaft oil slinger

Gear train timing marks

635

Kubota

drive gear. Draw out the gear by screwing in the center bolt.

26. Remove the four oil pump set bolts. Detach the oil pump.

27. Set a special puller on the crank gear.

28. Screw in the center bolt of the puller, and draw out the crank gear.

29. Installation is the reverse of removal. Note the following:

a. Heat the crank gear to about 210°F, and fit on the crankshaft. If there is a chance of the shaft being damaged, heat the gear more and fit.

b. Tighten the oil pump bolts to 7.233 to 8.68 ft.lb.

c. Apply a thin film of engine oil to each part, and reassemble so that the drive groove at the end of the camshaft engages with the driveshaft of the hydraulic gear pump. The balls to be contained in the governor ball case are thirty 0.1563 in. diameter ones and eight 0.5300 in. diameter ones.

d. Install the fork lever holder and crankcase after cleaning their contact surfaces.

e. Install the fork lever so that it will not hit the governor sleeve, and so that equal space is allowed on either side of the lever.

f. After installation, check to see that the fork lever is fixed to the fork lever shaft, and that it can turn smoothly in the holder.

g. Apply a thin film of engine oil to the camshaft before installation.

h. Check to see if each gear is aligned with its mark:
   1) Idle gear and crank gear
   2) Idle gear and camshaft gear
   3) Idle gear and injection pump gear

i. Tighten the idle gearshaft mounting bolt to 7.23 to 8.68 ft.lb.

j. Tighten the pulley nut to 101.26 to 115.73 ft.lb. Do not tighten the nut excessively or with an impact wrench; it may damage the oil slinger, causing oil leakage.

Piston and Crankshaft

REMOVAL AND INSTALLATION

1. Lay the engine cradle on its side.
2. Remove the eighteen oil pan mounting bolts with a T-wrench.
3. Detach the oil pan by lightly tapping the groove of the pan with a wooden hammer.
4. Remove the mounting bolt of oil filter.
5. Detach oil filter, being careful of the O-ring.
6. Remove the bolts from connecting rod.
7. Detach the splitmetal.
8. Remove the connecting rod bolts and large-end metal, turn the crankshaft by 180°, and bring the piston to top dead center.
9. Draw out the piston upward by lightly tapping it from the bottom of the crankcase with the grip of a hammer.
10. Draw out the other two pistons in the same manner as above.
11. Straighten the flywheel washers.
12. Remove the flywheel bolts, except for two which must be loosened and left as they are.
13. Set a flywheel puller, and remove the flywheel.
14. Remove the bearing case cover mounting bolts. First unscrew the eight bolts inside, and then work on the nine outside.
15. Screw the two bolts removed above into the two right and left holes for the bearing case cover, and pull off the cover by jacking it up.
16. Straighten the washers for the bearing case bolts.
17. Detach the bearing case bolts.
18. Draw out the crankshaft from the back of the crankcase by lightly tapping it with a copper hammer.
19. Remove the two mounting bolts for main bearing case assembly.
20. Detach the main bearing case, being careful with the side and crankshaft metals.
21. Detach the other bearing cases in the same manner. Be careful not to mix them up.
22. Grind the cylinder head mounting surface with an oil grinding stone.
23. Set a dry liner centering base adaptor on the frame head.
24. Set the liner centering base to the adaptor.
25. Set the bearing cradle to the liner centering base with the bearing on top.
26. Insert the pulling-out adaptor coupling, and fix the center bolt.
27. Contact the pulling-out adaptor with the bottom of the liner from the bottom of the frame.
28. Pull out the liner by turning the nut with a ratchet handle.

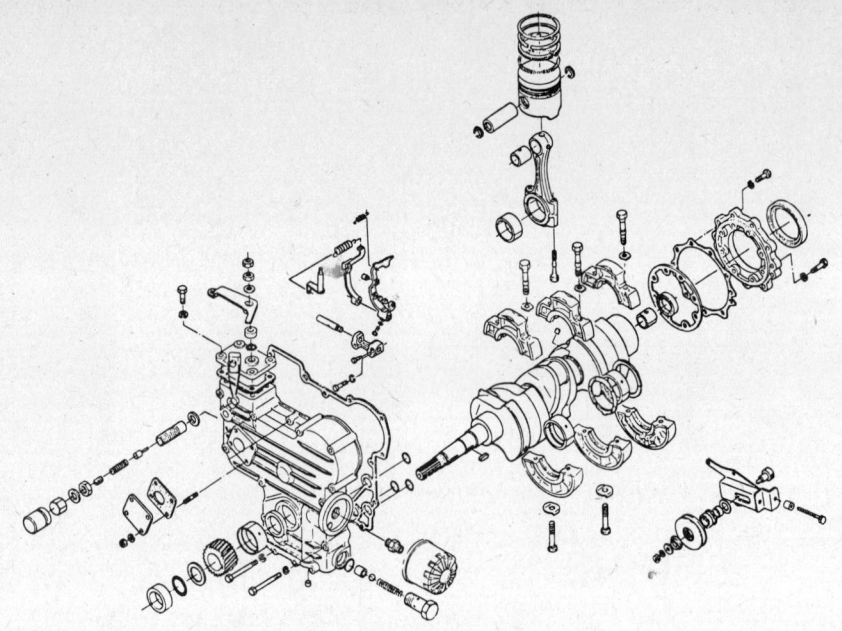

Piston and crankshaft

29. Installation is the reverse of removal. Note the following points:

a. Clean the frame holes where the liner fits, and apply oil.

b. Clean the surface of the liner, and apply oil.

c. Insert the liner first with the skirt.

d. After installation, insert a feeler gauge through the check window on the bottom of the liner centering base and check to see if the liner has been properly pressed in.
Tolerance: Liner projection ±0.0010 in.
Hone the cylinder liner: 1.2μRmax to 2μRmax.

e. Clean the oil holes in the main bearing case.

f. Install the main bearing case with their side marks toward the flywheel.
  Be sure to install main bearing with its oil groove facing outward.

g. Tighten the mounting bolts of the bearing case to 14.47 to 17.36 ft.lb., those for cases 2 and 3 to 21.70 to 25.31 ft.lb.

h. When tightening the main bearing case bolts, align the direction of the holes so that the crankshaft can be easily installed.

i. Tighten the bolts to 21.90 to 25.32 ft.lb.

j. When installing the main bearing case cover, check to see that there are no scratches on the oil seal lip. Then, apply a thin film of engine oil, and install. Be sure to check the top mark on the cover during installation.

k. Tighten the mounting bolts in diagonal order to 7.23 to 7.96 ft.lb.

l. Tighten the flywheel bolts to 39.78 to 43.40 ft.lb. in diagonal order.

m. Install the piston rings with their gaps making an angle of 120° or 180° to each other. (Place the top ring with their gaps on the opposite side of the combustion chamber.)

Kubota

n. Before inserting the pistons into the cylinders, apply enough engine oil to the pistons.
o. Tighten the connecting rod bolts to 19.53 to 22.42 ft.lb. If the bolts are not tightened uniformly, a connecting rod may deform or twist in a long run.

FUEL SYSTEM

Fuel Pump

REMOVAL AND INSTALLATION

1. Disconnect the fuel lines.
2. Remove the two fuel pump mounting nuts.
3. Detach the fuel pump.
4. Installation is the reverse of removal.

Injection Pump

REMOVAL AND INSTALLATION

1. Remove the four injection pump cover mounting bolts.
2. Remove the cover by lightly giving an impact.
3. Remove one end of the governor spring from the fork lever and leave it there.
4. Remove the nuts and bolts holding the injection pump.
5. Detach the injection pump. To prevent the pump rack from being caught, detach the pump along the removal groove.
6. In principle, the injection pump should not be disassembled.
7. Installation is the reverse of removal. Note the following:
a. Install the injection pump by aligning the control rack with the indicated position.
b. Addition or reduction of one shim changes the injection timing by 1.5°.
c. Install the injection pump shims after applying a non-drying adhesive.
d. After hooking the governor spring on the fork lever, bend the end of the spring so that it will go off easily.
e. One of the four injection pump cover mounting bolts has a holder which supports the release rod. Fit this bolt in the right position when returning the bolts.
f. Before installing the pump cover packing, apply a non-drying adhesive to it.

Model 175

FRONT WHEELS AND AXLE

Front Hubs

REMOVAL AND INSTALLATION

1. Remove the front wheel tire. Remove the front wheel cap. Remove the split pin exposed and then slacken the slotted nut.
2. Tap out the hub with a hammer (mallet, copper or plastic hammer) to outside.
3. Before assembling, apply sufficient wheel bearing grease (SAE multi-purpose type grease) to the bearings and oil seal. Also, fill the space inside the hub with EP multi-purpose grease up to ⅓ to ½ of the space.

CAUTION
Do not use grease contaminated with foreign matters. Be very careful to keep the hub free from mud, dirt, sand and other impurities.

4. After filling grease, proceed to assemble. Before tightening the slotted nut, make certain bearings are in correct place on the shaft.

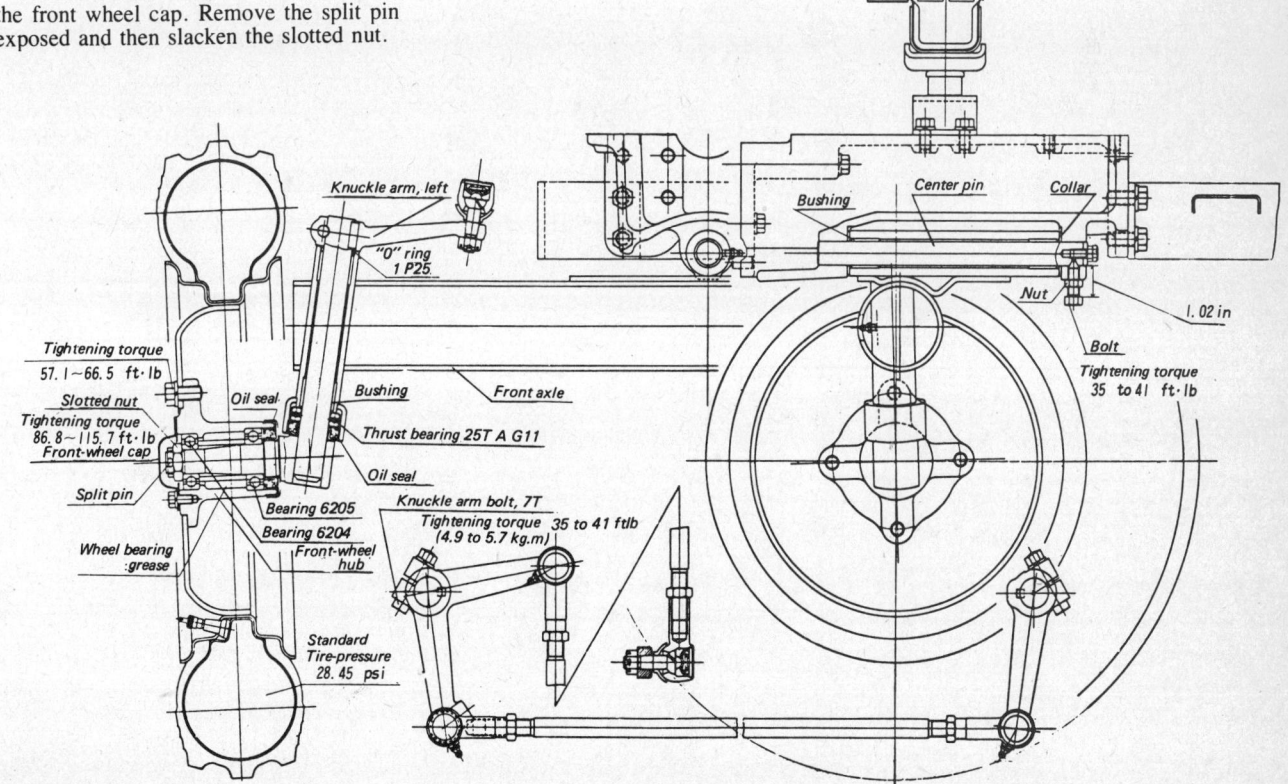

Front axle schematic

Kubota

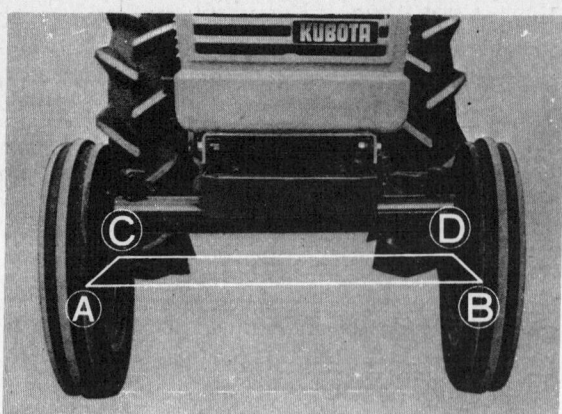

Toe-in adjustment

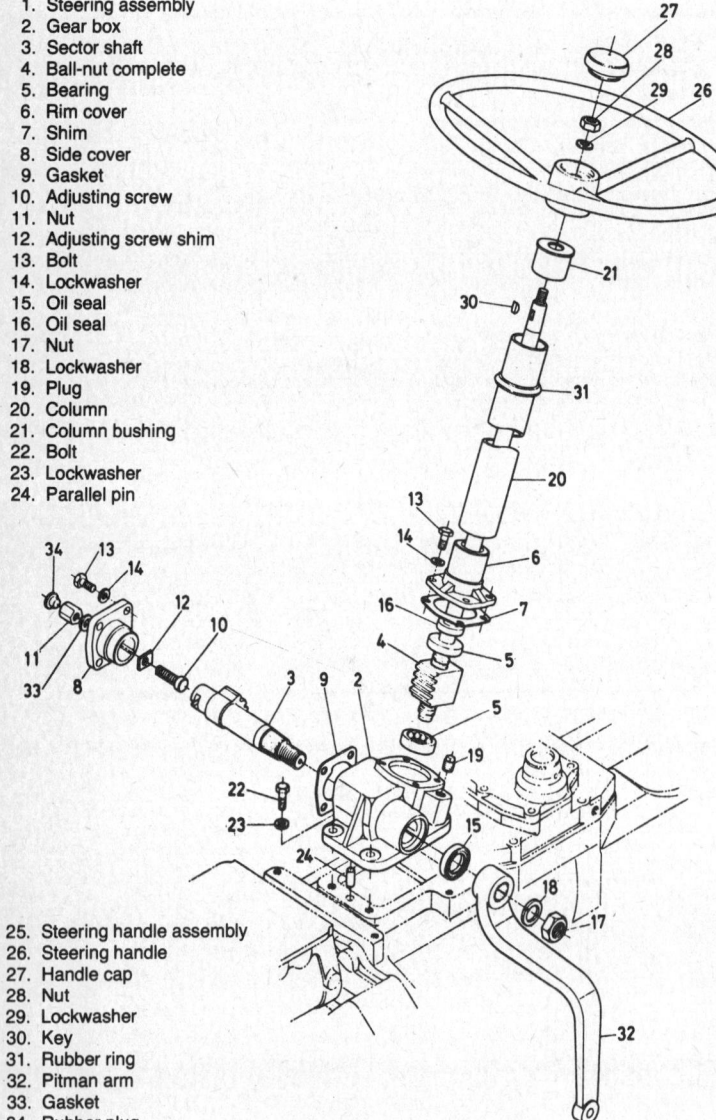

1. Steering assembly
2. Gear box
3. Sector shaft
4. Ball-nut complete
5. Bearing
6. Rim cover
7. Shim
8. Side cover
9. Gasket
10. Adjusting screw
11. Nut
12. Adjusting screw shim
13. Bolt
14. Lockwasher
15. Oil seal
16. Oil seal
17. Nut
18. Lockwasher
19. Plug
20. Column
21. Column bushing
22. Bolt
23. Lockwasher
24. Parallel pin
25. Steering handle assembly
26. Steering handle
27. Handle cap
28. Nut
29. Lockwasher
30. Key
31. Rubber ring
32. Pitman arm
33. Gasket
34. Rubber plug

Steering system components

Knuckle Shaft

REMOVAL AND INSTALLATION

1. Remove the knuckle arm and O-ring.
2. Lift up the front axle to the extent that the knuckle shaft can just be removed downward and then remove the knuckle shaft downward.
3. Before assembly, apply sufficient chassis grease to the thrust bearing and oil seal.

--- CAUTION ---
Pay attention to the thrust bearing because it is easily installed upside down. Assemble as shown in the component view.

Center Pin

REMOVAL AND INSTALLATION

Remove the bumper first and then the bolt which fixes the center pin. Lift up the front axle and pull out the center pin toward the front. The front axle will then be disconnected.

--- CAUTION ---
In assembling, check and make certain the bolt which fixes the center pin is locked securely with the locknut. The bolt protrusion is 1 in. If this bolt is loose, the center pin may come out.

Toe-in

ADJUSTMENT

Refer to the accompanying illustration. Loosen the locknut and turn the tie rod so that CD-AB = 0.08-0.31 in. To take measurements, use a toe-in gage. If one is not available, measure with a precise tape measure.

STEERING

Steering Gear

REMOVAL AND INSTALLATION

1. Using a Pitman arm puller, remove the Pitman arm first. Install the puller on the Pitman arm and screw in the center bolt while preventing the hooks from spreading with help of the stopper. The Pitman arm will then be pulled from the Pitman arm sector shaft.

NOTE: Pitman arm and sector shaft are marked with mating marks.

2. To remove the steering wheel, use a steering wheel puller. Loosen the nut at the center of the wheel, set the puller and screw in the center bolt. The wheel will

then be pulled from the steering shaft and column.

NOTE: Very tight wheel can be pulled off by tapping the center bolt with a hammer only if the bolt has been screwed into the normal torque.

3. Remove the fuel tank and instrument panel. The gearbox will then be exposed.
4. Remove the side cover of the box and take out the sector shaft.
5. Remove the bolts which fasten the rear cover to the box. Pull out the steering column and shaft as assembly. Note the number of shims used. Do not lose them.

CAUTION
The ball nut is a precision fit, so do not disassemble.

---

6. For assembly, wash the parts carefully with diesel fuel or kerosene and assemble in the reverse of disassembly.
7. Check and make certain the steering shaft turns lightly without chattering. Adjustment can be made by the rear cover shims. After confirming this, insert the sector shaft.
8. When the steering wheel is installed, turn the wheel gently and measure the free turn, which should lie between 0.79 in. and 1.97 in. on the circumference of wheel. Adjustment can be made with the lash adjust screw on the Pitman sector shaft.

NOTE: Counterclockwise turning of the screw removes overcenter load to increase the free turn of the wheel and clockwise turning of the screw decreases the free turn of the wheel.

9. Remove the oil inlet plug and fill the gearbox with SAE 90 EP gear oil. Capacity is 0.32 qt.

Tie Rods
REMOVAL AND INSTALLATION

There are four tie rod ends used in this tractor, two at the ends of the tie rod and the other two at the ends of the drag link. Each tie rod end is tapered into the knuckle arm or the Pitman arm and locked with a nut and split pin.

Loosen the nut and tap against both sides of the end at the same time with mallets or copper or a plastic hammer. The tie rod end tapered pin will then come out smoothly.

In case the end is stuck, the use of a tie rod pin puller is recommended. Attach the puller to the end in the manner shown.

BRAKES
REMOVAL AND INSTALLATION

1. Remove the brake connecting rod and brake cover. Then the anchor plate attached

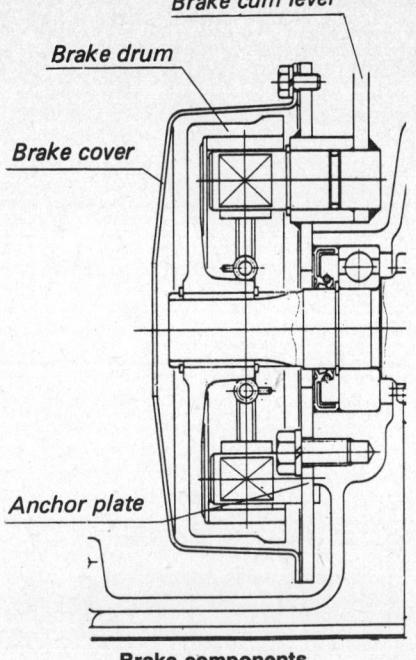

Brake components

to the rear axle case will be exposed with the brake cam lever and shoes.
2. Remove the circlip that retains the brake drum. Pull out the brake drum. The brake shoes can be checked or replaced if only springs are removed.
3. Loosen 6 bolts which fasten the anchor plate to the axle case. The anchor plate will then be detached outward.
4. When the anchor plate is removed, be sure to check the oil seal located behind the plate for leakage of lubricant.
5. Installation is the reverse of removal.

ADJUSTMENT

Travel of brake pedals changes due to wear on the brake linings. The adjustment of free travel is made by adjusting the length of each brake connecting rod with the turnbuckle used in the rod. When the free travel is excessive, shorten the rod; when the free travel is too little, lengthen rod.

Be sure that the total travel is the same between both pedals. Check and make certain of this by depressing both pedals linked together.

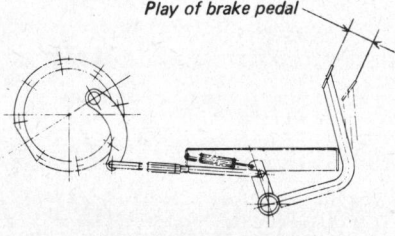

Brake pedal adjustment

Kubota

CLUTCH AND MAINSHAFT
REMOVAL AND INSTALLATION

To disassemble the clutch, the engine must be detached from the clutch housing first. The following are necessary precautions in disassembling and assembling the clutch.

1. Wiring of the alternator and starter should be handled only after the negative wire of the battery has been removed from the battery.
2. Before removing the hydraulic pipe from the pump, drain off the oil in the transmission case.

NOTE: Hydraulic pipe should be removed from the pump.

3. As an O-ring is used on the tip of hydraulic pipe, be careful not to lose or damage the ring.
4. Note that the engine is prone to fall or tip over when detached, so be very careful. Use a hoist or chain block for this removal.
5. It is preferable to have the clutch housing suspended with a hoist or chain block as well.
6. In case the rotary device is left attached to the tractor, be sure to lower this before starting the disassembly.
7. If the flywheel is rusted or scratched on the surface which contacts with the clutch the clutch performance will be affected. Polish the surface to clean with sand paper.
8. Do not apply any oil or grease to the contact surface of the flywheel, the surface of the clutch pressure plate or the surface of the clutch disc.
9. Check grease inside the clutch release hub. If its color has changed, replenish with SAE mulipurpose grease.
10. In attaching the clutch to the flywheel, be sure to make them concentric with each other. If misaligned, the mainshaft will not be inserted smoothly.

Insert a centering tool shaft into the center of the clutch assembly passing through the clutch disc, into the counterhole in the flywheel, to align the disc with the flywheel. Then tighten bolts which attach the pressure plate assembly.

After tightening securely, apply the arm gage to the surface of the flywheel, and adjust the heights of the release levers so that they are equal. After adjusting, remove the centering tool.

11. For aligning the clutch housing and engine, both should be moved on the same plane. Extreme care should be taken in this, because if not, it will be impossible to assemble even if the difference is very slight. Should you try to assemble, then the mainshaft will be very likely bent.
12. In assembling the engine and clutch housing, check the alignment at the spline of the mainshaft where it is to be fitted into the clutch disc. Make certain the deviation

639

Kubota

is within 0.004 in. If excessive, adjust by tapping lightly with a mallet or copper or plastic hammer.

ADJUSTMENT

Pedal Free Travel

Free travel of the clutch pedal is governed by the clearance of release thrust ball bearing to the release levers. This clearance is set to 0.098 to 0.12 in. when the tractor leaves the factory, but this changes with use.

To Adjust:

1. Open the clutch access window which is located on the righthand forward of the clutch housing.

2. Insert a 0.098 to 0.12 in. thickness gage between the release thrust ball bearing and the release levers. Adjust the clearance by the length of the clutch rod. Lengthen the rod when the clearance is narrower.

3. Depress the clutch pedal, and check and make certain that the free travel at the tip of the pedal is 0.8 to 1.2 in.

4. With the clutch pedal depressed to the lowest point, confirm that there is adequate gap between the clutch rod and safety starter switch.

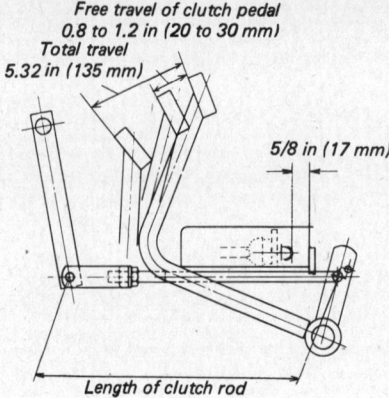

Clutch pedal adjustment

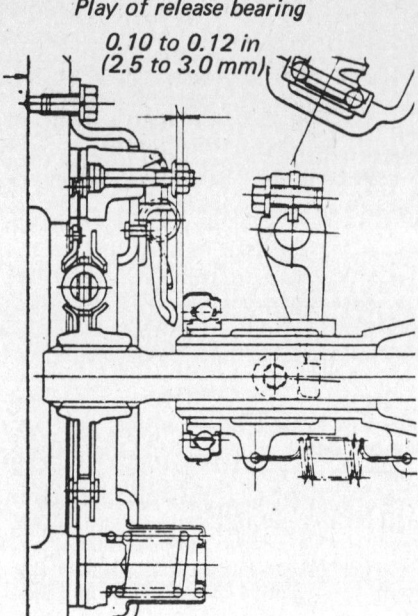

Clutch release bearing adjustment

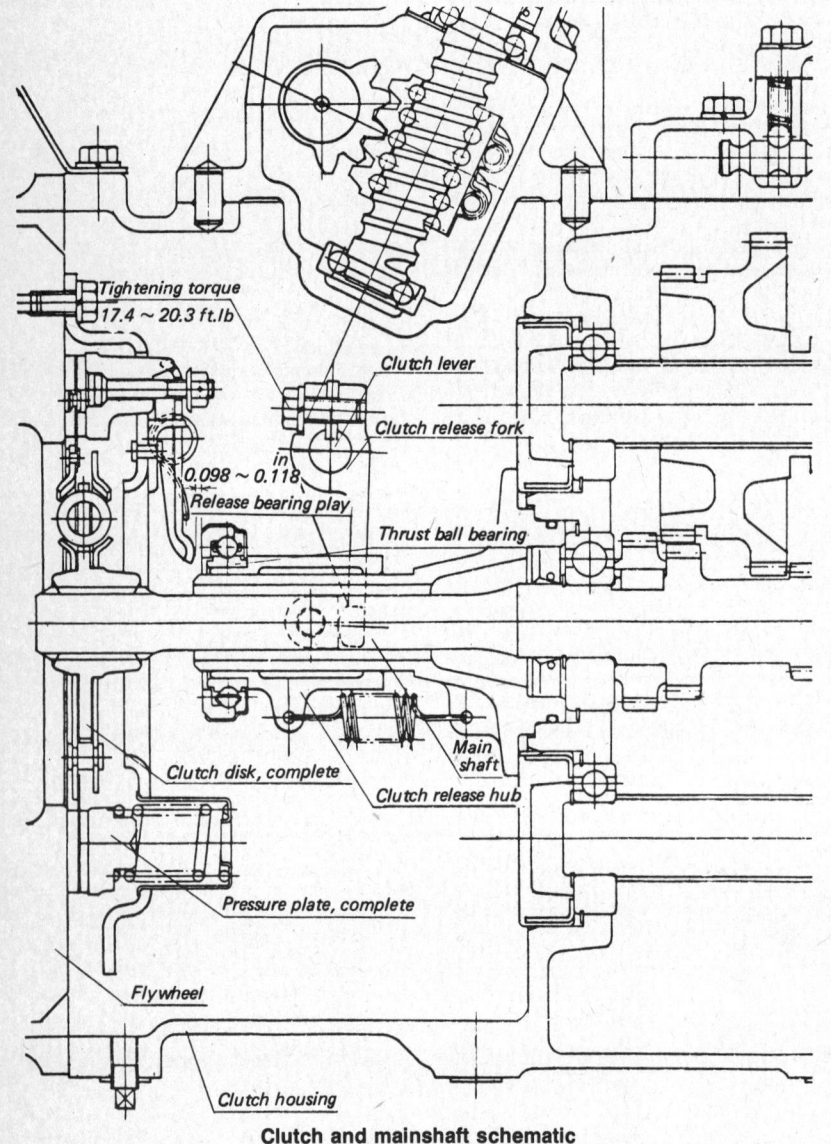

Clutch and mainshaft schematic

TRANSMISSION

NOTE: The transmission consists of two gear units, one on the clutch housing and one in the transmission case. It is generally not necessary to disassemble the transmission, the main exception being gear breakage.

Clutch Housing

DISASSEMBLY

1. It is not necessary to remove the instrument panel, fuel tank, steering unit, and clutch pedal, but separate the engine and then the transmission case from the clutch housing.

2. Put the main and PTO gearshifters in neutral and remove both upward together with the gearshifter cover.

3. Loosen 4 bolts which fasten the mainshaft case and pull out the mainshaft case forward.

4. Pull out the bearing cover at the front end of the countershaft, in such a manner as to tap the countershaft against the rear end. Do this to both of the countershafts.

5. Remove the circlip which retains the countershaft at the front end, and tap out the shaft backward. Do this to both of the countershafts.

6. Tap out the mainshaft frontward.

Transmission Case

DISASSEMBLY

1. Remove the engine from the clutch housing. Detach the fenders, rear wheels, seat, and hydraulic cylinder.

2. Separate the clutch housing from the transmission case.

Kubota

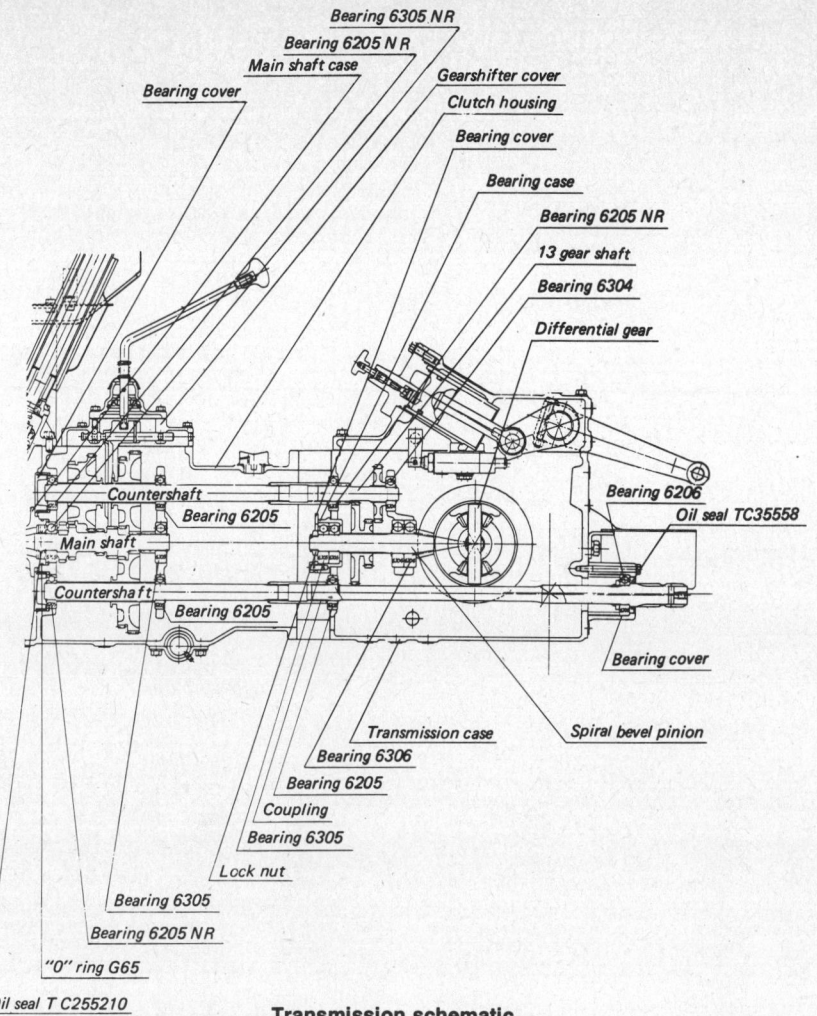

Transmission schematic

shaft at the rear end, and pull the PTO shaft backward.

NOTE: In case of removing the PTO shaft only, put the high-low gear shifter in high. The PTO shaft will then be extracted easily.

ASSEMBLY

Assembly is the reverse of disassembly. To insert the shift fork rod, use a shifter ball guide available from your Kubota dealer.

DIFFERENTIAL AND REAR AXLE

Differential

OVERHAUL

1. Remove the hydraulic cylinder located on top of the transmission case.
2. Remove both right and left halves of the rear axle. Remove differential bearing cases. Remove the differential upward.
3. Straighten washers which lock the set bolts for the spiral bevel gear, loosen the bolts and take out the spiral bevel gear. Then tap out the differential pinion shaft.
4. Next, move and take out pinion gears through the differential access window of the case.

Assembly is the reverse of disassembly.

CAUTION
Bolts tightening the spiral bevel gear are 7T. Tighten them to 35-41 ft.lb.

---

3. Remove the right and left rear axle cases. Take out the differential gears. Care is necessary in this process since the case may tilt and fall down if the work bench happens to be unstable.
4. Loosen 6 bolts which attach the bearing case and bearing cover and remove the bearing cover. Remove the nut which locks two bearings. These are all on the spiral bevel pinion shaft at the end opposite to the pinion. Tap out the pinion shaft backwad.
5. Drive out the 13-gear shaft forward. To effect this, remove circlips which retain the bearing supporting the shaft at the rear end and then the circlip which retains the bearing at the front end of the shaft.
6. Remove 4 bolts which fasten the cover for the bearing supporting the PTO

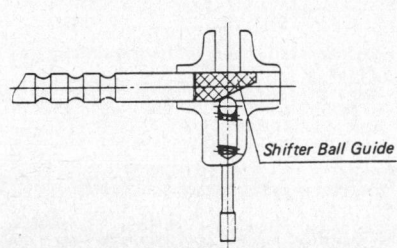

Insertion of shift fork rod

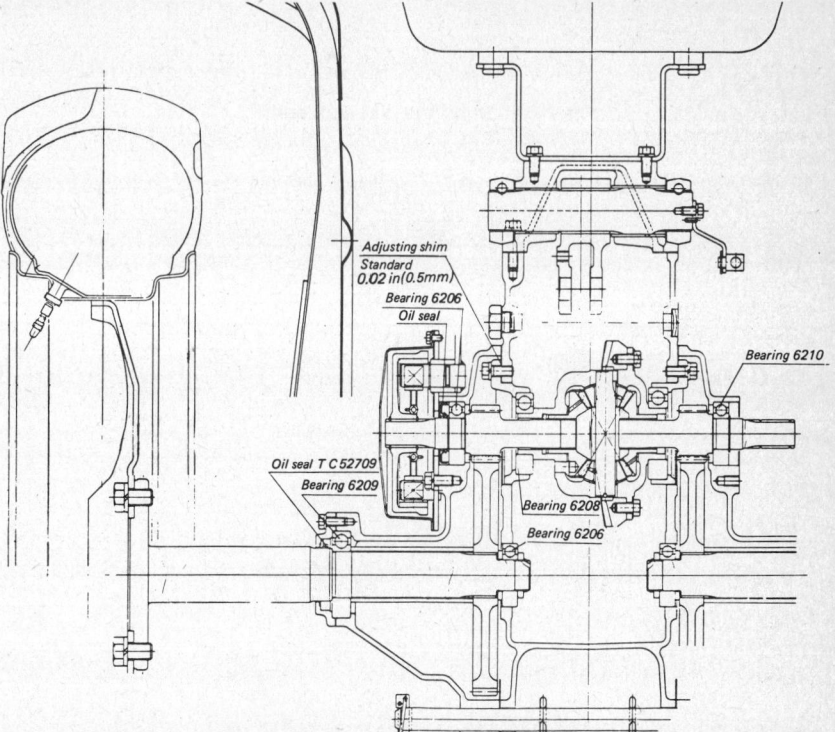

Differential and rear axle schematic

Kubota

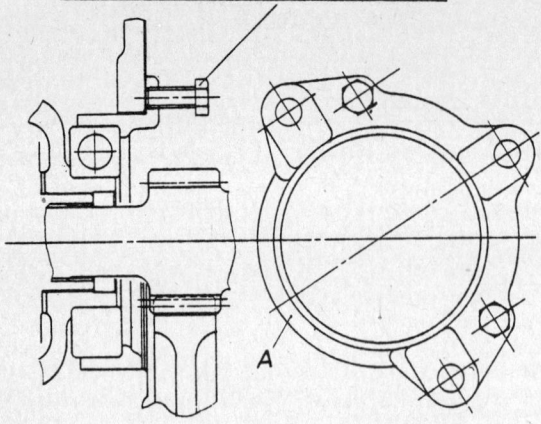

Removal of differential bearing case

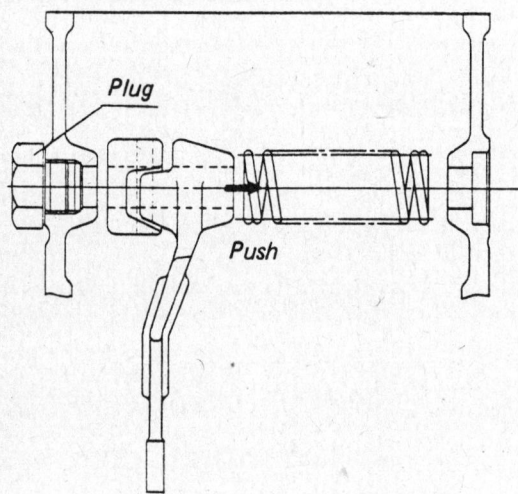

Inserting the differential lock spring

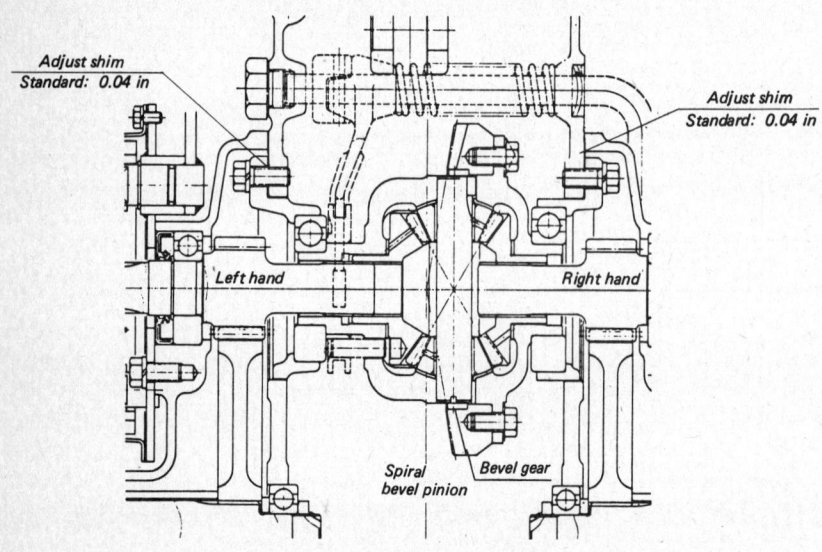

Spiral bevel gear and pinion adjustment

Differential Lock System

OVERHAUL

1. Remove the spring pin in the differential lock cam and pull out the differential lock pedal (rod). The differential lock cam, shift fork, and spring will then become loosened. The spring may jump out, so be careful.

---- **CAUTION** ----

Extreme care is necessary in pulling the pedal out of the case, so as not to damage lips of the oil seal.

2. For assembling, a jig may be used for installing the spring, but it can be placed without the jig if you will observe the following:
   a. Remove plug.
   b. Place spring between shift fork and transmission case inside surface opposite plug hole, and then place cam between the fork and plug hole.
   c. Pass differential lock pedal (rod) therethrough and then drive spring pin into the cam to fix the cam with the lock pedal (rod).
   d. Screw in the plug.

Rear Axle

OVERHAUL

1. Remove bolts and nuts which fasten the axle case to the bracket, and pull out the axle along with the case.
   Screw in M8 bolts little by little in turn while tapping lightly against the portion (A), so that the case can be extracted evenly.
2. Unlock the nut which locks the gear on the axle and remove the nut. Pull out the bearing and then the gear.
3. Unbolt the axle flange. Drive out the axle.

---- **CAUTION** ----

In driving out the axle, be careful not to damage the oil seal.

4. Be sure to bend the locknut securely into a notch in the axle end.
5. Do not fail to fill the oil seal with grease before assembly.

ADJUSTMENTS

Differential Clearance

Since the differential is supported on the right and left by ball bearings, an improper adjustment in shims used on the bearing cases will result in a differential which is too tight to turn smoothly or which rattles to right and left not allowing the pinion and gear to mesh properly. Gradually decrease or increase the number of shims, on both bearing cases, until the differential turns smoothly and freely.

Backlash

Fill and impress solder between teeth meshed and measure the thickness of im-

Kubota

pressed solder with a micrometer. If the measurement is 0.008 to 0.01 in., the adjustment is correct.

Tooth Contact

Apply a very thin coat of Prussian blue to contact surfaces of several teeth of the pinion. Mesh and turn the pinion with the gear lightly. Check the transferred material with the following diagrams.

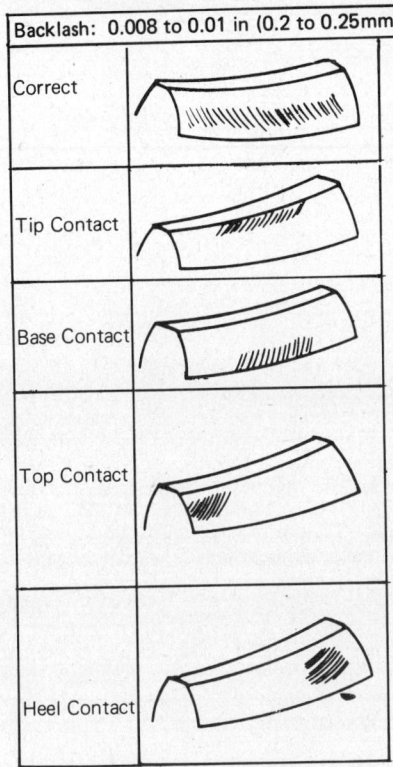

Tooth contact patterns

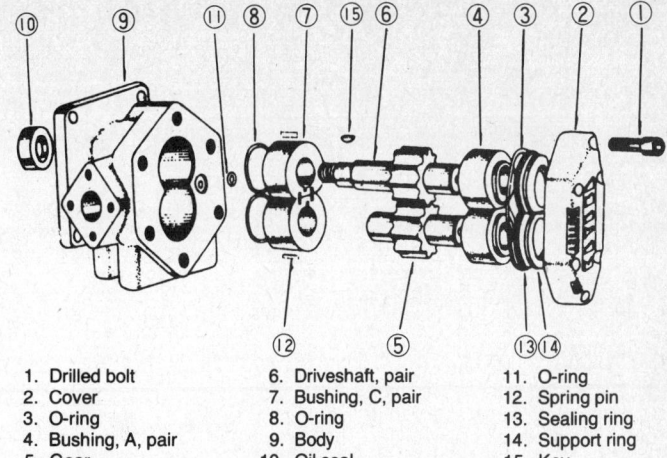

1. Drilled bolt
2. Cover
3. O-ring
4. Bushing, A, pair
5. Gear
6. Driveshaft, pair
7. Bushing, C, pair
8. O-ring
9. Body
10. Oil seal
11. O-ring
12. Spring pin
13. Sealing ring
14. Support ring
15. Key

Hydraulic pump components

HYDRAULIC SYSTEM

Pump

OVERHAUL

Disassembly

1. Loosen drilled bolt (1).
2. Remove end cover (2) and O-ring (3).
3. Remove the seal element which comprises two sealing rings and supporting rings (15, 16).

NOTE: Sealing ring (15) and supporting ring (16) can be removed as a unit.

4. Push driveshaft (6) into body (9). It will then be possible to lift up bushing (4) with the fingers. After removing the bushing, take out the driveshaft, gear (5), and the second bushing (7) to which O-ring (8) is attached.

CAUTION
Do not use a hammer to remove the bushing. If it is difficult to remove it, tap the body of the pump very lightly with something soft, like wood. This will make it possible to remove with ease.

Precautions

1. At the time of disassembly, arrange the disassembled parts in the order removed. Extreme care should be given to arrange the shaft and the bushing in their relative position so that there may be no mistakes at the time of assembling, such as to take the idler bushing for driveshaft bushing.
2. Refrain from removing oil seal (10) inside the casing if it is not damaged. In case the removal is necessary, then be very careful not to damage the inside surface of the casing while removing.

Assembly

Assembly is the reverse of disassembly.
1. Wash all parts clean in diesel fuel or kerosene with care so that no dirt and dust will stick to any of them.
2. Be sure to replace the parts into their original places.
3. In placing each pair of bushings, make certain the oil grooves in their surface are aligned with each other and not slanted.
4. After assembling, check and make certain that the pump has been reassembled properly. The pump should be turned with ease using something which is 4 in. in radius. If hard to turn, check to determine for the cause.

Testing

1. A new pump can be used as it is. So, attach to the engine, bleed air trapped, and put to use right away.
2. A reassembled pump should be put to break-in testing in the following manner:
   a. 10 minutes run under no load with engine at slow speed.
   b. 15 minutes run under 430 to 710 psi in pressure with engine at medium speed.
   c. 5 times of running, 5 seconds each, under relief pressure with engine at maximum speed.

Control Valve

OVERHAUL

Disassembly

1. Remove the control valve from the line and wash in diesel fuel or kerosene.

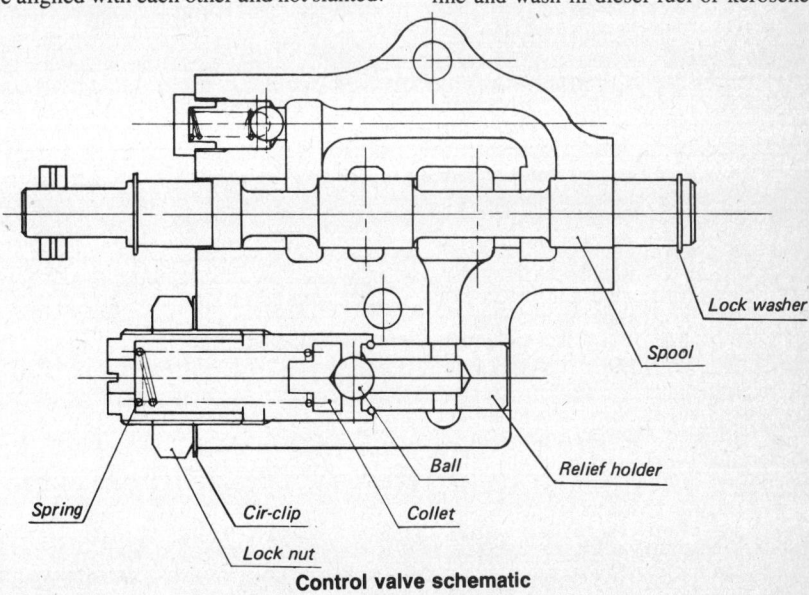

Control valve schematic

Kubota

2. Remove the retaining circlips and extract the spool.

3. Remove relief valve, unlock the washer and slacken the locknut. The spring, collet, and ball will then come out.

4. Extract the relief holder in the same manner as above.

Checking

1. Disassembled parts should be washed clean in diesel fuel or kerosene. Check the surface of each part to see if there are burrs, scratches, indents, or other defects. Light cases of scratches and burrs can be repaired by oil stone and lapping, but if there is any fear of oil leakage replace the part.

2. The spool and body are not interchangeable, so be sure to replace as a unit. The relief valve should also be replaced as assembly.

3. Each sliding part should work smoothly without any hitch. Each groove and passage should be free of impurities.

4. If the spring is deformed, bent or broken, replace.

Assembly

Assembly is in the reverse order of disassembly. Be very careful of the following points:

1. Wash all parts clean with diesel fuel or kerosene. All the O-rings should be renewed. Apply quality grease to the rings and assemble, with close care not to give any damage to them.

2. The relief valve should be adjusted.

3. The spring will be placed correctly if you hold the holder vertically and slip the spring in from the top.

ADJUSTMENT OF RELIEF VALVE SET PRESSURE

When a portion of the relief valve has been replaced or when the lifting force of the hydraulic system has deteriorated, it is necessary to confirm the set pressure of the valve, in the following manner:

1. Remove the plug in the cylinder cover and set a pressure gage instead. Use some bonding agent in installing the gage so that no oil will leak through the threaded part. Thread: PS⅜ in.

CAUTION
Do not adjust the pressure of relief valve without checking the pressure against a pressure gage. If the gage is not available replace the control valve as complete.

2. Remove the control rod from the lift arm unit, operate the engine at the maximum speed, and shift the control lever into "Lift". Read the pressure on the pressure gage at the moment the relief valve just starts working. The maximum preset pressure is 1700 psi.

3. In case the actual pressure differs from the specified, the adjustment can be made with the screw threads on the relief spring holder. Screwing in the nut will raise the pressure. Slackening the nut will lower the pressure.

4. After adjusting, be sure to bend the washer in correct manner to lock securely.

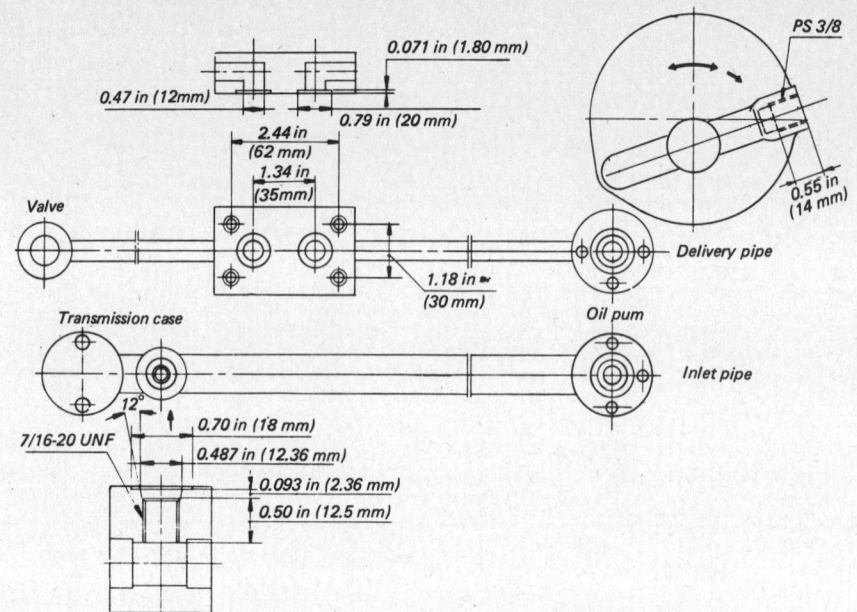

Relief valve set pressure adjustment

Hydraulic Cylinder

DISASSEMBLY

1. Remove the hydraulic cylinder assembly from the transmission case.

2. Remove 5 bolts (7T) which fasten the cylinder cover to the cylinder and push the lift arms down. The cover will then be pushed out. The cylinder liner may come out along with the cover.

3. Remove the bolt at each end of the hydraulic arm shaft and take out the lift arms. Drive out the arm shaft against the righthand end toward the lefthand side.

ASSEMBLY

Assembly is the reverse of disassembly, observing the following:

1. Wash all parts thoroughly in diesel fuel or kerosene care not to allow even a single dirt to enter into the assembly.

2. Coat each O-ring with grease before placing, with care not to scratch the surface. A worn ring or the one which has lost its elasticity should be replaced with new perfect one.

3. The hydraulic arm shaft and arm should be assembled according to the punch mating marks. In assembling the lift arm with the hydraulic arm shaft, be sure to align the bolt grooves.

CAUTION
If the punch mating marks are not lined up correctly, the feedback linkage and 3-point linkage will fail to work, so be very careful of these marks.

ENGINE

Cylinder Head

REMOVAL AND INSTALLATION

1. Remove the rocker arm cover.
2. Disconnect the fuel injection lines.

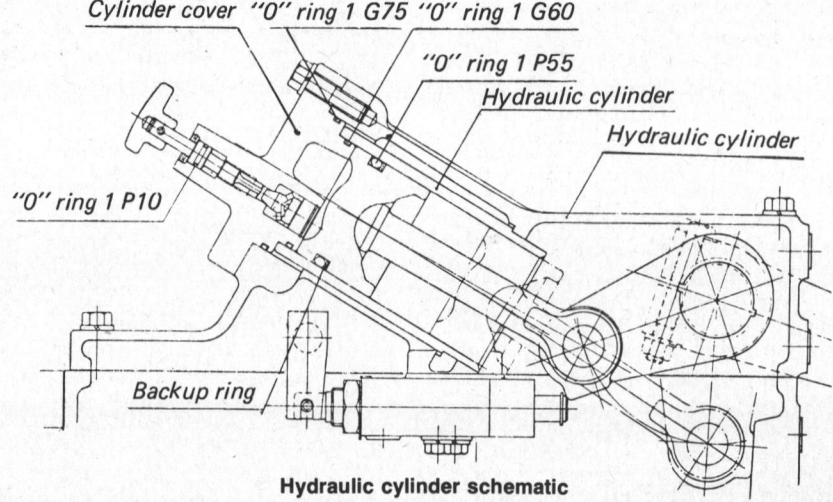

Hydraulic cylinder schematic

Kubota

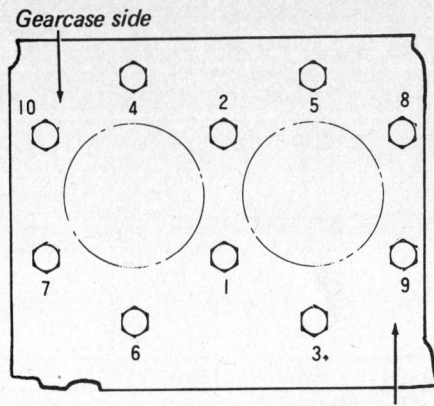

Cylinder head bolt tightening sequence

"IO"	Intake valve opens	TDC − 20°
"IC"	Intake valve closes	BDC + 45°
"EO"	Exhaust valve opens	BDC − 50°
"EC"	Exhaust valve closes	TDC + 15°

Valve timing chart

3. Remove the rocker arms and push rods, keeping the push rods in order.
4. Remove the head bolts and lift off the head.
5. Installation is the reverse of removal. Bleed the fuel system. Observe the cylinder head bolt torque diagram. Adjust the valves. Always use a new head gasket.

VALVE ADJUSTMENT

1. Remove cylinder head cover. With piston at compression top dead center, align the "TC" marked on the outside of flywheel with notched mark of peephole on the side of flywheel housing. This is the position where rocker arm is free, not pushing valve downward with its tip. Make adjustment. For this adjustment, engine should be cold.
2. Loosen locknut (1) and adjust clearance of rocker arm-to-valve stem to 0.007 to 0.009 in. After adjusting, tighten locknut securely.

VALVE TIMING

When the valve clearance has been adjusted the standard valve timing shown in the accompanying chart can be attained.

Timing Gears and Camshaft

REMOVAL AND INSTALLATION

1. Remove the crankshaft collar, the O-ring, oil slinger and the gear collar in that order.
2. Remove the external circlip and detach the idle gear.
3. Detach the idle gear collar.
4. Straighten the tang of the washer.
5. Remove the camshaft stopper bolt.
6. Detach the camshaft.
7. Detach the hydraulic pump and the hydraulic pump holder.
8. Remove the pump drive gear and the collar.

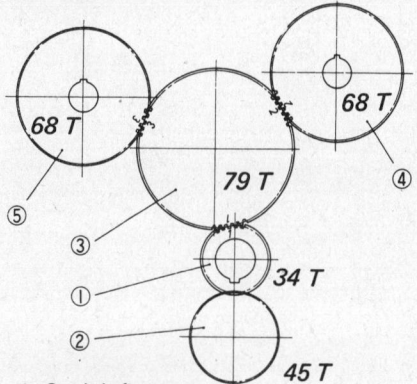

1. Crankshaft gear
2. Oil pump drive gear
3. Idle gear
4. Camshaft gear
5. Fuel injection pump drive gear

Engine gear train components, with timing marks aligned

1. Camshaft gear
2. Camshaft
3. Camshaft stopper
4. Circlip
5. Pin plug
6. Key

Attaching camshaft and gear

9. Remove three fork lever holder set bolts.
10. Remove the fuel camshaft stopper.
11. Remove the fuel camshaft and the fork lever shaft at the same time.
12. Detach the crank gear with a puller.
13. Remove the key.
14. Straighten the tang of the washer.
15. Detach the pump drive gear.
16. Detach the oil pump.
17. Installation is the reverse of removal.

Pistons and Connecting Rods

REMOVAL AND INSTALLATION

1. Detach the oil pan.
2. Remove the oil filter. Be careful of the O-ring.
3. Straighten the tang of the washer.
4. Detach the connecting rod bolt.
5. Remove the cap of the large end of the connecting rod.
6. Drive out the piston to the cylinder head side with a hammer handle.
7. After driving the piston out, attach a tag to each piston to indicate its number.
8. Remove the piston rings.
9. Straighten the tang of the washer.
10. Detach the flywheel bolts.
11. Remove the flywheel.
12. Straighten the tang of the washer.
13. Detach the bearing case bolt.
14. Straighten the tang of the washer. Remove the bolts.
15. Drive two M8 bolts into the bearing cover and then pull the cover out.
16. Tap the crankshaft until it comes out of the flywheel side; be careful not to scratch the crankshaft metal.
17. Straighten the tang of the washer.
18. Remove the bolts.
19. Remove the bearing case.
20. Remove the bearing case side metal on the flywheel side.

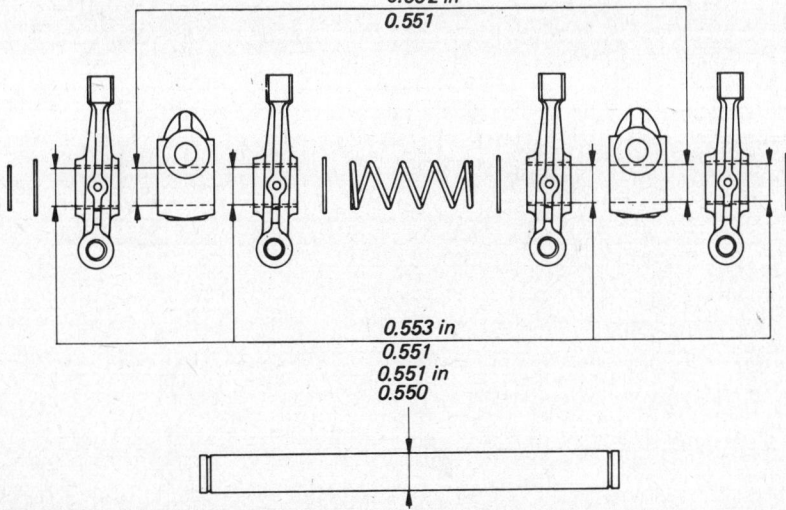

Standard relative position of rocker arms

645

Kubota

21. Attach the dry liner changer to the crankcase.
22. Draw out the liner.

FUEL SYSTEM

Injection Pump

TIMING

Fuel injection timing is adjusted by changing the number of shims used between the pump and the engine frame. A piece of the shim corresponds to 1.5°, approximately, in crank angle. Therefore, the injection will take place 1.5° later when a piece is added and 1.5° earlier when a piece is removed.

The timing is correct when the notch of the window on the side of flywheel housing aligns with "FI" marked on the flywheel outside.

NOTE: The manufacturer recommends that service to either the injection pump or the injectors be left to a qualified service technician.

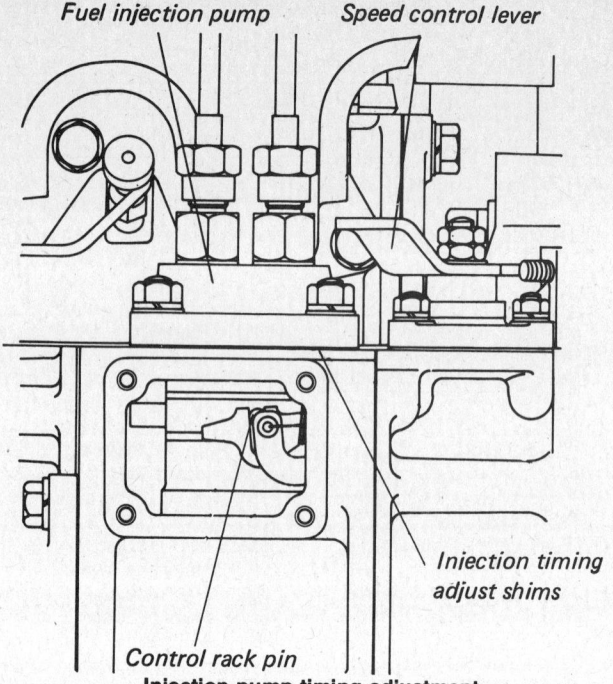

Injection pump timing adjustment

Model L225

FRONT WHEELS AND AXLE

Front Wheel Hub

REMOVAL AND INSTALLATION

1. Remove the front wheel tire. Remove the front wheel cap. Remove the split pin and then loosen the slotted nut.
2. Tap out the hub with a hammer (mallet, copper or plastic hammer).
3. Before assembling, apply sufficient wheel bearing grease (SAE multi-purpose type grease) to the bearings and oil seal. Also, fill the space inside the hub with grease up to ⅓ to ½ of the space.

— **CAUTION** —
Do not use grease contaminated with foreign matter. Be very careful to keep the hub free from mud, dirt, sand and other impurities.

4. After filling grease, proceed to assemble. Before tightening the slotted nut, make certain bearings are in correct place on the shaft.

Knuckle Shaft

REMOVAL AND INSTALLATION

1. Remove the knuckle arm and O-ring.

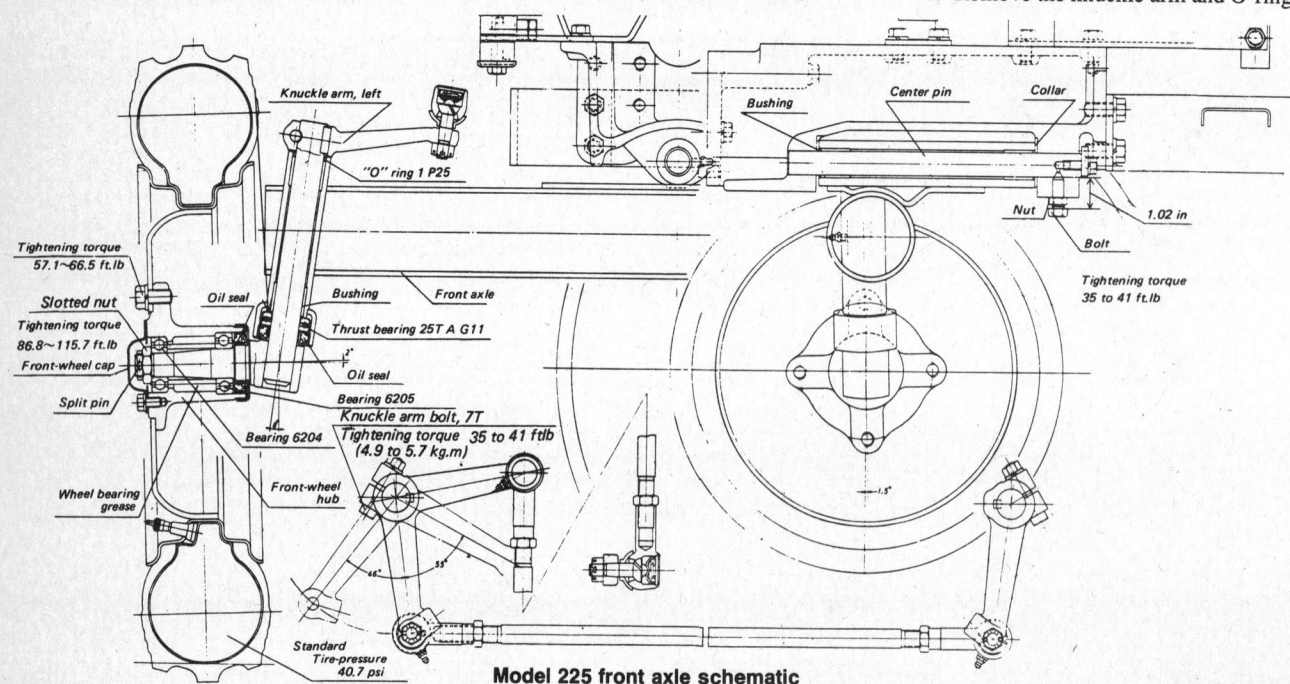

Model 225 front axle schematic

Kubota

2. Lift up the front axle to the extent that the knuckle shaft can just be removed downward and then remove the knuckle shaft.

3. Before assembly, apply sufficient chassis grease to the thrust bearing and oil seal.

Center Pin

REMOVAL AND INSTALLATION

Remove the bumper first and then the bolt which fixes the center pin. Lift up the front axle and pull out the center pin toward the front. The front axle will then be disconnected.

--- CAUTION ---
In assembling, check and make certain the bolt which fixes the center pin is locked securely with the locknut, with the bolt protrusion 1 in. If this bolt is loosened, the center pin may come out.

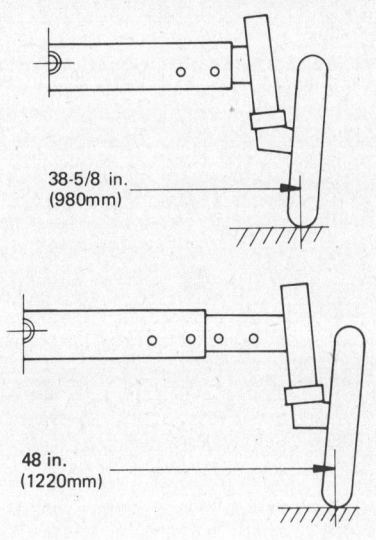

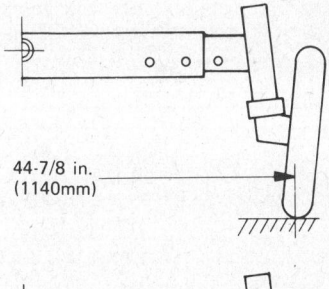

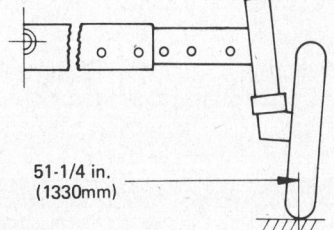

Model 225 front read adjustment

Model 225 toe-in adjustment

TOE-IN ADJUSTMENT

Loosen the locknut and turn the tie rod so that CD-AB = 0.08-0.31 in. To take measurements, use a wheel alignment tester or a toe-in gauge. If neither is available, measure with a precise tape measure.

FRONT TREAD ADJUSTMENT

Model L225FA

It is possible to change the front tread 4 stages, from 38⅝ in. to 51¼ in.

1. Loosen nut 2 of clamp 1 on the outer pipe of tie rod, and remove bolt 3.
2. Lift the front part of the tractor with a jack. Loosen nut 4 and remove bolt 5, four bolts. Then it is possible to separate the front axle (left) (right) and (center).
3. Insert bolt 5 into the hole of the desired width and tighten with nut 4. Insert bolt 3 into the inner pipe of the tie rod and tighten.
4. Select the bolt holes for the front axle (side), and (center) according to the diagram.

STEERING

Steering Gear

REMOVAL

1. Using a Pitman arm puller, remove the Pitmam arm first. Install the puller on the Pitman arm and screw in the center bolt while preventing the hooks from spreading with help of the stopper. The Pitman arm will then be pulled from the Pitman arm sector shaft.

NOTE: Pitman arm and sector shaft are marked with mating marks.

2. To remove the wheel, use a steering wheel puller. Loosen the nut at the center of the wheel, set the puller there and screw in the center bolt. The wheel will then be pulled from the steering shaft and column.

NOTE: A very tight wheel can be pulled off by tapping the center bolt with a hammer only if the bolt has been secured into the normal torque.

3. Remove the fuel tank and instrument panel. The gear box will then be exposed.
4. Remove the side cover of the box and take out the sector shaft.
5. Remove the bolts which fasten the rear cover to the box. Pull the steering column and shaft as assembly. Note the number of shims used. Do not lose them.

--- CAUTION ---
The ball nut is a precision fit, so do not disassemble it.

INSTALLATION

1. Prior to assembly, wash the parts carefully with diesel fuel or kerosene and assemble in the reverse of disassembly.
2. Check and make certain the steering shaft turns lightly without chattering. Adjustment can be made by the rear cover shims. After confirming this, insert the sector shaft.
3. When the steering wheel is installed, turn the wheel gently and measure the free turn, which should lie between 0.79 in. and 1.97 in. on the circumference of wheel. Adjustment can be made with the lash adjusting screw of the Pitman sector shaft.

NOTE: Counterclockwise turning of the screw removes overcenter load to increase the free turn of the wheel and clockwise turning of the screw decreases the free turn of the wheel.

4. Remove the oil inlet plug and fill the

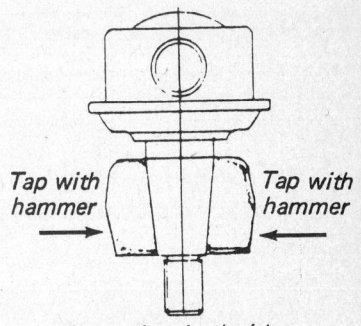

Tap against both sides at the same time.
Tie rod pin removal

647

Kubota

1. Steering assembly
2. Gear box
3. Sector shaft
4. Ball-nut complete
5. Bearing
6. Rim cover
7. Shim
8. Side cover
9. Gasket
10. Adjusting screw
11. Nut
12. Adjusting screw shim
13. Bolt
14. Lockwasher
15. Oil seal
16. Oil seal
17. Nut
18. Lockwasher
19. Plug
20. Column
21. Column bushing
22. Bolt
23. Lockwasher
24. Parallel pin
25. Steering wheel assembly
26. Steering wheel
27. Wheel cap
28. Nut
29. Lockwasher
30. Key
31. Rubber ring
32. Pitman arm
33. Gasket
34. Rubber plug

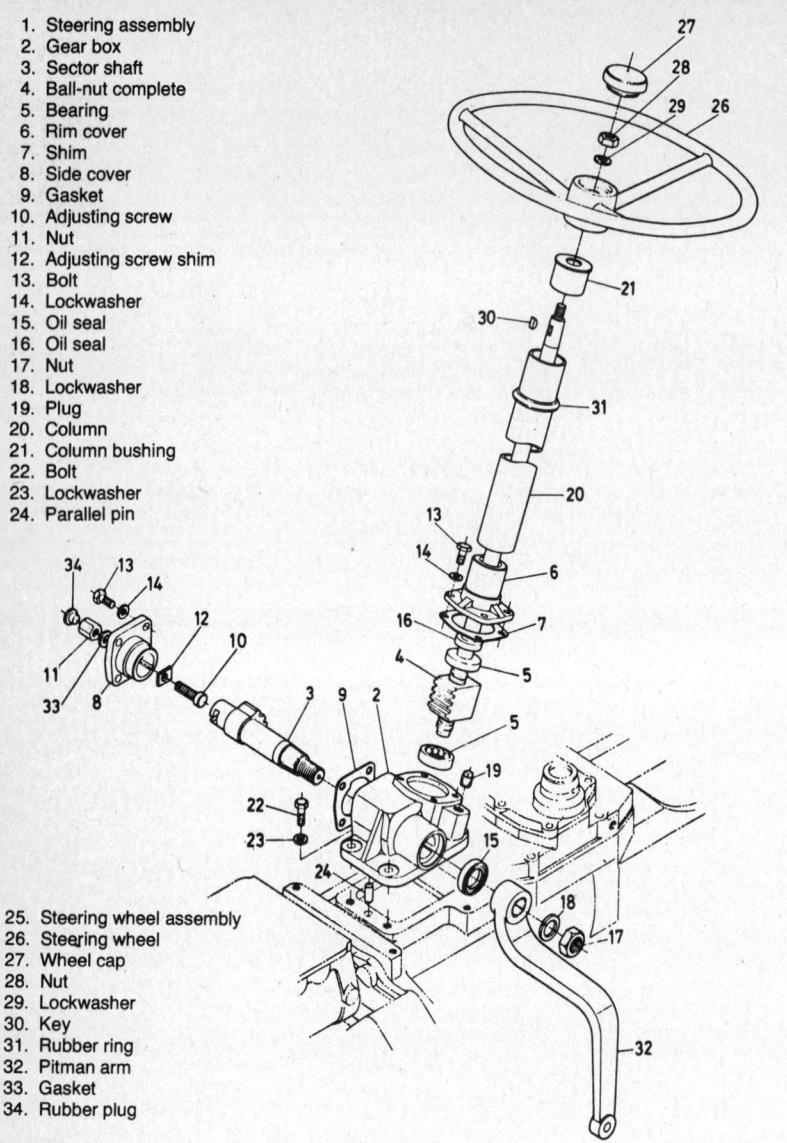

Steering system components

gear box with SAE 90 EP gear oil. Capacity amount is 0.32 qt.

Tie Rod Ends

REMOVAL AND INSTALLATION

There are four tie rod ends used in this tractor, two at the ends of the tie rod and the other two at the ends of the drag link. Each tie rod end is tapered into the knuckle arm or the Pitman arm and locked with a nut and split pin.

Loosen the nut and tap against both sides of the end at the same time with mallets or copper or plastic hammers. The tie rod end tapered pin will then come out smoothly.

BRAKES

For complete brake system service, see the model L175 section.

CLUTCH AND MAINSHAFT

REMOVAL

To disassemble the clutch, the engine must be detached from the clutch housing first. The following are necessary precautions in disassembling and assembling the clutch.

1. Wiring of the alternator and starter should be handled only after the negative wire of the battery has been removed from the battery.
2. Before removing the hydraulic pipe from the pump, drain off the oil in the transmission case.

NOTE: Hydraulic pipe should be removed from the pump in order to separate the engine.

3. Since an O-ring is used on the tip of hydraulic pipe, be careful not to lose or damage the ring.

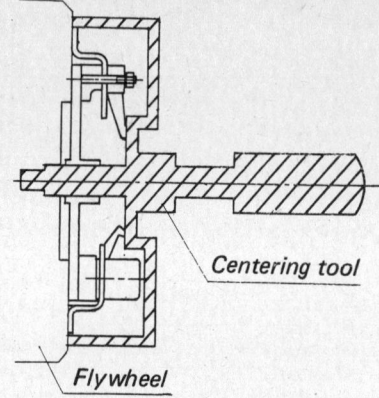

Clutch disc centering tool installed

4. Note that the engine is prone to fall or tip over when detached, so be very careful. Use a hoist or chain block for this removal.
5. It is preferable to have the clutch housing suspended with a hoist or chain block as well.
6. In case the rotary device is left attached to the tractor, be sure to lower this before starting the disassembly service.

NOTE: With the rotary device, the rear of the tractor will be heavier. Without the rotary device or with the rotary device lowered in the above manner, the front of the tractor will be heavier.

INSTALLATION

1. If the flywheel is rusted or scratched on the surface, which contacts with the clutch, the clutch performance will be affected. Polish the surface with sand paper.
2. Do not apply any oil or grease to the contact surface of the flywheel, the surface of the clutch pressure plate or the surface of the clutch disc.
3. Check grease inside the clutch-release hub. If its color has changed, replenish with SAE multi-purpose grease.
4. In attaching the clutch to the flywheel, be sure to make them concentric with each other. If misaligned, the mainshaft will not be inserted smoothly. Use a centering tool. Insert the centering tool shaft into the center of the clutch assembly passing through the clutch disc into the counterhole in the flywheel, to align the disc with the flywheel. Then tighten bolts

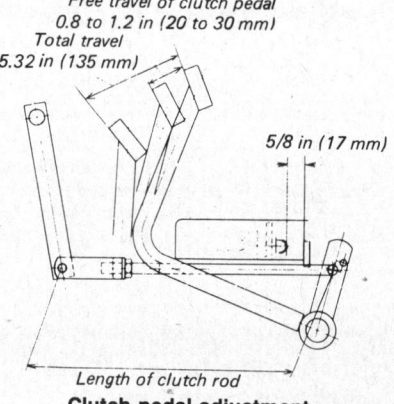

Clutch pedal adjustment

Kubota

Release Levers

1. If there is some difference in relative heights of release levers, remove the split pin, and turn the nut to adjust the difference to less than 0.02 in. This adjustment is made by measuring the clearance of the levers to the release bearing using a thickness gauge.

2. If the height of release levers is extemely low, remove the split pin and loosen the nut.

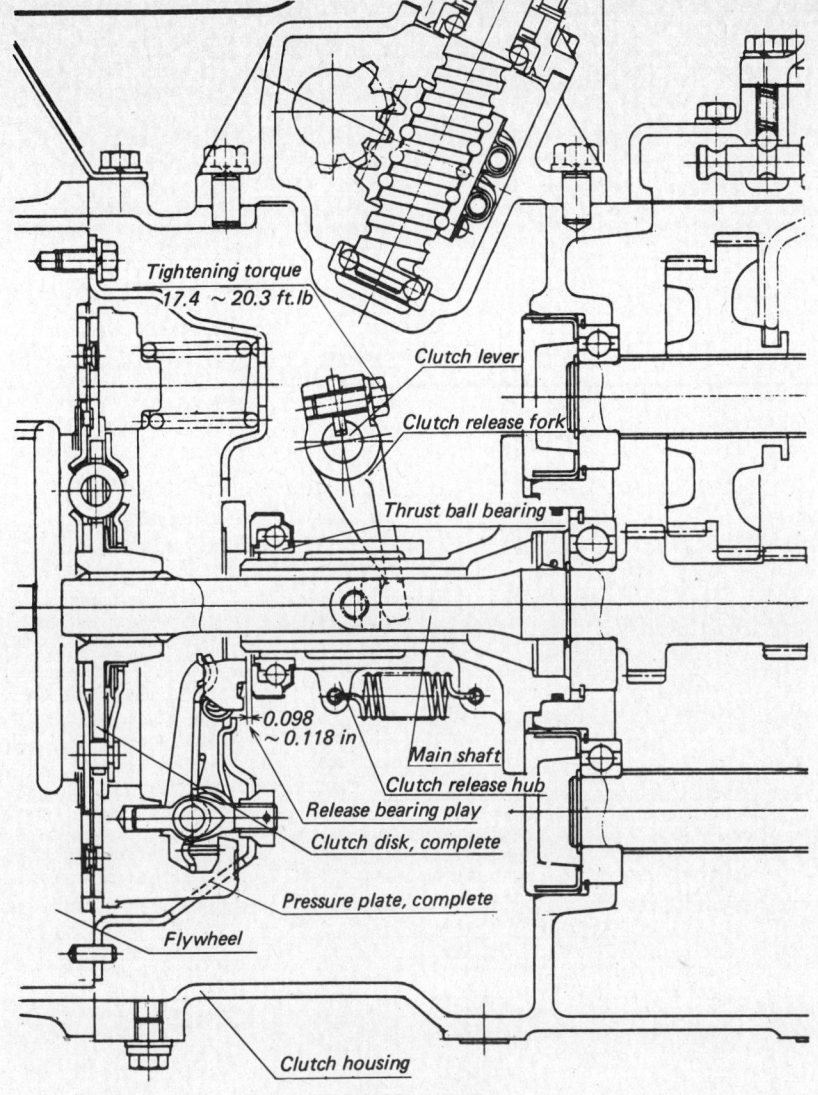

Clutch schematic

TRANSMISSION AND POWER TAKE-OFF

NOTE: This transmission consists of two gear units, one in the clutch housing and one in the transmission case.

Clutch Housing

DISASSEMBLY

1. It is not necessary to remove the instrument panel, fuel tank, steering unit, and clutch pedal, but separate the engine and then the transmission case from the clutch housing.

2. Put the main and PTO gearshifters in neutral and remove both upward together with the gearshifter cover.

3. Loosen 4 bolts which fasten the mainshaft case and pull out the mainshaft case forward.

4. Pull out the bearing cover at the front end of the countershaft, in such a manner as to tap the countershaft against the rear end. Remove the other two countershaft bearing covers in a similar manner.

5. Remove the circlip which retains the countershaft at the front end, and tap out the shaft backward. Remove the other two countershafts in a similar manner.

6. Tap out the mainshaft frontward.

Transmission Case

DISASSEMBLY

1. Remove the engine from clutch housing. Detach the fenders, rear wheels, seat, and hydraulic cylinder.

2. Separate the clutch housing from the transmission case.

3. Remove the right and left rear axle cases. Take out the differential gears. Take care in this disassembling process since the case may tilt and fall down if the work bench happens to be unstable.

4. Slacken 6 bolts which fix the bearing case and bearing cover and remove the bearing cover. Remove the nut which locks two bearings (6305). These are all on the spiral bevel pinion shaft at the end opposite to the pinion. Tap out the pinion shaft backward.

5. Drive out the 13-gear shaft forward. To effect this, remove circlips which retain

which fix the pressure plate assembly. After tightening securely, apply the arm gauge to the surface of the flywheel, and adjust the heights of the release levers so that they are equal. After adjusting, remove the centering tool.

ADJUSTMENTS

Pedal Free Travel

1. Open the clutch access window which is located on the righthand side, forward of the clutch housing.

2. Insert a 0.098 to 0.12 in. thickness gauge between the release thrust ball beard the release levers. Adjust the clearance by the length of the clutch rod. Lengthen the rod when the clearance is wide. Shorten the rod when the clearance is narrower.

3. Depress the clutch pedal, and check and make certain that the free travel at the tip of the pedal is 0.8 to 1.2 in.

4. With the clutch pedal depressed to the lowest point, confirm that there is adequate gap between the clutch rod and safety starter switch.

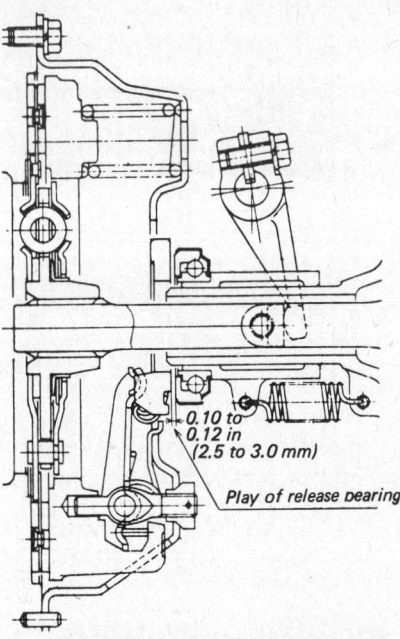

Clutch release lever adjustment

649

Kubota

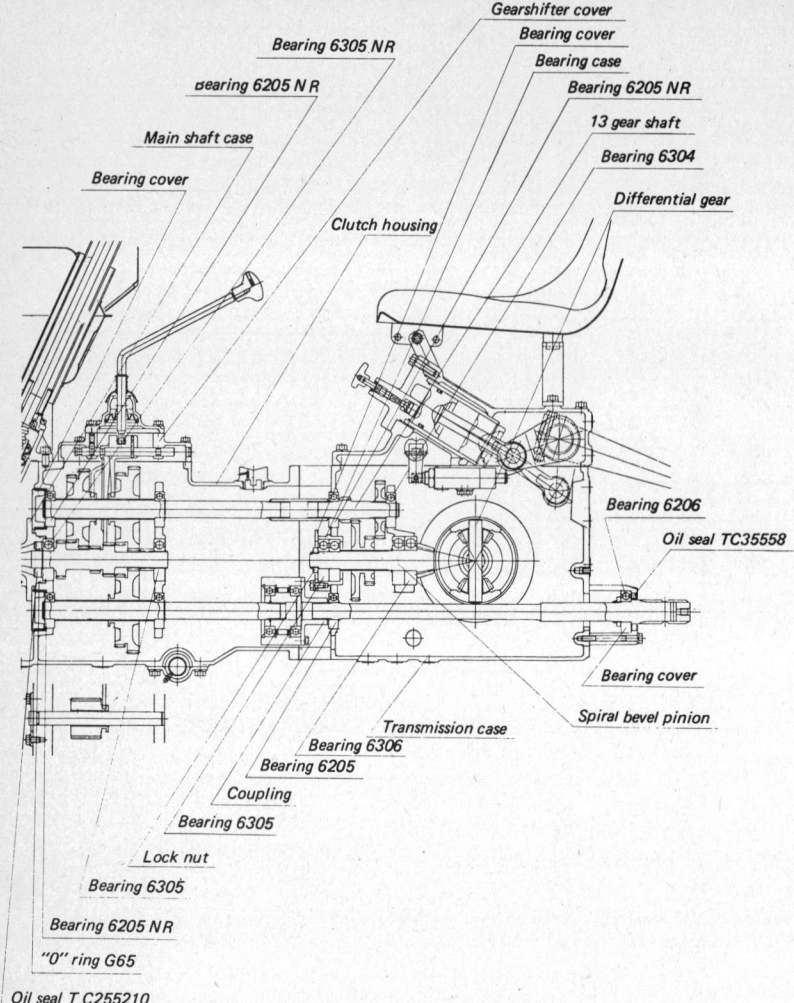

Transmission schematic

Differential and rear axle schematic

the bearing (6304) supporting the shaft at the rear end and then the circlip which retains the bearing (6205NR) at the front end of the shaft.

6. Remove 4 bolts which fasten the cover for the bearing supporting the PTO shaft at the rear end, and extract the PTO shaft backward.

NOTE: In case of removing the PTO shaft only, put the high-low gearshifter in high. The PTO shaft will then be extracted with ease.

Assembly of Transmission and Clutch

1. Wash the case, housing, and components in diesel fuel or kerosene.

CAUTION
Do not use gasoline for cleaning.
---

2. Next, check carefully to see if there might be any damaged parts. Slight defects can be repaired but severely damaged components should be replaced. Be especially careful to check rolling condition of oil seals.

3. Replace all components in the order removed.

4. To insert the shift fork rod into place, use a shifter ball guide available at your Kubota dealer.

Before inserting the rod, insert the guide to hold the ball. After holding the ball, push the rod into place and then pull out the guide.

DIFFERENTIAL GEAR AND REAR AXLE

Differential

OVERHAUL

1. Remove the hydraulic cylinder located on top of the transmission case.

2. Remove both right and left halves of the rear axle. Remove differential bearing cases. Remove the differential upward.

3. Straighten washers which lock set bolts for the spiral bevel gear, loosen the bolts and take out the spiral bevel gear. Then tap out the differential pinion shaft.

4. Next, move and take out pinion gears through the differential access window of the case.

ASSEMBLY

Assembly is in the reverse order to disassembly.

CAUTION
Bolts tightening the spiral bevel gear are 7T. Tighten them to the torque of 35-41 ft.lb.

Kubota

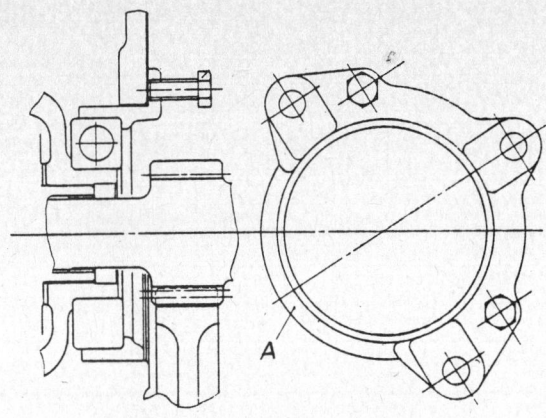

Differential bearing case removal

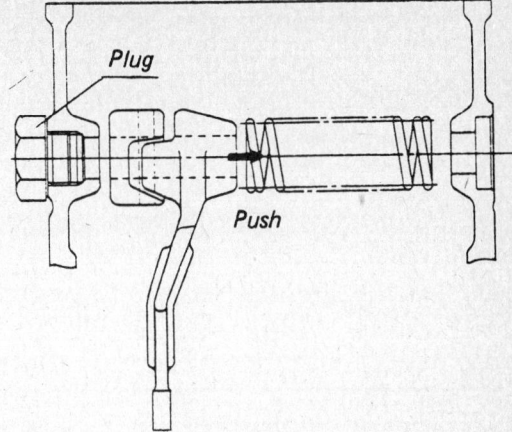

Insertion of differential lock spring

Differential Lock System

OVERHAUL

1. Remove the spring pin in the differential lock cam and pull out the differential lock pedal (rod). The differential lock cam, shift fork, and spring will then become loosened. The spring may jump out, so be careful.

CAUTION

Extreme care is necessary in pulling the pedal out of the case, so as not to damage lips of the oil seal.

---

2. For assembling, a jig may be used for placing the spring, but it can be installed without the jig if you will observe the following:
   a. Remove plug.
   b. Place spring between shift fork and transmission case inside surface opposite plug hole, and then place cam between the fork and plug hole.
   c. Pass differential lock pedal (rod) through and then drive spring pin into the cam to fix the cam with the lock pedal (rod).
   d. Screw in the plug.

Rear Axle

OVERHAUL

1. Remove bolts and nuts which fasten the axle case to the bracket, and pull out the axle along with the case.

NOTE: Screw in M8 bolts little by little in turn while tapping lightly against portion (A), so that the case can be extracted evenly.

2. Unlock the nut which locks the gear on the axle and remove the nut. Pull out the bearing and then the gear.

3. Unbolt the axle flange. Drive out the axle.

CAUTION

In driving out the axle, be careful not to damage the oil seal.

---

4. Be sure to bend the locknut securely into a notch in the axle end.

5. Fill the oil seal with grease between the lips before assembly.

ADJUSTMENT

Spiral Bevel Pinion and Gear

Adjustment of the spiral bevel pinion and gear is effected by shims used on the differential bearing cases. At the time of disassembly, note the shims used there and be careful not to lose them. During assembly return the shims to their original places, and then check meshing and backlash of the gear teeth. If both are correct, proceed with assembly.

However, if a tractor has been in a long period of service, it generally requires an adjustment in this unit:

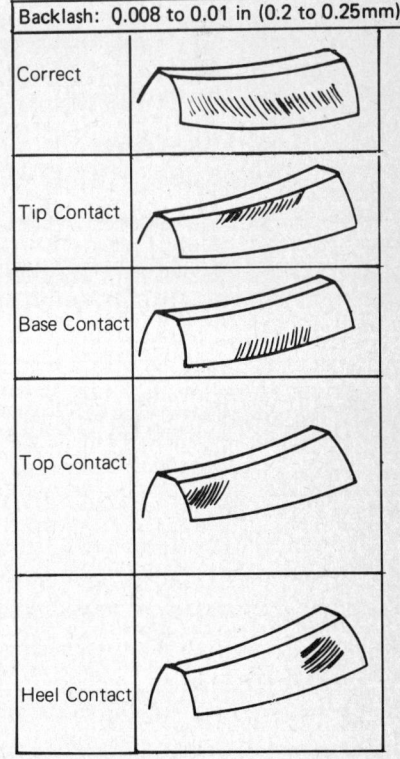

Differential tooth contact pattern

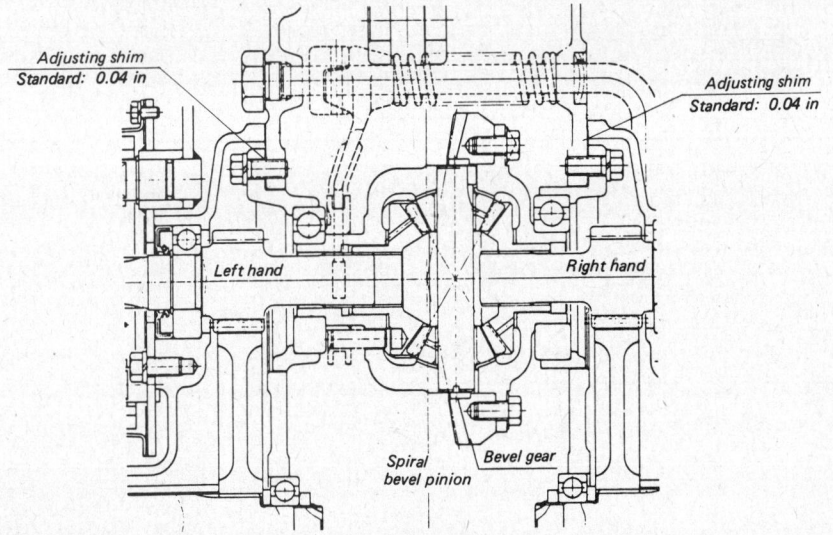

Spiral bevel pinion and gear adjustment

651

Kubota

NOTE: For easier adjustment, it is recommended that the spiral bevel pinion shaft be removed from the assembly.

Backlash and meshing of spiral bevel pinion and gear.

This adjustment is very important. Whether this has been performed correctly or not, seriously affects the life of the gears and consequently the durability of the tractor. Therefore, make the adjustment carefully. Simple ways of judging the results of adjustment are:

Backlash: Fill and impress solder between teeth and measure the thickness of impressed solder with a micrometer. If the measurement is 0.008 to 0.01 in., the adjustment is correct.

Contact of teeth: Apply a very thin coat of Prussian blue to contact surfaces of several teeth of the pinion. Mesh and turn the pinion with the gear lightly. Check the material transferred and stuck to the bevel gear and make certain that the wear patterns are in compliance with the following diagrams:

HYDRAULIC SYSTEM

For complete hydraulic system service, see the model L175 section.

ENGINE

For complete engine service, see the L185, 245, 295 section.

Model L285

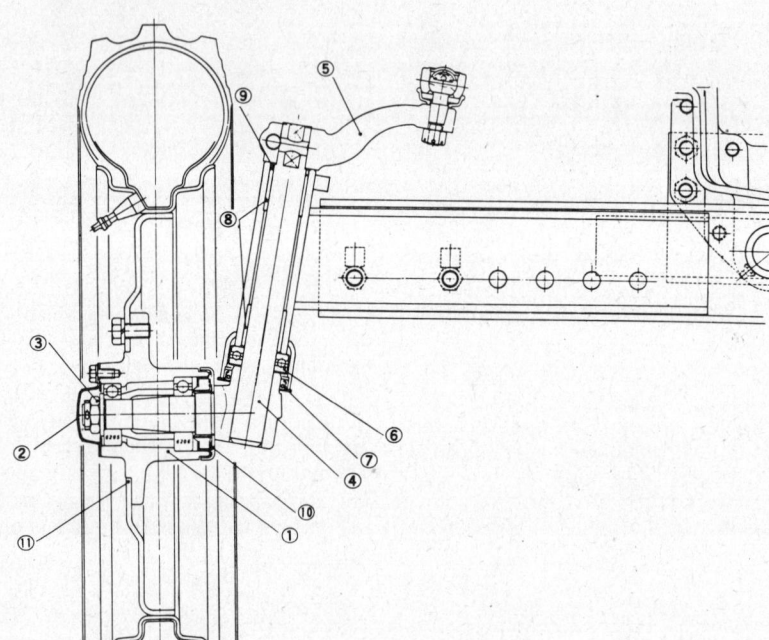

FRONT WHEELS AND AXLE

Front Wheel Hub

REMOVAL AND INSTALLATION

1. Remove the front wheel cap (2), and pull out the cotter pin from the grooved nut (3) to remove the nut.
2. Remove the four lock bolts for front wheel (11) and take off the front wheel tire assembly.
3. Tap out the front wheel hub with a hammer mallet, copper or resin hammer.
4. Before reassembly, apply 1.6 oz. of wheel bearing grease to the bearing, the special end oil seal (8) and the space inside of the hub.
5. Tighten the locknuts only after seeing that the bearing has been fully placed on the shaft.
6. Tighten the grooved nut to 87-116 ft.lb., and secure it with a cotter pin.

Knuckle Shaft

REMOVAL AND INSTALLATION

1. Remove the knuckle arm (5), then the O-ring.
2. Next, jack up the front axle to a height that allows downward removal of the knuckle shaft, then remove the knuckle shaft.
3. Before reassembly, apply ample chassis grease to the thrust bearing, the special end oil seal (7), and the bushing (8).

Center Pin

REMOVAL AND INSTALLATION

1. Remove the bumper and the center spring pin (13), jack up the front axle, and draw out the center pin frontward.
2. Apply ample grease to the bushing, tap it in from the front and set with a spring pin.

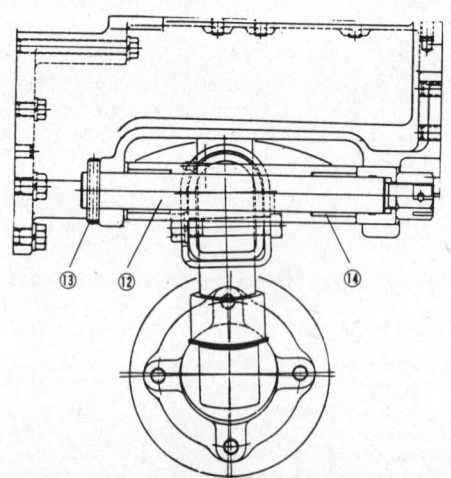

1. Front wheel hub
2. Front wheel cap
3. Nut
4. Knuckle
5. Knuckle arm
6. Thrust bearing
7. Oil seal
8. Bush
9. O-ring
10. Oil seal
11. Front wheel
12. Center pin
13. Spring pin
14. Bush

Model 285 front axle schematic

Kubota

STEERING

Steering Gear

REMOVAL AND INSTALLATION

1. Pull out the Pitman arm and the steering handle. To pull out the Pitman: Attach a Pitman arm puller to the Pitman arm, and drive in the center bolt, while preventing the claws from opening by the stopper.
(There are assembly marks on the Pitman arm and on the sector shaft.)
To pull out the steering wheel, loosen the nut in the center of the steering and set the steering wheel puller and drive in the center bolt. When the center bolt is tightened with a given load, the center bolt may be hammered for easier pulling.

2. Remove the fuel tank and the instrument panel to expose gear box.

3. Remove the side cover (6) and take out the sector shaft (3).

4. Remove the column (6) set bolts, and pull out the steering wheel (25), the column (14) and the ball nut (4). (Be careful of the shims.)

CAUTION
The ball nut (4), must not be disassembled.

5. When reassembling the gear box, wash carefully and reverse the disassembly procedure.

6. After seeing that the steering shaft turns without play, insert the sector shaft.

7. After installing the wheel, measure the amount of play, and adjust the wheel circumference play to 0.8-2.0 in., with the adjusting screw (10). (The amount of play is reduced by turning the adjusting screw (10) clockwise, and vice versa.)

8. Pour 0.32 qt. of SAE 90EP gear oil into the tapped hole.

Tie Rod Ends

REMOVAL AND INSTALLATION

1. The tie rod end is fixed to the knuckle and the Pitman arm with a taper. So loosen the nut a little, and tap with a mallet or a resin hammer. It will then come out easily.

2. The tie rod should be renewed as a complete unit. At the time of reassembly, be sure to insert a cotter pin in the nut and split the pin.

TOE-IN ADJUSTMENT

1. Loosen the locknut (2) and remove the bolt (3).

2. Turn the tie rod outer tube (1) and adjust by changing the length of the tie rod.

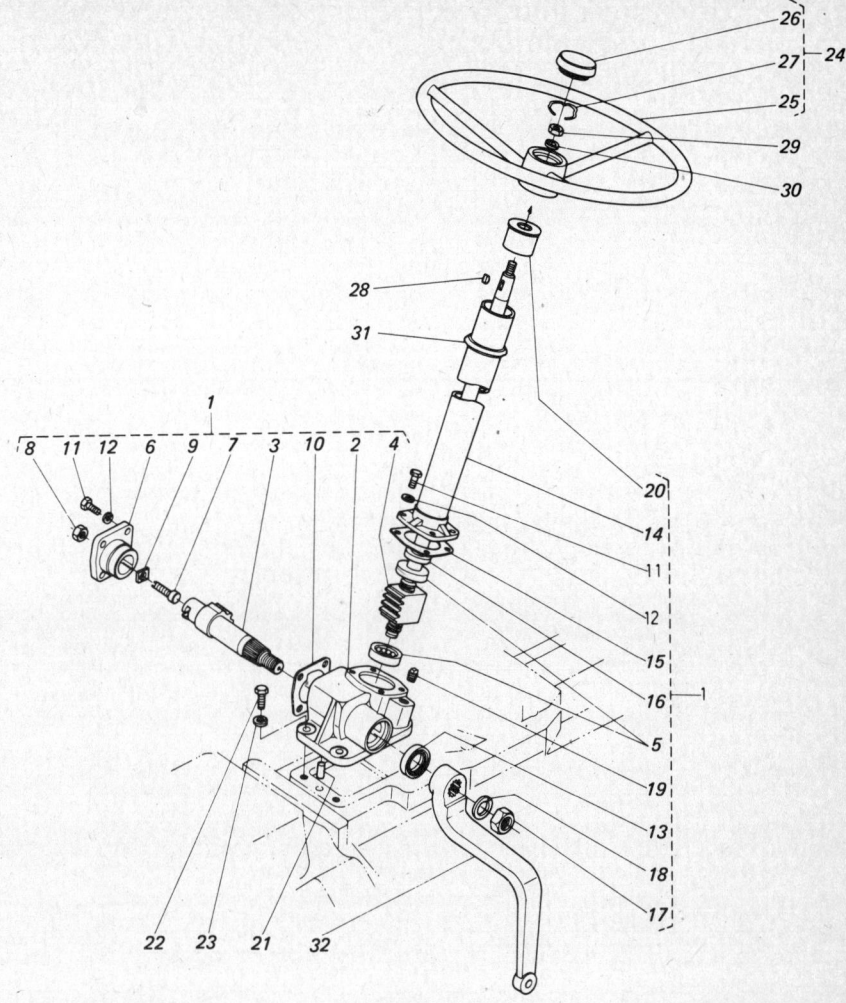

1. Steering assembly
2. Gear box
3. Sector shaft
4. Ball-nut complete
5. Bearing
6. Side cover
7. Adjusting screw
8. Nut
9. Adjusting screw shim
10. Packing
11. Bolt
12. Lockwasher
13. Oil seal
14. Column
15. Shim
16. Oil seal
17. Nut
18. Lockwasher
19. Plug
20. Column bushing
21. Parallel pin
22. Bolt
23. Lockwasher
24. Steering wheel assembly
25. Steering wheel
26. Wheel cap
27. Clip
28. Key
29. Lockwasher
30. Nut
31. Rubber ring
32. Pitman arm

Model 285 steering system components

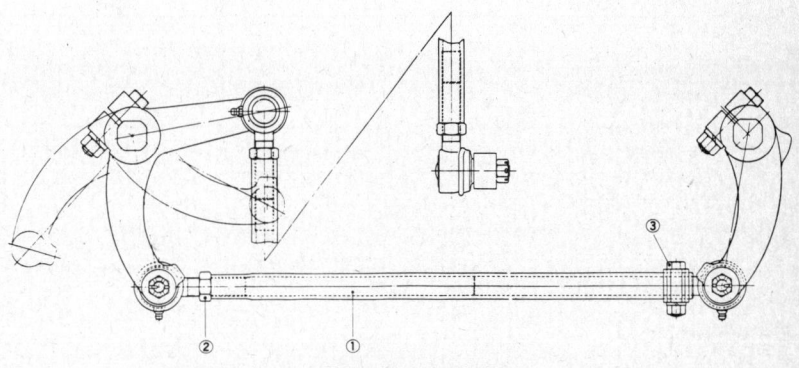

1. Exterior of tie rod pipe
2. Locknut
3. Bolt

Model 285 toe-in adjustment

653

Kubota

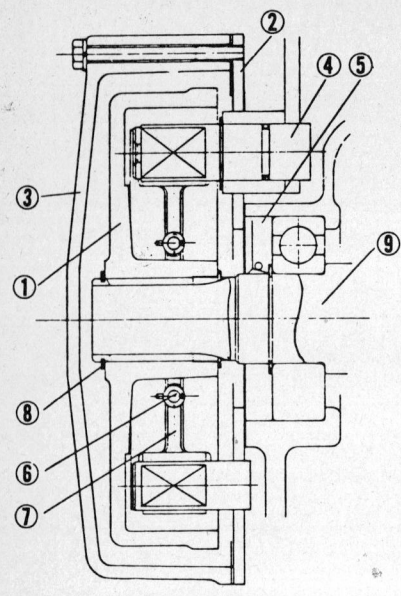

1. Brake drum
2. Anchor plate
3. Brake cover
4. Brake cam lever
5. Oil seal
6. Spring
7. Brake shoe
8. Set ring
9. Differential gear shaft

Brake components

CLUTCH

REMOVAL AND INSTALLATION

1. Separate the engine.
2. Remove the clutch assembly from the flywheel.
3. In attaching the clutch to the flywheel, align the clutch disc and flywheel centers for smooth insertion of propeller shaft. For alignment use a clutch disc centering tool. Insert the centering tool through the clutch disc into the center hole of the flywheel to center the clutch disc, and pull out the centering tool after fixing the pressure plate completely.

NOTE: Two reamer bolts are used to fix the pressure plate. Be sure of the bolt positions.

ADJUSTMENT

Remove the pin (3) that connects the clutch lever (1) and the clutch rod (2), loosen the locknut (5) and turn the joint (4) to adjust the rod length. Make it longer to reduce play; make it shorter to increase play.

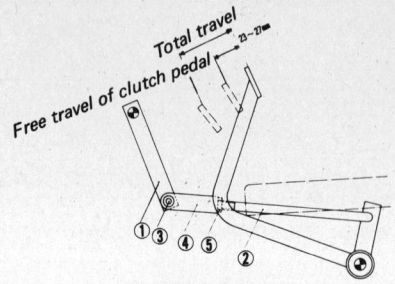

1. Clutch lever
2. Clutch rod
3. Pin
4. Joint
5. Locknut

Clutch pedal adjustment

TRANSMISSION

DISASSEMBLY

1. Drain out the oil from the clutch housing (1) and from the transmission case (13).
2. Transfer the creep speed change lever to "H" position.
3. Separate the clutch housing from the gear case (9).

BRAKES

OVERHAUL

1. Remove the brake rod and the brake cover, and the brake cam lever (4) and the brake shoe (7) will appear attached to the anchor plate (2).
2. Remove the stop ring (8) for the brake drum (1), and pull out the drum.
3. If the oil seal (5) is damaged and leaking, remove the axle case and tap the differential gear shaft (9) from inside to allow the oil seal to come off.

ADJUSTMENT

1. Loosen the nut on the turnbuckle, turn the turnbuckle to adjust the play to the pedal.
2. After the adjustment, securely lock the turnbuckle with the nut.

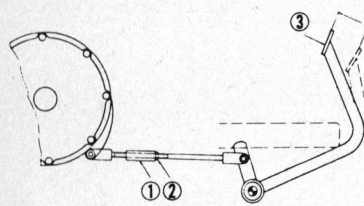

1. Turnbuckle
2. Nut
3. Brake pedal

Brake pedal adjustment

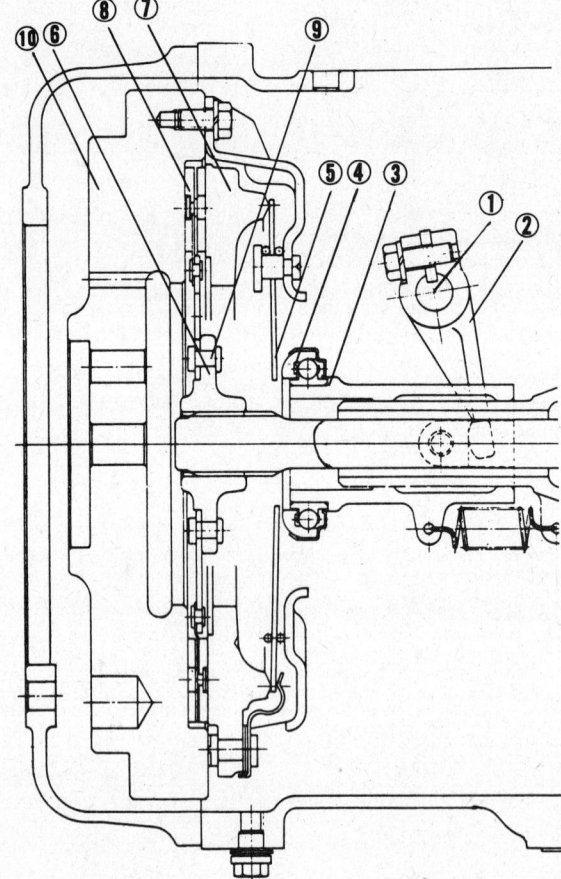

1. Clutch lever
2. Release fork
3. Bearing hub
4. Thrust bearing
5. Diaphragm spring
6. Clutch disc
7. Pressure plate
8. Facing
9. Rivet
10. Flywheel

Clutch schematic

Kubota

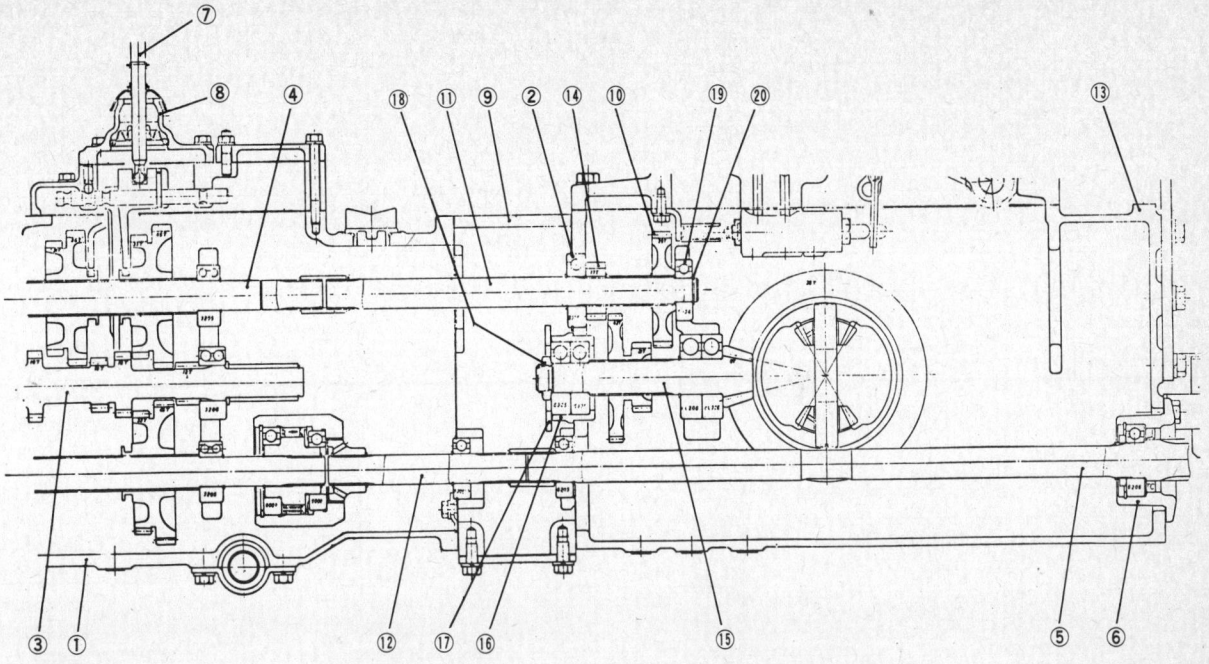

1. Clutch housing
2. Bearing
3. Mainshaft
4. Countershaft
5. PTO shaft
6. Bearing case
7. Gear shift lever
8. Gear shift cover
9. Gear case
10. 33 gear
11. Auxiliary change shaft
12. PTO transmission shaft
13. Transmission case
14. 17 gear
15. Spiral bevel pinion
16. Bearing case
17. Bearing cover
18. Nut
19. Bearing
20. Set ring

Transmission schematic

4. Pull off the gear case frontward while tapping with a mallet, for a knockpin is therein contained.
5. Remove the bolt from the propeller shaft case (2) and pull out the case frontward.
6. Pull off the bearing covers (5) frontward by tapping the countershafts (4), upper and lower, from the rear side.
7. Remove the snapring (6) from the countershaft bearing and tap out the countershaft from the front side.
8. Tap out the propeller shaft from the rear side.
9. Remove the stopring for the back gear shaft and tap out the shaft from the rear side.
10. Remove the six bolts from the bearing case (16) and from the bearing cover (17), and then the self-locking nut from the bevel pinion shaft (15).
11. Tap out the bevel pinion shaft rearward.
12. Remove the snapring (20) for the rear bearing (19) of 17 gear shaft (14), and the cap for the front bearing (21); tap out the shaft from the rear side.
13. Remove the bolts from the PTO shaft bearing case (23) and tap out the shaft rearward.

ASSEMBLY

1. Wash the parts in clean oil.
2. Replace broken gears and bearings.
3. Apply grease to O-rings and oil seals and place in position carefully so as not to scratch them.
4. Tighten the four bolts on the propeller shaft.

5. Tighten the self-locking nuts for the pinion shaft and the PTO shaft to 108-145 ft.lb.
6. Place the 33 gear (24) in the correct direction (its rear side has a higher boss).
7. Attach the propeller shaft case (2), with its spring fitting part down.
8. Attach the bearing cover (17), with its notch down.
9. Do not forget to place the collar between PTO shaft (22) bearing and oil seals.
10. Use needle bearings (17mm wide one at center and 20mm wide ones at front and rear) in 17-14 gear (25).
11. Use collars on both ends of 17-44 gear (25), with their grooved sides facing the gear.

DIFFERENTIAL

Differential Gear

OVERHAUL

1. Remove the hydraulic cylinder on the transmission case.
2. Remove the bolts (15) from the right and left shaft cases (14), take off the differential bearing cases (4) and (5), and then take off the differential from above.
3. Remove the bolts from the spiral bevel gears (6) and pull out the gears from the differential case.
4. Tap out the differential pinion shaft (7) from the side without key.

5. Turn the differential gear (1) by 90° and take it out through the window of the case.
6. Take out the differential gear (3).
7. Assembly is the reverse of disassembly.

Differential Lock

OVERHAUL

1. Draw out the spring pin (20) that connects the differential lock cam (8) and the differential lock pedal (9), and pull out the pedal while taking care not to scratch oil seals (21).
2. Remove the plug (12).
3. Place the spring (11) between the shift fork (10) and the mission case, then cam on the shift fork side.
4. Place the differential lock pedal, and

Inserting differential lock spring

655

Kubota

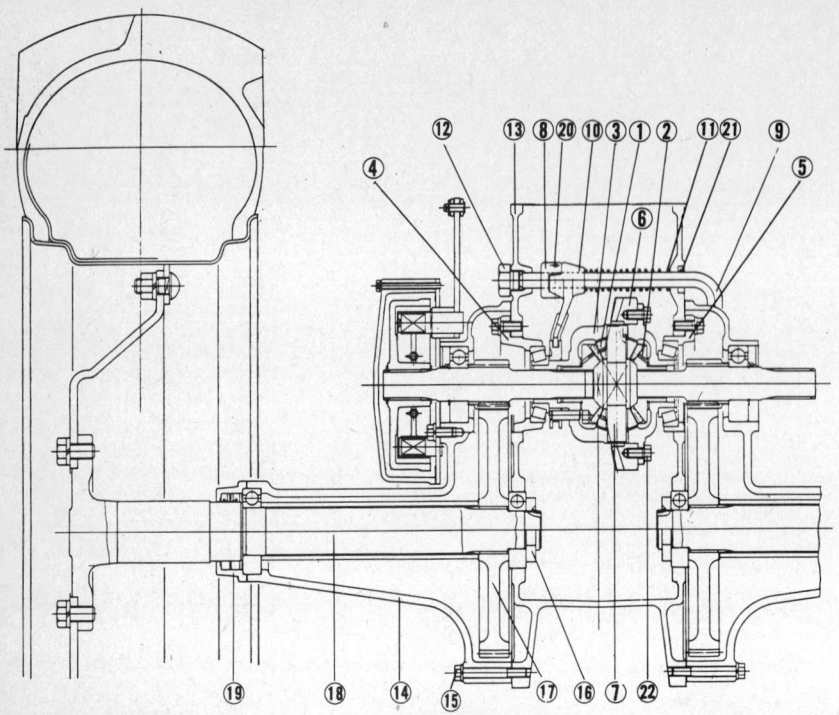

1. Differential pinion gear
2. Differential side gear
3. Differential case
4. Differential bearing case lefthand
5. Differential bearing case righthand
6. Spiral bevel gear
7. Differential pinion shaft
8. Differential lock cam
9. Differential lock pedal
10. Shift fork
11. Spring
12. Plug
13. Transmission case
14. Axle case
15. Bolt
16. Nut
17. 74 gear
18. Rear axle
19. Axle cover
20. Spring pin
21. Oil seal
22. Bevel gear bolt

Differential schematic

drive the spring pin into the differential lock cam to fix both the pedal and the cam.
5. Attach the plug.

ADJUSTMENT

Spiral Bevel Pinion and Gear

1. Set the number of right and left shims on the differential bearing case.
   a. Attach the differential bearing case without replacing the differential bearing case (3) shim.
   b. Select the shims that make a 0.04 in. thickness for the differential bearing case (4), and attach the differential gear.
   c. Turn the bevel gear by hand and adjust the number of shims so that it will not turn. (The rotation of the gear becomes heavier by adding shims and lighter by reducing them.)
   d. The number of shims thus obtained is the total number of the right and left shims.
2. Make the total thickness of the right shims on the differential bearing case equal to that of the left shims, and put the differential bearing case and the differential gear together.
3. Put together the spiral bevel pinions, and adjust the backlash to .007874-.008135 in. by changing the number of bearing case shims.
4. Apply oil-dissolved red lead thinly over the surfaces of the several pinions, and lightly rotate to check that over two thirds of the teeth surface comes into contact with the bevel gear.

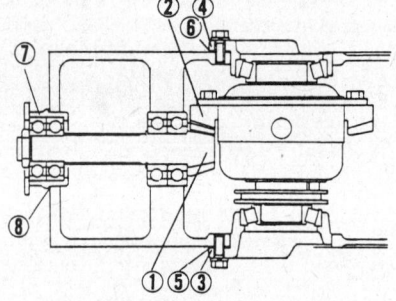

1. Spiral bevel gear pinion
2. Spiral bevel gear
3. Differential bearing case
4. Differential bearing case
5. Shim
6. Shim
7. Bearing case
8. Shim

Spiral bevel pinion adjustment

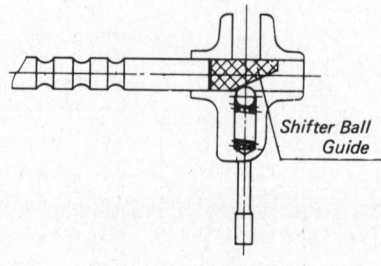

Inserting shift fork rod

Differential tooth contact patterns

Main Speed Change Gear and Power Take-Off Speed Change Gear

OVERHAUL

1. Separate the speed change cover (1) from the clutch housing.
2. Remove the speed change lever mounting from the speed change cover.
3. Draw out the shift fork spring pins.
4. Remove the setscrew (9) from the left side of the clutch housing, and tap out the PTO fork shaft (11) from the rear side.
5. Installation is the reverse of removal.

To insert a fork rod, use the shifter ball guide. As illustrated, insert the guide before shaft insertion and keep the ball to push in the fork rod; then tap out the guide by the fork rod.

When attaching the speed change cover, note that a placement of the cover from above the clutch housing will not enable the PTO speed change levers (5) and (6) to be installed on the pins (12) of the PTO speed change shift forks (7) and (8). The PTO speed change operation would, in that case, be impossible. So, separate the speed change cover from the speed change lever mounting (10), and lock from above to make sure that the PTO speed change levers are placed upon the pins. If it is hard to place them upon the pins, separate the PTO speed change levers from the speed change cover, and attach the cover to the clutch housing, then attach the PTO speed change levers one by one.

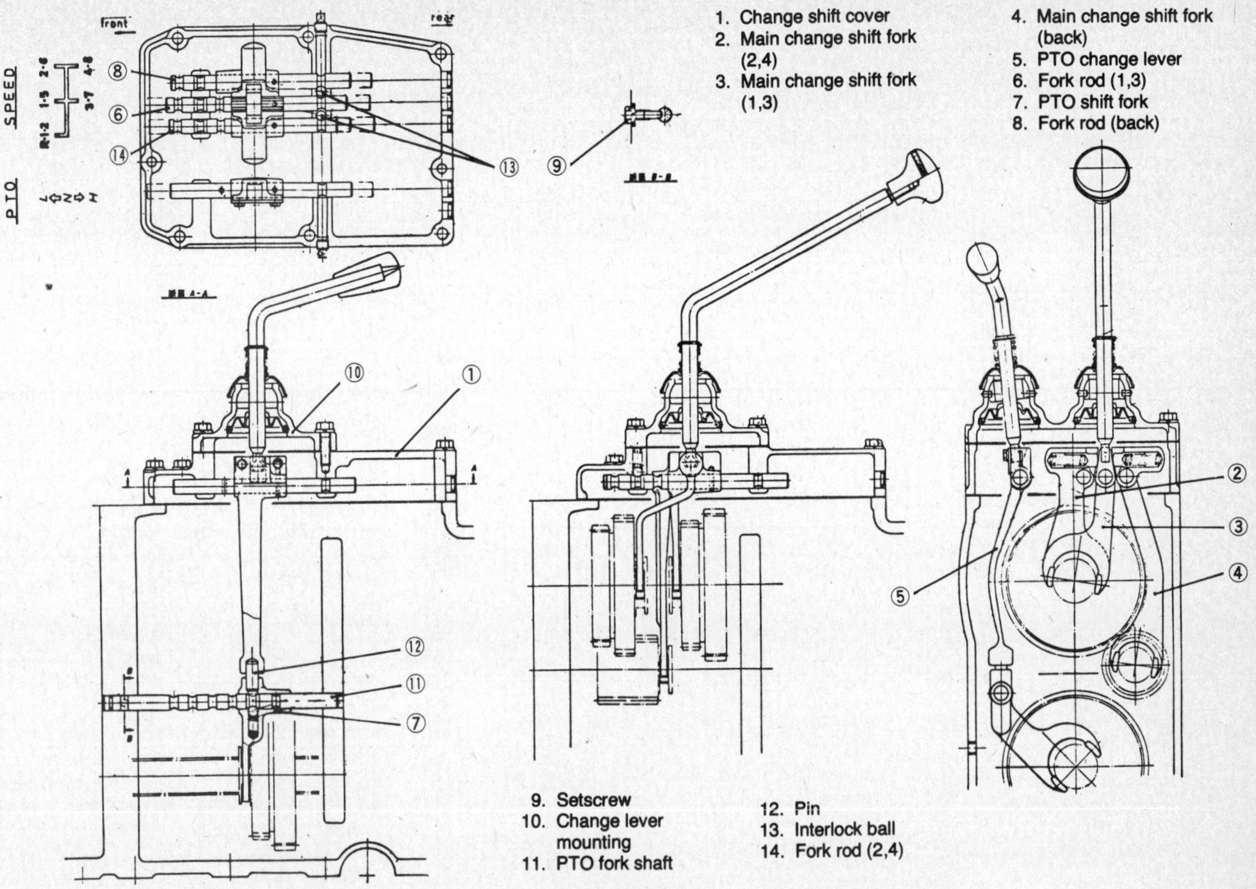

Kubota

1. Change shift cover
2. Main change shift fork (2,4)
3. Main change shift fork (1,3)
4. Main change shift fork (back)
5. PTO change lever
6. Fork rod (1,3)
7. PTO shift fork
8. Fork rod (back)
9. Setscrew
10. Change lever mounting
11. PTO fork shaft
12. Pin
13. Interlock ball
14. Fork rod (2,4)

Main speed change gear and PTO speed change gear components

HYDRAULIC SYSTEM

Hydraulic Cylinder

DISASSEMBLY

1. Remove the operator's seat and the tool box.
2. Remove the left rod of the 3-point hitch, then the delivery valve.
3. Remove the bolts and nuts from the hydraulic cylinder (1) and remove the entire hydraulic cylinder using a hoist.
4. Remove the feedback rod.
5. Remove the control valve (21).
6. Draw out the spring pin that connects the feedback arm (5) and the feedback lever shaft (2), and pull out the shaft.
7. Remove the lever guide (13).
8. Draw out the spring pin that connects the control (7) and the control lever shaft (6), and pull out the shaft outward.
9. Remove the six bolts from the cylinder cover (20), lower the lift arm (15) and push out the cover with the piston (19).
10. Tap out the cylinder liner (18) from the rear side using a mallet.
11. Remove the bolt from the left side of the lift arm and pull out the lift arm.
12. Pull out the hydraulic arm shaft (14) to the right.

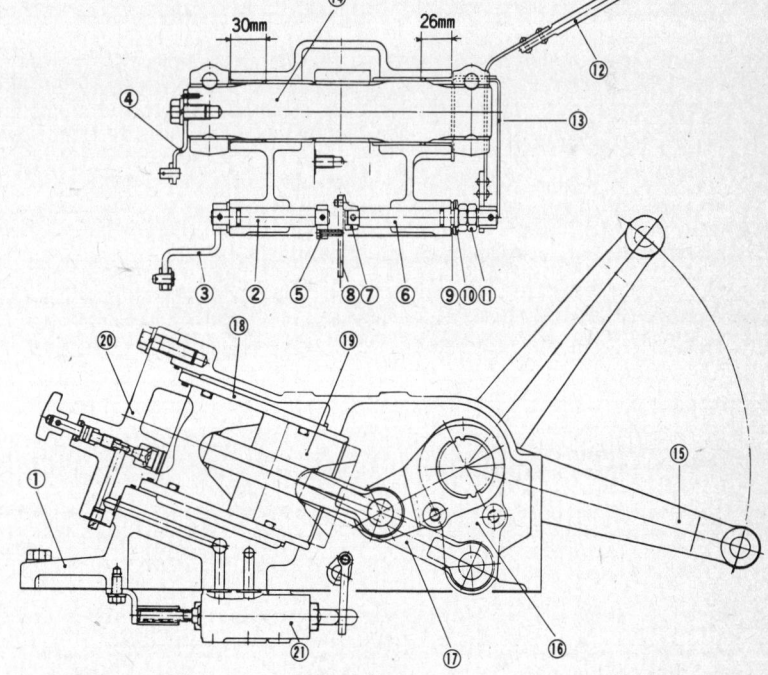

1. Oil pressure cylinder
2. Feedback lever shaft
3. Feedback lever—1
4. Feedback lever—2
5. Feedback arm
6. Control lever shaft
7. Control arm
8. Spool lever
9. Set plate
10. Plate spring
11. Nut
12. Control lever
13. Lever guide
14. Oil pressure arm shaft
15. Lift arm
16. Oil pressure arm
17. Rod
18. Cylinder liner
19. Piston
20. Cylinder cover
21. Control valve

Hydraulic cylinder components

657

Kubota

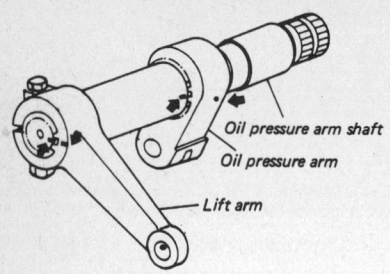

Aligning the assembly marks on the hydraulic arm shaft

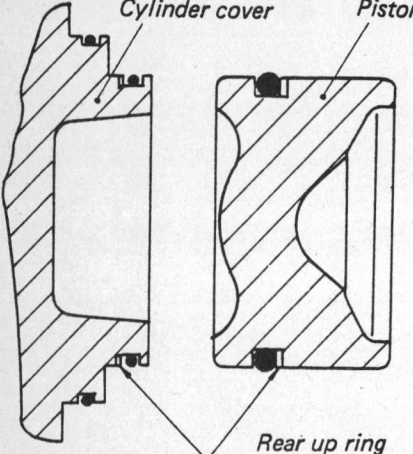

Backup ring installation

13. Remove the relief valve and the valve spacer.

---- **CAUTION** ----
Take care that dust does not enter the valve, piping, hydraulic cylinder and oil passages of cylinder cover.

ASSEMBLY

1. Wash parts thoroughly, and assemble using care no foreign substances get in anywhere.
2. Apply grease to O-rings and place in position with enough care to keep them free from damage.
3. At the time of renewing the hydraulic arm shaft bushing, drive in to specified dimension as shown in the schematic.
4. Align assembly marks on the hydraulic arm shaft, hydraulic arm, and lift arm.
5. Place the piston from the cylinder cover side.
6. Be sure the backup spring is placed in the right direction.

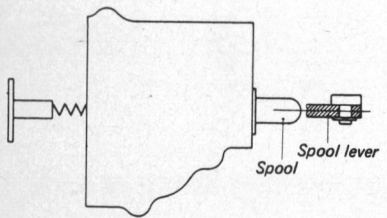

Divergence between spool and spool center

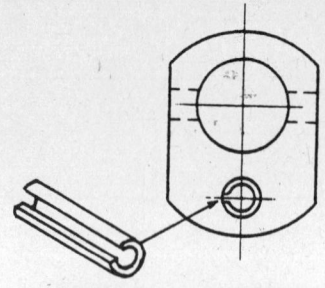

Spring pin installation

7. The edge gap of the spring pin of the feeback lever must be in the specified direction.
8. The divergence between the spool and spool centers should be below 0.04 in.
9. Adjust the position of the control lever (12) by the nut so that its tractive force at a position under the lever grip will be 11-25 ft.lb., and lock with locknut. The lever becomes heavy if the nut is tightened, and light if it is loosened.

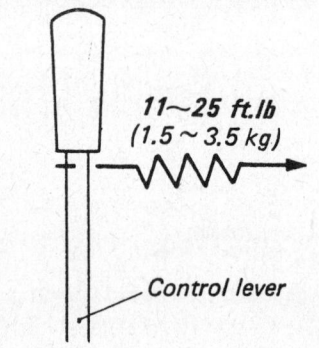

Adjustment of control lever tractive force

10. Confirm and place O-rings for control valve and valve spacer.
11. After reassembly and piping, pour transmission oil and start the engine. Move the control lever to the top position and adjust the length of the feedback rod that the relief valve may be immediately before operation when the lift arm has reached the top position.
(After the relief valve operates, turn the turnbuckle one revolution to make the rod shorter.)
Turn the turnbuckle in the direction for relief valve operation to make the rod longer; turn in the direction for relief valve non-operation to make it shorter.

Pump

DISASSEMBLY

1. Loosen the drilled bolt (1).
2. Remove the end cover (2) and O-ring (3).
3. Remove the seal element composed of two sealing rings (13) and two support rings (14). Each pair of the sealing rings (13) and support rings (14) can be removed as a unit.
4. Next, push the driveshaft (6) into the body, and lift up the bushing (4). Then take out the driveshaft (6), the gear (5) and the second bushing (7) with O-ring (8).

---- **CAUTION** ----
Do not use a hammer or the like to take out the bushings. If they are hard to remove, hit them lightly with something like wood.

---- **CAUTION** ----
Arrange the disassembled parts in the order disassembled. Use care that the shaft and the bushings are in their correct relative position. Take care not to mistake the idler side bushing for the driveshaft side bushing in reassembling the pump.
Refrain from removing the oil seal (10) if not damaged.
When removing the seal from inside of the casing, take care not to scratch the inside wall of the bushing.

REASSEMBLY

Reverse the disassembly procedure, paying special attention to the following:

1. Wash all parts clean with kerosene and be sure that dust will not stick to them.
2. Insert the bushing in such a way that the escape groove on the bushing will not tilt.
3. When parts are to be reused, be sure to replace them correctly.
4. After reassembly, check to see that the pump has been reassembled correctly. The pump ought to be turned lightly with a disc that is 4 in. in radius.

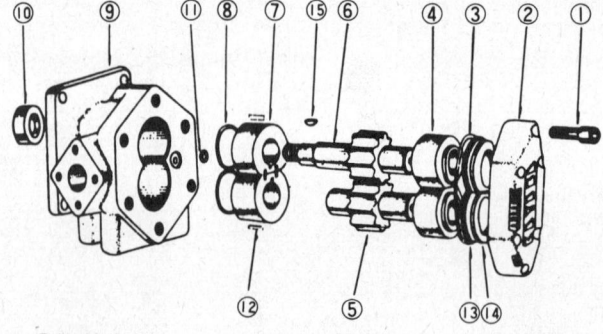

1. Drilled bolt
2. Cover
3. O-ring
4. Bushing, A, pair
5. Gear
6. Driveshaft, pair
7. Bushing, C, pair
8. O-ring
9. Body
10. Oil seal
11. O-ring
12. Spring pin
13. Sealing ring
14. Support ring
15. Key

Pump components

Kubota

Control Valve

DISASSEMBLY

1. Sleeve:
Remove the bolt (3) and push out the sleeve (2) from the side having no snap-ring. Two O-rings are inserted inside.
2. Unload valve:
Remove the plug (4). The spring (6) and the unload valve (5) will come out.
3. Seat:
Remove the plug (7). The spring (8) and the ball (9) will come out. Pull out the seat fitted with O-ring.

ASSEMBLY

1. Apply grease to O-rings and place them in position, being careful not to scratch them.
2. Secure the plugs (4) and (7) with a tightening torque of 43-51 ft.lb.
3. Check that the spool operates freely.

Relief Valve

PRESSURE ADJUSTMENT

1. Remove the plug of the relief valve and set a pressure gauge. (To prevent oil leakage from the tapped part, use packing material. Screw size: PS⅛.)
2. Set the engine to maximum speed and the control lever to "UP".
3. Turn the turnbuckle little by little to lengthen the feedback rod. Read the pressure gage when a beeping sound (relief valve operating sound) is heard.

   a. When the pressure is under 2000 psi: Increase the total thickness of shims placed inside of the plug of the relief valve. Determine the thickness of shims to be added by carrying out the following calculation: (Pressure increases by 2000 psi per 1mm total thickness of shims.)

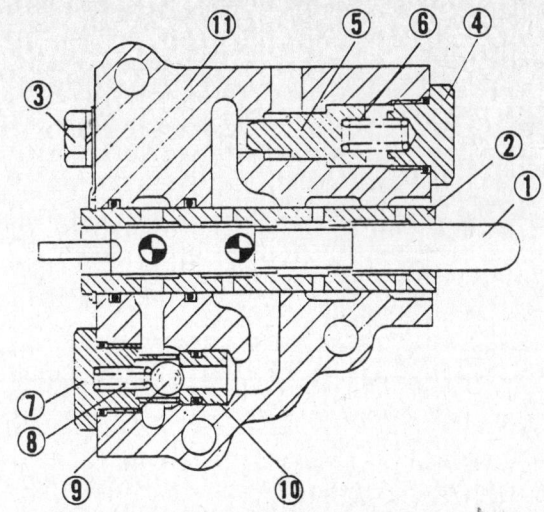

Shim thickness, mm = $\dfrac{\text{pressure gage reading psi} - 2000}{200}$

b. When the pressure is over 2000 psi: Reduce the total thickness of shims by determining the shim thickness as follows:

Shim thickness, mm = $\dfrac{\text{Pressure gage reading psi} + 2000}{200}$

NOTE: Use brass or copper sheet for shims.

Control valve components
1. Spool
2. Sleeve
3. Bolt
4. Plug
5. Unload valve
6. Spring
7. Plug
8. Spring
9. Ball
10. Sheet
11. Body

Relief valve components
1. Plug
2. Spring
3. Poppet
4. Sheet
5. Shim
6. Body

ENGINE

For complete engine service, see the L185, 245, 295 section.

Models L185, 245, 295

2-WHEEL DRIVE FRONT AXLE

REMOVAL AND INSTALLATION

1. Remove the set nuts from the inlet pipe holder.
2. Detach the battery support and the air cleaner.
3. Detach the front bumper.
4. Detach the radiator and the negative cable at the same time.
5. Remove the adjust nut.
6. Remove the center pin.
7. Remove the front axle bracket.
8. Installation is the reverse of removal.

Removing center pin

Kubota

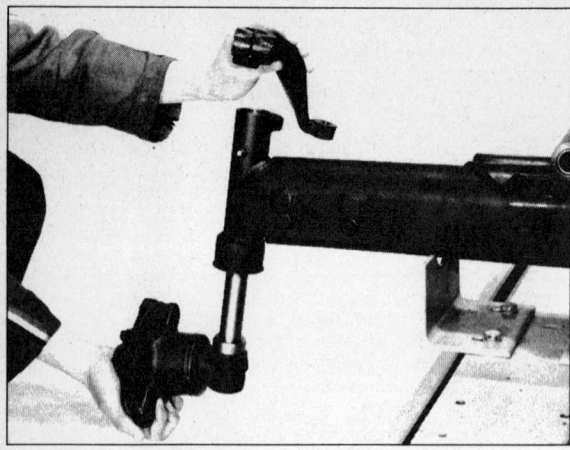
Removing knuckle arm and shaft

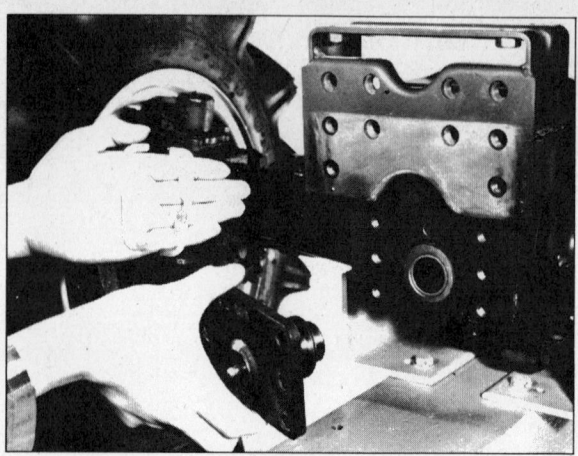
Removing front wheel drive axle center pin

Removing front axle bracket

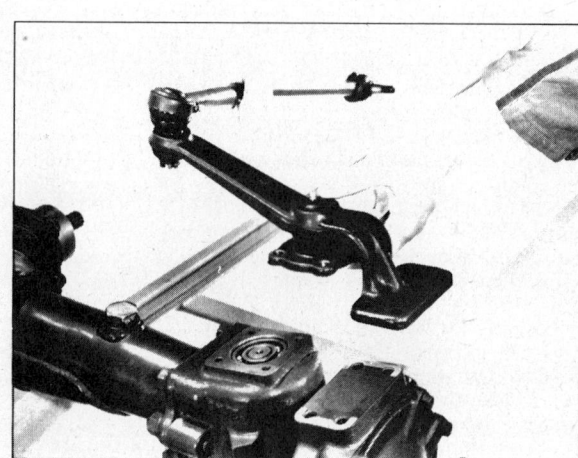
Removing front wheel drive axle drag link and knuckle pin

Removing front wheel drive axle knuckle pin

Removing front axle support from bevel gear case

Kubota

Removing differential gear

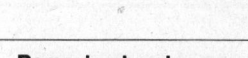

Removing bearing case

Knuckle Arm and Shaft

REMOVAL AND INSTALLATION

1. Remove the knuckle arm set bolt.
2. Remove the knuckle arm.
3. Remove the feather key.
4. Remove the O-ring.
5. Pull the knuckle shaft off.
6. Installation is the reverse of removal.

TOE-IN ADJUSTMENT

1. Adjust to the specified tire pressure (28.44 psi.).
2. Align the front wheel tires.
3. Measure the front and rear distances between the right and left front wheels, and obtain the difference.
4. Adjust by altering the length of the tie-rod.

4-WHEEL DRIVE FRONT AXLE

OVERHAUL

1. Remove the set bolts from the inlet pipe holder.
2. Dismantle the battery support and air cleaner as an assembly.
3. Remove the front bumper.
4. Dismantle the radiator.
5. Remove the center pin set bolts.
6. Remove the center pin.
7. Detach the front axle bracket by lifting its front up.
8. Remove the front wheels.
9. Remove the drag link end and knuckle pin as a set.
10. Remove the tie rod and knuckle arm as a set.
11. Detach the bevel gear case from the front axle support.
12. Detach the axle flange from the axle case.
13. Detach the bearing case from the axle case.
14. Tap the bevel gear shaft bottom off the axle case.
15. Remove the left front axle support.
16. Remove the set bolts from the differential bearing case.
17. Dismantle the differential bearing case and differential gear as a whole.

ADJUSTMENTS

Spiral Bevel Rotating Torque

1. Fit the spiral bevel pinion.
2. Measure the rotating torque by setting a torque wrench on the spiral bevel pinion nut.
3. Adjust by altering the torque on the spiral bevel pinion nut.

For L295DT, the correct rotating torque is obtained by tightening the spiral bevel pinion nut to 108.5-144.7 ft.lb.

Differential Gear Rotating Torque

1. Refit the spiral bevel pinion and the differential gear.
2. Set a torque wrench on the spiral bevel pinion nut and measure the rotating torque.

Adjust by increasing or reducing shims in the differential bearing case and differential gear case.

Shim thickness: 0.0039 in., 0.0079 in.

Spiral Bevel Pinion and Bevel Gear Backlash

1. Lock the bevel gear with a screwdriver. To gain access to the bevel gear, put the screwdriver through the drain plug hole on the differential gear case.
2. Set a lever-type indicator on the spiral bevel pinion.
3. Measure the backlash while turning the spiral bevel pinion by hand. Backlash should be 0.0079-0.0098 in.

For greater backlash, remove the shim from the differential gear case and add a shim of the same thickness to the differential bearing case.

For less backlash, remove the shim from

Measuring spiral bevel rotating torque

Measuring differential gear rotating torque

Kubota

Measuring the backlash between the spiral bevel pinion and the bevel gear

Measuring bevel gear in-case backlash

the diffferential bearing case and add a shim of the same thickness to the differential gear case.

Bevel Gear In-Case Backlash

1. Grip the differential york shaft in a vise.
2. Set a lever-type indicator on the bevel gear shaft.
3. Measure the backlash by turning the bevel gear shaft by hand.
4. Adjust by altering the shim on the front axle support. Backlash should be 0.0059-0.0118 in.

ASSEMBLING AXLE

Assembly is the reverse of disassembly.

STEERING

Steering Gear

REMOVAL AND DISASSEMBLY

1. Detach the Pitman arm from the sector shaft by using a Pitman arm puller.
2. Remove the side cover set bolts.
3. Remove the locknut on the center of the side cover and then detach the side cover by screwing the adjust screw in.
4. Tap the sector shaft off.
5. Remove the rear cover set bolts.
6. Pull the steering post off.

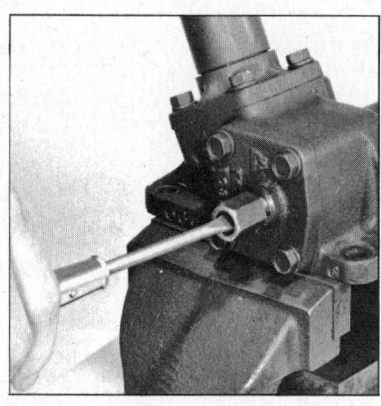

Steering gear adjustment

ASSEMBLY AND INSTALLATION

Installation is the reverse of removal. Note the following:
1. Place the sector shaft into the gear box and fully turn it to the right.
2. Place the ball nut assembly into the gear box and engage it with the sector gear.
3. Take care not to let the thrust bearing slip off the outer ring.
4. Line up the marks of the sector shaft and Pitman arm.
5. Tighten the nut to 86.8-115.8 ft.lb.

ADJUSTMENTS

1. Set a surface gauge on the bonnet and mark the steering wheel rim for free movement.
2. Measure the free movement of the rim with a rule. Play should be 0.78-2.00 in.
3. Loosen the adjust screw locknut on the steering gearbox.
4. Adjust by turning the adjust screw with a regular screwdriver.

Removing Pitman arm

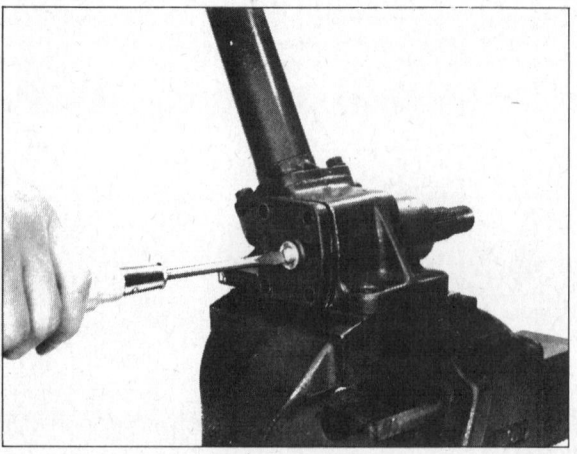

Removing side cover

Kubota

Sector shaft

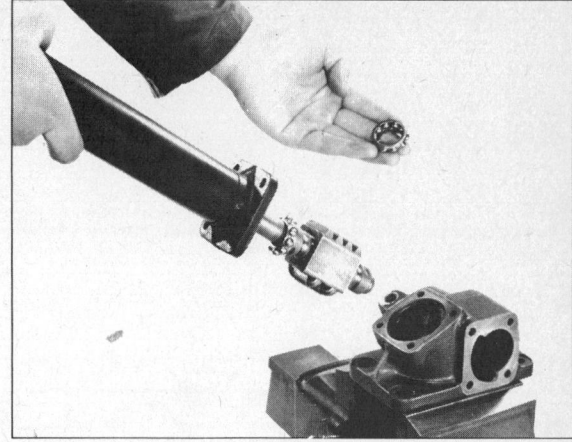

Removing the steering post

BRAKES

REMOVAL AND INSTALLATION

1. Remove the differential lock pedal bracket mounting bolts from the right axle case. Then remove the differential lock pedal.
2. Remove the axle case.
3. Disconnect the brake rod.
4. Detach the brake case.
5. When installing the brake case on the transmission case, fit the cam plates snugly onto the four projections of the differential bearing case. After installing, check to see that the brake functions well by moving the brake cam lever.
6. Place an O-ring snugly in the groove of the brake case.
7. Tighten first the bolts 1 and 2 of the largest pitch, then the rest in diagonal order.
8. Tighten the axle case mounting bolts and nuts to 57.1 to 66.5 ft.lb.

ADJUSTMENT

1. Step on the right and left brake pedals five times each with 33.0-66.0 lb.
2. Press the center of the pedal with 8.8-13.2 lb. and measure the free travel. Free travel should be 1.18-1.57 in. at the pedal end.
3. Adjust by altering the length of the brake rod.
4. Adjust the difference between the right and left brake pedals free travel to less than 0.1969 in.

CLUTCH

REMOVAL AND INSTALLATION

1. Separate the tractor.
2. Detach the pressure plate from the flywheel.
3. Remove the safety switch lever.
4. Remove the headed pin from the clutch rod.
5. Detach the clutch lever and fork.
6. Remove the hub return spring.
7. Draw the release hub off.
8. Installation is the reverse of removal.

TRANSMISSION

Clutch Housing Side

OVERHAUL

1. Detach the speed change base from speed change cover.
2. Remove the speed change cover.
3. Shift the PTO shift forks to neutral.
4. Shift the main speed change gear to F4 (F8)
5. Back off the reverse gear until it touches the 48T gear.
6. Shift the main speed change shift fork to F4 (F8).
7. Back off the reverse shift fork approximately 0.23262 in.
8. Shift the PTO shift forks to neutral.
9. Hold the three main shift forks with fingers to prevent the neutral piece from coming off.
10. Securely fit the three main shift forks into the gear shift fork grooves.
11. Fit the PTO levers onto the shift fork pins by slightly lifting the left side of the speed change cover.
12. Detach the safety switch lever.
13. Remove the head pin from the clutch rod.
14. Detach the clutch lever and fork.
15. Remove the hub return spring.
16. Draw out the release hub.
17. Detach the propeller shaft case.
18. Remove the two pieces of external circlips from the countershaft.
19. Tap the countershaft off the front of the bearing cover.
20. Remove the external circlip and pull the bearing off.
21. Remove the thrust collar.
22. Tap the countershaft off the rear of the gears.

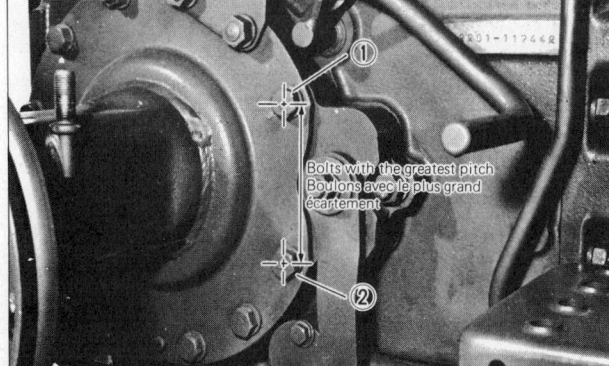

Brake unit

Clutch removal

663

Kubota

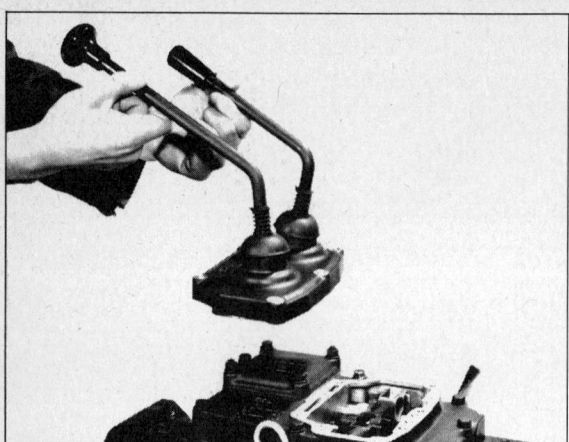
Removing speed change cover

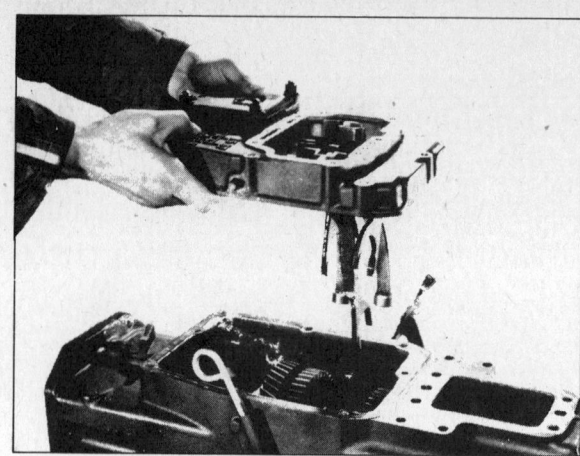
Removing speed change base

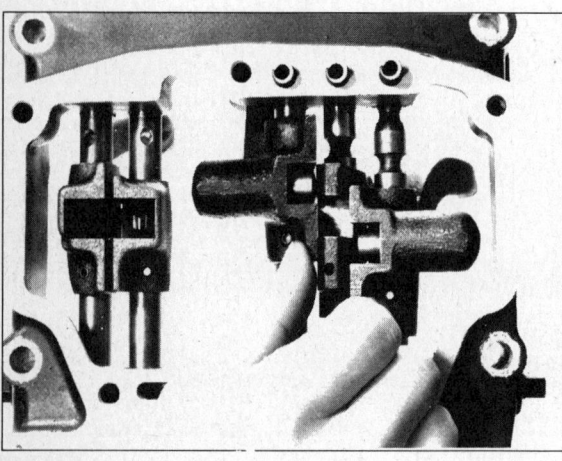
Shift fork removal

Removing clutch lever

Removing propellor shaft case

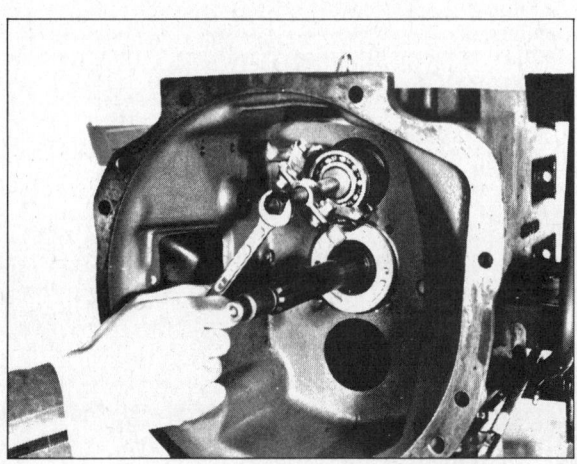
Removing main speed change countershaft bearing

Kubota

Removing main speed change countershaft

Removing mainshaft spacer

Removing mainshaft

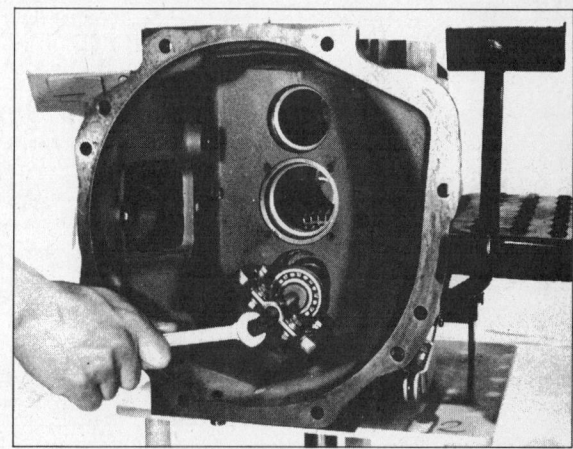
Removing PTO countershaft bearing cover

Removing PTO countershaft

Removing the 12T gear shaft

Kubota

Removing the spiral bevel pinion shaft nut

Removing differential bearing case

Removing differential gear

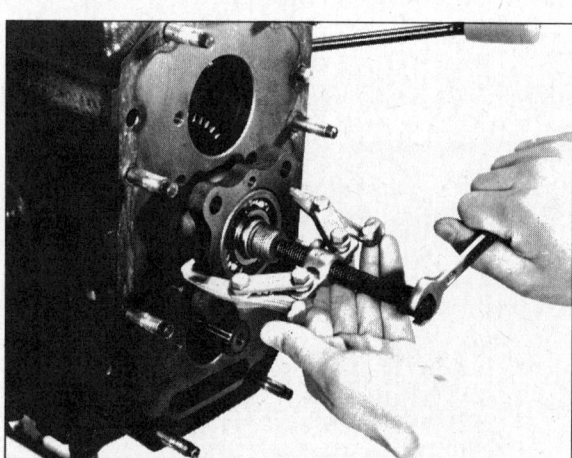
Removing spiral bevel pinion shaft bearing case

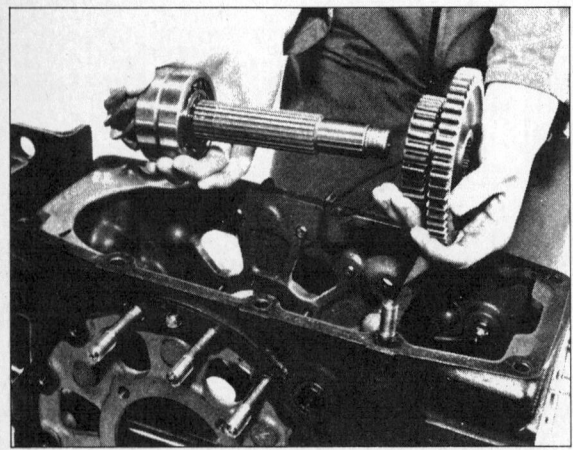
Removing spiral bevel pinion shaft

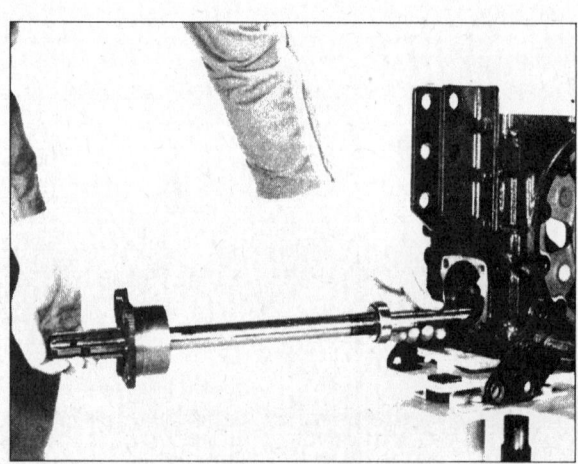
Removing PTO shaft

Kubota

23. Screw the two bolts into the spacer and pull it out.
24. Remove the external circlip.
25. Remove the collar.
26. Tap the mainshaft off the rear of the gears.
27. Remove the setscrew from the reverse shaft.
28. Draw out the reverse shaft and remove the gear.
29. Remove the fork shaft setscrew.
30. Tap the fork shaft off the front.
31. Remove the shift fork.
32. Remove two pieces of external circlip from the countershaft.
33. Tap the countershaft off the front of the bearing cover.
34. Remove the external circlip and then draw out the bearing.
35. Detach the thrust collar.
36. Tap the countershaft off the rear of the gears.
37. Installation is the reverse of removal. Note the following:
Face the grooves on the thrust collar toward the gear.
Face the outer raceway of the double-row bearing (3206) toward the differential gear.

Transmission Case Side

DISASSEMBLY

1. Remove the bearing retainer.
2. Shift the 12T gearshaft forward.
3. Remove the external circlip.
4. Pull the bearing, collar and 31T gear off the gearshaft.
5. Draw out the 12T gearshaft.
6. Lock the differential gear with a brass rod.
7. Unlock the nut and remove.
8. Remove the collar.
9. Remove the spring pin.
10. Pull the differential lock cam off.
11. Detach the shift fork.
12. Remove the differential bearing case mounting bolts.
13. Detach the case by screwing two M8 bolts in.

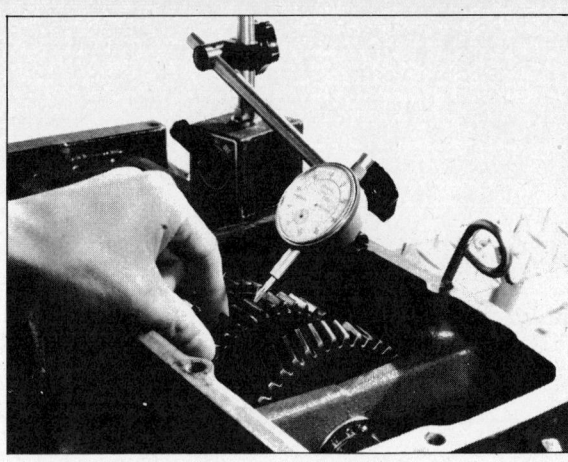

Measuring backlash

14. Take the differential gear out of the transmission case.
15. Remove the fork shaft setscrew.
16. Remove the fork shaft snappin.
17. Pull the fork shaft off the rear.
18. Remove the shift fork.
19. Remove the bearing cover.
20. Remove the bearing case mounting bolts.
21. Pull out the bearing case with a gear puller.
22. Pull the spiral bevel pinion shaft off the rear.
23. Pull out the 29–48T gears.
24. Remove the bearing case mounting bolts.
25. Pull the PTO shaft and bearing case off the rear.

ADJUSTMENTS

Gear Backlash

1. Set a dial gauge on one of the tooth faces.
2. Clamp the mating gear.
3. Measure backlash by turning the gear to be measured. Backlash should be 0.0039-0.0079 in.
4. If the reading exceeds the allowable limit (0.0157 in.), replace the gear.

Gear-to-Spline

1. Clamp the gear in a vise.
2. Set a lever-type indicator on the shaft.
3. Measure the clearance by turning the shaft. Clearance should be 0.0012-0.0039 in.
4. If the reading on the dial exceeds the allowable limits, replace the gear and spline.

Shift Fork-to-Gear

1. Place the shift fork in the shift gear groove and measure the clearance with a feeler gauge. Clearance should be 0.0059-0.0157 in.
2. If the measurement exceeds the allowable limit (0.0236) replace the shift fork.

ASSEMBLY

Assembly is the reverse of disassembly. Note the following points:
1. Drive the spring pin into the shift fork hole with its slot facing the direction of the force.
2. Replace the self-locking nut on the spiral bevel pinion.
3. Tighten the nut to 108.5-144.7 ft.lb.
4. Face the 31T gear chamfer toward the front.

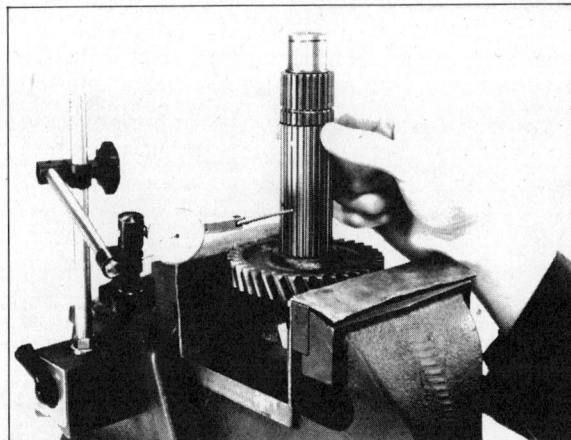

Measuring gear-to-spline clearance

Measuring shift fork-to-gear clearance

Kubota

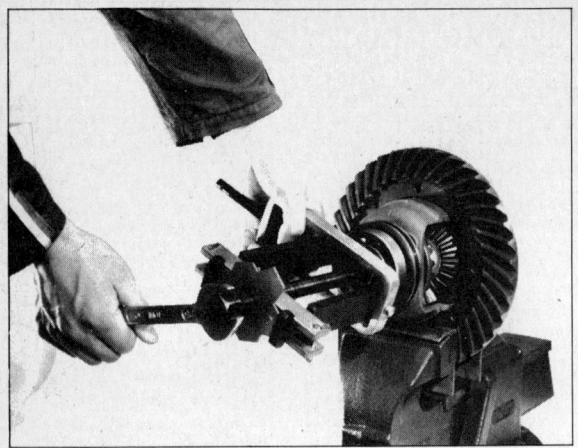

Pulling the 2-pinion differential gear bearing

Removing the 2-pinion differential spiral bevel gear

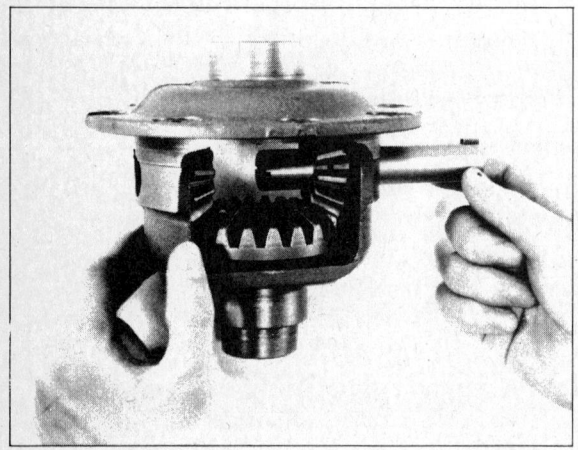

Removing the differential pinion gear and side gear

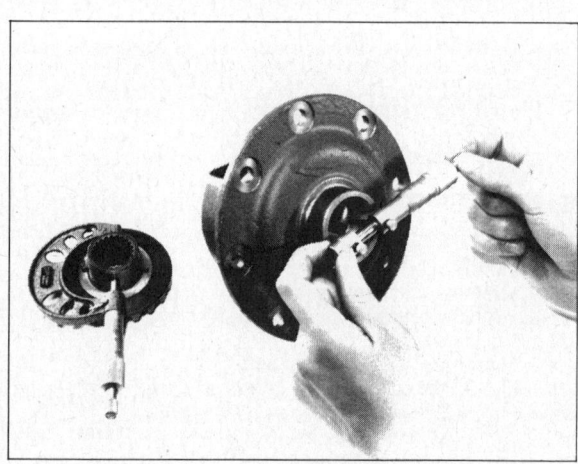

Measuring the case-to-side gear clearance

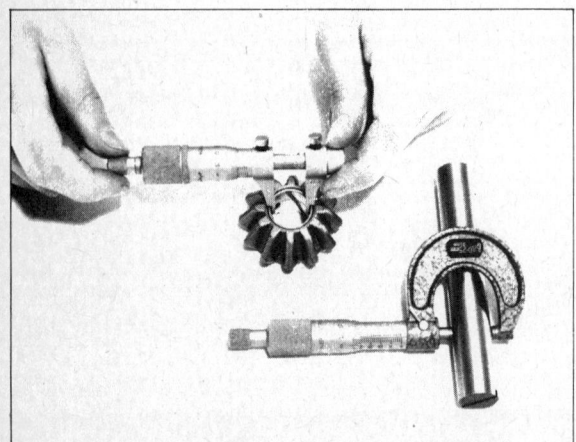

Removing pinion shaft-to-gear clearance

Measuring pinion-to-side gear clearance

Kubota

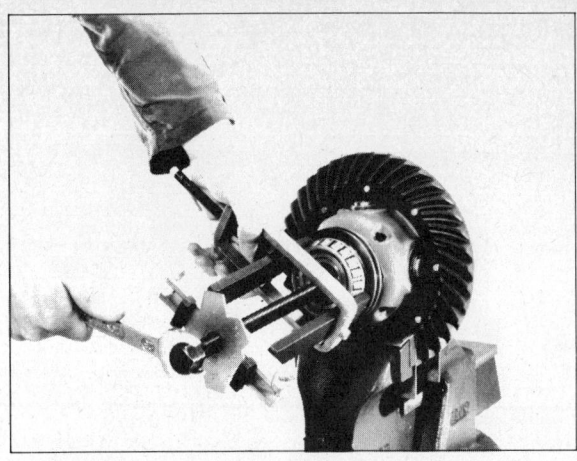
Removing the 4-pinion differential gear bearings

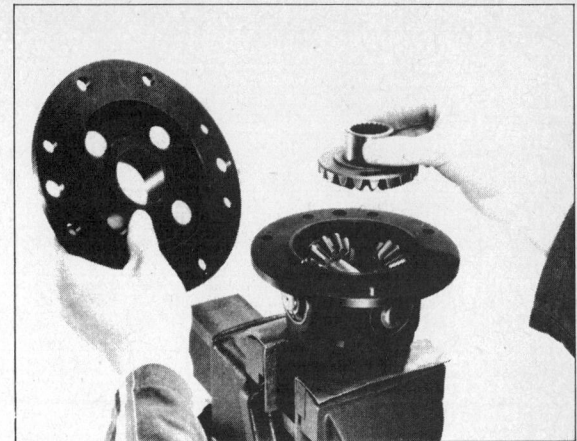
Removing differential case cover

DIFFERENTIAL

2-Pinion Differential Gear

DIASSEMBLY AND ASSEMBLY

1. Detach the bearing from each of the right and left differential cases.
2. Detach the differential lock shifter.
3. Remove bolts.
4. Detach the spiral bevel gear.
5. Push the differential pinion shaft out the key side.
6. Detach the differential pinion gear.
7. Detach the differential side gear.
8. Assembly is the reverse of disassembly. Torque the spiral bevel gear bolts to 45-59 ft.lb.

ADJUSTMENTS

Case-to-Side Gear
1. Measure the diameter of the differential side gear bushing with an outside micrometer.
2. Measure the inside diameter of the differential case bearing with an inside micrometer. Then take the difference for the clearance. Clearance should be 0.0020-0.0059 in.
3. If the clearance exceeds the allowable limit (0.0157 in.) replace.

Pinion Shaft-to-Gear
1. Measure the pinion shaft diameter with an outside micrometer.
2. Measure the inside diameter of the pinion gear with an inside micrometer. Then take the difference for the clearance. Clearance should be 0.0008-0.0024 in.
3. If the clearance exceeds the allowable limit (0.0118 in.) replace.

Pinion-to-Side Gear
1. Clamp the differential case in a vise.
2. Push the differential pinion and side gear toward the differential case.
3. Set a lever-type indicator on one of the tooth faces of the differential side gear.
4. Clamp the mating differential pinion gear.
5. Measure the backlash by turning the gear to be measured. Backlash should be 0.0059-0.0118 in.

Differential Gear Rotating Torque
Grip the spiral bevel pinion nut with a torque wrench and measure the rotating torque. Torque should be 0.3-0.7 ft.lb. (4-8 in.lb.).

Spiral Bevel Pinion-to-Bevel Gear Backlash
1. Clamp the spiral bevel pinion.
2. Set a dial gauge on one of the tooth faces of the bevel gear.
3. Measure the backlash by turning the bevel gear by hand. Backlash should be 0.0079-0.0098 in.

Tooth Contact
1. Place the differential gear under no load (remove the axle case).
2. Divide the bevel gear circumference into three equal parts and mark a few tooth faces around each of the three points with red lead.
3. Turn the pinion shaft while lightly braking the bevel gear circumference with a wood block.
4. Determine the amount of tooth contact.

4-Pinion Differential Case

DISASSEMBLY

1. Detach the bearings from the right and left differential cases.
2. Detach the differential lock shifter.
3. Remove the bolts.
4. Detach the spiral bevel gears.

Removing circlip and set collar

Removing the differential pinion bushing

Kubota

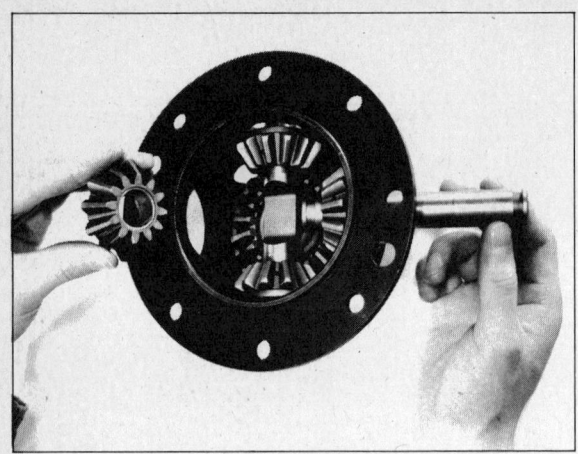
Removing differential pinion gear

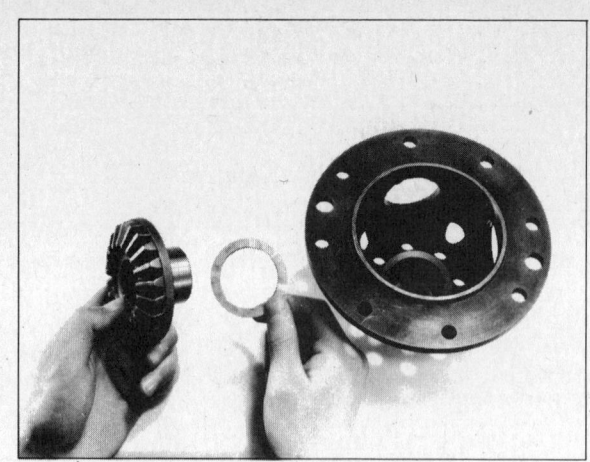
Removing differential side gear

Measuring pinion-to-side gear backlash

Removing axle case

Removing drive gear from case

Measuring rear wheel drive gear backlash

Kubota

5. Pull off the straight pin.
6. Remove the differential case cover.
7. Remove the differential side gear (right) and washer.
8. Remove the external circlip.
9. Remove the set collar.
10. Drive two M6 bolts into the bushing and then pull it out.
11. Remove the key.
12. Remove the differential pinion washer.
13. Slide the differential pinion shaft off the differential pinion gear.
14. Remove the differential side gear (left).
15. Remove the differential side washer.

ASSEMBLY

Assembly is the reverse of disassembly. Note the following:
1. Tighten the spiral bevel gear bolts to 44.8-52.1 ft.lb.
2. Fit a lock plate over the knock pin hole.

ADJUSTMENTS

Pinion-to-Side Gear Backlash

1. Put two pieces of soft lead on the differential side gear.
2. Turn the differential side gear with a regular screwdriver so that the lead is compressed.
3. Pick the lead pieces up and measure their thickness with a micrometer.
4. Backlash (in in.):
The sum of two lead pieces thicknesses
= $\dfrac{\text{(in in.)}}{2}$
Backlash should be 0.0059-0.0079 in.
Use a differential side gear washer and a set collar.

Washer thickness: 0.0591 in.
                              0.0630 in.
Set collar thickness: 0.1929 in.
                                0.1969 in.
                                0.2028 in.

Removing pump drive gear

REAR AXLE

REMOVAL AND INSTALLATION

1. Remove the axle case mounting bolts.
2. Tap the differential gear shaft off the axle case with a copper hammer.
3. Clamp the rear axle stud bolt in a vise. Then unfasten the self-locking nut and remove the bearing.
4. Remove the shims and collar.
5. Remove the differential gear shaft and rear wheel drive gear as an assembly.
6. Remove the shims and collar.
7. Tap out the rear axle.
8. Remove the bearing and oil seal.
9. Installation is the reverse of removal. Torque the drive gear nuts to 108-144 ft.lb. on 185 and 245 models; 180-217 ft.lb. on 295 models.

ADJUSTMENT

1. Set a lever-type indicator on the differential gear shaft.
2. Clamp the rear wheel drive gear.
3. Measure the backlash by turning the differential gear shaft. Backlash should be 0.0059 to 0.0079 in.
4. If the measurement exceeds the allowable limit (0.0197 in.) replace the gear.

HYDRAULIC SYSTEM

Pump

DISASSEMBLY

1. Clamp the gear pump body in a vise.
2. Straighten the lockwasher.
3. Remove the drive gear.
4. Remove the Woodruff key.
5. Remove the end cover.
6. Push the driveshaft in the direction of the end cover.
7. Remove the seal ring and support ring together.
8. Remove the "A" bushings together.
9. Pull out the driveshaft.

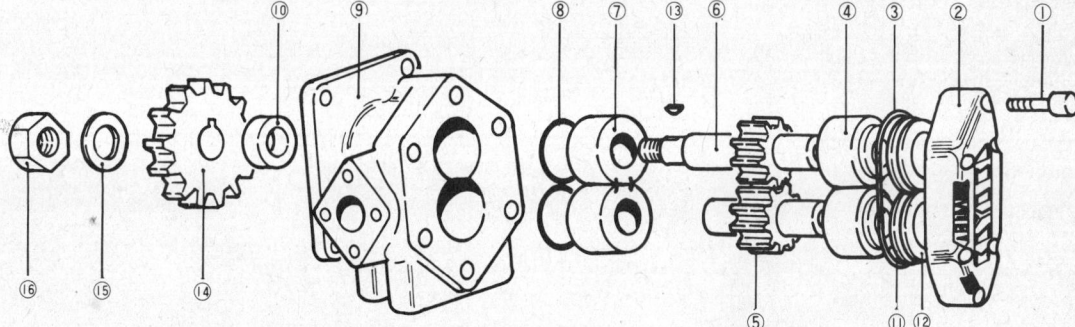

1. Bolt
2. End cover
3. O-ring
4. "A" bushing (pair)
5. Gear (idler)
6. Driveshaft (coupled with gear)
7. "C" bushing
8. O-ring
9. Body
10. Oil seal
11. Seal ring
12. Support ring
13. Woodruff key
14. 30T drive gear
15. Lockwasher
16. Nut

Hydraulic pump components

Kubota

Removing the "A" bushings

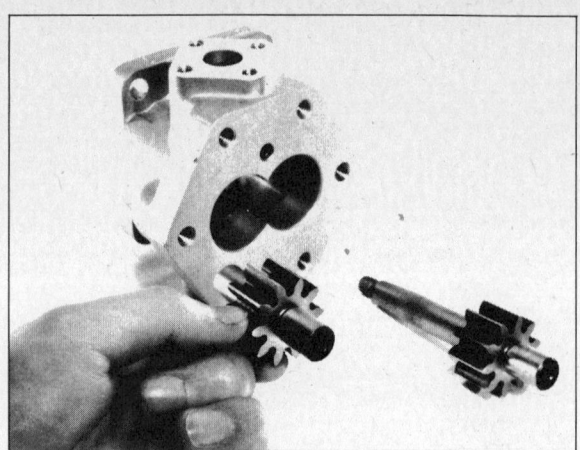
Removing pump gears

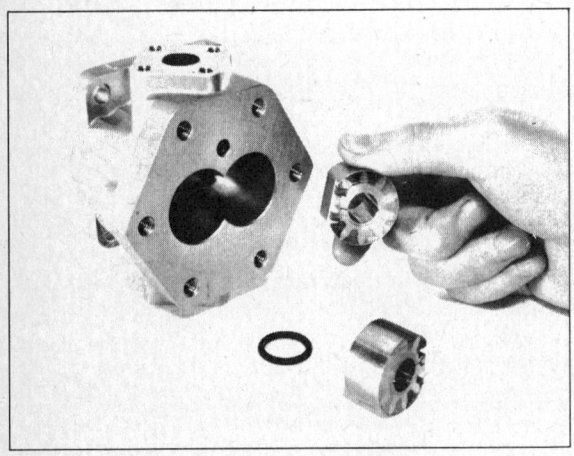
Removing the "C" bushings

Measuring gear tooth-to-body clearance

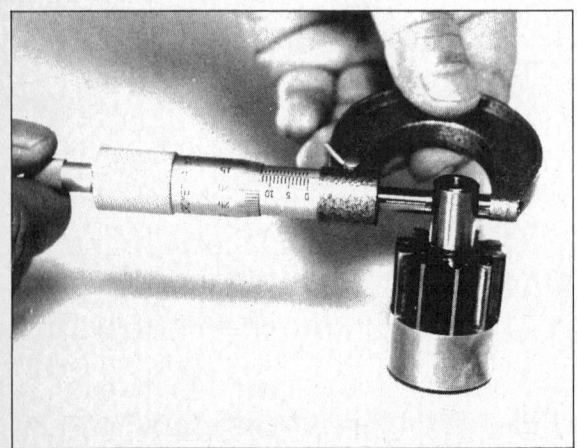
Measuring gear shaft wear

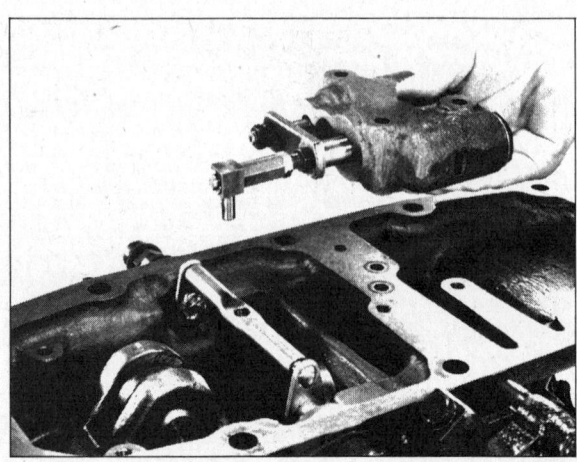
Removing the position control valve

Kubota

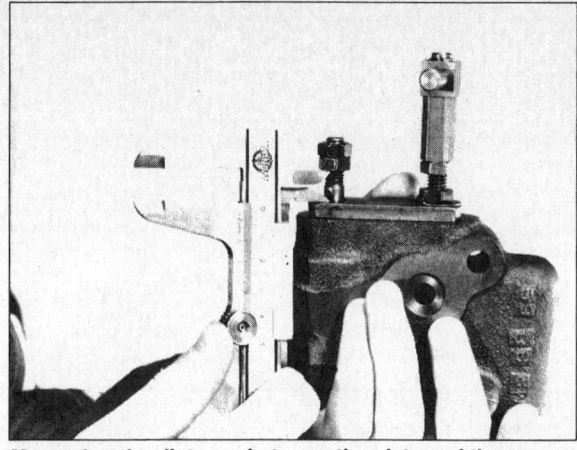

Measuring the distance between the plate and the poppet valve locknut

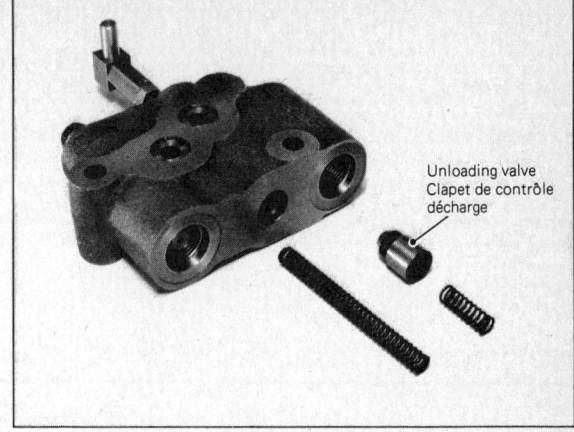

Removing the unloading valve and spring

10. Pull out the idle gear.
11. Knock the body on something soft until the two "C" bushings come out together.
12. Remove the O-rings.
13. Remove the oil seal.

INSPECTION

1. Insert the "C" bushings into the body.
2. Set the driveshaft and the idler in the body.
3. Measure the gap between the gear tooth crests and the inside surface of the body. The gap should be 0.0020 in.
4. Measure the shaft diameter with a micrometer. Shaft diameter should be 0.469 in.
5. Measure the inside diameter of the bushing with a micrometer.
6. Calculate the difference between the inside diameter of the bushing and the shaft diameter. The allowable limit is 0.0070 in.
7. Measure the length of the bushings with a micrometer. The bushing length is 0.7350 in.

ASSEMBLY

Assembly is the reverse of disassembly. Note the following points:
Be sure not to mistake the bushings on the driveshaft side for those on the idle gear side and not to mistake the direction of the idle gear. Refit them in the directions seen in the accompanying illustration.

Position Control Valve

DISASSEMBLY

1. Remove the bolts. Remove the control valve.
2. Bring the plate close to the body.
3. Measure the distance between the plate and the poppet valve locknut and write it down.
4. Remove the poppet valve seat plug.
5. Remove the unloading valve plug.
6. Remove the plate.
7. Remove the unloading valve spring.
8. Remove the unloading valve.
9. Remove the spool return spring.
10. Remove the poppet valve locknuts.
11. Draw out the spool.
12. Push the poppet valve to the seat plug side and pull it out.
13. Remove the seat.
14. Remove the check valve.
15. Remove the check valve spring.

ASSEMBLY

Assembly is the reverse of disassembly. Note the following points:
1. Be sure to stake the seat head with a punch.

2. Tighten the check valve nuts to 36.2 to 43.4 ft.lb.
3. Tighten the spool nuts to 13.0 to 15.9 ft.lb.
4. Stake the locknut with a punch so that it will not loosen.
5. Tighten the plug nut to 51.6 to 65.1 ft.lb.

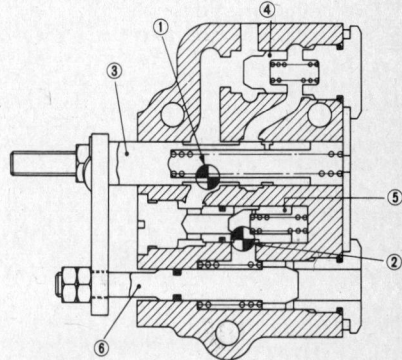

1. Gear pump port
2. Hydraulic cylinder port
3. Spool
4. Unloading valve
5. Check valve
6. Poppet valve

Position control valve schematic

Draft Control Valve

DISASSEMBLY

1. Disconnect the draft control rod from the top link holder.

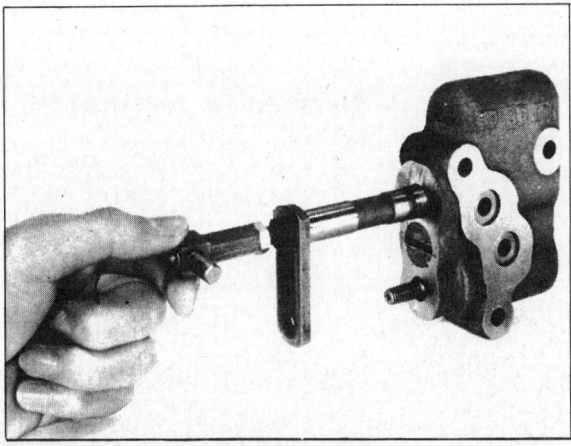

Removing the spool

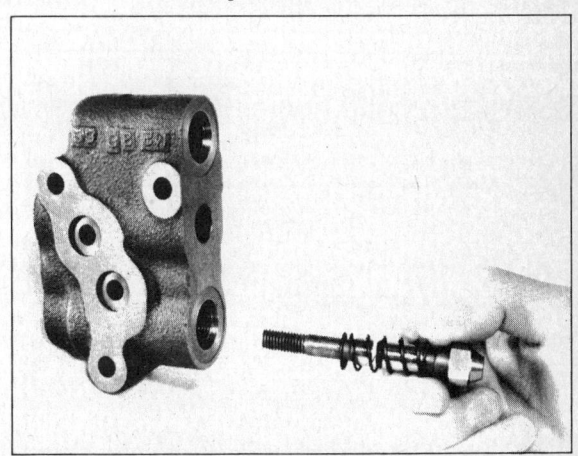

Removing the poppet valve

Kubota

Removing the check valve

Removing the draft control valve

2. Remove the draft control lever and the lever guide.
3. Remove the differential lock pedal and the pedal bracket.
4. Remove the control valve.
5. Remove the plug.
6. Remove the spool return spring.
7. Remove the bolts.
8. Separate the cap from the body.
9. Pull out the spool from the body.

ASSEMBLY

Assembly is the reverse of disassembly. Note the following point:
Check to see that the link and the arm are in the correct position.

Relief Valve

ADJUSTMENT

1. Remove the hydraulic cylinder cover plug, and connect a pressure gauge. (Prevent oil from leaking at the screw by applying seal tape.)
2. For L185 and L245
Remove the lever guide stopper, start and accelerate the engine to the maximum speed, and set the control lever in the up position. Read the pressure gauge while the relief valve is functioning, or while the buzzing sound can be heard.

3. For L295
Start and accelerate the engine to the maximum speed. Set the draft control lever at the up position, and the position control lever at the oil pressure take-off position, and read the pressure gauge.

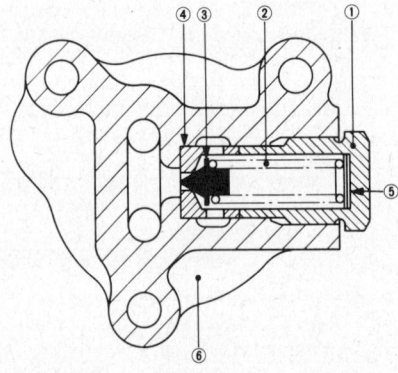

1. Plug 3. Poppet 5. Shim
2. Spring 4. Seat 6. Valve body

Relief valve schematic

4. Relief valve set pressure of model L185, L245 and L295 is 1991 to 2062 psi.
5. If the pressure is 1991 psi or less:
Remove the plug and add a shim (addition of a shim 0.0394 in. thick increases the pressure by approximately 199 psi).

6. If the pressure is 2062 psi or more:
Adjust by removing a shim. If the hydraulic control system is operated at more than 2062 psi relief valve set pressure, it may damage the gear pump.
7. Tighten the plug to 25.3 to 32.5 ft.lb.

Linkage

ADJUSTMENT

Position Control

1. Start the engine and run at low speed.
2. Set the control lever at the up position. (In case of the model L295, set the two levers at the up position.)
3. Shorten the feedback rod, and actuate the relief valve. Return the nut by two turns, and lock it there. (In case of the model L185 with turnbuckle, turn it by one turn.)
4. Set the control lever at the down position. Check to see that the relief valve does not function when the control lever is set at the up position.

Draft Control

1. Start and accelerate the engine to the maxium speed.
2. Apply a load of 15.4 to 22.0 lb. at the end of the lower link.
3. Fit a test bar to the top link holder,

Removing the spool return spring

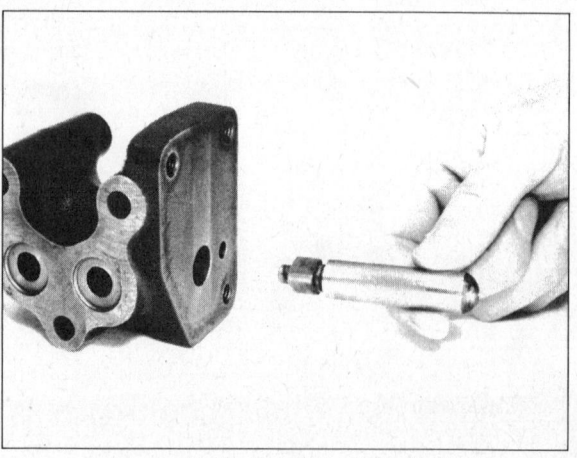
Removing the spool from the body

and push the top link holder all the way forward.

4. Set the position control lever at the up position.

5. Gradually move the draft control lever from the down position to the up position. Adjust the draft control rod so that the implement starts to lift up the moment the lever reaches the "1" on the lever guide.

Hydraulic Cylinder

DISASSEMBLY

1. Remove bolts and detach the control valve.
2. Remove bolts and nuts.
3. Detach the valve spacer.
4. Remove bolts.
5. Detach the cylinder and cylinder cover by tapping the cover with a plastic hammer.
6. Push the piston with a hammer handle to draw it out.
7. Remove the spring pin and detach the adjusting grip.
8. Remove the oil seal.
9. Remove the internal circlip.
10. Unscrew the adjusting screw by hand.
11. Remove the internal circlip.
12. Remove the adjust collar.
13. Remove the control link from the hydraulic arm shaft.
14. Remove lift arm mounting bolts.
15. Detach the lift arm.
16. Tap off the hydraulic arm shaft and the right lift arm as one assembly.
17. Remove the collar and the O-ring from the hydraulic arm shaft bearing.
18. Remove the set pin bolt.
19. Remove the set pin.
20. Disconnect the hydraulic rod.

INSPECTION

1. Measure the cylinder bore with a cylinder gauge.
2. Check to see if there are no scratches on the inside surface of the cylinder.
3. Replace if the measurement exceeds the allowable limit.

Models	Reference value	Allowable limit
L185	2.5591~2.5610 inches	+0.0059 inch
L245	2.9528~2.9547 inches	
L295	3.3465~3.3484 inches	

Cylinder wear dimensions

4. Measure the hydraulic rod (set pin hole) with an inside micrometer.
5. Measure the set pin outside diameter with an outside micrometer. Figure out the clearance between the two. Clearance should be 0.0008-0.0047 in.
6. If the measurement exceeds the allowable limit (0.0157 in.) replace.
7. Measure the hydraulic arm shaft outside diameter with an outside micrometer.
8. Measure the bushing inside diameter with an inside micrometer and figure out

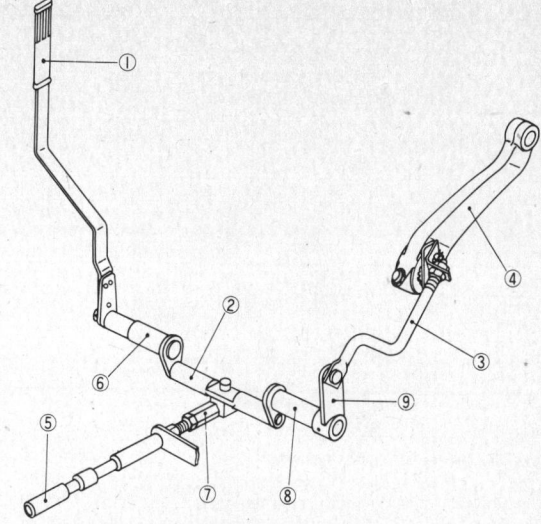

1. Position control lever
2. Spool drive lever
3. Feedback rod
4. Lift arm
5. Spool
6. Control arm
7. Spool joint
8. Lever shaft
9. Feedback lever

Linkage

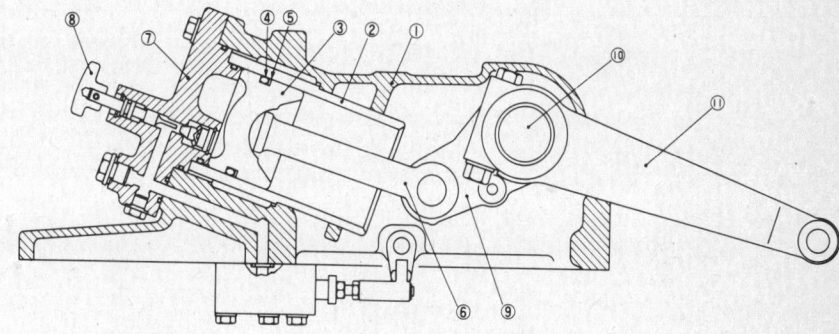

1. Hydraulic cylinder frame
2. Hydraulic cylinder liner
3. Hydraulic piston
4. O-ring
5. Backup ring
6. Hydraulic rod
7. Hydraulic cylinder cover, front
8. Grip
9. Hydraulic arm
10. Hydraulic arm shaft
11. Lift arm

Hydraulic cylinder schematic

Removing the relief valve

Kubota

Removing the hydraulic cylinder

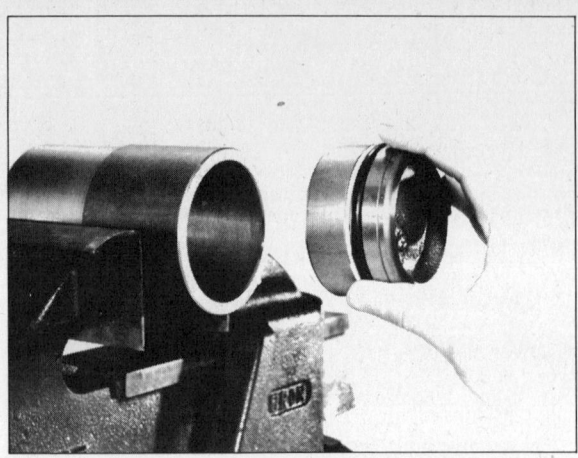
Separating the cylinder liner from the piston

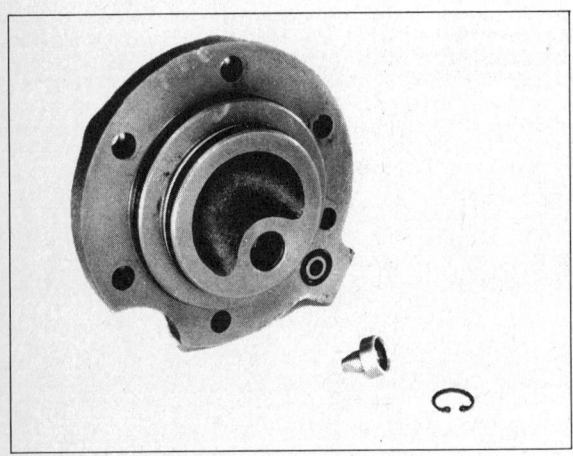
Removing the adjusting collar

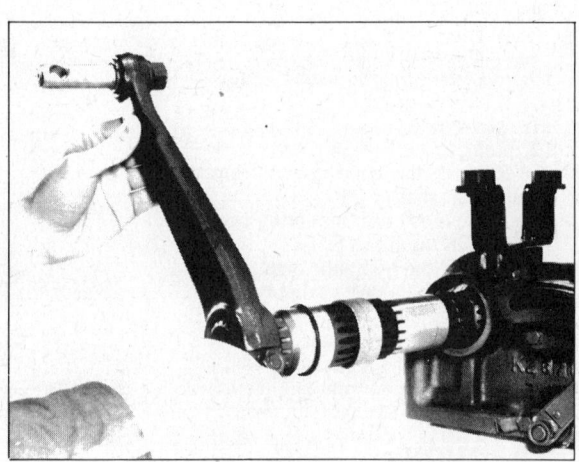
Removing the hydraulic arm shaft

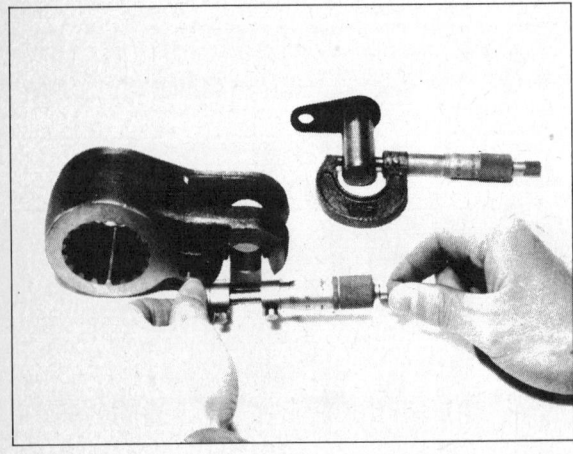
Measuring the hydraulic rod-to-set pin clearance

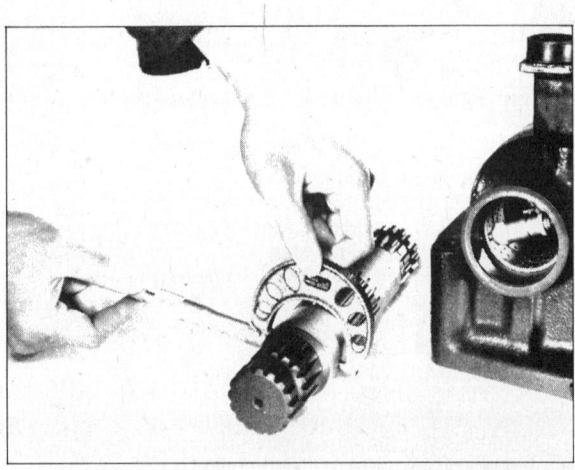
Measuring the arm shaft-to-bushing clearance

the clearance between the two. Clearance should be 0.0016-0.0037 in.

9. If the measurement exceeds the allowable limit (0.0197 in.) replace.

ASSEMBLY

Assembly is the reverse of disassembly. Note the following:

1. Grease the set pin and the tip of the hydraulic rod.
2. Grease the right and left bushings and the O-rings.
3. Line up the marks on the arm shaft, hydraulic arm, and lift arm.
4. Line up the marks on the arm shaft and lift arm.
5. To refit the left lift arm, bring it close to the hydraulic cylinder frame.
6. Insert the hydraulic piston from the cylinder cover side.

ENGINE

Cylinder Head

REMOVAL DISASSEMBLY AND INSTALLATION

1. Remove nuts from the cylinder head cover.
2. Remove the cylinder head cover and the gasket from the cylinder head.
3. Disconnect the injection pipe.
4. Remove the nozzle holder and the copper gasket.
5. Detach the inlet manifold.
6. Clamp the retaining ring nut in a vise.
7. Remove nut, eye joint and plain washer.
8. Remove the nozzle holder and take-out parts.
9. Remove the alternator.
10. Detach the fan belt.
11. Detach the rocker arm.
12. Remove the pushrods.
13. Detach the water return pipe.
14. Detach the cylinder head.
15. Remove the gasket and the O-ring.

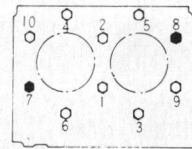

Head bolt torque sequence for 2 cylinder engines

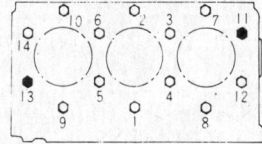

Head bolt torque sequence for 3 cylinder engines

Removing nozzle holder

16. Remove the valve cap and the valve spring collet.
17. Remove the valve spring retainer and valve spring.
18. Remove the valve stem seal and the valve.
19. Assembly is the reverse of disassembly. Note the following points:

a. Apply engine oil to each head bolt and nut; tighten them equally and in the order shown in the accompanying illustrations.
b. Tighten them to 54.2 to 57.9 ft.lb. regardless of engine models.
c. Tighten the nozzle retaining nut to 43.4 to 57.9 ft.lb.
d. Do not tighten it too much, or the needle valve will not slide easily and injection performance will be decreased.

INSPECTION

1. Clean the surface of the cylinder head.
2. Put a straight edge on the four sides and diagonal lines of the cylinder head to check the straightness of the surface.
3. Insert a feeler gauge between the straight edge and the cylinder head surface.
4. The maximum thickness inserted is the amount of distortion.
5. If the measurement exceeds the allowable limit, 0.0012 in., correct with a surface grinder.
6. Clean the valve seat surface.
7. Measure the width of the seat using a set of vernier calipers. Seat width should be 0.0827 in. Seat angle is 45°.
8. Apply red lead to the valve to check if the seat is not scratched or dented.
9. Measure the spring with a set of vernier calipers.

10. Replace it if it is not within the reference range of 1.6417-1.6535 in.
11. Put the spring on a surface plate, place a square on the side of the spring, and check to see if the entire side is in contact with the square.
12. Rotate the spring and measure the maximum B.
13. The flat surface at the end of the spring coil must exceed two-thirds of the full circumference.
14. Check all the surface of the spring for scratches.
15. If the measurement exceeds the allowable limit of 3%, replace the valve spring.
16. Place the spring on a tester, com-

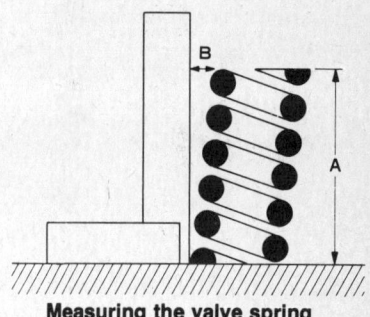

Measuring the valve spring

press it to the level to which the spring is actually compressed in the engine.

17. Read the compression load on the gauge. Load should be 26.4 lb. at 1.3839 in. If the measurement exceeds the allowable limit of 22.4 lb. at 1.3839 in., replace the valve spring.
18. Measure the inside diameter of the rocker arm bushing.
19. Measure the rocker arm shaft diameter.

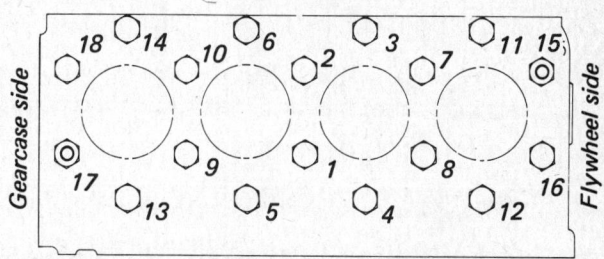

Head bolt torque sequence for 4 cylinder engines

Kubota

Measuring top clearance

Adjusting decompression clearance

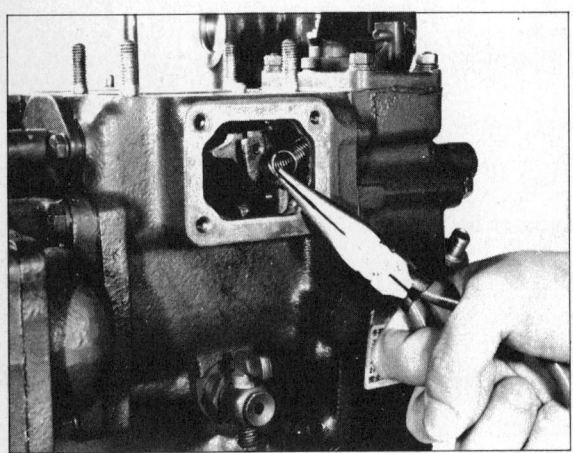
Removing the governor spring

Removing the speed control plate

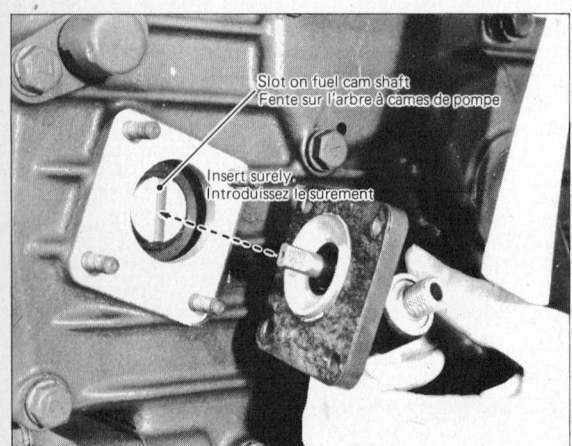
Removing the hour meter unit

Slot on fuel cam shaft
Fente sur l'arbre à cames de pompe
Insert surely.
Introduissez le surement

Removing the idler gear

Kubota

20. Calculate the clearance value. Clearance should be 0.0004-0.0028 in. If the measurement exceeds the allowable limit of 0.0059 in., replace.
21. Detach the nozzle holder.
22. Lower the piston in the cylinder to be measured.
23. Insert a length of high-quality solder through the nozzle holder hole. Be careful not to let the solder touch the valve surface.
24. Rotate the engine with your hand.
25. Take the solder out carefully.
26. Measure with a set of vernier calipers where the solder was crushed. If the measurement is not within 0.0276-0.0354 in., adjust by inserting a shim (0.0059 in. thick) between the cylinder head and gasket.
27. Measure the valve clearance with a feeler gauge after aligning each cylinder with the top dead center of compression. Adjust them in firing order. Clearance, cold, is 0.0071-0.0087 in.
28. Close the exhaust valve completely.
29. Remove the decompression adjust cover from the head cover.
30. Pull the decompression lever.
31. Reduce the valve clearance to zero by means of the decompression adjust bolt.
32. Gain access to the adjust bolt through the window. Then, screw in the bolt by 1 to 1.5 turns and tighten the locknut. Decompression clearance should be 0.0295-0.0443 in.

Gear Case

DISASSEMBLY

1. Detach the governor spring from the governor fork lever.
2. Remove the speed control plate and governor spring.
3. Remove the start spring from the gear case.
4. Straighten the tang of the washer.
5. Remove the fan drive pulley.
6. Remove the key.
7. Remove the hour meter unit.
8. Remove the gear case.
9. Remove the O-ring.
10. Detach the water pump from the gear case.

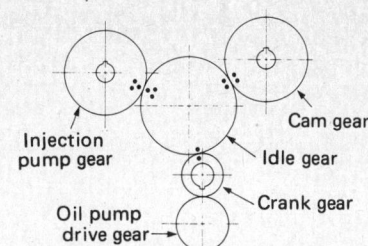

Aligning timing marks

11. Assembly is the reverse of disassembly.

Timing Gears, Camshaft, Oil Pump

REMOVAL AND INSTALLATION

1. Remove the crankshaft collar, the O-ring, oil slinger and the gear collar in that order.
2. Remove the external circlip and detach the idle gear.
3. Detach the idle gear collar.
4. Straighten the tang of the washer.
5. Remove the camshaft stopper bolt.
6. Detach the camshaft.
7. Detach the hydraulic pump and the hydraulic pump holder.
8. Remove the pump drive gear and the collar.
9. Remove three fork lever holder set bolts.
10. Remove the fuel camshaft stopper.
11. Remove the fuel camshaft and the fork lever shaft at the same time.
12. Detach the crank gear with a puller.
13. Remove the key.
14. Straighten the tang of the washer.
15. Detach the pump drive gear.
16. Detach the oil pump.
17. Installation is the reverse of removal. Align the timing marks on the gears.

INSPECTION

Measure the camshaft bearing in the crankcase with an inside micrometer. Measure the camshaft journal with an outside micrometer. Calculate the clearance. Clearance should be 0.0020-0.0036 in.

If the measurement exceeds the allowable limit of 0.0059 in., replace the camshaft.

Gently put the camshaft on V blocks.
Set a dial gauge on the journal.
While slowly rotating the camshaft, read the dial gauge. The camshaft flexure is indicated by half of the reading. The reading should be 0.0004 in.

If the measurement exceeds the allowable limit of 0.0008 in., replace the camshaft.

Measure the highest point of the cam with a micrometer.

If the measurement exceeds the allowable limit of 1.3114 in., replace.

Install a lever-type indicator between gear teeth.

Clamp one gear, rotate the other, and measure the backlash. Backlash should be 0.0016-0.0045 in.

Replace if the measurement exceeds the allowable limit of 0.0018 in.

Pistons and Crankshaft

REMOVAL AND INSTALLATION

1. Detach the oil pan.
2. Remove the oil filter 1. Be careful of the O-ring.
3. Straighten the tang of the washer.
4. Detach the connecting rod bolt.
5. Remove the cap of the large end of the connecting rod.
6. Drive out the piston to the cylinder head side with a hammer handle.
7. After driving the piston out, attach a tag to each piston to indicate its number.
8. Remove the piston ring.
9. Straighten the tang of the washer.
10. Detach the flywheel bolt.
11. Remove the flywheel.
12. Straighten the tang of the washer.
13. Detach the bearing case bolt.
14. Straighten the tang of the washer. Remove the bolts.

Removing the crank gear

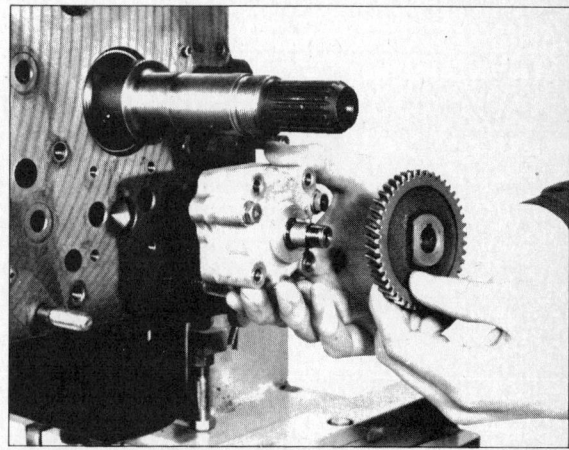

Removing the oil pump

Kubota

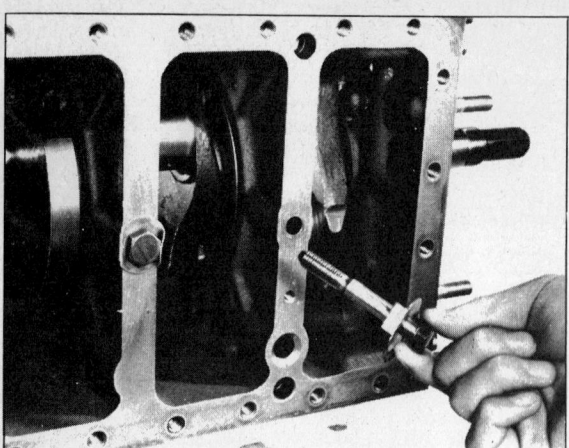
Removing bearing case bolt 2

Removing the bearing cover

Removing the crankshaft

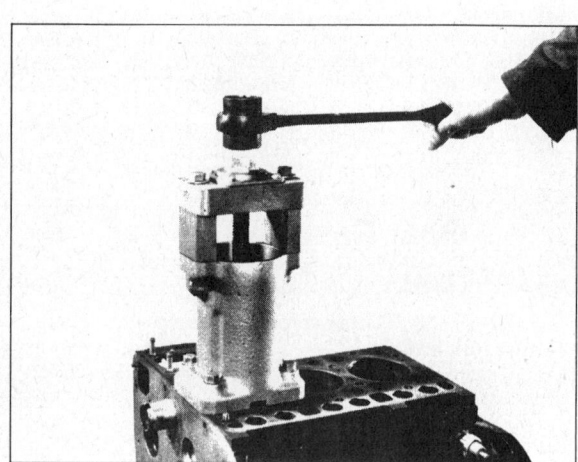
Pulling the cylinder liner

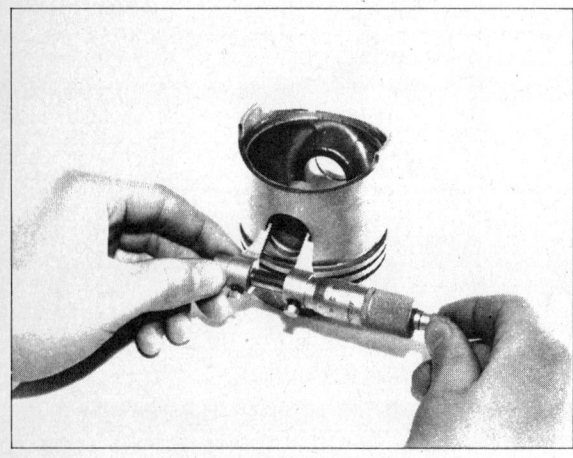
Measuring piston pin hole

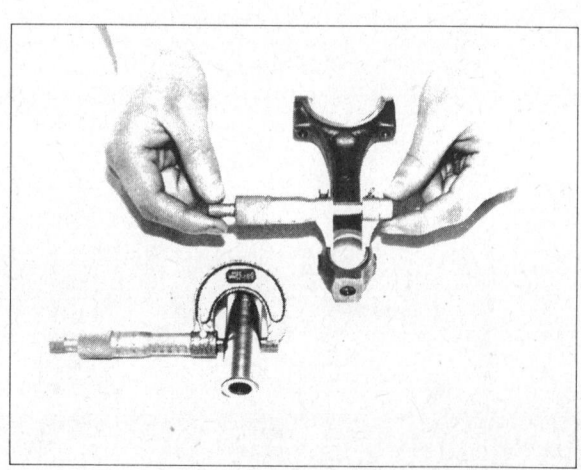
Measuring piston pin-to-bushing clearance

15. Drive two M8 bolts into the bearing cover and then pull the cover out.
16. Tap the crankshaft until it comes out of the flywheel side; be careful not to scratch the crankshaft metal 1.
17. Straighten the tang of the washer.
18. Remove the bolts.
19. Remove the bearing case.
20. Remove the bearing case side metal on the flywheel side.
21. Attach a liner puller to the crankcase.
22. Draw out the liner.
23. Installation is the reverse of removal. Note the following:
  a. Clean and oil the frame holes into which the liner is to be fitted.
  b. Clean and oil the outside surface of the liner to be force-inserted.
  c. Insert the liner with the most-chamfered end downward.
  d. After installation, bore and hone to the standard size.
  e. Regardless of engine models, tighten the main bearing cap nuts to 21.7 to 25.3 ft.lb.
  f. Face the oil groove of the side metal outward. As for bearing cases 1 and 2, line up their marks, and face フライホイール mark toward the flywheel side.
  g. Refit the bearing cases, starting with the one with the smallest outside diameter, as seen from the crank gear side.
  h. Regardless of engine models, tighten the bearing case bolts to 47.0 to 50.6 ft.lb.
  i. Align the bearing case hole with the crankcase hole.
  j. When installing a piston ring onto the piston, face the marked side of the ring toward the piston head.
  k. When installing the coil expander ring onto the piston, place the expander joint on the opposite side (180°) of the ring gap.
  l. Insert the piston into the cylinder liner with ring gaps marking a 120° angle to each other and do not face the ring gap toward the piston pin or the load of side pressure.
  m. Tighten the connecting rod cap bolts to 26.8 to 30.4 ft.lb. regardless of engine models.

INSPECTION

1. Measure the piston pin hole with an inside micrometer.
If the measurement exceeds the allowable limit of 0.9076 in., replace it.
2. Measure the piston pin with an outside micrometer.
3. Measure the inside diameter of rod small end bushing with an inside micrometer. Calculate the clearance. Clearance should be 0.006-0.0015 in.
If the measurement exceeds the allowable limit of 0.0059 in., replace it.
4. Put the piston ring in the cylinder.
5. Stand the piston upside down and push the ring into the cylinder with the piston head.
6. Insert a feeler gauge into the piston ring gap. If the measurement exceeds the allowable limit of 0.0492 in., replace it.

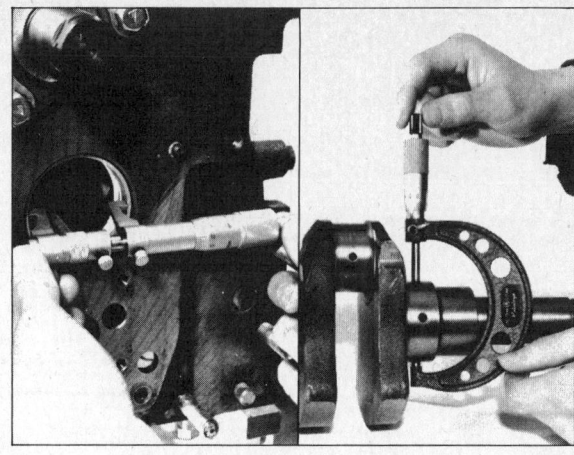

Measuring clearance between the journal and crankshaft metal 1

7. Remove the connecting rod crank pin metal and tighten the rod bolt.
8. Attach the connecting rod to the connecting rod aligner.
9. Place the gauge on the piston pin.
10. Measure the gap between the pin of the gauge and the flat surface of a straightedge. If the measurement exceeds the allowable limit of 0.0020 in., replace the rod.
11. Place V blocks on a surface plate, and put either end of the crankshaft journal on them.
12. Set a dial gauge on the center journal.
13. Read the dial gauge while rotating the crankshaft slowly. Crankshaft flexure is indicated by half of the reading.
If the reading is greater than 0.0008 in., replace the crankshaft.
14. Measure the crank journal (on the side of the crank pin metal 1) with an outside micrometer.
15. Measure the crank pin metal with an inside micrometer. Calculate the clearance. Clearance should be 0.0016-0.0046 in.
If the measurement exceeds the allowable limit of 0.0079 in., replace.
16. Place Plastigage® on the crankshaft metal.
17. Tighten the bearing case onto the crank journal to 21.7 to 25.3 ft.lb.
18. Detach the bearing case slowly, and measure the Plastigage®.
If the measurement exceeds the allowable limit of 0.0079 in., replace.
19. Place a Plastigage® on the crank pin metal.
20. Tighten the connecting rod cap onto the crank pin to 26.8 to 30.4 ft.lb.
21. Remove the large end cap carefully, and measure the Plastigage®.
22. If the standard size metals cannot be employed due to excessive wear of the crank pin, employ undersize metals. For undersize metal use, follow the precautions noted below:
  a. Cut corner radius of the crank pin to precisely 0.1378R ± 0.0079 in.
  b. Be sure to chamfer the oil hole circumference with an oil stone.
  c. Micro-finish the crank pin to higher than 0.000016 in. without fail.
23. Move the crankshaft to the crank gear side.
24. Set a dial gauge on the crankshaft.
25. Push the crankshaft toward the flywheel and measure the clearance. If the measurement is not within the reference range of 0.0059-0.0122 in., replace the side metal.
26. Adjust a cylinder gauge to a reference value of cylinder liner with an outside micrometer.

Size	Code number	Part name	Crank pin dia.	Metal grade mark
0.0079 in.	15221-2297-1	Crank pin metal 0.20 minus	1.7228 to 1.7234 in.	020US
0.0157 in.	15221-2298-1	Crank pin metal 0.40 minus	1.7149 to 1.7156 in.	040US

Crankshaft bearing dimensions

Table 1

Z751-A, DH1101-A	3.0118 to 3.0126 in.	Hone to 1.2–2 μ Rmax.
D1301-A	3.2480 to 3.2488 in.	Hone to 1.2–2 μ Rmax.

Table 2

	Oversize	Part name	Code number	Mark
Z751-A, DH1101-A	0.0197 in.	Piston 05	15221-2191-1	050S
		Piston ring 05 assembly	15221-2109-1	050S
D1301-A	0.0197 in.	Piston 05	15201-2191-1	050S
		Piston ring 05 assembly	15201-2109-1	050S

Cylinder liner refinishing dimensions

Kubota

	Reference value	Allowable limit
Z751-A DH1101-A	φ2.9921~2.9929 inch	+0.0059 inch
D1301-A	φ3.2283~3.2291 inch	

Cylinder liner reference dimensions

27. To find out the maximum wear, measure six points of cylinder diameters with the cylinder gauge, as shown.
28. When the cylinder liner has worn beyond the allowable limit, bore and hone the cylinder by 0.0197 in.
29. Finish the cylinder liner to the degree in Table 1.
30. The cylinder liner which has been oversized by 0.0197 in. should use a piston and ring of the same oversize. (See Table 2.)

FUEL SYSTEM

Injection Pump

REMOVAL AND INSTALLATION

1. Remove the injection pump cover.
2. Line up the control rack pin to the slot on the crankcase. Remove the injection pump.
3. Remove the injection pump shims. Take down the number of the shims for reference.
4. Installation is the reverse of removal.

Injectors

TESTING

1. Install the injector on a tester.
2. Move the tester handle up and down to prime fuel. Measure the pressure of fuel gushing out from the nozzle tip.

If the measurement is above or under the reference value of 1990.8-2133 psi, adjust with the adjust washer inside the nozzle holder.

3. An increase of every 0.0039 in. for washer thickness causes approximately 142.2 psi increase of fuel injection pressure.
4. Apply pressure 142.2 psi lower than the cracking pressure.
5. After keeping the nozzle under the specified pressure for 10 seconds, check to see that fuel does not leak from the nozzle valve seat.
6. If the valve seat should leak fuel, replace the nozzle piece.
7. Attach the nozzle to a nozzle tester and shoot it in the air. Check the shape of the fume. An even, fan-shaped spray should result.
8. If the shape is not acceptable, replace the nozzle piece.
9. Attach a pressure gauge to the pump.
10. Rotate the flywheel to increase the pressure to 8532 psi.
11. Align the plunger with the top dead center.
12. Measure the time needed to decrease the initial pressure 8532 psi to 7110 psi. If the measurement is not acceptable, replace the pump element. In this case, you should ask a repair shop for the replacement job because it needs a diesel fuel pump test stand. Be sure to give them adjustment reference data on the fuel injection pump.

Test Conditions
Nozzle ND-DN12SD12
Cracking Pressure .... 140 kg/cm²
Pipe .... φ6 x φ2 x 600 mm
Fuel Feed Pressure .... 0.03 kg/cm²
Cam profile .... See illustration
Prestroke .... 2.2 ± 0.05 mm
Test Fuel ..... JIS Diesel No. 2 light oil

13. Attach a pressure gauge to the pump.
14. Rotate the flywheel and increase the pressure to 1422 psi.
15. Align the plunger with the bottom dead center.
16. Measure the time needed to decrease to 71.1 psi from the initial pressure of 1422 psi.
17. If the measurement is not between 5 and 10 seconds, replace the delivery valve.
18. Start and run the engine at idle.
19. Attach a timing light to the injection pipe.
20. Check to see if the timing check window of the clutch housing is aligned with the F1 mark on the flywheel. If timing of the fuel injection is off, adjust with shims. Each shim changes crank angle by approximately 1.5°.

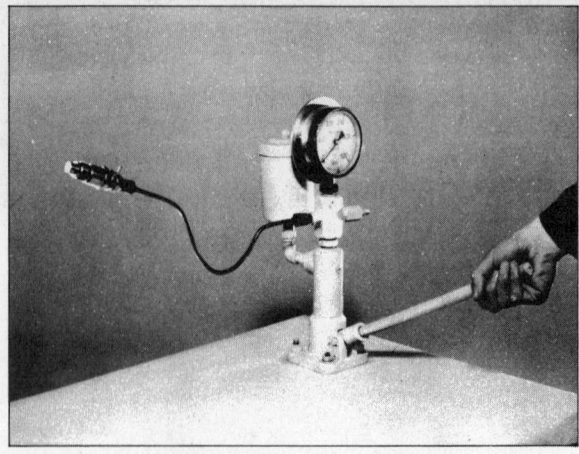

Injector installed on a tester

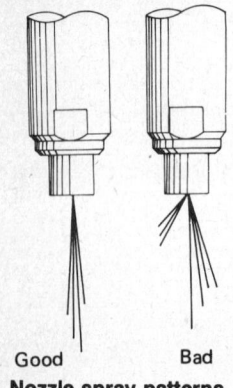

Good   Bad
Nozzle spray patterns

Control Rack Position *1 mm	Speed rpm	Amount of Injection mm³/st.	Allowance *3 mm³
9	1400	23 ± 1.5	less than ±1.5
8	1400	18.5 ± 7.5	less than ±3.8
7	1400	13.5 ± 7.5	less than ±3.8
0 ~ 3.5	1550	0 *2	

*1 Travel distance from non-injecting point of control rack.
*2 Zero cracking pressure and no injection.
*3 Allowance on the basis of standard cylinder.

Injection adjustment figures

Massey-Ferguson
INDEX

MF-85
Front Wheels and Axle........684
Steering.....................684
Clutch and Transaxle.........685
Engine.......................686

MF-1200
Front Axle and Wheels........687
Steering.....................687
Transaxle and Gearbox........688
Engine.......................689

MF-1450 & 1650
Front Axle and Wheels........689
Steering.....................689
Transmission.................690
Rear Axle....................692
Hydraulic System.............694
Engine.......................694

MF-1655 & 1855
Front Axle, Wheels and Mid-PTO 695
Steering.....................696
Transmission, Rear Axle and
Differential.................696
Hydraulic System.............696
Engine.......................697

MF-25 & 130
Brakes.......................697
Clutch.......................698
Transmission.................698
Differential and Rear Axle...701
Hydraulic System.............702
Engine.......................704
  R&R......................704
  Rocker Arm Shaft.........705
  Cylinder Head............706
  Valves...................707
  Piston and Connecting Rod...708
  Cylinder Liners..........710
  Front Cover..............710
  Timing Gears.............711
  Camshaft.................711
  Crankshaft and Main
  Bearings.................712
  Rear Main Seal...........713
Fuel System..................713
Lubrication System...........719

MASSEY-FERGUSON
MF-85

FRONT WHEELS AND AXLE

Axle Pivot Bushing

REPLACEMENT

1. Remove axle from tractor and remove old bushings with a sliding hammer puller.
2. Reinstall new bushings flush with machined surface of axle.
3. Reinstall axle on tractor.

Wheel Bearing

REPLACEMENT

1. Remove wheel and drive bearings out from inside of wheel with a suitable punch.
2. Install bearing with flange of bearing against outside of wheel hub and groove in bearing in alignment with grease fitting.

STEERING

Steering Gear

REMOVAL

1. Remove steering wheel.
2. Remove mower.
3. Disconnect drag link from steering arm.
4. Remove bolts securing steering gear to tractor frame.
5. Lower steering gear down and forward of mower lift linkage.

INSTALLATION

1. Raise steering gear up in front of mower lift linkage.
2. Continue raising steering gear into place and secure with retaining bolts.
3. Connect drag link to steering arm.
4. Install steering wheel.
5. Install mower.

DISASSEMBLY

1. Remove nut from Pitman arm and remove arm and sector gear as an assembly.
2. Remove nuts from shaft and remove pinion and shaft assembly from bracket.
3. Inspect all parts for wear or damage, replace as required.

REASSEMBLY

1. If required remove old bushings and install new bushings into place flush with machined surface.

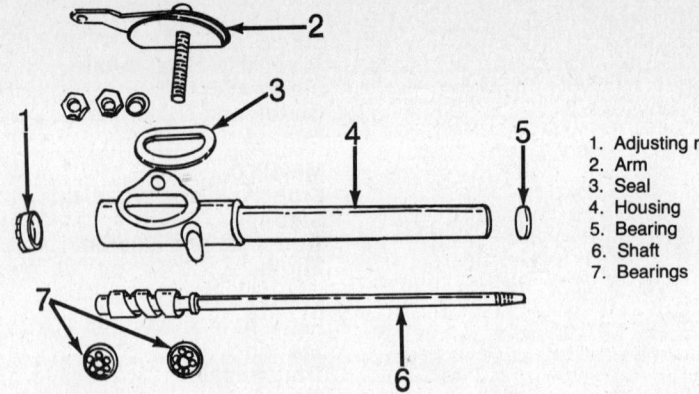

1. Adjusting nut
2. Arm
3. Seal
4. Housing
5. Bearing
6. Shaft
7. Bearings

Steering gear assembly showing the sequence of parts installation

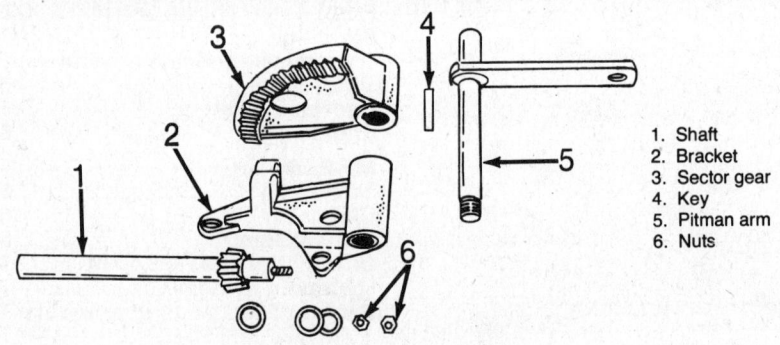

1. Shaft
2. Bracket
3. Sector gear
4. Key
5. Pitman arm
6. Nuts

Steering gear components

1. Shaft
2. Adjusting nut
3. Locknut
4. Adjusting nut
5. Pitman arm
6. Sector gear

Steering gear assembly

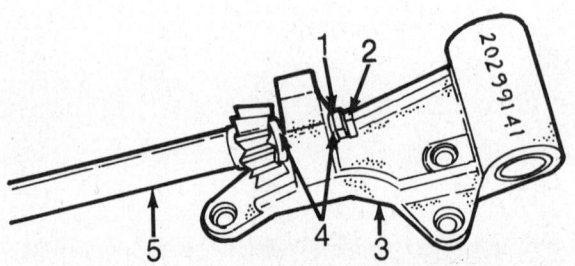

1. Adjusting nut
2. Locknut
3. Bracket
4. Thrust washers
5. Shaft

Shaft installed in bracket

Massey-Ferguson

2. Install shaft and pinion into bracket using a thrust on each side and secure with retaining nut.

3. Adjust nut at end of shaft to remove all end-play in shaft without any binding and secure with locknut.

4. Assemble sector gear on Pitman arm shaft and square key.

5. Install Pitman and sector gear into bracket and adjust gear backlash to obtain no binding by turning adjusting nut.

Steering Linkage

The steering linkage consist of a drag link and a tie rod. Each has replaceable ball socket ends.

SPINDLE ARM BUSHING REPLACEMENT

1. Disconnect tie rod and drag link from spindle arm.

2. Remove retaining ring from top of spindle and remove spindle.

3. Remove spindle bushing and install new bushing.

4. Reinstall spindle arm and secure with retaining ring.

5. Reinstall tie rod and drag link.

CLUTCH AND TRANSAXLE

REMOVAL

1. Remove console panel over drive belt idler pulleys and front of transaxle.

2. Shift transmission into 5th gear and disconnect shift linkage from transaxle.

3. Disconnect brake control linkage at brake.

4. Remove drive belt from transaxle pulley and remove pulley.

5. Remove bolts securing transaxle to frame, lift tractor and remove transaxle assembly.

NOTE: Righthand hanger bolt holds brake return spring bracket.

INSTALLATION

1. Move transaxle into position shift gears into 5th gear and lower tractor down on assembly.

2. Install retaining bolts with brake spring bracket on right hanger bolt and tighten securely.

3. Install pulley and drive belt on transaxle assembly.

4. Connect brake control linkage.

5. Connect shift linkage.

6. Install console over drive belt idlers.

DISASSEMBLY

The transaxle has five speeds forward and one reverse speed with a disc brake on output shaft.

1. Position shift lever in neutral and remove lever.

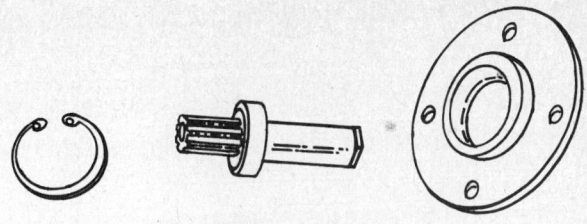

Shaft and bearing removed

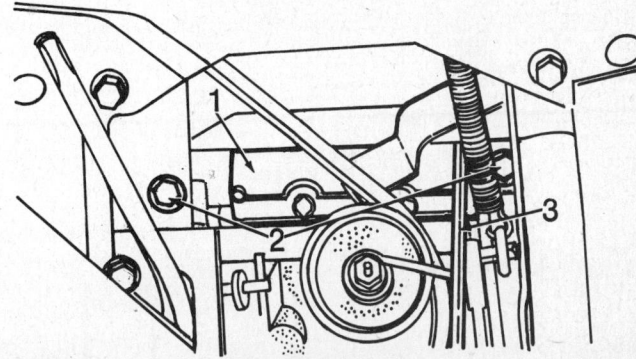

Top view of transaxle
1. Transaxle
2. Front mounting bolts
3. Shift linkage

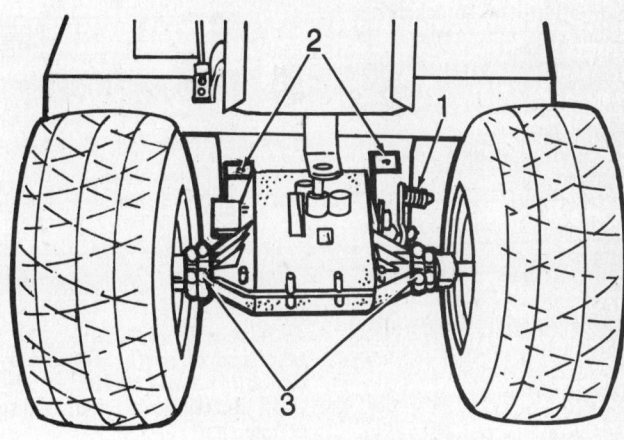

Removing the transaxle
1. Brake return spring
2. Front hanger brackets
3. Axle mountings

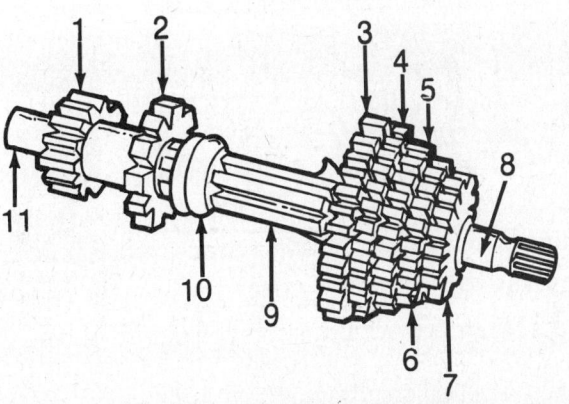

Output shaft and gears
1. Output gear
2. Sprocket
3. 1st gear
4. 2nd gear
5. 3rd gear
6. 4th gear
7. 5th gear
8. Bearing
9. Output shaft
10. Shift collar
11. Bearing

Massey-Ferguson

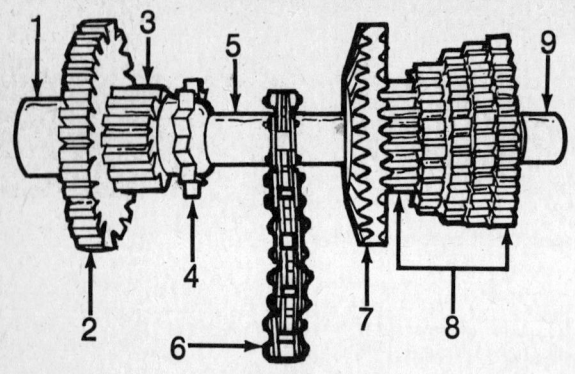

Countershaft and gears

1. Bearing
2. Output gear
3. Pinion
4. Sprocket
5. Countershaft
6. Chain
7. Bevel gear
8. Spur gears
9. Bearing

Differential components

1. Short axle
2. Thrust washers
3. Side gears
4. Long axle
5. Roll pin
6. Pinion pin
7. Pinion thrust washers (2)
8. Pinion gears
9. Differential gear assembly

2. Remove setscrew, spring and detent ball from top of cover.
3. Remove retaining screws and remove cover.
4. Remove shifter assembly.
5. Lift differential and axle assembly out.
6. Remove gear and shaft assemblies from case by lifting the two shafts out of bearing support taking care not to disturb drive chain relationship with sprockets.
7. Angle chain and sprocket ends of shaft together and remove chain from output shaft sprocket.
8. Remove splined bevel and spur gears from countershaft.
9. Remove bearing, output gear and sprocket from output shaft.
10. Remove five speed change gear, washers, shift collar and keys from output shaft.
11. Remove input shaft and pinion assembly.
12. Disassemble differential assembly.

ASSEMBLY

1. Reassemble differential assembly.
2. Install input shaft bearings in case.
3. Install gear and chamfered thrust washers on output shaft. Install all gears with cut out in gear toward outside and

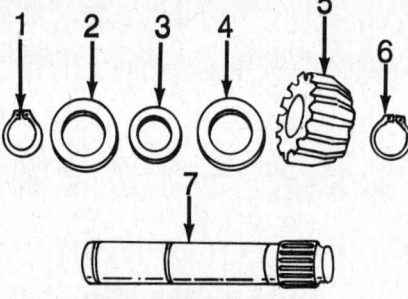

Input shaft components

1. Retaining ring
2. Washer
3. O-ring
4. Washer
5. Pinion
6. Retaining ring
7. Shaft

inside chamfer of washer toward largest gear.
4. Install shift collar and keys over other end of shaft, followed by sprocket, pinion, gear and bearing.
5. Install gears, chain and sprockets on countershaft.
6. Connect shafts together.
7. Position shafts and gears in case.
8. Install differential and axle assembly in case.

9. Install shift assembly.
10. Put 12 oz. of E.P. lithium grease around gears in case and install cover. Tighten cover retaining bolts 90-100 in.lb.
11. Install shift lever and detent ball and spring.

ENGINE

This tractor uses a Briggs & Stratton engine. For complete engine service, see the Briggs & Stratton engine part of the Engine Unit Repair section of this book.

REMOVAL

1. Remove hood assembly.
2. Disconnect and remove battery.
3. Disconnect from right side of engine.
   a. Fuel line at carburetor.
   b. Throttle control cable.
   c. Ignition cut-off wire.
4. Disconnect from left side of engine.
   a. Electrical wires.
   b. Starter cable.
5. Drain engine oil and remove oil drain tube.
6. Remove mower drive belt retainer from lower side of engine.
7. Depress clutch pedal and lock in park position.
8. Place mower control lever in the disengaged position (rearward).
9. Loosen traction drive belt guide at front idler pulley and slide belt off idler pulley.
10. Remove engine mounting bolts and lift engine out while working belts from engine pulleys.

INSTALLATION

1. Position idler pulley and belt.
2. Work engine into place and into belt.
3. Install mounting bolt and tighten securely.
4. Reconnect wires, line and cable.
5. Install oil drain tube and fill engine with oil.
6. Install hood assembly.

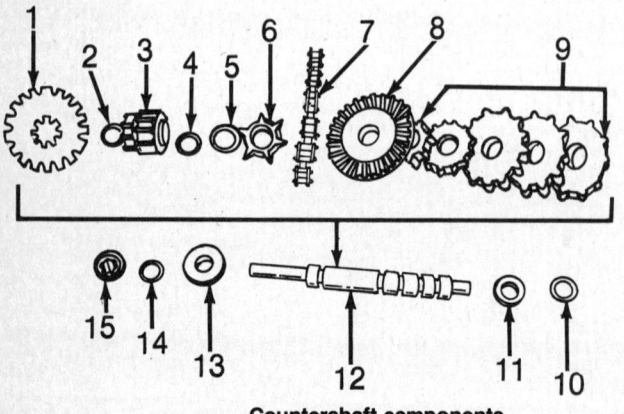

1. Output gear
2. O-ring
3. Pinion
4. O-ring
5. Washer
6. Sprocket
7. Chain
8. Bevel gears
9. Spur gears
10. Bearing
11. Washer
12. Countershaft
13. Washer
14. O-ring
15. Bearing

Countershaft components

Massey-Ferguson

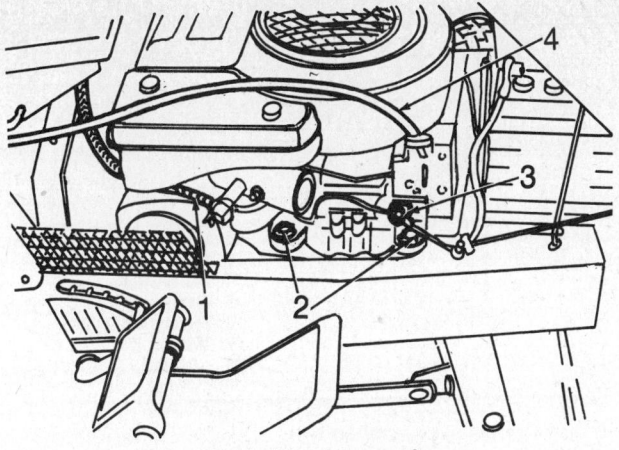

Right side of the engine
1. Fuel line
2. Mounting bolts
3. Ignition cut off wire
4. Throttle cable

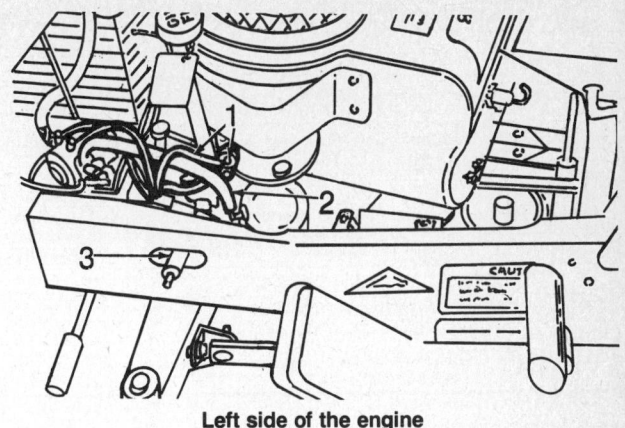

Left side of the engine
1. Electrical wires
2. Starter cable
3. Oil drain tube

MF-1200

FRONT AXLE AND WHEELS

Axle Pivot Bushing

REPLACEMENT

1. Remove mower drive idler at front of tractor.
2. Remove wheels and spindles.
3. Remove nut at rear of retaining bolt and remove bolt.
4. Lower axle and remove pivot pin.
5. Remove bushings from axle.
6. Install new bushings.
7. Reinstall axle onto tractor.

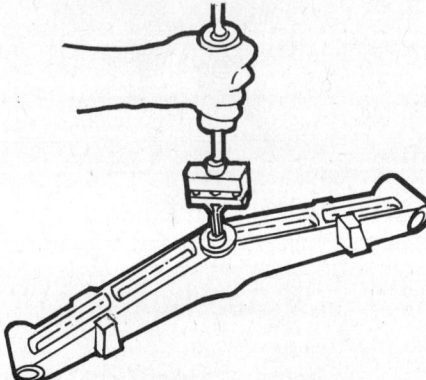

Removing axle pivot bushing

Wheel Bearing

REPLACEMENT

1. Remove wheel and drive bearings out from inside of wheel with a suitable punch.
2. Install bearings with flange of bearing against outside of wheel hub.
3. Reinstall wheel on axle and secure with retaining bolt.

CHECKING AND ADJUSTING TOE-IN

Before adjusting toe-in, be sure tie rods are of equal length.
Adjust toe-in so the distance between front of wheels measures ⅛ in. less than at back.

STEERING

Steering Gear

REMOVAL

1. Disconnect battery cables.
2. Remove fuel tank.
3. Remove cover over driveshaft and remove side shields.
4. Disconnect choke cable at engine.
5. Remove wire plug from ignition switch and PTO control switch.
6. Disconnect ammeter wires.
7. Remove steering wheel.
8. Remove retaining screws and lift instrument panel off tractor.
9. Remove retaining bolts and lift steering gear assembly out.

INSTALLATION

1. Position steering gear assembly in place and secure with retaining bolts.
2. Install instrument panel in place and secure with retaining screws.
3. Reconnect wires to ignition switch, PTO switch and ammeter.

NOTE: Green wire connects to + side of ammeter and red wire to − side.

5. Install steering wheel and fuel tank.
6. Install side shields and cover over driveshaft.
7. Connect battery cables.

DISASSEMBLY

1. Remove locknut and retaining nut from steering arm and remove arm, felt seal and retainer.
2. Remove adjusting nut and slide shaft and bearing from housing.
3. Remove upper bearing from housing.

REASSEMBLY

1. Lubricate bearings and insert one cup and bearing into gear housing.
2. Slide steering shaft into gear housing, followed with another bearing cone and cup.
3. Install adjusting nut and tighten until steering shaft has a slight bind when turned. At this point adjust nut until shaft has no end play with no binding when turned. Secure adjusting nut with cotter key.
4. Install bearing at upper end of shaft and tube.
5. Loosen adjusting screw in steering arm and install arm into housing with seal and retainer in place.
6. Install washer and one retaining nut on steering arm shaft. Tighten nut until steering arm has no binding or endplay. Secure retaining nut in this position with locknut.
7. Turn in on adjusting screw until a slight drag is felt on steering shaft at mid range of steering (this would be a straight ahead position of front wheels).
8. Back out adjusting screw to eliminate any binding of steering shaft and secure in this position with locknut.

Bellcrank Arm and Bearing

REPLACEMENT

1. Disconnect drag links from bellcrank.

Massey-Ferguson

2. Remove retaining bolt and slide bellcrank off.
3. Remove old bearings and install new bearings flush with outside surface.
4. Install bellcrank and secure with retaining bolt.
5. Connect drag links.

Spindle Arm Bushing

REPLACEMENT

1. Disconnect tie rod from spindle arm.
2. Remove retaining ring at upper end of spindle and remove spindle.
3. Remove spindle bushings.
4. Install new bushings.
5. Reinstall spindle arm and connect tie rod.

TRANSAXLE AND GEARBOX

The transmission and rear axle assembly (transaxle) is driven by a V-belt through a gear box. The transaxle has 4 speeds forward and 1 speed reverse.

REMOVAL

1. Disconnect wires and remove rear fender and seat assembly.
2. Remove drive belt from transaxle pulley and disconnect brake linkage at brake.
3. Attach a suitable lifting device to tractor frame.
4. Remove transaxle retaining bolts.
5. Raise tractor and remove transaxle assembly from tractor.

INSTALLATION

1. Position transaxle under tractor and lower tractor into place.
2. Install retaining bolts and tighten securely.
3. Install belt and reconnect brake linkage.
4. Install rear fender and seat assembly and reconnect wires.

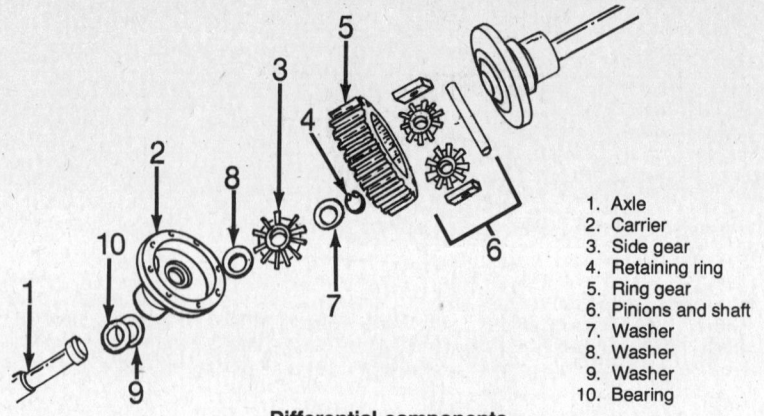

1. Axle
2. Carrier
3. Side gear
4. Retaining ring
5. Ring gear
6. Pinions and shaft
7. Washer
8. Washer
9. Washer
10. Bearing

Differential components

OVERHAUL

1. Shift transmission to neutral and remove shift lever.
2. Remove both axle housings.
3. Block up transmission so that no pressure is in brake shaft with case side facing up.
4. Cap dowel pins out of case into cover and remove bolts.
5. Lift case 1½ to 2 in. above cover, tilt so case will clear shift rods, rotate case and remove from cover and gears.
6. Remove the following:
   a. Thrust washer and 3 gear cluster from brake shaft.
   b. Reverse idler gear and shaft.
   c. Hold shifter assembly together and lift out.
   d. Low gear and shaft.
   e. 2 cluster gear from brake shaft.
   f. Differential assembly.
   g. Output Shaft.
   h. Brake shaft.
   i. Input shaft and bearing.
7. Inspect all parts for wear or damage, replace as required. Replace all seals and gaskets. If bearings require replacement, position bearing cage flush with machined surface of bearing bore and press against lettered side of bearing.
8. Install input shaft into case. Use a soft mallet to seat shaft completely in cover.
9. Install the following:
   a. Brake shaft and 2 cluster gear.
   b. Output shaft and gear.
   c. Differential assembly with bolt heads away from output gear.
   d. Shifter shaft.
10. Install low gear and shaft.
11. Hold shifter assembly together and install over shaft and into cover.
12. Install 3 gear cluster and reverse idler shaft and gear.
13. Work case around gears by tilting and rotating and install in place, and secure with retaining bolts.

NOTE: If case and cover does not go completely together, rotate input shaft to cause gears to mesh. Case should drop to about ¼ in. from cover.

14. Install axle housing and secure with retaining bolts.
15. Install shim washer and hub. Add or remove shims to eliminate axle end-play.
16. Fill transaxle to level plug with SAE-EP 90 transmission oil.

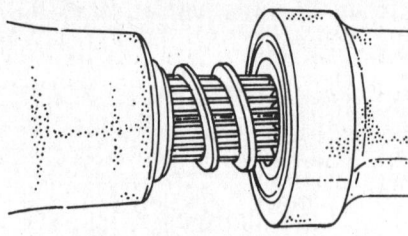

Hub and shim installation

Gearbox

REMOVAL AND INSTALLATION

1. Remove cover over driveshaft and remove belt from gear box pulley.
2. Remove bolts securing gear box to tractor frame.
3. Remove gear box from beneath tractor.
4. To reinstall gear box proceed as follows:
   a. Position gear box in place.
   b. Install bolts and tighten securely.
   c. Reinstall belt.

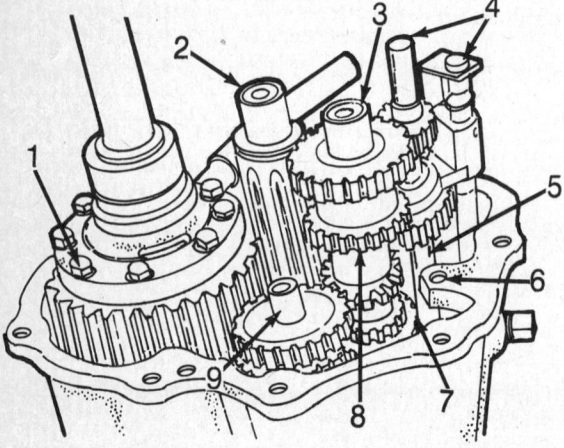

1. Differential assembly
2. Output shaft
3. Brake shaft
4. Shift rods
5. Shifter shaft
6. Reverse idler shaft (removed)
7. 2-cluster gear
8. 3-cluster gear
9. Low gear shaft

Case removed from transaxle

Massey-Ferguson

DISASSEMBLY

1. Remove gear box from tractor and remove pulley.
2. Punch two small holes in output shaft seal with a sharp punch as close to outer edge as possible. Do not drill these holes or bearing below seal could be damaged.
3. Install two metal screws into these holes.
4. Pry seal out of housing with a suitable bar or screwdriver.
5. Remove cover and seal from over input shaft.
6. Pull input shaft and bearing out of housing and gear.
7. Remove snapring from output shaft.
8. Remove output gear, bearing and shaft from housing.
9. Remove snapring and remove gear and bearing from output shaft.
10. If the other bearing for the input shaft requires removal, heat housing and remove bearing.

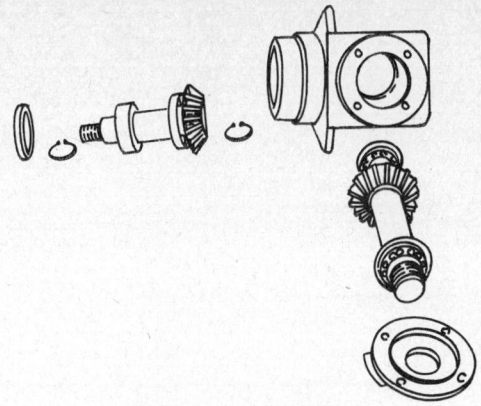

Gearbox components

ASSEMBLY

1. Heat housing and install input shaft bearing, if removed.
2. Assemble bearing and gear on output shaft and secure with retaining ring.
3. Install output shaft into housing.
4. Hold output shaft in place and install snapring.
5. Carefully slide seal over output shaft and install into housing. Drive seal into housing until it is flush with housing.
6. Install input shaft with bearing into housing and gear.
7. Install new seal into input shaft cover .040 in. to .050 in. below flush.
8. Install cover and seal with gasket over input shaft and secure with retaining bolts.
9. Fill case with approximately 4 oz. of moly EP lithium grease.

ENGINE

This tractor uses a Briggs & Stratton engine. For complete engine service, see the Briggs & Stratton engine part of the Engine Unit Repair section of this book.

REMOVAL

1. Remove hood and grille assembly.
2. Disconnect battery cables and electrical wires from engine.
3. Disconnect electrical clutch control wire.
4. Disconnect cable from starter motor.
5. Shut off fuel and remove line from carburetor.
6. Disconnect choke and throttle cables from engine.
7. Remove battery ground cable bolt from battery platform.
8. Remove pin from drive coupler and slide coupler sleeve rearward.
9. Remove engine mounting bolts and lift engine out.

INSTALLATION

1. Position engine in frame and secure with retaining bolts.
2. Reconnect cables, wires and fuel lines.
3. Slide drive coupler onto engine shaft and secure with pin.
4. Install hood and grille assembly.

MF-1450 and MF-1650

FRONT AXLE AND WHEELS

Axle Pivot Bushing

REPLACEMENT

1. Remove mower drive idler at front of tractor.
2. Remove wheels and spindles.
3. Remove nut at rear of retaining bolt and remove bolt.
4. Lower axle and remove pivot pin.
5. Remove bushings from axle.
6. Install new bushings.
7. Reinstall axle onto tractor.

Wheel Bearing

REPLACEMENT

1. Remove wheel and drive bearings out from inside of wheel with a suitable punch.
2. Install bearings with flange of bearing against outside of wheel hub.
3. Reinstall wheel on axle and secure with retaining bolt.

CHECKING AND ADJUSTING TOE-IN

Before adjusting toe-in, be sure tie rods are of equal length. Adjust toe-in so the distance between front of wheels measure ⅛ in. less than at back.

STEERING

Steering Gear

REMOVAL

1. Disconnect battery cables.
2. Remove fuel tank.
3. Remove cover over driveshaft and remove side shields.
4. Disconnect choke cable at engine.
5. Remove wire plug from ignition switch and PTO control switch.
6. Disconnect ammeter wires.
7. Remove steering wheel.
8. Remove retaining screws and lift instrument panel off tractor.
9. Disconnect hydraulic cylinder bracket and lower cylinder down as far as possible.
10. Remove retaining bolts and lift steering gear assembly out.

INSTALLATION

1. Position steering gear assembly in place and secure with retaining bolts.
2. Connect hydraulic cylinder bracket to frame with retaining bolts.
3. Install instrument panel in place and secure with retaining screws.
4. Reconnect wires to ignition switch, PTO switch and ammeter.

NOTE: Green wire connects to + side of ammeter and red wire to − side.

5. Install steering wheel and fuel tank.
6. Install side shields and cover over driveshaft.
7. Connect battery cables.

Massey-Ferguson

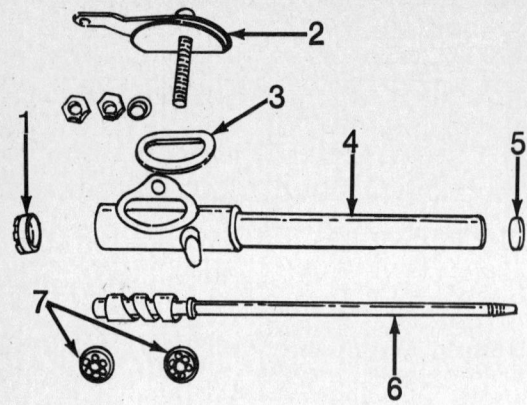

1. Adjusting nut
2. Arm
3. Seal
4. Housing
5. Bearing
6. Shaft
7. Bearings

Steering gear components

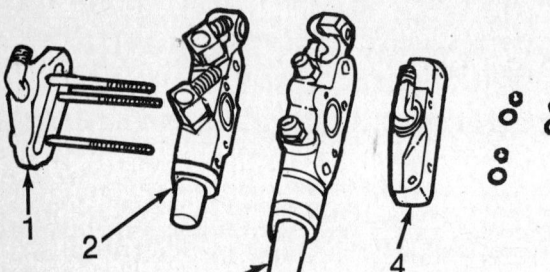

1. Outlet plate
2. 3-point hitch valve
3. Mid-mount linkage valve
4. Inlet plate

Control valve separated

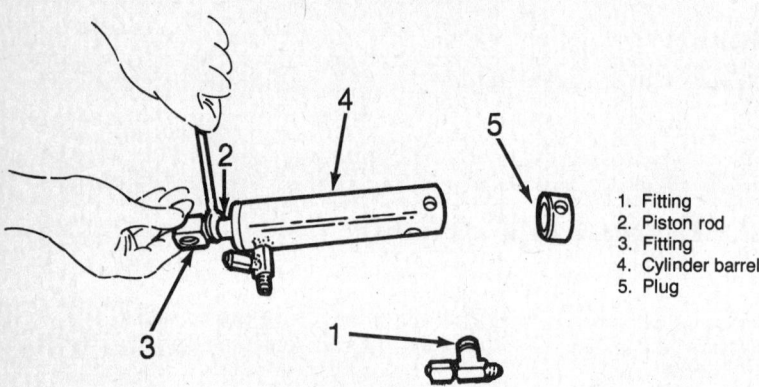

1. Fitting
2. Piston rod
3. Fitting
4. Cylinder barrel
5. Plug

Removing fitting from rod

DISASSEMBLY

1. Remove locknut and retaining nut from steering arm and remove arm, felt seal and retainer.
2. Remove adjusting nut and slide shaft and bearing from housing.
3. Remove upper bearing from housing.

ASSEMBLY

1. Lubricate bearings and insert one cup and bearing into gear housing.
2. Slide steering shaft into gear housing, followed with another bearing cone and cup.
3. Install adjusting nut and tighten until steering shaft has a slight bind when turned. At this point adjust nut until shaft has no endplay with no binding when turned. Secure adjusting nut with cotter key.
4. Install bearing at upper end of shaft and tube.
5. Loosen adjusting screw in steering arm and install arm into housing with seal and retainer in place.
6. Install washer and one retaining nut on steering arm shaft. Tighten nut until steering arm has no binding or endplay. Secure retaining nut in this position with locknut.
7. Turn in on adjusting screw until a slight drag is felt on steering shaft at mid range of steering (this would be a straight ahead position of front wheels).
8. Back out adjusting screw to eliminate any binding of steering shaft and secure in this position with locknut.

Bellcrank Arm and Bearing

REPLACEMENT

1. Disconnect drag links from bellcrank.
2. Remove retaining bolt and slide bellcrank off.
3. Remove old bearings and install new bearings flush with outside surface.
4. Install bellcrank and secure with retaining bolt.
5. Connect drag links.

Spindle Arm Bushing

REPLACEMENT

1. Disconnect tie rod from spindle arm.
2. Remove retaining ring at upper end of spindle and remove spindle.
3. Remove spindle bushings.
4. Install new bushings.
5. Reinstall spindle arm and connect tie rod.

TRANSMISSION

REMOVAL

1. Remove rear fender and seat assembly.
2. Remove three point hitch if equipped.
3. Remove PTO if equipped as follows:
   a. Remove rear cover shield over belts and remove belts.
   b. Remove retainer ring and upper pulley.
   c. Remove retaining bolts and remove pulley and bracket assembly.
4. Disconnect clutch/brake linkage at front.
5. Disconnect forward/reverse linkage at rear.
6. Disconnect wires from clutch/brake safety switch.
7. Disconnect hydraulic lines to auxiliary valve at transmission.
8. Attach a suitable lifting device to frame.
9. Remove bolts and plates securing transaxle to frame and maneuver assembly out.

INSTALLATION

1. Reposition transaxle into place.
2. Lower frame and secure assembly with retaining bolts and plates.

NOTE: Spacers between transmission and frame are different lengths, right-hand side is longest and can be identified by a groove around it. Two spacers are used between axle housing and frame on righthand side.

3. Connect forward/reverse linkage.
4. Connect clutch/brake linkage.
5. Reinstall PTO if equipped as follows:

Massey-Ferguson

a. Install pulley and bracket assembly.
b. Install upper pulley and secure with retaining ring.
c. Install belts and install rear cover shield.
6. Install three point hitch if equipped.
7. Install rear fender and seat assembly.

DISASSEMBLY

Before any servicing is done to the transmission it is suggested that a pressure check be made as follows:
1. Tilt seat forward and remove tool box.
2. Remove plug at top of transmission pump housing.
3. Install a 1000 psi gauge.
4. Start engine and observe reading on guage.
5. The reading on guage should be 70-150 psi.
6. If the pressure is not within this range, check pump charge relief valve located at upper righthand side of pump housing.
7. Shims are used to regulate pressure—add shims to increase pressure, remove to decrease.

Before disassembling the transmission, check and mark charging pump housing and motor housing positions. These two housings are directional according to pump shaft rotation.

If the motor housing is accidentally reversed, the tractor will go forward when the Hydra-Speed Lever is in the reverse position or back up when in forward position.

1. Separate transmission and rear axle assembly.
2. Remove bolts securing the charging pump housing and remove the housing.
3. Remove gerotor assembly and drive pin.
4. Remove the implement relief valve spring and shims from upper left side of the pump housing.
5. Remove the charge relief valve spring and shims from the upper right side of the pump housing.
6. Remove bolts securing motor housing and pump housing together and separate the units.
7. Remove cylinder block, thrust plate and pump valve plate.

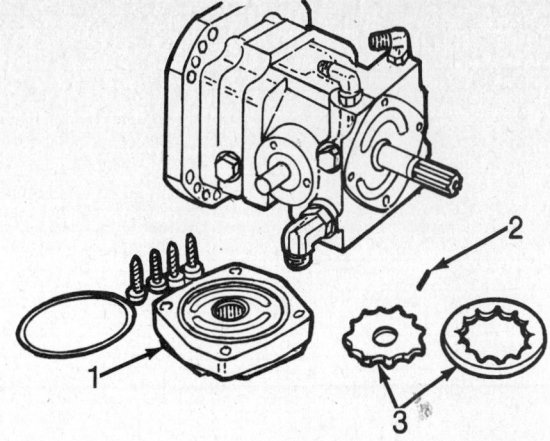

Charge pump components
1. Charging pump housing
2. Pin
3. Gerotor set

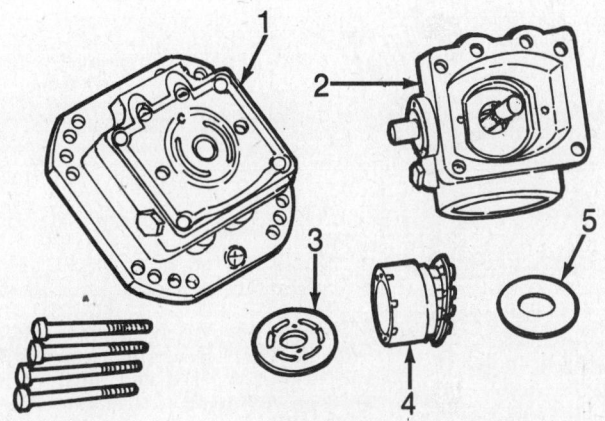

Pump and motor disassembled
1. Motor housing
2. Pump housing
3. Pump valve plate
4. Pump cylinder block
5. Thrust plate

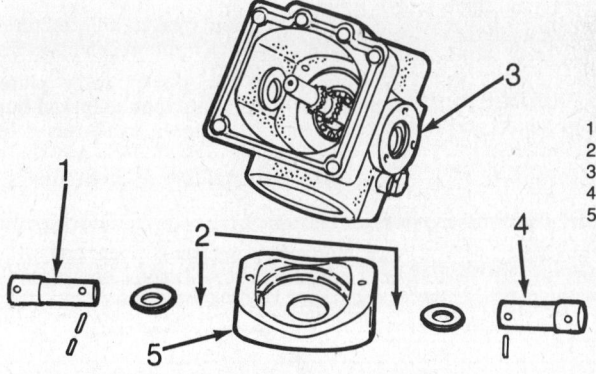

Swash plate and shafts
1. Control shaft
2. Seals (not shown)
3. Pump housing
4. Control shaft
5. Swash plate

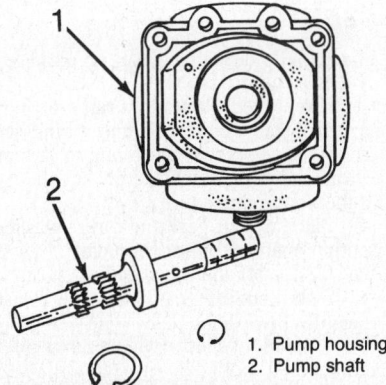

Pump shaft removed
1. Pump housing
2. Pump shaft

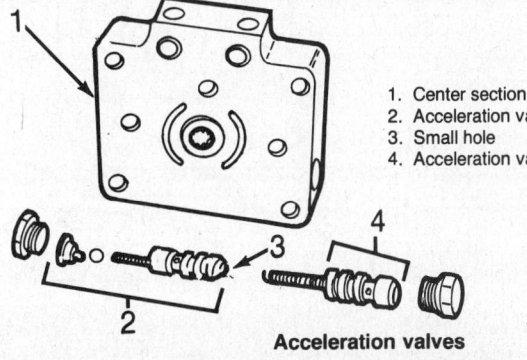

Acceleration valves
1. Center section
2. Acceleration valve (righthand side of tractor)
3. Small hole
4. Acceleration valve

691

Massey-Ferguson

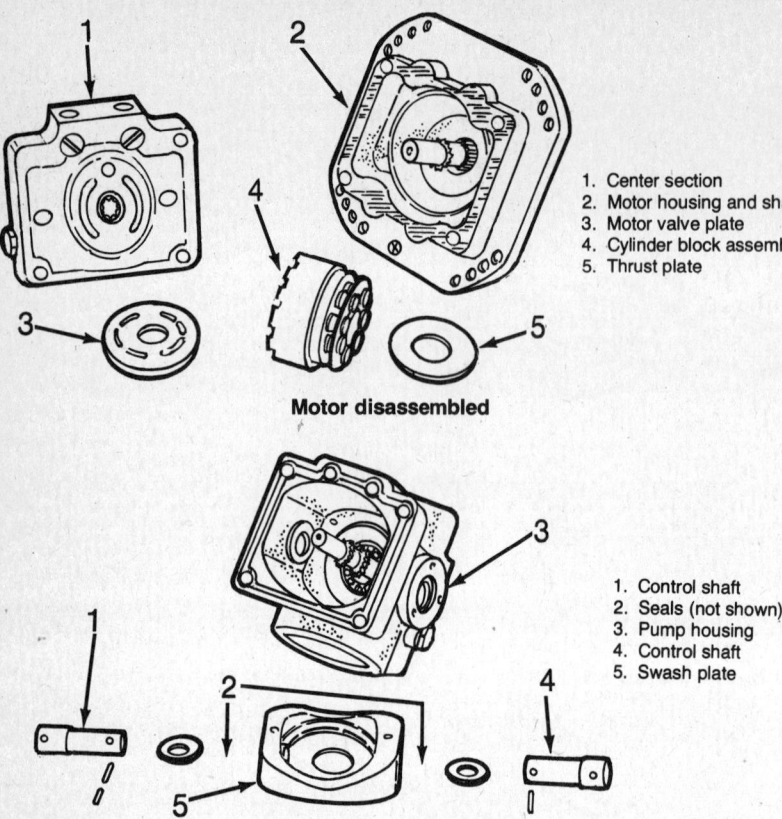

Motor disassembled

1. Center section
2. Motor housing and shaft
3. Motor valve plate
4. Cylinder block assembly
5. Thrust plate

Swash plate and shafts

1. Control shaft
2. Seals (not shown)
3. Pump housing
4. Control shaft
5. Swash plate

NOTE: The pistons may drop out of the cylinder block when the block is removed. The pistons are interchangeable and can be reinstalled into any hole.

8. Hold swashplate flat in housing and drive out roll pins. The lefthand shafts will have two roll pins. A hole is provided in the housing for the pins to be positioned flat in housing.
9. Remove pump shaft and bearing from housing. If bearing requires replacement, remove snapring and press shaft out of bearing.
10. Remove center section, valve plate, cylinder block and thrust plate from motor housing and shaft.
11. Remove acceleration valves from section.

NOTE: One of the acceleration valves has an extra hole in it. This valve must be installed on the righthand side of center section during reassembly.

12. Remove check valve assemblies from center section.
13. Remove shaft and bearing from motor housing.

REASSEMBLY

Lubricate all parts to aid in assembly.
1. Install motor shaft into motor housing and secure with retaining ring.
2. Install thrust plate.
3. Position motor housing with deep portion of fixed swash plate upward and slide cylinder block assembly onto shaft.
4. Install acceleration valves in center section.

NOTE: The acceleration valve with small hole at tapered end must be installed on righthand side of center section.

5. Install check valves in center section.
6. Position gasket and motor valve plate over dowel pin on center section.

NOTE: The pump valve plate and motor valve plate look identical but they are not. The motor valve has 4 lead-in chamfers at end of slots and the pump valve plate has 2 lead-in chamfers.

7. Position center section against motor housing with open end of check valves away from motor housing.

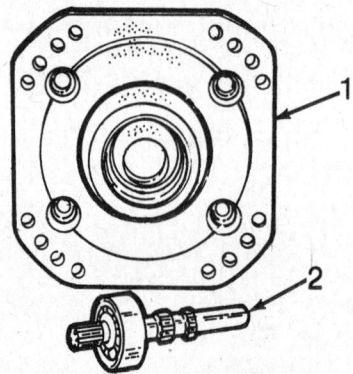

Motor shaft removed

1. Motor housing
2. Motor shaft and bearing

8. Install pump shaft and bearing into pump housing and install swash plate.
9. Position pump valve plate on center section.
10. Place the pump housing against center section and secure with retaining bolts.
11. Install implement relief valve into lefthand upper port on pump housing and charging relief valve on right side.
12. Replace oil seal in charging pump housing. Inspect bearing and replace if necessary.
13. Install drive pin, gerotor assembly and charging pump housing with lefthand on housing at top and secure with retaining bolts.
14. Connect transmission to rear axle assembly.

NOTE: When removing or installing oil filter be sure the oil filter retaining tube is turned into housing as far as possible.

15. Refer to appropriate heading and install transmission in tractor. Fill with M-1110 lubricant or a good grade of Type "A" transmission oil. Also fill suction line between reservoir and transmission with oil.

REAR AXLE

DISASSEMBLY

The rear axle assembly consists of a two-speed gear change with a differential assembly.
1. Remove wheel hubs and axle housings.
2. Remove housing side cover.
3. Remove thrust washer from output shaft and remove differential and axle shafts.
4. Remove retaining ring from shift rail and remove shift fork and sliding gear.

NOTE: It is not necessary to remove shift rail to disassemble the gear case. If it is desired to remove the shift rail, remove setscrew, spring and ball from outside of case before removing shift rail.

5. Remove remaining gears, shafts, thrust washers and spacers from the case.

ASSEMBLY

Inspect all parts for wear or damage. Replace as required.
1. If removed, insert shift rail into gear case and install detent ball and spring secured by a setscrew. Shift rail to detent position and turn detent screw in all the way then back out 1 turn.
2. Place output gear and thrust washer on output shaft and insert into case. Leave thrust washer off the other end of shaft.
3. Install sliding gear shaft with thrust spacer (thick washer).
4. Slide input shaft into low range gear (from beveled tooth side of gear) followed by a bevel gear, spacer and thrust bearing into the case.

Massey-Ferguson

5. Hold sliding gear with larger gear down, hold shift fork with long side of hub down and insert forks into slot in gear.
6. Position gear and fork in case followed by thrust washer.
7. Install spacer, high gear and thrust washer.
8. Install retaining ring on shift rail, differential assembly and output shaft thrust washer.

NOTE: The longest axle shaft of the differential assembly must be installed in the case side of unit.

9. Carefully install cover over axle and shift rail and secure with retaining bolts.
10. Install bearing and seal in axle housings and install over axles against thrust bearing and seal.
11. Secure housing to case with retaining bolts.

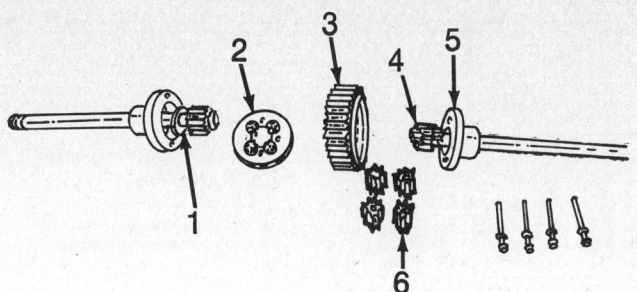

1. Thrust washer
2. Core and 4 pinions
3. Ring gear and core
4. Side gear
5. Carrier
6. Pinions

Differential components

Differential

OVERHAUL

1. Remove bolts securing differential together and separate the unit.
2. Inspect all parts for wear or damage. Replace as required.

The pinions, cores and side gears are interchangeable, the ring gear is nondirectional, the differential carriers and axle are different. During installation the carrier with the long hub must be installed on the longest axle shaft.

3. Install cores into ring gear with tabs on cores in same groove of ring gear.
4. Install pinions into the bores in the cores.
5. Install side gears on axle and secure with retaining rings.
6. Install thrust washers and differential carriers on axle shafts.

NOTE: Long hub on carrier must be on longest axle shaft.

7. Insert axle and side gears into pinions.
8. Bolt assembly together.

Transmission Adjustments

Before attempting to adjust transmission linkage, perform following checks:

1. Start tractor.
2. Place rear axle shift lever in low speed range.
3. Release neutral return and brake pedal.
4. Place right foot on heel and toe pedal.
5. Tilt right foot to rotate toe end of pedal downward. Tractor should travel forward.
6. Remove foot from pedal, tractor should stop and remain stopped.
7. Place right foot on pedal and tilt foot to rotate heel end of pedal downward. Tractor should travel rearward.
8. Remove foot from pedal, tractor should stop and remain stopped.
9. Drive tractor forward at maximum speed, then flip rear axle shift lever to neutral position and depress neutral brake pedal. Rear wheels should slide.

LINKAGE ADJUSTMENTS

1. Block tractor so that rear wheels are free to rotate.
2. Loosen nut on brake rod to end of rod.
3. Place speed control/parking brake lever in park position.
4. Depress clutch/brake pedal against step plate. Turn adjusting screw to obtain zero clearance between end of screw and selector link, then tighten locknut.
5. Raise tractor seat and clamp safety switch to hold circuit closed so engine will start.
6. Remove cotter pin and retainer from adjuster.
7. Start engine, shift speed control lever out of park and axle shift lever in gear.
8. Turn adjuster in one direction until wheels rotate then turn adjuster in opposite direction until wheels rotate in other direction. Place adjuster in center of these two turns. (This is a neutral position and wheels should not rotate.)
9. Make final check of adjustments as follows:
   a. Depress forward end of forward-reverse pedal. Wheels must rotate in forward direction. Remove pressure from pedal allowing it to return to neutral automatically. Wheels must stop within seven seconds.
   b. Depress reverse end of forward-reverse pedal. Wheels must rotate in rearward direction. Remove pressure from pedal allowing it to return to neutral automatically. Wheels must stop within seven seconds.
10. Readjust nut on brake rod as outlined under Brake Adjustment.
11. Adjust forward/reverse linkage to obtain pedal position to suit operator.

BRAKE ADJUSTMENT

1. Place speed control/parking brake lever in park position.
2. Loosen nut at end of brake rod.
3. Tighten nut to eliminate clearance between brake pucks and disc then back off one-half turn. This provides approximately .020 in. running clearance for brake disc.
4. With speed control lever in park position, pull brake rod forward with 10-20 lb. tension and tighten brake rod nut against tubular guide. Continue to tighten nut two additional turns.

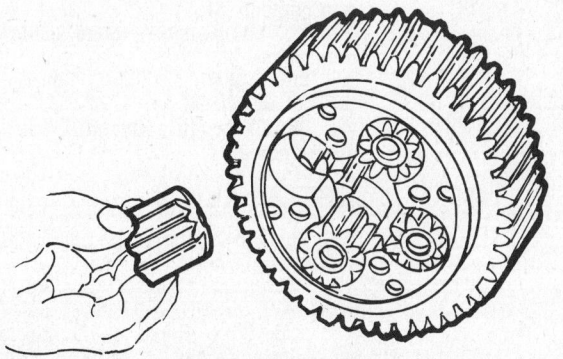

Installing pinion gears

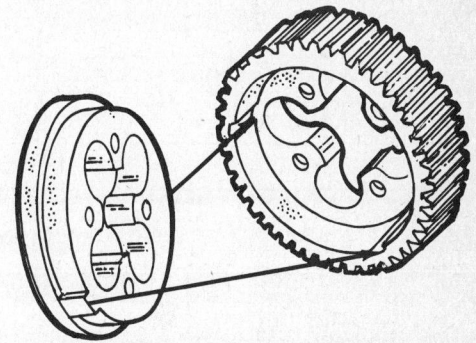

Positioning cores in ring gear

Massey-Ferguson

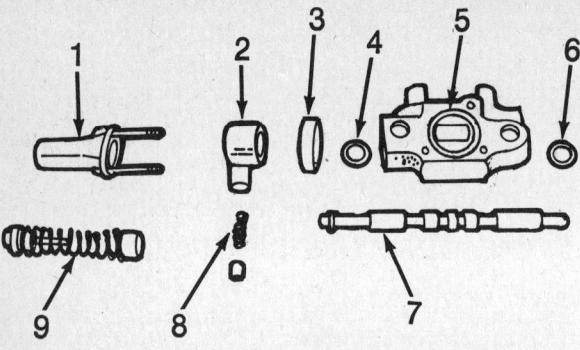

1. Cap
2. Detent housing
3. Spacer
4. O-ring
5. Body
6. O-ring
7. Spool
8. Detent assembly
9. Spring and retainers

Mid-mounted linkage control valve

1. Fitting
2. Fitting
3. Seal
4. Cylinder barrel
5. Piston and rod
6. Plug with seal

Hydraulic cylinder components

1. Cap
2. Spacer
3. O-ring
4. Body
5. O-ring
6. Spool
7. Spring and retainers

Point hitch control valve

HYDRAULIC SYSTEM

Auxiliary Control Valve

REMOVAL AND INSTALLATION

1. Remove rear fender and seat assembly.
2. Disconnect hydraulic lines.
3. Remove bolts securing valve and remove valve.
4. Reinstall valve and secure with retaining bolts.
5. Reconnect hydraulic lines.

Control Valve

REMOVAL AND INSTALLATION

The control valve may be either single or double. The second valve is used if tractor is equipped with 3-point hitch. Three tie bolts hold valve bodies together. If spool or body is damaged replace complete valve assembly. To replace O-ring seals valve may be disassembled as follows:

1. Remove tie bolt nuts and separate valve bodies.
2. Remove spool and O-rings.
3. Reassembly valve using new O-rings and assemble bodies together.

Hydraulic Cylinder

REMOVAL AND INSTALLATION

To service hydraulic cylinder proceed as follows:

1. Remove cylinder from tractor.
2. Remove fitting from head end of cylinder and bump plug and seal from barrel.
3. Remove fitting from end of piston rod.
4. Push piston and rod out of cylinder barrel.
5. Install new seals, lubricate with petroleum jelly and carefully insert rod and piston into barrel bore.
6. Push plug and seal into end of barrel.

NOTE: Use of shim stock will aid in protection of seals when piston and plug is installed.

7. Align hole in plug with hole in barrel and install fitting.
8. Install cylinder on tractor.

ENGINE

These tractors use Briggs & Stratton engines. For complete engine service, see the Briggs & Stratton engine part of the Engine Unit Repair section of this book.

REMOVAL

1. Remove hood and grille assembly.
2. Disconnect battery cables and electrical wires from engine.
3. Disconnect electrical clutch control wire.
4. Disconnect cable from starter motor.
5. Shut off fuel and remove line from carburetor.
6. Disconnect choke and throttle cables from engine.
7. Remove battery ground cable bolt from battery platform.
8. Remove pin from drive coupler and slide coupler sleeve rearward.
9. Remove engine mounting bolts and lift engine out.

INSTALLATION

1. Position engine on frame and secure with retaining bolts.
2. Reconnect cables, wires and fuel lines.
3. Slide drive coupler onto engine shaft and secure with pin.
4. Install hood and grille assembly.

Massey-Ferguson

MF-1655 & MF-1855

FRONT AXLE, WHEELS AND MID-PTO

Front Axle and Mid-PTO

REMOVAL

Mid power take-off is part of front axle pivot, therefore servicing of these units are together.

1. Remove front pulley shield and remove drive belts.
2. Remove retaining ring and remove pulley.
3. Unscrew large nut and bearing assembly off axle mounting sleeve and slide off.
4. Remove retaining ring from shaft.
5. Remove large nut from rear of axle mounting sleeve and slide PTO shaft with nut and bearing out rearward.
6. Disconnect drag link from steering pivot and attach a suitable lifting device to raise tractor.
7. Raise tractor enough to remove pressure on axle and remove mounting sleeve.
8. Continue raising tractor and remove axle, thrust washers and pivot tube.
9. Inspect bushings if worn, remove old bushings and install new ones flush with machined surface of axle.
10. Install bushings in bell crank arm flush with machined surface.

INSTALLATION

1. Place pivot tube in axle and position thrust washers on end of tube.
2. Lower tractor over axle pivot tube and thrust washers.
3. Install axle mounting sleeve with flat on sleeve down and rearward.
4. Install rear nut and bearing assembly on PTO shaft and install retaining ring in groove on shaft.
5. Install shaft into axle mounting tube at rear.
6. Secure shaft in place by turning nut and bearing onto threaded end of axle mounting tube.

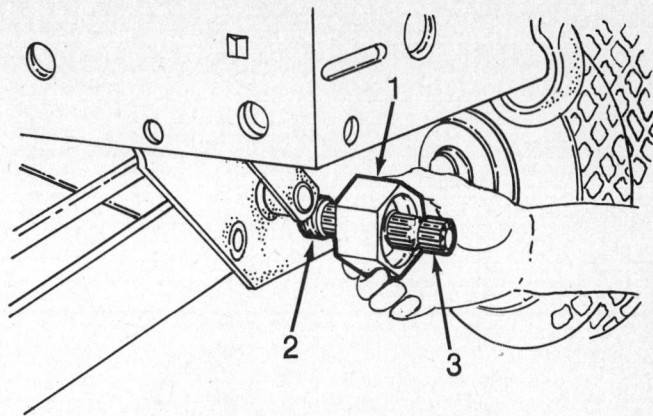

Removing nut and bearing assembly
1. Nut and bearing assembly
2. Axle mounting sleeve
3. PTO shaft

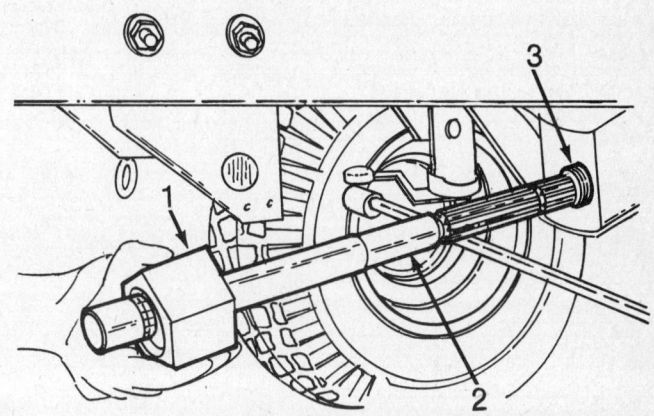

Removing PTO shaft assembly
1. Nut and bearing assembly
2. PTO shaft
3. Axle mounting sleeve

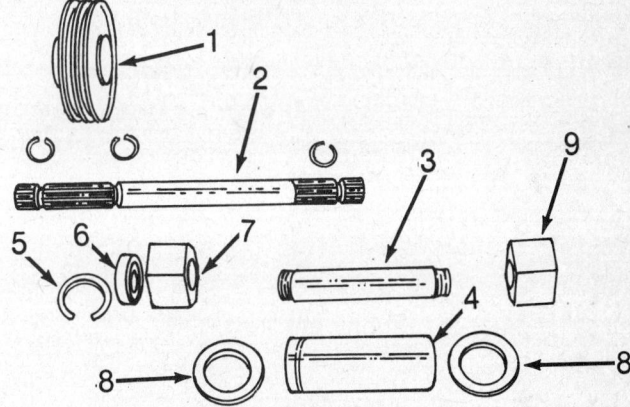

Mid-PTO and axle pivot components
1. Pulley
2. PTO shaft
3. Axle mounting sleeve
4. Axle pivot tube
5. Retaining ring
6. Bearing
7. Nut
8. Thrust washer
9. Nut and bearing assembly

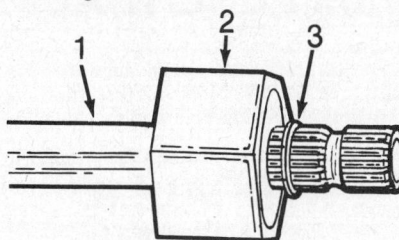

Rear nut and bearing installed on PTO shaft
1. PTO shaft
2. Nut and bearing assembly
3. Retaining ring

695

Massey-Ferguson

7. Position shaft so retaining ring can be installed at front.
8. Install nut and bearing over front end of PTO shaft and tighten both large nuts securely.
9. Install double pulley and secure with retaining ring.
10. Connect drag link to bellcrank and tighten nut securely.

Wheel Bearing

REPLACEMENT

1. Remove wheel and clean bearing with a suitable solvent. Remove old grease from inside of hub.
2. Inspect bearings and seals for wear. Replace as required.
3. Pack each bearing with lithium base grease. Install inner bearing in wheel and install new seal with lip of seal toward bearing.
4. Pack grease between lips of seal and bearing.
5. Install wheel on spindle and tighten retaining nut 15-20 ft.lb. while rotating wheel.
6. Install retainer on nut so that castellations are aligned with cotter pin hole.
7. Back off nut and retainer one castellation and secure with cotter pin.

CHECKING AND ADJUSTING TOE-IN

Adjust toe-in so the distance between front of wheels measures 1/8 in. less than at back. Adjust by loosening tie rod locknuts and turning tie rod.

STEERING

Steering Gear

REMOVAL

1. Disconnect battery and remove wire connector from ignition switch.
2. Use a suitable pulley and remove steering wheel.
3. Remove panel over driveshaft and side panels on each side of tractor.

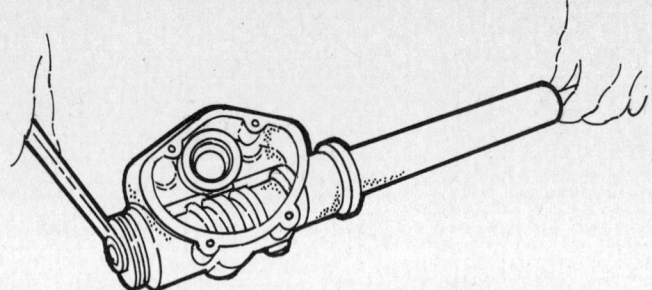

Setting preload on steering shaft bearings

4. Disconnect drag link and remove arm from steering sector shaft.
5. Remove bolts securing instrument panel in place.
6. Remove bolts securing steering gear and work assembly down and out toward right side of tractor.

INSTALLATION

1. Work steering gear up from right side of tractor and into place.
2. Secure steering gear assembly into place with retaining bolts.
3. Install steering arm while aligning timing marks. And secure with retaining washer and nut.
4. Install bolts securing instrument panel.
5. Install steering wheel, side panels and panel over driveshaft.
6. Install wire terminal on ignition switch and connect battery cables to battery.

DISASSEMBLY

1. Remove capscrews securing plate and remove plate.
2. Slide sector shaft from housing.
3. Remove lock and adjusting nut at end of shaft and remove shaft and bearings from housing.
4. Remove upper seal and bearings from housing.

ASSEMBLY

1. Lubricate both bearings and insert one cup and bearing cone into gear housing, followed with another bearing cone and cup.

2. Install adjusting nut and tighten until steering shaft has a slight bind when turned. At this point adjust nut until shaft has no end-play with no binding when turned. Secure adjusting nut with lock.
3. Install bearing, seal and retainer over upper end of shaft and into housing.
4. Fill cavity in housing with good grade of lithium base grease and install sector shaft.
5. Position gasket and plate over shaft and secure with retaining bolts.
6. Adjust steering gear backlash as follows:
   a. Turn in on adjusting screw until a slight drag is felt on steering shaft at mid range of steering (this would be a straight ahead position of front wheels).
   b. Adjust screw by turning in or out to eliminate any binding and secure in this position with locknut.

Spindle Arm Bushing

REPLACEMENT

1. Remove wheel and disconnect tie rod from spindle arm.
2. Remove retaining ring at upper end of spindle and remove spindle.
3. Remove upper bushing and reinstall new one flush with machined surface of axle.
4. Drive lower bearing out and reinstall new.
5. Reinstall spindle, connect tie rod and install wheel.

TRANSMISSION, REAR AXLE AND DIFFERENTIAL

For complete service on these components, see the MF-1450, 1650 section immediately preceding this section.

HYDRAULIC SYSTEM

For complete hydraulic system service, see the MF-1450, 1650 section immediately preceding this section.

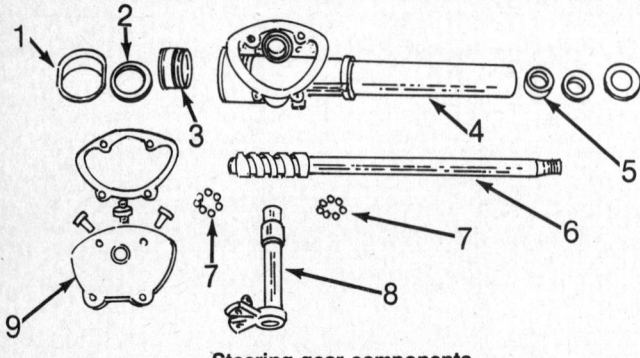

Steering gear components

1. Lock
2. O-ring
3. Adjusting nut
4. Housing
5. Bearing
6. Shaft
7. Bearings
8. Sector shaft
9. Plate

Massey-Ferguson

ENGINE

These tractors use Briggs & Stratton engines. For complete engine service, see the Briggs & Stratton part of the Engine Unit Repair section of this book.

REMOVAL

1. Remove hood and grille assembly.
2. Remove right side panel to gain access to tractor belt tension adjustment.
3. Disconnect battery cables.
4. Close fuel tank shut-off, disconnect fuel supply line from pump and remove fuel tank.
5. Disconnect electrical wiring (alternator wire) at right rear of engine and remove harness clamp bolt.
6. Loosen traction belt tension by turning nuts at lower portion of rod and remove pin from end of rod.
7. Remove traction belts.
8. Remove retaining clip from traction drive belt tension bracket.
9. Remove complete exhaust system and disconnect throttle and choke cables from engine.
10. Remove pin and slide coupling rearward on PTO shaft if equipped.
11. Disconnect battery cable from starter motor.
12. Disconnect ignition wire from ignition coil.
13. Disconnect electric clutch wire (blue wire) at right front of engine.
14. Remove front PTO drive belts.
15. Disconnect ground strap and remove engine mounting bracket nuts at front of engine.
16. Drain engine oil and remove drain pipe.
17. Remove nuts at rear of engine mounting brackets.
18. Attach a suitable lifting device and raise engine just enough to clear mounting studs.
19. Slide traction drive tension bracket from its pivot post while moving engine forward. Continue lifting engine clear of tractor.

INSTALLATION

1. Position engine in place and start tension bracket over pivot post at same time.
2. Continue aligning mounting brackets with studs and secure engine in place with retaining nuts.
3. Reconnect the following:
   a. Ground strap.
   b. Ignition wire to coil.
   c. Electric clutch wire.
   d. Cable to starter motor.
   e. Throttle cable.
   f. Alternator wire at right rear of engine.
4. Install choke cable.

MF-25 and MF-130

BRAKES

REMOVAL

1. Disconnect the brake rod.
2. Remove the three capscrews retaining the front brake shoe assembly and remove the complete assembly. The rear stationary shoe can be removed without removing the brake pulley. To do so, proceed, as follows:
   a. Remove the retaining screw.
   b. Tighten the adjusting screw and wedge the stationary shoe in the pulley groove. Then, remove the adjusting screw.
   c. Rotate the wheel until the shoe which is wedged in the pulley groove can be removed at the front opening in the axle housing.

INSTALLATION

When the front shoes are installed, the long side of the lining is toward the top. To remove the brake pulley, it will be necessary to remove the axle.

NOTE: **If the axle is to be removed, the housing will have to also be removed in order to protect the axle oil seals.**

If any indication of oil is found on the brakes, it will be necessary to replace the axle oil seal.

ADJUSTMENTS

To adjust the brakes, support the tractor on jack stands so the rear wheels can be rotated.

1. Make sure the retaining screw is tightened secure.
2. Turn the adjusting screw in until the stationary shoe contacts the brake pulley. (Wheel will be difficult to turn.)
3. Loosen the adjusting screw ½ turn and tighten locknut.
4. Adjust linkage rod to give a free pedal travel of 1½ in.
5. Adjust both pedals for balance of travel.

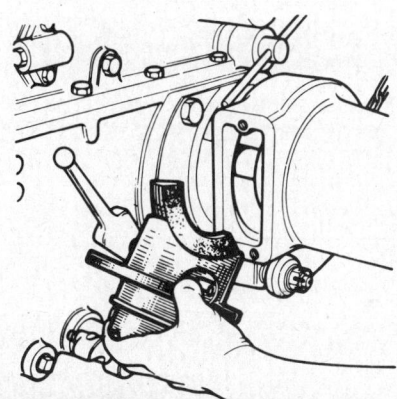

Installing primary brake shoe

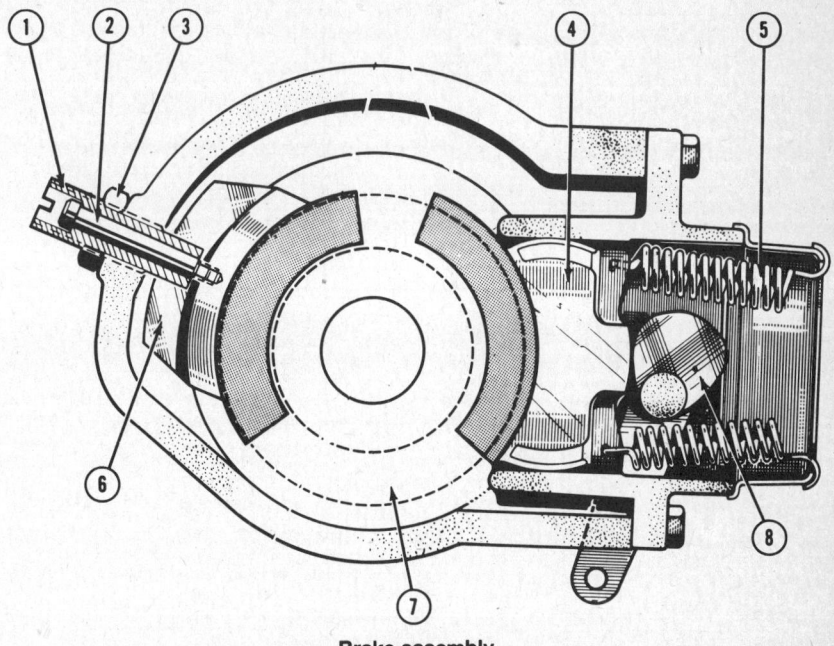

Brake assembly

1. Adjusting screw
2. Retaining screw
3. Locknut
4. Primary shoe
5. Retracting spring
6. Stationary shoe
7. Brake pulley
8. Cam

Massey-Ferguson

CLUTCH

REMOVAL AND INSTALLATION

1. Disconnect and remove the battery.
2. Remove the two nuts at the top of the clutch housing.
3. Loosen the four bolts holding the steering housing and instrument panel assembly in order to lift the assembly up far enough to clear the clutch housing.
4. Disconnect light wiring.
5. Disconnect the two wires from the starter safety switch (one from the starter switch on the instrument panel; the other from the starter solenoid).
6. Disconnect the oil feed and return lines from the hydraulic oil cooler at the connections under the starter.
7. Disconnect foot throttle linkage.
8. Remove the exhaust system.
9. Remove fuel lines from tank to auxiliary fuel pump and to thermostart.
10. Place a suitable stand under the frame and chock the front wheels.
11. Lift up the steering assembly, along with the rear of the fuel tank, approximately 1½ in. and block between the tank and the engine.
12. Place a floor jack under the transmission case.
13. Remove the three capscrews inside the frame on each side and one on the starter flange.
14. Remove the clutch housing cover and uncouple the clutch shaft sleeve and clamp. Notice the position of this clamp on the shaft. Long lip over snapring and toward the rear. If the ground speed PTO is now engaged, the rear of the tractor can be moved backwards by turning the PTO shaft.

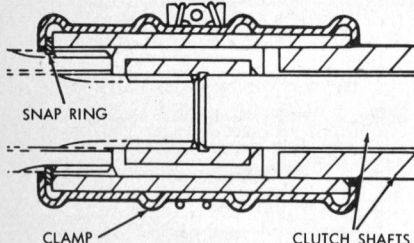

Clutch shaft sleeve and clamp

TRANSMISSION

Power Take-Off Shaft

DISASSEMBLY

The PTO shaft is 1⅜ in. in diameter and rotates 540 rpm at an engine speed of 1890 rpm.

If the PTO shaft is to be removed and replaced without removing the lift cover, the PTO shift lever must be placed in the engine PTO position. The lever must not be moved from this position until shaft is re-

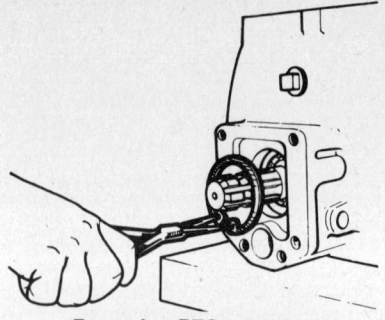

Removing PTO snapring

placed; otherwise, the gear will fall into the bottom of the transmission case, and cannot be installed without removing lift cover.

To remove the shaft assemblies, proceed, as follows:
1. Drain transmission oil.
2. Remove the PTO cover plate.
3. Remove oil seal in order to gain access to the snapring holding the bearing. (This seal must be renewed.)
4. Remove snapring and pull PTO shaft out with its bearing.

Transmission Case

DISASSEMBLY

The following can be done without splitting the transmission case from the clutch housing and engine.
1. Place suitable wedge blocks on each side between the front axle and engine.
2. Support engine and transmission on jacks or stands.
3. Disconnect rear light wires, remove seat, disconnect lift rods from lift arms, and remove the 14 capscrews securing lift cover.
4. Fit a suitable lifting plate to the seat mounting bolts.

Removing lift cover

5. Raise lift cover vertically, taking care not to hit and damage the hydraulic pump pinion.

NOTE: Do not lose shift rail detent springs. No hydraulic lines need to be disconnected unless external hydraulics are installed.

Power Take-Off
1. Drain oil in transmission case.

2. Remove cover plate at rear of transmission case and remove PTO shaft oil seal.

NOTE: Seal has to be removed in order to gain access to the snapring. A new seal must be installed when PTO shaft is assembled.

3. Remove snapring.
4. Remove PTO shaft with bearing.
5. Remove rear wheels.
6. Remove right and left axle housings, planetary units, and axles as a unit.
7. Remove the differential assembly out the top of the transmission case.
8. Unlock the castellated nut on the pinion shaft and unscrew it. Withdraw the pinion shaft rearward, catching nut, lock ring, washer and front bearing.

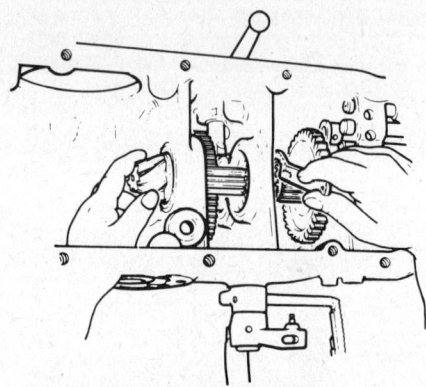

Removing pinion shaft nut

NOTE: There may possibly be a thrust collar between the nut and splined sleeve. If the drive pinion is renewed, this collar should not be used, the thrust of the splined sleeve being taken directly on the shoulder of the pinion shaft. If the pinion shaft is reused, the thrust collar should be installed to obtain a specified end float of .098 to .087 in. for the splined sleeve sliding pinion and splined sleeve.

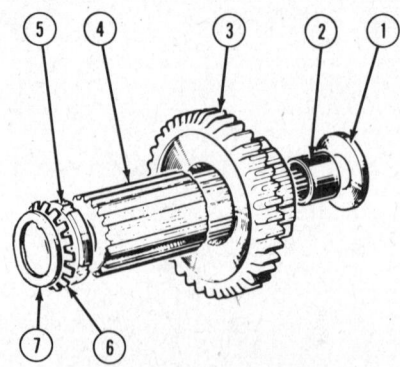

High-low shift assembly

1. Thrust washer
2. Bearing
3. Pinion
4. Splined sleeve
5. Nut
6. Lockwasher
7. Lip washer

Massey-Ferguson

Removing
High-Low Shift Assembly
1. Remove the splined sleeve.
2. Remove selector fork and pinion.
3. Remove thrust washer.

Removing Ground Speed PTO-Driven Pinion
1. Remove the differential assembly, pinion shaft, and PTO shaft.
2. Remove rear snapring on the PTO pinion sleeve and slide sleeve forward so the front snapring can be removed.
3. Remove sleeve from rear.
4. Remove gear and bearing.

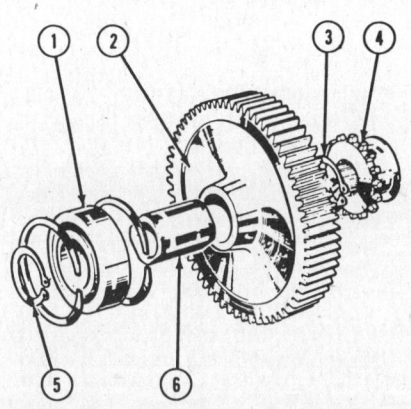

Ground speed PTO drive pinion assembly

1. Bearing
2. Pinion gear
3. Snapring
4. Shift collar
5. Snapring
6. Sleeve

SPLITTING THE TRANSMISSION AND CLUTCH HOUSING

The transmission case is held to the clutch housing by two large external studs at lower side of case and five smaller internal studs. To service the following items in the transmission, it will be necessary to split the transmission housing from the clutch housing and remove the differential assembly.

To split the transmission and the clutch housing, proceed, as follows:
1. Remove cover plate at top of clutch housing.
2. Remove clamp assembly holding transmission shaft and clutch shaft together.

NOTE: In assembly, install large lip over snapring.

3. Disconnect the two wires at the safety switch.
4. Remove the two external nuts from lower side and the five internal nuts through opening at top of clutch housing.
5. Use suitable hoist and remove transmission case from clutch housing.

DISASSEMBLY OF TRANSMISSION AFTER SPLITTING FROM CLUTCH HOUSING

1. Remove shift rail locking balls and groove pins.

NOTE: Install new groove pins in assembly procedure.

2. Remove right shift rail (selector lip forward, open side to left); shift fork pin hole forward.
3. Remove center shift rail and selector (notice alignment of selector hole to rail); shift fork pin hole rearward.
4. Remove left shift rail (selector with lip forward and open side to right); shift fork pin hole rearward.

NOTE: All parts can be removed, except shift rail, by sliding rail to rear. To remove shift rail, it will be necessary to remove expansion plug at front end opening.

5. Remove high-low shift rail (selector hole forward, shift fork flat side rearward).

Disassembly of Upper Transmission Shaft
1. Remove the oil seal in front of transmission case, which must be replaced on reassembly.
2. Remove snapring.
3. Remove PTO input shaft with bearing. Leave the pump drive pinion on the primary shaft. This pinion will be removed when the primary shaft is removed.
4. Disengage the snapring from its groove between the synchromesh and the second speed sliding gear.
5. Hold the snapring against the thrust washer and slide the shaft out of the bearings and gears.
6. Remove the pump drive pinion which was left on the shaft when the PTO input shaft was removed.
7. Remove the two gears at front of shaft (small gear, 15-tooth, at front and large gear, 18-tooth with both hubs to the rear).
8. Remove synchronizer assembly.

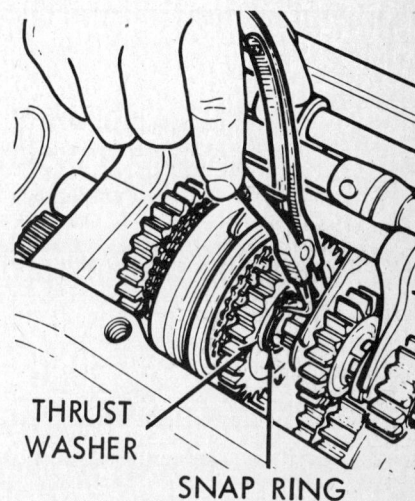

Disengaging snapring

Disassembly of Lower Transmission Shaft
1. Remove expansion plug from front of case.
2. Remove snapring, and gear from front end of front PTO shaft.
3. Remove oil seal, spacer, and snapring.
4. Remove the solid shaft with bearing forward, out of the case.
5. Remove snapring from rear of lower hollow shaft.
6. Remove shaft forward, collecting the pinions and spacers, as follows:

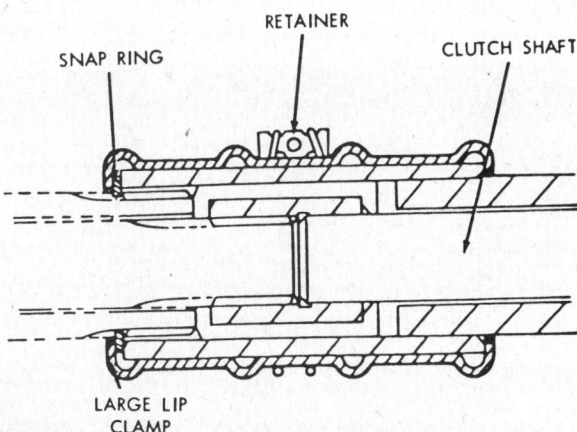

Transmission shaft clamp

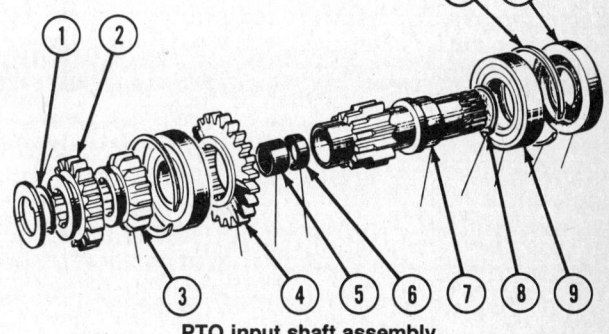

PTO input shaft assembly

1. Snapring
2. 18 tooth gear
3. 15 tooth gear
4. Pump drive gear
5. Needle bearing
6. Oil seal
7. PTO input shaft
8. Snapring
9. Bearing
10. Snapring
11. Oil seal

699

Massey-Ferguson

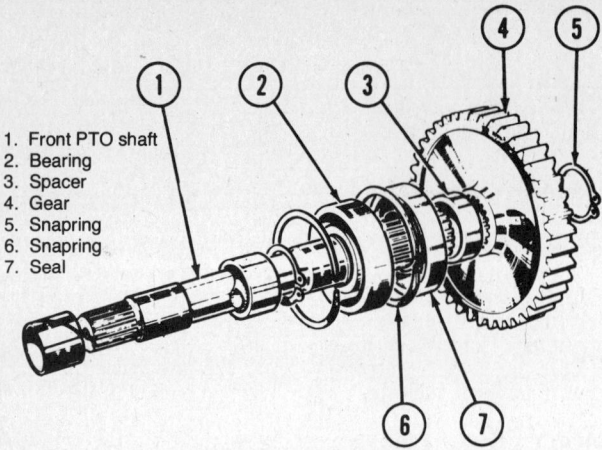

1. Front PTO shaft
2. Bearing
3. Spacer
4. Gear
5. Snapring
6. Snapring
7. Seal

Lower solid shaft assembly

a. Front gear: 36-tooth hub, rearward
   41-tooth hub, forward
   37-tooth hub, rearward
   26-tooth hub, forward
b. Long spacer: 32-tooth hub, rearward
   37-tooth hub, forward
c. Short spacer: 18-tooth long hub, forward

Lower the hollow shaft, assembled in the case.

REASSEMBLY OF TRANSMISSION

The reassembly of the transmission is the reverse of the disassembly procedure, however, to avoid mistakes, the following special points should be followed.

NOTE: The preload of the two pinion shaft bearings is the only adjustment to be made in the assembly of the transmission.

For the assembly of bearings and bushings, use suitable tool and avoid any impact stress while installing. When installing oil seals, use special tool and seal protector. Check all bearings before installing. Where a groove pin is used, such as in the shift forks, always install a new pin.

Assembly of Lower Shaft

Arrangement of gears on the shaft. This hollow shaft has two sets of splines of a different size and, in assembly, the larger gears may hang on the smaller splines of the shaft. (Do not force shaft through gears or damage may be caused to the splines.)

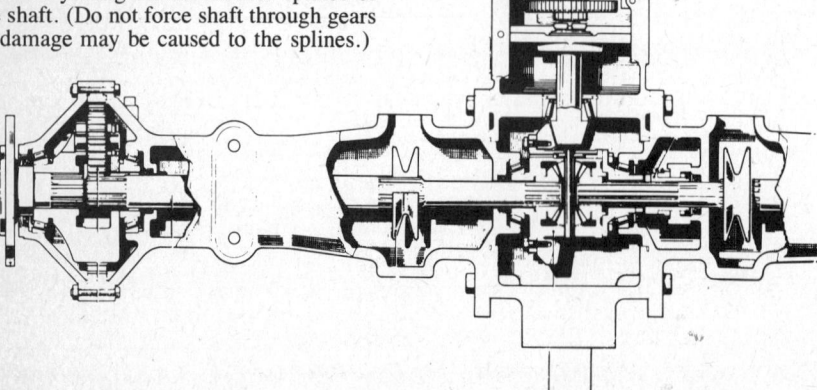

Differential and rear axle cross section

Assembly of Front PTO Shaft

This shaft rotates in the lower hollow shaft and is supported by a needle bearing in the front and a bushing in the rear. The clearance between the shaft and the bushing is .002 in. to .004 in.

1. Install shaft and bearing as an assembly and install snapring to secure in place.
2. Install spacer, gear, and snapring.
3. Install expansion plug with a sealing compound.

Assembly of Upper Transmission Shaft

1. Install the shaft in case and push as far to the rear as it will go.
2. Fit onto the shaft the synchromesh assembly (large gear 28 teeth forward and shifter hub taper rearward), pegged thrust washer, snapring, second speed sliding gear (18-tooth), first and reverse speed sliding gear (15-tooth); both shift hubs rearward. Slide through front bearing (this bearing is of a sealed type and seal should be to the rear) and, at the same time, position pump drive gear in place with hub rearward.
3. Position shaft by sliding forward and, at the same time, holding snapring with pliers in order to pass over grooves in the shaft. Position snapring in groove.
4. Check bearing and seal in primary PTO shaft. See procedure for installing seal and bearing. Use tool No. 267A to install seal bearing.
5. Install the primary PTO shaft, using seal protector tool to prevent damage to new seal. Use tool No. MFN 267 B to protect seal.
6. Install bearing, snapring, and oil seal. Use seal installing tool MFN 245.
7. After shift rails and forks have been installed, check the free clearance of the synchromesh ring at both limits of its full travel. If the cones are in good shape, this clearance should be .098-.118 in.

If the sliding jaws of the synchromesh bushing mesh with those of either the third or fourth speed pinions, there is an excessive clearance in the synchromesh assembly and the cones should be replaced.

Auxiliary Sliding Gear (High-Low)

Assemble the splined sleeve of the auxiliary sliding (high-low) gear with the thrust washer and gear on the rear of the upper transmission shaft (large gear to rear, inside taper of thrust washer toward shaft).

Assembly of the Ground Speed PTO Pinion

1. Install ground speed PTO pinion with its sleeve. The sleeve is splined internally at the forward end and must be pushed through the bearing from the rear far enough to be able to install the gear and snapring on the front end. It is then moved rearward so the rear snapring can be installed.
2. Install PTO sliding coupler on the splines of the lower transmission shaft and the PTO selector with snapring, retainer washer and O-ring.
3. Position lock ball and spring and install shift lever with new groove pin.
4. Install PTO shaft with bearing and snapring.
5. Install oil seal, using special tool. Use tool No. MFN 246.
6. Assemble the oil strainer with seal, plug, washer and spring.
7. Install rear cover plate on PTO shaft.

Drive Pinion and Differential

1. Install pinion shaft and bearings in housing. Fitting washer, locking ring, and nut also start shaft into splined sleeve.
2. Adjust preload on bearing to a rolling torque of 4-10 in.lb.
3. Lock the nut.
4. Install differential assembly, and axle housings.

Massey-Ferguson

ASSEMBLY OF TRANSMISSION TO CLUTCH HOUSING

1. After fitting the shift rails and forks, the transmission may be attached to the clutch housing.
2. Clean all surfaces, check expansion plugs, making sure they are sealed with a sealing compound.
3. Place new O-ring on auxiliary gearshift rail (longest rail).
4. Check the two oil cooler pipes on the right side of the transmission case. These pipes must not extend beyond the surface of the case. They may be recessed up to .004 in.
5. Place O-rings on the hydraulic oil cooler pipe openings in the clutch housing by the use of grease.
6. Attach transmission case to clutch housing and torque internal nuts: 55-60 ft.lb.; and external nuts: 100-110 ft.lb.

REPLACING THE HYDRAULIC LIFT COVER

1. Check the shift rail detent balls and springs and set sliding gears in a neutral position.
2. Install a new gasket and check O-ring on stand pipe.
3. Check the correct location of all transmission components.
4. Lower lift cover carefully into position, being careful not to damage the hydraulic pump pinion.
5. Tighten lift cover bolts evenly to a torque of 32-39 ft.lb.
6. Install lift arms and connect light wires.
7. Tighten drain plugs and install 5.48 U.S. gallons SAE 10w30 oil.

DIFFERENTIAL AND REAR AXLE

Differential and Pinion

REMOVING THE DIFFERENTIAL

In order to remove the differential assembly, it will be necessary to remove the lift cover.

1. Attach a lifting device to the seat mounting bolts, remove the fourteen capscrews holding cover to transmission housing and raise lift cover vertically.
2. Place a wedge on each side between the front axle and the engine.
3. Support the transmission on a jack or stand and remove the rear wheels.
4. Remove mounting capscrews on each axle housing and remove housings and planetary units as an assembly.
5. Remove the differential assembly through upper part of transmission case.

DISASSEMBLING THE DIFFERENTIAL

1. Mark the position of the differential lock coupling on the differential housing.
2. Remove the 8 bolts holding the assembly together.
3. Drive out the pins holding the differential pinion shaft.
4. Remove the shaft, the pinions, the thrust washers and the axle shaft pinion gears and their thrust washers.
5. Remove the 8 capscrews holding the ring gear to the differential cage, and remove ring gear.

REASSEMBLING THE DIFFERENTIAL

1. Install ring gear on differential cage, using capscrews and locks. Torque bolts 23-27 ft.lb. and secure locks.

NOTE: Shims may be used between ring gear and differential cage to obtain a clearance between ring and pinion of .004-.013 in.

2. Install axle shaft pinions, differential pinions and their thrust washers. Install pinion shaft and secure with a new lock pin.
3. Install differential lock coupling on the cage aligning the marks made in disassembly. Torque capscrews 23-27 ft.lb. and secure locks.

NOTE: The pre-load on the differential bearings is determined by manufacturing tolerances and cannot be adjusted. After reassembly of the differential assembly and the axle housing, the axle should be able to rotate under a rolling torque not to exceed 16.5 ft.lb. when the other axle is held stationary and the transmission in neutral.

Rear Axle Shafts and Housings

GENERAL DATA

The planetary gear reduction unit is located at the outer end of the axle housing, and has a gear ratio of 8.1-1.

The backlash between the pinion gears and the planetary ring gear is .003-.007 in. The backlash between center pinion teeth and planetary pinion gears is 0-.004 in. Bearing pre-load—.002-.010 in. Adjustment is made by use of shims .002-.005 in. thickness. Lubrication—.68 U.S. pint straight mineral oil, SAE 90 per unit.

DISASSEMBLING THE PLANETARY UNIT

1. Support the rear axle housing on a suitable stand.
2. Remove wheel and drain planetary unit.
3. Punch mark the relative position of the ring gear and cover to the axle housing.
4. Remove the 8 bolts attaching the planetary unit.
5. Use a screwdriver in notches provided and pry the unit apart, together with the shaft and carrier.
6. The outer ring gear can then be removed.

NOTE: Do not remove the axle shaft, since it would be necessary to remove the axle shaft housing to protect the oil seals in reassembly.

7. Drive groove pin securing the pinion shaft out.
8. Push the pinion shafts outward and remove the pinions with their needle bearings, two rows of 28 needles each and a spacer washer per pinion, also, a thrust washer on each side.

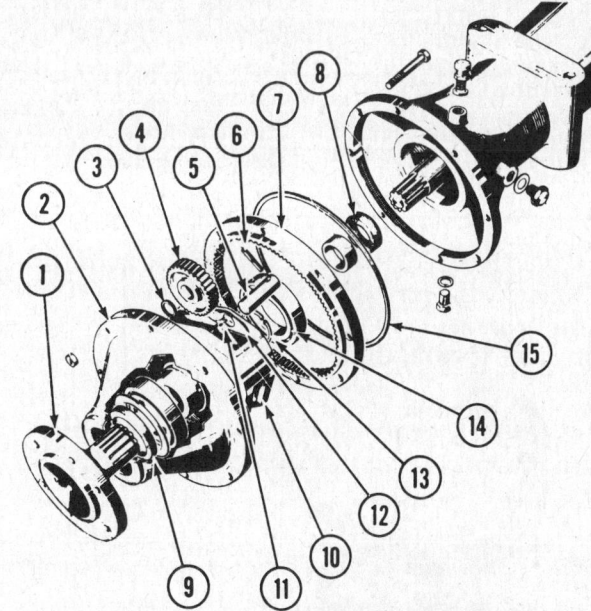

1. Axle
2. Housing
3. Spacer
4. Pinion gear
5. Pinion gear shaft
6. Groove pin
7. Ring gear
8. Oil seal
9. Oil seal
10. Carrier
11. Needle bearing
12. Thrust washer
13. Snapring
14. Bearing
15. Gasket

Axle planetary components

701

Massey-Ferguson

Installing pinions in planetary

9. Remove the snapring from the axle shaft.

10. Remove the carrier assembly from the axle by the use of a press with a capacity of at least 6 tons.

REASSEMBLING THE PLANETARY UNIT

1. Clean the grease from the splines and coat the edges of the planetary carrier splines with Locktite®, and press into place. The necessary pressure to do this is between 5,000 and 10,000 lb. Any assembly pressure less than 4,850 lb. is not permissible and it would be necessary to replace either the shaft or the planetary carrier, or both.

2. Replace the snapring on the axle shaft.

3. Install the pinion gears, bearings and shafts. This can be done by using grease to hold the needle bearings in place, and a thin piece of sheet metal. Leave the one thrust washer out until the pinion is located in position.

4. Install new groove pins in to secure the pinion shafts.

A bushing is fitted into the planetary carrier to support the axle shaft. This bushing should have an inside diameter after assembly of 1.693 to 1.692 in. The clearance of the shaft in this bushing should be .007 to .012 in.

5. Inspect the bearings, install a new gasket and assemble unit to axle housing. Torque bolts 19 to 22 ft.lb.

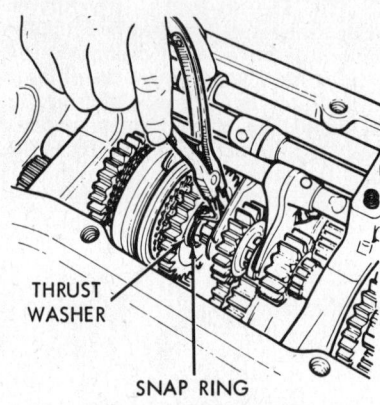

Disengaging snapring

Checking preload

ADJUSTING THE PLANETARY PRE-LOAD

The adjustment of the planetary bearing pre-load is determined by shims between the inner bearing outer race and the axle housing.

The initial pre-load is .002 to .010 in. and can be checked with a dial indicator.

After pre-load has been set and all bolts torqued, fill planetary with SAE 90 straight mineral oil.

REMOVING THE REAR AXLE AND HOUSING

To remove the axle shaft, it will be necessary to remove the axle housing as an assembly. The planetary assembly must be removed as described in paragraph Disassembly of the Planetary Unit.

To remove the right axle, proceed as follows:

1. Remove the differential lock control assembly by removing setscrew and sliding shaft out of the yoke, and remove yoke through opening at top of axle housing.

2. Put a light pressure against the spring, and remove the retaining snapring. The shift collar, spring and washer can then be removed. The previous procedure is necessary in order to gain access to the inner oil seal.

3. Install special tool No. MFN 690 on the inner end of the axle shaft to protect the inner oil seal.

4. Install seal protector collar No. MFN 689 in outer oil seal and unscrew driver hub from tool.

5. Remove the axle shaft through the outer end of axle housing while holding outer seal protector collar in place, and, at the same time, catch the brake hub as it slides off the end of the axle splines.

The removal of the left axle will be the same procedure as the right, with the exception of the differential lock.

ASSEMBLING THE REAR AXLE AND HOUSING

1. Check both oil seals and, it they are not damaged, use seal protector tools to install axle shaft. If new seals are to be installed, proceed as follows:

2. Install the inner seal, using tool No. MFN 690.

3. Install axle with brake hub in place.

4. After axle is in place, use tool No. MFN 689 and install the outer oil seal.

HYDRAULIC SYSTEM

Lift Cover

REMOVAL AND INSTALLATION

1. Disconnect light wires and remove seat.

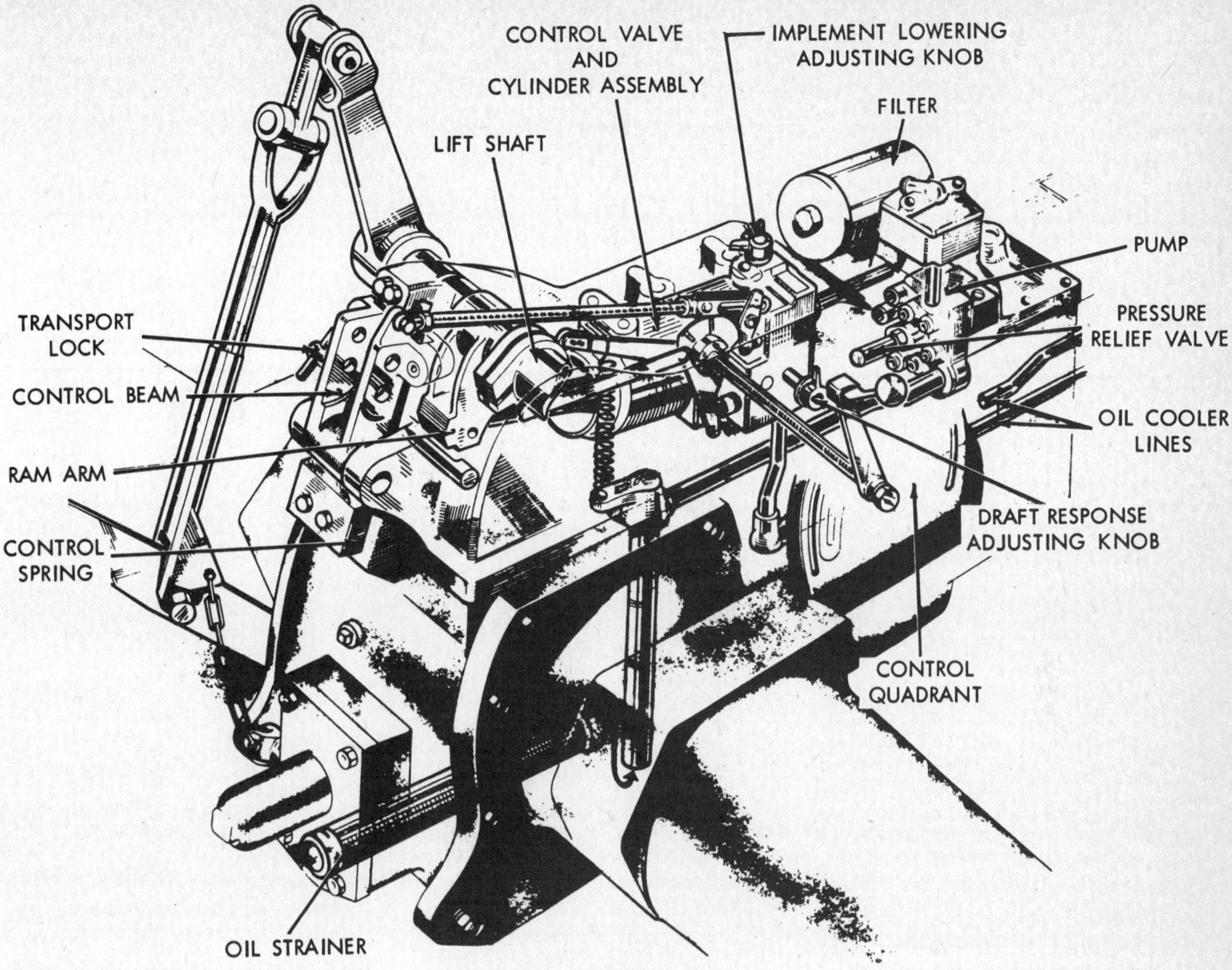

Massey-Ferguson

Hydraulic lift cover

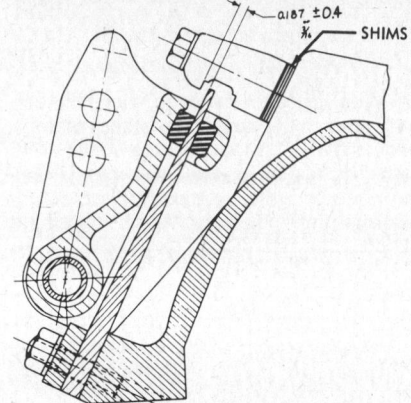

Control spring adjustment

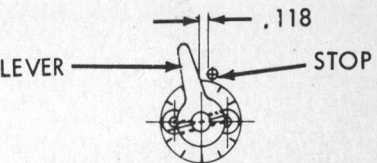

Draft response adjustment

2. Lock the lift arms in the raised position by positioning the transport lock.
3. Remove the capscrews securing the lift cover to the transmission case.
4. Fit two centering studs in the capscrew holes near the front side of the lift cover.
5. Attach the lifting tool MFN 298 to the seat mounting bolts and to the PTO shaft mounting bolts.
6. Raise lift cover and set on stand at rear of tractor.
7. Place a new gasket on the transmission housing and reinstall the two centering studs, if removed.
8. Raise lift cover from stand, rotate over the hinge and lower in place, being careful not to damage pump drive gear.
9. Install bolts, seat, lift links and light wires.

ADJUSTMENTS

No implement should be attached when making any adjustments on the hydraulic system. Two adjustments can be made with the lift cover installed on the tractor. They are, as follows:

Control Spring Adjustment

The stop bracket which limits the movement of the control spring is attached to the lift cover by two bolts. With the control spring in its neutral position, there should be a clearance of $3/16$ in. between the tip of the spring and the rear face of the stop. This clearance is obtained by shims between the housing and the stop. Shims are supplied in two thicknesses, .040 in. and .060 in.

Draft Response Adjustment

With the control lever removed, screw in the needle body until it bottoms on its seat. (Do not force needle against its seat.) Replace control lever and adjust to obtain a clearance of .118 in. on the left side of the stop, or, when lever is turned clockwise toward its stop, this clearance should be obtained.

The following adjustments require the removal of the lift cover.

Control Quadrant Position Adjustment

With lift cover removed, place the control in the rear notch of the quadrant, and loosen the two attaching bolts of the quadrant. Press the auxiliary valve in several times, making sure it is not binding. Turn the quadrant clockwise until the cam moves the roller into light contact with the auxiliary valve. Lock the two quadrant attaching bolts, making sure the quadrant doesn't move.

NOTE: Lift arms must be all the way toward the raised position before making this adjustment.

Massey-Ferguson

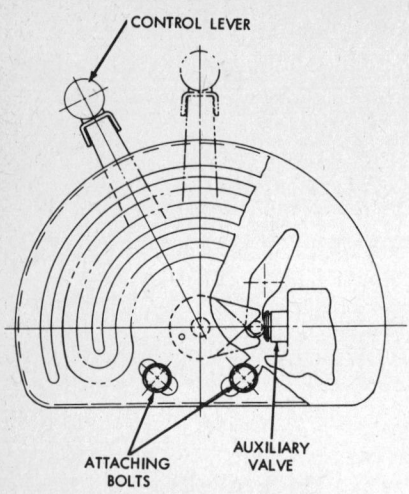

Control quadrant adjustment

Transport Adjustment

With lift cover removed, rotate lift arm toward raised position and place a .094 in. guage between the ram arm and the cast face of lift cover. Loosen cam and rotate until linkage pulls the secondary valve cam just enough over center of cam roller to allow the valve to click out. Tighten bolt on cam. To check this adjustment, lower lift arms and raise them very slowly past the ram arm lock. At the point where the ram arm reaches the .094 in. gauge setting, the secondary valve must click out.

This adjustment is critical because the ram arm could strike the rear of the lift cover and blow the relief valve or cause damage to the cover.

Position Control Adjustment

With lift cover removed, place the control lever in the rear notch of the quadrant.

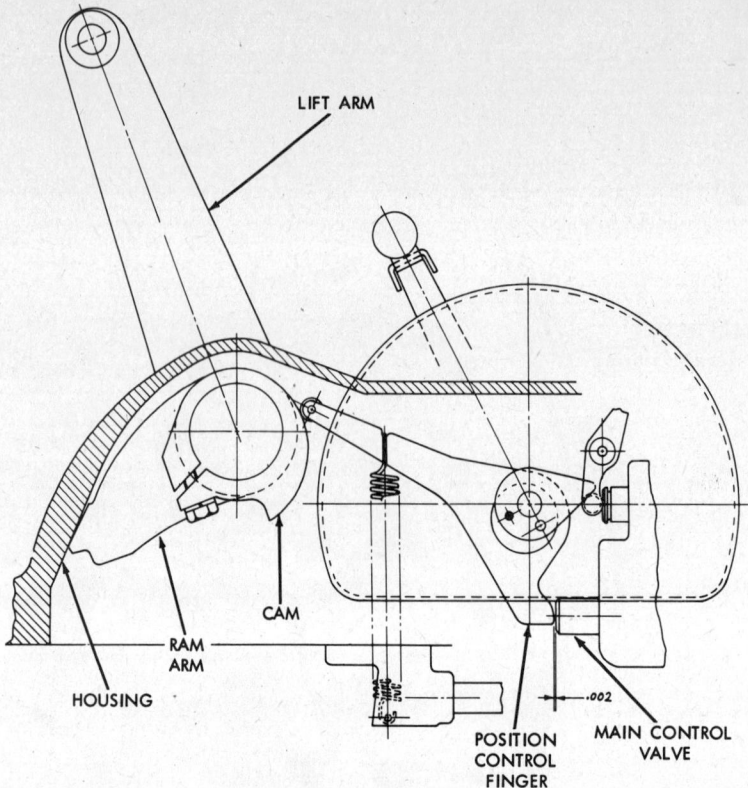

Position control adjustment

Loosen the position-control cam mounted on the lift shaft. (Check main control valve for freedom of movement.)

By means of the lift arms, keep the ram arm against the end of the housing, the position-control finger being held against the cam by its spring. Turn the cam so as to have a clearance of .002 in. between the finger and the main valve. In this position, tighten the cam clamping screws.

Draft Control Adjustment

With lift cover removed, place the control lever in the front notch of the quadrant. Force the control spring into contact with the rear face of its stop. This can be done by loosening the nuts holding the control spring in place and allowing this spring to contact the rear face of its stop.

The clearance between the draft control finger and the main valve should be .002 in. If necessary, correct this clearance by loosening the locknut and adjusting the link rod attached to the control beam.

ENGINE

REMOVAL

1. Drain the radiator and cylinder block, oil pan and fuel tank.
2. Remove the air cleaner bowl.
3. Disconnect and remove the battery.
4. Disconnect the air cleaner hose at the air cleaner body.
5. Disconnect the two rear bolts in the top section grille frame assembly which surrounds the air cleaner body.
6. Disconnect the hood support from the fuel tank.
7. Disconnect the lights to permit removal of hood.
8. Remove the screws securing the rear of the side panels.
9. Carefully support the hood assembly and remove the two bolts at the base of the grille in front of the front axle. Then, carefully spread the rear of the side panels and move the assembly forward to separate from tractor. Use care not to damage the sheet metal or to damage the rubber seals around the radiator.
10. Close the fuel shut-off valve at the sediment bowl, if the fuel was not drained earlier.
11. Disconnect the fuel supply line to the fuel lift pump.

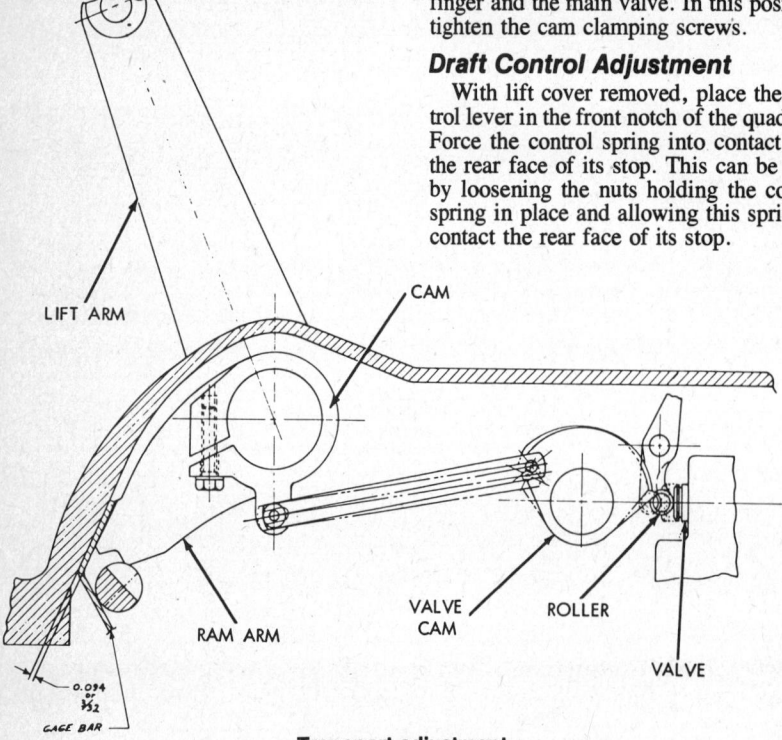

Transport adjustment

Massey-Ferguson

12. Disconnect and remove the filter leak-off line, the thermostart supply line, and the injector leak-off return line.
13. Disconnect the radiator support rod and loosen the two bolts securing the front of the fuel tank.
14. Remove the two bolts securing the rear of the fuel tank to the rear fuel tank bracket.
15. Move fuel tank rearward and carefully remove tank from tractor.
16. Disconnect the following items:
    - Upper and lower radiator hoses.
    - Cable from tractormeter drive.
    - Electrical wiring to starter and generator.
    - Electrical wiring to thermostart.
    - Temperature indicator sending unit.
    - Throttle control rod at the rear end.
    - Steering drag link at the rear steering arm.
    - Fuel shut-off control rod and slide rod rearward.
17. Remove the following items:
    - Oil pressure gauge line.
    - Fuel line, from injection pump to fuel filter.
    - Fuel line from fuel filter to injection pump.
    - Exhaust pipe and muffler assembly.
    - Starter motor, then, disconnect the hydraulic oil cooler lines beneath the starter motor.
    - Fuel line from fuel lift pump to fuel filter.
18. Install wooden wedges between the front axle and the engine frame to prevent the frame from rocking.
19. Support the engine and the transmission assembly in a suitable manner that will permit the units to be separated.
20. Unscrew and remove the following items:
    - The two nuts attaching the battery platform to the engine adaptor plate.
    - The two bolts at the bottom of the clutch housing.
    - Six long bolts attaching the engine frame and the adaptor plate, to the clutch housing.

NOTE: The two bottom bolts on each of the side frames are shouldered and act as pilot bolts. To avoid accidents, loosen all bolts before removing any of them.

21. Carefully separate the transmission assembly and the engine. To facilitate this operation, engage the ground driven PTO gear with the transmission gearshift levers in the neutral position, then, turn the PTO shaft. In separating the two assemblies, be careful not to damage the two oil cooler lines that pass through the adaptor plate.
22. Attach a chain, capable of lifting the engine (approximately 660 lb.), to the rear lifting bracket and the front fuel tank bracket.
23. Attach the chain to a crane and take up the slack to support the weight of the engine.
24. Support the side members of the engine frame in a suitable manner to allow removal of engine from frame.
25. Remove the bolts securing the two engine support brackets to the side members.
26. Lift the engine vertically from the frame and mount it on an engine stand or support it in an upright position.

INSTALLATION

Installing the Engine In Its Frame

1. Attach a chain to the engine, as described for its removal, and place the engine between the frame side members.
2. Slide the engine into place between the side members and align the attachment holes.

NOTE: If difficulty is encountered in alignment, insert a bolt or a tapered punch through one of the rear holes in the side members and into the engine adaptor plate.

3. Insert the setscrews in the front mounting pads, but do not tighten them.

Connecting the Engine and Engine Frame to the Transmission Assembly

1. Carefully align the rear of the engine with the clutch housing.
2. Remove any wires, lines or linkage that might interfere between the engine and the transmission and make sure that the two hydraulic oil cooler oil lines are arranged to pass easily through the hole in the engine adaptor plate.
3. Engage the transmission shafts with the clutch and carefully couple the engine and transmission together.

NOTE: The two bottom bolts in each side member are shouldered and act as pilot bolts.

4. Tighten all bolts, including those securing the engine to the front mounting pads.
5. Complete the reinstallation of the engine by reversing steps listed under the heading "Engine Removal."

NOTE: Be careful, when reinstalling the fuel lines, not to over-tighten them.

Before making any attempt to start the engine, make sure to bleed the fuel system of air.

Rocker Arm Shaft Assembly

REMOVAL

1. Remove fuel tank.
2. Remove the rear fuel tank bracket from the battery carrier.
3. Disconnect the breather pipe hose and remove the rocker cover.
4. Work from the center to both ends and gradually loosen the nuts securing the rocker arm support brackets. The rocker shaft oil line will pull out of the cylinder head at the cylinder head end.
5. Remove the nuts securing the brackets and remove the rocker assembly from the cylinder head.

DISASSEMBLY

1. Remove the retaining snapring from each end of the rocker shaft.
2. Remove the plain and spring washers, rocker arms, support brackets, spacer springs and rocker shaft oil line assembly from rocker shaft. Lay all parts in order, so they may be installed in their same location.
3. Remove the plugs from the rocker shaft so that the shaft may be thoroughly cleaned.

INSPECTION AND SERVICING

Clean all parts in a suitable solvent; make sure all oil holes are open, and proceed, as follows:

1. Inspect the rocker arm bearing surfaces of the shaft for scoring. Replace shaft, if scored.
2. Inspect for, and replace, any rocker arms that have seized and have loose bushings. Also, discard rocker arms with worn or pitted contact surfaces. The rocker arms should be an easy fit on the shaft without excessive side play.
3. Measure diameter of bearing surfaces on rocker shaft and inside diameter of rocker arm bushings. Proper clearance between rocker arm bushings and shaft is 0.0008-0.0035 in. Replace shaft and/or bushings, if clearance exceeds 0.0035 in.

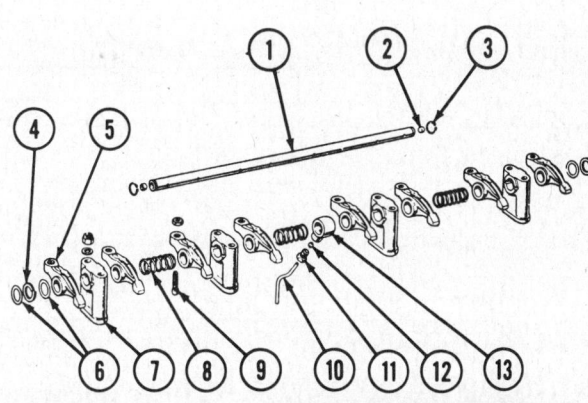

1. Rocker shaft
2. Rocker shaft end plug
3. Snapring
4. Spring washer
5. Rocker arm
6. Plain washers
7. Rocker shaft support bracket
8. Rocker shaft spring
9. Tappet adjusting setscrew
10. Oil line
11. Nut
12. Sleeve
13. Oil line connection

Rocker shaft components

705

Massey-Ferguson

When installing new rocker arm bushings, make sure oil hole in bushing is aligned with oil hole in rocker arm. Ream the new bushings, after they are installed, to an inside diameter of 0.6245-0.62575 in.

NOTE: Replacement rocker arms are supplied with the bushings already installed and reamed to size.

REASSEMBLY

1. Install all parts in their correct sequence.

NOTE: Check that the oil feed connection is in alignment with the large hole in the shaft.

2. Complete reassembly by installing a new snapring and plug at each end of the rocker shaft.

INSTALLATION

1. Inspect all of the push rods for straightness and install them in place, making sure each one is properly seated in its tappet.
2. Place the rocker arm shaft assembly over the studs and enter the oil feed line into the cylinder head.
3. Retain the assembly in position by installing the eight plain washers and new self-locking nuts. Tighten the two middle brackets down, first.
4. Adjust the valve clearance, as outlined under the heading "Adjusting the Valve Clearance".

Cylinder Head

The cylinder head is made of cast iron and is sealed with a copper asbestos steel gasket. The valve seats are machined directly in the cylinder head, but inserts are available, for service and can be installed for both the intake and exhaust.

The bottom face of the cylinder head contains replaceable combustion chamber caps which form the bottom half of the combustion chambers.

REMOVAL

NOTE: The fuel tank and cylinder head can be removed for minor servicing without the need of removing the hood and side panels. If major servicing is to be performed, however, remove the sheet metal to facilitate the repair.

1. Remove the battery and fuel tank.
2. Drain the radiator and cylinder block.
3. Disconnect the following items:
- Upper radiator hose.
- Exhaust pipe.
- Electrical wire to thermostart.
- Breather pipe hose connection.
- Temperature indicator sending unit from water outlet elbow.
- Rear of throttle control rod at bottom of bell crank.
- Clip retaining the return fuel line from injection pump to fuel filter.
- Disconnect front end of fuel shut-off rod and slide rod rearward.

4. Disconnect the clip retaining the fuel line going to No. 4 injector and remove all four of the high pressure fuel lines. Cap, or otherwise cover, all fuel line openings.
5. If the cylinder head is to be removed with the sheet metal in place, remove the screws securing the rear of the side panels.
6. Remove the two nuts securing the battery platform to the top of the clutch housing.
7. Disconnect the oil pressure gauge line at the gauge end, then, remove the two front bolts securing the steering gear housing. Loosen the two rear bolts, then, tip the housing and instrument panel assembly rearwards and retain it in this position by installing a block under the front end of the housing.

NOTE: The rear fuel tank bracket can be removed from the battery carrier to provide additional clearance.

8. Remove the rocker cover, rocker arm shaft assembly and push rods.
9. Remove the fuel leak off line assembly and remove the fuel injectors.
10. Progressively loosen and remove the cylinder head nuts in reverse order to the tightening sequence.

NOTE: The cylinder liners are a relatively loose fit in the cylinder block and might easily move, if the engine were turned with the cylinder head removed. To prevent such an occurrence, it is recommended that the liners be firmly located in position by placing suitable tubing over two of the cylinder head studs and securing them by means of washers and nuts.

INSPECTION

Disassemble the cylinder head and thoroughly clean it and the other connecting parts. Make sure all sealing surfaces are clean and not damaged. Clean all scale from the water passages in the cylinder head and inspect the head for cracks. Use a straight edge and determine if the cylinder head is straight and not warped due to overheating or extreme operation.

NOTE: Do not attempt to reface the cylinder head.

Examine the water jacket plugs on the cylinder head. Replace all damaged or defective parts.

ASSEMBLY AND INSTALLATION

All parts must be completely cleaned before replacement. Examine the water jacket plugs on the cylinder head and the studs on the top face of the cylinder block for looseness, damaged or stretched threads, etc. Cylinder head nuts should be examined to make sure they are not damaged.

1. Replace the valve assemblies, as outlined under the heading "Installing the Valves."
2. Remove the lengths of tubing securing the cylinder liners, if they were installed earlier, and carefully clean the faces of the cylinder head and cylinder block.
3. Use a new cylinder head gasket and coat both sides with a thin coating of aviation grade Permatex®, observing that the gasket is marked "front." Locate the gasket over the cylinder head securing studs.
4. With the bottom face of the cylinder head perfectly clean, it may be lowered into position on the studs and the securing nuts tightened down in the sequence. The nuts should be tightened to a torque of 40-42 ft.lb. The tightness of these nuts should again be checked after the engine has been run.

NOTE: The five long cylinder head nuts should be installed between the injectors.

5. Replace the generator adjusting link and adjust the fan belt.
6. Check the push rods for straightness and install them in place.
7. Loosen the tappet adjusting screws on the rocker assembly and install the rocker shaft onto the cylinder head, locating and tightening the oil feed pipe. Tighten down the rocker shaft, starting with the two middle brackets.
8. Adjust the valve tappet clearance to 0.012 in. "cold," for both intake and exhaust valves, by means of the adjusting screw and locknut.
9. To complete the assembly of the cylinder head to the cylinder block, reverse the procedure, as outlined under the heading "Cylinder Head Removal."

NOTE: When replacing the injectors, install new copper washers and tighten the nuts on the flange evenly to prevent the injectors from becoming cocked in the cylinder head. Check that the injec-

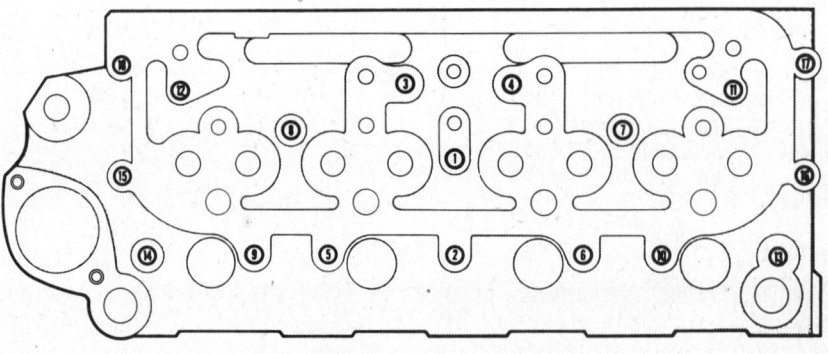

Cylinder head tightening sequence

tor flanges are positioned so that the cylinder head securing nuts are cleared by the scallops in the flange. Be careful, when tightening the connections on the high pressure fuel lines, not to overtighten them.

Combustion Chamber Caps

The combustion chamber caps are located in the bottom of the cylinder head and form the lower half of the combustion chambers. The cap contains the throat passage connecting the chamber to the cylinder. The combustion chamber inserts do not normally need any attention. They may be found cracked at the throat, but this is not detrimental, unless the cracks extend to beyond ¼ in. and start to move around to meet each other. An insert, excessively cracked, may be punched out through the injector bore. The bore in the cylinder head should be thoroughly cleaned and a new insert, which has a clearance fit of 0.001-0.003 in. dropped into place and secured with a new welch plug.

Before punching the center of the welch plug, check that the throat of the insert is in line with the cylinder head throat. The face of the insert, when installed, must be located within 0.001 in. below the cylinder head face to 0.002 in. above.

Valves

The valves are made of chrome steel and are retained in the head by caps and split-type cap retainers. Intake and exhaust valves are not interchangeable. Both valves are manufactured with 45 degree angle faces.

Valve specifications are, as follows:

Intake Valves
Face angle 45 degrees
Stem diameter 0.312-0.313 in.

Exhaust Valves
Face angle 45 degrees
Stem diameter 0.3115-0.3125 in.

REMOVAL AND INSPECTION

To remove the valves, use a valve spring compressor. Place valves in a numbered rack so that, if they are to be reused, they may be installed into their original valve guides. The original valves are numbered consecutively from the front of the engine.

Thoroughly clean and inspect all valves. If valves appear to be in good condition, measure the following:

1. Thickness of valve margin—discard valves with margins less than $1/32$ in.
2. Stem diameter—discard valves that have a stem diameter of 0.310 in. or less.

SERVICING

Reface new and old serviceable valves with a 44 degree face angle (one degree interference angle used). Discard any valves that are bent or have a margin thickness of $1/32$ in. or less, after refacing. Replacement valves should be numbered prior to installation.

INSTALLATION

1. Oil the valve stems and guides to provide initial lubrication.
2. Insert each valve into its correct port by making sure that the number on the valve head corresponds with the number stamped on the bottom face of the cylinder head adjacent to the valve seat.
3. Measure the valve head depth. Maximum depth must not exceed 0.048 in. for intake and exhaust valves, when measured using a new valve.
4. Locate the spring seats, springs, oil seal deflectors and spring caps. The damper coil of each spring should be toward the cylinder head.
5. Using a valve spring compressor, compress each valve, in turn, and install the retainers (locks). Tap each valve lightly with a rubber hammer to make sure that each assembly is correctly installed.
6. Replace the cylinder head assembly and adjust the valve tappet clearances, as outlined.

ADJUSTING VALVE TAPPET CLEARANCE

1. Run the engine until it is at its normal working temperature and remove the rocker cover. The engine firing order is 1-3-4-2.
2. Turn the crankshaft clockwise (as viewed from the front) until No. 1 piston is at top dead center on compression stroke with the valves of No. 1 cylinder both closed. The valves of No. 4 cylinder will be rocking up and down as a pair.
3. Using a 0.010 in. feeler gauge, check and, if necessary, adjust the clearance between the rockers and valve tips on No. 1 cylinder.
4. Turn the crankshaft 180° to the next compression stroke until the valves of No. 2 cylinder are on the rock and the valves of No. 3 cylinder are closed, and, if necessary, adjust the valve clearance of No. 3 cylinder valves to 0.010 in.
5. Turn the crankshaft 180° to the next compression stroke until the valves of No. 1 cylinder are on the rock and No. 4 cylinder valves are closed, and, if necessary, adjust the valve tip clearance of No. 4 cylinder to 0.010 in.
6. Turn the crankshaft another 180° to the next compression stroke until the valves of No. 3 cylinder are on the rock and No. 2 cylinder valves are closed, and, if necessary, adjust the valve tip clearance of No. 2 cylinder to 0.010 in.
7. Install the rocker cover and its gasket. Make sure the rocker cover gasket is correctly located.

NOTE: If the valve clearances are adjusted with the engine cold, then, the clearance should be set at 0.012 in.

Valve Guides

The valve guides are made of cast iron and are of the shoulderless type. Intake and exhaust valve guides are not interchangeable, as the exhaust guide is 0.310 in. longer than the intake guide.

The inside diameter of the valve guides is 0.314-0.3155 in. The operating clearance between the guide and valve stems is 0.001-0.0035 in. for the intake valves and 0.0015-0.004 in. for the exhaust valves. Replacement valve guides are presized and do not require reaming.

INSPECTION

Thoroughly clean and examine the guides. Replace guides that are scored, or otherwise damaged. Measure the inside diameter of the valve guides and replace guides, if diameter is more than 0.3155 in., or if valve guide-to-stem clearance exceeds 0.0045 in.

REPLACEMENT

Remove valve guides by pressing them out the top of the cylinder head. To install new valve guides, use a piloted drift 0.002 in. smaller than the inside diameter of the guides and press, do not drive, them into position from the top of the cylinder head.

When installing the shoulderless guides, insert the end with the long chamfer (both the ends are chamfered, but with different angles) from the top of the cylinder head. Press the valve guide into the head until the top end (short chamfered end) extends 0.820 in. (+ or −0.015 in.) above the top face of the head.

Valve Seats

Both the intake and exhaust valves seat directly in the cylinder head with no seat inserts being used.

INSPECTION

Thoroughly inspect each valve for pitting, burning or other evidence of leakage. If any of these conditions exist, or if new valve guides have been installed, the valve seats must be refaced.

SERVICING

Select the proper diameter stones and reface the intake and exhaust valve seats to a 45° angle (1° interference angle used). Remove only enough metal to clean up and true the face of the seat.

IMPORTANT: The head of the intake and exhaust valves, when installed in the cylinder head, must be between the limits of 0.028-0.048 in. below the cylinder head face. Do not reface the valve seat or valve to the point that the valve depth exceeds the amount specified.

After refacing the valve seats, install the valves and measure the valve head depth. Also, check width and location of valve seat contact area. Contact area should be $1/16$-$3/32$ in. wide and located in the middle of the valve face.

Massey-Ferguson

INSTALLING VALVE SEAT INSERTS

Valve seat inserts are not installed in production engines. It is possible, in most cases, however, to install inserts to service engines, if the existing valve seat is worn or damaged to the point where refacing would place the relationship of the valve head to the cylinder head face beyond the service limits of 0.028-0.048 in. This dimension applies to both the intake and exhaust valves.

1. Remove old valve guide and thoroughly clean the valve guide bore in the cylinder head.
2. Install a new valve guide into position.
3. Install a pilot into the new valve guide and machine the recess in the cylinder head face to the dimensions shown.
4. Remove ALL cuttings and thoroughly clean the insert recess; make sure that ALL burrs are removed and that the recess is completely clean.
5. Shrink the new seat insert by packing it in dry ice.
6. Select the correct sized driver extensions (slightly smaller than the diameter of the recess) and, using a pilot and driver, press the insert into position. Do not hammer the insert or use lubricant.
7. Make sure that the insert has been pressed in squarely and that it completely contacts the bottom of the recess.
8. Reface the valve seat insert, as in normal procedure.

NOTE: When refacing the insert, work as close as possible to the minimum valve depth in order to permit further refacing during subsequent overhauls.

Valve Springs

The valve springs, oil seal deflectors, caps and split-type retainers are interchangeable for the intake and exhaust valves. The valve springs have damper coils on one end which should be installed toward the cylinder head.

Discard any valve springs that do not have square ends, also, those that are discolored, damaged, or otherwise defective. If springs appear to be in good condition, check in a spring tester and compare the reading with the following specification:
Spring pressure @ 1-25/32 in.
compressed length 54-58 lb.

Piston and Connecting Rod Assembly

The pistons are cam ground and are made of a special light aluminum alloy. Five piston rings are installed to each piston; three compression rings and one oil ring above the piston pin and one oil ring below the pin.

All of the connecting rods are marked "front" and are fitted with replaceable piston pin bushings. The crank end of the rods are fitted with interchangeable precision-type bearing inserts. Replacement bearing inserts are available in standard size and 0.010 in., 0.020 in., and 0.030 in. undersize. The connecting rods and bearing caps are numbered from the front of the engine with the numbers toward the camshaft side of the engine. Connecting rods are graded for weight and are also numbered to indicate their weight. The rod weight numbers and their respective weights are, as follows:

Numbers	Rod Weight (includes bearing cap, cap setscrews and piston pin bushing)
No. 1	1 lb. 10 ozs.— 1 lb. 11 ozs.
No. 10	1 lb. 11 ozs.— 1 lb. 12 ozs.
No. 11	1 lb. 12 ozs.— 1 lb. 13 ozs.
No. 12	1 lb. 13 ozs.— 1 lb. 14 ozs.
No. 13	1 lb. 14 ozs.— 1 lb. 15 ozs.
No. 14	1 lb. 15 ozs.— 2 lbs.

The full floating-type piston pins are retained by snaprings and are available in standard size, only.

REMOVAL

Whenever it is necessary to service the piston and connecting rod assemblies, they should be removed, as follows:
1. Drain engine oil, also coolant, from radiator and cylinder block.
2. Remove the hood and side panel assembly, fuel tank, rocker cover, rocker assembly and cylinder head.
3. Remove oil pan.
4. Remove the oil pump.
5. Inspect the cylinder liners for wear. If cylinder taper exceeds 0.008 in., install new cylinder liners. If the cylinder liners are serviceable, install suitable tubing over two of the cylinder head studs and secrue the tubing with nuts and washers to prevent the liners from moving. Then, carefully remove the ridge from the top of the liners.
6. Remove the connecting rod caps and the lower bearing inserts and turn the crankshaft to bring the pistons to the top of their stroke. Then, raise the connecting rods to clear the crankshaft and remove the upper bearing inserts.
7. Remove the pistons out of the top of the cylinder block.
8. Reinstall the bearing inserts and caps to their original connecting rods to prevent them from becoming intermixed.

Removing Piston From Connecting Rod

1. Check that the pistons and connecting rods are correctly numbered to indicate their correct location in the engine.
2. Thoroughly clean and inspect the assemblies and, if the same pistons are to be used again, mark the pistons and connecting rods in such a way that they can be reassembled together the same way around.
3. Remove the snaprings from both ends of the piston pin holes.
4. Heat the pistons in hot water or oil and slip the piston pins out of the pistons.

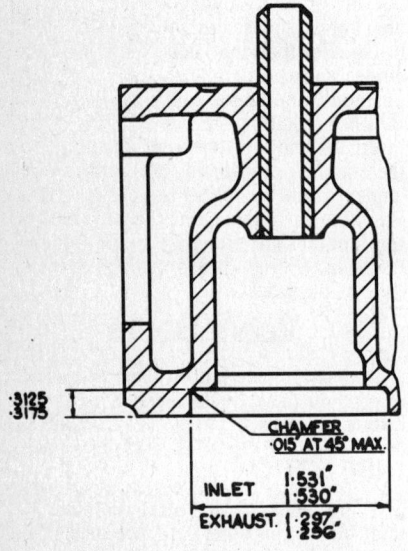

Machining dimensions for valve seat inserts

Homemade valve seat inserting tool. This tool should be made from EN.32A and case hardened and surface ground on the surface marked G. The shank of the tool is the pilot which slides into the valve guide

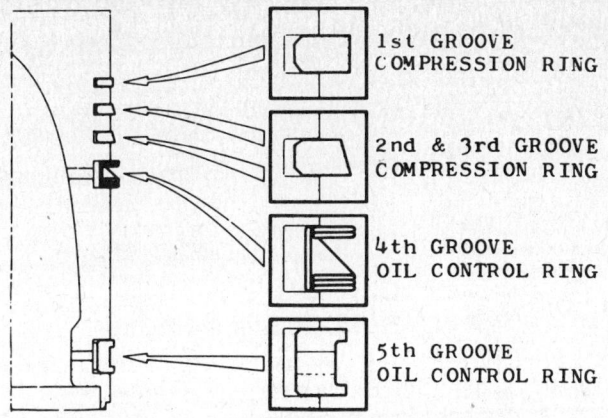

Piston ring arrangement, original equipment rings

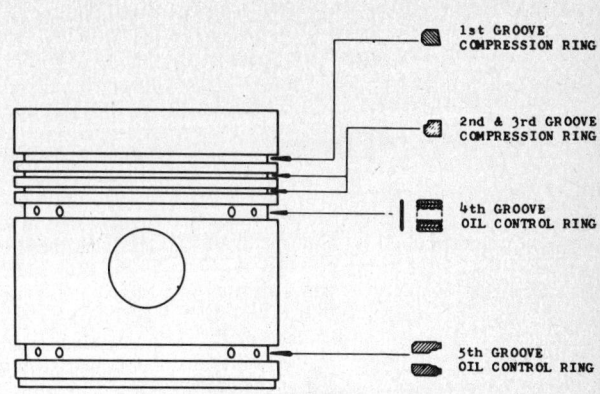

Piston ring arrangement, re-ring set

INSPECTION

Pistons and Rings
PISTONS

Discard pistons that are scored, or otherwise damaged. If pistons appear serviceable, proceed, as follows:

1. Measure the piston skirt diameter at a right angle and level to the piston pin.

NOTE: The piston is cam ground and is slightly tapered, being larger at the bottom than at the top. New piston skirt diameter (level with piston pin) is 3.120-3.121 in. Discard pistons that are collapsed or worn.

2. Inspect for worn ring grooves. Piston ring groove clearance specifications are, as follows:

No. 1 compression ring 0.002-0.004 in.
No. 2 compression ring 0.002-0.004 in.
No. 3 compression ring 0.002-0.004 in.
No. 4 oil ring   not applicable
No. 5 oil ring   0.0025-0.0045 in.

Discard pistons that have worn ring grooves or bent lands.

PISTON RINGS

Never install used rings onto a new or used piston. Before installing the piston rings, check the ring end gap in the lower, unworn portion of the cylinder sleeve. Proper ring end gap in a new sleeve is 0.012-0.014 in. for the top ring and 0.009-0.014 in. for all of the other rings, except the No. 4 oil control ring, which has no specific end gap.

When installing the piston rings, use a piston ring tool for all of the rings, except those used in the fourth ring groove. Install the lowest groove ring, first, then, the next lowest, etc., with the top groove ring installed last. Description of rings and the correct placement is, as follows:

ORIGINAL EQUIPMENT PISTON RINGS (USED WITH NEW PISTONS AND CYLINDER LINERS)

No. 1 compression ring is a chrome-plated, parallel-faced ring.

No. 2 compression ring is a cast iron, taper-faced ring; must be installed with side marked "T" toward top of piston.

No. 3 compression ring is a cast iron, taper-faced ring; must be installed with side marked "T" toward top of piston.

No. 4 oil control ring is a chrome-plated, Duaflex laminated ring (above the piston pin).

No. 5 oil control ring is a cast iron, slotted scraper ring.

PISTON RE-RING SET (USED WITH WORN CYLINDER LINERS)

No. 1 compression ring is a taper-faced ring; must be installed with side marked "T" toward top of piston.

No. 2 compression ring is an internally-stepped ring; must be installed with the step toward top of piston.

No. 3 compression ring is an internally-stepped ring; must be installed with step toward top of piston.

No. 4 oil control ring is a spring-loaded, laminated scraper ring, and should be installed as follows:

1. Install the internal expander in the ring groove.
2. Spiral two of the rail rings onto the piston and locate them at the bottom of the groove.
3. Install the spiral ring into the groove on top of the two rail rings.
4. Spiral two of the rail rings onto the piston and locate them in the top of the groove.
5. Turn the ring rails so that the gaps are staggered.

No. 5 oil control ring is a Microland, slotted scraper, and may be installed either side up.

Connecting Rods

Check connecting rods for proper alignment. Discard or straighten any rods that are misaligned.

NOTE: Replace all unserviceable rods with new ones of the same weight. Replacement rods are not numbered as to the correct cylinder location and should be marked prior to installation.

Install a new piston pin bushing, if inside bushing diameter exceeds 0.939 in.

Connecting rod bearing inserts are available in standard size or in 0.010 in., 0.020 in. and 0.030 in. undersize.
Specifications are, as follows:
Crankpin diameter   1.9995-2.000 in.
Recommended bearing clearance
                         0.0015-0.003 in.
Rod side play   0.003-0.009 in.

BUSHING REPLACEMENT

When replacing piston pin bushings, support the connecting rod in a press and press out the worn bushings. To install the new bushing, position the bushing on the connecting rod; make sure oil hole in bushing is in line with oil hole in connecting rod, then, use a bushing driver to press the new bushing into place. Ream the excess material from the bushing and, then, finish hone the bushing to an inside diameter of 0.93820-0.93875 in. Proper clearance between the piston pin and the piston pin bushing is 0.0005-0.00125 in.

INSTALLATION OF PISTON AND CONNECTING ROD ASSEMBLY

NOTE: If the oil cylinder liners are to be reused, deglaze them.

1. Make sure all parts are clean, then, remove the bearing cap and apply a liberal coating of oil to the cylinder liner bores and the piston assembly.

2. Stagger the ring gaps, then, using a piston ring compressor, start the assembly into the cylinder liner bore. Make sure that the side of the connecting rod marked "front" is installed, as indicated, and that they are installed in the proper cylinder location.

3. Rotate the crankshaft until the appropriate crankpin journal is at BDC, then, press the assembly into the cylinder bore.

4. Carefully clean the bearing bore in the connecting rod and the gearing insert and install the upper insert in place.

5. Lubricate the bearing insert and crankpin journal, then, pull the rod assembly into position onto the crankpin, making sure that the bearing insert is still correctly located.

6. Carefully clean the connecting rod cap and the bottom bearing insert and assemble it into the cap.

7. Lubricate the crankpin journal and the bottom insert.

8. With the cap location number toward the camshaft side of the engine, position the cap in place and install the bolts, using new locks. Tighten the connecting rod bolts to 30-35 ft.lb. torque.

Massey-Ferguson

NOTE: The top of the pistons, when at TDC, should be located within 0.0085-0.012 in. above the top of the cylinder block.

Cylinder Liners

The cylinder liners are of the wet type and are made of cast iron. The cylinder liners are manufactured with an inside diameter of 3.125-3.126 in. The top of the cylinder liners are flanged and, when installed, must be located within 0.001 in. below to 0.003 in. above the face of the cylinder block.

Each cylinder liner is sealed at the bottom by two synthetic rubber rings located in the cylinder block.

NOTE: The cylinder liners are a relatively loose fit in the cylinder block. If the original cylinder liners are to be reused, make sure to mark the liners and the cylinder block to allow them to be installed with the worm thrust side to the thrust side of the cylinder block.

REPLACEMENT

1. Remove the cylinder head, oil pan, oil pump and piston and connecting rod assemblies.
2. Use a sleeve puller, with a suitable adaptor, and remove the cylinder liners.

NOTE: The cylinder liners normally fit loose enough so that, by using a suitable brass plate across the bottom of the liner, it can be tapped upward until it is clear of the sealing rings.

3. After removing the cylinder liners, thoroughly clean the cylinder block bore, with particular attention given to the top recess for the flange of the liner and the two grooves in which the sealing rings are located.
4. On the injection pump side of the cylinder block, there are four $3/32$ in. diameter vent holes which should be cleaned and kept open at all times. Each hole is laterally aligned with the injection pump and is vertically aligned to the center of the cylinder. The holes are drilled through the side of the cylinder block and are located between the two sealing ring grooves. The object of the holes is to allow any coolant which might leak past the upper sealing ring to drain out, thus, relieving the lower sealing ring of any pressure above it. Make sure that these holes are kept open.
5. After carefully cleaning the cylinder block and cylinder liner, insert the new sealing rings in the two grooves in the cylinder block bore. Coat the sealing rings and the lower portion of the cylinder liner with soft soap or soapy water to enable the liner to slide freely into position.
6. Place the cylinder liner in position and press it fully into place by hand, using a brass plate and a hammer for the final stage. When fully located, the top of the flange must be within 0.001 in. below to 0.003 in. above the top face of the cylinder block. After installing the new cylinder liners, the block should be water tested with pressure of 20 psi, but without the water pump installed, i.e., the pump removed and the water passages well sealed.

NOTE: A slight oil leak from the vent holes may be observed with a new or newly overhauled engine. This is normal and should cease after a few days operation. A slight seepage of engine coolant is permissible.

Timing Gear Cover and Front Crankshaft Oil Seal

The front timing gear cover is made of pressed steel and contains the front crankshaft oil seal. The oil seal is of the spring-loaded type and is installed flush with the front face of the cover with the lip of the seal toward the engine.

REMOVING THE TIMING GEAR COVER

1. Remove the hood and side panels, also, the radiator and shroud.
2. Remove the generator, fan blade assembly and fan belt.
3. Remove the crankshaft pulley capscrew and remove the crankshaft pulley which is keyed to the crankshaft.
4. Remove the timing gear cover.

OIL SEAL REPLACEMENT

1. With the timing gear cover removed, press the old seal out the front face of the cover.
2. Carefully clean the oil seal recess in the cover and coat the outside diameter of the new oil seal with a sealing compound.
3. Using MFN 833 special tool, position the oil seal with the lip of the seal toward the engine and press it into the front of the cover until the tool locates the oil seal flush with the front of the cover.

INSTALLING THE TIMING GEAR COVER

To install the timing gear cover with the oil seal in place, proceed, as follows:
1. Install a new gasket to the timing gear cover, using sealing compound.
2. Locate the timing gear housing with four or five equally spaced setscrews. Do not tighten the setscrews at this time.
3. Install MFN 833 special tool in the timing gear cover to correctly locate the oil seal in the cover concentric with the crankshaft. Install the other setscrews securing the timing gear cover and tighten them securely.

NOTE: If there is any difficulty installing the setscrews, remove the timing gear cover and slightly loosen the four setscrews securing the front engine mounting plate, then, reinstall the cover with the aligning tool and align the

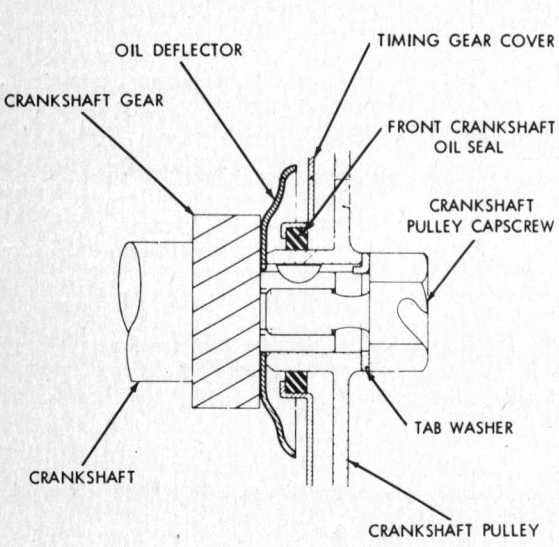

Timing gear cover installation

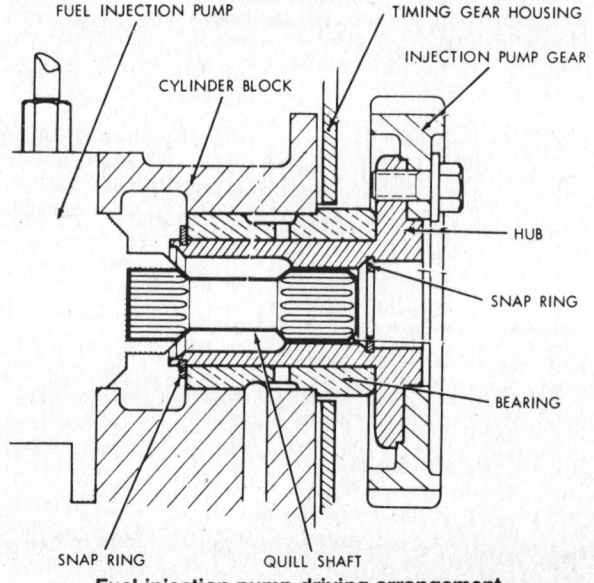

Fuel injection pump driving arrangement

Massey-Ferguson

mounting plate with the cover. Do not forget to retighten the four setscrews securing the front mounting plate.

Timing Gears

The engine timing gear train consists of the crankshaft gear, idler gear, camshaft gear and injection pump drive gear. The crankshaft gear drives the camshaft and the injection pump through the idler gear. All of the gears are of the helical type. The idler gear is mounted on a hub bolted to the front of the cylinder block. All of the timing gears are marked during production and, when reassembling the engine, all marks must align when No. 1 piston is at TDC on compression stroke.

All replacement timing gears have timing marks, except for the scribe lines on the inner face of the injection pump gear and driving hub. These scribe lines are marked in assembly. All gears are available in standard size, only.

NOTE: Due to the odd number of teeth in the idler gear, the complete set of timing marks will not align every time No. 1 piston is at TDC on compression stroke.

CHECKING TIMING GEAR BACKLASH

1. Remove the timing gear cover.
2. Check the backlash between the timing gears with the dial indicator or feeler gauge. The backlash between any two gears in the timing gear train must be between 0.0015-0.003 in.
3. The holes in the idler gear hub have sufficient clearance to allow movement of the hub on the setscrews to obtain the correct backlash. If the backlash of the timing gears is not within the limits of 0.0015-0.003 in., loosen the setscrew slightly and move the hub to obtain the desired clearance. Set the clearance between the idler and crankshaft gears and between the idler and the camshaft gears with a feeler gauge or a dial indicator.
4. Install the rocker arm shaft assembly and complete reassembly of tractor.

REMOVING THE TIMING GEARS

1. Remove the timing gear cover.
2. Remove the fuel tank, rocker arm cover and release the rocker arm shaft assembly.
3. Unlock and remove the two setscrews securing the ilder gear and its hub to the cylinder block. Then, remove the idler gear retaining plate, idler gear and hub. The idler gear hub will come away with the idler gear.
4. Remove the setscrews securing the camshaft gear and the injection pump gear and remove the gears.

REMOVING THE FUEL INJECTION PUMP HUB

1. With the injection pump gear removed, as described earlier, remove the fuel injection pump with its quill shaft.
2. To remove the hub, remove the snapring and slide the hub out through the front of the bearing. Another snapring is installed inside the hub to prevent the quill shaft from moving forward.

INSTALLING THE FUEL INJECTION PUMP HUB

1. Slide the hub into the bearing and secure it with a new snapring.
2. Rotate the hub to make sure it turns freely, then, measure the end play.

The hub end play should be within 0.003-0.011 in. Install a new snapring inside the hub.

INSTALLING THE TIMING GEARS

1. Observe on the hub of the camshaft and on the front of the camshaft gear the bolt holes with the letter "D" stamped next to them.
2. With the rocker arm assembly released to facilitate turning the camshaft, install the camshaft gear with the letters "D" in alignment. Do not attempt to install the gear using the slotted holes.
3. Observe the scribe lines on the injection pump gear and the drive hub and install the gear onto the hub with the scribe lines in alignment.

NOTE: If a new injection pump gear without the scribe line is installed, the injection pump must be retimed.

4. Turn the crankshaft and position it with the key straight up.
5. Turn the camshaft gear and the injection pump gear so that their timing marks will be approximately aligned with the center of the hub, when it is installed.
6. Use a new locking plate and install the idler gear hub, idler gear, retaining plate and setscrews with all of the timing marks in alignment. Tighten the setscrews securely, but do not lock them at this time.
7. Check the backlash between the timing gears. The backlash between any two gears in the timing gear train should be within the limits of 0.0015-0.003 in. If the backlash is not within these limits, adjust the backlash, as outlined under the heading "Checking Timing Gear Backlash".
8. Install the oil deflector onto the front of the crankshaft and complete reassembly of the engine.

Camshaft

The camshaft is situated on the righthand side of the cylinder block and operates the rocker assembly through tappets and pushrods. The tappets are a sliding fit in the cylinder block and are retained in position by the camshaft. The camshaft is driven by a helical type idler gear and is supported by three bearings which are machined directly in the cylinder block. Camshaft end play is controlled by two semi-circular thrust washers installed in a recess on the front of the camshaft and located in the front of the cylinder block. In addition to operating the tappets, the camshaft also operates the engine oil pump and fuel lift pump.

REMOVAL

To remove the camshaft, proceed, as follows:
1. Remove the timing gear cover.
2. Remove the rocker cover, rocker arm shaft assembly and push rods.
3. Disconnect the fuel lines at the fuel lift pump, and remove the tappet cover with the fuel lift pump attached. Remove the push rod which operates the pump from the eccentric on the camshaft.
4. Remove the idler gear, camshaft gear and injection pump gear.
5. Disconnect the tractormeter at the cylinder block, then, remove the snapring and remove the tractormeter driveshaft.
6. Remove the oil pan, oil pump and fuel injection pump.
7. Release the setscrew securing the front engine mounting plate and remove the mounting plate from the cylinder block.
8. Lift the tappets and secure them in the raised position.
9. With the tappets clear from the camshaft lobes, carefully withdraw the camshaft from the cylinder block. With the camshaft will come the camshaft thrust washers, which fit in the groove machined in the camshaft in front of No. 1 camshaft journal.

INSPECTION

Carefully clean and inspect the camshaft, as follows:
1. Inspect the camshaft bearing journals and cam lobes for wear or damage. If the journals are excessively scored, nicked or discolored by heat, or if the lobes are pitted, or rough, the camshaft must be replaced.
2. Measure the diameter of the bearing journals for wear and out-of-round with a micrometer. If the journal diameters are worn, or out-of-round, or if the camshaft lobes are worn, the camshaft must be replaced.

Specifications are, as follows:
Cam lobe lift   0.260-0.266 in.
Front bearing
  Front bearing journal 1.791-1.792 in.
  Front bearing bore 1.794-1.7955 in.
  Front bearing operating clearance
    0.002-0.0045 in.
Center bearing
  Center bearing journal 1.781-1.782 in.
  Center bearing bore 1.784-1.787 in.
  Center bearing operating clearance
    0.002-0.006 in.
Rear bearing
  Rear bearing journal 1.773-1.774 in.
  Rear bearing bore 1.776-1.778 in.
  Rear bearing operating clearance
    0.002-0.005 in.

INSTALLATION

1. Install the dowel in the thrust washer recess, then, make sure the tappets will clear the cam lobes and carefully slide the camshaft into the cylinder block until the

Massey-Ferguson

front journal just enters into the bearing in the cylinder block.

2. Install the thrust washers in the groove around the camshaft lining up the hole in the thrust washer with the dowel in the block recess.

3. Push the camshaft fully in place, being careful not to damage the thrust washers.

4. Using sealing compound, install a new front engine mounting plate gasket onto the front cylinder block and install the mounting plate.

5. Install the oil pump and oil pan.

NOTE: Install the tractormeter driveshaft and snapring after the oil pump has been installed.

6. Install the timing gears, oil deflector and the timing gear cover.

NOTE: Make sure to install the camshaft gear onto the camshaft with the letters "D" in alignment.

7. Install the fuel lift pump push rod, push rods and the rocker arm shaft assembly.

8. Complete reassembly by reversing the disassembly procedure.

NOTE: To facilitate installing the fuel lift pump when it is attached to the tappet cover, turn the engine and set the lift pump push rod in its lowest position.

Crankshaft and Main Bearings

The crankshaft is forged from chrome molybdenum steel and is supported by three main bearings. The crankshaft has four counter weights formed integrally with the crank webs. Crankshaft end play is controlled by three 180° thrust washers; one on each side of the rear main bearing cap and one to the rear of the cylinder block half of the bearing. The rear main bearing journal carries a shallow oil return helix which prevents oil seeping out past the rear oil seal. The main bearings are of the non-adjustable type and are steel-backed aluminum tin lined.

The main and connecting rod bearings are available in standard and 0.010 in., 0.020 in. and 0.030 in. undersizes.

The main bearing caps are made of heavy-duty cast iron and are located on dowels in the cylinder block. All bearing caps are specially dowelled and cannot be installed the wrong way around. In production, the main bearing bores are machined with the caps in position. If, for any reason, a main bearing cap becomes damaged, it will be necessary to replace the cylinder block complete with bearing caps.

NOTE: The front main bearing cap can be removed with the engine in place after removing the oil pan and the four bottom bolts in the timing gear cover. The rear main bearing cap cannot be removed without removing the flywheel.

Oil seals are provided at the front and rear ends of the crankshaft. The front oil seal is of the spring-loaded type and is contained in the timing gear cover. The rear oil seal is of the rubber core asbestos rope type and is contained in a two-piece sealed retainer.

CRANKSHAFT END PLAY

Crankshaft end play is controlled by three 180° thrust washers; one on each side of the rear main bearing cap and one to the rear of the cylinder block half of the bearing. The rear bottom thrust washer has a locating lug to prevent the rear thrust washer halves from turning out of position.

NOTE: Replacement thrust washers are available in standard size or 0.007 in. oversize.

To measure the crankshaft end play, force the crankshaft forward as far as it will go, then, measure the gap between the machined shoulder on the crankshaft web and the crankshaft thrust washer. End play should be within 0.003-0.009 in. Install new thrust washers, if end play exceeds 0.009 in.

REMOVAL

1. Remove the engine from the tractor and mount it in a suitable engine stand.
2. Remove the clutch and flywheel.
3. Remove the rear oil seal, oil pan and oil pump.
4. Remove the timing gear cover and idler gear.
5. Release the locks securing the connecting rod and main bearing setscrews and remove the setscrews.

NOTE: Measure crankshaft end play before removing crankshaft. If end play exceeds specifications, install new thrust washers when reassembling the engine.

6. Remove the connecting rod and main bearing caps, being careful not to damage the gasket between the front main bearing cap and the front adaptor plate.

NOTE: If this gasket does become damaged, it will be necessary to remove the camshaft and injection pump gears to allow the plate to be removed.

7. Carefully remove the crankshaft from the cylinder block.

NOTE: Identify and keep all parts in order so that, if serviceable, they may be installed in their original locations.

INSPECTION

Thoroughly clean and inspect the crankshaft. Specifications are as follows:

Crankshaft
  Main bearing journal diameters
    2.248-2.2485 in.
  Crankshaft end play   0.003-0.009 in.
  Crankpin journal diameters
    1.9995-2.000 in.
Main bearings
  Main bearing bore diameter
    2.250-2.2515 in.
  Bearing oil clearance 0.002-0.0035 in.

If any of the main or crankpin journals are worn beyond specifications, more than 0.0015 in. out-of-round, or tapered more than 0.001 in., the crankshaft should either be replaced, or reground and fitted with undersize bearings.

INSTALLATION

1. Make sure all parts have been carefully cleaned, then, locate the upper main bearing inserts into the cylinder block and lubricate them.

NOTE: Main bearing inserts are interchangeable.

2. Place the upper thrust washer in the block with the oil grooves outward and the steel face toward the bearing.

NOTE: A light coating of grease will assist in holding the thrust washer in place.

3. Carefully install the crankshaft in place.
4. Install the main bearing inserts into the bearing caps and lubricate them.
5. Coat the lower thrust washers with grease and install them into the rear main

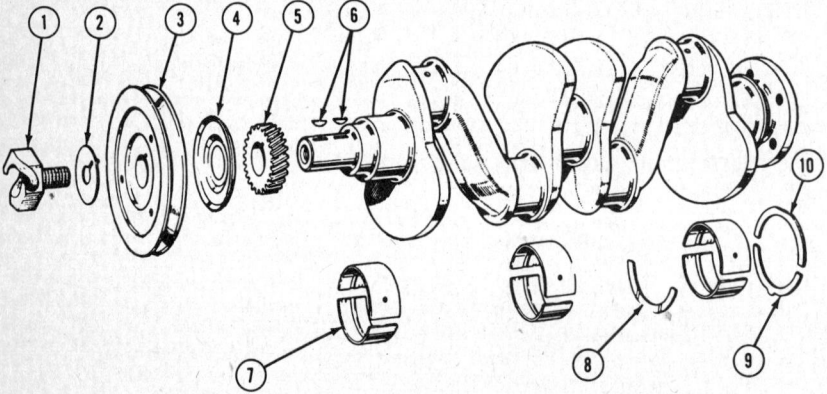

Crankshaft components

1. Crankshaft pulley capscrew
2. Tab washer
3. Crankshaft pulley
4. Oil deflector
5. Crankshaft gear
6. Crankshaft gear and pulley keys
7. Main bearing insert (6)
8. Front crankshaft thrust washer
9. Lower rear crankshaft thrust washer
10. Upper rear crankshaft thrust washer

bearing cap with the vertical oil grooves facing outward.

6. Using new locks on the setscrews, install the main bearing caps and tighten them to a torque of 80-85 ft.lb.

7. Measure crankshaft end play. End play should be within 0.003-0.009 in. Finish reassembling the engine by reversing the disassembly procedure.

Rear Main Oil Seal

The rear crankshaft oil seal is a two-piece rubber core asbestos rope type seal and is contained in a two-piece retainer housing which is bolted to the rear of the cylinder block and rear main bearing cap. New oil seals can be installed with crankshaft in position and the flywheel removed.

REMOVING THE OIL SEAL HOUSINGS

To remove the rear crankshaft oil seal, proceed, as follows:

1. Split the tractor and remove the clutch and flywheel.
2. Remove the transmission adaptor plate from the rear of the cylinder block.
3. Remove the six setscrews and spring washers securing the seal housings to the bearing cap and cylinder block.
4. Remove the two long setscrews and spring washers securing the housings together and remove the housings.

REPLACING OIL SEALS AND INSTALLING THE HOUSINGS

Before installing the new oil seals, make sure to carefully clean the seal grooves and old gasket material from the oil seal retainers.

To install the new oil seals, place each half of the seal retainer (one at a time) in a vise and proceed, as follows:

1. Press approximately one inch of the new asbestos seal into each end of the groove in the seal retainer, allowing the seal to project 0.010-0.020 in. beyond both ends of the joint face. This amount of projection assures correct contact between the end faces of the seals, when the retainer halves are installed around the crankshaft.

IMPORTANT: A projection exceeding that quoted must be avoided, because the excess may not settle in the retainer groove and may be forced over the joint face as the retainer halves are pulled together. Thus, the retainers may be held slightly apart and cause an oil leak.

Due to the interference between the seal and its groove, it is possible that the seal may not fully seat into the groove, giving the impression that it is too long. Each seal is of the correct length and must not be trimmed at any time.

2. The middle of the oil seal will bulge out of the groove and must be pushed in with the fingers, by working from the center, until well bedded in the groove. Use a round bar of metal and, by rolling and pressing, further bed the oil seal in place.

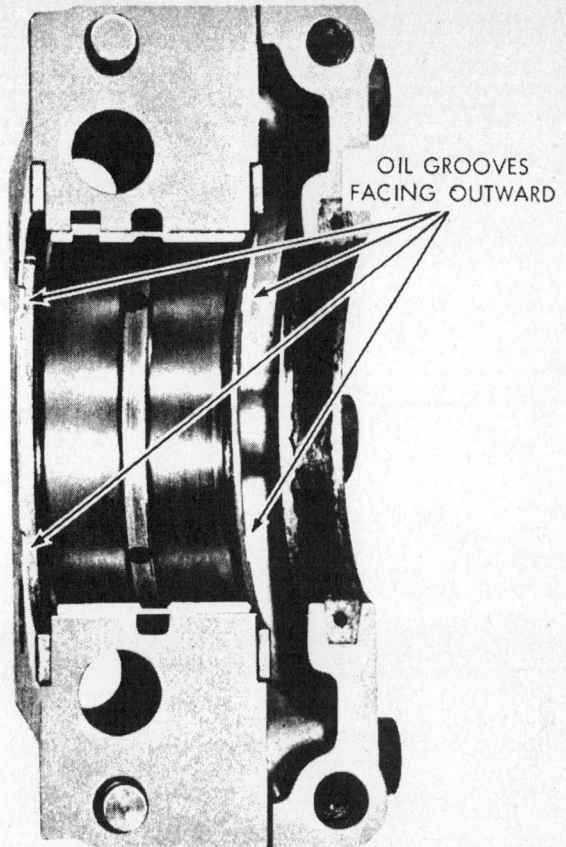

Thrust washer installed in rear main cap

Make sure that the seal extends 0.010-0.020 in. beyond both ends of the joint face. Any gap at this joint, when the seals are installed, will allow oil to leak past the seal.

3. Repeat the above steps for the other half of the seal retainer, then, proceed with step No. 4.

4. Remove all the old gasket material from the sealing faces of the cylinder block and rear main bearing cap, then, assemble the two oil seal housings together and tighten the two long bolts securing them together. Using sealing compound, place new gaskets on the back faces of the two housings, making sure that the butt faces of the gaskets meet and that all bolt holes align. Allow the sealing compound to dry to secure the gasket in position.

5. After the drying period, remove the two long bolts and separate the housings.

6. Lightly coat the butt faces of the two housings with sealing compound.

7. Liberally oil the oil return groove at the rear end of the crankshaft.

8. Liberally oil the exposed inside surface of the oil seals.

9. Assemble the two oil seal housings around the crankshaft over the oil return groove and assemble the housings together with the two long bolts.

10. Swivel the housings around the crankshaft to further bed in the seals and to make sure the assembly turns easily on the crankshaft.

11. Secure the housings to the cylinder block and main bearing cap with the six setscrews and spring washers.

12. Complete reassembly by reversing the disassembly procedure.

FUEL SYSTEM

BLEEDING THE SYSTEM

1. Open the fuel shut-off valve.
2. Loosen the banjo-bolt on the top of the fuel filter.
3. Pump the manual lever on the fuel pump until fuel, free of air, comes out the filter. Tighten the banjo bolt to 5-7 ft.lb. torque.

NOTE: In instances where the engine stops with the fuel pump rocker arm in the fully depressed position, it will be necessary to turn the engine slowly until the camshaft releases the arm and will allow full movement of the diaphragm in the fuel pump.

4. Loosen the lower vent plug on the injection pump body and operate the fuel pump manual lever until fuel, free of air, flows out of the vent plug.
5. Loosen the upper vent plug on the injection pump body and operate the fuel pump manual lever until fuel, free of air, flows from the port. Tighten the vent plug.
6. Loosen two of the high pressure fuel lines at the injector connections. Turn the

Massey-Ferguson

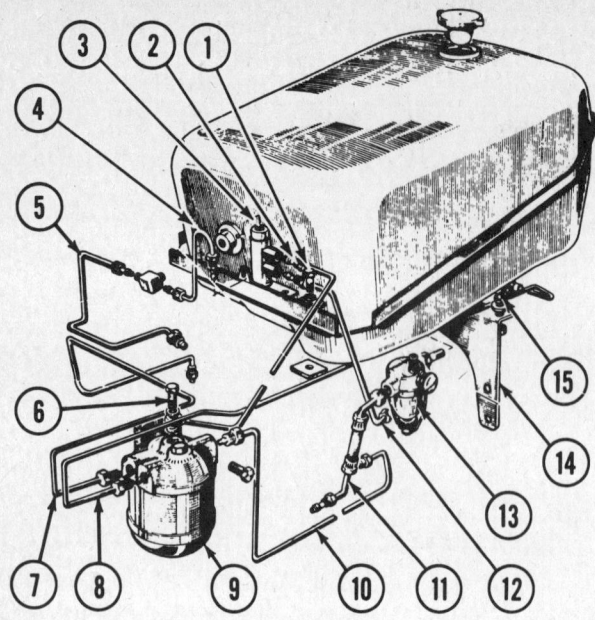

Fuel tank and lines

1. Rear fuel tank securing bolt
2. Fuel line, fuel filter to reservoir tank
3. Thermostart reservoir
4. Fuel line, tee to reservoir
5. Fuel line, leakoff line assembly to tee
6. "Banjo" bolt
7. Fuel line, fuel filter to injection pump
8. Fuel line, injection pump to fuel filter
9. Fuel filter
10. Fuel line, fuel lift pump to fuel filter
11. Fuel line, sediment bowl to fuel lift pump
12. Fuel line, reservoir tank to thermostart
13. Sediment bowl
14. Front fuel tank bracket
15. Front fuel tank securing bolt

engine over with starting motor until fuel, free of air, escapes from the high pressure line ends. Tighten the connections.

NOTE: If the fuel line from the thermostart reservoir to the thermostart has been disconnected, or removed, make sure to bleed it of air. If the thermostart is operated without fuel in the line, it may ruin the thermostart.

7. Turn the engine with the starter until the engine starts.

IMPORTANT: Do not operate the starter continuously for more than 15 seconds without allowing it to cool. The fuel system must be thoroughly bled of air or the engine will not operate properly.

Fuel Tank

REMOVAL

1. Disconnect or remove the battery.
2. Close the fuel shut-off valve at the sediment bowl.
3. Disconnect the main fuel line from the sediment bowl to fuel lift pump at sediment bowl end.
4. Remove fuel line from fuel filter to thermostart reservoir.
5. Remove fuel line from thermostart reservoir to thermostart.
6. Remove fuel line from injector leak-off line assembly to fuel tank tee.
7. Disconnect radiator support rod and hood support strap from fuel tank, then, carefully lay the hood down in its fully opened position.
8. Loosen the two bolts securing the front of the fuel tank.
9. Remove the two rear bolts securing the fuel tank, then, slide the tank rearward and lift it from the tractor.

INSTALLATION

To install the fuel tank, reverse the removal procedure.

NOTE: Bleed the fuel system of air before attempting to start the engine.

Fuel Lift Pump

REMOVAL

To remove the fuel lift pump, close the fuel shut-off valve and disconnect the fuel inlet and outlet lines from the pump. Remove the nuts securing the lift pump and remove the pump from the engine.

DISASSEMBLY

1. Clean the outside of the lift pump and make a file mark across the flange to facilitate the assembly of the pump.
2. Remove the six screws and washers securing the head of the pump to the pump body and separate the two halves.
3. Turn the diaphragm assembly 90° (¼ turn) in either direction and lift the diaphragm and pull rod assembly from the body.

The diaphragm and rod are serviced as an assembly and no attempt should be made to separate the layers. If any of the diaphragm layers are stuck together, or appear cracked, the assembly should be replaced. It must also be replaced if there is excessive wear in the link engagement slot of the pull rod. If the diaphragm return spring is corroded or distorted, it should be replaced by one of the same color.

4. Remove the valve retaining plate in the pump head and remove the inlet and outlet valves. Wash the valve assemblies in solvent and examine them. If the valves are damaged, or if there is any indication that the fuel had not been held by the valves, they should be replaced.

ASSEMBLY

1. Position the valves, then, place them in the pump head and install the valve retaining plate. Make sure that the valves are positioned as shown, with the smaller ends opposite the source of fuel.
2. Place the diaphragm assembly, complete with the return spring, metal washer and rubber washer, in the body. Press downward on the diaphragm and turn it 90° (¼ turn), so that the slots in the pull rod will engage the fork in the operating lever and align the holes in the diaphragm with those in the body.
3. Move the rocker arm upward until the diaphragm is level with the face of the pump body. Then, place the head of the pump into position, as indicated by the file marks on the flanges. Install the six screws and lockwashers and tighten only until the heads of the screws just touch the washers.
4. Release the pump rocker arm and move it downward to hold the diaphragm at the top of its stroke, then, alternately tighten the screws securing the head to the pump body.

NOTE: After securing the head to the pump body, as described above, the outside edges of the diaphragm should be about flush with the flanges. If the edges of the diaphragm extend beyond the flanges any appreciable amount, it is an indication of improper assembly, in which case, the screws should be loosened and retightened while holding the diaphragm at the top of its stroke.

INSTALLATION

1. Using a new gasket and sealing compound, position the pump up close to its studs, then, raise the pump rocker arm so that it will be on top of the push rod and slide the pump in place onto the studs.

NOTE: If there is any difficulty locating the rocker arm on top of the pump push rod, rotate the engine to position the push rod and its lowest level.

2. Install the nuts and lockwashers to secure the pump in place.
3. Install the fuel line, open the fuel shut-off valve and bleed the system of air.

Fuel Filter

The fuel filter is made by C.A.V. and is of the "FS" bowlless type. The filter uses

Massey-Ferguson

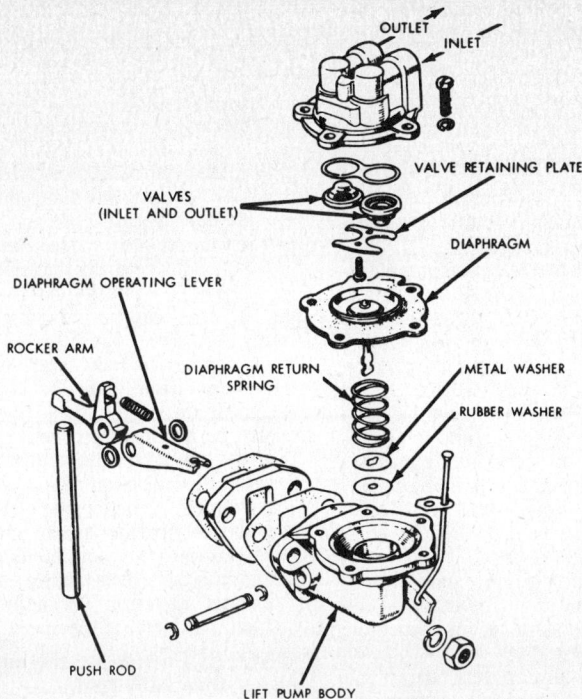

Fuel lift pump

a replaceable, throw-away, paper type element which should be replaced every 500 hours of operation.

REPLACING THE FUEL FILTER ELEMENT

1. Close the fuel shut-off valve at the fuel tank and thoroughly clean the outside of the fuel filter.
2. Unscrew the center retaining bolt located on top of the filter.
3. Remove the filter base, then, empty and thoroughly clean it.
4. Discard the old element and replace it with a new one.

NOTE: Use only genuine M-F fuel filter elements.

5. Using new seals, position the filter base and element in place and install the retaining bolt.
6. Open the fuel shut-off valve and bleed the fuel system of air.

Fuel Injection Pump

The fuel injection pump is of the DPA distributor type, and has a built-in mechanical governor with an automatic advance mechanism. The automatic advance provides a few extra degrees of advance at engine speeds higher than 1600 rpm.

The injection pump is flange-mounted on the lefthand side of the engine and is gear-driven through an idler gear by the crankshaft gear.

The fuel is pumped by a single element and the fuel charges are distributed in the correct firing order and at the proper timing intervals to each cylinder by means of rotary distributor integral with the pump. As a result, equal amounts of fuel delivery to each injector is an inherent feature of the pump and deliveries are not subject to maladjustment from one injector to another.

Accurate phasing is also a feature of the pump, as the timing intervals between injection strokes are determined by the accurate spacing of the distribution ports and the precision operating cams and are, therefore, not subject to adjustment.

The pump is a compact, oil-tight unit, completely lubricated by fuel oil, and requires no additional lubrication. Sensitive speed control is maintained by a governor of the mechanical fly-weight operated type, which is incorporated in the pump.

REMOVAL

To remove the fuel injection pump, thoroughly clean the ingine and proceed, as follows:

IMPORTANT: Before removing the injection pump to have it serviced, make sure that the injection pump really is at fault and that the difficulty is not caused by some other part of the fuel system.

1. Disconnect all linkage to the injection pump.
2. Remove the following parts:
   a. All high pressure fuel lines from injection pump to injectors.
   b. The low pressure fuel lines (fuel supply and return line) between fuel filter and injection pump.

NOTE: Cap, or otherwise cover, all open fuel lines and ports to prevent dirt from entering into the fuel system.

3. Remove the Allen setscrew and nuts securing the injection pump in place and carefully withdraw the pump and quill shaft.

NOTE: The quill shaft is a machine fit to the injection pump and should not be used with any other injection pump.

Make sure to cap, or otherwise cover, all fuel openings.

IMPORTANT: Do not disassemble the injection pump. Send faulty injection pumps to your nearest C.A.V. service station for repair. If the seals are broken on the pump, any warranty will become void.

INSTALLATION

1. Observe the end of the quill shaft and install it into the injection pump with the square end into the pump and the chamfered end toward the injection pump gear hub.
2. Make sure the snapring is installed inside the injection pump gear hub. Then, observe the master spline inside the hub and note where it it positioned.
3. Turn the quill shaft in the end of the injection pump so that its master spline will be approximately aligned with the one in the hub, then, carefully position the pump in place.
4. Rotate the pump until the scribe lines are aligned, then, secure the pump in place by installing the nuts, washers and lockwashers and Allen setscrew.
5. Install the low pressure fuel lines (fuel supply and return line) between the fuel filter and the injection pump. Also install all high pressure fuel lines from injection pump to injectors.

IMPORTANT: When connecting the fuel lines, make sure that they are not under stress. Fuel line stress will cause premature line breakage.

6. Connect the fuel shut-off and throttle control rods, then, air-bleed the fuel system.

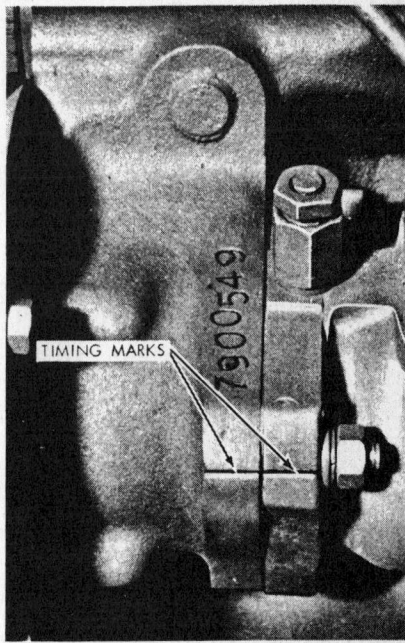

Injection pump timing marks aligned

715

Massey-Ferguson

FUEL INJECTION PUMP TIMING

The quill shaft connecting the injection pump gear hub to the injection pump has a master spline that will allow it to be installed in only one position. As long as the injection pump gear and hub are properly timed to the other timing gears, the injection pump may be removed and reinstalled without regard to the location of the crankshaft and the timing marks on the injection pump rotor. Whenever a new injection pump, pump gear, or hub is installed, or when incorrect pump timing is suspected, the timing must be checked, as follows:

NOTE: There are no timing marks on the flywheel.

Checking Injection Pump Timing

1. Remove the rocker arm cover and rotate the crankshaft in a clockwise direction until the valves of No. 4 cylinder are "on the rock" (No. 1 cylinder at approximately TDC on compression stroke).
2. With the crankshaft positioned as outlined in step No. 1, loosen one of the rocker arm adjusting screws for No. 1 cylinder (either valve). Remove the push rod and pivot the rocker arm out of the way to permit removal of the valve retainers, then, remove the retainers, spring cap, spring, oil deflector and spring seat.
3. Tie a string or a small wire around the groove in the valve stem and lower the valve onto the top of No. 1 piston.
4. Mount a dial indicator on the end of the valve stem.
5. Using the dial indicator, carefully position the piston at exactly TDC, then, zero the gauge.
6. Carefully turn the crankshaft backward (counterclockwise), in the opposite direction of engine rotation, until No. 1 piston is approximately 0.120 in. below TDC, as shown by the dial indicator, then, very carefully turn the crankshaft forward (clockwise) in its normal direction of rotation, moving the piston up in its cylinder until it is exactly 0.107 in. below TDC.

NOTE: The reason for turning the crankshaft back and then again forward, is to take up the backlash in the timing gears before reaching the desired position.

With the piston positioned at 0.107 in. below TDC (compression stroke), as outlined, the piston and crankshaft will then be positioned at 18° BTDC, which is the static point for the start of fuel injection to No. 1 cylinder.

7. Break the seal and remove the inspection cover from the side of the injection pump. Fuel will spill out when the inspection cover is removed from the pump.
8. With No. 1 piston positioned at 18° BTDC, as described in step No. 6, the scribe line on the rotor marked C must align with the scribe line across the lower eye of the snapring.
9. If the scribe line marked C on the rotor does not align with the scribe line on the snapring, then, the injection pump timing is incorrect and must be adjusted.

Adjusting the Fuel Injection Pump Timing

If the injection pump timing marks do not align after checking the timing as outlined, proceed, as follows:

1. To adjust the fuel injection pump timing, loosen the two nuts and Allen setscrew securing the pump in place, then, rotate the pump in the required direction to align the mark on the snapring with the C mark on the rotor.
2. Tighten the Allen setscrew and nuts securing the pump in place and again recheck the alignment of the timing marks.
3. If the timing marks are in alignment, replace the inspection cover, then, reassemble the tractor and bleed the fuel system of air.
4. If the injection pump cannot be rotated far enough to align the mark on the snapring with the C mark on the rotor, proceed, as follows:
   a. Remove the timing gear cover and loosen the three setscrews securing the injection pump gear to the hub, then, move the hub, in relation to the gear, in the required direction to align the timing marks inside the pump.
   b. Tighten the setscrews securing the gear to the hub and check that the mark on the rotor has not moved in the process of tightening the setscrews.
   c. If necessary, delete the pump timing marks on the injection pump gear and/or hub, and install new marks.
   d. Replace the inspection cover, then, reassemble the tractor and bleed the fuel system of air.

Adjusting Engine Speed

The fuel injection pump is provided with a low idle speed adjustment screw and a high idle speed adjustment screw to provide a means of regulating the engine speed within the proper limits.

The correct crankshaft speeds for the engine are listed, as follows:

Low idle speed .... 550-660 rpm
    (PTO speed of 157-171 rpm)
High idle speed (no load) hand throttle
    .... 2100-2150 rpm
    (PTO speed of 600-615 rpm)
Foot throttle .... 2400-2450 rpm
    (PTO speed of 685-700 rpm)

Check the throttle linkage, before adjusting the engine speed, to make sure that it is properly adjusted and not bent or damaged so that it restricts the full travel of the throttle lever on the injection pump.

Adjusting Low Idle Speed

1. Start the engine and allow it to reach normal operating temperature.
2. Position the hand throttle lever to the low idle position and check the idle speed.
3. To adjust the low idle speed, loosen the locknut and turn the adjusting screw "in" to increase speed, or "out" to decrease speed. Retighten the locknut after adjusting speed.

Adjusting High Idle Speed

1. With the engine running at normal operating temperature, position the hand throttle lever to the "wide open" position and check the engine speed.
2. If the throttle lever will not provide a high idle speed of 2100-2150 rpm, loosen the bolts securing the hand throttle quadrant to the steering post and rotate it to provide the desired speed. Before retightening the bolts securing the quadrant, move the quadrant vertically so that the hand throttle lever can be moved freely with a force of 2-5 lb.
3. Depress the foot throttle to the "wide open" position and check the engine speed. If an engine speed of 2400-2450 cannot be obtained, proceed, as follows:
   a. If the high idle adjustment screw from the injection pump restricts the movement of the throttle control rod and will not provide an engine speed of 2400-2450 rpm, then, break the seal and remove the cover sleeve. Loosen the locknut and turn the adjusting screw "out" to increase the speed.

NOTE: Do not set the engine speed higher than 2450 rpm.

Retighten the locknut after adjusting the engine, then, install and reseal the cover sleeve.

   b. If the injection pump throttle rod does not have sufficient travel to contact the high idle adjustment screw on the injection pump, adjust the linkage, as outlined.

Throttle Lever and Linkage Adjustment

1. Adjust the length of the injection pump throttle rod connected at the bottom of the bell crank and the link rod connected at the upper end of the bell crank, so that the stops on the lever stop plate, which is welded to the throttle lever, are the same distance from the stop in both minimum and maximum throttle positions.
2. Rotate and adjust the hand throttle quadrant horizontally, so that it acts as a stop for the hand throttle in the "wide open" position. Adjust the quadrant vertically so that the hand lever can be moved freely with a force of 2-5 lb.
3. Adjust the length of the foot throttle rod so that it is 2.2 in. long center-to-center (between the points of movement).
4. Adjust the length of the foot throttle control rod, so that the foot accelerator will touch its maximum stop position without exerting force on the high idle adjustment screw of the injection pump. This is adjusted by means of the nuts on each side of the adjusting block.

Fuel Injectors

REMOVAL

1. Carefully clean the engine and fuel lines. This is very important.
2. Loosen the high pressure fuel lne connections at both the injector and the injection pump ends.
3. After loosening the connections, dis-

connect the high pressure lines and the leak-off line assembly from the injectors.

IMPORTANT: Cap, or otherwise cover, all open fuel line fittings and connections. Also, do not bend the fuel lines when removing the injectors.

4. Remove the injector flange nuts and carefully remove the injectors from the cylinder head.

5. Make sure that the old injector washers (gaskets) are removed from the cylinder head.

INSTALLATION

1. Carefully clean each nozzle recess in the cylinder head. Use a piece of wood dowel or brass stock properly shaped to clean the recess.

NOTE: Do not use hard, sharp tools that would scratch the sealing surfaces.

2. Blow out all carbon and foreign material from each recess. Make sure all recesses are thoroughly cleaned.

3. Make sure the old injector washers (copper gaskets) have been removed and install a new washer in each recess.

4. Carefully install each injector into its recess with the scalloped portions of the securing flange toward the long cylinder head nuts. Install the nuts securing the flange and tighten them down evenly until the injector can just be turned by hand.

NOTE: Position the securing flange so that the scalloped portion will allow removal of the cylinder head securing nuts.

5. Turn the injector as required to position it for proper fuel line alignment, then, install and connect the high pressure fuel lines.

6. After the injectors have been correctly positioned, as outlined, to prevent bending the fuel lines, or placing the lines under stress, then, fully tighten the securing flange nuts to 21-24 ft.lb. torque.

NOTE: Tighten the nuts down evenly to prevent cocking the injector in the cylinder head.

7. Install the leak-off line assembly and carefully tighten the connections. The proper torque on the banjo-type leak-off bolts is 5-7 ft.lb.

NOZZLE TESTING

The main requirements of a good nozzle are as follows:
1. Correct valve opening pressure.
2. Pressure tight seats.
3. Good atomization and spray pattern.
4. Freedom from excessive back leakage.

IMPORTANT: Before attempting to test a nozzle, make sure you have a dependable nozzle tester and the necessary connectors. Also, use only clean, approved testing oil in the tester tank.

Injectors must be tested when performing the following operations:

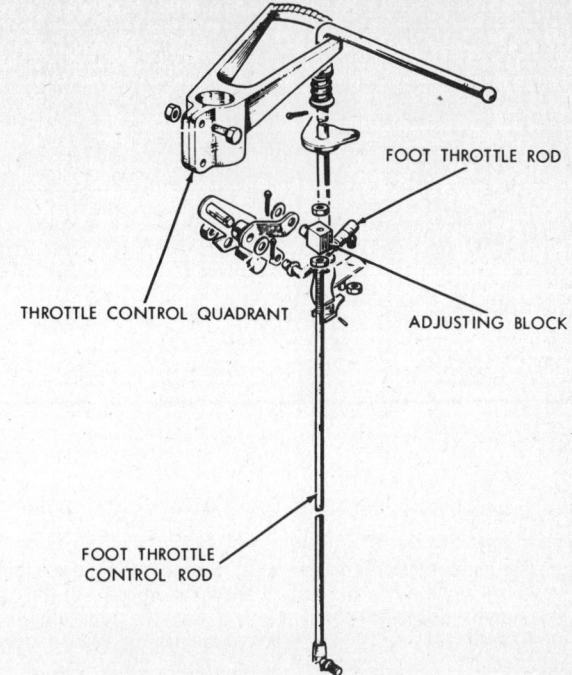

Foot throttle linkage

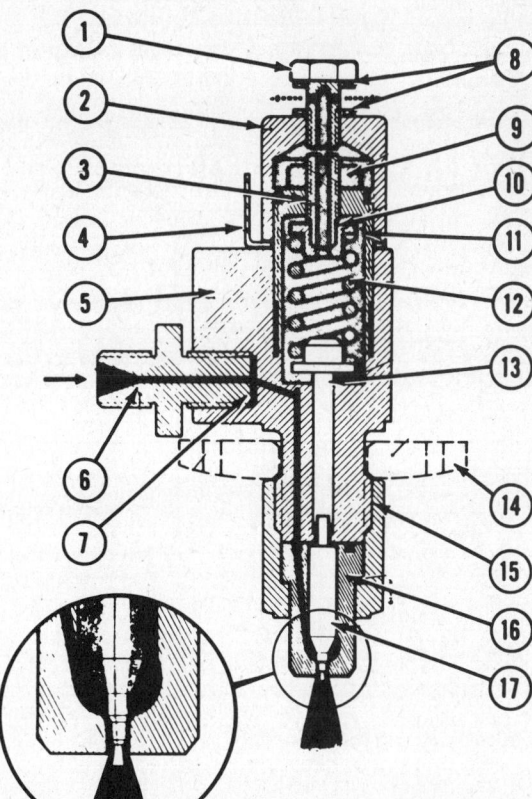

1. Leak-off banjo bolt
2. Nozzle holder cup
3. Adjusting screw
4. Cap washer
5. Nozzle holder body
6. Inlet connection
7. Copper washer
8. Leak-off washers
9. Locknut
10. Spring collar
11. Threaded cap
12. Pressure spring
13. Spring guide spindle
14. Injector securing flange
15. Nozzle cap
16. Nozzle body
17. Nozzle valve

Fuel injector

Massey-Ferguson

Correct positioning of the fuel injectors and securing flanges

1. When performing a complete engine overhaul.
2. When an injector is suspected of being faulty.
3. When an injector has been disassembled and serviced, prior to its reinstallation.
4. When replacing injectors or installing new nozzles, injectors must be cleaned and tested prior to installation in the engine.

To test the nozzles, follow the instructions for the care and operation of the particular nozzle tester being used and proceed, as follows:

1. Thoroughly clean the connector tube, then, connect it to the test stand and pump clean oil through it to flush out the tube.

NOTE: Practice extreme cleanliness to avoid pumping dirt into the injector.

2. Remove the nozzle holder cap, and connect the injector to the connector tube.
3. Close the pressure gauge valve (to avoid damaging it) and operate the hand lever rapidly for at least 10 long strokes to clear all air from the injector.

— CAUTION —
Do not allow the fuel spray from the nozzle to come in contact with the body. The spray has sufficient pressure to puncture the skin and should the fuel enter the blood stream, it may cause blood poisoning.

4. Testing for back leakage (leakage past the nozzle valve and body).

NOTE: A very small amount of back leakage is necessary for the lubrication of the nozzle valve and body. However, excessive back leakage caused by a worn or scored nozzle valve and/or body will affect the operation of the engine and must be corrected by installing a new nozzle.

To determine what is an excessive amount of back leakage, proceed, as follows:

1. Connect the injector to the test stand and open the pressure gauge valve.
2. Operate the hand lever to clear all air from the injector.
3. Set the nozzle opening pressure at 2350-2500 psi.
4. Pump up the pressure to just below the opening pressure, then, close the pressure gague valve and time the pressure drop from 1470 to 1000 psi. For a nozzle in good condition, this time must not be less than 6 seconds when using a dependable test oil at a timperature of 50-70°F. At higher temperatures, a period of time slightly less than 6 seconds may be considered satisfactory. Replace nozzles that do not meet these specifications. Be sure to reset serviceable injectors to their proper opening pressure after making this test.

IMPORTANT: When making this test, make sure that there is no leakage at the connections or the pressure faces of the nozzle holder and nozzle body, as any leakage of this type will cause erroneous test readings. Leakage past the pressure faces may be either external or internal. External leakage may be easily observed; however, internal leakage cannot easily be distinguished from back leakage past the nozzle valve and body.

If external leakage is apparent, or if internal leakage is suspected, do not attempt over-tightening the cap nut to stop such leakage. To correct the leakage, remove the nozzle and carefully examine the pressure faces to determine if they are dirty, scratched or otherwise damaged. If the pressure faces are not damaged, carefully clean them, then, reassemble the injector and retest it. If the pressure drop time is still low, this indicates excessive back leakage.

5. Testing for opening pressure.

Open the pressure gauge valve and operate the hand lever slowly until sufficient pressure is developed to lift the nozzle valve and spray the fuel. Observe the opening pressure as indicated on the pressure gauge.

Opening pressure on used injectors (more than 25 hours of operation) should be 1985 psi. New injectors should have an opening pressure of 2060 psi to allow the spring to "settle in". If the nozzle valve does not lift at the appropriate pressure, as given above, loosen the lock nut and turn the adjusting screw (turn in to increase pressure, or out to decrease pressure), until the correct opening pressure is obtained.

NOTE: Always recheck the opening pressure after retightening the locknut to make sure the pressure setting is not altered.

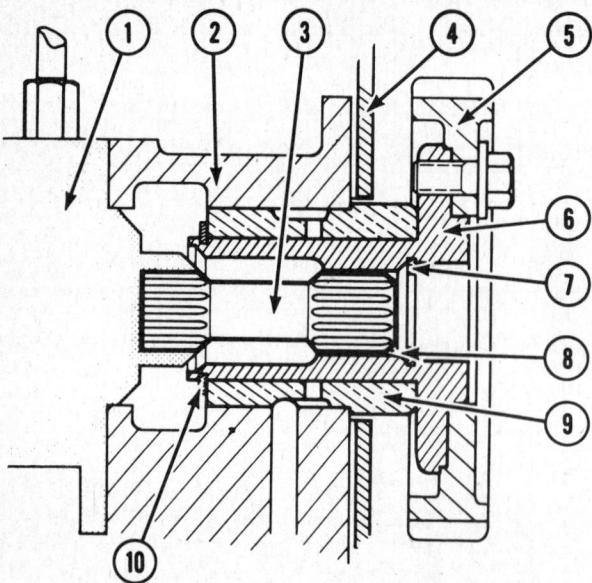

Injection pump hub assembly

1. Fuel injection pump
2. Cylinder block
3. Quill shaft
4. Engine adaptor plate
5. Injection pump timing gear
6. Injection pump hub
7. Snapring
8. Chamfered end
9. Bronze bearing
10. Snapring

Massey-Ferguson

6. Testing for valve seat leakage.

A used injector set at 1985 psi opening pressure must not leak at 1835 psi. A new injector set at 2060 psi opening pressure must not leak at 1910 psi.

To test new or used injectors for valve leakage, wipe the tip of the nozzle dry and slowly build up the pressure to 150 psi less than the opening pressure setting, then, hold this pressure for 10 seconds and observe the nozzle tip for leakage. If drops of fuel collect at this pressure, the nozzle valve is not seating properly and must be serviced and retested. A slight amount of dampness may be disregarded.

7. Testing for spray pattern.

Close the pressure gauge valve, then, operate the hand lever at a speed of 90-100 strokes per minute and observe the nozzle spray pattern. The spray pattern should be completely atomized without any irregular streaks. If the spray pattern is ragged, unduly wet, streaky, or not symmetrical, the nozzle must be serviced and retested.

8. Before removing the injector from the injector test stand, close the pressure gauge valve to prevent damage to the gauge, which may result from a sudden drop of pressure.

9. If the injector passes the back leakage, opening pressure, valve seat leakage and the spray pattern tests, reinstall the injector in the cylinder head. Do not disassemble injectors that pass all of the tests.

LUBRICATION SYSTEM

Oil Pan

REMOVAL

1. Drain the oil out of the engine.
2. Remove the sixteen (16) setscrews and the reinforcing plates around the flanges and remove the oil pan.
3. Remove the spring-loaded oil strainer from the end of the oil pump intake pipe.
4. Carefully clean the sealing faces on the oil pan and cylinder block. Soak the strainer in solvent and thoroughly wash the oil pan and strainer. Use compressed air to blow the strainer clean and free from solvent.

INSTALLATION

1. Using sealing compound, install the side gaskets onto the bottom face of the cylinder block, with the ends installed into the recesses of the front and rear main bearing caps.
2. Using sealing compound, install the front and rear oil pan gaskets into the recesses of the bearing caps with the ends contacting the side oil pan gaskets.
3. Install the oil strainer and spring, then, carefully position the oil pan in place and install the setscrews and reinforcing plates.

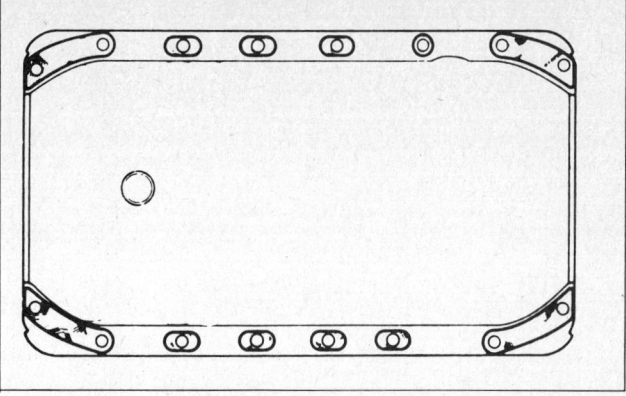

Oil pan with re-inforcing plates correctly installed

Oil Filter

REPLACEMENT

1. Remove the drain plug on the underside of the filter head casting and drain the oil out of the filter.
2. Unscrew the bolt in the center of the head casting and remove the filter bowl and element.
3. Remove the element sealing ring from the recess in the filter head. Discard the old element and clean the filter bowl.

NOTE: If the filter head is removed for major engine overhaul, wash out the by-pass chamber in the head and blow it clean with compressed air.

4. Install a new sealing ring in the recess of the filter head, then, install the new oil filter element in the filter bowl and install the bowl. Make sure the bowl seats properly within the

Oil Pump

REMOVAL

1. Remove the oil pan and oil strainer.
2. Disconnect the oil delivery pipe from the oil pump to the bottom face of the cylinder block.
3. Disconnect the tractormeter cable at the cylinder block end, then, remove the snapring and lift out the tractormeter driveshaft.

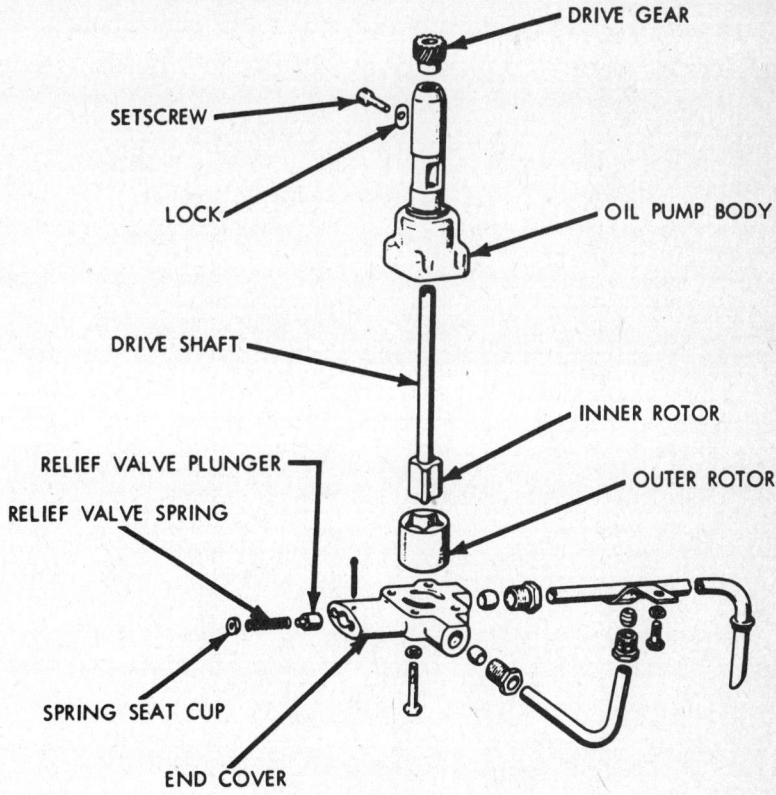

Oil pump components

4. Remove the setscrew on the outside of the cylinder block which secures the oil pump and remove the oil pump.

DISASSEMBLY

1. Using a suitable puller, pull the pump gear from the driveshaft. This gear is a press fit on the shaft and is not keyed.
2. Remove the four setscrews and washers securing the end cover to the pump body and remove the end cover. The relief valve body is integral with the end cover.
3. Disassemble the relief valve by pulling the cotter pin and removing the spring seat cap, spring and relief valve plunger.
4. Remove the outer rotor and the driveshaft with the inner rotor out of the pump body. The inner rotor is pressed onto the driveshaft and is secured with a pin.

INSPECTION

1. Thoroughly clean all parts in solvent.
2. Examine the rotors for cracks or scoring.

3. Install the rotors in the pump body.

NOTE: The chamfered end of the outer rotor must be installed inside the pump body.

4. Measure the operating clearance between the lobes of the inner and outer rotors. This clearance must not exceed 0.006 in.
5. Measure the clearance between the outer rotor and the pump body. Clearance must not exceed 0.010 in.
6. Measure the clearance between the top of the rotors and the face of the pump body with a straight edge and feeler gauge. This clearance must not exceed 0.003 in.

If the oil pump is considered defective in any way, it must be replaced as an entire assembly.

REASSEMBLY

1. Install the driveshaft and inner rotor into the pump body.
2. Install the outer rotor with the chamfered edge toward the inside of the pump body.
3. Install the end cover onto the body with the setscrews and spring washers. The end cover can only be installed one way, as the holes are offset.
4. Press the oil pump drive gear onto the driveshaft so that the maximum clearance between the inner face of the gear and the pump body is between 0.031-0.047 in.
5. Install the relief valve plunger, spring, cap and cotter pin.

INSTALLATION

Install the oil pump by reversing the removal procedure.

NOTE: If there is any difficulty installing the outlet pipe, due to misalignment, loosen the setscrews securing the pump end cover. Do not forget to retighten the setscrews.

Reinstall the tractormeter driveshaft and snapring after the oil pump is in place.

Satoh

INDEX

MODEL S-370
- 2-Wheel Drive Front Axle 722
- 4-Wheel Drive Front Axle 723
- Steering......................725
- Brakes........................726
- Clutch........................727
- Transmission..................727
- Front Power Take-Off..........731
- Hydraulic System..............731
- Engine........................734
- Fuel System...................737

MODELS S-373 BEAVER II & S-470 BUCK
- 4-Wheel Drive Front Axle 739
- 2-Wheel Drive Front Axle 742
- Steering, 4-Wheel Drive.......743
- Steering, 2-Wheel Drive.......744
- Brakes........................745
- Clutch........................745
- Transmission..................746
- Engine........................752
- Fuel System...................757

MODELS S-550G & 650G
- Front Axle....................759
- Steering......................760
- Brakes........................761
- Clutch........................762
- Transmission..................762
- Hydraulic System..............768
- Engine........................768
- Fuel System...................772

MODEL S-630
- 2-Wheel Drive Front Axle 775
- 4-Wheel Drive Front Axle 776
- Steering......................779
- Brakes........................781
- Clutch........................782
- Transmission..................782
- Rear Axle.....................788
- Hydraulic System..............789
- Engine........................789
- Lubrication System............795
- Fuel System...................795

SATOH
Model S-370

2-WHEEL DRIVE FRONT AXLE

REMOVAL AND INSTALLATION

1. Pull out the cotter pin from the center pin nut, and remove the castle nut.
2. Support the front axle, pull out the center pin, and remove the front axle.
3. Clean the center pin hole in the front axle and holes in the chassis, and install the front axle to the chassis with the longer boss facing forward.
Push the greased center pin to the rear, and tighten the castle nut to specification. Then back if off ¼ - ⅓ turn so that the cotter pin holes can be aligned.
Tightening torque: 108-123 ft.lb.

King Pin

REMOVAL AND INSTALLATION

1. Remove the cotter pin locking the castle nut on each end of the tie rod, loosen the castle nut, and remove the tie rod.

King pin installation

2. Loosen the knuckle arm bolt, and remove the knuckle arm while taking care so that the king pin does not fall.
Remove the drag link from the knuckle arm, as required.
3. By tapping the top end of the king pin, pull out the king pin.
4. Remove the righthand castle nut, and remove the king pin.
For the right side king pin, remove the castle nut and the king pin.
Before installing the front axle to the chassis, insert the bushing into the king pin.

1. Oil the king pin sparingly, and push the bushing into the king pin. Install the grease-coated O-ring.
2. Install the thrust bearing, and apply a liberal amount of grease to it, and install the king pin to the front axle.
3. Lock with the cotter pin.
4. Make sure the front axle swings smoothly without end play.
5. Insert the greased oil seal into the king pin, while taking care so that the king pin does not slip off.
6. Install the washer, and align the knuckle arm bolt hole with the cut on the king pin, and tighten the bolt.
Tightening torque: 15.2-21.7 ft.lb.
7. Make sure the gap between the knuckle arm lower side and the washer is 0.02 in. or less.
8. Install the washer to the righthand king pin, tighten the castle nut, and lock with the cotter pin.
Tightening torque: 32.5-39.7 ft.lb.
9. Make sure the king pin operates smoothly, and grease it as required.

Front Hub

REMOVAL AND INSTALLATION

1. Loosen the front wheel mounting bolt locknuts, and lightly loosen the three front wheel mounting bolts.
2. Jack up the front axle, loosen the front wheel mounting bolts, and remove the front wheels.
3. Remove the front wheel hub cap, straighten the tab of the lockwasher, and remove the sleeve nut.
4. Remove the front wheel hub using the gear puller, and remove the oil seal.
5. Remove the ball bearing from the front wheel hub.
6. Install the washer to the king pin. Install the grease-coated oil seal.
7. Install two circlips to the front wheel hub, and insert the ball bearing into the hub.

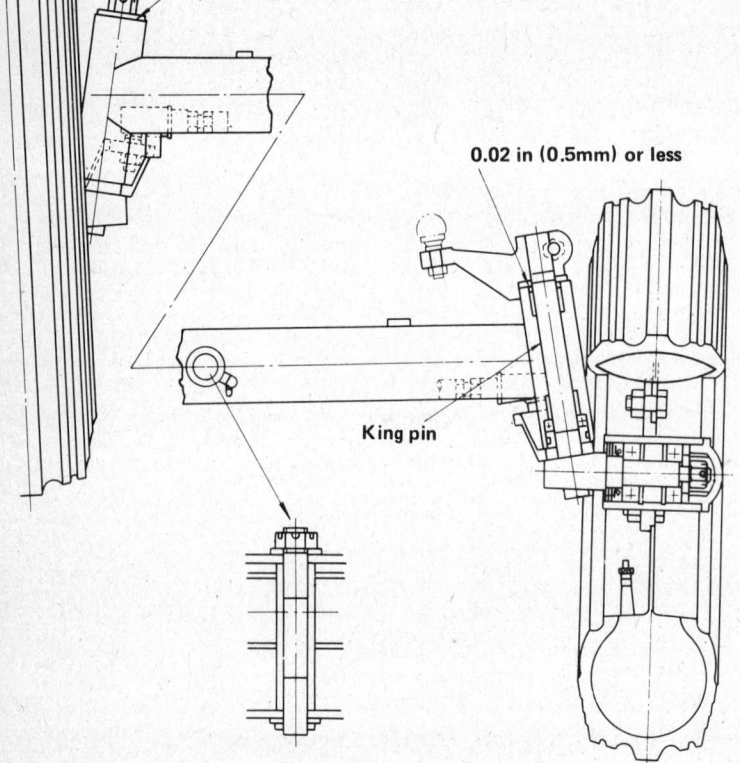

Knuckle arm-to-lower side gap

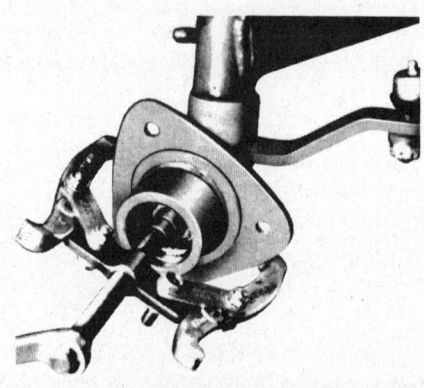

Hub bearing removal

8. Install the front wheel hub to the king pin, pack with Alvania grease, and insert the ball bearing into the hub.

9. Install the lockwasher, and tighten the sleeve nut.

Front wheel hub starting torque: 10.1-11.6 ft.lb.

10. After tightening the nut with starting torque, make sure that the front wheel hub rotates smoothly, and lock with the lockwasher. Apply a bond to the cap, and install it to the front wheel hub.

4-WHEEL DRIVE FRONT AXLE

Axle Housing

REMOVAL AND INSTALLATION

1. Remove the center pin grease nipple, and loosen the two bolts securing the center pin to the chassis.

2. Pull out the center pin, and remove the axle housing from the chassis.

3. Installation is the reverse of removal.

Knuckle

REMOVAL AND INSTALLATION

1. Jack up the front axle, and remove the front wheel.

2. Loosen the bolt securing the axle housing and knuckle assembly, and remove the knuckle assembly.

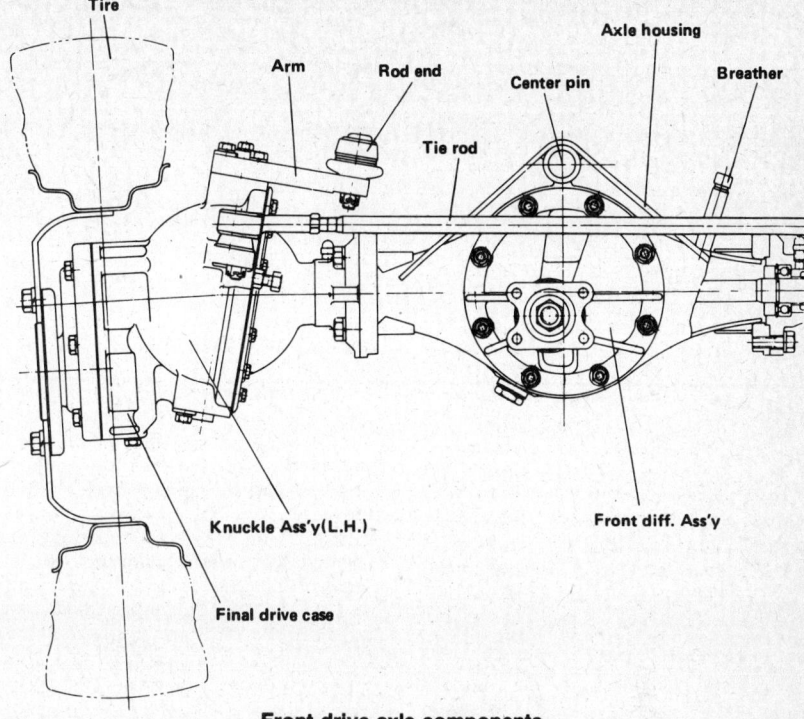

Front drive axle components

Front dive knuckle

NOTE: A long shaft is held to the knuckle assembly at one end. When removing the shaft, hold it carefully until it is completely pulled out.

3. Installation is the reverse of removal.

DISASSEMBLY

1. Loosen the oil seal retainer plate holding bolt on the spherical part, and remove the oil seal felt and oil seal.

2. Loosen the king pin bolt on both upper and lower parts of the knuckle arm, and remove the key plate.

Insert a screwdriver into the key plate, and pull out the king pin gently, and remove the king pin case and final gear case.

NOTE: There is a shim on the top of the king pin case and in the final gear case.

3. Remove the oil seal from the king pin case (the oil seal can not be reused), and remove the circlip on the king pin case side. Pull out the shaft with the ball bearing installed.

4. Remove the circlip from the shaft, and remove the ball bearing.

ASSEMBLY

1. Push the ball bearing into the differential shaft so that the sealed surface faces inward, set with the circlip, and install the king pin case.

2. Apply grease to the oil seal, and install the king pin case.

Final Gear Case

DISASSEMBLY

1. Loosen the bolt securing the final gear (B), and remove the final gear (B) from the final gear case (A).

2. Remove the ball bearing from the final gear case, and remove the circlip. Remove the gear, and by tapping the shaft, remove the yoke assembly from the final gear case.

NOTE: The yoke must be replaced as an assembly.

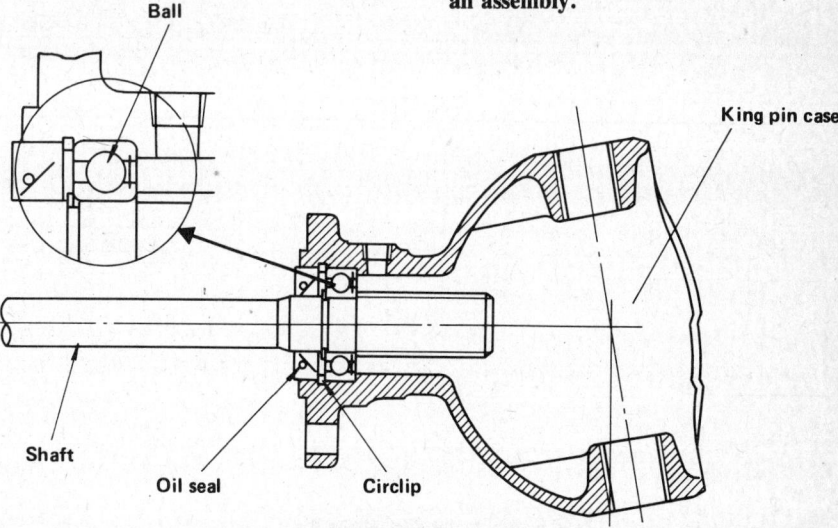

Knuckle cross section

Satoh

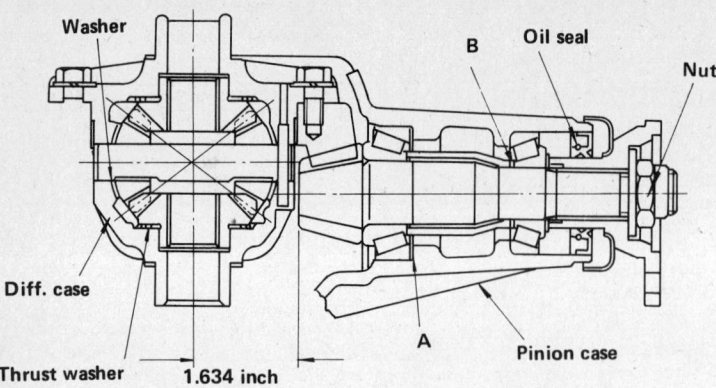

Final gear cross section showing measuring points

3. Remove the circlip from the wheel shaft, and by tapping the shaft, remove it from the final gear case (B).
Remove the ball bearing and oil seal.

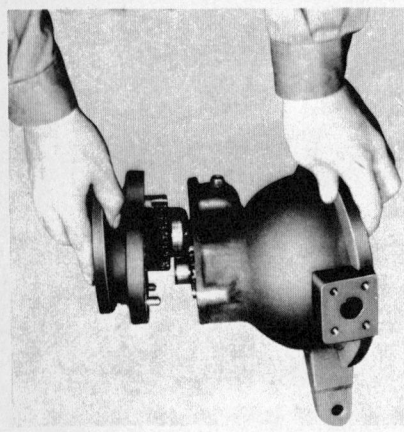

Separating the final gear from the final gear case

ASSEMBLY

1. Install the oil seal to the final gear case (B), and apply grease.
2. Install the wheel axle, and install the ball bearing, gear, and ball bearing, in that order. Set with circlips.

NOTE: The chamfered side must face the oil seal.

3. Install the thrust bearing in the king pin hole on the bottom (inside) of the final gear case (A) paying attention to the direction of the thrust bearing, as illustrated.
4. Adjust the clearance between the final gear case and king pin case to 0.01 in. or less using a shim.

Differential Gear Unit

DISASSEMBLY

1. Loosen the differential gear assembly mounting nut, and remove the differential gear assembly from the axle housing.
2. Loosen the bearing holder bolt, and remove the differential gear.
3. Remove the ball bearing from the differential case assembly, loosen the ring gear bolt, and remove the ring gear.

4. Pull out the pin locking the centerpin, and pull out the center pin, and remove the pinion gear and side gear as well as thrust washer.
5. Loosen the self-locknut, and pull out the pinion shaft while taking care so that the shim or spacer does not drop.
6. Remove the flange assembly, and remove the oil seal from the flange assembly.

ASSEMBLY

1. Always use new O-ring and oil seals.
2. Thoroughly wash all parts, and apply oil or grease to moving parts. Assemble them in the correct manner.
3. Using special tools, measure the pinion shaft cone center and preload correctly.
4. Install the thrust washer and side gear in the differential case, and set the pinion gear and thrust washer with the center pin, and measure the backlash between the side gear and pinion gear.
Backlash: 0.001-0.006 in.
5. Backlash adjustment should be made by changing the thickness of thrust washer installed, together with the side gear.
Thickness of thrust washer:
0.035 in.
0.04 in.
0.043 in.
0.047 in.
0.051 in.
6. When the backlash is correctly adjusted, lock the center pin with the lock pin, and futher clinch the lock pin with the center punch.
7. Install the ring gear to the differential case, and tighten it with the mounting bolt.
Tightening torque: 21.7-28.8 ft.lb.
8. Install the taper roller bearing collar and taper roller bearing to the pinion shaft, in that order and install the pinion shaft in the pinion gear case. Adjust the cone center to 1.634 in. (41.5 ± 0.03 mm) using a shim or shims of A portion.
Shim thickness: 0.002 in.
0.017 in.
0.004 in.
0.008 in.
0.012 in.
0.02 in.
9. After adjusting the cone center to 1:634 in., install the pinion shaft to the pinion shaft case, and install the flange. Tighten the self-lock nut to 72.2-86.7 ft.lb. and adjust the pinion shaft preload to 21.7-43.3 ft.lb. using the shim of the B portion.
Thickness of shim:
0.0669 in.	0.0740 in.
0.0681 in.	0.0752 in.
0.0693 in.	0.0764 in.
0.0705 in.	0.0776 in.
0.0717 in.	0.0787 in.
0.0728 in.	

10. Install the differential gear to the pinion case, and adjust the backlash be-

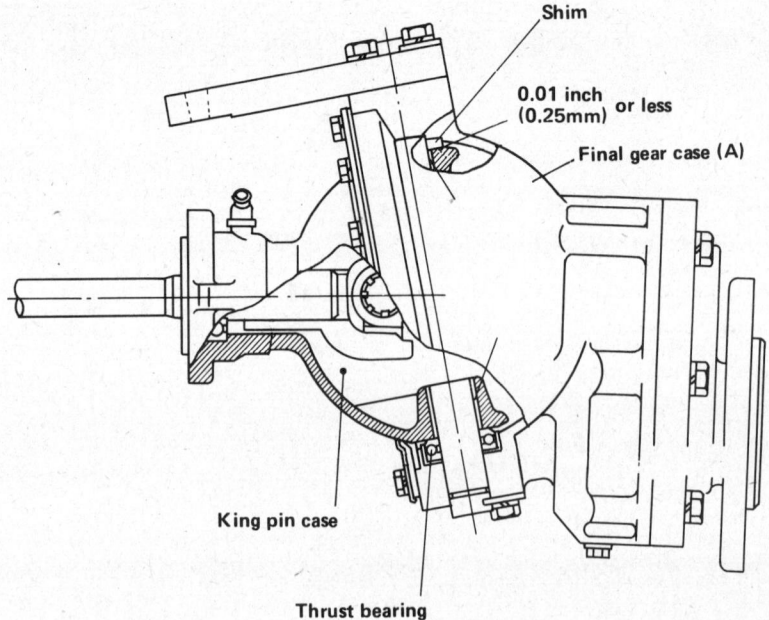

Shimming the thrust bearing

tween the ring gear and pinion gear to 0.004-0.006 in. using the shim of the C portion.

Backlash: 0.004-0.006 in.

Apply preload of 0.18 ft.lb. to the taper roller bearing.

Thickness of shim:
0.002 in.
0.0028 in.
0.004 in.
0.008 in.
0.012 in.
0.02 in.

11. Install the bearing metal, and tighten the bolts evenly. Make sure that the differential gear turns smoothly.

Tightening torque: 10.8-14.4 ft.lb.

12. Apply a sealing compound to the axle housing, and install the pinion gear case.

Tightening torque: 8.67-12.3 ft.lb.

13. Install the axle housing to the chassis, and adjust the clearance between the chassis and axle housing to 0-0.0077 in. using a shim, and lock with the center pin.

Tightening torque: 43.3-50.6 ft.lb.

Install the grease nipple to the center pin.

14. Make sure that the axle housing swings lightly.

15. Set the yoke assembly to the splined portion of the differential shaft in the king pin case, and pack the yoke with 0.44 lb. of grease.

16. Install the greased O-ring to the king pin. Install the shim (selected previously) into the hole on top of the final gear case. Install the king pin case and final gear case (A) while keeping the shim in its position with the king pin. Insert the ends of the yoke into the gear case (A) correctly.

17. Insert the key plate into each king pin.

18. Apply a sealing compound to the matching surfaces of the final gear cases (A) and (B), install the gasket, and install the final gear case (B) together with the dowel pin.

Tightening torque: 8.67-12.3 ft.lb.

19. Install the oil seal and oil seal felt (which is oiled) to the spherical part, hold them down with the oil seal retainer, and tighten the bolt.

Tightening torque: 3.61-5.06 ft.lb.

20. Install the steering lever to top of the left side final gear case, and install the key plate, then tighten the bolt.

Tightening torque: 18.1-21.7 ft.lb.

Make sure that the king pin case lightly moves and also the differential shaft turns lightly.

Tightening torque: 18.1-21.7 ft.lb.

NOTE: The king pin on the top of the left side final case is longer than the three other king pins since the steering lever is attached to it.

21. Push the ball bearing into the final gear case so that oil seal is on the inner side, and install the gear with the chamfered side on the inner side, and set with the circlip.

22. Install both right and left knuckle assemblies to the axle housing assembly.

Tightening torque is 43.3-50.6 ft.lb.

Front Axle Universal

REMOVAL

1. Loosen the universal joint cover, tightening bolts, and remove the universal joint cover.
2. Remove the universal joint flange yoke and flange assembly bolt, and remove the universal joint.
3. Remove the cotter pins from the castle nuts on the drag link and tie rod, loosen the castle nut, and remove the drag link and tie rod.

INSTALLATION

1. Apply oil to the mid PTO felt on the universal joint assembly, and install it to the PTO shaft, and fully push it downward.
2. Install the front part of the universal joint to the flange assembly of the axle housing.

Tightening torque: 14.4-18.1 ft.lb.

3. Make sure that the universal joint rotates smoothly.
4. Install the universal joint cover. Make sure that the universal joint does not make contact with the differential housing.
5. Install the tie rod, and set the castle nut with the cotter pin.
6. Install the drag link.
7. Adjust the toe-in to 0.23 in. with the tie rod, and lock the tie rod.
8. Adjust the steering angle of the front wheels to $44°{}^{-0°}_{+2°}$ by turning the stopper bolts.

STEERING

Steering Gear

REMOVAL AND DISASSEMBLY

1. Remove the steering column from the steering gear box, and remove the worm shaft.

NOTE: There is a shim for clearance adjustment between the steering gear box and steering column.
Take special care not to lose it.

2. Remove the Pitman arm from the sector shaft.
3. Remove the cover from the steering gear box, and remove the sector shaft.

NOTE: There is a collar, thrust liner, O-ring. Take special care not to lose them.

ASSEMBLY AND INSTALLATION

1. Push the two ball bearings into the worm shaft.
2. Install the steering column, and make an adjustment using a shim so that worm shaft end play is 0-0.079 in. (0-0.2 mm).

Worm shaft end play:
0-0.079 in.
(0-0.2 mm)

Adjusting shim:
1135-2120-000: 0.0039 in. (0.1 mm)
1135-2120-000: 0.0079 in. (0.2 mm)

NOTE: Because of improved accuracy of the worm at part B, the liner is not in use as from tractor serial numbers ST1300-700851 (2-wheel drive) and ST1300D-700641 (4-wheel drive).

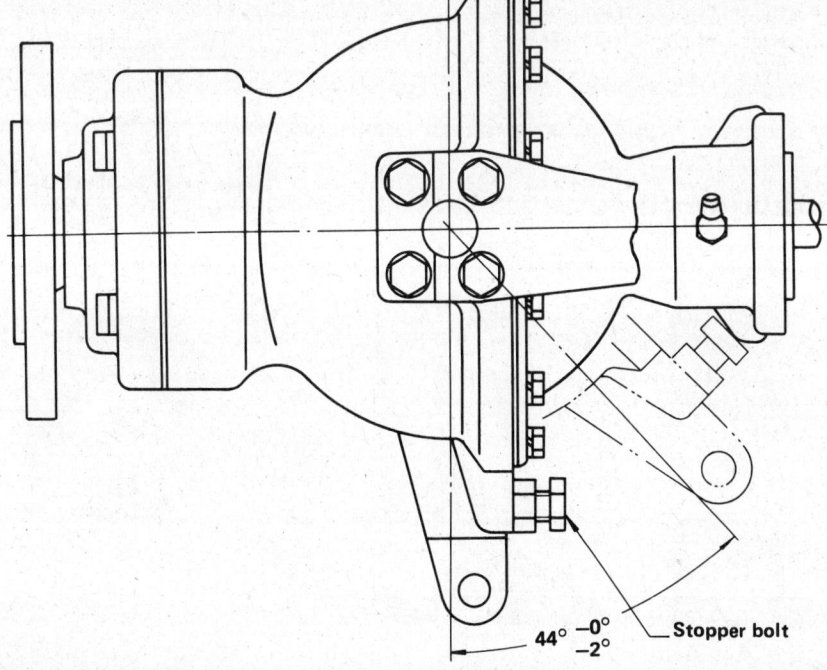

Front axle universal

Satoh

3. Incorporate the sector shaft in the steering gear box, install the collar and cover, and adjust the play of sector shaft end to 0.004 in. (0.1 mm) by inserting shims into the A and B portions.

Adjusting shim:
*1135-2109-000: 0.004 in. (0.1 mm)
 1135-2112-200: 0.004 in. (0.1 mm)
*1135-2110-000: 0.008 in. (0.2 mm)
 1135-2111-000: 0.008 in. (0.2 mm)
*1135-2112-100: 0.012 in. (0.3 mm)
 1135-2112-000: 0.016 in. (0.2 mm)

The shim marked * should be used at the B portion.

Select the proper shims for end play of the worm shaft and sector shaft, and proceed as follows:

4. Install the greased O-ring to the sector shaft, install the adjustment shims and collar, and install the cover with the gasket.
Tightening torque: 14.4-18.1 ft.lb. (2.0-2.5 kg.m)

5. Place the worm wheel in the case so it faces upward, push the worm shaft (to which the bearing is already installed), and engage the worm wheel with the worm gear.

6. Install the collar, shim, and install the steering column, together with the gasket, in the gear box.
Tightening torque: 8.67-12.3 ft.lb.

NOTE: In the machines starting with the below-indicated serial numbers, gaskets are eliminated between steering gear box and steering column and between box and cover: instead of gaskets, the sealant is used.

When reassembling this section of the linkage system, be sure to use gaskets if the machine is of previous production, or the sealant (THREE BOND No. 4) if the machine is of current production.
2-WD: C/#800001 and above
4-WD: C/#800001 and above

7. Align the punch mark on the sector shaft with the punch mark on the Pitman arm, and install the Pitman arm.
Tightening torque: 86.7-101 ft.lb.

NOTE: The Pitman arm used in some models is shaped as illustrated, but it can be installed in the similar way.

8. Install the steering gear box assembly to the clutch housing, while watching the dowel pin position.
Tightening torque: 14.4-18.1 ft.lb.

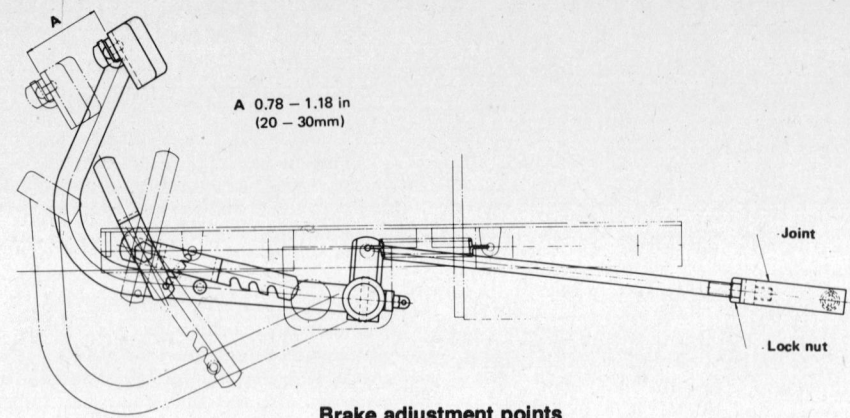

A 0.78 – 1.18 in (20 – 30mm)

Joint

Lock nut

Brake adjustment points

9. The fuel tank base, fuel tank, battery, instrument panel, steering wheel, etc. should be installed in the reverse order to the removal.

BRAKES

REMOVAL

1. Remove the brake rod from the brake cam arm, loosen the six M8x20 bolts securing the brake cover, and remove the brake cover.
2. Remove the circlip attached to the differential shaft, and remove the brake drum.
3. Remove the brake shoe from the brake cover, as required.

INSTALLATION

1. Install the greased O-ring to the cam arm, and apply grease to the grease groove. Install the cam arm to the brake cover, and set with the circlip.
2. Hook the spring to the brake lining, and install it to the brake cover.
3. Install the brake drum to the differential shaft, and set with the circlip.
4. Install the gasket to the brake cover, and install it to the final reduction case.
Tightening torque: 8.67-12.3 ft.lb.

ADJUSTMENT

1. Remove the joint from the brake cam arm, loosen the locknut, and adjust so that

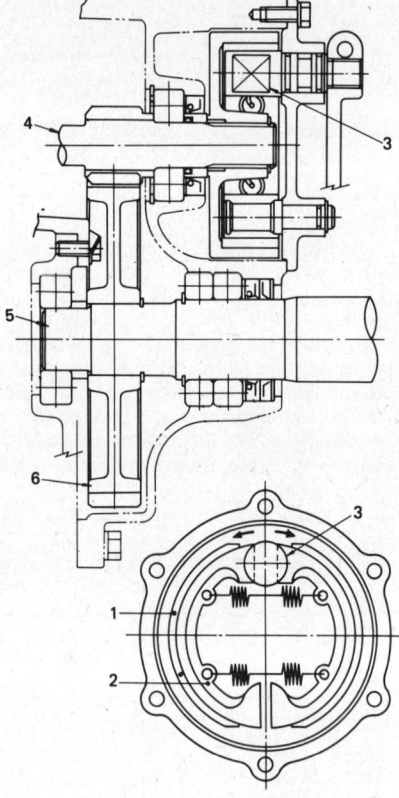

Brake cross section

1. Brake shoe
2. Brake lining
3. Brake cam
4. Diff. pinion shaft
5. Final shaft
6. Final gear

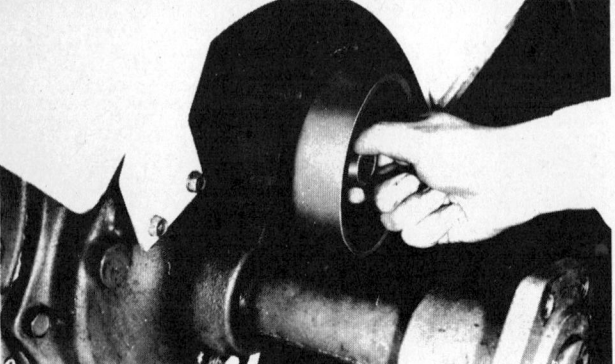

Brake drum removal

Brake shoe removal

the brake pedal free play becomes to be 0.78-1.18 in.

2. Make sure that both right and left wheels are braked evenly.

3. Make sure that the parking brake operates correctly.

CLUTCH

REMOVAL AND INSTALLATION

1. To remove the clutch, the engine must be removed first from the clutch housing.

2. After removing the engine from the clutch housing, loosen the six bolts securing the main clutch to the flywheel, and remove the pressure plate assembly and clutch disc from the flywheel. The bolts must be loosened evenly.

3. When loosening the bolts, care should be taken so that no extremely heavy load is imposed on a bolt.

4. Thoroughly wipe off the oil or grease on the flywheel or on the contact surface of the pressure plate assembly with the lining.

5. If the facings are worn down excessively so that the recession of the rivet heads from the facing is 0.0078 in., discard the disc and install a new one.

6. When a new disc is to be installed, make sure that the recession is 0.039-0.047 in.

7. Apply a thin coat of grease to the flywheel pilot bearing.

8. Place the clutch disc with the longer clutch disc splined boss on the transmission side, and center the clutch disc using the clutch disc center tool, and install the pressure plate to the flywheel.

9. While watching the two reamer bolt positions, tighten the six bolts evenly.
Tightening torque: 8.67-12.3 ft.lb.

ADJUSTMENT

Adjusting the Clutch Pedal Free Play

The clutch pedal free play should be adjusted from time to time to compensate for natural wear on the facings. If there is excessive pedal play, even full movement of the pedal to the floor board will not force the clutch release bearing in against the diaphragm spring. The correct pedal free play is 0.98-1.18 in.

Pedal Free Play Adjustment

Adjust the rod length so that the free play (A) becomes to be 0.98-1.18 in.

Pedal Effective Stroke Adjustment

1. Adjust the stopper bolt so that, when the pedal is depressed and the full effective stroke becomes 3.54-3.94 in., the release shaft lever firmly rests on the stopper head, and lock with the locknut.

2. After the adjustment is complete, make sure that the power from the engine is interrupted as the pedal is fully depressed

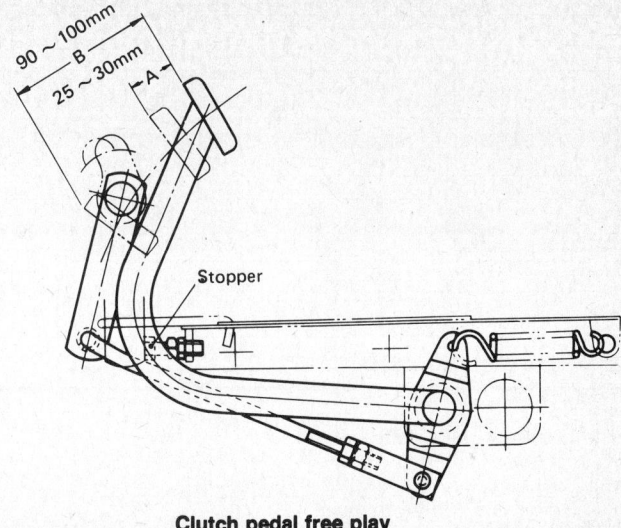

Clutch pedal free play

and that no abnormal noise is generated in the clutch system.

TRANSMISSION

REMOVAL

1. Remove the drain plug, and drain off the transmission case oil.

2. Remove the drag link from the Pitman arm.

3. Remove the front plate from the clutch housing.

4. Release the bonnet hook, open the bonnet, and disconnect the battery cables from the terminals.

5. Disconnect the wire connectors from the terminals.

6. Remove the safety guard screws.

7. Remove the fuel pipe from the fuel strainer, and make it so that fuel does not leak out. Remove the return pipe.

8. Pull out the cotter pin on the engine side, and remove the throttle rod.

9. Remove the bolts securing the suction pipe and delivery pipe from the hydraulic pump. Note that there is an O-ring installed.

10. Hang the engine with a chain block, and place the jack under the clutch housing. Loosen the bolts securing the engine and clutch housing, and demount the engine from the clutch housing by pulling it forward gently.

NOTE: On the 4-wheel drive tractor, remove the universal joint cover, loosen the universal joint bolt, and separate the engine.

11. Remove the driver's seat together with the tool box.

12. Jack up the rear wheel axle, and remove the rear wheels.

13. Remove the fenders.

14. Remove the brake rods and remove the steps.

15. Shift the jack under the clutch housing to under the transmission case, and separate the clutch housing from the transmission case.

16. Remove the hydraulic pipe, and bring the transmission case onto the work bench with the hydraulic case mounted.

17. Loosen the set bolt on the driveshaft coupling, and remove the mainshaft together with the coupling.

18. Loosen the hydraulic case bolt, and remove the hydraulic case.

19. Loosen the main shift cover bolt, and remove the cover.

20. Loosen the final reduction case bolt, and remove the case.

DISASSEMBLY

Disassembling the Driveshaft

1. Loosen the bearing holder bolt and nut, and remove the bearing holder together with the driveshaft.

Take care so that the needle bearing in the driveshaft does not fall off.

2. By tapping the end of the driveshaft, pull out the driveshaft from the bearing holder. Remove the driveshaft circlip, and remove the ball bearing.

Remove the oil seal and ball bearing from the bearing holder.

NOTE: On some early tractors, the bearing holder is secured by a bolt instead of a stud bold.

Disassembling the Select Shaft

1. Remove the circlip on the rear end of the select shaft, and by tapping the front end of the select shaft, pull it out so that the gear is removed.

2. Remove the select shaft circlip, and remove the ball bearing.

Removing the Differential Gear

Straighten the stopper washer on the bolt securing each bearing holder, and loosen the bolt.

1. Insert a screwdriver into the cut on the bearing holder and pry out the bearing holder or install the bolt to the bearing holder tap, and by tightening it, remove the bearing holder.

Satoh

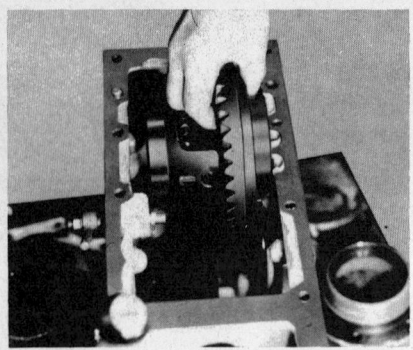

Removing the differential gear

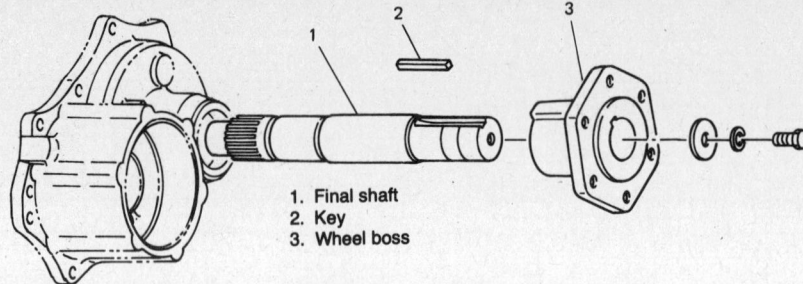

1. Final shaft
2. Key
3. Wheel boss

Wheel boss and axle

NOTE: When removing a differential gear which is in good condition, be sure to remember its thickness and quantity exactly. Otherwise, you will be confused when reinstalling.

2. Remove the differential gear from the transmission case.

Disassembling the Countershaft

1. Straighten the lockwasher, and loosen the sleeve nut, then remove the taper roller bearing.
2. Remove the circlip from the front end of the countershaft, and remove the gear collar and needle bearing, and thrust washer by gently pulling out the countershaft.
3. Remove the taper roller bearing from the countershaft.

Disassembling the Reverse Shaft

1. Remove the reverse shaft seal cap on the front end of the transmission case.
2. Pull out the reverse shaft lock's spring pin, and pull out the reverse shaft to the front by shifting the circlip, and remove the reverse gear, needle bearing and circlip.

Disassembly of the PTO Shaft

1. Loosen the bolt for tightening the PTO shaft plate at the rear side of the transmission case, and pull out the PTO shaft backwards.
2. Remove the plate from the PTO shaft, and remove the sleeve.
3. Remove the circlip and then remove the ball bearing from the PTO shaft.
4. Remove the collar from the PTO shaft joint, move the PTO shaft forward, and remove the ball bearing at the front of the PTO shaft by means of a gear puller.
5. Remove the PTO shaft by lifting the rear side of the shaft upward, remove the gear and the circlip.

NOTE: Although the 4-wheel tractor is equipped with an idler gear, the disassembly of the PTO shaft is the same as for a 2-wheel tractor. Remove the idler gear together with the needle bearing after removing the circlip.

Disassembly of the Differential Gears

1. Put reference marks on the differential gear case and differential gears.
2. Remove the ring gear by loosening the bolts after straightening the bent portion of a stopper washer.
3. Remove the pinion gear, side gear and thrust liner.

Disassembly of the Final Drive Case

Disassembly of the pinion shaft.
1. Remove the circlip, then remove the pinion shaft while lightly tapping it and remove the oil seal bush together with the ball bearing.
2. The righthand side drive case is equipped with a differential lock shifter. Remove the shifter after loosening the nuts.

Disassembly of the Rear Wheel Shaft

1. Remove the ball bearing and the collar final reduction gear.
2. Remove the 2 circlips, and remove the rear wheel shaft from the final gear case while lightly tapping it.

NOTE: Some early S-370 model tractors are equipped with separate wheel axles and wheel bosses, but the disassembly of such components is the same as above.

ASSEMBLY

Assembly of the Differential Gear

1. Install in the differential gear case the side gear on which an oil-coated thrust liner is mounted, assemble the pinion gear and the thrust liner and set them with a center pin.
2. Adjust the backlash of pinion gear and side gear to be 0.010-0.014 in. by means of a thrust liner.

Pinion gear and side gear backlash: 0.010-0.014 in.
Thickness of the adjusting shim:
1135-1408-001   0.047 in.
1135-1409-001   0.055 in.
1135-1411-001   0.063 in.

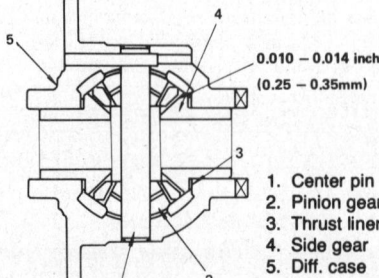

1. Center pin
2. Pinion gear
3. Thrust liner
4. Side gear
5. Diff. case

Pinion gear and side gear backlash

3. After adjusting the backlash properly, lock the center pin by driving a lock pin into it, install the pinion shaft and make sure that it rotates smoothly.
4. Install the ring gear together with a stopper washer in the differential gear case with their reference marks aligned, and tighten them with bolts.
Tightening torque: 21.7-25.3 ft.lb.

NOTE: The protruding portion of stopper washer A should be placed so that it covers the pin locking the center pin to prevent it from pulling out.

5. Install the ball bearing in the differential gear case.

Measuring the Cone Center of the Countershaft

1. Set a circlip in the center hole of the transmission case where the countershaft is placed, install circlip paying attention to the direction of the outer race of the taper roller bearing.
2. Install the taper roller bearing on the countershaft, paying attention to its direction.
3. Install the countershaft in the transmission case.

Then install a taper roller bearing at its front side and tighten it with a sleeve nut so as not to leave any end play. Then measure the cone center with a special tool.
Cone center of the countershaft: 3.031±0.002 in.

How to Use the Special Tool

Set jig A for measuring the countershaft at the portion of the transmission case where the differential gear case and bearing holder are installed, and insert jig B for measuring the countershaft between the top end of the countershaft pinion gear and jig A. Select shims which permit inserting the smaller diameter portion of jig B and which don't permit inserting the large portion, place these shims between the outer race of

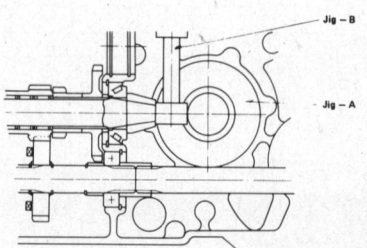

Countershaft measurement tool

the taper roller bearing and the circlip when assembling.

Countershaft Gear Clearance

Install on the countershaft on which the taper roller bearing is installed, the collar, gear 34T, gear 17T, liner gear 18-23T, gear 29T, gear 47T, and needle bearing collar liner.

With a circlip inserted, select the shim properly enough to provide a clearance of 0.004-0.016 in. between the circlip and the line;
Thickness of shim:
 1135-1315-011:0.004 in.
 1135-1316-011:0.008 in.
 1135-1317-011:0.020 in.

The shim thus selected should be inserted between the collar and the gear 34T when the countershaft is installed on the transmission case.
Gear clearance: 0.004-0.016 in.
Thickness of shim:
 1135-1314-011:0.008 in.
 1135-1318-000:0.016 in.

PTO Shaft

1. Install the PTO shifter in the transmission case without the O-ring. Install the grease coated O-ring on the shifter from outside of the transmission case, and tentatively install the guide plate for the PTO shift.
2. Install a circlip in the hole of the transmission case where the PTO shaft is placed.
3. Install a circlip at the rear of the PTO shaft and insert the collar and ball bearing.
4. Install a circlip, thrust liner, gear 30T, and liner and set them with the circlip, making sure the gear 30T is installed with its clutch facing forward.
5. Install the gear 20-26T on the PTO shaft. Install them on the transmission in the reverse order of disassembly. Install the PTO shaft on the transmission case by hammering the front side of the shaft.
6. Install the front side ball bearing.

NOTE: The disassembling and assembling procedures of a 4-wheel drive model is just the same as that for a 2-wheel drive model although at the rear of the gear 30T, the idle gear 32T, needle thrust liner, which are for the 4-wheel driving, are assembled.

7. Install the two ball bearings on the rear PTO shaft and set them with a circlip. Install the oil seal collar on the PTO shaft with the chamfered side of the collar facing rearward.
8. Install the joint collar on the PTO shaft inside the transmission. Mount the PTO shaft on the transmission case from the rear side. Install the grease-coated oil seal on the plate and install them together with a gasket on the rear of the transmission case.
Tightening torque: 36.1-43.3 ft.lb.

Assembling the Countershaft

1. Insert the shim selected in measurement of the cone center between the circular clip and taper roller bearing of the transmission case.

2. Push the countershaft to which the taper roller bearing is installed into the transmission case from the rear side and measure the collar gear clearance. Install shims to provide a clearance of 0.004-0.016 in., then install gear 34T, gear 17T, liner and gear 18-23T with gear 29T and 47T attached in the order listed. Insert a needle roller bearing collar and needle bearing in that order between gear 18-23T and the countershaft and set them with circlips. Attention should be paid to the direction of the gear installation.

3. Insert the taper roller bearing, install a tab washer and tighten them with a sleeve nut. Adjust the countershaft to eliminate end play by lightly tapping both ends and measuring the preload.
Countershaft preload: 0.57-0.72 ft.lb.
4. Lock the sleeve nut with a tab washer after setting the proper preload 0.57-0.72 ft.lb.

Adjustment of the PTO Gear

1. Put stopper spring, grease and the stopper ball on the PTO shift lever, install

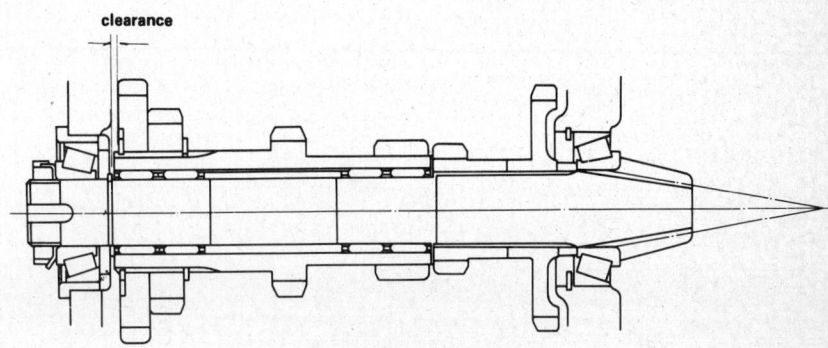

Countershaft gear clearance

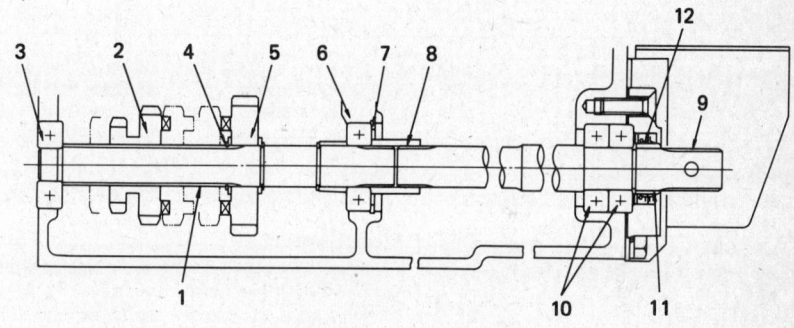

PTO shaft components

1. PTO shaft front
2. Gear 20-26T
3. Ball bearing
4. Circlip
5. Gear 30T
6. Ball bearing
7. Circlip
8. Collar
9. PTO shaft rear
10. Ball bearing
11. Oil seal bush
12. Oil seal

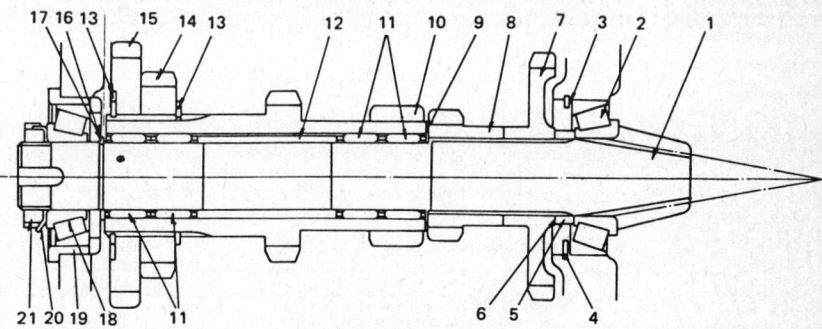

Countershaft components

1. Countershaft
2. Taper roller bearing
3. Shim
4. Circlip
5. Collar
6. Shim
7. Gear 34T
8. Gear 17T
9. Liner
10. Gear 18-23T
11. Needle bearing
12. Collar
13. Circlip
14. Gear 29T
15. Gear 47T
16. Liner
17. Circlip
18. Taper roller bearing
19. Holder
20. Tab washer
21. Sleeve nut

Satoh

the PTO shift lever and set it with a spring pin.
2. Align the gear teeth of the countershaft gears 29T, 23T and the PTO gear and fix the guide plate.
Tightening torque: 8.67-12.3 ft. lb. (1.2-1.7 kgm)
3. Install the knob on the PTO lever and lock it with a locknut.

Assembly of the Reverse Shaft
1. Insert the reverse shaft through the reverse shaft setting hole at the front side of the transmission case and insert a circlip, thrust liner, reverse gear with needle bearing installed, thrust liner and a circlip in this order, set the reverse shaft while moving it slowly, align the spring pin hole and set with a spring pin.

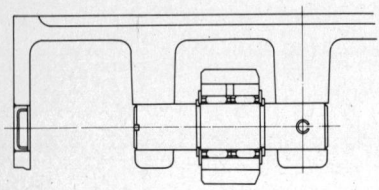

Reverse shaft

2. Drive the seal cap into the hole in the front of the transmission case. The cap should not protrude from the surface of the transmission case.

Assembly of the Select Shaft
Install the speed shifter in the transmission case.
Also install a grease coated O-ring from outside of the case and then install and tentatively tighten the sub-shift guide plate.
1. Install the ball bearing on the select shaft and set it with a circlip.
2. Set the circlips of the transmission case.

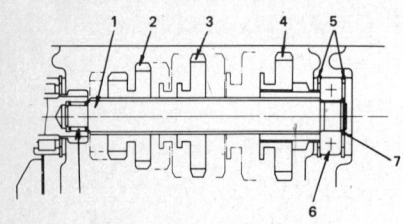

Select shaft components
1. Selector shaft
2. Gear 20-27T
3. Gear 32T
4. Gear 15-33T
5. Circlip
6. Ball bearing
7. Circlip
8. Needle bearing

3. Install the select shaft from the rear and set it with its side so that the shifting fork is placed facing forward together with gears 15T-33T and 32T, install gears 20T-27T with its smaller gear facing forward and set them with circlips.

NOTE: Gear should be correctly installed with great care.
If installed backwards it may become hard to change gears or may cause trouble.

Assembly of the Driveshaft
1. Install the roller bearing on the driveshaft and set it with a circlip. Install the roller bearing.

2. Install the ball bearing.
3. Install a grease coated oil seal on the bearing holder. Install the driveshaft in the bearing holder.

NOTE:
a. If there was a shim between the oil seal and ball bearing when the driveshaft was disassembled, assemble it as there was.
b. In case the end play of the driveshaft is excessive after all or part of the mechanical parts are replaced, adjust it using shims.
c. Check whether a circlip has been placed in the transmission case. Apply oil to the needle bearing, install it on the select shaft and assemble it in the bearing holder. Install the driveshaft combined with the bearing holder at the front of the transmission case together with a gasket coated with adhesive.

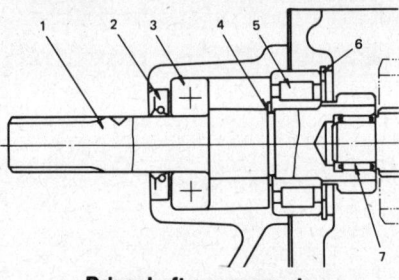

Driveshaft components
1. Driveshaft
2. Oil seal
3. Ball bearing
4. Circlip
5. Roller bearing
6. Circlip
7. Needle bearing

Tightening torque: 8.67-12.3 ft.lb.
d. Make sure that the shaft rotates smoothly.
e. Place the stopper spring, grease and stopper ball on sub-shift lever, install it on the shifter shaft and set it with a spring pin.
f. Align the teeth of gears 17T and 34T of the countershaft at their correct positions and fix the guide plate.
Tightening torque: 8.67-12.3 ft.lb.
g. Install the knob on the sub-shift lever and lock it with a locknut.

Assembly of the Differential Gear Unit
1. Install the differential gear with the ball bearing installed placing longer bearing holder on the right and the shorter one on the left and match the holder notch to that of the transmission case.
2. Temporarily tighten the bearing holders by bolts with the stopper washer.
3. Evenly place shims to provide the same thickness as the backlash adjusting shim for the circular area between the ball bearing holder on the left and the transmission and then tighten them.
4. Select shims to make a thickness so as to have a backlash of 0.010-0.014 in. in the ring gear.

NOTE: Be careful not to allow the stopper washer to protrude over the bearing holder notch.

Ring gear backlash: 0.010-0.014 in.
5. After a nominal backlash of 0.01-0.014 in. has been obtained, measure the clearance between the bearing holder on the ring and the transmission case. Select shims suitable for the clearance, place them evenly around the circumference and tighten the holder with stopper washers. Confirm that there is no end play by lightly tapping the differential gear case from both sides. Measure the backlash again to confirm that it is within the specified range, and finally tighten the bolts.
Tightening torque: 14.4-18.1 ft.lb.

Final Reduction
1. Push the ball bearing onto the pinion shaft. Apply grease to the O-ring, install it on the oil seal collar and install the collar so that the O-ring comes to the ball bearing side.
2. Install the pinion shaft in the final reduction case and set with a circlip.
3. Apply grease to the washers and oil seal and install on the pinion shaft. Install the brake drum, set it with a circlip and then install the brake cover.

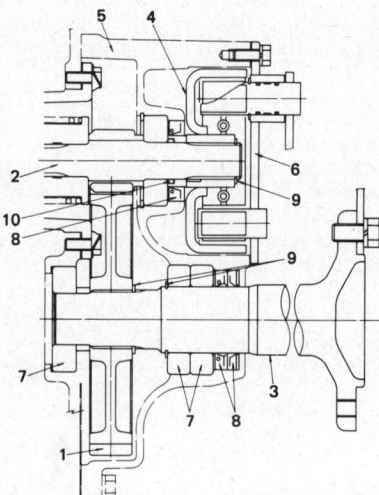

Final reduction unit components
1. Final gear
2. Diff. shaft
3. Final shaft
4. Brake drum
5. Final case
6. Brake cover
7. Ball bearing
8. Oil seal
9. Circlip
10. O-ring

Tightening torque: 8.67-12.3 ft.lb.
4. Apply grease to the washers and the two oil seals, and install them in the axle mounting portion.
5. Install the rear wheel shaft in the final drive case, push two ball bearings in and set them with circlips. Be sure to tap the inner ring when driving the ball bearing into position.
6. Place circlips to set the final drive gear, and install the reduction gear.
7. Put the collar into place and tighten the ball bearings.
8. Measure the height from the surface where the final drive case is installed on the transmission case to the surface of the ball bearing, and the distance from the said installation plane of the final drive case to the ball bearing housing.

Satoh

Removing the hydraulic case

Ram shaft removed

Select shims so that the end play of the rear wheel becomes 0.005-0.015 in. and place them in the ball bearing housing of the transmission case. In measuring dimensions, consider the thickness of the gasket.

NOTE: The gasket between transmission case and final drive is eliminated in the machines of recent production, starting with the below-indicated serial numbers: instead of the gasket, the sealant THREE BOND® No. 4 is used. When reassembling the transmission system of a machine of previous production (in which the gasket is used), *use the sealant instead of putting in the gasket.*
2-WD: C/#700661 and above
4-WD: C/#700641 and above

Shifter

DISASSEMBLY

After removing the speed shifter from the transmission case:
1. Loosen the shift rod set bolt after raising the bent portion of the stopper washer.
2! Pull the shift rod out forward and remove the shift fork, steel ball and stopper spring.
3. Remove the bracket after loosening bracket tightening bolt, and remove the shift lever with its knob removed.

ASSEMBLY

1. Place the shift lever and spring in the speed shift cover, install the bracket together with the stopper washer, tighten them with bolts and lock with stopper washers.
Tightening torque: 14.4-18.1 ft.lb.
2. Place the stopper spring and steel balls in the two shift forks and set with the jig.
3. The shift rod with the longest O-ring groove and steel ball groove is for the first speed and reverse; the shorter one is for the second and third speeds. Apply grease to the shift rod for the first speed and reverse and install it with the O-ring attached so that its shift fork faces the rear; then install the rod for the second and third speeds with a similar O-ring attached so that its shift fork faces forward. Next, tighten them with stopper bolts which are aligned with the corresponding holes, together with stopper washers.

NOTE: Transmission oil is easily filled before the cover is installed.

4. Apply adhesive to the surface of the transmission case where the cover is to be installed before installing the cover.
Tightening torque: 8.67-12.3 ft.lb.
5. The two bolts at the rear righthand side should be tightened at the position where the stopper plate is placed against the auxiliary speed shift lever to prevent it from being overshifted when shifting to a high speed.

FRONT POWER TAKE-OFF

DISASSEMBLY

1. In case the kit is equipped with an electric magnetic clutch, remove the guard and disconnect the connector of the electric cord. Remove the V-belt after loosening the tension pulley. Raise the bent portion of the tab washer at the top of the electric magnetic clutch and loosen the sleeve nut. Then remove the clutch after removing the bolts fixing the base.
2. Remove the front PTO by loosening the bolts fixing the bracket to the chassis.
3. Remove the circlips and pull out the shaft by tapping it forward with a plastic hammer.
4. Remove the oil seal collar, oil seal, and then remove the circlips.
5. Remove the ball bearing collar from the bracket.

ASSEMBLY

1. Apply grease to the oil seal and install it in the front of the bracket.
2. Install the ball bearing collar, ball bearing and oil seal collar on the PTO shaft, assemble it on the bracket and set it with circlips.
3. Apply grease to the oil seal and install it on the bracket.
4. Lightly apply grease to the spline portion of the PTO shaft, install it on the flange of the CG coupling, install the bracket on the chassis and tighten them taking care to avoid looseness of the PTO shaft.
Tightening torque: 36.1-43.3 ft.lb.
5. Assemble the electric magnetic clutch in the reverse order of the disassembly.

HYDRAULIC SYSTEM

Hydraulic Case

DISASSEMBLY

1. Remove the seat and tool box.
2. Remove the fender bracket.
3. Remove the delivery pipe tightening bolt.
4. Loosen the eight bolts (M10) tightening the hydraulic case assembly to the transmission case, remove the hydraulic case, and place it on the work bench.

NOTE: A bond is applied to the contact area of the hydraulic case with its mount. To remove the case, it is advisable to tap the case with a screwdriver.

5. Loosen the control valve mounting bolt lock plate, and loosen the bolt, then remove the control valve. Take care not to damage the O-ring.
6. Remove the auto-return feedback rod cotter pin from the end of the control lever. Loosen the bolt on the right side of the lift arm, and remove the lift arm bolt, then pull out the left side lift arm.
7. Remove the bushing set bolt, and by tapping the ram shaft from its both sides, remove the bushing and oil seal, then remove the ram shaft.

NOTE: When removing the ram shaft, note the lift fork position.

8. Remove the lift fork and connecting

Satoh

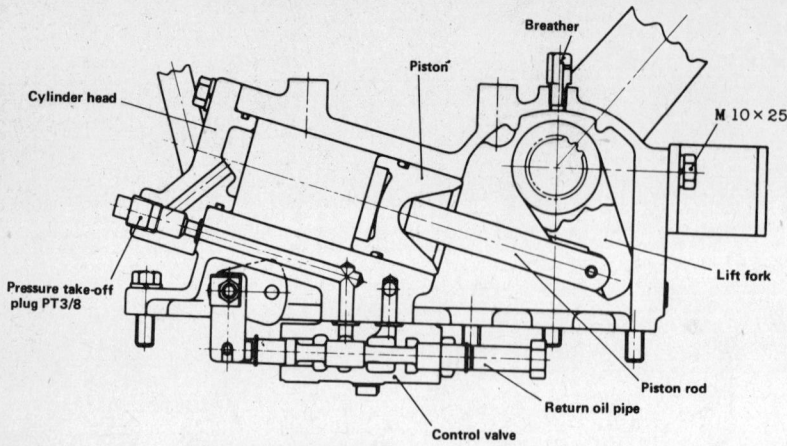

Hydraulic system cross section

Removing hydraulic cylinder head

Control lever removal

rod. Loosen the four M10 bolts holding the cylinder head, and remove the cylinder head.

9. Pull the piston out of the hydraulic case.
10. Loosen the control lever stopper ball holder, and remove the steel ball.
11. Loosen the arm tightening nut, remove the arm, collar and plate, and pull out the control lever.
12. Remove the flow control adjuster.
13. Remove the breather from the hydraulic case.
14. Remove the circlip on the opposite side to the spool valve spring pin, and remove the spool valve.
15. Loosen the pump relief valve locknut, and remove the locknut.
16. Loosen the relief adjuster, and remove the spring retainer, spring and ball.
17. By tapping the valve seat from outside, remove the valve seat.

ASSEMBLY

NOTE: Always use new O-rings coated with lithium based grease. Coat all mechanical parts with clean SAE 80 gear oil.

1. Install the connecting rod to the lift fork, and set with the cotter pin.
2. Place the lift fork in the hydraulic case, and align the punch mark on the ram shaft with the punch mark on the lift fork.
3. Hold the bushing with the chamfered side facing inward, and tap the set bolt into the ram shaft while paying attention to the set bolt position.

Tighten the ram shaft by tightening the set bolt (around which a seal tape is wound) from the top of the case.

Tightening torque: 43.3-50.6 ft.lb.

4. Grease the oil seal, and tap it in until it contacts the bushing.
5. Install the bake-up ring (fully damped with oil) and O-ring to the piston.

Oil the cylinder, and install the piston.

6. Install the larger. O-ring (fully

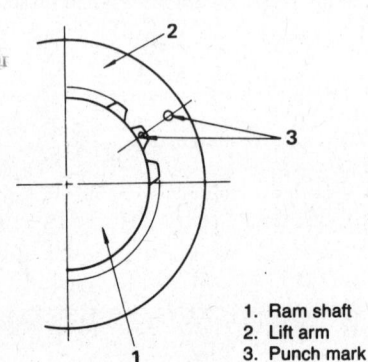

1. Ram shaft
2. Lift arm
3. Punch mark

Ram shaft alignment

damped with oil) to the cylinder head, and install the smaller O-ring to the cylinder.

Tightening torque: 43.3-50.6 ft.lb.

7. Install the greased O-ring to the control lever shaft, and install it to the case.
8. Install the plate (with the holed side facing rearward), collar and two arms to the control lever shaft, and tighten them with the nut and washer. Then lock the nut with the stopper washer.

Tightening torque: 8.67-12.3 ft.lb.

9. Install the O-ring to the control valve, and install it to the hydraulic case. Tighten it with the bolt, and lock with the stopper washer.

Tightening torque: 8.67-12.3 ft.lb.

NOTE: Make sure that the spring pin on the end of the control valve spool correctly fits in the cut on the arm installed to the control lever. Make sure that the flow control valve is installed to the control valve.

10. When reinstalling the pipe to the rear of the control valve, tighten it 5 or 6 turns so that the opened end of the pipe faces toward the center of the lift fork, then lock it with the locknut.
11. Apply Loctite® to the holder, and put the spring in two layers. While taking care so that the steel ball does not drop off, install the holder to the hydraulic case.

Make sure that the control lever stops at intervals of the same angle.

12. Matching the mark with the punch mark on the ram shaft, install the right and left arms. Install the auto-return arm to the right return arm, and tighten it with the bolt.

Tightening torque: 15.2-21.7 ft.lb.

13. Install the auto-return feedback rod to the control lever and auto-return arm, and set them with the cotter pins.
14. Install the breather.
15. Apply a sealing compound to the contact surface of the transmission case with the mating parts, and install the hydraulic case while taking care so that the dowel pin does not fall off. Then tighten it with the bolt.

Tightening torque: 43.3-50.6 ft.lb.

16. Lightly tighten the flow control valve adjuster from the transmission case side, and turn the lever 20°-25° forward, and secure with the nut.

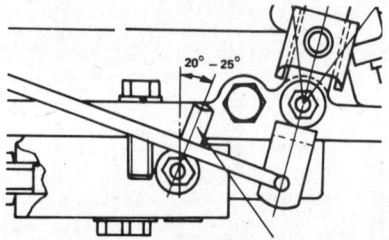

Control valave installation

17. Install the delivery pipe to the hydraulic case.

Tightening torque: 8.67-12.3 ft.lb.

18. Secure the pipe to the pump with the bolts.

Tightening torque: 5.78-7.22 ft.lb.

Satoh

Pump

REMOVAL

1. After removing the radiator, loosen the suction pipe and delivery pipe securing bolts, and remove the pipes from the pump.
2. Remove the pump from the engine timing gear case.

DISASSEMBLY

1. Loosen the bolt (K) with a hexagon hole, and remove the cover (J) and O-ring (I).
2. Remove the seal element consisting of two sealing rings and support rings (H and G).

NOTE: Both sealing ring (H) and support ring (G) are made into one piece assembly, and therefore, they cannot be separated.

3. Push the drive gear (C) into the body (A), and the bush (L) can easily be taken out by finger. Remove the bush (C), and remove the gear (D) and driven gear (E). The bush (F) can be taken out together with the O-ring (B).

INSPECTION

The disassembled parts should be arranged in order. As for the shafts and bushes, their positions should be correctly memorized.

When installing bushes, take care not to confuse the bush (C) for bush (F) in relation to their positions.

Each bush has a mark (A or B) stamped on its end. A bush having the same character should be installed on the cover side. (e.g. A indicates clockwise rotation, and C denotes counterclockwise rotation.) The oil seal (L) should not be removed unless otherwise broken. To pull out the oil seal (L) from the body, take special care not to scratch the housing bore.

Replacement of worn or damaged parts should be done with special care. In order to increase pumping efficiency, the gear rotates with its teeth in slight contact with the pump casing. The contact is evidenced on the low pressure side of the pump. After a long period of use, the oil will become dirty, and bush holes and journal bearings will be worn. As a result, the casing wall begins to show wear. If the amount of wear exceeds more than 0.001 in., the clearance between the gear teeth and the casing wall will be excessive, and oil leakage will increase. This will reduce the performance of the pump. Replacing the bush will not be effective to improve the performance. The use of a worn bush is not recommended except when the pump is operated under low pressure and with special care.

In general, working parts are subject to wear after a long period of use, and there will be not a big difference in wear between them. It is advisable, therefore, that when any component parts show an excessive wear, the pump itself should be replaced, instead of replacing worn parts. It will be more economical.

Each part should be washed with kerosene and blown with compressed air. All removed O-ring should also be replaced. Check the gears and gear shafts for scratches and broken teeth. The contact ratio in a pair of gears in mesh should be even. Measure the shaft diameter with an outside micrometer caliper (special tool No. 25MB). If the measurement shows a smaller value than 0.495 in., replace the shaft.

Check the bushes for deformed bore and scratches. If the discoloring of a gear can be considered to be related with a defective bush, check for the relief valve and related oil passages. If the length of a bush is shorter than the value as shown below, it should be replaced. 0.809 in.

Any discolored gear should also be replaced. Measure the clearance between the shaft and bush. If the measurement is larger than 0.006 in. the following value, the bush should be moved outward.

NOTE: If the clearance between the gear and the bush is more than 0.00019 in., both parts should be replaced together.

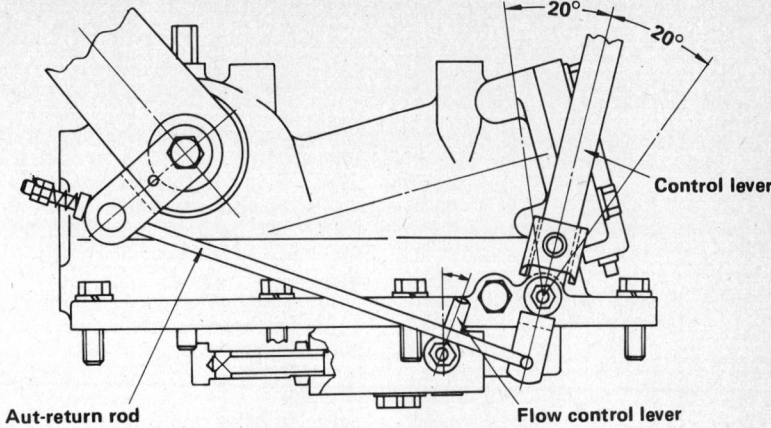

Hydraulic pump components

ASSEMBLY

Make sure that all parts are clean. If any oil seal (M) has been replaced, it should be replaced. To install the oil seal, use a press. (The seal lip must face inward.)

1. Place the O-ring (B) on the bush (F), and coat it with good quality of mineral grease. Insert it in the body (A) and push it in further. A pair of bushes must be at right angles to the bore in order to protect them against scratches. Tolerance is allowed for the bush so that it can be smoothly fitted in place without requiring force. If the bush has a scratch on its surface, it will not move smoothly. In this case, pull out the bush and smooth down the raised part with oil stone. Make sure that the surface of the bush is smooth. For this check, use a surface plate. After using oil stone, be sure to wash the bush. Oiling the bore will make it easy to install the bush. Make sure that after installation, the O-ring is in place.

The O-ring must be located between the bush and the body's bottom. The pressure balance type is greatly affected by the result of assembly. Make sure that the escape

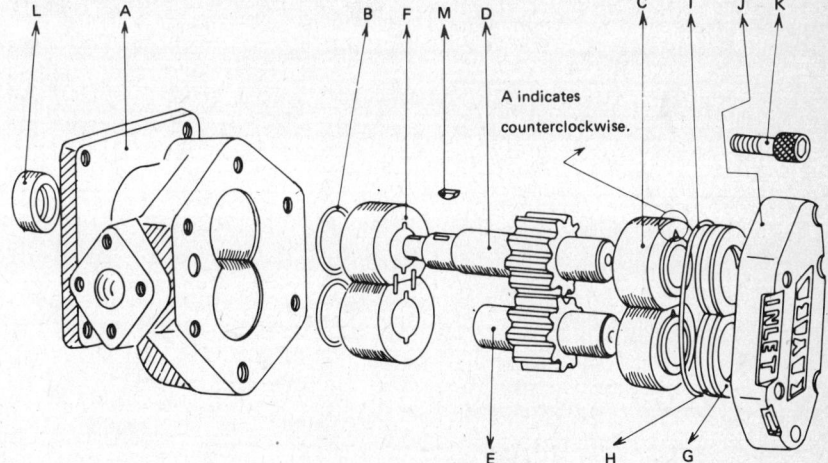

Hydraulic pump components

- A. Body (one)
- B. O-ring (two)
- C. Bushe A (one pair)
- D. Drive gear (pair with E)
- E. Driven gear (pair with D)
- F. Bush C (one pair)
- G. Support ring (two)
- H. Sealing ring (two)
- I. O-ring (one)
- J. Cover (one)
- K. Bolt with hex. hole (six)
- L. Oil seal (one)
- M. Key (one)

733

Satoh

groove in the bush surface is not inclined. Incorrect assembly will not provide the pressure balance effect, and as a result, the bush in the bore will be deformed by pressure.

2. When inserting the drive gear (D) into the body A, take care not to damage the oil seal with the stepped (machined) part of the gear shaft. To this end, wind a cellophane tape around the machined part or cover it with a specially designed sleeve.

The sleeve should be the same in outside diameter as the shaft (larger part in diameter), and its end is tapered. Install this sleeve firmly around the end of the drive gear (D), and insert it into the body (A).

The oil seal will not be damaged. If a gear which is once used is to be reinstalled, it should be positioned so that contact with the other gear in mesh will be exactly the same as before.

3. Fully grease the bushes (C) in pair, and insert them in the body. Place the seal ring (H) on the support ring (G), and set the position on the bush.

4. Place the O-rings (I) and (J) on the body, and place the cover. Finally tighten the bolt with a hole (K) by using a hexagon wrench (special tool AW-60-6 mm). Tightening torque is 5.77 ft. lb. Lock the bolt head with center punch, feed a small quantity of oil through the port.

5. Make sure that the assembly is done perfectly by turning the drive gear (D) with an open end wrench (special tool BT-9, 8x9). If the gear turns smoothly, the assembly is correct. If too tight, correct it.

INSTALLATION

1. Align the pump shaft with the pump driveshaft, and install it to the engine timing gear case.

2. Set the O-ring correctly, and install the suction pipe and delivery pipe to the pump.
Tightening torque: 5.78-7.22 ft.lb.

3. Install the radiator.

ENGINE

These tractors use Mitsubishi Ke55, Ke70 and Ke75 diesels.

REMOVAL AND INSTALLATION

1. Drain engine oil; it can easily be drained off when its temperature remains high. Drain cooling water completely from both the radiator and the cylinder block.

2. Remove the bonnet.

3. Remove the negative and positive battery cables from their respective terminals.

4. Disconnect lead wires from the alternator, oil switch, headlight, water temperature gauge, and starter switch.

5. Remove the light holder together with both the bonnet cover and air cleaner.

6. Loosen radiator hose clamps and disconnect the hoses. Then, remove the radiator.

7. Loosen the bolts securing hydraulic intake and exhaust pipes to the pump, and disconnect the pipes being very careful not to lose the O-rings.

8. Disconnect the throttle lever.

9. Disconnect the drag link from the Pitman arm.

10. Place a jack under the clutch housing. Keeping the housing slightly raised, remove the bolts securing the housing to the chassis and take out the chassis with the front axle installed.

11. Take out the engine rear plate.

12. Disconnect the fuel pipe; install the blind plug in the place of the pipe so that fuel will not flow out. Disconnect the return pipe.

13. Hoisting the engine, remove all clutch housing securing bolts. Gently move the engine backward to place it on the bench.

14. Remove six bolts securing the clutch housing to the transmission case. Take out the clutch housing.

15. Installation is the reverse of removal. Torque the engine-to-clutch housing bolts to 33-40 ft.lb.

Cylinder Head

REMOVAL AND DISASSEMBLY

1. Remove the air breather pipe.
2. Remove the inlet pipe and exhaust manifold.
3. Remove the oil pipe from the cylinder head.
4. Remove the alternator brace mounting bolts from the cylinder head.
5. Remove the fuel return pipe.
6. Remove the fuel injection pipe from the nozzle holder.
7. Remove the rocker cover.
8. Remove the cylinder head bolts as shown in the figure below.
9. Remove the cylinder head assembly. Scrape off the gasket stuck on the crankcase completely.

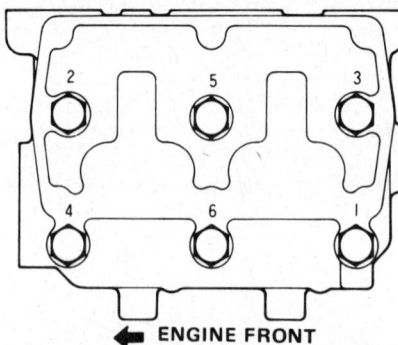

Head bolt loosening sequence

10. Remove the nozzle holder assembly.
11. Disconnect the glow plug lead wire, then remove the glow plug.
12. Loosen the rocker stay mounting bolts and remove the rocker shaft assembly.

13. Compress the valve spring with the valve lifter and remove the retainer, spring and valve.

Keep the valves, etc. for each cylinder so that none is lost.

Valve Guide

1. Check the valve stem-to-valve guide clearance. In case the service limit is exceeded, replace the valve guide and valve. When the valve is replaced due to worn valve guide, replace the valve guide also which must have been worn.

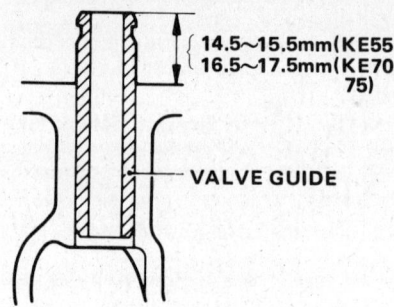

Valve guide dimensions

2. To replace the valve guide, press the old valve guide allway through the cylinder head from bottom to top using a press. Then press the new valve guide into position from the top of the cylinder head using a press as shown in the figure below. After the above press-in, check for proper valve guide-to-valve stem clearance. If the clearance is smaller than the standard, correct it with the reamer until the standard value is obtained.

ASSEMBLY AND INSTALLATION

1. Securely fit the valve stem on the valve guide.

2. Apply engine oil over the valve stem and insert it into the valve guide. Then assemble the valve springs and retainers into position, compress the spring with the valve lifter and install the retainer lock.

3. Assemble the rocker arms, rocker shafts and rocker stays as shown.

Note the direction of rocker shaft installation. Install the rocker shaft so that the identification mark at the front end of the shaft comes on the left side from front. Apply engine oil to the internal face of the rocker arm bush before installation.

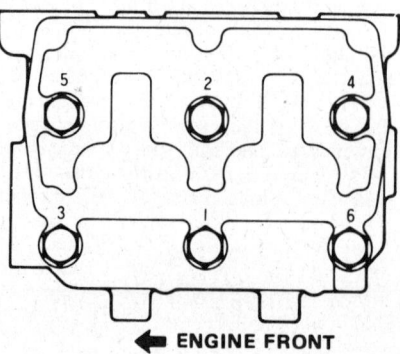

Head bolt tightening sequence

Satoh

4. Install the rocker arm shaft assembly on the top of the cylinder head and bolt it into place securely.
5. Install the glow plug. Tighten it by the specified torque.
6. Insert the nozzle holder gasket. Install the nozzle holder assembly. Tighten the mounting bolts by the specified torque.
7. Connect the flow plug lead wire.
8. Do not apply sealant over the cylinder head gasket since the sealant is already coated thereon.
9. Tighten the cylinder bolts in the order shown to 94-101 ft.lb. on engines number through 43399, and 101-108 ft.lb. on engines above number 43399.
10. Install new gaskets and packings. Apply sealant over the specified locations.

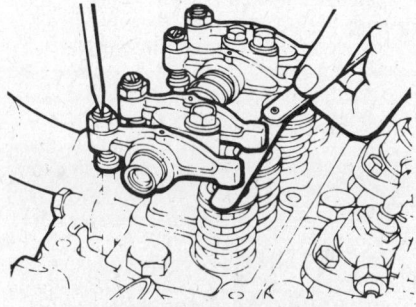

Valve clearance adjustment

11. Adjustment of valve clearance:
Insert the thickness gauge between the valve stem and the rocker arm with each cylinder at TDC on compression stroke and make adjustment by turning the adjusting screw until 0.014 in. clearance is obtained.

Crankcase

DISASSEMBLY

1. Refer to "Cylinder Head" for the removal of items belonging to the cylinder head.
2. Refer to Fuel System, for the removal of injection pump.
3. Remove the push rod. Extract the tappet upward from the crankcase.
4. Remove the oil pan and the gasket.

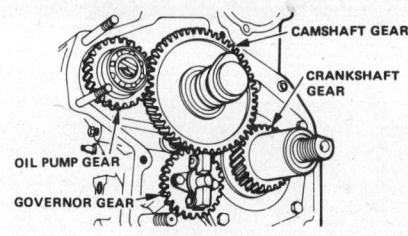

Timing gear train

5. Remove the flywheel.
6. Loosen crankshaft pulley nuts and remove the pulley and washer.
7. Remove the timing gear case and gasket.
8. Remove the oil pump gear.
9. After removal of the governor weight assembly, remove the snapring and then governor gear.

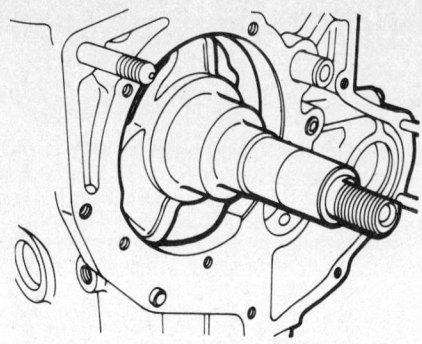

Crankshaft removal

10. Remove the camshaft and gears from the crankcase.
11. Remove the sleeve, crankshaft gear, plate and thrust washer.
12. Remove the connecting rod cap and remove the piston and connecting rod from the crankcase. Keep removed caps and bearings separately for each cylinder.
13. Remove the bearing housing from the crankcase.
14. Align the notch of the crankcase hole with the crankshaft counter weight and remove the crankshaft. Care should be taken not to damage the main bearing.
15. Disassembly of piston pin and connecting rod:
To disassemble the piston and connecting rod, use special tool piston pin setting tool XD998130. Set the piston and connecting rod assembly on the tool body, insert the tool push rod into the piston pin hole and remove the piston pin by pushing with a press.

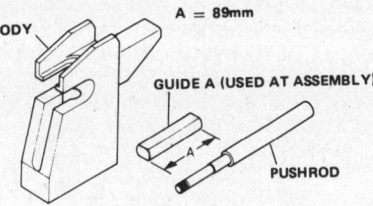

Piston pin setting tool

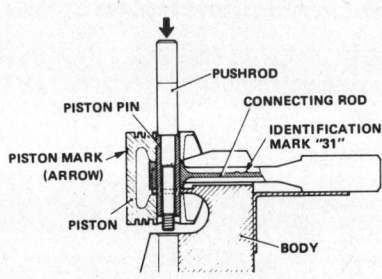

Removing the piston pin

NOTE: Always use a press for removing the piston pin. Never tap or strike the piston pin. When a great force is required to remove the piston pin seized, replace it with a new one. Do not apply a load exceeding 6,600 lb. to the pin setting tool.

For the proper setting of the piston and connecting rod on the tool body make sure the small end of the connecting rod securely rests on the tool body receiving part.

INSPECTION

Crankcase

1. Check the crankcase for possible cracks and damages. Replace defective crankshaft.
2. Check the water jacket for deposits, rust, etc. and correct as required.
3. Measure the cylinder liner ID at 3 points using the cylinder gauge. Correct the liner to oversize or replace as required.
4. In case the cylinder liner is still within the repair limit and only the piston ring is to be replaced, check for a ridge at the top of the cylinder liner. If there is a ridge, hone as required.

Piston, Piston Pin and Piston Ring

1. Check the piston for sticking, scores, wear, etc. Replace defective piston.
2. Measure the piston OD and replace excessively worn piston. If the piston-to-cylinder clearance is greater than normal, rebore and hone the cylinder bore to the oversize or replace the piston.
Measure the piston OD in the direction at right angle and parallel to the piston pin hole (thrust direction) at the bottom of the skirt.
3. When using the oversize piston, finish the cylinder bore to oversize. First measure OD of the oversize piston used (diameter in the direction of thrust at the bottom of the piston skirt), then bore the cylinder bore to obtain the cylinder bore-to-piston clearance of the specified value and hone the cylinder bore.
4. Measure the clearance between the piston ring groove and the piston ring (side clearance) and replace the piston ring as required. If the clearance is not corrected after replacement of piston ring, replace the piston.
5. Measure the piston ring gap clearance and replace the piston ring if the gap clearance is excessively great. For measuring the gap clearance, push the ring down to the point of the clearance with the thickness gauge.
6. When the clearance between the piston pin and piston or connecting rod small end is excessive, replace piston pin assembly or connecting rod assembly.

Connecting Rod

1. Measure the bend and distortion of the connecting rod using the connecting rod

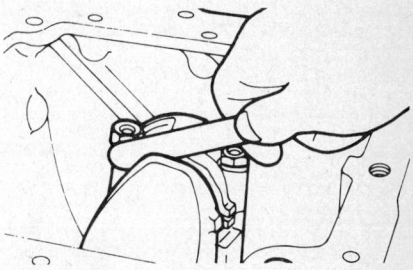

Measuring connecting rod thrust clearance

Satoh

aligner. If the bend of distortion is excessive, correct or replace the connecting rod.

2. Install the connecting rod on the crankshaft. Measure the thrust clearance. if the clearance is excessively great, replace the rod assembly.

Crankshaft

1. Check the periphery of the journal and pins for damages, sticking, etc. Measure OD. If wear is excessive, correct to an undersize and replace the main bearing and connecting rod bearing with those of the same undersize.

2. Measure the end play of the crankshaft. If the end play exceeds the specified value, replace the thrust bearing.

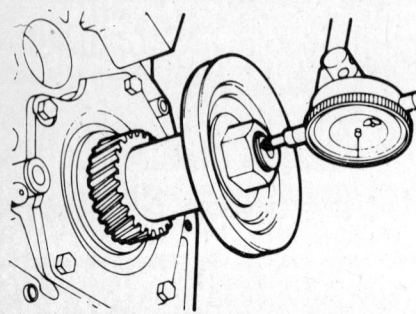
Measuring crankshaft end play

For measuring the end play, install the crankshaft and main bearing housing, then the gear, sleeve and pulley, and tighten the nuts. Apply the dial gauge to the tip of the crankshaft.

Main Bearing, Connecting Rod Bearing

1. Check the bearing surface for peeling, fusion, sticking and dents. Replace the defective parts.

2. Measure ID with the main bearing and the connecting rod bearing installed on the crankcase, bearing housing and connecting rod, respectively.

Then measure OD of the crankshaft journal and pin and calculate the oil clearance.

If the oil clearance is excessively great, replace the bearing. If the standard oil clearance is not obtained after bearing replacement, grind the crankshaft to an undersize and install the bearing of the same undersize.

3. Use special tool bearing installer for replacing the main bearing.

a. To remove the main bearing, place the installer body on the guidering as shown. To remove the main bearing from the bearing housing, push it all the way through from the front to the rear. To remove the main bearing at the back of the crankcase remove the oil seal, then push from the back of the crankcase to the case interior.

b. To press-fit the main bearing, assembly the installer body, guidering and main bearing to be pressed in as shown and press in from the same direction as for removal.

Install the main bearing by pressing so that the oil passage of the crankcase is aligned with that of the bearing. Set the main bearing so that the oil passage is aligned with the mark provided on the collar of the installer body and press it into position aligning with the oil passage of the crankcase. Once set, do not turn the tool or bearing. Press the main bearing so that its clench or clenched part comes above the center of the crankshaft.

Before installing the bearing, apply engine oil over the bearing surface. Always use a press. Never strike with a hammer, etc.

ASSEMBLY

NOTE: Coat all moving parts with clean engine oil, prior to assembly.

For assembling the piston and connecting rod, use the special tool piston pin setting tool MD998130 and follow the procedures steps below.

1. Insert the piston pin into the push rod of the tool and fully screw in the guide to the push rod.

2. Insert assembled push rod, piston pin and guide to the piston pin hole from the guide side. Assemble by passing them through the connecting rod small end hole.

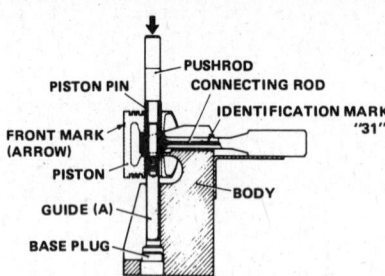
Piston pin installation

Assemble so that the front mark (arrow) at the top of the piston and the identification mark on the connecting rod face the same side. Sufficiently apply engine oil over the piston pin and the connecting rod small end bore.

3. Set the assembled piston, connecting rod and tool on the tool body. For the above setting, insert while aligning the notch of the guide and that of the body, then turn the guide 90°. At the end of setting, make sure that the connecting rod small end securely rests on the holding part of the body. Also, make sure that the front mark (arrow) at the top of the piston and the identification mark on the connecting rod point upward.

4. Install the piston pin using the press. Press-in load at this time will be 500-1500 kg. If other than the above, replace the connecting rod or piston pin assembly. The piston pin is pressed into place by the guide. At the end of installation, turn the push rod 90° to align notches of the guide and the body and remove the connecting rod assembly from the tool body.

5. Insert the crankshaft into the crankcase. Apply engine oil over the main bearing and journal. Be careful not to damage the main bearing.

6. Install the bearing housing. When installing, align the tag of the inner thrust bearing and the oil grooves of the bearing housing.

NOTE: The inner and outer thrust bearings are the same. Measure the thickness before installation and use only those of specified dimension. Thereby the end play of the crankshaft is controlled.

7. Install the outer thrust bearing in alignment with the bearing housing groove, then install the stopper plate, crankshaft gear and distance piece. Install so that the stopper plate has its chamfered side at the rear of the engine, the crankshaft has its ridged side at the rear of the engine and the distance piece of the larger ID goes first.

8. Tentatively install the crankshaft pulley and tighten the nut. Then check the

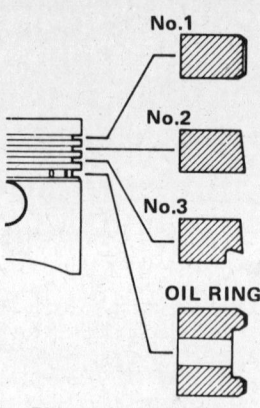
Piston ring positioning

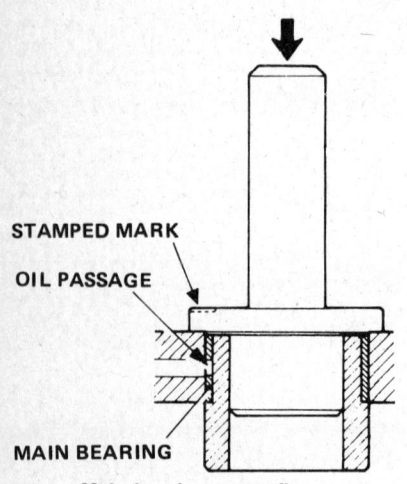
Main bearing press fit

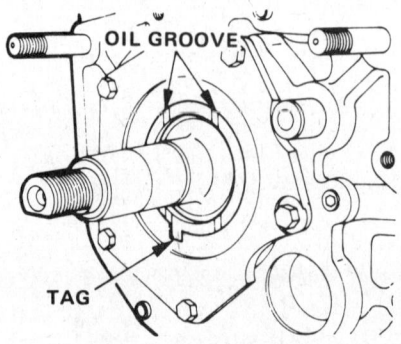
Thrust bearing installation

Satoh

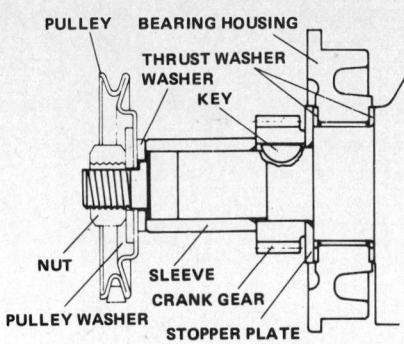

Crankshaft gear and sleeve installation

crankshaft for the end play. If the end play is far greater than the specified value, recheck the thrust bearing. Make certain that the crankshaft and sleeve piece have been installed securely. After checking the end play, loosen the nut and remove the pulley and washer.

9. Insert the piston and connecting rod assembly into the cylinder. Install the connecting rod cap, tighten it by the specified torque. Insert the bolt into the connecting rod side and tighten the nut from the cap side. When inserting the piston, firmly fasten the piston ring with the ring band and insert so that the front mark on the top of the piston and identification mark on the connecting rod point are directed to the engine front. Also, install so that the ring gap is neither in the direction of thrust nor in the direction of piston pin and the gaps of adjoining rings are spaced as far as possible.

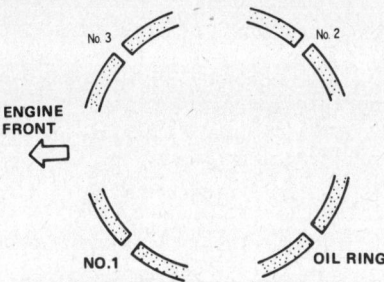

Positioning piston ring gaps

10. Key the camshaft to the cam gear and insert into the crankcase. At this time, align the alignment marks on the crankshaft gear with that on the camshaft gear as shown.
11. Install the governor gear and fit the snapring into the shaft. Then install the governor weight assembly and sliding shaft.
12. Install the oil pump drive gear having the ball bearing.
13. Install the timing gear case (the governor link is installed). Stick the gasket coated with the specified sealant on the gear case.
14. Install the key for pulley on the crankshaft and install the crankshaft pulley. Before fitting the nut, make sure that the crank pulley washer has been installed.
15. Install the flywheel and tighten the bolt by the specified torque.
16. Install the oil pan.
17. Insert the tappet.

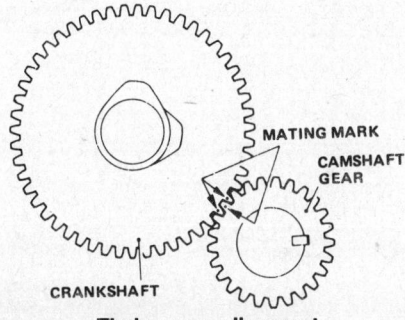

Timing gear alignment

FUEL SYSTEM

Nozzles and Holders

DISASSEMBLY AND ASSEMBLY

1. Remove the injection pipe from the nozzle holder.
2. Remove the overflow pipe from the nipple at the top of the nozzle holder.
3. Loosen the nozzle holder mounting bolt and remove the nozzle holder assembly. Disassemble the nozzle holder by the following procedure:
4. Separate the holder body from the retaining nut. To remove the nozzle from the retaining nut, gently tap the nozzle with a piece of wood. Be careful not to damage the nozzle.
5. Installation is the reverse of removal. Torque the holder nut to 57 ft.lb. for swirl chamber types; 47 ft.lb. for pre-combustion chamber types; 14 ft.lb. for the mounting nut.

TESTING

1. Injection start pressure test
Measure the injection start pressure with the nozzle tester. If the valve obtained does not correspond to the standard value, adjust to obtain the standard pressure by increasing or decreasing the thickness of the adjusting shim.

By changing the thickness of the adjusting shim by 0.1mm, the pressure changes by 10 kg/cm². When the nozzle holder or nozzle has been replaced with new one, the pressure more or less drops at the initial stage. Therefore, it is desirable to adjust after 30 to 50 hours operation.

Do not attempt to adjust unless the pump tester is available. In case of trouble, call the service factory for adjustment where the tester is on hand.

2. Injection test
a. Chattering test (small amount interrupted injection.)
Satisfactory if the fuel is injected interruptedly and positively as the tester lever is slowly operated (approx. 10 strokes/min). The fuel should be injected

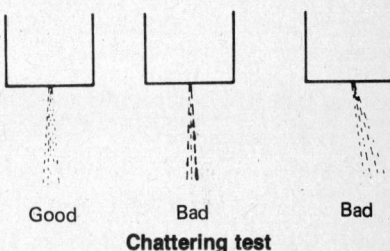

Chattering test

in the axial direction of the nozzle in a straight line. Unsatisfactory if the direction of injection varies excessively from time to time or the mist goes in several lines. Also unsatisfactory if the particles are large or granular.

b. Dripping
Unsatisfactory if in the above chattering test the fuel is collected at the bottom of the nozzle after injection and falls in drops. Replace the nozzle.

c. Manner of injection
The fuel should be injected in fine mists straightly in the axial direction of the nozzle when the tester lever is set to high speed (approx. 200 strokes/min). Bad if the particles are granular.

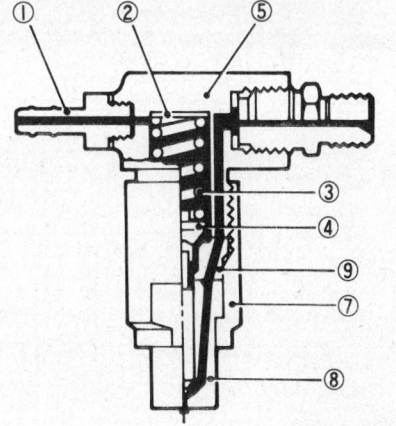
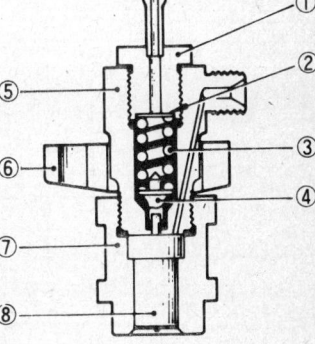

Nozzle components

1. Nipple
2. Adjusting shim
3. Pressure spring
4. Pressure pin
5. Body
6. Flange
7. Retaining nut
8. Nozzle
9. Distance piece

Satoh

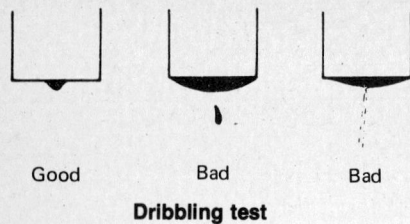

Good — Bad — Bad

Dribbling test

Governor

REMOVAL AND INSTALLATION

1. Remove the V-belt. (Refer to "Cooling System".)
2. Remove the crank pulley.
3. Remove the injection pump. (Refer to "Injection Pump".)
4. Remove the gear case mounting bolt and remove the gear case assembly.
5. Remove the governor spring.
6. Remove the governor lever.
7. Remove the speed control lever.
8. Remove the governor weight assembly and sliding shaft.
9. Installation is the reverse of removal.

Injection Pump

REMOVAL AND INSTALLATION

To remove the pump, disconnect and cap the fuel lines. Unbolt and remove the linkage, then dismount the pump.

1. When reinstalling the injection pump assembly, install the adjusting shim that has been removed.
2. When installing the new injection pump assembly, obtain the adjusting shim which can bring the dimension A shown (distance from the base circumference, the injection pump cam and the pump mounting face of the gear case) within 75.95-76.05 mm and install the pump assembly. Fit the projection of the control rack securely into the fork at the tip of the governor lever. Thickness of shims: 0.3, 0.5, 1.0 mm.
3. Connect the fuel feed hose. Loosen the injection pump air vent screw and vent air.
4. Check the injection timing. Remove the delivery valve holder, remove the delivery valve and spring, then install only the delivery valve holder.

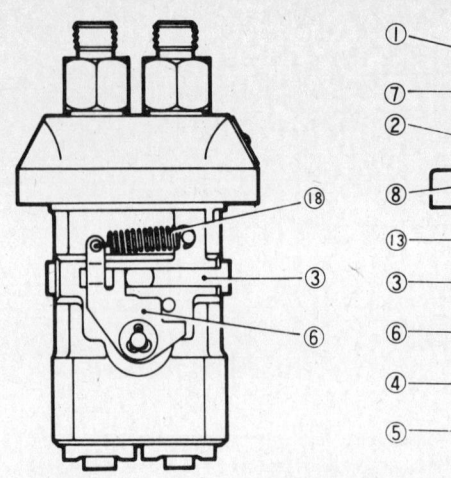

Injection pump components

1. Delivery valve holder
2. Delivery valve seat
3. Control rack
4. Pump body
5. Tappet
6. Stopper
7. Delivery valve spring
8. Delivery valve
9. Air vent screw
10. Plunger barrel
11. Lock pin
12. Control pinion
13. Plunger
14. Union bolt
15. Upper seat
16. Spring
17. Lower seat
18. Return spring

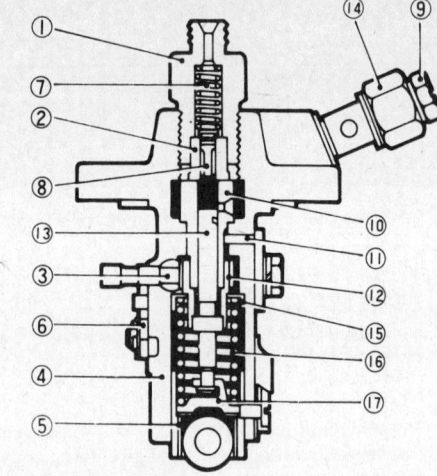

Turn the crankshaft to stop the flow of fuel from the discharge port of the delivery valve holder. The instant when fuel flow stops is the injection timing.

If the injection timing does not satisfy the standard value, adjust by increasing or decreasing the thickness of the adjusting shim (between the pump and gear case).

When the thickness of the shim is changed by 0.1mm, the injection timing changes by approximately 1 deg. In case the injection timing is to be adjusted outdoors or at a dusty place, perform adjustment with the delivery valve and spring being kept installed so as to prevent mission of parts and dust infiltration. In this

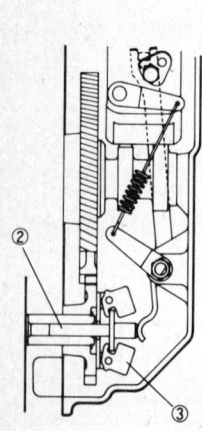

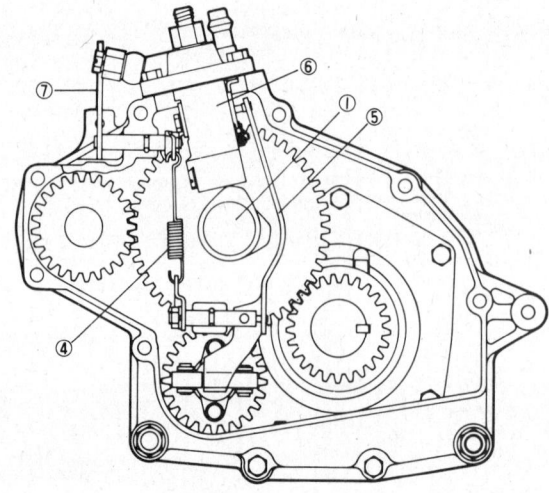

Injection pump mounting dimension point

case, remove the injection pipe No. 1 from the nozzle holder side, fit the socket wrench over the crank pulley nut and gradually turn the nut. The fuel starts flowing to the tip of the pipe.

The instant when the tip expands is the injection timing. This occurs at approximately 1 deg. behind the standard injection timing.

Governor components

1. Camshaft
2. Sliding shaft
3. Governor weight
4. Governor spring
5. Governor lever
6. Injection pump
7. Speed control lever

Satoh

Models S-373 Beaver III and S-470 Buck

FRONT AXLE, 4-WHEEL DRIVE

DISASSEMBLY

Removing the Universal Joint

To disconnect the propeller shaft from the front axle, remove the front universal joint, as follows:

1. Remove bolts securing the upper and lower covers, and take down these covers to expose the universal joint.
2. At the front end, remove the snap-rings and pull off pin from yoke. Push the joint backward, making its yoke slide off the pinion shaft extending out from the axle housing, and disconnect the joint from the shaft. Push the joint forward and remove the joint from the propeller shaft. (The rear joint can be similarly removed).

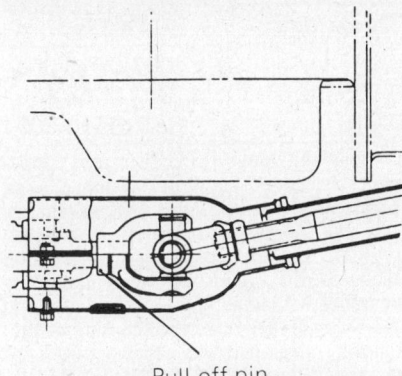

Rear joint removal

NOTE: When removing the snapring, be careful not to spread it out any more than is necessary to make it come off.

Disassembling the Yokes

Do not disassemble the yokes without valid reason. If it is necessary to disassemble them, remove circlips and carefully push the spider out of the yokes with a press.

To reassemble the yokes, use reverse of the disassembly procedure.

Disassembling the Front Axle

The two gear cases, right and left, are rigidly secured to the axle housing. These

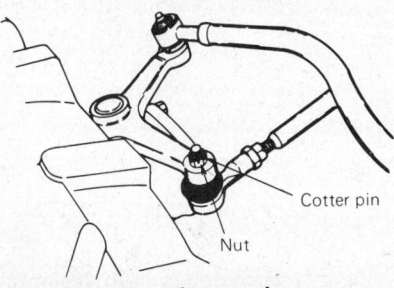

Tie rod removal

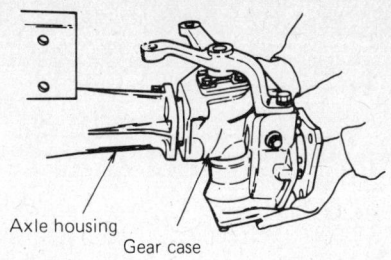

Removing the gear case from the axle housing

cases can be removed from the housing without requiring the whole front axle to be taken down.

The first step of front axle disassembly is to disconnect the propeller shaft in the manner already explained. After this step, proceed as follows:

1. Remove front tires. To raise the front end, push up the clutch housing with a jack.
2. On each side, remove tie rod and drag link by undoing respective ball socket joints. Each joint breaks apart when its cotter pin is pulled off locknut and the nut is then loosened and removed.
3. Remove bolts securing the gear case to the axle housing, and pull the gear case complete with the final case.

NOTE: Of the two bolts securing the gear case, two on the bottom side are reamer bolts.

Disassembling the Final Case

1. Remove bearing holder from the case. The holder will come out together with the wheel shaft.
2. Pull knuckle arm or tie rod arm off each final case. Be sure to recover the thrust liner.
3. Remove the bolts securing the holder to gear case, and take off the holder to expose the top end of final driveshaft. Drive lightly on the exposed shaft end so that the final case will slide off the gear case.
4. Remove the bottom cover from final case to expose the bottom end of the shaft. Draw the shaft off and take out the bevel pinion (15T).

Disassembling the Wheel Shaft

1. Using the gear puller, draw the bevel gear and ball bearing off the shaft.
2. Remove the shaft from bearing holder by lightly tapping on the shaft.
3. Remove circlip from the holder, and remove the bearing outer race out together with oil seal.

Disassembling the Axle Housings

The two axle housings connected together are heavy for handling. Exercise caution to avoid personal injury.

1. Place jacks under axle to take up the weight of the two housings. Loosen center pin securing bolts, thus severing the front axle from chassis, and transfer the whole axle to the work bench.

NOTE: Be sure to recover the shim between center pin and chassis.

2. Remove bolts securing the pinion case to the axle housing (A), take off the case.
3. Separate the axle housing (B) by removing bolts.
4. Pull out differential shaft from housing (B).
5. Remove differential gear assembly from axle housing (A).
6. Pull out differential shaft from housing (A).

Disassembling the Pinion Case

1. Straighten tap washer and remove sleeve nut after removing pinion case from axle housing (A).
2. Force pinion shaft out of the case by tapping lightly on shaft end.

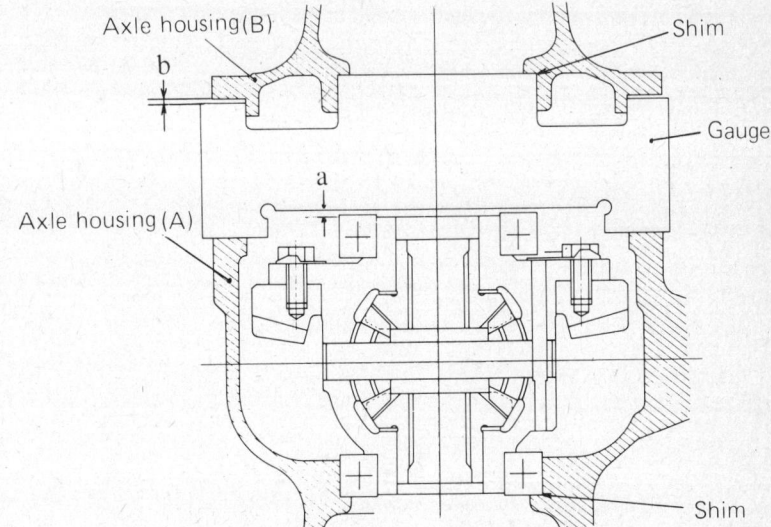

Axle housings, showing reference points for measurement

3. Remove oil seal and bearings, as necessary.

Disassembling the Differential
This differential is identical to the one located in the transmission. The procedures of disassembly and reassembly set forth for that differential apply to this differential.

Disassembling the Mid-PTO
Since this PTO is inside the clutch housing, its disassembly involves separation of clutch housing from transmission case. The procedure is outlined in the section for transmission.

INSPECTION
Clean the disassembled parts by washing. Inspect gears and pinions to be sure their teeth are in good condition. Check to be sure that each bearing is capable of smooth rotation. Examine each enclosure part (housing, case and the like) for evidence of cracking and repair or replace cracked parts, if any.

ASSEMBLY
Before commencing reassembly, be sure that necessary replacements and repair jobs have all been effected and that each part meets the dimensional and other specifications, if any. Fitting or installing rotating or sliding parts in dry condition is a bad practice: be sure to oil or grease their surfaces. Greasing is particularly needed for oil seals and O-rings being fitted.

Reassembling the Pinion Shaft
1. Fit the two outer races of tapered roller bearings into the pinion case, positioning each race as shown.
2. Fit the inner race of tapered roller bearing to gear side of pinion shaft, pushing it all the way against the pinion shoulder.
3. Insert pinion shaft into the case. Fit the other inner race to the shaft, and install oil seal bushing. Pinion shaft is now securely positioned inside the case.
4. Grease the oil seal, and force it into between case and bushing, making its outer end face flush with the mating face of pinion case.
5. Put on tab washer and run down sleeve nut. Tighten this nut to give the specified preload to the bearings. Pinion shaft bearing preload: 0.29-0.43 ft.lb.

NOTE: Be sure that the pinion shaft in place has no end play when checking its bearing preload.

6. After obtaining the specified preload, lock the sleeve nut by sharply bending tab washer.

Adjusting the Pinion Shaft Cone Center
Cone center specification: 1.69 ±0.002 in.

"Cone center" is the distance from the mating face of axle housing to the end face of pinion, and can be increased or decreased by decreasing or increasing the shim.

Shim stock for this adjustment is available in the following thicknesses:

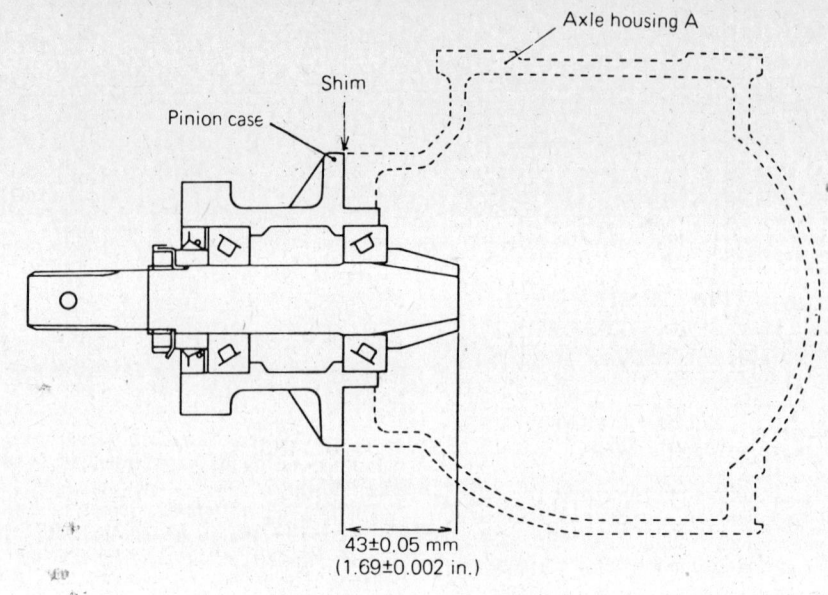

Shimming the pinion case

Thickness	Part No.
0.004 in.	1007 0605 000
0.008 in.	1007 0605 100
0.016 in.	1007 0605 200

Using the gauge for cone center adjustment, determine the required thickness of the shim by proceeding as follows:
1. Put the gauge alternately on the pinion shaft, and select the shim of such a total thickness that, with the short gauge, a clearance will occur between the gauge end and the case face but, with the long gauge, a similar clearance will occur between the gauge and the pinion's end face.
2. Fit O-ring (greased) to the mating face of pinion case, attach it to axle housing (A) and secure the case by tightening the bolts to this torque value: Tightening torque: 18-22 ft.lb.

Reassembling the Differential
Refer to the section for transmission, wherein the method of reassembly is set

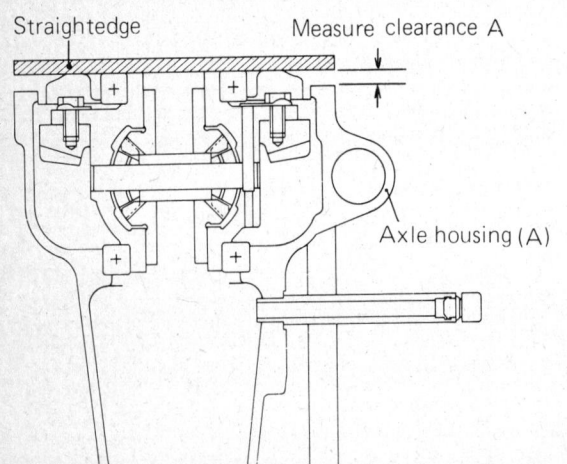

Measuring the clearance A when shimming between the differential and axle housing B

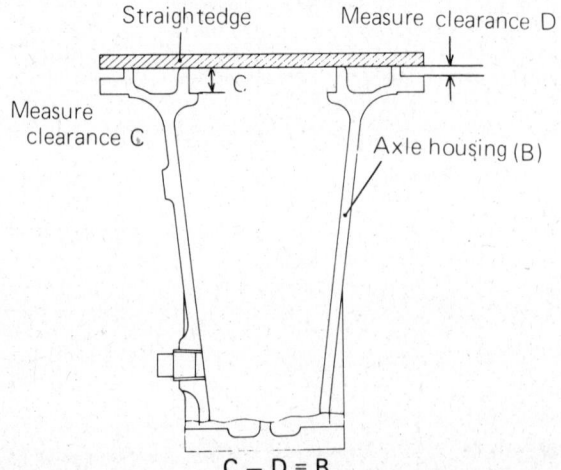

Measuring clearances C and D when shimming between the differential and axle housing B

forth. After reassembling, install the differential in the axle housing (A) by the same method.

Adjusting the Backlash

Backlash specification: 0.010-0.014 in.
Adjust the shim thickness (between bearing housing of axle housing (A) and bearing) to produce the specified backlash. Shim stock for this adjustment is available in these thicknesses:

Thickness	Part No.
0.004 in.	1007 0618 000
0.008 in.	1007 0118 100
0.016 in.	1007 0118 200

Shim Between Differential and Axle Housing B

After obtaining the specified backlash, determine the amount of shims needed, as follows:

1. Place the gauge on axle housing (A) and read the clearance (a) between ball bearing and gauge.
2. Place the other side of the gauge in the bearing housing of axle housing (B) and read the clearance (b) between housing flange and gauge.
3. Select from the shim stock mentioned above for backlash adjustment such an amount as will reduce the difference between the two readings (a − b) to anywhere between 0.004 in. a − b = 0-0.004 in.

If the gauge is not available, use a straightedge and proceed as follows:

1. After making sure that the backlash is correct, place the straightedge flat on the ball bearing.
2. Read the clearance A between axle housing (A) and straightedge.
3. Place the straightedge on axle housing (B) and read the clearances C and D. Subtract D from C, and call the difference B.
4. Select from the shim stock (mentioned above for backlash adjustment) such an amount as to obtain A-B = 0 to 0.004 in.

Reassembling the Axle Housing

1. Coat the mating faces of two housings (A) and (B) with the sealant, and attach the selected shim to the face of housing (B). Put the two housing together and fasten them by tightening the bolt to this torque value:
Tightening torque: 18-22 ft.lb.
2. Attach the axle housing to the chassis and insert the center pin to complete the axle-to-chassis connection. Be sure to apply the grease in the grease chamber.
3. Select such a shim as will produce 0 to 0.008 in. clearance between the axle housing and center pin. Insert the shim into between the chassis and center pin and tighten the bolts.
Tightening torque: 36-43 ft.lb.

Thickness	Part No.
0.008 in.	1119 0605 000
0.024 in.	1119 0605 100
0.04 in.	1119 0605 200

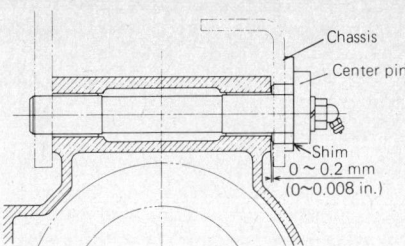

Shimming between the axle housing and center pin

4. Be sure that the axle housing as connected to the chassis is capable of smooth rocking motion.

NOTE: Apply THREAD LOCK® to the grease fitting, and run it into the center pin part, pointing it downward. Tighten the fitting good and hard.

5. Fit ball bearing to the outer end of each differential shaft, mount pinion (16T), and retain the pinion by installing circlip. The ball bearing has a groove cut in the end faces of its inner and outer races: be sure to position the bearing so that its grooved end comes on inner side, as shown.
6. The two differential shafts are now complete with bearings and pinions. Insert each into the axle housing.

Reassembling the Gear Case and Final Case

1. Insert bushing into final case.
2. Grease oil seal and fit it carefully to final case, making sure that the seal is trued up in place.

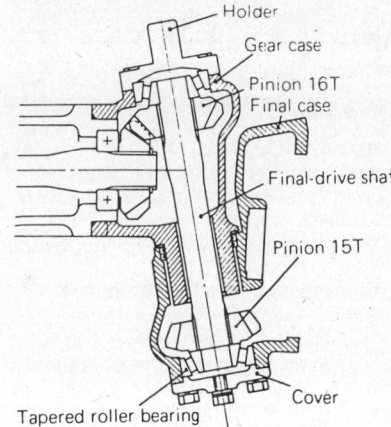

Final driveshaft components

NOTE: The steel part of this oil seal is so thin that the seal itself is prone to distort at the time of its installation. Apply a push evenly when fitting it.

3. Insert the final driveshaft into gear case, bringing its large-diameter end to top side and fitting pinion (16T) onto the splined end. Install inner race of upper tapered roller bearing, as shown.
4. Fit the bearing outer race into gear case. Apply the sealant to the mating face of the holder and secure it to case by tightening its bolts to this torque value:
Tightening torque: 18-22 ft.lb.

NOTE: Be sure to wrap each bolt with sealing tape before running it in.

5. Attach final case to gear case while fitting pinion (15T) onto the splined end of the shaft.
6. Install tapered roller bearing. Apply the sealant to the mating face of the cover, and secure the cover to final case, tightening its bolt to this torque value:
Tightening torque: 18-22 ft.lb.
7. Wrap the drain plug with sealing tape, and run it into the bottom cover. Tighten the plug good and hard.
8. Turn the final case around the gear case by hand to be sure the former is capable of smooth rotation.

Installing the Knuckle Arm (or Tie Rod Arm)

1. Insert bushing into knuckle arm carefully.
2. Position knuckle arm over the holder, fitting it to final case, and secure it tight to the case by bolting.
3. Check the thrust clearance, which is prescribed to be anywhere between 0.008 in.; if not, reduce the clearance to the specification by shimming. The shim stock for this adjustment is available in the following thicknesses:

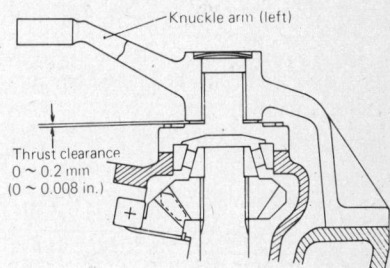

Measuring knuckle arm thrust clearance

Thickness	Part No.
0.03 in.	1438 1305 100
0.04 in.	3111 1302 100
0.05 in.	1438 1305 000
0.055 in.	1438 1304 000
0.063 in.	1468 3202 000
0.08 in.	1238 1357 000

4. After preparing the needed shim, take off knuckle arm. Apply lithium grease (heat- and water-resistant grease) to the OD part of the holder, put on knuckle arm with shim and secure it to final case by tightening the reamer bolts to this torque valve:
Tightening torque: 61-69 ft.lb.
5. Apply a bonding compound to the plug and fit it into the pivot bore of knuckle arm, thereby sealing off the top end of the holder.

Installing the Gear Case Assembly to Axle Housing

The gear cases, right and left, are now each complete with the final case and contains the final driveshaft.

1. In order to check the backlash be-

Satoh

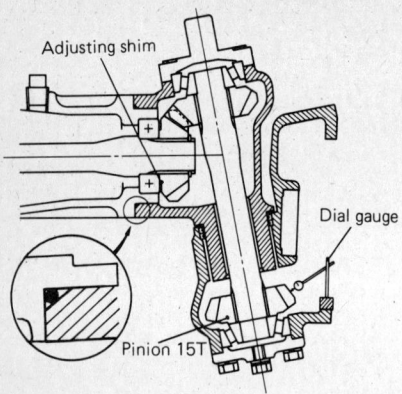

Gear case-to-axle housing clearance

tween the two pinions (16T) (one on final driveshaft and one on differential shaft), tentatively fit the gear case to the axle housing and tighten two bolts diametrically opposite.

2. Rig up a dial gauge on final case, putting the gauge spindle to a teeth of pinion (15T) (the pinion on the bottom end of final driveshaft). Take a backlash reading. This backlash is prescribed to be within this range: Pinion (16T) backlash: 0.008-0.016 in.

3. If the reading is off the specified range, take off the gear case and adjust the shim between the pinion (16T) on differential shaft and the ball bearing. The shim stock for this adjustment comes in these thicknesses:

Thickness	Part No.
0.004 in.	1135 3016 000
0.008 in.	0731 0002 502
0.016 in.	0731 0002 504

4. Having ascertained a proper pinion (16T) backlash, grease O-ring and fit it to the mating face of axle housing. Secure the gear case to the housing by tightening the bolts to this torque value: Tightening torque: 36-43 ft.lb.

Reassembling the Wheel Shaft

1. Grease oil seal and fit it to the bearing holder.
2. Insert the shaft into the holder, install ball bearing and retain the bearing by setting circlip.
3. Mount bevel gear (29T) on the splined end of the shaft.
4. Fit the bearing holder (complete with the shaft and bevel gear) to final case and tentatively secure it by bolting. Leave the bolts snug-tight.

NOTE: The inner ball bearing is left out at this time.

5. Check the backlash between bevel gear (29T) and pinion (15T). This is accomplished by putting the dial guage to the tip of wheel shaft. Adjust the shim between bevel gear (29T) and ball bearing if the reading is off the specification, which is: Bevel gear (29T) backlash: 0.008-0.016 in.
The shim stock for this adjustment comes in these thicknesses:

Thickness	Part No.
0.004 in.	1135 3016 000
0.008 in.	0731 0002 502
0.016 in.	0731 0002 504

6. Having obtained a proper backlash, remove the bearing holder, install the inner ball bearing on the shaft. Apply the sealant to the mating face of final case and attach the holder, tightening the bolts to this torque value: Tightening torque: 18-22 ft.lb.

7. Check the final case for torque needed to turn it around the gear case and be sure that no more torque than 0.22 ft.lb. is required. For this checking, use a torque wrench as shown.

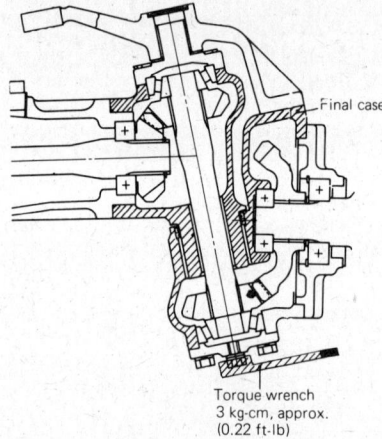

Final case torquing

Installing the Tie Rod and Drag Link

1. Clean the tapered portions of knuckle arm and tie rod arm, apply grease to the cleaned portions, and make connections, locking each nut by inserting cotter pin.
2. After installing the tie rod and drag link, mount front ties on wheel shafts, securing their hub bolts to: Tightening torque: 61-69 ft.lb.
Adjust the tie rod to obtain the specified "toe-in," which is: Toe-in specification: ¼ ±³/₃₂ in.
3. With the toe-in set properly adjust the inner-wheel steering angle to 53° on both sides, right and left, and secure each stopper bolt by tightening its locknut.

Installing the Universal Joints

Install the universal joints by reversing the order of removal.

FRONT AXLE, 2-WHEEL DRIVE

REMOVAL AND INSTALLATION

1. Pull out the cotter pin from the center pin nut, and remove the castle nut.
2. Support the front axle, pull out the center pin and remove the front axle.
3. Clean the center pin hole in the front axle and holes in the chassis, and install the front axle to the chassis with the longer boss facing forward.
4. Push in the greased center pin from the front side, and tighten the castle nut to specification. Then back it off ¼-⅓ turn so that the cotter pin holes can be aligned. Tightening torque: 108-123 ft.lb.
5. Lock with the cotter pin.
6. Make sure the front axle swings smoothly without end play.

Front Hubs

DISASSEMBLY

1. Loosen the front wheel mounting bolt lock nuts, and lightly loosen the three front wheel mounting bolts.
2. Jack up the front axle, loosen the front wheel mounting bolts, and remove the front wheels.
3. Remove the front wheel hub cap, straighten the tab of the lockwasher, and remove the sleeve nut.

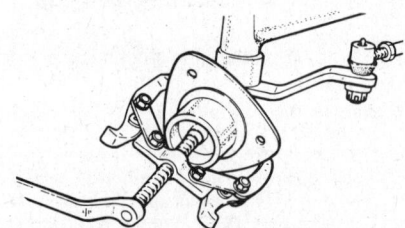

Pulling the front hub bearing

4. Remove the front wheel hub using the gear puller, and remove the oil seal.
5. Remove the ball bearing from the front wheel hub.

ASSEMBLY

1. Install the washer to the kingpin. Install the grease-coated oil seal.
2. Install two circlips to the front wheel hub, and insert the ball bearing into the hub.
3. Install the front wheel hub to the kingpin, pack with Alvania grease, and insert the ball bearing into the hub.
4. Install the lockwasher, and tighten the sleeve nut.
Front wheel hub starting torque: 10.1-11.6 ft.lb.
5. After tightening the nut with starting torque, make sure that the front wheel hub rotates smoothly, and lock with the lockwasher. Apply a bond to the cap, and install it to the front wheel hub.

Kingpins

REMOVAL

1. Remove the cotter pin locking the castle nut on each end of the tie rod, loosen the nut, and remove the tie rod.
2. Loosen the knuckle arm bolt, and remove the knuckle arm while taking care so that the kingpin does not fall.

3. Remove the drag link from the knuckle arm, as required.

4. By tapping the top end of the kingpin, pull out the kingpin.

5. Remove the rightside castle nut, and remove the rightside kingpin.

INSTALLATION

1. Oil the kingpin sparingly, and push the bushing onto the kingpin. Install the grease-coated O-ring.

2. Install the thrust bearing, and apply a liberal amount of grease to it, and install the kingpin to the front axle.

3. Fit the greased oil seal to the kingpin, while taking care so that the kingpin does not slip off.

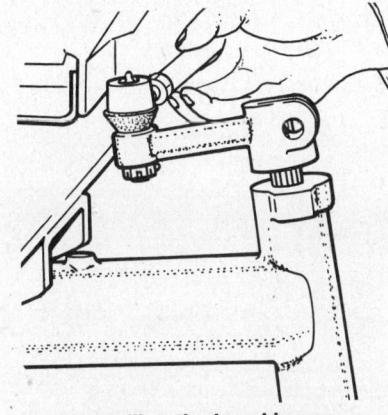

Installing the knuckle arm

4. Install the washer, and align the knuckle arm bolt hole with the cut on the left side kingpin, and tighten the bolt. Tightening torque: 15.2-21.7 ft.lb.

5. Make sure the gap between the knuckle arm lower side and the washer is 0.02 in. or less.

6. Install the washer to the rightside kingpin, tighten the castle nut, and lock with the cotter pin. Tightening torque: 32.5-39.7 ft.lb.

7. Make sure the kingpins operate smoothly, and grease them as required.

STEERING, 4-WHEEL DRIVE

Steering Gear

REMOVAL

1. Open hood to gain access to the battery.

2. Disconnect cables from the battery, and take it out.

3. Remove two safety guards, right and left.

4. From respective terminals, disconnect wires of glow plugs, oil pressure switch, water temperature switch, starting motor and alternator.

5. Disconnect the tractor meter cable.

6. Remove the cap on steering handwheel, loosen the nut and take off the handwheel.

7. Check to be sure, for safety's sake, that all electrical wires have been disconnected. From steering column, remove the two support panels.

8. Disconnect fuel return pipe from fuel tank.

9. Remove pipe between tank and fuel strainer. Plug up the pipe connection of the tank. Disconnect the pipe between strainer and injection pump, undoing the connection at the strainer.

10. Take down the fuel tank.

11. Disconnect engine control rod and foot accelerator rod from accelerator bracket.

12. Remove fuel tank bracket complete with speed control rod.

NOTE: It is not necessary to remove the plate located between engine and fuel tank.

13. Undo the ball socket joint between Pitman arm and drag link. The gear box is now accessible for its removal.

DISASSEMBLY

1. Remove the gear box from clutch housing.

2. Remove Pitman arm from sector shaft.

3. Loosen bolts securing the holder to the righthand side of gear box, turn adjusting screw clockwise, and take the holder off.

4. Reposition the sector shaft gear to the shape of gear box. Drive lightly on the Pitman armside end face of sector shaft to force it out.

5. From the sector shaft removed, separate adjusting screw and shim.

NOTE: If sector shaft is not to be replaced, preserve the shim for reuse. Keep the removed shim separate from others.

6. Loosen bolts securing the steering column, remove the column, and remove the ball nut subassembly from the gear box.

Reassembling the Steering Gear Box

Have all disassembled parts cleaned by washing. Be sure to oil moving parts before using them in reassembly.

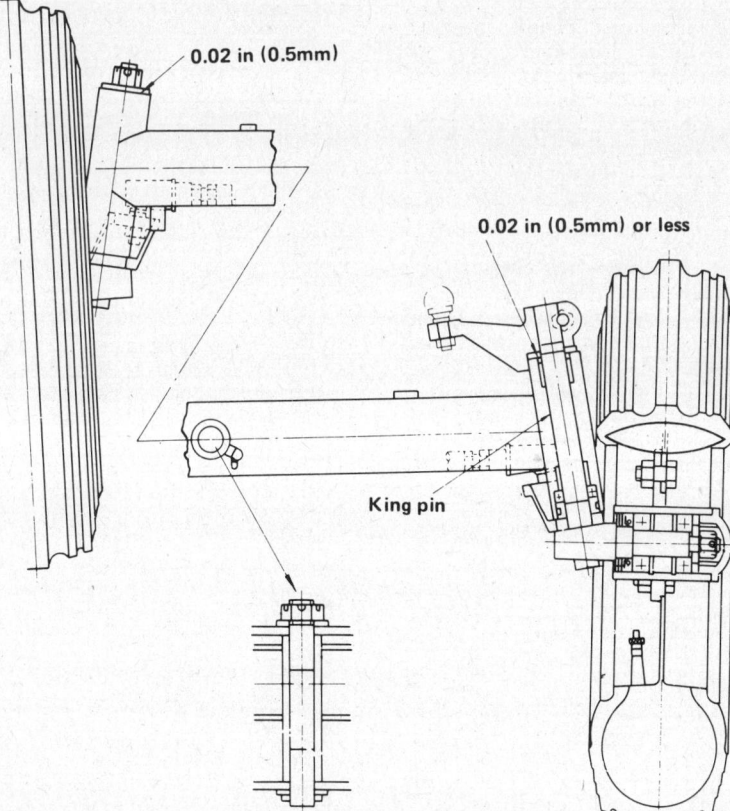

Kingpin measuring points

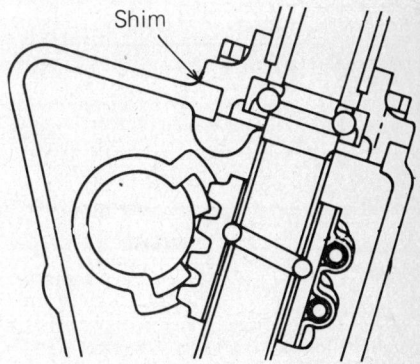

Bearing preload shim locaton

Satoh

1. Press bearing outer races into the bottom of the gear box and the column flange.
2. Fill the gear box with 0.63 pint of gear oil.
3. Fit bearing inner race to the ball nut subassembly, and set this subassembly in place inside the gear box.
4. Attach steering column to the gear box tentatively, making the bolts snug-tight.
5. Adjust the thickness of the shim (whose location is shown) in such a way as to give a bearing preload of 0.18-0.4 ft.lb. to the worm shaft in place. The shim stock for this adjustment is available in these thicknesses:

Thickness	Part No.
0.002 in.	1007 2109 000
0.0028 in.	1007 2110 000
0.003 in.	1007 2111 000
0.004 in.	1007 2112 000
0.008 in.	1007 2113 000

NOTE: This preload refers only to the worm shaft. This is the reason why the column is tentatively attached without its bushing.

6. Remove the column. Grease O-ring and fit it to steering column, and attach it to gear box by tightening to this torque value:
Tightening torque: 18-22 ft.lb.

NOTE: The end play for ball nut is prescribed to be from 0.0012 in. If this play is noted to be 0.1 mm or more, the ball subassembly must be replaced.
End play: 0.0012 in.

Reassembling the Sector Shaft

1. Make a shim adjustment to produce a clearance (T) of less than 0.002 in. between sector shaft and adjusting screw, using the shim stock available in the following thicknesses:

Thickness	Part No.
0.059+0.0022 +0.0006 in.	1007 2124 000
0.059+0.0026 +0.0018 in.	1007 2125 000
0.059+0.0037 +0.0030 in.	1007 2126 000
0.059+0.0049 +0.0041 in.	1007 2127 000
0.059+0.0061 0.053 in.	1007 2128 000

2. Install adjusting screw in sector shaft, together with the selected shim. Center ball screw and sector shaft in the gear box, as shown, and secure the shaft in place.
3. Coat the face of the gear box with the sealant. Secure the holder to gear box, tightening its bolts to this value:
Tightening torque: 18-22 ft.lb.
4. Grease oil seal, and fit the seal to gear box. The following specifications are to be met:
Worm shaft preload: 0.94 ft.lb.
Steering handwheel starting torque (as checked at the rim): 1.34 lb.

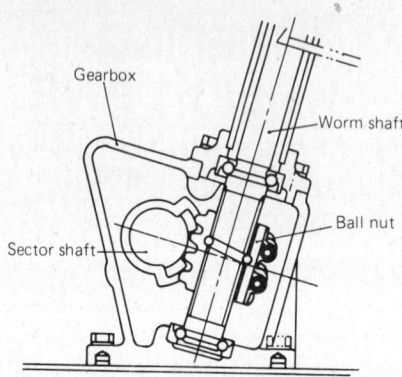

Sector shaft installation

5. Position adjusting screw in such a way that the backlash between ball nut and sector shaft will be anywhere between 0.4 in. as measured at the wheel rim. This much backlash corresponds to an angular play of 0° to 3° of sector shaft.
6. Having produced a proper backlash between sector shaft and nut, lock the adjusting screw by tightening its locknut.
7. Match marks are punched on sector shaft and Pitman arm. Fit the arm to the shaft, matching these marks, and secure the arm to the shaft by tightening its nut: be sure to use spring washer under this nut.

Installing the Steering Gear Box

1. Position gear box on clutch housing as dictated by dowel, and secure it to the housing by tightening its bolts to this value:
Tightening torque: 18-22 ft.lb.
2. Install all removed parts (such as fuel tank bracket, fuel tank, panel (A), panel (B), battery, etc., and reconnect all electrical wires and cables, in the sequential order which is reverse of disassembling and removing order.

STEERING, 2-WHEEL DRIVE

Steering Gear

REMOVAL AND DISASSEMBLY

Steering gear box removing procedure is the same with that of 4-wheel drive tractor.

1. Remove the steering column from the steering gear box, and remove the worm shaft.

NOTE: There are shims for clearance adjustment between the steering gear box and steering column.

2. Remove the Pitman arm from the sector shaft.
3. Remove the cover from the steering gear box, and remove the sector shaft.

NOTE: There are collars, thrust liner, and O-ring on the sector shaft. Take special care not to lose them.

ASSEMBLY AND INSTALLATION

1. Push and fit the two ball bearings to the wormshaft.
2. Install the steering column, and make an adjustment using a shim so that wormshaft end play becomes 0-0.079 in.

Thickness	Part Number
0.004 in.	1135 2120 000
0.008 in.	1135 2121 000

3. Incorporate the sector shaft in the steering gear box, install the collars and cover, and adjust the play of sector shaft end to 0.004 in. by inserting shims into the A portion.

Thickness	Part number
0.004 in.	1135 2112 200
0.012 in.	1135 2112 100
0.008 in.	1135 2112 000

After selecting the proper shims for end play of the wormshaft and sector shaft proceed as follows:

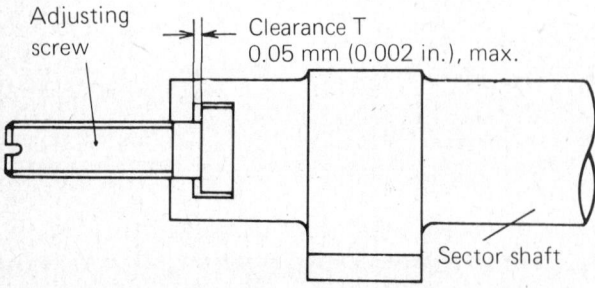

Sector shaft shim adjustment

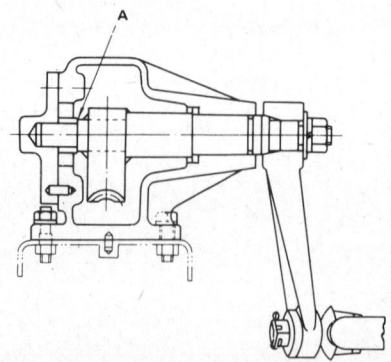

Sector shaft end play. Shims are inserted at point A

Satoh

4. Install the greased O-ring to the sector shaft, install the adjustment shims and collars, and install the cover.
Tightening torque: 14.4-18.1 ft.lb.

5. Place the sector shaft in the case so that its worm wheel faces upward, push in the worm shaft (to which the bearing is already installed), and engage the worm wheel with the worm gear of the worm shaft.

6. Fit the collar and shim to the worm shaft and secure the steering column, to the gear box.
Tightening torque: 8.67-12.3 ft.lb.

7. Align the punch mark on the sector shaft with the punch mark on the Pitman arm, and install the Pitman arm.
Tightening torque: 130 ft.lb.

8. Install the steering gearbox assembly to the clutch housing, while watching the dowel pin position. Tightening torque: 14.4-18.1 ft.lb.

9. To install fuel tank base, fuel tank, instrument panel A, instrument panel B, electrical wiring, etc., follow the reverse of the respective removal procedures.

BRAKES

DISASSEMBLY

Place a jack under the transmission case and take up the weight of the case with the jack to such an extent that the rear wheels will become slightly airborne. Apply parking brake, and loosen rear wheel bolts; remove the rear wheels and release the parking brake.

1. Remove brake rod from brake cam arm.

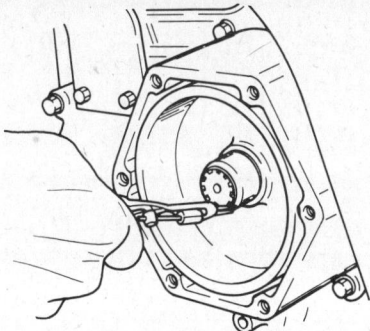
Removing circlip from differential shaft

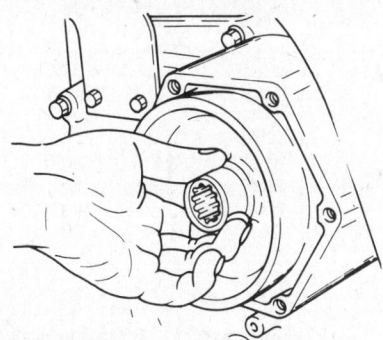
Removing brake drum

2. Loosen the six M8 x 20 bolts securing the brake cover, and remove the cover.
3. Remove circlip on differential shaft.
4. Remove brake drum.
5. Remove brake shoes from brake cover, as required.
6. Remove circlip on camshaft, and take off camshaft.

ASSEMBLY

1. Install greased O-ring to the camshaft, and apply grease to the grease groove. Install the cam to the brake cover, and set it with the circlip.
2. Hook the spring to the brake shoes and install them to the brake cover.
3. Install the brake drum to the differential shaft, and set it with the circlip.
4. Fit packing to brake cover; place the cover on final case; move cam arm to spread out brake shoes against drum; and, while centering brake cover, tighten bolts to this torque value: Tightening torque: 18-22 ft.lb.

ADJUSTMENT

1. Loosen locknut and rotate the joint piece to produce a free play of $13/16$ to $1\,3/16$ in. at the brake pedal. With this much play obtained, tighten the locknut.
2. Confirm that the right and left brakes operate simultaneously by running the tractor. If not, adjust both of them by means of the brake rod.
3. Make sure that the parking brake operates correctly.

CLUTCH

REMOVAL

1. Remove air cleaner cap, unhook hood.
2. Disconnect negative cord and positive cord from battery terminals in that order.
3. Remove safety guards, right and left.
NOTE: Remove the muffler as necessary.
4. Disconnect cords from respective terminals.
5. Turn off fuel filter cock, and disconnect and remove piping between fuel filter and injection pump.
6. Pull out snap pin on link pin by which governor lever is connected to engine control rod, and remove the control rod.
7. Remove bolts securing hydraulic pipes to hydraulic pump. Pay attention to O-rings. (Transmission need not be drained if the oil is up to but not above the prescribed level.)
8. Pull off cotter pin from the castle nut on the front end of drag link, loosen castle nut, and remove ball socket.
9. 4-WHEEL DRIVE TRACTOR
Remove joint covers, front and rear, take off circlips, draw out pins, and remove

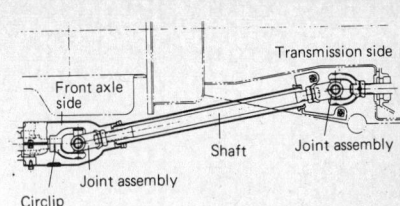
Front driveshaft

universal joints. (If the clutch housing is not to be removed, disconnect the front joint alone.)

10. Disconnect fuel return pipe from fuel tank. Loosen bolts on rear part of cylinder head cover.
11. Place a jack right under the bottom of clutch housing. Hitch lifting slings to engine hanger bolts and, by operating a chain block, take up the weight of engine, being careful not to lift the front tires off the floor.
12. Remove bolts securing clutch housing to engine, separate engine gently from clutch housing by inserting a screwdriver between clutch housing and engine rear plate. Pull out the engine toward the front in suspended state. (It is not necessary to remove bolts securing fuel tank bracket.)
13. Rest the engine, complete with chassis and front axle, on the work stand, keeping it in stable condition.
14. After removing the engine from the clutch housing, loosen the six bolts securing the main clutch to the flywheel, and remove the pressure plate assembly and clutch disc from the flywheel. The bolts must be loosened evenly.
15. When loosening the bolts, care should be taken so that no extremely heavy load is imposed on a bolt.

INSTALLATION

1. Apply a thin coat of grease to the flywheel pilot bearing.
2. Place the clutch disc with the longer clutch disc splined boss on the transmission side, and center the clutch disc using the clutch disc centering tool, and install the pressure plate to the flywheel.
3. Install the clutch cover to the flywheel. Pay attention to insert the two reamer bolts into correct positions and tighten the six bolts evenly to 18-22 ft.lb.
4. Assuming that the engine is ready for

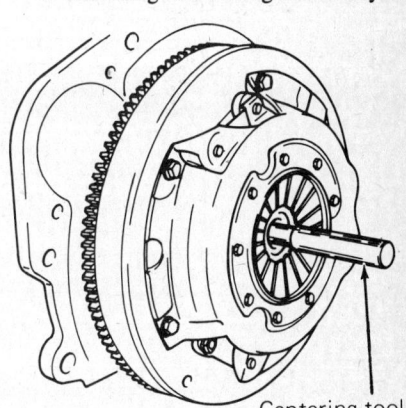
Installing the clutch

Satoh

mounting, coat the mating face of clutch housing with sealer, and give a thin coat of grease to the mainshaft splines. Bring in the engine in suspended state, level and steady it with clutch disc boss aimed squarely to the mainshaft, and move the engine toward the clutch housing while admitting the splined end of mainshaft smoothly into the boss. Make up the joint between clutch housing and engine by tightening the bolts to 61-69 ft.lb. for M12; 18-22 ft.lb. for M8.

5. Those parts removed for allowing the engine to be taken down must be restored to their positions in the sequential order that is reverse of removing sequence: they are 4-wheel drive universal joints and others. Installing a part the other way around, leaving out washers needed for bolts and similar mistakes could occur unless usual servicing precautions are exercised: this holds particularly true for wiring.

ADJUSTMENT
Free Play

The free play (A) in the clutch pedal should be between 1 and $1^{3}/_{16}$ in. and effective stroke (B) is $2^{13}/_{32}$-3 in.

After loosening the clevis locknut, pull the split pin out of the setting pin for the clutch pedal and joint, remove the pin and adjust the clutch pedal free play to be 1-$1^{3}/_{16}$ in. by turning the joint. After proper adjustment, set the cotter pin and lock the clevis securely with locknut.

NOTE: This adjustment is important for maintaining the clearance between the release lever and the release bearing to obtain a smooth gear shift and transfer all the driving power to the transmission.

Adjusting the Safety Starter Switch

After properly adjusting the clutch pedal free play and stroke, make this adjustment by the locknut so that the plunger of the switch is protruding 0.12 ± 0.04 in. out of the end of the safety starter switch body when the clutch pedal is fully depressed, and then lock it up.

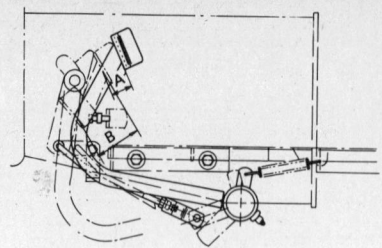

Pedal free play adjustment

Upon completion of adjustment, check to be sure that the starter motor does not rotate when the clutch pedal is released even if the starter switch is turned to start.

TRANSMISSION

REMOVAL

1. Remove drain plugs, right and left, on transmission case at its rear part. Be sure to drain the case completely.
2. Remove bolts securing the covers, front and rear, of 4-wheel drive universal joints, take off the covers, pick out circlips and pins, and remove the joint.
3. Remove tool box and seat together.
4. Place a jack under the rear part of transmission case, and take up the weight of the case with the jack. Loosen rear wheel bolts, jack up the rear axle at both ends, and take off the rear wheels. Take out the jack supporting the rear part of transmission case.
5. Remove fenders, right and left.
6. From each brake cam arm, disconnect brake rod.
7. Remove stepboards, right and left-, taking care not to injure the return springs for clutch pedal and brake pedals.
8. Remove hydraulic pipe clamp on the lefthand side of clutch housing, and disconnect suction pipe from the left part of transmission case. (The suction pipe is secured by bolts.) From the hydraulic case, remove banjo bolt, by which the pressure pipe is secured to this case.

NOTE: Be sure to recover the seal washers used on the banjo bolt.

9. Remove from the righthand side of clutch housing the shift lever for 4-wheel drive.
10. Remove 3-point link bracket.
11. Place two jacks under the machine to support clutch housing and transmission case, remove the bolts securing the two together, and separate them carefully.

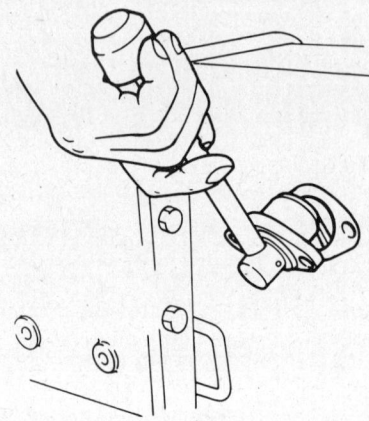

Removing the 4-wheel drive shift lever

12. Using a lifting sling and a hoist, lift the transmission out of the machine, and set it on the work stand.
13. Remove the bolts securing the hydraulic case to the transmission case, and take off the former case.
14. Remove the cover, in which the shift mechanism is built.

NOTE: This cover and also the hydraulic case are of an aluminum alloy: do not hammer them. They must be handled with care so as to avoid distortion, dent or any other damage.

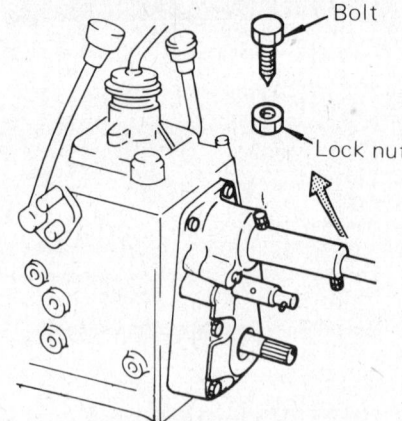

Loosening the driveshaft set bolt

15. Loosen set bolt on the driveshaft of the coupling between driveshaft and mainshaft, and separate and remove mainshaft, complete with the coupling, from driveshaft.
16. Loosen bolts securing the mid-PTO, and remove this PTO.
17. Loosen bolts securing each final case to the transmission case, and remove the two cases, right and left.

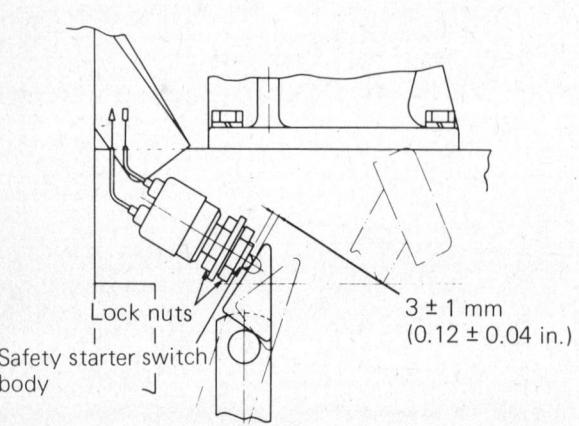

Safety starter switch adjustment

NOTE: Sealant is used in the joint between transmission case and final case. For this reason, the final cases may not separate easily from transmission case. Driving a chisel-like tool into the joint is sure to damage the case, which is made of an aluminum alloy. A soft-metal hammer may be used on the wheel boss in an attempt to sever it from the transmission case.

DISASSEMBLY

Disassembling the Select Shaft

1. Disassembly of this shaft can be effected with the transmission in place: the two parts to be removed in order to allow the shaft to be disassembled are the hydraulic case and the cover on shifting mechanism.

2. Pick out circlip on the rear portion of select shaft; drive on its forward end lightly; and pull it out with its rear portion foremost while taking out its gears. As necessary, remove ball bearing after picking out its circlip.

Disassembling the Driveshaft

Driveshaft comes out easily from the front end of transmission case when mid-PTO has been removed. Just pull out driveshaft. As necessary, remove roller bearing on driveshaft: a circlip must be taken out to allow this bearing to come off.

NOTE: Pay attention to the needle bearing fitted into driveshaft gear.

Removing the Differential Gear Assembly

1. Remove bearing holders, right and left, from transmission case. The bolts securing these holders are locked with stopper plates: be sure to straighten these plates and, after loosening the bolts, ease the holder off by putting the tip of a plain screwdriver to the notch formed of the holder.

NOTE: When taking out the differential gear assembly, be sure to recover the shim used for backlash adjustment and to check the shim thickness and the number of shim pieces used. The same shim must be reused in reassembly if the assembly has not been broken apart and is to be restored in its original condition.

2. Lift the differential gear assembly out of transmission case.

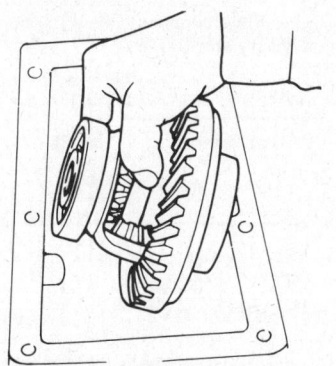

Removing differential gear

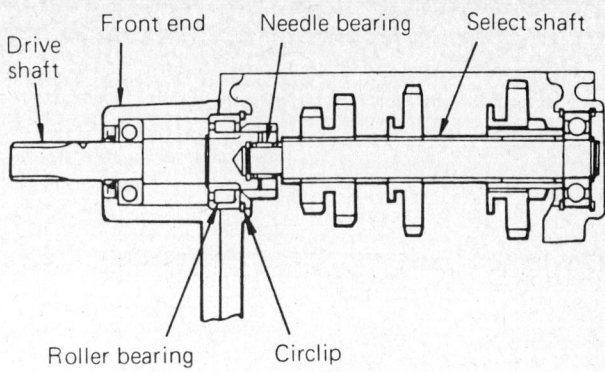

Driveshaft components

Disassembling the Countershaft

1. Straighten the tab washer under sleeve nut, loosen the nut and remove tapered roller bearing.

2. Pick out circlip on the forward part of countershaft and, while lightly driving on its forward end, force the shaft out to the rear, taking out the gears, collars, needle bearings and thrust washers, one by one, as the shaft comes out.

3. Draw the inner race of tapered roller bearing off countershaft.

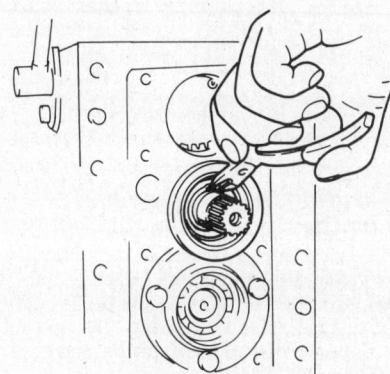

Removing countershaft circlip

Disassembling the Reverse Shaft

1. Remove the reverse shaft seal case from the front end of transmission case.

2. Pull out spring pin from reverse shaft lock and, while displacing the circlips, pull reverse shaft out from the front end. Remove circlips and needle bearings from reverse shaft.

Disassembling the PTO Shaft

PTO shaft can be removed with the transmission in place, provided that the hydraulic case be removed in advance. It is here assumed that this case has been taken down.

1. Loosen bolts securing the cover to the rear end of transmission case, and remove the cover.

2. Pick out circlip on the forward end portion of PTO shaft; lightly drive on the end face of PTO shaft and, while drawing it out, take out its gear. As necessary, remove ball bearings from PTO shaft, and oil seals from the cover.

Disassembling the PTO Driveshaft

This driveshaft is in two parts: rear side part and front side part. Removal of rear side shaft, too, can be effected with the transmission in place, after PTO shaft has been taken out.

REAR SIDE SHAFT REMOVAL

1. Ease out circlip on ball bearing inside transmission case.

2. Draw rear side shaft toward the rear, with a screwdriver hitched to ball bearing on the rear side.

3. When rear side shaft is halfway out, pick out circlip on the front side and the one already eased out of the groove next to ball bearing. With these circlips removed, draw out the shaft from transmission case.

FRONT SIDE SHAFT REMOVAL

To permit removal of this shaft, transmission case must be off clutch housing, and both countershaft and rear side part of PTO driveshaft must be out.

1. Remove bearing holder on the front end of transmission case.

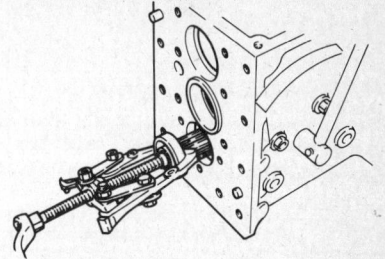

Removing front-side shaft ball bearing

2. Draw out the front side shaft from the front end.

3. Remove ball bearing (on the forward part of the front side shaft). Remove the shaft complete with gear 30T and gear 20T-26T from the upper side of transmission case.

Disassembling the Differential Gear Assembly

1. Give match marks to differential gear case and spiral gear.

2. Straighten the stopper washers under the heads of bolts securing the spiral gear.

3. Pull off locking pin from center pin, draw out the pin, and take out pinion gears, side gears and thrust liners.

Satoh

Disassembling the Final Cases

After removing the two final cases from transmission case and detaching brake cover and drum from each, proceed as follows:

DIFFERENTIAL SHAFT

1. Pick out circlip, remove pinion shaft (it may be necessary to lightly tap on this shaft), and draw out oil seal bushing together with ball bearing.
2. From the right final case, remove differential lock shifter: loosening the nut allows the shifter to come off.

REAR WHEEL SHAFT

1. Draw out ball bearing, and remove collar and final reduction gear.
2. Pick out two circlips and drive rear wheel shaft out of final case.

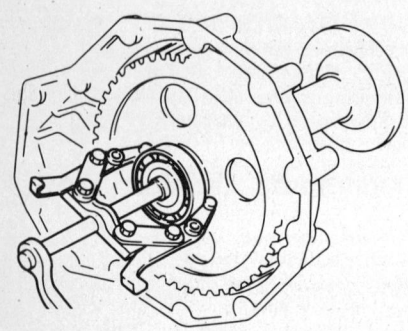

Removing the rear wheel shaft ball bearing

Disassembling the Mid-PTO

Draw out the mid-PTO shaft and remove ball bearings and gears together. There is no unusual step involved in this disassembly.

ASSEMBLY AND INSTALLATION

Reassembling the Differential Gear Assembly

1. Oil thrust liners, fit the liners to the differential gear case, install side gears, position pinion gears in place, together with liners, and insert center pin.
2. Adjust the thickness of thrust liners, as necessary, to secure a backlash of 0.010 to 0.014 in. between pinion gear and side gear.

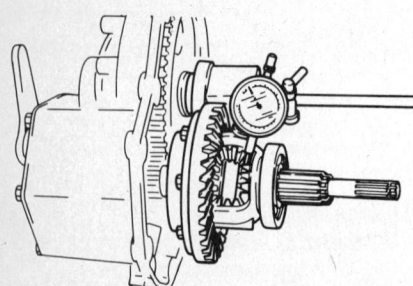

Measuring differential gear backlash

Backlash 0.010-0.014 in.
Shim stock for this backlash adjustment is available in the following thicknesses:

Thickness (in.)	Part No.
0.047	1135 1408 001
0.055	1135 1409 001
0.063	1135 1411 001

3. After producing the specified backlash, drive lock pin into center pin to lock the latter pin. Insert differential shaft and rotate the differential by hand to be sure that it rolls smoothly.
4. Fit spiral gear to the case, as guided by the match marks, setting the gear in its original position, put on stopper washers and bolt the gear to the case, tightening the bolts to 18-22 ft.lb.

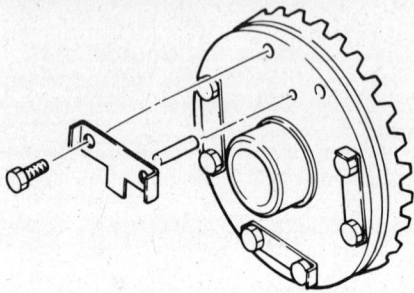

Stopper washer installation

NOTE: The stopper washer (part number 1135 1405 000) has a lug formed as shown. Position this washer in such a way that the lug will cover up the spring pin locking the center spring.

5. Install ball bearings in the differential case.

Measuring the Countershaft Cone Center for Shim Selection

1. In the bore provided in the transmission case wall, through which the countershaft is to enter, set circlip, and fit the outer race of tapered roller bearing. Be sure to discriminate between the inside end face and the outside end face of this race.
2. Mount the inner race of tapered roller bearing on countershaft.
3. Position countershaft in place, install the other tapered roller bearing on the front side, and secure it by tightening sleeve nut. Check to be sure that the countershaft so secured has no end play. Remember, cone center measurement with the use of a special tool presumes absence of end play on this shaft. Select the shim thickness, in the manner explained, to obtain countershaft cone center: 3.031 ±0.002 in.

How to Use the Two-Piece Special Tool

Set tool A on the seats (formed of transmission case) for the bearing holders of differential gear case. Position tool B between the end face of pinion (of countershaft) and tool A in place.
Select the shim thickness that permits the small-diameter end of tool B to enter freely and prevents the large-diameter end from entering. The shim so selected is to be inserted between the outer race of tapered roller bearing and the circlip at the time of reassembling the countershaft.

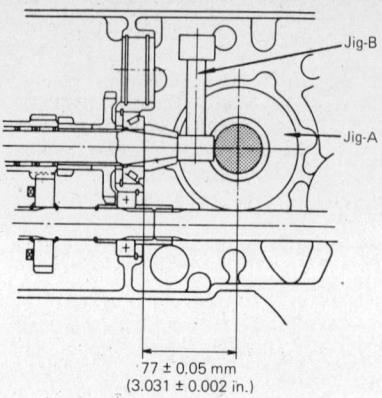

Using the two-piece special tool

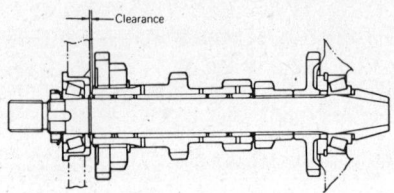

Measuring countershaft shim thickness

Shim stock for this selection is available in the following thicknesses:

Thickness (in.)	Part No.
0.004	1135 1315 011
0.008	1135 1316 011
0.020	1135 1317 011

The required shim thickness is equal to the clearance which occurs when the small-diameter tool B is placed between pinion gear and tool A.

Shim Selection for Countershaft Gear Clearance

The thickness of shim to be placed between the collar and gear 34T at the time of reassembling the countershaft must be determined in advance.
On the countershaft complete with tapered roller bearing, mount the collar, gear 34T, gear 17T, thrust liner, gear 18T-23T, gear 29T, gear 44T, needle bearings, collar, needle bearings, thrust liner—all in that order—and, as the last part, circlip next to the thrust liner. Under this condition, measure the clearance between the last circlip and the thrust liner, and on the basis of this reading, select the shim that will give the clearance to be anywhere between 0.004 and 0.016 in.
Shim stock for this adjustment is available in two thicknesses:

Thickness (in.)	Part No.
0.008	1135 1314 011
0.016	1135 1318 000

Reassembling the PTO Driveshaft

REAR SIDE SHAFT

1. Install circlip (N) in the bore for ball bearing (M). This bearing is the one at the middle of transmission case. Fit two circlip

(G) to the groove provided in rear side shaft (F).

2. Push ball bearing (H) onto rear side shaft, bringing the bearing all the way until it meets the circlip (G) on the rear end. Installing the shaft (F) in transmission case, mount collar (I) and gear 23T (P), with its claws coming on the rear side.

3. Fit the collar (O) to the shaft (F), making sure the splined fit is smooth.

FRONT SIDE SHAFT

1. Grease O-ring and fit it to PTO shifter. Install the shifter by bringing it into transmission case.

2. Install circlip (L) on the rear portion of front side shaft (J), and mount the inner race of ball bearing (M).

3. Set circlip (L) on that part of front side shaft where gear 30T takes its position. Mount thrust liner (E) and gear 30T (be sure to locate its clutching claws on the correct side), fit thrust liner, (E) and retain them by installing circlip (K).

4. Put on gear 20T-26T, locating the claws on the correct side, and feed the front side shaft into transmission case.

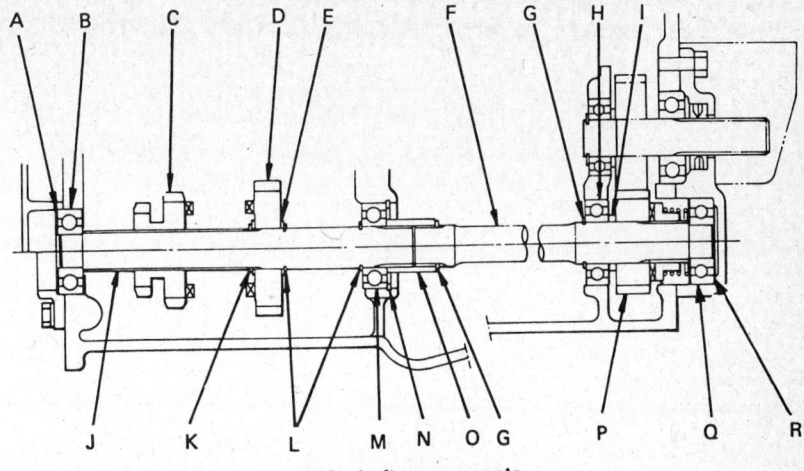

PTO shaft components

A. Liner
B. Ball bearing
C. Gear (20T-26T)
D. Gear (30T)
E. Liner
F. PTO driveshaft
G. Circlip
H. Ball bearing
I. Collar
J. Shaft
K. Circlip
L. Circlip
M. Ball bearing
N. Circlip
O. Collar
P. Gear (23T)
Q. Ball bearing
R. Circlip

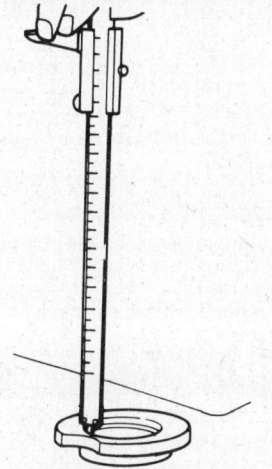

Front-side shaft bearing protrusion

5. Set the shifter gear 20T-26T, and position the shaft (J) and shifter in place, and install ball bearing (B) on front side.

6. Measure the amount by which the ball bearing (B) protrudes from transmission case and also the depth from bearing holder flange face to the bearing. On the basis of the readings, determine the liner (A) thickness necessary to reduce the clearance to less than .016 in., insert the needed liner between bearing holder and ball bearing, and secure these parts in place.

PTO driveshaft end play: 0.016 in., maximum.

Tightening torque: 18-22 ft.lb.

For this adjustment, use the liner (A) of the following thicknesses:

Thickness (in.)	Part No.
0.008	1135 1314 011
0.016	1135 1318 000

7. To the front end of PTO shaft, fit ball bearing and retain it by installing circlip.

8. Turn the shaft by hand to make sure that one-way clutch operates accurately and that the shaft rotates smoothly.

Reassembling the PTO Shaft

1. Install oil seal collar on PTO shaft, with the chamfered face of the collar coming on the front side.

2. Install ball bearing on PTO shaft, positioning the bearing in such a way that its circlip comes on the outer side and, through the rear end of transmission case, install the PTO shaft and then install collar and gear 27T on it.

3. Grease oil seal, and fit the seal to the cover.

PTO shaft installed

4. Through the rear end of transmission case, install one-way clutch on PTO driveshaft.

5. Insert spring and washer, and insert ball bearing.

6. Apply sealant to the cover, and install the cover from the rear end of transmission case.

NOTE: Wrap through-bolts with sealing tape.

Tightening torque: 18-22 ft.lb.

Reassembling the Reverse Shaft

1. Through the hole provided in the front end wall of transmission case, insert reverse shaft and, while pushing the shaft gradually, install it on the circlip, thrust liner, reverse gear complete with needle bearings, thrust liner and circlip, in that order. With the reverse shaft set in place, align the spring pin hole, and drive the pin into the hole to lock the shaft.

2. Spin the gear by hand and be sure that it is capable of smooth rotation.

3. Drive the seal cap into the hole in the front end wall of transmission case, making sure that the cap is flush with or down from the end surface.

Reassembling the Countershaft

1. Insert the shim (L) (which has been selected on the basis of cone center measurement) between the circlip (H) and the outer race of tapered roller bearing (K) in the wall of transmission case.

2. Insert countershaft (J) (to which tapered roller bearing (K) has been mounted) into transmission case through its rear end.

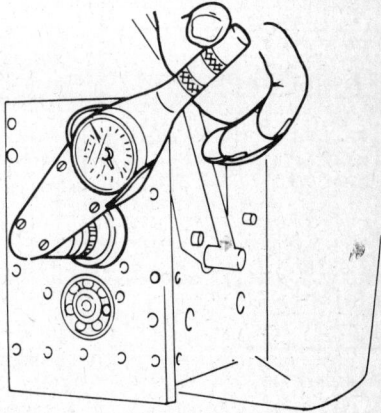

Checking countershaft bearing preload

Satoh

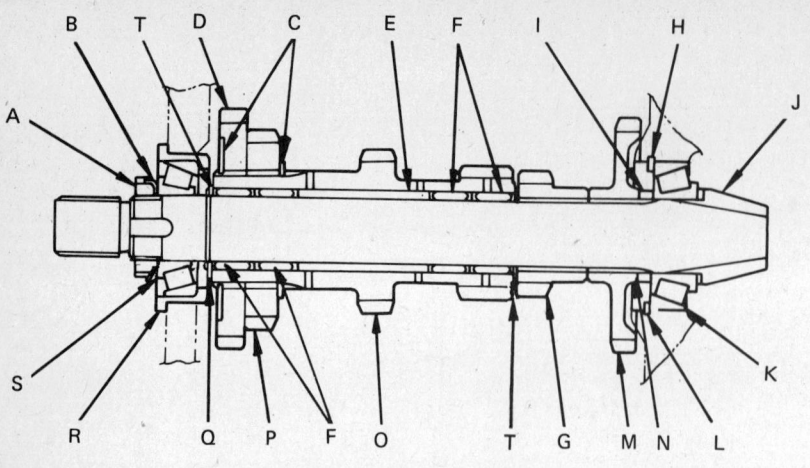

Assembling the countershaft

- A. Nut
- B. Washer
- C. Circlip
- D. Gear (44T)
- E. Collar
- F. Needle bearing
- G. Gear (17T)
- H. Circlip
- I. Collar
- J. Countershaft
- K. Tapered roller bearing
- L. Shim
- M. Gear (34T)
- N. Liner
- O. Gear (18T-23T)
- P. Gear (29T)
- Q. Circlip
- R. Holder
- S. Tapered roller bearing
- T. Liner

Place the shim (N) (which has been selected on the basis of clearance measured on countershaft gear) between collar (I) and gear 34T (M). Mount on the countershaft being inserted with the following parts: gear 17T (G), thrust liner (T), gear 18T-23T (O), gear 29T (P), gear 44T (D), needle bearings (F), collar (E), needle bearings (F), and thrust liner (T), in that order. Retain these parts by installing circlip (Q).

NOTE: These gears have their tooth ends chamfered. Be sure to bring the chamfered end to the side shown in the sectional view.

3. Install the outer race of tapered roller bearing (S) in the holder (R) fitted to the transmission case. Mount tapered roller bearing on countershaft and secure the countershaft in place by tightening its sleeve nut (A). Be sure to use tab washer (B) on this nut.

NOTE: When fitting the outer race, be sure to discriminate its inner end and outer end.

4. Tap lightly on each end of countershaft in place, and make sure that there is no end play on this shaft. Measure the preload of countershaft and, as necessary, adjust the tightness of its bearings to produce the specified preload: 0.58-0.73 ft.lb.

5. With a proper preload obtained, lock the sleeve nut by bending tab washer sharply.

PTO Gear Tooth Alignment and Adjustment

1. Tentatively secure guide plate to transmission case.

2. Fit stopper spring and stopper ball to PTO shift lever: grease the ball before inserting it. Attach shift lever to PTO shifter, and lock the lever by inserting spring pin.

3. Operate the shift lever to see if the teeth of PTO gear and counter gear are properly aligned; if not, align them by shifting the guide plate on transmission case. After making sure that the shifting action is satisfactory, secure guide plate permanently by tightening to 18-22 ft.lb.

Reassembling the Mid-PTO

1. Grease oil seal, and fit the seal to mid-PTO case.

2. Set 4-WD shifter-shaft stopper spring and stopper ball in mid-PTO case. Grease O-ring, and install O-ring and circlip on shifter shaft. Install the shaft in mid-PTO case, and drive spring pin into the forward end of shifter shaft.

3. Mount the roller bearing (C) inner race on driveshaft (J), bringing the flanged end of the race to the opposite side of the gear, and retain the inner race by fitting circlip (B).

4. Push in the ball bearing on mid-PTO shaft. Install gear with its longer boss coming on the front side. Push in the ball bearing on the shaft.

5. Install PTO shaft and driveshaft in the mid-PTO case.

6. Mount gear 15T, positioning it properly relative to the fork of shifter shaft. Apply sealant to the mating face of mid-PTO case and fit it to transmission case. Secure the mid-PTO case by tightening its bolts to 18-22 ft.lb.

Installing the Mainshaft

Fit the coupling to the mainshaft, and connect this shaft to driveshaft. Be sure to lock the mainshaft securely by means of set bolt. After tightening the set bolt, lock it positively.

Reassembling the Select Shaft

1. Install high-low shifter in transmission case. Grease O-ring, and fit it to shifter at its portion outside of the case while pushing it against the case.

2. Push in the shifter all the way, fit the plate to the case, and tighten its bolts tentatively, leaving them finger-tight.

3. Fit ball bearing (P) to select shaft (G), and put on circlip to retain the bearing (Q).

4. Install circlip (I) in the hole provided in the wall of transmission case for select shaft.

5. While inserting select shaft (G) through the rear end of transmission case, mount on this shaft the following parts: gear 15T-33T (H), gear 32T (F) (bring their shifter fork grooves on the front side) and gear 20T-27T (E) (bring gear 20T on the front side). Set shifter in the fork grooves of gear 15T-33T (H).

6. Fit collar (O) and needle bearing (N) to the forward part of the select shaft (G), and connect this part of select shaft to driveshaft (J). Be sure to oil the needle bearing (N) before installing it.

7. Install circlip (I) in the circlip groove provided in the wall of transmission case to retain the ball bearing on the rear part of select shaft and set the select shaft.

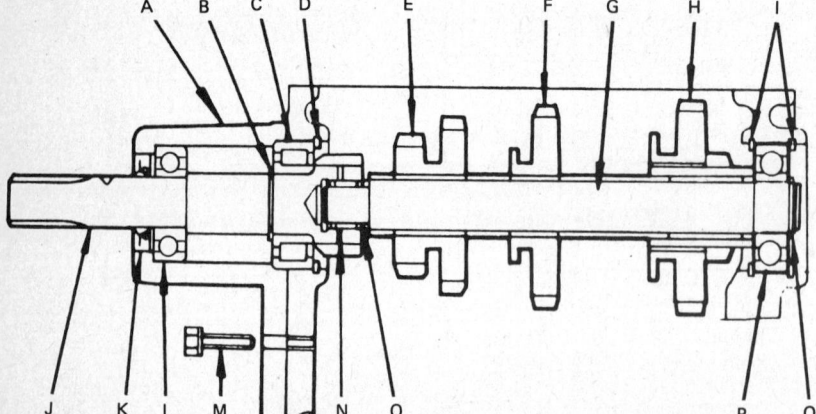

Select shaft components

- A. Bearing holder
- B. Circlip
- C. Roller bearing
- D. Circlip
- E. Gear (20T-27T)
- F. Gear (32T)
- G. Select shaft
- H. Gear (15T-33T)
- I. Circlip
- J. Driveshaft
- K. Oil seal
- L. Ball bearing
- M. Bolt
- N. Needle bearing
- O. Collar
- P. Ball bearing
- Q. Circlip

NOTE: Be sure to position the gears in place as shown. Discriminate the two ends of the gear: one end is chamfered but the other is not. "Hard shifting" is often due to these gears positioned the other way around.

8. Move the installed parts by hand to be sure that select shaft is capable of smooth rotation and that each gear is capable of smooth sliding motion.

9. Fit stopper spring and ball to high-low selector lever: grease ball before inserting it. Connect the lever to shifter shaft, and secure the connection by driving in spring pin.

10. Bring the gear 17T-34T on countershaft into correct positional relationship with the gear 33T-15T in reference to their faces, by adjusting the guide plate on transmission case. After this adjustment, secure the plate permanently by tightening its bolts to 18-22 ft.lb.

11. Attach the knob to high-low selector lever. Operate the lever in the usual manner to make sure that shifting action is smooth and positive.

Installing the Differential Gear Assembly

1. The differential gear assembly is already in built-up condition at this stage, complete with the ring gear. Lower it into transmission case, with the ring gear coming on the left side.

2. Position the two bearing holders (differing in length) in place, seating each holder correctly by matching its notch to the corresponding one formed of case. The long holder comes on the right side, and the short one on the left.

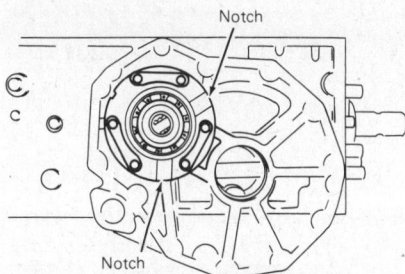

Aligning bearing holder notches

NOTE: Bear in mind that, if the holders are not positioned correctly as dictated by the matching notches, interference occurs between final gear and bearing holder. The final case might break if the case is secured under this condition (interference between holder and gear).

3. Tentatively tighten the two bearing holders in place, right and left, on transmission case, making their securing bolts snug-tight. Be sure that stopper washers are used on these bolts.

4. Using the below indicated sizes (thicknesses) of shim stock (liner), adjust the ring gear backlash to 0.01-0.014 in.

Shim stock for ring gear backlash adjustment

Thickness (in.)	Part No.
0.008	1135 1419 000
0.012	1135 1421 000
0.020	1135 1422 000

a. Insert the liner between transmission case and left side bearing holder. Be sure that the individual shims (each consisting of one or more sheets) of the liner are equal in thickness. This requirement is met by using the same number of shim sheets. The liner here is tentative and need not be exact in thickness, but make sure that the clearance is filled up uniformly all around.

b. Tighten the left side holder (which is now sided by the tentative liner) good and hard, but leave the right side holder in snug-tight condition. Take a backlash reading, as shown, to see if the backlash is within the specified range indicated above; if not, loosen the left side holder and increase or decrease the thickness of the liner there.

Differential gear backlash

c. A proper backlash having been secured, measure the clearance between the right side holder and case, and fill the clearance with another liner equal in thickness to the reading. This liner, too, is to be formed in the same way as above. Put on stopper washers and tighten the right side holder.

d. Tap lightly on the differential gear case in place, directing the tapping force leftward and rightward to be sure that this case has no end play. Re-check the backlash and, upon noting that it is within the specified range, lock the bolts securing the bearing holders, right and left, by bending stopper washers sharply. The holder bolts are to be tightened to 18-22 ft.lb.

Reassembling the Final Case

1. Grease oil seals, and fit the seals to those parts of final case admitting differential shaft and final shaft. For the final shaft, however, a washer must be installed before fitting the oil seal.

2. Press ball bearing onto differential shaft. Grease O-ring and fit it to oil seal collar. Feed the collar onto differential shaft, with its O-ring coming next to the ball bearing.

3. Insert differential shaft into final case, and set it in place by fitting circlip.

4. Position final shaft in final case, fit ball bearing by pressing, and retain the bearing by installing the circlip.

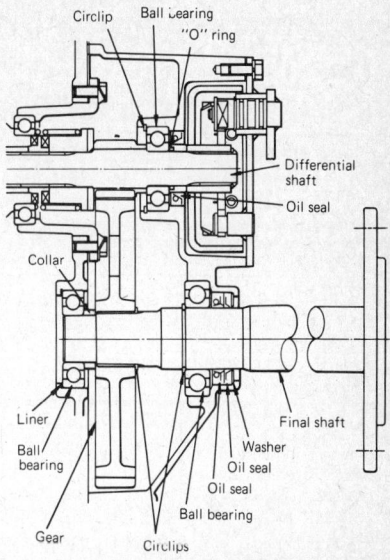

Final case assembly

NOTE: When forcing the bearing in, be sure to apply the push to its inner race.

5. Install circlip (for final gear stopper) on final shaft, mount final gear and collar, and press in ball bearing.

6. Take two measurements: 1) height of ball bearing face above the seat formed of transmission case for final case, and 2) distance from transmission case face to ball bearing holder. On the basis of these two measurements, determine the shim (liner) thickness necessary for giving an end play 0.005 to 0.016 in. to final shaft. The liner with the determined thickness is to be used on ball bearing holder. Final shaft end play is: 0.005-0.016 in.

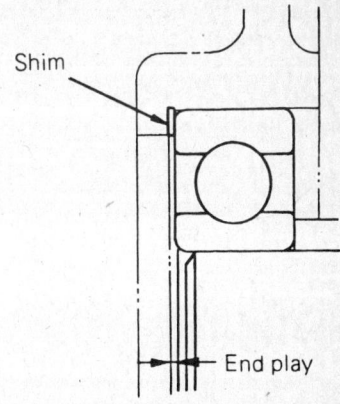

Shimming for end play

7. Apply sealant to the mating face of final case, fit the case to transmission case, and secure it by tightening its bolt to 36-43 ft. lb.

NOTE: The right side final case must be complete with differential lock shifter and lock sleeve; the shifter must be installed before inserting differential shaft into the case.

Satoh

Shift Mechanism

DISASSEMBLY

1. Loosen set bolts on the two shifter shafts, right and left Each set bolt is locked by stopper washer: straighten the locking portion of the washer and loosen the bolts.
2. Draw out each shifter shaft from the front side by pulling, and remove shift fork, steel ball and stopper spring.
3. Loosen bolts securing the bracket, take off bracket, remove knob from speed change lever, and take out the lever.

ASSEMBLY

1. Insert speed change lever into cover, attach spring, and position bracket in place. Put on stopper washer and tighten the bracket securing bolts to the torque value indicated below, and lock the bolts by bending the stopper washer. Tightening torque: 18-22 ft. lb.
2. Insert stopper springs and steel balls into the two shifter forks.
3. Grease O-ring and fit it to shifter shaft. Of the two shifter shafts, the distance between the two grooves (one for O-ring and the other for stopper ball) tells the difference between the two, 1st-reverse shaft and 2nd-3rd shaft. The distance is longer on 1st-reverse shaft.

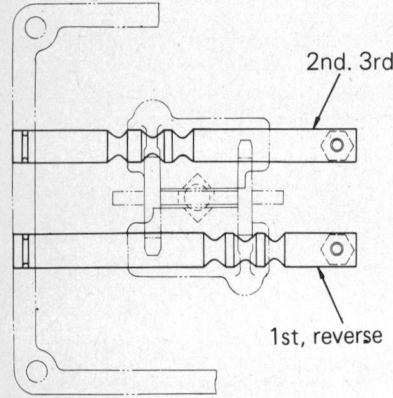

Shifter shaft installation

4. Mount shifter fork on the rear part of 1st-reverse shifter shaft, and install the shaft in the cover.
5. Mount shifter fork on the front part of 2nd-3rd shifter shaft, and install the shaft in the cover.
6. Adjust each shifter shaft, aligning the stopper hole through cover and shaft, put on stopper washer and lock after tightening.

NOTE: The transmission case may be filled with oil at this time, provided that the drain plug and hydraulic pipe connection are tight. Oil filling is easier at this stage.

7. Apply sealant to the mating face of cover, and secure it to transmission case by tightening its securing bolts to 18-22 ft. lb.

ENGINE

These tractors use Mitsubishi K3A-12 and K38-11 diesel engines.

NOTE: For engine removal procedures, see the Clutch section.

Cylinder Head

REMOVAL

1. Drain oil pan.
2. Drain the cooling system by opening the cock on the lower part of the radiator.
3. Remove air cleaner cap. Unlatch and open the bonnet.
4. Disconnect cables from the battery terminals, undoing the terminal connection of negative (−) cable first and that of positive (+) cable next.
5. Remove safety guards, right and left.
6. Loosen the clamp on air cleaner hose at manifold side, and disconnect the hose. Loosen the air cleaner band bolt, and take down the cleaner.
7. Remove the bolts securing air cleaner bracket in place, and take off the bracket.
8. Disconnect the radiator upper hose from the engine.
9. Disconnect the tractor meter wire at engine side.

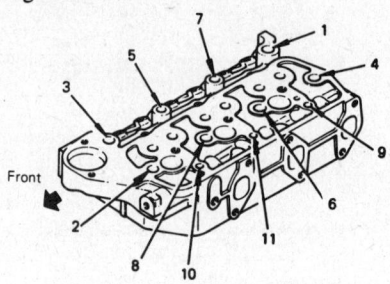

Head bolt loosening sequence

10. Disconnect the fuel return pipe from the fuel tank.
11. Undo the wire connectors of glow plug and water temperature wires.
12. Remove the bolts securing the alternator brace to cylinder head.
13. Disconnect the exhaust manifold (complete with muffler) from the cylinder head.
14. Disconnect and remove fuel injection pipes.
15. Remove the nuts fastening down the rocker cover, and take off the cover.
16. Remove the bolts securing the rocker stays, and take off the three stays complete with rocker arms and shaft.
17. Pull out the push rods one by one.
18. Loosen the cylinder head bolts sequentially in the ascending order of numbers as shown and remove the cylinder head.
19. Remove the cylinder head gasket.

DISASSEMBLY

1. Remove nozzle holders.
2. Disconnect glow plug lead wires, and remove the plugs.
3. Using the valve lifter, remove each valve in this manner: Compress the spring with the lifter; take off retainer locks; and pick out retainer, spring and valve in that order. Place the removed parts in trays or pans, separating them into three groups, one group for each cylinder. Be sure to identify each part for the cylinder it has been servicing.
4. Disconnect water bypass hose, and remove thermostat fitting.

INSPECTION

1. Wash the cylinder head clean. Before doint so, visually examine it for evidence of cracking, water leakage or any damage.
2. Check to be sure that the internal oil passages are all clear.
3. Using a straightedge and feeler gauge, check the gasketed surface for flatness.

Valve Guides

1. Take diameter measurements on valve guide and stem to find out the radial clearance by subtraction; if the determined clearance exceeds the limit, replace the guide or valve, or both.
2. To remove the guide for replacement, drive it out of the cylinder head by giving a push to the bottom side of the guide. Use the valve guide remover.

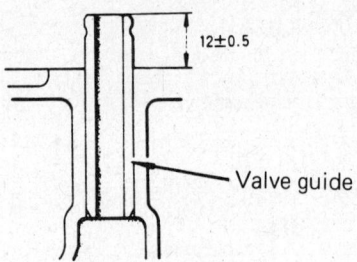

Pressing in of valve guide

Valve guide installation

To install the replacement guide, press it into the hole until its trailing portion measures 0.5±0.02 in. from cylinder head top. After pressing the guide in, check the radial clearance by inserting its valve stem just for a trial: if the clearance is too small, ream the guide to produce a proper radial clearance.

Valve Seats

1. Visually inspect each valve seat for seating contact pattern and for damage and, as necessary, repair it by lapping in the usual manner to the seat angle and diameter specified.
2. An insert-type valve seat in service is subject to beating action of its valve and might force itself, though very gradually, into the cylinder head, thereby presenting a phenomenon of "seat sinkage," which is primarily due to creeping effect of stressed metal. This sinkage shows up as an increment in the installed length of valve spring.

Measure the length of each valve spring in place and, if the increment (corresponding to the sinkage) is found to exceed the

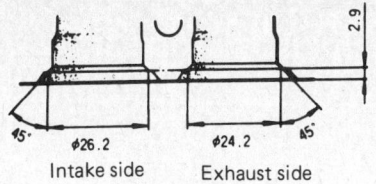
Valve seat repair

limit of 1.64 in., replace the whole cylinder head.

In measuring the installed length, it is permissible to read the distance between the bottom end of spring to the top face of spring retainer. In such a case, the thickness (measured in advance) of the retainer must be subtracted from the reading. The retainer (actually the flange) thickness is 0.079±0.010 in.

Valves

1. Visually inspect the seating face and stem of each valve for wear and damage, and repair or replace the valve, as necessary. Valve stem OD is 0.260 in.
2. A valve whose head is worn down to the limit of "T" value must be replaced.
3. Inspect the valve for localized wear at three places in particular, which are indicated in the illustration, and repair or replace the valve, as necessary. Make sure that the top end face and other surfaces of the stem as shown by arrows are smooth and that there are no dents nor groovy depressions on the stem.

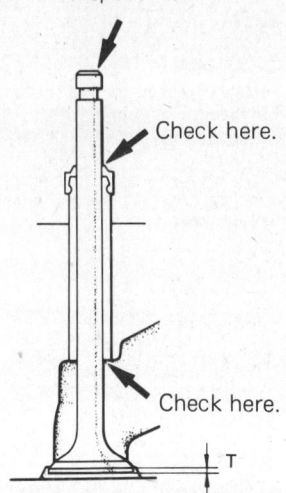
Checking valve dimensions

Valve Springs

1. Cracked, worn or otherwise damaged springs must be replaced.
2. A weakened valve spring, the weakness being evidenced by its reduced free length or spring rate, must be replaced.
3. Check each spring for squareness. Never re-use distorted valve springs.

Rocker Arms and Rocker Shaft

1. Take diameter measurement on rocker arms and rocker shaft to determine the clearance between the two. Replace the arms if the clearance exceeds the service limit of 0.0008 in.

2. A rocker arm whose end face for pushing down the valve stem is badly worn must be replaced. The same consideration is required for the adjusting screw, whose end face comes into contact with the top of the pushrod: if this face is found badly worn, replace the adjusting screw.

INSTALLATION

Installation is the reverse of removal. Observe the following points:

1. Install valve guides, making sure that each guide protrudes out of the top of the head to the specified height. (The method of installing them is explained in "Inspection," above.)
2. Fit the valve stem seal to each valve guide, making the seal settle snugly on the guide end.
3. Oil valve stems, and insert them into the guides. Put on valve springs, retainers and locks, in that order.
4. Build up the rocker mechanism by proceeding as follows: Hold the front stay with its mounting bolt hole coming on the righthand (nozzle) side. Insert the rocker shaft into the stay so that the identifying mark (0.1-in. dia. hole) on the end of the shaft faces the left-front side (the alternator). Mount the outer front rocker arm and secure it in place by fitting snapring; mount the inner front rocker, thus completing the reassembly of the first group.

Reassemble the second and third groups similarly, ending with the fitting of the rearmost snapring; set the whole mechanism on the cylinder head; secure it to the head. Be sure to use seat washers on the bolts for front and rear stays.

5. Run the glow plugs into the head, and tighten them to 18 ft.lb. (Gaskets are not required because of the tapered seal feature.)
6. Mount the nozzle holders. The mounting bolts must be tightened equally, to 47 ft.lb. Use new gaskets on the holders.
7. Connect lead wires to the glow plugs in place.
8. Do not apply any sealant to the cylinder head gasket: the replacement gasket comes with its surface coated with sealant.
9. When securing the cylinder head to the block, be sure to tighten its bolts sequentially to make sure that the pressure will be equalized. This is accomplished by running all bolts in till they become finger-tight, and then tightening them with a torque wrench gradually and in two or three steps, each time moving the wrench from one bolt to another in the sequential order indicated by the numbers and tighten to 58 ft.lb. for M10 bolts and 87 ft.lb. for M12 bolts.
10. Several kinds of gaskets and packings are used on the cylinder head. Be sure to use new gaskets and packings in reassembly. Also, be sure to use the prescribed sealant at the places specifically designated.
11. To adjust the valve clearance, proceed as follows:

   a. Valve clearance adjustment should be carried out with the piston in top dead center on compression stroke when the engine is cold.

Be sure to tighten the cylinder head bolts before adjusting the valve clearance.

   b. To bring the No. 1 cylinder piston to top dead center on compression stroke, align the timing (TDC) mark on the crank pulley with that on the gear case by turning the crankshaft in normal direction. Now, the intake and exhaust valves of the No. 1 cylinder are ready to be checked. Check the clearance and, if it is incorrect, adjust it by turning the adjusting screw. Valve clearance is 0.010 in. cold, for both intake and exhaust valves.

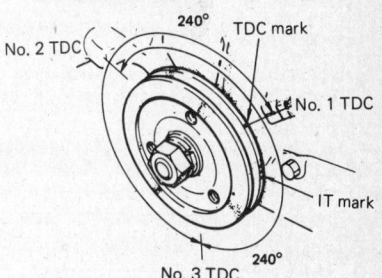
Crankshaft timing marks

Be sure to align the timing marks perfectly or the valves are moved away from the correction position, making it impossible to obtain the correct valve clearance.

   c. Next turn the crankshaft 240 degrees in normal direction to bring the No. 3 cylinder piston to top dead center on compression stroke. Having made sure that the timing marks are perfectly aligned, check and adjust the valve clearance on this cylinder.

   d. Further turn the crankshaft 240 degrees, and similarly adjust the valve clearance of No. 2 cylinder valves.

Pistons and Connecting Rods

REMOVAL AND INSTALLATION

1. Remove the head.
2. Disconnect tie rod from knuckle arm. 4-WD TRACTOR.
3. Remove the front and rear universal-joint covers; pick out the front circlip; pull off the pin; and disconnect the universal joint.
4. Remove the bolts securing the oil pan to the cylinder block, and take off the oil

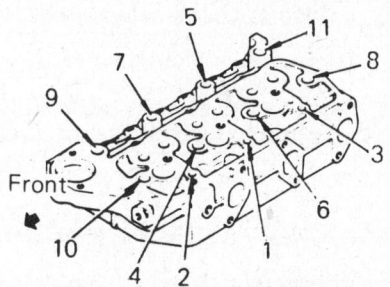
Cylinder head tightening sequence

Satoh

pan, taking care not to distort it. Remember, the sealing compound used in this joint may not permit the oil pan to separate easily.

5. Remove the cap bolts on each connecting-rod big end, take off the cap, and push out the piston assembly from block top.

6. Installation is the reverse of removal.

Crankcase

DISASSEMBLY

1. Take down the engine from the machine.

2. Remove the cylinder head assembly, as outlined in piston removal procedure, above.

3. Remove the water pump and electrical parts and components.

4. After pulling out the pushrods, draw out the tappets, taking them out from the cylinder block top.

5. Remove the speedometer driven unit.

6. Loosen the nut securing the crank pulley, and take off the pulley and washer.

7. Take down the flywheel: loosen the flywheel bolts just a little at a time.

8. Remove the rear plate and rear oil seal case.

9. Turn over the cylinder block upside down, and remove the oil pan and oil screen.

10. Remove the hydraulic pump gear bearing housing and the gear case. Just before detaching the gear case, be sure to remove the inspection peep hole cover and to disconnect the stopper spring and tie rod from the pump control.

NOTE: It is very important that, before the gear case is removed, the injection pump control rack should be disconnected from the tie rod. As mentioned previously, the front plate is bolted to the cylinder block, the bolts being run in from inside the gear case. Be careful not to remove the plate together with the gear case and also not to disturb the dowel pins.

11. Remove the fuel injection pump.

12. Remove the governor weight securing bolt, and take out the weights.

13. Remove the setscrew on pump camshaft.

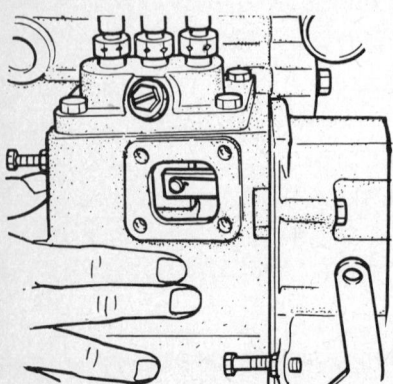

Injection pump removal

14. Remove oil filter and oil pump assembly, and draw out the pump camshaft.

15. Remove timing gears and detach the front plate from cylinder block.

16. The push rods, tappets and speedometer driven unit having all been removed, draw out the engine camshaft.

17. Open the big end of each connecting rod by removing the cap. Push out each piston assembly from block top.

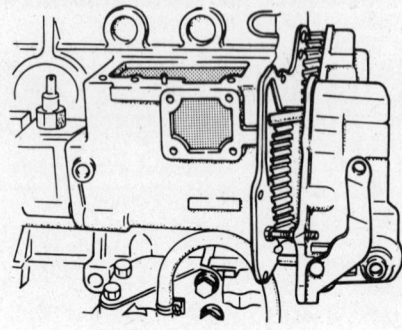

Removing gear case

NOTE: Have all removed parts laid out in groups, identifying each part for the cylinder from which it was taken. It is a standard practice to assign a reference number to each cylinder and use that number to refer to the group of parts, which of course include bearing shells, piston rings, etc. Never drive out the combination of piston and connecting rod: push on the mating face of big end with a wooden stick, as necessary, not to damage the bearing shell.

18. Use the piston pin setting tool (special tool) to separate piston pin from piston in the manner illustrated here: lay down the connecting rod on the tool body, fit the

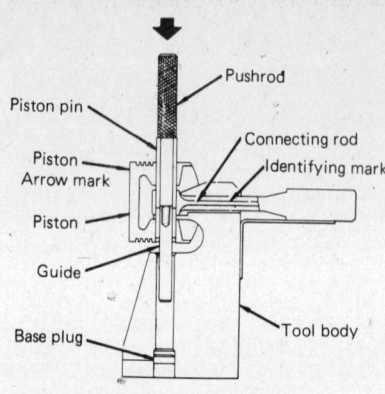

Removing piston pin

push rod tool to the piston pin in place, and press down the push rod to force the pin out. Never use a hammer to drive on the push rod tool.

NOTE: Never drive off the pin. If the pin is seized in the piston, it may be necessary to replace piston, pin and connecting rod with new ones. Do not apply a press force of more than 6615 lb. to the special tool.

19. Remove the main bearing caps. Set aside the removed caps and bearings separately in groups, each marked for its journal, so that the same combination as before can be reproduced at the time of reassembling. Before removing caps, read the crankshaft end play and write the reading down for reference.

20. Take off the crankshaft.

INSPECTION

Cylinder Block

1. Visually inspect the cylinder block before and after washing it clean. If any

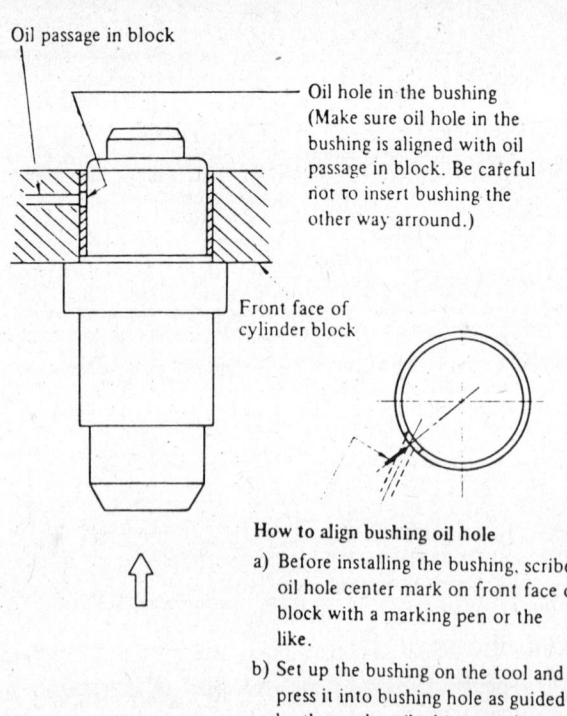

Checking camshaft front journal bushing

crack or serious damage is noted, replace the cylinder block.

2. Check the camshaft front journal bushing for wear. If the bushing is worn beyond the service limit or damaged beyond repair, remove it by using the special tool (installer) and replace it with new one. On the engines not equipped with the bushings, check for wear or damage the camshaft hole in the cylinder block. If this bore is worn or damaged, refinish it to the dimension shown in the illustration and install the bushing.

NOTE: When refinishing the bore to the new diameter: $1.89 ^{+0.0010}_{0}$ in., be sure that the refinished bore is parallel and concentric with the axis of the previous 1.77 in. bore within a tolerance of 0.0004 in.

3. Clean the water jackets of the cylinder block, removing water scales and rust, if any.

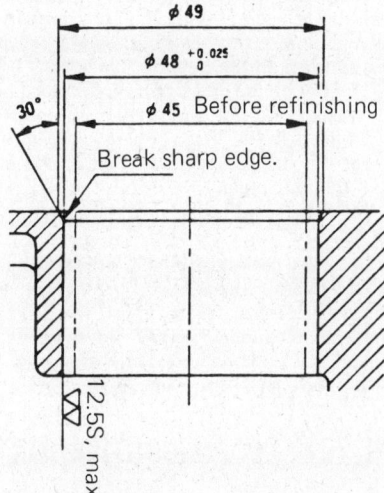

Camshaft bore refinishing dimensions

4. Check each cylinder bore for wear and inspect the bore wall for scoring, scuffing and other types of surface flaw, to determine whether repair by reboring or honing is necessary or not. To take micrometer readings for wear determination, be sure to measure at three levels, 1, 2 and 3, in two directions, thereby producing a total of six readings. Cylinder bore repair limit is 2.560 + 0.008 in. for the K3A engine and 2.680 + 0.008 in. for the K3B engine.

5. If the piston rings are the only parts to be renewed, there being no need of reboring or honing the cylinder bores, check the amount of "ridge" formed of the top portion of the bore and, as necessary, remove the ridge by reaming. The bore should be honed after this reaming.

Pistons, Piston Pins and Piston Rings

1. Burnt, grooved or badly scuffed pistons must be replaced.
2. Mike the piston at its skirt in the direction perpendicular to the piston pin to determine its radial clearance in the cylinder. The service limit is 0.010 in. If the

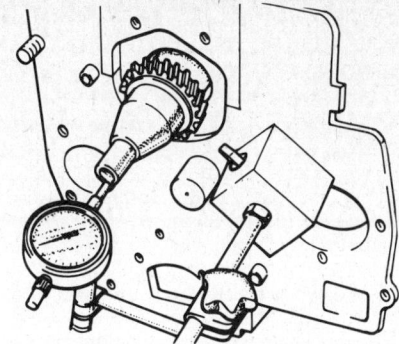

Checking crankshaft end play

piston is found excessively worn, replace it.

3. Check the side clearance of each ring in the groove and, as necessary, replace the ring. Side clearance of #1 is 0.0024-0.0047 in.; #2 is 0.0020-0.0035 in.; #3 is 0.0016-0.0031 in.; #4 is 0.0014-0.0028 in.

To measure the side clearance of No. 1 ring which is taper, hold the ring face flush with the land face, and insert a feeler gauge between the piston and the bottom (flat) side of the ring.

4. Piston rings whose joint gap is too large must be replaced. To check the tap, place the ring in the cylinder bore and push it down with the piston inserted upside down.

When the ring is located at the least worn part (lower section) of the cylinder, take out the piston and read the gap with a feeler gauge. Gap is 0.0060-0.0160 in.

Connecting Rods

1. To check each connecting rod for bend and twist, the connecting rod aligner must be used. Fit the connecting rod to the aligner and, if the rod is found to be excessively distorted, that is, bent or twisted, repair or replace it.

2. Check the big end for thrust clearance by fitting the big end to the crankpin and by using a feeler guage. Clearance is 0.004-0.014 in. If the clearance is noted excessively large, replace the connecting rod.

Crankshaft

1. A distorted crankshaft must be straightened or, if the distortion is too large to correct, be replaced. Runout is 0.002 in. Distortion here refers to the bow, if any, of the crankshaft, which can be measured with a dial indicator in the usual manner.

2. Inspect the surfaces of journals and crankpins for burning and damage and, as necessary, repair these surfaces by grinding to the next undersize. Grinding to an undersize necessitates replacement of the existing bearings by the corresponding undersize ones.

When grinding the journals and crankpins, be sure to finish the corner radii to 0.1 in.

3. Check the crankshaft end play and, if the measured play is in excess of the specification, replace the No. 3 main bearing. End play measurement is valid only when the crankshaft is set in place in the normal condition, with its main bearings fitted correctly and its bearing cap bolts tightened to 36-40 ft.lb. Use a dial gauge to read the end play. End play is 0.002-0.012 in.

Main Bearings and Connecting Rod Bearings

1. Inspect the bearing shells, paying particular attention to the tri-metal surface for evidence of flaking. Burnt, pitted or wiped shells and shells showing bad contact pattern must be replaced.

2. Mike the main bearings and connecting-rod bearings and also the crankshaft journals and crankpins to determine, on the basis of ID readings and OD readings, the amount of oil clearance available in each fit. (A press gauge can be used instead.)

Main bearing clearance: 0.004 in.; Rod bearing clearance: 0.006 in.

When reading the bearing ID, whether it is a main bearing or a crankpin bearing, be sure that the bearings shells are tight in the usual manner, with the cap bolts torqued to 23-25 ft.lb. If the clearance determined by computing with the ID and OD readings exceeds the limit, replace the bearings or, if mere bearing replacement does not produce the specified clearance, grind the crankshaft journals and crankpins to the next undersize and use the undersize bearings.

Timing Gears and Hydraulic Pump Gear

Inspect these gears for tooth contact pattern, tooth wear and damage and, as necessary, replace them. Inspect the Oldham coupling groove formed of the end of the pump gear; if this groove is disfigured or damaged, replace the gear.

CAMSHAFT

1. If the running clearance between the camshaft journal and its hole provided in the block is greater than 0.006 in, then either the camshaft or the block must be replaced. This clearance is to be determined by miking journal diameter and hole diameter. On the engines except K3B with E/#5181 and below, the front journal of the camshaft is supported by rolling bearing.

2. Visually inspect the cam faces for damage, and check each cam for cam height by miking. Replace the camshaft if any of the cams is in bad condition in regard to cam height and face.

Fuel-Injection-Pump Camshaft

Inspect and check this camshaft as in the

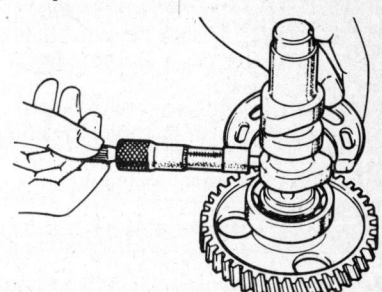

Measuring injection pump cam height

Satoh

case of engine camshaft. Additionally, inspect the shaft end, which is shaped for Oldham coupling: the camshaft must be replaced if the coupling end is disfigured.

ASSEMBLY

Assembly is the reverse of disassembly. Note the following points:

1. When installing the camshaft front journal bushing into the cylinder block, use the installer and press in the bushing so that the oil holes in the bushing and cylinder block are aligned. It is advisable to put a mark indicating oil hole position on the front face of cylinder block with a soft pen before installing the bushing for the convenience of aligning the holes and also of checking after the installation of the bushing is completed.

2. Fit the main bearing shells to the caps and to the half-bores formed of the block, making sure that each shell is correctly positioned.

3. Oil the crankshaft journals and crankpins, and set the crankshaft in place.

4. Put on main bearing caps and secure them by tightening their bolts to 36-40 ft.lb.

Each cap has an arrow mark and numeral cast out: refer to these marks and position the cap correctly. When installing Nos. 1 and 4 caps, be sure to apply sealant to their mating faces.

5. Check the end play of crankshaft.

6. Apply sealant to the periphery of the side seals, and push them into the front and rear caps. This completes the installation of the crankshaft.

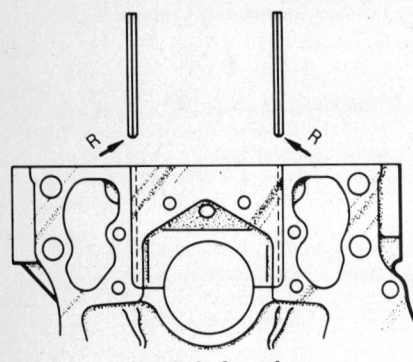

Crankshaft seals

7. To the rear oil seal case of crankshaft, fit the oil seal; and install the seal case with its gasket, securing the case fast to the cylinder block.

8. Using the piston pin setting tool (three-piece tool), combine piston with connecting rod in the following manner:

a. Fit piston pin to the pushrod (tool), and run guide (tool) all the way into pushrod.

b. Insert the combination of pushrod, pin and guide into the piston from the guide side, passing the pushrod through the small end of connecting rod, making sure that the front arrow mark (on piston crown) and the identifying mark on connecting rod come on the same side (top side).

Before inserting the pin, be sure to oil it as well as the small end.

c. Rest the whole work (piston, pin and connecting rod with pushrod and guide) on the tool body, bringing the notch of the guide into register with the notch provided in the tool body; and turn the guide by 90°, making sure that the small end is snugly settled in the recess of the body. Again, check to be sure that the front mark on piston crown and the identifying mark of the rod are both on top side and pointing upward.

d. Using a press and applying a force of anywhere between 1103 and 3308 lb., push the pin into the connecting rod. Should the pin go in with a push of less than 1103 lb. or greater than 3308 lb., the connecting rod or pin and piston must be replaced. The guide (tool) serves to locate the pin in the prescribed position. After pressing the pin in, turn the pushrod by 90° and take off the combination from the tool body.

NOTE: After combining piston with connecting rod, check to be sure that the pin is centered on the axis of connecting rod. If the pin is found displaced to one side, check the tool and, after correcting it as necessary, use it to push the pin back to the center position. As stated previously, restore all parts to their original positions in reassembly. Remember, piston and pin constitute a set and must not be interchanged. Be sure, too, that the three pistons are of the same size (same mark).

9. Fit the rings to the piston, discriminating the three compression rings, as shown, and distributing the gaps equiangularly. The side face of each ring with the maker and size marks comes on top side. When installing the oil ring with expander, be sure to position the expander tube opposite to the gap of the ring.

10. Insert the three combinations (piston and connecting rod) into the cylinders from the gasketed surface, using a ring band on each piston to embrace its rings. Make sure that the ring gaps are correctly distributed and that the arrow mark on piston crown points toward the front end of the engine. On the crankshaft side, connect the connecting rods to respective crankpins, with the bearings fitted properly, and secure the caps by tightening their bolts to 23-25 ft.lb.

11. Install the front plate, with its gasket properly positioned and doweling the plate securely.

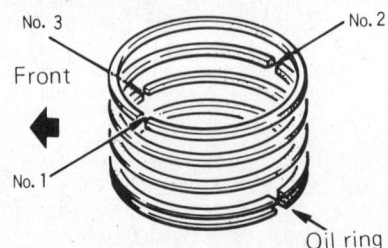

Piston ring gap positioning

12. Turn over the crankshaft to bring No. 1 piston to top dead center position.

13. Fit the key into the keyway formed of crankshaft, and install the crankshaft gear.

14. Set idle gear in place, matching its "I" mark to the "I" mark on crankshaft gear. If the crankshaft front bearing is already in place, the latter "I" mark may be hard to see and, in such a case, reference should be made to the engraved line, instead of the "I" mark, that is provided on the side face of gear boss. Fit the guide seals coated with sealant to the front and rear main bearing caps at this time by pushing the seals in, and have the cap bolts tightened to the torque limit. Insert the camshaft assembly into the cylinder block, positioning its gear in such a way as to bring its match mark "2" into register with the mark "2" on idler gear. Similarly install the injection pump camshaft, making the match mark "3" of its gear to the mark "3" of idle gear. Finally, install the hydraulic pump driveshaft, meshing its gear with camshaft gear.

15. Attach the governor weight assembly to the injection pump camshaft gear.

16. After installing governor parts, install the gear case, with its gasket properly set, while inserting the tie rod and its stopper spring into the cylinder block.

17. Put on the crank pulley, followed by its washer and nut, and tighten the nut to 145-181 ft.lb.

18. Fit the gasket to the block, and install the rear plate.

19. Attach the oil screen, and install the oil pan.

20. Turn over the cylinder block. While slowly rotating the camshaft or the speedometer driven gearshaft, install the driven unit, making sure to fit its O-ring correctly. Apply sealant such as Three-Bond #2® to the periphery of the sleeve.

To reassemble the driven gear unit, proceed as follows:

a. Fit O-ring in the groove in the sleeve.

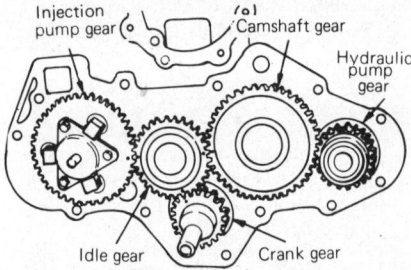

Installing timing gear

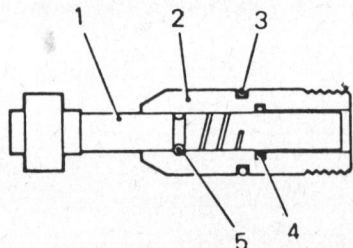

Speedometer driven gear

1. Driven gear shaft
2. Sleeve
3. O-ring (large)
4. O-ring (small)
5. Spring pin

Satoh

b. Apply EP (extreme-pressure) type grease to the periphery of driven gear shaft, especially to its O-ring surface, and insert the shaft.

c. Secure the shaft to the sleeve by inserting the spring pin, being sure that the groove of the pin faces outward and that the tip of the pin is not protruded beyond the periphery of the sleeve.

d. Put O-ring to the groove on the periphery of sleeve. After completing reassembly, check to be sure that the gear shaft rotates smoothly.

21. Oil the tappets liberally, insert them into the holes, and follow them with pushrods, making sure each pushrod fits snugly.

22. Install the cylinder head assembly on the block, as outlined in the preceding section.

23. Install the injection pump assembly: refer to the section dealing with the fuel system.

24. Install the oil pump and filter: refer to the section covering the lubrication system. Be sure to apply sealant to the screw threads of the oil pressure switch when installing this switch.

25. Install the fuel filter.

26. Install the water pump and cooling fan.

27. Install the starter and alternator. For this installation work, refer to the section dealing with the electrical system.

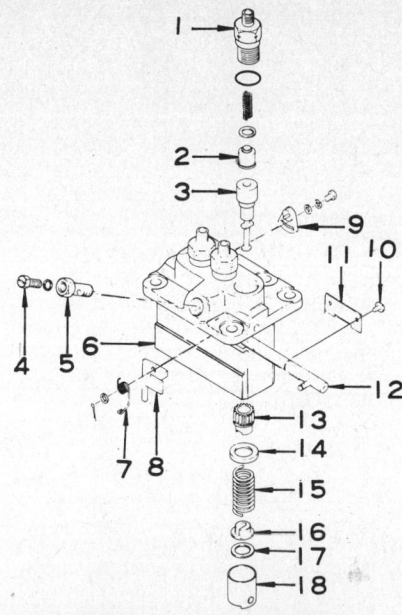

Injection pump components

1. Valve holder
2. Delivery valve
3. Plunger and barrell
4. Air vent screw
5. Hollow screw
6. Pump housing (body)
7. Return spring
8. Stopper
9. Adjusting plate
10. Tappet guide pin
11. Plate
12. Control rack
13. Pinion
14. Upper seat
15. Plunger spring
16. Lower seat
17. Adjusting shim
18. Tappet

9. Pick out spring (15) and upper seat (14).

10. Pull down and remove pinion (13).

11. Draw out plunger and barrel (3) from the delivery valve side of pump housing.

12. Be sure to group the delivery valve, plunger and barrel, so that these and related parts (pinion, spring, seats and shims) will be restored to the place to which they belong.

13. Remove the smoke-set stopper by pulling off split pin and taking off washer and return spring. Draw out control rack (12).

NOTE: Do not remove the injection-quantity adjusting plates since this removal makes it necessary to test the pump on a bench tester. If necessary to remove these plates, be sure to mark the plates and pump body to aid reassembly.

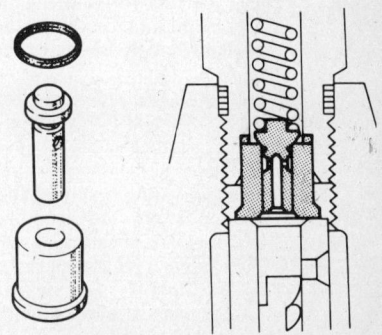

Removing delivery valve

FUEL SYSTEM

Injection Pump

REMOVAL AND DISASSEMBLY

1. Disconnect fuel injection pipes from delivery valve holders.

2. Removing the pump mounting bolts allows the pump to be taken down. Just before lifting the pump off, have the tie rod disengaged from the control rack inside: to gain access to this rod, the side cover must be removed.

3. Place the removed pump on the bench, and disassemble each pumping element in the following manner:

NOTE: During this disassembly, be sure to measure the thickness of the adjusting shims (indicated as (17) in the exploded view) and write down the reading and also the number of shims as reference data for reassembly.

4. Remove delivery valve holder (1).

5. Pick out delivery valve spring, valve (2) and O-ring.

6. Remove gasket and delivery valve seat.

7. Straighten the lock plate, which restrains the tappet guide pin; push in the tappet (18) just a little and pull off the guide pin with pincers. Take out the tappet.

8. Shims (17) and lower seat (16) will come out.

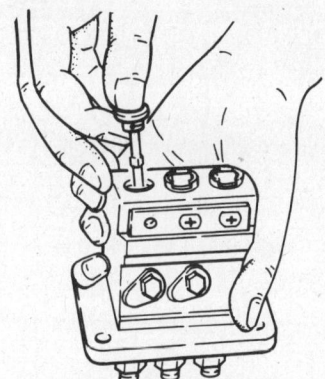

Removing plunger

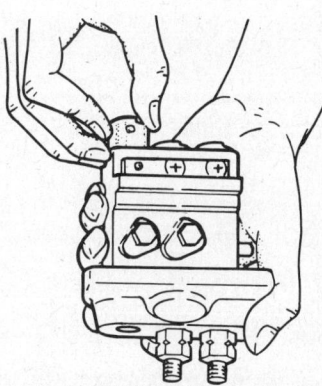

Removing tappet

ASSEMBLY AND INSTALLATION

1. Insert the barrel into the pump housing by aligning its notch with the dowel of adjusting plate.

2. Fit O-ring to valve holder.

3. Insert spring seat, gasket and valve assembly into the valve holder, and run the holder into the pump housing. With the wrench, tighten the holder in place to compress the O-ring fully.

4. Feed the control rack into the pump housing.

5. Install pinions, positioning each pinion in such a way as to index its tall tooth (sided by deep valleys) to the engraved line on the rack.

6. Insert the upper seat and its spring into each pumping element.

7. Combine plunger with lower spring seat, and insert the combination, bringing

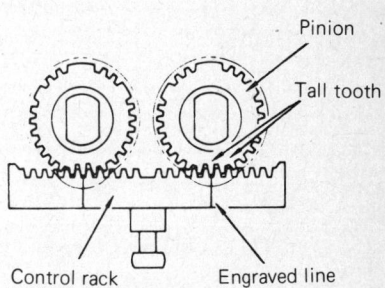

Meshing the rack with the pinions

757

Satoh

the "L" mark on plunger flange to control rack side.

8. Insert the tappets, each tappet being complete with shim. Be sure that guide pin holes in tappet and pump housing are aligned. Attach lock plate and, through the plate, insert guide pin. After installing the guide pins, lock the pins by bending the plate sharply.

9. Install the smoke-set device, positioning its return spring and washer correctly, and lock it by inserting split pin.

10. Adjust and set the reassembled injection pump in the following manner:

a. A proper amount of shim must be used on the mounting seat when positioning the pump in place. Engage the tie rod (linked to the governor lever) with the control rack, and fit the tie rod spring. Shim stocks are available in nine thicknesses: 0.01 to 0.04 in., the thickness increment being 0.004 in.

b. Reconnect the fuel feed line to the pump, admit fuel oil into the pump, and vent air out of this line by loosening the air vent screw.

c. In the present engine, fuel injection is prescribed to begin at 19° before top dead center. In other words, each pumping element of the pump is required to start delivering a slug of fuel when the piston in its corresponding cylinder comes to a position of 19° (crank angle) BTDC on compression stroke. This start, that is, injection timing, can be checked in the following way:

11. Take No. 1 cylinder as the reference. Remove the delivery valve holder, delivery valve and spring, and install the holder only, so that the fuel will continuously spill out from the holder.

12. Slowly turn over engine crankshaft by hand until the fuel ceases to overflow from the holder and, right then at the crank pulley, observe the timing mark to see if the piston (in no. 1 cylinder) is at 19° BTDC; if not, increase or decrease the thickness of the shim on the mounting seat.

13. The moment the fuel ceases to overflow corresponds to the start of injection. Increase the shim thickness to retard the timing and vice versa. Changing the thickness by 0.04 in. produces a change of about 1° in timing.

14. The start of injection can be checked at the end of injection pipe, with the delivery valve and spring in place and with the injection pipes properly installed: this is an alternate checking method. In this case, disconnect no. 1 pipe from its nozzle holder. Using a socket wrench at the crank pulley nut, gradually turn over engine crankshaft to let no. 1 pumping element force fuel out of the pipe. The moment the fuel starts swelling out of the pipe is the start of injection. This will occur approx. 1 degree behind the standard injection timing.

NOTE: After making sure that the injection timing is correct, install the tie rod cover. When installing the cover equipped with damper spring, keep the tie rod pushed in the direction of increasing the speed.

ADJUSTMENT

To set the high engine speed, proceed as follows:

1. With the damper spring in free state (the adjusting bolt backed), set the engine speed to 2840 rpm for the K3A-12 engine or 2740 rpm for the K3B-11 engine by means of the high-speed set bolt. After setting, lock the set bolt with locknut.

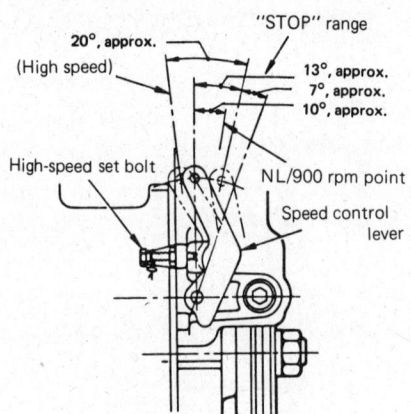
High speed adjustments

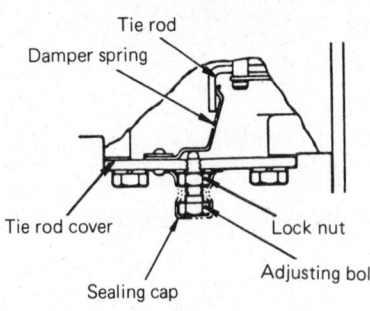
Damper spring adjustment

2. Run in the damper spring adjusting bolt to set the engine speed to 2850 rpm for the K3A-12 engine or 2750 rpm for the K3B-11 engine and lock the bolt with locknut. (Apply Super Three-Bond #20® to the threads of the bolt before locking.)

3. Seal the adjusting bolt with sealing cap.

4. Seal the high-speed set bolt with wire and cachet.

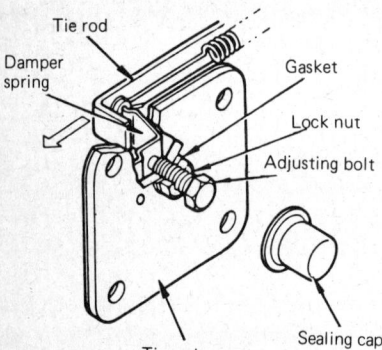
Installing tie rod cover

Nozzle Holders

DISASSEMBLY

1. From each nozzle holder, disconnect overflow pipe.
2. Similarly disconnect injection pipe.
3. Loosen nozzle holder securing bolts, and remove the holder assembly.
4. Break apart the nozzle holder assembly in the following manner:

a. Grip the holder body in the vise; put the wrench to the retaining nut and loosen the body. Use soft-metal pads (aluminum or copper) between vise jaws and holder body to protect the body when tightening the vise.

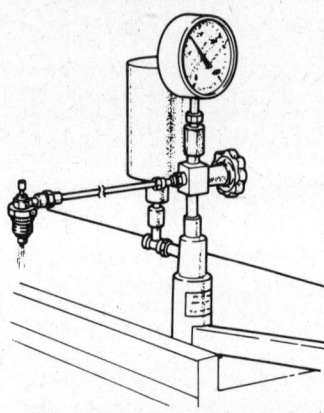
Start-of-injection pressure test

b. Take out shim washer, pressure spring, flange, pressure pin and distance piece in that order.

c. Take nozzle out of retaining nut. If the nozzle will not come easily, shake it loose by tapping on it lightly with a wooden mallet: be careful not to damage the needle valve part in the nozzle.

ASSEMBLY

1. Fit nozzle assembly, distance piece and pressure pin to retaining nut.
2. Fit shim, spring and flange to body, and tighten the body and retaining nut together by torquing to the prescribed value.

NOTE: The body may be gripped in the vise, but not the retaining nut. This is because the nozzle might suffer distortion if the nut is clamped between the vise jaws.

3. Fit gasket and nipple to the body.
4. When installing the nozzle holder assemblies, be sure to use new gaskets and tighten the securing bolts to the prescribed torque value.
5. Prior to installing the overhauled nozzle holder assemblies, test each for "start-of-injection" pressure, spray pattern, "after-injection" dribbling and fuel atomization.

a. Start-of-injection pressure test

A nozzle tester must be used to determine the pressure at which the nozzle starts spraying. If the pressure noted on the nozzle under test is at variance with the specification, increase or decrease the shim thickness. Changing the thick-

Satoh

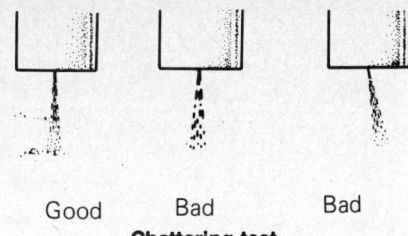

Good | Bad | Bad
Chattering test

ness by 0.04 in. changes the start-of-injection pressure by about 142 psi.

b. Spray pattern test

Operate the tester lever rather slowly to see if the nozzle shoots out fuel intermittently. A small amount of fuel is discharged as will allow the needle valve to chatter and, therefore, spray intermittently with some low-tone sound. The test is often called "chattering test." A good spray is characterized by fine atomization and straightforward jetting.

c. After-injection dribbling test

See if the nozzle dribbles after each injection. A dribbling nozzle must be replaced. In the chattering test, fuel might ooze out to form a globule of fuel at the nozzle tip but, since this is due to the chattering action of the needle valve, such a globule need not be taken as a cause of nozzle replacement.

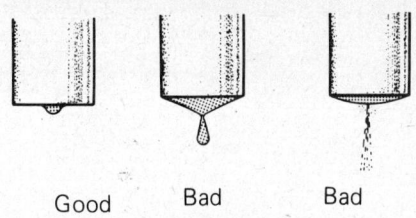

Good | Bad | Bad
Dribble test

d. Atomization test

Operate the tester lever rapidly, at a rate of about 800 strokes per minute to make the nozzle spray out with full force. Visually observe the spray to see if it consists of uniformly fine particles of fuel, straight in direction and having no fissures.

Governor

DISASSEMBLY AND ASSEMBLY

1. Remove fan belt.
2. Remove crank pulley nut, and take off the pulley.
3. Remove the fuel injection pump.
4. Remove the gear case.
5. Take out governor spring, taking care not to disfigure the spring.
6. Remove nut, washer and governor spring lever; and take out speed control lever from the gear case.
7. Remove nut, washer and spring lever; loosen the bolt securing the governor lever; and remove the lever.
8. Remove the governor weight assembly and sliding shaft from the pump camshaft.
9. From the governor lever, take off the tie rod and spring.
10. Assembly is the reverse of disassembly.

S-550G, S-650G

FRONT AXLE

REMOVAL

1. Raise the front of the tractor by placing a jack under the clutch housing: remove the front tire.
2. Remove the drag links and tie rod and then remove the steering levers. After removing the keys, take out the right and left king pins.
3. Pry off the cotter pin, remove the slotted nut and drive out the center pin while tapping it with a soft hammer. The front axle can then be taken out.

INSTALLATION

1. Apply grease to the bore in the front axle through which the center pin is inserted. Install the axle with the pin. Note the direction of the axle.
2. After making sure that the center pin is locked firmly, put the washer (2) and slotted nut (3) on the end of the pin and tighten the nut to specification. Secure the installation with the cotter pin (4).
Tightening torque: 32.5 ft.lb.
3. After the above steps have been completed, liberally apply grease to the pin and pin bore.

Front Hub

REMOVAL

1. In general, the front hub may be disassembled with the king pin installed on the front axle.
2. Raise the front of the tractor by placing a jack under the front axle. Raise the front just enough to remove load from the wheel tires. Without disturbing the above setup, remove the bolts which secure the wheels. Remove the tires.
3. Remove three M6 bolts (21) and take out the cap (20). Withdraw the cotter pin (19) and remove the nut (18).
4. With the use of a puller, draw out the front hub.
5. Remove the oil seal collar (12).

1. King pin
2. King pin bush (A)
3. Thrust bearing
4. M42 O-ring
5. O-ring bush
6. Key
7. Steering lever
8. M24 O-ring
9. King pin bush (B)
10. Front wheel hub
11. Ball bearing
12. Oil seal collar
13. Shim
14. Oil seal
15. Liner
16. Ball bearing
17. Washer
18. Slotted nut
19. Cotter pin
20. Cap
21. M6 screw
22. Gasket
23. Front axle

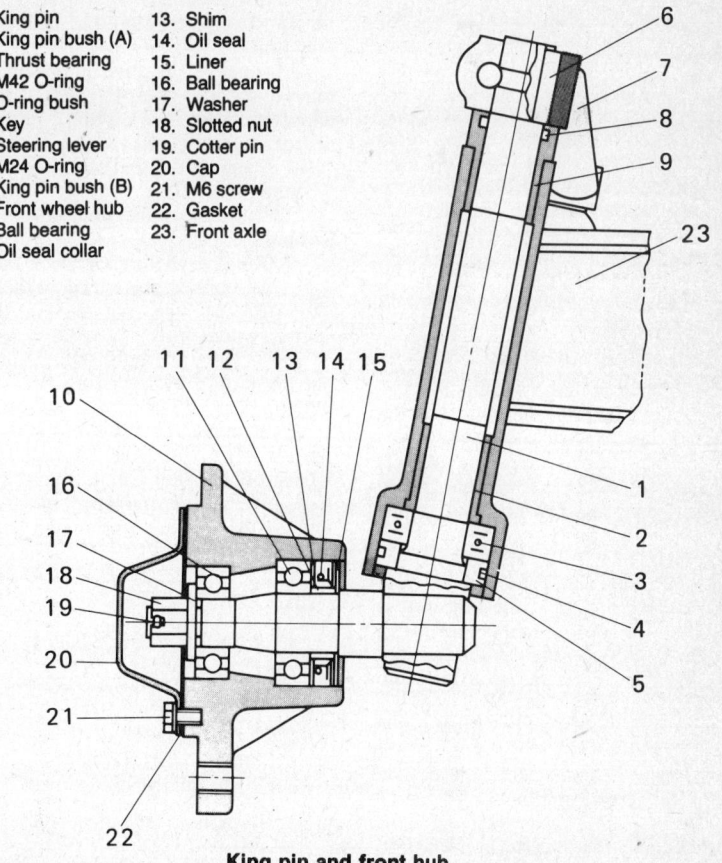

King pin and front hub

759

Satoh

INSTALLATION

1. Insert the liner (15) and oil seal collar (12) into place in the hub in this written order.
2. Assemble the ball bearing (11), shim (13) and oil seal (14) into the hub; install the above assembly to the king pin. Before installing, apply a coating or grease to the oil seal.
3. Hand pack all cavities in the hub with grease; insert fit the ball bearing (16).
4. Put the washer (17) on end of the wheel shaft; install the slotted nut (18) and torque to specification. Install the cotter pin (19) with the leg through the nut and bend the legs against the sides of the nut away from each other.
   Tightening torque: 32.5 ft.lb.
5. Place a gasket (22) on the cap (20) and tighten them to the hub with three M6 bolts (12).
   Tightening torque: 3.61 ft.lb.
6. Rotate the hub several times by hand to see that it rotates freely without binding.

King Pin

REMOVAL

1. Raise the front of the tractor by placing a jack under the front axle just enough to remove load from the axle. Without disturbing the above setup, remove the bolts that secure the front wheel. Further raise the axle and remove the front tire.
2. Pry off the cotter pin, remove the nut and, using a special tool tie rod end remover TRH-12, remove the ball socket. The drag link will then be disconnected from the tie rod.
3. Remove the steering lever attaching bolt; take out the lever.
4. Take out the key. While holding by hand the king pin from the bottom, give light hammer blows to the top of the king pin until it comes out thoroughly. Use a soft plastic hammer to drive the pin out.
5. When the king pin bushing is to be removed, use a special tool king pin bushing remover CS-5581ST.

INSTALLATION

1. Press the O-ring bushing (5) into place in the king pin (1) from the top.
2. Assemble the thrust bearing (3) in the king pin after coating it with grease.
3. Slide the king pin bushing (A) (2) into the king pin as far as it will go. Grease the M42 O-ring (4) and install it to the O-ring bushing.
4. Apply grease to the entire outer surface of the king pin and install it to the front axle. Use caution so that the key way is facing toward the rear.
5. Put the M24 O-ring (8) on the king pin bushing (B) (9); slide the bushing into the king pin. Install the steering lever (7) with the key way lined up with that in the king pin; install the key (6). Insert the bolt through holes in the steering lever and tighten to the correct torque.
   Tightening torque: 36.1 ft.lb.
6. Rotate the king pin several times to see if it moves freely without binding.

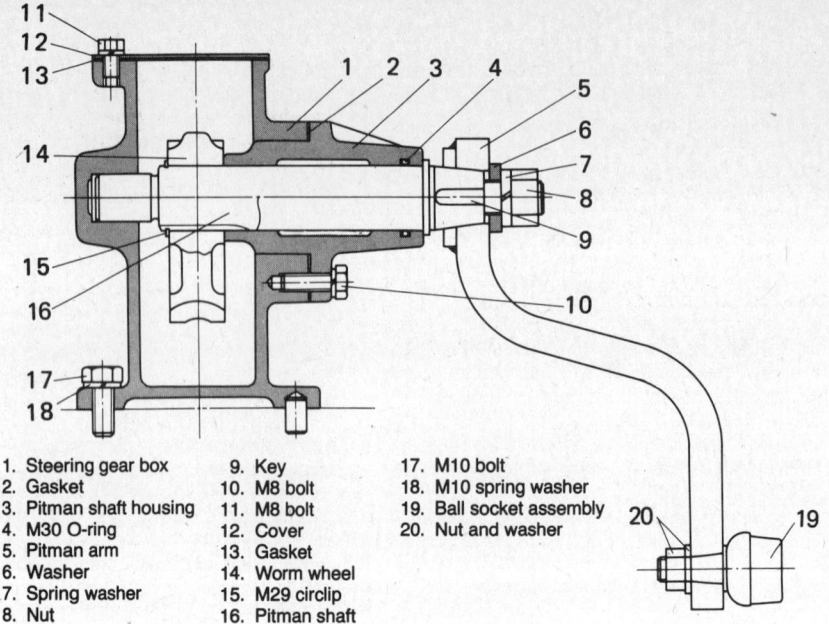

1. Steering gear box
2. Gasket
3. Pitman shaft housing
4. M30 O-ring
5. Pitman arm
6. Washer
7. Spring washer
8. Nut
9. Key
10. M8 bolt
11. M8 bolt
12. Cover
13. Gasket
14. Worm wheel
15. M29 circlip
16. Pitman shaft
17. M10 bolt
18. M10 spring washer
19. Ball socket assembly
20. Nut and washer

Steering gear components

STEERING

Steering Gear

REMOVAL

1. Disconnect the positive cord from the battery terminal.
2. Remove the drag link from the Pitman arm.
3. Remove the fuel pump circlip and remove the fuel pipe. Exercise care not to allow fuel from flowing out from the open end of the fuel pipe.
4. Remove the fuel tank and fuel tank bracket.
5. Disconnect the choke wire at the carburetor.
6. Remove the tractor meter cables.
7. Disconnect the wiring at the socket.
8. Remove the ring nut and then the starter switch.
9. Remove the light switch knob; take out the switch by removing the attaching nut.
10. Remove the cap from the steering wheel. After withdrawing the cotter pin, loosen off the slotted nut. Using a special tool universal steering wheel puller T-60B, remove the steering wheel.
11. Remove two bolts that secure the instrument panel; remove the panel.
12. Remove two bolts and separate the steering gear box from the clutch housing.

DISASSEMBLY

1. Remove the steering column support to the steering column support set bolt; take out the column and support.
2. Unscrew four M8 bolts securing the cover to the top of the gear box. Remove the cover.
3. Remove the Pitman arm by taking out the attaching nut. Loosen off four M8 bolts (10).
4. Pry off the circlip (15); take out the Pitman shaft from the gear box together with the Pitman shaft housing (3).
5. Pull the Pitman shaft off the Pitman shaft housing (3).
6. Take out the worm wheel from the gear box.

ASSEMBLY AND INSTALLATION

1. Set the O-ring in the Pitman shaft housing (H).
2. Apply a thin coating of grease to the inner surface of the Pitman shaft housing; install the shaft.
3. Apply grease to the bore in the steer-

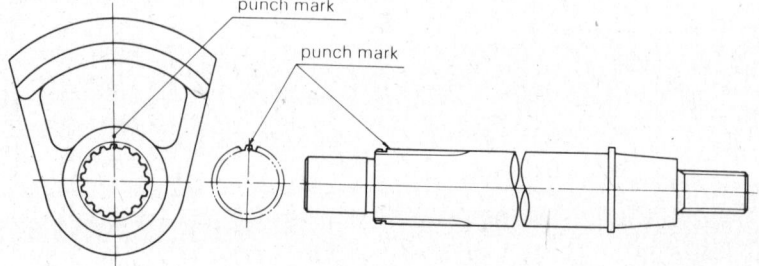

Aligning punch marks

ing gear box through which the Pitman shaft is inserted. Place a gasket on the Pitman shaft housing. Align the punch mark on the shaft and worm wheel in the steering gear box. Set the circlip (15) on the Pitman shaft to secure the Pitman shaft and worm wheel. Insert the above units to the bore on the steering gear box and attach the housing to the gear box with four bolts (10).

Tightening torque: 17.3 ft.lb.

4. Press the ball bearing into place on the steering column and assemble it into the gear box while engaging the worm with the worm wheel.

5. Place the collar (10) on top of the ball bearing. Attach the column support together with gasket to the steering gear box with the attaching bolts.

Tightening torque: 17.3 ft. lb.

6. Set the dowel pin on the clutch housing and install the steering gear box and tighten the attaching bolt together with fuel tank bracket to the correct torque.

Tightening torque: 36.1 ft. lb.

7. Pour 0.029 gal of SAE 80 gear oil into the gear box; install the cover (14) with the gasket under it.

8. Apply a coating of grease to the tapered end of the Pitman shaft and assemble the Pitman arm together with the key. Tighten the attaching nut to the correct torque.

Tightening torque: 110.1 ft.lb.

9. Install the instrument panel and then the starter switch and lighting switch. Connect the tractor meter cables.

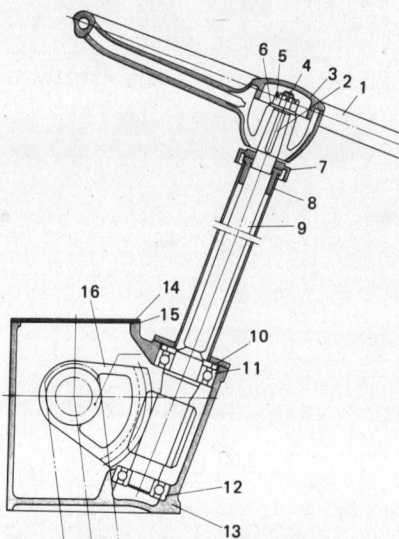

Worm wheel components

1. Steering wheel
2. Key
3. Steering wheel cap
4. Cotter pin
5. Slotted nut
6. Washer
7. End cap
8. Column
9. Steering Mast
10. Collar
11. Ball bearing
12. Ball bearing
13. Steering gear box
14. Cover
15. Gasket
16. Worm wheel

10. Install the steering column bushing and attach the cover.

11. Apply a thin coating of grease to the tapered end of the steering mast; Align keyway in the column with that in the wheel; insert the key in the key ways. Put the washer and castle nut on the end of the column and tighten the nut to specification. Insert and bend the cotter pin to secure the installation.

Tighten torque: 32.5 ft.lb.

12. Install the throttle control rod.

13. Install the fuel tank and connect the fuel pipe to the fuel pump. Secure the pipe with the pipe clamp.

14. Connect the choke wire.

15. Connect the drag link to the Pitman arm.

16. Test and adjust toe in: 0.236 in.

BRAKES

Brake Drum

REMOVAL

1. Loosen the rear axle wheel boss nut and jack up the rear. Remove the nut and the rear axle.

2. Remove the rear fenders. The battery should first be removed to remove the left fender.

3. Disconnect the brake adjust rod from the brake camshaft arm.

Removing brake drum

4. Unscrew the bolts that tighten the brake cover; remove the cover.

5. Remove the circlip and pull out the brake drum.

6. Remove the brake shoes.

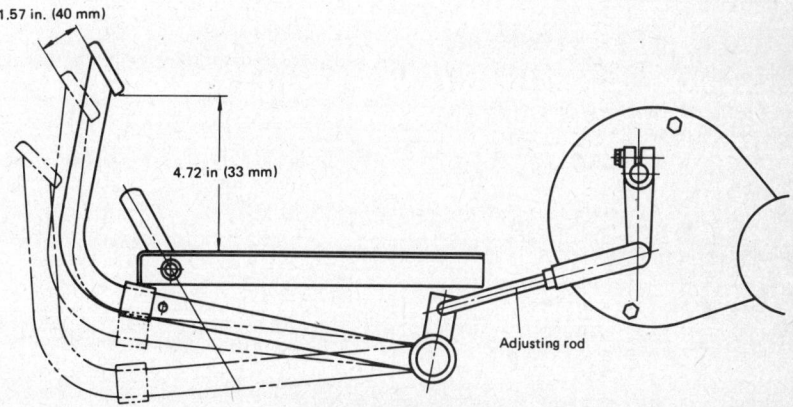

Brake adjustment

INSTALLATION

Installation is the reverse of removal. Note the following points:

1. Check the brake drum for any flaw on the surface, and if tpe flaw is so serious that the braking efficiency is reduced, grind the drum surface. If the grinded drum inner diameter exceeds 4.507 in. while the standard drum inner diameter is 4.488 in., replace the drum with a new one.

NOTE: The inside of the drum must be kept from oil, and grinded with a fine sandpaper in case it is again used.

Assemble the brake drum to the pinion shaft and secure it with the circlip.

2. The surface of the brake lining must be always clean and in even contact with the drum.

If it is found that the lining is applied to only a part of the drum, find the fault and correct it. The lining is 0.196 in. in width, and if it becomes less than 0.147 in. due to wear, replace it.

NOTE: When replacing any lining, always replace the left and right ones (four).

3. Put two M18 O-rings on the brake camshaft. Apply a thin coating of grease to the camshaft and install it in the brake cover.

4. Attach the brake return spring to the brake shoes and assemble them to the brake cover.

5. Install the brake camshaft arm to the brake camshaft and tighten with attaching bolt. Tightening torque: 17.3 ft.lb.

6. Install the brake camshaft so that the center of the brake arm is slightly behind the brake cover attaching bolt.

7. Install the brake cover to the brake housing with the gasket in between. Align the dowl with the dowel hole. Install the cover bolts and tighten to the specified torque:

Tightening torque: 17.3 ft.lb.

ADJUSTMENT

1. Adjust the right and left brake adjust rods so that both brakes are applied evenly.

2. Start the engine and rotate the rear wheels at 1,000 rpm at 5th gear, and depress the brake pedal. Make sure that each

rear wheel is locked at the same time, and tighten the locknut (F).

NOTE: When making the brake test after jacking-up, pay attention to car's surroundings and rotate the wheels gently.

3. Make sure that the parking brake lock plate operates smoothly.
4. As an overall test make a running test to see whether or not each wheel is braked evenly.

CLUTCH

REMOVAL AND INSTALLATION

1. Separate the tractor.
2. Remove the clutch by backing off the attaching bolts in criss-cross fashion to prevent warpage.
3. The disc should be carefully inspected to make sure that it is in good condition. Do not get any trace of oil or grease on the disc facings. Even small traces may cause clutch grabbing or slipping.
4. If the facings are worn down excessively so that the recession of the rivet heads from the facings is 0.0078 in. discard the disc and install a new one.
5. When a new disc is to be installed, make sure that the recession is 0.039-0.047 in.
6. Assemble the clutch disc to the flywheel center using a special tool. Note the pressure plate assembly and flywheel dowel.
7. Install and tighten the plate bolts in criss-cross pattern and in two or more steps to exert even pressure and avoid distortion. Tightening torque: 13.0 ft.lb.

ADJUSTMENT

1. The clutch pedal free play should be adjusted from time to time to compensate for natural wear on the facings. If there is excessive pedal play, even full movement of the pedal to the floor board will not force the clutch release bearing in against the diaphragm spring. The correct pedal free play is 0.80 in.
2. The free play can be adjusted by loosening the locknut (G) and turning the adjust bolt (D) either in or out as necessary.

The clutch stroke can be adjusted with the adjust bolt (E) so that it is 2.4 in. Tighten the locknut (H) to secure the adjustment.

TRANSMISSION

Gear Case

REMOVAL

NOTE: The steps to follow assume that the engine has been removed from the tractor chassis.

1. Loosen the rear wheel tightening bolts. Jack up the rear of the transmission and place a rack under the rear axle so that the rear wheels are clear of the ground (right and left).
2. Remove the battery and fenders.
3. Return the brake pedal; remove the spring. Remove the steps.
4. Remove the brake adjusting rod and withdraw the spring pin using the special tool spring pin rod guide CS55570. Remove the brake pedal and brake cross shaft.
5. Referring to Photo below, loosen the bolts, remove the pin and take out the operator's seat and tool box.
6. Unscrew the eight M10 bolts and remove the hydraulic case assembly. Use caution to avoid bending the upper cover excessively during operation.
7. The bolts are peculiar to their respective holes in length. When removing the case, be sure to keep the removed bolts in a part rack so that they can be placed back to their original locations from which they were removed.
8. Remove the upper cover complete with the select shifting mechanism.
9. Remove the thrust bearing carrier.
10. Jack up the clutch housing just enough to support its weight. Place the same rigid rack as was used in Engine Removal under the transmission case. Lower the jack carefully.
11. Again place the jack under the clutch housing to support its weight. Attach a lifting sling to the clutch housing and, using an overhead hoist, slightly lift the housing. Without disturbing the above setup, remove the eight clutch housing to transmission holddown bolts in criss-cross pattern. Slowly swing the housing away from the transmission case.

NOTE: Removal of the steering gear box should be determined according to the working condition.

12. Lift the transmission away from the tractor and lower it on the work bench or a clean surface.

DISASSEMBLY

1. Remove the final reduction case attaching bolts, starting from the lower bolt toward the one shown. Take out the reduction case.
2. The engine need not be removed from the tractor when the shaft alone is to be disassembled. To disassemble, remove the bolts securing the right side bearing holder in place to the transmission case. Move the select shaft toward the rear by applying careful hammer blows to the end as shown.

Removing final reduction case

Removing select shaft bearing holder stopper bolt

3. Take out the auxiliary transmission gear while moving the select shaft toward the rear.
4. Remove the circlip which retains the select shaft in place; take out the gear together with the shaft.

NOTE: Be sure that the ball bearing is not damaged while pulling out the select shaft.

Keep the auxiliary transmission shift lever in the neutral position. Neglecting this caution causes the shifter stopper to pop out since it is spring loaded. Do not drop the needle roller bearing from end of the select shaft during operation.

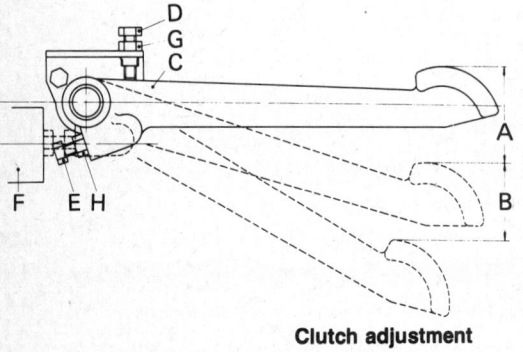

A. 0.80 in.
B. 2.40 in.
C. Clutch pedal
D. Clutch pedal stop bolt
E. Clutch stopper bolt
F. Clutch pedal stopper
G. Locknut
H. Locknut

Clutch adjustment

Satoh

Removing bearing holder

Removing PTO sub-shaft

5. Loosen there M8 bolts holding the bearing holder to the transmission case. Using a soft hammer, drive out the drive-shaft from inside the transmission case.

NOTE: The engine need not be removed when only the differential gear is to be removed.

6. Remove the final reduction case. Straighten the tabs of the lock plates and unscrew the bearing holder mounting M8 bolts.
7. Using forcing bolts, withdraw the holder by tightening the bolts.
8. Remove the differential gear assembly.
9. Remove the ball bearing at the side of the differential gear case.
10. Straighten the tabs of the lockwashers and loosen off the spiral bevel ring gear to the differential gear case bolts; take out the gear.

Differential gear center pin stopper screw

11. Unscrew the center pin securing screw and pull out the center pin. The differential gear can then be disassembled.
12. Remove the differential gear and then four M8 bolts that secure the countershaft end plate.
13. Straighten the tab of the lockwasher and remove the sleeve nut.
14. The countershaft can be taken out by lightly tapping it with a soft hammer. Take out the tapered roller bearing, thrust washer and gears while the shaft is being taken out.

PTO Shaft
NOTE: The PTO shaft can be taken out without removing the engine from the chassis.

1. Remove the hydraulic case assembly. Then, pry off the circlip at the differential gear side. In like manner as above, work the PTO gear retaining circlip out of place from the groove and slide it over the shaft slightly toward the front.
2. Remove the stopper ring retainer circlip; take out the stopper ring.
3. Drive the PTO shaft with a brass hammer from the front of the vehicle.
4. Remove the circlip at the differential gear side. The PTO shaft can be taken out by lightly tapping it with a shaft hammer complete with the ball bearing.
5. Remove the circlip, ball bearing and O-ring bushing from the PTO shaft in the order listed.

PTO Sub-Shaft
1. Remove the front and rear PTO shaft stopper plates from the transmission. This can be made by removing M8 bolts that secure them in place. Then, using a suitable wrench, back off four M10 bolts. The lower cover can then be taken out from the transmission case.
2. Reaching from the rear of the transmission, give light hammer blows to the PTO sub-shaft until it clears the case. When the bearing is to be removed, use a special tool bearing remover CS5575ST.
3. Working through the bottom opening, pry the circlip out of the groove. The PTO shaft can then be taken out by lightly tapping it with a brass hammer toward the rear. With the use of a gear puller, pull off the ball bearing together with the gear.

Reverse Gear
1. Remove four M10 bolts that attach the reverse gear on the right side of the transmission case.
2. Using a special tool "Spring Pin Rod Guide", pull off the spring pin and reverse gear shaft from the gear support in the following order.
3. Remove the band securing the dirt excluder; lift the excluder upward. Pull the shift lever straight-up after removing the M40 circlip.
4. The shift forks can be taken out easily by removing four nuts that tighten the cover to the transmission.
5. Pry off the spring pin from the shift lever; remove the shift forks from inside the transmission case.
6. Be sure that the shifter stop pin and stopper spring are not scattered and lost.

ASSEMBLY
PTO Sub-Shaft
1. Press the bushing (5) into place in the ball bearing (7) and collar (6) on the PTO sub-shaft (1).
2. Put the T18 gear (2) in the sub-shaft with the toothed end facing the inside.
3. Fit the ball bearing (4) and bushing (5) assembly to the sub-shaft, being sure that the flange of the bushing (5) is facing toward the T18 gear (2).
4. Install the M25 circlip (8) to the inside of the sub-shaft. With the ball bearing, collar and gear assembled on the sub-shaft, insert the shaft from the rear of the transmission case.
5. Place the T20 gear (3) on the sub-shaft; secure the installation with the M25 circlip (8).
6. Put the roller bearing (9) on the sub-shaft. Note the direction of the bearing. Install the cap (10) and secure the plate (11) (thick one) with M8 bolt (12).

Tightening torque: 17.3 ft.lb.

7. Make sure that the sub-shaft is set in the transmission case properly.

NOTE: Care should be taken when performing Step 9 below when the sub-shaft is not installed properly.

8. Install the roller bearing in the direction as shown.
9. Seat the shaft in the bores in the transmission case by lightly tapping it from both ends. The operating position of the sub-shaft requires the use of an adjustment shim (14) between the holder (14) and bearing (4) so that the end float is 0-0.0157 in.
   a. Measure the distance from the rear end of the transmission to the ball bearing (4).
   b. Measure the thickness of the holder (14).
   c. Adjust by shim (15) so that the transmission end float is 0-0.015 in. The thickness of the shim is 0.0157 in.
10. Enter the shim (15) in the bore in the

Satoh

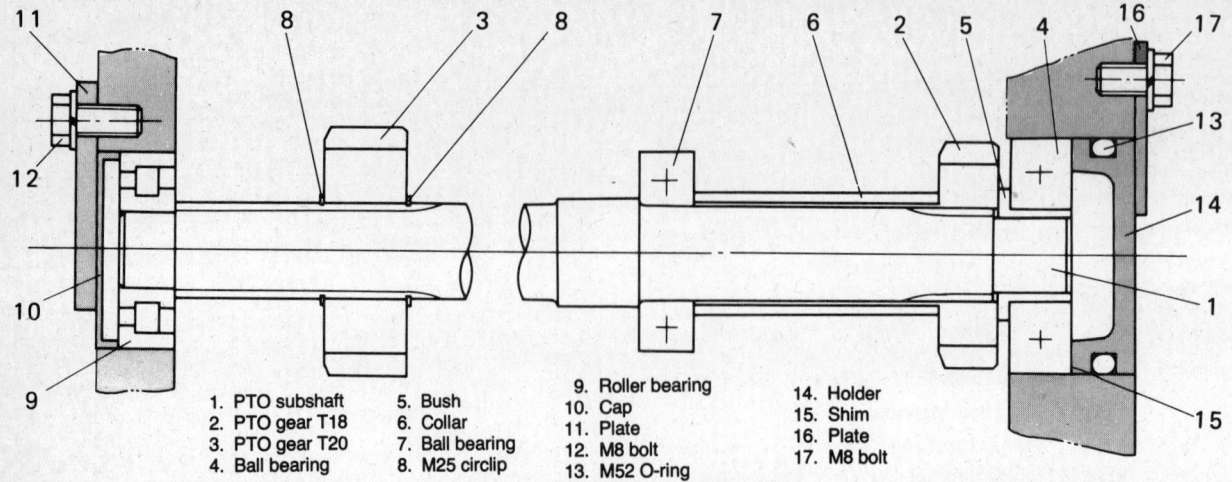

PTO sub-shaft components

1. PTO subshaft
2. PTO gear T18
3. PTO gear T20
4. Ball bearing
5. Bush
6. Collar
7. Ball bearing
8. M25 circlip
9. Roller bearing
10. Cap
11. Plate
12. M8 bolt
13. M52 O-ring
14. Holder
15. Shim
16. Plate
17. M8 bolt

transmission case. Liberally apply grease to the M52 O-ring (13) and put it in the groove in the outerpriphery of the holder (14). After placing the holder on the bearing, secure with the plate (16) and M8 bolts (17).

Tightening torque: 17.3 ft.lb.

PTO Shaft

1. Install the M62 circlip (5A) inside the transmission case.

2. The first operation is to install the oil seal collar (2) on the PTO shaft (1). Slide the M30 O-ring (3) over the shaft so that it is in the clearance between the collar (2) and shaft (1). Press the bearing (4) into place on the shaft (1) and secure with the stopper ring (6) and M36 circlip (7). Put the M30 circlip (8) in the groove in the shaft (1).

3. After installing the PTO shift mechanism, enter the PTO shaft in the transmission case from the rear. Put the gear on the PTO shaft while installing the T38 gear (9) and PTO shift fork.

4. Reaching through the rear of the transmission case, install the M62 circlips (5B). Put the ball bearing (1) on the PTO shaft from inside the transmission case; secure the installation with the M20 circlip.

5. Fit the oil seal (12) using a special tool oil seal installer CS5572ST.

Countershaft

1. Put the ball bearing (2), and liner (3) on the countershaft (1) in the order listed.

2. Press the outer races of the tapered roller bearings (21) and (22) into place in the bearing holder (20). Install the holder (20) under it on the transmission case, but do not tighten at this stage of assembly. Use two M8 bolts to temporarily secure the holder.

3. Assemble the liner (17) and tapered roller bearing (21) on the countershaft. Reaching through the inside of the transmission case, install the countershaft; press the tapered roller bearing (22) on end of shaft and secure with the nut (24). Do not tighten the nut (24) at this point of assembly.

4. Measure the centering distance to see if it is 4.1732 in. ±0.00197 in. Use a special tool counter gauge set CS5526T.

5. Adjustment requires the use of a shim between the transmission case and bearing holder. Shims (19) are available in 0.00393 to 0.019 in. thickness.

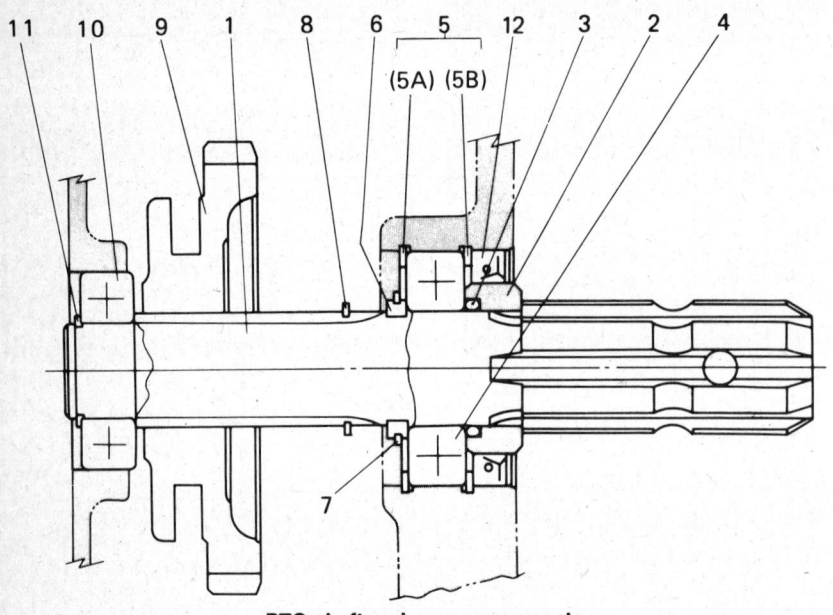

PTO shaft and gear components

1. Power take-off shaft
2. Oil seal collar
3. M30 O-ring
4. Ball bearing
5. M62 circlip
6. Stopper ring
7. M36 circlip
8. M30 circlip
9. T38 PTO gear
10. Ball bearing
11. M20 circlip
12. Oil seal

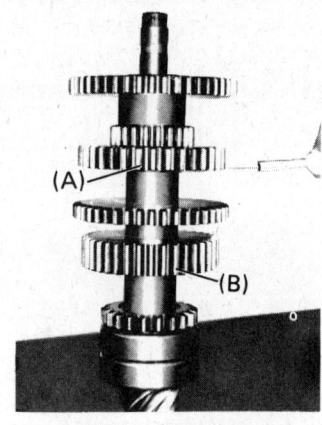
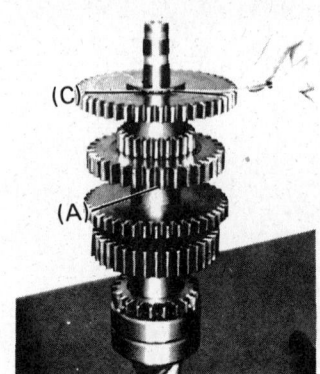

Measuring countershaft shim thicknesses

Shims available: 0.00393 in.
0.00984 in.
0.01968 in.

Countering distance: 4.1732 in. ± 0.00197 in.

6. Remove the nut (24) and pull out the countershaft.

7. Using a thickness gauge, measure the clearance at point "A" and note the required thickness of shim (6) to obtain 0-0.0098 in. clearance. The shim should be installed at the point "B" at assembly.

Shim thickness: 0-0.0098 in.
End float: 0-0.0098 in.

9. Measure the clearance at the point "C" to see if it is from 0.0039 to 0.0118 in. Three liners are available:
Liner thickness: 0.079 in
0.090 in.
0.102 in.
End float: 0.0039-0.0118 in.

10. Insert the countershaft from inside the transmission case. Put the gear (4), collar (5), shim (6), gear (7) and collar (8) in this written order.

Note the directions of the roller bearings and gears.

11. Press the outer race of the tapered roller bearing (21) into place in the bearing holder (20) until it bottoms. Apply a thin coating of grease of the M70 O-ring (18) and seat it in the groove in the outer-periphery of the holder (20).

12. After installing the centering adjust shim, press the roller bearing (22) into place on the countershaft (1). Install the washer (23) and nut (24) on the threaded end of the shaft and tighten so that the bearing preload is 3.972-4.333 ft.lb.

Countershaft bearing preload: 3.972-4.333 ft.lb.

13. Install the cover (25) on the transmission case with the gasket (27) under it. Install M8 bolts and tighten to 17.3 ft.lb.

14. Make sure that the countershaft and three front gears rotate freely.

Differential Gear

1. Set the differential side gears in the differential gear case with 0.0472 in. thrust liner (thinnest). Install the thrust liner to the pinion gear and set the center pin.

2. After two pinion gear align, pull out the center pin and then insert it in alignment with the center of the differential case. Use caution not to allow the thrust liner to come off during operation.

3. Measure the backlash if it is 0.0059 in.-0.0098 in. If the backlash is out of specification, correct by shim settled at the side gear to obtain a satisfactory unit.

Thrust liner availble: 0.0472 in.
0.0551 in.
0.0629 in.
Backlash: 0.0059-0.0098 in.

4. After ajustment has been made, set the center pin with the stopper screw.

5. Position the spiral bevel ring gear in place on the differential gear case with holes properly aligned. Install M8 high tension bolts in the holes with the lockwasher under them. Tighten the bolts to specification and bend the washers to prevent the bolts from being turned out during operation.

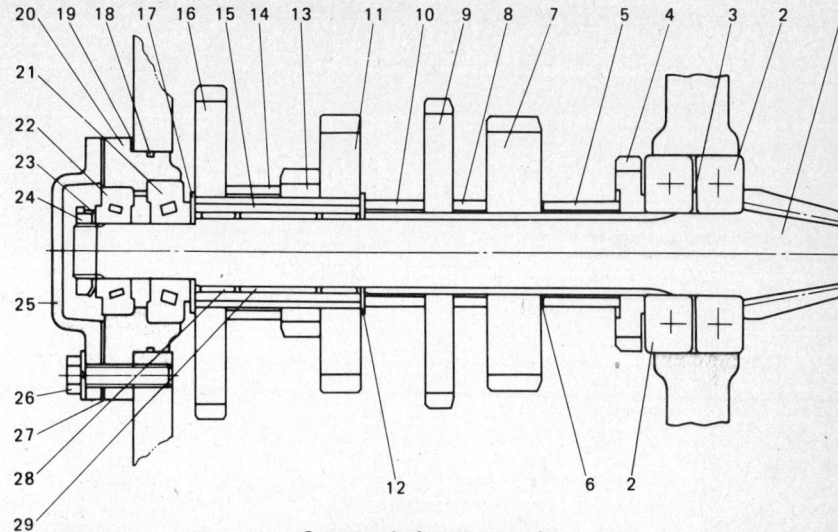

Countershaft components

1. Countershaft
2. Ball bearing
3. Liner
4. T21 gear
5. Collar
6. Shim
7. T37A gear
8. Collar
9. T42 gear
10. Collar
11. T33 gear
12. Thrust liner
13. T22 gear
14. Collar
15. Boss
16. T46 gear
17. Liner
18. M70 O-ring
19. Shim
20. Bearing holder
21. Taper roller bearing
22. Taper roller bearing
23. Tab washer
24. Locknut
25. Cover
26. M8 bolt
27. Gasket
28. Needle roller bearing
29. Collar

Tightening torque: 17.3 ft.lb.
Be sure to lock the center pin stop screw.

6. Install the ball bearing in the differential gear case.

7. Insert the spiral bevel gear in the left side of the transmission case; install the bearing holders from both sides of the case.

8. Align holes in the left bearing holder with those in the transmission case. Install M8 bolts in the holes with the lockwashers under them.
Tighten the bolts securely using a suitable wrench.

9. The right bearing holder uses special bolts equipped with a spring. Do not tighten these bolts so that the springs are fully compressed.

10. Adjustment of the spiral bevel gear backlash requires the use of an adjustment shim between the bearing holder and transmission case. To measure backlash, set up a dial indicator as shown, after tightening the bearing holder with bolts. Adjust by shim so that the backlash is within 0.0059 in.-0.0100 in.

Shims available: 0.0100 in.
0.01181 in.
0.01968 in.
0.003937 in.

NOTE: After backlash has been adjusted, tighten M8 bolts and lock with the lockwashers. With the use of a thickness gauge, measure the clearance between the right bearing holder and transmission case. A slightly thicker shim must be selected to eliminate undue load on the ball bearing.

11. Remove the special bolts; install the shim selected during the previous step between the bearing holder and transmission case. Install the M8 bolts with the lockwasher under them. After tightening the bolts to the specified torque, bend the lockwashers to prevent turning during operation.

Tightening torque: 17.3 ft.lb.

12. Clean the bevel gear and the transmission shaft pinion gear and apply red lead uniformly to 8-9 teeth on the bevel gear.

13. Giving resistance to the bevel gear, rotate the transmission shaft in the forward direction, thus making a pattern on the gear teeth.

14. Check the pattern on the bevel gear teeth, and if the teeth meshing is bad, make the following adjustments.

a. Heel meshing
Increase the thickness of the transmission shaft adjusting liner and allow the pinion to come near the bevel gear.
Keep the gear off the pinion and adjust the backlash.

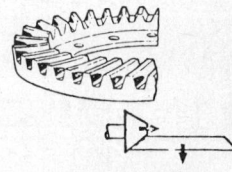

Heel meshing

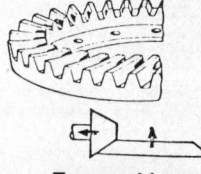

Toe meshing

Satoh

b. **Toe meshing**

Decrease the thickness of the transmission shaft adjusting liner and keep the pinion off the bevel gear.

Then allow the bevel gear to come near the pinion and adjust the backlash.

c. **Flank meshing**

Make the same adjustment as of toe meshing.

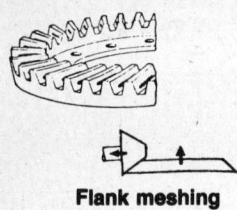

Flank meshing

d. **Face meshing**

Make the same adjustment as of heel meshing.

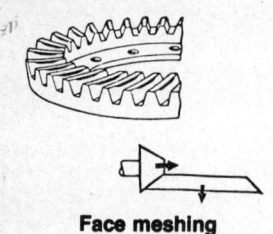

Face meshing

NOTE: One of the above will never occur independently; usually heel meshing and face meshing combined or toe meshing and flank meshing combined.

Repeat the above-mentioned adjustments until good teeth meshing can be obtained.

When the correct teeth meshing is obtained, make sure again of the bevel gear and the transmission shaft pinion backlash.

Backlash: 0.00787-0.00984 in.

NOTE: After adjustment, clean the teeth with gasoline.

Driveshaft

1. Put the ball bearing (2), liner (4) and ball bearing (3) on the driveshaft (1) in the

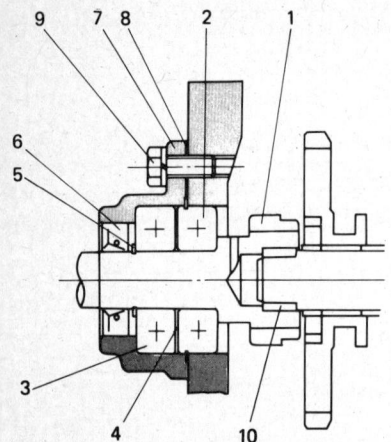

Driveshaft components

1. Driveshaft
2. Ball bearing
3. Ball bearing
4. Liner
5. M20 circlip
6. Oil seal
7. Bearing holder
8. Gasket
9. M8 bolt
10. Needle roller bearing

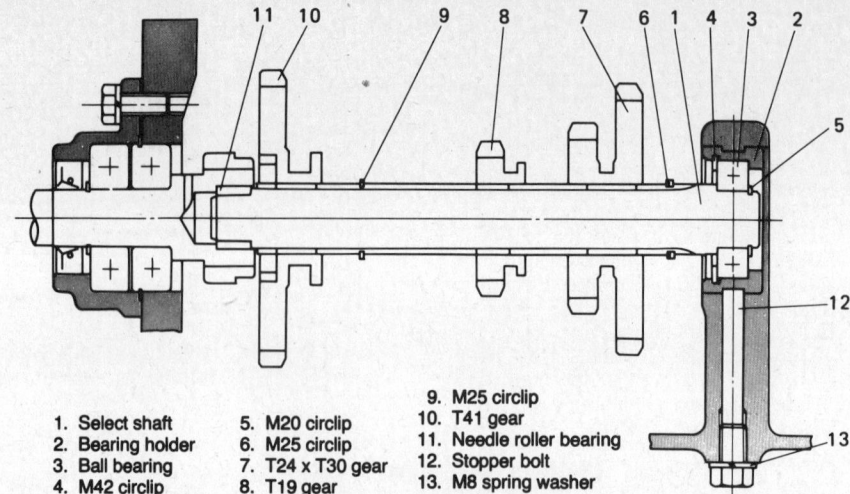

Select shaft components

1. Select shaft
2. Bearing holder
3. Ball bearing
4. M42 circlip
5. M20 circlip
6. M25 circlip
7. T24 x T30 gear
8. T19 gear
9. M25 circlip
10. T41 gear
11. Needle roller bearing
12. Stopper bolt
13. M8 spring washer

order listed; seat the circlip M20 in the groove in the shaft. Set the above assembly in the transmission case.

2. Apply a thin coating of grease to the oil seal (6) and install the seal in the bore in the bearing holder (7). Position the holder in place on the transmission case with holes properly aligned and tighten with M8 bolts (9). Be sure to install the gasket under the holder, exercising care not to damage the sealing lip during installation.

3. Grease the needle roller bearing (10) and install the needles around the shaft end.

Tightening torque: 17.3 ft.lb.

Select Shaft

1. After assembling the sub-shaft shift fork, press the ball bearing (3) in the bore in the bearing holder (2). Install the M42 circlip (4), being certain that it seats in the groove snugly.

2. Put the bearing holder on the end of the select shaft (1) with the bearing (3) inside. Install the M20 circlip (5) so that the holder and bearing are retained in place on the select shaft. Fit the M25 circlip (6) to the shaft.

3. Set the select shaft in the transmission case from the inside. Slide the T24 x T30 gear (7) over the shaft with the T24 gear facing the inside.

4. Fit the T19 gear (8) to the shaft and set the M25 circlip (9).

5. Dock the select shaft on the end of the driveshaft, paying attention to the T41 gear (10) and sub-shaft fork.

6. From the right side of the transmission, install and tighten the stopper bolt (12) with the M8 spring washer under it. Before installing, wrap the thread with a sealing tape.

Tightening torque: 36.1 ft.lb.

7. Check to be certain that the end of the stopper bolt falls in the groove in the bearing holder.

Reverse Gear

1. Put the needle roller bearing (3) in the T22 reverse gear (10) assemble the gear to the support (5) with the thrust liners (4) on both sides. Lock the shaft (2) in the holes in

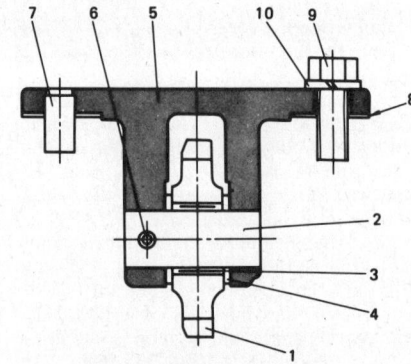

Reverse gear components

1. T22 reverse gear
2. Shaft
3. Needle roller bearing
4. Thrust liner
5. Support
6. Spring pin
7. Dowel
8. Gasket
9. M10 bolt
10. M10 spring washer

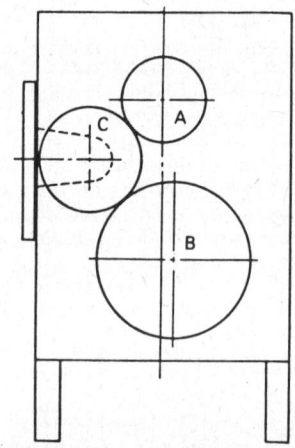

Backlash measurement points

the support (5) with the spring pin (6). Be sure that the pin and chamfered end of the gear (1) are on the front of the unit when installed on the transmission case.

2. With the chamfered end of the gear (1) facing the front, install the gear assembly on the case with the gasket (8) placed in

Satoh

between. Measure the backlash in gears using the setup shown.

3. Select the gasket which gives 0.0039-0.1181 in. backlash.
Tightening torque: 36.1 ft.lb.

4. Select the correct thickness of gasket so that the backlash is 0.0039-0.1181 in. between the gears A and C, and B and C.
Backlash: 0.0039-0.1181 in.
Gaskets available: 0.0118 in.
0.0236 in.
0.0393 in.

Shift Unit

1. Screw in M8 stud bolts (9) in the selector brackets (11) and (15).

2. Set the holder (17) on the select fork (12) with the pin (18); then, install the M14 circlip (19) to the select fork.

3. Set the holder (16) on the select fork (13) with the spring pin (18). Install the M14 circlip (19) in the groove in the select fork.

4. Enter the stopper spring (21) and stopper steel ball (20) in the hole in the bracket (11). Assemble the forks (12) and (13).

5. Install the selector bracket (15) and then the gear rocker (14). Bring the transmission into Neutral.

6. Apply sealant to both sides of the gasket (8) and place it on the upper cover (7) so that the selector bracket is facing the front. Tightening torque: 17.3 ft.lb.

Selector Lever

1. Set the selector lever (2) in the ball housing (1). Install the support plate (3) in the housing and secure with the M40 circlip (4).

2. With the oil grooves toward the rear, assemble the ball housing to the upper cover (7) with the gasket (6) under it. Tighten M6 screws (5) to 3.61 ft.lb.

3. Install the dirt excluder (26) and then the clamp (27).

4. Run down the M10 locknut (29) and screw in the knob (28) until it will no longer go.

Sub-Shift Unit

1. Assemble the selector fork (C) (2) into bore in the transmission case. Put the M16 O-ring (3) in groove in the fork (2).

2. After putting the plate (9) over the fork (2), assemble the range lever (1) to the fork shaft and secure with the spring pin (4).

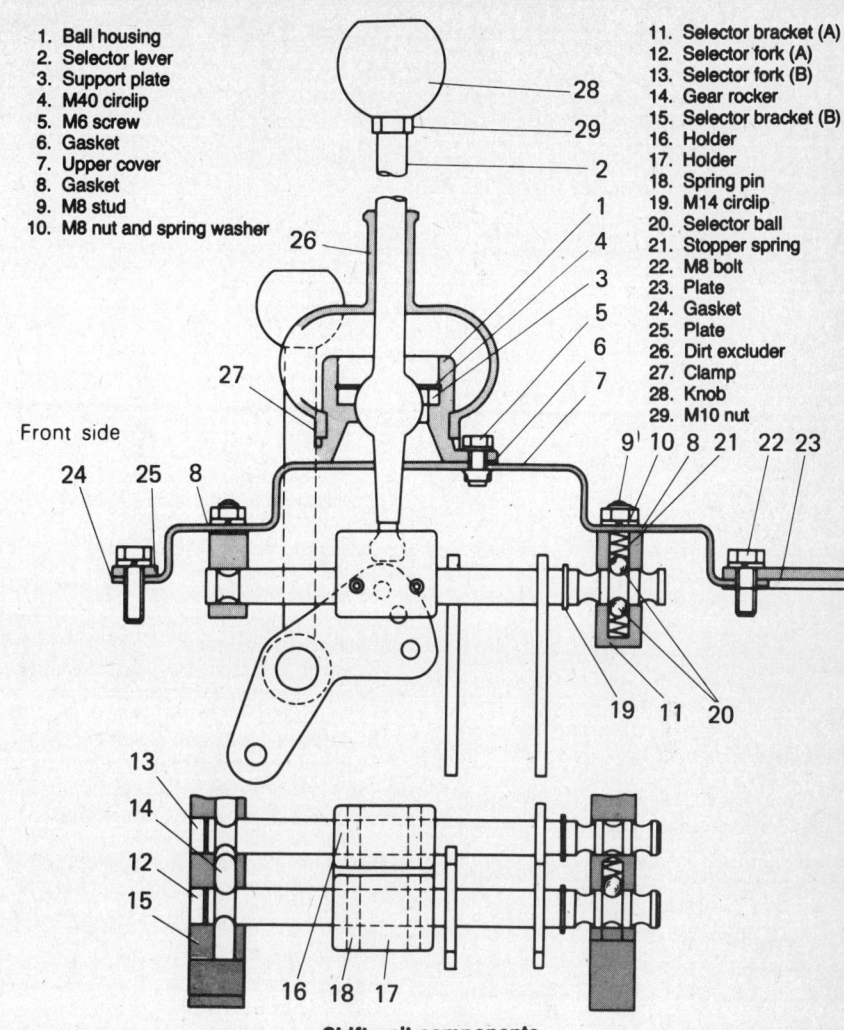

1. Ball housing
2. Selector lever
3. Support plate
4. M40 circlip
5. M6 screw
6. Gasket
7. Upper cover
8. Gasket
9. M8 stud
10. M8 nut and spring washer
11. Selector bracket (A)
12. Selector fork (A)
13. Selector fork (B)
14. Gear rocker
15. Selector bracket (B)
16. Holder
17. Holder
18. Spring pin
19. M14 circlip
20. Selector ball
21. Stopper spring
22. M8 bolt
23. Plate
24. Gasket
25. Plate
26. Dirt excluder
27. Clamp
28. Knob
29. M10 nut

Shift unit components

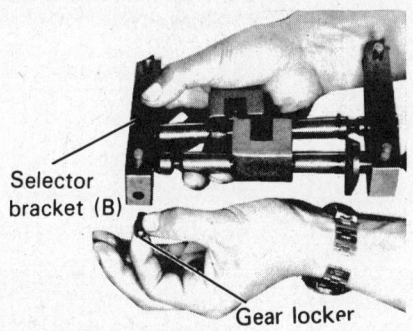

Gear locker

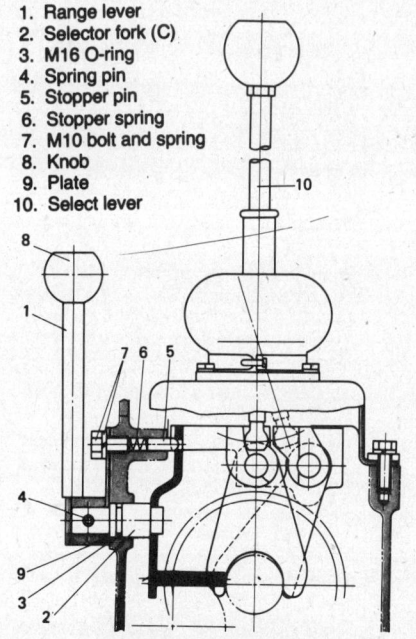

1. Range lever
2. Selector fork (C)
3. M16 O-ring
4. Spring pin
5. Stopper pin
6. Stopper spring
7. M10 bolt and spring
8. Knob
9. Plate
10. Select lever

Sub-shift components

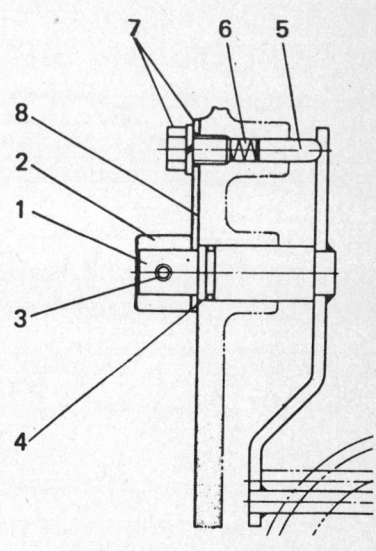

1. PTO selector fork
2. Selector lever
3. Spring pin
4. M16 O-ring
5. PTO select pin
6. Stopper spring
7. M10 bolt and spring
8. Plate

PTO shift unit components

Satoh

3. Install the stopper pin (5) and spring (6) in this written order. Wrap the threads of the M10 bolts (7) with vinyl tape and install them in holes in the plate (9) and case. Tightening torque: 36.1 ft.lb.

4. Install the knob (8).

PTO Shift Unit

1. Assemble the PTO shift fork (1) to the transmission case, install the M16 O-ring (4).

2. Set the selector lever (2) with the spring pin (3). This also holds the plate (8) in its place.

3. Install the PTO select pin (5) and stopper spring (6). Tighten the plate (8) with the M10 bolt (7). Before installing the bolt (7), wrap a sealing tape around the threads to prevent leaks. Tightening torque: 36.1 ft.lb.

Final Reduction Unit

DISASSEMBLY

1. Work the M40 circlip (3) out of place from the groove in the final driveshaft (1); remove the gear (2).

2. Using a brass or copper drift, drive out the axle shaft from the final reduction case.

3. When the ball bearing (7) and oil seals (10) and (11) are to be removed, use a special tool "Bearing Installer CS5550ST".

4. Using a suitable wrench, loosen off M8 nuts (11); remove the brake cover (9). Pry off the M26 circlip (8) and remove the brake drum (7).

5. Take out the M72 circlip (6) from inside the final reduction case. Remove the M72 circlip (6) and then the pinion shaft (1) by lightly tapping it with a soft hammer. Remove the oil seal (5) by removing the M72 circlip (6).

6. Remove the ball bearing (2) and oil seal bushing (3) from the pinion shaft (1).

ASSEMBLY

1. Install the M90 circlip (6) in groove in the final reduction gear case. Press fit the ball bearing (7), liner (13) and oil seals (10) and (11) on the outside of the gear case. Note the direction of the oil seals.

2. Press the plate (12) and oil seal collar (9) in the axle (1); install the M40 O-ring (8).

3. Assemble the axle shaft to the final reduction gear case.

4. Press fit the inner ball bearing (5), install the gear (2) and set the M40 circlip. Measure the clearance between the gear and circlip. If the clearance exceeds 0.0157 in. insert a liner (4) between the bearing (5) and gear (2).

End float: 0.0157 in.
Liner thickness: 0.0157 in.

5. Press the ball bearing (2) into place on the pinion shaft (1) unit it will no longer go. Assemble the M28 O-ring (4) and oil seal collar (3) in this written order.

6. Fit the M72 circlip (6) to the outside of the final reduction gear case; install the pinion shaft in position.

7. After installing the oil seal (5), set the brake drum (7) with the M26 circlip (8).

8. Place a gasket on the brake cover (9). Align the knock bolt with the hole in the cover. Install the attaching bolts (10) and tighten the nuts (11) to 17.3 ft.lb.

9. Attach the final drive case to the transmission case with the dowels and a gasket in between and plate at outside. Tighten the attaching bolts to the following torque:

Tightening torque:
M 8 bolt: 17.3 ft.lb.
M10 bolt: 36.1 ft.lb.
M14 bolt: 84.8 ft.lb.

10. Be sure that certain bolts should be wrapped with a sealing tape to prevent oil leaks through the threads.

1. Final drive shaft
2. Final drive gear
3. M40 circlip
4. Shim
5. Ball bearing
6. M90 circlip
7. Ball bearing
8. M40 O-ring
9. Oil seal bush
10. Oil seal
11. Oil seal
12. Washer
13. Liner
14. Final reduction case

Final driveshaft

HYDRAULIC SYSTEM

For complete service procedures, see the S-370 section.

ENGINE

NOTE: These tractors use the Mazda PB engine.

REMOVAL

1. Disconnect the positive and negative cables at the battery terminals.

2. Disconnect the head light cords and remove the bonnet.

3. Remove the bolts securing the radiator in place; loosen the water hose clip and pull out the water hose from the radiator. Remove the radiator and radiator water hose.

4. Remove all cables from the oil pressure switch (A), cooling water thermostat switch (B); alternator (C), ignition coil (D) and starting motor (E) at the terminals or cable joints or connections.

5. Disconnect the choke wire at the carburetor.

6. Remove the cover from the governor by backing off the attaching bolt; remove the throttle control rod.

7. Raise the fuel pump fuel pipe clip and remove the fuel pipe. Bend the open end of the pipe and tie with a string to prevent fuel from leaking from the end.

8. Remove the fuel tank by loosening the bolts which tighten the fuel tank bands.

9. Remove the fuel tank brackets.

10. Disconnect the meter cables at the tractor instrument panel.

11. Pull out the cotter pin from the ball socket connecting the Pitman arm and drag link and the slotted nut. Remove the ball socket by using the special tool (No. TRH-12 tie rod end remover).

12. Loosen the four nuts holding the silencer pipe to the exhaust manifold. Remove the bolt which secures the silencer in place to the brake housing. The silencer pipe and silencer can then be taken out.

13. Remove the M6 bolt securing the pipe to the hydraulic pump, exercising care not to damage the O-ring during the operation.

14. Loosen off the three M8 bolts holding the discharge pipe in place to the hydraulic lift case; take out the pipe in place to the hydraulic lift case; take out the pipe. Avoid damaging the O-ring when removing the pipe. Remove the pipe clamp and then take out the pipe.

15. Remove the suction pipe from the oil filter case. This can be done by removing the two bolts.

16. Remove the alternator strap bolt. Remove the alternator bracket and the bolt attaching the alternator. Then remove the alternator.

Satoh

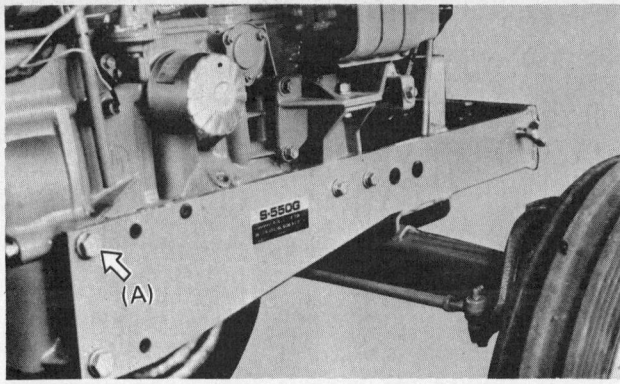

The arrow indicates the last chassis bolt to be removed prior to engine removal

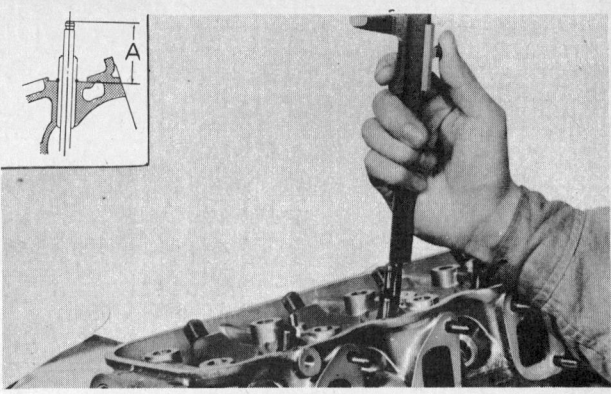

Valve spring seat lower measurement

17. Install the chain block on the engine hanger on the manifold on both exhaust and intake side. Do not hoist the engine but pull the wire so that it becomes slightly tight.

18. Place a floor jack under the clutch housing. Do not jack up the tractor. Just raise the jack so that it is just tightly locked under the tractor.

19. Loosen a total of eight M16 bolts holding the chassis to the clutch housing. (Remove the bolt (A) last and carefully swing the frame horizontally toward the front.)

20. Remove the bolt securing the engine support in place. Remove the bolts which hold the engine support to the governor bracket. Also remove the alternator bracket mounting bolt.

21. Remove the engine support after the chassis has been taken out. Remove the rear plate.

22. Remove the starter motor mounting bolt and remove the starter motor.

23. Remove the six nuts, mounting the engine and clutch housing. Next, adjust the chain block suspending the engine so that the stud bolt will not be under excessive load. Then pull the engine toward the front.

24. Remove the three mounting nuts and remove the alternator bracket (B) from the cylinder block.

25. Installation if the reverse of removal.

Cylinder Head

REMOVAL

1. Remove the pump.
2. Remove the intake and exhaust manifolds
3. Remove the distributor.
4. Remove the rocker arm cover.
5. Unscrew ten cylinder head nuts.
6. Pull upward rocker arm assembly.
7. Pull upward eight push rods.
8. Remove the cylinder head assembly from the cylinder block.
9. Remove the cylinder head gasket.

INSPECTION

1. To check the cylinder head bottom for bend, measure the points (1)-(6) of the cylinder head bottom, using a thickness guage and a surface gauge. If the measured value is beyond the limit of 0.006 in., correct it with a surface grinder.

2. Check the valve seat for wear and damage, and if it is defective, correct it with a special tool (No. 0490140 valve seat cutter). Besides, if the sinking of the valve seat is more than 0.059 in. replace the cylinder head with a new one.

Seat Cutter Pilot:
45° cutter (for IN side sheet)
45° cutter (for EX side sheet)
15° cutter (for IN side port)
15° cutter (for EX side port)
75° cutter (for IN side face)
75° cutter (for EX side face)

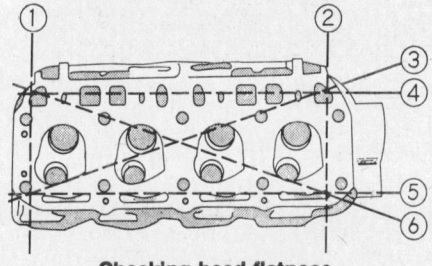

Checking head flatness

3. Check the valve face for wear and damage, and if it is defective, correct it with a valve face grinder. If the valve head is less than 0.039 in. thick, replace it with a new one.

NOTE: After grinding the valve sheet and valve face, insert and adjust the valve spring seat lower measurement (A) 1.596 in.

The valve spring seat lower is **0.020 in. thick.**

4. Check the diameter of the valve stem with a micrometer. If the valve stem is worn more than the limit of 0.2758 in., replace the valve with a new one.

5. Check the clearance between the valve stem (A) and valve guide (B). If the clearance exceeds the limit of 0.008 in., replace the valve and valve guide.

6. Check the valve spring for free length (1.596 in.) spring pressure (43.6 lb., closed) and squareness in the procedure outlined below and if it is defective, replace it with a new one.

a. Measure squareness of the valve spring with a level block and a square. Replace the spring if out of squareness is 0.118 in. per 3.937 in.

b. Measure the free length of the valve spring. Replace the spring if it decreases more than 3% of the standard dimension.

c. Measure the spring pressure with a valve spring tester. Replace the spring if the spring pressure under set condition reduces more than 15% of the standard dimension.

INSTALLATION

1. Place the cylinder head gasket on the cylinder block and install the cylinder head assembly.

2. Insert eight push rods into the tappet-floor with the longer ones into the EX side.

3. Assemble the rocker arm, rocker arm

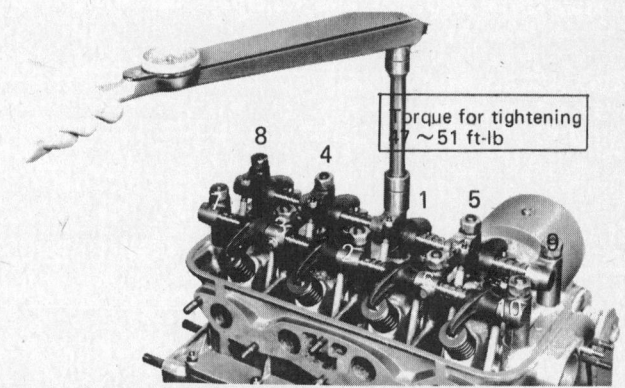

Head bolt tightening sequence

Satoh

shaft, rocker arm supporter, lock bolts, and washers.

4. Install the rocker arm assembly on the cylinder head assembly.

NOTE: Check the adjusting screws of the push rod and rocker arm.

5. Place ten cylinder head holding nuts and adjust the EX rocker arm supporter so that the center of the valve stem is by 0.04 in. shifted from that of the rocker arm.

6. Tighten the cylinder head holding nuts to 47-51 ft.lb. in the order shown.

Intake Manifold

REMOVAL AND INSTALLATION

1. Unscrew eight nuts and remove the engine hanger. Then remove the intake manifold from the cylinder head, together with the governor cover stay and the air cleaner stay mounted thereon.

2. Installation is the reverse of removal. Torque the nuts to 12-20 ft.lb.

Exhaust Manifold

REMOVAL AND INSTALLATION

1. Unscrew eight nuts, remove the engine hanger, and then remove the exhaust manifold from the cylinder head.

2. Installation is the reverse of removal. Torque the nuts to 7-14 ft.lb.

Timing Chain Cover, Adjuster and Chain

REMOVAL AND INSTALLATION

1. Undo nine nuts and remove the timing chain cover.

2. Undo two nuts and remove the chain adjuster assembly.

3. Remove the oil baffle plate from the crankshaft.

4. Remove the key from the crankshaft. Undo two nuts and remove the vibration damper assembly.

5. Undo two nuts securing the camshaft.

6. Remove the camshaft and cam sprocket wheel and the timing sprocket wheel together with the timing chain from the crankshaft and the cylinder block.

7. Set the first cylinder at top dead center.

8. Fit the camshaft and cam sprocket wheel and the timing sprocket wheel to the timing chain.

9. There are eleven lanes between both match marks.

10. When installing the sprocket wheel and timing chain to the cylinder block and crankshaft, be careful not to go wrong with the meshing of them.

11. Turn the cam sprocket wheel to match the key grooves of the crankshaft and the timing sprocket wheel, and fit the key.

12. Secure the camshaft and cam sprocket wheel to the cylinder block with two nuts.

13. Fit the oil baffle plate.

14. Secure the vibration damper to the front plate with two bolts.

15. Straighten the bent edges of the lockwasher and remove the plug (A) from the body (C).

16. Insert the spring (E) and inner cylinder (?) into the slipper head (F), and fit a hexagonal wrench (?) to the inner cylinder and rotate the wrench clockwise until the inner cylinder enters the slipper head completely.

17. Insert the slipper head into the body.

18. Secure the chain adjuster assembly to the cylinder block with two bolts.

19. Rotate the inner cylinder in the direction of the arrow, with the hexagonal wrench and loosen the slipper head.

20. Secure the lockwasher and the plug (A) to the body and bend the lockwasher around the plug.

21. Apply a bit of graphite grease to the oil seal (A) and secure the timing chain cover together with the packings to the cylinder block temporarily with nine nuts.

22. Rotate the crankshaft and ensure that it runs smoothly, and then tighten the timing chain cover holding nuts securely.

Pistons, Connecting Rod, and Crankshaft

REMOVAL

1. Undo twenty-two nuts securing the oil pan (B) and lift off the oil pan from the cylinder block.

2. Unfasten the lock bolts and remove the crankshaft pulley (A) from the crankshaft.

3. Remove the timing chain.

4. Loosen the connection nuts and detach the oil pipe from the cylinder block.

5. Unscrew two nuts and remove the oil strainer from the oil pump.

6. Unscrew four nuts and lift up the oil pump.

7. Remove the cylinder head.

8. Removing the connecting rod cap.

9. Loosen eight connecting rod lock bolts and remove four connecting rod caps (A) together with the connecting rod bearing metal and the lock bolts.

NOTE: The connecting rod caps are marked with cylinder numbers (1, 2, 3 and 4), respectively.

10. Straighten the bent edges of the lockwasher and ten nuts securing the main bearing caps.

11. Remove five main bearing caps together with the main bearing metal.

12. Remove the crankshaft from the cylinder block.

13. Remove the five main bearing metals from the cylinder block.

14. Lift up four piston and connecting rod assemblies from the cylinder block.

15. Remove three piston rings and the expander.

16. Remove the clips at both ends of the piston pin.

17. Heat the piston in an oven, to 250-300°F and pull out the piston pin by means of the special tool (No. 0490070 piston pin installer) and detach the piston from the connecting rod.

INSPECTION

1. Check the piston for crank and damage. If it is defective, replace it.

2. Measure the (A) thrust diameter just below the oil ring and (B) thrust skirt diameter.

Standard diameters of (A) and (B) are as follows:

(A) 2.6756 ± 0.0004 in.
(B) 2.6779 ± 0.0004 in.

3. Measure the allowance between the piston and piston pin. If the measured value

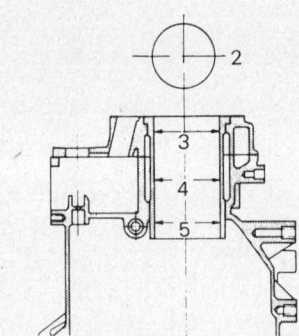

Checking the cylinder liner for wear at 6 points as shown

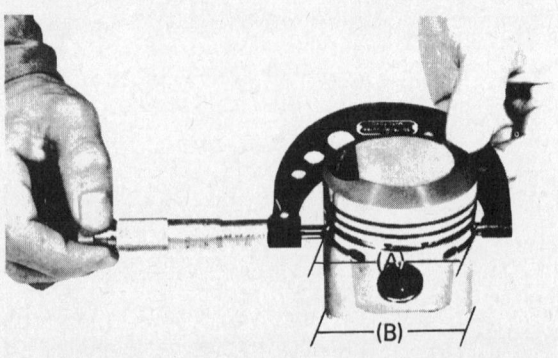

Measuring thrust diameter A and skirt diameter B

is not 0.0002-0.0009 in., replace the piston and piston pin.

NOTE: When fitting and removing the piston and piston pin, heat the piston to 250-300°F.

4. Measure the clearance between the piston ring and ring groove with a special tool (No. N026 standard feeler gauge). If the measured value exceeds the clearance limit, replace the piston ring.

Parts	Standard Clearance	Clearance Limit
Top ring	0.001-0.003 in.	0.006 in.
Second ring	0.001-0.003 in.	0.006 in.
Oil ring	0.001-0.002 in.	0.006 in.

5. Measure the piston ring end gap with a feeler gauge.
6. If the measured value exceeds the standard end gap of 0.008-0.016 in., replace the piston ring.
7. Check the inside of the cylinder for scratching, and if it is defective, replace the cylinder liner.
8. Check the inside of the cylinder liner for wear.
9. If the measured value is 0.006 in. more than the standard value below, replace the cylinder liner.

Cylinder Mark	Standard Value	Wear Limit
A	2.6772	0.006 in.
No Mark	2.6772	0.0006 in.
C	2.6772	0.006 in.

10. When removing and fitting the cylinder liner from and to the cylinder block, use a special tool (No. 0490090 cylinder liner puller).
11. When fitting the cylinder liner and top deck, fit them together bearing same marks in order to keep the amount of protrusion uniform (marks are classified into two, A and no mark, as shown in the following table). Marks are on the cylinder liner front and rear and on the top of the top deck.
12. The standard clearance between the cylinder and piston is 0.001-0.007 in.
13. The cylinders and pistons are classified into three (A, no mark, and C) and those with same marks are fitted together.
14. Check the clearance between the connecting rod small end bush and piston pin. If the clearance exceeds 0.0018 in., replace the piston pin and small end bush.
15. When pulling and press-fitting the small end bush, use a bushing tool and a press.

NOTE: When press-fitting the small end bush to the connecting rod, align the oil aperture of the small end bush with that of the connecting rod.

16. Measure the connecting rod large end play with a special tool (No. N026 standard feeler gauge), and it is good if the measured value is in the range of 0.004-0.008 in.
17. Check the connecting rod deflection with a conrod aligner, and if the measured value is not satisfactory, correct the connecting rod with a press or replace it.

Connecting rod deflection: less than 0.002 in. per 3.94 in.

Distance between the large end and small end: 4.84 ± 0.002 in.

18. Check the oil clearance between the connecting rod bearing metal and crank pin with a plastic guage in the following procedure. Clear the bearing metal and crank pin of dust, oil, etc. Place the plastigauge on the crank pin. Fit the bearing cap to the connecting rod and secure it to the crank pin with cap bolts at the specified torque of 25.3-28.9 ft.lb.

Loosen the cap bolts and remove the connecting rod and place the plastigauge. Oil clearance is 0.001-0.003 in.

19. Measure the diameter of the crank pin and main journal with a special tool (No. 75MB outside micrometor caliper).

In case wear is more than 0.002 in., correct the crankshaft by grinding to the undersize of 0.01, 0.02 or 0.03 in. with a crankshaft grinder.

Mark	Top Deck	Cylinder Liner	Protrusion
A	0.3346 $^{+0.0012}_{-0}$ in.	0.3346 $^{+0.0024}_{+0.0016}$ in.	0.0004-0.0024 in.
No Mark	0.3346 $^{+0}_{-0.0012}$ in.	0.3346 $^{+0.0016}_{+0.0006}$ in.	0.0006-0.0028 in.

Mark	Cylinder Liner Standard Value	Piston Standard Value	Clearance
A	2.6772 $^{+0.0008}_{+0.0005}$ in.	2.6756 $^{+0.0004}_{+0.0001}$ in.	0.0017-0.0022 in.
No Mark	2.6772 $^{+0.0005}_{+0.0002}$ in.	2.6756 $^{+0}_{-0.0001}$ in.	0.0018-0.0022 in.
C	2.6772 $^{+0.0002}_{+0}$ in.	2.6756 $^{-0.0001}_{-0.0004}$ in.	0.0017-0.0022 in.

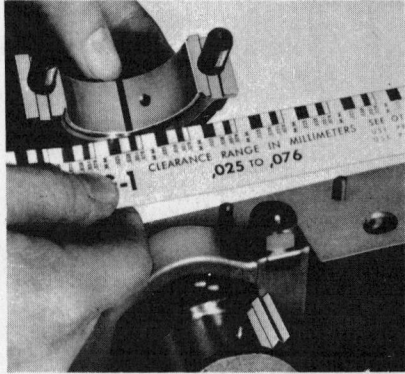

Using Plastigage® to check bearing clearance

20. Measure the standard oil clearance between the main bearing metal and main journal with a plastigauge.

Standard oil clearance: 0.0007-0.0029 in.

Main bearing cap torque: 43-47 ft.lb.

Undersize metal: 0.01 in., 0.02 in., 0.03 in.

21. Measure the crankshaft end play with a special tool (No. N026 standard feeler gauge). If the measured value ex-

	Internal Diameter (after press-fitted)	External Diameter
Small end bush	0.787φ $^{+0.0006}_{+0.0001}$ in.	0.9055φ $^{+0.0034}_{-0.0022}$ in.

	Standard Diameter	Wear Limit	Grinding Limit
Crank pin	1.7717φ $^{-0.0018}_{-0.0024}$ in.	0.002 in.	0.03 in.
Main journal	2.2048φ $^{-0.0018}_{-0.0024}$ in.	0.002 in.	0.03 in.

Satoh

ceeds the limit of 0.0023 in., replace the thrust washer inserted into the main bearing cap rear with an oversize washer in order to obtain the standard end play.

INSTALLATION

1. Fit a clip to the clip groove at one end of the piston.
2. Heat the piston with the piston heater 250-300° F and insert the piston pin quickly into the contact point with the clip by means of the special tool (No. 0490070 piston pin installer).

NOTE: When fitting the piston pin to the connecting rod, the mark F (B) on the piston must be positioned in relation to the oil jet hole (C) of the connecting rod.

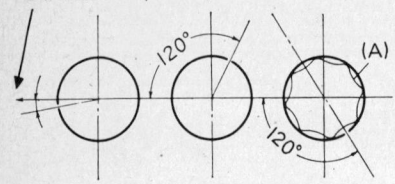

Positioning piston ring gaps

3. Fit a clip to the clip groove at the other end of the piston.
4. Fit each piston ring following the procedure outlined below:
   a. Fit the expander, to the lowest groove on the piston and next the oil ring (A).
   b. Fit the second ring (B) and top ring (C).

NOTE: Be sure to fit each piston ring with the mark R (D) upwards. Besides, do not expand rings more than necessary to install.

   c. Place eash piston ring at about 120° apart and alternate the gap of the oil ring and that of the expander (A).
5. Using a special tool (No. 0490080 piston inserting guide) (B), install the piston and connecting rod large end with "F" mark on the piston toward the front of the engine, tapping the head of the piston with a plastic hammer.

NOTE: Prevent the gap of the piston ring from facing the thrust side and the piston pin side.

6. Turn the engine upside down and now the crankshaft is upward.
7. Fit the main bearing metals to the cylinder block and the main bearing cap.

NOTE: Fit the main bearing metals with the same mark at the disassembly.

8. Fit the block thrust washer to both ends of the cylinder block rear with the oil groove outside.
9. Install the crankshaft on the main bearing metals.

NOTE: Be careful not to drop the thrust washer.

10. Apply graphite grease to the lip of the oil seal and fit it to the rear of the crankshaft.
11. Insert the side seals (?) into the grooves at both sides of the no. 5 main bearing cap. Then fit the thrust washers (?) to the cap and install the cap to the cylinder block and crankshaft.

NOTE: Apply the special tool (No. TL-05 THREE BOND® No. 5) to the side seal and fit it with the wider lip of the side seal toward the flank of the engine.

12. Fit the main bearing caps and tighten them to 43–47 ft.lb.

NOTE: Fit the main bearing caps in accordance with the marks showing the arrangement order (A, B, A and D from the front of the engine).
Bend the lockwasher around the nut.

13. Fit the piston and connecting rod large end, and the connecting rod caps to the crankshaft.

NOTE: When fitting them, be sure to match the mark of the cap with that of the large end.

14. Tighten each connecting rod cap to the specified torque of 25.3-28.9 ft.lb. with eight cap bolts.
15. Insert the oil pump from above the cylinder block and secure it with four nuts. Then, fit the oil pipe to the cylinder block and oil pump.
16. Secure the oil strainer to the oil pump with two nuts.

NOTE: Insert the O-ring into the oil strainer.

17. Apply a thin coat of the special tool (No. TL-05 THREE BOND® No. 5) to the parts on the cylinder block where the No. 5 main bearing cap and the timing chain cover are fitted.
18. Secure the oil pan together with the packings to the cylinder block with twenty-two nuts.

Oil Pump

DISASSEMBLY

1. Remove the two nuts, and remove the oil strainer from the pump cover.
2. Remove the pump cover (B), outer rotor (C) and pump driveshaft assembly (D) from the pump body (A).
3. Remove the split pin (H). Pull out the spring seat (E), pressure control spring (F) and pressure control plunger (G) from the pump body.

NOTE: Take care so that the spring seat and pressure control spring will not spring out when the split pin is removed.

INSPECTION

1. Measure the clearance between the oil pump body and outer rotor with a feeler gauge. Replace both if the measurement exceeds 0.0118 in.
2. Measure the clearance between the oil pump cover and outer rotor with a straight scale and a feeler gauge. Replace both parts if the measurement exceeds the maximum limit of 0.0059 in.
3. Measure the clearance between the oil pump drive shaft assembly (1) and the pump body (2) and pump cover (3) bearings (A) and (B). If the measurement exceeds the specified maximum limit of 0.0039 in., replace the parts.

ASSEMBLY AND INSTALLATION

Reverse the above procedures.

FUEL SYSTEM

Fuel Pump

REMOVAL

1. Remove the clip and disconnect the fuel hoses (?) from the fuel pump (?) at the both inlet and intake sides.
2. Remove the two nuts, and remove the fuel pump.

DISASSEMBLY

1. Remove the five set screws, and separate the housing (A) from the body (B).
2. Remove the two screws, and remove the valve (C) from the housing.
3. Remove the diaphragm (D) and diaphragm spring (E) from the body.
4. Pull out the pin, and remove the rocker arm (G) and rocker arm spring (F) from the body.

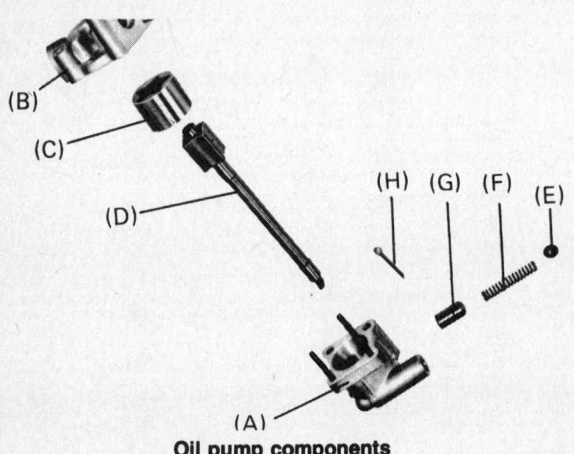

Oil pump components

Satoh

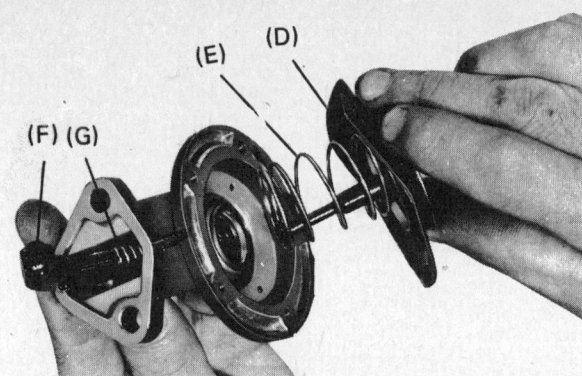

Removing pump internal components

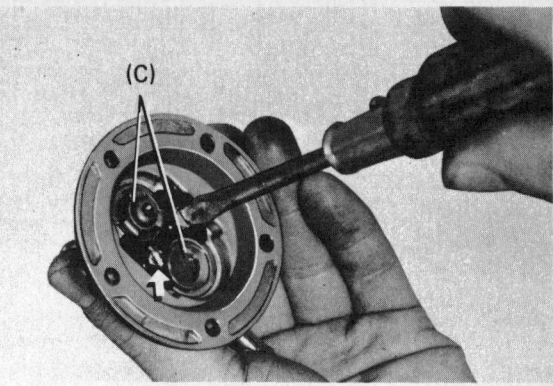

Removing pump valve

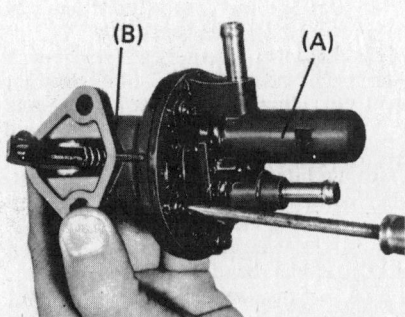

Separating pump components

ASSEMBLY AND INSTALLATION

Reverse the above procedures. Fuel pump discharge pressure should be 1.5-2.0 psi.

Carburetor

REMOVAL

1. Remove the air funnel by loosening the clamp band.
2. Remove the connecting bolt (B) and remove the fuel hose. Pull out the vacuum advance hose (C) and remove the rod assembly (?).
3. Remove the two mounting nuts (E) and remove the carburetor from intake manifold.

DISASSEMBLY

1. Pry off the cotter pin and separate the choke connecting rod (A) from the throttle lever.
2. Pry off the cotter pin and remove the pump connecting rod (B) from the pump arm.
3. Remove the return spring. Remove the five screws and take out the air horn.
4. Remove the gasket.
5. Pull off the float pin and take out the float.
6. Remove the float valve spring and retainer from the float valve seat.
7. Removing the accelerator pump.
8. Remove the accelerator pump from the carburetor body.

9. Remove the slow air bleed (A), slow jet (B), main air bleed (C), pump weight (D), steel ball (under the pump weight) and inlet check valve (E) from the carburetor body.
10. Remove the plug (F) and take out the main jet (G).
11. Loosen the two screws and remove the body from the flange.
12. Remove the gasket.

ADJUSTMENTS

Throttle Opening

When the choke valve is fully opened, the clearance (A) of the throttle valve on the primary side should be 0.036 in. If not, adjust it by bending the choke connecting rod. Opening angle for starting is 13-17°.

Float Level

1. Remove the air horn from the carburetor, and raise the float (A).
2. Slowly lower the float (A) and stop it when the float seat (C) contacts the needle valve stem.
3. Measure H, and if the measurement is 0.16 in., the float level is correct. If not, adjust it by bending the float seat.

Idle Speed

1. Connect a tachometer to the engine.
2. Fully turn in the idle adjust screw, and back it off 3 turns. Then screw in the throttle adjust screw 2 or 3 turns, and start the engine.

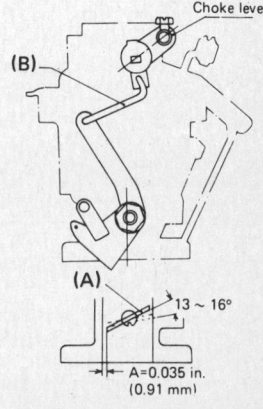

Throttle opening adjustment

3. Slowly back out the throttle adjust screw, and the engine speed will slow down. Stop turning the screw just before the engine begins to run roughly.
4. Then slowly turn in the idle adjust screw. When the engine begins to run smoothly at maximum speed, stop turning the screw.
5. Slow down the engine speed by turning out the throttle adjust screw. Repeat this operation so that the engine smoothly run at 700-800 rpm.

NOTE: Do not run in the idle adjust screw too hard, otherwise, the screw end will be damaged.

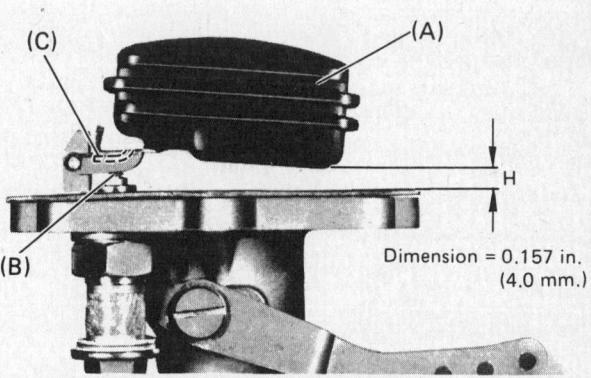

Float level adjustment

Satoh

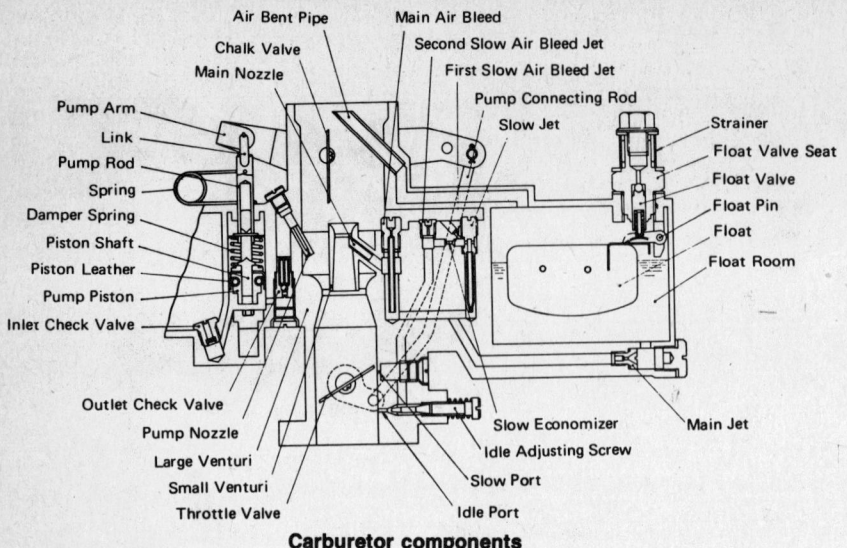

Carburetor components

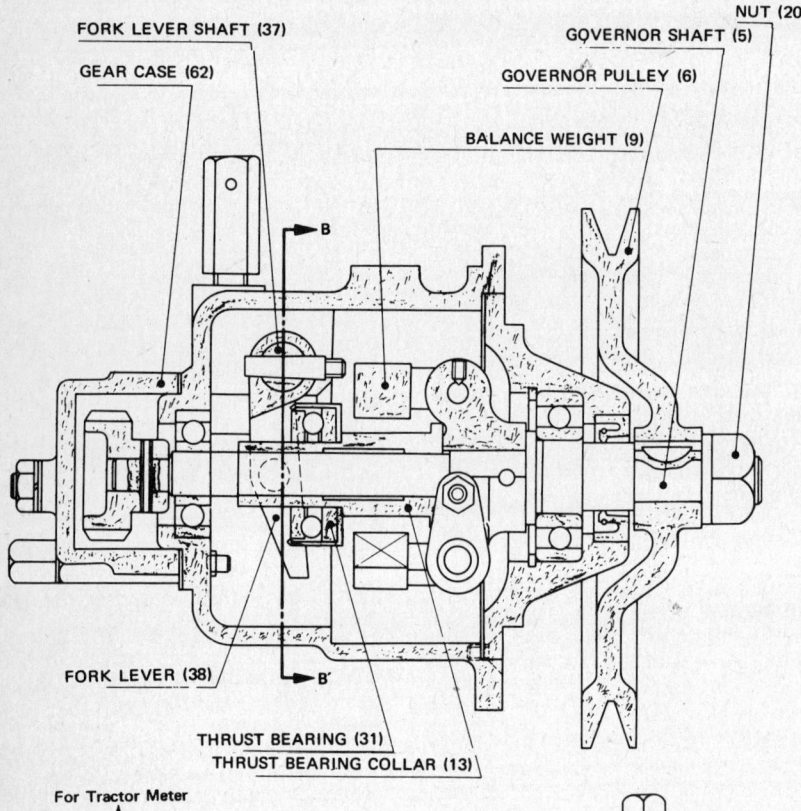

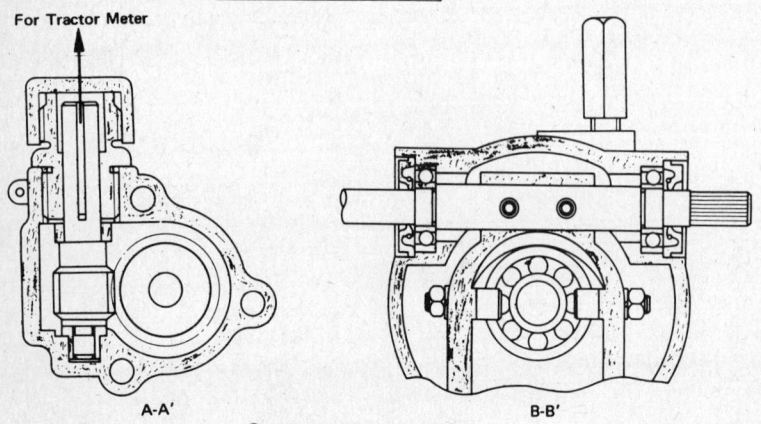

Governor components

ASSEMBLY AND INSTALLATION

Reverse the above procedures.

Governor

DISASSEMBLY

1. Remove the ball joint at the end of the governor.
2. Remove the nut at the front of the pulley, and remove the pulley.
3. Remove the bracket mounting bolts, and remove the governor assembly.
4. Remove the gear case stud nut, and remove the gear case kid.
5. Pull out the drive gear roll pin, and remove the drive gear.
6. Remove the counter-sunk screw at the front of the governor case.
7. Hold the governor case flange with your hand, and lightly strike the governor shaft projecting to the governor case rear side with a hammer. The governor shaft kid and bearing case kid can be removed.
8. Pull out the thrust collar. The bearing case kid can be removed.
9. Remove the bayonet setscrew on the rear of the weightholder. The governor shaft kid can be disassembled into the weight set pins, weights, and weight pins, in that order.
10. Remove the taper pin, and the weight holder can be removed with ease.
11. Remove the snapring, and disassemble the bearing case kid in order of the oil seal and bearing.
12. Pull out the lever joint cotter pin located on the bottom end of the adjust lever.
13. Remove the adjust lever snapring. The governor case kid can be disassembled into the adjust lever, governor spring, spring bolt, and lever rod, in that order.
14. Remove the spring bolt holder taper pin, and the bolt holder can be pulled out with ease.
15. Loosen the governor lever fitting bolt, and remove the governor lever.
16. Remove the fork lever fitting ten pin in the case.
17. Lightly strike the fork shaft projecting into the case (on the bolt holder side) with a hammer, and the fork lever can be removed. One half of the lever is attached with the oil seal and bearing.
18. Remove the sleeve, and the driven gear can be removed.

ASSEMBLY

Reverse the disassembly procedure. Note the following points:
1. Thoroughly wash all disassembled parts with light oil.
2. Be sure to replace all gaskets and oil seals with new ones. Otherwise oil leakage may result.
It is also advisable to replace bearings as a rule.
When press-fitting the bearing case kid into the governor shaft, grease the oil seal inner surface and bearing side surface.
3. When installing thrust collars, apply

Satoh

a small quantity of mobile oil to contacting surfaces.

4. Feed a plenty of fiber grease into the gear case. Tighten the pulley with a torque of 36.12 ft.lb.

5. When replacing the balance weights, make sure that new weights are the same in mass as old ones.

6. Make sure that the governor lever is at an angle of 4° to the horizontal line.

7. Set the spring bolt so that the distance from the center of the spring bolt holder to that of the spring hole is 1.1024 in.

8. Make sure that the runout of the pulley V groove is 0.0098 in. or less measured from the governor case mounting surface, after assembling is completed.

ADJUSTMENTS

Interlock Adjustment between Carburetor and Governor

1. The interlock adjustment is required only when the governor is found to be faulty because of instable engine speed, fluctuations in specified engine speed and increases in speed variation.

2. Turn in the throttle adjusting screw to bring the throttle valve opening to zero.

3. Pull out the adjusting lever fully, and loosen the governor lever not so that the throttle valve opening will be between 70° and 75°.

4. Return the adjusting lever to its original position, and adjust the fork lever stopper so that the throttle adjusting valve will return to the zero position.

5. By using the throttle adjusting screw and the idle adjusting screw, perform idle speed adjustments.

• Idle adjusting screw . . . Adjust so that the vacuum gauge reading will be at maximum.

• Throttle adjusting screw . . . Use a tachometer.

6. Adjust the adjusting lever position by using the stopper bolt so that the engine speed will be at $2.650 ^{+0}_{-100}$ rpm with no load and with full throttle.

Engine Speed

Adjustment can be done by turning the adjust lever stopper bolt and by changing the governor spring tension. Increasing the length of the spring installed causes the engine speed to increase, and shortening causes the speed to decrease.

Speed Sluctuation

Adjustment can be done by changing the distance from the center of the spring bolt holder to that of the spring hole. When the distance is longer, the coefficient is larger, and vice-versa.

Periodic Speed Variation

If hunting (to slight degree) is found, it should be adjusted by the damper spring attached to the lever rod end.

If hunting is great, governor adjustment is incorrect. Readjustment should be made by adjusting the specific speed and fluctuation coefficient, alternately.

Ignition System

Adjustments

IGNITION TIMING

1. Connect the timing light.
2. Connect the tachometer.
3. Start the engine to reach the ordinary idling revolution of 700-800 rpm.
4. Adjust the timing so that the notch on the crankshaft pulley is aligned with the indicator pin end.
5. To adjust the ignition timing, loosen the distributor locknut and turn the distributor.
6. Retighten the locknut and again inspect the ignition timing with the timing light.

POINT GAP ADJUSTMENT

1. Set the contact arm heel to the high part of the cam so that the clearance between points becomes widest.
2. Insert a feeler guage into the point gap and measure the point gap. Gap should be 0.016-0.020 in.

Timing marks

3. To adjust the point gap, loosen two screws and adjust the point gap and retighten the screws.
4. Dwell angle should always be checked when the points are serviced. Dwell should be 58°±3°.

S-630

2-WHEEL DRIVE FRONT AXLE

REMOVAL

1. Pull out the cotter pin from the ball socket nut securing the tie rod and drag link, loosen the nut, and remove the tie rod and drag link.

2. Place a jack under the clutch housing and jack up so that the front tires are slightly off the ground.

3. Loosen the axle outer bolts and remove both axle outers.

4. Loosen the center pin set bolt, and tapping on the center pin, pull out the center pin and remove the front axle. Care should be used so that O-rings are not lost.

INSTALLATION

1. Apply grease or oil to the center pin holes in the chassis and front axle center.

2. Apply grease to the O-rings and install them on both sides of the front axle center.

3. Install the front axle center to the chassis and install the center pin.

4. Install the liner between the rear part of the front axle and the chassis, and then install the center pin.

NOTE: The liner should be so set that the end play of the front axle is 0.008 in.

5. Make sure the front axle functions smoothly.

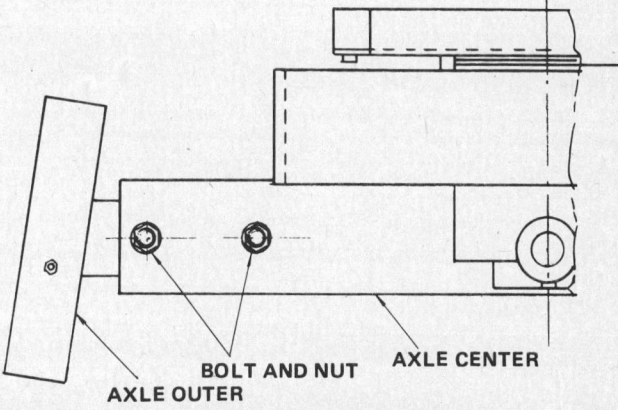

Axle center

Satoh

6. Install the bumper on the chassis end. Tightening torque: 68.64-75.86 ft.lb.

NOTE: When the chassis is reinstalled, the radiator, battery, hood, etc. should be installed in the reverse order of removal.

Front Hub

REMOVAL

1. Jack up the front axle with a jack, loosen the front wheel bolts, and remove the front tires.
2. Remove the front wheel hub cap bonded to the front wheel hub, and pull out the cotter pin. Loosen the nut, and remove the front wheel hub from the king pin knuckle using a gear puller.
3. Remove the ball bearing oil seal from the king pin knuckle.

NOTE: Always use a new oil seal.

INSTALLATION

1. Install the oil seal and ball bearing to the king pin knuckle.
2. Install the front wheel hub and pack with grease.

Drive in the bearing, install the nut and tighten the nut.

Tighten the slotted nut until a preload of 0.072-0.108 ft.lb. is obtained when measured at a hub bolt hole.

3. Pack the cap with grease, and apply bond to the front wheel hub, and install the cap taking care not to be deformed.
4. Check to see that the front wheel hub rotates smoothly.

King Pin

REMOVAL

1. Loosen the steering lever bolt, and remove the steering lever.

NOTE: If necessary, remove the tie rod and drag link ball socket.

2. Taking care so that the king pin will not drop off, remove the key and pull out the king pin downward.

INSTALLATION

1. Install the O-ring (greased) in the O-ring groove in the lower part of king pin.
2. Install the thrust bearing and grease the king pin.
3. Install bushes to both ends of the axle outer and grease the bushings.
4. Insert the king pin into the axle outer, and install the front tire taking care so that the king pin will not drop off.
Tightening torque: 86.7-97.5 ft.lb.
5. Install the O-ring (greased) to the top end of the king pin, drive in the key, and install the steering lever.
6. Tighten the bolt so that the clearance between the top end of the axle outer and bottom end of the steering lever is 0-0.02 in. Tightening torque: 21.68-28.9 ft.lb.

NOTE: If the clearance between the axle outer and steering lever becomes 0.04 in. or more, it should be adjusted to be 0-0.02 in.

7. Install the axle outer to the front axle center and adjust the tread as required. Tightening torque: 86.7-97.5 ft.lb.
8. After cleaning the tapered ball socket surfaces of the drag link and tie rod, apply grease to the surfaces and install the drag link and tie rod on the steering lever.

Secure them with nuts and lock with cotter pins. Tightening torque: 32.5-39.7 ft.lb.

9. Check to make sure each installed system is functioning properly.

4-WHEEL DRIVE FRONT AXLE

REMOVAL AND INSTALLATION

1. Loosen the center pin set bolt holding the axle housing. Remove the center pin, and then remove the axle housing.
2. Installation is the reverse of removal.

Front Axle Universal

REMOVAL

1. Loosen the universal joint cover bolt, and remove the cover.
2. Loosen the bolt securing the universal joint and flange, and remove the universal joint.

Remove the knuckle assembly, remove the axle housing final drive case drain plug, and drain the oil.

INSTALLATION

1. Apply grease to the universal joint splines and install the joint to the PTO shaft. Align the front end with the flange fitted to the drive pinion shaft and lock it with the bolt and nut.
Tightening torque: 14.45-18.06 ft.lb.
2. Install the universal joint cover, and tighten the bolt.
Tightening torque: 14.45-18.06 ft.lb.

NOTE: Take care so that the universal joint does not contact the cover.

3. Clean the tapered portion of the tie rod ball socket and apply grease. Install the socket to the steering lever, the washer together with spring washer, and the castle nut, and lock with the cotter pin.

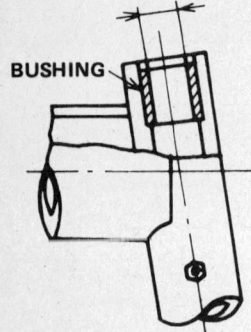

STANDARD I.D.
1.02401 – 1.02901 in
(26.010 – 26.137 mm)

Knuckle fit in bushing

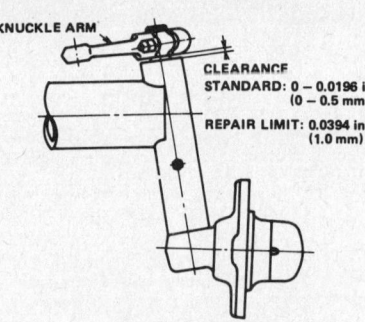

KNUCKLE ARM
CLEARANCE
STANDARD: 0 – 0.0196 in (0 – 0.5 mm)
REPAIR LIMIT: 0.0394 in (1.0 mm)

King pin measurements

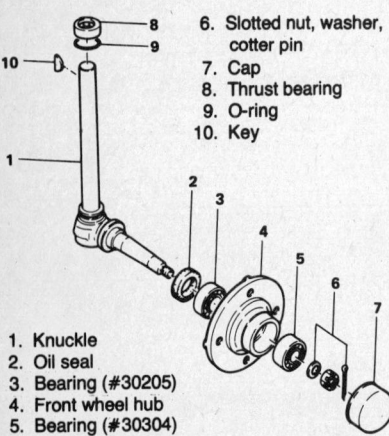

1. Knuckle
2. Oil seal
3. Bearing (#30205)
4. Front wheel hub
5. Bearing (#30304)
6. Slotted nut, washer, cotter pin
7. Cap
8. Thrust bearing
9. O-ring
10. Key

Knuckle components, 2-wheel drive

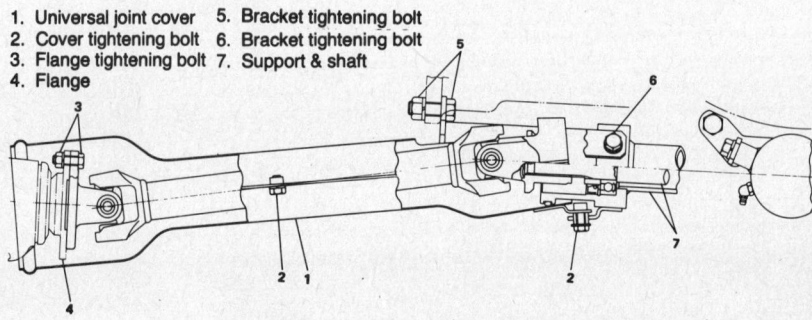

1. Universal joint cover
2. Cover tightening bolt
3. Flange tightening bolt
4. Flange
5. Bracket tightening bolt
6. Bracket tightening bolt
7. Support & shaft

4-wheel drive front axle universal

Tightening torque: 86.7-97.5 ft.lb.

4. Install the drag link so that the curved-inward side faces toward the engine, and apply grease to the tapered portion of the ball socket. Install the washer, spring washer, and castle nut, and lock with the cotter pin.

Tightening torque: 32.5-39.7 ft.lb.

5. Install the front tires.

Tightening torque: 86.7-97.5 ft.lb.

6. Adjust the toe-in to be 0-0.2 in. (0-5 mm) by turning the tie rod and lock the turnbuckle with the locknut.

7. After adjusting the toe-in correctly, adjust the inner front wheel toe-out to be 45° using the stopper bolt, and lock the bolt with the nut.

8. Pour oil into the final gear case and axle housing.

Pack the king pin case with enough grease.

Oil capacity final gear case:
Right 0.079 U.S. gal.
Left 0.079 U.S. gal.
Axle housing
0.264 U.S. gal.
King pin case 12.2 cu. in.

Knuckle and King Pin Case

REMOVAL

1. Removing the cotter pins from the castle nuts on the drag link and tie rod loosen the castle nut and remove the drag link and tie rod.

2. Jack up the front axle and remove the front wheel.

3. Loosen the bolt securing the axle housing and knuckle assembly and remove the left and right knuckles.

DISASSEMBLY

1. Loosen the oil seal retainer plate holding bolt on the spherical part and remove the retainer plate oil seal felt and rubber seal.

2. Loosen the steering lever bolt, and remove the steering lever together with the king pin installed in the final gear case.

3. Loosen the king pin stopper plate bolt under the final gear case, remove the plate, and pull out the king pin using a screwdriver.

Separate the king pin case from the final gear case.

NOTE: There are shims on the top of the king pin case and in the final gear case.

4. Remove the yoke assembly, then the circlip from the final driveshaft in the king pin case, and pull out the shaft.

5. Remove the oil seal from the king pin case. (No reuse of the oil seal is allowed).

NOTE: As required, remove the king pin case bushing.

ASSEMBLY AND INSTALLATION

NOTE: After inserting the bushing, install the final drive shaft ball bearing in the king pin case, and lock with the circlip (ϕ1.85 in.). Grease the oil seal, and install it in the king pin case. Install the final driveshaft, and lock with the circlip (ϕ0.984 in.).

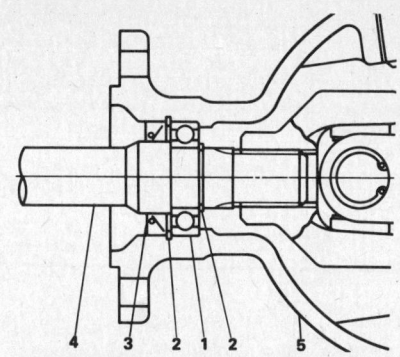

Inserting the final driveshaft ball bearing into the king pin case
1. Ball bearing
2. Circlip
3. Oil seal
4. Final drive shaft
5. Kingpin case

1. Install an oiled oil seal to the contact area of the final gear case B with the wheel axle.

2. Install the wheel axle to the final gear case B, and drive in the ball bearing.

3. Install the ball bearing to the pinion shaft, then in the final case B, and then install gear 49T.

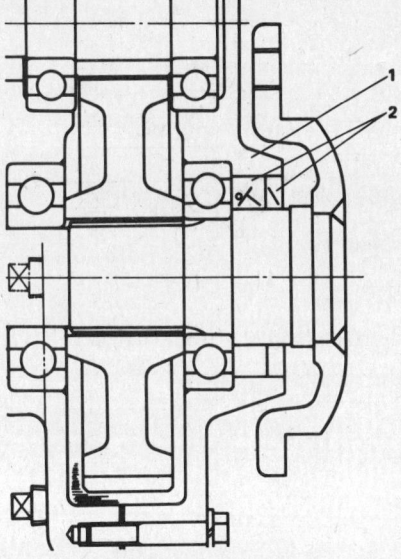

Final gear case
1. Final drive case B
2. Oil seal

4. Drive the ball bearing into the wheel axle, install the washer, and lock with the circlip.

5. Drive the ball bearing into the pinion shaft.

6. Apply grease to the gear case A and install the oil seal.

7. Apply sealant (Three-Bond® No. 4 or Helmeseal® 201) to both faces of the gasket, and install the dowel pin to the gear case A. Align the gear case B with the gear case A and tighten the bolts.

Tightening torque: 18.1-21.7 ft.lb.

8. Install the circlip to the pinion shaft, and make sure that the wheel axle rotates smoothly.

Final Gear Case

DISASSEMBLY

1. Remove the knuckle assembly from the final gear case and remove the circlip from the pinion shaft.

2. Loosen the eight bolts securing the final gear case A and final gear case B, and split the gear cases into A and B.

3. Remove the wheel shaft circlip and remove the ball bearing.

4. Remove the final gear and pinion shaft, the final case oil seal, and then the pinion shaft ball bearing.

5. Remove the wheel shaft from the gear case and remove the ball bearing oil seal.

ASSEMBLY

1. Place the final gear case under the king pin case, and install the yoke assembly.

Apply grease to the contact areas of the moving parts.

2. Hold the yoke assembly in a vertical position, and bring the splines of the final driveshaft (installed in the king pin case) to align with splines of the yoke, and install the king pin case in the final drive case.

3. Apply grease to the contact surfaces of the king pin case with the final gear case, and install the thrust bearing.

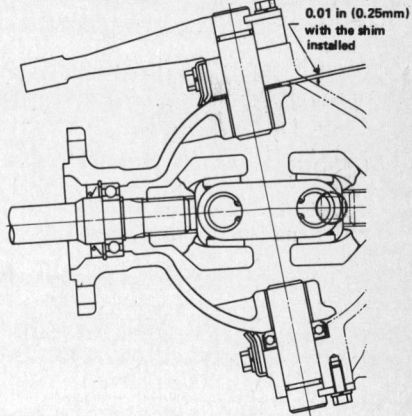

Shimming between the king pin case and the final drive case

4. Install the O-ring (greased) to the king pin on the bottom side, install the final gear case, and set the king pin case.

5. Align the plate with the king pin groove, and tighten the bolt.

Tightening torque: 8.67-12.28 ft.lb.

6. Select the shim so that the clearance between the top of the king pin case and gear case is 0.01 in. or less.

Thickness of shim:
0.03937 in.
0.04724 in.
0.05512 in.

7. Place a shim between the king pin case and the final gear case, and make sure

Satoh

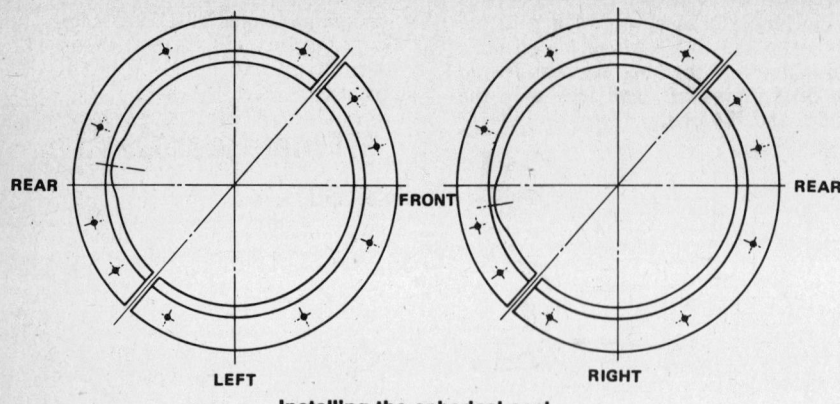

Installing the spherical seal

that the end play of the king pin is 0.01 in. Grease the O-ring and install it to the king pin, insert it into the steering lever, and install the key plate.

8. Insert the king pin (which is installed to the steering lever) into the final gear case from its bottom, and lock with the bolt.

Tightening torque: 28.9-36.13 ft.lb.

NOTE: The drag link is connected to the left side steering lever.

9. As illustrated, install the spherical seal to the sphere and tighten it, together with the felt and retainer plate.

Tightening torque: 8.67-12.28 ft.lb.

NOTE: Check to see that the king pin case and final case move smoothly.

10. Apply sealant (Three-Bond® No. 4 or Helmeseal® 201) to the gasket, and install it between the axle housing and the king pin case. Install the knuckle assembly to the axle housing.

Tightening torque: 28.9-36.13 ft.lb.

Differential

DISASSEMBLY

1. Loosen the nut securing the differential gear assembly to the axle housing.

2. Put a match mark to the bearing cap, remove the nut, and remove the differential gear.

3. Remove the taper roller bearing from the differential gear case assembly, loosen the ring gear bolt, and remove the ring gear from the differential gear case.

4. Pull out the differential pinion center lock pin, pull out the center pin and remove the pinion gear, side gear, thrust washer and pinion gear seat.

Pinion Gear

1. Remove the locknut at the end of the pinion shaft, and remove the pinion shaft

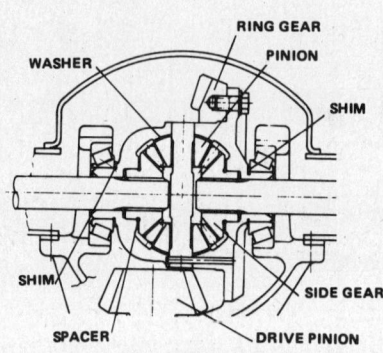

Measuring differential backlash

taking care so that the shim, washer and spacer will not drop off.

2. Remove the flange assembly, and remove the oil seal.

ASSEMBLY

1. Install the spacer and side gear in the differential case, and set the pinion gear and thrust washer with the center pin, and measure the backlash between the side gear and pinion gear.

Backlash: 0 – 0.002992 in.

Spacer 0.03937 $\begin{matrix} 0 \\ -0.002755 \end{matrix}$ in.

0.03937 $\begin{matrix} -0.003149 \\ -0.006692 \end{matrix}$ in.

0.03937 $\begin{matrix} -0.007086 \\ -0.009842 \end{matrix}$ in.

Shim 0.031496 in.

When the backlash is correctly adjusted lock the center pin with the lock pin and further clinch the pin with the center punch.

2. Install the ring gear to the differential gear case and tighten it with the mounting bolt.

Tightening torque: 25.287-28.90 ft.lb.

Pinion Shaft

1. Install the outers of the taper roller bearings, #32207 and #30305, in the carrier case.

2. Install the shim (A) and taper roller bearing, #32207, in the drive pinion shaft, and install this assembly in the carrier case. Install the spacer and shim (B), and insert the taper roller bearing, #30305.

3. Install the washer (A), and bring the flange splines to mesh with the pinion gear shaft, and tighten the nut to 115.6-158.95 ft.lb.

4. Adjust with the spacer and shim (B) so that the preload is 5.058-7.225 ft.lb.

Spacer: 2.259 in.
         2.272 in.

Shim (B): 0.0787 in.   0.0858 in.
          0.0799 in.   0.0870 in.
          0.0811 in.   0.0882 in.
          0.0823 in.   0.0894 in.
          0.0835 in.   0.0906 in.
          0.0846 in.   0.0917 in.

5. Measure the preload of the drive pinion shaft, and by changing the thickness of the shim located between the taper roller bearing, #32207, and pinion gear, adjust the distance between the end of the drive pinion and the center of the ring gear to be 2.04 in.

Distance between ring gear
and drive pinion: 2.04 in.

Thickness of shim A:
  0.0543 in.   0.0614 in.
  0.0555 in.   0.0626 in.
  0.0567 in.   0.0638 in.
  0.0579 in.   0.0650 in.
  0.0591 in.   0.0118 in.
  0.0602 in.

6. After adjusting the distance between the ring gear and the drive pinion, remove the flange again and grease the taper roller. Also grease both faces of the washer A and install the oil seal in the carrier case.

7. Bring the drive pinion shaft splines

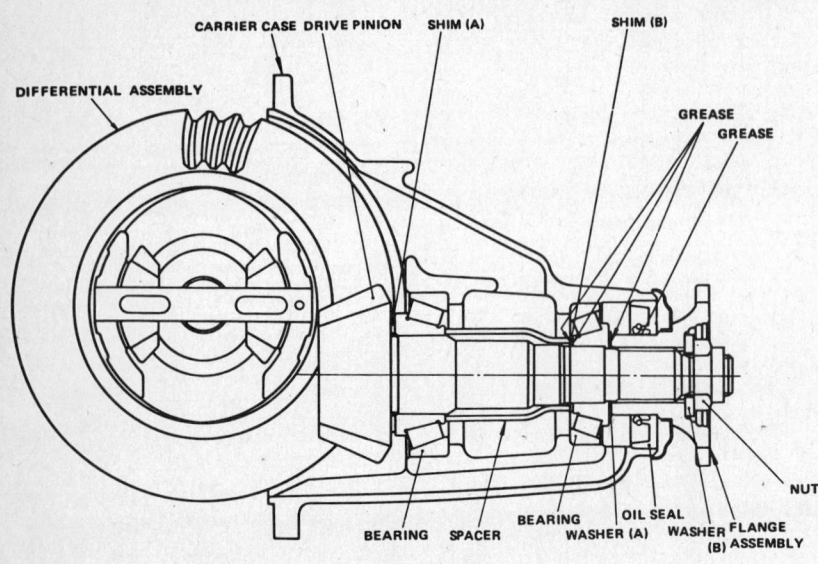

Pinion assembly

to align with the flange splines, install the flange, install the washer B and tighten the nut.

Tightening torque: 115.6-158.95 ft.lb.

8. Set the ring gear in the differential gear case, install a proper shim, and install the taper roller bearing.

9. Install the differential gear assembly in the carrier case, and taking care not to allow both right and left caps to drop off, tighten the bolts.

10. By changing the thickness of the shims between the differential case and the taper roller bearing, adjust the backlash between the drive pinion shaft and the ring gear to be 0.0051-0.007 in.

End play at both sides: 0.002 in.
Thickness of shim:
0.00197 in.   0.00787 in.
0.00276 in.   0.01181 in.
0.00394 in.   0.02756 in.

11. After adjusting the backlash and preload correctly, tighten the bearing cap.

Tightening torque: 25.29-28.9 ft.lb.

12. Apply sealant (Three-Bond® No. 4 or Helmeseal® 201) to both faces of the gasket, and install it in the axle housing. Install the differential carrier case in the axle housing and tighten the nut.

Tightening torque: 18.06-21.67 ft.lb.

13. Apply grease to the axle housing center pin mounting hole, and make sure that the center pin can move into the hole. Install the axle housing on the chassis, together with O-rings and thrust washers, insert the center pin and lock it with the set bolt.

14. Swing the center pin gently, and make sure there is no play back and forth.

Mid-Power Take-Off

REMOVAL

1. Loosen the oil filter cover bolt, and remove the cover, and drain the transmission oil.

Remove the MID PTO gear case drain plug, and drain oil.

2. Remove the universal joint cover, loosen the bolt securing the flange to the universal joint, and remove the universal joint.

3. Loosen the bolt securing the countershaft support to the bracket, and remove the countershaft together with the support. Remove the joint.

4. Loosen the MID PTO gear box setting bolt, and remove the MID PTO from the transmission case. Be careful so that the dowel pins are not lost.

DISASSEMBLY

1. Pull out the spring pin securing the gear 30T to the shaft, and remove the shaft from the gear case, and remove the gear 30T.

If necessary, remove the circlip from the gear 30T, and remove the needle roller bearing.

2. Remove the MID PTO gear case cap, and the circlip. Remove the PTO shaft by

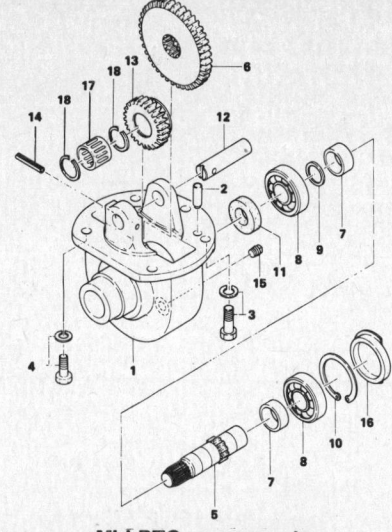

Mid-PTO components

1. Housing
2. Dowel pin
3. Washer bolt
4. Washer bolt
5. Shaft
6. Gear
7. Spacer
8. Ball bearing
9. Liner
10. Circlip
11. Oil seal
12. Shaft
13. Gear
14. Spring pin
15. Plug
16. Rubber cap
17. Needle bearing
18. Circlip

tapping it with a wooden hammer. Remove the gear 55T, spacer, ball bearing and liner.

As required, remove the oil seal.

When the cap has been removed, discard it.

ASSEMBLY

1. Grease the oil seal, install it to the MID PTO gear case and install the ball bearing.

2. Install the gear, spacer and liner to the shaft in that order and drive the same shaft. Then install another spacer, drive the ball bearing and lock it with the circlip.

The gear should be installed so that the chamfered faces toward the front.

Adjust the liner so that the shaft end play is 0.016 in. or less.

3. Apply sealant (Three-Bond® No. 4 or Helmeseal® No. 201) to the outer surface of the new cap, and install it to the gear case.

4. Install the circlip in the gear 30T.

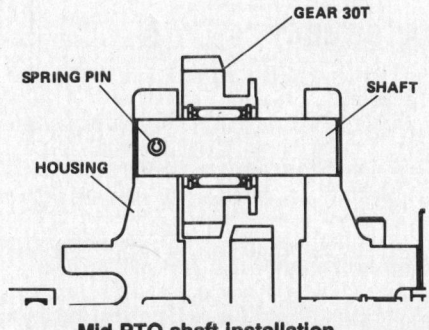

Mid-PTO shaft Installation

Apply oil to the needle roller bearing, install it to the gear, and lock with the circlip.

5. Insert the shaft so that the shift groove of gear 30T faces backward, and drive the spring pin, and install the shaft in the case. Care should be taken so that the spring pin position is not reversed.

INSTALLATION

Shift the sub-shift lever to Low speed.

1. Apply sealant (Three-Bond® No. 4 or Helmeseal® No. 201) to the mating surfaces of MID PTO gear case and transmission case.

2. Align the 4-wheel shift pin with the gear 30T shifter groove, install the MID PTO gear case (attached with dowel pins) to the transmission case, and secure with the bolt around which sealing tape is wound.

Tightening torque: 36.13-43.35 ft.lb.

3. By operating the front drive shift lever, check to see that the front drive is shifted properly.

STEERING

Steering Gear

REMOVAL

1. Open the hood and disconnect the wires from the instrument panel at the terminals.

2. Loosen the steering wheel cap, and remove. Loosen the nut, and using the steering wheel puller, remove the steering wheel.

NOTE: If the steering wheel is hard to pull out, give impacts to the puller center so it can be removed.

3. Pull out the cotter pin in the bottom part of the throttle lever, loosen the nut, and remove the spring washer.

4. Pull out the spring pin and pull out the accelerator lever upward.

5. Loosen the meter panel mounting bolts, and remove the meter panel. Loosen the bolts in the upper part and on both right and left sides of the instrument panel, and remove the instrument panel.

NOTE: Make sure that wires and tractor cables are removed.

6. Disconnect the fuel return pipe from the fuel tank side, remove the fuel pipe from the filter side, and plug up so that no oil will leak out.

7. Loosen the fuel tank band nut, remove the band, and demount the fuel tank.

8. Loosen the instrument panel support bolts, and remove the instrument panel support.

9. Loosen the Pitman arm nut, and remove the Pitman arm from the steering gear box.

As required, remove the drag link from the Pitman arm.

10. Loosen the steering gear box bolts, and remove the steering gear box.

Satoh

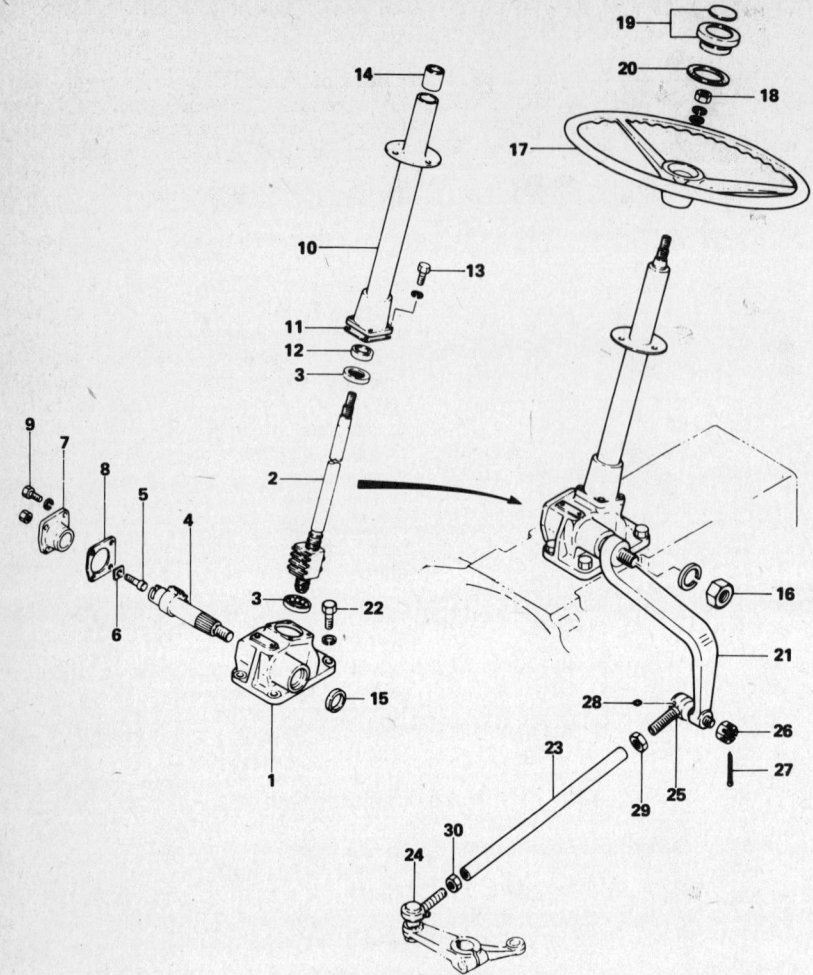

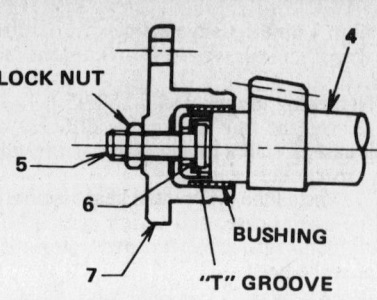

T-groove clearance adjustment

4. Sector shaft
5. Adjusting screw
6. Shims
7. Side cover

Steering system components

1. Steering gear box
2. Ball-and-nut assembly
3. Bearing
4. Sector shaft
5. Adjusting screw
6. Adjusting screw shims
7. Side cover
8. Gasket
9. Bolt (M8 x 22)
10. Steering column assembly
11. Rear cover shims
12. Oil seal
13. Bolt (M8 x 22)
14. Bushing
15. Oil seal
16. Nut (M20)
17. Steering wheel
18. Nut (M14)
19. Steering wheel cap
20. Gasket
21. Pitman arm
22. Bolt (M10 x 25)
23. Drag link tube
24. Socket set
25. Socket set
26. Nut
27. Split pin
28. Grease fitting
29. Nut
30. Nut

DISASSEMBLY

NOTE: Prior to disassembly, measure the preload of the ball-and-nut assembly to help make shim adjustment properly at the time of assembly.

1. Remove the steering column assembly as described below.
   a. Unscrew nuts (5) securing steering column assembly (2) to gear box, and remove the assembly.
   b. Remove oil seal (4).
   c. Remove bushing (6).
   d. Remove bearing (1).

NOTE: Be careful not to damage shims (3) when removing them.

2. Remove the sector shaft and ball-and-nut assembly as described below:
   a. Unscrew bolts (9) securing side cover (7), and pull out the sector shaft (4) complete removing the cover.
   b. Pull ball-and-nut assembly (2) from gear box (1).
   c. Remove oil seal (10) and bearing (3).
   d. Remove the locknut from adjusting screw (5), and screw the adjusting screw to remove the side cover.

NOTE: Take care not to damage shims (6) when removing them.

ASSEMBLY

1. Install the sector shaft and ball-and-nut assembly as described below:
   a. Press bearing (3) into gear box (1), and place oil seal (5) in the box.
   b. Place ball-and-nut assembly (2) in the gear box, and insert sector shaft (4) from the right side of the box.
   c. Adjust the "T" groove clearance between sector shaft (4) and adjusting screw (5) by means of shims (6). Clearance is 0-0.001968 in.; tighten the adjusting screw.
   d. Attach the gasket to the side cover, and install the cover in position with bolts (9). Torque the bolts to 20 ft.lb. Temporarily lock adjusting screw (5).

2. Install the steering column assembly to the gear box as described below:
   a. Install the oil seal (4) and bearing (1) to the column assembly (2).
   b. Place shims (3) in position, and secure the column assembly (2) with bolts (5). Torque the bolts to 20 ft.lb.

NOTE:
a. On a tractor which has been in service longer than 100 hours, add a 0.00197 in. shim when installing the column assembly temporarily.
b. It is unnecessary to add such a shim when the ball nut assembly is replaced.

3. Adjust the backlash between the ball-and-nut assembly and sector shaft.

SPECIFICATIONS

Item	Standard
Backlash measured at steering shaft	0-3°
Backlash measured at steering wheel rim	0-0.3937 in.
Backlash measured at drag link side end of Pitman arm	0-0.00866 in

To adjust, turn adjusting screw (5). After adjusting, lock the adjusting screw with locknut.

4. Adjust the preload of the steering shaft (ball-and-nut assembly) by adding or removing shims (11) as outlined below:

NOTE: The steering column assembly consists of rear cover and column.

a. The bearing preload measured at the center of the steering shaft should be as specified below:

SPECIFICATIONS

Item	Standard
Preload measured w/o sector shaft	0.144-0.36 ft.lb.
Preload measured on assembled gear box	0.144-0.397 ft.lb.

b. The bearing preload measured at the rim of the steering wheel shaft should be as specified below:

SPECIFICATIONS

Item	Standard
Preload measured w/o sector shaft	0.2646-0.6615 lb.
Preload measured on assembled gear box	0.2646-0.7277 lb.

NOTE: When adjusting the bearing preload, be sure to check the backlash, making sure it is correct.

To measure the preload, hook a spring balancer to one of the steering wheel spokes as close to the rim as possible.

5. Install the steering gear box assembly to the clutch housing.

BOLTS, TOOL AND TIGHTENING TORQUE

Qt.	Diameter	Length	Tool	Torque
Bolt 4	10mm	0.9843 in.	14mm offset wrench	21.67-28.90 ft. lb.

6. Mount the fuel tank, meter panel and steering wheel.
7. Fill the steering gear box with an approximately 200 cc of the recommended grease as described below:
   a. Remove the bolts from the inspection hole and filler.
   b. Apply grease through the filler until it appears at the inspection hole.

INSTALLATION

Installation is the reverse of removal. Note the following points:
1. Steering gear box tightening torque: 24.57-31.68 ft.lb.
2. Instrument panel support tightening torque: 21.67-28.90 ft.lb.
3. Pitman arm tightening torque: 86.70-115.60 ft.lb.
4. Instrument panel bolt 7.22-8.67 ft.lb.
5. Force to be required to operate the throttle lever 6.615-8.820 lb.

Tighten the nut so that force of 3 or 4 kg is obtained at the end of the throttle lever to operate it, but if the throttle lever is moved by force of the governor at full-throttle operation, tighten the nut further.

6. Steering wheel tightening torque: 36.12-57.80 ft.lb.

BRAKES

The brake is installed in the rear axle housing. To disassemble the brake, the rear axle housing must be removed. Since the rear axle housing is heavy, care should be taken when it is removed.

Rear Axle Housing

REMOVAL

1. Loosen the oil filter cover bolt on the right side of the transmission case, remove the cover, and drain off the transmission case oil.
After draining, remove the oil filter.
2. Place a jack under the transmission case, and jack up. Loosen the rear wheel bolts, and remove the rear wheels.

NOTE: When loosening the rear wheel bolts with the rear wheels jacked up, be sure to apply the parking brake.

3. Remove the wires of working lights and turn signals at connectors.
4. Remove the fender bolts at the rear axle housing and step, and remove the fender.
5. Remove the brake pedal return spring, and remove the brake rod.
6. Loosen the rear axle housing bolt, and remove the rear axle housing with special care. Be careful so that the dowel pins are not lost.

INSTALLATION

1. Apply sealant (Three-Bond ® No. 4 or Helmeseal ® No. 201) to the rear axle housing. Install the dowel pins to the transmission case, and install the rear axle housing.
Tightening torque: 57.8-65.0 ft.lb.
2. Install the fender to the rear axle housing, and install the rubber cushion, plate and lock plate, and tighten the bolts. Lock the bolts with the lock plate.
Tightening Torque:
  On the rear axle side: 57.8-65.0 ft.lb.
  On the step side: 12.28-14.45 ft.lb.
3. Thoroughly wash the oil filter, and install the cover together with the gasket.
Tightening torque: 7.22-8.6 ft.lb.
4. Install the rear wheels.
Tightening torque: 72.25-86.7 ft.lb.

Brake Drum and Shoes

DISASSEMBLY

1. Remove the circlip from the brake drum installed on the final driveshaft, and remove the brake drum.
2. Remove the brake shoes from the anchor pin brake camshaft.
3. Straighten the lockwasher at the end of the brake camshaft, and loosen the nut.

Remove the brake cam lever, and pull out the brake camshaft.
4. As required, remove the anchor pin, brake camshaft bushing, and O-ring.

INSPECTION

Measure the inside diameter of the brake drum. Replace the drum if worn beyond the service limit.

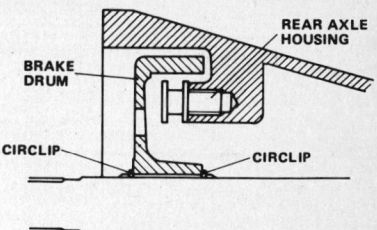
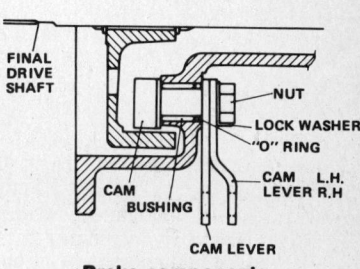

Brake components

SPECIFICATIONS

Item	Standard	Service Limit
Brake drum I.D.	6.3-6.35 in.	6.38 in.

3. Measure the thickness of brake shoe linings. Replace the linings as pairs if worn beyond the service limit.

SPECIFICATIONS

Item	Standard	Service Limit
Brake shoe lining thickness	0.177 in.	0.05 in.

4. Check the fit of the brake camshaft in the bushing. Replace the bushing if the fit exceeds the service limit.

SPECIFICATIONS

Item	Standard	Service Limit
Fit of brake camshaft in bushing	0.014 in.	0.027 in.

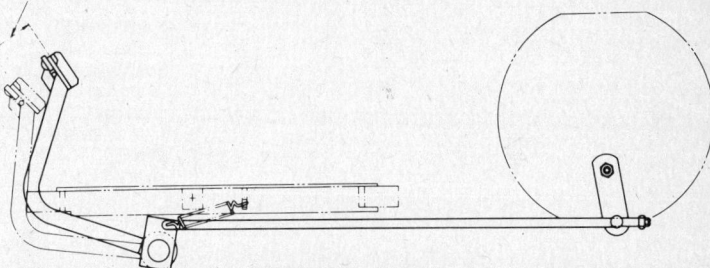

Brake adjustment

Satoh

ASSEMBLY

1. Drive the bushing into the rear axle housing, and install the anchor pin.
Tightening torque: 86.7-94.5 ft.lb.
2. Apply grease to the inner surface of bushing, and install the brake camshaft. Grease the ring and install it.
3. Install the brake rod so that the straight brake camshaft lever is on the inner side. Be sure that the right and left brake camshaft lever position is not reversed. Install the lockwasher and tighten the nut.
Tightening torque: 21.68-28.90 ft.lb.

NOTE: Be sure to bend the lockwasher to lock the nut.

4. Install the spring to the brake shoe and install the brake shoe.
5. Install the brake drum and set it with the circlip.

ADJUSTMENTS

1. Adjust the brake rod adjusting nut so that the pedal free play will be 0.3943-0.787 in.
2. Confirm that the right and left brakes operate simultaneously by running the tractor. If not adjust both of them by means of the rod.

CLUTCH

REMOVAL

1. Remove the hood catch and stopped rod cotter pin on the hood side, and then loosen the front bolt and take out the hood.
2. Remove the battery ground wire from the battery terminal.
3. Disconnect the wires coming from the instrument panel.
4. Remove the leak-off pipe on the fuel tank side. Disconnect the fuel pipe on the filter side, remove the fuel pipe on the filter side, and secure the pipe to the fuel tank.
5. Remove the left side safety guard, the throttle control rod snap pin, and then the control rod.
6. Loosen the hydraulic pipe tightening bolt, and disconnect the suction pipe and delivery pipe from the pump. Take care so that O-ring is not lost.
7. Pull out the cotter pin from the drag link nut and loosen the nut. Using the ball socket remover, remove the drag link from the steering lever.
Then place a jack under the clutch housing and jack up. Connect a wire rope to the hanging bolt on the cylinder head, and suspend the engine with a chain block so that the front is slightly off the ground.
8. Remove the four bolts securing the rear plate and clutch housing.

NOTE: It is unnecessary to remove the rear plate bolt on the engine side. The mating surfaces of the rear plate and clutch housing are coated with sealant.

9. Loosen the six bolts securing the engine and clutch housing alternately, and gently demount the engine toward the front, with the front axle installed.

NOTE: On 4-wheel drive models, loosen the U-joint cover bolt, and remove the cover. Remove the flange bolt.

10. Insert the clutch center tool into the flywheel pilot bearing. Loosen the clutch bolts evenly, and remove the clutch pressure plate assembly and clutch disc.

INSTALLATION

1. Apply grease to flywheel pilot bearing.
2. Install the clutch disc to the clutch center tool so that the longer boss is facing outward, and install them to the flywheel pilot bearing.
3. Secure the pressure plate assembly to the flywheel with M6 x 16 bolts. The bolts should be tightened evenly.
Tightening torque: 3.6-5.1 ft.lb.
4. Turn the nut in and out so that the height of release lever is 0.799-0.854 in. above the flywheel.

For assembly of the engine and clutch housing, reverse the procedure for disassembly paying attention to the following points:
a. Hold the tractor body and engine in a horizontal position, and proceed to assembly with special care.
Tightening torque: 39.68-47.62 ft.lb.
b. Be sure to apply bond evenly to the surfaces of the rear plate to be installed to the lower side of the clutch housing so that no water enters the clutch housing.
c. Check to see that the hydraulic pipe flange O-ring is correctly set in the groove, and install the flange to the pump.
Tightening torque: 3.6-5.1 ft.lb.
d. Connect each connector to the wire correctly, and tighten them firmly.

4-Wheel Drive

1. Join the universal joint to the flange, and secure them with four bolts.
Tightening torque: 14.45-18.06 ft.lb.
2. Install the universal joint cover.

ADJUSTMENT

Adjust the clutch pedal rod clevis pin so that the free play of clutch pedal is 0.4-0.8 in.

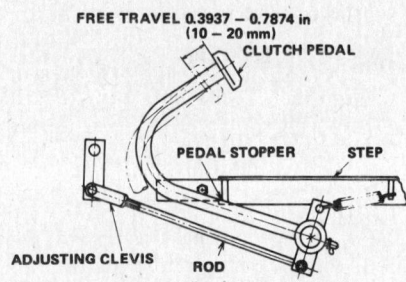

Clutch release adjustment

TRANSMISSION

NOTE: Transmission removal is not necessary for disassembly.

REMOVAL

1. Loosen the oil filter cover setting bolt, remove the cover, and drain off the transmission oil. After draining, remove the hydraulic oil filter.

NOTE: For a 4-wheel drive tractor, remove the MID PTO drain plug and drain oil.

2. Disconnect working light and turn signal wires at the connectors, and move them to the clutch housing side.

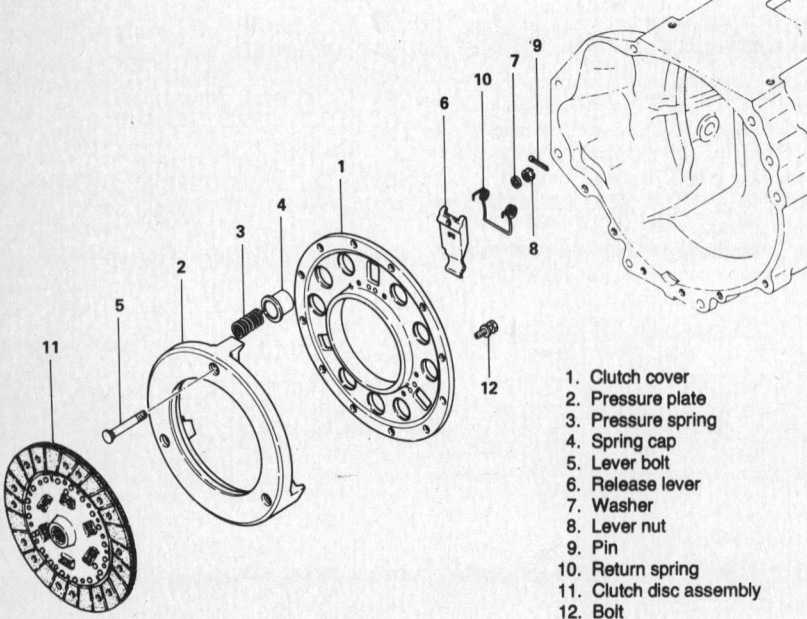

1. Clutch cover
2. Pressure plate
3. Pressure spring
4. Spring cap
5. Lever bolt
6. Release lever
7. Washer
8. Lever nut
9. Pin
10. Return spring
11. Clutch disc assembly
12. Bolt

Clutch components

Satoh

3. Remove the bolts from both right and left fenders and steps.
4. Remove the bolts behind the fenders.

NOTE: It is unnecessary to remove the bolts in front of steps.

5. Remove the return spring from the brake rod, and remove the brake rod from the brake pedal.
6. Remove the seat from the hydraulic case, with the seat bracket installed.
7. Remove the delivery pipe and suction pipe from the transmission case and the hydraulic case. Take care so that O-ring and seal washers are not lost.
8. Now, place a jack under the clutch housing (as much backward as possible), and place a garage jack under the front part of the transmission case.
9. Remove the bolts securing the clutch housing and the transmission case, and separate the clutch housing from the transmission case taking care not to bend the hydraulic pipe.

NOTE: Remove the one-way clutch or coupling.

NOTE: For a 4-wheel drive tractor, remove the universal joint, support and shaft for the front axle drive, and separate the clutch housing from the transmission case.

INSTALLATION

1. Apply sealant (Three-Bond® No. 4 or Helmeseal® 201) to the gasket between the clutch housing and the transmission case, and install it with dowel pins.
2. Install the over-running clutch spring and overrunning clutch to the PTO countershaft on the clutch housing, and install the coupling to the upper countershaft.
3. Install the over-running clutch (A) to the end of the transmission PTO shaft, and aligning the coupling splines, install the transmission case to the clutch housing. Tightening torque: 57.8-65.0 ft.lb.

NOTE: On the 4-wheel drive tractor, the mid pto countershaft, support, etc. have to be installed.

DISASSEMBLY

MAIN AND PTO Shifters

The main shifter and PTO shifter are positioned on the clutch housing side of the transmission case.

1. Loosen the bolts setting the clutch housing cover in which the main shift lever and PTO shift lever are assembled, and remove the cover.

NOTES: Sealant is applied to the cover. Remove the cover tapping with a screwdriver. Check to see that the shift rail stopper spring is in position.

2. Pull out the shift fork installed to the shift rail and adaptor set spring pin, and remove the shift rail. Remove the shift fork only after the countershaft is pulled out.

NOTES: When pulling out the shift rail, take care so that the stopper ball is not lost. Separately place the shift rail, shift fork and adaptor so that you will not be confused when reassembling.

3. Loosen the bearing retainer plate bolt and remove the plate.
4. Pull out the countershaft backward, on which the main shifter gear is mounted, and remove the ball bearing at the front of countershaft. While removing gears, 32T, 36T, 33T and 30T, remove the countershaft at the same time.
5. Remove the PTO shifter fork.

Mainshaft

1. Pull out both countershaft (installed with PTO gear) together with mainshaft backward. Make sure that the ball bearing at the front end of the countershaft is removed from the case, and pull out the mainshaft taking care so that the roller bearing at the front end of mainshaft does not contact the PTO gear.

NOTE: If only the mainshaft is removed, the roller bearing at the front end of the mainshaft will be damaged.

2. As required, remove the mainshaft bearing.

PTO Countershaft

Remove the bearing in the front of PTO countershaft. Remove the thrust liner, gear 39T, needle roller bearing, thrust liner, gear 32T, gear 36T, and gear 33T, and then remove the countershaft.

Remove the ball bearing, as required.

Important Notes

1. Don't attempt to disassemble the main shifter countershaft and PTO countershaft blind plug, unless oil leakage is noticed.
2. If the blind plug is leaky, it should be replaced. To replace the blind plug, the engine and clutch housing must be separated.

Reverse Gear

Pull out the 3 spring pins from the reverse shaft, remove the reverse shaft and remove the gear.

Differential Gear

1. Flatten the stopper washer on the differential gear support securing bolt. Loosen the bolt and remove the differential support.
2. Insert a screwdriver between the differential gear case and the ball bearing to pry out the ball bearing, and then take out the differential gear assembly.

Sub-Shifter Shaft

Remove the circlip, pull out the shaft backward, and remove the ball bearing, gear and collar. As required, remove the ball bearing and circlip from the shaft.

Drive Pinion Shaft

Straighten the tab washer at one end of the drive pinion shaft, and remove the sleeve nut.

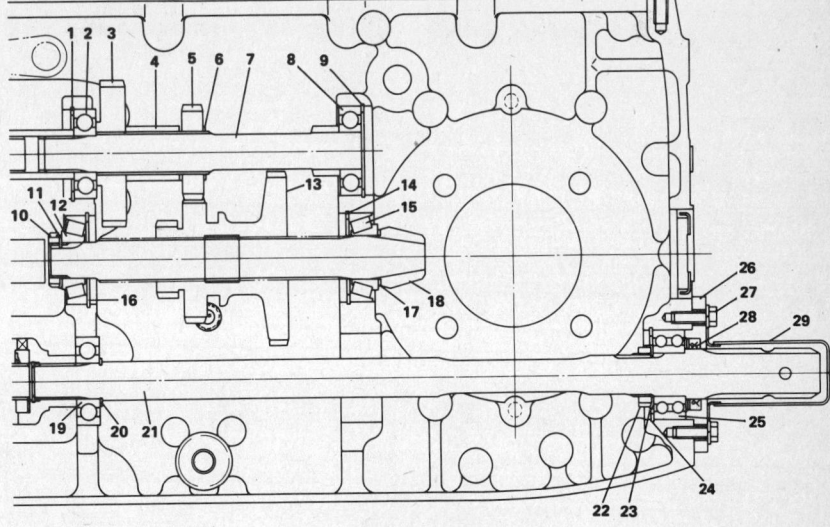

Sub-shifter shaft assembly

1. Circlip
2. Ball bearing #6305
3. Gear 49T
4. Collar
5. Gear 31T
6. Circlip
7. Shaft transfer
8. Ball bearing #6305
9. Circlip
10. Nut
11. Tub washer
12. Taper roller bearing
13. Gear sub change
14. Circlip
15. Taper roller bearing
16. Circlip
17. Shim
18. Pinion drive
19. Ball bearing
20. Circlip
21. PTO shaft
22. Nut
23. Tub washer
24. Circlip
25. Ball bearing
26. Cap PTO shaft bearing
27. Bolt
28. Oil seal M8 x 25
29. PTO cap

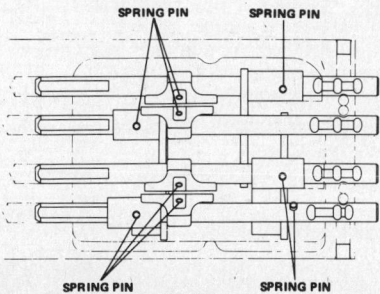

Main and PTO shifters

Satoh

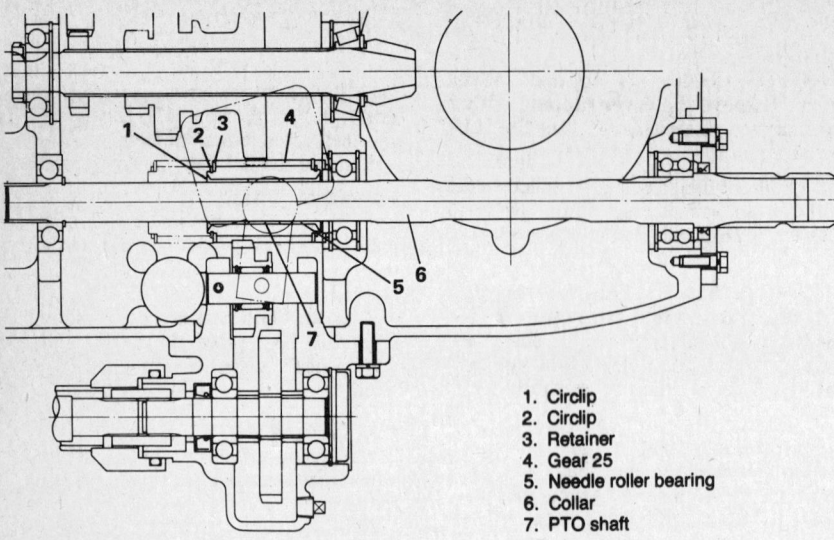

4-wheel drive gear unit

1. Circlip
2. Circlip
3. Retainer
4. Gear 25
5. Needle roller bearing
6. Collar
7. PTO shaft

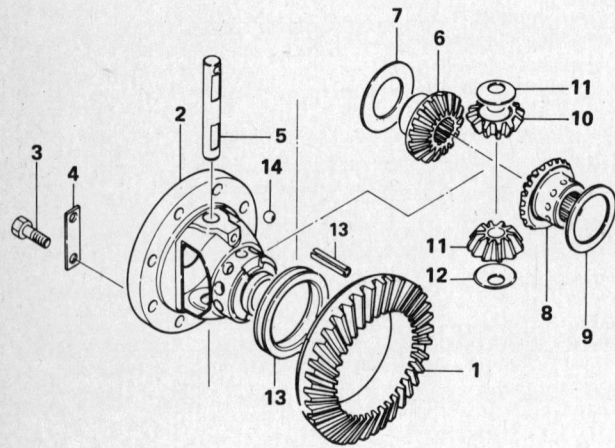

Differential gear components

1. Ring gear
2. Differential gear case
3. Bolt (10 x 25)
4. Lock plate
5. Shaft
6. Differential side gear
7. Thrust washer
8. Differential side gear
9. Thrust washer
10. Differential pinion
11. Thrust washer
12. Spring pin
13. Differential lock hub
14. Steel ball

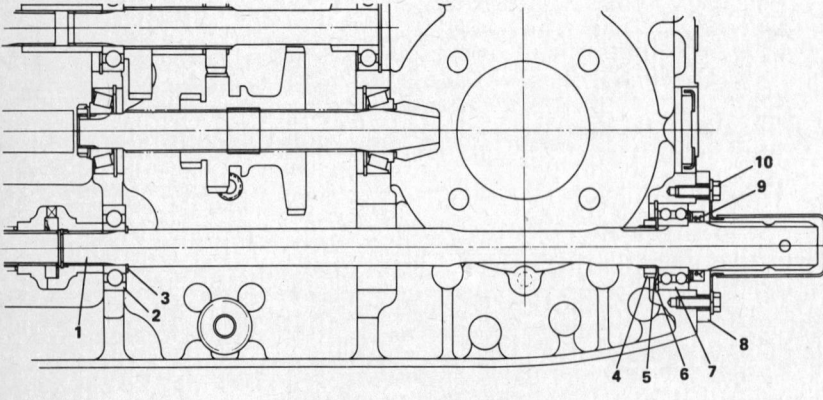

PTO shaft, 2-wheel drive

1. PTO shaft
2. Ball bearing
3. Circlip
4. Sleeve nut
5. Tub washer
6. Circlip
7. Ball bearing
8. PTO cap
9. Oil seal
10. Bolt

2. By tapping one end of the drive pinion shaft, pull out the shaft backward, and remove the gear.

As required, remove the shim and circlip for taper roller bearing adjustment.

PTO Shaft

1. Loosen the PTO shaft cap bolt on the rear part of the transmission case, and remove the cap.
2. By tapping the PTO shaft, pull it out backward. As required, remove the ball bearing and oil seal.

4-Wheel Drive Gear

Remove the circlip locking the 4-wheel drive gear, and pull out the gear circlip, liner, needle roller bearing, and collar while pulling out the PTO shaft.

4-Wheel Lever

1. Pull out the spring pin locking the 4-wheel shift lever, and remove the shift lever.

Remove the stopper ball and stopper spring.

2. Loosen the plate bolt, and remove the plate.
3. Remove the 4-wheel shift arm.

Sub-Shift Lever

1. Pull out the spring pin setting the sub-shift lever, and remove the sub-shift lever. Take care so that the stopper ball and stopper spring will not be lost.
2. Loosen the plate bolt, and remove the plate.
3. Remove the sub-shift arm.

ASSEMBLY

Differential Gear

1. Place thrust washers (7) & (9) on the side gears (6) & (8) respectively.
2. Place the side gears in differential case (2).
3. Place thrust washers (11) complete with differential pinions (10) in the differential case, and drive in the shaft (5). Measure the backlash, and adjust it to be 0.004-0.012 in. by changing side gear thrust washers.
4. Drive in the spring pin (12) to secure the shaft (5).
5. Place ring gear (1) on the differential case, and secure it with lock plates (4) attaching bolts (3).
Tightening torque: 21.67-28.9 ft.lb.

2-Wheel Drive PTO Shaft

1. Install the ball bearing to the PTO shaft, fasten it with washer and sleeve nut, and then lock with a tab washer.
Tightening torque: 28.9-50.6 ft.lb.
2. Fit the circlip in the circlip groove at one end of the PTO shaft.
3. Fit the circlip in the circlip groove in the rear of transmission case.

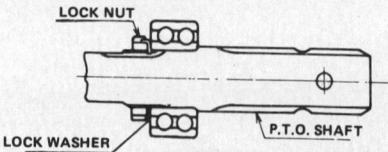

Installing the bearing on the PTO shaft

4. Apply grease to the PTO shaft cap, and set the oil seal.

5. Insert the PTO shaft into the transmission case from its rear side, install the gasket to the PTO cap, and install the cap to the transmission case.
Tightening torque: 7.22-8.67 ft.lb.

6. Fit the ball bearing to the front end of PTO shaft by striking with a hammer.

4-Wheel Drive PTO Shaft

1. Install the ball bearing to the PTO shaft, install the lockwasher and sleeve nut, and bend the lip of washer over the sleeve nut.
Tightening torque: 28.9-50.6 ft.lb.

2. Fit the circlip in the circlip groove at the rear end of the transmission.

3. Install the circlip to the area where the ball bearing at the center of the transmission case is installed, drive in the ball bearing, and lock with the circlip.

4. Fit a greased oil seal to the PTO shaft cap.

5. Install the circlip in the bore of the front drive counter gear the needle roller bearing, collar, and needle roller bearing (all of which should be oiled preliminarily), in that order, and lock with circlips.

6. Insert the PTO shaft into the transmission case from its rear end, and install the front drive counter gear. Install the gasket to the PTO cap and further install it in the transmission case.

7. Fit the circlip in the circlip groove at the front end of PTO shaft, and drive in the ball bearing.

4-Wheel Drive Shift Lever

1. Install a greased O-ring to the 4-wheel drive shift arm, and install it by inserting from inside of the transmission.

2. Install the plate in the transmission case.
Tightening torque: 7.22-8.67 ft.lb.

3. Install the stopper spring and stopper ball to the 4-wheel shift lever, and install the lever to the shift arm while taking care so that the spring and ball will not come off. Align the spring holes, and drive in the spring pin.

Assembling the Sub-Shift Lever

1. Fit the O-ring (oiled) in the O-ring groove on the sub-shift arm, and install the sub-shift arm from the inside of the transmission case.

2. Install the plate in the transmission case.
Tightening torque: 7.22-8.67 ft.lb.

3. Install the stopper spring and stopper ball to the sub-shift lever, and the sub-shift lever to the shift arm taking care so that the stopper spring and stopper ball do not fall off.
Align the spring pin with the spring pin hole and drive it in.

Drive Pinion Shaft

1. Adjust the cone center with shims (5).

a. Formula for determining the thickness of shims without using a special tool:

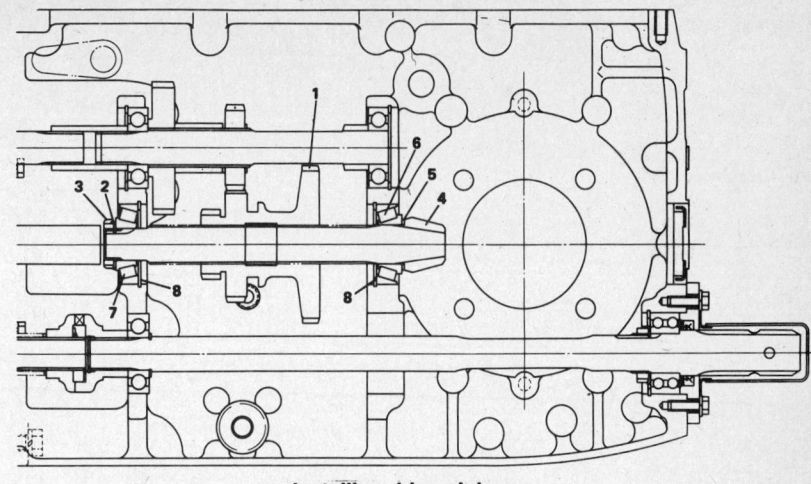

Installing drive pinion

1. Sub change gear
2. Lockwasher
3. Nut
4. Drive pinion
5. Shim
6. Tapered roller bearing (#32206)
7. Ball bearing (#6305)
8. Snapring

$$t \text{ (mm)} = 2.0 - (D) - \frac{X}{100}$$

where
t = thickness of shims (mm)
2.0 = factor (mm)
D = value indicated on the inner wall of the differential housing (mm)
X = value stamped on the end face of the pinion

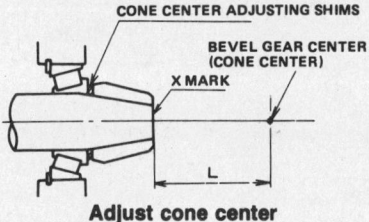

Adjust cone center

Example—1
Where the value (D) indicated on the inner wall of the differential housing is +1.75 and the value (X) stamped on the end face of the pinion is −9, we obtain

$$t = 2.0 - (D) - \frac{X}{100}$$
$$= 2.0 - (+1.75) - \left(\frac{-9}{100}\right)$$
$$= 2.0 - 1.75 - (-0.09)$$
$$= 2.0 - 1.75 + 0.09$$
$$= 0.34$$

Example—2
Where the value (D) is −0.26 and the value (x) is +, we obtain

$$t = 2.0 - (-0.25) - \frac{+15}{100}$$
$$= 2.0 + 0.25 - (+0.15)$$
$$= 2.10$$

b. Formula for determining the thickness of shims, utilizing a special tool:

$$t \text{ (mm)} = (L - 64) - \frac{X}{100}$$

where
t = thickness of shims (mm)
L = value measured (mm)
X = value stamped on the end face of the pinion (mm)
64 = designed value (mm)
$(L - 64) = 2.0 - (D)$

REFERENCE

Shim	Thickness (in.)
67700-30300	0.00276
67700-30400	0.00512
67700-30500	0.00984
67700-30600	0.02992
67733-01700	0.05906

2. Place shims (5) of the thickness determined as above on the drive pinion, and press bearing (6) to the housing.

3. Install snapring (31) in the housing.

4. Insert the drive pinion shaft from the rear, align the sub-shift lever pin with the sub-change gear shifter groove, and then install the gear to the driver pinion shaft. Next, drive in the bearing (30).

5. Place lockwasher (2), and tighten the locknut (3) until the specified preload is obtained.

6. Adjust the preload of the tapered roller bearing supporting the drive pinion shaft as described below:

a. Adjust the preload on a tractor which has been in service not exceeding 100 hours, or after the tapered roller bearing is replaced on a tractor which has been in service for longer than 100 hours.

• The preload should be 0.4335-0.8670 ft.lb. as measured at the center of the pinion shaft or 2.205-4.410 lb. as measured at the sub-change gear (59T) in use of a wire.

b. Adjust the preload if the tapered roller bearing is not replaced on a tractor which has been in service for longer than 100 hours.

• The preload should be 0.2168-0.2529

Satoh

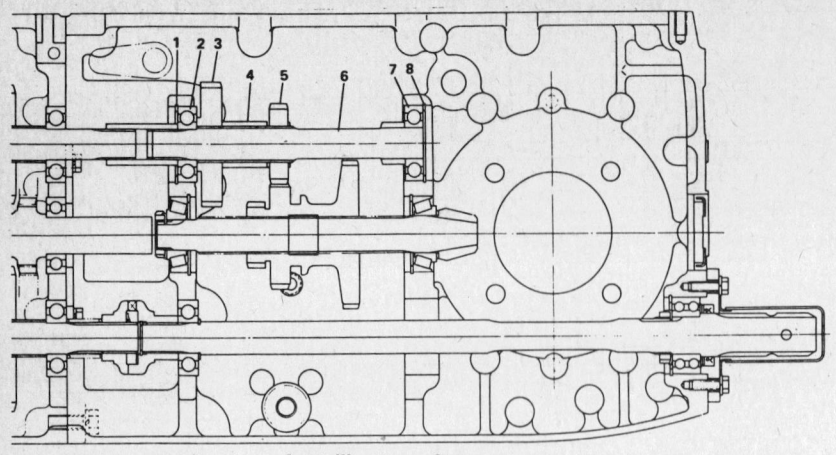

Installing transfer case
1. Circlip
2. Ball bearing
3. Gear 49T
4. Collar
5. Gear 31T
6. Shaft transfer
7. Ball bearing
8. Circlip

ft.lb. as measured at the center of the pinion shaft or 1.103-1.323 lb. as measured at the sub-change gear (59T) in use of a wire.

NOTE: When measuring the preload, be sure to give blows of a soft-faced hammer or a mallet to the pinion.

c. After adjusting the preload, be sure to lock the adjusting nut by bending the lockwasher (2) properly.

7. Install the straight PTO shift fork in the PTO gear (42T) groove in the shift fork so that the fork is on the rear side.

8. Install the shift fork, whose shift rail side is bent, to the PTO gear so that the fork is on the rear side.

Transfer Shaft
1. Install the ball bearing to the gear side at the rear end of the transfer shaft.
2. Install the circuit in the groove at the center of the transfer shaft.
3. Insert the transfer shaft from the rear side, and install the gear 31T, collar, and gear 49T, in that order. Lock the ball bearing with the circlip.
4. Drive in the ball bearing from the front end of the shaft, and install the circlip.
5. Shift the sub change gear, and check the gear teeth for defects.

Differential
1. Make sure the differential lock ball is firmly installed, and insert the differential assembly into the transmission case.
2. Install the ball bearing to the left side of the differential gear case by striking it.
3. Apply Heltite® to the O-ring groove on the screw, install the O-ring, and set the differential support temporarily.
4. Install the differential support (with the setscrew to the transmission case, and the O-ring together with the lockwasher.
5. Drive the taper roller bearing inner into the differential gear case.
6. Install the taper roller bearing outer to the differential support (with the setscrew), and install the differential support to the right side of the transmission case, together with the lockwasher, in a similar way to that for the left side differential support.
Tightening torque: 21.73-28.9 ft.lb.

BACKLASH ADJUSTMENT
1. Place a dial gauge contact point against the ring gear tooth surface end, and adjust gear backlash by turning the screw installed to the differential support.

NOTE:
1. To decrease backlash, loosen the lefthand adjusting nut and tighten the righthand adjusting nut.
2. To increase backlash, loosen the righthand adjusting nut and tighten the lefthand adjusting nut.

Measuring the Preload on the Differential Gear Taper Roller Bearing
Preload 1.0-2.5 ft.lb. (without pinion shaft)

Adjust the preload of the tapered roller bearing. To determine whether or not the preload is correct, add the standard value to the preload measured on the drive pinion. The procedure for measuring the preload is as follows:

a. Measure the preload of the drive pinion proper (with the differential assembly off the pinion).
b. Add 1.0-2.5 ft.lb. to the value obtained in the step (a).
Example—1
If the preload of the drive pinion is 0.5780 ft.lb., we obtain 0.578 ft.lb. + [0.0361-0.1084 ft.lb., = 0.6141-0.6864 ft.lb.
Example—2
If the preload of the drive pinion proper is 0.867 ft.lb., we obtain 0.8670 ft.lb. + [0.0361-0.1084 ft.lb. =0.9031-0.9754 ft.lb.
c. The preload must be measured at the transfer 3rd-speed gear. The formula for computing the preload from the measurement made by using a wire is

[Preload with drive pinion and differential assembly installed]

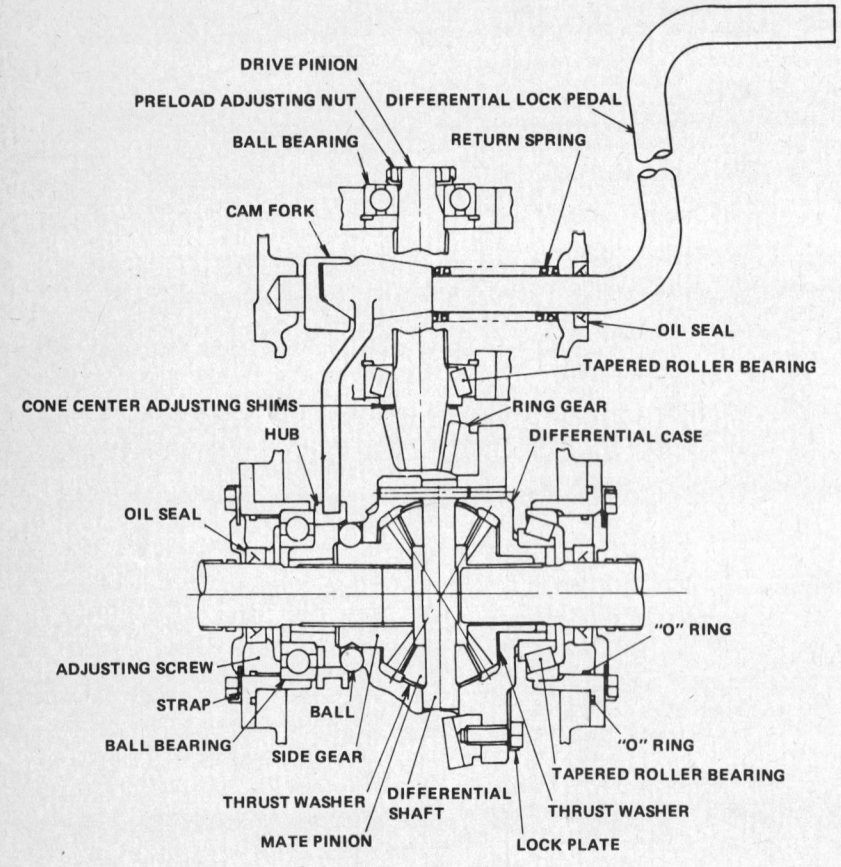

Assembling the differential

Preload of drive pinion = $\dfrac{0.0361\text{-}0.1084 \text{ ft.lb.}}{1.9291 \text{ in. (radius of transfer 3rd-speed gear)}}$

d. Adjust backlash properly and the preload, and lock the differential support with the lockwasher.

CHECKING THE RING GEAR TOOTH CONTACT

Clean the bevel gear and the transmission shaft pinion gear and apply red lead uniformly to 8-9 teeth on the bevel gear.

Giving resistance to the bevel gear, rotate the transmission shaft in the forward direction, thus making a pattern on the gear teeth.

Check the pattern on the bevel gear teeth, and if the teeth meshing is bad, make the following adjustments.

1. Heel meshing

Increase the thickness of the transmission shaft adjusting liner and allow the pinion to come near the bevel gear.

Keep the bevel gear off the pinion and adjust the backlash.

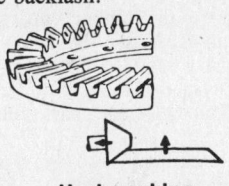

Heel meshing

2. Toe meshing

Decrease the thickness of the transmission shaft adjusting liner and keep the pinion off the bevel gear.

Then allow the bevel gear to come near the pinion and adjust the backlash.

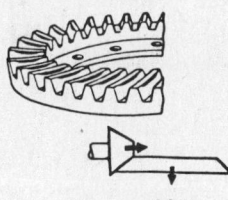

Toe meshing

3. Flank meshing

Make the same adjustment as of toe meshing.

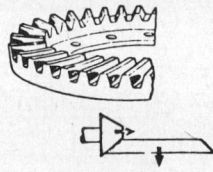

Flank meshing

4. Face meshing

Make the same adjustment as of heel meshing.

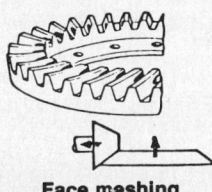

Face meshing

NOTE: One of the above will never occur independently; usually heel meshing and face meshing combined or toe meshing and flank meshing combined.

Repeat the above-mentioned adjustments until good teeth meshing can be obtained.

When the correct teeth meshing is obtained, make sure again of the bevel gear and the transmission shaft pinion backlash.

Backlash 0.006-0.012 in.

NOTE: After adjustment, clean the teeth.

Differential Lock Pedal

1. Apply grease to the oil seal, and fit it to the differential lock pedal on the right side of the transmission case.

2. Install the spring, differential lock and shift fork cam, in that order, and align the spring pin with the spring pin hole in the differential pedal, and drive it in.

NOTE: Firmly install the shifter fork to the differential lock hub.

NOTE: When the transmission case is disassembled, install the clutch housing first, and install the rear axle housing, hydraulic case, fender and rear tires in that order.

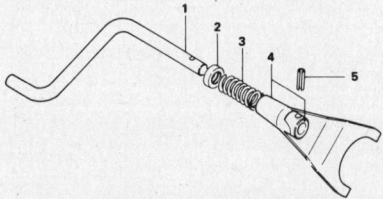

Differential lock pedal components
1. Differential lock pedal
2. Oil seal
3. Spring
4. Fork and cam
5. Spring pin

Installing the Rear Axle Housing

1. Apply sealant (Three-Bond® No. 4 or Helmeseal® No. 201) to the rear axle housing, install the dowel pins, and install the oil seal to the transmission case. During the operation, hold the rear axle housing in a horizontal position and take care so that the oil seal is not damaged.
Tightening torque: 57.8-65.0 ft.lb.

2. Install the bolt in the rear part of the step.

3. Apply sealant (Three-Bond® No. 4) to the transmission case, install the dowel pins, and tighten the hydraulic case bolt.

4. Install the delivery pipe to the hydraulic case taking care not to loosen the washer.

5. Install the O-ring to the suction pipe flunge, and then install the suction pipe to the transmission case.

6. Wash the oil filter clean, and install the cover together with the gasket.

7. Install the fender to the rear axle housing, the rubber cushion, plate, and lock plate, and secure them with bolts. Lock the bolt with the lock plate.

NOTE: After tightening the fender setting bolts, be sure to lock bolts.

8. Install the brake pedal rod, and install the return spring.
9. Install the rear wheel.
10. Install the seat to the hydraulic case.
11. Connect wires to the working lights and turn signals.
12. Install the 3-point link drawbar, if it has been removed.

Assembling the Main Shifter and PTO Shifter (Clutch Housing)

Thoroughly wash the interior of the clutch housing, and apply oil or grease to the moving parts. Assemble the shifter in the correct manner.

NOTE: If the engine is not demounted, care should be taken so that the blind plug will not come off.

Reverse Gear

1. If a reverse gear bushing is worn, it should be replaced. Apply oil to the inner surface of the bushing and shaft, and insert the shaft from the rear side. Install the collar and reverse gear so that its boss faces rearward as illustrated, and install the shaft.

2. Align the spring set holes in the shaft, case shaft and collar, and drive the spring pin into the holes.

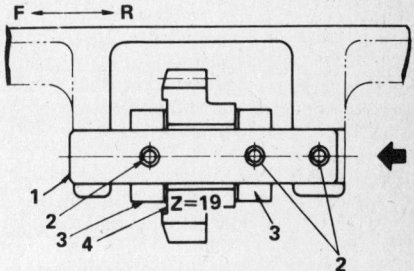

Assembling the reverse gear
1. Reverse shaft   3. Collar
2. Spring pin   4. Reverse gear

PTO Countershaft

1. Install the circlip to the clutch housing, and drive in the ball bearing.
2. Install the ball bearing to the rear end (splined) of PTO countershaft.
3. Insert the shaft from the rear end of the clutch housing and install gears 39T-36T so that 36T is on the rear side.
4. Install gear 42T to the countershaft so that the shifter fork groove is facing rearward. Install the liner, two needle roller bearings (oiled) and liner, and drive the countershaft into the bearing.

NOTE: Make sure that the PTO countershaft turns smoothly, and also check to see that the gears slide smoothly.

Mainshaft

1. Drive the ball bearing into the rear end of the mainshaft.
2. Insert the mainshaft from the rear end of the clutch housing, bring it to engage with the PTO shifter gear, and install the mainshaft.
3. Make sure that the mainshaft turns smoothly.

Satoh

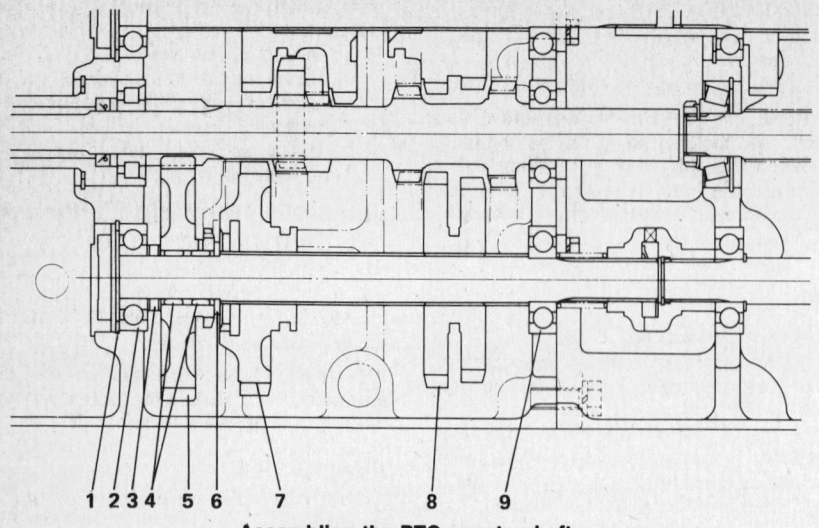

Assembling the PTO countershaft

1. Circlip
2. Ball bearing
3. Liner
4. Needle roller bearing
5. Gear 45T
6. Liner
7. Gear 42T
8. Gear 39T, 36T
9. Ball bearing

Main Shifter Countershaft

1. Fit the circlip to the clutch housing, and drive in the ball bearing.
2. Drive in the ball bearing from the rear end (splined) of the countershaft.
3. Insert the countershaft from the rear end of the clutch housing, and install gears 33T-30T to the shaft with the shifter fork groove facing forward. Install the countershaft with gear 36T on the rear side.

Shift Fork and Rail

1. Install one of the two shift rails having a spring pin hole at center and rear end to the bent one of the PTO shift forks (3rd and 4th). Install the adaptor and lock with the spring pin.
2. Install the two steel balls to the gear locker, and insert the shift fork having three spring pin holes. Install the adaptor, install the shift rail to the PTO shift fork (1st and 2nd), and lock with the spring pins.
3. Install the 1st reverse shift fork (with a shorter boss) to the gear so that the boss is on the front side. Insert the shift rail (having two spring pin holes at its center) from the rear side, install the adaptor and shift fork (1st and reverse), and lock with the spring pins.
4. Install the 3rd and 4th shift forks (with a longer boss) to the gear so that the bosses are on the rear side.

Install two gear locker steel balls, and insert the rail of the same type as the PTO shift rail from the rear side. Install the adaptor and lock with the spring pin.

NOTE: Take care so that the steel ball for the gear locker does not fall. Drive in the spring in the correct direction.

5. Operate the shift rail, and make sure that the gear will not mesh with the rail simultaneously.
6. Install the shift rail stopper ball and stopper spring to the respective shift rails.
7. Apply sealant to the clutch housing cover, and install it to the clutch housing.
   Tightening torque: 21.681-28.90 ft.lb.
8. By operating the shift lever, make sure that it moves smoothly.

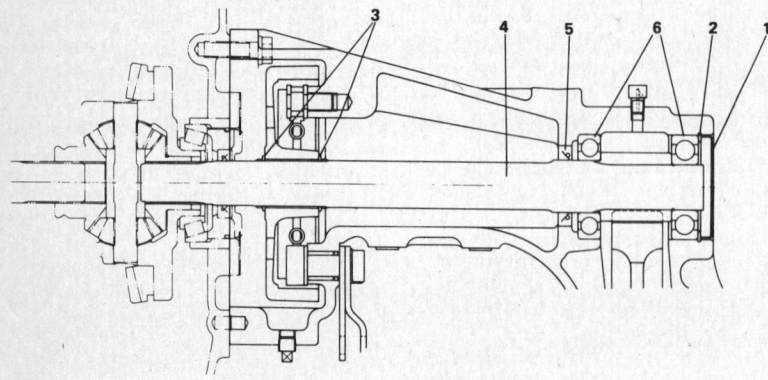

Rear axle components

1. Rubber cap
2. Snapring
3. Snapring
4. Final drive shaft
5. Oil seal (SC405211)
6. Bearing (#6208)
7. Bearing (6307)

REAR AXLE

DISASSEMBLY

After removing the rear axle housing from the transmission case, proceed as follows:

1. Remove the final gear case drain plug, and drain off the gear oil.
2. Remove the circlip and the brake drum.
3. Remove the brake shoes and the circlip from the final driveshaft.
4. Remove the rubber cap and remove the circlip from the rear axle housing.
5. By tapping on the final driveshaft with a soft-faced hammer, pull it out from the transmission case side.

NOTE: As required, remove the ball bearing and oil seal.

Wheel Shaft

1. Loosen the rear axle housing oil pan bolt, and remove the oil pan.
2. Remove the rubber cap, straighten the lockwasher, loosen the sleeve nut, and pull out the wheel shaft by tapping on it with a soft-faced hammer.
3. Remove the final gear, collar A and collar B.

As required, remove the taper roller bearing, oil seal and circlip.

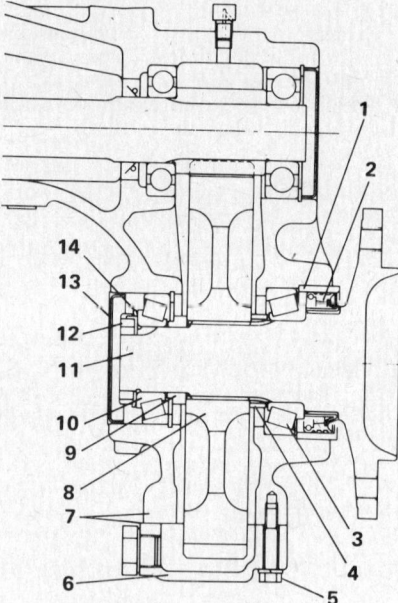

Wheel shaft components

1. Bolt (M8 x 40)
2. Bolt (M8 x 20)
3. Oil pan
4. Gasket
5. Rubber cap
6. Lockwasher
7. Nut
8. Axle shaft
9. Final gear
10. Collar A
11. Collar B
12. Bearing (#30208)
13. Oil seal
14. Bearing (#32210)
15. Collar

INSPECTION

1. Using a wire gauge, measure the backlash between the splines of the final driveshaft and brake drum. If the backlash

exceeds the service limit, replace the shaft and drum as an assembly.

2. Using a calipers, measure the displacement over two teeth at the pitch circle of the pinion of the final driveshaft. If the displacement exceeds the service limit, replace the final driveshaft.

SPECIFICATIONS

Item	Standard	Service Limit
Backlash between final drive shaft and brake drum splines	0.0035 in.	0.012 in.

SPECIFICATIONS

Item	Standard	Service Limit
Displacement over two teeth of final drive pinion	0.77-0.771 in.	0.763 in.

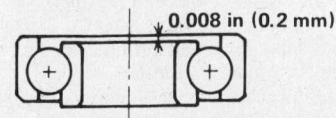

Checking ball bearing axial clearance

3. Measure the displacement over nine teeth at the pitch circle of the final driven gear. If the displacement exceeds the service limit, replace the gear.

SPECIFICATIONS

Item	Standard	Service Limit
Displacement over nine teeth of final driven gear	4.0573-4.060 in.	4.0533 in.

4. Check the ball bearing for axial clearance. Replace the bearing if the clearance exceeds 0.0079 in.

SPECIFICATIONS

Item	Standard	Service Limit
W7 x 24 disc wheel (for 9-24 tires) Lateral runout and radial runout	0.118 in.	0.197 in.

5. Check the disc wheel for runout and face runout. Replace the wheel if the radial and/or lateral runout exceeds the service limit.

ASSEMBLY

1. Install the circlip to the rear axle housing and install the taper roller bearing outers in their respective housings.
2. Install the collar for oil seal mounting in the housing so that its chamfered area is on the outer side.
3. Install the oil seal (greased) to the final driveshaft and drive it in so that the oil seal end is flush with the housing surface.
4. Drive in the ball bearing.
5. Put the final gear in the housing, and align it with the wheel shaft splines, and install.
6. Install the collar, drive in the taper roller inner race, install the tub-washer, and tighten the sleeve nut.

Tighten sleeve nut to preload the bearing as described below:

a. Tighten the nut so that the torque for starting rotation of the wheel shaft is increased by 1.445 to 2.168 ft.lb. than is before giving preload to the bearing.
b. To measure the starting torque, hook a spring balancer to a hub bolt hole and pull the balancer squarely.

7. Apply sealant to the outer surface of the rubber cap, and drive it into the housing.

NOTE: Be sure to always use a new rubber cap.

8. Apply sealant to the oil pan gasket and install the oil pan in the rear axle housing.

Tightening torque: 12.28-13.72 ft.lb.

Final Driveshaft

1. Install the final driveshaft to the ball bearing, drive in the ball bearing, and install the circlip in the housing.
2. Apply sealant to the outer surface of the rubber cap, and install it in the housing. Always use a new rubber cap.
3. Install the brake drum lock circlip to the final driveshaft.
4. Install the brake shoes and lock with the circlips.
5. Install the seal washer to the drain plug and tighten it to the housing firmly.

HYDRAULIC SYSTEM

For complete service, see the S-370 section.

ENGINE

The engine used in this tractor is the Mitsubishi KE130 diesel.

REMOVAL

1. Place the tractor on a flat place, and drain off the engine oil and cooling water (or coolant).
2. Place a garage jack under the clutch housing, and jack up the tractor a little so that the front tires are still in contact with the ground.
3. Remove the cotter pin beside the bonnet and then take out the bonnet stopper rod.
4. Disconnect the minus cable (earth) from the battery terminal.
5. Disconnect the wire of the instrument panel from each connecting part.
6. Remove the safety guard and take out the throttle control lever.
7. Loosen the hydraulic pipe tightening bolt and remove the bracket and the suction pipe and delivery pipe from the pump. In this case care should be taken of O-ring.
8. Remove the leak-off pipe beside the fuel tank. Then remove the fuel pipe beside the filter and lock the pipe to the fuel tank taking care so that any fuel may not flow out.
9. Remove the cotter pin of the drag

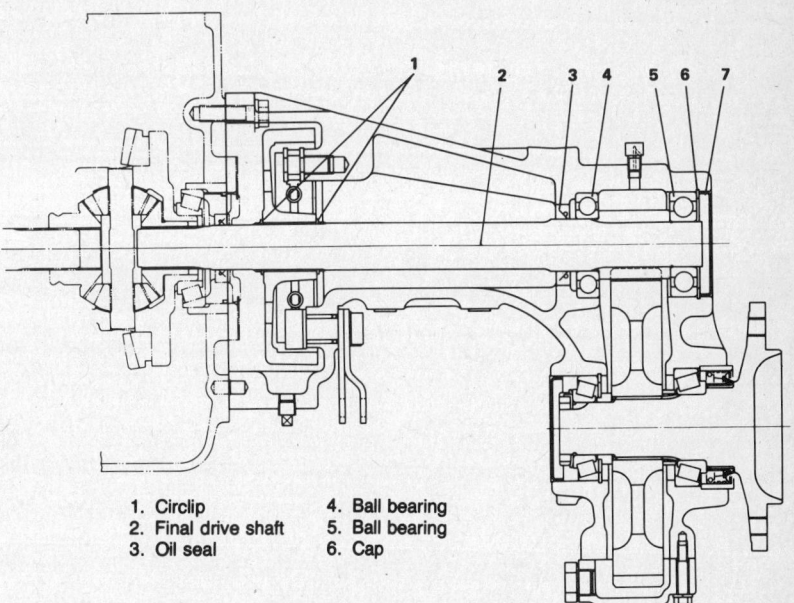

1. Circlip
2. Final drive shaft
3. Oil seal
4. Ball bearing
5. Ball bearing
6. Cap

Final drive shaft components

Satoh

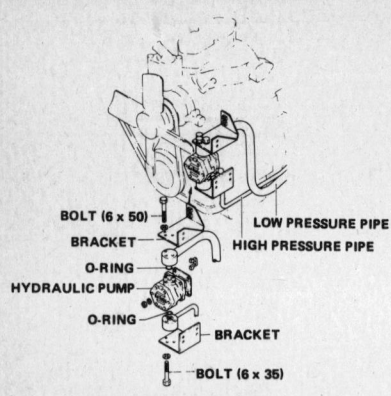

Hydraulic pump removal

link ball socket tightening nut, loosen the nut and remove the drag link from the steering lever with the ball socket remover.

10. Loosen the radiator hose clamp beside the engine and remove the radiator hose.
11. Remove the air cleaner hose from the intake manifold setting part.
12. Loosen the chassis tightening bolt on the both sides of the chassis and at the front end and then carefully demount the chassis pulling forward with the front axle radiator skirt installed.
13. In the above-mentioned status check and see that jack under the clutch housing is properly and securely supporting the clutch housing, fasten wire rope to the engine cylinder head hanging bolt and then suspend the engine with the chain block as much as the wire is stretched.
14. Remove four bolts installing the rear plate and the clutch housing.

NOTE: It is not necessary to remove the rear plate setting bolt beside the engine. Take special care not to cause any bending when demounting the engine because sealant is applied between the plate and the clutch housing.

15. Make uniform tightening and removal of six bolts tightening the engine and the clutch housing, take out the engine slowly and carefully from the clutch housing and then place it on the work bench.

Cylinder Head

REMOVAL

1. Remove the air breather pipe.
2. Remove the intake manifold, intake pipe and exhaust manifold.
3. Remove the lube oil pipes from the cylinder head.
4. Remove the alternator brace mounting bolts.

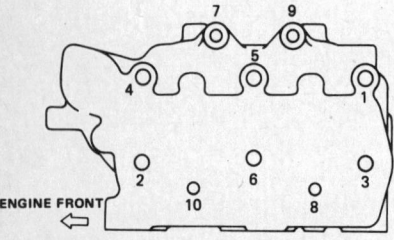

Head bolt loosening sequence

5. Disconnect the fuel injection pipe from each nozzle holder.
6. Remove six rocker cover securing bolts and remove the rocker cover.
7. Loosen the cylinder head bolts in the sequence shown in the figure, and take them off.
8. Remove the cylinder head assembly.

DISASSEMBLY

1. Remove the nozzle holder assemblies being careful of the copper packing.
2. Remove the flow plug lead wires, and remove the glow plugs.
3. Loosen off the rocker stay mounting bolts, and remove the rocker shaft assembly.
4. Using a valve lifter, compress each valve spring, and remove the valve cotter, and then remove the spring retainer and valve spring.

INSPECTION

1. Check the clearance between the valve stem and valve guide. If the clearance is in excess of the service limit of 0.006 in. (in.) or 0.008 in. (ex.), replace the valve guide and valve.

When replacing a valve whose stem is excessively worn, replace the matched valve guide with new one.

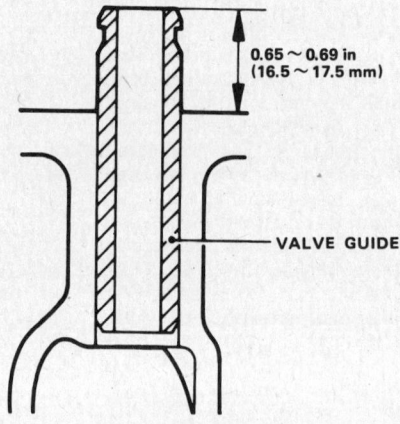

Valve guide replacement

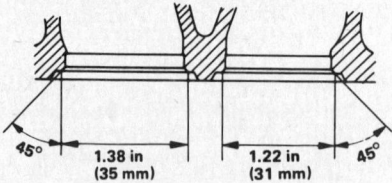

Valve seat dimensions

2. When replacing a valve guide, press out the old valve guide from the lower side of the cylinder head to the upper side using a press. Then, press a new valve guide into the cylinder head from the upper side until the specified height above the cylinder head is obtained. After installing the valve guide, check the clearance between the valve guide and valve stem. If the clearance is too small, ream the valve guide until the specified clearance is obtained.
3. Check each valve seat for damage or improper contact. Correct the valve seat, if necessary, as shown in the figure. After refacing the valve seat, lap the valve against the valve seat insert by using a proper grade of lapping compound.

NOTE: Refacing of the valve seat should be made after valve guide is checked for wear. If the valve guide must be replaced, reface the valve seat after replacing the guide.

SPECIFICATIONS

Item		Standard	Service Limit
Valve seat width	Intake	0.051-0.071 in.	0.089 in.
	Exhaust	0.051-0.071 in.	0.089 in.
Valve head thickness	Intake	0.006 in.	0.02 in.
	Exhaust	0.06 in.	0.02 in.

4. Determine the amount of valve seat sinkage by measuring the installed length of the valve spring as shown here. If the sinkage is in excess of the service limit of 0.04 in., replace the cylinder head.
5. Check the valve stem and seat for wear, damage, or distortion. Repair or replace the valve if necessary.
6. Check the edge thickness (T) of the valve head. If this thickness is in excess of the service limit, replace the valve.
7. Check the valve stem end (A) (con-

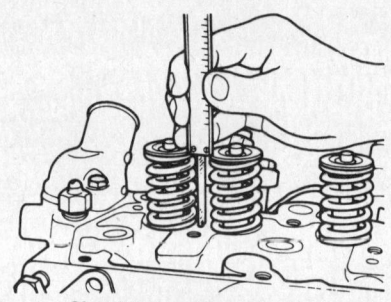

Checking valve seat sink

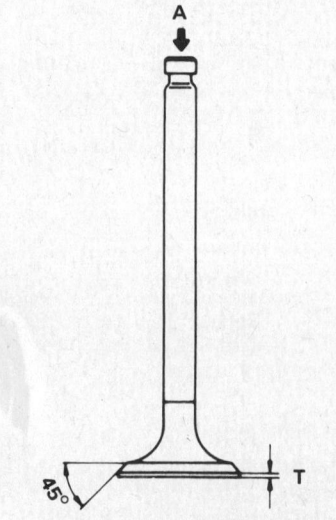

Checking valve face dimensions

tacting surface with rocker arm) for wear or galling. Repair or replace the valve if necessary.

8. Check each valve spring for crack or damage.

9. Measure the free length and as-installed tension of each valve spring. If the spring is weakened, replace the valve.

10. Check each valve spring for squareness, and replace if necessary.

SPECIFICATION

Item	Standard	Service Limit
Free length	2.28 in.	
Tension	46.74 lb. at 1.83 in.	15% decrease
Installed height	1.83 in.	15% decrease
Squareness	2°, max.	2°, max.

11. Check the valve contacting surface of each rocker arm for wear or damage. Replace any unserviceable rocker arm with new one. Check the adjusting screw on each rocker arm. If excessive wear or damage is found on the push rod contacting surface, replace the screw with a new one.

12. Measure the inside diameter of the rocker arm bushing and the outside diameter of the rocker shaft at the rocker arm bearing portion, and replace the rocker arm assembly if the clearance exceeds the service limit.

ASSEMBLY AND INSTALLATION

NOTE: Fit the new valve stem seal to each valve guide securely.
Do not reuse any valve stem seal that has been used.

1. Coat the valve stem with clean engine oil, and insert the stem into the valve guide.
Assemble the valve spring and retainer, and fit the retainer lock while compressing the spring with a valve lifter.

2. Assemble the rocker arms, rocker shaft and rocker stay as shown.

3. Be sure to install the rocker shaft in the correct direction. The direction is correct if the identification mark on the shaft front end is facing right (toward the push rod) as viewed from the engine front side.

4. Install the rocker arm and shaft assembly to the cylinder head, and tighten the mounting bolts securely.

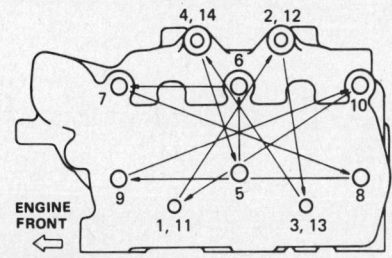

Cylinder head tightening sequence

5. Install the glow plugs, and tighten them to 18 ft.lb.
NOTE: This glow plug has a tapered end construction requiring no gasket as leakage of gases can be prevented by tight contact of this end against the cylinder head.

6. Place the nozzle holder gasket, and install the nozzle holder assembly. Tighten the mounting bolts to 57 ft.lb.
NOTE: Discard any nozzle holder gasket that has been used.
The cylinder head gasket has a coating of sealant by dipping method. Do not apply any sealing agent to the gasket.
Discard the gasket that has been used.

7. With the model number side of the gasket facing the cylinder head, install the push rod. The cylinder head bolts should be tightened to the specified torque in the sequence shown in the figure.
Tighten the cylinder head bolts in two or three steps gradually increasing the torque up to 101-115 ft.lb.

Valve Clearance Adjustment

With the piston positioned at the top dead center on compression stroke, insert a 0.014 in. thickness gauge between the valve stem end and the rocker arm, and turn the adjusting screw until the specified clearance (0.014 in.) is obtained.

Cylinder Block

DISASSEMBLY

1. Pull out the push rods, and take out the tappets using a magnet.
2. Turn the cylinder block upside down and remove the oil pan.
3. Remove the oil screen and oil pump.
4. Loosen off the crankshaft pulley mounting nut, and remove the pulley.
5. Remove the fuel injection pump being careful of the shim between the pump and the crankcase.
NOTE: Be very careful not to let dust particle enter the fuel injection line.
6. Remove the gear case with great care.
7. Remove the injection pump camshaft complete with the ball bearings from the crankcase.
8. Remove the governor weight assembly from the camshaft.
9. Remove one bolt and plain washer sucuring the camshaft bearing, and pull out the camshaft.
10. Remove the crankshaft gear.
11. Straighten the lockwashers on the connecting rod cap tightening bolts, and remove the rod cap.
12. Then, push out the piston complete with the connecting rod to the cylinder head side.
NOTE: Identify the pistons, connecting rods, bearings and caps (of the same number) to aid reassembly. In pushing out the piston and connecting rod, be sure to use the handle (wooden) of a hammer.
13. To separate the piston and connecting rod, first remove the snaprings at both ends of the pin, then pull off the piston pin.
14. After removing the main bearing caps, remove the crankshaft.
NOTE: Arrange the removed main bearing with the respective bearing cap from which it has been removed.

INSPECTION

Cylinder Liners

1. Check the inside surface of each cylinder liner for scratches, damage, or wear; replace the liner if faulty.
2. Measure the inside diameter of the cylinder liner using a cylinder gauge, and take readings in two directions, "A" and "B", at three elevations as shown here. Rebore the liner to an oversize, or replace it, as necessary.

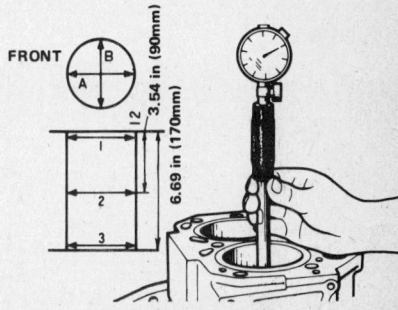

Checking liner dimensions

Specifications

Item	Standard	Repair Limit	Service Limit
ID of cylinder	3.62 in.	0.008 in.	0.047 in.
Out of roundness	0.0004 in.	—	—
Protrusion above top face of crankcase	0.0012-0.006 in.	—	—

3. When replacing the cylinder liner, be sure to replace the O-ring with a new one.
Before inserting a cylinder liner into the crankcase, fit the O-ring into the groove provided in the outer lower portion of the cylinder liner, and coat the O-ring lightly with engine oil.

4. When the cylinder liner is not worn beyond the service limit, but the piston rings alone are to be replaced, check the upper end of the cylinder bore for ridge. Correct ridge by using a ridge reamer and refinish the liner by honing, if necessary.

Piston, Piston Pins and Piston Rings

1. Check the piston for sticking, scratches and wear. Replace it if found defective.

Satoh

2. Measure the piston diameter at its skirt in the direction transverse to the pin hole, using a micrometer. Replace the piston if worn excessively. If the clearance between the piston and cylinder is excessive, replace either the liner or the piston whichever is worn badly, or replace both of them.

SPECIFICATIONS

Item	Standard	Service Limit
Piston OD (at skirt)	3.54 in. (90 mm)	
Cylinder-to-piston clearance	0.0045-0.006 in.	-0.012 in.

3. When using an oversize piston, rebore the cylinder liner to the oversize as follows:
Measure the diameter of an oversize piston to be used (the diameter at piston skirt in the direction transverse to the pin hole), and rebore the cylinder liner so that the specified cylinder-to-piston clearance (0.0045 to 0.006 in.) can be obtained, then finish the liner by honing. The oversizes are available in four sizes: 0.0098, 0.02, 0.03, 0.04 in.

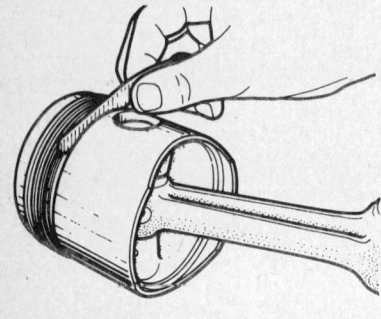

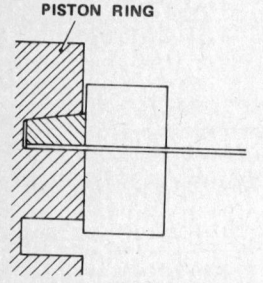

Measuring piston ring side clearance

4. Measure the piston ring side clearance; if found excessive, replace the ring. If the clearance is still excessive with a new piston ring installed, replace the piston.
When measuring the side clearance of the No. 1 compression ring, use a straight edge in such a manner that the ring face is flush with the piston side face.

SPECIFICATIONS

Item		Standard	Repair Limit
Width of ring	Compression ring	0.098 in.	
	Oil control ring	0.16 in.	
Piston ring side clearance	No. 1 ring	0.012 in.	
	No. 2 and No. 3 rings and oil control ring	0.008 in.	

5. Measure the piston ring end gap; if the gap is in excess of the service limit, replace the ring. To measure the ring gap, fit the piston ring into the cylinder, and press down the ring with a piston head to the lower portion of the cylinder where the wear is the smallest, and measure the gap between the ends of ring using a thickness gauge.

SPECIFICATIONS

Item		Standard	Service Limit
Piston ring gap	No. 1 thru No. 3 rings	0.012-0.02 in.	0.06 in.
	Oil control ring	0.008-0.016 in.	0.06 in.

6. Check the clearance between the piston pin and the piston, and the clearance between the piston pin and the connecting rod small end bushing. Replace the piston pin assembly or connecting rod assembly if the clearance is excessive.

SPECIFICATIONS

Item	Service Limit
Piston pin-to-piston clearance	0.0031 in.
Piston pin-to-connecting rod bushing clearance	0.004 in.

Connecting Rods

1. Measure the bend or twist of each connecting rod, and repair the rod with a press if the repair limit is not exceeded. Use a connecting rod aligner for checking.

SPECIFICATIONS

Item	Repair Limit
Connecting rod bend or twist	0.006 in.

2. Measuring connecting rod thrust clearance: Secure the rod to the crankpin by tightening the cap bolts to the prescribed torque, push the rod to one side, and measure the thrust clearance of the rod by

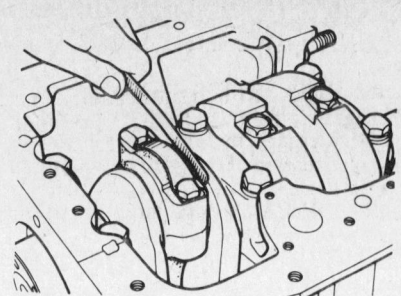

Measuring connecting rod thrust clearance

inserting a thickness gauge on the other side of the rod. If the clearance is too great, replace the rod assembly.

SPECIFICATIONS

Item	Service Limit
Connecting rod big end thrust clearance	0.02 in.

Crankshaft

1. Measuring crankshaft bend:
Hold the crankshaft with V-blocks or supports as shown here, and measure the runout by putting a dial gauge to the journal and turning the shaft slowly. The amount by which the shaft is bent is one half of the runout (as read on the dial gauge). When the bend is within the repair limit, straighten the crankshaft with a power press.

SPECIFICATIONS

Item	Standard	Repair Limit
Crankshaft bend	0.0006 in.	0.002 in.

2. Check the crankshaft journals and crankpins for damage and sticking; replace the shaft if faulty.

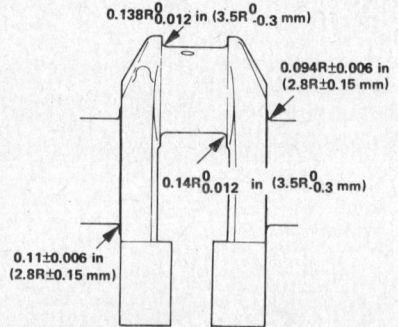

Measuring crankshaft fillet radius

3. Measure the crankpin and journal diameters with a micrometer. If wear exceeds the limit, regrind them to the next undersize, and replace the main bearings and connecting rod bearings to the same undersize.

SPECIFICATIONS

Item		Standard	Service Limit
Journal		2.32 in.	−0.037 in.
Crankpin OD		2.07 in.	−0.047 in.
Under-sizes	Journals	0.0098 in.	2.3108-2.3114 in.
		0.02 in.	2.3010-2.3016 in.
		0.03 in.	2.2911-2.2917 in.
	Crank-pins	0.01 in.	2.0761-2.0768 in.
		0.02 in.	2.0663-2.0670 in.
		0.03 in.	2.0565-2.0571 in.
		0.04 in.	2.0466-2.0472 in.

NOTE: When grinding the crankpin, be sure to remove the counterweight. After grinding, install the counterweight to the original position, and tighten the mounting bolts to the specified torque. Regrinding of the crankshaft should be made after correction of crankshaft bend. Finish the fillets to the specified radius, as shown.

4. Measure the crankshaft end play; if the end play is excessive, replace the rear main bearing upper.

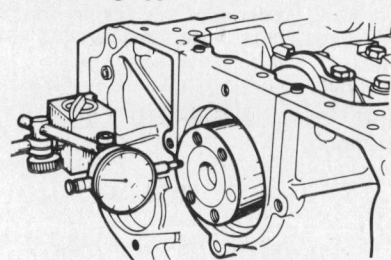

Measuring crankshaft end play

To measure the end play, assemble the main bearings, crankshaft and main bearing caps in position, and tighten the bolts to the specified torque, and put the gauge spindle to the rear end face of the crankshaft as shown. Force the crankshaft toward the dial gauge and then force it in the opposite direction, and note the amount of end play on the gauge shown as the deflection of pointer.

SPECIFICATION

Item	Standard
Crankshaft end play	0.0026-0.0079 in.

Main Bearings and Connecting Rod Bearings

1. Check each bearing surface for flaking, corrosion, sticking and the sign of uneven contact; replace the bearing if found faulty.

2. Install the main bearings and the connecting rod bearings to the crankcase and connecting rods, respectively, and tighten the cap bolts to the prescribed torque, 65.0 to 72.2 ft.lb. for main bearing cap bolts, and 36.1 to 39.7 ft.lb. for connecting rod bearing caps, and measure the inside diameter of each bearing. Next, measure the crankshaft journal OD and the crankpin OD, and compute the oil clearance. If the oil clearance is excessively large, replace the bearing. If the specified oil clearance cannot be obtained even by the replacement of the bearing, regrind the crankshaft to the next undersize, and assemble the bearing of the same undersize. The undersizes of the bearings are 0.0098, 0.019 and 0.0295 in. and 0.0394 in. for connecting rod bearings only.

SPECIFICATION

Item		Service Limit
Oil Clearance	Connecting rod bearings	0.006 in.
	Main bearings	0.004 in.

Timing Gears

Check the tooth face of each gear for pitting, scarring or wear, and replace a defective gear.

Camshaft

1. Check the clearance between the camshaft rear journal and the bore in the crankcase. If the clearance is excessive, replace the camshaft or crankcase.

SPECIFICATION

Item	Repair Limit
Clearance between rear journal and crankcase	0.006 in.

2. Check the cam lobe for wear or damage; replace the camshaft if the cam lobe has been worn excessively or damaged.

SPECIFICATIONS

Item	Standard	Service Limit
Cam height (for both intake and exhaust valves)	1.605 in.	−0.039 in.

Injection Pump Camshaft

Check the cam lobe for wear or damage; replace the camshaft if the cam lobe has been worn excessively or damaged.

SPECIFICATIONS

Item	Standard	Service Limit
Cam height	1.220 in.	−0.039 in.

Ball Bearings

1. Check the ball bearing for wear or damage, or abnormal noise or binding in rotation; replace the ball bearing if found defective.

2. When press-fitting a ball bearing to the gear or shaft, be sure to apply load to the inner race only.

Tappets

1. Check the bottom face (cam contacting face) of each tappet for cracks, flakes, or scratches; replace the tappet if the face is defective.

2. Check the clearance between the tappet and bore in the crankcase; replace the tappet if the clearance is excessive.

SPECIFICATIONS

Item	Standard	Repair Limit
Tappet OD	1.024 in.	−0.039 in.

Push Rods

1. Check each end of the push rod for wear; replace the push rod if wear is excessive.

2. Place the push rod on a surface plate, and measure the bend at the center portion. If the bend is excessive, correct or replace the push rod.

SPECIFICATION

Item	Standard
Push rod bend	0.012 in.

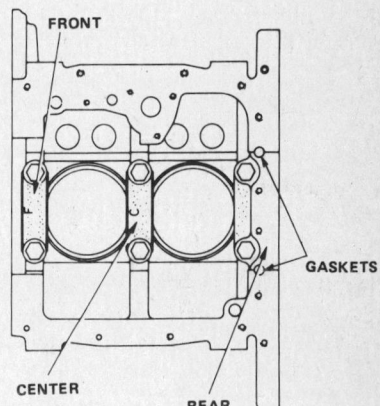

Installing main bearing caps

ASSEMBLY

1. Install the cylinder liners into the crankcase. When inserting a cylinder liner, replace the O-ring in the groove at the lower portion with new one, and coat the O-ring with engine oil sparingly. Push the cylinder liner straight down to the bottom. After installation, check the protrusion above the gasket surface of the crankcase.

2. Install the main bearing upper shells to the crankcase, and the lower shells to the respective bearing caps. Fit the flanged bearing shell to the rear bearing saddle in the crankcase. Be sure to install the bearing shells to their original positions.

3. Install the crankshaft. Apply a coat of engine oil to the journals and pins.

4. Install the main bearing caps to the respective positions, and tighten the bearing cap bolts to the specified torque.

Satoh

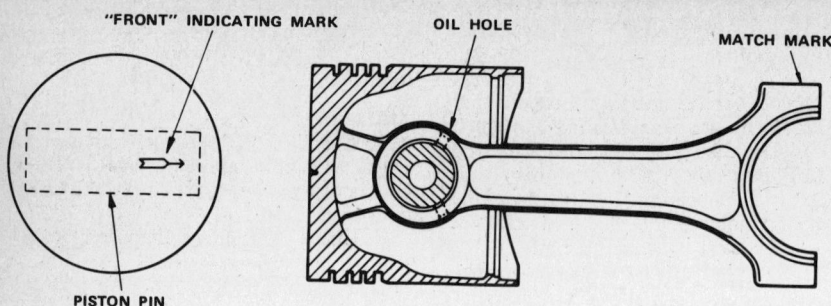

Piston and rod positioning

SPECIFICATION

Item	Standard
Main bearing cap bolt tightening torque	65.0-72.2 lb.ft.

The front and center main bearing caps are marked "F" (front) and "C" (center), respectively, as viewed from the flywheel side. Install them as shown below. After installation, rotate the crankshaft to see if it can be turned smoothly. Apply Three-Bond® No. 20 to the upper face (mating surface to the crankcase) of the rear bearing cap.

5. Apply Three-Bond® No. 2 to the outside surface of the gasket, and insert it into position from the bottom side.

6. Press-fit the crankshaft rear oil seal to the crankshaft from the rear side of the crankcase.

NOTES:
a. Apply a thin coat of engine oil to the lip of the oil seal.
b. Apply Helmeseal® 52B to the outside peripheral portion of the oil seal.
c. Use care not to damage the lip portion when fitting the seal.

7. Check the crankshaft end play.

8. Install the piston rings to the piston. Each piston ring should be installed to the piston so that the side having the manufacturer's mark and size mark faces upward (piston top side).

9. Install the piston and connecting rod assembly into the cylinder from the cylinder top, while compressing the piston rings. Fit the big end of the connecting rod to the crankpin.

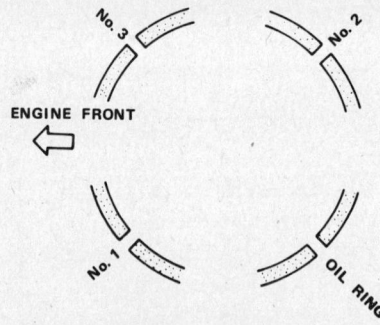

Ring gap arrangement

Before compressing the piston rings for installation, position them so that their gaps are away from the piston pin axis and thrust direction and are staggered.

Make sure the arrow mark on the piston head faces the engine front.

Assemble the connecting rod caps to their respective connecting rods, with the matching marks on them matched, and tighten the cap mounting bolts to the specified torque.

SPECIFICATION

Item	Standard
Connecting rod cap mounting bolt tightening torque	36.1-39.7 lb.ft.

10. Rotate the crankshaft until the No. 1 cylinder piston is at the top dead center.
11. Install the crankshaft gear.
12. Install the injection pump gear by aligning its matching mark with that on the crankshaft gear.
13. Install the camshaft and gear assembly to the crankcase, aligning the matching mark on the camshaft gear with that on the injection pump gear. Then, install the camshaft bearing stopper.
14. Install the governor weight assembly to the camshaft gear.
15. Install the pressure oil pump gear on which the ball bearing has been installed.
16. Install the gear case assembly.

NOTE: Apply Helmeseal® 52B to both sides of the gear case gasket.

17. Install the fuel injection pump.
18. Install the crankshaft pulley, and tighten the nut.
19. Install the oil pump and oil screen.
20. Install the oil pan to the crankcase.

NOTE: Coating of any sealant to the oil pan gasket is unnecessary as sealant is applied to both sides of the gasket in advance.

21. Insert the tappets, and install the push rods.

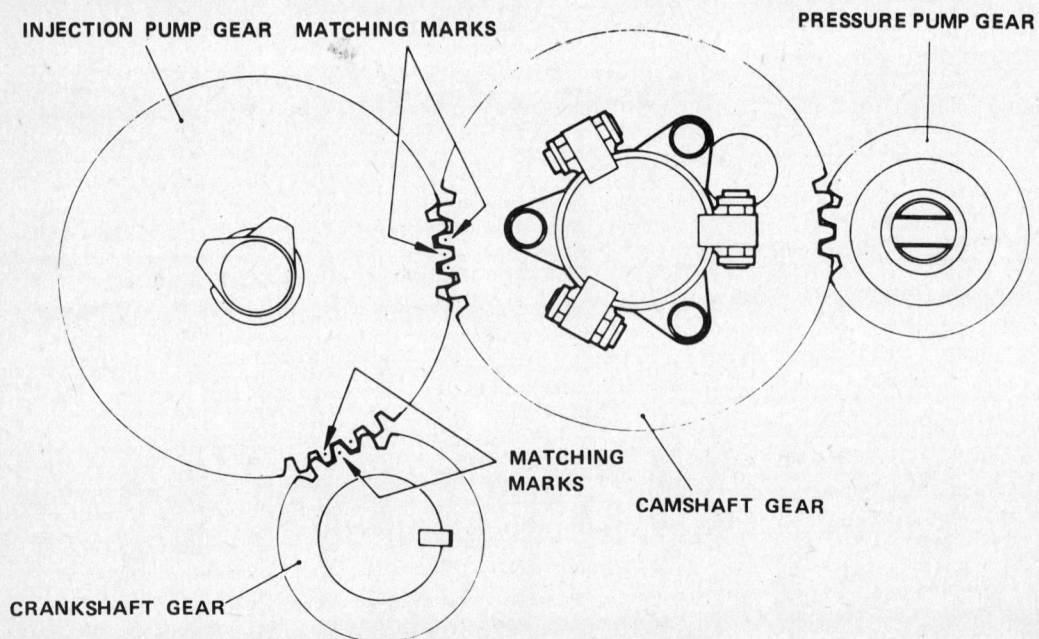

Engine gear train timing mark alignment

Satoh

LUBRICATION SYSTEM

Oil Pump

REMOVAL

1. Loosen off the drain plug, and drain thoroughly the engine oil pan.
2. Remove the oil pan.
3. Remove the oil screen, and then remove the oil pump.

INSPECTION

1. Clearance between outer rotor and body.
   Measure the clearance between the outer rotor and body with a thickness gauge. Replace the rotor assembly or the body if the clearance is excessive.

Measuring outer rotor-to-body clearance

SPECIFICATION

Item	Standard
Clearance between outer rotor and body	0.006-0.008 in.

2. Clearance between inner rotor and outer rotor: Measure the clearance between the outer rotor and inner rotor with a thickness gauge. Replace the rotor assembly if the clearance is excessive.

SPECIFICATION

Item	Standard
Clearance between outer rotor and inner rotor	0.002-0.0047 in.

3. Clearance between outer rotor and cover: Insert the outer rotor into the pump

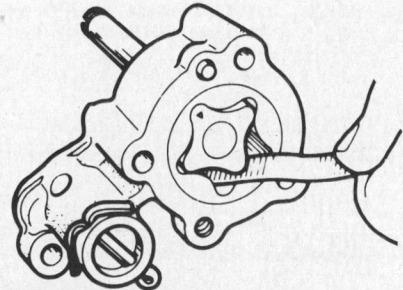
Measuring rotor-to-rotor clearance

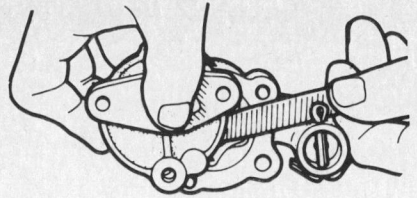
Measuring rotor-to-cover clearance

body and, using a straightedge and a thickness gauge, measure the clearance between the straightedge and the outer rotor. If the clearance is out of the specified range, replace the rotor assembly or body.

SPECIFICATION

Item	Standard
Clearance between outer rotor and cover	0.002-0.0038 in.

INSTALLATION

1. Install the oil pump.
   Be sure to mesh the groove on the oil pump shaft end with the lug provided on the lower end of the gear shaft.
   Replace the oil pump gasket with a new one.
2. Install the oil screen.
3. Install the oil pan.
   Replace the existing oil pan gasket with a new one. Coating of sealant to the new oil pan gasket is unnecessary as both sides of the gasket are factory-coated with sealant.
4. Refill the oil pan with the specified quantity of engine oil.

FUEL SYSTEM

Injection Pump

REMOVAL

1. Disconnect the fuel feed hose at the fuel injection pump.
2. Remove the injection pump.
3. Loosen off the injection pump mounting bolts and dismount the pump assembly.

Record the thickness and number of adjusting shims removed for correct reinstallation.

INSTALLATION

1. When reinstalling the injection pump assembly, also reinstall the adjusting shims to their original positions.
2. When installing a new injection pump assembly, select the adjusting shim thickness so that the dimension A shown (the distance from the injection pump cam base to the gear case pump mounting surface) is 3.26 to 3.27 in. Be sure to fit the lug of the control rack correctly to the forked portion at the end of the governor lever. Shims are available in 0.012, 0.020 and 0.039 in.
3. Connect the fuel feed hose to the injection pump, and loosen the air vent screw of the pump to bleed air.
4. Check the injection timing:
   a. Remove the delivery valve holder, and take off the delivery valve and spring from the holder. Then, install the delivery valve holder on the pump.
   b. Turn the crankshaft while observing fuel flowing out of the outlet of the delivery valve holder. The moment at which the fuel stops flowing is the injection timing. If the injection timing is out of the specification, adjust it by changing the thickness of the adjusting shim (between pump and gear case). A change of 0.0039 in. in the thickness of shim varies the injection timing by about 1 deg.
   c. When examining the injection timing in dusty condition, it is a good practice to keep the delivery valve and its spring unremoved from the valve holder to prevent dirt from getting inside the pump. To check the timing, disconnect the No. 1 injection pipe at the nozzle holder side, and rotate the crankshaft, noting the disconnected end of the pipe. The injection timing is the moment the fuel starts flowing out of the disconnected end of the pipe. This movement, however, is retarded about one deg. and therefore be taken as 22 deg. before top dead center.

DISASSEMBLY

1. Remove the delivery valve holder.

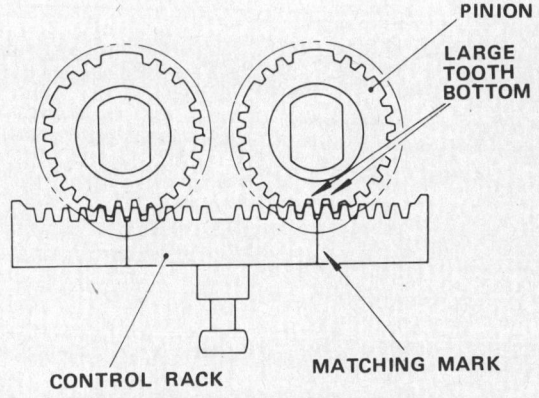
Assembling control rack and pistons

Satoh

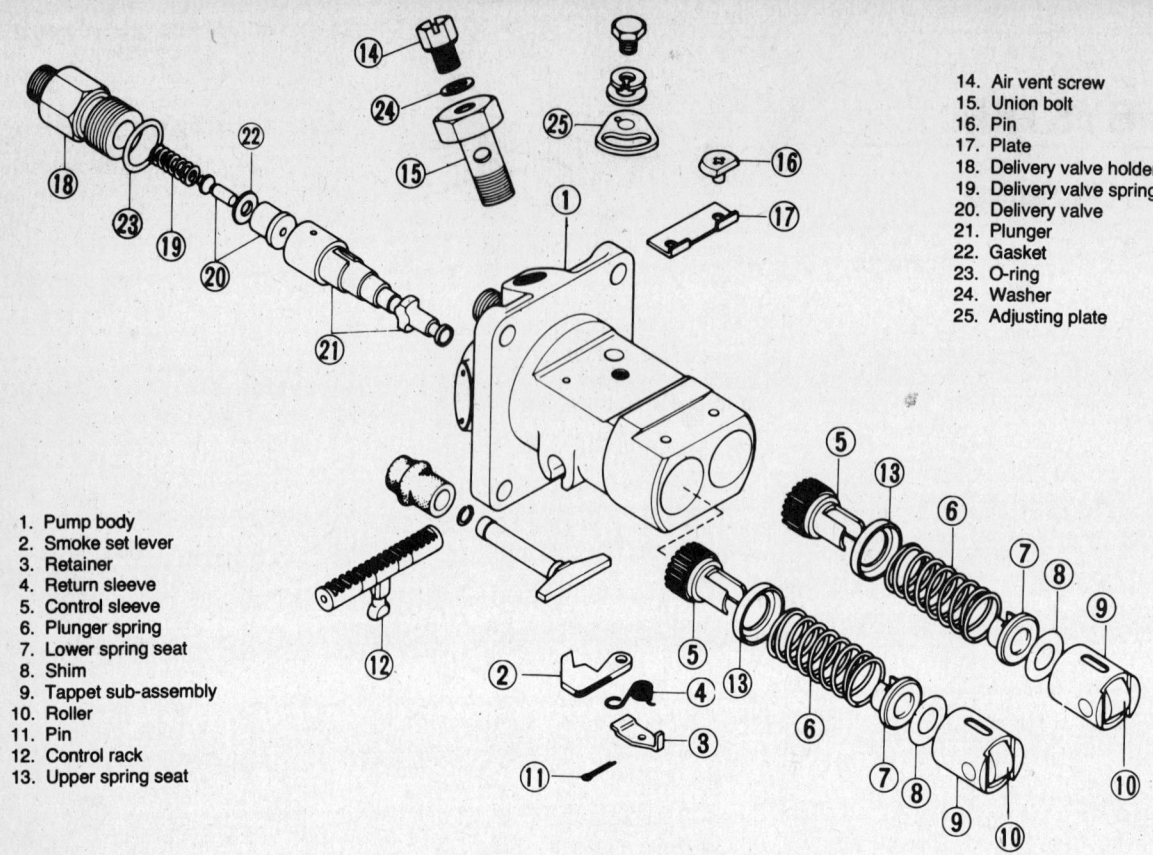

14. Air vent screw
15. Union bolt
16. Pin
17. Plate
18. Delivery valve holder
19. Delivery valve spring
20. Delivery valve
21. Plunger
22. Gasket
23. O-ring
24. Washer
25. Adjusting plate

1. Pump body
2. Smoke set lever
3. Retainer
4. Return sleeve
5. Control sleeve
6. Plunger spring
7. Lower spring seat
8. Shim
9. Tappet sub-assembly
10. Roller
11. Pin
12. Control rack
13. Upper spring seat

Injection pump components

2. Take out the delivery valve spring and delivery valve.
3. Take out the gasket and delivery valve seat.
4. Straighten the lock plate of the tappet guide pin, push in the tappet and pull out the tappet guide pin, then remove the tappet.
5. Remove the lower spring seats and plungers.
6. Remove the springs and upper seats.
7. Remove the pinions.
8. Pull out the plunger barrels toward the top of the pump housing.
Keep the removed plungers and plunger barrels arranged as matched for each cylinder.
9. Pull out the split pin, remove the washer, return spring and smoke set stopper, and remove the control rack.

INSPECTION

1. Delivery valve seat:
a. Check the delivery valve seat for ridges, scores and scratches. Replace if found faulty.
b. Oiltightness test for delivery valve:
No pressure drop should occur in 5 seconds when a pressure of 2132.4 psi is applied to the delivery valve. The service limit is reached if the pressure drops from 2132.4 psi to 1848.1 psi in 5 seconds.
2. Plunger and barrel:
a. Oiltightness test for plungers:

Raise the oil pressure up to 4264.8 psi, using an oiltightness tester, and measure the length of time for the pressure to drop from 2843.2 psi to 1421.6 psi. If this pressure drop takes longer than 6 seconds, the plunger is in good condition.
b. Plunger inspection:
Check the plunger for sticking, scratches and rust. Put it back into the barrel and check for smooth movement. The plunger and barrel may not be reused by lapping. Replace with new ones as a matched set.
3. Control rack and pinions:
Check the tooth face of the control rack and pinions for wear and damage; replace if found faulty.
4. Tappets:
Check the tappets, tappet rollers and shafts for wear or plunger damage; replace the tappets if found faulty.

ASSEMBLY

1. Insert the plunger barrel into the pump body with its notch fitted to the dowel on the body.
2. Fit the O-ring to the delivery valve holder.
3. Fit the delivery valve spring, delivery seat gasket and delivery valve assembly to the delivery valve holder, and tighten the holder to the pump body with the specified torque. Make sure the O-ring is installed correctly.

SPECIFICATION

Item	Standard
Delivery valve holder tightening torque	28.9-32.5 lb.ft.

4. Assemble the control rack.
5. Engage the control sleeve with the center gear tooth (the both sides of which are grooved deeper) of the pinion aligned with the scribed line on the control rack.
6. Install the upper spring seat and spring.
7. Install the plunger and lower spring seat, and install the plunger with the "R" mark on the plunger flange facing the rack side.
8. Align the tappet guide hole with the dowel pin hole in the pump body, and install guide pin by pushing the tappet lightly by hand. Before installing the tappet guide pin, attach the lock plate. Bend the lock plate for locking after installing the guide pin.
9. Install the smoke set lever, return spring and washer, and attach the split pin.

Injection Nozzles

REMOVAL

1. Remove the fuel injection pipe from the nozzle holder.

Satoh

2. Disconnect the overflow pipe from the overflow nipple provided at the top of the nozzle holder.

3. Loosen off the nozzle holder mounting bolt, and remove the nozzle holder assembly.

DISASSEMBLY

1. Hold the retaining nut in a vise, and remove the nozzle holder by using a wrench. Take out the shim, pressure spring, flange and pressure pin.

When holding the nozzle holder in a vise, be sure to place metal such as copper or aluminum between vise jaws and the nozzle.

2. Remove the nozzle from the retaining nut. The nozzle may be pulled out easily if tapped lightly with a wooden hammer.

Use extreme care not to cause any damage to the nozzle.

INSPECTION

1. Check each nozzle for damage; replace as an assembly if found faulty.
2. Check the pressure spring for damage.

ASSEMBLY

NOTES:
a. Clean each part in clean diesel fuel before assembly. Do not wipe it with cloth or the like.
b. Be sure to tighten the nozzle holder body to the retaining nut with the specified torque. Poor tightening will cause insufficient compression, and overtightening will result in unsmooth movement of needle in the nozzle, leading to poor injection performance.

1. Fit the nozzle into the retaining nut.
2. Tighten the nozzle holder body to the specified torque.

SPECIFICATION

Item	Standard
Nozzle holder body tightening torque	43.3-57.8 lb.ft.

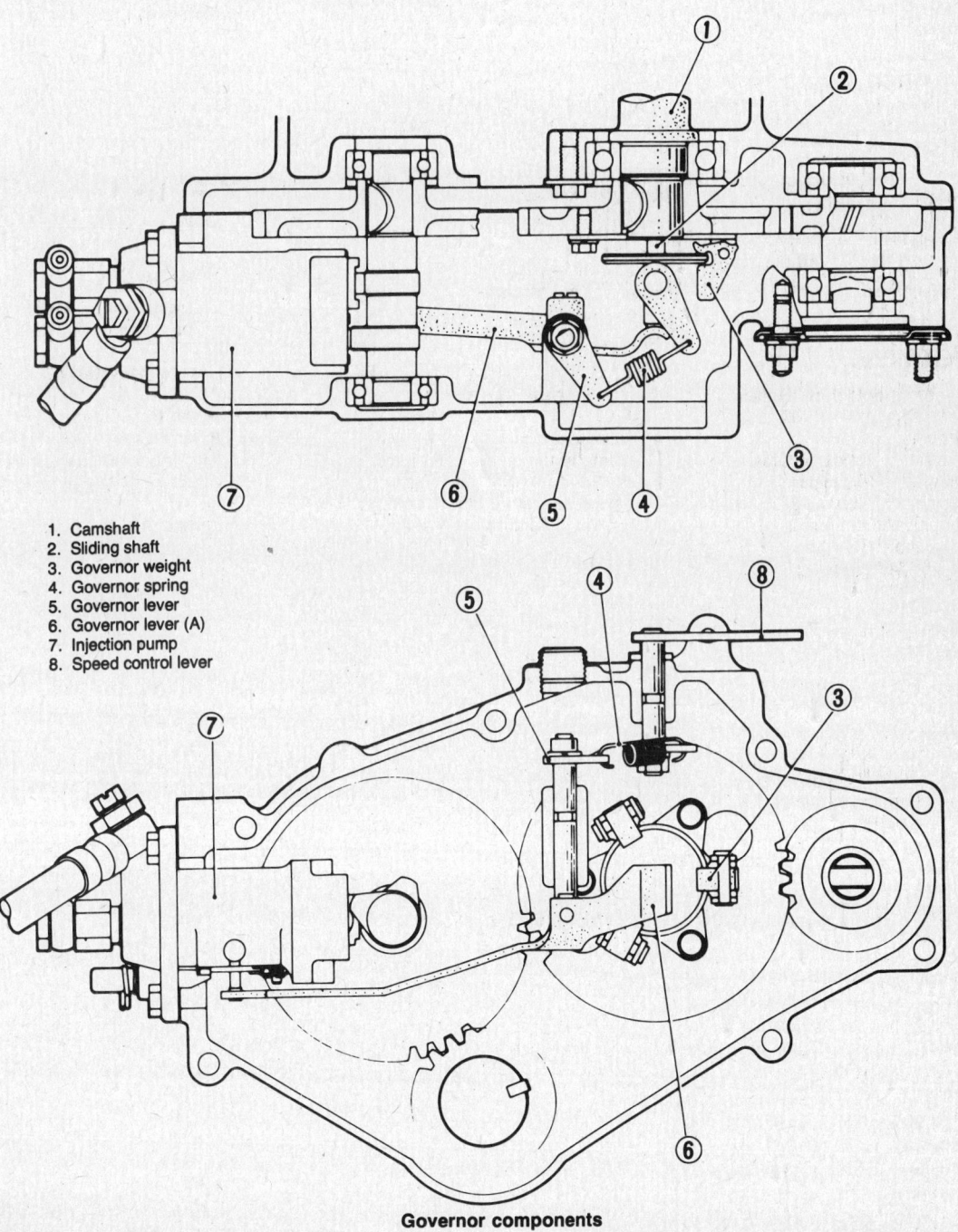

1. Camshaft
2. Sliding shaft
3. Governor weight
4. Governor spring
5. Governor lever
6. Governor lever (A)
7. Injection pump
8. Speed control lever

Governor components

Satoh

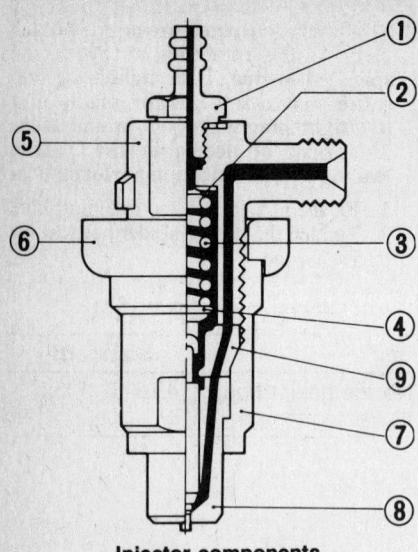

Injector components
1. Overflow nipple
2. Adjusting shim
3. Pressure spring
4. Pressure pin
5. Nozzle holder body
6. Flange
7. Retaining nut
8. Nozzle
9. Distance piece

3. Install the pressure pin, pressure spring, and shim, and then install the nipple assembly.

INSTALLATION AND ADJUSTMENT

1. Install the nozzle holder on the head. Torque the mounting bolts to 11-14 ft.lb.
2. Valve opening pressure
Measure the initial injection pressure of each nozzle using a nozzle tester. Adjust the pressure to the specified value, if necessary, by changing the thickness of the adjusting shim. A change of 0.0039 in. in shim thickness varies injection pressure by about 142.2 psi. To replace the shim, secure the retaining nut in a vise and loosen off the nozzle holder body using a wrench.

Be sure to tighten the retaining nut to the specified torque.

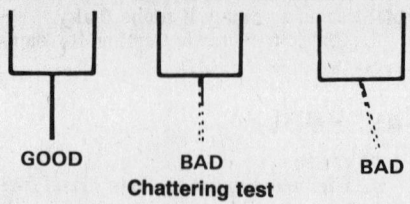
Chattering test

3. Chattering test
Move the tester lever slowly up and down with about 10 strokes per minute. The nozzle is normal if the fuel is injected sharply on each stroke with intermittent sound.

The fuel should be injected in the axial direction of nozzle in a straight line. If the fuel is injected in a wrong direction, or branched into several lines, the nozzle is not in good condition.

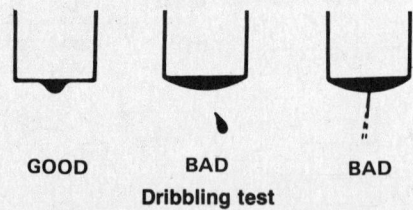
Dribbling test

4. Dribbling
Dribbling at the end of each injection, if noted in the course of the chattering test above, is an indication that the nozzle is at fault. However, a tiny drop of fuel found on the nozzle tip at the end of injection is due to chattering and is not a matter of concern.

5. Spray condition
When the lever is moved faster (with about 200 strokes per minute), fuel should be finely atomized and sprayed in a conical pattern without droplets.

Governor

DISASSEMBLY AND ASSEMBLY

1. Remove the V-belt.
2. Remove the crank pulley.
3. Dismount the fuel injection pump. Refer to "Fuel injection pump."
4. Remove the gear case mounting bolts, and remove the gear case assembly.
5. Remove the governor spring.
6. Remove the governor lever.
7. Remove the speed control lever.
8. Remove the governor weight assembly and sliding shaft from the camshaft.

INSPECTION

1. Governor assembly
a. Check the pawl of the governor weight for wear or damage. Replace if found unserviceable.
b. Check the sliding shaft for smooth operation. Repair or replace if found faulty.
2. Governor levers
Check the sliding shaft contacting portion and the forked portion of the governor lever A for wear.
Replace the lever if worn excessively.
3. Governor spring
Check the spring for breakage and fatigue; replace if found faulty.

Wheel Horse
INDEX

B, C, AND D SERIES
Front Axle and Steering 800
Brakes 800
Transaxle 801
   B Series 801
   C Series 802
   D Series 810
Differential 811
Power Take-Off 812
Engine 813

D-250
Front Axle 814
Steering 814
Brakes 815
Clutch 815
Transaxle 815
Axle Tube 819
Hydraulic System 820
Engine 820
   R&R 820
   Cylinder Head 821
   Cylinder Block 822
   Pistons and Liners 824
Cooling System 825
Fuel System 825
Electrical System 826

WHEEL HORSE
B, C, and D Series

FRONT AXLE AND STEERING

Steering Gear

REMOVAL

1. Remove steering wheel and foam dust cover.
2. Disconnect the drag link from steering gear lever.
3. Cut any wire ties holding electrical wiring to steering column.
4. Remove through bolts holding gear housing to steering gear bracket.
5. Remove steering gear assembly from the bottom.

INSTALLATION

1. Line up steering post and jacket.
2. Push steering gear up through the grommet in the console.
3. Insert the two lever shaft nuts through the large hole in the bracket.
4. Align and install the three bracket to steering gear bolts.
5. Center the jacket in the console grommet and tighten the three bolts evenly.
6. Connect the drag link to the steering arm.
7. Install the steering wheel on the splined upper shaft.
8. Check steering alignment.

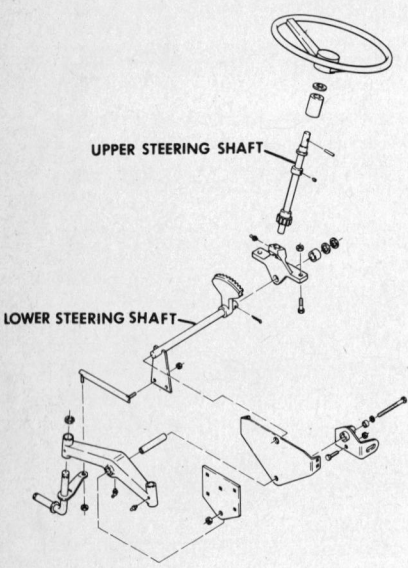

B series steering components

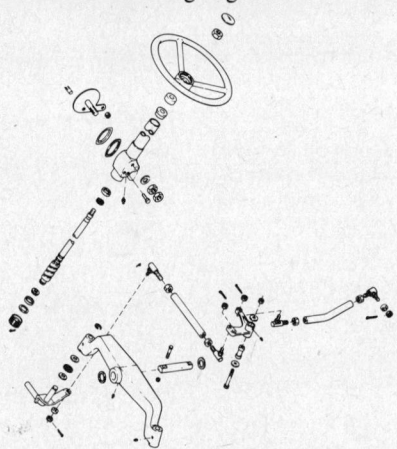

D series steering components

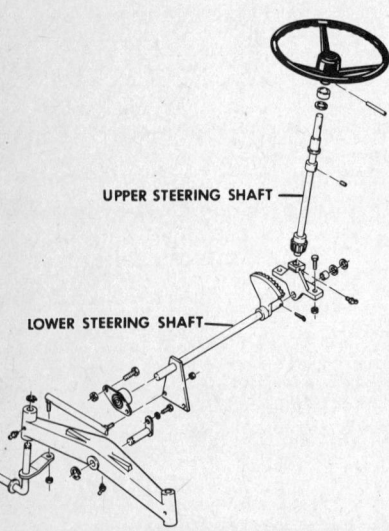

C series steering components

OVERHAUL

1. Place the steering gear assembly in a vise.
2. Back off the locknut and loosen the lever stud adjusting screw.
3. Remove both large nuts and washer and remove the lever and shaft assembly.
4. Remove the cotter pin and adjusting plug from the end of the steering gear housing.
5. Remove the steering post, worm, and bearings from the housing.
6. Place the main housing firmly in a vise.
7. Lubricate and place the upper and lower ball bearings and races on the wormshaft.
8. Insert the worm and bearings into the housing.
9. Install the spring washer in the adjusting plug and thread the plug into the housing.
10. While turning the post and worm, screw the adjusting plug in place until all end play is removed.
11. Tighten until a slight drag is felt while turning the steering shaft.
12. Install the cotter pin to lock the adjusting plug in place.
13. Back off the lever adjusting stud and insert the lever shaft into the housing assembly.
14. Install the washer and two nuts but do not tighten.
15. Install $3/32$ in. thick washer as a guage between the lever and housing.
16. Tighten the inside nut until the lever is just against the gauge washer.
17. Tighten the locknut against the inside nut and remove the $3/32$ in. gauge washer.
18. Turn the steering shaft from full left to full right.
19. Use a screwdriver to tighten the lever stud until the high spot (slight drag) is just felt as you pass through the center of travel.
20. Hold the stud and tighten the locknut securely. Check for proper operation and make sure the adjustment did not move when locking the nut.
21. Check steering alignment.

ADJUSTMENTS

Upper Adjustment

1. Release the setscrew on the collar.
2. Press down on the steering wheel.
3. Slide collar up against bushing and tighten setscrew.

Lower Adjustment

1. Remove cotter pin from lower steering shaft.
2. Add or remove enough shims so that 0 to .015 in. end play remains in shaft.

BRAKES

B Series

ADJUSTMENT

1. Depress the foot brake and set parking brake in the first notch.
2. Loosen locknut and tighten adjusting nut on brake cam lever.
3. Push or pull tractor by hand. Proper brake adjustment is achieved when rear wheels skid across floor.
4. Release brakes and check that brake disc turns freely.
5. Tighten the locknut and recheck brake adjustment.

C Series

ADJUSTMENT

1. Depress the brake pedal and engage parking brake.
2. With the parking brake engaged, adjust the nut on the end of the rod until the

Wheel Horse

B series brake adjustment

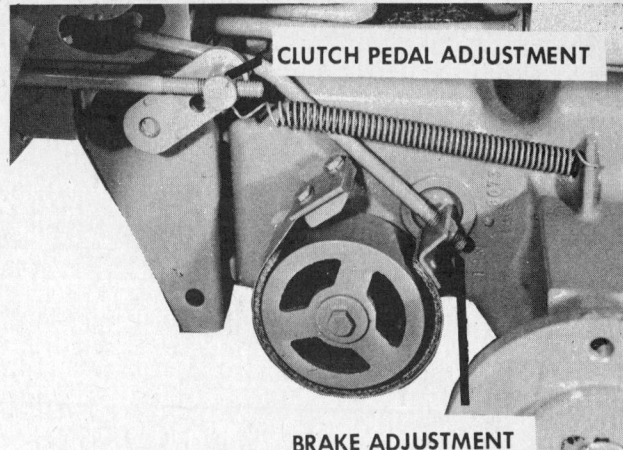

C series clutch and brake adjustment

brake band is tight enough to skid both rear wheels when the tractor is pushed.

3. Tighten the nut another ½ turn.

D Series

ADJUSTMENT

1. Release both foot and park brake.
2. Tighten the band adjustment nut until the brake band can no longer be slid from side to side on the brake drum.
3. Back off the nut until the brake band is again free to slide from side to side.
4. Front brake adjustment is only required if the linkage is removed. The jam nuts are adjusted so the spacer just contacts the spring, and the spring remains free to turn.

TRANSAXLE

B Series

REPLACEMENT

1. Lift tractor to relieve weight from transaxle.
2. Place gearshift lever in neutral.
3. Depress clutch and slip drive belt off transaxle pulley.
4. Disconnect brake lever and return spring at brake.
5. Remove 3 bolts from transaxle support bracket.
6. Remove 6 bolts from axle support brackets at frame.
7. Raise tractor frame and remove transaxle while guiding gearshift lever through access hole and the belt.
8. Reverse above procedure for replacement.

DISASSEMBLY

1. Remove axle hubs and input pulley.
2. Place the transmission assembly in a vise or suitable holding fixture.
3. Place the gear shift lever in neutral and pull the gear shift lever rubber boot away from the transmission boss. Loosen the gear shift lever retaining locknut and setscrew and remove the gear shift lever.
4. Remove the two (2) brake band retaining bolts and remove the brake band.
5. Remove the brake drum retaining bolt and washer, and remove the brake drum.
6. Remove the six (6) ⅜ in. x 16 bolts and nuts that retain the righthand and lefthand cases.
7. Remove paint and burrs from both axle shafts, brake shaft, and the input shaft.
8. Remove the lefthand case.
9. Lift out the axle and differential assembly.

NOTE: If only the differential is to be overhauled the transmission need not be further disassembled.

10. Shift the 2nd and high shift fork assembly up into the 2nd speed position.
11. Remove the splined pinion shaft and gear. At the same time the cluster gear and brake shaft assembly may be removed.
12. Remove the large pinion and reduction gear assembly, (and the thrust washer if used) and the reverse idler gear.

NOTE: Prior to 1973 models the pinion reduction gear and thrust washer was welded together and was serviced as one part. Starting with the 1973 models the pinion, reduction gear and thrust washer are serviced separately. The thrust washer goes on the stub shaft next to the big gear. The pinion may be pressed in and out of the gear with a suitable arbor press.

13. Remove the 2nd and high sliding gear and the low and reverse sliding gear.
14. Place the shift forks in neutral and remove first the low and reverse shift rail and fork assembly, then the 2nd and high shift rail and fork assembly, being careful not to lose the two stop balls, spring and stop pin as they are released.

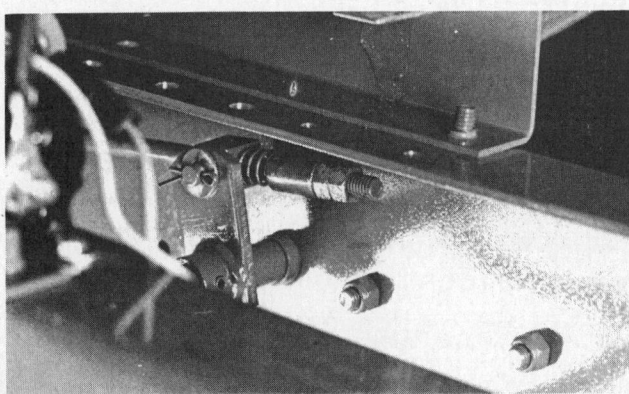

D series brake band adjustment, front

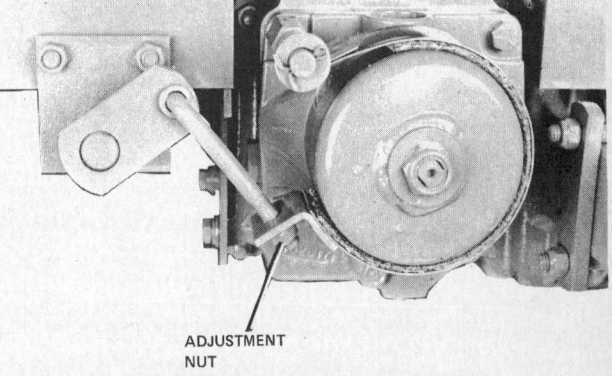

D series brake band adjustment, rear

Wheel Horse

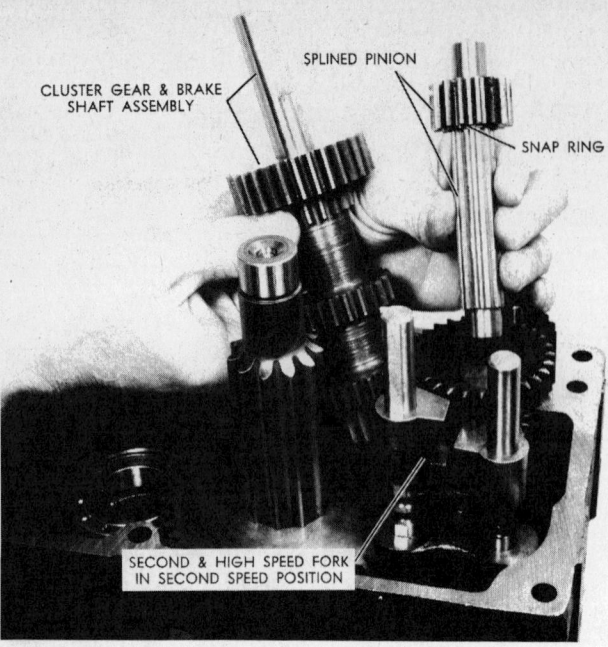

Transmission gear shaft removal

15. Remove the reverse idler shaft and the input shaft and gear.

NOTE: The reverse idler shaft may sometimes stick in the lefthand case.

16. If it is desired to replace the case bearings they may easily be removed and installed with suitable arbors. When replacing bearings they should be pressed in so that they are flush with the inside of the case, as shown.

17. The cluster gear and brake shaft assembly may be disassembled, as shown. Disassemble and assemble carefully to avoid damage to the Woodruff key and bushings. The cluster gear bushings may be removed and installed using suitable arbors.

18. With a screwdriver or other suitable tool remove the plug from the end of the detent passage.

ASSEMBLY

1. Install the input shaft and gear, and reverse idler shaft.
2. Install the 2nd and high shift rail and fork assembly (the one with the unequally spaced detent grooves.)
3. Install one stop ball, spring and stop pin.
4. Install the 2nd stop ball and using a punch push the ball in against the spring. Move the 2nd & high shift rail into the neutral position and continue to push on the punch until the ball is flush with the shift rail bore. At the same time insert the low and reverse shift rail and fork assembly, removing the punch, as the shift rail passes the ball.
5. Shift the 2nd and high shift fork into the 2nd speed position and install the 2nd and high sliding gear, and the low and reverse sliding gear in position on their respective forks. Note that the shift fork grooves of the sliding gears face each other, and that the smaller of the two gears is the 2nd and high gears which goes on the bottom.
6. Install the reverse idler and the large pinion and reduction gear assembly (and the thrust washer if used), making sure that the flanged end on the reverse idler goes down.
7. Install the splined shaft and gear through the two splined sliding gears seating the bottom end of the shaft into the input gear bearing. Install the cluster gear and brake shaft assembly carefully seating it in its bearing and meshing it with its related gears.

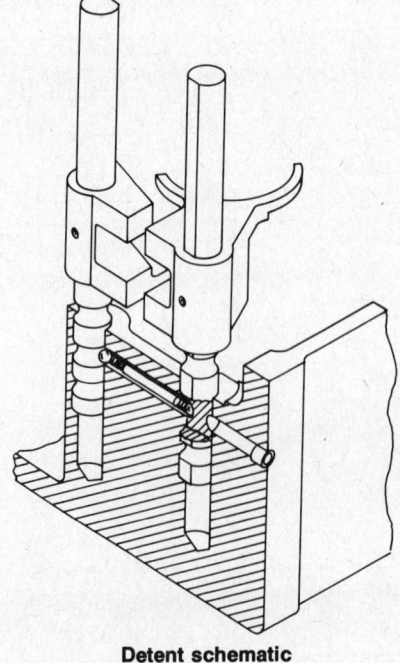

Detent schematic

8. Install the differential and axle assembly, making sure that four (4) nuts are up toward the lefthand case.
9. Place shift forks in neutral position and check for a free neutral and proper operation in all gears.
10. Install a new case gasket holding it in place with gasket sealer and install the lefthand case.
11. Install the six (6) ⅜ in. x 16 bolts and nuts and tighten all bolts evenly to avoid distortion.
12. Install the shift lever assembly and retain with the setscrew and locknut. Install gear shift lever rubber boot.
13. Check again for a free neutral and operation in all gears.
14. Install the brake drum and secure with the bolt and washer.
15. Install the brake band and secure with the two retaining bolts.
16. Install a new plug to seal the detent passage and drive it in flush with the case.

C Series

REMOVAL AND INSTALLATION, 8-SPEED

1. Drain the transmission oil.
2. Remove gear shift and park brake knobs and shift cover plate.
3. Depress clutch pedal and slip belt off drive pulley.
4. Remove seat hinge brackets and seat.
5. Unbolt seat pan and disconnect seat switch wires and tail light wires if applicable.
6. Remove seat pan.
7. Remove two (2) bolts holding seat pan support bracket to top of transmission. Remove fuel line and conduit clamps, close fuel valve, and disconnect fuel line at tank.
8. Remove two (2) bolts and nuts holding seat pan support bracket (not the fuel tank forward bolts).
9. Lift off fuel tank along with both brackets, and set tank aside.
10. Drive spirol pin from range selector lever.
11. Remove clutch return spring from transmission casting.
12. Disconnect rear brake rod at bellcrank.
13. Support frame of tractor and remove four (4) bolts holding transmission to frame.
14. Disconnect range lever as transmission is pulled away from frame.
15. To replace transmission, reverse above procedure. Refill with oil.

DISASSEMBLY, 8-SPEED

1. Remove axle hubs and input pulley.
2. Place the transmission assembly in a vise or suitable holding fixture (shown with levers, brake band, brake drum, hubs, and input pulley removed).
3. Place the shift lever in neutral and pull the shift lever rubber boot from the transmission boss. Loosen the shift lever retaining locknut and setscrew and remove the shift lever.

Wheel Horse

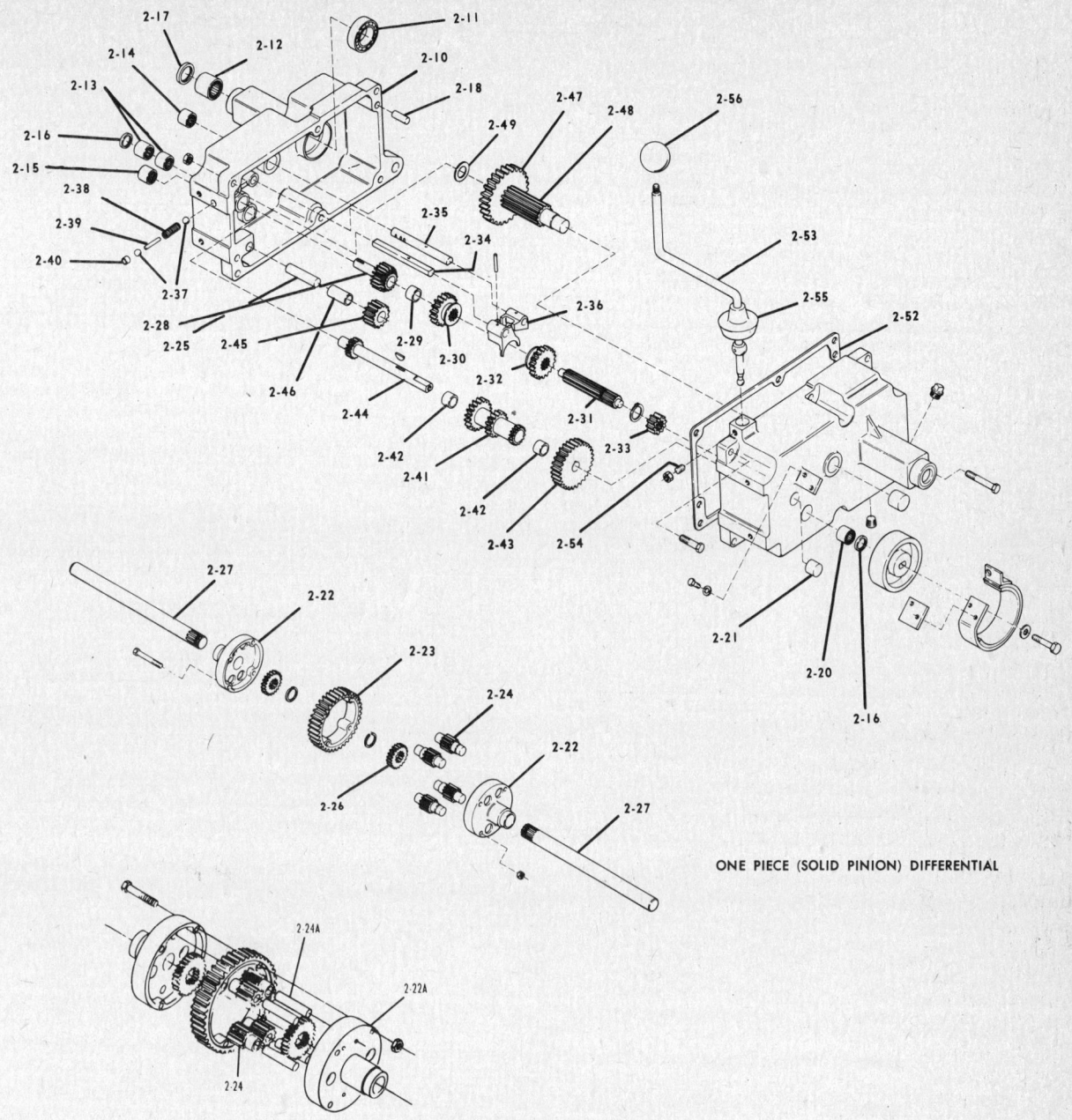

ONE PIECE (SOLID PINION) DIFFERENTIAL

B series 4-speed transmission components

2-10	Righthand case	2-24A	Differential pinion gear	2-38	Shift stop spring
	Lefthand case	2-25	Shaft	2-39	Shift stop pin
2-11	Ball bearing, 1½ in. I.D.	2-26	Axle gear	2-40	Cup plug
2-12	Needle bearing, 1 in. I.D.	2-27	Axle	2-41	Cluster gear
2-13	Needle bearing, ¾ in. I.D.		Axle snaping	2-42	Bronze bearing, ¾ in. I.D.
2-14	Needle bearing, 1 in. I.D.	2-28	Input gear	2-43	Reduction gear
2-15	Needle bearing, ¾ in. I.D.	2-29	Needle bearing, ⅝ in. I.D.	2-44	Cluster gear/brakeshaft
2-16	Oil seal, ¾ in. shaft	2-30	High and 2nd gear	2-45	Reverse idler gear
2-17	Oil seal, 1 in. shaft	2-31	Spline shaft	2-46	Bronze bushing, ½ in. I.D.
2-18	Dowel pin		Snapring, ¾ in.	2-47	Gear, 44 teeth
2-20	Needle bearing, ¾ in. I.D.	2-32	Low and reverse gear	2-48	Gear, 11 teeth
2-21	Needle bearing, ⅝ in. I.D.	2-33	Splined pinion gear	2-49	Thrust washer, 1 in. I.D.
2-22	Differential case	2-34	Front shift rail	2-52	Case gasket
2-22A	Differential case	2-35	Rear shift rail	2-53	Shift stick assembly
2-23	Differential bull gear	2-36	Shift fork	2-54	Setscrew, ¼-20 x ¾ in.
2-24	Differential pinion gear	2-37	Shift stop ball	2-55	Boot
				2-56	Knob

803

Wheel Horse

4. Remove the roll pin and remove the high-low range lever.

5. Remove the two (2) brake band retaining bolts and remove the brake band.

6. Remove the brake drum retaining bolt and washer, and remove the brake drum using a suitable puller as required.

7. Remove paint and burrs from both axle shafts, brake shaft and the input shaft.

8. Remove the six (6) ⅜ in. x 16 bolts and nuts that retain the righthand and lefthand cases.

9. Remove the lefthand case leaving all gears intact.

10. Lift out the axle and differential assembly.

NOTE: If only the differential is to be overhauled, the transmission section need not be further disassembled.

11. Shift the 2nd and high shift fork assembly up into the 2nd speed position.

12. Remove the cluster gear and brake shaft assembly, and the reverse idler gear and shaft.

13. Remove the large reduction gear assembly.

NOTE: Prior to 1973 models the pinion reduction gear and thrust washer was welded together and was serviced as one part. Starting with the 1973 models the pinion, reduction gear and thrust washer are serviced separately. The thrust washer goes on the stub shaft next to the big gear. The gear and pinion may be pressed in and out of the gear with a suitable arbor press.

14. Remove the splined pinion gear from the input shaft.

15. With a screwdriver or other suitable tool remove the plug from the detent passage.

16. Place the shift forks in neutral and remove the low and reverse shift rail and fork assembly (the one nearest the end of the case) together with the low and reverse sliding gear, making sure to catch the stop ball and other detent parts as the shift rail is removed.

17. Remove the 2nd and high shift rail and fork assembly together with the 2nd and high sliding gears.

18. Remove the remaining detent parts—ball, spring and stop pin from the detent passage.

19. Remove the gear and spline assembly from the input shaft.

20. Remove the input shaft.

21. Remove the sliding gear from the high-low range shift fork assembly.

22. Remove the input shaft to case thrust washer.

23. Remove the high-low range detent bolt and shift fork assembly.

24. Remove the detent bolt from the shift fork being careful to catch the stop ball and spring.

25. Remove the reduction gear shaft by driving it out toward the inside of the case, and remove the reduction gear.

26. Remove the high-low range shift lever and shaft assembly from the inside of the case.

ASSEMBLY, 8-SPEED

1. Install the high-low range shift lever in the hole of the right case below the shift rail supports.

2. Position the reduction gear assembly in place with the large gear down and insert the reduction gear shaft, driving it in place from the inside of the case.

3. Assemble the high and low range shift fork assembly by inserting the spring and ball in the detent hole. Using a ³⁄₁₆ in. punch, press on the ball, compressing the spring. At the same time insert the detent bolt, withdrawing the punch as the bolt slides in place.

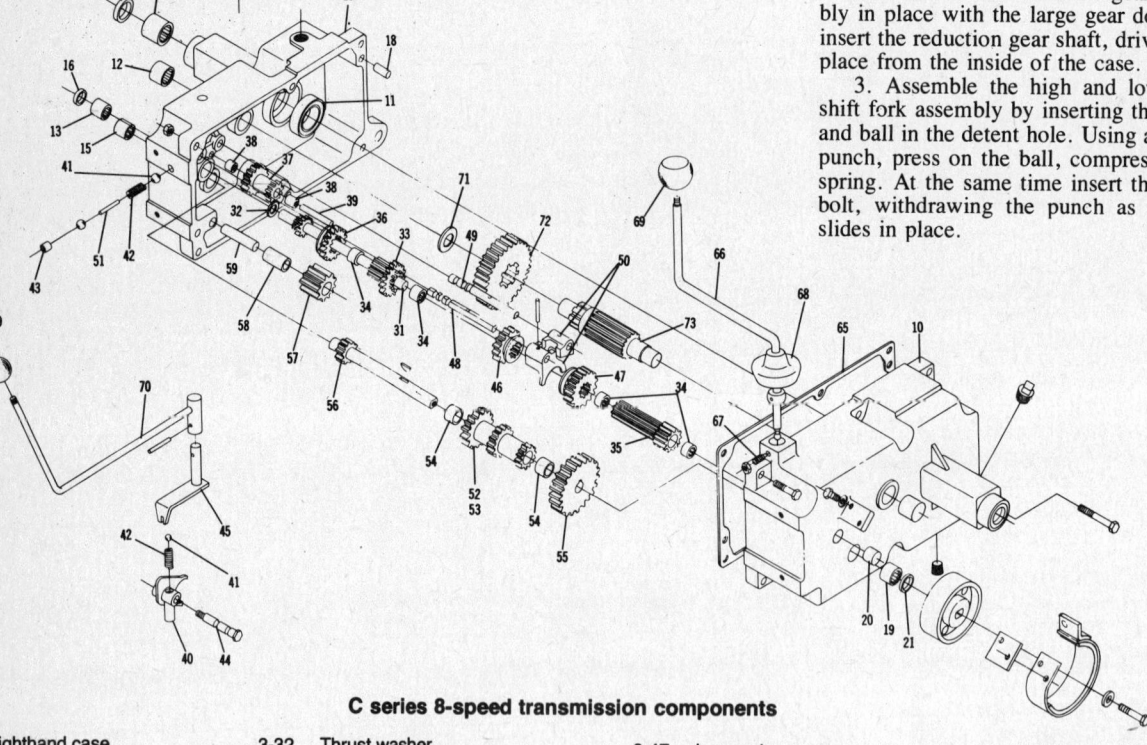

C series 8-speed transmission components

3-10	Righthand case	3-32	Thrust washer	3-47	Low and reverse gear	3-61	Nipple
	Lefthand case	3-33	Gear and spline	3-48	Front shift rail	3-63	Dipstick
3-11	Ball bearing 1½ in. I.D.	3-34	Needle bearing ⅝ in. I.D.	3-49	Rear shift rail	3-65	Case gasket
3-12	Needle bearing 1 in. I.D.	3-35	Gear and spline pinion	3-50	Shift fork	3-66	Shift stick
3-13	Needle bearing ⅝ in. I.D.	3-36	Sliding gear		Roll pin, ⅛ x 1 in.	3-67	Setscrew—dog point
3-14	Needle bearing 1⅛ in. I.D.	3-37	Reduction gear	3-51	Shift stop pin		—¼-20 x ¾ in.
3-15	Needle bearing ¾ in. I.D.	3-38	Needle bearing ⅜ in. I.D. x ½ in.	3-52	Gear and bearing, complete		Nut, ¼-20 in.
3-16	Seal—⅝ in. shaft	3-39	Reduction shaft	3-53	Cluster gear	3-68	Boot
3-17	Seal—1⅛ in. axle	3-40	Shift fork	3-54	Bronze bearing ¾ in. I.D.	3-69	Knob
3-18	Dowel pin	3-41	Stop ball	3-55	Reduction gear	3-70	Shift handle
3-19	Needle bearing ¾ in. I.D.	3-42	Stop spring	3-56	Cluster gear and brakeshaft		Roll pin ³⁄₁₆ x ¾ in.
3-20	Needle bearing ⅝ in. I.D.	3-43	Plug		#9 Woodruff key	3-71	Thrust washer
3-21	Seal—Brakeshaft	3-44	Detent bolt	3-57	Reverse idler gear	3-72	Gear assembly
	Differential case	3-45	Shift lever	3-58	Bronze bushing ½ in. I.D.	3-73	Gear 11T
3-31	Input shaft	3-46	High and 2nd gear	3-59	Shaft		Gear 44T

Wheel Horse

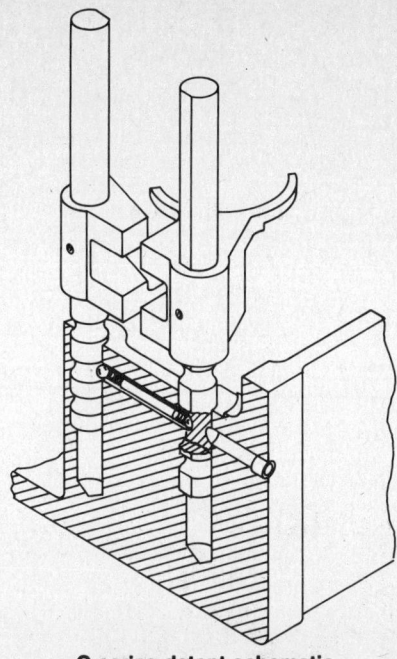

C series detent schematic

4. Position the detent bolt and shift fork so that the slot of the high and low range shift lever engages the shift fork and thread the bolt in place until it seats securely against the shoulder. Torque the detent bolt to 12-17 ft.lb.

5. Using a 3/16 in. punch to move the high-low range shift lever, check the operation of the detent and fork assembly.

6. Lift up on the reduction gear and slide the input shaft thrust washer into position.

7. Install the sliding gear into the high-low range fork with the groove down.

8. Install the input shaft, sliding it down through the high-low range sliding gear and thrust washer.

9. Install the gear and spline assembly over the input shaft, sliding the spline down through the sliding gear.

10. Install the 2nd and high shift rail and fork assembly (the one with the unequally spaced detent grooves) together with the 2nd and high sliding gear (the smaller of the two sliding gears). The sliding gear fits down over the input shaft with the shift fork groove up.

11. Install one of the stop balls, stop spring and stop pin into the detent passage.

12. Install the 2nd stop ball and using a punch, push the ball in against the spring. Move the installed shift rail into the neutral position and continue to push on the punch until the ball is flush with the shift rail bore. At the same time insert the low and reverse shift rail and fork assembly together with the low and reverse sliding gear, removing the punch as the shift rail passes the ball.

NOTE: The low and reverse gear fits down over the input shaft with the shift fork groove down.

13. Install the splined pinion gear.

14. Shift the 2nd and high shift rail and fork assembly into 2nd speed position.

15. Install the large pinion and reduction gear assembly (and the thrust washer if used).

16. Install the reverse idler gear shaft and gear assembly making sure the shoulder of the gear goes down.

17. Install the cluster gear and brake shaft assembly.

18. Install the differential and axle assembly with the retaining bolt nuts **up** toward the lefthand case.

19. Recheck for proper assembly, a free neutral and proper operation in all gears in both ranges. Place the shift forks in neutral position.

20. Install a new case gasket holding it in place with gasket sealer and install the lefthand case.

21. Install the six (6) 3/8 in. x 16 bolts and nuts and tighten all bolts evenly to avoid distortion.

22. Install the shift lever assembly and retain with the setscrew and locknut. Install gear shift lever rubber boot.

23. Install the high-low range selector lever and retain with the 3/16 in. roll pin.

24. Recheck for a free neutral and proper operation in all gears and ranges.

25. Install a new plug to seal the detent passage and drive it in flush with the case.

26. Install the brake drum and secure with the retaining bolt and washer.

27. Install the brake band and secure with the two retaining bolts.

REMOVAL AND INSTALLATION, AUTOMATIC

1. Drain the transaxle oil.
2. Remove knob from drive clutch handle and remove control cover plate.
3. Disengage transmission drive clutch and slip belt off drive pulley.
4. Remove seat hinge brackets and seat.
5. Unbolt seat pan and disconnect seat switch wires and tail light wires (if applicable).
6. Remove seat pan.
7. Remove two (2) bolts holding seat pan support bracket to top of transaxle. Remove fuel line clamp and conduit clamp.
8. Close fuel valve and disconnect gas line at tank.
9. Remove two (2) bolts and nuts holding seat pan support bracket.
10. Lift off fuel tank along with both brackets and set tank aside.
11. Remove nut holding pin to rear of motion control lever assembly.
12. Remove left side panel.
13. Remove brake rod adjusting nut.
14. Models with hydraulic lift: disconnect two hydraulic lines at pump and mark for replacement.
15. Support frame of tractor and remove 4 bolts holding transaxle to frame.
16. Remove transaxle assembly.
17. To replace transaxle, reverse above procedure. Refill with oil. Adjust brake band.

OVERHAUL
Gear Case

1. Remove both wheel hubs by loosening the locknuts and setscrews. Slide the hubs from the axle. A special wheel puller may be required.

2. Remove the Woodruff keys and file any rough edges from the key sots and the ends of the axle.

3. Remove the six bolts holding the transmission case halves together.

4. Lightly tap the half to be removed with a plastic or rawhide mallet to break the seal. Slide the case-half off the axle carefully to avoid damaging the oil seals.

5. Remove the old gasket and carefully clean the sealing surfaces.

6. Carefully slide the differential assembly out of the case.

7. Remove gear assembly consisting of Nos. 103395, 101887 and 102781. Also remove large gear No. 101885 intact.

8. Slide gears off of the No. 5965 shaft.

9. Remove the four nuts and bolts from the differential assembly. The unit may now be separated allowing the replacement of the bull gear, pins, pinion gears or the differential cases. Note the position of the pinion gears. Adjacent pinions are installed in opposite directions (teeth up or teeth down). Therefore, the positions of diagonally opposite pinions are the same.

10. The axle gears may be separated from the axle by removing the snapring.

NOTE: During reassembly, tighten all bolts securely in sequence.

Pump

1. Remove the four 3/8-16 socket-head capscrews that hold the finned aluminum pump housing to the pump end cap.

2. With the pump assembly held horizontally, carefully remove the finned aluminum pump housing, together with the input shaft, swash plate and cylinder block assembly.

NOTE: Make sure the cylinder block and piston assembly does not drop off the input shaft. The valve plate may stick to the cylinder block and come out with it, or it may stick on the charge pump housing.

3. Remove the pump housing/pump end cap gasket.

4. Carefully slide off the cylinder block and piston assembly from the pump shaft.

IMPORTANT: If any of the pistons slip out, return them to their original cylinder bores. Place the cylinder block and pistons on a lint free towel so the valving surface and slippers will not be damaged. If the pump housing is to be disassembled to remove the shaft, trunnion or swash plate, refer to the Pump Housing Disassembly and Assembly section.

5. Remove the valve plate from the charge pump housing, noting that the steel surface fits against the charge pump housing and over a dowel pin to keep it from rotating.

6. Using a 5/16 in. 12-point socket, remove the two short and two long 5/16-24 charge-pump-housing-to-end-cap bolts.

805

Wheel Horse

7. Remove the charge pump assembly from the pump end cap, being careful to keep the gerotor set together. Carefully note the position (dowel pin down) of the charge pump assembly in relation to the end cap.

8. Remove the 5/8 in. hex plug from the top left corner of the pump end cap. (Top left means viewed from the motor side of the end cap). If the unit is equipped for hydraulic lift operation, remove the charge ball valve and spring from the pump end cap (top view). If the unit is **not** equipped for hydraulic lift operation, there will be no spring and ball valve in this area (bottom view).

9. Remove the 5/8 in. hex plug from the top right corner of the pump end cap. (Viewed from the motor side of the pump end cap.) If the unit is equipped for hydraulic lift operation, there will be shims located in the spring cavity of the plug. Do not lose these shims as they determine the amount of implement pressure. Remove the spring and cone valve. Do not mix up this implement valve spring with the charge valve spring removed from the top left corner.

10. Remove the two slotted pump check valve plugs together with their respective check valve balls and springs. (These valves are located in the passage just below the charge and implement valve passage.)

11. Remove the push or free-wheeling valve by unscrewing it from the housing.

12. Remove the split back-up ring and the O-ring seal from the valve.

13. If the needle bearing in the charge pump housing is damaged, replace as follows:

    a. Using a $13/16$ O.D. flat washer with two opposite edges ground to a width of

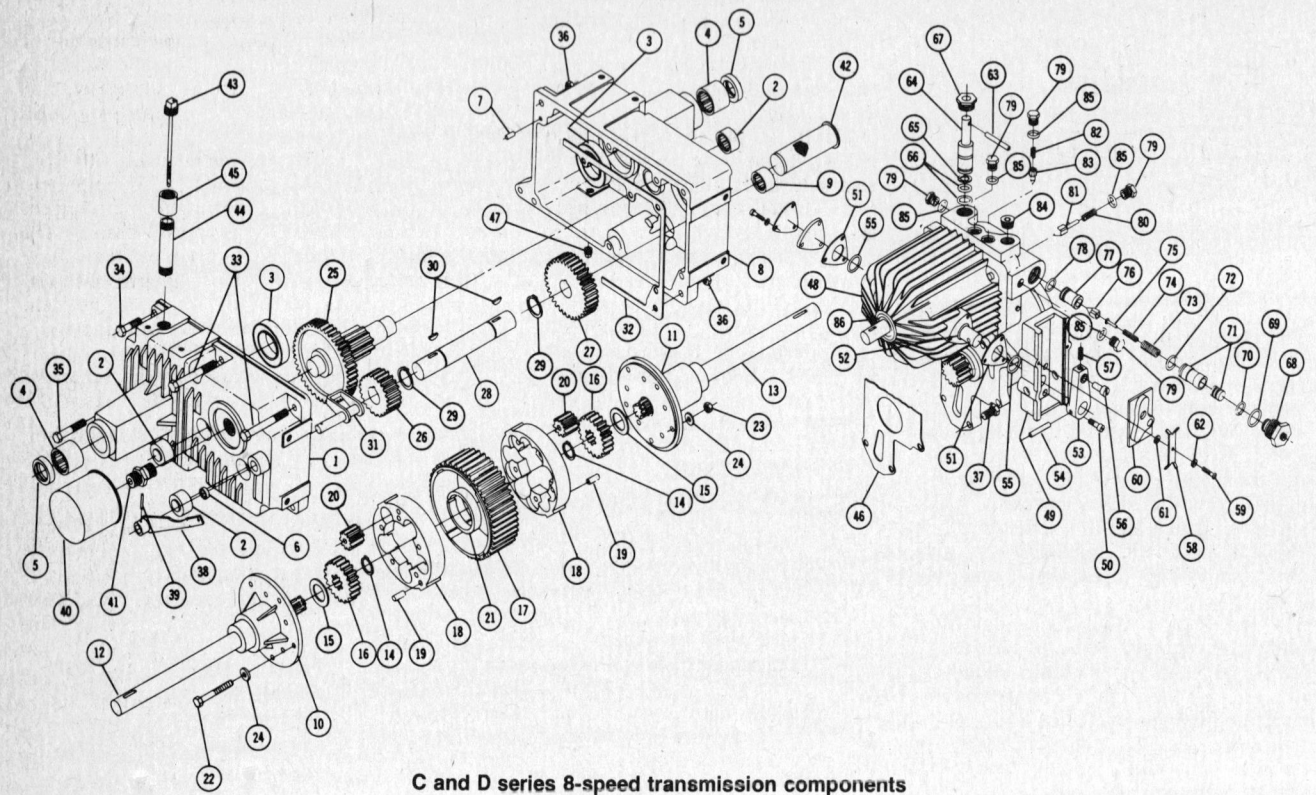

C and D series 8-speed transmission components

1. Righthand case
2. Needle bearing 1in. I.D.
3. Ball bearing 1½ in. I.D.
4. Needle bearing 1⅛ in. I.D.
5. Seal 1⅛ in. I.D.
6. Brakeshaft seal ½ in. I.D.
7. Dowel pin
8. Lefthand case
9. Needle bearing 1 in. I.D.
10. Righthand end cap
11. Lefthand end cap
12. Righthand axle rear
13. Lefthand axle rear
14. Snapring
15. Thrust washer
16. Axle gear
17. Differential ring gear
18. Body
19. Roll pin ¼ x ½ in.
20. Differential pinion gear
21. Cylindrical spring
22. Bolt ⅜-16 x 3¼ in.
23. Elastic stop ⅜-16 nut
24. Flat washer, ⅜ SAE
25. Gear—44 teeth—11 teeth
26. Gear—22 teeth
27. Gear—33 teeth
28. Shaft 1 in. dia.
29. Snapring 1 in. shaft
30. Woodruff key #9
31. Park brake assembly
32. Case gasket
33. Hex bolt ⅜-16 x 5 in.
34. Hex bolt ⅜-16 x 2 in.
35. Hex bolt ⅜-16 x 3½ in.
36. Nylok nut ⅜-16 in.
37. Nylok hex bolt ⅜-16 x 1 in.
38. Brake assembly lever
39. Roll pin ⅛ x ¾ in.
40. Filter assembly
41. Fitting
42. Strainer
43. Dipstick and filter assembly
44. Nipple ½ in. pipe x 3½ in.
45. Coupling ½ in. pipe
46. Pump gasket
47. Plug ½ in. pipe
48. Hydro gear unit (complete) assembly
49. Cam block support
50. Socket head screw ¼-20 x ½ in.
51. Shim
52. O-ring
53. Cam follower arm
54. Roll pin ¼ x 1¼ in.
55. O-ring
56. Cam follower eccentric
57. Nylok setscrew ¼-20 x 5/16 in.
58. Tension plate
59. Round head screw #8-32 x ½ in.
60. Cam
61. Washer
62. Lockwasher #8
63. Roll pin 3/16 x 2 in.
64. Valve
65. Back up ring
66. O-ring
67. Plug
68. Dampening valve assembly (with O-rings)
69. O-ring
70. O-ring
71. Assembly piston and sleeve (matched assembly)
72. O-ring
73. Relief valve spring
74. Relief valve spring
75. Spring seat
76. Relief valve cone
77. Sleeve
78. O-ring
79. Plug
80. Check valve spring
81. Check valve valve
82. Charge pump relief valve spring
83. Relief valve cone
84. Pipe hex socket ½-14 in. plug
85. O-ring
86. Seal

Wheel Horse

1. Pump shaft seal
2. Snapring
3. Ball bearing
4. Snapring
5. Pump shaft
6. Variable swash plate
7. Thrust plate
8. Pump housing
9. Stub trunnion shaft
10. Control trunnion shaft
11. Needle bearing
12. Seal
13. Washer
14. Retaining ring
15. Spirol pin
16. Capscrew
    Cylinder block kit—consists of nine pistons and a cylinder block (not available separately) plus the correct number of items 18, 20, 21, 22 and 23.
18. Slipper retainer
20. Washer
21. Spring
22. Front washer
23. Retaining ring
24. Valve plate
25. Locating pin
26. Needle bearing
27. Charge pump housing
28. Gerotor assembly
29. O-ring, large
30. Back-up ring
31. Short capscrew
32. Long capscrew
33. Socket head plug
34. Pump gasket
35. Pump end cap
36. Socket head plug
37. Socket head plug
38. Pipe plug
39. Pipe plug, ¼ in.
40. Short capscrew
41. Ball
42. Check valve spring
43. O-ring
44. Check valve plug
45. O-ring
46. Back-up ring
47. Push valve
48. Ball
49. Charge relief valve spring
50. O-ring
51. Plug
52. Plug
53. Acceleration valve body
54. Shim set
55. Implement relief valve spring
56. Relief valve cone
57. Hex plug
58. O-ring
59. Metering plug
60. O-ring
61. Ball
62. Spring
63. Acceleration valve assembly
64. O-ring
65. Back up ring
66. O-ring
67. Motor gasket
68. Motor end cap
69. Valve plate
    Cylinder block kit—consists of nine pistons and a cylinder block (not available separately) plus the correct number of items 70, 71, 72, 73 and 86.
70. Retaining ring
71. Washer
72. Spring
73. Retainer
74. Retaining clip
75. Acceleration valve spring
76. Pipe plug
77. Motor cover plate
78. Capscrew
79. Motor shaft
80. Capscrew
81. Centering pilot
82. O-ring
83. Seal retainer with O-ring
84. Needle bearing
86. Slipper retainer
88. Thrust plate
90. Ball bearing
91. Retaining ring
92. O-ring
93. Copper washer
94. Socket head capscrew
95. Hex head capscrew

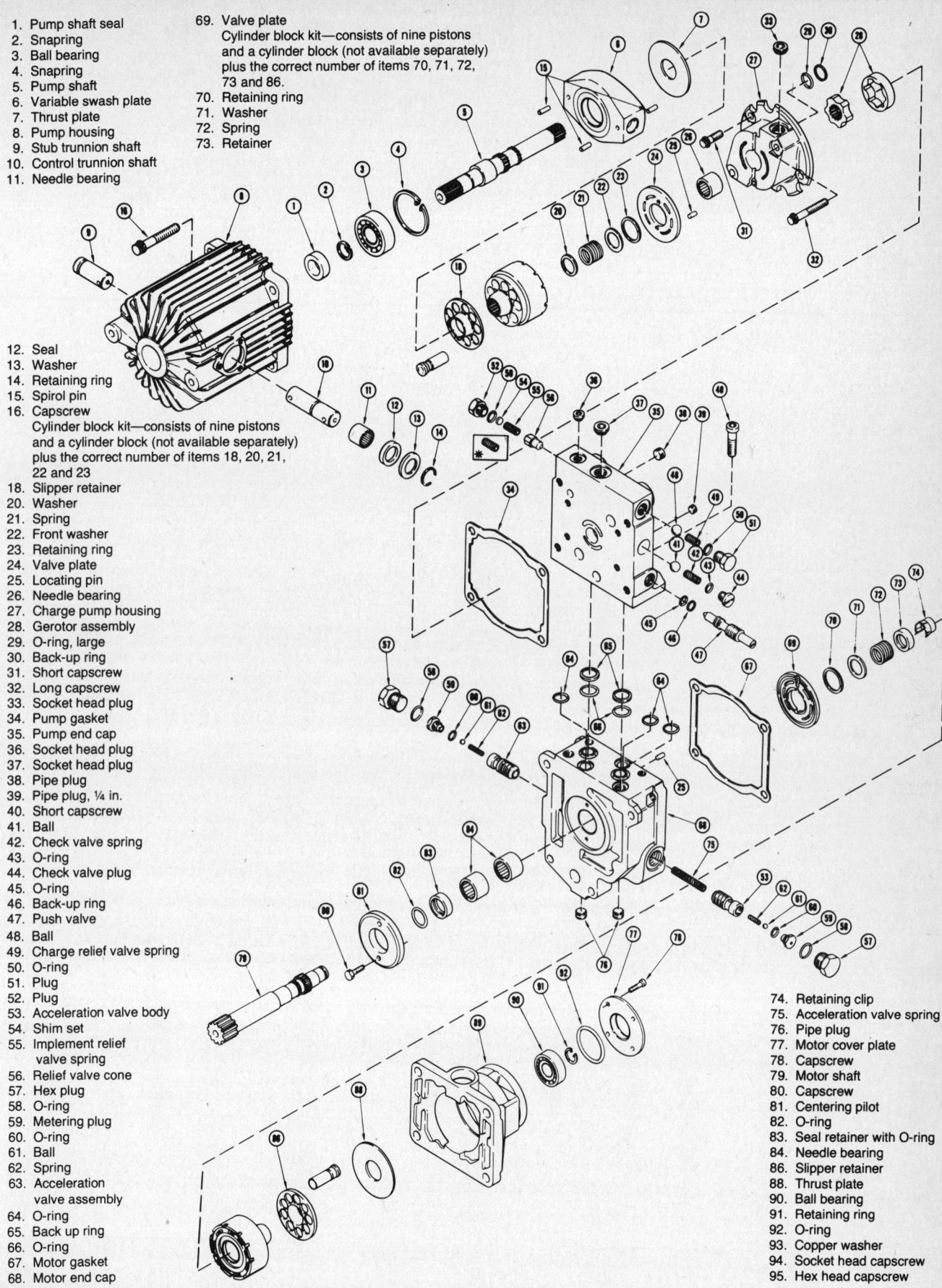

Piston-Piston hydrostatic transmission components

$^{11}/_{16}$ in. as a tool, insert it from the gerotor side against the inside of the needle bearing. Then, pressing against the washer with an arbor, remove the needle bearing.

b. To install the needle bearing, press it in place from the front side.

NOTE: Always install needle bearings pressing on the heavy end, which may be easily identified. It is the end where the identification numbers are found. Press bearing to the correct depth: .100 in. should be left out of the bore.

Pump Housing

1. Remove the thrust plate from the swash plate assembly.
2. Remove the pump shaft seal.
   a. Use a sharp awl and puncture the seal retainer.
   b. Pry out the seal, being careful not to scratch or damage the shaft seal surface or the aluminum housing.
3. Using snapring pliers remove the retaining ring from the shaft.
4. Remove the shaft by tapping on the input (both ends are splined) end and pushing it through the bearing.
5. Using a $^3/_{16}$ in. punch, drive the swash-plate-to-trunnion-shaft roll pins out toward the closed end of the case.

NOTE: One roll pin is used at the short stub shaft and two roll pins are used at the control shaft.

6. Remove the trunnion shaft retaining rings and washers. Drive the stub trunnion shaft in from the outside and remove. Drive the control shaft out from the inside using a long punch inserted through the removed trunnion shaft hole.
7. Remove the swash plate housing.
8. Pry out the trunnion seals, being careful not to damage the housing.

NOTE: If examination of the trunnion shaft needle bearings and the pump shaft ball bearing shows them to be in good condition, steps 9 and 10 should not be performed.

9. Using a suitable arbor such as a socket, press or drive needle bearings out of the housing from the inside.
10. With snapring pliers, remove the internal retaining ring that retains the pump shaft ball bearing in the pump housing, then remove the bearing by pressing it toward the inside of the housing.
11. Install pump shaft ball bearing in the front of the pump housing and retain with the internal tru-arc retaining ring.
12. **Note:** Only if needle bearings have been removed, install the two trunnion shaft needle bearings from the outside, pushing on the lettered end of the bearing. The bearings should be pressed in flush with the bottom of the seal bore so they will not interfere with the seal installation.
13. Install new trunnion shaft seals, pressing them in until they touch bottom. Oil the seal lips with 10W30 engine oil.
14. Insert the swash plate assembly into the case. Slide the trunnion shafts into each side, and into the swash plate housing.

Line up the roll pin holes and install the roll pins. One roll pin is used at the short trunnion shaft and two roll pins are used at the control shaft. Drive the dirst pin in so it enters into the far side of the swash plate. Then drive the second roll pin down against the first pin until it is ¼ in. below the surface of the swash plate. Drive the single roll pin at the short trunnion side so that it is ¼ in. below the surface.

15. Install the trunnion seal washers and the retaining rings.
16. Install the pump shaft by tapping it through the bearing from the inside and retain it with the external snapring.
17. Lubricate the pump shaft seal and install over the pump shaft with the lip side toward the pump. Press it in place so that it is flush with the outside of the housing.
18. Install a new O-ring seal and new back-up ring on the free wheeling valve, making sure the seal is toward the pump housing and the split back-up ring is toward the outside. Install the valve assembly.
19. Install both pump check valve balls and springs, together with their slotted plugs and O-ring seals.
20. Viewed from the motor side of the pump end cap, install charge relief valve and implement valve parts.
   a. Models not equipped for hydraulic lift:
      1. Install the hex head plug and new O-ring seal in the top left corner of the pump end cap.
      2. Install the cone valve, charge valve spring together with the hex head plug, and new O-ring seal in the top right corner of the pump end cap.
   b. Models equipped for hydraulic lift:
      1. Install the $^7/_{16}$ in. ball valve and spring together with the hex plug and O-ring in the to left corner of pump end cap.
      2. Install the cone valve and spring in the top right corner of the pump end cap, then, making sure that the original shim pack is in the spring cavity of the plug, install the plug, O-ring and shim assembly.
21. Install the gerotor set in the charge pump housing and install the pump housing and gerotor assembly on the pump end cap using new seal rings (O-rings and back up rings) with the valve plate dowel pin toward the bottom. Install the two long and two short bolts and tighten evenly.
22. Install the valve plate on the charge pump housing, centering it around the needle bearing with the steel face of the plate against the charge pump housing and the slot over the dowel pin to keep it from turning. Mounted properly, the plate should be flush against the housing so the cylinder block will operate on the bronze surface.
23. Install a new pump-housing-to-end-cap gasket on the pump end cap.
24. Apply oil to the thrust plate and place it on the swash plate.
25. Lay the pump end cap on a flat surface and install the cylinder block with the piston and slipper assembly on the valve plate.
26. Center the charge pump gerotor

drive so the pump shaft and spline can enter it.

27. Making sure the control shaft is on the correct side, carefully install the aluminum pump housing and shaft so the shaft spline enters the cylinder block and gerotor drive. Install the four pump housing capscrews to align the pump, gasket and end cap. Push the pump housing in place (it may be necessary to turn the pump shaft to align it with the charge pump spline). After the housing and shaft are pushed in place, tighten the four bolts evenly.

Acceleration Valves

1. Remove the ⅞ in. hex head plug from each side of the motor end cap.
2. Remove the first valve by pulling it out. Remove the second valve by pushing it out with the spring and remove the spring.

NOTE: To simplify removal of the forward valve, located at the rear of the end cap housing, move the tractor about an in. and oil pressure will force the valve to pop out.

3. The acceleration valves may be disassembled for inspection and cleaning.
4. Install the forward acceleration valve assembly (the one with the wider relief flats) in the bore at the rear of the end cap housing.

Insert the reverse acceleration valve assembly (the one with the narrow relief flats) together with the spring into the bore at the front of the end cap housing.

Make sure the spring seats in the spring cavity of each valve. When properly seated each valve will have approximately four or five threads exposed so the ⅞ in. hex plugs may be easily installed.

5. Install both plugs using new O-ring seals and tighten securely.

Hydrostatic Motor

1. Remove the four 10-24 Allen head capscrews that retain the cover plate, and remove the cover plate.
2. Remove the large O-ring seal from around the ball bearing.
3. Remove the snapring from the end of the motor shaft.
4. Remove the four ⅜-16 capscrews that retain the aluminum housing to the motor end cap (main iron housing) and remove the aluminum housing. Note the position of the housing. It must be installed with the webbed section up. If the cover is reversed, the unit will operate in the opposite direction.
5. Remove the housing-to-end-cap gasket.
6. Remove the thrust plate from the fixed swash plate in the housing.
7. Remove the ball bearing from the aluminum housing.
8. Remove the two ¼ in. capscrews that fasten the aluminum pilot (centering ring) to the motor end cap and remove the pilot.
9. Place the end cap, cylinder block and motor shaft assembly in a press, making sure the cylinder block and valve plate are seated on the end cap housing.
10. Press on the motor shaft until the

Wheel Horse

motor shaft retaining spring clip pops loose from the retainer (in the center of the cylinder block). This pressing operation only requires moving the shaft a short distance.

--- **CAUTION** ---
Do not press the shaft through the cylinder block at this time.

11. Remove the assembly from the press and remove the cylinder block.
12. Remove the spring retaining clip from the motor shaft.
13. Remove the motor shaft from the end cap housing.
14. Remove the motor valve plate.
15. Remove the O-ring from the seal retainer insert at the output end of the motor end cap.
16. Bearing removal and replacement:
    A. **Removal:**
    If the motor end cap bearings are to be removed, they may be pressed out by using a suitable arbor placed against the seal retainer. Then press the seal retainer and both needle bearings out the cylinder block side of the end cap.
    B. **Installation:**
    Install the first needle bearing into the output end of the motor end cap with the lettered end of the bearing out. Press the bearing in to the flush point. Then press the seal retainer insert in until it is flush with the end cap.*
    Install the second needle bearing into the cylinder block end of the end cap and, pressing against the lettered end, press it into the housing leaving .100 in. protruding from the face of the end cap. To obtain the .100 in. protrusion try using a $3/32$ in. cotter pin or washer as a stop for the press.
    *For convenience, the seal retainer may be pressed in with the bearing.
17. Install a new O-ring seal in the recess of the output shaft seal retainer.
18. Install the aluminum centering pilot on the output end of the motor end cap and retain with the two ¼ in. capscrews.
19. Install the valve plate with the steel side toward the motor end cap, making sure it is centered over the needle bearing and over the end cap dowel pin.
20. Apply 10W30 oil to the bearing surfaces of the shaft and install the motor shaft with the gear at the output end of the motor end cap.
21. Install the motor shaft spring clip with the prongs toward the output gear.
22. Apply 10W30 oil on the valve plate surface, the cylinder block valving surface and through the valving parts into the cylinder bores.
23. Slide the cylinder block, pistons, slippers and slipper retainer assembly on the motor shaft. Push it in place until the spring retaining clip seats.
24. Install the ball bearing in the end of the aluminum housing with lettered end out.
25. Install the four ⅜-16 in. capscrews through the aluminum housing and install a new gasket over the bolts.
26. Install the thrust plate on the swash plate surface of the housing.
27. With the webbed side of the housing to the top, using the screws as a pilot, install the housing over the shaft and push it in enough to start the screws. Snug up the screws evenly until the housing is approximately $3/16$ in. away from the end cap. Do not tighten at this time.
28. Install the bearing retaining ring on the end of the pump shaft.
29. Install the large O-ring seal in the housing recess around the outside of the bearing.
30. Install the end cover and secure with the four Allen head screws and tighten evenly.
31. Push the housing up flush to seat against the end cap and tighten all four screws evenly.

Cylinder Block and Pistons

1. Remove the pistons, slippers and slipper retainer assembly from the cylinder block.
2. Make up a special spring compressor made from a ⅜ in. x 3 in. long hex head bolt, nut, ⅜ SAE washer and a ⅜ in I.D. x $15/16$ in O.D. washer.
3. Center the $15/16$ in. O.D. washer on the cylinder block spring retainer. Insert the ⅜ in. x 3 in. bolt through the washer and on through the cylinder block. Place the ⅜ SAE washer and nut on the end of the bolt and tighten the nut until the spring is compressed.
4. With the spring compressed, remove the spiroloc retaining ring.
5. Loosen and remove the compressor and remove the outside retaining washer, spring, and inside spring seat.

NOTE: The pump cylinder block incorporates a flat washer for the inside spring seat. The motor cylinder block incorporates a special spring seat retainer which also serves as the stop for the motor shaft spring clip.

6. Install the inside spring retainer in the cylinder block (washer, if a pump cylinder block—special retainer if a motor cylinder block). When installing the special retainer in the motor cylinder block, make sure that the large end is placed toward the spring.
7. Install the spring in the bore of the cylinder block.
8. Place the outside retainer washer on the coil spring.
9. Using the special spring compressor tool, compress the spring and install the spiroloc retaining ring.

Hydrogear
REMOVAL

1. If the transmission has been contaminated either with metal filings or other foreign matter, the fluid must be changed and the oil filter replaced. The transaxle must also be flushed thoroughly before a new Hydrogear assembly is installed.
2. Support rear of tractor with rear wheels off the floor. Remove the left rear wheel. If tractor has tail light, disconnect wire and bring it forward of the tool box. If the tractor has a hydraulic lift, disconnect the hoses from the hydrogear and move them forward out of the way. Remove the instruction plate (located in front of the seat) by removing four mounting screws.
3. Remove the four tool box to transmission bolts and remove the fender, seat and tool box as an assembly.

NOTE: On 1965, 1966 and 1967 models it is also necessary to remove the fan guard. The 1968 and 1969 models will also require the removal of the two belt guard-to-tool box bolts.

4. Remove the belt guard and drive belt. To do this on 1968 and 1969 models, the righthand foot rest must be removed. Remove the cooling fan from the drive pulley, if so equipped, and remove the pulley.
5. Clean all dirt and clippings off the hydrogear and surrounding area with solvent and compressed air to keep the interior of the transaxle free of foreign matter when the hydrogear is separated from the transaxle.
6. Place a drain pan beneath the hydrogrear.
7. Remove three capscrews and two nuts which secure the hydrogear to the transaxle. Slide the hydrogear rearward to disengage the cam block pin from the cam block and lift the assembly off the transaxle.
8. Remove the strainer and clean with solvent and compressed air. The strainer contains a magnet to retain metal particles circulated by the fluid.

INSTALLATION

1. Clean the gasket surface on the side of the transaxle. Make sure the strainer is in place in the transaxle. Coat the gasket surface with petroleum jelly to hold the gasket in place and position a new gasket on the hydrogear.
2. If a new hydrogear assembly is being installed, remove the plastic plugs from the inlet and outlet parts of the new unit.
3. Position the hydrogear assembly on the transaxle with the cam block pin engaged in the cam block. Start three capscrews and two nuts on the through bolts and tighten until snug. Then, in rotation, tighten the nuts and screws securely.
4. Reinstall and align the pulley, fan if so equipped), and drive belt.
5. Fill the reservoir to proper level with fresh, type "A" automatic transmission fluid. Run the engine and hydrogear at low speed to circulate fluid throughout the system and to check for leaks. Recheck the fluid level and add as necessary.
6. Install belt guard, righthand foot rest (1968 and 1969 models), and the fender, seat, and tool box assembly. Connect tail light if so equipped.

NOTE: To facilitate installation of hydraulic lift hoses (see following paragraph) leave the tool box assembly loose enough to be shifted at this time.

7. If a new hydrogear assembly is being installed on a tractor which is equipped with a hydraulic lift, transfer the implement relief valve from the replaced hydrogear to the new assembly and connect the hydraulic hoses. Complete the tightening of tool box assembly.
8. Install the left rear wheel, adjust

Wheel Horse

neutral position at the cam block, and install the instruction plate.

9. Lower the rear of the tractor to the floor and test operation of the hydrogear.

D-Series

REMOVAL

1. Drain the transaxle oil.
2. Block up the tractor with jack stands adjacent to the rear foot rest.
3. Remove both rear wheels.
4. Remove axle housing links, or dozer blade hitch if so equipped.
5. Remove brake rod from brake band.
6. Remove two ½ in. and two ⅜ in. capscrews that retain the hydraulic manifold to the hydrostatic motor.
7. Tie up the rear of the manifold so it is held clear of the motor pad.
8. Remove dipstick and tube assembly from transaxle and install a ½ in. pipe plug to keep out dirt.
9. Place a floor jack under transaxle assembly to support it for remaining bolt removal.
10. Remove four front butt plate-to-case bolts.
11. Remove four top frame plate to transaxle bolts.
12. Lower the transaxle, keeping it aligned between the butt plate and the frame angles to prevent binding.

INSTALLATION

1. Install new seal rings on the motor manifold mounting pad. The two large high pressure ports each require an O-ring with a square section backup ring centered on top of each O-ring.
2. Apply a small amount of grease in the recesses and on the rings to hold them in place.
3. To keep out dirt during installation, cover the motor manifold mounting pad and seals with a piece of cardboard and hold with a piece of wire.

NOTE: After transaxle installation, cut retaining wire and remove cardboard. Lift manifold to make sure seals are still in place.

4. Place transaxle squarely in the jack, and jack the unit up into place. Prevent binding between the front plate and the angle brackets.
5. Start the top four case bolts using a punch for alignment. Leave them slightly loose at this time.
6. Align the front case holes with the butt plate using a punch.
7. Install the four bolts and washers, making sure the transaxle is squarely against both plates.
8. Tighten all eight bolts.
9. Install the axle links or hitch as applicable.
10. Connect the brake rod to the brake band.
11. Install the dipstick and tube assembly.
12. Carefully align and lower the mainifold pad in place and install the two ⅜ in. bolts and two ½ in. bolts.

NOTE: Make sure the special seal ring is used under the head of the front ½ in. bolt.

13. Tighten all four bolts evenly and securely.
14. Install both rear wheels.
15. Fill the transaxle with oil, start the engine, and check for any leaks and proper operation. Adjust brake band.

PUMP REMOVAL

1. To remove the hydrostatic pump see Engine Removal. The engine may be moved forward on the frame for access to the pump, or removed from the frame, as desired.
2. Disconnect and remove the battery.
3. Remove the right console panel, together with the battery supports.
4. Disconnect all three control rods connected to the pump linkage.
5. Remove the four capscrews holding the front manifold pad to the pump pad. The two rear screws have elastic stop nuts on the top. An oil drain pan should be placed under this area, since there will be some oil loss.
6. Disconnect the hydraulic lift tubes located at the top of the pump.
7. Disconnect the temperature sending unit wire.
8. Remove the two $5/16$ in. bolts and nuts that hold the rear pump bracket to the steering gear support.
9. Remove the bolt that attaches the front pump bracket to the left side panel.
10. Remove the pump and brackets.

PUMP INSTALLATION

1. Place a small amount of clean grease in the manifold plate seal ring recesses and on the seal rings.
2. Place a small O-ring in each of the low pressure port recesses and a large O-ring in each of the large high pressure port recesses.
3. Place a square section type backup ring over the top of each of the large O-rings, centering them.
4. To facilitate alignment of the pump to the manifold pad, fabricate two aligning pins. Cut the heads off two ⅜-16 x 2 bolts to make two studs. Saw a screwdriver slot and taper the ends.
5. Screw the aligning pins into the two front threaded holes in the pump pad.
6. Install the pump by placing it carefully in position.
7. Insert the aligning pins into the front manifold holes as the pump is lowered into place.
8. Allow the pump bracket to rest on the steering gear brace. Install punches through the side panel holes and into the front bracket to hold the pump in position.
9. Install the two $5/16$ in. bolts and nuts and secure the rear pump bracket to the steering gear brace.
10. Secure the left panel to the front bracket with the bolt and nut.
11. Connect the hydraulic tubes to the pump fittings.

12. Connect the temperature indicator wire to the sending unit.
13. Install the two rear manifold-to-pump capscrews from underneath. Place the ⅜ in. elastic stop nuts on top, but do not tighten at this time.
14. Remove the two aligning studs and install the two front capscrews and tighten all four capscrews evenly.
15. Connect the three control rods to their respective levers.
16. Install the right side panel and battery supports.
17. Lubricate the pump spline and splined washer with "moly" grease and slide the engine carefully to the rear.
18. Engage the pump spline with the pump coupling; center the engine on the pump shaft.
19. Bolt the engine securely to the frame.
20. Connect engine controls as required.
21. Install and connect battery.
22. Install the grille shroud and hood.
23. Test the unit for proper operation.
24. Check the oil level, filling as required.

MOTOR REMOVAL

1. Jack the tractor up under the frame and remove the left rear wheel.
2. Place an oil drain pan under the motor and transaxle to catch oil as the motor is removed.
3. Remove the four capscrews that hold the manifold pad to the motor.
4. Tie up the manifold so it will clear the motor pad.
5. Disconnect brake rod from brake band.
6. Remove the nuts from the two bolts securing the top section of the motor to the transaxle.
7. Remove the two lower motor-to-case capscrews which thread into the case.
8. Remove the motor.

MOTOR INSTALLATION

1. Apply a small amount of grease to the motor manifold pad recesses and the O-rings to hold them in place.
2. Install two small O-rings in the two lower pressure ports.
3. Install a third small O-ring around the right front bolt hole, located at the top left corner of the manifold pad.
4. Center a square section backup ring on top of each of the large O-rings and install them in the two large high pressure ports.

NOTE: When properly installed, the bottom of the backup rings will be just below the top of the recess. Be careful not to get dirt on the pad surface.

5. Install a new gasket on the transaxle.
6. Line the motor up on the two top case bolts and install the nuts. Install the two lower capscrews.
7. After the motor has been secured to the transaxle, check the seal rings to make sure they are in position.

Wheel Horse

8. Release the back of the manifold if it was tied up during the motor removal.
9. Line up the manifold and install the four (4) bolts. Make sure the special seal washer is under the head of the right front bolt.
10. Connect brake rod to brake band.
11. Test the unit for proper operation.
12. Check the oil level, filling as required. Adjust the brake band.

MANIFOLD REMOVAL

1. Remove two ⅜ in. Allen head screws and two ½ in. capscrews holding the manifold pad to the motor.
2. Lift the manifold and remove the O-rings.
3. Remove the two capscrews and two hex head bolts and nuts holding the manifold to the pump.
4. Remove these O-rings being careful to keep dirt away from the open oil ports.
5. Remove the manifold by turning it past the steering gear bracket and pulling it out towards rear of tractor.

NOTE: Cover all open oil ports in both motor and pump to keep loose dirt from entering.

MANIFOLD INSTALLATION

Install new seal rings on the manifold pad as follows:
1. Place a small amount of grease in the seal ring recesses and on the seal rings.
2. Place a small O-ring in each of the low pressure port recesses and a large O-ring in each of the high pressure port recesses.
3. On top of each large O-ring, place a square section type backup ring. Center it exactly on top of each of the O-rings.
4. Place a protective cardboard cover over the seal rings to hold them in place and keep the area free from dirt during installation.

NOTE: Two ⅜ in. bolts and washers are used at the left side of the pad, and two ½ in. bolts at the right side. Also note that the rear ½ in. bolt does not use a washer. The front ½ in. bolt, however, requires a special seal washer. An O-ring is used between the manifold and the motor pad at this location.

These extra seals are required since this bolt goes down into a pressure area. If it is not sealed, there will be a major oil leak.

5. Carefully hold the manifold in place.
6. Align the bolt holes in the rear manifold pad with the bolt holes in the motor and install all four bolts.
7. Remove the protective cover from the front manifold pad and check to make sure all seal rings are in place. This is done by flexing the tubes down just enough to feel if all the O-rings and backup rings are in place.
8. Position the manifold pad so the bolt holes line up.
9. Install the two short front hex screws. Leave them loose at this time.

10. Install the two longer bolts in the two rear holes, with the elastic stop nuts on top.
11. Tighten all four bolts evenly, holding the nuts on the rear bolts as required.
12. Tighten all four of the rear manifold pad-to-motor bolts.

OVERHAUL

For complete service, see the C Series Overhaul section.

DIFFERENTIAL

4 Pinion Model Used in B Series

DISASSEMBLY

1. Remove the four (4) retaining bolts and nuts.
2. Lift off the differential side case together with the axle shaft and gear.
3. Remove the axle shaft and gear from the case.
4. Remove the differential bull gear.
5. Note position of the four (4) pinions. Adjacent pinions are installed in opposite directions, which make the position of the diagonally opposite pinions the same.
6. Remove the pinions (and shafts on the early type), and remove the remaining axle shaft and gear from the case.
7. The axle shaft gears may easily be removed by driving out the roll pins or removing the snaprings as required.

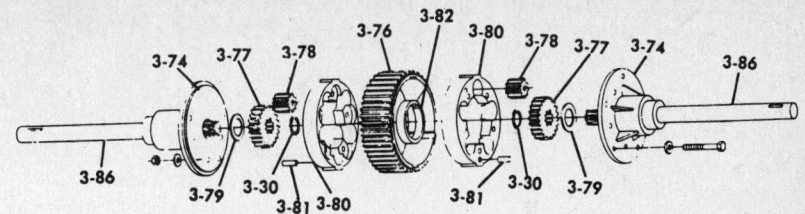

8-speed, 10 pinion differential components

3-30 Snapring	3-77 Axle gear	3-81 Roll pin
3-74 Righthand end cap	3-78 Differential pinion gear	3-82 Spring
3-74 Lefthand end cap	3-79 Thrust washer	3-86 Righthand axle
3-76 Ring gear	3-80 Body	3-86 Lefthand axle

ASSEMBLY

1. If the axle gears were removed install them on the axles and install the retaining roll pins or snaprings as required.

NOTE: Install the snaprings with the sharp side of the snapring toward the inner end of the axles.

2. Install one of the axle and gear assemblies in the differential case and on the early type install the four (4) pinion shafts.
3. Install two (2) pinions with their teeth up diagonally opposite each other, and install the other two (2) pinions with their teeth down diagonally opposite each other.
4. Install the differential bull gear.
5. Install the remaining axle and gear assembly in the remaining differential case.
6. Position the axle and gear, and differential case over the pinion shafts, mesh the gears and seat the case against the ring gear.
7. Center the bull gear and install the four (4) retaining bolts, and nuts. Tighten securely and evenly to avoid distortion.

8 Pinion Model Used on C Series

DISASSEMBLY

1. Remove the four (4) retaining bolts and nuts.
2. Lift off the differential side case together with the axle shaft and gear.
3. Remove the axle shaft and gear from the case.
4. Remove the differential bull gear.
5. Note position of the eight (8) pinions.

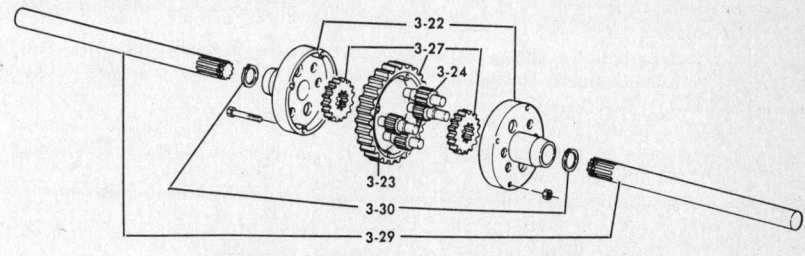

8-speed, 4 pinion differential components

3-22 Differential case	3-27 Axle gear
3-23 Ring gear	3-29 Axle
3-24 Differential pinion gear	3-30 Snapring

811

Wheel Horse

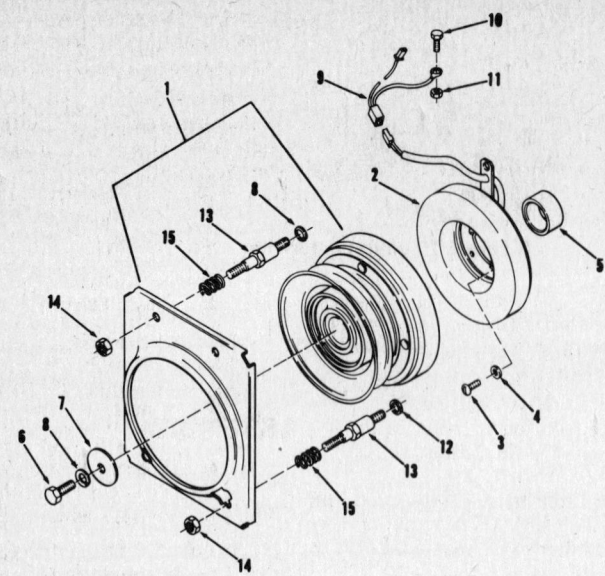

D-160 electric PTO clutch components (no keys)

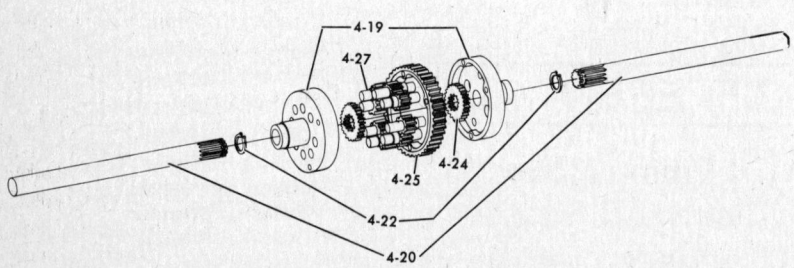

Planetary gear differential components

4-19 Differential case	4-22 Snapring
4-20 Axle	4-24 Axle gear
	4-25 Ring gear
	4-27 Differential pinion gear

Adjacent pinions are installed in opposite directions, which make the position of the diagonally opposite pinions the same.

6. Remove the pinions and remove the remaining axle shaft and gear from the case.
7. The axle shaft gears may easily be removed by removing the snaprings.

ASSEMBLY

1. If the axle gears were removed, install them on the axles and install the snaprings with the sharp side of the snapring toward the inner end of the axle.
2. Install one of the axle and gear assemblies in the differential case and install the eight (8) pinions. Properly installed each axle gear will mesh with four pinions.
3. Install the differential bull gear.
4. Install the remaining axle and gear assembly in the remaining differential case.
5. Position the other axle, gear, and differential case over the pinion shafts, mesh the gears and seat the case against the ring gear.
6. Center the bull gear and install the four (4) retaining bolts and nuts. Tighten securely and evenly to avoid distortion.

POWER TAKE-OFF

B-Series

CLUTCH REPLACEMENT

1. Separate PTO linkage under frame by removing clevis pin and hairpin cotter at PTO clutch turnbuckle.
2. Remove PTO brake adjustment screw and bracket.
3. Remove PTO clutch cone assembly.
4. Remove crankshaft bolt and bearing race.
5. Relieve drive belt tension by depressing clutch pedal and remove PTO clutch housing from driveshaft.
6. Reverse above procedure for installation of PTO clutch.

ADJUSTMENT

1. Engage PTO clutch.
2. Loosen the locknut on the PTO brake adjustment screw.
3. Turn the adjustment screw so there is a .010 in. gap between the brake pad and the clutch pulley face.
4. Tighten the locknut securely.
5. Check that the gap between the washer on the PTO rod and the clutch/brake bracket is ¼ in.
6. If clutch adjustment is necessary, change the length of the PTO rod using the turnbuckle on the rod. Loosen the locknut behind turnbuckle and remove the hairpin cotter and clevis pin from the opposite end of the turnbuckle.

C Series

CLUTCH REPLACEMENT

1. Disconnect PTO clutch rod and spring at trunnion.
2. Remove yoke pin and swing yoke aside.
3. Remove PTO brake.
4. Slide PTO hub off shaft.
5. Remove 4 bolts holding locking plates onto clutch plate and face.
6. Slide bearing race off shaft.
7. Remove clutch plate and facing.
8. Reverse above procedure for installation of PTO clutch.

ADJUSTMENT

Adjustment is required only when the PTO clutch slips noticeably. If clutch slippage is apparent, turn the trunnion toward the rear of the rod in one-turn intervals until clutch slippage is eliminated.

PTO Brake

1. Engage the PTO clutch.
2. Loosen the two bolts that hold the brake pad bracket to the support bracket.
3. Place a .012 in. feeler gauge between the brake pad and the clutch pulley.
4. While holding the brake pad against the feeler gauge and pulley, tighten the two brake bracket bolts.

D Series

CLUTCH/BRAKE REPLACEMENT, D-160

1. Disconnect the PTO wire connector plug No. 9.
2. Remove four nuts No. 14 holding clutch housing to engine block.
3. Remove bolt No. 6, lockwasher No. 8, and large special washer No. 7 from center of shaft.
4. Slide clutch off crankshaft and remove springs No. 15.
5. Remove four machine screws No. 3, remove stator No. 2, and spacer No. 5.
6. To replace PTO clutch, reverse above procedure and adjust PTO brake.

CLUTCH REMOVAL, D-200

1. Disconnect the clutch rod trunnion from the clutch bars.
2. Loosen the two brake bracket bolts, allowing the brake pad and bracket to hang down to clear the pulley.

Wheel Horse

3. Remove the clutch capscrew and special capwasher.
4. Remove the front plate and pulley assembly from the crankshaft.
5. Remove the 3/8 in. bolt and nut holding the clutch bars together.
6. Remove the cotter pin and washer from the pivot hook.
7. Remove the two clutch bars from the hook and the clutch collar.
8. Slide the clutch plate and the clutch collar off the crankshaft.

CLUTCH INSTALLATION, D-200

1. Check and adjust PTO clutch hook so that the hook centerline is $1^{13}/_{16}$ in. from the engine face.
2. Clean the crankshaft and key.
3. Apply a small amount of "moly" grease on the crankshaft and key.
4. Line up the key way in the rear clutch plate with the key and slide the clutch plate and collar assembly onto the crankshaft.
5. Position the bottom end of the two clutch bars over the pivot hook and over their respective clutch collar guide pins.
6. Install the 3/8 in. bolt and nut to hold the clutch bars together.
7. Before tightening the bolt, install the clutch hook washer and cotter pin.
8. Make sure of the alignment of the two bars and then tighten the bolt and nut.
9. Place the spacer washer inside the front hub next to the internal snapring. This determines the position of the hub on the crankshaft.
10. Line up the key way of the front cluch plate with the crankshaft key.
11. Slide the plate and pulley assembly in position on the crankshaft.
12. Install the retaining bolt, washer, and end cap and tighten securely.
13. Line up the clutch rod trunnion and clutch bars.
14. Attach the trunnion to the clutch bars and insert the washer and cotter pin to retain it.
15. Operate the clutch control lever and check the rear clutch plate and clutch bars for possible interference.
16. Adjust the clutch pivot hook in or out to eliminate any interference.
17. Adjust PTO clutch and brake.

ADJUSTMENTS

PTO Brake, D-160 Models

1. Remove the grille.
2. Position a piece of .012 in. shim stock in each of the four slots in the brake flange.
3. Turn ignition switch to Run position and PTO switch to On position.
4. Tilt the tractor's seat forward and place a weight on the seat switch button, sufficient to actuate it.
5. Loosen the four locknuts holding the brake flange.
6. Push on the brake flange until it bottoms out and retighten the four locknuts. Do not over-tighten.
7. Turn PTO switch to Off position and remove ignition key.
8. Recheck the gap at all four positions with a feeler gauge. The gap should be between .010 and .015 in.

PTO Clutch, D-160 Models

The electric PTO clutch on the D-160 is self-compensating for wear and dows not require adjustment. If clutch problems are encountered, check the stator, PTO switch and wiring for continuity. Replace the rotor if the clutch friction surface is excessively worn.

PTO Brake, D-200 Models

1. Remove the grille and engage the PTO clutch.
2. Loosen the two brake bracket bolts.
3. Place a .012 in. feeler gauge between the brake pad and the clutch pulley.
4. While holding the brake pad against the feeler gauge and pulley, tighten the two brake bracket bolts.

PTO Clutch, D-200 Models

1. Remove the grille and disengage the PTO clutch.
2. Loosen the two brake bracket bolts.
3. Loosen the locknuts and adjust the turnbuckle on the clutch rod so that the rear clutch plate facing just clears the pulley, allowing it to be turned freely by hand.
4. Engage PTO clutch and check tension. The washer at the front end of the clutch rod should just clear the trunnion so it can be turned by hand.
5. Adjust PTO brake.

ENGINE

The B series tractors use various Briggs & Stratton engines, the C series use various Kohler engines, the D series use both Briggs and Stratton and Onan engines. See the appropriate part of the Engine Unit Repair section.

REMOVAL AND INSTALLATION

B Series

1. Remove negative battery cable.
2. Disconnect choke and throttle cables, fuel line, and all four electric lines: magneto ground (green), alternator/lights (yellow); alternator/charge (red), starter (black).
3. Separate PTO linkage under frame by removing clevis pin and hairpin cotter at PTO clutch turnbuckle.
4. Remove PTO brake adjustment screw and bracket.
5. Remove PTO clutch cone assembly.
6. Remove crankshaft bolt and bearing race.
7. Relieve drive belt tension by depressing clutch pedal and remove remaining PTO clutch housing from crankshaft.
8. Remove four (4) engine mounting bolts and lift engine off frame.
9. For engine installation, reverse removal procedure.

C-Series

1. Remove negative battery cable.
2. Disconnect choke and throttle cables, fuel lines, and all electric lines:
   * red—2 wires to solenoid
   * battery cable to solenoid
   * black—coil
   * rectifier regulator plug
   * orange—rectifier plug
3. Remove the belt guard.
4. Remove the hairpin cotter from the PTO clutch rod.
5. Remove the PTO shaft clevis pin, and pivot yoke away from shaft.
6. Remove the PTO brake.
7. Disengage the transmission drive clutch on automatic models or depress clutch on 8-speed models.
8. Remove drive belt from the transmission drive pulley and slide belt forward to clear engine pulley.
9. Remove through bolts from front isomounts.
10. Remove nuts securing rear isomount block to frame.
11. Remove engine with rear block still attached.
12. Align front isomount through bolt with front block, but do not tighten.
13. Maneuver rear block so studs drop through holes in frame and secure with nuts.
14. Tighten all mounting bolts.

C-161 Twin

1. Remove negative battery cable.
2. Disconnect choke and throttle cables.
   * battery cable to solenoid
   * black—magneto ground
   * red and yellow—alternator plug
3. Remove the belt guard.
4. Remove the hairpin cotter from the PTO clutch rod.
5. Remove the PTO shaft clevis pin, and pivot yoke away from shaft.
6. Remove the PTO brake.
7. Disengage the transmission drive clutch on automatic models or depress clutch on 8-speed models.
8. Remove drive belt from the transmission drive pulley and slide belt forward to clear engine pulley.
9. Remove through bolts from front isomounts.
10. Remove nuts securing rear isomount block to frame.
11. Remove engine with rear block still attached.
12. Align front isomount through bolt with front block, but do not tighten.
13. Maneuver rear block so studs drop through holes in frame and secure with nuts.
14. Tighten all mounting bolts.

D-160

1. Remove negative battery cable.
2. Disconnect choke and throttle controls, and fuel line from fuel pump.
3. Disconnect electric wires to coil, starter, orange wire to rectifier plug, headlights, and PTO clutch. Mark wire colors for replacement.
4. Remove grille.

Wheel Horse

5. Remove upper muffler clamps, and four (4) bolts on grille shroud.
6. Remove mufflers and complete hood and grille shroud assembly.
7. Remove four (4) bolts holding engine to frame and slide engine forward until the flex coupling slides off pump drive shaft.
8. Remove the engine from the frame.
9. For installation, reverse removal procedure.

D-200

1. Remove negative battery cable.
2. Disconnect choke and throttle controls and fuel line from fuel pump.
3. Disconnect electric wires to coil, starter, headlights, and orange wire to rectifier plug. Mark wire colors for replacement.
4. Remove mufflers and complete hood and grille shroud assembly.
5. Disconnect the PTO rod trunnion from the clutch bar.
6. Remove four (4) bolts holding engine to frame and slide engine forward until the flex coupling slides off pump drive shaft.
7. Remove the engine from the frame.
8. For installation, reverse removal procedure.

D-250

FRONT AXLE

REMOVAL

1. Disconnect the drag link from the Pitman arm.
2. Remove the front PTO belt guard and slide the drive belts off the PTO pulley.
3. Bend the lock tabs back on the front axle mounting bolts and remove the four nuts.
4. Raise the chassis and roll out the front axle assembly.

INSTALLATION

1. Roll the front axle assembly under the chassis, lower the chassis on the axle and install the mounting nuts.
2. Slide the PTO drive belts on the pulley.
3. Reinstall the PTO belt guard.
4. Reconnect the drag link to the Pitman arm.

OVERHAUL

1. Remove the front axle assembly.
2. Disconnect an end of the tie rod from one of the steering spindle assemblies.
3. Remove the spindle/wheel assemblies from the front axle.
4. Remove the front PTO pulley.
5. Remove the snapring and washers on the front of the PTO shaft.
6. Loosen the setscrew on the locking collar and turn collar by tapping the other hole with a punch until loose.
7. Remove the locking collar on the other side of the PTO shaft in the same manner as described above.
8. Slide out the PTO shaft.
9. Remove the bearings from the axle by tapping on the outer race with a punch.
10. Remove the inner C clips from the axle and install on the new axle.
11. Lubricate the axle bearings and install in the axle.
12. Inspect the bearings in the axle support. Replace by removing the bearing cap, turning the bearing 90°, aligning with the slots and pulling out.
13. Place the front axle in the axle support and install the PTO shaft with shims. Make sure the larger shims are in the front.
14. Install the locking collar, washers and snapring on the back of the PTO shaft.
15. Tap the PTO shaft toward the front and tighten the locking collar.
16. Install the locking collar on the front and tighten.
17. Shim the front of the PTO shaft up to the snapring to prevent axial play and install the snapring.
18. Adjust the axle axial play by screwing the adjusting bolt in until the play is eliminated. Be careful not to overtighten and cause binding.
19. Install the PTO pulley.
20. Reinstall the spindle/wheel assemblies on the front axle.
21. Reconnect the tie rod to the steering spindle.
22. Reinstall the front axle assembly.

STEERING

Spindle
REMOVAL

1. Elevate the front of the tractor.
2. Disconnect the steering spindle from the tie rod.
3. Remove the nut from the steering spindle shaft and slide out the steering spindle.

INSTALLATION

1. Grease the steering spindle shaft and slide it into the front axle.
2. Install the nut and cotter pin on the steering spindle shaft.
3. Reconnect the steering spindle arm to the tie rod.

Steering Gear
REMOVAL

1. Remove the steering wheel, using a wheel puller if necessary.
2. Remove the coolant overflow bottle.
3. Remove the upper front steering gear bracket bolt.
4. Remove the upper rear steering gear bracket bolt.
5. Jack up the rear of the tractor about 10 in.
6. Remove the drag link from the steering gear Pitman arm.

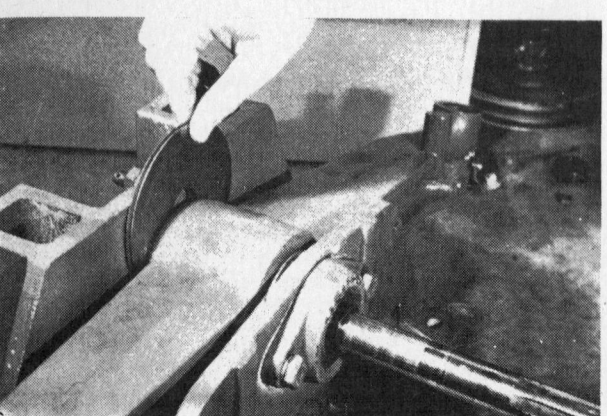

Shimming the front axle

Wheel hub removal

Wheel Horse

7. Remove the driveshaft flange from the clutch shaft, and move the driveshaft aside.
8. Remove the lower steering gear bracket bolt and slide the steering gear assembly out from bottom.

INSTALLATION

1. Position the steering gear and install the lower mounting bolt.
2. Install the two upper mounting bolts.
3. Reconnect the driveshaft to the clutch shaft.
4. Reconnect the drag link to the steering gear Pitman arm.
5. Reinstall the coolant overflow bottle.
6. Reinstall the steering wheel.

Front Wheel Bearings

REPLACEMENT

1. Tap the bearings out to remove them.
2. Grease and install the inner bearing, with the sealed side of the bearing facing out.
3. Grease and install the spacer.
4. Grease and install the outer bearing with the sealed side facing out.

BRAKES

REMOVAL

1. Elevate the tractor chassis.
2. Remove the appropriate wheel.
3. Pull off the brake drum.
4. Remove the wheel hub with a puller.
5. Remove the brake shoes by pulling out.

INSTALLATION

1. Install the new brake shoes on the axle tube.
2. Reinstall the wheel hub.
3. Place the brake drum on the wheel hub.
4. Reinstall the wheel.

ADJUSTMENT

1. Loosen the locknut on the turnbuckle.
2. Rotate the turnbuckle to achieve a uniform grabbing effect at each wheel. The adjustment should not be so tight that the brake shoes rub against the brake drum without the brake pedal depressed.
3. Retighten the turnbuckle locknut.

CLUTCH

REMOVAL

1. Disconnect the following items from the tractor.
   a. Drag link from Pitman arm.
   b. Hydraulic lift cylinder.
2. Drive out the roll pin on the right side of the rear brake linkage shaft and rotate the shaft down.
3. Remove the driveshaft flange bolts from the clutch shaft and lower the drive shaft.
4. Disconnect the clutch adjusting rod from clutch fork.
5. Remove the clutch housing mounting bolts and lift out the clutch housing and the clutch housing plate together with the clutch fork.

NOTE: If reusing the clutch, mark its position on the flywheel.

6. Remove the clutch mounting bolts and lift out the clutch disc and clutch pressure plate.

INSTALLATION

1. Using special tool (part no. 90.45.269) to center the clutch disc, install the clutch disc and clutch pressure plate. Make sure the protruding side of the clutch disc faces away from the engine.
2. Check the clutch shaft bearings. If necessary, press the shaft out and replace any worn parts.
3. Check the release bearing (throwout bearing) on the clutch housing plate and replace if necessary by disconnecting the spring clips.
4. Place the clutch housing and the clutch housing plate together on the engine.
5. Reinstall the clutch housing mounting bolts.
6. Reconnect the clutch adjusting rod to the clutch fork. Install the special washer and nut and adjust clutch play.
7. Install the driveshaft flange on the clutch shaft. Make sure the alignment arrows on the splined shafts of the driveshaft are matched up.
8. Rotate the rear brake linkage shaft up and install the roll pin.
9. Reconnect the drag link and the hydraulic lift cylinder.

TRANSAXLE

REMOVAL

1. Put the speed range in high and the shifter in 2nd gear.
2. Remove the seat.
3. Remove the two mounting bolts for the seat fender pan from each side of the tractor and lift off the seat fender pan.
4. Remove the two rear mounting brackets for the seat from the bottom.
5. Remove the front mounting bolts for the gas tank.
6. Remove the rear mounting brackets for the gas tank.
7. Disconnect the rubber gas line from the metal line (running up the left channel to the engine) and plug the rubber hose.
8. The gas tank **must** be raised in the rear of the tractor before sliding it out to avoid damaging the plastic fittings. Elevate the rear of the tank and slide it out.
9. Disconnect the items listed below on the tractor.
   a. Main shift shaft
   b. Right foot brake
   c. Rear axle lock
   d. Rear PTO linkage
   e. Parking brake linkage
   f. Left foot brake
   g. Parking brake indicator light switch
   h. Return spring
10. Remove the cotter pin from the speed range shifter linkage, so that it will release as the transaxle is removed.
11. Raise the rear of the tractor about 1 in. off the ground and place jack stands under the frame.
12. Remove the flange between the universal joint and the transaxle.
13. Remove the two front transaxle mounting stud nuts.
14. Remove the two side mounting brackets for the transaxle.
15. Elevate the rear of the tractor another 4 to 5 in.
16. Walk the transaxle out from under the tractor by rotating the transaxle input shaft.
17. Use a jack stand to support the front end of the transaxle.
18. Drain the transaxle fluid.

INSTALLATION

1. Add transaxle fluid.
2. Place the speed range in high and the shifter in 2nd gear.
3. Walk the transaxle into position under the tractor by rotating the transaxle input shaft.
4. Lower the tractor chassis.
5. Install the two side mounting brackets for the transaxle.
6. Install the two front mounting bolts for the transaxle.
7. Reinstall the flange between the universal joint and the transaxle.
8. Reconnect the speed range shifter linkage.
9. Reconnect the items listed below:
   a. Main shift shaft
   b. Right foot brake
   c. Rear axle lock
   d. Rear PTO linkage
   e. Parking brake linkage
   f. Left foot brake
   g. Parking brake indicator light switch
   h. Return spring
10. Replace the gas tank, taking care not to damage the plastic fittings.
11. Reconnect gas line.
12. Reinstall rear gas tank mounting brackets.
13. Reinstall the front mounting bolts for the gas tank.
14. Reinstall the two rear mounting brackets for the seat.
15. Reinstall the seat fender pan.
16. Reinstall the seat.

DISASSEMBLY

1. Remove the tranaxle from the vehicle.
2. Remove the left wheel and tire only (determined by sitting in the seat).
3. Stand the transaxle up on the right tire.

Wheel Horse

PTO removal

Countershaft removal

NOTE: Do not attempt to remove the front cover plate before removing the rear cover plate. Removing the front plate first will cause parts to fall out of place in the rear portion of the transaxle.

4. Remove the flange from the front of the transaxle.

Rear PTO

1. Remove the cover plate on the rear of the transaxle.
2. Take care not to lose the shims on the PTO input shaft.
3. Remove the spacer plate.
4. Tap out the PTO shaft/bearing assembly with a plastic hammer.
5. Remove the two needle bearings and place them inside the PTO shaft until you are ready to reassemble.
6. If the bearing or sleeve seal on the PTO shaft assembly need replacing, remove the snaprings and press the shaft out.
7. Remove the large snapring from the circumference of the housing.
8. Slide off counter gear.
9. Remove snapring from PTO countershaft.
10. Remove the change gear and spacer washer.

Reduction Gear

1. Pull the speed range shift lever out to high position.

NOTE: The lever must be in the high position to remove the reduction gear cover.

2. Remove mounting bolts for front cover plate.
3. Tap the front cover off with a soft hammer and remove the reduction gear assembly.
4. Remove the two needle bearings and thrust washer from the intermediate shaft.
5. Remove capscrews holding reduction gear housing to transaxle case.
6. Tap reduction gear housing off with a plastic hammer. Take care not to lose shims between housing and two roller bearings.

Gearshift Housing

1. Remove the three upper bolts for the gearshift housing.
2. Loosen the three lower mounting bolts for gearshift housing, but do not remove them.

Left Gear Housing Removal

1. Remove ten bolts which hold the two halves of the transaxle case together. The two larger bolts are locating bolts, and usually have to be tapped out with a punch.
2. Remove the two bolts screwed from the right half of the housing up into the left half of the housing (from the bottom up).
3. Tap the two halves of the housing apart with a plastic hammer while separating. The halves may be difficult to separate due to the sealer used between the two halves.

Countershaft

1. Remove the gearshift housing and assembly.
2. Remove the countershaft, retaining ring and shim from the gear housing.

Worm Wheel/Wormshaft

1. Pry the worm gear up a little with a screwdriver or tire iron to provide clearance, being careful not to damage the bearing bores in the housing.
2. Lift out the wormshaft assembly.
3. Pry the worm gear the remainder of the way out of the casing. If the bearings remain in the casing, be careful not to lose the shims.

Reverse Idler Gear

Remove the reverse idler gear and idler gear shaft.

Rear PTO Shift Lever

1. Loosen the nut on the rear PTO shift lever.
2. Tap the retaining pin out of the shift lever.
3. Tap the shift shaft (held by pin) out and remove the assembly.
4. Replace any defective parts of the PTO shift lever and reinstall the assembly in the casing.

Differential

NOTE: The differential housing and worm gear can be changed independently. If both are to be reused, mark the position of the worm gear on the differential housing before disassembling.

Wormshaft removal

Wormshaft components

Wheel Horse

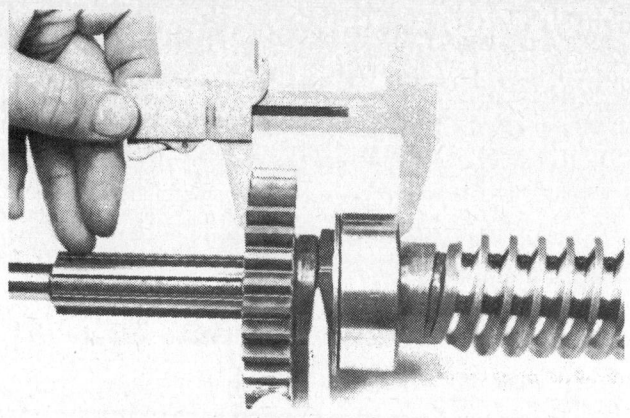

Measuring bearing-to-idler gear clearance

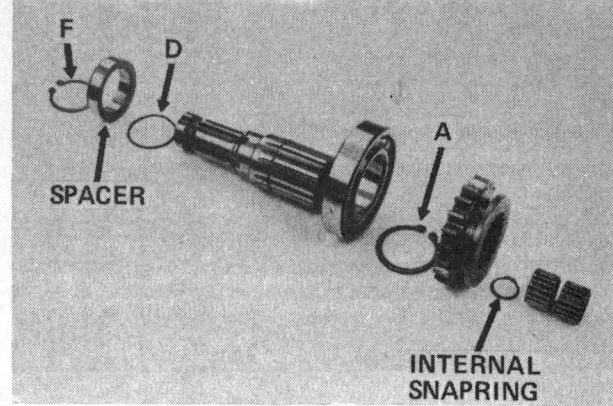

PTO shaft components; A and F are snaprings, D is an O-ring

1. Bend back locking tabs and remove eight bolts on differential housing.
2. Remove the differential end caps from the worm gear.
3. Tap out the differential shaft.
4. Inspect all parts and replace any if necessary.
5. Reinstall the large pinions in each of the differential end caps with their thrust washers.
6. Install the smaller pinions and the differential axle.
7. Assemble the worm gear between the two differential end caps, and align the marks previously made.
8. Install the eight bolts with locking tabs and torque evenly to 47 ft.lb. The head of the bolt should be on the side with the smallest clearance to prevent bolts from turning.

NOTE: Use only bolts and nuts of 10K class.

Wormshaft

1. Slide the change gears and intermediate shaft from the wormshaft.
2. Inspect the gears and bearings and replace any if necessary.
3. If replacing the wormshaft bearing, it is necessary to remove the idler gear.
4. Remove the snapring holding the idler gear to the shaft. Be careful not to lose the thrust washers on both sides of the gear.
5. Remove the snapring holding the parking brake disc to the shaft and remove the disc.
6. Bend up the lock tabs and remove the nut in front of the bearing by turning it clockwise. This nut has lefthanded threads.
7. Press the bearing off the shaft.
8. Heat the new bearing in an oil bath 175-195°F and press on the shaft.
9. Reinstall the bearing retaining nut and tighten to 72 ft.lb.
10. Bend the locking tabs over the nut.
11. Reinstall the locating ring, parking brake disc and snapring.
12. Reinstall the locating ring, idler gear with thrust washers and the snapring.
13. Using a caliper, measure the distance between the outside of the bearing and the idler gear.
14. The distance should be 2.226-2.338 in.

Adjust distance if necessary by placing shims between the bearing and the idler gear (where pencil points).

15. Install the change gears on the wormshaft and check to see that they move freely.
16. If necessary, replace the bearing on the intermediate shaft with the aid of a press.
17. Check to see that the snapring (N) that positions the needle bearing is installed within the shaft.

Rear PTO Drive

1. If it is necessary to replace the bearing, disassemble the rear PTO shaft.
2. Press off the old bearing.
3. Install snapring "A" on the shaft.
4. Press on new bearing against "A".
5. Install O-ring "D" and the seal spacer (with the bevel toward the O-ring).
6. Install snapring "F". If there is a space between the O-ring seal, spacer and snapring, tighten with shims. Insert the shims between the snapring and the seal spacer.
7. Install internal snapring and needle bearings.
8. Place the thrust washer and PTO shaft on the wormshaft.
9. Install the intermediate shaft on the wormshaft.

Parking Brake

1. Mark the position of the parking brake lever on the shaft.
2. Remove the lock bolt on the parking brake lever.
3. Remove the lever and cam.
4. Replace the O-ring in the cover cap.
5. Replace any parts if necessary.

Parking brake lever removal

6. Install the brake lever on the shaft and align marks.
7. Reinstall lock bolt on parking brake lever.

ASSEMBLY

Adjusting the Worm Wheel

1. Insert the bearing calibration block (special tool no. 00.40.286/7) in the right half of the transaxle case.
2. Measure the distance between the top of the case and the bearing calibration block with a depth gauge.

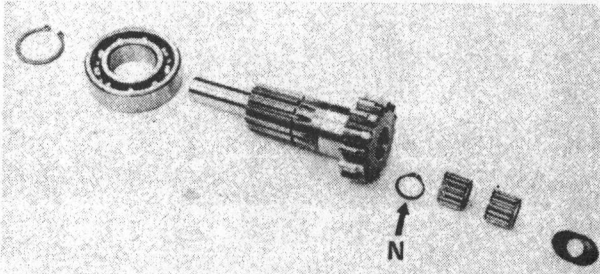

Intermediate shaft components; N is the snapring that positions the bearing

Wheel Horse

3. Insert the bearing calibration block (special tool no. 00.40.282/11) into the left case and measure the distance with a depth gauge.

4. Measure the worm gear with an outside caliper then, using inside calipers accurately determine the outside caliper opening.

5. Calibrate the necessary shims (See example below).

EXAMPLE:

Right casing (Step 2)	1.594 in.
Left casing (Step 3)	1.705 in.
Worm gear (Step 4)	3.150 in.
Shims For Right Casing	
Right casing	1.594 in.
−One half the worm gear	−1.575 in.
	.019 in.
−Allowable Axial play	− .004 in.
*Shims for Right Casing	.015 in.
Shims For Left Casing	
Left casing	1.705 in.
−One half the worm gear	−1.575 in.
	.130 in.
−Allowable Axial Play	−.004 in.
*Shims for Left Casing	.126 in.

6. Install the required shims on each side of the differential assembly, between the differential and the bearing. *The right side has the large I.D.*

7. Replace the bearings on the differential assembly.

8. Install the worm gear and wormshaft assembly.

9. Check the meshing of the differential gear. If possible, paint the gears with machinist's ink (Prussian blue), and bolt the cases together. Place the shift lever in fourth gear and turn the input shaft several revolutions in both directions to ensure that the worm gear has completed at least one revolution. Reopen the case and check the gear contact.

Reverse Idler Gear

1. Check the condition of reverse idler needle bearings and replace if necessary.

2. Reinstall reverse idler and shaft in casing.

Countershaft Assembly

1. Replace the countershaft assembly in casing.

2. Measure the distance between the inside face of the front cluster gear and the outside of the case.

3. The distance should be .787 in. +.008 in. If necessary, install shims to area shown to adjust.

4. Check the meshing of the countershaft and wormshaft gears.

5. Mount the gear shift housing with a new gasket on the right casing.

6. Make sure the shift forks are meshed in the grooves of the gears.

7. Test the meshing by shifting into all the speeds. Make sure all gears turn freely and do not rub against other gears or bearings.

Left Gear Housing

NOTE: Before the left casing is installed, it is necessary to check the intermediate shaft axial play.

1. Tap the end of the intermediate shaft in with a hammer and measure the distance between the intermediate shaft bearing and the casing.

2. Measure the centering protrusion on the reduction gear housing with a depth gauge.

3. Subtract the measurement in step 2 from the measurement in step 1.

Step 1 measurement
−Step 2 measurement
Axial play

The answer is the axial play, which should be .008 − .012 in.

4. If the axial play is too large, adjust by installing shims behind the intermediate shaft. If it is too small, remove shims.

5. Coat each half of the housing with sealing compound.

6. Install two bolts in opposite corners to use for locators.

7. Lower the left casing down and tap the two together with a plastic hammer.

8. Reinstall the remaining bolts and torque to 18 ft.lb.

Reduction Gear

1. Push the high-low range shift shaft in and lift out the change gear.

2. Pull out the PTO countershaft. Take care not to lose the shim washers.

3. Pull out the cluster gear assembly.

4. If bearing is to be replaced, remove with puller and take care not to lose shims.

5. To remove driveshaft input gear, detach the snapring and drive the shaft from the outside to inside with a plastic hammer.

6. If necessary, replace the oil seal on drive shaft input gear.

7. If PTO countershaft needle bearings are to be replaced, the new bearing must be flush with the housing.

8. If the shift lever needs to be serviced, remove the shaft setscrew.

9. Drive out the retaining pin on the shift fork.

10. Slide the shaft out, taking care not to lose the spring or detent ball.

11. Replace the rubber oil seal and any other worn parts.

12. When reinstalling the shaft, hold the spring and ball in place with a screwdriver while sliding the shaft in.

13. Check to make sure the lever can be easily shifted.

14. Install the countershaft in place and measure the distance from the top of the largest gear to the top of the casing with a depth gauge.

NOTE: If the bearing is not pressed on the shaft, the special calibration bearing (Part No. 00.41.283/5) may be used when taking the measurement.

15. The distance should be .429 in. ± .008 in. If necessary, shim between the bearing and larger gear.

16. Install the thrust washer, shim washer and PTO countershaft.

17. Check the mesh of the gears. The top of the gear on the PTO shaft should be level with cluster gear. Adjust with shims if necessary.

18. Push the high-low shift lever to low position and install the change gear.

19. Pull the lever to high position and check the mesh of the gears.

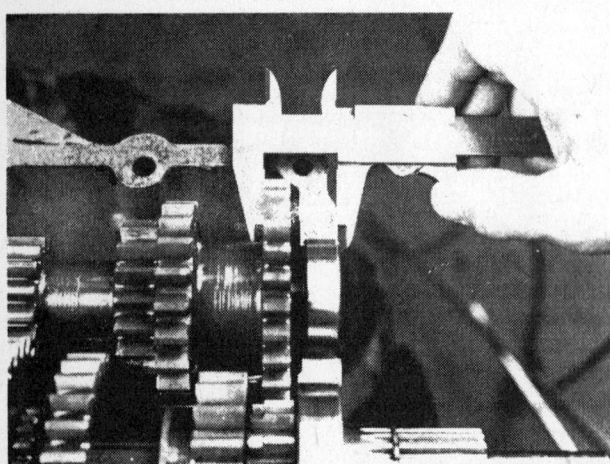

Countershaft axial play measurement. This is the area to be shimmed

PTO output shaft measurement

Wheel Horse

20. Spread grease on the shim washer for the countershaft bearing and press it into place.
21. Spread sealing compound on the reduction gear housing.
22. Mount the housing to tranaxle with the six hexagon bolts and special washers.
23. Grease the thrust washer and needle bearings and install.
24. Spread sealant on reduction gear cover contact surfaces.
25. Slide the reduction gear cover assembly on the transaxle.

Rear PTO Cover

NOTE: If you are replacing the PTO cover, perform the following procedures.

1. Install oil seal and fill with grease.
2. Press in needle bearing.
3. Install dowel pin.
4. Press in screws.
5. Install the snapring on the PTO countershaft.
6. Grease thrust washer and needle bearings and place on wormshaft.
7. Replace PTO change gear, countergear and snapring.
8. Slide the PTO output shaft assembly on and tap in with a plastic hammer.
9. Reinstall PTO spacer plate and gasket.
10. Measure the distance between the end of the bearing and the spacer plate with a depth gauge.
11. Measure the distance on the PTO cover (with gasket on) between the bearing contact surface and the gasket surface.
12. Subtract the measurement of step 10 from the measurement of step 11 to calculate axial play.

Step 11
−Step 10
Axial Play

13. The axial play should be .016 in.-.020 in. Adjust with shims if necessary.
14. Tap the PTO countershaft gear in with a plastic hammer and measure the distance between the gear and the spacer plate.
15. Add shims equal to the measurement obtained in step 10 to end of the PTO countershaft gear. Axial play is provided by cover gasket.
16. Install the PTO cover and gasket.

AXLE TUBE

NOTE: The axle tubes may be removed and repaired with the transaxle mounted in the vehicle.

REMOVAL

(With transaxle in tractor)
1. Support the tractor on jack stands.
2. Remove the wheel and brake drum on the axle tube being removed.
3. Remove the wheel hub with a wheel puller.
4. Remove the brake backing plate mounting bolts and lift off.
5. Remove the axle tube mounting bolts and lift off the axle.

NOTE: Be careful not to lose shims between the bearing and the worm wheel.

Differential Lock (Right Tube Only)

REMOVAL

1. Remove the locking bolt and the detent ball from the differential lock lever.
2. Slide off the differential locking gear.
3. Pull out the differential lock lever.

DISASSEMBLY

1. Remove the oil seal by tapping a punch through its side and prying off.
2. Remove the snapring retaining the bearing in place.
3. Tap the half axle toward the wheel end from the differential end of the case with a plastic hammer.
4. Remove the bearing in a press.

ASSEMBLY

1. Install the bearing in the casing, make sure a snapring is on each side of the bearings.
2. Press the half axle into the bearing.
3. Install a new oil ring.
4. Slide on the oil seal race with the bevel side toward the bearing.
5. Install the new oil seal.

INSTALLATION

1. Install the differential lock lever in the housing.
2. Slide the differential locking gear into position.
3. Install the detent ball, spring and locknut.
4. Check operation of the lever.
5. Coat the contact surfaces with sealer.
6. Bolt the axle tube to the transaxle.

Gearshift Box

DISASSEMBLY

1. Loosen the locking nuts and remove the adjusting screws from each side.
2. Remove the mounting screws for the selector shaft cover plates from each side.
3. To remove selector shaft and fork, tap the shaft with a punch. Take care not to lose detent springs or balls.
4. If replacing shift lever assembly, remove the two remaining shafts.
5. Remove the shift gate mounting bolts and slide out the shift gate.
6. Loosen the Allen bolt holding the selector lever to the selector shaft, and tap out the selector shaft.

ASSEMBLY

1. Replace both O-rings on the selector shaft.
2. Install the selector shaft in the housing, being careful not to reverse direction.
3. Spread Loctite® on the selector shaft setscrew and install with an Allen wrench.
4. Install the shift gate assembly, taking care to get it in the *proper direction*.
5. In order to install the various detent balls and washers, it is *absolutely necessary* to fabricate a special tool.

Make the tool from a steel rod ⅜ in. in diameter. The longer piece should be 5½ in. long, and the shorter 1½ in. long. Cut the rod at approximately the angle indicated. Slot each end.

6. Install the ½ or center selector shaft first.
7. Place the spring and ball bearing in the detent hole and hold with a screwdriver while installing the longer piece of the fabricated tool.
8. Rotate the long portion of the tool 180°. Drive in the short part of the tool and rotate the short tool 180°.
9. Place the selector fork in the gear housing and push the ½ selector shaft into position.
10. Put the selector fork in neutral position.
11. Install the other two selector forks in the same manner making sure to insert detent balls between the shift forks also.
12. Install the side covers with gaskets on the shift housing.
13. Install the adjusting screws and locknuts.
14. Turn the screws snugly against both ends of the rods and adjust so that each shaft is centered. This can be determined by making sure the screws protrude the same amount beyond the locknut at each end of each of the rods.
15. Tighten the locknuts.

NOTE: After the gearshift box assembly is installed on the transaxle, it may be necessary to make minor adjustments to ensure that the sliding gears in the transaxle are positioned properly in all speeds.

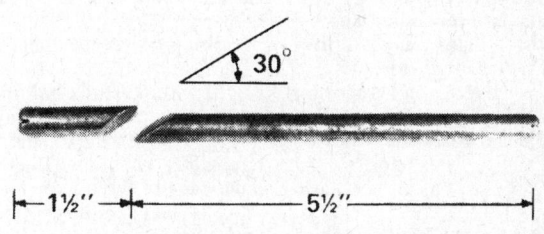

Shift installation special tool

Wheel Horse

HYDRAULIC SYSTEM

Pump

REPLACEMENT

1. Remove the drive belt cover on the hydraulic pump.
2. Loosen the tension on the drive belt and remove the belt from the pulley.
3. Disconnect the shift rod from the hydraulic pump.
4. Remove the shift shaft from the hydraulic pump by loosening the locking nut and stud.
5. Remove the hydaulic pump mounting bolts.
6. Remove the hydraulic lines from the old pump and install on the new one.
7. Place the new pump on the mounting bracket and install the mounting bolts.
8. Reinstall the shift shaft, with retaining stud and locknut.
9. Reconnect the shift rod to the hydraulic pump.
10. Place the drive belt on the pulley and adjust the tension.
11. Reinstall the drive belt cover on the hydraulic pump.
12. Bleed the hydraulic system.

Lift Cylinder

REPLACEMENT

1. Remove the retaining pins on each end of the lift cylinder and remove the cylinder.
2. Remove the hydraulic line from the old cylinder and install on new one.
3. Install the new cylinder.
4. Bleed the hydraulic system.

ENGINE

The engine used in these tractors is made in France by Renault, the automobile company.

REMOVAL

1. Remove the hood covering the engine.
2. Disconnect the battery.
3. Remove the bolt from both sides of the tractor with a 13mm wrench.
4. Remove the two remaining nuts.
5. Remove the radiator water overflow hose from the top of the radiator.
6. Remove the cover from the radiator water reservoir bottle (compensating tank).
7. Disconnect and mark the wires on the ignition coil and the generator.
8. Remove the two nuts on the dashboard panel.
9. Remove the plastic wraps which hold the hydraulic line to the grille assembly.
10. Remove the clamp holding the hydraulic line to the radiator.
11. Lift the grille assembly up and rotate it back to rest on the seat.
12. Drain the coolant from the radiator.
13. Remove the water hoses from the radiator.
14. Remove the radiator mounting bolts which are labeled "A", remove the radiator.
15. Remove the water hoses from the water pump.
16. Disconnect the muffler from the exhaust manifold.
17. Disconnect the muffler mounting brackets from the engine block and remove the muffler.
18. Disconnect and mark the wires on the starter, the temperature sending unit and the oil pressure sending unit.
19. Remove the air horn from the carburetor.
20. Remove the air cleaner assembly by loosening the two bolts mounted on the engine block and sliding the mounting bracket apart.
21. Remove the throttle cable mounting bracket from the carburetor.
22. Remove the hand and foot throttle cables from the governor lever.
23. Disconnect the choke cable from the carburetor.
24. Disconnect the clutch rod from the clutch lever.
25. Loosen the bolts on the belt guard for the twin V-belt PTO pulley and slide the guard out of the way.
26. Slip the drive belts off the PTO pulley.
27. Remove the hydraulic control rod (shift rod) from the hydraulic pump.
28. Remove the hydraulic hose labeled "A" from the hydraulic pump and screw the fitting in to prevent contamination of the fluid in the pump.

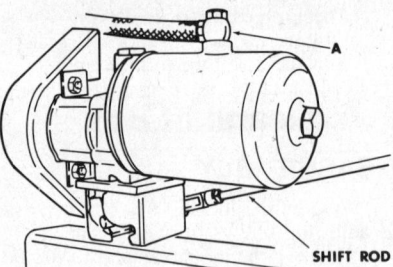

Hydraulic pump; A is the hydraulic hose

29. Remove the cotter pin on the front brace of the hydraulic lift cylinder, lower the cylinder and disconnect the hydraulic line.
30. Remove the fuel line from the fuel pump.
31. Remove the bolt mounting the horn to the engine block and lay the horn aside.
32. Loosen the top mounting bolt on the regulator and rotate the regulator toward the back of the tractor for clearance.
33. Remove the bolt from the front engine bracket.
34. Remove the bolts from the two side engine brackets.
35. Disconnect the linkage between the governor and the carburetor.
36. Mount the lifting hoop (special tool No. 90.45.289) on the cylinder head.
37. Remove the engine.

INSTALLATION

1. Remove the driveshaft coupling from the engine before reinstalling.
2. Lower the engine in place on the tractor.
3. Remove the lifting hoop from the cylinder head.
4. Reconnect the linkage between the governor and the carburetor.
5. Install the bolts in the two side engine mounting brackets.
6. Install the bolt in the front engine mounting bracket.
7. Replace the top mounting bolt on the regulator.
8. Reinstall the horn to the engine block.
9. Install the fuel line on the fuel pump.
10. Reinstall the hydraulic line on the hydraulic lift cylinder and remount the cylinder.
11. Reinstall the hydraulic hose labeled "A" on the hydraulic pump.
12. Install the shift rod to the hydraulic pump.
13. Replace the drive belts on the PTO pulley.
14. Position the PTO belt guard and tighten the mounting bolts.
15. Reconnect the clutch rod to the clutch lever.
16. Connect the choke cable to the carburetor.
17. Install the hand and foot throttle cables on the governor lever.
18. Install the throttle cable mounting bracket on the carburetor and adjust the tension of the cables.
19. Reinstall the air cleaner assembly on the block.
20. Reconnect the air horn to the carburetor.
21. Reconnect the wires to the starter, temperature sending unit and oil pressure sending unit.
22. Remount the muffler mounting brackets on the engine block.
23. Reconnect the muffler to the exhaust manifold.
24. Install the water hoses on the water pump.
25. Place the radiator on the chassis and install mounting bolts.
26. Reinstall the water hoses on the radiator.
27. Rotate the grille assembly off the seat and back to its normal position on the tractor.
28. Reinstall the clamp holding the hydraulic line to the radiator.
29. Fix the hydraulic line to the grille assembly with plastic tie wraps (or wire).
30. Reinstall the two nuts on the dashboard panel.
31. Reconnect the wires to the ignition coil and the generator.
32. Reinstall the cover on the radiator

Wheel Horse

water reservoir bottle (compensating tank).

33. Reinstall the radiator water overflow hose on the top of the radiator.
34. Reinstall the two remaining nuts.
35. Reinstall the two remaining bolts on both sides of the tractor.
36. Reconnect battery cables.
37. Fill the engine with oil.
38. Fill the cooling system with coolant.
39. Check the hydraulic fluid and bring it up to level.
40. Reinstall the hood on the vehicle.
41. Place the driveshaft flange on the driveshaft, making sure to align the positioning arrows.

 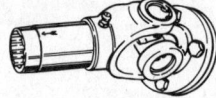

Driveshaft aligning arrows

---- **WARNING** ----

The arrows on the driveshaft must be aligned for proper balance of the driveshaft. Failure to align the arrows will result in damage to the bearings.

42. Bolt the driveshaft flange to the rear of the engine.

Cylinder Head

REMOVAL

1. Disconnect the battery.
2. Drain the cooling system.
3. Remove the air horn from the carburetor.
4. Disconnect throttle and choke linkages from carburetor.
5. Remove drive belt from water pump.
6. Remove the clamp between exhaust manifold and exhaust pipe.
7. Disconnect and mark ignition wires from spark plugs.
8. Remove the valve cover.
9. Remove the head bolts.
10. Lift off the cylinder head with carburetor, water pump, etc. attached.
11. Remove the push rods and mark them for reinstallation.

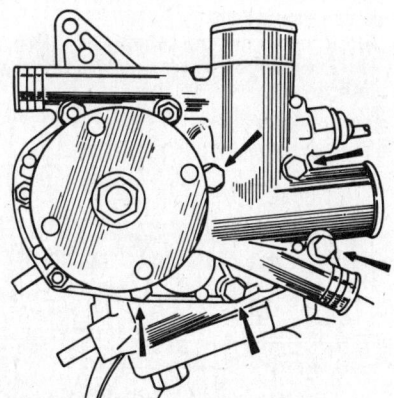

Water pump removal

DISASSEMBLY

1. Remove the intake and exhaust manifolds from the head.
2. Remove the pulley from the water pump.
3. Remove the generator, water pump and spacer plate from the cylinder head.
4. Remove the four retaining clips and springs from the rocker arm shaft.
5. Remove the rocker arm shaft locking bolts.
6. Pull out the rocker arm shaft with special tool (part no. 90.45.266).
7. Compress the valve springs with special tool (part no. 90.45.260) and remove valve keepers.
8. Remove valve spring compressor tool.
9. Lift out valve springs, washers and valves.
10. Remove the cover plate in the rear of the cylinder head.
11. Clean the cylinder head with a solvent.

ASSEMBLY

1. Check gasket surface of the cylinder head with a straight-edge and feeler gauges. The maximum deformation should be .002 in.
2. Reface the cylinder head if necessary. The normal height of the combustion chamber is 3.73 in. and the *minimum height* is 3.71 in.

NOTE: If the cylinder head height is shaved below 3.72 in., the volume of the combustion chambers should be readjusted to 1.68 in. (27.3 cc). This is done by removing metal from the sides of the combustion chamber. The spark plugs and valves should be in position when measuring volume of combustion chamber.

3. Press the valve guides out with special tool (part no. 90.45.291).
4. Ream out valve guide hole with special tool (part no. 90.45.290).
5. Replacement valve guide should be the next largest size. Use chart below.

	Inside Diameter	Outside Diameter	Identification Markings
Original	.2756 in.	.4331 in.	No Rings
1st oversize	.2756 in.	.4370 in.	1 ring
2nd oversize	.2756 in.	.4429 in.	2 rings

6. Grease the new valve guide and press in with special tool (part no. 90.45.291) until flush with cylinder head.
7. Ream out valve guide bore with special tool (part no. 90.45.290).
8. Regrind valve seats if necessary. The valve seat angle is 45° for both.
   Inlet Valve Seat Width—.060 in.
   Exhaust Valve Seat Width—.071 in.

9. Clean the head with solvent before installing valves.
10. Lubricate valves and reinstall in cylinder head in their correct locations.
11. Place the valve springs and washers on the valves.
12. Install the valve spring compressor tool and compress the springs.
13. Reinstall the valve keepers on the valve stems. Note that the keepers for the intake valves differ from exhaust valves.
14. Install the rocker arm shaft, placing the rocker arms and springs on the shaft as you slide it in.

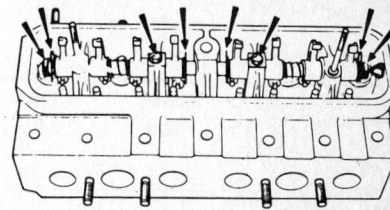

Rocker arm shaft assembly

15. Reinstall the mounting bolts on the rocker arm shaft.
16. Reinstall the retaining clips and springs on the rocker arm shaft.
17. Reinstall the plug and rubber seal in the hole which the rocker arm shaft is removed from.

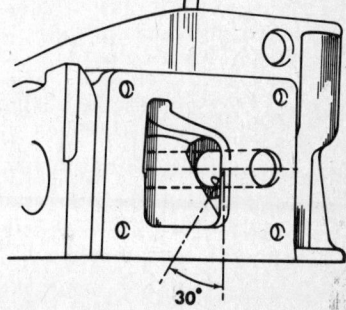

Water distributor tube installation

18. Replace the water distributor tube if necessary. Mount the new tube so that the water holes are at an angle of 30° from vertical of valves.
19. Reinstall water pump and spacer plate with a new gasket on the cylinder head.
20. Reinstall the pulley on the water pump.
21. Reinstall the intake and exhaust manifold with a new gasket on the cylinder head.

INSTALLATION

1. Reinstall the push rods in the block.
2. Place the cylinder head on the block with a new gasket, taking care to get push rods positioned in the lifters.
3. Torque the head bolts, in a circular pattern, starting from the center, rotating outwards, to 44 ft.lb. using special socket (part no. 90.45.263).
4. Adjust the intake valves to .006 in. and exhaust valves to .008 in. See below for details on valve adjustment.
5. Reinstall the valve cover with a new gasket.

Wheel Horse

6. Reinstall ignition wires on spark plugs.
7. Reinstall the clamp between the exhaust manifold and exhaust pipe.
8. Reinstall the generator on the engine.
9. Reinstall drive belt on water pump and adjust tension.
10. Connect throttle and choke linkage on carburetor.
11. Place air horn on carburetor.
12. Refill the cooling system.
13. Add oil to engine.
14. Reconnect the battery cables.

VALVE ADJUSTMENT

	Exhaust Valve Gap	Intake Valve Gap
Engine cold	.008 in.	.006 in.
Engine warm (wait until engine has cooled for 50 min.)	.010 in.	.007 in.

1. Remove the valve cover.
2. For identification purposes number the valves 1 to 8, starting with No. 1 at the flywheel end of the engine.
3. Rotate the engine until No. 1 valve is completely open (depressed by the rocker arm to the lowest position).
4. Adjust valve No. 8 to the proper valve gap specifications. Special tool (part no. 90.45.265) makes the adjustment procedure easier to accomplish.
5. Repeat steps 3 and 4 to adjust the remaining valves with the information listed below. Note that the numbers of two valves in each process always add up to 9.

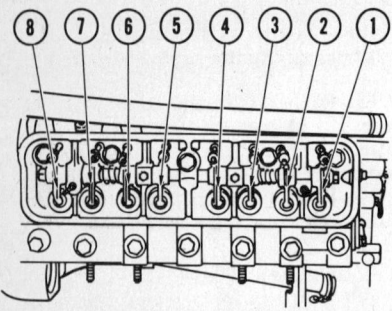
Numbering the valves

Rotate Engine Until This Valve Is Completely Open	Adjust This Valve To Specifications
1—Exhaust	8—Exhaust
3—Intake	6—Intake
5—Exhaust	4—Exhaust
2—Intake	7—Intake
8—Exhaust	1—Exhaust
6—Intake	3—Intake
4—Exhaust	5—Exhaust
7—Intake	2—Intake

6. Reinstall valve cover with new gasket.

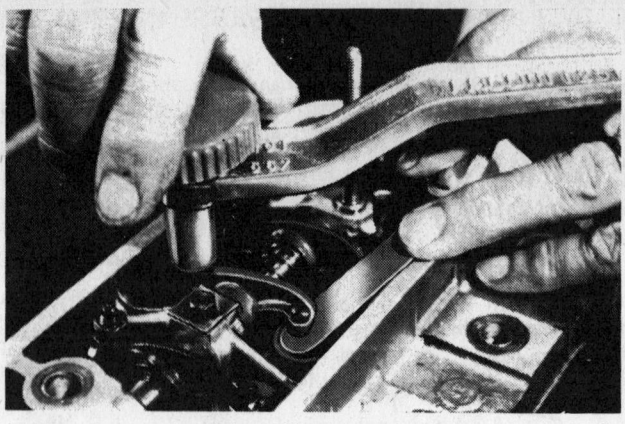
Valve adjusting tool

Cylinder Block

DISASSEMBLY

1. Remove the engine from tractor.
2. Remove the hydraulic pump/ignition coil bracket by removing the three nuts from the studs and remove the governor.
3. Remove the engine mounting bracket on the left side of the engine.
4. Mount the block on an engine stand if available.
5. Remove the distributor.
6. Remove the fuel pump.
7. Remove the oil dipstick.
8. Remove the oil pressure sending unit.
9. Remove the clutch housing and clutch assembly.
10. Remove the generator.
11. Remove the valve cover.
12. Remove the cylinder head bolts. Special socket (part no. 90.45.263) may be used for this operation.
13. Remove the cylinder head from the block, leaving the water pump, manifolds and carburetor attached to the head.
14. Remove the push rods and mark them if reusing so they may be reinstalled in the same order.
15. Remove the cylinder head gasket.
16. Install special tools (part no. 90.45.264) to prevent cylinder liners from falling out if the engine is turned over.
17. Remove the starter.
18. Remove the front PTO drive.
19. Remove the hex bolts and Allen bolts on the timing chain cover, and lift off the cover.
20. Pull out the distributor drive pinion using special tool (part no. 90.45.261).
21. Remove the generator (alternator) drive pulley from the rear of the camshaft.
22. Remove the valve lifters.
23. Remove the camshaft flange bolts and slide out the camshaft.
24. Remove the oil pan, oil pump and gasket.
25. Remove the timing chain tensioner.

NOTE: It is only necessary to remove the sprocket on the crankshaft if the sprocket or the crankshaft are being replaced.

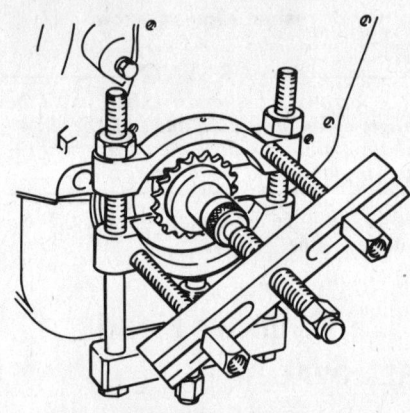
Crankshaft sprocket removal

26. Pull off the crankshaft using special tools (part no. 90.45.313/312).

NOTE: If is not necessary to remove the flywheel from the crankshaft in order to remove the crankshaft from the block. However, if the flywheel and the crankshaft are both being removed and reused, mark the position of the flywheel on the crankshaft.

27. Number the connecting rods and main bearing caps with a punch for later identification. Start no. 1 at the flywheel (back) end of engine.
28. Remove the connecting rod cap nuts, caps and bearing shells.
29. Remove crankshaft main bearing caps and bearing shells.
30. Remove the crankshaft and fillets.
31. Remove the upper halves of the main bearings.
32. Turn the block over and remove the

Connecting rod cap numbering

Wheel Horse

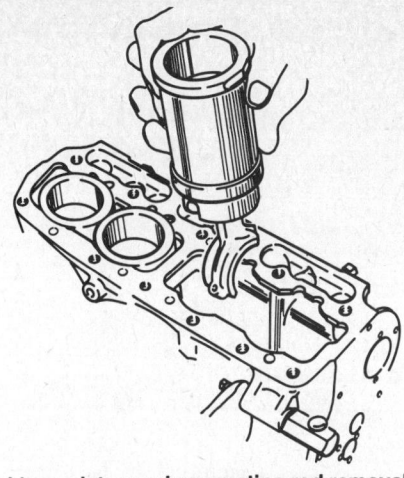

Liner, piston and connecting rod removal

tools holding the cylinder liners onto the block.

33. Remove the liner/piston/connecting rod assemblies as a unit from the top.

ASSEMBLY

1. Remove the spacer plate and oil strainer.
2. Clean the block.
3. Drill holes $^{17}/_{64}$ in. in diameter in oil passage plugs.
4. Install threaded 8 x 125 mm bolts in the holes, and use them to remove the oil plugs.
5. Clean the oil passages by passing copper wire through them.
6. Install new oil plugs and locking them with a punch and reinstall the spacer plate.
7. Clean the crankshaft and pass copper wire through the oil holes.
8. If the bronze clutch shaft pilot bushing is damaged, remove it with a tap and replace.
9. Check the connecting rod and main bearing journal diameters with a micrometer (See table below for tolerances).

	Nominal Diameter	Regrind Diameter	Regrinding Tolerances
Connecting rod journals	1.496 in.	1.486 in.	.00098-.00160 in.
Main bearing journals	1.575 in.	1.565 in.	.00035-.00099 in.

NOTE: When regrinding the connecting rod journals, the roll hardening must remain intact over a 140° section, facing the rotational center line of the crankshaft.

10. Remount the cylinder block on engine stand, if available.
11. Perform the following tasks as you reassemble the remainder of the block.
    a. Replace all oil seals and lockwashers.
    b. Make sure all oil passages are clean and oil plugs are tight.

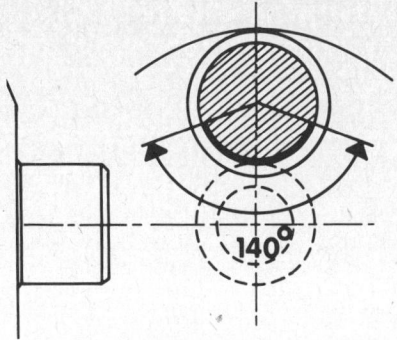

Connecting rod journal regrinding

12. Mount the main bearings in the block.
13. Lubricate the main bearing journals on the crankshaft and lay the crankshaft in place.
14. Reinstall the crankshaft fillets.
15. Mount the lower halves of the main bearings in the main bearing caps.
16. Mount the main bearing caps according to the marks previously made.
17. Torque the main bearing cap bolts to 43 ft.lb.
18. Check the crankshaft to make sure it turns freely.
19. Measure the axial play of the crankshaft with a feeler gauge or a dial gauge.
20. The axial play of the crankshaft should be .002-.010 in.
21. Adjust the axial play by changing the fillet size, if necessary. Fillets are available in .080 in., .081 in., .083 in. and .085 in. sizes.
22. Lubricate the main bearing oil seal and mount on the crankshaft.
23. Mount the flywheel on the crankshaft with *new* bolts. If using both the original crankshaft and flywheel, be sure to align the marks made previously.
24. Torque the flywheel bolts to 36 ft.lb.
25. Check the flywheel runout with a depth gauge. Maximum runout allowable is .003 in.
26. Insert the cylinder liners into the block without seals and press down *by hand* to ensure proper seating.
27. Check the protrusion of each liner from the block with a feeler gauge and straightedge or using a dial indicator and special tools (part no. 90.45.293/294).

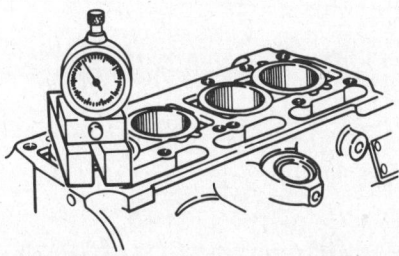

Measuring the liner clearance with a dial gauge

28. The correct protrusion height is .002 in. to .005 in. (.04 to .12 mm), with the maximum difference between any of the four liners being .001 in. (.02 mm).

29. Adjust the cylinder liner heights with seals, which are available in the sizes shown below.

Thickness of Seal	Color Code
.003 in.	Blue
.004 in.	Red
.006 in.	Green

30. Remove the liners after completing adjustment.
31. If the pistons have been removed from the liners, reinstall them using special tool (part no. 90.45.292).
32. Lubricate the connecting rod inserts and place them in the connecting rods.
33. Insert liners with pistons and connecting rods in block, taking care to get them in their original location.
34. Install special tool (part no. 90.45.264) to hold liners in place.
35. Install the connecting rod caps with bearing inserts on the crankshaft.
36. Install new connecting rod nuts and torque to 25 ft.lb.
37. Check the crankshaft to see that it rotates freely.
38. Mount the oil pump with a new gasket on the block.
39. Insert the key in the crankshaft sprocket and mount the sprocket on the crankshaft. Be sure the timing mark on the sprocket is on the outside.
40. Lubricate the bearings and install the camshaft in the block.
41. Install the timing chain with the yellow link placed on the marked tooth of the crankshaft sprocket, and the other marked link aligned with the marked tooth on the camshaft sprocket.
42. Install the camshaft flange mounting bolts.

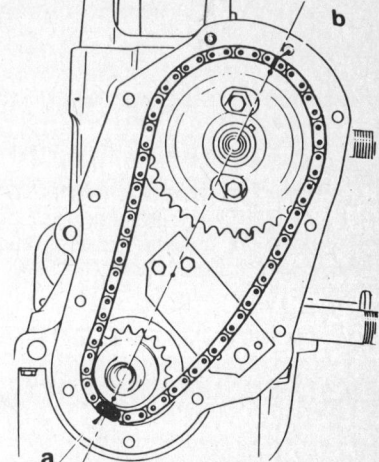

Timing chain installation

43. Install the chain tensioner with spacer plates.
44. Use a ⅛ in. Allen wrench to adjust the tensioner.
45. Grease the rear oil seal for the camshaft and press into place.
46. Install the timing chain cover with a new gasket.
47. Install the oil pan with new gaskets.

Wheel Horse

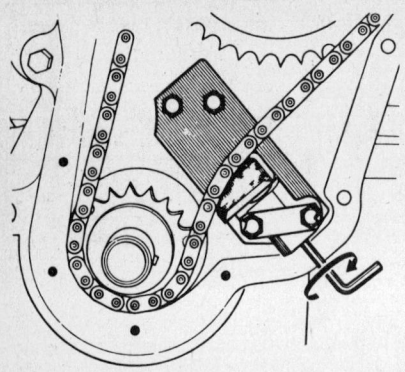

Installing the timing chain tensioner and spacer plates

48. Remove the cylinder liner holders.
49. Reinstall the valve tappets and the push rods.
50. Reinstall the cylinder head and adjust valves.
51. Install the valve cover.
52. Install the starter.
53. Install the generator.
54. Install the clutch assembly.
55. Install front PTO drive.
56. Install the oil pressure sending unit.
57. Install the fuel pump.
58. Install distributor drive pinion and distributor.
59. Install left engine mount.
60. Install the hydraulic pump bracket, ignition coil bracket and governor.
61. Replace the engine in the tractor.

Pistons/Liners

NOTE: The engine does not have to be removed from the vehicle to change piston/liner assemblies.

1. Disconnect the battery.
2. Drain the cooling system.
3. Drain the engine oil.
4. Remove the air horn from carburetor.
5. Disconnect throttle and choke linkage from carburetor.
6. Remove the drive belt from water pump.
7. Remove the clamp between the exhaust manifold and the exhaust pipe.
8. Disconnect ignition wires from spark plugs.
9. Remove the valve cover.
10. Remove the head bolts.
11. Lift off the cylinder head with carburetor, water pump, etc., attached.
12. Remove push rods and mark for reinstallation.
13. Remove the oil pan.
14. Remove the oil pump.
15. Install special tool (part no. 90.45.264) to hold cylinder liners.
16. Mark the connecting rods and connecting rod bearing caps so they can be replaced in their original position. The No. 1 piston is at the flywheel end of the engine.
17. Remove the connecting rod cap nuts.
18. Remove the connecting rod caps and bearing inserts.
19. Remove the special tool holding the piston liners.
20. Remove the liner-piston-connecting rod assemblies from the top of the engine.

Servicing

1. Remove the piston pin by pressing it out using special tool (part no. 90.45.925).
2. When reassembling piston and connecting rod, make sure they are positioned properly.
   a. Pistons with an arrow should be positioned so that the arrow points downward.

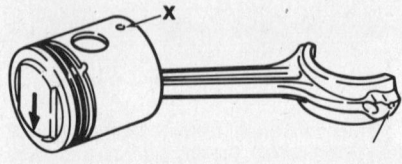

Piston and connecting rod alignment marks

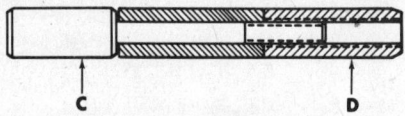

Piston pin centering tool; C is the punch and D is the centering sleeve

   b. Pistons with a hole in the side should be placed with the hole on the top.
3. Slip the piston pin on the punch ("C") and centering sleeve ("D").

NOTE: The small end of the connecting rod must be heated to 475°F (250°C) to install piston pin.

4. Lubricate the piston pin and push in the piston and connecting rod by hand.
5. Check to make sure the piston pin does not protrude from the circumference of the piston, and that the piston rotates freely on the pin.
6. The piston rings are pregapped, and require no adjustment.

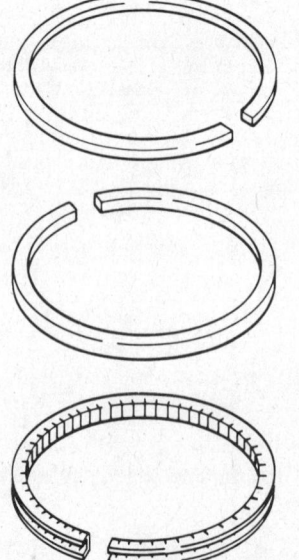

Piston ring installation showing gap arrangement

7. If replacing rings on a piston, install as follows:
   a. Combustion ring on top
   b. Compression ring in middle
   c. Oil scraper ring on bottom

NOTE: Be sure the side on the compression ring marked "O" or "Top" is facing toward the cylinder head.

8. Stagger the ring gaps equally around the piston.

REASSEMBLY

1. Clean the inside of the block, cylinder liners, cylinder liner base seal locations and the crankshaft.
2. Insert cylinder liners into the block and press down by hand to ensure proper seating.
3. Check the protrusion of each liner from the block using a feeler gauge and straight edge or a dial indicator and special tools (part no. 90.45.293/294).
4. The correct protrusion height of the liners is .002 in. to .005 in., with the maximum difference between any of the four liners being .001 in.
5. Adjust the cylinder liner heights with seals, which are available in sizes shown below.

Thickness of Seal	Color Code
.003 in.	Blue
.004 in	Red
.006 in.	Green

6. Remove the liners after completing adjustment, and keep them in order for reinstalling.
7. Lubricate the pistons and install them in liners using special tool (part no. 90.45.292).
8. Insert liners with pistons and connecting rods in block:
   a. Making sure to get them in proper location.
   b. With arrow on piston pointing to flywheel or hole in piston body on the timing chain side of the engine.
9. Install special tool (part no. 90.45.264) to hold liners in place.
10. Install the connecting rod caps with bearing inserts on the crankshaft.
11. Install *new* connecting rod bolts and torque to 25 ft.lb.
12. Check the crankshaft to see that it rotates freely.
13. Mount the oil pump with a new gasket on the block.
14. Reinstall the oil pan with new gaskets.
15. Check the gasket face on the cylinder head for warpage, and reface if necessary.
16. Grind the valves and recut the valve seats if necessary.
17. Remove cylinder liner holders.
18. Reinstall the cylinder head with a new gasket.
20. Adjust the valves.
21. Install the valve cover.
22. Reinstall the ignition wires on spark plugs.
23. Reinstall the clamp on the exhaust pipe.

Wheel Horse

24. Reinstall drive belt on water pump.
25. Reconnect throttle and choke linkages on carburetor.
26. Refill engine with oil.
27. Refill cooling system.
28. Reconnect battery.

COOLING SYSTEM

Water Pump

REPLACEMENT

1. Disconnect the battery.
2. Drain the cooling system.
3. Loosen belt tensions and remove the drive belts from the water pump.
4. Remove the hoses and the thermostat from the water pump.
5. Remove the large pulley from the water pump.
6. Remove the water pump mounting bolts and lift off the water pump and spacer plate.
7. Remove the small pulley, and the rubber caps from the old water pump, and install on the new one.
8. Install the water pump with spacer plate and new gaskets on the engine. Torque the mounting bolts to 8 ft.lb.
9. Reinstall the large pulley on the water pump.
10. Place the thermostat in the water pump so that the hole "K" in the water pump is not covered by the hoop "L" on the thermostat.
11. Reinstall the hoses on the water pump.
12. Place the drive belts on the water pump and adjust tension.
13. Refill the cooling system.
14. Reconnect the battery.

Radiator

1. Drain the cooling system.
2. Remove the water hoses from the radiator.
3. Remove the front guard plate.
4. Remove the top mounting nuts for the radiator.
5. Remove the fan shroud mounting screws and slide it back.
6. Remove the two mounting nuts on the bottom of the radiator and lift off the radiator.

NOTE: Take care not to damage cooling cores and hose fixtures on the radiator when installing.

7. Place the new radiator on the tractor and install the bottom mounting nuts.
8. Reinstall the fan blade guard mounting screws on the radiator.
9. Reinstall the top mounting nuts for the radiator.
10. Reinstall the front guard.

11. Reinstall the water hoses on the radiator.
12. Refill the cooling system.

FUEL SYSTEM

Carburetor

REMOVAL

1. Remove the governor linkage and choke cable from carburetor.
2. Disconnect the fuel line from the carburetor.
3. Remove the air horn from carburetor.
4. Remove the mounting nuts and lift off the carburetor.

INSTALLATION

1. Place the carburetor with throttle plate assembly and gasket on the manifold.
2. Install the mounting nuts.
3. Install the air horn on the carburetor.
4. Connect the fuel line to the carburetor.
5. Install the governor linkage and choke cable on the carburetor.

OVERHAUL

1. Disassemble the carburetor.
2. Check the metal tab on the float (E) for wear or indentation and replace if necessary.
3. Check the float to see if liquid has leaked inside and replace if necessary.
4. Clean the fuel filter screeen (K) on the gas intake fitting.
5. Remove the jets (B), (F) and (G) and blow them out with compressed air.
6. Blow out the air passages with compressed air.
7. Reassemble the carburetor.

CHOKE ADJUSTMENT

1. Remove the carburetor from the manifold.
2. With the choke plate completely closed, the throttle plate should be open .0433 in. Use a drill bit to measure the throttle plate opening.
3. Adjust the throttle plate opening by bending the rod between the choke and throttle plates.
4. Reinstall the carburetor on manifold.

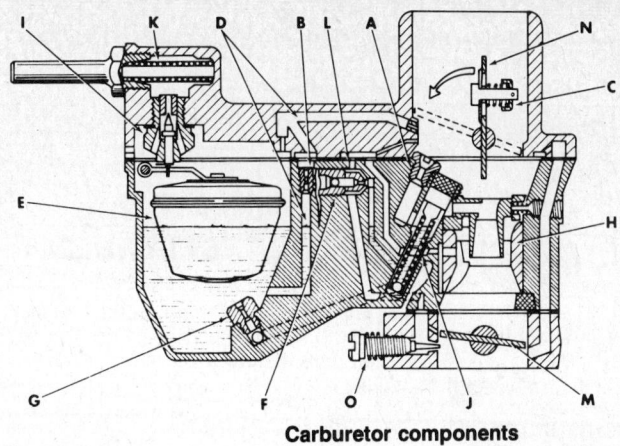

A. Air adjustment jet
B. Partial load jet
C. Air valve
D. Partial load channel
E. Float
F. Idle jet
G. Main jet
H. Venturi
I. Needle valve
J. Mixture tube
K. Fuel filter screen
L. Idle air bore
M. Throttle plate
N. Choke plate
O. Idle screw

Carburetor components

ADJUSTMENT

Adjusting Engine Idle

1. The engine idle speed should be 800 rpm.
2. If the idle speed is incorrect, adjust the idle speed screw.

Adjusting Air/Fuel

1. Turn the air/fuel metering screw in and out to obtain the smoothest idle. This is usually the highest speed obtainable if using a tachometer.
2. Reset the engine idle speed to 800 rpm.

NOTE: If the generator light remains on with the idle adjusted to 800 rpm, increase the idle speed until the generator light goes out.

Adjusting Maximum rpm

1. The maximum rpm should be 3000 rpm. Check to see that both the hand throttle and the foot pedal achieve the proper maximum engine speed.
2. If there is a problem, first check to see that the throttle cables are adjusted properly.
3. Finally, adjust the maximum rpm if necessary, by turning the nut on the governor.
4. Turning the nut clockwise increases the maximum rpm, and counterclockwise decreases it.
5. Adjust the maximum speed to 3000 rpm.

— CAUTION —
Setting the maximum engine speed above 3000 rpm can cause permanent damage to internal engine parts.

Governor

REMOVAL

1. Disconnect the throttle cables and carburetor linkage from the governor.
2. Loosen the tension on the governor drive belt.
3. Remove the drive belt cover from the hydraulic pump.
4. Slide the drive belt off the hydraulic pump pulley and over the front of the governor fan.
5. Remove the lower portion of the oil bath filter and the filter cartridge.

6. Remove the air horn hose from the oil filter.
7. Remove the two mounting bolts and lift the governor assembly straight up and out of the engine.
8. Remove the air filter bath and mounting bracket from the governor.

INSTALLATION

1. Install the air filter bath and mounting bracket on the governor.
2. Slide the governor into position and install the mounting bolts.
3. Install the air horn hose on the air filter.
4. Reinstall the lower portion of the oil bath filter and the filter cartridge.
5. Slip the drive belt over the fan and onto the hydraulic pump pulley.
6. Reinstall the drive belt cover on the hydraulic pump.
7. Adjust the tension of the drive belt.
8. Reconnect the throttle cables and carburetor linkage on the governor.

GOVERNOR LINKAGE ADJUSTMENT

1. Push the upper governor lever against the valve cover.
2. Measure the distance between the tab extending down from the upper governor lever and the lower governor lever.
3. The distance should be .04 in. Adjust with the linkage rod between the lower governor lever and the carburetor.
4. Recheck the distance after making adjustment.

GOVERNOR/THROTTLE LINKAGE ADJUSTMENT

1. With hand and foot throttles closed, the cables should have .04 in. free play at the upper governor lever.
2. Adjust the tension of the cables by moving the cable housing at the clamp.
3. Check choke adjustment.

GOVERNOR SPEED ADJUSTMENT

1. Adjust the maximum rpm if necessary by turning the nut on the governor.
2. Turning the nut clockwise increases the maximum rpm, and counterclockwise decreases it.
3. Adjust maximum speed to 3000 rpm.

---— CAUTION ———
Setting the engine speed above 3000 rpm can cause permanent damage to internal engine parts.

ELECTRICAL SYSTEM

Points and Condenser
REPLACEMENT

1. Remove the condenser and replace with a new one.
2. Remove the set of points from the distributor.
3. Install the new set of points.
4. Adjust the point gap .016 in-(.018 in)-.020 in. with feeler gauges, making sure that a high point on the distributor cam is aligned with the rubbing block of the points. New points should always be set to the larger dimension.

Ignition Timing
MARKING THE FLYWHEEL

Since several distributors are available for this engine, it is first necessary to identify the distributor and mark the flywheel accordingly before proceeding to time the engine.

1. Locate the identification numbers on the body of the distributor.
2. Use the chart below to determine the ignition advance or the distance the timing mark should be from the Top Dead Center mark (All flywheels are only notched at Top Dead Center, which is not always the timing mark reference).

Distributor I.D.	Ignition Advance on Flywheel
A46	.400 in. ± .04 in., 3/8 in. ± 1/16 in.
R252	0 in. ± .04 in., 0 in. ± 1/16 in.
R284	.470 in. ± .04 in., 1/2 in. ± 1/16 in.
R285	.320 in. ± .04 in., 5/16 in. ± 1/16 in.

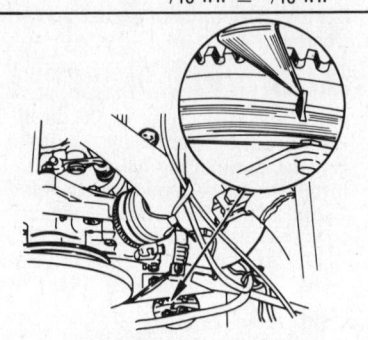
Flywheel timing marks

3. The flywheel timing marks are observed at the rear of the engine.
Example:
For an engine with an A46 distributor, you would draw the timing mark .400 in. ± .04 in. to the right of the Top Dead Center mark with white paint. The R252 distributor would be timed at the Top Dead Center mark.

STROBE LIGHT METHOD

This method incorporates adjusting the engine timing while the engine is running. It is therefore necessary to set the idle speed low enough to ensure that the centrifugal advance mechanism is not operative.
1. Connect a tachometer and timing light to the engine according to the instructions accompanying the instruments.
NOTE: No. 1 cylinder is closest to the flywheel, or at the back of the engine.
2. Loosen the distributor retaining clamp nut until the distributor can be rotated by hand.
3. Aim the timing light at the timing mark reference point.
4. Rotate the distributor until the timing mark (See previous section for determining timing mark location) aligns with the pointer.
5. Tighten the distributor retaining clamp nut, and recheck the timing to make sure it hasn't changed.
6. Disconnect the timing light and adjust the idle speed to 800 rpm.
7. Proceed to Carburetor Adjustment section.

TEST LIGHT METHOD

1. Rotate the engine by hand in the same direction as it travels when running until the timing mark (See previous section for correct timing mark location) aligns with the pointer.

NOTE: The engine *must* be rotated in the same direction as it operates for accurate adjustment.

2. It is now necessary to determine whether the number 1 piston is on a compression stroke. This can be done two ways.
   a. If the rotor is pointing toward the number 1 ignition wire on the distributor cap, the cylinder is on a compression stroke.
   b. Remove the valve cover and check to see that both valves are closed for number 1 piston (closest to flywheel). This assures you that the piston is on a compression stroke. If the valves are not closed, rotate the flywheel around another revolution.
3. Remove the distributor cap.
4. Connect one lead of a 12 volt test light to ground and the other to the primary distributor terminal.
5. Push the ignition key into the "On" position.
6. Loosen the distributor clamp retaining nut.
7. Rotate the distributor in the direction of rotor rotation (clockwise) until the light is off.
8. Then rotate the distributor in the direction opposite of rotor rotation (counterclockwise) slowly until the light just begins to illuminate.
9. Tighten the distributor clamp retaining nut.
10. Turn the key off.
11. Replace the distributor cap.

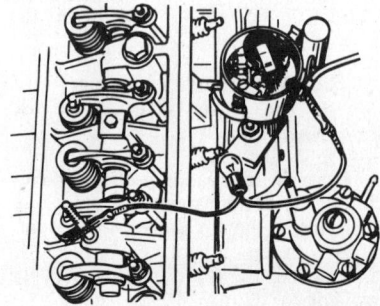
Connecting the test light

White

INDEX

- **Front Axle** .... .828
- **Steering** .... .828
- **Brakes** .... .830
- **Clutch** .... .830
- **Transmission** .... .830
  - Single Speed Reversing Unit .. .830
  - Model 200 2-speed .... .832
  - Model 200 3-speed .... .832
  - Model 700 .... .833
  - Models 350 and 400 .... .834
  - Model 500 .... .834
  - Model 600 .... .835
  - Models 1200, 1400, 1700, 2000 .837
  - Peerless 4-speed .... .838
- **Reduction Units** .... .838
- **Differential** .... .840
- **Final Drive** .... .841
- **Right Angle Drive** .... .842
- **T Drive** .... .842
- **Shifter Units** .... .842
- **Engine** .... .844

WHITE
ALL MODELS

FRONT AXLE

REMOVAL AND INSTALLATION

1. Raise and support the tractor on jack stands.
2. Disconnect the steering linkage.
3. Disconnect the tie rods and Pitman arm.
4. Remove the center pin noting the presence of any shims.
5. Roll the axle from under the tractor.
6. Installation is the reverse of removal.

STEERING

Steering Wheel, Shaft and Gear

REMOVAL AND INSTALLATION

1. Remove the capscrew and hex nut holding the steering wheel to the steering shaft assembly.
2. Remove hood (two thumbscrews).
3. Remove battery from dash assembly.
4. Drain and remove fuel tank from tractor.
5. Disconnect and mark for reassembly all electrical wiring under the dash assembly. Disconnect the choke and throttle cables at the engine.
6. Remove the capscrews holding the dash assembly to the frame of the tractor. Remove the retaining ring from the bushing on the steering shaft assembly. Remove the dash assembly. Remove bushing and lower retaining ring from steering shaft.
7. Remove locknut, flat washer, and spacer securing the fork and pinion assembly to the frame. Remove shaft with fork and pinion from tractor.
8. Remove capscrew and hex nut securing universal joint pin to steering shaft. Remove steering shaft.
9. Remove two capscrews, lockwashers, flat washers, and spacers attaching universal joint pin to fork and pinion. Remove universal joint, capscrew, flat washer, and bushing from fork and pinion assembly.
10. Remove locknut, flat washer, spacer and capscrew, holding drag link to steering gear.
11. Remove capscrew, two special washers, spacer, flat washer, lockwasher and hex nut attaching steering gear to frame. Remove steering gear.
12. Installation is the reverse of removal.

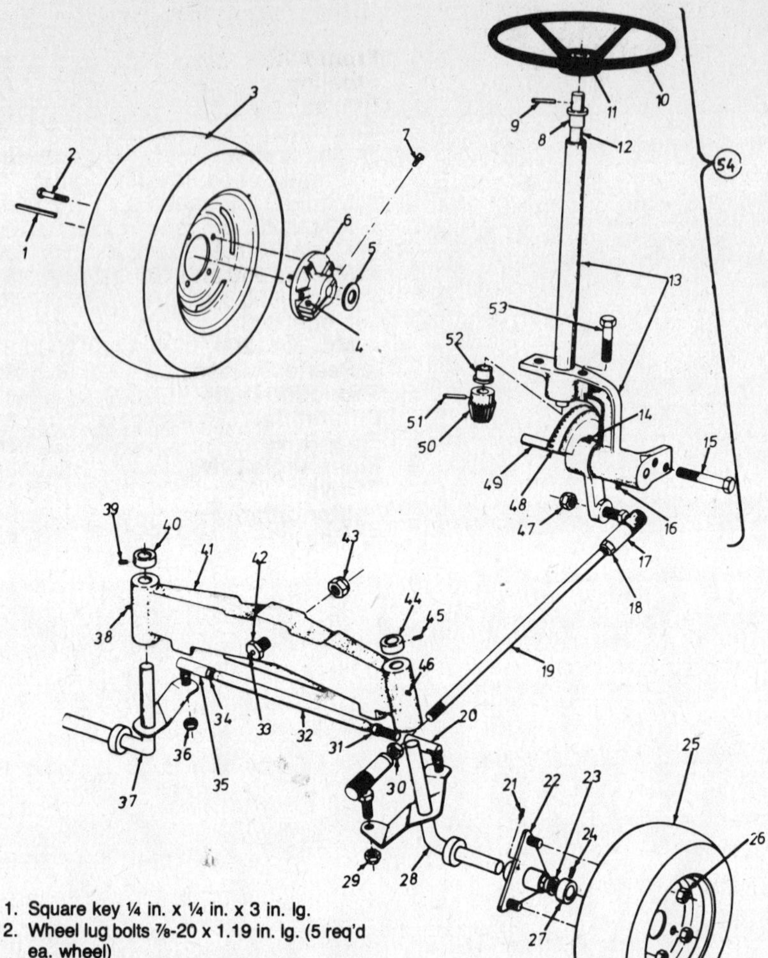

1. Square key ¼ in. x ¼ in. x 3 in. lg.
2. Wheel lug bolts ⅞-20 x 1.19 in. lg. (5 req'd ea. wheel)
3. Rear wheel assembly comp. 23.0 x 9.50
   Rim assembly only—rear wheel
4. Hex head capscrew ⅜-16 1.25 in. lg.
5. Flat washer 1.03 I.D. x 1.62 O.D. x .03 hardened
6. Hub assembly—rear wheel
7. Square head setscrew ⅜-16 x .75 in. lg.
8. Flanged bushing—steering tube
9. Spring pin spiral ¼ in. dia. x 1.50 in. lg.
10. Steering wheel
11. Steering wheel cap
12. Steering rod
13. Steering tube and segment support assembly
14. Square head setscrew 5/16-18 x .75 in. lg.
15. Hex head capscrew ⅜-16 x 1.00 in. lg.
16. Grease fitting
17. Drag link joint assembly
18. Hex jam nut ½-20 in. thd.
19. Drag link
20. Ball joint assembly ⅜-24 thd. (tie rod end)
21. Grease fitting
22. Hub assembly—front wheel (includes bearings and grease fitting)
23. Ball bearings
24. Square head setscrew 5/ 6-18 x .50 in. large cup point
25. Front wheel assembly
    Front wheel rim assembly only
26. Tire only 16.0 x 6.50—front wheel
26. Cone nut ⅜-24 thd.
27. Collar ¾ in. I.D. (axle)
28. Axle assembly
29. Hex inserted locknut ½-20 thd.
30. Hex jam nut ½-20 thd.
31. Hex jam nut ⅜-24 thd.
32. Tie rod (threaded both ends)
33. Grease fitting
34. Hex jam nut ⅜-24 thd.
35. Ball joint assembly ⅜-24 thd. (tie rod end)
36. Hex inserted locknut ⅜-24 thd.
37. Axle assembly
38. Grease fitting
39. Square head setscrew 5/16-18 x .50 in. lg. (cup point)
40. Collar ¾ in. I.D.
41. Front axle support
42. Pivot bolt ¾-10 x 4.00 (special)
43. Hex inserted locknut ¾-10 thd.
44. Collar ¾ in I.D. (axle assembly)
45. Square head setscrew 5 /16-18 x .50 in. lg. (cup point)
46. Grease fitting
47. Hex inserted locknut ½-20 thd.
48. Gear segment
49. Segment shaft
50. Pinion gear
51. Spring pin spiral ¼ in. dia. x 1.50 in. lg.
52. Flanged bearing—steering tube
53. Hex head capscrew ⅜-16 x .75 in. lg.
54. Steering assembly complete
    Snapring (outside of rear wheel) (not shown)

Front axle and steering components

White

CLEANING, INSPECTION AND REPAIR

1. Clean all parts thoroughly using a suitable cleaning solvent to remove all grease and foreign matter.
2. Inspect teeth of steering gear and fork and pinion assembly. Small nicks and burrs on gear teeth may be removed by using a slip stone or hone. If damage is severe, replace part.
3. Check bushing and spacers for damage and replace as required. Lubricate bushing with a light oil before assembly.

Tie Rod and Spindle

REMOVAL

1. Raise tractor front wheels off ground and block up axle.
2. Remove hex nut and lockwasher attaching ball joint to steering lever. Remove ball joint from steering lever and move drag link away from tractor.
3. Loosen setscrews on steering arm and remove steering lever and keys from spindle.
4. Pry off hub caps from both front wheels.
5. Loosen set collar setscrews and slide set collar, washer, and wheel and tire assembly from both spindles. Remove remaining washer.
6. Remove both capscrews, flat washers, spacers, lockwashers, and hex nuts attaching tie rod to spindles. Remove tie rod from tractor.
7. Tap lefthand spindle assembly from front axle. Remove one washer and, using a screwdriver and hammer, tap two bushings from axle. Remaining washer should be removed from spindle.
8. Remove righthand spindle retaining ring and tap righthand spindle from front axle. Remove one washer and, using a screwdriver and hammer, tap two bushings from axle. Remaining washer should be removed from spindle.

CLEANING, INSPECTION AND REPAIR

1. Thoroughly clean all parts.
2. Inspect tie rod for unusual bending or damage. Replace as required.
3. Examine each of four bushings for nicks, scratches, or unusual damage. Replace as required.
4. Inspect spindle for wear, damage, or weak or broken weldments. Check key way in spindle to see if its sides are square with its bottom and not rounded. If possible, file key way square. Replace spindle damaged beyond immediate repair or if weldments are weak or broken.
5. Replace broken or bent keys.

INSTALLATION

1. Using a rubber mallet, tap four bushings in position in front axle.
2. Slide one washer on each key way end of spindle.
3. Insert righthand spindle (key way end) into its respective front axle and secure it in place with a washer and retaining ring.
4. Insert lefthand spindle (key way end) into front axle and slide washer over key way end of spindle.
5. Position keys in key ways. Install steering lever on lefthand spindle and secure it by tightening two setscrews on keys.
6. Secure drag link ball joint on steering lever with lockwasher and hex nut.
7. Install both wheel and tire assemblies on spindles, placing one washer on each side of wheel and securing the assembly with a set collar and setscrew. Be sure setscrew is tight and wheel turns freely.
8. Lower tractor and tap on hub caps.

LUBRICATION

1. Using a lithium base automotive type grease, lubricate both front wheels and spindles.
2. A few drops of engine oil should be applied to lubrication points. Do not allow oil to get on belts and pulleys.

Drag Link and Steering Lever Assembly

INSPECTION

Check drag link, ball joint, and steering lever for damage. If repair is necessary, remove as follows:

REMOVAL

1. Position front wheels straight forward.
2. Remove locknut, flat washer, spacer and capscrew attaching drag link to steering gear.
3. Remove hex nut and lockwasher attaching ball joint to steering lever.
4. Remove drag link and ball joint from tractor.
5. Note location of jam nut on drag link and separate ball joint from drag link.
6. Loosen two setscrews on steering lever and pull lever straight up from spindle, being careful not to damage both keys in spindle key ways.
7. Remove two keys.

CLEANING AND REPAIR

1. Clean all parts with a suitable cleaning solvent.
2. Replace drag link with a new one if it is broken or bent beyond normal straightening.
3. Inspect key ways in spindle and keys for damage. Replace bent or broken keys and attempt to file key ways straight if their corners are rounded. If key ways are damaged severely, replace spindle according to the procedure in this section.
4. Check setscrews for stripped threads or unusual wear. Replace as required.
5. Be sure grease fitting in axle is clean.

INSTALLATION

Install the steering lever and drag link in reverse order of removal noting the following:

1. Keys must be inserted completely into key ways and setscrews must be secured tightly against them.
2. Ball joint must be screwed on drag link and locked in position by the jam nut the same distance on drag link as when it was removed.

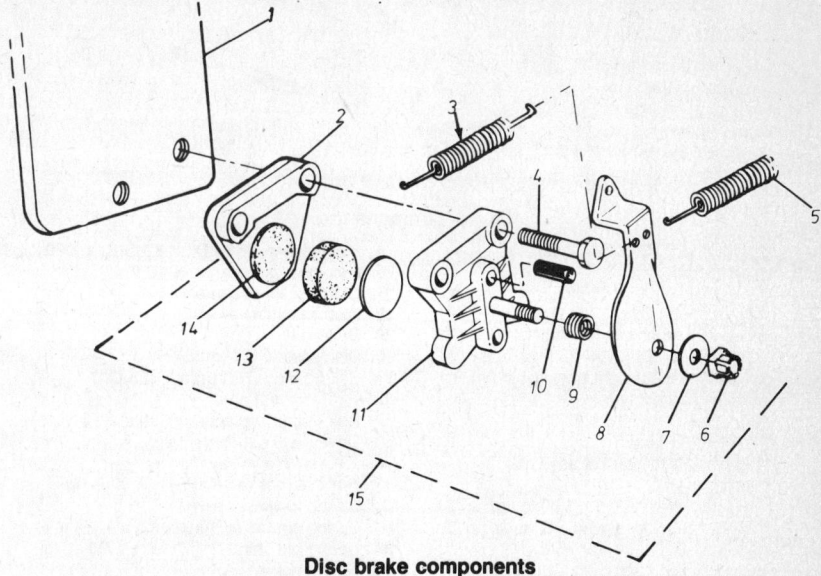

Disc brake components

1. Transaxle mounting bracket assembly
2. Casting-carrier side
3. Spring (brake)
4. Hex head capscrew 5/16-18 x 1½ in. lg.
5. Brake return spring
6. Locknut
7. Thrust washer
8. Cam lever
9. Compression spring
10. Push pin
11. Casting cam side
12. Back-up disc.
13. Friction pad 1.110 in. dia. x .370 thk.
14. Friction pad 1.110 in. dia. x .245 thk.
15. Caliper disc brake assembly comp.

White

BRAKES

ADJUSTMENT

The disc brake is located on the left side of the transaxle. To adjust, tighten the locknut. The locknut should be tightened just enough to stop the tractor when the brake pedal is depressed, allowing the disc to turn freely between the brake pads when the pedal is released.

CLUTCH

REMOVAL AND INSTALLATION

1. Remove the spark plug from the engine and turn the steering wheel all the way to the left. Place the PTO lever in the disengaged position.
2. Remove the support bracket behind the clutch.
3. Loosen the leveler screw on the idler so the belt is slack.
4. Remove the four screws and the four-hole washer holding the rear of the mounting bracket in place.
5. Remove the belt.
6. Remove the elastic locknut and shouldered pivot bolt.

NOTE: Use a ¼ in. Allen wrench to loosen the shoulder bolt.

7. Push the clutch assembly to the rear of the tractor until the pins clear the clutch disc, then move the clutch assembly towards the left side of the tractor.
8. Remove the elastic locknut and all the washers from the stud. The large cup washers and spring can be set aside.
9. Remove the two screws holding the "L" bracket to the frame.
10. Remove the clutch assembly from the tractor.
11. After the clutch has been repaired or replaced, assemble in reverse order of this assembly.

ADJUSTMENT

CAUTION
Remove the spark plug from the engine when adjusting the clutch.
---

1. Rotate the clutch disc by hand and tighten the elastic locknut until the bearing stops rotating.

NOTE: All the play has been removed from the throwout assembly when the bearing stops turning.

2. Loosen the elastic locknut just enough for the bearing to rotate freely as you rotate the clutch disc.

TRANSMISSION

Single Speed Reversing Unit

DISASSEMBLY

1. Remove retaining clip and drive sprocket from output shaft.
2. Remove retaining clip and flat washer from opposite end of output shaft.
3. Remove six (6) ¼-28x⅝ transmission case capscrews and hex nuts. Two of six capscrews are located in lower area of transmission case.
4. Carefully remove case half without shift fork assembly from case half having shift fork assembly. Do not remove shift fork at this time.
5. Remove large bevel gear from output shaft.
6. Remove pinion shaft assembly. Do not disturb shift collar or fork assembly.
7. Remove shift fork and collar by pulling shift fork shaft from shaft bore in case. Use care not to lose detent ball and spring.
8. Remove remaining large bevel gear and output shaft. Keys need not be removed from output shaft.
9. Output shaft bushings are pressed from exterior of case to interior of case. In most cases bushings can be tapped out with hammer and punch.
10. Remove retaining ring from pinion shaft and press pinion gear from pinion shaft.
11. Remove bushings from pinion shaft.

ASSEMBLY

1. Press case bushing into case from inside of case. Place large bevel gear in case half having shifter assembly. Place output shaft in position.
2. Place clutch collar and shift fork in position and begin to position shift fork shaft into case bore.
3. Use small screwdriver to depress detent ball and spring and push shift fork shaft into case until it seats.
4. Place bushings on pinion shaft and press pinion gear onto shaft. Install retaining ring.
5. Place pinion shaft assembly in case and align protrusions on bushings with slots in transmission case.
6. Place large bevel gear on output shaft.
7. Pack case with 5 oz. of WHITE 32-000 8311 High Temperature Grease.
8. Place remaining half of transmission case on assembled half of transmission case. Install six (6) ¼-28x⅝ capscrews and hex nuts into transmission case.
9. Install sprocket with widest flange toward transmission case. Install sprocket retaining clip on output shaft.
10. Install spacer washer and retaining clip to opposite end of output shaft.

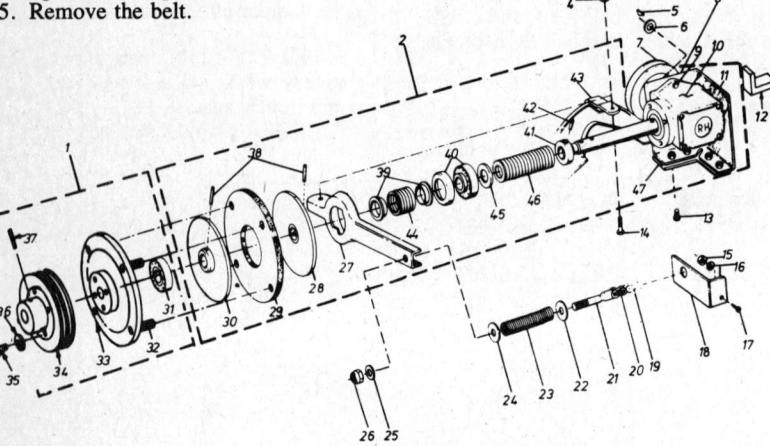

Clutch components

1. Drive plate and PTO pulley assembly
2. Clutch assembly
3. Hex inserted locknut ⅜-16 thd.
4. Flat washer .385 I.D. x .87 O.D. x .060
5. #9 Hi-Pro-key ¾ in. x ¾ in. dia.
6. Flat washer .938 I.D. x 1.47 O.D. x .100
7. Sheave assembly for right angle gear box
8. Gear box mounting plate
9. Hex head capscrw ⅜-16 x 1.25 in. lg.
10. Right angle gear box
11. Gear box mounting bracket assembly
12. Brace assembly
13. Hex head capscrew ⅜-16 x 1.00 in. lg.
14. Socket head shoulder screw .50 in. dia. x .75 in.
15. Hex nut ¼-20 thd.
16. Spring lockwasher ¼ in.
17. Hex head capscrew ¼-20 x .62 in. lg.
18. Spring retainer
19. #48 Chain ½ in. pitch x 19 links
20. #48 Master link ½ in. pitch type 1
21. Clutch adjusting rod
22. Belleville washer .535 I.D. x 1.50 O.D. X .052
23. Yoke return spring
24. Belleville washer .535 I.D. x 1.50 O.D. x .052
25. Flat washer .385 I.D. x .87 O.D. x .060
26. Hex inserted locknut ⅜-16 thd.
27. Clutch yoke assembly
28. Disc assembly—clutch
29. Drive disc
30. Disc assembly—clutch
31. Ball bearing .750 bore 1.625 O.D.
32. Replacement stud
    Hex nut for replacement stud
33. Drive plate sub assembly
34. PTO engine pulley
35. Hex head capscrew 5/16 x 2.00 in. lg.
36. Spring lockwasher 5/16 in.
37. Spring pin spiral 5/16 in. dia. x 1.75 in. lg.
38. Spring pin spiral ¼ in. dia. x 2.00 in. lg.
39. Spring guide
40. Bearing assembly
41. Spring collar
42. Spring pin spiral ¼ in. dia. x 1.50 in. lg.
43. Clutch yoke pivot bracket assembly
44. Clutch helper spring
45. Flat washer .885 I.D. x 1.37 O.D. x .060
46. Clutch spring
47. Gear box mounting washer

White

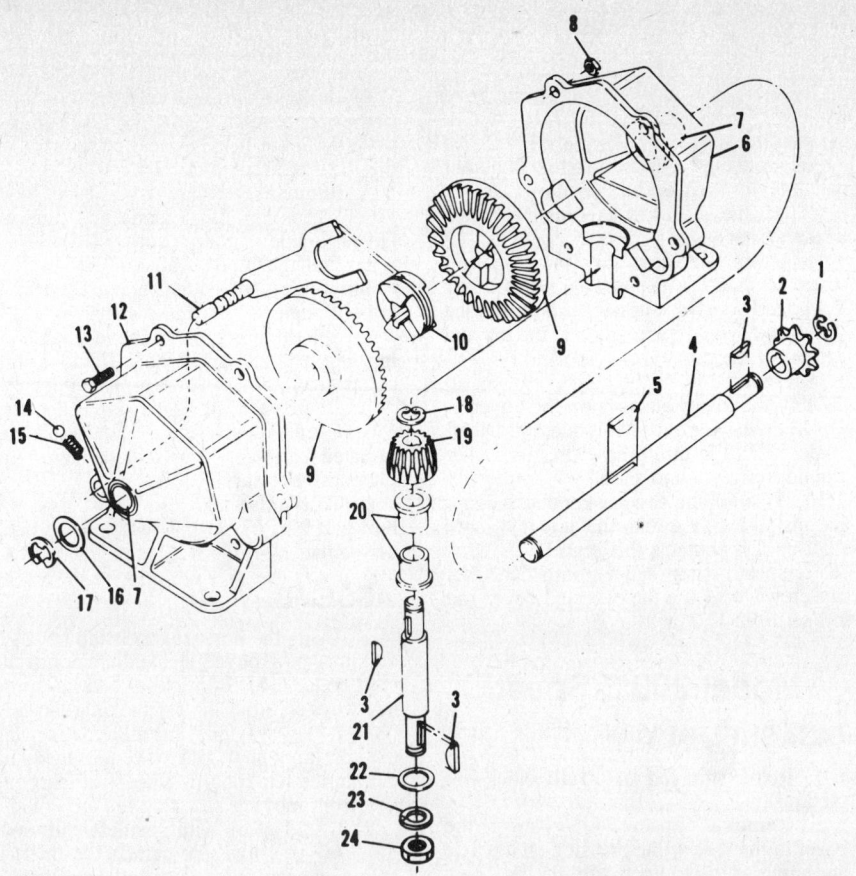

1. E-ring for .500 in. dia. shaft
2. #41 sprocket center 8 tooth
3. #41 Hi-Pro key 3/32 x 5/8 in. dia.
4. Output shaft
5. #9 Hi-Pro key 3/16 x 3/4 in.
6. Transmission case—lefthand
7. Flange bearing
8. Hex centerlock 1/4-28
9. Bevel gear
10. Clutch collar
11. Shift yoke assembly
12. Transmission case—righthand
13. Hex head capscrew 1/4-28 x .62 in.
14. Detent ball
15. Detent spring
16. Fl-wash. .635 I.D. x .93 O.D.
17. E-ring for .625 in. dia. shaft
18. Snapring for .500 in. dia. shaft
19. Pinion gear
20. Bearing .627 I.D.
21. Pinion shaft
22. Fl-wash. .531 I.D. x .93 O.D.
23. Spring L-wash. 1/2 in.
24. Hex jam nut 1/2-20 thd.
   Grease—high temp. (5 oz.)

Single speed transmission components

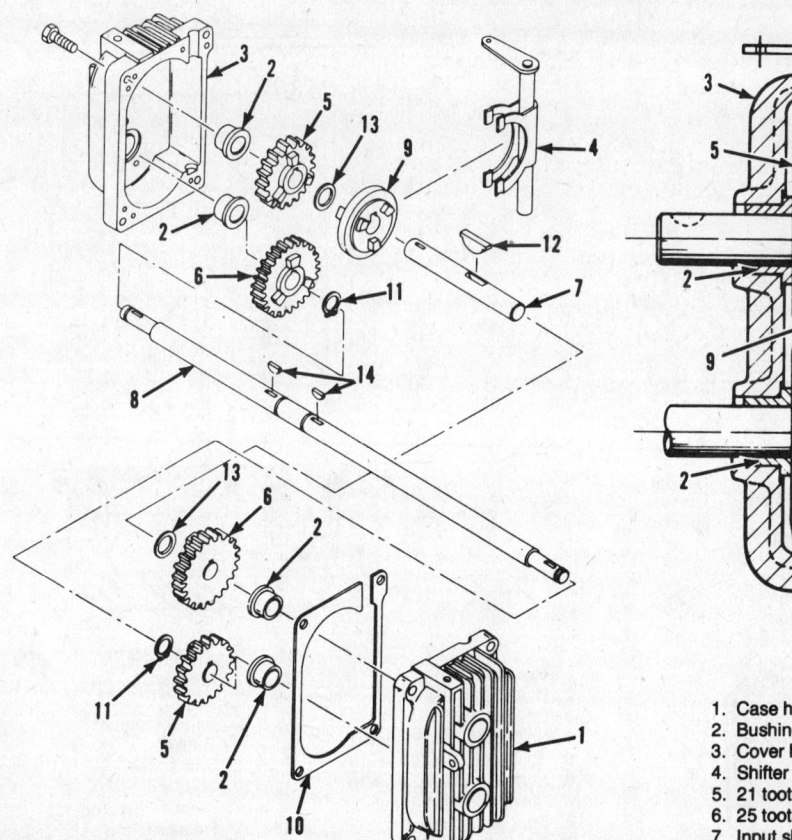

1. Case half
2. Bushing
3. Cover half
4. Shifter rod, fork, and lever
5. 21 tooth gear
6. 25 tooth gear
7. Input shaft
8. Axle (output) shaft
9. Shifter lug
10. Gasket
11. Snapring
12. Shouldered key
13. Thrust washer
14. Woodruff key

Model 200 components

White

Model 200 2-Speed

DISASSEMBLY

1. Remove the capscrews retaining the housing halves together. Lift the case from the cover while pressing on the axle and input shafts to keep those parts remaining with the cover.
2. Clean excess grease from the cover and from around the transmission internal parts.
3. Lift out the axle and gears. Note that the axle extension from the gears is of different length.
4. Remove the input shaft, drive lug and gears, and shifter shaft as a unit; then separate parts.
5. Degrease all parts.
6. Check keys and keyways for wear, galling, or breaks.
7. To remove and install bushings, use tool 670210. Use sizing ball and driver 670212.

ASSEMBLY

1. Position snaprings and keys on axle. Install 25 tooth gear to the short side of the axle and one 21 tooth gear to the longer side.
2. Smear E.P. lithium grease around the bushings in the case and cover.
3. Press the shouldered key into the large keyway on input shaft, then slide the shifter lug onto the shaft over the key.
4. Position a thrust washer on each side of the key.
5. Install the remaining 21 tooth gear on the Woodruff keyway side of the shaft and the 25 tooth gear on the smooth side of the shaft.
6. Hold the input shaft so that the other parts stay in their correct position.
7. Fit the shifter forks over the flange of the shifter lug.
8. Install the assembly into the unit cover so that the exposed Woodruff key slot goes through the upper bushing and the shifter rod lays in its recesses at the top and bottom of the cover.
9. Insert the short axle extension through the lower bushing on the cover.
10. Smear 2 oz. of Molybdenum Disulphide E.P. lithium grease in the cover around the gears and shafts.
11. Position the new gasket on the cover and install the case onto the axle and input shaft until it contacts the gasket.
12. Install four self-tapping ¼-20x¾ capscrews to secure the case and cover and torque to 90-110 in.lbs.

Model 200 3-Speed

DISASSEMBLY

1. Rotate shift rod lever fully clockwise (neutral).
2. Remove capscrews securing the cover to the case while pressing on the axle and input shafts to keep parts in the case.
3. Remove the gasket and clean any grass, dirt, or grease from around the external surfaces of the transmission.
4. Remove shift rod lever from its location in the case by lifting the lever up and away from the case by clearing the pins on the rod from the shift collar.
5. Remove the axle from case, with gears. Remove gears, items 8, 9 and 10 keyed to axle. Remove spacer (16).
6. Pull the input shaft out of the case. Remove the shift spur gears. (If the keys are not engaged, the input shaft will come out without the gears). Remove the retaining ring from the input shaft. Remove shift collar (15) and key (12) from input shaft.
7. Degrease all the disassembled parts.
8. Check keys, key ways, gears and associated parts for wear and damage. Replace as necessary.
9. To remove and install bushings, use tool part No. 670210. Use sizing ball and driver part No. 670212.

ASSEMBLY

1. Using the input shaft, install key (12) end into the grooved inside diameter of the shift collar (15). Pilot collar and keys into key ways on input shaft with collar locating opposite key way end. Install retaining ring (18) on input shaft in groove provided. Be certain the retaining ring keeps at least one key from moving past ring toward collar.
2. Install gears with smallest shift spur gear toward collar. Be certain the internal spline on each gear faces away from retain-

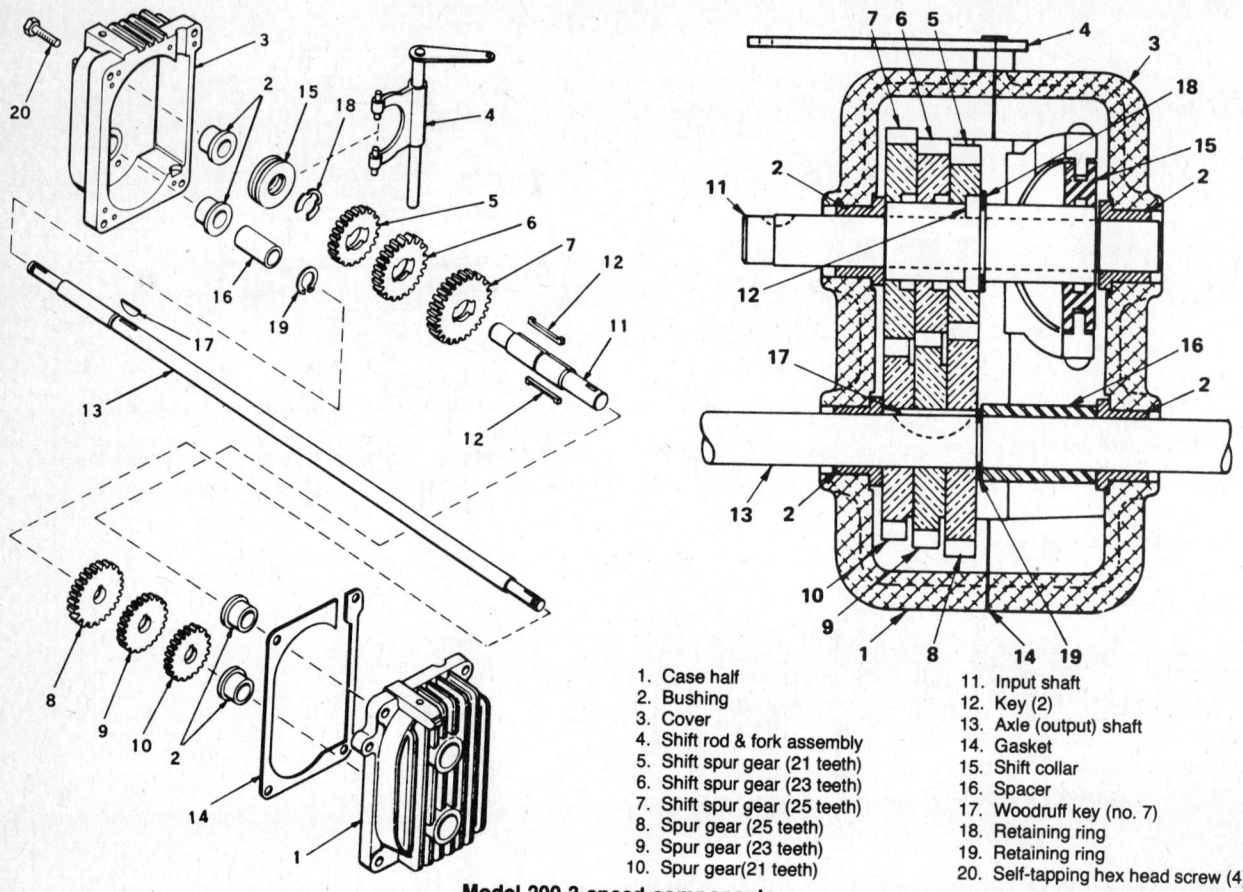

1. Case half
2. Bushing
3. Cover
4. Shift rod & fork assembly
5. Shift spur gear (21 teeth)
6. Shift spur gear (23 teeth)
7. Shift spur gear (25 teeth)
8. Spur gear (25 teeth)
9. Spur gear (23 teeth)
10. Spur gear (21 teeth)
11. Input shaft
12. Key (2)
13. Axle (output) shaft
14. Gasket
15. Shift collar
16. Spacer
17. Woodruff key (no. 7)
18. Retaining ring
19. Retaining ring
20. Self-tapping hex head screw (4)

Model 200 3-speed components

ing ring side. Install input shaft with gear (5) into the case half.

3. Install retaining ring (19) on axle. Install spacer on axle shaft.

4. Apply 2 ounces of extreme pressure (E.P.) lithium grease in case around gears.

5. Install largest to smallest spur gears on the key on the axle. Be certain the machined spacer side of gears face the cover half. Install axle with gears into case half. All six gears must be in mesh.

6. Install shift rod and fork assembly, making sure pins on fork locate in groove on collar.

7. Install new gasket on the case and install cover on case until both halves lock gasket in place.

8. Install self-tapping screws to secure the cover.

9. Torque the case to 90-110 in.lbs.

Model 700

DISASSEMBLY

1. Clean outside surface of transmission. Position shift lever is neutral position as indicated by the shift pattern. Remove shift lever. If installed, remove neutral start switch.

2. Remove setscrew, spring and index ball from transmission cover.

3. Remove six capscrews that maintain cover to case. Remove cover.

4. Remove shifter assembly (includes shaft, pins and fan) from transmission case by lifting shaft out of case.

5. Remove gear and shaft assemblies from case half of the transmission by lifting the two shafts out of the bearing supports taking care not to disturb drive chain relationship with sprockets.

6. Angle chain and sprocket ends of shaft toward each other, removing the bearing and sprocket from the countershaft. Note the collar on the sprocket faces the bevel gear. Remove chain.

7. Remove bevel spur gear combination and spur gears from the countershaft; these gears are splined to the countershaft.

8. Remove the output sprocket and brake disc from the output shaft. Remove the bushings, shift spur gears, chain sprocket, collar and keys.

9. Remove snapring from input shaft, remove bevel gear and pull shaft through case.

10. Input shaft needle bearings should be installed flush to .005 below bearing bore surfaces from inside and outside case.

ASSEMBLY

1. Install and secure the input shaft and bevel gear in the case. See step 9 under "Disassembly" and reverse the order.

2. Install collar and keys on output shaft. Thick side of collar MUST face shoulder on shaft.

3. Install thrust washers and shifting gears on output shaft as shown. The 45° chamfer in the inside diameter of the thrust washers MUST face the shoulder on the output shaft. The flat side of the shifting gears ALWAYS face the shoulder on the output shaft.

NOTE: The thrust washer on the shift gear end of the output shaft does not have a chamfer on the inside diameter and must be positioned as shown. It is thicker than the other thrust washers separating the gearworks from the bearings.

4. Install bevel spur gear and smallest to largest spur gears to the splined end of the countershaft.

5. Install chain over two shafts registering chain on output shaft sprocket and

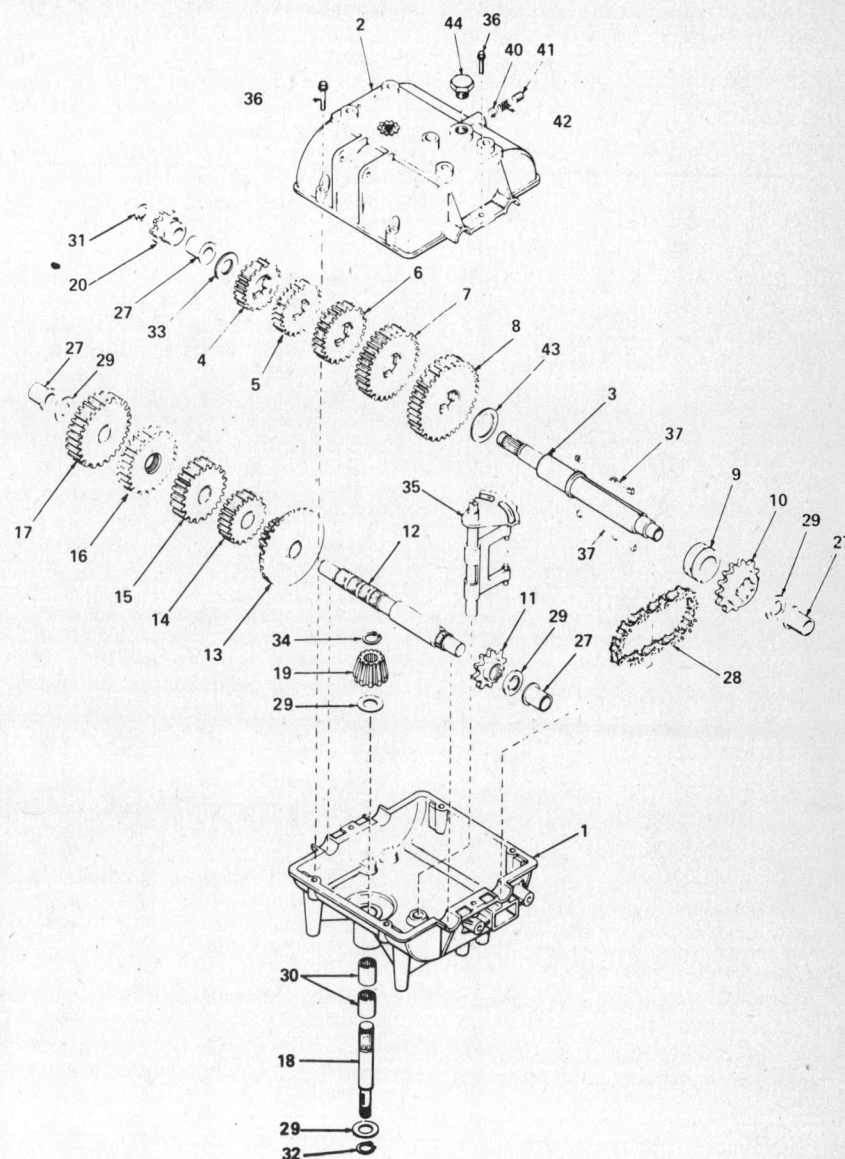

Model 701 components

1. Transmission case
2. Transmission cover
3. Output and brake shaft
4. Spur gear (20 teeth)
5. Spur gear (22 teeth)
6. Spur gear (25 teeth)
7. Spur gear (30 teeth)
8. Spur gear (35 teeth)
9. Shift collar
10. Sprocket (14 teeth)
11. Sprocket (10 teeth)
12. Countershaft
13. Bevel gear (42 tooth and 15 tooth spur gear)
14. Spur gear (20 teeth)
15. Spur gear (25 teeth)
16. Spur gear (28 teeth)
17. Spur gear (30 teeth)
18. Input shaft
19. Input bevel pinion
27. Flanged bushing
28. Roller chain (no. 41 chain, 22 links)
29. Thrust race
30. Needle bearing
32. Retaining ring
33. Washer
34. Retaining ring
35. Rod and fork assembly
36. Screw ¼-20 x 1¼ in. tap-tite
37. Key
40. Ball, 5/16 in.
41. Setscrew, ⅜-16 × ⅜ in.
42. Spring
43. Thrust washer
44. Plug

White

in-line with serration on countershaft. Be sure collar on sprocket faces shifting keys and collar.

6. With collar on countershaft sprocket facing the bevel spur gear install sprocket onto serration and install chain. Install all (4) thrust washers to shafts. The thickest thrust washer MUST be positioned on the shifting gear end of the output shaft. Install bronze bearings and disc and sprocket.

7. Install shaft assemblies into case utilizing piloting locators on bearings to properly align notches in case.

---- CAUTION ----

Be sure bearing locators are seated in transmission case.

---

8. Install shifter assembly (shaft, pins and fan), 12 oz. E.P. lithium grease around gearing, and reinstall cover on case. Torque capscrews 90-100 in.lbs.

9. Install index ball, spring and setscrew in that order into cover and slowly tighten the screw 2 turns below flush.

10. Check for binding by turning input shaft. Install transmission on equipment. Install brake and linkage and adjust to disengage when clutch is engaged. Consult equipment owners manual.

NOTE: If adjusted incorrectly the brake will do one of two things:

1. It will not brake (stop) the vehicle when the brake is applied, or

2. If adjusted too tight a drag or continued braking effect will be evident until the brake wears out.

Models 350 and 400

DISASSEMBLY

1. Clean the outside surface of the transmission. Position shift fork in neutral position. Remove screws (3) holding shift lever and shift lever housing. Remove shift lever housing.

2. If a brake shaft is on the unit, remove snapring from shaft. Clean shaft of dirt or burrs.

3. Remove snapring holding sprocket to the output shaft and remove sprocket. Clean shaft of dirt and burrs.

4. Remove two dowel pins by tapping out with metal punch. Remove cover screws (8).

NOTE: Wipe grease from parts as they are removed.

5. Prior to lifting off the cover, install seal protector 670182 to protect seal for output shaft on units having it. Remove cover and discard gasket.

6. Remove reverse idler gear, shaft, and spacer. Install gear with bevel out toward spacer.

7. Remove shift fork, gears and shaft assemblies. Grasp shifter forks, gears and shaft and raise up while tapping shifter shaft bevel gear with handle of hammer to separate from shaft splines. If no service is required, put unit aside for easy reassembly.

8. Remove clutster gears and shaft. To separate gears, remove outer gears first, then slide off middle gear and key. For replacement, note that the larger gear belongs on the short key way opposite the output end of the shaft. Note the bevel edge position of the gears.

9. Remove the shifter shaft bevel gear, and the thrust bearing and washers.

10. To remove the input shaft oil seal in the 400 series, use metal screws to puncture the seal casing and lift out seal. Seal must be replaced. Clean the input shaft of scratches and sharp edges. Remove the snaprings and thrust washer and press or tap the input shaft into the case. A thrust washer (Model 350) or a thrust washer and a thrust bearing (Model 400) should be on the shaft.

ASSEMBLY

1. Install and secure the input shaft.

2. Install cluster gears and shaft with a thrust washer between the large gear and case. The small and middle gear bevel faces down, the large gear bevel faces up.

3. Install the shifter bevel gear. For 400 series, be sure thrust washers and bearings are between gear and case. Align the gear with the center of the hole.

4. Install the shift mechanism. Try to align the gear and shaft splines before inserting the shaft. To do so, visualize the shifter mechanism in position and note the position of one spline. Position a spline on the gear so that its relationship is the same. Carefully guide the shifter shaft through the gear, disturbing the gear and thrust washers as little as possible. The shifter mechanism must be held firmly to keep parts from changing position. Install shifter stop.

5. Install reverse idler shaft, gear and spacer. Install gear with bevel out toward spacer.

6. Install washers on the cluster gear and shifter shafts. Coat 12 oz. of E.P. lithium grease around gearing if unit uses grease.

7. Install gasket to case.

8. Install cover to case and secure by cross-tightening eight capscrews to 90-110 in.lbs. If cover does not close, use needle-nosed pliers to reposition shifter components until cover seats. Do not force cover on.

9. Install oil seals on 400 series. Use oil seal sleeve #670143 and oil seal driver #670203 on the brakeshaft oil seals. Use oil seal sleeve #670102 and oil seal driver 670209 on the input shaft oil seal.

10. Install new gasket, shift lever and housing with three socket head capscrews.

11. Install sprocket and snapring on output shaft.

12. Turn input shaft to check for binding. Check for correct shifting pattern.

ASSEMBLY

1. Install and secure the input shaft and bevel gear in the case.

2. Install collar and shifter keys to output and brake shaft and follow with largest to smallest shift spur gears. Install chain sprocket, bearing, thrust washer and output sprocket.

3. Install bevel spur gear and smallest to largest mating spur gears to splined end of countershaft. Install chain over two shafts registering chain on output shaft sprocket and in-line with serration on countershaft. Slip sprocket onto serrations and install chain, thrust washers and remaining bearings to shafts.

4. Install shaft assemblies into case utilizing piloting locators on bearings to properly align notches in case.

---- CAUTION ----

Be sure bearing locators are seated in transmission case.

---

5. Install shifter assembly (shaft, pins and fan), 12 oz. E.P. lithium grease around gearing, and reinstall cover on case. Torque capscrews 90-110 in.lbs.

6. Install index ball, spring and setscrew in that order into cover and tighten the screw two full turns from flush. Install shift lever.

7. Check for binding by turning input shaft. Install transmission on equipment. Install and adjust brake linkage to disengage when clutch is engaged.

NOTE: If adjusted incorrectly the brake will do one of two things: It will not brake (stop) the vehicle when the brake is applied, or if adjusted to too tight a drag, a continued braking effect will be evident until the brake wears out.

Model 500

DISASSEMBLY

1. Clean outside surface of transmission. Position shift lever in neutral position as indicated by shift pattern. Remove shift lever. If installed, remove neutral start switch.

2. Remove setscrew, spring and index ball from transmission cover.

3. Remove six capscrews that maintain cover to case. Remove cover.

4. Remove shifter assembly (includes shaft, pins and fan) from transmission case by lifting shaft out of case.

5. Remove gear and shaft assemblies from case half of the transmission by lifting the two shafts out of the bearing supports taking care not to disturb drive chain relationship with sprockets.

6. Angle sprocket ends of shafts towards each other, removing the bearing and sprocket from the countershaft. Note the collar on the sprocket faces away from the bevel gear. Remove chain.

7. Remove bevel spur gear combination and spur gears from countershaft; these gears spline to countershaft.

8. Remove output sprocket from output shaft and remove shift spur gears, keys and collar and chain sprocket.

9. Remove snapring from input shaft, remove bevel gear and pull shaft through case.

10. Input shaft needle bearings should be installed flush to .005 below bearing bore surfaces from inside and outside case.

White

holding the case and cover together. Drive out the dowel pins used for alignment of the case and cover.

4. Lift off the cover assembly. Use a seal protector on axle shaft and lift off transaxle cover assembly. Because this seal is a single lip type, it may be reused, if care is taken to see that it isn't scratched or cut. Discard gasket.

5. To remove differential assembly, it may be necessary to replace two or three screws to hold center plate assembly down. Pull assembly straight up. If tight, tap on lower axle with soft mallet.

─── **CAUTION** ───
Do not use steel hammer. Remove gear on top of shifter shaft.

6. Remove temporary holding screws, if used, and lift off center plate assembly. Discard gasket.

7. Remove complete shifter assembly by grasping shifter gears, shaft and both shifter rods as a unit.

NOTE: Examine assembly carefully; if no service is required, retain assembly as a unit for easy reassembly.

8. Remove reverse idler shaft and spacer, cluster gear assembly and thrust washer. For removal and replacement of gears on cluster, see paragraph (12) on the next page.

9. Lift idler gear assembly out of case.

NOTE: Caution required as needles from shifter and brake shaft bearing may fall out.

10. Remove input shaft oil seal to allow access to snapring. Remove snapring and input shaft will slide out. A removed seal must be replaced by a new seal.

11. One model (612) has a sealed ball bearing instead of an oil seal. To remove this unit, remove snapring inside the case and drive out. On model 612-A, remove the oil seal in the normal manner.

12. Cluster Gear Sub-Assembly
    a. The cluster gear can be disassembled. All gears are replaceable if damaged or worn. Preferably use a press to drive the gears squarely.
    b. The small and middle gear bevel faces down, there is no beveled edge on large gear. Shorter section between middle and large gear.
    c. Key edge ends must align with shaft ends.

13. Shifting Assembly
    The shifting assembly is usually removed from and installed into the transaxle as a unit. The assembly is removed and replaced by grasping the shifting rods firmly. This will cause the binding necessary to hold the assembly together. Before removal or installation of the shifting assembly, notches in the shifter forks should be aligned with notches in the shifter stop. This indicates that shifting assembly is in a neutral position. The shifter stop must be so positioned that the notch aligns with notches in shifter forks.

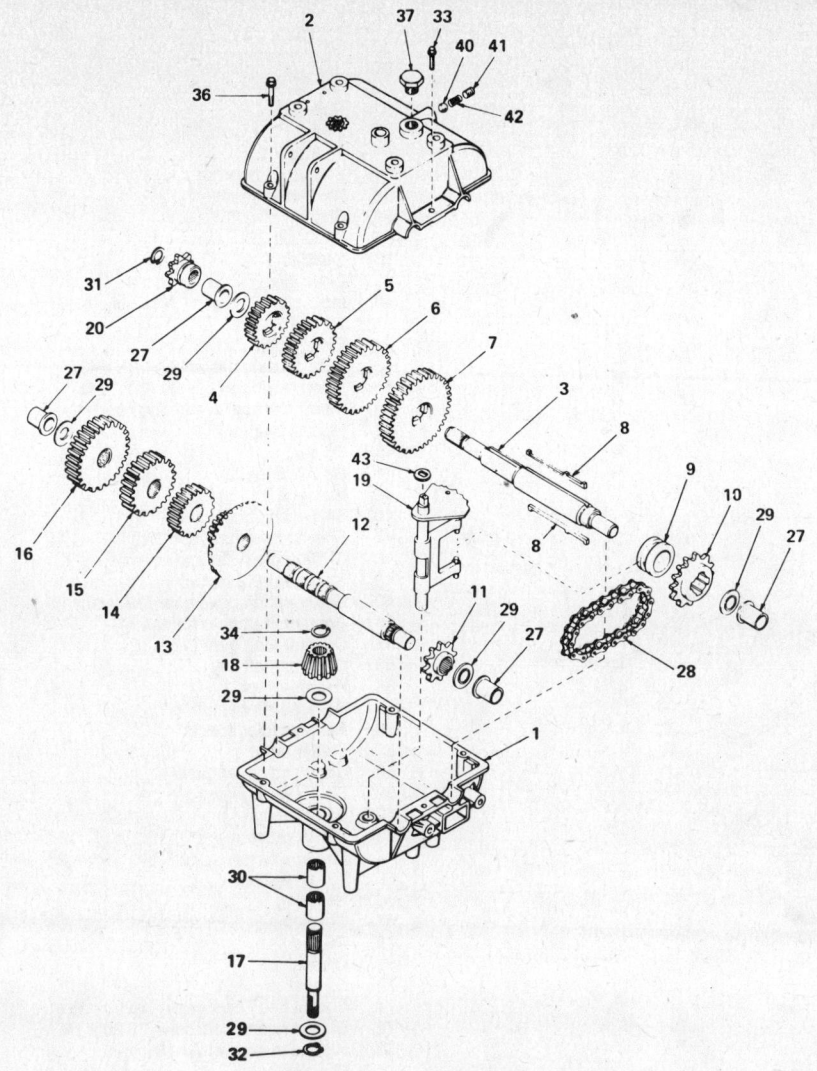

Model 501 components

1. Transmission case
2. Transmission cover
3. Output and brakeshaft
4. Spur gear (20 teeth)
5. Spur gear (25 teeth)
6. Spur gear (30 teeth)
7. Spur gear (35 teeth)
8. Key
9. Shift collar
10. Sprocket (14 teeth)
11. Sprocket (10 teeth)
12. Countershaft
13. Bevel gear (42 tooth & 15 tooth spur gear)
14. Spur gear (20 teeth)
15. Spur gear (25 teeth)
16. Spur gear (30 teeth)
17. Input shaft
18. Input bevel pinion
19. Rod and fork assembly
20. Sprocket (8 teeth)
27. Bushing
28. Roller chain (#41 chain, 22 links)
29. Thrust race
30. Needle bearing
31. Retaining ring
32. Retaining ring
33. Screw, 1/4-20 x 3/4 tap tite
34. Retaining ring
36. Screw, 1/4-20 x 1 1/4 tap tite
37. Plug
40. Ball, 5/16 in. steel
41. Setscrew, 3/8-16 x 3/8 in.
42. Spring
43. Washer

Model 600

DISASSEMBLY

1. Clean the outside surface of the transaxle, away from the area where disassembly will take place. Position shift lever in neutral position to help disassembly. Remove screws (3) holding shift lever opening. Remove all keys from key ways, remove all burrs and dirt from shafts. On hardened shafts, use a stone to remove burrs. All seals should be replaced whenever a shaft is pulled through a seal. Always use a new gasket whenever the gasket surfaces have been separated.

2. After removing axle housings, place the unit in a receptacle, bench or clamp the transaxle in a soft jaw vise. Position the transaxle so that the socket head capscrews are facing up.

3. Remove the socket head capscrews

White

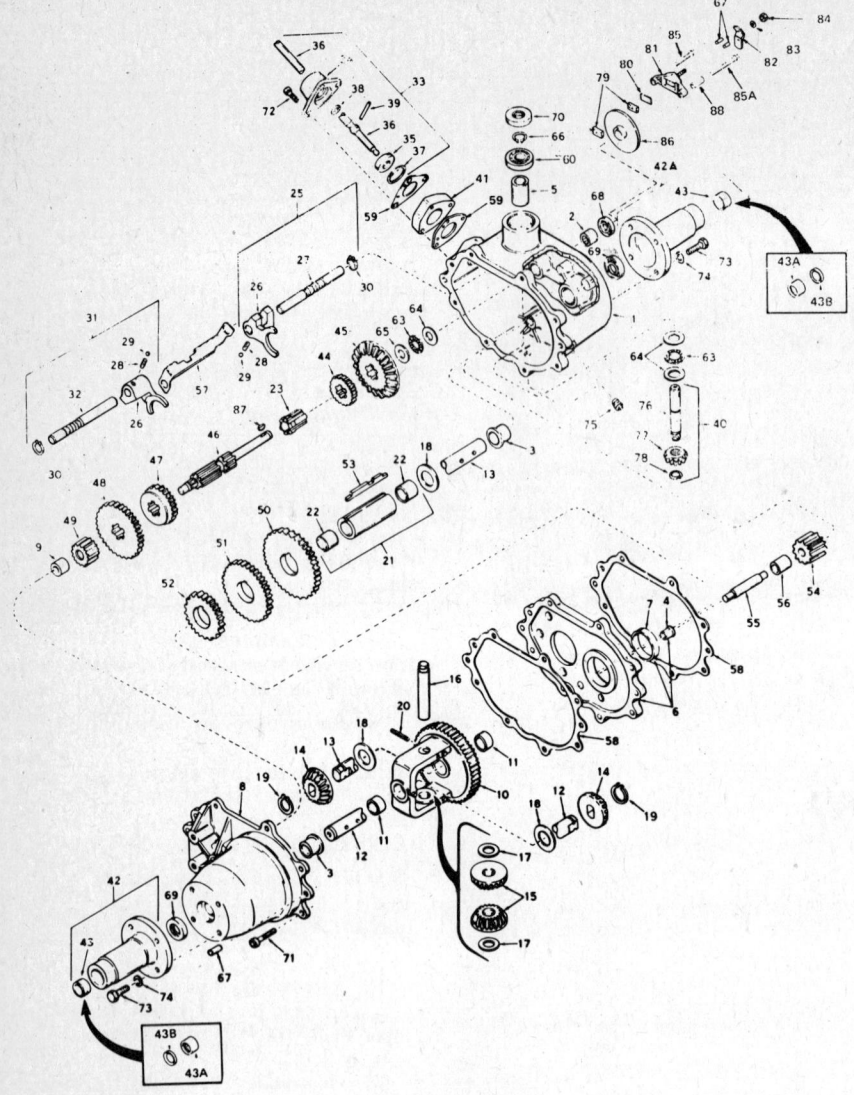

29. Steel ball
30. Snapring
31. Shift rod assembly (2nd and 3rd) (incl. nos. 26, 28, 29 30, 32)
32. Shift rod
33. Shift lever and housing assembly (incl. nos. 34 thru 39)
34. Shift lever housing
35. Shift lever keeper
36. Shift lever
37. Snapring
38. Quad ring
39. Drive pin
40. Input shaft and gear assembly (incl. nos. 76, 77 and 78)
41. Riser block
42. Axle housing assembly (incl. #43)
42A. Axle housing assembly (inlc. #43)
43. Bearing & seal assembly, needle
43A. Needle bearing
43B. Oil seal
44. Gear (16 teeth)
45. Bevel gear (33 teeth)
46. Shifter and brake shaft
47. Shifting gear (2nd and 3rd)
48. Shifting gear (1st and reverse)
49. Spur gear (12 teeth)
50. Countershaft drive gear (39 teeth)
51. Countershaft gear (34 teeth)
52. Countershaft gear (25 teeth)
53. Countershaft key
54. Reverse idler
55. Reverse idler shaft
56. Reverse idler spacer
57. Shifter stop
58. Case and cover gasket
59. Shift lever housing gasket
60. Ball bearing
63. Thrust bearing
64. Thrust washer
65. Thrust washer
66. Snapring
67. Dowel pin
68. Oil seal
69. Oil seal
70. Oil seal
71. Socket head capscrew, ¼-20 x 1½ in.
72. Socket head capscrew, ¼-20 x 1¾ in.
73. Hex head screw, 5/16-18 x 1 in.
74. Lockwasher 5/16 in.
75. Pipe plug ⅛ in.
76. Input shaft
77. Input pinion
78. Retaining ring
79. Brake pad
80. Brake pad plate
81. Brake pad holder
82. Brake lever
83. Flat washer
84. Locknut
85. Hex head capscrew, ¼-20 x 1¼ thd. forming
85A. Hex head capscrew, ¼-20 x 2¼ thd. forming
86. Brake disc
87. Woodruff key #61
88. Spacer

1. Case assembly
2. Needle bearing
3. Bronze bearing
4. Bronze bearing
5. Bronze bearing
6. Plate assembly
7. Bronze bearing
8. Cover assembly
9. Needle bearing
10. Differential gear assembly
11. Bronze bearing
12. Lefthand axle
13. Righthand axle
14. Bevel gear
15. Bevel pinion
16. Drive pin
17. Thrust washer
18. Thrust washer
19. Snapring
20. Roll pin
21. Countershaft sleeve assembly (incl. no. 22)
22. Bronze bearing
23. Idler shaft
25. Shift rod assembly (1st and rev.) (incl. nos. 26 thru 30)
26. Shift fork
27. Shift rod
28. Spring

Model 654 components

ASSEMBLY

1. Install thrust washers and bearing on input shaft.

2. Install input shaft into case assembly. Lock on with snapring retainer. Install oil seal.

3. Set case assembly open side up. Insert the idler shaft gear assembly, thrust washers and bearing. Note sequence of washers and bearings.

NOTE: Place reverse idler shaft into bearing to aid in holding washers, thrust bearing, idler shaft and gear assembly prior to installing shifter assembly.

4. Insert the washer and then the three gear cluster assembly.

5. Insert shifter assembly. Check that rods are seated properly.

NOTE: Reverse idler shaft will be pushed out at this time.

6. Install reverse idler. Make sure beveled edge is up. Spacer on top of gear.

7. Place new gasket on case and install center plate.

8. Place new gasket on center plate and install differential assembly, longer axle in down position. Be sure gear on shifter shaft is on shaft.

9. Install gear case dowel pins. Leave dowel pins slightly exposed on top to locate cover assembly.

10. Install transaxle cover assembly, and secure with eight (8) capscrews.

11. Install bearings and/or bushings, if necessary, using bearing driver and bushing tool. See bearing chart below.

12. Install axle housing assembly. Fill with 1½ pints SAE EP 90 oil.

White

1. Lever and housing assembly (incl. nos. 2 thru 7)
2. Snapring
3. Quad ring
4. Drive pin
5. Shift lever housing
6. Shift lever keeper
7. Shift lever
8. Shift rod assembly (incl. nos. 9 thru 12)
9. Spring
10. Steel ball
11. Shifter fork
12. Shifter rod
13. Shift rod assembly (incl. nos. 9, 10, 11, 14 and 55)
14. Shifter rod
15. Righthand axle
15A. Lefthand axle
16. Thrust washer
17. Hex head capscrew, ¼-20 x 2¼ in.
18. Lockwasher, ¼ in.
19. Ring gear
20. Drive pin
21. Drive block
22. Bevel pinion
23. Idler pinion and bushing assembly
25. Shifter shaft and bearing assembly (incl. no. 26)
26. Bearing
27. Idler gear
28. Idler shaft
29. Shifter stop
30. Case to cover gasket
31. Shift lever housing gasket
32. Shifting gear
33. Shifting gear
34. Spur gear (26 teeth)
35. Spacer
36. Spur gear (22 teeth)
37. Spacer
38. Spur gear (16 teeth)
39. Input shaft
40. Input shaft spur gear
41. Washer
42. Spacer
43. Output gear
44. Output pinion
45. Oil seal
46. Housing and bushing assembly (incl. no. 51)
46A. Axle housing and bushing assembly (incl. no. 51)
47. Transaxle cover assembly (incl. nos. 57 and 58)
48. Transaxle case assembly (incl. nos. 56 and 57)
49. Socket head capscrew, ¼-20 x ¾ in.
50. Dowel pin
51. Bushing
52. Pipe plug
53. Bevel gear
54. Snapring
55. Snapring
56. Bearing
57. Bearing
58. Bearing
60. Oil seal
64. Reverse idler shaft
65. Reverse idler spacer
66. Reverse idler
68. Washer
69. Differential carrier
70. Differential carrier

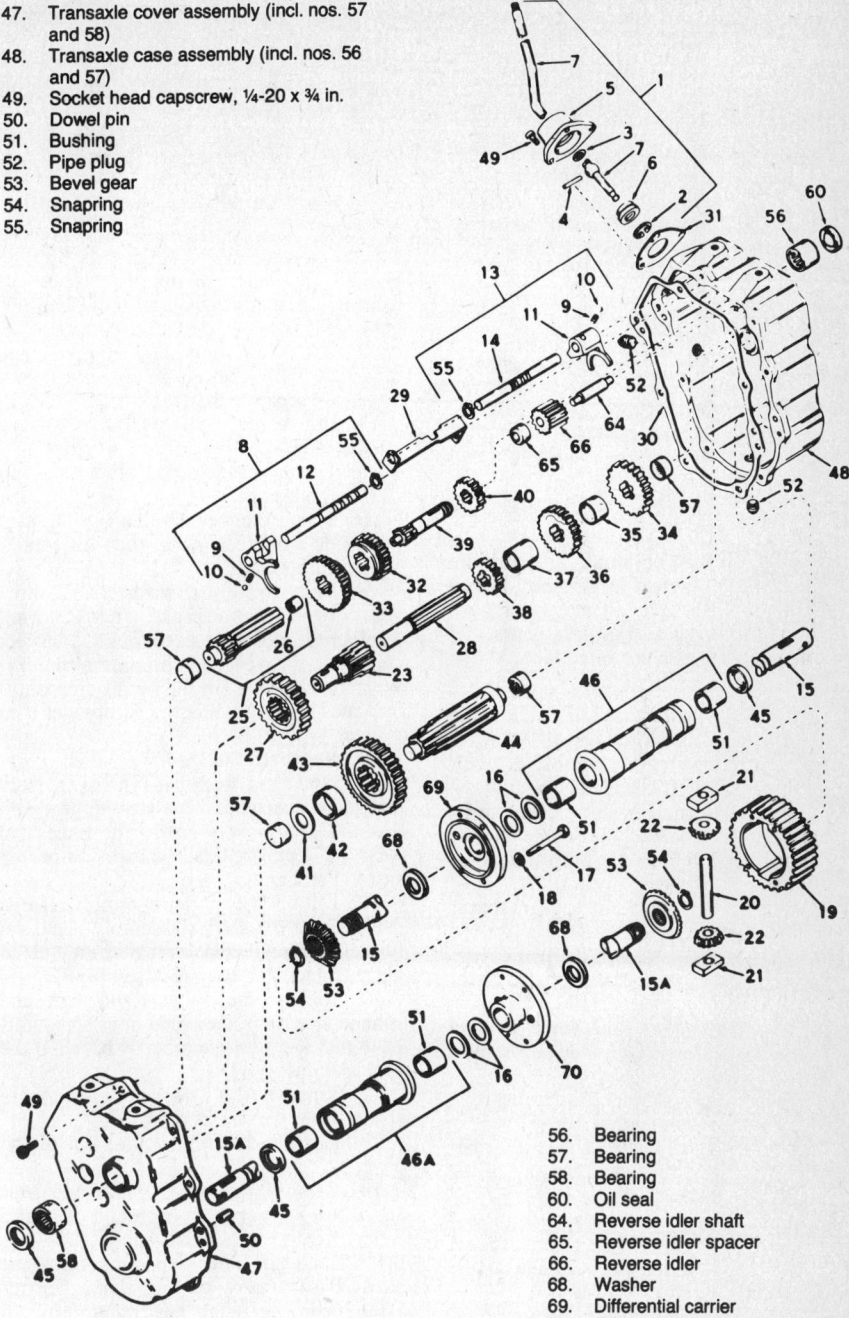

Model 1218 components

NOTE: For a neutral position, shift notches in forks and notch in shifter stop must be aligned and centrally located.

Part Location	Tool Remove & Replace	Tool Bearing Sizer
Brake Shaft	670210	(27 needles)
Axle	670204	670214
Input Shaft*	670207*	670208
Center Plate	670205	
Brake Shaft	670213	(30 needles)
Differential	670204	670214
Cluster Gear	670204	670214
Idler Gear	670210	670212
Axle Housing Reverse Idler	670204	670214

*Bearing must be flush with top of case. Secure with Loctite®.

Models 1200, 1400, 1700, 2000

DISASSEMBLY

1. Position the shifter forks in neutral.
2. On 1700 and 2000 series transaxles, remove the axle supports. On the 2000 series units, use the axle as a ram to press out the seal retainers.
3. When disassembling the rest of the unit, it should be held so that:
   a. It lies on the case, properly blocked so that no weight rests on the input shaft or differential, yet the case is rigid.
   b. It can be worked on without the chance of falling, or causing injury.

NOTE: On some 1700 series, the screw heads are on the case side of the unit. Take screws out from below.

4. Oil seals have a double lip so seal sleeves do not offer much protection during removal. Upon replacement, new seals should be used.
5. Tap dowel pins into the case and remove socket head capscrews.
6. NOTE: Some units have a threaded cover, so capscrew removal will be awkward.

Lift the cover off from case. Discard gasket.

White

Remove output gear and shaft. Note that the 2000 series has a spacer and thrust washer.
7. Remove the shifting assembly as one unit.
8. Remove the reverse idler shaft, spacer and gear.
9. Lift out the three gear cluster.
10. Remove the differential.
11. Tap the input shaft out of the case.

ASSEMBLY

1. Install input shaft in case. Use a soft mallet to seat shaft and gear completely. Binding can occur if the shaft is driven in only part way.
2. Install the differential assembly. The four capscrew heads should go down into the case.
3. Install the three gear cluster, with the smallest gear up.

NOTE: Bevels of small and middle gear go down toward large gear. Large gear bevel is up. The short spacer goes between the large and middle gears.
The 1400 series has a one piece, 3 gear cluster.

4. Position the reverse idler shaft in the unit, then install gear and spacer.
5. Install the shifter assembly as a unit into the case. When installed correctly, the neutral square formed by the shifting forks should appear through the case opening for attaching the shift housing. Both shift gears should be out of mesh.
6. Install the output shaft, gear, spacer and thrust washer.
7. Install brake shaft in the unit cover.
8. Position a new gasket on the cover mounting surface, then install cover.
9. Align cover with the dowel pin and secure with the socket head capscrew. Torque to 10 ft.lbs.
10.a. On 1200 series units, install axle seals using sleeve and driver.
   b. On 1700 series, install axle seals before installing axle supports using sleeve and driver.
   c. On 2000 series, press seals into seal retainers, then use sleeve to protect seal when installing into the case and cover. Install O-ring seal.
11. Install axle supports (1700 and 2000 series). Be sure that the mounting pad position is correct before tightening down capscrews to 13 ft.lb.
12. Install a new gasket and shift lever housing. Torque screws to 10 ft.lb. Be sure the shift lever is in the proper position to allow shifting.

Peerless 4-Speed Transaxles

DISASSEMBLY

1. Remove the transaxle.
2. Position the shifter forks in neutral before disassembly.
3. Remove both axle housings and use the exposed axle as a ram to separate the seal retainers from the case and cover.
4. When disassembling the rest of the unit, it should be held so that:
   a. It lies on the cover, properly blocked up, so that no weight rests on the brake shaft.
   b. The cover should sit rigidly so that removal of parts can be done in a systematic step by step procedure.
   c. It will not fall causing an accident or injury.
5. Oul seals are of the double lip type so sleeve protectors do not offer much protection when removing them. Upon replacement, new seals should be used.
6. Tap dowel pins into the cover and remove eight socket head capscrews.
7. To separate the case from the cover:
   a. Lift the case 1½ to 2 inches above the cover.
   b. Tilt the case so that shift rods will clear edge.
   c. Rotate the case so that boss hidden inside will clear gears, then lift free of the differential.
8. Remove thrust washer and three gear cluster from brake shaft, noting whether the cluster has a sloppy fit. Inspect gear teeth for wearing, chipping or breaks. Wear or chipping on the bevel area only, indicates shifting while the equipment is in motion.
9. Remove the reverse idler gear, spacer, and shaft from boss in cover. Note that the spacer goes between the gear and that the gear bevels go down. Excessive wear on teeth bevels indicates improper shifting technique.
10. Lift out the shifter assembly. If it is evident that the shifter assembly needs no further teardown, place it aside, in a clean place, intact, for easy re-assembly.
11. Remove the low gear and shaft, and splined spur gear. Separate gear and shaft. Note that NO thrust washer is between the gear and case.
12. Remove the two gear cluster and spacer from the brake shaft.
13. Lift the differential unit out of the cover.
14. Remove the output shaft and gear and thrust washer from each end of shaft.
15. Remove the brake shaft. Note that the brake shaft idler separates from the shaft. If separated, be sure that when reassembled, the idler gear chamfers are away from the cover.
16. Remove input shaft from case by tapping with a non-metallic hammer.

ASSEMBLY

1. Install input shaft in case. Use a soft mallet to seat shaft and gear completely. Often, binding in the assembled unit can be traced to a partially installed input shaft.
2. Center one $1/32$ in. thick by 1 in. I.D. thrust washer on the cover brake shaft needle bearing, then install the brake shaft and gear (chamfer side away from cover).
3. Install the output shaft and gear after centering a $1/16$ in. thick by $15/16$ in. I.D. thrust washer on each end of the shaft.
4. Insert the differential assembly in the cover. Note that the four bolt heads should be out away from the output gear.
5. Install the two gear cluster and spacer on the brake shaft.
6. Install a $1/16$ in. thick by ¾ in. I.D. thrust washer, gear, and low gear idler shaft in cover. Do not put a thrust washer on the exposed end of this shaft. Be sure the small gear meshes with the larger gear of the two gear cluster.
7. Center one $1/32$ in. thick by ⅞ in. I.D. thrust washer on cover shifter shaft bearing.
8. Install shifter assembly as a unt into the cover.
9. With the small gear of the three gear cluster toward the spacer, install the three gear cluster and other $1/32$ in. thick by ⅞ in. I.D. thrust washer on the brake shaft.
10. Install the reverse idler shaft, spacer, and gear into the cover. The beveled side of the idler gear should be down into the cover.
11. Position the gasket on the cover sealing surface, then install case over the differential shaft. Be sure the boss goes under gears and that edge of the case goes over the shaft rods in the opposite manner from which it was removed.
12. Once in position, if case hangs ½ to 1 in. high, turn the input shaft to get gears to mesh. The case should drop to about ¼ in. from closing.
13. Use a pair of needle nose pliers on the shifter stop on each shifter fork to agitate the shifter rod ends into their machined recesses in the case.
14. Align the case and cover with the two dowels, then install and tighten the eight socket head capscrews. Torque screws to 10 ft.lb. Unit can now be placed flat on the work bench.

Position seal retainers and new seals in position.

---- **CAUTION** ----
Sleeves must be used to protect seals, especially axle ends or where wheels attach.

15. Install new O-rings on seal retainers and position axle supports to case and cover. Be sure mounting pads face in same position as when removed. Install capscrews and torque to 13 ft.lb.
16. Install shift lever housing and new gasket.

REDUCTION UNITS

1300 Series

DISASSEMBLY

1. Remove lockscrews and tap dowel pins out of cover. Lift off cover and discard gasket.
2. Lift out brake shaft, gear, and thrust washers on each side of gears.
3. Lift output shaft, gear, spacer, and thrust washer from case. At the same time, lift out the differential assembly.

White

NOTE: No thrust washer is located between the output shaft and case.

4. To separate axle supports from the case and cover, use an arbor or hydraulic press. A piece of bar stock should be used to protect the support from the press ram.

ASSEMBLY

1. When installing axle support, be sure case and cover alignment is true with the press. Press supports in until flanged surfaces contact case and cover.
2. Install differential and output shaft simultaneously. Position gear ¾ in. I.D. spacer, and thrust washer on shaft.
3. Center one ¾ in. I.D. thrust washer over case needle bearing then install brake shaft gear, and other 1⅛ in. I.D. thrust washer.
4. Position a new gasket on the mounting surface of the case, then install cover. Align cover and case by tapping dowel pins into cover and secure with lockscrews torqued to 10 ft.lb.
5. Install new brakeshaft oil seal using sleeve number 670179 and driv driver number 670180.
6. Install new axle support oil seals using sleeve number 670179 and driver number 670180.
7. Add 2¾ (44 oz.) pts. oil (SAE EP 90) before securing hydrostatic drive to the 1300 series unit. Clean mounting surfaces and use a new gasket between the units. Torque 4 mounting bolts to standard torque for bolt used.

2400 Series

DISASSEMBLY

1. Clean axles of burrs, rust and sharp edges.
2. Remove axle supports. Be sure to note in which position and to which side they attach.
3. Drain oil from unit.
4. Remove seal retainers and O-rings.
5. Remove eight socket head capscrews securing case to cover. Drive dowel pins out of case into cover, then lift case off of the cover.
6. Before removing differential unit,

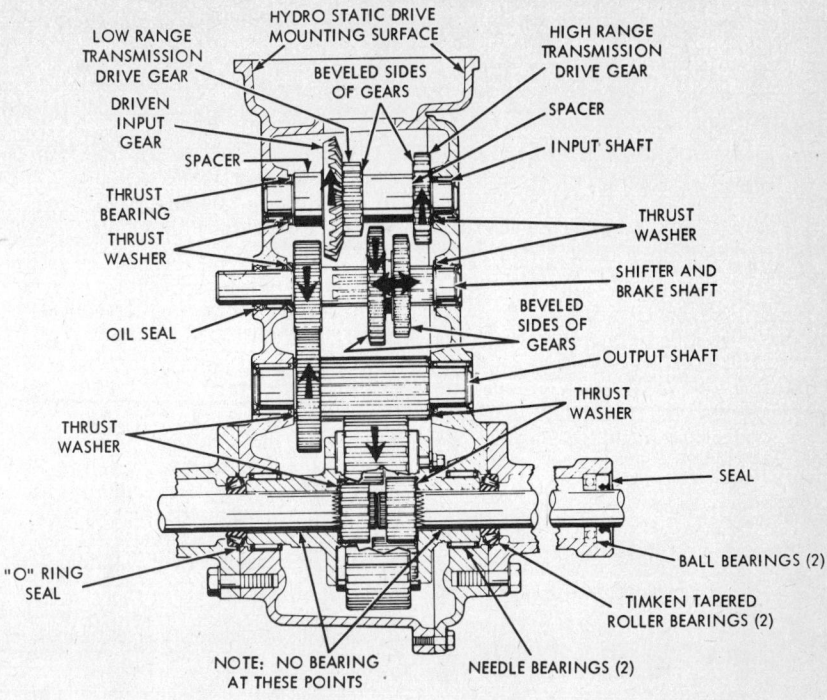

2500 series components

note the position of the capscrew heads. Replace the unit in the same way to insure that axle lengths will be correct when installing axle supports.

7. Remove the input shaft and gear assembly. Note that thrust washers are on both ends of shaft.
8. Remove output pinion and gear splined to it. Note position of thrust washers on each end of pinion.
9. To remove the brake shaft, the shaft must be tapped from the gear splined to it. Use a soft hammer. Note that both ends of the shaft have thrust washers.

ASSEMBLY

1. Install thrust washer in cover for input, output and brake shafts.
2. Install brake shaft into gear in cover.
3. Install input and output shaft (with gear) and position spacers on brake shaft. Install thrust washers on all shafts.

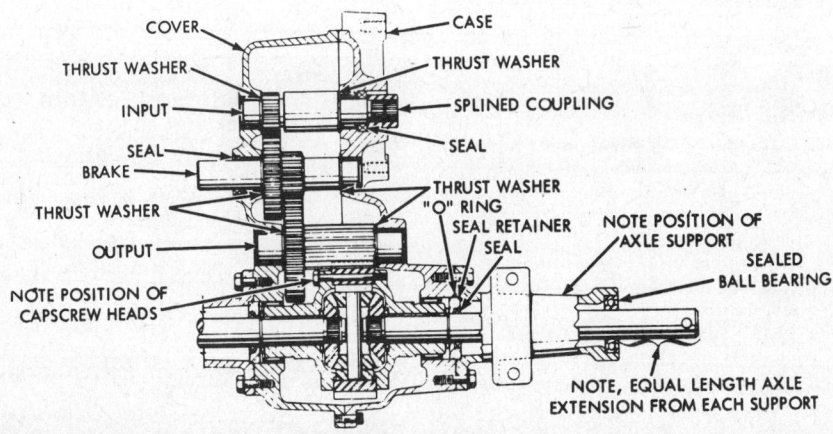

2400 series reduction unit

4. Install differential unit with capscrew, facing up.
5. Position new gasket on slightly raised dowels and install case. Turn down eight socket head capscrews lightly. Tap dowel pins in and tighten capscrews to 10 ft.lb.
6. Use seal sleeve and install seal and retainer assemblies. Position new O-ring on seal retainer. Be sure the mounting pad is in correct position. Install each axle support and bolt down. Torque the capscrews to 13 ft.lb.
7. Using proper seal sleeve and driver, replace input and brake shaft seals. Rotate shafts to check for binding or noise.
8. Add lubricant (2 pints SAE EP 90 oil) to level of fill plug with unit in normal mounted position.

2500 Series

DISASSEMBLY

1. Remove the axle supports. If supports are different or if there is a chance for confused reassembly, scribe alignment marks.
2. Remove and discard square O-ring seal.
3. If the tapered roller bearings are loose, remove them.
4. Position the unit on the "cover up" side, then remove the dowels and screws. Lift off the cover and discard the gasket.
5. Note whether the unit is as shown. Disassembly and reassembly will have to follow one view or the other.
6. Because the hydrostatic can be driven either way and input rotation direction is a matter of application, the driven bevel gear must be changeable to accept either rotation. The gears and shafts also are

839

White

"flopped" depending upon input rotation direction.

7. Disassemble as called out in Step (6) or (7).

(Counterclockwise input for forward)
a. Remove the output shaft, thrust washer and gears.
b. Lift out the differential assembly.
c. Lift out the input bevel, gear and shaft as an assembly.
d. Work the shifter shaft and gears out of mesh with the shifter fork and rod and remove.
e. To remove the shifter rod and fork, the setscrew, spring and ball should be removed at the outside of the case.

(Clockwise input for forward)
a. Lift out the differential assembly.
b. Remove the output gear and shaft.
c. On the outside of the case remove the setscrew, spring and ball.
d. Remove both the input and shifter gear and shaft assemblies, along with the shifter rod and fork.

ASSEMBLY

1. After shifter rod is positioned, install ball, then spring and setscrew. Turn setscrew in slowly while raising and lowering rod—until ball stops rod movement.

2. Be sure thrust washers and spacers are between every shaft and case and cover.

3. Install new gasket. It may be helpful to dampen the gasket to get it to lie flat.

4. Be sure differential bolt heads go opposite output gear (large gear).

5. Use seal sleeve 670245 to protect axle support oil seals during installation.

6. Install axle supports correctly. Do not rotate the support to a new position when pressed tight against the square O-ring seal or the seal may be cut.

7. To install brake shaft oil seal, use seal sleeve 670179 and drive 670180.

8. To install shifter rod oil seal, use seal sleeve 670206 and driver 670211.

9. Lubrication for all modesl is 7 to 8 pints of SAE Type A Automatic transmission fluid. When filling with fluid, allow fluid to settle behind the taper bearings into the axle supports. This may necessitate filling, checking and adding. The lubrication fill and check hole is located in the case. If the shrouding on the vehicle does not allow access to the fill and check hole when the 2500 is installed in the chasis, lubrication should be filled prior to installation on vehicle. If filling takes place before installing 2500 in vehicle the unit must be oriented as shown in figure 7-18 or at a 90° angle to the axles. There should be no fluid leak.

10. Install drive assembly according to manufacturers instructions.

DIFFERENTIAL

Model 100

DISASSEMBLY

1. Clean outside of differential. Remove all keys, pins, etc. Remove all burrs from key ways and holes. Use stone on hardened shaft.

2. Remove 4 locknuts, bolts and sprocket. Separate differential carrier housings.

3. Remove drive pin, pinion gears and thrust washers as a unit.

4. Remove snapring, bevel gear and thrust washer. Slide axle from differential carrier housing.

NOTE: Bushings are replaceable in the differential carrier housing. To replace bushing, use bushing tool 670204.

ASSEMBLY

1. Slip axle in differential housing carrier. Place thrust washer and bevel gear on axle and secure with snapring.

2. Place pinion gears and thrust washers on drive pin and insert assembly into either differential housing carrier.

3. Use 1 oz. SAE EP 90 lithium grease as lubricant.

4. Assemble differential carrier housings and sprocket with 4 bolts and locknuts.

NOTE: No oil seals or gaskets are required in this unit.

Model 600

DISASSEMBLY

1. Drive out roll pin that secures drive pin with suitable driver.

2. Remove drive pin.

3. Thrust washers must be removed before attempting to remove the pinions. Remove bevel pinions simultaneously by rotating the gears in opposite directions; gears will move out of position.

4. Drive out double roll pin and slide axle out. On roll pin drive types, drive the bevel gears from the axle.

5. On double "D" type drives, remove snapring, bevel gear and thrust washer. Slide axle out.

6. Inspect bushings and gears for wear and replace when necessary.

ASSEMBLY

1. Place axles (left and right) into differential gear assembly. Install thrust washers.

NOTE: The axles differ in length so select the proper axle.

2. On roll pin drive models, install double roll pins into holes in each shaft. Place bevel gears on shaft. Roll pins fit into the recess in back of the gears, bevel gears must be seated tightly on the roll pins or binding will occur.

3. On double "D" type drives, place bevel gears on the shaft and install snapring in groove on the shaft.

4. Install bevel pinions SIMULTANEOUSLY FROM OPPOSITE SIDES by rotating pinions in opposite directions while sliding into position in gear assembly. Check alignment by inserting fingers into drive pin holes. If not aligned, drive pin cannot be inserted. Remove and replace bevel pinions as only one tooth out of position will cause misalignment.

5. After aligning, insert thrust washers behind each pinion. Insert drive pin and secure with roll pin.

Models 1200, 1300, 1700, 2000, 2300, 2400

DISASSEMBLY

NOTES:

1. The 1200 series differential carrier is supported directly on the axle (1). Roller thrust bearings (2) are used between the bevel gear (3) and the differential carrier (4). This illustration shows axles with snapring (5) retainers, some earlier production had rolled over axle ends to secure the assembly. Thrust washers (6) are used at the ends of the differential carriers and case/cover thrust face. The drive pin (7) and drive blocks (8) are similar to those used in the accompanying illustration. Replace the differential carrier if worn in excess of .878 at point A.

2. The 1700 series differential has rolled on the ends (1) to retain the bevel gear to the axle. The 1700 differential is also made with snapring retainer on the axle. In event it is necessary to replace parts, the new axles will be snapring type and the spacer (2) will be eliminated. Replace differential carriers worn in excess of 1.004 at point A.

3. Roller thrust bearings (3) and (4) are used between the carrier and case/cover and between the bevel gear and carrier. The bushings (5) support the axles.

4. The 2000 series, three speed and the 2300 series, four speed differential. Examine the external bearing race on the differential carriers (1) for wear, pitting, replace if evident. The differential carriers in this assembly have replaceable bushings (2) replace if worn in excess of .878, point A. These differentials have been built with rolled axle ends and also snaprings (3).

DISASSEMBLY

1. Clean the differential assembly, then check and note the axle lengths and their relation to the heads of the four hex head bolts.

2. If the unit will not turn freely, note where the unit binds. Check and replace those parts.

3. Place the differential in a large vise with soft jaws (hex head bolts up). Do not clamp the vise on the bearing race of a differential carrier.

a. Remove the four hex head bolts and the upper axle and differential carrier. Remove the drive blocks, pinions, drive pin and thrust spacer if used, by lifting out of the ring gear. Tap the ring gear lightly with a mallet to loosen from the differential carrier.

b. If a snapring is used, the axle assembly may be disassembled. If the axle end has been rolled, do not attempt to break the rolled retaining edge. The parts are to be replaced as an assembly.

c. Remove the snapring and the thrust washer, if used. Separate the bevel gear and differential carrier from the axle.

ASSEMBLY

1. Select the correct axle for the side of the differential opposite the hex head bolts. If the wrong axle is used, it will require complete tear down of the differential, or possibly the entire transaxle if the error is not detected until later.
2. Clamp the axle, in a soft jaw vise (not bearing or oil seal surfaces). The differential carrier with threaded holes is assembled to this axle.
3. Torque the four hex head bolts to 7 ft.lb.

Duo-Trak® Limited Slip Model

DISASSEMBLY

1. Remove four through-bolts.
2. Separate axle assemblies from body cores.
   To disassemble axles, remove snapring and retained parts. Be sure that flanged thrust washer goes toward hub end of axle upon reassembly.
3. Use a pair of large 90° tip snapring pliers and remove the cylindrical spring putting tension on the ten pinion gears. Once the spring is removed, the gears can be removed.
4. Separate the two body cores from the ring gear.

ASSEMBLY

1. Install body cores to ring gear so that pockets in one core are out of alignment with pockets in the other core.
2. Re-assemble thrust washers, bearing, carrier and side gear to axle and secure with the snapring.
3. Install pinion gears on one side, then use the differential carrier and axle to hold them from falling out when the unit is turned over. The side gear must mesh with the five pinions.
4. Install pinions in other side to mesh with previously installed pinions.
5. Insert the cylindrical spring with a pair of large 90° tip snapring pliers so that it bottoms on the side gear. Most of the ten pinions should be in contact with the spring.
6. Install other axle and secure assembly with four through bolts. Torque to 7-10 ft.lb.

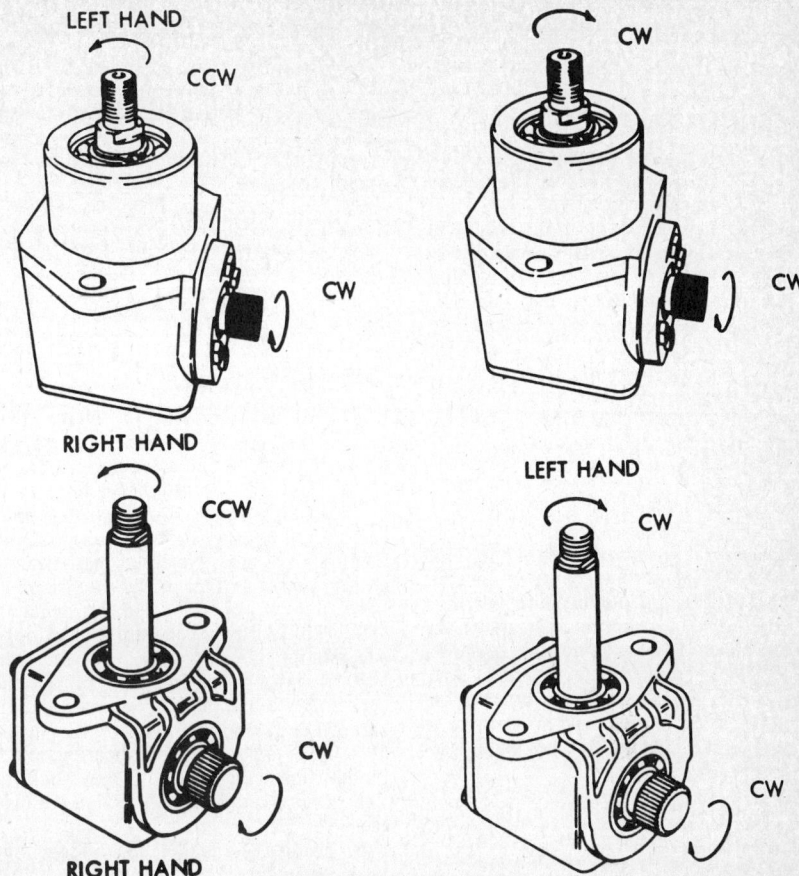

Final drive rotation patterns

FINAL DRIVE

DISASSEMBLY

1. Remove the cover and gasket and remove the lubricant.
2. Different output shafts achieve a different rotation by mounting a gear at the top or bottom of the shaft.
   a. If the driven bevel gear is on the bot-

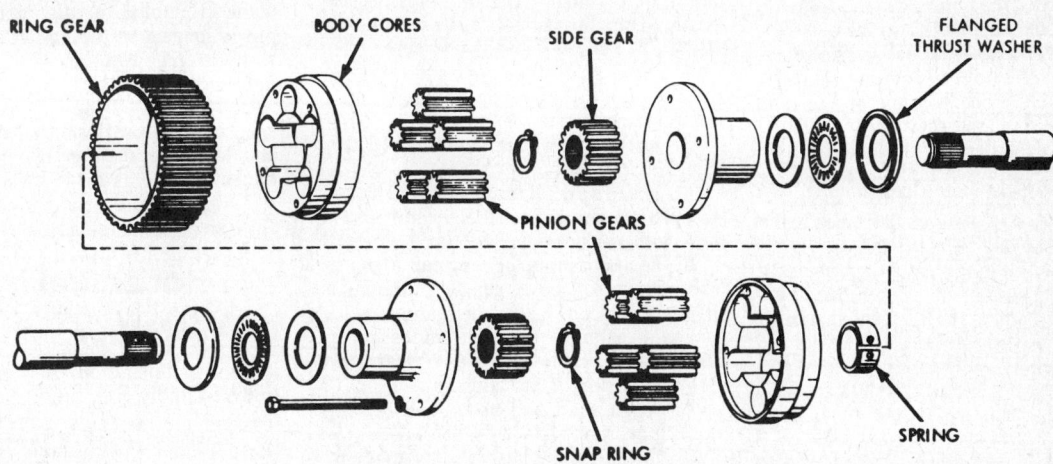

Duo-Trak limited slip differential

tom remove the snapring on the output shaft and pull out the output shaft.

b. Remove the output shaft bearing by using a large machine bolt, washer and nut.

c. The input shaft bearing, input shaft and bevel gear may now be driven out of the housing. Use tool 27569. Press bearing into housing to remove.

d. The top ball bearing in the blind end of the housing is removed by either heating the housing on a hot plate or tapping the housing on the outside with a mallet.

e. Right angle units with the driven bevel gear at the top will have to be disassembled. The output shaft is larger in diameter inside than outside. When the output shaft is pulled out the bearing will also be removed.

ASSEMBLY

Assemble in reverse of the disassembly instructions. It is important that only ball bearings with snaprings be used for the input shaft and top output shaft bearing. Use tool 27569 for driving in bearings. In the event bearing bore is tight in housing, heat housing on hot plate. Place a steel plate on the hot plate and rest housing on steel plate. Do not overheat. Work rapidly after removing housing from hot plate. Use care to prevent burns.

RIGHT ANGLE DRIVE

DISASSEMBLY

1. Remove the cover, gasket and lubricant.
2. If the unit is built with the bevel gear on the input shaft toward the cover, remove the snapring on the input shaft from the groove.
3. Remove four screws and the cover and seal assembly.
4. Remove the seal at the output shaft.
5. Remove the snapring on the output shaft and tap it with a mallet on the outside end to drive it and the inner ball bearing out of the housing. Remove the output end ball bearing by driving out from the inside.
6. Remove the input shaft from the ball bearing with an arbor press. Always support the inner race when removing and replacing the input shaft into the ball bearing. Use tool 670185 to support the inner ball bearing race.
7. The ball bearing that supports the inner end of the input shaft is removed by holding the housing in the hand and tapping the housing sharply on the outside with a soft mallet behind the bearing. It may at times be necessary to heat the housing to remove this bearing.

ASSEMBLY

Assemble the unit in reverse of the disassembly. When building up the units it is important to install the correct input shaft and identify it with the correct cover if there is any identification on the cover, either righthand or lefthand, see the accompanying illustration, to determine rotation. Use tools 28679, 670158, and 27569 to support bearings being pressed on shafts or into housing.

T DRIVE

DISASSEMBLY

1. Before removing T Drive from equipment, be sure to scribe marks at one mounting hole to insure correct reassembly.
2. To disassemble T Drive self-tapping screws and housing cover and clean grease from internal area. Note and mark near the casting gasket surface on the side where the beveled input gear is located. To switch the "T" 180° will result in output shaft opposite rotation.
3. Remove capscrews and both retainer caps and seal assemblies. Separate and discard oil seals and gaskets.
4. Press input shaft ball bearings out of the housing with fingers. If they stick, tap lightly using a drift punch around the outer race.
5. Using a soft mallet, separate the input shaft and gear. A slight press fit holds a keyed surface on the shaft in a groove of the gear.
6. Remove and discard the output shaft oil seal. Do not scratch shaft. Remove and discard snapring. Tap the shaft into the housing, using a soft mallet.
7. If necessary, remove the snapring to separate the gear and bearing from the shaft. Hold the gear and bearing in one hand and tap the end of the shaft vigorously with a soft mallet.

ASSEMBLY

1. If separated, install inner ball bearing and bevel gear on output shaft. Be careful of alignment. Use of a press is preferable to tapping parts together with a mallet.
2. Press shaft, bearing, and gear assembly into housing until outer bearing race bottoms in retaining cavity.
3. Install outer bearing and new snapring.
4. Install new oil seal using seal sleeve No. 670185 and driver 28679 until seal is flush with housing.
5. Position input bevel gear in mesh with output shaft bevel gear. Tap the input shaft into place with a soft hammer. Use one hand to hold the gear and shaft to dampen tapping blows. Be sure gear is on the marked side of the housing.
6. Align shaft and insert ball bearing on each bearing surface by hand.
7. Install new seals in retainer caps, using driver 28679.
8. Using seal sleeve 670185 over the shaft serrated ends, install new gaskets and bearing cap. Tighten retaining capscrews to 8-11 ft.lb.
9. Fill housing with 4 oz. of E.P. lithium grease.
10. Install gasket and cover and secure with self-tapping screws. Torque to 20-24 ft.lb.
11. Align scribe marks and install T Drive on equipment.

SHIFTER UNITS

Vertical Input

DISASSEMBLY

1. Place the shift lever in a vise so that the shift lever housing is at least one inch from the top of the vise jaws.
2. **Dowel Pin Type** Locate the dowel pin holding the retainer in the housing from the outside. Place a ¼ in. flat face punch on the gasket surface directly over the dowel pin. Strike the punch sharply but lightly with a hammer to dislodge the retainer from the shift lever housing. Always use a new dowel pin for reassembly.
3. **Snapring Type** Use the proper compressing type tool for removing the snapring. Loosen the vise and disassemble the pieces.
4. Remove the shift lever from the shift lever housing. Examine the roll pin in the ball of the shift lever, if bent or worn, replace. When inserting a new roll pin in the ball, position so that equal lengths protrude from both sides of the ball.
5. Oil leakage past the point where the shift lever enters the shift lever housing will require replacement of the quad ring seal in the shift lever housing.
6. Prior to reassembly, be sure that bends in the shift lever correspond to the mounting on the vehicle.

ASSEMBLY

1. **Dowel Pin Type** Secure with a new dowel pin. A second dowel pin is used in some assemblies for alignment. This dowel pin is located in the gasket surface of the shift lever housing and fits into a mating hole in the transaxle.
2. **Snapring Type** Secure parts with the snapring. Before installing the shift lever and housing to the transaxle housing, check the shifting forks for Neutral position.
3. Always use new gaskets between the shift lever housing and the transaxle.

Horizontal Input

DISASSEMBLY

Follow illustrations A-F in order. Prior to disassembly compare the assembly with the illustrations. This will aid during the reassembly.

ASSEMBLY

1. Reassemble the shifting assembly by

following illustrations F-A. Pay particular attention to the annular grooves in the shifter rods and the snapring.

a. Assemble the shifter forks to the shifter rods as illustrated in Figure F. The shifter forks are interchangeable.

b. Refer to Figure F. Slide the shifter fork onto the shifter rod until it comes to the hole with the indexing ball and spring. With a flat blade screwdriver press the indexing ball into the hole and move the shifting fork completely onto the shifter rod.

c. Move the shifting fork to the Neutral position. The neutral groove is the center groove. If the shifter rod has four grooves, the neutral groove is the second groove from the shortest end. This neutral groove can be seen through the hole in the shifter fork.

d. When the shifter forks are properly assembled to the shifter rods and positioned in neutral, the ends of the notches in the shifter forks are in alignment.

2. Assemble the two flanged gears onto the shifter shaft. Note that the large gear is placed on the shaft first with the flange side toward the needle bearing in the end of the shifter shaft. Slide on the smaller gear with the flange toward that of the larger gear.

3. When assembling the shifter forks and rod to the flanged gears on the shifter shaft, that shifter fork which is on shifter rod "A" always engages in flange in the larger gear. To determine which is shifter rod "A" compare the parts to illustrations. Hold the shifter shaft in the hand as illustrated during assembly.

4. After the shifter fork and rod assemblies have been engaged with the flanged gears allow the shifter rods to lay open in the hand and position the shifter stop. The notch in the shifter stop is the guide for correct positioning. Align this notch with the corresponding notches in the shifter forks and insert the shifter stop. Move the shifter rods together, and insert into the transaxle. Remember to squeeze

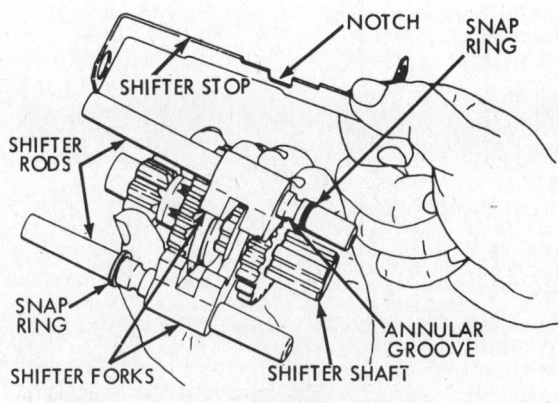

Figure A

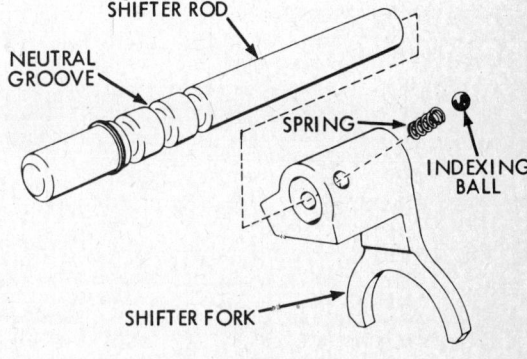

Figure B

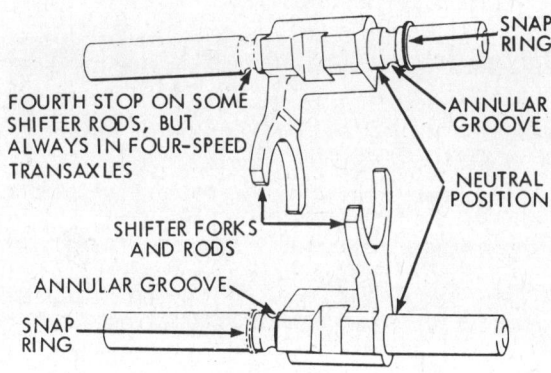

Figure C

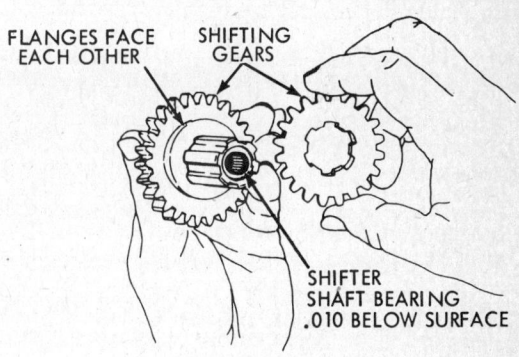

Figure D

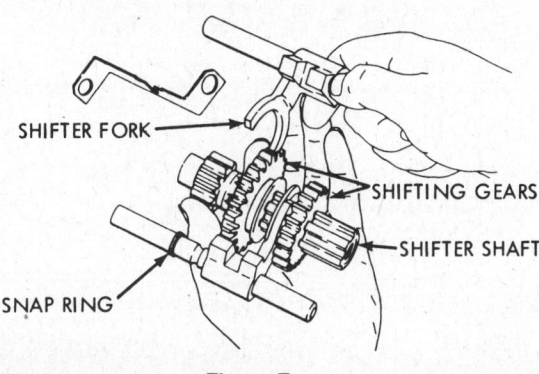

Figure E

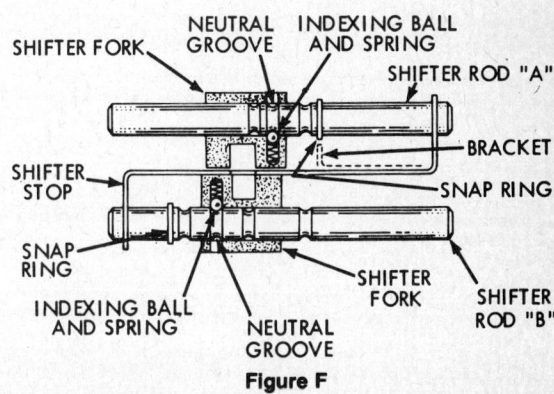

Figure F

the ends of the shifter rods to cause the assembly to bind and stay together.

5. In three speed transaxles the needle bearing end is inserted first into the case to engage the end of input shaft.

6. When placing the shifting assembly into the four speed transaxle be sure the thrust washer is on the bearing. Place the assembly into the transaxle with the needle bearing end of the shifter shaft up. Allow the end of the shifter shaft to protrude below the ends of the shifter rods, this will ease the alignment of the assembly.

7. The shifter assembly is correctly installed in the transaxle if the notches in the shifter forks are just about in the center of the opening in the case or cover of the transaxle.

ENGINE

Various Briggs and Stratton engines are used in these tractors. For complete engine service, see the Unit Repair section of this book.

Yanmar

INDEX

Front Axle 846
Steering..................... 846
Brakes...................... 846
Clutch 846
Transmission and Differential...846
Hydraulic System............. 851
Engine 855
Fuel System 856

YANMAR
All Models

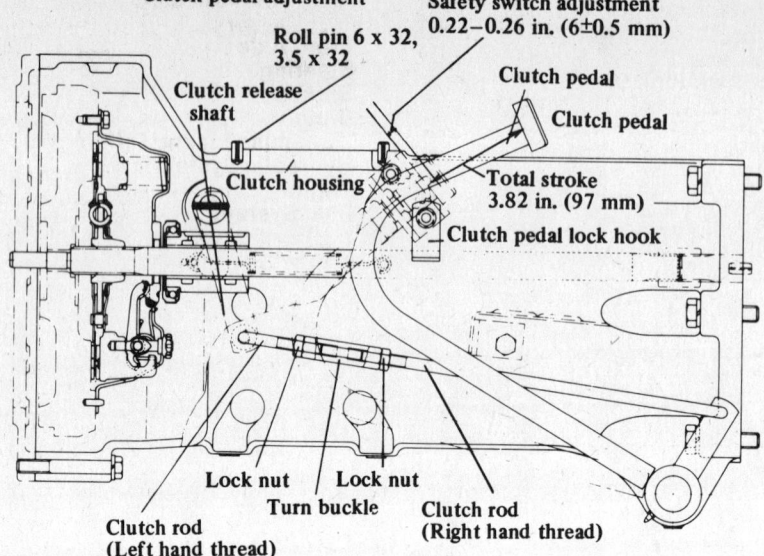

FRONT AXLE

REMOVAL AND INSTALLATION

1. Raise and support the tractor on jack stands.
2. Disconnect the steering linkage.
3. Support the axle and remove the center pin.
4. Roll the axle from under the tractor.
5. Installation is the reverse of removal.

STEERING

Steering Gear

REMOVAL AND INSTALLATION

1. Remove the steering wheel.
2. Remove the instrument panel assembly.
3. Disconnect the Pitman arm. Make alignment marks for assembly.
4. Remove the steering column, noting the number and location of the adjusting shims.
5. Remove the steering gear box side cover.
6. Remove the sector shaft by knocking it out towards the direction of the side cover.
7. Remove the ball-nut assembly.
8. Installation is the reverse of removal.
Note that the ball-nut is installed with the threaded portion in the center engaging the mid-section of the gear.

ADJUSTMENT

Adjust the steering wheel play by turning the adjusting screw on the side of the gear box. Steering wheel play should be 0.8–2.0 in. measured at the wheel rim.

BRAKES

Brake Drum

REMOVAL AND INSTALLATION

1. Raise and support the rear of the tractor on jack stands.
2. Remove the wheel.
3. Remove the rear axle housing.
4. Unbolt and remove the brake drum. The brake components may now be serviced.
5. Installation is the reverse of removal. The drive pinion nut is torqued to 58-72 ft.lb.; the drum nuts to 17-22 ft.lb.

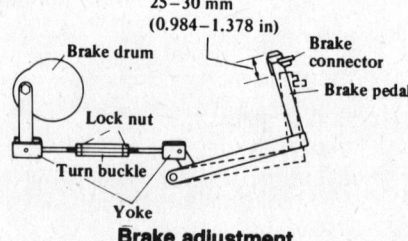

ADJUSTMENT

Adjustment is made by adjusting the turnbuckle on the brake rod. Brake pedal play should be 1.2-1.6 in.

CLUTCH

REMOVAL AND INSTALLATION

1. Separate the engine and transmission.
2. Remove the clutch assembly from the flywheel.
3. Installation is the reverse of removal. Torque the pressure plate bolts alternately and evenly to 17-22 ft.lb.

TRANSMISSION AND DIFFERENTIAL

DISASSEMBLY

1. Separate the engine and transmission.

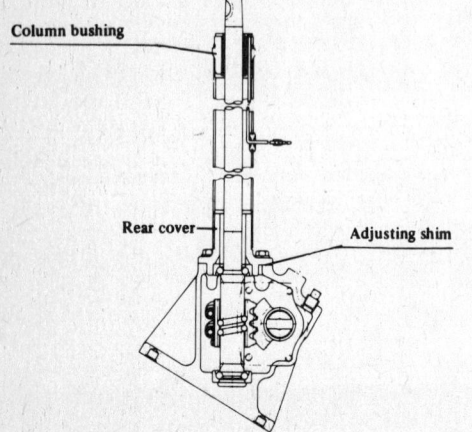

Steering gear

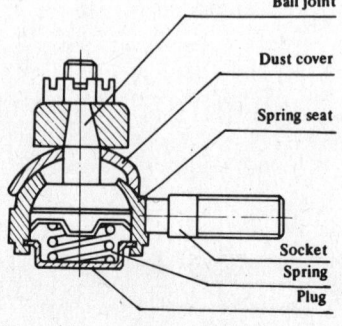

2. Remove the PTO gear assembly.
3. Remove the transmission case cover.
4. Remove the axle housings.
5. Remove the brake drums, noting the ring gear adjusting shims.
6. Remove the differential lock shaft.
7. Remove the differential assembly.
8. Remove the reverse shaft.
9. Remove the mainshaft.
10. Remove the countershaft.

INSPECTION

Use the following table to make replacement determinations for the transmission components. Note that items 19 through 26 refer to the differential. Item 20 is carried out at a point 90° from the axial direction of the input shaft. Also note:

1. When using a new ring gear and pinion
   a. Use a mandrel instead of the differential carrier (mandrel dimensions: diameter 4.3305-4.3301 in. and use shims to adjust the reading for A to 0.612-0.608 in.
   b. Fit the ring gear in the differential case, and insert shims (four sizes available: 0.004, 0.008, 0.012, 0.039 in. between the left differential carrier and the transmission case so that the backlash (Item 19) is 0.005-0.007 in.
   c. At this time, take care to see that there is no space between the differential carrier and the ring gear, and the ring gear and the differential pinion.
   d. When assembling the righthand differential case, use shims so that the total thickness of the shims is 0.002-0.004 in. more than the thickness of the shims used in the righthand differential case.
   e. After assembling the left and right sides, inspect the bearing of the teeth visually, or coat the teeth with red lead and turn the drive pinion to check the bearing.

NOTE: If the adjustments have been made correctly, about half of the width of the ring gear should be bearing on the drive pinion, somewhat towards the small ends.

Item	Standard Dimension	Replacement Limit	Testing Equipment
Shaft axial play			
Main shaft	0.0047-0.0177	0.0315	Dial gauge
PTO pinion shaft	0.0244-0.0457	0.059	Dial gauge
PTO shaft	0.0059-0.0220	0.0315	Dial gauge
Reduction gear	0.0075-0.0201	0.0315	Dial gauge
PTO 1st gear	0.0059-0.0228	0.0315	Dial gauge
Reduction gear (39T/20T)			
Outer diameter of mainshaft	1.1811-1.1806		Micrometer
Inner diameter of needle bearings	1.1821-1.1814		Cylinder gauge
Clearance	0.0003-0.0015	0.0039	
Counter gear (34T/38T/28T/23T/17T)			
Outer diameter of countershaft	1.1811-1.1806		Micrometer
Inner diameter of needle bearings	1.1821-1.1814		Cylinder gauge
Clearance	0.0003-0.015	0.0039	
Reverse gear (22T)			
Outer diameter of reverse shaft	0.7474-0.7869		Micrometer
Inner diameter of bushings	0.7910-0.7882		Cylinder gauge
Clearance	0.0008-0.0041	0.0079	
PTO 1st gear (46T)			
Outer diameter of PTO shaft	1.7725-1.7713		Micrometer
Inner diameter	1.7753-1.7746		Cylinder gauge
Clearance	0.0020-0.0040	0.008	
Over-running clutch			
Outer diameter of shaft	0.7067-0.7060		Micrometer
Inner diameter of clutch	0.7089-0.7069		Cylinder gauge
Clearance	0.0001-0.0010	0.012	

Play in direction of rotation between each of sliding gears and shaft spline sections
  Mainshaft (18T/23T/28T/34T)
  Countershaft (41T)
  PTO pinion (17T)
	0.0021-0.0058	0.008	Dial gauge

Backlash for all gears	0.0047-0.0106	0.02	Dial gauge
Deflection for all shafts	—	0.002	

Yanmar

Item	Standard Dimension	Replacement Limit	Testing Equipment
Shifter assembly			
Width of groove for shift forks for sliding gears	0.287-0.280	—	Calipers
Thickness of shift fork tip	0.272-0.264	—	Calipers
Clearance	0.008-0.024	0.040	Clearance gauge
Shaft forks			
Outer diameter of fork shaft	0.5899-0.5892		Micrometer
Inner diameter of shift fork	0.5922-0.5906		Cylinder gauge
Clearance	0.0006-0.0030	0.008	
Low gear			
Diameter of low gear (41T) shift arm hole in housing	0.7895-0.7882		Cylinder gauge
Outer diameter of low gear (41T) shift arm	0.7874-0.7854		Micrometer
Clearance	0.0008-0.0041	0.020	
Fork lock assembly			
Free length of fork lock spring	1.33	1.15	Calipers
Fitted load of fork lock spring	1.781-16.14 lbs	13.67 lbs	
Fitted length	1.04	1.04	
Backlash of drive pinion and ring gear [3]	0.005-0.007	0.012	
Clearance between drive pinion and ring gear [1] [2] [3]			
Backlash of differential pinion and differential side gears [3]	0.008-0.012	0.020	Lead wire Micrometer
Differential pinion			
Inner diameter of differential pinion [3]	0.7898-0.789		Cylinder gauge
Outer diameter of differential pinion [3]	0.7866-0.7858		Micrometer
Clearance [3]	0.024-0.0040	0.016	
Play in direction of rotation between differential side gear and final pinion spline [3]	0-0.0035	0.012	
Thickness of differential pinion thrust liner [3]	0.041-0.037	0.024	Calipers
Thickness of differential side gear thrust liner [3]	0.041-0.037	0.024	Calipers
Backlash for final gears	0.0063-0.0122	0.024	Lead wire Micrometer
Play in direction of rotation between final gears and rear axle splines	0-0.0035	0.012	Dial gauge
PTO gear backlash	0.0047-0.0106	0.020	
PTO shaft deflection (at tip)		0.012	

Unit: Inch

[1] Check to see that there is no looseness between the inner and outer races of the ball bearings and needle bearings. If this is the case, or if they emit abnormal noises when turned, or do not turn smoothly, replace. Particular care should be taken in the case of needle bearings to see that the rollers are not cracked or chipped.

[2] If the gears show any uneven wear, galling, or scoring, they should be repaired or replaced.

[3] Refer to measurements for the differential assembly; drive pinion clearance is carried out at a point 90° from the axial.

Yanmar

2. If the drive pinion is bearing on the small ends of the ring gear teeth, remove the shims inserted between the differential case, and move the ring gear towards the drive pinion.

3. If the drive pinion is bearing on the base of the teeth, insert additional shim(s) between the differential case, and move the ring gear away from the drive pinion.

4. If the ring gear is bearing on the small ends of the drive pinion teeth, remove the shim(s) between the countershaft bearing retainer to move the drive pinion towards the ring gear.

5. If the ring gear is bearing on the base of the teeth, the countershaft is too close to the ring gear, so add shim(s) between the countershaft bearing retainer, and move the drive pinion away from the ring gear.

ASSEMBLY

Transmission Case

1. Service transmission case.
2. Fit M8 x 20 mm (3 pcs) and 8 x 32 mm (4 pcs) stud bolts into front part of transmission case.
3. Fit 10 x 30 mm (2 pcs) and 10 x 25 mm (6 pcs) (indicated by ⊙ mark) into right side of transmission case.
4. Drive in 10 x 20 spring pins (2 pcs).
5. Fit stud bolts into left side of case, in same manner as indicated in 2 and 3 above.
6. Fit 12 x 32 mm (2 pcs) stud bolts into rear part of transmission case.

NOTE: Coat all stud bolts with a bonding adhesive to prevent them from getting loose.

Countershaft Assembly

1. Press-fit type 5206 ball bearing onto countershaft. The side of the bearing on which the number is inscribed should be facing the teeth. Check the surfaces of the teeth to ascertain that there are no mars or other damage.
2. Fit the type 30 bearing retainer ring onto the shaft.

NOTE: Ring is fitted firmly and positively into the groove.

Counter Gears Assembly

1. Fit the type 41 circlip on the counter gear proper, press-fit the 5th speed counter gear on, and then fit the type 45 circlip.
2. Press-fit the counter gear spacer, the 3rd speed counter gear, and the 1st speed counter gear in that order, and secure them by type 41 circlip.

NOTE: When press-fitting the gears, coat the counter gear with oil to minimize peeling. Any peelings must be removed completely. Make sure that the gears face in the proper direction. Check the faces of gear teeth to make sure that there are no mars resulting from severe impacts or flash.

3. Press-fit the needle bearings. Use the special tool available for this purpose (note that they are to be fitted so that they are sunk to a depth of 0.04 in.). The side on which the bearing type is inscribed must face the ends of the counter gear.

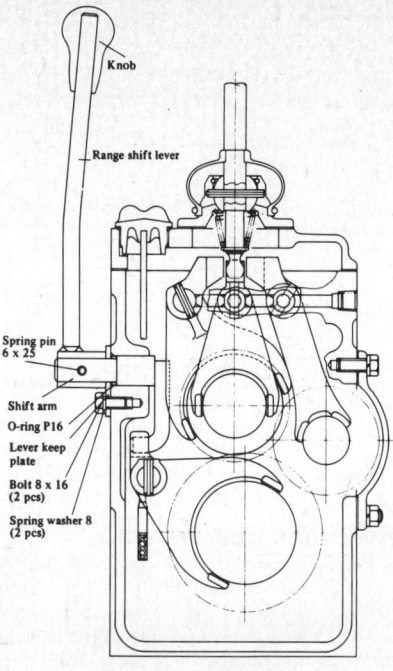

Range shift system

Reverse Gear Assembly

Press-fit reverse gear wound bushings—20 x 1.5 x 15 mm (2 pcs). Use the special tool available for this purpose. See that the bearing does not project beyond the gear.

Range Shift System

1. Insert 5/16 steel ball and fork lock spring A in low-high fork shaft, and secure by 3 x 28 mm cotter pin.

NOTE: Be sure to turn back ends of split pin positively.

2. Drive in 4 x 18 mm spring pin into low-high fork shaft. Amounts of upper and lower projection are to made the same. Split down spring pin is positioned as shown.
3. Fit O-ring on low-high shift arm, and apply thin coating of grease.
4. Fit nut M12 on range shift lever.

NOTE: Bevelled side of nut should be facing upwards.

Counter Gears and Shaft Assembly

1. Insert countershaft gear assembly in following order.
Fit range shift gear, and counter gear assembly. Ascertain that range shift gear slides smoothly. Take care not to damage the counter gear assembly's needle bearing ring.
2. Fit the countershaft shims (0.04 in.) in position provisionally, insert the jig, tighten, and then tighten the countershaft by the M20 nut.
3. Measure and adjust position of countershaft (remove jig). Pass a mandrel through the 4.3307 in. diameter portion and adjust distance to tip of countershaft to 0.6067-0.6102 in.
4. Fit in bearing retainer assembly and 20 x 42 x 7 mm spacers. Make sure that the shims do not drop out at this time.
5. Press-fit the ball bearing and the bearing sleeve.

NOTE: The side of the bearing on which the number is stamped should face outwards.

6. Fit the washer and tongued lockwasher in position, fit and tighten the M20 nut, and then bend up the tongues of the lockwasher. Ascertain that the tongues of the lockwasher are not severed. Tightening torque; 58-72 ft.lb.
7. Fit the packing of the countershaft cap and the countershaft cap itself, and secure the cap with the 8 spring washers (4 pcs) and nuts M8 (4 pcs). Coat the packing with liquid sealant before installing it. Tightening torque; 17-22 ft.lb.

Range Shift System and Reverse Gear Assembly

1. Low-high shift fork assembly is fitted in the groove of the range shift gear, and the low-high fork shaft assembly is fitted in position.

NOTE: See that the end surfaces of the fork shaft do not project beyond the front face of the transmission case.

2. Fit in the low-high shift arm assembly, and secure it in position by the shift lever retainer plate with the spring washers (8) and 8 x 16 mm nuts. The retainer plate should be fitted so that its curved section faces outwards. Tightening torque; 17-22 ft.lb.
3. Fit the reverse gear assembly and reverse shaft in position, and secure it with the 6 x 14 spring pin.

NOTE: Take care to see that the split portion of the spring pin is positioned correctly.

Right Differential Carrier Assembly

1. The bearing 6306UU is fitted onto the final pinion R, and the O-ring S25 is fitted in position.

NOTE: When installing the O-ring, coat it with grease, and be careful not to tear or cut it.

2. Press-fit the assembly referred to in Item (1) above into the right differential assembly, and secure it with two spring pins (double).

NOTE: The head of the spring pin should be driven in to a depth of approximately 0.04 in. from the face of the carrier.

3. Insert the seal collar into the oil seal TC385811, and at the same time, press-fit it into the final pinion. Use the special tool available for this purpose. Coat liberally with grease, and make sure that there is no peeling or cutting of the lip section.
4. Fit the G105 O-ring in position. Coat with grease.
5. Screw in the tapered plug 1/8 PT. It should be tightened with a wrench.

Yanmar

Left Differential Carrier Assembly

1. The bearing 6306UU is fitted onto the final pinion L, and the O-ring S25 is fitted in position.

NOTE: When fitting the O-ring, coat it with grease, and be careful not to tear or cut it.

2. Press-fit the assembly (referred to in the previous step) into the left differential assembly and secure it with two spring pins (double).

NOTE: The head of the spring pin should be driven in to a depth of approximately 0.4 in. from the face of the carrier.

3. Insert the seal collar into the oil seal, and at the same time, press-fit it onto the final pinion. Use the special tool available for this purpose. Coat liberally with grease, and make sure that there is no cutting of the lip.

4. Fit the G105 O-ring in position. Coat with grease.

5. Screw in the ⅛ tapered plug. It should be tightened with a wrench.

6. Fit the differential lock slider to the differential lock sleeve, and insert it in the final pinion. All splined sections should move with ease.

Differential Gears

1. Insert thrust washer, side gear, pinion liners (2) and pinions (2) in the differential case, and pass differential pinion shaft through them. Ascertain that the pinions and side gear will turn smoothly. Completely remove the "edges" formed on the pinion shaft by machining.

2. Drive in the spring pin, and secure it to the shaft.

NOTE: Drive in to a depth of approximately 0.04 in.

3. Insert the thrust washer and side gear in the ring gear.

4. Place the items assembled in (1) and (2) above onto the top of the ring gear assembly, and secure them with bolts M10 x 26.5 mm (6 pcs). Apply adhesive to the bolts. Tightening torque: 58-72 ft.lb.

Front Oil Seal Case Assembly

Press-fit oil seal into front oil seal case. Coat with grease, and press-fit using jig; make sure that there is no peeling. Be careful not to cut the lip section nor let the spring fall out of position.

Reduction Gear Assembly

Press-fit needle bearing in reduction gear. Be sure to use special tool available for this purpose (conform to 0.24 in. and 0.04 in. clearances. The side of the bearing on which the number is stamped must face outwards when press-fitted in position

Mainshaft Assembly

Press-fit ball bearing 6304 onto mainshaft.

Differential Device Assembly

1. Fitting of righthand differential assembly (differential carrier shims should be fitted provisionally).

2. Secure using 10 x 20 mm bolts (2 pcs), M10 nuts (2 pcs), and spring washers 10 (4 pcs). Tighetning torque: 33-43 ft.lb.

3. Insert differential lock slider in final pinion. Make sure that it slides with ease.

4. Fit differential assembly in position.

5. Attach lefthand differential carrier assembly, and secure provisionally with light torque.

6. Measure and adjust, if necessary, backlash between countershaft and ring gear. Backlash: 0.005-0.007 in.

7. Measure clearance between transmission case and lefthand differential carrier, and insert shims to provide an additional 0.002-0.004 in.

NOTE: Dimension at * position in the accompanying illustration. When con-

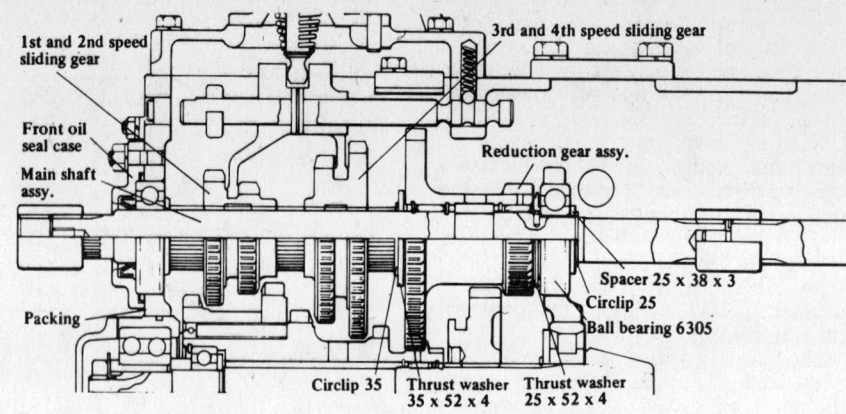

Main gears and shaft assembly

ducting measurements, make sure you are hitting bottom.

8. Final tightening of lefthand differential carrier assembly:
   10 x 30 bolts (2 pcs)
   M10 nuts (2 pcs)
   Spring washers, 10 (4 pcs)

NOTE: Tightening torque for bolt-nuts: 33-43 ft.lb.

Main Gears and Shaft Assembly

1. Insert mainshaft assembly, and then assemble thrust washer, shaft snapring, reduction gear assembly, thrust washer, 3rd to 5th speed sliding gears, and 1st and 2nd speed sliding gears, in that order.

NOTE: The curved side of the snapring should face the rear.

2. Drive in spring pins, 6 x 14 mm and 3.5 x 16 mm (double). The split in the 6 x 14 spring pin should be opposite the bearing. The depth to which the pins should be driven in should be about the same as the outer ring (inside) of the bearing.

3. Attach plugs with hexagonal heads M8 and copper packing. Tightening torque: 9-12 ft.lb.

4. Fit snapring referred to in step (1) above in groove.

5. Press-fit bearing 6304.

6. Attach front oil seal case packing and oil seal case, and secure in position with 8 spring washers (3 pcs) and nuts M8 (3 pcs). Coat packing with liquid sealant. Tightening torques; 17-22 ft.lb. The mainshaft must turn freely.

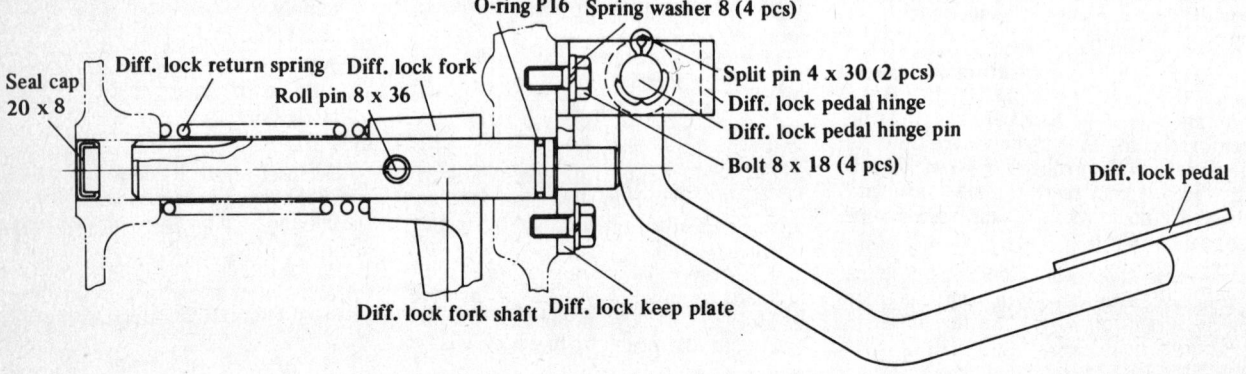

Differential lock shaft

7. Fit air breathers (left and right). Use adhesive. Breathers should face rear.
 * Amount of end float:
   mainshaft—0.0012-0.02 in.
   reduction—0.0094-0.022 in.

Differential Lock Pedal Assembly
DIFFERENTIAL LOCK SHAFT INSTALLATION
1. Install spring.
2. Tighten the nut by hand.
3. Fit fork as illustrated.
4. Remove stopper.
5. After checking the position of center of fork, remove the installing shaft.
6. Insert differential lock shaft, set the position of the hole of the shaft with the groove of the fork, and drive in spring pin.

Gear Shift Lever Assembly
1. Insert the shift lever ball in the gear shift lever, and secure with spring pin 5 x 40 mm.

NOTE: The spring pin should be driven in so that the split faces upwards.

2. Fit the assembly referred to in Item (1) above into the retainer of the gear shift lever and secure by fitting circlip 30.
3. Fit the shaft lever spring and washer 13.2 x 19 x 2 mm, and the circlip (12).

Transmission Case Cover Assembly
1. Fit the fork lock spring A and steel ball 5/16 on the 1st-2nd shift fork, and then fit the 1st-2nd fork shaft (equipped with O-ring) in.

NOTE: Do not forget to fit the steel ball. Coat the O-ring with grease.

2. Fit the fork shaft (equipped with O-ring) into the 3rd-5th shift fork. Coat the O-ring with grease.
3. Fit the fork shaft (equipped with O-ring) into the reverse shift fork. Coat the O-ring with grease.
4. Secure each of the fork shafts with spring pins, 4 x 10 mm (3 pcs).

NOTE: Drive in to depth of 0.04 in. from face of case.

5. Insert the steel ball and fork lock spring into the 3rd-5th speed fork and reverse fork, and secure them with spring pins, 3 x 20 mm.

NOTE: The split of the spring pin should face the opposite direction from the fork lock spring.

6. Press-fit the seal caps, 15 x 5 mm (3 pcs).
7. Fit shift lever guides on the grooves of each of the forks.

NOTE: Make sure that the guides can function freely.

8. Fit shift lever retainer packing and shift lever assembly, and secure with spring washer 6 (4 pcs) and bolts, 6 x 16 mm (4 pcs). Coat packing with liquid sealant. Coat bolts with adhesive. Tightening torque: 5.8-8.7 ft.lb.

PTO Shift Lever Assembly
1. Insert shift plate in PTO shift lever, fit plain washer (10), and secure with split pin, 2.5 x 18 mm.

NOTE: Bend back the ends of the split pin firmly.

2. Insert PTO shift lever in the shift plate, fit washer, and secure with split pin 2.5 x 18 mm.

Transmission Case Cover and Oil Port Cap Assembly
1. Fit transmission case cover packing and transmission case cover assembly, secure with spring washer (10) (4 pcs) and bolts 10 x 55 mm (4 pcs).
2. Set oil port cap assembly to transmission case. Coat packings with liquid sealant. Make sure that all forks are fitted properly into the groove of gears. Tightening torque: 33-43 ft.lb.

Rear Axle Housing Assembly
1. Insert circlip 44.5 on rear axle shaft, and press-fit ball bearing.

NOTE: The side of the bearing with the number showing should face outwards.

2. Press-fit the assembly referred to in Item (1) above into the rear axle housing.
3. Fit the oil seal case packing and oil seal case assembly, and secure with the spring washers (10) (4 pcs) and bolts, 10 x 35 mm (4 pcs). Coat the packing with liquid sealant. Tightening torque: 33-43 ft.lb.
4. Press-fit seal collar.

NOTE: Take care to see that the oil seal is not scraped or damaged while being inserted; to prevent damage, press-fit while rotating shaft.

5. Press-fit rear wheel hub.

NOTE: Remove any snaprings.

6. Insert the spacer, and press-fit the bearing.

NOTE: The side of the bearing on which the number is stamped should face outwards.

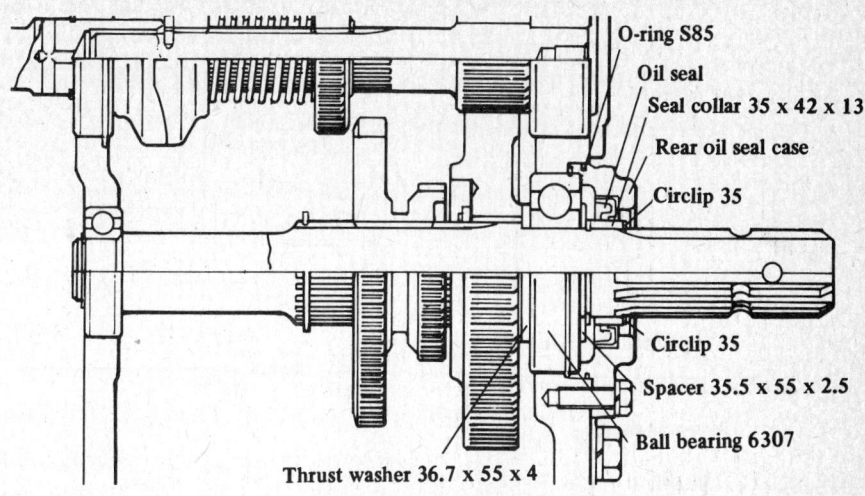

PTO Shaft assembly

HYDRAULIC SYSTEM

The manufacturer recommends that all system repair be left to trained, qualified technicians. If the system must be serviced, do so in a clean, lint free area and with great care. Note that the system uses transmission fluid and prior to assembly, all parts must be coated with clean oil.

OVERHAUL
Whatever the components being worked on, remove and disassemble it or them noting the arrangement of all parts. Lay out the parts in the order disassembled.

Oil Filter and Hydraulic Cylinder Case Assembly
1. Fit oil filter cover packing to oil filter assembly, and secure using 8 spring washers (3 pcs) and bolts, 8x20 mm (3 pcs). Make sure that there is no debris such as swarf or bits of rubber inside the transmission case. Tightening torque: 9-12 ft.lb. The packing should be coated with liquid sealant.
2. Fit the hydraulic cylinder case packing to the hydraulic mount and insert leader bolts (2 pcs).

NOTE: The packing should be coated with liquid sealant.

3. Mount the hydraulic cylinder case, and secure it with spring washers (10) (6 pcs) and bolts, 10 x 32 mm (6 pcs). Tightening torque: 33-43 ft.lb.

Piston Assembly
Fit the backup ring for P55 and O-ring P55 into the piston. Check the ends of the backup ring to make sure they conform to the illustration. Note the positions of the backup ring and O-ring. Make sure that there are no projections on the curved section of the piston (indicated by the arrow).

Yanmar

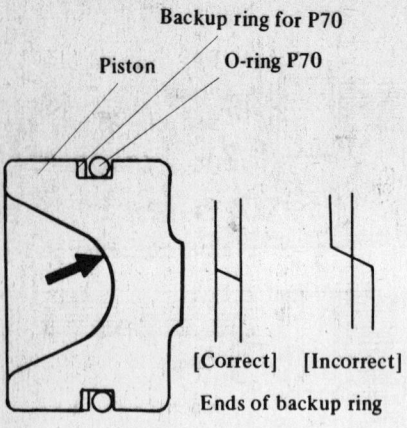

Piston assembly

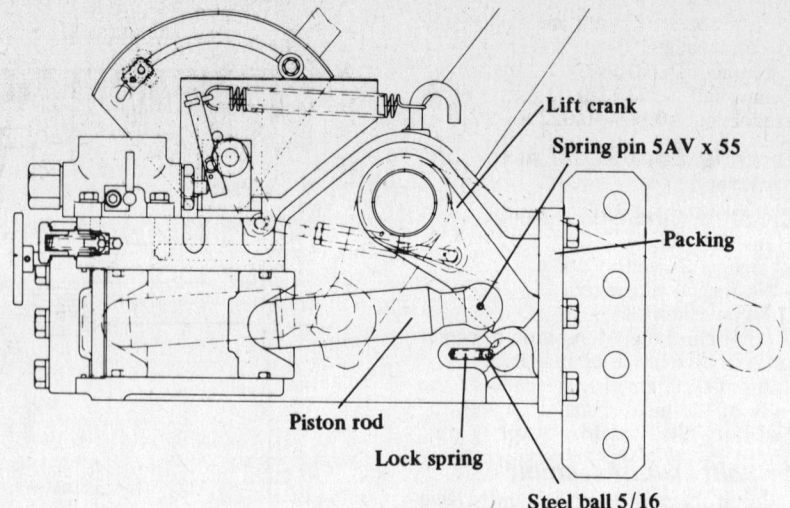

Lift crank assembly

Lift Crank Assembly

Couple the piston rod and the lift crank by driving in the spring pin 5AWx55 mm. Make sure that there are no cavities or projections on the curved surface of the lift crank. Note the direction of the split of the spring pin. Make sure that the piston rod will slide effortlessly.

Lift Shaft Assembly

Drive the spring pin 4AWx14 mm into the lift shaft.

NOTE: Check the fitting of the pin.

Lift Arm Assembly

Fit the lift arm pin in the tapered hole of the lift arm, and secure it with the spring washer 16 and nut M16. Tightening torque: 123-152 ft.lb.

Feedback Arm Assembly

1. Press-fit the lever knob onto the feedback rod. Use adhesive to secure.
2. Drive the spring pin 4AWx18 mm into the feedback rod. The spring pin must project equally on both sides.
3. Fit the plain washer (10), the feedback arm, the plain washer (6), the spring, and the plain washer (6), in that order. Apply grease to the surfaces of the feedback arm and feedback rod which move and rub against each other.
4. Fit the cotter pin 2.5 x 15 mm in position and bend the ends to secure the feedback arm and feedback rod together.
5. Pass the feedback rod through the adjuster, and then fit the nut M8 and the locknut M8 on the rod and secure it at a suitable position so that the adjuster does not fall off.

NOTE: The position of the adjuster will be adjusted later.

Control Lever Shaft Assembly

1. Fit the plain washer (8) and the feedback lever A on the control shaft, and secure feedback lever with the E-shaped retainer ring. Apply grease to the surfaces which move and rub against each other. Make sure that the feedback lever turns smoothly.
2. Fit the split pin 4x30 mm in the feedback lever A. Bend the ends back firmly. Note the position of the head of the split pin.
3. Fit the shaft hinge on the control lever shaft, and secure it by driving in the spring pin 4AWx18 mm. Apply grease to the surfaces which move and rub against each other. Be sure the shaft hinge is fitted in the proper direction.

Control Lever Guide Assembly

1. Fit the spacer 6.5x17x7 mm on the bolt M6x8 mm, insert it in the control lever guide hole; then from the opposite direction, fit on the plain washer and the spring washer (6); secure with the nut M6; and secure the spacer on the control valve guide.
2. Drive the spring pin 4AWx10 mm into the locknuts 6, place it in the control lever guide groove, fit the plain washer 6, tighten the knob screw, and secure the locknut on the control lever guide.

NOTE: Note the relative positions of each of the components. Adjustments will be made later.

Control Lever Assembly

Press-fit the control lever knob onto the control lever.

NOTE: Use adhesive

Control Bracket Assembly

1. The left and right cushion plates are secured to the control bracket by the bolt M8x20 mm, spring washers (8), and nut M8.

NOTE: Do not get the left and right plates mixed up. Tightening torque: 17-22 ft.lb.

2. Fit two plain washers 13 on the feedback shaft, pass it through the hole at the

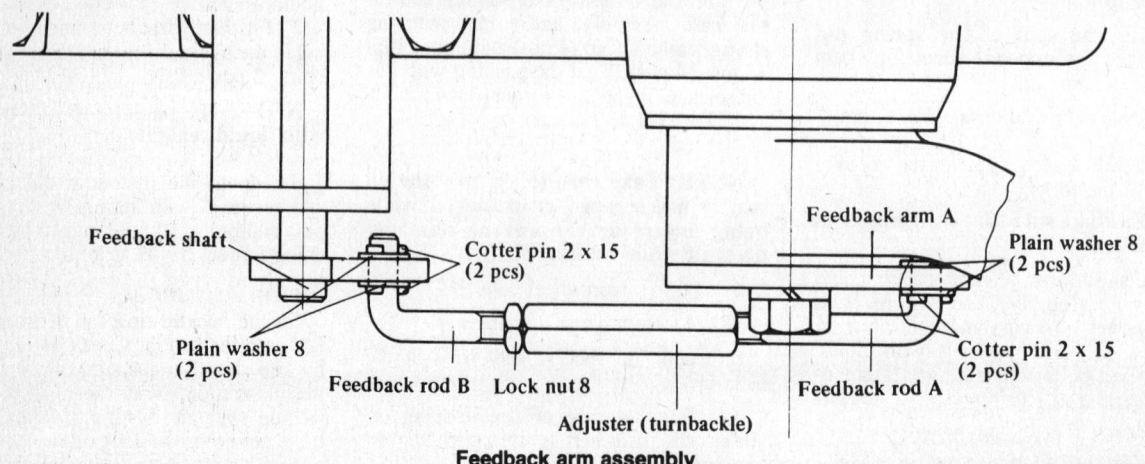

Feedback arm assembly

Yanmar

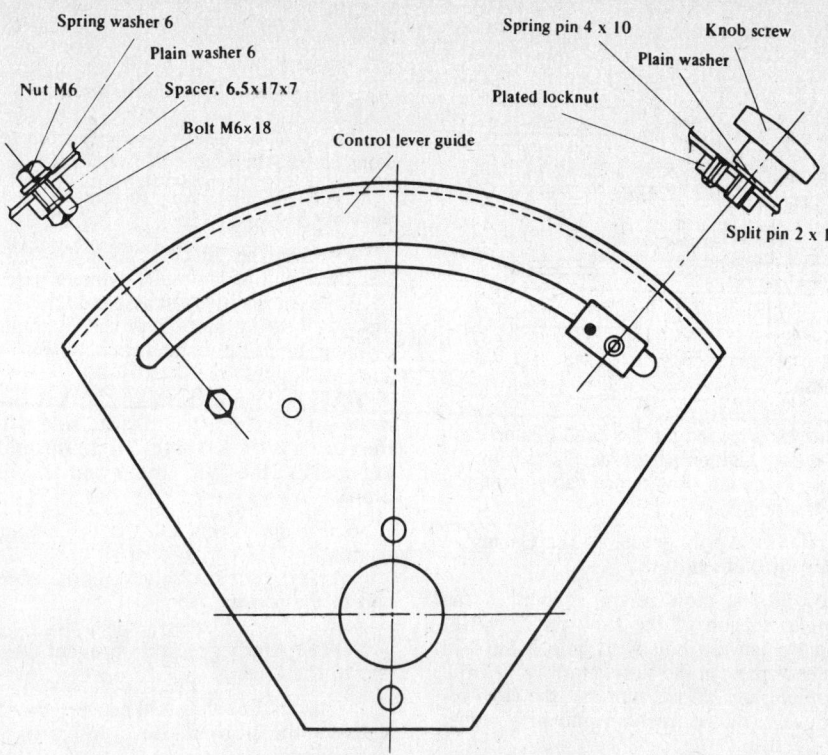

Control lever guide assembly

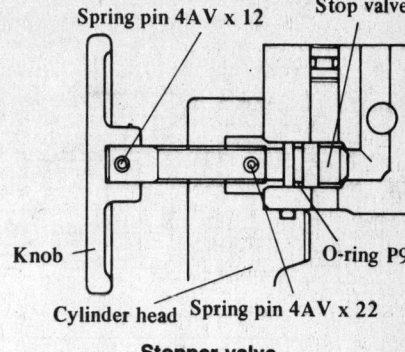

Stopper valve

left end of the control bracket, put the feedback lever into the end of the feedback shaft, drive in the spring pin 4AVx25 mm, and secure the feedback lever B on the feedback shaft; also secure the feedback shaft on the control bracket. Apply grease to the surfaces of the feedback shaft and control bracket which move and rub against each other. Make sure that the feedback shaft turns smoothly. When driving in the pin, make sure that the relative positions of the feedback lever and the feedback shaft are correct.

3. The control lever shaft assembly is positioned by driving the spring pin 4AVx10 into the control bracket. Make sure that the relative positions of the feedback lever A and feedback lever B are correct. Apply grease to the surfaces of those components that move against each other, as mentioned earlier.

4. Secure the control lever guide assembly to the control bracket, using spring washers (6) and nuts M6. Tightening torque: 5.8-8.7 ft.lb.

5. Put the control lever assembly into the control shaft and secure by driving in the spring pin 4AWx18 mm.

NOTE: In driving in the pin, make sure that the relative positions of the lever and control lever shaft arm are correct.

6. Insert the friction plate between the control lever and the control lever guide, then insert the bolt M6x50 mm on which has been fitted the plain washer (6), the spring, the plain washer 6, and the friction plate, in that order; secure from the outside using the nut M6 and the locknut M6 to secure the control lever. Control lever operating force: 6.5-13 lb.

Stop Valve Retainer Assembly

Screw the stop bolt to the stop valve retainer and drive the spring pin 4AWx12 mm to secure the stop bolt.

NOTE: Make sure that the stop bolt turns smoothly.

Cylinder Head Assembly

1. Fit the O-ring P9 in the groove on the stop valve, and insert into the cylinder head. Apply grease to the O-ring to prevent damage to it.

2. Secure the stop valve retainer assembly to the cylinder head using the M8x25 mm bolts and spring washers (8), fitting it in from the upper part. Tightening torque: 17-22 ft.lb.

3. Insert the O-ring P12, insert the relief valve seat, and tighten the valve seat screw to secure the relief valve seat the the cylinder head. Apply grease to the O-ring to prevent damage to it.

4. Insert in the upper part of the above the relief valve holder, the relief valve spring, the spring seat, and the nozzle adjusting plate, and fit and tighten the relief valve that has been fitted with the copper packing (20). Increase or reduce the number of nozzle spring adjusting plates to adjust the relief pressure to 2,060 psi. Tightening torque for relief valve spring: 87-108 ft.lb.

5. Fit the copper packing (18) on the plug with head M18, and fit and tighten in

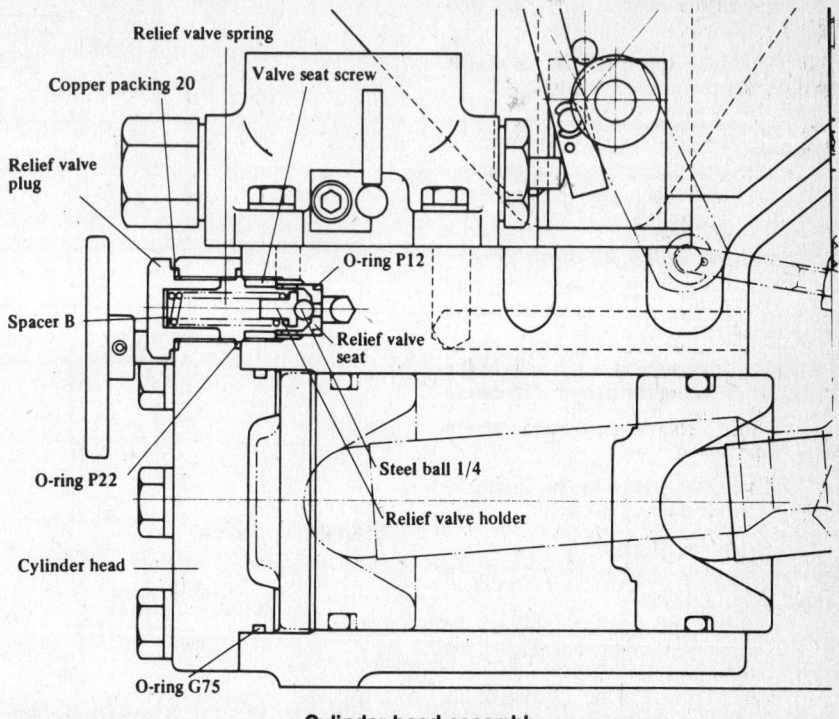

Cylinder head assembly

Yanmar

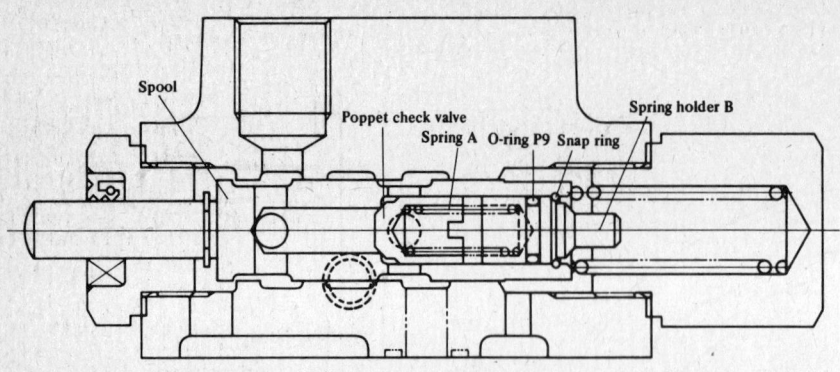

Spool assembly

cylinder head. Tightening torque: 72-94 ft.lb.

6. Fit the copper packing (10) on the plug with hexagonal head M10, and fit and tighten in cylinder head. Tightening torque: 33-43 ft.lb.

7. Fit the O-ring P5 on the blind cover, and press into port at upper part of cylinder head.

NOTE: Apply grease to the O-ring in order to prevent damage to it.

8. Fit the O-ring G60 into the groove in the cylinder head shaft section.

9. Fit the O-ring S14 into the groove on the inner surface of the cylinder head.

Spool Assembly

Press into the upper part of the above, an assembly comprising the poppet valve fitted with spring A, and O-ring P9. Fit in the snapring to prevent the poppet valve from dropping out.

NOTE: Apply grease to the O-ring in order to prevent damage to it.

Pushrod Assembly

1. Press-fit the oil seal in the pushrod holder.

NOTE: Make sure that the oil seal is fitted facing the correct direction.

2. Fit the E-shaped retainer ring (7) on the pushrod.

3. Insert the above into the pushrod holder. Make sure there is no peeling at the lip section. Apply grease.

Slow Return Valve Assembly

1. Drive the spring pin, 4x40 mm, into the slow return valve and encase the spring pin with the vinyl tubing, 4x30 mm.

NOTE: Make sure that the split of the spring pin is facing the proper direction.

2. Fit the O-ring P7 in the slow return valve groove.

NOTE: Apply grease to the O-ring in order to avoid damage to it.

Control Valve Assembly

1. Fit spool assembly in position. It should move freely.

2. Fit the copper packing (24) on the push rod assembly, and tighten. Tightening torque: 9-12 ft.lb.

3. Fit the spring B in the aperture of the spring holder and the copper packing (24) onto the threaded portion, and tighten each of them. Tightening torque: 9-12 ft.lb.

4. Place the slow return valve assembly in position.

NOTE: Apply grease to the O-ring to prevent damage to it.

5. Fit the slow return support in the groove section of the slow return valve, and tighten the bolt with hole M8x14 in order to prevent the slow return valve from dropping out. Make sure that the slow return valve moves freely. Tightening torque: 17-22 ft.lb.

6. Fit the O-ring P12 in the O-ring groove at the bottom.

Hydraulic Cylinder Case Assembly

1. Clean the hydraulic cylinder case thoroughly.

2. Press-fit the wound bushings 50x55x40 mm in the left and right sides. Make sure that the difference in height between the end face of the boss and the end face of the wound bushings conforms to the figures indicated in the illustration. After press-fitting them into position, make sure that the inner diameter of the wound bushings conforms to the figures indicated in the illustration.

3. Insert the piston assembly from the front part of the hydraulic cylinder case.

NOTE: Apply grease to the O-ring to prevent damage to it.

4. Insert the lift crank assembly from the lower part of hydraulic cylinder case.

5. Insert the lift shaft assembly through the side bearing section of the hydraulic cylinder, and fit it on the lift crank spline.

NOTE: Do not mistake the left and right side of the lift shaft (the side with the spring is the left side). Align the mating marks (the "O" mark) on the lift crank.

6. Fit the O-ring G50 in the bearing section.

7. Fit the lift shaft sleeve in the upper part of the above.

NOTE: Apply grease to prevent damage to the O-ring.

8. Insert the spline seal between the lift shaft and the lift shaft sleeve. Insert firmly into position.

9. Fit lift arm assembly onto lift arm shaft splines. Left side: Fit lift arm assembly, feedback arm assembly, and spring washer (12), and secure with bolt M12x25 mm. Right side: Fit lift arm assembly, washer, and spring washer, and secure with bolt M12x25 mm.

a. Align mating marks of lift arm and lift shaft (the "O" marks)

b. Make sure that the lift arm can be raised and lowered with little effort. (Force required to raise the lift arm

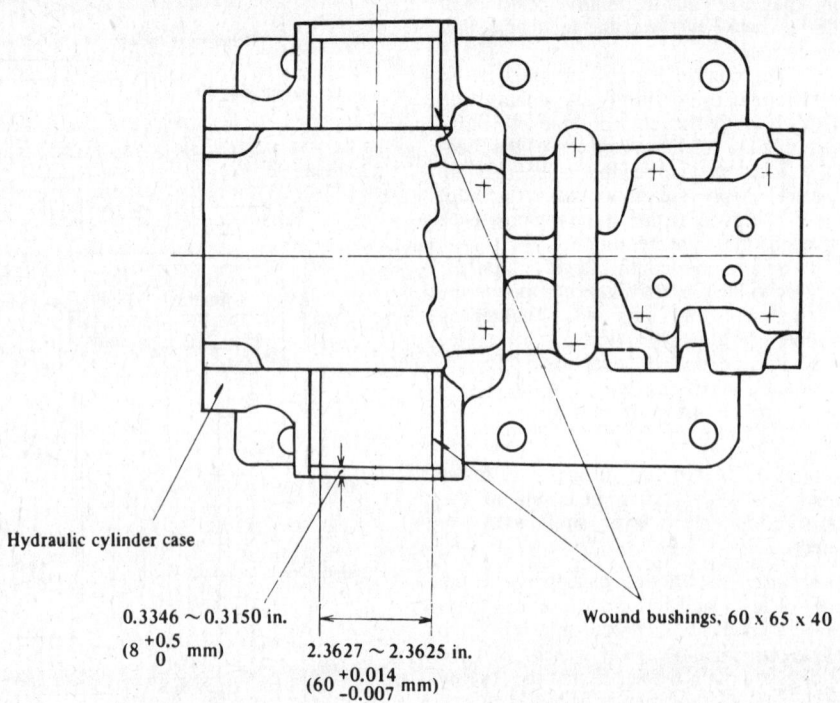

Hydraulic cylinder case assembly

from the horizontal should be less than 17.5 lb. at the tip of the arm).

c. When inserting the feedback arm assembly, align the hole of the feedback arm with the spring pin of the lift arm.

d. The washer should be fitted in position with the curved surface facing outwards.

e. Tightening torque: 58-72 ft.lb.

10. Fit the cylinder head assembly in the opening at the forward part of the hydraulic cylinder case, and secure the cylinder head assembly by fitting and tightening using bolts M12x35 mm fitted with the spring washers (12). Apply grease to prevent damage to the O-ring. Tightening torque: 58-72 ft.lb.

11. Mount the control valve assembly upon the control valve mount on the upper part of the hydraulic cylinder case, inserting a valve packing in between, and secure it in position using bolts M8x25 mm fitted with spring washers (8). Make sure that the O-ring P12 is fitted correctly in the O-ring groove. Tightening torque: 17-22 ft.lb.

12. Secure the top link hinge to the port at the rear of the hydraulic cylinder case using bolts M12x30 mm fitted with spring washers and inserting a hinge packing in between. Coat the hinge packing with liquid sealant. Tightening torque: 58-72 ft.lb.

13. The port on the right side of the hydraulic cylinder case is plugged with the M10 plug, with the copper packing (10) fitted on the plug. Tightening torque: 33-43 ft.lb.

14. The control bracket assembly is fitted and secured to the control bracket mount on the upper part of the hydraulic cylinder case, with spacer A in between, using the bolts M8x22 mm and M8x28 mm, with the spring washers 8 fitted on the bolts.

a. Apply grease between the feedback lever A and the pushrod.

b. The bolts M8x28 mm are secured on the right side together with the shaft hinge.

c. Tightening torque: 17x22 ft.lb.

15. Spacer B is inserted between the bracket stay and the control bracket, and the bolt M8x22 mm, the spring washer (8), and nut M8 are used to secure the bracket stay in position. The hole on the bracket stay is lined up with the transmission case mounting hole above the hydraulic cylinder case. Tightening torque: 17x22 ft.lb.

16. Couple the adjuster of the feedback rod and the feedback shaft, insert the plain washer (8), and secure with the E-shaped retainer ring (6). Apply grease to the connection.

17. Screw in the air breather into the threaded hole on the upper part of the hydraulic cylinder case. Coat the threaded portion with thread lock adhesive. The curved portion should face ahead.

18. Screw the bolt M8x35 mm into the threaded hole on the upper part of the hydraulic cylinder case. Screw in until bolt strikes bottom. Tightening torque: 17x22 ft.lb.

19. Attach the return spring between the bolt mentioned above and the head of the split pin on the feedback lever A.

Engine

The following engines are used:
135 series: 2T73A
155 series: 2TR13A
195 series: 2T84A
240 series: 2TR20A-X
330 series: 3T84A

All are diesel engines, 4-cycle configuration, overhead valve design and all are 2-cylinder except the 3-cylinder 3T84A.

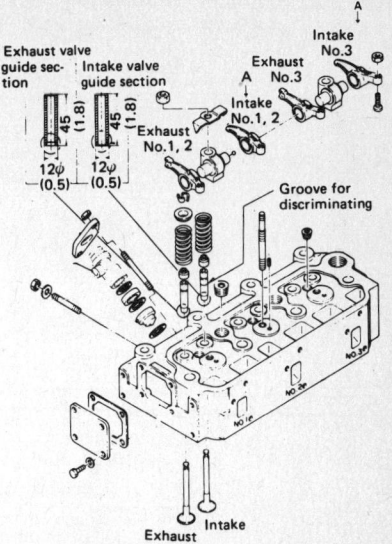

Cylinder head components

Cylinder Head

REMOVAL

1. Drain lube oil.
2. Remove fan belt.
3. Detach generator and starter motor.
4. Remove air cleaner and rocker arm cover.
5. Detach fuel and lube oil pipes.
6. Detach rocker arm assembly.
7. Remove push rods.
8. Remove cylinder head nuts. (Do not forget the nut in the intake manifold.)
9. Remove cylinder head gasket.

INSPECTION

Use the following table to check the head components:

CYLINDER HEAD COMPONENTS INSPECTION DIMENSIONS

Item	Standard Value	Replacement Limit	Testing Equipment
Inner diameter of valve guides (after assembly)	0.3159-0.3154 ($8^{+0.025}_{+0.010}$)	0.3181 (8.08)	Wire gauge
Outer diameter of valve stems	0.3138-0.3134 ($8^{-0.030}_{-0.040}$)	0.3110 (7.9)	Micrometer
Clearance between valve guide and valve stem	0.0018-0.0028 (0.045-0.070)	0.0059 (0.15)	

Valve seat width
  Intake
  Exhaust
	0.0835 (2.12)	0.0984 (2.5)	Calipers

Amount of valve sinkage	—	0.0197 (0.5)	Calipers

Valve springs
  Inclination
  Free length
	—
	1.5748 (40)
		2 degrees
		(39.7)
			Calipers

Valve seat angle, intake/exhaust	90 degrees	—	

Unit: Inch (mm)

Yanmar

INSTALLATION

1. Fit gasket, taking care to see that right side is up (side with stripes should be visible).
2. Install push rods.
3. Install cylinder head. Torque the bolts to 126-134 ft.lb. in a circular pattern starting at the center.
4. Install rocker arm assembly.
5. Install lube oil pipe.

VALVE CLEARANCE ADJUSTMENT

1. No. 1 cylinder (flywheel side cylinder) is set to the TDC where there is compression.
(TDC is the point where when the marking marks of the crankshaft/V-pulley are aligned with the pointer of the gear case, neither intake nor exhaust valves move.)
2. The adjusting screw is loosened, clearance is adjusted to 0.0059 in. (for both intake and exhaust), and after adjusted, tightened once again.
3. The valves for No. 2 cylinder are also adjusted to clearances of 0.0059 in. after turning the engine exactly 180° from the TDC for No. 1 cylinder where there is compression.

DECOMPRESSION STROKE ADJUSTMENT

1. Set the decompression lever at engine-operating position.
2. Set the piston at top (TDC) as the inlet and exhaust valves are closed.
3. Loosen the adjusting screw of the decompression lever shaft.
4. Turn the adjusting screw clockwise until the tip of it touches the exhaust valve arm.
5. Then turn the adjusting screw again 360° and lock it by locknut.
The amount of decompression lift at this time will be about 0.0314 in.
6. Repeat the same adjustment for each.

Gear Case

DISASSEMBLY

1. Remove fuel injection pump room cover; detach governor linkage.
2. Remove crankshaft pulley, using the gear puller.
3. Detach gear case.

NOTE: The hydraulic pump and tachometer gear unit are not detached from the gear case but are removed as a unit.
When assembling, remove the hydraulic pump and tachometer gear unit to make fitting easier.

ASSEMBLY

1. Fit in gear case while leading regulator spring and governor linkage towards fuel injection pump compartment.
2. Coat oil seal with lube oil, and push-fit crankshaft V-pulley, taking care not to damage lip section of oil seal.

3. Install lube oil pressure control valve and lube oil filter.
4. Install crankshaft pulley.

Pistons and Connecting Rods

REMOVAL

1. Remove the head.
2. Turn cylinder block on its side (with the fuel pump below.)
3. Remove under cover of cylinder block.
4. Detach lube oil inlet pipe.
5. Remove rod bolts.

---- CAUTION ----
Be careful not to drop the crankpin bearings.

6. Remove piston assembly.

---- CAUTION ----
For later reference clearly mark which is the No. 1 piston and which is the No. 2 piston.

DISASSEMBLY

1. Detach the two piston pin retaining rings;
2. Heat up the piston to 176°-212°F.
3. Remove the piston pin by tapping it with a lead bar and a hammer.

ASSEMBLY

1. Fit in one of the piston pin retaining rings;
2. Heat up the piston to 176°-212°F.
3. Fit the piston pin in position.
4. Fit in the remaining piston pin retaining ring.

INSPECTION

Measure the outer diameter of the piston at the piston skirt section at right angles to the axis of the piston pin. Diameter should be 3.3029-3.3040 in.
2. Measure the clearance between the piston and the cylinder liner by comparing the outer diameter of the piston and the inner diameter of the cylinder liner. The clearance is the difference between the minimum clearance figure obtained and the outer diameter of the piston. Clearance should be 0.0030-0.0056 in.
3. To measure the piston ring end clearance, the cylinder liner is placed on a surface table, the ring is fitted in the skirt section, and after making sure that the ring is straight, the end clearance is measured with a thickness gauge. End clearance for all rings should be 0.0118-0.0197 in.
When the piston rings are being fitted in the piston, be sure that the marking on the ring is facing up.
4. To replace the small end bushings, the new bushings are press-fitted in position. In doing so, take care to see that there is no peeling or contraction of the bushings. Measure the inner diameters after the new bushings have been press-fitted in po-

sition. If the diameter is too small, it should be corrected, using a reamer. ID should be 1.826-1.821 in. When press-fitting new bushings into position, be sure to align the oil passages of the bushings with those of the connecting rod.

INSTALLATION

Installation is the reverse of removal. Torque the connecting rod caps to 33-36 ft.lb. Stagger the ring gaps 90° apart.

Crankshaft

REMOVAL

1. Remove the pistons.
2. Remove the main bearing caps.
3. Lift out the crankshaft.

INSPECTION

See the table below for determining component replacement.

INSTALLATION

Installation is the reverse of removal. Torque the main bearing caps to 22-29 ft.lb.

FUEL SYSTEM

Injection Pump

DISASSEMBLY

Disassemble the pump by removing parts in the following order:
1. Plunger guide roller pin
2. Plunger guide roller
3. Plunger guide
4. Plunger and plunger spring retainer
5. Plunger position adjusting shim
6. Plunger spring
7. Plunger spring retainer
8. Fuel control pinion
9. Delivery valve holder
10. O-ring
11. Delivery valve spring
12. Delivery valve
13. Delivery valve seat (including packing)
14. Plunger barrel
15. Plunger barrel packing
    [No. 2 cylinder]
16. Fuel control pinion screw (check scribed line)
17. Plunger guide roller pin
18. Fuel control pinion
19. Fuel control pinion
20. Rack

ASSEMBLY

Assemble the parts in the following order:
1. Plunger barrel packing
2. Plunger barrel
3. Delivery valve seat and packing
4. Delivery valve
5. Delivery valve holder O-ring

Yanmar

CRANKSHAFT INSPECTION DIMENSIONS

Item	Standard Value	Replacement Limit	Testing Equipment
Outer diameter of crank pin	2.1246-2.1240 ($54^{-0.036}_{-0.050}$)	2.1224 (53.91)	Micrometer
Outer diameter of crank journal	2.7545-2.7539 ($70^{-0.036}_{-0.050}$)	2.7520 (69.9)	Micrometer
Distortion of crankshaft	Less than 0.0006 (0.015)	0.0059 (0.15)	Dial gauge
Inner diameter of main bushings	2.7559-2.7578 ($70^{+0.0419}_{+0}$)	2.7598 (70.1)	Cylinder gauge
Clearance between outer diameter of crank journal and inner diameter of main bushings	0.0014-0.0039 (0.036-0.099)	0.0079 (0.2)	
Thickness of thrust bearings	0.1161-0.1142 ($2.95^{0}_{-0.05}$)	0.1083 (2.75)	Micrometer
Crankshaft end play	0.004-0.008 (0.1-0.2)	0.0157 (0.4)	Dial gauge
Roundness of main bushings	0.0004 (0.01)		

Unit: Inch (mm)

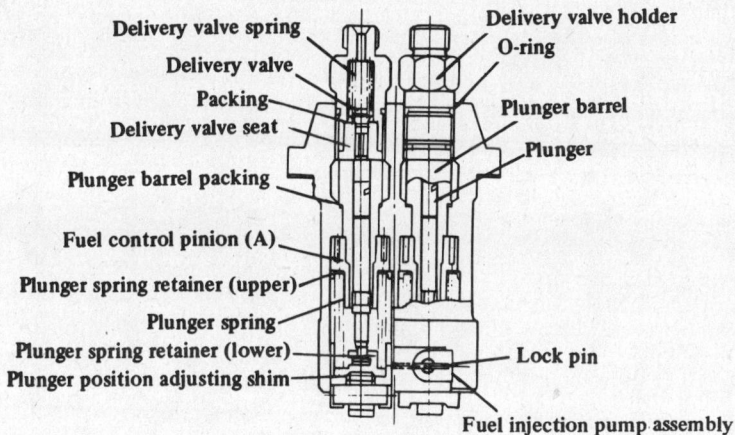

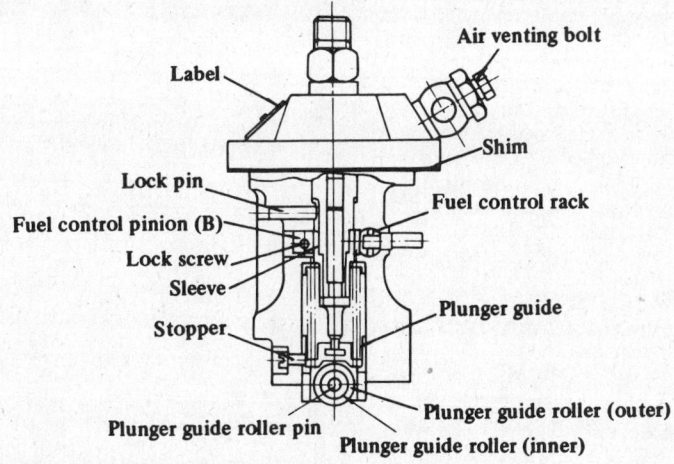

Injection pump components

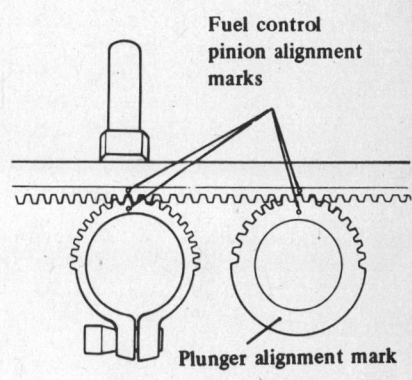

Rack alignment marks

Tightening torque: 30-32.6 ft.lb.
   6. Delivery valve holder
   7. Rack. Make sure that rack is fitted in correct direction.
   8. Fuel control pinion. Line up alignment marks.
   9. Plunger spring retainer. Check which side should be up.
  10. Plunger spring
  11. Plunger. Line up alignment marks.
  12. Plunger spring lower retainer
  13. Plunger position adjusting shim. Adjust the plunger position to close the port of barrel by the shim(s) in which pre-lift volume is 0.098 in.
  14. Plunger guide
  15. Plunger guide roller
     [No. 2 cylinder]
  16. Fuel control pinion. Line up alignment marks on rack and control pinion.
  17. Plunger barrel
  18. Insert control pinion of fuel control sleeve. Line up scribed lines on control pinion and sleeve.

Yanmar

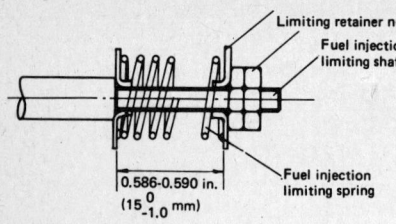

Limiting shaft adjustment

19. Secure plunger guide roller pin

ADJUSTMENT

Limiting Shaft

1. Remove the fuel pump chamber cover.
2. Remove the cap nut and loosen the locknut.
Turn the injection control to the right (to screw in the shaft).
3. Move the governor lever as far as it will go towards the gears (to the left).
4. Loosen the governor linkage adjustment screw, move the governor linkage to the maximum output side (towards the fuel pump side), and tighten the governor linkage adjustment screw.
5. Turn the injection control shaft to the left (so that the shaft emerges from the gear case side).
6. Lock the control shaft when the injection control strikes the governor lever, the rack of the fuel pump starts to move to the right (in the deceleration direction), and when the center of the punched mark on the rack is lined up with the datum level of the pump.
7. Restore the fuel pump chamber cover and the lube oil filler port cover to their respective positions.

Injection Timing

1. The No. 1 cylinder is set to the TDC position at which there is compression.
2. The flywheel is moved back and forth in small increments.
3. The flywheel is stopped at the very moment that fuel is discharged from the delivery valve.
4. The angle between the stamped mark on the crankshaft V-pulley and the indicating pointer on the gear case at the time fuel is discharged is checked.
5. Correct fuel injection delivery timing is when the stamped marking and the pointer are lined up, which is 22° degrees before TDC.
6. If the fuel injection timing is lagging,

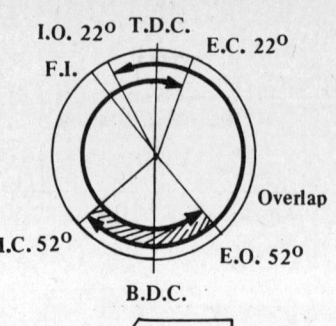

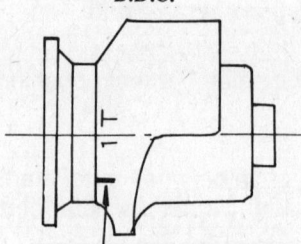

This stamped marking shows the injection timing.

Injection timing adjustment

the timing is adjusted by removing the shim(s), and if it is ahead, add the shim(s).
Every 0.0039 in. shim provides approximately 1° of variation.

Engine Overhaul

INDEX

ROUTINE CARE, MAINTENANCE, STORAGE, AND OPERATING PRECAUTIONS
- **Operating Precautions** 860
- **Routine Maintenance** 861
- **Storage** 862
- **Troubleshooting** 862

BRIGGS AND STRATTON
- **Engine Identification** 864
- **Maintenance** 865
- **Lubrication** 866
- **Tune-Up** 867
  - Spark Plugs 867
  - Breaker Points 867
  - Mixture Adjustments 873
  - Governor Adjustments 874
  - Choke Adjustments 874
  - Compression Checks 875
- **Fuel System** 875
  - General 875
  - Pulsa-Jet and Vacu-Jet Series 82000 & 92000 only 877
  - All Pulsa-Jet exc. 82000 & 92000 878
  - All Vacu-Jet exc. 82000 & 92000 878
  - Two-piece Flo-Jet 879
  - One-piece Flo-Jet 880
  - Governors 881
- **Engine Overhaul** 882
  - Cylinder Head and Valves 882-885
  - Pistons, Rings and Connecting Rods 885
  - Crankshaft and Camshaft 887
  - Cylinders 889
- **Oiling System** 890

KOHLER
- **Engine Identification** 891
- **Maintenance** 892
- **Tune-Up** 893
  - Spark Plugs 893
  - Breaker Points 893
  - Mixture Adjustments 894
  - Governor Adjustments 895
  - Choke Adjustments 896
  - Valve Adjustments 896
  - Compression Check 896
- **Fuel System** 896
  - Carburetors 896
  - Fuel Pump 897
- **Engine Overhaul** 897
  - Disassembly 898
  - Cylinder Block 898
  - Crankshaft 899
  - Connecting Rod 899
  - Pistons and Rings 899
  - Cylinder Head and Valves 900
  - Assembly 901

ONAN
- **Engine Identification** 905
- **Maintenance** 905
- **Tune-Up** 906
  - Spark Plugs 906
  - Breaker Points 906
  - Ignition Timing 906
  - Mixture Adjustments 907
  - Governor Adjustments 908
  - Choke Adjustments 909
  - Valve Adjustment 909
  - Compression Test 909
- **Fuel System** 910
  - Carburetors 910
  - Mechanical Fuel Pumps 912
  - Electric Fuel Pumps 913
  - Governors 913
- **Engine Overhaul** 914
  - Valves 914
  - Gear Cover 915
  - Timing Gears 916
  - Piston and Rings 916
  - Connecting Rods 917
  - Block 917
  - Crankshaft and Bearings 918
- **Lubricating System** 918

WISCONSIN 921
- **Engine Identification** 921
- **Maintenance** 921
- **Tune-Up** 922
  - Spark Plugs 923
  - Breaker Points 923
  - Ignition Timing 923
  - Mixture Adjustments 924
  - Governor Adjustments 925
  - Compression Check 925
- **Fuel System** 925
  - Carburetors 925
  - Fuel Pump 928
  - Governor 928
  - Fuel Tank 928
- **Engine Overhaul** 928
  - Cylinder Head 929
  - Valves 929
  - Flywheel 930
  - Gear Cover 930
  - Piston, Rings and Connecting Rod 931
  - Camshaft and Tappets 933
  - Breaker Push Pin 933
  - Crankshaft 933
- **Lubrication System** 934

TECUMSEH 934
- **Engine Identification** 934
- **Maintenance** 934
- **Tune-Up** 936
  - Spark Plugs 936
  - Breaker Points 936
  - Ignition Timing 936
  - Mixture Adjustments 936
  - Governor Adjustments 937
  - Compression Check 937
- **Fuel System** 937
  - Carburetor, General 937
  - Float Feed Carburetor 938
  - Tecumseh Automatic, Non-adjustable Float Feed Carburetor 940
  - Walbro and Tillotson Float Feed Carburetors 940
  - Tillotson E Float Feed Carburetor 941
  - Walbro Carburetors for V80, VM80, H80 and HM80 Engines 941
  - Diaphragm Carburetors 941
  - Craftsman Carburetors 942
  - Governors 944
- **Engine Overhaul** 945
  - Timing Gears 945
  - Crankshaft 945
  - Pistons 945
  - Piston Rings 945
  - Connecting Rods 946
  - Camshaft 946
  - Valve Springs 947
  - Valve Lifters 947
  - Cylinder Reboring 948
  - Bearings 949
- **Lubrication System** 952

Engine Overhaul
ROUTINE CARE, MAINTENANCE, STORAGE AND OPERATING PRECAUTIONS

OPERATING PRECAUTIONS

General

1. Before starting the engine, make sure it is *full* of the proper grade of lubricating oil. Some sort of record of the viscosity of the oil in use should be kept so that the proper grade may be added when necessary and the oil may be changed to keep the viscosity in conformance with recommendations for the outside air temperature range that is prevalent. If the engine is run in hot weather with an oil of too low a viscosity in the crankcase, very serious damage can result. The same sort of damage (from overheating) can occur if the oil level is too low. Remember that all engines must consume some oil to run properly and that, even if the engine is in continuous service, it should be stopped every few hours for an oil level check, and the crankcase refilled.

In the case of two-stroke engines, a recommended two-stroke oil must be used, and it must be mixed in the proportions recommended *for that engine* by the manufacturer. If several two-strokes using different fuel mixes are in use at the same location, each fuel container should be marked so that only the mix recommended for each engine is used in it. Do not simply use a pre-mixed batch of fuel—there is no standard mix. Make sure what you're using is in the exact proportion recommended by the manufacturer. Also, fuel and oil must be uniformly mixed. Usually, this is done by filling the container about 25% of the way with gas, pouring in the oil, and then completing the filling job. After this, the container should be shaken vigorously with the cap applied tightly to complete the mixing process. When using fuel that has been setting for several weeks, shake up the can before refilling the engine fuel tank to ensure uniform mixing.

2. Check that all cooling system air inlets are clear before starting the engine, and check occasionally to make sure nothing has clogged the air intake, especially if a screen is used. Check the condition of fins, fan blades, and thermostatic airflow controls as described above in the description of air cooling systems, periodically.

3. Watch the engine during operation to ensure it is running smoothly at governed speed. If operation is sluggish, stop it immediately and check for possible overloading, overheating, or lack of oil or two stroke lubrication. If the engine lacks power and overheats, check fuel air mixture adjustments (especially on two-stroke engines) and ignition timing.

4. Check for exhaust smoke during operation. If the engine smokes, check first for excessive oil consumption. If this is not due to running too hot or with the crankcase too full, the crankcase breather or engine must be repaired. Smoke can also occur due to a very rich mixture. If necessary, service the air cleaner and check for any restrictions in the air intake. Make sure the choke is fully opened if it is a manual design, and that it is opening properly if it is an automatic design. Adjust the carburetor mixture, if necessary. Smoke may also accompany severe engine knock due to very advanced ignition timing. Check the timing if there is knock.

5. Make sure to use only clean, fresh fuel. Fuel can begin deteriorating only a month after it is purchased. Fuel containing water or rust or other dirt should be discarded. Use leaded or unleaded fuel as per manufacturer's recommendations. Make sure the fuel also meets octane requirements.

6. Set the throttle only slightly above idle in starting (unless this conflicts with specific recommendations for starting), and idle the engine for several minutes before putting it to work. This will allow oil to become thin enough to reach all moving parts of the engine before they begin carrying too much load. It's also a good idea to idle the engine for a minute or two before shutdown, to cool the hottest parts more gradually.

7. Make sure the engine remains tightly mounted, as vibration can severely damage the engine or other parts, or cause potentially dangerous mechanical damage.

Safety Precautions

1. Never refuel the engine unless it is stopped. Hot exhaust system parts or an electrical spark could ignite the fuel. Also, avoid spillage of fuel, especially when the engine is hot. If fuel spills, make sure it is completely removed before starting the engine. Check the engine occasionally for fuel leaks and repair them immediately. Also, keep ignition high tension wiring in top shape. Brittle insulation can crack, causing a spark which could ignite spilled fuel.

2. Keep all sources of ignition away from batteries, especially when they are being rapidly charged or caps are off. When installing jumper cables between batteries, remove caps and place a rag over open vent holes. Make positive connections first (make sure you connect positive to positive); make negative connections by first connecting the cable to the negative side of the good battery, and making the final connection to a bare spot on the frame of the piece of equipment with the dead battery.

3. Be careful not to come in contact with output terminals on an electric generator, which are often located externally. Remember that metal tools are excellent conductors, and that they too must be kept away from electrical terminals. Even if you do not become exposed to electrical shock via tools, they can serve as a conductor and become hot enough to cause serious burns. This can happen if they touch positive and negative terminals of a battery.

Many electric generators are used only in case of a failure in commercial power. In these cases, they are often connected into a circuit through a transfer switch. Remember that the transfer switch is energized even when the generator is not running. Remember, too, that the output terminals of a generator are hot once the generator is running and electrically excited—it does not need to be carrying a load.

4. Make sure that exhaust gases are properly vented. If you are working near an engine which is enclosed, leaks in the exhaust system can prove dangerous even when the bulk of the exhaust is being carried to the outside air.

5. If your small engine is utilized in a marine engine compartment, remember that the compartment must be thoroughly aired out before starting the engine, or explosion of accumulated fumes may result.

6. Remember that many pieces of equipment driven by small engines are too heavy to be safely carried by a single person. Often, even though the total weight may be within reasonable limits, you may strain yourself because of the difficulty of handling the unit's bulk. Get help!

7. Keep the operating area clean. It should be wiped clean of fluids which could catch fire, and kept free of debris which might be drawn into a fan or blower and thrown out at high speed. This, of course, includes removing debris from a lawn which is to be mowed with a rotary mower or blown clean by a rotary blower.

8. When working on any engine driven accessory, disconnect the spark plug wire to avoid possible accidental starting of the motor if it should be turned over.

9. Keep all safety guards tightly in place, and replace them should they become damaged. Always be fully aware of all moving parts and the possibility of coming into contact with them even if guards are in place.

10. Keep governors in good operating condition, and do not reset for a speed higher than the recommendation of the manufacturer of the equipment.

11. Periodically tighten mounting bolts of both the engine and any machinery driven off it, especially that which turns at crankshaft rpm, such as rotary blades.

Hot and Cold Weather Operation

There are a few things that must be done when a small engine is to be operated in hot or cold weather, that is, when temperatures are either below 30°F, or above 75°F.

Engine Overhaul

In hot weather, do the following:

1. Keep the cooling fins of the engine block clean and free of all obstructions. Remove all dirt, built up oil and grease, flaking paint and grass.
2. Air should be able to flow to and from the engine with no obstructions. Keep all fairing and cover openings free from obstructions.
3. During hot weather service, heavier weight oil should be used in the crankcase. Follow the manufacturer's recommendations as to the heaviest weight oil allowed in the crankcase.
4. Check the oil level each time the fuel tank is filled. An engine will use more oil in extremely hot weather.
5. Check the battery water level more frequently since, in hot weather, the water in the battery will evaporate more quickly.
6. Be on the lookout for vapor lock, which occurs within the carburetor.
7. Use regular grade gasoline rather than premium.
8. Use unleaded gasoline if possible.
9. The most important thing to remember is to keep the engine as clean as possible. Blow it off with compressed air or wash it as often as possible.

CAUTION
Wash the engine only after it has had sufficient time to cool down to ambient temperatures. Avoid getting water in or even near the carburetor intake opening.

Precautions for operating a small engine in cold weather are as follows:

1. A lightweight oil should be installed in the crankcase when operating in cold weather. Consult the manufacturer's recommendations.
2. If the engine is filled with summer weight oil, the engine should be moved to a warm (above 60°F) location and allowed to reach ambient temperature before starting. This is because a heavy summer weight oil will be even thicker at cold temperatures. So thick, in fact, that it will be unable to sufficiently lubricate the engine when it is first started and running. Damage could occur due to lack of lubrication.
3. Change the oil only after the engine has been operated long enough for operating temperatures to have been reached. Change the oil while the engine is still hot.
4. Use fresh gasoline. Fill the gas tank daily to prevent the formation of condensation in the tank and fuel lines.
5. Keep the battery in a fully charged condition, since cold weather infringes upon a battery's maximum current output capabilities.
6. If the engine is run only for short periods of time, have the battery charged every so often to ensure maximum power output when it is needed most.

ROUTINE MAINTENANCE

Care of a small engine is divided into the following five categories: lubrication; filter service; tune-up; carburetor overhaul and fuel pump repair/replacement; combustion chamber deposit removal and valve repair; complete overhaul. Routine maintenance consists of the first three categories. These are described below.

Lubrication

This category includes simple replenishment of lost fluids, and, in part, is the responsibility of the operator. Operators must be aware of not only the need to run the engine only when it is adequately lubricated, but of the need to cease operation and perform required maintenance, even if the actual work is done by a mechanic.

Before starting the engine, fill the crankcase and the air cleaner with the proper oil and fill the gasoline tank. Never try to fill the fuel tank of an engine that is running, and if the engine is still hot from running, allow it to cool down before refueling it.

Use a good grade, clean, fresh, lead free or leaded regular grade automotive gasoline. The use of highly leaded gasoline (high octane) should be avoided, as it causes deposits on the valves and valve seats, spark plugs, and the cylinder head, thus shortening engine life.

Any high quality detergent oil having the American Petroleum Institute classification "For Service SC or SD or MS" can be used. Detergent oils keep the engine cleaner by retarding the formation of gum and varnish deposits. Do not use any oil additives. In the summer (above 40°F) use SAE 30 weight oil. If that is not available, use SAE 10W-30 or SAE 10W-40 weight oil. In the winter (under 40°F) use SAE 5W-20 or SAE 5W-30 weight oil. If neither of these is available, use SAE 10W or SAE 10W-30 weight oil. If the engine is operated in ambient temperatures that are below 0°F, use SAE 10W or SAE 10W-30 weight oil diluted 10% with kerosene.

The oil should be changed after each 25 hours of service or engine operation, and more often under dirty or dusty operating conditions, or as the manufacturer specifies. In normal running of any engine, small particles of metal from the cylinder walls, pistons and bearings will gradually work into the oil. Dust particles from the air also get into the oil. If the oil is not changed regularly, these foreign particles cause increased friction and a grinding action which shorten the life of the engine. Fresh oil also assists in cooling the engine, for old oil gradually becomes thick and cannot dissipate the heat fast enough. Old oil will also gradually lose its lubricating properties.

In two-stroke engines, lubrication consists of ensuring the engine runs on a mix of fuel and oil which is in the proper proportion, and that the oil used meets the specifications of the manufacturer. Since running a two-stroke engine on straight gasoline or an improper mix is very much like operating a four stroke engine without oil, it must be seen that proper preparation of fuel/oil mix is literally a life and death matter for the engine—failure to provide the proper mix may result in immediate engine failure. Always observe the following points:

1. Use an oil specifically designed for two-stroke engines, and of the viscosity recommended by the manufacturer. The wrong oil may solidify in many different parts of the engine, may leave ash deposits in the combustion chamber, or foul the spark plug. Don't forget that the oil must not only lubricate well, but burn well.
2. Measure the oil accurately into the fuel container in the exact proportion recommended. Mix thoroughly according to the directions on the can (see "Two-Stroke Lubrication" above). Do not simply use a standard, pre-mixed fuel unless you can determine that it is in the correct proportion. Where engines requiring different mixes are used at a common site, label fuel cans with the fuel/oil mixture ratio contained.
3. Remember that available lubrication in a two-stroke also depends on fuel/air ratio. Ensure that carburetors are properly adjusted, and that there are no air leaks so that sufficient lubrication will always be available. Watch, too, for clogged air cleaners, partially closed chokes, or too rich an adjustment, as these will lead to plug fouling. Correct immediately any conditions causing four-cycling or misfire, as gasoline may dilute oil lubing pistons and rings under these conditions.

Filter Service

Air filter service is usually performed at the time of oil change, but may be performed at a longer interval—check specific recommendations. The air cleaner must be serviced much more frequently if the engine is operated in dusty conditions. Check specific recommendations here, also.

Oil type air cleaners require draining of old oil; a thorough cleaning of oil bowl and element with solvent; oiling of the element; and refilling of bowl *to the specified level* with new oil of the type used in the engine.

Most dry element air cleaners require that the element be replaced—they usually cannot be cleaned with compressed air. Some also employ a swirl chamber to remove large dust particles before they reach the main element. This chamber must be thoroughly cleaned out and, in some cases, a dust catching bowl must be emptied and cleaned.

Other types of dry element type air cleaners may require cleaning in soap and water, thorough drying and, in some cases, oiling.

Fuel sediment bowls and strainers, or filters are used on many engines to ensure that the use of dirty fuel or the entrance of dirt into the gas tank will not cause dirt to get into the carburetor. Since only dirt that enters with the fuel or works its way into the tank reaches the filter, it should be obvious that the first step in fuel filter maintenance is the use of clean gas, and the second, proper maintenance of the tank filler cap and gasket. The filter is usually serviced at the same time the oil is changed

Engine Overhaul

or at twice that interval. Fuel tank valves are turned off, and the bowl or filter housing is removed and cleaned. Strainers are cleaned in solvent and dried, and pleated paper type elements are replaced.

Oil filters are replaced at every oil change or every other change—consult specific recommendations. Throw-away type filters usually require the use of a strap wrench for ready removal. Wipe the filter base clean, lubricate the seal on the filter with *clean* oil, and tighten only by hand, or the amount specified on the filter. In the case of cartridge type filters, clean the housing with solvent and dry. Make sure to replace seals both at the filter base and around the mounting bolt, as applicable.

Filters deserve the same consistency of attention to recommended service intervals as oil changes. Oil change intervals are determined by the ability of the filter to prolong the life of the oil directly in mind. If the filter is allowed to accumulate dirt to the point where it bypasses due to loss of oil pressure, the oil will be subjected to a much greater than normal amount of material to keep in suspension. This shortens the potential life of the oil drastically, greatly increases the chances of clogging engine oil passages, and may allow abrasive particles large enough to be trapped between moving parts to circulate with the oil. Remember, too, that operation in dusty areas can cause a filter to become clogged and bypassed very quickly. Follow manufacturer's recommendations for more frequent changes under these conditions.

Tune-Up

The following list of procedures is rather extensive for a simple tune-up. Normally one would just check the condition of the spark plug, points, condenser, and wiring, make the necessary adjustments to these components and the carburetor, maybe change the oil, and service the carburetor if needed.

However if the following is performed, you will either be sure that the engine is functioning properly or you will know what major repairs will be made. In other words the engine is going to run well or you will find the cause of any problems.

1. Remove the air cleaner and check for the proper servicing.
2. Check the oil level and drain the crankcase. Clean the fuel tank and lines if separate from the carburetor.
3. Remove the blower housing and inspect the rope, rewind assembly and starter clutch of the starter mechanism. Thoroughly clean the cooling fins with compressed air, if possible, and check that all control flaps operate freely.
4. Spin the flywheel to check compression. It should be spun in the direction opposite to normal rotation, and as rapidly as possible. A sharp rebound indicates good compression. Four-stroke engines may also be checked with a compression gauge in place of the spark plug—consult manufacturer's specifications in the individual repair section.
5. Remove the carburetor and disassemble and inspect it for wear or damage. Wash it in solvent, replace parts as necessary, and assemble. Set the initial adjustments.
6. Inspect the crossover tube or the intake elbow for damaged gaskets.
7. Check the governor blade, linkage, and spring for damage on wear; if it is mechanical, check the linkage adjustment.
8. Remove the flywheel and check for seal leakage, both on the flywheel and power take off sides. Check the flywheel key for wear and damage.
9. Remove the breaker cover and check for proper sealing.
10. Inspect the breaker points and condenser. Replace or clean and adjust them. Check the plunger or the cam. Lubricate the cam follower.
11. Check the coil and inspect all wires for breaks or damaged insulation. Be sure the lead wires do not touch the flywheel. Check the stop switch and the lead.
12. Replace the breaker cover, using sealer where the wires enter.
13. Install the flywheel and time the ignition if necessary. Set the air gap and check for ignition spark.
14. Remove the cylinder head, check the gasket, remove the spark plug, clean off the carbon, and inspect the valves for proper seating.
15. Replace the cylinder head, using a new gasket, torque it to the proper specification, and set the spark plug gap or replace the plug if necessary.
16. Replace the oil and fuel and check the muffler for restrictions or damage.
17. Adjust the remote control linkage and cable, if used, for correct operation.
18. Service the air cleaner and check the gaskets and element for damage.
19. Run the engine and adjust the idle mixture and high speed mixture of the carburetor.

STORAGE

If an engine is to be out of service for more than 30 days, the following steps should be performed:

1. Run the engine for 5 to 10 minutes until it is thoroughly warmed up to normal operating temperatures.
2. Turn off the fuel supply while the engine is still running, and continue running it until the engine stops from lack of fuel. This procedure removes all fuel from the carburetor.
3. Drain the oil from the crankcase while the engine is still warm.
4. Fill the crankcase with clean oil and tag the engine to indicate what weight oil was installed.
5. Remove the spark plug and squirt about an ounce of oil into the cylinder. Turn the engine over a few times to coat the cylinder wall, the top of the piston, and the head with a protective coating of oil. Reinstall the spark plug and tighten it to the proper torque.
6. Clean or replace the air cleaner. Refer to the manufacturer's recommendations.
7. Clean the governor linkage, making sure that it is in good working order and oiling all joints.
8. Plug the exhaust outlet and the fuel inlet openings. Use clean, lintless rags.
9. Remove the battery and store it in a cool place where there is no danger of freezing. Do not store any wet cell battery directly in contact with the ground or cement floor, as it will establish a ground and discharge itself. A completely discharged battery will never be able to be brought back to its original output capacity. Store the battery on a work bench or on blocks of wood on the floor.
10. Wipe off or wash the engine. Wash only after the engine has had time to cool down to ambient temperature and avoid getting water in the carburetor intake port.
11. Coat all parts that might rust with a light coating of oil. Paint all non-operating parts with a rust inhibiting paint.
12. Provide the entire unit with a suitable covering. Plastic is good where the application and removal of sunlight will not promote the formation of condensation under the plastic covering. If this is the case, use a covering that is able to "breathe," such as a canvas tarpaulin.

TROUBLESHOOTING

HOW TO GO ABOUT IT

Start with the simplest, most obvious causes first—many engine mechanics and operators have difficulty identifying trouble because they start out assuming everything that is obvious has already been checked. Check to see that there is fuel in the tank, and that it is clean, that the tank is properly vented, and that the fuel filter or sediment bowl is not full of dirt. Check to see that the spark plug wire is connected and that the spark plug is not fouled. If the cause of the trouble is not immediately obvious, use your basic knowledge of how the engine works. For example, if the engine runs fine but is very hard to start, you might conclude that the choke does not close, since its function is confined, mainly, to engine starting.

The guide below will point out many possible causes of the most basic problems. Find the "PROBLEM" which matches the

Engine Overhaul

engine's behavior, and then check out the possibilities listed under "CAUSES AND REMEDIES." Refer to the manufacturer's section which pertains to your engine, if necessary, in making repairs.

Troubleshooting Guide

PROBLEM: The engine does not start or is hard to start.
CAUSES AND REMEDIES:
1. The fuel tank is empty.
2. The fuel shut-off valve is closed; open it.
3. The fuel line is clogged. Remove the fuel line and clean it. Clean the carburetor, if necessary.
4. The fuel tank is not vented properly. Check the fuel tank cap vent to see if it is open.
5. There is water in the fuel supply. Drain the tank, clean the fuel lines and the carburetor, and dry the spark plug. Fill the tank with fresh fuel. Check the fuel supply before pouring it into the engine's fuel tank. Chances are it might be the source of the water.
6. The engine is over choked. Open the choke and throttle wide on manual choke engines. On engines with automatic chokes, close the throttle. Then, turn the engine over with several pulls of the starter rope. If engine does not start, set throttle to just above idle, close choke again, and again attempt to start the engine. If one or two pulls does not make engine fire, try cranking with the choke closed only half way. If engine still fails to start, remove the spark plug and dry it, and spin the engine over several times to clear excess fuel out of the engine. Replace the spark plug and perform the normal starting procedure. Over choking is most often due to *continued cranking* with the choke fully shut.
7. The carburetor is improperly adjusted; adjust it to the standard recommended preliminary settings. See the carburetor section.
8. Magneto wiring is loose or defective. Check the magneto wiring for shorts or grounds and repair it, if necessary.
9. No spark. Check for spark, and if there is none, check and, if necessary, replace the contact points, and set contact gap and timing. If there is still no spark, replace further magneto parts (especially coil and high tension wire) as necessary.
10. The spark plug is fouled. Remove, clean, and regap the spark plug.
11. The spark plug is damaged (cracked porcelain, bent electrodes etc.). Replace the spark plug.
12. Compression is poor. The head is loose or the gasket is leaking. Sticking or burned valves or worn piston rings could also be the cause. In any case, the engine will have to be disassembled and the cause of the problem corrected.

PROBLEM: The engine misses under load (if a two-stroke, it may "four-cycle.")
CAUSES AND REMEDIES:
1. The spark plug is fouled. Remove, clean, and regap the spark plug.
2. The spark plug is damaged. Replace the spark plug.
3. The spark plug is improperly gapped. Regap the spark plug to the proper gap.
4. The breaker points are pitted or improperly gapped. Replace the points, or set the gap.
5. The breaker point's breaker arm is sluggish. Clean and lubricate it.
6. The condenser is faulty. Replace it.
7. The carburetor is not adjusted properly. Adjust it.
8. The fuel system is partly clogged, or the fuel shut-off valve is partly closed. Open the valve and check the fuel filter/strainer, tank, lines, and carburetor for dirt. Clean all parts as necessary.
9. If the engine is a two-stroke, the exhaust ports may be clogged. Remove the exhaust manifold and inspect the ports. If they are clogged with carbon, clean them with a soft tool such as a wooden stick. Check also for bad crankshaft seals.
10. The valves are not adjusted properly. Adjust the valve clearance.
11. The valve springs are weak. Replace them.

PROBLEM: The engine knocks.
CAUSES AND REMEDIES:
1. The magneto is not timed correctly. Time the magneto.
2. The carburetor is not properly adjusted (may be too lean). Adjust the carburetor for best mixture.
3. The engine has overheated. Stop the engine and find the cause of overheating.
4. Carbon has built up in the combustion chamber, resulting in retention of excess heat and an increase in compression which causes pre-ignition. Remove the cylinder head, and remove the carbon from the head and top of the piston.
5. The connecting rod is loose or worn. Replace it.
6. The flywheel is loose. Check the flywheel key and key way and the end of the crankshaft. Replace any worn parts. Tighten the flywheel nut to the specified torque.
7. The cylinder is worn. Rebuild/replace parts as necessary.

PROBLEM: The engine vibrates excessively.
CAUSES AND REMEDIES:
1. The engine is not mounted securely to the equipment that it operates. Tighten any loose mounting bolts.
2. The equipment that the engine operates is not balanced. Check the equipment.
3. The crankshaft is bent. Replace the crankshaft.
4. The counter balance shaft is improperly timed (recent reassembly) or broken. Disassemble the crankcase, inspect, and replace or repair parts as necessary.

PROBLEM: The engine lacks power.
CAUSES AND REMEDIES:
1. The choke is partially closed. Open the choke.
2. The carburetor is not adjusted correctly. Adjust it.
3. The ignition is not timed correctly. Time the ignition.
4. There is a lack of lubrication or not enough oil in the crankcase. fill the crankcase to the correct level.
5. The air cleaner is fouled. Clean it.
6. The valves are not sealing. Do a valve job.
7. Ring seal is poor. Repair/replace rings, piston, or cylinder/cylinder liner.
8. If the engine is a two stroke, the exhaust ports may be clogged with carbon. Remove the exhaust manifold and inspect. Clean with a soft instrument such as a wooden stick, if dirty. Ports may clog frequently if the carburetor mixture is adjusted too rich, or if there is excessive oil or oil of the wrong type in the fuel.

PROBLEM: The engine operates erratically, surges, and runs unevenly.
CAUSES AND REMEDIES:
1. The fuel line is clogged. Unclog it.
2. The fuel tank cap vent is clogged. Open the vent hole.
3. There is water in the fuel. Drain the tank, the carburetor, and the fuel lines and refill with fresh gasoline.
4. The fuel pump is faulty. Check the operation of the fuel pump if so equipped.
5. The governor is improperly set or parts are sticking or binding. Set the governor and check for binding parts and correct them.
6. The carburetor is not adjusted properly. Adjust it.

PROBLEM: Engine overheats.
CAUSES AND REMEDIES:
1. The ignition is not timed properly. Time the engine's ignition.
2. The fuel mixture is too lean. Adjust the carburetor.
3. The air intake screen or cooling fins are clogged. Clean away any obstructions.
4. The engine is being operated without the blower housing or shrouds in place. Install the blower housing and shrouds.
5. The engine is operating under an excessive load. Reduce the load and check associated equipment.
6. The oil level is too high. Check the oil level and drain some out if necessary.
7. There is not enough oil in the crankcase. Check the oil level and adjust accordingly.
8. The oil in the crankcase is of too low a viscosity or is excessively contaminated with fuel (four stroke). If the engine is a two-stroke, check for adequate fuel/oil mix—oil must be mixed with the fuel in proper proportions and be fully mixed. Check condition of crankcase oil (four-stroke) and if it appears very dirty, or there is doubt about proper viscosity, replace it.
9. The valve tappet clearance is too close. Adjust the valves to the proper specification.
10. Carbon has built up in the combustion chamber. Remove the cylinder and clean the head and piston of all carbon.
11. An improper amount of oil is mixed with the fuel (two stroke engines only). Drain the fuel tank and fill with correct mixture.

Engine Overhaul

PROBLEM: The crankcase breather is passing oil (four stroke engines only).

CAUSES AND REMEDIES:

1. The crankcase is substantially overfilled with oil. Check oil level several minutes after engine has stopped. Wipe the dipstick clean before checking the level. If the crankcase is too full, drain oil as necessary until oil level is at or slightly below the upper mark.

2. The engine is being operated at too high rpm. Slow it down by adjusting the governor.

3. The oil fill cap or gasket is missing or damaged. Install a new cap and gasket and tighten it securely.

4. The breather mechanism is damaged. Replace the reed plate assembly.

5. The breather mechanism is dirty. Remove, clean, and replace it.

6. The drain hole in the breather is clogged. Clean the breather assembly and open the hole.

7. The piston ring gaps are aligned. Disassemble the engine and offset the ring gaps 90° from each other.

8. the breather is loose or the gaskets are leaking. Tighten the breather to the crankcase.

9. The rings are not seated properly or they are worn. Install new rings.

PROBLEM: The engine backfires.
CAUSES AND REMEDIES:

1. The carburetor is adjusted so the air/fuel mixture is too lean. Adjust the carburetor.

2. The ignition is not timed correctly. Time the engine.

3. The valves are sticking. Do a valve job.

BRIGGS AND STRATTON

ENGINE IDENTIFICATION

The Briggs and Stratton model designation system consists of up to a six digit number. It is possible to determine most of the important mechanical features of the engine by merely knowing the model number. An explanation of what each number means is given below.

1. The first one or two digits indicate the cubic inch displacement (cid).

2. The first digit after the displacement indicates the basic design series, relating to cylinder construction, ignition and general configuration.

3. The second digit after the displacement indicates the position of the crankshaft and the type of carburetor the engine has.

4. The third digit after the displacement indicates the type of bearings and whether or not the engine is equipped with a reduction gear or auxiliary drive.

5. The last digit indicates the type of starter.

The model identification plate is usually located on the air baffle surrounding the cylinder.

BRIGGS AND STRATTON MODEL NUMBERING SYSTEM

Cubic Inch Displacement	First Digit After Displacement Basic Design Series	Second Digit After Displacement Crankshaft, Carburetor Governor	Third Digit After Displacement Bearings, Reduction Gears and Auxiliary Drives	Fourth Digit After Displacement Type of Starter
6	0	0-	0-Plain Bearing	0-Without Starter
8	1	1-Horizontal Vacu-Jet	1-Flange Mounting Plain Bearing	1-Rope Starter
9	2			
10	3	2-Horizontal Pulsa-Jet	2-Ball Bearing	2-Rewind Starter
13	4			
14	5	3-Horizontal Flo-Jet (Pneumatic Governor)	3-Flange Mounting Ball Bearing	3-Electric-110 Volt, Gear Drive
17	6			
19	7	4-Horizontal Flo-Jet (Mechanical Governor)	4-	4-Elec. Starter-Generator-12 Volt, Belt Drive
20	8			
23	9			
24		5-Vertical Vacu-Jet	5-Gear Reduction (6 to 1)	5-Electric Starter Only-12 Volt, Gear Drive
30				
32		6-	6-Gear Reduction (6 to 1) Reverse Rotation	6-Wind-up Starter
		7-Vertical Flo-Jet	7-	7-Electric Starter, 12 Volt Gear Drive, with Alternator
		8-	8-Auxiliary Drive Perpendicular to Crankshaft	8-Vertical-pull Starter
		9-Vertical Pulsa-Jet	9-Auxiliary Drive Parallel to Crankshaft	

Engine Overhaul

GENERAL ENGINE SPECIFICATIONS
Aluminum Engines

Model	Bore Size (in.)	Horsepower
140000	2.750	6
170000, 171700	3.000	7
190000, 191700	3.000	8
251000	3.4375	10

GENERAL ENGINE SPECIFICATIONS
Cast Iron Engines

Model	Bore Size (in.)	Horsepower
19, 190000, 200000	3.000	8
23, 230000	3.000	9
243000	3.0625	10
300000	3.4375	13
320000	3.5625	16

MAINTENANCE

Air Cleaners

A properly serviced air cleaner protects the engine from dust particles that are in the air. When servicing an air cleaner, check the air cleaner mounting and gaskets for worn or damaged mating surfaces. Replace any worn or damaged parts to prevent dirt and dust from entering the engine through openings caused by improper sealing. Straighten or replace any bent mounting studs.

SERVICING

Oil Foam Air Cleaners

Clean and re-oil the air cleaner element every 25 hours of operation under normal operating conditions. The capacity of the oil-foam air cleaner is adequate for a full season's use without cleaning. Under very dusty conditions, clean the air cleaner every few hours of operation.

The oil-foam air cleaner is serviced in the following manner:
1. Remove the screw that holds the halves of the air cleaner shell together and retains it to the carburetor.
2. Remove the air cleaner carefully to prevent dirt from entering the carburetor.
3. Take the air cleaner apart (split the two halves).
4. Wash the foam in kerosene or liquid detergent and water to remove the dirt.
5. Wrap the foam in a clean cloth and squeeze it dry.
6. Saturate the foam in clean engine oil and squeeze it to remove the excess oil.
7. Assemble the air cleaner and fasten it to the carburetor with the attaching screw.

Oil Bath air Cleaner

Pour the old oil out of the bowl. Wash the element thoroughly in solvent and squeeze it dry. Clean the bowl and refill it with the same type of oil used in the crankcase.

Dry Element Air Cleaner

Remove the element of the air cleaner and tap (top and bottom) it on a flat surface or wash it in non-sudsing detergent and flush it from the inside until the water coming out is clear. After washing, air dry the element thoroughly before reinstalling it on the engine. NEVER OIL A DRY ELEMENT.

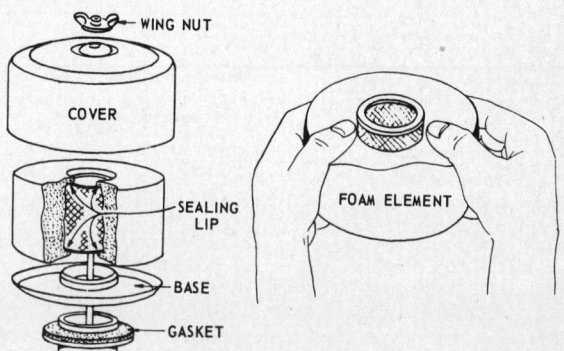
Oil foam air cleaner

Oil foam air cleaner

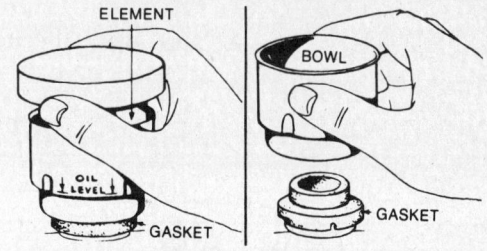
Oil bath air cleaner

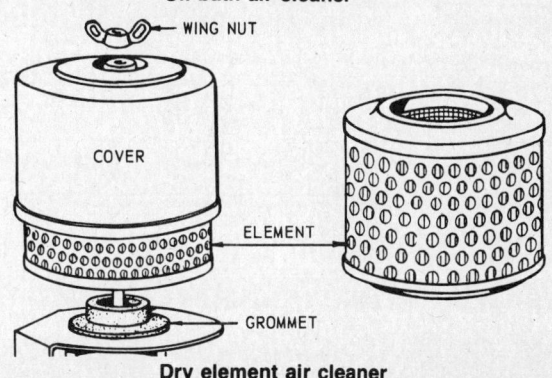
Dry element air cleaner

Engine Overhaul

Heavy Duty Air Cleaner

Clean and re-oil the foam pre-cleaner at three month intervals or every 25 hours, whichever comes first.

Clean the paper element every year or 100 hours, whichever comes first. Use the dry element procedure for cleaning the paper element of the heavy duty air cleaner.

Use the oil foam cleaning procedure to clean the foam sleeve of the heavy duty air cleaner.

If the engine is operated under very dusty conditions, clean the air cleaner more often.

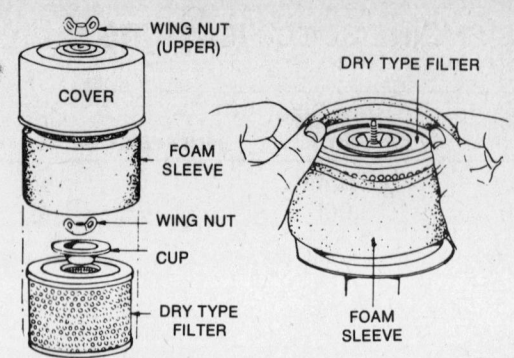

Briggs and Stratton heavy duty air cleaner

LUBRICATION

Oil and Fuel Recommendations

Briggs & Stratton recommends either a leaded regular grade gasoline or low-lead or unleaded fuel. Low-lead or unleaded fuel is preferable because of the reduction in deposits that results from its use, but its use is not required. Nor is it necessary to use leaded fuel occasionally in an engine which runs primarily on unleaded fuel, as is the case with some older automobile engines. Premium fuel is not required, as regular or unleaded will have sufficient knock resistance if the engine is in proper condition. The factory recommends that fuel be purchased in lots small enough to be used up in 30 days or less. When fuel is older than that, it can form gum and varnish, or may be improperly tailored to the prevailing temperature.

You should use a high quality detergent oil designated "For Service SC, SD, SE, or MS." Detergent oil is recommended because of its important ability to keep gum and varnish from clogging the lubrication system. Briggs & Stratton specifically recommends that no special oil additives be used.

Oil must be changed every 25 hours of operation. If the atmosphere in which the engine is operating is very dirty, oil changes should be made more frequently, as often as every 12 hours, if necessary. Oil should be changed after 5 hours of operation in the case of brand new engines. Drain engine oil when hot.

In hot weather, when under heavy load, or when brand new, engines may consume oil at a rate which will require you to refill the crankcase several times between oil changes. Check the oil level every hour or so until you can accurately estimate how long the engine can go between refills. To check oil level, stop the engine and allow it to sit for a couple of minutes, then remove the dipstick or filler cap. Fill the crankcase to the top of the filler pipe when there is no dipstick, or wipe the dipstick clean, reinsert it, and add oil as necessary until the level reaches the upper mark.

On cast iron engines with a gear reduction unit, crankcase and reduction gears are lubricated by a common oil supply. When draining crankcase, also remove drain plug in reduction unit.

On aluminum engines with reduction gear, a separate oil supply lubricates the gears, although the same type of oil used in the crankcase is used in the reduction gear cover. On these engines, remove the drain plug every fourth oil change (100 hours), then install the plug and refill. The level in the reduction gear cover must be checked during the refill operation by removing the level plug from the side of the gearcase, removing the filler plug, and then filling the case through the filler plug hole until oil runs out the level plug hole. Then, install both plugs.

On 6-1 gear reduction engines (models 6, 8, 8000, 10000, and 13000), no changes are required for the oil in the reduction gear case, but level must be checked and the case refilled, as described in the paragraph above, every 100 hours. Make sure the oil level plug (with screwdriver slot and no vent) is installed in the hole on the *side* of the case.

ENGINE OIL CAPACITY CHART

Basic Model Series	Capacity Pints
Aluminum	
6, 8, 9, 11 Cu. in. Vert. Crankshaft	1¼
6, 8, 9 Cu. in. Horiz. Crankshaft	1¼
10, 13 Cu. in. Vert. Crankshaft	1¾
10, 13 Cu. in. Horiz. Crankshaft	1¼
14, 17 Cu. in. Vert. Crankshaft	2¼
14, 17, 19 Cu. in. Horiz. Crankshaft	2¾
25 Cu. in. Vert. Crankshaft	3
25 Cu. in. Horiz. Crankshaft	3
Cast Iron	
9, 14, 19, 20 Cu. in. Horiz. Crankshaft	3
23, 24, 30, 32 Cu. in. Horiz. Crankshaft	4

OIL VISCOSITY RECOMMENDATIONS

Winter (under 40° F.)	Summer (above 40° F.)
SAE 5W-20, or SAE 5W-30	SAE 30
If above are not available: SAE 10W or SAE10W-30 (under 0° F.) Use SAE10W or SAE10W-30 in proportions of 90% motor oil/ 10% kerosene	If above is not available: SAE10W-40 or SAE10W-30

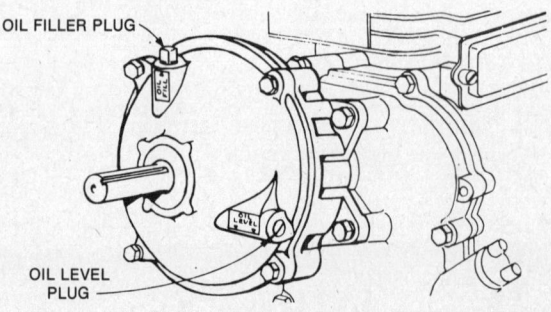

Location of oil fill plug on 6:1 gear reduction units

Engine Overhaul

TUNE-UP

Spark Plugs

Remove the spark plug with a ¾" (1½ in. plug) or a ¹³⁄₁₆" (2" plug) deep well socket wrench. Clean carbon deposits off the center and side electrodes with a sharp instrument. If possible, you should also attempt to remove deposits from the recess between the insulator and the threaded portion of the plug. If the electrodes are burned away or the insulator is cracked at any point, replace the plug. Using a wire type feeler gauge, adjust the gap by bending the side electrode where it is curved until the gap is .030 in.

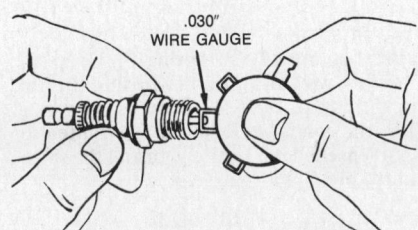

Checking spark plug gap

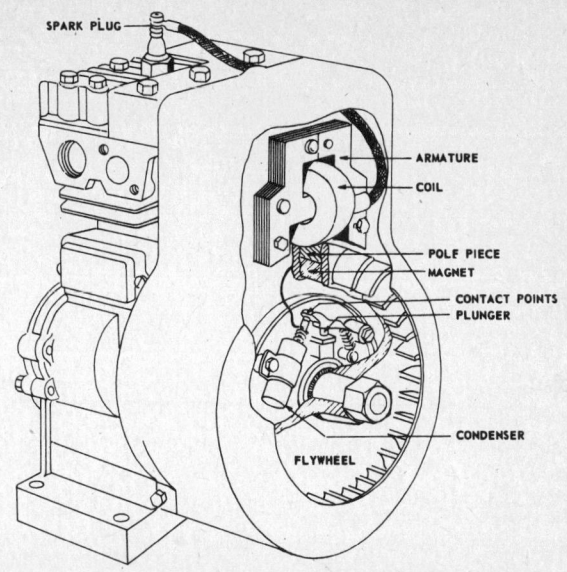

Flywheel ignition internal breaker

When installing the plug, make sure the threads of the plug and the threads in the cylinder head are clean. It is best to oil the plug threads very lightly. Be careful not to over-torque the plug, especially if the engine has an aluminum head. If you use a torque wrench, torque to about 15 ft.lb.

Breaker Points

All Briggs and Stratton engines have magneto ignition systems. Three types are used: Flywheel Type—Internal Breaker Flywheel Type—External Breaker, and Magna-Matic.

FLYWHEEL TYPE— INTERNAL BREAKER

This ignition system has the magneto located on the flywheel and the breaker points located under the flywheel.

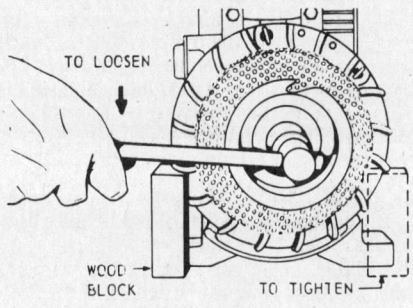

Loosening large flywheels

The flywheel is located on the crankshaft with a soft metal key. It is held in place by a nut or starter clutch. The flywheel key must be in good condition to insure proper location of the flywheel for ignition timing. Do not use a steel key under any circumstances. Use only a soft metal key, as originally supplied.

The key way in both flywheel and crankshaft should not be distorted. Flywheels are made of aluminum, zinc, or cast iron.

Flywheel, Nut and/or Starter Clutch
REMOVAL AND INSTALLATION

Place a block of wood under the flywheel fins to prevent the flywheel from tunring while you are loosening the nut or starter clutch. Be careful not to bend the flywheel. There are special flywheel holders available for this purpose; Briggs & Stratton recommends their use on flywheels of 6¾" diameter or less.

TUNE-UP SPECIFICATIONS

Model	Plug Type	Plug Gap (in.)	Point Gap (in.)	ARMATURE GAP 2 Leg	ARMATURE GAP 3 Leg	Idle Speed
Aluminum Block						
140000, 170000, 190000, 251000	①	.030	.020	.010-.014	.016-.019	1750
Cast Iron Block						
19, 190000, 200000	①	.030	.020	.010-.014	.022-.026	1200
23, 230000	①	.030	.020	.010-.014	.022-.026	1200
243400, 300000, 320000	①	.030	.020	.010-.014	—	1200

①Manufacturer's Code

1½ in. plug	2 in. plug	Manufacturer
CJ-8	J-8	Champion
RCJ-8	RJ-8	Champion (resistor)
A-7NX	A-71	Autolite
AR-7N	AR-80	Autolite (resistor)
CS-45	GC-46	A.C.
—	R-46	A.C. (resistor)

Engine Overhaul

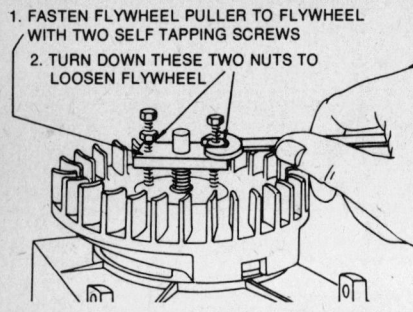
Removing the flywheel with a puller

On rope starter engines, the ½ in. flywheel nut has a left-hand thread and the ⅝ in. nut has a right-hand thread. The starter clutch used on rewind and wind-up starters has a right-hand thread.

Some flywheels have two holes provided for the use of a flywheel puller. Use a small gear puller or automotive steering wheel puller to remove the flywheel if a flywheel puller is not available. Be careful not to bend the flywheel if a gear puller is used. On rope starter engines leave the nut on for the puller to bear against. Small cast iron flywheels do not require a puller.

Install the flywheel in the reverse order of removal after inspecting the key and key way for damage or wear.

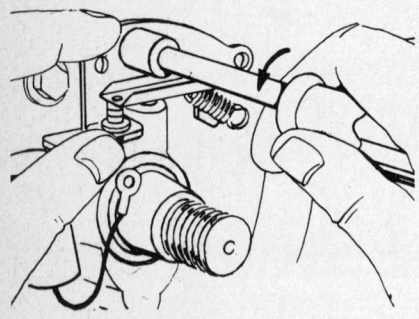
Breaker point assemblies

Breaker Point Removal and Installation

Remove the breaker cover. Care should be taken when removing the cover, to avoid damaging it. If the cover is bent or damaged, it should be replaced to insure a proper seal.

The breaker point gap on all models is 0.020 in. Check the points for contact and for signs of burning or pitting. Points that are set too wide will advance the spark timing and may cause kickback when starting. Points that are set too close will retard the spark timing and decrease engine power.

On models that have a separate condenser, the point set is removed by first removing the condenser and armature wires from the breaker point clip. Loosen the adjusting lockscrew and remove the breaker point assembly.

On models where the condenser is incorporated with the breaker points, loosen the screw which holds the post. The condenser/point assembly is removed by loosening the screw which holds the condenser clamp.

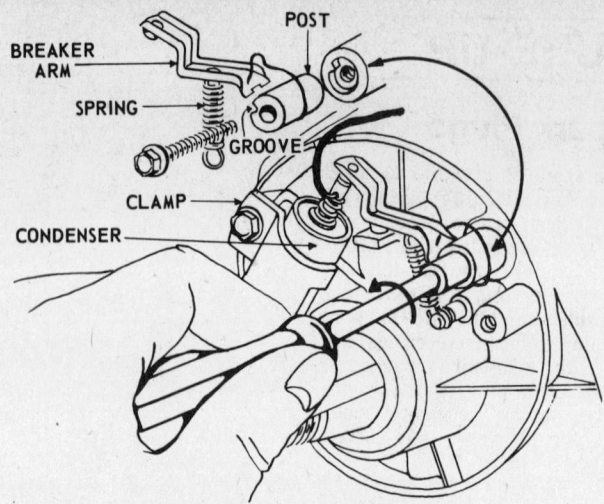
Breaker point assemblies

When installing a point set with the separate condenser, be sure that the small boss on the magneto plate enters the hole in the point bracket. Mount the point set to the magneto plate or the cylinder with a lock screw. Fasten the armature lead wire to the breaker points with the clip and screw. If these lead wires do not have terminals, the bare end of the wires can be inserted into the clip and the screw tightened to make a good connection. Do not let the ends of the wire touch either the point bracket or the magneto plate, or the ignition will be grounded.

To install the integral condenser/point set, place the mounting post of the breaker arm into the recess in the cylinder so that the groove in the post fits the notch in the recess. Tighten the mounting screw securely. Use a ¼ in. wrench. Slip the open loop of the breaker arm spring through the two holes in the arm, then hook the closed loop of the spring over the small post protruding from the cylinder. Push the flat end of the breaker arm into the groove in the mounting post. This places tension on the spring and pulls the arm against the plunger. If the condenser post is threaded, attach the coil primary wire and the ground wire (if furnished) with the lockwasher and nut. If the primary wire is fastened to the condenser with a spring fastener, compress the spring and slip the primary wire and ground wire into the hole in the condenser post. Release the spring. Lay the condenser in place and tighten the condenser clamp securely. Install the spring in the breaker arm.

Point Gap Adjustment

Turn the crankshaft until the points are open to the widest gap. When adjusting a breaker point assembly with an integral condenser, move the condenser forward or backward with a screwdriver until the proper gap is obtained (0.020 in.). Point sets with a separate condenser are adjusted by moving the contact point bracket up and down after the lockscrew has been loosened. The point gap is set to 0.020 in.

Breaker Point Plunger

If the breaker point plunger hole becomes excessively worn, oil will leak past the plunger and may get on the points, causing them to burn. To check the hole, loosen the breaker point mounting screw and move the breaker points out of the way. Remove the plunger. If the flat end of the #19055 plug gauge will enter the plunger hole for a distance of ¼ in. or more, the hole should be rebushed.

To install the bushing, it is necessary that the breaker points, armature, and

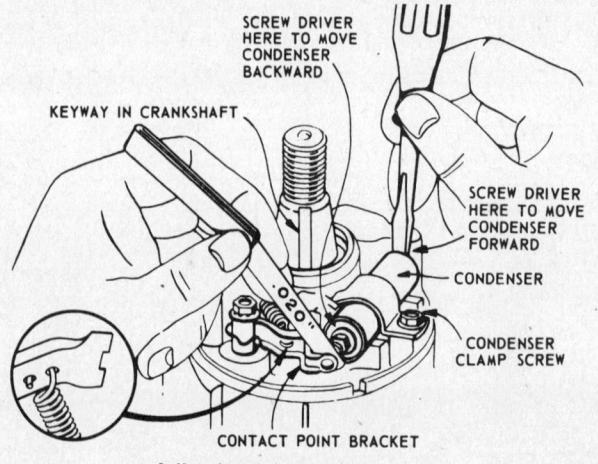
Adjusting breaker point gap

Engine Overhaul

crankshaft be removed. Use a #19056 reamer to ream out the old plunger hole. This should be done by hand. The reamer must be in alignment with the plunger hole. Drive the bushing, #23513, into the hole until the upper end of the bushing is flush with the top of the boss. Remove all metal chips and dirt from the engine.

If the breaker point plunger is worn to a length of 0.870 in. or less, it should be replaced. Plungers must be inserted with the groove at the top or oil will enter the breaker box. Insert the plunger into the hole in the cylinder.

Armature Air Gap Adjustment

Set the air gap between the flywheel and the armature as follows: With the armature up as far as possible and just one screw tightened, slip the proper gauge between the armature and flywheel. Turn the flywheel until the magnets are directly below the armature. Loosen the one mounting screw and the magnets should pull the armature down firmly against the thickness gauge. Tighten the mounting screws.

FLYWHEEL TYPE— EXTERNAL BREAKER

Breaker Point Set Removal and Installation

Turn the crankshaft until the points open to their widest gap. This makes it easier to assemble and adjust the points later if the crankshaft is not removed. Remove the condenser and upper and lower mounting screws. Loosen the locknut and back off the breaker point screw. Install the points in the reverse order of removal.

To avoid the possibility of oil leaking past the breaker point plunger or moisture entering the crankcase between the plunger and the bushing, a plunger seal is installed on the engine models using this type of ignition system. To install a new seal on

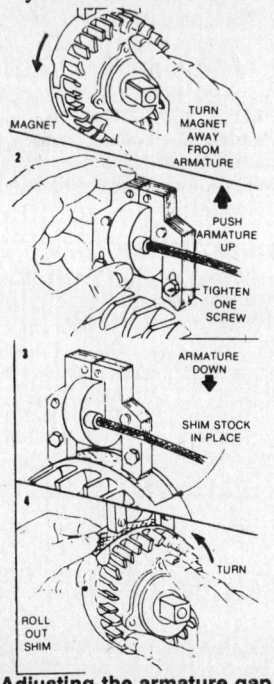

Adjusting the armature gap

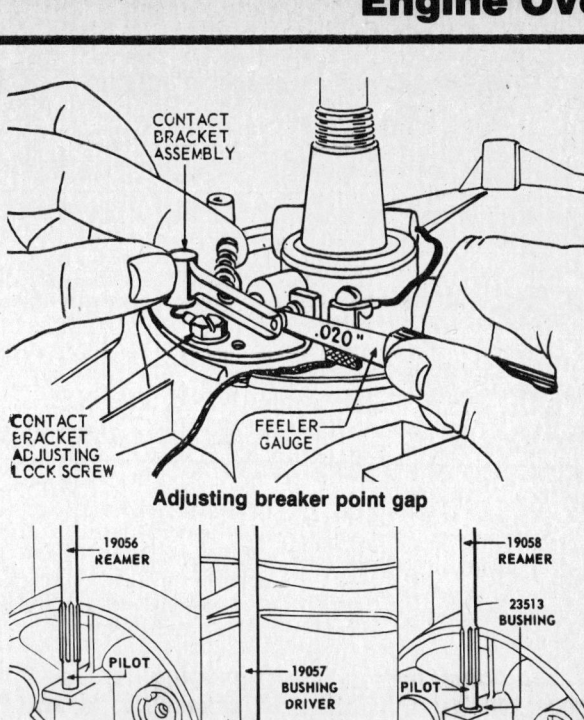

Adjusting breaker point gap

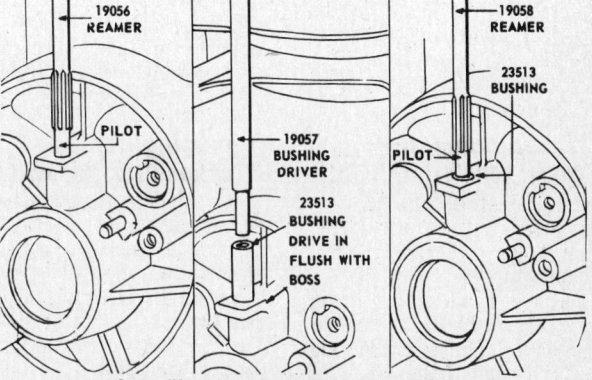

Installing breaker plunger bushing

Variations in armature positioning

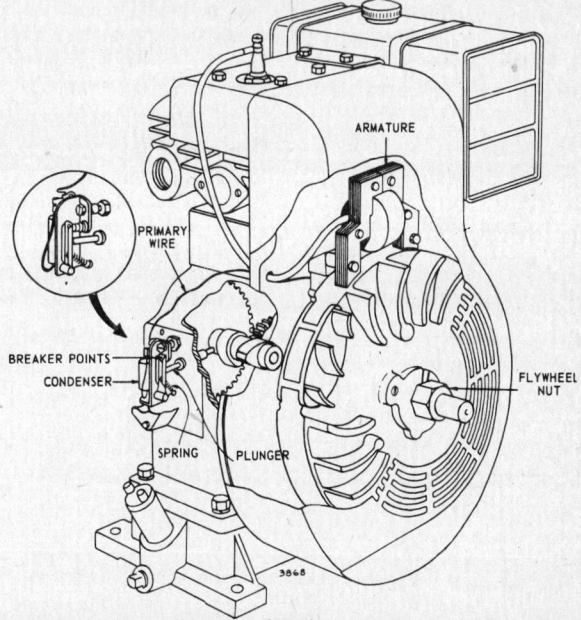

Flywheel ignition external breaker, models 193000, 200000, 233000, 243000, 300000, 320000

Engine Overhaul

the plunger, remove the breaker point assembly and condenser. Remove the retainer and eyelet, remove the old seal, and install the new one. Use extreme care when installing the seal on the plunger to avoid damaging the seal. Replace the eyelet and retainer and replace the points and condenser.

NOTE: Apply a small amount of sealer to the threads of both mounting screws and the adjustmet screw. The sealer prevents oil from leaking into the breaker point area.

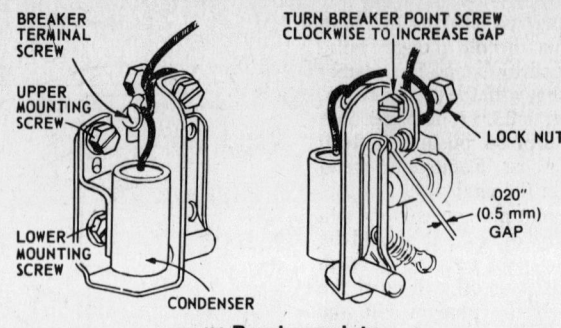
Breaker points

Point Gap Adjustment

Turn the crankshaft until the points open to their widest gap. Turn the breaker point adjusting screw until the points open to 0.020 in. and tighten the locknut. When the cover is installed, seal the point where the primary wire passes under the cover. This area must be resealed to prevent the entry of dust and moisture.

Armature Timing Adjustment
MODELS 193000, 200000, 230000, 243000, 300400, 320400

Using a puller, remove the flywheel. Set the point gap at 0.020 in. Position the flywheel on the crankshaft taper. Slip the key in place. Install the flywheel nut finger tight. Rotate the flywheel and the crankshaft clockwise until the breaker points are just opening. Use a timing light. When the points just start to open, the arrow on the flywheel should line up with the arrow on the armature bracket.

If the arrows do not match, slip off the flywheel without disturbing the position of the crankshaft. Slightly loosen the mounting screw which holds the armature bracket to the cylinder. Slip the flywheel back onto the crankshaft. Insert the flywheel key. Install the flywheel nut finger tight. Move the armature and bracket assembly to align the arrows. Slip off the flywheel and tighten the armature bracket bolts. Install the key and flywheel. Tighten the flywheel nut to 110 to 118 ft.lb. on the 193000 and 200000 series. On all the rest, tighten to 138 to 150 ft.lb. Set the armature gap at 0.010 to 0.014 in.

Armature Timing Adjustment
MODELS 19D, 23D

With the points set at 0.020 in. and the flywheel key screw finger tight together with the flywheel nut, rotate the flywheel clockwise until the breaker points are just opening. The flywheel key drives the crankshaft while doing this. Using a timing light, rotate the flywheel slightly counterclockwise until the edge of the armature lines up with the edge of the flywheel insert. The crankshaft must not turn while doing this. Tighten the key screw and the flywheel nut. set the armature air gap at 0.022 to 0.026 in.

Replacing Threaded Breaker Plunger and Bushing

Remove the breaker cover and the condenser and breaker point assembly.

Place a thick ⅜ in. inside diameter washer over the end of the bushing and screw on the ⅜-24 nut. Tighten the nut to pull the bushing out of the hole. After the bushing has been moved about ⅛ in., remove the nut and put on a second thick washer and repeat the procedure. A total stack of 3 ⅜ in. washers will be required to completely remove the bushing. Be sure the plunger does not fall out of the bushing as it is removed.

Place the new plunger in the bushing with the large end of the plunger opposite the threads on the bushing. Screw the ⅜-24 in. nut onto the threads to protect them and insert the bushing into the cylinder. Place a piece of tubing the same diameter as the nut and, using a hammer, drive the bushing into the cylinder until the square shoulder on the bushing is flush with the face of the cylinder. Check to be sure that the plunger operates freely.

Replacing Unthreaded Breaker Plunger and Bushing

Pull the plunger out as far as possible and use a pair of pliers to break the plunger off as close as possible to the bushing. Use a ¼-20 in. tap or a #93029 self threading screw to thread the hole in the bushing to a depth of about ½-⅝ in. Use a ¼-20 x ½ in. Hex. head screw and two spacer washers to pull the bushing out of the cylinder. The bushing will be free when it has been extracted 5/16 in. Carefully remove the bushing and the remainder of the broken plunger. Do not allow the plunger or metal chips to drop into the crankcase.

Correctly insert the new plunger into the new bushing. Insert the plunger and the bushing into the cylinder. Use a hammer and the old bushing to drive the new bushing into the cylinder until the new bushing is flush with the face of the cylinder. Make sure that the plunger operates freely.

PLUNGER SEAL

Later models with Flywheel Type External Breaker Ignition feature a plunger seal. This seal keeps both oil and moisture from entering the breaker box. If the points have become contaminated on an engine manufactured without this feature, the seal may be installed. Parts, part numbers, and their locations are shown in the illustration. Install the seal onto the plunger very carefully to avoid fracturing it.

MAGNA-MATIC IGNITION SYSTEM

Removing the Flywheel

Flywheels on engines with Magna-Matic ignition are removed with pullers similar to factory designs numbered #19068 and 19203. These pullers employ two bolts, which are screwed into holes tapped into the flywheel. The bolts are turned until the flywheel is forced off the crankshaft. Only this type of device should be used to pull these flywheels.

Armature Air Gap

The armature air gap on engines equipped with Magna-Matic ignition system is fixed and can change only if wear

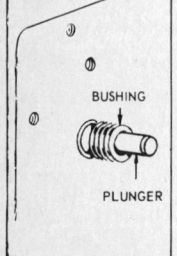
Illus. 1

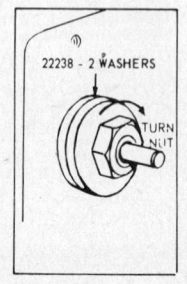
Illus. 2

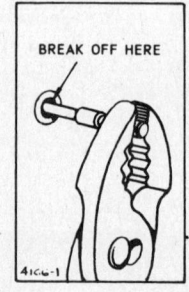
Illus. 3

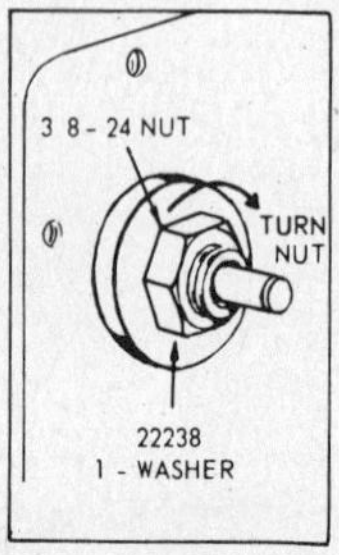

Removing plunger and threaded bushing

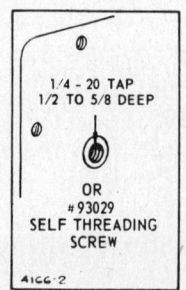

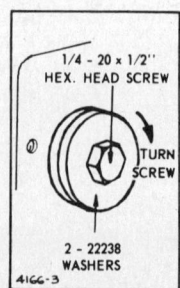

Removing bushing and plunger

Engine Overhaul

occurs on the crankshaft journal and/or main bearing. Check for wear by inserting a ½ in. wide feeler gauge at several points between the rotor and armature. Minimum feeler gauge thickness is 0.004 in. Keep the feeler gauge away from the magnets on the rotor or you will have a false reading.

Rotor Removal and Installation

The rotor is held in place by a Woodruff key and a clamp on later engines, and a Woodruff key and set screw on older engines. The rotor clamp must always remain on the rotor, unless the rotor is in place on the crankshaft and within the armature, or a loss of magnetism will occur.

Loosen the socket head screw in the rotor clamp which will allow the clamp to loosen.

It may be necessary to use a puller to remove the rotor from the crankshaft. On older models, loosen the small lockscrew, then the setscrew.

To install the setscrew type rotor, place the Woodruff key in the key way on the crankshaft, then slide the rotor onto the crankshaft until the setscrew hole in the rotor and the crankshaft are aligned. Be sure the key remains in place. Tighten the setscrew securely, then tighten the lockscrew to prevent the setscrew from loosening. The lockscrew is self-threading and the hole does not require tapping.

To install the clamp type rotor, place the Woodruff key in place in the crankshaft and align the keyway in the rotor with the Woodruff key. If necessary, use a short length of pipe and a hammer to drive the rotor onto the shaft until a 0.025 in. feeler gauge can be inserted between the rotor and the bearing support. The split in the clamp must be between the slots in the rotor. Tighten the clamp screws to 60 to 70 in.lb.

Rotor Timing Adjustment

The rotor and armature are correctly timed at the factory and require timing only if the armature has been removed from the engine, or if the cam gear or crankshaft has been replaced.

If it is necessary to adjust the rotor, proceed as follows: with the point gap set at 0.020 in., turn the crankshaft in the normal direction of rotation until the breaker points close and just start to open. Use a timing light or insert a piece of tissue paper between the breaker points to determine when the points begin to open. With the three armature mounting screws slightly loose, rotate the armature until the arrow on the armature lines up with the arrow on the rotor. Align with the corresponding number of engine models, for example, on Model 9, align with #9. Retighten the armature mounting screws.

Coil and/or Armature Replacement

Usually the coil and armature are not separated, but left assembled for convenience. However, if one or both need replacement, proceed as follows: the coil primary wire and the coil ground wire must be unfastened. Pry out the clips that hold the coil and coil core to the armature. The

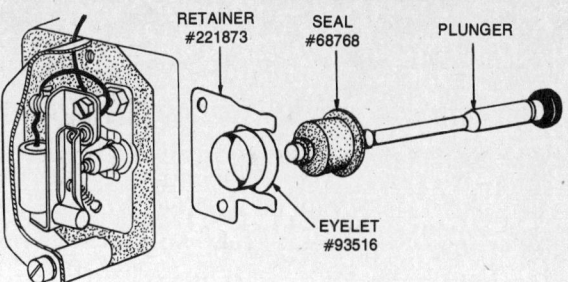
Plunger seal used on later model engines

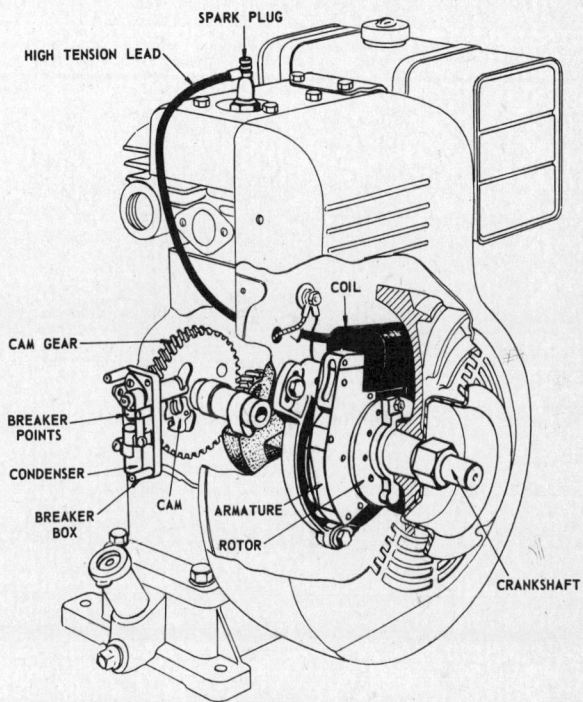
Magna-Matic system

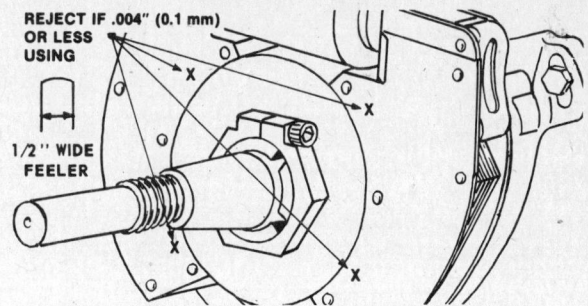
Checking armature gap on Magna-Matic systems

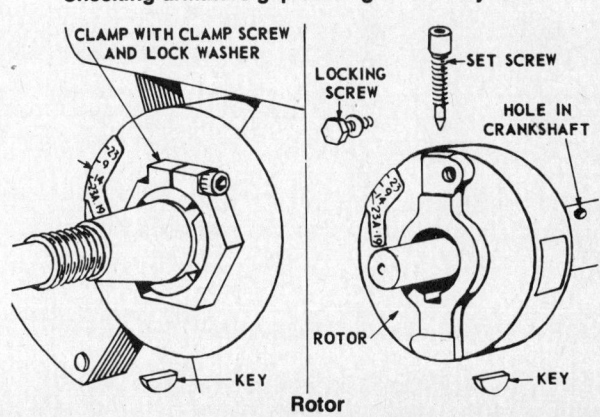
Rotor

Engine Overhaul

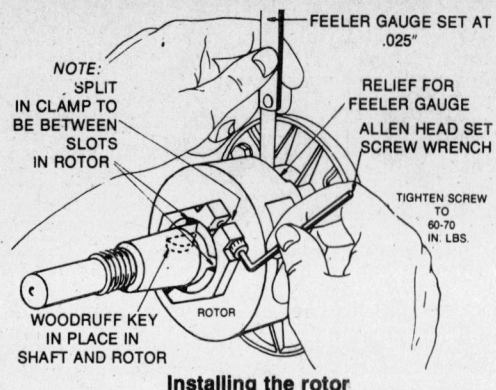
Installing the rotor

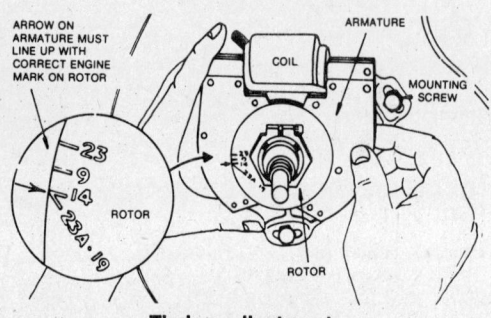
Timing adjustment

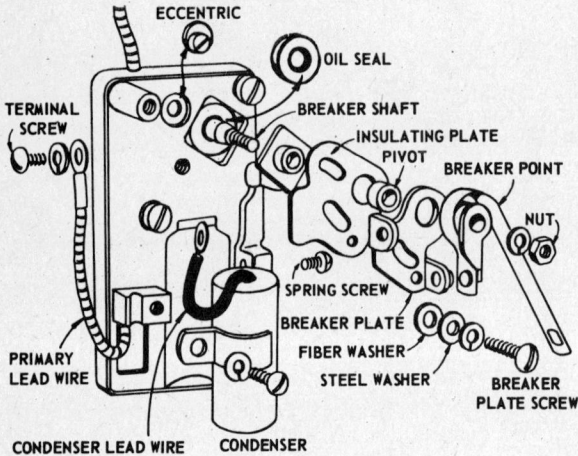
Breaker box assembly

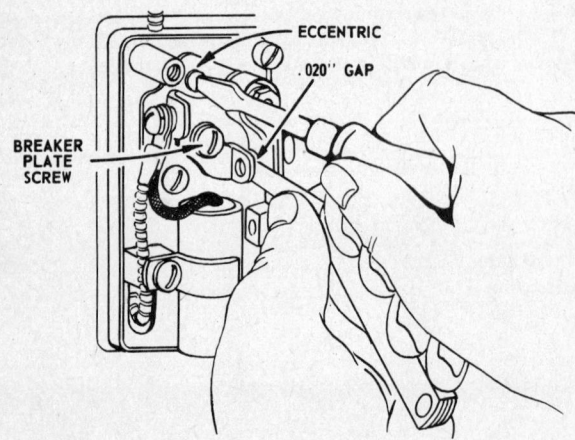
Adjusting breaker points

coil core is a slip fit in the coil and can be pushed out of the coil.

To reassemble, push the coil core into the coil with the rounded side toward the ignition cable. Place the coil and core on the armature with the coil retainer between the coil and the armature and with the rounded side toward the coil. Hook the lower end of the clips into the armature, then press the upper end onto the coil core.

Fasten the coil ground wire (bare double wires) to the armature support. Next, place the assembly against the cylinder and around the rotor and bearing support. Insert the three mounting screws together with the washer and lockwasher into the three long oval holes in the armature. Tighten them enough to hold the armature in place but loose enough so the armature can be moved for adjustment of the timing. Attach the primary wires from the coil and the breaker points to the terminal at the upper side of the backing plate. Push the ignition cable through the louvered hole at the left side of the backing plate.

NOTE: On Model 9 engines, knot the ignition cable before inserting it through the backing plate. Be sure all wires are clear of the flywheel.

Breaker Point
Removal and Installation

Turn the crankshaft until the points open to the widest gap. This makes it easier to assemble and adjust the points later if the crankshaft is not removed. With the terminal screw out, remove the spring screw. Loosen the breaker shaft nut until the nut is flush with the end of the shaft. Tap the nut to free the breaker arm from the tapered end of the breaker shaft. Remove the nut, lockwasher, and breaker arm. Remove the breaker plate screw, breaker plate, pivot, insulating plate, and eccentric. Pry out the breaker shaft seal with a sharp pointed tool.

To install the breaker points, press in the new oil seal with the metal side out. Put the new breaker plate on the top of the insulating plate, making sure that the detent in the breaker plate engages the hole in the insulating plate. Fasten the breaker plate screw enough to put a light tension on the plate. Adjust the eccentric so that the left edge of the insulating plate is parallel to the edge of the box and tighten the screw. This locates the breaker plate so that the proper gap adjustments may be made. Turn the breaker shaft clockwise as far as possible and hold it in this position. Place the new breaker points on the shaft, then the lockwasher, and tighten the nut down on the lockwasher. Replace the spring screw and terminal screw.

Breaker Box
Removal and Installation

Remove the two mounting screws, then remove the breaker box, turning it slightly to clear the arm at the inner end of the breaker shaft. The breaker points need not be removed to remove the breaker box.

To install, pull the primary wire through the hole at the lower left corner of the breaker box. See that the primary wire rests in the groove at the top end of the box, then

Engine Overhaul

tighten the two mounting screws to hold the box in place.

Breaker Shaft Removal and Installation

The breaker shaft can be removed, after the breaker points are removed, by turning the shaft one half turn to clear the retaining spur at the inside of the breaker box.

Install by inserting the breaker shaft with the arm upward so the arm will clear the retainer boss. Push the shaft all the way in, then turn the arm downward.

Breaker Point Adjustment

To adjust the breaker points, turn the crankshaft until the breaker points open to the widest gap. Loosen the breaker point plate screw slightly. Rotate the eccentric to obtain a point gap of 0.020 in. Tighten the breaker plate screw.

Mixture Adjustment

920000 ENGINES BUILT SINCE 1968 (AUTOMATIC CHOKE)

1. Start the engine and run it long enough to reach operating temperature. If the carburetor is so far out of adjustment that it will not start, close the needle valve by turning it clockwise. Then open the needle valve 1½ turns counterclockwise.

2. Move the control so that the engine runs at normal operating speed. Turn the needle valve clockwise until the engine starts to lose speed because of too lean a mixture. Then slowly turn the needle valve counterclockwise and out past the point of smoothest operation until the engine just begins to run unevenly because of too rich a mixture. Turn the needle back clockwise to the midpoint between the rich and lean mixture extremes. This should be where the engine operates smoothest. The final adjustment of the needle valve should be slightly on the rich side (counterclockwise) of the mid-point.

3. Move the engine control to the slow position and turn the idle adjusting screw until a fast idle of about 1750 rpm is obtained. If the engine idles at a speed lower than 1750 rpm, it may not accelerate properly. It is not practical to attempt to obtain acceleration from speeds below 1750 rpm, because the mixture which would be required would be too rich for normal operating speeds.

4. To check the idle adjustment, move the engine control from slow to fast speed. The engine should accelerate smoothly. If the engine tends to stall or die out, increase the idle speed or readjust the carburetor, usually to a slightly richer mixture.

Flooding can occur if the engine is tipped at an angle for a prolonged period of time, if the engine is cranked repeatedly with the spark plug wire disconnected, or if the carburetor mixture is too rich.

In case of flooding, move the governor control to the stop position and pull the starter rope at least six times.

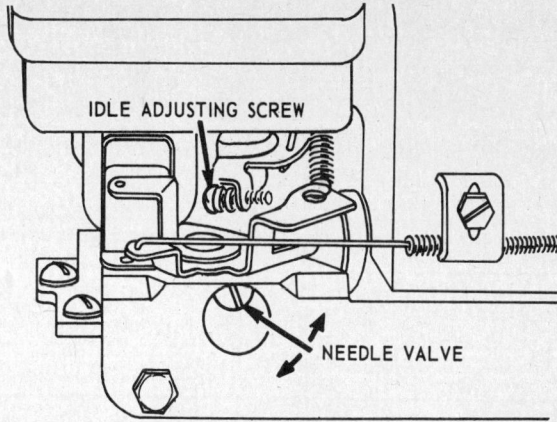

Carburetor adjustment

When the control is placed in the stop position, the governor spring holds the throttle in a closed idle position. Cranking the engine with a closed throttle creates a higher vacuum which opens the choke rapidly, permitting the engine to clear itself of excess fuel.

Then move the control to the fast position and start the engine. If the engine continues to flood, lean the carburetor needle valve by about ⅛–¼ of a turn clockwise.

PULSA-JET AND VACU-JET (MODEL SERIES 82000, 92000 ONLY)

Models 82500 and 92500 have a Vacu-Jet carburetor and Models 82900 and 92900 have a Pulsa-Jet carburetor.

Adjust the carburetor with the air cleaner installed and the fuel tank half full.

Turn the needle valve clockwise to close it. Then open it about 1½ turns. This will permit the engine to be started and warmed up before making the final adjustment.

With the engine running at normal operating speed (about 3000 rpm without a load) turn the needle valve clockwise until the engine starts to lose speed because of a too lean mixture.

Then slowly turn the needle valve counterclockwise past the point of smoothest operation, until the engine just begins to run unevenly. This mixture will give the best performance under a load.

Hold the throttle in the idle position. Turn the idle speed adjusting screw until a fast idle is obtained (about 1750 rpm).

Test the engine under full load. If the engine tends to stall or die out, it usually indicates that the mixture is slightly lean and it may be necessary to open the needle valve slightly to provide a richer mixture. This slightly richer mixture may cause a slight unevenness in idling.

The breather tube and fuel intake tube thread into the cylinder on the model 82500 and 82900 engines. The fuel intake tube is bolted to the cylinder on the model 92500 and 92900 engines. Check for a good fit to prevent any air leaks or dirt entry. The fuel intake tube must not be distorted at the point where the carburetor O-ring fits or air leaks will occur.

TWO PIECE FLO-JET

1. Start the engine and run it at 3,000 rpm until it warms up.

2. Turn the needle valve (flat handle) both extremes of operation noting the location of the valve at both points. That is, turn the valve inward until the mixture becomes too lean and the engine starts to slow, then note the position of the valve. Turn it outward slowly until the mixture becomes too rich and the engine begins to slow. turn the valve back inward to the mid-point between the two extremes.

3. Install a tachometer on the engine. Pull the throttle to the idle position and hold it there through the rest of this step. Adjust the idle speed screw until the engine idles at 1,750 rpm if it's an aluminum engine, or 1,200 rpm, if it's a cast iron engine. Then, turn the idle valve in and out to adjust mixture, as described in Step 2. If

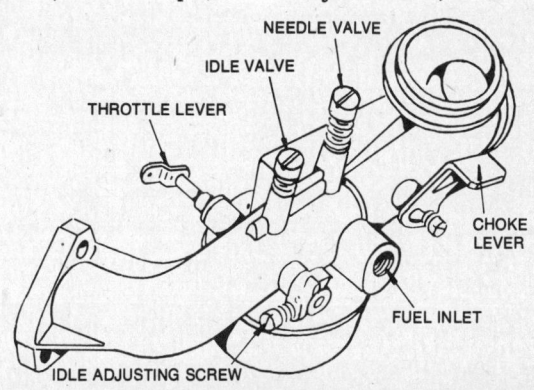

Idle valve and needle valve locations on the one-piece Flo-Jet

Engine Overhaul

idle valve adjustment changes idle speed, adjust speed to specifications.

4. Release the throttle and observe the engine's response. The engine should accelerate without hesitation. If response is poor, one of the mixture adjustments is too lean. Readjust either or both as necessary. If idle speed was changed after idle valve was adjusted, readjust the idle valve first.

ONE PIECE FLO-JET

Follow the instructions for adjusting the Two Piece Flo-Jet carburetor (above). On the large, One Piece Flo-Jet, the needle valve is located under the float bowl, and the idle valve on top of the venturi passage. On the small One Piece Flo-Jet, both valves are adjusted by screws located on top of the venturi passage. The needle valve is located on the air horn side, is centered above the float bowl, and uses a larger screw head.

Governor Adjustments

SETTING MAXIMUM GOVERNED SPEED WITH ROTARY LAWNMOWER BLADES

NOTE: Strict limits on engine rpm must be observed when setting top governed speed on rotary lawnmowers. This is done so that blade tip speeds will be kept to less than 19,000 feet per minute. Briggs & Stratton suggests setting the governor 200 rpm low to allow for possible error in the tachometer reading.

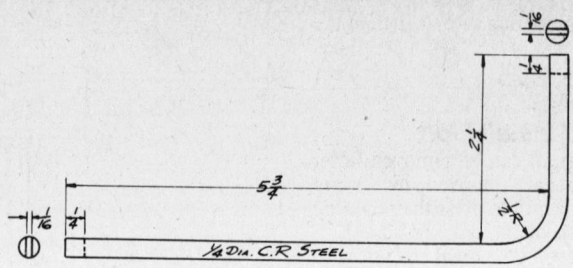

Spring anchor tab adjusting tool

These figures below, based on blade length, must be strictly adhered to, or a serious accident could result!

Blade Length (in.)	Max. Governed Speed (R.P.M.)
18	4032
19	3820
20	3629
21	3456
22	3299
23	3155
24	3024
25	2903
26	2791

MODEL 140000

Loosen the screw which holds the governor lever to the governor shaft. Turn the governor lever counterclockwise until the carburetor throttle is wide open. With a screwdriver, turn the governor shaft counterclockwise as far as it will go. Tighten the screw which holds the governor lever to the governor shaft.

CAST IRON MODELS 19, 190000, 200000, 23, 230000, 240000, 300000, 320000

Loosen the screw which holds the governor lever to the governor shaft. Push the lever counterclockwise as far as it will go. Hold it in position and turn the governor shaft counterclockwise as far as it will go. This can be done with a screwdriver. Securely tighten the screw that holds the governor lever to the shaft.

ALUMINUM MODELS 140000, 170000, 190000, 251000

Vertical and horizontal shaft engine governors are adjusted by setting the control lever in the high speed position. Loosen the nut on the governor lever. Turn the governor shaft clockwise with a screwdriver to the end of its travel. Tighten the nut. The throttle must be wide open. Check to see if the throttle can be moved from idle to wide open without binding.

Adjusting Top No Load Speed

Set the control lever to the maximum speed position with the engine running. Bend the spring anchor tang to get the desired top speed.

Choke Adjustment

CHOKE-A-MATIC— PULSA-JET AND VACU-JET CARBURETORS

To check the operation of the choke linkage, move the speed adjustment lever to the choke position. If the choke slide does not fully close, bend the choke link. The speed adjustment lever must make good contact against the top switch.

Install the carburetor and adjust it in the same manner as the Pulsa-Jet carburetor.

TWO-PIECE FLO-JET AUTOMATIC CHOKE

Hold the choke shaft so the thermostat lever is free. At room temperature (68°F),

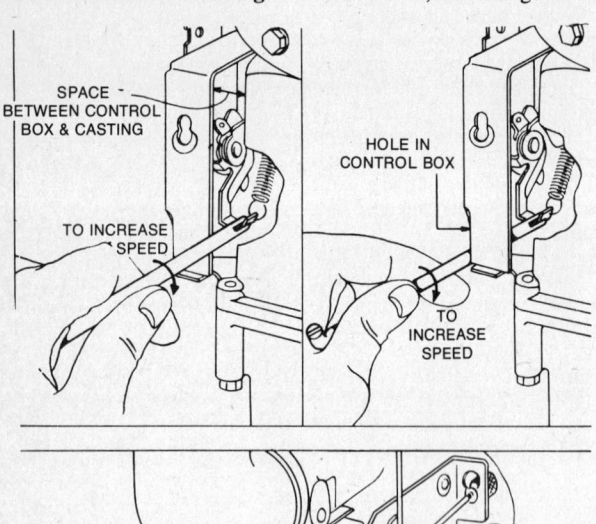

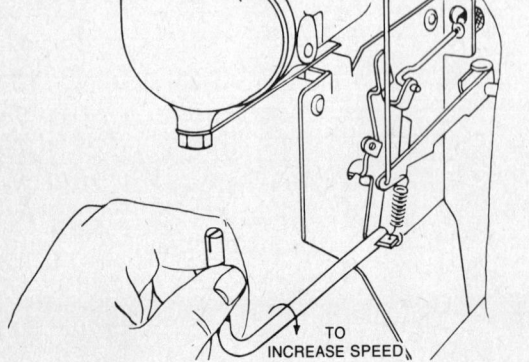

Bending the spring anchor tab to get the desired top speed

Engine Overhaul

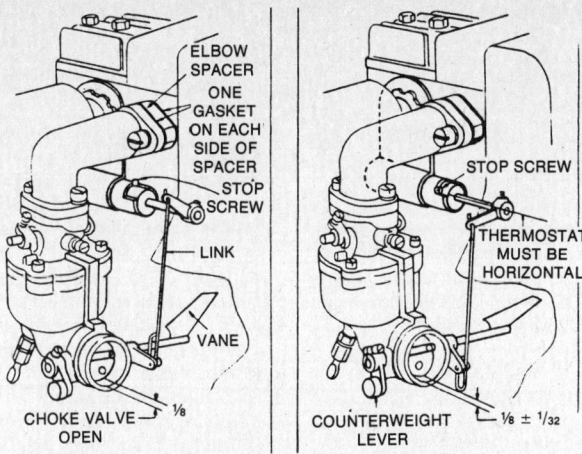

Adjusting automatic choke on two-piece Flo-Jet

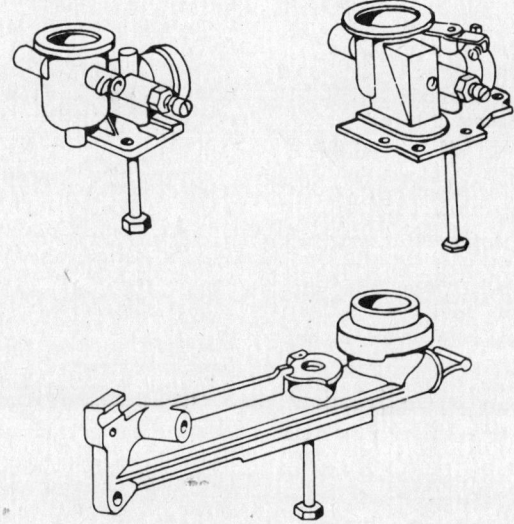

Three types of Vacu-Jet carburetors

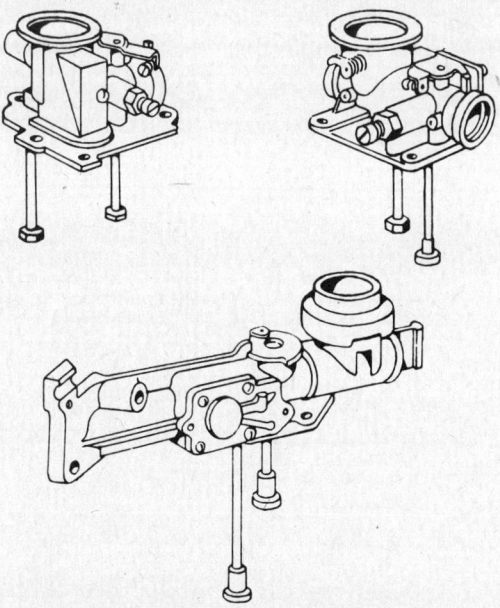

Three types of Pulsa-Jet carburetors

the screw in the thermostat collar should be in the center of the stops. If not, loosen the stop screw and adjust the screw.

Loosen the set screw on the lever of the thermostat assembly. Slide the lever to the right or left on the shaft to ensure free movement of the choke link in any position. Rotate the thermostat shaft clockwise until the stop screw strikes the tube. Hold it in position and set the lever on the thermostat shaft so that the choke valve will be held open about 1/8 in. from a closed position. Then tighten the set screw in the lever.

Rotate the thermostat shaft counterclockwise until the stop screw strikes the opposite side of the tube. Then open the choke valve manually until it stops against the top of the choke link opening. The choke valve should now be open approximately 1/8 in. as before.

Check the position of the counterweight lever. With the choke valve in a wide open position (horizontal) the counterweight lever should also be in a horizontal position with the free end toward the right.

Operate the choke manually to be sure that all parts are free to move without binding or rubbing in any position.

Compression Checking

You can check the compression in any Briggs and Stratton engine by performing the following simple procedure: spin the flywheel counterclockwise (flywheel side) against the compression stroke. A sharp rebound indicates that there is satisfactory compression. A slight or no rebound indicates poor compression.

It has been determined that this test is an accurate indication of compression and is recommended by Briggs and Stratton. Briggs and Stratton does not supply compression pressures.

Loss of compression will usually be the result of one or a combination of the following:

1. The cylinder head gasket is blown or leaking.
2. The valves are sticking or not seating properly.
3. The piston rings are not sealing, which would also cause the engine to consume an excessive amount of oil.

Carbon deposits in the combustion chamber should be removed every 100 or 200 hours of use (more often when run at a steady load), or whenever the cylinder head is removed.

FUEL SYSTEM

Carburetors

There are three types of carburetors used on Briggs and Stratton engines. They are the Pulsa-Jet, Vacu-Jet and Flo-Jet. The first two types have three models each and the Flo-Jet has two versions.

Before removing any carburetor for repair, look for signs of air leakage or mount-

Engine Overhaul

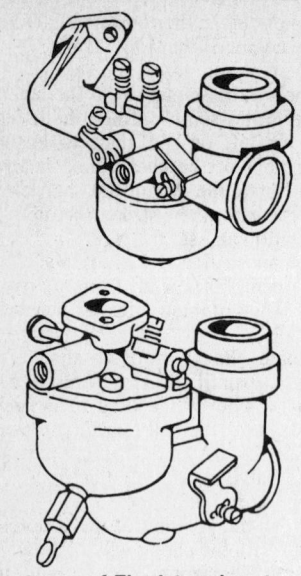

Two types of Flo-Jet carburetors

ing gaskets that are loose, have deteriorated, or are otherwise damaged.

Note the position of the governor springs, governor link, remote control, or other attachments to facilitate reassembly. Be careful not to bend the links or stretch the springs.

AUTOMATIC CHOKE

All 92000 model engines built since August 1968 have an automatic choke system.

The automatic choke operates in conjunction with engine vacuum, similar to the Pulsa-Jet fuel pump.

A diaphragm under the carburetor is connected to the choke shaft by a link. A calibrated spring under the diaphragm holds the choke closed when the engine is not running. Upon starting, vacuum created during the intake stroke is routed to the bottom of the diaphragm through a calibrated passage, thereby opening the choke.

This system also has the ability to respond in the same manner as an accelerator pump. As speed decreases during heavy loads, the choke valve partially closes, enriching the air/fuel mixture, thereby improving low speed performance and lugging power.

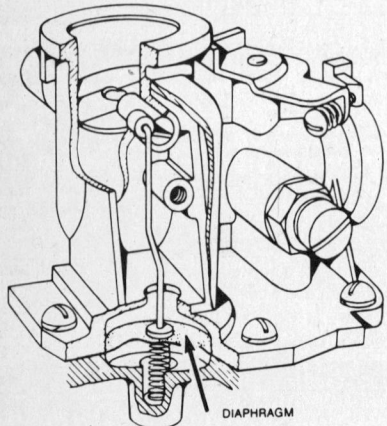

Automatic choke system

To check the automatic choke, remove the air cleaner and replace the stud. Observe the position of the choke valve; it should be fully closed. Move the speed control to the stop position; the governor spring should be holding the throttle in a closed position. Give the starter rope several quick pulls. The choke valve should alternately open and close.

If the choke valve does not react as stated in the previous paragraph, the carburetor will have to be disassembled to determine the problem. Before doing so, however, check the following items so you know what to look for:
Engine is over-choked.
1. Carburetor is adjusted too lean.
2. The fuel pipe check valve is inoperative (Vacu-Jet only).
3. The air cleaner stud is bent.
4. The choke shaft is sticking due to dirt.
5. The choke spring is too short or damaged.
6. The diaphragm is not preloaded.
Engine is over-chocked.
1. Carburetor is adjusted too rich.
2. The air cleaner stud is bent.
3. The choke shaft is sticking due to dirt.
4. The diaphragm is ruptured.
5. The vacuum passage is restricted.
6. The choke spring is distorted or stretched.
7. There is gasoline or oil in the vacuum chamber.
8. There is a leak between the link and the diaphragm.
9. The diaphragm was folded during assembly, causing a vacuum leak.
10. The machined surface on the tank top is not flat.

REPAIRING THE AUTOMATIC CHOKE

Inspect the automatic choke for free operation. Any sticking problems should be corrected as proper choke operation depends on freedom of the choke to travel as dictated by engine vacuum.

Remove the carburetor and fuel tank assembly from the engine. The choke link cover may now be removed and the choke link disconnected from the choke shaft. Disassemble the carburetor from the tank top, being careful not to damage the diaphragm.

CHECKING THE DIAPHRAGM AND SPRING

The diaphragm can be reused, provided it has not developed wear spots or punctures. On the Pulsa-Jet models, make sure that the fuel pump valves are not damaged. Also check the choke spring length. The Pulsa-Jet spring minimun length is $1\frac{1}{8}$ in. and the maximum is $1\frac{7}{32}$ in. Vacu-Jet spring length minimum is $\frac{15}{16}$ in., maximum length 1 in. If the spring length is shorter or longer than specified, replace the daiphragm and the spring.

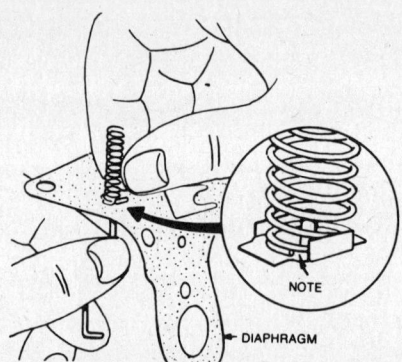

Installing the spring on a new diaphragm

CHECKING THE TANK TOP

The machined surface on the top of the tank must be flat in order for the diaphragm to provide an adequate seal between the carburetor and the tank. If the machined surface on the tank is not flat, it is possible for gasoline to enter the vacuum chamber by passing between the machined surface and the diaphragm. Once fuel has entered the vacuum chamber, it can move through the vacuum passage and into the carburetor. The flatness of the machined surface on the tank top can be checked by using a straightedge and a feeler gauge.

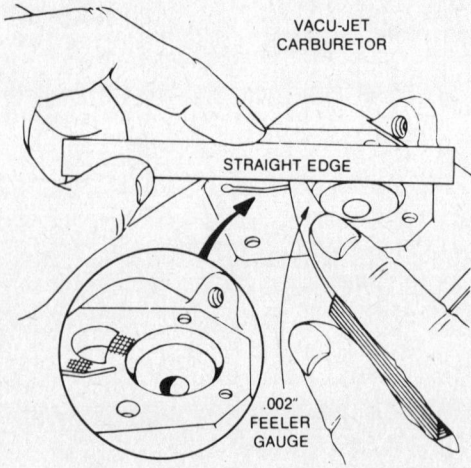

Checking the tank top for warpage

Engine Overhaul

The surface should not vary more than 0.002 in. Replace the tank if a 0.002 in. feeler gauge can be passed under the straightedge.

If a new diaphragm is installed, assemble the spring to the replacement diaphragm, taking care not to bend or distort the spring.

Place the diaphragm on the tank surface, positioning the spring in the spring pocket.

Place the carburetor on the diaphragm ensuring that the choke link and diaphragm are properly aligned between the carburetor and the tank top. On Pulsa-Jet models, place the pump spring and cap on the diaphragm over the recess or pump chamber in the fuel tank. Thread in the carburetor mounting screws to about two threads. Do not tighten them. Close the choke valve and insert the choke link into the choke shaft.

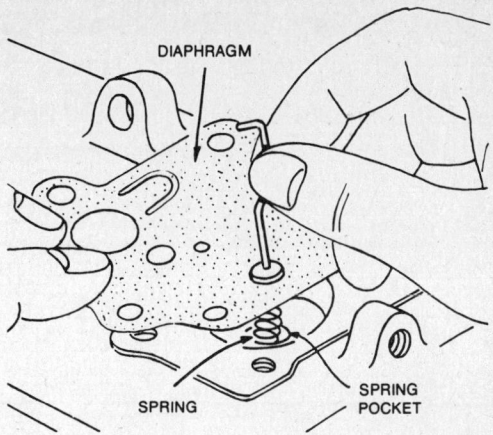

Installing the diaphragm and spring into the spring pocket

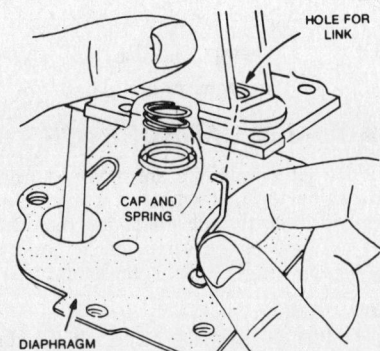

Positioning the diaphragm on top of the tank

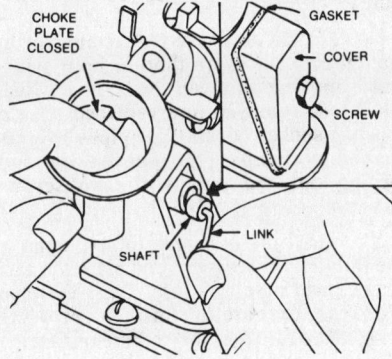

Inserting the choke link into the choke shaft

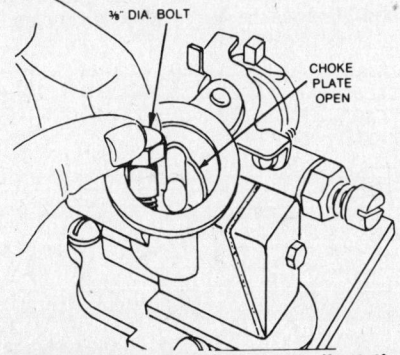

Preloading the diaphragm to adjust the choke

Remove the air cleaner gasket, if it is in place, before continuing. Insert a ⅜ in. bolt or rod into the carburetor air horn. With the bolt in position, tighten the carburetor mounting screws in a staggered sequence. Please note that the insertion of the ⅜ in. bolt opens the choke to an over-center position, which preloads the diaphragm.

Remove the ⅜ in. bolt. The choke valve should now move to a fully closed position. If the choke valve is not fully closed, make sure that the choke spring is properly assembled to the diaphragm, and also properly inserted in its pocket in the tank top.

All carburetor adjustments should be made with the air cleaner on the engine. Adjustment is best made with the fuel tank half full. See the Tune-Up section.

PULSA-JET AND VACU-JET (MODEL SERIES 82000, 92000 ONLY)

Models 82500 and 92500 have a Vacu-Jet carburetor and Models 82900 and 92900 have a Pulsa-Jet carburetor.

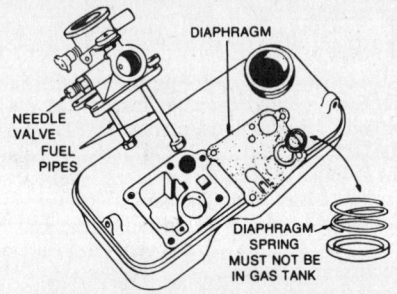

Removal of the Pulsa-Jet diaphragm

Rebuilding

1. Remove the carburetor and fuel tank assembly from the engine by removing the two attaching bolts.
2. Disconnect the governor link at the throttle, leaving the governor link and the governor spring hooked to the governor blade and control lever.
3. Slip the carburetor and tank assembly off of the engine.
4. Remove the carburetor from the tank. Always remove all nylon and rubber parts if the carburetor is soaked in solvent.
5. Remove the O-ring and discard it. Remove and inspect the needle valve, packing and seat.
6. Metering holes in the carburetor body should be cleaned with solvent and compressed air. Do not clean the holes with a pin or a length of wire because of the danger of altering their size.
7. Remove the choke parts on models 82500 and 82900 by pulling the nylon choke shaft sideways to separate the choke shaft from the choke valve. On the 92500 and 92900, remove the choke parts by first disconnecting the choke return spring at the pin in the carburetor body. Then pull the nylon choke shaft sideways to separate the choke shaft from the choke valve.
8. If the choke valve is heat-sealed to the choke shaft, loosen it by sliding a sharp pointed tool along the edge of the choke shaft. Do not re-seal parts on assembly.
9. When replacing the choke valve and shaft, install the choke valve so the poppet valve spring is visible when the valve is in full choke position.

On these models, the nylon fuel pipe is threaded into the carburetor body. Use a socket to remove and replace it. Be careful not to overtighten it and do not use any sealer.

The Pulsa-Jet diaphragm also serves as a gasket between the carburetor and the tank. Inspect the diaphragm for punctures, wrinkles, and wear. Replace it if it is damaged in any way.

To assemble the carburetor to the tank, first position the diaphragm on the tank. Then place the spring cap and spring on the diaphragm. Install the carburetor, tightening the mounting screws evenly to avoid distortion.

To install the carburetor and tank assembly onto the engine, make sure that the governor link is hooked to the governor blade. Connect the link to the throttle and slip the carburetor into place. Align the carburetor with the intake tube and breather tube grommet. Hold the choke lever in the open position so it does not catch on the control plate. Be sure the O-ring in the carburetor does not distort when fitting the carburetor to the intake tube. Install the mounting bolts. Adjust the carburetor as described in the Tune-Up section.

Engine Overhaul

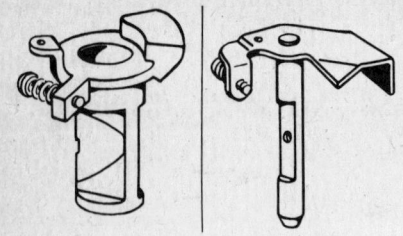

The two types of throttle shafts

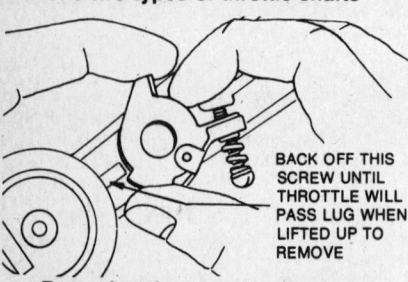

Removing the cast throttle shafts

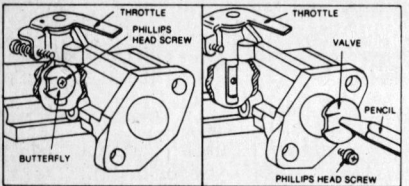

Removing the throttle plate

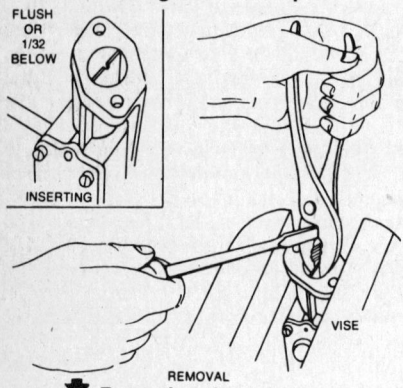

Removing the spiral

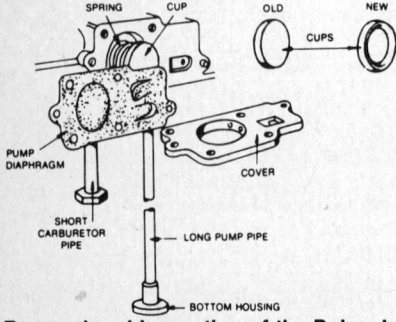

Removal and inspection of the Pulsa-Jet pump cover diaphragm

PULSA-JET CARBURETOR

Throttle Plate Removal

Cast throttle plates are removed by backing off the idle speed adjustment screw until the throttle clears the retaining lug on the carburetor housing.

Stamped throttles are removed by using a phillips screwdriver to remove the throttle valve screw. After removal of the valve, the throttle may be lifted out. Installation is the reverse of removal.

Some carburetors may have a spiral in the carburetor bore. To remove it, fasten the carburetor in a vise about ½ in. below the top of the jaws. Grasp the spiral firmly with a pair of pliers. Place a screwdriver under the edge of the pliers. Using the edge of the vise, push down on the screwdriver to pry out the spiral. When installing the spiral, keep the top flush, or $1/32$ in. below the carburetor flange, and parallel with the fuel tank mounting face.

Fuel Pipe

Check balls are not used in these fuel pipes. The screen housing or pipe must be replaced if the screen cannot be satisfactorily cleaned. The long pipe supplies fuel from the tank to the pump. The short pipe supplies fuel from the tank cup to the carburetor. Fuel pipes are nylon or brass. Nylon pipes are removed and installed by using a socket, or open-end wrench.

NOTE: Where brass pipes are used, replace only the screen housing. The housing is driven off the pipe with a screwdriver with the pipe held in a vise. The new housing is installed by lightly tapping it onto the pipe with a soft hammer.

Needle Valve and Seat

Remove the needle valve to inspect it. If the carburetor is gummy or dirty, remove the seat to allow better cleaning of the metering holes. Do not insert pins or wires in the metering holes. Use solvent or compressed air.

Pump

Remove the fuel pump cover, diaphragm, spring, and cup. Inspect the diaphragm for punctures, cracks, and fatigue. Replace it if damaged. On early models, the spring cap is solid; on later models, the cap has a hole in it. The new style supersedes the old style. When installing the pump cover, tighten the screws evenly to insure a good seal.

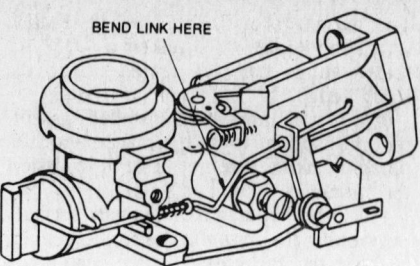

Choke-a-Matic choke adjustment

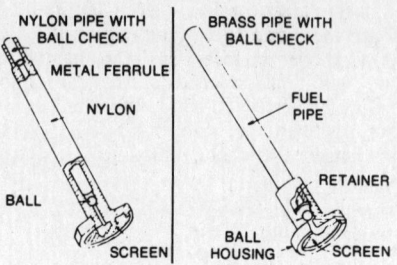

The two types of fuel pipes

VACU-JET CARBURETORS

Vacu-Jet carburetors are removed from the engine together with the fuel tank as one unit. The throttle plates are removed and installed in the same manner as the throttles in the Pulsa-Jet carburetors.

Fuel Pipe

The fuel pipe contains a check ball and a fine mesh screen. To function properly, the screen must be clean and the check ball free. Replace the pipe if the screen and ball cannot be satisfactorily cleaned in carburetor cleaner.

NOTE: Do not leave the carburetor in the cleaner for more than ½ hour without removing all nylon parts. Nylon fuel pipes are removed and replaced with a $9/16$ in. socket. Brass fuel pipes are removed by clamping the pipe in a vise and prying out the pipe with two screwdrivers.

To install the brass fuel pipes, remove the throttle, if necessary, and place the carburetor and pipe in a vise. Press the pipe into the carburetor until it projects $2^{9}/_{32}$–$2^{5}/_{16}$ in. from the carburetor face.

Needle Valve and Seat

Remove the needle valve assembly to inspect it. If the carburetor is gummy or dirty, remove the seat to allow better clean-

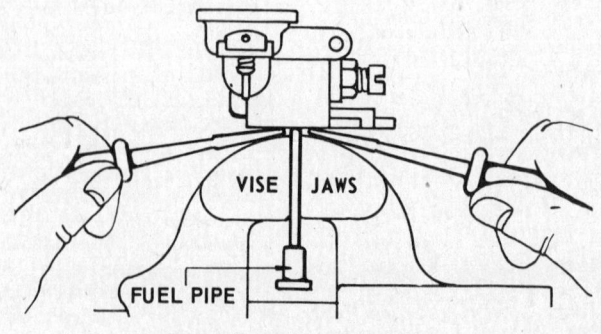

Removing a brass fuel pipe

Engine Overhaul

ing of the metering holes. Do not clean the metering holes with a pin or a length of wire.

Choke-A-Matic Linkage

To remove the choke link, remove the speed adjustment lever and the top switch insulator plate. Work the link out through the hole in the choke slide.

Replace all worn or damaged parts. To assemble a carburetor using a choke slide, place the choke return spring and three washers on the choke link. Push the choke link through the hole in the carburetor body, turning the link to line up with the hole in the choke slide. The speed adjustment lever screw and the stop switch insulator plate should be installed as one assembly after placing the choke link through the end of the speed adjustment lever.

TWO PIECE FLO-JET CARBURETORS (LARGE AND SMALL LINE)

Checking the Upper Body for Warpage

With the carburetor assembled and the body gasket in place, try to insert a 0.002 in. feeler gauge between the upper and lower bodies at the air vent boss, just below the idle valve. If the gauge can be inserted, the upper body is warped and should be replaced.

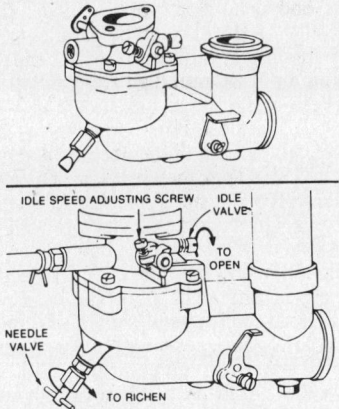

Two-piece Flo-Jet carburetor

Checking the Throttle Shaft and Bushings

Wear between the throttle shaft and bushings should not exceed 0.010 in. Check the wear by placing a short iron bar on the upper carburetor body so that it just fits under the throttle shaft. Measure the distance with a feeler gauge while holding the shaft down and then holding it up. If the difference is over 0.010 in., either the upper body should be rebushed, the throttle shaft replaced, or both. Wear on the throttle shaft can be checked by comparing the worn and unworn portions of the shaft. To replace the bushings, remove the throttle shaft using a thin punch to drive out the pin which holds the throttle stop to the shaft; remove the throttle valve, then pull out the shaft. Place a ¼ in. x 20 tap or an E-Z Out®

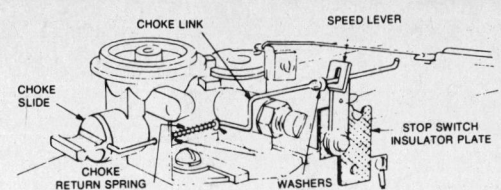

Adjustment of the Choke-a-Matic choke linkage on Vacu-Jet carburetors

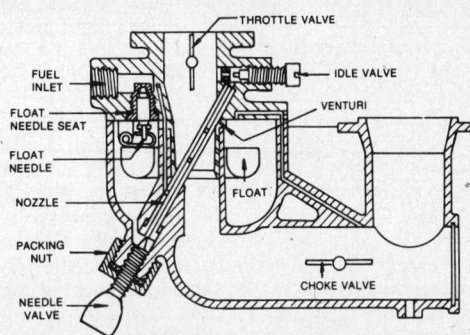

Cutaway view of the Flo-Jet

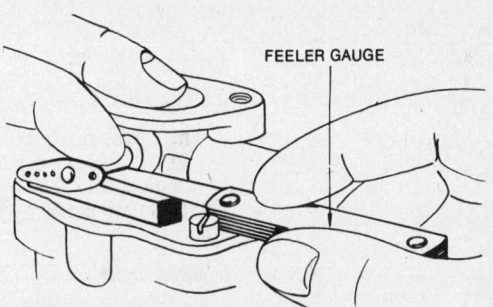

Checking throttle shaft wear

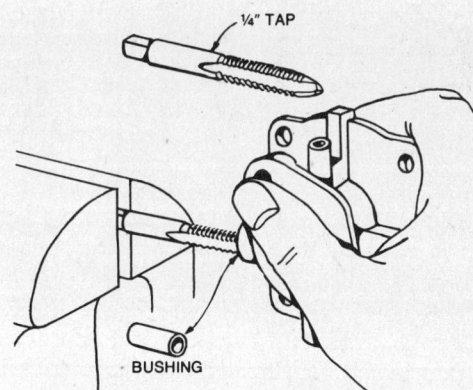

Removing the throttle shaft bushing

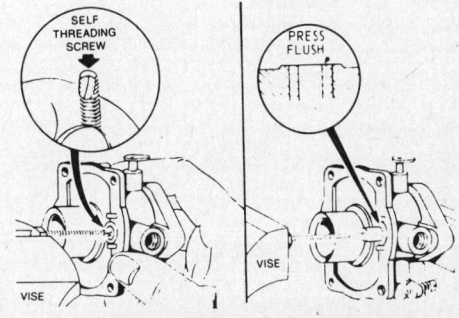

Replacing the float valve seat

879

Engine Overhaul

in a vise. Turn the carburetor body so as to thread the tap or E-Z Out® into the bushings enough to pull the bushings out of the body. Press the new bushings into the carburetor body with a vise. Insert the throttle shaft to be sure it is free in the bushings. If not, run a size 7/32 in. drill through both bushings to act as a line reamer. Install the throttle shaft, valve, and stop.

Disassembly of the Carburetor
1. Remove the idle valve.
2. Loosen the needle valve packing nut.
3. Remove the packing nut and needle valve together. To remove the nozzle, use a narrow, blunt screwdriver so as not to damage the threads in the lower carburetor body. The nozzle projects diagonally into a recess in the upper body and must be removed before the upper body is separated from the lower body, or it may be damaged.
4. Remove the screws which hold the upper and lower bodies together. A pin holds the float in place.
5. Remove the pin to take out the float valve needle. Check the float for leakage. If it contains gasoline or is crushed, it must be replaced. Use a wide, proper fitting screwdriver to remove the float inlet seat.
6. Lift the venturi out of the lower body. Some carburetors have a welch plug. This should be removed only if necessary to remove the choke plate. Some carburetors have nylon choke shaft.

Repair
Use new parts where necessary. Always use new gaskets. Carburetor repair kits are available. Tighten the inlet seat with the gasket securely in place, if used. Some float valves have a spring clip to connect the float valve to the float tang. Others are nylon with a stirrup which fits over the float tang. Older float valves and engines with fuel pumps have neither a spring nor a stirrup.

A viton tip float valve is used in later models of the large, two-piece Flo-Jet carburetor. The seat is pressed into the upper body and does not need replacement unless it is damaged.

Replacing the Pressed-In Float Valve Seat
Clamp the head of a #93029 self threading screw in a vise. Turn the carburetor body to thread the screw into the seat. Continue turning the carburetor body, drawing out the seat. Leave the seat fastened to the screw. Insert the new seat #230996 into the carburetor body. The seat has a starting lead.

NOTE: If the engine is equipped with a fuel pump, install a #231019 seat. Press the new seat flush with the body using the screw and old seat as a driver. Make sure that the seat is not pressed below the body surface or improper float-to-float valve contact will occur. Install the float valve.

Checking the Float Level
With the body gasket in place on the upper body and the float valve and float installed, the float should be parallel to the body mounting surface. If not, bend the tang on the float until they are parallel. Do not press on the flat to adjust it.

Assembly of the Carburetor
Assemble the venturi and the venturi gasket to the lower body. Be sure that the holes in the venturi and the venturi gasket are aligned. Some models do not have a removable venturi. Install the choke parts and welch plug if previously removed. Use a sealer around the welch plug to prevent entry of dirt.

Fasten the upper and lower bodies together with the mounting screws. Screw in the nozzle with a narrow, blunt screwdriver, making sure that the nozzle tip enters the recess in the upper body. Tighten the nozzle securely. Screw in the needle valve and idle valve until they just seat. Back off the needle valve 1½ turns. Do not tighten the packing nut. Back off the idle valve ¾ of a turn. These settings are about correct. Final adjustment will be made when the engine is running. See the Tune-Up section for mixture and choke adjustments.

ONE-PIECE FLO-JET CARBURETOR
The large, one-piece Flo-Jet carburetor has its high speed needle valve below the float bowl. All other repair procedures are similar to the small, one-piece Flo-Jet carburetor.

Disassembly
1. Remove the idle and needle valves.
2. Remove the carburetor bowl screw. A pin holds the float in place.
3. Remove the pin to take off the float and float valve needle. Check the float for leakage. If it contains gasoline or is crushed, it must be replaced. Use a screwdriver to remove the carburetor nozzle. Use a wide, heavy screwdriver to remove the float valve seat, if used.

If it is necessary to remove the choke valve, venturi throttle shaft, or shaft bushings, proceed as follows.
1. Pry out the welch plug.
2. Remove the choke valve, then the shaft. The venturi will then be free to fall out after the choke valve and shaft have been removed.
3. Check the shaft for wear. (Refer to the "Two-Piece Flo-Jet Carburetor" section for checking wear and replacing bushings.)

Repair of the Carburetor
Use new parts where necessary. Always use new gaskets. Carburetor repair kits are available. If the venturi has been removed, install the venturi first, then the carburetor nozzle jets. The nozzle jet holds the venturi in place. Replace the choke shaft and valve. Install a new welch plug in the carburetor body. Use a sealer to prevent dirt from entering.

A viton tip float valve is used in the large, one-piece Flo-Jet carburetor. The seat is pressed in the upper carburetor body and does not need replacement unless it is damaged. Replace the seat in the same manner as for the two-piece Flo-Jet carburetor.

Checking the Float Level
With the body gasket in place on the upper body and float valve and the float installed, the float should be parallel to the body mounting surface. If not, bend the tang on the float until they are parallel. Do not press on the float.

Install the float bowl, idle valve, and needle valve. Turn in the needle valve and the idle valve until they just seat. Open the needle valve 2½ turns and the idle valve 1½ turns. On the large carburetors with the needle valve below the float bowl, open the needle valve and the idle valve 1⅛ turns.

These settings will allow the engine to start. Final adjustment should be made when the engine is running and has warmed up to operating temperature. See the "Two-Piece Flo-Jet Carburetor" adjustment procedure.

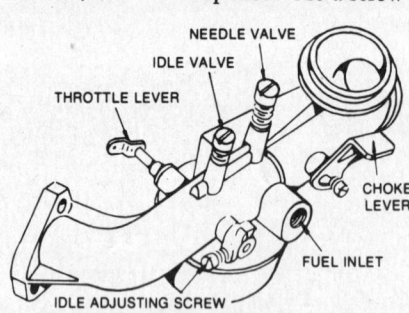

Small one piece Flo-Jet

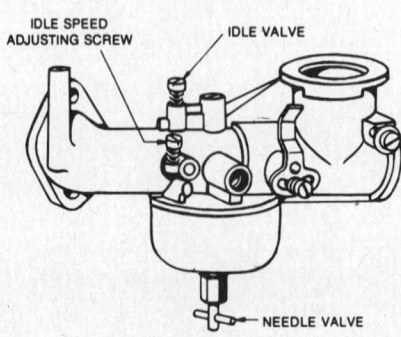

Large one-piece Flo-Jet

FLOAT LEVEL CHART

Carburetor Number	Float Setting (in.)
2712-S	1 9/64
2713-S	1 9/64
2714-S	¼
2398-S*	¼
2336-S	¼
2336-SA	¼
2337-S	¼
2337-SA	¼
2230-S	1 7/64
2217-S	11/64

*When resilient seat is used, set float level at 9/32 ± 1/64.

Engine Overhaul

Governors

The purpose of a governor is to maintain, within certain limits, a desired engine speed even though the load may vary.

AIR VANE GOVERNORS

The governor spring tends to open the throttle. Air pressure against the air vane tends to close the throttle. The engine speed at which these two forces balance is called the governed speed. The governed speed can be varied by changing the governor spring tension.

Worn linkage or damaged governor springs should be replaced to insure proper governor operation. No adjustment is necessary.

MECHANICAL GOVERNORS

The governor spring tends to pull the throttle open. The force of the counterweights, which are operated by centrifugal force, tends to close the throttle. The engine speed at which these two forces balance is called the governed speed. The governed speed can be varied by changing the governor spring tension.

GOVERNOR REPAIR

The procedures below describe disassembly and assembly of the various kinds of mechanical governors. Look for gears with worn or broken teeth, worn thrust washers, weight pins, cups, followers, etc. Replace parts that are worn and reassemble.

MODEL 140000

Disassembly

1. Loosen the governor lever mounting screw, and pull the lever off the shaft.
2. Remove the two housing mounting screws. Carefully pull the housing off the block, being careful to catch the governor gear, which will slip off the shaft. Pull the steel thrust washer off the shaft.
3. Remove the governor lever roll pin and washer. Unscrew the governor lever shaft by turning it clockwise and remove it.

Assembly

1. Push the governor lever shaft into the crankcase cover, threaded end first. Assemble the small washer into the inner end of the shaft, and then screw the shaft into the governor crank follower by turning it counterclockwise. Tighten it securely.
2. Turn the shaft until the follower points down slightly, in a position where it would press against the cup when the housing is installed.
3. Place the washer on the outside end of the shaft. Install the rollpin, so the leading end just reaches the outside diameter of the shaft and the back end protrudes.
4. Install the thrust washer and the governor gear on the shaft in the housing (in that order).
5. Hold the crankcase cover in a vertical (the normal) position and install the housing with the gear in position so the point of the steel cup on the gear contacts the fol-

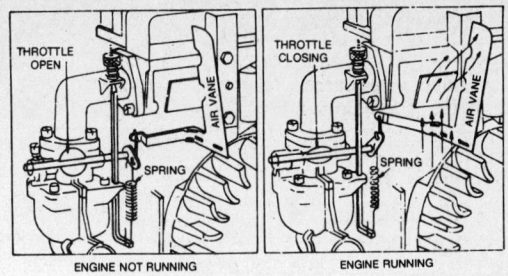
Air vane governor installed on horizontal crankshaft engines

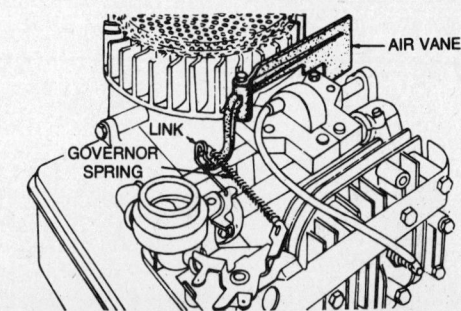
Air vane goveror installed on vertical crankshaft engines

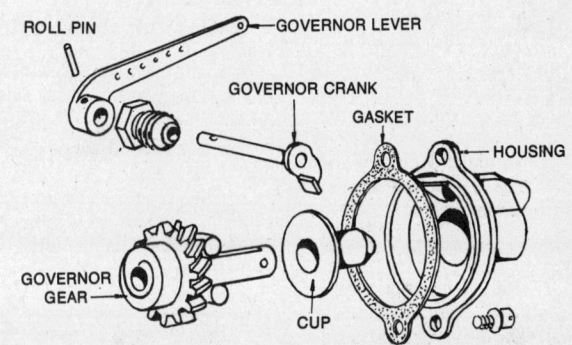
Governor housing and gear assembly

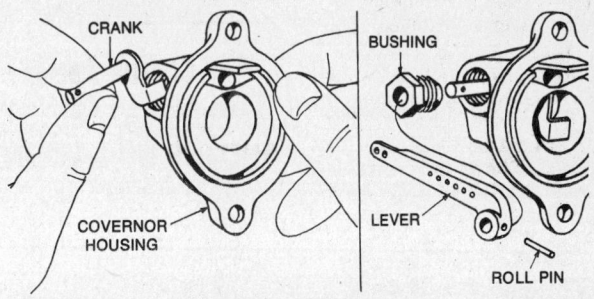
Installing crank and lever

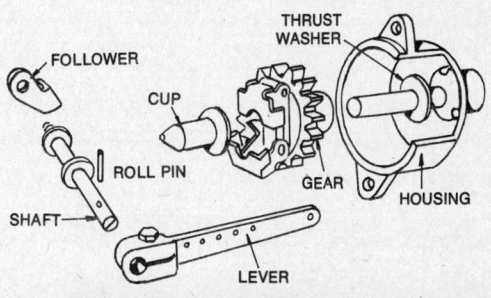
Mechanical governor exploded view

Engine Overhaul

lower. Install and tighten the housing mounting screws.

6. Install the lever on the shaft pointing downward at an angle of about 30 degrees. Adjust as described in the Tune-Up section.

CAST IRON MODELS 19, 190000, 200000, 23, 230000, 240000, 300000, 320000

Disassembly

1. Remove the cotter key and washer from the outer end of the governor shaft. Remove the governor crank from inside the crankcase.
2. Slide the governor gear off the shaft.

Assembly

1. Install the governor gear onto the shaft inside the crankcase. Then, insert the governor shaft assembly through the bushing from inside the crankcase.
2. Install the governor lever to the shaft loosely, and then adjust it as described in the Tune-Up section.

ALUMINUM MODELS 140000, 170000, 190000, 251000

Disassembly

On horizontal shaft models: Remove the governor assembly as a unit from the crankcase cover.

On vertical shaft models: Remove the entire assembly as part of the oil slinger (see the Overhaul Section).

Assembly

1. Assemble governors on horizontal crankshaft models with crankshaft in a horizontal position. The governor rides on a short stationary shaft which is integral with the crankcase cover. The governor shaft keeps the governor from sliding off the shaft after the cover is installed. The governor shaft *must* hang straight down, or it may jam the governor assembly when the crankcase cover is installed, breaking it when the engine is started. The governor shaft adjustment should be made (see the Tune-Up section) as soon as the crankcase cover is in place so that the governor lever will be clamped in the proper position.

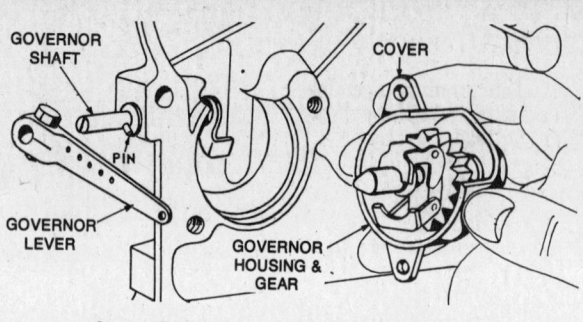

Assembling the mechanical governor

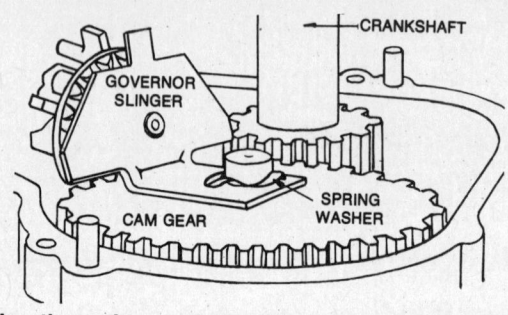

Installing the spring on the camshaft on models 100900 and 130900

2. On both horizontal and vertical crankshaft models, the governor is held together through normal operating forces. For this reason, the governor link and all other external linkages must be in place and properly adjusted whenever the engine is operated.

3. On vertical shaft models 100900 and 130900, be sure the spring washer is in place on the camshaft after the governor is in position.

ENGINE OVERHAUL

Cylinder Head

REMOVAL AND INSTALLATION

Always note the position of the different cylinder head screws so that they can be

CYLINDER HEAD BOLT TORQUE SPECIFICATIONS

Basic Model Series	In. lbs Torque
Aluminum Cylinder 140000, 170000, 190000, 251000	165
Cast Iron Cylinder 19, 190000, 200000, 23, 230000, 240000, 300000, 320000	190

properly reinstalled. If a screw is used in the wrong position, it may be too short and not engage enough threads. If it is too long, it may bottom on a fin, either breaking the fin, or leaving the cylinder head loose.

Remove the cylinder screws and then the cylinder head. Be sure to remove the gasket and all remaining gasket material from the cylinder head and the block.

Assemble the cylinder head with a new gasket, cylinder head shield, screws, and washers in their proper places. Graphite grease should be used on aluminum cylinder head screws.

Do not use a sealer of any kind on the head gasket. Tighten the screws down evenly by hand. Use a torque wrench and tighten the head bolts in the correct sequence.

Valves

REMOVAL AND INSTALLATION

Using a valve spring compressor, adjust the jaws so they touch the top and bottom of the valve chamber, and then place one of

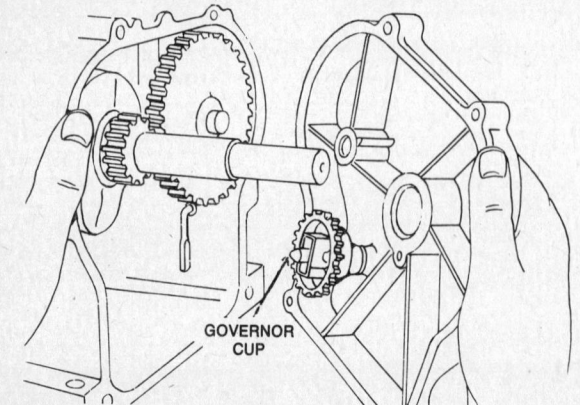

Positioning the governor and shaft for assembly on horizontal crankshaft engines

Engine Overhaul

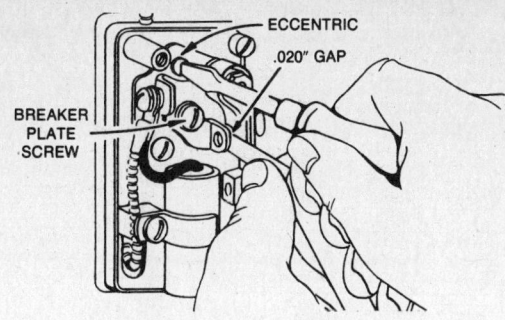

Adjusting the breaker point gap

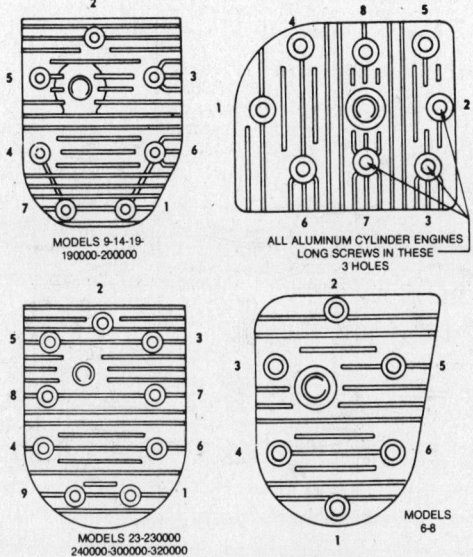

Head bolt torque sequences

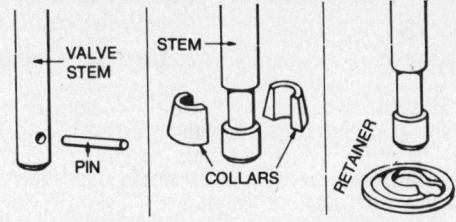

The three types of valve spring retainers

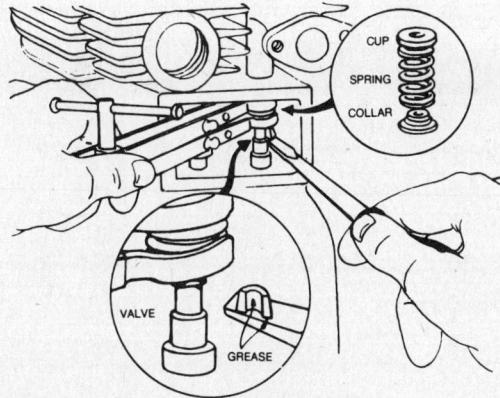

Using a valve spring compressor

the jaws over the valve spring and the other underneath, between the spring and the valve chamber. This positioning of the valve spring compressor is for valves that have either pin or collar type retainers. Tighten the jaws to compress the spring. Remove the collars of pin and lift out the valve. Pull out the compressor and the spring.

To remove valves with ring type retainers, position the compressor with the upper jaw over the top of the valve chamber and the lower jaw between the spring and the retainer. Compress the spring, remove the retainer, and pull out the valve. Remove the compressor and spring.

Before installing the valves, check the thickness of the valve springs. Some engines use the same spring for the intake and exhaust side, while others use a heavier spring on the exhaust side. Compare the springs before installing them.

If the retainers are held by a pin or collars, place the valve spring and retainer and cup (Models 9-14-19-20-23-24-32) into the valve spring compressor. Compress the spring until it is solid. Insert the compressed spring and retainer into the valve chamber. Then drop the valve into place, pushing the stem through the retainer. Hold the spring up in the chamber, hold the valve down, and insert the retainer pin with needle nose pliers or place the collars in the groove in the valve stem. Loosen the spring until the retainer fits around the pin or collars, then pull out the spring compressor. Be sure the pin or collars are in place.

To install valves with ring type retainers, compress the retainer and spring with the compressor. The large diameter of the retainer should be toward the front of the valve chamber. Insert the compressed spring and retainer into the valve chamber. Drop the valve stem through the larger area of the retainer slot and move the compressor so as to center the small area of the valve retainer slot onto the valve stem shoulder. Release the spring tension and remove the compressor.

Valve Guides

Removal and Installation

Models 19, 23, 140000, 170000, 190000, 200000, 230000, 240000, 251000, 300000, 320000

First check valve guide for wear with a plug gauge, Briggs & Stratton part #19151 or equivalent. If the flat end of the valve guide plug gauge can be inserted into the valve guide for a distance of $5/16''$, the guide is worn and should be rebushed in the following manner. See the illustration.

Procure a reamer #19183 and reamer guide bushing #19192, and lubricate the reamer with kerosene. Then, use reamer and reamer guide bushing to ream out the worn guide. Ream to only $1/16''$ deeper than valve guide bushing #230655. BE CAREFUL NOT TO REAM THROUGH THE GUIDE.

Press in valve guide bushing #230655 until top end of bushing is flush with top end of valve guide. Use a soft metal driver

Engine Overhaul

(brass, copper, etc.) to make sure the top end of bushing is not peened over.

The bushing #230655 is finish reamed to size at the factory, so no further reaming is necessary, and a standard valve can be used.

CAUTION
Valve seating should be checked after bushing the guide, and corrected if necessary by refacing the seat.

REFACING VALVES AND SEATS

Faces on valves and valve seats should be resurfaced with a valve grinder or cutter to an angle of 45°.

NOTE: Some engines have a 30° intake valve and seat.

The valve and seat should then be lapped with a fine lapping compound to remove the grinding marks and ensure a good seat. The valve seat width should be 3/64-1/16 in. If the seat is wider, a narrowing stone or cutter should be used. If either the seat or valve is badly burned, it should be re-

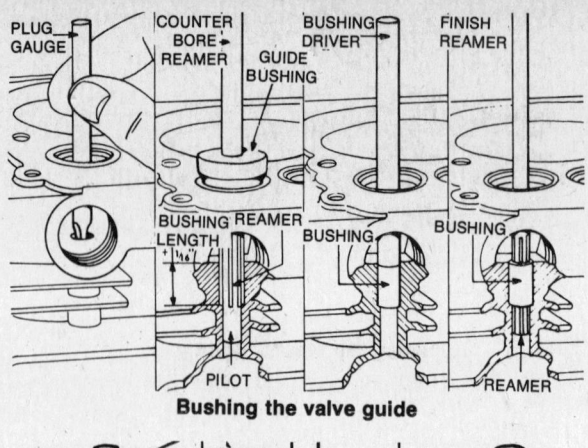
Bushing the valve guide

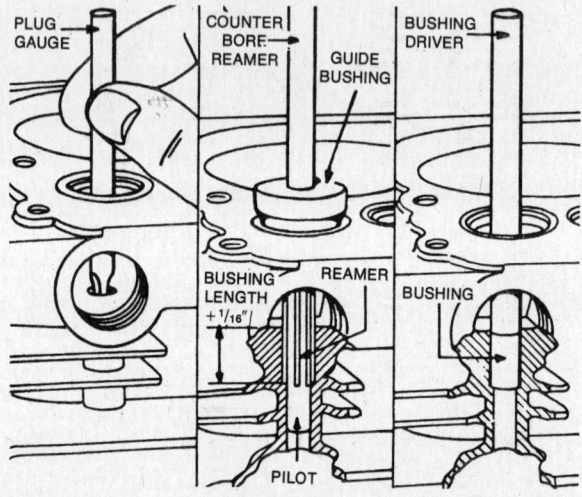
Bushing the valve guide

placed. Replace the valve if the edge thickness (margin) is less than 1/64 in. after it has been resurfaced.

CHECK AND ADJUST TAPPET CLEARANCE

Insert the valves in their respective positions in the cylinder. Turn the crankshaft until one of the valves is at its highest position. Turn the crankshaft one revolution. Check the clearance with a feeler gauge. Repeat for the other valve. Grind off the end of the valve stem if necessary to obtain proper clearance.

NOTE: Check the valve tappet clearance with the engine cold.

Valve Seat Inserts

Cast iron cylinder engines are equipped with an exhaust valve insert which can be removed and replaced with a new insert. The intake side must be counterbored to allow the installation of an intake valve seat insert (see below). Aluminum alloy cylin-

VALVE TAPPET CLEARANCE CHART

Model Series	INTAKE Max.	INTAKE Min.	EXHAUST Max.	EXHAUST Min.
Aluminum Cylinder 140000, 170000, 190000, 251000	.007	.005	.011	.009
Cast Iron Cylinder 19, 190000 200000	.009	.007	.016	.014
23, 230000, 240000, 300000, 320000	.009	.007	.019	.017

VALVE SEAT INSERTS CHART

Basic Model Series	Intake Standard	Exhaust Standard	Exhaust Stellite	Insert Puller Assembly[1]	Puller Nut
Aluminum Cylinder 140000, 170000, 190000	211661	211661	210940[2]	19138	19141
250000	211661	211661	210940	19138	19141
Cast Iron Cylinder 19, 190000	21880	21880	21612	19138	19141
200000, 23, 230000	21880	21880	21612	19138	19141
240000	21880	21612	21612	19138	19141
300000, 320000		21612	21612	19138	19141

[1] Includes puller and #19182, 19141, 19140 and 19139 nuts [2] Before code #7101260 use #211892

Engine Overhaul

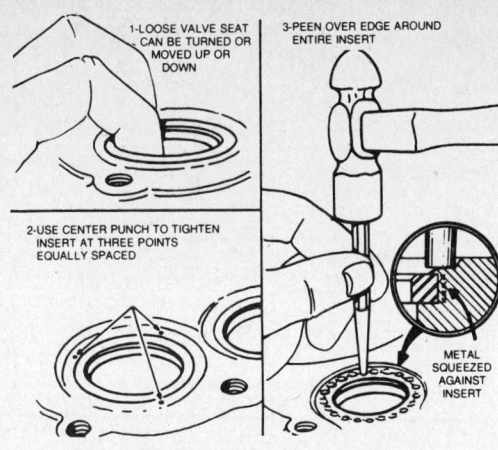
Installing valve seat inserts

der models are equipped with inserts on both the exhaust and intake valves.

REMOVAL AND INSTALLATION

Valve seat inserts are removed with a special puller.

NOTE: On Aluminum alloy cylinder models, it may be necessary to grind the puller nut until the edge is $1/32$ in. thick in order to get the puller nut under the valve insert.

When installing the valve seat insert, make sure that the side with the chamfered outer edge goes down into the cylinder. Install the seat insert and drive it into place with a driver. The seat should then be ground lightly and the valves and seats lapped lightly with grinding compound.

NOTE: Aluminum alloy cylinder models use the old insert as a spacer between the driver and the new insert. Drive in the new insert until it bottoms. The top of the insert will be slightly below the cylinder head gasket surface. Peen around the insert using a punch and hammer.

NOTE: The intake valve seat on cast iron cylinder models has to be counterbored before installing the new valve seat insert.

COUNTERBORING CYLINDER FOR INTAKE VALVE SEAT ON CAST IRON MODELS

1. Select the proper seat insert, cutter shank, counter bore cutter, pilot and driver from the table. These numbers refer to Briggs & Stratton parts—you may get equivalent parts from other sources if available.
2. With cylinder head resting on a flat surface, valve seats up, slide the pilot into the intake valve guide. Then, assemble the correct counterbore cutter to the shank with the cutting blades of the cutter downward.
3. Insert the cutter straight into the valve seat, over the pilot. Cut so as to avoid forcing the cutter to one side, and be sure to stop as soon as the stop on the cutter touches the cylinder head.
4. Blow out all cutting chips thoroughly.

Pistons, Piston Rings, and Connecting Rods

REMOVAL

To remove the piston and connecting rod from the engine, bend down the connecting rod lock. Remove the connecting rod cap. Remove any carbon or ridge at the top of the cylinder bore. This will prevent breaking the rings. Push the piston and rod out of the top of the cylinder.

Pistons used in sleeve bore, aluminum alloy engines are marked with an "L" on top of the piston. These pistons are tin plated and use an expander with the oil ring. This piston assembly is not interchangeable with the piston used in the aluminum bore engines (Kool bore).

Pistons used in aluminum bore (Kool bore) engines are not marked on the top.

To remove the connecting rod from the piston, remove the piston pin lock with thin nose pliers. One end of the pin is drilled to facilitate removal of the lock.

Remove the rings one at a time, slipping them over the ring lands. Use a ring expander to remove the rings.

INSPECTION

Check the piston ring fit. Use a feeler gauge to check the side clearance of the top ring. Make sure that you remove all carbon from the top ring groove. Use a new piston ring to check the side clearance. If the cylinder is to be resized, there is no reason to check the piston, since a new oversized

CONNECTING ROD BEARING SPECIFICATIONS

Basic Model Series	Crank Pin Bearing	Piston Pin Bearing
Aluminum Cylinder		
140000, 170000	1.095	.674
190000	1.127	.674
251000	1.252	.802
Cast Iron Cylinder		
19, 190000	1.001	.674
200000	1.127	.674
23, 230000	1.189	.736
240000	1.314	.674
300000, 320000	1.314	.802

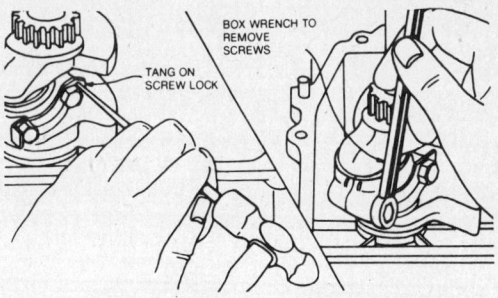
Removing the connecting rod caps

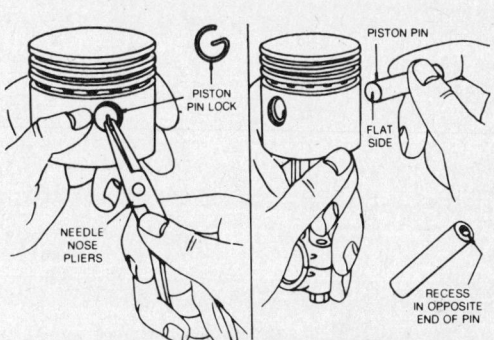
Removing the wrist pin and connecting rod

Engine Overhaul

piston assembly will be installed. If the side clearance is more than 0.007 in., the piston is excessively worn and should be replaced.

Check the piston ring end gap by cleaning all carbon from the ends of the rings and inserting them one at a time 1 in. down into the cylinder. Check the end gap with a feeler gauge. If the gap is larger than recommended, the ring should be replaced.

NOTE: When checking the ring gap, do not deglaze the cylinder walls by installing piston rings in aluminum cylinder engines.

Chrome ring sets are available for all current aluminum and cast iron cylinder models. No honing or deglazing is required. The cylinder bore can be a maximum of 0.005 in. oversize when using chrome rings.

If the crankpin bearing in the rod is scored, the rod must be replaced. 0.005 in. oversize piston pins are available in case the connecting rod and piston are worn at the piston pin bearing. If, however, the crankpin bearing in the connecting rod is worn, the rod should be replaced. Do not attempt to file or fit the rod.

If the piston pin is worn 0.0005 in. out of round or below the rejection sizes, it should be replaced.

PISTON RING GAP SPECIFICATIONS

Basic Model Series	Comp. Ring	Oil Ring
Aluminum Cylinder 140000, 170000, 190000, 251000	.035	.045
Cast Iron Cylinder 19, 190000, 200000, 23, 230000, 240000, 300000, 320000	.035	.035

WRIST PIN SPECIFICATIONS

Basic Model Series	Piston Pin	Pin Bore
Aluminum Cylinder 140000, 170000, 190000	.671	.671
251000	.799	.801
Cast Iron Cylinder 19, 190000	.671	.673
200000	.671	.673
23, 230000	.734	.736
240000	.671	.673
300000, 320000	.799	.801

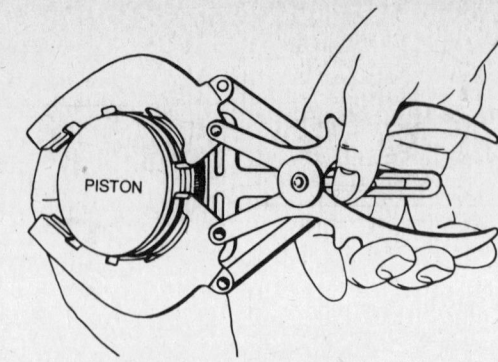

Replacing the piston rings

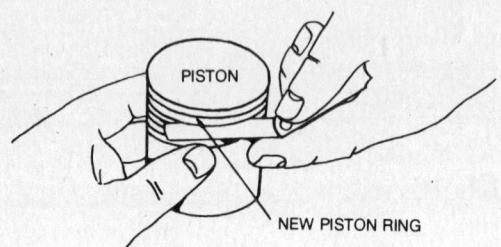

Measuring the ring side gap

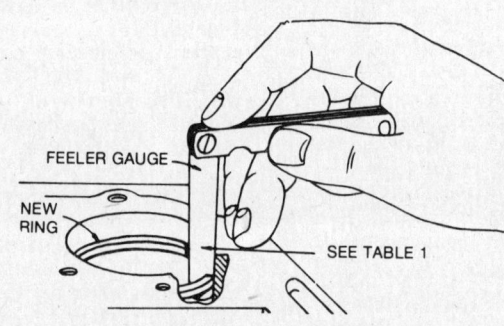

Measuring the ring end gap

INSTALLATION

The piston pin is a push fit into both the piston and the connecting rod. On models using a solid piston pin, one end is flat and the other end is recessed. Other models use a hollow piston pin. Place a pin lock in the groove at one side of the piston. From the opposite side of the piston, insert the piston pin, flat end first for solid pins; with hollow pins, insert either end first until it stops against the pin lock. Use thin nose pliers to assemble the pin lock in the recessed end of the piston. Be sure the locks are firmly set in the groove.

Install the rings on the pistons, using a piston ring expander. Make sure that they are installed in the proper position. The scraper groove on the center compression ring should always be down toward the piston skirt. Be sure the oil return holes are clean and all carbon is removed from the grooves.

NOTE: Install the expander under the oil ring in sleeve bore aluminum alloy engines.

Oil the rings and the piston skirt, then compress the rings with a ring compressor. On cast iron engines, install the compressor with the two projections downward; on aluminum engines, install the compressor with the two projections upward. These instructions refer to the piston in normal position—with skirt downward. Turn the piston and compressor upside down on the bench and push downward so the piston head and the edge of the compressor band are even, all the while tightening the compressor. Draw the compressor up tight to fully compress the rings, then loosen the compressor very slightly.

CAUTION
Do not attempt to install the piston and ring assembly without using a ring compressor.

Place the connecting rod and piston assembly, with the rings compressed, into the cylinder bore. Push the piston and rod down into the cylinder. Oil the crankpin of the crankshaft. Pull the connecting rod against the crankpin and assemble the rod cap so the assembly marks align.

NOTE: Some rods do not have assembly marks, as the rod and cap will fit

Engine Overhaul

together only in one position. Use care to ensure proper installation. On the 251000 engine, the piston has a notch on the top surface. The notch must face the flywheel side of the block when installed. On models 300000 and 320000, the piston has an identification mark "F" located next to the piston pin bore. The mark must appear on the same side as the assembly mark on the rod. The assembly mark on the rod is also used to identify rod and cap alignment. Note, on these pistons, that the top ring has a beveled upper surface on the outside, while the center ring has a flat outer surface. The "F" mark or notch must face the flywheel when the piston is installed.

Where there are flat washers under the cap screws, remove and discard them prior to installing the rod. Assemble the capscrews and screw locks with the oil dippers (if used), and torque to the figure shown in the chart to avoid breakage or rod scoring later. Turn the crankshaft two revolutions to be sure the rod is correctly installed. If the rod strikes the camshaft, the connecting rod has been installed wrong or the cam gear is out of time. If the crankshaft operates freely, bend the cap screw locks against the screw heads. After tightening the rod screws, the rod should be able to move sideways on the crankpin of the shaft.

CONNECTING ROD CAPSCREW TORQUE

Basic Model Series	Inch lbs Avg. Torque
Aluminum Cylinder 140000, 170000, 190000	165
251000	185
Cast Iron Cylinder 19, 190000, 200000	190
23, 230000	190
240000, 300000, 320000	190

Crankshaft and Camshaft Gear

REMOVAL

Aluminum Cylinder Engines

To remove the crankshaft from aluminum alloy engines, remove any rust or burrs from the power take-off end of the crankshaft. Remove the crankcase cover or sump. If the sump or cover sticks, tap it lightly with a soft hammer on alternate sides near the dowel. Turn the crankshaft to align the crankshaft and camshaft timing marks, lift out the cam gear, then remove

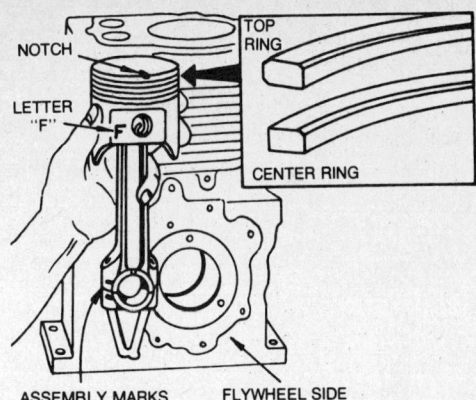
Assembling the piston and connecting rod

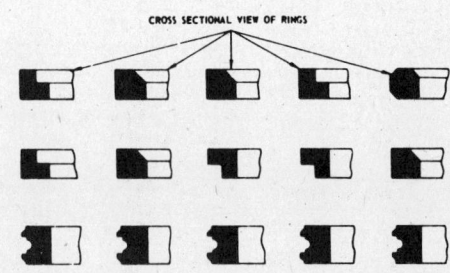
Various types of Briggs and Stratton piston rings

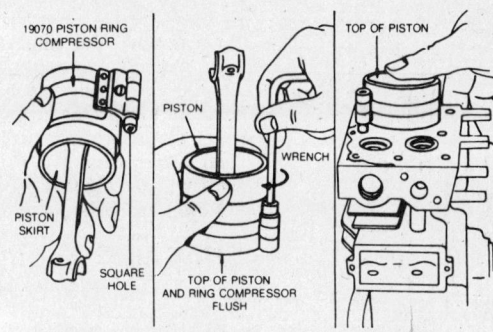
Installing the piston and rod assembly

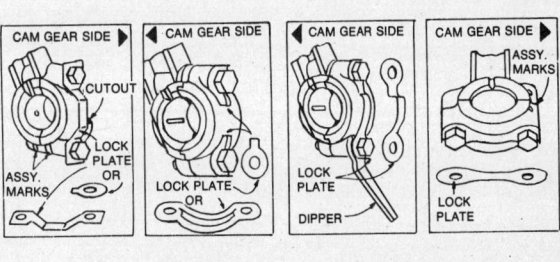

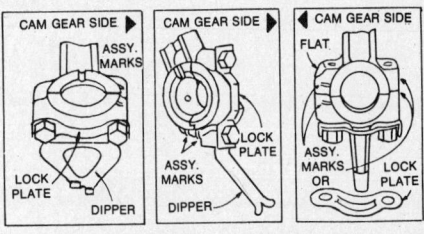
Connecting rod installation

Engine Overhaul

the crankshaft. On models that have ball bearings on the crankshaft, the crankshaft and the camshaft must be removed together with the timing marks properly aligned—see illustration.

Cast Iron Cylinder Models

To remove the crankshaft from cast iron models (19-190000-200000-23-230000-240000-300000-320000), remove the crankcase cover. Revolve the crankshaft until the crankpin is pointing upward toward the breather at the rear of the engine

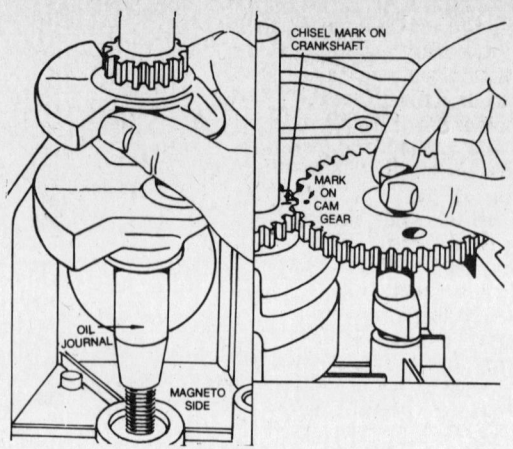

Timing mark alignment

CRANKSHAFT SPECIFICATIONS

Basic Model Series	PTO Journal	Mag. Journal	C Crank-pin
Aluminum Cylinder			
140000, 170000	1.179	.997①	1.090
190000	1.179	.997①	1.122
251000	1.376	1.376	1.247
Cast Iron Cylinder			
19, 190000	1.179	1.179	.996
200000	1.179	1.179	1.122
23, 230000②	1.376	1.376	1.184
240000	Ball	Ball	1.309
300000, 320000	Ball	Ball	1.309

①Synchro balanced magneto bearing reject size—1.179
②Gear reduction P.T.O.—1.179

CAMSHAFT SPECIFICATIONS

Basic Model Series	Cam Gear or Shaft Journals	Cam Lobe
Aluminum Cylinder		
140000, 170000, 190000	.498	.977
251000	.498	1.184
Cast Iron Cylinder		
19, 190000	.497	1.115
200000	.497	1.115
23, 230000	.497	1.184
240000	.497	1.184
300000	①	1.184
320000	①	1.215

①Magneto side—.8105, P.T.O. side—.6145

(approximately a 45° angle). Pull the crankshaft out from the drive side, twisting it slightly if necessary. On models with ball bearings on the crankshaft, both the crankcase cover and bearing support should be removed.

On cast iron models with ball bearings on the drive side, first remove the magneto. Drive out the camshaft. Push the camshaft forward into the recess at the front of the engine. Then draw the crankshaft from the magneto side of the engine. Double thrust engines have capscrews inside the crankcase which hold the bearing in place. These must be removed before the crankshaft can be removed.

To remove the camshaft from all cast iron models, except the 300400 and 320400, use a long punch to drive the camshaft out toward the magneto side. Save the plug. Do not burr or peen the end of the shaft while driving it out. Hold the camshaft while removing the punch, so it will not drop and become damaged.

CHECKING THE CRANKSHAFT

Discard the crankshaft if it is worn beyond the allowable limit. Check the key ways for wear and make sure they are not spread. Remove all burrs from the key way to prevent scratching the bearing. Check the three bearing journals, drive end, crankpin, and magneto end, for size and any wear or damage. Check the cam gear teeth for wear. They should not be worn at all. Check the threads at the magneto end for damage. Make sure that the crankshaft is straight.

NOTE: There are 0.020 in. undersize connecting rods available for use on reground crankpin bearings.

CHECKING THE CAMSHAFT GEAR

Inspect the teeth for wear and nicks. Check the size of the camshaft and camshaft gear bearing journals. Check the size of the cam lobes. If the cam is worn beyond tolerance, discard it.

Check the automatic spark advance on models equipped with the Magna-Matic ignition system. Place the cam gear in the normal operating position with the movable weight down. Press the weight down and release it. The spring should lift the weight. If not, the spring is stretched or the weight is binding.

REMOVAL AND INSTALLATION OF THE BALL BEARINGS

The ball bearings are pressed onto the crankshaft. If either the bearing or the crankshaft is to be removed, use an arbor press to remove them.

To install, heat the bearing in hot oil (325° F maximum). Don't let the bearing rest on the bottom of the pan in which it is heated. Place the crankshaft in a vise with the bearing side up. When the bearing is quite hot, it will slip fit onto the bearing

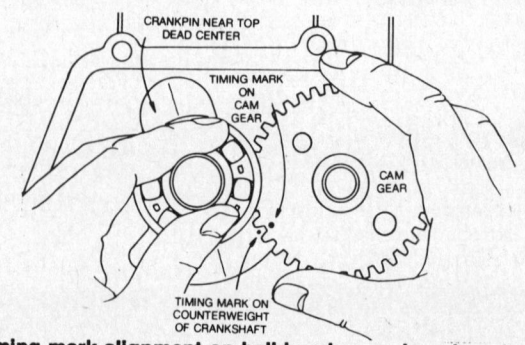

Timing mark alignment on ball bearing equipped engines

Engine Overhaul

journal. Grasp the bearing, with the shield down, and thrust it down onto the crankshaft. The bearing will tighten on the shaft while cooling. Do not quench the bearing (throw water on it to cool it).

INSTALLATION

Aluminum Alloy Engines—Plain Bearing

In aluminum alloy engines, the tappets are inserted first, the crankshaft next, and then the cam gear. When inserting the cam gear, turn the crankshaft and the cam gear so that the timing marks on the gears align.

Aluminum Alloy Engines—Ball Bearing

On crankshafts with ball bearings, the gear teeth are not visible for alignment of the timing marks; therefore, the timing mark is on the counterweight. On ball bearing equipped engines, the tappets are installed first. The crankshaft and the cam gear must be inserted together and their timing marks aligned.

Crankshaft Cover and Crankshaft

INSTALLATION

Cast Iron Engines—Plain Bearings

Assemble the tappets to the cylinder, then insert the cam gear. Push the camshaft into the camshaft hole in the cylinder from the flywheel side through the cam gear. With a blunt punch, press or hammer the camshaft until the end is flush with the outside of the cylinder on the power takeoff side. Place a small amount of sealer on the camshaft plug, then press or hammer it into the camshaft hole in the cylinder at the flywheel side. Install the crankshaft so the timing marks on the teeth and on the cam gear align.

Cast Iron—Ball Bearings

Assemble the tappets, then insert the cam gear into the cylinder, pushing the cam gear forward into the recess in front of the cylinder. Insert the crankshaft into the cylinder. Turn the camshaft and crankshaft until the timing marks align, then push the cam gear back until it engages the gear on the crankshaft with the timing marks together. Insert the camshaft. Place a small amount of sealer on the camshaft plug and press or hammer it into the camshaft hole in the cylinder at the flywheel side.

Crankshaft End-Play Adjustment

The crankshaft end-play on all models, plain and ball bearing, should be 0.002 in. to 0.008 in. The method of obtaining the correct end-play varies, however, between cast iron, aluminum, plain, and ball bearing models. New gasket sets include three crankcase cover or bearing support gaskets, 0.005 in., 0.009 in., and 0.015 in. thick.

The end-play of the crankshaft may be checked by assembling a dial indicator on the crankshaft with the pointer against the crankcase. Move the crankshaft in and out. The indicator will show the end-play. Another way to measure the end-play is to assemble a pulley to the crankshaft and measure the end-play with a feeler gauge. Place the feeler gauge between the crankshaft thrust face and the bearing support. The feeler gauge method of measuring crankshaft end-play can only be used on cast iron plain bearing engines with removable bases.

On cast iron engines, the end-play should be 0.002 in. to 0.008 in. with one 0.015 in. gasket in place. If the end-play is less than 0.002 in., which would be the case if a new crankcase or sump cover is used, additional gaskets of 0.005 in., 0.009 in., or 0.015 in. may be added in various combinations to obtain the proper end-play.

Aluminum Engines Only

If the end-play is more than 0.008 in. with one 0.015 in. gasket in place, a thrust washer is available to be placed on the crankshaft power take-off end, between the gear and crankcase cover or sump on plain bearing engines. On ball bearing equipped aluminum engines, the thrust washer is added to the magneto end of the crankshaft instead of the power take-off end.

NOTE: Aluminum engines never use less than the 0.015 in. gasket.

Cylinders

INSPECTION

Always inspect the cylinder after the engine has been disassembled. Visual inspection will show if there are any cracks, stripped bolt holes, broken fins, or if the cylinder wall is scored. Use an inside micrometer or telescoping gauge and micrometer to measure the size of the cylinder bore. Measure at right angles.

If the cylinder bore is more than 0.003 in. oversize, or 0.0015 in. out of round on lightweight (aluminum) cylinders, the cylinder must be resized (rebored).

NOTE: Do not deglaze the cylinder walls when installing piston rings in aluminum cylinder engines. Also be aware that there are chrome ring sets available for most engines. These are used to control oil pumping in bores worn to 0.005 in. over standard and do not require honing or glaze breaking to seat.

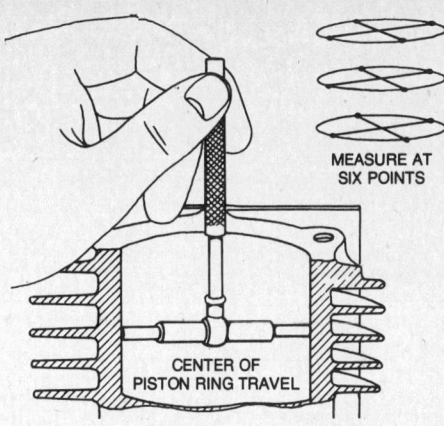

Checking the cylinder bore

RESIZING

Always resize to exactly 0.010 in., 0.020 in., or 0.030 in. over standard size. If this is done accurately, the stock oversize rings and pistons will fit perfectly and proper clearances will be maintained. Cylinders, either cast iron or lightweight, can be quickly resized with a good hone. Use the stones and lubriation recommended by the hone manufacturer to produce the correct cylinder wall finish for the various engine models.

If a boring bar is used, a hone must be used after the boring operation to produce the proper cylinder wall finish. Honing can be done with a portable electric drill, but it is easier to use a drill press.

1. Clean the cylinder at top and bottom to remove all burrs and pieces of base and head gaskets.

2. Fasten the cylinder to a heavy iron plate. Some cylinders require shims. Use a level to align the drill press spindle with the bore.

3. Oil the surface of the drill press table liberally. Set the iron plate and the cylinder on the drill press table. Do not anchor the cylinder to the drill press table. If you are using a portable drill, set the plate and the cylinder on the floor.

4. Place the hone driveshaft in the chuck of the drill.

5. Slip the hone into the cylinder. Connect the driveshaft to the hone and set the stop on the drill press so the hone can only extend ¾ in. to 1 in. from the top or bottom of the cylinder. If you are using a portable drill, cut a piece of wood to place in the cylinder as a stop for the hone.

6. Place the hone in the middle of the cylinder bore. Tighten the adjusting knob with your finger or a small screwdriver until the stones fit snugly against the cylinder wall. Do not force the stones against the cylinder wall. The hone should operate at a speed of 300-700 rpm. Lubricate the hone as recommended by the manufacturer.

Engine Overhaul

NOTE: Be sure the cylinder and the hone are centered and aligned with the driveshaft and the drill spindle.

7. Start the drill and, as the hone spins, move it up and down at the lower end of the cylinder. The cylinder is not worn at the bottom but is round so it will act to guide the hone and straighten the cylinder bore. As the bottom of the cylinder increases in diameter, gradually increase your strokes until the hone travels the full length of the bore.

NOTE: Do not extend the hone more than ¾ in. to 1 in. past either end of the cylinder bore.

8. As the cutting tension decreases, stop the hone and tighten the adjusting knob. Check the cylinder bore frequently with an accurate micrometer. Hone 0.0005 in. oversize to allow for shrinkage when the cylinder cools.

9. When the cylinder is within 0.0015 in. of the desired size, change from the rough stone to a finishing stone.

The finished resized cylinder should have a cross-hatched appearance. Proper stones, lubrication, and spindle speed along with rapid movement of the hone within the cylinder during the last few strokes, will produce this finish. Cross-hatching provides proper lubrication and ring break-in.

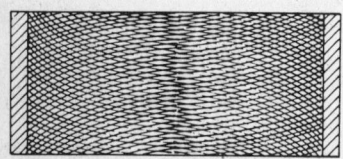

Cross hatch pattern after honing

NOTE: It is EXTREMELY important that the cylinder be thoroughly cleaned after honing to eliminate ALL grit. Wash the cylinder carefully in a solvent such as kerosene. The cylinder bore should be cleaned with a brush, soap, and water.

CYLINDER BORE SPECIFICATIONS

Basic Engine Model or Series	Std. Bore Size Diameter	
	Max.	Min.
Aluminum Cylinder		
140000	2.750	2.749
170000, 190000	3.000	2.999
251000	3.4375	3.4365
Cast Iron Cylinder		
19, 23, 190000 200000	3.000	2.999
230000	3.000	2.999
243400	3.0625	3.0615
300000	3.4375	3.4365
320000	3.5625	3.5615

CRANKSHAFT BEARING SPECIFICATIONS

Basic Engine Model or Series	PTO Bearing	Bearing Magneto
Aluminum Cylinder		
140000, 170000	1.185	1.004
190000	1.185	1.004
251000	1.383	1.383
Cast Iron Cylinder		
19, 190000, 200000	1.185	1.185
23, 230000	1.382	1.382
240000, 300000	Ball	Ball
320000	Ball	Ball

Bearings

INSPECTION
Plain Type

Bearings should be replaced if they are scored or if the plug gauge will enter. Try the gauge at several points in the bearing.

REPLACING PLAIN BEARINGS
Models 9-14-19-20-23

The crankcase cover bearing support should be replaced if the bearing is worn or scored.

REPLACING THE MAGNETO BEARING
Aluminum Cylinder Engines

There are no removable bearings in these engines. The cylinder must be reamed out so a replacement bushing can be installed.

1. Place a pilot guide bushing in the sump bearing, with the flange of the guide bushing toward the inside of the sump.
2. Assemble the sump on the cylinder. Make sure that the pilot guide bushing does not fall out of place.
3. Place the guide bushing into the oil seal recess in the cylinder. This guide bushing will center the counterbore reamer even though the oil bearing surface might be badly worn.
4. Place the counterbore reamer on the pilot and insert them into the cylinder until the tip of the pilot enters the pilot guide bushing in the sump.
5. Turn the reamer clockwise with a steady, even pressure until it is completely through the bearing. Lubricate the reamer with kerosene or any other suitable solvent.

NOTE: Counterbore reaming may be performed without any lubrication. However, clean off shavings because aluminum material builds up on the reamer flutes causing eventual damage to the reamer and an oversize counterbore.

6. Remove the sump and pull the reamer out without backing it through the bearing. Clean out the remaining chips. Remove the guide bushing from the oil seal recess.
7. Hold the new bushing against the outer end of the reamed out bearing, with the notch in the bushing aligned with the notch in the cylinder. Note the position of the split in the bushing. At a point in the outer edge of the reamed out bearing opposite to the split in the bushing, make a notch in the cylinder hub at a 45° angle to the bearing surface. Use a chisel or a screwdriver and hammer.
8. Press in the new bushing, being careful to align the oil notches with the driver and the support until the outer end of the bushing is flush with the end of the reamed cylinder hub.
9. With a blunt chisel or screwdriver, drive a portion of the bushing into the notch previously made in the cylinder. This is called staking and is done to prevent the bushing from turning.
10. Reassemble the sump to the cylinder with the pilot guide bushing in the sump bearing.
11. Place a finishing reamer on the pilot and insert the pilot into the cylinder bearing until the tip of the pilot enters the pilot guide bushings in the sump bearing.
12. Lubricate the reamer with kerosene, fuel oil, or other suitable solvent, then ream the bushing, turning the reamer clockwise with a steady even pressure until the reamer is completely through the bearing. Improper lubricants will produce a rough bearing surface.
13. Remove the sump, reamer, and the pilot guide bushing. Clean out all reaming chips.

REPLACING THE PTO BEARING
Aluminum Cylinder Engines

The sump or crankcase bearing is repaired the same way as the magneto end bearing. Make sure to complete repair of one bearing before starting to repair the other. Press in new oil seals when bearing repair is completed.

REPLACING OIL SEALS

Note the following points:
1. Assemble the seal with the sharp edge of leather or rubber toward the inside of the engine.
2. Lubricate the inside diameter of the seal with Lubriplate® or equivalent.
3. Press all seals so they are flush with the hub.

LUBRICATION

The primary purpose of oil is, of course, lubrication. However, oil performs three very other important functions as well: cooling, cleaning, and sealing. Oil absorbs

Engine Overhaul

and dissipates heat created by combustion and friction. Oil cleans by trapping and holding dirt and by-products of combustion. This dirt is held in suspension by the oil until it is drained. Oil seals the combustion chamber by coating the rings, thus helping increase and maintain compression.

Briggs and Stratton engines are lubricated with a gear driven splash oil slinger or a connecting rod dipper.

Extended Oil Filler Tubes and Dipsticks

When installing the extended oil fill and dipstick assembly, the tube must be installed so the O-ring seal is firmly compressed. To do so, push the tube downward toward the sump, then tighten the blower housing screw, which is used to secure the tube and bracket. When the dipstick assembly is fully depressed, it seals the upper end of the tube.

A leak at the seal between the tube and the sump, or at the seal at the upper end of the dipstick can result in a loss of crankcase vacuum, and a discharge of smoke through the exhaust system.

Breathers

The function of the breather is to maintain a vacuum in the crankcase. The breather has a fiber disc valve which limits the direction of air flow caused by the piston moving back and forth in the cylinder. Air can flow out of the crankcase, but the one-way valve blocks the return flow, thus maintaining a vacuum in the crankcase. A partial vacuum must be maintained in the crankcase to prevent oil from being forced out of the engine at the piston rings, oil seals, breaker plunger, and gaskets.

INSPECTION OF THE BREATHER

If the fiber disc valve is stuck or binding, the breather cannot function properly and must be replaced. A 0.045 in. wire gauge

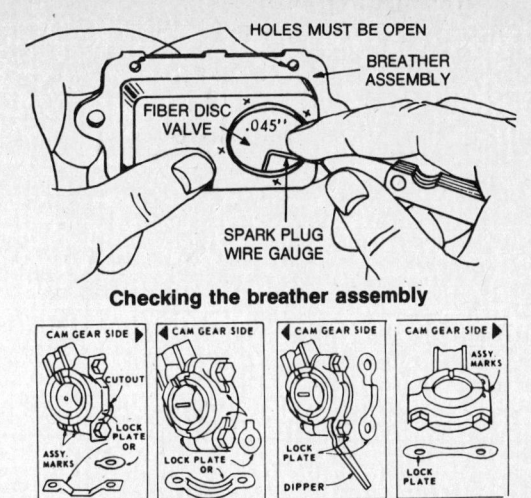

Checking the breather assembly

Installing the connecting rod in a horizontal crankshaft engine

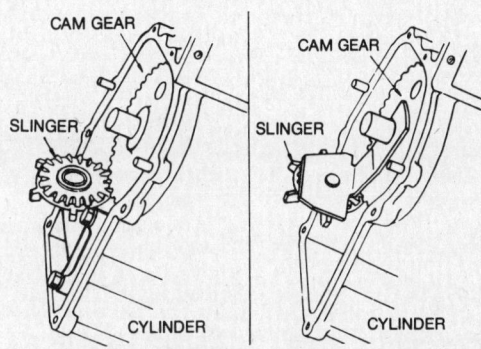

Oil slinger installation

should not enter the space between the fiber disc valve and the body. Use a spark plug wire gauge to check the valve. The fiber disc valve is held in place by an internal bracket which will be distorted if pressure is applied to the fiber disc valve. Therefore, do not apply force when checking the valve with the wire gauge.

If the breather is removed for inspection or valve repair, a new gasket should be used when replacing the breather. Tighten the screws securely to prevent oil leakage.

Most breathers are now vented through the air cleaner, to prevent dirt from entering the crankcase. Check to be sure that the venting elbows or the tube are not damaged and that they are properly sealed.

Oil Dippers and Slingers

Oil dippers reach into the oil reservoir in the base of the engine and splash oil onto the internal engine parts. The oil dipper is installed on the connecting rod and has no pump or moving parts.

Oil slingers are driven by the cam gear. Old style slingers using a die cast bracket assembly have a steel bushing between the slinger and the bracket. Replace the bracket on which the slinger rides if it is worn to a diameter of 0.490 in. or less. Replace the steel bushing if it is worn. Newer style oil slingers have a stamped steel bracket.

KOHLER 4-STROKE

ENGINE IDENTIFICATION

An engine identification plate is mounted on the carburetor side of the engine blower housing. The numbers that are important, as far as ordering replacement parts is concerned, are the model, serial, and specification numbers.

The model number indicates the engine model series. It also is a code indicating the cubic inch displacement and the number of cylinders. The model number K181, for instance, indicates the engine is 18 cu. in. in displacement and that it has 1 cylinder. The letters following the model number indicate that a variety of other equipment is installed on the engine. The letters and what they mean are as follows:

C	Clutch model
G	Housed with fuel tank
H	Housed less fuel tank
P	Pump model
R	Reduction gear
S	Electric start
T	Retractable start

NOTE: A model number without a suffix letter indicates a basic rope start version.

The specification number indicates model variation. It indicates a combination of various groups used to build the engine. It may have a letter preceding it which is sometimes important in determining superseding parts. The first two numbers of the specifications number is the code designating the engine model; the remaining numbers are issued in numerical sequence as each new specification is released, for example, 2899, 28100, 28101, etc. The

Engine Overhaul

current specification number model code is as follows:

K141-29
K161-28
K181-30
K241-46
K301-47
K321-60
K341-71

The serial number lists the order in which the engine was built. If a change takes place to a model or a specification, the serial number is used to indicate the points at which the change takes place. The first letter or number in the serial number indicates what year the engine was built. The letter prefix to the engine serial number was dropped in 1969 and thereafter the prefix is a number. Engines made in 1969 have either the letter "E" or the number "1." The code is as follows:

A—1965
B—1966
C—1967
D—1968
E—1969

First Digit Numbers
1—1969
2—1970
3—1971
4—1972
5—1973
6—1974

GENERAL ENGINE SPECIFICATIONS

Model	Bore & Stroke (in.)	Displacement	Horsepower
K141 (—29355)	2⅞x2½	16.22	6.25
K141 (29356—)	2¹⁵⁄₁₆x2½	16.9	6.25
K161 (—281161)	2⅞x2½	16.22	6.25
K161 (281162—)	2¹⁵⁄₁₆x2½	16.9	6.25
K181	2¹⁵⁄₁₆x2¾	18.6	8.0
K241	3¼x2⅞	23.9	10.0
K241A	3¼x2⅞	23.9	8.0
K301	3⅜x3¼	29.07	12.0
K301A	3⅜x3¼	29.07	12.0
K321	3½x3¼	31.27	14.0
K321A	3½x3¼	31.27	14.0
K341	3¾x3¼	35.89	16.0
K341A, AS	3¾x3¼	35.89	16.0

MAINTENANCE

Air Cleaners

A dirty air cleaner can cause rich fuel/air mixture and consequent poor engine operation and sludge deposits. If the filter becomes dirty enough, dirt that otherwise would be trapped can pass through and may wear the engine's moving parts prematurely. It is therefore necessary that all maintenance work be performed precisely as specified.

DRY AIR CLEANERS

Clean dry element air cleaners every 50 hours of operation, or every 6 months (whichever comes first) under good operating conditions. Service more frequently if the operating area is dusty. Remove the element and tap it lightly against a hard surface to remove the bulk of the dirt. If dirt will not drop off easily, replace the element. Do not use compressed air or solvents. Replace the air cleaner every 100-200 hours, under good conditions, and more frequently if the air is dusty.

Observe the following precautions:
1. Handle the element carefully—do not allow the gasket surfaces to become bent or twisted.
2. Make sure the gasket surfaces seal against back plate and cover.
3. Tighten wing nut only finger tight—if it is too tight, cleaner may not seal properly.

If the dry type air cleaner is equipped with a precleaner, service this unit when cleaning the paper element. Servicing consists of cleaning the precleaner in soap and water, squeezing the excess out, and then allowing it to air dry before installation. Do not oil!

OIL BATH AIR CLEANERS

This type of unit may be used to replace the dry type in applications where very frequent replacement of the element is required. The conversion is simple and requires the use of an elbow to fit the oil bath unit onto the engine in a vertical position.

Service the unit every 25 hours of operation under good conditions and, under dusty conditions, as often as every 8 hours of operation. Service as follows:
1. Remove cover and lift element out of bowl.
2. Drain dirty oil from bowl, and then wash thoroughly in clean solvent.
3. Swish the element in the solvent and then allow it to drip dry. *Do not dry with compressed air.* Lightly oil the element with engine oil.
4. Inspect air horn, filter bowl, and cover gaskets, and replace as necessary (if grooved or cracked).
5. Install filter bowl gasket on air horn, then put the bowl into position. Fill bowl to indicated level with engine oil.
6. Install element, put the cover in position, and then install copper gasket (if used) and wingnut. Tighten wingnut with fingers only to avoid distorting housing. Make sure all joints in the unit seal tightly.

Lubrication

CRANKCASE

Oil level must be maintained between F and L marks—do not overfill. Check every day and add as necessary. On new engines, be especially careful to stop engine and check level frequently. When checking, make sure regular type dipstick is inserted fully. On screw type dipstick, check level

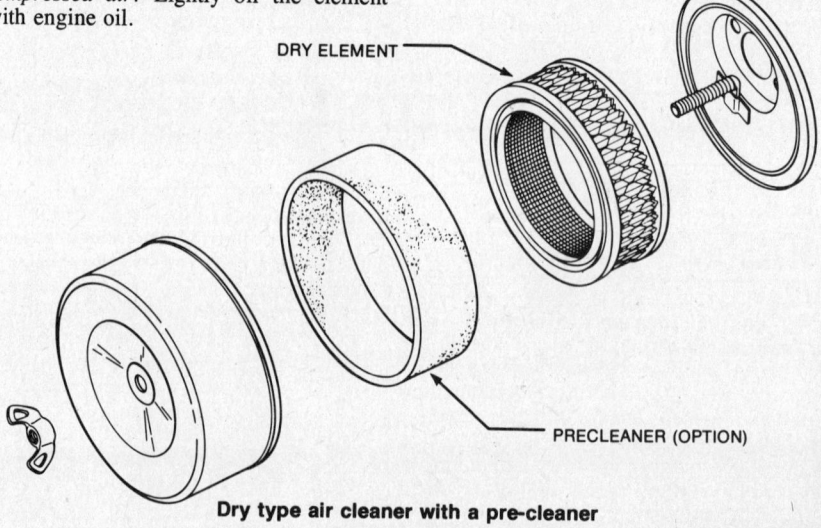

Dry type air cleaner with a pre-cleaner

Engine Overhaul

with dipstick inserted fully but *not* screwed in. On this type, however, make sure to screw dipstick back in tightly when oil level check is completed.

Use SC type oils meeting viscosity specificatons according to the prevailing temperature as shown in the chart below.

OIL VISCOSITY CHART

Air Temperature	Oil Viscosity	Oil Type
Above 30° F	SAE 30	API Service SC*
30° to 0° F	SAE 10W-30	API Service SC*
Below 0° F	SAE 5W-20	API Service SC*

*SC standard recommendation—CC (MIL 2104B) and SD class oils may also be used.

Change initial fill of oil on new engines after five hours of operation. Then, change oil every 25 hours of operation. Change oil when engine is hot. Change more frequently in dusty areas. If the engine has just been overhauled, it is best to fill it initially with a non-detergent oil. Then, after 5 hours, refill with SC type oil.

Oil capacities are:
K141, K161, K181—1 qt.
K241, K301, K321—2 qts.
On K241A, K301A, K321A, K341A, AS, install 1 qt., then fill to F mark on dipstick.
K331—3 qts.

REDUCTION GEAR UNITS

Every 50 hours, remove the oil plug on the lower part of the reduction unit cover to check level. If oil does not reach the level of the oil plug, remove the vented fill plug from the top of the cover and refill with engine oil until level is correct. This oil need not be changed unless unit has been out of service for several months. In this situation, remove the drain plug, drain oil, then replace plug and fill to proper level as described above.

FUEL RECOMMENDATIONS

Use either leaded or unleaded regular grade fuel of at least 90 octane. Unleaded fuel produces fewer combustion chamber deposits, so its use is preferred.

Purchase fuel from a reputable dealer, and make sure to use only fresh fuel (fuel less than 30 days old). If the engine is stored, drain the fuel system or use a fuel stabilizer that is compatible with the type of fuel tank the engine is equipped with.

TUNE-UP

Spark Plugs

SERVICE

The spark plug should be removed and serviced every 100 hours of engine operation. The plug should have a light coating of light gray colored deposits. If deposits are black, fuel/air mixture could be too rich due to improper carburetor adjustment or a dirty air cleaner. If deposits are white, the engine may be overheating or a spark plug of too high a heat range could be in use.

Kohler reommends that the plug be replaced rather than sandblasted or scraped if there are excessive deposits. Torque plugs to 18-22 ft.lbs.

TESTING

To test a plug for adequate performance, remove it from the engine, attach the ignition wire, and then rest the side electrode against the cylinder head. Crank the engine vigorously. If there is a sharp spark, the plug and ignition system are allright, although ignition timing should be checked if the engine fires irregularly.

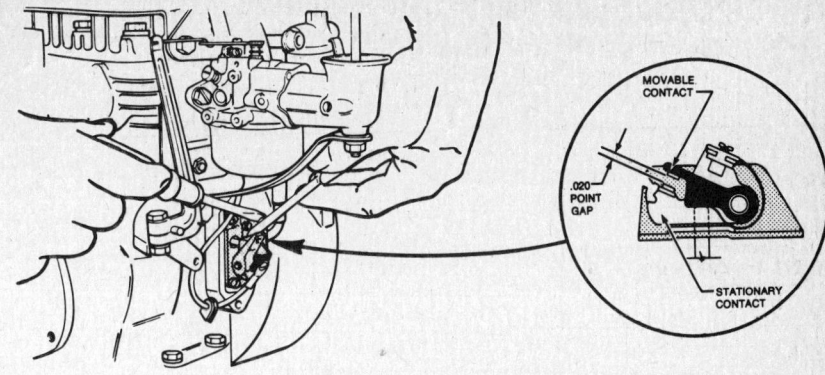

Adjusting the breaker gap

Breaker Points

INSPECTION

Remove the breaker cover and inspect the points for pitting or buildup of metal on either the movable or stationary contact every 100 hours of operation. Replace the points if they are badly burned. If there is a great deal of metal buildup on either contact, the condenser may be faulty and should be replaced.

To replace points, remove the primary wiring connector screw and pull off the primary wire. Then, remove contact set mounting screws and remove the contact set. Install the new set of points in reverse order, leaving upper mounting screw slightly loose. Then set point gap and timing as described below.

SETTING BREAKER GAP AND TIMING

1. Remove the breaker cover and disconnect the spark plug lead. Rotate the engine in direction of normal rotation until the points reach the maximum opening.

2. Using a clean, flat feeler gauge of .020 in. size, check the gap between the points. Gauge should just slide between the contacts without opening them when flat between them. If the gap is incorrect, loosen the upper mounting screw (if necessary), and shift the breaker base with the blade of the screwdriver, until gap is correct.

3. There is a timing sight hole in either the bearing plate or the blower housing. If there is a snap button in the hole, pry it out with a screwdriver.

4. While observing the sight hole, turn the engine slowly in normal direction of rotation. When the S or SP mark (engines with Automatic Compression Release) or the T mark on engines without ACR appears in the hole, the points should just be beginning to open. If timing is incorrect, breaker gap will have to be reset slightly (.018-.022 in.). If the points are not yet opening when the timing mark is centered in the hole, make the point gap wider. If points open too early, narrow it. Recheck the setting after tightening the upper breaker mounting screw by turning the engine in normal direction of rotation past the firing point and checking that the points open at just the right time.

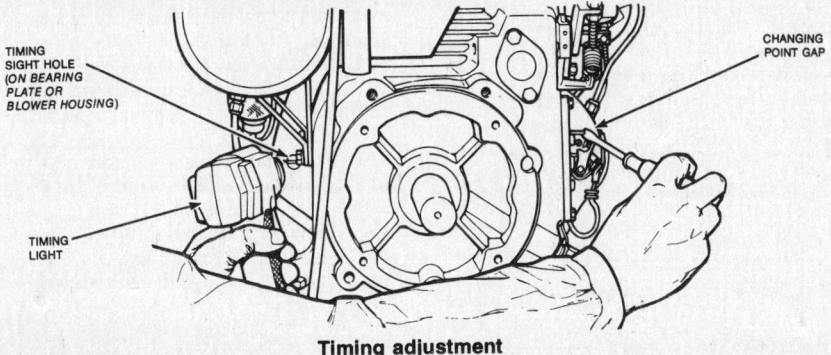

Timing adjustment

Engine Overhaul

TUNE-UP SPECIFICATIONS

Model	Plug Gap (in.)	Breaker Point Gap (in.)	Trigger Air Gap (in.)	Normal Timing (deg)	Retard Timing (deg)
K141 (small bore)	.025①	.020	.005-.010	20	3B
K141 (large bore)	.025①	.020	.005-.010	20	—
K161 (small bore)	.025①	.020	.005-.010	20	3B
K161 (large bore)	.025①	.020	.005-.010	20	—
K181	.025①	.020	.005-.010	20	3B
K241	.025①	.020	.005-.010	20	3A
K301	.025①	.020	.005-.010	20	3A
K321	.025①	.020	.005-.010	20	—
K341	.025①	.020	.005-.010	20	—

B—Before
A—After
① Shielded plug gap—.020 in.

SPARK PLUG SPECIFICATIONS

Engine Model	Plug Size	Hex Size	Plug Reach	STANDARD PLUGS Solid Post		STANDARD PLUGS Knurled Nut		RESISTOR PLUGS Non-Shielded		RESISTOR PLUGS Shielded	
K141	14 mm	13/16"	3/8"	J-8	270321-S	J-8	220040-S	XJ-8	232604-S	XEJ-8	220258-S
K161	14 mm	13/16	3/8"	J-8	270321-S	J-8	220040-S	XJ-8	232604-S	XEJ-8	220258-S
K181	14 mm	13/16"	3/8"	J-8	270321-S	J-8	220040-S	XJ-8	232604-S	XEJ-8	220258-S
K241	14 mm	13/16"	7/16"	H-10	235040-S	Not Available		XH-10	235041-S	XEH-10	235259-S
K301	14 mm	13/16"	7/16"	H-10	235040-S	Not Available		XH-10	235041-S	XEH-10	235259-S
K321	14 mm	13/16"	7/16"	H-10	235040-S	Not Available		XH-10	235041-S	XEH-10	235259-S
K341	14 mm	13/16"	7/16"	H-10	235040-S	Not Available		XH-10	235041-S	XEH-10	235259-S

Gap Setting—gasoline .025" (shielded .020") tightening torque—All plugs 18 to 22 foot lbs.
(Champion plugs listed—use Champion or equivalent plugs.)

NOTE: This procedure may be performed with the engine running at 1,200-1,800 rpm if a timing light is available. Connect the timing light according to manufacturer's instructions. You may have to chalk the timing mark to see it adequately.

TRIGGER AIR GAP

Trigger air gap is set within the range .005-.010 in. As long as the gap falls within this range, the ignition system should perform adequately. Optimum ignition performance during cold weather starting is provided if the gap is adjusted to .005 in. If you wish to adjust this or to ensure that the gap falls within the proper range, rotate the flywheel until the flywheel projection is lined up with the trigger assembly. Then, loosen the trigger bracket capscrews and slide the trigger back and forth to get the proper gap, as measured with a flat feeler gauge. Then, retighten capscrews.

IGNITION COILS

Coils do not require regular service, except to make sure they are kept clean, that the connections are tight, and that rubber insulators are in good condition (replace if cracked). If you suspect poor performance of a breakerless type ignition system and trigger air gap is correct, check resistance with an ohmmeter. To do this, disconnect the high tension lead at the coil and connect the meter between coil terminal and coil mounting bracket. If resistance is not about 11,500 ohms, replace the coil. Also, check the reading with the meter lead going to the coil terminal pulled off and connected to the spark plug connector of the high tension lead. If there is continuity here, replace the coil.

PERMANENT MAGNETS

These may be checked for magnet stength by holding a screwdriver (non-magnetic) blade within one inch of the magnet. If the magnetic field is good, the blade will be attracted to the magnet. Otherwise, replace it.

Mixture Adjustments

NOTE: Before making any adjustments, be sure that the carburetor air cleaner is not clogged. A clogged air cleaner will cause an over-rich mixture, black exhaust smoke, and may lead you to believe that the carburetor is out of adjustment when, in reality, it is not. The carburetor is set at the factory and rarely needs adjustment unless, of course, it has been disassembled or rebuilt.

1. With the engine stopped, turn the main and idle fuel adjusting screws all the way in until they bottom *lightly*. Do not force the screws or you will damage the needles.
2. For a preliminary setting, turn the main fuel screw out 2 full turns and the idle screw out 1¼ turns.
3. Start the engine and allow it to reach operating temperatures; then operate the engine at full throttle and under a load, if possible.
4. For final adjustment, turn the main fuel adjustment screw in until the engine slows down (lean mixture), then out until it slows down again (rich mixture). Note the positions of the screw at both settings, then set it about halfway between the two positions.
5. Set the idle mixture adjustment screw in the same manner. The idle speed (no-

Engine Overhaul

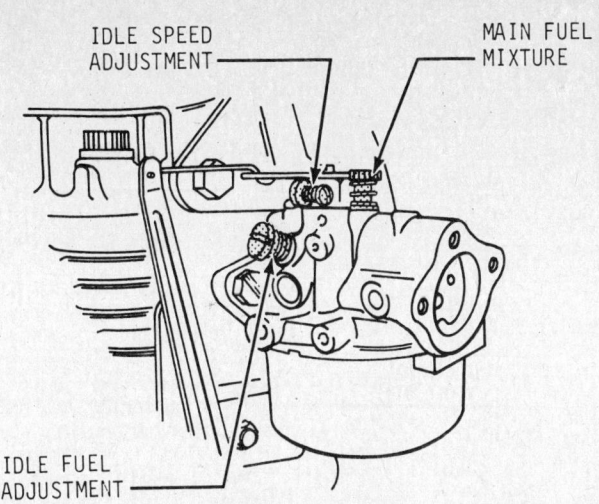

Side draft carburetor

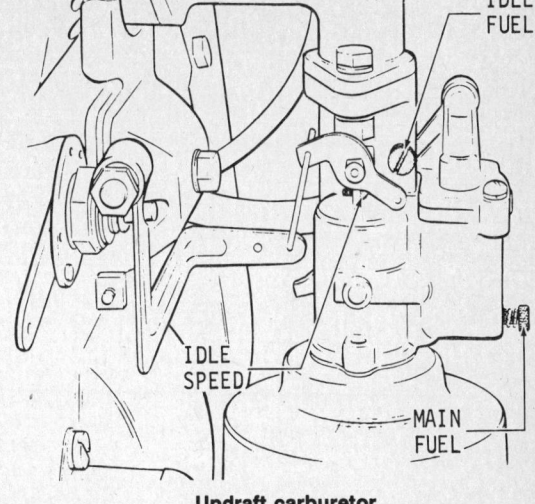

Updraft carburetor

load) on most engines is 1200 rpm; however, on engines with a parasitic load (hydrastatic drives) the engine idle speed may have to be increased to as much as 1700 rpm for best no-load idle.

Governor Adjustment

All Kohler engines use mechanical, camshaft driven governors.

INITIAL ADJUSTMENT

1. Loosen, but do not remove, the nut that holds the governor arm to the governor cross shaft.
2. Grasp the end of the cross shaft with a pair of pliers and turn it in counterclockwise as far as it will go. The tab on the cross shaft will stop against the rod on the governor gear assembly.
3. Pull the governor arm away from the carburetor, then retighten the nut which holds the governor arm to the shaft. With updraft carburetors, lift the arm as far as possible, then retighten the arm nut.

FINAL ADJUSTMENT

K141–K181

After making the initial adjustment and connecting the throttle wire on the variable speed applications, start the engine and check the maximum operating speed with a tachometer. If adjustment is necessary:

1. Loosen the bushing nut slightly.
2. Move the throttle bracket in a counterclockwise direction to increase speed, or in a clockwise direction to decrease engine speed. Maximum speed is 3600 rpm.
3. With the speed set to the proper range, tighten the bushing nut to lock the throttle bracket in position.

K241

Engine must be adjusted to 3,600 rpm.

1. Start the engine and measure the speed with a tachometer.
2. If the speed is incorrect, adjust as follows:
   a. *Constant Speed Governor*—Tighten the governor adjusting screw to increase speed, or loosen to decrease speed until the correct speed is attained.
   b. *Variable Speed Governor*—Loosen

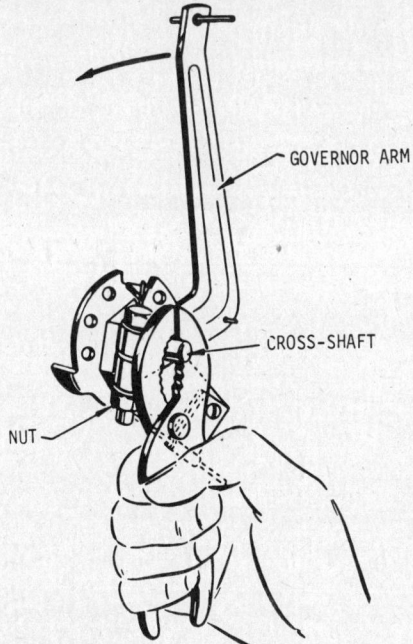

Initial governor adjustment on K141, 161, 181 engines

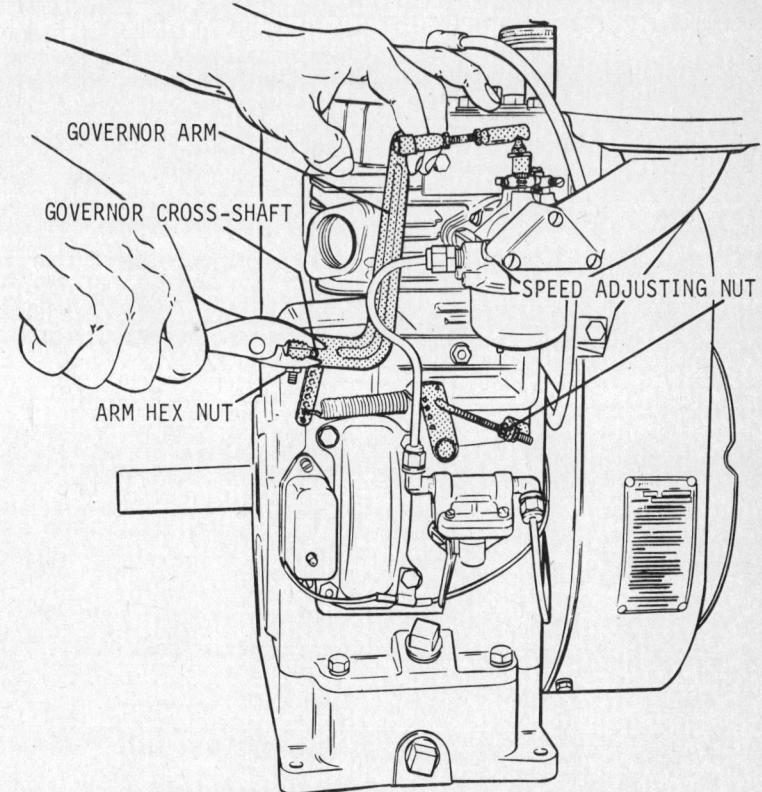

Initial governor adjustment on K241 through 341 engines

895

Engine Overhaul

the capscrew, move the high speed stop bracket until the correct speed is attained, and then retighten the capscrew.

If the governor is too sensitive (causing hunting or surging), or not sensitive enough (causing too great a drop in speed when load is applied), the governor sensitivity should be adjusted. Make the governor more sensitive by moving the spring to holes further apart. Make it less sensitive by moving it to holes that are closer together. Standard setting is the third hole from the bottom on the governor arm and second hole from the top on the speed control bracket.

Choke Adjustment

THERMOSTATIC TYPE

If the engine does not start when cranked, continue cranking and move the choke lever first to one side and then to the other to determine whether the setting is too lean or too rich. Once the direction in which lever must be moved has been determined, loosen the adjusting screw on the choke body. Then, move the bracket downward to increase choking or upward to decrease it. Then, tighten the lockscrew. Try starting it again and readjust as necessary.

ELECTRIC-THERMOSTATIC TYPE

Remove the air cleaner from the carburetor and check the position of the choke plate. The choke should be fully closed when engine is at outside temperature and the temperature is very low. In milder temperatures, slightly less closure is required.

If adjustment is required, move the choke arm until the hole in the brass shaft lines up with the slot in the bearings. Insert a #43 (.089 in.) drill through the shaft and push it downward so it engages the notch in the base of the choke unit. Then, loosen the clamp bolt on the choke lever and push the arm upward to move the choke plate toward the closed position. When the desired position is obtained, tighten the clamp bolt. Then, remove the drill.

Remount the air cleaner, and then check for any binding in the choke linkage. Correct as necessary. Finally, run the engine until hot, and make sure the choke opens fully. If not, readjust it toward the open position as necessary.

Valve Adjustment

On K241, K301, K321, K341 engines, adjustable valve tappets are provided. With the engine cold, turn crankshaft until it reaches Top Dead Center timing mark. If valves are slightly open, turn the crankshaft another turn until valves are closed and engine is again at Top Center. Check valve clearances with a flat feeler gauge. Note that exhaust and intake clearances are different, and make sure you're using the right gauge for each valve. If the valve clearance is correct, a gauge can just be inserted between tappet and valve stem. A slight pull is required to bring it back out. If clearance is incorrect, loosen the locking nut and turn the adjusting nut in or out to get the proper clearance. Hold the adjusting nut while tightening the locknut and recheck clearance.

Compression Check

Compression is checked by removing the spark plug lead and spinning the flywheel forward against compression. If the piston does not bounce backward with considerable force, checking with a gauge may be necessary. On Automatic Compression Release engines, rotate the flywheel backward against power stroke—if little resistance is felt, check compression with a gauge.

The compression gauge check requires rapid motoring (spinning) of the crankshaft, at about 1,000 rpm. Install the gauge in the spark plug hole and motor the engine. Gauge should read 110-120 psi. If reading is less than 100 psi, the engine requires major repair to piston rings or valves.

FUEL SYSTEM

Carburetor

If a carburetor will not respond to mixture screw adjustments, then you can assume that there are dirt, gum, or varnish deposits in the carburetor or worn/damaged parts. To remedy these problems, the carburetor will have to be completely disassembled, cleaned, and worn parts replaced and reassembled.

Parts should be cleaned with solvent to remove all deposits. Replace worn parts and use all new gaskets. Carburetor rebuilding kits are available.

DISASSEMBLY

Side Draft Carburetors

1. Remove the carburetor from the engine.
2. Remove the bowl nut, gasket, and bowl. If the carburetor has a bowl drain, remove the drain spring, spacer and plug, and gasket from inside the bowl.
3. Remove the float pin, float, needle, and needle seat. Check the float for dents, leaks, and wear on the float lip or in the float pin holes.
4. Remove the bowl ring gasket.
5. Remove the idle fuel adjusting needle, main fuel adjusting needle, and springs.
6. Do not remove the choke and throttle plates or shafts. If these parts are worn, replace the entire carburetor assembly.

Updraft Carburetors

1. Remove the carburetor from the engine.
2. Remove the bowl cover and the gasket.
3. Remove the float pin, float, needle

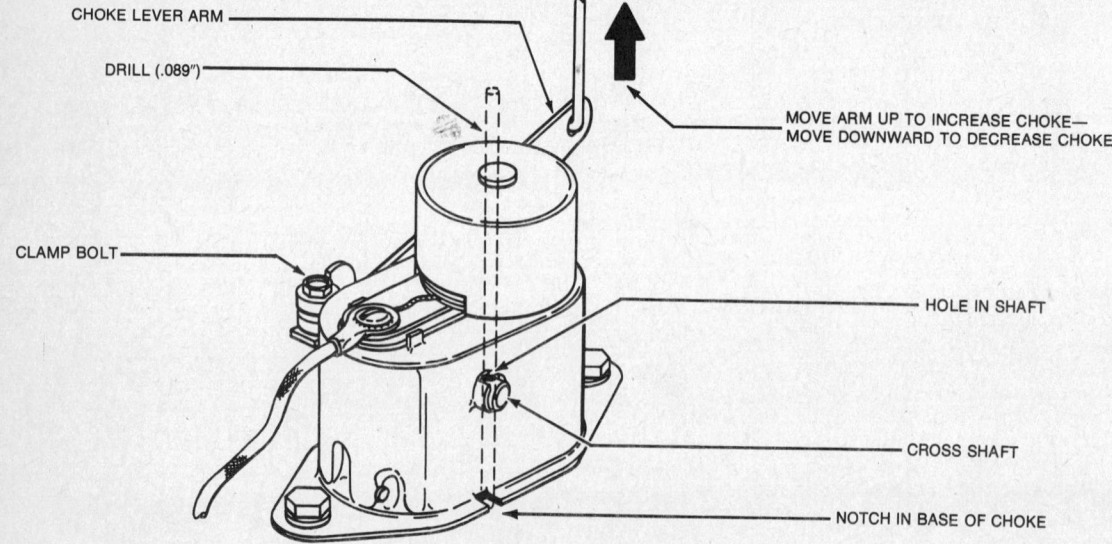

Adjusting the electric-thermostatic choke

Engine Overhaul

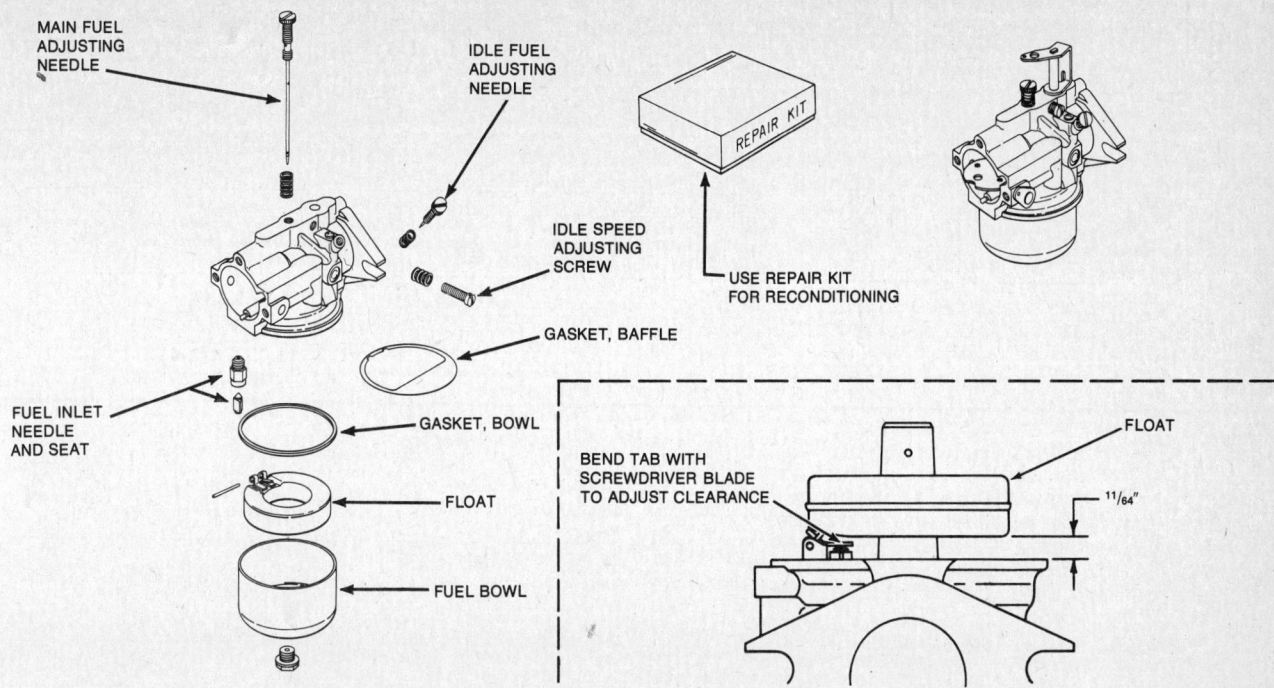

Side draft carburetor exploded view with the insert showing adjustments

and needle seat. Check the float pin for wear.

4. Remove the idle fuel adjustment needle, main fuel adjustment needle, and the springs. Do not remove the choke plate or the shaft unless the replacement of these parts is necessary.

ASSEMBLY

Side Draft Carburetor

1. Install the needle seat, needle, float, and float pin.

2. Set the float level. With the carburetor casting inverted and the float resting against the needle in its seat, there should be $11/64$ in. plus or minus $1/32$ in. clearance between the machined surface of the casting and the free end of the float.

3. Adjust the float level by bending the lip of the float with a small screwdriver.

4. Install the new bowl ring gasket, new bowl nut gasket, and bowl nut. Tighten the nut securely.

5. Install the main fuel adjustment needle. Turn it in until the needle seats in the nozzle and then back out two turns.

6. Install the idle fuel adjustment needle. Back it out about 1¼ turns after seating it lightly against the jet.

7. Install the carburetor on the engine.

Updraft Carburetor

1. Install the throttle shaft and plate. The elongated side of the valve must be toward the top.

2. Install the needle seat A $5/16$ in. socket should be used. Do not over-tighten.

3. Install the needle, float, and float pins.

4. Set the float level. With the bowl cover casting inverted and the float resting lightly against the needle in its seat, there should be $7/16$ in. plus or minus $1/32$ in. clearance between the machined surface casting and the free end of the float.

5. Adjust the float level by bending the lip of the float with a small screwdriver.

6. Install the new carburetor bowl gasket, bowl cover, and bowl cover screws. Tighten the screws securely.

7. Install the main fuel adjustment needle. Turn it in until the screw seats in the nozzle and then back it out 2 turns.

8. Install the idle fuel adjustment needle. Back it out about 1½ turns after seating the screw lightly against the jet.

Install the idle speed screw and spring. Adjust the idle to the desired speed with the engine running.

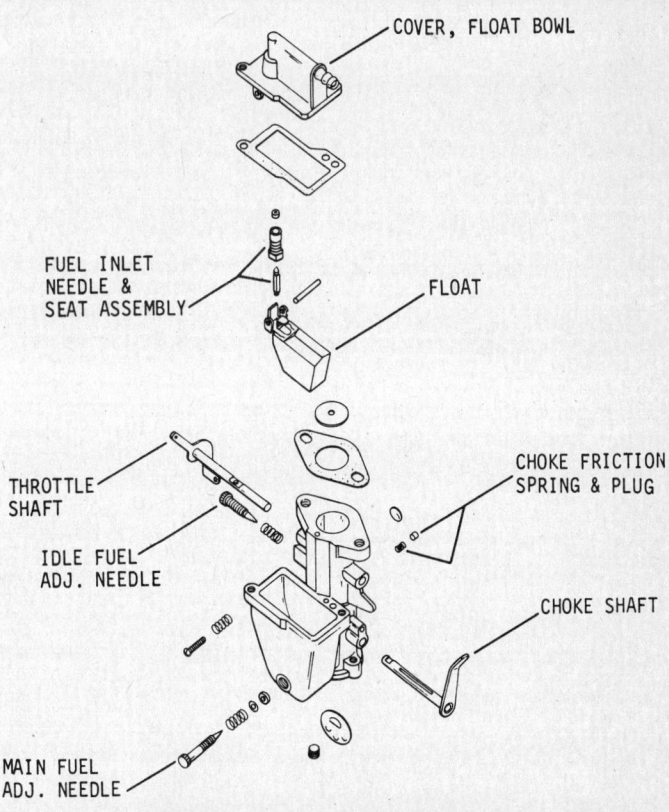

Exploded view of the updraft carburetor

Engine Overhaul

9. Install the carburetor on the engine.

Fuel Pump

Fuel pumps used on single cylinder Kohler engines are either the mechanical or vacuum actuated type. The mechanical type is operated by an eccentric on the camshaft and the vacuum type is operated by the pulsating negative pressures in the crankcase. The k91 vacuum type pump is not serviceable and must be replaced when faulty. The mechanical pump is serviceable and rebuilding kits are available.

1. Disconnect fuel lines, remove mounting screws, and pull the pump off engine.
2. File a mark across some point at the union of pump body and cover. Remove the screws and remove the cover.
3. Turn the cover upside down and remove the valve plate screw and washer. Remove the valve retainer, valves, valve springs, and valve gasket, after noting the position of each part. Discard the valve springs, valves and valve retainer gasket.
4. Clean the fuel head with solvent and a soft wire brush. Hold the pump cover with the diaphragm surface upward; position a new gasket into the cavity. Put the valve spring and valves into position in the cavity and reassemble the valve retainer. Lock the retainer into position by installing the fuel pump valve retainer screw.
5. Rebuild the lower diaphragm section.
6. Hold the mounting bracket and press down on the diaphragm to compress the spring underneath. Turn the bracket 90 degrees to unhook the diaphragm and remove it.
7. Clean the mounting bracket with solvent and a wire brush.
8. Stand a new diaphragm spring in the casting, put the diaphragm into position, and push downward to compress the spring. Turn the diaphragm 90 degrees to reconnect it.
9. Position the pump cover on top of the mounting bracket with the indicating marks lined up. Install the screws loosely on mechanical pumps; on vacuum pumps, tighten the screws.
10. Holding only the mounting bracket, push the pump lever to the limit of its travel, hold it there, and then tighten the four screws.
11. Remount the fuel pump on the engine with a new gasket, tighten the mounting bolts, and reconnect the fuel lines.

ENGINE OVERHAUL

Disassembly

The following procedure is designed to be a general guide rather than a specific and all inclusive disassembly procedure. The sequence may have to be varied slightly to

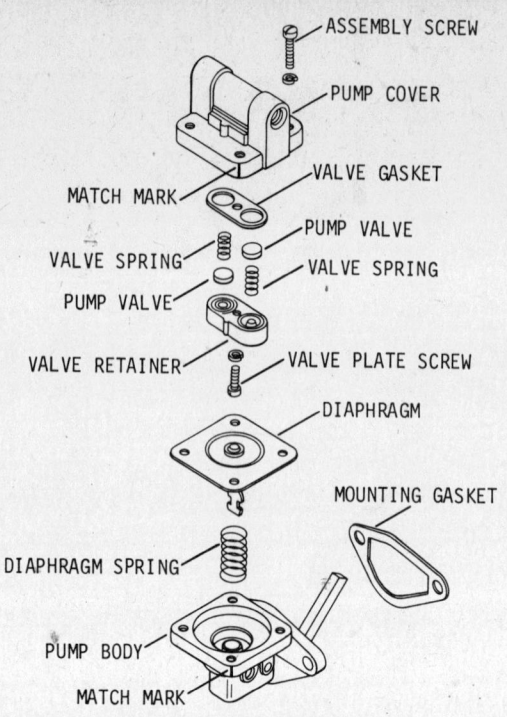

Mechanical type fuel pump

allow for the removal of special equipment or accessory items such as motor/generators, starters, instrument panels, etc.

1. Disconnect the high tension spark plug lead and remove the spark plug.
2. Close the valve on the fuel sediment bowl and remove the fuel line at the carburetor.
3. Remove the air cleaner from the carburetor intake.
4. Remove the carburetor.
5. Remove the fuel tank. The sediment bowl and brackets remain attached to the fuel tank.
6. Remove the blower housing, cylinder baffle, and head baffle.
7. Remove the rotating screen and the starter pulley.
8. The flywheel is mounted on the tapered portion of the crankcase and is removed with the help of a puller. Do not strike the flywheel with any type of hammer.
9. Remove the breaker point cover, breaker point lead, breaker assembly, and the push-rod that operates the points.
10. Remove the magneto assembly.
11. Remove the valve cover and breather assembly.
12. Remove the cylinder head.
13. Raise the valve springs with a valve spring compressor and remove the valve spring keepers from the valve stems. Remove the valve spring retainers, springs, and valves.
14. Remove the oil pan base and unscrew the connecting rod capscrews. Remove the connecting rod cap and piston assembly from the cylinder block.

NOTE: It will probably be necessary to use a ridge reamer on the cylinder walls before removing the piston assembly, to avoid breaking the piston rings.

15. Remove the crankshaft, oil seals and, if necessary, the anti-friction bearings.

NOTE: It may be necessary to press the crankshaft out of the cylinder block. The bearing plate should be removed first, if this is the case.

16. Turn the cylinder block upside down and drive the camshaft pin out from the power take-off side of the engine with a small punch. The pin will slide out easily once it is driven free of the cylinder block.
17. Remove the camshaft and the valve tappets.
18. Loosen and remove the governor arm from the governor shaft.
19. Unscrew the governor bushing nut and remove the governor shaft from the inside of the cylinder block.
20. Loosen, but do not remove, the screw located at the lower right of the governor bushing nut until the governor gear is free to slide off of the stub shaft.

Engine Rebuilding

CYLINDER BLOCK SERVICE

Make sure that all surfaces are free of gasket fragments and sealer materials. The crankshaft bearings are not to be removed unless replacement is necessary. One bearing is pressed into the cylinder block and the other is located in the bearing plate. If there is no evidence of scoring or grooving and the bearings turn easily and quietly it is not necessary to replace them.

The cylinder bore must not be worn, tapered, or out-of-round more than 0.005 in. Check at two locations 90 degrees apart and compare with specifications. If it is, the cylinder must be rebored. If the cylin-

Engine Overhaul

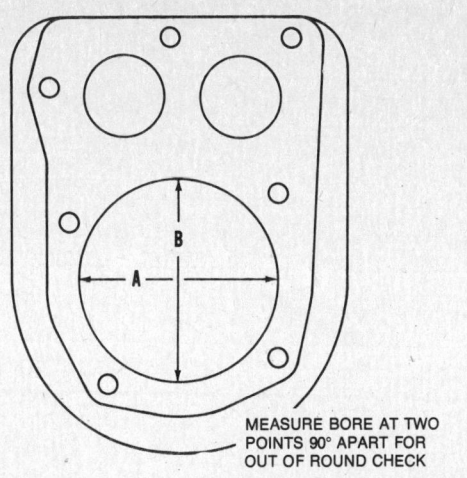
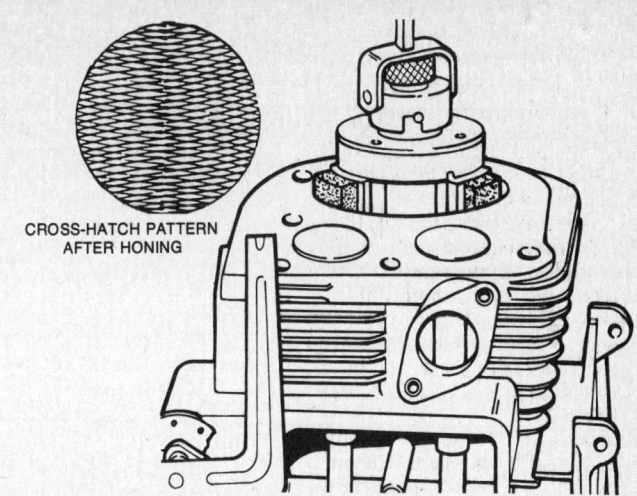

Left side: measuring the cylinder bore; right side: the cross hatch after honing

der is very badly scored or damaged it may have to be replaced, since the cylinder can only be rebored to either 0.010 in. or 0.020 in. and 0.030 in. maximum. Select the nearest suitable oversize and bore it to that dimension. On the other hand, if the cylinder bore is only slightly damaged, only a light deglazing may be necessary.

HONING THE CYLINDER BORE

1. The hone must be centered in relation to the crankshaft crossbore. It is best to use a low speed drill press. Lubricate the hone with kerosene and lower it into the bore. Adjust the stones so they contact the cylinder walls.
2. Position the lower edge of the stones even with the lower edge of the stones even with the lower edge of the bore, hone at about 600 rpm. Move the hone up and down continuously. Check bore size frequently.
3. When the bore reaches a dimension .0025 in. smaller than desired size, replace the coarse stones with burnishing stones. Use burnishing stones until the dimension is within .0005 in. of desired size.
4. Use finishing stones and polish the bore to final size, moving the stones up and down to get a 60 degree cross-hatch pattern. Wash the cylinder wall thoroughly with soap and water, dry, and apply a light coating of oil.

CRANKSHAFT SERVICE

Inspect the keyway and the gears that drive the camshaft. If the keyways are badly worn or chipped, the crankshaft should be replaced. If the cam gear teeth are excessively worn or if any are broken, the crankshaft must be replaced.

Check the crankpin for score marks or metal pickup. Slight score marks can be removed with a crocus cloth soaked in oil. If the crankpin is worn more than 0.002 in., the crankshaft is to be either replaced or the crankpin reground to 0.010 in. undersize. If the crankpin is reground to 0.010 in. undersize, a 0.010 in. undersize connecting rod must be used to achieve proper running clearance.

CONNECTING ROD SERVICE

Check the bearing area for wear, score mark and excessive running and side clearance. Replace the rod and the cap if they are worn beyond the limits allowed.

PISTON AND RINGS SERVICE

Production and Service Type

Rings are available in the standard size as well as 0.010 in., 0.020 in., and 0.030 in. oversize sets.

NOTE: Never reuse old rings.

The standard size rings are to be used

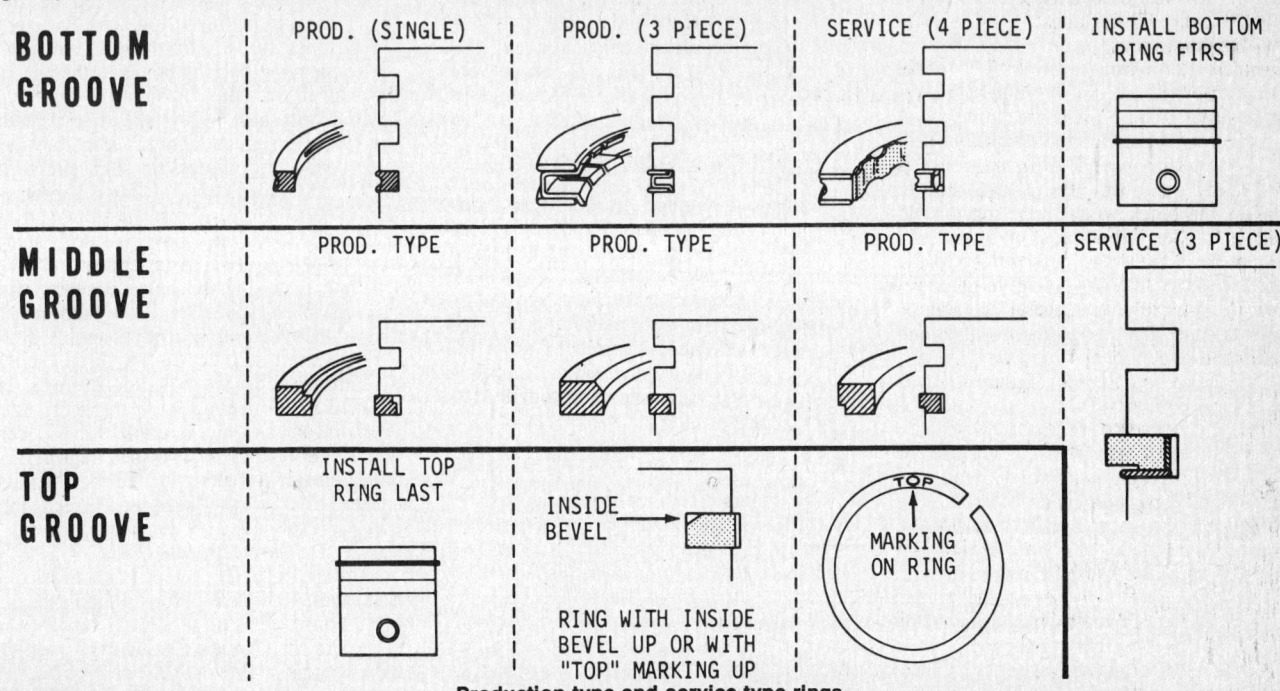

Production type and service type rings

Engine Overhaul

when the cylinder is not worn or out-of-round. Oversize rings are only to be used when the cylinder has been rebored to the corresponding oversize. Service type rings are used only when the cylinder is worn but within the wear and out-of-round limitations; wear limit is 0.005 in. oversize and out-of-round limit is 0.004 in.

The old piston may be reused if the block does not need reboring and the piston is within wear limits. Never reuse old rings. After removing old rings, thoroughly remove deposits from ring grooves. New rings must each be positioned in its running area of the cylinder bore for an end clearance check, and each must meet specifications.

The cylinder must be deglazed before replacing the rings. If chrome plated rings are used, the chrome plated ring must be installed in the top groove. Make sure that the ring grooves are free from all carbon deposits. Use a ring expander to install the rings. Then check side clearance.

PISTON AND ROD SERVICE

Normally very little wear will take place at the piston boss and piston pin. If the original piston and connecting rod can be used after rebuilding, the piston pin may also be used. However if a new piston or connecting rod or both have to be used, a new piston pin must also be installed. Lubricate the pin before installing it with a loose to light interference fit. Use new piston pin retainers whether or not the pin is new. Make sure they're properly engaged.

VALVES AND VALVE MECHANISM SERVICE

Inspect the valve mechanism, valves, and valve seats or inserts for evidence of wear, deep pitting, cracks or distortion. Check the clearance between the valve stems and the valve guides.

Valve guides must be replaced if they are worn beyond the limit allowed. K91 model engines do not use valve guides. To remove valve guides, press the guide down into the valve chamber and carefully break off the protruding end until the guide is completely removed. Be careful not to damage the block when removing the old guides. Use an arbor press to install the new guides. Press the new guides to the depth specified, then use a valve guide reamer to gain the proper inside diameter.

Make sure that replacement valves are the correct type (special hard faced valves are needed in some cases). Exhaust valves are always hard faced.

Intake valve seats are usually machined into the block, although inserts are used in some engines. Exhaust valve seats are made of special hardened material. The seating surfaces should be held as close to $1/32$ in. in width as possible. Seats more than $1/16$ in. wide must be reground with 45° and 15° cutters to obtain the proper width. Reground or new valves and seats must be lapped in for a proper fit.

After resurfacing valves and seats and lapping them in, check the valve clearance.

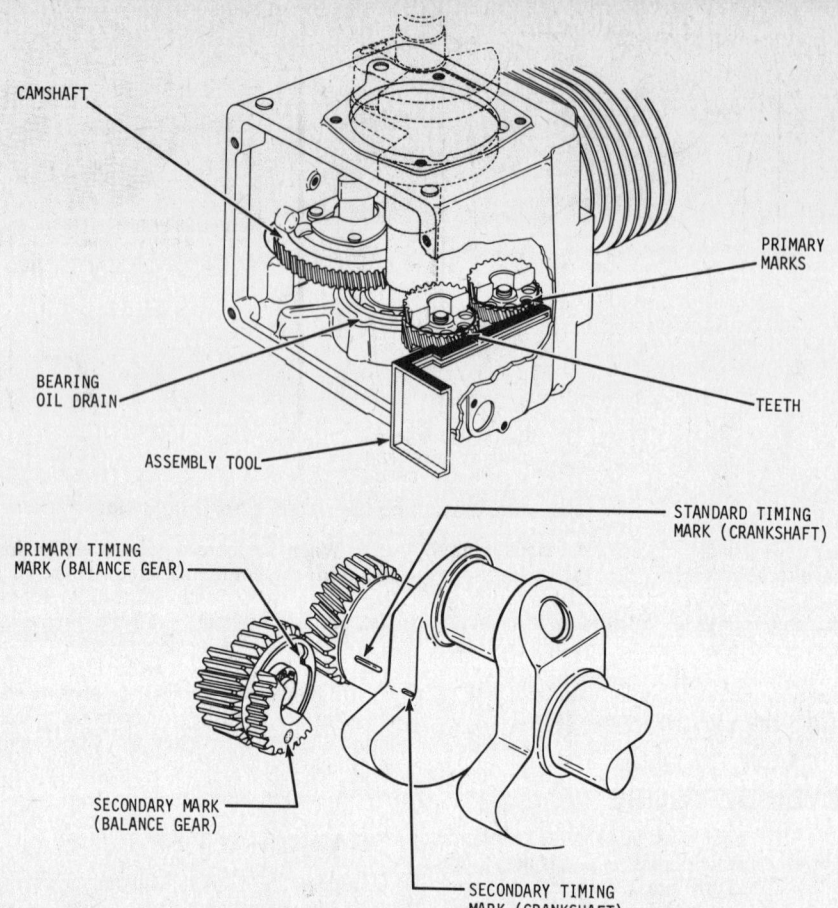

Timing marks for the dynamic balance system

Hold the valve down on its seat and rotate the camshaft until it has no effect on the tappet, then check the clearance between the end of the valve stem and the tappet. If the clearance is not sufficient (it will always be less after grinding), it will be necessary to grind off the end of the valve stem until the correct clearance is obtained. This is necessary on all engines except the K241, K301, K321 and K341 engines which all have adjustable tappets.

CYLINDER HEAD SERVICE

Remove all carbon deposits and check for pitting from hot spots. Replace the head if metal has been burned away because of head gasket leakage. Check the cylinder head for flatness. If the head is slightly warped, it can be resurfaced by rubbing it on a piece of sandpaper placed on a flat surface. Be careful not to nick or scratch the head when removing carbon deposits.

DYNAMIC BALANCE SYSTEM SERVICE

The dynamic balance system consists of two balance gears which run on needle bearings. The gears are assembled on two stub shafts that are pressed into special bosses in the crankcase. Snaprings hold the gears and spacer washers are used to control end-play. The gears are driven off of the crankgear. The dynamic balance system is found on special versions of K241 and K301 models and is standard equipment on K321 engines.

If the stub shaft is worn or damaged, press the old shaft out. The new shafts must be pressed in a specified distance which depends upon the distance between the stub shaft boss and main bearing boss. Measure the distance the stub shaft boss protrudes above the main bearing boss and then press the shaft in for a protrusion of the shaft end beyond stub shaft boss as specified. If stub shaft boss protrudes about $7/16$ in. beyond main bearing boss, press the shaft in until it is .735 in. above stub shaft boss. If protrusion is about $1/16$ in., press the stub shaft in until it is 1.110 in. above the stub shaft boss, and then use a $3/8$ in. spacer.

When installing the balance gears, slip one 0.010 in. spacer onto the stub shaft, then install the gear/bearing assembly onto the stub shaft with the timing marks facing out. Proper end-play of 0.002–0.010 in. is attained with one 0.005 in. spacer, one 0.010 in. spacer, and one 0.020 in. spacer which are all installed on the snapring retainer end of the shaft. Install the thickest spacer next to the retainer. Check the end-play and adjust it by adding or subtracting 0.005 in. spacers.

To time the balance gears, first press the crankshaft into the block and align the primary timing mark on the top of the balance gear with the standard timing mark next to the crankgear. Press the shaft in until the crankgear is engaged $1/16$ in. into the top

Engine Overhaul

gear (narrow side). Rotate the crankshaft to align the timing marks on the crankgear and camgear. Press the crankshaft the remainder of the way into the block.

Rotate the crankshaft until it is about 15° past BDC and slip one 0.010 in. spacer over the stub shaft before installing the bottom gear/bearing assembly.

Align the secondary timing mark on this gear with the secondary timing mark on the counterweight of the crankshaft and then install the gear on the shaft. The secondary timing mark will also be aligned with the standard timing mark on the crankshaft after installation. Use one .005 in. spacer and one .020 in. spacer (with larger spacer next to retainer) to get proper end play of .002–.010 in. Install the snapring retainer, then check and adjust the end-play.

ENGINE ASSEMBLY

Rear Main Bearing

Install the rear main bearing by pressing it into the cylinder block with the shielded side toward the inside of the block. If it does not have a shielded side, then either side may face inside.

Governor Shaft

1. Place the cylinder block on its side and slide the governor shaft into place from the inside of the block. Place the speed control disc on the governor bushing nut and thread the nut into the block, clamping the throttle bracket into place.
2. There should be a slight end-play in the governor shaft and that can be adjusted by moving the needle bearing in the block.
3. Place a space washer on the stub shaft and slide the governor gear assembly into place.
4. Tighten the holding screw from outside the cylinder block.
5. Rotate the governor gear assembly to be sure that the holding screw does not contact the weight section of the gear.

Camshaft

1. Turn the cylinder block upside down.
2. The tappets must be installed before the camshaft is installed. Lubricate and install the tappets into the valve guides making sure that the short tappet is installed in the exhaust valve guide on the K141, K161, K181 ACR engines. All other tappets are the same size.
3. Position the camshaft inside the block.

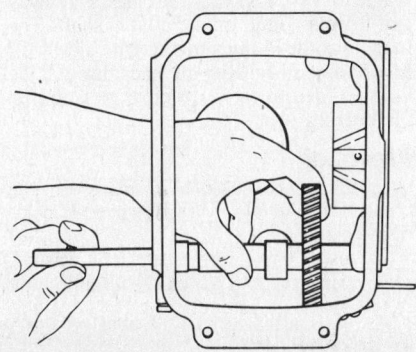
Installing the camshaft pin

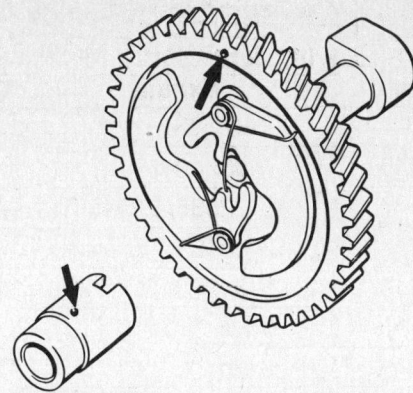
Timing marks for the automatic advance system

NOTE: Align the marks on the camshaft and the automatic spark advance, if so equipped.

4. Lubricate the rod and insert it into the bearing plate side of the block. Install one 0.005 in. washer between the end of the camshaft and the block. Push the rod through the camshaft and tap it lightly until the rod just starts to enter the bore at the PTO end of the block. Check the end-play and adjust it with additional washers if necessary. Press the rod into its final position.
5. The fit at the bearing plate for the camshaft rod is a light to loose fit to allow oil that might leak past to drain back into the block.

Crankshaft

1. Place the block on the base of an arbor press and carefully insert the tapered end of the crankshaft through the inner race of the anti-friction bearing, or sleeve bearing on the K141.

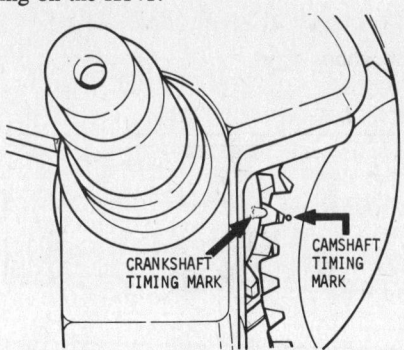
Alignment of the timing marks on the crankshaft and camshaft

2. Turn the crankshaft and camshaft until the timing mark on the shoulder of the crankshaft lines up with the mark on the cam gear.
3. When the marks are aligned, press the crankshaft into the bearing, making sure that the gears mesh as it is being pressed in. Recheck the alignment of the timing marks on the crankshaft and the camshaft.
4. The end-play of the crankshaft is controlled by the application of various thickness gaskets between the bearing plate and the block. Normal end-play is achieved by installing 0.020 in. and 0.010 in. gaskets, with the thicker gaskets on the inside.

Bearing Plate

1. Press the front main bearing into the bearing plate. Make sure that the bearing is straight.
2. Press the bearing plate onto the crankshaft and into position on the block. Install the cap screws and secure the plate to the block. Draw up evenly on the screws.
3. Measure the crankshaft end-play, which is very critical on gear reduction engines.

Piston and Rod Assembly

1. Lubricate the pin and assemble it to the connecting rod and piston. Install the wrist pin retaining ring. Use new retaining rings.
2. Lubricate the entire assembly, stagger the ring gaps and, using a ring compressor, slide the piston and rod assembly into the cylinder bore with the connecting rod marks on the flywheel side of the engine.
3. Place the block on its end and oil the connecting rod end and the crankpin.
4. Attach the rod cap, lock or lockwashers, and the cap screws. Tighten the screws to the correct torque.

NOTE: Align the marks on the cap and the connecting rod.

5. Bend the lock tabs to lock the screws.

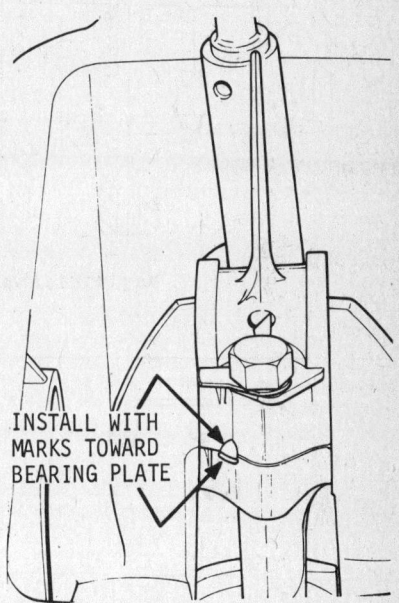
Connecting rod and cap alignment marks

Crankshaft Oil Seals

Apply a coat of grease to the lip and guide the oil seals onto the crankshaft. Make sure no foreign material gets onto the knife edges of seal, and make sure the seal does not bend. Place the block on its side and drive the seals squarely into the bearing plate and block.

Oil Pan Base

Using a new gasket on the base, install pilot studs to align the cylinder block, gasket, and base. Tighten the four attaching screws to the correct torque.

Engine Overhaul

VALVE SPECIFICATIONS

Dimension		MODEL K141, K161, K181 Intake	Exhaust	MODEL K241, K301, K321, K341 Intake	Exhaust
A	Seat Angle	89°	89°	89°	89°
B	Seat Width	.037/.045	.037/.045	.037/.045	.037/.045
C	Insert OD	—	1.2535/1.2545	—	1.2535/1.2545
D	Guide Depth	1.312	1.312	1.586	1.497
E	Guide ID	.312/.313	.312/.313	.312/.313	.312/.313
F	Valve Head Diameter	1³⁄₈	1¹⁄₈	1.370/1.380	1.120/1.130*
G	Valve Face Angle	45°	45°	45°	45°
H	Valve Stem Diameter	.3105/.3110	.3090/.3095	.3105/.3110	.3084/.3091

*2.125" on all K341 and K321 engines with spec suffix "D" and later.

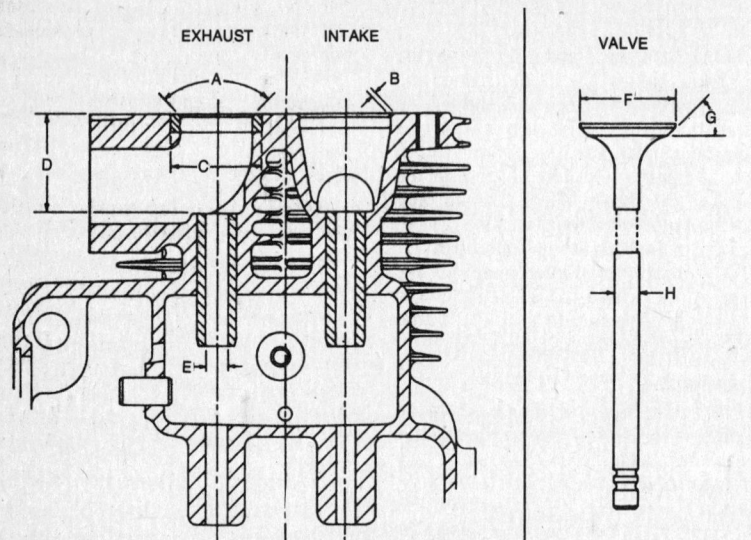

Key to the valve specifications chart

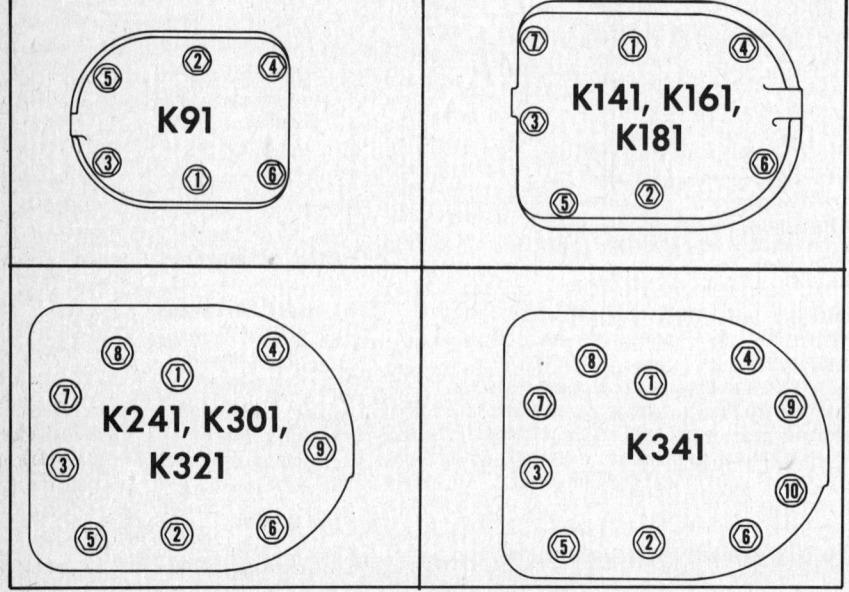

Cylinder head torque sequences

Valves
1. See the engine rebuilding section of the Briggs and Stratton chapter for details concerning installation of the seats and guides. Clean the valves, seats, and parts thoroughly. Grind and lap-in the valves and seats for proper seating. Valve seat width must be ¹⁄₃₂ in.–¹⁄₁₆ in. After grinding and lapping, slide the valves into position and check the clearance between stem and tappet. If the clearance is too small, grind the stem ends square and remove all burrs. On engines with adjustable valves, make the adjustment at this time.
2. Place the valve springs, retainers, and rotators under the valve guides. Lubricate the valve stems, and then install the valves down through the guides, compress the springs, and place the locking keys or pins in the grooves of the valve stems.

Cylinder Head
1. Use a new cylinder head gasket.
2. Lubricate and tighten the head bolts evenly, and in sequence, to the proper torque.
3. Install the spark plug.

Breather Assembly
Assemble the breather assembly, making sure that all parts are clean and the cover is securely tightened to prevent oil leakage.

Magneto
On flywheel magneto systems, the coil-core assembly is secured onto the bearing plate. On magneto-alternator systems, the coil is part of the stator assembly, which is secured to the bearing plate. On rotor type magneto systems, the rotor has a keyway and is press fitted onto the crankshaft. The magnet rotor is marked "engine-side" for proper assembly. Run all leads through the hole proided at the 11 o'clock position on the bearing plate.

Flywheel
1. Place the washer in place on the crankshaft and place the flywheel in position. Install the key.
2. Install the starter pulley, lock washer, and retaining nut. Tighten the retaining nut to the specified torque.

Breaker Points
1. Install the pushrod.

Engine Overhaul

2. Position the breaker points and fasten them with the two screws.

3. Place the cover gasket into position and attach the magneto lead.

4. Set the gap and install the cover.

Carburetor

Insert a new gasket and assemble the carburetor to the intake port with the two attaching screws.

Governor Arm and Linkage

1. Insert the carburetor linkage in the throttle arm.

2. Connect the governor arm to the carburetor linkage and slide the governor arm into the governor shaft.

3. Position the governor spring in the speed control disc on the K141, K161, and K181.

4. Before tightening the clamp bolt, turn the shaft counterclockwise with pliers as far as it will go; pull the arm as far as it will go to the left (away from the carburetor), tighten the nut, and check for freedom of movement. Adjust the governor.

Blower Housing and Fuel Tank

Install the head baffle, cylinder baffle, and the blower housing, in that order. The smaller capscrews are used on the bottom of the crankcase. Install the fuel tank and connect the fuel line.

Run-In Procedure

1. Fill the crankcase with a *non-detergent* oil and run it under load for 5 hours to break it in.

2. Drain oil and refill crankcase with the recommended detergent type oil. Non-detergent oil must not be used except for break-in.

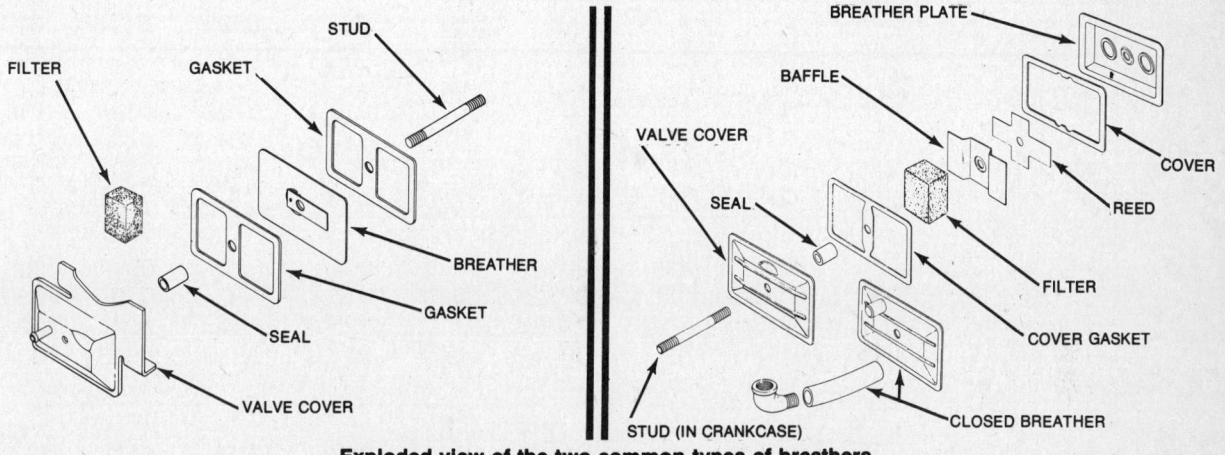

Exploded view of the two common types of breathers

ENGINE REBUILDING SPECIFICATIONS

Specification	K141 2⁷⁄₈" Bore	K141 2¹⁵⁄₁₆" Bore	K161 2⁷⁄₈" Bore	K161 2¹⁵⁄₁₆" Bore	K181	K241	K301	K321	K341
Displacement									
Cubic Inches	16.22	16.9	16.22	16.9	18.6	23.9	29.07	31.27	35.89
Cubic Centimeters	265.8	276.99	265.8	276.99	304.8	391.65	476.37	528.46	588.24
Horsepower (Max RPM)	6.25	6.25	7.0	7.0	8.0	10.0	12.0	14.0	16.0
Cylinder Bore									
New Diameter	2.875	2.9375	2.875	2.9375	2.9375	3.251	3.375	3.500	3.750
Maximum Wear Diameter	2.878	2.9405	2.878	2.9405	2.9405	3.2545	3.3785	3.503	3.753
Maximum Taper	.0025	.0025	.0025	.0025	.0025	.0015	.0015	.0015	.0015
Maximum Out of Round	.005	.005	.005	.005	.005	.005	.005	.005	.005
Crankshaft									
End Play (Free)	.002/.023	.002/.023	.002/.023	.002/.023	.002/.023	.003/.020	.003/.020	.003/.020	.003/.020
Crankpin									
New Diameter	1.186	1.186	1.186	1.186	1.186	1.500	1.500	1.500	1.500
Maximum Out of Round	.0005	.0005	.0005	.0005	.0005	.0005	.0005	.0005	.0005
Maximum Taper	.001	.001	.001	.001	.001	.001	.001	.001	.001
Camshaft									
Run Clearance on Pin	.0005/.003	.0005/.003	.0005/.003	.0005/.003	.0005/.003	.001/.0035	.001/.0035	.001/.0035	.001/.0035
End Play	.005/.010	.005/.010	.005/.010	.005/.010	.005/.010	.005/.010	.005/.010	.005/.010	.005/.010
Connecting Rod									
Big End Maximum Diameter	1.1885	1.1885	1.1885	1.1885	1.1885	1.5025	1.5025	1.5025	1.5025
Rod-Crankpin Max Clear	.0035	.0035	.0035	.0035	.0035	.0035	.0035	.0035	.0035

Engine Overhaul

ENGINE REBUILDING SPECIFICATIONS

Specification	K141 2⁷⁄₈" Bore	K141 2¹⁵⁄₁₆" Bore	K161 2⁷⁄₈" Bore	K161 2¹⁵⁄₁₆" Bore	K181	K241	K301	K321	K341
Small (Pin) End-New Dia	.62565	.62565	.62565	.62565	.62565	.85975	.87585	.87585	.87585
Rod to Pin Clearance	.0006/.0011	.0006/.0011	.0006/.0011	.0006/.0011	.0006/.0011	.0003/.0008	.0003/.0008	.0003/.0008	.0003/.0008
Piston									
Thrust Face-Max Wear Dia ①	2.866	2.9305	2.866	2.9305	2.9305	3.2445	3.3625	3.4945	3.7425
Thrust Face Bore Clearance ①	.006/.0075	.006/.008	.006/.0075	.006/.008	.006/.008	.0075/.0085	.0065/.0095	.007/.010	.007/.010
Ring-Max Side Clearance	.006	.006	.006	.006	.006	.006	.006	.006	.006
Ring-End Gap in New Bore	.007/.017	.007/.017	.007/.017	.007/.017	.007/.017	.010/.020	.010/.020	.010/.020	.010/.020
Ring-End Gap in Used Bore	.027	.027	.027	.027	.027	.027	.030	.030	.030
Valve-Intake									
Valve-Tappet Cold Clear	.006/.008	.006/.008	.006/.008	.006/.008	.006/.008	.008/.010	.008/.010	.008/.010	.008/.010
Valve Lift (Zero Lash)	.2778	.2778	.2778	.2778	.2778	.324	.324	.324	.324
Stem to Guide Max Wear Clear	.0045	.0045	.0045	.0045	.0045	.0045	.0045	.0045	.0045
Valve-Exhaust									
Valve-Tappet Cold Clear	.015/.017	.015/.017	.015/.017	.015/.017	.015/.017	.017/.020	.017/.020	.017/.020	.017/.020
Valve Lift (Zero Lash)	.2542	.2542	.2542	.2542	.2542	.324	.324	.324	.324
Stem to Guide Max Wear Clear	.006	.006	.006	.006	.006	.0065 ②	.0065 ②	.0065 ②	.0065 ②
Tappet									
Clearance in Guide	.0005/.002	.0005/.002	.0005/.002	.0005/.002	.0005/.002	.0008/.0023	.0008/.0023	.0008/.0023	.0008/.0023
Ignition									
Spark Plug Gap-Gasoline	.025	.025	.025	.025	.025	.025	.025	.025	.025
Spark Plug Gap-LP Gas	.018	.018	.018	.018	.018	.018	.018	.018	.018
Spark Plug Gap (Shielded)	.020	.020	.020	.020	.020	.020	.020	.020	.020
Breaker Point Gap	.020	.020	.020	.020	.020	.020	.020	.020	.020
Trigger Air Gap (Breakerless)	.005/.010	.005/.010	.005/.010	.005/.010	.005/.010	.005/.010	.005/.010	.005/.010	.005/.010
Spark Run ° BTDC	20°	20°	20°	20°	20°	20°	20°	20°	20°
Spark Retard	3° BTDC ③ (ACR-NONE)	ACR ONLY (No Retard)	3° BTDC ③ (ACR-NONE)	ACR ONLY (No Retard)	3° BTDC ③ (ACR-NONE)	3° ATDC ③ (ACR-NONE)	3° ATDC ③ (ACR-NONE)	ACR ONLY (No Retard)	ACR ONLY (No Retard)
Torque Values									
Spark Plug (foot lbs)	18-22	18-22	18-22	18-22	18-22	18-22	18-22	18-22	18-22
Cylinder Head	15-20 ft lbs	15-20 ft lbs	15-20 ft lbs	15-20 ft lbs	15-20 ft lbs	25-30 ft lbs	25-30 ft lbs	25-30 ft lbs	25-30 ft lbs
Connecting Rod	200 in. lbs	200 in. lbs	200 in. lbs	200 in. lbs	200 in. lbs	300 in. lbs	300 in. lbs	300 in. lbs	300 in. lbs
Flywheel Nut	50-60 ft lbs	50-60 ft lbs	50-60 ft lbs	50-60 ft lbs	50-60 ft lbs	60-70 ft lbs	60-70 ft lbs	60-70 ft lbs	60-70 ft lbs

①Measured just below oil ring and at right angles to piston pin
②Measured at top of guide with valve closed
③Engines built before automatic compression release (ACR)

Engine Overhaul

Onan

ENGINE IDENTIFICATION

All Onan engines have an identification plate attached to the left side of the cooling shroud (facing the flywheel) on one cylinder engines, and on the right side of the cooling air duct (facing the flywheel) on four cylinder engines. The Model and Specification number shown there may be interpreted as follows:

1. Factory code for general engine identification.
2. Engine specific type:
S-Manual starting with stub shaft power takeoff.
MS-Electric starting with stub shaft, starter, and generator.
MV-Vacu-Flo cooling. Similar to MS, but with cooling air drawn in through a front end duct.
3. Factory code for optional equipment on the engine.
4. Specification letter. Letter advances alphabetically with production modifications.

Serial numbers consist of a two digit month of manufacture indicator and a six digit engine number.

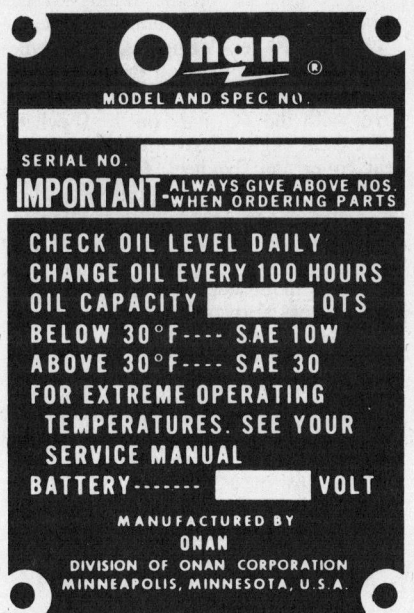
Onan engine identification plate

MAINTENANCE

Air Cleaner Service

OIL BATH AIR CLEANER

The oil bath air cleaner uses engine oil to collect dirt. Dirt is deposited in a sump full of oil in the bottom of the unit, and on a

GENERAL ENGINE SPECIFICATIONS

Model	Bore & Stroke (In.)	Displacement (cu. in.)	Horsepower @ RPM
LKB	3.25 x 3.00	24.9	8.5 @ 1800
NB	3.562 x 3.00	30.0	12.0 @ 3600
CCK	3.25 x 3.00	49.8	12.9 @ 2700
CCKA	3.25 x 3.00	49.8	16.5 @ 3600
CCKB	3.25 x 3.00	49.8	20.0 @ 3900
BF	3.125 x 2.615	40.3	16.0 @ 3600

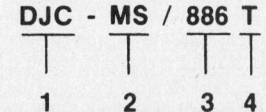

Model and specification number breakdown

screen contained in the upper portion. Dirt sticks to the screen because the incoming air pulls some of the oil out of the sump, deposits it on the screen, and this wetting action keeps the screen sticky. After each 100 hours, disassemble the unit, drain the oil out of the sump, clean the dirt out of the sump with a solvent soaked rag, and refill the sump with clean oil of the type used in the engine. If the air around the engine is very dusty, inspect the air cleaner more frequently. Make sure to change the oil in the sump before dirt contained in the sump reaches the ring which is slightly above the bottom. In extreme cases, it may be necessary to clean the screen with solvent and coat it with clean oil as part of the service.

MOISTENED FOAM AIR CLEANER

This cleaning medium in this type of cleaner is an oil soaked sponge mounted around a screen. After each 100 hours, wash the sponge in soap and water or solvent, dry it, soak it in engine oil, and then squeeze excess oil out of it before installation.

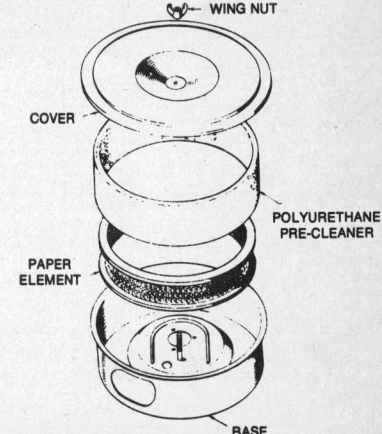
Dry paper type air cleaner with polyurethane pre-cleaner

DRY PAPER AIR CLEANER

Dry paper air cleaners use a replaceable paper element. All that is required in servicing is disassembly of the unit, replacement of the element, and a brief removal of dust from inside the unit with a rag. Some of these units also employ a polyurethane precleaning element, which is also replaced at time of service, or a dust cup, which is cleaned with a solvent soaked rag and dried. Service standard units every 100 hours; units with precleaner at 500 hours.

Lubrication

OIL AND FUEL RECOMMENDATIONS

Onan gasoline engines may use either regular grade leaded, unleaded, or low lead gasoline. The use of unleaded or low lead fuels is strongly recommended in order to reduce combustion chamber deposits, especially on governor controlled engines which run at a fairly steady load. If you are switching from leaded to unleaded fuel, combustion chamber deposits should be removed before beginning operation on the unleaded fuel because of its lower octane

Oil bath air cleaner

905

Engine Overhaul

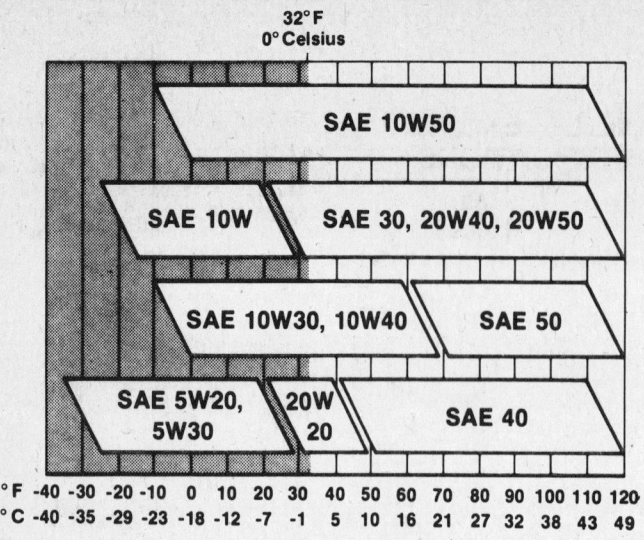

Oil viscosity chart

CRANKCASE CAPACITY CHART

Model	Pints
LKB	2.0
NB	2.0
CCK	4.0
CCKA	3.0
CCKB	4.0
BF	2.0

rating and consequent potential for damage due to engine knock.

It is best to use an SE oil, although oils not so well equipped to resist sludge formation at both high and low temperatures may be used in very moderate service—i.e. long operating periods under light loads. Consult the chart for recommended viscosity ranges for the prevailing temperature.

Spark Plugs

Spark plugs should be replaced every 200 operating hours. Set the gap with a

TUNE-UP SPECIFICATIONS

Model	Spark Plug Gap (in.)	Ignition Point Gap (in.)	Ignition Timing (Deg. BTDC @ RPM)
LKB	.025	.020	24
NB	.025	.020	22
CCK	.025	.020	19
CCKA	.025	.020	24
CCKB	.025	.020	24
BF	.025	.025	21

wire gauge to .025 in. It is best to torque the plugs with a torque wrench to avoid cylinder head damage, distorted gap dimension, or leakage. Use new gaskets and torque to 15-20 ft.lbs. on NB and BF engines., 25-30 ft.lbs. on other engines. Plugs should also be inspected and cleaned with a wire brush once or twice, depending on type of fuel and service, during their service life. Use new plug gaskets even when reinstalling old plugs.

Breaker Points

Breaker points must be inspected and cleaned up and gapped or replaced as required. On BF engines with side adjust points, they should be replaced every 200 hours. Gain access to the points as described below under "Ignition Timing". Inspect the points for burning and pitting. If only slightly worn or damaged, points may be dressed with an abrasive stone and regapped. Otherwise, they must be replaced.

To replace points, disconnect the primary electrical connection, remove the mounting screws and remove the old set. Install the new set with mounting screws slightly loose. Turn the engine over slowly until points reach their maximum gap. Then, using a flat feeler gauge of proper dimension (see the tune-up chart), shift the contact set base plate as necessary, or adjust the Allen screw, to get the proper gap, tighten the mounting screw, and recheck. In some cases, a slight further adjustment may be needed to get the proper timing. Make sure to lube the cam follower with high temperature grease or oil, as necessary, when replacing or servicing points.

Ignition Timing
CCK, CCKA, CCKB ENGINES

1. Remove breaker box cover, turn engine until points are wide open, and set point gap to .020 in. with a flat feeler gauge.
2. Once point gap is correct, timing is adjusted by shifting the entire breaker box.

Note that timing is set at 19 degrees on engines running at 2,400 rpm and below, and at 25 degrees on engines running at 2,500 and above. On all engines with automatic spark advance, make this preliminary setting at 5 degrees. See Step 4 for the location of the timing marks. A preliminary check of the timing may be made by turning the flywheel backwards and then turning it forward with engine on the compression stroke until the timing marks just line up. If the points open just when the timing marks line up, the timing is o.k. If the timing is too early, shift the breaker box toward #1 cylinder to retard the timing; if too late, shift the box in the opposite direction.

3. Tighten the breaker box mounting screws, and check the timing at the suggested rpm as specified in the next step. Shift the breaker box position and recheck until the timing is correct.
4. Timing mark locations and rpm at which timing should be checked are as follows:

Vacu-Flo Engines Without Spark Advance: Remove the dot button from the top of the blower housing. Run the engine at 1,400-1,600 rpm. Viewing through the round hole in the blower housing, the TC flywheel mark should line up with the 25 degree mark on the gear cover.

Vacu-Flo Engines With Spark Advance: Follow the procedure above. After timing at 1,400-1,600 rpm is set, slow the engine to less than 800 rpm and make sure the timing retards. If not, the timing advance mechanism may need cleaning or repair.

Other CCK Series Engines: Some types have the timing marks on the gear cover. On these, align the correct mark on the gear cover with the TC mark on the flywheel. Other CCK engines have marks on both the gear cover and flywheel. On these, align the TC flywheel mark with the correct timing mark on the gear cover, or align the correct timing mark on the flywheel with the TC mark on the gear cover. On CCK engines without Vacu-Flo cooling, timing is 24 degrees at over 1,100 rpm.

BF ENGINES
Side Adjust Breaker Points

1. Remove two screws and remove the breaker cover.
2. Remove the air intake hose that connects to the blower housing. On Power Drawer units, remove the dot button on the blower housing to see the timing marks.
3. Loosen the mounting screw on the points. Rotate the crankshaft clockwise until the timing mark on the gear cover aligns with the mark on the flywheel. Turn another 90 degrees.
4. Insert a screwdriver into the notch on the points and set the gap to .025 in. with a flat feeler gauge. Tighten the mounting screw.
5. The points should open when the crankshaft passes the 25 degree mark (26 degrees on Power Drawer engines). Timing may be checked more precisely by connecting a test lamp between the breaker box terminal and a ground on the engine. The lamp should go out just as timing mark

Engine Overhaul

lines up. If necessary, change point gap slightly to get timing to occur at the right point. Timing may also be checked on either spark plug with a timing light for greatest accuracy.

Top Adjust Breaker Points

Follow the procedure above for "Side Adjust Breaker Points", but set point gap to .021 in. by turning the point gap adjusting screw with an Allen wrench.

NB ENGINES

1. Remove the breaker box cover and crank the engine slowly by hand until the maximum point opening is achieved.
2. Adjust the breaker gap to .020 in. with a flat feeler gauge. To adjust the breaker base, loosen the lower mounting screw.
3. Check the timing by turning the engine backward, and then going forward slowly as the 22 degree mark on the flywheel passes the mark on the gear cover. Points should open just as the timing marks line up. If the timing is incorrect, loosen the breaker box mounting screws (located inside the breaker box) and slide the box upward to retard the timing or downward to advance it. Tighten the mounting screws securely and recheck timing.
4. A more accurate check may be made with a test lamp connected from the breaker box terminal to an engine ground. The light should go out when the timing marks align as the engine is turned slowly forward.

Mixture Adjustments

Factory mixture adjustments should not be disturbed unless the engine clearly is running too rich or too lean. Standard settings are: Main Jet—1¼ turns open; Idle Jet—1 turn open. Do not force the needle against the seat by turning it in hard when making these settings. Turn the screw in slowly until it just bottoms very gently, then turn it out the specified amount.

If it is necessary to adjust further, run the engine until it is hot and remove all load. Run the engine at idle speed and adjust the idle mixture screw in and out slowly until highest speed and smoothest running are obtained. Adjust the main jet mixture screw similarly, with the engine running at normal speed without load. Test the response of the engine when accelerating from idle and open the main jet ¼ turn more if acceleration response is poor. If the

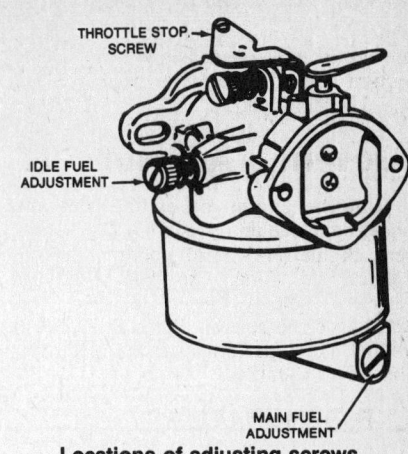

Locations of adjusting screws

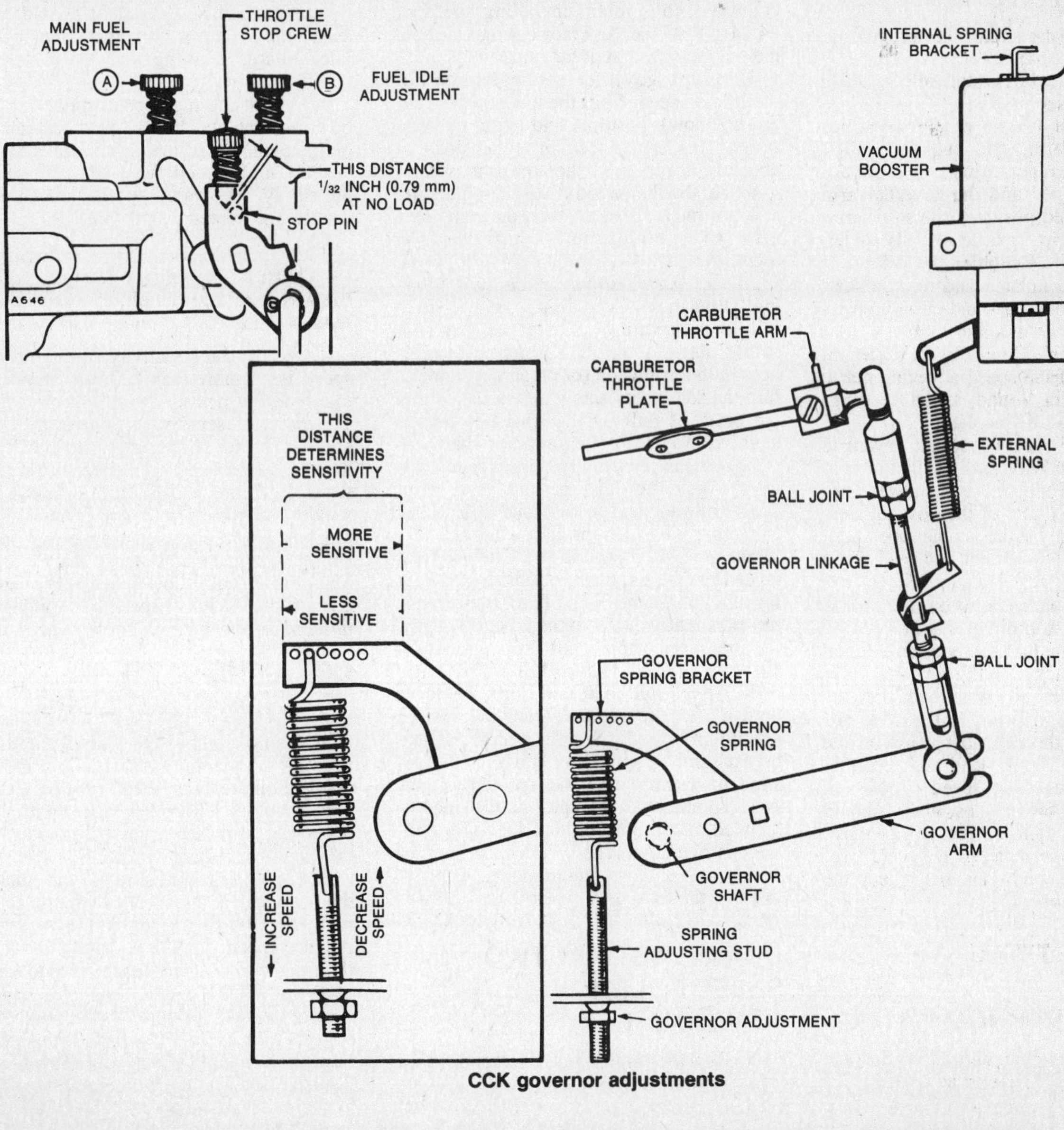

CCK governor adjustments

Engine Overhaul

governor "hunts" open and closed, the main jet may be opened as ¼ turn more or ½ turn beyond point of highest operating speed without load.

Governor Adjustments

Run the engine or engine/generator under light load until it is hot. On generator sets, engine rpm controls both output frequency and voltage, so both speed and sensitivity settings are critical. Use of an accurate tachometer is the best means of getting good performance without overspeeding the engine.

LKB GENERATORS

First adjust the speed adjusting nut (clockwise to increase) to the setting called for on the nameplate. If the voltage drop with increase in load is too great, loosen the locknut and turn the sensitivity adjusting screw inward, or toward the shaft, retighten the locknut, and recheck speed. Repeat adjustments as necessary.

CCK GENERATORS

1. Adjust the carburetor for best mixture with the engine at full load.
2. Adjust the carburetor idle mixture with all load disconnected.
3. Adjust the length of the governor linkage by rotating the ball joint (first loosen the locknut). Adjust, so that with the engine stopped and the governor arm held in the closed position, the stop screw on the carburetor throttle lever is 1/32 in. from the stop pin. Retighten the locknut.
4. Check the linkage and throttle shaft for binding or looseness and clean up parts or replace them as necessary.
5. Disconnect the booster external spring, and turn the speed adjusting nut to obtain the voltage and speed readings shown on the unit nameplate.
6. Move the governor spring inward to get greater sensitivity until a hunting condition occurs. Then, move it outward (toward the outward end of the govenor arm) as necessary to ensure stable operation. Recheck speed adjustment and readjust as necessary.
7. Set the distance between the throttle stop screw and stop pin at 1/32 in.
8. Connect the vacuum booster external spring to the bracket on the governor link. With the unit operating with load, slide the bracket on the governor link just to the position where there is no tension on the external spring. Apply load and carefully watch the cycles. The speed should not drop more than four cycles, and it should recover rapidly. If it is necessary to make the speed booster more or less sensitive, change the cotter pin to another hole in the return spring strap.

CCK INDUSTRIAL VARIABLE SPEED GOVERNOR

1. Run the engine until it is hot and make necessary carburetor adjustments.

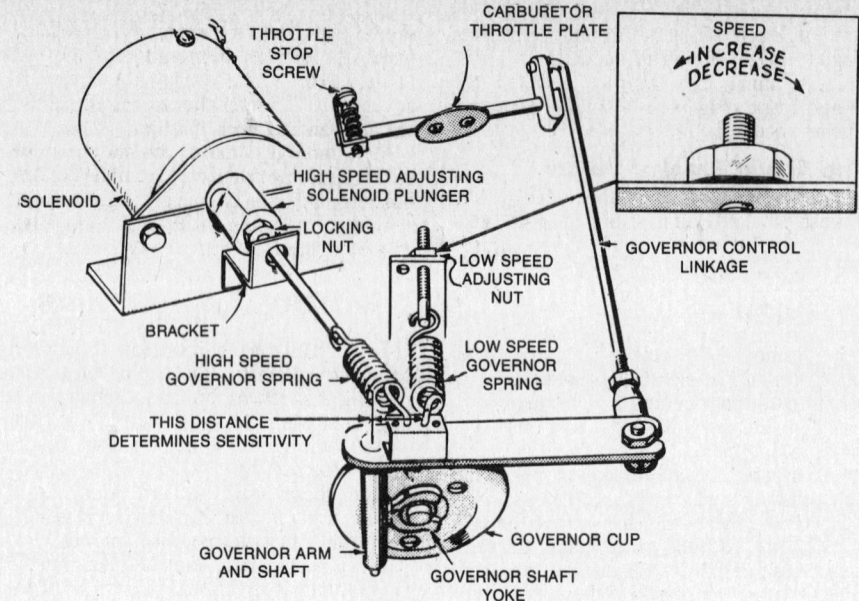

Two-speed solenoid type goveror adjustments

2. Adjust the throttle stop screw on the carburetor to a minimum idling speed of 1450 rpm so the governor spring can hold the engine speed at 1500 rpm.
3. Adjust governor spring tension for minimum speed. Shift the lever to the minimum (slow) position and with no load connected, adjust the spring tension for about 1500 rpm with the adjusting nut.
4. Adjust the sensitivity while operating at a minimum speed to attain the smoothest yet quickest no load to full load operation as follows:

To decrease sensitivity and prevent hunting (allow more speed drop from no load to full load operation): Move the governor spring outwards into a different groove or hole in the extension (or on earlier models, turn the sensitivity screw outwards) so that the point of pull by the spring is moved slightly away from the governor shaft.

To increase sensitivity (closer regulation by the governor that permits less speed drop from no load to full load operation): Move the governor spring inward to a different groove or hole in the extension (or on earlier models, turn the sensitivity screw inward) so that the point of pull by spring is moved slightly closer to the governor shaft. Engine speed should not drop more than 100 rpm from no-load to full-load.
5. Apply full load and shift the lever until the engine speed reaches the desired maximum speed. Set the screw in the bracket slot to stop lever travel at the desired maximum full load speed position. Approximately 3000 rpm is the recommended maximum full load speed for continuous operaton. The speed must be consistent with the load requirements. Adjust it lower if the load must be driven at a lower rpm.

CCK INDUSTRIAL TWO SPEED SOLENOID TYPE GOVERNOR

1. Run the engine until hot and adjust the carburetor for best operation.

2. Adjust the low speed spring tension for the desired low speed (it should be at least 1,500 rpm for units that carry load at idle). Increase spring tension to increase speed.
3. Adjust the high speed spring tension by loosening the locknut and turning the plunger on the adjusting stud to give the desired high speed when the plunger is pulled all the way into the solenoid (it should not exceed 3,000 rpm).

NOTE: If the plunger does not pull all the way into the solenoid due to excessive tension, electrical damage will occur.

4. Adjust the sensitivity so that no load speed is no more than 100 rpm above full load speed (measure with a tachometer). To increase sensitivity, move the high speed governor spring inward to a different hole of the bracket, or, to decrease it, outward. If, in order to adjust high speed sensitivity, you need to use the hole in the bracket that is occupied by the low speed spring, you can simply move that spring as required. Usually, low speed sensitivity is not critical. If it is necessary to adjust it, proceed as for the high speed spring.

CCKA AND CCKB TRACTOR GOVERNORS

1. Using a tachometer, adjust the throttle stop screw for 1,000 rpm on CCKA applications, and to 1,200 rpm on CCKB governors. Readjust the idle mixture as necessary, and then bring the closed throttle speed back to specification.
2. On CCKA units, adjust the nuts on the low speed (smaller) adjusting spring with the control in the "slow" position. Adjust speed to 1,200 rpm, and then tighten the nuts against each other to lock.
3. On CCKB units, move the speed control to the "fast" position, and then turn the high speed (larger) adjusting spring adjusting nuts so the engine runs at 3,800-3,850 rpm. Then tighten the adjust-

Engine Overhaul

ing nuts against each other to lock, making sure rpm does not exceed 3,850.

BF TRACTOR GOVERNOR

1. Disconnect the throttle linkage from the governor arm. Then, hold both linkage and governor arm towards the carburetor, and note which hole in the arm is closer to the position of the linkage.
2. Connect the linkage to the hole which more nearly lines up with the position of the link.
3. If it is necessary to increase the sensitivity, move the spring loop into the governor arm hole nearest to the governor. If it is necessary to decrease sensitivity, go in the opposite direction.
4. Adjust the low speed with the adjusting screw on the control wire bracket.

BF Power Drawer

1. Run the unit for 15 minutes under light load to warm it up.
2. Adjust the length of the linkage by rotating the ball joint on the governor arm. With the engine stopped and slight tension on the governor spring, adjust so that the stop on the carburetor lever just touches the carburetor bowl.
3. Adjust with an accurate tachometer and voltmeter. Check voltage and speed at no load and at full load, and compare with specifications in the charts.
4. If it is necessary to change sensitivity, shift the spring toward the outer end of the governor arm to decrease it, or toward the inner end to increase it.
5. After adjusting and checking sensitivity, adjust speed by tightening or loosening the speed adjusting nut at the end of the spring.

Automatic Choke Adjustment

CCK

1. Loosen the choke cover screws (2) and rotate the cover as necessary to get the following dimension between the inner edge of the choke plate and the wall of the carburetor air horn: At 58°F—¼ in. open; at 66°F—½ in. open; at 76°F—¾ in. open; at 82°F—fully open.
2. Tighten the choke cover screws.

VOLTAGE AND SPEED REGULATION LIMITS FOR BF POWER DRAWER GOVERNOR

Voltage Chart for Checking Governore Regulator	120 Volt 1 Phase 2 Wire
Maximum No-Load Voltage	126
Minimum Full-Load Voltage	110

Speed Chart for Checking Governor Regulation	
Maximum No-Load Speed (RPM)	1890
Hertz (Current Frequency)	63
Minimum Full-Load Speed (RPM)	1770
Hertz	59

THERMAL-MAGNETIC CHOKE

NOTE: Make sure the engine has been off for at least one hour.

1. Loosen the screw which secures the choke body.
2. Turn the choke body clockwise to enrichen the setting, or counterclockwise to lean it out until measurement of the dimension between the choke and carburetor air horn conforms to specifications shown in the illustration. Tighten the choke body screw.

ELECTRIC SOLENOID CHOKE

1. With the engine cold, disconnect the linkage to the carburetor choke shaft. Rotate the choke lever in the closed direction until the hole in the shaft is aligned with the notch in the shaft bearing.
2. Insert a 1/16 in. diameter rod through the shaft hole, engaging the rod in the notch of the mounting flange to lock the shaft in place.
3. Loosen the choke lever clamp screw enough to permit moving the lever on the shaft. Remove the air cleaner and verify that the linkage to the choke lever is properly in place. Adjust the choke assembly lever so the choke is from just closed to not more than 1/16 in. open.
4. Tighten the choke lever clamp screw and remove the locking rod from the shaft.
5. Press downward on the choke lever to the limit of its travel and make sure the choke opens completely. If not, adjust the position of the choke shaft lever as necessary.
6. Make sure that when the engine is hot, the choke is wide open.

BF POWER DRAWER CHOKE

1. Remove the clip and bushing, and loosen the choke lever clamp screw.
2. With the lever fully forward (or, away from carburetor), adjust so the choke valve is completely closed.
3. Tighten the clamp screw, and replace the bushing and clip.

Valve Adjustment

Adjust the valve clearance with the engine cold and in TDC position with the spark plug about to fire. Adjustment is checked by sliding a flat feeler gauge of the proper dimension (see specifications, noting that clearances are different for exhaust and intake valves) between the tappet and head of the valve stem. If necessary, turn the adjusting nut on the tappet while holding tappet with a second wrench to get a slight pull on the gauge.

Testing Compression

1. Run the engine until hot. Stop and remove spark plugs.
2. Insert the compression gauge in one of the spark plug holes and crank the engine with the starter. Record the reading.
3. Squirt a small amount of SAE 30 oil into the cylinder and repeat the check.
4. Repeat steps 2 and 3 for each cylinder. Compression specifications are listed below. If compression is low, but increases substantially when oil is squirted into the cylinder, the compression problem is probably with the pistons, rings, and cylinders. Compression pressures are: (psi)
LKB—100–120
NB—105–115
CCK—90–110
CCKA, CCKB—100–120
BF—110–120

AMBIENT TEMP. (°F)	60	65	70	75	80	85	90	95	100
CHOKE OPENING (Inches)	1/8	9/64	5/32	11/64	3/16	13/64	7/32	15/64	1/4

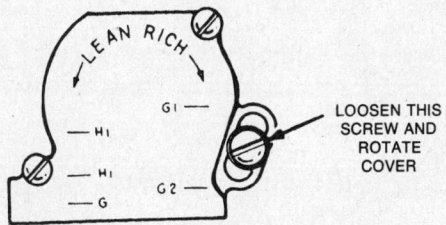
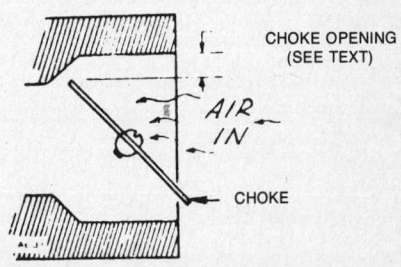

Thermal-magnetic choke adjustments

Engine Overhaul

FUEL SYSTEM

Carburetion

The carburetor is a side draft or horizontal float type with two adjusting screws for the idle needle and the main fuel nozzle needle.

BASIC DOWNDRAFT AND SIDEDRAFT CARBURETORS

Removal and Installation

1. Remove air cleaner, fuel line, governor linkage and choke apparatus from carburetor.
2. Remove two carburetor mounting nuts and pull off carburetor. On CC Engines, first remove the intake manifold; then remove the carburetor from the manifold.
3. Reverse above steps to install carburetor on engine.

Overhaul

1. Remove air cleaner adapter and choke from carburetor.
2. Remove main fuel adjustment needle and needle retainer.
3. Remove top of carburetor from carburetor base.
4. Remove carburetor float, lift out float valve and unscrew and remove its seat.
5. Remove no load adjusting needle.
6. Remove throttle plate and throttle shaft.
7. Remove choke plate and choke shaft.
8. Remove nozzle assembly.
9. Soak all components thoroughly in carburetor cleaner, following cleaner manufacturer's instructions. Clean all carbon from carburetor bore, especially in the area of the throttle valve. Blow out passages with compressed air. Avoid using wire to clean out passages.
10. Check adjusting needles and nozzle for damage. If the float is loaded with fuel or damaged, replace it. The float should turn freely on its pin without binding. Invert the carburetor body and measure float level.
11. To adjust float level, bend small lip that needle valve rides on.
12. Check choke and throttle shafts for excessive side play and replace if necessary.
13. Install throttle shaft and valve, using new screws. The bevel on the throttle plate must fit flush with carburetor body. On valve plates marked with a "C," install them with the mark on the side toward idle port as viewed from flange end of carburetor. To center throttle valve (Bendix/Zenith Carburetor) back off stop screw, close throttle lever, and seat valve by tapping it with a small screwdriver, then tighten the two throttle plate screws.
14. Install choke shaft and choke plate. Center choke plate in same manner as throttle valve. Always fasten plate in position with new screws.
15. Install main nozzle. Make sure it seats in body casting.
16. Install main fuel adjustment needle and its retainer.
17. Install no load adjusting needle.
18. Install intake valve seat and intake valve.
19. Install float and float pin. Center pin so float bowl doesn't ride against it.
20. Check float level. Adjust if necessary.
21. Install a new body-to-bowl gasket and secure the two sections together.
22. Reinstall choke.
23. Install air horn assembly.
24. Install carburetor on engine.
25. Adjust both main and idle fuel mixture needles, as described in the Tune-Up section.

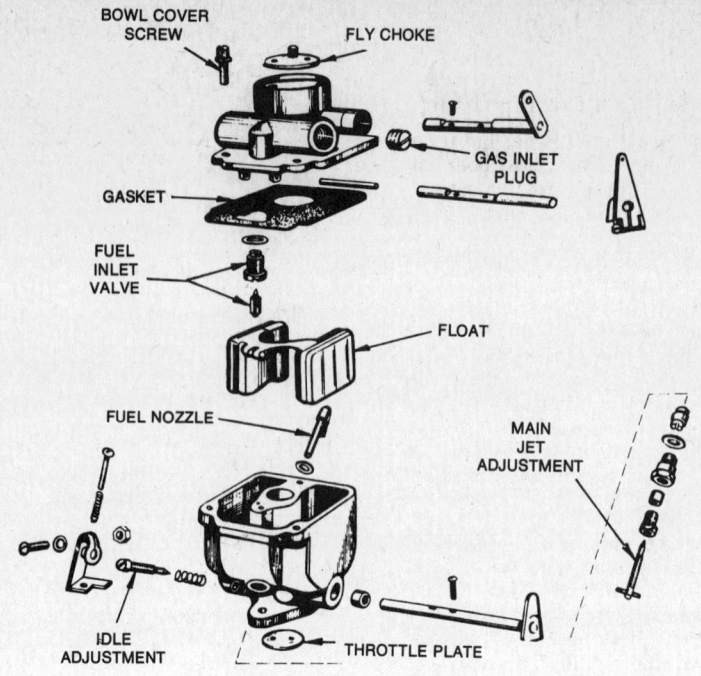

Downdraft type carburetor

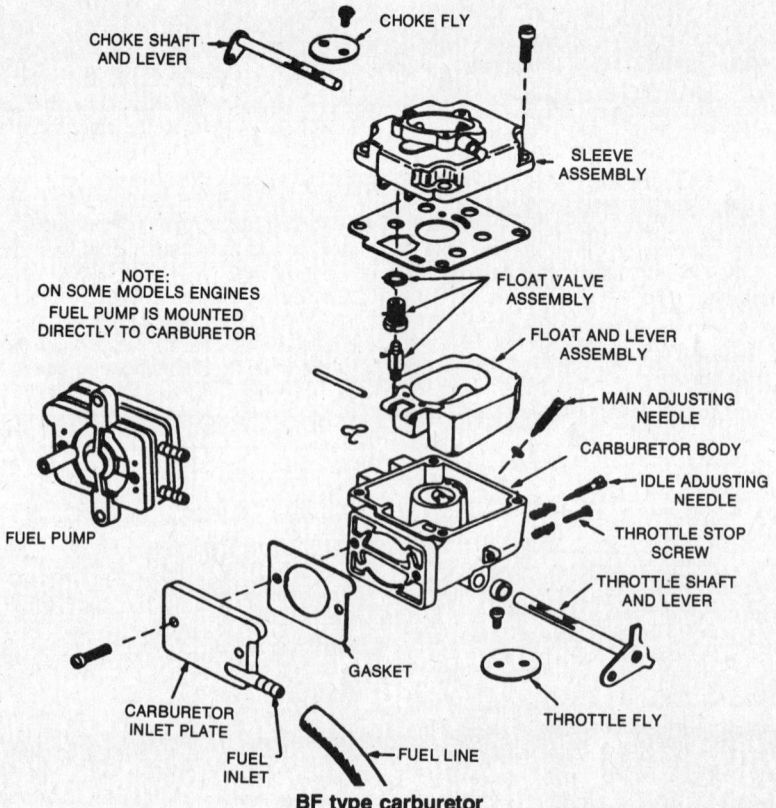

BF type carburetor

Engine Overhaul

BF ENGINE CARBURETOR

Removal and Installation
1. Remove air cleaner and hose.
2. Disconnect governor and throttle linkage, choke control and fuel line from carburetor.
3. Remove four intake manifold capscrews and lift complete manifold assembly from engine.
4. Remove carburetor from intake manifold.
5. Installation is the reverse of the removal procedure.

Overhaul
Generally follow the procedures applying to the downdraft and sidedraft carburetors as described above. Refer to the two items below for needle and seat replacement and float adjustment.

Replacing Needle and Seat
1. Remove four screws from top of carburetor and lift off float assembly.
2. Invert float assembly as shown.
3. Push out pin that holds float to cover.
4. Remove float and set aside in a clean place. Pull out needle and spring.
5. Remove valve seat and replace with a new one, making sure to use a new gasket.
6. Install new bowl gasket.
7. Clip new needle to float assembly with spring clip. Install float.

Float Adjustment
1. Invert float assembly and casting.
2. With float resting lightly against needle and seat, there should be 1/8-inch clearance between bowl cover gasket and free end of float.
3. If it is necessary to reset float level, bend float tangs near pin to obtain a 1/8-inch clearance, as shown.

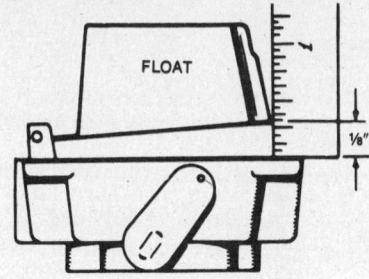

BF type float adjustment

BF POWER DRAWER CARBURETORS

Removal and Installation
1. Disconnect fuel inlet hose, crankcase breather hose and air inlet hose.
2. Disconnect governor, throttle linkage, and choke control.
3. Remove two hold-down nuts and lift carburetor from intake manifold.
4. Installation is the reverse of the removal procedure.

Overhaul
Generally follow the procedures applying to the downdraft and sidedraft carburetors as described above, referring to the exploded view. Refer to the two items

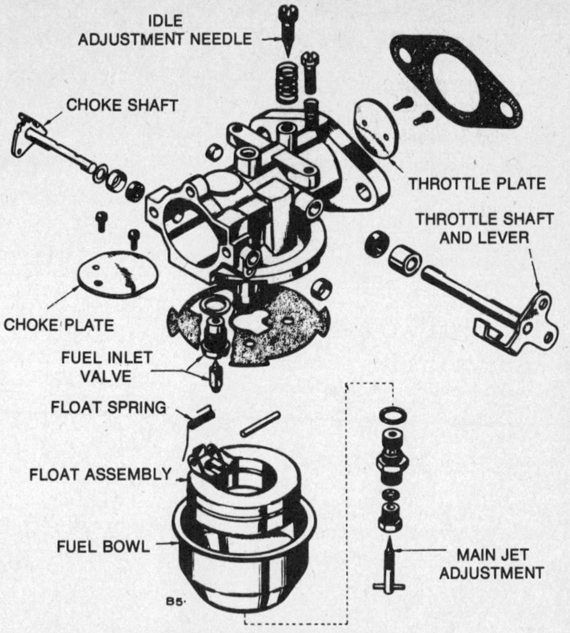

Side draft carburetor

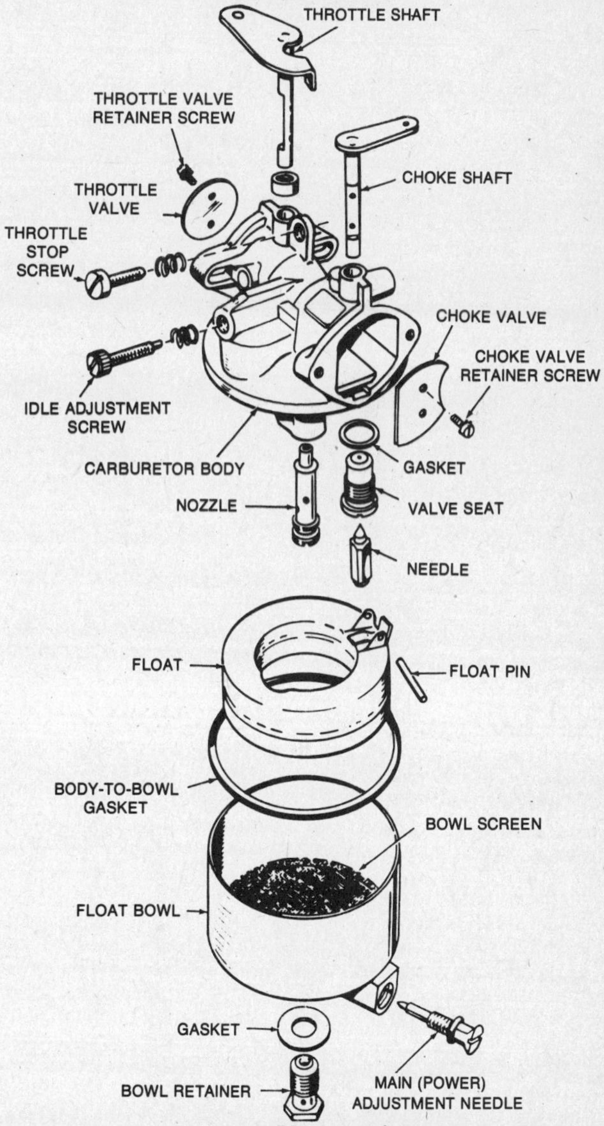

BF Power-Drawer carburetor

Engine Overhaul

below for needle and seat replacement and float level adjustment.

Replacing Needle and Seat

1. Remove 7/16 in. hex at base of fuel bowl and lift bowl from carburetor.
2. Push out pin that holds float to carburetor body.
3. Remove float and set aside in a clean place. Pull out needle and using a large screwdriver remove needle valve seat.
4. Install new valve seat and needle and replace float.

Float Adjustment

1. Invert float and casting.
2. With float resting lightly against needle and seat, there should be .07 in. to .11 in. clearance between base of float and carburetor casting.
3. If it is necessary to reset float level, remove float from carburetor and bend float tang near pin to obtain correct float level.

CAUTION

Do not bend the float when installed; doing so may cause deformation of needle or seat.

---

4. Check float carefully for signs of leakage. Repair or replace float if damaged or filled with gasoline.
5. Before assembling carburetor, remove filter screen from float bowl and clean both screen and base of float bowl.
6. Install new gaskets when reassembling.

MECHANICAL FUEL PUMPS

Removal and Installation

1. Remove the fuel lines.
2. Remove the two mounting capscrews, and pull the pump off the engine. Remove and discard the gasket.
3. Clean all gasket material from both mounting pad and pump flange. Apply an oil resistant sealer to both sides of a new gasket and to the threads of the attaching bolts.
4. Position the new gasket on the pump flange and position the pump and gasket on the mounting pad. To check their position, make sure the rocker arm rides on the pump cam lobe. Turn the crankshaft until the rocker arm is at the low point of its stroke.
5. Position the pump tightly against the pad and install the mounting bolts, torquing them alternately in several stages to specifications.
6. Connect the fuel lines and then operate the unit to check for leaks.

Service

1. Scribe a mark across the flanges of the pump body and valve housing so these parts can be assembled in original positions.
2. Remove the valve housing from the body of the pump. Tap pump body with a screwdriver to do this.
3. Remove both valves and their gaskets from valve housing. Note position of valves in their housing so new valves can be correctly installed.

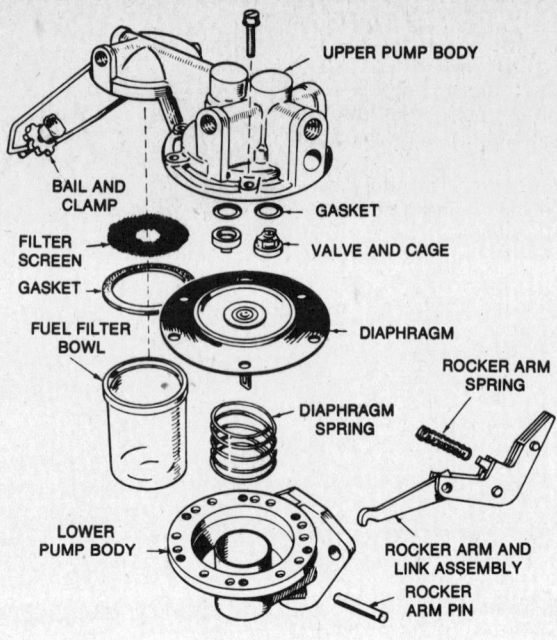

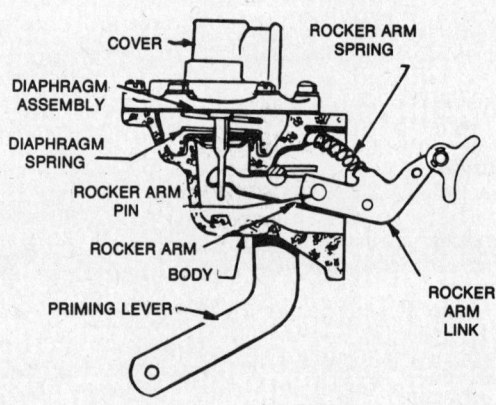

CCK mechanical fuel pump

4. Using a blunt punch, drive rocker arm pin out of pump.
5. Press diaphragm into fuel pump body and then pull rocker arm outward to unhook diaphragm actuating rod from rocker arm link assembly.
6. Remove diaphragm and diaphragm return spring, rocker arm and link assembly, and rocker arm return spring from pump body.
7. Remove diaphragm actuating rod oil seal from pump body.
8. Clean and inspect all fuel pump components and replace all unserviceable parts.
9. Install inlet and outlet valves and their gaskets in their respective positions. Seat valves firmly.
10. Lubricate diaphragm actuating rod.
11. Position fuel pump diaphragm and spring assembly into pump body as shown.
12. Hold diaphragm assembly in pump body and position pump body so mounting flange faces up. Apply slightly more pressure to lower edge of diaphragm and insert rocker arm link assembly.
13. Hook rocker arm link to diaphragm actuating rod.
14. Install rocker arm return spring and hold it in place by cocking rocker arm slightly.
15. Install the rocker arm pin in the pump body.
16. Position the valve body and pump body so the two previously scribed marks align. Install all screws and lockwashers until they just engage the pump body, being careful not to tear the diaphragm fabric.
17. Alternately and evenly tighten all screws.

BF ENGINE PULSATING DIAPHRAGM FUEL PUMP

Service

1. Disconnect vacuum and fuel lines. Inspect lines for cracks and replace as necessary.

NOTE: On some engines, the pump is mounted on the side of the engine and has an obvious fuel discharge line. On other applications, the pump is an integral part of the carburetor and only a suction line is present.

Engine Overhaul

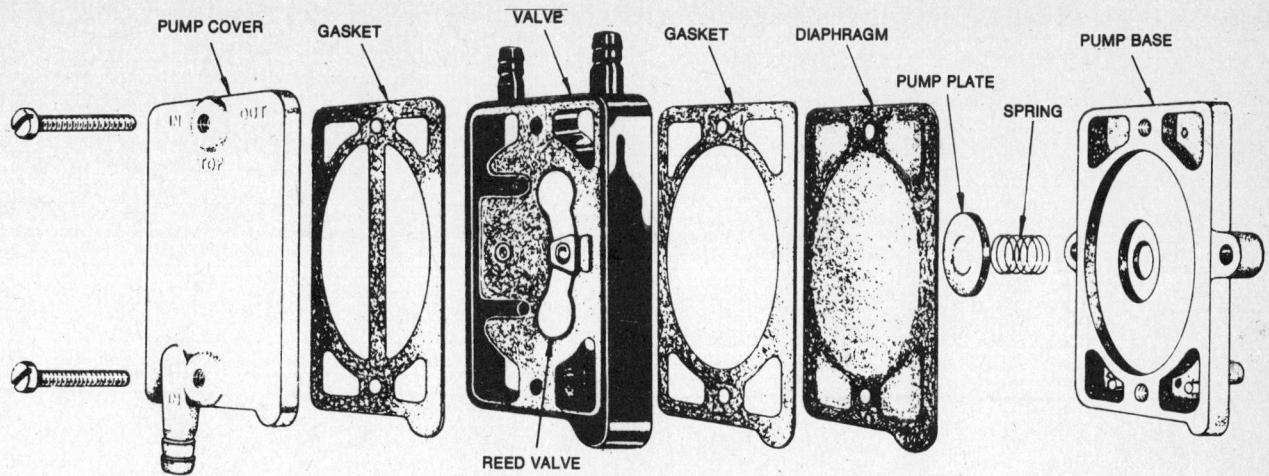

BF pulsating diaphragm fuel pump

2. Scribe two lines (one each on opposite ends of the pump) across the pump parts. This will ensure correct alignment of the pump parts with each other and the carburetor when the pump is reassembled.

3. Remove the fuel pump attaching screws.

4. Holding the pump carefully, pull the sections of the pump apart. The diaphragm, plunger, return spring and plate, pump body and gaskets will now be loose.

5. Check parts for wear and damage. Replace them with new parts where necessary.

6. Unclog the pump air bleed hole in the pump to allow unrestricted movement of pump diaphragm.

7. Replace the gaskets and reassemble the pump.

NOTE: That all parts must be perfectly aligned or there will be leakage and a consequent fire hazard.

8. Install the pump according to the marks scribed above. Reconnect the fuel lines, with the clamps tightly in their proper positions.

BENDIX ELECTRIC FUEL PUMPS

Service

1. Release the bottom cover (1) from the bayonet fittings. Twist the cover by hand to remove it from the pump body.

2. Remove the filter (4), magnet (3) and cover gasket (2) (see appropriate illustration). Wash the filter in cleaning solvent and blow out dirt and cleaning solvent with air pressure. Check the cover gasket and replace if deteriorated. Clean the cover.

3. Remove the retainer spring (5) from plunger tube (11), using thin nose pliers to spread and remove the ends of the retainer from the tube. Then remove the washer (6), O-ring seal (7), cup valve (8), plunger spring (9) and plunger (10) from the tube (11), (see appropriate illustration).

4. Wash all parts in cleaning solvent and blow out with air pressure. If the plunger does not wash clean or if there are any rough spots, gently clean the surface with a crocus cloth. Slosh the pump assembly in cleaning solvent. Blow out the tube with air pressure. Swab the inside of the tube with a cloth wrapped around a stick.

5. Insert the plunger assembly (10) in the tube with the buffer spring end first. Check the fit by slowly raising and lowering the plunger in the tube. It should move fully without any tendency to stick. If a click cannot be heard, the interrupter assembly is not functioning properly in which case pump should be replaced.

6. To complete the assembly, install the plunger spring (9), cup valve (8), O-ring seal (7) and washer (6) as shown. Compress the spring (9) and assemble the retainer (5) with ends of the retainer in the side holes of tube (11).

7. Place the cover gasket (2) and magnet (3) in the bottom cover (1) and assemble the filter (4) and cover assembly. Twist the cover by hand to hold it in position on the pump housing. Securely tighten the bottom cover.

ONAN ELECTRIC FUEL PUMP

Service

Clean the filters every 100 operating hours. Remove the four Phillips screws from the top, and lift off the filter assembly. Clean the two screen-type filters in a safe solvent. Install, ensuring the gasket is in the proper position to prevent leaks.

Governor Repair

1. Draw a sketch of the governor linkage or lay the parts out in position as you disassemble them. Disassemble the linkage. Remove the gear cover.

2. Remove the snapring that holds the governor cup to the camshaft gear, being ready to catch the flyballs, which will come out as the governor cup is removed.

3. Clean all parts thoroughly in a safe solvent. Inspect as follows and replace parts which are found defective:
   a. flyballs for grooves or flat spots.
   b. ball spacers for arms with noticeable wear or damage.
   c. governor cup with a rough or grooved race surface.
   d. governor cup which does not have a free spining fit on the camshaft center pin, or which is loose and wobbles.

4. To install the governor cup, tilt the engine to make the gear face upward. Space the flyballs at equal distances on the gear and then install the cup and snapring on the camshaft center pin.

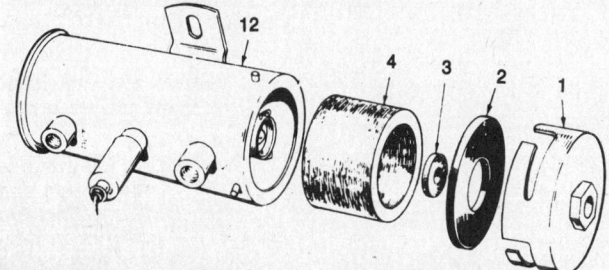

Bendix electric fuel pump cover, gasket and filter

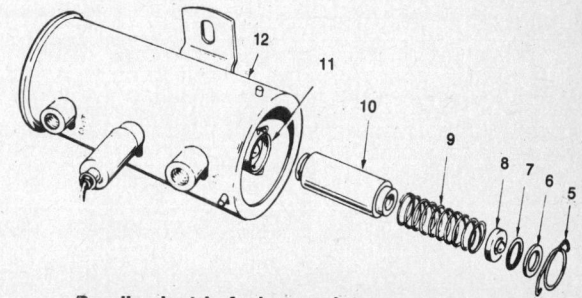

Bendix electric fuel pump internal components

Engine Overhaul

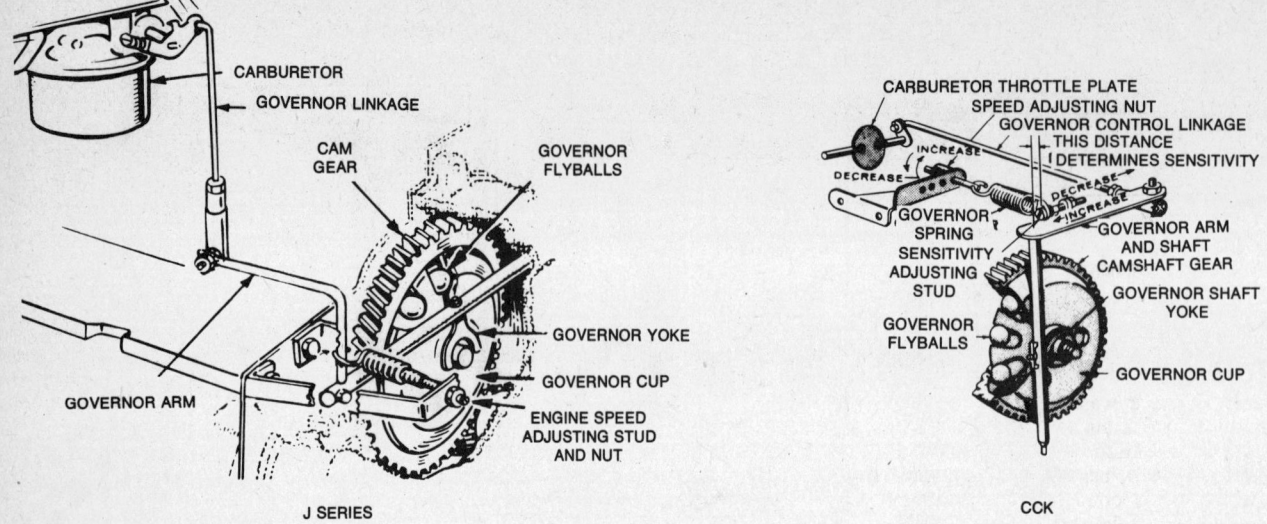

Two centrifugal type governors

5. On CCK engines, measure the distance the center pin extends outward—it should be ¾ in. On J series this dimension should be $^{25}/_{32}$ in. Hold the cup against the flyballs to make this measurement. If the distance is not correct, pull out the center pin and replace it, pressing the new pin in just the required amount. If the pin extension is o.k., grind off the hub of the cup as required. In and out travel distance must be $^{7}/_{32}$ in. or the engine will race.

ENGINE OVERHAUL

Valves

SERVICE

In order to remove the valves, the cylinder head must be removed. Remove the cylinder head screws and then remove the head from the engine. If the head sticks to the block, hit it lightly with a soft hammer, being careful not to damage any of the cooling fins. Remove the cylinder head gasket and discard it.

Use a conventional type valve spring compressor to compress the valve springs so that the spring retainer can be removed. The retainers are the split, tapered type and will most likely fall out when the valve spring is compressed. After removing the retainers, lift the valve out through the top of the valve guide. Clean the valves of all carbon deposits and inspect them, looking for warpage, worn stems, and burned surfaces that are partially destroyed. Determine whether or not the valve can be reused, whether it can be reground, or if the valve has to be replaced.

Check out the valve stem-to-guide clearance and, if it is too large, the guides must be replaced. They can be removed through the valve chamber. The valve tappets are also replaceable from inside the valve chamber, once the valves have been removed.

In removing the valve guides, first wire brush carbon and other deposits from the top guide surfaces, or the guide bores may be damaged during guide removal. Note that a gasket must be used on the intake valve guides for the LK, LKB, and BF engines. Where used, place the gasket on the intake guide and install the intake and exhaust guides from within the valve chamber. Before installing the guides, run a small polishing rod with a crocus cloth through the guide holes to clean out deposits.

Valve Seat Insert
Removal and Installation

Use a solvent to clean carbon or stuck gasket material from cylinder head and block surfaces. If necessary, follow up using a metal scraper. Inspect the head gasket surfaces for cracks, nicks, or burrs. Check the head for cracks, and replace if any are present. An oil stone may be used to remove burrs or nicks.

Use a straightedge at three angles to check the head for flatness. Try to insert a .003 in. flat feeler gauge under the straightedge at any point, and replace the head if it is not flat.

Replace the valve seats if cracked or loose, or excessively worn. On cast iron engines, the seat may be driven out using a knockout tool as shown. The tool must be inserted under the port side of the valve seat with the square end extending over the cylinder bore. With the sharp edge of the tool at the joint between the seat and its recess, strike a sharp blow on the end of the tool with a light hammer. This will crack the insert and permit removal.

CAUTION
Since this may shatter the relatively brittle material of the seat, you should wear goggles when performing the procedure.

If the engine has an aluminum block, use a ¾ in. or one inch pipe tap to suit the seat

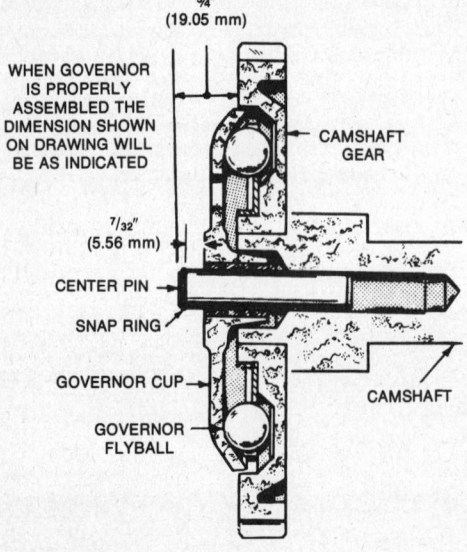

Measuring governor travel

Engine Overhaul

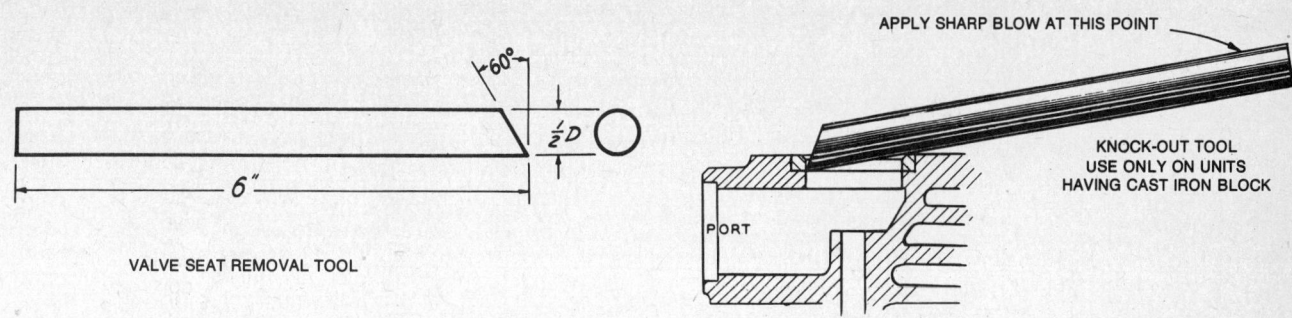

Special tool for removing valve seats from cast iron engines

diameter. Place a washer on top of the valve guide for the tap to bottom against. Turn the tap in until the seat begins to turn. As it starts to turn, begin pulling outward on the tap while continuing the turning motion to pull out the seat. Make sure the valve guide is not pushed downward by the tap or, if it is, that it is pressed back into position. Use an oversize replacement seat on aluminum engines.

Clean any carbon or burrs from the insert recess, and install a new seat as follows:

1. Gradually heat the block to 325 degrees F. Place the new seat in dry ice until thoroughly chilled.
2. Insert the pilot of an appropriate special tool in the valve guide hole in the block and quickly and evenly drive in the seat insert. It must seat on the bottom of the recess.

The valve face angle is 44°. The valve seat angle is 45°. The 1° interference angle assures a sharp seating surface between the valve and the seat and good sealing characteristics. The valve seat width must be between $1/32$ in. and $3/64$ in. Valves should not be hand lapped if at all possible. This is especially important if stellite valves are used.

To check the valves for a tight seal, make pencil marks around the valve face, then install the valve and rotate it a part of a turn. If the marks are all rubbed off uniformly, then the seal is good.

1. Insert the tappets in the crankcase holes.
2. Install the valves, springs and guides.
3. Using a valve spring compressor, compress each valve spring and insert the valve spring retainer and retainer locks.
4. Set the valve clearance to the specifications listed at the back of the chapter.
5. Install the heads and gaskets to the cylinder block.
6. Tighten the head bolts to the correct torque following the sequence in the appropriate illustration.
7. Install the exhaust manifold, oil lines, spark plugs and carburetor.

Gear Cover

REMOVAL AND INSTALLATION

In order to gain access to the camshaft gear and other internal components of the engine, it is necessary to remove the gear

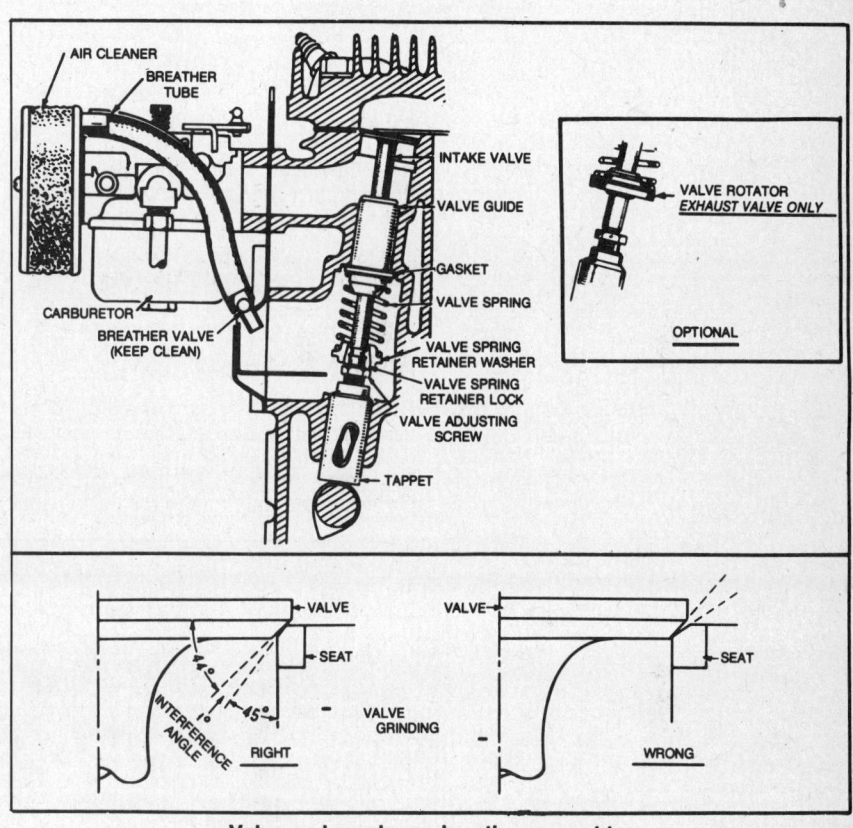

Valve and crankcase breather assembly

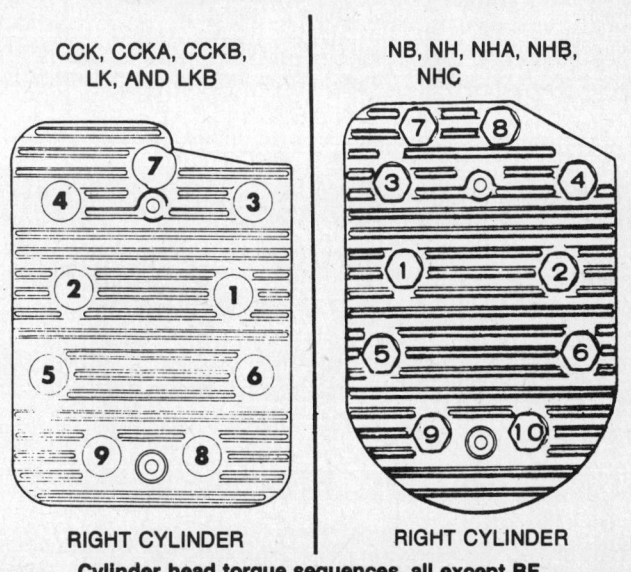

Cylinder head torque sequences, all except BF

915

Engine Overhaul

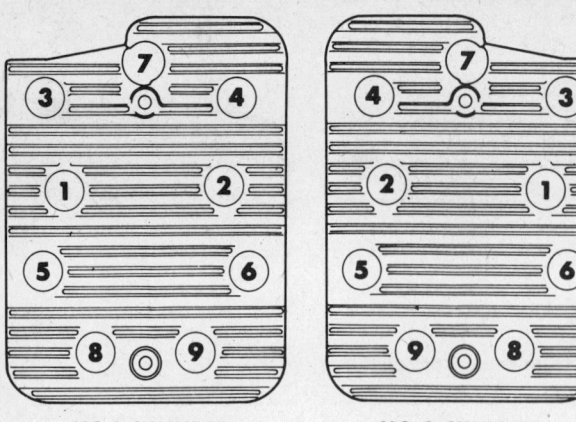

BF cylinder head torque sequences

cover. In order to remove the gear cover, the magneto assembly must be removed. Disconnect the spark plug wire at the spark plug and disconnect the stop wire. Remove the attaching screws that hold the magneto assembly to the gear cover and remove the magneto.

When the gear cover is removed, the governor shaft disengages from the governor cup, which is part of the camshaft.

During the installation of the cover, be sure to engage the pin on the cover, with the chamfered hole located in the governor cup. To do this, turn the governor cup so that the hole is located at the top or in the 12 o'clock position. Turn the governor shaft clockwise as far as it will go and hold it there until the cover is installed. Position the cover on the engine and make sure that it fits flush against the engine. Be careful of the gear cover oil seal during installation. Use a new gasket if the old one is damaged.

Timing Gears

REMOVAL AND INSTALLATION

If it becomes necessary to replace either the crankshaft or camshaft gears because of broken teeth, extreme wear, cracks, etc., both gears must be replaced as a pair. Never replace only one of the gears. Both gears are pressed onto their respective shafts.

To remove the crankshaft gear, insert two #10-32 screws into the threaded holes in the gear and tighten the screws alternately a little at a time. The screws will press up against the crankshaft shoulder and force the gear off the end of the crankshaft.

To remove the camshaft gear, it is necessary to remove the entire camshaft assembly from the engine. First remove the crankshaft gear lock ring and washer. Remove the cylinder head, valve assemblies, fuel pump (if so equipped), and the valve tappets. Remove the governor cup assembly and then remove the camshaft and gear assembly from the engine. The camshaft gear may now be pressed off the camshaft. Do not press on the camshaft center pin as it will be damaged. The governor ball spacer is press fit into the camshaft gear.

When the camshaft gear is replaced on the camshaft, be certain that the gear is properly aligned and the key properly positioned before beginning to press the gear onto the camshaft.

Install the governor cup before replacing the camshaft and gear assembly back into position in the engine.

There are two stamped 'O' marks, one on each gear, near the gear teeth. When the crankshaft and camshaft gears are meshed, these marks must be exactly opposite each other. When installing the camshaft gear assembly, be sure that the thrust washer that goes behind the camshaft gear is installed. Replace the retaining washer and lockwasher on the crankshaft.

Cylinder Bore, Piston, and Piston Rings

The cylinder can be rebored if it becomes heavily scored or badly worn. If the cylinder bore becomes cracked, replace the cylinder block.

The cylinder can be bored out to 0.010 in., 0.020 in., or 0.030 in. oversize.

There are pistons and rings in the above oversizes available to accommodate an oversized cylinder bore. If the cylinder bore has to be bored out only 0.005 in. to remove the damage in the cylinder, use standard size parts.

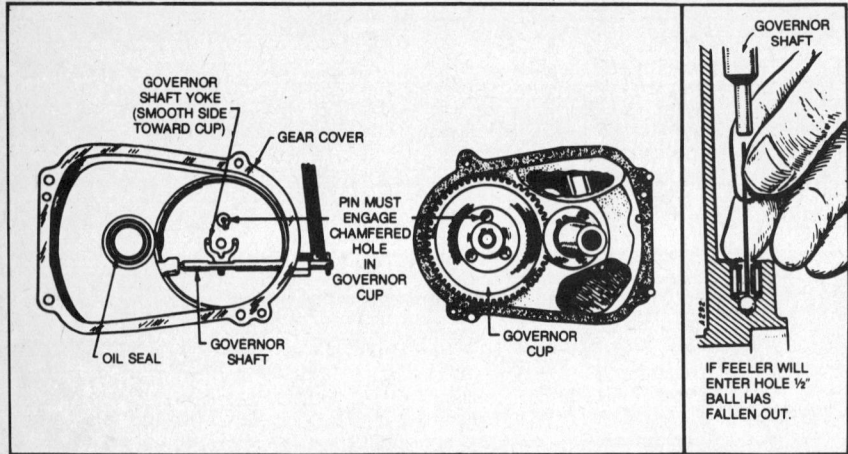

Gear cover assembly

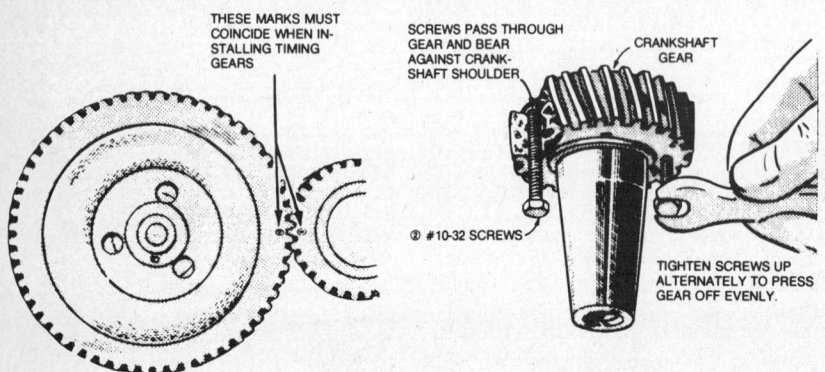

Aligning the timing marks on the timing gears and removing the crankshaft gear

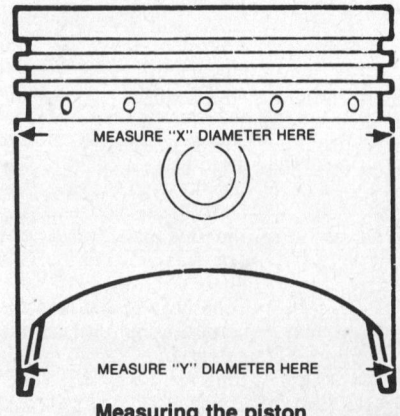

Measuring the piston

Engine Overhaul

Use a ridge reamer to remove the ridge that may be present at the top of the cylinder bore to avoid damaging the piston rings when the piston and connecting rod assembly is removed. Hone the cylinder and create a cross hatch pattern on the cylinder walls if new rings are being installed. Clean the cylinder with SAE 10 engine oil after honing.

Some engines were originally built with 0.005 in. oversize pistons and are so indicated by a letter 'E' following the serial number stamped on the identification plate and on the side of the crankcase.

The piston is fitted with two compression rings and one oil control ring. When the piston assembly is removed from the engine, clean off all carbon deposits and open all of the oil return holes in the lower ring groove. Before installing new rings, check the ring gap by installing the rings squarely in the cylinder bore and measuring the gap between the two ends of the rings. If the gap is too small, it is possible to file the ends to obtain the proper size gap.

Tapered type rings are usually marked with the word 'TOP' on one side. This side must be installed facing toward the top or closed end of the piston. Position the ring gaps evenly around the circumference of the piston with no ring gap over the piston pin.

The piston pin is held in place by two lock rings, one at each end. Make sure that the lock rings are properly installed before installing the piston assembly in the cylinder. Be sure to check the size of the piston, piston pin, piston pin bore, and the size of the cylinder before installing any of these parts back into the engine. Replace any parts that are worn beyond the maximum allowed specification. The piston should be replaced when the side clearance of the top ring reaches .008 in., or if there are signs of scuffing, scoring, worn ring lands, fractures, or preignition damage. Measure the piston dimensions at points shown in the illustration.

In fitting rings, install each ring in the cylinder bore, invert the piston and push the ring to the end of the ring travel—about halfway to the end of the bore. With the ring at exactly 90 degrees to the centerline of the bore, measure the end gap with a feeler gauge. Fit rings by choosing the right size for the bore—do not file the ends.

Connecting Rod

Before removing the connecting rod from the crankshaft, mark the cap and rod so they can be installed in exactly the same position from which they are removed.

If abnormal bearing wear (worn on one side more than the other) is noticed, this would indicate that the connecting rod is bent. It is possible to have the connecting rod straightened, but this should be done at a machine shop or small engine service shop.

Measure all of the bearing surfaces for size, including the piston pin hole and the crankpin bearing. Inspect the bearings for burrs, breaks, pitting and wear. Replace if scored or if the overlay is wiped out. Scratching is also reason for replacement. If bearings look all right, check bearing clearances. Place a piece of Plastigage® of the proper width in the bearing cap about ¼ in. off center. Rotate the crankshaft about 30 degrees from bottom center, and install the bearing cap. Torque to specification. Without turning the crankshaft, remove the bearing cap, leaving the flattened Plastigage on the bearing or journal. Compare its widest point with the scale on the Plastigage envelope to determine clearance. If clearance is excessive, replace the connecting rod bearings.

Use a new piston pin to check the pin bushing for war. A push fit clearance is required. If a new pin will fall through a dry rod bore of its own weight, replace the rod bushing.

To replace the bushings, press them out using a press and proper driving tool. Press in the new bushings so that the ends are flush with the sides of the rod and a $1/16$ in. oil groove is formed in the center. If there are oil holes, make sure they are at least half way open. Make sure the pin is a push fit after the bushing is installed.

Replace rod nuts or bolts if the threads are damaged. Replace rods which are nicked or fractured, or which have bores which are out of round more than .002 in. Straighten or replace rods which are twisted more than .012 in. or out of line more than .005 in.

Be sure to reinstall the oil dipper to the connecting rod cap when assembling the piston and connecting rod assembly to the crankshaft.

Cylinder Block

INSPECTION

Check the entire block thoroughly for cracks, and check the cylinder bore for scoring. Measure the bores for out-of-round at points indicated. A is the point where greatest ring wear occurs, and B is the bottom of ring travel. C and D are at the same heights as A and B respectively, but

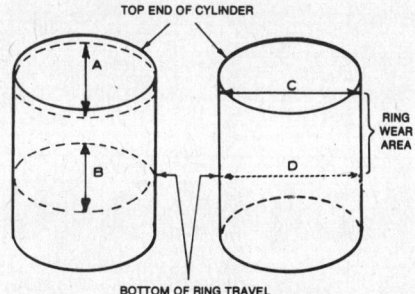

Measuring the cylinder

at 90 degrees. Compare (subtract the smaller from the larger) A and B and then compare C and D to determine taper. Cylinder must be rebored for the next oversize piston if taper exceeds .005 in. Comparing A to C and B to D indicates out-of-round. If the cylinder is out-of-round .002 in., it must be rebored for the next oversize piston.

HONING CYLINDERS TO OVERSIZE DIMENSIONS

1. Anchor the block solidly for either vertical or horizontal honing. Use either a drill press or heavy-duty drill which operates at about 250 to 450 rpm.
2. Lower the hone into the cylinder until it protrudes ½ to ¾ inch past the end of the cylinder. Rotate the adjusting nut until the stones just come in contact with the cylinder wall at its narrowest point.
3. Loosen the adjusting nut until the hone can be turned by hand.
4. Connect the drill to the hone and start the drill. Move the hone up and down in the cylinder about 40 times per minute. Usually the bottom of the cylinder must be worked out first because it is smaller. When the cylinder takes a uniform diameter, move the hone up and down all the way through the bore. Follow the hone manufacturer's recommendations for wet or dry honing and oiling the hone.

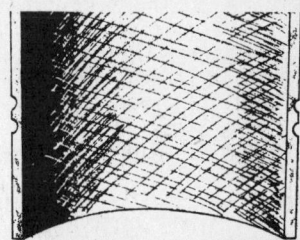

PRODUCE CROSS HATCH SCRATCHES FOR FAST RING SEATING

AVOID THIS FINISH

Cross hatch patterns

5. Check the diameter of the cylinder regularly during honing. A dial bore gauge is the easiest method but a telescoping gauge can be used. Check the size at six places in the bore; measure twice at the top, middle and bottom at 90-degree angles.
6. When the cylinder is within about 0.002 inch of the desired bore, change to fine stones and finish the bore. The finish should not be smooth but crosshatched as shown. The crosshatch formed by the scratching of the stones should form an angle of 32 degrees. This can be achieved by moving the hone up and down in the cylinder about 40 times per minute.
7. Clean the cylinder block thoroughly with soap and water and clean rags. A clean white rag should not be soiled on the wall after cleaning is complete.

─────── **CAUTION** ───────

Do not use a solvent or gasoline since they wash the oil from the walls but leave the metal particles.

Engine Overhaul

8. Dry the crankcase and coat it with oil.

Crankshaft and Bearings

CRANKSHAFT REMOVAL

1. Remove the lock ring and retainer washer from the crankshaft gear, and pull off gear with a puller.
2. Remove the oil pan, head(s), pistons and rods. See the previous section on piston and ring service.
3. Remove the rear bearing plate. Remove the crankshaft through the rear opening in the crankcase.

INSPECTION

Clean the crankshaft thoroughly, including blowing out all oil passages with compressed air. Check the journals for out-of-round, taper, grooves, and ridges, and place on V-blocks and rotate to measure runout. It should not exceed .003 in. Compare all dimensions with specifications: out-of-round should not exceed .005 in.; staper should not exceed .001 in.; wear should not exceed .002 in. If the limits are exceeded, regrind the shaft for the use of undersize bearings.

Replace bearings that are warped, scored, or have been overheated. If such damage is present, check the bearing bores in the block for excessive size. A new rear main bearing seat can be installed by replacing the rear main bearing plate. After doing this, check the main bearing bore alignment on a line boring machine.

The main crankshaft bearings are precision bearings. They are available in the standard size as well as 0.002 in., 0.010 in., 0.020 in., and 0.030 in. undersize. The precision type bearings are *NOT* to be line reamed.

The bearings are press fit into the cylinder block. Before trying to install the bearing into the cylinder block or bearing plate, heat the plate or block by running hot water over them or placing them in an oven heated to 200°F. This will cause the block or bearing plate to expand and facilitate the installation of the sleeve bearing.

The oil hole in the bearing and the oil hole in the bearing bore must be aligned when the bearings are installed. On pressure lubricated engines, the hole should be opposite the crankshaft. On engines that are splash lubricated, the hole should be upward.

Install rear bearings in the rear bearing plate using a special driver. Install them to 1/64 in. below the end of the bore. If a special tool is not available and the lock pins must be removed with side cutters or a screw extractor, install new lock pins.

CRANKSHAFT INSTALLATION

1. Oil the bearing surfaces thoroughly. Install the crankshaft from the rear of the crankcase (through the rear bearing plate hole).

2. Put the rear bearing plate gasket in place and lubricate the rear end plate bearing. Slide the thrust washer, with grooves toward the crankshaft and the bearing plate, over the end of the crankshaft.

NOTE: Line up the thrust washer notches with the lock pins before tightening the end plate bolts, or the lock pins and thrust washer will be damaged.

If you have trouble getting the thrust washer to stay in place, it may be lubricated with a light coating of oil.

3. Torque the bearing plate bolts.
4. Heat the timing gear to 350°F. Install a new crankshaft key and drive the gear into position. Install the washer and lock ring.
5. Adjust the crankshaft end play as described below.
6. Complete the reassembly of the engine.

CHECKING END PLAY

Check the end play with the rear bearing plate bolts properly torqued. If end play is excessive, remove the end plate and install a shim between the thrust washer and plate. When installing the plate, line up the notches in the thrust washer and shim with the lock pins. Torque the end plate and recheck end play. If total gasket and shim thickness required is more than .015 in., use a steel shim of the proper thickness and two thin gaskets, or gasket compression and consequent loose bolts may result.

Crankshaft Oil Seals

The crankshaft oil seals are installed with the open sides facing toward the inside of the engine. To replace the rear oil seal the rear bearing plate must be removed. To replace the front oil seal, the front gear cover must be removed. Be careful not to damage the oil seal during installation or to turn back the edge of the seal lip.

Lubrication

Onan engines are either splash lubricated or pressure lubricated.

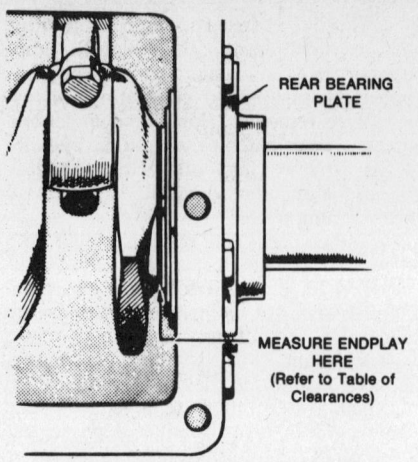

Measuring crankshaft end play

The splash lubrication system consists of an oil dipper attached to the connecting rod cap and various oil passages to catch and channel the oil that is splashed up by the oil dipper.

The pressure lubrication system consists of a gear type oil pump, an oil intake cup, a non-adjustable pressure relief valve and the various oil passages and channels.

OIL PUMP SERVICE

1. Drain the crankcase oil. Then, remove the gear cover and oil base.
2. Unscrew the intake cup from the oil pump. Loosen the two capscrews holding the pump and remove it.
3. Disassemble the pump by removing the two cap screws holding the cover in position. Inspect for excessive wear of gears and shafts a replace the pump if any parts are badly damaged. Gears may be measured for excess wear by positioning them in the housing, running a straightedge across the sides of the housing, and inserting a flat feeler gauge between the straightedge and side of gear. If either this clearance or the clearance between oil pump teeth is excessive, replace the unit.

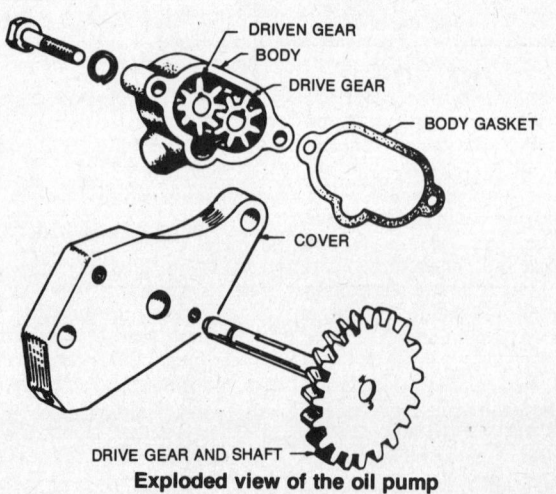

Exploded view of the oil pump

Engine Overhaul

4. In reassembling the pump, side clearance between gears and housing can be adjusted by using a thinner gasket. Use the thinnest gasket that permits freedom in operation of the pump.

5. CAUTION: *Fully prime the pump with oil before assembly.* Mount the pump into position on the engine and adjust for a clearance of .005 in. between the pump and crankshaft gears.

6. Install the intake cup on the pump parallel with the bottom of the crankcase.

7. Install the gear cover and oil base, fill the crankcase with oil, and run engine, checking for good oil pressure.

CRANKSHAFT AND CONNECTING ROD SPECIFICATIONS
All measurements are given in inches

Engine Model	CRANKSHAFT				CONNECTING ROD		
	Main Brg Journal Dia	Main Brg Oil Clearance	Shaft End-Play	Thrust on No.	Journal Diameter	Oil Clearance	Stem Clearance
LKB	1.9992-2.0000	.0020-.0030	.006-.0012	Rear	1.6252-1.6260	.0020-.0030	.002-.016
NB	1.9992-2.0000	.0025-.0038	.006-.0012	Rear	1.6252-1.6260	.0020-.0030	.002-.016
CCK	1.9992-2.0000	.0025-.0038	.006-.0012	Rear	1.6250-1.6252	.0005-.0023	.002-.016
CCKA	1.9992-2.0000	.0025-.0038	.006-.0012	Rear	1.6250-1.6252	.0005-.0023	.002-.016
CCKB	1.9992-2.0000	.0025-.0038	.006-.0012	Rear	1.6250-1.6252	.0005-.0023	.002-.016
BF	1.9992-2.0000	.0025-.0038	.006-.0012	Rear	1.6250-1.6252	.0020-.0033	.002-.016

VALVE SPECIFICATIONS

Engine Model	Seat Angle (deg)	Face Angle (deg)	Spring Test Pressure (lbs @ in.)	Spring Installed Height (in.)	GUIDE DIAMETER (in.)		STEM DIAMETER (in.)	
					Intake	Exhaust	Intake	Exhaust
LKB	45	44	71-79 @ 1.375	—	.3440-.3460	.3440-.3460	.3425-.3430	.3425-.3430
NB	45	44	71-79 @ 1.375	—	.3440-.3460	.3440-.3460	.3425-.3430	.3410-.3415
CCK	45	44	71-79 @ 1.375	—	.3440-.3460	.3440-.3460	.3425-.3430	.3410-.3415
CCKA	45	44	71-79 @ 1.375	—	.3440-.3460	.3440-.3460	.3425-.3430	.3410-.3415
CCKB	45	44	71-79 @ 1.375	—	.3440-.3460	.3440-.3460	.3425-.3430	.3410-.3415
BF	45	44	71-79 @ 1.375	—	.3440-.3460	.3440-.3460	.3425-.3430	.3410-.3415

PISTON CLEARANCE

Engine Model	Piston to Bore Clearance (in.)	Engine Model	Piston to Bore Clearance (in.)
LKB	.0005-[1].0015	CCKB	.0015-[2].0035
NB	.0025-[2].0045	BF	.0010-[2].0030
CCK	.0015-[2].0035		
CCKA	.0015-[2].0035		

[1] Measure piston diameter across dimension Y
[2] Measure piston diameter across dimension X

Engine Overhaul

RING GAP
All measurements are given in inches

Engine Model	Top Compression	Bottom Compression	Oil Control
LKB	.010-.023	.010-.023	.010-.023
NB	.013-.023	.013-.023	.013-.023
CCK	.010-.023	.010-.023	.010-.023
CCKA	.010-.023	.010-.023	.010-.023
CCKB	.010-.023	.010-.023	.010-.023
BF	.010-.020	.010-.020	.010-.020

RING SIDE CLEARANCE
All measurements are given in inches

Engine Model	Top Compression
LKB	.002-.008
NB	.002-.008
CCK	.002-.008
CCKA	.002-.008
CCKB	.002-.008
BF	.002-.004

TORQUE SPECIFICATIONS

		USE ENGINE LUBRICATING OIL AS A THREAD LUBRICANT									DO NOT USE ANY LUBRICANT ON THESE THREADS			
Engine Series		Cylinder Head (Cold)	Conn Rod	Rear Bearing Plate	Main Bearing (4 Cyl)	Flywheel To Crankshaft	Oil Base	Exhaust Manifold (Tighten Evenly)	Intake Manifold	Damper Flywheel Assy Nut (4 Cyl)	Rocker Arm Stud in Head	Armature Thru Stud Revolving Armature Units	Revolving Field Units	Spark Plugs	Injection Nozzle
LKB	lb-ft	29-31	26-28	20-25	—	35-40	25-30	—	—	—	—	35-40	—	25-30	—
CCK, CCKA, CCKB, MCCK, RCCK	lb-ft	29-31	①	20-25	—	35-40	43-48	—	15-20	—	—	35-40	—	25-30	—
BF, BG, BFA, BGA	lb-ft	14-16	14-16	25-27	—	35-40	18-23	6-10	—	—	—	45-50	—	15-20	—
B43M, M48M	lb-ft	16-18	14-18	25-27	—	35-40	18-23	9-11	6-10	—	—	35-40	—	15-20	—
NB	lb-ft	29-31	①	30-35	—	30-35②	38-43	—	—	—	—	35-40	—	15-20	—
NH	lb-ft	22-25	27-29	25-27	—	30-35②	18-23	—	—	—	—	35-40	—	15-20	—
NHA, NHB, NHC, NHAV, NHBV, NHCV, NHP, NHPV, N52M ⑤	lb-ft	17-19	27-29	20-23	—	35-40	18-23	10-12	18-20	—	—	45-50	—	15-20	—
JA	lb-ft	28-30③	27-29	40-45④	—	65-70	32-38	13-15	13-15	—	25-30	30-40	—	25-30	—
JB	lb-ft	28-30③	27-29	40-45④	—	65-70	45-50	13-15	13-15	—	25-30	—	55-60	25-30	—
JC	lb-ft	28-30③	27-29	40-45④	97-102	65-70	45-50	13-15	13-15	—	25-30	—	55-60	25-30	—
MJA	lb-ft	44-46③	27-29	40-45④	—	65-70	32-38	13-15	13-15	—	35-40	30-40	—	25-30	—
MJB	lb-ft	44-46③	27-29	40-45④	—	65-70	45-50	13-15	13-15	—	35-40	—	55-60	25-30	—
MJC	lb-ft	44-46③	27-29	40-45④	97-102	65-70	45-50	13-15	13-15	17-21	35-40	—	55-60	25-30	—
MDJA	lb-ft	44-46③	27-29	40-45④	—	65-70	32-38	13-15	13-15	—	35-40	30-40	—	—	20-21
DJA	lb-ft	37-40③	27-29	40-45④	—	65-70	32-38	13-15	13-15	—	35-40	30-40	—	—	20-21
MDJB	lb-ft	44-46③	27-29	40-45④	—	65-70	45-50	13-15	13-15	—	35-40	—	55-60	—	20-21
DJB, DJE	lb-ft	37-40③	27-29	40-45④	—	65-70	45-50	13-15	13-15	—	35-40	—	55-60	—	20-21
MDJE	lb-ft	44-46③	27-29	40-45④	—	65-70	45-50	13-15	13-15	—	35-40	—	55-60	—	20-21

Engine Overhaul

TORQUE SPECIFICATIONS

USE ENGINE LUBRICATING OIL AS A THREAD LUBRICANT | **DO NOT USE ANY LUBRICANT ON THESE THREADS**

Engine Series		Cylinder Head (Cold)	Conn Rod	Rear Bearing Plate	Main Bearing (4 Cyl)	Flywheel To Crankshaft	Oil Base	Exhaust Manifold (Tighten Evenly)	Intake Manifold	Damper Flywheel Assy Nut (4 Cyl)	Rocker Arm Stud in Head	Armature Thru Stud Revolving Armature Units	Revolving Field Units	Spark Plugs	Injection Nozzle
MDJC	lb-ft	44-46③	27-29	40-45④	97-102	65-70	45-50	13-15	13-15	17-21	35-40	—	55-60	—	20-21
DJC	lb-ft	37-40③	27-29	40-45④	97-102	65-70	45-50	13-15	13-15	17-21	35-40	—	55-60	—	20-21
MDJF	lb-ft	44-46③	27-29	40-45④	97-102	65-70	45-50	13-15	13-15	17-21	35-40	—	55-60	—	20-21
RDJE, RDJEA	lb-ft	44-46③	27-29	40-45④	—	65-70	45-50	13-15	13-15	—	35-40	—	—	—	20-21
RJC	lb-ft	44-46③	27-29	40-45④	97-102	65-70	45-50	13-15	13-15	17-21	35-40	—	55-60	—	20-21
RDJC	lb-ft	44-46③	27-29	40-45④	97-102	65-70	45-50	13-15	13-15	17-21	35-40	—	55-60	—	20-21
RDJF	lb-ft	44-46③	27-29	40-45④	97-102	65-70	45-50	13-15	13-15	17-21	35-40	—	55-60	—	20-21

① Aluminum rods 24-26 lb-ft, forged rods 27-29
② Zinc or aluminum wheel. Cast iron wheel 40-45 lb-ft
③ Use NEVER-SEEZE® or equivalent when torquing to this value.
④ Use LOCTITE® when torquing bolts.
⑤ When using compression washers torque should be 13-15 lbs.

Wisconsin

ENGINE IDENTIFICATION

There is a Wisconsin name plate attached to the blower housing of the engine on which is stamped the model number, serial number, and specification number along with the size and rpm rating. The model, serial, and specification number must be given when obtaining replacement parts for any of the engines. Make certain that the identification plate remains with the engine on which it was originally installed.

Name plate

MAINTENANCE

Air Cleaner Service

DRY ELEMENT TYPE

If the unit is operated in a very dusty atmosphere, remove the element by unscrewing the wingnut and removing the cover. Shake out accumulated dirt (do not tap) once each day. Under normal operating conditions, the most frequent service required is a weekly washing of the element. Rinse the element under cold water and then dip repeatedly into a solution of

GENERAL ENGINE SPECIFICATIONS

Model	Bore & Stroke (in.)	Displacement (cu in.)	Horsepower @ RPM
S-7D	3 x 2⅝	18.6	7.25 @ 3600
S-8D	3⅛ x 2⅝	20.2	8.25 @ 3600
TR-10D	3⅛ x 2⅝	20.2	—
TRA-10D	3⅛ x 2⅞	22.05	10.1 @ 3600
TRA-12D	3½ x 2⅞	27.66	12.0 @ 3600
ALN	2⅝ x 2¾	14.9	6.0 @ 3600
BKN	2⅞ x 2¾	17.8	7.0 @ 3600
AEN	3 x 3¼	23.0	9.2 @ 3600
AENL	3 x 3¼	23.0	9.2 @ 3600
AENS	3 x 3¼	23.0	9.2 @ 3600

Engine Overhaul

mild, non-sudsing detergent and warm water. Rinse in cold water and allow to dry overnight. Avoid freezing temperatures until the element is dry.

After five washings, or one year of service, whichever comes first, replace the element.

OIL BATH TYPE

All Series Except ACN, BKN

If the engine is run in a very dusty atomosphere, service the cleaner daily according to the instructions printed on the unit.

Ordinarily, remove the wingnut and pull off the cover and filter element and wash it in solvent. Drain oil from the filter bowl and clean it with a clean rag. Fill the filter bowl with engine oil to the line, and reassemble the unit. Install the wing nut.

ACN, BKN

The air cleaner must be serviced frequently: weekly in a clean atmosphere, and as often as daily in a dusty atmosphere. Ordinary service consists of snapping off the spring bail and removing the bowl from the bottom of the unit for service. Clean out the cup and baffle, and then refill with about ¼ pint of engine oil.

The filter element does not ordinarily need service and may be left on the engine. However, if extreme conditions have made it dusty, remove it from the engine bracket and wash it in solvent.

Lubrication

OIL AND FUEL RECOMMENDATIONS

Oils of grades MS, SD, or SE may be used in Wisconsin engines. Viscosity recommendations are as follows:

Above +40° F	SAE 30
+15° - +40° F	SAE 20-20W
0° - +15° F	SAE 10W
Below 0° F	SAE 5W-20

Fuel should be regular grade of 90 octane or above. Fuel should be of known quality to provide adequate protection against gum formation, and adequate assurance that it will be free of moisture and sediment. Remember that fuel of too low an octane rating may cause engine knock and severe damage.

Check the oil level every 8 hours and

CRANKCASE CAPACITY CHART

S-7D, S-8D, TR-10D, TRA-10D, TRA-12D	1 qt.
AEN, AENL, AENS	3 pts.
ACN, BKN	2 pts.
Clutch Unit Housing— ACN, BKN	½ pt.
Reduction Unit Housing— ACN, BKN	1 pt.

Dry type air cleaner

Oil bath air cleaner

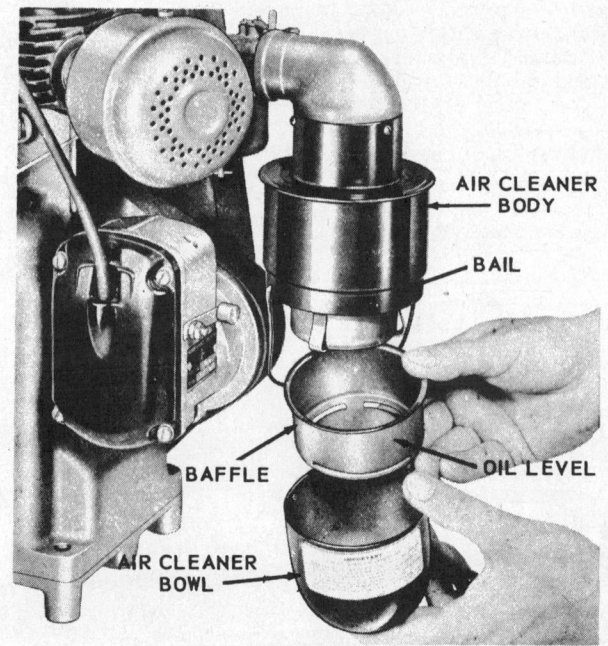

ACN and BKN air cleaner

Engine Overhaul

replenish. Check more frequently when the engine is new. Drain the old oil and replace it every 50 operating hours. Always change the oil when the engine is hot, by removing the crankcase drain plug. Fill crankcase to the level of the filler plug hole.

TUNE-UP

Spark Plugs

The outside of the plug, the electrodes and insulator on the underside should be kept clean. As often as significant deposits form, remove the plug and wire brush deposits away. Set the spark plug gap by bending the side electrode to get a gap of .030 in. as measured by a wire type feeler gauge. Clean the threads on the plug and in the cylinder head before installing the plug. Use a new gasket and torque to 25-30 ft.lb. If the plug has deposits that cannot be removed, or if the electrodes are badly burned or there is evidence of cracking of the insulator, either inside or outside, replace the plug.

Breaker Points

REMOVAL AND INSTALLATION

1. Remove the breaker box cover.
2. Disconnect the terminal strip by loosening the screw and pull it off the contact set.
3. Remove the point attaching screws and remove the contact set.
4. To install, first position the points, noting that on some types a prong located on the underside of the contact set must fit into a hole in the breaker box. Install the mounting screw or screws just tightly enough to hold the contacts in place.
5. Set the gap and time the engine as described in the procedures below.

SETTING BREAKER GAP

1. If necessary, loosen the breaker mounting screws so the point gap can be changed. Turn the engine flywheel back and forth until the contacts are as far apart as they can be.
2. Place the screwdriver in the adjusting slot and slide a flat feeler gauge of the proper dimension (see tune-up chart) between the contacts.
3. Adjust the gap with the screwdriver until the gauge has a very slight pull when sliding straight through the point gap.
4. On AEN, AENL, AENS, ACN, and BKN, tighten the contact mounting screw. Then, recheck the gap. Reset if necessary. On these engines, timing need not be reset after contact gap adjustment unless the magneto or timer position has been disturbed.
5. On all other engines, leave the contact mounting screw(s) only slightly tight, and proceed to the engine timing procedure below.

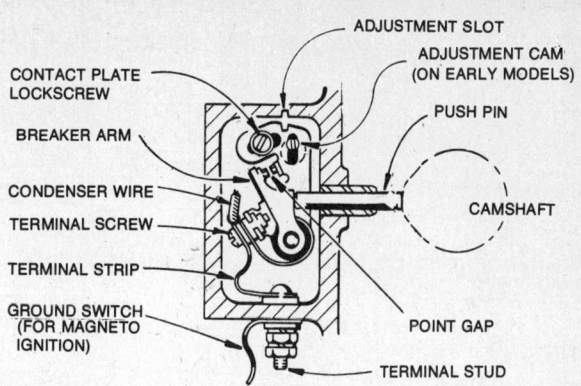

Typical breaker box

IGNITION TIMING ADJUSTMENT

S-7D, S-8D, TR-10D, TRA-10D and TRA-12D

1. Remove the breaker box cover. Disconnect the coil primary wire at the bottom of the breaker box.
2. Line up the flywheel timing mark and the pointer with the engine on compression stroke. The timing mark on the flywheel can be seen through the opening on the right side of the flywheel shroud. The engine is on the compression stroke if the breaker arm push pin is moving as the timing marks approach alignment.
3. Connect a self powered test lamp or timing light between an engine ground and the terminal stud on the bottom of the breaker box. If necessary, slightly loosen the contact set mounting screw so the gap can easily be changed.
4. Close the points slowly with a screwdriver in the adjusting slot, just until

TUNE-UP SPECIFICATIONS

Model	Plug Type	Plug Gap (in.)	Point Gap (in.)	Idle Speed
S-7D	AC-C86, Champion D-16J	.030	.020	①
S-8D	AC-C86, Champion D-16J	.030	.020	①
TR-10D	AC-C86, Champion D-16J	.030	.020	②
TRA-10D	AC-C86, Champion D-16J	.030	.020	②
TRA-12D	AC-C86, Champion D-16J	.030	.020	②
ACN	AC-C86, Champion D-16J	.030	.020 ③	④
BKN	AC-C86, Champion D-16J	.030	.020 ③	④
AEN	AC-C86, Champion D-16J	.030	.020 ③	④
AENL	AC-C86, Champion D-16J	.030	.020 ③	④
AENS	AC-C86, Champion D-16J	.030	.020 ③	④

① Throttle screw 2 turns open, or lowest smooth speed
② Throttle screw 1¼ turns open, or lowest smooth speed
③ Applies to battery ignition—with magneto, point gap is .015
④ Lowest smooth speed

Engine Overhaul

the light goes out. Tighten the mounting screw.

5. Turn the flywheel counterclockwise until the light goes on and then rotate it slowly forward and stop just as light goes out. At this point, the timing marks should be lined up. If necessary, readjust the gap slightly. Widen the gap if the light goes out too early; narrow the gap if the light goes out too soon.

6. Install the breaker cover and reconnect the primary wire to the terminal stud.

AEN, AENL, AENS, ACN and BKN with Magneto

1. The magneto need not be timed unless it is removed from the engine. On ACN and BKN engines, take off the shroud and remove the timing inspection hole plug. On AEN, AENL, and AENS engines, simply remove the plug, which is located near the magneto mounting. Remove the spark plug. Then, turn the engine over until the piston is coming up on compression stroke (air will be expelled from the spark plug hole) and the D/C and X marked vane on the flywheel lines up with the mark on the vertical centerline of the cooling shroud. On AEN, AENL, and AENS engines, remove the plug from the hole in the shroud to see the flywheel marks.

2. When installing the magneto on ACN and BKN engines, mesh the magneto and camshaft gears so the two timing marks line up. They are visible through the inspection hole located to the left side of the flywheel. On AEN, AENL, and AENS engines, mesh the magneto gears to so that the X marked gear tooth is visible through the inspection hole.

3. Timing may be checked by slowly rotating the engine past the point where the flywheel D/C and X marked vanes pass the vertical centerline mark. The impulse coupling will snap when the marks are lined up, if the timing is correct.

AEN, AENL, AENS, ACN, and BKN
ENGINES WITH BATTERY IGNITION

1. Set the engine flywheel at the position described in step 1 of the procedure above, in the same way.

2. If the timer unit has been removed from the engine, turn the timer cam counterclockwise, using the gear on the back of the unit, until the points just begin to open (you will feel increased friction). Then, mount the timer to the engine.

3. Loosen the clamp lever screw which keeps the unit from rotating. On ACN and BKN engines, turn the timing unit clockwise 3/64 in. as measured on the circumference of the timer body, to get 2 degrees of spark advance. On AEN, AENL, and AENS engines, rotate the unit clockwise 1/8 in. to get 5 degrees of advance.

4. Mark the timing marks with chalk, install the spark plug, and connect a timing light.

5. Start the engine and run it at 1,800 rpm or higher, as measured with a tach. On ACN and BKN engines, turn the unit as

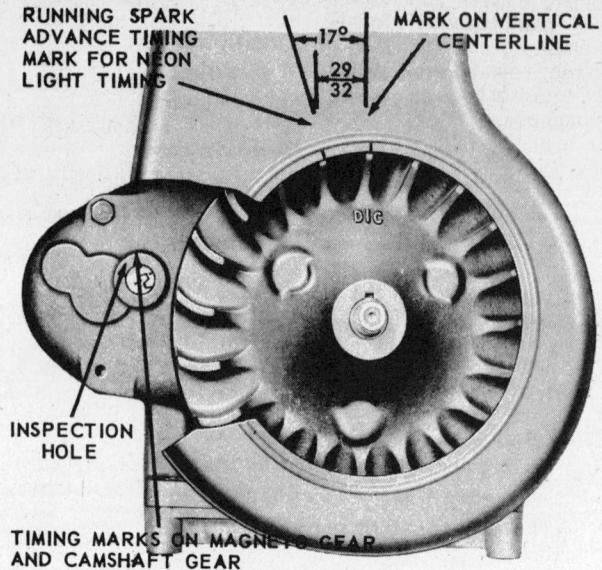

ACN and BKN magneto timing marks

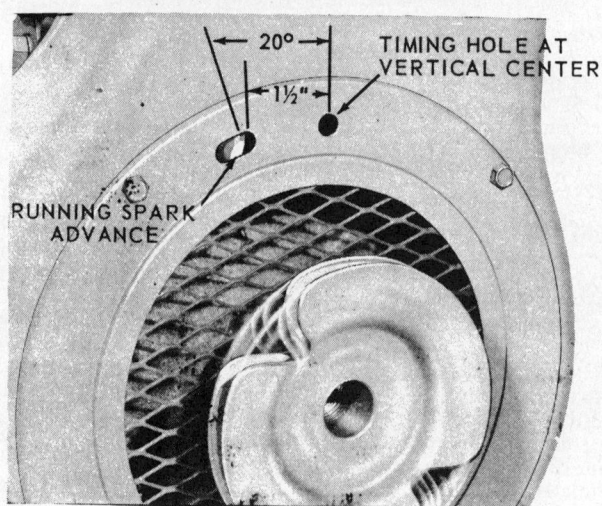

AEN, AENL and AENS timing marks

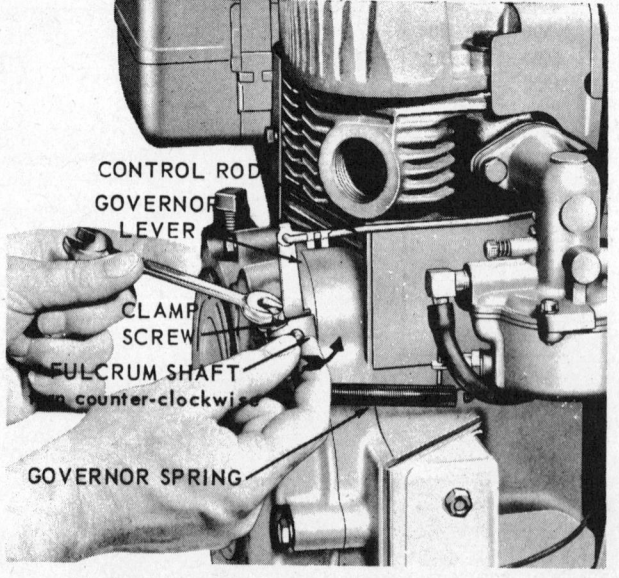

S-7D and S-8D governor adjustments

Engine Overhaul

required to align the flywheel mark and the running advance timing mark located on the shroud to the left (counterclockwise) of the centerline mark. On AEN, AENL, and AENS engines, rotate the unit as necessary to align the marked vane of the flywheel with the running advance timing hole in the shroud. Tighten the clamp screw.

Carburetor Mixture Adjustments

1. If the engine seems to be running very poorly due to improper fuel/air mixture, or it will not start, make the following preliminary settings. Make the settings by turning the mixture screw in until it seats *only very gently,* then outward the required number of turns:
S-7D, S-8D—Turn main jet adjustment out 1-1¼ turns.
AEN, AENL, AENS—Turn main jet adjustment out 1¼ turns.
ACN, BKB—Turn main jet adjustment out 1¼ turns.
TRA-12D, TRA-10D, TR-10D—Turn main jet open 1¼ turns, idle jet open 1 turn.

2. After making preliminary settings, run the engine until hot and check acceleration. If engine stumbles, open the main jet ¼ turn at a time until response is smooth.

3. Slow the engine down to idle speed and adjust idle mixture screw in or out for the smoothest idle.

Governor Adjustment

S-7D, S-8D, TR-10D, TRA-10D, TRA-12D

1. Loosen the governor lever clamp screw so that the fulcrum shaft can be turned independently of the governor lever and the lever moves to full throttle position. Then, turn the shaft counterclockwise until the internal governor vane stops against the flywheel thrust pin.

2. Tighten the clamp screw. See "Speed Adjustment" below.

ACN, BKN, AENL, AEN, AENS

1. Disconnect the rod at the governor lever. Move the rod as far as possible toward the carburetor.

2. Move the governor lever as far as possible in the same direction.

3. Hold both parts in this position and turn the rod in or out of the swivel block until the hole in the lever indexes with the end of the rod. On ACN and BKN engines, turn the rod two more turns inward. Install the rod in the lever and install the cotter pin. See "Speed Adjustment" below.

SPEED ADJUSTMENT

The governor lever is provided with a number of holes so that the engine can be operated at different speeds. If the governor spring has been removed from the hole, or if the speed range of the engine is to be changed, the proper hole in the lever must first be selected, and the adjusting screw must then be turned for fine adjustment.

1. Run the engine until hot, and connect a tachometer. Open the throttle control and install the governor spring into each of the holes in the lever to get as close as possible to the desired rpm (holes further away from the fulcrum of the lever give more speed). Once you've found the hole nearest the desired speed, note which hole you are using.

2. Loosen the locknut, disconnect the spring, and turn the screw for more tension to increase speed, or for less tension to decrease it.

Compression Check

No precise method of checking compression is required. However, on engines without compression release, compression may be checked by spinning the engine in the normal direction of rotation and checking for a substantial increase in resistance when the piston begins coming up on the compression stroke.

Generally, when compression is poor, the engine requires disassembly and major work. However, compression can be low because a long period without operation has permitted oil to drain off the cylinder walls. If this is suspected, remove the spark plug and squirt a small quantity of engine oil into the combustion chamber to seal it.

FUEL SYSTEM

Carburetor

DISASSEMBLY AND REASSEMBLY

NOTE: See item below for inspection and cleaning procedure.

Zenith 87, Wisconsin L-51

1. Remove the three bowl assembly screws (37 & 38) and lockwashers (36) and separate fuel bowl (30) from throttle body (9).

2. Remove the main jet adjustment (34) and fiber washer (33), using a 9/16" open end wrench.

3. Remove the main jet (32) and fiber washer (31), using Zenith Tool No. C161-83 main jet wrench or equivalent.

4. Remove the idle jet (29), using a small screwdriver.

5. Remove the bowl drain plug (35).

6. Remove the float axle (26) by pressing against the end with the blade of a screwdriver.

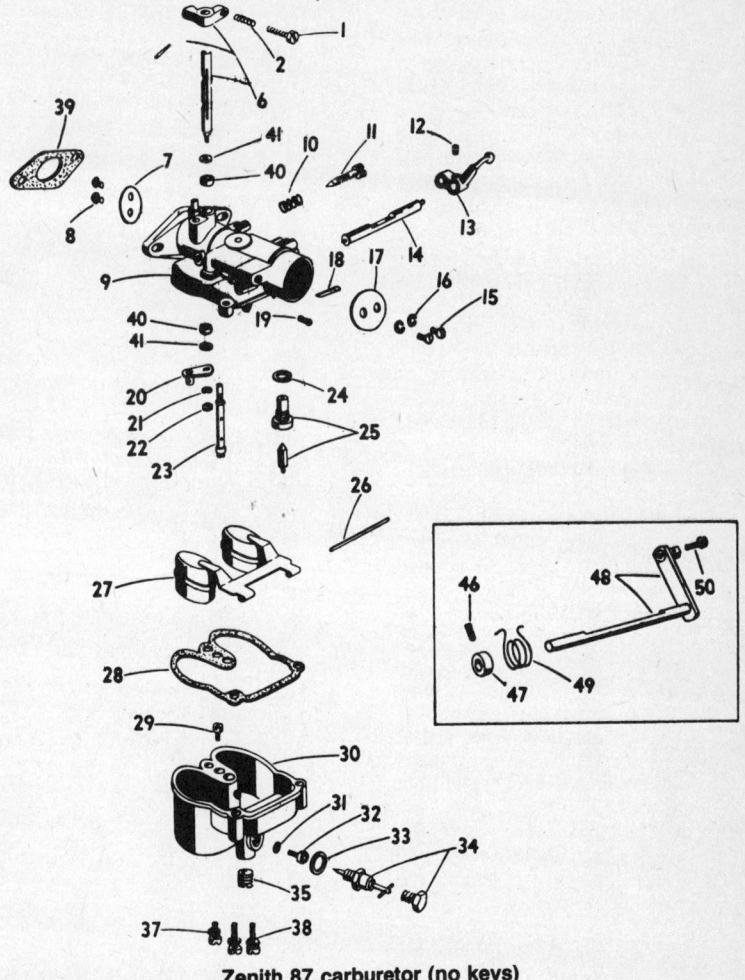

Zenith 87 carburetor (no keys)

Engine Overhaul

7. Remove the float (27).
8. Remove the fuel valve needle (25), with your fingers.
9. Remove the fuel bowl to throttle body gasket (28).
10. Remove the main discharge jet (23), using a small screwdriver.
11. Remove the fuel valve seat (25) and fiber washer (24), using Zenith Tool No. C161-85 or equivalent.
12. Remove the idle adjusting needle (11) and spring (10).
13. Install the fuel valve seat (25) and fiber washer (24), using Zenith Tool No. C161-85 or equivalent.
14. Install the main discharge jet (23), using a small screwdriver.
15. Install fuel valve needle (25) in seat (25), followed by float (27). Insert tapered end of float axle (26) into float bracket on side opposite slot and push through the other side. Press float axle (26) into slotted side until the axle is centered in bracket.
16. Check position of float assembly for correct measurement to obtain proper fuel level using a depth gage.

NOTE: Do not bend, twist, or apply pressure on the float body.

With bowl cover assembly in an inverted position, viewed from free end of float, the float body must be centered and at right angles to the machined surface. The float setting is measured from the machined surface (no gasket) of float bowl cover to top side of float body at highest point. This measurement should be $61/64$ in., plus or minus $1/32$ in. To increase or decrease distance between float body and machined surface use long nosed pliers and bend lever close to float body. Replace with new float if position is off more than $1/16$ in.

17. Install throttle body to fuel bowl assembly gasket (29) on machined surface of throttle body (9).
18. Install the idle adjusting needle (11) and spring (10).
19. Install the main jet (32) and fiber washer (31), using Zenith Tool No. C161-83 main jet wrench or equivalent.
20. Install the main jet adjusting needle assembly (34) and fiber washer (33), using a $9/16$ in. open end wrench.
21. Install the idle jet (29), using a samll screwdriver.
22. Install the bowl drain plug (35).
23. Install the three bowl assembly screws (38) and lockwashers (36) through the fuel bowl and into the throttle body and draw down firmly and evenly.

Zenith 72Y6

1. Remove the three assembly screws (2) that hold the bowl cover to the bowl.
2. Separate the bowl cover assembly (1) from the bowl assembly.
3. Remove the float axle (13) and float (12).
4. Remove the bowl cover gasket (16).
5. Remove the fuel valve needle and seat (15) with the gasket (14). Remove the fuel valve seat. Use tool C161-85 or equivalent.
6. Remove the idle adjusting screw needle (3) and spring (4).

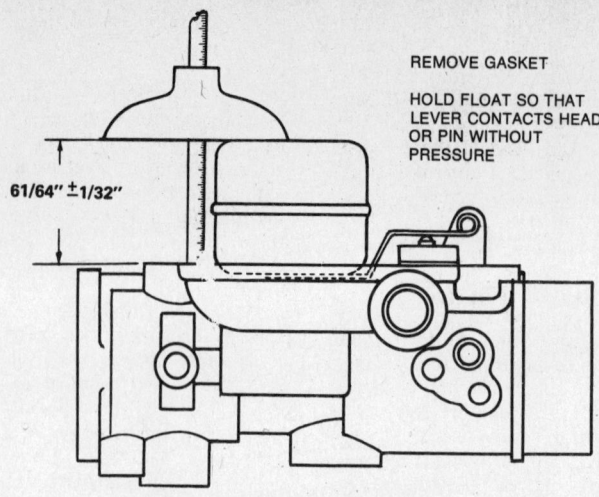

Zennith 87 float level adjustment (no keys)

7. Remove the throttle stop screw (5) and spring (6).
8. Hold the bowl cover inverted with the mounting flange to your right. Note that the closed throttle plate slopes down and away from you and that there is a mark stamped on the high side of the throttle plate. It is important that upon reassembly this same relationship is retained.
9. Hold the throttle shaft in the closed-position, remove the throttle plate screw and lockwasher (11) and throttle plate (10).
10. Remove the throttle shaft and lever (9).
11. With a screwdriver, or similar tool, remove the throttle shaft dust seal retainer (8) and rubber shaft seal (7).

12. Remove the venturi (17) and idle tube (18) by inverting bowl cover.

NOTE: Do not attempt to remove the main discharge jet. This part is pressed in and is not a serviceable item.

13. Remove the lower main jet plug if a plug is used or remove the main jet adjustment (20) and gasket (21). Then remove the main jet (22). Use tool C161-83 or equivalent.
14. Hold the bowl (19) in a vertical position with the air intake and bottom of bowl next to you and note that the closed choke plate slopes down and away. Observe the marking on the choke plate. It is important that upon reassembly, this same relationship is retained.

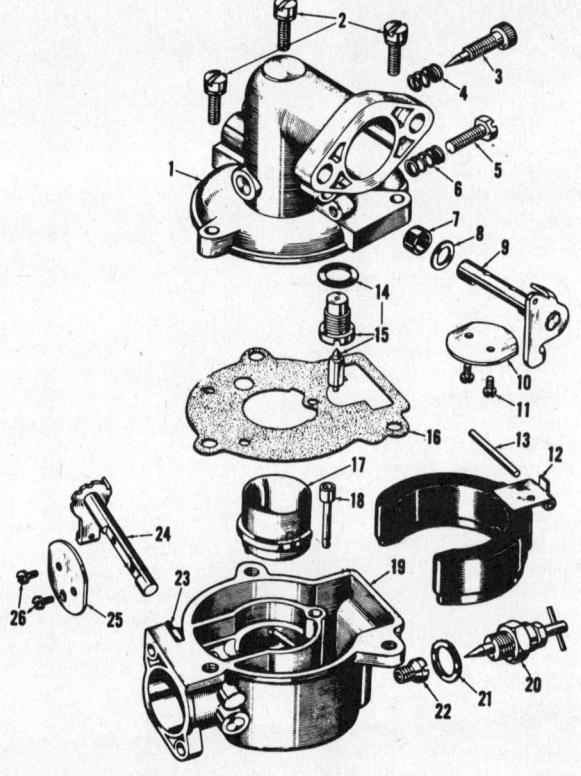

Zenith 72Y6 carburetor (no keys)

Engine Overhaul

15. Remove the choke plate screws and lockwashers (26), choke plate (25), shaft and lever (24).

NOTE: During cleaning, *do not* soak the float in solvent.

16. Hold the bowl (19) vertically (with air intake upward), and insert the choke shaft and lever with the lever pointing downward.
17. Install the choke plate (25) with the letter 'Z' toward the bottom of bowl.
18. Start, but do not tighten both the choke plate screw and the lockwasher (26).
19. Center the choke plate in the air intake bore by lightly tapping the choke plate on the high side. Hold it in this position with a finger and tighten the choke plate screws.
20. Install the main jet (22) using tool C161-83 or equivalent.

NOTE: Before installing the main jet (20) and gasket (21), turn the adjusting needle several turns to the left (counter-clockwise) to avoid damage to the main jet orifice during assembly.

21. Install the main jet plug with a new fiber washer (21).
22. Hold the bowl in the operating position and install the idle jet (18), tube end down.
23. Install the venturi with the key at the lower edge of the venturi, in the matching slot at the choke valve side of the bowl.
24. Assemble a new rubber dust seal (7) against the throttle shaft bearing, with the lips of the seal toward the outside.
25. Install and stake the seal retainer washer (8).
26. Assemble the throttle shaft and lever (9) with the wide open stop lug (narrow lug) on shaft lever in contact with the stop on the casting when the shaft is in the wide open throttle position.
27. Hold the bowl cover inverted, with the mounting flange toward your right.
28. Install the throttle plate (10) with the mark stamped on the throttle plate on the high side of the plate and toward you.
29. Start, but do not tighten both throttle plate screw and lockwasher (11).
30. Gently tap the high side of the throttle plate to center the plate. Hold in this position with a finger and tighten the throttle plate screws.
  Install the throttle stop screw (5) and spring (6).
32. Install the idle adjusting needle (3) and spring (4).
33. Install the fuel valve seat (15) with a fiber washer (14). Use tool C161-85 or equivalent.
34. Assemble the fuel valve needle and bowl cover gasket (16).
35. Carefully examine the float assembly (12) for evidence of wear or damage. This type of float is not adjustable and wear in any part of the fuel valve and float hinge assembly will raise the fuel level.
36. Install the float and float axle pin (13). Insert the bowl cover and check the float in the closed position. The float setting will be within limits if the float is parallel to the gasket seating surfaces of the bowl cover. Any necessary float correction should be made by replacing worn parts. DO NOT attempt to bend the float bracket.
37. Attach the bowl cover assembly (1) to the bowl using the three assembly screws (2).

Zenith 68-7

Use the detailed exploded view to guide you in disassembling and reassembling the carburetor. Clean and inspect all parts as described in the section below. When the float has been reassembled to the throttle body, invert the throttle body and support the float so that the lever contacts the head of the float pin *without pressure*. Measure from the surface of the casting (without gasket) to the top surface of the float (which is the bottom surface during normal operation). The distance should be $15/32$ in. plus or minus $1/32$ in. If distance is incorrect, bend the float lever close to the float body with long nose pliers.

CLEANING AND INSPECTION

1. Clean all of the metal parts in a suitable solvent, removing all carbon deposits from the throttle bore and idle discharge passages. To ensure that all dirt is removed, blow compressed air through all passages in the throttle body and fuel bowl in the reverse direction from normal flow.

NOTE: Never use wire or a drill to clean jet orifices or idle port openings.

2. Check the float and make sure that it is not soaked with gasoline. Also check for wear on the float hinge and where the float contacts the inlet needle. Replace the float if any of the above conditions exist.
3. Inspect the main jet adjustment needle and the idle adjusting needle tapered ends to make sure that they are smooth and not grooved from being seated too hard. If there is a groove around the end of the taper, or if it is pitted, replace the needle.
4. Check the fuel inlet valve and seat for wear or damage. Replace the entire assembly as a unit if it doesn't look like new.
5. All gaskets, seals, retainers, and rubber O-rings must be replaced every time the carburetor is overhauled, with the pos-

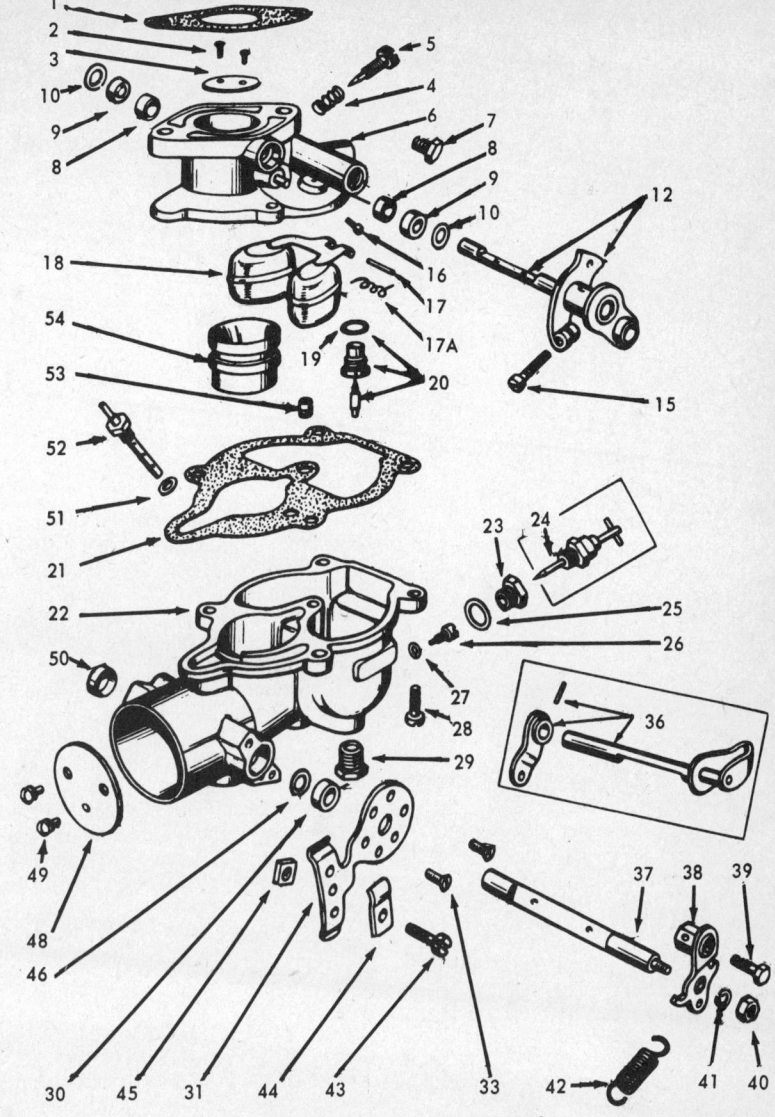

Zenith 68-7 carburetor (no keys)

927

Engine Overhaul

sible exception of the rubber O-rings which can be retained if they are in good condition.

6. Make preliminary adjustments of the main and idle jet adjusting needles before remounting the carburetor on the engine.

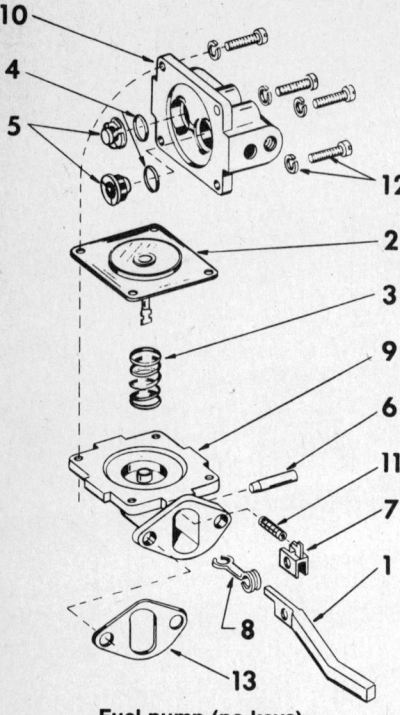

Fuel pump (no keys)

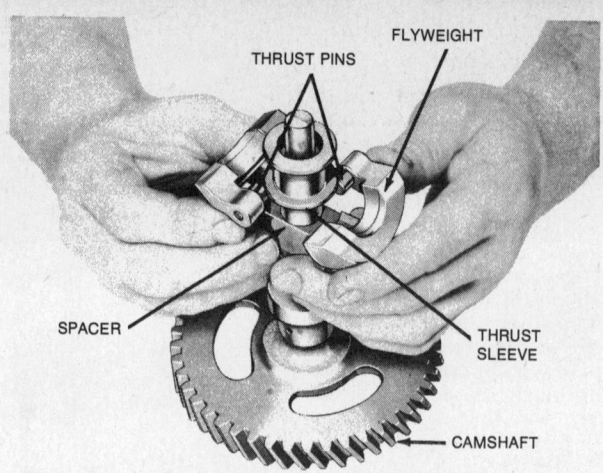

Assembling the governor

Fuel Pump

These instructions refer to overhaul of the LP-62 Series fuel pumps used on some Wisconsin engines. The pump requires rebuilding sometime after 500 hours of operation.

1. Disconnect the fuel lines and, if so equipped, remove the fuel strainer.
2. Scribe a mark across the two halves of the body. Use this mark to positively indicate fuel line inlet and outlet positions. Then, remove the fuel head-to-bracket screws (12), and remove the fuel head (10).
3. Turn the fuel head over, note the positions of the valve assemblies, and then discard them. Clean the fuel head thoroughly in kerosene using a fine wire brush.
4. Hold the head with the diaphragm surface upward, and evenly press in new valve gaskets. Carefully press in new valve assemblies evenly and without any distortion. Make sure each assembly faces in the proper direction—they are check valves.
5. Remove the rocker arm spring (11) from the lower diaphragm section by inserting a screwdriver between coils and prying it out.
6. Hold the mounting bracket (9) in your left hand with the rocker arm toward you and your thumbnail on the end of the link (8). Compress the diaphragm spring (3) by placing the heel of your other hand on the diaphragm (2), and then rotate your hand 90 degrees clockwise to unhook the diaphragm from the link. Remove the diaphragm.
7. Clean the mounting bracket in the same way you cleaned the fuel head.
8. Install the new diaphragm spring onto the bracket (9). Reconnect the new diaphragm to the link by reversing the removal procedure (Step 6). Replace the rocker arm spring (11).
9. Mount the completed mounting bracket assembly (9) onto the engine, using a new gasket (13).
10. Crank the engine over until the diaphragm is laying flat on the mounting bracket. Remount the fuel head (10) with match marks aligned, tightening the screws only three turns. Crank the engine over until the diaphragm is pulled down to its lowest position. Fully tighten screws.
11. Remount the strainer (if so equipped) and install fuel lines to proper connections.

Governor

The governor consists of hardened parts which are only slightly stressed, so repair is rarely necessary. If parts must be replaced, however, the following procedure is useful in reassembling these heavily sprung parts:

Slip the spacer onto the camshaft first. Then, separate the flyweights far enough to permit the thrust sleeve to pass between them. Slide the thrust sleeve back so the flyweights will be closed down between the two flanges of the thrust sleeve.

ENGINE OVERHAUL

The disassembly, inspection, and assembly of each component of the engine is discussed separately, because many times it is not necessary to disassemble the entire engine. The order in which the disassembly procedures are given may be changed to suit the job.

Whenever the engine is either partially or completely disassembled, all of the parts removed should be thoroughly cleaned. Be sure to use new gaskets when reassembling the engine and to lubricate all bearings.

If the engine is to be completely overhauled, remove the engine from the machinery it drives or operates and remove any accessories. If an external component is to be removed, or a minor adjustment made, it may not be necessary to remove the engine from the equipment it powers.

Fuel Tank

Close the fuel tank outlet valve and remove the fuel line. Unscrew the nuts or bolts that retain the tank to the cylinder head bolts and crankcase. The tank and bracket may then be removed as a complete unit. Replace the tank in reverse order, torquing cylinder head in sequence.

Air Cleaner and Carburetor

Unscrew the wing nut and remove the air cleaner. Remove the breather line at the inspection cover, the throttle rod clip at the governor lever, and the fuel line. Unscrew the bolts which hold the carburetor bracket and manifold to the engine and remove the carburetor and air cleaner bracket and the manifold as one. Replace in reverse order.

Starter Sheave and Flywheel Shroud

Remove the starter sheave by removing the three screws and washers which retain it to the flywheel. Remove the top cover and the cylinder side shroud. Disconnect the governor spring and remove the four screws that hold the flywheel shroud to the back plate. The entire flywheel shroud may now be removed. The back plate can be removed, if necessary, only after the flywheel is removed. Reassemble in the reverse order. Apply a ¼ in. long bead of #271 Loctite® to the thread ends of the capscrews for mounting the sheet metal starter sheave. Use plain washers and lockwashers in place of the rubber washers used previously, and torque to 9-10 ft.lbs.

Engine Overhaul

Rope Starter Sheave

Loosen and remove the rope starter sheave by installing a wrench on the hexagonal hub of the sheave and striking a sharp blow in the proper direction. Install in reverse order.

Air Shroud

Remove the cylinder head capscrews and, in cases where so equipped, the crankcase capscrews, and remove the air shroud. Usually, the fuel tank must be removed to remove this shroud, and in some cases, common mounting bolts may be used.

Cylinder Head

Remove the spark plug and unscrew the five capscrews that attach the cylinder head. Remove the cylinder head and gasket. Clean the carbon from the combustion chamber and all dirt from the cooling fins. Use a new gasket when installing the head. If screws of different lengths are used, judge their locations in reassembly from the lengths of the bosses on the head.

Torque the bolts precisely according to instructions below:

ACN, BKN—Torque bolts to 14–18 ft.lbs. TR-10D, TRA-10D, TRA-12D, S-7D, S-8D—torque to 10 ft.lbs. all around; then to 14 ft.lbs.; and finally to 18 ft.lbs.

AEN, AENL, AENS—Coat screw threads with a mixture of oil and graphite and torque to 32 ft.lbs.

Valves and Valve Seat Inserts

Remove the valve inspection cover which is also the breather assembly. Use a valve spring compressor to compress the valve springs. On TR and TRA Model engines, be careful not to damage the breather reed in the valve spring compartment in compressing valve springs. Remove the valve spring retainers, the compressor, and the valve springs; take the valve out from the top of the cylinder block. Clean all carbon deposits from the valves, seats, ports, and guides. Inspect the condition of the valves, stems, guides, and seats, looking for burned, pitted, scored, or warped surfaces.

The exhaust valve and seat are made of stellite. A valve rotator is used on the exhaust valve only. Clean the valve rotator and make sure that it operates properly.

Both intake and exhaust valves have removeable seat inserts on AEN, AENL, and AENS models. On all other models, only the exhaust seat insert is replaceable. Valve seats are removed by means of a special puller. After the new seats are installed, they should be ground to the proper angle.

Before grinding the seats or valves, check the valve-to-guide clearance. The illustration shows specifications for TR, TRA, and S series engines. For AEN, AENL, AENS, ACH, stem-to-guide clearance is .003–.005 in. initially, and the limit is .007 in. Valve and seat angles are 45 degrees for these engines, also. Try replacing the valve to get the proper clearance. If clearance is still excessive, the guides can be pressed out and new ones installed (pressed in). A special tool, Wisconsin DF-72 driver or equivalent is required in installation of new guides. The guide must go in with the internal chamfer downward (towards the camshaft). All guides are pressed in with the top surface flush with the guide boss except for exhaust valve guides on TR and TRA series engines. On these models, the exhaust guide must extend 1/32 of an inch above the guide boss.

On TR and TRA series engines *only*, valve guides must be reamed to the dimensions shown in the illustration *after* they have been pressed into the guide bosses.

The valves should be ground (machined) at an authorized Wisconsin engine service outlet or other qualified machine shop to the specifications shown in the illustration. Then, they must be lapped, using a valve grinding compound by turning them back and forth from above with light downward pressure. Check the effectiveness of the lapping process by putting a dye such as "Prussian blue" or a similar product on the valve sealing surface and seating the valve. The dye will show the pattern of the effective contact between valve and seat on the seat. The pattern shown must be a wide, uniform ring.

Finally, clean the valves and block with soap and water, rinse and wipe thoroughly and then apply a coating of light oil to the cylinder walls to prevent rust.

Valve tappet clearance must be checked before the springs and keepers are reassembled except on engines with adjustable tappets. Install the valves into the guides and seat them. Turn the camshaft as necessary until the cam for the valve to be checked points downward (the tappet is at the lowest possible position). If the engine uses a compression release, make sure the tappet is not riding on the compression release spoiler cam. Check the clearance between the head of the valve stem and the tappet. Clearance should be:

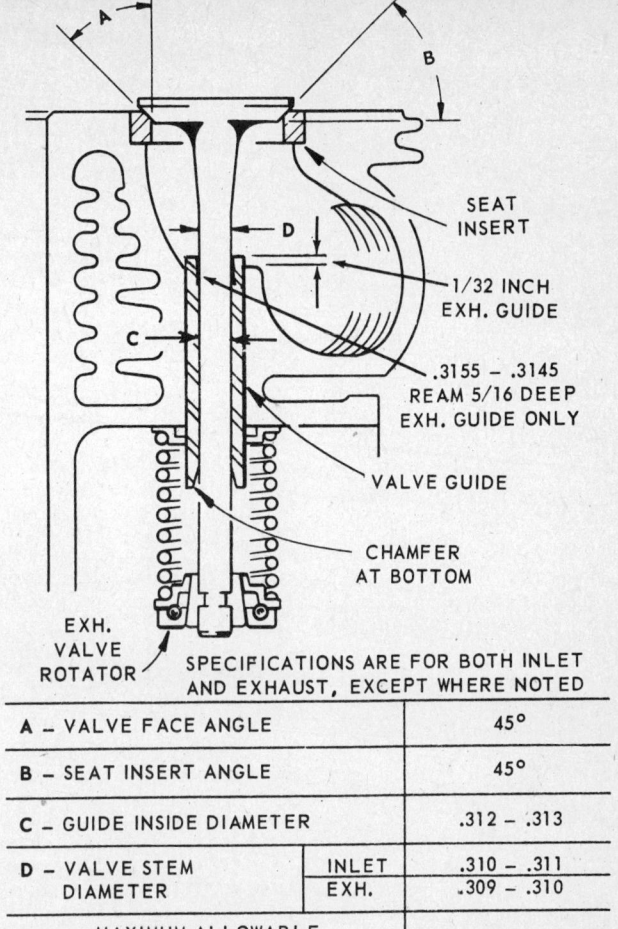

Valve and guide measurements for TR, TRA, S-7, and S-8 engines

A – VALVE FACE ANGLE		45°
B – SEAT INSERT ANGLE		45°
C – GUIDE INSIDE DIAMETER		.312 – .313
D – VALVE STEM DIAMETER	INLET	.310 – .311
	EXH.	.309 – .310
MAXIMUM ALLOWABLE CLEARANCE BETWEEN C AND D		.006

	Intake	Exhaust
TR and TRA Series Engines	.006	.015
AEN, AENL, AENS Series Engines	.008	.016
S-7D, S-8D Engines	.006	.012
ACN, BKN Engines	.008	.014

Engine Overhaul

On engines with adjustable tappets, loosen the locknut with an open end wrench and turn the adjusting nut with another open end wrench until the gauge fits between the tappet and valve stem and can be pulled between the two with a slight amount of effort. On engines with plain tappets, if the clearance is smaller than specification, so that gauge cannot be inserted without lifting the valve off the seat, remove the valve and grind a small amount off the end of the stem. Recheck the clearance until it is adequate. Make sure the stem end is ground absolutely flat (parallel to the valve face) and that all grinding chips, etc. are removed from the valve stem before installation.

Assemble the springs, spring and spring seats or rotators, compress the springs, and install the retainer locks. Make sure the springs are seated properly in the locator cups. If they are not properly seated, they could cock to one side and cause the valve to stick.

Valve Seat and Face Angle	VALVE-TO-GUIDE CLEARANCE	
	Inlet	Exhaust
45°	0.001-0.003	0.003-0.005

Flywheel

ALL TR, TRA, AND S ENGINES

If the flywheel is to be removed, loosen the retaining nut before the gear cover on the opposite end is removed.

NOTE: Do not try to loosen the flywheel after the gear cover is removed. Do not strike the crankshaft when it is not supported by the gear cover.

To remove the flywheel, first straighten the tab of the washer under the flywheel retaining nut. Place the correct size wrench on the flywheel retaining nut and strike the wrench sharply with a hammer to loosen the nut. Do not remove the nut completely, just unscrew it until it is flush with the end of the crankshaft. Turn the crankshaft until the keyway is at 10 o'clock. Pry outward on the flywheel with the outer end of the prybar at the 10 o'clock position on the flywheel. At the same time strike the end of the crankshaft with a soft hammer. This will loosen the flywheel from the tapered end of the crankshaft. Loosen the flywheel, but do not remove it at this point. It is necessary for the flywheel to remain on the crankshaft and support it while the gear cover and connecting rod are removed. Remove the flywheel only after the piston and connecting rod are removed.

When reassembling the engine, install the flywheel after the crankshaft is installed. Make sure that the woodruff key is in place before positioning the flywheel onto the crankshaft. Do not drive the flywheel onto the crankshaft by striking it with a hammer. Place a small length of pipe against the hub of the flywheel and tap the end of the pipe with a soft hammer until the flywheel is seated on the crankshaft taper. Assemble the washer and nut to the crankshaft with the tab of the washer inserted into the keyway of the flywheel. Tighten the nut only enough to hold the flywheel in place. Only after the crankshaft endplay has been adjusted is the flywheel nut to be tightened by sharply striking the wrench with a soft hammer. Bend the tab of the washer up against the nut.

AENL, AEN, AENS, ACN, BKN ENGINES

1. Remove the four air intake screen mounting screws, and remove the screen.
2. Pull outward on the flywheel air fins, and gently tap on the end of the crankshaft with a soft hammer (do *not* use an ordinary, hard hammer) until the flywheel slides off the crankshaft taper.
3. To install the flywheel, first put the crankshaft key into position in the crankshaft keyway. Then, line up the keyway in the flywheel with the key, and slide the flywheel into position on the crankshaft taper. Finally, position a piece of pipe around the crankshaft and against the hub of the flywheel and strike the end of it sharply with the hammer.

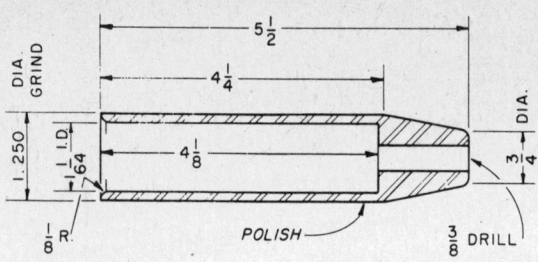

Dimensions of a suitable oil seal installation sleeve

Gear Cover

TR, TRA, AND S ENGINES

To remove the gear cover, unscrew the cover capscrews and remove the governor lever. Tap the two dowel pins lightly from the crankcase side to break the cover loose from the crankcase.

NOTE: A steel ball for the end thrust of the camshaft will most likely fall out when the cover is removed. Remove the spring from the end of the camshaft so it won't be lost.

To reassemble the gear cover to the engine, position the spring into the end of the camshaft and mount the governor flyweight assembly. Lubricate the bearings, gears and tappets. Tap the dowel pins into the crankcase until they protrude about 1/8 in. from the mounting flange face. Place a finger full of grease into the hole in the cover to retain the camshaft spring and ball in place. Lubricate the lip of the oil seal with engine oil. Lubricate the gear cover

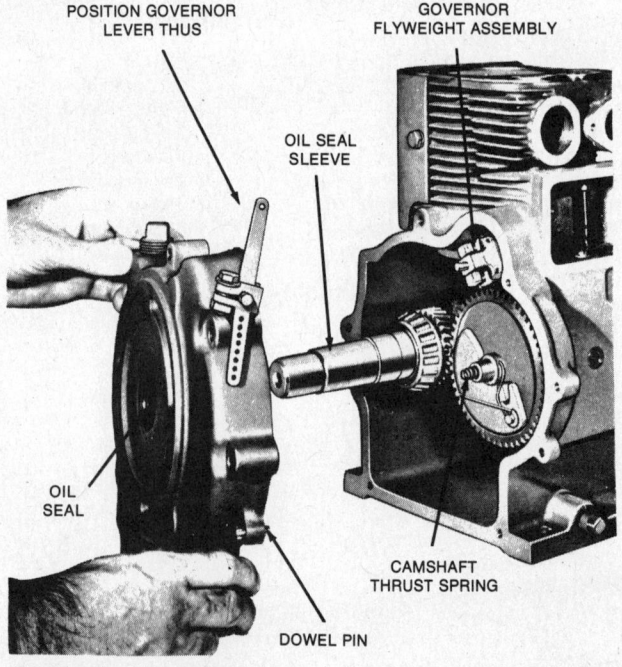

Positioning the governor lever during gear cover installation

Engine Overhaul

face with a light film of oil to hold the gasket in place. The best means of getting the oil seal onto the crankshaft is to make a tapered installation sleeve such as that shown in the illustration. If such a sleeve is available, or you can make one, install it onto the crankshaft.

Position the governor lever as shown in the illustration. Then, gently locate the cover around the crankshaft. If the seal sleeve is being used, it can simply be pushed into position. If a seal sleeve is not available, press the cover into position very carefully. It may be necessary to hold the crankshaft still and very gently rotate the cover back and forth in order to get it over the crankshaft sealing surface without damaging the seal.

Finally, remove the seal sleeve (if used) and torque the cover capscrews to 8 ft.lbs. Tap the dowel pins into place.

Connecting Rod, Piston and Piston Rings

TR, TRA, AND S SERIES ENGINES

Unscrew the two cap bolts which hold the connecting rod cap to the connecting rod. The oil dipper will come off with the cap screws. Tap the ends of the bolts to loosen the connecting rod cap.

Remove all deposits from the cylinder that might hinder the removal of the piston. This is done with a ridge reamer.

Turn the crankshaft until the piston is at the top of the cylinder and push the connecting rod and piston assembly up and out of the engine.

The piston skirt is elliptical in shape. When measuring the piston-to-cylinder wall clearance, you must take the measurement at the bottom of the piston skirt thrust face. The thrust faces of the piston skirt are located at a 90° angle from the piston pin hole axis.

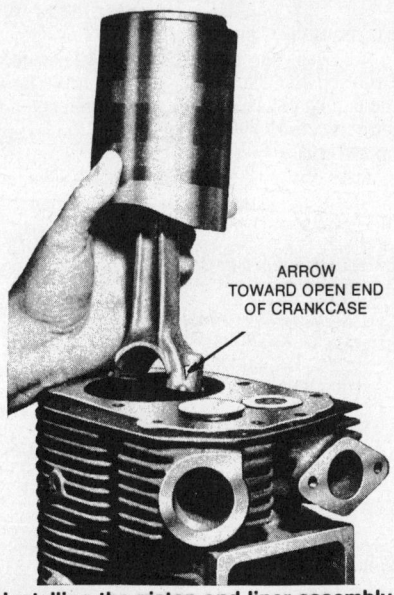

ARROW TOWARD OPEN END OF CRANKCASE

Installing the piston and liner assembly

PISTON TO CYLINDER AT PISTON SKIRT THRUST FACES		.004 to .0045"
PISTON RING GAP		.010 to .020"
PISTON RING SIDE CLEARANCE IN GROOVES	TOP RING	.002 to .0035"
	2nd RING	.001 to .0025"
	OIL RING	.002 to .0035"
CONNECTING ROD TO CRANK PIN	DIAMETER	.0015 to .0005"
	SIDE	.009 to .016"
PISTON PIN TO CONNECTING ROD		.0002 to .0008"
PISTON PIN TO PISTON		.0000 to .0008" tight

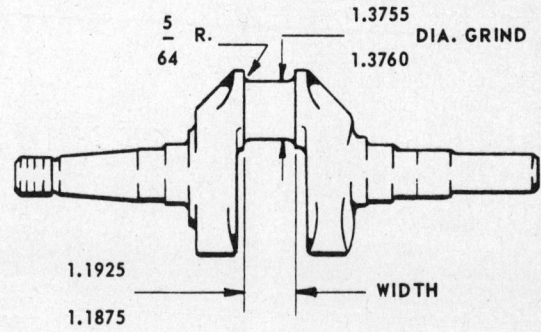

STANDARD CRANK PIN DIMENSIONS

Specifications for S, TR and TRA engines, except that the piston-to-cylinder clearance for TRA-12D engines is .0025-.0030 in.

CYLINDER BORE		3.0005 to 2.9995
PISTON TO CYLINDER AT PISTON SKIRT (THRUST FACE)	CAM-GROUND	SPLIT-SKIRT
	.003 to .0035"	.0045 to .005"
PISTON RING GAP		.010 to .022"
PISTON RING SIDE CLEARANCE IN GROOVES	TOP RING	.002 to .0035"
	2nd, 3rd RING	.001 to .0025"
	OIL RING	.0025 to .004"
PISTON PIN TO CONNECTING ROD BUSHING		.0005 to .0011"
PISTON PIN TO PISTON		.0000 to .0008" tight
CONNECTING ROD TO CRANK PIN – SIDE CLEARANCE		.009 to .018"
CONNECTING ROD SHELL BEARING TO CRANK PIN DIA. (VERTICAL)		.0011 to .0030"
CONNECTING ROD BABBITT BEARING TO CRANK PIN		.0007 to .0020"

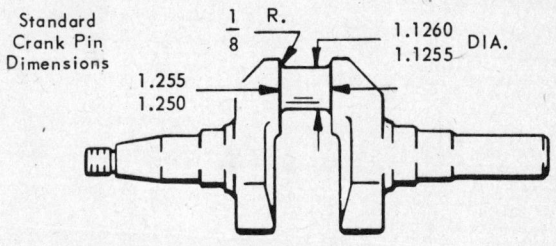

AEN, AENL and AENS specifications

Engine Overhaul

Install the piston rings so that the ring gaps are 90° apart around the circumference of the piston. A ring expander tool should be used to remove and install piston rings. If the tool is not available, the rings can be installed by placing the open end of the ring into the appropriate groove and working the ring down over the piston. Install the bottom oil control ring first, the scraper ring second and the compression ring last. Be careful not to bend or distort the rings in any way. A notch mark or the word "top" will be stamped on each ring so as to identify which side of the ring should face the top of the piston. Before installing the piston assembly into the cylinder, oil the rings, cylinder wall, rod bearings, wrist pin and the piston itself. Use a ring compressor to install the piston assembly into the cylinder bore.

The piston and rod are mounted with the arrows on the connecting rod bolt boss and on the cap matched up and facing toward the open end of the crankcase. The oil hole in the dipper, which is integral with the cap, will be toward the camshaft side of the engine.

If the cylinder is worn more than .005 in. beyond the standard size, you should have it reground at an authorized Wisconsin shop or other reputable machine shop. It might be wise to consult with the machinist as to whether or not the rings should be replaced with a set of chromium rings. Rotate the crankshaft until it is at the bottom of its stroke. Tap the piston down so that the connecting rod seats onto the crankpin. Tighten the cap screws to 18-22 ft.lbs.

AEN, AENL, AENS, ACN, BKN ENGINES

Drain the oil from the crankcase, and then place the engine on its side. Remove the base capscrews and washers, and remove the base and gasket. On AEN, AENL, and AENS, remove the two capscrews which hold the oil pump to the crankcase and remove it.

Use a ½ in. socket wrench to remove the hex locknuts from the rod bolts. If there are lockwasher tabs, these must be straightened first. Tap the ends of the rod bolts lightly to free the cap, and remove it.

Use a ridge reamer to remove all carbon deposits from the cylinder wall above the piston. Turn the crankshaft until the piston is at the top of the cylinder. Push the rod and piston out through the top of the cylinder from below.

NOTE: Do not let the rod bolts come in contact with the crankpin!

AEN, AENL, and AENS engines were originally furnished with babbit cast connecting rod bearings. The shell bearing type rods are now used, and these are interchangeable with the older type rod for service replacement. In reassembling shell bearings, make sure the locating lug for both bearing halves are on the same side of the rod—the side on which numbers are stamped. Fit the bearings according to the specifications shown.

In installing rings, use an expander, or,

PISTON TO CYLINDER AT PISTON SKIRT	**MODEL ACN** Up to 3000 R.P.M. .005 to .0055" 3000 R.P.M. & above .006 to .0065"	
	MODEL BKN Up to 3000 R.P.M. .0055 to .006" 3000 R.P.M. & above .006 to .0065"	
PISTON RING GAP	.012 to .022"	
PISTON RING SIDE CLEARANCE IN GROOVES	TOP RING	.002 to .0035"
	2nd, 3rd RING	.001 to .0025"
	OIL RING	.0025 to .004"
CONNECTING ROD TO CRANK PIN – SIDE CLEARANCE	.009 to .016"	
CONNECTING ROD **SHELL BEARING** TO CRANK PIN DIA. (VERTICAL)	.0009 to .0032"	
CONNECTING ROD **BABBITT BEARING** TO CRANK PIN	.0007 to .002"	
PISTON PIN TO CONNECTING ROD	.0001 to .0007"	
PISTON PIN TO PISTON	.0000 to .0008" tight	

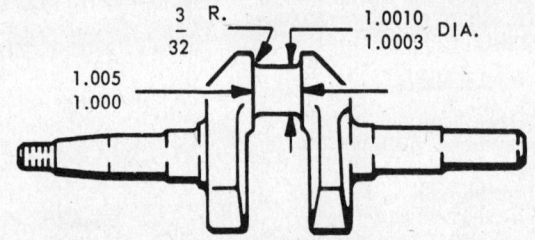

ACN and BKN specifications

if none is available, install the rings open end first. Be careful to open the ring *only* far enough to get it onto the piston. Install the rings so the gaps are 90 degrees apart. Go from bottom to top. *Make sure* the oil scraper ring is mounted as shown, with the scraper edge down, or severe oil pumping will result.

If the cylinder is worn more than .005 in. beyond the standard size, you should have it reground at a Wisconsin authorized shop or other reputable machine shop. It might be wise to consult with the machinist on whether or not the rings should be replaced with a set of chromium rings.

On the AENL engine, if the split skirt type piston originally used is to be re-used, be sure to install it with the split toward the manifold side of the engine. In the case of cam ground pistons used on AEN, AENL, and AENS engines, install the piston with the wide section of the skirt (wide thrust face) toward the fuel tank. Piston-to-cylinder clearance is measured at the center of the thrust face, at the bottom of the skirt.

When installing the piston into the cylinder, oil the rings, piston pin, rod bearings and cylinder wall. Use a ring compressor to hold the rings compressed while sliding the piston into the cylinder. On AEN, AENL, and AENS engines, the numbers stamped on the rod and cap must be on the same side and the oil hole in the cap must face toward the oil pump.

On ACN and BKN engines, the arrow cast onto the connecting rod bolt boss must face toward the take-off end of the crankcase and the oil hole in the rod must face the camshaft. The rod cap must be installed with the cast arrow lining up with the arrow on the rod.

Turn the crankshaft to Bottom Center position, and insert the piston into the cylinder, using a ring compressor until after

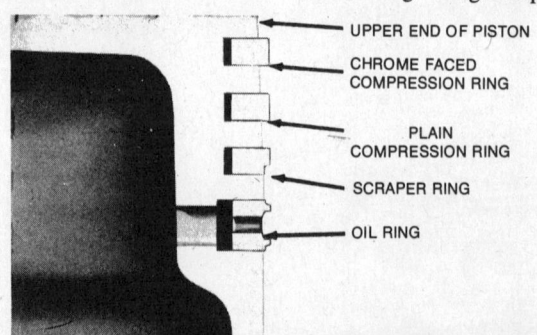

Piston ring installation for AEN, AENL, AENS, ACN engines

Engine Overhaul

the rings enter cylinder. Tap the piston down (with the rod hanging straight down) until the rod contacts the crank pin. Install the cap in the proper position as described above, and install the bolts and nuts (use new nuts on AEN, AENL, and AENS). Torque to 14-20 ft.lbs. on ACN and BKN engines, 18-20 ft.lbs. on AEN, AENL, and AENS engines. Fold the lockwasher tabs over hex head and bolt boss, if so equipped.

Install the oil pump on AEN, AENL, and AENS engines. Install the engine base using a new gasket. Torque the bolts to 6–8 ft.lbs. on ACN and BKN engines, and to 7–9 ft.lbs. on AEN, AENL, and AENS engines.

Camshaft and Valve Tappets

S, TR AND TRA SERIES ENGINES

When removing the camshaft, turn the engine over on its side and push the tappets away from the camshaft so that they will clear the camshaft lobes when the camshaft is removed. The valves must be removed for this operation. After the camshaft is removed, mark the tappets as to location and then remove the valve tappets and inspect them for wear. The tappet stem diameter must be .309-.310 in., and the clearance in the guide hole must be .002-.006 in.

Install tappets into their original guide holes before installing the camshaft. Install the camshaft with the timing mark on the camshaft gear located between the two marked teeth on the crankshaft gear. Put the camshaft thrust spring into the end of the camshaft before installing the gear cover. Adjust or check the valve tappet clearance as described above.

ACN, BKN, AEN, AENL, AND AENS ENGINES

To remove the camshaft, first raise the tappets until they clear the cam lobes. Pry out the expansion plug from the flywheel end of the crankcase. With a drift punch, drive out the camshaft pin from the flywheel end of the crankcase until it emerges from the opposite end of the crankcase. The camshaft should drop down inside the crankcase, with the expansion plug emerging in front of the camshaft pin.

On installation, align the timing marks as shown. Use new expansion plugs and make sure to drive in the camshaft support pin from the takeoff end of the crankcase.

Breaker Push Pin and Bushing

TR, TRA, AND S SERIES ENGINES

Remove the breaker arm push pin and inspect it for wear. Replace parts as necessary. If you're replacing the pin, install the assist spring, small end toward the groove in the tapered end, from the plain end. The pin goes into the guide hole with the plain end toward the camshaft. If there is excessive clearance between pin and bushing, replace the bushing and then ream it to an inside diameter of .2785—.2790 in. Bushings are pressed in. Loctite® may be used to fasten them in place if there is excessive clearance between outside of the bushing and the crankcase.

Crankshaft

TR, TRA, AND S SERIES ENGINES

The crankshaft is removed after the gear cover has been removed, the connecting rod disconnected and raised up out of the way. Remove the flywheel nut, flywheel, and the Woodruff key. The crankshaft may now be pulled out of the open end of the crankcase. When reinstalling the crankshaft, mount the flywheel after the crankshaft is inserted into the crankcase. The flywheel supports the crankshaft while the connecting rod is attached. The flywheel nut is tightened only enough to hold the flywheel during end-play adjustment.

Stator Plate and End-Play Adjustment

The end-play of the crankshaft is ad-

Timing mark alignment on S, TR and TRA engines

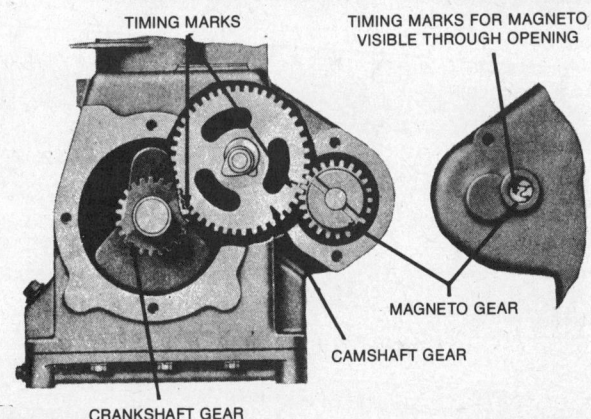
Timing mark alignment on ACN, BKN, AEN, AENL and AENS engines

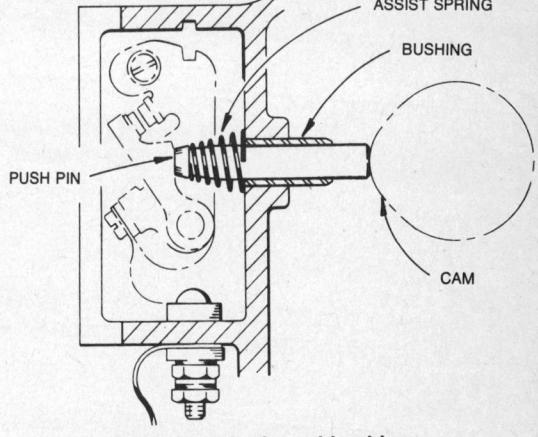
Breaker push pin and bushing

Engine Overhaul

justed by the application of various size gaskets behind the stator plate, which doubles in function as the front bearing support and an adaptor for the magneto coil. The stator plate should not be removed from the crankcase unless it has to be replaced.

To remove the stator plate, remove the four retaining screws and tap the plate from the inside until it falls off. Reassemble the stator plate to the crankcase using new gaskets with the same total thickness as those originally installed. The stator plate mounting screws are to be torqued 8 ft.lbs. on all S, TR, and TRA series engines except the TRA-12D engine; tighten them on TRA-12D engines to 20–22 ft.lbs. End play is checked after the crankshaft, gear cover, and flywheel are mounted. End play is .002–.005 in. on S-7D and S-8D engines, and .001–.004 on TR-10D, TRA-10D, and TRA-12D.

Crankshaft end play is measured with a dial indicator mounted on the PTO side of the crankshaft and a lever prying behind the flywheel. If new crankshaft roller bearings have been installed, they must be properly seated by tapping the ends of the crankshaft with a lead hammer before measuring the crankshaft end-play.

Crankshaft and End Play

Remove the four main bearing plate capscrews at the power take-off end. Pry off the plate and pull the crankshaft out.

On installation, use the same thickness of gaskets initially, and torque the mounting bolts to 10–12 ft.lbs. on ACN and BKN engines and 20–22 ft.lbs. on AEN, AENL, and AENS engines. Check the end play as described at the end of the section above. It should be .001–.003 in. on AEN, AENL, and AENS engines, and .002–.005 in. on ACN and BKN engines. Change the gasket thicknesses in order to correct improper end play.

Make sure to align the punch mark on the front face of the crankshaft between two marked teeth of the camshaft gear.

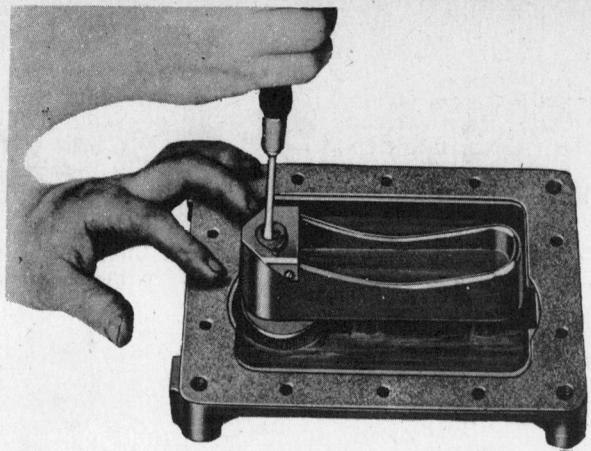
Priming the ACN and BKN oil pump

Oil Pump

ACN, BKN ENGINES

Drain the crankcase, place the engine on its side, and remove the engine base. Carefully note the order of disassembly of the check balls, springs, and other parts.

When assembling the pump, tap the check ball at the bottom of the pump very lightly with a punch and hammer to seat it. After the pump is assembled, fill the engine base with oil and work the pump plunger up and down with a screwdriver in order to check the pump's operation and fill the oil trough. Use a new base gasket and torque the bolts to 6–8 ft.lbs.

AEN, AENL, and AENS Engines

Remove the engine base by draining oil placing engine on its side, and then removing the capscrews and washers. Remove the two oil pump mounting capscrews and remove the oil pump. The main pump plunger, springs, and check balls come out the top, once the pump is away from the drive pushrod. The plug on the side of the discharge tube is removed to gain access to the discharge check ball and spring. Wash all parts in a good solvent.

New plunger-to-bore clearance is .003–.005 in. The limit is .008 in. The pump should be replaced if the clearance is greater than the limit. Inspect the check ball seat in the bottom of the pump cylinder for wear, pitting, or dirt. Clean or replace the pump as necessary.

On reassembly:

1. Drop the intake check ball into the bottom of the pump cylinder and tap it very lightly in order to seat it. Insert the retainer, spring, and plunger into the bore. Install the discharge check ball and spring into the discharge tube.
2. Fill the engine base and put the oil pump into position. Operate the plunger with your finger to prime the pump and check operation. Install the oil pump mounting bolts.
3. Make sure the oil pump pushrod makes good contact with the plunger and the strainer screen is in good condition and properly mounted.
4. Install the engine base, using a new gasket. Torque the mounting bolts to 7–9 ft.lbs.

Break-In

An overhauled engine should be operated at 1600–1800 rpm with no load for one-half hour. It should be operated at normal operating rpm, but still without load, for an additional four hours.

TECUMSEH

ENGINE IDENTIFICATION

Tecumseh-Lauson 4 cycle engines are identified by a model number stamped on a nameplate. The nameplate is located on the crankcase of vertical shaft models and on the blower housing of horizontal shaft models.

A typical model number appears on the illustration showing the location of the nameplate for vertical crankshaft engines. This number is interpreted as follows:
V—vertical shaft engine
60.—6.0 horsepower

70360J—the specification number. The last three numbers (360) indicate that this particular engine is a variation on the basic model line.
2361J—serial number
2—year of manufacture
361—the calendar day of manufacture
J—line and shift location at the factory.

MAINTENANCE

Air Cleaner Service

Service all of the oil/foam polyurethane and oil bath air cleaner elements in the

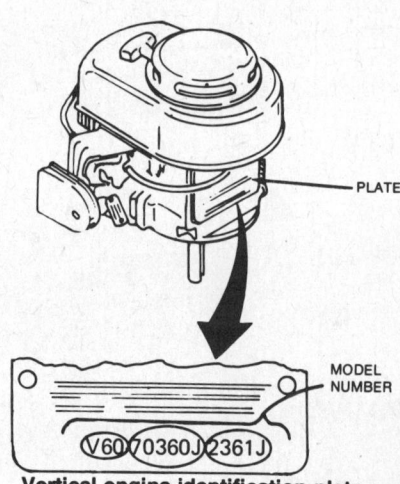
Vertical engine identification plate

Engine Overhaul

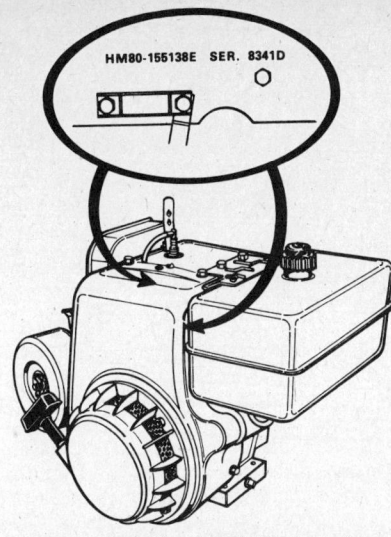

Horizontal engine identification plate

GENERAL ENGINE SPECIFICATIONS HORIZONTAL CRANKSHAFT ENGINES

Model	Bore & Stroke (in.)	Displacement (cu in.)	Horsepower
H60	2.6250x2.5000	13.53	6
HH60	2.6250x2.5000	13.53	6
H70	2.7500x2.5313	15.0	7
HH70	2.7500x2.5313	15.0	7
HM70	2.9375x2.5313	17.16	7
H80	3.0620x2.5313	18.65	8
HM80	3.0620x2.5313	18.65	8
HM100	3.1870x2.5313	20.2	10
ECH90	2.5000x1.8438	9.06	—

GENERAL ENGINE SPECIFICATIONS VERTICAL CRANKSHAFT ENGINES

Model	Bore & Stroke	Displacement	Horsepower
V60	2.625x2.5000	13.53	6
VH60	2.625x2.5000	13.53	6
V70	2.750x2.5313	15.0	7
VH70	2.750x2.5313	15.0	7
VM70	2.750x2.5313	15.0	7
V80	3.062x2.5313	18.65	8
VM80	3.125x2.5313	19.41	8
VM100	3.187x2.5313	17.16	8
ECV100	2.625x1.8438	10.0	—
TNT100	2.625x1.8438	20.2	—
ECV105	2.625x1.9375	10.5	—
ECV110	2.750x1.9375	11.5	—
ECV120	2.812x1.9375	12.0	—
TNT120	2.812x1.9375	12.0	—

same manner as the Briggs and Stratton components.

The Tecumseh treated paper element type air cleaner consists of a pleated paper element encased in a metal housing and must be replaced as a unit. A flexible tubing and hose clamps connect the remotely mounted air filter to the carburetor.

Clean the element by lightly tapping it. Do not distort the case. When excessive carburetor adjustment or loss of power results, inspect the air filter to see if it is clogged. Replacing a severely restricted air filter should show an immediate performance improvement.

Check the oil level in the oil bath type air cleaners regularly to make sure the level is correct. To add oil, unscrew the wingnut, pull off the filter element and add oil along the side of the filter until the level is correct. Use the same type and viscosity oil used in the engine.

If the filter is dirty, remove it and wash it in solvent. Also remove the filter bowl, drain the oil, and wash the filter in solvent. Refill the bowl with clean oil after putting it into postion on the air horn.

A plain paper element is also used. It should be removed every 10 hours, or more

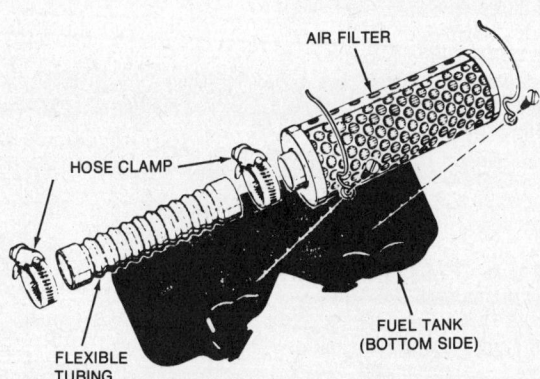

Treated paper air cleaner element used on Craftsman engines

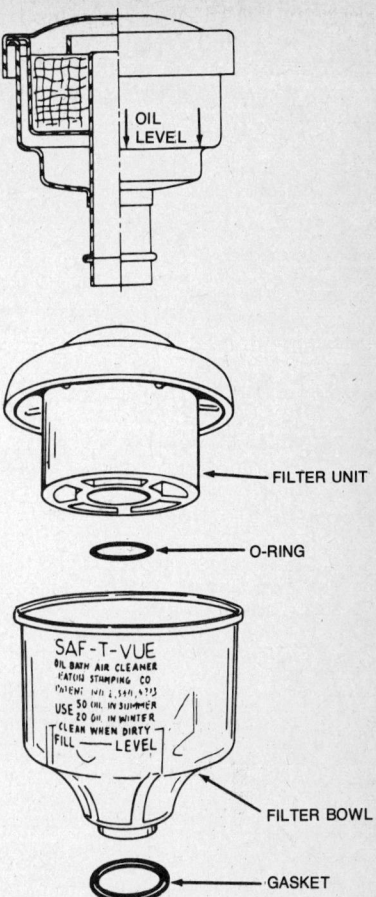

Oil bath air cleaner

often if the air is dusty. Tap or blow out the dirt from the inside with low pressure air. This type should be replaced at 50 hours. If clogged sooner, it may be washed in soap and water and rinsed by flushing from the inside until the water is clear. Blow dry with low pressure compressed air.

To service the KLEEN-AIRE® system, remove the element, wash it in soap and mild detergent, pat dry, and then coat with oil. Squeeze the oil to distribute it evenly and remove the excess. Make sure all mounting surfaces are tight to prevent leakage.

Lubrication

OIL AND FUEL RECOMMENDATIONS

Use fresh (less than one month old) gasoline, of "Regular" grade. Unleaded fuel is preferred, but leaded fuel is acceptable.

Use oil having MS, SC, SD, or SE classification. Use these viscosities for aluminum engines:

Summer—above 32°F.—S.A.E. 30 (S.A.E. 10W30 or 10W40 are acceptable substitutes).

Winter—Below 32°F.—S.A.E. 5W30 (S.A.E. 10W is an acceptable substitute). (Including Snow King Snow Blower Engines)

Winter—Below 0° only S.A.E. 10W diluted with 10% kerosene is an acceptable

substitute. (Including Snow King Snow Blower Engines)
Use these viscosities for cast iron engines:
Summer—Above 32°F.—S.A.E. 30
Winter—Below 32°F.—S.A.E. 10W

TUNE-UP

Tune-Up Specifications

The following basic specifications apply to all the engines covered in this section:
Spark Plug Gap: .030 in.
Ignition Point Gap: .020 in.
Valve Clearance: .010 in. for both intake and exhaust
For timing dimension, which varies from engine to engine, see the complete specifications at the rear of this section.

Spark Plug Service

Spark plugs should be removed, cleaned, and adjusted periodically. Check the electrode gap with a wire feeler gauge and adjust the gap. Replace the plugs if the electrodes are pitted and burned or the procelain is cracked. Refer to the Tecumseh master parts manual for the correct replacement number. Apply a little graphite grease to the threads to prevent sticking. Be sure the cleaned plugs are free of all foreign material.

Breaker Points

ADJUSTMENT

1. Disconnect the fuel line from the carburetor.
2. Remove the mounting screws, fuel tank, and shroud to provide access to the flywheel.
3. Remove the flywheel with a puller.
4. Remove the dust cover and gasket from the magneto and crank the engine over until the breaker points of the magneto are fully opened.
5. Check the condition of the points and replace them if they are burned or pitted.
6. Check the point gap with a feeler gauge. Adjust them, if necessary, as per the directions on the dust cover. Refer to the specifications chart at the end of this section for point gap.

Replacement

1. Gain access to the points and inspect them as described above. If the points are badly pitted, follow the remaining steps to replace them.
2. Remove the nuts that hold the electrical leads to the screw on the movable breaker point spring. Remove the movable breaker point from stud.
3. Remove the screw and stationary breaker point. Put a new stationary breaker point on the breaker plate; install the screw, but do not tighten. This point must be moved to make the proper air gap when the points are adjusted.
4. Position a new movable breaker point on the stud.
5. Adjust the breaker point gap with a flat feeler gauge and tighten the screw.
6. Check the new point contact pattern and remove all grease, finger-prints, and dirt from contact surfaces.
7. Adjust the timing as described below.

Ignition Timing Adjustment

1. Remove the cylinder head bolts, and move the head (with gasket in place) so that the spark plug hole is centered over the piston.
2. Using a ruler (through the spark plug hole) or special plunger type tool, carefully turn the engine back and forth until the piston is at exactly Top Dead Center. Tighten the thumbscrew on the tool.
3. Find the timing dimension for your engine in the specifications at the rear of the manual. Then, back off the position of the piston until it is about halfway down in the bore. Lower the ruler (or loosen the thumbscrew and lower the plunger, if using the special tool) exactly the required amount (the amount of the timing dimension). Then, hold the ruler in place (or tighten the special tool thumbscrew) and, finally, carefully rotate the engine forward until the piston just touches the ruler or tool plunger.
4. Install a timing light or place a very thin piece of cellophane between the contact points. Loosen and rotate the stator just until the timing light shows a change in current flow or the cellophane pulls out of contact gap easily. Then, tighten stator bolts to specified torque.
5. Install the leads, point cover, flywheel and shrouding.

SOLID STATE IGNITION SYSTEM CHECKOUT

The only on-engine check which can be made to determine whether the ignition system is working, is to separate the high tension lead from the spark plug and check for spark. If there is a spark, then the unit is alright and the spark plug should be replaced. No spark indicates that some other part needs replacing.
Check the individual components as follows:
High Tension Lead—Inspect for cracks or indications of arcing. Replace the transformer if the condition of the lead is questionable.
Low Tension Leads—Check all leads for shorts. Check the ignition cut-off lead to see that the unit is not grounded. Repair the leads, if possible, or replace them.
Pulse Transformer—Replace and test for spark.
Magneto—Replace and test for spark.
Time the magneto by turning it counter-clockwise as far as it will go and then tighten the retaining screws.
Flywheel—Check the magnets for strength. With the flywheel off the engine, it should attract a screwdriver that is held 1 in. from the magnetic surface on the inside of the flywheel. Be sure that the key locks the flywheel to the crankshaft.

CARBURETOR MIXTURE ADJUSTMENTS

1. If the carburetor has been overhauled, or the engine won't start, make initial mixture screw adjustments as specified in the chart.
2. Start the engine and allow it to warm up to normal running temperature. With the engine running at maximum recommended rpm, loosen the main adjustment screw until engine rpm drops off, then tighten the screw until the engine starts to cut out. Note the number of turns from one extreme to the other. Loosen the screw to a point midway between the extremes.

NOTE: Some carburetors have fixed jets. If there is no main adjusting screw and receptacle, no adjustment is needed.

CHART OF INITIAL CARBURETOR ADJUSTMENTS

Adjustment	For Engines Built Prior to 1977	For Engine Built After 1977
Main Adjustment Up to 7HP	V50-60-70-1¼ H50-60-70-1¼	Same
Main Adjustment VM70-80-100 & HM70-80-100	1¼	1½
Idle Adjustment Up to 7 HP	V50-60-70-1 Turn H50-60-70-1 Turn	Same
Idle Adjustment VM70-80-100 & HM70-80-100	1½	1¼
Idle Speed (Top of Carburetor) Regulating screw		Back out screw, then turn in until screw just touches throttle lever and continue 1 turn more (if idle RPM is given set final idle speed with a tachometer)

Engine Overhaul

3. After the main system is adjusted, move the speed control lever to the idle position and follow the same procedure for adjusting the idle system.

4. Test the engine by running it under a normal load. The engine should respond to load pickup immediately. An engine that "dies" is too lean. An engine which ran roughly before picking up the load is adjusted too rich.

Governor Adjustment

AIR VANE TYPE

1. Operate the engine with the governor adjusting lever or panel control set to the highest possible speed position and check the speed. If the speed is not within the recommended limits, the governed speed should be adjusted.

2. Loosen the locknut on the high speed limit adjusting screw and turn the adjusting screw out to increase the top engine speed.

MECHANICAL TYPE

1. Set the control lever to the idle position so that no spring tension affects the adjustment.

2. Loosen the screw so that the governor lever is loose in the clamp.

3. Rotate both the lever and the clamp to move the throttle to the full open position (away from the idle speed regulating screw).

4. Tighten the screw when no end-play exists in the direction of open throttle.

5. Move the throttle lever to the full speed setting and check to see that the control linkage opens the throttle.

Compression Check

1. Run the engine until warm to lubricate and seal the cylinder.

2. Remove the spark plug and install a compression gauge. Turn the engine over with the pull starter or electric starter.

3. Compression on new engines is 80 psi. If the reading is below 60 psi, repeat the test after removing the gauge and squirting about a teaspoonful of engine oil through the spark plug hole. If the compression improves temporarily following this, the problem is probably with the cylinder, piston, and rings. Otherwise, the valves require service.

FUEL SYSTEM

Carburetor

NOTE: Four-cycle Tecumseh engines use float or diaphragm type carburetors.

REMOVAL AND INSTALLATION

1. Drain the fuel tank. Remove the air cleaner and disconnect the carburetor fuel lines.

2. If necessary, remove any shrouding or control panels to provide access to carburetor.

3. Disconnect the choke or throttle control wires at the carburetor.

4. Remove the capscrews, or nuts and lockwashers that hold the carburetor to the engine; remove the carburetor.

5. Secure the carburetor on to engine.

6. Install the shrouding or control panels. Connect the choke and throttle control wires.

7. Position the control panel to carburetor. Connect the carburetor fuel lines.

8. Install the air cleaner.

9. Adjust the carburetor as described above.

GENERAL OVERHAUL INSTRUCTIONS

1. Carefully disassemble the carburetor removing all non-metallic parts, i.e.; gaskets, viton seats and needles, O-rings, fuel pump valves, etc.

NOTE: Nylon check balls used in some diaphragm carburetor models may or may not be serviceable. Check to be sure of serviceability before attempting removal.

2. Clean all metallic parts with solvent.

NOTE: Nylon can be damaged if subjected to harsh cleaners for prolonged periods.

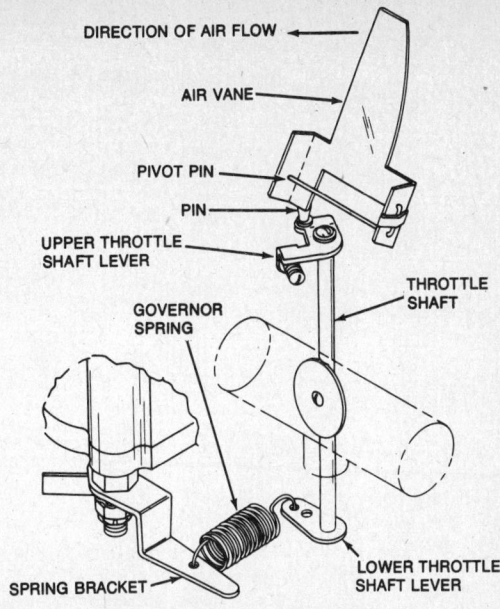

Air vane governor

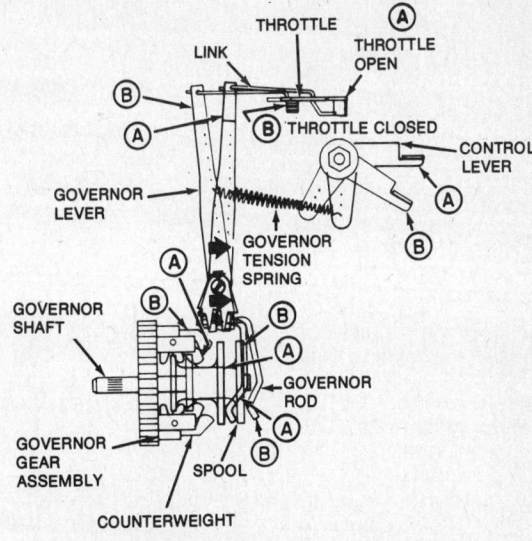

Mechanical governor

Engine Overhaul

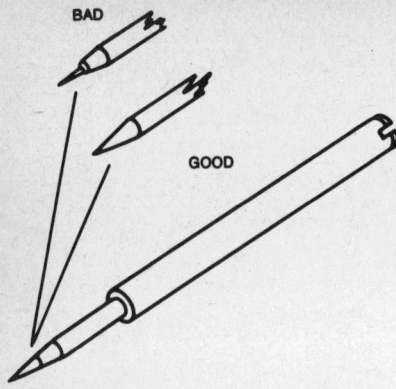

Mixture adjusting screw inspection

3. The large O-rings sealing the fuel bowl to the carburetor body must be in good condition to prevent leakage. If the O-ring leaks, interfering with the atmospheric pressure in the float bowl, the engine will run rich. Foreign material can enter through the leaking area and cause blocking of the metering orifices. This O-ring should be replaced after the carburetor has been disassembled for repair. Lubricate the new O-ring with a small amount of oil to allow the fuel bowl to slide onto the O-ring properly. Hold the carburetor body in an inverted position and place the O-ring on the carburetor body and then position the fuel bowl.

4. The small O-rings used on the carburetor adjustment screws must be in good condition or a leak will develop and cause improper adjustment of carburetor.

5. Check all adjusting screws for wear. The illustration shows a worn screw and a good screw. Replace screws that are worn.

6. Check the carburetor inlet needle and seat for wear, scoring, or other damage. Replace defective parts.

7. Check the carburetor float for dents, leaks, worn hinge or other damage.

8. Check the carburetor body for cracks, clogged passages, and worn bushings. Clean clogged air passages with clean, dry compressed air.

9. Check the diaphragms on diaphragm carburetors for cracks, punctures, distortion, or deterioration.

10. Check all shafts and pivot pins for wear on the bearing surfaces, distortion, or other damage.

NOTE: Each time a carburetor is disassembled, it is good practice to install a repair kit.

11. Where there is excessive vibration, a damper spring may be used to assist in holding the float against the inlet needle thus minimizing the flooding condition. Two types of springs are available; the float shaft (hinge pin) type and the inlet needle mounted type.

12. Float shaft spring positioning:
a. The spring is slipped over the shaft.
b. The rectangular shaped spring end is hooked onto the float tab.
c. The shorter angled spring end is placed onto the float bowl gasket support.

13. Note that on late model carburetors the spring clip fastened to the inlet needle has been revised to provide a damping effect. The clip fastens to the needle and is hooked over the float tab.

FLOAT FEED CARBURETOR

Note the following points when rebuilding these carburetors.

Removal

Remove the carburetor from the engine. It is easier to remove the intake manifold and carburetor assembly from the engine, disconnect the governor linkage, fuel line, and grounding wire and then disassemble the carburetor from the intake manifold on a work bench. Be sure to note the positions of the governor and the throttle linkage to facilitate reassembly.

Throttle

1. Examine the throttle lever and plate prior to disassembly. Replace any worn parts.

2. Remove the screw in the center of the throttle plate and pull out the throttle shaft lever assembly.

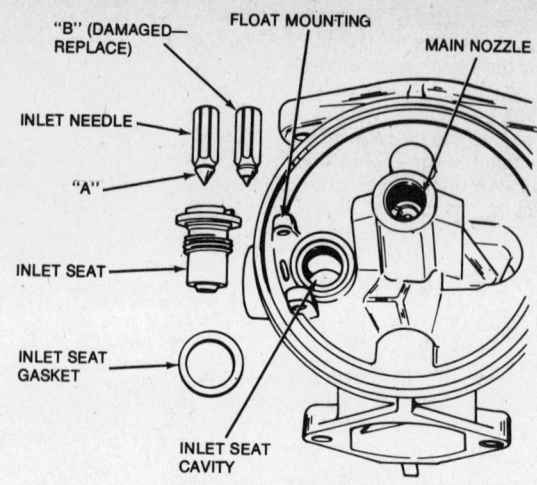

Inlet needle inspection

3. When reassembling, it is important that the lines on the throttle plate are facing out when in the closed position. Position the throttle plates with the two lines at 12 and 3 o'clock. The throttle shaft must be held in tight to the bottom bearing to prevent the throttle plate from riding on the throttle bore of the body which would cause excessive throttle plate wear and governor hunting.

Choke

Examine the choke lever and shaft at the bearing points and holes into which the linkage is fastened and replace any worn parts. The choke plate is inserted into the air horn of the carburetor in such a way that the flat surface of the choke is toward the fuel bowl.

Idle Adjusting Screw

Remove the idle screw from the carburetor body and examine the point for damage to the seating surface on the taper. If damaged, replace the idle adjusting needle. Tension is maintained on the screw with a coil spring and sealed with an

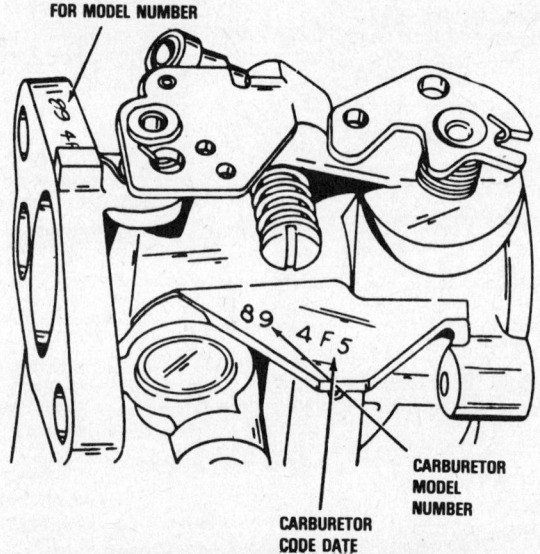

Float feed carburetor identification

O-ring. Examine and replace the O-ring if it is worn or damaged.

High Speed Adjusting Jet

Remove the screw and examine the taper. If the taper is damaged at the area where it seats, replace the screw and fuel bowl retainer nut as an assembly.

The fuel bowl retainer nut contains the seat for the screw. Examine the sealing O-ring on the high speed adjusting screw. Replace the O-ring if it indicates wear or cuts. During the reassembly of the high speed adjusting screw, position the coil spring on the adjusting screw, followed by the small brass washer and the O-ring seal.

Fuel Bowl

To remove the fuel bowl, remove the retaining nut and fiber washer. Replace the nut if it is cracked or worn.

The retaining nut contains the transfer passage through which fuel is delivered to the high speed and idle fuel system of the carburetor. It is the large hole next to the hex nut end of the fitting. If a problem occurs with the idle system of the carburetor, examine the small fuel passage in the annular groove in the retaining nut. This passage must be clean for the proper transfer of fuel into the idle metering system.

The fuel bowl should be examined for rust and dirt. Thoroughly clean it before installing it. If it is impossible to properly clean the fuel bowl, replace it.

Check the drain valve for leakage. Replace the rubber gasket on the inside of the drain valve if it leaks.

Examine the large O-ring that seals the fuel bowl to the carburetor body. If it is worn or cracked, replace it with a new one, making sure the same type is used (square or round).

Float

1. Remove the float from the carburetor body by pulling out the float axle with a pair of needle nose pliers. The inlet needle will be lifted off the seat because it is attached to the float with an anchoring clip.

2. Examine the float for damage and holes. Check the float hinge for wear and replace it if worn.

3. The float level is checked by positioning a #4 (0.209 in.) twist drill across the rim between the center leg and the unmachined surface of the index pad, parallel to the float axle pin. If the index pad is machined, the float setting should be made with a #9 (0.180-0.200 in.) twist drill.

4. Remove the float to make an adjustment. Bend the tab on the float hinge to correct the float setting.

NOTE: Direct compressed air in the opposite direction of normal flow of air or fuel (reverse taper) to dislodge foreign matter.

Inlet Needle and Seat

1. The inlet needle sits on a rubber seat in the carburetor body instead of the usual metal fitting.

2. Remove it, place a few drops of heavy engine oil on the seat, and pry it out with a short piece of hooked wire.

3. The grooved side of the seat is inserted first. Lubricate the cavity with oil and use a flat faced punch to press the inlet seat into place.

4. Examine the inlet needle for wear and rounding off of the corners. If this condition does exist, replace the inlet needle.

Fuel Inlet Fitting

1. The inlet fitting is removed by twisting and pulling at the same time.

2. Use sealer when reinstalling the fitting. Insert the tip of the fitting into the carburetor body. Press the fitting in until the shoulder contacts the carburetor. Only use inlet fittings without screens.

Carburetor Body

1. Check the carburetor body for wear and damage.

2. If excessive dirt has accumulated in the atmospheric vent cavity, try cleaning it with carburetor solvent or compressed air. Remove the welch plug only as a last resort.

NOTE: The carburetor body contains a pressed-in main nozzle tube at a specific depth and position within the venturi. Do not attempt to remove the main nozzle.

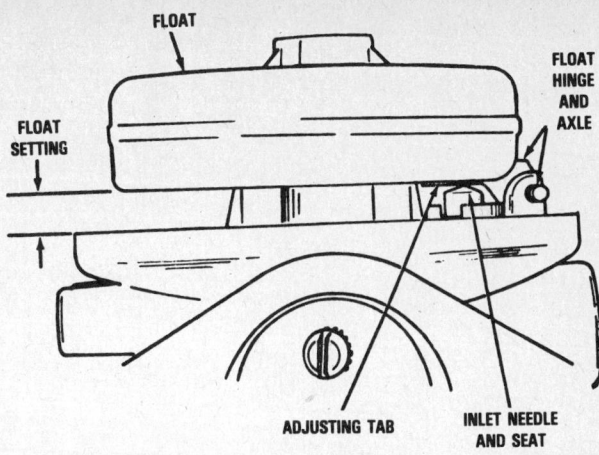

Float adjustment

Any change in nozzle positioning will adversely affect the metering quality and will require carburetor replacement.

3. Clean the accelerating well around the main nozzle with compressed air and carburetor cleaning solvents.

4. The carburetor body contains two cup plugs, neither of which should be removed. A cup plug located near the inlet seat cavity, high up on the carburetor body, seals off the idle bleed. This is a straight passage drilled into the carburetor throat. Do not remove this plug. Another cup plug is located in the base where the fuel bowl nut seals the idle fuel passage. Do not remove this plug or the metering rod.

5. A small ball plug located on the side of the idle fuel passage seals this passage. Do not remove this ball plug.

6. The welch plug on the side of the carburetor body, just above the idle adjusting screw, seals the idle fuel chamber. This plug can be removed for cleaning of the idle fuel mixture passage and the primary and secondary idle discharge ports. Do not use any tools that might change the size of the discharge ports, such as wire or pins.

Resilient Tip Needle

Replace the inlet needle. Do not attempt to remove or replace the seat in the carburetor body.

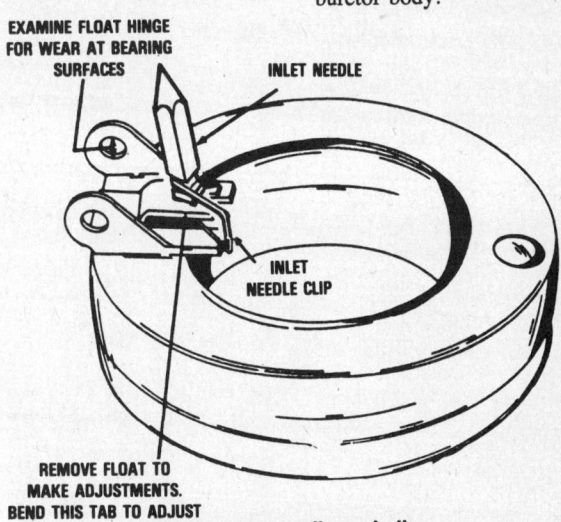

Float, inlet needle and clip

Engine Overhaul

Viton Seat

Using a 10-24 or 10-32 tap, turn the tap into the brass seat fitting until it grasps the seat firmly. Clamp the tap shank into a vise and tap the carburetor body with a soft hammer until the seat slides out of the body.

To replace the viton seat, position the replacement over the receptical with the soft rubber like seat toward the body. Use a flat punch and a small hammer to drive the seat into the body until it bottoms on the shoulder.

TECUMSEH AUTOMATIC NON-ADJUSTABLE FLOAT FEED CARBURETOR

This carburetor has neither a choke plate nor idle and main mixture adjusting screws. There is no running adjustment. The float adjustment is the standard Tecumseh float setting of 0.210 in. (#4 drill).

Cleaning

Remove all non-metallic parts and clean them using a procedure similar to that for the other carburetors. Never use wires through any of the drilled holes. Do not remove the baffling welch plug unless it is certain there is a blockage under the plug. There are no blind passageways in this carburetor.

Some engines use a variation on the Automatic Nonadjustable carburetor which has a different bowl hold—on nut and main jet orifices. There are two main jet orifices and a deeper fuel reserve cavity, but service procedures are the same.

WALBRO AND TILLOTSON FLOAT FEED CARBURETORS

Procedures are similar to those for the Tecumseh float carburetor with the exceptions noted below.

Main Nozzle

The main nozzle in Walbro carburetors is cross drilled after it is installed in the carburetor. Once removed, it cannot be reinstalled, since it is impossible to properly realign the cross drilled holes. Grooved service replacement main nozzles are available which allow alignment of these holes.

Float Shaft Spring

Carefully position the float shaft spring on models so equipped. The spring dampens float action when properly assembled. Use needle-nosed pliers to hook the end of the spring over the float hinge and insert the pin as far as possible before lifting the spring from the hinge into position. Leaving the spring out or improper installation will cause unbalanced float action and result in a touchy adjustment.

Float Adjustment

1. To check the float adjustment, invert the assembled float carburetor body. Check the clearance between the body and the float, opposite the hinge. Clearance should be ⅛ in. ±¹⁄₆₄ in.
2. To adjust the float level, remove the float shaft and float. Bend the lip of the float tang to correct the measurement.
3. Assemble the parts and recheck the adjustment.

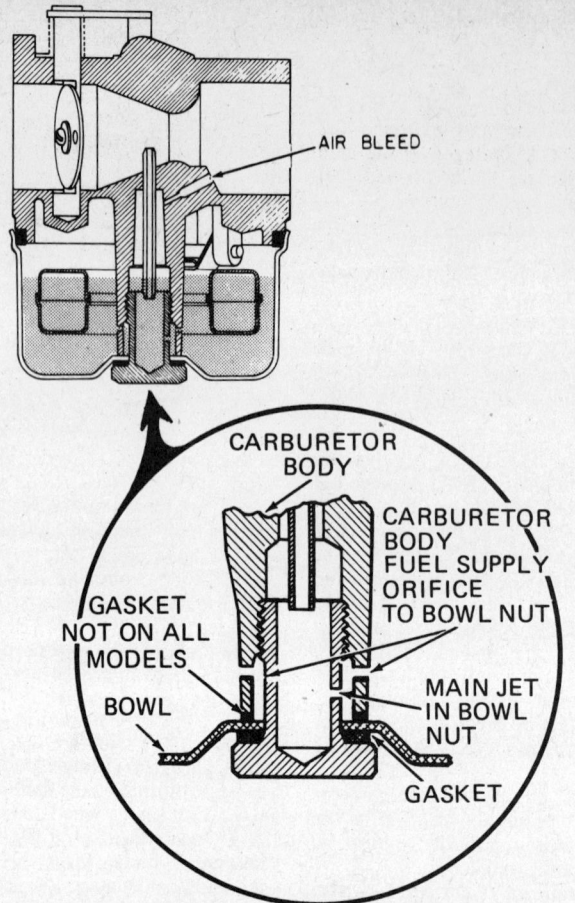

Variation on non-adjustable float feed carburetor

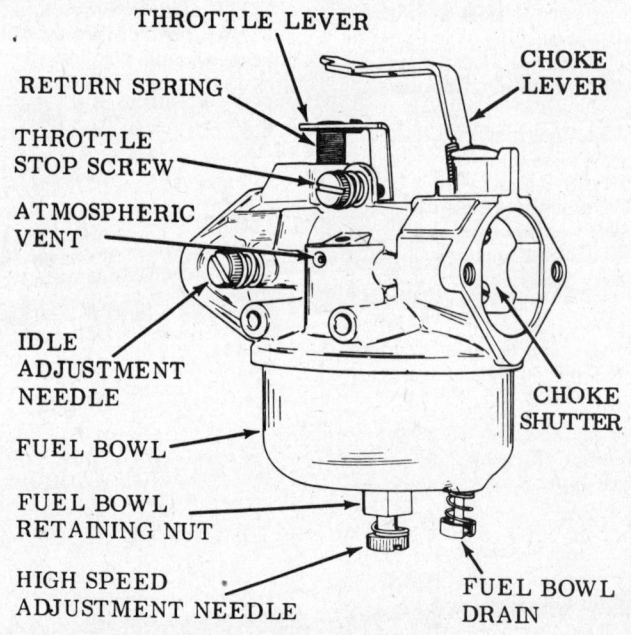

Walbro float feed carburetor

Engine Overhaul

TILLOTSON E FLOAT FEED TYPE CARBURETOR

The following adjustments are different for this carburetor.

Running Adjustment

1. Start the engine and allow it to warm up to operating temperatures. Make sure the choke is fully opened after the engine is warmed up.
2. Run the engine at a constant speed while slowly turning the main adjustment screw in until the engine begins to lose speed; then slowly back it out about 1/8-1/4 of a turn until maximum speed and power is obtained (4000 rpm). This is the correct power adjustment.
3. Close the throttle and cause the engine to idle slightly faster than normal by turning the idle speed regulating screw in. Then turn the idle adjustment speed screw in until the engine begins to lose speed; then turn it back 1/4 to 1/2 of a turn until the engine idles smoothly. Adjust the idle speed regulating screw until the desired idling speed is acquired.
4. Alternately open and close the throttle a few times for an acceleration test. If stalling occurs at idle speeds, repeat the adjustment procedures to get the proper idle speed.

Float Level Adjustment

1. Remove the carburetor float bowl cover and float mechanism assembly.
2. Remove the float bowl cover gasket and, with the complete assembly in an upside down position and the float lever tang resting on the seated inlet needle, a measurement of $1^{5}/_{64}$ in. should be maintained from the free end flat rim, or edge of the cover, to the toe of the float. Measurement can be checked with a standard straight rule or depth gauge.
3. If it is necessary to raise or lower the float lever setting, remove the float lever pin and the float, then carefully bend the float lever tang up or down as required to obtain the correct measurement.

WALBRO CARBURETORS FOR V80, VM80, H80, AND HM80 ENGINES

Adjustment

The following initial carburetor adjustments are to be used to start the engine. For proper carburetion adjustment, the atmospheric vent must be open. Examine and clean it if necessary.

1. Idle adjustment—1¼ turn from its seat.
2. High speed adjustment—1½ turn from its seat.
3. Throttle stop screw—1 turn after contacting the throttle lever.

After the engine reaches normal operating temperature, make the final adjustments for best idle and high speed within the following ranges. Recommended speeds: Idle: 1800-2300 rpm. High speed: 3450-3750 rpm.

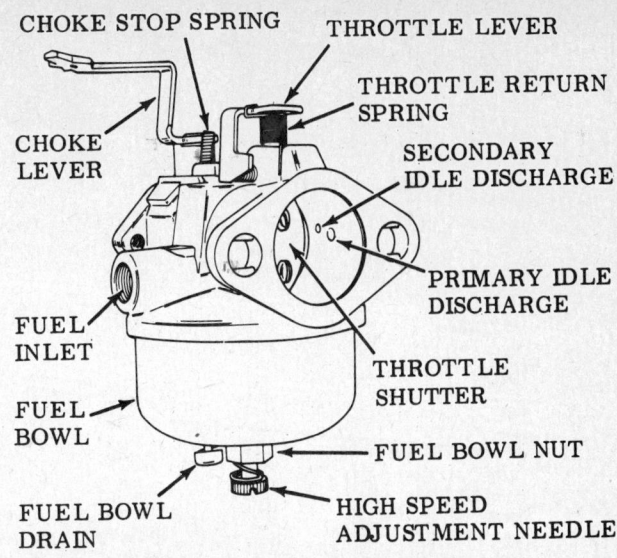

Walbro carburetor #631635, engine side

Rebuilding Notes

1. The throttle plate is installed with the lettering (if present) facing outward when closed. The throttle plate is installed on the throttle lever with the lever in the closed position. If there is binding after the plate is in position, loosen the throttle plate and reposition it.
2. Before removing the fuel bowl nut, remove the high speed adjusting needle. Use a $7/_{16}$ in. box wrench or socket to remove the fuel bowl nut. When replacing the fuel bowl nut, be sure to position the fiber gasket under the nut and tighten it securely.
3. Examine the high speed needle tip and, if it appears to be worn, replace it. When the high speed jet is replaced, the main nozzle, which includes the jet seat, should also be replaced. The original main nozzle cannot be used. There are special replacement nozzles available.
4. The inlet needle valve is replaceable if it appears to be worn. The inlet valve seat is also replaceable and should be replaced if the needle valve is replaced.

Float Adjustment

1. The float setting for this carburetor is 0.070 to 0.110 ($5/_{64}$ to $7/_{64}$ in.).
2. The float is set in the traditional manner, at the opposite end of the float from the float hinge and needle valve.
3. Bend the adjusting tab to adjust the float level.

DIAPHRAGM CARBURETORS

Diaphragm carburetors have a rubber-like diaphragm that is exposed to crankcase pressure on one side and to atmospheric pressure on the other side. As the crankcase pressure decreases, the diaphragm moves against the inlet needle allowing the inlet needle to move from its seat which permits fuel to flow through the inlet valve to maintain the correct fuel level in the fuel chamber.

An advantage of this type of system over

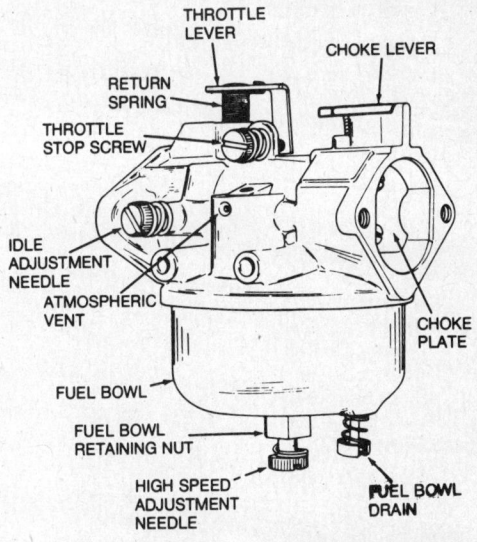

Walbro carburetor #631636, intake side

Engine Overhaul

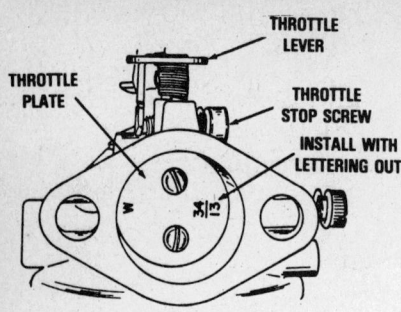

Installation of the throttle plate

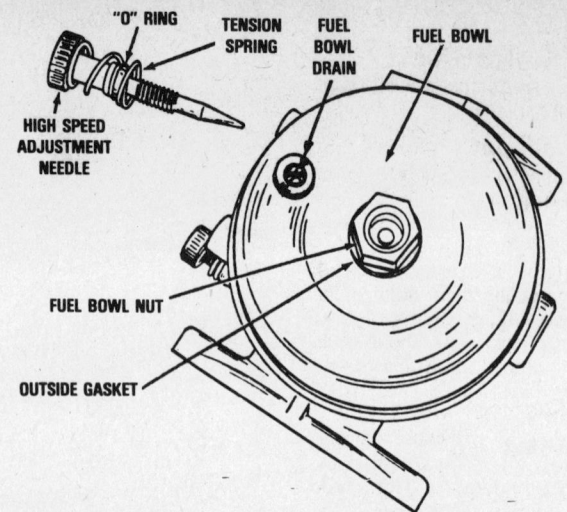

Installation of the fuel bowl and high speed adjustment needle

the float system, is that the engine can be operated in any position.

NOTE: In rebuilding, use carburetor cleaner only on metal parts, except for the main nozzle in the main body.

Throttle Plate
Install the throttle plate with the short line that is stamped in the plate toward the top of the carburetor, parallel with the throttle shaft, and facing out when the throttle is closed.

Choke Plate
Install the choke plate with the flat side of the choke toward the fuel inlet side of the carburetor. The mark faces in and is parallel to the choke shaft.

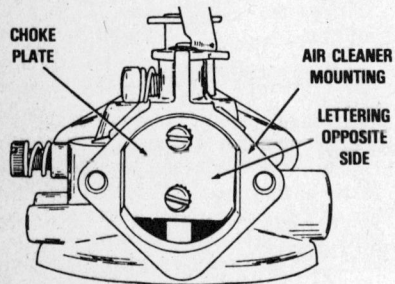

Installation of the choke plate

Idle Mixture Adjustment Screw
There is a neoprene O-ring on the needle. Never soak the O-ring in carburetor solvent. Idle and main mixture screws vary in size and design, so make sure that you have the correct replacement.

Idle Fuel Chamber
The welch plug can be removed if the carburetor is extremely dirty.

Diaphragms
Diaphragms are serviced and replaced by removing the four retaining screws from the cover. With the cover removed, the diaphragm and gasket may be serviced. Never soak the diaphragm in carburetor solvent. Replace the diaphragm if it is cracked or torn. Be sure there are no wrinkles in the diaphragm when it is replaced. The diaphragm rivet head is always placed facing the inlet needle valve.

Inlet Needle and Seat
The inlet seat is removed by using either a slotted screwdriver (early type) or a $^9/_{32}$

in. socket. The inlet needle is spring loaded, so be careful when removing it.

Fuel Inlet Fitting
All of the diaphragm carburetors have an integral strainer in the inlet fitting. To clean it, either reverse flush it or use compressed air after removing the inlet needle and seat. If the strainer is lacquered or otherwise unable to be cleaned, replace the fitting.

CRAFTSMAN FUEL SYSTEMS

Changes in Late Model Carburetors
The newest Craftsman carburetors incorporate the following changes:
a. The cable form of control is replaced by a control knob.
b. The fuel pickup is longer and has a collar machined into it which must be installed tight against the carburetor body.

c. The fuel pickup screen is pressed onto the ends of the fill tubes on both models, but the measured depth has changed.
d. The cross-drilled passages have been eliminated, as has the O-ring on the body. There are no cup plugs.
e. The fuel tank and reservoir tube have been revised—the reservoir tube being larger.

Disassembly and Service
1. Remove the air cleaner assembly and remove the four screws on the top of the carburetor body to separate the fuel tank from the carburetor.
2. Remove the O-ring from between the carburetor and the fuel tank. Examine it for cracks and damage and replace it if necessary.
3. Carefully remove the reservoir tube from the fuel tank. Observe the end of the tube that rested on the bottom of the fuel tank. It should be slotted.

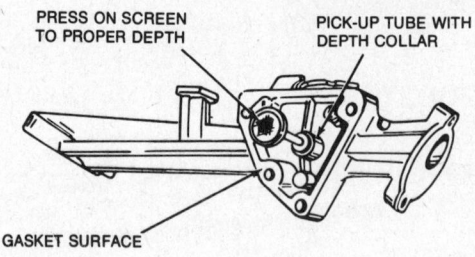

LATE MODEL

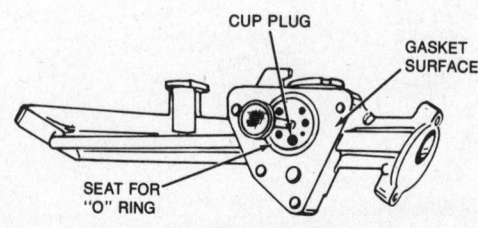

EARLY MODEL

Details of the pick-up tube used on Craftsman carburetors

Engine Overhaul

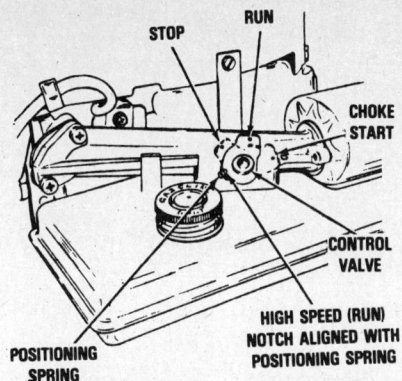

Craftsman fuel tank mounted carburetor

4. Remove the control valve by turning the valve clockwise until the flange is clear of the retaining boss. Pull the valve straight out and examine the O-ring seal for damage or wear. If possible, use a new O-ring when reassembling.

5. Examine the fuel pick up tube. There are no valves or ball checks that may become inoperative. These parts can normally be cleaned with carburetor solvent. If it is found that the passage cannot be cleared, the fuel pick up tubes can be replaced. Carefully remove the old ones so as

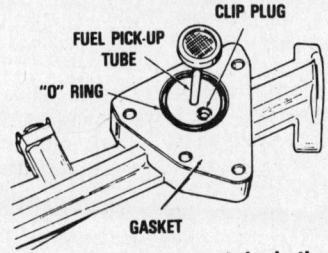

Positioning of the reverse tube in the fuel tank

not to enlarge the opening in the carburetor body. If the pick-up tube and screen must be replaced, follow the directions shown in the illustration for the type of carburetor (early or late model) on which you are working.

6. Assemble the carburetor in reverse order of disassembly. Use new O-rings and gaskets. When assembling reservoir tube, hold the carburetor upside down and place the reservoir tube over the pickup tube with the slotted end up. Make sure the intake manifold gasket is correctly positioned—it can be assembled blocking the intake passage partially.

Adjustments

1. Move the carburetor control valve to the high speed position. The mark on the

The fuel pick-up tube and O-ring in early Craftsman carburetors

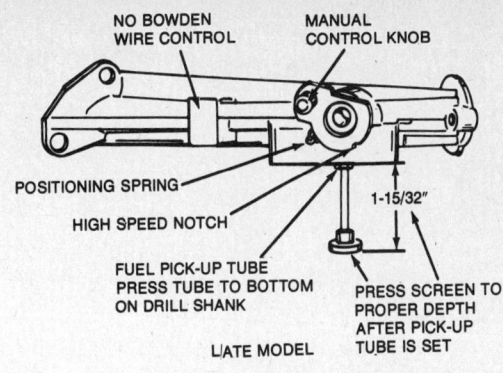

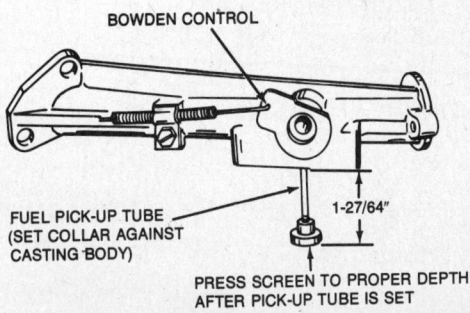

Details of the manual control knob and other changes incorporated in Craftsman carburetors

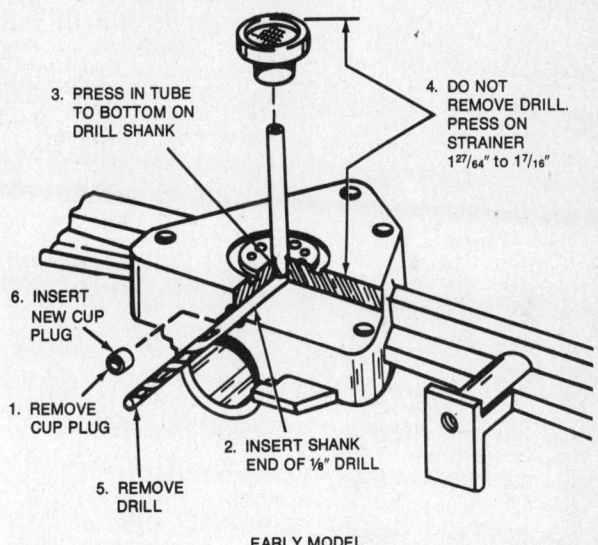

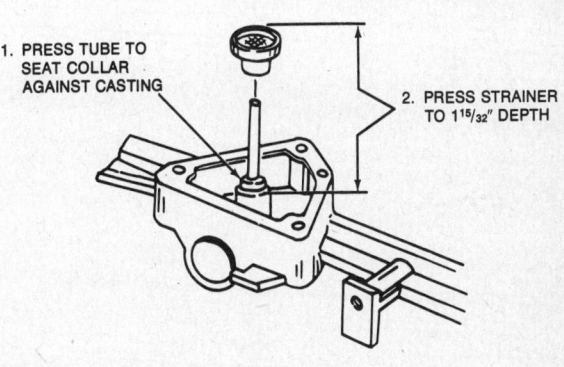

Pick-up tube replacement

943

Engine Overhaul

face of the valves should be in alignment with the retaining boss on the carburetor body.

2. Move the operator's control on the equipment to the high speed position.

3. Insert the bowden wire into the hole of the control valve. Clamp the bowden wire sheath to the carburetor body.

NOTE: If the engine was disassembled and the camshaft removed, be sure that the timing marks on the camshaft gear and the key way in the crankshaft gear are aligned when reinserting the camshaft. Then lift the camshaft enough to advance the camshaft gear timing mark to the right (clockwise) ONE tooth, as viewed from the power take-off end of the crankshaft.

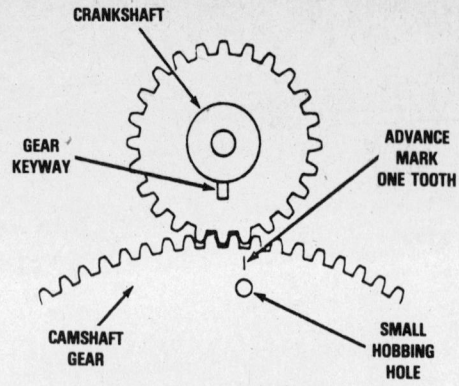

Engine timing on Craftsman engines with fuel tank mounted carburetors

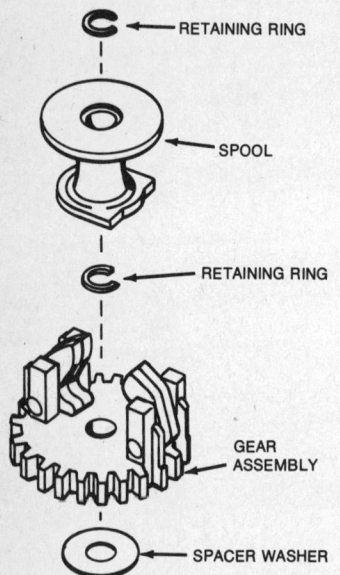

Standard mechanical governor

SHAFT INSTALLED DIMENSION/ENGINE MODEL CHART

Engine Model	"A" Exposed Shaft Length
LAV30-50 H25-35 HS40-50 TNT100-120 ECV100-105-110-120 ECH90 TVS75-90-105-120	1⁵⁄₁₆"
V50-60 VM70-80-100	1¹⁹⁄₃₂"
H50-60 HH40-70	1⁷⁄₁₆"
HM70-80-100	1¹³⁄₃₂"

CRAFTSMAN FLOAT TYPE CARBURETORS

Craftsman float type carburetors are serviced in the same manner as the other Tecumseh float type carburetors. The throttle control valve has three positions: stop, run, and start.

When the control valve is removed, replace the O-ring. If the engine runs sluggishly, consider the possibility of a leaky O-ring.

To remove the fuel pickup tube, clamp the tube in a vise and then twist the carburetor body. Check the small jet in the air horn while the tube is out. In replacement, position the tube squarely and then press in on the collar until the collar seats.

In assembly, install the gasket and bolt through the bowl, position the centering spacer onto the bolt, and then attach the parts to the carburetor body.

The camshaft timing mark must be advanced one tooth in relation to crankshaft gear timing mark, as shown in the illustration above. However, when this type carburetor is used with a float bowl reservoir and variable governor adjustment, time it as for other Tecumseh engines—with the camshaft and crankshaft timing marks aligned.

Governor

The mechanical governor is located inside the mounting flange. See engine disassembly instructions, below. To disassemble the governor, see the illustration, and: remove the retaining ring, pull off the spool, remove the second retaining ring, and then pull off the gear assembly and retainer washer.

Check for wear on all moving surfaces, but especially gear teeth, the inside diameter of the gear where it rides on the shaft, and the flyweights where they work against the spool.

If the governor shaft must be replaced, it should be started into the boss with a few taps using a soft hammer, and then pressed in with a press or vise. The shaft *can* be installed by positioning a wooden block on top and tapping the upper surface of the block, but the use of a vise or press is much preferred.

The shaft must be pressed in until just the required length is exposed, as measured from the top of the shaft boss to the upper end of the shaft. See the chart.

The governor is installed in reverse of the removal procedure. Connect the linkage and then adjust as described in the Tune-Up section.

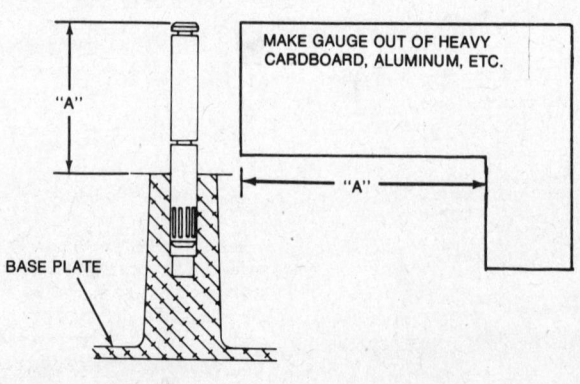

Governor shaft installed dimension

Engine Overhaul

Timing Gears

Correctly matched camshaft gear and crankshaft gear timing marks are necessary for the engine to perform properly.

On all camshafts the timing mark is located in line with the center of the hobbing hole (small hole in the face of the gear). If no line is visible, use the center of the hobbing hole to align with the crankshaft gear marked tooth.

On crankshafts where the gear is held on by a key, the timing mark is the tooth in line with the keyway.

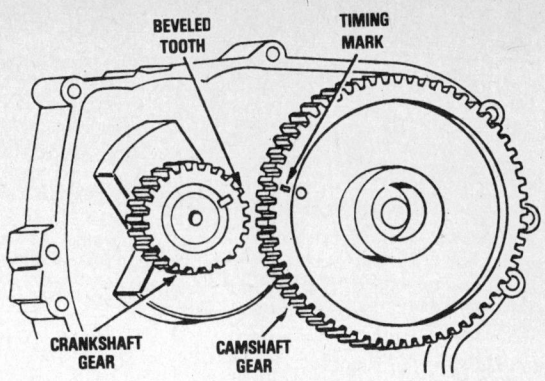

Timing marks

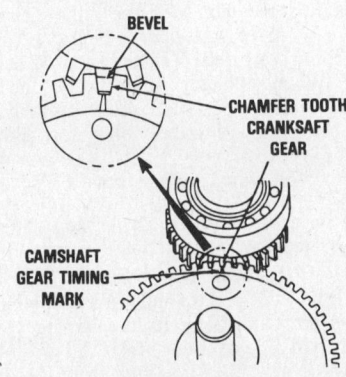

Timing marks

On crankshafts where the gear is pressed onto the crankshaft, a tooth is bevelled to serve as the timing mark.

On engines with a ball bearing on the power take-off end of the crankshaft, look for a bevelled tooth which serves as the crankshaft gear timing mark.

If the engine uses a Craftsman type carburetor, the camshaft timing mark must be advanced clockwise one tooth ahead of the matching timing mark on the crankshaft, the exception being the Craftsman variable governed fuel systems.

NOTE: If one of the timing gears, either the crankshaft gear or the camshaft gear, is damaged and has to be replaced, both gears should be replaced.

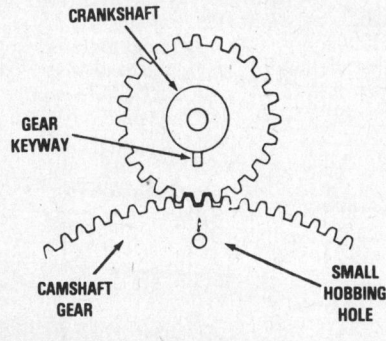

Timing marks

Crankshaft

INSPECTION

Inspect the crankshaft for worn or crossed threads that can't be redressed; worn, scratched, or damaged bearing surfaces; misalignments; flats on the bearing surfaces. Replace the shaft if any of these problems are in evidence—do not try to straighten a bent shaft.

In replacement, be sure to lubricate the bearing surfaces and use oil seal protectors. If the camshaft gear requires replacement, replace the crankshaft gear, also.

LAV35-LAV50 and H35-HS50 crankshafts have a press fit gear. If the camshaft gear requires replacement on these engines, the crankshaft must be replaced, as the gear cannot be replaced separately.

Pistons

When removing the pistons, clean the carbon from the upper cylinder bore and head. The piston and pin must be replaced in matched pairs.

A ridge reamer must be used to remove the ridge at the top of the cylinder bore on some engines.

Clean the carbon from the piston ring groove. A broken ring can be used for this operation.

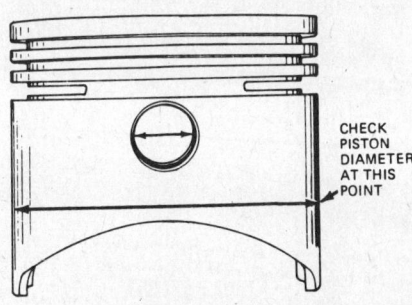

Checking piston dimensions

Some engines have oversize pistons which can be identified by the oversize engraved on the piston top.

There is a definite piston-to-connecting rod-to-crankshaft arrangement which must be maintained when assembling these parts. If the piston is assembled in the bore

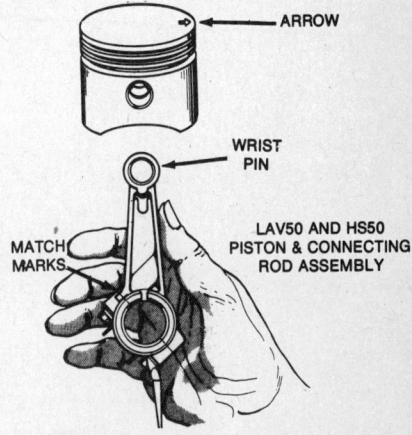

Piston and rod, LAV50 engines

180° out of position, it will cause immediate binding of the parts.

Piston Rings

Always replace the piston rings in sets. Ring gaps must be staggered. When using new rings, wipe the cylinder wall with fine emery cloth to deglaze the wall. Make sure the cylinder wall is thoroughly cleaned after deglazing. Check the ring gap by placing the ring squarely in the center of the area in which the rings travel and measuring the gap with a feeler gauge. Do not spread the rings too wide when assembling them to the pistons. Use a ring leader to install the rings on the piston.

The top compression ring has an inside chamfer. This chamfer must go UP. If the

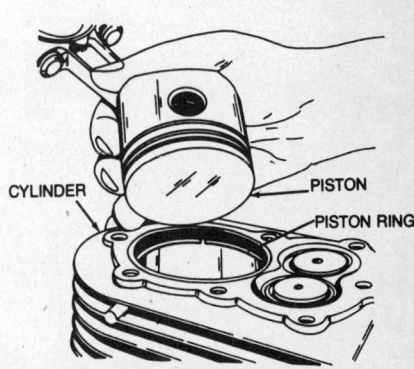

Squaring the ring in the bore

Engine Overhaul

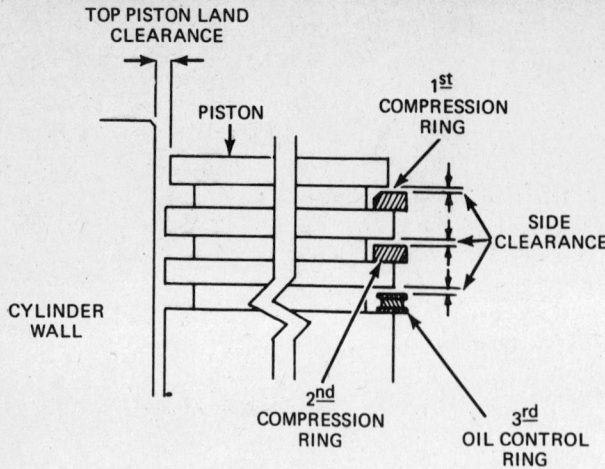

Ring arrangement and dimensions

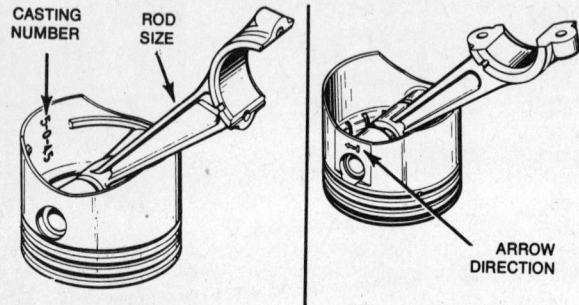

VM70, V80, VM80, HM70, H80, HM80, VM100 & HM100 PISTON AND CONNECTING ROD ASSEMBLY

Piston-to-rod relationship

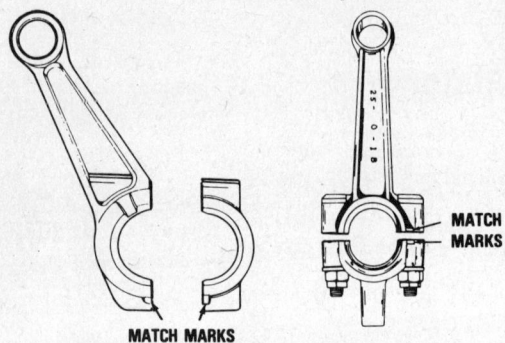

LAV40 and HS40 connecting rod assembly

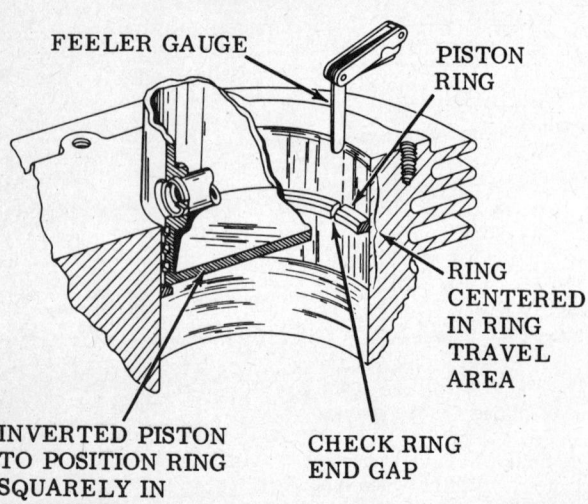

Checking ring end gap

second ring has a chamfer, it must also face UP. If there is a notch on the outside diameter of the ring, it must face DOWN.

Check the ring gap on the old ring to determine if the ring should be replaced. Check the ring gap on the new ring to determine if the cylinder should be rebored to take oversize parts.

NOTE: Make sure that the ring gap is measured with the ring fitted squarely in the worn part of the cylinder where the ring usually rides up and down on the piston.

Connecting Rods

Be sure that the match marks align when assembling the connecting rods to the crankshaft. Use new self-locking nuts. Whenever locking tabs are included, be sure that the tabs lock the nuts securely. NEVER try to straighten a bent crankshaft or connecting rod. Replace them if necessary. When replacing either the piston, rod, crankshaft, or camshaft, liberally lubricate all bearings with engine oil before assembly.

The following engines have offset connecting rods: LAV40, LAV50, HS40, HS50, V70, VH70, VM70, V80, VM80, H70, HH70, HM70, H80, HM80. The LAV40, LAV50, HS40, HS50, ECV105, ECV110, ECV120, and TNT120 engines have the caps fitted from opposite to the camshaft side of the engine. The following engines have the cap fitted from the camshaft side: V70, VH70, VM70, V80, VM80, H70, HH70, HM70, H80, VM100 and HM100.

On H70, HH70, HM70, H80, HM80, V70, VH70, VM70, V80, VM80, VM100 and HM100 engines, a dipper is stamped into the lockplate. Use a *new* lockplate whenever the rod cap is removed.

On engine with Durlock rod bolts, torque the bolts as follows: LAV25-50, H25, 35, HS40-50, TVS75, 90, 105, 120, ECH90, TNT100, 120, ECV100, 105, 110 and 120—110 in.lbs. V50-80, H50-80, VH50-70, HH50-70, VM70-100 and HM70-100—150 in.lbs.

NOTE: Early type caps can be distorted if the cap is not held to the crank pin while threading the bolts tight. Undue force should not be used.

Later rods have serrations which prevent distortion during tightening. They also have match marks which must face out when assembling the rod. On the V80 and H80 engines, the piston and rod must fit so that the number inside the casting is on the rod side of the rod/cap combination. On the LAV50 and HS50 engines, the piston must be fitted to the rod with the arrow on the top of the piston pointing to the right and the match marks on the rod facing you when the piston is pinned to the rod.

Camshaft

Before removing the camshaft, align the timing marks to relieve the pressure on the valve lifters, on most engines. On VM80

Engine Overhaul

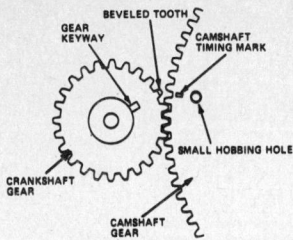

VM80 and VM100 timing marks

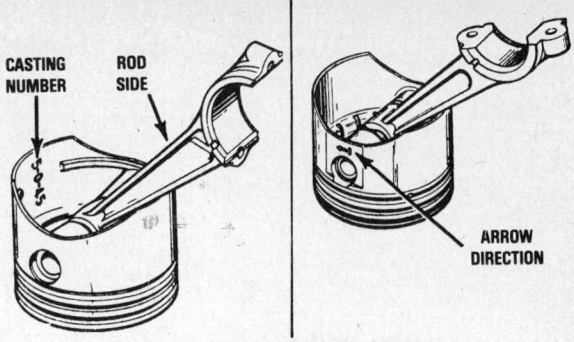

VM70,V80,VM80,HM70,H80,HM80,VM100 & HM100 PISTON AND CONNECTING ROD ASSEMBLY.

V70, V80, H70, H80, HM80, VM100 and HM100 piston and connecting rod

and VM100 engines, while the basic timing is the same, the crankshaft must be rotated so that the timing mark is located 3 teeth further counterclockwise (referring to *crankshaft* gear rotation). This clears the camshaft of the compression release mechanism for easier removal.

In installation, align the gears for this type of camshaft as they were right before removal. After installation, turn the crankshaft gear clockwise in order to check for proper alignment of timing marks.

Clean the camshaft in solvent, then blow the oil passages dry with compressed air. Replace the camshaft if it shows wear of evidence of scoring. Check the cam dimensions against those in the chart.

If the engine has a mechanical fuel pump, it may have to be removed to properly reinstall the camshaft. If the engine is equipped with the Insta-matic Ezee-Start Compression Release, and any of the parts have to be replaced due to wear or damage, the entire camshaft must be replaced. Be sure that the oil pump (if so equipped) barrel chamber is toward the fillet of the camshaft gear when assembled.

NOTE: If a damaged gear is replaced, the crankshaft gear should also be replaced.

Valve Springs

The valve springs should be replaced whenever an engine is overhauled. Check the free length of the springs. Comparing one spring with the other can be a quick check to notice any differences. If a difference is noticed carefully measure the free length, compression length, and strength of each spring. See the specifications chart at the end of this section.

Some valve springs use dampening coils—coils that are wound closer together than most of the coils of the spring. Where these are present, the spring must be mounted so the dampening coils are on the stationary (upper) end of the spring.

Valve Lifters

The stems of the valves serve as the lifters. On the 4 hp light frame models, the lifter stems are of different lengths. Because this engine is a cross port model, the shorter intake valve lifter goes nearest the mounting flange.

The valve lifters are identical on standard port engines. However once a wear pattern is established, they should not be interchanged.

Valve Grinding and Replacement

Valves and valve seats can be removed and reground with a minimum of engine disassembly.

Remove the valves as follows:
1. Raise the lower valve spring caps while holding the valve heads tightly against the valve seat to remove the valve spring retaining pin. This is best achieved by using a valve spring compressor. Remove the valves, springs, and caps from the crankcase.
2. Clean all parts with a solvent and remove all carbon from the valves.
3. Replace distorted or damaged valves. If the valves are in usable condition, grind the valve faces in a valve refacing machine and to the angle given in the specifications chart at the end of this section. Replace the

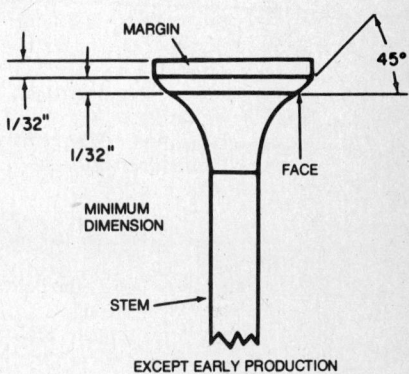

Valve face dimensions

valves if the faces are ground to less than 1/32 in.
4. Whenever new or reground valves are installed, lap in the valves with lapping compound to insure an air-tight fit.

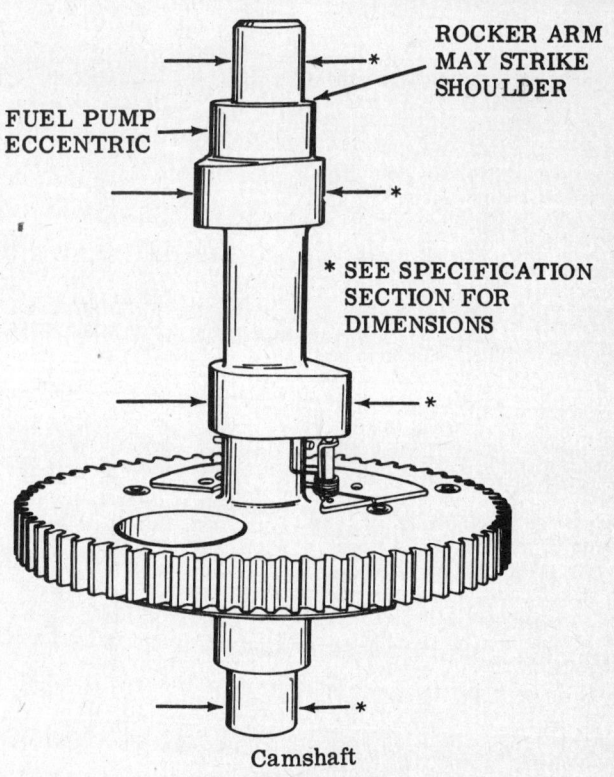

Camshaft
Checking the camshaft dimensions

947

Engine Overhaul

NOTE: There are valves available with oversize stems.

5. Valve grinding changes the valve lifter clearance. After grinding the valves check the valve lifter clearance as follows:
   a. Rotate the crankshaft until the piston is set at the TDC position of the compression stroke.
   b. Insert the valves in their guides and hold the valves firmly on their seats.
   c. Check for a clearance of 0.010 in. between each valve stem and valve lifter with a feeler gauge.
   d. Grind the valve stem in a valve resurfacing machine set to grind a perfectly square face with the proper clearance.
6. Install valves as follows on Early Models:
   a. Position the valve spring and upper and lower valve spring caps under the valve guides for the valve to be installed.
   b. Install the valves in the guides, making sure that the valve marked "EX" is inserted in the exhaust port. The valve stem must pass through the valve spring and the valve spring caps.
   c. Insert the blade of a screwdriver under the lower valve spring cap and pry the spring up.
   d. Insert the valve pin through the hole in the valve stem with a long nosed pliers. Make sure the valve pin is properly seated under the lower valve spring cap.

Install the valves as follows on Later Models:
   a. Position the valve caps and spring in the valve compartment.
   b. Install the valves in guides with the valve marked "EX" in exhaust port. The valve stem must pass through the upper valve cap and spring. The lower cap should sit around the valve lifter exposed end.
   c. Compress the valve spring so that the shank is exposed. DO NOT TRY TO LIFT THE LOWER CAP WITH THE SPRING.
   d. Lift the lower valve cap over the valve stem shank and center the cap in the smaller diameter hole.
   e. Release the valve spring tension to lock the cap in place.

REBORING THE CYLINDER

1. First, decide whether to rebore for 0.010 in. or 0.020 in.
2. Use any standard commercial hone of suitable size. Chuck the hone in the drill press with the spindle speed of about 600 rpm.
3. Start with coarse stones and center the cylinder under the press spindle. Lower the hone so the lower end of the stones contact the lowest point in the cylinder bore.
4. Rotate the adjusting nut so that the stones touch the cylinder wall and then begin honing at the bottom of the cylinder. Move the hone up and down at a rate of 50 strokes a minute to avoid cutting ridges in the cylinder wall. Every fourth or fifth stroke, move the hone far enough to extend the stones 1 in. beyond the top and bottom of the cylinder bore.

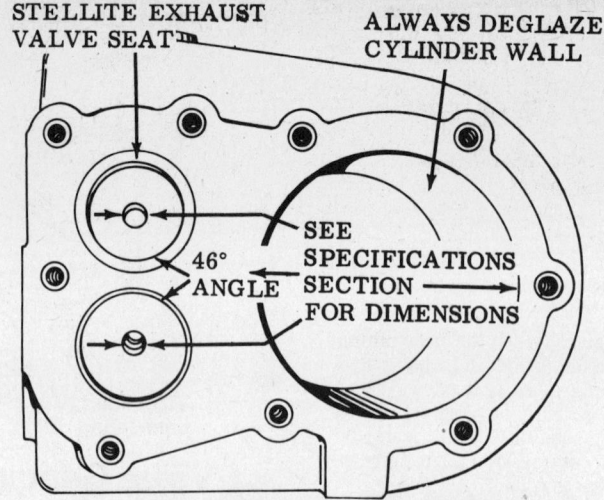

Valve seats, guides and cylinder dimensions

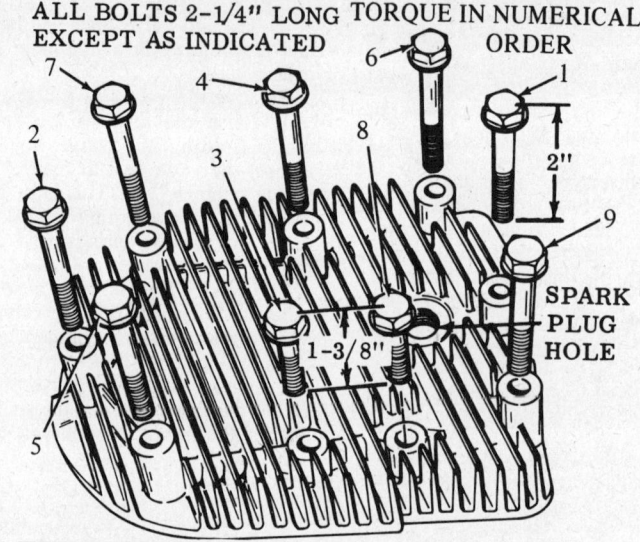

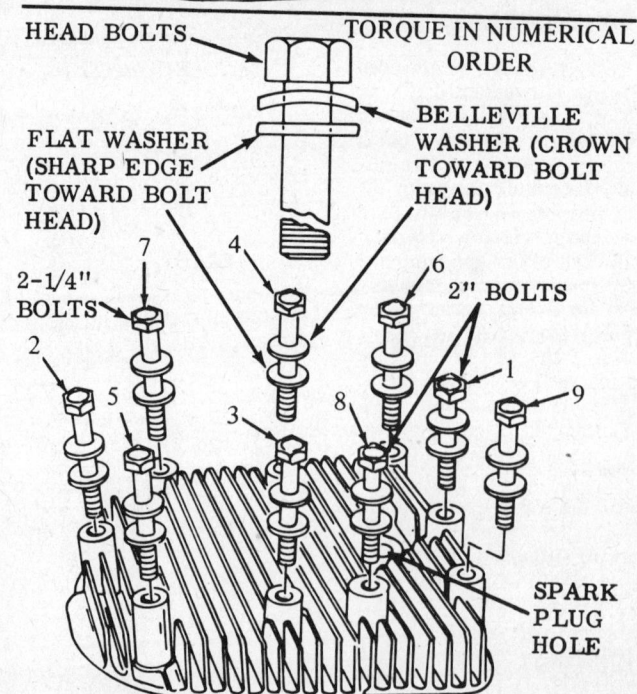

Cylinder head bolt torque sequences; early types (top) and later types (bottom)

Engine Overhaul

5. Check the bore size and straightness every thirty or forty strokes. If the stones collect metal, clean them with a wire brush each time the hone is removed.

6. Hone with coarse stones until the cylinder bore is within 0.002 in. of the desired finish size. Replace the coarse stones with burnishing stones and continue until the bore is to within 0.0005 in. of the desired size.

7. Remove the burnishing stones and install finishing stones to polish the cylinder to the final size.

8. Clean the cylinder with solvent and dry it thoroughly.

9. Replace the piston and piston rings with the correct oversize parts.

Reboring Valve Guides

The valve guides are permanently installed in the cylinder. However, if the guides wear, they can be rebored to accommodate a 1/32 in. oversize valve stem. Rebore the valve guides in the following manner:

1. Ream the valve guides with a standard straight shanked hand reamer or a low speed drill press. Refer to the specifications chart at the end of this section for the correct valve stem guide diameter.

2. Redrill the upper and lower valve spring caps to accommodate the oversize valve stem.

3. Reassemble the engine, installing valves with the correct oversize stems in the valve guides.

Regrinding Valve Seats

The valve seats need regrinding only if they are pitted or scored. If there are no pits or scores, lapping in the valves will provide a proper valve seat. Valve seats are not replaceable. Regrind the valve seats as follows:

1. Use a grinding stone or a reseater set to provide the proper angle and seal face dimensions.

2. If the seat is over 3/64 in. wide after grinding, use a 15° stone or cutter to narrow the face to the proper dimensions.

3. Inspect the seats to make sure that the cutter or stone has been held squarely to the valve seat and that the same dimensions has been held around the entire circumference of the seat.

4. Lap the valves to the reground seats.

Torquing Cylinder Head

Torque the cylinder head to 200 in.lbs. in 4 equal stages of 50 in.lbs. Follow the sequence shown in the appropriate illustration for each tightening stage.

Bearing Service

LIGHTWEIGHT ALUMINUM BEARING REPLACEMENT

The aluminum bearing must be cut out using the rough cut reamer and the proce-

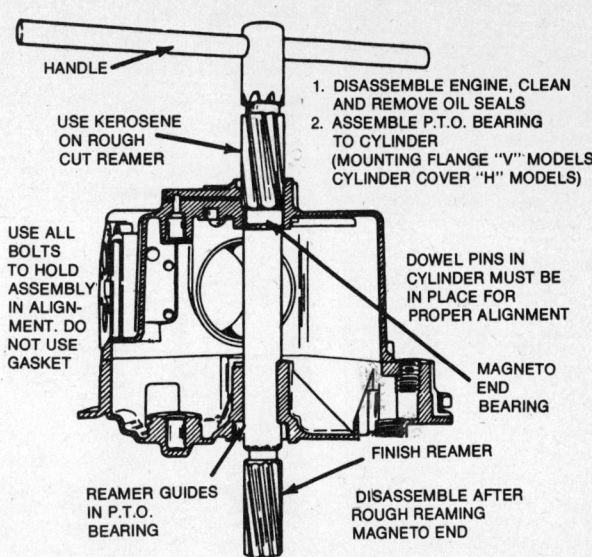

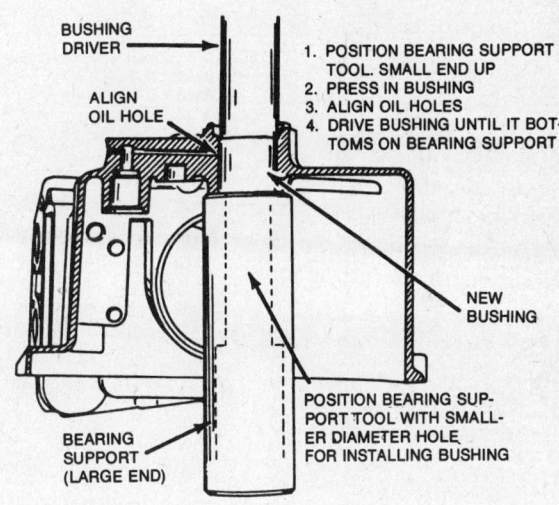

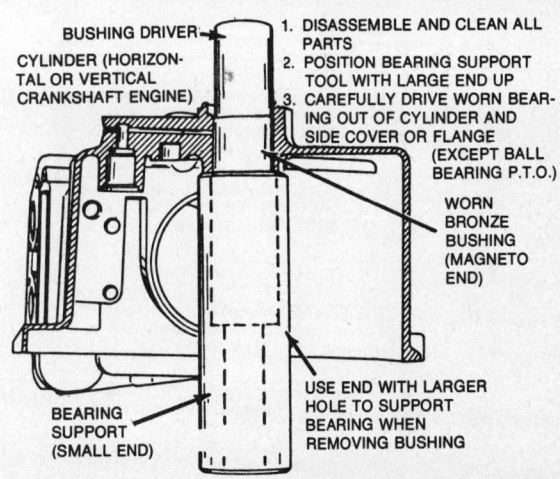

Engine Overhaul

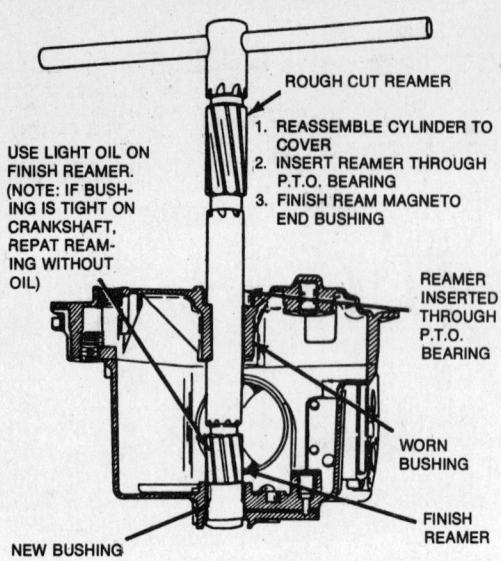

FINISH REAMING NEW MAGNETO BUSHING

- ROUGH CUT REAMER
- USE LIGHT OIL ON FINISH REAMER. (NOTE: IF BUSHING IS TIGHT ON CRANKSHAFT, REPAT REAMING WITHOUT OIL)
  1. REASSEMBLE CYLINDER TO COVER
  2. INSERT REAMER THROUGH P.T.O. BEARING
  3. FINISH REAM MAGNETO END BUSHING
- REAMER INSERTED THROUGH P.T.O. BEARING
- WORN BUSHING
- FINISH REAMER
- NEW BUSHING

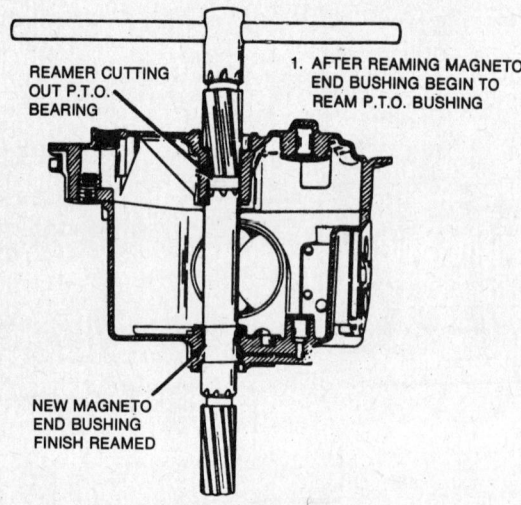

ROUGH REAMING WORN ALUMINUM BEARING (P.T.O. END) FOR ALUMINUM BEARING ONLY

- REAMER CUTTING OUT P.T.O. BEARING
  1. AFTER REAMING MAGNETO END BUSHING BEGIN TO REAM P.T.O. BUSHING
- NEW MAGNETO END BUSHING FINISH REAMED

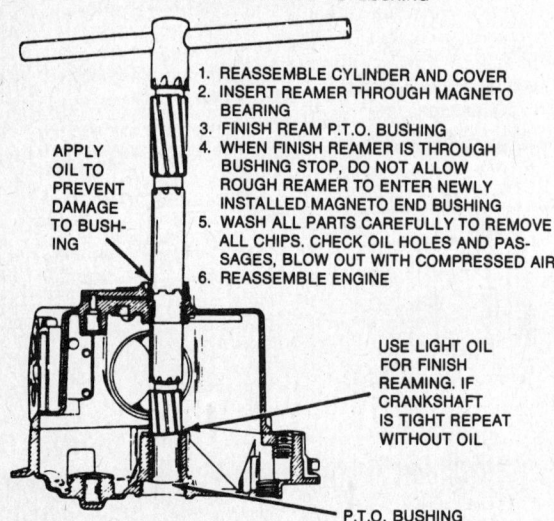

FINISH REAMING P.T.O. BUSHING

- APPLY OIL TO PREVENT DAMAGE TO BUSHING
  1. REASSEMBLE CYLINDER AND COVER
  2. INSERT REAMER THROUGH MAGNETO BEARING
  3. FINISH REAM P.T.O. BUSHING
  4. WHEN FINISH REAMER IS THROUGH BUSHING STOP, DO NOT ALLOW ROUGH REAMER TO ENTER NEWLY INSTALLED MAGNETO END BUSHING
  5. WASH ALL PARTS CAREFULLY TO REMOVE ALL CHIPS. CHECK OIL HOLES AND PASSAGES, BLOW OUT WITH COMPRESSED AIR
  6. REASSEMBLE ENGINE
- USE LIGHT OIL FOR FINISH REAMING. IF CRANKSHAFT IS TIGHT REPAT WITHOUT OIL
- P.T.O. BUSHING

magneto end bearing must be rebushed. The PTO bearing should also be rebushed to assure proper alignment.

4. Oil should be used to finish—ream the bushings. In the event the crankshaft does not rotate freely repeat the finish—reaming operation without oil.

5. Kerosene should be used as a cutting lubricant while rough—reaming.

6. Be sure that the dowel pins are in the cylinder block when assembling the mounting flange or cylinder cover. Use all bolts to hold the assembly together.

7. Remove the reamer by rotating it in the same direction as it is turned during the reaming operation. DO NOT TURN THE REAMER BACKWARDS.

dures shown in steps 1A and 4A. Follow illustrated steps 1A, 2, 3, 4A, 5 and 6 to install bronze bushing in place of the aluminum bearings.

LONG LIFE AND CAST IRON ENGINES WITHOUT BALL BEARINGS

The worn bronze bushings must be driven out before the new bushing can be installed. Follow illustrated steps 1B, 2, 3, 4B, 5 and 6, to replace the main bearings on these units.

LONG LIFE AND CAST IRON ENGINES WITH BALL BEARING ON THE PTO (POWER TAKE-OFF) SIDE OF THE CRANKSHAFT

The side cover containing the ball bearing must be removed and a substitute cover with either a new bronze bushing or aluminum bearing must be used instead. Follow illustrated steps 1B, 2 and 3 only.

SPECIAL TOOLS

The task of main bearing replacement is made easier by using one of two Tecumseh main bearing tool kits. Kit 670161 is used to replace main bearings on the lightweight engines except HS, LAV40 and 50 models. Kit 670165 is used to replace main bearings on the medium weight engines except HS, LAV40 and 50 models.

GENERAL NOTES ON BUSHING REPLACEMENT

1. Your fingers and all parts must be kept very clean when replacing bushings.
2. On splash lubricated horizontal engines, the oil hole in the bushing is to be lined up with the oil hole that leads into the slot in the original bearing.
3. In the event it is necessary to replace the mounting flange or cylinder cover, the

REMOVAL AND INSTALLATION OF CRANKSHAFT BUSHING FOR 8 AND 10 HORSEPOWER ENGINES

The illustrations show the mounting

Engine Overhaul

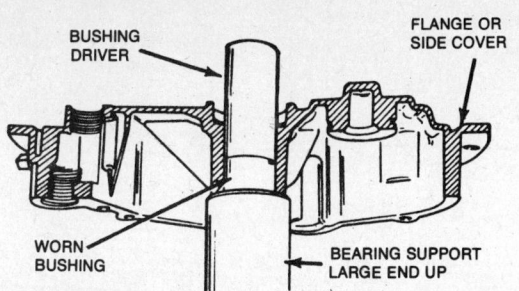

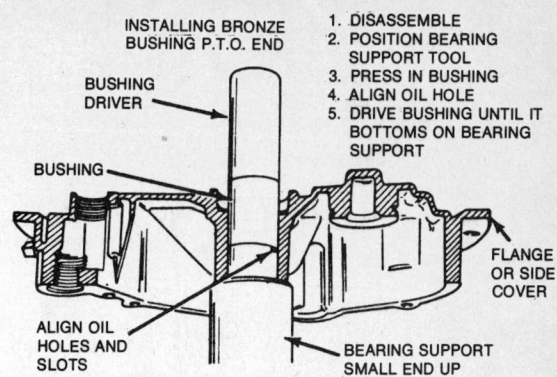

flange for a vertical engine. Procedures also apply to the cylinder cover for a horizontal engine. Use tool No. 670247 removal end and arbor press to press bushing from PTO bearing end. Note the position of oil slots in the bushing which must align with the oil slots in the mounting flange.

To install, insert a new bushing on the installation end of tool No. 670247. Position the slots so they properly align with the oil slots in the cover and press the bushing in with an arbor press.

After the new bushing is installed, use a light coating of oil and finish reaming with reamer, part No. 670248 (handle 690160). Assemble the PTO mounting flange to the cylinder. Use all bolts with dowel pins to hold the assembly in alignment. Insert the tool through the bushing and cylinder crankshaft magneto end bearing as shown. Rotate the cutting edge clockwise in the PTO bushing. Remove the tool in the same direction of rotation. Do *not* allow the cutting edge of reamer to touch the magneto end of the cylinder.

BALL BEARING SERVICE— H20 THROUGH HS50 H.P. HORIZONTAL CRANKSHAFT ENGINES

1. Remove the crankshaft PTO end oil seal. Drive an awl or similar tool into the metal seal body and pry out.
2. Use snap ring pliers to remove the snapring.
3. Reassembly is in reverse order. Secure the cylinder cover, install the snap ring and oil seal. Protect the oil seal to prevent damage during installation.

BALL BEARING SERVICE— H40 THROUGH HM100 H.P. HORIZONTAL CRANKSHAFT ENGINES

1. Prior to attempting removal of the cylinder cover, observe the area around the crankshaft PTO oil seal. Compare it with the illustration, and if there are bearing locks, follow instructions below:
a. Remove the locking nuts using the proper socket wrench. Note fiber washer located under nut; this must be reinstalled. Lift side cover from cylinder after removing the side cover bolts.
b. Install the bearing retainer bolts, fiber washer and locking nuts in the proper sequence in the cover.
2. Also note the following points:
a. On some engines, a locking type retainer bolt is used. To release the bolt, merely loosen the locking nut and turn the retainer bolt counterclockwise to the unlocked position with needle nose pliers to permit the side cover to be removed. Note that the flats on the retainer bolts must be turned so they face the crankshaft to be relocked upon installation. Don't force them! Torque the locking nuts only to 15-22 in.lbs.
b. The ball bearing used in horizontal crankshaft engines has a restricted fit. The bearing is heated and put onto the cold crankshaft. As the bearing cools in grasps the crankshaft tightly and must be removed cold. Remove the ball bearing with a bearing splitter (separator) and a puller. The bearing may be heated by placing it into a container with a sufficient amount of oil to cover the bearing. The bearing should not rest on the bottom of the container. Suspend the bearing on a wire or set the bearing onto a spacer block of wood or wire mesh. Heat the oil and bearing carefully until the oil smokes, quickly remove the bearing and slide it onto the crankshaft.
c. The bearing must seat tightly against the thrust washer which in turn rests tightly against the crankshaft gear.
d. When a ball bearing is used it is not possible to see the key way in the crankshaft gear which is normally used for timing. Because of this, one tooth of the crankshaft gear is chamfered. This chamfered tooth of the crankshaft gear is

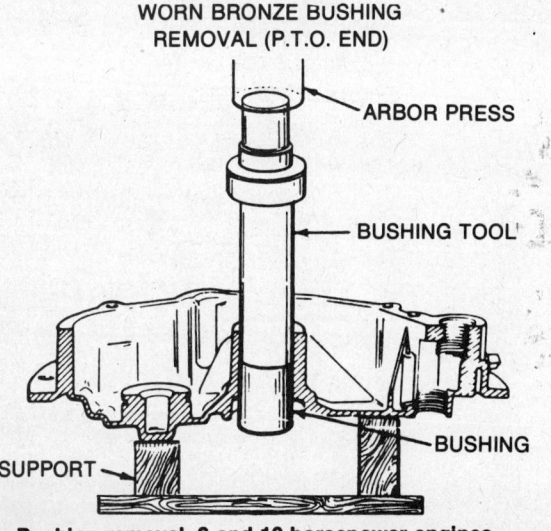

Bushing removal, 8 and 10 horsepower engines

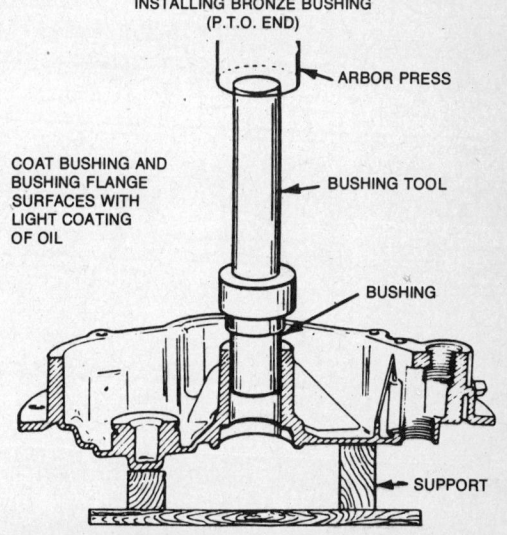

Bushing installaton, 8 and 10 horsepower engines

Engine Overhaul

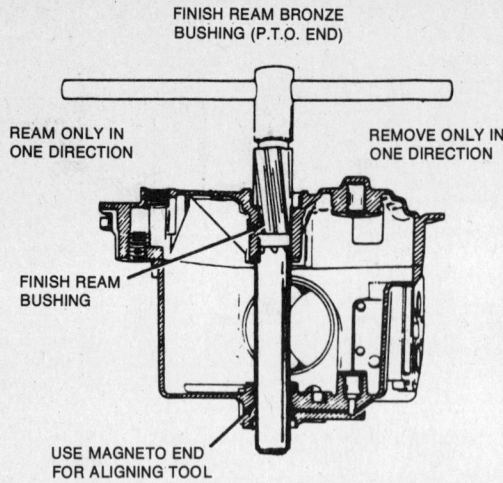

Finish-reaming the bushing for 8 and 10 horsepower engines

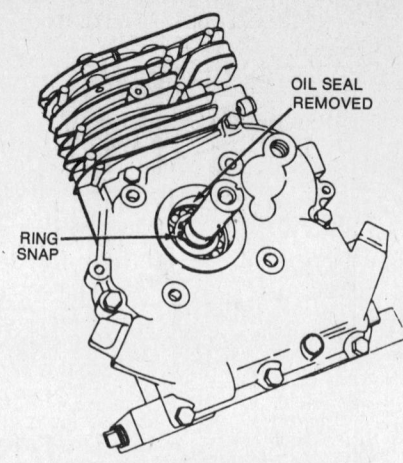

On H20-HS50 H.P. horizontal crankshaft engines, remove the oil seal and snapring to remove the cylinder cover

positioned opposite the timing mark on the camshaft gear. The use of a ball bearing requires the removal of the crankshaft when it is necessary to remove the camshaft. When replacing the crankshaft and camshaft, mate the timing marks and insert it into the cylinder block as an assembly.

Lubrication

BARREL AND PLUNGER OIL PUMP SYSTEM

This system is driven by an eccentric on the camshaft. Oil is drawn through the hollow camshaft from the oil sump on its intake stroke. The passage from the sump through the camshaft is aligned with the pump opening. As the camshaft continues rotation (pressure stroke), the plunger forces the oil out. The other port in the camshaft is aligned with the pump, and directs oil out of the top of the camshaft.

At the top of the camshaft, oil is forced through a crankshaft passage to the top main bearing groove which is aligned with the drilled crankshaft passage. Oil is directed through this passage to the crankshaft connecting rod journal and then spills from the connecting rod to lubricate the cylinder walls. Splash is used to lubricate the other parts of the engine.

A pressure relief port in the crankcase relieves excessive pressures when the oil viscosity is extremely heavy due to cold temperatures, or when the system is plugged or damaged. Normal pressure is 7 psi.

Service

Remove the mounting flange or the cylinder cover, whichever is applicable. Remove the barrel and plunger assembly and separate the parts.

Clean the pump parts in solvent and inspect the pump plunger and barrel for rough spots or wear. If the pump plunger is scored or worn, replace the entire pump.

Before reassembling the pump parts, lubricate all of the parts in engine oil. Manually operate the pump to make sure the plunger slides freely in the barrel.

Lubricate all the parts and position the barrel on the camshaft eccentric. If the oil pump has a chamfer only on one side, that side must be placed toward the camshaft gear. The flat goes away from the gear, thus out to work against the flange oil pickup hole.

Install the mounting flange. Be sure the plunger ball seats in the recess in the flange before fastening it to the cylinder.

SPRAY MIST LUBRICATION

Late model LAV40, LAV30, and LAV35 engines have a spray mist lubrication system. This system is the same as the barrel and plunger oil pump system except that (1) the pressure relief port is changed to a calibrated spray mist orifice and (2) the crankshaft is not rifle-drilled from the top main to the crank pin. Lubrication is sprayed to the narrow rod cap area through the spray mist hole.

SPLASH LUBRICATION

Some engines utilize the splash type

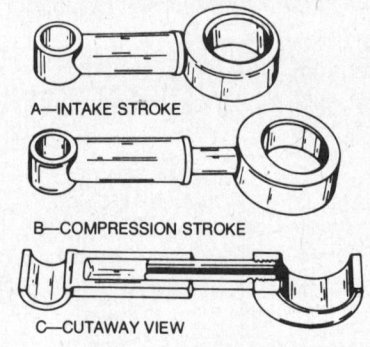

Barrel type oil pump

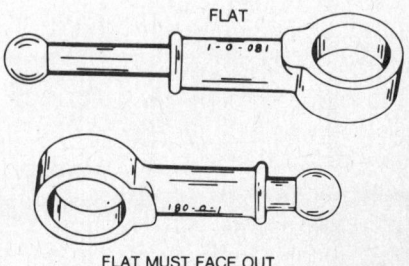

Barrel type oil pump installation

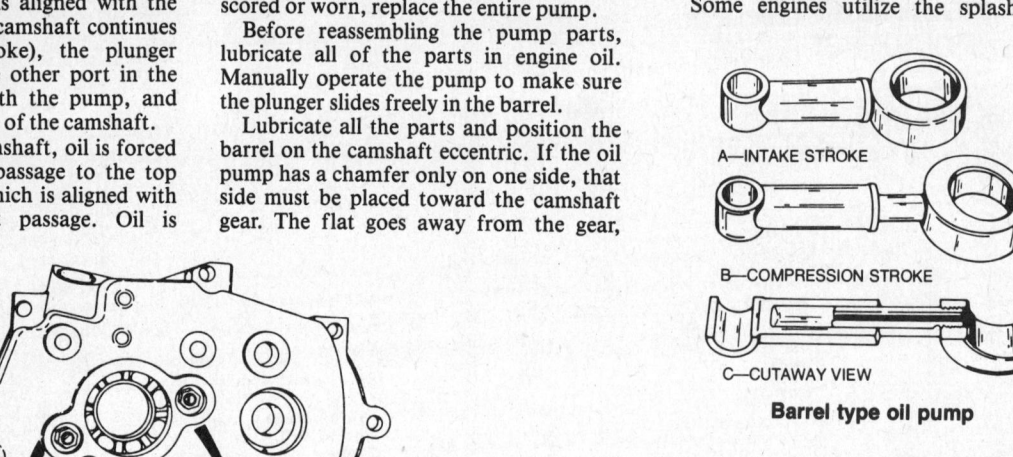

H40-HM100 H.P. horizontal crankshaft engines: locking and unlocking the bearing from the outside

Engine Overhaul

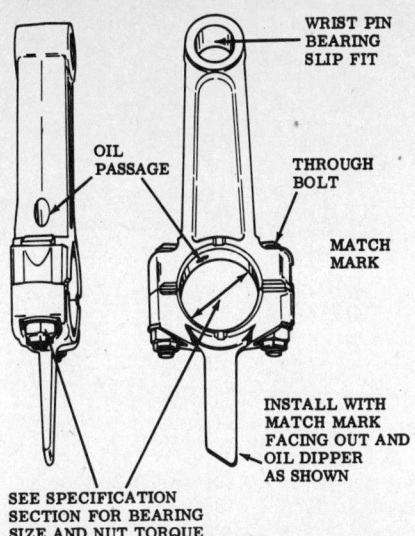

Splash type lubrication connecting rod

GEAR TYPE OIL PUMP SYSTEM

The gear type lubrication pump is a crankshaft driven, positive displacement pump. It pumps oil from the oil sump in the engine base to the camshaft, through the drilled camshaft passage to the top main bearing, through the drilled crankshaft, to the connecting rod journal on the crankshaft.

Spillage from the connecting rod lubricates the cylinder walls and normal splash lubricates the other internal working parts. There is a pressure relief valve in the system.

Service

Disassemble the pump as follows: remove the screws, lockwashers, cover, gear, and displacement member.

Wash all of the parts in solvent. Inspect the oil pump drive gear and displacement member for worn or broken teeth, scoring, or other damage. Inspect the shaft hole in the drive gear for wear. Replace the entire pump if cracks, wear, or scoring is evident.

To replace the oil pump, position the oil pump displacement member and oil pump gear on the shaft, then flood all the parts

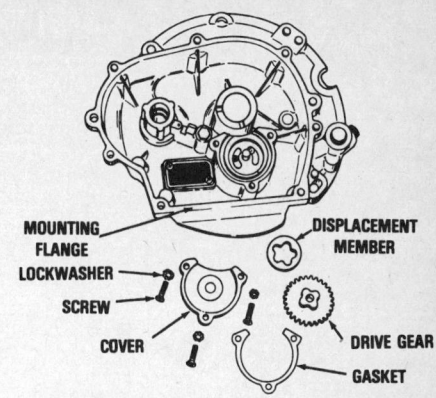

Gear type lubrication system

with oil for priming during the initial starting of the engine.

The gasket provides clearance for the drive gear. With a feeler gauge, determine the clearance between the cover and the oil pump gear. The clearance desired is 0.006 to 0.007 in. Use gaskets, which are available in a variety of sizes, to obtain the correct clearance. Position the oil pump cover and secure it with the screws and lockwashers.

lubrication system. The oil dipper, on some engines, is cast onto the lower connecting rod bearing cap. It is important that the proper parts are used to ensure the longest engine life.

CRAFTSMAN ENGINES CROSS REFERENCE CHART

Craftsman Engine Models	See Column	Craftsman Engine Models	See Column	Craftsman Engine Models	See Column
143.50040	7	143.176012-143.176092	8	143.197082	5
143.50045		143.177012-143.177072	7	143.201032-142.203012	1
143.131022-143.131102	1	143.181042-143.183042	1	143.204022	4
143.135012-143.135112	9	143.184012-143.184212	2	143.204032-143.204052	2
143.136012-143.136052	8	143.184232-143.184252	4	143.204062	4
143.137012	7	143.184262-143.184402	2	143.204072-143.204092	2
143.137032		143.185012-143.185052	9	143.204102	4
143.141012-143.143032	1	143.186012	8	143.204132	
143.145012-143.145072	9	143.186022-143.186042	10	143.204142-143.204192	2
143.146012	8	143.186052	8	143.204202	4
143.146022		143.186062		143.205022	9
143.147012-143.147032	7	143.186072-143.186112	10	143.206012	8
143.151012-143.153032	1	143.186122	8	143.206022	10
143.154012-143.154142	2	143.187022-143.187102	3	143.206032	8
143.155012-143.155062	9	143.191012-143.191052	1	143.207012-143.207052	3
143.156012	8	143.194012-143.194052	2	143.207062	5
143.156022		143.194062	4	143.207072	3
143.157012-143.157032	7	143.194072-143.194092	2	143.207082	5
143.161012-143.163062	1	143.194102	4	143.213012-143.213042	1
143.164012-143.164202	2	143.194112-143.194142	2	143.214012-143.214032	2
143.165012-143.165052	9	143.195012	9	143.214042-143.214072	4
143.166012-143.166052	8	143.195022		143.214082-143.214252	2
143.167012-143.167042	7	143.196012-143.196032	8	143.214262-143.214282	4
143.171012-143.171172	1	143.196042-143.196072	10	143.214292	2
143.171202	2	143.196082	8	143.214302	
143.171212-143.173042	1	143.197012	3	143.214312	4
143.174012-143.174292	2	143.197022	5	143.214322	
143.175012-143.175072	9	143.197032		143.214332	2
		143.197042-143.197072	3	143.214342	

Engine Overhaul

CRAFTSMAN ENGINES CROSS REFERENCE CHART

Craftsman Engine Models	See Column
143.214352	4
143.216012-143.216032	10
143-216042-143.216062	3
143.216072-143.216092	10
143.216122	8
143.216132	11
143.216142	8
143.216152	11
143.216162	
143.216172	
143.216182	8
143.217012-143.217032	5
143.217042-143.217072	3
143.217092	5
143.217102	3
143.223012-143.223052	1
143.224012	2
143.224022	
143.224032	4
143.224062	2
143.224072	4
143.224092-143.224132	2
143.224142	1
143.224162-143.224222	2
143.224232	4
143.224242	
143.224252-143.224282	2
143.224292	4
143.224302	
143.224312-143.224342	2
143.224352	4
143.224362	
143.224372-143.224422	2
143.224432	4
143.225012	13
143.225022	
143.225032-143.225052	9
143.225062	13
143.225072	
143.225082-143.225102	9
143.226012	8
143.226032	
143.226072	10
143.226082	
143.226092-143.226122	11
143.226132-143.226182	8
143.226192	11
143.226202	10
143.226212	
143.226222-13.226262	8
143.226272	10
143.226282	
143.226292	11
143.226302	10
143.226312	11
143.226322	8

Craftsman Engine Models	See Column
143.226332	
143.226342	10
143.226352	11
143.227012-143.227072	12
143.233012	1
143.233032	
143.233042	
143.234022-143.234052	2
143.234062-143.234092	4
143.234102-143.234162	2
143.234192	
143.234202	
143.234212-143.234232	4
143.234242-143.234262	2
143.235012	13
143.235022	
143.235032	6
143.235042	13
143.235052	
143.235062	9
143.234072	6
143.236012	8
143.236022-143.236042	11
143.236052	8
143.236062	11
143.236072	
143.236082	8
143.236092	10
143.236102	8
143.236112	
143.236122	10
143.236132	8
143.236142	11
143.236152	8
143.237012	12
143.237022	
143.237032	5
143.237042	3
143.244032	2
143.244042	4
143.244052	
143.244062	
143.244072-143.244112	2
143.244122-143.244142	4
143.244202	2
143.244212	4
143.244222	2
143.244232	
143.244242	4
143.244252	4
143.244262-143.244282	2
143.244292-143.244332	4
143.245012	6
143.245042	9
143.245052-143.245072	13

Craftsman Engine Models	See Column
143.245082	
143.245092	6
143.245102-143.245132	13
143.245142	6
143.245152	
143.245162	13
143.245172	6
143.245182	
143.245192	13
143.246012	8
143.246022	11
143.246032	
143.246042	8
143.246052-143.246072	10
143.246082	11
143.246092	
143.246102	10
143.246112	
143.246122	11
143.246132	10
143.246142	
143.246152-143.246212	11
143.246222	10
143.246232	11
143.246242	10
143.246252	11
143.246262	10
143.246272-143.247292	11
143.246302	10
143.246312	
143.246322	11
143.246332	
143.246342	10
143.246352	8
143.246362	16
143.246382	
143.246392	8
143.254012-143.254052	2
143.254062	4
143.254072-143.254122	2
143.254142-143.254192	4
143.254212	2
143.254222	
143.254232-143.254292	4
143.254302	2
143.254312	
143.254322	4
143.254332	2
143.254342	4
143.254352	
143.254362	2
143.254372	4
143.254382	
143.254392	2
143.254402	4
143.254412	

Engine Overhaul

CRAFTSMAN ENGINES CROSS REFERENCE CHART

Craftsman Engine Models	See Column
143.254432	2
143.254442	4
143.254452	2
143.254462	4
143.254472	2
143.254482	
143.254492	4
143.254502-143.254532	2
143.255012-143.255112	6
143.256012	11
143.256022	8
143.256032	10
143.256042	11
143.256052	8
143.256062	11
143.256072	
143.256082	8
143.256092	
143.256102	10
143.256112	11
143.256122	8
143.256132	10
143.257012-143.257072	3
143.264012-143.264042	2
143.264052-143.264082	4
143.264092	2
143.264102	4
143.264232-143.264342	2
143.264352-143.264372	4
143.264382	2
143.264392-143.264412	4
143.264422	2
143.264432-143.264482	4
143.264492	2
143.264502	
143.264512	4
143.264522	2
143.264542	
143.264562-143.264672	4
143.264682	2
143.265012-143.265192	6
143.266012	11
143.266022	
143.266032	8
143.266042	10
143.266052	
143.266062	8
143.266082	
143.266092-143.266132	10
143.266142-143.266242	11
143.266252	8
143.266262	11
143.266272-143.266302	10
143.266312	11
143.266322	
143.266332	10
143.266342	11
143.266352	10
143.266362	11
143.266372-143.266412	8
143.266422	11
143.266432-143.266452	8
143.266462	16
143.266472	
143.266482	11
143.267012-143.267042	3
143.274022-143.274072	4
143.274092-143.274132	2
143.274142	4
143.274152	
143.274162-143.274182	2
143.274192-143.274242	4
143.274252	2
143.274262	4
143.274272-143.274322	2
143.274402-143.274482	4
143.275012-143.275052	6
143.276022	4
143.276032	11
143.276042	
143.276052	16
143.276062-143.276162	11
143.276182	8
143.276192	10
143.276202	8
143.276222	10
143.276242	11
143.276252	8
143.276262	11
143.276272	11
143.276282	10
143.276292	11
143.276302	11
143.276322-143.276342	10
143.276352	11
143.276362	16
143.276372-143.276392	10
143.276402	16
143.276412	8
143.276422	10
143.276432-143.276472	11
143.276482	16
143.277012	3
143.277022	
143.284012	2
143.284022	1
143.284032	2
143.284042	4
143.284052	2
143.284062	
143.284072	4
143.284082	2
143.284092	
143.284102	4
143.284112	2
143.284142	
143.284152	
143.284162	
143.284182	
143.284212	4
143.284312	2
143.284322	
143.284332	4
143.284342	
143.284352	
143.284372	16
143.284382	4
143.284402	2
143.284412	
143.284432	4
143.284362	
143.284392	2
143.284442	
143.284482	
143.284452	4
143.284472	
143.285012	6
143.285022	
143.285032	
143.286102	16
143.286022	17
143.286032	10
143.286072-143.286092	11
143.286112	
143.286122	
143.286132	10
143.286142	11
143.286152	
143.286162	
143.286172	
143.505010	8
143.505011	
143.521081	9
143.525021	9
143.526011	
143.526021	8
143.526031	8
143.531052	1
143.531082	
143.531122	
143.531132	
143.531142	2
143.531152	1
143.531172	
143.531182	
143.534012-143.534072	2

955

Engine Overhaul

CRAFTSMAN ENGINES CROSS REFERENCE CHART

Craftsman Engine Models	See Column
143.535012-143.535062	9
143.536012-143.536062	8
143.537012	7
143.541012	1
143.541042-143.541062	1
143.541112-143.541152	1
143.541172-143.541202	1
143.541222	1
143.541282-143.541302	1
143.544012-143.544042	2
143.545012-143.545042	9
143.546012-143.546022	8
143.547012-143.547032	7
143.551012	1
143.551032	
143.551052-143.551192	1
143.554012-143.554082	2
143.555012-143.555052	9
143.556012-143.556282	8
143.557012-143.557082	7
143.565022	9
143.566002-143.566202	8
143.566212	9
143.566222-143.566252	8
143.567012-143.567042	7
143.571002-143.571122	1
143.571152	2
143.571162	1
143.571172	
143.574022-143.574102	2
143.575012-143.575042	9
143.576002-143.576202	8
143.581002-143.581102	1
143.584012-143.584142	2
143.585012-143.585042	9
143.586012-143.586042	8
143.586052-143.586062	10
143.586072-143.586082	8
143.586112-143.586142	10
143.586152	8
143.586162	10
143.586172-143.586242	8
143.586252	10
143.586262-143.586282	8
143.587012-143.587042	3
143.591012-143.591142	1
143.594022-143.594082	2
143.594092	2
143.594102	
143.595012	9
143.595042	
143.596012	10
143.596022	
143.596042	8

Craftsman Engine Models	See Column
143.596052	10
143.596072-143.596122	8
143.597012-143.597032	3
143.601022-143.601062	1
143.604012	2
143.604022	4
143.604032	2
143.604042	
143.604052	4
143.604062	2
143.604072	
143.605012	9
143.605022	
143.605052	
143.606012-143.606052	10
143.606092	8
143.606102	10
143.607012-143.607032	3
143.607042-143.607062	3
143.611012-143.611112	1
143.614012-143614032	4
143.614042	2
143.614052	4
143.614062-143.614162	2
143.615012-143.615092	9
143.616012	10
143.616022-143.616112	8
143.616122	10
143.616132	8
143.616142	
143.617012-143.617182	3
143.621012-143.621092	1
143.624012-143.624112	2
143.625012-143.625132	9
143.626012	10
143.626022	8
143.626032	10
143.626042	8
143.626052-143.626122	10
143.626132	8
143.626142	10
143.626152	
143.626162	8
143.626172	10
143.626182	8
143.626192	10
143.626202	8
143.626212	10
143.626222-143.626262	8
143.626282	11
143.626292	10
143.626302	8
143.626312	10
143.626322	
143.627012-143.627042	3

Craftsman Engine Models	See Column
143.631012-143.631092	1
143.634012	2
143.634032	
143.635012	9
143.635022	
143.635032	6
143.635052	9
143.636012	11
143.636022	
143.636032	10
143.636042	11
143.636052	8
143.636062	10
143.636072	11
143.637012	3
143.641012-143.641062	1
143.641072	2
143.644012-143.644082	2
143.645012-143.645032	6
143.646012-143.646032	10
143.646042	11
143.646052	
143.646072-143.646102	10
143.646112	8
143.646122	10
143.646132	
143.646142	11
143.646152	10
143.646162	11
143.646172	10
143.646182	
143.646192	8
143.646202	10
143.646212-143.646232	11
143.647012-143.647062	3
143.651012-143.651072	1
143.654022-143.654322	2
143.655012	6
143.655032	
143.656012-143.656052	8
143.656062	10
143.656082	11
143.656082	8
143.656102	10
143.656112	8
143.656122-143.656152	10
143.656162-143.656182	8
143.656192	10
143.656202	8
143.656212	11
143.656222	
143.656232	10
143.656242	11
143.656252	8
143.656262	10
143.656272	

Engine Overhaul

ENGINE SPECIFICATIONS

Reference Column	1	2	3	4	5	6	7	8	9	10	11	12	13	14	15	16	17
Displacement	7.75	9.06	10.5	10.0	10.5	12.0	11.04	13.53	12.17	15.0	18.65 See Note A	11.5	12.0	10.0	12.0	20.2	17.16
Stroke	1²⁷⁄₃₂″	1²⁷⁄₃₂″	1¹⁵⁄₁₆″	1²⁷⁄₃₂″	1¹⁵⁄₁₆″	1¹⁵⁄₁₆″	2¼″	2½″	2¼″	2¹⁷⁄₃₂″	2¹⁷⁄₃₂″ See Note B	1¹⁵⁄₁₆″	1¹⁵⁄₁₆″	1²⁷⁄₃₂″	1¹⁵⁄₁₆″	2¹⁷⁄₃₂″	2¹⁷⁄₃₂″
Bore	2.3125 / 2.3135	2.5000 / 2.5010	2.625 / 2.626	2.625 / 2.626	2.525 / 2.626	2.812 / 2.813	2.5000 / 2.5010	2.625 / 2.626	2.625 / 2.626	2.750 / 2.751	3.062 / 3.063	2.750 / 2.751	2.812 / 2.813	2.625 / 2.626	2.812 / 2.813	3.187 / 3.188	2.9375 / 2.9385
Timing Dimension Before Top Dead Center for Vertical Engines	V.060 / .070	V.065	V.035	.035	.035	V.040 / .060	V.050	V.050	H.050	V.050	V.070	V.035	V.035	V.035	V.035		V.070
Timing Dimension Before Top Dead Center for Horizontal Engines	H.060 / .070	H.030 / .040	H.035			H.055	H.050	H.050	H.050	H.050	H.070					H.070	H.070
Point Setting	.020	.020	.020	.020	.020		.020	.020	.020	.020	.020	.020	.020	.020	.020	.020	.020
Spark Plug Gap	.030	.030	.030	.030	.030	In. .280 Ex. .278	.030	.030	.030	.030	.030	.030	.030	.030	.030	.030	.030
Valve Clearance	.010 Both	.010 Both	.010 Both	.010 Both	.010 Both	.010 Both	.010 Both	.010 Both	.010 Both	.010 Both	.010 Both	.010 Both	.010 Both	.010 Both	.010 Both	.010 Both	.010 Both
Valve Seat Angle	46°	46°	46°	46°	46°	46°	46°	46°	46°	46°	46°	46°	46°	46°	46°	46°	46°
Valve Spring Free Length	1.135″	1.135″	1.135″	1.135″	1.135″	1.135″	1.562″	1.462″	1.462″	1.462″	1.462″	1.135″	1.135″	1.135″	1.135″	1.462″	1.462°
Valve Guides Over-Size Dimensions	.2805 / .2815	.2805 / .2815	.2805 / .2815	.2805 / .2815	.2807 / .2817		.3432 / .3442	.3432 / .3442	.343 / .344	.3432 / .3442	.3432 / .3442	.2805 / .2815	.2805 / .2815	.2805 / .2815	.2805 / .2815	.3432 / .3442	.3432 / .3442
Valve Seat Width	.035 / .045	.035 / .045	.035 / .045	.035 / .045	.035 / .045	.035 / .045	.042 / .052	.042 / .052	.042 / .052	.042 / .052	.042 / .052	.035 / .045	.035 / .045	.035 / .045	.035 / .045	.042 / .052	.042 / .052
Crankshaft End Play	.005 / .027	.005 / .027	.005 / .027	.005 / .027	.005 / .027	.005 / .027	.005 / .027	.005 / .027	.005 / .027	.005 / .027	.005 / .027	.005 / .027	.005 / .027	.005 / .027	.005 / .027	.005 / .027	.005 / .027
Crankpin Journal Diameter	.8610 / .8615	.8610 / .8615	.9995 / 1.0000	.8610 / .8615	.9995 / 1.0000	.9995 / 1.0000	1.0615 / 1.0620	1.0615 / 1.0620	1.0615 / 1.0620	1.1865 / 1.1870	1.1865 / 1.1870	.9995 / 1.0000	.9995 / 1.0000	.8610 / .8615	.9995 / 1.0000	1.1865 / 1.1870	1.1865 / 1.1870

Note A. For VM80 & HM80 engines only—Displacement is 19.41″.
Note B. For VM80 & HM80 engines only—Bore is 3.125″ (3⅛″). 3.126″
Note C. For VM80 & HM80 engines only—Piston Diameter is 3.1205″ 3.1195″

Note A. For VM80 & HM80 engines only—Displacement is 19.41″.
Note B. For VM80 & HM80 engines only—Bore is 3.125″ (3⅛″). 3.126″
Note C. For VM80 & HM80 engine only—Piston Diameter is 3.1205″ 3.1195″

Engine Overhaul

ENGINE SPECIFICATIONS

Reference Column	1	2	3	4	5	6	7	8	9	10	11	12	13	14	15	16	17
Cylinder Main Bearing Dia.	.8755/.8760	.8755/.8760	1.0005/1.0010	.8755/.8760	1.0005/1.0010	1.0005/1.0010	1.0005/1.0010	1.0005/1.0010	1.0005/1.0010	1.0005/1.0010	1.0005/1.0010	1.0005/1.0010	1.0005/1.0010	.8755/.8760	1.0005/1.0010	1.0005/1.0010	1.0005/1.0010
Cylinder Cover Main Bearing Dia.	.8755/.8760	.8755/.8760	1.0005/1.0010	.8755/.8760	1.2010/1.2020	1.0005/1.0010	1.0005/1.0010	1.0005/1.0010	1.0005/1.0010	1.0005/1.0010	1.1890/1.1895	1.0005/1.0010	1.0005/1.0010	.8755/.8760	1.0005/1.0010	1.1890/1.1895	1.1890/1.1895
Conn. Rod. Dia. Crank Bearing	.8620/.8625	.8620/.8625	1.0005/1.0010	.8620/.8625	1.0005/1.0010	1.0005/1.0010	1.0630/1.0635	1.0630/1.0635	1.0630/1.0635	1.1880/1.1885	1.1880/1.1885	1.0005/1.0010	1.0005/1.0010	.8620/.8625	1.0005/1.0010	1.1880/1.1885	1.1880/1.1885
Piston Diameter	2.3090/2.3095	2.4950/2.4955	2.6200/2.6205	2.6200/2.6205	2.604/2.608	2.8070/2.8075	2.492/2.4945	2.6210/2.6215	2.6210/2.6215	2.7450/2.7455	3.0575/3.0585 See Note C	2.7450/2.7455	2.8070/2.8075	2.6200/2.6205	2.8070/2.8075	3.1817/3.1842	2.9325/2.9335
Piston Pin Diameter	.5629/.5631	.5629/.5631	.5629/.5631	.5629/.5631	.5631/.5635	.5629/.5631	.6248/.6250	.6248/.6250	.6248/.6250	.6248/.6250	.6248/.6250	.5629/.5631	.5629/.5631	.5629/.5631	.5629/.5631	.6248/.6250	.6248/.6250
Width of Comp. Ring Groove	.0955/.0977	.0955/.0975	.0925/.0935	.0955/.0975	.0955/.0975	.0955/.0975	.0955/.0975	.0955/.0975	.0955/.0975	.0795/.0805	.0955/.0975	.0795/.0815	.0955/.0975	.0955/.0975	.0955/.0975	.0955/.0975	.0975/.0955
Width of Oil Ring Groove	.125/.127	.125/.127	.156/.158	.156/.158	.156/.158	.156/.158	.156/.158	.156/.158	.156/.158	.188/.189	.188/.190	.1565/.1585	.1565/.1585	.1565/.1585	.1565/.1585	.188/.190	.188/.190
Side Clear- ance of Ring Groove (Top)	.002/.005	.002/.003	.002/.004	.002/.005	.002/.005	.003/.004	.002/.003	.002/.004	.002/.004	.002/.003	.003/.004	.002/.004	.003/.004	.003/.0045	.0028/.0039	.0020/.0050	.0028/.0051
Side Clear- ance of Ring Groove (Bot.) Oil			.001/.004	.001/.004	.001/.004	.002/.003		.002/.004	.002/.004	.001/.003	.002/.003	.001/.002	.001/.002	.0010/.0030	.0018/.0038	.001/.004	.0018/.0029
Ring End Gap	.007/.020	.007/.020	.007/.020	.007/.020	.007/.020	.007/.020	.007/.020	.007/.020	.007/.020	.007/.020	.007/.020	.007/.020	.007/.020	.007/.020	.007/.020	.007/.020	.007/.020
Top Piston Land Clearance	.0015/.0145	.015/.018	.0165/.0215	.017/.022	.017/.022	.017/.022	.015/.018	.017/.020	.017/.020	.023/.028	.031/.034	.024/.027	.018/.021	.017/.022	.017/.022	.029/.034	.030/.035
Piston Skirt Clearance	.0025/.0040	.0045/.0060	.0045/.0060	.0045/.0060	.0050/.0065	.0045/.0060	.0055/.0070	.0035/.0050	.0035/.0050	.0045/.0060	.0035/.0055	.0045/.0060	.0045/.0060	.0045/.0060	.0045/.0060	.0028/.0063	.004/.006
Camshaft Bearing Dia.	.4975/.4980	.4975/.4980	.4975/.4980	.4975/.4980	.505/.513	.4975/.4980	.6230/.6235	.6230/.6235	.6230/.6235	.6230/.6235	.6230/.6235	.4975/.4980	.4975/.4980	.4975/.4980	.4975/.4980	.6230/.6235	.6230/.6235
Dia. of Crankshaft Mag. Main Brg.	.8735/.8740	.8735/.8740	.9985/.9990	.8735/.8740	.9985/.9990	.9985/.9990	.9985/.9990	.9985/.9990	.9985/.9990	.9985/.9990	.9985/.9990	.9990/.9995	.9990/.9995	.8735/.8740	.9985/.9990	.9985/.9990	.9985/.9990
Dia. of Crankshaft P.T.O. Main Brg.	.8735/.8740	.8735/.8740	.9985/.9990	.8735/.8740	.9985/.9990	.9985/.9990	.9985/.9990	.9985/.9990	.9985/.9990	.9985/.9990	1.1870/1.1875	.9985/.9990	.9985/.9990	.8735/.8740	.9985/.9990	1.1870/1.1875	1.1870/1.1875

Engine Overhaul

CRAFTSMAN ENGINES CROSS REFERENCE CHART

Craftsman Engine Models	See Column
143.656282	11
143.657012-143.657052	3
143.661012-143.661062	1
143.664012-143.664332	2
143.665012-143.665082	6
143.666012	10
143.666022	
143.666032	11
143.666042-143.666072	10
143.666082	11
143.666092	
143.666102-143.666142	8
143.666152	11
143.666162	
143.666172	8
143.666202	
143.666222	10
143.666232	11
143.666242	8
143.666252	10
143.666272	8
143.666282	10
143.666292	8
143.666302	10
143.666312	
143.666322	11
143.666332	16
143.666342	10
143.666352	11
143.666362	16
143.666372	8
143.666382	10
143.667012	3
143.667022	
143.667032	6
143.667042-143.667082	3
143.674012	2
143.675012	6
143.675022	
143.675032	9
143.675042	6
143.676012	11
143.676022	
143.676032	10
143.676042	11
143.676052	
143.676062	16
143.676072	
143.676102	10
143.676112	8
143.676122	10
143.676132	8
143.676142	11
143.676152	16
143.676162	
143.676172	10
143.676182	11
143.676192	10
143.676202	11
143.676212	16
143.676222	11
143.676232	8
143.676242	8
143.676252	11
143.676262	16
143.677012	3
143.677022	
143.686012	17
143.686022	
143.686032	11
143.686042	
143.686052	
143.686062	10
143.686072	9
143.687012	3
143.694126	4
143.694132	2
143.694134	11

TORQUE SPECIFICATIONS

Model/Part	Inch Pounds	Ft. pounds
Cylinder Head Bolts	160-200	13-16
Connecting Rod Bolts	65-75	5.5-6
ECH90, ECV100, TNT100	75-80	6.2-6.7
6 H.P. Medium Frame	86-110	7.1-9.1
6 H.P. Medium Frame (Durlok Rod Bolts)	130-150	10.8-12.5
ECV105, ECV110, ECV120, TNT120	80-95	6.6-7.9
7, 8 & 10 Medium Frame	106-130	8.8-10.8
7, 8 & 10 Medium Frame (Durlok Rod Bolts)	150-170	12.5-14.1
Cylinder Cover or Flange-to-Cylinder	65-110	5.5-9
Cylinder Cover 6-7 H.P. Medium Frame, H Models	100-140	8.3-11.6
Flywheel Nut	360-396	30-33
Spark Plug	180-360	15-30
Magneto Stator to Cylinder	40-90	3.3-7.5
Starter to Blower Housing or Cylinder	40-60	3.5-5
Housing Baffle to Cylinder	48-72	4-6
Breather Cover (Top Mount ECV)	40-50	3.3-4.1
Breather Cover	20-26	1.7-2.1
Intake Pipe to Cylinder	72-96	6-8
Carburetor to Intake Pipe	48-72	4-6

Engine Overhaul

TORQUE SPECIFICATIONS

Model/Part	Inch Pounds	Ft. Pounds
Air Cleaner to Carburetor (Plastic)	8-12	1
Tank Plate to Bracket (Plastic)	100-144	9-12
Tank to Housing	45-65	3.7-5
Muffler Bolts to Cylinder 6-10 H.P. Medium Frame	90-150	8-12
6:1 Gear Reduction Housing to Cylinder	100-144	8.5-12
Gear Reduction Cover to Housing	65-110	5-9
Oil Drain Plug		
1/8-27	35-50	1.1-4.1
1/4-18	65-85	4.5-7
3/8-18	80-100	6.6-9
5/8-18	90-150	7.5-12.5
1/2-14	80-100	6.6-9
Ball Bearing Retainer	15-22	1.5
Craftsman Exclusive Fuel System to Cylinder	72-96	6-8
Electric Starter-to-Cylinder	50-60	4-5

CROSS REFERENCE FOR VERTICAL CRANKSHAFT ENGINES

Model	Column
10.0 CI	
ECV100	4
TNT100	14
10.5 CI	
ECV105	5
11.0 CI	
ECV110	12
12.0 CI	
ECV120	13
TNT120	15
6 HP	
V60	8
VH60	8
7 HP	
V70	10
VH70	10
VM70	17
8 HP	
V80	11
VM80	11
10 HP	
VM100	16

CROSS REFERENCE CHART FOR HORIZONTAL CRANKSHAFT ENGINES

Model	Column
9.0 CI	
ECH90	2
6 HP	
H60	8
HH60	8
7 HP	
H70	10
HH70	10
HM70	17
8 HP	
H80	11
HM80	11
10 HP	
HM100	16

Acknowledgements

The Chilton Book Company wishes to express its sincere appreciation to the following manufacturers and suppliers for their technical assistance and original artwork, without which the preparation of this manual would have been impossible:

Allis-Chalmers Corp., Milwaukee, Wisconsin
Bolens-FMC Corp., Fort Washington, Wisconsin
J.I. Case Co., Racine, Wisconsin
John Deere Co., Moline, Illinois
Ford Motor Co., Troy, Michigan
Gilson Brothers Co., Plymouth, Wisconsin
Gravely Corp., Clemmons, North Carolina
International Harvester Co., Westmont, Illinois
Jacobsen Div., Textron, Inc., Racine, Wisconsin
Kubota Tractor Corp., Compton, California
Massey-Ferguson, Inc., Des Moines, Iowa
Satoh Agricultural Machine Mfg. Co., New York, New York
Wheel Horse Products, Inc. South Bend, Indiana
White Farm Equipment Co., Oak Brook, Illinois
Yanmar Diesel Division, Mitsui Co., Bensenville, Illinois